ENVIRONMENTAL PHYSIOLOGY OF ANIMALS

Environmental Physiology of Animals

SECOND EDITION

Pat Willmer
School of Biology
University of St Andrews
St Andrews

Graham Stone
Institute of Cell, Animal, and Population Biology
University of Edinburgh
Edinburgh

Ian Johnston
School of Biology
University of St Andrews
St Andrews

BLACKWELL PUBLISHING
350 Main Street, Malden, MA 02148-5020, USA
108 Cowley Road, Oxford OX4 1JF, UK
550 Swanston Street, Carlton, Victoria 3053, Australia

First edition published 2000 by Blackwell Publishing Ltd
Second edition published 2005

Library of Congress Cataloging-in-Publication Data
Willmer, Pat, 1953–
Environmental physiology of animals / Pat Willmer, Graham Stone, Ian Johnston.– 2nd ed.
p. cm.
Includes bibliographical references (p.).
ISBN 1-4051-0724-3 (hardback : alk. paper)
1. Adaptation (Physiology) 2. Physiology, Comparative.
3. Ecophysiology. I. Stone, G. II. Johnston, Ian A. III. Title.

QP82.W48 2004
571.1–dc22 2003023170

A catalogue record for this title is available from the British Library.

Set in 9/11.5pt Minion
by Graphicraft Limited, Hong Kong
Printed and bound in the United Kingdom
by William Clowes Ltd, Beccles, Suffolk

The publisher's policy is to use permanent paper from mills that operate a sustainable forestry policy, and which has been manufactured from pulp processed using acid-free and elementary chlorine-free practices. Furthermore, the publisher ensures that the text paper and cover board used have met acceptable environmental accreditation standards.

For further information on
Blackwell Publishing, visit our website:
http://www.blackwellpublishing.com

Contents

Preface to Second Edition

The first edition of this book found a useful place in many libraries and on the shelves of teachers and students as a text for a wide range of undergraduate courses, and we are glad that our targeted market found it readable, interesting and suitably up-to-date in its approach. Our general aim of integrating animal physiology into a more holistic approach, one that includes both an ecological setting and an appreciation of the range of behavioral responses open to individual animals before specific physiological responses need to come into play, has clearly met a need and found a receptive audience.

However, we met with two general criticisms that, although not aimed at the content or the written style, were enough to put off some of the potential users. Firstly, many readers while appreciating the consistent style of illustration found the book visually rather dull. We have been able to address that in this second edition by adding a colour to the production, making the figures both clearer and we hope more attractive. Secondly, and perhaps more importantly, many lecturers reported that they could not adopt this book as their core text for physiology courses because of the omission of material on excitable tissues and control systems. Leaving these topics out of the original book was a conscious decision designed to retain the strongly environmental flavour of the treatment—after all nerves, muscle and hormones are not inherently very variable between different habitats. Furthermore there are excellent treatments of nerve and muscle available at suitable levels in existing books on neurobiology. However, many students are unable or unwilling to purchase more than one book in a particular area of their studies, so there is clearly a need to offer a fully comprehensive coverage in one volume. Here then we have added in the two new chapters that were requested: Chapter 9 covers nerves, sense organs, muscles and the integration of all of these components into functional sensory and motor systems, while Chapter 10 covers the hormonal control systems operating across the animal kingdom. We hope that these provide coverage at the level needed for any comparative physiology course. Chapter 9 is unusually long, reflecting the enormous advantages made in the last two decades to bring the realms of neurobiology into the molecular age; it should in itself also prove suitable for many junior courses on excitable tissues.

Other chapters in the book retain the same format but have been updated where appropriate with recent insights, and additional molecular and evolutionary evidence has been included as it emerges. The suggested reading has been enhanced with recent reviews.

This new edition will therefore provide a core text for students taking a wide range of physiological modules in colleges and universities, and it should also serve as a useful reference for many lecturers striving to keep updated on the interactions of animal physiology with their own related areas of ecology, behavior and environmental biology.

Pat Willmer

Preface to First Edition

There are many books dealing with physiological functioning and the comparative adaptations of animals. Why should one more be added to the list?

We hope that there are three key reasons why this book is different.

1 Above all, we seek to be novel by setting the issues in a strongly environmental context; the largest part of this book is devoted to analyzing the problems for each kind of environment in turn. In one chapter you can find out how different kinds of seashore animal cope with the ionic problems of salinity change, the thermal problems of both cold sea water and warm drying aerial exposure, and the respiratory hazards of alternate aerial and underwater breathing. Inevitably we include more varied material, from a wider range of animal types; you will find less of a focus on terrestrial reptiles, birds, and mammals than in most other books (after all, they represent less than 1% of all animals!).

2 We try to put together issues other than just conventional physiology, to look at mechanisms and responses with perspectives from the related and overlapping fields of ecology, behavior, and evolutionary biology. Thus you can find out how the seashore animals also deal with the mechanical problems of tides and waves, how their senses may need to be modified, the difficulties of finding a mate and producing viable offspring, and some of the imposed (anthropogenic) problems that man has introduced to different habitats. The book thus has more environmental biology, more ecology, and a greater evolutionary perspective than many others.

3 Wherever possible we include modern molecular biological insights into adaptive problems, and take the level of analysis to the cell, the membrane, and the enzyme.

In essence we want to look at how animals cope with particular kinds of environment, putting together the biochemical, physiological, behavioral, and ecological adaptations that allow animals to survive there. We believe that this will be a useful contrast to current texts; most of these instead take each system of the animal body in turn (such as ionic and osmotic issues, energy balance, temperature, circulating fluids, nerves, hormones) and then give examples of how they operate in different kinds of animals or in different kinds of habitat. We believe that our approach will prove useful for a much broader range of environmental biology courses, as well as more traditional environmental physiology courses.

Some readers will have little physiological background initially, and will want to understand physiological processes in detail, so in Section B we also offer an updated review of comparative physiology, looking at mechanisms and functions system by system, but incorporating new information on how adaptation is operating at a molecular level. We leave out most of the "examples" here, and concentrate on principles. This section provides most of the material needed for understanding comparative physiology, though it does not deal with processes largely unaffected by environment (the basic mechanisms of neural, hormonal, and muscle physiology, for example, which are broadly conserved in all animals and where excellent texts for specialist courses already exist).

Other readers may already know a good deal about physiological principles, and can use Section B merely as a reference point and reminder when considering particular environments. For them, Section C provides all the key material on environmental adaptation, relevant to a wide range of environmental biology issues. It covers far more than traditional animal physiology textbooks. Extensive cross-references to principles in Section B give the reader a chance to check their understanding of the core physiology.

Thus we hope this book presents material that will be useful to a wide range of readers, in varying kinds of environmental biology courses and at different levels. Above all we hope it will help provide an integrated understanding of environmental adaptation.

Acknowledgments

In this textbook we give further reading sections at the end of each chapter, mostly to modern review material. We have cited sources from which material in the illustrations has been drawn wherever a figure has been used in unmodified form, but we do not cite every source of information in the text. At this level, copious references break up the flow and markedly lengthen the text, making it harder to read. However, we realize that this carries two important risks: firstly for the reader, who might not be able to follow up on points of interest and find out more; and secondly for the scientists who carried out the work and might feel their contributions are going unrecognized. We hope that both of these risks are reduced by modern information technology, making it easier to access literature searches and track down the primary references. Thus we expect that the work we quote from, whether a classic paper or a very recent new insight, can usually be traced quickly and its authorship established. However, we take this opportunity at the outset to offer an enormous general acknowledgment and a debt of thanks to all those whose studies have contributed to this text.

More specifically, we offer personal and very grateful thanks to those who have read and commented on the manuscript for us. Simon Maddrell (University of Cambridge, UK) and Jim Childress (University of California, Santa Barbara, USA) took on the heroic task of reading the whole book, and gave us tremendously helpful inputs. Others read parts of the text as follows: Jon Harrison (Arizona State University, USA) (Chapters 3–8); David Scholnick (University of Colorado at Boulder, USA) (Chapters 3–8); John Spicer (University of Sheffield, UK) (Chapters 4–8); John McLachlan (University of St Andrews, UK) (Chapter 2); Lynn McIlroy (University of St Andrews, UK) (Chapters 4, 5, 6, and 8); Clare Maynard (University of St Andrews, UK) (Chapter 12); Alex Rowe (University of Edinburgh, UK) (Chapter 17); David Paterson (University of St Andrews, UK) (Chapter 12); and Ray Huey (University of Washington, USA) (Chapters 15 and 16).

We also thank those who have provided invaluable assistance in locating sources of both information and illustrations over the course of several years. Here Sandy Edwards has been an invaluable and endlessly patient colleague, and Christina Lamb has also offered enormous help with seeking out references, while Peter Slater, Gordon Cramb, Mike Fedak, Anne Magurran, Su Bryan, Jim Aiton, and Susie Whiten have helped out beyond the call of duty. Finally, colleagues at Blackwell Science, especially Ian Sherman and Katrina McCallum, have had a huge and greatly valued impact on the successful production of this book.

To all these, and to many other colleagues and students over the years who have encouraged us in thinking about and writing this book, we are very grateful.

Acknowledgments for second edition

Many readers of the first edition, including our own students, have offered comments that we have tried to address or incorporate here. PGW is especially grateful to those friends and colleagues who have read all or part of the new chapters or who have loaned their books and academic papers or made suggestions for new material that might be added. Special mention should be made of Peter Slater, Keith Sillar, and Susie Whiten for their help and commentaries, and we also thank those who reviewed Chapters 9 and 10 for us: Lovise Milligan (University of Western Ontario), Guillermo Paz-y-Miñoc (University of Nebraska), Michaela Hau (Princeton University), and Carol Lee (University of Wisconsin).

Once again, we thank Sandy Edwards who acted as a tireless gopher and aide-de-camp on many occasions, and our colleagues at Blackwell Publishing, especially Sarah Shannon, Rosie Hayden, and Jane Andrew who have once again made the production a relatively painless affair!

Abbreviations

ACE	angiotensin-converting enzyme
ACh	acetylcholine
ACTH	adrenocorticotropic hormone
ADH	antidiuretic hormone
ADP	adenosine diphosphate
AFGP	antifreeze glycoprotein
AFP	antifreeze protein
AKH	adipokinetic hormone
ALH	ascending loop of Henle (of vertebrate kidney)
AMP	adenosine monophosphate
ANP	atrial natriuretic peptide
AP	action potential
AQP	aquaporin
ATP	adenosine triphosphate
ATPase	adenosine triphosphatase
AVP	arginine vasopressin
AVT	arginine vasotocin
BAT	brown adipose tissue ('brown fat')
BMR	basal metabolic rate
BOD	biological oxygen demand
BP	boiling point
CaM	calmodulin
CAM	cell adhesion molecule
CaMK	calmodulin kinase
cAMP	cyclic adenosine monophosphate
CCK	cholecystokinin
CD	collecting duct
CEH	critical equilibrium humidity (above which water vapor is absorbed)
CEWL	cutaneous evaporative water loss
CFC	chloroflurocarbon
cGMP	cyclic guanosine monophosphate
CNS	central nervous system
CO	carbon monoxide
CoA	coenzyme A
CPG	central pattern generator
CPK	creatine phosphokinase
CREB	cAMP response element binding protein
CRH	corticotropin-releasing hormone
DAG	diacylglycerol
DCT	distal convoluted tubule (in vertebrate kidney)
DH	diuretic hormone
DLH	descending loop of Henle (in vertebrate kidney)
DMSP	dimethylsulfoniopropionate
2,3-DPG	2,3-diphosphoglycerate
DVC	discontinuous ventilation cycle
E	electric membrane potential
E_X	equilibrium potential due to a particular ion X
ECF	extracellular fluid
ECV	extracellular volume
EDNH	egg development neurosecretory hormone
EH	eclosion hormone
EHWN	extreme high water, neap (of tide)
EHWS	extreme high water, spring (of tide)
ELWN	extreme low water, neap (of tide)
ELWS	extreme low water, spring (of tide)
Epo	erythropoietin
EPP	end-plate potential
EPSP	excitatory postsynaptic potential
ER	endoplasmic reticulum
ETS	electron transfer system (in mitochondria)
EWL	evaporative water loss
FAD	flavin adenine dinucleotide
FAP	fixed action pattern
F-1,6-BP	fructose 1,6-bisphosphate
FP	freezing point
FPD	freezing point depression
FSH	follicle-stimulating hormone
GABA	γ-amino butyric acid
GDP	guanosine diphosphate
GH	growth hormone
GHIH	growth hormone-inhibiting hormone
GHRH	growth hormone-releasing hormone
GIP	gastric-inhibitory peptide
GIT	gastrointestinal tract
GnH	gonadotropic hormone
GnRH	gonadotropin-releasing hormone
GTP	guanosine triphosphate
H	hydrogen
Hb	hemoglobin
HBTS	high body temperature setpoint
Hc	hemocyanin
Hif-1	hypoxia-inducible factor 1
HLW	high-level waste

HO	heme oxygenase
HSP	heat shock protein
5-HT	5-hydroxytryptamine
HTH	hypertrehalosemic hormone
HVA	homeoviscous adaptation
HWN	high water, neap (of tide)
HWS	high water, spring (of tide)
ICF	intracellular fluid
Ig	immunoglobulin
IGF	insulin-like growth factor
IL	interleukin
ILW	intermediate-level waste
IMP	inosine monophosphate
INA	ice-nucleating agent
INP	ice-nucleating protein
IP_3	inositol triphosphate
IPPC	Intergovernmental Panel on Climate Change
IPSP	inhibitory postsynaptic potential
JH	juvenile hormone
K_m	Michaelis–Menten constant
LAP	L-aminopeptidase
LBTS	low body temperature setpoint
LC	light chain
LCT	lower critical temperature
LDH	lactate dehydrogenase
LGN	lateral geniculate nuclei
LH	luteinizing hormone
LHRH	luteinizing hormone-releasing hormone
LLW	low-level waste
LPH	lipotropin
LTD	long-term depression
LTP	long-term potentiation
LVP	lysine vasopressin
LWN	low water, neap (of tide)
LWS	low water, spring (of tide)
M_b	body mass
MAO	monoamine oxidase
MAP	mitogen-activated protein
MDH	malate dehydrogenase
MDOC	minimum depth of occurrence
MEPP	miniature end-plate potential
MF	methyl farnesoate
MHC	major histocompatibility complex
MHWN	median high water, neap (of tide)
MHWS	median high water, spring (of tide)
MIH	molt-inhibiting hormone
MLR	mesencephalic locomotor region
MLWT	median low water (of tide)
MMR	maximum metabolic rate
mPSP	miniature postsynaptic potential
MR	metabolic rate
mRNA	messenger RNA
MSH	melanocyte-stimulating hormone
MWL	median water level (of tide)
mya	million years ago
nAChR	nicotinic acetylcholine receptor
NAD	nicotinamide adenine dinucleotide
NADH	reduced nicotinamide adenine dinucleotide
NIS	non-indigenous species
NK	natural killer (cell)
NMJ	neuromuscular junction
NO	nitric oxide
NOS	nitric oxide synthase
NST	nonshivering thermogenesis
O	oxygen
OML	oxygen minimum layer
OMP	ovary maturing parsin
ORC	olfactory receptor cell
P_b	barometric pressure
P_c	critical partial pressure
P_i	inorganic phosphate
P_{Na}	permeability to sodium
P_{O_2}	oxygen partial pressure
P_{osm}	osmotic permeability of a membrane
P_w	permeability to water
P_X	permeability for a particular ion *X*
PAH	*para*-aminohippuric acid
PAL	present atmospheric level
PAr	phosphoarginine (arginine phosphate)
PCB	polychlorinated biphenyl
PCr	phosphocreatine (creatine phosphate)
PCT	proximal convoluted tubule (in vertebrate kidney)
PDE	phosphodiesterase
PEP	phosphoenolpyruvate
PFK	phosphofructokinase
PIP2	phosphatidylinositol biphosphate
PK	protein kinase
PKA	protein kinase A
PKC	protein kinase C
ppb	parts per billion
ppm	parts per million
ppt	parts per thousand
PRL	prolactin
PSD	postsynaptic density
PSMO	polysubstrate mono-oxygenase
PSP	postsynaptic potential
PTH	parathyroid hormone
PTTH	prothoracicotropic hormone
Q_{10}	temperature coefficient
RBC	red blood cell
REWL	respiratory evaporative water loss
RH	relative humidity
RI	refractive index
RIA	radioimmunoassay
RMR	resting metabolic rate
RP	resting potential
RPD	redox potential discontinuity
RQ	respiratory quotient
rRNA	ribosomal RNA
RVD	regulatory volume decrease
SA	sinoatrial
SCP	supercooling point
SDA	specific dynamic action (heating effect during digestion)

SDE specific dynamic effect
SIN sensorimotor integration neuron
SMR standard metabolic rate
SR sarcoplasmic reticulum
STT shivering threshold temperature
T3 triiodothyronine
T4 thyroxine
T_a ambient temperature
T_{abd} abdominal temperature
T_b body temperature
T_s surface temperature of body
T_{pref} preferred body temperature
T_{th} thoracic temperature
TBT tributyl tin
Tc cytotoxic T (cell)
TEA tetraethylammonium (potassium channel blocker)
Th T-helper (cell)
THP thermal hysteresis protein
TMAO trimethylamine oxide
TNZ thermoneutral zone
T6P trehalose 6-phosphate
TRH thyroid-releasing hormone
tRNA transfer RNA
TSD temperature-dependent sex determination
TSH thyroid-stimulating hormone
TTX tetrodotoxin (sodium-channel blocker)
UCP uncoupling protein
UCT upper critical temperature
UT urotensin
$\dot{V}O_2$ rate of oxygen consumption
VIP vasoactive intestinal peptide
VMI vasomotor index
VNO vomeronasal organ
VP vapor pressure
WLR water loss rate

PART 1 Basic Principles

1 The Nature and Levels of Adaptation

1.1 Introduction: comparative, environmental, and evolutionary physiology

This book is about how animals cope with the problems posed, and exploit the opportunities offered, by their particular environments. Traditionally the mechanisms for coping with the environment have been treated as issues of **comparative physiology**, which is concerned with investigating both general principles of organismal function (the similarities that exist between all organisms) and the exceptions to the general rules. Comparative physiologists may be interested to look for different ways in which animals solve particular challenges of living in different habitats. Do these different solutions also depend on factors such as animal size or design, or the biological materials used? Much of this traditional physiology relies on laboratory study of either classic laboratory animals (mainly mammals, and certainly with vertebrates very dominant and only a sporadic use of other taxa) or of extremely specialized animals pushed to their extreme performance. This can tell us a great deal about mechanisms, since organisms living in extreme environments illustrate the range of evolutionary possibilities amongst living animals, so that species with an extreme development of a particular physiological property are therefore often useful as model systems. But this study of extremes should not make us lose sight of the "norm" of performance or of more "generalist" animals. We also need to be wary of the tendency of laboratory-maintained animals to become rather unrepresentative, as there are famous cases where within a few months or a few generations after "domestication" such animals have dramatically changed their physiological performance.

For these reasons it is very important to study animals in the context of their own habitat and their real needs, so that "comparative physiology" has tended to be replaced by **environmental physiology** or **ecophysiology** or **physiological ecology**, which in a real sense add traditional natural history to the study of comparative physiology. The primary aim of ecological or environmental physiologists is to understand how animals function in and respond to their natural environments, at all stages of their life cycles. Indeed it quickly becomes apparent that many animals, for most of their lifetime, do not need extreme physiological adaptation and rely instead on behavioral strategies to avoid the worst of their difficulties.

However, ecophysiology has in turn become subject to an increase in conceptual rigor, partly as a result of important critiques of the story-telling nature of the "adaptationist program" (the assumption that everything has an adaptive function). We need to move beyond anecdotal science, where after collecting lots of examples of particular features their functions are to be gleaned merely by their apparent correlations with environmental features. Understanding the processes by which particular ecophysiological features arose, and the values of physiological parameters as outcomes of natural selection, may be termed **evolutionary physiology**. Evolutionary physiology is a discipline still somewhat in its infancy, but learning rapidly from other areas of evolutionary biology and from the analytical techniques of population biology and (especially) molecular biology. It involves a more explicit attempt to integrate both short-term and long-term genetic perspectives into physiological ecology. Within species, understanding the genetic basis of physiological traits, and the magnitude and causes of physiological variation, may reveal how they can be shaped in the relatively short term by natural selection. Evolutionary physiology also examines the evolution of traits over longer time periods, across species or higher taxa. Here the crucial point is that the traits we see represent an interaction between ancestral traits and selection, set in the context of ancestral environments. We can only understand this by following the evolution of these traits through a family tree or phylogeny. A phylogenetic analysis allows the evolutionary physiologist to address two crucial issues: firstly, how does the physiology of an ancestral species affect what is possible in its descendants, and secondly, how rapidly can physiological traits evolve? The recent explosion in our knowledge of the relationships between organisms, much of it based on DNA or RNA sequence analysis, provides a wealth of opportunity in this area. Thus, not only can we use explicitly evolutionary analyses to help us understand patterns of mechanistic physiology, but we can now also track the evolution of molecular components of physiological adaptation as a short cut for many years of complex laboratory analysis of comparative species differences. Evolutionary physiology may develop all the more quickly set in a strong environmental context, taking both past and present environments into account; for unless there has been substantial climatic or geological upheaval, or migration, animals tend to inherit their environment as well as their genes from their immediate ancestors. Furthermore, organisms from more extreme environments are particularly likely to show clear examples of evolutionary adaptation as a result of intense selective pressures.

This book is written in the conviction, increasingly common in all kinds of biological literature, that the trend needs to be taken yet

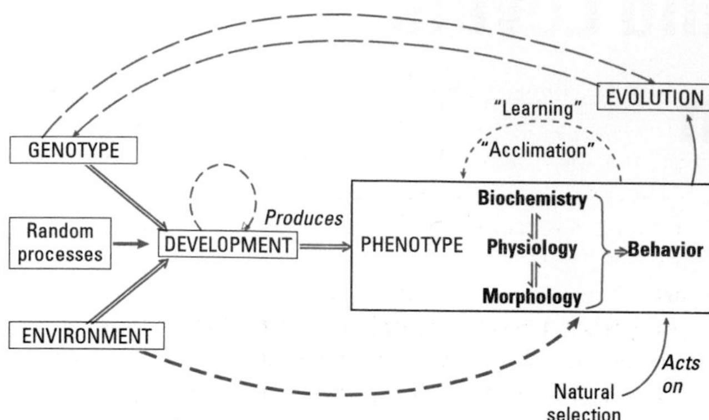

Fig. 1.1 The genotype and environment interact though developmental processes to give a particular phenotype, a suite of biochemical, physiological, and morphological traits.The phenotype also includes behavior, which is limited by all the other phenotypic characters. Selection acts at the whole organism level, and therefore mainly on behavior. Biochemical and physiological traits are normally only subject to selection if they have detectable and stable effects at the level of behavioral performance (e.g. escape speed, reproductive behaviors, food-gathering efficiency, etc.) that in turn affect reproductive fitness.

further; that physiology cannot be isolated from behavioral strategies and from ecological and life-history strategies, while also taking account of evolutionary and molecular studies. Thus, environmental adaptation has to be an interdisciplinary and fully integrated study as biological sciences come of age. A schematic view of the interrelationships of genotype, environment, and phenotype is shown in Fig. 1.1, stressing this wider viewpoint; note especially the intervening effect of developmental processes, and the key role of behavior, both as the outcome of phenotypic characters and as the level at which natural selection often acts most strongly. Genotypic variation (long-term evolutionary change) can only be subject to selection if it affects reproductive success, mediated by changes in performance and behavior, though the more immediate changes may concern biochemical, physiological, or morphological features, which are what we normally see as "adaptations". All these issues are discussed further in this chapter.

1.2 The meaning of "environment"

The two basic concepts in natural selection are fitness and environment. Fitness is strongly linked to adaptedness and therefore to adaptation, and all these concepts are subject to intensive analysis by ecologists and evolutionary biologists (see section 1.3). However, the concept of the environment is usually largely ignored, perhaps because it is seen as unproblematic. Nevertheless it is worth considering just what we do mean by an animal's environment.

At an obvious level, the environment means the kind of habitat in which an animal lives—the deep sea, a tropical forest, a hot desert, or whatever. In this sense an environment is equivalent to the concept of a **biome**, with grossly similar kinds of living space having similar physical characteristics all lumped together. Knowing whether an animal lives in water, in air, or amphibiously, or in cold, seasonal, or warm latitudes, certainly tells us a lot about the problems it will encounter and the kinds of design and strategy it is likely to show.

Environments are obviously rather more complex than this though, and each species of animal has a more precisely defined environment within a biome, perhaps in the deep-sea benthos, or predominantly arboreal in a forest, or in the litter layer of a freshwater pond. By considering the environment at this level, with some elements of biotic interaction coming in as well as mere physical factors, we gain rather more knowledge about the animal's requirements. Indeed for the parasitic animals considered in Chapter 17, properties of the host organism and of other parasites may be far more important than abiotic factors.

At a third level, each individual animal has its own environment: the totality of all the external factors it experiences, both biotic and physical. This environment is commonly modified by its own behavioral choices and indeed by its very presence. At this level we are really considering microenvironments, or microhabitats, or (on land) microclimates. This is also the level that really matters in terms of environmental and evolutionary physiology. Just because an animal lives in a "type" of environment as perceived and classified by humans, it does not necessarily experience that environment in the way that we see and experience it. Each animal chooses where to spend its time, to forage or to rest, or to seek mates, from a range of possible options. These options change on a very fine spatial scale and a very rapid temporal scale for small animals, but much more coarsely for large animals (for them the environment is sometimes described as "coarse-grained"). In many cases an animal will choose the least stressful microhabitat, whether in and amongst the ameliorating effects of vegetation, or within a burrow or nest, or merely within the boundary layer of relatively still fluid above a substratum. The presence of the animal will modify local conditions, by adding excretory products (including CO_2) or depleting oxygen, or modifying humidity and temperature; an environment containing an animal is no longer the same as the environment before the animal moved in. Thus the environment we want to know about is that of the organism itself and its immediate surroundings, measured on a temporal and spatial scale appropriate to that animal. An animal then becomes part of an interactive system that includes its own internal physicochemical state and the physical and chemical conditions of its immediate surroundings.

At whatever level we consider them, environments can be helpfully categorized in terms of three important interacting parameters: the basic stress intensity, the magnitude and timescale of fluctuations, and the energy or resource availability.

1.2.1 Environmental stress

Environments are enormously variable in relation to the stress that they impose on their inhabitants, and this stress may be both **abiotic** (physical and chemical factors) and **biotic** (direct and indirect effects of other organisms, including competition and habitat modification). Since life on Earth evolved in seas that were thermally and osmotically relatively stable, and all cellular machinery was fundamentally selected from its inception to work best in stable, rather cool marine conditions, it is often useful to see the abiotic stress as being dependent on how much conditions have diverged from those starting points. Thus life in cool sea water is relatively "easy", life in a seasonal pond is somewhat tricky, and life in hot desert conditions is spectacularly difficult. Other extreme environments would include polar lands, mountain peaks, hypersaline lakes, hot springs, and deep-sea thermal vents. But biotic stress may sometimes work in the opposite direction, in that "easy" habitats may also have large and diverse populations of other organisms and so impose more competition and predation pressure. Either of these kinds of stress may put an organism at a disadvantage in having to expend more of its own energy to survive, whether in physiological regulation, or avoidance tactics, or competitive or defensive activities.

1.2.2 Magnitude of fluctuations

Environments may be very stable on all timescales relevant to living organisms, the classic case perhaps being deep seas. Or they may vary on an evolutionary and geological timescale of tens or hundreds or thousands of years as land masses move, sea levels rise and fall, materials erode and deposit elsewhere, and rivers change their courses. There may also be changes with a regular annual, lunar, or daily cyclicity. Finally, there are changes on a much shorter timescale of hours or minutes or seconds, as the weather changes. Note again that the magnitude of the change is a relative phenomenon, particularly linked to the size of the perceiver. Short-term changes are especially important in relation to very local microenvironments and therefore to very small animals; the difference between the environment above a leaf and the environment below it may be profound, and both may change within seconds in relation to varying solar radiation (insolation), air movements, and rainfall. Changeable environments put a high selective premium on versatility or tolerance in animals, rather than on precise adaptations to particular conditions. This may be particularly true where man has intervened in the natural ecosystem to put new stresses on animals, whether from habitat destruction, climate modification, or the introduction of many kinds of toxic chemicals.

1.2.3 Energy or resource availability

Energy is rarely freely available to animals (as it may be to plants), but it is certainly more easily obtained in some habitats than in others. Traditionally it has been thought that where energy is severely restricted, as it may be in deserts and in polar regions, the results are simple communities with short food chains, largely made up of animals with low metabolic demands.

By contrast, in rather stable environments with high primary productivity and rapid energy flow (exemplified by tropical rain forests or coral reefs), species may develop more specific habitat preferences, leading to specialization and producing diverse and complex animal communities. The corollary of all this is that stable high-energy communities would favor adaptive radiation, whereas rates of evolutionary change may be rather low in low-energy communities.

Varying resource availability between environments is also important in leading to **resource polymorphism** (or trophic polymorphism), where differences in behavior, life-history strategy, or morphology occur within a single species. Striking polymorphisms may occur where interspecific competition is relaxed, and/or there are unfilled niches that can be rapidly exploited, and we will meet many examples of this in Part 3.

1.2.4 Selection and the environment

Interactions between all three of these components of an environment tend to determine the kinds and diversity of animals that occur, and the type of selection that operates. Traditionally two main types of selection are recognized, representing either end of a continuum: ***r*-selection**, which occurs in unpredictable environments, and ***K*-selection**, which occurs in more predictable environments (Table 1.1). The prefix *r* refers to the rate of population increase, which tends to be maximized in *r*-selected species, while *K* represents the carrying capacity of the environment. Typically, *r*-selected animals are the small, rapidly reproducing, early maturing, and short-lived species, producing large numbers of relatively low-investment progeny, many of which will not survive, often reproducing just once and then dying (semelparity) and with potentially wide swings in population size; while *K*-selected animals are large, slowly reproducing, and long-lived, producing just a few young and investing heavily in each, often reproducing repeatedly (iteroparity), and with relatively stable population sizes. The *r*-selected animals should live in disturbed habitats as early successional species, have large geographic ranges and relatively nonspecialist interactions with other species; they are colonists and opportunists, in environments of high stress and high levels of fluctuation. *K*-selected animals should be more common in climax communities, having complex coevolved relationships with other organisms; they do best with low abiotic environmental stress (though often high biotic stress) and low levels of fluctuation. These designations therefore relate to features both of the environment and of the organisms inhabiting those environments, especially their life-history strategies.

This dichotomy is useful as a way of thinking about different kinds of lifestyle and the kinds of environments and features that might be expected to go together. However, it has many imperfections, and many animals do not fit even into a continuum between the two extremes, having instead some very *r*-selected features and some very *K*-selected features. Some authors prefer to use a three-way model, where *K*-selection is the norm in predictably favorable habitats but is replaced by ***A*-selection** (adversity selection) in more extreme environments where conditions are predictably unfavorable. *A*-selection occurs in habitats that are of high environmental stress but with a low magnitude of fluctuation, and with low energy availability. Animals here have high stress resistance, low fecundity, late maturity, and long lifespans, with very low levels of biotic interaction. *A*-selection might be expected in many kinds of extreme environment: deserts, polar regions, montane habitats, caves, anoxic muds, etc. This may also link to the argument about rates of speciation in different environments. In practice maximum adaptive radiation tends to be exhibited, not in thoroughly *K*-selected environments, but in areas of abiotic stability where energy is moderately available but not unrestricted (e.g. we find tens or hundreds of closely related species in African lakes, in some areas of deep sea, and in some patches of tropical forest). Here organisms do not become too specialist, but they can speciate rather rapidly by switching between a wide range of possible resources. The opposite environmental combination of abiotic instability with high resources (perhaps exemplified by a temperate estuary) may produce an alternation between periods of relatively low diversity when species are coping with fluctuating adverse conditions, and periods

Table 1.1 Types of selection on animals.

	r-selection	*K*-selection	*A*-selection
Environment			
Stability	Low	High	High
Abiotic stress	High	Low	High
Energy	Low	High	Low
Individuals			
Body size	Small	Large	Small *or* large
Lifespan	Short	Long	Long
Maturity	Early	Late	Late
Reproduction			
Pattern	Semelparous	Iteroparous	Either
Generation time	Short	Long	Either
Fecundity	High	Low	Low
Offspring	Many, small	Few, large	Either
Parental care	Absent	Common	Possible
Populations			
Density	Fluctuating	High	Low, or fluctuating
Stability	Fluctuating	Steady	Fluctuating
Range	High	Low	Either
Competition	Low	High	Low
Biotic interactions	Few, simple	Many, complex	Few, simple
Overview	Small	Large	Very varied
	Rapid reproductive output	Slow reproductive output	Usually slow
	Colonists	Climax communities	Simple climax
	Generalists	Specialists	Specialists

of abrupt change when abiotic conditions alter too drastically, potentially leading to a phase of rapid evolution and the generation of whole new groups of related species (cladogenesis).

1.3 The meaning of "adaptation"

Adaptation is a central concept in biology and one that attracts enormous controversy. It is often used in several different senses, to describe both a pattern and a process; and more often than not it is used rather loosely.

1 Firstly, adaptation is often used as a term for the **characters** or **traits** observed in animals that are the result of selection; for example, the presence of hemoglobin might be said to be an adaptation to allow a greater oxygen carriage in blood.

2 Alternatively, and most "correctly", adaptation might be defined as a **process**; the means by which natural selection adjusts the frequency of genes that code for traits affecting fitness (most simply, the number of offspring surviving in succeeding generations). For example, increasing hemoglobin concentrations within a taxon might be seen as an adaptation to potentially low oxygen (hypoxic) environments. Evolutionary adaptation then becomes almost synonymous with natural selection itself—necessary attributes for both include variability, repeatability, heritability, and differential survivorship of offspring. Adaptation in this sense is a process that normally occurs extremely slowly, over hundreds or thousands of generations, and is not usually reversible. However, in extreme environments or (as we shall see in Part 3) where selective pressures from human interference are strong, it can sometimes occur very quickly.

3 "Adaptation" is also used to describe short-term **compensatory changes** in response to environmental disturbance. This kind of change is the outcome of **phenotypic plasticity**, where pre-existing traits are differentially expressed as appropriate to the local conditions. Here the terms **acclimation** or **acclimatization** are technically more correct (see section 1.5); evolutionary biologists would not use the term adaptation at all in this context.

Great caution must be exercised in using the terms "adaptation" or "adaptive", and in the simplistic interpretation of individual traits at the molecular, cellular, tissue, or organ levels, in relation to particular environmental factors. A trait should only be considered an adaptation if there is some evidence that it has evolved (has been changed through its evolutionary history) in ways that make it more effective at its task, and thus enhance **fitness**. In other words, a trait should only be deemed to be an adaptation if it is a consequence of selection for the task it performs; an incidental ability to perform the task is not enough, and nor is the mere existence of a good general fit between organism and environment. Evidence that a characteristic is an adaptation is quite hard to come by, and may be of three kinds:

1 Correlation between a character and an environment or use. This is the commonest approach, and where it involves interspecific comparisons of extant species it forms the core of comparative physiology. However, it has inherent dangers, which are discussed in more detail in the next section.

2 Comparisons of individual differences within a species. Geographic patterns in the frequencies of different gene variants (alleles) in natural populations can be mapped over environmental gradients.

3 Observation of the effects of altering a character. An organ can be experimentally altered, or a behavior prevented, and the effects on the efficiency of some particular function in some particular environment observed. In modern physiology we can improve matters by either knocking out or overexpressing a specific gene (rather than a phenotypic character) and observing the effects; indeed, if we know enough about the evolution of a given gene we may be able to

piece together the changing physiology that goes with it. This has been possible with a few very well-studied systems such as the lactate dehydrogenase (LDH) gene alleles that are involved in key anerobic adaptations and that show clines with latitude and temperature (see Chapter 8). But the danger remains of interfering with other processes and thereby breaking the link between cause and effect.

These approaches generally involve **correlative evidence**, which is never really conclusive. Clear evidence for adaptation is not always possible, and the literature is still dominated by inference; nevertheless this is reasonably well founded by sheer weight of evidence under criterion 1 above. Most of the material in Part 3 of this book presents adaptations supported by this kind of inference, which may be logically imperfect but is nevertheless very probably correct. In fact, unlike many of the characters that evolutionary biologists deal with, many physiological traits have direct and clearly quantifiable effects in terms of the survival of organisms, and many would argue that it is not necessary to show "adaptation" (as enhanced survivorship in the field) to make clear inferences about physiological adaptedness.

However, it is important to realize at the outset that not all differences in physiology between species are adaptive. Remember that traits do not evolve "for a reason"; they evolve entirely by chance and are selected after they arise if they enhance the success (fitness) of their possessor. They may persist merely because they do no harm or are linked with some other beneficial trait. They may not have been selected because of the advantage they now seem to endow, but perhaps for some other and now quite unapparent reason; these kinds of traits are sometimes called "exaptations" rather than adaptations (e.g. Darwin noted that the sutures between bones in mammalian skulls may now be adaptive for allowing the birth of relatively large-brained offspring in some species, but they were selected for long ago in evolutionary history for quite different reasons). In addition, many traits were fixed because of natural selection in an ancestral species but are now present for reasons that have little to do with the current selective regime on the species being considered; this phenomenon is termed "phylogenetic inertia", and may be attributable to the fundamental conservatism of development, of physiological mechanisms, and of underlying genes.

Remember also that natural selection does not necessarily provide maximal or even optimal solutions: **optimality** is a relative concept, and a particular design need only be sufficient to do better than the existing alternatives. Therefore it is impossible to say that any one design is "optimal", or to predict whether a better optimum might exist. The use of optimality in adaptation debates has therefore led to considerable confusion and criticism. It is likely that there will be some sort of match between biological structures and functional requirements, but how good this match should be is not clear. A great many features of animal design, including physiological mechanisms and pathways, have to meet more than one, and sometimes several, conflicting functional demands. They are therefore very likely to represent a working compromise between these competing demands. Adequacy or sufficiency in one function, without unduly disrupting others, may well be the most common evolutionary outcome.

Even where design features appear to serve only one function we should not expect perfect or optimal form or performance. Natural selection filters phenotypic success through genotypic transmission only in relation to differential survivorship, relative to other possible phenotypes and genotypes. Optimal construction and performance are therefore not the inevitable outcomes of selection, and the match of optimality to actual design may be largely a reflection of the strength of the selective force. Genetic drift may also tend to reduce the rate of improvement in characters, especially in small populations with low immigration. The potential for optimal adaptation may also be limited by developmental constraints, and by genetic constraints where traits are correlated or linked.

This is why it is so helpful to consider the whole organism and its integrated functioning within its habitat, rather than individual characters or particular physiological systems in isolation. The approach adopted in this book, looking at all aspects of coping with a particular type of environment, may be more useful as a background to modern evolutionary physiology than the "systems" approach more commonly presented, that deals with each system in turn and only adds in the other bodily systems and the environment as afterthoughts.

1.4 Comparative methods to detect adaptation

Comparisons between and amongst sets of species are the most common ways of looking for adaptations; indeed, it is fair to say that adaptations are essentially "comparative" phenomena and can only be measured comparatively between or within species. Many studies choose two or more species differing in behavior or ecology and compare them, to determine whether they show phenotypic differences that could be interpreted as adaptations to their different selective regimes. In a very loose sense, this is called the "comparative method", and it has served biology well in giving "evolutionary" explanations of why things are the way they are—in the biological sciences, experiments repeating actual evolutionary events are usually impossible!

However, there is an inherent problem in looking at patterns across species. As soon as two species have diverged from a common ancestor, differences will appear between them for many traits. Some of these changes will be due to chance shifts in gene frequencies (genetic drift). This is particularly true if one or both species experience low population size (a "genetic bottleneck"), in which drift can be especially significant. Differences in a given trait therefore do not represent proof of selective modification in either of the species. So what do we need if we want to link differences in environment with differences in physiology? Imagine one species from a warm environment and one from a cold environment, differing in their physiologies. The difference may have nothing to do with environment but could just be due to genetic drift. However, if selection has modified one of the species, how can we tell which has acquired a novel physiology, and thus determine the direction of evolutionary change? We can only work this out if we know the physiology of the common ancestor, usually inferred using a technique called **outgroup comparison** (Box 1.1). Then we can say which species has changed, and by how much, giving us a correlation between temperature and physiology.

While one pair of species may show a change in one direction with temperature, another (phylogenetically distinct) pair may show the opposite change; thus we still do not have a general rule. In

Box 1.1 Statistics and biological comparisons

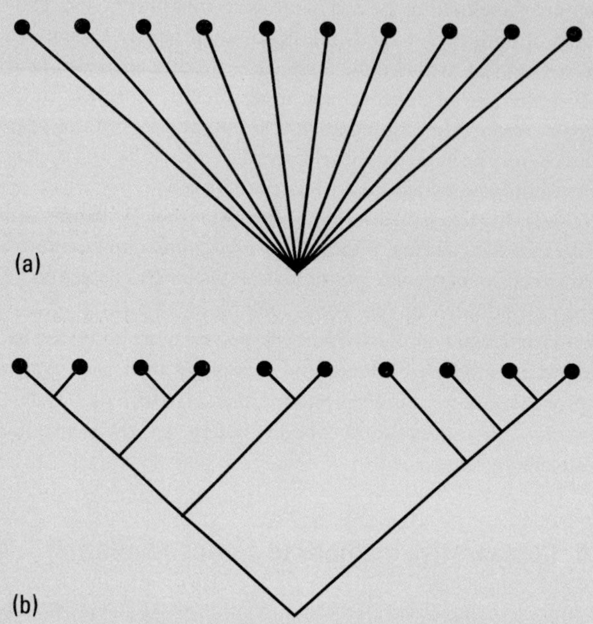

(a)

(b)

Typically, statistical methods assume that data points are independent, but this would be the case only if we studied 10 "ideal" species that were genuinely unrelated or were absolutely equally related, as shown in the top figure; here, instantaneous speciation resulted in 10 independent lineages leading to 10 living species.

Thus, if we were to test for a correlation between the mean values for these species for two phenotypes (e.g. relative heart mass and maximal oxygen consumption), or perhaps one phenotype and an environmental factor (e.g. blood hemoglobin concentration and altitude), we could claim the nominal $n-2=8$ degrees of freedom for hypothesis testing.

If, instead, the 10 species were actually related as shown in the lower figure, with a hierarchical evolutionary relationship following descent with modification, we would have something fewer than 8 degrees of freedom available for hypothesis testing. The trait of interest might only have arisen or changed once or twice, and then been inherited by several descendant species. These could not be treated as separate examples of adaptive acquisition of that trait. Various analytical methods now exist that explicitly use the phylogenetic topology and branch lengths to allow valid hypothesis testing.

fact a single two-species comparison never gives sufficient evidence for adaptation (statistical tests of correlation always require a minimum of three data points); but unfortunately such comparisons are remarkably common as a way of proceeding in comparative physiology. However, if many distinct evolutionary radiations show the same change in physiology linked to the same difference in environment then there is evidence for a general correlation between the two. The more radiations there are that show the pattern, the stronger the case. Crucially, because each radiation starts with a different common ancestor, each represents an independent test of whether trait and environment are related. Without a phylogenetic perspective, we make the error of assuming that all species represent statistically independent data points in our analysis, even though we know that the more closely related ones are more likely to share common features and common past environments (see example in Box 1.1).

Taking a phylogenetic perspective also helps us to avoid two common errors made in comparative physiology. First, knowledge of phylogeny may reveal that we have less support for a relationship than we thought (Fig. 1.2). Second, it may show that there are significant associations present, where a naive analysis would miss them (Fig. 1.3).

So, in addition to physiological data, a **phylogenetic comparative analysis** requires an accurate phylogeny to work with. For an increasing diversity of animal groups, reasonably reliable phylogenetic trees have been generated, increasingly based on DNA sequence data. Using outgroup comparison we can than infer the phenotypes of ancestors for each radiation—the branch points or "nodes" on the phylogenetic tree. Once we have worked out the characters at the "nodes", then the magnitude and direction of changes that have occurred along each branch segment of the phylogenetic tree can be calculated. If we also have independent information on divergence times, available traditionally from the fossil record or more recently by applying the idea of a molecular clock to the sequence analyses, then the rates of evolution of characters can also be studied. This approach also permits statistical testing for correlations in the changes of two or more characters, which may allow us to see how the sequence of changes that occurs during the evolution of a complex character may predispose some other trait to change in a particular direction. If associations between particular characters and particular environmental factors are revealed, then justified and sensible inferences about adaptation are possible.

It must be noted, however, that this is an idealized program. For the majority of animals we do not yet have broadly agreed phylogenetic trees, nor is there a complete consensus on how these can be achieved. An incorrect phylogeny can lend statistical support to bogus relationships or can obscure genuine ones; thus the comparative method can also appear to support physiological correlations that may yet turn out to be wrong. Phylogenies based on morphological data are particularly prone to errors resulting from convergent evolution, and this may be rife in animals that are adapting to similar environmental constraints, where physiological convergence is especially likely and may be especially revealing from a functional viewpoint. Thus comparative evolutionary physiology has some substantial hurdles to overcome; and the core of our knowledge of physiological function must not be underrated just because it does not measure up to "phylogenetic correctness". Nevertheless, understanding evolutionary ecophysiology at this kind of level is certainly an outcome much to be desired, and the limitations of more traditional and "anecdotal" species comparisons should be borne in mind throughout the reading of this book.

1.5 Physiological response on different scales

1.5.1 Different timescales

We considered the nature of adaptation in section 1.3 above, but we also need here to distinguish it from other kinds of physiological

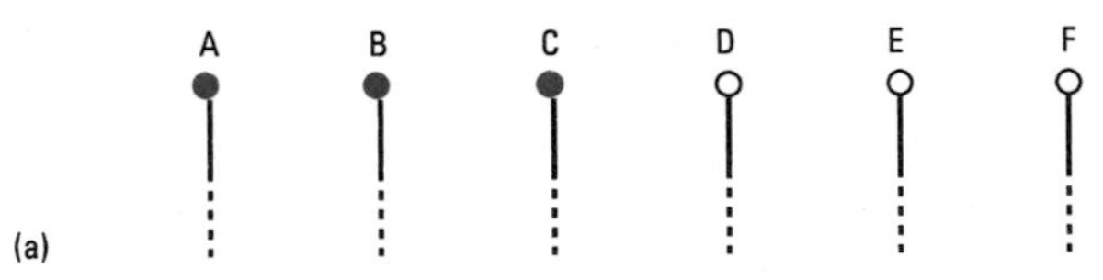

Tree unknown. Inference: apparent correlation of Hb presence and inshore life (apparently six independent data points). *But* these species may be linked by evolutionary history and therefore not independent. The inference is untrustworthy

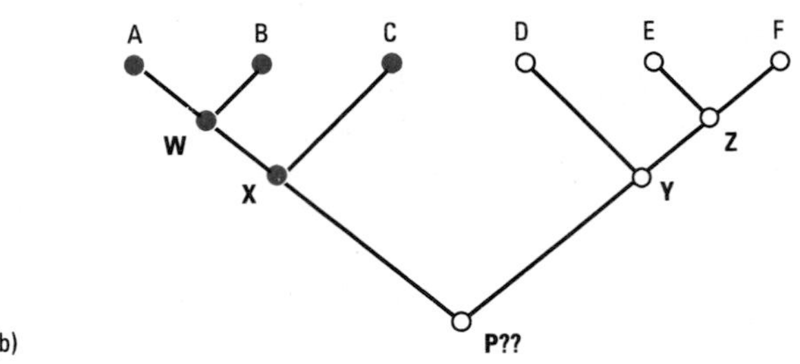

Tree known. ABC form one group, DEF another. Ancestral species WXYZ are included, but their character states cannot be measured directly, though they may be reconstructed (e.g. using parsimony which assumes that changes in state are rare and minimizes the number of transitions required to give the observed pattern). Here W and X are reconstructed as deep sea/Hb$^-$, and Y and Z as inshore/Hb$^+$. This means that the six data points only give *one* real comparison, and no real evidence for a link between Hb and environment. Also since ancestor P is uncertain we do not know which state is ancestral and which is derived

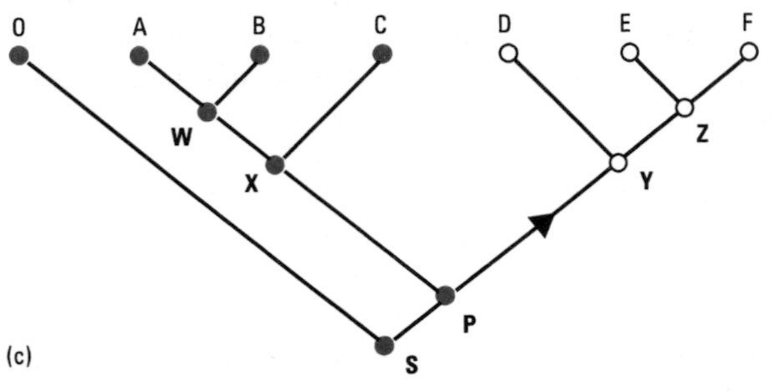

Tree known *and* condition of an outgroup (O) is known. Relationships and ancestral states as in (b) but now the inference (by "outgroup comparison") is that deep sea/Hb$^-$ represents the ancestral condition (as in ABC and O). However we still have only one comparison (ABC against DEF) and no real evidence for a link between Hb and environment

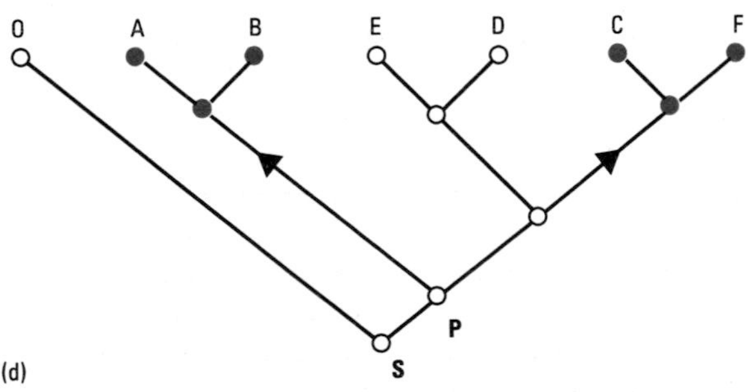

Tree known and outgroup known, but now suggesting that ancestors S and P were inshore/Hb$^+$. Now there are *two* independent transitions from this ancestral state to the derived deep sea/Hb$^-$ state. This strengthens the inference that Hb and environment are causally linked. Including more species in the phylogeny, with more independent transitions, may eventually give statistical support for such a link

Fig. 1.2 The effects of phylogenetic trees on possible interpretations of a single character distribution (presence or absence of hemoglobin, Hb), showing the utility of outgroup comparisons. Green circles show deep-sea taxa with Hb absent; open circles show inshore taxa with Hb present. Triangles show the sites of changes of state.

response: acclimation, acclimatization, and acute adjustments such as changing heart rate. The process of adaptation is usually a long and slow one occurring over generations, and is rarely reversible. In contrast, **acclimatization** is a more rapid phenomenon whereby a physiological or biochemical change occurs within the life of an individual animal, resulting from exposure to new conditions in the animal's environment. Thus migration up a mountain may lead to acclimatization to low oxygen and low pressure; movement southwards out of Arctic areas may force acclimatization to warmer temperatures; and year-long survival in one place may require acclimatization to alternating summer and winter conditions. The term **acclimation** is normally used for similar processes occurring in the laboratory, in response to experimentally imposed changes in conditions. Both acclimatization and acclimation may be reversible. Thus a polar bear is presumed to be adapted to polar temperatures; a human polar explorer may become (at least partly) acclimatized to them; and a laboratory mouse may be forced to acclimate to them. The human and the mouse are both likely to revert gradually to "normal" when conditions change again.

In practice the terms acclimatization and acclimation are often used interchangeably in modern literature, especially since the term "acclimatory" is a useful shorthand for a short-term nongenotypic

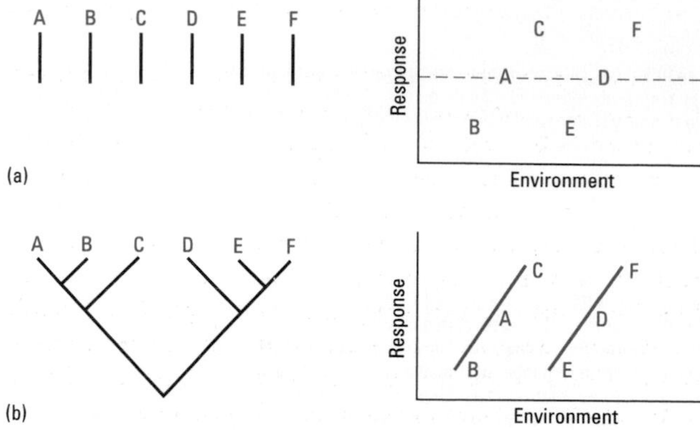

Fig. 1.3 The importance of understanding phylogenetic relationships in making inferences about adaptation and environmental effects on continuous variables. (a) Six species are considered as if they are independent entities; there appears to be no consistent effect of environment. (b) The phylogenetic tree is known showing that ABC form one radiation and DEF another. Now we can see that *within each radiation* there is a trend for the physiological response to increase in relation to position on the environmental axis. Given more radiations showing similar patterns we could reasonably analyze the effect of the environmental parameter in bringing about a specific adaptive response. (From Huey 1987, courtesy of Cambridge University Press.)

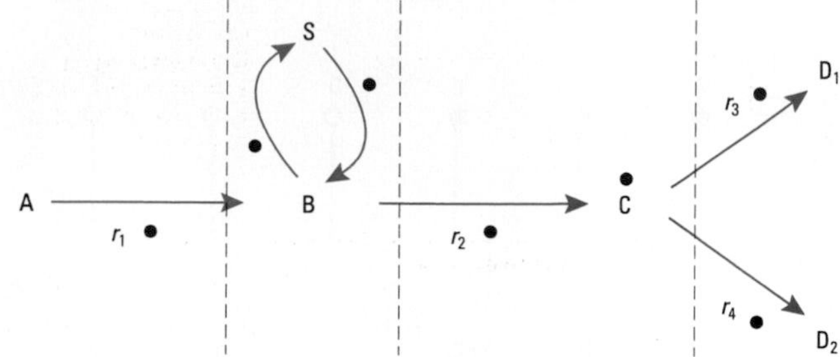

Fig. 1.4 Developmental plasticity and its interaction with the environment. The normal developmental program of a hypothetical animal runs irreversibly through immature stages A→B→C, and then may switch to alternative adult phenotypes D_1 or D_2. The environment (•) acts at several stages. Firstly, it controls the rates of transition between stages (r_1–r_4). Secondly, it determines whether an optional developmental arrest (S) occurs at stage B. Thirdly, it determines which of the adult phenotypes is produced. (Adapted from Smith-Gill 1983.)

response that has no equivalent derived from the word "acclimatization". Thus we will often speak of acclimatory responses in animals within their natural environments.

Short-term changes

Very short-term changes in physiological state, such as an increase in heart rate or ventilation or urine flow, are usually acute responses following some behavioral effect such as exercise or a bout of feeding. On a slightly longer timescale, it follows from the discussion above that responses to environmental change occurring within hours or days or weeks are normally acclimatization rather than adaptation. But changes occurring regularly and repeatably, on a seasonal, monthly, or daily basis, are also often acclimatizations rather than adaptations. Remember, though, that these changes in phenotype may be underpinned by genotypic change, in the sense of differential expression of particular genes, even if they are not brought about by any irreversible or heritable change in the genotype itself.

Developmental effects

Embryonic, larval, and juvenile phases in a life cycle may occupy very different environments from adults, and may have very different environmental responses. Such changes of phenotype during development and ontogeny operate on a slightly longer timescale and are more permanent than the acclimatory responses dealt with above. They are of course ubiquitous, and are commonly linked to changes in the environment of the embryo and/or the juvenile that affect its form and function. For example, it is possible to raise littermates of desert rodents either without drinking water or with free access to it; the drought-raised adults have increased relative medullary thickness (the ratio of the cortex to the medulla within the kidney, where it is the medulla that generates the urine-concentrating mechanism) and significantly higher urine concentrations. This kind of phenomenon is usually termed **phenotypic** or **developmental plasticity,** and again operates through differential gene expression.

The environment of an organism interacts with its developmental program to play a major role in determining the expressed phenotype (see Fig. 1.1). A whole range of environmental factors can act upon development, from the first meiotic division through to the later assumption of juvenile and then adult form. These factors may be abiotic, such as temperature, pressure, pH, humidity, salinity, and photoperiod. But there are also biotic factors acting, and these may be external, such as resource availability or population density, or internal, such as hormones. Many of these factors may interact to affect the phenotype produced, usually in a highly complex fashion (Fig. 1.4).

Environmental cues most conspicuously work by activating switches in the patterns of gene expression that make up the developmental program. This may lead either to the initiation of a developmental arrest, or to the production of alternative phenotypes differing in morphological features, behaviours, and life-history characters. Diapause in insects (a resting stage permitting survival through a regularly recurring season of adverse conditions) provides a good example of a triggered developmental arrest; many other animals in more hostile environments also show periods of arrested development. A striking example of a developmental switch to produce alternate phenotypes is temperature-dependent sex determination (TSD) in reptiles, dealt with in some detail in Chapter 15. Species with TSD are generally found in thermally patchy environments, which allows for the production of both sexes. The sex of the embryo is determined by the cumulative effects of the nest temperature in the period from shortly after egg laying through the first half of embryonic development.

Phenotype variation also arises due to the effects of environmental factors on the rates and degrees of expression of the developmental program. In contrast to developmental switches, which as we have seen produce discontinuous phenotypes, environmental effects on the rate of development produce a continuum of phenotypes. This kind of phenotypic plasticity has been called "continuous lability",

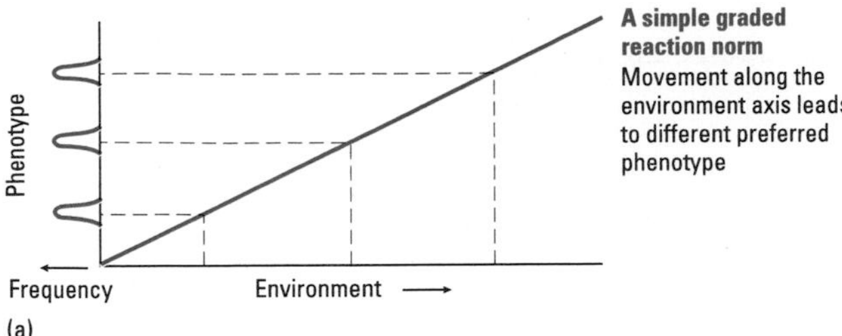

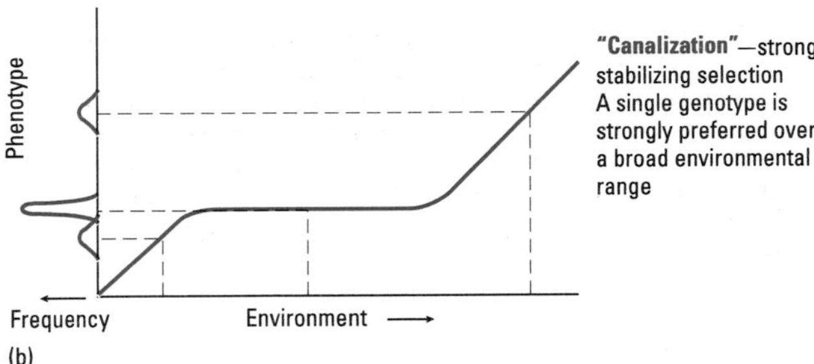

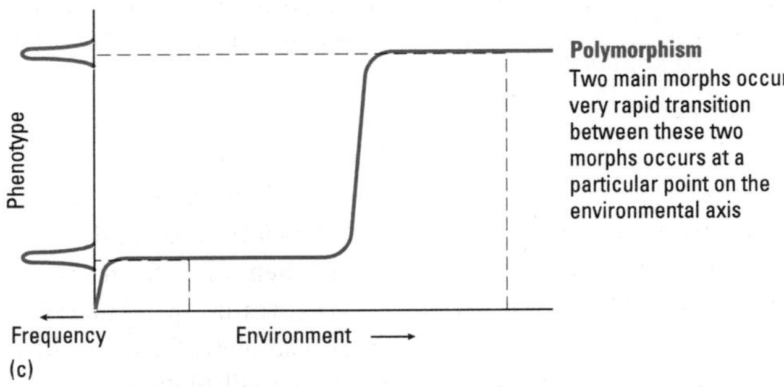

Fig. 1.5 "Reaction norms" as a way of understanding the effects of environment on phenotype.

or "phenotypic modulation". The effects of environment on development vary between traits and at different embryological stages. For example, temperatures close to the upper and lower thermal tolerance limits of a species in the first quarter of the developmental period produce a high probability of abnormal phenotypes, but later on may have little effect. Similarly, at any given point during embryogenesis the organs and tissues will have reached different points in their developmental programs and may therefore exhibit somewhat different environmental sensitivities. In some cases, differential sensitivities of tissues manifest as a change in the relative timing of growth and maturation of different parts of the body, and this is referred to as "developmental heterochrony".

The profile of phenotypes produced by a particular genotype in different environments is usually called the "norm of reaction" or **reaction norm** (Fig. 1.5), and this can clearly be of very variable shape. Morphogenetic processes which show little or no developmental plasticity are said to be "canalized", i.e. they are fixed at a given level by strong stabilizing selection and produce a very narrow range of phenotypes. Characters that show discontinuous plasticity (and thus appear polymorphic) might be explained by a continuous underlying response to the environment at the cellular or genetic level, which has a distinct threshold for phenotypic expression; in this case, the reaction norm would have steeply S-shaped regions (Fig. 1.5c).

There is some indication that the phenotypic variation induced by variable environments can be subject to strong selection, i.e. that there may be selection for plasticity itself as an adaptation. Perhaps the most compelling evidence for this comes from studies of temperature effects in amphibian life cycles. Field and laboratory experiments have shown that temperature can account for most of the observed variation in growth and differentiation in populations of the frog *Rana clamitans* along an altitudinal gradient. Growth in montane frog populations has a reduced temperature sensitivity relative to lowland populations, which serves to offset the effect of low temperature. Lowland frogs taken to high altitude have their growth retarded by low temperature to such an extent that the tadpole stage is extended by a whole season relative to the resident montane tadpoles. Such transplant experiments indicate that the phenotypic modulation of this particular trait is under genetic control. Another striking example of plasticity as adaptive in itself comes from Galápagos iguanas, where body size is linked to algal food supply, which in turn varies with environmental factors, notably El Niño (ENSO) events. During an ENSO cycle, large males die quickly, but many individuals respond with a reduction in size (shrinking by up to 20% of body length), thus significantly improving their chances of survival.

Longer term genotypic/evolutionary effects

This is the most important timescale of adaptive effects, and perhaps the only timescale where we should strictly use the term adaptation. Natural selection acts on phenotypes, regardless of their genetic basis, so there can be selection for the results of phenotypic plasticity, as well as for the results of genes producing discrete phenotypes. But the evolutionary response to selection is always at the level of the genotype, with genetic change occurring from one generation to the next. Therefore intraspecific genetic variability and hence heritable variation in fitness are essential prerequisites for long-term evolutionary change. We take a more detailed look at the mechanisms of adaptation at this level in Chapter 2.

1.5.2 Different functional levels

Avoidance, conformity, and regulation

When an animal is confronted with changes in its environment, it normally shows one of three categories of response: avoidance, conformity, or regulation. Traditionally physiology has largely concerned itself with the last of these, dissecting the mechanisms and the underlying biochemistry of the regulation of cells, tissues, and the whole body. Thus **homeostasis** (the maintenance of a constant internal environment) always takes an extremely prominent place in physiology textbooks, together with the regulatory systems needed to achieve it. But homeostasis can often be achieved more cheaply by avoidance and behavioral tricks; and it may not need to be achieved at all, with many animals surviving and flourishing with a conforming lifestyle, involving much less energy and resource expenditure. The distinction of these different kinds of adaptive and acclimatory responses constitutes a key message early on in this

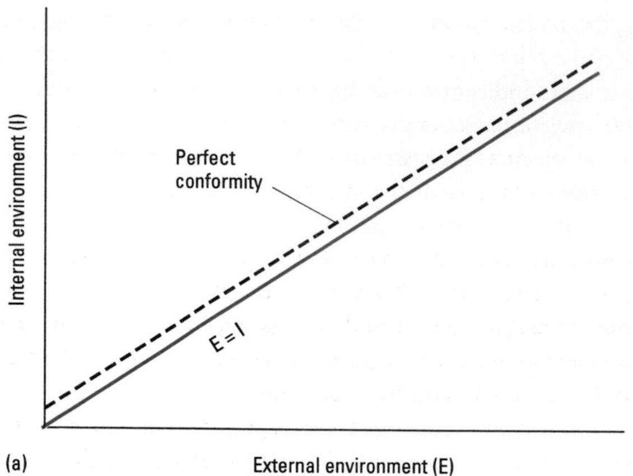

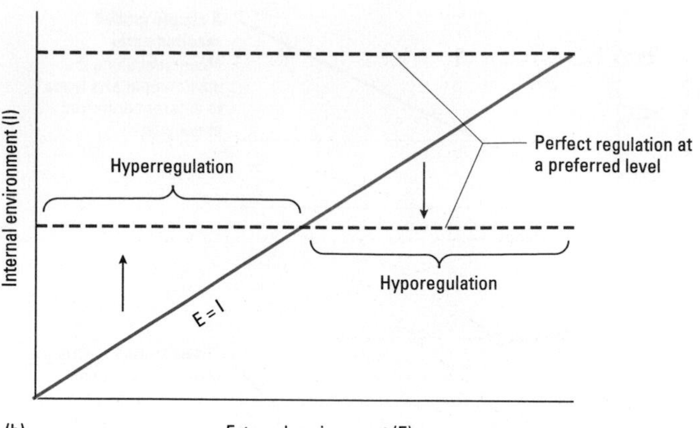

Fig. 1.6 The relation between the internal environment of an animal and the external conditions, showing the basic principles of (a) conforming (E = I) and (b) regulating (I = constant).

book. In the context of environmental adaptation it is appropriate to treat the other levels of response as being at least as important, and homeostasis merely as one of several options.

The strategies accompanying these three responses may be summarized as follows:

1 **Avoiders** have some mechanism for getting away from an environmental problem either in space (e.g. seeking unstressed microhabitats in crevices or burrows, or larger scale migration) or in time (using torpor or diapause, or producing a resistant egg, pupa, or cyst to survive difficult times).

2 **Conformers** undergo changes of internal state similar to the changes of state imposed externally. (They are therefore sometimes termed "tolerators", although this is a little confusing since regulating animals are also "tolerating" the external conditions in the sense that they are surviving in them.) Conformers do not attempt to maintain a homeostatic condition for the whole body.

3 **Regulators** maintain some or all of the components of their internal environment close to the original or "normal" level, irrespective of external conditions.

Figure 1.6 compares the general patterns of conformity and regulation. This kind of analysis may apply to a whole range of environmental and internal state variables, including temperature, osmotic concentration, oxygen levels, and pH. Thus we can identify animals that are ionoconformers, osmoconformers, thermoconformers, or oxyconformers, and plot similar kinds of diagrams to those in Fig. 1.6a, and other animals that are osmoregulators, etc., as in Fig. 1.6b. There will be many such examples later in this book, with hyperregulation being rather common and hyporegulation rather rare for most of the key variables.

However, it should be stressed that these categories are not absolute and do merge into each other; commonly we find that there are limits to both regulating and conforming (Fig. 1.7) and that there are no such things as perfect regulators or perfect conformers. For example, osmoconforming animals tend to show some regulation at extremely low salinities, so that their blood is not so dilute that the cells are irreversibly damaged by excessive swelling;

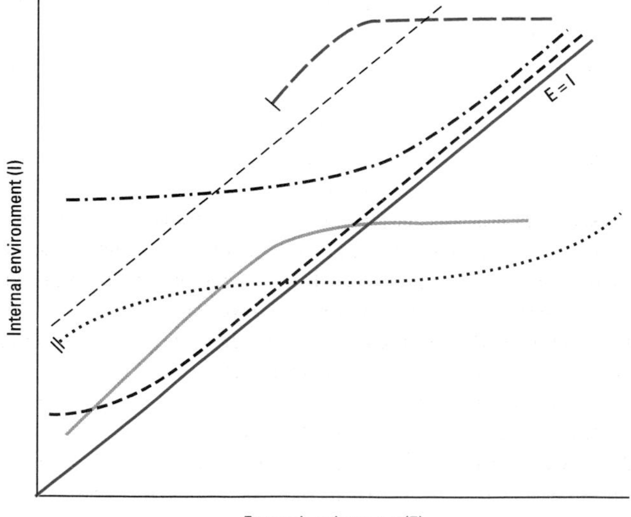

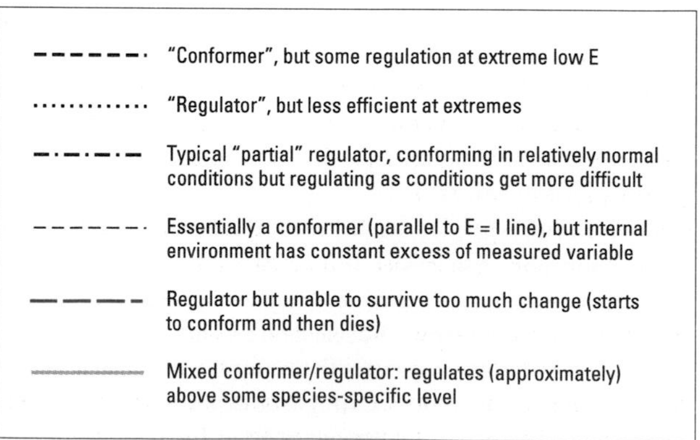

Fig. 1.7 Conformity and regulation in the real world: a variety of options for partial conforming and partial regulating.

Box 1.2 Levels at which adaptive responses can occur

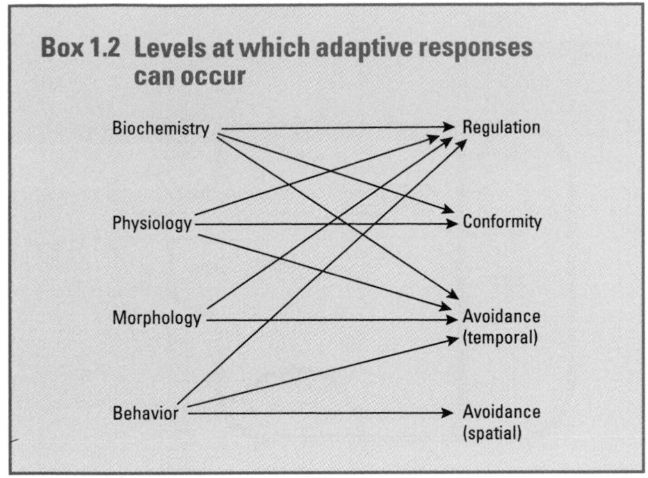

while osmoregulators sometimes lose an ability to regulate at lower salinities and become conformers. Oxyregulators tend to have a critical (species specific) point beyond which they revert to conforming. Thermoregulators may lose an ability to control their body temperature at both high and low ambient temperatures. Particular species may regulate for some parameters and conform for others.

Behavior, physiology, biochemistry, and morphology

There are four major different but interacting functional levels at which avoidance, conformity, or regulation can be effected (Box 1.2), corresponding to the four subdivisions of phenotypic traits in Fig. 1.1. These levels are often seen as being hierarchical, though from a modern perspective, in which all changes are viewed as being fundamentally mediated by genes and biochemistry, this hierarchy is somewhat artificial.

Avoidance in space is primarily an attribute brought about by behavior. For a small animal it might involve a search for an appropriate habitat, using phototactic or chemotactic responses. The "microhabitat" effect is central here, animals making choices in favor of less stressful local conditions within an apparently harsh macroenvironment. For a larger migratory species the behavioral component might need to be supplemented with physiological adjustments, for example accumulation of food reserves. Avoidance in time may require more complex responses at all levels: an animal entering torpor may accumulate food, construct or find a refuge (thus also avoiding in space), then huddle in a ball to reduce its exposed surface area; it may reduce its core temperature and lower its metabolic rate; it may acquire a thicker insulating layer; and it may mobilize or generate new forms of enzymes and new components in its membranes.

Conforming is largely concerned with changes at the physiological and biochemical levels. If the internal conditions are allowed to vary markedly, whether in terms of temperature or salinity or oxygen supply, then tissues and cells will need to have biochemical systems in place that can continue to function in the new conditions, especially in terms of appropriate enzymes and stabilized membranes. Just enough must be done to keep the animal functional (though usually at a very low level) in extreme conditions, avoiding potentially irreversible damaging effects of freezing or hypoxia or osmotic water loss (and also the damaging effects of coming out of these states, since, for example, reoxygenation has hazards of its own). In general the physiological and biochemical changes will be small and cheap to institute, so that conformity (while having the cost of reduced activity, growth, or reproduction) may have the benefit of economy.

Regulating, by contrast, may require substantial and expensive changes at all these hierarchical levels. Behavior, as we shall see repeatedly throughout Part 3 of this book, may remain as the first line of defense; even the best mammalian endotherms continue to use a whole range of behaviors (basking, burrowing, wallowing, huddling, erecting or concealing appendages, etc.) to regulate their temperature. However, behavior will be augmented by substantial physiological and biochemical adjustments; in the case of thermoregulation these may include changes of blood flow or respiratory rate, an increase in shivering or nonshivering thermogenesis, or the production of heat-shock proteins or antifreeze molecules.

It is impossible to give unconditional generalizations about patterns of avoidance, tolerance, and regulation across the animal kingdom. But to a first approximation it would be fair to say that different strategies do tend to be found in different phyla, with different body designs, and in different habitats, as follows:

1 Smaller and soft-bodied animals are more likely to be avoiders and conformers. They can use microhabitats more effectively, with concealment in protected crevices or burrows, or on and in other organisms. Because they have a high surface area to volume ratio, they will experience relatively rapid fluxes (whether the flux is of water, ions, thermal energy, or respiratory gas) across their surfaces, and working to restore a status quo against these fluxes would be very expensive. They also have little inbuilt protection against swelling and shrinking, and lack the complex outer layers that can be modified to give some insulation or impermeability. In predictably variable habitats, such as estuaries, cyclic avoidance is common (burrowing in mud, hiding in a crevice) and conforming is also often a good solution, with the cyclic environmental changes somewhat smoothed out. But in terrestrial habitats where there is both continuous high environmental stress and high fluctuation, conforming may not be an option and exceptional strategies for avoidance (torpor, estivation, encystment, cryptobiosis, etc.) are more common.

2 Animals with hard outer layers (exoskeletons), of small and medium size, may have better options for some regulation and a greater independence of their environments. Arthropods of all kinds are more likely to show partial regulation of osmotic concentrations, their exoskeletons giving them an inevitable "built-in" resistance to shrinkage or swelling. The outer surfaces can have very much reduced permeability, and may be partly thermally insulated by the addition of fine cuticular hairs, so that all fluxes are slowed and a degree of regulation becomes economically feasible. But behavioral avoidance, aided by the efficient limbs (and sometimes wings) that can be built from an exoskeleton, remains a major part of the overall strategy for coping with environmental change, especially in the more rapidly changing terrestrial habitats.

3 Large animals are much more likely to be regulators in all environments, with the important exception of the relatively equable and unchanging open oceans (where only vertebrates have a serious problem and need to regulate, due to their history of secondary invasion as discussed in Chapter 11). Larger animals operate in a larger scale (coarse-grained) environment, where rapid changes (due to local water or air movements, or localized patches of sunshine) are relatively unimportant. They have lower surface area to volume ratios so that rates of change of state internally are much slower, giving them an "inertia" effect that smoothes out the fluctuations and gives time for regulatory mechanisms to operate. They may have better opportunities for energy storage (and indeed storage of other resources such as water and even thermal energy). They may also have "room" for more complex internal regulatory centers, both neural and hormonal. In terrestrial habitats where environmental changes are inherently faster, all of these factors may work together to make regulation the only real option for a large animal. Again, though, remember that regulation does not just mean physiological and biochemical effects; behavior often still forms the first line of regulatory response.

There are very different costs and benefits of each strategy in terms of energy usage and lifestyle. Avoidance by shutting down in time is cheap but effectively causes the animal to opt out of the race for a while and achieve no growth or reproductive output. Avoidance in space by migration may be transiently very expensive but allows the animal to keep on increasing the species biomass in another environment. Avoidance of poor physical environments by either means may give additional benefits, for example by also avoiding predation or competition. Conformity at the extremes of temperature, salinity, or hypoxia that are experienced may allow only a minimal "ticking over" lifestyle, but over a broad range of less extreme variation it is a cheap way of insuring a reasonably productive lifestyle most of the time. Regulation is usually rather expensive; osmotic regulation underpinned by ionic pumping takes a moderate proportion of the total energy budget of estuarine and freshwater animals, while thermal regulation in terrestrial endotherms may take as much as 90% of the total budget. The pay-off comes primarily in the level of performance and greatly extended activity periods during times and at places of environmental adversity; food can be gathered more or less continuously, and all the avoiders and conformers that are relatively inactive become potential prey. With all this extra food, regulators can grow and reproduce faster and/or more reliably and be such successful competitors that despite high costs they may become dominant in many ecosystems.

1.5.3 Different spatial levels

Adaptive responses may occur fundamentally at a molecular level, but they "appear" at various different spatial levels in the whole animal. Some responses are essentially subcellular, and others affect the morphology or activity of whole cells. Yet others manifest as effects on entire tissues or organs, for example changes in muscle size, heart volume, or arrangements of vascularization.

However, there is another sense in which different spatial scales of response are important. Animals are made up of several distinct compartments (Fig. 1.8), each of which may show different kinds of adaptive/acclimatory response. The individual cells contain their own fluid environment (**intracellular fluid**, ICF), and they in turn are directly bathed in tissue fluid or **extracellular fluid** (ECF). In many animals the ECF is distinct from a circulating fluid termed

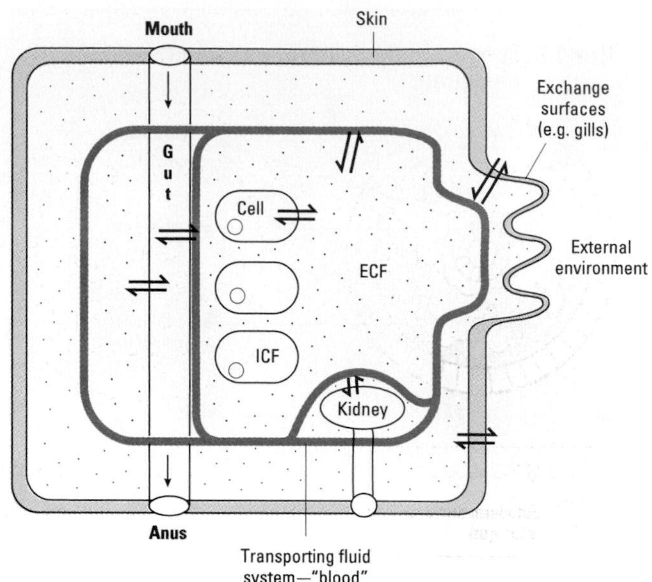

Fig. 1.8 A model of the major body compartments in an animal and exchange routes between them. ECF, extracellular fluid; ICF, intracellular fluid.

blood or **hemolymph**, and this may be chemically different from the ECF. (In fact, the relations between these fluids and animal design and body cavities are quite complex and there are other kinds of possible arrangement; see Box 1.3. However, this simple version of a three-compartment system is adequate for the kinds of animals we are normally concerned with.)

In terms of the whole animal, adaptations may therefore occur at several sites:

1 At the outside surfaces to maintain differences between the outside world and the circulating blood. Here the adaptive features tend to relate to the "skin", whether this is a relatively unspecialized epidermal cell layer or a complex multilayered structure with keratinous, chitinous, or lipid-containing elements. Sometimes the adaptations relating to exchange processes will be concentrated in or even confined to particular parts of the skin, such as the gill surfaces, with other areas of skin relatively inert and impermeable.

2 Between the circulating fluid and the ECF. This mainly occurs in vertebrates, where some constituents of blood pass out from capillaries into the ECF and others may be returned to the blood system via the lymph vessels.

3 Between the ECF and the cells. Here the adaptive surface is the cell membrane itself, controlling exchanges between the ECF and ICF. Total concentrations inside cells must be similar to those outside or the cell will shrink or swell osmotically; cell membranes cannot withstand substantial pressure differences. However, the exact makeup of the cell fluid is very different from that of the ECF, as we shall see in Chapter 4.

4 Within cells. Cells themselves are, of course, strongly compartmentalized and internal membranes may be responsible for regulating exchanges between the cytoplasm and the nucleus, mitochondria, or endoplasmic reticulum.

Adaptation to the environment may require modification at all these spatial levels. For many invertebrate marine animals the ECF

Box 1.3 Patterns of body fluids and body cavities

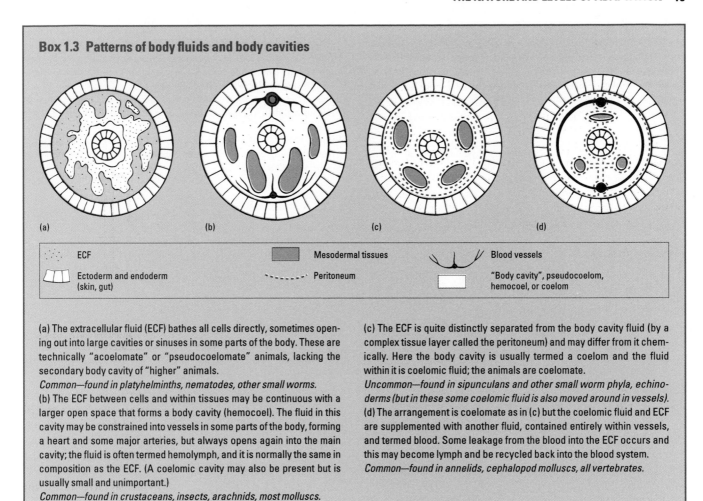

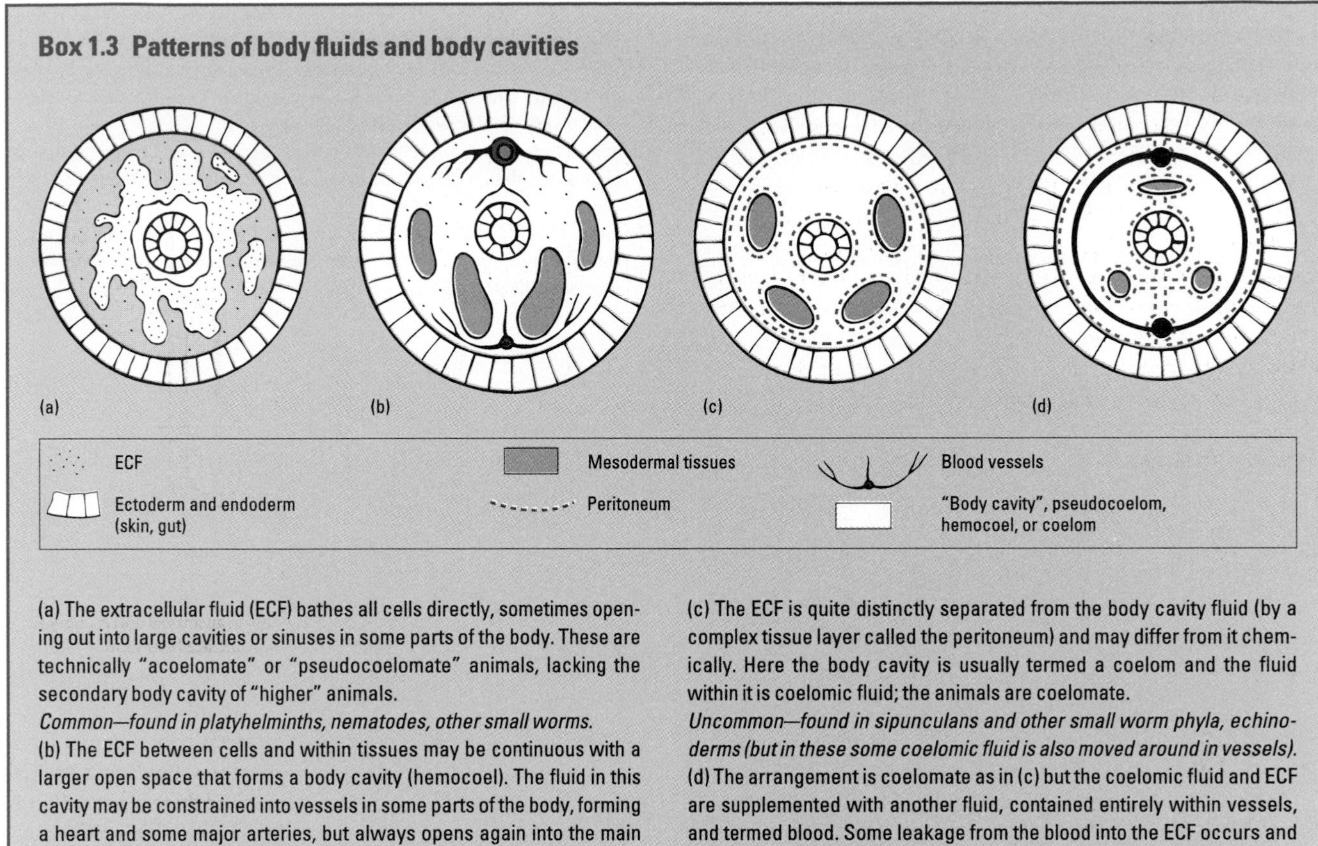

(a) The extracellular fluid (ECF) bathes all cells directly, sometimes opening out into large cavities or sinuses in some parts of the body. These are technically "acoelomate" or "pseudocoelomate" animals, lacking the secondary body cavity of "higher" animals.
Common—found in platyhelminths, nematodes, other small worms.

(b) The ECF between cells and within tissues may be continuous with a larger open space that forms a body cavity (hemocoel). The fluid in this cavity may be constrained into vessels in some parts of the body, forming a heart and some major arteries, but always opens again into the main cavity; the fluid is often termed hemolymph, and it is normally the same in composition as the ECF. (A coelomic cavity may also be present but is usually small and unimportant.)
Common—found in crustaceans, insects, arachnids, most molluscs.

(c) The ECF is quite distinctly separated from the body cavity fluid (by a complex tissue layer called the peritoneum) and may differ from it chemically. Here the body cavity is usually termed a coelom and the fluid within it is coelomic fluid; the animals are coelomate.
Uncommon—found in sipunculans and other small worm phyla, echinoderms (but in these some coelomic fluid is also moved around in vessels).

(d) The arrangement is coelomate as in (c) but the coelomic fluid and ECF are supplemented with another fluid, contained entirely within vessels, and termed blood. Some leakage from the blood into the ECF occurs and this may become lymph and be recycled back into the blood system.
Common—found in annelids, cephalopod molluscs, all vertebrates.

and blood are effectively identical with sea water and adaptation concerns only levels 3 and 4. But for nonmarine animals there is always extensive regulation both at level 1, i.e. the skin of the whole animal, and at the cellular levels.

1.6 Conclusions

Environmental adaptation is a complicated business, integrating all aspects of animal biology. It requires an understanding of animal design and animal physiology above all, but this must be put in context with a detailed understanding of the environment (measured on a suitable temporal and spatial scale), and with an appreciation of ecological and evolutionary mechanisms. There is a need to move away from purely descriptive accounts and oversimplified comparisons, and to set the encyclopedic knowledge already accumulated about what animals can do into a realistic framework of why and how they came to be as they are. Equally there is a need to look beyond the confines of traditional isolated physiological "systems" (circulation, excretion, respiration, etc.) and to see the whole picture of what is needed in order to live in a particular environment: the physiological needs of course, but also the mechanical, sensory, reproductive, and life-history adaptations that together make up a successful fully functional animal.

FURTHER READING

Books

Feder, M.E., Bennett, A.F., Burggren, W.W. & Huey, R.B. (1987) *New Directions in Ecological Physiology*. Cambridge University Press, Cambridge, UK.

Harvey, P.H. & Pagel, M.D. (1992) *The Comparative Method in Evolutionary Biology*. Oxford University Press, Oxford.

Hochachka, P.W. & Somero, G.N. (2002) *Biochemical Adaptation: Mechanism and Process in Physiological Evolution*. Oxford University Press, Oxford.

Louw, G.N. (1993) *Physiological Animal Ecology*. Longman, Harlow, UK.

Sibley, R.M. & Calow, P. (1986) *Physiological Ecology of Animals: an Evolutionary Approach*. Blackwell Scientific Publications, Oxford.

Spicer, J.I. & Gaston, K.J. (1999) *Physiological Diversity and its Ecological Implications*. Blackwell Science, Oxford.

Withers, P.C. (1992) *Comparative Animal Physiology*. Saunders College Publishing, Fort Worth.

Reviews and scientific papers

Angiletta, M.J., Wilson, R.S., Navas, C.A. & James, R.S. (2003) Tradeoffs and the evolution of reaction norms. *Trends in Research in Ecology & Evolution* **18**, 234–240.

Calow, P. & Forbes, V.E. (1998) How do physiological responses to stress translate into ecological and evolutionary processes? *Comparative Biochemistry & Physiology* A **120**, 11–16.

Felsenstein, J. (1985) Phylogenies and the comparative method. *American Naturalist* **125**, 1–15.

Garland, T. & Carter, P.A. (1994) Evolutionary physiology. *Annual Review of Physiology* **56**, 579–621.

Harvey, P.H., Read, A.F. & Nee, S. (1995) Further remarks on the role of phylogeny in comparative ecology. *Journal of Ecology* **83**, 733–734.

Hochachka, P.W. (1998) Is evolutionary physiology useful to mechanistic physiology? The diving response in pinnipeds as a test case. *Zoology, Analysis of Complex Systems* **100**, 328–335.

Piersma, T. & Drent, J. (2003) Phenotypic flexibility and the evolution of organismal design. *Trends in Research in Ecology & Evolution* **18**, 228–233.

Somero, G.N. (2000) Unity in diversity; a perspective on the methods, contributions, and future of comparative physiology. *Annual Review of Physiology* **62**, 927–937.

2 Fundamental Mechanisms of Adaptation

2.1 Introduction: adaptation at the molecular and genome level

In Chapter 1 we considered adaptation in relation to selection on phenotypes, as determined by genes and their constituent DNA. But adaptation in practice is essentially a phenomenon associated with proteins. Proteins make up most of the dry matter of cells, and underlie all cellular processes; they form the structural components of the process, they catalyze and control the process, or as enzymes they are responsible for manufacturing the nonprotein components that carry out the process. DNA may store the information needed to make a cell, but it has little other function, and cannot "do" anything for itself. RNA passes this same information to the cytoplasm outside the nucleus, and has some role in cell regulation, but in essence its function is to make proteins. Proteins by contrast very obviously "do things"; they serve as the main agents of molecular recognition, and are precision machines, with moving parts that perform and integrate chemical and mechanical functions. They can transduce motions across membranes, integrate several distinct signals into a single outcome, control rates of reaction of other molecules, and control gene expression itself. They are, however, only moderately stable, their structures relatively easily perturbed by chemical or physical factors in their environment, so that environmental adaptation has much to do with preserving protein function in the face of a changing world.

Proteins are the product of the genetic code in the DNA of a cell, and sequence changes in a length of protein-coding DNA often (though not always, due to code redundancy) result in changes in amino acid sequence. While DNA has very limited structural diversity and an essentially linear character, the resultant proteins have enormous diversity, being made up from assorted chains of 20 different amino acids, each of which has its characteristic three-dimensional structure and thus its own biochemical "personality".

Genomes are regularly altered by point mutations, especially by single base deletions and duplications, but they are more substantially remodeled by recombination events, which may produce larger transpositions, deletions, and duplications in sections of chromatids. It is these recombinations that are more likely to give a change of physiological performance or of morphology at the cell and the organism level, and thus the phenotypic variation on which selection will act. Adaptation occurs when any change at the DNA level becomes expressed, via protein change, as a trait that is beneficial in a particular environment and so persists due to selective advantage, eventually spreading through a population.

Anything that controls proteins is therefore at the core of adaptation. It is conventional to look at proteins in a "temporal sequence" of how they are made (transcription, translation), then how they are modified (post-translational changes), and then how their production is controlled by other factors. But since our main perspective here is environmental adaptation, we look first at how the functioning of proteins can be controlled in general terms to alter their function, and then at how proteins are made and destroyed in cells, how processing at the nucleic acid level affects them, and thus how proteins might change through evolutionary time. Then we consider how their functioning may also be controlled at a wider level by signals between cells and within whole organisms. Key phenomena that facilitate protein evolution and organismal adaptation are thus highlighted.

2.2 Controlling protein action

Potential protein activity in cells is set by a balance of synthesis and degradation, processes which are examined later, but achieved protein activity needs more precise spatial and temporal control. The most important system of control is the activity of **ligands**, small molecules that bind onto proteins and alter their conformation, which have an **allosteric** effect. A simple example is the binding of glucose onto the enzyme hexokinase, which then facilitates the binding of ATP (resulting in the glucose being transformed into glucose 6-phosphate in the first step of glycolysis, dealt with in Chapter 6). We understand this interaction fairly clearly at the molecular level. The hexokinase consists of two separate but linked structural sections ("domains"; see section 2.3.1), and the binding sites for both glucose and ATP lie in the cleft between the two domains; when glucose binds onto its site, deep in the cleft, it helps to pull the two domains close together and so creates a more precise fit for the ATP site on the edge of the cleft (Fig. 2.1a). Binding of glucose causes a 50-fold increase in the affinity of the hexokinase for ATP. In other proteins the sites at which two ligands bind may not be so close together but they can still affect each other's binding, and are said to be **allosterically coupled** (Fig. 2.1b). Proteins in which conformational changes can affect two widely separated binding sites are likely to have been selected for in evolution because they

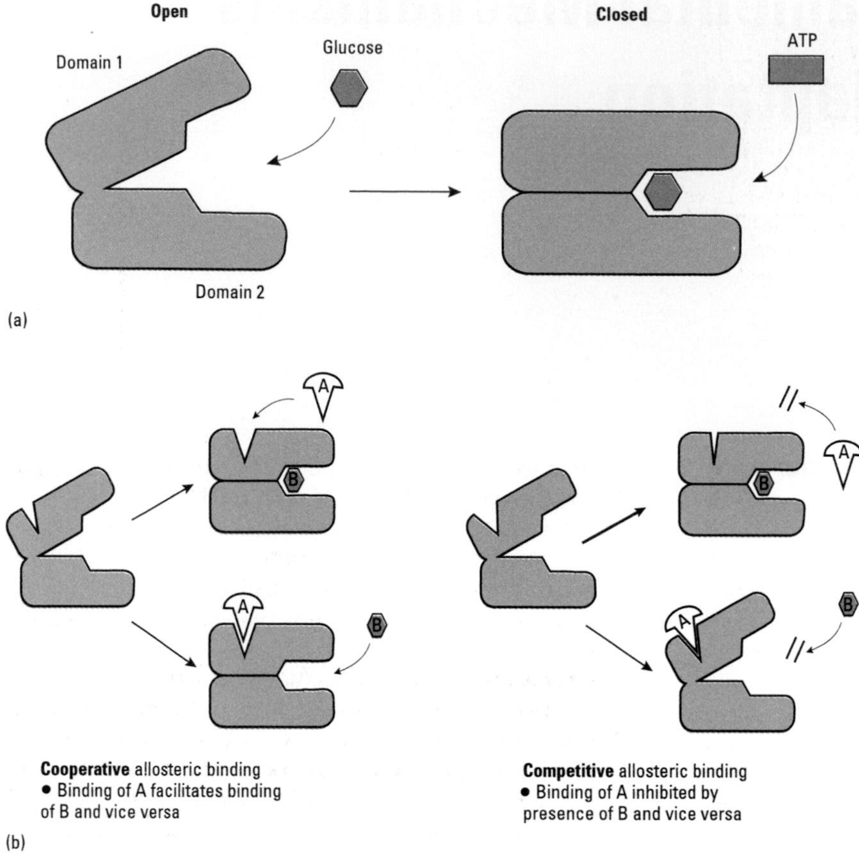

Fig. 2.1 (a) An allosteric effect: the binding of glucose between the two domains of hexokinase, allowing ATP to bind more easily. Glucose binding effectively changes the enzyme from an *open* to a *closed* conformation. (b) Models for two-ligand allosteric effects, either cooperative or competitive.

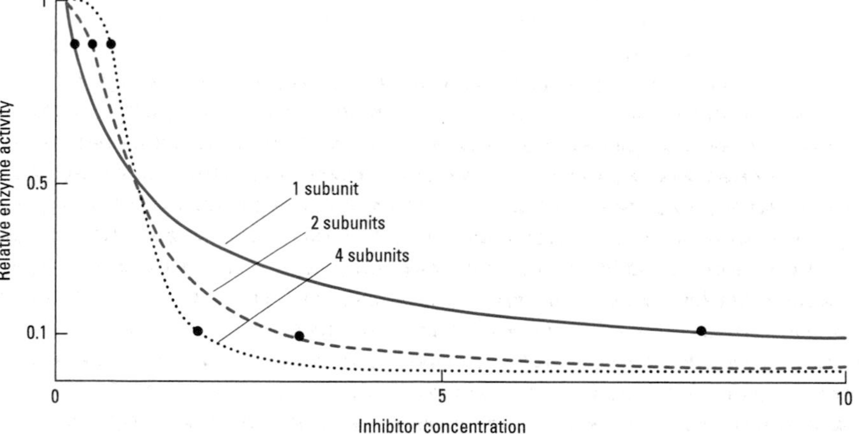

Fig. 2.2 The effects of inhibitory ligand binding on enzyme activity for monomeric and multimeric allosteric enzymes. With only one subunit, effects are relatively slow and high concentrations are needed for 90% inhibition; while for a tetrameric (four subunit) enzyme inhibition is faster and lower concentrations are needed for 90% inhibition. (From Alberts *et al.* 1994, Fig. 5.7.)

allow the cell to link together the fate of two molecules. Allosterically controlled proteins are therefore very widespread.

The link between the two ligands may be positive, as in the example of hexokinase. But it may also be negative; here the two ligands bind preferentially to two different conformations of the protein, thus one ligand may effectively "turn on" or "turn off" the enzyme in relation to the other ligand. These kinds of allosteric interaction are crucial in regulating metabolic pathways, because it is often the end-product of one pathway that becomes the regulatory ligand for an early step in the path, so providing negative feedback. Allosteric interactions can also be used to give a positive feedback though, where the ligand is something that accumulates when the cell is short of an end-product of the pathway.

Regulation of enzymes by simple negative feedback may be rather imprecise, and in many cases it is refined by having enzymes made up of several different subunits where the regulatory sites are interactive. The binding of a ligand to a single site can then trigger an allosteric change in other subunits, helping them in their turn to bind the same ligand (Fig. 2.2). A relatively small change in ligand concentration can thus switch the whole protein assembly from

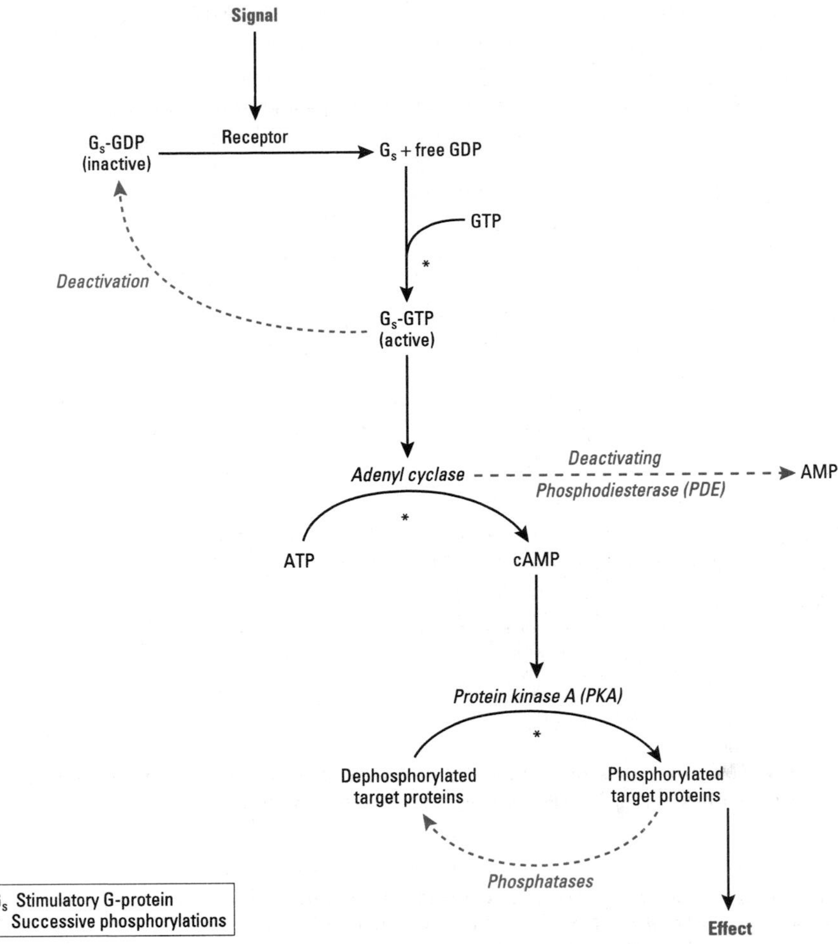

Fig. 2.3 A general model of G-protein cascade effects.

inactive to fully active or vice versa. This complex on–off switching is understood in precise molecular detail for some enzymes, with the key amino acids in the polypeptide chains identified.

One of the commonest forms of allosteric control in animal cells is that of reversible **phosphorylation** of the protein. A phosphate group carries two negative charges (PO_4^{2-}) and so can cause a structural change in a protein to which it attaches, perhaps by attracting one or several positively charged amino acid side chains. This can in turn alter the protein's configuration elsewhere, allosterically affecting the active binding sites. At any one time at least 10% of the proteins in a cell are phosphorylated, with the phosphate being added from an ATP molecule by a **protein kinase**, and removed by a **protein phosphatase**. These two types of enzyme are therefore key players in the regulation of cell functioning and physiology.

Protein kinases (PKs) all belong to a large protein family containing a similar stretch of about 250 amino acids (the "kinase domain"). Every cell has hundreds of different kinases. There are two main groups: one group works by transferring a phosphate group from ATP onto the hydroxyl group of a serine or threonine side chain in a protein; the other group works by similarly transferring phosphate onto a tyrosine side chain. These reactions are not spontaneously reversible, because of the large amount of chemical energy released when ATP is converted to ADP; but they can be reversed when a protein phosphatase specifically removes the phosphate again, giving a fine-control system. Kinases work in teams, phosphorylating each other in integrated cascade systems where some steps amplify the preceding effect. Figure 2.3 shows an example for G-protein activation (discussed below). Thus the cell contains a series of **switches**, each of which may itself integrate two or more signals. For example, there are "cyclin-dependent protein kinases" that only work when another particular protein (cyclin) is present *and* one specific phosphate-binding site is occupied *and* another such site is empty (with phosphate having been removed by a phosphatase); three signals are being integrated. Kinases may also have an indirect switch effect by phosphorylating an inhibitory protein that normally binds to another protein, allowing the last of these to be unmasked and to become active. The kinases are therefore extraordinarily abundant and variable, with over 1000 isoforms of protein kinase known in eukaryotic cells, reflecting the importance of their roles (especially in signal transduction and transmembrane information flow) in the adaptive evolution of organisms and the generation of physiological diversity.

There is another group of proteins that are important in regulating cell activity and are active in cascades of cellular protein phosphorylations; these are the **GTP-binding proteins** (or GTPases). The kinases discussed above achieved phosphorylation by catalyzing the transfer of a phosphate group from ATP. The GTP-binding proteins instead incorporate the whole nucleotide, GTP, at a binding site within their structure. When this GTP is hydrolyzed to GDP the protein becomes inactive, and it is probably so most of the time. This inactivation is brought about by a GTPase-activating protein, which binds to the protein and forces the GTP to hydrolyze to GDP. It must then exchange its GDP for a GTP molecule to become active again, which requires the action of a guanosine nucleotide-releasing protein, causing release of the GDP; the space is immediately filled by a GTP as this is in excess over GDP in the cell cytoplasm.

This activating effect only occurs in response to various specific extracellular signals. Simple GTP-binding proteins have also been built up through evolution into large versions composed of many subunits (multimeric), the most famous of which are the **G-proteins**, which are associated with receptors on cell membranes. These control important biological activities by the hydrolysis of GTP molecules, leading to changes of protein conformation; we will meet them again below.

Proteins that hydrolyze ATP or GTP can be used to perform unusually complex actions. Some of them become directional **motor proteins**, bringing about various kinds of movement. This is achieved because of the high free energy released in the hydrolysis reaction, so that it becomes effectively irreversible. Thus if a protein has three structural conformations and the transitions 1→2, 2→3, and 3→1 are all driven by ATP hydrolysis, then this series of changes will proceed automatically, without reversing. This means that a protein undergoing small shape changes can be made to "move along" the substrate it is attached to (Fig. 2.4). For example, a protein can move along a DNA strand in one direction (allowing "reading" of the genetic code), or a myosin filament can move along an actin filament (producing muscle contraction); in the latter case the physical movement produced by one ATP-linked conformation change is about 5 nm.

Another complex action that ATP-driven proteins can perform is that of pumping, most commonly pumping ions across a membrane, and these **pump proteins** are particularly important to physiologists. The most famous example is the sodium pump, or Na^+/K^+-ATPase, which normally (as we will see in detail in Chapter 4) moves three sodium ions out of the cell and two potassium ions into the cell for each cycle of conformational changes.

Membrane-bound allosteric pumps that are linked to ATP hydrolysis can also be made to work in reverse; they can employ an existing ionic gradient to synthesize ATP molecules from ADP plus phosphate. The energy from the hydrogen ion (H^+) gradient across the inner mitochondrial membrane is used by such an enzyme, called **ATP synthase**, to make most of the ATP that animals need (see section 6.4.2).

The ability of cells to achieve phosphorylations by harnessing the energy in nucleotide triphosphates (ATP or GTP) to drive allosteric changes in proteins has therefore been central to all evolutionary processes. It provides the fundamental machinery of cells, including switches, motors, and transducers (energy converters). Proteins that function in this manner can be assembled into complex arrays that function as multistep machines. Tiny changes in protein structure, produced by the alteration of a single key amino acid in the protein chain (and thus the result of as little as one base pair change in the DNA), can produce greatly amplified effects at the functional level; thus small variation at the molecular level can be strongly selected for or against in the organism. In practice, many of the processes occurring in cells are so intimately linked to each other that there must be strong stabilizing selection to prevent change. However, any small genetic change that is nonlethal but produces protein conformation change can clearly have quite dramatic benefits on a much larger scale. We will meet below a number of reasons why such nonlethal or nonharmful changes are rather likely to occur.

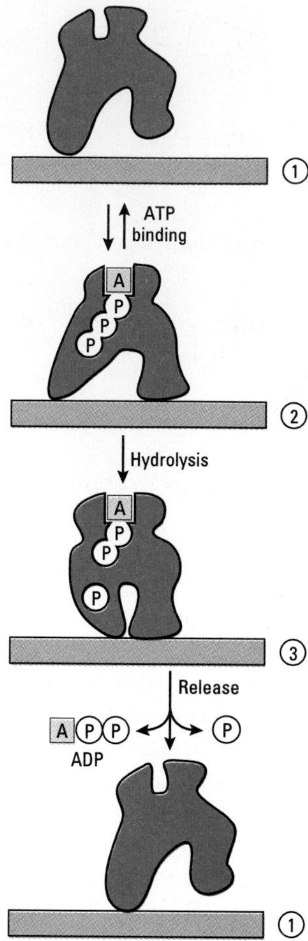

Fig. 2.4 A model of an allosteric "motor protein". The protein has three possible shapes (1, 2, and 3). ATP binding causes the shape change 1→2, and ATP hydrolysis causes the change 2→3, so that the two binding sites effectively "walk" along a substrate. Release of ADP and phosphate cause the final change 3→1, completing one "step" or "stride". Reversals of movement are normally precluded because one of the stages involves splitting ATP which is essentially energetically irreversible. The protein therefore moves continuously to the right along the substrate. (From Alberts *et al.* 1994, Fig. 5.22.)

2.3 Control of protein synthesis and degradation

We have seen how protein activity can be controlled by other intracellular molecules (often themselves proteins) that bind to them. However, the availability of proteins to be so controlled is in its turn determined by the more fundamental cellular processes of protein synthesis (from DNA via messenger RNA molecules) and of protein degradation. These too must be subject to elaborate control to produce an appropriate intracellular protein pool.

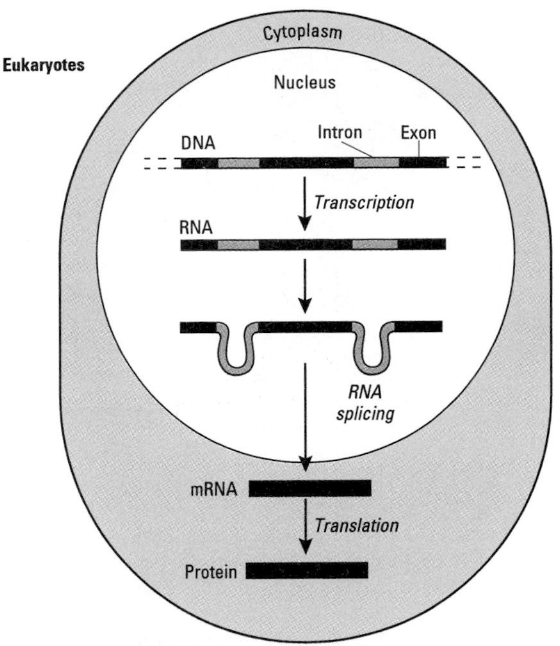

Fig. 2.5 The transfer of information from DNA in the nucleus to protein in the cytoplasm via transcription, splicing, and translation. See text for more details. (From Alberts *et al.* 1994, Fig. 3.15.)

2.3.1 Protein synthesis and morphology

The basic process by which DNA structure is repeatedly transformed into protein sequences, via DNA replication, transcription of DNA into RNA, and translation of the RNA into protein (Fig. 2.5), does not need to be rehearsed here, having become the starting point for nearly all of modern biology. The nucleotide sequence of a gene determines the amino acid sequence of the corresponding protein, via the classic codon–anticodon system in which a triplet of consecutive bases encodes a single amino acid. Remember, though, that DNA is transcribed into single-stranded RNA only in short sections; some of the resultant pieces (the messenger RNAs or mRNAs) are the protein coders, corresponding to just one or a few proteins, while others are transfer RNAs (tRNAs) or ribosomal RNAs (rRNAs) that will take part in the translational process.

Proteins once synthesized fold up into their correct configuration. The polypeptide chain automatically forms a complex three-dimensional shape determined by the amino acid side branches and their tendency to form weak bonds with each other. In general the hydrophobic side chains tend to be pushed to the interior of the protein molecule and the hydrophilic side chains protrude outwards. This folding is partly controlled by special proteins termed "molecular chaperones" whose main function is to facilitate protein folding (we will meet these again in their role as the "heat shock proteins" in Chapter 8).

It has become clear that there are several structural patterns of folding that occur repeatedly. Two are particularly common: the **α-helix** and the **β-sheet**. The α-helix often forms short lengths of protein that cross membranes; the configuration is not very stable in aqueous environments, though two helices may coil together (the "coiled coil") to give a particularly stable structure common in fibrous proteins (e.g. collagen, keratin). The β-sheet folding system may form large parts of the framework of globular proteins (e.g. lysozyme).

At the next level of complexity, there are only a limited number of ways of combining the α-helices and β-sheets, so certain patterns are commonly repeated and are known as **motifs**. An example is the "hairpin beta motif", consisting of two antiparallel β-sheets joined at a sharp angle by a loop of polypeptide chain. Combinations of motifs of α-helices and β-sheets may together make up compactly folded globular units within proteins that are known as **domains**, usually containing 50–300 amino acids. A typical domain is therefore made up of a compact core of motifs plus some uniquely shaped linking sections, with the surface covered by protruding loop regions; these loops, or the pits between them, commonly form the binding sites for other molecules.

The number of possible proteins that could theoretically be formed by the normal 20 amino acids transcribed in animals is staggering (20^n, where n is the chain length; this gives a value of 160,000 for just four amino acids, and 10^{390} for a more typical chain of 300 amino acids). Yet very few of these possibilities exist naturally. This is probably largely because very few of them would be stable in three dimensions; many would form straggly molecules with many different possible conformations, which would be impossibly difficult to control. The proteins that are selected for in evolution are those that can reliably and repeatably fold up into stable forms, where small conformational changes can be controllably achieved to perform specific structural or catalytic functions.

2.3.2 Controlling protein synthesis

In any cell at any particular time some of the DNA is being used to make very large amounts of RNA and other parts are not being transcribed at all. Transcription is controlled by an **RNA polymerase** enzyme, which makes RNA copies of DNA molecules (though in fact the full "transcription initiation complex" comprises the polymerase and at least seven other proteins). A polymerase appears to act by colliding randomly with DNA molecules but only sticking tightly when it contacts a specific region of DNA termed a **promoter**, which contains the "start sequence" signaling where RNA synthesis should begin. It opens up the DNA double helix here and then moves along the DNA strand in the 5′ to 3′ direction, at any one time exposing just a short stretch of unpaired DNA nucleotides. Eventually it meets a stop signal, and releases both the DNA template and the new RNA chain. Figure 2.6 shows the organization of a promoter and the five genes that it controls, all of them involved in a particular pathway (in this case tryptophan synthesis in the bacterium *Escherichia coli*). This arrangement means that a set of related genes pertinent to a particular metabolic pathway or morphological change can be controlled as one unit.

In animals there are three kinds of RNA polymerase, designated I, II, and III, and each is responsible for transcribing different sets of genes. They are structurally similar, all being large multimeric proteins with masses in excess of 500,000 Da. RNA polymerase II is responsible for transcribing genes into the RNAs that will make most of the cell's proteins.

Each active gene may make thousands of copies of mRNA in each cell cycle, and the resultant mRNA strand may translate into many

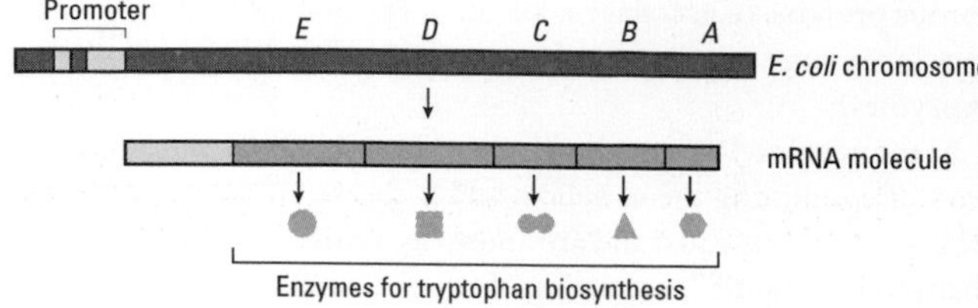

Fig. 2.6 The cluster of genes (*A–E*) in *Escherichia coli* that code for enzymes involved in the manufacture of the amino acid tryptophan. All five are transcribed as a single mRNA molecule controlled by a single promoter. The cluster is called an operon. (From Alberts *et al.* 1994, Fig. 9.24.)

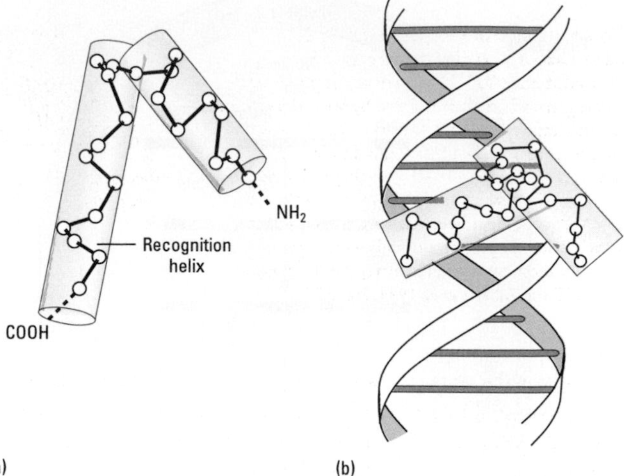

Fig. 2.8 (a) The "helix-turn-helix" motif in a protein; and (b) the way in which the motif binds to the outside of the DNA double helix. (From Alberts *et al.* 1994, Fig. 9.19.)

thousands of polypeptide chains in the same period, so that amplification factors of millions may be involved. The amount of RNA made from a particular region of DNA must therefore be very tightly controlled, and this is partly determined by the goodness of match between the promoter region sequence and the RNA polymerase. Genes that produce large amounts of mRNA thus have strong promoters and genes that produce mRNA only in small amounts have weak promoters.

Further control of protein synthesis is needed, however, and there are at least five more stages at which this can occur (Fig. 2.7). The first of these, and for many genes the most crucial, is that of transcriptional control. This is the function of a special class of **gene regulatory proteins**, which bind to specific sites on the DNA close to the gene coding sequences. Their manufacture takes up several percent of the capacity of the genome in higher animals. Their effect is to insure that the right genes are switched on in each cell type at the right time, leading to the differentiation between, for example, nerve cells, gland cells, or epidermal cells. All these cells in any one animal contain the same DNA, and all of them have some proteins in common (key metabolic enzymes, cytoskeletal proteins, etc.). However, there are some proteins that are only ever expressed in one cell type (e.g. vertebrates express hemoglobin only in red blood cells), and many other intermediate protein types that have a differential intensity of expression between cell types.

The important fact in gene regulation by proteins is that the DNA helix can be "read" by proteins from the outside, i.e. without splitting the double helix. There are structural patterns (of the hydrogen bonding sites and hydrophobic patches) inherent to the external grooving on the helix that are unique for each of the base pairs. Thus a protein can recognize and bind to a particular set of nucleotides. Very short DNA sequences, often only 6–15 base pairs long, act as genetic switches; for example in *Drosophila* the regulatory protein product from the gene *bicoid* recognizes the sequence GGGATTAGA (the matching bases are excluded here for simplicity). Thus specific sequences are read by particular motifs on the regulatory proteins. There appear to be rather few such motifs. The most common one is the "helix-turn-helix" motif (two α-helices separated by a short amino acid chain, creating a fixed angle). The longer carboxyl-terminal α-helix of this motif fits neatly into the groove of one turn of the DNA helix (Fig. 2.8). The now famous "homeodomain" proteins that serve as vital switches in the development of animals are a subset of the helix-turn-helix regulatory proteins, having one extra helix region added. Another common motif in gene regulatory proteins is the so-called "zinc finger" motif, where areas of α-helix or β-sheet are linked by a zinc molecule. A third important type is the "leucine zipper motif", differing in that it works via an interaction between α-helices on two separate monomers of the protein. There are several different monomers, each recognizing a different DNA binding site, and so able to interact with each other to give more sophisticated control (i.e. specific combinations of zipper proteins

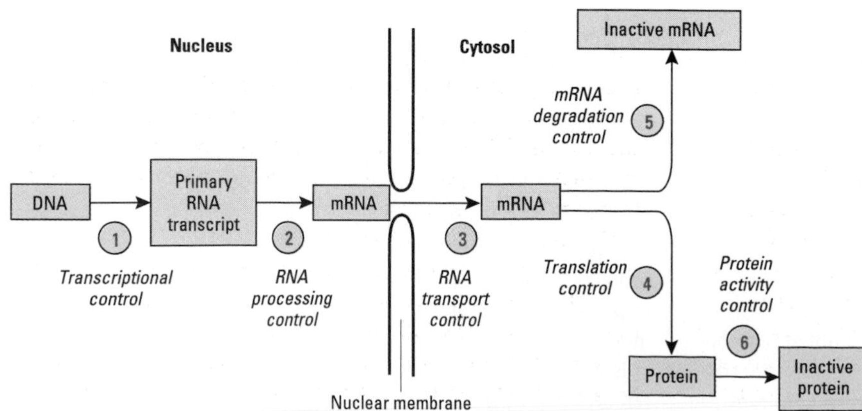

Fig. 2.7 Six steps at which gene expression can be controlled. (From Alberts *et al.* 1994, Fig. 9.2.)

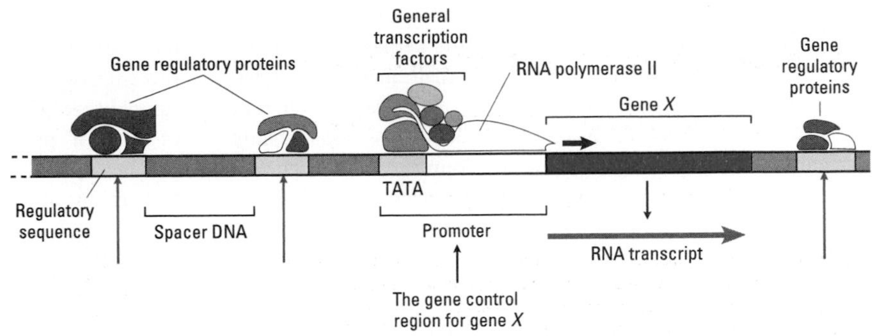

Fig. 2.9 A model of a single gene (X) and all its associated regulatory proteins and transcription factors. The promoter region is next to the gene, where the crucial RNA polymerase and the general transcription factors assemble. The promoter region always contains a short section of TATA repeats, recognized by one key transcription factor. More distantly there are several regions of DNA sequence (arrowed) that are recognized by gene regulatory proteins, usually working in clusters; these are separated by regions of spacer DNA. (From Alberts *et al.* 1994, Fig. 9.34.)

controlling the gene regulation process). As might be expected, the match required between DNA helix and regulatory protein is very precise; a protein once in place may be making up to 20 bonds of various kinds with the DNA helix structure.

How do regulatory proteins work? At their simplest they recognize a region within the promoter sequence of a DNA strand (see above and Fig. 2.6). This small section within the promoter is termed the **operator**. Promoter and operator regions are arranged so that when the operator is occupied by the regulatory protein, the binding of RNA polymerase to the promoter is affected. One possibility is that the polymerase cannot then bind properly to the promoter so transcription cannot occur. This gives a negative control. In the case of the tryptophan synthesis system shown in Fig. 2.6, the regulatory protein is only able to bind to its operator site once it has already bound two molecules of tryptophan. Thus the whole system (one **operon**) can be switched on or off according to the level of free tryptophan in the cell; when there is plenty of tryptophan, the repressor is activated and therefore bound to the operator, so the RNA polymerase cannot bind to the promoter, and enzymes to produce more tryptophan will not be transcribed. A simple on–off switch with automatic negative feedback is created. Positive control systems are also possible. Gene regulatory proteins that are activating rather than inhibiting generally work by binding to operator regions in such a way that the RNA polymerase is contacted and its own binding to the promoter region improved. The two types of gene regulatory protein, having positive and negative effects, are usually termed **transcriptional activators** and **transcriptional repressors** (or gene activator proteins and gene repressor proteins), respectively. More complex switches can be produced by combining the two, and many genes probably have both repressor and activator proteins, which can be derived from more than one chromosome.

In animal cells, additional complexity of control is achieved in a number of ways. One important mechanism is the existence of **general transcription factors**, which must be bound to the promoter site in a complex array before any transcription by RNA polymerase II can occur. Often, one of the components is a protein kinase, which phosphorylates the polymerase and allows it to disengage partly from the transcription factor and begin its own work of transcription. Another important regulatory system involves **enhancers**, sequences of DNA at some distance from the promoter region that can nevertheless influence transcription, seemingly by loops or folds occurring in the actual DNA to allow direct contact of the enhancer at the promoter site. Enhancer regulatory proteins are probably widespread in animals, and most of them exert positive rather than negative control. The "regulatory complex" of transcription controllers that pertain to any one particular gene can therefore be extremely intricate, and may be spread out over as much as 40,000–50,000 base pairs, including large areas of material that is no more than spacer (Fig. 2.9).

An individual gene regulatory protein may take part in more than one gene regulatory complex, and it might even be an activator in one situation and a repressor in another. Each functions as a regulatory unit that is used to assemble regulatory complexes in different places at different times. This gives an enormous range of possibilities for the control of gene transcription. A major way of setting up positional information in animal embryos is by the uneven distribution of gene regulatory proteins along the length of the embryo, so that chromosomes in cells or nuclei at one end of the system are exposed to a different regulatory environment from those at the other end. Where the embryo is fully cellular (that is, in most animals other than insects, which are unusual in having a mosaic embryo with many nuclei sharing a common cytoplasm), setting up these gradients must also require regulation via messengers across cell boundaries, so that the regulatory proteins are themselves properly regulated. Some of the general ways in which proteins can be controlled from outside the cell, discussed in section 2.2 above, must then be operative.

At a yet higher level of organization, there are also some critical gene regulatory proteins (encoded by **master control genes**) that serve as key switches for a whole set of other regulator complexes across a range of cell types. In many cases such genes work via methylation of some part of the genome, or condensation of the chromatin, both of which limit access of the RNA transcribing enzymes. The homeobox genes are preeminently of this kind, determining large-scale patterning of the developing animal body. Another example relates to the whole process of muscle cell differentiation, which is triggered by a group of "myogenic regulatory factors" expressed strongly only in muscle cells and integrating the action of many other regulatory systems relating to muscle tissue structure and function. Genes that determine sex could also be included in this category, with a whole complex of characters related to expression of a key control gene.

One final point here is that regulation of gene expression can sometimes be passed on between cells through time as well as through space. Most obviously, once a cell has switched to become a particular cell type, with the appropriate genes turned on and off, it passes on this suite of gene expression to its daughter cells through mitotic divisions. This **cell memory** allows the formation of tissues.

Clearly the control of gene expression can be extremely complicated, but it is based on simple principles and perhaps on relatively few molecules with complex interactions. Small changes in regulatory proteins are likely to have large effects, yet these will often be well-controlled effects, for example producing positional change without functional changes during development. One important way to get major adaptive change in animals at a physiological level is therefore via mutation in the DNA coding for these regulatory proteins. Another must be the possibility of spatially reshuffling the various regulatory centers on the genome adjacent to the actual gene, thus creating a new regulatory circuitry and modifying existing controls.

2.3.3 Controlling protein degradation

Another crucial aspect of good cellular functioning is the ability to "shut off" proteins completely. Intracellular proteolytic mechanisms exist in all animal cells. These mechanisms confer an automatic short half-life on certain normal proteins, as well as recognizing and eliminating abnormal proteins that have not folded up correctly, or disposing of proteins that have been damaged during their working life. Some proteins are degraded continuously, and others may be stable for some time but then get degraded rapidly when a particular point of the cell cycle is reached. Degradation may occur in the cytoplasm or in the endoplasmic reticulum, and in special cases it also occurs in lysosomes.

Cytoplasmic degradation is mainly the work of large protein complexes, sometimes given the special name **proteasomes**. Each of these is shaped like a hollow cylinder formed from many distinct proteinase enzymes with their active sites facing inwards. The entrance to the core of the cylinder is guarded by another large protein complex, which probably selects the proteins destined for destruction and lets them into the inner chamber. Then the proteinases do their work and a multitude of short peptide chains are released.

Proteins due for destruction are recognized by being tagged with a particular marker molecule, **ubiquitin**. This is a small protein that links covalently (thus rather firmly) to the destruction-bound proteins, and serves as a degradation signal. Linking it onto a target protein is achieved by a ubiquitin-conjugating enzyme, which presumably recognizes a signal on a misfolded protein. Recognition of "normal" proteins destined for a short half-life is a more complicated matter, and seems to reside largely in the nature of the N-terminus of the peptide chain. Certain amino acids confer protection, while others (such as arginine, aspartate, and glutamate) provoke proteolytic attack. The sequence of this region can thus be rather easily selected over evolutionary time to give an appropriate protein half-life.

2.3.4 Control via nucleic acid processing systems

DNA organization

Most of the chromosomal DNA in any animal cell does not code for RNA or for protein; it is "excess DNA" and does not form genes. The size of the genome, and the proportion of it that codes for RNA, varies between animal groups, and even more widely between the kingdoms of living organisms (Fig. 2.10). The mammalian genome in principle contains enough DNA to make about 3 million average proteins, but probably only constructs about 60,000 proteins in reality. Much of the DNA lies between genes, and a large part of it consists of repeated nucleotide sequences; some of this is tandemly repeated "satellite DNA" (with very large numbers of repeats of small sequences, perhaps only 2–10 nucleotides in length), and the rest is interspersed repeated DNA. Remember also that even within coded sequences there are introns that are spliced out and discarded after transcription (see below). Thus only about 1% of the DNA sequence is transcribed into functional RNA sequences. A schematic view of the organization of a chromosome is shown in Fig. 2.11.

Another feature of DNA organization is the existence of **transposable elements** ("jumping genes" or transposons). These are short pieces of DNA (a few hundred to 10,000 nucleotide pairs) that have the remarkable property of being able to move around the

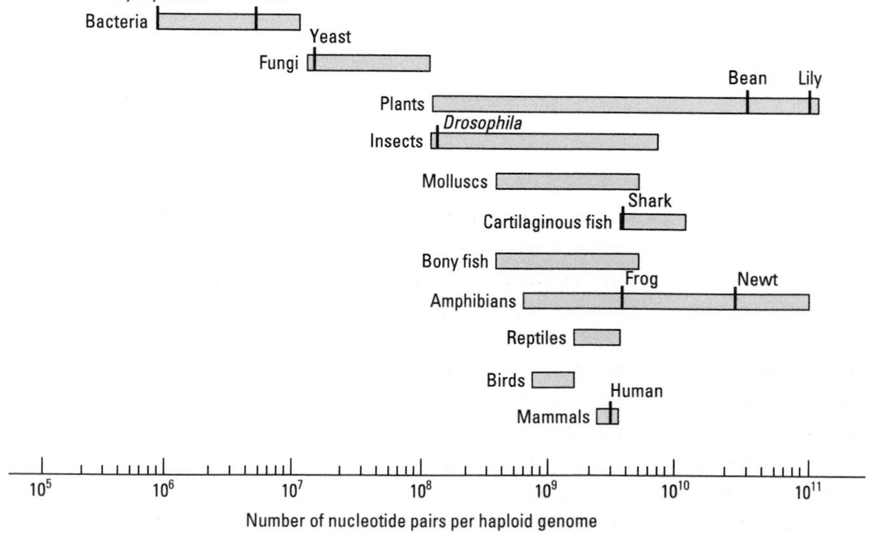

Fig. 2.10 The amount of DNA in the haploid genome of a range of organisms. Note that there is no particular relation with "complexity" within the eukaryotes: certain amphibians and plants have notoriously large amounts of DNA. (From Alberts *et al.* 1994, Fig. 8.6.)

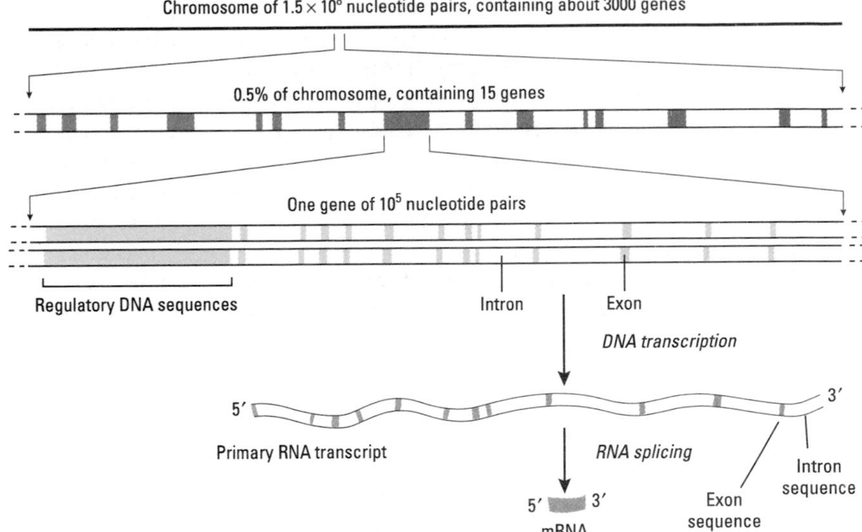

Fig. 2.11 The organization of genes on a typical animal chromosome. The number of genes per chromosome is highly variable, as is the length of each and its exon–intron structure. (From Alberts *et al.* 1994, Fig. 8.7.)

genome, either as DNA or via an RNA intermediate. Some merely carry the information for transposition; these may accelerate mutation at the sites where they insert and exit, inducing localized base pair mutation or some mispairing and thus sequence duplication; they can also multiply and spread within the genome, causing more substantial sequence shifts. Others (transposons) carry both the instructions for transposition and some other gene as well; the best known cases are genes for antibiotic resistance in bacteria. Thus transposable elements may disproportionately affect the rate of evolutionary change.

DNA repair

DNA and RNA are complex molecules made from simple subunits, and change in DNA is the fundamental mechanism of evolution. Not surprisingly, cells have elaborate mechanisms to limit DNA damage and alteration and so eliminate a high proportion of mutations before they lead to any expression of "damaged" and useless proteins. Damage arises quite independently of replication for a number of reasons: thermal degradation of purine bases is extremely common, while less frequent changes involve spontaneous deamination of cytosine to uracil, the effects of ultraviolet radiation, and the influence of certain metabolites such as oxygen radicals. Each day thousands of bases may be altered at random in any animal cell. But fewer than one in a thousand alterations causes a heritable mutation; all the rest of them are removed by repair mechanisms (representing the most fundamental kind of "stabilizing selection" to prevent inappropriate functional change).

Several different kinds of repair occur, each catalyzed by a series of enzymes; all rely on the fact that there are two complementary DNA strands and a damaged strand will not align properly with its undamaged complementary copy. **DNA repair nucleases** recognize altered portions of DNA strands and hydrolyze the phosphodiester bonds that join a damaged nucleotide to its adjacent undamaged nucleotides. **DNA polymerase** then binds to the 3′ end of the cut DNA and helps the synthesis of a new copy of the good strand. Finally, **DNA ligase** seals the original gap in the damaged strand. Some or all of these repair enzymes can be induced in response to severe DNA damage, providing an important safety mechanism against environmental hazards.

The basic mechanisms of DNA repair are an essential component of the process of genetic recombination in meiosis, where chromosomes undergo crossing-over. Maternal and paternal sequences may be slightly different, so that in the crossing region there will be some mismatched base pairs. This minor problem can be corrected by the DNA repair machinery, giving rise to a major gene conversion where, for example, a copy of the paternal allele has been changed to a copy of the maternal allele (Fig. 2.12). However, where there is a particularly poor match between the recombining strands the pairing is usually aborted and no recombination occurs. Excessive scrambling of the genome, which might be detrimental to physiological function, is thus averted.

RNA processing

Selectivity in the synthesis of RNA from DNA occurs at two levels. Only a part of the DNA is transcribed to produce RNA at all; and only a minor proportion of this RNA survives to be exported into the cytoplasm to take part in protein synthesis (see Fig. 2.5). We dealt with the first of these processes in considering the control of protein synthesis, but we still need to consider the second. In eukaryotic cells there is the additional post-transcription stage of **RNA splicing** or editing, where some sections of the transcribed mRNA are removed and take no further part in the translation into protein. **Introns** are the regions of DNA that are edited out from the primary transcript, and **exons** are the areas that are spliced back together to form the secondary transcript, which is then encoded (often with one exon corresponding to one protein domain). RNA processing therefore involves the removal of long nucleotide sequences from the middle of RNA molecules. This RNA splicing of introns occurs in the nucleus, out of reach of the ribosomes.

The numbers of introns are very variable (Table 2.1). There may be none at all in some genes, but as many as 75 introns occur within a few mammalian genes. Introns range in size from about 80 to

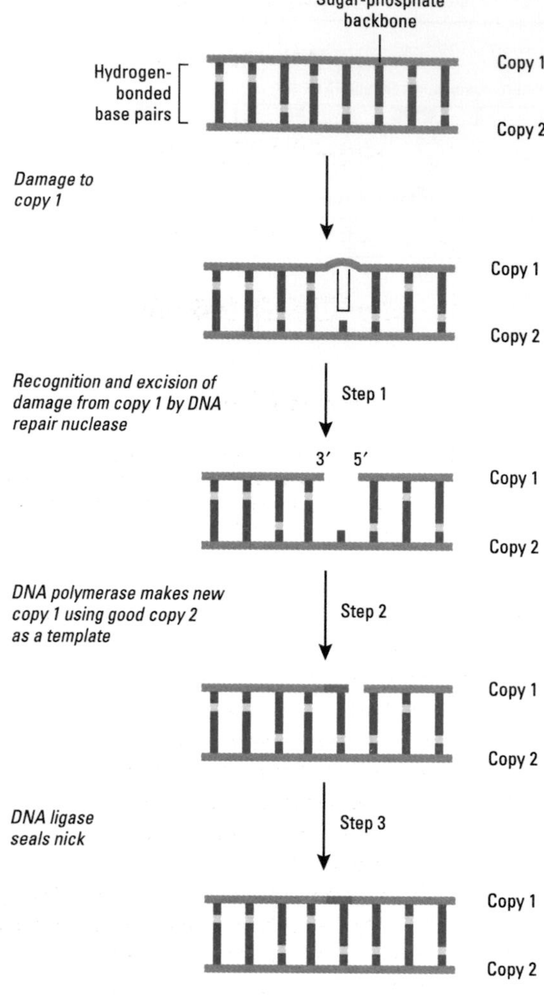

Fig. 2.12 The role of three kinds of DNA repair enzymes in recognizing and correcting damage (here a single point error) to the DNA double helix. (From Alberts *et al.* 1994, Fig. 6.66.)

Table 2.1 The sizes of genes before and after editing to mRNA, and the numbers of introns they may contain, from the human genome.

Gene	Gene nucleotides	Resultant mRNA nucleotides	Number of introns
β-globin	1500	600	2
Insulin	1700	400	2
Protein kinase C	11,000	1400	7
Albumin	25,000	2100	14
Clotting factor VIII	186,000	9000	25
Thyroglobulin	300,000	8700	36

more than 10,000 nucleotides. Their precise sequence is probably irrelevant and they may contain many accumulated mutations; they are sometimes described as "junk" RNA. The only highly conserved sequences they possess are those required for their removal, at the 5′ splice site and the 3′ splice site. Splicing is achieved by small complexes of RNA and protein termed **small nuclear ribonucleoproteins**; a number of these together assemble into a **spliceosome**, and collectively they effect removal of the intron.

Single nucleotide changes may be enough to inactivate a splice site, and when this occurs the splicing mechanism may stop or start at the wrong point. Alternative RNA splicing patterns can produce different forms of a protein from the same gene, either at random or more commonly in a regulated pattern. This may involve the production of a functional protein rather than a nonfunctional one, or the production of different morphs of the protein in different cell types. RNA processing therefore becomes a major way of producing these alternative protein morphs. There is also some evidence that changes in the splicing patterns could be an important source of variation in the overall evolution of genes and proteins, and thus of organisms.

There are other ways in which RNA is processed and controlled, usually less important in animals but contributing to overall gene expression control. These include systems that: alter the 3′ end of the RNA (often by poly A additions); regulate movement from nucleus to cytoplasm; localize mRNAs to certain parts of a cell; or regulate mRNA degradation. These controls are probably mainly effected by further regulatory proteins, and offer yet more scope for relatively controlled and potentially adaptive change.

We should also note here that there are some systems where RNA is itself the controlling agent, particularly during development. Nontranslated RNA (including in some instances antisense RNA strands and even mitochondrial ribosomal RNA) is known to play some role in particular aspects of patterning (e.g. the development of polarity, and induction of muscles) in embryos of *Drosophila* and *Xenopus*, and possibly also in tunicates and mammals. In this context, remember also that the promoter and enhancer sections of DNA are also crucial in development and evolution, though never normally transcribed and translated; it is not *just* proteins that have controlling roles.

2.4 Protein evolution

Point mutation is the most familiar means of genetic change, yet because of the repair processes outlined above only about one nucleotide in 200,000 is randomly (heritably) changed by this process every 1000 years. Thus mutation probably serves as no more than "fine-tuning" for most evolutionary change in proteins. The more important system promoting change is that of recombination, providing the "scissors and paste" of genetic change, especially the general recombination that occurs between homologous chromosomes during the crossing-over phase of meiosis, giving quite radical genomic change with duplications, deletions, or transpositions of whole sections of the nucleotide sequence (Fig. 2.13).

There are several features of protein synthesis and regulation that interact with recombination events to have a major effect on protein evolution:

1 Duplication resulting from recombination is particularly important (Fig. 2.14) because it means there are often multiple copies of genes available. Often these occur as multiple "tandem repeats" adjacent to each other on a chromosome, and while they stay adjacent there are repair mechanisms that tend to limit mutation in any

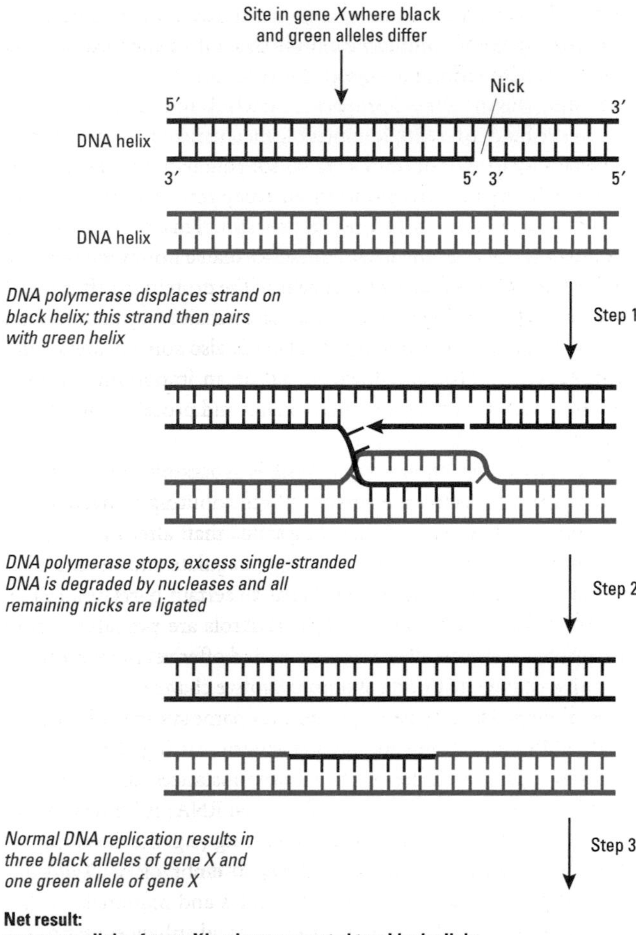

Fig. 2.13 A general recombination event causing gene conversion.

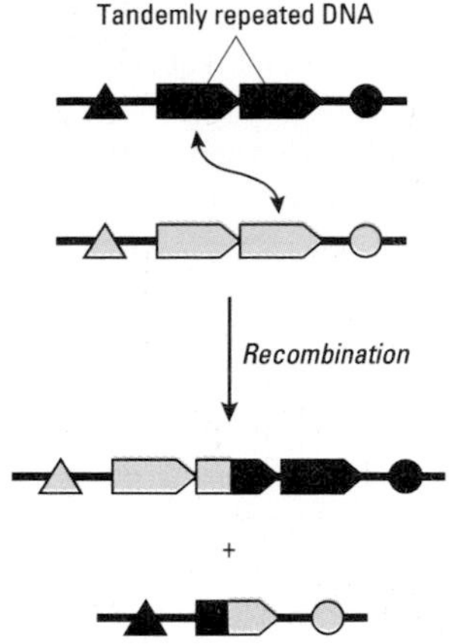

Fig. 2.14 Where there are tandemly repeated genes in a genome, copies are easily gained or lost due to unequal crossing-over between the sister chromosomes. The long regions of homologous sequence easily align with each other to produce this "error". (From Alberts *et al.* 1994, Fig. 8.73.)

one copy. Multiple identical copies may be very useful where large amounts of the transcript are needed; for example, ribosomal RNA genes in vertebrates may occur as hundreds of repeats, many of them adjacent. Heterozygote advantage may also bring about selection for duplications. Possession of two copies of the same gene allows an individual to be homozygous for different alleles at each of them but still effectively heterozygous. But if one copy of a gene is translocated away by further recombination then mutation becomes more likely, and mutational changes in any one copy are not too problematic for the organism. Some altered copies may never be of any use (they are then termed **pseudogenes**), but others may accumulate mutations and eventually acquire new functions. This gradually leads to the evolution of **gene families**, with related proteins having new functions.

2 The intron–exon system is another important factor, since it means that a "gene" is actually a series of separate bits of nucleotide sequence, including protein-coding sections and regulatory sections. It is presumed to have persisted in eukaryotes because it makes protein synthesis in a multicellular organism much more versatile. Because of RNA processing steps the transcripts of genes can be spliced together in various ways, perhaps with different combinations of regulatory systems, depending on the cell type and developmental stage, so that in practice different but related proteins can be produced from the same gene.

3 The system of incorporating introns must also have led to extra possibilities of nondamaging recombination events between exons, speeding up the process whereby organisms can evolve entirely new proteins with new functions from parts of preexisting proteins. Further families of genes result. Classic examples of gene groups resulting from exon recombinations are the immunoglobulins and the albumins, discussed in Chapter 17.

4 At least half of all spontaneous mutations observed in *Drosophila* are thought to be due to transposable element effects. Sometimes this is a direct effect on the gene, but often it is because a transposable element affects the timing or the spatial pattern of expression of a nearby gene, having subtle effects on animal development. There is some evidence that transposable elements undergo long periods of quiescence in the genome and then have sudden bursts of activity, known as "transposition bursts". These may be triggered by environmental change, giving an obvious link between environment and adaptation. The bursts seem likely to result in increased biological diversity, as the transpositions will potentially bring together two or more new traits that were of little value alone but that become very useful when working together; thus these transposition bursts can produce randomly modified progeny. If it is true that they can do so particularly at times of environmental stress, then the transposable elements may be acting not as mere parasites in the genome but as useful symbiotic factors, generating diversity just when it is most needed.

5 Another possible source of gene families is the joining together by recombination of DNA sequences coding for stable domains, making entirely new proteins but with some shared ancestry; many proteins do show signs of being assembled from shuffling together of preexisting domains in new combinations.

6 There may be post-translational joining of existing proteins into even larger "protein complexes", multimeric forms with molecular weights of 10^5–10^6.

7 Both duplication and recombination may be involved in putting together new genes and new proteins with multiple internally repeated sequences; the collagen family, for example, always has multiple repeats of an ancestral sequence that gives a particularly stable fibrous structure.

8 A different sort of factor that may help explain why new proteins largely evolve by alteration of old ones is the very low probability (referred to earlier) of generating a stable functional new protein from scratch.

For any or all of these reasons, there may be many variants of any one protein, termed **protein isoforms**; these may be the product of different but related genes, or may be the product of one gene modified by RNA processing. Furthermore, genes themselves can clearly be united into gene families and these may show quite high levels of sequence homology, usually also retaining very high levels of three-dimensional structural homology. In most animals multiple variant forms of almost every gene exist: different opsin genes for each kind of photoreceptor, different collagen genes for each kind of connective tissue, or for each stage of the developmental process from embryo to adult, and so on. Families of genes can also be traced between species and often between taxa at much higher levels. Gene families such as globins, collagens, and actins occur throughout the animal kingdom in multifarious forms. As an example, the evolutionary history of the globin genes is traced in Fig. 2.15.

Increasingly the evidence points to many multigene families being the results of ancient introns splitting up genes and providing new sites where the genetic material could be shuffled into new arrays. As we have seen, some motifs and domains also occur very widely indeed, shared between otherwise unrelated proteins. Protein evolution can, then, be viewed largely as a trial-and-error system of recombining separate exons coding for separate protein domains. This will regularly produce new proteins, and there is a high probability of these retaining a good degree of functional integrity but exhibiting slightly altered properties on which selection can act.

The existence of multiple copies of genes means that every animal may have a range of options as to which proteins it uses at any one time or in any particular place. This can happen in two ways:

1 Different variants of a protein (or specifically of an enzyme) occur within one organism, and these are termed **isozymes**; thus we may speak of a muscle dehydrogenase, a liver dehydrogenase, etc., all expressed within one individual. There may also be more than one such variant expressed in one cell type, and this opens up considerable scope for acclimatory change. For example, an animal may express one isozyme in its muscles at low temperatures and switch this off and another on (or at least change their proportionate expression) under different environmental conditions with higher temperatures. One isozyme may have greater thermal stability than another, or better performance at a particular pH, and so on.

2 Genetically distinct variants (alleles) may occur at a single locus, and where the resultant proteins have different properties they are termed **allozymes**. Populations in different environments commonly have different allele frequencies. Here we have scope for truly adaptive change, with one allozyme being selected in one part of a species' range and another in some other part of its range, leading to gradual diversification and potential speciation. Geographic trends in allele frequencies linked to a selective gradient are termed **clines**.

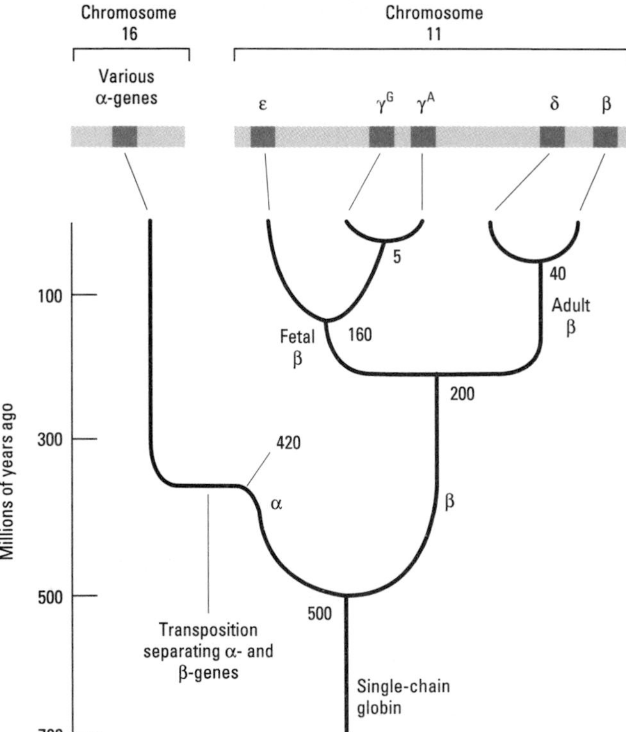

Fig. 2.15 Evolution of the globin gene family in animals. The location of these genes on human chromosomes is also shown. Numbers indicate calculated divergence times (millions of years).

2.5 Physiological regulation of gene expression

We have considered thus far the ways in which proteins may be synthesized, controlled, and degraded within cells largely in relation to proteins as the immediate products of genes. However, proteins are also the main targets of messages from other possible sources, both from within the cell and from the outside environment; this kind of regulation of proteins is the process underlying most of what we conventionally call physiology. The proteins concerned in this control and signaling system may be at the cell surface or intracellular, the latter including agents we have already met (protein kinases, protein phosphatases, GTP-binding proteins) as well as receptor proteins and all the other proteins with which all these key agents interact. There are also some smaller nonprotein components of intracellular signaling, however, and these are ubiquitous in all living cells. Together these provide the systems whereby gene expression is communicated to other cells and to all parts of a functioning animal.

2.5.1 Receptors

All signaling molecules, to be effective, have to be recognized by a specific receptor in or on their target cell. The signaling molecules are released by exocytosis or diffusion, and they are extremely varied, including proteins, polypeptides, small peptides, or single amino acids, as well as nucleotides, fatty acid derivatives, steroids, and even strange molecules such as nitric oxide. But regardless of their chemical nature nearly all these signals act rapidly on their target cells via specific receptor proteins (the exception is nitric oxide which can diffuse through membranes and act directly on target enzymes). The signals are commonly at extracellular concentrations of only 10^{-6} to 10^{-9} M and the receptors have very high affinities for them. The signal molecules act as ligands to the receptors so that the receptor protein undergoes a conformational change, usually across the span of the cell membrane, and triggers a series of intracellular signals. We know much more about the nature of receptors since we have been able to target them with recombinant DNA technology and produce large amounts of purified receptor protein.

Receptor proteins are of four main type: one type is intracellular, while the other three main families are cell surface receptors.

Intracellular receptors

Steroid hormones, together with thyroid hormones, vitamin D, and the vitamin A-related retinoids, share a hydrophobic character and act by diffusing through the cell membrane to act directly on intracellular receptors (see section 10.1.2). These receptors belong to the steroid hormone–receptor superfamily, characterized by the zinc finger motif in a key site. Some of these steroid hormones can persist in the blood for hours or days, having long-lasting effects. The receptors to which they link all bind to DNA sequences termed **hormone response elements** to activate gene transcription. Often one or more small groups of genes are activated quickly and the products of these genes in turn activate more genes, giving a long and complex response pattern of gene expression. For example, Fig. 2.16 shows the time course of various egg-white protein mRNAs after the application of estrogen. At the same time certain other target genes may be repressed by the hormone–receptor complex; for example, prolactin genes are repressed by glucocorticoid action.

Enzyme-linked receptors

These receptors either function directly as enzymes themselves or are closely linked to enzymes; mostly they are themselves, or are linked to, a protein kinase. The ligand-binding site is outside the cell and the catalytic site inside.

Ion channel-linked receptors

These perform the function of both channels and receptors, and are involved in rapid signaling between excitable cells, mediated by fast-acting locally produced neurotransmitters. The signal molecule causes a transient opening or closing of a channel through the protein structure (see section 9.2.3).

G-protein-linked receptors

A G-protein is a trimeric GTP-binding regulatory protein, as we met before. G-proteins sit in membranes adjacent to both a receptor protein and another target protein, which may belong to either of the two categories already described (channel-linked or enzyme-linked). This is therefore a slightly less direct form of receptor system, but it is much the most common. Its core structure includes a long extracellular binding domain and a chain that folds back and forth across the membrane seven times in all; intracellularly there are G-protein-binding domains and phosphorylation sites (Fig. 2.17).

For all these last three types of cell surface receptors, signals received at the surface must be relayed to the interior of the cell, and often to the nucleus so that the expression of particular genes can be affected. Ion channel-linked systems are a rather special case dealt with in Chapter 9, but in most examples of the other two categories the immediate result of receptor activation is a cascade of phosphorylations mediated by a second messenger.

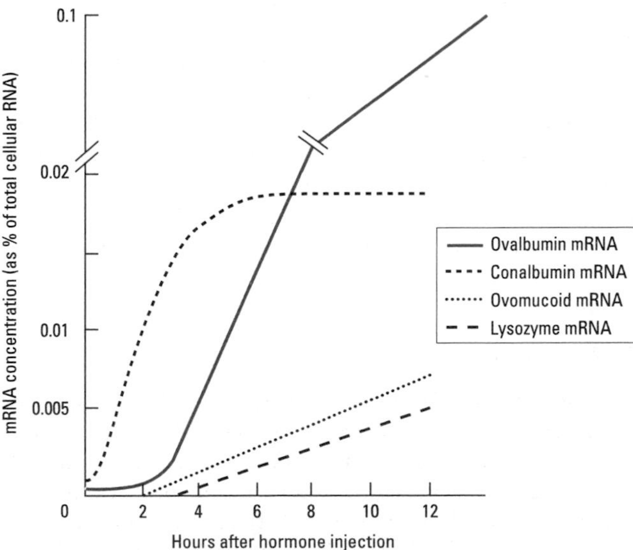

Fig. 2.16 Time courses of the appearance of various "egg white" protein mRNAs, after induction by the hormone estrogen. (Adapted from Palmiter *et al.* 1977.)

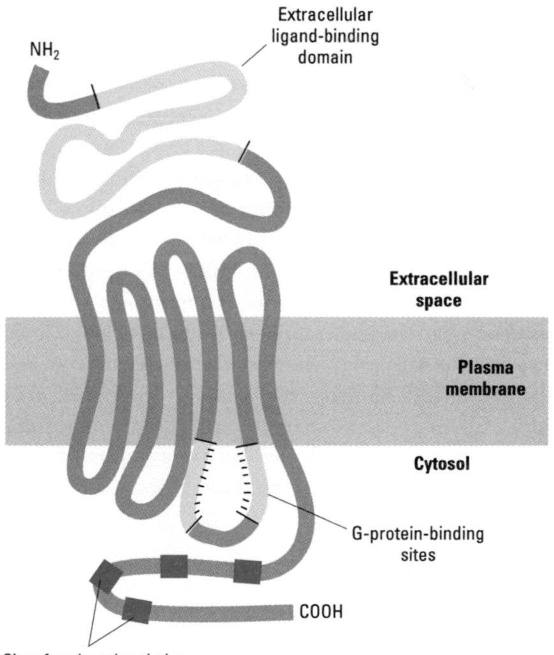

Fig. 2.17 Schematic view of a single G-protein-linked receptor, showing the typical seven folds across the membrane.

2.5.2 Intracellular mediators and second messengers

Molecules that help to control major pathways and systems within cells are those that link between protein receptors and their target enzymes or structural proteins or motor proteins or ion channels, or of course their target genes. The receptors are most commonly G-protein-linked receptors, as described above, so the linking stage is usually a G-protein in the cell membrane. The messengers produced are often small enough (unlike proteins or polypeptides) to move through gap junctions between adjacent cells, so that signaling can be shared amongst cell communities. They allow extracellular signals to be greatly amplified and also to be integrated together. Two of these "second messengers" are particularly common—**cAMP** and ionic calcium; these interact with each other and also with other small messenger molecules that are produced in related reactions, as we will see.

Cyclic AMP (cAMP)

In animal cells cAMP is a ubiquitous transmitter between signal receptor sites on the exterior surface of cells and the regulatory enzymes of metabolism. It is synthesized from ATP by a membrane-bound enzyme **adenyl cyclase**, and rapidly destroyed by **cAMP phosphodiesterase** enzymes. When a signal is received that activates the cAMP system, the cytoplasmic level of cAMP can increase at least five-fold in a few seconds; this is achieved by modification at the synthesis end (via adenyl cyclase) rather than at the breakdown end.

It works primarily via the amplifier system known as "cAMP-dependent protein kinase" (or **kinase A**), being linked to the receptor via one or two G-proteins (Fig. 2.18). There may be two G-proteins (G_s and G_i, indicating stimulatory and inhibitory, respectively) because usually each kinase molecule can be activated or inhibited by ligands binding to separate receptor sites. For example, activation may be by a β-adrenergic agonist or by adrenocorticotropic hormone, and inhibition by an α-adrenergic agonist, muscarine or an opiate. Inside the nucleus, the eventual product of cAMP-induced phosphorylations may bind to a cAMP-response element, adjacent to the transcribed gene, to produce the final response.

The action of 30 or more peptide and catecholamine hormones, as well as some prostaglandins, proceeds via pathways involving elevated cAMP concentrations and the activation of kinase A systems (see Chapter 10). Yet individual hormones trigger distinct and diverse physiological responses in the same cell. How is this achieved? A probable mechanism is that individual kinase pools are compartmentalized at their site of action, with specific anchoring proteins, and only become active when the cAMP in the immediate microenvironment becomes elevated by a particular hormone. In addition, individual hormones activate specific protein kinase subtypes.

Calcium

Calcium is normally regulated by specific ion pumps at very low levels (around 10^{-7} M) intracellularly, but higher (usually > 10^{-3} M) extracellularly and in the endoplasmic reticulum (ER). The large transmembrane gradient means that calcium can serve as a messenger simply by opening Ca^{2+} channels and producing an automatic inward flux with a rise of cytoplasmic calcium concentration. When a particular G-protein (G_q) is activated by an appropriate signal it in turn activates a membrane-bound enzyme, **phospholipase C**. This then hydrolyzes a specific phospholipid in the cell membrane to produce a soluble intermediate, **inositol triphosphate** (IP_3) (Fig. 2.19). The IP_3 in its turn travels through the cytoplasm and releases calcium (via IP_3-gated Ca^{2+} channels) from internal membrane stores. These are formed from modified portions of ER, which also contain calcium-binding proteins (such as calsequestrin and calreticulin) and pumps to resequester the calcium so that it can be used again. Calcium thus released from the ER serves as an intracellular messenger and as such is even more widespread than cAMP. Calcium levels in the cytoplasm of a stimulated cell often show marked oscillations persisting as long as the receptors are activated, suggesting complex positive and negative feedback systems are in operation.

Calcium ions may then activate their targets by directly binding to enzymes as an allosteric ligand. Examples of enzymes that directly bind calcium ions are the Ca^{2+}-inhibited guanyl cyclase of rod cells (see section 9.8.4) and protein kinase C. Alternatively, calcium may act indirectly by binding to **calmodulin** (CaM). CaM is a protein that has four Ca^{2+}-binding sites, and a relative molecular mass of about 16,000; it may make up as much as 1% of the total protein content of an animal cell. As more Ca^{2+} is bound so more and more hydrophobic regions are exposed in the molecule. Ca^{2+}–CaM itself acts directly by binding to enzymes, or indirectly through a CaM-

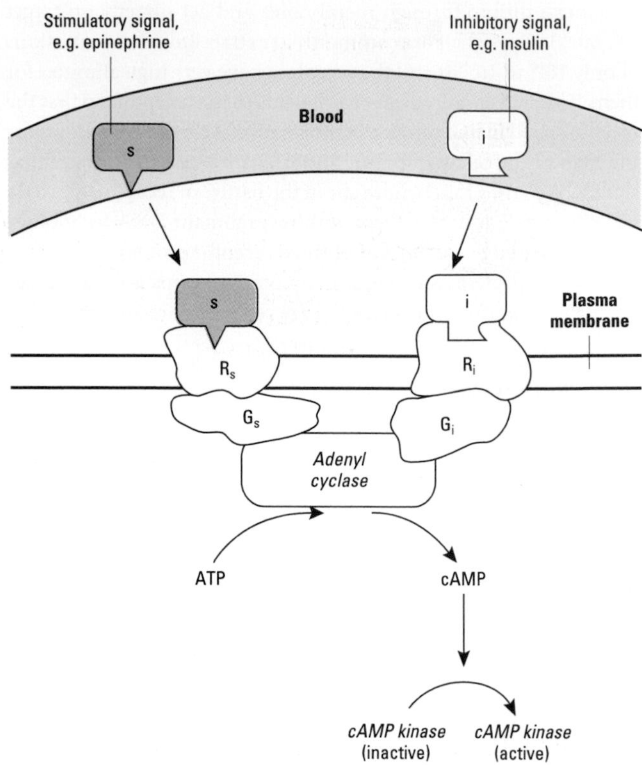

Fig. 2.18 Interaction of two G-proteins with a single cAMP-producing adenyl cyclase, giving both stimulatory(s) and inhibitory(i) pathways. G, G-proteins; R, receptors.

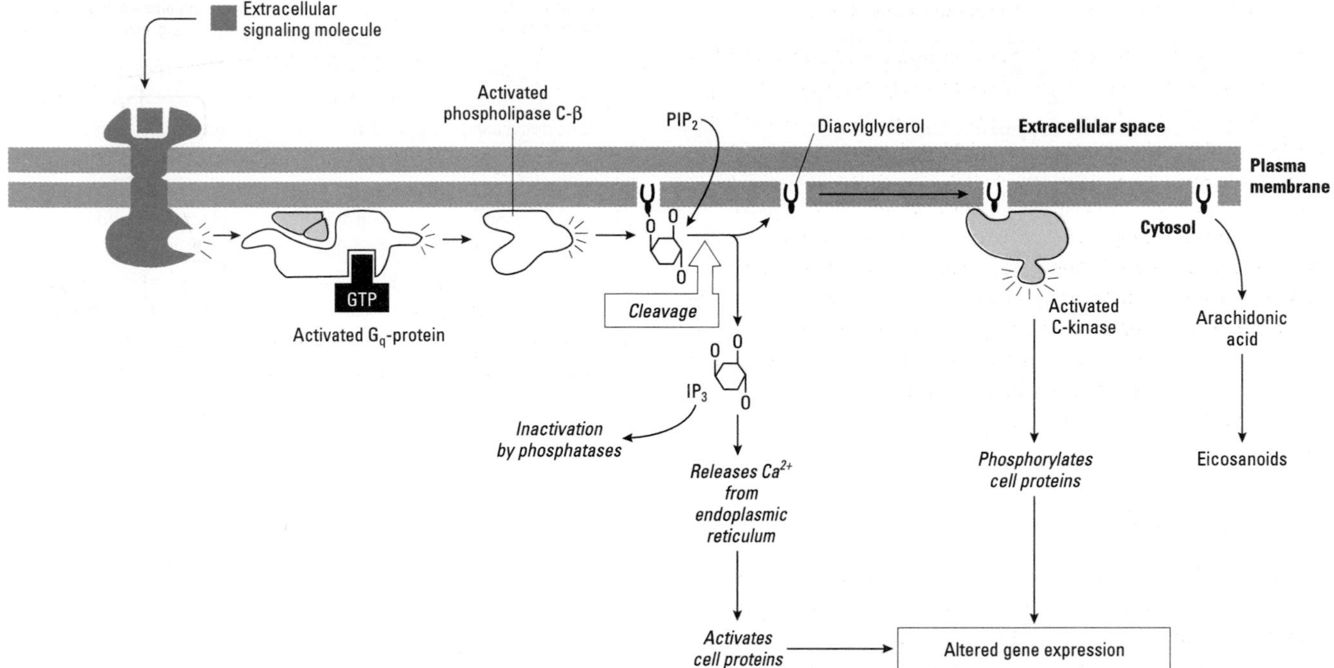

Fig. 2.19 The roles of inositol triphosphate (IP_3) and diacylglycerol as intracellular messengers. Both are produced from cleavage of a specific membrane phospholipid following activation of a G-protein by an extracellular signal. PIP_2, phosphatidylinositol biphosphate. See text for more details. (From Alberts *et al.* 1994, Fig. 15.13.)

dependent multifunctional kinase (PKII or CaM-kinase II). This kinase is particularly important in neural cells, and because it can phosphorylate itself it remains active even after the Ca^{2+} has been sequestered, giving it a role in certain kinds of memory (see section 9.5.4). Examples of enzymes mainly regulated by calmodulin are Ca^{2+}-ATPase, brain adenyl cyclase, myosin light chain kinase, and phosphorylase kinase.

In all animals calcium is especially important as the immediate trigger for muscle contraction, working via a sarcoplasmic reticulum (SR) receptor similar to that on the ER, and called a ryanodine receptor (see section 9.11.2). In some invertebrate muscles and in all vertebrate striated muscles, control is exerted via released Ca^{2+} binding to troponin C, which is one of the thin (actin) filament proteins. By contrast, in numerous invertebrate muscles and in vertebrate smooth muscles calcium control of contraction is exercised at the level of the thick (myosin) filament (see section 9.13). In most of these cases a cAMP-dependent protein kinase/phosphatase pair is involved, which changes the phosphorylation state of a small subunit of myosin thus turning contraction on and off. Relaxation occurs because calcium is sequestered back into the SR by specific pumps. Thus very localized changes in intracellular Ca^{2+} concentration can have marked physiological effects.

cAMP and calcium interactions

The cAMP and Ca^{2+} intracellular pathways interact in several important ways. Enzymes that influence cAMP levels (adenyl cyclase and phosphodiesterase) are regulated by Ca^{2+}–CaM complexes. Enzymes regulated by cAMP and by Ca^{2+} can also influence each other, or may have interacting effects on shared target molecules. The phosphorylase kinase of skeletal muscle is a classic example, responsible for phosphorylating the key enzyme that mobilizes glycogen.

Phosphorylase kinase has four subunits, one of which is catalytic, one which is regulated by Ca^{2+}–CaM, and two which are regulated by cAMP-mediated systems. Since calcium also controls muscle contraction as we saw above, coordination of crucial ATP-utilizing (myosin ATPase) and ATP-generating (glycolysis) pathways is assured.

Inositol triphosphate, diacylglycerol, and protein kinase C

When IP_3 is released by phospholipase C causing hydrolysis of a membrane phospholipid (see above and Fig. 2.19), the other cleavage product is **diacylglycerol** (DAG). The production of these two second messengers is initiated by a variety of signals, including neurotransmitters, hormones and growth factors. IP_3 and DAG have roles in the control of numerous cellular processes, including gametogenesis, fertilization, cell growth, secretion, smooth muscle contraction (as above), sensory perception, and neuronal signaling.

IP_3 is at the focal point of two major control systems, one involving G-protein-linked receptors and the other involving tyrosine-linked receptors. These signaling pathways show a diversity of individual components, enabling precise control of the function of a huge variety of specialist cells. Diversity is generated by the stimuli that act through separate receptor types, and by heterogeneity in the G-proteins, phospholipase C, intracellular IP_3 receptors, and protein kinase C. The latter is a serine/threonine kinase, particularly common in brain tissues; it starts a phosphorylation kinase cascade that leads to at least two important changes in gene regulatory proteins.

DAG has two important roles in cell signaling. Firstly, it directly activates protein kinase C thus interacting with the IP_3 pathways. Secondly, it is a precursor for **arachidonic acid**, which is a messenger in its own right and also serves in turn as the precursor for **eicosanoids** (which include the **prostaglandins** and are important effectors in pain, inflammation, and fever responses). Chapter 10 considers these issues in more detail.

Cyclic GMP (cGMP)

Cyclic GMP is generated directly from an enzyme-linked receptor; this receptor is guanyl cyclase, and there is no intervention of G-proteins. The role of cGMP is at present less well understood than that of cAMP, although an increasing number of cGMP-sensitive protein kinases have been characterized. One important role for cGMP is in rod vision in vertebrates. Light energy falling on the receptors in rod cells of the retina is transformed into a chemical change and amplified, closing Na^+ channels in a membrane to induce hyperpolarization, thus initiating a nerve impulse (see section 9.8.4).

Calcium and cGMP are also linked together in some situations by a nitric oxide synthase, where **nitric oxide** (NO) itself is an intermediate signaling molecule. The complex interactions of calcium and cGMP with the NO system are intricate and somewhat variable between tissues, with effects in neural cells and arterial smooth muscle proving to be particularly important in vertebrates. Further examples of the role of NO in cell–cell interactions are described for parasitic environments in Chapter 17.

The major target and effect of all these second messengers is, ultimately, altered gene expression, giving a host of ways in which genes can be turned on or off as needed in response to signals perceived at the cell surface. Now we need to consider what those signals may be.

2.5.3 Extracellular control signals: growth factors, hormones, neurotransmitters

The major extracellular signals for activating intracellular systems are local mediators between cells (paracrine signals), neurotransmitters delivered from nerve cells via chemical synapses, and hormones, which may be either local protein "hormones" (which we often term growth factors) or more conventional hormones delivered at longer range and transmitted through the blood circulation. All of these bind to the receptors, usually in cell membranes, which transduce information into the cell, to start a cascade of intracellular events affecting channels, pumps, enzymes, the cytoskeleton, and numerous other proteins and genes.

These signaling molecules have varying half-lives and thus work on very different timescales. In relation to adaptive change, it is the hormones that are the major controllers of genetic expression and thus of protein synthesis. They can influence differentiation of dividing cells, and change the patterns of gene expression in already differentiated cells. They may activate master switches that can trigger expression of whole suites of genes, particularly where they are involved in developmental programs (the juvenile hormone in insects is a familiar example) or have trophic effects on the growth of whole tissues. They may mediate exchanges between the whole animal and the outside world, via effects on excretory and secretory processes and on behavior. But they may also have rather more subtle effects, producing different outcomes on different cells. For example, acetylcholine causes contraction of skeletal muscles, but produces decreased rate and force of contraction in heart muscle cells and causes secretion in yet other nonmuscular cells. Dopamine plays a role in numerous aspects of behavior (sexual responses, hunger, thirst, vigilance) as well as influencing lactation and immune responses. Hormones largely achieve this diversity of response by being effective with several different G-protein-linked receptors; for example, epinephrine is known to link with at least nine such receptors, and dopamine with at least five.

Although these issues are dealt with fully in Chapter 10, a good example of multiple hormonal effects for now is the adrenergic stimulation of a vertebrate heart muscle cell (Fig. 2.20). Norepinephrine binds to β-adrenergic receptors on the cell membrane, which in turn interact with G-proteins complexed with GDP. The GDP is exchanged for cytoplasmic GTP to produce activated forms of the G-proteins. The activated G-proteins interact with adenyl cyclase in the membrane and produce an increase in the intracellular cAMP as described above. The cAMP activates protein kinase A, which can then phosphorylate a range of target proteins. It may convert the inactive β-dimer of phosphorylase to the active α-dimer. At the same time the kinase A activates Ca^{2+} channels leading to an increase in calcium concentration in the cytoplasm; this further activates glycogen phosphorylase, and also overcomes the inhibition of myosin ATPase activity, leading to muscle contraction. There is enormous amplification of the initial signal, and coordination of a variety of enzyme sequences that produce and then utilize ATP, giving an immensely complex but organized response to the hormone.

As a second illustrative example we might consider the action of **growth factors**, small extracellular proteins acting through specific surface receptors and of immense importance in developmental programs. They have local effects, not involving the circulatory system, and they mediate crucial processes like cell division and cell differentiation. Their action does not involve G-proteins or cAMP, but instead a series of intermediates triggered via a tyrosine-specific protein kinase. Under the influence of this kinase, the protein P60 becomes phosphorylated to PP60, and a chain reaction results, the last product acting on gene expression in the nucleus.

Additional sophistication in control systems at the whole organism level (reflecting control at the protein synthesis level) comes about due to the pulsatility of many hormones. Secretion of mammalian growth hormone is cyclic, due to an underlying rhythm in the release of two hypothalamic peptides (growth hormone-releasing hormone and somatostatin); one or both of these peptides also leads to a cyclicity in the release of the main gonadotropic hormones, luteinizing hormone (LH), and follicle-stimulating hormone (FSH) (see Chapter 10). This appears to result in turn from a pulse generator in cells of the basal hypothalamus, possibly due to reciprocal inhibition circuits in two sets of neurons (see Chapter 9). The frequency of hormonal release can thus be controlled in its turn by neural signals related to intracellular signals, giving fine-tuning to the system and a new layer of regulation at which adaptation could operate.

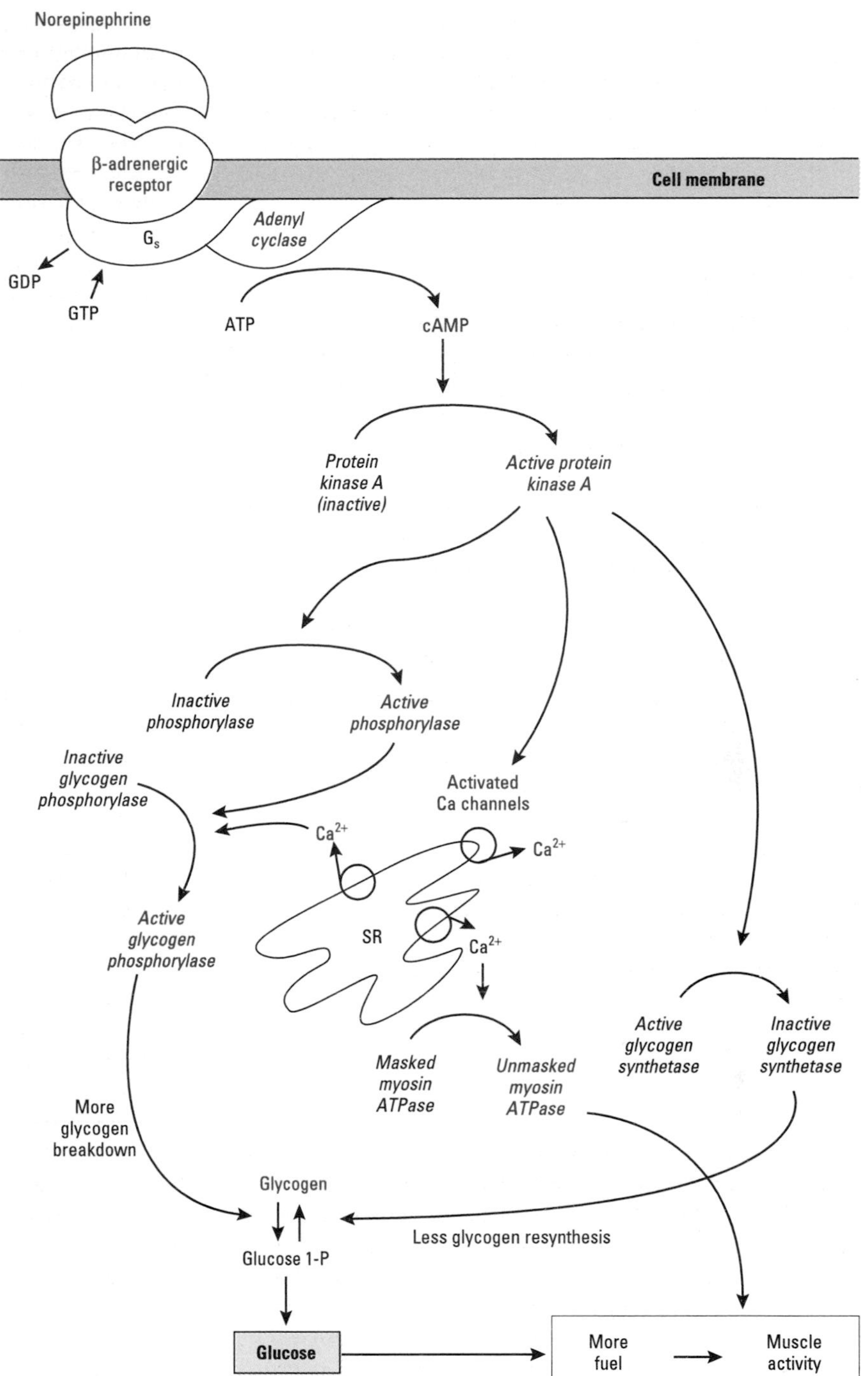

Fig. 2.20 The action of a stimulatory hormone, norepinephrine (noradrenaline), via cell membrane receptors, on various intracellular processes, leading to increased fuel supply and increased muscle activity. SR, sarcoplasmic reticulum.

2.5.4 Feedback systems and metabolic control

Repeatedly through the discussions in this chapter we have met examples of feedback regulation of one protein or molecule by another. This is a fundamental process in all organismal function, and it is worth dealing with the concept here as it will recur throughout other chapters of this book. The essential principles are shown in Fig. 2.21.

We can most easily examine it by looking at specific cases of enzyme regulation. Enzymes, of course, are protein catalysts that convert one kind of molecule to another. The flow of carbon through complex metabolic pathways in animals (discussed fully in Chapter 6) is controlled by enzymes, which are subject to all the general kinds of control outlined in the previous sections. The control systems operating are generally reversible, and types with and without some kind of amplification are found.

Where no amplification is present control is usually by a simple feedback; an end-product accumulates or disappears and this provides the **positive feedback** or **negative feedback** by acting allosterically on one of the key enzymes earlier in the pathway to change its

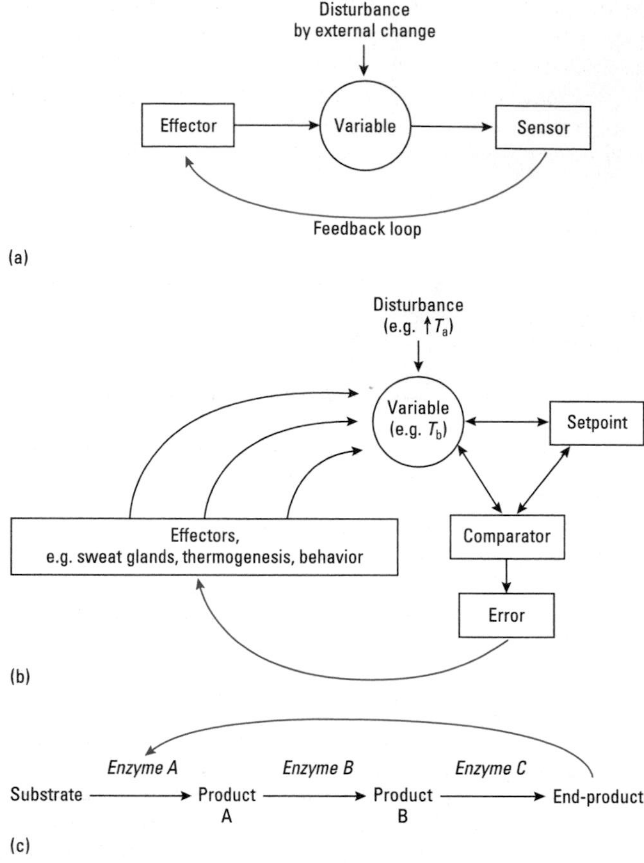

Fig. 2.21 (a) Principles of feedback control. (b) A more complex system where there is a setpoint and a comparator detecting any mismatch with the setpoint and producing an error signal, as in many physiological systems. T_a, ambient temperature; T_b, body temperature. (c) The commonest type of negative feedback at the biochemical level, underlying many enzyme regulation systems; accumulating end-product here serves as the "error" signal and feeds back to an earlier enzyme as the "effector".

activity. Negative feedback is extremely common, and most often involves the final product of a pathway feeding back as an allosteric inhibitor onto the first enzyme in the pathway; thus if large quantities of the product accumulate any wasteful entry of precursors into the path is automatically prevented. An example of the rather less common phenomenon of positive feedback occurs where ADP, the precursor of ATP, is a ligand for several enzymes involved in the pathways of catabolism of foodstuffs that produce ATP. Here ADP is purely regulatory, taking no part in the actual reaction, but serving to insure an adequate rate of synthesis of ATP. Another example is that of fructose 1,6-bisphosphate (F-1,6-BP) in the liver; this is a product of an early stage of glycolysis (see Fig. 6.6) that activates the terminal enzyme in glycolysis *before* the "slug" of metabolites reaches it. This insures that the central part of the glycolytic chain can work at maximum throughput, limited only by the supply of ADP and phosphate. This control mechanism is stable and works well because as the concentration of F-1,6-BP falls so the later acting pyruvate kinases also lose activity. Thus, although positive feedback sounds potentially risky, stability will always be gained if the response times of an early and a late enzyme in the sequence are different.

In more complex enzyme-control systems that have built-in amplification there may be very high gains, i.e. a single signaling molecule can bring about the activation of a large number of enzyme molecules. Systems with built-in amplification typically have a cascade mechanism involving reversible phosphorylations, as in Fig. 2.3. How big the amplification is depends on a number of factors, including the speed of dephosphorylations (related in part to the concentration of the phosphatase enzymes) and the number of steps in the cascade. For example, it has been demonstrated experimentally that for glutamine synthase the amplification is 67 for a single stage, and depending on the conditions it may be 250–1500 for a two-step cascade. Control steps work on different timescales, and this is very important for the integrated functioning of metabolism, as it provides great flexibility. For example, the immediate conformational responses of enzymes to ligand binding can occur in milliseconds, whereas covalent modifications such as phosphorylation are rarely complete in less than a second and can take many minutes. The effects can be seen in the multistep cascade leading to the activation of glycogen phosphorylase, shown in Fig. 2.20, which has four steps or cycles. This shows the utility of having control steps with different time courses: because the activation of glycogen synthase is much slower than the activation of the phosphorylase, the futile cycling of glucose residues (repeatedly adding and removing phosphate groups) is avoided.

Regulated enzymes are often controlled by more than one feedback mechanism. Usually, the important flux-controlling enzymes have at least two different control systems. For example, the membrane-bound enzyme hydroxymethylglutaryl-CoA reductase is subject to feedback control by cholesterol and to allosteric regulation by other steroid metabolites. Its activity is also regulated by a reductase kinase and by a phosphatase. Furthermore, because the enzyme has a short half-life before being dealt with by proteasomes, changes in concentration also modify its activity. A final control is exerted by a marked diurnal rhythm in its rate of synthesis.

The complex patterns of enzyme regulation observed in present-day animals to some extent reflect a "fossil record" of the control mechanisms that have evolved. There are probably many examples where the importance of a particular control feature has diminished over time, perhaps due to a changing environment or to some evolutionary innovation. Unless it is positively harmful in the changed circumstances it is likely that the original regulatory domain on the enzyme will remain to supplement the new control system or will gradually assume a new function over time. There may thus be considerable hidden variation in the signaling pathways of cells, and recent evidence has indicated that this can lead to rather rapid change in morphology during periods of environmental stress. Here the heat-shock proteins (HSPs, covered in detail in Chapter 8) are implicated, these being molecules that bind and stabilize proteins. In periods of environmental stress the common form HSP90 is fully occupied with protecting the unfolded proteins in a cell and may therefore be unavailable to monitor and protect the signaling proteins directing development. In *Drosophila*, in the absence of HSP90 a small number of viable mutants therefore appear with a range of developmental malformations that were previously masked. HSP90 thus normally serves as a buffer against existing variation in morphogenesis; environmental stress enables this cryptic variation to be expressed, giving a mechanism whereby phenotypic change can be accelerated just when it is needed as the environment changes.

The need to have redundancy, cryptic variation, or "fail-safe" mechanisms in important control systems, together with the accumulated variety of past evolutionary experiments, probably accounts for the apparent complexity of the present-day regulation of enzyme activity.

2.6 Conclusions

It is the essential character of proteins that underlies biological adaptation. Small changes in protein shape govern most of what goes on in a cell, by producing switching effects or motor effects or by channeling other molecules across membranes and between cell compartments. These effects can be subject to extremely complex regulations and amplifications to produce controlled effects on whole organisms. Permanent changes in protein structure, brought about either by point mutations or more commonly by recombinational changes in the DNA sequence of a cell, can therefore produce subtle changes in enzymic activity, signaling activity, and subcellular morphology, and perhaps, above all, effects on the expression of other proteins. All of these can in turn lead to permanent heritable change in the development of organisms. This is what we see outwardly as change, at a series of levels. At the biochemical level there may be increased thermal tolerance of an enzyme, or expression of a more pH-stable allozyme; morphologically there may be developmental changes of gene expression leading to altered positioning of muscle cells, nerves, blood vessels, or even completely modified appendages; physiologically there might be heart rate increments under the influence of calcium and cAMP signaling regimes, or sodium pumping rate changes as more or different channel proteins are synthesized; and at the behavioral level there may be increased attack speed, greater sensitivity to a sex pheromone, or even "conscious" changes in response (the ultimate adaptive trick). All of these protein-induced changes provide the raw material for adaptation.

However, when all this is said, it remains the case that from a physiologist's viewpoint the organism is in some senses more than the sum of its molecular or genetic parts. Genes can only function, and can only be selected for, as part of a fully functional reproducing organism. Reducing the scale of analysis to molecules, even when these can now be experimentally manipulated, does not in the end "explain" what is going on at the whole animal level. That is why the rest of this book concentrates on adaptation at a higher level of organization, from cell and tissue to organism and environment, using a specific molecular understanding of adaptation only in those relatively rare instances where it has been elucidated. We hope, however, that this chapter has set the process of adaptation in a suitable molecular context, so that new information on the molecular interactions and genomic changes underlying ecophysiological modification can be easily assimilated as it becomes available.

FURTHER READING

Books

Alberts, B., Bray, D., Lewis, J., Raff, M., Roberts, K. & Watson, J.D. (1994) *Molecular Biology of the Cell*. Garland, New York.

Hardie, D.G. (1991) *Biochemical Messengers*. Chapman & Hall, London.

Hochachka, P.W. & Somero, G.N. (2002) *Biochemical Adaptation: Mechanism and Process in Physiological Evolution*. Oxford University Press, Oxford.

Storey, K.B. & Storey, J.M. (eds) (2000) *Environmental Stressors and Gene Responses*. Elsevier, Amsterdam.

Stryer, L. (1988) *Biochemistry*. Freeman, New York.

Watson, J.D., Hopkins, N.H., Roberts, J.W., Steitz, J.A. & Weiner, A.M. (1987) *Molecular Biology of the Gene*. Benjamin-Cummings, Menlo Park, CA.

Wolfe, S.L. (1993) *Molecular and Cellular Biology*. Wadsworth, Belmont, CA.

Reviews and scientific papers

Berridge, M.J. (1993) Inositol triphosphate and calcium signalling. *Nature* **361**, 315–325.

Borgese, F., Sardet, C., Cappadoro, M., Pouyssegur, J. & Motais, R. (1992) Cloning and expression of a cAMP-activated Na^+/H^+ exchanger; evidence that the cytoplasmic domain mediates hormonal regulation. *Proceedings of the National Academy of Sciences USA* **89**, 6765–6769.

DeCamilli, P., Emr, S.D., McPherson, P.S. & Novick, P. (1996) Phosphoinositides as regulators in membrane traffic. *Science* **271**, 1533–1539.

Divecha, N. & Irvine, R.F. (1995) Phospholipid signalling. *Cell* **80**, 269–278.

Gilman, A.G. (1987) G proteins: transducers of receptor-generated signals. *Annual Review of Biochemistry* **56**, 615–649.

Hunter, T. (2000) Signaling—2000 and beyond. *Cell* **100**, 113–127.

Pigliucci, M. (1996) How organisms respond to environmental changes: from phenotypes to molecules (and vice versa). *Trends in Ecology and Evolution* **11**, 168–173.

Rutherford, S.L. & Lindquist, S. (1998) Hsp90 as a capacitor for morphological evolution. *Nature* **396**, 336–342.

Snyder, S.H. (1985) The molecular basis of communication between cells. *Scientific American* **253**, 132–140.

Tanaka, C. & Nishizuka, Y. (1994) The protein kinase C family for neuronal signalling. *Annual Review of Neuroscience* **17**, 551–567.

Wurgler-Murphy, S.M. & Saito, H. (1997) Two-component signal transducers and MAPK cascades. *Trends in Biochemical Science* **22**, 172–176.

3 The Problems of Size and Scale

3.1 Introduction

Living organisms occur in a wide range of sizes (Table 3.1), spanning 21 orders of magnitude (10^{21}). The smallest living things are the viruses, which contain only genetic material, all their other functions being undertaken by the host cell. Mycoplasmas (also known as pleuropneumonia-like organisms) are only slightly larger, weighing less than 0.1 pg, and are the smallest things that can lead an independent existence and survive and reproduce in an artificial medium. The smallest animals (i.e. multicellular organisms in the kingdom Metazoa) are considerably larger at around 2–10 μg. At the other end of the scale the largest extant animals are mammals: the blue whale weighing in excess of 100,000 kg in the seas, and the African elephant at 5000 kg on land. Each taxon of animals has its own particular size range, largely dependent on inbuilt design constraints; some phyla such as Bryozoa or Platyhelminthes may contain species (both living and extinct) all within two or three orders of magnitude of mass, while others such as Crustacea and Mollusca, or the single class Mammalia, are hugely variable in size. For example, the smallest mammal is a shrew weighing only about 1 g when fully grown, thus being about 100 million times less heavy than the blue whale.

Just the fact of being a particular size has profound consequences for structural and functional relationships in animals. Most structures and physiological variables change in a predictable way with increasing body size. The study of these size-related effects is known as **scaling**. Knut Schmidt-Nielsen defined scaling as dealing with "the structural and functional consequences of changes in size or scale among otherwise similar organisms"; evolutionary biologists may think of it in terms of a major constraint on the adaptation of certain characters. Mass is most commonly used to measure the size of animals, although length is also useful in a few circumstances, for example in swimming fish. As with manmade structures, three parameters can be changed as the size of an organism is increased, namely the **dimensions**, the **materials** used for its construction, and the **design**. Changing dimensions or proportions as size increases is an obvious solution, widely used in the animal kingdom. A familiar example is that bones are disproportionately thicker in vertebrates as the mass of an animal increases, to support the extra weight and the forces occurring during locomotion without breaking. But the use of new materials, or an innovative design, are also often selected for to overcome some fundamental constraint associated with size or dimensions. There are many examples of this in the chapters that follow. For example, gas transport by simple diffusion is fast over very short distances but soon becomes limiting as body size increases. The evolutionary solutions to this problem have been mass transport by convection, either on the outside of the animal (ventilation), or inside the animal (circulation), or both. A whole range of specialized ventilated respiratory surfaces have been developed together with pumped circulatory systems to enable animals to overcome the constraints to maximum size posed by diffusion (see section 7.2).

Table 3.1 Ranges of organism and animal sizes.

Organism	Mass: Grams	Mass: Units to scale
Mycoplasma	10^{-13}	<0.1 pg
Typical bacterium	10^{-10}	0.1 ng
Tetrahymena	10^{-7}	0.1 μg
Amoeba	10^{-4}	0.1 mg
Rotifer	10^{-4}	0.1 mg
Aphid	10^{-3}	1 mg
Bee	10^{-1}	100 mg
Pygmy shrew	10^{0}	1 g
Hamster	10^{2}	100 g
Human	10^{5}	100 kg
Elephant	5×10^{6}	5000 kg (5 tonnes)
Blue whale	10^{8}	100,000 kg (100 tonnes)

3.2 Principle of similarity: isometric scaling

Relatively simple rules govern the relationships between geometrically similar objects. For example, the linear dimensions of isometric triangles are shown in Fig. 3.1. Geometrically similar objects are sometimes called **isometric**. The surface areas (SA) and volumes (V) of isometric objects are related to their linear dimensions to the second and third power, respectively. Since surface areas are $(\text{length})^2$ and volumes are $(\text{length})^3$, it follows that surface area is $(\text{volume})^{2/3}$, or volume to the power 0.67. If the surface area of a cube is plotted against its volume on logarithmic coordinates a straight line is obtained that corresponds to the equation:

$$\log SA = \log 6 + 0.67 \log V \quad \text{(see Fig. 3.2)}$$

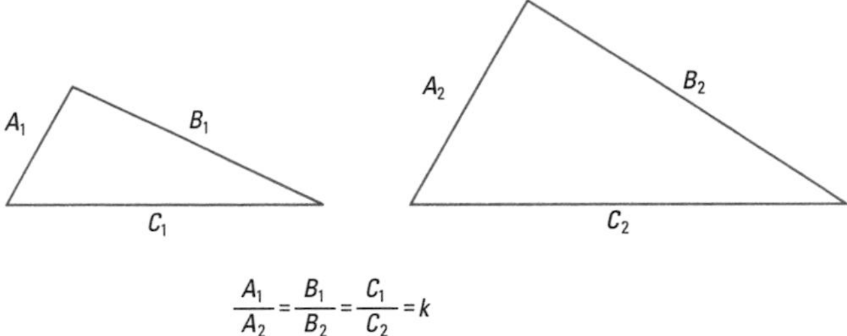

Fig. 3.1 Isometric triangles, showing that all corresponding linear dimensions, in any isometric figures, always have the same relative proportions.

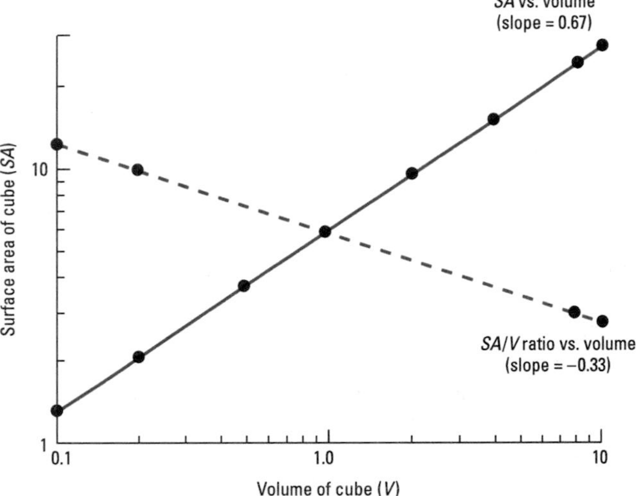

Fig. 3.2 Double logarithmic plot of the surface area (*SA*) vs. the volume (*V*) of a cube (solid line). The regression line has a slope of 0.67. The dashed line shows a plot of the surface area per unit volume against volume, with a slope of −0.33, showing that the relative surface area (the *SA* / *V* ratio) decreases with increasing size. (From Schmidt-Nielsen 1984.)

For other shapes the intercept 6 will change but the **isometric scaling exponent** of 0.67 remains the same; in other words the surfaces of isometric bodies are always related to their volumes by the power 0.67. Expressed another way, if the surface area per unit volume is plotted against volume a straight line is produced with a slope of −0.33. Thus smaller bodies have larger surface areas relative to their volumes than larger objects of the same shape. The surface area to volume ratio of cells, tissues and organs is often a major consideration in physiological comparisons.

In scaling, the units in which measurements are made are not in themselves important; scaling deals with dimensionless ratios. Box 3.1 gives an overview of dimensions and units as used in this book.

Box 3.1 Dimensions and units

If we write an equation using an equals sign, both the magnitude and the dimensions of variables on the two sides of the equation must be identical. The units used to describe the magnitude of a variable must always correctly reflect its dimensions in the three aspects of mass (M), length (L), and time (T).

According to Newton's second law of thermodynamics, **force** is the product of **mass** and **acceleration**:

$$F = ma$$

Acceleration is the rate of change of velocity, and velocity has the dimensions of length per unit time or LT^{-1}; thus acceleration must have the dimensions LT^{-1}/T, or LT^{-2}. Hence we know that force must have the dimensions MLT^{-2}.

Units represent an arbitrary but self-consistent system to represent the magnitude of a variable, and **SI units** (Système International) are now almost invariably used, where the standard units are **kilograms, meters,** and **seconds**. The unit for acceleration is therefore meter per second squared (m s^{-2}), and force is measured in kg m s^{-2}, also known as Newtons and symbolized as N. On average objects fall with an acceleration due to gravity of 9.81 m s^{-2}, and each kilogram of the object is subject to a gravitational force of 9.81 N.

By considering the dimensions of variables we can appreciate that the mass and the weight of an object are not the same thing, and should not be used interchangeably. It is nearly always appropriate to use the mass of biological materials.

The dimensions of some important physical, physiological, and ecological variables are as follows:

Variable	Description	Dimensions
Fundamental variables		
Mass		M
Length		L
Time		T
Derived variables		
Density	Mass/volume	ML^{-3}
Velocity	Distance/time	LT^{-1}
Acceleration	Velocity/time	LT^{-2}
Force	Mass × acceleration	MLT^{-2}
Stress	Force/area	$ML^{-1}T^{-2}$
Work	Force × distance	ML^2T^{-2}
Power	Work/time	ML^2T^{-3}
Mass specific power	Power/mass	L^2T^{-3}
Biomass density	Mass/area	ML^{-2}
Production	Mass/time	MT^{-1}
Productivity	Mass/(time × area)	$ML^{-2}T^{-1}$

3.3 Allometric scaling

If the relationship is plotted between some physiological or ecological function (e.g. oxygen consumption or gonad mass or reproductive output) and some measure of size (e.g. body mass or length), then there are three possible outcomes as shown in Fig. 3.3. There may be a straight line relation passing through the origin; then the variable changes in direct proportion or isometrically with body size, as outlined above, and the ratio between that variable and body size is always the same. Alternatively there may be a straight line relation that does not pass through the origin, or there may be a curvilinear relation; in either of these cases there is no simple ratio between the two measurements (thus a doubling of size does not mean a doubling of oxygen need, or gonad size, etc.) and the function varies **allometrically** with body size (from the Greek *allos* meaning different). In practice, even in animals of the same general

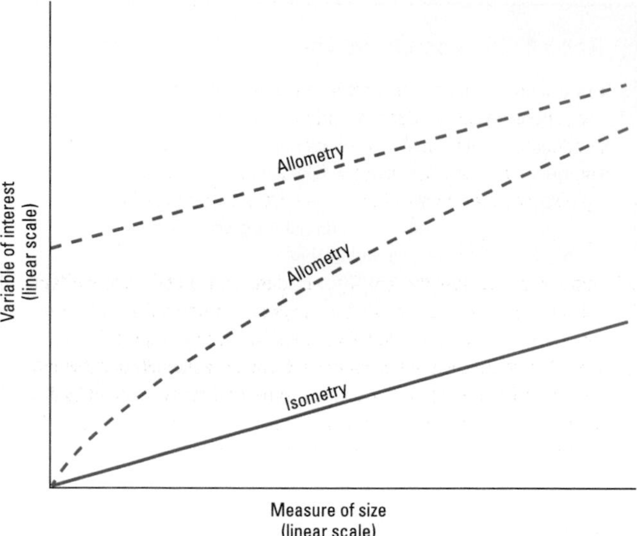

Fig. 3.3 Allometric and isometric variation. *Isometric* variation occurs when a particular variable (Y) is related to body size (X) by a straight line through the origin, so that the ratio Y/X does not vary with changing body size. *Allometric* variation occurs when the same plot intersects the y-axis at some point other than zero, *or* when the plot is a smooth curve; in either case the ratio Y/X varies with body size.

body plan, characters do not scale isometrically and biologists must deal with this nonisometric, or allometric, scaling.

By far the most common independent variable considered in issues of biological scaling is that of body mass (M_b). A great range of parameters of interest to physiologists has been found to vary with body mass according to allometric equations of the general form:

$$Y = aM_b^b$$

In its logarithmic form this becomes:

$$\log Y = \log a + b \log M_b$$

This is described as a "power formula" because the dependent variable Y changes as some power of the independent variable (body mass). Since the equation is of the general form for a straight line ($y = mx + c$), it follows that plots of Y against M_b on logarithmic axes always produce a straight line, with a slope of b and an intercept of $\log a$. The calculated intercept at unity body mass (a) is then known as the **proportionality constant**, and the slope or gradient b is the **mass exponent**. Box 3.2 provides some background on dealing with logarithmic and exponential relationships and on handling the simple mathematics involved.

The proportionality constant can be useful in comparisons between data sets, looking for differences in the level or setting of a particular parameter between animal groups. For example, marsupial mammals have a lower metabolic rate overall (about 30% lower) than do eutherian mammals, with the slope having a lower intercept (lower value of a); however, the relation between metabolic rate and size is the same for both groups (i.e. they have the same value of b).

The mass exponent b is usually the parameter of most interest. If Y increases in simple proportion to body mass then the slope b will have a value of 1.0. Examples of this are blood volume or lung volume in mammals, both of which are a constant fraction of body mass. If a parameter does not change with body size then the slope b will be zero; a good example here is the concentration of hemoglobin in vertebrate blood, which is broadly invariant across species. If Y increases less than would be expected by simple proportionality then b will have a positive value of less than 1.0, as with the surface area to volume ratio already mentioned; here another well-known example is the resting metabolic rate of mammals (see below). If Y increases more than expected, b will have a positive value greater than 1, as with the bone thickness of vertebrates already mentioned, where for mammals the exponent is 1.08. There may even be negative values for b, a good example being the relation between mammalian heart rate and size (25 beats per minute in an elephant, compared with up to 1200 per minute in a pygmy shrew), the exponent being −0.25.

The value of b provides information on how the variable of interest changes with body size. Similar values of b may indicate that similar principles are operating to determine the scaling of this variable. If b values are the same, then values of a are also directly comparable and can be used to make comparisons between different groups of animals (as above for marsupials and placentals). Note that values for the slope and intercept in real populations are not fixed values, but means about which natural variation occurs. This variation is usually quantified using confidence limits (Fig. 3.4). For a linear relationship the best fit line and the confidence limits about that line are calculated using various possible kinds of regression analysis. Confidence limits in Y are not a constant value above and below the line, but get wider towards the ends of the sampled variation in X. Confidence limits can also be calculated for the slope and for the intercept of the best fitted line, and are particularly affected by

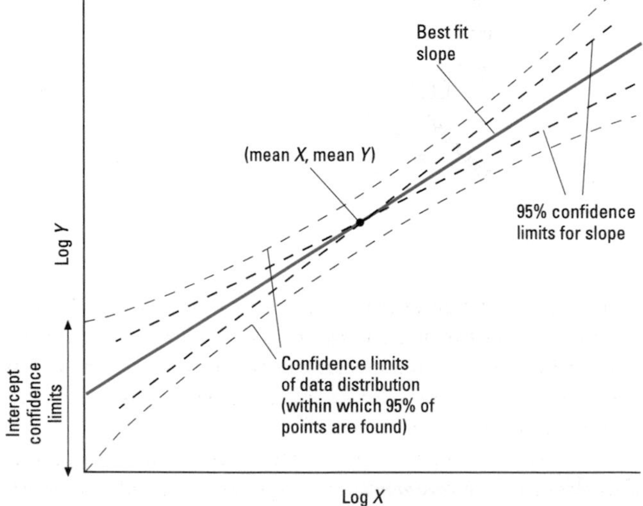

Fig. 3.4 A regression line calculated by the least squares method. Note that the 95% confidence limits for the regression get wider towards the upper and lower extremes of the range. The slope will be affected more by an outlier here than by a point near the middle of the line. 95% confidence limits for the slope make the line more or less steep by rotation about (mean X, mean Y) (dashed lines).

Box 3.2 Use and handling of logs and exponents

Why use logs?

Taking logs is useful in biological comparisons for several reasons. One is illustrated below: on a log scale it is much easier to compare diagrammatically across a very large size range. With a base 10 scale and exponents, the size axis here can be more easily expressed.

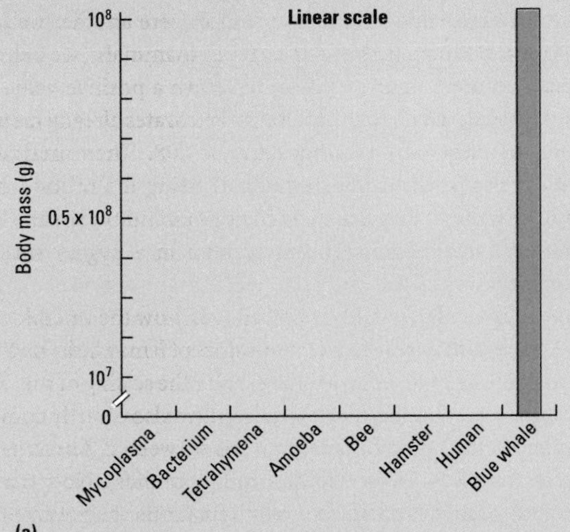

(a)

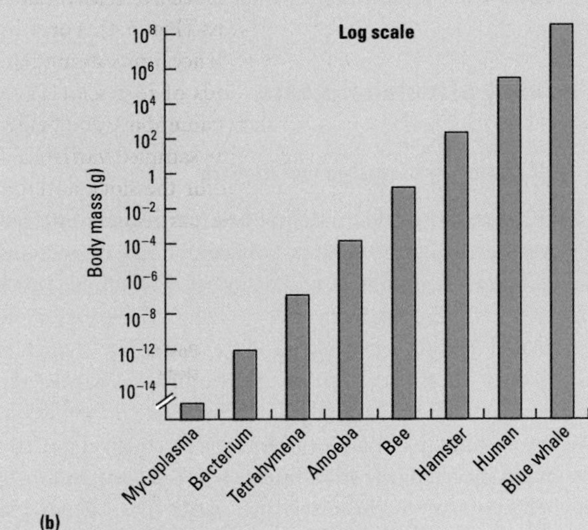

(b)

Taking logarithms is also useful because it commonly transforms curvilinear biological relationships into a linear form, making statistical analysis much easier. The curvilinearity very often results from a relationship between two variables that is a power function (see text) and taking logs produces an equation that represents a straight line:

$$\log Y = \log a + b \log X$$

It is tedious to calculate the logs for each data point, but the same effect is achieved by plotting untransformed data on log–log paper. Note that the intercept occurs where $\log X = 0$, i.e. $X = 1$. Linear regression then gives a value of the correlation coefficient r, or more usefully of r^2, which describes the fraction of the variation in Y-values that is explained by the variation in X-values. It also gives the standard errors or the 95% confidence limits for the slope and the intercept of the straight line (see Fig. 3.4).

Handling logs

Logarithms represent the power to which some base must be raised, so a change of 1 log unit represents a change of an order of magnitude in the antilog. Thus an arithmetic change (e.g. 1 to 2 to 3) in the log corresponds to a multiplicative change in the antilog (e.g. 1 to 10 to 100). Handling logarithms is not difficult—they can be added, subtracted, multiplied, and divided according to simple rules:

$$\log (1/X) = -\log X$$
$$\log (XY) = \log X + \log Y$$
$$\log (X/Y) = \log X - \log Y$$
$$\log X^a = a \log X$$
$$\log \sqrt[a]{X} = 1/a \log X$$

Remember also: log 1 in base 10 is zero, and positive numbers below 1 have negative logs (e.g. $\log 0.1 = -1$, $\log 0.001 = -3$); there is no log for zero itself or for negative numbers.

There are also simple rules for operating with powers, analogous to those for logs:

$$X^a X^b = X^{a+b}$$
$$X^a / X^b = X^{a-b}$$
$$(X^a)^b = X^{ab}$$
$$\sqrt[b]{X^a} = X^{a/b}$$
$$X^a Y^a = (XY)^a$$
$$X^a / Y^a = (X/Y)^a$$
$$X^a / X = X^{a-1}$$
$$1/X^a = X^{-a}$$
$$X^0 = 1$$

points with very high or low X values (outliers). Notice also that when comparing two or more good data sets even quite small differences in b are potentially important; remember that the log-log plot inevitably appears to reduce differences that would look substantial on a nonlogarithmic graph.

Analysis of scaling can be applied at many levels. Differing allometries may emerge when analyzing differences within a species:

1 During the growth of individual animals.
2 Across individuals of the same age or developmental stage.
3 Across a range of ages or stages.

It is therefore important to use the right species value when making larger interspecific comparisons. Given this precaution, variations in physiological parameters can be assessed on the basis of allometric effects at almost any taxonomic level, up to and including the whole animal kingdom. Such analyses have shown that many of the differences between species once attributed to adaptation can largely be explained by scaling effects.

Allometric analysis has had a major impact on environmental physiology; for example, the way in which energy assimilation, respiration, and mortality vary with body mass for each species can be described by three allometric equations, and the slopes and intercepts of these have been described as the "six facts of life", with various combinations of these six facts being possible. However, the very importance of scaling prompts some further words of caution.

Box 3.3 Problems with scaling and adaptive interpretations

Large male deer have extremely long, heavy and cumbersome antlers; these were up to 3.3 m across in the extinct Irish elk, and many explanations for this have been put forward. But antler size is strongly correlated with body size, and once it was realized that the antlers of the larger species had the same relation to body weight as those of the smaller species some authors maintained that no further search for an adaptive explanation for the huge ornaments on the heads of large deer was necessary.

This cannot be true; the antlers represent a great cost in materials to their possessor (especially as they are renewed every year), and can be a significant handicap in normal daily activities. Deer use their antlers in male/male fighting, and it transpires that the size of the antlers is proportionately greater in species with polygynous mating systems, where male/male competition is greatest. Hence there is clearly selection for larger antler size operating. Probably the two characters of body weight and antler size are under strongly correlated selection pressures, the forces selecting for antler size being the same as those selecting for large bodies.

Hence when we see strong allometric trends we should still be looking for the selective forces responsible for them, just as we would seek for the explanation for any single character.

Allometric equations are purely descriptive and do not represent laws; it is therefore important that they should not be used for extrapolating beyond the range of data on which they are based. Allometric relations do not indicate some kind of unbreakable body-weight-related constraint on evolution, since they may be the end-product of several interacting selection pressures. Above all, they do not remove the need for adaptive explanation, as has sometimes been maintained or implied. We may be able to *predict* some character fairly well from a knowledge of another parameter and of the equation relating the two, but this in itself does not *explain* the character or the selective pressures that gave rise to it. In other words, allometric trends are not "nonadaptive". The best known example to throw light on this is probably the story of deer antlers (Box 3.3).

Allometric relations can also be of use in another way. Single observations may not necessarily fall on the calculated allometric regression line, and they may be smaller or larger than expected. Assuming that no errors have been made in making the measurements (or in the calculations or statistics), then further insight may be obtained from the data points that significantly deviate from the general allometric line. Comparative biologists are always intrigued by species that do not fit with the pattern of their near relatives, or with individuals that seem to differ from the rest of their species. We noted above that blood volume in mammals is a constant proportion of body mass, but the striking exceptions are the marine diving mammals, many of which have much larger blood volumes than expected, which turns out to be because the blood is used as a significant oxygen store during a dive. In the case of metabolic rate, we might also predict some deviations from a simple body size relationship: sloths (if they live up to their reputation) should fall below the line and small rodents and small passerine birds might lie above it. For brain size against body size, we might hope that primates, and *Homo* in particular, would lie well above the line (this does turn out to be true, but is somewhat confounded by an interaction with unusual longevity in some primates and man).

Numerous examples of such scaling are given in this chapter and throughout the book; in particular see section 6.12 on energy balance, sections 7.2–7.4 on respiration and the circulatory system, and sections 8.7 and 8.8 on temperature regulation. Here we pick out a few examples to show some of the principles and effects of scaling.

3.4 Scaling of metabolic rate

3.4.1 The allometry of resting metabolism

The scaling of metabolism is one of the best known and most intensively studied examples. An animal's metabolic rate determines its food intake, growth, excretion, and many other functions, which as a consequence will also vary with body size. The terminology of metabolic rates and their measurement is fully explained in Chapter 6.

To compare the effects of some environmental variable on the metabolic rate of different populations or species it is first necessary to take into account that fraction of the variation in the data that can be attributed more simply to differences in the body mass of the groups being compared. The importance of this can be illustrated by an example: in a recent study of the routine oxygen consumption of 143 species of zooplankton covering eight different phyla, which were apparently highly variable, it was found that as much as 84–96% of this variation in metabolic rate could be explained simply by differences in body mass and temperature. Thus it is essential to start by examining the underlying scaling relationships.

Hundreds of scientific papers have been published on the scaling of basal or resting metabolic rate (see Chapter 6) in birds and mammals over the last 60 years. If basal metabolic rates (BMRs) are plotted against body mass on logarithmic coordinates a straight line is obtained, often known as the "mouse to elephant curve" or the Kleiber relation. Max Kleiber, in 1932, surveyed the metabolic rate of mammals varying in body mass from 0.15 kg to 679 kg. Metabolic rate (watts or kcal day^{-1}) was related to body mass (M_b) by the allometric equation:

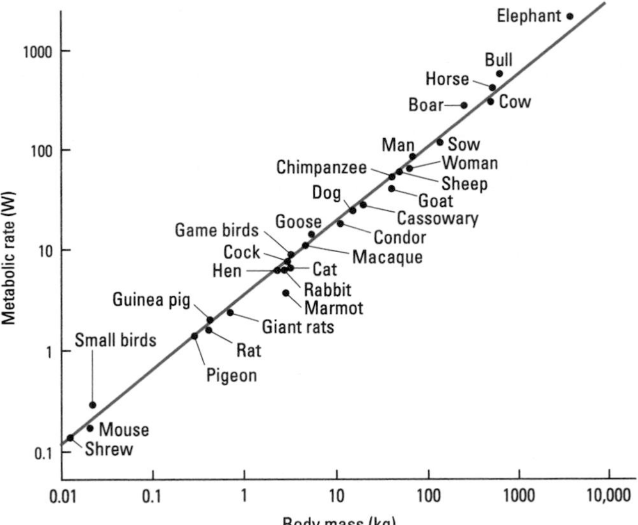

Fig. 3.5 Metabolic rates for a variety of birds and mammals plotted against mass on a double logarithmic scale. (From Schmidt-Nielsen 1984, courtesy of Cambridge University Press.)

$$\text{Metabolic rate} = 73.3\ M_b^{0.74}$$

Statistically the intercept was not different from 70 and the slope was not different from 0.75 (Fig. 3.5), so we can rewrite this as:

$$\text{Metabolic rate} = 70\ M_b^{0.75}$$

The reasons for this exponent of 0.75 are considered further in Chapter 6, but for the present we need to consider what use can be made of this allometric relation. Within a particular taxonomic group, variation in body mass can in itself explain at least 90% of the variation in metabolic rate. In practice, there are considerable differences in the BMRs of mammals and birds even when they are of the same size; but only when we have factored out the size effect can we begin to examine the causes of the remaining size-independent variation.

This raises a second important issue that has only recently been properly taken into account, which is the role of taxonomy. With size taken out, it commonly transpires that the next most important factor is what taxon the animals belong to. For example, we have seen that marsupial mammals generally have about 30% lower BMRs than eutherian mammals of the same size.

Several authors have also suggested that environmental or life-history features affect BMR: for example that BMR may be low in desert species, or low in the various groups of ant-eating mammals. This can only be analyzed properly once the effects of scale and of taxonomy have been taken out, and the latter is particularly difficult because species with similar lifestyles are often taxonomically related. Some recent studies have begun to disentangle such problems, usually by using the techniques of modern comparative analysis (see Chapter 1). A good example comes from a recent analysis of BMR in muroid rodents (the rats and mice and their relatives). Here the species could be split into three broad strategies of living, the "vole type", the "mouse type", and the "hamster type". With size and taxonomy taken into account, it turned out that the hamster strategy (rich diet and warm, dry open habitat) was associated with the lowest BMR and the mouse strategy (rich diet and warm, wet dense habitat) with the highest. This may be partly explained by disproportionately large organs in the high BMR groups, particularly the organs such as heart and kidney that have intrinsically high metabolic demands. These larger organs may in turn be needed to maintain relatively higher daily energy budgets, and the two factors may therefore be "coevolved" by natural selection in relation to particular species' environments. This kind of analysis allows us to see the real and quantifiable effects of the evolutionary shaping of physiological variables in step with environmental variation—but the first step was to understand the overriding scaling effect.

3.4.2 The allometry of specific metabolic rate and metabolic scope

Metabolic rate per unit body mass is referred to as the **specific metabolic rate** (see Chapter 6), in other words:

$$\text{Specific metabolic rate} = \text{metabolic rate}/M_b$$

Substituting from our earlier equation we get:

$$\text{Specific metabolic rate} = 70\ M_b^{0.75}/M_b$$

which can be rearranged as:

$$\text{Specific metabolic rate} = 70\ M_b^{-0.25}$$

So specific metabolic rate decreases with increasing body size, the slope of the regression equation being −0.25. This equation tells us that energy metabolism per unit mass is higher in small than in large mammals; in fact small animals like mice degrade chemical energy into heat at about 10 times the rate of an equal mass of much larger mammals like moose. This is at least partly because the smaller mammals have relatively larger surface areas and heat will dissipate more readily from them. If a moose had a specific metabolic rate comparable to that of a mouse, other things being equal it would be liable to cook its own tissues from within!

This kind of analysis also provides information on the constraints on energy metabolism that may be imposed by changes in scale. Again, some species have been able to overcome these constraints following the evolution of different designs. A good example comes from analysis of scaling in mammalian respiration. An important comparison here is the **metabolic scope** of a particular species (this is the ratio of resting rates and maximum rates of metabolism; see Chapter 6). To examine this the oxygen consumptions of animals that have been trained to run on treadmills can be measured. Oxygen consumption ($\dot{V}O_2$) increases with speed of running up to a maximum ($\dot{V}O_{2max}$), after which further increases in speed require anaerobic metabolism and result in the build-up of lactic acid in the blood (see Chapter 6). Some of the results obtained for wild and domesticated mammals are plotted in Fig. 3.6. For the combined group of 21 species shown, the best fit equation is:

$$\dot{V}O_{2max} = 1.92\ M_b^{0.81}$$

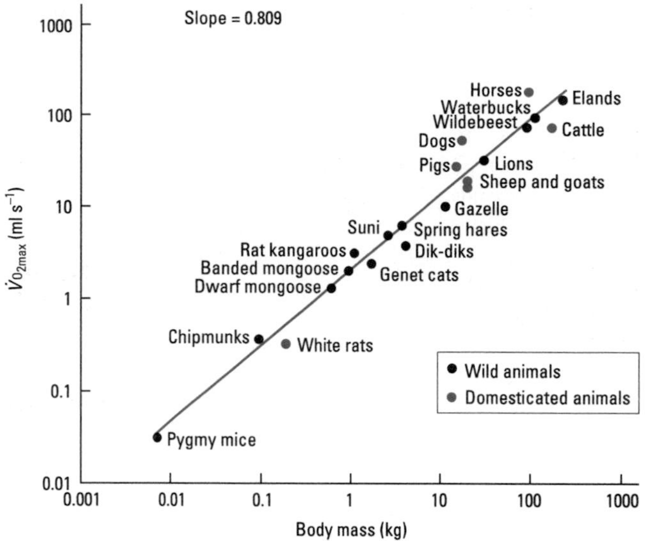

Fig. 3.6 Double logarithmic plot of the maximal rates of oxygen consumption when running on a treadmill, for 21 species of African mammals (mass range 0.007–263 kg). (Reprinted from *Respiration Physiology* **44**, Taylor, C.R. *et al.*, Design of the mammalian respiratory system. III. Scaling of maximal aerobic capacity to body mass: wild and domestic animals, pp. 25–37, copyright 1981, with permission from Elsevier Science.)

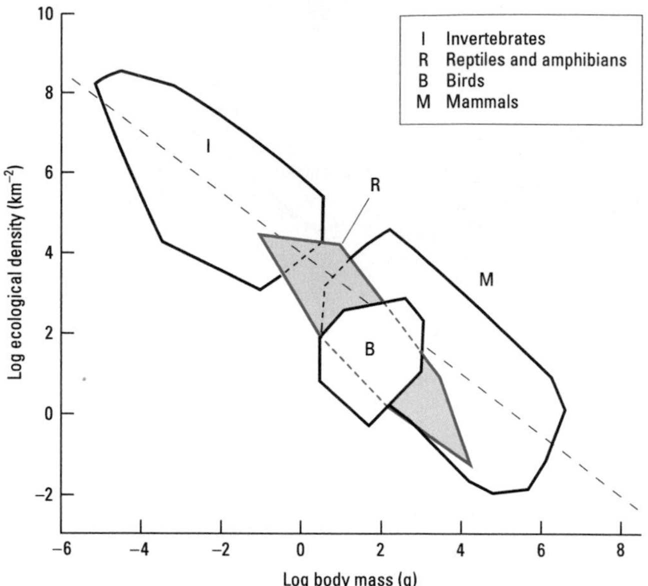

Fig. 3.7 The scaling of ecological densities with body mass for terrestrial animals. Polygons enclose all the data points for particular taxa. Density values for ectotherms were adjusted to correct for the differences in average metabolic rates from endotherms. (Adapted from Damuth 1991.)

and the 95% confidence limits on the mass exponent b are 0.75–0.87. Other studies have come up with similar values, using a wide range of mammals.

For most mammals, the aerobic metabolic scope values turn out to be around 10, meaning that the maximum metabolic rate is about 10 times the resting or basal metabolic rate. But this generalization is again helpful in revealing anomalies for individual species; for example, dogs have aerobic scopes of about 30. Again this raises questions about the particular features of dogs compared with other mammals.

3.4.3 Environmental implications of the scaling of metabolic rate

Because small mammals degrade more energy per unit mass than large mammals, it should follow that a given energy supply will run out more quickly for small animals; or, looked at per unit time, that a given energy supply will support a much smaller biomass of mice than of large moose or elephants. What about actual numbers of animals for a particular energy supply or for a given area? The abundance of animals in particular habitats is described in terms of population density. Studies spanning many taxa and a large size range of species have repeatedly shown that the population density for each species decreases approximately as the −0.75 power of body mass. This is the reciprocal of the scaling of individual metabolic requirements discussed above, which means that the overall amount of energy each species uses per unit area of its habitat is roughly independent of body mass (Fig. 3.7).

In practice, guilds of species that include some medium to large mammals usually do show slopes for population density against body mass equal to or more steeply negative than −0.75. This will indeed cancel out the effects of individual metabolic rate scaling, so that from the point of view of energy control at the ecological level there may be no particular advantage to being a large or a small animal. But a different pattern is found when ecological densities are compared within guilds made up of species of small birds or small mammals that are closely related. In these cases, the slopes on double-logarithmic plots are often less steep than −0.75, which means that the somewhat larger species are better at getting and using energy than smaller species from similar taxa (Fig. 3.8); in ecological terms they are likely to be better competitors. Again the importance of considering taxonomic relations is underlined.

3.5 Scaling of locomotion

3.5.1 Scaling of running, swimming, and flying

Locomotion is one of the most obvious activities in which animals use up metabolic substrates. The mechanics of different types of locomotion are dealt with in Chapter 9 (see section 9.15), but here we consider specifically how scaling equations can be very helpful in showing up the relative costs of different kinds of locomotion at different velocities and for different body sizes.

For running animals on land, oxygen consumption increases with speed of running as would be expected, and the slope of this increase is greater for small animals (Fig. 3.9; note that this is a linear plot, not a logarithmic one). It is more useful for interspecific comparisons to look at the "net cost of transport", C, this being the additional energy cost (over and above resting metabolic rate) per unit distance moved, or even better to look at the "specific net transport cost", i.e. the additional energy an animal uses in moving 1 kg

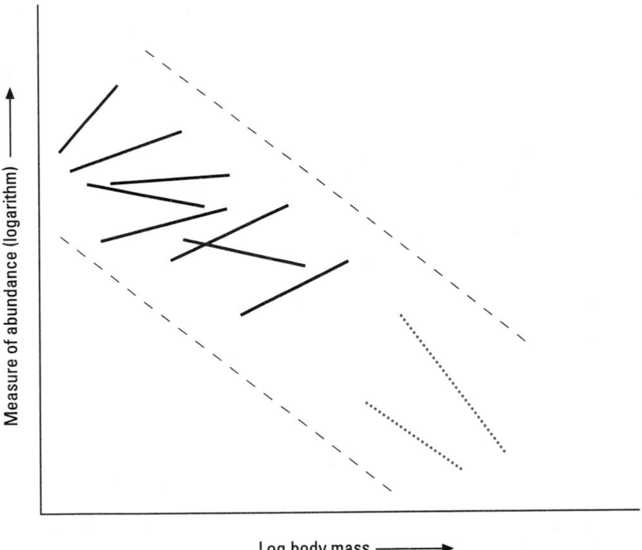

Fig. 3.8 The idealized empirical relationships between abundance and body size at different scales of analysis. For both ecological and regional densities, the overall relationship across a diverse array of species is negative, falling between the dashed lines and giving a slope of about −0.75. For guilds of larger species the individual slopes probably fit well with this, as in the lower right-hand corner (dotted green lines). But within guilds of closely related small species (solid black lines) the slopes are often shallower than −0.75 or even positive, showing that the larger guild members control more of the energy. (From Damuth 1991.)

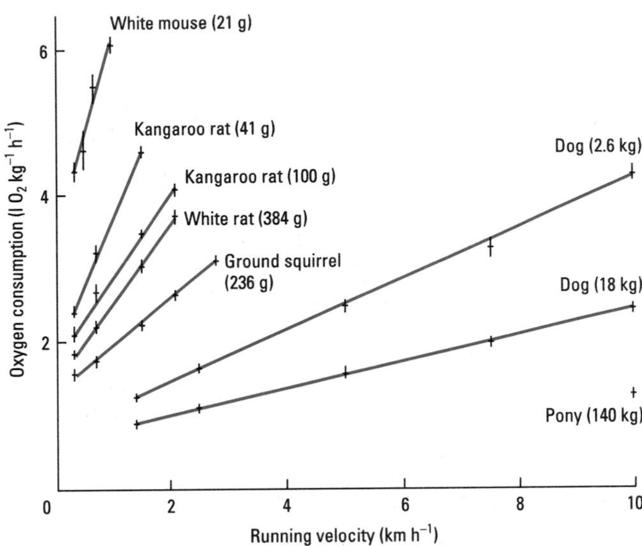

Fig. 3.9 Running velocity plotted against $\dot{V}O_2$ for various different size running mammals. Oxygen usage increases with running speed, and for each species the increase is linear, but oxygen usage increases more steeply for a small species than for a large one. (Adapted from Taylor *et al.* 1970.)

of its body mass over a specified distance (C/M_b). The relationship between specific net transport cost and body mass is shown in Fig. 3.10; the mass exponent is −0.28 for the entire data set, telling us that the cost of traveling a unit distance is relatively smaller for larger animals. The figure also illustrates running in reptiles, birds, and mammals, and it is evident that birds (mass exponent −0.32) have rather higher transport costs than mammals, and larger ectothermic reptiles usually have slightly lower costs. For land animals as a whole, this parameter (C/M_b) also turns out to be unaffected by the velocity of movement; in other words, the greater cost of running faster is almost exactly offset by the shorter time needed to cover the distance.

Specific total transport costs are rather higher than the figures so far discussed, because there are other costs involved, such as the need to maintain the correct (usually rather upright) posture while running. These secondary costs are relatively higher for larger animals; for example, because of scaling effects on the skeleton, elephants have to keep their legs rather straight and their body carefully upright to avoid excess stress on their limbs and potential bone breaks. Similarly, large animals require more energy to start and to stop, and take longer to change velocity, because of the greater difficulty of overcoming their own inertia.

The muscles of small animals consume much more energy per gram during steady running than do the muscles of large ones. This is partly due to the general phenomenon of raised specific metabolic

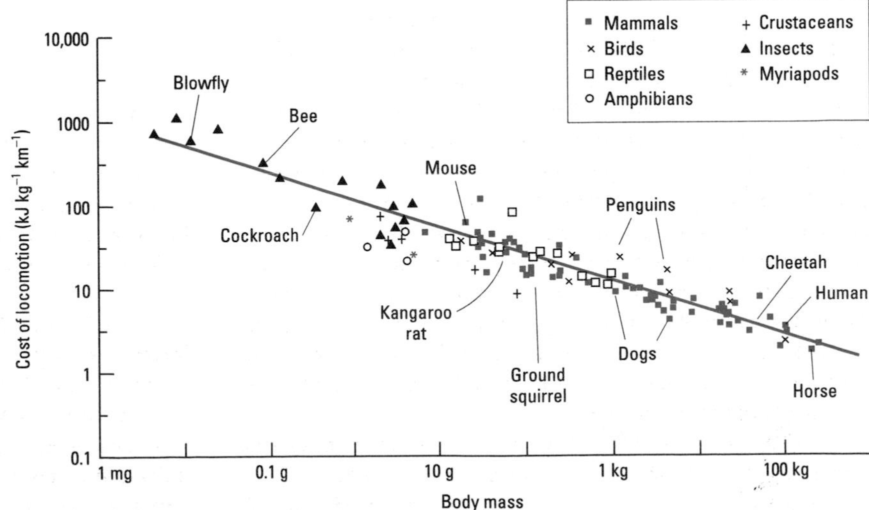

Fig. 3.10 The specific cost of transport in running mode for animals of various kinds, on a double logarithmic plot. To move a unit mass of their tissue a given distance is more expensive for small animals regardless of their taxon or the number of legs they may have. The differences in slope between taxa are slight.

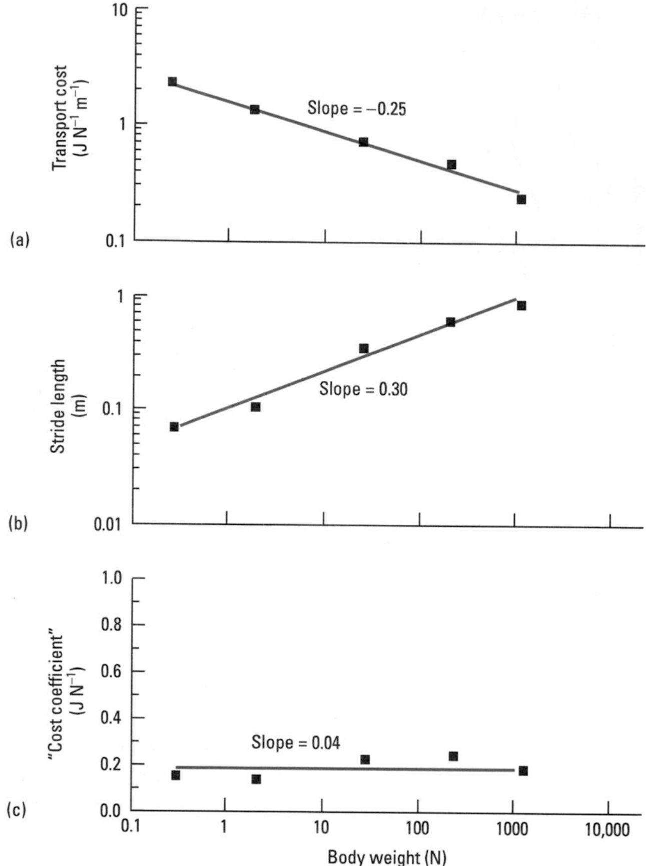

Fig. 3.11 Scaling of strides and the specific cost of transport, based on studies of five species of mammal ranging from horses to kangaroo rats. In (a) the specific cost of transport is shown to decrease with weight; (b) shows stride length increasing with weight; and (c) shows the product of these two as an almost constant "cost coefficient" independent of body size. (Adapted from Kram & Taylor 1990.)

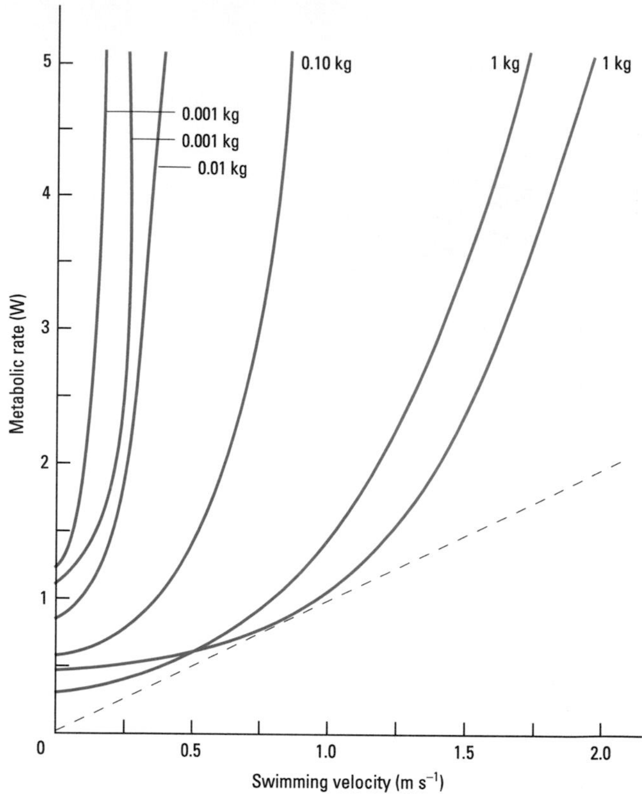

Fig. 3.12 The effects of swimming velocity on metabolic rate for a range of different sizes of fish, showing the sharp rise in costs above a particular size-dependent velocity. For each curve, the intersection with a straight line through the origin (as shown for a 1 kg fish) gives the velocity for minimum total cost of transport. (From Peters 1983, courtesy of Cambridge University Press.)

rate in smaller animals, but also probably because of the increased stride frequency for small animals, so that they must turn their muscles on and off at a higher rate to generate the forces required to move at an "equivalent speed". If the cost of each stride in a land mammal is multiplied by the stride period, effectively normalizing for these differences, then the mass-specific cost of locomotion per stride is similar (slope = 0.04) over four orders of body mass (Fig. 3.11).

As a result of the various factors discussed above, almost 85% of the variation in the costs of terrestrial locomotion can be explained by variations in body mass. This turns out to be true not just for mammals but also for a range of other land vertebrates (including even some limbless examples such as snakes), and for many invertebrates with radically different skeletal structures, leg architectures, and numbers of legs.

The discussion so far has dealt with terrestrial locomotion—walking, running, hopping, or even the sinusoidal movements of snakes. When it comes to aquatic and aerial locomotory patterns the scaling relationships become more complex. Swimmers expend little energy in postural control, but do need to do work against the resistance of the medium. The metabolic rate of swimming animals increases with speed, but the increase is not linear; above a particular size-dependent velocity, metabolic rate increases sharply (Fig. 3.12). This effectively puts an upper limit on swimming speeds, and the range of achievable speeds is particularly narrow for small animals. Fish normally swim at the speed at which total transport costs are minimal, but as expected this is more flexible for larger fish than for small ones. Once again, in spite of the different designs and lifestyles among fish, the energetic cost of swimming per unit body mass over unit distance appears to be fairly similar for all species studied, though declining somewhat with increased size.

For flying animals (birds, bats, and insects), there is the additional complication that stopping all forward motion does not save energy but rather increases the need for it, as the animal then has to hover. Hovering may be at least twice as expensive as forward flight, and therefore for some birds the metabolic rate against velocity plot is strongly curved (Fig. 3.13a) with a clear optimal velocity where costs are minimized. However, the figure also shows that for other fliers, including some birds and all the bats and insects so far tested, the relation between velocity and metabolic rate is fairly flat; for example, the bumble-bee *Bombus terrestris* uses about 350 W kg^{-1} at all speeds. Note also (Fig. 3.13b) that the specific costs per unit of body mass decrease with mass, as discussed earlier. However, the minimum total transport costs for flight, like those for running and swimming, increase with increasing body size. Figure 3.14 shows

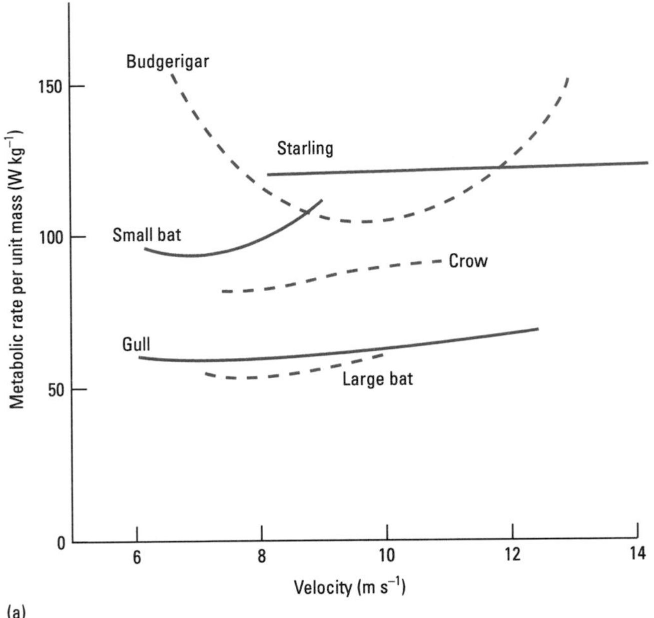

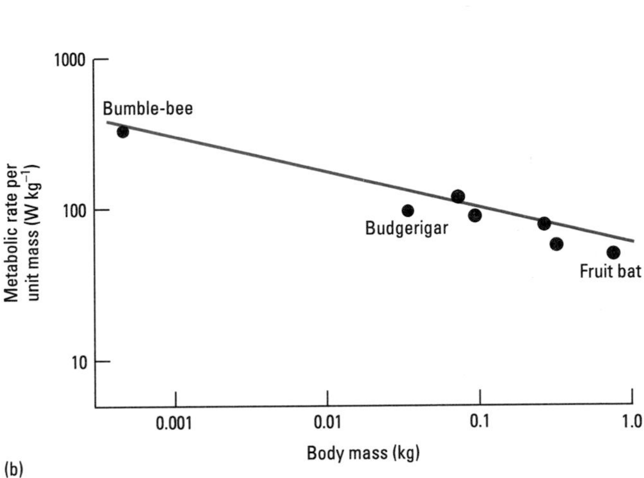

Fig. 3.13 (a) The relation between flight velocity and specific metabolic rate. Initial experiments (as with the budgerigar) showed a strongly U-shaped curve; but not all animals show this, and in many the relation is almost flat. (b) Shows increasing specific metabolic rate with body mass. (Adapted from Alexander 1999.)

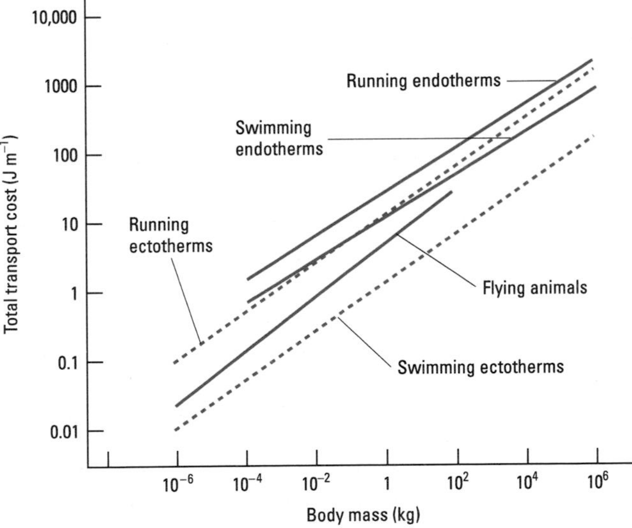

Fig. 3.14 Total transport costs associated with swimming, running, and flying in ectotherms (dashed lines) and endotherms (solid lines). (From Peters 1983, courtesy of Cambridge University Press.)

that for any given body size, the lowest total costs are associated with swimming and the highest costs with walking or running, with flight being intermediate.

The effect of body size on locomotory velocity can also be estimated, as shown in Fig. 3.15, and (as expected) maximum speed does increase with increasing size. The exponents are 0.14 for flying, 0.35 for swimming, and 0.38 for running, indicating the lesser sensitivity of flight speeds to body size. For animals of similar size, flying is fastest and swimming slowest. For the largest animals in each category, though, the maximum speeds are not very different for the three locomotory modes, with flight, running, and burst swimming all achieving a peak of around 20 m s^{-1} (calculated for a 1 kg flier, 10 kg runner, and 100 kg swimmer, respectively). Measurements of the speeds that animals move at in their normal daily routines indicate that land animals and aquatic animals rarely approach their maximum velocities, whereas flying animals generally do operate closer to their maximum flight velocities (Fig. 3.16).

3.5.2 Environmental implications of locomotory scaling

Given the known maximum and average speeds for animals of particular body sizes, it is feasible to estimate the range over which they can operate, either on a daily basis or as the distance from a central point such as a nest. In the latter case a "foraging radius" can be assessed (Fig. 3.16). The resultant foraging area is clearly largest for flying animals, and increases with size within each locomotory mode. This will have implications for territory sizes, and for the habitat (and hence diet) quality required by particular species.

Similarly, different sizes and locomotory modes affect the ability to migrate effectively. For any given migration time, flying animals will be able to complete a much longer distance than walkers or swimmers of similar size; or, to achieve the same distance in the same time, walkers or swimmers would have to be much larger. Global terrestrial or marine migrations are therefore rather slow and prolonged, often taking many months (as in salmonid fish), whereas aerial migration by birds can be completed in a matter of days or weeks. This picture is clearly complicated by the need for food, as few animals undertake nonstop migrations relying entirely on their food stores, though many do accumulate fat stores of 25–50% body weight before migrating. Because fish are generally ectothermic, with roughly 10-fold lower metabolic rates than similar

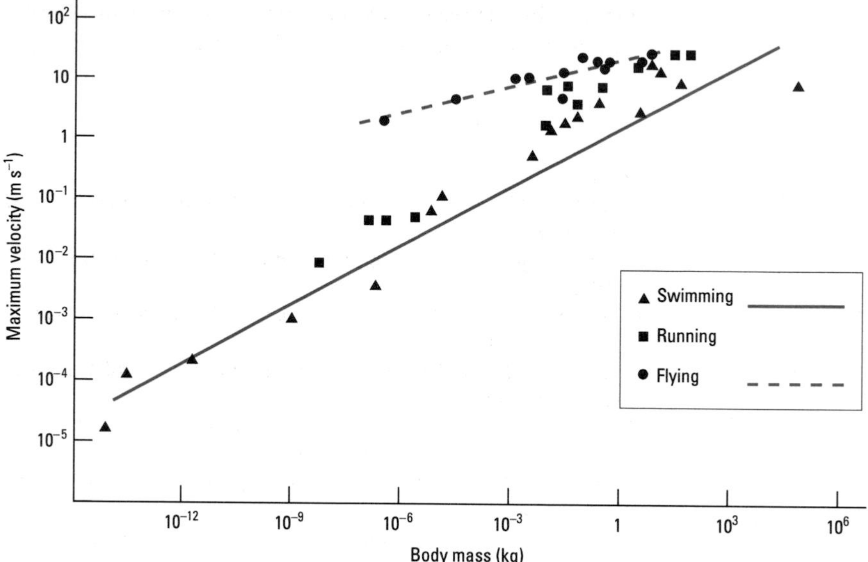

Fig. 3.15 Maximum velocity against body size; the top speed of an animal increases with body size for all locomotory modes. For animals of similar size, flying is the fastest mode and swimming the slowest (though flight velocity is relatively insensitive to size and the *difference* in speed is therefore smaller for large animals). (From Peters 1983, courtesy of Cambridge University Press.)

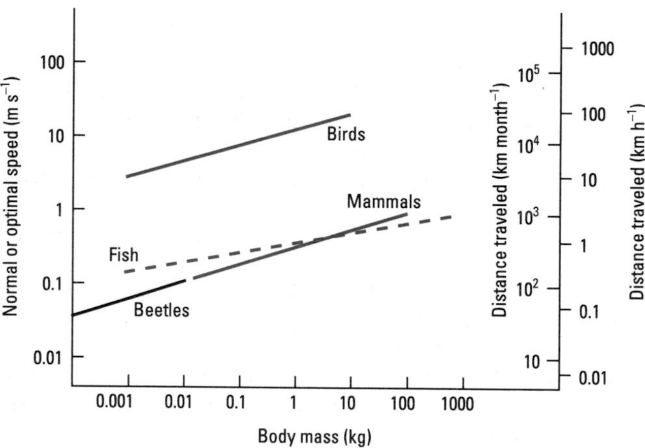

Fig. 3.16 Normal or optimal speeds measured for beetles, fish, birds, and mammals, showing that only birds normally approach their maximum velocities of about 20 m s^{-1}. The right-hand axis shows how this influences ecologically relevant factors: the effective foraging radius that results (km traveled in 1 h spent foraging), and the effective migration distance (km traveled in 1 month, moving for 12 h per day). (From Peters 1983, courtesy of Cambridge University Press.)

sized birds and mammals (see Chapter 6), they can in practice migrate much further without feeding, and have the greatest overall migratory capability.

The scaling effects on migration may have important implications for the "avoidance strategy" of choice in times of environmental stress: birds are more likely to migrate, whereas terrestrial animals, especially smaller ones, are often more inclined to escape by hibernation and torpor. The smallest migrating mammals have a body mass of about 20 kg (some African antelope), whereas very small birds and even the1 g monarch butterfly are regular migrators. Many kinds of aquatic animals of all sizes migrate, whether latitudinally or vertically through the water column.

3.6 Conclusions: is there a right size to be?

Scaling and size dependency are crucial factors in all comparative physiology, and the examples above amply demonstrate the importance of thinking quantitatively and of avoiding oversimplistic interpretations of trends and patterns in animal adaptation. We will meet many other examples, from all environments, where scaling effects must be incorporated in the analysis to understand what is really going on and must be "taken out" before any other more specific adaptive mechanisms can be properly revealed.

This still leaves us with the question of what really determines the size of a particular species of animal. Clearly there are crucial interacting constraints on size due to various factors:

1 Phylogenetic inheritance (e.g. insects are essentially "small", vertebrates are "large", and the two groups only just overlap in body mass).

2 Basic physiological design (e.g. animals with open blood systems are larger than those with no circulation, and those with closed blood systems may be larger again).

3 Basic mechanical design (e.g. animals with hydrostatic skeletons are usually relatively small, as are those with exoskeletons, while those with tubular endoskeletons are relatively large).

4 Habitat (e.g. any given design may be larger in aquatic habitats giving inherent support than on land where self-weight is a problem).

But within any one design and habitat, what determines species and individual sizes? Again very obvious selective pressures that relate to physiological balances may be invoked, and we discuss in Chapter 16 why either small or large (but not medium) sizes may be favored in various extreme environments. The best studied examples of size limits and size distributions are the mammals, and this is therefore an instructive example to end with. Mammals show a very clear-cut pattern that also appears in other taxa, that of "right-skewed" distribution of body masses so that the modal mass is well below the mean mass (Fig. 3.17). Put simply, there are unexpectedly large numbers of rather small species. This observation

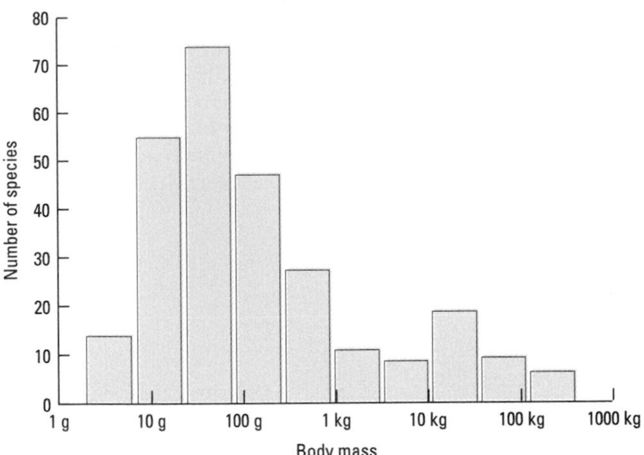

Fig. 3.17 Patterns of body sizes of mammals, for terrestrial Palaearctic species, showing the typical "right-skewed" distribution and the apparent "optimum" at about 100 g. (From Purvis & Harvey 1997.)

seems to suggest that there is an optimum body size for a given taxon, which for mammals is set at about 100 g. It may be that there are evolutionary reasons for this, relating to the time of origin of a taxon and a putative tendency to increase in average size through time. Ecological reasons are also feasible. One view might relate to the carrying capacity of habitats for rather a lot of small species but only a few large ones; this is given some support by the observation that within mammals larger land masses support both larger "large species" and smaller "small species", giving a greater overall size range. It may also be linked to the ecological pyramid or food web concepts where only a few large (carnivorous) species can be adequately fed by a larger number of small species as prey. One recent analysis suggested yet another reason, which is primarily a function of life histories, with mammals having a maximum "reproductive power" (the rate at which acquired energy is channeled into offspring) at about 100 g. However, this would imply that the scaling of reproductive power with size should change sign at this optimum size, and a test of the model using large numbers of species of bat has failed to find any support for this prediction. Another analysis of data for the past 65,000 years and covering continents and islands, suggests that body mass of the "top" species in any one habitat is positively related to land area, and for a given area the size of this top species decreases in the sequence ectothermic herbivore → endothermic herbivore → ectothermic carnivore → endothermic carnivore. These patterns neatly explain the size of the largest known (extinct) land mammals, though not that of the largest dinosaurs.

For both mammals and for other taxa it seems that the jury is still out on just what is the right size to be, but undoubtedly there is an interaction of physiological, ecological, and evolutionary factors here. An interesting consequence may be that interspecific allometries for key factors such as assimilation, respiration, and mortality may be "by-products" of body size optimization, with natural selection acting to switch allocations of resources to production and to growth independently in different species. In that case, many of the interspecific comparisons that produce clear allometric relationships may be somewhat artefactual, and should not be used uncritically in considering the "big questions" of ecology and life history.

FURTHER READING

Books

Alexander, R.McN. (1996) *Optima for Animals.* Princeton University Press, Princeton, NJ.

Alexander, R.McN. (1999) *Energy for Animal Life.* Oxford University Press, Oxford.

Brown, J.H. & West, G.B. (eds) (2000) *Scaling in Biology.* Oxford University Press, New York.

Calder, W.A. (1984) *Size, Function and Life History.* Harvard University Press, Cambridge, MA.

Harvey, P.H. & Pagel, M.D. (1992) *The Comparative Method in Evolutionary Biology.* Oxford University Press, Oxford.

Pennycuick, C.J. (1992) *Newton Rules Biology.* Oxford University Press, Oxford.

Peters, R.H. (1983) *The Ecological Implications of Body Size.* Cambridge University Press, Cambridge, UK.

Schmidt-Nielsen, K. (1984) *Scaling. Why is Animal Size so Important?* Cambridge University Press, Cambridge, UK.

Reviews and scientific papers

Atkinson, D. & Sibly, R.M. (1997) Why are organisms usually bigger in colder environments? Making sense of a life history puzzle. *Trends in Ecology & Evolution* **12**, 235–239.

Biewener, A. (1989) Scaling body support in mammals; limb posture and muscle mechanics. *Science* **245**, 45–48.

Blackburn, T.M. & Gaston, K.J. (1996) On being the right size: different definitions of "right". *Oikos* **75**, 551–557.

Burness, G.P., Diamond, J. & Flannery, T. (2001) Dinosaurs, dragons and dwarfs: the evolution of maximal body size. *Proceedings of the National Academy of Sciences USA* **98**, 14518–14523.

Chai, P. & Millard, D. (1997) Flight and size constraints: hovering performance of large hummingbirds under maximal loading. *Journal of Experimental Biology* **200**, 2757–2763.

Damuth, J. (1991) Ecology—of size and abundance. *Nature* **351**, 268–269.

Damuth, J. (1993) Copes Rule, the Island Rule, and the scaling of mammalian population density. *Nature* **365**, 748–750.

Damuth, J. (1994) No conflict among abundance rules. *Trends in Ecology & Evolution* **9**, 487.

Elgar, M.A. & Harvey, P.H. (1987) Basal metabolic rates in mammals: allometry, phylogeny and ecology. *Functional Ecology* **1**, 25–36.

Full, R.J. & Tu, M.S. (1991) Mechanics of six-legged runners. *Journal of Experimental Biology* **148**, 129–146.

Gillooly, J.F., Brown, J.H., West, G.B., Savage, V.M. & Charnov, E.L. (2001) Effects of size and temperature on metabolic rate. *Science* **293**, 2248–2251.

Hedenstrom, A. & Alerstam, T. (1995) Optimal flight speed of birds. *Philosophical Transactions of the Royal Society B* **348**, 471–487.

Jones, K.E. & Purvis, A. (1997) An optimum body size for mammals? Comparative evidence from bats. *Functional Ecology* **11**, 751–756.

Kozlowski, J. & Weiner, J. (1997) Interspecific allometries are by-products of body size optimisation. *American Naturalist* **149**, 352–380.

Kram, R. & Taylor, C.R. (1990) Energetics of running: a new perspective. *Nature* **346**, 265–267.

Marquet, P.A. & Taper, M.L. (1998) On size and area: patterns of mammalian body size extremes across landmasses. *Evolutionary Ecology* **12**, 127–139.

Packard, G.C. & Boardman, T.J. (1988) The misuse of ratios, indices and percentages in ecophysiological research. *Physiological Zoology* **61**, 1–9.

Pennycuick, C.J. (1997) Actual and "optimum" flight speeds: field data reassessed. *Journal of Experimental Biology* **200**, 2355–2361.

Purvis, A. & Harvey, P.H. (1997) The right size for a mammal. *Nature* **386**, 332–333.

Ritchie, M.E. & Olff, H. (1999) Spatial scaling laws yield a synthetic theory of biodiversity. *Nature* **400**, 557–560.

Taylor, C.R., Maloiy, G.M.O., Weibel, E.R. *et al.* (1981) Design of the mammalian respiratory system III. Scaling maximum aerobic capacity to body mass: wild and domestic mammals. *Respiration Physiology* **44**, 25–37.

West, G.B., Brown, J.H. & Enquist, B.J. (1997) A general model for the origin of allometric scaling laws in biology. *Science* **276**, 122–126.

PART 2

Central Issues in Comparative Physiology

4 Water, Ions, and Osmotic Physiology

4.1 Introduction

The interactions of all life on this planet with the most common molecule found here, water, are probably the most crucial aspects of any review of physiological adaptation. It is a truism to say that water is essential for life, since to a large degree water has been the determinant of life. All living processes on Earth have evolved around the very peculiar properties of this simple molecule, so that retrospectively we are often somewhat mistakenly led to think of water as being serendipitously "ideal" for supporting life.

All living organisms are in essence a series of interconnected aqueous solutions, contained within bags of phospholipid membranes. Water constitutes 60–90% of the total body mass of animals, this proportion being greater in soft-bodied invertebrates and lower in animals with substantial amounts of stiff skeletal materials such as vertebrates and arthropodan groups. The water is distributed between the cells themselves and the various extracellular fluids (Fig. 4.1). Again, soft- and hard-bodied animals have different patterns, with higher extracellular volumes (ECVs) in soft animals arising from their spacious body cavities that serve a hydrostatic function; for worms and slugs, ECVs can be as much as 70% of total water, compared to only 20% in many vertebrates. These volumes must be maintained for the animal to function properly; but even more crucially the concentration of the fluids, particularly intracellularly, has usually to be kept reasonably constant. Yet the concentrations of many components of the internal fluids invariably differ from concentrations in the environment. This chapter is concerned with the mechanisms for keeping volumes and concentrations of biological solutions under control, and thus keeping animal tissues operative, in the face of this fundamental challenge.

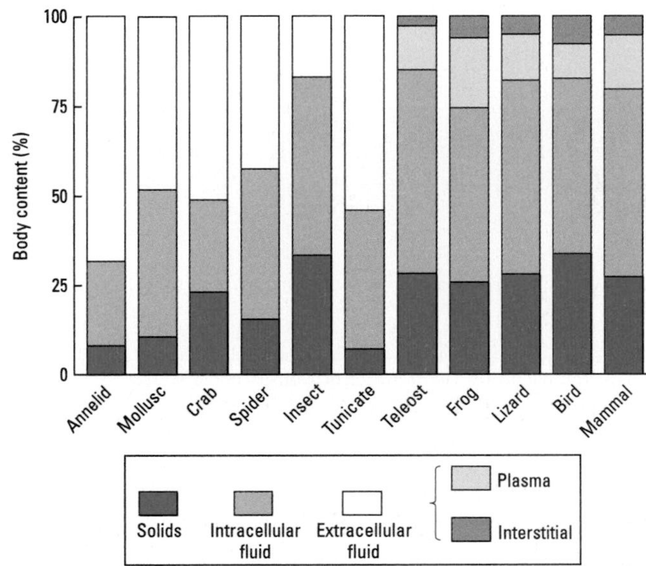

Fig. 4.1 Proportions of body solids, and the distribution of body water, in various fluid compartments for a range of animal taxa. (From *Comparative Animal Physiology*, 1st edn, by Withers © 1992. Reprinted with permission of Brooks/Cole, a division of Thomson Learning: www.thomsonrights.com. Fax 800 730-2215.)

4.2 Aqueous solutions

4.2.1 Properties and advantages of water as a basis for life

Water, because of its molecular architecture, has a special set of properties. It consists of two hydrogen atoms covalently bonded to a central oxygen atom; but because the O atom is quite strongly electronegative the bonds have some characteristics of ionic bonds and the water molecule exhibits a dipole effect, with each H atom bearing a slight positive charge and the O atom bearing a double negative charge. This in turns distorts the molecule away from linearity, giving a bond angle of 104.5°.

The presence of this dipole effect confers on water its ability to form secondary weak bonds (hydrogen bonds) with other molecules (Fig. 4.2) via its two H atoms (essentially protons, almost bare of electrons since these have been attracted closer to the parent oxygen atom). Most crucially, one water molecule can bond transiently with two others, the H atoms of one water molecule being attracted to O atoms of two other water molecules; this gives a tetrahedral three-dimensional organization to liquid water. Such organization is very transitory, as it takes only 20 kJ mol^{-1} to break the hydrogen bonds (compared with about 110 kJ mol^{-1} for the H–O covalent bonds within the water molecule), so that each hydrogen bond lasts only about 10^{-10} to 10^{-11} s. But at any one moment, a significant proportion of the water molecules are hydrogen bonded together. Therefore the **melting point** and **boiling point** of water are markedly higher than would be expected for similar kinds of molecule (such

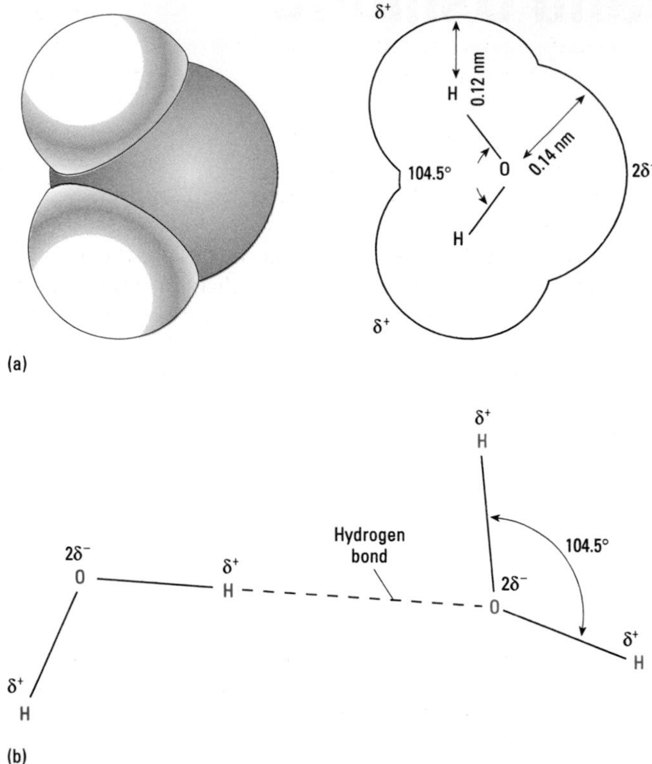

Fig. 4.2 (a) The structure of the water (H_2O) molecule, with bond angle and charge distribution. (b) The formation of hydrogen bonds between adjacent water molecules.

as NH_3 or H_2S). This means that water will occur both as a solid and as a gas, but by far most commonly as a liquid, at the temperatures experienced at the Earth's surface. As a liquid it has a relatively chaotic structure (Fig. 4.3a), with the hydrogen-bonded molecules in rapid motion (described as the "flickering cluster" structure). Due to the hydrogen bonding, water also has curious packing properties such that it achieves its **maximum density** at 4°C, a little above its freezing point. Once frozen as ice, the water molecules are packed in one of at least nine possible conformations, the commonest being a regular open hexagonal lattice (Fig. 4.3b), and the material is somewhat less dense. Hence ice forms at the surface of water bodies, and floats on top of such water bodies; if water were a more conventional liquid, with maximum density at its freezing point, water bodies would freeze suddenly throughout their volume. In reality, liquid water below ice is insulated from further excessive temperature change, allowing freshwater organisms to survive harsh winters while still living in a fluid medium. Sea water, with its highest density and its freezing point both at about −1.9°C, could in theory freeze to a substantial depth, but in practice it only freezes at the surface when the oceans encounter very low air temperatures, the resulting sea ice having lower density than the sea water and so floating upon it. In fact this freezing usually involves the water component only, leaving salts in solution and thus raising the salinity in the water immediately below the sea ice. However, it should be noted here that most of the ice we see at sea is actually derived from continental and glacial icebergs (a freshwater source) rather than as a result of the sea itself freezing.

The formation of ice crystals requires the presence of a nucleus of molecules onto which the water molecules can condense. As the temperature of fresh water is lowered the probability of this happening increases, and at 0°C at atmospheric pressure ice crystals can form and exist in equilibrium with the liquid phase. Almost all water contains impurities (microcrystalline dust particles, often called "motes") that absorb water molecules and act as catalysts for

Fig. 4.3 (a) Diagrammatic representation of the structure of bulk water, bound water, and membrane-linked water in a cell (dotted lines are hydrogen bonds). (b) The structure of ice.

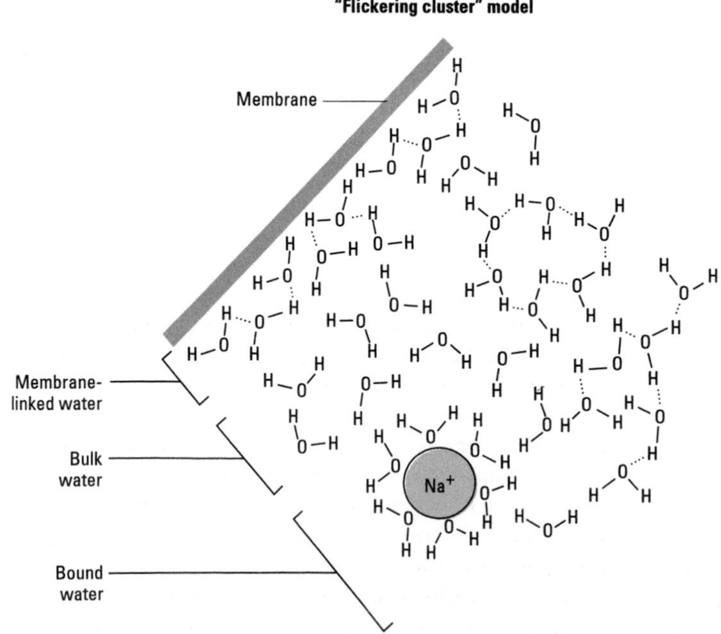

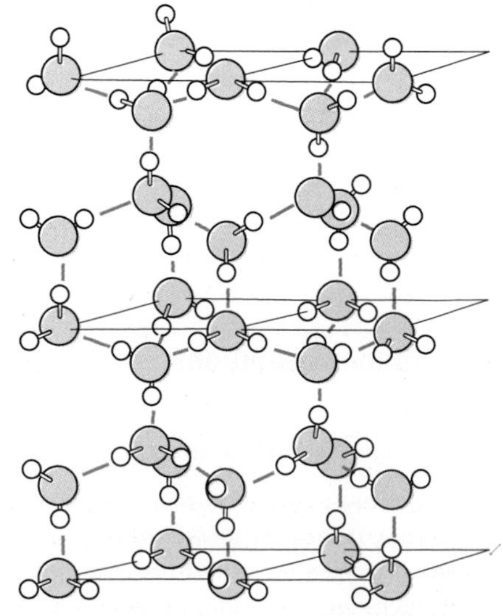

freezing by increasing the surface area for nucleation. It is possible to make ultrapure water which does not freeze at normal pressures until about –40°C. Water cooled to below 0°C without freezing is referred to as "undercooled" or "supercooled", and many animals rely on supercooling to survive freezing conditions. It is also worth noting that water in animal tissues that is trapped between macromolecules behaves differently from water in the bulk cytoplasm, since it is unable to assume a stable lattice structure and hence is unfreezable; around 35% of the water in a cell is in this so-called "bound state".

The hydrogen bonding in water also acts as an energy store (again 20 kJ mol^{-1}), and when ice forms at low ambient temperatures and the crystalline structure of Fig. 4.3b results, almost all the water is hydrogen bonded. Hence the melting of ice, which involves breaking a proportion of these bonds, requires a substantial **latent heat of fusion** (6 kJ mol^{-1}). Similarly, vaporizing water involves breaking all the remaining hydrogen bonds, giving an even greater **latent heat of vaporization** (40–44 kJ mol^{-1} depending on the temperature at which the vapor phase forms—equivalent to about 2.5 kJ g^{-1} or 580 cal g^{-1}). Many terrestrial animals exploit this property during evaporative cooling, as an excellent means of controlling body temperature.

This internal bonding in water also results in a substantial **surface tension**, exploited by many small animals living at aquatic interfaces. Small animals with hydrophobic cuticles are able to live supported on the water surface of ponds and rock pools, because penetrating the water surface in either direction is quite difficult. Special arrangements of hydrophobic surfaces or hairs may be required for submerged animals such as fly larvae to breathe air through siphons, or for terrestrial animals such as damselflies to lay eggs below the water surface. Internal bonding in liquid water also results in a very low **compressibility**, so that water is an excellent hydraulic medium to use in hydrostatic skeletons for locomotion in the many groups of worms and other flexible-bodied creatures.

Similarly, water may develop weak bonds with macromolecules and with biological membranes, to give relatively organized **boundary layers** within cells. It may also bind quite strongly to ionic moieties, so that small cations and anions always carry with them a substantial "shell" of bound water, affecting their transit across membranes and their rates of movement through cytoplasm.

Water also has a small inherent **capacity to ionize.** As the internal hydrogen bonds break and reform every so often three H atoms will be left associated with one molecule and only one with another, forming H_3O^+ and OH^- ions (hydronium and hydroxyl ions, respectively). The former moiety is often written as H^+, as if it were just a proton, but in fact it always exists as the hydronium ion. Only a very small proportion of water molecules are ionized at any one time, the concentration of the two ions being around 10^{-7} mol l^{-1}. However, an existing hydronium ion can readily donate its spare proton to another water molecule, or displace one of that recipients' protons, and protons may thus appear to "cascade" through an aqueous system even though any one proton may move only a short distance. Proton cascade conduction through water may play an important part in some biochemical reactions.

The ionization property of water has two other important consequences. Firstly, we can see that technically H_3O^+ is an acid (i.e. it can donate a proton) and OH^- is a base (it can receive a proton). Water can thus act as either an acid or a base, i.e. it is **amphoteric**. It can therefore neutralize charged groups dissolved in it. This is especially important for dissolved amino acids and proteins, as in solution these contain both positively charged amino residues ($-NH_3^+$) and negatively charged carboxyl residues ($-COO^-$); which form predominates will depend upon the acidity of the overall solution, and at normal cellular pH most proteins bear a net negative charge.

The second major consequence of water being ionized is that its **electrical conductivity** is somewhat greater than would be expected, as H_3O^+ and OH^- ions can act as carriers of charge. The conductivity of pure water is fairly low, but when any electrolytes are present it is greatly enhanced.

This brings us to the final point: above all, water acts as a superb **universal solvent**. It can dissolve crystalline ionic compounds such as NaCl, because the dipole on the water molecules can overcome the electrostatic attractions between individual Na^+ and Cl^- ions. Anything ionic will dissolve in water (although certain ions such as carbonate and silicate dissolve only very slightly and are thus often used for making shells and other "permanent" biological structures). It can also dissolve a range of nonionic and organic compounds such as sugars and alcohols, as these also have polar properties; and can react with partially polar, technically "amphipathic", molecules such as soaps (sodium oleate and related compounds), which have hydrophilic heads and hydrophobic tails, dispersing the solute into tiny droplets or "micelles". Only completely nonpolar molecules fail to dissolve in water. In a variety of ways, all of these solvent properties are exploited in living cells.

4.2.2 Concentrations of aqueous solutions: definitions and measurements

Aqueous solutions have to be described in terms of the concentrations of their various solutes, and here there are several options encountered in the biological literature (Box 4.1). Although **molality** is technically a more "correct" term, in practice it is most convenient to use **molarity**, as it is more easily measured and molar solutions are easy to prepare volumetrically. The differences between molar and molal measurements are nearly always small enough to be ignored.

A 1 molar (1 M) solution contains 1 mol of solute (i.e. its molecular weight in grams) dissolved in 1 liter of solution. Thus a 1 molar solution of sodium chloride contains 58.4 g NaCl per liter of solution. Most biological solutions are 1 molar or less, so that for individual component concentrations it is much more convenient to use the millimolar range. For example the concentration of Na^+ ions in human blood is about 140 mM.

Molarity is a very commonly used measurement in both chemical and biological science; but biological studies also use another kind of concentration measure, that of osmolality or more usefully **osmolarity**, as this is a particularly simple way of expressing the overall concentration of a fluid, and of considering how water movements may occur between biological fluid compartments. Osmolarity measures how many solute particles are present in a fluid in total, so that all the constituent solutes contribute to it. For a simple mixture where 100 mM glucose solution is added to 100 mM sucrose solution, the concentration of the resulting solution can be given as 200 milliosmolar or 200 mOsM. Hence the overall

Box 4.1 Measurements of concentration in biological systems

The chemical properties of a molecule in solution are usually adequately described by its **concentration** (C), although in concentrated media molecular interactions can alter its behavior so that its **activity** (A) is slightly less than its concentration ($A = \delta C$, where δ is the activity coefficient).

Concentration can be measured in a variety of units:
molarity—moles of solute per liter of solution;
molality—moles of solute per kilogram of solvent.

Osmotic concentration (Π) has the equivalent measures:
osmolarity—total moles of all solutes per liter of solution;
osmolality—total moles of all solutes per kilogram of solvent.

Molal and osmolal solutions in biology always relate to water as the solvent, so that they can be determined as moles per liter of solvent.

In practice when making up artificial solutions the difference is between weighing out the solutes and making them up to a total volume of 1 liter in a volumetric flask, or weighing them out and then adding a premeasured 1 liter of water, giving a somewhat greater final volume.

Chemical potential (μ) can also be used: this is related to concentration in a logarithmic fashion, as follows:

$$\mu = \mu^\circ + RT \ln C$$

μ° is used to specify a standard chemical potential, R is the gas constant, and T the absolute temperature. For most biological systems these can be regarded as constants, and μ is taken as proportional to $\ln C$.

However, chemical potential also depends on factors other than concentration. Pressure affects it, as does gravity, and an electrical potential can also have an effect in biological systems. The overall equation becomes:

$$\mu = \mu^\circ + RT \ln C + VP + zFE + mgh$$

where V is the molar volume of the molecule, P the pressure, z the charge on the ion, F the Faraday constant, E the electrical potential, m the mass of the molecule, g the gravitational constant, and h the height above ground level.

This more complex equation is normally needed only to compare the chemical potential between two compartments of a system and thus determine which way the molecule of interest will move. Many of the constants therefore cancel out, while P, E, and h can all be expressed as differences (ΔP, etc.) between the two regions.

Water potential (ψ) is often particularly useful when looking at movements in different phases (e.g. from liquid to vapor). This is the water chemical potential, as above, divided by the partial molar volume of water, and is given by:

$$\Psi = P - \Pi + \rho_w gh$$

where ρ_w is the density of water and overall units are as a pressure (kPa or Torr, etc.). For a liquid it is a composite of osmotic potential, pressure potential, and hydrostatic potential.

For water vapor, the equation is:

$$\Psi = RT/V \ln (\mathrm{RH}/100) + \rho_w gh$$

where RH is relative humidity. This is a composite of a vapor pressure potential and a gravitational potential.

concentration of a cellular or extracellular fluid can be given as a single value of osmolarity. This overall **osmotic concentration**, when compared for different adjacent solutions separated by a permeable surface, will indicate which way water will move through the system. It will always flow into the osmotically more concentrated fluid, and out of the more dilute fluid. More concentrated fluids are termed **hyperosmotic**, and more dilute fluids are hypo-osmotic or **hyposmotic**; where two fluids have the same osmotic concentration they are iso-osmotic or **isosmotic**.

Osmolarity is therefore a very useful concept, but it is not especially easy to measure directly. Another way of describing concentration in aqueous solutions is to give a related parameter. Particularly useful here are the **colligative properties**, a series of properties of fluids all of which are related specifically to the number of particles present in a solution but *not* to the nature, size, or charge of these particles. Colligative properties include osmotic concentration itself, already referred to, where as we have seen the concentrations of different kinds of molecule can be treated additively irrespective of their nature. But they also include depression of the freezing point (FP), raising of the boiling point (BP), vapor pressure (VP) reduction, and increasing of the refractive index (RI or μ). Any solute particles present in any solution will depress the freezing point and reduce the vapor pressure, while raising the boiling point and the refractive index of that solution, and always in a predictable manner. This means that any of these secondary characteristics of a solution can be used to measure its overall osmotic concentration. Simple refractometers, which measure the bending of light caused by a thin film of the solution introduced between two glass prisms, are a convenient way of checking concentrations in many areas of biological fieldwork. Instruments that measure melting points of fluids (rather than freezing points, which may be affected by supercooling) are even more frequently encountered, and this principle forms the basis of many kinds of laboratory osmometer. Any salt dissolved in fresh water lowers its freezing point, hence the common practice of putting salt on roads during cold nights. Theoretically a 1 molal aqueous solution of an ideal dissolved nonionic molecule has a freezing point of $-1.86°C$, and this factor can be used to convert measured freezing point depressions (FPD) into milliosmolar osmotic concentrations.

All that we have said so far is true for nonelectrolytes, where molarity and osmolarity are effectively equivalent. But since all the colligative properties depend on the actual number of dissolved particles, electrolytes that by definition dissociate into more than one particle have a much greater effect than their molarity predicts, and this is another reason why the concept of osmolarity is often particularly useful. Sodium chloride, for example, will dissociate into Na^+ and Cl^- ions, giving *two* moles of particles for each mole of solute, and thus giving roughly twice the FPD expected. Calcium chloride, $CaCl_2$, with three ions when dissolved, will give roughly three times the FPD. Hence freezing point measurements for ionic solutions can still be directly converted into osmotic concentrations and could also be used to calculate molar concentrations. In practice, the conversions do not quite work. This is because solutions of electrolytes do not behave "ideally" by dissociating fully into their constituent ions. Instead, the various ions within the solution show a degree of interaction, cations and anions attracting each other and also attracting the water molecules, so that the solution is incompletely dissociated. Each ionic molecule therefore has its own **osmotic coefficient** (sometimes described as the "activity coefficient"), expressing the extent to which it does dissociate. To complicate matters further, this coefficient also varies with concentration, with increasing interaction and a lower coefficient as concentration increases. For most biologically important electrolytes the osmotic coefficient is high, usually above 0.8 (80% dissociation) in the normal range of concentrations, so that the osmolarity can be roughly estimated as the expected multiple of the molarity. For example, if a 1 molar aqueous solution of NaCl is made up its osmolarity should be about 2 OsM, and the expected FPD will be about 2 × 1.86°C. In fact the measured FP is only −3.38°C, indicating an actual osmolarity of 1.82 OsM, so that the osmotic coefficient here is 0.91.

Box 4.2 Fick's law of diffusion

Following the argument in Box 4.1 on measures of concentration, we know that the overall **chemical potential** of a molecule in solution is given by:

$$\mu = \mu^\circ + RT \ln C + VP + zFE + mgh$$

A gradient of chemical potential is the driving force for diffusion, so **diffusional flux** (J) over a distance x will be directly proportional to the chemical potential gradient $\Delta\mu/\Delta x$.

Given two regions (concentrations C_1 and C_2) separated by a surface of area A over which exchanges can occur, the flux in one direction can be calculated from the concentration, the gradient of chemical potential, and the ability of the molecule to move (its mobility u):

$$J_1 = uAC_1(-\Delta\mu/\Delta x)$$

The flux in the opposite direction (J_2) can be calculated similarly:

$$J_2 = uAC_2(-\Delta\mu/\Delta x)$$

Since A and $\Delta\mu$ and Δx are the same in both directions, the net flux is therefore:

$$J_{net} = J_2 - J_1 = uA(C_2 - C_1)(-\Delta\mu/\Delta x)$$

which can be simplified (again see Box 4.1) as:

$$J_{net} = -uRTA\,\Delta C/\Delta x$$

The various constants (u, R, T) can be replaced by a single **diffusion coefficient** (D, which varies with temperature and with molecular weight), giving Fick's law:

$$J_{net} = -DA\,\Delta C/\Delta x$$

i.e. net rate of diffusion is proportional to concentration gradient and to exchange area.

This is of central importance in physiology in considering the diffusion of solutes in solution. The same treatment can be extended to describe diffusion of heat through an animal, and diffusion of electric charge in excitable tissue.

4.3 Passive movements of water and solutes

4.3.1 Movements through solutions

The movements of both water and solutes through living tissues and between different compartments of living systems are very complex, but can be seen as following a few relatively simple rules and principles.

The most fundamental rules are those governing **diffusion**, described by Fick's law (Box 4.2) as depending upon the diffusion area, the concentration gradient of the substance moving, and a constant known as the diffusion coefficient (which in practice is often substituted by a parameter such as permeability, P). No energy is required or released, the process depending entirely on the inherent kinetic energies of the moving molecules. All diffusional movement therefore usually takes place down a gradient, whether of chemical concentration or heat or electric potential.

But in the special case of ions, diffusive movement is sometimes against a chemical gradient, because ions may also be driven by differences in electric charge, and most biological membranes have a resting potential across them, with the cell contents negatively charged relative to the extracellular fluids (see below). Thus positive ions may move into a cell, or negative ions move out of it, against their chemical gradients; but in all cases of diffusional passive movement these ions are moving *down* a combined electrochemical gradient.

4.3.2 Passive movements through membranes

Membrane structure and properties

The structure and chemistry of cell membranes have been intensively studied since their essential **lipid bilayer** character was established early in the twentieth century. The bilayer is clearly visible in electron micrographs (Fig. 4.4a), and the familiar Davson–Danielli phospholipid bilayer model of its constitution explains most of its properties. However, the current more complex model, the **fluid mosaic** structure (Fig. 4.4b), proposed by Singer and Nicholson, allows a better understanding of some kinds of membrane permeation. The membrane appears to be a somewhat disorderly lipid bilayer with fluid properties, having an internal lipid hydrocarbon zone, which is made up of the hydrophobic lipid tails, and superficial polar regions where the hydrophilic lipid molecule heads orientate outwards to the surrounding aqueous solution. The lipid components (25–50% of the dry weight) are predominantly phospholipids, especially glycerophosphatides and sphingolipids, with a small percentage of cholesterol in the outer layer acting as a structural stabilizer. Examples of key lipid structures are shown in Fig. 4.5.

There are additional orderly arrays of proteins and lipids in some parts of the membrane surface. Complex proteins straddle the structure completely, forming channels and carriers. Other proteins project only on the inside of the cell, where they form part of internal messenger systems; or they project only on the outside, many of the latter having short carbohydrate tails that are negatively charged and act to control cell adhesion and separation (forming a 5–7 nm "thatch" on the outside of the cell), or act as part of external receptor mechanisms. These integral proteins generally have their nonpolar portions buried in the hydrocarbon core of the membrane, and their polar regions projecting. They have a globular structure, and can be "seen" as 5–8 nm particles in electron micrographs of freeze-etched cell membranes (Fig. 4.6); these particles are progressively

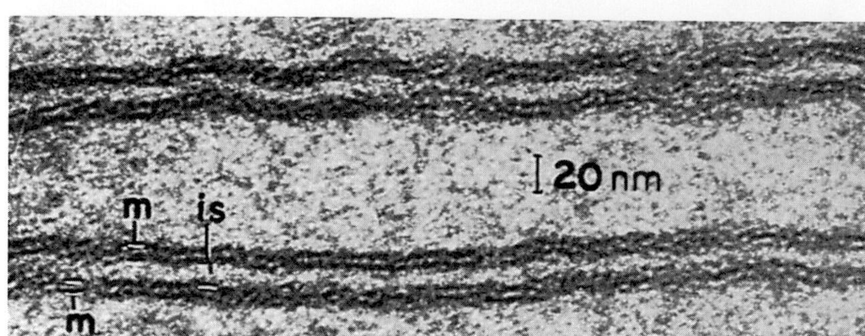

(a)

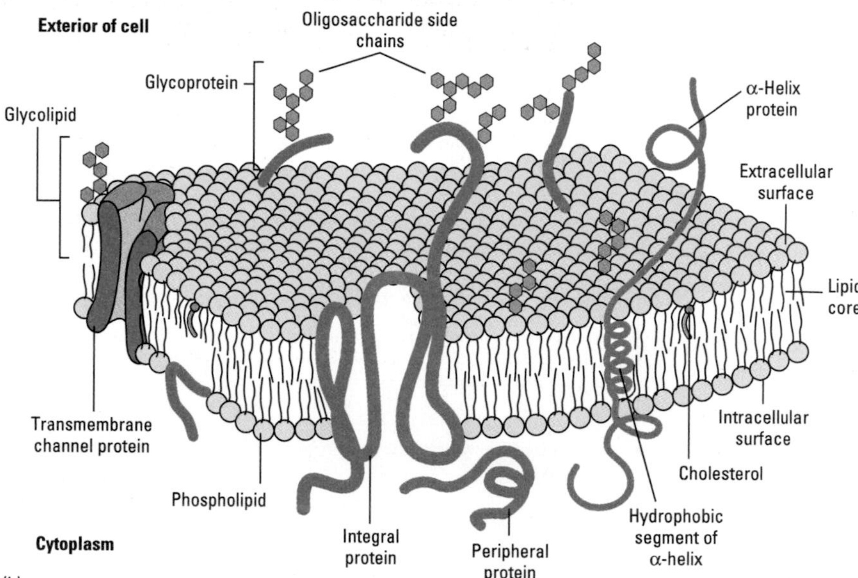

(b)

Fig. 4.4 (a) Electron micrograph of a cell membrane indicating the bilayer structure. m, membrane; is, intercellular space. (From Villee 1977.) (b) Diagrammatic view of the "fluid mosaic" model of membrane structure, showing the lipid constituents arranged as a bilayer, together with the various protein constituents lying in and crossing the membrane.

destroyed when the membranes are treated with proteolytic agents. Electron microscopy has also revealed that the inner side of the membrane is linked to microtubules and microfilaments, which may provide lateral movements and rearrangements of membrane regions according to the cells' needs.

This membrane structure allows considerable possibilities of variation, both within and between cells, so that it is quite incorrect to think of a "typical" cell membrane as if it covered all possibilities. Within cells, there may be areas of local concentration of particular lipids, creating especially fluid or especially "solid" zones, or areas where transport proteins are concentrated to create zones of rapid permeation and hence local ion concentration gradients. Even more variation can occur between cells, since specific lipid constituents may be quite different. This can lead to cells with highly insulated and inert surfaces, such as the myelin sheath of some nerve cells; or to specifically active membranes such as those of mitochondria, where protein subunits are tightly packed together, linked to high enzymatic activity. It can also allow the production of membranes that are resistant to environmental change; for example, membranes may be altered at low temperatures or at high pressures to have lipid constituents that are less saturated and/or of shorter chain length, decreasing the melting point and so maintaining membrane fluidity (see Chapter 8). Cholesterol content may also be increased in some cell membranes under various kinds of environmental stress, producing the curious effect of making lipids more mobile when in the solid phase but less so in the liquid phase, so smoothing out the process of lipid transition and preventing sudden changes in membrane performance.

MEMBRANE PERMEATION

The permeability of any solute in any biological system, whether a whole animal, a tissue, an epithelial layer, or a single cell, depends upon passage through these lipid bilayer membranes. For lipid-soluble solutes—especially those organic molecules with few or no –OH residues (potential hydrogen bonds) and for which there are no special channels or carriers—passage through cell membranes is essentially diffusional and occurs at a rate dependent simply on their lipid solubility. Simple diffusion also occurs for some small water-soluble solutes such as CO_2, which may pass readily through membranes, although water itself may move mainly through small channels. Larger water-soluble molecules need some other way of getting across membranes, and probably use specific larger diameter channels. For example, vertebrate bladder membranes appear to have many small channels for water movement, and rather fewer larger channels permitting urea movement. Membrane permeability to water can be readily measured using radioisotopes of water (deuterium and tritium oxides), and is usually expressed in micrometers per second ($\mu m\ s^{-1}$). The biological range for water **diffusion**

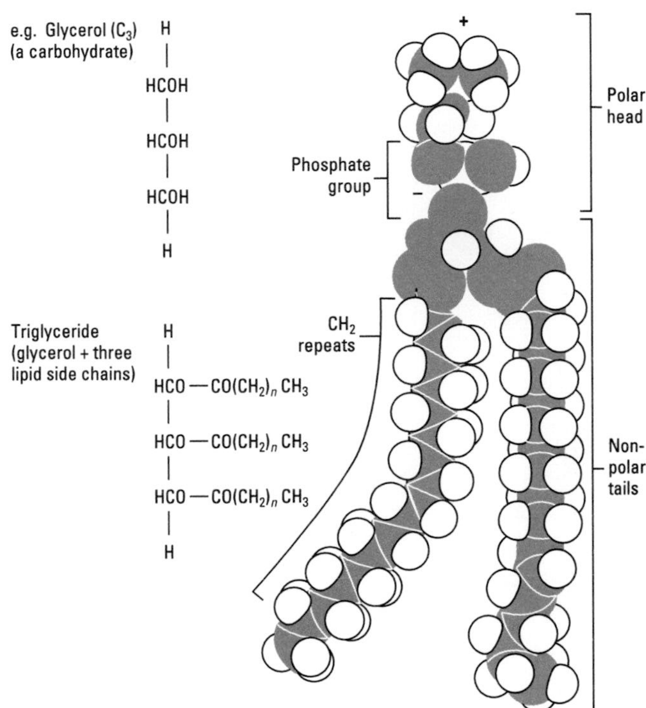

Fig. 4.5 Structures of typical biological lipids, and a pictorial view of a glycerophosphatide, phosphatidyl choline, showing the charges that give the head region its polar character.

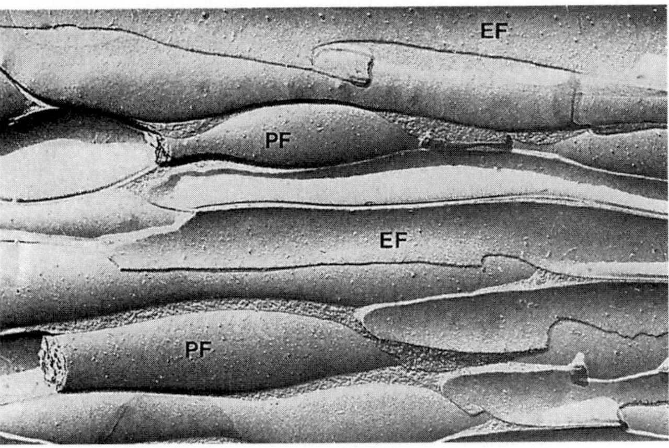

Fig. 4.6 Scanning freeze–fracture electron micrograph of the membrane surfaces of small axons in the neural connective of a mussel. Both the protoplasmic (PF) and the external (EF) faces show protruding membrane proteins as discrete particles 6–10 nm in diameter.

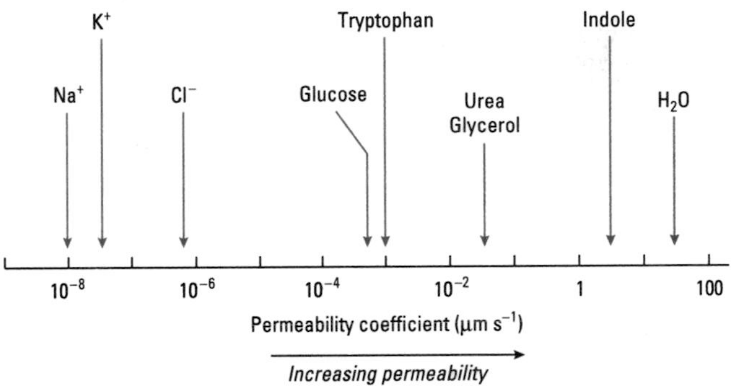

Fig. 4.7 A comparison of diffusional permeabilities through biological membranes for key ions and molecules. (From Alberts *et al.* 1994, Fig. 11.2.)

permeability across membranes is around 0.1–1 µm s^{-1}, and rarely up to 10 µm s^{-1}.

The process of **osmosis** can be seen as a special case of diffusion across membranes in which only the solvent (water, in all biological systems) is moving across the membrane (i.e. through the water channels), the solute molecules being restrained by the membrane (Fig. 4.7), which is therefore described as semipermeable. Hence the values of cell water permeability actually measured are often called **osmotic permeability**. This varies dramatically for different biological membranes (Table 4.1). Although values for some skins and cuticles are very low, the values for unprotected membranes are clearly generally higher than the diffusional permeability values mentioned above. In some cases this may be partly because water movements are being affected by concurrent salt transport (see below), but it also indicates that actual osmotic flow of water across membranes is *not* purely diffusional. The detailed physics of osmosis remain controversial; put simply, the water molecules in the more dilute (hyposmotic) of two solutions could be thought of as randomly colliding with and passing through the membrane more frequently than those on the other side where water concentration is lower, giving rise to a net flow across the membrane from dilute to concentrated solutions. This movement could in theory be through natural transient pores created in the membrane by the random thermal motion of the hydrocarbon "tails" of the membrane lipids, which might create gaps as large as 0.8 nm, large enough even for hydrated Na^+ and K^+ ions. However, this model does not fully explain the high rate of many osmotic processes, nor does it explain the action of certain hormones that can increase the rate of water flow across membranes (e.g. antidiuretic hormone in vertebrate kidneys). Recent work shows that osmosis partly and perhaps predominantly occurs via specific and controllable membrane **water channels** instead, controlled by proteins known as **porins** or aquaporins (AQPs), having a central water-filled pore; these are now known to be present in all living organisms. Their presence explains the fairly common observation of higher osmotic permeability in one direction for animal cells, usually with inward osmosis happening faster than outward osmotic loss (though linked salt transport may again be confusing the issue here).

Table 4.1 Osmotic permeabilities for biological membranes, skins, and cuticles.

	P_{osm} (μm s^{-1})	
Source/species	**Membranes**	**Skins and cuticles**
Artificial collodion membrane	1200	
Protozoans		
Marine ciliate	1.0	
Zoothamnium	2.6	
Amoeba	0.5	
Soft-bodied invertebrates		
Asterias (echinoderm)		32.2
Hydra (cnidarian)		9.4
Squid axon membrane	10.8	
Arthropods		
Carcinus (crustacean)		1.4
Potamon (crustacean)		0.6
Astacus (crustacean)		0.3
Uca (crustacean)		0.24
Artemia (crustacean)		0.1
Sialis (insect)		0.05
Aedes (insect)		0.005
Crustacean axon membrane	32.0	
Crustacean muscle membrane	65.4	
Vertebrates		
Salmo (salmon)		0.6
Anguilla (eel)		
freshwater		0.8
saltwater		0.2
Rana (frog)		5.6
Bufo (toad)		23.6
Lizard		0.006
Snake		0.0008
Frog muscle membrane	221.6	
Mammal cell membrane	142.8	

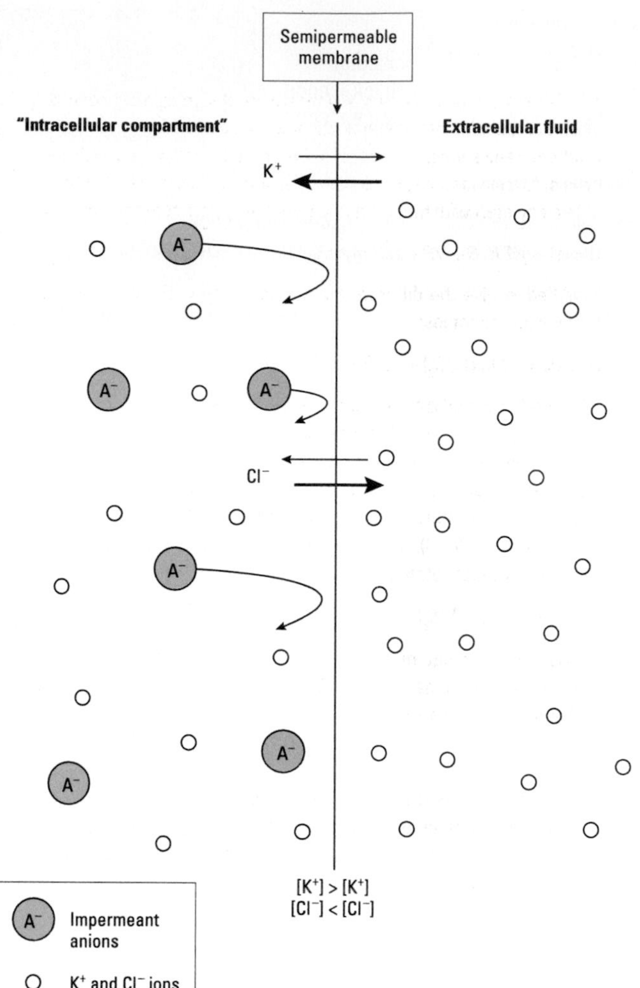

Fig. 4.8 Osmosis and the Donnan equilibrium across a semipermeable membrane. With only KCl present and both the anion and cation freely diffusible, both sides of the membrane would equilibrate at identical concentrations. But here, on one side of the membrane (equivalent to the intracellular compartment) there are additional impermeant anions; K^+ diffuses in and Cl^- diffuses out of this compartment until electrochemical equilibrium is reached, when the concentration gradients for the ions (thin arrows) are balanced by an electrical gradient (thick arrows).

Clearly the osmotic movement of water across membranes is not entirely straightforward, and its rate varies depending upon the osmotic gradient created by transport of solutes such as NaCl, a phenomenon often described as "bulk flow". It is worth noting that where water is moving fast down an osmotic gradient by this kind of bulk flow system, it may sweep other small solutes along in its path, by a process of **solvent drag**, so speeding up other movements across membranes.

Osmotic movements can be prevented by applying a physical pressure to the more dilute solution, exactly counteracting the tendency for the water molecules to move across the membrane into it; this is why an osmotic difference between two solutions is often given as an **osmotic pressure** (Π).

4.3.3 Electrochemical balance across membranes

Donnan equilibrium

A further complication arises in real biological solutions because of the presence of some solutes within cells that cannot pass across membranes but which bear a charge: proteins occurring in the cell in colloidal suspension are the obvious example, commonly bearing a net negative charge at the usual pH of the cell contents. Imagine a cell with a membrane permeable to all ions, where the intracellular solution contains such proteins and all other electrolyte concentrations are equal intracellularly and extracellularly (Fig. 4.8). If the system is allowed to equilibrate, some positive ions from the extracellular fluid will move into the cell to counteract the negative charge there, despite the fact that there is no concentration gradient. Similarly, some diffusible anions such as chloride will leave the cellular compartment because of the electrical gradient, despite this resulting in a slight chloride concentration gradient in the opposite direction. Both anions and cations will continue to redistribute themselves until their overall electrical *and* chemical gradients are in balance and the tendency to move in either direction is matched. At this point there will be a small net charge across the membrane, inside negative, resisting the tendency of any ions to move back down their concentration gradients. This equilibrium situation, naturally occurring in all cells simply because their membranes are

Box 4.3 The Nernst equation

The Nernst equation is very important to physiologists because it allows calculation of the electrical potential of ions from their measured concentrations, or in reverse the calculation of concentration gradients from measured voltage difference across a membrane.

We can start with the formula for chemical potential from Box 4.1:

$$\mu = \mu^\circ + RT \ln C + VP + zFE + mgh$$

Modified to give the difference in chemical potential between two regions this becomes:

$$\mu_1 - \mu_2 = RT \ln (C_1/C_2) + V\Delta P + zF\Delta E + mg\Delta h$$

where ΔP, ΔE, and Δh are the pressure difference, electrical potential difference, and height difference between the two regions.

An ion in equilibrium across a membrane has the same chemical potential on both sides ($\mu_2 = \mu_1$ so $\mu_2 - \mu_1 = 0$), and we can assume the same temperature and pressure and gravitational state on both sides (ΔP and $\Delta h = 0$). So we can rearrange and calculate the difference in electrical potential as:

$$\Delta E = (RT/zF) \ln (C_1/C_2)$$

This is the Nernst equation; it is more familiar and more useful in base 10 log form, where the value of RT/F (for a univalent ion where $z = 1$) is 58 mV. Thus we can write:

$$\Delta E = 58 \log(C_1/C_2)$$

The electrical potential (membrane equilibrium potential, E) produced by a particular ion is therefore directly proportional to the log of the concentration gradient for that ion. An example is given in the text; see also Table 4.3.

semipermeable and they contain nondiffusible charged moieties, is called the **Donnan equilibrium**. All cells, even without any of the more complex active transport systems considered later, would thus have an unequal distribution of diffusible ions, and a small resultant electric charge, across their membranes.

When any ion is in equilibrium across a biological membrane, the electric charge that results across that membrane can be calculated; this most important relationship is described by the **Nernst equation** (Box 4.3). The **electric membrane potential** (E) is related to the logarithm of the concentration gradient. To take a specific example, for the monovalent ion K^+ a mammalian cell might contain a 140 mM solution with the extracellular fluid at 20 mM, so that the membrane potential can be calculated as $58 \times \log(140/20)$, or −49 mV (by convention given as the internal potential relative to the outside, hence inside negative).

The Donnan equilibrium is one of the causes of electrical potentials existing across cell membranes, and in a hypothetical cell that was only permeable to one cation, *or* that was equally permeable to all ions, it would on its own give a substantial membrane potential. However, it usually only contributes a small part of the total potential recorded in real living cells, because several ions can leak across the membrane at different rates. There are in fact two other major reasons why cells have a so-called "resting potential": the existence of substantial diffusion potentials resulting from differential permeability of the membranes, dealt with next; and the existence of active transport mechanisms across membranes, covered in section 4.4.

Selective permeability and ion channels

All ions have a particular mobility when in solution; some move readily, others rather slowly. A major determinant of this is frictional drag, which in turn is related to ionic size. But it is not necessarily the case that smaller ions are more mobile: as we have seen, ions carry a shell of water molecules bound to them, the "hydration layer", so that quite small ions may become relatively less mobile, while larger ions may carry a lesser hydration shell (they have a bigger surface area and thus a lower charge density, so bind less tightly to water molecules) and may be quite rapidly mobile. Values of important hydration numbers, and both unhydrated and hydrated ionic radii are given in Table 4.2 (though these values are averages, as the water shell is itself continually exchanging with the surrounding free water). Mobilities are also given, showing that the mobility is greater as the hydration shell gets smaller, and cations of larger atomic mass are in practice *more* mobile.

Differential ionic mobility is one reason for differential ionic permeability in cell membranes. The mechanism of differentiation resides in **ion channels**, though, which are formed by pores within integral proteins in the membrane. All membranes appear to have these channels in them, and the channels are relatively specific for a particular type of ion of a particular size, charge distribution, and mobility. Thus we speak of Na^+ channels, K^+ channels, Ca^{2+} channels and so on. These cationic channels will accept other cations, but will permit their transit across the membrane only slowly. For example, the Na^+ channel in vertebrate nerve axons is also reasonably permeable to Li^+ ions, but rather impermeable to K^+ or NH_4^+. The K^+ channel in the same cell membrane is very impermeable to Na^+ ions. Values of relative permeability are also given in Table 4.2; note that they are not directly related to ionic radius, since charge

Table 4.2 Hydration states, ionic radii, and mobilities for important biological ions, together with relative permeabilities through the Na^+ channel and K^+ channel in frog axons.

Ion	Atomic mass	Hydration number	Ionic radius (nm)		Mobility (μm s^{-1}/volts cm^{-1})	Na^+ channel (P/P_{Na})	K^+ channel (P/P_K)
			Crystal	Hydrated			
Li^+	6.94	>6	0.060	–	4.01	0.930	<0.010
Na^+	22.99	4.5	0.095	0.512	5.20	1.000	<0.010
K^+	39.10	2.9	0.133	0.396	7.64	0.086	1.000
Ca^{2+}	40.08		0.099	–	–	0	0
NH_4^+	18.04		–	–	–	0.160	0.130
Rb^+	85.47		0.148	–	–	<0.012	0.910
Cl^-	35.45	2.9	0.181	–	7.91	–	–
Br^-	79.70	2.4	0.195	–	8.28	–	–

distribution matters as well as size. In certain other cells there are more general cationic or anionic channels, with reasonable permeability to any ion of the correct charge. Most animal cells in their resting state have quite high K^+ permeabilities (e.g. P_K 0.001–0.03 μm s^{-1}), moderate Cl^- permeability, and relatively low Na^+ and Ca^{2+} permeabilities. (However, it is worth noting that all these values of ion permeability are around 5–8 orders of magnitude lower than the values of cell water permeability, P_{osm}, in Table 4.1, so that any active movement of ions will result in a rapid osmotic water movement, faster than the ion gradient can run down.)

This kind of variable selectivity must be attributable to rather specific pore structures, resulting from membrane proteins with a very precise configuration, and incorporating a "pore" that discriminates between the hydrated ions according to their size and charge. Investigation of pore structure and selectivity has been greatly aided by the discovery of particular toxins that will block certain kinds of channel and not others. The most famous of these is tetrodotoxin (TTX), derived from the Japanese puffer fish and its relatives, which will quite specifically block the axonal sodium channel when applied to the outside of nerve cells by binding to a site on the outer part of the channel protein. Some local anesthetics are also specific sodium channel blockers. Other toxins, derived from snakes, scorpions, and sea anemones, are known to bind to other parts of the sodium channel structure, causing it to stay abnormally open and thus increasing sodium permeability.

It is worth adding here that certain channels may also include a "gating" mechanism, the protein undergoing some kind of conformational change to open or close the pore to ion flow. This gating system may be sensitive to ion concentrations, or to certain hormones and intracellular messengers (ligand-gated channels). It may also be voltage sensitive, so that the channel opens or closes as the membrane itself is depolarized or hyperpolarized. This system of altering permeabilities via alterations to the channel protein forms the basis of excitability—the action potential—in nerve and muscle cells.

In recent years channel proteins have begun to be sequenced so that we can recognize families of related membrane channels. The familiar ion channels best known from excitable tissues probably form one family, and the voltage-gated channels and ligand-gated channels are rather different from these. There is also another category of ion channel, however, strongly selective for sodium and blocked by amiloride, and this has a wide and puzzling distribution including mammalian kidney tubules (where its frequency is increased by the hormone aldosterone), certain snail sensory neurons, some nematode cells that are probably mechanosensors, and parts of the vertebrate brain.

Resting membrane potential

Living animal cells, then, have a membrane that is selectively permeable to a few ions (especially potassium), but rather impermeable to others. This has very important consequences for ionic distributions and electrical potentials developed across the membrane. If we consider an artificial situation, with a membrane that is selectively permeable to potassium and which has a simple concentration gradient across it resulting from 1 M KCl on one side, and 0.1 M KCl on the other, we can visualize what will happen. K^+ ions will tend to diffuse down their concentration gradient to the more dilute solution, and in doing so will set up a small positive potential on the more dilute 0.1 M side and an equal small negative potential on the 1 M side. However, this process will not go on unchecked until the two sides have equal potassium concentrations, because the electrical potential it creates (as the charged particles move) will lead to an opposing force tending to move potassium ions in the other direction. Therefore diffusion will go on until the electric forces and the chemical forces balance each other, and at this point there will be a potential difference across the membrane given by the Nernst equation discussed above. This calculated value will be the **potassium equilibrium potential** (E_K). It can be calculated from the Nernst equation for any particular living cell given the internal and external potassium concentrations. Similarly, the **sodium equilibrium potential** (E_{Na}) could be calculated; it would be the achieved membrane potential for a system where sodium was the only mobile ion.

If we look at the actual equilibrium potentials for individual ions in real living cells, with high internal potassium and low sodium and calcium levels, the values obtained are strikingly different from each other. Table 4.3 includes the values for different kinds of axon (including the squid axon, the first cell for which these measurements were made) and for various muscle cells. E_K values are large and negative (i.e. potassium drives the cell to a negative potential internally, because positively charged potassium ions tend to move outwards along their concentration gradient). E_{Cl} levels are moderately to highly negative, but E_{Na} (and E_{Ca}) levels are relatively high and positive, because of the tendency of both these ions to move into the cell down their concentration gradients and so produce a small internal positive charge.

But in reality, of course, membranes are exposed to solutions containing complex mixtures of ions, with both cations and anions able to move across the barrier. It is quite easy to extend the Nernst relation to consider several ions at once, and the more complex resultant equation is known as the **Goldman–Hodgkin–Katz relation** (Box 4.4). The only additional parameters needed are the

	Intracellular fluid (mM)			Extracellular fluid (mM)			Equilibrium potentials (mV)			Resting potential (mV)
	Na	K	Cl	Na	K	Cl	E_K	E_{Na}	E_{Cl}	
Squid nerve	49	410	90	440	22	560	−74	+54	−46	−68
Crab nerve	52	410	26	510	12	540	−87	+57	−76	−80
Insect nerve	86	555		212	124		−37	+23		−25
Frog muscle	10	124	2	110	2	78	−102	+59	−92	−90
Insect muscle	22	86		117	11		−51	+41		−40
Mammal muscle	12	155	4	145	4	123	−98	+67	−91	−90

Table 4.3 Equilibrium potentials in a range of nerve and muscle tissues, with the internal and external ion concentrations. Note that resting membrane potentials are generally close to E_K. (Calculated at 15°C, except for mammalian muscle which was calculated at 37°C.)

Box 4.4 The Goldman–Hodgkin–Katz relation

In Box 4.3 the derivation of the important Nernst equation was examined, giving the relation between the electrical potential between any two compartments and the concentration difference between them for a particular molecule. This same equation describes the equilibrium potential (E, in millivolts) across a cell membrane due to a Donnan equilibrium (see text), as follows:

$$E = (RT/zF) \ln (C_1/C_2)$$

or:

$$E = 2.303\,(RT/zF) \log (C_1/C_2)$$

or:

$$E = 58 \log (C_1/C_2)$$

In the simple Donnan equilibrium described there were just two permeant ions, K^+ and Cl^-; in a real cell there may be several. But the principle remains the same: we can calculate the final membrane potential (E) from all the notional contributing potentials *but* we must take into account the relative permeabilities of the ions. The net contribution to E of any ion species by its movement from a compartment across a membrane is the product of its concentration in the original compartment and its permeability. The Goldman–Hodgkin–Katz equation simply puts all these contributions, for each ion in each direction, together:

$$E = \frac{RT}{F} \ln \frac{[\mathrm{K}]_o P_K + [\mathrm{Na}]_o P_{Na} + [\mathrm{Cl}]_i P_{Cl} + [\mathrm{Ca}]_o^{1/2} P_{Ca}}{[\mathrm{K}]_i P_K + [\mathrm{Na}]_i P_{Na} + [\mathrm{Cl}]_o P_{Cl} + [\mathrm{Ca}]_i^{1/2} P_{Ca}}$$

The extracellular cation concentrations (e.g. for potassium this is $[\mathrm{K}]_o$) are placed in the numerator by convention, and the intracellular cation concentrations (e.g. $[\mathrm{K}]_i$) in the denominator, so that we calculate the potential of the inside relative to the outside. Note also that the valency or charge term z has been incorporated into the initial constant group as if it were one, which has two consequences: for the only divalent ion (calcium) the terms are raised to the power $1/z$, or 1/2, and for the only ion where $z = -1$ (chloride) the numerator and denominator must be reversed.

For the squid giant axon, the full equation above gives $E = -73.6$ mV.

In practice, calcium ions are not very permeable across cell membranes, and chloride ions are distributed entirely passively, so the equation can be simplified to contain only the sodium and potassium terms. E is then calculated as −73.7 mV.

Furthermore, as P_K is usually much greater than P_{Na}, we can see from Table 4.3 that the resting potential E for many cells is close to E_K (in this case −74 mV). Only at very low levels of external potassium concentration, in laboratory manipulations with cells, does E differ significantly from E_K, as the contribution of sodium ions becomes more important.

relative permeabilities of each of the ions involved: P_K, P_{Na}, and so on. Freely permeable ions will move readily across the membrane and contribute substantially to the membrane potential E, whereas impermeant ions will not move and can have no effect on E. The net contribution of any one ion is therefore the product of its concentration gradient and its permeability. Using the full equation, the value of E for a living cell (sometimes called its **resting potential**) can be calculated; it generally works out at about −30 to −90 mV, in close agreement with the values measured using microelectrodes. In practice the equation is useful in reverse: measured resting potentials can be a tool to calculate ionic permeabilities for particular cells.

In practice the Goldman–Hodgkin–Katz equation can be simplified to take account of only sodium and potassium ions (see Box 4.4), since calcium ions have a very low permeability, and chloride ions are nearly always distributed entirely passively across cell membranes. Using this simple version with the squid axon, E is calculated as −73.7 mV, in reasonable agreement with the measured value of −68 mV.

But a quick examination of the actual membrane potential in these cells shows that E is always very close to E_K. For the squid axon E_K is −74 mV with E being −68 mV, whilst for typical vertebrate muscle E_K is −98 mV and E is −90 mV. So it is often quite acceptable to use a very simple version of the original equation (see Box 4.4) and calculate E simply as being equivalent to E_K. This is because in nearly all cells, the permeability to K^+ ions is much greater than to Na^+ ions (often at least 100-fold greater), and potassium is therefore by far the most important determinant of the resting membrane potential.

To summarize then, many ions move passively across membranes, through selectively permeable pores. Some like K^+ and Cl^- move relatively freely having a high permeability, others like Na^+ and Ca^{2+} have much lower permeability. These differences lead to a membrane potential (E) that is always quite close to the potassium equilibrium potential (E_K). This small electrical potential across living cell membranes is immensely important for many aspects of cell functioning, from simple cell adhesion and processes of secretion through to more dramatic effects such as excitability and contractility (see Chapter 9). We know that it is brought about by passive ion movements, determined by the varying concentrations of ions inside and outside cells, and the properties of the membrane and of the ion channels therein. But we also need to know how the cell concentration initially came to be so different from the extracellular fluids, setting up these ionic gradients upon which the membrane potential depends.

This brings us to the topic of active ion movements. There are ion "pumps" in all cell membranes, which are the major reason for the unequal distribution of Na^+ and K^+ between cells and extracellular fluids. These pumps are continuously moving potassium ions into the cell and sodium ions out to produce the high-potassium cellular milieu. These pumps are therefore, indirectly and in the long term, the single most important reason for the membrane potential, by causing K^+ to be the predominant intracellular cation. In the following section we examine how active pumps work and what other effects they have in cells.

4.4 Nonpassive solute movements

Diffusion and osmosis are responsible for the bulk properties and equilibrium compositions of most of the fluids found in the world. The additional type of process that controls the composition of biological aqueous solutions is mediated transport, whereby substances are moved across membranes by interaction with protein **transporters**. This may be against an electrical and/or chemical gradient, or at unexpectedly high rates that are unrelated (or nonlinearly related) to concentration differences. Such movements are also characterized by being susceptible to chemical inhibition.

Mediated transport is fundamentally of two kinds. This field has for 30 years been dogged with alternative and conflicting terminologies, but the terms used here are those coming into generally agreed use now. **Primary transport processes** include all those directly dependent on coupling with metabolic reactions in the cell. The best known form of this is primary solute translocation, commonly known simply as **active transport**, where an enzyme within a membrane catalyzes the movement of a solute up an electrochemical gradient. Group translocations are a special type of active transport where the transported solute is simultaneously transformed by the transporter enzyme and is released on the other side of the membrane as a different molecular species. Electron translocations are also a form of active transport, where electrons are transported between coupled membrane enzymes; these are sometimes termed "redox potential pumps".

These forms of active transport are all distinct from **secondary transport processes**. These include **symport** and **antiport** (collectively secondary active transport), where two solutes are transported together in the same (symport) or opposite (antiport) directions, with one of the solutes moving up an electrochemical gradient. Secondary transport processes also include the important category of **uniport**, where a single solute is moved down its electrochemical gradient essentially passively, needing no metabolic energy and no other solute gradient, but with specific mediated interactions with the membrane components involved so that the process does not have the kinetics of simple diffusion; this process has traditionally been termed **facilitated diffusion**.

All these processes involve a membrane-bound protein molecule that binds reversibly to the solute molecule at a specific site. The protein–solute complex then undergoes a change in conformation that effectively translocates the solute to the other side of the membrane, where the solute and protein disengage. The protein molecule may actually move across the membrane, like a ferry, or it may span the membrane and effectively act as a gate (Fig. 4.9a), guarding a pore in the membrane. Both kinds of system are known to occur, and they are quite difficult to distinguish in practice.

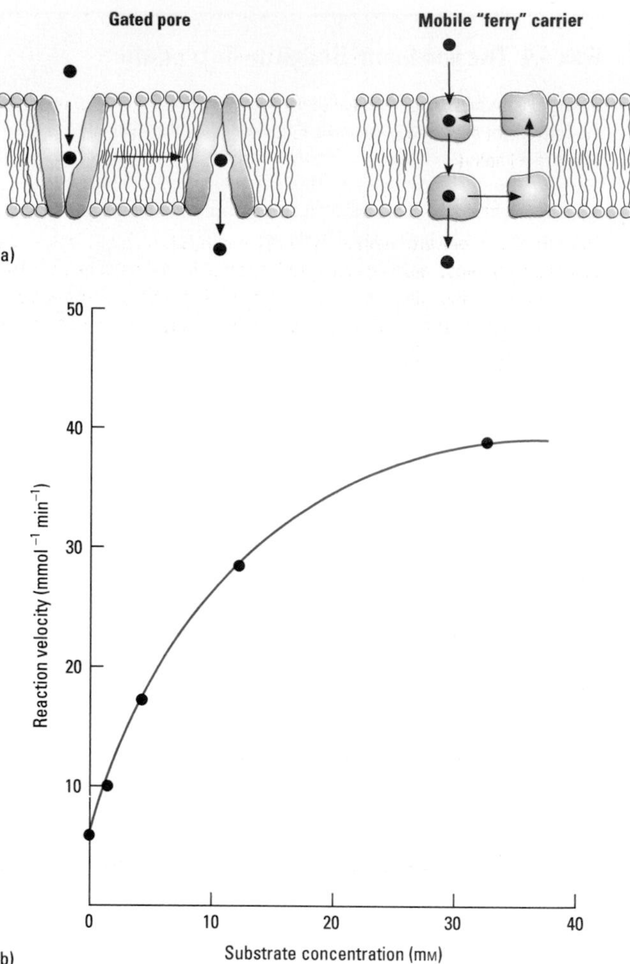

Fig. 4.9 (a) "Gate" and "ferry" models for protein-mediated movements across cell membranes. (b) With such carrier-mediated transport systems there is a nonlinear relationship between rate of transport and substrate concentration, giving typical saturation kinetics.

4.4.1 Facilitated diffusion (uniport)

Uniporters are the simplest of the membrane transport protein systems, and the term may sometimes be used to include both channels (dealt with above) and facilitated diffusion carriers (or "facilitated transporters"); here we use it only in the latter sense. The carrier proteins are distinguished from channels by having substrate-binding sites that are only accessible from one side of the membrane at any one time.

Uniport-facilitated diffusion is thus a protein-mediated, but still "downhill", process. Anything moving across a biological membrane by this means will not obey Fick's law but instead will show saturation kinetics, where transport rate increases hyperbolically with solute concentration and thus approaches saturation, where transport rate cannot increase further (Fig. 4.9b). A classic example is that of glucose uptake into human erythrocytes (red blood cells). The protein carrier for this has a carbohydrate tail on the outside of the cell membrane, but the active site for glucose binding is within the protein and appears to be exposed alternately to the outside and then the inside of the membrane, so shuttling the glucose across. Both sugars and amino acids are moved across eukaryotic cell membranes by uniporter systems, though both are also handled against their concentration gradients by symporters, discussed below. Recently it has also been shown that urea movements in vertebrate kidneys involve uniporters (facilitated transporters), especially in the terminal part of the collecting duct (see Chapter 5).

4.4.2 Primary active transport: membrane ion pumps

Active transport of any solute involves protein-mediated "uphill" processes, and therefore ultimately requires a continuous expenditure of energy by the splitting of ATP. Energy from a readily reduced, oxidized, or hydrolyzed chemical bond is converted into the energy of an electrochemical gradient (usually of cations or of protons). In its most familiar form as primary solute translocation, active transport is usually mediated by a membrane protein molecule that is linked directly to ATP splitting (so that the protein

Box 4.5 Enzyme kinetics and the K_m value

Biological reactions commonly occur under the influence of a catalyst in the form of an enzyme (E), with the overall reaction from substrate (S) to product (P) being:

$$E + S \rightleftharpoons ES \rightleftharpoons E + P$$

The rate of the forward reaction (V_f) is then k_1[S][E], where k_1 is the rate constant; for most enzyme-catalyzed reactions the reverse reaction (with rate k_2) is negligible and can be ignored. The relationship between reaction velocity and substrate concentration [S] is then hyperbolic as shown below:

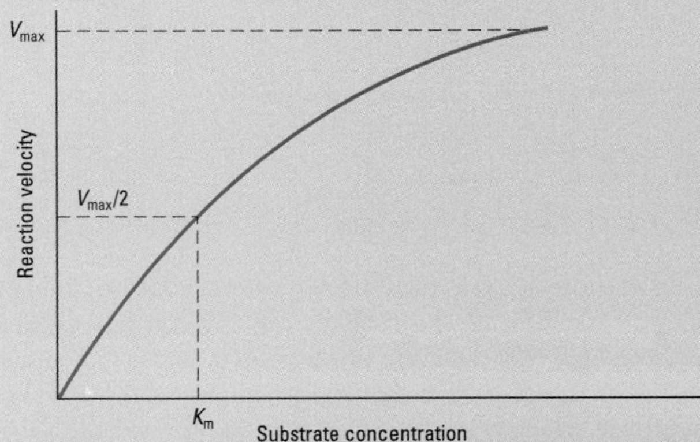

The **Michaelis–Menten equation** describes this hyperbolic curve as follows:

$$V_f = V_{max}[S]/(K_m + [S])$$

Maximal reaction velocity (V_{max}) occurs at infinitely high substrate concentrations, so that in practice biochemists normally think in terms of the substrate concentration at half maximal velocity; this is what is termed the **Michaelis–Menten constant, K_m**.

It is difficult to estimate K_m and V_{max} from a hyperbolic plot as above, so reciprocal plots (1/V against 1/[S]) are normally used instead, given that the equation above can be rearranged as follows:

$$1/V_f = K_m/V_{max}[S] + 1/V_{max}$$

The reciprocal plot (known as a Lineweaver–Burk plot) gives a straight line that has an intercept on the 1/V axis at $1/V_{max}$, and an intercept on the 1/[S] axis at $-1/K_m$.

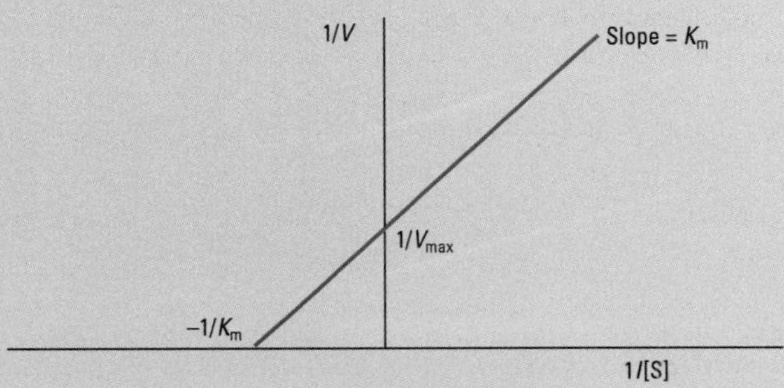

Values for K_m and for V_{max} can then be fairly readily found and compared between reactions.

These values are also useful to physiologists in allowing us to compare between tissues and animals such phenomena as rate of ion pumping (equivalent to rate of an ATPase-catalyzed reaction).

- K_m values (usually in mM) can be seen as reflecting the "affinity" of the pump for the ion that it moves; they are often termed K_t values instead ("transport constants").
- V_{max} values (in s^{-1}) indicate the maximum rate of pumping (ion flux) that is possible; again they are often termed J_{max} for pumps and carriers.

carrier is an ATPase enzyme) and that combines reversibly with the substance being moved across the membrane. Active transport by ion pumps is therefore subject to typical enzyme-mediated reaction kinetics (Box 4.5), and the affinity of the pump (its K_m value) and its maximum rate (its V_{max} value) can be calculated in a standard fashion from the Michaelis–Menten equation as shown in Box 4.5.

At present it is still unclear how many primary ion pumps (ATPases) have evolved in animal systems. Initial work indicated many different kinds operating, but recent studies have tended to reduce this profusion and indicate that relatively few basic mechanisms are at work, though they may handle ions differently in different cells according to circumstance. By far the best studied example of this kind of transport is the **Na^+/K^+-ATPase** or **sodium pump**, which is an ATP-splitting enzyme that works only in the presence of both Na^+ ions in the cell and K^+ ions outside it. This pump now appears to be common to all animal cells, usually localized in the apical membrane (the side away from the basement membrane on which the epithelium rests) but sometimes in the basal and lateral membranes (e.g. in the vertebrate kidney). In many cells this pump exchanges three Na^+ "out" for every two K^+ "in". This unequal exchange clearly produces a net movement of charge, and the pump is therefore described as **electrogenic**, or more strictly **electrophoretic**. But sometimes the same pump appears more straightforwardly to exchange one Na^+ for one K^+ (or to operate 1 : 1 Na^+/Na^+ or K^+/K^+ exchanges), becoming **electroneutral**. In many cells the electrogenic effect of the pump contributes directly to the membrane potential, although this direct effect is usually only a few millivolts. The two roles that the sodium pump plays in setting the resting membrane potential can be demonstrated using inhibitors, since the Na^+/K^+-ATPase has a specific reaction to the drug ouabain. Inhibition of a cell by ouabain will often bring about a rapid depolarization of 2–10 mV as the electrogenic "topping up" effect of the pump is blocked. Then, over a much longer time course, the membrane depolarizes more substantially and the membrane potential, *E*, declines towards zero, as the ionic gradients upon which the passive equilibrium potential depends are gradually dissipated without the continuing action of the pumped Na^+/K^+ exchanges.

The Na^+/K^+-ATPase pump has been extremely well studied from many different tissues, and is now being characterized at the sequence level. It is known to be a four-part protein, or tetramer, with two α polypeptide subunits (molecular weight 110 kDa) that span the cell membrane and contain all the catalytic sites, and two

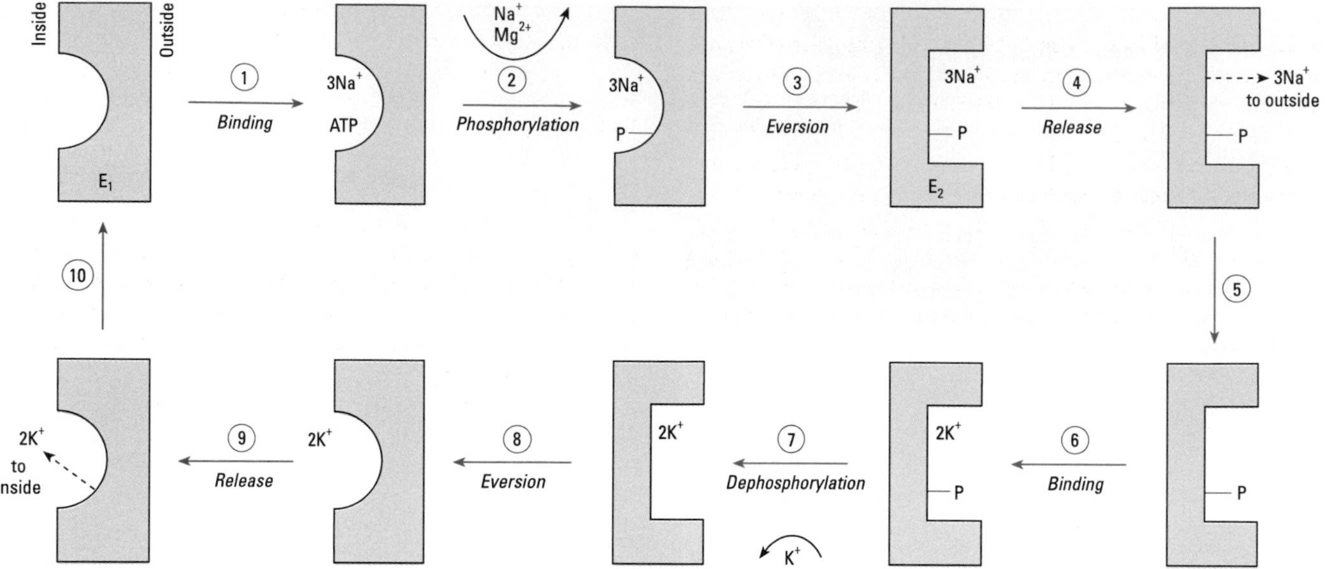

Fig. 4.10 A model for the series of changes taking place when the Na^+/K^+-ATPase binds with and hydrolyzes ATP and simultaneously exchanges Na^+ and K^+ across the cell membrane.

smaller β-subunits (35 kDa). At present there are three identified variants (isoforms) of the α-unit and two isoforms of the β-unit, expressed differently according to species, tissue, and developmental stage. These polypeptides have distinct regions, in sequence, that are Na^+ sensitive, ouabain sensitive, and K^+ sensitive, and other zones concerned with phosphorylation. These are arranged in the membrane by complex molecular folding such that there are binding sites for Na^+ and for ATP on the inside of the cell, and for K^+ and ouabain on the outside. The cycle of events that occurs when the pump is operative and ATP is being hydrolyzed is summarized in Fig. 4.10.

A second ion pump that seems to be ubiquitous in animal cells is a calcium-activated enzyme, the Ca^{2+}-ATPase, also usually localized in the apical plasma membrane and found at particularly high concentration in the sarcoplasmic reticulum of striated muscles. Again, it may operate electrogenically, or in partial exchange for H^+ ions, according to circumstance. This pump is also made up of α- and β-dimers (94 kDa and 34 kDa, respectively), and these have quite substantial sequence homology with the Na^+/K^+-ATPase α- and β-subunits.

The third type of pump that appears to be universal in cells is a proton (H^+) pump. Proton gradients may have been the first system developed in living systems to interconvert chemical energy from energy-rich compounds into electrochemical energy as directional ion gradients; they may date back over 3 billion years. Proton gradients arise in two ways: firstly from "redox potential pumps", where the electron transfer systems in cell membranes are coupled to proton translocation, and secondly from reversible ATP-dependent proton pumps. Both are present in the same membranes, and in mitochondria they both pump protons from the center of the organelle into the lumen between the double membrane. This mitochondrial proton pump belongs to a gene family known as the **F-ATPases**.

In all eukaryotic cells there are also two additional kinds of proton pumps moving H^+ ions out of the cell. One of these is usually an electroneutral H^+/K^+-ATPase (made up of α- and β-subunits, and again with strong sequence homology with the Na^+/K^+-ATPase); it appears to belong to a single gene family together with the Na^+/K^+-ATPase and the Ca^{2+}-ATPase, and they are collectively designated the **P-ATPase** family. In vertebrates the P-ATPase proton pump operates in the stomach epithelium to acidify the contents in exchange for K^+ ions, and is capable of operating against a 4 million-fold proton gradient. The other well-established proton pump is an H-ATPase belonging to a third important gene family, the **V-ATPases**. This pump is often electrogenic and is particularly responsible for acidification in various tissues, for example in urinary compartments such as the reptile bladder and the mammalian nephron; in insects it seems to drive most of the epithelial transport systems. It is worth noting that proton pumps cannot act entirely alone as the proton gradients would be electrically damaging to cells. There is normally some linked movement of an anion as well as a proton, though not always in a 1 : 1 stoichiometric fashion.

The three classes of ATPase have very different structures and operate very differently. The P-ATPases have a phosphorylated enzyme (phosphoenzyme) as an intermediate and are characterized by sensitivity to vanadate; they can exist as a single unit of 94–110 kDa, or may have two kinds of subunit as described for the sodium and calcium pumps above. The V-ATPases and F-ATPases are more like each other; they do not use phosphoenzyme intermediates, and both require several cooperating subunits for full functioning. The V-ATPase may be the primordial proton pump, found in archaebacteria and operating as a hexamer of a single gene product. F-ATPases evolved from this in the eubacteria, as oxygen levels in the Earth's atmosphere increased (see Chapters 6 and 7). Thus modern eukaryotic cells have V-ATPases in their plasma membranes but F-ATPases in their inner mitochondrial membranes (reflecting the descent of these intracellular organelles from eubacterial endosymbionts). Some V-ATPases developed important cation-handling functions in eukaryotes, still seen today in many invertebrates; the family may include sodium-dependent pumps that are insensitive to K^+ (unaffected by ouabain but blocked by the drug ethacrynic acid), and K-dependent electrogenic anionic (Cl^-) pumps, which dominate the process of fluid resorption in the insect rectum.

Apart from the V and F families, the P-ATPases developed somewhat later and have proved much more versatile, radiating into many distinct versions and handling a variety of cations, with the Na^+/K^+-ATPase becoming the dominant membrane pump in most animal cells. This is especially the case in vertebrates, where this pump underlies most ionic distributions, with movements of Cl^- and K^+ ions nearly always explained by symport (cotransport) mechanisms.

4.4.3 Secondary active transport: symport and antiport

This raises the issue that, in many cases of active transport of ions or small solutes, the linkage to ATP splitting is often much less direct than we find with the primary ion pumps, and the active transport is driven instead by the "energy store" represented by an existing electrochemical gradient, most commonly the Na^+ gradient in vertebrates (though often a proton gradient in prokaryotes, plants, and insect epithelia). This may be termed **secondary active transport**, and it is responsible for redistributing actively transported ions and many organic and inorganic solutes such as sugars, amino acids, and neurotransmitters in cells. Substances are moved in coupled pairs, so that the system is sometimes called **cotransport**; one of the cotransported substances (A) moves in a direction that dissipates its electrochemical gradient, and the other (B) is thereby moved against its gradient (so the direction of movement can change if the relative gradient changes). If A and B both move in the same direction the process is termed **symport** and the protein carrier is a **symporter** (or simply a **cotransporter**). If B moves in the opposite direction to A then the terms **antiport** and **antiporter** are used (or the carrier may simply be termed an exchanger). In either case the cotransport is dependent on the ATP used to maintain the initial ionic gradient. Overall, then, secondary active transport involves a net loss of the energy stored in the gradients set up by primary active transport.

Given the variations in terminology, inevitable confusion arises in the literature, particularly since the genes for individual transporter proteins have become amenable to study. Thus for example we used to speak of a Na^+/glucose cotransporter, and the gene for it became known as *SGLT*, but we would now call this transporter the glucose/Na^+ symporter (specifying the driven solute first and the driver solute second), and thus the accepted gene abbreviation looks less relevant. Similarly the "Na^+/H^+ exchanger" in older literature should now ideally be termed the H^+/Na^+ antiporter and its relation to the "*NHE* gene" becomes a little mystifying.

Symporters

Some of the clearest examples of symport occur in marine invertebrates (e.g. annelid worms) that can take up amino acids and monosaccharides from sea water against enormous chemical gradients. This uptake occurs in the presence of extracellular sodium, but almost ceases if there is no extracellular sodium. The process is driven by the standard Na^+/K^+-ATPase exchange pump on the basal membrane of the epithelial cells, creating low intracellular sodium and a substantial electrochemical gradient for sodium uptake from the sea water. As a result there is an inward leak of sodium ions, and these are coupled by the specific membrane symporter (protein carrier) to an equal inward movement of the amino acid or sugar (Fig. 4.11). The presence of Na^+ has thus increased the rate of uptake of the required solute, which (continuously driven by the continuously renewed Na^+ gradient) may occur "uphill" into the animal's cells against 10- or 20-fold concentration gradients. This kind of symport is blocked by application of the Na^+/K^+-ATPase blocker, ouabain, because of the indirect effect of blocked sodium pumps. It operates in an essentially similar manner in vertebrate intestinal and renal cells, using the symporter products of the *SGLT* genes mentioned above. Most of the animal Na^+-coupled symporters work with ion : substrate ratios of at least 2 : 1 (e.g. two Na^+ per single sugar molecule moved).

Symport is also enormously important in the flow of anions across membranes. In many systems concerned with salt transport or with volume regulation, chloride ions are handled by a specific symporter system. The apical membrane may bear sodium pumps, and the basal and lateral membranes house the symporter, so that when sodium ions leak in basally along the gradient created by the pump, chloride ions are also moved. The net result is usually transport of both sodium and chloride across the cell or epithelium. The mechanism is very commonly a $Na^+/K^+/2Cl^-$ symporter, and therefore electroneutral. Some tissues reportedly show different Na^+/Cl^--only, or K^+/Cl^--only, systems, though it may be that all these will eventually be revealed as having a similar underlying mechanism. A common feature is that the Cl^- symporters are all sensitive to the diuretic drug furosemide, and all have a very high chloride specificity, handling bromide only partially and ignoring all other chloride substitutes. Other anionic symporters do exist, however; in vertebrate intestine and kidney there are Na^+/phosphate symporters and Na^+/sulfate symporters, together responsible for a large part of the animal's phosphate and sulfate transmembrane transport.

Antiporters

Antiport is involved in the maintenance of pH in all cells and in the maintenance of very low intracellular calcium concentrations in many cells. Na^+/H^+ antiporters operate to regulate the internal cell pH (electroneutrally in vertebrates, but apparently electrophoretically with a 2 : 1 ratio in most invertebrate epithelia). At the same time, $[Ca^{2+}]_i$ is usually less than 10^{-6} mM, and must be kept at these very low levels by a continuous outward flux (efflux) of calcium. When extracellular Na^+ is removed, so that there is no inward sodium leak, efflux of calcium slows very substantially. This suggests that the two ions may be linked via an exchanging carrier (Na^+ leaking in and Ca^{2+} leaking out), and this must be a Na^+/Ca^{2+} antiporter. The immediate source of energy for the exchange is again the classic sodium pump, as this has set up the sodium gradient on which the Na^+/Ca^{2+} exchange depends.

Antiporters also operate in electrogenic K^+ transport in some invertebrate sensilla and in the insect midgut, salivary glands, and Malpighian tubules. In the lepidopteran midgut K^+ is secreted into the lumen from special "goblet cells" by an electrogenic $K^+/2H^+$ antiporter, driven by the gradient resulting from the proton V-ATPase.

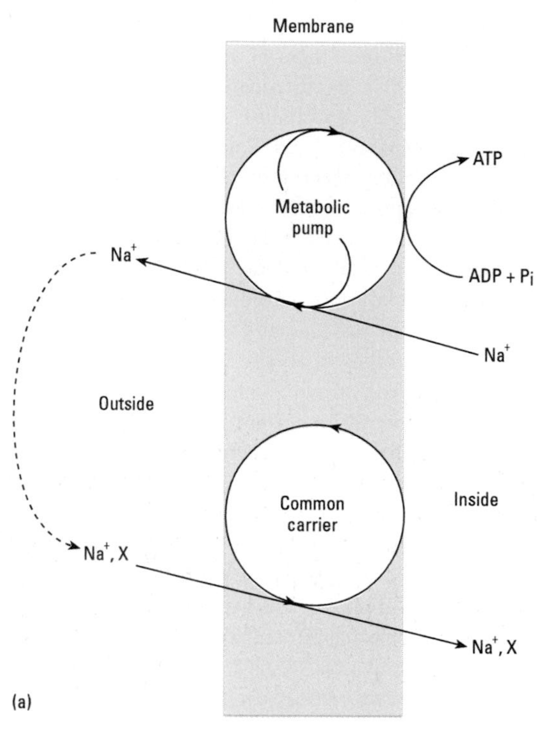

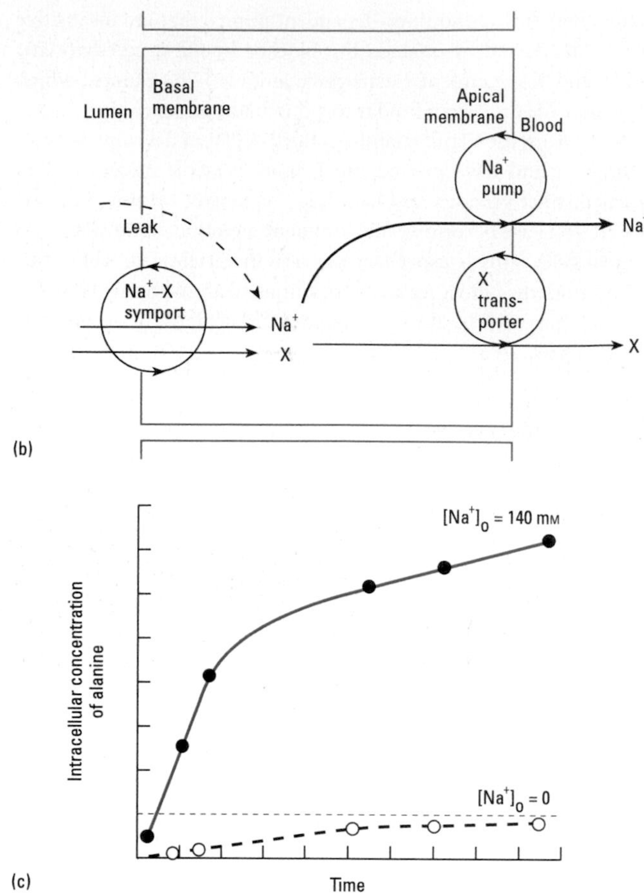

Fig. 4.11 Models of Na^+-linked symport of amino acids or sugars (X) across (a) a single membrane or (b) an epithelium. The chemical energy of ATP is used to drive the apical "sodium pump" so that Na^+ leaves the cell. In (a) the Na^+ then leaks back in passively and carries X with it; in (b) Na^+ leaks in across the other (basal) side of the epithelium and moves X into the cell with it, with a separate X transporter possibly handling movement of X out of the cell apically. In (c) the effects of Na^+ concentration on intracellular concentrations of the amino acid alanine are shown.

This generates a high (lumen-positive) electrical gradient, which in turn drives K^+ along with amino acids back across the normal lumen cells using a symporter, giving an overall effect of amino acid uptake. In this system, carbonate ions are also secreted into the lumen, making it highly alkaline (pH 8–12).

It is important to remember that water itself is *never* handled by primary or secondary active transport; the process only applies to solutes. In all situations where water may appear to move against the osmotic gradient, closer analysis has always revealed an underlying active transport of ions. We will meet many examples of this in ion and water regulatory systems.

4.4.4 Transepithelial solute movements

We have seen that active transport mechanisms operating across the cell membrane essentially move a substance either in or out of a particular cell. But because animal cells are usually arrayed in organized and polarized sheets as **epithelia**, with different transport properties at the apical and basal membranes, it is possible to use ATPases, both as primary and secondary active transport, to move substances across cells and across sheets of epithelia, and thus to vary concentrations and conditions between compartments of the animal body. This is achieved by salt-transporting epithelia (arranged in spherical or tubular fashion around a lumen, or as multilobular glands) in which the cells always have a very characteristic appearance (Fig. 4.12). The apical side (often termed luminal when referring to a glandular or tubular system with a clear lumen) has a dense coating of microvilli, giving a very large surface area, while the basal side has less dense but often longer indentations. One or both sides may have heavy endowment with mitochondria, and this serves as a good clue to the site of the transporters. In some transporting tissues the active transporters are apical, and basal movements are passive or involve secondary transport systems, but in others most of the active transporting is basal and apical flow is passively down the gradient created. Active processing at both membranes is not necessary; moving ions across either one is sufficient to create the gradients that will draw ions passively across the other.

In summary, both organic and inorganic solutes can be transported across biological membranes and thus across cells, against their electrochemical gradients, by coupling to the downhill movements of other molecular species. In animals the coupling is almost always

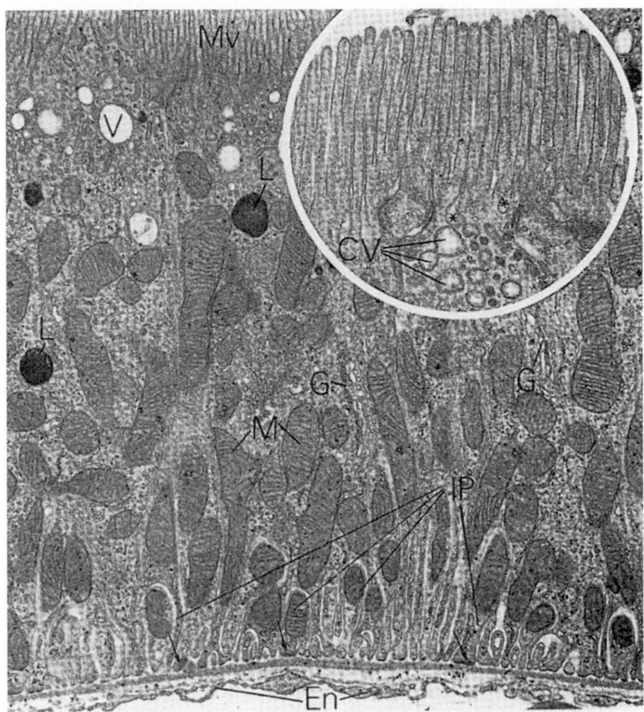

Fig. 4.12 Electron micrograph of a typical salt-transporting epithelial cell (from the proximal convoluted tubule of a vertebrate kidney). Inset shows the inner (luminal) margin. Mv, microvilli (at inner edge, forming a "brush border"); V, vesicle; M, mitochondria; G, Golgi bodies; IP, interdigitating processes of outer cell margin; L, lysosomes. (From Ross & Romrell 1989.)

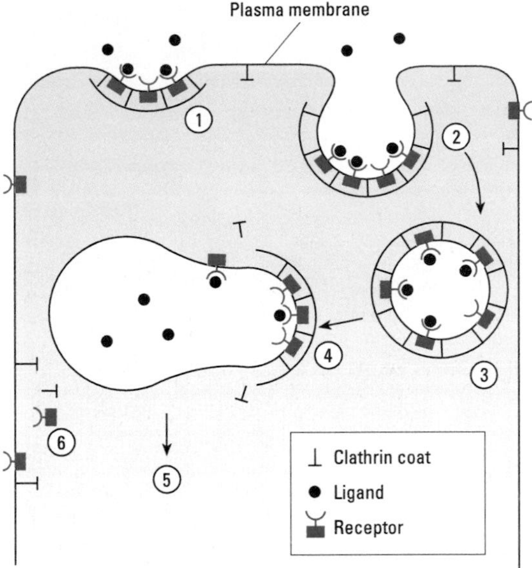

Fig. 4.13 The formation of coated vesicles during receptor-mediated endocytosis: (1) ligand molecules bind to surface receptors in coated pits formed by clathrin molecules bound to the surface membrane; (2) invagination of the coated pit so that (3) a coated vesicle is formed; (4) fusion with an existing vacuole and (5) further processing, with (6) clathrin and receptor molecules recycled. (Reprinted from *Trends in Biochemical Sciences* **5**, Pearse, B., Coated vesicles, pp. 131–134, copyright 1980, with permission from Elsevier Science.)

symport or antiport with either sodium or hydrogen ions, the energy being provided by P-ATPase or V-ATPase enzymes and transiently stored as an electrochemical gradient of Na^+ or H^+.

4.4.5 Other active processes at cell surfaces

There are several other ways in which materials can be moved across membranes, cells, and epithelia faster than by simple diffusion. Endocytosis and exocytosis are well-known methods for moving small bulk quantities into and out of cells, respectively. **Endocytosis** involves trapping material lying outside the plasma membrane within a small surface invagination, which is pinched off as an internal vesicle. This is then lysed, to release the contents into the cytoplasm. The process can be split into pinocytosis, where fluid is engulfed, and phagocytosis, where solid matter is involved. The best known mechanism (though there may be several different kinds) is that of **receptor-mediated endocytosis** (Fig. 4.13). Here receptor molecules in the cell membrane detect and bind to certain ligands in the extracellular medium; these may be viruses or toxins that the cell will destroy, or plasma proteins, hormones, or immunoglobulins to which it will produce a specific response (see Chapter 17). The receptor–ligand complexes can diffuse laterally in the membrane, and once bound they tend to accumulate in specific areas termed "coated pits", where there are many clathrin molecules; clathrin is a specific protein organized in complex tightly packed membrane arrays. The coated pit invaginates to become a "coated vesicle", which then fuses with pre-existing vacuoles within the cell. Then the clathrin-coated piece of membrane, with its receptors, is recycled to the surface to attract further ligand–receptor complexes.

Exocytosis is the reverse process, where a small portion of the intracellular medium is organized into a vesicle, which is then extruded at a surface. It is very important in neurosecretory and hormonal systems, with calcium ions playing an important controlling role in vesicle release, but it is less involved with cell transport systems in general. However, some cells retain, in reserve just below the transporting apical membrane, a population of vesicles whose membranes contain large quantities of the proteins that are specific ionic channels or pumps. These can be moved into position in the surface membrane using exocytosis controlled by microtubules and microfilaments, to change the pattern or rate of ion movements in response to hormonal triggers; an increase in cell surface area is sometimes detectable. In *Xenopus*, for example, up to 50% of the pumps in developing oocytes are within the cell, and these are continuously exchanged with the surface pumps; as the embryonic blastomeres form, sodium pump distribution becomes polarized and directional Na^+ transport occurs in concert with the formation of the primary embryonic cavity (blastocoel). Pumps and channels can also be mobilized from this vesicle reserve by the action of various hormones.

Another category of movements we should consider briefly are those that occur through membrane junctions. Cells are organized into sheets by junctions; these are of two types: tight junctions and gap junctions. **Tight junctions** (Fig. 4.14) serve to bind cells tightly in place to each other so that they can act as organized epithelia, the outer surface of two neighboring cells making very close contact and leaving no extracellular space. The linked area may encircle each cell

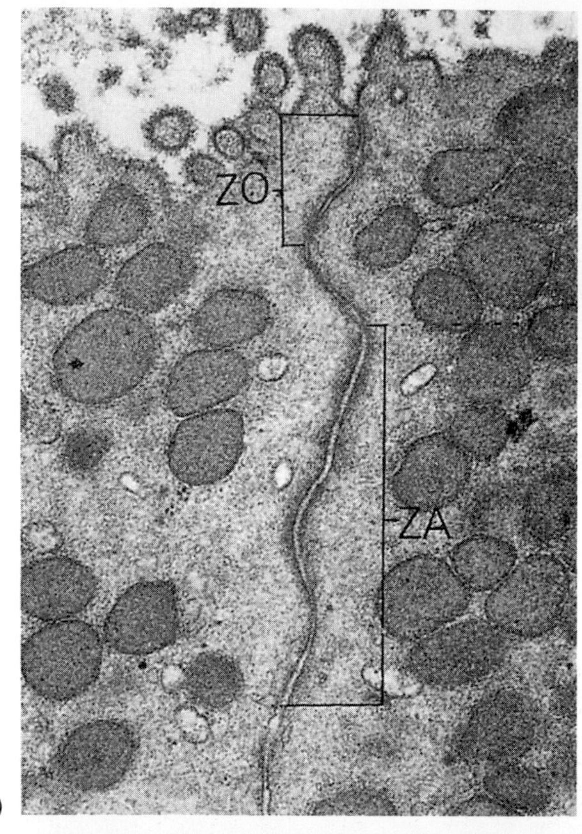

(a)

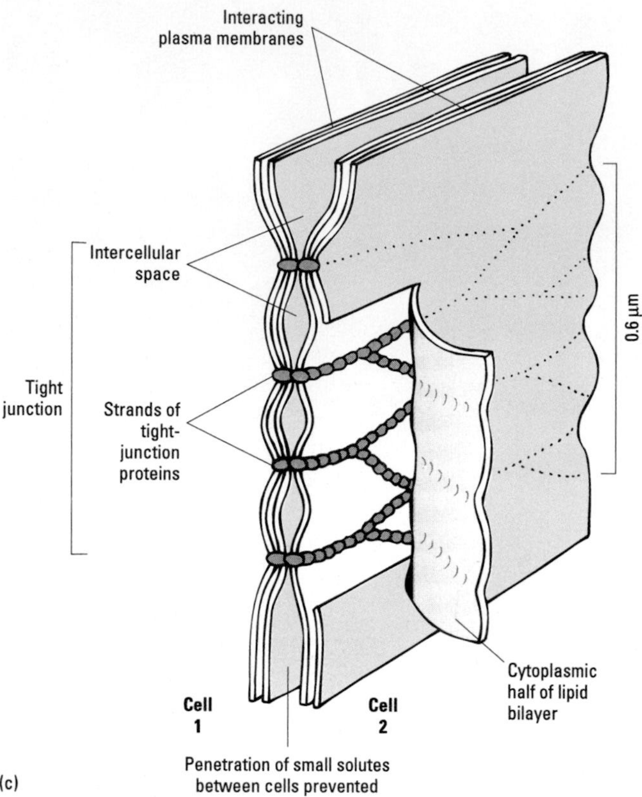

(c)

(b)

Fig. 4.14 (a) Electron micrograph of a typical tight junction between epithelial cells. (b) The structure in a freeze–fracture view in epithelial cells. Mv, microvilli; arrows show the ridge of the tight junctions. (c) Diagrammatic view of a tight junction complex. (a and b, From Ross & Romrell 1989; c, from Alberts *et al.* 1994, Fig. 19.5.)

in the epithelium, acting as a barrier against any back leakages *between* cells of substances that have been transported through the cell cytoplasm. Some epithelia may be described as quite "leaky" and others as quite "tight" (though the distinction is not very clear-cut). **Gap junctions** are more structured and may have a more active role in cellular transport. They involve a close proximity of two cell membranes, but retain a small intercellular gap of about 2 nm in which a highly organized array of hexagonal units can be detected (Fig. 4.15). Each unit is made largely of protein, with six subunits structured like a doughnut, leaving a central pore through which the two cells can communicate. Molecules as large as certain dyes are able to pass through, and ions, amino acids, nucleotides, and sugars can certainly pass through. Passage through these cell–cell links can be stopped by varying the intracellular $[Ca^{2+}]$ or $[H^+]$, probably by effects on the spacing of the subunits around each pore.

Finally, we should mention **extracellular pathways** for solute movements. Here we are dealing with transport across an epithelium that occurs *between* the cells rather than through them—the route is sometimes termed **paracellular** to distinguish it from random extracellular diffusion. Clearly, paracellular movement could not

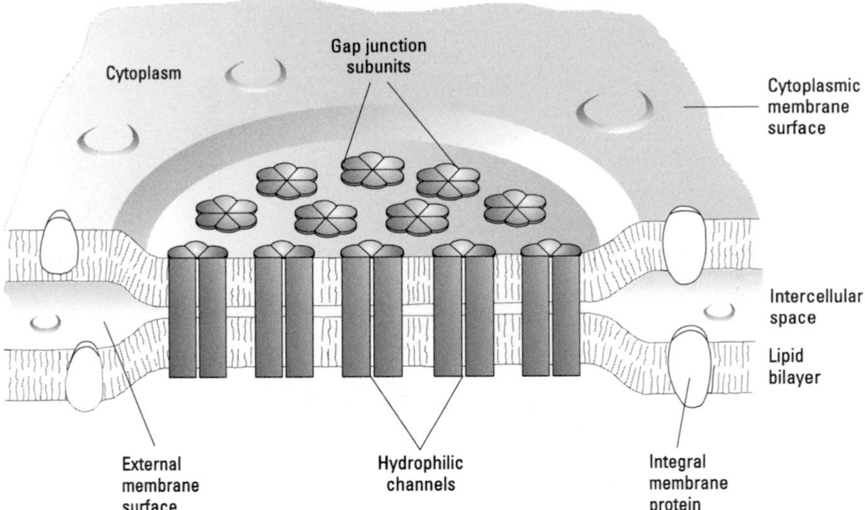

Fig. 4.15 Structure of a gap junction, where neighboring cells are coupled through an array of hexagonal subunits on the two closely apposed membranes, forming through-channels. The passage of molecules larger than about 2 nm (such as proteins and nucleic acids) is prevented, but ions, amino acids, sugars, and nucleotides can pass from one cell to the other. (From Staehelin 1974.)

occur across an epithelium that had nonleaky tight junctions around every cell, but it could nevertheless be an important route in many tissues. Water and various solute molecules could pass between cells down their respective gradients, and in some tissues the lateral gaps between cells could be used specifically to create local high concentrations to facilitate such flow. This is the basis for several models of flow across epithelia to explain rapid transport of water in a variety of tissues. Figure 4.16 shows how such a system might work. Salts (e.g. Na^+, with Cl^- following) are actively transported by classic ion pumps into the intercellular clefts, where the salt concentration builds up rapidly in the restricted volume available. Thus there will be a "standing gradient", with a high concentration deep in the cleft and a lower concentration luminally, which is eventually indistinguishable from the actual lumen concentration. Water is therefore drawn into the cleft osmotically from the apical side, either through the leaky junction area (which may be permeable to water but not to ions) or through the cytoplasm, and then fluid moves on down the cleft towards the lumen. The steady pumping of salt at a localized area of membrane thus results in a steady osmotic flow of water across the cell epithelium—so-called "solute-coupled water transport".

Evidence for this kind of transport is patchy, and quite how important these paracellular routes are remains controversial. In some tissues, such as the mammalian gall bladder, there is a detectable distension of the intercellular clefts during rapid secretion, indicating that at least part of the bulk fluid flow is passing by this route; this epithelium is described as "leaky", as is that of the mammalian small intestine (jejunum) and the insect Malpighian tubule. Such epithelia can produce net fluid transport providing the active transport is fast relative to the "leak". But recent evidence has tended to support older ideas that most movement of water in most epithelia is actually transcellular after all, involving simple osmosis and very small concentration gradients at the apical and basolateral cell borders, which have a high water permeability. Thus epithelia

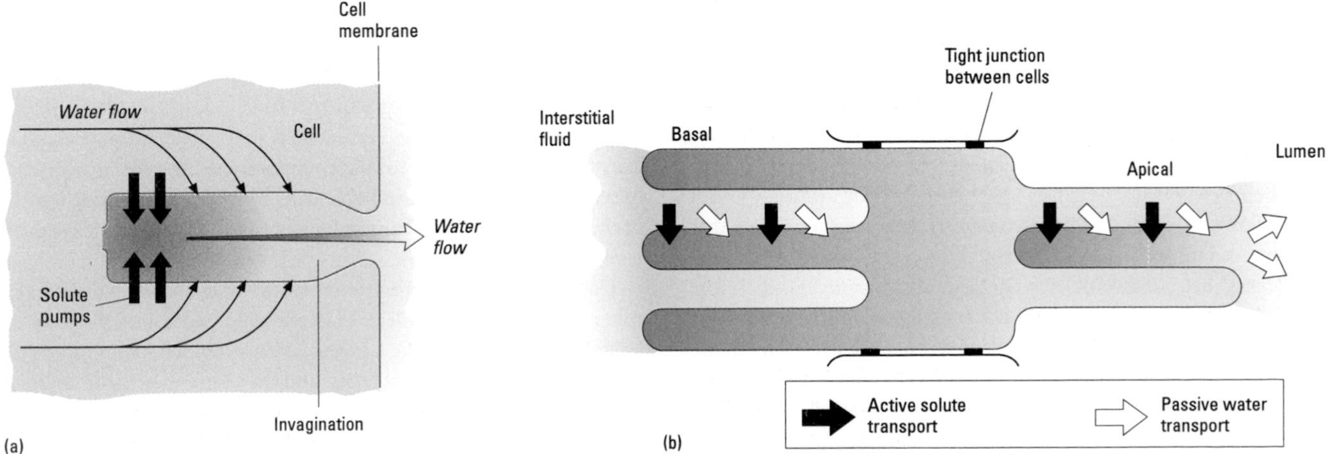

Fig. 4.16 The standing gradient model for solute-linked water transport. (a) In the simple version salts are actively transported out of the cell into the deep clefts (either invaginations into the cell, or clefts between cells), and water follows passively, giving a net flow of fluid. (b) In a more complicated version for a whole epithelium salt is pumped into cell invaginations basally and out to extracellular clefts apically, so that the whole epithelium moves fluid from, for example, its bathing blood to its lumen. The transported solution here is isosmotic, but it can be hyperosmotic if there is solute recycling within the cell (see Chapter 5).

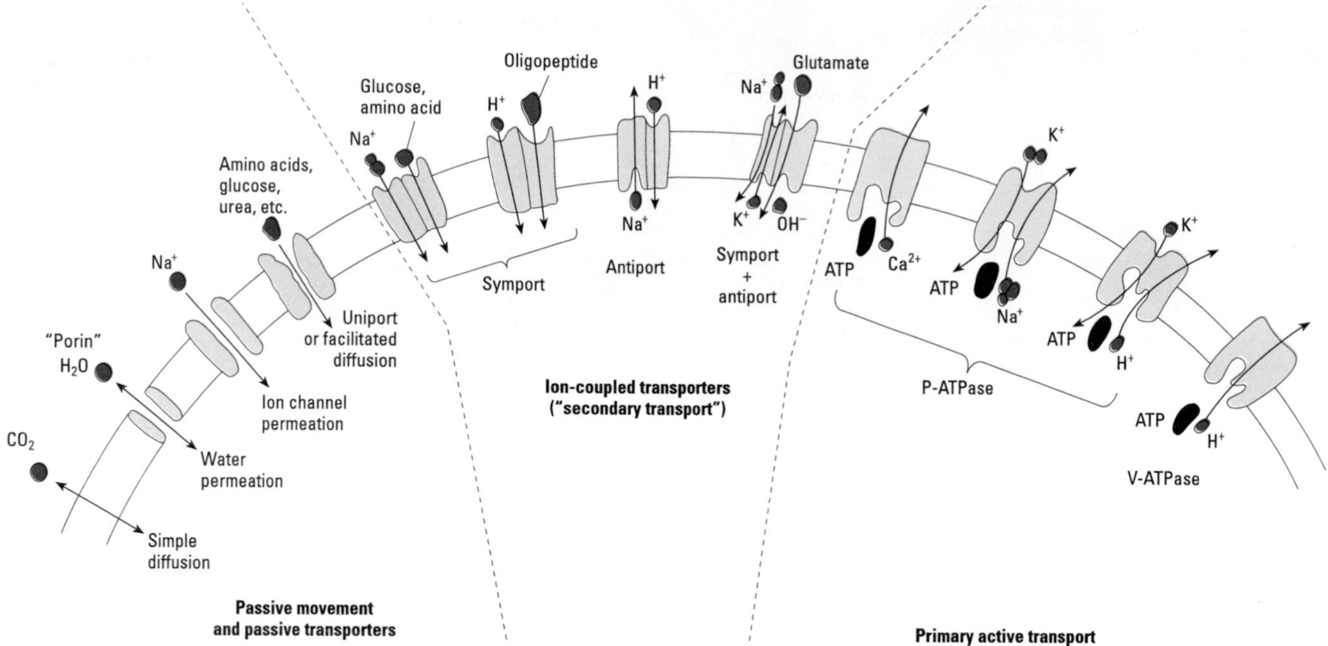

Fig. 4.17 A pictorial summary of the variety of possible methods for moving molecules across membranes.

such as those of the mammalian and insect rectums are "tight" (in the sense of being nonleaky) and this may well be the norm.

In any event, it is clear that complex patterns of transepithelial transport can be achieved by a combination of the relatively simple subcellular transport systems and variable permeation mechanisms, which operate in particular localities on cell membranes. An overview of possible transport mechanisms across membranes and epithelia is shown in Fig. 4.17.

4.5 Concentrations of cell contents

We should turn now to look at the actual concentrations of ions and other solutes found inside cells. The earliest living organisms, enclosed by a biological membrane, evolved perhaps 3 billion years ago in the Precambrian, probably in the fringes of shallow seas. The nature of those seas is somewhat uncertain (see Chapter 11), but we can be reasonably confident that the waters were moderately salty, and there is a reasonable consensus that their composition was not too different from that of modern oceans. It seems very likely that cells evolved in, and have run through almost their entire evolutionary history in, a condition that was essentially marine. It is therefore no surprise to discover that the intracellular contents of the vast majority of cells still reflect this ancestry; most phyla of animals are predominantly marine, and most have intracellular concentrations roughly isosmotic with sea water. It is only the relatively few groups that have successfully invaded freshwater and terrestrial habitats that diverge from this pattern, commonly having both cells and extracellular fluids that are one-quarter to one-third the concentration of sea water. Examples of the ionic and osmotic concentrations of cells from a range of animals are given in Table 4.4.

As is clear from discussions earlier in this chapter, the dominant intracellular cation in animals is always potassium, rather than the sodium that dominates most extracellular fluids and the marine and fresh waters in which animals may live. This preferential accumulation of K^+ (compared to Na^+) may have arisen from its lower destabilizing effects on proteins, and/or from its lower charge density, which makes it tend to break up the interactions between water molecules and so allow other solutes to be dissolved within the cell more readily. But even for animals living in very concentrated media such as salt lakes, the total intracellular cation concentration ($[Na^+]_i + [K^+]_i$) does not normally exceed about 300 mM (rarely up to 440 mM as in squid giant axons). This relatively low overall intracellular ionic concentration is probably again related to the destabilizing effects of salts on protein structure, a process sometimes called "salting in" since it leads to proteins solubilizing into the aqueous phase, and beginning to unfold. Animal cells also differ from their environments in having extremely low calcium concentrations; the values given in Table 4.4 include a large proportion of calcium that is effectively bound, and the true free $[Ca^{2+}]_i$ is normally below 10^{-6} mM. Hence even the earliest cells may have had in place the transporter mechanisms for pumping cations across the membrane to set up ion gradients, thus benefitting from the various side effects of these gradients but of course having to pay the energetic costs of continuous ion regulation.

The other major point to notice from Table 4.4 is the important role of small organic molecules inside cells. Most animals have at least 30% of their osmotic concentration provided by amino acids and other small organic moieties, and the figure may be up to 70% in certain cell types. These organic components make up for the low ion levels discussed in the last paragraph, to give an overall osmotic concentration close to that of extracellular fluid and of the environment. These intracellular **osmolytes** play an important role in osmotic adaptation and volume regulation (see section 4.6), being moved in and out of the cell as its surroundings change. Many of the

Table 4.4 Intracellular solute compositions for muscle or nerve tissues from a range of animals from different habitats.

	Solute concentration (mM)							
Animal group	**Na^+**	**K^+**	**Cl^-**	**Ca^{2+}***	**SO_4^{2-}**	**PO_4^{3-}**	**Amino acids**	**Total NPN†**
Marine								
Annelids								
Neanthes (ragworm)	125	195	124	14				412
Molluscs								
Mytilus (mussel)	79	152	94	7	9	39	289	
Sepia (cuttlefish)	31	189	45	2				678
Crustaceans								
Nephrops (prawn)	24	188	53		1	164	476	602
Carcinus (crab)	54	146	53	5				617
Chelicerates								
Limulus (horseshoe crab)	29	129	43	4	1	96	136	
Vertebrates								
Myxine (hagfish)	32	142	43		104		290	
Chimaera (ratfish)	28	120	37	3	189		378	767
Salmo (salmon)	21	264	3		46		49	
Freshwater								
Molluscs								
Anodonta (mussel)	5	21	2	12		20	11	
Crustacea								
Astacus (crayfish)	11	122	14	9				153
Insects								
Sialis (alder fly)	1	135	1					
Land								
Vertebrates								
Rana (frog)	15	126	1	5		10		68
Rattus (rat)	16	152	5	2				70
Homo (human)	14	140	4		4		8	

* Values for calcium are total concentrations; most is sequestered and free calcium levels are generally much less than 1 mM, often less than 1 μM.
† For some species total nonprotein nitrogen (NPN) is given rather than amino acid concentrations, yielding a higher value.

organic molecules used intracellularly instead of ions are negatively charged at normal pH, and they contribute substantially to the anionic fraction, so that the major inorganic anion, chloride, is also markedly less concentrated inside animal cells than in extracellular fluids, and $[Cl]_i$ is usually less than the total internal ionic cation concentration. As mentioned earlier, this chloride distribution arises entirely passively in most cells. Other inorganic anions are numerically relatively insignificant, with sulfate at low concentrations and bicarbonate and phosphate ions often even lower in concentration (though functionally very important as buffers of cell pH levels).

4.6 Overall regulation of cell contents

In the previous section we looked at cell contents in a range of animals. Clearly these are average conditions, measured in resting animals under "normal" conditions, and most of the problems of water and ion balance arise in relation to maintaining these concentrations in the face of environmental challenges. Cells are rather permeable to water, and external changes in medium concentration therefore tend to cause cells to either shrink or swell quite quickly and concentration gradients to run down. The next chapter will deal with the problems of water and ion balance at the level of the whole animal, but we need first to summarize how control of concentration and volume is manifested at the level of individual cells.

Cell concentration regulation depends on uptake and exchange mechanisms whose activity is modified in relation to perceived change, either directly or via intervening sensors and second messengers, and which are crucial to maintaining cell homeostasis. The sodium pump Na^+/K^+-ATPase system, already dealt with, is the single most important effector system for the major cationic Na^+ and K^+ concentrations, keeping internal potassium levels high and sodium levels low. Sodium levels are also in part regulated by pH levels in the cell, via H^+ extrusion and Na^+/H^+ exchange; pH itself depends on a complex acid–base regulation system linked with CO_2 and HCO_3^- levels.

Calcium levels depend on at least two processes already discussed. There are Na^+/Ca^{2+} antiporters, powered by the sodium gradient created by the Na^+/K^+-ATPase system, but this is often the less important system for extruding calcium from cells and maintaining a low $[Ca^{2+}]_i$ because if extracellular Na^+ is entirely removed, calcium efflux persists, albeit at a reduced level. This is due to the specific "calcium pump" or Ca^{2+}-ATPase membrane protein. At normal calcium levels, this pump may be responsible for nearly all the active extrusion of calcium from cells, with the Na^+/Ca^{2+} antiport exchange system only operating when $[Ca^{2+}]_i$ gets abnormally high.

Much of the magnesium present in cells is chelated and bound internally to other molecules; for example, most of the ATPases also have Mg^{2+}-binding sites. Thus free intracellular magnesium is at

Box 4.6 The van't Hoff relation

Cells tend to shrink in hyperosmotic media due to osmotic loss of water and swell up in hyposmotic media due to osmotic water influx. In a simple physical system the relation between cell volume and concentration would be given by the **van't Hoff equation**:

$$\Pi_1(V_1 - b) = \Pi_2(V_2 - b)$$

where Π is the osmotic pressure, V the cell volume, and b the total content of solids in the cell (for two sets of conditions 1 and 2, initial and final). Thus $(V_1 - b)$ is the initial solvent volume and $(V_2 - b)$ the final solvent volume in the cell.

Plotting cell volume against the ratio of the two osmotic conditions (Π_1/Π_2) should give a straight line with a slope of $(V_1 - b)$, *if* the system behaves as a perfect osmometer. This is rarely the case, however, for reasons given in the text.

An indication of the degree of volume regulation can therefore be obtained by comparing the slope value $(V_1 - b)$ with the known water content (W) of the cell (or of a whole organism).

This ratio $(V_1 - b)/W$ is termed **Ponder's *R*-value**; it is about 0.87 for mammalian kidney cells, 0.94 for vertebrate fibroblasts, and very close to 1 for human red blood cells, which are nearly perfect osmometers. Many invertebrates have values around 0.8.

very low levels. Regulation probably involves symport and antiport with other transport systems, but detailed information is lacking.

Chloride levels are determined largely passively in most vertebrate cells that have been fully analyzed, though the anion is symported with Na^+ and K^+ as described above, often extensively so in invertebrate epithelia. Other anions involve more complex regulatory patterns: phosphate has complex interactions with many other transport systems, and undergoes much binding within cells, and the same is probably true of sulfate and bicarbonate ions.

Overall, for the great majority of cells it would be fair to say that only Na^+, K^+, and Cl^- are really important and variable osmotic components of the cytoplasm. All other ions are tied up in more complex interactions, affecting each other's concentrations in feedbacks and balanced transport systems, with relatively low and relatively constant unbound intracellular levels.

Cell volume regulation depends on all these mechanisms. Cells have ions both inside and outside their membranes, but organic concentrations are much higher inside. In isosmotic conditions the cells would therefore tend to swell, and outward ion pumping is a key factor in counteracting this. The ubiquitous Na^+/K^+-ATPase is crucial here, since Na^+ pumped out will stay out and K^+ pumped in will leak back out again as P_K is higher. The net effect is sodium extrusion, hence the common name of "sodium pump", for which prevention of cell swelling could well be the ancestral function in the most primitive of cells.

Cells are many times more permeable to water than they are to any ions, so that the immediate challenge to the cell in a nonisosmotic environment is usually to limit the damage due to water movements—to maintain its volume and avoid swelling or shrinking. In practice, cells readily act as osmometers, shrinking quite rapidly in hyperosmotic solutions and swelling in hyposmotic solutions (strictly speaking the terms "hypertonic" and "hypotonic" are more correct here, referring directly to swelling and shrinking effects, taking into account membrane permeabilities as well as osmotic gradients). A perfect osmometer would behave according to van't Hoff's equation (Box 4.6), so that a graph of cell volume as a function of initial osmotic concentration/experimental osmotic concentration would be a straight line. But although living cells often show an initially straight-line reaction, the response soon deviates from this ideal line and volume changes are less than expected. A number of causes probably underlie this effect. Firstly, a significant portion of internal cell water is bound to membranes, proteins, and other constituents and is not free to move. Secondly, neither solutes nor solvents behave "ideally" except when the solutions are very dilute. Thirdly, some solutes will pass across the membranes too, because real cells are not perfect osmometers. But finally, and most importantly, cells have the specific capacity to regulate their volume, and these regulatory mechanisms will be switched on soon after the initial phase of swelling or shrinking begins. Hence most cells, and indeed many whole animals with soft bodies, when transferred to nonisosmotic media will exhibit the kind of volume response shown in Fig. 4.18, with an initial perturbation of volume and then a return to the original value. Commonly the volume stabilizes within a few minutes or hours for cells, but full regulation may take several days of exposure for whole animals such as marine worms (see Chapters 11 and 12). The capacity to volume regulate can be easily expressed as the ratio of predicted weight (or volume) change to actual weight (or volume) change; this ratio would be 1.0 for a perfect osmometer, and 0 for a perfect volume regulator.

Cells have at least four mechanisms that contribute to volume regulation. Firstly, as we discussed earlier, they may have membranes that are differentially permeable to water movements (inward osmotic gain being faster than outward osmotic loss). Secondly, there may be adaptive reduction in the cell's permeability to water, either as a direct response of the postulated membrane channels or sometimes in response to hormones such as antidiuretic hormone (ADH); this slows water movement though it does not in itself bring about volume regulation. Thirdly, a number of secondary transport systems seem to be switched on in response to cell volume change: $Na^+/K^+/2Cl^-$ symport into the cell occurs during shrinkage, whilst K^+/Cl^- exit occurs during swelling, bringing about regulatory volume decrease by parallel activation of K^+ and Cl^- channels. These secondary transport systems are strongly linked to the fourth mechanism, that of changes in the intracellular solute composition, involving ionic and nonionic solutes acting as **osmotic effectors**. Such changes may be the most important in longer term volume regulation for the majority of cells. For example, some vertebrate epithelial cells, and many tissue-culture cells, will volume regulate in hyperosmotic media by taking up solutes (especially NaCl), while in hyposmotic media they show loss of solute, in this case mostly KCl. Inevitably the membrane Na^+/K^+-ATPase (sodium pump) is heavily involved in this, so that we could list volume regulation as another of its functions. A vertebrate red blood cell treated with ouabain, or with a metabolic poison that shuts off the ATP supply to the pump, loses most of its ability to control its volume.

But adjustment via ionic solutes risks undue changes to cell concentrations, with potential effects on many enzyme reaction rates,

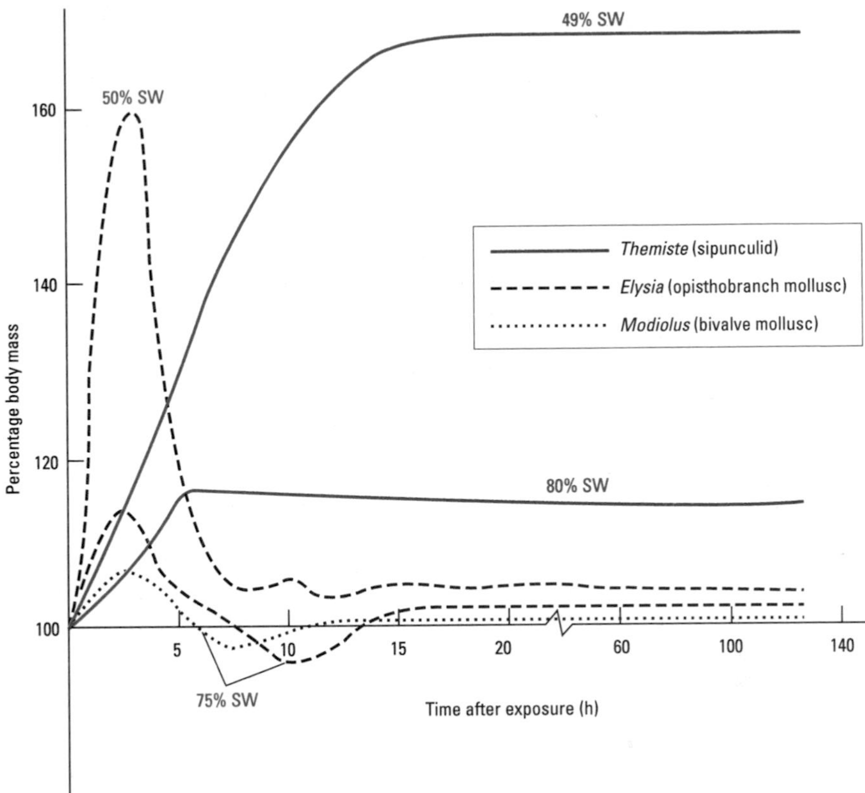

Fig. 4.18 Volume changes in three different invertebrates on transfer to dilute sea water (SW). All show initial volume increments; the sipunculid shows very poor volume regulation, whereas both molluscs are able to regulate their volume back to near normal levels after 5–10 h exposure. (Data from Oglesby 1968; Pierce 1971; Pierce *et al.* 1983.)

protein stability, and other aspects of normal cell functioning. Perhaps for this reason, many cells show a different mechanism of long-term volume regulation, by alterations to their nonionic solutes or **compensatory osmolytes**, which have far fewer effects on enzyme activities. By far the most important compensatory osmolytes in many invertebrate cells are the amino acids, which may make up a large percentage of the normal intracellular osmotic concentration. For example, during acclimation from fresh water to sea water (i.e. to hyperosmotic conditions), most cells from marine invertebrates show a gradual increase of amino-nitrogen in their cells, and an equivalent rise can be detected in the blood of the whole animal (Fig. 4.19a). During hyposmotic adaptation, a decrease in amino acid levels occurs in the cells; the resting level of amino acids closely correlates with the medium salinity (Fig. 4.19b). Both these effects are commonly due to changes in the levels of *specific* amino acids, rather than the amino acid pool as a whole. Some amino acids are especially suitable, since they are stabilizing with respect to protein structure and function, opposing the "salting out" effect of ions—glycine, alanine, and serine are particularly effective here. Muscles, and more especially nerves, tend to use some glutamate and aspartate too, as these help to maintain the important intracellular negative charge. However, in many animal cells less familiar methylamine compounds such as betaine and taurine are partly responsible, whilst some insects use proline or carbohydrates like trehalose and the polyol sugars such as glycerol or sorbitol, and a few animals use metabolic end-products such as trimethylamine oxide (TMAO) or urea. All of these osmotic effectors are highly soluble, lack a net charge or are slightly negative overall, and also lack strongly hydrophobic regions in their structures; hence they can occur at high concentrations without much interaction with charged intracellular macromolecules, and can counteract the effects of ionic changes. They are described as "kosmotropic" due to their strong indirect stabilizing influence on membranes; methylamines coordinate the water molecules around them, so they are excluded from the hydration layers around peptides, making protein unfolding energetically unfavorable. Thus having relatively high levels of such compounds is both a general protection for macromolecules and membranes during the initial shock phase of osmotic exposures and ionic changes, and a compensatory protection for the osmotic gap left by the maintenance of ion levels in the longer term.

The compensatory changes in these intracellular solutes are brought about by the changing levels of extracellular sodium, causing concomitant small changes in intracellular sodium. In animals acclimating to reduced concentrations (e.g. to lowered salinity), it is the lowered $[Na^+]_i$ that acts as the trigger within the cell, and it may bring about several effects (Fig. 4.20). Firstly, it changes the activity of one or several enzymes (especially glutamate dehydrogenase) that control amino acid synthesis and/or degradation, so leading to altered amino acid levels. Secondly, it alters transmembrane movements of amino acids or sugars, so that there is a rapid efflux and a direct reduction of overall osmotic concentration in the cell. Thirdly, it may work indirectly on gene expression via MAP kinases (which are implicated as osmotic sensors or signal transducers). The resultant changes of concentrations of amino acids or sugars are such as to decrease the osmotic stress on the cell, and so limit water movements and volume changes, without compromising ion levels.

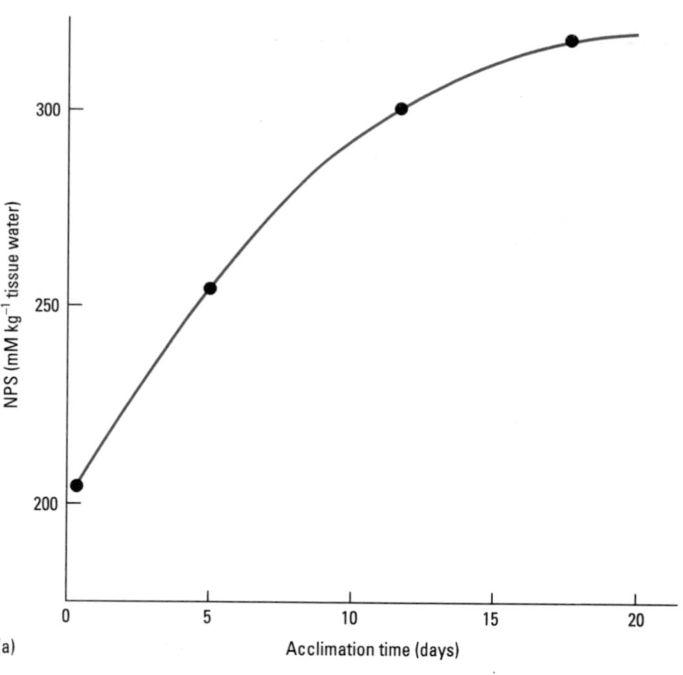

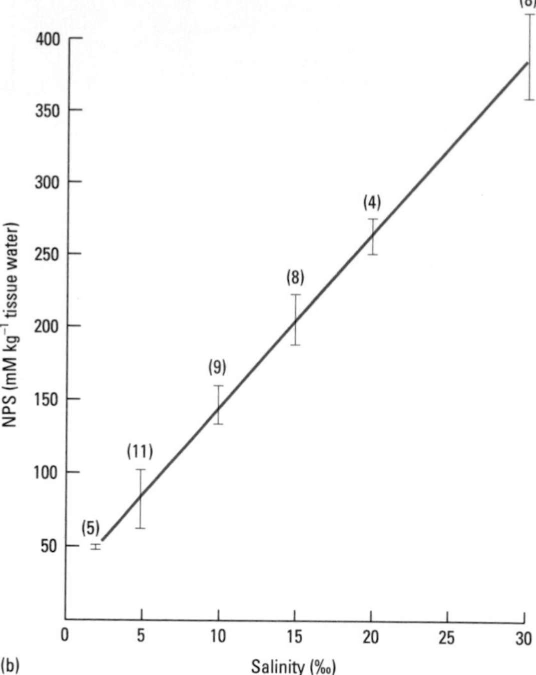

Fig. 4.19 Changes in amino-nitrogen (NPS = ninhydrin positive substances, ninhydrin reacting with amino groups) in tissues of animals exposed to varying osmotic media. (a) The increase in amino acids in *Eriocheir sinensis* (a crab) during acclimation to sea water from fresh water. (b) The equilibrium levels of amino acids in muscles of *Mya arenaria* (a clam) in relation to salinity of the medium. Numbers in parentheses are sample size, vertical bars show standard error of the mean. (a, Data from Gilles 1979; b, data from Virkar & Webb 1970.)

This use of small intracellular organic solutes as osmotic effectors was first studied in marine and estuarine invertebrates (see Chapter 12, where we will meet specific examples), but similar phenomena are now well known in many aquatic vertebrates' cells. They may be less common in the cells of terrestrial birds and mammals, which are highly regulated in other ways, but comparable systems certainly operate in the highly stressed renal medulla cells of mammals, where betaine and certain polyols like sorbitol seem to operate to maintain intracellular osmotic balance. In the excretory systems of insects similar systems have been found, and the proteins coded by the *inebriated* (*ine*) gene have been identified as the osmolyte transporters that confer this osmotic tolerance; this gene family is likely to be important in all animals showing osmotic stress responses.

Cellular adaptation to osmotic stress must involve further steps beyond osmolyte adjustments. The induction of stress proteins as chaperones to protect key proteins is certainly important (see section 8.2.3), and DNA repair systems may be required. If some cells are irreversibly damaged then activation of apoptosis (organized cell death) may also ensue.

4.7 Conclusions

The movements of water, ions, and other small solutes within and between cells, and thus within biological systems in general, can be attributed to the fundamental structure and properties of the cell membrane. The simple characteristics of cell permeation are mediated through a small range of selective ion channels and passive transporters, and of selective ion "pumps" operating active transport processes that may be coupled with transporters of other ions and small solutes. These processes can be linked together in cells and epithelia, and coupled with osmotically induced water flow. Various combinations of these simple processes turn out to underlie most of the larger scale and apparently complex processes of osmoregulation, salt regulation, excretion, and overall animal water balance, which will be covered in the next chapter.

FURTHER READING

Books

Andreoli, T.E., Hoffman, J.F., Fanestil, D.D. & Schultz, S.G. (eds) (1987) *Membrane Physiology*. Plenum Medical, New York.

Bonting, S.L. & dePont, J.J.H.H.M. (eds) (1981) *Membrane Transport*. Elsevier/North Holland, Amsterdam.

Harvey, W.R. & Nelson, N. (eds) (1994) *Transporters*. Company of Biologists Ltd, Cambridge, UK.

Hille, B. (1984) *Ionic Channels of Excitable Membranes*. Sinauer, Sunderland, MA.

Reuss, J.M., Russell, J.M. & Jennings, M.L. (1993) *Molecular Biology and Function of Carrier Proteins*. Rockefeller University Press, New York.

Somero, G.N., Osmond, C.B. & Bolis, C.L. (eds) (1992) *Water and Life*. Springer-Verlag, Berlin.

Stein, W.D. (1986) *Transport and Diffusion across Cell Membranes*. Academic Press, San Diego, CA.

Strange, K. (1994) *Cellular and Molecular Physiology of Cell Volume Regulation*. CRC Press, Boca Raton, FL.

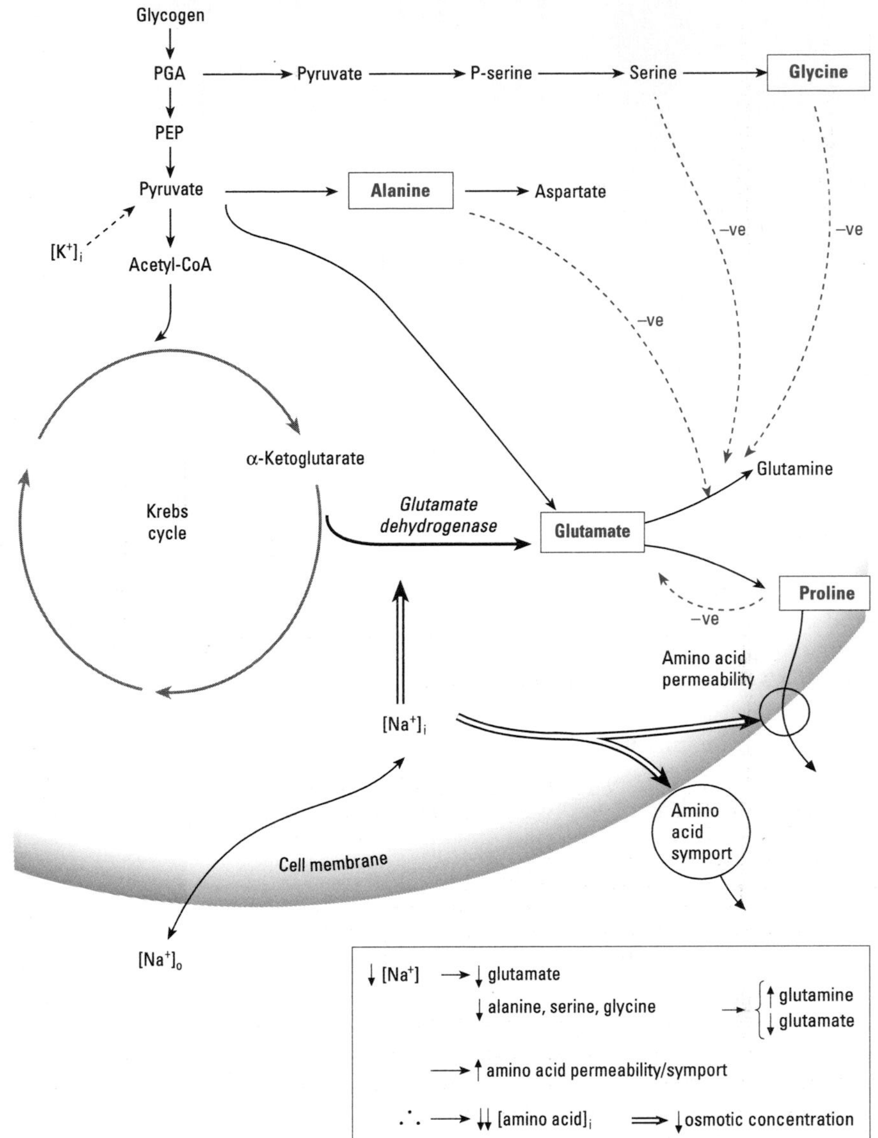

Fig. 4.20 Influences of ion concentrations on key points in the glycolytic and Krebs cycle pathways linking to amino acid synthesis and transaminations, and on transmembrane amino acid movements. Changes in $[Na^+]_o$ can thus bring about changes in intracellular amino acid levels and achieve osmotic compensation. CoA, Coenzyme A; PEP, phosphoenolpyruvate; PGA, phosphoglycerate.

Reviews and scientific papers

Ahearn, G.A., Zhuang, Z., Duerr, J. & Pennington, V. (1994) Role of the invertebrate electrogenic $2Na^+/1H^+$ antiporter in monovalent and divalent cation transport. *Journal of Experimental Biology* **196**, 319–335.

Chamberlin, M.E. & Strange, K. (1989) Anisosmotic cell volume regulation: a comparative view. *American Journal of Physiology* **257**, C159–C173.

Eveloff, J.L. & Warnock, D.G. (1987) Activation of ion transport systems during cell volume regulation. *American Journal of Physiology* **252**, F1–F10.

Galinski, E.A., Stein M., Amendt, B. & Kinder, M. (1997) The kosmotropic (structure-forming) effect of compensatory solutes. *Comparative Biochemistry & Physiology A* **117**, 357–365.

Gilles, R. (1997) "Compensatory" organic osmolytes in high osmolarity and dehydration stresses: history and perspectives. *Comparative Biochemistry & Physiology A* **117**, 279–290.

Heymann, J.B. & Engel, A. (1999) Aquaporins: phylogeny, structure and physiology of water channels. *News in Physiological Science* **14**, 187–193.

Houssier, C., Gilles, R. & Flock, S. (1997) Effects of compensatory solutes on DNA and chromatin structural organisation in solution. *Comparative Biochemistry & Physiology A* **117**, 313–318.

Kinne, R.K.H. (1993) The role of organic osmolytes in osmoregulation; from bacteria to mammals. *Journal of Experimental Zoology* **265**, 346–355.

Okada, Y., Maeno, E., Shimizu, T., Dezaki, K., Wang, J. & Morishima, S. (2001) Receptor-mediated control of regulatory volume decrease (RVD) and apoptotic volume decrease (AVD). *Journal of Physiology* **532**, 3–16.

Pequeux, A., Gilles, R. & Marshall, W.S. (1988) NaCl transport in gills and related structures. *Advances in Comparative Environmental Physiology* **1**, 1–73.

Shuttleworth, T.J. (1989) Overview of epithelial ion-transport mechanisms. *Canadian Journal of Zoology* **67**, 3032–3038.

Smith, P.R. & Benos, D.J. (1991) Epithelial Na^+ channels. *Annual Review of Physiology* **53**, 509–530.

Yancey, P.H. (2001) Water stress, osmolytes and proteins. *American Zoologist* **41**, 699–709.

5 Animal Water Balance, Osmoregulation, and Excretion

5.1 Introduction

5.1.1 Problems of water balance

Maintaining a reasonably constant water balance is a straightforward business for the majority of animals. As we saw in the last chapter, the cells and body fluids of marine animals are in osmotic equilibrium with the surrounding medium, and numerically marine animals dominate the planet (see Chapter 11). However, when living organisms began to colonize more osmotically challenging habitats—the seashores, estuaries, and rivers, and eventually the land—the problems of water balance became paramount. As living cells moved out of the sea, they faced real problems: water always has a tendency to move into them (from dilute surroundings such as estuaries and fresh water) or move out of them (into the surrounding dry air of terrestrial zones), and neither of these fluxes can be left uncorrected if the cells are to continue functioning.

For nonmarine multicellular life, continuing healthy existence depends upon maintaining the cells at a fairly constant water balance, i.e. in reasonable osmotic stasis, and this usually involves using the extracellular fluids to buffer the cells against excessive environmental stress. Controlling this cellular and extracellular osmotic balance ultimately depends on principles outlined in the preceding chapter: the permeation characteristics of membranes and associated surfaces throughout the organism, active processes regulating ion levels, and the control of intracellular osmotic effector levels. But for whole animals these processes have additive and interactive effects between different tissues and organs, with particular epithelia carrying out regulatory functions to maintain blood composition and thus providing osmotic stasis for the rest of the body. Therefore the problem of animal water balance can best be examined in terms of the actions, and the control, of particular effector organs.

5.1.2 Components of water balance

Figure 5.1 summarizes the various routes through which water and ions can be gained and lost in idealized animals. For an aquatic animal (Fig. 5.1a), water and ions may be exchanged through the general skin surface and at specialized areas of the skin such as the gills. They may be taken in through the mouth and gut, in the food, or via drinking; and they may be lost through the feces or in urine. For terrestrial animals (Fig. 5.1b), there is normally no opportunity for skin uptake of water or ions (exceptions do occur, and will be covered later), so that the skin and the respiratory system serve only as avenues of ion and water loss; again, losses occur through fecal and urinary deposition. In addition to these routes, all animals may incur small losses of ions and water associated with reproduction, and unpredictable losses from injury, while some may also lose water during molting processes or from secretions such as slime, mucus, saliva, defensive fluids, or toxins. Furthermore, all will have some gain of water from their own internal metabolic oxidative processes, of which water is nearly always a by-product, and this "metabolic water" may be a very significant component of water balance for small terrestrial animals.

Some of these routes of water and ion gains and losses are unavoidable, continuous, and essentially passive, and their effects will be determined by the same factors of concentration gradient, surface area (or surface area to volume ratio), and permeability as were involved in the passive movements across cell membranes discussed in Chapter 4. These routes may be partly controlled, via changes in permeability and manipulation of gradients, but movements of water or ions remain passive. Other components of overall water balance are discontinuous routes that may be operative or switched off according to need, and these generally depend upon the principles of active transport also considered in Chapter 4, and therefore depend ultimately on metabolic factors dealt with in Chapter 6.

The principles and mechanisms governing all of these potential gains and losses need to be considered at the level of epithelia, organs, and whole animals, in relation to the habitat and design of different animals. First, though, we should look at broad strategies for coping with ionic and osmotic problems and at the water balance problems encountered in different kinds of environment.

5.1.3 Avoidance, tolerance, conformity, and regulation in osmotic biology

As we discussed in the first chapter of this book, there are distinct general patterns that animals show in dealing with environmental changes. They may **avoid** them by behavioral tricks of various kinds, or they may allow their bodies to be exposed to the changes and thus **tolerate** the conditions. The tolerators may be split into two groups, since some animals may **conform** to changes by letting their own internal conditions follow those of the outside world, while others may **regulate** their own body to maintain reasonable independence

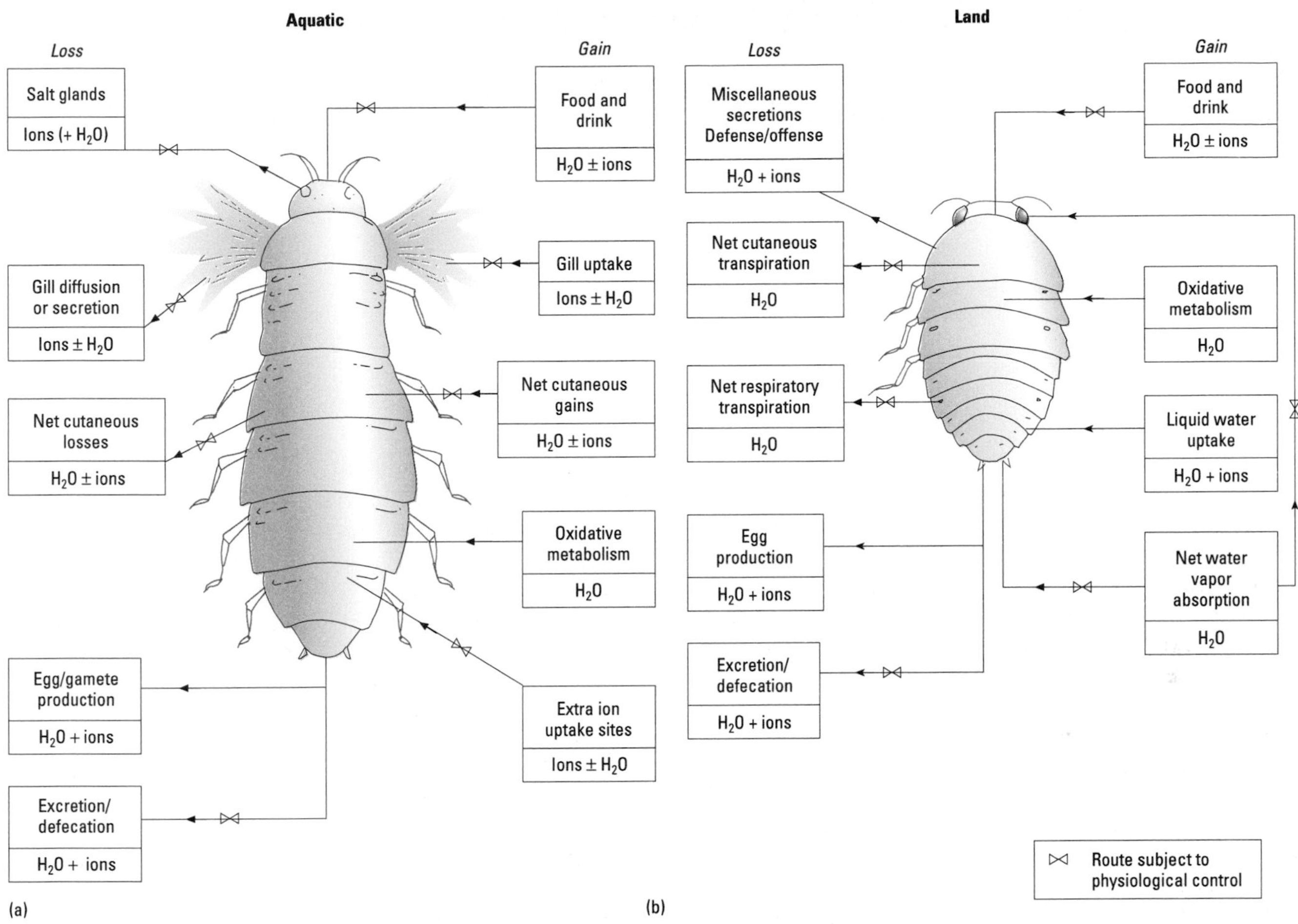

Fig. 5.1 Summary of avenues of water and ion gains and losses for (a) an aquatic animal and (b) a terrestrial animal. (b, Adapted from Edney 1980.)

of the environment. In the context of water balance, all these options may be available, depending on the habitat, but we also need to distinguish the responses to ionic variance (**ionoconforming** and **ionoregulating**) and to osmotic variance (**osmoconforming** and **osmoregulating**). Table 5.1 summarizes some of the ionic and osmotic conditions found in the body fluids of animals from different habitats, together with the composition of sea water and fresh water, as a background to what follows.

For marine animals, avoidance is usually neither a need nor an option. Conditions are similar throughout, so that an animal cannot "avoid" by moving to a less stressful part of the environment. For most marine animals, including virtually all invertebrates and a few vertebrates, conditions are "easy", since the cells and extracellular fluids (ECFs) are in osmotic equilibrium with the medium and the ECF is only marginally different ionically from sea water (see Chapter 11). All these marine animals are therefore conformers, but most of them cannot survive very much change in the concentration of their surroundings; their conforming tolerance *range* is small, and they may be described as **stenohaline.** In practice the lack of environmental variation means that an ability to regulate would be quite unnecessary, so the great majority of animals in the sea are osmoconformers and ionoconformers, and water balance poses no problem.

For many vertebrates living in the sea, different patterns occur because the taxon has an evolutionary history in fresh water, with rather dilute blood, and modern fish and tetrapods are secondary invaders of the oceans. Most of the cartilaginous fish show complete osmoconformity but also strong ionic regulation; their blood is osmotically in equilibrium with sea water due to the presence of various dissolved organic compounds exerting an osmotic effect (**osmolytes**, such as urea and other nitrogenous compounds), while the blood ion levels are kept low, similar to the levels found in freshwater animals. Again this means that water balance is not a significant problem, with no significant osmotic influx or efflux. But in the marine teleost fish (and in nearly all the secondarily marine tetrapods such as whales and seals), the blood is both ionically and osmotically dilute, so that water always has a tendency to move out of the animal and ions to move in, and regulation must be continuous. This is by far the most "difficult" physiological strategy for a marine animal, and it is considered in some detail as a special topic at the end of Chapter 11.

In aquatic habitats other than the sea, osmotic and ionic problems inevitably arise. In estuaries and littoral habitats, many invertebrate animals can operate an avoidance strategy, as conditions follow a tidal cycle and temporary evasion is a good short-term trick when there is a guarantee that conditions will return to the preferred

Table 5.1 Extracellular fluid compositions in a range of animals from different habitats.

Habitat/animal group	Genus	Ionic concentration (mM)							Osmolarity (mOsm)
		Na	K	Ca	Mg	Cl	SO_4	HPO_4	
Marine									
Seawater		*470*	*10*	*10*	*55*	*570*			*1100*
Cnidaria	*Aurelia*	454	10	10	51	554	15		
Mollusca	*Aplysia*	492	10	13	49	453	28		
	Loligo	419	21	11	52	522	7		
	Sepia	465	22	12	58	591			1160
	Eledone	432	14	11	54	516			1061
	Mytilus	490	13	13	56	573	29		1148
	Strombus	503	11	11	60	577			
	Nerita	484	11	11	55	554			
Annelida	*Arenicola*	459	10	10	52	527	24		
	Neanthes	483	14	13	44	545			1108
Crustacea	*Homarus*	472	10	16	7	470			
	Carcinus	525	13	14	21	502	17		1100
	Pachygrapsus	465	12	11	29				
	Nephrops	512	9	16	10	527			1108
	Ligia	586	14	36	21	596			
Chelicerata	*Limulus*	445	12	10	46	514			1042
Echinodermata	*Asterias*	428	10	12	49	487	27		
	Parastichopsis	473	10	10	53	548			1117
	Echinus	444	10	10	51	519	28		1065
Chordata									
Chondrichthyes	*Scyllium*	269	4	3	1	258	1	1	1075
	Squalus	296	7	3	4	276			1096
Teleostei	*Paralichthys*	180	4	3	1	160	<1		337
	Gadus	174	6	7	3	150			308
Freshwater									
Soft freshwater		*<1*	*<0.01*	*<0.1*	*<0.1*	*<1*			*1–2*
Mollusca	*Anodonta*	16	<1	8	<1	12	<1		66
	Theodoxus	45	2	2	3	33			
	Viviparus	34	1	6	<1	31			
Crustacea	*Cambarus*	146	4	8	4	139			
	Potamon	259	8	13		242			522
	Astacus	208	5	14	1	250			477
	Asellus	137	7			125			
Chordata									
Teleostei	*Carassius*	142	2	6	3	107			
	Salmo	161	5	6	1	120			
Amphibia	*Rana*	92	3	2	2	70			210
Reptilia	*Alligator*	140	4	5	3	111			278
Aves	*Anas*	138	3	2		103		2	294
Terrestrial									
Mollusca	*Poteria*	31	2	5	2	25			
	Pomatias	110	6	16	3	106			
Annelida	*Lumbricus*	76	4	3		43			
Crustacea	*Oniscus*	230	8	17	9	236			
	Porcellio	227	8	15	11	279			
	Gecarcinus	468	12	17	8				
	Holthusiana	270	6	16	5	266			
Insecta	*Locusta*	60	12	17	25				
	Periplaneta	161	8	4	6	144			
Chordata									
Mammalia	*Homo*	142	4	5	2	104	1	2	295
	Rattus	145	6	3	2	116			

range on a regular basis. Thus aquatic estuarine animals may burrow, or seal their shells, when the tide goes out, while terrestrial animals may climb vegetation, or retreat into sealed crevices and burrows, when the tide comes in. For those animals that remain active and exposed all the time, though, a broad tolerance range is essential, so these are **euryhaline** animals coping with a broad range of salinities. Many species conform to changes, with the osmotic and ionic concentrations of their ECF following those of the outside fluids. Others are regulators to varying degrees, keeping their body fluids more concentrated (**hyperosmotic regulation**) when the tide goes out and freshwater inputs produce dilution of the habitat. More rarely, some regulators can also keep their body fluids more dilute than their surroundings (**hyposmotic regulation**) by excreting or sequestering osmolytes when they are exposed to salinities at or above that of normal sea water. There may be three broad groups in this hyporegulating category, though numerically they are all

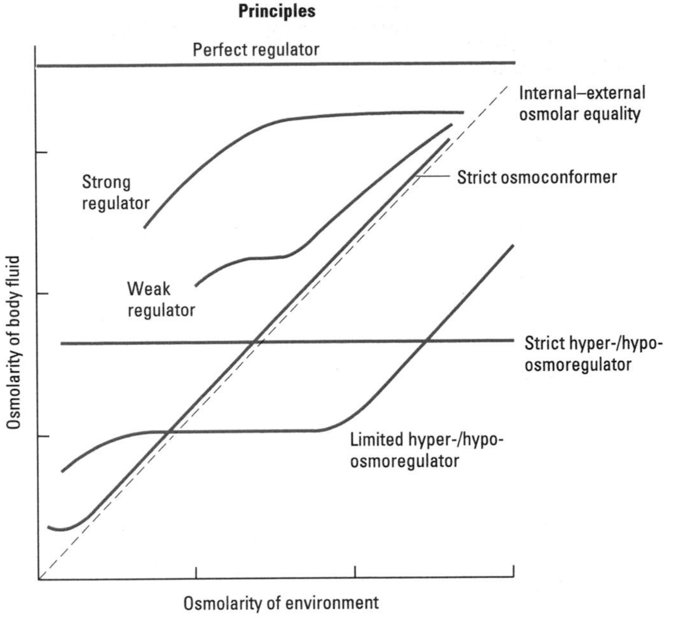

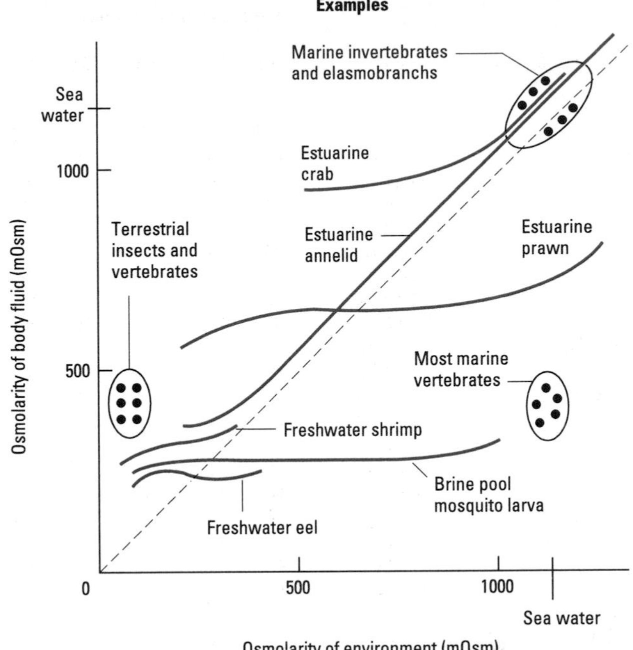

Fig. 5.2 General categories of responses of animal body fluids to variations in external concentrations.

insignificant: the extremely tolerant euryhaline animals that can live in salt lakes, such as the brine shrimp *Artemia* and to a lesser extent some larval mosquitoes (see Chapter 14); others like *Palaeomonetes* and *Leander* (two of the common prawns) that may have had a freshwater evolutionary history but now live littorally; and a few animals that live high up on rocky shores where pools may become substantially concentrated by evaporation on hot days. Animals that can achieve hyposmotic regulation can generally also manage hyperosmotic regulation, though the reverse is not true. These patterns are compared in Fig. 5.2 with respect to osmotic concentrations; we will meet many specific examples on similar figures in Chapters 12 and 13. It is worth noting now that there are no *perfect* regulators, and no *perfect* conformers—all animals lie between these two extremes, and all experience some problem with the maintenance of water balance. In reality some animals do cope with a very wide range of environmental concentration, while others are relatively restricted either to saline or to fresh waters (the points where a line cease reflects the lower and upper tolerance limits). The level of body fluid concentration maintained in fresh water (itself commonly 0.1–5 mM) is obviously very variable, and shows no very clear phylogenetic pattern. Despite this variation, all freshwater animals (and all terrestrial animals making a secondary return to fresh water) are clearly regulators to some degree, and all have continuous water balance problems because water will always tend to flow into their bodies from the hyposmotic surrounding fluid.

On land, there is again no alternative to continuous osmotic and ionic regulation, but here maintaining water balance is a struggle to resist water losses to the surrounding drying air. Air does not have an "osmotic concentration", and the normal terminology cannot be applied; but as a rough guide, all terrestrial animals' body fluids will lose water across any available permeable surface to air at anything less than 99.4% relative humidity—the kind of almost saturated air that can normally only be found in deep, wet caves and crevices, or in air gaps amongst very wet litter and soil. Thus resisting, or compensating for, the efflux of water becomes a dominant factor in the environmental physiology of most terrestrial animals.

5.1.4 Tolerance of water loss

There is another aspect that we should consider in relation to overall water balance, and that is the varying tolerance that animals show to actual loss of body water. Uncontrolled dehydration inevitably leads to a proportional increase in osmolyte concentration; if there is no osmoregulation, a loss of x% of body water will lead to osmotic concentrations increasing by a factor of $100/(100 - x)$, corresponding to a factor of 1.33 for 25% water loss or 2.0 for 50% water loss. All animals are inherently 60–90% water, but this is distributed between the various compartments of the body (see Fig. 4.1), so that many have inbuilt stores or buffers against this effect. Some vertebrates lose water preferentially from the gut when water stressed, worms lose coelomic fluid from the main body cavity, while insects lose hemolymph water, but nearly all animals have mechanisms to avoid excessive water loss from the cells. Most animals can therefore tolerate an overall water loss of a few percent without undue osmotic problems. However, many animals can tolerate substantially greater loss than this. Clearly for aquatic animals this ability will correlate to a substantial degree with the tolerance of osmotic variation in the body fluids; animals that can cope with high salinities are necessarily tolerating considerable water loss from their bodies. For terrestrial and shore animals, where evaporative water loss is a problem, the tolerance of such loss is extremely variable (Table 5.2). Some slugs and snails, and littoral molluscs such as limpets and chitons, can tolerate 35–80% water loss, causing

Table 5.2 Tolerance of water loss (as maximum % weight loss tolerated) in a variety of animals from terrestrial or semiterrestrial habitats.

Annelids	
Allolobophora (earthworm)	75
Molluscs	
Patella (limpet)	35–60
Chitons	75
Helix (snail)	45–50
Limax (slug)	80
Sphincterochila (desert snail)	50–55
Crabs	
Gecarcinus	15–18
Uca	18
Insects	
Temperate beetles	25–45
Temperate roaches	25–35
Desert cicada	25
Desert ants, grasshoppers	40–70
Desert tenebrionid beetles	60–75
Frogs	
Rana	28–35
Hyla	35–40
Bufo	42–45
Scaphiopus	45–48
Birds and mammals	
Small birds	4–8
Rat	12–15
Human	10–12
Camel	30

substantial shrinkage of the body. Amphibious animals such as anurans (frogs and toads) can survive around 30–40% loss, with values up to 48% in the spadefoot toad *Scaphiopus*. Many insects will survive 30–50% loss of body water, and those from arid habitats commonly cope with 40–60%; global comparative analyses show a strong relation between insect water loss rate (WLR) and precipitation regime. At the other extreme, most terrestrial birds and mammals cannot survive more than a few percent loss of water (in humans 3–4% is no problem, 5–8% causes dizziness, and over 10% leads to real physiological deterioration with death at 15%); the camel is exceptional in tolerating 30%. These vertebrates are very much more dependent on high blood pressure and constant blood composition to maintain their respiratory pigments in working order, so insuring adequate oxygen uptake to support their high metabolic rates. Clearly the terrestrial vertebrates need very sophisticated controls to maintain water balance, even though it is the smaller terrestrial animals such as insects, arachnids, and land snails that will suffer potentially faster fluxes due to their very high surface area to volume ratio.

5.2 Exchanges occurring at the outer body surface

5.2.1 Passive exchanges

If the surroundings of an animal are at a different osmotic concentration than its interior, gains or losses of water and ions across the animal's outer surface are inevitable; there is no such thing as an "impermeable" biological surface. We are therefore only discussing *differences* in permeability to water or ions, in considering how animals may control their cutaneous osmotic fluxes. Water permeability and ion permeability are rather separate issues, just as they were distinct in relation to passage across cell membranes, as discussed in Chapter 4. However, for whole animals, the surfaces across which such fluxes occur are usually structurally and chemically much more complicated than simple biological membranes. In some marine animals there may be little more than a surface epithelium, with a thin and loosely organized extracellular coating; at the other extreme, land arthropods are covered in a dense and usually rather rigid cuticle of very complex structure with intricate waterproofing properties, while land tetrapods have a thick outer covering of dead skin cells forming a keratinized layer. Clearly the varying structures and properties of skins will have very different consequences for rates of water and ion flux.

Water permeability

Water permeability in itself is a somewhat complicated issue. For whole animals, P_w is commonly measured from rates of *weight* loss (or gain), which are considered equivalent to rates of water loss (or gain) after correction for any other observed water outputs or inputs such as defecation, urination, or feeding. More recently water loss has been assessed using electronic water vapor sensors, or from rates of flux of labeled water (tritiated water, ^3HOH), often expressed as daily turnover of water. Such measures can yield a "permeability" in units of weight of water lost per unit surface area per unit time, and permeability has therefore been expressed traditionally (but rather loosely) in units such as mg cm^{-2} h^{-1}.

For terrestrial animals, the evaporative water loss (EWL) varies with the dryness (water vapor density) of the surrounding air, and so is more properly expressed as water lost per unit of water vapor pressure deficit, although there are substantial problems in controlling temperature and humidity adequately to insure correct assessment of this. But this figure of total WLR is still potentially misleading for two reasons; firstly it takes no account of the changing conditions very close to the animal's surface (boundary layers of unstirred, more humid air); and secondly it almost inevitably contains a substantial component of respiratory evaporative water loss (REWL) in addition to the cutaneous evaporative water loss (CEWL). The first problem is nearly insuperable, as correct modeling of water movements just above and just below an irregular evaporating surface is notoriously complex. The second problem is also difficult to get around. The relationship between the two components (REWL and CEWL) varies with activity, and even in a "resting" animal is highly variable between taxa (Fig. 5.3), so it is not a simple matter to make corrections to the EWL and thereby calculate CEWL. Therefore many recent studies instead express the data on cutaneous loss or gain for all animals as a resistance, *R*, measured in seconds per centimeter (Box 5.1); on this scale a free water surface has a resistance of zero. Resistance can be determined from experiments on isolated pieces of tissue, removing the respiratory water loss element. But it can also be calculated from whole animal rates of water loss according to the formula given in Box 5.1, and for quiescent animals with enclosed (or artificially blocked) respiratory surfaces the errors will be small. Cuticle permeability is then effectively the reciprocal of resistance ($1/R$), in centimetres per second.

Whatever the measurement used, there is a huge range of permeabilities to be found in "skins". Table 5.3 shows values of rate of loss

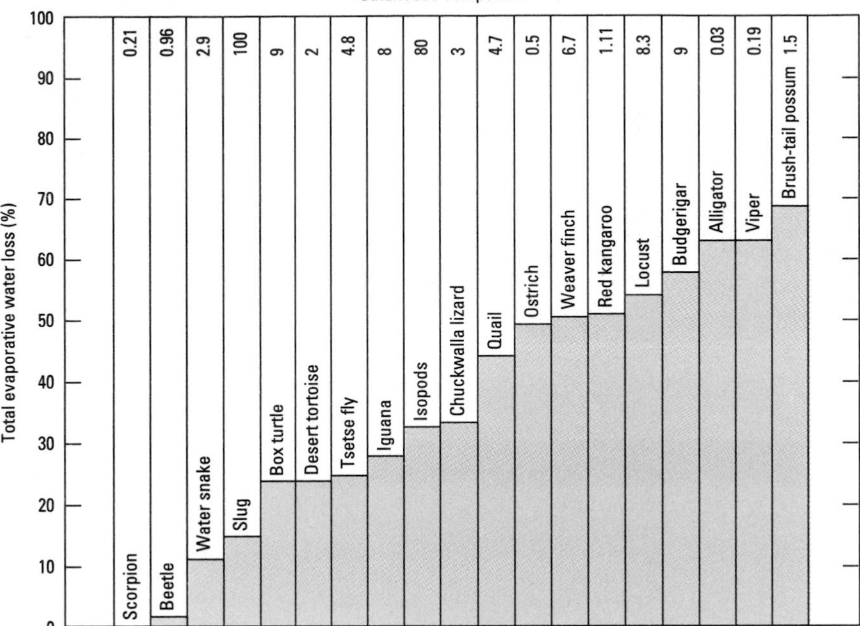

Fig. 5.3 Proportion of total evaporative water loss (EWL) attributable to respiratory loss (REWL, shaded) and to cutaneous loss (CEWL, white) for a variety of animals in their resting state. The total resting EWL is given (mg g^{-1} h^{-1}) for each species. (From *Comparative Animal Physiology*, 1st edn, by Withers © 1992. Reprinted with permission of Brooks/Cole, a division of Thomson Learning: www.thomsonrights.com. Fax 800 730-2215.)

Box 5.1 Resistance, permeability and evaporative water loss rates

Evaporative water loss (EWL) depends on the difference in water vapor density χ (in mg water per liter) between the animal and the surrounding air, and by analogy with Fick's law we can derive:

$$EWL = -D\,\Delta\chi/\Delta d$$

where D is the diffusion coefficient for water vapor in air, and d is the diffusion distance (the boundary layer thickness). χ for the animal is taken as the χ for saturated air in equilibrium with the body fluids. In practice the diffusion coefficient and the diffusion distance can be replaced by a single factor, R, the resistance, giving:

$$EWL = -\Delta\chi/R$$

Values of R can be compared between animals, assuming a constant difference in water vapor density applies (roughly true for a particular vapor pressure level, given the similarity of all terrestrial animals' body fluids).

Alternatively, water loss data (EWL) expressed in conventional units as μg cm^{-2} h^{-1} $mmHg^{-1}$ can be converted into permeability coefficients (P, with units of cm s^{-1}) using the equation:

$$EWL \times 2.94 = P \times 10^4$$

This is accurate at 20°C but slightly inacccurate at other temperatures (e.g. the conversion factor is 2.91 at 30°C). This value of permeability P in cm s^{-1} is then the reciprocal of resistance (R) in s cm^{-1}.

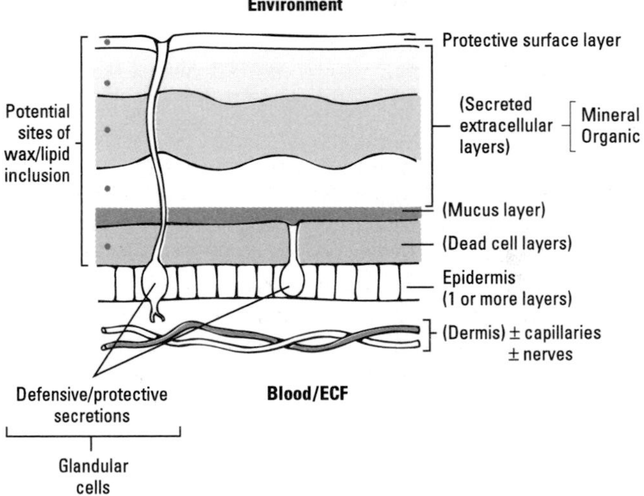

Fig. 5.4 A generalized view of "skin" and its possible components, in relation to cutaneous water loss. ECF, extracellular fluid.

(or gain) of water, together with resistance or turnover rate where possible, recorded from a wide spectrum of animals from different habitats. Although the values vary phylogenetically, with some taxa always more permeable than others, there is a clear-cut superimposed habitat effect, with members of the same phylum or class that come from aquatic, mesic, and xeric habitats having increasingly impermeable skins. Note that there may also be effects of diet; for example, nectar-feeding birds that get plentiful water supplies are decidedly less impermeable than similar-sized birds that feed on seeds or insects.

Skin surfaces clearly vary from epithelia with nearly unrestricted permeation to very resistant nearly "waterproof" cuticles. A model of the various layers that may occur in a skin and its cuticular covering is shown in Fig. 5.4. Several features potentially contribute to low P_w. Firstly, the epidermis itself may exert some control over P_w, and hormones may operate here to affect water loss. Secondly, the

Table 5.3 Comparative measures of "permeabilities" across the skins of animals.

Habitat	Animal group	Resistance (s cm^{-1})	Water flux (mg cm^{-2} h^{-1})	Transpiration rate (μg cm^{-2} h^{-1} Torr^{-1})	Water turnover (ml g^{-1} day^{-1})	P_{osm} (μm s^{-1})
Marine	Crabs	6–14	14–25			
	Echinoderms					30
	Reptiles	33	0.5			0.002
	Fish	7–35				0.2
Hypersaline	Crustaceans	20–100				
Estuarine/littoral	Amphipods and isopods	3–6		200–300		
	Crabs	4–40		80–200		1.4
Freshwater	Insect larvae	10				0.01
	Insect adults	200				0.05
	Crustaceans	11–46				0.3
	Fish	2–8				
Amphibious	Frogs	3–100	5–40	2400		5–25
	Crocodile	5	70	400		
Terrestrial	Snails active (inactive)	2 (46)		2500 (16)		
	Earthworms	9				
	Isopods	3		80–160		
	Crabs	30–75	1–3			
	Scorpions			8–80		
	Spiders and mites	80		30–60		
	Myriapods		2–4	40–270		
	Collembolans			700		
	Termites			28–37		
	Orthopterans	200	1–2	20–70	0.1–0.4	
	Caterpillars			190		
	Dipteran flies			50–76		
	Beetles			24–50		
	Ants			25		
	Frogs	1.5		2000–2500		
	Lizard	198	0.2			0.005
	Birds	50–70	0.7–2.7			
	Mammals		1–10			
Arid/desert	Isopods			14–30		
	Scorpions, spiders, mites	1300–4000		0.6–2.0		
	Millipedes	430		8		
	Apterygotes			15		
	Hemipteran bugs			12–14		
	Cockroach, cicadas			12–100		
	Beetles	5030	0.1–0.2	3–15	0.05	
	Beetle pupa			1		
	Tsetse fly			8		
	Tsetse fly pupa			0.3		
	Caterpillars			40		
	Ants and wasps			4–26		
	Frog, cocooned	457				
	Tortoises	120	1.6		0.003	
	Birds	158	1.7		0.09	
	Lizards	1360	0.1		0.03	
	Small mammals (rodents)		0.5–0.7		0.03–0.13	
	Large mammals		2–6		0.03–0.09	

bulk cuticle material, whatever its constituents (e.g. chitin, proteins, keratin, calcium salts) will usually be inherently relatively impermeable, and its water resistance may alter according to the density of its molecular packing as it becomes more or less hydrated. Thirdly, and usually most important of all, there may be a specific layer conferring even greater impermeability due to a high lipid content; this is commonly at the outer surface, but could be anywhere within the cuticle.

Relatively simple skins are found in many marine soft-bodied animals such as sea anemones, annelid worms, and echinoderms. There is often no secreted layer except perhaps a loose extracellular matrix external to the epidermis, often composed of mucopolysaccharides. Some consolidation of this matrix may occur in essentially "soft-bodied" animals and may confer a degree of water-flux resistance. Freshwater animals generally have lower permeabilities than related marine species, and brackish-water species may be even less permeable. This may initially seem surprising, but of course most brackish waters change concentration cyclically, and low permeability is a particularly good strategy to smooth out these external changes. In many of the more euryhaline aquatic invertebrates P_w

can be acclimated to some degree, over a time course of a few hours or day, according to environmental exposure (see Chapter 12).

The really impermeable surfaces are, not surprisingly, found in animals with complete coverings of exoskeletal materials, especially where the habit is terrestrial. Land animals possessing some kind of relatively impermeable complete outer covering include inactive molluscs, where only the calcareous shell and a small area of mantle tissue are exposed. The other invertebrate groups that show a near complete and highly impermeable covering are the insects and arachnids from drier habitats. These arthropods provide the very best examples of protective surfaces that minimize CEWL. The main chitin + protein cuticle layers are of relatively low permeability, containing some lipid and with a variable hydration state. But the particular asset of land insects is an additional epicuticular layer of high lipid content. This is made up of C_{20}–C_{37} fatty acid derivatives plus some branched hydrocarbons and some C_{40} wax esters, which may appear as a specific wax layer on the surface of the epicuticle. At one time all these epicuticular lipids were thought to be organized in a single orientated monolayer at the epicuticular boundary, but this theory has now been replaced by a more realistic view (one better able to account for the amount of lipid present and its actual chemistry), which interprets the epicuticular lipid as present in bulk and in a less organized "liquid crystal" form, able to change phase from more liquid to more pseudocrystalline solid according to conditions (especially temperature). This explains the transition in cuticle permeation (increasing P_w) that often occurs when an insect is warmed (Fig. 5.5), and which is even sharper if related to temperature of the cuticle rather than to air temperature.

In the most xeric insects and arachnids the presence of a waxy epicuticle insures much slower water loss (R up to 5000) than in similarly sized crustaceans (R = 15–75), with the same general cuticle design but little or no waxy component. This combination of epidermis, chitin–protein cuticle, and waxy epicuticle provides an almost complete body covering that is sufficiently "waterproof" to allow survival in a desert with no drinking water at all.

Vertebrate integuments also constitute "complete outer coverings", but their P_w varies quite widely. At one extreme ($R < 50$) are the soft and nonkeratinized skins of amphibians, used for ionic exchanges and even more crucially for a large part of the respiratory exchange of the animal; these skins have little or no embellishment external to the epidermis itself, apart from secreted mucus. However, in a few relatively xeric amphibian species, such as *Phyllomedusa*, P_w can be further reduced by a waxy secretion from glands that the frog spreads over itself, when R may be raised to 400–500.

Intermediate permeabilities are found in many fish, where keratin scales overlie the softer dermal layers, and the skin itself has little or no respiratory role. As with the invertebrates, species from brackish waters and the very euryhaline species are often the least permeable. Many highly tolerant fish show a hormonally mediated decrease in P_w in response to saltwater exposure.

But the amniotic vertebrates—reptiles, birds, and mammals—have distinctly more impermeable skins, with no respiratory function to compromise their osmoregulatory function. The skin is formed largely from layers of dead keratinized cells overlying the living epidermis, often thickened into plates and scales in reptiles, or modified into hair or feathers in the mammals and birds. Lipid material is included in a dispersed fashion in mammalian skin,

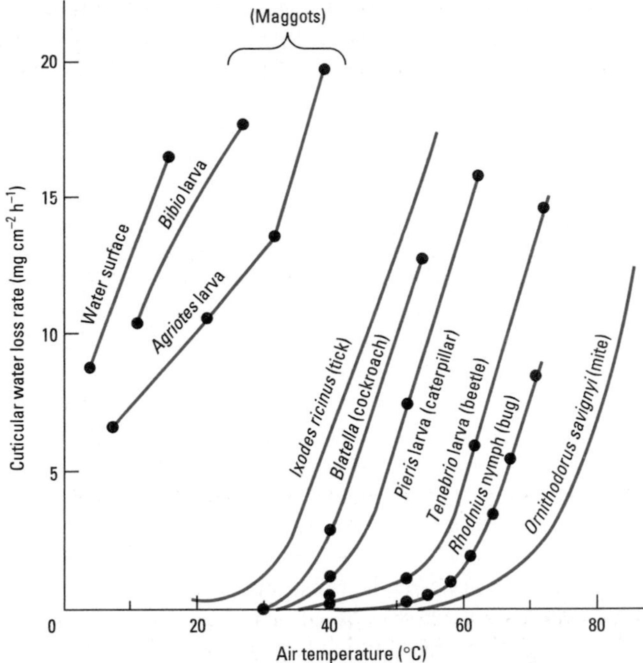

Fig. 5.5 The phenomenon of "transition temperatures" in a variety of arthropods; cuticular water loss increases with temperature, but begins to increase particularly sharply above some species-specific transition temperature. Note that the transition is not very sharp, and commonly occurs well above normal environmental temperatures. (Data from Wigglesworth 1945; Lees 1947.)

while in reptile skin the stratum corneum specifically has a rather high phospholipid content. All these animals are distinctly less prone to water loss than most fish and amphibians; aquatic reptiles by only a small degree (R = 5–50), but terrestrial reptiles (R = 200–500) and all birds and mammals (R = 50–200) by a very substantial amount. These vertebrates still do not approach the degree of waterproofing found in the arthropodan groups. However, the greater surface area (SA) to volume (V) ratio of desert reptiles and rodents compared with much smaller insects often means that they too lose water so slowly as a percentage of their body weight that they can also survive without access to drinking water.

To complicate the picture, cutaneous water loss in some of these groups can be altered more drastically by controlled and "planned" losses related to cooling: the phenomenon of sweating, dealt with in detail in Chapter 8. Many mammals have a background low level of water loss from the sweat glands superimposed on the loss through the rest of the skin, with a much faster loss that is turned on during exercise and in hot climates. Similarly, the more waterproof species of frogs effectively sweat through their epidermis when heat stressed; and a few insects also "sweat" when overheated.

Ionic permeabilities

The ion permeation properties of skins are due to ion channels in underlying membranes (varying with their number and their properties), and to the structure and electric charge distributions in overlying extracellular matrices or cuticular layers. Where substantial extracellular material is present the situation is somewhat complicated by ion-binding effects; for example, mucus is often a good ionic binding matrix, and the rate of movement of Na^+ or H^+ ions through it is unexpectedly hard to predict. Certain reagents have traditionally been used to stain sites of high salt permeability in aquatic animals: silver nitrate and potassium permanganate are the most widely used.

Again comparisons are somewhat complicated by varied methods of measurement and different ways of expressing salt movements; some studies determine labeled sodium loss rates directly through pieces of isolated skin or cuticle, while others consider overall sodium loss from whole animals. Data therefore have to be treated with some care. However, for most aquatic invertebrates salts certainly move relatively freely across the surfaces, and again P_{Na} in particular tends to be much lower in brackish and freshwater forms. Specific loss rates of sodium (in μmol g^{-1} h^{-1}) when exposed to dilute media vary from over 1000 in fully marine crabs, through 800–900 in marine/littoral crabs, 100–200 in estuarine/freshwater crabs, to just 5 in the fully freshwater crayfish. Similar effects can be seen even within a genus, as we will see in Chapter 12.

Unlike P_w, there is little evidence for acclimation or variation of P_{Na} in invertebrates, although it can vary genotypically within a species. However, many aquatic vertebrates can vary their sodium permeability individually according to need, and it is directly under the control of the pituitary hormone prolactin in fish and in frog skin.

5.2.2 Acclimatory changes of permeability

We have mentioned a number of cases where permeabilities to ions or to water can be altered in the short term—within the life of an individual. How can this be achieved? There are two main possibilities. Firstly, at the level of cell membranes, acclimation involves adding or destroying ion channels (and probably also water channels), and there are two possible timescales for this, either recruiting pre-existing channels from vesicles within the cells as described in Chapter 4, or actually synthesizing new channel proteins. Secondly, permeabilities may be altered on a larger scale by alterations to the extracellular components; again this may involve spatial rearrangements, involving phase changes in lipid layers, for example, or it may involve synthesis of new material added to the surface. Insects show these effects particularly clearly, since the permeability of successive exuviae (cast-off skins) can be measured for an individual; they contain lipids of significantly higher melting point when the individual is acclimated to higher temperatures, largely due to incorporation of more straight-chain hydrocarbons relative to branched forms.

5.3 Osmoregulation at external surfaces

A substantial proportion of aquatic animals are always either hyperosmotic (most brackish and freshwater animals) or hyposmotic (most marine vertebrates) to the medium in which they live, as we have seen. In both situations, some of the regulation to maintain this situation is done at the external surfaces of the animal, either through the skin as a whole (which thus becomes an important organ actively participating in water balance), or through the more permeable areas such as gills, or via particular modified glands or appendages. External "skin" surfaces are thus a very important component of overall osmoregulation, and in a few animals they may be the only regulatory system in operation.

5.3.1 Hyperosmotic regulation and salt uptake

All the inhabitants of fresh waters, and most of those in brackish waters, are hyperosmotic regulators, continuously having to counter the twin problems of water tending to flow into the animal, and solutes tending to be lost. The water that enters the body has to be excreted, and this process inevitably entails yet more solute lost in the urine. Decreasing the permeability of the skin will help, as discussed in section 5.2, but the respiratory surface of the skin must remain large enough and permeable enough to give adequate gaseous diffusion, so that large gills could be rather an osmotic liability. However, some part of the body must be used to regain solute and redress the balance against the passive fluxes, and the obvious surface to use is that of the permeable respiratory surface itself, localizing all exchange systems there. Hence many aquatic animals, both invertebrates and fish, show **active uptake of ions at the gills**; they may also use their gills for nitrogenous excretion and to regulate acid–base balance. Freshwater animals may take up Na^+ and other ions against at least a 100-fold gradient, for example from a medium at < 5 mM to their own blood at perhaps 500 mM.

The respiratory site is nearly always the site for this uptake, so for animals that normally rely on cutaneous respiration, such as freshwater worms, uptake is directly through the skin. In amphibians salt uptake also occurs through the general skin surface (though anuran amphibians also have lungs), and frog skin is a classic physiological preparation for studying the properties of salt-absorbing epithelia. In most other freshwater animals, salt uptake is localized at the gills. In some larval insects salt uptake occurs in appendages at the tail end, the anal papillae, also inaccurately known as "anal gills". But in most molluscs and crustaceans, and in adult insects, this uptake can be shown to be localized in cells of the genuine gill epithelium. Often the whole gill surface is osmoregulatory, but in some animals there is a division of labor; for example, in the mitten crab *Eriocheir* some posterior gills are primarily osmoregulatory while the anterior ones are mainly respiratory.

The standard Na^+/K^+-ATPase operates at the basolateral cell membranes in all these sites, with Na^+ symport/antiport systems on the apical membranes. The kinetic properties of the overall sodium transport are highly variable, with K_m values (the concentration for half maximal transport, so that low K_m indicates high affinity; see Box 4.5) dependent on the medium in which an animal lives (Table 5.4). Thus K_m is 6–20 mM Na^+ in marine invertebrates and only 0.04–0.2 mM Na^+ in freshwater and amphibious animals, reflecting a high-affinity pump suited to the lower sodium concentrations in their environment. There is still some uncertainty as to whether chloride movements are always dependent on the sodium pump, or whether there is a separate and independently active mechanism for handling Cl^- ions, as reported for some brackish-water and freshwater invertebrates. Certainly in many epithelia it has been shown that the two ions appear to move into cells independently of each other, each being coupled to the efflux of an ion of like charge (Na^+ for H^+ or NH_4^+, Cl^- for HCO_3^-; see Chapter 4).

Table 5.4 Values of K_m for sodium transport in various aquatic animals.

Habitat/species	Animal group	K_m (mM Na^+)
Marine		
Marinogammarus finmarchicus	C	6–10
Carcinus maenas	C	20
Brackish		
Cyathura carinata	C	6
Mesidotea entemon	C	9
Gammarus duebeni	C	1.5–2.0
G. zaddachi	C	1–1.5
Rana cancrivora	V	0.40
Freshwater		
G. lacustris, G. pulex	C	0.1–0.15
Lymnaea stagnalis	M	0.25
Margaritana margaritana	M	0.04
Hirudo medicinalis	A	0.14
Salmo gairdneri	V	0.46
Carassius auratus	V	0.26
Xenopus laevis	V	0.05
Amphibious (freshwater/land)		
Rana pipiens	V	0.20
Bufo boreus	V	0.25
Hyla regilla	V	0.14

A, annelid; C, crustacean; M, mollusc; V, vertebrate.

In freshwater teleosts, salt uptake occurs across the integument and gills, as well as in particular cells of the gill epithelium—these cells are conspicuously thicker than the main respiratory epithelium, with the deep basal infoldings and high density of mitochondria characteristic of salt-transporting cells. These so-called **chloride cells** (Fig. 5.6) (also known as "mitochondrion-rich cells") primarily function to secrete excess salt (NaCl) in marine fish, so their mechanism is covered in the next section; they are less active in freshwater species. However, other ions (especially Ca^{2+}, H^+, HCO_3^-, and NH_4^+) appear to be handled by these cells according to circumstance, and in fresh water substantial chloride–bicarbonate exchange is recorded.

5.3.2 Hyposmotic regulation and salt secretion

Coping with environments that are more concentrated than body fluids is a relatively uncommon problem given the numerical dominance of marine invertebrates (i.e. animals isosmotic with the medium), but it is very important because it occurs in all marine teleosts and tetrapods, and also in the invertebrates that can live in hypersaline seas. For all these animals, water will rapidly efflux from the body, potentially causing tissue shrinkage, and salts will tend to leak in.

For invertebrates faced with these problems (see Chapter 14), the drinking rate is increased to acquire the essential water, but this produces a greater salt load, which must be excreted via the gills, excretory organ, or rectum.

Hyporegulation in marine vertebrates occurs in lampreys, teleosts, and all tetrapods examined to date except the crab-eating frog. In general these animals all have blood concentrations about one-third that of the sea in which they dwell, and their particular physiological problems are dealt with in detail in Chapter 11. Again, drinking sea water is the common response, with water entering the body via the gut. In teleosts, drinking rate measurements indicate that only a proportion of the salts come via this route, with substantial amounts also entering through the relatively permeable gills. Monovalent ions are the main input at both sites; divalent ions in the drinking water are largely retained in the gut lumen and

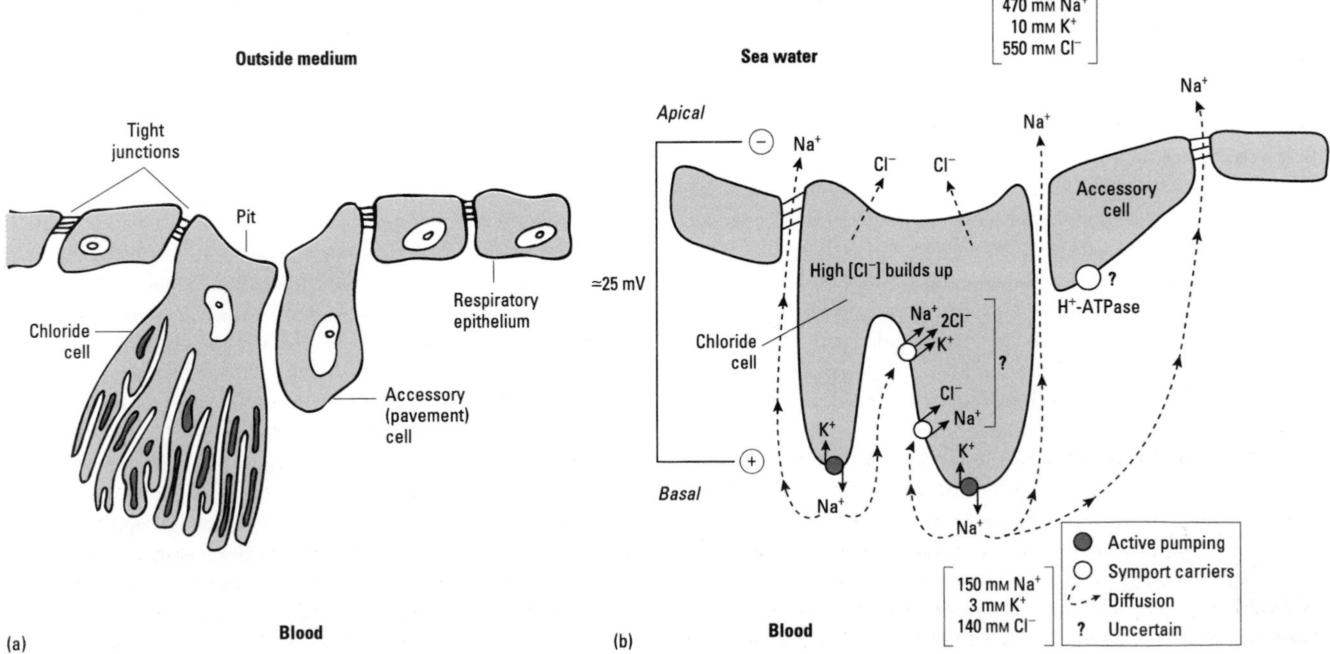

Fig. 5.6 The structure and function of "chloride cells" occurring in fish gills and salt glands. (a) General structure, thickened cells within the respiratory epithelium. (b) Model of salt transport across the cells in a marine teleost.

excreted with the feces, while any leaking in at the gills are dealt with by the kidneys. The excess monovalent salt load is secreted largely by the gill epithelia at the **chloride cells** already referred to (Fig. 5.6). These have high levels of chloride ions, especially near their external borders, but present evidence indicates that the main ion pumps are actually, once again, the classic Na^+/K^+-ATPase "sodium pumps", since ouabain blocks salt transport (see section 4.4.2). The sodium pumps are basal, and seem to pump in the "wrong" direction. How can this work? In effect, they produce a steep inward Na^+ gradient basally, so that symporters (either Na : Cl or more probably Na : K : 2Cl) can draw ions into the cell from the blood, across the basal membrane. Na^+ continues to be pumped out again, and K^+ will diffuse back out, so the net effect is a pile-up of chloride, which then leaves apically via chloride channels, producing a net electrical gradient. This in turn draws sodium across to the apical side, mainly by the paracellular route (chloride probably cannot pass back by this route due to a net negative charge within the cell junctions). Chloride cells are surrounded by accessory "pavement cells", which may be involved in sodium uptake coupled to an H^+-ATPase. They also exert a special kind of morphological control over the chloride cells, being able to move in and occlude them or move back and expose them as conditions dictate (see Chapter 12).

The gills are thus the main osmoregulatory organs for the major osmotic effectors in a marine teleost. In contrast, for marine elasmobranches, rectal glands are the main site of salt secretion, with a rather similar mechanism to that of the teleost chloride cells.

Other marine vertebrates (the tetrapods) are numerically insignificant in comparison to teleosts, and they do *not* normally use the body surface, or any particular parts thereof, for osmoregulation. Most of them are essentially terrestrial air-breathing animals that spend long periods in and around salt waters, feeding on marine life with a high salt content, and drinking marine waters, so that their water balance problems are not unlike those of the teleosts. However, they are much more physiologically isolated from sea water in that they do not need large surface areas of permeable gill, since they are periodically breathing air at the surface of the sea; the salt load problem is therefore rather less severe for them. Within this assemblage of tetrapods, two distinct osmoregulatory strategies occur:

1 The reptiles and birds, endowed with kidneys incapable of producing concentrated urine (with the exception of a few birds where the urine is slightly hyperosmotic to the blood), swallow some sea water and gain more salt via the general body surface, and then must excrete the excess salt load via structures known as **salt glands**, functioning in a similar fashion to the teleost chloride cells by pumping Na^+ in the "wrong" direction to produce a steep gradient back again. The fluid produced is highly concentrated, made up almost entirely of NaCl at a particular concentration for each species, and it is only produced sporadically when the animal concerned is particularly osmotically stressed. The salt gland system is thus an all-or-nothing osmoregulatory effector (unlike the tetrapod kidney, which can produce varying volumes and concentrations and works all the time). It is also an extremely powerful system, producing higher rates of ion transport per gram of tissue than almost any other regulatory organ (see Chapter 11).

2 Marine mammals are rather different, since like all mammals they can produce hyperosmotic urine, and thus they can regulate using the kidney, leaving even less role for the external surfaces. Whales and seals can produce urine that is hyperosmotic not just to their own blood but also to the surrounding sea water, so ridding themselves of a substantial salt load. Most marine mammals do not drink sea water to any great extent anyway, though they may take in water incidentally when feeding. There are indications that some marine mammals can hold food in the esophagus while using a reversed peristalsis to pump sea water back to the mouth where it is spat out.

5.3.3 Terrestrial regulation and water uptake

There are only three possible avenues of water gain in a terrestrial animal: production of oxidative water by metabolic processes, uptake of liquid water, and uptake of water in the vapor phase. All animals inevitably use the first mode, simply by virtue of being alive, but this has only limited scope for regulation. This topic is dealt with in more detail in section 5.8, as it is relevant to the water balance of all animals. Most land animals in practice acquire liquid water by drinking or in combination as food; a few have special modes of acquiring water from puddles and pools, from condensing dew, or from interstitial water held in the substrate. A very small range of animals have special mechanisms to extract water from the vapor phase—from subsaturated air. Chapter 15 gives a more detailed treatment of these purely terrestrial strategies.

Liquid uptake

For the majority of land animals the most important sources of liquid water are drink and food. However, there are a number of other possibilities used in special circumstances.

1 Use of dew. Dewdrops form by condensation, usually at dawn, and may be drunk as they form on plants or on the soil; but some desert species "collect" dew specifically, on their own bodies or in constructed condensation traps. Spider silk is hygroscopic and dew that collects on webs at dawn or whenever the humidity is high may be drunk by the resident spider.

2 Osmotic uptake of standing water. This is usually only an option for animals that retain fairly permeable skin surfaces. Some crabs can take up water at their gills by immersing themselves in freshwater pools, and some frogs achieve the same thing by squatting in puddles.

3 Osmotic uptake of water from wet soils and surfaces. Quite a range of insects and myriapods (millipedes and centipedes) can use eversible structures at the top of the legs called "coxal sacs" to absorb water, while some woodlice (isopod crustaceans) use the brood pouch tissues for this purpose. The ventral tube of springtails (collembolan insects) bears eversible vesicles that take up water osmotically, by a process involving active transport of sodium.

4 Suctorial and capillary uptake from soils. Use of interstitial liquid water from less saturated soils generally requires either a very good muscular pump, or structural modifications of the skin surface to produce a strong capillarity that will overcome the capillary attraction of water to soil. Some spiders can extract water from soils by suction, and some ghost crabs can suck water from sand into their gill chambers using muscles acting on the chamber walls. Some littoral amphipod crustaceans have a capillary channel formed by

Table 5.5 Critical equilibrium humidity (CEH) values for various nonflying arthropods able to take up water vapor from unsaturated air. Where a range of values is given for a species, the extremes represent variation with temperature.

Taxon	CEH (as % relative humidity)
Arachnida	
Acari—mites and ticks	
Ornithodorus	94
Ixodes	92
Echinolaelaps	90
Acarus	70
Dermatophagoides	52–69
Amblyomma	87–89
Insecta	
Thysanura—apterygotes	
Ctenolepisma	48
Thermobia	44–47
Mallophaga/Siphonaptera—fleas	
Ceratophyllus	82
Goniodes	60
Xenopsylla	50
Liposcelis	54
Phthiraptera—lice	43–52
Psocoptera—booklice	
Liposcellis knulleri	70
L. bostrychophilus	60
L. rufus	58
Orthoptera—roaches + grasshoppers	
Arenivaga	80–83
Chortophaga nymphs	92
Lepidoptera	
Tinea larva	95
Coleoptera—beetles	
Tenebrio larva	88
Lasioderma larva	43
Onymacris larva	84
Crustacea	
Isopoda—woodlice	
Porcellio	87–91

the uropods (terminal posterior appendages) that can conduct water to the ventral surface and thus to the mouth. Many semiterrestrial crabs can take up water from sand using a ventral patch of hairy cuticle; these hairs draw up water by capillarity, and conduct it to the leg bases, from where it is drawn into the gill cavity.

Vapor uptake

A small and exclusive range of flightless arthropods has developed the ability to take up water from nonsaturated air (Table 5.5); at present the list includes some 70 species, ranging through mites and ticks, some of the apterygote insects, and a range of pterygotes including lice, fleas, and orthopterans, plus a few beetles and caterpillars. Recently, examples have been added from amongst the terrestrial woodlice. Each species has a characteristic **critical equilibrium humidity** (CEH; or sometimes equilibrium relative humidity), the lowest air humidity at which it can extract water (Table 5.5). For most species this value is above 80% relative humidity; in arachnids the total range known is currently 52–94%, and in insects it is 43–95%. An example of the pattern of responses seen during water vapor uptake is shown in Fig. 5.7.

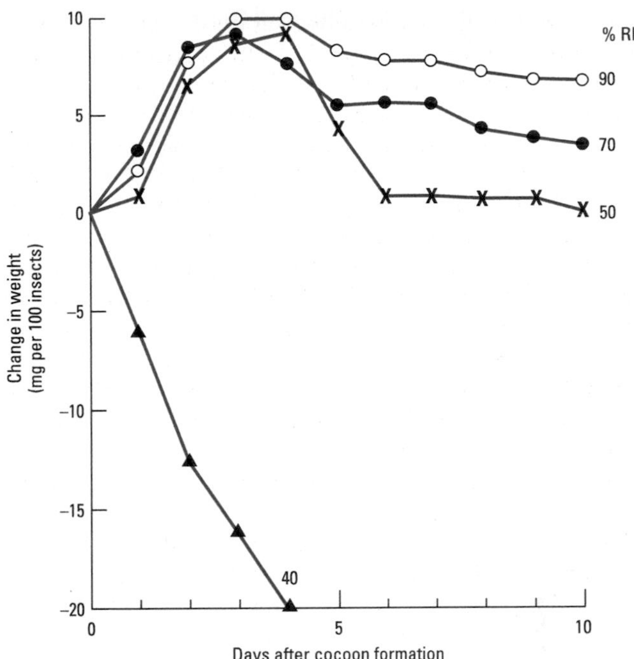

Fig. 5.7 In prepupae of the flea *Xenopsylla*, water vapor uptake (seen as an increase in weight) occurs at humidities above 50% relative humidity (RH) but not at 40% RH, which is below the critical equilibrium humidity value for this species. (Adapted from Edney 1947.)

The mechanisms of vapor uptake differ, but always serve to produce a large water activity gradient between the body fluids and the uptake surface. This may be achieved in different sites:

1 In some of the mites and in the desert cockroach, uptake occurs at the mouthparts, in association with salivary glands. Two main principles appear to be involved: the secretion (or perhaps muscular extrusion) of high concentrations of hygroscopic solutes, on which water will condense, and/or the use of structurally complex exposed surfaces to hold this concentrated fluid and draw in fluid by capillarity.

2 In most other cases uptake is via the rectal or anal surfaces, which are already specialized for water absorption. Air is drawn into the anus and exposed to the absorptive surfaces, which in turn may be in close approximation to the ends of the Malpighian tubules where KCl concentrations may reach > 4000 mOsm (see below). There may also be special hygroscopic materials, or muscular pumping to create pressure cycles, to facilitate water uptake at such sites in species with very low CEH values; examples are considered in more detail in Chapter 15.

In addition to these localized uptake systems, there are suggestions in the literature of a more general occurrence of surface uptake of water vapor in some insects, including pupal stages where there is no other source of water and where no salivary or anal openings occur; but these have never been fully substantiated and no widely accepted mechanisms have been proposed. The physiology of the uptake of water vapor is an area where details of the physical and chemical processes remain substantially unresolved.

5.4 Osmoregulatory organs and their excretory products

Some animals can do all of their osmotic regulation and control across all or part of the general body surface, by the mechanisms considered above. However, most animals also have some kind of internal structure that is usually described as their excretory organ, and often the term "kidney" is used generically to include all these structures, which range from subcellular and unicellular structures such as contractile vacuoles and flame cells, through to complex organs with several different constituent tissues. In practice the primary function of these organelles and organs is almost always that of osmoregulation rather than excretion. The nitrogenous waste from the body that needs to be eliminated is added to the osmoregulated urine almost incidentally (and sometimes not at all). It might therefore be more correct to call these structures osmoregulatory organs, although the distinction becomes somewhat blurred if excretion is allowed a broader definition in terms of removing all excess waste metabolites (thus including excess water or excess ions).

Despite the huge variation in origin, size, and apparent complexity in osmoregulatory organs, there are certain common architectural and physiological principles operating in all of them (Fig. 5.8) so that they can be treated together. Nearly all of them are made up of one or many tubular structures, and most include an initial collecting area where the primary urine is formed, then one or more areas where it is modified by the addition or removal of particular solutes. Many of the organs include an even more distal area where the urine as a whole can be made more concentrated (hyperosmotic) or more dilute (hyposmotic) than the body fluids.

5.4.1 Production of primary urine

There are only two basic ways of making the initial urinary product in animals (Fig. 5.9). Most taxa use a system based on **ultrafiltration**, where hydrostatic pressure forces blood (or coelomic fluid, or some other component of the body fluids) through a semipermeable membrane into the lumen of the osmoregulatory organ. The membrane acts as a filter, holding back blood cells, proteins, and other large molecules, but allowing water, ions, and small solutes such as amino acids and sugars to pass through. The proportion of a solute that passes into the urine (its filtration factor) is roughly determined by its hydrated size; this proportion is generally 1.0 for water, urea, glucose, and sucrose, and for mammals it is 0.75 for myoglobin, 0.03 for hemoglobin, and less than 0.01 for other serum proteins. The product is thus a simple filtrate of the body fluid, with very little change in osmotic concentration or composition except that cells and proteins are lacking. Most components of the body fluids would therefore be ejected from the body if no further modification of the urine were to take place. This makes the system ("eject everything and then take back what is wanted") rather safe, in that any small toxic molecules entering the body will be automatically eliminated at the "kidney" by ultrafiltration. However, it requires large volumes of fluid to be handled; in some animals as much as 99% of the initial filtrate is taken back into the body, involving substantial and expensive active transport.

Ultrafiltration is usually demonstrated by injecting a passive marker unfamiliar to the animal's cells and small enough to go through the filter; its presence can then be detected in the urine. The most common marker is inulin, a soluble polysaccharide that is apparently completely inert in biological systems and thus with no side effects on cell functioning, and which has a filtration factor in the vertebrate kidney of about 0.98. Presence of inulin in primary urine is a useful (but not infallible) indication of a filtration-based system.

By far the best known ultrafiltration system is that of the glomerulus in the vertebrate kidney (Fig. 5.10), where a network of blood capillaries is contained within the Bowman's capsule of each nephron and a filtrate of the blood plasma is forced out under

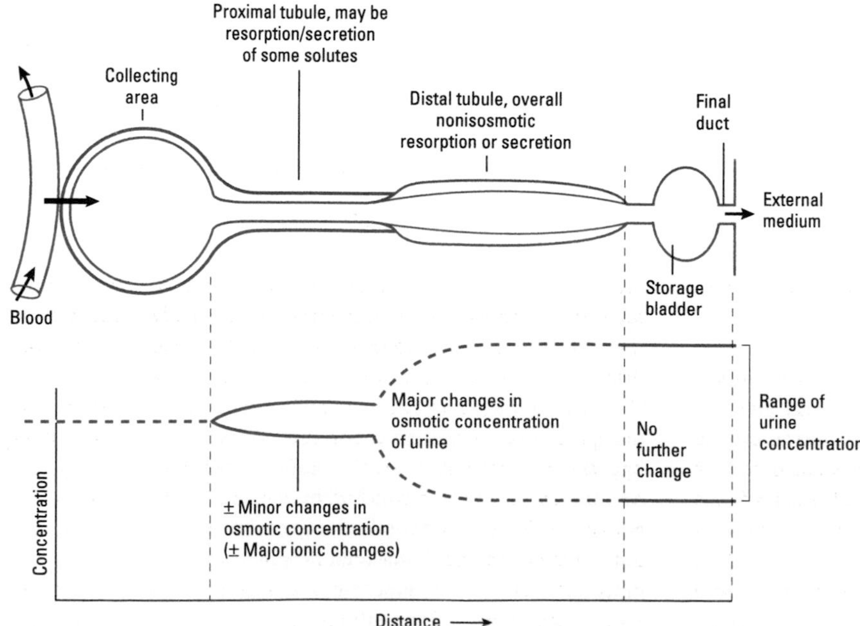

Fig. 5.8 A schematic basic design of osmoregulatory/excretory systems, and the patterns of change of concentration commonly found in each region.

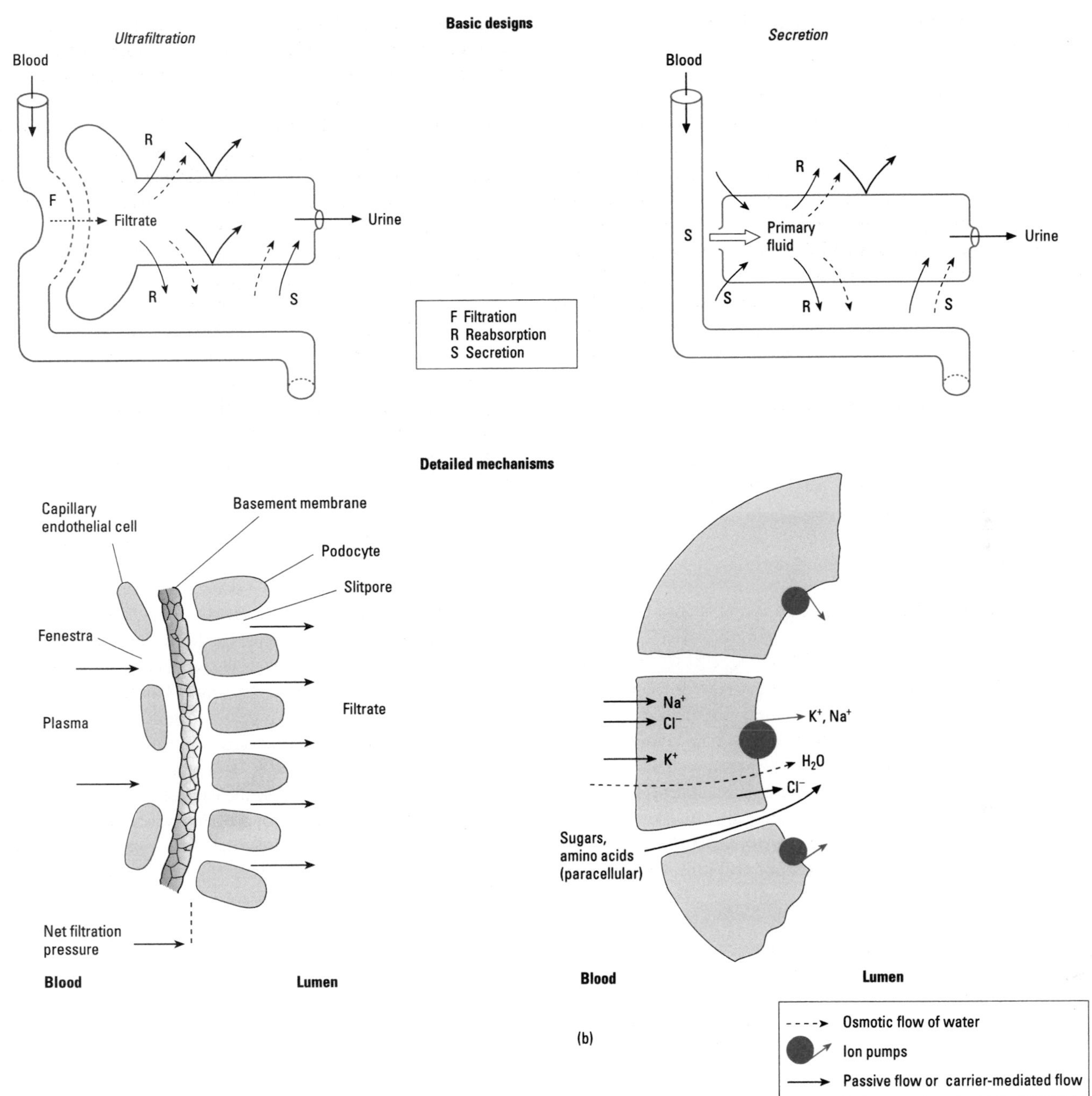

Fig. 5.9 Urine production by (a) ultrafiltration and (b) secretion. Basic principles and examples of mechanisms.

pressure through the capillary walls. However, filtration systems also occur in most invertebrate excretory organs (Fig. 5.11), including nephridia in worms, the antennal glands of crustaceans, and the molluscan coelomoduct kidney where the filtrate is taken directly from the heart into the pericardial cavity. The closed end of a flame cell in some simple freshwater animals almost certainly operates by ultrafiltration as well (see Chapter 13).

A few groups of animals, notably insects but also leeches and some species of teleost fish, use a different system where the primary urine is produced by an **active secretion** process (see Fig. 5.9). Membrane pumps move ions from the general body fluids into the lumen of the organ concerned. A model for a single tubule cell (based on the Malpighian tubule in insects) is shown in Fig. 5.12. The membrane pumps will handle most cations, and principally transport whichever is most common (varying with diet). Other oppositely charged ions may then follow down the electrical gradient created, or may be handled by separate pumps, while water follows readily down the osmotic gradient and may carry with it other small molecules in the bulk flow set up through the leaky cell junctions. (Note that these other solutes are effectively filtered across a semipermeable membrane again, with anything of relative

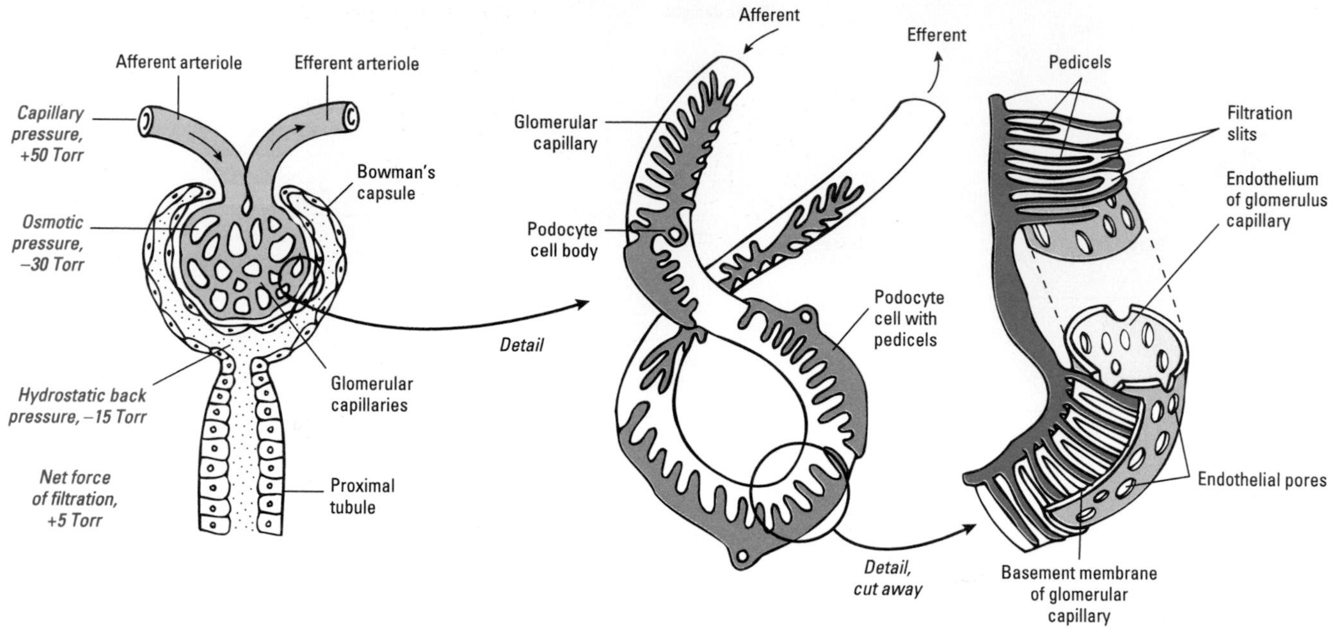

Fig. 5.10 The structure of the vertebrate kidney glomerulus where ultrafiltration of the blood occurs.

molecular mass less than about 10,000 getting across, so the two methods of urine formation are not quite as distinct as at first appears.) Again, the primary product may well be isosmotic to the blood; but if the pumping of ions is very intense, or if the product is rapidly removed from the lumen to other areas of the organ, or if the membranes where the pumps are operating are rather impermeable to water or to particular ions, then the primary urine may potentially differ quite substantially in osmotic and ionic concentration from the main body fluids.

Active secretion of primary urine has the advantage of being much more controllable. The rate of secretion can be varied according to need and circumstance—for example, being accelerated just after a large fluid meal in sap or blood feeders. On the other hand the system is potentially much less "foolproof" than a filtration-based mechanism, since only molecules that are either specifically pumped, or are of the right size and charge to follow passively, will be passed into the urine; hence the need for a very "leaky" secretory epithelium (see Chapter 4). Thus inulin injected into an insect does turn up in the urine via the leak route. But there is always a risk that moderately large toxic molecules that cannot leak through and are not "recognized" by pumps in the tubule walls will be left circulating in the body fluids. Thus it may be that some cold-water marine teleosts, living in a relatively "safe" and nonvariant habitat, can get away with a nonglomerular, secretion-based kidney. However, the argument hardly applies to insects, frequently encountering "foreign" chemicals in their variable terrestrial world.

5.4.2 Modification of urine solutes

After production of the primary urine, nearly all animals have mechanisms operating further along the tubes or ducts of the osmoregulatory organs that will retrieve particularly useful solutes, or add extra unwanted components, before the urine leaves the body (see Fig. 5.8). In particular, glucose and other small sugars, and crucial amino acids, need to be resorbed into the body fluids. In animals with highly regulated body fluids and a strong homeostatic system, especially the terrestrial vertebrates and insects, the excretory system may be responsible for regulating water and certain ion levels. Potassium regulation and the control of pH via bicarbonate concentration are common properties of excretory tubules, and in many cases there is also active transport of magnesium into the urine (with sulfate often linked to this) and selective resorption of calcium. In marine elasmobranch fish, where urea is used to increase the osmotic concentration of the blood, there must be a urea-resorbing system in the kidney tubules to recover this compound after it has been ultrafiltered.

The site of these active transport mechanisms is usually at the proximal end of the system, soon after the primary urine is formed. For example, in insects the lower third or so of the Malpighian tubule and the first part of the hindgut (the ileum) often have a resorptive function, taking back KCl and other ions, and leaving a reduced volume of a slightly hyposmotic urine due to a reduced P_w here. Similarly in the vertebrate kidney, the volume of the primary urine is reduced greatly in the proximal convoluted tubule (see Fig. 5.16) by resorption of NaCl, with water following; also, all the filtered glucose is actively resorbed here. However, regulation of potassium and bicarbonate levels in mammals is a property of the more distal parts of the tubules and the collecting ducts, and in other taxa a degree of resorption and secretion occurs in the distal bladder or the cloaca, just before the urine is excreted (see below).

Certain other molecules are specifically secreted into the urine in many animals. In fact kidneys have a surprising ability to secrete quite "strange" larger molecules into the lumen, and so protect the body from toxins and drugs. In vertebrates this seems to be largely because the liver initially modifies such molecules, or parts of them,

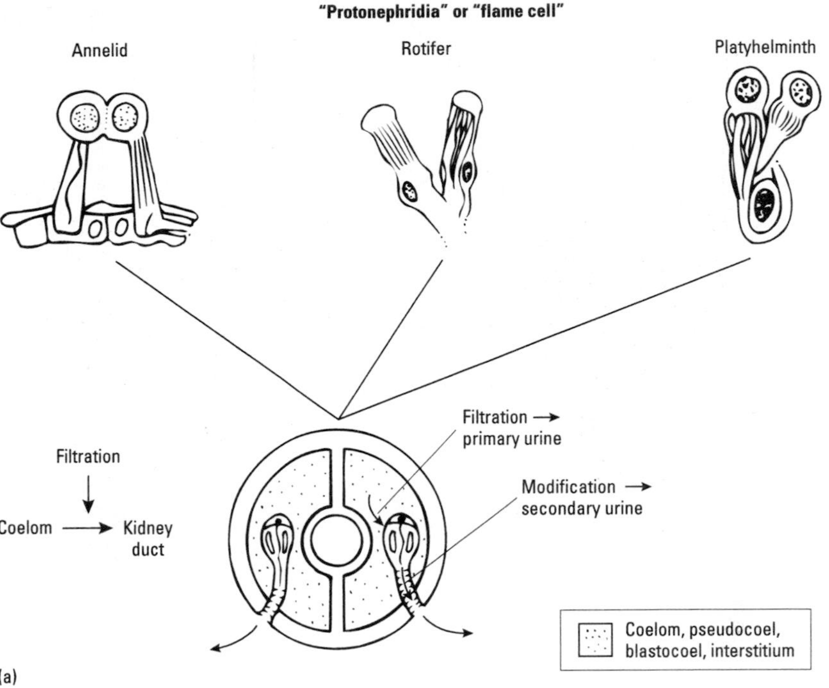

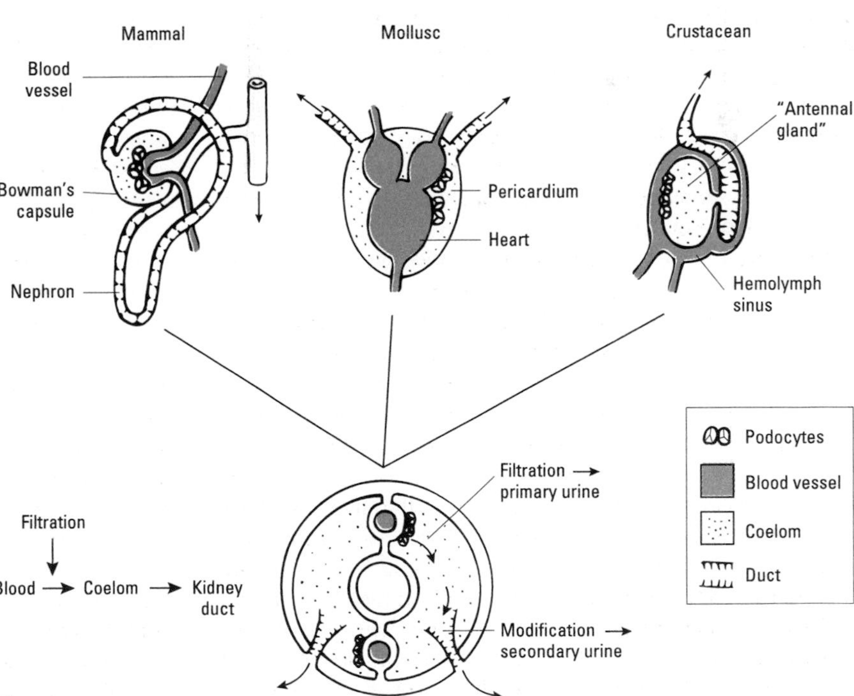

Fig. 5.11 Variety and occurrence of excretory systems based on ultrafiltration: (a) in flame cells; and (b) in more complex nephridial structures. (Adapted from Ruppert & Smith 1988.)

into forms that can be recognized by kidney transport systems. Some molecules have become widely used as markers to study these secretory processes, two of the most useful being *para*-aminohippuric acid (PAH) and the dye phenolsulphonphthalein (phenol red). Both of these are actively secreted into the tubules of a wide range of osmoregulatory organs. PAH is especially useful because it is secreted very rapidly indeed, with none remaining in the efferent blood supply from a healthy mammalian kidney. Phenol red is useful because it colors the urine, and its concentration can be determined colorimetrically. As a result we know that active secretion

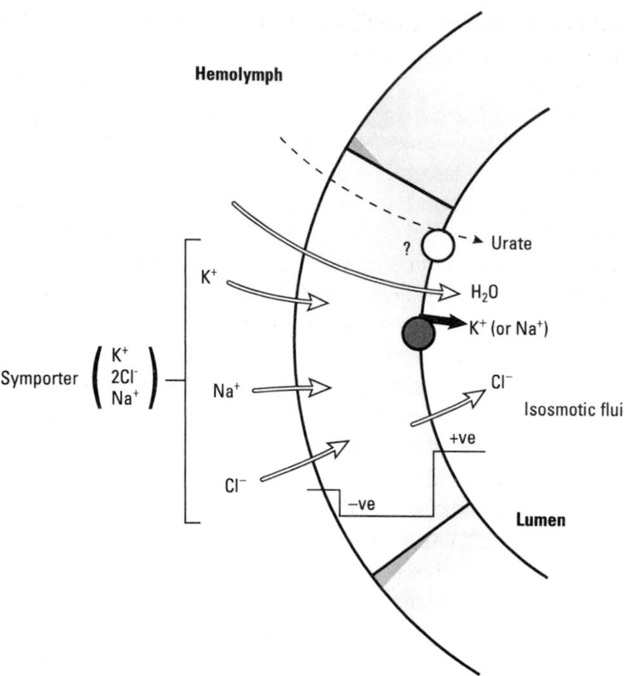

Fig. 5.12 Secretory urine production across cells of Malpighian tubules in insects, driven by an apical cation pump. (Simple model; more details in Chapter 15, see Fig. 15.13.)

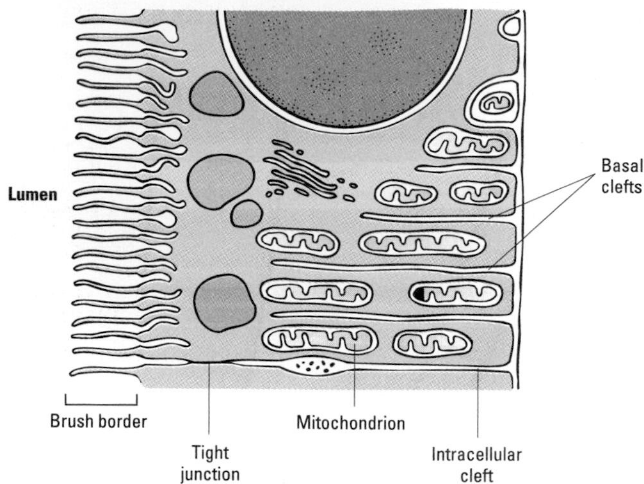

Fig. 5.13 Ultrastructure of a kidney cell from the proximal convoluted tubule where isosmotic resorption occurs. The luminal side has conspicuous projections ("brush border") and the basal side has deep clefts between which mitochondria are tightly packed.

certainly occurs into the urine in many molluscan kidneys and in crustacean antennal glands. Active secretion of other toxins such as nicotine, atropine, and morphine, often derived from plant diets, has also been identified in insects. Secretion also occurs in the proximal convoluted tubule of the tetrapod kidney, where PAH, certain phenolic compounds, and the antibiotic penicillin are known to be secreted. Again, the distinction between filtration-based and secretion-based kidney systems becomes somewhat blurred.

5.4.3 Controlling urine volume and concentration

Even when an animal excretes isosmotic urine, as in most marine invertebrates, some further modification of the primary urine is required before it passes out of the body. Almost always a large part of the volume initially produced must be recovered, and this can be simply achieved by active transport of salts out of the lumen, with water following passively. This allows the waste products and any unfamiliar compounds to be excreted in the smaller volume of (still isosmotic) fluid that remains. The functioning of ion pumps in resorbing ions from the lumen essentially follows the principles outlined in the previous chapter; most commonly, sodium pumps (transmembrane Na^+/K^+-ATPases) operate to extract Na^+ from the urine into the cells, and chloride ions follow. The pump is normally sited on the basal side of classic transporting epithelial cells; the general type was shown in Fig. 4.12, and a specific example from the kidney tubule of a vertebrate is shown in Fig. 5.13. This kind of highly differentiated cell, with luminal microvilli (the "brush border") giving a vast increase in surface area for water to follow the osmotic gradients created, and with basal indentations housing the pumps and linked with numerous mitochondria, can be found in osmoregulatory organs throughout the animal kingdom.

Hyposmotic urine, which rids the body of excess water, is a characteristic product of freshwater invertebrates and also of most freshwater vertebrates such as frogs or crocodiles. Its production again requires the active resorption of salts from the osmoregulatory tubules, but in this case as ion pumps remove salts from the urine it becomes progressively more dilute, since the P_w of the luminal membranes is kept quite low so that water is less able to follow osmotically. A hyposmotic fluid is either excreted continuously or gathers in a bladder for periodic ejection.

Animals on land generally need to conserve as much water as possible and therefore many of them have the opposite problem of producing a **hyperosmotic urine**. Secondarily marine vertebrates also need to conserve water, but as we saw earlier (section 5.3.2) only the marine mammals and birds can do this by producing a hyperosmotic excretory fluid. How can animals apparently remove water from their urine, back into their own more dilute body fluid compartment, when we have already established that water itself cannot be actively transported? The mechanisms involved turn out to be various modifications of the simple model shown in Fig. 5.14: a system of cyclical active transport of ions, combined with particular kinds of compartmentalized cellular and tissue architecture in association with differential membrane permeabilities. In some cases this is improved by the addition of a countercurrent flow component.

Concentration by compartmentalization

The model in Fig. 5.14 can operate on a subcellular or on a multicellular level to achieve apparent removal of water from a system and hence produce a hyperosmotic fluid in one compartment by secreting a hyposmotic fluid into another. The cells form part of an epithelium between the lumen of an organ and the blood. Each cell has three essential constituents: a large intracellular compartment

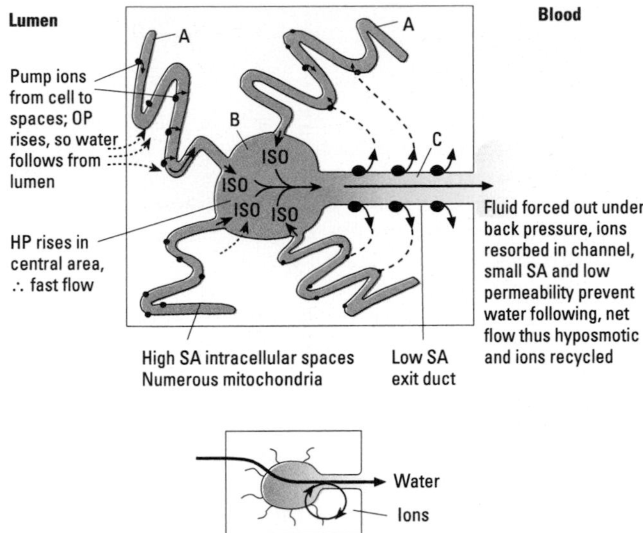

Fig. 5.14 A model for "water transport" across a cell within a resorptive epithelium, where solute recycling within the cell occurs and a fluid hyposmotic to the lumen is discharged into the blood, representing net water resorption. HP, hydrostatic pressure; ISO, isosmotic fluid; OP, osmotic pressure (or osmotic concentration); SA, surface area. See text for further details.

(B) that is continuous with the blood side via a relatively broad duct (C), and a series of fine convoluted blind-ending channels (A) near the luminal side of the cell, all of which lead into compartment B (Fig. 5.14). Cation pumps are concentrated in the membranes lining these fine channels, and pump Na^+ or K^+ into their narrow lumens, with anions following. Water is drawn osmotically out of the intracellular fluid into the channels, since their walls have a high P_w, and ultimately water is drawn down a gradient from the nearby organ lumen contents to balance this ion pumping. This creates an isosmotic fluid in A, which is continuously forced along the channels to accumulate in the intracellular compartment B. Both B and C have much lower P_w values. Fluid accumulating in B has nowhere to go except out to the blood via C, and the flow rate through C will be high because of hydrostatic pressure resulting from the much larger total surface area of the channels (A) producing all the fluid. As the isosmotic fluid flows down C further Na^+ or K^+ ion pumps operate, this time to remove ions from the lumen back into the cell. Because of the high flow, the much reduced surface area, and the relative impermeability to water in this part of the system, water has no chance to follow the ions isosmotically, and so the fluid in C becomes progressively more dilute, and by the time it discharges into the body fluid it is very severely depleted of ions. Thus the cell has in effect transported water from the organ and deposited it in the blood. The organ lumen contents become more and more concentrated. The ions that are pumped into intracellular channels at A are recycled back into the cytoplasm at C and very little net movement of ions occurs.

A specific example of cellular architectural compartmentalization is shown in Fig. 5.15, based on the resorptive cells found in the rectum of many insects, where the excretory fluid from the Malpighian tubules is finally concentrated. Here the cells remove water from the gut back into the body fluids, to produce a very concentrated excreta. A similar combination of architectural compartmentalization, where different parts of the system have very different surface areas and very different water permeabilities, can be found on a more complicated multicellular scale in the rectal papillae and pads of some insects, covered in Chapter 15. Mechanisms based on this general pattern are at work in other systems that appear to transport water "actively", including parts of the gut of most other terrestrial animals, but at present the insect rectal cells system is the only system where net cellular production of a hyposmotic fluid is really understood. Clearly water itself is not being actively moved at all; ions are pumped and water follows passively as always.

Additional concentration by countercurrent

A relatively small number of animals are able to achieve very considerable concentration of their urine by supplementing subcellular and cellular architecture with a pattern of gross anatomy whereby

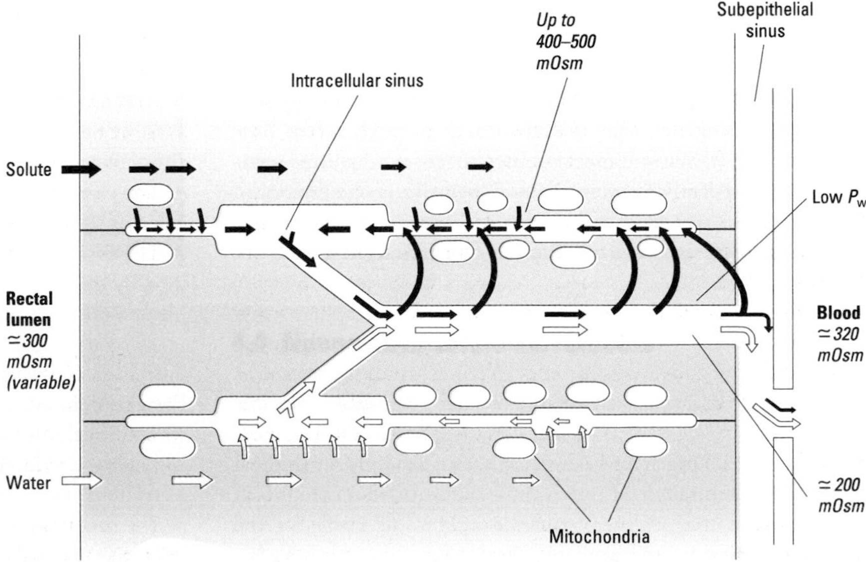

Fig. 5.15 A real example of a resorptive epithelium in the insect hindgut (the rectal cells of the cockroach *Periplaneta*) illustrating the principles of intracellular architecture and differential permeability outlined in Fig. 5.14. Note that solute and water movements take place in the *same* intracellular channels in reality. (Adapted from Wall & Oschman 1970.)

Box 5.2 Countercurrent exchangers and countercurrent multipliers

Arranging two flowing parallel systems so that the direction of flow in one is opposite to the direction of flow in the other is the most effective and simple way of insuring efficient exchange between the two, whether of **water, ions, gases,** or **heat.**

Initially (when flows are first "turned on") there is a large gradient of component X from fluid A to fluid B at all points in the system; but even after flow has continued for some time there will still be a small gradient from A to B throughout the length of the system. At the left-hand side, B has gained a large concentration of component X but still cannot be as concentrated as the incoming fluid A, and will continue to pick up more of X. Almost complete transfer of X from A to B occurs, at all but the highest of flow rates.

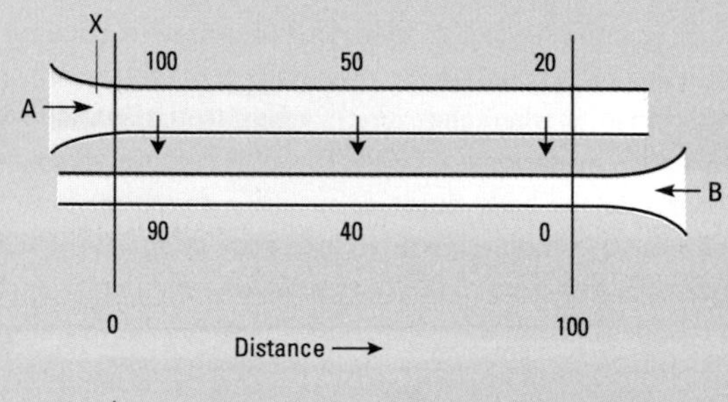

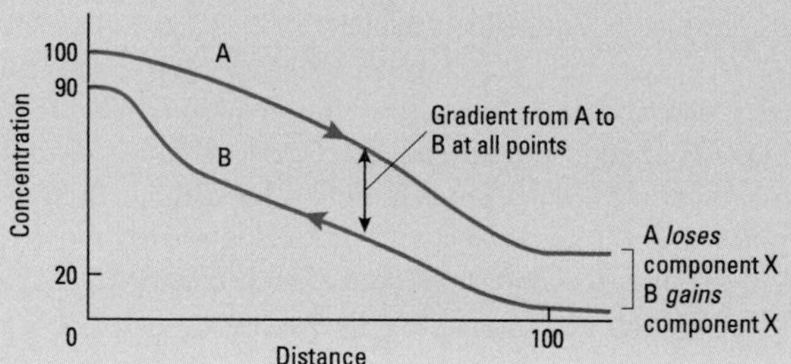

Exchangers

These work "downhill", acting as passive devices to achieve maximum uptake or exchange. They work for anything diffusible: whatever can pass across the walls will be transferred.

The fluid in the two channels may be:

1 The same, for example cold blood and warm blood, in many rete systems that keep parts of the body cooler or warmer than the rest (e.g. whale flipper, tuna swimming muscle).

2 Different, for example oxygenated sea water and deoxygenated blood (many fish gills); oxygenated air and deoxygenated blood (bird lung).

The two channels may be:

1 Quite separate, as in the gills and lungs.

2 Arranged as a loop, as in the whale flipper, where warm blood moving distally meets cooled blood coming back towards the body in the same continuous flow system.

The exchange may operate:

1 Spatially, as in all the above examples with two distinct channels.

2 Temporally, where there is just one channel but flow alternates in different directions (e.g. tidal flow in and out of vertebrate lungs). Cool air is warmed as it is drawn in, then loses most of its heat to the cooled passages as it is expired.

Multipliers

These work "uphill", needing input of energy, acting to concentrate a substance (which may be relatively nondiffusible) in some part of the body.

Therefore they *always* require a loop structure, *always* with the same fluid (blood) in both channels, and they are *always* spatial in character:

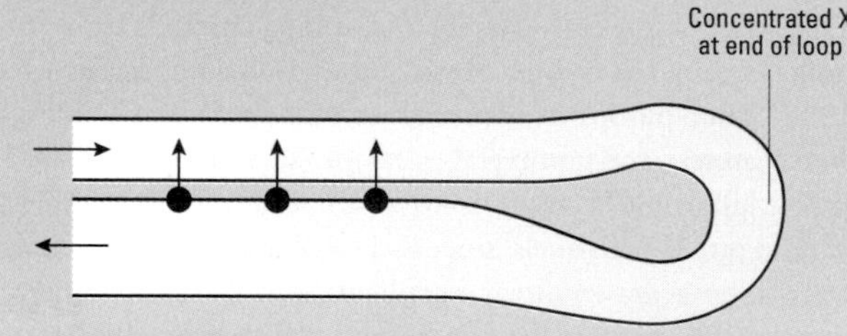

Examples include the loop of Henle in the mammalian kidney, and the gas-accumulating rete of the fish swim-bladder.

two parts of the tubule system run close together and in opposite directions, setting up what is known as a **countercurrent flow** system (Box 5.2). We will meet countercurrent mechanisms regularly in the field of environmental physiology, so it is very important to read this box and get the principles and advantages of **countercurrent exchange** and **countercurrent multiplication** clear from the outset.

In the case of osmoregulation, countercurrent multipliers based on active transport are set up in the excretory/osmoregulatory organs using the constituent tubules to create standing concentration gradients, and so resorb as much water as possible from the urine.

It is worth looking at an example in some detail. By far the best known, architecturally relatively simple but somewhat complex in mechanism, is that of the vertebrate kidney, or rather of the nephrons from which it is made up (Fig. 5.16).

VERTEBRATE KIDNEY FUNCTION

First we need to describe the main functions of the various parts of the nephron.

1 The glomerulus in Bowman's capsule produces the ultrafiltrate, filtering up to 25% of the blood per minute in an active mammal.

2 The primary urine is almost immediately reduced in volume by 75% by the proximal convoluted tubule (PCT), which has a classic "leaky" salt-transporting epithelium. This takes out NaCl isosmotically by the usual combination of ion pumping and passive water entrainment, the ions and water passing back into the efferent renal blood system. As it reaches the edge of the cortex the urine is therefore still at about 300 mOsm, isosmotic with the blood. But the 75% reduction in volume has led to a four-fold increase in concentration of relatively impermeant solutes such as urea.

3 The descending loop of Henle (DLH) contains simple epithelial cells with no specializations for transport, having very low P_{Na} and

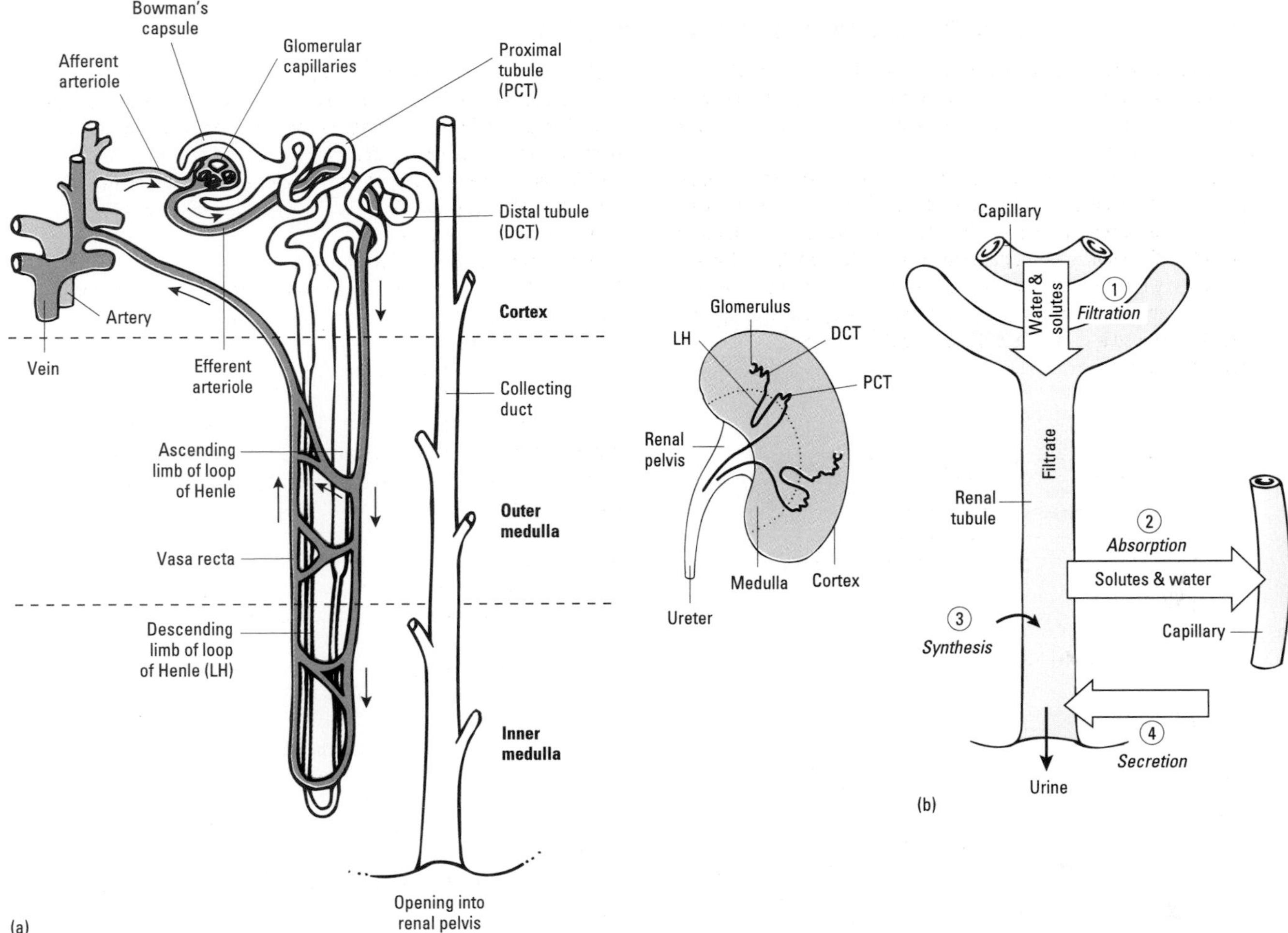

Fig. 5.16 (a) The basic structure of a mammalian nephron, and (b) a simple model of nephron function.

P_{Cl} and P_{urea}, but a high P_w. The first part of the ascending limb likewise has no indications of active transport but the cell membranes have high P_{Na} and P_{Cl}, low P_{urea}, and very low P_w. However, the upper ascending limb is thick walled with actively transporting cells moving Na and Cl from the tubule lumen to the surrounding interstitial space, in a manner similar to that described for the chloride cells earlier (see section 5.3.2). The effect is to make the interstitial space hyperosmotic. This region also has very low P_w so water does not move rapidly by osmosis to reduce the hyperosmoticity.

4 The distal convoluted tubule (DCT) is rather complex; it is concerned with the detailed modification of ion balance in the urine, usually transporting K^+, H^+, and NH_3 into the lumen and resorbing more Na^+, Cl^-, and HCO_3^-. The DCT is under precise endocrine control according to the animal's immediate osmotic and ionic needs (see Chapter 10); aldosterone from the adrenal glands increases sodium uptake here in particular. Water follows passively here, so again the fluid in the tubule is isosmotic with the surrounding cortex tissue.

5 The collecting duct (CD), into which many nephrons feed, has very low P_{Na} and P_{Cl}, but a variable permeability to urea. Some NaCl is actively resorbed here, but by far the most important effect is that as the urine descends through the medullary interstitial space, water leaves the duct osmotically down the steep concentration gradient created through the thickness of the medulla. To allow this the CD normally has a high P_w (although this is under the control of antidiuretic hormone (ADH), discussed below).

This brief review of the sequence of events brings us back to the question of how the medullary concentration gradient was set up. The medulla is made up of all the loops of Henle (which may vary considerably in length) from the thousands or millions of individual nephrons, and their common collecting ducts. In effect it is an assemblage of countercurrent multipliers. Microsampling of both ascending loop of Henle (ALH) and DLH fluids, as well as the interstitial fluids, shows a standing gradient of concentration (Fig. 5.17), increasing from the edge of the cortex towards the inner medulla, where fluids at the hairpin end of the loops can reach 1000 mOsm in normal mammals and more than 3000 mOsm in some species, especially those from arid habitats (see below).

This high concentration in the deeper medulla results from the permeability properties of the loop together with the ion transport in the ascending limb. Ion pumping from the upper ALH makes the interstitial fluids (ECF) more concentrated and thus also makes the fluid in the descending loop gradually more concentrated as water

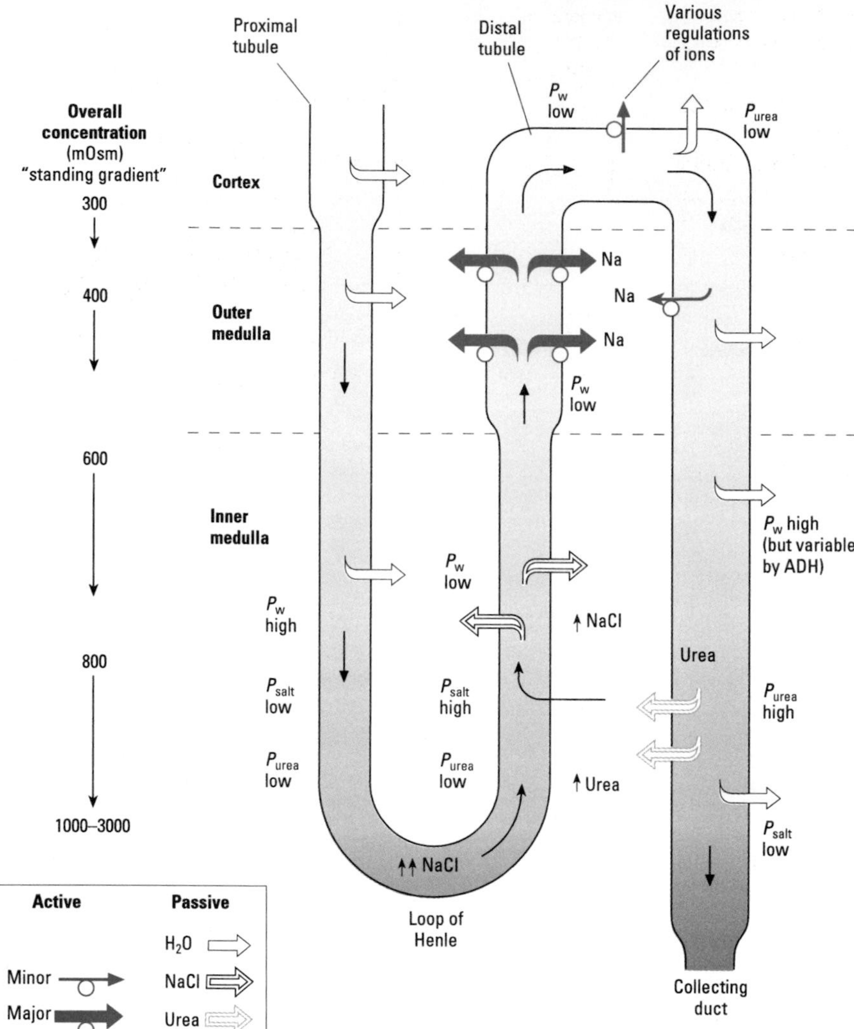

Fig. 5.17 The renal countercurrent multiplier system in mammals; how the standing osmotic gradient across the kidney medulla is set up and maintained by ion fluxes and differential permeabilities. Colored arrows indicate active transport and white arrows indicate passive flow. This is the state in a dehydrated mammal. In a hydrated mammal, the supply of antidiuretic hormone (ADH) from the pituitary declines and the distal convoluted tubule and collecting duct become almost impermeable to water so that a copious dilute urine leaves the kidney.

passes out of its relatively high P_w membranes to the more concentrated interstitial area. The counterflow of the two limbs of the loops insures that the hairpin area becomes and remains the most concentrated zone. As the very concentrated fluid ascends the loop again it will supply the pumps in the upper ALH with yet more Na^+ and Cl^- to work with, so that they can pump more ions out, get more water withdrawn, and so on, so that the single effect of the ion pump is multiplied by the anatomy of the system. Fluid ascending the ALH thus becomes gradually less concentrated, and will pass on to the cortex and the DCT at around 300 mOsm again.

This high medullary osmotic concentration is significantly supplemented by urea leaking out of the lower collecting duct cells as these ducts traverse the medulla. Until then, urea is retained within the urine by consistently low P_{urea} values, but the deeper areas of the collecting ducts are at last more permeable to this solute and it follows its concentration gradient into the medullary interstitial spaces. This movement of urea is now known to be facilitated by specific protein uniporters in the collecting duct membranes (see Chapter 4).

The high concentration resulting from the ionic concentration in the loops of Henle, and the urea concentration, inevitably has the desired effect of drawing water out of the urine in the collecting ducts and into the medullary tissue, just before the completed urine is due to leave the kidney, yielding the hyperosmotic product characteristic of most mammals and birds. Furthermore, the extent of this water loss from the collecting ducts can be simply controlled by regulating the P_w of the lower collecting duct cell membranes, via ADH levels in the blood. The increase in P_w seems to occur both by opening existing water channels and by incorporating subapical cytoplasmic vesicles that are particularly well endowed with water channels into the apical cell membrane; these channels have now been isolated and sequenced. ADH thus keeps the water channels numerous and open. When ADH is withdrawn water cannot leave the duct, and diuresis (copious dilute urine flow) results.

The water that is drawn into the kidney medulla from the collecting ducts is prevented from destroying the crucial concentration gradients by a parallel system of medullary blood vessels (the vasa recta), which are freely permeable to water and solutes and are also arranged as countercurrents (see Fig. 5.16). Thus the concentration of the blood follows that of the tissues of the medulla. However, the capillary walls do not let proteins pass, so that the blood (already relatively enriched in protein after passage through the glomerulus) has a colloid osmotic pressure that is sufficient to draw water into the bloodstream.

The net effect of all this structural complexity, ionic pumping, and variable permeation is that each nephron produces a small volume of urine at a higher concentration than the body fluids, and often even more concentrated than sea water. At the same time the efferent renal blood has become more dilute than the afferent supply and takes the recovered water away from the kidney. Once again, water has been removed from a biological fluid without there being any active handling of water itself.

Regulation of the **acid–base balance** (**pH**) is also achieved in the tetrapod kidney. Chemical buffering in the blood (by bicarbonate, phosphate, and protein buffer systems) is the major pH control system; but the kidney plays an additional role on a longer timescale of hours or days, and it is the *only* organ in a land-living vertebrate that can remove some particular acidic products generated by metabolic activities, notably phosphoric, uric, and lactic acids, and some ketones. The main strategy used is to vary the level of excretion, resorption, or generation of bicarbonate ions. These processes all depend on the rate of secretion of protons (H^+ ions) into the kidney filtrate, mainly in the proximal convoluted tubule (Fig. 5.18a). For each hydrogen ion excreted, one sodium ion and one bicarbonate ion are resorbed. The secreted H^+ combines with bicarbonate present in the tubular lumen to form carbonic acid (H_2CO_3), which then dissociates to CO_2 and water; the CO_2 diffuses back into the cells and will trigger further H^+ secretion. As long as the filtered bicarbonate is indirectly reclaimed in this manner, the secreted H^+ is not really "lost", because having turned into water it will mostly be resorbed later in its transit through the kidney. In effect, the rate of H^+ secretion varies with the CO_2 levels in the ECF of the kidney, these CO_2 levels being directly related to blood pH. The system can respond to either rising or falling pH levels. Only when the body has excess H^+ (usually derived from the diet) will the filtered bicarbonate all get used up, and then the H^+ that is secreted starts to appear in the urine, where it binds with buffers (mostly phosphates, especially HPO_4^{2-}) to prevent the urine becoming too acid itself. This part of the process happens mostly in the collecting ducts (Fig. 5.18b), where H^+-ATPases add hydrogen ions to the lumen, at the same time generating "new" bicarbonate ions that are added (by cotransport with chloride) to the blood in the capillaries. Thus when the animal's body is suffering acidosis, the kidneys generate new bicarbonate to alkalinize the blood as well as adding an equal amount of H^+ to the urine and thus acidifying the urine.

At first sight there is an enormous interspecific diversity in the structure and function of mammalian kidneys, and it would appear that the presence and the length of the loops of Henle in a vertebrate kidney should be related to an ability to produce concentrated urine. The first assumption is broadly true: groups that do not produce a hyperosmotic product usually lack the loops and have much simpler nephron structures than the xeric terrestrial birds and mammals (Fig. 5.19). The second assumption (that length is related to concentration) is not strictly true. Small desert mammals do produce very hypertonic urine, approaching 6000 mOsm in some species, and even in excess of 9000 mOsm in the Australian hopping mouse *Notomys*, but they do not have unusually long loops of Henle within the kidney. Filtration pressure in the glomerulus of a mammalian kidney does not increase with increasing size, and total nephron length is also relatively constant (limited by this constant filtration driving force). However, the filtration demand does increase in direct proportion to metabolism, which scales positively with body size. Hence the *number* of nephrons must increase as body mass increases, which results in changes in the geometry of the kidney, increasing the amount of cortex (where most of the nephron is sited) at the expense of the relative thickness of the renal medulla. Urine-concentrating ability should be a function of relative medullary thickness because of the mechanisms already described, and Fig. 5.20 confirms this point. So it might follow that larger mammals should inevitably be incapable of producing urine as hypertonic as that of small mammals. Thus body size alone could have major implications for kidney capability, which may explain why the ubiquitous house mouse, *Mus musculus*, can produce as concentrated a urinary product as some of its desert relatives.

This is a somewhat oversimplified view, because it confuses

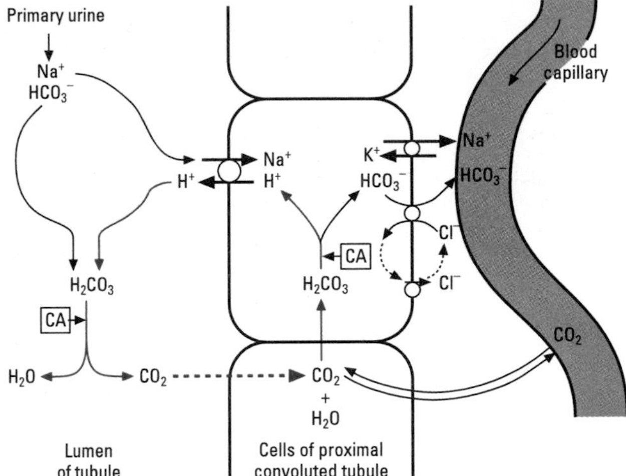

(a) Normal: HCO_3^- present in filtrate

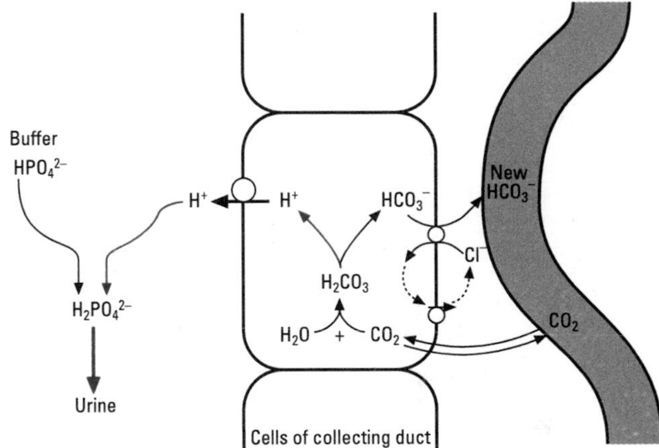

(b) Excess H^+: Filtrate HCO_3^- all used up

Fig. 5.18 The regulation of pH in the tetrapod kidney. (a) Resorption of filtered bicarbonate is coupled with the secretion of hydrogen ions into the filtrate. HCO_3^- in the filtrate is thereby in effect transferred into the capillaries. (b) Where H^+ ions are in excess they are buffered by phosphate in the urine and "new" bicarbonate is added to the blood capillaries. CA, carbonic anhydrase enzyme.

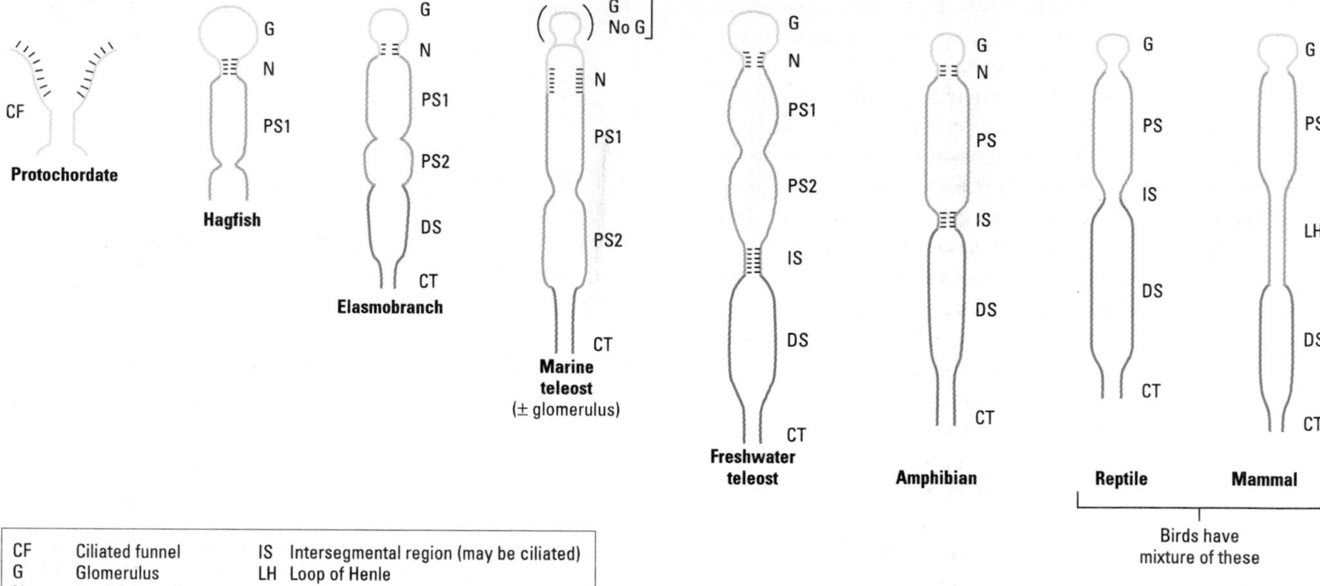

Fig. 5.19 A comparison of the structure of kidney tubules in different taxa of vertebrates. In general, the more "difficult" the environment the more complex the tubule becomes.

relative size and absolute size. Other things being equal, it should be the absolute length of a tubule loop that matters, not its relative length. *Notomys* has a loop length of only 5.2 mm, whereas a horse has loops seven times longer yet produces urine only up to 1900 mOsm. In fact it turns out that there is no real relationship between absolute loop length and urine concentration for mammals; we need to direct our analysis at a more cellular level to understand urine concentration. Small mammals have higher mass-specific metabolic rates (see Chapter 6), and within their kidneys the tubule cells have more densely packed mitochondria, with denser internal cristae, and more elaborate lateral membrane foldings. All this indicates that small mammals should have a much more intense capacity for the active transport mechanisms on which the urine-concentrating system depends. Thus a mouse kidney could maintain a much stronger osmotic gradient along the loop of Henle than a horse kidney, achieving higher urine concentration with a smaller kidney.

Nevertheless, Fig. 5.20 does reveal that many small rodents living in arid (xeric) habitats do have relatively thicker medullas than other mammals, which correlates with their concentrated urine, and which could be viewed as a "desert adaptation" superimposed on a basic body size-dependent pattern. Desert rodents tend to have almost exclusively long loops and very few short ones, with a high capacity for active transport per unit length, explaining the greater gradients and higher urine concentrations produced. Aquatic mammals have very few long loops, and temperate mammals may have only 25–50% of their loops dipping deep into the medulla. Thus the relations between medulla or loop dimensions, body size, and habitat are not nearly as clear-cut as elementary texts would have us believe—this serves as an object lesson in letting enthusiasm for scaling-based explanations (see Chapter 3) exceed their legitimate limits.

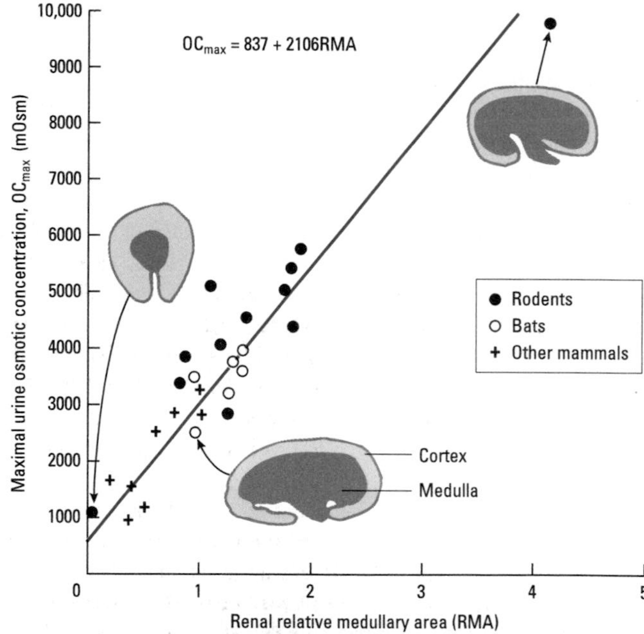

Fig. 5.20 The relationship between the relative medullary area in the mammalian kidney (taken at the midline in sagittal section) and the maximal urine concentration that can be produced. In general, proportionally larger medullas produce a more concentrated product; note that the relatively largest medullas and very concentrated urines mostly occur in rodents, which are of small body size and are common in arid habitats. Insets show diagrammatic cross-sections for three kidneys (not to scale). (From *Comparative Animal Physiology*, 1st edn, by Withers © 1992. Reprinted with permission of Brooks/Cole, a division of Thomson Learning: www.thomsonrights.com. Fax 800 730-2215.)

RECTAL COUNTERCURRENT CONCENTRATION SYSTEMS

One further example of concentration of excreted fluids by countercurrent occurs in the cryptonephridial system found in the rectum of certain xeric insects. The system is shown in diagrammatic form

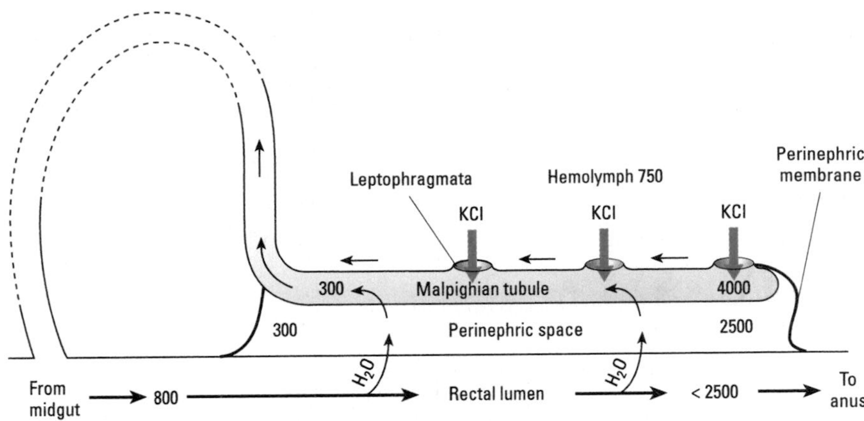

Fig. 5.21 The essential principles of concentration by countercurrent in the cryptonephridial system of certain insects. Values shown are in mOsm.

in Fig. 5.21. Essentially the Malpighian tubules, which in most insects lie freely in the hemolymph, are here folded back to lie against the rectal walls, with fluid flowing forward in them counter to the rearward fluid flow in the rectum. The whole array is enclosed in a very impermeable "perinephric membrane". At a few sites, this membrane is broken by specialized cells (leptophragmata) on which all the KCl-transporting mechanism of normal Malpighian tubules is centered, secreting the primary urine constituents into the tubules. But the impermeable perinephric membrane prevents the usual entrained water flow from the hemolymph, and KCl accumulates at high concentrations. Thus it is able to withdraw water osmotically from the rectal contents instead, even when these get quite concentrated, and in effect a standing gradient is set up along the complex such that water is withdrawn along a small osmotic gradient at all points in the rectum. Even close to the anus, the highly concentrated fluid in the gut still loses water to the even more concentrated fluids in the perinephric space and tubule, and ultimately to the blood. More details of this system in particular insects are given in Chapter 15.

5.4.4 Final urine composition

After all the processes described above, urine is passed from the body, often via a storage bladder in amphibious and terrestrial animals—partly because for many of these animals intermittent urination takes place only in prescribed places (middens) or as a territory marking system. Occasionally the bladder is used as a site of final modification; for example, acidification in the reptile bladder by the unusual V-ATPase H^+ pump (see Chapter 4) is regulated in part by CO_2 levels, which cause vesicles bearing the ATPase to move to the apical membrane of the bladder epithelium.

Table 5.6 gives examples of the range of (and/or maximum) urine concentrations achieved in different taxa living in different habitats, including the net urine : plasma (U : P) ratio to clarify the concentration relative to the blood. The best examples of the insect osmoregulatory system are capable of producing urine at around 7000–9000 mOsm; and in some cases where the rectum is used for vapor uptake it is at least notionally resorbing at around 30,000 mOsm concentrations.

Table 5.6 Urine concentrations and urine : plasma (U : P) ratios in animals from different taxa and habitats. For land animals the values are generally maximal rates from dehydrated animals.

Taxon	Habitat	Urine concentration (mOsm)	U : P ratio
Crustaceans			
Lobster	Marine	1050	1.0
Crab	Marine	1020	1.02
Crab	Freshwater	550	1.04
Crab	Littoral	1200	1.17
Molluscs			
Cuttlefish	Marine	1100	1.02
Elasmobranchs			
Dogfish	Marine	800	0.8
Teleosts			
Goby	Marine	406	0.9
Flounder	Marine	375	0.96
Reptiles			
Crocodile	Amphibious	267	0.8
Tuatara	Mesic	270	1.0
Iguana	Mesic/xeric	362	1.0
Desert tortoise	Xeric	337	1.0
Birds			
Pigeon	Mesic	655	1.7
Pelican	Marine/mesic	700	2.0
Ostrich	Xeric	900	2.7
Savanna sparrow	Xeric	2020	5.8
Small mammals			
Rat	Mesic	2900	9
Domestic cat	Mesic	3100	10
Vampire bat	Mesic/xeric	4650	14
Kangaroo rat	Xeric	5500	16
Large mammals			
Beaver	Amphibious	520	1.7
Man	Mesic	1400	4–5
Porpoise	Marine	1800	5
Eland	Xeric	1880	6

Excretory product	Formula	M_r	Solubility (mM)	Diffusion coefficient	H : N ratio	C : N ratio	Toxicity
Ammonia	NH_3	17	52.4	High	3	0	High
Urea	$CO(NH_2)_2$	60	39.8	Moderate	2	0.5	Moderate
Allantoin	$C_4H_6O_3N_4$	158	0.016	Low	1.5	1.0	Low
Guanine	$C_4H_5ON_5$	151	0.0013	Low	1.25	0.8	Low
Uric acid	$C_5H_4O_3N_4$	168	0.0015	Low	1.0	1.25	Low

Table 5.7 Nitrogenous excretory products and their characteristics.

5.4.5 Nitrogenous excretion in "excretory" systems

Comparative biology of nitrogenous excretory products

There is a relatively small range of nitrogen-containing excretory products in animals, with three principal molecules dominating the field. These are **ammonia**, **urea**, and **uric acid**. Their chemical formulae and properties, together with those of a couple of much less common waste products, are shown in Table 5.7.

When proteins and their constituent amino acids are metabolized, the amino group that is removed forms ammonia. Ammonia is the ideal waste product for most aquatic animals, with a high nitrogen content making it the least wasteful of useful organic carbon structures; its high toxicity does not matter in aquatic habitats as there is plenty of water available to dissolve it and disperse it away from the body. Often the osmoregulatory organ is not involved at all, with ammonia diffusing away from any permeable surface (very often this will mainly be the gills). Animals that use ammonia as their main excretory product are termed **ammonotelic**. They include most aquatic invertebrates, including the aquatic larval phases of many insects, plus the cyclostome and teleost fish. In many of these animals the ammonia combines with H^+ to form NH_4^+, and this is used as an exchange for Na^+ pumped in at the gills, reducing the electrical work required. In many fish up to 90% of the excreted ammonia leaves via the gills rather than the kidneys. Aquatic representatives of other normally terrestrial groups may also excrete a high proportion of ammonia, for example most aquatic insect larvae, the tadpole larvae of frogs and toads, and the crocodiles. Amphibians that remain aquatic throughout their life, such as *Xenopus* and the axolotl, are persistently ammonotelic. Most of the terrestrial woodlice also excrete ammonia, releasing urine from the maxillary glands (where urea is stored as glutamate, which is deaminated just before release to form NH_3) and spreading it over the flattened appendages (pleopods) on the ventral surface of their body, allowing gaseous ammonia to diffuse away. In some species the watery urine is then resorbed in the rectum. Just occasionally, truly terrestrial animals still produce gaseous ammonia: guano bats eliminate a moderate percentage of their excess nitrogen as NH_3 through the lungs, and some hummingbirds can switch to ammonotely in cool conditions when dilute nectar is freely available to give a plentiful water supply.

Urea is something of a compromise product, synthesized from the initial ammonia in combination with carbon dioxide via the urea/ornithine cycle (Fig. 5.22). It has a moderate nitrogen content and intermediate solubility and toxicity, so it can be dissolved at fairly high concentrations and thus requires little water for its excretion. It is commonly used by larger land vertebrates (mammals, amphibians, and some reptiles) and by cartilaginous fish, and it occurs sporadically elsewhere in the animal kingdom, for example in intertidal fish like blennies, in some earthworms and land snails, in freshwater turtles, and in some animals from stagnant freshwater habitats. All these animals are described as **ureotelic**. Some of the ureotelic vertebrates use the synthesis of urea as a means of generating protons that contribute to pH homeostasis, and in these cases (notably the ruminant mammals) up to 50% of the liver ATP function may be devoted to urea production.

Uric acid is a purine, synthesized from various amino acids and from the general carbon pool of a cell. It is extremely insoluble and can be produced as an almost solid paste without toxic effects on adjacent living tissues. It is therefore ideal for land animals that are smaller and often particularly water stressed, including the insects, the terrestrial snails, the birds and most smaller reptiles, and a few arid frogs; these are described as **uricotelic**. Where the excretory product is nominally uric acid, an extra bonus is achieved in that all urate salts are also very insoluble, so that cations can be added to the precipitated paste of excreta and the salts are lost with almost no associated water loss. Some insects exploit this by producing NH_4^+-urate, thus unloading ammonia without any necessary water loss. Terrestrial vertebrates can also lose less water by excreting Na^+ and K^+ as urates from the kidney than they would by using their salt glands. This is true for reptiles; indeed, some xeric reptile species that excrete urates have dispensed with salt glands altogether. However, in birds the usual excretory product is a colloidal gel with a rather high water content, in which uric acid represents little saving over the use of urea. Urate is useful for birds at the embryonic stage, however—it is an ideal excretory product for any animals with relatively yolky and impermeable eggs, where a solid paste of excreta can be left behind within the eggshell so that it is not an encumbrance to the hatchling.

The distinction between these three excretory modes is very far from absolute, and many anomalies occur. It is evident from the distribution of these products that the use of particular excretory products is largely related to habitat (water availability) and not to phylogeny. For example, different species of frog can exhibit any of the three modes, with the xeric genera *Chiromantis* and *Phyllomedusa* (both almost independent of water, with special reproductive tricks to avoid the need to return to ponds) producing substantial amounts of uric acid. The same versatility is found in turtles, with the more terrestrial species producing at least 50% uric acid, and with switching possible even within one individual according to its state of hydration. However, even the correlation between habitat and excretory product cannot be taken too far: uric acid turns up in some entirely marine species, for example some prosobranch snails, and also in some freshwater members of the same taxon, whilst some terrestrial snails can excrete gaseous ammonia through their mantle epithelium.

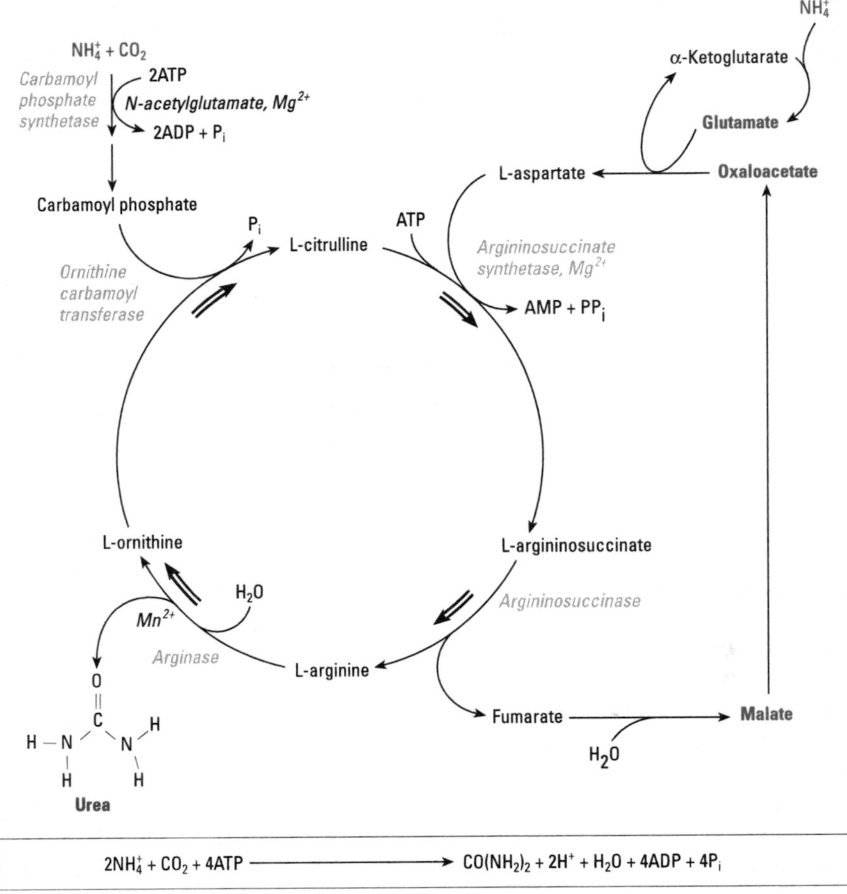

Fig. 5.22 The urea cycle, or ornithine cycle, whereby urea is synthesized from ammonia and carbon dioxide by condensation with ornithine, which is then reformed by a cyclical process.

A few other nitrogenous residues occur less commonly in animal excreta. Some are breakdown products of nucleic acids (the purine and pyrimidine components), rather than of protein. Creatinine is one example, occurring in small quantities in vertebrate urine. Allantoin and allantoic acid also result from purine degradation, and occur in many molluscs and insects, as well as in most mammals except primates. In freshwater insects these products may become the major excretory output, having a greater solubility than uric acid. Another important excretory product is guanine, accounting for up to 90% of the nitrogen excreted in spiders, and occurring in crystalline form in the scales and swim-bladders of some fish, where it reduces the surface permeability. Terrestrial snails also excrete some guanine. Finally, some animals produce small amounts of trimethylamine as a breakdown product of choline metabolism, and trimethylamine oxide (TMAO) is then retained in the blood of some elasmobranch fish (sharks and rays) as an important osmolyte (see Chapter 11).

Excretion of these products

In general, the main nitrogenous excretory product of an animal gets into its urine passively, either in the ultrafiltrate or during bulk flow of water in a secretory urine production system. This is normally true whether that product is ammonia, urea, or uric acid (though urates are at least partly actively transported in insect systems). Final concentration of the waste product in the urine is largely determined by the permeability properties of different parts of the tubule system.

However, there are several ways of increasing or of regulating the flow of nitrogenous wastes. One of the most important involves urea; in some animals that can survive prolonged dehydration when urine flow almost ceases, some urea may still be added into the osmoregulatory/excretory tubules by secretion. This is known to occur in some frogs. Shark kidney tubules can also handle urea actively, but in this case it is active resorption that occurs, to keep the high urea blood level used as an osmotic balancing strategy in these fish.

In some snails there is a zone of specialized secretory cells at the distal end of the kidney, and here active uric acid secretion can occur during periods of dehydration. Other water-stressed terrestrial animals, particularly some insects with relatively short adult lives, give up excreting the uric acid at all and resort to depositing it in crystalline form in various parts of their body ("storage excretion"), sometimes using it as a cuticular pigment.

In some animals, tubular filtration and secretion systems for ridding the body of nitrogenous waste are supplemented by tubular synthesis. Some excreted compounds are actually made within the epithelium of the kidney, often by deamination processes resulting in the formation of ammonia (NH_3), which then diffuses into the urine and combines with H^+ to form NH_4^+ ions ready for excretion. This is an important process in both mammals and insects, where it is particularly noticeable in the hindgut of carrion-eating larvae such as the flesh-fly *Sarcophaga*.

Sequestration of these products

Some animals, particularly arthropods, are able to cope with varying states of desiccation by regulating the patterns of sequestration of their excretory products. The cockroach *Periplaneta* responds to dehydrating environments by sequestering sodium ions together with uric acid as sodium urate crystals in the fat body, with chloride apparently stored in the hindgut; thus no water need be wasted in excreting these surplus and accumulating osmolytes. Urate crystals are accumulated by a wide range of insects—this strategy is particularly common in juvenile stages, as the excess urate can then be "dumped" with the old cuticle at the next molt. Some animals also accumulate and store their own urine for prolonged periods; pulmonate snails can do so in the mantle cavity, and some anuran amphibians use the bladder for similar purposes. Such stores may then become sources of water during subsequent drought, particularly in desert species.

5.5 Water regulation via the gut

In those aquatic animals where drinking large volumes of the medium is a major part of their osmoregulatory strategy, the gut clearly becomes a frontline osmoregulatory organ. The same is true, though less immediately obvious, for all noncarnivorous animals; for whereas "meat" is roughly the same in composition as the tissues of the animal eating it, all other food sources are dissimilar from animal tissue. Since the food intake is therefore not in osmotic and ionic equilibrium with the body tissues, the gut lining may be used as a barrier against excessive ionic stress. Guts may therefore play a crucial role in osmotic and ionic regulation.

In the simplest invertebrate animals, such as sea anemones, hydras, and flatworms, the primitive gut (enteron) is the only fluid-filled cavity within the body and inevitably has a vital role in the regulation of salt and water balance, especially for those species that live in estuarine or freshwater habitats. The gut is a blind-ending sac enclosed by the body wall, with a single opening; thus there is little opportunity for compartmentalization or specialization. The primary function of digestion is inevitably compounded with that of ion and water regulation, and compromises have to be reached in a multifunctional epithelium.

In all other animal groups a separate anus is added to the system so that the gut becomes tubular, and this opens up new possibilities for modifying the contents differentially in successive areas. It also allows for altering conditions within the gut by adding secretions, so that enzyme functioning can be tuned to local conditions. Digestion can occur at a specific pH in certain zones, and ion regulation or water resorption can be localized at a different pH in other areas. In the more complex invertebrates, a variety of ceca and "digestive glands" are frequently added as side branches of the gut, which becomes an integrated part of the iono/osmoregulatory system of the whole animal. This is seen most clearly in insect herbivores and in terrestrial herbivorous vertebrates, where the gut becomes extraordinarily complex (see Chapter 15). In all these animals with through-guts, a series of functions are performed from anterior to posterior, by specialized areas that show a degree of similarity throughout the animal kingdom.

5.5.1 Water balance in various gut regions

Salivary glands

In most animals, salivary products serve to lubricate the food as it passes through the mouth and esophagus, and often begin the process of food breakdown with one or more enzymes. Thus the products of the salivary glands represent no net loss to the animal as they can be recycled through succeeding parts of the gut. However, in some cases the saliva also has an excretory and osmoregulatory role, particularly for animals with a blood-feeding habit. Ixodid ticks, for example, eliminate most of their excretory water, Na^+, and Cl^- by injecting it into the host as they feed; sodium from the blood meal is rapidly taken up from the midgut to the hemolymph, and is then pumped into the salivary lumen (Fig. 5.23). In other arthropods, water stored in the salivary glands may be used as an emergency resource when an individual is water stressed; termites and some fly larvae appear to be able to mobilize water from this source when needed.

Midgut and gut ceca

Midgut epithelia (equivalent in vertebrates to the stomach and a large part of the intestine) are the most important sites of digestive enzyme secretion, and of food absorption, in nearly all animals. But they are also important centers for ion and water absorption, and ion regulation, in many taxa. In most animals at least 75% of ingested water, and more than 95% of ingested monovalent cations, are taken into the body here.

Complex ion cycling occurs in the midgut, especially in the pterygote insects (Fig. 5.24), with cycles of solute absorption being coupled by symporters to ion movements, and the ions being returned to the system at a different point in the cycle, often in a blind-ending gut cecum. In fact the midgut serves as the main excretory/osmoregulatory organ in many apterygotes; these primitive insects lack Malpighian tubules and use the gut in concert with labial glands opening near the mouthparts and acting as "kidneys".

Systems to bypass the midgut absorptive epithelium are often found in animals that are at risk of getting too much water from their diet. This is commonly the case in plant sap feeders taking in xylem fluids. Many xylem-feeding insects and mites therefore have gut bypass arrangements (Fig. 5.25), where a "filter chamber" brings about close contact (in countercurrent) between the anterior midgut, Malpighian tubules, and anterior hindgut. Much of the water taken into the gut therefore bypasses the main midgut epithelium and passes directly from the anterior midgut to the rectum, leaving a more concentrated fluid in the rest of the midgut from which digestion and absorption can occur.

Rectum

In most animals, and especially in terrestrial species, the hindgut or rectum becomes a major site of water balance regulation, serving as a site for water resorption from the feces (and, in many arthropods, from the excretory fluid that has been added to the gut as well). Solute-linked water resorption, described above, is a primary function here in all terrestrial animals. In birds and reptiles, part of the

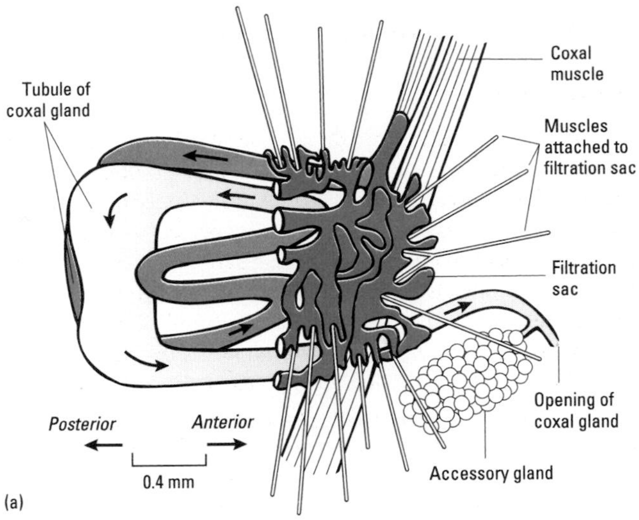

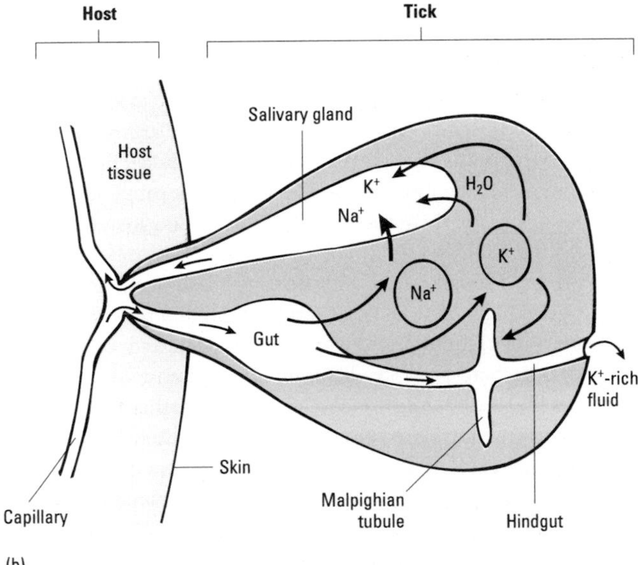

Fig. 5.23 (a) Structure and (b) function of coxal salivary glands in a tick; about 96% of sodium and 74% of water is excreted by this route during feeding. (a, From Little 1983, courtesy of Cambridge University Press; b, from Kaufmann & Phillips 1973, courtesy of Company of Biologists Ltd.)

hindgut (the "integrative segment") is a major resorptive area, so that the resultant precipitation of uric acid further reduces water loss.

In insects even more efficient solute-linked water resorption is possible in the rectum, by mechanisms discussed above. This may be facilitated by the presence of a cuticular lining in the insect hindgut, protecting the rectal cells from the high concentrations of larger fecal toxins that will result from such an extreme degree of concentration as water is removed. In some cases the rectum can even be used for absorbing water vapor from air. However, many arthropods have a very different strategy and use their hindguts as a water store for emergency use. In isopods, semiterrestrial crabs, millipedes, and many juvenile and adult insects the gut contents are usually rather dilute, and become significantly more concentrated during times of water stress as the stores are used up.

Integration of gut and excretory functions

The most xeric animals, especially the insects and arachnids from desert habitats, have achieved their success at least partly by integrating most of their salt and water regulation together, with the excretory tubules disgorging their watery product into the gut and letting the hindgut epithelium do most of the resorption. Some relatively xeric earthworms (Australian megascolecids) have independently arrived at the same solution. Most birds and reptiles use their cloacal region (the site where reproductive, rectal, and urinary outputs come together) rather similarly to make further changes to the excreta, being able to reflux urine through the cloaca and back into the integrative segment of the hindgut for further modification. In the avian cloaca, ions present as urate salts may be released by the action of bacteria, with Na^+ and NH_4^+ resorbed into the tissues in exchange for K^+ ions. Remember also the woodlouse trick of allowing urine to flow over the ventral pleopods and then resorbing the fluid via the rectum.

5.6 Regulation of respiratory water exchanges

Respiratory surfaces always have gases diffusing across them, and water therefore inevitably also moves across these surfaces in any situation where there is a water gradient—i.e. in *all* animals except the isosmotic marine invertebrates. Even in the most xeric animals, the respiratory surfaces are therefore always moist; this has sometimes been described as a necessity for gas exchange but actually represents an increased barrier to rapid gas exchange. In nonisosmotic aquatic animals, the solution has been to make the best of the situation and use the permeable gill surfaces as the main site of ion and water regulation, taking up or secreting salts there. The general evolutionary trend in nonaquatic animals has been to replace external and exposed respiratory structures with internalized ones where water balance can be better controlled. In terrestrial animals from very moist habitats (e.g. earthworms and many amphibians), cutaneous respiration is sometimes an adequate solution. But in more xeric forms either gills are concealed within enclosed chambers or they are replaced with either tracheal systems or lungs, entirely within the body and with very restricted openings. Details of these respiratory systems are covered in Chapter 7; here we are concerned only with their role in water loss.

Oxygen uptake, carbon dioxide excretion, and evaporative water loss are all essentially diffusive processes. The ratio of water loss to oxygen uptake for any exposed respiratory surface operating in air can therefore be relatively easily calculated from the ratios of the diffusion coefficients for the two compounds (which works out at about 1.3), multiplied by the ratios of the gradients for the two compounds. Figure 5.26 shows the situation for a cutaneous surface, a lung, and a tracheal tube. At a simple cutaneous surface exposed to air the value obtained would be about 2.1 mg H_2O lost per ml O_2 taken up. Any terrestrial system of respiration that can improve on this ratio will be selected for. Invaginated systems achieve this by greatly reducing the partial pressure of oxygen (Po_2) at the final

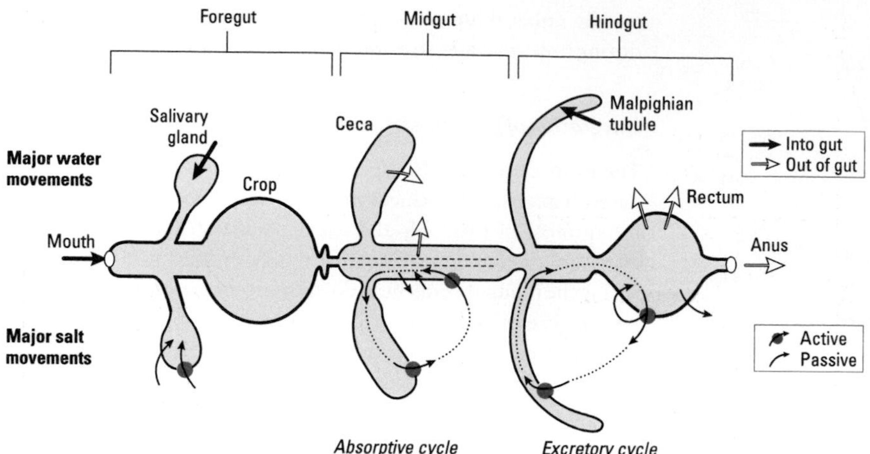

Fig. 5.24 A summary of water (above) and solute (below) movements in the insect gut.

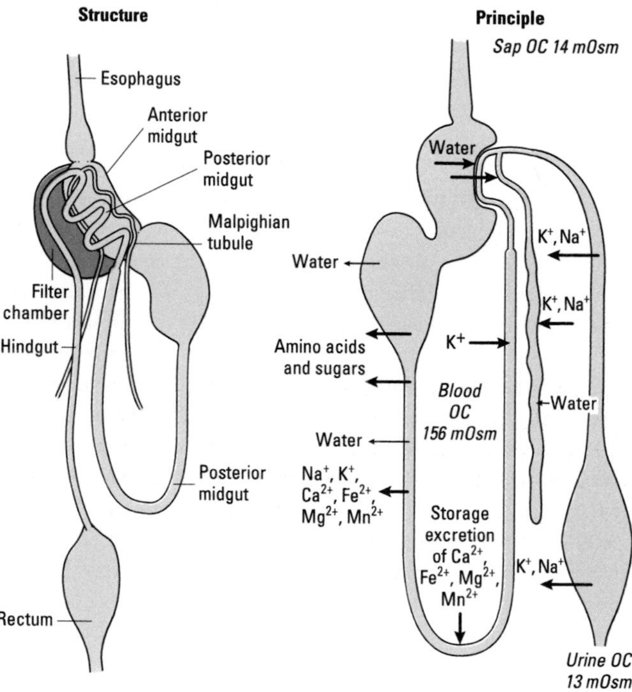

Fig. 5.25 The structure of a "gut bypass" system in a sap-feeding insect, whereby much of the fluid taken in is filtered directly to the hindgut, bypassing the midgut resorptive areas. OC, osmotic concentration. (Adapted from Cheung & Marshall 1973.)

air–fluid interface, so making the O_2 gradient steeper and increasing the difference between the water gradient and the oxygen gradient.

Lungs and tracheae should therefore both represent a major advantage in terms of water conservation (Fig. 5.26b,c). Most of the simple lungs found in invertebrates achieve some Po_2 reduction and thus some water saving, but in vertebrate alveolar lungs this effect is particularly striking, as the Po_2 at the internal alveolar surfaces is reduced to around half of its atmospheric level, and the ratio of water loss to oxygen uptake is thus considerably improved compared with cutaneous breathing, to about 0.2–0.3 mg H_2O lost per ml O_2 taken up (a 90% saving). The essential difference between lungs and tracheae is that in the latter the air is delivered directly to the tissues in the gaseous phase, and therefore the Po_2 deep within the tracheoles can fall very low, especially when the spiracles are closed, giving a further improvement in the relative water and oxygen gradients (Fig. 5.26c). Theoretically this would allow a water loss as low as 0.1 mg per ml O_2 taken up. A recent analysis showed that the interspecific variation in WLRs from arthropods with tracheal systems was strongly related to global rainfall; but also that WLR and metabolic rate were much less tightly coupled in desert species, so that for them respiratory transpiration losses becomes relatively more important.

Inevitably, though, water loss from the lungs and tracheae is always quite high, because large surface areas and slow flow internally are always required, and equilibration of air and water occurs at the deep exchange surfaces. Thus moist air, almost invariably saturated, is breathed out. More xeric land animals, therefore, have to operate further controls over the respiratory openings, and may add condensers and countercurrent exchange systems to the respiratory flow, to regain as much of the water as possible from the expired air.

The main water-related adaptations of terrestrial vertebrate respiratory systems are clearly concerned with recovering some of the expired water, and the obvious way to do this is to cool the exhalant air. Any cooling of the expired air relative to its temperature at the respiratory surface of the lung will cause it to give up moisture as it leaves the body, since saturated cold air holds less water vapor than saturated warm air. Nasal countercurrents to achieve this can be found in almost all birds and mammals (the rare exceptions are animals with unusually modified "large" noses such as elephants and pelicans). At their best, where the nasal passage is elongated and the internal anatomy highly specialized, the nasal **respiratory turbinals** of vertebrate endotherms can substantially reduce respiratory water loss. The principle and the results are shown in Fig. 5.27. Incoming air is warmed in the nose before it reaches the alveolar surfaces, and this draws some heat out of the nose, leaving its surfaces cooler. Expired air is subsequently cooled by heat exchange with the cooled nasal mucosa, to an extent depending upon the anatomy of the system, so that the final expired air temperature can be considerably lower than the animal's core body temperature. Hence water condenses in the nose, which is a site of countercurrent recovery both for water and for heat. We will therefore meet this system again

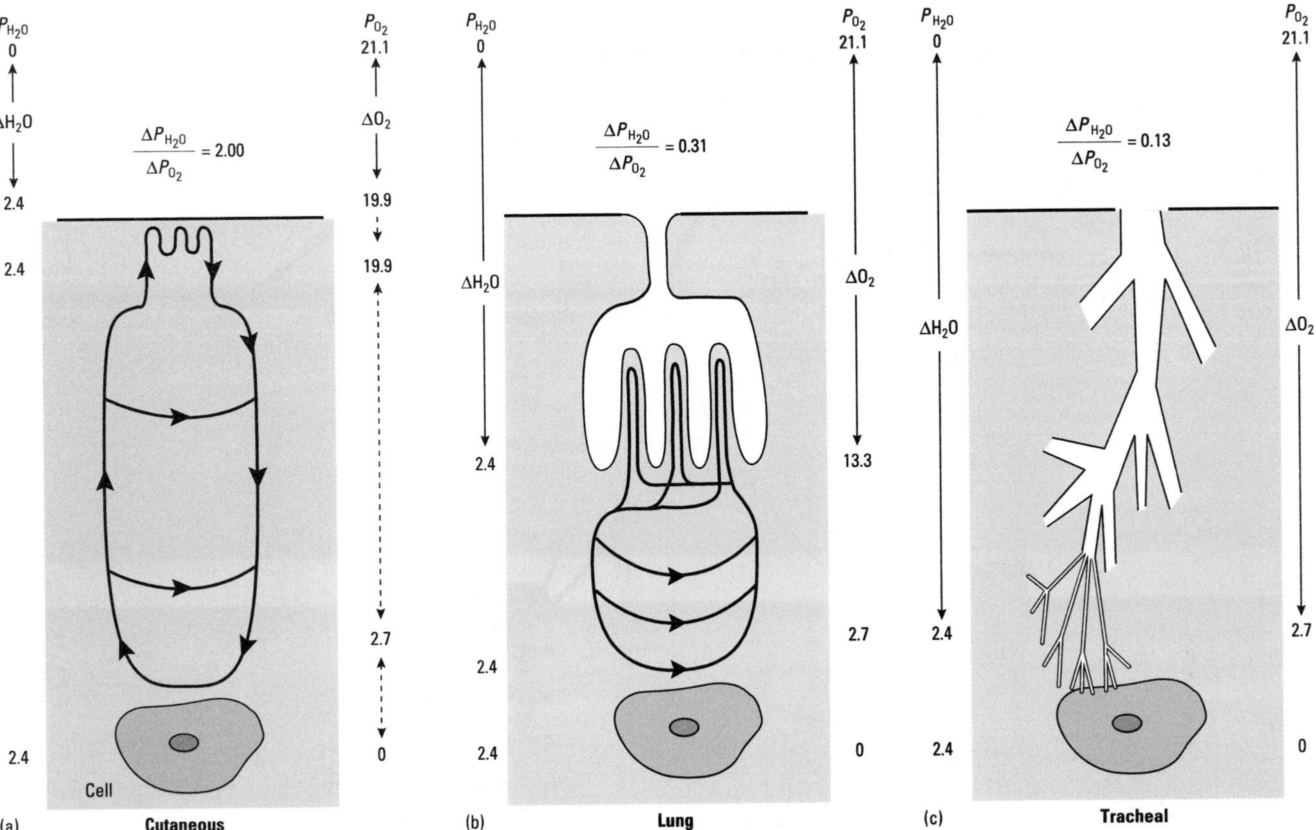

Fig. 5.26 The relationship between diffusive oxygen uptake and water loss across respiratory systems. (From *Comparative Animal Physiology*, 1st edn, by Withers © 1992. Reprinted with permission of Brooks/Cole, a division of Thomson Learning: www.thomsonrights.com. Fax 800 730-2215.)

in considering thermal biology in Chapter 8, and specific cases of it in operation in terrestrial animals in Chapters 15 and 16.

In tracheal systems air is supplied in the vapor phase via finely divided tubules, almost to the point of uptake in the tissues. However, the final blind endings of the tracheoles are fluid-filled, and at this point gases go into solution and the final uptake of oxygen is largely in the fluid phase. But there are some more sophisticated features that can be added to improve control and limit water loss from tracheal systems:

1 Directional control of ventilation. Considerable heat is generated in a flying insect, principally in the thorax as this is the site of the main locomotory muscles, so the air there is warmed in passing over the thoracic tracheolar surfaces. As it exits rearwards the air will then encounter cooler tissues, so that some water from the tracheal system should be recondensed in the abdomen. This is analogous to the nasal cooling system discussed above for vertebrates, though the evidence for it is less secure.

2 Control of spiracular openings. These can be controlled in response to levels of respiratory gas as a means of regulating for respiratory need, or in response to osmotic concentration or ambient relative humidity, both of which impact more directly on water balance. In flight, the metabolic demand for oxygen and the need to conserve water are strongly in conflict and the spiracle opening has to be finely adjusted. In flying *Drosophila* careful spiracle control results in a 23% saving of water compared with having spiracles fully opened. In resting insects though (and also in some chelicerates and centipedes with tracheal respiration), reduced spiracular access can be taken to extremes by making breathing discontinuous, with very occasional spiracle gaping and thus significant water savings.

Examples of these mechanisms are dealt with in detail in Chapter 15.

5.7 Water loss in reproductive systems

Only in terrestrial animals is water loss associated with reproduction a severe problem, and there is a standard solution in all taxa that are "properly" terrestrial and breed away from water. They all adopt effectively internal fertilization, either using spermatophores or direct internal insemination. In the first case, relatively impermeable spermatophores containing a little fluid and a package of sperm are left by males in some suitable spot and are then picked up by females; or the spermatophore package may be inserted into the female (often into her reproductive system, but sometimes at any point on her body wall), where it breaks down and releases the sperm. In most terrestrial animals the second system is used. There is a switch to full copulation and direct insemination, with a small volume of isosmotic fluid containing the sperm released directly inside the female's vagina.

Eggs and larvae of course have their own water balance problems. In aquatic animals these are essentially an exaggerated version of the

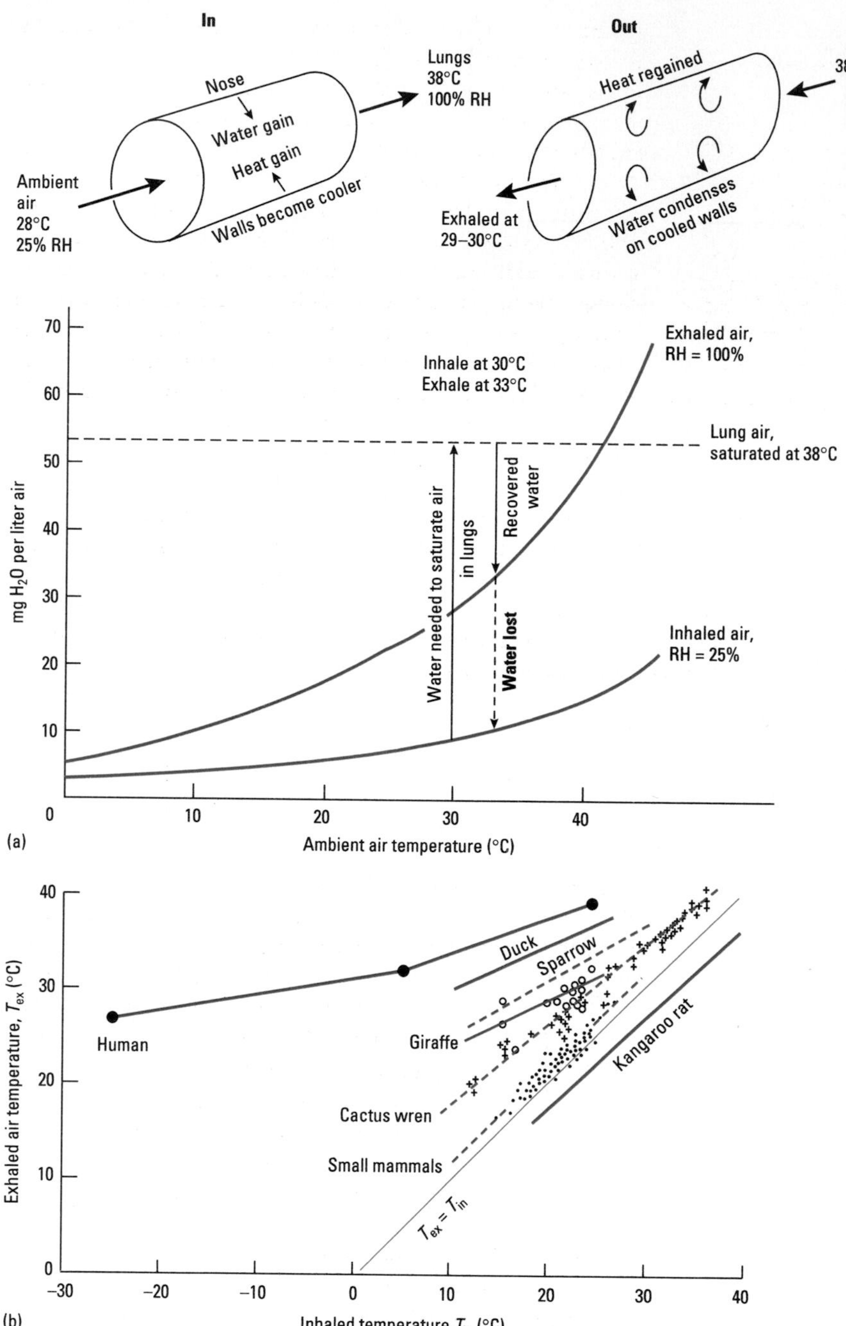

Fig. 5.27 (a) The principle of nasal countercurrent systems that recover heat and/or water from exhaled air; and (b) the varying degrees of lowering of exhaled air temperature in different mammals and birds. Notice that the kangaroo rat exhales air not only well below its body temperature but even below that of the inhaled air. RH, relative humidity. (b, Adapted from Schmidt-Nielsen *et al.* 1970b.)

adult's problems, compounded by the much larger surface area to volume ratio of the small juvenile stages. Hence the osmoregulatory organs commonly make up an even larger percentage of the body weight in larvae. Many littoral and freshwater animals shorten or entirely suppress the larval stage, and use direct development from well-endowed yolky eggs instead, which reduces these osmotic problems. So too do most of the terrestrial animals, although only the placental mammals and a few isolated examples from other groups have switched to full viviparity, where the offspring's osmotic problems are entirely dealt with by the maternal body. In terrestrial animals that retain egg-laying habits, use of microclimates chosen by the mother in finding an oviposition site very often ameliorates the problems, the extreme case being an endoparasitic habit where the host provides a stable environment. Eggs may be laid in carefully constructed nests in species that use maternal aftercare, or in very humid sites on (or inside) a food source, where water and temperature balance are relatively easy to achieve for the egg itself and then for the hatchling.

Nevertheless the egg stage has to undergo a prehatching period of respiration through the surface coating, without losing too much water. This problem is particularly acute for terrestrial insects, reptiles, and birds. Insect eggs, like the adults, contain waxy layers lying beneath the chorion of the egg "shell", and show temperature transitions in EWL rate similar to those of adult insects. The chorion

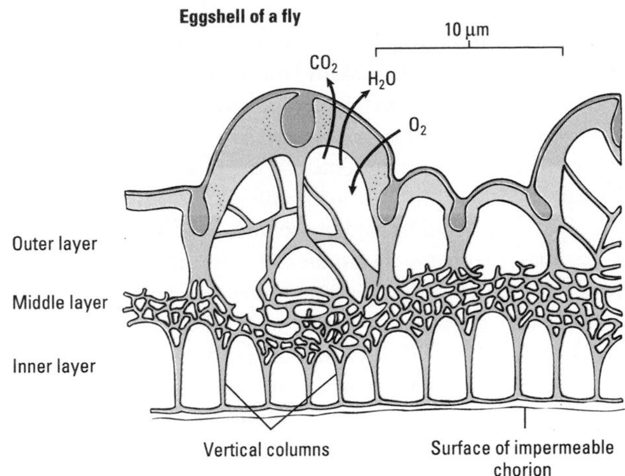

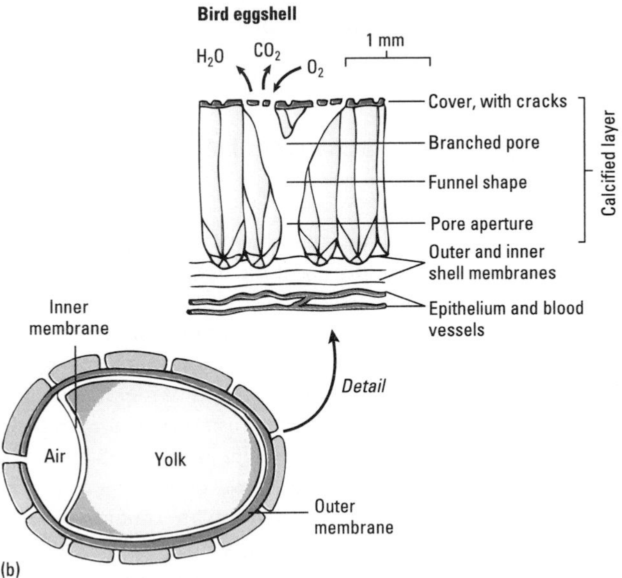

Fig. 5.28 Water balance in eggs of (a) insects and (b) birds, both having shells with a relatively impermeable inner layer (chorion) to control water loss and above this a surface "plastron" layer of highly reticulate air spaces (aeropyle) to allow oxygen uptake. The smaller size of insect eggs necessitates a less permeable surface and a more elaborate aeropyle. Many birds' eggs have an internal air store within the egg. (a, Reprinted from Hinton, H.E., *Biology of Insect Eggs*, copyright 1981, with permission from Elsevier Science.)

itself is a complex coating (Fig. 5.28a) with an "aeropyle" layer of air-filled channels so that air can diffuse into the egg. In some species there is hydrophobic material above the chorion to prevent these air spaces getting flooded, and in others there may be hygroscopic material in parts of the chorion forming a "hydropyle" layer that draws water into the egg.

The amniotic egg in reptiles and birds has a fairly standard structure but its surface may have very different properties. Some reptiles (lizards and snakes) have leathery eggs only partially impregnated with $CaCO_3$, and these swell and shrink quite substantially, normally increasing in volume as they develop because they are laid in damp soils where their hydration is assured. Hard-shelled eggs occur in crocodiles and most turtles, and these are unable to take up water and change volume. Some of the hard-shelled reptilian eggs share with birds' eggs the property of only allowing the passage of water vapor, being nearly impermeable to liquid water. These shells always have pores for respiratory exchange (Fig. 5.28b), whose size and properties are related to the crystalline structure of the mineral shell material and are characteristic for particular species. But in all cases, water is inevitably lost; this can only be controlled after laying by control of the humidity and temperature of the nest site, which is achieved by parental nesting in birds but usually only by initial site selection in most reptiles. Even in a carefully constructed and tended hen's nest, it can be calculated that an egg incubated for a standard 21 days takes up 5 l of oxygen, gives off 4 l of carbon dioxide, and loses 10 l of water vapor (though this is equivalent to just a few grams of water).

5.8 Water gain

5.8.1 Metabolic water

Respiration of foodstuffs, whether carbohydrate, lipid, or protein, ultimately yields carbon dioxide and water, so that metabolic water is invariably a component of water gain in animals. Metabolic water yields for particular substrates are shown in Table 5.8, indicating a greater water output for lipids per gram of substrate, but of glucose per unit of energy yield (because fats require much more oxygen uptake for their complete oxidation). However, for many animals, even in relatively dry terrestrial habitats, metabolic water is only a small component of their overall water balance compared with inputs from food and drink; and while metabolic water production may be highest in absolute terms in summer when activity is greater, its percentage contribution to overall water inputs may be greater when an animal is in a resting state in winter.

For small xeric animals, though, metabolic water can be the principal contributor to achieving net water balance. Clothes moths (*Tineola*) kept at very low humidities may use as much as 70% of their food intake primarily for water production. Desert tenebrionid beetles in the dry desert winter may get a large input from metabolic water, whereas in summer they are more active and seek out any alternative external water sources.

Table 5.8 Metabolic water yields during the oxidation of different food types.

Substrate	Water formed (g water per g food)	Metabolic energy yield (kJ per g food)	Water/energy (g water per kJ)
Starch	0.56	17.6	0.032
Lipid	1.07	39.3	0.027
Protein (with urea excreted)	0.39	17.0	0.022
Protein (with uric acid excreted)	0.50	18.4	0.027

Table 5.9 Water contents of various types of food.

Food type	Water content (% by weight)
Leaves	75–90
Xylem sap	>95
Fruits	60–90
Most seeds	5–10
Rice grains	12
Nectar	30–90
Milk	87
Meat	70–90
Whole animals	60–80

Can metabolic water be regulated? In animals with simple diffusive respiration, such as many nonflying insects, extra food does seem to be eaten at low humidities, with a rise in metabolic rate, to give extra water; this occurs in flour-moth larvae (*Ephestia*), for example, and can best be interpreted as a means of improving the animal's water balance. But in animals with ventilated respiration this would not normally work as the extra respiration needed to metabolize the extra food leads to a net water loss. Hence, of course, the old idea that camels can metabolize their fat store to gain water is a fallacy, as more water would be lost (in extra respiration to provide the necessary oxygen) than could be gained by the metabolic processes. However, recent work shows that in rapidly flying insects water is indeed lost more slowly than it is generated; some large bees produce excess water in flight and have to void it.

5.8.2 Food and drink

Drinking and eating ("preformed water" sources) are far more commonly the main component of water gain and the main avenue for regulating the input side of water balance. The water content of food (Table 5.9) is high for carnivores and insectivores, but extremely variable for plant feeders, ranging from dry seeds to exceptionally watery sap (many xylem tubes contain fluid that is 99.8–99.9% water), which may put the consumer into water overload and necessitate gut bypass mechanisms (see section 5.7). Blood-sucking animals and dilute nectar feeders may have a limited version of the same problem.

The nature of the food also affects the nature of the animal's salt problems. Eating animal flesh, or blood, naturally yields principally Na, K, and Cl in roughly the correct proportions for the consuming animal. But eating plants may give a high potassium diet with inadequate sodium (probably because plants have lower ECF volumes), and eating certain plant fluids (nectar, and many xylem and phloem saps) may give almost no salt input.

Many animals therefore have sophisticated systems for regulation of food choice, mediated by chemoreceptors (especially for salts and sugars) on food-handling appendages or in the mouth region. Direct stimulation from such sites is usually integrated with inhibition from gut stretch receptors, to regulate the occurrence of feeding. Control from higher centers (hypothalamic nuclei in vertebrates) has a separate role in determining food preferences. Here nitrogen contents, calorific values, or water contents may be key factors in many species, but specific attractants may also operate. Modified amino acids are often the key triggers; for example, glutathione is a potent feeding attractant for many cnidarians (hydras and sea anemones), and taurine has a similar effect in shrimps and some other marine filter feeders. Many terrestrial insects prefer to feed on plants with specific secondary compounds, often using these same compounds as part of their defensive repertoire (see Chapter 15). Herbivorous mammals are generally less fussy and tend to maximize nutrient intake, though many will select legumes (with higher nitrogen, higher sodium and phosphorus, and lower fiber) in preference to most other plant families. In more difficult environments, however, food choices may be more refined. Many nectar feeders will only take certain nectar concentrations (rather dilute in desert bees, compared with a fairly broad concentration range preferred in temperate species); and primates, such as baboons, from arid areas become much more selective for plants of high moisture content.

In general, the limited volume of the gut means that eventually feeding may cease even in the presence of preferred foods, and before the needs of water balance are fully met. This is usually brought about by the inhibitory effect of stretch receptors in the gut wall.

5.9 Costs and energetics of regulating water and ion balance

The energy required for ionic and osmotic regulation cannot be assessed accurately just by looking at the oxygen consumption or metabolic rate of an animal when exposed to different media. The analysis of metabolic rates is fraught with problems anyway (see Chapter 6); and in this situation too many factors are changing, and the varying costs of locomotion or buoyancy, or of respiration itself, will be compounded with those of actual osmoregulation. In practice, many euryhaline animals, such as the mitten crab *Eriocheir*, show little or no difference in metabolic rate in sea water or in very dilute brackish waters, and results from fish are confusingly variable. The cost of osmoregulating must therefore be calculated from simple thermodynamic considerations, though this can only give the minimum costs involved. Comparisons for different kinds of animal can then show the relative costs of different osmotic strategies such as high impermeability (hence low urine volume), or of localized ion uptake areas, or of producing isosmotic as against hyposmotic urine. Some examples for invertebrates, including both hyporegulators and weak and strong hyperosmotic regulators, are shown in Table 5.10, and several points are apparent:

1 It may be quite surprisingly "cheap" to be an osmoregulator. The minimum cost in fresh water or brackish water of hyperosmotic regulation is rarely more than 10% of the total metabolic output for an aquatic animal, and may be estimated as less than 1%. (However, real costs are probably markedly higher than these estimates, given the density of mitochondria found in all osmoregulatory tissues.)

2 The cost of osmoregulation is lower for the freshwater and estuarine crustaceans shown (values for the shore crab *Carcinus* are affected by a low measured metabolic rate) than for the soft-bodied animals such as the freshwater clam *Anodonta*. The latter is only a very weak regulator, maintaining a very low blood concentration (one-tenth that of the crayfish *Potamobius*), but because it has relatively permeable tissues it takes in water quite rapidly and has a high urine volume. Animals with relatively high permeability are therefore

Table 5.10 Examples of the costs of osmoregulating.

Species	Concentration of medium (mOsm)	Blood concentration (mOsm)	Urine concentration (mOsm)	Osmotic work ($J\ g^{-1}\ h^{-1}$)			% of metabolic rate
				Ion uptake	Urine	Total	
Potamobius (crayfish)	12	840	240			0.002	0.3
Eriocheir (mitten crab)	12	640	640			0.005	0.5
Anodonta (freshwater clam)	12	85	50			0.001	1.3
Nereis (ragworm)	3	200	200	0.110	0.003	0.113	2.7
Carcinus (shore crab)	370	600	600	0.301	0.008	0.309	11
Aedes (mosquito larva)	1000	330	High	0.29	1.07	1.36	22
Artemia (brine shrimp)	6000	350	High	1.92	?	(1.92)	>33

constrained to have very low blood concentrations; if they tried to osmoregulate more strongly, the costs would escalate substantially as the ionic gradients increased. This may explain why costs for regulating fish, also relatively permeable, are often in the range 2–4% of total metabolic rate. The advantages of the arthropodan design, with "built-in" low permeability and also a degree of inherent volume control, are evident.

3 There is an inherent advantage to producing hyposmotic urine, since the crayfish and mollusc, which recover up to two-thirds of the ions from their urine, have lower osmotic costs than the mitten crabs and ragworms, which produce isosmotic urine. Thus it would seem to be cheaper to osmoregulate (take up ions) via the kidney rather than via the skin or gills, presumably because the ionic gradients are lower there. It may also be cheaper to produce urine in the kidneys by secretion (as insects do) rather than by ultrafiltration (as most other animals do). The latter system requires a very large volume of fluid to be sent through the kidney, and a great deal of active resorption is therefore needed; probably more active transport is required in total than in a system where the primary urine is made by active salt transport.

4 Hyposmotic regulation in the brine shrimp and the mosquito larva, which involves very hyperosmotic urine production, is relatively much more costly (22–33% of the metabolic budget) than hyperosmotic regulation in any animal yet investigated.

Water vapor uptake, in those few animals that can do it, is certainly thermodynamically difficult to achieve, since coupling energy expenditure to actual water absorption is tricky, but the energetic cost again turns out to be quite low. Ticks and insects that are able to absorb water vapor use only a few percent of their daily metabolic budgets in this way, and there is no detectable increase in metabolic rate when vapor uptake occurs.

5.10 Roles of nervous systems and hormones

Many epithelia involved in osmoregulation are under hormonal control, either to regulate permeability to particular ions or water, to control the rate of transport of one or more ions, or even to change the direction of this transport. Detailed accounts are given in Chapter 10, section 10.3.

Permeability changes are best known in the vertebrate kidney, where aldosterone increases the P_{Na} of the apical membrane in the distal convoluted tubule, while ADH causes an increase in epithelial water permeability in the collecting duct. Similar mechanisms operate in invertebrates; in Malpighian tubule cells peptide hormones can raise the chloride permeability or cause changes in P_w in the lower tubules.

Controlling the rates of secretion of ions generally involves activation of secretory tissues, principally via their ion pump ATPases. For example in insect tubules, neuropeptides activate the cation-transporting apical V-ATPase, and in the salt gland of elasmobranch fish a peptide hormone from the gut activates the Na^+/K^+-ATPase and also enhances the symport pathway handling Na^+ and Cl^-.

Changes of transport direction are well known for NaCl absorption and secretion in the mammalian lower gut, in teleost gills, and in the hindgut of the larval stages of some euryhaline insects. In at least some of these cases the same cells are involved in both directions.

Such individual control systems are almost invariably integrated into multiple feedback control systems, especially in insects and mammals. Several hormones interact to control permeability values, whether of skin, gill, or renal surfaces, and another suite of hormones act to modulate salt uptake or loss rates and the overall directions of ion movements. A single example may serve to show the complex interactions that can operate: Fig. 5.29 summarizes hormonal control mechanisms in avian water balance.

5.11 Conclusions

Life on Earth has evolved around the properties and peculiarities of water, and all life remains dependent upon water. In aquatic ecosystems, and in most temperate and tropical terrestrial habitats, water is abundant and its conservation within animal bodies is not too difficult, but maintaining a precise osmotic equilibrium may nevertheless be a perpetual task, requiring some proportion of the

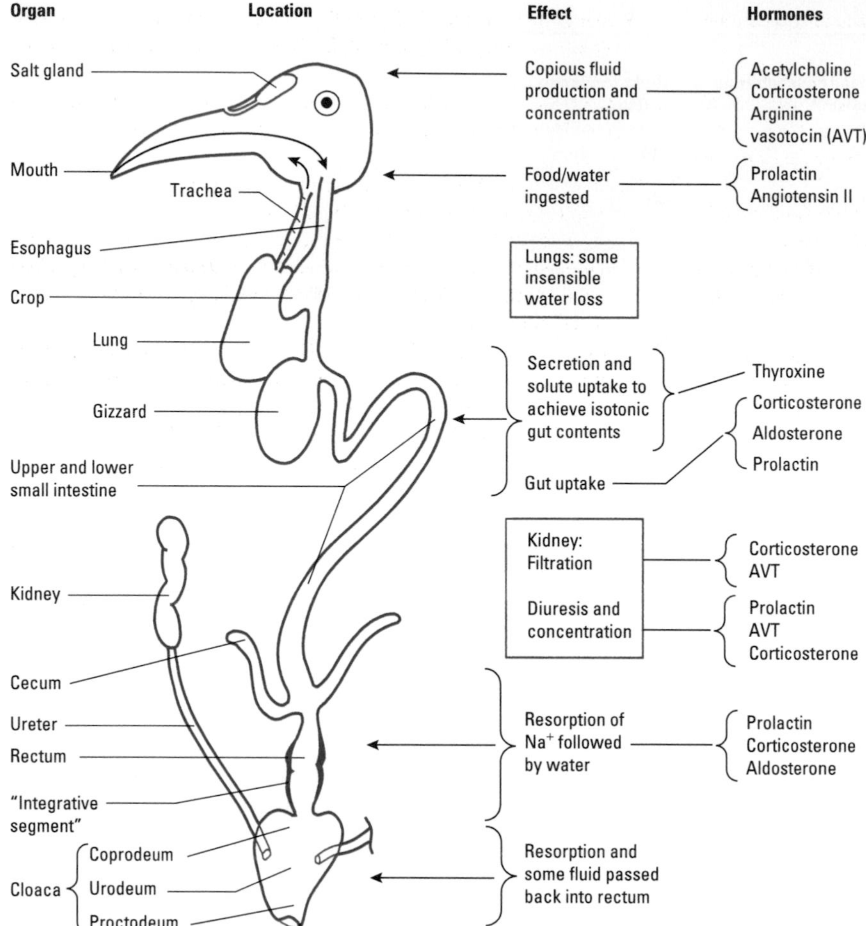

Fig. 5.29 A summary of the organs involved in osmoregulation and excretion in a terrestrial avian vertebrate, and the hormones that have a regulatory effect at each site.

metabolic efforts of all cells or of some specialized subset of body cells. In hyposmotic habitats it is excess water uptake that must be avoided, and the proportion of the metabolic effort devoted to this may be increased. In many land habitats, dehydration is a permanent danger, and sophisticated structural, physiological, and behavioral adaptations are necessary to prevent fatal water losses. It is probably fair to say that controlling water balance is even more essential than controlling temperature stresses, and (together with acquiring a food supply) is the single most important factor in the lives of many animals.

FURTHER READING

Books

Brown, J.A., Balmet, R.J. & Rankin, J.C. (eds) (1993) *New Insights in Vertebrate Kidney Function*. Cambridge University Press, Cambridge, UK.

Dejours, P., Bolis, L., Taylor, C.R. & Weibel, E.R. (eds) (1987) *Comparative Physiology: Life in Water and on Land*. Liviana Press, Padova, Italy.

Gilles, R. & Gilles-Baillien, M. (1985) *Transport Processes: Iono- and Osmoregulation*. Springer-Verlag, Berlin.

Greger, R. (1988) *Advances in Comparative and Environmental Physiology: NaCl Transport in Epithelia*. Springer-Verlag, Berlin.

Hadley, N.F. (1994) *Water Relations of Terrestrial Arthropods*. Academic Press, San Diego, CA.

Kinne, R.K.H. (1990) *Urinary Concentration Mechanisms*. Karger, Basel.

Somero, G.N., Osmond, C.B. & Bolis, C.L. (eds) (1992) *Water and Life*. Springer-Verlag, Berlin.

Reviews and scientific papers

Addo-Bediako, A., Chown, S.L. & Gaston, K.J. (2001) Revisiting water loss in insects: a large scale view. *Journal of Insect Physiology* **47**, 1377–1388.

Atkinson, D.A. (1992) Functional roles of urea synthesis in vertebrates. *Physiological Zoology* **65**, 243–267.

Bayley, M. & Holmstrup, M. (1999) Water vapor absorption in arthropods by accumulation of myo-inositol and glucose. *Science* **285**, 1909–1911.

Beuchat, C.A. (1990) Body size, medullary thickness and urine concentrating ability in mammals. *American Journal of Physiology* **258**, R298–R308.

Foskett, J.K. & Schaffey (1982) The chloride cell: definitive identification as the salt-secretory cell in teleosts. *Science* **215**, 164–166.

Gaede, K. & Knülle, W. (1997) On the mechanism of water vapour sorption from unsaturated atmospheres by ticks. *Journal of Experimental Biology* **200**, 1491–1498.

Gibbs, A.G. (1998a) The role of lipid physical properties in lipid barrier functions. *American Zoologist* **38**, 268–279.

Gibbs, A.G. (1998b) Waterproofing properties of cuticular lipids. *American Zoologist* **38**, 471–482.

Gibbs, A. & Mousseau, T.A. (1994) Thermal acclimation and genetic variation in cuticular lipids of the lesser migratory grasshopper (*Melanoplus sanguinipes*): effects of lipid composition on biophysical properties. *Physiological Zoology* **67**, 1523–1543.

Gilles, R. (1987) Volume regulation in cells of euryhaline invertebrates. *Current Topics in Membrane Transport* **30**, 205–247.

Greenwald, L. (1992) Concentrating mammalian urine. *Nature* **356**, 295.

Handler, J.S. & Moo Kwon, H. (1997) Kidney cell survival in high tonicity. *Comparative Biochemistry & Physiology A* **117**, 301–316.

Jørgensen, C.B. (1997) Urea and amphibian water economy. *Comparative Biochemistry & Physiology A* **117**, 161–170.

Karnaky, K.J. (1986) Structure and function of the chloride cells of *Fundulus heteroclitus* and other teleosts. *American Zoologist* **26**, 209–224.

Kültz, D. & Burg, M.B. (1998) Evolution of osmotic stress signalling via MAP kinase cascades. *Journal of Experimental Biology* **201**, 3015–3021.

Lockey, K.H. (1988) Lipids of the insect cuticle: origin, composition and function. *Comparative Biochemistry & Physiology B* **89**, 595–645.

Rasmussen, A.D. & Anderson, O. (1996) Apparent water permeability as a physiological parameter in crustaceans. *Journal of Experimental Biology* **199**, 2555–2564.

Ruppert, E.E. & Smith, P.R. (1988) The functional organisation of filtration nephridia. *Biological Reviews of the Cambridge Philosophical Society* **63**, 231–258.

Schmidt-Nielsen, B. (1988) Excretory mechanisms as examples of the principle "The whole is greater than the sum of its parts". *Physiological Zoology* **61**, 312–321.

Shoemaker, V.H. & Nagy, K. (1977) Osmoregulation in amphibians and reptiles. *Annual Review of Physiology* **39**, 449–471.

Somero G.N. & Yancey, P.H. (1997) Osmolytes and cell volume regulation: physiological and evolutionary principles. In: *Handbook of Physiology, Section 14, Cell Physiology, II* (ed. W. Dantzler), pp. 1445–1477. Oxford University Press, New York.

Treherne, J.E. & Willmer, P.G. (1975) Hormonal control of integumentary water loss: evidence for a novel neuroendocrine system in an insect. *Journal of Experimental Biology* **63**, 142–159.

Wright, P.A. (1995) Nitrogen excretion: three end products, many physiological roles. *Journal of Experimental Biology* **198**, 273–281.

6 Metabolism and Energy Supply

6.1 Introduction

Metabolic reactions within cells are the source of all the macromolecules within a living body, synthesizing these large end-products by **anabolic** processes. At the same time other large macromolecules are being broken down to produce usable energy, by **catabolic** processes. The free energy (ΔG^o) produced by catabolism is involved in so many physiological processes that it is essential to understand the source of this energy and to consider the effects of the environmental variables that will influence its rate of production, i.e. the "metabolic rate" of an animal. The main energy-consuming processes in animals are protein turnover, the sodium and calcium pumps, myosin ATPases involved in muscle contraction, and gluconeogenesis; together these account for over 45% of the metabolic rate of a mammal.

The phosphagens are the high-energy phosphate compounds that mediate energy transfers for all these processes in living organisms. **Adenosine triphosphate (ATP)** is the best known phosphagen and the common currency of energy metabolism for animals, providing the link between energy-yielding and energy-requiring reactions. ATP may be produced independently of oxygen (anaerobic metabolism) or through pathways which require oxygen (aerobic metabolism). Life on Earth certainly evolved in a reducing anaerobic environment, perhaps 3500 million years ago (mya). Free oxygen would be created in such an environment in small quantities by the photolysis of water, but would have been rapidly taken up by reduced ferrous rocks, which acted as an oxygen sink for at least another 1500 million years. Even after some of the prokaryotic heterotrophs acquired chlorophyll (probably around 3000 mya) and became photosynthetic "plants" that released oxygen as a by-product, free oxygen levels stayed very low because the oceans and vast beds of reduced minerals acted as sinks. Thus the first prokaryotic organisms were necessarily dependent on anaerobic metabolism; oxygen was toxic to them.

Perhaps around 2300–2000 mya the ocean and mineral sinks were becoming filled and free oxygen levels began to rise slowly (Fig. 6.1). This change may have taken place in three stages, from total planetary anoxia, through atmospheric and surface ocean oxidization, to deep oceanic oxidization and the formation of the first geological oxidized "red beds" rich in ferric oxides (dated 2000 mya). All organisms would have had to develop O_2 detoxification systems to survive, and at this point aerobic metabolism also probably began to evolve. Current thinking suggests that this important evolutionary innovation may have had multiple origins in different groups of prokaryotes, and that it became possible at about 1% of present atmospheric levels (PAL) of oxygen. Gradual further rises in oxygen level accompanied the subsequent evolution of eukaryotic cells and then (perhaps 1000–800 mya) of the first multicellular plants and animals.

Most metazoan animals now use aerobic metabolism for most of their energy requirements, reaping the benefits of pathways that are many times more efficient (generating 36–38 moles of ATP for each mole of substrate broken down aerobically, instead of just 2–6 with different anaerobic routes). However, many organisms still thrive in totally anaerobic conditions, for example in marine muds and sediments, in bogs and wetlands, or in the guts of other animals. Furthermore, nearly all organisms commonly revert to using anaerobic pathways during temporary periods of reduced oxygen availability (hypoxia) or when their rate of energy utilization is greater than can be met by the slower aerobic pathways, for example during strenuous exercise.

6.2 Metabolic intermediaries

6.2.1 Adenosine triphosphate (ATP)

Adenosine triphosphate is the classic "high-energy molecule", and interconversions between it and the other adenylates (the diphosphate and monophosphate forms, ADP and AMP) are the key to many metabolic processes. ATP consists of a purine base (adenine) linked to a five-carbon sugar (ribose) to form adenosine, and then three phosphate groups linked on by high-energy esteric bonds. Hydrolysis of the terminal phosphate group (Fig. 6.2) releases about 30.5 kJ mol^{-1} of ATP under normal cell conditions; remember that 1 mol of ATP is equivalent to 6×10^{23} molecules of ATP (Avogadro's number). ATP acts to transfer chemical energy to other molecules during energy-requiring interactions, yielding ADP and inorganic phosphate (P_i) in the process; this ADP is then recharged with a phosphate group, again by cellular metabolism. As a result energy is constantly cycling through ATP, and the individual molecules have a relatively short lifetime.

Around 60% of the energy that is used to recharge ADP appears as heat, due to inefficiencies in the reactions involved. Hence

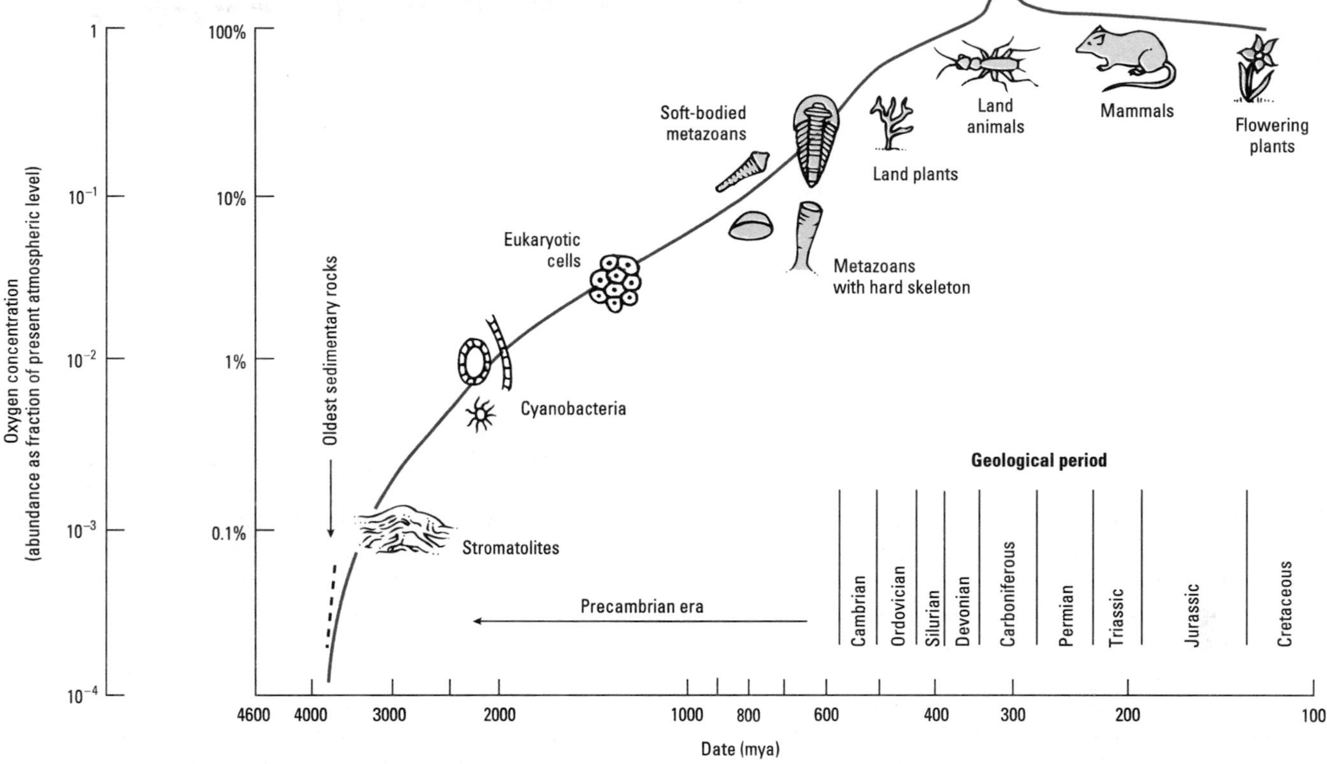

Fig. 6.1 Atmospheric oxygen levels through the history of Earth, with key events shown. (Adapted from Wayne 1991, courtesy of Oxford University Press.)

metabolic activity always generates heat, referred to as "metabolic heat" when considering problems of animal temperature balance (see Chapter 8). With the exception of muscles, the rate of ATP turnover is relatively low in most tissues. During the transition from rest to maximal activity, the ATP turnover (or cycling, or flux) in muscle can increase by one or more orders of magnitude, representing a massive heat production in a very active animal. However, the rates of ATP turnover also vary systematically between large and small animals so that mass-specific metabolic rates are higher in smaller animals (see Table 6.1 and Chapter 3).

Fig. 6.2 Structure and hydrolysis of adenosine triphosphate.

H_2O H^+

Adenosine triphosphate (ATP) ⇌ **Adenosine diphosphate (ADP)** + **Inorganic phosphate (P_i)**

6.2.2 Phosphagens: rapidly mobilized fuels

At any one time an animal does not contain a great amount of ATP, with values of only 2–8 μmol ATP g^{-1} being common; evidently it is not normally used to store much energy. Other **phosphagens** are used for this storage, the commonest being creatine phosphate (PCr) and arginine phosphate (PAr), which are phosphorylated derivatives of guanidinium compounds (Fig. 6.3). These may be up to 10 times as concentrated as ATP in tissues such as muscle and brain, and they are the immediate short-term source of ATP in most animals. They may be used to provide the necessary ATP under both aerobic and anaerobic conditions; thus they act in effect as a "buffering" system for ATP, insuring that its concentration is low and the ATP : ADP ratio is always kept favorable for ATP synthesis.

Phosphagens are the simplest and most rapid precursors for generating ATP as only one enzymic step is involved, catalyzed by a

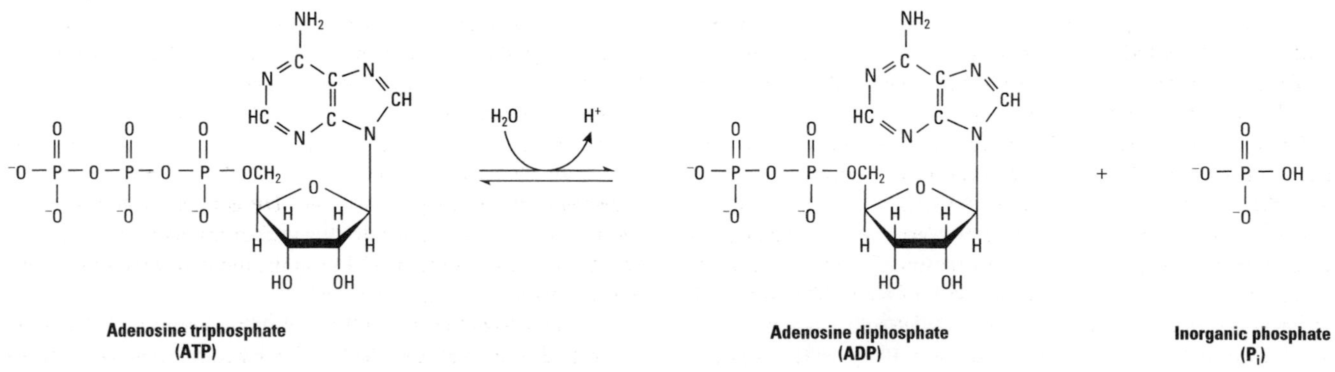

Table 6.1 ATP turnover rates during peak exercise in animals of varying size.

Species	Body mass	Muscle ATP turnover (μmol ATP g^{-1} min^{-1})
Insects		
Locust	1 g	5400
Birds		
Hummingbird	2.5 g	600
Mammals		
Mouse	7.2 g	227
Ferret	500 g	82
Wallaby	4.8 kg	57
Goat	20 kg	48
Human	70 kg	30
Eland	210 kg	32

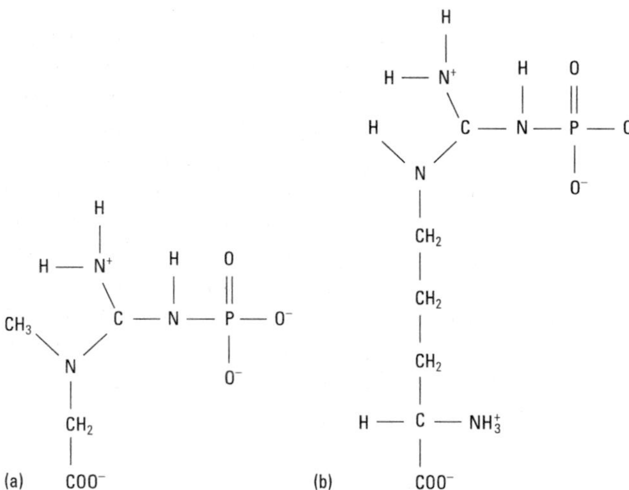

Fig. 6.3 Structures of the phosphagens (a) phosphocreatine (PCr) and (b) phosphoarginine (PAr).

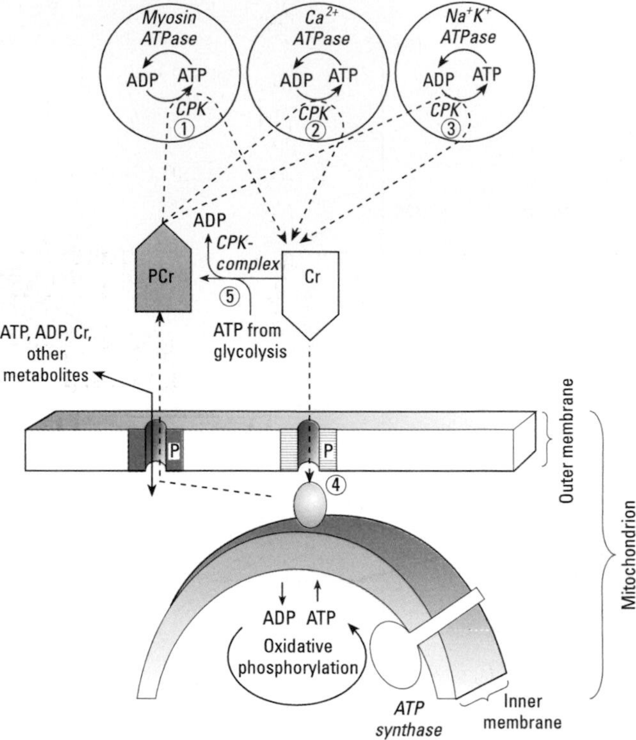

Fig. 6.4 The role of creatine phosphokinase (CPK) in controlling the flow of adenylates between sites of ATP synthesis and utilization in muscle. Different isozymes of CPK occur in at least five sites where they facilitate the supply of ATP from phosphocreatine (PCr). Three of these (1–3) are associated with ATPases. A fourth (4) is linked to porins (P) in the outer mitochondrial membrane, using the ATP from oxidative phosphorylation, while a fifth (5) occurs in the cytosol and uses ATP from glycolytic processes. (Adapted from Wallimann *et al.* 1992.)

kinase enzyme (e.g. creatine phosphokinase, CPK). Kinases belong to a class of enzymes that transfer the high-energy phosphate bond of ATP to another molecule:

$$\text{PCr} + \text{ADP} + \text{H}^+ \rightarrow \text{ATP} + \text{creatine}$$

$$\text{PAr} + \text{ADP} + \text{H}^+ \rightarrow \text{ATP} + \text{arginine}$$

For a phosphagen to be a useful source of ATP, it must be able to transfer its high-energy phosphate bond (~P) to ADP at high rates and at the correct time. This is achieved in most cells by having large amounts of the right kinase, with the right kinetic properties, present in the right locations in the cytoplasm. These kinases therefore occur in high concentrations in the cell as a whole, and are present as different isozymes (see Chapter 2), close to their sites of utilization and often bound to other proteins. For example, in muscles different isozymes occur, some bound to myosin ATPase, and others within the sarcolemma and sarcoplasmic reticulum with easy access to Na^+/K^+-ATPase and Ca^{2+}-ATPase, respectively (Fig. 6.4). The overall concentration of kinases is highest in tissues with high ATP demands, such as the skeletal muscles of animals capable of very rapid movements (see section 9.12.3). These same tissues usually have high concentrations of the phosphagens. Thus the white muscle of the rainbow trout contains around 30 mmol PCr kg^{-1} wet mass, and the adductor muscle of the scallop contains about 50 mmol PAr kg^{-1} wet mass. Both of these muscles are involved in escape behaviors. In slower muscles and in tissues such as neural ganglia the PCr levels may be only one-tenth of these values.

At equilibrium the enzymes catalyzing the reactions shown above are driven to the right (i.e. in the direction of ATP synthesis) by a number of mechanisms, including the phosphagens' high affinity for ADP (at least during the initial stages of work). The equilibrium constants for these kinases at pH 7.0 are of the order 2×10^9, which means that the kinases work extremely effectively and the phosphagen stores can be largely used up without affecting ATP concentrations (Fig. 6.5).

PCr and PAr stores are sufficient to support a modest amount of work lasting just a few seconds (about 2–5 s in mammalian muscle). In fish and many invertebrates the energy required for capturing prey or escaping from predators can often be supplied entirely from the breakdown of phosphagens, which are subsequently recharged using aerobic pathways. However, further amounts of ATP can be obtained by a pathway that results initially in the formation of AMP:

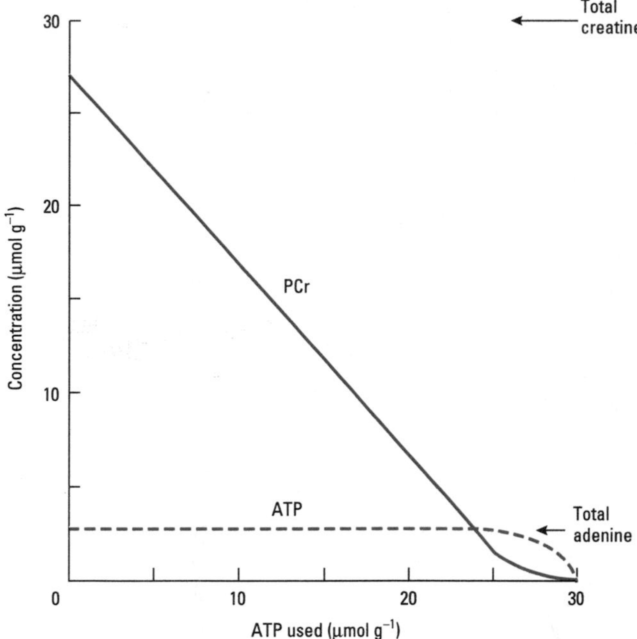

Fig. 6.5 The concentration of phosphocreatine (PCr) showing how it is used to maintain ATP levels constant despite varying ATP usage regimes. Total creatine and adenine levels are also shown. (From Woledge *et al.* 1985.)

$$2ADP \rightarrow ATP + AMP$$

This reaction is at equilibrium and can only proceed in the direction of further ATP synthesis if the AMP is removed in another reaction, catalyzed by the enzyme AMP deaminase and yielding inosine monophosphate (IMP) and ammonia:

$$AMP \rightarrow IMP + NH_3$$

Molluscs and crustaceans lack the AMP deaminase enzyme and therefore cannot exploit this source of ATP. Those animals that do have it may show a depletion of their total adenylate pool through the breakdown of AMP when exercised to exhaustion.

In addition to being fuels in anaerobic metabolism, phosphagens (particularly PCr) play an important part in aerobic pathways, by shuttling P_i between sites of ADP formation and utilization, and by involvement in the facilitated diffusion of ATP. Hummingbird flight muscles are among the most aerobic tissues known and have very high levels of PCr.

In summary, phosphagens are a useful way of supporting cellular metabolism at high power outputs while buffering the cell's ATP levels, especially at localities close to vitally active ATPase enzymes.

6.3 Anaerobic metabolic pathways

6.3.1 Uses and significance of anaerobic metabolism

Most animals use anaerobic pathways as part of their repertoire in maintaining their energy budgets, and in certain circumstances where oxygen is unavailable the entire metabolism may be dependent on anaerobic processes. We will meet examples in more detail when looking at animals from marine, freshwater, and semiterrestrial sediments, and at parasites living in the vertebrate gut. In many environments the need for anaerobic metabolism is transient, though. This may occur on a regular cycle for animals in intertidal and estuarine environments, for example, where anoxia may persist for hours or even for days according to the state of the tides. Or it may occur more randomly where spurts of exercise are needed (e.g. for prey capture or escape responses) or where an air-breathing diving animal spends time under water, suffering bouts of oxygen shortage. All these animals are very tolerant of hypoxia or anoxia (see Chapter 7). Their metabolic rates usually decline during anaerobic respiration, and in intertidal invertebrates the reduction may be to only a fraction of the normal resting metabolic rate (see below). In vertebrates tolerance of hypoxia is much less profound, except in a few special cases where, for example, fish can survive in ice-bound ponds for several weeks. Higher vertebrates only use anaerobic pathways to any great extent during short bursts of intense activity. Here the rapidity of ATP generation is an advantage; the reaction is entirely in the cell cytoplasm, and nothing need be transported in or out of the cell to support the catabolic reactions. However, the rapid build-up of end-products soon inhibits further anaerobiosis, and the amount of energy that can be generated before this occurs is severely limited. Anaerobic energy yields are therefore only really important in activities lasting up to a few minutes. Longer term submaximal work by mammalian muscles does involve some anaerobic component, but the end-product (lactate) is removed by the liver and other aerobic tissues at about the same rate that it is formed.

6.3.2 Basic glycolytic pathways

The process of **glycolysis** takes place in the cytoplasm and if oxygen is limited it can be an entirely anaerobic process. Written at its simplest it involves the reaction:

$$Glucose + 2ADP + 2P_i \rightarrow 2\ lactate + 2ATP$$

However, the glycolytic pathway is in fact a complex fermentation reaction, consisting of 10 enzyme-catalyzed steps (Fig. 6.6). Because it is a very ancient pathway, many of its features, such as the relative proportions of the different enzymes, are highly conserved between animals, and its dual roles in providing ATP and providing carbon molecules for synthetic (anabolic) reactions are similar throughout the metazoan kingdom. However, as we will see there are a number of ways in which its rate and outcome can be modified according to environmental circumstance.

Glucose usually enters cells by means of specific transport proteins (see Chapter 5). It is initially phosphorylated by ATP to form glucose 6-phosphate, a reaction catalyzed by the enzyme hexokinase:

$$Glucose + ATP \rightarrow glucose\ 6\text{-}phosphate + ADP$$

Under anaerobic conditions, 1 mol of glucose can be fermented to 1 mol of anaerobic end-product, yielding 2 mol of ATP (Fig. 6.6). If

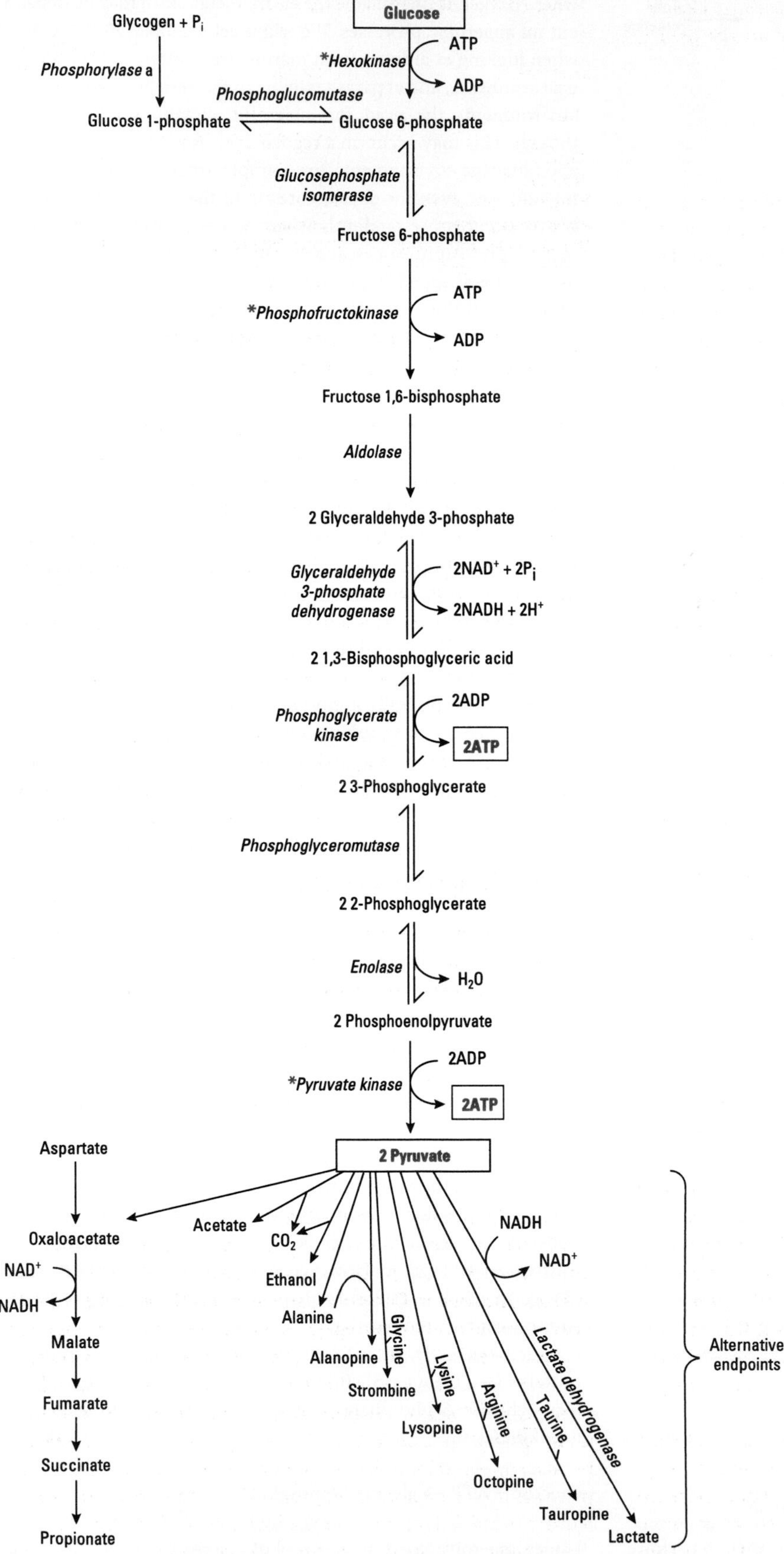

Fig. 6.6 Glycolysis pathways and key side paths, also showing the main points of control (*).

the ATP is coupled to an ATPase, such as the myosin ATPase in muscle, then 2 mol of H^+ are also generated, which are accepted by the coenzyme NAD:

$$\text{Glucose} + 2\text{ADP} + 2\text{P}_i + 2\text{NAD} \rightarrow 2\ \text{pyruvate} + 2\text{ATP} + 2\text{NADH} + \text{H}^+$$

In practice at physiological pH the net generation of protons is less than this reaction would suggest, as some are consumed in converting the pyruvate to lactate, a common end-product of glycolysis. Nevertheless, the generation of H^+ cannot proceed indefinitely as it leads to acidosis, and this is one factor limiting the duration of anaerobic glycolysis that can be sustained. Protons are buffered or exported from cells to minimize the impact of their accumulation. Lactate concentrations can reach 40 mmol kg^{-1} wet tissue mass in mammalian muscle, and 200 mmol kg^{-1} wet tissue mass in diving turtles where survival of hypoxia is crucial.

The reducing equivalents (NADH) formed by the action of glyceraldehyde phosphate dehydrogenase during glycolysis (Fig. 6.6) are oxidized to regenerate NAD in the terminal step of the pathway, which is catalyzed by another dehydrogenase enzyme. This enzyme is lactate dehydrogenase (LDH) when lactate is the endpoint of glycolysis acting as the terminal electron acceptor (note that LDH is so ubiquitous and so relatively easy to study and sequence that it has become an important source of comparative and evolutionary studies).

6.3.3 Alternative endpoints of glycolysis

Arthropods, echinoderms, and vertebrates all rely mainly on classic glycolysis for anaerobic energy production, resulting in the accumulation of large amounts of lactate and sometimes small amounts of alanine. In most vertebrates too, glycolysis to lactate is the only anaerobic pathway of any real importance, and is used in exercise and also in more chronic anoxia such as during diving in air-breathing tetrapods. But a few invertebrates, especially but not exclusively in parasitic groups, have ethanol as an endpoint; parasite anaerobic pathways are discussed further in Chapter 17. Ethanol is also an endpoint in some fish, of which the goldfish is the best studied. In these animals a complete absence of oxygen can be tolerated for long periods at low temperatures, with CO_2 and ethanol as end-products. Glycogen is converted to pyruvate by anaerobic glycolysis in the cytoplasm, and pyruvate is then decarboxylated to acetaldehyde in the mitochondria; the acetaldehyde is subsequently reduced to ethanol in the cytoplasm by alcohol dehydrogenase (note that this is the same pathway that is quite common in plants). Ethanol is excreted, allowing glycolysis to proceed without the accumulation of an acid end-product. Goldfish using this pathway can survive through several days of anoxia at 0–4°C because their metabolic rate is reduced by more than 80%, sparing their glycogen stores.

In many invertebrate taxa a different strategy is used, with lactate dehydrogenase replaced by a functionally analogous imino dehydrogenase, having an identical ATP yield per glycosyl unit. Here the final step is the condensation of an amino acid with the pyruvate produced by glycolysis, allowing the regeneration of the electron acceptor coenzyme, NAD:

Table 6.2 ATP yield from different fermentation routes.

Substrate	Endpoints	mol ATP per mol substrate
Glucose	Lactate, octopine, alanopine or strombine	2
Glucose	Succinate	4
Glucose	Propionate	6
Aspartate or glutamate	Succinate	1
Aspartate or glutamate	Propionate	2
Branched-chain fatty acids	Volatile fatty acids	1

$$\text{Pyruvate (RCO} - \text{R}') + \text{R}'' - \text{NH}_2 + \text{NADH} + \text{H}^+ \rightarrow \text{RCHR}' - \text{NHR}'' + \text{H}_2\text{O} + \text{NAD}$$

Particular amino acids such as alanine, glycine, lysine, and arginine are used, according to species, to remove the pyruvate; the end-products are the imino acids alanopine, strombine, lysopine, and octopine, respectively. In effect, oxygen acts as the terminal electron acceptor to form water. Octopine formation appears to occur when there are high resting levels of phosphoarginine in the muscles, and is common in swimming cephalopods and certain bivalves.

Aspartate and certain branched-chain amino acids can also be fermented directly in some invertebrates (see Fig. 6.6). For example, in intertidal molluscs, where hypoxia can be a problem, the aspartate is deaminated and then fed on via parts of the Krebs cycle (see below) to produce a higher ATP yield and a variety of end-products such as succinate and propionate; the amino group is transferred into glutamate and then alanine, so may again lead to the formation of alanopine. Alternatively, in some metazoan parasites in profoundly hypoxic conditions other glycolytic products occur (see Chapter 17), especially oxaloacetate derived from pyruvate. These feed on into the Krebs cycle, and are again converted through succinate and propionate, sometimes eventually leading to methylated products such as 2-methylbutyrate. Such processes may yield an extra 4–6 ATP per glucose molecule. The ATP yields of the different substrates of fermentation are shown in Table 6.2. Pathways producing succinate or propionate from aspartate or glutamate, or those using branched amino acids, are not favored for anaerobic muscular work, because they cannot sustain a high enough flux of ATP.

6.3.4 Control of glycolysis

Control of glycolysis is achieved in two important ways: firstly by exogenous activating signals (hormones or neurotransmitters, acting through the G-protein/adenyl cyclase/cAMP/protein kinase systems discussed in Chapter 2); and secondly through the regulatory properties of glycolytic enzymes themselves where products act as negative feedbacks on earlier stages in the process (see Chapter 2).

There are tissue-specific isozymes of almost every glycolytic enzyme, and most of the steps in glycolysis are readily reversible under physiological conditions. The exceptions are those catalyzed by isoenzymes of hexokinase, phosphofructokinase (PFK), and pyruvate kinase (starred in Fig. 6.6). These three enzymes act in a coordinated fashion to control a major part of the flux through the pathway. PFK is the most important control step; this enzyme is inhibited by high levels of ATP and citrate, and is activated by

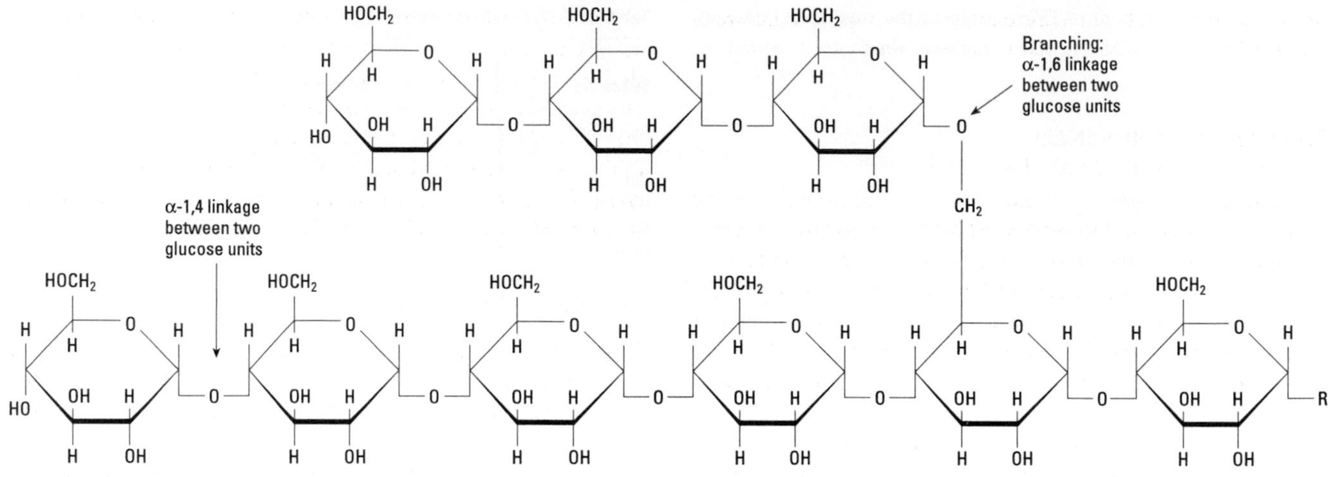

Fig. 6.7 Glycogen structure and bonding.

AMP and fructose 1,6-bisphosphate. The presence of fructose 1,6-bisphosphate in the vertebrate liver signals that glucose is abundant. This pattern of regulation insures that the activity of PFK matches the need either for generating ATP or for providing the building blocks for synthetic reactions. When PFK is inhibited, glucose 6-phosphate accumulates, and inhibits hexokinase, the first enzyme in the pathway (Fig. 6.6). Pyruvate kinase is the last step in the pathway. It is allosterically inhibited by ATP and alanine, and is activated by fructose 1,6-bisphosphate (Fig. 6.6). The feed-forward activation of pyruvate kinase by glycolytic intermediates insures that its activity increases in parallel with PFK when the demand for ATP is high (see Chapter 2). Pyruvate kinase occurs in a number of different forms (isoenzymes) providing variation in the control of the terminal step of glycolysis between tissues. The isoenzyme present in mammalian liver is controlled by a reversible phosphorylation. The hormone glucagon (which in concert with insulin controls blood sugar levels in vertebrates) triggers a cascade that leads to an increase in the less active phosphorylated form of pyruvate kinase. This insures that the liver does not consume glucose for synthetic reactions when it is more urgently needed for ATP generation in other tissues such as muscles and brain.

However, a puzzling feature of glycolytic control is that it appears to be "more than the sum of its parts": manipulation of gene expression in yeast has shown that changes in overall rates of glycolysis are greater than can be accounted for by changes in individual enzyme steps. There is recent evidence that in muscle some fraction of the key enzymes may be present but inaccessible or unreactive, providing a latent pool that can be called upon to increase rates of flux without substrate concentrations changing much. This may be the main control mechanism, with varying substrate or product or modulator levels providing the fine tuning.

6.3.5 Fuel sources for glycolysis

Glucose is the primary fuel source for glycolysis, but it is stored as the polysaccharide glycogen in many tissues, especially muscle (and the liver of vertebrates). Glycogen can enter glycolysis either via conversion to glucose or indirectly by conversion to glucose 6-phosphate (Fig. 6.6). Glycogen concentrations are particularly high in species that have a well-developed tolerance for hypoxia or anoxia. For example, in species of marine bivalves found in the intertidal zone stored glycogen concentrations can exceed 1% of wet body mass (see Chapter 12).

Glycogen is a large branched polymer where glucose molecules are linked together by α-1,4-glycosidic bonds in the straight parts of the polymer, and the branches are created by α-1,6-glycosidic bonds (Fig. 6.7). Glycogen is often present in the cytoplasm in the form of granules (readily visible with the electron microscope) that also contain the enzymes for glycogen synthesis and degradation as well as some of the enzymes that control these processes. The synthesis and degradation of glycogen may be hormonally regulated via different enzyme pathways involving glycogen phosphorylase and glycogen synthase. Both of these enzymes exist in active and inactive forms. The reaction cascades that convert the inactive to the active forms are controlled by phosphorylation reactions which have cAMP as a common second messenger (see Chapter 2). The fact that the glycogen polymer is branched increases the solubility of glycogen molecules and creates a large number of terminal residues that are the sites of action for glycogen phosphorylase and synthase. Thus branching increases the maximum rate at which glycogen synthesis and breakdown can occur.

Other sugars, including fructose and galactose, can also enter the glycolytic pathway. Galactose is converted in four steps to glucose 6-phosphate, which can then enter glycolysis. In mammals, the lactose present in milk is hydrolyzed to fructose and glucose. Fructose (which is present in many plant foods, and therefore also in honey) can enter the glycolytic pathway of animals via a fructose 1-phosphate pathway, which is particularly important in liver cells. Fructose can also be phosphorylated to fructose 6-phosphate by hexokinase. This reaction only predominates in tissues such as white fat cells, which have higher concentrations of fructose than glucose. In other tissues the formation of fructose 6-phosphate is competitively inhibited because the affinity of hexokinase for glucose is much greater than for fructose.

We saw above (section 6.3.2) that amino acids can also feed into glycolytic pathways. Thus carbohydrates and amino acids from proteins can be used as fuel for glycolysis, giving a range of end-products. However, the one fuel that cannot be used is lipid, which is normally in too reduced a state to be fermented.

6.4 Aerobic metabolism

In fermentation, organic molecules serve as terminal electron and proton acceptors. Since life evolved in an essentially anaerobic atmosphere these pathways must be among the most ancient means of energy production. However, carbon skeletons are only partially degraded in the process, and despite a roughly 35–50% efficient process (depending on the end-products) only about 4.7% of the total energy of the glucose molecule is released in producing pyruvate. Whenever possible then, animals proceed further with the breakdown of glucose, insuring that it is more completely oxidized with O_2 acting as the terminal electron acceptor. These additional steps represent the evolutionarily "later" reactions of **aerobic** metabolism. The net reaction for complete glucose oxidation in aerobic conditions can be written as:

$$\text{Glucose} + 36(\text{ADP} + \text{P}_i) + 6O_2 \rightarrow 36\text{ATP} + 6CO_2 + 6H_2O$$

6.4.1 Krebs cycle

The classic way of achieving further carbon compound breakdown and better energy yield is to convert the glycolytic end-product pyruvate not into lactate but (by condensation with coenzyme A) into acetyl-CoA. This then serves as the entry substrate into the series of reactions known as the Krebs cycle (also known as the citric acid cycle or tricarboxylic acid cycle), occurring in the mitochondrial matrix (Fig. 6.8). The Krebs cycle is both the major pathway for ATP production and a source of intermediates for the synthesis of a whole range of other molecules, including proteins, as the figure shows.

The acetyl-CoA (a two-carbon fragment, C_2) that starts the cycle is combined with oxaloacetate (also C_2) to form citrate (C_4). This then passes on through a series of C_6, C_5, and C_4 intermediates, the last of which is again oxaloacetate. The two carbons that entered as acetyl-CoA are released as $2CO_2$ in each turn of the cycle, at steps 5 and 6 in Fig. 6.8. At the same time, one molecule of the high-energy phosphate compound GTP is formed by the succinyl-CoA synthetase reaction (step 7 in Fig. 6.8) per turn of the Krebs cycle; this is then transferred to ATP. Simultaneously, H^+ is transferred to the proton acceptors NAD and FAD (flavin adenine dinucleotide) at four stages in the cycle, with an additional proton transfer to NAD at the pyruvate–acetyl-CoA stage. The net reaction per turn of the cycle is therefore:

$$\text{Pyruvate} + 4\text{NAD} + \text{FAD} + \text{GDP} + \text{P}_i$$
$$\rightarrow 2CO_2 + 4\text{NADH} + \text{FADH}_2 + \text{GTP}$$

(Remember that as each glucose yields two pyruvate molecules to feed into the cycle, the yield per molecule of glucose is twice this.)

Any Krebs cycle intermediates that are drawn off for biosynthesis must be replenished by a series of "anaplerotic pathways" (from the Greek meaning "to fill up"). For example, oxaloacetate converted into amino acids for protein synthesis is subsequently replaced by the carboxylation of pyruvate using the enzyme pyruvate carboxylase.

The CO_2 formed in the Krebs cycle is a major waste product of aerobic metabolism, and it is generated by the production of ever shorter carbon fragments, from the carboxyl groups of the cycle intermediates. At the same time molecular oxygen is used up, and this will appear as water after combining with hydrogen molecules during oxidative phosphorylation, so that water becomes another end-product (the so-called "metabolic water" referred to in considering animal water balance in Chapters 4 and 5).

6.4.2 Electron transfer system

During the Krebs cycle, the two carbon atoms in acetyl-CoA have appeared as CO_2 with the formation of NADH and $FADH_2$. Each glucose molecule produces a total of 10 NADH and two $FADH_2$. These represent the "fuel" for the major energy-storing process of aerobic respiration, as they are reoxidized via the **electron transfer system** (ETS) located in the inner mitochondrial membranes.

The reduced molecules NADH and $FADH_2$ each contain a pair of electrons with a high transfer potential that can be donated to molecular oxygen via a series of intermediates. In the process free energy is liberated, which can be used to generate ATP by a process called **oxidative phosphorylation** (Fig. 6.9). Oxidative phosphorylation is carried out by molecular complexes termed respiratory assemblies, located on the folded cristae of the inner mitochondrial membranes. NADH itself cannot easily cross the membranes, but electrons are transferred from NADH via carriers and then passed on to O_2 via the respiratory assemblies. These comprise three large protein complexes: NADH-Q reductase, cytochrome reductase, and cytochrome oxidase (Fig. 6.9a,c). The electron-carrying groups on these proteins are flavins, iron–sulfur clusters, hemes, and copper ions. Electrons are carried from NADH-Q reductase to cytochrome reductase by the reduced form of a hydrophobic quinone called ubiquinone (Q in Fig. 6.9). Ubiquinone also functions to carry electrons from $FADH_2$ to cytochrome reductase. The protein cytochrome *c* then shuttles electrons from cytochrome reductase to cytochrome oxidase (note that cytochrome *c* has been another good source of comparative sequence studies). Cytochrome oxidase at the end of the ETS chain catalyzes the transfer of electrons to molecular oxygen, the final electron acceptor.

The step-by-step transfer of electrons from NADH or $FADH_2$ to O_2 through these carriers does not directly produce any ATP, but results in the pumping of protons out of the mitochondrial matrix to the space between the inner and outer mitochondrial membranes, generating a proton-motive force consisting of a pH gradient and a transmembrane electrical potential. ATP is then synthesized when protons flow back to the mitochondrial matrix through an **ATP synthase**, a large protein complex straddling the inner membrane and made up from F_1 and F_0 subunits (Fig. 6.9b). The precise details of this process remain unclear, though it is known that the proton gradient is needed not to form the ATP but to release it from the enzyme. Whatever the detail, it is evident that oxidation and phosphorylation are coupled by the large proton electrochemical gradient across the inner mitochondrial membrane.

As we have seen, the H^+ that were transferred from intermediates to coenzymes in the Krebs cycle produce a total of four reduced coenzyme molecules for each two-carbon fragment entering the Krebs cycle. The three molecules of NADH donate their hydrogen

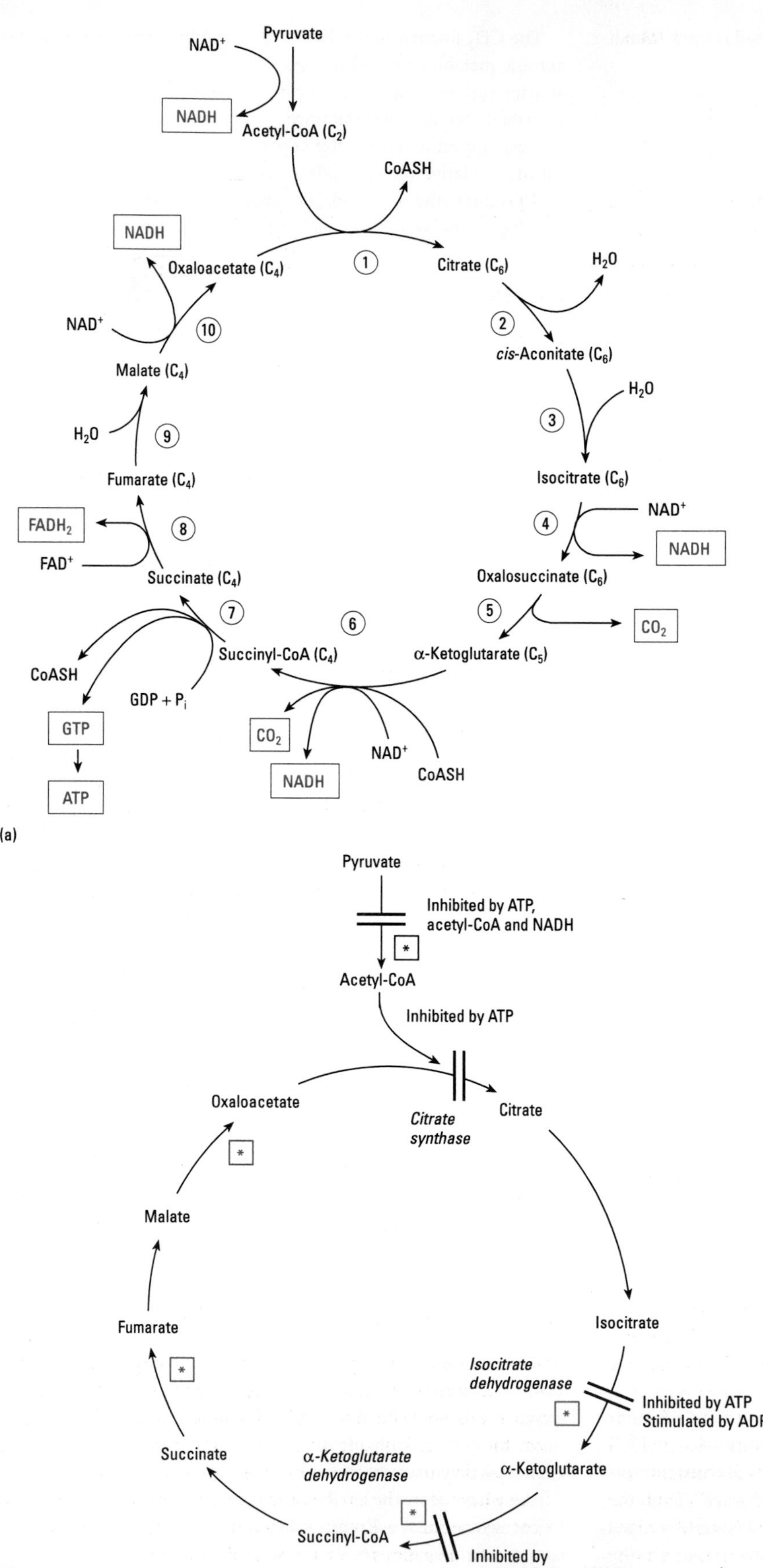

Fig. 6.8 The Krebs cycle (a) in detail, showing the carbon content of each molecule through the cycle and (b) showing the main enzymes that function as control points (note that * indicates steps also controlled by availability of the electron acceptors NAD+ or FAD that are regenerated by oxidative phosphorylation).

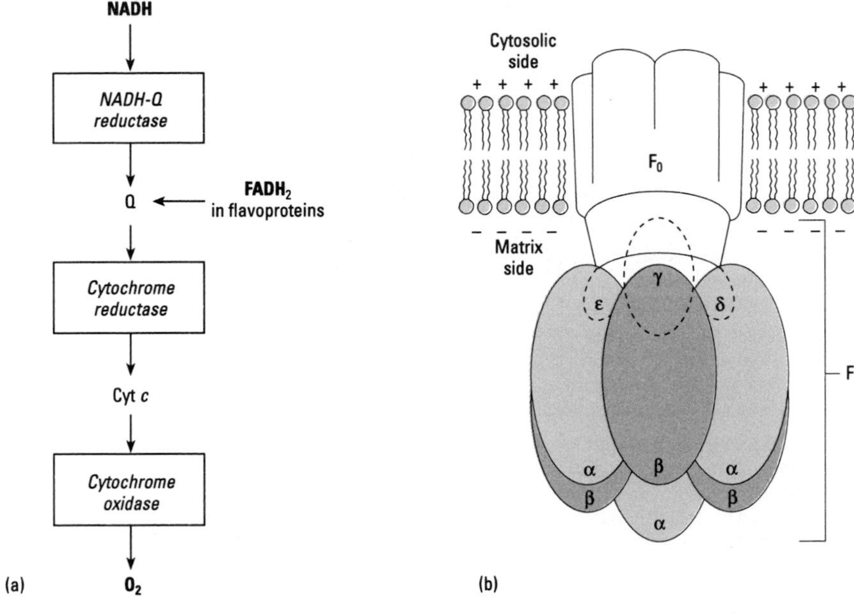

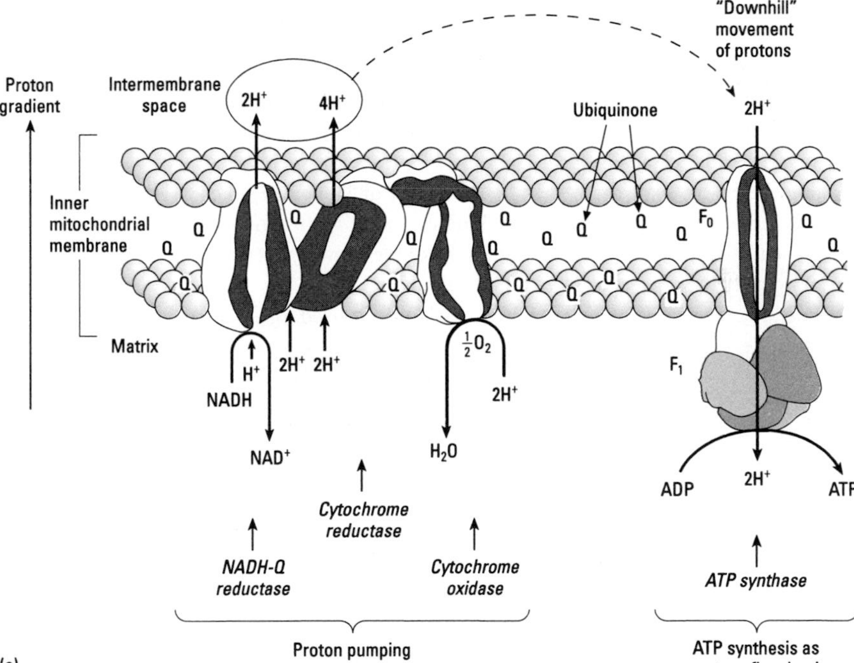

Fig. 6.9 Oxidative phosphorylations by the electron transfer system (ETS) in mitochondrial membranes. (a) The basic sequence of electron carriers, protons being "pumped" by the three complexes shown in boxes. (b) The ATP synthase, where a positive membrane potential allows proton flow through the F_0 subunit activating the F_1 ATP-synthesizing unit. (c) The arrangement of these major proteins in the inner mitochondrial membrane and a view of the overall process.

atoms to cytochromes at the beginning of the electron transfer chain forming a total of nine ATP molecules. The reduced coenzyme $FADH_2$ enters at a later point in the ETS chain producing a further two ATP molecules. Thus each glucose precursor produces $2 \times (9 + 2)$ ATP via the Krebs cycle and ETS, plus further ATP from the NADH generated in glycolysis and in the pyruvate → acetyl-CoA step. The net reaction for complete aerobic glucose oxidation produces 36 molecules of ATP (36 mol ATP mol^{-1} glucose) (Table 6.3). Where glycogen is the starting point the net yield is slightly higher. The conversion efficiency is about 41%, with the remaining 59% of available energy released as heat.

6.4.3 Fuels for aerobic metabolism

Glucose and related sugars, derived from carbohydrates, are not the only fuel sources for aerobic carbon breakdown. In fact the Krebs cycle is extremely flexible in allowing other potential fuels to be fed into the system, some of which need to be considered here as they have particular consequences for the overall energetic gain. Fuels from within the cell may be used for short-term activity, but long-term aerobic exercise requires muscles and other tissues to be supplied with exogenous fuels, for example from the liver in vertebrates or from the fat body in insects.

Table 6.3 Overview of ATP production in aerobic metabolism.

Reaction sequence	ATP yield per glucose
Glycolysis (in cytoplasm)	
Phosphorylation of glucose	−1
Phosphorylation of fructose 6-phosphate	−1
Dephosphorylation of 2 molecules of 1,3-bisphosphoglycerate	+2
Dephosphorylation of 2 molecules of phosphoenolpyruvate	+2
(2 NADH are formed in the oxidation of 2 molecules of glyceraldehyde 3-phosphate)	
Conversion of pyruvate into acetyl-CoA feeding into Krebs cycle (in mitochondria)	
(2 NADH are formed)	
Krebs cycle (in mitochondria)	
2 molecules of GTP are formed from 2 molecules of succinyl-CoA	+2
(6 NADH are formed in the oxidation of 2 molecules each of isocitrate, α-ketoglutarate, and malate)	
(2 $FADH_2$ are formed in the oxidation of 2 molecules of succinate)	
Oxidative phosphorylation (in mitochondria)	
2 NADH formed in glycolysis; each yields 2 ATP	+4
2 NADH formed in the oxidative decarboxylation of pyruvate; each yields 3 ATP	+6
2 $FADH_2$ formed in the Krebs cycle; each yields 2 ATP	+4
6 NADH formed in the Krebs cycle; each yields 3 ATP	+18
Net yield per glucose	+36

For abbreviations, see Abbreviations list (p. xi).

Lipids

Lipids in the form of fatty acids are one of the most common inputs to metabolism in most animals. In general, the fatty acids present in animal cells contain an even number of carbon atoms, often between 14 and 24 (see Fig. 4.5). The most common fatty acids are C_{16} and C_{18}. The alkyl chain may be saturated, or it may be unsaturated (i.e. containing one or more double bonds); 16 : 0 denotes a fatty acid with 16 carbon atoms and no double bonds, whereas 16 : 2 signifies the presence of two double bonds.

Note that at physiological pH fatty acids are ionized (giving, for example, palmitate, stearate, or oleate). However, the fatty acids are stored in an uncharged form in combination with glycerol as "triglycerides", more correctly known as triacylglycerols (see Fig. 4.5). The triacylglycerols are highly concentrated fuels because they are both highly reduced and very nonpolar, enabling then to be stored in an anhydrous state. One gram of nearly anhydrous triacylglycerol stores more than six times as much energy as 1 g of hydrated glycogen (note that glycogen cannot be stored unhydrated, and 1 g anhydrous glycogen binds about 2 g water). Many animals contain cells that are specialized for the synthesis, storage, and mobilization of triacylglycerols, for example the fat body in insects and white adipose tissue in vertebrates (Fig. 6.10). In vertebrates triacylglycerol is hydrolyzed back to fatty acids and glycerol by lipases; the resultant process is termed **lipolysis** and is upregulated by a variety of hormones, including epinephrine, norepinephrine, glucagons, and adrenocorticotropic hormone (ACTH) (see Chapter 10). These hormones commonly stimulate adenyl cyclases to cause an increase in cAMP concentration, which then stimulates a protein kinase that phosphorylates the lipase and so activates it. Lipolysis in vertebrates is inhibited by the hormone insulin.

The glycerol produced by lipolysis can be converted to glyceraldehyde 3-phosphate and either metabolized to pyruvate by glycolysis or to glucose by "gluconeogenic" pathways. Before the fatty acids released by lipolysis can be used as fuels they must first be activated to acyl-CoAs on the outer mitochondrial membrane. Since the inner mitochondrial membrane is relatively impermeable to long-chain acyl-CoA molecules, a special transporting system is needed to get them into the matrix. The acyl-CoA molecules combine with

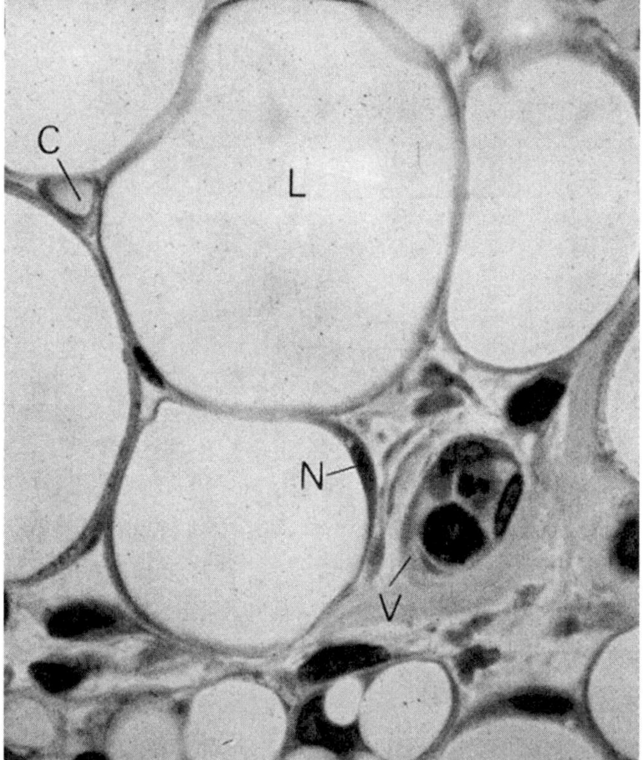

Fig. 6.10 Micrograph of adipose tissue in a mammal. The cytoplasm of the cells is largely occluded by huge lipid deposits (L), with a nucleus (N) sometimes visible. A capillary (C) and a venule (V) are also present. (From Ross & Romrell 1989.)

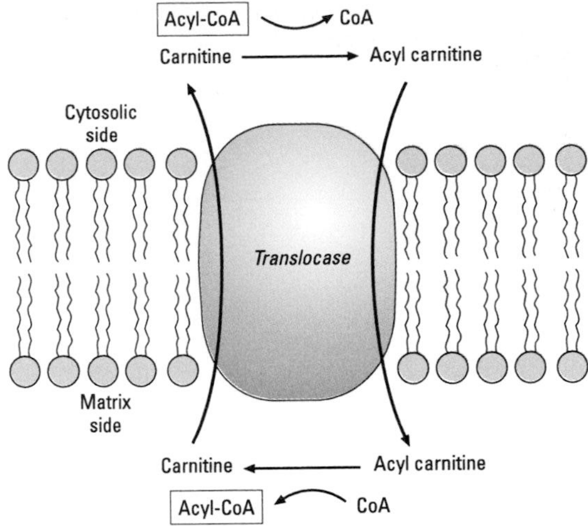

Fig. 6.11 Mechanisms for acyl coenzyme A (CoA) crossing a membrane; acyl carnitine is formed and translocated into the mitochondrial matrix, then the carnitine returns to the cytoplasmic side in exchange for the acyl carnitine.

carnitine and are shuttled across the inner mitochondrial membrane by a translocase enzyme (Fig. 6.11). Saturated acyl-CoA is then degraded by the so-called β-oxidation spiral, involving four basic steps generating acetyl-CoA (Fig. 6.12), which is further metabolized by the Krebs cycle and ETS as before (Figs 6.8 and 6.9). With 1 mol of the fatty acid palmitate as a starting substrate, a total of 129 mol ATP are generated; starting with a triacylglycerol (glycerol + 3 palmitates) the yield is 403 mol ATP. These values compare very favorably with the yields of 39 ATP from glycogen and 18 ATP from lactate (moles per mole of substrate), underlining the value of fats in energy storage.

Amino acids

Amino acids, ultimately derived from protein breakdown, are also metabolized by pathways feeding into the Krebs cycle at various different points (Fig. 6.13). The ATP yields are relatively modest; with alanine, glutamate, and proline as substrates they are 15, 27, and 30 mol per mole of substrate, respectively. Nevertheless cephalopods, and some insects, rely heavily on proline as a fuel during exercise. Note that where proteins or amino acids are fed into the oxidative cycle there will also be production of bicarbonate and ammonium ions as wastes.

Other fuels

In the tissues of some animals acetoacetate and 3-hydroxybutyrate are important alternative fuels. For example, the heart and renal cortex of mammals use acetoacetate in preference to glucose, whilst the brain can adapt to acetoacetate during periods of starvation. The utilization of acetoacetate as a fuel is shown in Fig. 6.14.

Lactic acid

The lactic acid formed in glycolysis (see above) is another important fuel for aerobic metabolism in some animals. In many vertebrates lactate produced in muscle fibers specialized for glycolysis is transported to more aerobic tissues by the circulation where it is oxidized. Even within mammals the types and sequence of fuels used vary between species. For example, the average human contains sufficient glycogen to sustain a lactate turnover rate of 500 mmol kg^{-1} min^{-1}, which is sufficient for about 3 h of moderate intensity exercise, while in llamas lactate flux only makes a contribution to work for a few minutes at the start of exercise. The shuttling of

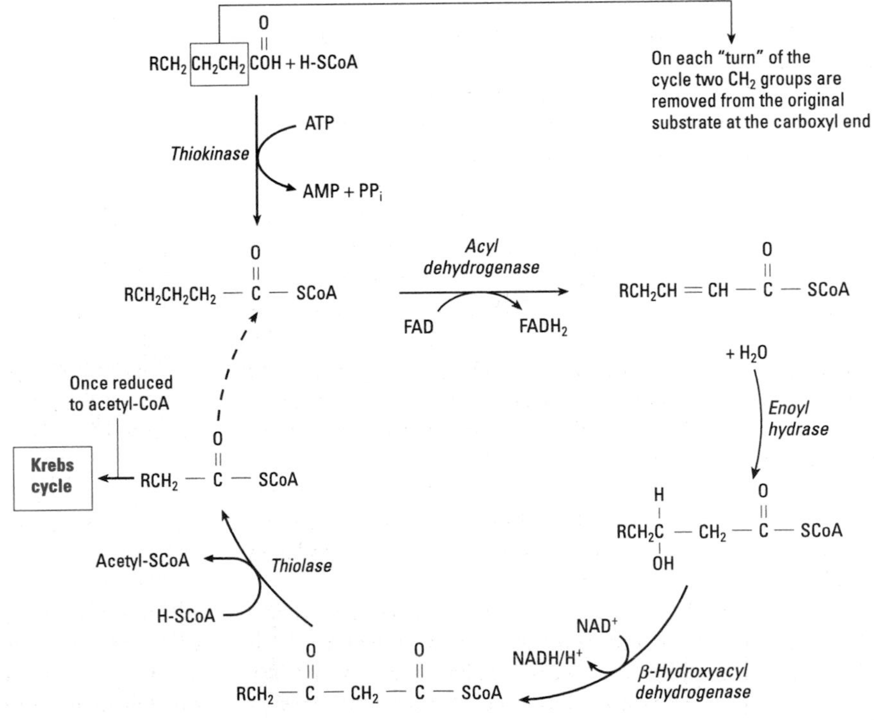

Fig. 6.12 The β-oxidation spiral.

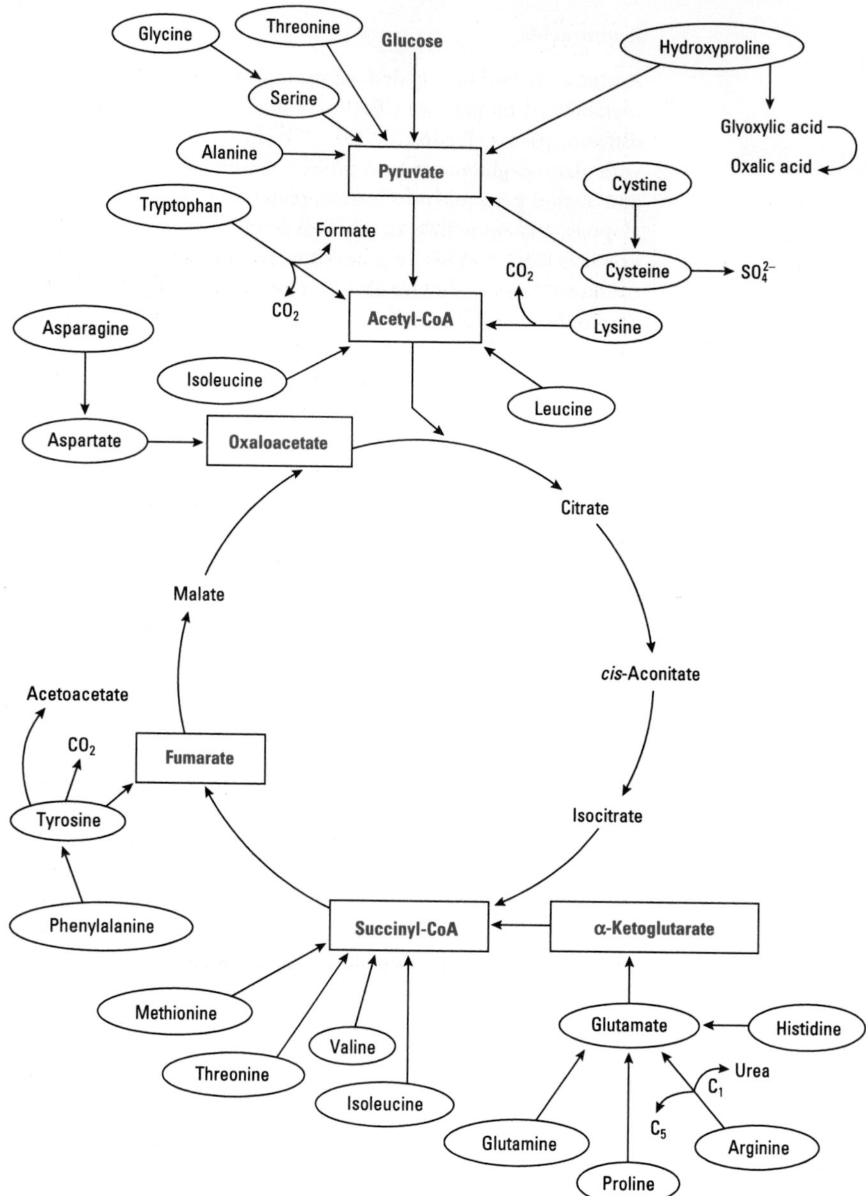

Fig. 6.13 Amino acids feeding into the Krebs cycle as alternative fuels.

lactate between sites of formation and utilization is also minimized or absent in fish and insects. Although lactate concentrations can reach 40 mmol kg^{-1} in the white muscle of salmonid fish following strenuous exercise (and even higher levels in some diving vertebrates), its clearance requires many hours of recovery. Some of the lactate is oxidized and some is converted back to glucose in the kidney and the liver where it is stored as glycogen.

Each of these different potential fuels yields a different net ATP output. These can perhaps best be compared by the amount of each fuel needed for a particular level of ATP generation. Thus an ATP turnover rate of 20 mmol g^{-1} min^{-1} could be supported by a palmitate flux (turnover rate) of 0.15 units, a glucose flux of 1.06 units, a lactate flux of 1.2 units, and an oxygen flux of 3.3 units. Since the supply of oxygen needed is so high, the maximum rate of aerobic work in many animals may in practice be limited by the capacity of the heart and circulatory system to deliver oxygen-loaded blood to the tissues.

To summarize this review of fuels, it should be noted that many animals use different fuels at different times, usually mobilizing them in a clear hierarchical order. This order of recruitment is species specific but is also modified by fuel availability and by current functional need, such as duration and intensity of exercise. This in turn also links to the muscle type involved (see Chapter 9), with red muscle mostly catabolizing fats and white muscle using carbohydrate and protein. A clear example of these effects is seen in migrating fish such as salmon, which use fat reserves for the first 200–300 km, and then start to metabolize their protein. At any time when sudden fast swimming is required, for an escape response or to swim up a fast-flowing river, some carbohydrate metabolism is added; but as

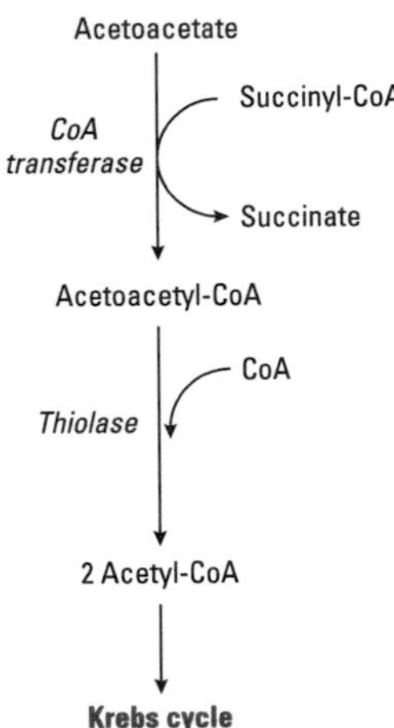

Fig. 6.14 Acetoacetate as a fuel.

far as possible the liver and muscle glycogen stores are conserved until the end of migration when they power the flurry of activity needed for spawning.

6.4.4 Control of aerobic metabolism: metabolic depression

Most of the enzymes in the Krebs cycle, the β-oxidation spiral, and the ETS do not seem to occur as distinctly different isozymes, and control is exerted mainly by enzyme–substrate interactions and feedbacks. The major control points of the Krebs cycle are illustrated in Fig. 6.8. Highly phosphorylated molecules such as ATP can reduce the rate of formation of acetyl-CoA by diminishing the activities of citrate synthase, isocitrate dehydrogenase, and α-ketoglutarate dehydrogenase. The activity of the pyruvate dehydrogenase complex is also regulated, by a combination of reversible phosphorylation, feedback regulation by nucleotides, and product inhibition.

Enzyme phosphorylation is an important mechanism for coping with hypoxia and anoxia. Key glycolytic enzymes can be "turned down" by phosphorylation, with their substrate and activator affinities altered, and the effect can be magnified by changing the concentration of the allosteric activators such as fructose 1,6-bisphosphate. By such means, a coordinated **metabolic depression** right through the glycolysis/Krebs cycle/ETS pathway results. This "hypometabolic state" is used in many animals faced with poor oxygen supply, dehydration, increased acidity or alkalinity, or freezing. In most species the reduced metabolic rate may be 0.05–0.4 normal levels, while in some dehydrated cryptobiotic animals it may be less than 0.05 of normal, and in brine shrimp eggs in diapause there is (uniquely) effectively zero metabolism. In all cases the metabolic depression is related to oxygen sensing (see Chapter 7), and is accompanied by reduced pH, the presence of latent mRNA, changed protein phosphorylation levels, downregulated protein synthesis, but reasonably well-maintained ion pumping levels.

6.4.5 Aerobic metabolism in insects

Insects are a notable exception to some of the patterns of metabolic fueling outlined above. Insect flight is a very energy-demanding process, and one which is supplied almost entirely by aerobic metabolic pathways. The maximum rate of oxygen consumption increases 50-fold or more between rest and take-off, and magnetic resonance spectroscopy in resting and flying locusts has shown a 600-fold increase in ATP production. However, relatively little energy is stored in the flight muscles themselves. The most important energy store is the fat body, which contains fat (triacylglycerol) and glycogen, and these are broken down and delivered to the muscles via the hemolymph. The direct substrates for aerobic respiration are the disaccharide trehalose, the lipid diacylglycerol, and the amino acid proline. Carbohydrate is used exclusively in many Diptera and Hymenoptera, whereas lipids (the fatty acid chains of diacylglycerol) are the main fuels in Orthoptera and Lepidoptera. Proline is used as a fuel in insects such as the tsetse fly (*Glossina*) and the Colorado potato beetle (*Leptinotarsa*), where it is partly oxidized to alanine (Fig. 6.15). The alanine is transported to the fat body for the resynthesis of proline, and the acetyl-CoA required is derived from fatty acids. Thus proline-supported insect flight can also deplete lipid stores. Insect flight muscles contain very low concentrations of lactate dehydrogenase, and lactate is of minimal importance as a fuel. The flight muscles are also unusual in containing only low levels of phosphagens such as phosphoarginine. This is because insect flight muscle can shift between its resting and maximal rates of ATP production more rapidly than the tissues of any other animals, due to its radically different oxygen transport system (see Chapter 7). It therefore does not require significant buffering of ATP.

As a rule of thumb insects that are oxidizing exclusively carbohydrates usually only make short flights; for example, honey-bees (*Apis*) rarely travel more than 3 km from their hives. Lipid stores contain more energy per unit mass than carbohydrate stores and are therefore preferable for long flights, as in migrating monarch butterflies. In many cases carbohydrate is used at and soon after take-off, followed by lipid during prolonged flight. For example, during a continuous flight lasting 10 h, which requires a metabolic rate of 280 J g^{-1} body mass h^{-1}, a locust would have to oxidize 500 mg of glycogen, equivalent to 35% of its body mass. In contrast, a similar flight requires the oxidation of only 70 mg lipid. In other words, on average, 1 mg of lipid delivers as much energy as about 8 mg of glycogen. This difference is related to the higher energetic value of lipid relative to carbohydrate and to the fact that glycogen is stored in cells in association with water.

The mobilization of fuels in insects is under the control of hormones secreted from the glandular lobe of the corpus cardiacum (part of a neurosecretory complex just behind the brain). These hormones are octapeptides and decapeptides, with sequences that are highly conserved between different insect species. The hormones activate a lipid shuttle whereby high-density lipoproteins are used to transport digested lipid from the gut to tissues and to and from storage sites. In addition octopamine (a short neurosecretory peptide) is probably involved in the rapid mobilization of diacylglycerol during the first few minutes of insect flight. The whole system is supported by oxygen delivery via the tracheoles (see Chapter 7), which in insects is so effective that the *in vivo* cytochrome oxidase turnover rates approach those of the theoretical maximum (whereas in vertebrates the cytochrome oxidase only works at about 10% of its theoretical maximum).

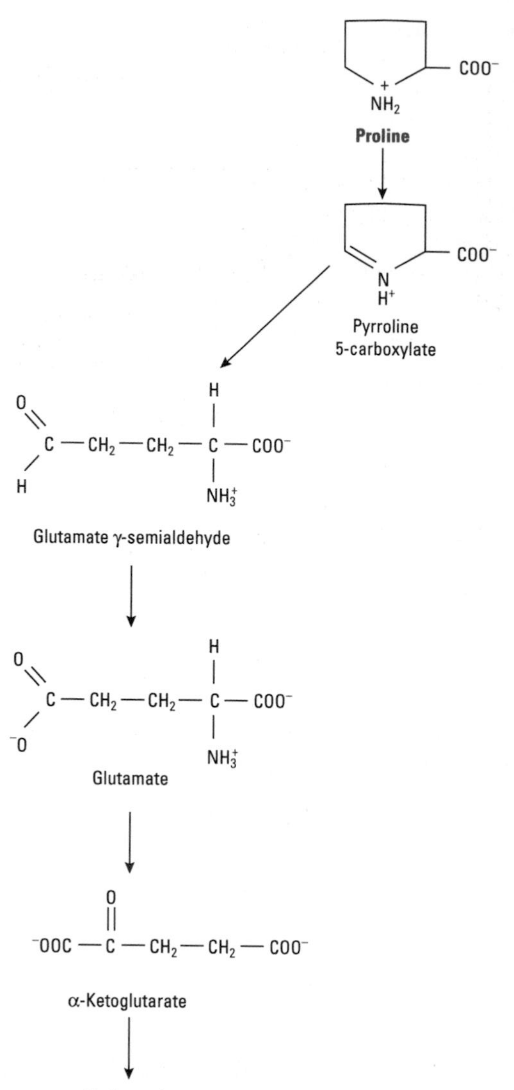

Fig. 6.15 Proline as a fuel.

6.5 Metabolic rates

Metabolic rate is one of the most commonly measured physiological variables, and is a measure of the total energy metabolized by an animal in unit time. It has proved immensely useful in comparative studies of animal adaptation and performance. But despite its wide usage, its value can be determined only indirectly, from any one of four kinds of analysis:

1 The energy value of food taken in minus that of the wastes excreted.
2 The amount of oxygen used up (or carbon dioxide produced).
3 The amount of heat produced.
4 The amount of metabolic water produced.

Of these, the usage of oxygen is most easily measured, and rate of **oxygen consumption** ($\dot{V}O_2$) has therefore been the routine measure of metabolic rate in physiological studies. It does not of course take

Table 6.4 Energy content, respiratory quotient (RQ), and metabolic water production of foods.

	Heat production			
Nutrient	**kJ per g consumed**	**kJ per l O_2 consumed**	**RQ (CO_2 formed /O_2 used)**	**Metabolic water (g per g food)**
Carbohydrates	17.4	20.9	1.00	0.56
Lipids	39.3	19.6	0.71	1.07
Proteins	17.8	18.6	0.80	0.4–0.5

account of any anaerobic metabolism, so can be very misleading for some kinds of animal. However, oxygen consumption does correlate closely with **heat production**, whatever the kind of foodstuff being metabolized (Table 6.4), with an average value of about 20 kJ of heat per liter of oxygen. Conversions between oxygen usage and heat production are therefore relatively simple. Heat production can be assessed by whole animal calorimetry (Fig. 6.16), and calorimeters can be calibrated with some fuel of known energy content (Table 6.4).

Conversions between these two parameters and the first measure given above, the **food value taken in**, are less straightforward because we know that different foodstuffs have different calorific values (see above); fat yields more than twice as much energy per gram used up, in comparison with carbohydrates or proteins (Table 6.4). This makes fat a very good energy store for mobile animals where body weight matters, as it provides greater energy per unit weight of body tissue (though of course this stored fat still requires as much oxygen to release the energy as any other store would do). The difficulty of using food intake as a measure of metabolic rate in situations where diet composition may not be known in great detail is partly overcome by use of the **respiratory quotient** (RQ). This is the ratio (fairly simple to measure) of carbon dioxide released to oxygen used up, and it differs for each of the main nutrient categories (Table 6.4). An RQ value close to 0.7 indicates that fat is principally being metabolized, whereas an RQ approaching 1.0 indicates carbohydrate-based metabolism. Intermediate values are more confusing, with several possible combinations of substrate, but even these can be disentangled if the amount of nitrogen excretion (ammonia, urea, or uric acid) is also measured, as this gives a value for protein metabolism; with this taken out, the proportions of the remainder that are due to fat and carbohydrate, respectively, can be calculated.

The final technique for assessing metabolic rates, from **metabolic water production**, is in practice rather difficult, since a small amount of metabolic water is diluted into the large water content of the whole animal. However, improvements in isotope trace techniques (e.g. using doubly labeled water) have facilitated these kinds of analyses, and they are increasingly being used. Again the release of water depends on the nature of the substrate being metabolized, so that corrections as outlined above must be made.

In theory then, and to a large extent in practice, metabolic rates can be measured for any aerobically respiring animal. There are, however, a number of problems of definition and measurement inherent in making any comparative analyses of metabolic rates in animals.

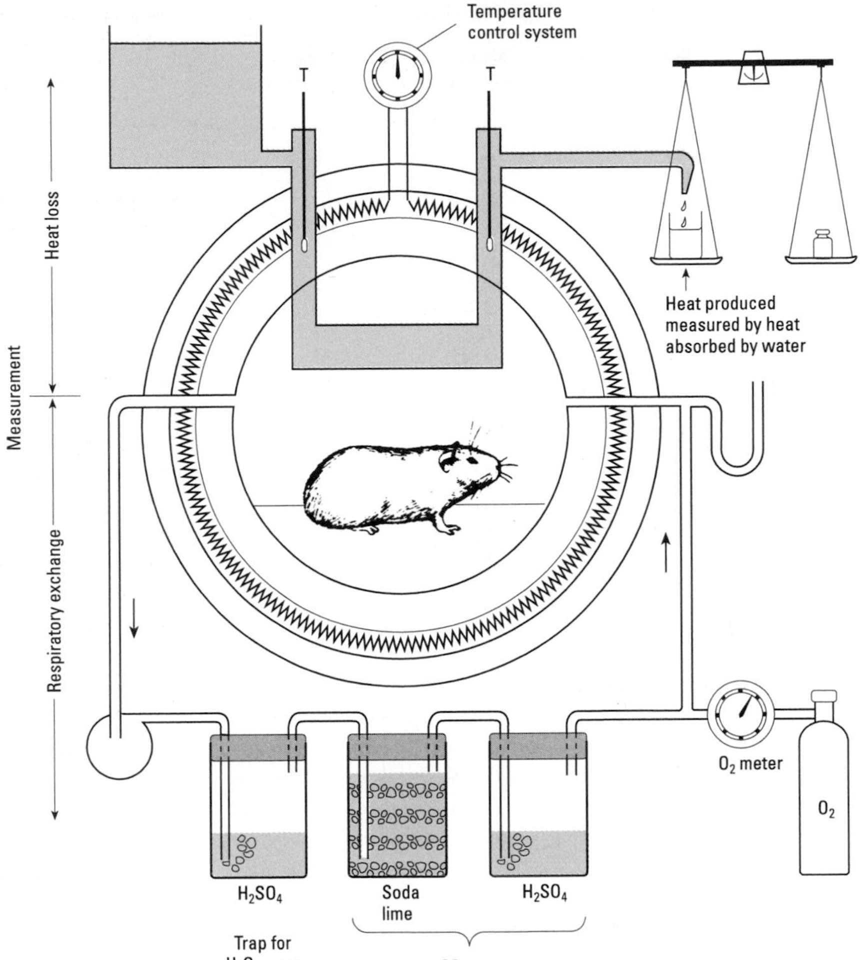

Fig. 6.16 Basic whole-animal calorimetric apparatus. (Adapted from Kleiber, M. *The Fire of Life: an Introduction to Animal Energetics*, copyright 1961, reprinted by permission of Wiley-Liss, Inc., a subsidiary of John Wiley & Sons, Inc.)

6.5.1 Metabolic rate and activity

The first complication in measuring an animal's metabolic rate is that it is substantially affected by levels of rest and activity. An active fish or crustacean may have a metabolic rate several times that of a resting individual, whilst for flying insects the change may be more than 100-fold (partly because body temperature rises sharply during this activity). To complicate matters further, in certain kinds of activity such as diving, metabolic rates may actually be temporarily reduced (see Chapter 11).

Therefore several levels of metabolism (and metabolic rates) are commonly recognized:

1 **Standard metabolic rate** (SMR) is the level required for a minimal resting lifestyle with no spontaneous activity, no digestion of food, and no physical, thermal, or psychological stress. In ectotherms this is temperature dependent. In an endotherm, where this standard metabolism is not varied with temperature (see later), the term **basal metabolic rate** (BMR) may be used instead, and must be measured in the thermoneutral zone (see Chapter 8).

2 **Routine metabolic rate** (or "resting metabolic rate") is the level of metabolism for minimal (but normal and unrestrained) activity, and is rather loosely defined in practice.

3 Average daily metabolic rates may be worked out for the routine activities of a 24 h period and this is sometimes termed **field metabolic rate**; or a metabolic rate may be calculated in laboratory studies over a long enough time period to insure that body mass (M_b) remains constant, food inputs balancing metabolic outputs, giving the **sustained metabolic rate**. Field and sustained metabolic rates are probably the most useful parameters to characterize an animal's normal energy expenditure.

4 **Active metabolic rate** is the level required for specified levels of exercising activity (forced if necessary), up to the maximum possible level of **maximum metabolic rate** (MMR).

Measuring standard metabolism directly is usually impractical, as even in a darkened chamber most animals show some spontaneous activity. Therefore it is necessary to record a series of metabolic rates along with an estimate of the activity level during the recording, and extrapolate back from these to the SMR at zero activity. For laboratory measurements of routine or sustained or active metabolism the actual level of activity during the recording has to be carefully specified, and interspecific comparisons are therefore particularly difficult. It is crucial to control for stress, feeding history, acclimation temperature, activity, and body size (some of these aspects are considered in more detail below) and to compare like with like in terms of developmental stage or age, and lifestyle.

Table 6.5 Resting and peak metabolic rates (MR) and factorial aerobic scopes.

Species	Resting/standard MR (ml O_2 g^{-1} h^{-1})	Active/peak MR (ml O_2 g^{-1} h^{-1})	Factorial aerobic scope
Invertebrates			
Barnacle	14	58	4
Periwinkle	0.2	0.4	2
Fruit fly	2.3	30	13
Butterfly	0.6	100	170
Fish, reptiles			
Sunfish	0.03	0.29	9
Salmon	0.08	0.60	8
Monitor lizard	0.08	0.38	5
Desert iguana	0.05	0.89	17
Turtle	0.03	0.64	20
Birds, mammals			
Budgerigar	1.7	18.9	11
Hummingbird	2.8	42.0	15
Mouse	2.5	20.0	8
Dog	0.33	4.0	12
Human	0.23	3.2	14
Elephant	0.11	1.2	11
Whale	0.04	0.6	16

In mammals, the range of possible metabolic rates, from basal to maximum, may be very large (Table 6.5). The ratio between maximum and basal rates of oxygen consumption is termed the **factorial aerobic scope**; the difference between the two levels is the **absolute aerobic scope**. Factorial aerobic scopes are easier to use in comparisons between species, and are usually in the range 5–12 for vertebrates, though larger and more athletic mammals may have somewhat higher values. Invertebrates tend to have values in the range 2–10, the lower values occurring in smaller and less active species and in cool aquatic habitats where body temperatures during activity are lower; some terrestrial invertebrates (especially flying insects) may have much larger aerobic scopes. However, insects are very different in design and in metabolic responses from most other animals, and they may have very large factorial aerobic scopes during flight, with values commonly in the range 80–120 for large bees and moths, whose body temperature (T_b) is considerably elevated when flying. Even when temperature corrected, these factorial scopes may be about 30–50, reflecting the huge metabolic rates of insect flight muscles.

Comparisons between sustained metabolic rates and BMRs are also of interest, and in mammals this value (the **sustained metabolic scope**) ranges between 1.3 and 7.0, with a value of 4.3 in the best highly trained human athlete. The apparent ceiling of about 7 on this parameter may be due to constraints in the energy-consuming machinery (mainly the muscles) or in the energy-supply machinery (gut, liver, circulation, or lungs perhaps). Or it could be that animals have evolved with the capacities of some or all of these limiting factors being matched to each other. The fact that sustained metabolic scope may differ for different kinds of energy expenditure in the same animal (e.g. in mice it is 3.6 for physical exercise, 4.8 for heat production, and 6.5 for lactation) suggests that more peripheral limitations on the energy-consuming systems may be most important. However, the observation that sustained metabolic rate tends to be coupled with routine metabolic rate (perhaps because increases in the former demand increased masses of energy supply organs, which in turn require high maintenance costs that disproportionately raise routine levels) complicates the whole issue.

6.5.2 Metabolic rate and temperature

It is unusually difficult to measure the metabolic rate of an animal in relation to changing thermal conditions, not least because the animal itself continually alters the imposed experimental regime. Rates of oxygen consumption are the most commonly measured variable, and these will certainly show variation in response to a sudden shift of temperature. But the response will be erratic initially (as the animal's behavior alters) and will often overshoot and then oscillate for a period of several hours or even days; even in quiescent stages such as eggs or pupae, or in anesthetized animals, these overshoots occur, suggesting that they reflect perturbations to an underlying cellular steady state.

To be useful, then, experimental analyses involving an acute change of temperature must carefully control for handling and shock effects, and must last long enough to reach the new stabilized condition, which may be particularly difficult for sudden large temperature changes. This immediately introduces a further problem, as animals will commonly acclimate to a changed thermal regime by changing their own metabolic rate in compensation for the direct rate-changing effects of temperature; and the extent to which they can do this may be variable through a season, or may change in relation to the immediate history of the particular animals tested. As these acclimatory changes may take only a few days to implement, they can counteract the effect being studied and reduce the slope of the rate–temperature curve that we are trying to determine. Any study is therefore a compromise between avoiding the influence of the initial shock reaction, and avoiding the onset of physiological adjustments by the experimental animal.

Metabolic rate commonly has a Q_{10} of about 2–3 for the majority of animals; i.e. a 10°C increase of ambient temperature produces a 2–3-fold increase in metabolic rate (Fig. 6.17; see also Chapter 8).

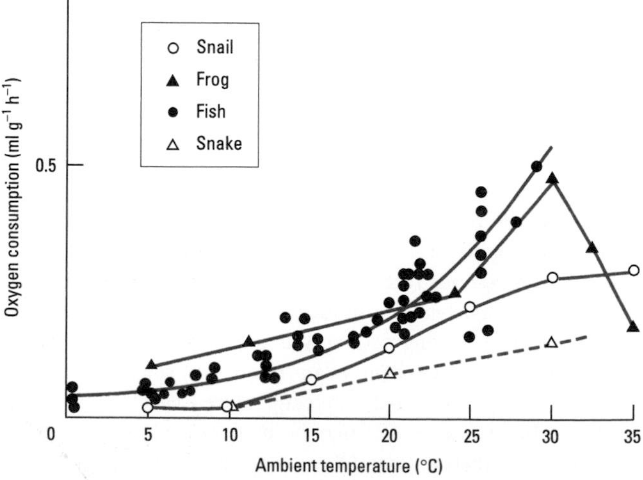

Fig. 6.17 Relationship between ambient temperature and metabolic rate (as oxygen consumption) for four ectotherms; the Q_{10} is commonly 2–3. (Data from Haugaard & Irving 1943; Aleksiuk 1971; Jusiak & Poczopko 1972; Santos *et al.* 1989.)

Values may vary depending on whether the 10°C range studied is distant from the normal environmental temperatures of the species studied; some fish and crustaceans from polar seas appear to have a relatively high Q_{10} for a temperature range around 0°C but a reduced Q_{10} as the test temperatures approach their lethal limits (which may be only 4–8°C).

For endotherms (primarily the mammals and birds) the normal Q_{10} value of 2–3 is obscured because these animals control their body temperature so precisely. In such cases, only when the animals are in states of metabolic depression, either torpid, hibernating, or estivating, does the underlying effect of changing temperature on metabolism become apparent (Fig. 6.18).

Overall, though, when resting metabolic rates are mass and temperature compensated, they are quite similar across all organisms, with a factor of only 20 separating the lowest (unicells and plants) from the highest (endothermic vertebrates).

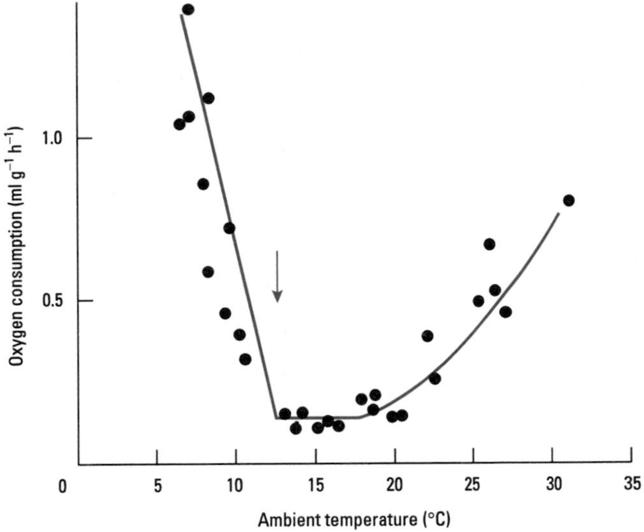

Fig. 6.18 The relationship between ambient temperature and metabolic rate (MR) (as oxygen consumption) for an endotherm that undergoes torpor. MR declines to a new level during torpidity, but increases sharply again if ambient temperature falls below a species-specific minimum (arrow).

6.5.3 Metabolic rate and size

Many of the component processes that contribute to metabolic demand in a resting animal (e.g. ventilation rate, heart rate) are related to body weight. Therefore, as the last section implied, estimates of SMR and of active metabolic rate strictly have to be compared between individuals of the same size. In other words, as in so many areas of comparative biology, scaling effects must be very carefully considered, and we touched on this issue in relation to metabolic rate in Chapter 3.

The relationship between metabolic rate and body mass, both interspecifically and within a species, remains one of the most important and controversial issues in comparative animal physiology. If we look initially at the pattern just for the mammals, a group that has inevitably attracted most attention, it is obvious that a large elephant has a higher oxygen consumption than a small mouse. The problem concerns the proportionality of this relationship, which takes the general form of a power curve:

$$\text{Metabolic rate} = a\ \text{mass}^b$$

On a simple plot, a curve results with a crowded left-hand side (Fig. 6.19) where b is the mass exponent of the curve. It is more useful in practice to take logarithms of the metabolic rate and body mass values, so that the relationship becomes:

$$\text{Log metabolic rate} = \log a + b \log \text{mass}$$

The persistent controversy concerns whether or not this log plot is a straight line, and what the value of b, the gradient, might be. An example of the log plot is shown in Fig. 6.20. Most of the analyses that have been done for mammals give a value for b of about 0.75 (Table 6.6).

Whatever the true relation here, this kind of plot has its uses in revealing the "aberrant" physiological performances of atypical mammals. Any species whose oxygen consumption is notably greater or less than that for other animals of the same size becomes of special interest to the comparative biologist, and reasons for the anomaly can be sought. The same is true intraspecifically; a value close to (often rather lower than) 0.75 is usually obtained within any

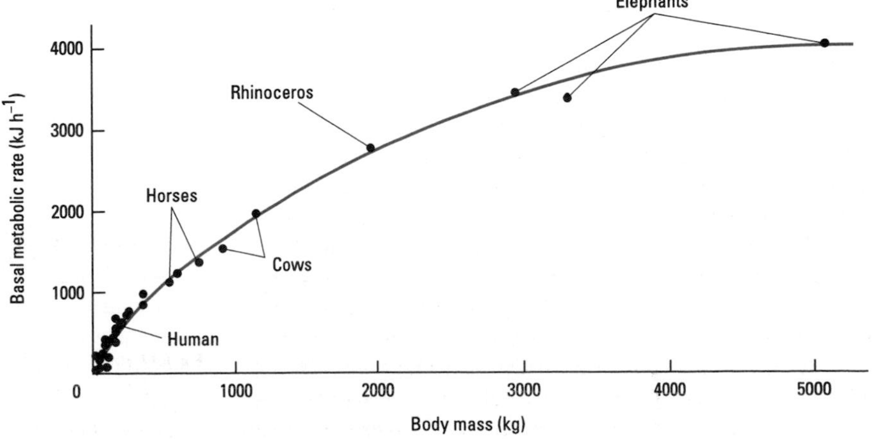

Fig. 6.19 Linear plot of metabolic rate against mass.

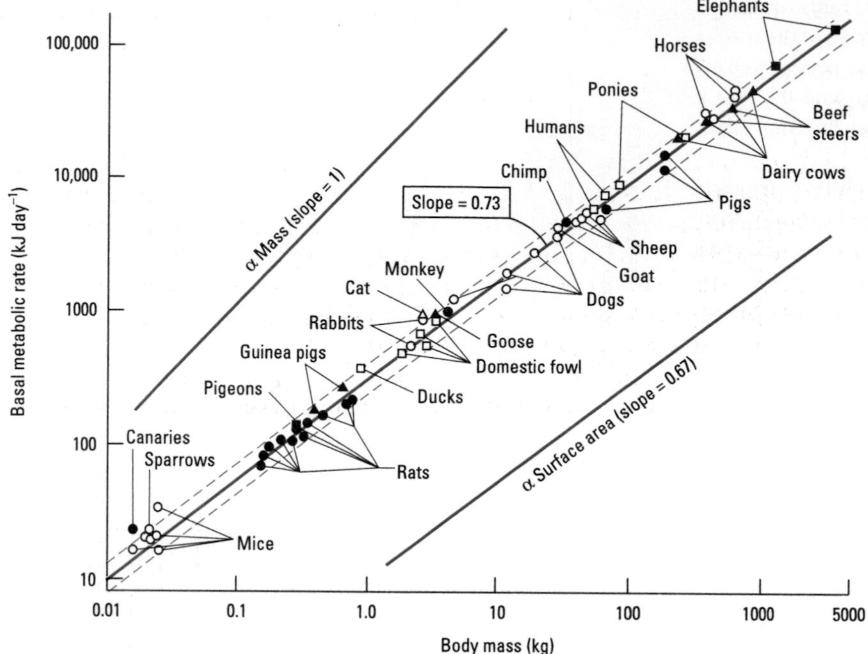

Fig. 6.20 Log plot of metabolic rate against mass for birds and mammals. (Data from Kleiber 1932; Heusner 1982.)

Table 6.6 Mass exponents for the relationship between mass and metabolic rate.

Animal group	Temperature (°C)	Mass exponent (*b*)
Ectotherms		
All combined	20	0.80–0.88
Nematodes		0.72
Annelids		0.61–0.82
Molluscs		0.75
Echinoderms		0.65–0.80
Crustaceans		0.81
Spiders		0.65
Insects		0.62–0.67
Reptiles		0.77–0.80
Snakes		0.81–0.98
Amphibians		0.75–0.86
Fish		0.70–0.88
Endotherms		
All birds	41	0.69–0.75
Passerines		0.72–0.73
Honey-eaters		0.68
Ratites		0.73
All mammals	39	0.73–0.76
Marsupials		0.75
Edentates		0.66
Bats		0.72
Rodents		0.72
Shrews		0.75

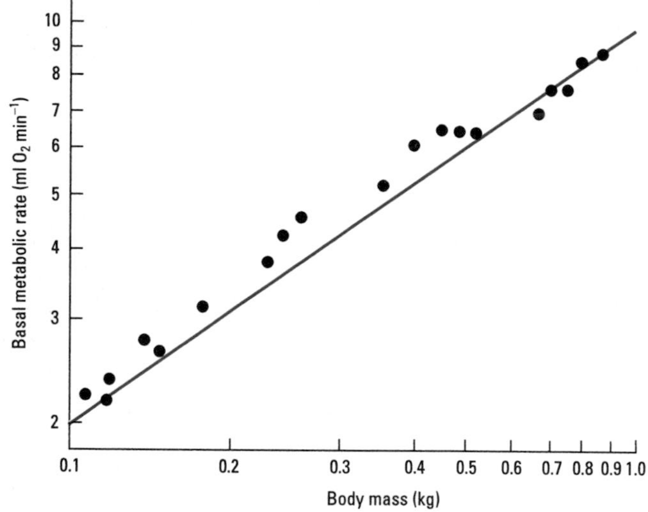

Fig. 6.21 Log plot of metabolic rate vs. mass for a single species, the guinea pig. The line drawn has a slope of 0.67. (From Wilkie 1977.)

one species (Fig. 6.21), and outlying individuals can again provide interesting insights.

Moving to a broader scale, we can do the same kind of comparison for the whole of the animal kingdom, and an example of this kind of plot was used in Chapter 3 when we considered scaling problems in general. It turns out that all organisms have a similar relationship between body mass and metabolic rate, though the slope is set within somewhat narrower bounds for birds and mammals (0.70–0.75) than it is for other animals (0.6–0.9 range, with many invertebrate groups having a mass exponent of 0.66–0.72), and it is somewhat lower again for unicellular organisms than for the multicellular animals (Table 6.6). The intercept values, *a*, are also of course lower for the majority of animals that cannot generate much heat internally (the ectotherms; see Chapter 8), with mean values about 3 J h^{-1}, than they are for endothermic mammals and birds (mean 116 and 139 J h^{-1}, respectively) (Fig. 6.22).

The metabolic differences between ectotherms and endotherms seem to be explained by a number of small structural and physiological differences rather than by any one factor. For example, cells in endotherms have somewhat greater mitochondrial volumes, and slightly greater membrane areas per unit of mitochondrion, with

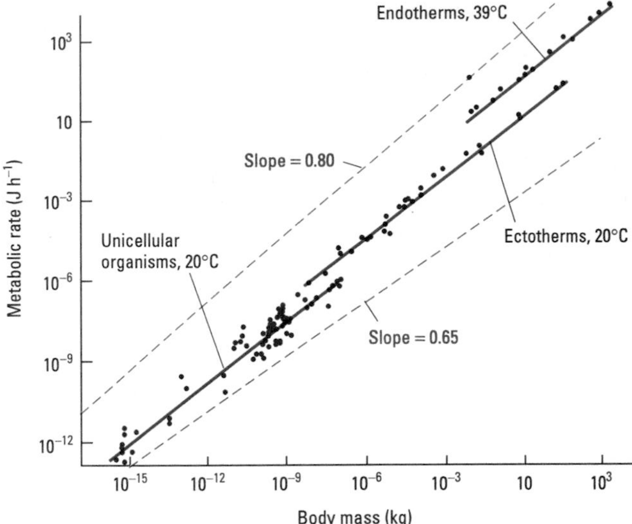

Fig. 6.22 Log plot of metabolic rate vs. body mass for unicells and for both ectothermic and endothermic animals. (From Phillipson 1981.)

raised enzyme activities, and rather more "leaky" membranes around both the cytoplasm and mitochondria (especially the inner mitochondrial membrane). They also have proportionately larger organs (liver, kidneys, and brain in particular).

As to why metabolic rates always show a mass exponent of 0.75 in relation to body mass, there are several hypotheses but no clear-cut answers. Metabolic rate could theoretically be proportional to surface area ($mass^{0.67}$) because crucial metabolic processes and exchanges are occurring across surfaces, but it could also be proportional to volume ($mass^{1.0}$) since catabolic reactions occur and heat is generated within all the three-dimensional volume of the animal. Perhaps the exponent 0.75 represents a balance of these two effects. If so, it is a little unclear why the value of 0.75 should be found in ectotherms as well as endotherms, since the balances of body temperature, heat exchange, and surface area differ substantially for the two strategies (see Chapter 8). However, support for this explanation based on a balance of surface-related and volume-related factors comes from considering the scaling of purely anaerobic metabolism; here processes occur entirely within the cytoplasm, requiring no surface-related exchanges, and the mass exponent for a range of animals is close to 1.0, as would be expected for purely volume-related effects.

However, other explanations for the aerobic mass exponent have been offered, bringing in mechanical considerations, fractal scaling factors in capillaries and respiratory surfaces, and the effects of four-dimensional rather than three-dimensional scaling (perhaps with time as the fourth dimension). These have not yielded any real new insight into the problem, but the debate continues. It may be that there are different explanations for the 0.75 figure applying between thermal strategies, between taxa, and between individuals, and that the "average" value of 0.75 is a bit misleading (bear in mind that we are looking at a log-log plot, where a several-fold deviation from the "norm" for any one species is scarcely detectable when both the axes have many orders of magnitude). Many species do in fact violate the "0.75 law".

Table 6.7 Mass-specific oxygen consumption in mammals.

Species	Mass (kg)	O_2 consumption ($ml\ g^{-1}\ h^{-1}$)
Shrew	0.005	7.40
Harvest mouse	0.009	1.50
Kangaroo mouse	0.015	1.80
House mouse	0.025	1.65
Ground squirrel	0.095	1.03
Rat	0.290	0.87
Cat	2.5	0.68
Dog	11.7	0.33
Sheep	43	0.22
Lion	50	0.23
Human	70	0.22
Eland	240	0.17
Elephant	3850	0.07
Blue whale	100,000	0.04 (est.)

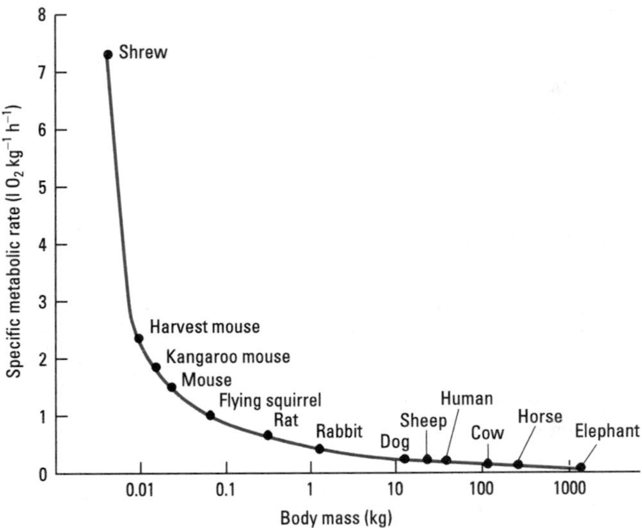

Fig. 6.23 Semilogarithmic plot of specific oxygen consumption against body mass for mammals—the "mouse to elephant curve". (From Schmidt-Nielsen 1997, courtesy of Cambridge University Press.)

Given these controversies about metabolic rate, it is clearly often more useful to compare the metabolic rate of different animals per unit of body mass by determining the **mass-specific oxygen consumption** (mass-specific metabolic rate). Table 6.7 shows that the small mammal actually has tissues that are very much more metabolically active than a large mammal; the rate of oxygen consumption per gram of tissue *decreases* consistently as body weight increases (Fig. 6.23). This semilogarithmic plot indicates that mouse tissue has about 100 times the oxygen consumption of elephant tissue (around 7 ml O_2 g^{-1} h^{-1} for mice and shrews, and only 0.07 ml O_2 g^{-1} h^{-1} for an adult elephant). This may initially seem odd—why should the homologous tissues in two quite closely related animals have such different metabolic rates just because the animals are different sizes? Why does cellular metabolism decline as organism size increases? There may be several factors operating here, including the relatively larger membrane leaks and inner mitochondrial proton leaks in small animals, requiring more active compensating pump

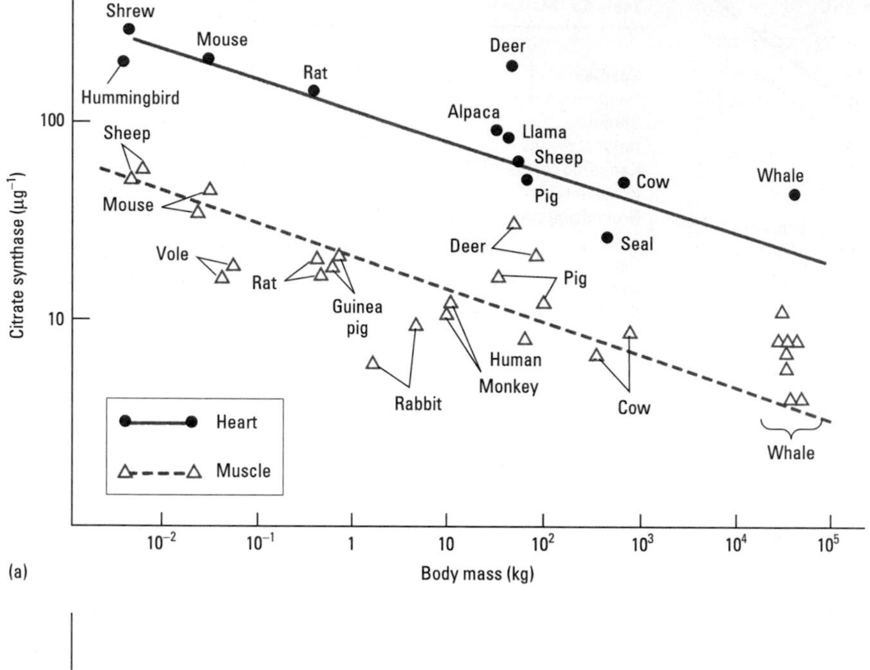

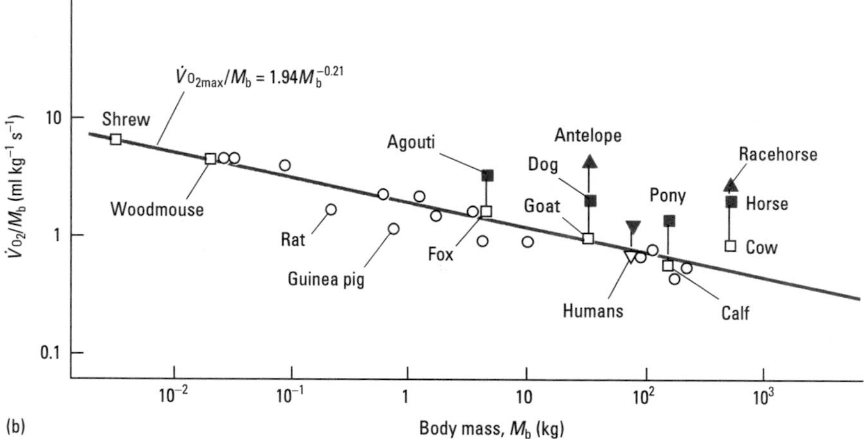

Fig. 6.24 Scaling of (a) mass-specific enzyme activities in bird and mammal skeletal muscles and heart muscles, and (b) maximal oxygen demands, with body mass. In (b) the more athletic species (green symbols) always have values of $\dot{V}o_{2max}$ well above those for sedentary species of similar body mass (for humans a sedentary adult and a highly trained athlete are compared). (a, From Hochachka *et al.* 1988; b, from Hoppeler & Weibel 1998, courtesy of Company of Biologists Ltd.)

activity. The higher leakage pathways in cells of small bodies link in turn to lower phospholipid saturation in their cellular and intracellular membranes, so that biological membranes ultimately serve as the "metabolic pacemakers".

Because of the size/metabolic rate relationship between mouse and elephant, the blood flow supplying the oxygen must be about 100 times greater per unit of tissue in the mouse, and all related variables of the cardiovascular systems, such as heart function and respiratory rate, must be similarly scaled up for the mouse. Even more critically, the specific food intake of a mouse must also be 100 times greater, whence the familiar observation that small mammals almost never stop eating during their waking hours.

The scaling of the MMR in mammals suggests that mass-specific oxygen consumption scales to body mass to the power of –0.25. Extrapolation of the data to blue whales weighing 100,000 kg would suggest a mass-specific oxygen consumption of only 0.02–0.04 ml O_2 g^{-1} h^{-1} (though such extrapolations are not strictly valid, as we discussed in Chapter 3). What are the biochemical consequences of these large differences? Are there upper and lower limits to energy metabolism that are related to body size?

It has been shown for fish, birds, and mammals that mass-specific enzyme activities in heart and in skeletal muscle scale in much the same fashion as the mass-specific oxygen consumption of the whole animal, varying with body mass to the power of –0.21. Data for birds and mammals are shown in Fig. 6.24a. The roughly parallel scaling of maximal oxygen demands (Fig. 6.24b) and maximal enzyme activities insures that the carbon flux through an animal can be increased during intense activity, in all species of all sizes. Substrate concentrations are thus essentially scale independent, and are kept within the range for maximum sensitivity of regulation at all body sizes.

The cellular units of aerobic metabolism are the mitochondria (see section 6.3). The fraction of muscle fiber volume (the volume density) that is occupied by mitochondria also scales with a slope of –0.21 on double-logarithmic plots. For most species it would appear that 1 ml of muscle mitochondria consumes about 200 μmol O_2 min^{-1} during maximal exercise, regardless of the oxygen source. Since specific oxygen consumption increases with decreasing size, this implies that the volume density of muscle mitochondria also increases systematically with decreasing body size. However, there is also an effect of lifestyle, in that for a given body size more athletic species have a higher volume of mitochondria per unit muscle mass. For example, the ratios of maximum oxygen consumption to body mass in dogs and horses are 2.5 times greater than in goats and cattle, respectively, matched by 2.5 times more mitochondria per unit mass in their muscles.

Clearly, in the tiniest mammals and birds there must be some limit to the number of mitochondria that can be packed into a muscle fiber without compromising its function for doing external mechanical work (see section 9.12). The packing of mitochondria in the flight muscles of hummingbirds (birds that can hover) appears to be close to the upper limit possible in vertebrates, at around 50–55%. Volume densities of mitochondria higher than this would probably be counterproductive, since this would decrease the space available for myofibrils and sarcoplasmic reticulum, the structural elements required for force generation and calcium cycling within the muscle fibers. Hovering flight is energetically expensive, requiring wing beat frequencies of 60–80 Hz (see section 9.13.2), and metabolic rate is very high due to the small size of the birds (1–2 g), so it is logical that small hovering hummingbirds should be at the upper limit of mitochondrial packing.

Double-logarithmic plots of the surface density of the inner cristae membranes of mitochondria (which contain the electron transport chain assemblages; see section 6.3) against body mass also scale to about the power of –0.21, but there are some very interesting exceptions, again including hummingbirds. Mitochondria from hummingbird flight muscles have around twice the surface density of cristae and hence twice the rate of oxygen consumption of other birds of similar size. This can be viewed as an evolutionary design innovation to overcome the general constraints imposed by the scaling of the structural elements responsible for oxygen uptake. The increased density of cristae effectively doubles the maximal rate of oxygen consumption that can be achieved with a volume density of 50% mitochondria.

However, there is evidence that the real limit to aerobic capacity in hummingbirds is not at the enzyme level but is related to the cardiovascular system's ability to deliver oxygen-loaded blood to the capillary circulation. In mammals the rate of oxygen delivery per milliliter of capillary is practically invariant, but capillary density scales negatively with size (and thus positively with mass-specific metabolic rate) over the size range 20 g (mice) to 500 kg (horses). Insects, with their unique tracheal system for delivering oxygen, have overcome these constraints in an entirely novel way, and achieve much higher levels of mass-specific oxygen consumption than birds and mammals.

The limits set on muscle activities can be overcome in other ways, but only at some cost. Some animals have muscles that contract much faster than hummingbird flight muscles, but at the expense of the work done per contraction cycle (see section 9.12). For example, the tymbal (sound-producing) muscles of cicadas operate at a contraction frequency of 550 Hz. The volume densities of mitochondria, sarcoplasmic reticulum, and myofibrils are in the ratio of approximately 1 : 1 : 1 (i.e. the mitochondrial volume density is only about 33%). A very high frequency, required to produce the necessary sound, has been achieved, but because there are more of the energy-supply systems there are fewer myofibrils, and the gain in frequency of contraction is at the expense of the muscles' ability to do mechanical work.

6.5.4 Metabolic rate and oxygen availability

Given that metabolic rates are often assessed by measuring oxygen consumption, it is important to understand the relation between metabolism and oxygen supply (though here again we are really only considering aerobic animals). There are two main patterns to be found, corresponding to the distinctions between regulators and conformers made in Chapter 1. **Metabolic conformers** have standard (basal) metabolic rates that are directly proportional to ambient O_2. By contrast the **metabolic regulators** maintain their standard (basal) metabolic rates even when Po_2 is reduced, at least down to some critical value (below which their metabolism slows down markedly and they enter a hypometabolic state, as discussed in section 6.4.4). There is no obvious phylogenetic or environmental pattern to the occurrence of this strategy; we will look at examples in more detail in Chapter 7 when we consider respiratory adaptations.

It is worth noting that metabolic rates also vary with other variables, including developmental stage and age (usually increasing through development, and declining in old age), food availability, diet and exercise regime, photoperiod (with diel patterning), salinity, and hormonal balance. In each chapter in Part 3 of this book we will meet examples of these kinds of interactions. Indeed there is a common theme of **hypometabolism** associated with various physiological problems; animals may adopt a hypometabolic state ("turning down the metabolic burners") when they enter torpor daily or seasonally, when they acclimate to unusual drought or cold or heat, when they dive, when they burrow, or when they starve. The machinery underlying this response, at the biochemical level, is common to all these conditions.

6.6 Energy budgets

All the energy acquired by animals through the ingestion of food is ultimately either used in metabolic processes, or deposited as new body tissue (growth or reproduction), or lost as waste products. Energy budgets are used to describe the partitioning of the ingested energy into these various components. The simplest form of balanced energy budget can be represented as:

$$E_{in} = E_{out} + E_p$$

where E_{in} is the energy ingested as food, E_{out} is the sum of the energy losses, and E_p is the energy associated with production (Fig. 6.25). The equation is usually expanded to further subdivide each of the terms on each side:

$$E_{food} + E_{drink} = E_{feces} + E_{urine} + E_m + E_p$$

where E_m represents the energy losses in metabolism. Metabolism can be further subdivided into basal metabolism (energy used to maintain bodily functions), active metabolism (energy used in locomotion and other activities), and energy used in the digestion, absorption, and processing of food. The E_p term in the equation is also often partitioned into somatic growth and reproductive growth (energy expended on the production of gametes).

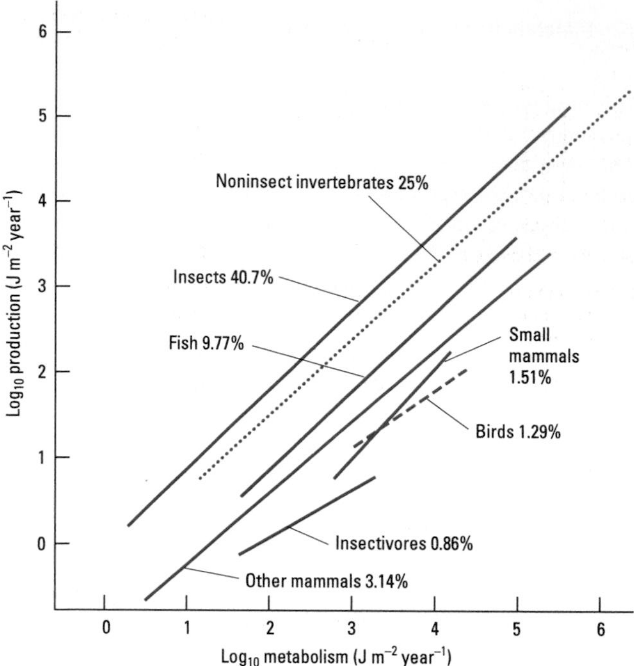

Fig. 6.25 Energy losses in metabolism: the relation between production and metabolic rate, with the mean for net production efficiency shown for each taxon. (Adapted from Humphreys 1979.)

All components of energy budgets must be expressed in the same units, usually joules or kilojoules. Clearly the ingestion rate sets the upper limit to all other variables in the equation, so ingestion must be considered first.

6.6.1 Foodstuffs and ingestion

Ingested foodstuffs, whether as pure compounds or as the mixtures provided by natural foods, provide varying levels of energy content (Table 6.8). The biological materials that most animals eat have energy values in the range 15–30 kJ g^{-1}, with those of high lipid content being most valuable if seen solely as a source of energy. However, foodstuffs may not be utilized with complete efficiency by an animal, and their **apparent digestible energy** must also be considered: this is the difference between the energy that they could provide and the energy from them that is lost in the feces. The percentage of the energy consumption that is actually assimilated is often termed the **assimilation efficiency**. This is usually much lower for herbivores than for carnivores due to the indigestibility of cellulose; by contrast it may be very high for fluid feeders such as blood-suckers, phloem feeders, and suckling mammals, as their liquid diets are very readily assimilated (Table 6.9).

A power function provides a good description of the variation in ingestion rate in relation to body mass. Figure 6.26 provides an example for ectothermic carnivorous tetrapods, where the slope of the regression line is 0.82 and the ingestion rate of a 1 kg animal (obtained from the intercept) is 0.78 W. This implies that ingestion rate varies with body size but is relatively independent of diet. The same conclusion is reached from a comparison of the scaling of ingestion rates in endothermic herbivores and carnivores (Fig. 6.26). In this case ingestion rates scale to $M_b^{0.70}$, but the intercept is much higher at 10.7 W. The mass exponents obtained in this kind of analysis do not differ much from the 0.75 slope describing the effects of size on respiration rate (see above). The similarities of these scaling relationships suggest that species that deviate significantly from the general constraints of body size on respiration rate will show corresponding deviations in ingestion rate. For example, mammals with unusually low metabolic rates, such as sloths, should eat more slowly, and metabolically active animals, such as weasels, should eat more quickly than predicted on the basis of body mass alone.

Numerous factors influence ingestion rate, including seasonal cycles of growth, reproduction, and the accumulation of fat stores, as well as many environmental variables such as day length, oxygen availability, and temperature. Acute changes in body temperature also have a major effect on the amount of food consumed in ectotherms. For example, increases in temperature initially lead to

Table 6.8 Energy contents and composition of different diets and foodstuffs.

Foodstuff	Energy: KJ g^{-1} wet wt	Energy: KJ g^{-1} dry wt	Protein (%)	Lipid (%)	Carbohydrate (%)
Pure carbohydrate		17	0	0	100
Pure protein		23	100	0	0
Pure lipid		38	0	100	0
Phytoplankton	~1	15–20	3	1	4
Macroalgae	~1		1	<1	4
Pulses	2–8	17–21	10	2	70
Plant sap	~0.1		0	0	2–5
Leaves	~1		2	<1	1
Fruits	1–2		~1	0	6–15
Earthworms	3		11	1	1
Crabs	4–5		13	3	2
Insects (larvae)	3–6		10–16	1–5	2
Fish (oily)	6–8	23–25	14–18	10	1
Fish (nonoily)	5–6		17	3	1
Meat (vertebrate muscle)	8–10	23–28	20	20–30	0
Milk (cow's)	2–3		3	4	5
Egg		25–28			

Table 6.9 Assimilation efficiencies for different diets.

Feeding habit	Consumer	Food	Assimilation/consumption (% absorption efficiency)
Herbivory	Ectotherms, aquatic	Algae	30–70
	Ectotherms, aquatic	Macrophytes	30–60
	Ectotherms, land		40–50
	Endotherms, land		60–70
Granivory	Ectotherms, land		75–80
	Endotherms, land		70–77
Nectarivory	Ecto- or endotherms, land		95+
Carnivory	Ectotherms, aquatic	Invertebrates	65–85
	Ectotherms, aquatic	Fish	80–90
	Ectotherms, land	Flesh	85
	Endotherms, land	Flesh	85
	Endotherms, land	Milk	95
	Various	Blood	85+
Detritivory	Ectotherms, aquatic		40–45
	Ectotherms, land		10–20
Endoparasitism	Ectotherms		70–80

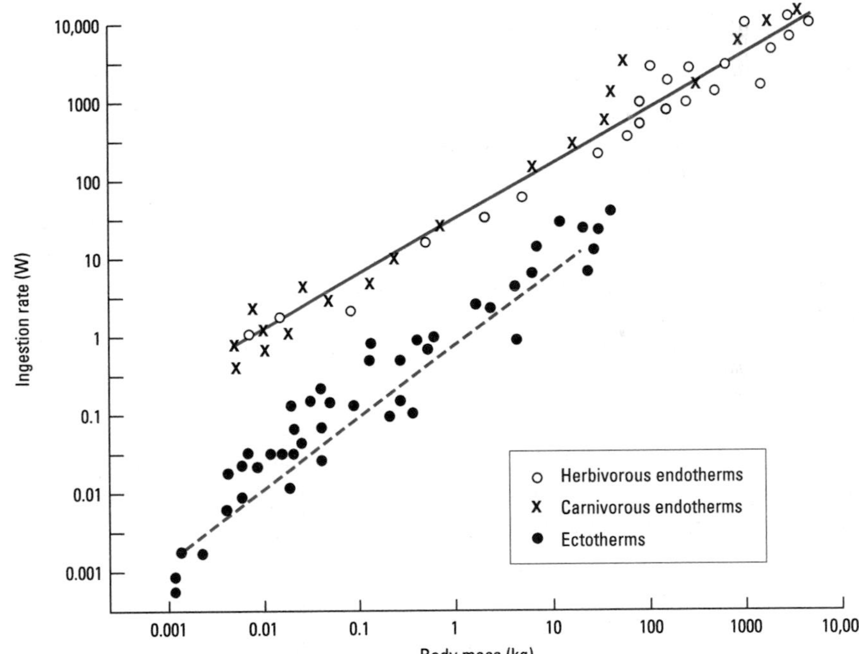

Fig. 6.26 Ingestion rate (IR) as a function of body mass for vertebrates. In ectotherms the regression line for the data is $IR = 0.78\ M_b^{0.82}$, while for endotherms the ingestion rate is substantially higher and $IR = 10.7\ M_b^{0.70}$. (Data from Farlow 1976.)

an increased rate of ingestion in fish (Fig. 6.27). Food intake peaks at some intermediate temperature, before declining markedly as the temperature approaches the upper thermal tolerance limit of the species. The energy available for growth, E_p, initially increases with temperature. However, at high temperatures the combination of a declining ingestion rate and an increasing metabolic rate markedly reduces the scope for growth. Seasonal temperature changes may serve to modify the responses shown in Fig. 6.27.

Terrestrial herbivores have a special problem in relation to ingestion and metabolism. Plant cell walls contain cellulose, which is not normally broken down by digestion. This material, referred to as roughage or fiber in the human diet, increases gut transit times and the volume of the feces. The material in plant cell walls can be digested by symbiotic microbes, and animals that eat mostly plant material have evolved a variety of behaviors and structures to promote microbial fermentation; these are dealt with in Chapter 15 when we look at land animals in detail. The gastrointestinal tract is an anoxic environment and may house symbiotic bacteria (in ruminants and other fermenters) that convert monosaccharides, mixed amino acids, and fats into volatile fatty acids, methane, and hydrogen. Volatile fatty acids represent a good source of energy (about 20% less than glucose on a molar basis).

The metabolic cost of processing ingested food is usually referred to as the **specific dynamic action** (SDA) or specific dynamic effect; it leads to a raised metabolic rate after feeding, sometimes termed "postprandial thermogenesis" (Fig. 6.28). In fasted animals, oxygen consumption rises markedly following a satiating meal; 30–40% is not uncommon in mammals, and in very infrequent feeders such as

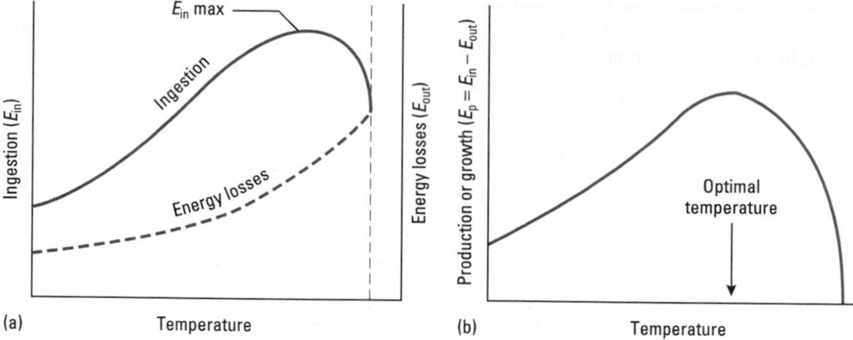

Fig. 6.27 The influence of temperature on (a) the rates of ingestion and metabolism (energy loss) in fish, where the vertical dashed line shows the upper thermal tolerance for the species; and (b) the consequences for the available energy remaining for growth, which has a clear optimum. (From Jobling 1993, Fig. 1.2, with kind permission from Chapman & Hall.)

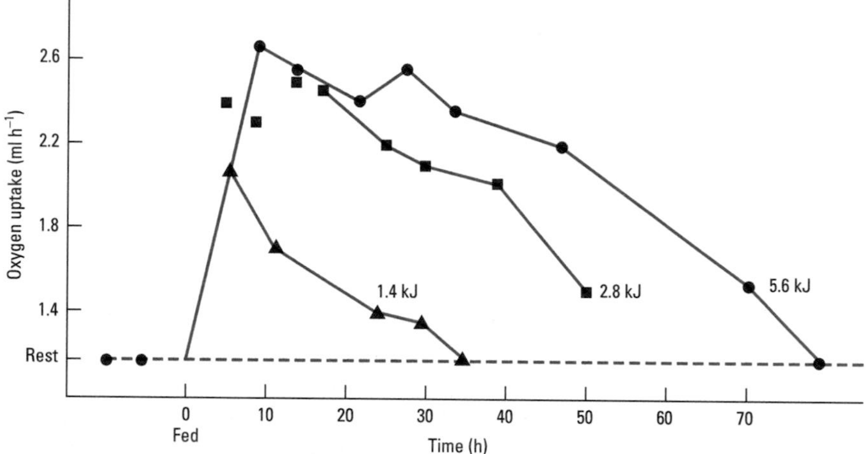

Fig. 6.28 The specific dynamic action in terms of increased oxygen uptake after feeding at three different levels in the plaice. The energy content (in kJ) of each of the meals is shown. The dashed line shows the routine oxygen consumption before feeding. (From Jobling & Davis 1980.)

boas and pythons the metabolic rate during meal processing may be as much as 20–40 times the normal resting rate, more extreme than the change incurred in a mammalian sprinter and sustained for much longer. This partly reflects the low fasting metabolic rate in pythons, which is 13 times less than in a mammal of equivalent size (after correcting for differences in body temperature). The very large increase in metabolic rate arises from the costs of regulatory processes including the upregulation of enzymes, transporters, and gastrointestinal secretions, and the rapid growth of organs that have atrophied in fasting periods. For example, within 24 h of feeding the python's small intestine doubles in dry mass as a result of a six-fold increase in microvilli length; and within 3 days there are also 50–100% increases in the masses of stomach, heart, liver, lung, and kidney.

The duration of the SDA varies with the size and composition of the meal, but it appears that the metabolic costs of processing particular classes of foodstuffs are relatively independent of taxon. The ingestion of 1 kJ of protein requires approximately 310 J of additional metabolic power, regardless of whether the protein was consumed by an invertebrate, a fish, a bird, or a mammal. The specific dynamic action associated with various diets varies considerably, and is generally lower for most plant foods, with the exception of nectar, pollen, and seeds, which contain little or no cellulose.

In fasted individuals, the relative increase in oxygen consumption after a single satiating meal is independent of body temperature. However, the duration of the SDA is markedly temperature dependent. For example, in teleost fish the duration of the SDA is around four times shorter for tropical than Antarctic species (Fig. 6.29). Note that, after correcting for differences in body mass and meal size, the area under the curves in Fig. 6.29 is found to represent a similar proportion of the energy ingested for both the warm-water and cold-water fish; it simply reflects the composition of the meals, which was the same in both cases. The slow rate at which ingested food can be metabolized in Antarctic fish limits the rate at which energy can be made available for all other purposes, providing a general constraint to somatic and reproductive growth.

The increase in metabolic rate with feeding is a reflection of numerous metabolic processes, including the digestion, absorption, and storage of nutrients, the deamination of amino acids, the synthesis of excretory products, and the increased synthesis of protein and other macromolecules associated with growth. The relative contribution of each of these processes is very difficult to quantify with any accuracy. Intravenous infusion of amino acids in vertebrates also results in an increase in oxygen consumption, which is a significant fraction of the maximum SDA observed with feeding. This suggests that the SDA largely reflects postabsorptive metabolism, and that costs associated with processing the food in the gut are relatively small. Tissue components are constantly being synthesized and degraded and it has been suggested that as much as 60–80% of the increase in metabolic expenditure associated with the ingestion of food may result from increased protein and lipid synthesis.

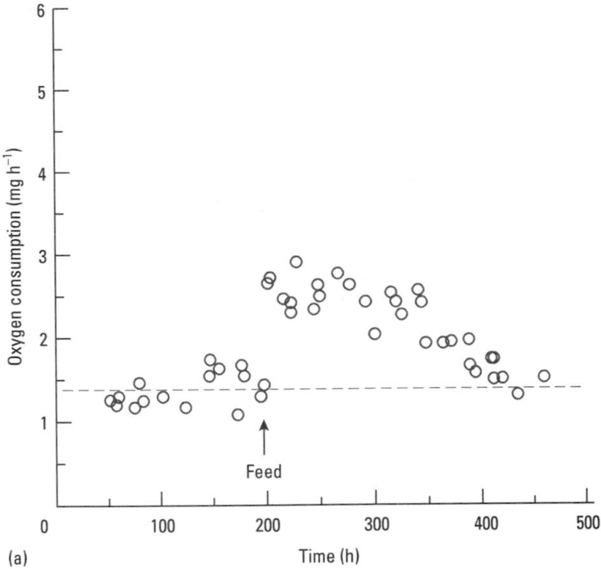

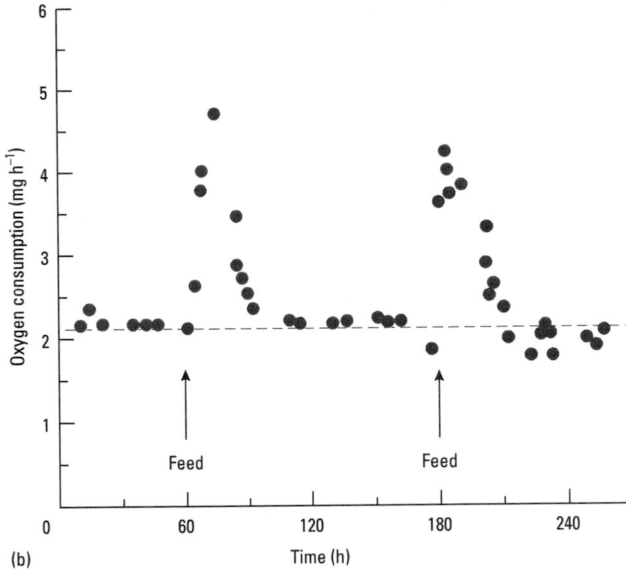

Fig. 6.29 The specific dynamic action effect in fish given a satiating meal of shrimp: (a) is the polar fish *Notothenia* at −0.9°C, and (b) the tropical fish *Cirrichthys* at 25°C. The dashed line shows the fasting oxygen consumption in each case. (From Johnston & Battram 1993.)

6.6.2 Energy losses in excreted products

Some of the food eaten by animals is lost as a waste product, usually as a component of the feces (though some simpler animals lack through-guts and lose the waste material in other ways). In addition to unabsorbed food, the feces may also contain mucus, cells sloughed off from the wall of the gut, bile salts, digestive enzymes, and bacteria derived from the gut microflora. In some animals, particularly insects, reptiles, and birds, the nitrogenous excretory products are also passed into the gut and exit the body together with the feces; in these cases, E_{urine} and E_{feces} cannot be distinguished.

There are a variety of other exit routes for excreted materials, including gills, skin, and kidney tubules, and any or all of these may excrete forms of nitrogenous products, including amino acids, creatine, creatinine, choline, betaine, and glycine. However, the most important nitrogen waste products are ammonia, urea, and uric acid (see Chapter 5). Even fasted animals excrete nitrogenous end-products resulting from the transamination and deamination of amino acids that arise as a result of basal protein turnover (the endogenous nitrogen excretion). Exogenous excretion results from the direct deamination of amino acids ingested and absorbed in the food and is influenced by factors such as feeding rate, the protein content of the food, and the amino acid composition of the diet.

6.6.3 Energy, growth, and production

Ingested energy that it is not lost as fecal or excretory products or used for metabolism is available for growth and production. Biosynthesis is relatively cheap in energy terms, and may be around 95% efficient in principle, though this is somewhat reduced in practice by the losses due to ongoing metabolism and excretion. It can be directly assessed where embryos grow rapidly from the closed supply of stored yolk in the egg, and values for efficiency in such studies are around 50–70%.

The partitioning of ingested energy within an animal may vary with the composition and mineral content of the food. In insects, for example, the utilization of ingested energy has been shown to increase with the nitrogen content of the food. Energy partitioning changes during ontogeny as somatic growth slows, and the production term in the energy budget is then largely accounted for by the production of gametes and of secretions associated with reproduction. Figure 6.30 illustrates the energy budget derived for the meadow spittlebug (*Philaenus*). Rapidly growing nymphs have a production of about 3 J day^{-1}, whereas adult stages have a negligible production, mostly associated with the energy costs of eggs, sperm, and reproductive secretions. In other terrestrial animals much greater amounts of energy are invested in reproduction (see Chapter 15).

The energy that females invest in reproduction (E_{rep}) is usually assessed indirectly by some index of **reproductive effort.** E_{rep} scales with a mass exponent, b, between 0.5 and 0.9, which means that larger species tend to invest relatively less in their offspring per unit time. But larger animals also tend to live longer: **lifespan** scales with a mass exponent of 0.15–0.29. Thus over a female's lifetime, these two factors combine, with the result that the energy invested in reproduction varies more or less isometrically with respect to body mass ($b = 1.0$). This is also a reflection of the fact that lifetime energy assimilation and energy expenditure also scale isometrically.

In species in which body mass does not change much during the reproductive phase of their lives, the energy invested in reproduction (E_{rep}) will equal the total energy assimilated minus the individual's energy requirements excluding reproduction and growth (E_{req}):

$$E_{rep} = E_{in} - E_{req}$$

However, if females reproduce over a wide range of body mass then the mass exponent will be larger, as some of the available energy in smaller individuals will go to growth (E_{gro}):

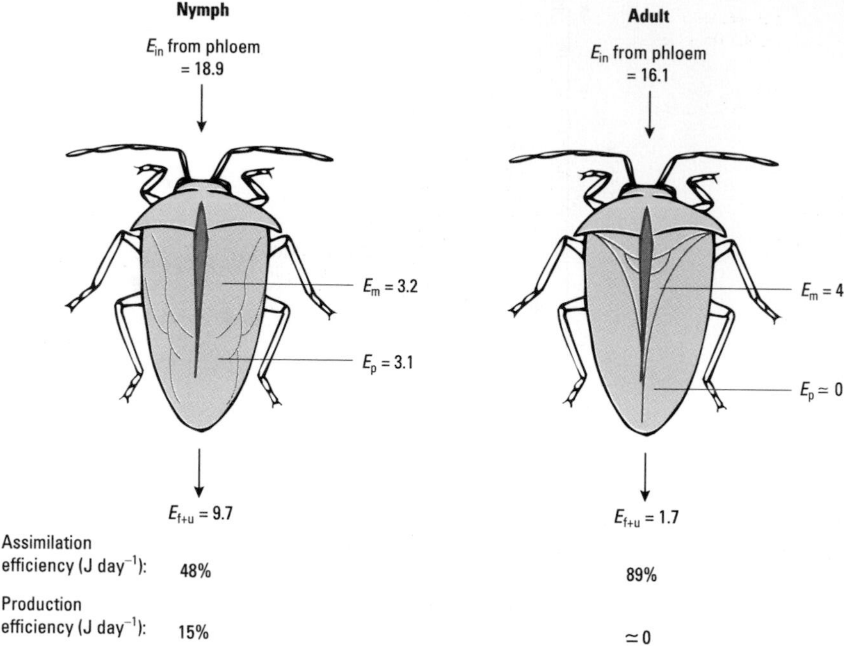

Fig. 6.30 Pictorial summaries of energy (E) budgets for nymphs and adults of the common meadow spittlebug insect (*Philaenus*). Terms are explained in the text, see p. 133. (Data from Weigert 1964.)

$$E_{gro} + E_{rep} = E_{in} - E_{req}$$

It is often of great interest to measure the growth rate of individual species in ecology, agriculture, and fisheries.

Age at maturity can be related to **growth rate** and **generation time**. Since growth scales to $M_b^{0.7}$ it follows that age at maturity would be expected to scale between species according to the relationship:

$$M_b/M_b^{0.7} = M_b^{0.3}$$

Studies of generation time and age at maturity in types of organism as diverse as viruses and giant sequoia trees, and including a wide range of metazoans, have found that, as expected, these variables scale with body mass to the power of 0.21–0.33.

Mortality, **production**, and **life-history theory** are also interrelated because of links with energy budgets. The growth and reproductive characters of animals show genetic variation within populations and can be subject to selection. However, any gains in fitness achieved by increasing rates of growth and reproductive output may be at the expense of an increased risk of mortality. It would appear that there is some kind of trade-off between personal survival and production, such that it is not advantageous to select for maximum production. For example, it would be possible to maximize production by more intense feeding, but such a change in behavior may bring increased risks of predation leading to increased mortality. Interspecific studies have shown positive correlations between growth rate and mortality for echinoderms and for fish. Intraspecific correlations for these variables have also been demonstrated for aphids and springtails.

It is also worth considering the influence of **genotype** on maintenance energy requirements and production. Many enzymes occur in a number of polymorphic forms within populations (see section 2.4). By scoring the number of individuals in a population that are heterozygous for polymorphic alleles, an index for heterozygosity can be calculated. Positive correlations between maintenance energy requirements and heterozygosity have been demonstrated in molluscs, salamanders, and fish, particularly under conditions of energy limitation.

Marine bivalves show particularly high levels of genetic variation on both micro- and macrogeographic scales, with between 12 and 35% of enzyme loci showing allelic variation. The common edible mussel, *Mytilus edulis*, has been the subject of many studies on the relationship between genotype and energy budgets. Changes in the growth rate and mortality of *Mytilus* following reciprocal transfer between populations have shown that genetic differences can

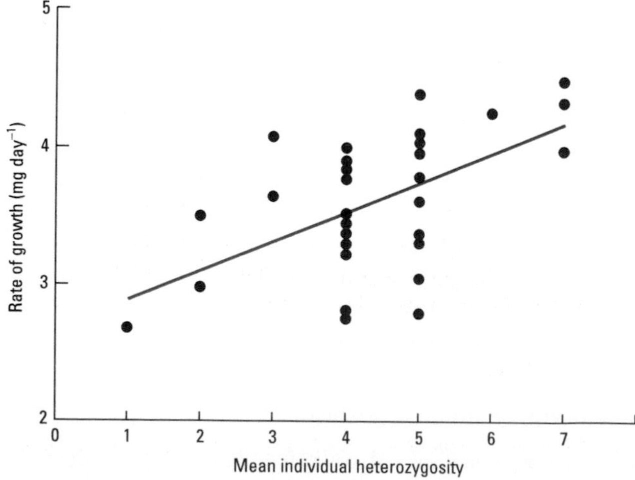

Fig. 6.31 The relationship between heterozygosity (mean number of heterozygotes per individual at seven sites) and growth rate for the mussel *Mytilus galloprovincialis*. (From Bayne & Hawkins 1997, courtesy of University of Chicago.)

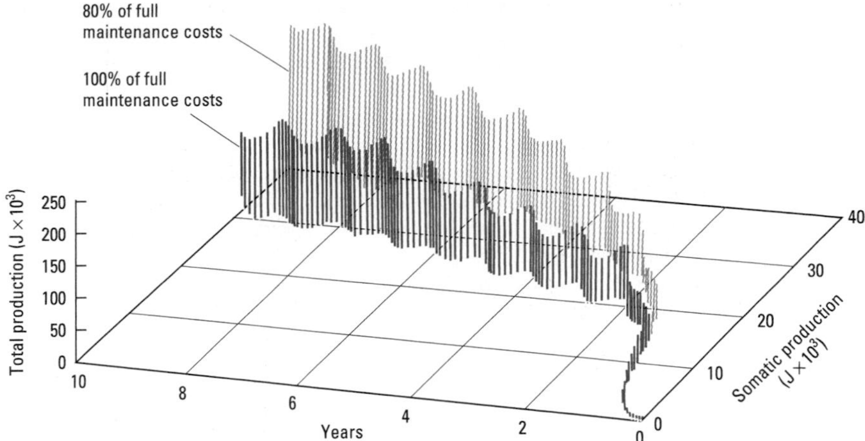

Fig. 6.32 Simulated effects of a 20% reduction in energy requirements for maintenance in *Mytilus edulis* (the common mussel) upon its performance over 10 years after first settlement. Energy is partitioned between growth, reproduction, and storage depending on the amount already stored and current energy balance. After 10 years the mussels with only an 80% energy requirement have achieved a one-third increment in somatic production and almost double the total production. (From Hawkins *et al.* 1989.)

account for up to 27% of the variation in production. Several studies have shown positive correlations between growth rate and heterozygosity at glycolytic and lysosomal enzyme loci. Typically, heterozygosity at the polymorphic enzyme loci can explain around 10–30% of the variation in individual growth rates (Fig. 6.31). Body proteins are continuously being synthesized and degraded, both energy-consuming processes. Protein turnover costs are thought to represent about 20–30% of the total maintenance energy requirements in ectotherms. There is a trend for the more heterozygous individuals to have lower rates of protein turnover and hence reduced energy requirements for maintenance, which means that more of the ingested energy is available for growth. In fact relatively small differences in maintenance energy requirements can have a major impact on energy balance. Simulations have shown that a 20% decrease in maintenance energy requirements relative to a "normal" mussel could theoretically double the cumulative production over a 10-year period (Fig. 6.32). Mussels with the lowest rates of protein turnover also have the lowest rates of both protein synthesis and protein degradation, and show the best rates of survival under a range of stressful environmental conditions studied. It is possible that faster rates of protein turnover in more homozygous mussels may be advantageous under other conditions (e.g. by facilitating metabolic regulation, detoxification, and/or replacement of damaged proteins), although this has yet to be established.

The genetic basis for these correlations between heterozygosity and individual differences in protein turnover, energy expenditure, and growth are unknown, although there is evidence that it reflects functional differences in the studied genes rather than merely indicating linked loci. Here again we have an area where molecular and genetic analyses may soon give us a better handle on understanding the nature of environmental adaptations.

FURTHER READING

Books

Alexander, R. McN. (1999) *Energy for Animal Life.* Oxford University Press, Oxford.

Blaxter, K. (1989) *Energy Metabolism in Animals and Man.* Cambridge University Press, Cambridge, UK.

Hochachka, P.W. & Guppy, M. (1987) *Metabolic Arrest and the Control of Biological Time.* Harvard University Press, Cambridge, MA.

Hochachka, P.W. & Somero, G.N. (1994) *Biochemical Adaptation.* Princeton University Press, Princeton, MA.

Hochachka, P.W. & Somero, G.N. (2002) *Biochemical Adaptation: Mechanism and Process in Physiological Evolution.* Oxford University Press, Oxford.

Martin, B.R. (1987) *Metabolic Regulation: a Molecular Approach.* Blackwell Scientific Publications, Boston, MA.

Prosser, C.L. (1991) *Environmental and Metabolic Physiology.* Wiley-Liss, New York.

Salway, J.G. (1994) *Metabolism at a Glance*, 2nd edn. Blackwell Science, Oxford.

Stryer, L. (1988) *Biochemistry.* Freeman, New York.

Townsend, C.R. & Calow, P. (1981) *Physiological Ecology: an Evolutionary Approach to Resource Use.* Blackwell Scientific Publications, Oxford.

Reviews and scientific papers

Balban, R.S. (1990) Regulation of oxidative phosphorylation in the mammalian cell. *Journal of Applied Physiology* **73**, 737–743.

Brookes, P.S., Buckingham, J.A., Tenreiro, A.M., Hulbert, A.J. & Brand, M.D. (1998) The protein permeability of the inner membrane of liver mitochondria from ectothermic and endothermic vertebrates and from obese rats: correlations with standard metabolic rate and phospholipid fatty acid composition. *Comparative Biochemistry & Physiology B* **119**, 325–334.

Dobson, G.P & Headrick, J.P. (1995) Bioenergetic scaling: metabolic design and body size constraints in mammals. *Proceedings of the National Academy of Science USA* **92**, 7317–7321.

Elgar, M.A. & Harvey, P.H (1987) Basal metabolic rates in mammals: allometry, phylogeny and ecology. *Functional Ecology* **1**, 25–36.

Fothergill-Gilmore, L.A. (1986) The evolution of the glycolytic pathway. *Trends in Biochemical Science* **11**, 47–51.

Gillooly, J.F., Brown, J.H., West, G.B., Savage, V.M. & Charnov, E.L. (2001) Effects of size and temperature on metabolic rate. *Science* **293**, 2248–2251.

Guppy, M., Fuery, C.J. & Flanigan, J.E. (1994) Biochemical principles of metabolic suppression. *Comparative Biochemistry & Physiology B* **109**, 175–189.

Guppy, M. & Withers, P. (1999) Metabolic depression in animals: physiological perspectives and biochemical generalizations. *Biological Reviews* **74**, 1–40.

Hawkins, A.J.S. (1995) Effects of temperature change on ectotherm metabolism and evolution: metabolic and physiological interrelations underlying the superiority of multi-locus heterozygotes in heterogeneous environments. *Journal of Thermal Biology* **20**, 23–33.

Hochachka, P.W. (1997) Oxygen—a key regulatory metabolite in metabolic defence against hypoxia. *American Zoologist* **37**, 595–603.

Hochachka, P.W., Buck, L.T., Doll, C.J. & Buck, L.T. (1996) Unifying theory of hypoxia tolerance: molecular/metabolic defense and rescue mechanisms for surviving O_2 lack. *Proceedings of the National Academy of Science USA* **93**, 9493–9498.

Hulbert, A.J. & Else, P.L. (1999) Membranes as possible pacemakers of metabolism. *Journal of Theoretical Biology* **199**, 257–274.

Hulbert, A.J. & Else, P.L. (2000) Mechanisms underlying the cost of living in animals. *Annual Review of Physiology* **62**, 207–235.

Jobling, M. (1993) Bioenergetics—food intake and energy partitioning. In: *Fish Ecophysiology* (ed. J.C. Rankin & F.B. Jensen), pp. 1–144. Chapman & Hall, London.

Lovegrove, B.G. (2000) The zoogeography of mammalian metabolic rate. *American Naturalist* **156**, 201–219.

Poppitt, S.D., Speakman, J.R. & Racey, P.A. (1993) The energetics of reproduction in the common shrew (*Sorex araneus*): a comparison of indirect calorimetry and the doubly labeled water method. *Physiological Zoology* **66**, 964–982.

Secor, S.M. & Diamond, J. (1998) A vertebrate model of extreme physiological regulation. *Nature* **395**, 659–662.

Speakman, J.R., McDevitt, R.M. & Cole, K.R. (1993) Measurement of basal metabolic rates: don't lose sight of reality in the quest for comparability. *Physiological Zoology* **66**, 1045–1049.

Storey, K.B. & Storey, J.M. (1990) Metabolic rate depression and biochemical adaptation in anaerobiosis, hibernation and estivation. *Quarterly Review of Biology* **65**, 145–174.

van Waarde, A. (1991) Alcoholic fermentation in multicellular organisms. *Physiological Zoology* **64**, 895–920.

West, G.B., Brown, J.H. & Enquist, B.J. (2000) The origin of universal scaling laws in biology. In: *Scaling in Biology* (ed. J.H. Brown & G.B. West), pp. 87–112. Oxford University Press, New York.

7 Respiration and Circulation

7.1 Introduction

Life is thought to have originated in an anoxic (reducing) environment, with simple anaerobic heterotrophs present in the oceans within 1 billion years of the formation of the Earth. Anaerobic life has therefore been in existence for at least 3500 million years, utilizing the pathways outlined in the previous chapter. Even small amounts of oxygen would produce real problems for any existing life in such an environment. It produces some highly reactive reduction products, including the free radicals superoxide (OO^-) and hydrogen peroxide (H_2O_2). These are more reactive than "normal" molecular oxygen (O_2) because they have unshared electrons in their outer shells, and they are effectively poisonous to living cells because of their potential to damage macromolecules. It seems likely that early microorganisms had to develop methods of detoxification to deal with oxygen so that these free radicals were not generated. In modern organisms, enzymes such as superoxide dismutases, peroxidases, and catalases break down free radicals to give harmless products, and antioxidant organic molecules such as vitamin E, vitamin C, and β-carotene will scavenge any free radicals that have otherwise escaped attention. Any or all of these safeguards may have appeared in the early Precambrian prokaryotes. The complex cytochrome systems in modern mitochondria may have originally developed in their prokaryotic ancestors as oxygen detoxification mechanisms, working alongside simple anaerobic fermentation pathways; later they were selected for their more lucrative role in aerobic respiration, producing far more energy per unit of carbon substrate than the original fermentation pathways.

Once molecular oxygen became more readily available (see section 6.1), its potential advantages could begin to override the problems it caused, and organisms would have had the chance to exploit these advantages. Rising atmospheric oxygen permitted the formation of an ozone layer, and ozone is the only atmospheric gas that is a good absorber of harmful UV-B radiation from the sun, so its accumulation meant that nucleic acids would become less prone to radiation damage. Increased oxygen may also have allowed the synthesis of certain crucial biological macromolecules, such as collagen (and the plant equivalents including lignin), that permit the assembly and mechanical integration of multicellular bodies. Thus planetary oxygen levels came to permit the evolution of metazoan life—animals—that could make use of this gas in oxidative cellular metabolism (or "internal respiration"). As the size of such early metazoans increased and their surface area to volume ratios decreased they came to require external respiration, the exchange of gases between the surrounding medium and the tissues of a living organism, with substantial oxygen uptake and approximately equivalent carbon dioxide excretion.

As plants continued to diversify and spread, giving net oxygenation, animals radiated to use the gas in their own metabolic processes. Atmospheric levels of O_2 gradually rose towards present-day levels, and although the pattern of this atmospheric accumulation remains controversial, a useful "best guess" is shown in Fig. 7.1. Current evidence suggests that the level may have been about 10% of the present atmospheric level (10% PAL) at the Cambrian boundary, not settling at around 100% PAL (i.e. to 21% of the atmospheric gases) until after the end of the Palaeozoic (> 65 mya). There may, however, have been a period of unusually high oxygen levels (perhaps 30–40%, approaching twice PAL) in the Carboniferous era (around 350–300 mya).

Oxygen availability in the modern world varies in both time and space (Fig. 7.2). In the seas it is reasonably constant at 4–6 ml l^{-1} through much of the water, but decreases in the oceanic "oxygen minimum layer" (see Chapter 11) and in deep water in the Pacific basin. It is also drastically reduced in the benthic sludges and in deep marine vents. In the terrestrial world the air contains a standard 21% or 210 ml l^{-1}, but microhabitats in burrows and nests, within the litter layer, or air pockets underground or underwater may be substantially less oxygen-rich (as low as 10% O_2 and up to 5–10% CO_2), and any habitat at altitude also experiences lowered oxygen availability. Where animals live at the interface of water and air (e.g. intertidally) but can only breathe from one or other of these media, they may suffer periodic withdrawal of available oxygen. Furthermore, in any of these habitats, an animal undergoing vigorous exercise may be unable to acquire oxygen quickly enough to offset its consumption, as we saw in Chapter 6. Many animals are therefore subject to temporary or permanent low oxygen (**hypoxia**), and some will experience a complete absence of oxygen (**anoxia**). They may also experience accompanying rises in carbon dioxide levels, or **hypercapnia**.

The respiratory responses that animals show when subject to hypoxia can be placed in one of two categories, as with so many response systems. Some animals are **oxygen conformers**, where oxygen consumption decreases in proportion to decreasing ambient oxygen concentration. These include a few protozoans, most

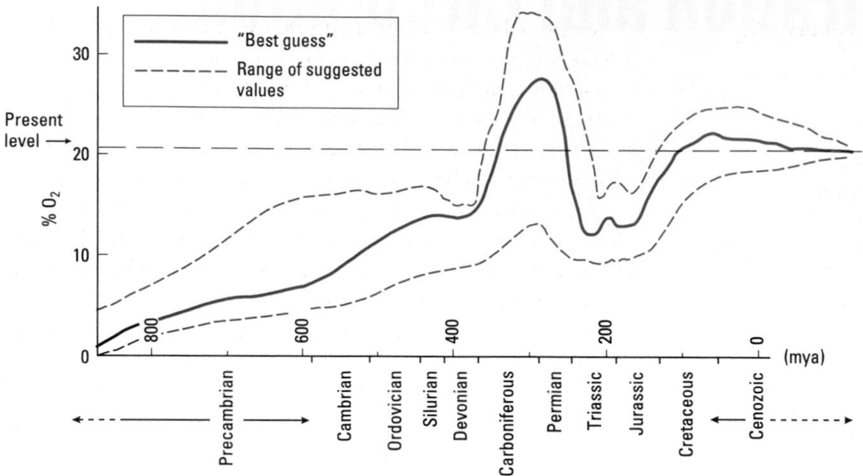

Fig. 7.1 Patterns of atmospheric oxygen from the late Precambrian, through the Palaeozoic and Mesozoic to the present. Details are controversial, but there appears to have been a steady rise, with a "burst" around the Carboniferous period.

cnidarians, most marine worms and most parasitic worms, some molluscs and crustaceans, most echinoderms, and a few aquatic insects. Generally, oxygen conformity is common in marine invertebrates, especially in large sedentary forms or smaller but less active forms that lack highly specialized respiratory organs and circulatory systems. In contrast, other animals will maintain their own oxygen consumption at ambient oxygen partial pressures (Po_2) well below normal. These are termed **oxygen regulators**, and metabolism is maintained at normal levels down to some species-specific Po_2 level termed the **critical partial pressure**, P_c, values of which are shown in Table 7.1. Regulators include most protozoans, many freshwater and terrestrial annelids, many molluscs and crustaceans, some aquatic and probably all terrestrial insects, and almost all vertebrates.

Fig. 7.2 The general pattern of availability of oxygen in different habitats in the modern world. (Adapted from Dejours 1981.)

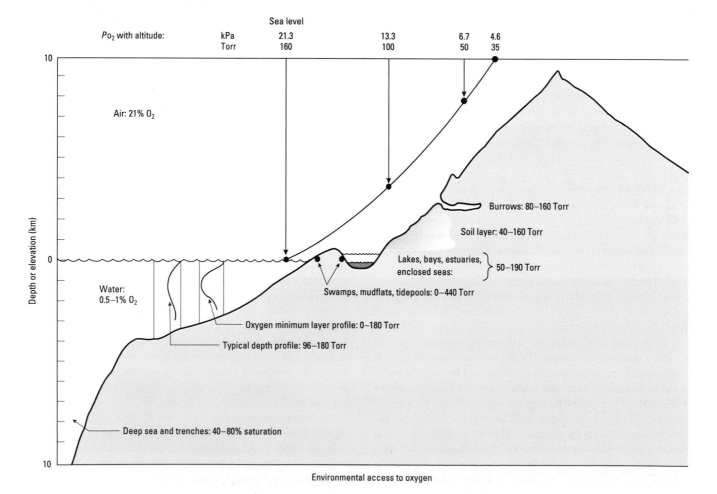

Table 7.1 Values for P_c, the critical partial presure at which regulation of O_2 uptake ceases and animals become O_2 conformers.

Organism	P_c (kPa)	Habitat
Aurelia (cnidarian)	16	SW
Mytilus (bivalve)	10	SW/shore
Loligo (squid)	6	SW
Gnathophausia (crustacean)	0.8	SW (O_2 minimum layer)
Uca (crab)	1	SW/land
Pelmatohydra (cnidarian)	8	FW
Tubifex (oligochaete)	3	FW
Cambarus (crustacean)	5	FW
Cloeon (mayfly)	4	FW
Lumbricus (oligochaete)	10	FW/terrestrial
Helix (pulmonate)	10	Terrestrial
Hyalophora (caterpillar)	3	Terrestrial

FW, fresh water; SW, sea water.

Some examples of oxygen usage in different animals are shown in Fig. 7.3.

In this chapter we look at the fundamental design of respiratory systems whereby aerobically respiring animals take up the oxygen they require (and may also release carbon dioxide). These systems always involve an **exchange site**, with a large surface area where the epithelial layers in contact with the medium are of minimal thickness. For larger and/or more active animals, respiratory efficiency also requires **ventilatory systems** to insure that sufficient volumes of the external medium pass across the exchange site. Such animals also commonly have internal perfusion of the exchange site linked to internal **circulatory systems** to move the dissolved oxygen from the exchange site to the respiring tissues. This is essential because while diffusion is very effective at very short range it is exceedingly slow over even moderate distances (e.g. only about 30% of diffusing O_2 molecules arrive at a point 1 m from their source in 4 years!). These circulating fluids are known in more complex animals as cardiovascular systems, and in most cases serve many other functions such as moving nutrients and wastes around the body.

The presence of adequate ventilation and/or transport systems may be the most important determinant of whether an animal is a conformer or a regulator, since conformers seem to be limited mainly by an inability to transport oxygen quickly enough to the metabolizing tissues. Evidence of this is seen with isolated tissue slices from such animals, which may have higher rates of oxygen consumption than the whole animals.

The distinction between oxygen conformity and oxygen regulation is therefore not a sharp one, particularly when the critical partial pressure (P_c) of a species is high. Moreover, both invertebrates and ectothermic vertebrates may be oxygen regulators at low temperatures when oxygen consumption is quite low, but conformers at high temperatures when their metabolic rates are much higher. The value of P_c may also change during acclimation; it decreases with previous exposure to hypoxia, reflecting acclimation in one or more of the supply components (respiratory system, blood properties, or cardiovascular functioning). Thus the metabolic response

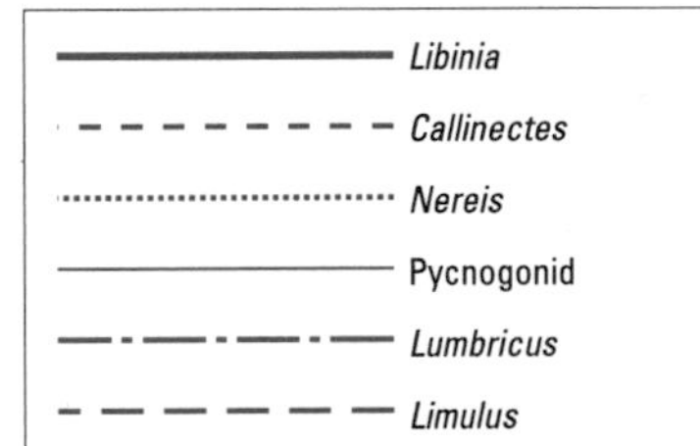

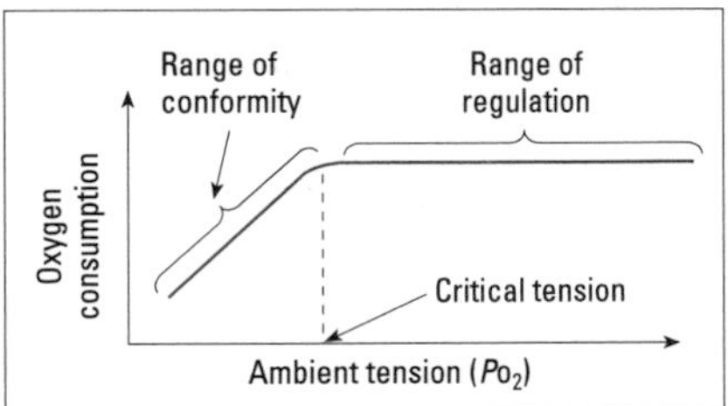

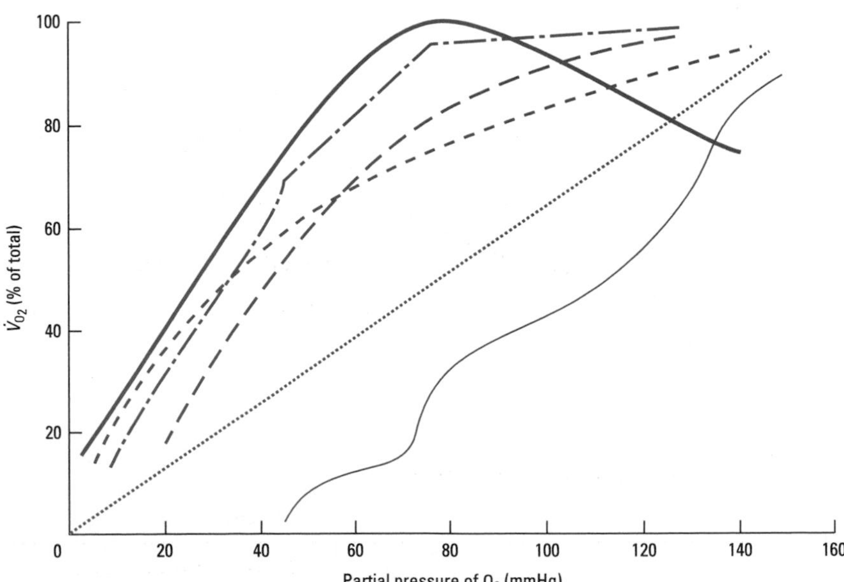

Fig. 7.3 Oxygen conformers and regulators: O_2 consumption expressed as percentage of total O_2 consumption is shown in relation to ambient Po_2 for six species of invertebrate. The pycnogonid, annelid worm (*Nereis*), crab (*Callinectes*) and horseshoe crab (*Limulus*) are conformers, while the earthworm *Lumbricus* and the crab *Libinia* are regulators but only above 70–80 mmHg. (Data from de Fur & Mangum 1979; Davenport *et al.* 1987.)

to oxygen availability is not absolutely fixed, but can range from conformity to regulation depending on external conditions and individual experience.

7.2 Uptake and loss of gases across respiratory surfaces

7.2.1 Diffusion, partial pressures, gas concentrations, and gas solubilities

Fundamentally, the uptake and loss of gas by animals is always achieved by simple diffusion. Diffusion does not rely on metabolic energy but is merely the result of the random thermal motion of molecules, at a velocity proportional to the absolute temperature. Although individual molecules move at random, where there is a concentration gradient there will be a net movement of the gas down the gradient; from an oxygen-rich environment into the respiring (oxygen using) animal, and from the increasingly carbon dioxide-rich animal back into the environment. At this level of analysis, respiratory exchange is very simple. However, the time taken for a randomly moving molecule to move in a particular direction increases with the square of the distance moved. Thus diffusional transport becomes extremely slow over large distances.

In a mixture of gases such as air, the total pressure exerted is the sum of the pressures exerted by each gas (Dalton's law). The **partial pressure** (*P*) exerted by a gas can be calculated from its fractional content of the total gas (*F*) and the total barometrically measured pressure (P_b):

$$Po_2 = Fo_2\, P_b$$

$$Po_2 = Fo_2\, P_b$$

For normal dry air at sea level, P_b is of course "1 atmosphere", or 760 mmHg (760 Torr), or 101 kPa. Under the same conditions, the partial pressure of oxygen is 21.2 kPa, and that of carbon dioxide is 0.03 kPa.

Uptake of respiratory gases always depends eventually on the gas (oxygen) being dissolved in a liquid phase, though this may only occur deep within the tissues in some terrestrial animals. In an idealized system at equilibrium, the gas pressure in the two phases (gaseous and liquid) will be equal. This remains true when the gaseous phase is a mixture of several gases, such as air; each of the components will have equal partial pressures in the gaseous and liquid phases. However, to calculate the amount of gas dissolved in a volume of liquid we also need to take solubility into account, given by Henry's law as follows:

$$Q = \alpha\, p$$

where *Q* is the molar concentration of dissolved gas, *P* is the partial pressure of that gas and α is the solubility coefficient (in mol l^{-1} kPa^{-1}). Water and saturated air in equilibrium with each other will have the same gas partial pressures, but may have very different molar concentrations of the gases. In practice, gases can therefore diffuse "up" a concentration gradient because they are diffusing down a partial pressure gradient. Most obviously this occurs when a gas dissolves into a liquid in which it is very soluble, even if already present there in high concentration, because this reduces the partial pressure difference between liquid and gas.

Table 7.2 Respiratory gas solubilities (in μmol l^{-1} kPa^{-1}) and effects of temperature and salinity.

Temperature (°C)	Oxygen FW	Oxygen 50% SW	Oxygen SW	Carbon dioxide FW	Nitrogen FW
0	21.7	18.9	16.6	767	–
10	16.9	–	–	531	–
20	13.7	12.2	10.8	386	6.8
30	11.6	–	–	294	–
40	10.2	9.3	8.3	235	5.5

FW, fresh water; SW, sea water.

In general, of the three important atmospheric gases, oxygen and nitrogen have fairly low solubility but carbon dioxide has high solubility; therefore O_2 and N_2 are present in water at much lower molar concentrations than in air, while CO_2 is present at about the same concentration in the two media. But solubility for any gas is affected by the nature of the solvent (water in biological systems), and also by temperature and ionic strengths in the solution (Table 7.2). Thus at higher temperatures there is less CO_2 in water than in air, whereas at low temperatures there will be more CO_2 present in the water. Similarly at normal biological temperatures more oxygen will dissolve in solutions of low ionic concentration; at 10°C, the solubility coefficient for O_2 (in μmol l^{-1} kPa^{-1}) varies from 16.9 (fresh water) to 13.4 (sea water). However, within animal tissues the solubility of the important gases is fairly insensitive to the presence of other gases or to the nature of the electrolytes present (although as we will see gases and ions may critically affect the binding of oxygen to carrier pigments).

7.2.2 Major respiratory structures in animals

In the simplest and smallest animals, exchange of gases can take place over the general body surface, and this can be shown by simple consideration of diffusion rates (the relationship of time taken to the square of the distance moved, discussed above) to be adequate for animals up to about 1 mm in diameter, or 1 mm in thickness if they are of elongated flattened form, as in the platyhelminths, or elongated tubes, as in nematodes. For a body of given shape, a two-fold increase in diameter increases the surface area by a factor of 4 and the volume by a factor of 8. Any tissue mass greater than a 1 mm sphere would have insufficient surface to supply the oxygen needed by the contained volume of metabolizing tissue, and would thus have an effectively anoxic zone at its core, in the absence of any extra specialization. Since metabolic rate, and therefore oxygen consumption, scale with body mass with an exponent of only 0.75, as discussed in Chapter 6, there is a small mitigating factor here in that larger bodies need proportionally less oxygen; but this effect is far too small to compensate for the surface area to volume effect for oxygen uptake. In fact the limit may be set well below the theoretical 1 mm level because of the existence of unstirred oxygen-depleted boundary layers around the surface of the living organism, delaying

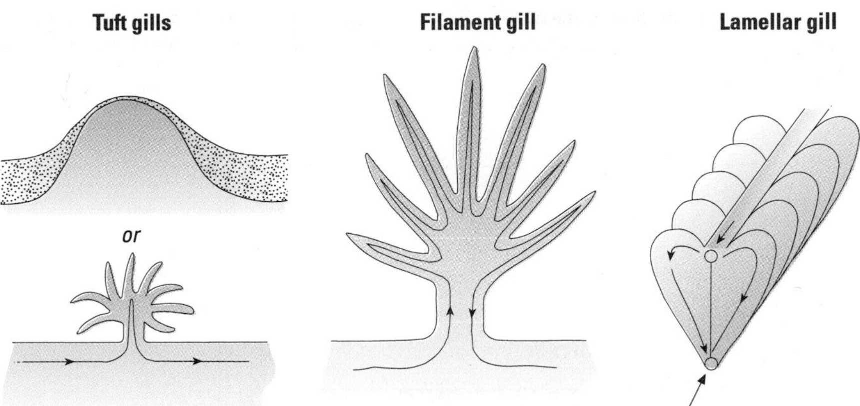

Fig. 7.4 A schematic view of the three main types of gill. Arrows show blood flow.

diffusion even further. Most single cells are considerably less than 1 mm in diameter, and where they do exceed this they are usually of irregular shape to keep the maximum diffusion path small. Most animals bigger than 1 mm in thickness, and also many smaller animals that regularly experience lowered external oxygen concentrations or that are relatively active, have evolved specialized gas exchangers, at the tissue or organ level.

Commonly this occurs at the outside of the animal, and where the structure is a protruding evaginated surface it is usually called a **gill**; most aquatic animals have one or more pairs of gills. Three main designs of gill can be identified, with increasing three-dimensional complexity (Fig. 7.4). **Tuft gills** can be little more than a raised thinned area of the general skin, as seen in the papulae and modified podia (tube-feet) of some echinoderms (starfish and sea urchins), but range through to the segmental tufts and the elaborate terminal plume-like structures common in many aquatic worms. **Filament gills** are best exemplified by aquatic arthropods, where a series of feather-like structures occur, supported by a thin external cuticle, and with a more elaborate internal flow system than is seen in any of the tuft arrangements. Filament gills also occur in some vertebrates, for example in embryonic elasmobranch fish, in some salamanders and tadpoles, and on the pelvic fins of some adult lungfish as a supplementary breathing mode. **Lamellar gills** are found as flat platelets extending out from a central strut or "gill arch", with a specific orientation to water flow. The pleopods (modified abdominal limbs) of most of the isopod crustaceans take this form, and beat gently against the ventral surface of the body. Lamellar gills also occur in the larger and more active decapod crustaceans (crabs, etc.), and littoral species often use such gills for air breathing: the shore crab *Carcinus maenas* has gill lamellae that are rather more sturdy than in fully aquatic crabs, and can be ventilated in air as well as in water; other species will allow water from their branchial chambers to flow out over a rather hairy region of their carapace, where it becomes oxygenated, subsequently trickling back into the gill chamber through the inhalant openings at the leg bases. Lamellar gills of particularly elaborate form also occur in both elasmobranch and teleost fish, where there may be 4–14 pairs of gills along the axis of the body, often protected by a flap-like covering (the operculum).

Horseshoe crabs (not actually crabs, but related to the arachnids) and some isopods have respiratory structures that are gill-like, though somewhat enclosed and ventilated by being waved through water, and these tend to be termed **gill books** or book gills. In arachnids, such as spiders, a design of respiratory organ that appears rather gill-like also occurs, but it is well protected within a chamber having only limited access, and functions in air rather than water. Here the lamellae may be fused more securely to the chamber wall for extra rigidity, held apart by cuticular struts to stop the lamellae collapsing on to each other, and supported by internal props to keep the blood space open. This arrangement is termed a lung book or **book lung**.

In relatively few animals the respiratory surface is completely internalized and forms by invagination rather than evagination of the body surface, and it is then usually termed a **lung**. This kind of structure occasionally occurs in marine animals, notably in the somewhat enigmatic "respiratory tree" of holothurian echinoderms (sea cucumbers). But lungs are particularly associated with partially or fully terrestrial animals, where water loss from a superficial structure would be an overwhelming disadvantage. Lungs may vary from simple sacs to complicated subdivided structures with internal alveoli (Fig. 7.5).

Simple lungs occur in at least one earthworm, where the rear portion of the body is tucked back on itself to form a simple vascularized chamber; and in some land-dwelling molluscs and crustaceans. Sac lungs are achieved in some terrestrial crabs where the gill lamellae are lost and the branchial chamber becomes vascularized; ocypodids (ghost crabs) can pump air through the water in these chambers. Crabs that are more fully terrestrial have air-filled and fully vascularized branchial chambers, losing their gills either partially or fully; some soldier crabs are obligate air breathers and drown if their mud-walled burrows do not keep out the incoming tide. At the extreme, the Trinidadian mountain crab *Pseudothelphusa* has a highly vascular branchial wall surrounding a network of interconnected airways, approaching the complexity of many vertebrate lungs. Molluscs show a similar trend towards elaboration of the lung. In intertidal prosobranch snails, such as *Pomacea*, we find the intermediate situation where half the branchial chamber is filled with gill filaments and the other half is a vascularized simple lung. Most pulmonate snails and slugs have quite simple lungs, albeit with highly complex vascularization of the mantle wall; but some slugs have finely divided cavities with air diverticula interspersed with blood sinuses.

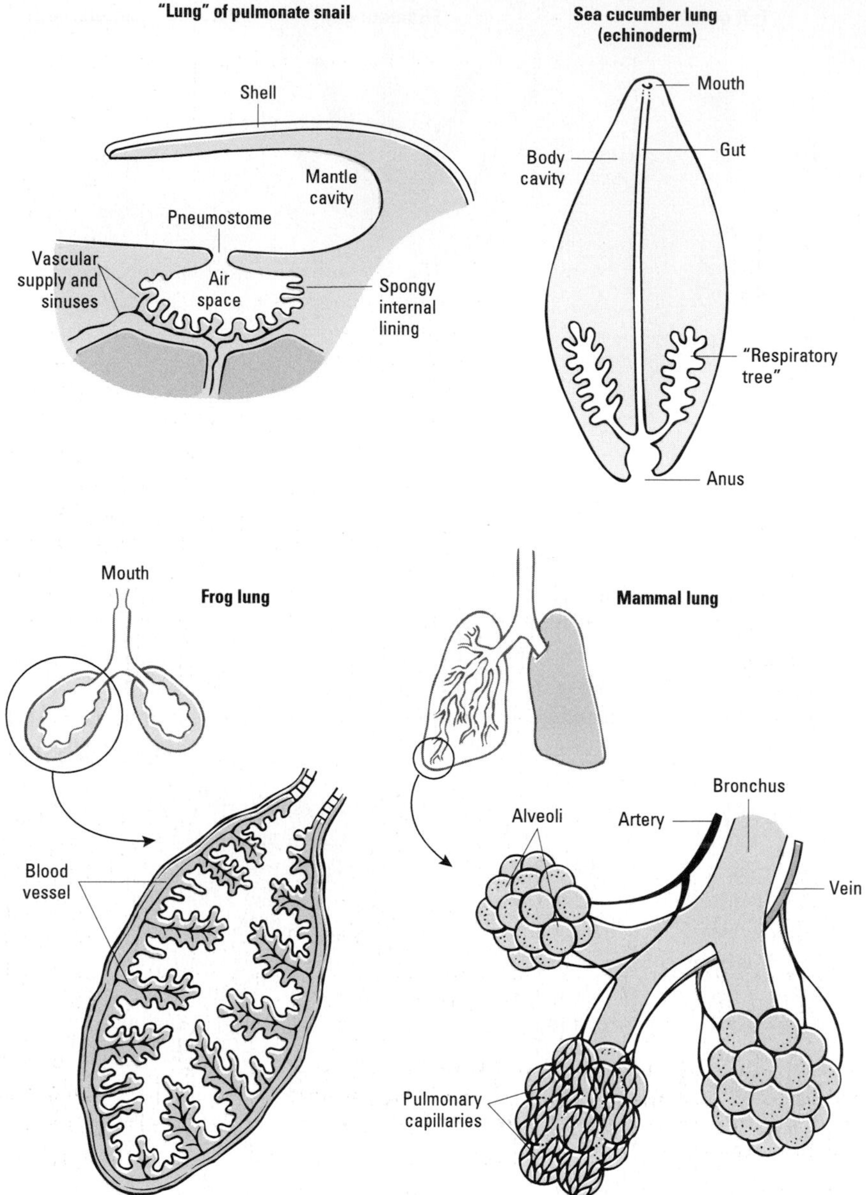

Fig. 7.5 Different types and complexities of lung.

Invaginated sacs for breathing also occur in a range of fish. In many relatively primitive fishes that live in swamps or ponds the swim-bladder is used as a lung, while in some of the perch family a "suprabranchial pouch" is developed as a lung-like structure, and in a few species from various families parts of the gut are used for breathing. The lungs of modern air-breathing fish are relatively little subdivided, whether they occur in chondrosteans (e.g. *Polypterus*), holosteans (gars and bowfins), or dipnoans (true lungfishes).

"True" lungs, pouching off from the lower pharynx, are most obviously associated with the tetrapods: amphibians, reptiles, birds, and mammals. Amphibian lungs exhibit the whole range from secondary absence (some salamanders), through simple sacs, to finely alveolar structures, though in all cases a substantial part of the oxygen requirement is supplied via the cutaneous route (see below). In reptiles, lungs are of course always present but their complexity is very variable.

In the birds and mammals the demands of endothermic life always require a high oxygen consumption and thus a highly alveolar lung, giving a dense spongy tissue; 1 cm^3 of mammalian lung has at least 600 cm^2 of alveolar surface, and the two lungs together may have 300 million alveoli, each only 150–300 μm in diameter, the overall internal surface area scaling in simple proportion with metabolic rate (Fig. 7.6). Air enters the paired bronchi from the trachea before branching into ever finer airways (Fig. 7.5); in mammals, ventilation largely ceases beyond the tertiary branches and air merely diffuses to the final alveolar exchange sites. However, in birds the lungs are additionally ventilated by air sacs with a two-stroke one-way through-flow system (see below and Fig. 7.11), and ventilation extends throughout the bronchial network down to parabronchi of only 500 μm diameter, the final passages before the alveoli (often termed "air capillaries" in birds). Lung volume, pulmonary capillarity, and surface area of the blood–gas barrier all scale

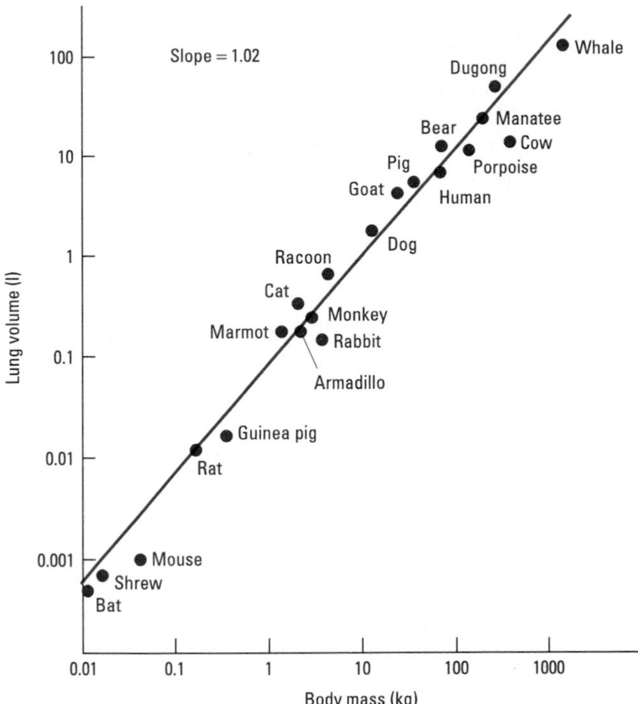

Fig. 7.6 Lung volume in mammals, scaling in simple proportion to body mass. (Data from Tenney & Temmers 1963.)

linearly with body weight in birds, while the weight-specific blood–gas surface area scales negatively, indicating that the smallest birds (with very high metabolic rates) have the most specialized and highly adapted of all lungs.

One other respiratory exchange system is found in the animal kingdom, the **trachea** system of insects and some other terrestrial arthropods (myriapods, onychophorans, some spiders, and many other arachnids). This again involves invagination into the body, but of many separate tubular arrays rather than a single pair of sacs from a single opening as with lungs. Spiders may have a combination of book lungs and simple "sieve tracheae" (branching in bundles from a book lung-like cavity); but in more advanced spiders and in other arachnids only true "tube tracheae" are present. Some terrestrial isopod crustaceans have a combination of gills and tracheae, in that their abdominal pleopods contain internal air-passages called "pseudotracheae". The pterygote insects and most myriapods always have tube tracheae. There the system starts with a series of spiracles, primitively arranged with one pair per segment, but some pairs often being lost, and closable (certainly in insects and in at least some centipedes too). Each spiracle leads to an air-filled primary tracheal tube up to 1 mm in diameter, then branches into finer secondary and tertiary tracheae ramifying to all the tissues and down to about 2 μm in diameter. These retain a moderately thick and probably rather impermeable cuticular lining, and are not sites for respiratory exchange. They lead ultimately to the very finely divided **tracheoles**, blind endings that may be 400 μm long but only 0.6 μm across, penetrating next to and even within individual cells, with a permeable cuticle only 0.01 μm thick. In a very large insect flight muscle cell there may be 5–10 million effectively intracellular tracheolar indentations. Air is delivered directly to the tissues without ever being pumped around the body dissolved in liquid (blood); if a blood system is also present in the animal, it tends to have no respiratory role. In a simple tracheal system flow is essentially tidal, but in larger more active insects (and probably in all flying insects) the tracheae may fuse together into a complex network throughout the body (Fig. 7.7), with air sacs at intervals for storage purposes or to act as bellows.

A tracheal system may also be modified for aquatic respiration (Fig. 7.8; see also Chapter 13). Some aquatic insects merely restrict their spiracular openings to their posterior end, make the cuticle and hairs there hydrophobic, and poke the tip of the body through the surface film to breathe from air. But many freshwater insect adults and larvae have other solutions: there may be extensions of the body (commonly laterally or terminally on the abdomen) containing dense arrays of tracheae (tracheal gills); or "plastrons" of air contained within a dense fur on the cuticular surface; or tufted extensions from the spiracular openings. In dragonflies water is drawn in and out of the rectum and oxygen diffuses into tracheoles there, giving the additional advantage of using the expelled water for jet propulsion! Yet other insects carry air bubbles under water with them when they dive (see Chapter 13).

No listing of respiratory structures would be complete without mentioning the **skin** itself, however. A surprising range of animals possess no special "structures" for gas exchange, and get most of their oxygen supply through their skin (cutaneous exchange). Even quite large and complex animals use the cutaneous route for much of their oxygen supply, without external ventilation. They can do this because they get over the 1 mm maximum diffusion problem by having efficient internal perfusion systems, with a flow of deoxygenated blood or hemolymph passing close to the respiratory surface. This combination of direct cutaneous diffusion externally and efficient perfusion internally works perfectly well in a range of small aquatic invertebrates (including most larval forms and many adult marine worms as well as freshwater leeches), and in semiterrestrial earthworms and tropical slugs. It also occurs in some aquatic vertebrates, including the early larval stages of many fish, adult eels, catfish, salamanders, and sea snakes, and most notably in the more terrestrial anuran amphibians (frogs and toads). In all these vertebrates it is probably a secondary development as the immediate ancestors were more highly dependent on gills or lungs. All these skin breathers, where the skin is permeable to oxygen, are inevitably at risk from desiccation and the phenomenon therefore only occurs in aquatic animals or those from rather moist terrestrial microhabitats; in fact in the few modern frogs that live permanently in water, respiration is entirely cutaneous.

7.2.3 Diffusion barriers

All the respiratory exchangers discussed above have in common the presence of a very thin gas-exchange barrier (Table 7.3). In many animals this diffusional barrier will be no more than a single epithelial cell. In birds and mammals the diffusion barrier across the alveolar surface may be as thin as 0.1 μm, and in fish gills it may appear equally thin but is often in practice augmented by a protective mucus layer, which produces a somewhat thicker boundary layer.

In arthropods there will inevitably be an additional layer of chitinous cuticle at the gill surface. In crab gills this can vary from

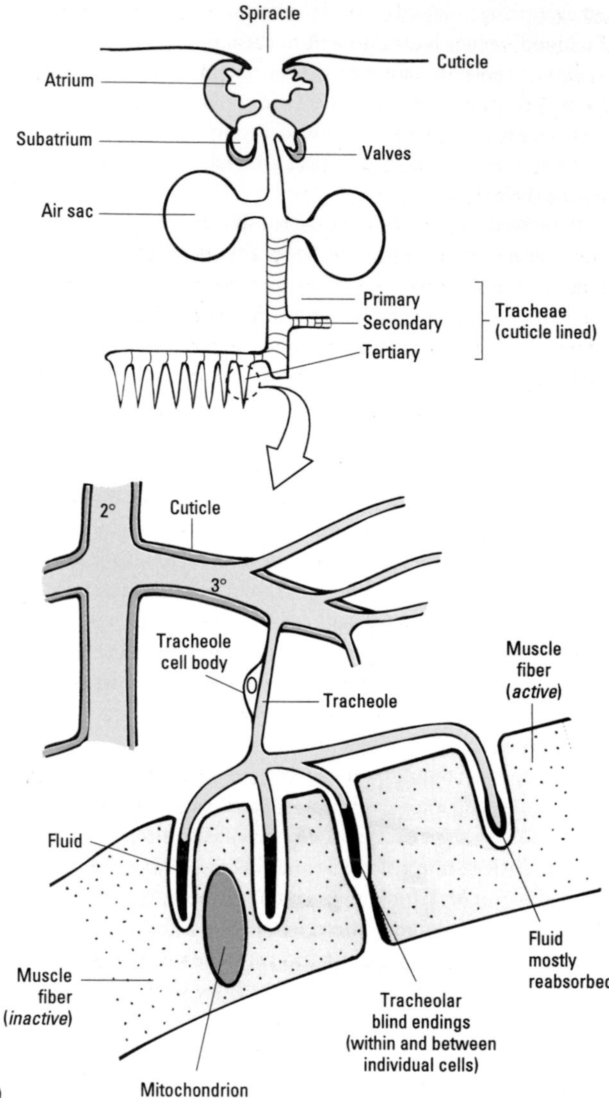

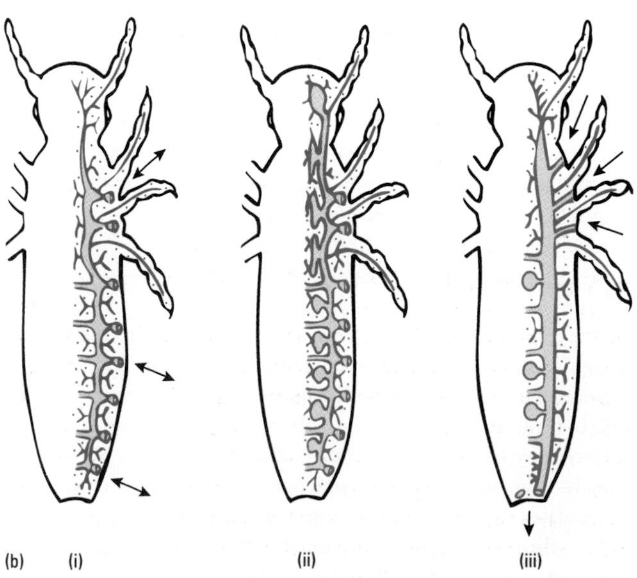

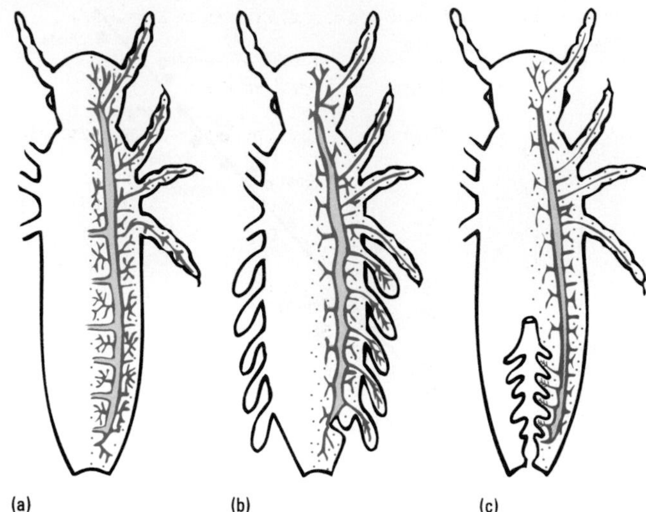

Fig. 7.8 Aquatic tracheal systems with (a) no spiracles, cuticular diffusion only; (b) abdominal gills; and (c) rectal gills.

Table 7.3 The thickness of respiratory gas-exchange barriers (ambient medium to capillary or cell) in air-breathing animals.

Animal group/species	Diffusion distance (μm)
Crab	0.2–0.4
Insect	0.1–0.3
Fish (air-breathing)	
Haplochromis	
Gill	0.3–2.0
Saccobronchus	
Gill	3.6
Air sac	1.6
Anabas	
Gill	10
Branchial wall	0.2
Amphibian	
Toad	1.3–3.0
Birds	
Pigeon	0.1–1.4
Swallow	0.09
Mammals	
Rat	0.13–0.26
Human	0.36–2.5

2 μm in active shore-dwelling species to 15 μm in some deep-sea low-metabolic-rate species, though in land crabs the barrier within the lung tissues may be only 0.2–0.4 μm. Since chitin has a much lower diffusion coefficient than cytoplasm and vertebrate connective tissues (Table 7.4), the actual barrier this imposes will be substantial. In terminal insect tracheoles the whole diameter may be only 0.1–0.2 μm so the cuticular barrier is evidently very thin.

However, insects may be unusual in being able to alter the depth of their diffusion barrier according to need, by variation in the

Fig. 7.7 (*left*) (a) Basic tracheal respiratory systems in insects. The ultimate blind-ending tracheoles supply individual cells with inspired air. They are fluid-filled at rest but during activity fluid is withdrawn (see text) and there is a higher surface area for O_2 uptake. (b) Varying patterns of tracheal network: (i) with many spiracles and only partly interlinked tracheae; (ii) with air sacs added for ventilation; and (iii) with air sacs and reduced numbers of functional tracheae giving unidirectional flow from front to back.

Table 7.4 Diffusion coefficients for gases in air, in water, and in biological tissues.

Medium	Diffusion coefficient ($cm^2\ s^{-1}$)		
	Oxygen	Carbon dioxide	Water vapor
Air			
0°C	0.18	0.14	0.24
20°C	0.20		
Water			
20°C	20×10^{-6}	18×10^{-6}	
37°C	33×10^{-6}		
Tissues			
Human lung	23×10^{-6}		
Muscle	14×10^{-6}		
Amphibian skin	14×10^{-6}		
Amphibian egg	10×10^{-6}		
Fish egg capsule	6×10^{-6}		
Eel skin	2.4×10^{-6}		
Insect chitin	0.7×10^{-6}		

amounts of fluid within these tracheolar endings. The system appears to operate automatically in that when an insect is very active and its metabolic rate rises, the cells around the tracheolar endings are breaking down stored sugars into smaller metabolites so that the osmotic concentration in the cells rises. This draws water out of the tracheolar fluid, so that more of the space is air filled rather than fluid filled, with just a thin surface coating on the tracheolar wall in the aqueous phase, and gaseous exchange can therefore occur more quickly to supply the needs of the active animal.

7.3 Ventilation systems to improve exchange rates

Ventilation involves convective flow of the external medium, maintaining a movement of fluid over the respiratory exchange site, reducing unstirred boundary layers, and replenishing the source of oxygen more or less continuously. The rate of convective transport of oxygen (Q) from the medium (high concentration, C_2) into the animal (low concentration, C_1) is then given by:

$$Q = V_w (C_2 - C_1)$$

where V_w is the rate of mass flow of the medium. This convective transport rate is always much higher than for diffusional transport alone, but there is the disadvantage that energy is expended in creating the ventilatory current, especially for aquatic animals where the fluid to be moved is relatively dense and viscous water.

7.3.1 Mechanisms of ventilation

In some animals ventilation can be achieved quite passively by exploiting the properties of moving fluids at interfaces. Where a fluid flows past a surface there is usually a velocity gradient (slower nearer the surface), and pressure is therefore automatically higher at the surface and lower further up (Bernoulli's theorem). Thus for a perforated sessile animal attached to the surface, fluid will inevitably flow in at the base and out at the top, explaining why a dead sponge still gets ventilated. The same principle applies to ventilation of flow-through burrows, if one end of the burrow is raised above the surface pressure is lower at that end and water or air will flow through the burrow automatically (as in prairie dogs' terrestrial burrows, and in termite nests).

Ventilatory mechanisms where some force is generated by the animal range from ciliary systems to complex muscular machinery; all these in effect function either as pumps or as paddles. Cilia are the most common way of achieving a "paddle" effect, especially in aquatic invertebrates. The sponges (Porifera), which are amongst the simplest and probably most primitive of invertebrates, can attain very large body masses with a design based on water-filled channels, where sea water is propelled along with the aid of cilia borne on the specialized choanocyte cells. Here the cilia are effecting both ventilation and perfusion and, as in many aquatic animals, the current they generate serves both respiratory and filter-feeding purposes. In more complex marine animals such as molluscs, coordinated cilia on the gills produce substantial bulk-flow water currents, which are particularly obvious when channeled through the siphons of larger bivalves, for example clams.

Muscle-related respiratory ventilation at its simplest merely involves moving the animal's body in a suitable manner. Swimming or crawling movements in a benthic animal may stir up the water enough to provide adequate oxygenation. Positioning the body across an existing water current is a more direct way of achieving ventilation, exhibited in several freshwater aquatic insects such as stoneflies and blackflies. However, muscles may be used much more specifically for ventilation where the whole gill structure is moved around, as with the pleopods of isopod crustaceans. Muscles more commonly produce a pumping effect; this may be seen at its simplest in tube dwellers (marine annelids such as the lugworm *Arenicola*) where peristalsis of the body wall draws water in and out of the tube, and in a somewhat more complex form in the related annelid *Chaetopterus*, where the appendages of two segments are modified as fans to draw water through the tube (Fig. 7.9a). Pumps are more common where the respiratory surfaces are enclosed, however, as in the gills of crustaceans and prosobranch and cephalopod molluscs. Most larger crustaceans, and especially the decapods such as crabs and lobsters, have a paddle-shaped "baler" (technically the scaphognathite) that acts as a pump to force water in and out of the branchial chamber (Fig. 7.9b), with a fairly fixed stroke volume but a variable frequency. The smooth muscle layers of the mantle in *Nautilus* and in octopuses contract rhythmically to provide a pulsatile flow over the gills within the mantle cavity (but note that here again the respiratory current can have an additional function, this time in locomotion). In most fish, the muscular pumping effect is more complex again, with the mouth and the gill cover (operculum) coordinated to produce a two-phase, pumped, one-way flow over the intervening gills (Fig. 7.10), with both stroke volume and frequency capable of modification according to metabolic demand. However, in certain actively swimming fish, such as sharks and the tuna and mackerel family, the mouth is kept open and water is forced through the gills continuously by the passage of the animal through the water, a system known as "ram ventilation". Note that in the gills of an active animal, whether decapod or fish, the optimal ratio of ventilation to perfusion (water flow to blood flow) is rather high, often 10–20. This is because of the low oxygen content of

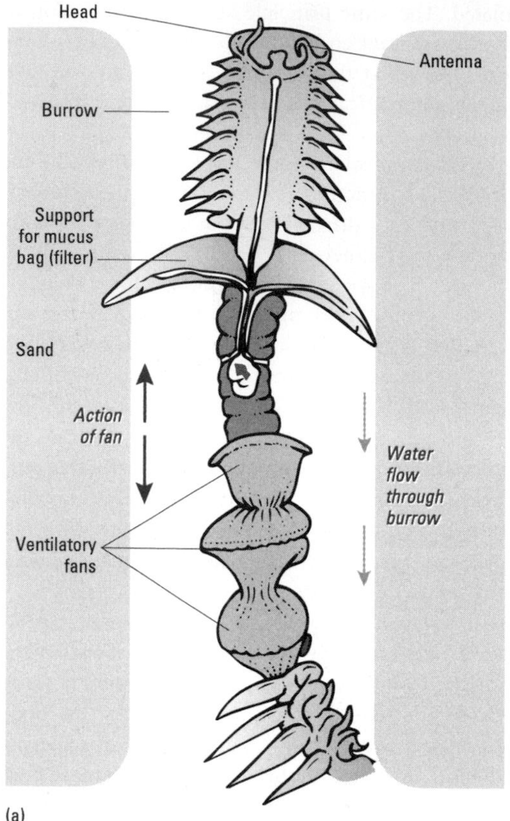

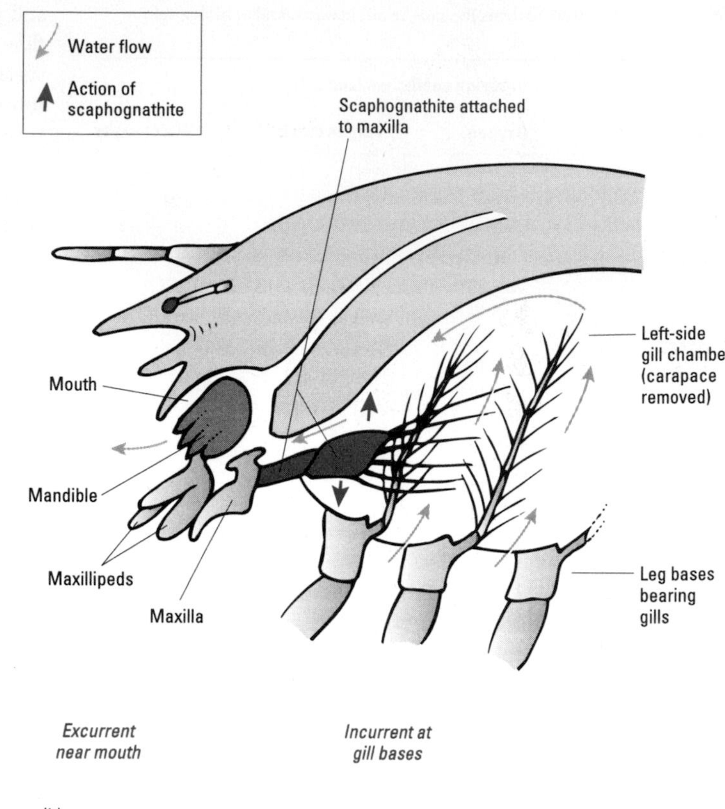

Fig. 7.9 Gill ventilation systems. (a) The parapodial "fans" in the tube-dwelling annelid *Chaetopterus*. (b) The scaphognathite, or "baler", in a decapod crab (longitudinal section).

water relative to the oxygen-carrying capacity of blood, so that ventilation has to be highly organized and rather rapid to support the animal's oxygen demands. Ventilation becomes a significant cost in animals with high metabolic rates, and ram ventilation is a way of saving energy in an active fish, though it means the animal has to keep moving indefinitely.

Whereas gills are usually ventilated unidirectionally, most of the lungs and tracheae of terrestrial animals have to be tidally ventilated with alternate filling and emptying along common channels. Tracheal systems in small insects are entirely diffusional, but in many larger forms ventilation may be assisted by muscular contractions (e.g. of the flight muscles) adjacent to air sacs, allowing them to act as positive-pressure bellows. Movements of the abdomen associated with air pumping are easily seen in larger bees and locusts, and the rate of pumping increases with metabolic rate, so that supply tracks demand. Air movements through the system may be organized into unidirectional pathways, usually with air entering at the thoracic spiracles and leaving through the abdominal ones. In fact in large flying insects air may be forced into the anterior spiracles in a manner analogous to the ram ventilation of swimming fish, and in some beetles the spiracles are shaped so that air is "scooped in" at the thorax in flight and automatically flows through the animal from front to back. The tracheae may thus have unidirectional air flow, though the tracheoles are always blind-ending with simple steady-state diffusion.

Lung books in arachnids appear to be simple diffusion systems, with little or no ventilation. "True" lungs may be ventilated by positive-pressure inflation as in the systems described above, using the cloacal muscles in holothurians, the scaphognathite in crabs, or the buccal muscles or opercular muscles (or both, in some lungfish) as a pump, or again by utilizing air sacs as bellows in birds. Alternatively, they can be ventilated by aspiration breathing, where negative pressure is created to suck air in. In *Pomacea*, the prosobranch snail with both gills and lungs in the branchial chamber, a flap of mantle is modified as a siphon which can pump air into the "lung" section from above the water surface when the water becomes oxygen depleted. In the common snail, *Helix*, air is drawn into the lungs by raising and lowering the mantle floor. Aspiration ventilation in vertebrates may be achieved by ribs alone, for example in most reptiles where the rib cage muscles (intercostals) cause the expansion of the thorax and hence lung inflation. Alternatively it can be achieved by a diaphragm alone, though this system is described in only one fish (the Amazonian *Arapaima*), and in effect in a few reptiles, including crocodilians, where the liver is moved back and forth as a functional diaphragm. In tortoises (where the ribs are fused to the carapace and cannot help to inflate the lungs) the limbs are extended to decrease the intracarapace pressure, so pulling down the diaphragm and inflating the lungs. In most cases, though, ribs are accompanied by a muscular diaphragm to isolate the thoracic space and enhance the negative pressure resulting from rib movements, as in birds and mammals.

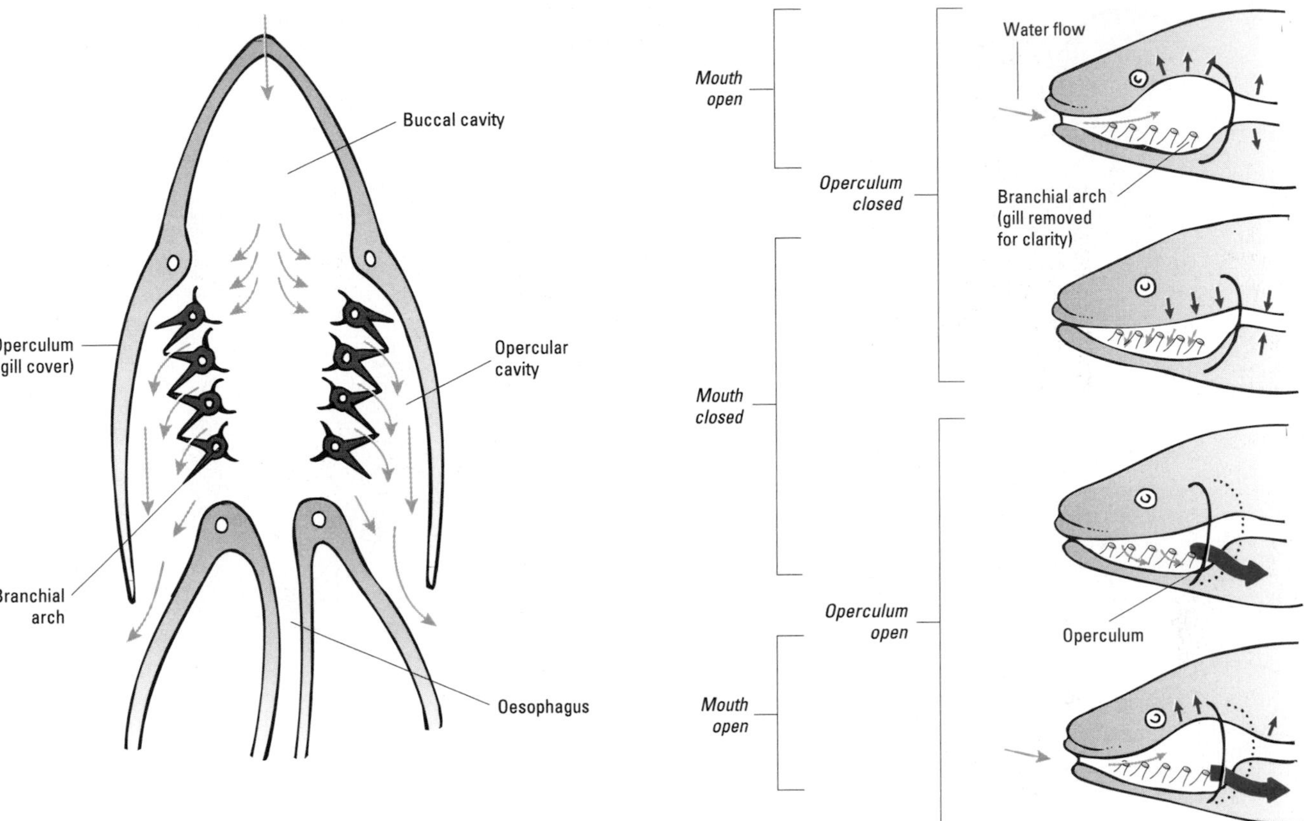

Fig. 7.10 Ventilation of teleost fish gills, also showing the patterns of buccal and opercular movements to produce a two-stroke pump action.

In birds, the extra demands of flight are met by adding air sacs to the system; these work with the ventilatory musculature to achieve a one-way fully ventilated flow through much of the system (Fig. 7.11). The lungs remain at a fairly constant volume but are ventilated by a bellows effect from the neighboring nonalveolar air sacs which surround the viscera and even extend into some of the bird's bones. Expansion of the anterior sacs draws air in through the trachea and mesobronchi, and on to the posterior air sacs via the dorsal bronchi, parabronchi, and ventral bronchi of the lung. Collapse of the air sacs in expiration forces air from the posterior sacs and back out by the same route of dorsal bronchi, parabronchi, and ventral bronchi. Thus flow through the parabronchi adjacent to the alveoli is continuous and always in the same direction, quite unlike the tidal flow of the mammalian lung. Parabronchial flow is roughly countercurrent to the main blood flow (see below), and in the alveoli, which are radially arranged around the parabronchi, it is roughly crosscurrent or at best nearly countercurrent to the finest capillaries, allowing bird lungs to be proportionally smaller and thus lighter than in other tetrapods.

Since air is both far less dense than water and has a much higher O_2 content, ventilation with air is significantly cheaper than aquatic ventilation for an equivalent oxygen demand, though always with the drawback of inescapable evaporative water loss. For land animals, some further savings of energy are made by insuring that the ventilation rate is tied in with other patterns, particularly of locomotion; birds ventilate at the same rate as they flap their wings, or in a strict ratio to wingbeat, and many mammals also breathe in synchrony with their gait. This probably provides greater mechanical efficiency, and may even assist ventilation, as for example in jumping kangaroos where the rising and falling viscera act as an extra pump against the elastic diaphragm and so reduce the cost of lung deflation.

7.3.2 Cocurrents, crosscurrents, and countercurrents

Ventilation systems can be usefully characterized by the relative orientation of the flow of the external medium and of the internal perfusing fluid (Fig. 7.12; see also Box 5.2 for a review of general principles). If the two flows are parallel to each other the flow is termed **cocurrent** and the gas content (e.g. the partial pressure of oxygen) of the internal perfusate leaving the exchanger cannot exceed that of the external medium leaving the exchanger; at best the two fluids will equilibrate, with 50% exchange occurring. However, if the two flows are arranged as a **countercurrent** the internal perfusing fluid may reach almost the same level of Po_2 as is present in the bathing medium. Here the effluent perfusate is in contact with the incoming and maximally oxygenated external medium, so that even if the perfusate has already picked up a good loading of oxygen within the exchanger the gradient is still such that it will continue to load up, leading to almost complete exchange between the two. The **crosscurrent** exchanger has properties lying between these two extremes, and the effluent perfusate will load to a Po_2 somewhere between those of the incoming and outgoing external media.

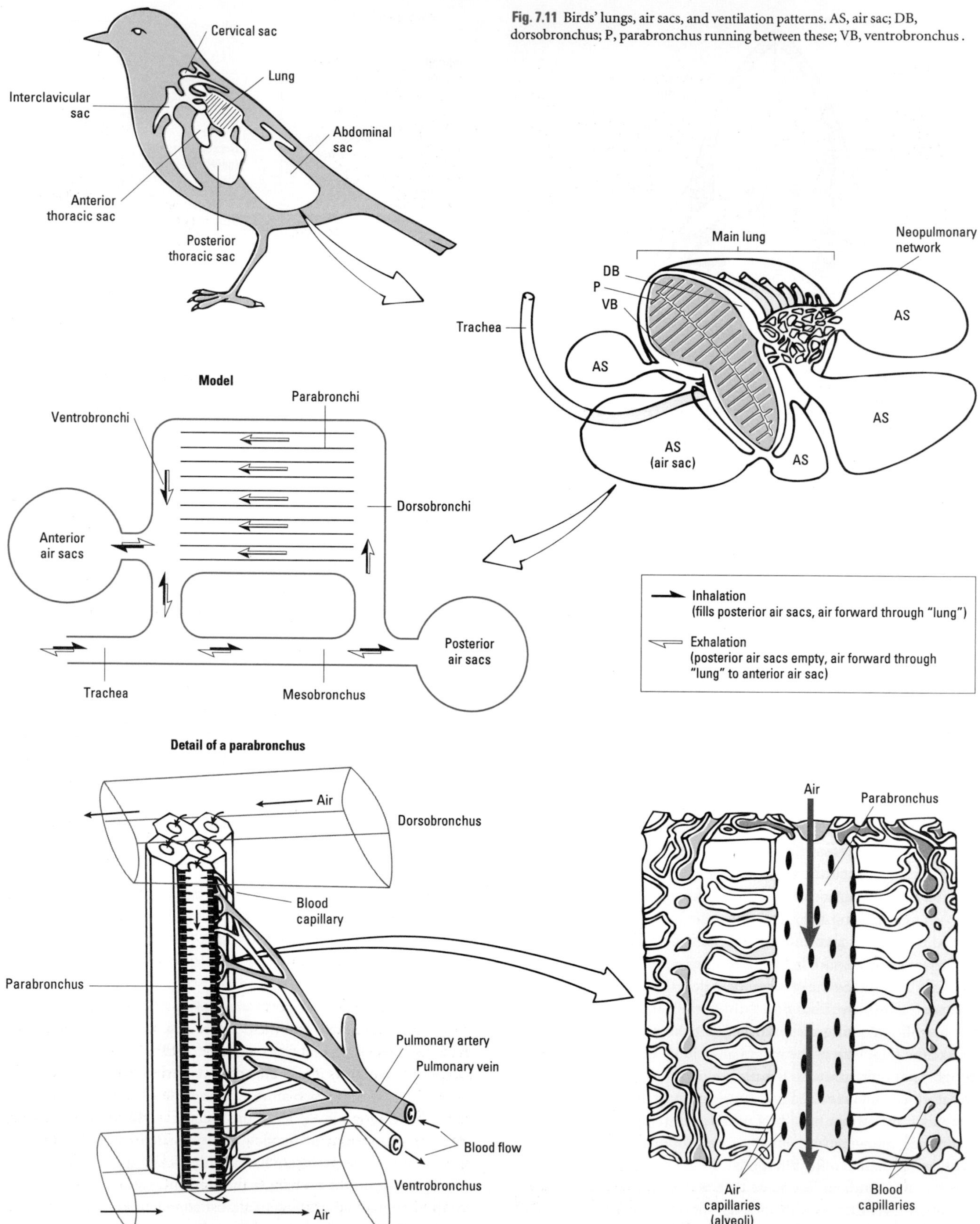

Fig. 7.11 Birds' lungs, air sacs, and ventilation patterns. AS, air sac; DB, dorsobronchus; P, parabronchus running between these; VB, ventrobronchus .

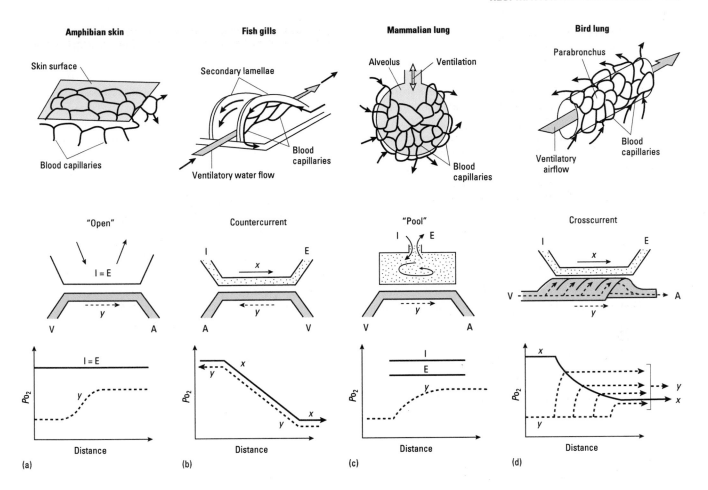

Fig. 7.12 Simple models for respiratory gas exchange. For the blood (green, y), V = venous, A = arterial; for the water or air flow (stippled, x), I = inhalant, E = exhalant. (Adapted from Piiper & Scheid 1982.)

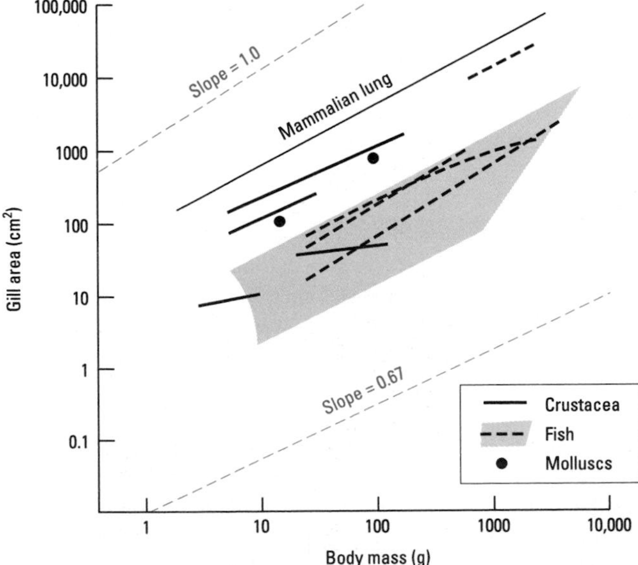

Fig. 7.13 Relationship between gill surface areas and body masses for various taxa, shown in relation to mammalian lung surface area.

Many of the more complex gills in animals achieve countercurrent flow, and in filamentous and lamellar gills there are normally parallel vascular channels within the gill to insure that the blood flows in the right directions. In fish and the larger more active invertebrates such as cephalopods, countercurrent flow is almost essential to insure that oxygen uptake is adequate to compensate for the difficulty and energetic expense of ventilating with a viscous medium. In practice, the advantages of full countercurrent flow (and the generally lower metabolic rates of most aquatic animals) offset the difficulty of ventilation, so that aquatic animals generally require less respiratory surface area than lung breathers (Fig. 7.13), which may be helpful in limiting their osmotic and ionic exchanges with the water.

Crosscurrent flow approximates to the situation found in many bird lungs, discussed above, and is adequate in a terrestrial animal even given the high metabolic demands of flight. But mammalian lungs, with a strictly tidal flow and effectively unventilated alveoli, cannot be characterized in these terms, and are sometimes described as "uniform pool" gas exchangers (Fig. 7.12c), with the alveolar air reaching a steady state by diffusion. The same term could be applied to cutaneous respiration. It may also be applicable to many insects, though in the species that have unidirectional flow with air sacs and controlled ventilation there may be an approximation to countercurrent flow in parts of the system such as the flight muscle, where air flows primarily front to back and may be pulsed with the muscle contractions and resultant wingbeats.

7.4 Circulatory systems

Virtually all metazoan animals have some form of circulatory system, designed to move fluids in bulk within the body and thus reduce diffusion distances by fluid perfusion of all the exchange sites where gases, and also foods and wastes, must be taken up or disposed. In most larger animals this system serves to reduce all diffusion paths to 0.1 mm or less, insuring very rapid diffusive transfer at the tissue level. In effect, a circulatory system becomes a system of forced convection of respiratory gases around the body, so that only the final exchanges at the point of uptake and at the point of delivery to the tissue mitochondria are reliant on the slow process of diffusion. Using such forced convection, a large mammalian body can move a molecule of oxygen from the uptake point in the lung over a distance of about 2 m to peripheral capillaries in a limb in about 1 min.

Circulatory systems, once evolved, can serve a multiplicity of roles. They will transport oxygen and carbon dioxide, water and salts, food components such as glucose, fatty acids, and amino acids, vitamins and trace elements, and waste products such as ammonia, urea, and creatine. Hormones may be distributed, so that the circulatory system becomes a key player in maintaining homeostasis; the cellular and chemical components of bodily defense systems are also distributed (see Chapter 17). The circulated fluid may also have a hydraulic role, either as a hydrostatic skeleton for the whole body (in many marine worms) or as a means of protruding tissues and organs under pressure in special areas of the body. Examples here include echinoderm tube-feet, legs in spiders (which have retractor muscles but no extensors), gills ("branchial crowns") in many annelids, and erectile tissues such as throat display pouches and penile tissues in many animals including vertebrates. Thus the circulatory system can come to play a role in locomotion, in reproduction, and in behaviors associated with aggression, dominance, and mating.

7.4.1 Open-plan and closed plumbing

The circulation in an animal can range from a slightly controlled pumping of the external medium through the body, to continuous circulation of a contained regulated internal fluid endowed with specialized cellular components through highly organized closed conduits.

Use of the external medium is generally associated with smaller animals, but can be effective in larger species with relatively sedate or sessile lifestyles, such as sponges or cnidarians. It is mainly found in marine animals, where the sea water provides a suitable osmotically balanced and well-oxygenated medium to bathe the tissues, but also occurs in some freshwater species; in both cases, the circulating medium may also be used as a source of filterable food.

Circulation of internal fluid can be achieved by a movement of the general extracellular fluid, so that the extracellular spaces of all tissues are confluent with the circulating fluid, which moves through loose sinuses within tissues and larger lacunae or cavities between the tissues. In this situation the circulatory fluid is usually termed **hemolymph** and the system is described as an **open circulation**. Open circulation occurs in most of the smaller animals that have bodies designed around the acoelomate or pseudocoelomate grades, where fluid in the blastocoel bathes the tissues directly (see Box 1.3). No specific pump is needed in these animals, as the movements of the body wall maintain an adequate circulation. Such a system is also found in a more sophisticated version in most of the arthropodan groups (Fig. 7.14d,e), in many molluscs (Fig. 7.14f), and in less well-known invertebrates such as tunicates. Here the hemolymph is partially directed around the body through vessels and may flow through short sections of finely divided networks of thin-walled vessels in key sites such as excretory organs, gills, or central ganglia. It then passes into the general open hemocoel, which lacks any lining separating it from the muscles and other tissues, before returning to one or more simple hearts. In a few cases there are specific vessels leading from the respiratory system back to the heart; for example, some spiders have "lung veins" draining the book lungs.

Alternatively, the circulating fluid may be highly specialized and largely separated from the rest of the extracellular fluid, and it is then usually termed **blood**. This separation results in a **closed circulation**, where the blood travels in vessels lined with endothelial cells, and only comes into contact with the tissues when it traverses the walls of the finest vessels, known as **capillaries**. Within such a system there may be areas of apparently "open" circulation where larger blood sinuses occur, but these will always be endothelium-lined. Clearly in this situation, where the whole circuit is in narrow vessels, one or more powerful pumps will be needed to force the blood around the conduits of the body; in larger and more active animals the pump is a highly muscular heart (see below), which itself needs a substantial blood supply.

The distinction between open and closed circulatory systems is clearly not absolute. The presence of capillaries is sometimes said to be the defining difference, but capillaries can occur in the most finely divided reticulate sections of the essentially "open" systems in large crustaceans, and within the phylum Mollusca the complete range from open to closed systems occurs, with cephalopods having fully formed endothelium-lined capillaries. Open systems are also usually characterized by large blood volumes (at least 30% of body weight is commonplace, compared with about 8% in most mammals; see Fig. 4.1), by low pressures relative to the values of 100–200 mmHg found in many vertebrates, and by long circulation times. But in terrestrial crabs the peak ventricular pressures can be as high as 50 mmHg, although once the blood has passed beyond the gills the pressure falls almost to zero, with virtually no pulse detectable. In terrestrial pulmonate molluscs, too, pressures of 20 mmHg have been recorded. Perhaps the best way to distinguish open and closed systems is by the presence or absence of an identifiable extracellular space separate from the blood; note that the distinction is not then between invertebrates and vertebrates, since the cephalopods at least would qualify as having fully closed circulations. This allows them to generate pressures of over 100 mmHg during activity, comparable to vertebrate levels.

In both open and closed blood systems, the carriage and distribution of blood to tissues is carefully regulated, with fluid pressure and fluid volume monitored by central or peripheral receptors (see section 7.7). Closed systems are in no way inherently more complex or more highly regulated than open ones. However, in the higher vertebrates the fluid composition of the blood is more stable, and perhaps more rigorously monitored than in other animals, due to their high oxygen demand coupled with the susceptibility of their

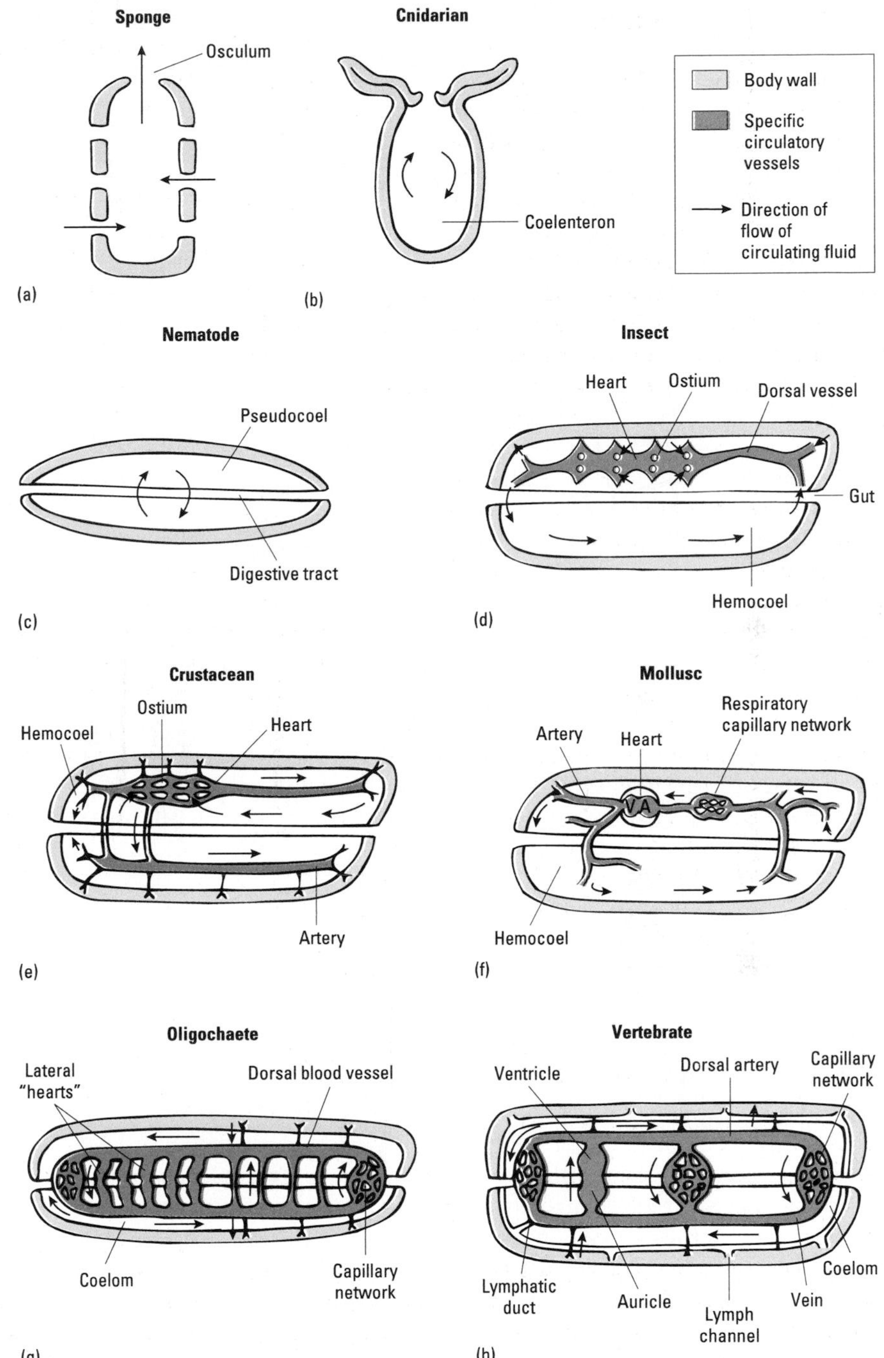

Fig. 7.14 Diagrammatic views of circulatory patterns in a range of animal taxa.

hemoglobin properties to environmental variability (see section 7.5.2).

7.4.2 Pumping stations: hearts

In animals of moderate activity and/or large size, some form of pumping station is needed to keep the blood or hemolymph circulating adequately, and this pump is conventionally termed a **heart**. In essence the heart is a section of one of the principal blood vessels where the vessel wall is thickened and muscularized, and this ancestral design can be traced even in the complex multichambered hearts of tetrapods. To qualify as a fully functional heart, a pumping structure should also have a degree of nervous synchronization of its contractions, and be equipped with internal valves to ensure one-way flow.

Invertebrate hearts

Many invertebrate groups are rather flexible in their arrangement of hearts, with one or many being present and sited according to need, so that there may be accessory "branchial hearts" near the gills, for example. Hearts appear at their simplest in annelid worms, where sections of the main dorsal vessel, carrying blood forwards along the

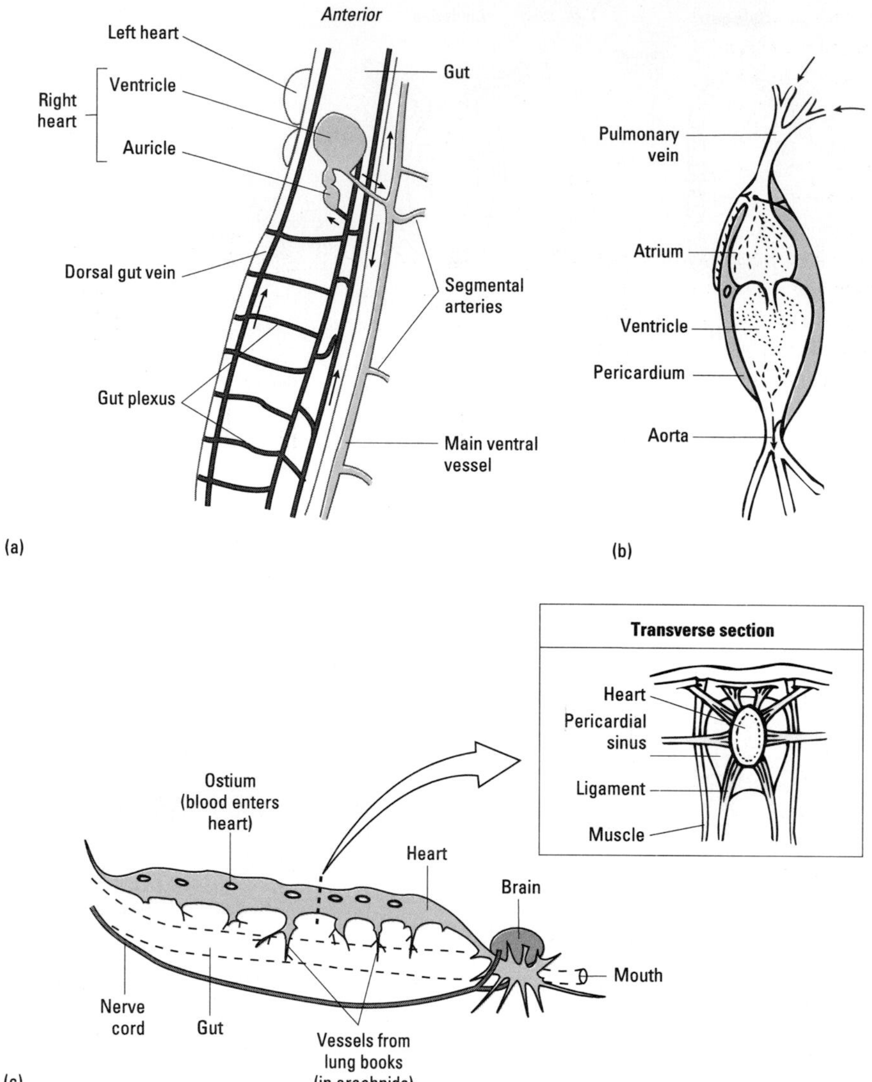

Fig. 7.15 Structure of the heart in (a) earthworms, (b) molluscs, and (c) arthropods.

body, are thickened and show rhythmic contractility. In some of the oligochaete annelids (earthworms and their relatives) these dorsal hearts are supplemented or even supplanted by a series of paired hearts on the lateral vessels that link the dorsal and ventral longitudinal vessels (Fig. 7.15a). In the giant megascolecid earthworms of Australasia, which are up to 3 m long and weigh 0.5 kg, these lateral hearts can generate pressures of up to about 20 mmHg.

Molluscs show a greater diversity of hearts, with a two-chambered arrangement being common (Fig. 7.15b). In most classes of mollusc there is a thin-walled **auricle** (or atrium) collecting blood from venous vessels and sinuses, then a thick-walled muscular **ventricle** as the main pump, supplying several main arteries leading in parallel to the important regions of the body (the kidney, the visceral hump where digestion and absorption occur, and the head region). The heart is enclosed in a fairly rigid sac, the **pericardium**, so that the sequential contractions of the auricle and then the ventricle provide suction pressures to speed up the refilling process. In cephalopod molluscs this same structure exists but is proportionately larger and thicker walled, and the pumping action is supplemented by a pair of branchial hearts at the base of the gills, providing a pressure boost in the venous system as it enters the complex gill vasculature where viscous resistance must be overcome.

Rather different kinds of hearts are found in arthropod groups, where the absence of a piped venous return makes the return of blood to the heart more problematic. Most crustaceans solve this with a series of valved holes or "ostia" in the heart itself (Fig. 7.15c), taking in blood from the pericardial cavity during cardiac expansion; as the heart then contracts with the ostial valves closed, blood is expelled forwards into the arteries and the negative pressure created in the rigid pericardium draws further venous blood in. A similar system operates in insects, and in all these arthropod groups the degree of vascular development (particularly of arteries) is strongly positively correlated with size and activity.

Vertebrate hearts

Fish hearts represent only a slight increase in complexity on the molluscan and crustacean hearts already described. In addition to the **auricle** (or atrium) and **ventricle**, two further sections are found in series to the system. The **sinus venosus** precedes the auricle and is the principal convergence point of the great veins, which empty into it via valves; neither the sinus nor the auricle are particularly muscular, and their weak contractions generate only small pulse pressures. By contrast, the ventricle (lying ventrally to the auricle) is usually extremely muscular. In less active fish it may become reticulate and spongy with only a small open lumen, but in fast-swimming fish there is a good-sized lumen and an outer layer of circumferential muscle with its own coronary blood supply, producing a fast and forceful contraction. In both elasmobranch and teleost fish the ventricle leads into a fourth chamber, either a **bulbus arteriosus** (a thickened elastic structure in teleosts) or a **conus arteriosus** (both elastic and muscular in elasmobranchs). The main function of this extra section to the heart is to smooth out the pressure changes created by the ventricle.

In a few fish where lungs have been added to the respiratory system, notably the dipnoan lungfish, we see the first indications of a separation of the oxygenated and deoxygenated bloods in vertebrates. The return vessel from the lungs, carrying oxygenated blood, enters the sinus venosus separately from the other veins, and an **intra-auricular septum** keeps this oxygenated blood largely separate from the deoxygenated systemic venous blood as it traverses the auricle. The ventricle is also partially divided by a plug of tissue, and the conus arteriosus is modified with a spiral twist and complex valves. The net effect is that the two bloods are only partially mixed and the lungs get a relatively deoxygenated supply compared to the oxygen-rich blood entering the systemic arterial loop. This partial separation is continued in the amphibians, with an improved distinction of the pulmonary and systemic return circuits brought about by a full separation of the auricle into right and left chambers, and a more effective partial **ventricular septum**. Differential timings of the flows from each return system, and channeling of flows by mechanical obstacles (ridges within the heart), seem to insure the supply of nearly fully oxygenated blood to the systemic circuitry. However, the amphibian heart, at least when compared with those of many active fish, is relatively weak, with a spongy poorly muscularized ventricle and no independent coronary blood supply.

In the amniote vertebrates there is increasing separation of both the venous and the arterial circuits, with septation inside the heart increasing and separation of the auricles evident externally as the sinus venosus is progressively lost. In most reptiles the ventricle is not quite fully divided, but the sinus has been largely incorporated into the right auricle, while the bulbus has been divided into two separate channels opening via valves right back at the ventricular wall. It should be noted that partially separated circulations in the "lower" vertebrates are not necessarily inferior, and allow the individual to choose the pattern of blood oxygenation in different parts of the body, which is particularly useful in diving animals.

In adult crocodilians, and in all birds and mammals, the ventricular septation is complete, with independent right and left ventricles opening separately to the pulmonary arteries and the aorta, respectively. The detailed anatomy of the system differs somewhat in each of these three groups, but all have achieved the pattern of two side-by-side pumps, with the right auricle and ventricle handling deoxygenated blood for transit on to the lung, and the left auricle and left ventricle handling the oxygenated blood from the lung for transit on to the rest of the body (Fig. 7.16). All these amniote vertebrates also have highly muscular ventricles and independent coronary arteries supplying the heart muscles. Birds have the most powerful of all hearts, coping with the 10–15-fold increase in oxygen demand during flight and giving a 5–7-fold higher rate of tissue perfusion compared with a similarly sized mammal.

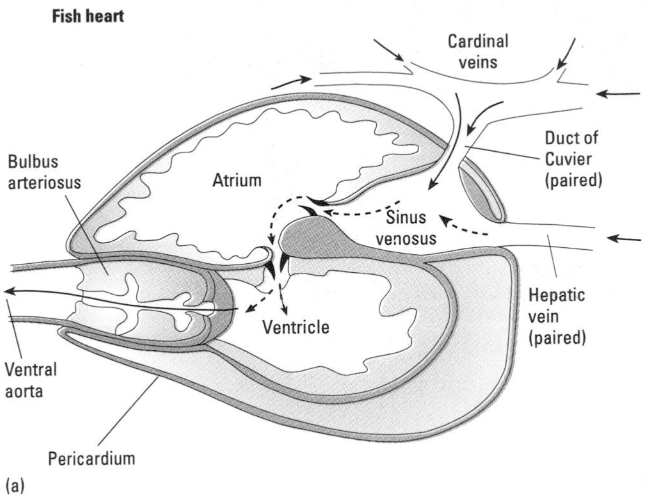

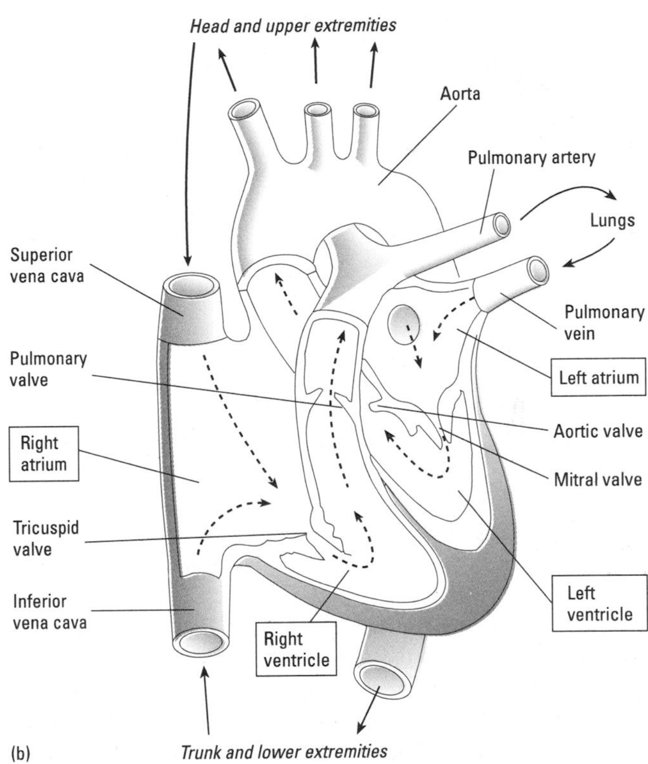

Fig. 7.16 Vertebrate heart complexity in (a) fish and (b) mammals.

In the avian and mammalian heart, the muscular ventricle fills with blood while it is relaxed (**diastole**), and then contracts to force blood out into the arteries (**systole**). The terms diastolic pressure and systolic pressure can also be applied to less sophisticated circulations, with the highest pressures always recorded at peak systole.

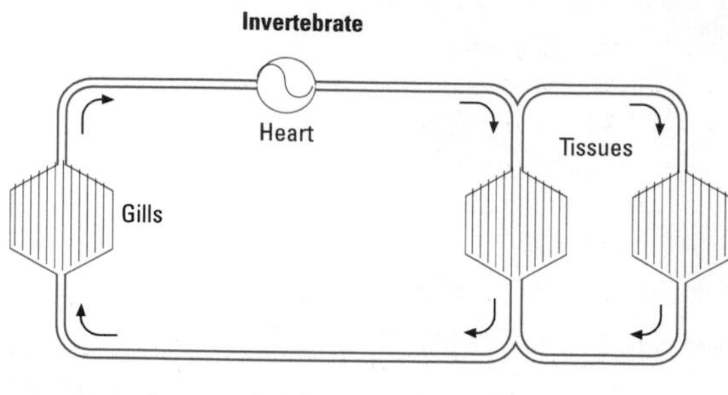

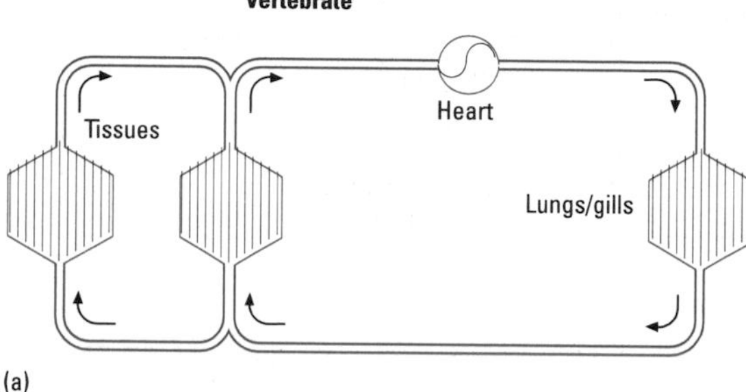

(a)

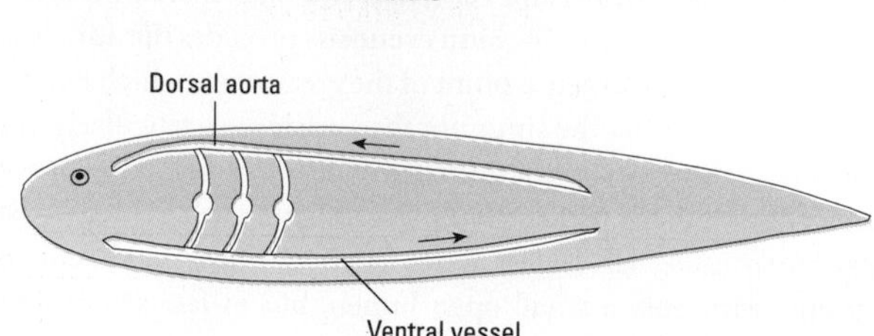

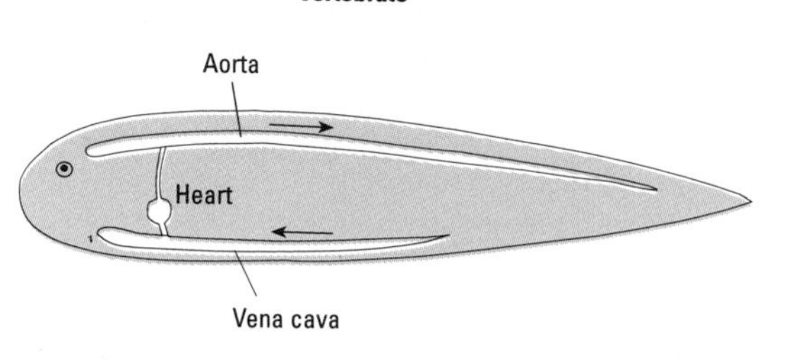

(b)

Fig. 7.17 (a) Differing patterns and (b) directionalities of flow in major vessels in invertebrates and vertebrates. In invertebrates the respiratory organs are downstream of the tissues, whereas in vertebrates (until two separate circuits are established in reptiles) the respiratory organs are served directly from the heart. Dorsal vessels also carry blood forwards in almost all invertebrate groups and backwards in vertebrates.

7.4.3 Circuitry

With invertebrates generally, there is a single circuit of blood around the body, and the circuit runs from heart to systemic system to gills or lungs and then back to the heart (Fig. 7.17a). In vertebrates, however, even where there is only a single-circuit system there are two differences in the pattern of the circuitry. Firstly, in vertebrates such as fish the blood from the heart goes first to the respiratory exchange site (gills or lungs) and thence to the main systemic organs before returning to the heart (Fig. 7.17b). Secondly, all invertebrates with vascular systems have a dorsal vessel where the flow runs from posterior to anterior and a ventral vessel with flow in the opposite (backwards) direction, whereas in all vertebrates in the main ventral vessel blood flows anteriorly and in the dorsal vessel it flows posteriorly (Fig. 7.18).

Where dual circuitry is found, in birds and mammals (and crocodiles), the two halves of the circulation have very different physiological properties. The **pulmonary circuit** is low pressure (Table 7.5), but fairly high volume, driven by the right ventricle. In mammals the average pressure here is only about 13 mmHg, peaking at 25 mmHg. The system is very compliant to changes in output from the heart, and there is little capacity for control (except in crocodiles, where the unusual circuitry allows flow to be diverted away from the lungs by increases in vascular resistance). By contrast, the **systemic circuit** is highly variable in its needs, with the demands of muscles, kidney, and skin varying with activity levels, and with water and temperature balance. Thus the pressure is always relatively high, but pressures and volumes in different parts of the system may be varied and controlled on a minute-to-minute basis.

Table 7.5 Typical systemic and pulmonary pressures in animals with closed circulatory systems.

Animal group	Systolic systemic pressure (kPa)	Systolic pulmonary pressure (kPa)
Cephalopod	6.0	–
Lungfish	3.5	2.8
Teleost	3.9	–
Amphibian	4.4	3.2
Reptile	5.5	3.5
Mammal	15.5	2.5
Bird	21.0	2.7

In all vertebrates, the blood supply to most organs arises directly from the dorsal aorta, so that each organ is supplied in parallel and at full pressure. However, a few organs are supplied in series with other organs, receiving the venous outflow of another organ, giving a **portal circulation**. This entails a delivery of partially deoxygenated blood, but it permits particular metabolites to be localized in the circulation, so that they do not reach the body as a whole. This is an obvious advantage in some circumstances, for example where the liver receives digestion products directly from the intestinal and splenic circulation; this is the hepatic portal system, responsible for 75% of the supply to the liver in mammals. The liver can then deal with any toxins present in the food before they reach other tissues. Portal systems also operate in the vertebrate kidney, where it is effluent blood from the glomeruli that supplies the nephron tubules (see Chapter 5), and in the brain where a portal vein carries blood leaving the hypothalamus (and thus carrying several key peptide hormones) directly to the anterior pituitary gland.

Fig. 7.18 *(opposite)* An overview of circulation patterns in various vertebrate taxa.

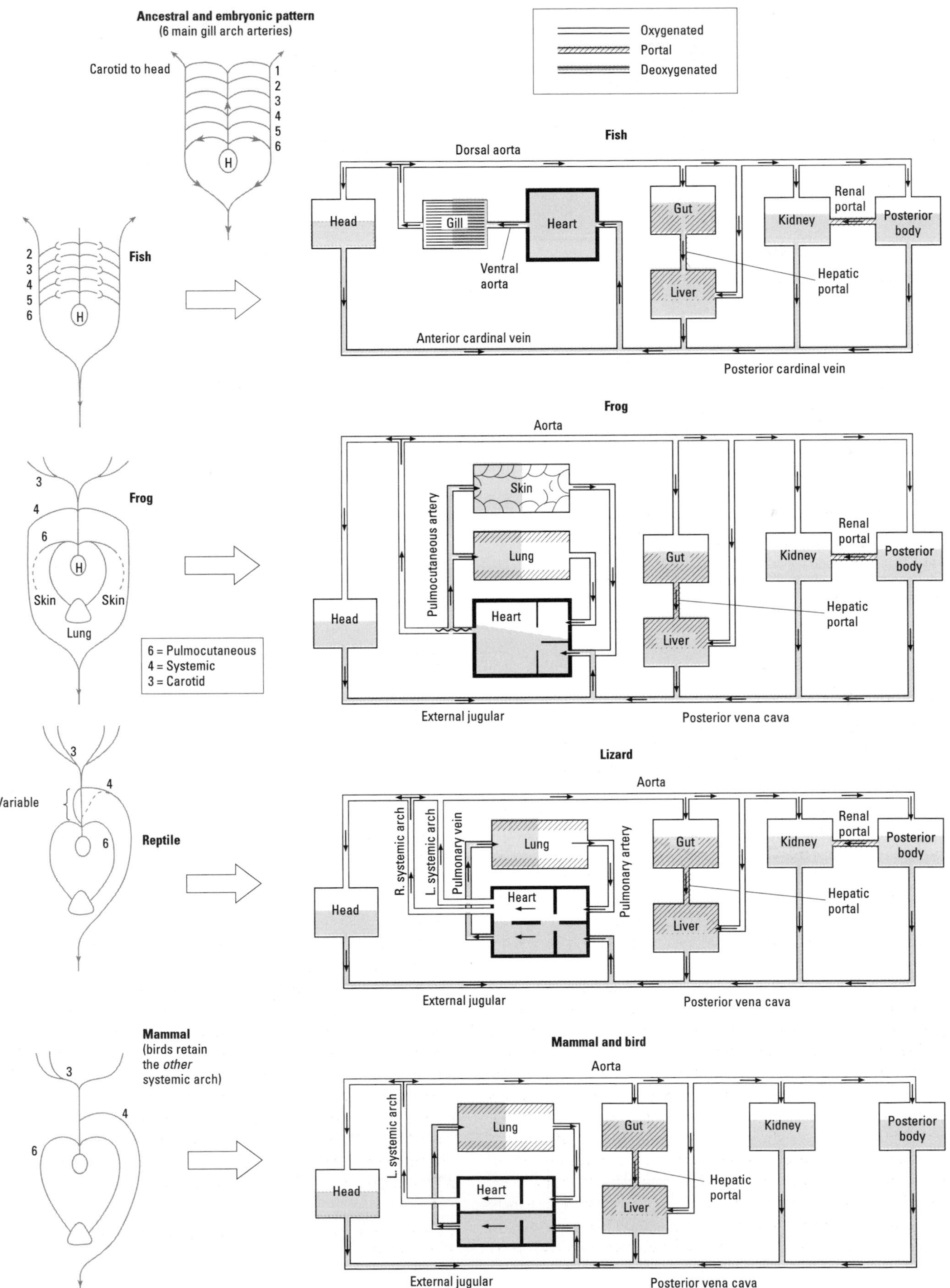

Ancestral and embryonic pattern
(6 main gill arch arteries)
Carotid to head
1
2
3
4
5
6
H
Oxygenated
Portal
Deoxygenated
Fish
2
3
4
5
6
H
Fish
Dorsal aorta
Head
Gill
Heart
Ventral aorta
Gut
Liver
Kidney
Renal portal
Posterior body
Hepatic portal
Anterior cardinal vein
Posterior cardinal vein
Frog
3
4
6
H
Skin
Skin
Lung
6 = Pulmocutaneous
4 = Systemic
3 = Carotid
Frog
Aorta
Skin
Lung
Heart
Pulmocutaneous artery
Head
Gut
Liver
Kidney
Renal portal
Posterior body
Hepatic portal
External jugular
Posterior vena cava
Variable
3
4
6
Reptile
Lizard
Aorta
R. systemic arch
L. systemic arch
Pulmonary vein
Lung
Heart
Pulmonary artery
Head
Gut
Liver
Kidney
Renal portal
Posterior body
Hepatic portal
External jugular
Posterior vena cava
Mammal
(birds retain the *other* systemic arch)
3
4
6
Mammal and bird
Aorta
L. systemic arch
Lung
Heart
Head
Gut
Liver
Kidney
Posterior body
Hepatic portal
External jugular
Posterior vena cava

Table 7.6 Properties of different kinds of blood vessels and distribution of blood in mammals (total volume = 8–9% body volume).

	Typical diameter	Wall thickness	Composition of vessel wall (% diameter)				% of total blood volume contained	
			Endothelium	Muscle	Collagen	Elastin	Systemic	Pulmonary
Aorta	25 mm	2 mm	5	22	33	40	–	–
Artery	4 mm	1 mm	10	35	25	30	12	7
Arteriole	30 μm	20 μm	10	50	20	20		
Capillary	8 μm	<1 μm	100	0	0	0	11	1
Venule	20 μm	2 μm	40	10	40	10	41	14
Vein	20 mm	1 mm	8	32	40	20		
Heart							14	

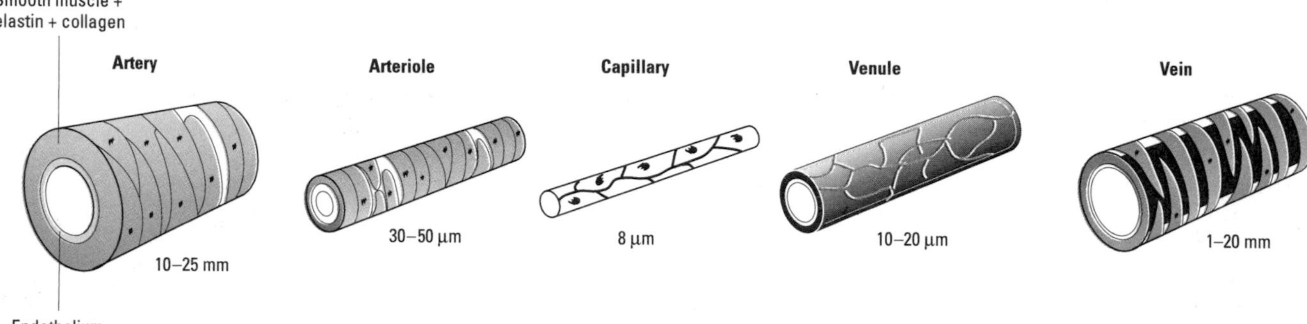

Fig. 7.19 Transverse sections of vertebrate blood vessels, showing typical dimensions.

7.4.4 Blood flow and pressure

Systolic pressures close to the heart in any animal are always well above atmospheric pressure, whilst other parts of the circulatory system may be at almost zero pressure. It is this pressure gradient that drives the blood around the circuit and from arteries to veins via the capillaries. Because hearts beat intermittently, the systemic arterial pressure is not constant but decays between heartbeats; in a human it decays from around 120 mmHg to about 80 mmHg. This gives a pulsatile driving force and a pulsatile flow.

The relationship between pressure and flow is therefore rather complex. But we can model it for a more simple situation with flow along a rigid tube at constant pressure, where the flow, Q, is directly proportional to the difference between the inlet pressure (P_1) and the outlet pressure (P_2):

$$Q \propto P_1 - P_2$$

Note that flow is by definition a rate (mass or volume per unit time), so that it is unnecessary to refer to a "flow rate".

If a constant is inserted into the above equation, it is transformed into a simple expression describing flow:

$$Q = K(P_1 - P_2)$$

K is called the hydraulic conductance of the tube, and its reciprocal is called the resistance to flow or hydraulic resistance (R):

$$Q = (P_1 - P_2)/R$$

Thus the flow in the tube is proportional to the "driving pressure", i.e. the pressure differential ($P_1 - P_2$), and is inversely related to the hydraulic resistance. This is a form of Darcy's law of flow (which is in turn analogous to Ohm's law relating resistance, voltage, and electric current). It tells us that in any animal's circulatory system, flow can be altered either by changing the driving pressure exerted by the pump, or by altering the vascular resistance.

In practice, blood pressure tends to be kept relatively constant (often by hormone actions, discussed in Chapter 10), and flow is regulated by resistance changes brought about by the contraction and relaxation of vessels—**vasoconstriction** and **vasodilation**.

7.4.5 Blood vessels

There are various kinds of blood vessel in any animal's circulatory system, but those in the vertebrates are best known and may be used as a general model. The pattern is certainly fairly constant: an aorta leading from the heart to one or several major arteries, progressively breaking into narrower, high-resistance vessels (arterioles) then branching into a network of capillaries that converge again into venules that unite into the veins. The characteristics of the various types of vessel in a mammal are given in Table 7.6.

With the exception of the capillaries, all blood vessels are made up of three layers of tissue (inner two shown on Fig. 7.19). An inner (intimal) layer of flat endothelial cells rests on a middle (medial) layer of circular smooth muscle cells embedded in a matrix of elastin and collagen fibers. Outside these is a connective tissue layer (the adventitial layer), containing nerves and often (in the large arteries in particular) some small blood vessels to provide circulation to the vessel wall.

Within this general structure, certain distinct types of vessel can be distinguished. **Elastic arteries** include the pulmonary artery, aorta, and major limb arteries; these have very extensible walls due to the presence of large amounts of elastin fibers in the medial layer (elastin being six times more extensible than rubber). These major arteries therefore expand during ventricular emptying and recoil again later, thus having the effect of converting the intermittent pumping of the heart into a relatively continuous flow through the more distal vessels. They also have a good proportion of collagen fibers in the medial layer; these fibers are stiffer than elastin and probably insure that the vessels do not get overdistended to cause aneurysms or bursts.

Muscular arteries are less important in smoothing out blood flow but have a rich autonomic nerve supply and can contract. Examples include many smaller arteries such as the cerebral and coronary arteries. They have thick walls to prevent arterial collapse at sharp bends such as the knee and elbow joints. In diving animals they contract sharply during a dive to limit the blood supply to the limbs and insure that the brain and heart are well supplied.

Arterioles have the highest wall thickness to lumen ratio of any vessels. The larger arterioles are richly innervated by vasoconstrictor neurons, while the smaller ones are poorly innervated and have a particularly narrow lumen. The pressure drop here is greater than at any other point in the circulatory system, indicating that the systemic arterioles represent the chief resistance to flow (Fig. 7.20). The smaller arterioles are therefore referred to as **resistance vessels**, and by changes in their luminal diameter (vasoconstriction and vasodilation) they control the blood flow to organs and tissues, mediated by a variety of chemical and neural signals. Full contraction of the terminal arterioles can shut off the blood flow to a particular region of capillaries (a capillary "bed") entirely.

Capillaries have the smallest diameter of any vessels (only 3–100 μm) and are generally very short (250–1000 μm). They are present in very large numbers, reducing the diffusion distances from circulation to cell. Their thin walls consist of a single layer of flattened endothelium, through which gases and metabolites can permeate. Blood passes along capillaries in short, smoothed pulses ("bolus flow"), and because of this and their very high density and thus large total cross-sectional area they offer rather little resistance to flow. The transit time of a red blood cell in mammalian muscle capillaries is around 1–2 s, allowing sufficient time for the unloading of oxygen and uptake of carbon dioxide.

Certain tissues, such as the skin and the mucosa of some chemosensory organs, possess shunt vessels, 20–130 μm in diameter and called **arteriovenous anastomoses**. These enable blood to pass directly from arterioles to venules without going through the capillary bed. Such an anastomosis has a rich sympathetic innervation, and in many mammals the anastomosing vessels have an important role as bypasses in regulating blood flow to the surface of the body as part of the process of temperature regulation.

Venules and small **veins** are more numerous than arterioles and arteries (see Table 7.6), and hence they provide a low resistance to flow. For example, in humans a pressure differential of only 10–15 mmHg is sufficient to drive the cardiac output from the venules back to the vena cava (the great vein leading to the heart). In addition to returning blood to the heart, the venous system of vertebrates serves as a variable reservoir for blood; the vessels have thin walls that are readily distended or collapsed, having only low elastin content, enabling their blood content to vary enormously according to their degree of vasoconstriction. The venules and veins are therefore referred to as **capacitance vessels**. In the limbs of vertebrates, the veins have pairs of semilunar valves that prevent the backflow of blood even when operating at high capacity.

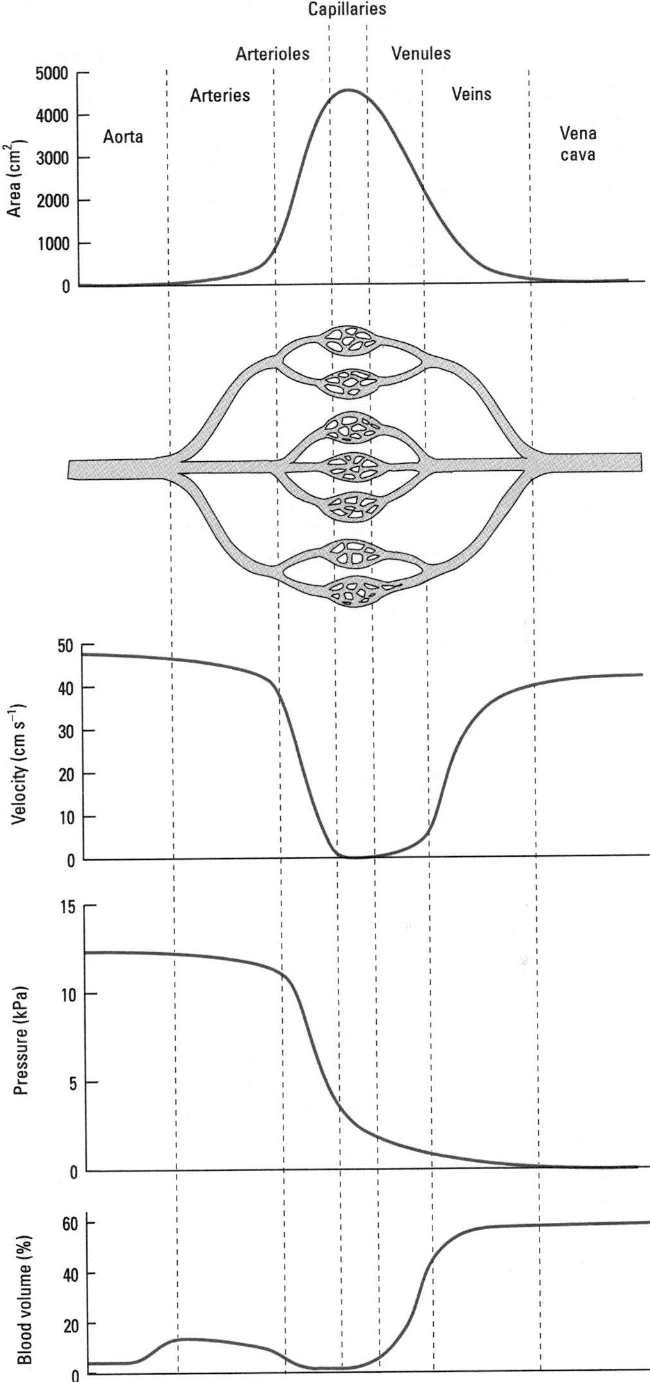

Fig. 7.20 Pressure variation, in relation to surface area and velocity, through the mammalian circulatory system. Area peaks in the capillary beds, where velocity is lowest; the major drop in pressure occurs through the arterioles.

Table 7.7 Protein contents of plasma in different taxa.

Animal group	Plasma protein (g l^{-1})
Most marine invertebrates (e.g. cnidarians, sipunculans, echinoderms)	0.2–2.0
Marine crustaceans	10–90
Freshwater crustaceans and molluscs	10–50
Cephalopod molluscs	80–110
Insects (varies with molt cycle)	20–100
Vertebrates	30–80

7.4.6 Blood and hemolymph

All animal circulatory fluids are primarily water, with low and variable concentrations of ionic solutes, waste products, dissolved gases, foodstuffs, some proteins, and some cells. The solute composition is largely related to osmotic regulation and habitat, and was dealt with in Chapters 4 and 5, where waste products were also discussed. Concentrations of food breakdown products (especially glucose, amino acids, and free fatty acids) are variable with an individual's state of activity and recency of feeding, so that systematic differences between taxa or between habitats are very hard to detect. However, the protein and cellular components are more relevant here in the context of respiration and gas transport. Many marine invertebrates have very low levels of protein in the blood or hemolymph or coelomic fluid (Table 7.7), though the level is higher in decapod crustaceans and very high in cephalopods. Most vertebrates have moderately high plasma protein levels.

These high blood protein levels are responsible for a raised **colloid osmotic pressure** within the blood vessels, since the proteins are too large to permeate the vessel walls. A few large proteins have less effect than many smaller protein molecules, and animals with high concentrations of small blood proteins have a substantial Donnan effect influencing the distribution of other charged solutes in the body fluids (see section 4.3.3). These proteins are predominantly the respiratory blood pigments, dealt with below.

Most animal bloods also contain cells, generically termed **hemocytes**. These have many functions (Table 7.8), particularly involving defensive systems (e.g. clotting, repair, phagocytosis, and chemical immune responses; see Chapter 17) and storage (of foods, excretory products, or blood pigments). In some animals a subpopulation of the blood cells are specialized as containers of the respiratory pigment, and these may be anucleate; they are commonly given the special name of **erythrocytes**.

Blood cells make up a highly variable proportion of blood volume, which is usually expressed as the **hematocrit** (packed cell volume as a percentage of total volume). This ranges from about 20–30% in most amphibians and reptiles to about 30–45% in birds and mammals (up to 50–55% in some very small mammals and in some diving marine mammals).

7.5 Delivering and transferring gases to the tissues

7.5.1 Oxygen carriage: pigments and their properties

We have seen that when animals become too large to rely on diffusion alone, the solutions they have generally adopted include ventilation systems externally, and circulatory systems internally to carry respiratory gases to and from the metabolizing tissues. Further sophistication can be added to this. The simplest way to carry gases in the blood is by dissolution, since, for example, oxygen in the blood will move down the diffusion gradient into any tissue where oxygen is being consumed. However, the quantities of gas that can be moved around in this manner will be limited by the solubility coefficients of the gas concerned (see the equation for Henry's law, section 7.2.1). The blood of a marine animal, having roughly the ionic composition of sea water, will hold about 5.5 ml l^{-1} of oxygen at 20°C when exposed to normal air. Under the same conditions this blood will hold 159 ml l^{-1} of carbon dioxide, a gas with much greater solubility (see Table 7.2), giving about a 29-fold increase in capacity. Oxygen limitation will therefore become a serious problem for larger animals with a high metabolic rate, especially where raised body temperature further compromises the blood plasma's ability to hold oxygen. Therefore in nearly all of the more complex animals there are special components of the blood that enhance its oxygen-carrying capacity—the **blood pigments**, or **respiratory pigments**.

Pigment types

These compounds (Table 7.9) are highly pigmented because they are metalloproteins, containing metallic ions bound to one or more polypeptide groups, so that the metal as it undergoes partial changes in its valency state confers varying color on the pigment as a whole.

Table 7.8 Functions of blood cells in different animal groups.

Function	Annelids, sipunculans, echiurans	Molluscs	Crustaceans, myriapods, insects	Echinoderms	Vertebrates
Phagocytosis	✓	✓	✓	✓	✓
Encapsulation	✓	✓	✓	✓	
Wound plugging	✓	✓	✓	✓	✓
Clotting	✓		✓	✓	✓
Immune responses	✓	✓	✓	✓	✓
Glycogen storage	✓	✓		✓	
Lipid storage			Some	Some	
Hb storage (few)	Few	Some	Some	Few	✓
Storage excretion	✓		Some	✓	
Nacrezation (pearl formation)		Some			

Table 7.9 Respiratory pigments.

Pigment	Structure	Color (+ change)	Oxygen capacity (ml g^{-1})	Molecular weight (kDa)	Cells or solution
Hemocyanin	Protein + Cu^{2+}	Blue (colorless)	0.3–0.5	25–7000	Solution
Hemoglobin	Protein + heme + Fe^{2+}	Red (purple/blue)	1.2–1.4	16–2000	Either
Chlorocruorin	Protein + heme + Fe^{2+}	Green	0.6–0.9	3000	Solution
Hemerythrin	Protein + Fe^{2+}	Violet (colorless)	1.6–1.8	16–125	Either

The **hemoglobins** (Hbs) are closely allied to the noncirculating tissue myoglobins found in many organisms (including fungi, some plants, and many protoctists as well as all animals). They all have iron as their metallic component, in its ferrous state (Fe^{2+}), and bound in the center of a porphyrin ring (Fig. 7.21) to give a functional heme group. There are four such groups in a typical mammalian hemoglobin molecule. Oxygen is bound reversibly to the iron by a partial transfer of one electron to the oxygen, producing a transient ferric state in the iron and a superoxide of oxygen (OO^-); the result is oxyhemoglobin HbO_2. The iron-porphyrin functional group is loosely bound to proteins, which are of variable form. Hemoglobins are very diverse, with subunits ranging from monomers of low molecular weight to huge aggregates of up to 3 million daltons in some annelids (where they are often green and are termed erythrocruorins or chlorocruorins). Hemoglobins can be found in some nematodes and nemertines, many annelids, a few molluscs and insects, some crustaceans, and virtually all vertebrates (the exceptions being a few icefish and a few amphibians). They are normally extracellular, but occur within **red blood cells** (RBCs) in at least seven different animal phyla, mostly marine and intertidal in habitat. Vertebrate hemoglobins are always intracellular and of relatively low molecular weight (~17 kDa).

Hemerythrins are less common, occurring in three phyla, which from recent molecular evidence may have a common phylogenetic origin (sipunculids, priapulids, and brachiopods), as well as in one family of marine annelids. Their most common forms are trimeric or octomeric, each polypeptide containing an iron-binding site within a fold of its tertiary structure where two Fe^{2+} ions are located. The probable binding system involves forming a peroxo compound (O_2^{2-}) of ferric iron. **Hemocyanins** also involve direct reversible binding of a metal to protein structures, but in this case the metal is copper and a peroxo compound of cupric (Cu^{2+}) ions occurs, the active site involving six histidine residues. Hemocyanins are very large proteins, occurring in multimeric forms with up to 48 of the original polypeptide units. They occur in molluscs (some

Fig. 7.21 The structure of heme (porphyrin ring plus a ferrous ion), of myoglobin, and of tetrameric hemoglobin.

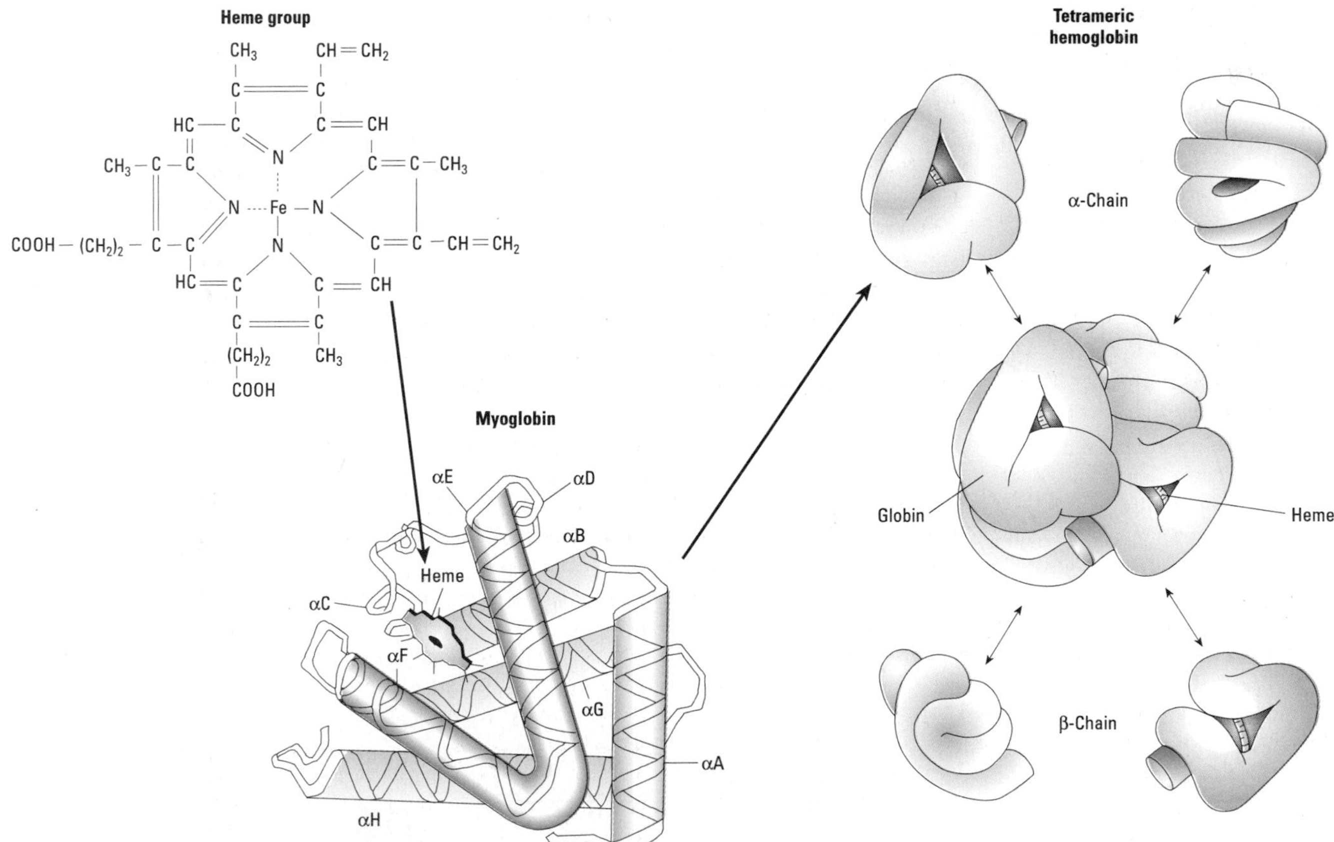

gastropods and all cephalopods, plus chitons and a few bivalves) and in some arthropods (crustaceans, the horseshoe crab, arachnids, and centipedes), but the protein components in molluscs and arthropods are sufficiently distinct to suggest independent origins.

Most animals, whatever their pigment, use it in an extracellular form. However, hemoglobin as we saw above is sometimes packaged into blood cells instead, and hemerythrin is also intracellular in most animals where it occurs. Traditionally it was argued that intracellular pigments allowed higher pigment concentration without compromising the colloid osmotic pressure of the blood or making the blood too viscous, but neither of these arguments holds up well. Alternatively, it might be that an intracellular site allows closer control of oxygen binding by allosteric modulators. However, most of the invertebrates with pigments contained in cells do not have complex modulatory systems. It seems likely that the vertebrate RBC was simply an inherited phenomenon, to which allosteric modulators were added as respiratory demands became more sophisticated.

Oxygen binding

The degree of oxygenation of a pigment depends on the partial pressure of oxygen, Po_2; the pigment is fully saturated at some specific Po_2 where all its oxygen-binding sites are occupied. The extent of saturation can be calculated as:

$$\% \text{ saturation} = \frac{100 \times [HbO_2]}{[Hb] + [HbO_2]}$$

The relation between this value and Po_2 is traditionally used to show pigment properties, and is termed the **oxygen equilibrium curve** (also known as the loading curve or dissociation curve).

To be physiologically useful, any respiratory pigment must bind oxygen at relatively high partial pressures, and release it at the kind of partial pressures found in metabolizing tissues. In most animals, the oxygen equilibrium curve is sigmoid (Fig. 7.22). Nonlinearity is to be expected; at low Po_2 and low pigment saturation, any oxygen molecule is likely to hit an unoccupied site on a deoxygenated Hb molecule, while at higher Po_2, its chances of hitting a free site are lower and increasingly more oxygen molecules are needed to increase the pigment loading by a particular increment. This might produce a hyperbolic curve. But in addition, the **oxygen affinity** of the pigment as a whole (at least of hemoglobins and hemocyanins, which have been best studied) is a function of its precise molecular shape (its tertiary and quarternary structure), and binding oxygen to any one heme group slightly affects the shape of the rest of the molecule. In fact all the four heme components in vertebrate hemoglobins interact, so that the binding of oxygen to the first heme increases the affinity of the second one, and so on, giving the sigmoid effect. A useful quick comparison between such curves is given by the P_{50}, the partial pressure at which 50% (half) saturation occurs; a high-affinity pigment has a low P_{50} and a curve well to the left, whereas a low-affinity pigment has a high P_{50} and a curve shifted to the right.

When oxygen-binding curves are expressed as percentage saturation the relative shapes of the curves in different species can be readily compared (as in Fig. 7.22), and this kind of comparison will often be used in later chapters. Invertebrate Hbs commonly have lower P_{50} values than vertebrate Hbs, and Hbs from more active species amongst the vertebrates (carnivores and insectivores) have particularly low affinities. But note that where the vertical axis of a dissociation curve is the actual oxygen content then differences not only in shape (affinity) but in overall capacity are more clearly revealed (Fig. 7.23). On any such plot, the concentrated pigments of mammals turn out to be very effective indeed. Arterial mammalian blood is usually at least 95% saturated, whilst venous blood may be less than 75% saturated (but it is by no means fully unloaded in the tissues and does not therefore normally turn blue in the veins).

A summary of oxygen-carrying capacity for fully saturated bloods of a range of animals is shown in Table 7.10, again emphasizing the

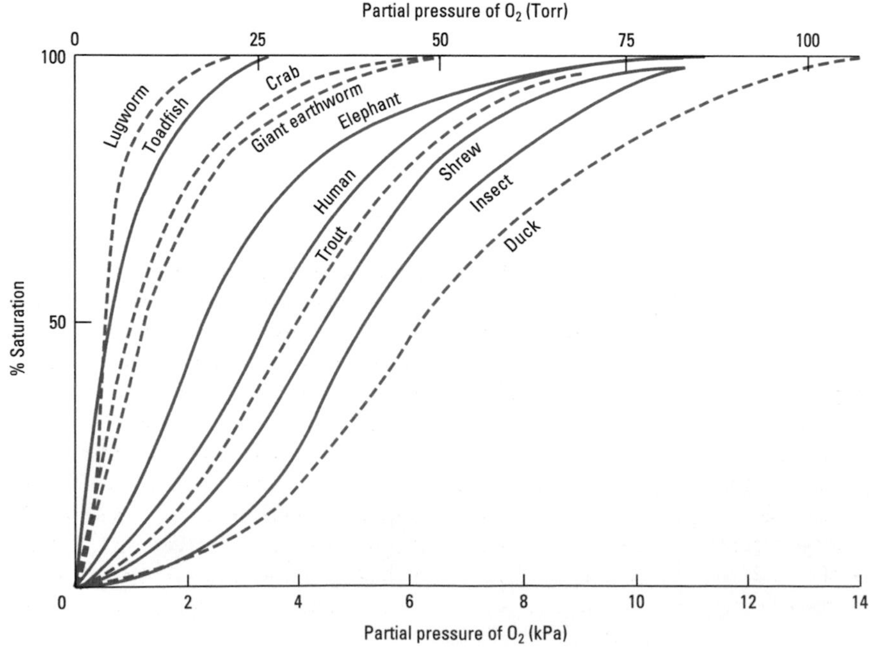

Fig. 7.22 Oxygen equilibrium curves as percentage saturation.

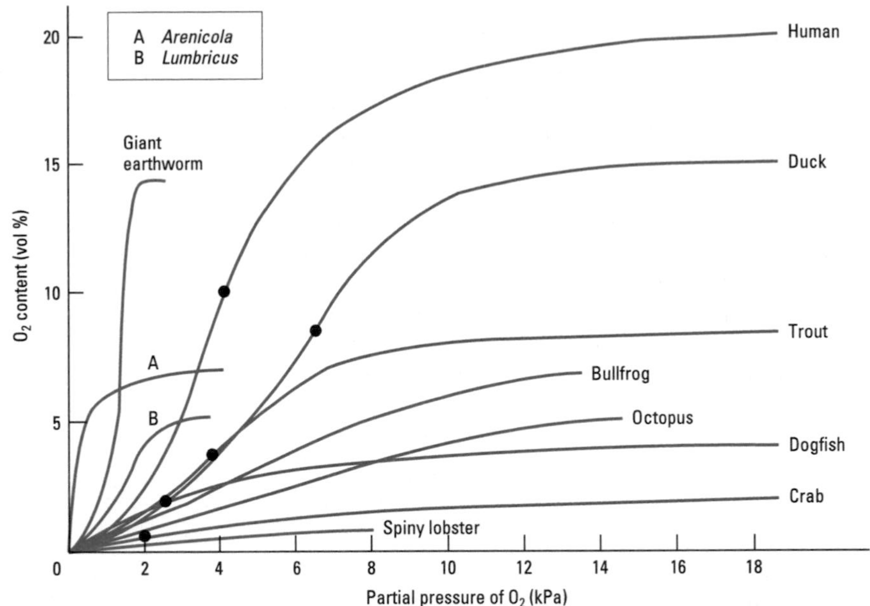

Fig. 7.23 Oxygen equilibrium curves as actual O_2 content, stressing differences in O_2 capacity.

Table 7.10 Oxygen-carrying capacity of bloods with pigments from a range of animals.

Animal group	Pigment	ml O_2 100 ml^{-1}
Nematode	Hb	1–3
Annelid	Hb	0.1–20
	He	3–6
Echiuran	Hb	4–5
Sipunculan	He	2–3
Cephalopod mollusc	Hc	3–4
Gastropod mollusc	Hc	1–3
Crustacean	Hb	2–3
	Hc	1–4
Insect	Hb	5–12
Elasmobranch	Hb	4–5
Teleost	Hb	4–20
Amphibian	Hb	6–10
Reptile	Hb	6–12
Bird	Hb	10–22
Mammal	Hb	14–32
Water	–	0.65

Hb, hemoglobin; He, hemerythrin; Hc, hemocyanin.

considerable augmentation produced by pigments, especially in the birds and mammals.

7.5.2 Modulating oxygen-carrying capacity

Oxygen-carrying capacity can be altered either by altering the amount of respiratory pigment or by altering its properties. As we saw in Chapter 2, the functional properties of a protein can be changed either by the action of extrinsic cofactors or by internal changes in molecular structure. Both mechanisms do occur, but while both are common for hemocyanins the latter is rather rare in hemoglobins, especially in endothermic vertebrates. Here we briefly survey factors that control pigment function, and the control of pigment production.

Temperature

Nearly all hemoglobins and hemocyanins suffer decreasing oxygen affinity (raised P_{50} values) as temperature increases (Fig. 7.24). Though irrelevant for bird and mammal hemoglobins (where T_b is constant), this is probably advantageous to most ectothermic animals, since raised temperature will be associated with higher tissue respiration rates and unloading will be assisted. In a range of crustaceans from different thermal niches, it seems that the hemocyanin P_{50} is fairly similar when measured at the normal ambient condition: the values for the Antarctic isopod *Glyptonotus* at 0°C, for the shore crab *Carcinus* at 10°C, and for the terrestrial crab *Pseudothelphusa* at 25°C are all in the range 15–25 mmHg, with some seasonal variation in the latter two species. For species living at lower temperatures, a higher proportion of the oxygen transport can take place in simple solution in the blood, so that a lower hematocrit but higher blood volume is sometimes seen (in *Glyptonotus* and in some Antarctic fish).

pH levels and carbon dioxide levels

Hemoglobins and hemocyanins are almost invariably affected by altered pH levels, usually with a shift of the oxygen equilibrium curve to the right as the pH decreases (Fig. 7.25). This effect seems to be universal in vertebrates, and almost so in invertebrates. It is again due to subtle changes in the higher order structure of the protein, with the pH of the solution altering the charge characteristics of the amino acid residues and thus the precise shape of the polypeptide chains. Most animals therefore have mechanisms to buffer pH change; buffering involves weak acids and alkalis, and the most important components in animals are bicarbonate, phosphate, and weak protein anions, especially histidine residues intracellularly. Thus animal bloods show a surprising degree of uniformity in their pH and also in the response of pH to body temperature (Fig. 7.26), insuring a constant "relative alkalinity" (and hence protein charge) as temperature changes.

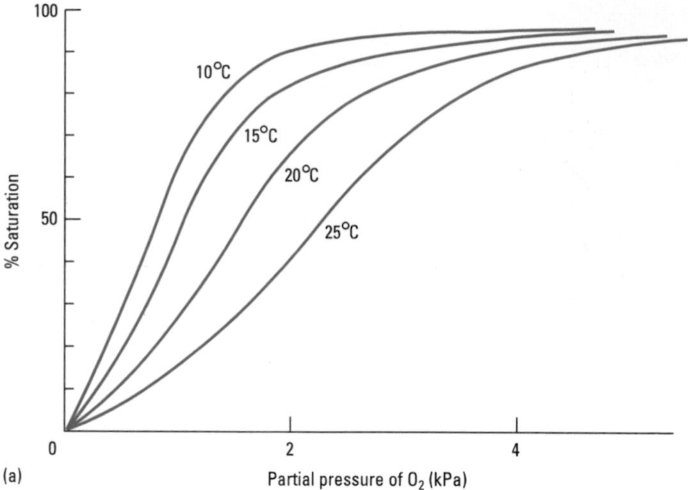

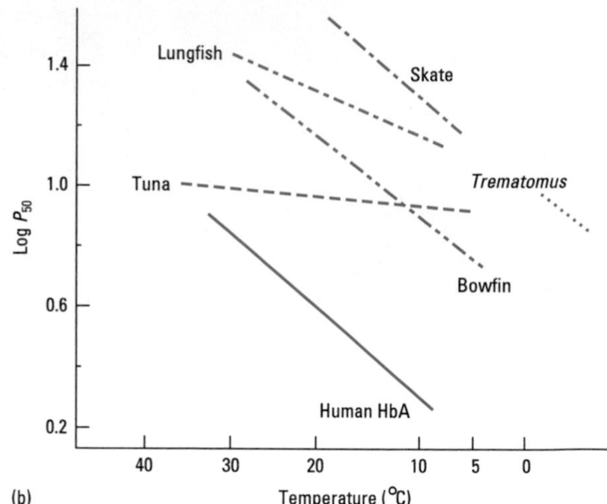

Fig. 7.24 The decreasing affinity of blood pigments with increasing temperature: (a) for hemocyanin in the spiny lobster; (b) as a semilog plot (P_{50} against temperature) for various vertebrates (the thermal insensitivity in tuna is probably related to the rapid changes in T_b occurring in its swimming muscles; see Chapter 11). (a, Data from Redmond 1955; b, data from Hochachka & Somero 1973.)

Since pH is decreased by increasing concentrations of CO_2, it follows that raised carbon dioxide levels will also shift the oxygen equilibrium curve to the right. This is the so-called **Bohr shift**, a term now used for both pH and Pco_2 effects (Fig. 7.25). This shift can be very pronounced with both hemoglobins and hemocyanins, and is useful in accentuating unloading of oxygen on the venous side of capillary networks. In the hemocyanins of some molluscs and of the horseshoe crab (*Limulus*), however, a reverse Bohr shift has been reported with both pH and CO_2. Carbon dioxide also seems to have a more direct effect on oxygen binding to blood pigments, independent of its pH effect, by forming carbamino compounds ($NHCOO^-$ groups) with the constituent proteins.

Carbon dioxide may affect the maximum oxygen binding, as well as the shape of the equilibrium curve, particularly in the hemoglobins of some fish species. High CO_2 levels decrease the total bound oxygen, a phenomenon known as the **Root effect** (Fig. 7.27). This is used to good effect in controlling the activities of fish swimbladders (see section 11.6.2). Root effects have not been observed with hemocyanins.

Ionic levels

Since hemoglobin has weakly dissociated groups in its own structure, it is affected by overall ionic strength in the blood, and in general any increase in inorganic ion concentrations will lead to decreased oxygen affinity (increased P_{50}). This is probably not very important in vertebrates, where the Hb is packaged inside tightly regulated cells, but it may be very significant in many euryhaline and osmoconforming invertebrates, where the blood concentration varies with that of the medium. More specific effects also occur, with

Fig. 7.25 The Bohr shift due to the effects of pH or Pco_2 on respiratory pigments: (a) in the hemocyanin of the mangrove crab with varying Pco_2; (b) in the hemoglobin of the dog with varying pH. (Data from Roughton 1964.)

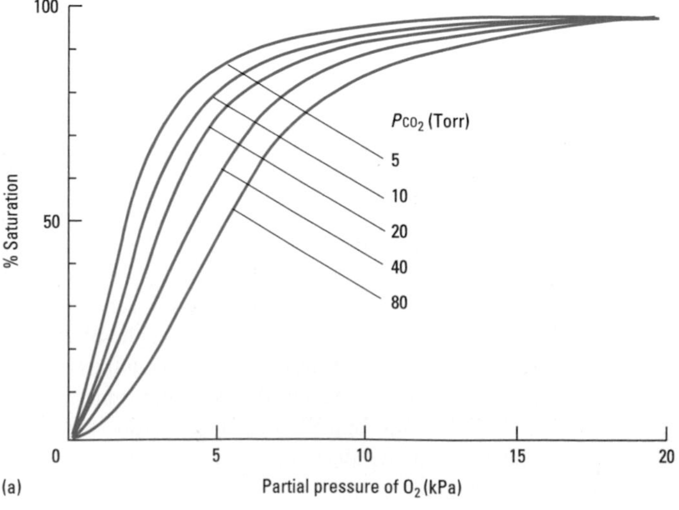

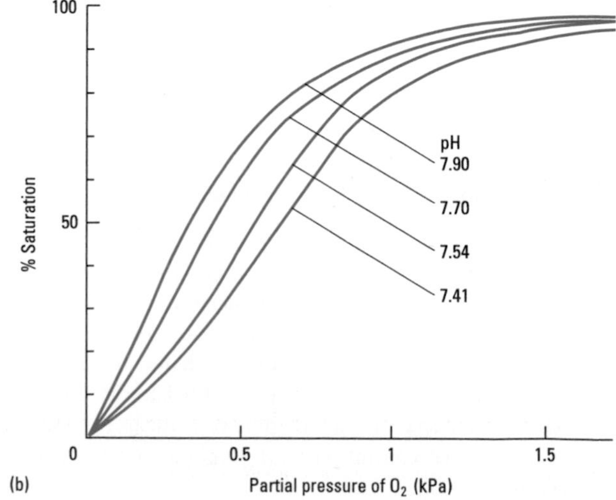

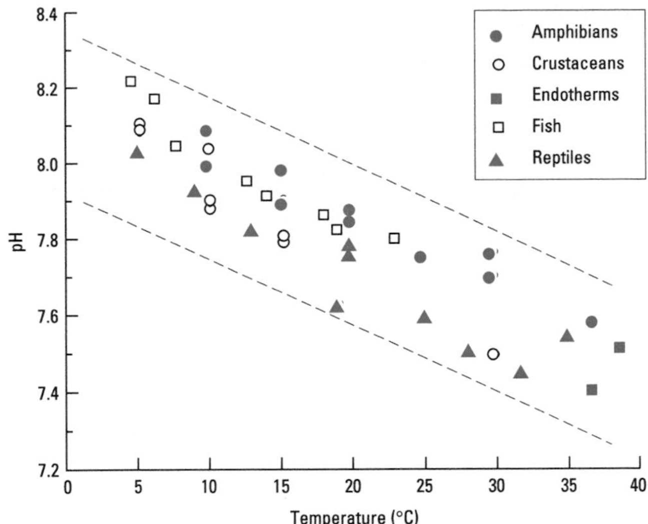

Fig. 7.26 The relation between blood temperature and pH across a range of taxa, suggesting that animals regulate their blood at a fairly constant "relative alkalinity" at which there is a constant protein net charge. (From Feder 1987, courtesy of Cambridge University Press.)

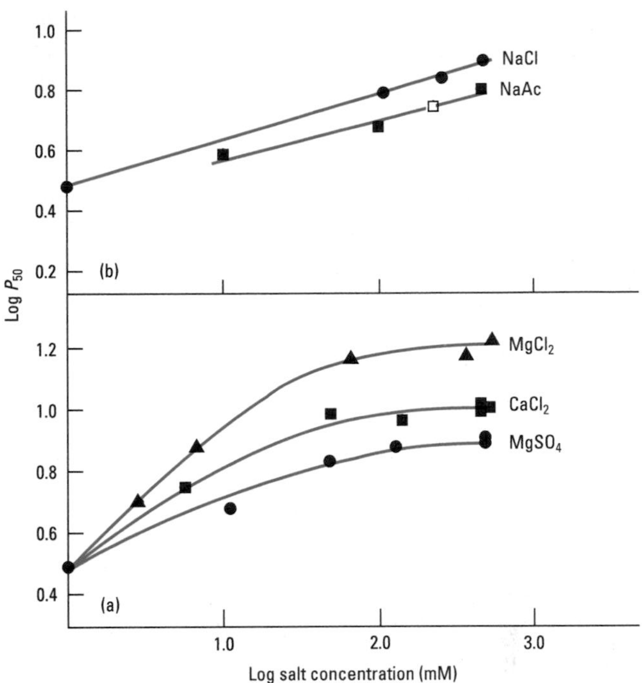

Fig. 7.28 The effects of specific ions on O_2 dissociation in crustacean hemocyanin; the log P_{50} is shifted up as salt concentrations increase. NaAc, sodium acetate.

calcium and magnesium both affecting Hb and Hc binding directly (Fig. 7.28).

Organic modulators

Various small organic compounds are known to be modulators of hemoglobin affinity in animals. In vertebrate water breathers the main strategy during hypoxia is the release of an organic modulator (phosphate) which alters the P_{50}. In some aquatic animals, including many fish, ATP itself seems to regulate oxygen affinity. For air breathers one of the most important modulators is 2,3-diphosphoglycerate (2,3-DPG), the addition of which decreases the oxygen affinity of the blood of most mammals. This may be significant in high-altitude adaptation (see Chapter 16), and it perhaps also serves to enhance fetal–maternal gas exchange. It is also used in developing bird and reptile eggs.

Pigment production

Many species show varying patterns of expression of oxygen

Fig. 7.27 The Root effect, a right shift in the O_2 equilibrium curve due to increased P_{CO_2} or pH, with a decrease in the O_2-binding capacity. (a) For fish hemoglobin and (b) for octopus hemocyanin, showing the nonlinearity of the effect.

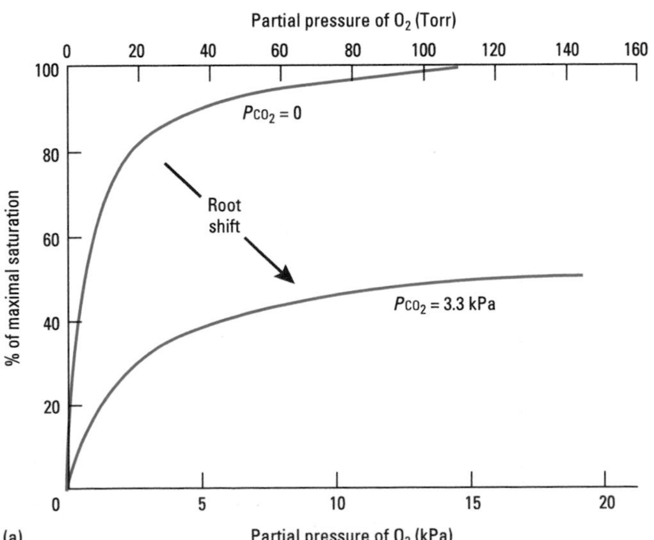

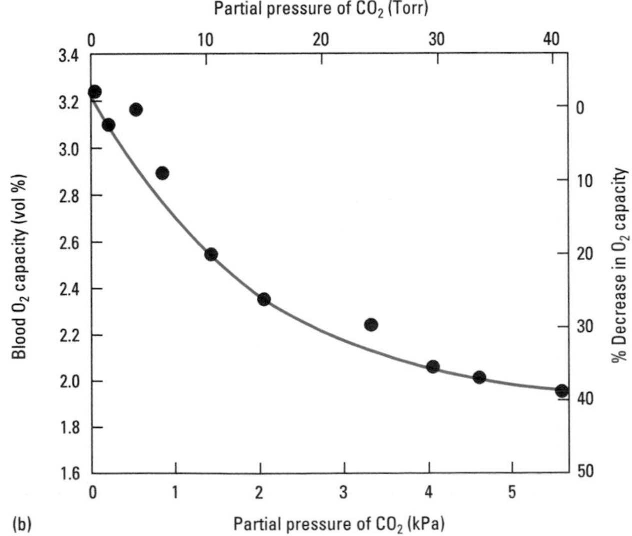

transport proteins during ontogeny. For example, larval chironomid flies express a hemolymph Hb that has very different properties and Bohr effects from the adult form. More rapid and reversible changes in oxygen-carrying capacity are also possible in relation to need, and are often brought about by increments of the pigment concentration, which in vertebrates means production of more RBCs (**erythropoiesis**). This is controlled by the hormone erythropoietin. In many vertebrates, especially strong runners such as dogs and horses and also aquatic species, some RBCs are stored in the spleen and can be quickly mobilized when needed.

7.5.3 Oxygen storage

Respiratory pigments can transfer oxygen between themselves, and this frequently occurs in invertebrates, where there may be separate pigments in cells, in plasma and in the coelomic fluids. In such cases the intracellular pigment may serve as a short-term store during transient anoxia. A similar phenomenon occurs in many animals with muscle **myoglobins**, where the relatively small and mobile myoglobin facilitates diffusion across the cell membrane.

Pigments have a particularly important role as oxygen stores in animals that burrow, such as benthic annelid worms, many littoral species, and certain freshwater fly larvae. Here the pigment-bound oxygen is generally insufficient to support aerobic respiration for more than about an hour, so that it could not be used for a full intertidal cycle in the littoral zone; but it is enough to support aerobic metabolism between periodic bouts of gill ventilation in a benthic burrower, or during an escape response.

Stores are also important in air-breathing animals that dive (see Chapters 11 and 13). In aquatic mammals the blood is the most important store of O_2 during the period of apnoea, but muscle myoglobin is a very important secondary source; the muscle myoglobin store in seals is particularly high relative to that in similar-sized non-diving mammals such as man.

7.5.4 Carbon dioxide transport

Carbon dioxide is a relatively soluble gas, and its transport in solution presents no particular problems for most animals. It will diffuse into the blood from active tissues, and be disposed of into the surrounding medium at the external exchange surfaces. Most of the CO_2 in an animal's fluids is actually chemically combined with water to form carbonic acid and thus bicarbonate ions, as follows:

$$CO_2 + H_2O \rightarrow H_2CO_3 \rightarrow H^+ + HCO_3^-$$

The first of these reactions is relatively slow but the second very fast, so that under normal conditions the formation of the carbonic acid would be the rate-limiting step. However, the enzyme **carbonic anhydrase** catalyzes the reaction, and is present within vertebrate red cells. Thus CO_2 is very readily taken up into such cells at its sites of formation in the tissues (Fig. 7.29), and the resulting HCO_3^- then diffuses out into the plasma in exchange for chloride ions (the **chloride shift**, which slightly increases the osmotic concentration of the RBCs so that water enters and they swell a little; thus venous blood has a detectably higher hematocrit than arterial blood). The remaining H^+ is effectively buffered by the hemoglobin inside the blood cells, and the acid hemoglobin (HHb) has a lower affinity for oxygen, allowing O_2 to be unloaded to the tissues. About 65% of the carbon dioxide ends up in the plasma in the bicarbonate form, and no carriers of any kind are involved. The carriage of CO_2 is therefore not saturable, showing a rapid increase at low P_{CO_2} and thereafter a slow steady increase (Fig. 7.30).

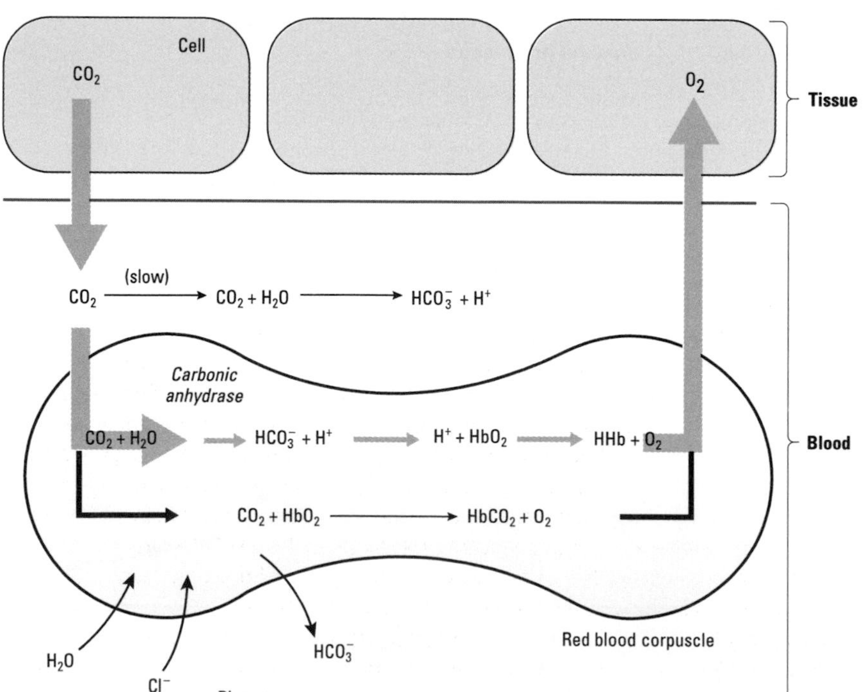

Fig. 7.29 Carbon dioxide carriage in red blood cells (details in text).

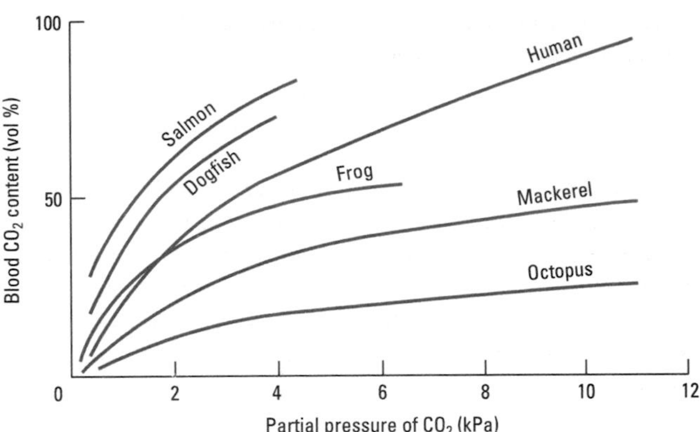

Fig. 7.30 Carbon dioxide equilibrium curves in vertebrates.

In invertebrates CO_2 transport has been less well studied but shows similar patterns. Carbonic anhydrase is sometimes involved, and carriage as bicarbonate is the norm, with Hb or Hc acting as efficient H^+ buffers.

However, the issue of carbon dioxide distribution does become somewhat complicated because of the effects of CO_2 on acid–base balance in the animal as a whole, so that CO_2 levels in the circulating fluid have knock-on effects on various other physiological systems.

Hypoxia *per se* has little effect on acid–base relations, but since in most animals it leads to increased ventilation more CO_2 is lost from the body and **respiratory alkalosis** can set in. However, where the hypoxia was itself a result of exercise, there will also be an increase in metabolism and of metabolic end-products that are acidic, so that respiratory alkalosis is countered by a degree of induced **acidosis.**

Hypercapnia (an increase in $P\text{CO}_2$ in the blood) may be caused independently of hypoxia by an inadequate excretion of CO_2 or an increase in the environmental $P\text{CO}_2$. It has a rapid effect on animal body fluids, with pH decreasing sharply. In air-breathing animals this acidosis will normally be quickly rectified by increased ventilation, since blood $P\text{CO}_2$ directly triggers the ventilatory control centers (see section 7.7); and in the longer term the kidney will excrete an excess acid load. However, in aquatic animals, including fish and crabs, blood $P\text{CO}_2$ and hence HCO_3^- levels are routinely low, and regulation of pH cannot be achieved by varying $P\text{CO}_2$; thus ventilation is less important as a response to carbon dioxide levels or acidosis. Here the gills are used as an ion regulatory exchange site, performing the classic Na^+/H^+ and Cl^-/HCO_3^- exchanges (see Chapter 5), though this may leave a degree of alkalosis due to accumulated bicarbonate, so that complete compensation is rare.

7.5.5 Other functions of pigments

Oxygen-transporting proteins have taken on several additional roles in animals. In deep-sea vent animals they can transport and detoxify hydrogen sulfide, as we will see in Chapter 14. In vertebrates the RBCs are involved in the regulation of blood pressure because the hemoglobin is able to bind nitric oxide, an important signal molecule (see Chapter 2) acting as a relaxing factor in blood vessel endothelium (see Chapter 10).

7.6 Coping with hypoxia and anoxia

Animals in many different kinds of habitat may have to deal with periods of hypoxia or anoxia; examples include aquatic animals in benthic muds or on shores when the tide goes out or the pond dries up or freezes over, and terrestrial animals in underground burrows or at altitude. Thus nematodes, many bivalves and annelid worms, and some amphibious teleosts are particularly hypoxia tolerant. Many other situations lead animals to enter an induced **hypometabolic state**, or **metabolic depression**, as briefly discussed in Chapter 6. All such animals commonly show varying degrees of "hypoxia adaptation", depending upon downregulation of energy turnover, and upregulation of energetic efficiency. Different combinations of these two strategies may be used, but it appears that suppression of energy turnover (often by an order of magnitude) supplies the best protection against severe hypoxia, and the process has been likened to a boiler that has its main burners switched off and is "turned down to the pilot light". The oxygen sensor at the cell level may well be a heme protein, and from this a transduction process ultimately leads to the activation of some genes and suppression of others, with enzyme phosphorylations as a major outcome so that metabolic depression results.

The changes to fundamental cellular metabolism that accompany hypoxia and anoxia, including switching to various anaerobic pathways, were covered in Chapter 6 (especially section 6.4.4). As a result of these changes, protein metabolism, glucose synthesis, and maintenance of electrochemical gradients are all depressed. Cellular permeabilities are also reduced, so that the lower sodium pumping rate can still keep excitable membranes functioning, albeit at a lower firing frequency. In the brain of vertebrates, hypoxia tolerance is geared to maintaining the crucial ionic gradients, so that increased blood supply, reduced electrical activity, increased release of inhibitory transmitters and neuromodulators, and downregulation of excitatory ion conductances are all coordinated. In hypoxia-intolerant species (e.g. most mammals) neurons lose ATP progressively, depolarizing and suffering toxic increases in intracellular calcium levels, with cell swelling and membrane damage. However, in hypoxia-tolerant vertebrates, such as diving turtles, the neurons are able to decrease their energy usage and ion channel activity in a coordinated fashion, regulating intracellular calcium at a new setpoint.

The best known molecular mechanism of hypoxia-dependent regulation involves the action of **hypoxia-inducible factor 1** (Hif-1). This is a helix–loop–helix transcription factor, and binds to various sequences on the regulated genes whenever hypoxia occurs (e.g. in humans in response to high altitude, anaemia, or severe wounding). Hif-1 is seen as the master regulator of oxygen homeostasis in mammals, with particular effects on erythropoietin (controlling RBC production) and on vascular endothelial growth factor (controlling capillary density), as well as an upregulating effect on glycolytic enzymes. Long-term hypoxic adaptation also requires binding of inducing regulators such as Hif-1 to other sections of mRNA to increase their stability, as well as the binding of repressing regulators, which may include the transcription factor Sp3. Coupling between the oxygen sensors and these regulatory systems is still being explored; it excites interest in mammals because of the role of hypoxia in mortality from heart attacks and strokes.

The further problem posed by bouts of hypoxia or anoxia is that of the recovery phase when cells and organs become reoxygenated. This phase may be accompanied by high levels of particularly reactive oxygen species, notably the free radicals superoxide and hydrogen peroxide referred to in the introduction to this chapter. These radicals can damage macromolecules, especially lipids; they produce conjugated dienes, and lipid hydroperoxides, which are useless and potentially damaging in themselves. Animals therefore need protection by the use of detoxifying **antioxidants** during this phase. Species that regularly experience bouts of anoxia therefore tend to have particularly high levels of the key enzymes such as superoxide dismutases, peroxidases, and catalases. Levels of these are generally higher in ectothermic animals than in endotherms, and higher in animals from variable habitats, indicating a good match between antioxidant defenses and the probability of experiencing temperature change and hypoxia. Nonenzymic antioxidants also occur: invertebrates often use the small antioxidant glutathione, and vertebrates sometimes show high levels of the antioxidants vitamin E and vitamin C. During anoxia the levels of these may decrease, but they rapidly increase (especially in tissues such as the brain, heart, and kidney) during aerobic recovery.

7.7 Control of respiration

Since environmental conditions vary so markedly, and with the additional problem of varying animal activity levels, all animals require some further system for controlling their gas exchange and respiratory rates. Metabolic demand for oxygen, and (secondarily perhaps) the need for CO_2 excretion, are the major determinants of the rate needed; Po_2, Pco_2, and pH are likely to be the main feedback triggers. To regulate the needs of the organism as a whole, an animal is likely to need both internal and external receptors, and may also need specialized "defense receptors" to insure appropriate protective responses to unexpected noxious stimuli in the environment.

7.7.1 Aquatic animals

In most aquatic animals the respiratory surface is a rather delicate site of high surface area at the periphery of the body, even protruding into the medium, so that it is peculiarly vulnerable to physical or chemical damage. Receptors to detect damaging conditions—**nociceptors**—may therefore be sited close to the inhalant openings. Such receptors have been identified in the nares (nasal passages) of certain fish, and in the anterior segments of tube-dwelling annelids. They may elicit an inhibition of gill ventilation, or even a transient reverse ventilation; in fish this results in a "cough" response.

Many aquatic animals must also have external receptors sensitive to Po_2, though these are rarely identified. Some teleost fish gill nerves clearly respond to oxygen levels, and in elasmobranchs oxygen-sensitive receptors seem to be widely distributed over the branchial cavity, though their precise role remains controversial. Some crustaceans and the horseshoe crab *Limulus* also appear to have oxygen-sensitive areas in the gill filaments. In molluscan gills the cilia have an intrinsic beat, though their frequency may be altered by the external oxygen concentration.

However, **internal chemoreceptors** are probably more important in determining the overall respiratory response in most aquatic animals. Fish can detect blood Po_2 with chemoreceptors in regions just distal to the ventral aorta, where venous oxygen tension is readily monitored. In aquatic animals the primary response is nearly always to oxygen rather than to carbon dioxide, as Pco_2 is low and changes only slightly with varying metabolic rate because of its high solubility. However, osmotic or mechanical stresses, and a range of external chemical stimuli, will also affect respiratory patterns.

All these inputs are integrated centrally, and in many animals the organizing center then feeds its integrated output to a pacemaker system affecting both heart and breathing rates. The central integration of respiratory systems in invertebrate aquatic animals is almost completely unexplored. In the lugworm *Arenicola* and in some freshwater insects, respiratory integration is known to involve a group of neurons in the ventral nerve cord. In decapod crustaceans the pacemaker neurons are in the subesophageal ganglion; these are nonspiking oscillatory cells, providing the basic frequency signal, and are modified by neurons that are fed by the gill receptors. The oscillator cells link to motor neurons in two groups, serving as depressors and levators to the scaphognathites (see Fig. 7.9) that act as the balers to ventilate the branchial chambers on each side of the body. The principal response shown is an increased ventilation rate when the animal is hypoxic; there is little response to environmental hypercapnia, and CO_2 regulation is simply dealt with by excretion at the body surface.

In fish the integration of respiration occurs in the medulla of the brain, the respiratory control input signals arriving via the vagus nerve (cranial nerve X) and to a lesser extent via nerves V, VII, and IX. Efferent motor pathways also originate at the medullary level, and the fore- and midbrains do not appear to affect respiratory rhythms. The efferent signals control three principal factors in fish:

1 **Ventilation volume**, by adjustments to the mouthparts, operculi, and gill lamellae.

2 **Cardiac output**, with the heartbeat pattern determined largely by the intrinsic rhythmic discharge of the sinoatrial node (SA node), inhibited to varying degrees by the vagus nerve.

3 **Perfusion pattern** within the gills.

In a fish faced with a hypoxic environment, the commonest responses (i.e. the result of integrating all these inputs and outputs) are increased locomotion to move it out of the hypoxic area, and increased ventilation. The magnitude of each response is species specific, as fish vary greatly in their tolerance of reduced oxygen. Many fish also show a slowing of the heart rate (bradycardia), but with an increase in stroke volume to maintain blood flow, and changes in blood distribution within the gill, so that transit time and gas exchange are increased. As with crustaceans and other aquatic animals, respiratory rhythms are not very responsive to hypercapnia.

7.7.2 Amphibious animals

Amphibious animals, in terms of respiration, come in two categories. Firstly, there are water dwellers with additional air-breathing organs. These include a variety of fish, including not only the lungs of lungfish but also modifications of stomach areas in catfish, of buccal areas in the electric eel, of part of the gill chamber in gouramis, and even of parts of the skin or fins in a few species. This category also includes a range of invertebrates, such as land crabs and some littoral molluscs. Secondly, there are animals of terrestrial origin that live temporarily or permanently in water: here there are many insects and an occasional spider, plus a range of vertebrates including amphibians, some reptiles, and some mammals.

Again we need to use the better studied vertebrates as models for how control systems might be organized. Air-breathing fish do not generally monitor the gas concentrations in the air itself, as that is effectively invariant, but instead monitor gas using **venous chemoreceptors** in the veins draining the air-breathing organ. In addition they may have **gill arch chemoreceptors** to monitor hypoxia in the aquatic environment and initiate the switch to air breathing. Most of the fish that breathe air lose carbon dioxide primarily through the gills and skin rather than from the lungs, and so lack significant CO_2 sensitivity in the vicinity of the air-breathing organ. The other major control of aerial respiration is from **mechanoreceptors**, detecting lung distension with both tonic and phasic receptors. All these inputs are integrated in the medulla of the brain, but (unlike fish) there appears to be no central pattern generation for air breathing; instead a variety of reflex responses occur, giving an on-demand intermittent breathing pattern.

Within the amphibians and reptiles there is an almost complete spectrum from fully aquatic forms (some frogs, forms such as *Xenopus* and many neotenous salamanders, sea snakes, and a few turtles) to fully terrestrial forms (desert frogs and toads, most reptiles). In the middle of this spectrum are animals with genuinely bimodal breathing, including many frogs and toads and some turtles. Despite this diversity, the respiratory controls found are fairly similar, and indeed match those found in nearly all other tetrapods. Submergence usually invokes closure of the mouth and nares, with receptors around these openings triggering the **apnoeic response** (breath holding). A back-up is provided by water-sensitive taste receptors in the tongue. In animals that dive for significant periods these back-up circuits become the main ones, as the mouth is often left open. The receptors of the tongue, larynx, and posterior nares then become the first part of a reflex arc that produces apnoea and may cause lung contraction and reduced pulmonary blood flow. The lung itself again has stretch receptors with tonic and phasic responses. To monitor oxygen levels, the internal chemoreceptors of the **carotid arch** blood vessel are particularly important, often elaborated into a carotid labyrinth. Denervation of this area produces reduced breathing and a lack of response to hypoxia or hyperoxia. Other areas of chemosensitivity also exist, in the vicinity of the pulmonary (pulmocutaneous) arch or the aorta, but these are more taxon specific and probably affect blood flow distribution more than ventilation.

Most amphibians and reptiles breathe only intermittently, with bursts of breathing separated by apnoeic periods. This contrasts both with the majority of fish and with birds and mammals. The overall control of this pattern is complex and relatively poorly understood, not least because of the variety of breathing musculature found, varying from buccal breathing in frogs to aspiration breathing in turtles (by limb and carapace movements) and in crocodiles (having a liver–diaphragm action). With the added complication of extensive cutaneous respiration in many of these animals, itself subject to controls by modifications to cutaneous blood flow and also by ventilating the skin through bodily movements, it is clear that amphibious respiration in vertebrates is one of the more complicated breathing control systems found in animals.

Table 7.11 Blood P_{CO_2} and HCO_3^- concentrations for a range of water-breathing (W) and air-breathing (A) animals.

Species	Breathing	P_{CO_2} (kPa)	HCO_3^- (mM)
Tadpole	W	0.3	4
Trout	W	0.35	5
Crab	W	0.4	7
Dogfish	W	1.0	6
Salamander	W/A	0.8	9
Garfish	W/A	1.2	8
Toad	W/A	1.6	22
Bullfrog	W/A	1.8	31
Lungfish	W/A	2.2	33
Electric eel	W/A	4.1	30
Lizard	A	2.3	13
Turtle	A	2.0	40
Human	A	5.4	24
Ground squirrel	A	5.7	23

The same may well be true of amphibious invertebrates, but too little is known to generalize here. Some land crabs have been investigated, and in parallel with the fish, they do show a response to hypercapnia in addition to the normal aquatic response to hypoxia; but the receptors and the circuitry of this are unknown.

7.7.3 Terrestrial breathers

As a general principle, control of breathing in land animals is more commonly related to **perception of hypercapnia** than to hypoxia, continuing the trend seen in amphibious animals as compared to aquatic ones. This difference may be partly because of the increasing importance of acid–base balance; certainly blood P_{CO_2} levels are usually markedly higher in amphibious and land animals (Table 7.11), so that changes in this value should be more easily detected. The switch may also be partly because blood P_{O_2} will be an increasingly less reliable predictor of available O_2 as pigments that store oxygen become more important.

Invertebrates on land breathe with either simple lungs, with book lungs or gill books, or most commonly with tracheae. Only for the last of these do we have significant understanding of the control mechanisms involved (though it is at least clear that land crabs have made the switch to relying on P_{CO_2} for ventilatory control). In insects there are very local controls at a cellular level, since epidermal cells that are short of oxygen are able to send out processes to "capture" and drag back a nearby tracheole. There are also controls of ventilation pattern in larger insects where the tracheal system forms a single interconnected network, with air sacs acting as bellows. But the major controls for the whole animal lie at the level of the **spiracles**, which can be fully closed to prevent water loss and where the timings and durations of closure are significant controllers of respiratory rate. In most insects the spiracles open to a small atrium and have both opener and closer muscles (Fig. 7.31). Rhythmic bursts from the ventral nerve cord may innervate these and give an intrinsic rhythm of opening and hence of ventilation. Each segment may have its own respiratory center, and these may be differentially sensitive to P_{CO_2}, with trigger levels varying from 0.2% CO_2 to 15% CO_2. To a lesser extent the tonic activity in the nerves is also

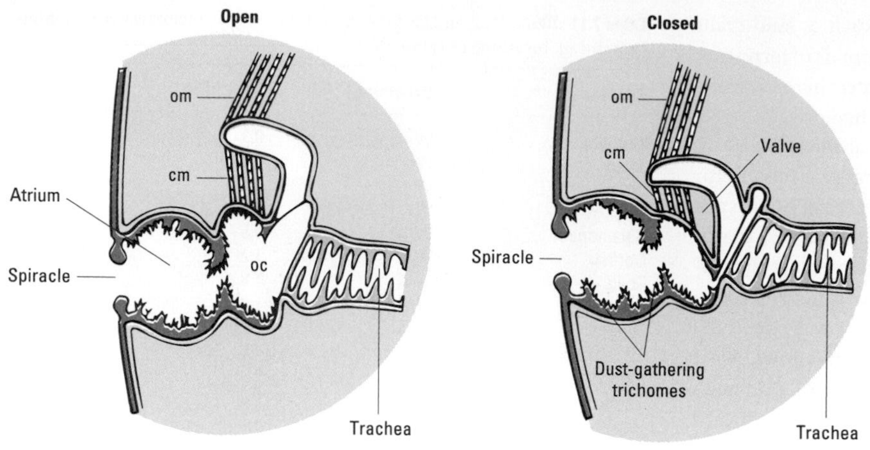

Fig. 7.31 Spiracle opening and closing mechanisms in the abdomen of an ant. The internal occluding chamber (oc) is constricted when the closer muscle (cm) contracts, and opened again by the opener muscle (om) contracting. (Adapted from Richards & Davies 1977, with kind permission from Kluwer Academic Publishers.)

modulated by O_2. However, the spiracular muscles can also be directly affected by CO_2; in locusts the spiracle closer muscle is relaxed by a direct application of CO_2, so that the spiracle opens due to its own elasticity. In some insects local pH of the tissues (affected by activity levels and lactate build-up) may also have an effect. These controls in turn may interact with a direct effect of humidity in at least some insects, insuring that respiratory needs do not excessively dominate the need for water conservation.

All land vertebrates breathe with fully alveolar lungs, and all have a common form of control (Fig. 7.32). The most important element on the receptor side is provided by chemoreceptors in the **carotid bodies**, adjacent to the carotid artery and innervated by the vagus nerve. These primarily transmit information about arterial Po_2, and will invoke responses to both hypoxia and hyperoxia. They may also respond to hypercapnia, but this role is less clear and varies between taxa. Further receptors occur in scattered locations along the aorta and major arteries, apparently mostly concerned with pressure regulation. Within the lungs, chemoreceptors are commonplace in birds, and also in some lizards, snakes, and crocodilians, with a primary response to Pco_2. This is in contrast with the lungs of the water-dwelling amphibians and reptiles, which have only stretch receptors. However, this difference is not habitat related, as mammals too lack chemoreceptors and rely on stretch receptors. It appears to reflect phylogenetic history, as the mammals are descended from a stem reptilian lineage predating the divergence of most modern reptile and bird groups. Probably both stretch receptors and carbon dioxide receptors in the lungs are modifications of a common ancestral mechanism, and either will serve to monitor the efficacy of ventilation.

Central regulation in birds and mammals is primarily related to Pco_2 and pH levels, again with little oxygen sensitivity. The response to Pco_2 and/or pH (it is not clear which is the primary trigger) is centered in the **medulla** (see Chapter 9), and is much slower than the response to peripheral changes, probably giving a low-level

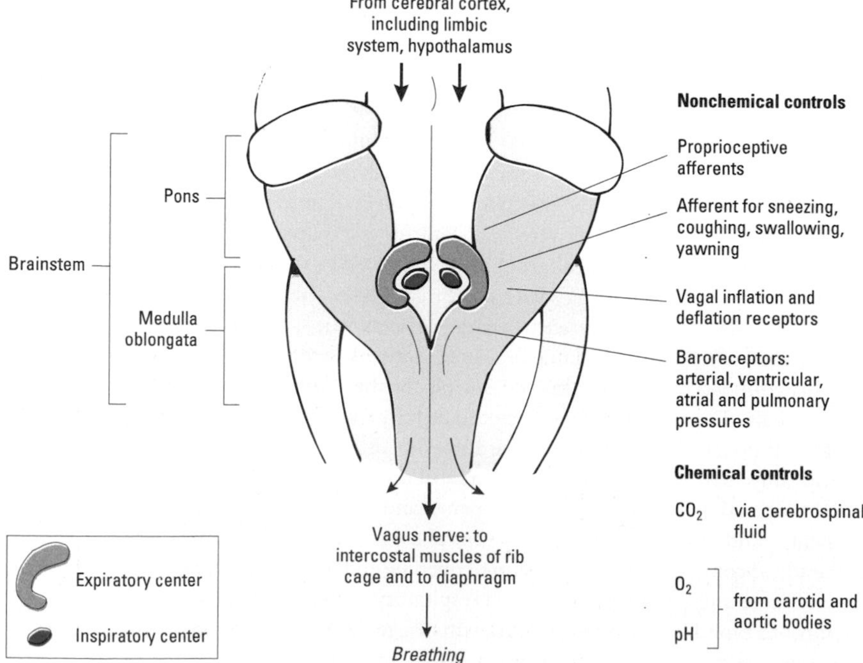

Fig. 7.32 Control of respiration in a terrestrial vertebrate.

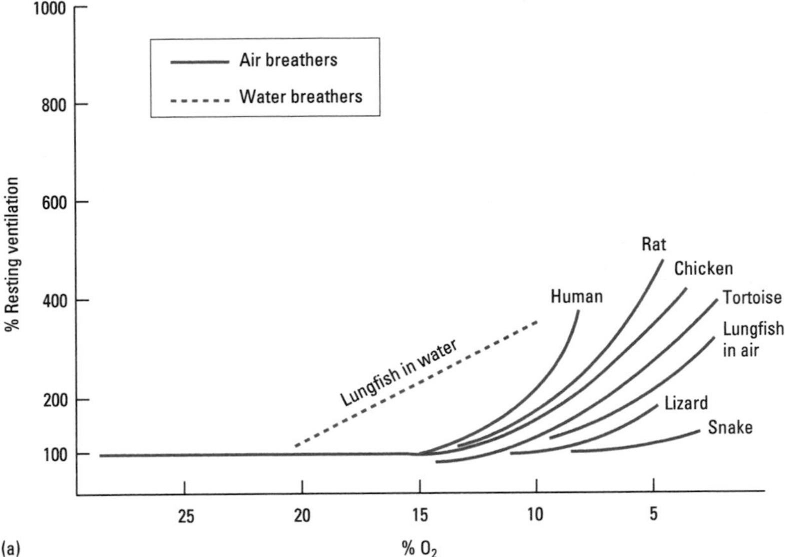

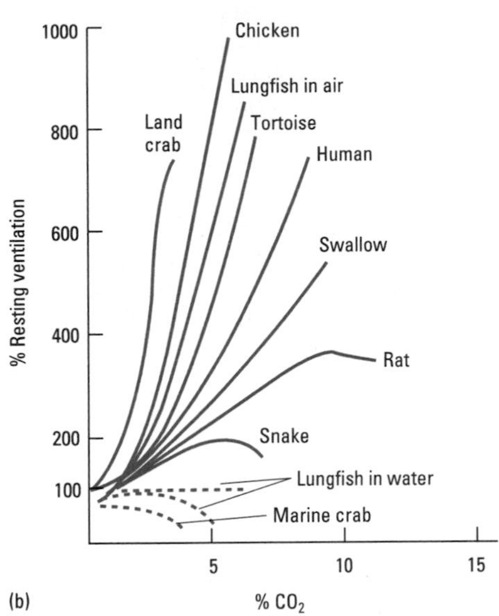

Fig. 7.33 The roles of O_2 and CO_2 in the control of respiration. In aquatic animals, falling O_2 has a large effect and rising CO_2 is relatively unimportant. In land animals (air breathing), decreasing Po_2 has relatively little effect but increasing Pco_2 causes a massive increase in ventilation.

background regulation feeding in to the central integration processes. This integration reverts to the use of a rhythmic pattern generator (as in aquatic fish), involving areas of the brain cortex, pons, and medulla, and producing oscillatory sequences in the intercostal and diaphragm muscles controlling inspiration and expiration. Both hypoxia and hypercapnia produce very rapid increases in the rate of this ventilatory pattern, the former via the carotid bodies and the latter probably triggered from the lungs, the carotid receptors, and the brain medulla. Animals and birds that normally experience raised CO_2 in their environments—for example, species that burrow, and some that dive—show reduced CO_2 sensitivity.

Viewing the progression from aquatic to air breathing in the whole range of animals that have been studied, the overriding change is from reliance on control by Po_2 levels to control by hypercapnia via Pco_2 levels. Figure 7.33 shows some comparative examples of Po_2 and Pco_2 responses for a variety of species to underline this point.

FURTHER READING

Books

Boutilier, R.G. (ed.) (1990) *Vertebrate Gas Exchange; from Environment to Cell.* Advances in Environmental Physiology 6. Springer-Verlag, Berlin.

Bryant, C. (ed.) (1991) *Metazoan Life Without Oxygen.* Chapman & Hall, London.

Cameron, J.N. (1989) *The Respiratory Physiology of Animals.* Oxford University Press, New York.

Dejours, P. (1988) *Respiration in Air and Water.* Elsevier, Amsterdam.

Gilles, R. (1985) *Circulation, Respiration and Metabolism.* Springer-Verlag, Berlin.

Hochachka, P.W. (1980) *Living Without Oxygen: Closed and Open Systems in Hypoxia Tolerance.* Harvard University Press, Cambridge, MA.

Houlihan, D.F., Rankin, J.C. & Shuttleworth, T.J. (1982) *Gills.* Cambridge University Press, Cambridge, UK.

Johansen, K. & Burggren, W.W. (1985) *Cardiovascular Shunts.* Munksgaard, Copenhagen.

Lamy, J., Truchot, J-P. & Gilles, R. (1985) *Respiratory Pigments in Animals: Relations, Structure and Function.* Springer-Verlag, Berlin.

Prange, H.D. (1996) *Respiratory Physiology: Understanding Gas Exchange.* Chapman & Hall, New York.

Vogel, S. (1993) *Vital Circuits: on Pumps, Pipes and the Workings of Circulatory Systems.* Oxford University Press, New York.

Reviews and scientific papers

Boutilier, R.G. & St.Pierre, J. (2000) Surviving hypoxia without really dying. *Comparative Biochemistry and Physiology* A **126**, 481–490.

Bramble, D.M. & Jenkins, F.A. (1993) Mammalian locomotor–respiratory design; implications for diaphragmatic and pulmonary design. *Science* **262**, 235–240.

Dejours, P. (1994) Environmental factors as determinants in bimodal breathing: an introductory review. *American Zoologist* **34**, 178–183.

Feder, M.E. & Burggren, W.W. (1985) Cutaneous gas exchange in vertebrates: design, patterns, control and implications. *Biological Reviews of the Cambridge Philosophical Society* **60**, 1–45.

Gracey, A.Y., Troll, J.V. & Somero, G.N. (2001) Hypoxia-induced gene expression profiling in the euroxic fish *Gillichthys mirabilis. Proceedings of the National Academy of Science USA* **98**, 1993–1998.

Hadley, N.F. (1994) Ventilatory patterns and respiratory transpiration in adult terrestrial insects. *Physiological Zoology* **67**, 175–189.

Hochachka, P.W. (1997) Oxygen—a key regulatory metabolite in metabolic defence against hypoxia. *American Zoologist* **37**, 595–603.

LaBarbera, M. (1990) Principles of design of fluid transport systems in zoology. *Science* **249**, 992–1000.

Lehmann, F-O. (2001) Matching spiracle opening to metabolic need during flight in *Drosophila. Science* **`94**, 1926–1929.

Lighton, J.R.B. (1994) Disconti..ıuous ventilation in terrestrial insects. *Physiological Zoology* **67**, 142–162.

Mangum, C.P. (1994) Multiple sites of gas exchange. *American Zoologist* **34**, 184–193.

Mangum, C.P. (1998) Major events in the evolution of the oxygen carriers. *American Zoologist* **38**, 1–13.

Paul, R.J., Bihimayer, S., Colmorgen, M. & Zahler, S. (1994) The open circulatory system of spiders: a survey of functional morphology and physiology. *Physiological Zoology* **67**, 1360–1382.

Smatresk, N.J. (1994) Respiratory control in the transition from water to air breathing in vertebrates. *American Zoologist* **34**, 264–279.

Storey, K.B. & Storey, J.M. (1990) Metabolic rate depression and biochemical adaptation in anaerobiosis, hibernation and estivation. *Quarterly Review of Biology* **65**, 145–174.

Van Holde, K. (1998) Respiratory proteins of invertebrates: structure, function and evolution. *Zoology (Jena)* **100**, 287–297.

Wasserthal, L.T. (1997) Interaction of circulation and tracheal ventilation in holometabolous insects. *Advances in Insect Physiology* **26**, 298–351.

Weber, R.E. & Jensen, F.B. (1988) Functional adaptations in hemoglobins from ectothermic vertebrates. *Annual Review of Physiology* **50**, 161–179.

Wenger, R.H. & Gassmann, M. (1997) Oxygen(es) and the hypoxia-inducible factor-1. *Biological Chemistry* **378**, 609–616.

Zhu, H. & Bunn, H.F. (2001) How do cells sense oxygen? *Science* **292**, 449–451.

8 Temperature and its Effects

8.1 Introduction

The lowest temperature yet recorded in the Earth's biosphere is –89.2°C, in Antarctica, whilst the highest temperatures range from 80°C in large deserts to 100°C in certain geothermal hot springs and to over 350°C at high hydrostatic pressures in deep-sea hydrothermal vents. Prokaryotic life can be found over much of this temperature range, but active animal life is restricted to a relatively narrow range of thermal conditions. For example, in hot springs there may be thermophilic bacteria living at temperatures in excess of 90°C, but for metazoans the upper limits are set well below this: for land animals the upper limits are attained by certain desert insects and reptiles, in which body temperature may sometimes exceed 50°C; for aquatic invertebrates maximum heat tolerance occurs in some specialized ostracod crustaceans, which can tolerate 49°C for short periods; and among the vertebrates a few species of fish can thrive at 44°C. The bottom end of the range for aquatic animals is inevitably set by the freezing point of sea water at –1.86°C, but on land large polar mammals and birds can tolerate ambient temperatures of –60°C. Some indications of the broader range of ambient thermal conditions survived by active terrestrial animals are shown in Fig. 8.1. Some animals survive in inactive states somewhat outside these limits, most spectacularly the nematodes, tardigrades, and eggs of insects that can withstand freezing in liquid helium at –269°C (see Chapter 14).

Despite the rather narrow range of temperatures most animals

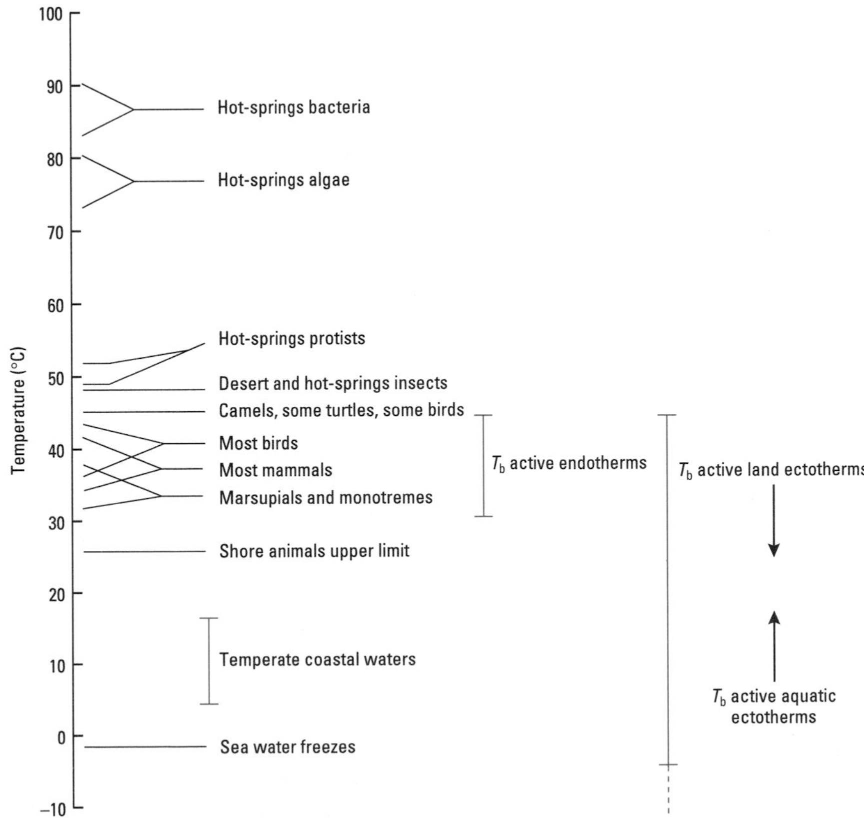

Fig. 8.1 The range of temperatures occurring naturally at the surface of the Earth, and thermal limits to animal life there. T_b, body temperature.

encounter, they are drastically affected at all levels of their biological organization by any change in their thermal surroundings. Thus temperature changes pose a real challenge to animals. Climate change over geological time periods is thought to have played a significant part in evolution, influencing adaptive radiations and in some cases leading to mass extinctions. Temperature may also change relatively dramatically over a few hundreds or thousands of years, in cyclical patterns related to astronomical anomalies and to continental (plate tectonic) movements. Moreover, temperatures may alter very widely even during the course of a year, especially at moderate and high latitudes. However, on annual and circadian timescales we cannot easily make broad generalizations, because of the major differences between the thermal characteristics of terrestrial and aquatic habitats. Water is a reasonably good conductor of heat and it has a much higher specific heat capacity than air. As a result, the temperature of relatively large bodies of water is highly buffered and a large heat input/output is required to modify the temperature. In contrast, air is an effective insulator and it has a low specific heat, such that relatively minor inputs of solar radiation can produce dramatic rises in temperature. Thus the surface air temperature just north of the Arctic circle in Siberia can change from −70°C in winter to +36.7°C in summer (more than 100°C of temperature change); while close to the Antarctic continent the temperature of the ocean water varies by less than 0.2°C through the entire year. Similarly in low-latitude deserts there may be exceptionally high daylight temperatures but very cold nights, while just offshore the sea temperatures show no daily variation. Thus dramatic daily temperature changes are the norm for many terrestrial animals, but (with the exception of species living in small bodies of water such as puddles and rockpools) are relatively rare for aquatic organisms.

In this chapter we review the effects of temperature on animals, and the kinds of adaptation they show to withstand or to counter temperature change.

8.2 Biochemical effects of temperature

Temperature is a measure of the average molecular motion within a material—at higher temperatures, there is more molecular vibration. In fact the mean kinetic energy of molecules is proportional to the absolute temperature. However, particular molecules within a material have differing kinetic energies about this mean value, and at any particular temperature a certain proportion of the more active molecules in a system are moving fast enough to react with other molecules when they collide, i.e. their energy is above the required activation energy ($\Delta G°$) for a given reaction. As the temperature rises, the number of molecules whose activity exceeds this threshold also rises, so that any chemical reaction becomes faster at higher temperature. In fact, while a rise of only 10°C may raise the kinetic energy by just a few percent, it may increase the reaction velocity by several hundred percent by pushing a large proportion of molecules across this activation energy barrier (Fig. 8.2). The same is true, within limits, of all biochemical reactions catalyzed by enzymes. The reaction rate may rise very sharply with temperature due to thermal effects on the reaction of substrates with enzyme catalysts and the consequent faster release of products. The energy barrier of enzyme-catalyzed reactions is itself subject to evolutionary modification. However, there is normally an upper thermal limit beyond which reactions are drastically reduced or cease altogether as the biochemical structures involved are damaged.

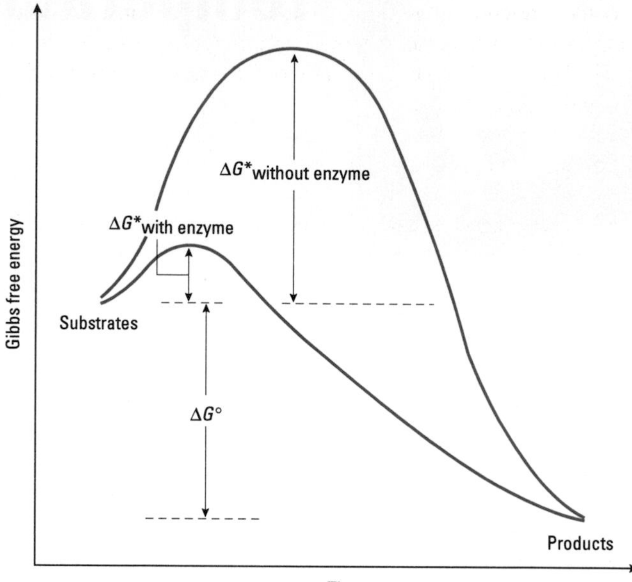

Fig. 8.2 Free energy changes during a chemical reaction, showing the decreased activation energy (ΔG^*) for an enzyme-catalyzed reaction where the enzyme–substrate complex facilitates breakdown of substrate(s) to product(s).

Many biological functions are made up of lots of steps, the individual components of which cannot easily be measured. In such cases it is not appropriate to derive an activation energy and instead **Q_{10} values** are used to describe the effects of temperature on the rate of the function being studied (Table 8.1).

Table 8.1 Q_{10} values for a variety of reactions involved in biological processes.

	Temperature (°C)	Q_{10}
Physical reactions		
Diffusion	20	1.03
Biochemical reactions		
Cytochrome reductase (possum)	>20	1.5
	<20	2.5
Pyruvate kinase V_{max} (rat)	>25	1.7
	<25	3.2
Hemoglobin coagulation	60	13.8
Physiological reactions		
Anemone oxygen consumption	5	2.0
	15	2.3
Crayfish heart rate	5	2.4
	15	1.6
	25	0.8
Crustacean gill movements	5	3.8
	15	1.7
Beetle oxygen consumption	10	2.4
	20	2.1
Insect thermal induction of diapause	20	1.4
	10	3.7
Torpid mammal oxygen consumption	20	4.1

Usually reaction rates are compared at two or more different temperatures to show how thermally dependent a particular function is. The Q_{10} value is then the ratio of the velocity constants k_1 (at t°C) and k_2 (at t + 10°C):

$$Q_{10} = k_{2(t+10°C)}/k_{1(t°C)}$$

or more generally, the ratio between k_1 and k_2 corrected for a 10°C difference, i.e.:

$$Q_{10} = (k_2/k_1)\ (10/(t_2 - t_1))$$

In practice the rate of most processes changes as a logarithmic function of temperature. This means that Q_{10} values are not constant for different 10°C increments (e.g. 10–20°C and 30–40°C), but instead normally decrease at higher temperatures. The value therefore has to be quoted for a *particular* 10°C interval. This may not be easy, as effects are rarely measured exactly 10°C apart, so the Q_{10} value is in practice extrapolated from observations at any two experimental temperatures.

The Q_{10} values for physical processes such as diffusion, and for many inorganic chemical reactions, are normally about 1. Values for photochemical reactions (as in photosynthetic systems) are often only a little above 1. However, Table 8.1 shows that for enzyme-mediated biochemical reactions, and for physiological processes, such as heart rate or metabolic rate, that rely on underlying enzymic processes, the Q_{10} is normally higher, usually in the range 2–3 (i.e. a doubling or tripling of reaction rate per 10° temperature change, so that the rate plot is an exponential upward curve). This could mean that for an animal able to survive over a 30°C temperature range, the metabolic rate might be raised at least 27-fold at the upper temperatures. In reality, since we know that the Q_{10} values tend to be variable in different parts of the temperature range, such a spectacular rise in metabolic rate or heart rate probably will not occur. For marine invertebrate heart rates, the Q_{10} values are commonly over 3 below 10°C, around 2 between 10 and 20°C, and less than 2 when the temperature exceeds 20°C. Marine animals will not encounter such a range (except in the special case of hydrothermal vent animals considered in Chapter 14), but simple calculation shows that the increase in heart rate incurred might be up to 12-fold at the higher temperatures. In some cases, for processes such as ventilation rate, for example, there may be extra complications due to sharp transitions in Q_{10} values at particular ambient temperatures, suggesting rather fundamental underlying changes in the materials or tissues or mechanisms involved. Indeed the Q_{10} for oxygen consumption may show a reduced value in the region of maximum activity temperatures for some animals, which has been interpreted as an adaptive means of controlling energy expenditure ("metabolic homeostasis"). There is good reason to suppose that Q_{10} values are under strong selective pressure with regulatory processes in place, given that phenomena that need to be temperature independent (e.g. the functioning of biological rhythm controllers or "clocks") are commonly found to have a Q_{10} of 1.0.

Thus, although Q_{10} values are often helpful and illuminating, and are frequently quoted in relation to biochemical and physiological adaptations (see Table 8.1 for examples), they clearly have to be treated with care. We need a more detailed understanding of the processes and effects at these levels to appreciate how temperature changes really influence animals.

8.2.1 Enzymic adaptation to changing conditions

Enzymes are proteins, and their performance is inevitably susceptible to thermal changes, but as temperatures rise or fall their performance can be modified in a variety of ways that may be adaptive, extending the thermal range of the species. The nature of these modifications varies with the timescale of the temperature change.

Short-term changes

The activity of many enzymes is controlled directly or indirectly via hormones and/or the nervous system. These provide mechanisms for modulating enzyme activity over the timescale of a few seconds to several hours, compensating for changes in temperature. There are at least four possible types of change that could be instituted quickly by hormonal or neurotransmitter intervention:

1 The enzyme's effective concentration or activity could be altered.
2 The concentration of the relevant substrate could be altered.
3 The energy supply for the reaction being catalyzed could be increased.
4 The intracellular environment could be altered to improve the enzyme's effects.

A great many animals have been said to use the first of these adaptive responses, by increasing effective enzyme concentrations in colder conditions. This could occur by direct modifications to the enzyme (such as phosphorylation to an active form), or by changed compartmentalization within the cell (such as mobilization from membrane-bound areas into the cytoplasm) or by altered activity due to allosteric modulation by activators or inhibitors.

In practice, though, it is usually impossible to disentangle mechanisms 1–3 given above, as all of them interact. A neat example of hormonal control of enzyme concentrations, substrate concentrations, and energy supply is provided by cold-tolerant insects, where a falling temperature triggers synthesis of cryoprotectant molecules (these are dealt with specifically in section 8.3.1). The first step is the release of a hormone that binds to receptors on the fat body cell membranes, causing an increase in intracellular cAMP and Ca^{2+} levels via a classic protein kinase cascade (see Chapter 2). The result is that inactive phosphorylase b becomes phosphorylated to the active form, phosphorylase a, giving a net increase in the cellular concentration of phosphorylase a. This in turn allows the breakdown of more glycogen, which is the "fuel" to provide the extra substrates for the synthesis of the cryoprotectants.

Enzyme performance can be modulated in a less direct manner by changing the intracellular environment in which the enzyme works. Thus ionic concentration within the cell commonly changes slightly during short-term thermal exposures, and both pH and potassium concentration are known to alter enzyme activity. In particular, pH is usually observed to increase (i.e. become more alkaline) in animal cells as the temperature drops, and this provides a significant stabilization of protein structure and thus reduced enzyme temperature dependence (sometimes termed "alphastat regulation"). Since this effect could influence the thermal responses of large numbers of proteins, it should represent a particularly efficient way of dealing with acute thermal stress. Certain low molecular weight solutes also affect protein stability; the sugar trehalose is a powerful stabilizer of both protein and membrane structures, and is synthesized in some yeasts after heat shock. It also appears in many metazoan organisms entering cryptobiotic states (see Chapter 14), where it protects against extreme dehydration as well as temperature changes.

Medium-term or acclimatory changes

Over the timescale of days to weeks, temperature change often influences the concentration of enzymes more fundamentally, via differential effects on protein synthesis and protein degradation. Normally this involves just a few of the enzymes that govern a complex reaction sequence—those that are rate-limiting—and it may be reflected in a rise in mitochondrial protein content as the temperature falls, by as much as 50% in some animal tissues. Over a similar timescale, new enzymes may also be synthesized that can regulate other enzymes' activities. Here the best known examples are enzymes that influence the fatty acid composition of membrane phospholipids (see section 8.2.2), resulting in improved performance of other membrane-bound enzymes following a period of low temperature acclimation.

However, since most enzymes exist as a variety of isozymes with subtle differences in kinetic properties, the best and the commonest solution for maintaining the catalytic efficiency of an enzyme may be to change the suite of isozymes present. The net effect is often to alter the fate of metabolites rather than overall reaction rates; for example, in frogs and fish low temperatures favor the synthesis of the "heart-type" H_4 lactate dehydrogenase (LDH) isozyme in skeletal muscles and thus the oxidation of pyruvate, at the expense of its reduction to lactate by the more normal "muscle-type" M_4 isozyme.

In practice it seems to be quite rare for organisms to carry a range of isozymes that are specific to particular parts of their thermal range. Perhaps the best documented example is myosin ATPase. In goldfish and common carp subjected to several weeks of low temperatures, an altered pattern of expression of the genes for the heavy and light myosin chains is found, so that ATPase activity is increased and muscle contraction speed and force are improved. This contributes to an increased maximum swimming speed at low temperature, having obvious survival value in escaping from predators; but it occurs at the expense of a trade-off in performance at warm temperatures in the cold-acclimated fish.

Clearly there must be a reasonable time period for switching on the new isozyme locus and synthesizing the new protein. For example, trout living at 17°C apparently have an acetylcholinesterase enzyme quite distinct from the acetylcholinesterase found in trout living at 2°C, and a fish moved from one habitat to the other is not immediately able to mobilize the new form of its enzyme to cope. However, trout living at 12°C seem to have copies of both isozymes and need very little time to adapt to a change in thermal conditions. This is perhaps an example of "on–off" gene expression, where one or other isozyme may be switched off according to the thermal history. In other cases, for example malate dehydrogenase (MDH) of the goby *Gillichthys*, regulation of expression is more subtle, with adjustments of the ratio of expression of two isozymes occurring.

Although this strategy of having a suite of isozymes with different temperature optima is actually not very common, nevertheless quite large numbers of animals express different isozymes in different tissues (such as liver, heart, and muscle). The ability to alter gene expressions and thus switch between protein components in response to seasonal (or even diurnal) change is a key factor in the adaptive repertoire of some animals with particularly wide temperature tolerances.

Long-term or evolutionary changes

Natural selection offers the possibility of evolving enzymes tailored to particular thermal environments both in terms of maximal activities and regulatory properties. In some cases the mechanisms are similar to those observed in acclimatory changes, with a quantitative increase in enzyme concentrations. For example, individual mitochondria in Antarctic fish have relatively low rates of oxygen uptake, but their concentration is greatly increased: they occupy 30–60% of the fiber volume of red muscles in Antarctic fish, compared with 15–25% in tropical fish, and only 8–10% in terrestrial reptiles.

Other possibilities include genotypic changes in enzyme sequence and structure to give altered properties, resulting from the combined effects of mutation and selection to give more suitable protein variants for particular temperature regimes. The ligand-binding site is usually highly conserved, but adjacent areas are altered to change the thermal stability and/or flexibility of the enzyme. In general, early studies have indicated that enzymes of animals in thermally stable environments have lower thermal stabilities and less resistance to thermal perturbation of substrate binding than the homologous enzymes of an animal from a thermally variable habitat; and that enzymes of warm-adapted species tend to show more thermostability than the homologs from cold-adapted species.

Adaptive changes in the thermal properties of homologous enzymes from different populations of a species reflect the underlying allelic variation maintained by selection. The enzyme variants are then termed allozymes (or orthologous homologs), as discussed in Chapter 2. A classic example of these multiple allele effects is the muscle M_4 LDH enzyme, which in southern populations of the fish *Fundulus* is predominantly the LDH_a type, this being progressively replaced by the LDH_b type in more northern populations. Measured under certain conditions *in vitro* the kinetic properties of the enzymes appear to be optimal at 20°C for the cold-adapted northern fish, but optimal at 30°C for the southern fish living in warmer seas. It has been suggested that these differences are maintained by strong selection, although nonadaptive explanations (such as random genetic drift) are possible.

Analysis of enzymic function from animals that have evolved to cope with different thermal regimes has become more sophisticated in recent years. Firstly, thermal denaturation temperature has been studied, and this correlates closely with adaptation temperature for the M_4 LDH enzyme (Fig. 8.3). The homolog from polar fish species starts to unfold at a much lower temperature than the homolog from thermophilic species such as desert lizards. Work on this and other enzymes indicates that only tiny changes in protein structure—perhaps only a few residues—are enough to affect the enzymes' thermostability. It has been suggested that enzymes from

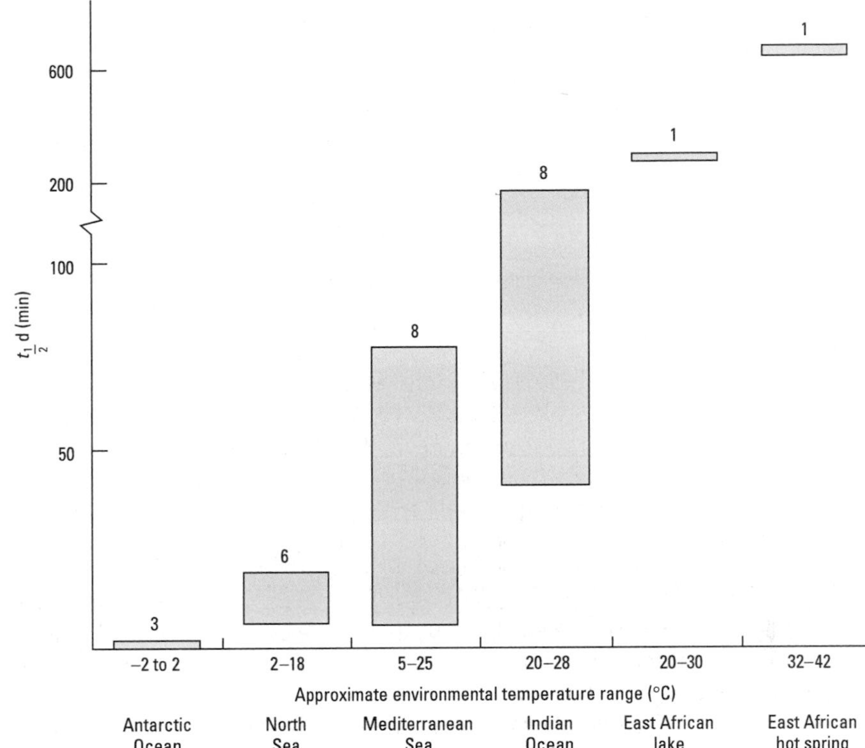

Fig. 8.3 Thermal stability of enzymes in marine animals is related to their thermal habitat. The thermal denaturation half-times at 37°C for myofibrillar ATPase from fish species taken from six habitats are compared. Those from cold seas denature within a few minutes, while those from the Indian Ocean (normally 20–28°C) have half-times of up to 3 h. Note that enzymes in fish from warm freshwater lakes and springs are even more thermally stable. Numbers above bars indicate numbers of species tested. (Adapted from Johnston & Walesby 1977.)

Table 8.2 Substrate turnover rates (K_{cat}) for lactate dehydrogenase (LDH) from barracuda (see also Fig. 11.9), showing the similar (conserved) values for different species when measured at normal operating temperatures.

Species (distribution)	Normal mean temperature of habitat (T_m) (°C)	K_{cat} at 25°C (s^{-1})	K_{cat} at T_m (s^{-1})
Sphyraena argentea (northern)	18	893	667
Sphyraena lucasana (intermediate)	23	730	682
Sphyraena ensis (southern)	26	658	700

animals with low body temperatures have a more "open" structure (stabilized by fewer and weaker bonds) than those from animals with warmer body temperature (T_b) values, and that (while reducing thermostability at low temperatures, as described above) this permits more efficient conformational changes during catalysis. Enzymes from animals with higher body temperatures need more compact and rigid enzyme structures because of the higher kinetic energy of the cell.

Secondly, enzymes from cold-bodied species show major differences in substrate turnover numbers (K_{cat}, the moles of substrate converted to product per mole of enzyme per second). In general, enzymes from species with lower T_b values have higher K_{cat} values than the homologous enzymes from warmer bodied species (Table 8.2). Thirdly, it has been observed that the activation energy of homologous enzymes across a range of species from different habitats is such as to counter the effect of temperature. That is, cold-adapted enzymes have a lower activation energy than warm-adapted enzymes. Thus they can avoid the reduction in reaction velocity that would otherwise take place, and in some cases even allow an increase in reaction rate in cold conditions.

Enzyme affinities present a somewhat more complicated picture. We now know that substrate concentrations in cells are generally rather similar at all body temperatures. The ability of an enzyme to bind its substrate (its "affinity") can often be approximated by the apparent Michaelis–Menten constant (K_m) of the reaction (see Box 4.5 for a reminder). Cold-adapted isozymes also tend to show an increased affinity (lower K_m value) compared to warm-adapted ones when tested at the same temperature (Fig. 8.4). But this also shows that cold-adapted enzymes have rather similar K_m values to warm-adapted isozymes when each is determined at its relevant working temperature, i.e. under physiological assay conditions. These careful interspecific comparisons of enzyme activities measured *in vitro* indicate that ligand-binding characteristics are also highly conserved. Evolution seems to have favored the conservation of the important kinetic properties of enzymes within fairly narrow ranges, so that enzymes that differ within species and even within individuals retain maximal responsiveness to changes in the normal cellular concentrations of substrates and allosteric modulators.

Differences in K_{cat} and K_m may occur in closely related species with only a small difference in mean habitat temperature, with enzymes differing by only a few amino acids, indicating the potent effects that can be achieved with tiny selective adjustments in tertiary protein structure. Indeed there are indications that virtually *no* changes in structure are needed to produce changes in thermal stability for LDH. The enzymes from different species of goby (*Gillichthys*) are indistinguishable in most respects with cDNAs

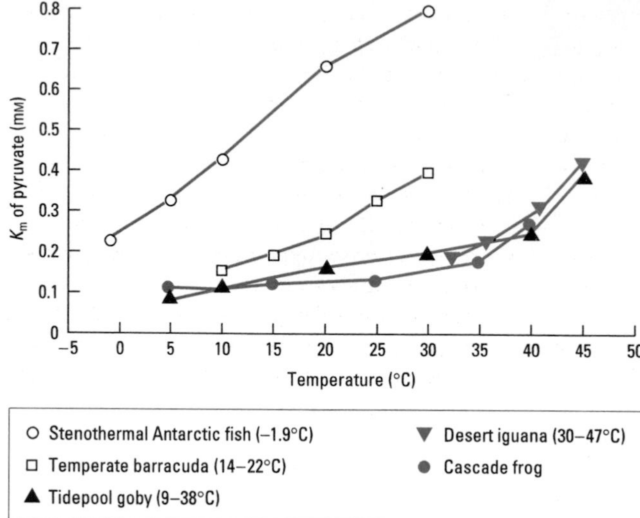

Fig. 8.4 The effects of temperature on the K_m values for pyruvate breakdown by lactate dehydrogenase (M_4-LDH) from vertebrates. For each species the normal body temperature (T_b) is shown. K_m values vary widely, but when comparisons are made at the animal's preferred T_b all values are in a similar range (around 0.1–0.3 mM). (From Somero *et al.* 1996.)

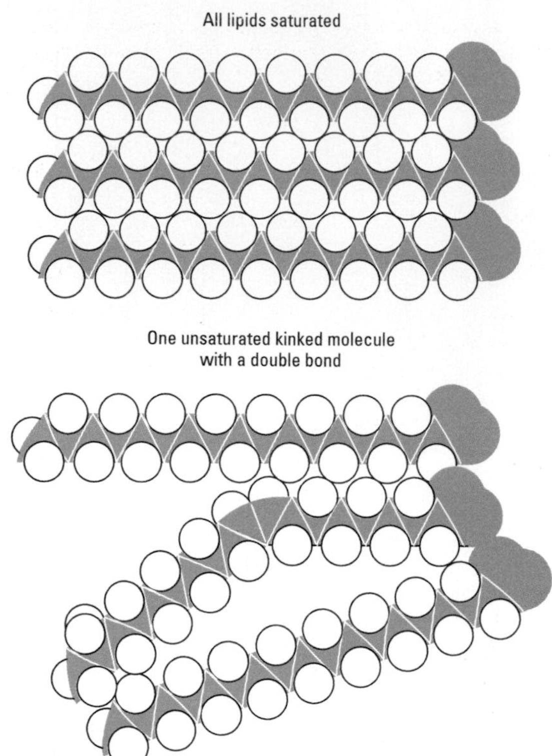

Fig. 8.5 Double bonds in membrane lipids producing a less densely packed structure.

differing only by four synonymous nucleotides, yet have different temperature sensitivities, suggesting that in some cases the same isozyme may exist in different folded conformations and have different functional properties.

8.2.2 Thermal effects on membranes and cellular structures

The physical properties of biological membranes, and especially of their lipid components (which may be 25–50% of the membrane dry weight), are markedly influenced by temperature. The lipids normally exist in a "liquid crystal" state that is intermediate between a highly fluid material and a rather rigid structure. This delicately balanced structural state can be disrupted rather easily, and this is reflected in substantial and abrupt changes in viscosity in preparations of membrane phospholipids exposed to changing temperature *in vitro*. Heat death may result from disruptions to membrane function, especially at very sensitive sites, such as synapses, where any malfunction would obviously produce immediate behavioral effects. In living cells, however, a phenomenon known as **homeoviscous adaptation** (HVA) occurs to keep the membrane in its normal moderately fluid state with its protein components properly functioning.

The most important contributor to this acclimatory or adaptive process appears to be variation in the fatty acid composition of the lipids. Shorter chain fatty acids, and those that are unsaturated (with some carbon-carbon double bonding), are inherently more fluid than longer chain or fully saturated fatty acids. This is because a long molecule with a saturated and therefore straight molecular backbone can achieve more stable bonding with adjacent molecules compared with the rather kinked unsaturated fatty acid molecules (Fig. 8.5). The introduction of the first unsaturated bond has the greatest effect on membrane physical properties. If this first unsaturation is at the central 9–10 position from the carboxyl terminus then the largest increases in fluidity are observed. Cholesterol also has some effect on HVA, increasing membrane order and affecting phase behavior and permeability. In general more cholesterol is present in membranes from endotherms than from ectotherms, but changes in cholesterol content with temperature may be in either direction, probably reflecting its multiple roles.

Temperature-induced changes in membrane fluidity can be detected by relatively new techniques such as electron spin resonance spectroscopy or fluorescence polarization spectroscopy. In the light of the resulting new understanding of membrane structure and membrane domains (see Chapter 4), the use of terms such as viscosity or fluidity in relation to membranes has been partially superseded by motional rate parameters and measures of static molecular ordering. The concepts of changing "fluidity" or of changing states of "order" and "disorder" are both useful ways of envisaging the effect of lipid alterations on the properties of the membrane.

The membrane lipids of a wide range of animals show varying fatty acid compositions at varying ambient temperature that are just as we would expect for adaptive maintenance of the homeoviscous state (Table 8.3). The exact nature of the fatty acids that are varied differs according to taxon, and is complicated because membrane composition also depends upon diet. Indeed a diet artificially high in polyunsaturated lipid (resulting in membranes of lower viscosity at a given temperature) has been shown to lead to a lowered preferred T_b in lizards, with a reduction of as much as 5°C. This may indicate that seasonal changes in diet can automatically bring about

Table 8.3 The ratio of saturated to unsaturated fatty acid residues in three phospholipids from cell membranes in brain tissues of various vertebrates acclimated to different temperatures.

Species	Body temperature (°C)	Phospholipid Choline	Ethanolamine	Serine inositol
Arctic sculpin	0	0.59	0.95	0.81
Goldfish	5	0.66	0.34	0.46
	25	0.82	0.51	0.63
Desert pupfish	34	0.99	0.57	0.62
Rat	37	1.22	0.65	0.66

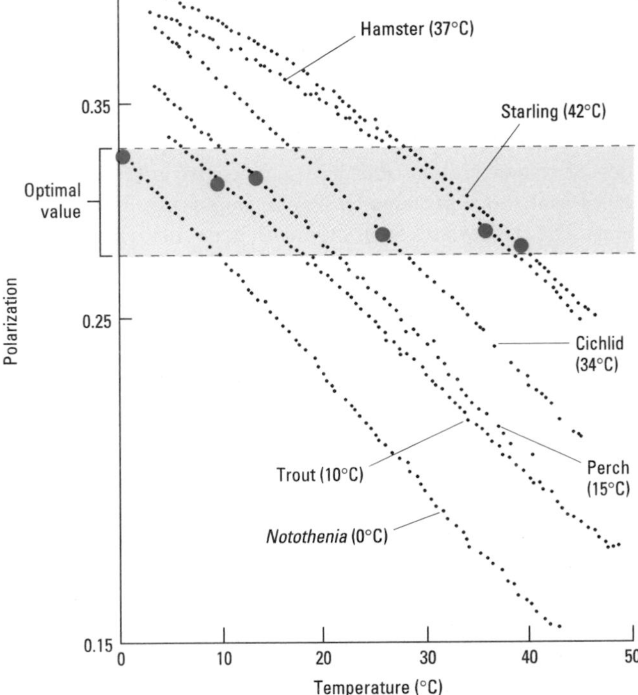

Fig. 8.6 Homeoviscous adaptation (HVA) seen in brain membranes, for various vertebrates. HVA is measured as the polarization of fluorescence (see text for details: decreased polarization indicates increased "fluidity" and decreased "order"). Values at the normal body temperature for each species (given in parentheses) are shown as large green circles. (From Behan-Martin *et al.* 1993.)

adjustments of thermal physiology. Certainly both the type of fatty acids and the proportions of pre-existing saturated and unsaturated forms are varied naturally in most species that have been studied. This also means that direct inference of adaptation by changing fatty acid saturation is quite difficult.

Perhaps the most impressive evidence of thermal adaptation of biomembranes comes from studies of species that have evolved in particular thermal habitats or maintain elevated and constant body temperatures, as in birds and mammals. Figure 8.6 compares the brain membranes isolated from a variety of teleost fish species and from a representative mammal and bird. In this study the measure of membrane physical condition was the fluorescence anisotropy of two standard fluorescence probes, a technique which provides information on the degree of hindrance to the free rotational motion of the rod-shaped probe. There is a clear relationship between the position of the curve and the temperature at which the species lives, so that at a given temperature membranes from the Antarctic species, *Notothenia*, were more fluid, and less ordered, than equivalent membranes from the perch, *Perca fluviatilis*, and so on. The mammalian and avian species possess relatively ordered membranes, which matches their need to maintain stable membranes at high body temperatures of 37°C and 41°C, respectively. The figure also indicates the value of anisotropy at the respective body temperature for each species, showing that these were always very similar. Thus it appears that selection designs the brain membranes of vertebrates to give equivalent physical properties under *in vivo* conditions of environmental or body temperature, the implication being that the condition is in some way favorable for the functional properties of that membrane system. Evolutionary responses to cold regimes therefore combine a higher proportion of unsaturated fatty acids with a marked increase in the number of unsaturated bonds in each unsaturated fatty acid, to maintain overall membrane fluidity.

All these studies on HVA are correlative, and do not establish causal relationships between increased phospholipid unsaturation and decreased membrane order. Molecular genetics offers some alternative and less equivocal approaches for establishing a degree of experimental control over membrane adaptation, by intervening in the expression of enzymes involved in the response. The overexpression or abolition of a specific enzyme, or its induced expression in the absence of temperature change, allow its role in the adaptive regulation of membrane structure to be demonstrated more confidently. The changes in fatty acid composition are now known to be brought about by a group of **desaturase enzymes**; for example the Δ^9-desaturase enzyme incorporates the first double bond into stearic or palmitic acid. The activity of desaturases is directly related to temperature, so that following cold transfer in fish this activity is induced, with enzyme activity increasing up to 30-fold. Diet also affects this, so that feeding on more saturated lipids causes an up-regulation of desaturase activity, even at high temperatures.

It is possible to make a chemical probe that binds to the mRNA for the carp liver Δ^9-desaturase, and then look for the induction of the desaturase transcript relative to that of some other gene not thought to be involved in temperature adaptation. Figure 8.7 shows the variation in Δ^9-desaturase transcript levels relative to 18S ribosomal RNA from carp acclimated to 30°C, and at different periods through a progressive 3-day cooling regime to 10°C. Transcript levels were low in warm-acclimated carp and after 24 h of cooling, but increased 8–10-fold after 48 h. Levels peaked at 2–3 days and then quickly subsided. The transcript levels after long periods in the cold remained about double those observed in warm-acclimated carp liver. Thus although cooling led to a dramatic increase in transcript levels, the effect was transient. Evidently the change in lipid saturation requires only a transient increase in desaturase expression, and low levels of desaturase activity are then sufficient to maintain the *status quo* on long-term exposure to cold.

In prokaryotic cells and in plants the genetic manipulation of unsaturation of fatty acids in membrane lipids is easier to perform and is known to influence growth and chilling tolerance. For example, the introduction of cytoplasmic DNA for glycerol 3-phosphate acyltransferase from a squash plant (which is chill sensitive) or from

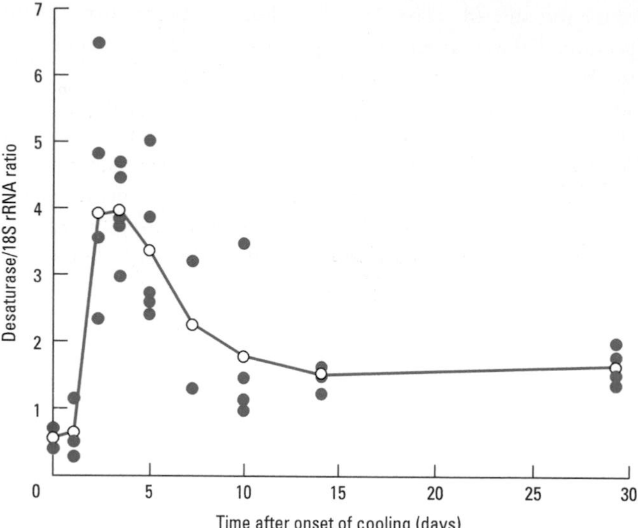

Fig. 8.7 The effect of cold acclimation in carp on the levels of desaturase enzyme (measured as Δ^9-desaturase mRNA transcript from liver, relative to a standard 18S rRNA signal). Green symbols represent data from individual animals, and open symbols represent mean values, showing the clear increase in the first few days of cooling. (From Gracey *et al.* 1996, courtesy of Cambridge University Press.)

Arabidopsis (which is chill resistant) into *Nicotiana* (tobacco, which has intermediate chill sensitivity) results in altered chilling sensitivity and resistance, respectively. In the former case, the level of unsaturated fatty acids in glycerophosphatides falls significantly and chilling sensitivity is markedly increased, whilst in the latter there is an increase in both the unsaturation of fatty acids in glycerophosphatides and in chilling tolerance.

In terrestrial animals that experience rapid temperature fluctuations these changes can be extremely rapid, with a time course of just a few hours. Recent evidence also shows that the changes vary in different tissues within an individual; the degree of compensation of membrane order (the "homeoviscous efficacy") varies from 75% in the carp intestinal mucosa membranes to 25% in brain myelin and to almost zero in the sarcoplasmic reticulum of skeletal muscle. More spectacularly, it is now also clear that homeoviscous efficacy can vary within one cell. Some neurons have high saturation in the axon membrane close to the cell body in the core of the animal, but fewer saturated lipids where the axon reaches the periphery, so that the nerve functions properly along its entire length even in the cold.

At present we know relatively little about the nature of the temperature sensor in what may be a long and complex feedback pathway. Some transcriptional activation seems to be linked to changes in the fatty acid saturation and/or the physical properties of the plasma membrane, so that the ultimate sensor may be some aspect of membrane physical condition. In at least some cases the thermally induced change in enzyme activity is probably due to configurational changes within the membrane, such that the active site of the desaturase is masked in relatively fluid/disordered membranes but becomes exposed as the temperature falls and the membrane becomes less fluid and more ordered.

This neatly indicates the complex interrelation of membrane effects and enzymic effects in a thermally stressed animal. Indeed a similar process of membrane fluidity/enzyme masking may be part of the explanation for the widespread changing of activation energies of membrane-bound enzymes as temperature changes, discussed above.

In addition to membrane components, other subcellular components will be adversely affected by temperature changes, and the best studied examples are microtubules. In mammals the constituent tubulins are cold-labile, depolymerizing at 0–4°C, but the tubulins from polar fish are much more stable with temperature and will self-assemble *in vitro* at –2°C. These differences seem to be intrinsic to the tubulin subunits themselves, and arise from relatively minor variations in amino acid sequences, largely in the α-chains of the tubulin structure at a domain responsible for contact between tubulin dimers. This produces more hydrophobic interactions in the microtubules of the polar fish.

8.2.3 Thermal induction of stress proteins

A third effect of temperature at the biochemical level is the induction of certain proteins that have a protective effect, commonly called **heat shock proteins** (HSPs) or more generally **stress proteins**. The HSPs were a focus of interest in the 1980s and 1990s, as they were proved to play a key role in regulating protein folding and in coping with proteins affected by heat and other stresses.

HSPs belong to several families, defined by their sizes (molecular weights, in daltons). In animals the relatively large protein families dominate (HSP60, HSP70, HSP90, HSP100, and Lon), whereas in plants there are many more of the small HSPs (including HSP10, HSP27, and ubiquitin). The structures of these are highly conserved even across kingdoms, and their action also seems to be highly conservative. They can be induced by many stresses: natural changes such as hypoxia or hyperoxia, osmotic shock and pH change, and less natural stresses such as the presence of alcohols, ionizing radiation, heavy metals, toxins, and free radicals. Hence they can bring about "cross-tolerance", i.e. once induced in relation to a particular stress, they make the organism more tolerant of other stresses. Some of the smaller HSPs appear to be developmentally patterned in animals, with HSP27 regulated in part by ecdysone in at least some insects. The higher molecular weight HSPs are more commonly the forms that are inducible by temperature change, though the 70 kDa family includes both inducible (HSP70) and noninducible or "constitutively expressed" (HSC70) members.

Induction of HSP expression in relation to temperature is related once again to normal thermal regimes, so that species from warm environments undergo a stress response at a substantially higher threshold temperature than those from cold environments (indeed some Antarctic fish such as *Trematomus*, living in permanently stable cold waters, show virtually no heat shock response). Induced HSP concentrations also differ substantially between tissues within an individual. It is evident that the ability to synthesize HSPs is strongly correlated with thermal tolerance, in all animal taxa.

Expression of the proteins and the resultant increased thermal tolerance occurs within minutes to hours, substantially faster than the acclimatory changes of lethal limits normally recorded in animals. Molecular studies using gene deletions or insertions of extra copies show that there is a direct match between protein expression and this inducible thermotolerance, both in the laboratory and in

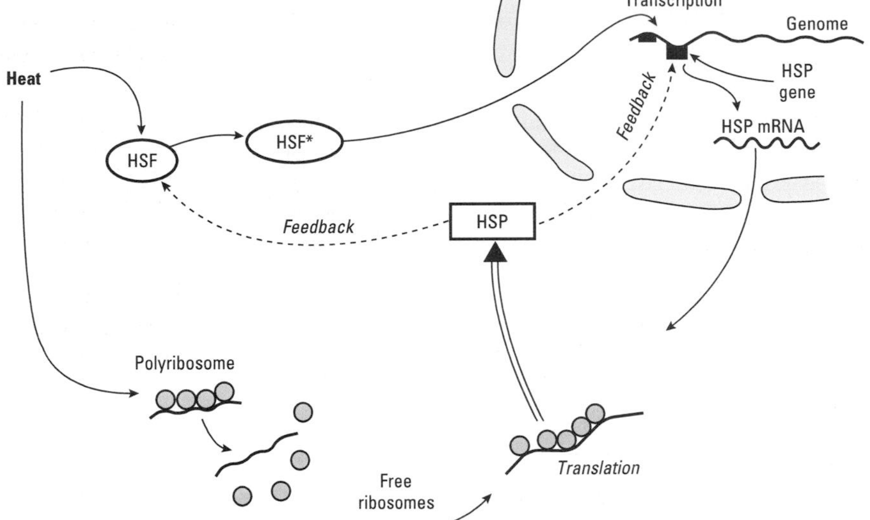

Fig. 8.8 A model for regulatory systems during heat shock. Changing temperature induces a cytoplasmic factor, HSF, which then changes from an inactive monomer to an active trimeric form, HSF*. This has a regulatory effect on the promoter region of the heat shock protein (HSP) gene, initiating production of HSP mRNAs. Since existing ribosome complexes have also been disrupted by the thermal shock, RNAs that were bound to them are released or degraded, and the new mRNA is preferentially translated, the HSPs then being distributed to the most sensitive nuclear and cytoplasmic sites. There is also a feedback interaction between some HSPs and HSF (see text).

wild populations of fruit fly larvae (*Drosophila*) living in the stressful alcoholic environment of rotting fruits.

The mechanisms of HSP action are now understood in terms of "molecular chaperonage". Proteins are normally folded into specific tertiary configurations, requisite for their proper functioning; but they may be unfolded during the early stages of their synthesis, during intracellular transport across membranes, or as a response to various kinds of stress. Such unfolded proteins may refold wrongly, and may be susceptible to interactions with other cellular components, including other unfolded proteins, rendering them useless. Molecular chaperones are used to limit these interactions, by recognizing and binding to the exposed side groups of the unfolded proteins, stabilizing the unfolded state. When appropriate (and often in association with ATP breakdown), the chaperone molecule may release its bound protein to allow correct refolding and a resumption of normal function; or alternatively the chaperone may release its bound protein for degradation or export from the cell. HSPs are seen as the primary molecular chaperones synthesized in response to any stress that causes protein unfolding. In some cases HSPs from different families work together as co-chaperones.

The regulation of the heat shock response is now known to be mediated via a cytoplasmic detector, the heat shock factor (HSF), which exists as a monomer until heat shock occurs, when it trimerizes and moves into the nucleus. There it binds to a particular site on the promoter region of HSP genes, initiating transcription. One of the main HSPs, HSP70, appears to have a feedback interaction with HSF, since it acts to inhibit trimerization. Thus unfolded proteins and HSF monomers may compete for interactions with HSF—if there is an excess of unfolded protein, HSP70 will be mopped up and the HSF will be free to trimerize and bring about more HSP transcription (Fig. 8.8).

8.3 Physiological effects of temperature

Biological processes generally exhibit a biphasic effect of temperature (Fig. 8.9). The first phase is a consequence of the rate-enhancing

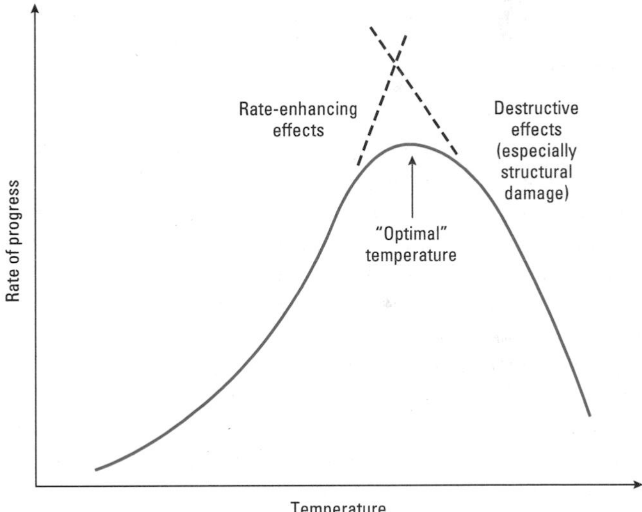

Fig. 8.9 Biphasic effects of temperature increase on the rate of biological processes.

influence of temperature on enzyme function discussed in section 8.1, so that rates of activity rise as temperature increases up to a certain point. But in a second phase beyond this (and normally beyond the natural range of temperatures experienced), the destructive effects of temperature supersede and rates of activity decline. The rate–temperature curve therefore reveals an "optimum" temperature for any particular activity.

Determining the rate–temperature curve is relatively simple for systems of no greater complexity than single-enzyme reactions, but it becomes more difficult where tissues are used, and nearly impossible for whole animals whose own reactions continually affect the conditions of the experiment. It is therefore appropriate to move from the effects of temperature on cellular activities first, to the integrated effects on whole animals.

8.3.1 Low temperature effects on cells and organisms

We have seen that enzymes are intrinsically less effective at low temperature, and that membranes might become more viscous, but that cells show a range of adaptive mechanisms to maintain reasonable activity. If enzymes and membranes are so adaptable, what then actually brings about cell death at low temperature? Some animal cells can tolerate extensive freezing, even down close to absolute zero in a few very special cases, yet others (e.g. cells from tropical fish) are so sensitive to cold that the cells are dead well before 0°C is reached on a cooling cycle.

Cells and animals that can survive low temperatures are exhibiting the general phenomenon of "cold hardiness", and most of them have some ability to survive subzero temperatures, raising the general problem of the potential freezing of their contained water. Surveying the animal kingdom as a whole there appear to be two broad strategies for dealing with this problem. Some animals show **freeze tolerance**, and can cope with extensive freezing with ice formation occurring within the body. Others show **freeze intolerance**, surviving at –40 to –50°C without ice forming in their bodies, but usually dying very quickly if any ice crystals do begin to form; they survive by avoiding freezing, and are therefore also described as showing **freeze avoidance**. However, the two different options are not entirely exclusive (e.g. a species can show tolerance as a juvenile and intolerance as an adult, or vice versa), and they show no clear phylogenetic distributions. Indeed, animals may be killed by **chill injury** before they show any signs of freezing, and within species there is often marked variation in the extent of chill injury and resultant mortality.

Any solutes present in water lower its freezing point in a predictable manner: remember that freezing point is one of the colligative properties, from which the concentration of dissolved solutes can be determined experimentally (see Chapter 4). Most marine animals are isosmotic and therefore have blood that freezes at a similar temperature to sea water, –1.86°C, whilst vertebrate body fluids (around 300 mOsm) have a freezing point of about –0.6 to –0.7°C. Invertebrates living in sea water therefore have no particular likelihood of ice formation, as the high latent heat of fusion of water coupled with the thermal inertia of the marine system makes it most unlikely that the sea itself will freeze; but the hyposmotic vertebrates in sea water may be at risk. Freshwater animals have body fluids more concentrated than their surroundings, and hence a lower freezing point, and so will not be at risk of tissue freezing unless their entire habitat is frozen (see Chapter 13). But intertidal invertebrates may be exposed at high tide to very cold air and will rapidly freeze, many of them undergoing a regular twice daily freezing and thawing cycle in winter (see Chapter 12). Small terrestrial animals such as insects may also experience exceptionally cold temperatures on a seasonal basis, while some desert forms such as scorpions only encounter cold temperatures at night; these too will undergo internal freezing.

Freeze-tolerant animals

These perhaps represent the lesser puzzle. When any animal does freeze, ice crystals can potentially grow very rapidly indeed if there are any nuclei present on which the water molecules can condense, and the crystals may severely damage cells in the process. In a biological system there are bound to be particles present in the extracellular and intracellular solutions to act as nuclei. Ideally, therefore, ice crystals should be allowed to form rather slowly at relatively high subzero temperatures, perhaps –2 to –10°C, and they should be confined to the extracellular fluids, leaving the cellular machinery undamaged down to much lower lethal temperatures, even when as much as 65% of the body water is frozen. This can be achieved, somewhat unexpectedly, by having **ice-nucleating agents** (INAs) within the extracellular body fluids to encourage ice to form there; at the same time, much of the intracellular water is kept highly ordered by reactions with macromolecules and is relatively unfreezable.

Nucleating agents in freeze-tolerant invertebrates appear to be proteins (**ice-nucleating proteins** or INPs), with a very hydrophilic character. Their structure provides many hydrophilic sites that order the water molecules into small crystals, reducing the energy barrier for nucleation. They thus allow a slow controlled ice formation that involves only the actual water content of the fluids, and only the extracellular compartment normally freezes. The solutes (ions and metabolites) therefore concentrate in the cells and in the limited body fluids that remain unfrozen, giving membranes time to undergo controlled shrinkage. Once a high percentage of the body water is frozen the cells must be tolerating very high osmotic concentrations, which incidentally render them even less likely to freeze intracellularly. The tissues become very distorted by the surrounding ice formation, with cells becoming shrunken and very changed in morphology.

Protecting the cells from injury during all these drastic changes is largely achieved by a second group of molecules, the **cryoprotectants**. These are of two kinds. **Colligative cryoprotectants** accumulate in high concentrations (0.2–2 M) and raise the osmotic concentration of the body fluids so that only a limited percentage of total body water can turn into extracellular ice. The commonest such molecules are polyhydric alcohols (polyols) or sugars, which are nontoxic even at high concentrations and are metabolically inert, being the end-products of biochemical pathways rather than intermediate stages. Glycerol (C_3) is the most commonly encountered (at up to 30% of body weight in some insects), but other polyols such as sorbitol (C_6) and ribitol (C_5) are also used. **Noncolligative cryoprotectants** occur at lower concentrations (usually <0.2 M) and are membrane protectants, binding in place of water and thus preserving the subcellular structure from long-term damage; trehalose and proline are common examples in insects. Many freeze-tolerant animals have multiple cryoprotectants. They are synthesized from glycogen reserves in the liver or fat body (Fig. 8.10), and must be in place before the first freezing exposure of winter.

The cells of freeze-tolerant animals must survive and maintain reasonable homeostasis in the face of long-term lack of blood flow (ischemia) imposed by the surrounding ice. Their enzymes need to be able to function at high salt concentrations and very low temperatures, with a reduced metabolic rate and little or no respiration or circulation of body fluids. In the frozen state energy production often switches over to a phosphagen-hydrolysis system (using phosphoarginine or phosphocreatine), followed by anaerobic glycolysis, so that lactate and alanine accumulate as end-products.

Freeze tolerance is quite common in many groups of invertebrates including some insects, bivalves, gastropods, annelids, and

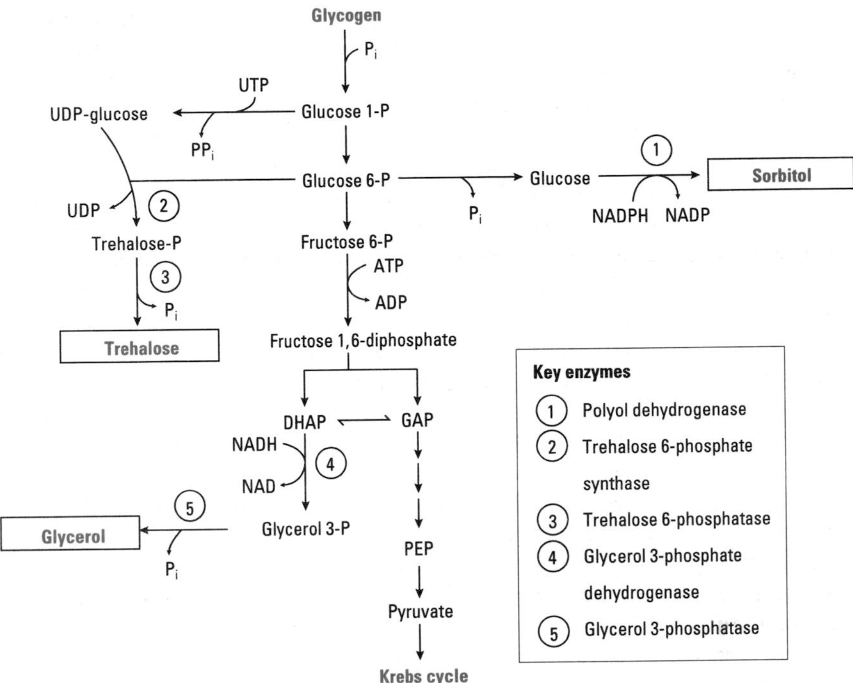

Fig. 8.10 Synthesis of the major cryoprotectants in insects.

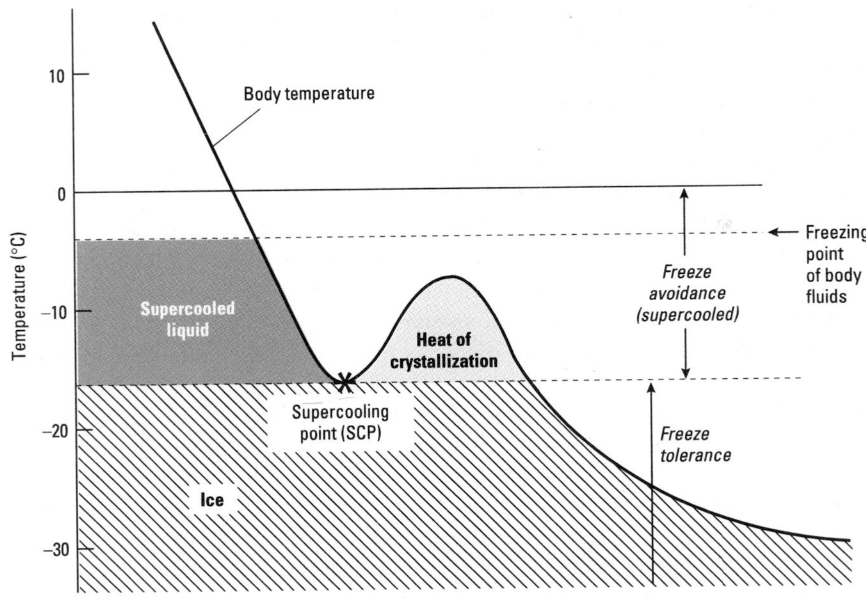

Fig. 8.11 Supercooling in aqueous fluids: freeze-avoiding animals may spend long periods in the supercooled state.

nematodes. Freezing survival to −25 or −35°C for many weeks is common, with a number of Arctic species having lower lethal temperatures at −55 to −70°C. This strategy is rare in the vertebrates, where it apparently only occurs in a few specialized amphibians and reptiles that overwinter in cooler terrestrial habitats (see Chapter 16).

Freeze-intolerant animals

Animals whose strategy is freeze avoidance are decidedly puzzling. They live with their body fluids well below 0°C but with no ice formation. For example, many Antarctic mites can survive unfrozen at −10 to −20°C, and some species may remain active at these temperatures. This situation arises for two main reasons: because of the peculiar phenomenon of supercooling, and because of the additional presence of specific antifreezes.

Any liquid can undergo **supercooling**, i.e. being cooled below its freezing point without actually solidifying, but water is particularly good at supercooling (Fig. 8.11). The actual freezing process depends on three main variables: temperature, time, and the presence of nuclei on which crystals of ice can begin to form. If foreign nucleating agents, such as dust particles, are carefully excluded, pure water can readily be supercooled to −20 or even −40°C without any ice appearing. If dissolved substances are present in the water we know

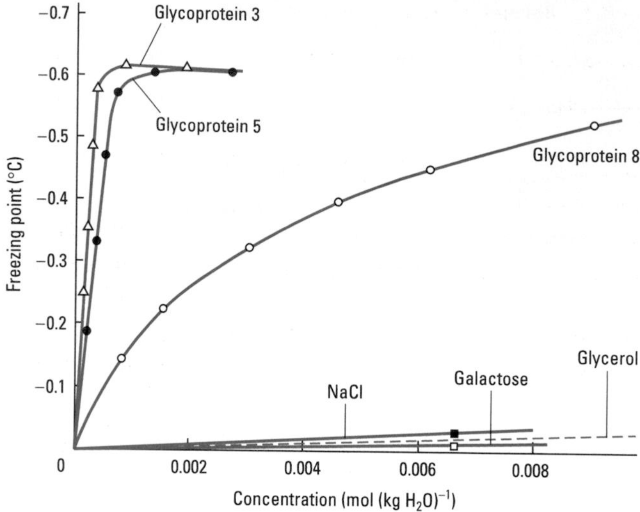

Fig. 8.12 Freezing points of aqueous solutions of important sugars, salts, and glycoproteins used as antifreezes, the last of these being much more effective mole for mole and nonlinear (suggesting a noncolligative action). (Data from de Vries 1980, 1988.)

Table 8.4 Freezing points and melting points of body and blood fluids from some cold-hardy animals showing thermal hysteresis.

Genus	Freezing point (blood) (°C)	Melting point (blood) (°C)	Supercooling temperature (°C)
Spiders			
Philodromus	−5.2	−2.7	−26.2
Clubiona	−4.8	−2.9	−15.4
Insects			
Meracantha	−5.0	−1.3	−10.3
Dendroides	−7.0	−3.4	−9.5
Fish			
Gadus (cod)	−1.1	−0.7	
Rhigophalia (eel pout)	−2.0	−0.9	
Myxocephalus (sculpin)	−2.0	−1.1	
Eleginus	−2.2	−1.2	
Pagothenia	−2.7	−1.1	

that the freezing point is always lowered anyway. But the presence of dissolved solutes also generally leads to the supercooling point (SCP) being lowered, so that body fluids have an inherently greater ability to supercool than pure water.

Animals that routinely encounter low temperatures (but are not continuously exposed to them) commonly enhance their own supercooling abilities by voiding their guts (where food particles could act as ice nucleators) and by adding extra solutes—known as **antifreeze** compounds—to their blood. Any solute, such as a sugar or salt, will have a limited antifreeze effect in accordance with its colligative properties (Fig. 8.12). But certain kinds of low molecular weight osmolytes are particularly effective antifreezes: glycerol and its relatives are, once again, the most famous examples in small terrestrial animals such as insects and arachnids. Glycerol occurs in high concentrations in many cold-hardy invertebrates that do not actually freeze, some of which can survive winter temperatures of −60°C. But these polyols still only have a simple colligative effect in lowering the freezing point. As we saw above they occur in some freeze-tolerant species too, probably to prevent the many small ice crystals recrystallizing into more stable larger crystals over time, which would be inherently dangerous. Thus animals with this strategy tend to show accumulation of polyols or sugars in the fall, lowering the supercooling point; additional antifreezes are then synthesized at the onset of winter to stabilize the supercooled state.

In polar marine fish there is a more severe problem to be overcome, because for them the threat of ice formation may be continuous. The animal's fluids are more dilute than the surrounding medium and thus can be below their own natural freezing point of about −0.7°C. They are in part protected by higher plasma NaCl levels than are normal for teleosts, but these animals also need compounds that will actually prevent existing ice crystals from growing. At least 11 different families of teleost fish have a range of **antifreeze peptide** (AFP) or **antifreeze glycopeptide** (AFGP) molecules that can do exactly this (see Fig. 8.12). They bring about a protective phenomenon called "thermal hysteresis", such that the freezing point of the solution is somewhat lower than its melting point (Table 8.4); hence they are also sometimes called "thermal hysteresis proteins", or THPs. On a molar basis, these antifreezes are far more effective than expected, reaching peak effectiveness at only 4% by mass, and thus being about 200 times as effective as the equivalent concentrations of NaCl. Again, it has recently been shown that they may also occur in freeze-tolerant species, and in both types of animal they confer protection against sudden unexpected freezing. Since their main effect is in marine fish, more details are given in Chapter 11. The AFGP and AFP molecules act by linking to the edges of ice crystal lattices and preventing new water molecules adding on there. They probably attach by their hydroxyl groups, and once in place they break up the ice front into small highly curved minifronts (Fig. 8.13), increasing the surface area in relation to the volume. Water molecules cannot add on to these curved ice fronts unless the liquid temperature is much lower; in other words, the freezing point of the solution is decreased.

A summary of the characteristics of cold survival shown in the two kinds of animals is given in Table 8.5; there are obvious overlaps, but the essential difference is the synthesis of ice-nucleating agents in freeze-tolerant species which override the supercooling effects of polyols. Freeze tolerance and freeze intolerance have different advantages and disadvantages. Freeze-tolerant animals have a lower lethal temperature for particularly high levels of blood osmolality. Freeze tolerance could be regarded as the "cheaper" strategy, as with so many kinds of environmental tolerance; it insures a sharp decline in metabolic rate in cold spells, and avoids the need to synthesize large quantities of polyols or other antifreeze molecules. It is perhaps most useful in areas where there are consistently severe winter climates. However, freeze intolerance (freeze avoidance, with supercooling) may be the commonest overall strategy in animals, and may in particular be the best strategy in environments that are thermally quite variable. Here switching rapidly from inactive to active states during winter thaws or summer frosts is advantageous. Freeze intolerance may also be linked with lower rates of cuticular permeability, and hence reduced problems with water loss. Some animals, notably larval beetles, can actually switch from freeze tolerance to freeze intolerance, giving yet greater flexibility.

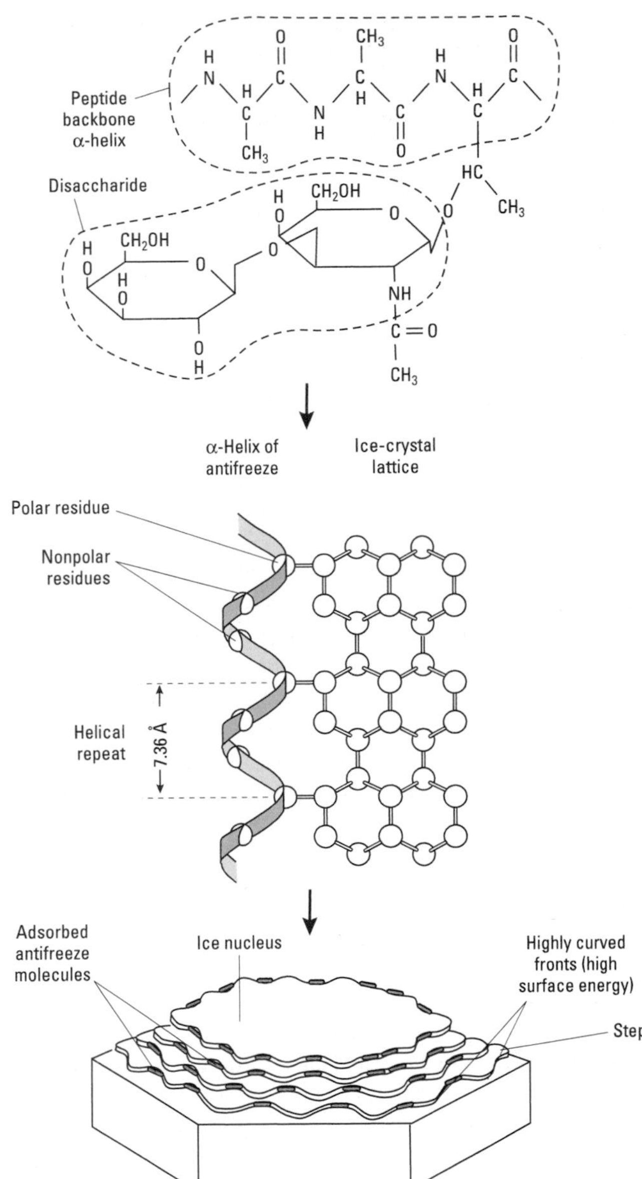

Fig. 8.13 The repeating structure of glycopeptide antifreezes: a model of how this may allow adsorption to ice crystals, preventing growth at the sites of attachment and forcing intermediate sites to grow out into a series of curved ice fronts of high surface energy, which require lower temperatures for further growth.

However, we should also remember that at an individual level many animals (particularly overwintering insects) are not really either tolerant or intolerant because they die from chill injury well above freezing point.

8.3.2 High temperature effects on cells and organisms

In general terms, enzymes and membranes are active and functional over a broader range of temperature than is the case for individual cell organelles, which are active over a broader thermal range than whole cells, and organisms in their turn have a narrower thermal tolerance zone than do cells. There is an upper thermal limit for any

Table 8.5 Freeze tolerance and freeze avoidance.

Characteristic	Freeze avoiders	Freeze tolerators
Ice formation	Lethal	Extracellular ice tolerated
LCT	–5 to –20°C (rarely –60°C)	–20 to –70°C
Supercooling capacity	High	Low
Supercooling point	Close to LCT	Well above LCT
Ice-nucleating agents	Absent or masked	Present and active (proteins)
Antifreezes	Polyols ± peptides in fish	–
Cryoprotectants	–	Polyols, ± trehalose, ± proline
Occurrence	Many invertebrates	Common in invertebrates
	Some vertebrates, notably polar fish	A few frogs

LCT, lower critical temperature.

given cell, beyond which cell death occurs; and a lower maximum for an animal as a whole. Clearly lethal limits are only partly determined by temperature effects on protein denaturation and coagulation so far described, though these are almost certainly part of the story. Protein (enzyme) inactivation is a less drastic but potentially equally lethal effect, when inactivation rates exceed the rate at which new proteins can be formed via the ribosomes. However, both these effects really only apply in those cells whose upper lethal temperature is in the range 30–55°C, and cannot account for heat death at 6°C in some polar fish. Some other mechanisms must also be involved, and three are particularly important:

1 There may be a disruption of the balance between reactions of differing thermal sensitivity, so that metabolic pathways are subverted. Several hundred enzymes contribute to cellular intermediary metabolism, and each has its own particular thermal properties. If one part of a pathway is greatly slowed the pathway it contributes to must also slow down for lack of appropriate substrates, affecting the balance in the system. Similarly, where particular steps are accelerated the immediate end-product of that step will build up and may itself affect the operation of steps elsewhere in the system, again distorting the balance between pathways. Thus small changes of temperature can have drastic effects, and the cells within an animal may run out of energy supplies.

2 There may be effects on membrane structures (lipid–lipid and lipid–protein interactions), so that the various transport systems into and between cells (discussed in Chapter 4) become out of balance with each other (or even entirely inoperative), again upsetting the intracellular conditions on which enzyme function depends. In crayfish the first sign of heat death is a breakdown of the normal permeability of the gills, so that hemolymph Na^+ concentration decreases as gradients run down. Abnormal neural activity soon occurs and the animal becomes severely uncoordinated just before death. Many other animals also show extracellular decrease of Na^+ levels and increments of K^+.

3 Animals may also be forced into death by the upper thermal limit for particular tissues, with nerves and heart muscle often showing a lower thermal tolerance than skeletal muscles, for example, so that

Table 8.6 Upper critical temperature (UCT) for a variety of animals from different habitats.

Group	Example (habitat)	UCT (°C)
Prokaryotes	Bacteria (aquatic)	70–75
	Bacteria (thermophilic)	90–91
	Cyanobacteria	75
Molluscs	*Modiolus* (SW bivalve)	38
	Nassa (SW gastropod)	42
	Clavarizona (SW gastropod)	43
Annelids	*Lumbricus* (land earthworm)	29
Echinoderms	*Asterias* (SW starfish)	32
	Ophioderma (SW brittlestar)	37
Crustaceans	*Palaeomonetes* (SW/littoral prawn)	34
	Porcellio (SW crab)	39–41
	Uca (littoral/land crab)	39–45
	Armadillidium (land woodlouse)	41–42
Insects	*Lepisma* (land springtail)	36
	Thermobia (land firebrat)	40+
	Sphingonotus (land moth)	41
	Bembex (land sandwasp)	42
	Onymacris (desert beetle)	49–51
	Dasymutilla (land sandwasp)	52
	Ocymyrmex (desert ant)	51.5
	Melophorus (desert ant)	54
Arachnids	*Buthotus* (land scorpion)	45
	Leiurus (land scorpion)	47
Vertebrates		
Fish	*Pagothenia* (polar SW)	6–10
	Fundulus (cold SW)	35
Amphibians	Salamanders (FW/land)	29–35
	Anurans (FW/land)	36–41
Reptiles	Alligators (land/FW)	38
	Turtles (SW/land)	41
	Lizards (land/desert)	40–47
	Snakes (land)	40–42
Birds	Passerines	46–47
	Nonpasserines	44–46
Mammals	Monotremes	37
	Marsupials	40–41
	Placentals	42–44

FW, fresh water; SW, sea water.

uncoordinated movement may persist for a while after the heart and central nervous system have succumbed.

For some or all of these reasons, there will be a distinct thermal maximum for a particular organism. This critical temperature may be 70–90°C for bacteria and other prokaryotes, but is rarely above 50°C for animals (where to avoid complications due to varying tissue sensitivities it is often measured as the temperature eliciting loss of neuromotor control). In fact this **upper critical temperature** (UCT) in animals commonly lies between 30 and 45°C for all but deep-sea animals and polar aquatic species (Table 8.6). The UCT is reasonably consistent within taxonomic groups, but with variation superimposed through ontogeny (caterpillars are more sensitive than butterflies, and tadpoles more sensitive than adult frogs and toads). Variation also occurs with habitat; for example, temperate lizards have lower UCTs than tropical and desert lizards, and on a smaller scale within a particular habitat lizard species from open grassy areas are more tolerant than nearby forest dwellers.

8.3.3 Overview

Biochemical and physiological effects clearly underlie the thermal tolerances, preferences, responses, and susceptibilities of animals. The thermal phenotype of a metazoan comprises many hundreds of traits; it is important not merely to describe these traits but to understand their ensemble activities for the organism as a whole. Some traits will covary in evolutionary terms, while others will act almost independently but may on their own exert major effects on evolutionary fitness. Unambiguous attribution of a difference in thermal phenotype to any particular genetic change (or group of changes) is always going to be problematic.

It is also worth noting here that biological processes do not commonly show anything like complete **compensation** for the effects of temperature—they seem to be adapted to keep going, but not to keep going at the same rate irrespective of thermal regime. Thus, for example, there appears to be little or no temperature compensation in the rate of development of embryos for a wide range of marine invertebrates and fish. Figure 8.14 shows the relationship between temperature and development time for fish and for sea urchins with planktonic larvae, and it is striking that polar, temperate, and tropical species all fit to the same relationship. The same lack of compensation is true for the resting and maximum metabolic rates in fish discussed in Chapter 7, and it is likely a widespread phenomenon (perhaps not surprising, since maintaining high rates at low temperatures would be very costly). This underlines again the caution needed in evolutionary physiology—adaptation and acclimation do not produce what we may see as theoretically ideal solutions, merely solutions that are "good enough" and reasonably cheap.

8.4 Terminology and strategies in thermal biology

Different animals have very different ranges of body temperatures in their natural surroundings, and differing degrees of thermal constancy. Unfortunately, though, this is an area of comparative biology where the terminology used has often tended to obscure rather than clarify the underlying patterns and processes. This section therefore needs to begin with a review of the various categories that have been used, and of their meanings.

8.4.1 Patterns of body temperature

Initially, the terms **warm blooded** and **cold blooded** were coined to describe animal temperature patterns, and they are still used today in general conversation. The terms stress the perceived difference between animals that are always warm to the touch—the birds and mammals—and those that commonly feel rather cool, a grouping that includes all other animals but which has particularly been applied to other vertebrates (the fish, amphibians, and reptiles). However, these terms never enjoyed much scientific favor, since it is evident that many supposedly "cold-blooded" animals can become distinctly warm, for example after a period of basking; reptiles and many insects regularly achieve an elevated T_b of the same order as that found in birds and mammals. Moreover, a hibernating mammal may have a T_b as low as 5°C. Such expressions as warm and cold blooded are therefore clearly best avoided.

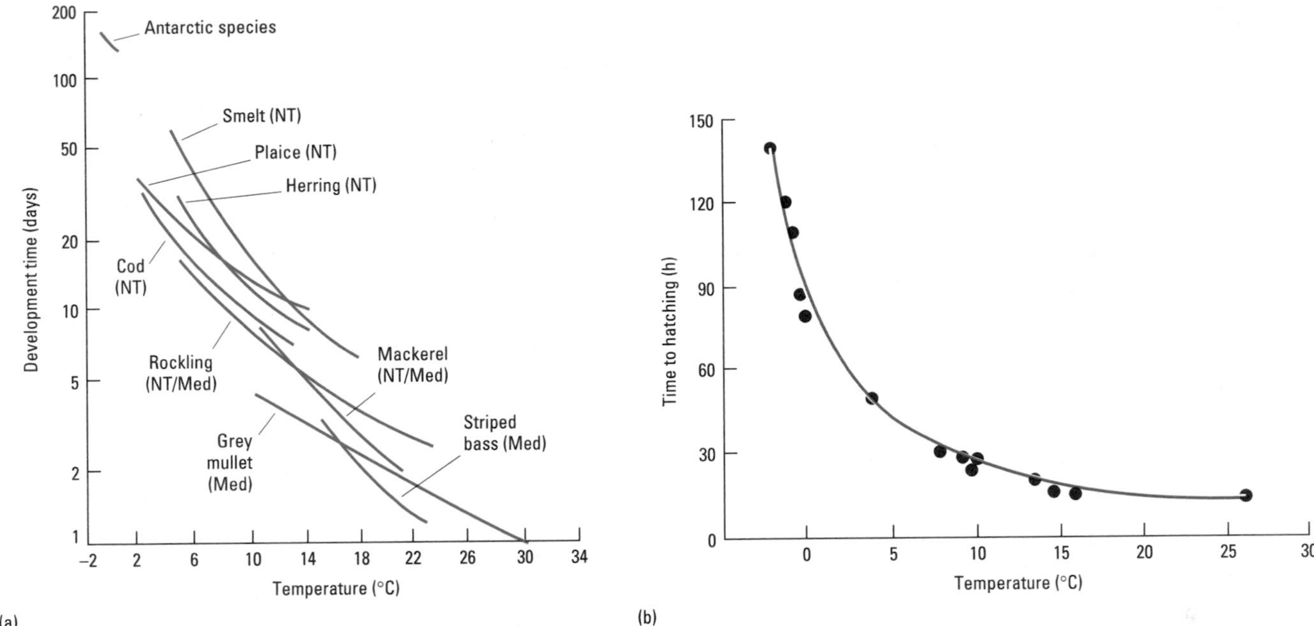

Fig. 8.14 Temperature and development time for (a) various marine fish and (b) sea urchin planktonic larvae. Note in (a) the similar form of relationship for very different latitudinal populations. Med, Mediterranean; NT, north temperate. (a, From Johnston 1990; b, from Bosch *et al.* 1987.)

The more favored terms that were in general use for much of the twentieth century were **poikilothermic** and **homeothermic**, referring to the constancy of the body temperature rather than its actual setting. Poikilotherms (from the Greek *poikilos* meaning changeable) have a variable T_b, whereas in homeotherms it is rather constant. In principle this makes perfect sense, except that the terms have been regularly misused and misunderstood. In particular, "homeothermic" has often been seen as a synonym of "warm blooded" and applied in an exclusive sense to the birds and mammals, where T_b is nearly always around 37–40°C. All other animals are thus seen as poikilotherms. Yet in this usage the term is clearly misleading; a fish or a worm living in cold, deep water at 5°C is not a poikilotherm at all, but is every bit as homeothermic as a hamster or a human. An insect, or treefrog, in the litter layer of a tropical forest may also be close to homeothermy. In other words, homeothermy may be achieved physiologically in some specialist animals (the sense in which the term was primarily intended to be used), but it may also be brought about quite passively by environmental constancy. To categorize such different conditions together seems unhelpful. The concept of poikilothermy can be similarly muddling —a hibernating mammal or a small tropical hummingbird with nightly torpor may have a body temperature almost as variable as that of a reptile living in similar conditions.

A more recent and more sensible terminology refers to **endothermic** and **ectothermic** animals. These terms emphasize the heat sources used, rather than the settings or constancies of achieved T_b, i.e. they refer to mechanisms in thermal biology rather than to conditions. As such, they are unambiguous:

- **Ectotherms** have a body temperature principally dependent on external heat sources, almost always ultimately the sun, either directly ("heliothermy") or from a heated substrate ("thigmothermy"). Geothermal heat may also be available in some cases.
- **Endotherms** have a body temperature principally dependent on their internally generated metabolic heat.

Again the distinction is largely between birds and mammals, being endothermic, and the rest of the animal kingdom, using ectothermic strategies. But on a broader view the distinction is actually independent of that between poikilotherms and homeotherms (Fig. 8.15). This arises because our new terminology is still not without its own problems. The main complication is that in practice *all* animals generate some internal heat, as a natural consequence of being alive and having metabolic conversions going on in their constituent cells (especially their muscles) that are not perfectly efficient. And nearly *all* animals also use external heat sources; most terrestrial birds or mammals will choose to bask, for example, maintaining their body temperatures as cheaply as possible, when solar radiation is available. However, the two terms remain readily distinct, for the principal *mechanisms* of heat gain do not merge as a continuum. While all animals may use external heat sources and will generate some internal heat, a few groups of animals are quite distinctly endothermic in that they have an additional special ability to raise their metabolic rate markedly (especially in the liver and viscera) in excess of all their normal needs, specifically benefitting from the resultant heat generation. In fact a useful complementary terminology exists, with the endotherms referred to as **tachymetabolic** (having a fast metabolism) and the ectotherms being **bradymetabolic** (slow metabolism). For a given body size, an endotherm will normally have at least five times the metabolic rate of an ectotherm.

Although ectothermy and endothermy are thus distinct and nonoverlapping, there are some intermediate conditions, which are sometimes collectively called **heterothermy**, although they occur for a variety of different reasons and would be better treated separately.

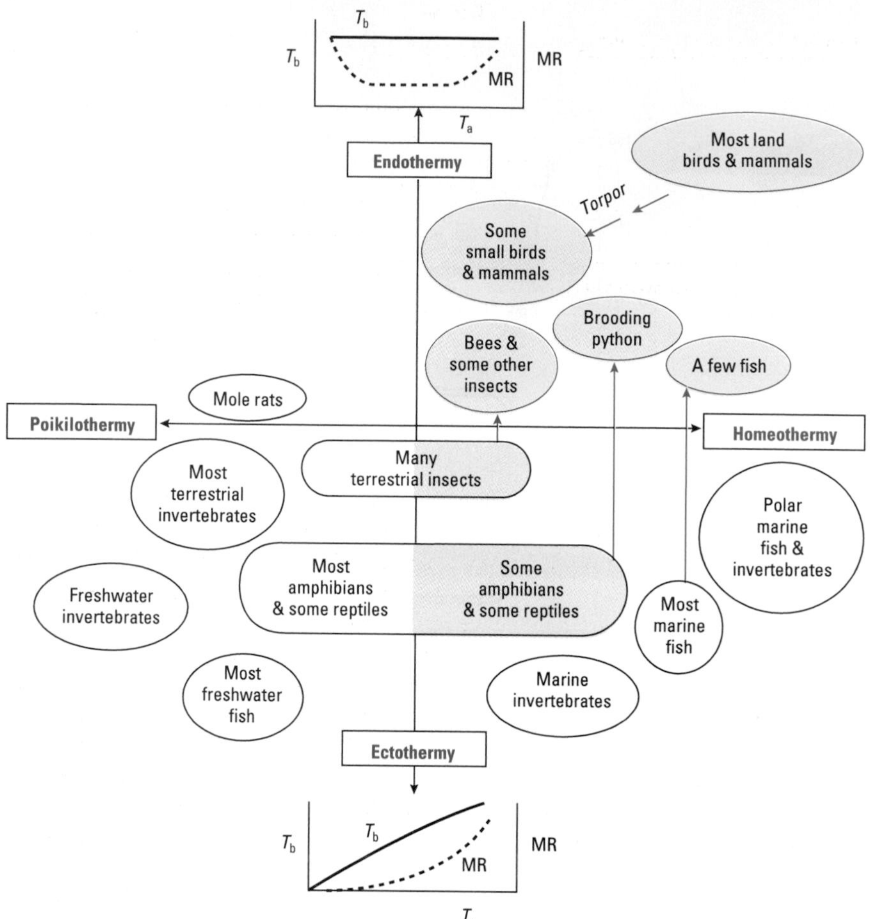

Fig. 8.15 Terminology in thermal biology–how the terms used are distinctive or overlapping. Green shading shows thermoregulators; white shows thermoconformers; arrows indicate temporal heterothermy. MR, metabolic rate; T_a, ambient temperature; T_b, body temperature.

1 Partial endothermy occurs in some of the very smallest birds and animals (hummingbirds, tiny rodents, etc.), where the normally high metabolic rate is turned down seasonally or even each night, to reduce energy expenditure in conditions of cold or of low food and water supply. A temporary torpor results, with the body temperature falling substantially (see section 8.9 and Chapter 16). At the extreme, the curious burrowing naked mole rats of southern Africa, and also the golden moles living in similar habitats, appear to be almost entirely "turned off", and operate more like nonmammalian ectotherms, perhaps because they would lose too much heat by conduction to the burrow walls if they attempted continuously strongly elevated and regulated body temperatures. (Here is a further reason why the term homeotherm should be avoided; some traditionally homeothermic birds and mammals just aren't!)

2 Facultative endothermy is increasingly recognized in a variety of animals that, while normally ectothermic, can "turn on" an endothermic heat generation system in some parts of their body when they need to become active at low ambient temperatures. Insects such as bumble-bees are a classic case, being essentially ectotherms but with considerable scope for raised metabolic rates leading to temporary endothermy. Most bees, and quite a wide range of other insects, have this ability, and more detail of the mechanisms involved will be found in Chapter 15.

Note that strategies 1 and 2 are sometimes collectively called **temporal endothermy** or **temporal heterothermy**, as they involve periods of endothermy alternating with a slow metabolic state closer to ectothermy. Figure 8.16 compares the relative body size ranges in which these strategies occur in vertebrates and in insects.

3 Regional endothermy occurs in a variety of fish and some reptiles. In these cases, localized areas of the musculature routinely operate at much higher temperatures than the rest of the body, allowing faster or more sustained locomotory activity, or maintained sensory abilities, in cold environments. Some of these cases, involving specialized muscle cells that have lost their contractility, are described in Chapter 11.

Here we should also mention **regional heterothermy**, having somewhat similar effects but for very different reasons. It occurs in many birds and mammals in cold environments, where it is common for the extremities to be much cooler. This is usually the result of cardiovascular adjustments serving to reduce heat loss (see section 8.8.2).

4 Inertial homeothermy/endothermy should also be dealt with here. Animals that have no specific strategies for raised metabolic rate and are essentially ectothermic and bradymetabolic can still end up with a rather high and rather constant T_b if they are particularly large bodied. They have a very high thermal inertia because of scaling effects, the animal's surface area being relatively too small to dissipate even the slowly generated internal metabolic heat from their comparatively enormous volumes, so that internally generated heat is the major contributor to their body temperature as with an

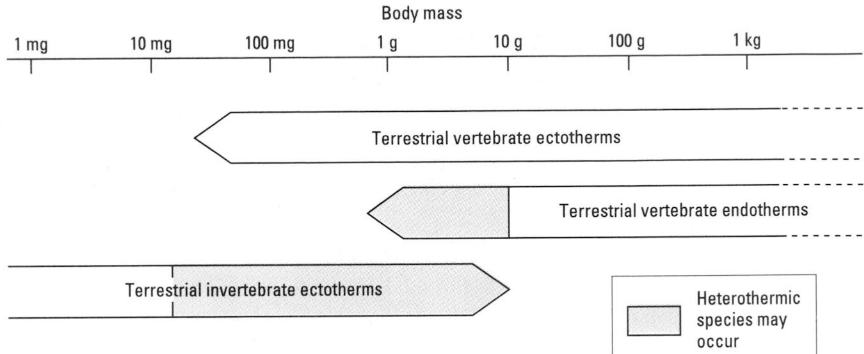

Fig. 8.16 Relative sizes of ectotherms and endotherms, showing the size ranges for the occurrence of temporal heterothermy.

endotherm. Inertial homeothermy is the main reason why the largest dinosaurs, though reptilian and perhaps lacking any specific endothermic capacity or tachymetabolism, must have been effectively "warm blooded" all the time.

Despite these four kinds of equivocal case, there is no doubt that the ectotherm/endotherm terminology is the most satisfactory of those available to us. In this book we will therefore use only the terms ectotherm and endotherm to describe basic thermal adaptive strategy.

These terms separate essentially distinct thermal strategies, while all other systems, by concentrating on the result rather than the mechanism, involve the inappropriate bracketing together of quite different kinds of thermal strategy (see Fig. 8.15). This is because the *result* of all thermal strategies is rather often a relatively constant body temperature, achieved by whatever means are available. For most aquatic animals T_b is close to that of the medium, whilst for aquatic mammals and for most active land animals T_b during activity is commonly in the range 25–40°C (Table 8.7; cf. Fig. 8.1).

Table 8.7 Body temperatures (T_b) during normal activity in ectothermic and endothermic animals.

	Ectotherms		Endotherms	
Habit	Group/species	T_b	Group/species	T_b
Aquatic	Marine deep-water fish	4–6	Fur seal	38
	Marine surface fish	5–10	Whale	36
	Temperate riverine fish	5–15		
	Reef fish	20–28		
	Tropical turtles	29–35		
	Warm temperate turtles	20–25		
Amphibious	Alligator in air	32–35		
	Alligator in water	20–35		
	Bullfrog tadpoles	22–30		
	Bullfrog adults in air	22–28		
Terrestrial	Blowfly	22–29	Man	37
	Flour beetle	25–30	Small rodents	35–37
	Bumble-bee (flying)	35–39	Bats	35–39
	Carpenter bee (flying)	38–42	Echidna	31
	Housefly (flying)	30–33	Armadillo	34–36
	Housefly larva	30–37	Chicken	40
	Chicken louse	38–42	Dove	39–42
	Temperate caterpillar	20–28	Owl	38
	Semitropical caterpillar	25–35	Zebra finch	41–42
	Tropical lizard	28–36		
	Desert iguana	36–41		

8.4.2 Patterns of temperature maintenance

So far we have dealt only with the terminology of overall thermal strategies. There is a further complication pertaining to the terminology of body temperature. It is very often implied that some animals have a special ability as **thermoregulators**, while most do not bother and are **thermoconformers**, at the mercy of the thermal variation of the environment. This is fundamentally untrue, for most animals do indeed thermoregulate. Probably the only "conformers" are those animals that live in environments, such as the deep sea, that have no thermal variability, where body temperature is virtually identical to the invariant ambient temperature at all times (it will in practice be detectably elevated by internal metabolic processes, but normally by less than 1°C). The misconception reflects once again the longstanding overemphasis on physiological mechanisms as the keystone of adaptation. It results from the view that only **physiological thermoregulation** is "real" thermoregulation, whereas in practice **behavioral thermoregulation** is at least as crucial to many animals, and can be equally effective.

The muddled thinking about thermoregulation probably results from the association of the term with the "homeothermic" endotherms. Physiological regulation is certainly used by most endotherms to help achieve their state of T_b constancy. They may have physiological mechanisms to insure a high heat production or heat gain, and also mechanisms to reduce or to hasten heat loss. But they also use a great many behavioral tricks, such as basking, huddling, and burrowing. Nor is physiological regulation their exclusive prerogative; for whilst ectotherms may have a very obvious repertoire of behavioral thermoregulatory options, many of them also have a suite of physiological strategies, as we will see.

To summarize:

1 Ectotherms can be regulators or conformers, and use both physiological and behavioral mechanisms to regulate.

2 Endotherms are always regulators, but also use both kinds of regulatory mechanism to achieve constant T_b.

8.4.3 Patterns of temperature tolerance

There is a third and final problem with terminology that needs to be sorted out, but which is fortunately relatively straightforward. This concerns the familiar usage of the terms **eurythermal** and **stenothermal**. As with the equivalent terms euryhaline and stenohaline used in the discussion of water balance and osmotic problems (see

Chapter 1 and section 5.1.3), these concern the range of tolerance an animal will show. A eurythermal animal tolerates, and is active with, a rather wide range of T_b. For example, many temperate insects and reptiles will continue feeding and locomoting with body temperatures of perhaps 8–38°C, tolerating a range of 30°C, and similar ranges of tolerance exist in some intertidal rock pool fish. By contrast, a stenothermal animal only operates when its T_b is within a narrow range. Most mammals and birds are stenothermal, but some lizards and even a few insects are too, and many aquatic animals are very strict stenotherms, with some polar fish tolerating only about 6°C in total range (from the freezing point of sea water to roughly 4°C).

Linking this back to our discussion of temperature effects at the biochemical level, the terminology can be applied at that level too. It is generally true that stenotherms have proteins and membranes that are themselves tolerant of only narrow temperature changes, i.e. are stenothermal, while eurytherms have biochemical components with wider (eurythermal) tolerances; the differences are in large part genetically fixed.

8.5 Thermal environments and thermal exchanges

The thermal exchange between any two objects is proportional to the difference in their temperatures, and clearly the same will apply to exchange between an animal and the components of its environment. Temperature control in animals is largely concerned with manipulating the various avenues of heat exchange with that environment in the animal's favor, depending on its immediate needs and circumstances. Since the thermal environment of many animals, especially those that are terrestrial, is very complex, these patterns of exchange control can be remarkably elaborate.

Before considering the details of these control systems, we need to deal with the basic principles of heat exchange, which as in any physical system will involve **conduction**, **convection**, and **radiation**. In addition, since animals are made up of watery tissues those that live on land may sometimes have the option of heat control by **evaporation** or **condensation**.

8.5.1 Conduction

Conduction is the direct flow of heat (kinetic energy) between two materials in contact, whether those materials are solid or fluid. Heat flows from the warmer to the cooler material, by a direct transfer of the kinetic energy as molecules collide. The rate of heat transfer therefore depends upon the temperature difference (or temperature gradient), the area of contact, and the conductive properties of the two materials (Box 8.1). Clearly an animal in direct contact with its environment has control over some of these parameters. Any motile animal can exercise choice over the nature of the substrate on which it rests, and thus choose an appropriate temperature difference. It can also vary the area of contact rather easily by changing its posture, from minimal contact at the tips of appendages to a full exposure of up to half of all its surface area to the substrate.

Since these parts of the heat conduction relationship are variable in the short term by behavioral means, thermal conductivity is often the most important long-term adaptive element in determining conductive loss and gain of heat. A variety of ways of expressing this and related concepts are used (Box 8.1), of which the **insulation** value is perhaps the most familiar. Insulation may be provided by internal air sacs or fat layers, or externally by cuticular bristles of chitin or by various hairs and other structures made of keratin (Table 8.8). Internal insulators are much less effective: a layer of blubber 60–70 cm thick has about the same insulating properties as 2 cm of effective mammalian fur. In practice, most external biological insulating materials, including the familiar furs and feathers, have an insulation value fairly similar to that of still air (Table 8.8), which is not surprising as they nearly all rely on trapping a layer of still air close to the animal's surface. Heat transfer through bird plumage is about 95% by convection through the trapped air and conduction along the feather structure combined, with only 5% of total heat flow via radiation. Of course the insulation provided depends rather directly on the thickness of this trapped layer (Fig. 8.17), so that a large mammal able to carry a thick fur layer (e.g. about 6 cm thick in a polar bear) has much better insulation than a tiny mouse with only a 0.2 cm fur layer, on which any thicker fur would severely restrict movement. Nevertheless, per unit thickness the mouse fur turns out to be a somewhat better insulator, as might be expected.

Box 8.1 Conduction, conductivity, and insulation

Conduction involves direct transfer of heat between two solid materials, and rate of conductive heat transfer (Q_c) in joules per second (or watts) can be calculated by a modified version of Fick's law as follows:

$$Q_c = -kA(T_2 - T_1)/d$$

where A is the area of contact and d the distance between the two measured temperatures T_1 and T_2. Here the constant k is the **thermal conductivity**, with units of J s^{-1} °C^{-1} cm^{-1}.

Since A and d are often difficult to measure, the equation is often rewritten as:

$$Q_c = C\Delta T$$

where ΔT is the thermal gradient; now C is the **thermal conductance** (or conductive heat transfer coefficient) with units of J s^{-1} °C^{-1}.

Conductivity (k) is a fixed property of materials, and is directly related to **insulation**, which is the reciprocal of the total heat flux per unit area per unit of temperature difference, and therefore with units of °C cm^{-2} s J^{-1}. In practice it is usually measured per meter squared, and a common unit given is the "clo" (clothing unit), where 1 clo = 0.16 °C m^2 W^{-1}.

For biological use, an insulative layer is often described by a **resistance** value (r, measured in s cm^{-1}), where the resistance takes into account both the thermal conductivity and the thickness (l) of the layer:

$$r = \rho C_p l/k$$

Thus there is a reciprocal relation between resistance and thermal conductivity.

Here ρ C_p is the specific heat capacity of the medium (air or water).

Table 8.8 Insulation values of biological and other materials.

Material	Insulation (°C m^2 W^{-1})
Silver	0.015
Steel	0.14
Ice	2.9
Water	11
Human tissue	14
Dry soil	20
Natural rubber	38
Fat	38
Cattle fur	50
Pigeon feathers (flat)	99
Sheep wool	102
Goose-down feathers	122
Husky dog fur	157
Lynx fur	170
Still air	270

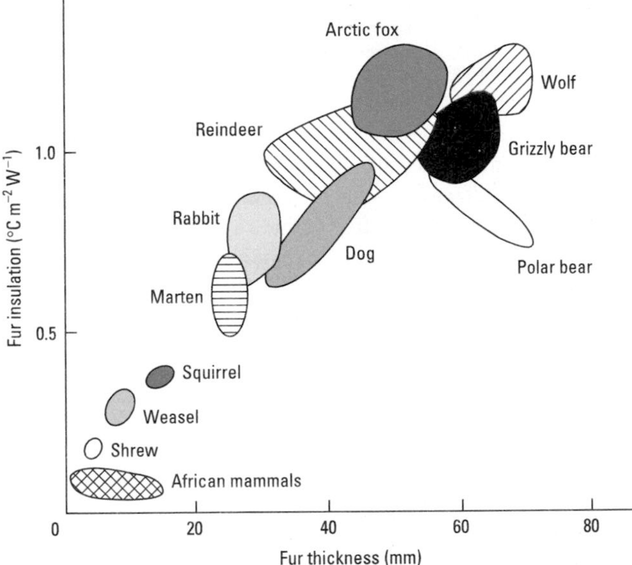

Fig. 8.17 Fur thickness and fur insulation values; small mammals of necessity have short fur, of relatively poor insulation value. Note that these examples are cool temperate and polar species; the fur in African mammals rarely exceeds 15–20 mm in length, and this decreases rather than increases with body size, being only 0.5 mm in the eland (the largest antelope) (see Chapter 16). (Data from Scholander *et al.* 1950; Hofmeyr & Louw 1987.)

Mammals are also adept at *varying* fur thickness and therefore conductivity over the body, with denser fur layers commonly around the trunk and rather scanty fur on distal parts of the limbs (this may sound counteradaptive at first, allowing huge heat loss from large surface areas, but linked to blood and heat distribution patterns, discussed in section 8.7, it becomes more intelligible). Both birds and mammals naturally also make use of extrinsic insulators, in their choice of nesting materials.

For large terrestrial mammals there may be the opposite problem of eliminating excess metabolic heat, especially after periods of activity in warmer climates. This problem is compounded because total heat production increases with $M_b^{0.75}$ (see Chapters 3 and 6), whilst surface area for heat dissipation only increases with $M_b^{0.67}$. In the tropics, very large mammals such as the elephant and rhinoceros are therefore largely hairless, to aid heat loss and keep them in thermal balance.

8.5.2 Convection

Convection is the flow of heat between two bodies by the mass movement of an intervening fluid, whether gas or liquid. Convection currents may be "free" or "natural", caused by inherent temperature differences (since warm air or warm water will rise), or may be imposed by outside mechanical factors such as wind, water currents, or movements by the animal itself ("forced convection").

Convection is naturally much faster than conduction. For animals, convective heat movements often occur continuously within the body, when hot blood is moved from the core to the periphery. In vertebrates with complex circulations, any disorder that reduces this convective flow may cause an explosive rise in body temperature, since purely conductive heat transfer to the periphery is much too slow to cope.

However, the convective heat exchange of animals with their environment is the key issue here, and is essentially the process of conductive heat exchange through the boundary layer of fluid that separates that animal from (and differs in temperature from) the free mass of fluid constituting the environment, whether air or water. This boundary layer is extremely complex, modified on a very small spatial and temporal scale by water or air flow, and assessment of convective temperature exchange across animal skins is notoriously difficult. In both air and water, the temperature difference between the animal and its environment is an important determinant of heat exchange, as are the animal's dimensions, the curvature of its surfaces, and the fluid's thermal expansion coefficient, density, and viscosity. But even when all these are known, the exact pattern of convective exchange across and just above the skin may defy accurate description or prediction. It is therefore common, and not too inaccurate, to treat conduction and boundary-layer convection together when dealing with animal heat balance.

In water, convective heat transfer is very rapid and it is very costly for aquatic animals to maintain temperatures above that of their surroundings; few do so. In air, convection is slower and large temperature gradients between the body and its surroundings become feasible, again relying on the effects of insulating layers that effectively expand the boundary layer.

8.5.3 Radiation

Radiation travels with the speed of light and is emitted by all bodies whose temperature is above absolute zero (−273°C). The wavelength of this radiation depends on the surface temperature (T_s) of the emitting body, the predominant wavelength (l_{max}) being inversely proportional to T_s (Wien's law). For a living organism, which in absolute terms is relatively cool, the l_{max} is around 9–10 μm, which is rather long-wave infrared radiation (Fig. 8.18). The total rate at which such radiation is emitted from any object is, according to Stefan's law, proportional to the fourth power of T_s, so it rises very rapidly with small increases in surface temperature. It is also dependent on the radiative surface area and its **emissivity**.

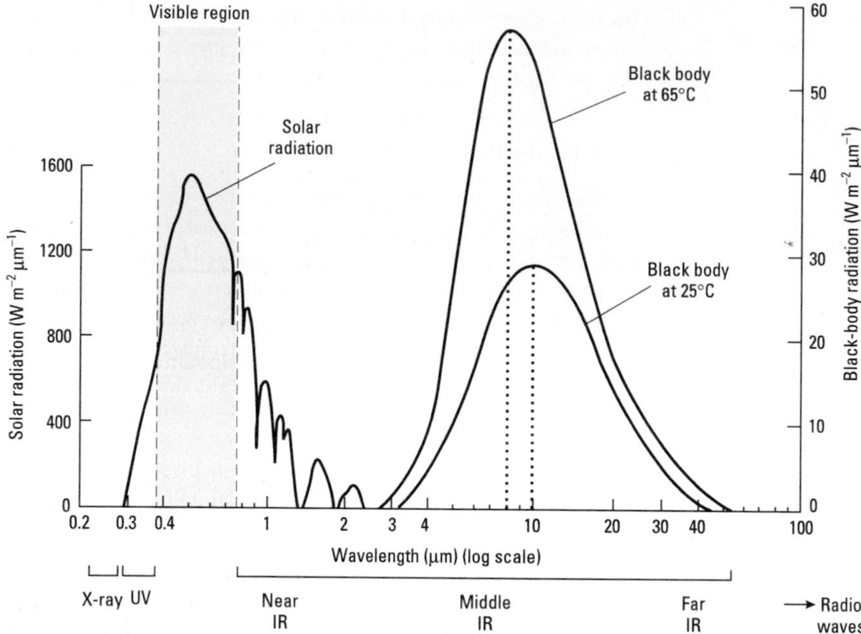

Fig. 8.18 Thermal radiation from bodies of different temperatures: the higher the surface temperature, the shorter the wavelength and the higher the intensity. Note especially the spectral distribution from the sun (6000+°C), from a sun-heated rock or animal surface at 65°C, and from a mammalian body surface at about 25°C. (Dotted lines show peak wavelengths.) (Adapted from Hardy 1949; Bond *et al.* 1967.)

This latter parameter is a dimensionless constant for any material and is related to the familiar concepts of **color** and shininess of the surface. Emissivity for a particular surface actually varies for different wavelengths, which is what normally gives rise to perceived color, a red object having a high reflectivity of (and low emissivity for) incident red light. For most biological materials, the emissivity at biological wavelengths in medium to long infrared radiation is between 0.90 and 0.99, where a perfect "black body" has an emissivity of 1.00. Some external skeletal materials, such as insect cuticle or snail shell, can be white and shiny, though, and are of very much lower emissivity, although in practice this is commonly recorded as a high **reflectance** (Table 8.9). In cuticles high reflectance is often due to tiny air bubbles, while in shells the properties of calcium carbonate are often exploited.

Table 8.9 Emissivities and reflectances of various materials.

Material	Emissivity	Reflectance (%)
Water	0.95	
Cork	0.95	
Glass	0.94	
Wood	0.90	
Ice	0.92	
Steel	0.07	
Aluminum	0.20	
Copper	0.03	
Insect cuticle		
Black matt		2–5
Black shiny		4–23
Pale colors		5–25
White		10–35
Snail shell		
Dark		2–30
White		95
Human skin		
"Black"		5–9
"White"		5–10

However, color as perceived in relation to visible light wavelengths is not necessarily a good guide to emissivity for the important biological infrared radiation; black and white skins in humans, though they appear very different, have very similar properties in relation to emission of thermal radiation, and also differ little in their absorption of long-wave radiation that is re-emitted from the ground and other solids. The difference, as with most differences in skin or fur colors, lies in their absorption of the visible components (almost 50%; see Fig. 8.18) of direct solar radiation. Dark-colored skins and furs do indeed absorb more of this incident light than pale skins and furs. The situation is further complicated by the "penetrance" of the radiation, which varies with the microstructure and hence microoptics of the feathers or hairs, and with their "packing". To give an example, black pigeons absorb a greater radiative heat load than white pigeons in still air, but this difference becomes reversed at high wind speeds because radiation penetrates more deeply into white plumage and therefore the white birds' heat gain is less affected by ruffling of the feathers in windy conditions (Fig. 8.19). Note that confusion can also arise due to structural colors (often seen as iridescence), which can be generated by the scattering of small particles, by multilayered reflectors, or by diffraction gratings, and which do not relate closely to thermal properties.

In general, though, color bears some relation to habits and habitats; animals have pale surfaces where they need to stay cool in hot climates, and dark surfaces if they need to warm up by basking, etc. However, there are exceptions in every environment (black desert goats, white polar bears) which will be dealt with in the relevant chapters of Part 3. For now, we should note that it is always important to avoid a simple-minded view of color in relation to temperature.

It must also be stressed that emissivity, color, or reflectance, or any other similar parameter, affects only the rate of change of temperature and cannot alter the equilibrium temperature that an object will finally reach. For example, a white ball and a black ball will both eventually reach the same temperature if left in the sun,

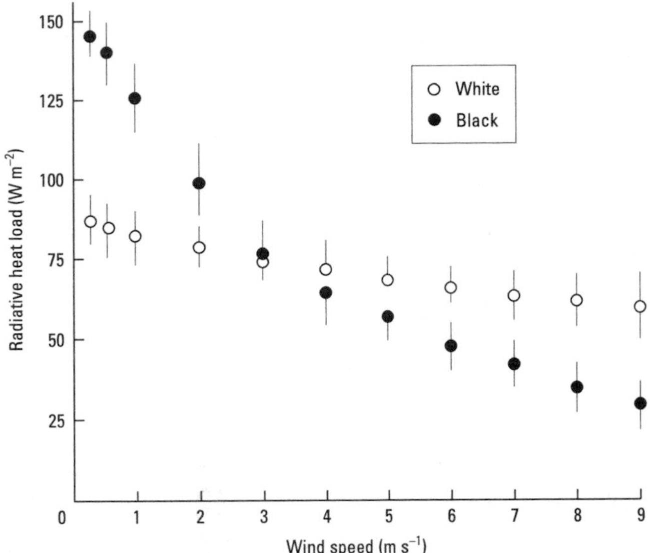

Fig. 8.19 Color effects in the pigeon. Radiant heat reaching the skin under a high insolation regime is greater for black plumage at low wind speeds but greater for white plumage at high wind speeds. (Adapted from Walsberg & King 1978.)

but the highly reflective white ball will take substantially longer to reach this equilibrium.

Objects at temperatures in the biological range can be calculated to emit radiant energy in the range 300–500 W m^{-2}. This is a very low emission compared to hot bodies such as a light bulb or an electric heating element, but represents a substantial loss relative to energy production from metabolism. For a resting 10 kg mammal with about 1 m^2 of radiating surface the emissive loss would be 300–500 W, whilst metabolic production would be only 20 W. In practice, of course, such an animal is also gaining radiative heat all the time from components of its environment. Solar irradiance perpendicular to the Earth's surface may exceed 1000 W m^{-2}, which is much greater than the metabolic heat production of a typical mammal, while surrounding surfaces (soil, rocks, vegetation) of similar temperature (T_{sur}) will also be radiating heat to the animal at roughly the same rate as the animal is emitting it (Fig. 8.20).

8.5.4 Evaporation

Evaporative loss of water is an excellent way of dissipating heat; the latent heat of the vaporization of water is 2500 J g^{-1} at 0°C, and is still 2400 J g^{-1} at 40°C (or about 580 cal g^{-1}). The rate of evaporation depends not only on the surface temperature, but also on the difference in water vapor density between the animal's surface and the environment, and on the resistance to water loss from that surface (measured as permeability or some related parameter).

Air at the immediate surface of the animal is usually assumed to be saturated, at 100% relative humidity (RH), and to decrease in saturation in a regular manner away from that surface through the boundary layer. As with convection, though, the actual situation in this layer is complex and difficult to model accurately, being modified by very small-scale fluid movements. This makes rates of

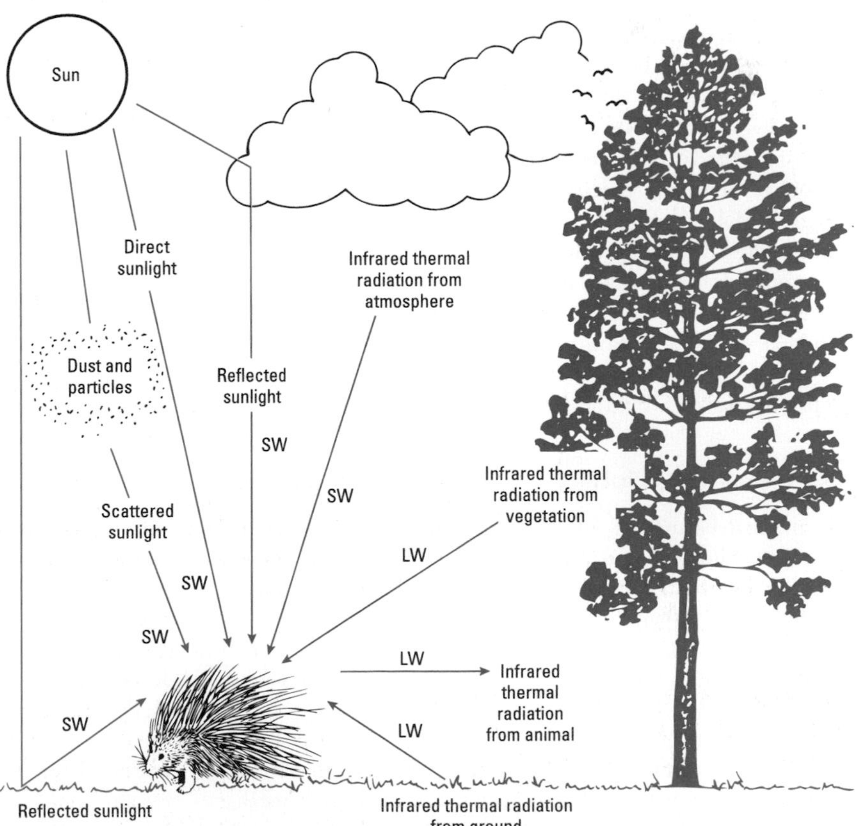

Fig. 8.20 Radiation exchanges for a land animal, emphasizing direct shorter wavelength inputs and reflected longer (infrared) wavelength inputs. LW, longwave; SW, shortwave.

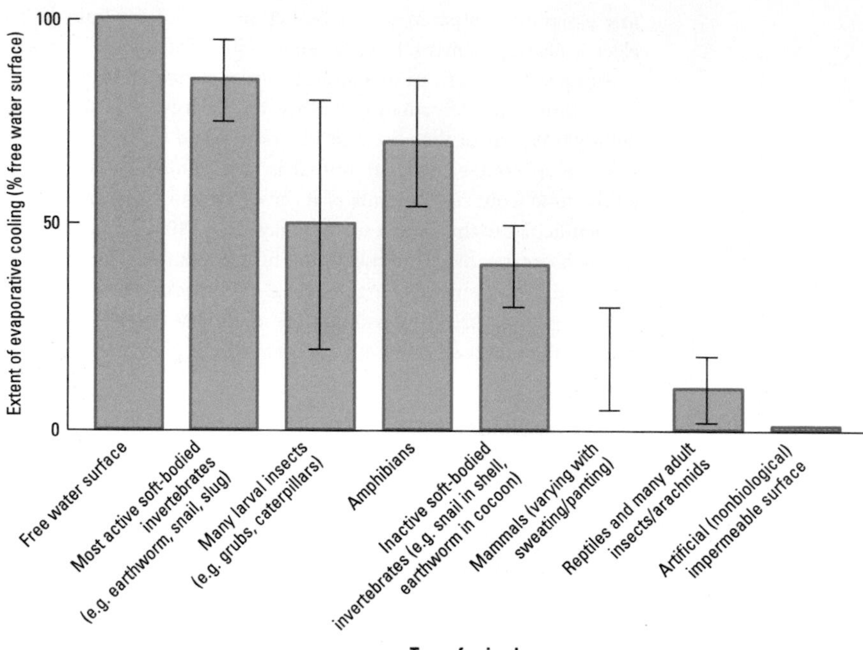

Fig. 8.21 Relative extents of evaporative cooling in different types of terrestrial animals (compared with free water and artificial impermeable membranes). Bars indicate ranges.

water loss and thus evaporative heat loss difficult to measure (note that it also complicates the issue of determining the permeabilities of biological surfaces, as Chapter 6 showed that these are generally estimated indirectly from rates of water loss!). Evaporation in relation to the water content of air also has to be very carefully expressed, in view of the interactive effects of temperature and water vapor concentration discussed in Chapter 5.

Evaporation as a means of temperature regulation is unique amongst the methods discussed here, as it can *only* be used by animals living in air. It is thus a fairly common technique for animals on seashores and in humid terrestrial zones, where heat stress can be a problem but water is reliably available. However, evaporative cooling declines in importance in most terrestrial zones, and it is rarely used in deserts, even though these may be very hot and will put animals under considerable thermal stress, since the water loss that it entails cannot then be tolerated. Only a handful of moderately large species, away from the depths of deserts, use it extensively (see Chapter 16). Figure 8.21 summarizes the extent of evaporative cooling in a variety of land animals.

8.5.5 Overall thermal balance

The thermal balance of any animal is determined by the net exchanges of heat through all the avenues outlined above, together with metabolic heat production (M). The general equation describing heat balance is therefore:

$$M = h_{cond}\,(T_b - T_a) + h_{conv}\,(T_s - T_a) + h_{ad}\,(T_s - T_{sur}) + E + S$$

where h values are the various heat transfer coefficients, which are in practice rather similar to the three main temperature exchange routes. E is any evaporative heat loss, while S is any heat storage that occurs, dependent on the change of temperature, the mass of the body, and the specific heat capacity of the body tissues. A pictorial summary of these complex inputs and outputs for a terrestrial animal such as a tortoise is shown in Fig. 8.22.

This equation can be greatly simplified for a steady state, with no change of temperature in the animal, no heat storage, and no significant difference between T_s and T_{sur}, as:

$$M = h\,(T_b - T_a) + E$$

where h is the total heat transfer coefficient.

For a thermoconforming ectotherm in a thermally stable environment, E and M are negligible and the equation simplifies readily to $T_b = T_a$. However, for more complex thermal strategies, particularly in land animals, the balancing of this equation is clearly going to be a more difficult procedure. A resting human in still air loses about 60% of body heat by radiation and 20% by conduction and convection, but this balance changes drastically during exercise and in dry windy conditions. Hence a wide range of adaptations to control different elements of the animal's heat-exchange systems have evolved.

8.6 Avoidance, tolerance, and acclimation in thermal biology

Now that the avenues of heat exchange have been considered, we need to examine the specific strategies that animals use to reduce, accentuate, or control these exchanges. As with so many other aspects of comparative physiology, these can be considered under the general categories of avoidance strategies, tolerance strategies, and regulatory mechanisms, the latter including both acclimatory responses and a variety of strictly regulatory systems. Thermal regulation is the most complex of these topics because so many different

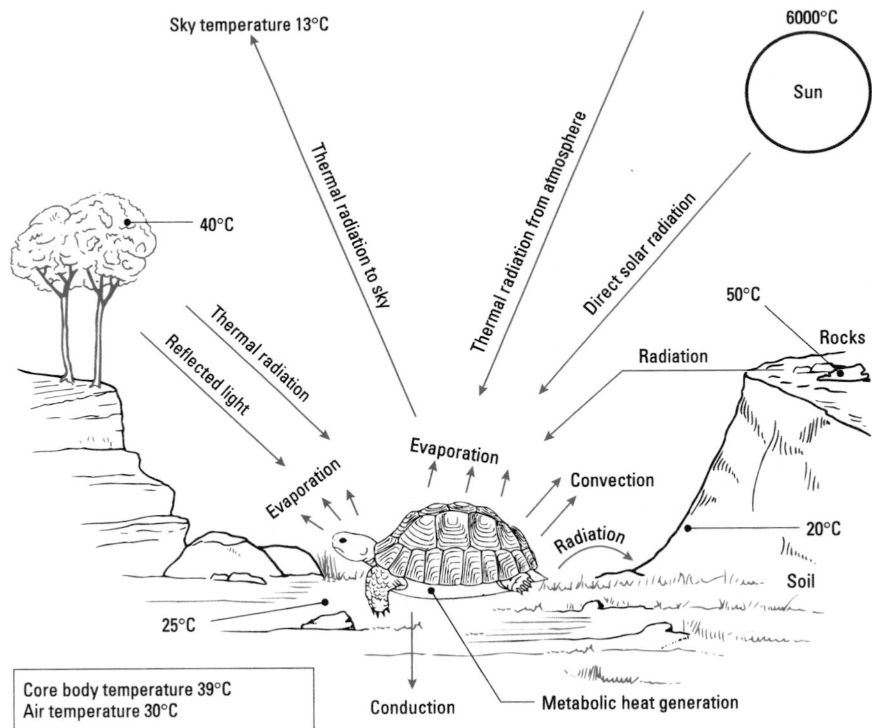

Fig. 8.22 An overview of the thermal exchanges between a terrestrial animal and its environment.

possible techniques exist in animals. It is therefore covered in depth in sections 8.7 and 8.8, dealing separately with the regulation of heat gain and heat loss.

8.6.1 Avoidance

Avoidance of environmental temperature extremes is an important part of the behavioral repertoire of a huge range of animals; virtually all land animals use avoidance as a first line of defense and thus as a key part of their thermal biology, as do many littoral, estuarine, and freshwater animals. The strategy is less obvious in truly marine animals as the temperature variation in their environment is rather limited, but thermal avoidance behaviors certainly do operate in marine animals that encounter permanent or seasonal thermoclines (steep environmental temperature gradients; see Chapter 11).

Avoidance in partly or fully terrestrial animals may take many forms, but can be generally viewed as the location and use of appropriate **climatic conditions** in time and space. In this context an animal's size becomes a crucial variable influencing its thermal balance, in two key respects. Firstly, cooling and heating rates will be faster for a small animal than for a large one given the same thermal gradient; the larger animal has a larger **thermal inertia**. The time taken for thermal equilibrium to be reached increases with body size and with the initial temperature differential (Fig. 8.23). Within each taxon the time to equilibration increases around 10-fold for a 100-fold increase in body mass. Large insects and reptiles can make use of their relatively large thermal inertia, for example, by heating their bodies in a warm microhabitat and then foraging in a previously inhospitable cooler one. Conversely, species that forage in warm environments may gain too much heat, which then has to be lost by returning to the shade. In general the activity bouts and cooling-off bouts will have to be shorter for smaller species. Such observations are usually linked to "Bergmann's rule", which states that in colder climates animals (and their constituent parts) are larger, which seems to hold true in relation to latitude for most endotherm taxa and for many ectotherms too (although not perhaps for some insect groups).

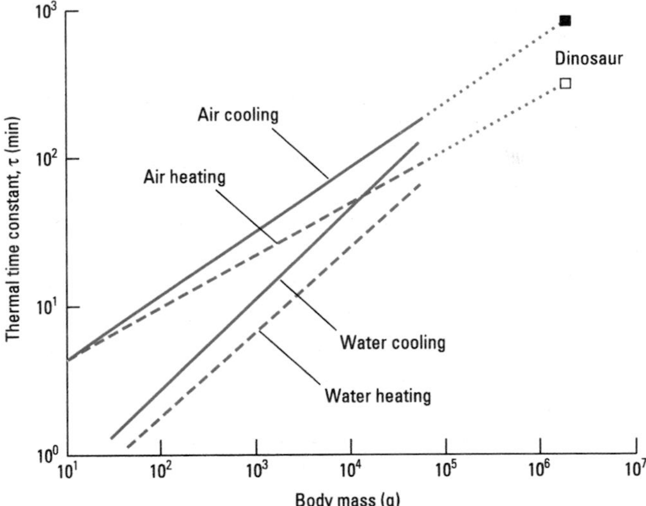

Fig. 8.23 The relationships between the thermal time constant (τ) and size in reptiles. The time constant scales with mass over four orders of magnitude for extant species, varying according to whether the animal is heating or cooling and in water or in air. Warming occurs more quickly (i.e. lower time constant) than cooling, especially in water (having a higher conductance). Extrapolation to reptiles the size of a large dinosaur indicates the very substantial thermal stability that these animals must have had, with a time constant for cooling of up to 16 h.

Size is obviously crucial in a second respect in that the concepts of microclimates and microhabitats are very labile depending on the size of the animal experiencing the environment. For organisms with dimensions in millimeters the environment is experienced as a very "fine grained" one, with conditions varying widely in time and space. It is vital to appreciate the huge range of microclimates available to a small terrestrial invertebrate; conditions vary enormously on a scale of millimeters or centimeters, and change from second to second when there is air movement. Familiar examples of microclimate exploitation by smaller animals include the occupation of thermally equitable refugia such as burrows or nests, which on a seasonal or diurnal or moment-to-moment basis can make the difference between survival and death for animals in very hot or very cold climates (see Chapter 16). Less familiar, but equally dramatic in effect, are the subtle movements of insects or spiders up and down a grass blade, around a plant stem, or in and out of the boundary layer above hot sand by standing up on the tips of long limbs (stilting), movements which can allow the animal to avoid highly adverse conditions and achieve an entirely equitable microclimate within seconds. Many other kinds of thermal avoidance strategies will be met in Chapters 15 and 16, dealing with terrestrial adaptations; their importance for small animals can hardly be overstated.

But large animals do not have this option of avoidance to anything like the same extent. They physically do not fit within the varied microhabitats open to a small creature, and their thermal inertia would be too large to allow them to take advantage of these temporally variable microhabitats anyway.

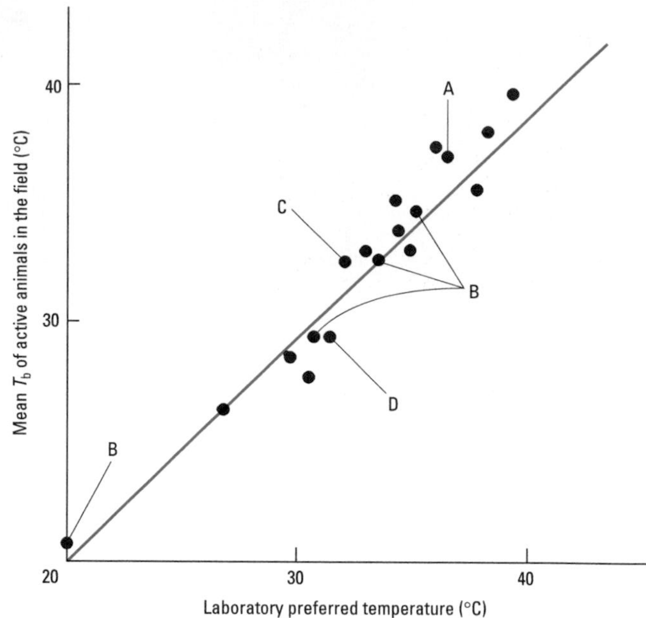

Fig. 8.24 The match between preferred body temperatures as measured in the laboratory and observed body temperatures in the field, for reptile species ranging from monitor lizards (A) through small lizards (B), marine iguanas (C), and snakes (D). (From Cossins & Bowler 1987, Fig. 3.4, with kind permission from Kluwer Academic Publishers.)

8.6.2 Tolerance

Animals vary substantially in their tolerance of varying body temperatures, and thus in the temperatures that elicit acclimatory or regulatory responses. We saw in Table 8.6 that some polar marine animals have an upper lethal temperature as low as 6°C, while for desert terrestrial animals this value may be close to 50°C. Looked at more closely, these kinds of data show clear-cut phylogenetic and geographic patterns. Certain kinds of animal tend to have broader tolerance ranges than others; for example, insects on land, and crustaceans and molluscs in aquatic habitats, are generally rather catholic in their thermal preferences, whilst vertebrates in all habitats commonly have more restricted tolerance zones. Within a taxon, temperate animals will usually show broader tolerances than either polar or tropical forms.

Most animals have a **preferred body temperature** (T_{pref}) at which they operate, which is usually rather close to the "optimal" temperature measured for many of their physiological processes. They are often very good at maintaining their T_b very close to their measured preferred temperature during activity (Fig. 8.24). In many terrestrial ectotherms, though, the T_{pref} for different activities varies. For example, in snakes the digestive processes are more sensitive to temperature than are the locomotory speeds, so there can be no single optimal T_b for prey capture. It is often more useful therefore to consider a **thermal performance breadth**, a range of T_b values for which performance is still high (a value of 80% of maximum performance value is often used to define the limits of this term). Beyond this again is the **thermal tolerance range**, the full range of body temperatures over which an animal can survive indefinitely. Finally there is the even broader **thermal survival zone**, between the upper critical temperature (UCT) and the lower critical temperature (LCT). As a broad generalization it would be fair to say that these limits are roughly matched to the environmental temperatures experienced, thus varying with latitude. Tolerance ranges are also generally broader, and critical temperatures spread further apart, for larger animals, so that the lower critical temperatures of birds and of mammals are clearly size dependent (Fig. 8.25).

However, measuring any of these parameters for a particular species is fraught with difficulty. There is inherent variability between individuals, as with so many biological issues. Therefore, critical temperatures are often defined in terms of 50% effects; for example, the lethal temperature for a species is that temperature at which 50% of the animals die (T_{L50}). This would be tedious and dreadfully wasteful to determine exactly in practice, but if desired it can be established graphically by plotting percentage survival against temperature for a range of experimental conditions and reading off the predicted value for 50% survival. However, we have already noted that the destructive effects of temperature extremes are time dependent, so that a temperature that can be tolerated for a few minutes may not be survivable over a longer period. The point at which exposure time begins to have an effect is sometimes termed the **incipient lethal temperature** (Fig. 8.26), and can be used to distinguish the "zone of resistance" from the "zone of tolerance". A better analysis of thermal survival is therefore given by temperature against time to 50% mortality (Fig. 8.27a), which commonly gives a straight line on a semilogarithmic plot (Fig. 8.27b).

An even more complete picture of a species' thermal tolerance is illustrated by the **tolerance polygon** (Fig. 8.28). The animals are

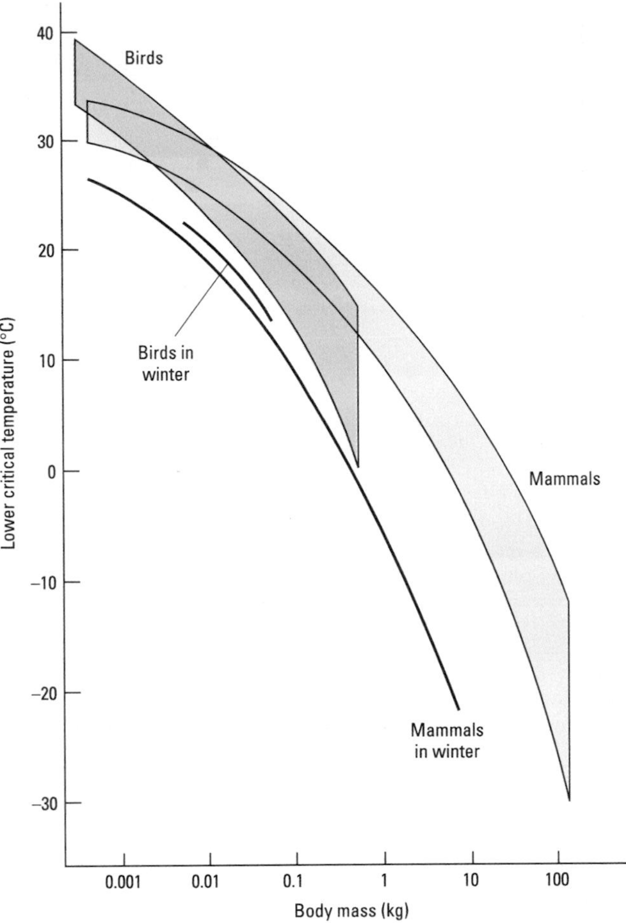

Fig. 8.25 The relation between lower critical temperature (LCT) and body mass for mammals and birds, showing the lower temperatures survived by larger species, and the seasonal effects particularly in mammals (lowered LCT in winter). (From Peters 1983, courtesy of Cambridge University Press.)

given sufficient time to acclimate to the maximum possible range of temperatures, and the lethal temperature at each acclimation temperature is then determined as before. The area between the two curves represents the combination of acclimation and lethal temperatures that 50% of the population can withstand for an indefinite period. The greater the enclosed area (measured as °C^2), the greater that species' thermal tolerance range. Table 8.10 shows this kind of analysis for a range of aquatic organisms, revealing the generally higher values for freshwater forms over their seawater relatives.

In practice, it is still difficult to quote a single figure for thermal tolerance for a species, even using the tolerance polygon, for a number of reasons:

1 Plots of percentage mortality versus temperature are usually sigmoidal, i.e. most individuals in a population die over a restricted range of temperatures but some are more and some are less susceptible.

2 Tolerance of high and low temperatures is rarely fixed in a species and may be modified by conditioning. For example, many perennial animals are more cold tolerant and heat susceptible in the winter. Birds that are artificially exposed to hot environments can increase their T_b by as much as 3–4°C.

3 Evolution may produce marked differences in thermal tolerance throughout a species' geographic range.

4 Any or all of the commonly quoted thermal parameters may be variable through the life of an animal (Fig. 8.29). In particular, thermal performance breadth, and thermal tolerance polygons tend to be reduced in young animals, often substantially so in metamorphic invertebrate animals where larval stages are especially vulnerable to thermal stress.

5 Temperature tolerance may be modified by interactions with other factors, such as salinity in aquatic organisms and humidity in terrestrial organisms.

6 In mammals and birds, higher core temperatures may be tolerated with internal heating (exercise) than with external environmental heating.

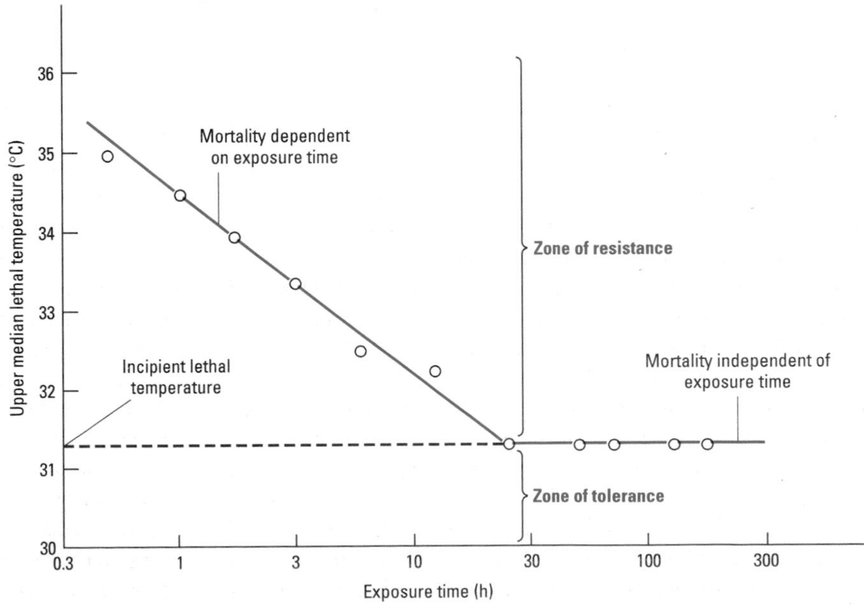

Fig. 8.26 Interactions of exposure time and temperature on mortality. The median lethal temperature (T_{L50}) can be established from standard mortality curves; it decreases as exposure time is increased, and reaches a plateau where mortality is independent of time, known as the "incipient lethal temperature", separating the zone of resistance and the zone of tolerance. (From Cossins & Bowler 1987, Fig. 6.4, with kind permission from Kluwer Academic Publishers.)

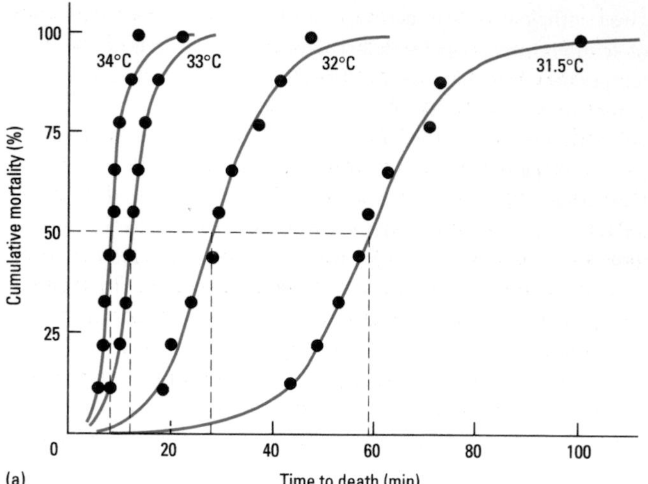

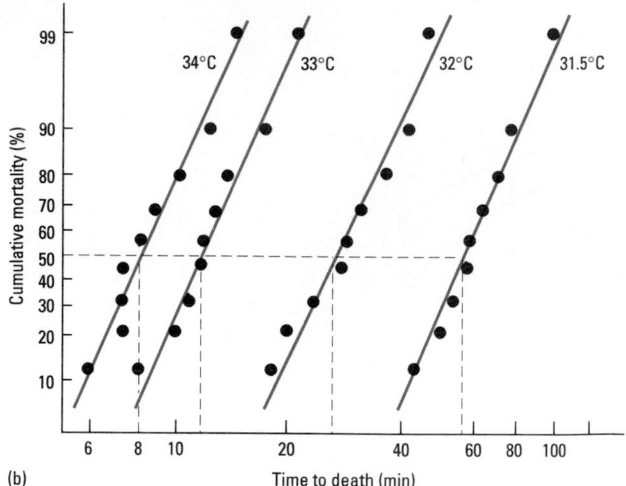

Fig. 8.27 (a) Determinations of the lethal times for 50% mortality at raised temperatures in the freshwater crayfish *Austropotamobius pallipes*; in (b) the same data are plotted logarithmically. (From Cossins & Bowler 1987, Fig. 6.2, with kind permission from Kluwer Academic Publishers.)

Table 8.10 Thermal tolerances of fish and crustaceans, as given by the area of tolerance polygons (units of °C^2). Compare Fig. 8.28.

Species	Tolerance	Habitat
Goldfish	1220	Fresh water, widespread
Bullhead	1162	Fresh water, widespread
Lobster	830	Marine, widespread
Greenfish	800	
Silverside	715	Mostly marine, widespread
Flounder	685	Marine, temperate
Trout	625	Fresh water/marine, temperate
Puffer fish	550	Marine, north temperate
Chum salmon	468	Marine/fresh water, north temperate
Rock perch	47	Antarctic

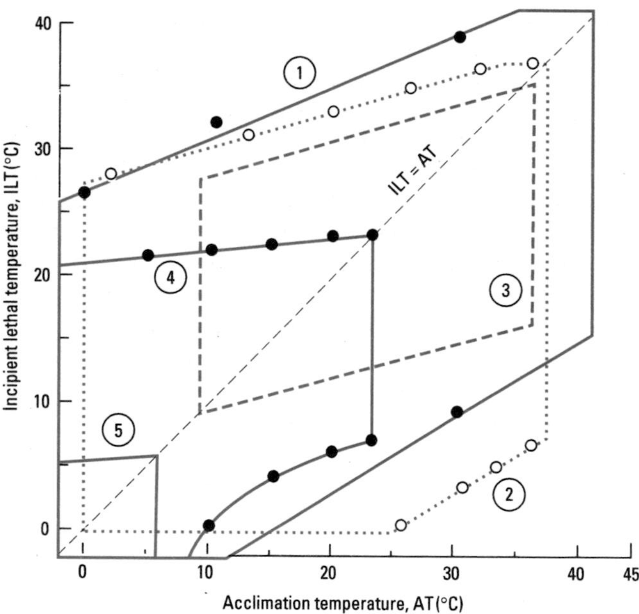

Fig. 8.28 Tolerance polygons for five species of fish. At any particular acclimation temperature, the upper critical temperature and lower critical temperature can be read off. 1, goldfish, eurytherm; 2, bullhead, eurytherm; 3, puffer fish, temperate stenotherm; 4, chum salmon, cold stenotherm; 5, Antarctic rock perch, extreme cold stenotherm.

7 Finally, of course, thermal tolerance may be modified within a rather short timescale for a particular individual due to its own adaptive responses, discussed below.

8.6.3 Thermal acclimation

Thermal tolerance as discussed above is partly determined by an animal's ability to acclimate to temperature change, and in this sense acclimation is a response intermediate between tolerance and thermoregulation. Although most biochemical and physiological processes have a Q_{10} value between 2 and 3 (see section 8.2), the rates of many such processes can be gradually adjusted in living animals to compensate for variations in the thermal environment, whether on daily, monthly, or annual timescales. As we have seen before, this adjustment tends to be called acclimatization when it occurs in nature, but acclimation when it is induced in laboratory studies; but since all reliable studies are based on laboratory analyses the effects are generally called acclimatory responses. Thus many animals have lower T_{pref} in winter than in summer—they have acclimatized to the extent that in winter they may even be killed by raised temperatures at which they would have been normally active in summer. The measured underlying acclimatory responses may involve heart rate, ventilation rate, metabolic rate, neural conduction persistence, and enzyme reaction rates. The effect may be expressed not just in T_{pref} but also in thermal tolerance range and in LCT and UCT.

As a general rule, acclimatory responses act to keep biological processes operating at roughly the same rate whatever the temperature; the changes in enzymes and membranes discussed in section 8.2 are clearly examples of acclimatory responses. Perfect acclimation is rare, however, and numerous attempts have been made to classify the exact nature of acclimatory response. One useful scheme (based on the Prosser scheme) is shown in Fig. 8.30. Type 1 involves

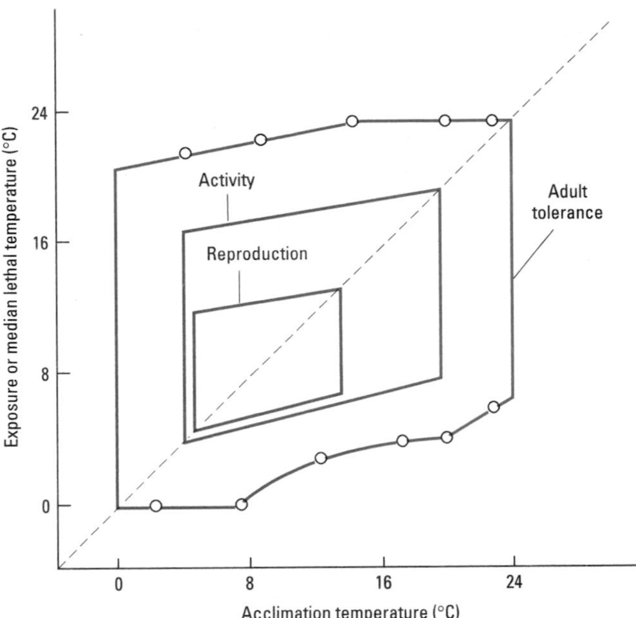

Fig. 8.29 The varying polygons for particular activities in the sockeye salmon. (Adapted from Fry & Hochachka 1970.)

no acclimation of rate with temperature; type 2 moves the rate response up or down in a translational movement, while type 3 involves a rotational movement of the rate curve (implying a change in Q_{10} value as well as in rate). The more complex type 4 includes both translational and rotational effects.

Another widely used classification uses the terms **capacity acclimation** (or capacity adaptation) for a change in the rate of some function during long-term exposure, and **resistance acclimation** (adaptation) for a change in upper or lower tolerance limits. Both kinds of response are widespread, and they very often occur simultaneously. The usage of this terminology can therefore be confusing, and has probably been overworked in the literature.

Examples of these varying kinds of thermal acclimation occur in all kinds of animal and in virtually all habitats. In invertebrates and ectothermic vertebrates, acclimation occurs in response to varying T_b, and examples of acclimation of LCT, UCT, T_{pref}, and metabolic rate can all be found for creatures ranging from sea anemones to desert lizards. Some of the "lower" organisms perhaps show relatively little compensation. Some littoral invertebrates, for example, really do become quiescent in winter with little or no acclimation of their thermal properties, and are sluggish unless artificially warmed (type 1 response); at best they may show resistance acclimation of their lower thermal limits. Ectothermic vertebrates and terrestrial invertebrates are more likely to show the more complex acclimatory responses, types 3 and 4, with an upward translation to a higher rate and a clockwise rotation to a lower Q_{10} value, i.e. with both capacity and resistance acclimation occurring. But there are certainly no clear-cut rules here. It is also particularly important to recall the points made in Chapter 1 about adaptation and acclimation, and realize that a changed response is not necessarily evidence for a beneficial adaptive response. It is possible to think up "reasons" and "advantages" for each of the types of acclimatory response described above, but this does not prove that they are specific adaptations of benefit to the animals exhibiting them.

From ectothermic vertebrates, littoral invertebrates, and the insects, particularly from the well-studied *Drosophila* fruit flies, come some of the best known cases of naturally occurring acclimatory gradients, along latitudinal or altitudinal clines. For example, amphibians show clear clines of LCT and UCT with latitude; southern populations of oysters (*Crassostrea*) spawn at higher water temperatures than more northern populations; and southern populations of *Drosophila funebris* are more resistant to high temperatures than are northern (cooler climate) populations. These effects could be due to acclimation, or could represent genotypic physiological races of the animals. Both situations do occur, but in *Drosophila* at least some of these effects can be directly correlated with chromosomal differences or even with precise genetic loci effects. This is an area where real and significant advances are being made in our understanding of comparative adaptation.

Fig. 8.30 Four types of acclimation as defined by Prosser. The solid line indicates the initial rate–temperature curve, and the dashed line indicates this curve after thermal acclimation (to higher temperatures). C, cold acclimatized; W, warm acclimatized.

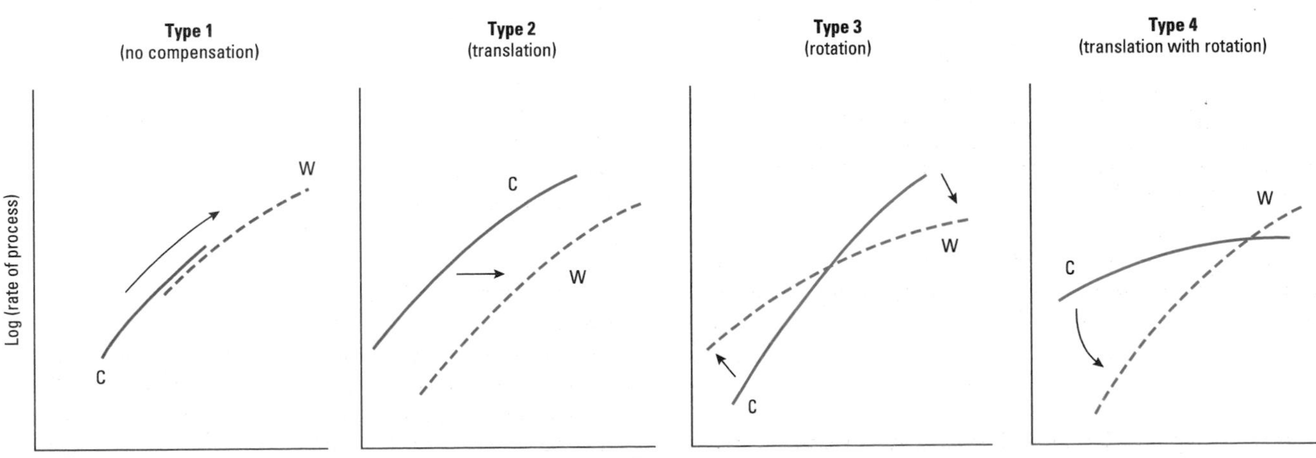

Endotherms show acclimation in a similar fashion to the ectotherms thus far considered, but here the effects are, of course, in response to variation in T_a rather than T_b. The acclimatory effects in endothermic birds and mammals may be very elaborate, ranging from behavioral patterns (e.g. reduced activity outside a burrow, or gathering of extra nest insulation) through morphological features (e.g. fur or feather characteristics, or patterns of fat deposition) and physiological features (e.g. rates of thermogenesis, or nerve conduction velocity), to biochemical features such as the nature and properties of membrane constituents, or the concentrations of enzymes.

8.6.4 Rates of acclimation and effects of history

Rates of acclimation generally follow a hyperbolic curve, with complete acclimation in a laboratory setting taking from a few hours to several days depending on the kind of animal involved. This is most easily expressed as the time for 50% acclimation. In small terrestrial animals, such as insects and woodlice, acclimation is often particularly fast. Acclimation times again have to be expressed rather carefully, as for any given species they tend to vary with temperature and can be drastically affected if oxygen supplies are less than optimal. Furthermore, acclimation to a particular increment of temperature change tends to be quicker from low to high than from high to low.

Acclimation rates also depend on age, size, and previous thermal history. Even for animals kept in constant conditions for very long periods, and thus lacking any apparent thermal cues, acclimation can occur, albeit slowly. For example, thermal preferences may change to a colder range in winter for a laboratory goldfish, in the absence of any water temperature change; the fish is apparently responding to a photoperiod cue.

8.7 Regulating heat gain and keeping warm

8.7.1 Metabolic heat and body temperature

Heat is produced by all metabolic activities, and in a tachymetabolic (endothermic) animal it is produced in substantial quantities in the core of the body. The tissues of endotherms have substantially higher mitochondrial densities and higher activities of mitochondrial enzymes than do similarly sized ectotherms (Fig. 8.31), reflecting this ability to produce metabolic heat. For birds and mammals at rest, the thoracic and abdominal organs (gut, liver, kidney, heart, lungs, etc.) produce up to three-fourths of the metabolic heat, whereas when active, with the metabolic rate perhaps elevated 10-fold (see Chapter 6), it is principally the muscles that produce extra heat. The core temperature of the animal therefore varies according to activity, so that quoting a standard core temperature for a species is difficult. This is particularly so for some large desert mammals, such as the camel, that has spectacular daily variations in body temperature as part of their thermal strategy (see Chapter 16). The core temperature of most birds and mammals also varies a little with a diurnal periodicity (Fig. 8.32) according to the light cycle, even when an individual is kept under constant thermal conditions. Diurnal species are warmer by day, and nocturnal species warmer by night, with a mean difference of up to 2°C, and in extreme cases (such as the tiny tree shrews) of about 5°C. Even when light cues are removed the diurnal effect persists, with a slightly different periodicity, indicating an endogenous rhythm of heat production. Further longer term complications arise in relation to estrous cycles in mammals, where females may be 0.5–1°C warmer at ovulation.

Table 8.11 shows the core body temperatures (allowing for the problems outlined above in determining this mean value) for a range

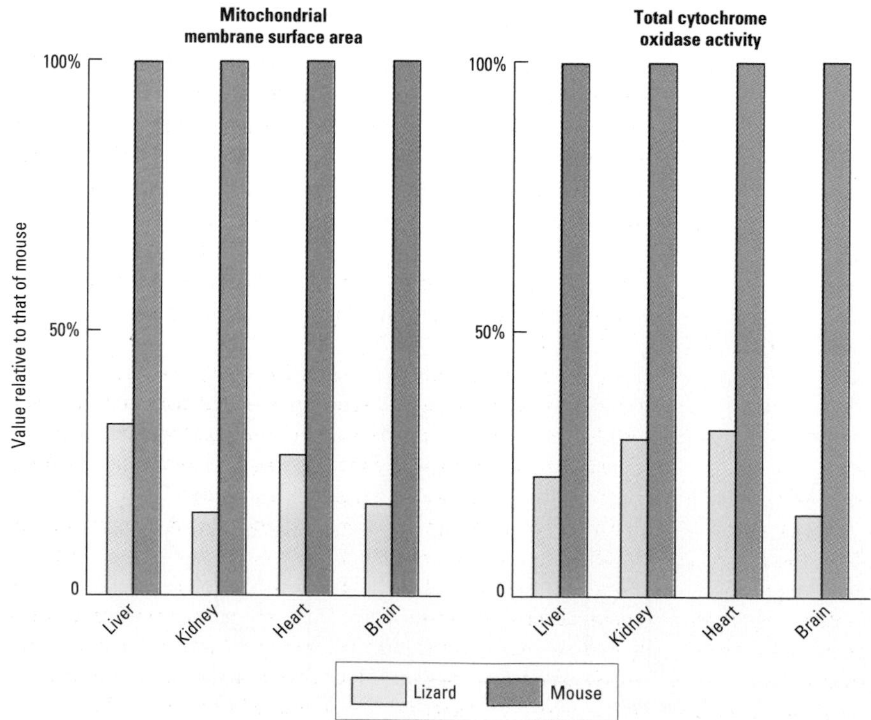

Fig. 8.31 A comparison of mitochondrial density and cytochrome oxidase activity in the tissues of an ectotherm (lizard) and endotherm (mouse) of similar body mass. Values are given relative to the mouse as 100%, and are roughly one-third to one-fourth as high for the lizard. (From Else & Hulbert 1985.)

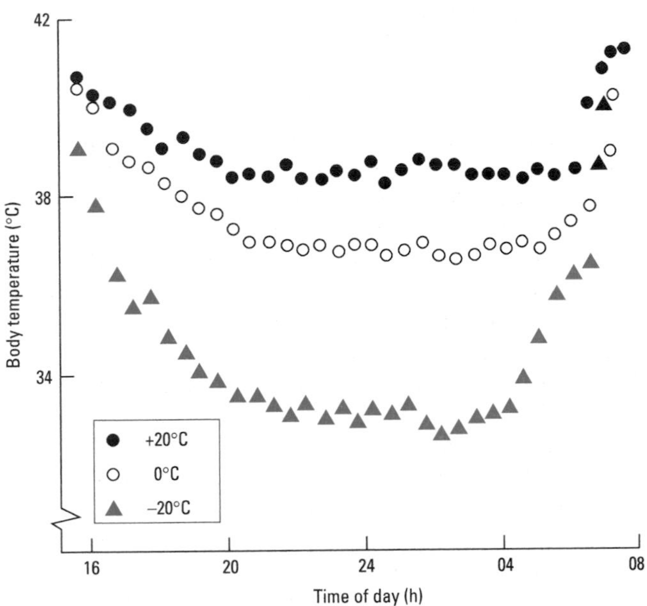

Fig. 8.32 Diurnal variation in body temperature for a small bird (the willow tit) during the night at three different ambient temperatures. (From Reinertsen & Haftorn 1986.)

Table 8.11 Core body temperatures (T_b) characteristic of endothermic vertebrates.

Taxon	Common names	T_b (°C)
Mammals		
Monotremes	Echidna, platypus	30
Edentates	Anteaters, etc.	33–34
Marsupials	Possums, kangaroos, etc.	36
Insectivores	Hedgehogs, moles, etc.	36
	Shrews	37–38
Chiropterans	Bats	37
Cetaceans	Whales, etc.	37–38
Pinnipeds	Seals, etc.	
Rodents	Mice, rats, etc.	
Perissodactyls	Tapir, rhinoceros, horse	
Primates	Monkeys, humans, etc.	38–39
Carnivora	Dogs, cats, etc.	
Artiodactyls	Cow, camel, pig, etc.	
Lagomorphs	Rabbits	
Birds		
	Penguins	38
	Ostrich, petrels, etc.	39–40
	Pelicans, parrots, ducks, gamebirds	41–42
	Passerine songbirds	42

of birds and mammals, whether measured in laboratory studies or by implanted thermistors transmitting telemetric records of T_b in natural surroundings. Birds generally maintain a T_b of about 40°C, with the temperature of smaller species (especially passerines) somewhat higher than larger species, and that of ratites (flightless birds) particularly low. In mammals, clear phylogenetic effects can be seen, with the monotremes at 30–31°C, marsupials at 36°C, and eutherians at about 38°C, but with the T_b of "primitive" groups somewhat lower than more "advanced" groups (e.g. insectivores, such as shrews, have T_b values of 34–38°C, slightly varying with ambient temperature). The reasons for lower body temperatures in some groups are not at all clear, and the apparent evolutionary trends are puzzling. Certainly the earlier groups in evolutionary terms (e.g. monotremes) are every bit as good at thermoregulating at their specific preferred T_b as the evolutionarily later groups that have "chosen" a higher T_b, so the differences are not due to any primitive failing in the ability to control thermal balance.

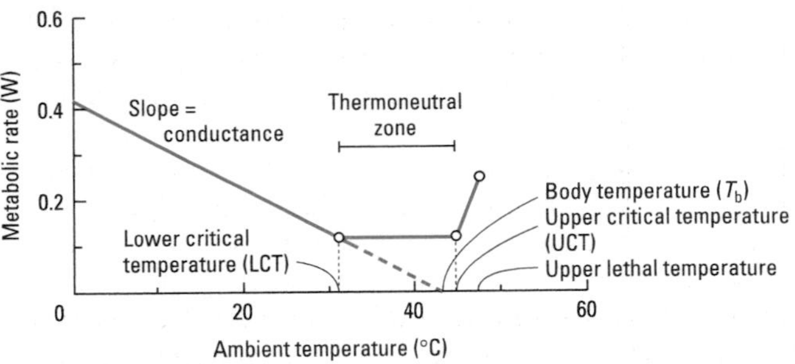

Fig. 8.33 The influence of ambient temperature on the metabolic rate (MR) of a small temperate endotherm, showing the "thermoneutral zone" where MR is constant.

To maintain a constant body temperature, at whatever level, there must be a steady-state condition in which the rate of heat production (M) equals the overall rate of heat loss (Q, where $Q = h\,(T_b - T_a) + E$; see section 8.5). It follows that if T_b is to be constant then the only factors that can be regulated are M (the metabolic heat production, and the distribution of this heat to the surface), or the conductance factor (represented by h), or the evaporative loss E.

8.7.2 Heat production: shivering and nonshivering thermogenesis

Heat production in animals is continuous and inevitable; the important distinction is that in endotherms it occurs at a higher rate (4–8 times that of ectotherms at the same T_b) because of a greater concentration of mitochondria and perhaps also a different kind of mitochondria in which oxidative phosphorylation is always partially uncoupled so that more heat is released for a given level of ATP generation. Thermoregulatory heat production is extremely expensive, and makes up the single largest component of the energy budget in endotherms, reducing the energy available for growth and reproduction.

For such endothermic animals the **thermoneutral zone** (TNZ) is a useful concept, being the range of ambient temperatures over which the animal's metabolic rate and thus heat production is not varied. A plot of ambient temperature against metabolic rate for a mammal or bird is shown in Fig. 8.33, and within its TNZ any particular endotherm keeps its energy metabolism fairly constant. It is in this range that the resting metabolic rate (see Chapter 6) should always be measured. Below some specific lower temperature, a mammal will show a rise in metabolic rate with decreasing temperature in a linear fashion. Above some specific higher temperature, metabolic rate may again rise due to the direct effects of temperature on metabolic processes and because of additional metabolic costs

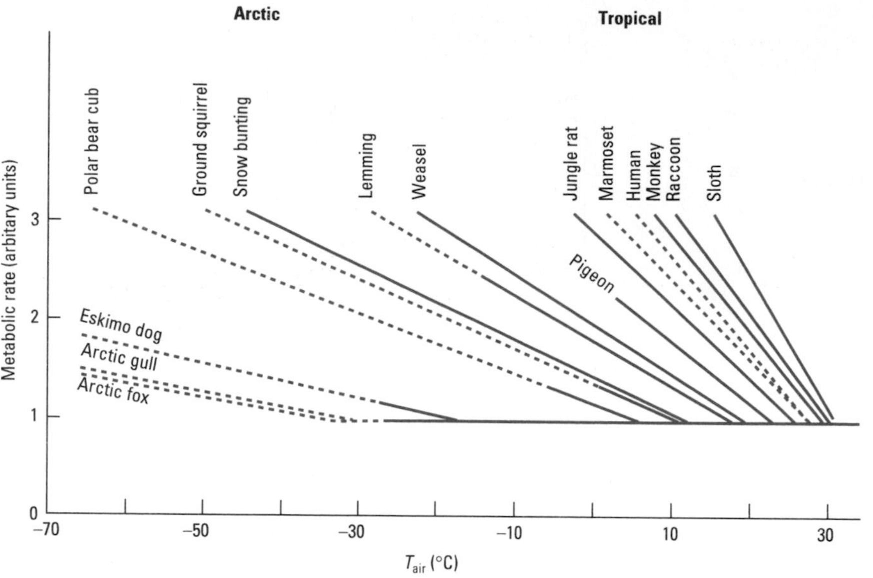

Fig. 8.34 Plots of metabolic rate (MR) against ambient temperature for a variety of tropical and polar mammals and birds. The plots are equivalent to the left-hand side of Fig. 8.33, with the MR in the thermoneutral zone (TNZ) taken as unity to allow direct comparisons. Polar animals have a much wider TNZ, surviving to lower ambient temperatures before MR is elevated: this is mainly due to their much better insulation. (Adapted from Scholander *et al.* 1950.)

associated with thermoregulation (such as increasing the blood flow to the skin—see below); eventually these processes run out of control (before declining finally as the cell machinery is irreversibly damaged and death ensues). The upper lethal temperature is often only 3–6°C above the normal body temperature.

Comparing TNZs between animals is difficult unless the resting metabolic rate of each is normalized as 100% (Fig. 8.34). A comparative plot then reveals the much greater width of TNZs in cold-adapted animals. An Arctic mammal does not need to increase its metabolic rate to keep warm unless the external temperature is well below 0°C (e.g. −40°C for an Arctic fox), and even then the metabolic rate/ambient temperature plot is very shallow. In contrast, a tropical mammal may show an increased metabolism if the ambient temperature only drops to 25°C, and the rate of increase will be much sharper with declining temperature.

These differences in TNZ are largely due to reductions in the thermal conductance of the cold-adapted endotherms—they have thick insulating pelts (see sections 8.5 and 8.7.4) that greatly increase their thermal tolerance. Larger mammals and birds have thicker fur or feather layers, and/or subcutaneous fat layers, and therefore are better protected against heat loss; specific heat conductance (C^*, heat flow per unit body mass) varies with $M_b^{-0.50}$ for birds, marsupials, and eutherian mammals. Specific heat conductance decreases with body size more rapidly than does specific heat production, so that the temperature an animal can support without increasing its heat production also increases with body mass. This mismatch between heat production and conductance as size decreases means that a large endotherm can tolerate a much greater degree of cooling without increasing its heat production; it will inevitably have an extended TNZ at the lower end.

Nevertheless, once conductive heat loss has been decreased to an irreducible minimum, a further fall in ambient temperature will necessitate an increase in heat production. Maximum heat production is generally around five times the basal metabolic rate. Heat production can be increased according to need in three ways. Muscular activity by physical exercise greatly raises the metabolic rate as we have seen, and could be used to warm the body. Involuntary muscular activity could also be initiated, this being what we perceive as shivering. As a third possibility, thermogenesis could be achieved without muscular contractions, by a variety of chemical means, so-called nonshivering thermogenesis.

Voluntary muscular activity to achieve warm-up is not uncommon in animals, and is of course used by humans. Some insects may use short periods of flight for no other purpose than to warm their bodies up, and recent evidence suggests that some can alter their heat production in flight by alterations of wingbeat frequencies and wing kinematics to help maintain a constant T_b. However, activity is not a simple solution to being cold for most animals, as exercise may in itself accelerate some avenues of heat loss: it tends to involve high surface area positions, reduces the thickness of boundary layers, and most seriously may reduce the value of insulation layers by wetting them (in sweating mammals) or making them dishevelled (as with bird plumage).

Shivering is used by nearly all adult endotherms, and by many ectotherms as well. It is a high-frequency reflex operating via efferent nerves to the muscle spindles, to produce oscillatory contraction of mutually antagonistic skeletal muscles, giving little net movement. Thus the ATP hydrolyzed to provide energy for contraction is producing minimal physical work and being diverted into heat instead. The intensity of shivering is linearly related to oxygen consumption. Shivering is mediated via the normal neuromuscular junctions in the nervous system (largely controlled by the cerebellum), and it is therefore susceptible to neuromuscular blocking drugs such as curare.

The use of such drugs gives a convincing demonstration that other forms of thermogenesis are commonly present; curarized laboratory animals can still exhibit an increase in metabolic rate when cooled. This results from **nonshivering thermogenesis** (NST), a contentious but exciting phenomenon in the physiological repertoire of many classic endotherms, including placental mammals (especially juveniles and small species), some marsupials, and rather few birds. It certainly occurs during arousal from torpor in specific

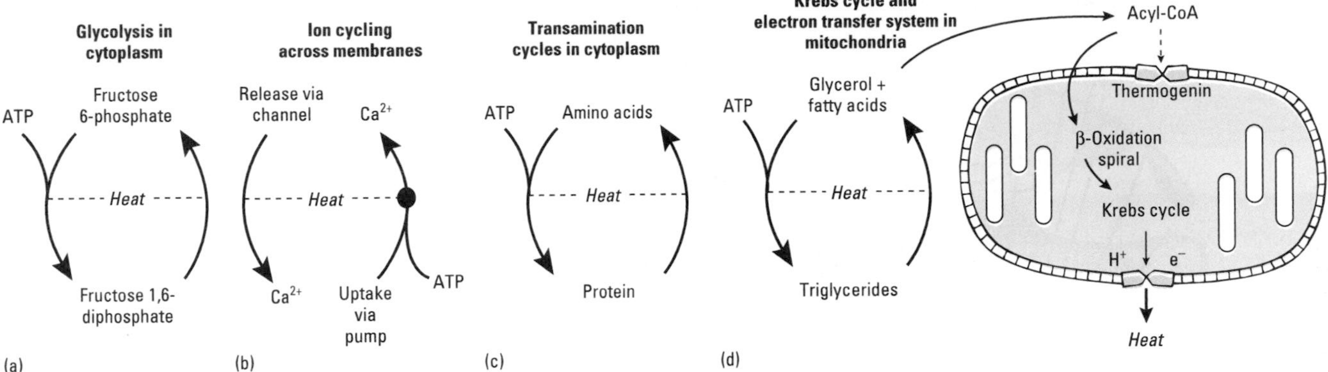

Fig. 8.35 Possible heat-generating "futile" biochemical cycles: (a) is used in some insects with endothermy, (b) in some fish heater cells, and (d) is probably the route in tetrapod brown adipose tissue.

mammalian tissues such as brown fat (**brown adipose tissue**, or **BAT**), which consists of fat cells with a very high mitochondrial content (with highly proliferated inner mitochondrial membranes), highly vascularized with a direct venous return to the heart, and with sympathetic nervous innervation. It probably also happens in the liver, kidneys, and elsewhere in chilled mammals. Some semi-aquatic mammals also use it to warm up after diving or prolonged swimming.

Nonshivering thermogenesis equally clearly occurs in at least some ectotherms, including leatherback turtles and a few insects, where for example warm-up can occur in immature forms (social wasps and others) lacking any proper skeletal muscle to shiver with. Its occurrence in adult insects such as bees has been disputed, but now seems incontrovertible, and will be discussed in Chapter 15.

The actual mechanisms and fuels are somewhat variable (Fig. 8.35), but the primary heat allowing warm-up before muscle activity begins is generated from futile cycling of substrates through a series of linked anabolic and catabolic reactions simply to generate the by-product of heat. The best understood mechanism is that in the brown adipose tissues of mammals (see Fig. 6.10). Under resting conditions the BAT mitochondria function in the normal way, protons flowing back into the mitochondrial matrix through the ATP synthase enzyme, generating ATP and some heat (see Chapter 6). But when stimulated by norepinephrine from the sympathetic nerve endings the BAT mitochondria switch to a different proton route (Fig. 8.35d), due to the activity of **uncoupling proteins** (UCPs) on the mitochondrial inner membranes. This route is normally inhibited in the resting state by purine nucleotides. Once this inhibition is released, the protons flow back into the matrix via proton translocase channels ("thermogenin" sites, now identified as one isoform of UCP), and since there is no associated ATP synthase all the energy is liberated as heat. In effect, proton fluxes are uncoupled from the normal link to ATP synthesis. When this proton shunt pathway is operating, respiration no longer depends on ADP supplies, and the release of heat is only limited by carbohydrate and lipid substrates. Heat production can be as high as 500 W kg^{-1}, almost 10 times higher than in normal mammalian skeletal muscle. It is worth noting here that since UCPs were discovered in the BAT of mammals their homologs have been reported in fish, birds, insects, and even plants, and there is currently great interest in elucidating their wider physiological roles.

Other heat sources may play a minor role in heat generation in animals. Digestion certainly produces some heat, and in ruminants the additional heat from fermentation can substitute for shivering in the overall thermal repertoire. We can assume that large herbivorous dinosaurs gained significant heat in this way, adding to the effect of their large thermal inertia to insure that they were effectively "warm blooded".

8.7.3 Heat distribution: exchangers and countercurrents

Heat is produced in all body tissues, but particularly in the core organs of a resting animal, and in muscles during activity. This must be distributed around the body, principally in the blood or other body fluids, and dissipated suitably at the surface according to need. Clearly this need will vary; sometimes an animal will have to offload the excessive heat produced in a bout of activity, at other times it will be vital to conserve as much of the metabolic heat as possible. Most terrestrial animals, and many in aquatic environments, therefore have adaptations to insure appropriate blood flow patterns in relation to thermal needs.

The simplest form of control is to manage the rate and volume of the flow of blood to the surface relative to the core. In a cool and resting animal where most of the heat is generated in the core it can be effectively retained there with minimal peripheral vasodilation, keeping most of the circulatory flow in arterioles and venules and away from superficial capillary beds (Fig. 8.36). There may be direct arteriole to venule pathways (known as arteriovenous anastomoses) in vertebrates, as we saw in Chapter 7, allowing this bypassing of capillaries and keeping heat in the core. In an active or hot animal, peripheral arterioles dilate, sometimes with specific valve actions, and the capillary beds are incorporated into the circulation.

This technique in effect involves regulation of surface temperature, T_s, and it might be thought that it would inevitably become less effective for larger mammals and birds (with relatively less surface area) and especially for those with a substantial external insulating layer. However, the ability to control T_s can be measured in mammals as the **vasomotor index** (VMI), and this has been shown to scale positively with body weight with an exponent of 0.28. This means that large mammals use vasomotor control and regulation of T_s very intensively as a thermoregulatory strategy, whereas smaller mammals (with an intrinsically high level of thermogenesis

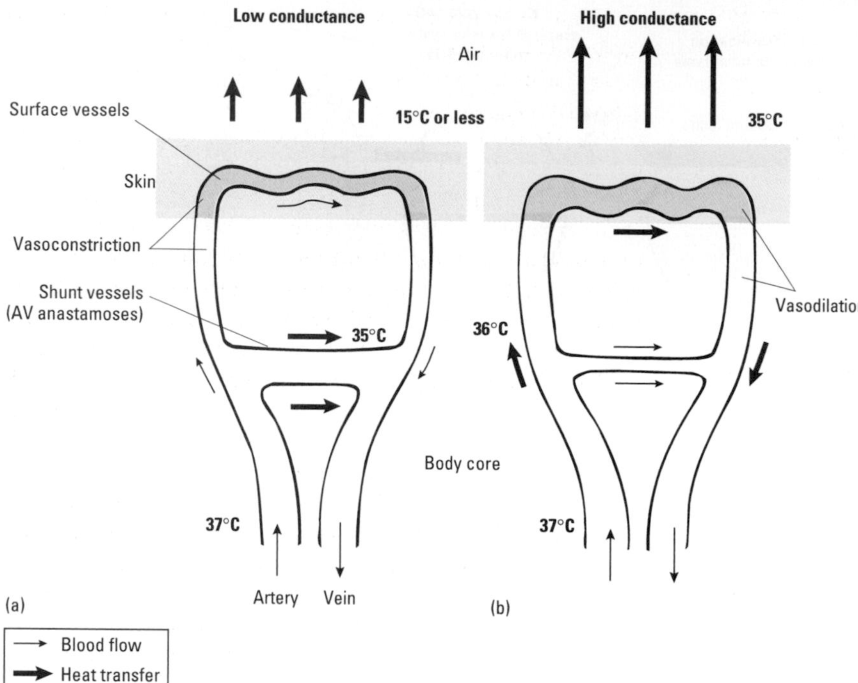

Fig. 8.36 The role of blood flow to the skin in regulating the heat conductance of the body surface and hence heat loss. Vasomotor control of peripheral arterioles and arterio-venous (AV) anastomoses is mainly responsible for shunting blood either (a) to or (b) away from the skin.

and limited absolute surface area) instead have to rely more on reducing their metabolic rate when heat stressed. Indeed, conspicuous vasodilation is most frequently described in large and endothermic animals.

Nevertheless, physiological thermoregulation by peripheral blood flow control is certainly also present in many smaller and ectothermic animals, such as lizards and even insects. It is a major part of the explanation for the common observation that heating rates in animals can be substantially faster than cooling rates (see Fig. 8.23). During solar heating in a reptile, for example, the heart rate is kept high and blood flow to the skin is high, whereas during cooling the heart rate drops very quickly and little of the hot blood flows peripherally, so controlling and reducing heat loss (see Fig. 15.25).

A more subtle control of heat distribution comes from the addition of **countercurrent heat exchangers** to the blood circuitry. These involve the principles of countercurrent flow discussed in Chapter 5 (see Box 5.2) and Chapter 7, and a model of their function in regulating thermal balance (together with examples) is shown in Fig. 8.37. Countercurrent heat exchangers acting to conserve core heat are commonly found in the extremities of animals living in cold habitats, particularly in the legs or flippers of polar and cool temperate birds and mammals (e.g. whales and porpoises, seals, mink, platypus, gulls), the horns of some ungulates, the tails of beavers and other rodents, and the ears of Arctic hares. In all these cases, warm blood in the arterioles runs parallel and very close to the venous return, so that heat can be exchanged effectively and returned to the body, leaving the relatively poorly insulated appendage perpetually cool. This may work in conjunction with peripheral vasodilation and constriction patterns (Fig. 8.38). The result is a magnified thermal gradient along the limb or other extremity, giving rise to **regional heterothermy** (Fig. 8.39).

This kind of heat exchanger also occurs in some tropical and temperate animals. To a limited degree the legs of a human have an artery–deep vein exchange system, while some sloths and armadillos and some of the more primitive tropical forest primates (lemurs and lorises) exhibit limb exchangers too, probably to counter the cooling that would occur for an inactive creature on a wet and windy night.

Heat exchangers to conserve warmth certainly achieve significant thermal benefits for the animal, but they produce very cold extremities that could run the risk of freezing and tissue damage. Just enough heat must be added to the peripheral tissues to prevent this, and many birds and mammals will begin to pass occasional pulses of warm blood to their feet as the ambient temperature is lowered beyond 0°C. This is reflected in a slight increment in the metabolic rate curve at this same temperature (Fig. 8.40).

Heat exchangers can also be used to conserve heat within specific parts of the body, rather than just in the trunk as a whole relative to extremities. Examples occur in some of the fish that show regional endothermy, maintaining the central parts of the swimming musculature at an elevated temperature in an otherwise essentially ectothermic animal (see Chapter 11). In such fish there are parallel arrangements of arterial and venous pathways at the outer edge of the trunk myotomes, so that heat is effectively retained within the musculature.

In certain circumstances heat exchangers can be used to keep cool as well, as section 8.8 will show.

8.7.4 Insulation effects

Changing insulative properties is an obvious way of regulating heat gain, for example by an ability to vary the thickness of an insulating layer whether this be fur, feathers, or a layer of fatty tissue.

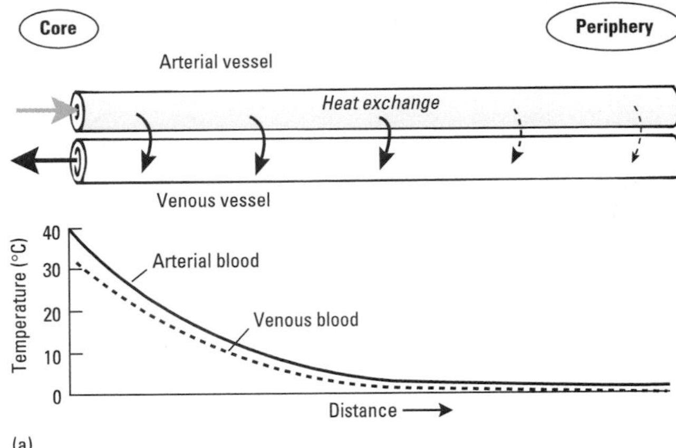

1 **Vena comitantes:** 2–3 veins surrounding central artery

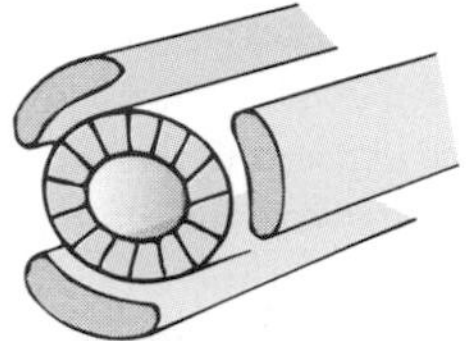

2 **Central rete:** central artery and many small veins

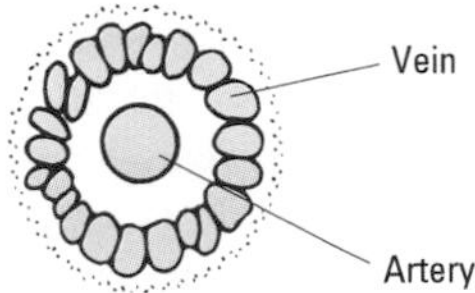

3 **Artery–vein rete:** many interdigitating vessels

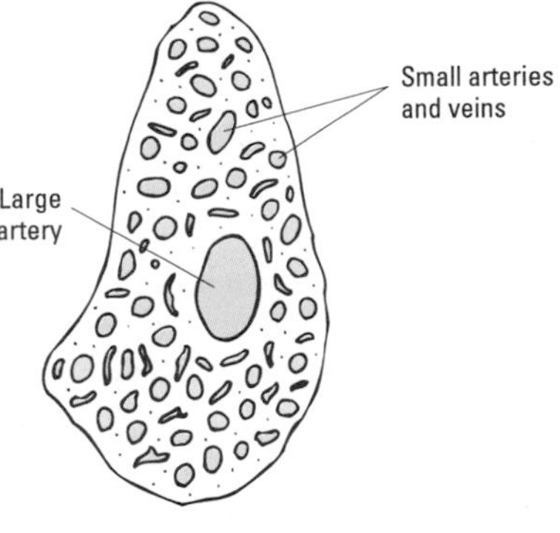

(b)

Fig. 8.37 Thermal countercurrent systems. (a) A model of flow and heat exchange. (b) Examples of three kinds of vascular arrangement.

For most mammals insulation change can be achieved in the short term by piloerection, which can decrease thermal conductance very substantially, retaining more heat in the body. In effect, the angle of each hair to the skin surface is changed by the contraction of muscles attached to the hair root, maximal contraction bringing the hair perpendicular to the skin. However, while partial erection thickens the insulating hair layer and helps to conserve heat, full erection on the dorsal surfaces of the body may increase heat loss by opening up routes for convective transfer of the warm air rising away from the skin surface. Some animals may piloerect more fully on their ventral surface when overheated, while keeping the dorsal fur flattened.

Birds achieve the same kind of heat-loss effect as mammals by the equivalent process of ptiloerection ("fluffing up" of their feathers), but since feathers are structurally more complex, and their position can be controlled and changed more radically, birds often have better insulation control than mammals. The soft down feathers closest to the skin provide the best insulation, but interlocking layers of contour and flight feathers above this add extra dimensions of insulating variability. All birds also possess a uropygial gland ("preen gland") near the base of the tail that provides an oily liquid to clean and lubricate the feathers. This may have little insulation value in itself, but by protecting the feathers from becoming waterlogged it insures that air remains trapped within the feathers so they retain their insulating effect even in heavy rain or after immersion; hence the gland's output is particularly copious in aquatic birds.

For aquatic mammals using the third main insulation system (i.e. subcutaneous fat deposits), changed insulation value may be effectively achieved by shunting the blood away from the surface, below the adipose layer (Fig. 8.41). This adds sophistication to the general strategy of shunting blood away from radiating surfaces, and it is a major advantage of having the main insulative layer internal to the heat-dissipating surface, rather than externally as with the fur and feathers of terrestrial birds and mammals, where no bypass is possible.

The strategies described so far may reduce heat loss in the short term. In the longer term, many birds and mammals grow much denser winter coats, often with a change of hair structure and color, while aquatic mammals may deposit thicker adipose layers as winter approaches (though of course this also serves the additional major function of providing a food store during migration between feeding grounds in whales or during beached periods for mating and giving birth in many seals). These winter strategies for retaining heat in endotherms are dealt with in Chapter 16.

Invertebrates and ectotherms in general lack any ability to change their heat conduction by varying the effective depth of insulative layers. However, some insects do have quite sophisticated insulation from hairs or from internal air sacs. In bumble-bees there are outer guard hairs and underlying more feathery down hairs, a striking parallel with the situation in birds and mammals. These hairs may differ in density over the body, and the ventral part of the abdomen often serves as a "thermal window", lacking the fur covering of the rest of the body. If the bee becomes overheated pulses of hot blood flow close to this uninsulated ventral abdominal cuticle to enhance heat loss (see Chapter 15).

8.7.5 Behavioral aspects

There are a great many ways of regulating heat gain (and heat loss) by behavioral means, but most of them can be viewed in general terms as ways of locating an appropriate microclimate, and ways of changing effective surface area. Techniques mainly operate for an individual, but in some cases group behaviors can help in thermal regulation.

Choice of microclimate is clearly going to be most important for smaller terrestrial animals, where the possibilities are so much greater. The issue of microclimates on land is dealt with more thoroughly in Chapter 15, but clearly intermittent basking in sunlit spots

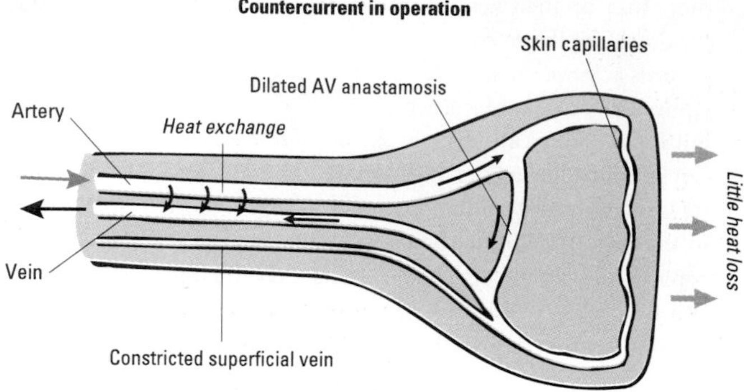

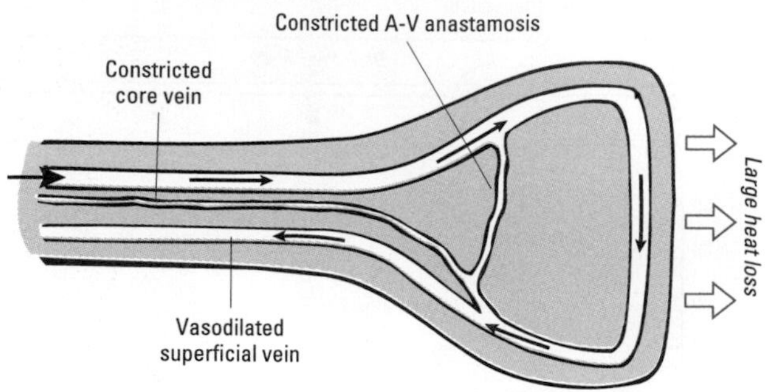

Fig. 8.38 How a countercurrent system can supplement the peripheral dilation and constriction shown in Fig. 8.36.

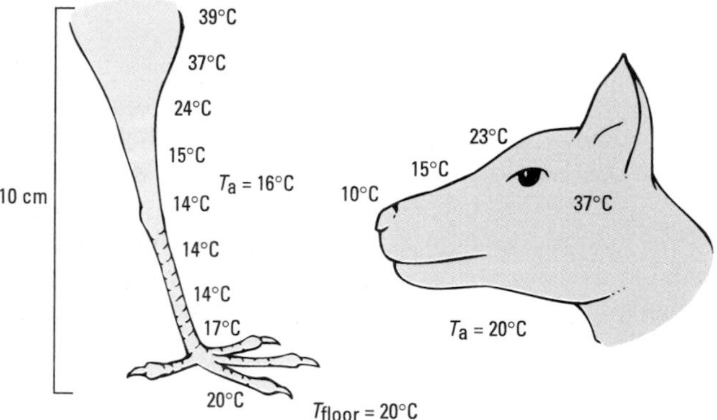

Fig. 8.39 Regional heterothermy resulting from peripheral blood flow management and countercurrents, in limbs and noses.

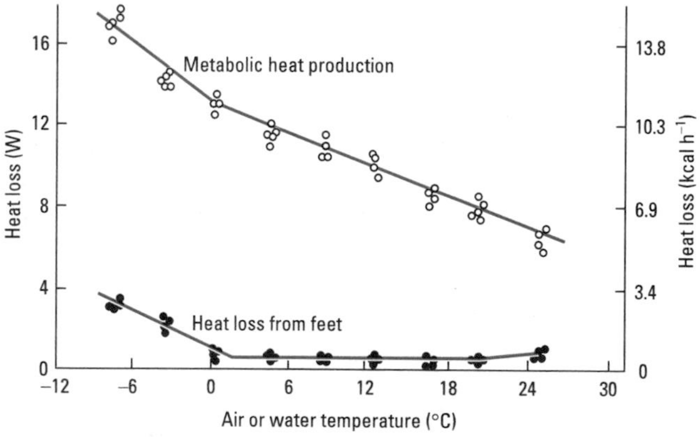

Fig. 8.40 At and below 0°C, intermittent pulses of increased blood flow to extremities are used to prevent freezing and tissue damage, reflected in increased heat loss from the feet of a duck and increased metabolic rate. (From Kilgore & Schmidt-Nielsen 1975.)

underlies many strategies: sun/shade shuttling behaviors, rotational movements around an area as the sun moves across the sky, vertical movements up and down surfaces projecting through the boundary layer, and the use of hide-outs, nests, and burrows are all of relevance here. Even nocturnal animals such as geckos and other reptiles can regulate their body temperatures quite substantially by choosing appropriate microclimates amongst available retreat sites.

Individual behaviors often involve general posture, and thus control the amount of surface exposed to the sun in basking. Basking is particularly useful for radiative warm-up in winged insects with large exposable surfaces. Many of the crests, sails, and ridges on early reptiles, such as *Dimetrodon* and the stegosaurs, were probably used as radiant heat gatherers in basking. In mammals, changing posture substantially affects the thermal balance by altering the insulation of the receiving surfaces—densely furred backs absorb very differently from nearly naked ventral and limb surfaces. Using posture to vary the surface exposed to conduction from heated substrata is another possibility; for example, many animals warm up by crouching against hot rocks, or in the heated sand at a burrow entrance.

Group behaviors to regulate warming are perhaps most obvious in the social insects, particularly those with colonies that overwinter such as hive-dwelling honey-bees in the genus *Apis*. Here a tight cluster of bees share their slowly generated body heat to raise the whole cluster to a temperature far above that attainable by any one individual. This can even be used as a form of aggression, since *Apis* will huddle around an invading hornet to overheat it. Huddling also occurs in many other terrestrial animals, and is therefore described more fully in Chapters 15 and 16.

8.8 Regulating heat loss and keeping cool

Keeping cool is not usually an enormous problem for the majority of animals. Aquatic animals rarely meet high temperatures, and cooling is only likely to be critical for some freshwater pond dwellers, and for some rock-pool animals, in high summer. Most terrestrial animals behave in such a way as to avoid high ambient temperatures, and smaller ones can use the microclimates around plants, or underground, to great effect. Only the larger land animals, unable to evade or avoid the midday heat of the sun, are obviously going to have a problem with cooling; though for them the high radiation loads around midday may pose an overheating problem even in polar latitudes.

However, all thermal issues have to be seen in relation to a particular animal's habitat and experience, and many other animals may in practice have to regulate the heat-loss side of their thermal

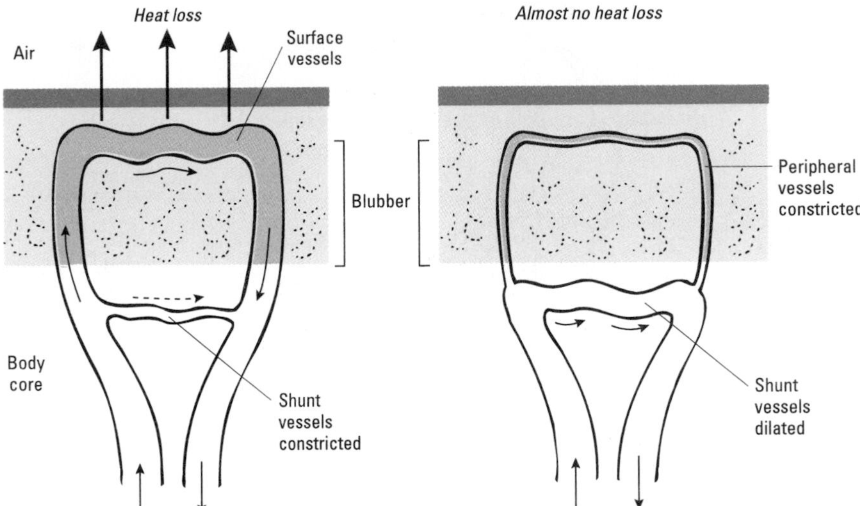

Fig. 8.41 The use of shunt vessels in aquatic endotherms to send blood either above the blubber layer for heat loss, or to keep it below the blubber where it is insulated.

balance to maintain their own T_{pref}. They also have to achieve appropriate activity levels at appropriate times of day and in the right places. After all, avoidance may be the easiest strategy, but at some point every individual has to feed, meet others of its species, and reproduce, and during these periods heat may inevitably be gained from the environment or from the metabolic effort supporting the activity, so that regulating heat loss may need attention.

8.8.1 Heat storage

Many animals use transient hyperthermia (where T_b exceeds the normal regulated value) as a way of managing exposure to hot conditions or heat load due to exercise. They thereby retain a gradient for shedding heat slowly to the environment without having to employ specific heat-loss mechanisms that may have other associated costs. The larger the animal, the longer it can tolerate a degree of hyperthermia. For example, small desert rodents (100–250 g) may tolerate brief overheating to a T_b of 43°C during foraging bouts lasting 15–30 min, shedding this heat load on return to inactivity in the cool burrow; whereas a 250 kg camel may take all day to warm from its overnight T_b of 35°C to an evening temperature of 40°C, shedding the excess heat overnight again.

Having a high body temperature has two advantages. Firstly, it increases the thermal gradient for heat loss from the body, or at least (if T_a exceeds T_b) it reduces the heat gain. Secondly, for each 1°C increase in T_b around 3.4 kJ of heat is retained per kilogram body mass, and this can be dissipated when heat production returns to the resting level or ambient temperatures drop (and without wasting water in evaporative heat loss).

Heat storage is clearly a useful strategy for small animals with cool refugia, or for large animals in climates where night temperatures are much lower, but it is only a stop gap solution for most animals—at some point, actual heat-loss regulatory systems must be engaged.

8.8.2 Heat exchangers and blood distribution

One way for an animal to lose excess heat is to shunt more of the hot blood within the core to the surface, effectively increasing the thermal conductance of the outer layers; this is essentially the inverse of the blood-shunting techniques mentioned in section 8.7. Peripheral vasodilation is an essential component of homeostasis during exercise, and occurs in many situations of temporary overheating. It forms part of a general suite of blood-shunting strategies, with flow increased to muscles, heart, and brain during moderate exercise, and then with blood shunted even more peripherally to the skin, with concomitant reductions of flow to the gut and viscera, as the body warms further (Table 8.12; note that in mammals, birds, and reptiles the increased flow is largely due to an increased heart rate, whilst in fish it is mainly brought about by an increased stroke volume of the heart).

Table 8.12 Cardiovascular changes during exercise in vertebrates. Numerical values shown are for a maximally exercising adult.

	Human male		Pigeon		Salmon	
Factor	**Resting**	**Exercise**	**Resting**	**Exercise**	**Resting**	**Exercise**
Heart rate (min^{-1})	75	180	100	600	38	51
Stroke volume (ml)	77	130	1.0	1.0	0.5	1.1
Cardiac output (ml min^{-1})	5700	25,000	–	–	18	53
Cardiac output per kg	85	360	–	–	18	53

Blood shunting during exercise

Blood flow increments:	Skeletal muscle, brain, heart, skin
Blood flow reductions:	Nonskeletal muscle, gut, kidney, liver, spleen

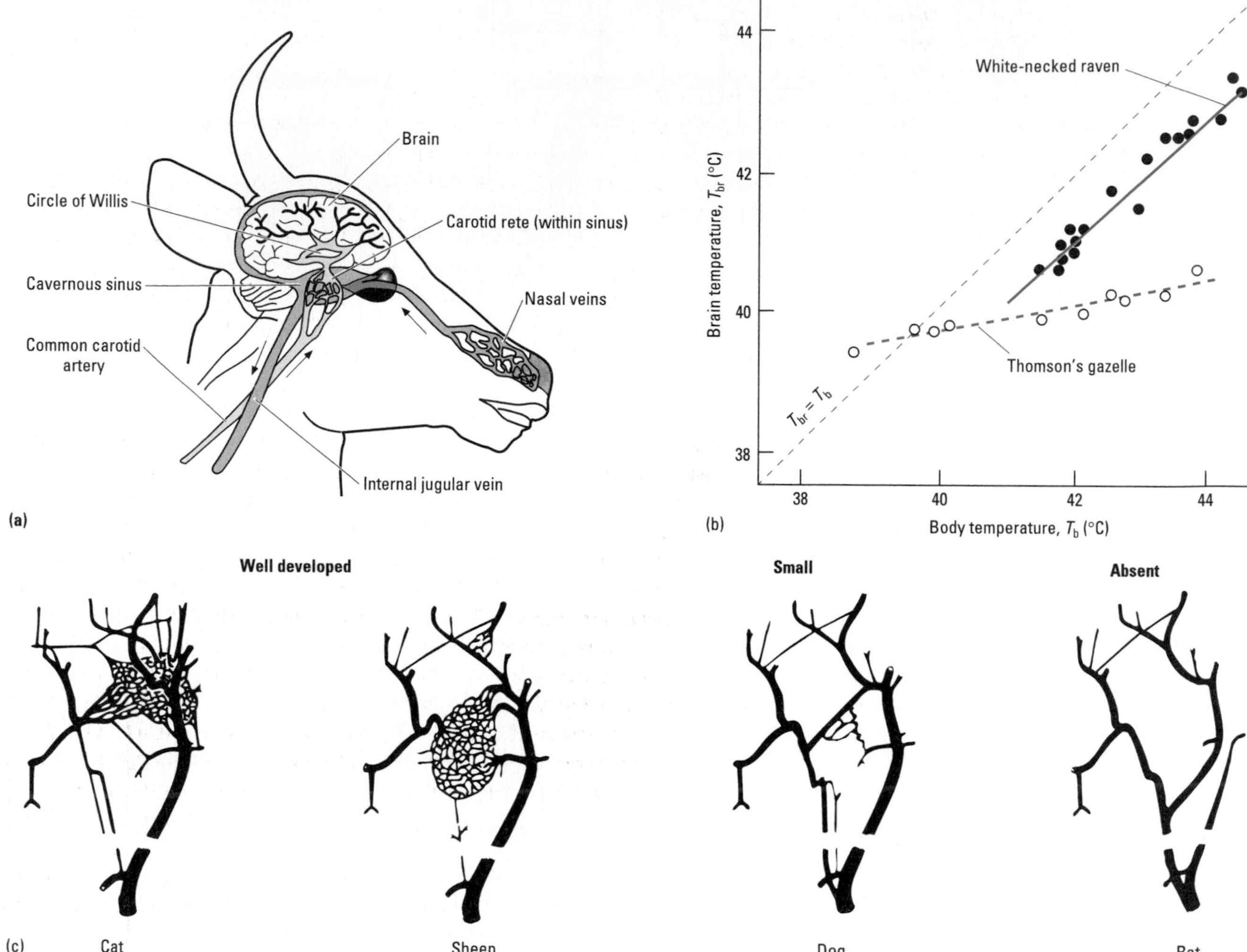

Fig. 8.42 The carotid rete to give brain cooling, as found in many mammals and birds. (a) Basic design, (b) effect on brain temperature in two species, and (c) patterns of rete blood flow in various mammals. (b, Adapted from Taylor & Lyman 1972; Kilgore *et al.* 1976; c, from Schmidt-Nielsen 1972, after Daniel *et al.* 1953.)

More specific regulation of heat loss can again be achieved with heat exchangers. The single most vital area that cannot be allowed to overheat in active animals is the brain, as neurons are extremely sensitive to thermal disturbance. Yet at the same time the brain requires a continuing high blood flow. The commonly adopted mechanism to keep the brain cool is to install a heat exchanger on the supply side to the brain, so that the arterial input (the carotid artery in tetrapods) breaks up into a capillary network and runs parallel to a cooled blood supply just before it enters the brain itself. Such a structure is known as a carotid rete (Fig. 8.42), and may occur in mammals and birds from hot environments (see Chapter 16) and in those that use intense bursts of activity for prey capture, as well as in some lizards. It has the additional advantage of keeping the thermal receptors in the brain below the level at which they might trigger additional cooling by sweating or panting, which could allow excessive evaporative water loss in a desert species. A few insects have an alternative approach to brain cooling; nectar feeders may achieve cooling of the head by regurgitating nectar onto the tongue and allowing this to evaporate.

It is also worth noting that when animals from cool environments do have overheating problems, for example after prolonged activity, they may use most of the peripheral heat-exchange systems mentioned previously as heat dissipators rather than as conservation systems. Ears, flippers, horns, and the like may become temporary thermal windows through which the excess heat is off-loaded, by sending arterial blood peripherally instead of through the deep arteries adjacent to the venous return path.

8.8.3 Changing surface properties

Another possible way to regulate heat loss is to modify the actual properties of the heat-dissipating surfaces of the body. Changing the insulative value of a furred or feathered surface is an obvious example, and this is taken to extremes in humans, who can take off layers of clothing. The commoner response is to reverse any piloerection, flattening the fur or plumage and reducing the trapped air layer. Wetting the insulation by a brief immersion is even more effective, and is often employed by birds and mammals in hot weather, largely for thermal purposes. Two effects are achieved: there is heat loss by evaporation, and this is supplemented by extra heat loss through the greatly enhanced conductance of the wet surface layer. For example, the insulation of eider duck feathers is about 1.6°C m^2 s J^{-1} in air, and only 0.8 in water. This is principally because the trapped water is a better conductor than air, but also because the layer is compressed into a reduced thickness by the water anyway.

Occasionally an apparently unlikely response occurs, when animals in hot environments partially raise their dorsal fur or plumage. This is observed when the air temperature exceeds the animal's T_b, such that a thick insulation layer will reduce conduction, convection, and radiation into the body, thus reducing heat uptake. This also explains why some desert dwellers, such as ostriches, goats, sheep, and even camels (at least dorsally), have surprisingly dense pelage.

Coloration also affects radiative exchanges as we have seen, and some animals can change skin or cuticle color in ways that are adaptive for keeping cool. Some lizards and a few frogs can do it; in lizards this may be due to "iridophore" cells, where quite rapid thermally induced color change arises from alterations in the packing between layers of reflective crystalline platelets (which are organelles containing crystals) inside the cells. Insect and crustacean examples are perhaps best studied, though. In insects color change generally involves phase changes of lipid or other material within the cuticle, and tends to produce a change from a nearly black surface appearance in an insect that is warming up to a shiny almost metallic blue or green in the same insect with a raised T_b, with a time course of minutes or hours. This sort of adaptive response is known in some grasshoppers, beetles, and dragonflies. In littoral and terrestrial crustaceans, particularly some isopods and smaller crabs, color change is brought about instead by dispersion or contraction of pigment-containing chromatophores, and may be a balanced response reflecting both camouflage and thermal needs.

8.8.4 Evaporation

Beyond the rather limited regulative possibilities already discussed, there is no doubt that evaporative heat loss is the key to keeping cool for most larger terrestrial animals, and for small animals in humid semiterrestrial niches such as the littoral zone and tropical forest litter.

There are various ways of achieving evaporative cooling, as different kinds of body fluid can be used. Some insects and reptiles indulge in urinating or in regurgitating gut fluids over their own surfaces, and we have just seen that some nectar-feeding animals, like bees and butterflies, will regurgitate nectar and saliva onto their elongated tongues and wave the droplet about to achieve some cooling ("tongue lashing"). Many animals will lick themselves to spread saliva over the chest, limbs, or flanks to achieve a cooling effect. But most obvious are the two techniques used by birds and mammals: **sweating** from special glands at various points over the body surfaces, and **panting**, a rapid breathing out through the mouth (rather than the nose, which is commonly designed to conserve water—see Chapter 5) to lose water from the respiratory surfaces. In effect these two mechanisms involve supplementary cutaneous evaporative water loss (CEWL) and supplementary respiratory evaporative water loss (REWL), respectively (see Chapter 5). Most species use only one of these possible avenues of evaporative heat loss, though some can switch between alternative mechanisms.

The amount of heat needed to evaporate a fixed amount of water is the same for all these sources, so it is not immediately clear as to why one technique might be preferred over another. Sweating is virtually exclusive to mammals, though something similar happens in a few frogs and in one or two insects such as cicadas (see Chapter 16), and some CEWL is reported in heat-stressed birds where epinephrine causes increased blood flow to the skin. Since heat load is proportional to body surface area, and water available for evaporative cooling is proportional to body mass or volume, small mammals with a high surface area to volume ratio are at a disadvantage and do not normally sweat because of the danger of dehydration. Mammalian panting is therefore commoner in smaller herbivores and in nearly all carnivores; and panting is ubiquitous in birds (which lack sweat glands). Licking is frequently observed in Australian marsupials, and in many rodents, but is relatively ineffective and laborious as a means of evaporating water. Some mammals choose different methods in different circumstances; for example, kangaroos will sweat during exercise, but pant when stationary while subjected to a radiative heat load.

Sweating is clearly most effective when the skin is not hairy or feathered, and in relatively hairless mammals it can bring about a 100-fold increase in CEWL; perhaps not surprisingly it is more effective in humans than in almost any other mammal, with the sweating rate being highly attuned to small changes in internal temperature (Fig. 8.43). Yet many larger furred mammals, such as antelopes and camels, also rely on it to lose heat, perhaps because they live in such dry atmospheres that water evaporates readily even through their fur.

An obvious advantage of panting over all the other systems mentioned is that it involves the loss of only water, in the vapor form from the lungs, whereas other mechanisms involve at least some loss of solutes as well (heavily sweating mammals notoriously run the risk of salt depletion). A second advantage is that panting allows the body surface to remain hot, whereas sweating cools the skin so that heat gain may increase and loss by radiation is reduced. Finally, panting creates its own airflow and can thus speed up the evaporative process.

However, evaporation by panting does require increased ventilation above that which the animal would otherwise need. This can lead to two problems. Firstly it requires muscular work and so increases the heat production that it is designed to control. Secondly it can lead to an unusually rapid loss of carbon dioxide and hence a degree of alkalosis in the tissues. Humans can easily become dizzy and nauseous if they deliberately hyperventilate, because of this raised tissue pH. Both of these effects can be ameliorated by a commonly observed adaptation during panting—a switch to a much shallower and higher frequency respiration (Fig. 8.44a), often up to 10 times the normal respiratory rate. This reduces the muscular work substantially by utilizing the intrinsic elastic properties of the lungs, inflating them at about their own natural oscillation frequency (resonant frequency). Inhalation stretches the elastin elements in the alveolar walls, which return quickly to their unstretched

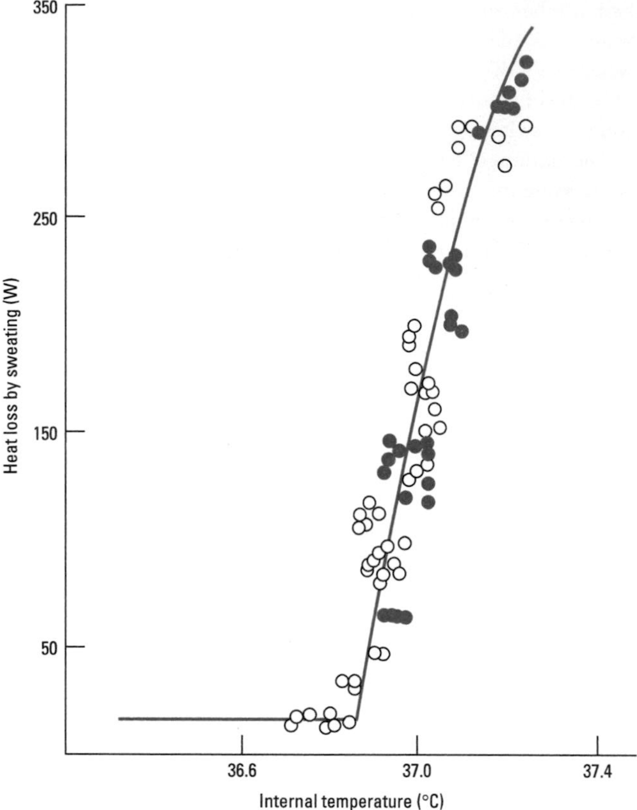

Fig. 8.43 Heat loss by sweating in a human, showing the small changes in body temperature (T_b) that elicit very substantial sweating rates. Core T_b was elevated either by exercise (green circles) or by increased ambient temperature (open circles). (From Benzinger 1961.)

state when exhalation begins, with little additional muscular effort. Shifting to a shallower respiration with a small tidal volume (inevitable when the respiratory rate is so much increased) also aids the alkalosis problem as much of the ventilation then involves only the upper dead space of the lungs, perhaps only 15% of the normal breath volume, and so does not remove carbon dioxide from the deeper alveoli. In birds this shallow, fast respiration is even more specialized as a "gular flutter" (Fig. 8.44b), which largely involves rapid (> 600 min^{-1}) vibration of the floor of the mouth by muscles pulling on the hyoid apparatus, with little lung ventilation; this can be used with or without simultaneous panting.

Above and beyond all these advantages and disadvantages of varying methods of evaporative cooling, it must be stressed that though the techniques are highly efficient for regulating heat loss they are also potentially highly dangerous. This is because of the close interrelation of thermal balance and water balance; very few terrestrial animals can afford too much use of evaporative cooling, as they lack the necessary water reserves. For many animals it is a "last resort" strategy when the risk of overheating overrides all other considerations.

8.8.5 Behavioral aspects

As with heat gain, the behavioral regulation of heat loss may operate on an individual or a group basis, and many of the techniques employed are the inverse of those dealt with in section 8.7.5 above. Many small-scale movements of invertebrates, ectothermic vertebrates, and even of smaller endotherms are primarily thermally related behaviors. Postures where minimal surfaces are exposed to the sun, and/or maximum surfaces to a breeze, are common in overheated insects, and endotherms may also "shut" their thermal windows by altering their posture. The opposite response to crouching for maximal heat gain is that of stilting for heat loss (or at least for minimal heat gain), most effective in long-legged insects and lizards and in some long-legged mammals.

On an individual basis there may be extra possibilities that involve exposing or erecting parts of the body that can act as radiators. These will be areas, such as ears, that are thin, vascularized, and of large surface area, so that heat can be lost quickly from the peripheral blood supply. "Pointing" this hot part of the body at a cool part of the sky may give particularly effective cooling, so that small mammals may sit in shade and point their noses or ears at the sky away from the sun. Examples of appendages as radiators are most obvious in large animals, particularly endotherms, and are dealt with in Chapter 16.

Group behaviors pertinent to heat loss most obviously involve huddling, a phenomenon widespread in terrestrial animals from many habitats but perhaps most spectacularly effective in the tundra and polar animals considered in Chapter 16. Other huddling animals include many young birds and mammals within their nests, where they can achieve an energy saving of up to 30% at low temperatures. Aggregation behavior akin to huddling is also used by many ectothermic animals, including some insects and littoral invertebrates such as isopods, to reduce heat loss.

8.9 Opting out: evasion systems in space or time

In seasonal climates at moderate and high latitudes, many animals adopt the simplest strategy of all when thermal problems become too severe, and effectively opt out of attempting to maintain a suitable body temperature. For a great many animals, particularly for small invertebrates, this involves a seasonal life cycle, where the periods of the year that are too hot or too cold are spent as a resistant stage, i.e. egg, cyst, or (in holometabolous insects) pupa. For many other animals there is a period of juvenile or adult quiescence or torpor. In insects this is determined intrinsically by endocrine changes and is termed diapause. For adult endotherms there may be periods of torpor, either on a diurnal cycle or through the winter ("hibernation") or through the summer ("estivation"). All of these options represent a way of evading thermal problems in time. There is a further alternative of evasion in space, which can be seen most spectacularly in the massive seasonal migrations undergone by some species, at least partly as a way of escaping adverse thermal conditions.

8.9.1 Torpor and diapause in ectotherms

Many aquatic animals show marked seasonal changes in activity that are related to water temperature. Some fish become relatively inactive and cease feeding, entering a state described as torpor or

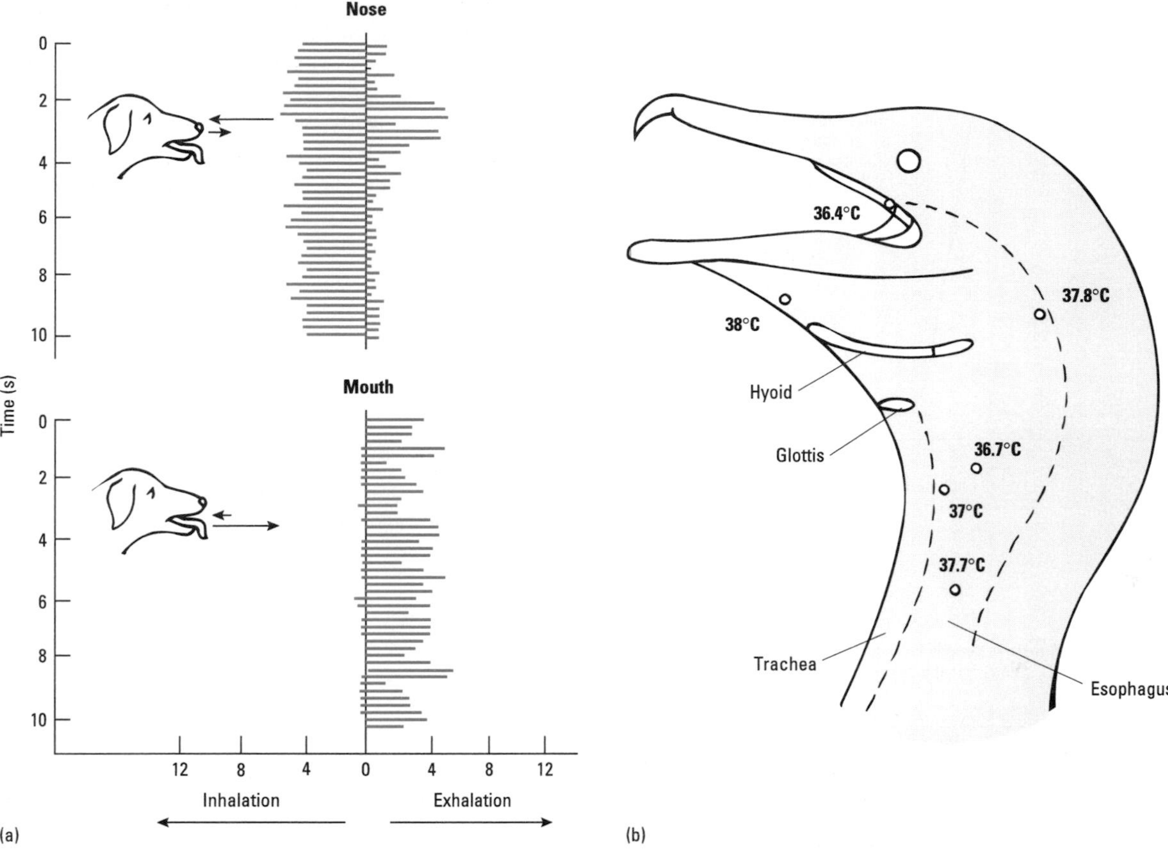

Fig. 8.44 (a) Respiratory patterns during panting in a dog: inhalation is predominantly through the nose and exhalation through the mouth, so avoiding heat retention in the nasal turbinals. Note that the respiratory rate is elevated to about 5 s^{-1}. (b) Temperatures in parts of the head of a bird during a period of gular fluttering, when the floor of the mouth and throat area is vibrated rapidly by movements of the hyoid apparatus, giving rapid evaporative heat loss and up to 2°C cooling in the mouth (T_b = 38°C). (a, From Schmidt-Nielsen *et al.* 1970a; b, from Lasiewski & Snyder 1969.)

dormancy. This strategy spares their food reserves until prey become more abundant again, and it is more common in freshwater species than in marine taxa. On land, amphibians and reptiles that show winter torpor are described as hibernating, whilst insects go into diapause in the egg, pupal, or adult stages. In all these cases, metabolic rates are lowered, and there is little or no activity or feeding. Fat development in subcutaneous layers of tetrapods, or in the insect fat body, is usually pronounced. Blood composition and muscle tone may be highly modified.

Ectotherms such as lungfish, frogs, and desert snails may show the "opposite" response of estivation in summer, with a substantially reduced metabolic rate despite the high temperatures. These animals often maintain their tissues in an unusually dehydrated state, and have various ways of resisting the arid conditions they would normally encounter by burrowing, sealing the skin and/or orifices with relatively waterproof membranes, or using air layers to buffer the tissues from excess heat; details are considered in Chapter 16. Estivation also occurs in a range of insects, where it is again termed diapause.

Diapause in insects is a rather special case of arrested development, and it may occur in any part of the life cycle according to the species (Fig. 8.45c). It usually involves an "anticipation" of forthcoming adverse conditions, and is generally triggered by changing day length. There may be a specific sensitive period, with the photoperiodic signals acting via the neuroendocrine system in several different fashions, suggesting several independent evolutionary origins. Before its onset, the larval or adult animal may accumulate extra nutrients and seek a sheltered microhabitat, often spinning a cocoon or constructing a shelter from leaves or litter or other extrinsic materials. Insect diapause can normally only be broken after a suitable period of cooling, again by an endocrine signal, this time from the brain to the prothoracic gland, normally serving to reinstate a molting program (Fig. 8.45d). In some species it has a strong genetic component in that certain populations are much more prone to diapause than others (with evidence of maternally controlled inheritance via the egg cytoplasm). Diapause is commonest in eggs and pupae, which are inherently inactive, and it is often triggered in the previous phase of the life cycle, so that larval instars respond to the triggers promoting pupal diapause and the diapause of eggs may be predetermined in the mother (e.g. in crickets and in the bog-dwelling dragonfly *Somatocholora alpestris*) thus providing an example of a cross-generational effect of temperature on phenotype. Diapause serves to synchronize the whole life cycle with the environment, insuring that further growth and development, and eventual reproduction, happen at a time when they will have the

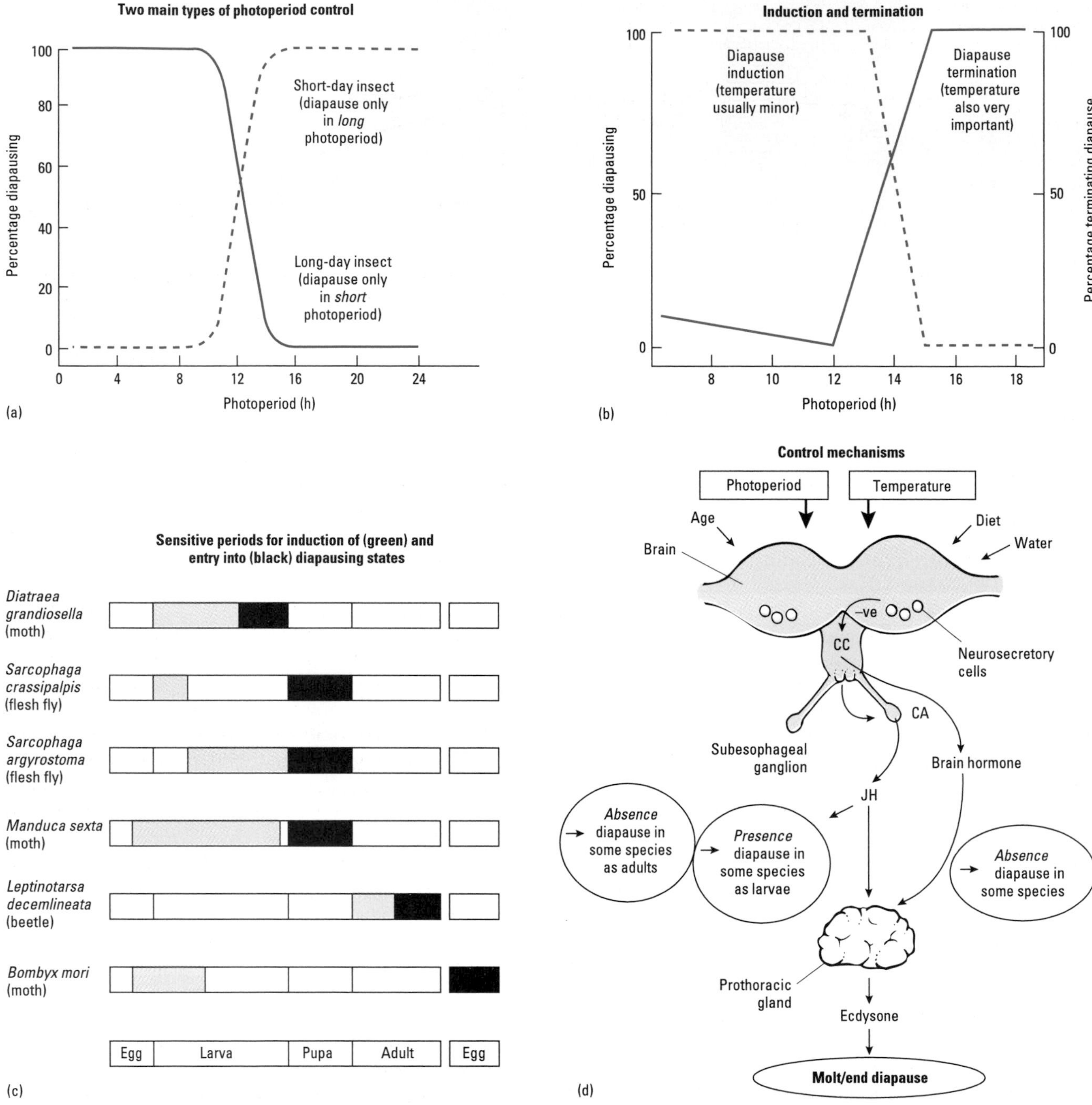

Fig. 8.45 Control of diapause in insects. CA, corpus allatum; CC, corpus cardiacum; JH, juvenile hormone. (c, From Kerkut, G.A. & Gilbert, L.I. *Comprehensive Insect Physiology, Biochemistry and Pharmacology*, Vol. 8. *Endocrinology II*, copyright 1985, with permission from Elsevier Science.)

best possible outcome. Insects in the diapausing state have extremely reduced respiratory rates, with very viscous body fluids and (for winter diapause) a high level of antifreezes such as glycerol. Some herbivorous species may spend as much as 10 months of the year in diapause, only emerging to be active at the time when their particular host plant is flourishing.

8.9.2 Hypothermia, torpor, and hibernation in endotherms

The terminology of hypothermia is very confused, and some clarification is needed at the outset. Many mammals, and relatively few birds, undergo regular winter periods of reduced metabolic activity, when the T_b drops substantially. This **torpor** occurs primarily in small endotherms such as rodents, insectivores, and bats, the small marsupial possums, and the monotreme echidnas, and in some hummingbirds and swifts. In advance of seasonal torpor substantial deposits of fat are usually laid down, giving a fuel reserve. Then metabolic rate, heart and ventilation rates, and most other

Table 8.13 Body temperatures (T_b) and metabolic rate (MR) changes in mammals and birds showing torpor. Basal MRs are similar to nontorpid mammals; here the ratio of torpid MR to normal basal MR is given. (Data from Geiser 1988.)

Species	Normal T_b (°C)	Torpid T_b (°C)	Ratio of torpid MR to normal MR
Monotremes			
Echidna	32.2	5.7	0.44
Marsupials			
Pygmy possum	33.7	10.1	0.22
Insectivores			
Shrew	34.7	14.0	0.10
Rodents			
Mouse	37.4	19.0	0.44
Dormouse	37.7	7.0	0.35
Ground squirrel	37.1	5.0	0.15
Carnivores			
Badger	37.0	28.0	0.50
Birds			
Poorwill	37.0	10.0	0.17
Hummingbird	40.0	21.0	0.13

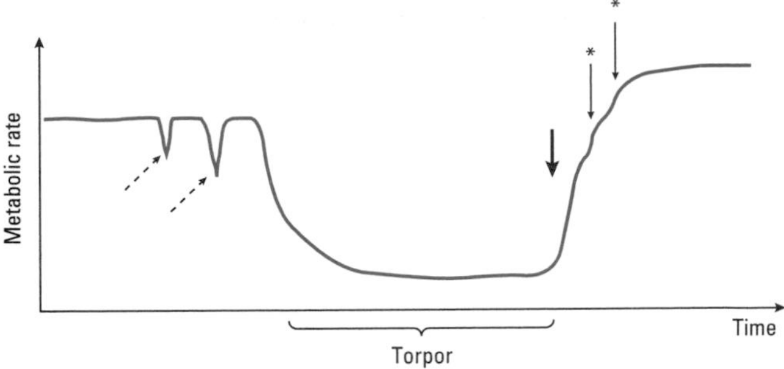

Fig. 8.46 The pattern of changes in metabolic rate during entry into and exit from torpor in a small mammal. Dashed arrows show preliminary "trial runs" when metabolism transiently slows; then entry into torpor occurs smoothly but relatively slowly. Torpor ceases abruptly (solid arrow) when the brown adipose tissue (BAT) is activated, and metabolic rate may subsequently show further brief periods of increment as shivering occurs (*).

physiological processes are greatly reduced, and the animal becomes quiescent and unresponsive to stimulation. Some examples of animals that use torpor, and the percentage savings of energy over maintained basal metabolism, are shown in Table 8.13. Most of the animals that undergo true winter torpor weigh less than 200 g, and the largest animals in this category are marmots, weighing up to 8 kg. For these small animals, the energy savings of torpor are clearly potentially huge.

True torpor is *not* a state in which endothermy and thermoregulation are abandoned; instead, it involves regulation at a new level, with a new critical minimum temperature being maintained. If ambient temperature falls below this, the metabolic rate will be increased to maintain the critical T_b. In many very small endothermic animals from temperate climates (such as hummingbirds and shrews), using torpor as an escape from transient seasonal or even nightly cold, the critical T_b is around 15°C, but in somewhat larger species from similar habitats (e.g. poorwills and honeypossums) it may be as low as 5°C. The terms "shallow" and "deep" torpor are sometimes used to distinguish these, though in practice there is a continuum of states. Many animals that undergo torpor on a daily basis use a shallow type (defined merely as T_b dipping below the normothermic range, as recorded by small implantable data loggers in free-ranging animals), and those with longer term and seasonal torpor normally use a deeper type, but there are many exceptions to this (see Chapter 16, p. 658).

Torpor is not easy to distinguish with clarity from "winter sleep", the latter often occurring in much larger animals such as skunks, beavers, or bears, where there is a relatively mild degree of torpidity, with the metabolic rate dropping a little and T_b only partially reduced, yet with many physiological functions continuing fairly normally. Torpor and winter sleep together are often loosely termed "hibernation". Nor is torpor entirely distinct from the kind of physiological shut-down and inactivity that may occur in many invertebrates in winter. Furthermore, many animals that show a deep winter torpor also show a degree of much shorter term nightly torpor, again with decreased metabolic rate and body temperature (e.g. hummingbirds and some bats, including many tropical species).

However, in the laboratory it transpires that the same rather common physiological principles probably underlie all these phenomena and they can best be treated together as variations on the theme of controlled hypothermia.

There is no single trigger for going into a torpid hypothermic state; entry may vary with food supply and with the imposed ambient temperature regime, though where there is a yearly cycle the onset of torpor is generally regulated through photoperiod cues. Torpor onset largely results from failure to increase the metabolic rate as T_a declines, because the setpoint for regulation has been changed. This leads to an inevitable decrease in T_b and a further decrease in metabolism, so that the animal slides rather gently into a torpid state (Fig. 8.46). In the fully torpid state, the metabolic rate may be only 1–20% of its normal resting value.

Arousal from torpor is a specific and very energy-demanding process, which may occur quite rapidly (1–6 h) relative to the process of entering torpor. Each species has a minimum temperature from which it can arouse, the **critical arousal temperature**. Arousal in most mammals is a two-stage process. At first it is not accompanied by obvious shivering or muscular activity, because a large part of the thermogenesis is achieved by nonshivering thermogenesis (NST) in the brown adipose tissue. Later on, shivering also occurs as the muscles become activated (Fig. 8.46).

Since these phenomena of torpor and arousal are particularly associated with land animals in cold environments they are dealt with in some detail in Chapter 16.

8.9.3 Migration

Migration is an alternative way of escaping seasonally adverse environmental conditions, although it is often triggered by dwindling food supplies, and/or by photoperiod, rather than being a direct response to temperature. It can occur on various spatial scales. Many animals show fairly local movements to more equable winter refugia. Littoral and terrestrial invertebrates commonly retreat deep into crevices in rocks or tree bark; freshwater invertebrates, and many frogs and turtles, seek refugia in the deeper waters of ponds

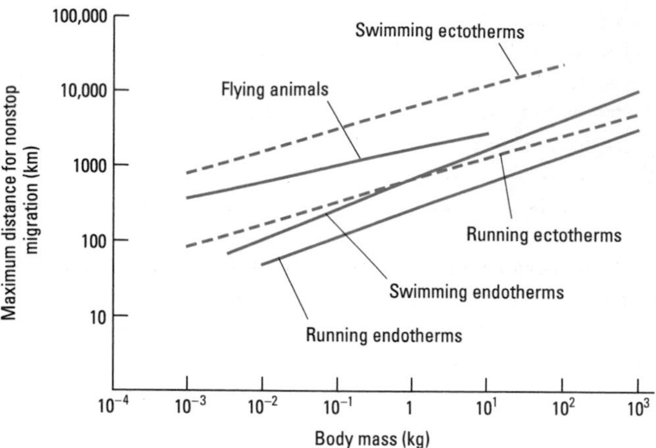

Fig. 8.47 Maximum possible migration distance in relation to size. Fish and flying animals can achieve long distances relatively cheaply, and larger animals can achieve much longer distances than small ones (compare Figs 3.14 & 3.16). (From Peters 1983, courtesy of Cambridge University Press.)

and streams, where the water will remain at 4°C below ice cover; a whole variety of land animals use caves (ranging from spiders, woodlice, and hoverflies to snakes, bats, and bears), or dig down 1 m or more to insure that they stay below the frostline, often combining this with some form of controlled hypothermia as discussed above. Summer migrations may also occur to seek out damper and cooler refugia, often to the deeper soils in deserts.

On a grander scale, migration involves a translocation to entirely new biomes, by long-distance migrants. We discussed briefly in Chapter 3 (see section 3.5.2) the effects of scaling on migration, and while net costs of transport are lowest for swimming, especially in ectotherms (see Fig. 3.14), so that ectothermic fish can undertake prolonged seasonal migrations due to their low feeding requirements, it is flight that permits much longer range movement for most animals over a wide range of sizes (Fig. 8.47). Migration by swimming or walking is only possible for endotherms if they are relatively large, and takes many weeks or months to complete. Amongst endotherms only the birds are conspicuous and common migrators, together with some marine mammals; a few large land mammals do use this strategy (perhaps rather fewer now than in the recent past, due to man's impact on traditional migration routes) but most high-latitude mammals use an *in situ* torpor as the escape strategy of choice.

All long-range movements require specialized navigational skills, with a need to detect the spatial relationships of the starting point and the expected endpoint. Many animals use the stars as navigational aids, and this may be backed up by a magnetic sense (at least in bees and some birds). Landmark learning and olfactory cues are also important parts of the navigational repertoires. Figure 8.48 shows major routes taken over land and through the oceans by migrating species.

Aquatic migrations by fish are particularly well studied where they involve a movement between marine and fresh waters. Eels, for example, move out to deep oceanic waters to breed, returning to fresh water for most of their life cycle (a **katadromous** migration). A number of tropical fish species show similar behavior. In contrast, many high-latitude salmonids show **anadromous** migration, where

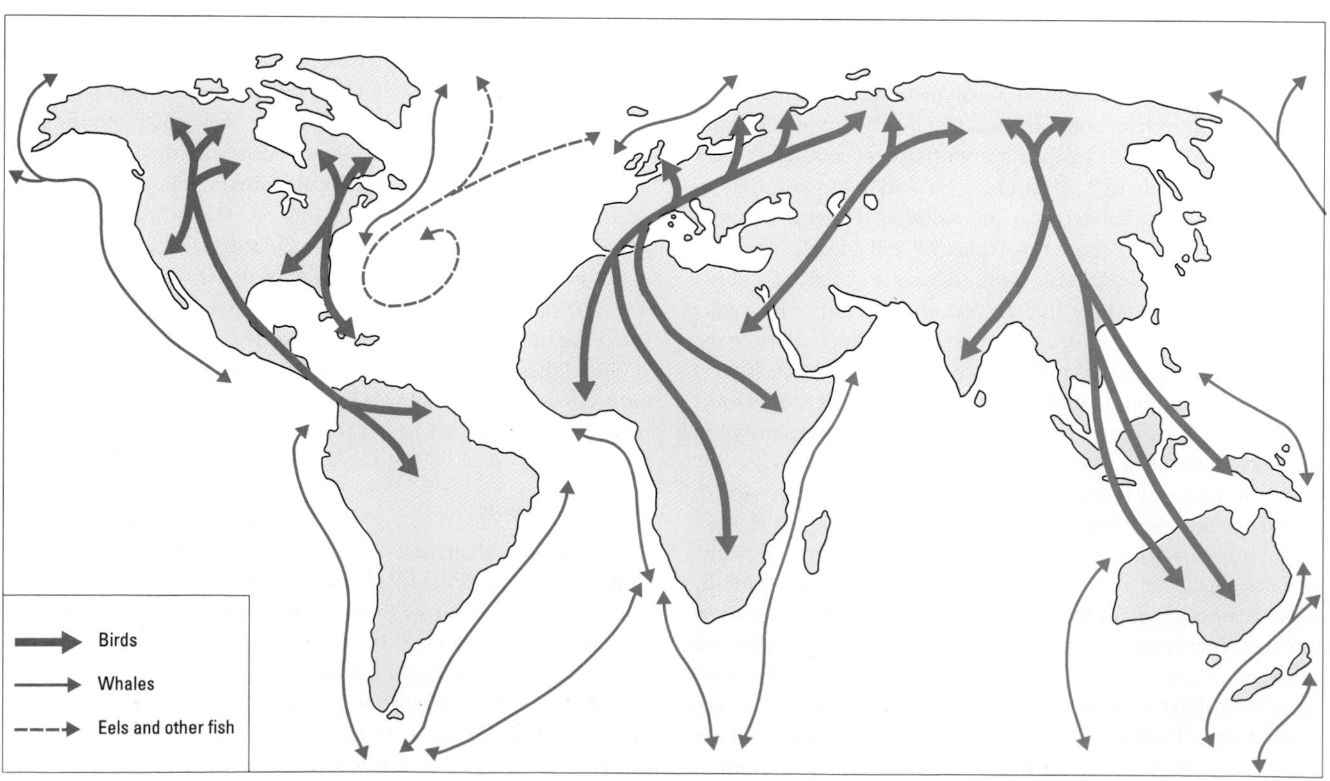

Fig. 8.48 Major marine and terrestrial migration routes.

breeding occurs in the headwaters of rivers and the young then migrate out to sea to feed, often for several years. Olfactory cues seem to be the major basis for "homing" behavior, allowing the returning fish to locate the river system they originated from.

Cetaceans also show marine migrations, with various species of *Balaenoptera* (blue, right, humpback, and minke whales) particularly well studied. Some of these baleen whales spend the months from April to September feeding on krill (*Euphausia*) in the Arctic and move to tropical waters around November; others spend November to April feeding similarly in the Antarctic, and move to the tropics around May. During breeding in the tropical waters there is little food intake and body weights may fall by 25–40%. Few baleen whales cross between hemispheres, and the populations rarely intermingle. Amongst the toothed whales (such as the sperm whale *Physeter*) migrations are less extensive, with populations shifting between subtropical and tropical waters of either hemisphere. Feeding is more or less continuous and changes in blubber thickness are much less severe than in the baleen whales.

Migration by flight can be equally spectacular. Amongst insects, long-range movement is relatively uncommon, but the most familiar examples are the monarch butterflies (*Danaus plexippus*) that migrate from North America to southern California and to Mexico each autumn (up to 3000 km), spending the winter in enormous aggregations in the warmer forests they encounter. Some bats also migrate over perhaps 100–1000 km, usually moving from northern coniferous forests to more southerly Mediterranean or chaparral habitats. Bird migrations are far more spectacular and can encompass almost the entire globe, with many species moving annually between hemispheres. Several hundred Palaearctic-breeding species (perhaps 5–6 billion individual birds) move southwards to feeding grounds in Africa each year, many of them using the narrowest water gap routes to get maximum uplift from thermals over land, and thus concentrating over southern Spain, through Italy, and across the Middle East and the Sinai Peninsula; similar numbers from the eastern Palaearctic move into Southeast Asia and Australasia. At the same time a rather high percentage of birds breeding in northern North America migrate to the forested areas of Florida, Mexico, and northern parts of South America for the winter.

On land, two groups of ungulates provide the major examples of long-distance migration. In the far north, elk and caribou (reindeer) migrate northwards into the tundra in spring, returning to the coniferous forests of the taiga belt in the autumn. In the savanna biomes of the world a variety of ungulates, including zebra, wildebeest, and several gazelle species, move regularly between grazing areas as the pattern of precipitation changes with the season, covering perhaps 700–1000 km in the course of a whole year. Both these migratory systems are governed largely by food availability, though linked, respectively, with snowfall and rainfall patterns; neither involves a great alleviation of thermal problems.

8.10 Regulating thermal biology: nerves and hormones

Clearly animals must have a range of thermal effectors used to control thermal exchanges, and hence T_b, by modifying a great variety of processes, from cellular enzyme concentrations to the behavior of the whole body. Equally clearly, there must be sensors within the body, and control centers to integrate the various thermal responses of each individual. This has been studied most intensively in mammals and birds, where central sensing and processing occur predominantly in interneurons of the preoptic center of the hypothalamus, with additional input from sensors in the skin, spinal cord, and abdomen. Similar principles of thermal coordination are likely to apply in other animals.

Current evidence indicates that there are dual setpoint regulatory systems in mammalian and reptilian nervous systems, with separate neural thermostats for controlling heat gain and heat loss. Some neurons ("warm receptors"; see Chapter 9) increase firing when the hypothalamic temperature rises, and probably link to neurons that activate heat-dissipation systems such as shade seeking, sweating, or vasodilation. Other neurons ("cold receptors") decrease in firing under the same stimulus, and may initiate vasoconstriction, basking, and other behavioral effects. Still others only increase in firing when the hypothalamic temperature falls below a specific setpoint, and these are likely to trigger heat-production mechanisms (shivering, etc.) and conservation mechanisms (piloerection, huddling behaviors, etc.). A highly schematic view of the possible control systems is shown in Fig. 8.49.

In many animals, including mammals and lizards from a range of aquatic and terrestrial habitats, the resultant T_b shows a negatively skewed distribution around T_{pref}. Normally this involves substantially more regulatory precision for the high setpoint. The desert iguana has a trigger point for mechanisms to increase T_b, such as shuttling into the sun, at 36.4°C (also known as the **low body temperature setpoint**, LBTS); and a trigger point for cooling mechanisms (**high body temperature setpoint**, HBTS) at 41.7°C. Thus an iguana may sit in the sun until T_b reaches 41.7°C and then shuttle into the shade until it cools to 36.4°C. In practice as these are averages each behavioral thermal shuttling cycle varies a little; the LBTS is generally more variable than the HBTS. This is likely to be adaptive, as T_b rising towards the critical thermal maximum is inherently more dangerous to the animal than a T_b falling moderately below the preferred levels.

In all animals studied to date, the role of the peripheral thermoreceptors is very much secondary to that of the core receptors and brain regulation center (Fig. 8.50). Clearly reflex adjustments of local vasodilation or piloerection, when an extremity is close to a hot rock or dipped into cold water, cannot be allowed to override central needs determining the whole body's core temperature.

However, in some of the smaller endotherms the peripheral sensors are used in a key role to reset the setpoints of the central thermoregulating system, so that as the weather cools the central setpoint rises to allow for greater heat production in the difficult time ahead. For example, in lizards the reception of light of different wavelengths by the parietal eye (a photoreceptive structure on the dorsal surface of the head) affects the thermoregulatory setpoints. Similar but more drastic adjustment of the neural setpoints must underlie the torpor and hibernation responses discussed above.

Hormones seem only rarely to have direct controlling effects on animal temperature balance, and these processes are covered in detail in section 10.6. A few of the vertebrate hormones (some steroids, and epinephrine) have effects on glycogen breakdown as we saw earlier, and hence may raise metabolic rate; the thyroid

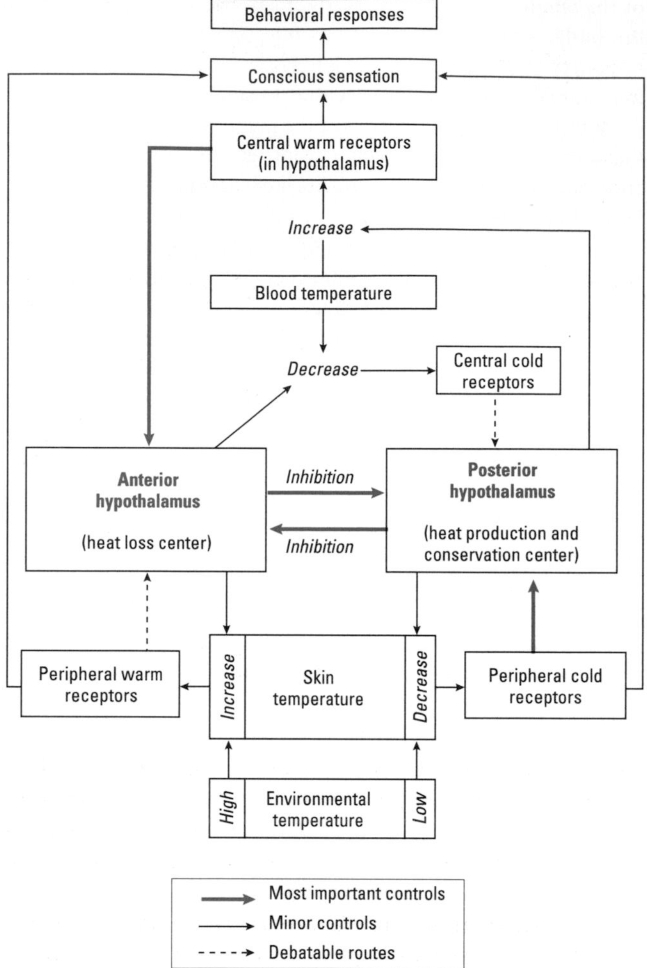

Fig. 8.49 Control systems regulating body temperature in mammals. (From Davenport 1992, Fig. 4.6, with kind permission from Kluwer Academic Publishers.)

hormones (thyroxine) also have direct effects in most endotherms in causing an increase in metabolic rate, and may act similarly in some lizards. Prostaglandins also have some regulatory effects, and are particularly involved in the fever response. Effects similar to those of thyroxine appear to operate at the fat body in insects and crustaceans, and prostaglandins may again be the main controlling factors triggering physiological and behavioral hyperthermia.

8.11 Evolution and advantages of varying thermal strategies

8.11.1 Relative merits of ectothermy and endothermy

Ectotherms have an approach to living that essentially involves low energy flows, and this can work in a very wide range of body sizes. Terrestrial ectotherms also usually have less need for water, therefore a whole range of arid microniches are available to them that endotherms cannot use (though ectotherms cannot normally occupy a nocturnal niche, except in tropical and summer temperate climates). Ectotherms can also devote a larger proportion of their energy budget to reproduction, with less investment in day-to-day maintenance, so that they are more likely to be good colonizers of poor-quality environments.

However, they are less able to sustain high activity bursts, running a risk of oxygen debt. They tend to use periods of anaerobic metabolism for activities beyond relatively slow movement; therefore, although briefly active, they fatigue rapidly due to lactate accumulation. Many of them are therefore ambush predators themselves, but are susceptible to intense sustained predation by endotherms.

In contrast endotherms use high energy flow systems, and they require larger body sizes, with relatively low surface area to volume ratios, to avoid excessive rates of heat loss and to enable them to eat enough to keep up with their metabolic needs. They have a continuous need for high-quality and/or high-quantity food, often coupled with a high water need, so that their numbers are severely limited in very poor-quality habitats. The extra respiration they require also gives them an increased risk of disrupted acid–base balance.

They can undergo much longer periods of intense aerobic activity than ectotherms, including prolonged nocturnal activity in all habitats and an ability to exploit colder environments at extreme latitudes or altitudes. Because they can maintain high-level activity for long periods, they are able to forage widely and migrate over long distances. Thus they have become the dominant terrestrial animals at all higher latitudes.

It may also be the case that a high and constant body temperature at all stages of the life cycle allows the development of more sophisticated neural processing in endotherms, and a greater complexity of learned behaviors; but this is allied with, and difficult to disentangle from, a longer contact period with the mother necessitated by the reproductive constraints on endotherms.

8.11.2 Evolution of endothermy

Endothermy has clearly evolved a number of times in animals, and some of the points made above probably indicate why this should be so in general, given its advantages in particular kinds of habitat. But much speculation has also centered on the particular issue of why and when endothermy evolved in the birds and the mammals, two different offshoots from a reptilian stock. There are two possible theories currently in play.

1 **Thermoregulatory advantage**, with maintenance of a steady T_b being primarily selected for. Such maintenance would be facilitated when the T_b level was rather high (see section 8.11.3), leading to selection for endothermic heat generation and a high resting metabolic rate.

A major difference between ectotherms and endotherms resides in their cellular machinery, with endotherms having increased net mitochondrial membrane density and rather "leaky" membranes surrounding both mitochondria and cytoplasm. The leaky plasma membranes (to Na^+ and K^+ especially, and also to protons) require extra metabolic energy to be generated to maintain normal solute concentration gradients, as discussed in Chapter 4. This is associated with increments in tissue protein levels and phospholipid concentrations in endotherms, and perhaps with increased unsaturation of the plasma membranes. Similarly the mitochondrial

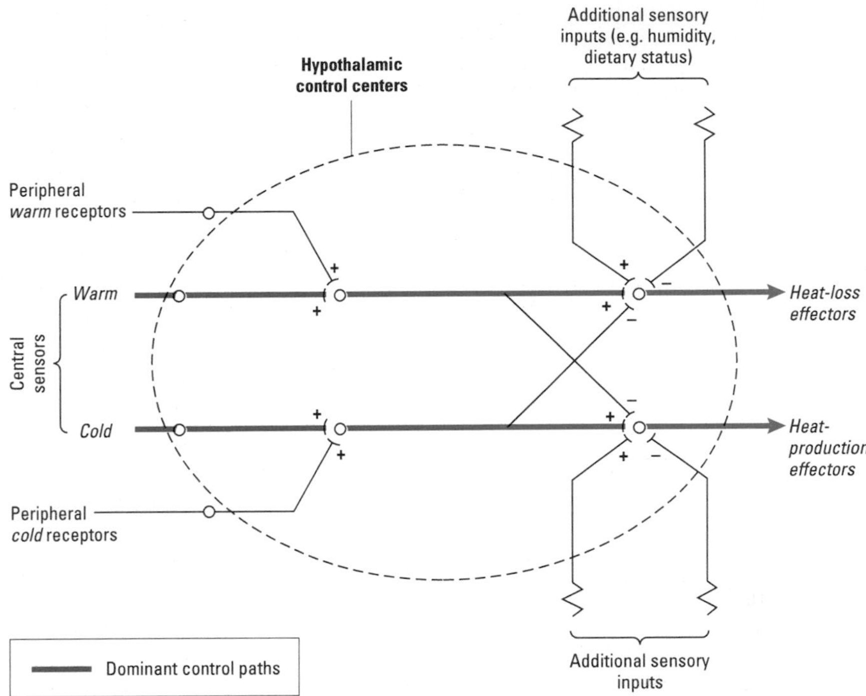

Fig. 8.50 The role of peripheral thermoreceptors.

"leak" is reflected in the increased rate of futile proton cycling in endotherm mitochondria. It has been proposed that these differences are linked to a need for increased thermogenesis in endotherm cells; i.e. a need for increased heat and heat regulation *per se*, supporting this thermoregulatory theory of endothermy origins.

There is an important argument against this theory, however. The evolution of increased metabolism on thermoregulatory grounds alone is difficult to explain in view of its high energetic cost. Small increments in maintenance metabolism would not have been sufficient to establish a steady T_b, but would have added to energetic demands with little apparent benefit. Thus intermediate states may have had no benefits in terms of fitness, if thermoregulation alone was the selective driving force. It is also worth noting that endothermy has evolved in plants, insects, fish, and possibly dinosaurs *without* being linked to routine thermoregulation; only in birds and mammals do the two go together. The "proton leak" may therefore be an unavoidable cost (potentially up to 20% of the basal metabolic rate of a mammal) of trying to maintain a high and steady body temperature.

2 **Aerobic capacity advantage**, with selection for high rates of aerobiosis allowing sustainable physical activity—effectively, selection for high maximum metabolic rates (MMRs). The net costs of transport for terrestrial walkers and runners are substantially greater than those for swimmers of equal body size (see Chapter 3). However, the aerobic capacities of amphibians and reptiles are not greatly expanded relative to those of fish of the same body size and temperature. Thus the ectothermic land tetrapods reach the limits of their endurance at much lower speeds and levels of exertion than their aquatic ancestors, and there might well have been strong selection to improve their endurance. Increased levels of maximum oxygen consumption would result in enhanced performance in a variety of activities that would influence fitness, such as prey pursuit, predator avoidance, territorial defense, and courtship. Ancestral animals in both the mammalian and avian lineages do seem to have been progressively more capable of vigorous activity, as shown by general skeletal arrangements such as changes in limb suspension and reduction of skeletal mass.

On this theory, elevated MMRs led secondarily to elevated resting metabolic rates (RMRs). There may be some fixed relation between them, and with carefully controlled comparisons the correlation between RMR and MMR is certainly rather tight. It may be that intense aerobic activity leads to increased permeability of membranes to supply oxygen, thereby requiring extra sodium pump activity all the time and linking to a permanently increased T_b. This might lead ultimately to metabolic rates high enough to achieve continuous endothermy and homeostasis. RMRs in birds and mammals are both higher (by a factor of 5–15) and more constant than those in most ectothermic vertebrates. However, this maintenance heat is not myogenically derived in endothermic vertebrates but is mainly generated by visceral organs and the central nervous system (e.g. in man 70% of resting metabolic heat is generated by the metabolism of the heart, kidneys, brain, liver, and intestines, even though they only constitute 7% of the body mass). It is not easy to see how this could have arisen as a way of supporting sustained muscular activities, although it is possible that once the advantages of "leaky" membranes had been acquired they could be spread through the body. Supporters of the aerobic capacity theory claim that the leakiness of membranes in endotherms might also be linked to increased rates of protein and phospholipid turnover. Increased levels of sustainable activity, on this model, would require extra provision of high-endurance oxidative-type muscle fibers, with increased demand for support and resupply to these from the digestive and excretory systems, linking in turn to enhanced resting as well as maximum metabolic rates.

Perhaps the "price" of increased maximal oxygen consumption was the disadvantage of inevitably but expensively increased RMRs. Thus it may be that the early stages in the selection for endothermy were indeed dominated by selection for increased endurance, and the later stages by selection for thermoregulatory homeostasis to insure maximum benefit from the inevitably raised resting metabolism.

Most of the paleontological evidence supports the aerobic capacity theory, particularly since incipient endothermy probably appeared in moderately sized predatory therapsids (mammal-like reptiles) at a time when they lived in warm equable climates and were likely to be moderate "inertial homeotherms" anyway, and capable like modern reptiles of highly efficient behavioral thermoregulation. However, it is difficult to test these theories directly by looking for the real origins of endothermy in the fossil record. We can rarely be sure whether particular fossil groups (such as dinosaurs, therapsids, pterosaurs, or *Archaeopteryx*) were endothermic or not, since endothermy is largely manifested in aspects of the soft tissues, of which no trace remains. Several possible lines of evidence for endothermy from paleontological studies have been proposed, and most remain controversial. These include: the gait and posture; the nature of the bones; the proportions of predators and prey in particular populations; the size of the braincase; the apparent growth rates; and the oxygen isotope ratios in bones. More recently the development of the nasal respiratory turbinals (especially the maxilloturbinate, bearing respiratory epithelium) has been advanced as evidence of endothermy, this being a more direct correlate of endothermic ability as it reflects a high ventilation rate and the capacity for conservation of water (see Chapter 6) and heat (this chapter) in the expired air. Complex turbinates occur in xeric and mesic mammals, and (convergently) in all birds, but are absent in reptiles. These structures, or at least the ridges to which they attach, are readily preserved in fossils, and thus may give a clue to the presence of endothermic respiratory rates. Such ridges are found throughout many of the mammal-like reptiles, even including pelycosaurs; endothermy seems to have accompanied or even predated the other aspects of "mammalness". However, in birds there is only a rather weak crest for turbinate attachment, and evidence from fossils is therefore less compelling, though it would indicate an absence of turbinates in *Archaeopteryx* and its immediate dinosaur ancestors. Thus it is inferred that endothermy postdated the origin of birds and avian flight by some 50 million years. Hence the selective regime favouring endothermy in birds was perhaps rather different from that in mammals, though in both cases a selection for increased aerobic scope was probably a major component.

8.11.3 Evolution of high body temperatures

Irrespective of the question of why endothermy might have evolved, there is a separate question as to why endotherms almost invariably settle for a body temperature of 35–40°C (see Table 8.11). These temperatures are also similar to those maintained by quite a range of active terrestrial ectotherms, notably reptiles and insects. This might suggest separate convergent evolution on this thermal range. Why should this be so?

One explanation would be the need for body temperature to be set rather high because biochemical reactions and particularly enzyme functions are then faster. Organisms can therefore respond more rapidly to any changes in cell conditions and thereby keep their own cellular environment more stable, allowing selection for enzymes with narrow thermal preferenda and for maximal efficiency within that narrow temperature range. Faster and more constant catalytic rates should also permit faster and more forceful muscle action.

Another point commonly made is that T_b needs to be above the highest T_a likely to be experienced, because then the animal can lose heat in a regulated manner by conduction, convection, and radiation. If the T_b is below ambient temperature, then the body can only cool by evaporation, and that is usually a dangerous strategy in terms of water balance. On this argument, a routine T_b approaching 40°C would be adequate in many terrestrial habitats.

Relatively high body temperatures will also favor faster neurological and hormonal function, hence allowing selection for quicker integration and behavioral sophistication.

All these theories would predict a selection for higher temperatures, but may not explain why the selected T_b is not higher still (remember that proteins can certainly be constructed to function at much higher temperatures if necessary, e.g. as in hot-springs prokaryotes). An alternative explanation for the choice of 35–40°C comes from recent evidence showing that the physical properties of water are such that the ideal balance of viscosity, specific heat, and state of ionization in living tissues is achieved at around 35–37°C; this input from physical chemistry may be all the explanation we need, though bird T_b values of 42°C remain somewhat problematic.

We may be able to see reasons for the selection for higher body temperature in those cases where animals exhibit unusually raised T_b values. An important example of this is the fever response in endotherms; controversy continues as to whether this is an adaptive response of benefit to the host (helping to disable the pathogen), or an effect induced by the pathogen and deleterious to the host. Table 8.14 shows examples of fever responses. There is evidence that in some cases the host actually regulates T_b slightly higher than normal, and that the survival of a host with a fever reaction is enhanced, the higher temperature helping to overcome the pathogen. This view would be supported by observations that an equivalent effect also occurs in ectotherms; leeches, grasshoppers, and some reptiles and amphibians will choose warmer habitats during an infection. *Anolis* lizards appear to show adaptive responses in that some species exhibit this form of "fever", whilst others with lower preferred T_b values do not. (However, the issue is complicated because some parasites can apparently also "control" whether they cause fever. Parasites with vectors are especially likely to cause fever, perhaps because it is then in the parasite's interest to have the host prostrate and more likely to be bitten by further vectors, whereas parasites with direct transmission are favored by keeping the host active, so fever is then less prevalent.)

There are also a few documented cases of animals selecting lower temperatures for particular reasons relating to interactions with parasites or prey items. Certain bumble-bees "choose" lower temperatures to reduce the effects that parasitic nematodes have upon them, and will stay out of the nest at night to this end. Spider-hunting wasps have also been shown to store their paralyzed prey items in chambers at lower temperatures than their normal living quarters, to slow deterioration and deter fungal growths; and some spiders

Table 8.14 Body temperatures, T_b (°C), during fever in invertebrate and vertebrate taxa.

Species	Normal (or preferred) T_b	Fever T_b	Difference
Invertebrates			
Crayfish	22.1	23.9	1.8
Shrimp	31.0	35.5	4.5
Lobster	16.0	20.7	4.7
Cockroach	32.3	35.9	3.6
Desert beetle	33.0	34.5	1.5
Horseshoe crab	27.0	33.0	6.0
Leech	20.5	30.0	9.5
Scorpion	25.0	40.0	15.0
Fish			
Sunfish	30.1	32.3	2.1
Bass	29.6	31.9	2.3
Amphibians			
Frog	25.5	27.9	2.4
Reptiles			
Desert iguana	41.0	42.7	1.7
Birds			
Pigeon	39.7	41.5	1.8
Mammals			
Dog	38.2	39.4	1.2
Rabbit	39.5	40.8	1.3
Monkey	38.9	40.1	1.2
Man	37.4	41.3	3.9

will also move prey around their nests or temporarily shift them out of the nest to keep the prey storage temperature within a narrow optimal range. Thus the selection of particular temperatures by animals is not a simple matter, but reflects a balance of biochemical, physiological, behavioral, and coevolutionary needs.

FURTHER READING

Books

Cossins, A.R. & Bowler, K. (1987) *Temperature Biology of Animals.* Chapman & Hall, London.

Davenport, J. (1992) *Animal Life at Low Temperature.* Chapman & Hall, London.

Heinrich, B. (1993) *The Hot-Blooded Insects.* Springer-Verlag, Berlin.

Johnston, I.A. & Bennett, A.F. (eds) (1996) *Animals and Temperature: Phenotypic and Evolutionary Adaptation.* Cambridge University Press, Cambridge, UK.

Lyman, C.P., Willis, J.S., Malan, A. & Wang, L.C. (1982) *Hibernation and Torpor in Mammals and Birds.* Academic Press, New York.

Morimoto, R., Tissieres, A. & Georgopoulos, C. (1994) *The Biology of Heat Shock Proteins and Molecular Chaperones.* Cold Spring Harbor Laboratory Press, Plainview, NY.

Reviews and scientific papers

Argyropoulos, G. & Harper, M.E. (2002) Uncoupling proteins and thermoregulation. *Journal of Applied Physiology* **92**, 2187–2198.

Atkinson, D. (1995) Effects of temperature on the size of aquatic ectotherms: exceptions to the general rule. *Journal of Thermal Biology* **20**, 61–74.

Atkinson, D. & Sibly, R.M. (1997) Why are organisms usually bigger in colder environments? Making sense of a life history puzzle. *Trends in Ecology & Evolution* **12**, 235–239.

Baker, M.A. (1982) Brain cooling in endotherms in heat and exercise. *Annual Review of Physiology* **44**, 85–96.

Bakken, G.S. (1992) Measurement and application of operative and standard operative temperatures in ecology. *American Zoologist* **32**, 194–216.

Barclay, R.M.R., Lausen, C.L. & Hollis, L. (2001) What's hot and what's not: defining torpor in free-ranging birds and mammals. *Canadian Journal of Zoology* **79**, 1885–1890.

Bennett, A.F. (1987) Evolution of the control of body temperature: is warmer better? In: *Comparative Physiology: Life in Water and on Land* (ed. P. Dejours, L. Bolis, C.R. Taylor & E.R. Weibel), pp. 421–434. Liviana Press, Padova, Italy.

Block, B.A. (1994) Thermogenesis in muscle. *Annual Review of Physiology* **56**, 535–577.

Boyer, B.B. & Barnes, B.M. (1999) Molecular and metabolic aspects of mammalian hibernation. *BioScience* **49**, 713–724.

Bozinovic, F. (1992) Scaling of basal and maximum metabolic rate in rodents and the aerobic capacity model for the evolution of endothermy. *Physiological Zoology* **65**, 921–932.

Brock, T.D. (1985) Life at high temperatures. *Science* **230**, 132–138.

Caceres, C.E. (1997) Dormancy in invertebrates. *Invertebrate Biology* **116**, 371–383.

Cannon, B. & Nedergaard, J. (1985) Biochemical mechanisms of thermogenesis. In: *Circulation, Respiration and Metabolism* (ed. R. Gilles), pp. 502–518. Springer-Verlag, Berlin.

Crockett, E.L. (1998) Cholesterol function in plasma membranes from ectotherms: membrane specific roles in adaptation to temperature. *American Zoologist* **38**, 291–304.

Detrich, H.W. (1997) Microtubule assembly in cold-adapted organisms: functional properties and structural adaptations of tubulins from Antarctic fishes. *Comparative Biochemistry & Physiology A* **118**, 501–513.

Duman, J.G. (2001) Antifreeze and ice-nucleator proteins in terrestrial arthropods. *Annual Review of Physiology* **63**, 327–357.

Else, P.L. & Hulbert, A.J. (1985) An allometric comparison of the mitochondria of mammalian and reptilian tissues: implications for the evolution of endothermy. *Journal of Comparative Physiology* **156**, 3–11.

Else, P.L. & Hulbert, A.J. (1987) Evolution of mammalian endothermic metabolism: 'leaky' membranes as a source of heat. *American Journal of Physiology* **253**, R1-R7.

Feder, M. & Hofmann, G.E. (1999) Heat shock proteins, molecular chaperones and the stress response; evolutionary and ecological physiology. *Annual Review of Physiology* **61**, 243–282.

Franks, F., Mathias, S.F. & Hartley, R.M.H. (1990) Water, temperature and life. *Philosophical Transactions of the Royal Society of London* **326**, 517–534.

Gibbs, A. & Mousseau, T.A. (1994) Thermal acclimation and genetic variation in cuticular lipids of the lesser migratory grasshopper: effects of lipid composition on biophysical properties. *Physiological Zoology* **67**, 1523–1543.

Hazel, J.R. (1984) Effects of temperature on the structure and metabolism of cell membranes in fish. *American Journal of Physiology* **246**, R460–R470.

Himms-Hagen, J. (1985) Brown adipose tissue metabolism and thermogenesis. *Annual Review of Nutrition* **5**, 69–94.

Huey, R.B. & Kingsolver, J.G. (1993) Evolution of resistance to high temperature in ectotherms. *American Naturalist* **142**, S21–S46.

Hughes, L. (2000) Biological consequences of global warming: is the signal already apparent? *Trends in Research in Ecology & Evolution* **15**, 56–61.

Johnston, I.A. & Walesby, N.J. (1977) Molecular mechanisms of temperature adaptation in fish myofibrillar adenosine triphosphatases. *Journal of Comparative Physiology* **119**, 195–206.

Kearney, M. & Predavec, M. (2000) Do nocturnal ectotherms thermoregulate? A study of the temperate gecko *Christinus marnoratus. Ecology* **81**, 2984–2996.

Kuhnen, G. (1997) Selective brain cooling reduces respiratory water loss during heat stress. *Comparative Biochemistry & Physiology A* **118**, 891–895.

Laszlo, A. (1988) The relation between heat shock proteins, thermotolerance and protein synthesis. *Experimental Cell Research* **178**, 401–408.

Leroi, A.M., Bennett, A.F. & Lenski, R.E. (1994) Temperature acclimation and competitive fitness; a test of the beneficial acclimation assumption. *Proceedings of the National Academy of Sciences USA* **91**, 1917–1921.

Logue, J.A., DeVries, E., Fodor, E. & Cossins, A.R. (2000) Lipid compositional correlates of temperature-adaptive interspecific differences in membrane physical structure. *Journal of Experimental Biology* **203**, 2105–2115.

Lovegrove, B.G., Heldmaier, G. & Ruf, T. (1991) Perspectives of endothermy revisited: the endothermic temperature range. *Journal of Thermal Biology* **16**, 185–197.

Lowell, B.B. & Spiegelman, B.M. (2000) Towards a molecular understanding of adaptive thermogenesis. *Nature* **404**, 652–660.

McNab, B.K. (1983) Energetics, body size and the limits to endothermy. *Journal of Zoology* **199**, 1–29.

Morrison, R.L., Sherbrooke, W.C. & Frost-Mason, S.K. (1996) Temperature-sensitive, physiologically active iridophores in the lizard *Urosaurus ornatus*: an ultrastructural analysis of color change. *Copeia* **1996**, 804–812.

Parker, A.R. (1998) The diversity and implications of animal structural colours. *Journal of Experimental Biology* **201**, 2343–2347.

Partridge, L. & Coyne, J.A. (1997) Bergmann's rule in ectotherms: is it adaptive? *Evolution* **51**, 632–635.

Pough, F.H. (1980) The advantages of ectothermy for tetrapods. *American Naturalist* **115**, 92–112.

Ruben, J. (1995) The evolution of endothermy in mammals and birds: from physiology to fossils. *Annual Review of Physiology* **57**, 69–95.

Somero, G.N. (1995) Proteins and temperature. *Annual Review of Physiology* **57**, 43–68.

Storey, K.B. (1997) Organic solutes in freezing tolerance. *Comparative Biochemistry & Physiology A* **117**, 319–326.

Storey, K.B. & Storey, J.M. (1996) Natural freezing survival in animals. *Annual Review of Ecology & Systematics* **27**, 365–386.

Turner, J.S. (1987) The cardiovascular control of heat exchange: consequences of body size. *American Zoologist* **27**, 69–79.

Turner, J.S. (1988) Body size and thermal energetics: how should thermal conductance scale? *Journal of Thermal Biology* **13**, 103–117.

Van Voorhies, W.A. (1997) On the adaptive nature of Bergmann size clines: a reply. *Evolution* **51**, 635–640.

Whiteley, N.M., Taylor, E.W. & El Haj, A.J. (1997) Seasonal and latitudinal adaptation to temperature in crustaceans. *Journal of Thermal Biology* **22**, 419–427.

Willmer, P.G. & Unwin, D.M. (1981) Field analysis of insect heat budgets: reflectance, size and heating rates. *Oecologia* **50**, 250–255.

Wolf, B.O. & Walsberg, G.E. (2000) The role of plumage in the heat transfer of birds. *American Zoologist* **40**, 575–584.

Excitable Tissues: Nervous Systems and Muscles

9.1 Introduction

Nervous systems can be seen as having three basic functions: the **processing of inputs** (sensory information from receptors), the **organization of motor activity** (outputs to, and thus control of, the effectors that are primarily muscles), and (intervening between these two) the **process of decision-making**. At its simplest, a nervous system includes sensory neurons and motor neurons peripherally, performing the first two functions, and interneurons in the central nervous system (CNS) that assume the decision-making role. Understanding nervous systems is one of the greatest challenges to physiologists, and requires integration of analyses at all levels from molecules within cells to behavior of the whole animal.

The interactions of excitable tissue activities and environmental problems can be found at many levels. Above all, nerves and their associated receptors and effectors are responsible for detecting external (environmental) change and initiating responses to it. They are thus in a crucial sense the single most important factor in determining any animal's ability to cope with differing environmental conditions, and it has been estimated that up to half of the genome of a mammal or bird is involved in the differentiation and functioning of the nervous system. At the same time, the nervous system itself must be able to adapt to environmental variation; a neural pathway must continue to function whatever the outside world is doing, without altering (or at least only with predictable variations in) the way it encodes receptor information. The same resilience is required of the other major excitable tissue, muscle, if coordinated movement is to be maintained. Furthermore, excitability resides in a large degree in the differential gradients of small cations across membranes, and any environmental factor that could affect these gradients (e.g. changing salinity) or could affect membrane integrity or permeability (e.g. changing temperature or pressure) is going to be especially problematic for excitable cells.

In this chapter we therefore examine the basic functioning of excitable tissues, and how they permit detection of environmental change, response to it, and indeed learning about it; but we also examine the environmental adaptability of individual nerves, receptors, and muscles as they face different kinds of environmental stress. Because the topic is enormous, this chapter is divided into three main sections: the nerves, the muscles, and then the interaction between nerves and muscles.

SECTION 1: NERVES

9.2 Neural functioning

9.2.1 Neural tissues: neurons and glia

Nervous systems are uniquely and invariably constructed from special types of cell termed **neurons**. These are usually highly branched, the branches being termed neurites and including several incoming **dendrites** and (usually) one outgoing **axon** (Fig. 9.1). Neurons make contact with each other at specific points termed **synapses**. Collections of neurons, communicating with each other at synapses, are the key elements of structure and function in all nervous systems, and understanding the special functioning of neurons is therefore the key to neurobiology.

Neurons are in many respects merely an elaboration of "normal" single cells. They have all the key constituents of cells that allow metabolic and mechanical stability, and exchanges with the extracellular fluid (ECF). They are, of course, constructed from the same genes and gene products as all other cells in an animal's body. They differ only in the manner and extent of their communication with other cells, and hence in their capacity to receive and process information. Thus they invariably have a number of key specializations:

1 At least part of their surface is elaborated into a branching network of dendrites, with points of synaptic contact with adjacent (and also sometimes very distant) nerve cells; they may have as many as 100,000 of these fine synapse-bearing branches.

2 The neuronal cell body is unusually active in the synthesis of macromolecules such as membrane phospholipids, intrinsic membrane proteins (ion pumps and channels; see Chapter 4), cytoskeletal proteins necessary to support the elaborate dendritic structure, chemical transmitters for use at the synapses, and receptors for other incoming transmitters. Hence the cell body has a particularly high concentration of endoplasmic reticulum where these macromolecules are synthesized and packaged.

3 There must be special intracellular transport systems to move the macromolecules quickly and reliably to the points where they are needed.

4 The cell membrane must have a high level of electrical excitability. All cells have a degree of excitability and can, of course, communicate with each other since a degree of organized cell-to-cell communication

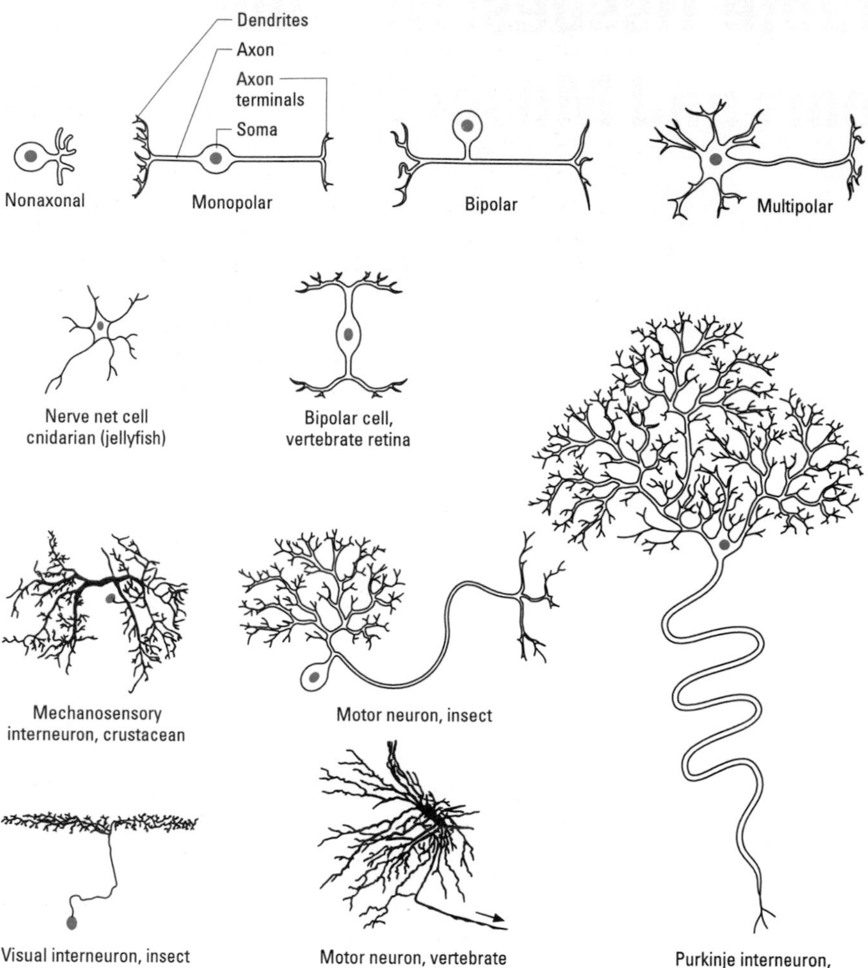

Fig. 9.1 Basic neuron types in animals; a simple classification above, and some examples below.

is a prerequisite for multicellular life. As we saw in Chapter 4, all cells have an electrical potential (the **resting potential**, RP) across their membranes, and if very strongly stimulated by physical distortion this potential can alter, with some lateral conduction of the potential change. Even prokaryotic cells are therefore excitable and have some elements of receptor function. However, the neuron is very specialized in its ability to use these fundamental cellular properties in much more elaborate forms of information processing. Neurons with these characteristics evolved very early on in animal life, being present in all known animal phyla except the sponges (Porifera), and probably dating from the late pre-Cambrian, perhaps 800–1000 mya.

Nervous systems always contain a second type of cell, known as glia. **Glial cells** surround and protect neurons, commonly outnumbering them (10–50-fold in some nervous systems, and up to five-fold even in the human brain with its 10–50 billion neurons). They tend to be small and highly ramifying (Fig. 9.2), wrapping around both axons and cell bodies, and also tightly linked with blood capillaries. In the vertebrate brain and spinal cord some of the glial cells are specialized as **oligodendrocytes**, small but very flattened cells that wrap around the axons many times over, giving a lamellate covering to the axons and thus providing important electrical insulation. In the peripheral nervous system, the same property is conferred by glia known as **Schwann cells**, resulting in the formation of a **myelin** layer. The glial cells have several other key functions: they provide support and protection; they communicate with the neurons via their own sensitivity to electrical potential changes and sensitivity to transmitters; they help to control the ECF around the neurons and the exchanges of metabolites between ECF and neurons; and they have also been shown to act as synaptic regulators, able to mask and unmask certain synaptic surfaces. They also help to guide growing neurons in the correct pathways, since (unlike most neurons) glial cells continue to grow and divide throughout the life of an animal. All the types of glia so far mentioned are formed from ectoderm (as are neurons); however another kind, the so-called "microglial" cells, are mesodermal in origin, invading the nervous system as it develops, and these seem to have a defensive function, removing debris resulting from neuronal damage or degeneration.

9.2.2 Generating a signal: electrotonic potentials

In Chapter 4 we looked in detail at how an electrical membrane potential is set up across the surface of most living cells; this is based on the inequality of ion concentrations on the inner and outer sides of that membrane and its differential permeability, especially to Na^+ and K^+ ions, as determined by the characteristics of **ion channels** in

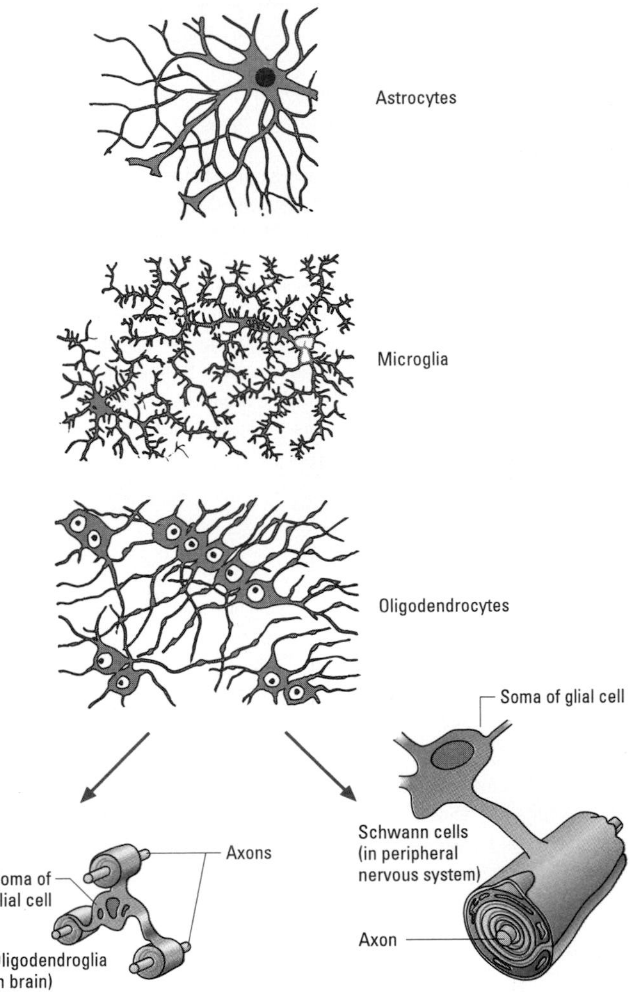

Fig. 9.2 Glial cells from various sites in vertebrates. The oligodendroglia provide the "wrapping" for many axons, and in the Schwann cell this is exaggerated to form the myelin sheath.

the membrane. The resulting electrochemical equilibrium leads to a **resting potential** (abbreviated here as RP, but symbolized as E) that is commonly between −50 and −80 mV, with the inside of the cell being negatively charged relative to the outside. This can be approximately calculated using the Nernst equation (see Box 4.3), and is close to the potassium equilibrium potential, E_K, for reasons explained in the earlier chapter. It results primarily from a small proportion of the K^+ ions leaving the inside of the cell and accumulating on its outside; for a "normal"-sized neuron this might be 30–40 million K^+ ions, far too small a number to affect the overall internal potassium concentration $[K]_i$. Similarly, any change in membrane potential induced externally results in K^+ ions moving across the membrane to re-establish the equilibrium potential. For an electric change of 100 mV around 50 million potassium ions will flow across the membrane. It is possible to induce many thousands of these small electrical perturbations and promote these tiny ion fluxes without any detectable effect on the cell's potassium levels. This is the key to signal generation and transmission in nerves.

In practice the neuronal membrane is permeable to most ions to some degree, and a fuller description of the RP is given by the Goldman equation (known in full as the Goldman–Hodgkin–Katz equation, see Box 4.4), taking Na^+ and Cl^- ions into account as well. There is never a fully stable thermodynamic equilibrium potential across a membrane because some Na^+ will always leak into the cell from the ECF, and some K^+ leaks out. These leakages are compensated for by the normal functioning of the "sodium pump" (the Na^+/K^+-ATPase; see section 4.4), which in all cells continuously pumps Na^+ out and K^+ in, thus indirectly contributing to the resting potential. However, it also plays some direct electrogenic role, as its normal stoichiometry involves moving 3 Na^+ outwards for every 2 K^+ moved in, giving a net electrical effect of one positive charge outwards for each "turn" of the pump. In many neurons, up to 20% of the resting potential is due to this direct electrogenic effect.

How can a neuron, with a resting potential across its membranes, generate an electrical signal? The answer is to have a very transient *change* of potential across the membrane acting as the signal, and this can be produced by a very brief ionic current flowing into or out of the cell. Box 9.1 sets the background to this, showing an electrical analog of the membrane as a combined resistive and capacitative system, and the time course associated with its excitation.

9.2.3 Long-range signals: action potentials

Neuronal conduction

Neurons conduct electrical signals "electrotonically" by currents spreading through the material of which they are made, in much the same way that metal wires or cables conduct electrical signals. The major differences are of scale and of time course. Firstly, neurons are roughly 10 million times lower in conductance than typical metals, and are so thin that they have a relatively high internal resistance to ionic currents. Secondly, neuronal currents are carried by ions, which are relatively large and slow moving (and much less concentrated) compared with the electrons that carry currents in metals. Thirdly, the neuron membrane is a poor insulator and relatively leaky to ions, so a proportion of every ionic current will leak out sideways rather than passing longitudinally down the neuron. Any sustained (tonic) voltage signal is therefore quite rapidly attenuated as it passes along the fiber. Both reduction and distortion of the signal inevitably result, and both will increase with distance from the source of the signal, especially for faster signals.

This should make a neuron an unpromising long-range communication system. However, it works well as long as the length of a particular neuron is small in relation to its own "length constant", and the signal is slow in relation to its "time constant" (Box 9.1 explains these terms). This is commonly the case for short interneurons in the brains and other parts of the CNS. Such neurons are "nonspiking", and can function entirely by the spread of electrotonic potentials. Therefore only the longer axons in the body need to use the more specialized neurons that generate active all-or-nothing "spikes" of potential change, and use them to transmit signals over relatively large distances. These much larger transient potential changes are what we record as **action potentials**.

Action potentials

To achieve long-range signal transmission, an electrical signal that is

Box 9.1 Electrical analog of an axon membrane

The axon can be modeled very much like a cable, if its membrane is seen as a capacitor (C_m) because of the insulating lipid bilayer, and a resistor (R_m) depending on the presence of conducting ion channels; together these affect the passive change of potential when a current is applied across the membrane:

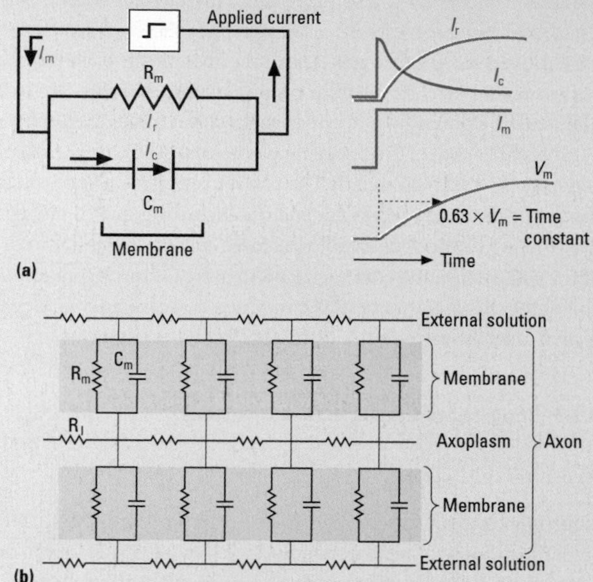

The total current that flows can then be divided into:

1 A capacitative current I_c, peaking quickly then decaying once the capacitor is fully charged.

2 A resistive current I_r, building up after the capacitor is fully charged and then persisting until the potential change ends.

I_r and I_c summed together give the total current across the membrane, I_m.

The development of the membrane potential V_m largely depends on the time course of the resistive current I_r, and both are exponential (related to the natural log e). In effect, the membrane capacitance distorts the time course of the rise and fall of potential.

Comparisons of timings between membranes are thus usually expressed as the **time constant**, which is the product of C_m and R_m. This is proportional to the time required for the changing potential to reach V/e (63%) of its value.

To analyze electrotonic spread along an axon it can best be modeled as a series of these capacitative + resistive elements, connected by "internal" resistors (R_i) corresponding to the axoplasm, as shown in figure (b) in the first column.

If current injection is used to give a steady potential difference E applied at position x, the membrane depolarization falls off with distance from this point, and reaches a steady state with exponential decay. The **length constant** that describes this decay is given by the square root of R_m/R_i, and this corresponds to the distance required for the potential to decay by 1/e (by 63%, i.e. to roughly one-third):

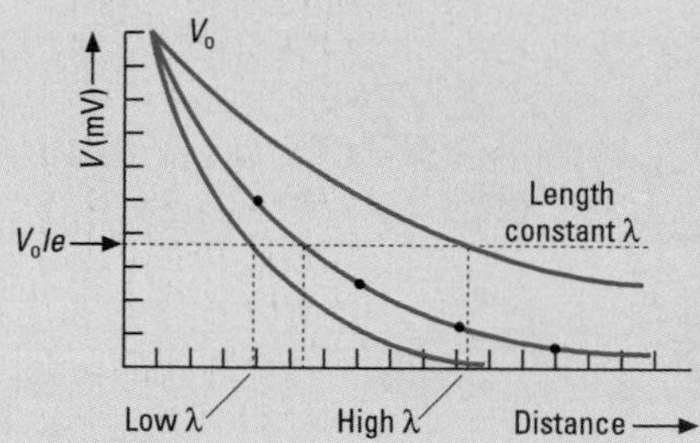

Note that the time constant depends on relative membrane resistance and capacitance, while the length constant depends on the balance of membrane and axoplasm resistance. The main influence on the length constant is therefore the size (diameter) of the axon.

Some examples are given in the table below:

	Diameter (μm)	Length constant (mm)	Time constant (ms)
Squid axon	500	5	0.7
Lobster axon	75	2.5	2
Crab axon	30	2.5	5
Frog muscle	100	2	8
Crab muscle	330	1.6	14.5

(Figures adapted from Aidley 1998.)

larger and not susceptible to decrement or distortion is required. Thus a passively spreading electrotonic potential has to be converted into an action potential (AP) at some point in the neuron. An AP is an actively evoked and rapid potential change; it is "all-or-nothing", of constant stereotyped size and shape in a particular neuron, and conducts quickly along the axon. It appears as a rapid and very transient reversal of the membrane potential (from the negative resting state to slightly positive), which sweeps along a neuron. Generation of an AP is a very special excitatory phenomenon, restricted to neurons (and then only to certain kinds of neurons) and absent in any other cell type in animals.

An AP results from a rapid and regenerating depolarization of the cell membrane, in response to a local electrotonic depolarization of the normal membrane RP. To evoke an AP, this initial depolarization must be greater than a particular **threshold** value, the threshold differing for different neurons. Once evoked, an AP has the classic generalized shape and time course shown in Fig. 9.3. There is a very **rapid depolarization** so that the membrane polarity is briefly reversed (RP positive, usually +20 to +50 mV), followed by a slightly **slower repolarization**, often with an "undershoot" to a potential value a little below the normal RP. The rise and fall of the AP are normally completed in less than a millisecond (0.5–1 ms) but the undershoot or **negative after potential** may persist for another 2–3 ms.

The AP is fundamentally due to activities going on at a molecular level, involving the ion channels that we considered in Chapter 4. Some of the protein-based ion channels in a cell membrane have the particular property of being **voltage-gated**; that is, the proteins change their configuration as the membrane potential changes. At one potential the channels are effectively "closed" to the passage of

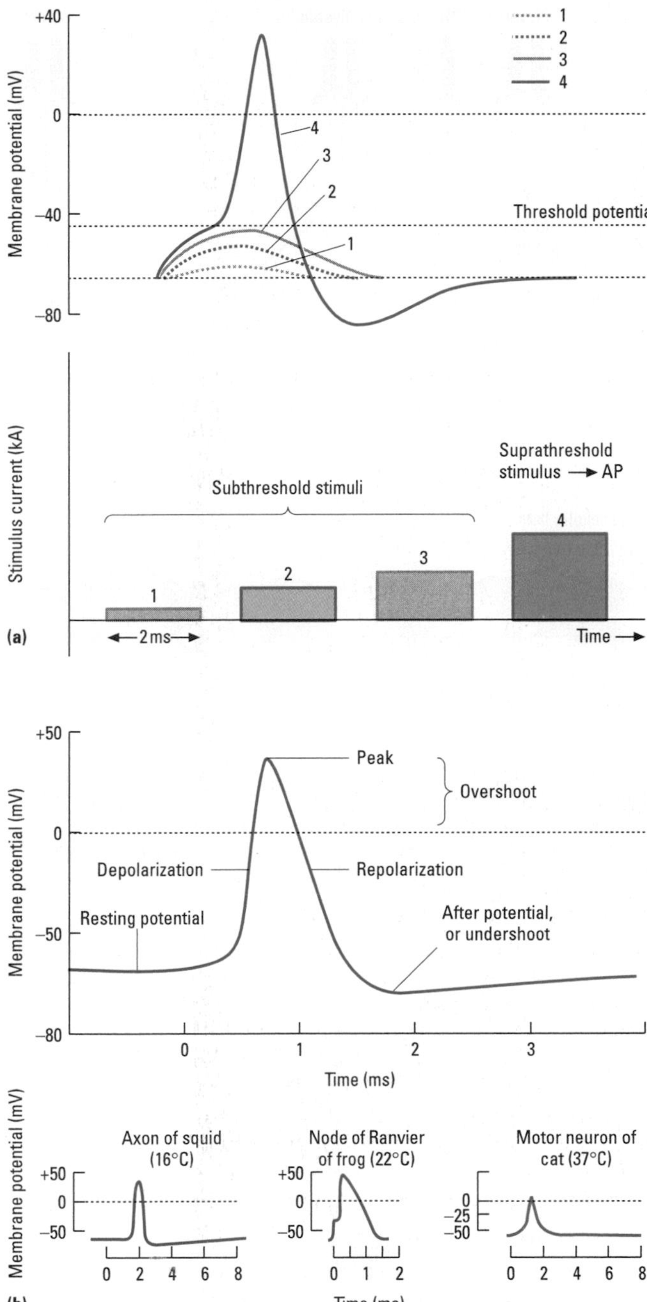

Fig. 9.3 (a) The effects of varying levels of stimulus current on a neuronal membrane. Small and moderate voltage steps produce passive, graded, and voltage-dependent depolarizations, but at some threshold level an active all-or-nothing response—the action potential (AP)—is produced. (b) The classic shape of an AP, together with the AP recorded from several named sources.

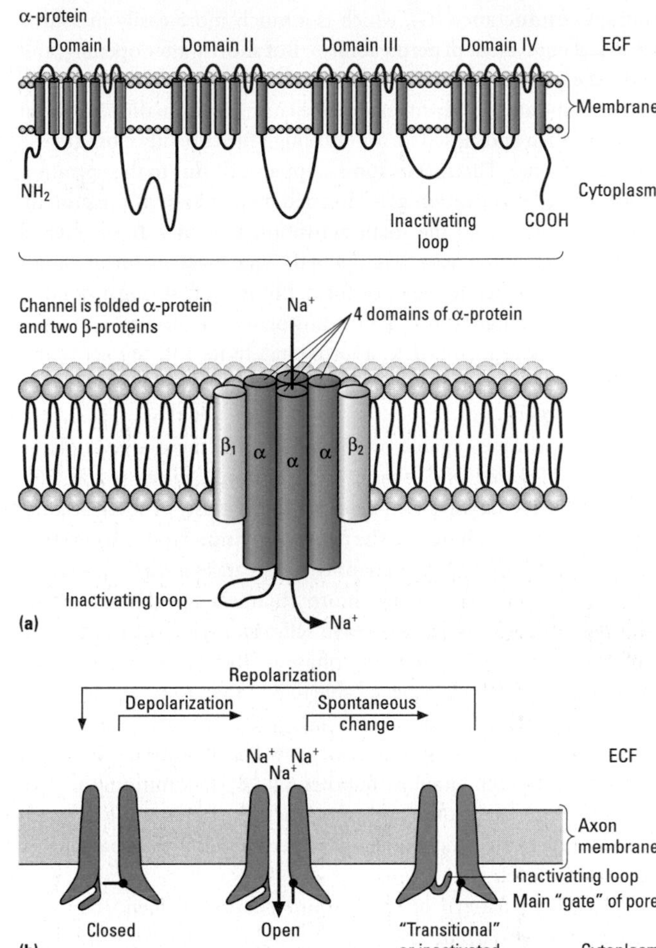

Fig. 9.4 (a) The molecular structure of a voltage-gated sodium channel, with (above) a large α-subunit composed of four homologous repeat protein domains each of which has six transmembrane helices, and which fold together (below) to enclose the conducting pore, also linking with two smaller β-proteins to form the whole channel complex. The voltage-sensing region is in domain I, and the inactivating effect resides in the internal peptide loop between domains III and IV. (b) Representations of the closed, open, and transitional or inactivated states of the channel; at the normal resting potential the closed state is favored, but during a strong depolarization the open state is the most energetically favorable.

the ions that can normally pass through them, and at some other potential value they switch to an "open" configuration. The open and closed configurations are the only two stable configurations of the protein, so the protein in effect becomes a "gate". In Chapters 2 and 4 we looked in some detail at how changes of protein configuration can be brought about, and how they affect cellular processes. Sometimes changed configuration is due to ligand binding, or to changes in ion concentrations, pH, etc. Here we have the special case of configurational change due to electrical change. The key performer in an AP is a **voltage-gated Na^+ channel**, with a secondary role played by a **voltage-gated K^+ channel.**

The neuronal sodium channel is formed from a glycoprotein that spans the membrane and has a molecular weight of 260 kDa. Most of the length of this protein is coiled up into four main transmembrane domains, interlinked by short lengths of the polypeptide chain. The four domains are essentially identical, and are probably arranged symmetrically around a central pore (Fig. 9.4). When the cell membrane is at its RP, the four domains produce a closed structure with no functional pore. As the neuron is depolarized the open form is energetically favored and progressively more of the channels switch to the open mode, selectively allowing Na^+ ions to diffuse inwards through their pores, down the sodium concentration gradient. In effect, the channel is acting as a sensor for local changes in the electric field, and producing an increase in sodium permeability (or sodium **conductance** (G), which is a much more easily measured electrical equivalent of permeability). But shortly after opening, and with the membrane still depolarized, the Na^+ channel protein converts spontaneously to an inactive configuration, different from both the open and closed forms (although functionally closed to the passage of ions). This inactivation is apparently due to the operation of a specific "inactivation gate" formed by one loop of the protein, which swings across the main activation gate at a fixed interval after the activation gate has opened. The inactive form cannot immediately switch to the open form, but must first transform into the closed configuration. This transformation occurs when the membrane is repolarized to approximately its RP; the activation gate then closes in its own right, and the inactivation gate opens again (Fig. 9.4b).

In a membrane capable of AP generation, there are very large numbers of gated Na^+ channels, commonly measured as 35–500 μm^{-2}. Given a threshold depolarization many of these become open, and Na^+ ions flow into the neuron, giving a "**sodium current**" of a few picoamps (pA) per channel. This causes a slight additional depolarization so that a few more channels open; thus we get positive feedback and an explosive self-amplifying depolarization, which can be seen as the rising phase of the AP. At its peak, the membrane is much more permeable to Na^+ than to K^+ (quite unlike the normal situation; see section 4.3) and experiences an inward sodium current of around 1 mA cm^{-2}. The membrane is driven close to its sodium equilibrium potential (E_{Na}), commonly around +50 mV. It normally does not quite reach E_{Na} because the channels start to inactivate, and the sodium permeability falls back to its normal low level.

The overall pattern of the sodium channel behavior and the change of potential it produces are shown in Fig. 9.5. Note that there is a threshold of depolarization necessary for the self-amplifying process to take off and generate an AP. If the initial depolarization is too small, the few Na^+ channels that do open are more than offset by the background level of potassium conductance (permeability) and the K^+ ions moving outwards.

In most neurons, the duration of the AP is reduced below that shown in Fig. 9.5 by the addition of the potassium-gated channel. This again is a glycoprotein in the membrane and has both open and closed configurations, and again switches from closed to open in response to depolarization. However, it does so significantly more slowly than the Na^+ channel, and it has no spontaneous inactive form during extended depolarization. Instead it only converts back to a functionally closed form when the membrane is repolarized. Thus, when a neuronal membrane is depolarized above threshold a chain of events is automatically triggered (Fig. 9.6):

1 Sodium channels open, and an inward sodium current flows, giving further depolarization and a rapid rise towards the sodium equilibrium potential E_{Na}; the membrane potential rapidly swings from −70 to +50 mV.

2 Before maximum depolarization is reached, the sodium channels start to inactivate and the potassium channels increasingly open, giving an outward potassium current and a move back towards the potassium equilibrium potential E_K.

3 The membrane repolarizes and may undershoot below the RP to a value very close to E_K.

4 The potassium channels start to close and initial conditions are re-established so that the membrane potential returns to its resting value.

5 During phases 3 and 4 it is difficult to re-excite the membrane, and there is a **refractory phase**. Initially this refractoriness is absolute, as too many Na^+ channels are still inactive, but later it is only relative, and enough Na^+ channels are amenable to opening to give a

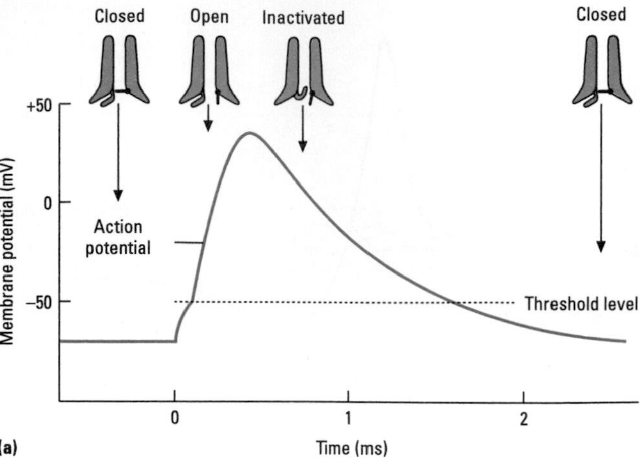

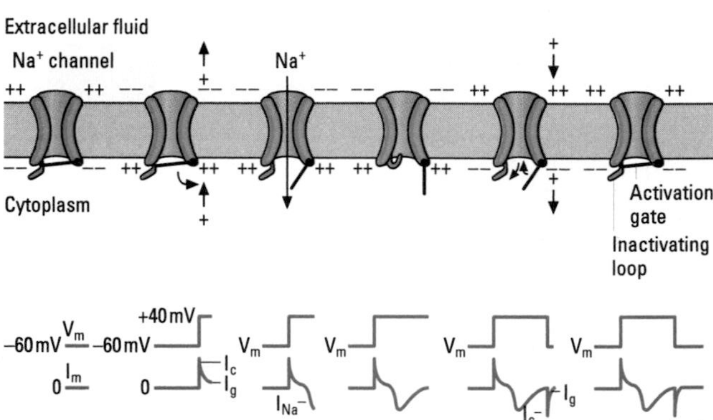

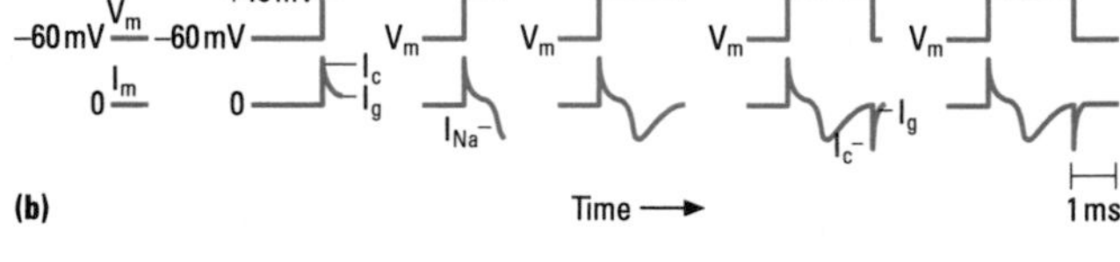

Fig. 9.5 (a) The time course of an action potential and the associated behavior of the voltage-gated sodium channel in relation to measured potential change. (b) The currents recorded during a voltage clamp experiment in relation to channel behaviour: I_c and I_g are related to the electrical behavior of the membrane, and the important extra component is I_{Na}, the sodium current carried by Na^+ ions moving through the open channel.

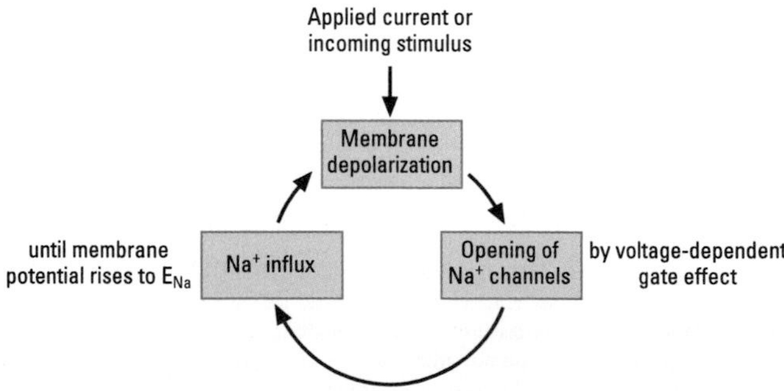

Fig. 9.6 A summary of the self-amplifying chain of events occurring during an action potential.

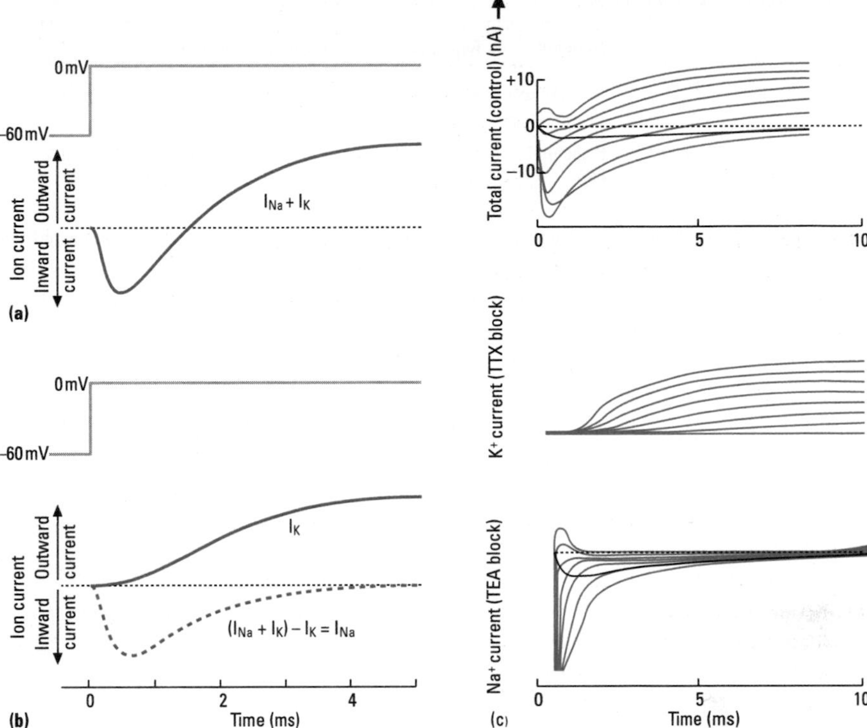

Fig. 9.7 The currents recorded in voltage clamp experiments. (a) The total current recorded in normal conditions with a substantial imposed step depolarization. (b) The potassium current (I_K) recorded when the sodium current is abolished, either by using a low Na^+ bathing saline (so that $Na_o = Na_i$), or by blocking with a sodium channel blocker such as tetrodotoxin (TTX; see text), and also shows the difference between the total current and I_K, which must be the sodium current I_{Na}. Note that the experiment can be reversed by blocking the potassium current with tetraethylammonium (TEA), thus recording the sodium current directly, and then calculating the potassium current. Hence in (c) the values of the Na^+ and K^+ currents for different step depolarizations are both shown by using the selective blocking agents. (Adapted from Hodgkin & Huxley 1952.)

further AP (albeit an unusually small one) if a strong enough stimulus occurs.

It is possible to measure the ion current flowing during an AP using "**voltage clamp**" techniques. Fine microelectrodes are inserted into the neuronal surface and are used to record electrical changes as usual, but are also used to supply a fixed step voltage to the system and then record the "extra" current that needs to be applied to counter the ion currents and keep the membrane potential constant. Figure 9.7 shows the current recorded at different levels of applied voltage change in a typical experiment. This current can be separated into its two main components using selective channel blockers: the classic examples are tetrodotoxin (TTX) derived from the puffer fish, which is highly selective for sodium channels, and tetraethylammonium (TEA), which blocks potassium channels. Using these, we can see the shape and time course of the Na^+ and K^+ currents (conductances) separately (Fig. 9.8). With these techniques it can be shown that the potassium current will persist as long as a depolarization is applied—the channels do not spontaneously close in the manner of the Na^+ channels. "**Patch clamp**" recordings allow an even smaller scale measurement of the tiny piece of membrane clamped to the end of a microelectrode, and can give recordings of the flow through single channels. Such measurements confirm this difference between the two voltage-gated channels, and can be used to analyze the detailed molecular functioning of the ion channels.

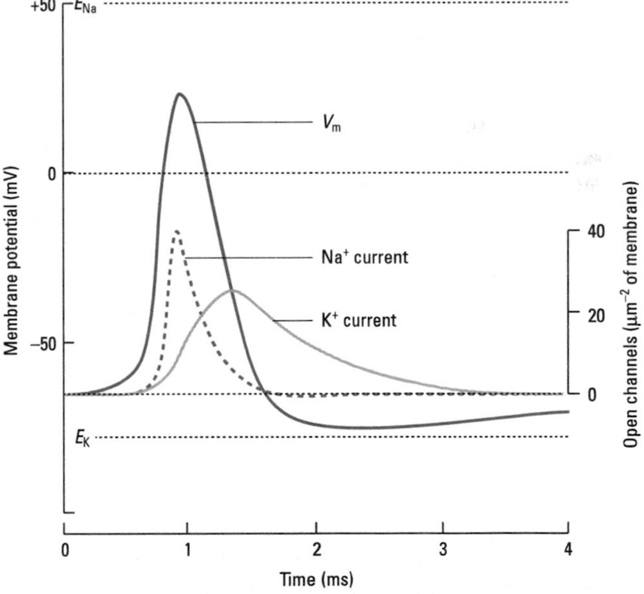

Fig. 9.8 From experiments similar to those shown in Fig. 9.7, the sodium and potassium conductances can be plotted, and are here converted into estimates of the number of ion channels of each type opening in the axon membrane, through the time course of an action potential. E_{Na}, sodium equilibrium potential; V_m, maximum velocity.

Thus far we have described the "classic" AP, as first detected and analyzed by Hodgkin and Huxley in the 1950s using the useful "giant axons" of the squid. We now know that other neurons have variants on the same theme. Most axons have sodium currents through sodium channels, and these are very uniform in performance and in molecular subunit structure. However, many dendrites and cell bodies (as opposed to axons) use an inward current partly or mainly based on a calcium channel rather than a sodium channel, yielding a longer duration AP. Potassium channels are commonly used to restore the RP in all neurons, but are much more variable (at least 50 types are now known), with differences in triggering, timing, and

Table 9.1 Voltage-gated ion channels.

	Notation	Response to depolarization	Function	Blocking agents
Sodium channel	Na channel	Activated	Fast depolarizing phase of action potential (AP)	Tetrodotoxin (TTX) Saxitoxin (STX)
Potassium channel (delayed rectifier K channel)	K channel (K_V)	Activated	Repolarizing phase of AP, giving maximum firing rate	Tetraethyl ammonium ions (TEA) 4-amino pyridine Calcium
Fast early potassium channel	K_A channel	Activated	Inactivated by prolonged depolarization, regulates slow firing in pacemaker neurons	TEA (low concentration) 4-amino pyridine
Inward rectifying potassium channel	K_{ir} channel (K_x)	Inactivated	Generation of plateau potentials (regulates heart muscle)	Rubidium, caesium TEA
Calcium-dependent potassium channel	K_{Ca} channel	Activated by $[Ca]_i$	Affects intracellular Ca^{2+}, slow firing rates	Barium TEA, apamin
Calcium channel	Ca channel	Activated by strong depolarization	Slower repolarization in some long-lasting APs. Main depolarizing phase of AP in some muscles	Nickel, cobalt Verapamil
Chloride channel	Cl channel	Activated in some cases	Stabilize resting potential (AP in plant cells)	Thiocyanate

Also other variations on potassium channels: K_M, K_s, and K_{ACh} channels, opening partly related to neurotransmitters.

interactions with other ions. The important voltage-gated channels, and their roles in signaling processes, are shown in Table 9.1. Note that a single neuron may have regions with different channels and thus different excitable properties.

It is also important to realize that some neurons fire spontaneously, producing APs without any external stimulus. These are the so-called **pacemaker** neurons, sometimes in the CNS and sometimes in organs such as the heart that must be active continually. In vertebrates they normally fire because they have an unusually high inward sodium current, which by itself can continually depolarize the membrane towards threshold. Different kinds of potassium channel serve to speed up or slow down the rate of pacemaker firing. In molluscs and other invertebrates there are pacemaker neurons that fire due to the influx of calcium into the cells, exceeding the rate at which it can be pumped out. This influx depolarizes the cell to give a train of spikes, but then the $[Ca^{2+}]_i$ rises enough to activate the calcium-dependent potassium channels, giving a restoring hyperpolarization and a cessation of spiking, during which the excess calcium can be pumped out to restart the cycle.

Signal propagation

It should be evident from the discussion above that an AP in one part of a neuron is not only a self-amplifying system at that site, but is **self-propagating** along the membrane. The local currents that underlie the AP spread out electrotonically to the adjacent membrane, where they are of sufficient magnitude to set off depolarization and a further AP response. Thus a "wave" of AP moves along the axonal membrane. This movement will be in one direction only, as the parts of the cell that have just experienced an AP will still be refractory; the AP can only move "forward" and excite the zones "beyond" itself that have not yet produced an AP (Fig. 9.9). The AP therefore propagates along the axon in one direction only, regenerating continuously, and there is no attenuation of the signal whatever the distance traveled. The amplitude of an AP is 4–5 times greater than the depolarization threshold for the opening of Na^+ channels, so there is a considerable built-in safety factor insuring propagation. A useful analogy is that of a fuse, or more spectacularly a child's "sparkler" firework: there, a certain input (temperature) is required to ignite it, but once lit the burning area of the sparkler ignites the next region in front of it and the zone of ignition progresses smoothly and with no diminution along the wire without any further inputs; and the burning cannot pass backwards because there the flammable material has been used up, giving a refractory effect.

The velocity of AP propagation is a function of the membrane properties, essentially the rate at which the local currents can depolarize the membrane in front of the active zone. This again depends on the neuron's length and time constants, and on the two key factors of r_i (internal longitudinal resistance) and r_m (membrane resistance). Faster velocity results from decreased r_i or increased r_m. The larger an axon the lower its internal longitudinal resistance to current flow: thus larger axons usually have faster conduction velocities (Table 9.2). The squid epitomizes this solution to the problem of maximizing conduction velocity; its giant axons are formed by the fusion of many smaller axons, and have a conduction velocity of up to 20 m s^{-1}, so it can achieve a synchronous signal to all the muscles of its mantle for efficient jet propulsion (see section 9.15.1). There is an alternative way to increase conduction velocity where large axons are inappropriate (for example in advanced brains where having a great many very small neurons is preferable); in such cases, the membrane resistance can be increased. This is achieved by the special trick of **insulation**, relying on the insulative properties of additional biological membranes. Many axons in the invertebrate nervous systems are therefore enclosed in a loose sheath formed by glial cells, but in vertebrate nervous systems the axons are wrapped in a dense sheath of glial myelin. Peripherally this is achieved by Schwann cells, and centrally (in the vertebrate brain) by oligodendrocyte glial cells. In either case, the glial cells wrap tightly and repeatedly around the neuron to give a dense layer of doubled membranes as a coating, with nearly all the cytoplasm excluded (Fig. 9.10). **Myelin** is therefore essentially stacks of double membrane, and like all membranes is composed of about 70% phospholipids, glycolipids, and cholesterol (see section 4.3), making it an

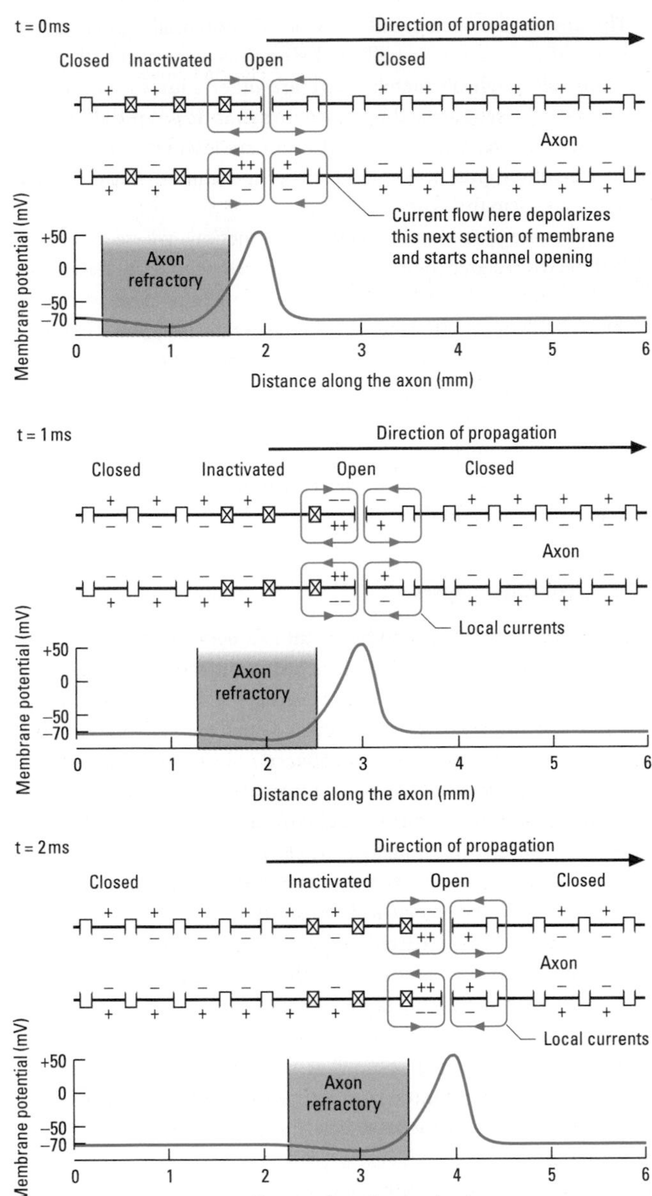

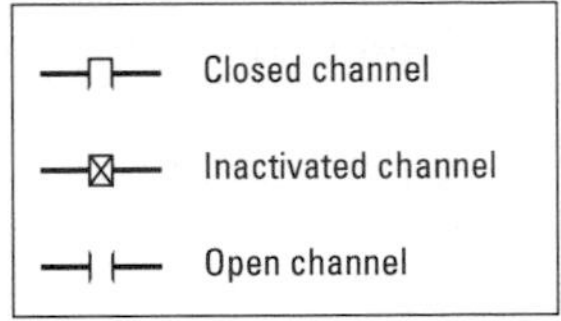

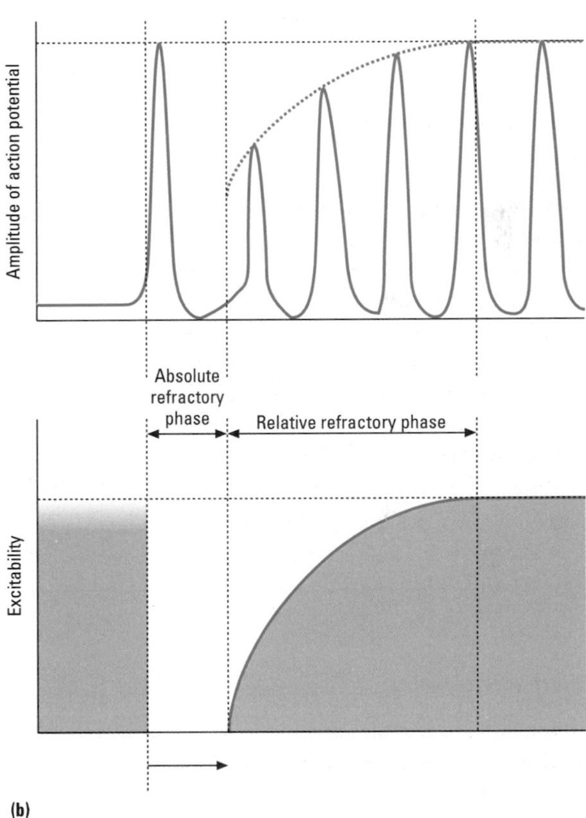

Fig. 9.9 (a) Forward propagation of an action potential (AP) along an axon membrane, showing three successive time frames and the associated currents and ion channel states, together with the refractory periods that result. (b) The refractory phases in relation to excitability and the form of an AP that can be produced. (Adapted from Alberts *et al.* 1994.)

excellent insulator. Each glial cell may wrap up about 1 mm of axon in myelin, and each axon has a series of such cells along its length. Between the glial cells the axon membrane is briefly exposed, at the **nodes of Ranvier**. The stretches of myelinated axon (internodes) have a greatly increased membrane resistance, and the axonal membranes in these regions have very few ion channels, with transmembrane currents virtually precluded. At the nodes, however, the ion channels are densely clustered (often in excess of 1000 μm^{-2}) and the currents can flow freely. In effect, the whole system has been transformed into a cable with a very good length constant, so that local currents (and therefore APs) "jump" from one node to another largely unattenuated (Fig. 9.11). The local depolarizations do not have to spread to, and initiate APs in, every successive soft membrane. This "jumping" or **saltatory propagation** in myelinated axons is significantly faster than in a similar-sized unmyelinated axon; some vertebrate myelinated axons only10 μm in diameter can reach a conduction velocity of 100 m s^{-1}. At that velocity, a single AP of 1 ms duration may affect 10 cm of axonal membrane, and involve around 100 nodes, at one time.

Myelination allows rapid accurate performance in a limited volume of nervous tissue. Without it, the vertebrate spinal cord would need to be meters thick to work at the same speed. Damage to myelin is therefore a very serious event, and diseases that affect myelination, such as multiple sclerosis, can lead to an irreparable loss of motor function.

Table 9.2 Diameters and conduction velocities for motor axons, including giant axons (*) from invertebrates and fish, and mammalian myelinated axons (**).

	Diameter (μm)	Conduction velocity (m s^{-1})
Cnidaria		
Anemone	1–2	0.1
Nanomia	30*	3
Aglantha	30–40*	3–4
Mollusca		
Helix	2	0.8
Loligo	5–7	4
	450*	30–35
Annelida		
Nereis		
medial	9*	2–3
lateral	35*	5
Lumbricus	5	0.6
medial	90*	25
lateral	60*	11
Myxicola	1000*	20
Crustacea		
Cambarus	150*	15
Homarus	125*	18
Insecta		
Periplaneta	5	2
	40*	10–12
Vertebrata		
Fish		
Enterosphenus	50*	5
Cyprinus	65*	55
Mammal		
Frog**	3–10	7–25
Snake**	5–10	10–35
Cat**	4	20
	10	50
	20	120

Information content

Although there is some variation in the shape and the ionic basis of APs between and within neurons, and considerable variation in the speed of conduction of the signal along neurons, nevertheless at any one time and place the AP is still all-or-nothing. Thus its mere presence cannot carry significant information; the information can only be coded into the frequency of APs, and not their magnitude. Fortunately the frequency and pattern of APs is highly variable, limited only by the capacity of the membrane to regenerate open sodium channels and emerge from the refractory period; the frequency can be as high as 500–1000 s^{-1}. We will return to the problem of coding information in considering aspects of sensory activation later in the chapter.

9.3 Synaptic transmission

We have seen how a small change of voltage can be transformed into a larger AP in a nerve membrane, and how this AP can be transmitted along the nerve rapidly and reliably without attenuation. But how is the original change of voltage that starts the whole process generated? Where does the initial input come from?

The answer lies with small electrotonic potentials generated at the ends of the neurons, usually at their points of contact with other neurons (although ultimately there are other sources, in sensory receptors). The synapse is the site where **synaptic potentials** are induced, which spread electrotonically and are integrated with many other synaptic inputs to produce either a summated electrotonic signal or an AP in the axon.

Synapses are specialized intercellular junctions where this summated electrical signal in one neuron is passed, with high spatial and temporal precision and high speed, to another neuron (or some other excitable cell). The synapse is normally the only point of communication between two neurons, and somehow the signal from one cell has to pass to the other across the discontinuity of an intercellular gap. There are two ways in which this can happen: either by direct electrical signaling, or by a chemical message crossing the gap. Both systems occur, though the latter is more common and we will therefore consider it first.

9.3.1 Chemical synapses

For chemical transmission to be useful in a nervous system it must have three characteristics: it must be fast, it must be point-to-point (rather than diffuse), and it must be simple (excitatory or inhibitory, but not complex or ambiguous). Chemical synapses are the main system for communicating between cells in a nervous system. They may occur between any two neurons, or occasionally between different parts of one neuron. A chemical synapse (Fig. 9.12) is a zone where two neuronal surfaces lie close together (commonly the axon terminals of one neuron abutting onto the dendrites or cell body of another) and a chemical transmitter diffuses across the gap between them. It is constructed by the close apposition of specialized regions of the plasma membranes of the two participating neurons, forming a clearly defined synaptic interface with a narrow intervening cleft. The presynaptic cell contains a cluster of transmitter-filled vesicles and may have a somewhat thickened membrane, while the postsynaptic membrane is usually clearly thickened due to an accumulation of receptor sites and the associated submembrane "scaffolding".

In essence, the functioning of a synapse requires that:

1 The spreading depolarization at the presynaptic terminal of neuron A brings about a specific release of chemical.

2 The chemical diffuses from the membrane into the synaptic cleft.

3 The diffusing chemical arriving at receptors on the postsynaptic terminal in neuron B initiates a new depolarization.

Chemical synapses have the general advantage over electrical synapses of being **polarized**, with information flowing in only one direction from the presynaptic cell to the postsynaptic cell (although if bidirectional information flow is wanted it can be obtained with reciprocal chemical synapses, having two adjacent synapses transmitting chemically in opposite directions). However, chemical synapses have the potentially major disadvantage of being relatively **slow**, and where maximum speed is required, a series of chemical synapses between a series of neurons is not ideal.

Presynaptic structure

The presynaptic terminal is a site for the conversion of the electrical potential in the axon into a secretory process. This is achieved

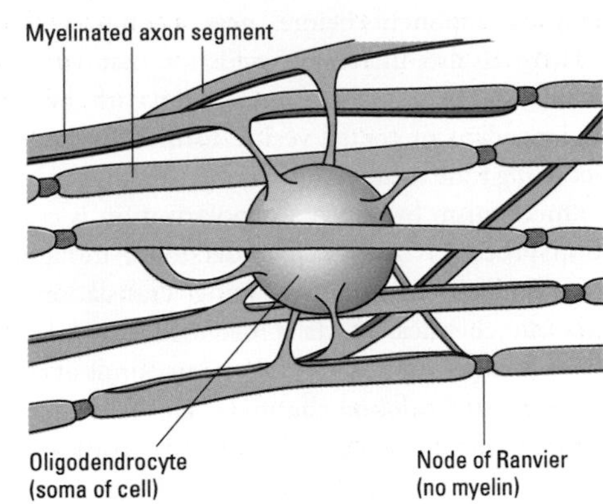

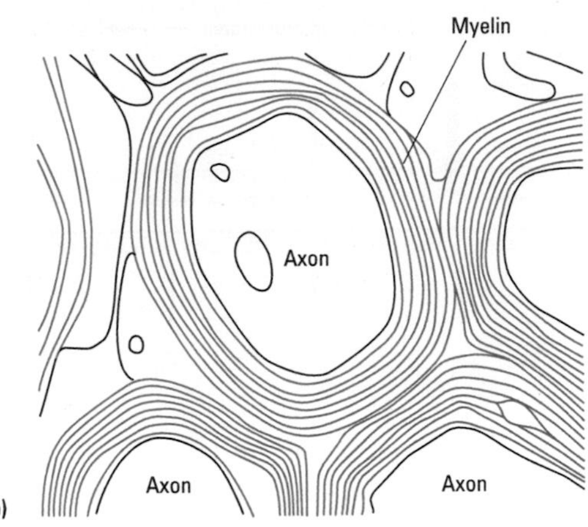

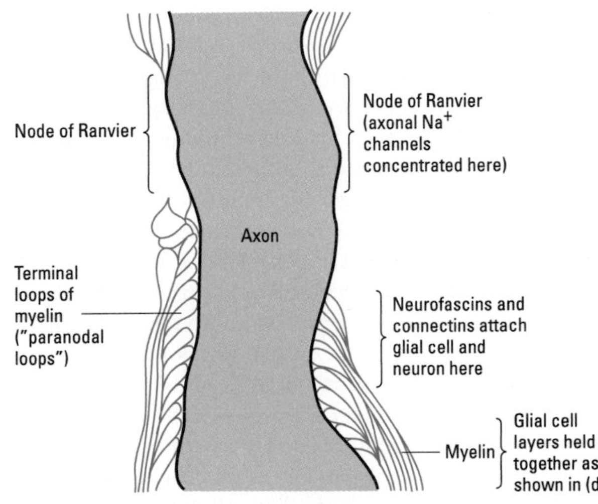

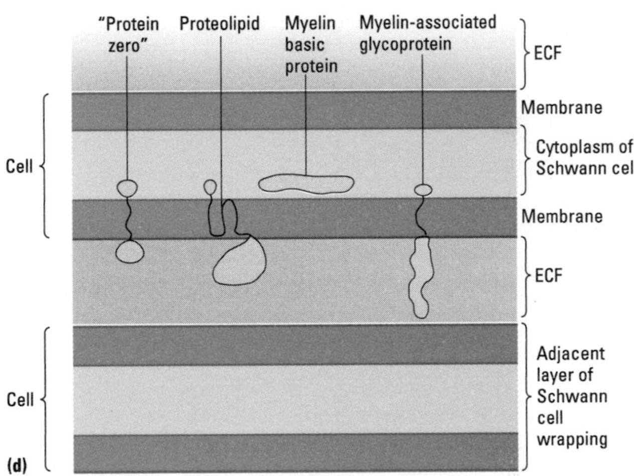

Fig. 9.10 (a) Schematic view of how a glial oligodendrocyte (Schwann cell) wraps around several adjacent axons and myelinates them, leaving the axons exposed only at the nodes of Ranvier. (b) Cross-section through a Schwann cell and several axon, showing how the glial membranes wrap around to produce the myelin sheath, with tight layering along the length of the resulting "internode" and with membrane loops at the node itself (c). Various proteins are associated with the myelin sheath, shown diagrammatically in (d). (Adapted from Reichert 1992, and other sources.)

through a sequence of events relying on a particular intracellular architecture (Fig. 9.12). The presynaptic zone is usually a slightly swollen knob-like structure, within which lie many secretory vesicles, densely packed together; the numbers per terminal normally vary from around 50 to several hundred, but may reach thousands in some specialized synapses. The vesicles are normally 30–50 nm in diameter, or rarely up to 150 nm, and they are spherical. Vesicles can be readily isolated, and their membrane proteins have been particularly well studied as a tool for understanding membrane trafficking processes. One vesicle contains only about 10,000 molecules of phospholipid, and around 200 of proteins with a combined molecular weight of 5–10,000 kDa. These include at least nine families of protein concerned with membrane traffic (including synapsins, synaptobrevins, and rab proteins) (Table 9.3), many of them derived from the Golgi apparatus; however their functions are only just being established in any detail.

The adjacent cell membrane appears dense in electron micrographs due to the presence of an organizing submembrane scaffold. Vesicles are localized close to the synapse by interaction with specific proteins in this scaffolding, notably further synapsins that bind both vesicles and actin filaments (polymerized actin, more familiar as a major muscle component, is abundant in the cytoskeleton at presynaptic terminals). The membrane itself also houses a series of intramembrane particles that are functionally calcium channels, and which are related to specialized "active zones" where vesicle attachment occurs. At any one time a few vesicles (often 5–10, but sometimes only one) are attached to each active zone, ready for release.

To support all the requisite molecular events, many of which require ATP inputs and the management of complex protein arrays, presynaptic terminals must also include some mitochondria and a moderate complement of endoplasmic reticulum. Some synapses also contain large densely staining vesicles about 200 nm in diameter, thought to be involved in amine and peptide supply.

Presynaptic events

The complete "**synaptic vesicle cycle**" is summarized in Fig. 9.13. Vesicles become filled with neurotransmitter by H^+-linked active transport, and are translocated by diffusion through the scaffold

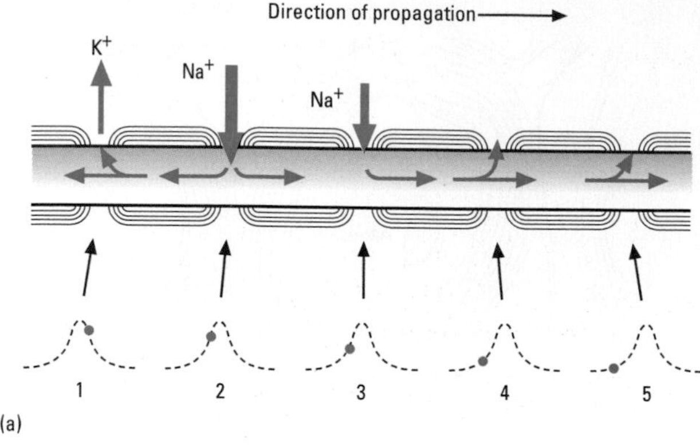

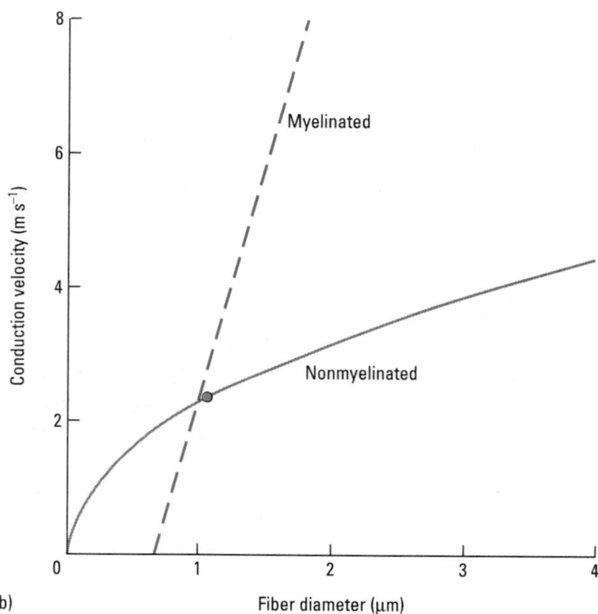

Fig. 9.11 Saltatory conduction between the nodes of Ranvier in a myelinated axon. (a) The longitudinal current flow between the nodes, and the Na^+ and K^+ currents at successive nodes. Node 1 is in the falling phase of the passing action potential (AP); node 2 in the rising phase, with a peak sodium current; and nodes 3 and 4 are in successively earlier phases of the approaching AP, which has not yet had any effect at node 5. The dots here indicate the value of the membrane potentials all recorded at the same instant at the five nodes. (b) The relative speed of conduction of myelinated and nonmyelinated axons of different diameters.

elements to lie alongside the presynaptic membrane at the active zones where they become "docked vesicles". These then undergo several pre-fusion processes (sometimes termed "priming", requiring ATP and perhaps involving partial membrane fusion). The next steps only occur when a calcium influx occurs at the terminal, triggering full fusion and transmitter release. The vesicles are recycled into the cell almost immediately after emptying. This can be achieved by endocytosis (see Chapter 4 and Fig. 4.13), the vesicles becoming coated with clathrin and the coated vesicles then budding from the plasma membrane to the interior of the nerve terminal where the clathrin is lost again. The empty vesicles become acidified via proton pump activity (thus generating the gradient on which transmitter uptake depends), and either refill immediately or fuse with elements of the endoplasmic reticulum (endosomes) that eliminate any damaged components before "new" synaptic vesicles are regenerated. There is also increasing evidence that vesicles can engage in a much faster "kiss and run" interaction with the synaptic membrane, independent of coated vesicle formation, and with a much faster recycling time course.

The precise mechanism by which a depolarization is converted into a secretion process is now well understood, though many details at the molecular level are still emerging. Translation of electrical excitation into chemical effects is mediated here (as in other situations such as muscle contraction and various kinds of chemical release) by **voltage-gated calcium channels**. The calcium channel proteins in synaptic membranes are very similar chemically to the axonal sodium channels (about 60% amino acid sequence homology), and very similar in structure, with four homologous transmembrane domains that form the main "pore" region. Potential change opens the pore or channel by altering the protein configuration, and the resultant inward flow of Ca^{2+} ions serves as the intracellular messenger to activate further change. This presynaptic calcium current is only evoked above some specific threshold of incoming potential change, but above this threshold its magnitude is strongly voltage dependent, and its duration depends on the duration of the depolarizing phase of the signal that elicits it. However, the calcium current flow is somewhat delayed, and only begins towards the end of the presynaptic AP (or in some cases the presynaptic electrotonic potential change). This delay is related to a rather slow channel-opening process (much slower than the opening of a sodium channel, for example) (Fig. 9.14), and is the prime reason for so-called "**synaptic delay**", whereby a time lapse of around 0.5 ms occurs at typical synapses. Once open, the calcium channels also inactivate rather slowly.

The inward flow of calcium ions as the channels open produces a localized increase in $[Ca^{2+}]$ to about 10–100 μm (way above the resting value of 50–100 nm), and it is this that leads to vesicle fusion and transmitter release. The process has been demonstrated with voltage clamp techniques, and can be verified with local injection of calcium ions. An increase in $[Ca^{2+}]$ elicits fusion of vesicles with the presynaptic membrane, and thus the release of the vesicle contents into the synaptic cleft. In fact calcium entry presynaptically is related to voltage or current detected postsynaptically by a power relation, with an exponent of 2–4, giving a marked amplification effect (i.e. doubling the Ca^{2+} influx produces a 4–16-fold increase in response).

The fusion of docked vesicles with the presynaptic membrane is clearly the key process for transmitter release. But whereas at any one time 5–10 vesicles may be docked and "ready" at each active zone, only one vesicle normally discharges. Furthermore, release only occurs with a fairly low probability of about once every 5–10 APs. In other words, fusion occurs for only one in every 5–10 ready vesicles and only once every 5–10 calcium signals; the system has been described as "reliably unreliable". This may be a key to the high level of plasticity or regulation of response that can occur, discussed later in more detail.

Fusion of a single vesicle involves at least three or four calcium ions binding to a low affinity protein site, and it occurs remarkably quickly, within about 0.1 ms (probably many more complex stages are involved in the pre-fusion phase, which takes much longer). The

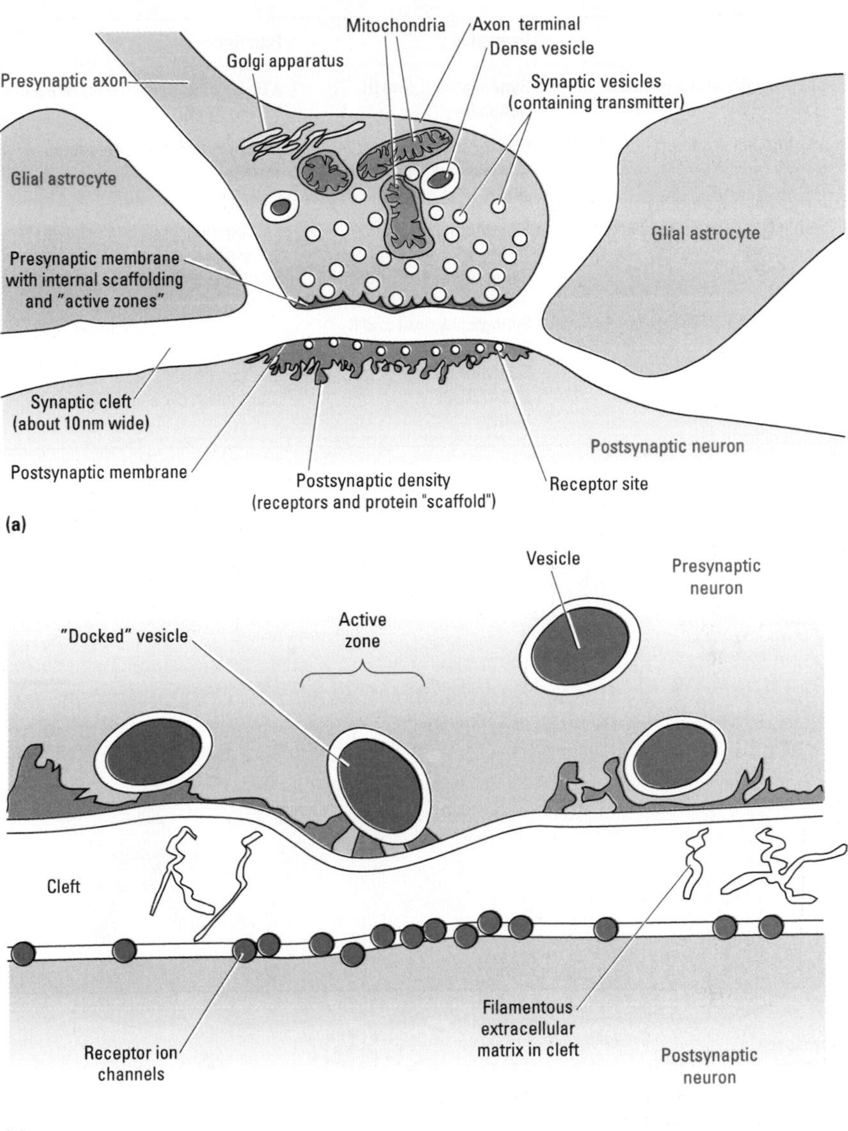

Fig. 9.12 (a) Diagrammatic view of the basic structure of a chemical synapse between two nerves. (b) Details of presynaptic structure and the synaptic cleft.

response to the calcium influx may be a direct effect, or may be mediated through calcium-binding proteins; calmodulin, rabphilin, and synapsin 1 have all been implicated, but the best candidates may be **synaptotagmins,** which bind to various targets and have several calcium-binding domains. Work with knock-outs and mutants has shown that synaptotagmins are crucial for the fast Ca^{2+}-triggered vesicle release but not for any other stage of exocytosis, and so act as important regulators of synaptic function.

Thus we may have identified the "calcium sensor" part of the process, but the specific molecular stages of vesicle fusion are still being worked out. They include formation of **SNARE complexes,** acting just before or in concert with the calcium-dependent step. SNARE proteins ("soluble NSF attachment receptors") are abundant both in the target cell membrane (t-SNAREs, including **syntaxin** and **SNAP-25**) and on the vesicles (v-SNAREs, including **synaptobrevin** also known as VAMP). These fuse together during exocytosis to form a "core complex", with four α-helix regions from the various proteins uniting into a stable coiled-coil bundle that brings the membranes into very close proximity, rather like a "zipper" action. A current model of SNARE action is shown in Fig. 9.15, and similar models may account for other instances of membrane fusion, with specificity conferred by a requirement for "cognate SNAREs". Recent evidence shows that synaptobrevin may be the limiting component at synapses, having only a sparse availability on the synaptic vesicles; whenever it meets a syntaxin–SNAP complex on the nearby membrane, a very rapid protein folding into the coiled-coil arrangement is induced that results in the two membranes fusing. Incidentally, the toxins associated with botulinum and with tetanus both act on this SNARE-binding system and thus can fatally block synaptic function.

A further part of the vesicle fusion control system may be regulation by **rab3 proteins**. These are GTP-binding proteins (GTPases) that are membrane bound but can be removed from the vesicle membranes after hydrolysis to GDP occurs. They interact with rabphilins, and together seem to limit the amount of transmitter release (per calcium signal) from the synapse, giving a negative

Table 9.3 Protein components of synaptic vesicles.

Type	Examples	Functions
Peripheral membrane proteins	Synapsins I, II, and III Rabphilin	ATP-binding, essential for NT release, ?enzymatic
Lipid-attached proteins	CSP Rab proteins (esp rab 3)	?Chaperone for a target protein
Transmembrane proteins		
Single transmembrane region	Synaptotagmins I and II	Ca^{2+} binding triggering exocytosis, ?role in endocytosis
	Synaptobrevins I and II (also termed VAMPs)	Binding to syntaxin and SNAP-25 on the plasma membrane
4 transmembrane regions	Synaptophysins I and II Synaptogyrin SCAMPs 1 and 4	
12 transmembrane regions	SV2A, 2B, 2c SVOP	?Regulation of calcium levels

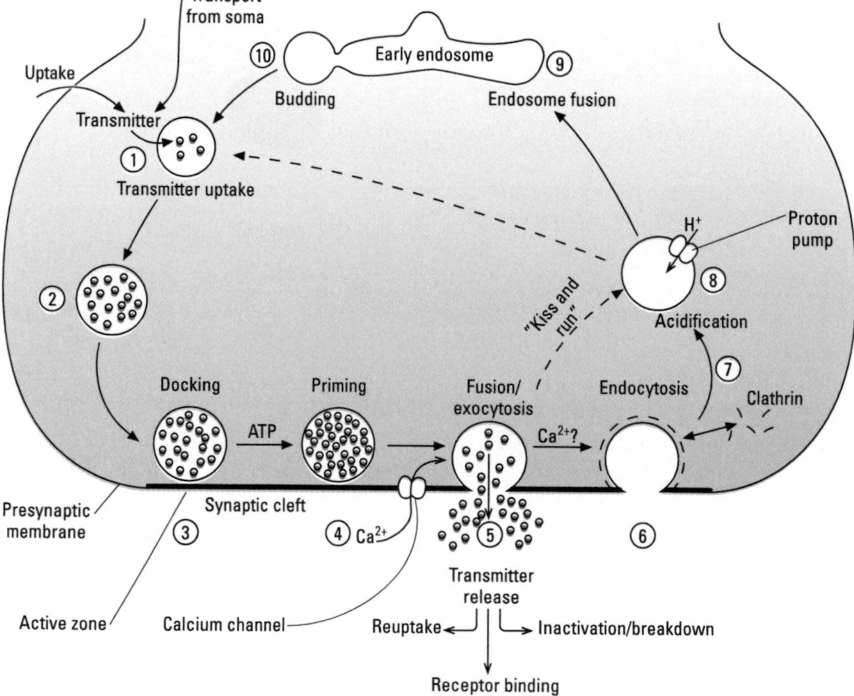

Fig. 9.13 The vesicle cycle at a typical synapse. 1, Synaptic vesicles are filled with transmitter. 2, Vesicles are moved to the active zone, and become docked there (3). 4, A complex energy-requiring process of "prefusion" or "priming" occurs at the active zone. 5, An incoming stimulus causes calcium influx through the voltage-gated channels, triggering fusion of the vesicle membrane and presynaptic cell membrane and exocytosis of the neurotransmitter. 6, The empty vesicle becomes coated with clathrin, then fuses together again (endocytosis). 7, The coated vesicle moves off the active zone and back into the cell, shedding its clathrin covering. 8, The vesicle contents become acidified by a proton pump. 9, The vesicle fuses with an endosome that eliminates damaged proteins. 10, Budding from the endosome produces a fresh vesicle. Note that steps 6–7 may be bypassed in an alternative fast "kiss and run" fusion process, and steps 9–10 may be bypassed for undamaged vesicles. (Adapted from Südhof 1995.)

control system. In the absence of rab3, presynaptic long-term plasticity (see below) is greatly reduced or lost.

Finally, there is accumulating evidence that many or all of the molecular interactions and protein complexes involved in vesicle fusion are protected and stabilized by the action of **molecular chaperones**, with different chaperones responsible for the regulation of SNARE and SNAP and of the uncoated vesicles. Such tight regulation insures the precise and crucial sequencing of events needed for reliable synapse functioning.

Transmitter release

The fusion of a vesicle with the plasma membrane inevitably leads to release (secretion) of the contained transmitter. The resultant transmitter release, crucially, is **quantal** in nature—it involves discrete units or packages. The idea of a "quantum" of transmitter can be readily related to the morphological feature of vesicles already described; put simply, one quantum represents a single vesicle fusing with the presynaptic membrane. Each vesicle contains a few thousand transmitter molecules (with a low variance of perhaps only 10%), and these become a single quantum of transmitter release.

It should be evident that there is no direct relation between the change in concentration of calcium in the presynaptic cell and the size of the quanta released; rather, it is the probability of a particular number of quanta being released that varies. The number of quanta released per normal presynaptic AP is normally quite low, in the range of 1–10 (although at vertebrate neuromuscular synapses it may be up to 200).

Transmitter release occurs into a specific synaptic cleft, often delimited laterally by glial cells (see Fig. 9.12) to keep the transmitter

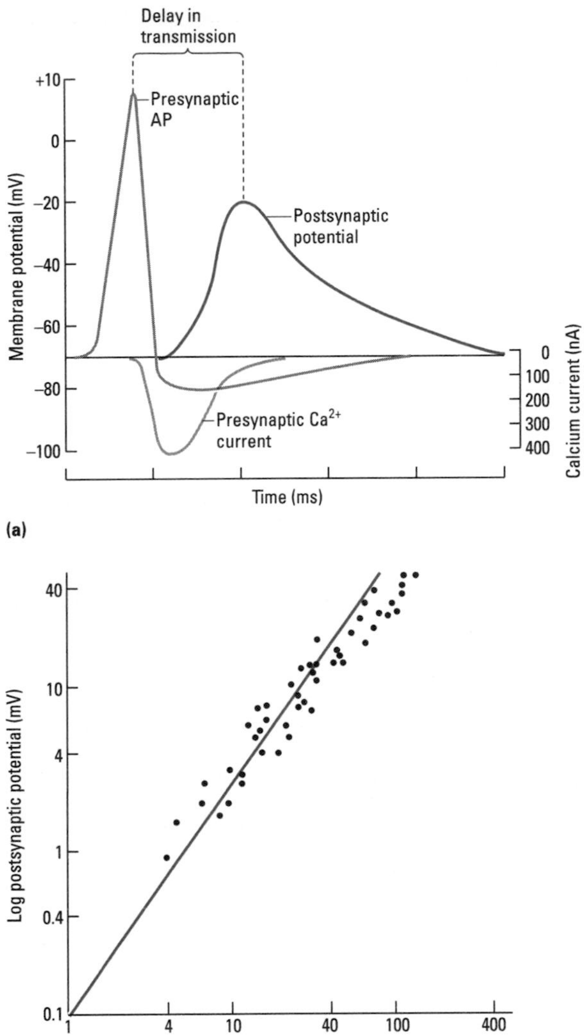

Fig. 9.14 At most chemical synapses a delay in transmission occurs mainly because the Ca^{2+} channel opening in response to the incoming action potential is relatively slow, and hence the postsynaptic effects of the transmitter released by the calcium influx are delayed. However, as shown in (b), the resulting postsynaptic potential is precisely (logarithmically) determined by the magnitude of the presynaptic calcium influx, insuring a "quantitative" onward transmission of the message. (b, Adapted from Reichert 1992.)

localized. Morphologically the synapse and its cleft are very like normal tight junctions occurring in other cell types, though they differ in being highly polarized. The cleft is narrow (~20 nm) and it is not occluded by any specific structure (except in the special case of neuromuscular junctions where there is a basal lamina between the two cell membranes). However, some filamentous or amorphous material does appear to cross the gap in most electron micrograph images. This is now characterized as an array of protein filaments, mechanically connecting the two membranes, and a diversity of **cell adhesion molecules** (CAMs, including cadherins, integrins, neurexins, and neuroligins), which are glycoproteins in the cell membrane, and which bind to each other from one cell to another, acting like glue. At synapses they help to maintain synaptic size and integrity, without directly affecting transmitter movement across the gap.

The synaptic cleft is also very commonly contacted intimately by at least one glial cell (astrocytes in the vertebrate CNS), and these have some role in mopping up transmitter from the cleft and so limiting the time course of transmitter actions. In many excitatory synapses the majority of transporters for the glutamate transmitter (see below) are on the astrocytes.

Transmitters

The neurotransmitters released from presynaptic sites are synthesized in the neuron itself, and are aggregated at high concentrations in the presynaptic vesicles. They can be applied experimentally to nerves and give the same effects as *in vivo* transmission, and they can be blocked by specific pharmacological agents. They are packed into the vesicles as soon as they are synthesized, so they become protected from enzymic degradation. In this form they are then easily transported through the neuron to their site of action.

The best known of all excitatory transmitters is acetylcholine (ACh), which is synthesized from acetyl-CoA, a crucial molecule in the intermediary metabolism of all cells (see Chapter 6, Fig. 6.8). The enzyme cholinacetyltransferase directs this ACh synthesis, using the synthesized acetyl-CoA plus choline from the diet. ACh operates at vertebrate neuromuscular junctions and at many other excitatory sites.

In chemical nature, other transmitters are many and varied, and it is often hard to decide which are truly transmitters and which are more properly described as "neuromodulators". This problem has in part been caused by discoveries that most neurons contain more than one transmitter, and these can act together as "cotransmitters". However, the majority of known transmitters can be broadly classified into two types: the amines and the peptides. Details of these and other transmitters are shown in Table 9.4. Most transmitters occur across the animal kingdom, though as yet octopamine and tyramine have been found only in invertebrates.

The **amine transmitter** group (Fig. 9.16) includes norepinephrine, dopamine, 5-HT (5-hydoxytryptamine, or serotonin), octopamine, and γ-amino butyric acid or GABA, this last being by far the most important inhibitory transmitter. The known classic amine transmitters in nervous systems are all small charged molecules, usually bearing a positive charge. All of them are metabolized directly from amino acids, and this can be achieved locally in the presynaptic terminal as well as in the neuron cell body. Most of these transmitters are monoamines, and many are in the group known as catecholamines with a 3,4-dihydrobenzole core structure. They are synthesized from the amino acid tyrosine, and are inactivated by monoamine oxidases (MAOs) or by catechol-*O*-methyltransferase. Note that a few (sometimes given their own separate group) are simple unmodified **amino acids**. Of these, by far the most important is glutamate, now confirmed as the main excitatory transmitter between neurons, at least in vertebrates, despite the fact that it is also an important component in intermediary metabolism and occurs at high concentrations (~20 mM) throughout animal brains.

Neuroactive **peptide transmitters** (Fig. 9.17) are somewhat larger molecules and far more numerous than amines, though their effects are generally less well known. Many of them merge into the category of neurosecretory hormones dealt with in Chapter 10. Over 100 neuroactive peptides are now known, but there may be

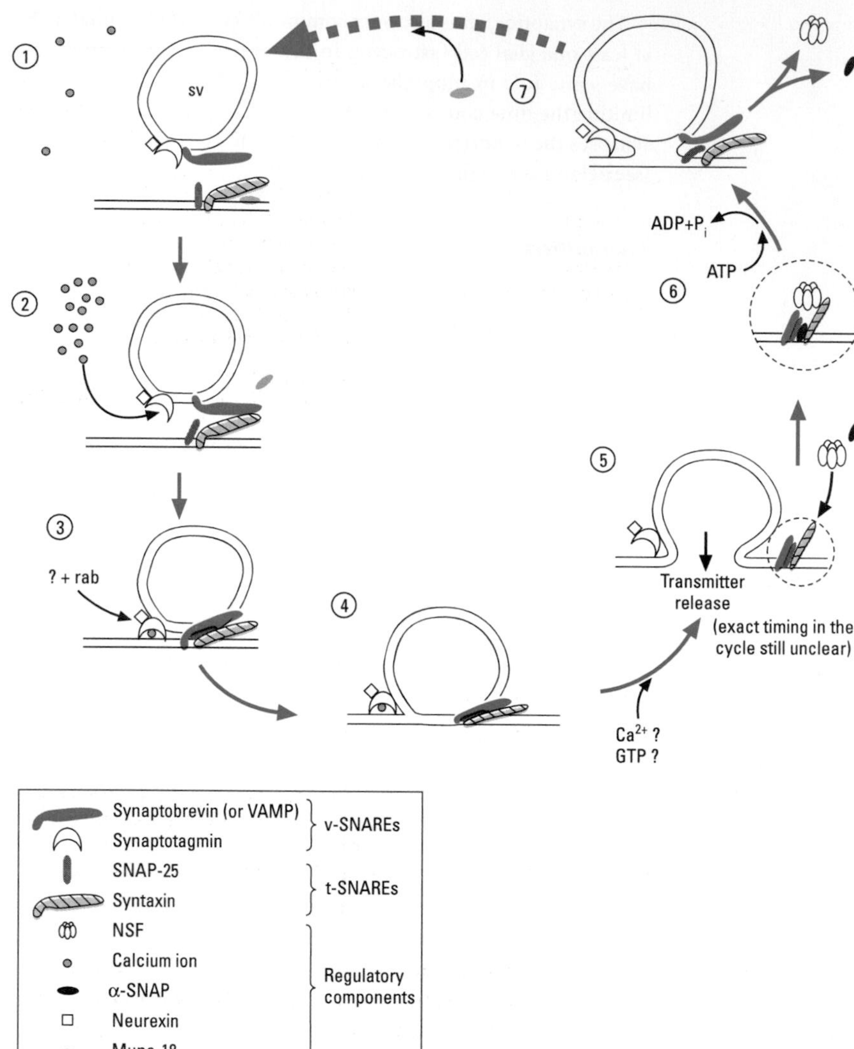

Fig. 9.15 Diagrammatic summary of our current understanding of SNARE action whereby synaptic vesicles fuse with the presynaptic membrane. 1, The resting state, with v-SNAREs and t-SNAREs well separated. 2, Following a calcium influx, Ca^{2+} binds to the synaptotagmin so that its binding to synaptobrevin is weakened; synaptobrevin is activated, losing its link to the inhibitory protein Munc18 and able to bind instead with SNAP-25 and syntaxin on the synaptic membrane. 3, The helices of synaptobrevin, SNAP-25, and syntaxin rapidly intertwine with each other and flip into a new orientation, forming the SNARE complex, and begin a "zippering" process that fuses the two membranes together (4). 5, Membrane fusion (probably requiring GTP input) leads to pore opening and transmitter release. 6, The stable SNARE complex binds to NSF and α-SNAP from the cytoplasm, which then (7) dissociates the complex using ATP and allows the v-SNAREs and t-SNAREs to become localized, respectively, in the vesicle and synapse membrane again. Various other inhibitory and regulatory proteins are also known to be involved.

hundreds more to be discovered. They are short chains of amino acids, many only 2–10 units long (though some are up to 50). They include endorphins and enkephalins, acting as neuromodulators in vertebrate brains where they bind to at least three kinds of opioid receptors, and mimicking the effects of natural plant opiates in producing euphoria and analgesia (pain suppression). They also include a good many compounds that were already well known as peptide hormones, acting over much longer distances; for example, vasopressin, GnRH (gonadotropin-releasing hormone), and gastrin, which as hormones are released respectively from the pituitary, the hypothalamus, and the intestine (see Chapter 10).

When acting as neurotransmitters, these small peptides are often derived from a single strand of mRNA which encodes for a polyprotein that may then be broken down into multiple active peptides, including multiple copies of the same peptide. In this way some long-term amplification can be built in, as one stimulus at the genomic level yields several active peptides.

Additional types of transmitter also occur in nervous systems that cannot be fitted into the binary classification above, as summarized in Table 9.4. The most unusual of these are **neurotransmitter gases**, nitric oxide (NO) and carbon monoxide (CO). NO was first noticed for its effects on blood vessels, where it mediates the action of ACh on relaxation processes in smooth muscle. ACh triggers the endothelium to release NO, which then enters the muscle cells. NO is produced from arginine, and was soon also identified as having transmitter-like roles in other situations where arginine had been known to be important—for example in macrophage attack on tumor cells, and in the cerebellum. Since as a gas this transmitter is peculiarly elusive, the occurrence and effects of NO are now largely studied by seeking out the enzyme nitric oxide synthase (NOS). NOS-containing neurons represent only about 1% of the mammalian brain, and are highly localized, but these neurons ramify so extensively that many or perhaps most brain cells will contact a NO nerve terminal. Whether NO is truly a "transmitter" is debatable, since as a gas it cannot be stored in vesicles or be released by exocytosis, instead diffusing freely and rapidly between cells. Nor is it detected by specific receptors, instead binding postsynaptically to the rather common guanylyl cyclase to affect cGMP levels. However, NO does form in neurons in response to calcium influx, and it is very tightly regulated so it is produced only in specific

Table 9.4 Neurotransmitters and their properties (those marked * are usually neuromodulators).

Type	Transmitter	Source	Locations
Cholinergic	Acetylcholine	Acetyl CoA + choline	Vertebrate parasympathetic ANS, CNS, and PNS. Many invertebrate neurons
Biogenic amines	Dopamine	Tyrosine	CNS and ANS of vertebrates. Many invertebrate neurons
	Norepinephrine (noradrenaline)	Tyrosine	CNS and ANS of vertebrates. Many invertebrate neurons
	Octopamine	Tyrosine	Common in invertebrate neurons
	Serotonin (5-HT)	Tryptophan	Vertebrate CNS
	Histamine		Vertebrate PNS
	Tyramine		Some invertebrate neurons
Amino acids	γ-amino butyric acid (GABA)		Inhibitory synapses
	Glutamate		Vertebrate CNS, excitatory. Arthropod NMJ
	Aspartate		Excitatory synapses
	Glycine		Vertebrate spinal cord
Peptides*	Endorphins		
	Enkephalins		
	Various hormones (angiotensin II, carnosine, glucagon, neurotensin, oxytocin, vasopressin, (somatostatin, gastrin, antidiuretic hormone)		
	Various hormone-releasing factors		
	Substance P		
Purines*	Adenosine and ATP	Ribose-P + purine	CNS
Fatty acid derivatives*	Arachidonic acid, prostaglandins, thromboxanes	Membrane phospholipids	CNS
Neurotrophins*	Nerve growth factor (NGF) NT3, NT4, NT6, etc.	Transcription from mRNA	Developing CNS
Gases*	Nitric oxide (NO)	Arginine	Vertebrate and invertebrate brains and ANS
	Carbon monoxide (CO)	Heme protein	Localized parts of vertebrate brain

ANS, autonomic nervous system; ATP, adenosine triphosphate; CNS, central nervous system; CoA, coenzyme A; 5-HT, 5-hydroxytryptamine; NMJ, neuromuscular junction; PNS, peripheral nervous system.

circumstances. It has a clear role as a transmitter in blood vessels, especially in the gut, and in the brain it has a major function in inhibiting aggressive and sexual behaviors (NOS knockout mice display an excess of both traits!). It may also play a role in learning and memory, since chemicals that release NO also facilitate long-term potentiation or LTP (see section 9.5).

CO may be a second gaseous neurotransmitter. It is produced in the body by breakdown of heme proteins, by the enzyme heme oxygenase (HO). HO is a stress-induced protein (formerly classified with the heat shock proteins; see Chapter 8), and again plays a role in smooth muscle, CO often being a cotransmitter with NO. CO also appears to regulate some olfactory neurons, probably via cGMP.

As is clear from the above outlines, new insights into transmitter function and diversity are still emerging. However, it seems safe to conclude that while any one neuron may release more than one transmitter (e.g. there are many GABA + 5-HT and GABA + dopamine neurons), so far as is known one neuron always produces the same transmitter, or combination of transmitters, at all of its terminals.

Postsynaptic structure

The postsynaptic membrane is characterized by a densely staining area known as the "postsynaptic density" or PSD, which contains an array of receptor protein sites and an underlying matrix of signal-transducing and scaffolding proteins (with different elements at excitatory and inhibitory synapses). The receptor sites are sometimes heterogeneous and randomly distributed, but in other synapses they are organized into zones; for example glutamatergic synapses often have an outer ring of NMDA receptors around a core of AMPA receptors (see below). The postsynaptic membrane also contains **signal transduction proteins** that couple receptors to the intracellular second messenger pathways, and the insertions of **adhesion proteins** (neuroligins, cadherins, etc.) that as we have seen anchor the membrane to similar presynaptic adhesion proteins.

In the vertebrate cortex, most of the excitatory postsynaptic sites are on so-called "dendritic spines", 0.2–2 μm long, giving the dendrites a thorny appearance with a much increased surface area. These spines have a swollen head and a rather narrow neck, limiting the diffusion of ions and chemical messengers; they contain a prominent actin–myosin-based cytoskeleton, and recent evidence shows that the diameter of the neck may be altered (by the "contractile" interaction of actin and myosin, discussed fully in section 9.10) to give a means of regulating the flow of signals from the spine to the main dendritic shaft.

Postsynaptic events

At the postsynaptic membrane, it was noted early in the era of cellular neurobiological studies that tiny spontaneous potential changes can occur (Fig. 9.18a). These are termed **miniature postsynaptic**

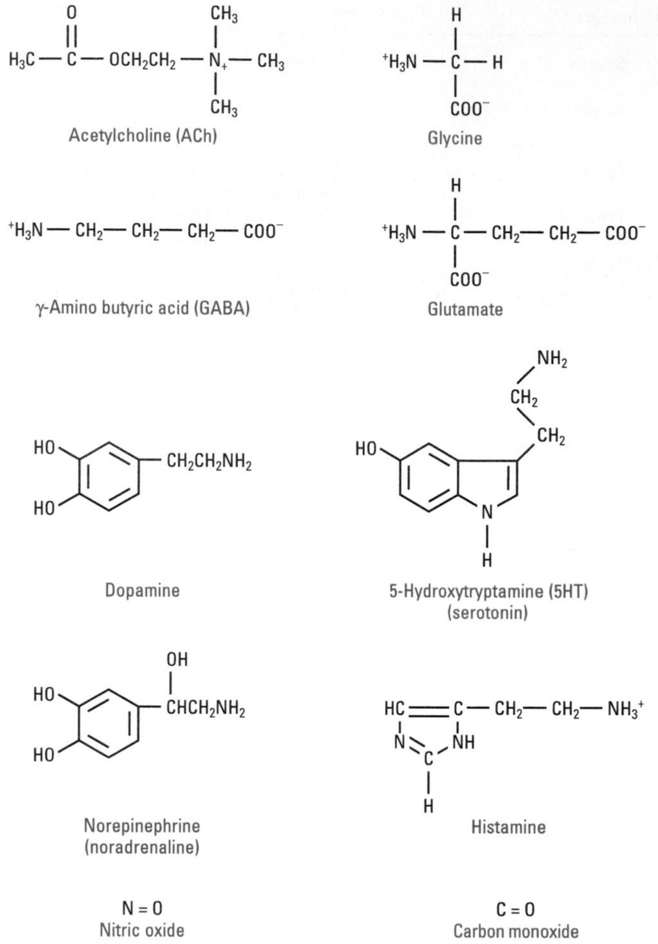

Acetylcholine (ACh)

Glycine

γ-Amino butyric acid (GABA)

Glutamate

Dopamine

5-Hydroxytryptamine (5HT) (serotonin)

Norepinephrine (noradrenaline)

Histamine

N = O
Nitric oxide

C = O
Carbon monoxide

Fig. 9.16 Amine neurotransmitter structures.

potentials, or mPSPs, and are random in frequency, with an amplitude of less than 1 mV. These tiny potential changes are a manifestation of the spontaneous release of a single quantum of transmitter; that is, the discharge of single presynaptic vesicles, following their random fusion with the presynaptic membrane. The contents of these vesicles elicit a subthreshold effect on the postsynaptic membrane via the receptor molecules that are sited there. When a presynaptic AP leads to the discharge of a number of vesicles the effect is simply a multiplication of this, leading to a summed **postsynaptic potential** (PSP) (Fig. 9.18b).

The underlying mechanism for this involves the tight and highly specific binding of the transmitter molecule, released into the synaptic cleft, onto receptor sites on the outward-facing postsynaptic membrane. For every neurotransmitter there are specific receptor molecules (proteins) with binding sites of high affinity. In most central synapses there may be around 1000 receptors per square micrometer (μm^2), giving estimates of total receptor number per synapse of 15–450 (GABA synapses) or 11–300 (glutamate synapses); though we should note here that receptor densities and numbers are perhaps 10–20 times greater at neuromuscular junctions. Fusion of one presynaptic vesicle may give a response in 20–200 receptors (i.e. nearly all of those present at one synapse).

The transmitter molecule induces allosteric changes in the receptor (see Chapter 2), which leads to further postsynaptic events. The effect that is seen depends on the receptor, and not on the transmitter. Thus the same transmitter can have quite different effects at different synapses, even to the extent of being excitatory at some and inhibitory at others (Fig. 9.19 gives an illustration of this). It is possible to distinguish two quite different kinds of acetylcholine-mediated synapses by the pharmacology of the two different receptor sites that are in action. At one kind of synapse, the ACh action is also induced by nicotine, and blocked by α-bungarotoxin, while at the other kind muscarine is an activator and atropine is a blocker. Thus we speak of nicotinic and muscarinic ACh synapses. Similarly, there are at least three main kinds of GABA receptor (given subscripts A, B, and C) distinguished by their pharmacology. Perhaps the most complex are the glutamate receptors (Table 9.5), occurring in at least four varieties (including the important NMDA and AMPA types, named in accordance with acronyms for specific molecules that can block them). At least 22 different protein subunits for these receptors have been identified in mammals alone.

The effect of the receptor-binding phenomenon is apparently always one of two types, often termed **ionotropic** or **metabotropic** (Fig. 9.20). In the first case the consequence of transmitter binding is a change of ionic permeability in the postsynaptic membrane involving the direct opening of a channel, and hence a transmembrane flow of ions (ionic current). In the second case, transmitter binding has an indirect effect, via the G-protein/second messenger system discussed in Chapter 2.

The first type (channel-mediated or ionotropic effects) involves proteins that are "**transmitter-gated ion channels**", a special case of ligand-gated channels. These are often simply termed "receptor channels", since they combine the properties of channels and receptors in one membrane-spanning molecule. They switch between different "closed" and "open" configurations at the central pore within the protein molecule, in response to transmitter binding at an adjacent site. This transition is usually very fast, so that a postsynaptic potential is generated within milliseconds. Like most ion channels, the transmitter-gated channels are usually very specific (to sodium, or potassium, or calcium, or chloride). But in contrast to normal ion channels, they are not normally very sensitive to voltage change. Most of the receptor channels so far investigated seem to belong to a single protein family, and share large degrees of sequence homology. They are constructed from several homologous polypeptide units (often with α and β versions, sometimes four different versions), each spanning the membrane several times in an α-helix conformation (see Chapter 2). Each polypeptide is around 40–65 kDa, and the complex of polypeptides forms a loosely cylindrical shape protruding just a few nanometers into the extracellular space. One or more of the subunits incorporate extracellular receptor sites, and when occupied these induce configurational change through the whole complex. Receptors that involve direct transmitter-gated channels include the nicotinic ACh receptor (especially well studied in vertebrate neuromuscular junctions), the GABA receptor, and most of the glutamate and glycine receptors.

The other (and less common) category of receptors are metabotropic, with an indirect action, and do not immediately open a channel in the postsynaptic membrane (though this may be their

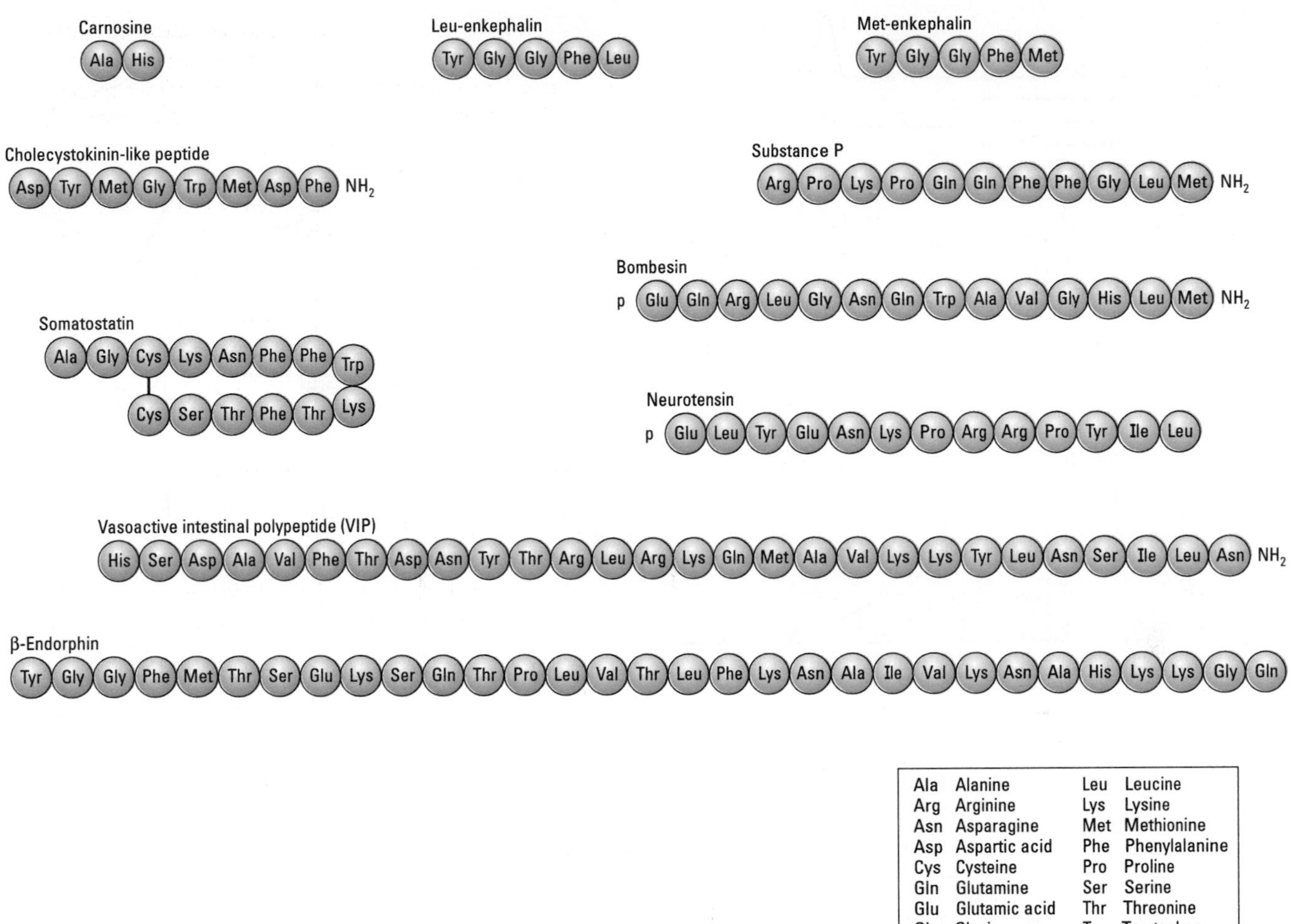

Ala	Alanine	Leu	Leucine
Arg	Arginine	Lys	Lysine
Asn	Asparagine	Met	Methionine
Asp	Aspartic acid	Phe	Phenylalanine
Cys	Cysteine	Pro	Proline
Gln	Glutamine	Ser	Serine
Glu	Glutamic acid	Thr	Threonine
Gly	Glycine	Trp	Tryptophan
His	Histidine	Tyr	Tyrosine
Ile	Isoleucine	Val	Valine

Fig. 9.17 The amino acid sequences of common peptide transmitters. Most are short and single-branched, though somatostatin is two-branched via a dicysteine bridge. (Somatostatin is a neurohormone, this and others being dealt with in Chapter 10.)

ultimate effect). Instead, when transmitter molecules bind to the receptor the effect is a biochemical cascade on the intracellular side, involving the kind of amplification process we looked at in Chapter 2. The postsynaptic receptor changes its configuration and initiates further change in a **coupling protein** (**G-protein**) nearby, such that GDP exchanges for GTP at that protein (see Fig. 2.3) and it becomes activated. This is the trigger either for change in a nearby channel (in a relatively few simpler cases) or for further more complicated signal-amplifying effects. The latter operate either via cAMP and protein kinases, or via phosphodiesterases, diacylglycerol (DAG), and inositol triphosphate (IP_3) which in turn affect kinases and/or calcium levels. These alternative pathways were covered in detail in Chapter 2. This kind of receptor system allows for complex effects, extending even as far as the regulation of gene activity. The end effect can be an allosteric change in an ion channel, or a covalent binding to such a channel; either way, the electrical properties of the postsynaptic membrane are altered. Inevitably, this kind of transmitter-mediated effect is rather slower than the direct ionotropic effect described in the previous paragraph. The signal processing is more complex, and the ion-gating effect takes at least a few hundred milliseconds, with after effects lasting potentially minutes or hours. The receptors involved are large transmembrane proteins with significant homologous sections, each with seven hydrophobic α-helix regions that span the membrane (cf Fig. 2.17). Examples of metabotropic postsynaptic receptors are the muscarinic ACh receptor (Fig. 9.21), the β-adrenergic receptor and the serotonin (5-HT) receptor, as well as one group of the glutamate receptors. These kinds of neurotransmitters do not necessarily obey the three criteria of "fast, point-to-point, and simple" communication set out earlier, and are therefore often called **neuromodulators** instead. (However there is no exact match between neuromodulation and metabotropic processes, **neuromodulation** being a broader term encompassing many heterosynaptic situations where one neuron can modify the responses of other synapses. Indeed some neurons are specialized for modulation, having diverse projections to many areas of the brain and thus modulating overall neural activity.)

All transmitter binding interactions with receptors ultimately lead to a potential change in the postsynaptic membrane. However, the released transmitter must somehow be removed from the

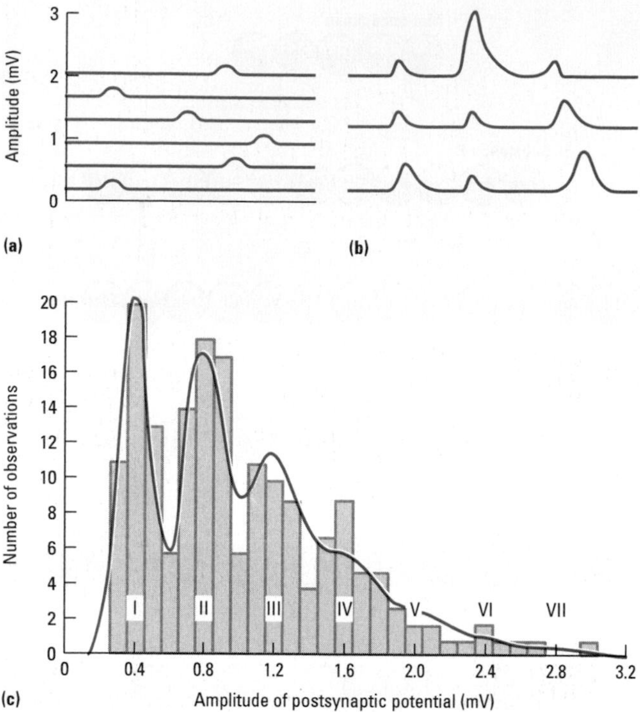

Fig. 9.18 (a) Microelectrodes recording from a postsynaptic neuron will pick up spontaneous potential changes that are of a constant size—the miniature postsynaptic potential, or mPSP, usually less than 0.5 mV in amplitude, and corresponding to the random release of the contents of a single synaptic vesicle. (b) Certain kinds of stimulation and environmental changes will elicit a few larger potential changes, but these are always exact multiples of the mPSP, indicating the release of two, three, or four vesicles. Hence (c) shows these evoked small potentials plotted as a histogram with peaks at whole number multiples (up to seven) of the mean amplitude of the mPSP. The full postsynaptic potential resulting from excitation of the synapse is usually at least an order of magnitude larger than this. (c, Adapted from Boyd & Martin 1956.)

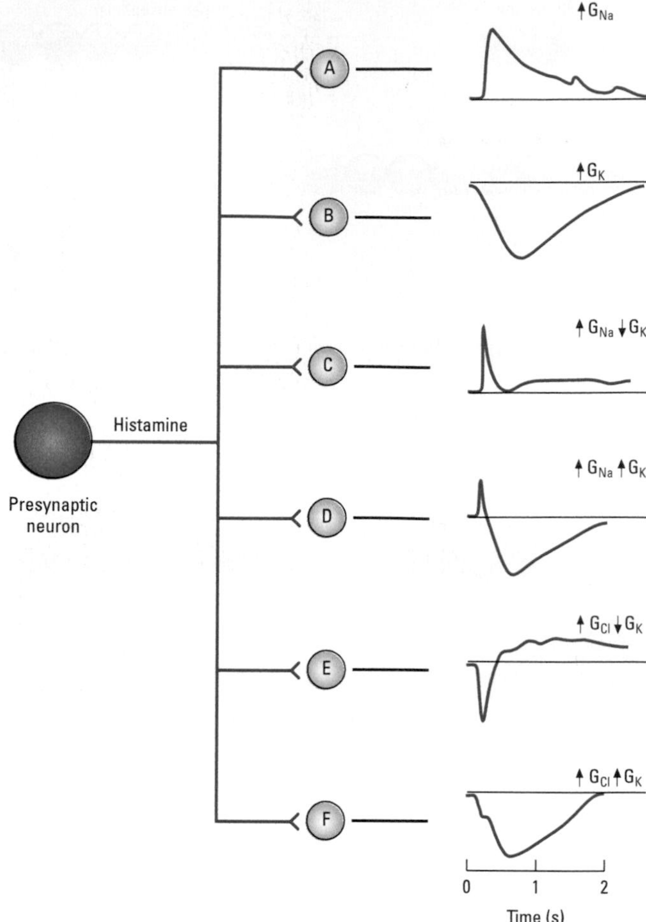

Fig. 9.19 The same neurotransmitter—even when released from the same neuron—may have very different effects at different synapses. Here histamine from a single neuron in the sea slug *Aplysia* has at least six contrasting effects in terms of the ion currents induced and the postsynaptic potentials recorded in different target neurons. (G refers to the conductance to different ions in each case.)

Table 9.5 Diversity of GABA and glutamate receptors.

GABA receptors	
Ionotropic receptors	Chloride channels, with subunits from at least six gene families; anchored via GABA-RAP protein
$GABA_A$	
$GABA_B$	
$GABA_C$	
Glutamate receptors	
Ionotropic receptors (glutamate-gated ion channels)	
NMDA	Postsynaptic excitatory, heteromeric, long cytoplasmic tail linking with many signaling and scaffold proteins; permeable to calcium, involved in calcium signaling
AMPA	As above, but lacking calcium permeability and often sited on intracellular membranes
Kainite	Pre- or postsynaptic, otherwise as above
γ-type	Postsynaptic, mostly in Purkinje cells of cerebellum
Metabotropic receptors	
mGluR 1 and 5	Postsynaptic, operate through PLC and intracellular calcium
mGluR 2–4, 6–8	Pre- or postsynaptic, operate through adenylyl cyclase

AMPA, α-amino-3-hydroxy-5-methyl-isoxazolepropionic acid; GABA, γ-amino butyric acid; NMDA, *N*-methyl-D-aspartate.

synaptic cleft to turn off the signal and allow new inputs of information. This can be achieved in several ways. The simplest is for the transmitter to diffuse away, and this need only happen over a very short distance since the receptors are normally only found in the small areas of membrane immediately beneath the presynaptic terminals. Another possibility is for the transmitter to be degraded by the action of specific enzymes; this is particularly well known for acetylcholine, split by the action of acetylcholinesterase, located in the extracellular matrix at the synaptic cleft, from where the breakdown product choline is taken up and reused presynaptically. The third possibility is that the whole transmitter molecule is taken up and recycled, and this is commonly the case for the neuroactive amines. Either the presynaptic terminal, or nearby glial cells, may perform this function. If the former, there must be receptor sites on the presynaptic membrane itself.

In addition to all the relatively fast events described in this section, it is becoming clear that neurotransmitters may also have much longer term effects when they are detected postsynaptically. They can affect neuronal differentiation and survival, axonal growth, and

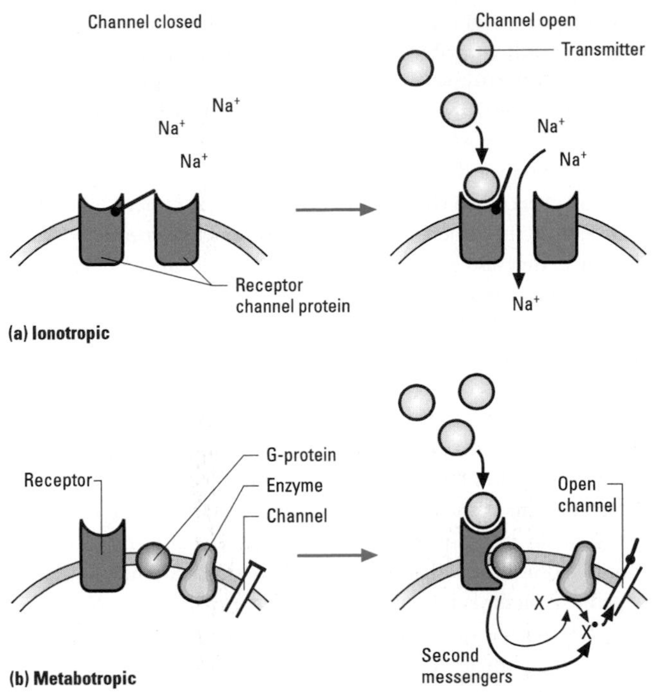

Fig. 9.20 (a) Ionotropic and (b) metabotropic receptors. The former operate directly as ligand-gated channels and the latter via coupling to G-proteins that, once activated, can alter ion channels directly or initiate second messenger cascades operating via enzymes, but again usually with a changed channel conductance as the final outcome.

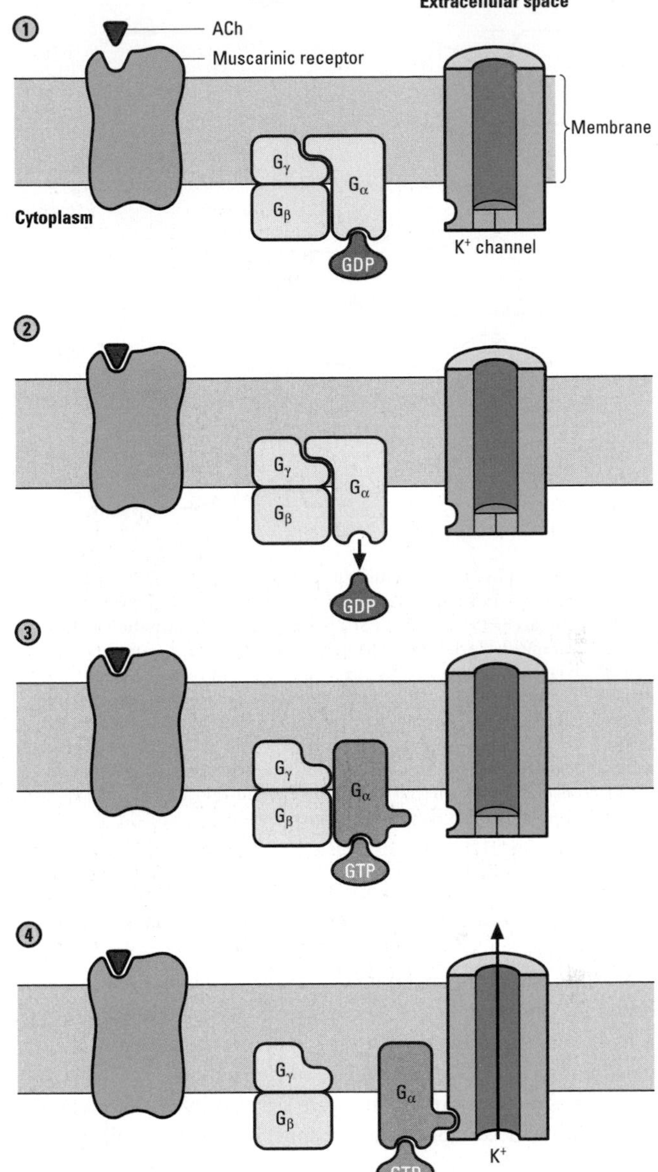

Fig. 9.21 The acetylcholine (ACh) muscarinic receptor as an example of a simple metabotropic receptor structure and its operation. Here there are no second messengers; when ACh binds externally the G-protein changes conformation slightly, releases a bound GDP and takes up a GTP. This in turn activates the G_α-subunit, which binds to a neighboring potassium channel causing it to open.

synaptic strength. The best known case is glutamate, which is not only an inhibitory transmitter but which, by bringing about calcium influx through transmitter-gated and voltage-gated channels, plays a role during embryogenesis in the survival of developing neurons, and a role in the adult in neuronal adaptation and hence information storage. These issues are discussed in later sections.

EPSPs and IPSPs

Postsynaptic potentials (PSPs) generated by the transmitter action can be either excitatory or inhibitory, and are therefore termed either **EPSPs** or **IPSPs** respectively. EPSPs help to evoke APs in the postsynaptic cell, or, if it is nonspiking, help to induce electrotonic potentials that in turn lead to transmitter release (usually at the cell's onward synapses, or occasionally in a regulatory action at the same synapse where the PSP is generated). IPSPs act in the opposite fashion, preventing the development of an AP or onward transmitter release. Both EPSPs and IPSPs tend to be quite small (often around 1 mV) in magnitude, and they can summate with each other. They are not self-regenerating in the same manner as an AP. Their properties are largely determined by the characteristics of the postsynaptic ion channels through which the current flows.

If the ion channel that is opened by a particular transmitter is permeable to just one ion, then the PSP will normally be seen as a potential change away from the RP to (or close to) the equilibrium potential of that ion, here often termed the **reversal potential**. If more than one kind of ion can flow through the channel, then the reversal potential can be calculated using the Goldman equation (see Box 4.4). The current will be inward (depolarizing if as usual it is a cation current) when the RP is more negative than the reversal potential; but outward (hyperpolarizing for a cation) when the RP is more positive than the reversal potential. Thus it is crucial to realize that the effect produced by any one transmitter and channel can vary according to the RP of the postsynaptic cell. At most RPs the effect may be depolarizing, but at very positive RPs the effect will be hyperpolarizing (Fig. 9.22). The total synaptic current, made up of all the tiny currents of less than a picoamp flowing through individual channels, may be up to about 0.1 mA, and can produce a postsynaptic potential change of up to 60 mV. The full postsynaptic current, and the full PSP, show exactly the same dependence on

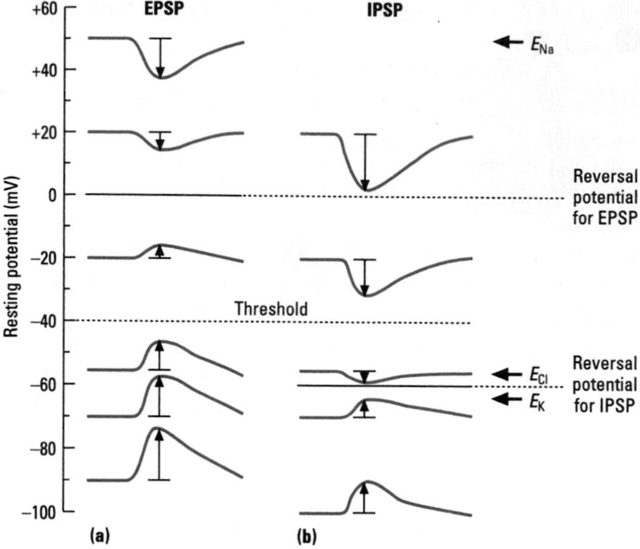

Fig. 9.22 The inhibitory and excitatory postsynaptic potentials (IPSPs and EPSPs) depend on the state of the receiving membrane. (a) The EPSP at different resting potential (RP) values: the EPSP appears as a hyperpolarization if the membrane is unusually depolarized, but in the normal RP range (around −60 to −70 mV) it appears, as expected, as a depolarization. The reversal potential for this EPSP is around 0 mV, well above the threshold for an action potential (AP) to be produced (around −40 mV). (b) The same principles apply in reverse for an IPSP, which (given a reversal potential of about −60 mV) will appear at the normal RP as a small depolarization, but which causes hyperpolarization near the AP threshold and so will tend to inhibit AP production. (E_K, E_{Na}, and E_{Cl} are the equilibrium potentials for the three principle ions involved.)

RP as do the individual currents and potential changes that are contributory.

A PSP is effective as an excitatory event (i.e. it is an EPSP) when it drives the postsynaptic cell to further activity, such as AP production. It does this normally when its own reversal potential is more positive than the cell's threshold for AP generation, so that it gives a suprathreshold depolarization. Again the nicotinic ACh receptor is the best known example; its ion channel is roughly equally permeable to Na^+ and K^+ ions, so its reversal potential is between 0 and −10 mV, roughly the midpoint between E_{Na} (around +50 mV) and E_K, (around −70 mV). When the RP is more negative than this, as is of course usual, the Na^+ inward current predominates and we see a depolarizing EPSP. But it would in theory be possible to see a hyperpolarizing EPSP if the RP were to be artificially reset at a much less negative starting point (Fig. 9.22a).

An IPSP requires that the reversal potential for the current that carries it is more negative than the AP generation threshold. IPSPs can be produced by either cation or anion channels. Many are due to Cl^- channels, including the best known example, the $GABA_A$ receptor–ion channel complex, probably the most important inhibitory system in the vertebrate brain. Others are due to K^+ channels. In either case, the reversal potentials (typically −60 mV for chloride and −70 mV for potassium) are normally below the RP, and the effect of channel opening is a hyperpolarization (Fig. 9.22b), opposing the effect of any EPSPs that are also being received by the neuron.

There are also instances where PSPs are produced by the closure of ion channels rather than by their opening. Some neurons have postsynaptic membranes in which the main ion channels are held open at the RP, contributing to the value of that RP. These can then be closed by transmitter–receptor interactions. Some K^+ channels act like this under the influence of muscarinic ACh receptors, and the receipt of an ACh signal leads to channel closure and thus a decrease in outward potassium current, hence a depolarization and an EPSP effect. Such channel closures tend to be rather long acting, and because they lead to decreased membrane permeability (conductance) they also enhance the amplitude of all other EPSPs (from Ohm's law: current flow produces a greater potential change when resistance is higher). Synapses using this mechanism can therefore mediate heightened excitability over relatively long timescales in some parts of animal nervous systems.

Most neurons receive hundreds or thousands of synaptic contacts, from different presynaptic cells, producing a variety of PSPs each generally less than 1 mV in magnitude. To produce an AP in the neuron, many such EPSPs must summate (and must counter the effects of any incoming IPSPs). The response of any one neuron to inputs from chemical synapses is therefore extremely variable, depending on the subset of inputs active at any one time. **Neuronal integration** is the process whereby all these inputs are spatially and temporally summated, and is a topic we will return to later (see section 9.5). This summation or integration often occurs at a particular site in the neuron where the membrane threshold for AP initiation is especially low, a zone termed the "axon hillock" (see Fig. 9.1). Here the amplitude and time course of all the integrated PSPs are encoded into the frequency and pattern of APs propagating along the axon.

Synaptic plasticity

There are three crucial variables in synaptic transmission: the number of release sites (N), the probability of a quantal release (P), and the size of the quantal response (q). Controlling these three variables underlies much of what goes on at a synapse. All three have morphological correlates that can be relatively easily studied: N is the number of active zones in the synapse, P is linked with the number of docked vesicles, and q is the size or molecular content of a single vesicle (although q is also modified by postsynaptic variables). Any factors that alter one or more of these parameters will alter the outcome of synaptic transmission, and so lead to synaptic plasticity (in effect, modification by past use). Thus synaptic transmission is a dynamic process, with the postsynaptic responses waxing and waning as presynaptic activity changes through time.

Various kinds of "autoregulation" are possible within a chemical synapse, as a result of which transmission can be modified by a factor of 10-fold or more. This regulation is termed **homosynaptic plasticity**, or **short-term plasticity** (in contrast to longer term processes affecting more than one synapse—heterosynaptic—which are dealt with in section 9.5 below). Short-term plasticity can most simply be classified by effect, as either synaptic depression or synaptic facilitation. Both phenomena are widespread in animal nervous systems, and the same neuron may show quite different expressions of each at synapses with different target cells.

1 **Synaptic depression** (sometimes more loosely termed synaptic fatigue) is seen as a gradual decrease in the amplitude of PSPs when a neuron has been in use for long periods (Fig. 9.23a). The effect is

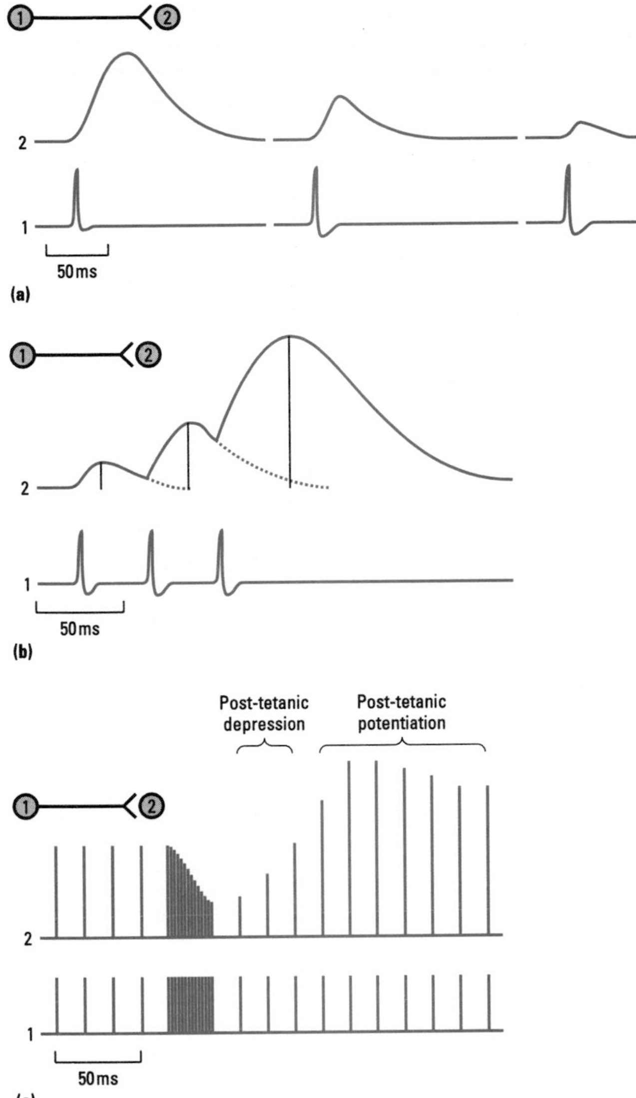

Fig. 9.23 (a) Synaptic depression: the postsynaptic potential (PSP) is decreased with repetitive stimulation. (b) Synaptic facilitation: the PSPs increase (and may summate) with repetitive stimulation. (c) Post-tetanic effects at neuromuscular synapses, where high-frequency stimulation leads to a much longer lasting increase (post-tetanic potentiation) in the PSP.

commonly localized to the presynaptic site, where it is due to a decrease in the number of quanta of transmitter (i.e. the number of vesicles) released per AP; the protein rab3 is strongly implicated, with synapsins also involved. Depression may also be a postsynaptic phenomenon, arising by modification of the responses of the receptors (by phosphorylations, or more subtle molecular desensitizations); and it may also be brought about by neuronal–glial interactions in some instances.

2 **Synaptic facilitation** may be less common, but is well known in vertebrate neuromuscular synapses, where rapid repeated stimulation of the motor neuron leads to larger amplitude PSPs in the muscle cells (Fig. 9.23b). This effect lasts only a few hundred milliseconds. A longer term effect ("augmentation") may occur over a few seconds, and if the incoming APs are at particularly high frequency then "tetanic stimulation" may occur. A brief phase of reduced transmission (post-tetanic depression) is followed by a phase of increased PSPs, termed post-tetanic potentiation (Fig. 9.23c), and lasting for several minutes. The major cause of facilitation lies with an increase in the calcium concentration in the presynaptic terminal, each transient evoked increase in $[Ca^{2+}]$ failing to be corrected before the next round of calcium channel opening, so allowing summation with previous events. The increased residual calcium concentration (referred to as $[Ca^{2+}]_{res}$) then leads to a considerable increment in the evoked transmitter release, so that calcium in effect is acting as a second messenger. It is probably operating, at least partly, via calcium/calmodulin-dependent kinases (CaMKI and CaMKII), which interact with synapsins on vesicle surfaces. At postsynaptic sites facilitation may be achieved through similar calcium-mediated effects, and/or by adding or perhaps unmasking new receptors.

As may be seen, synaptic plasticity can also be classified in terms of its source of induction, site of expression, and molecular basis (summarized in Fig. 9.24). For homosynaptic plasticity the source is within the same synapse, whereas for heterosynaptic plasticity the source is a third neuron. The site of expression in either case may be pre- or postsynaptic, the former involving changes to the amount of transmitter released and the latter changes to the response to a fixed amount of transmitter. The molecular basis is more varied, with many mechanisms now known, but all involving second messengers from cell surfaces to interior effectors. In short-term plasticity the messenger is usually calcium.

Thus synapses do not just transmit information, they also process it in various ways according to their own time-varying filters, and thereby perform the pivotal role in neural functioning.

Variation in synapses

Analysis of synapses, first with electron microscopy and more recently with molecular probes, has led to an increasing appreciation of their diversity. An early structural distinction between "type I" and "type II" synapses has turned out to correspond rather well (though not perfectly) with excitatory glutamate and inhibitory GABA synapses, respectively. The former have wider synaptic clefts, thicker submembranous scaffolds, and spherical rather than ovoid presynaptic vesicles. Some synapses (especially in sensory organs) have ribbon-like processes presynaptically, or berry-like clusters of polyribosomes.

The dimensions of presynaptic sites also vary markedly, and this links to their specificity. In some cases the vesicles are highly clustered at active zones and the transmitter is released very locally. In other neurons, for example in parts of the vertebrate autonomic nervous system, the presynaptic membranes are not locally specialized and vesicles are present throughout the terminal, with transmitter being delivered over a large postsynaptic target area. This variation can be viewed as part of a continuum that at its extreme merges into neurosecretion processes, where a transmitter is released so diffusely that it enters the bloodstream and can travel to quite distant targets to act as a hormone (see Chapter 10).

But perhaps most crucially, synapses vary greatly in their "normal" values of N, P, and q. Some synapses have hundreds of active sites, a few have only one; recent evidence indicates that the size

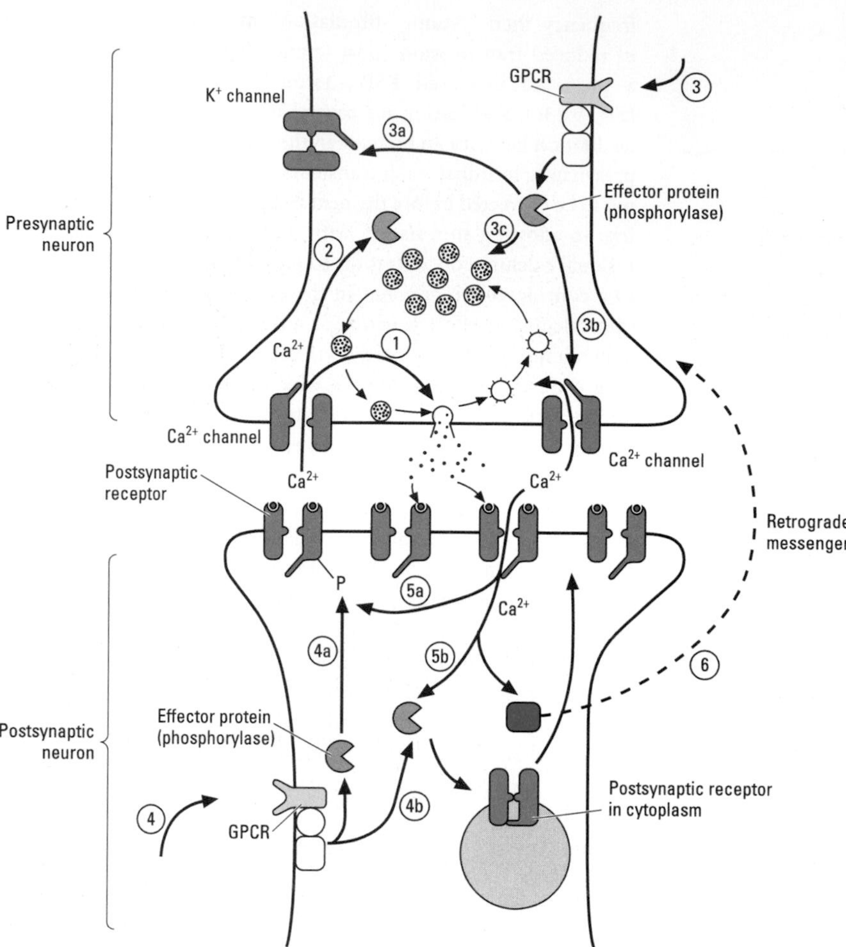

Fig. 9.24 Various types of presynaptic and postsynaptic plasticity. 1, Variations in calcium entry trigger normal vesicle release. 2, In homosynaptic plasticity, calcium can act as a second messenger affecting synaptic proteins that modulate the vesicle release. 3, In heterosynaptic plasticity, external signals trigger G-protein-coupled receptors (GPCRs) that act via phosphorylases protein kinases A and C to (3a) alter the potassium channels, or (3b) feedback onto the calcium channels, or (3c) directly affect proteins involved in the synaptic vesicle cycle. 4, Postsynaptic plasticity commonly also acts via GPCRs, either (4a) via effects on phosphorylations of the postsynaptic ionotropic receptor channels, or (4b) by affecting the cycling of these receptors to and from the cytoplasmic pool. 5, Postsynaptic changes may also act via calcium influx, via the same two routes (5a and 5b). 6, The postsynaptic neuron may produce signals that act on the presynaptic cell (retrograde messengers). (Adapted from Malenka & Siegelbaum 2001.)

of the postsynaptic receptor cluster correlates inversely with distance from the soma, which might be a way of compensating for the attenuation that would occur for more distant synaptic inputs. Some synapses have an almost 1 : 1 link between incoming APs and docked vesicle fusion, while for others an AP rarely triggers fusion. Furthermore, for some synapses each quantal release produces a large depolarization, whereas for others depolarization is almost undetectable. Together these differences help to determine the extent to which each synapse is plastic in its response, and therefore affect all the "higher" neural processes described in later sections.

9.3.2 Electrical synapses

If two neurons are close enough to each other at a synapse, there is always the possibility that some of the current flow in the first neuron could reach the second one and cause a potential change there. However, this is inherently unlikely in most situations, because the resistance of the path from one cell through the ECF and across the second cell's membrane is much greater than the resistance of the path back to the original cell's active site. For this reason we do not normally see direct electrical signaling between neurons; in particular, there is no lateral electrical spread between closely apposed unmyelinated axons running side by side, which is fortunate as it allows neighboring axons to carry quite separate signals.

The most common way to achieve an effective electrical synapse is to overcome this problem of path resistance by having intercellular connections or "junctional elements", high conductance bridges linking the cytoplasm of the two neighboring neurons at a specific synaptic zone (Fig. 9.25). Here, instead of the intercellular gap that is normally around 10–20 nm, the gap between the two membranes is reduced to only about 2 nm, forming a typical "gap junction" (see Chapter 4), and the gap itself is bridged by proteins here termed "**connexons**". To form a functional intercellular canal two connexons (one from each membrane) must meet, effectively forming a channel from one cell to another through their core. At an electrical gap junction there are normally many such paired protein connections. This allows almost perfect electrical communication and electrotonic transmission between the pre- and postsynaptic cells.

Such synapses usually have no asymmetry and can conduct a signal equally well in either direction—they are "nonrectifying". However, in some situations rectification is useful, with signals only allowed in one direction, and this can be achieved by incorporating voltage dependency into the connexons, so that they switch between open and closed configurations according to the direction of the depolarizing current. Changes in intracellular pH or [Ca^{2+}] may also alter the directional properties of such a synapse, again by effects on the connexon proteins' configuration.

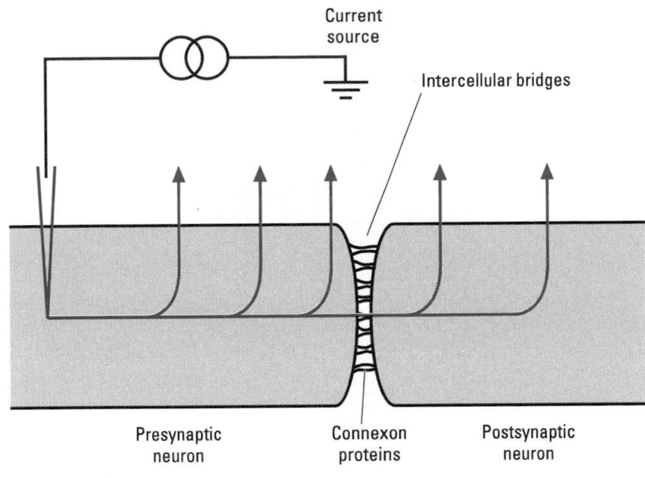

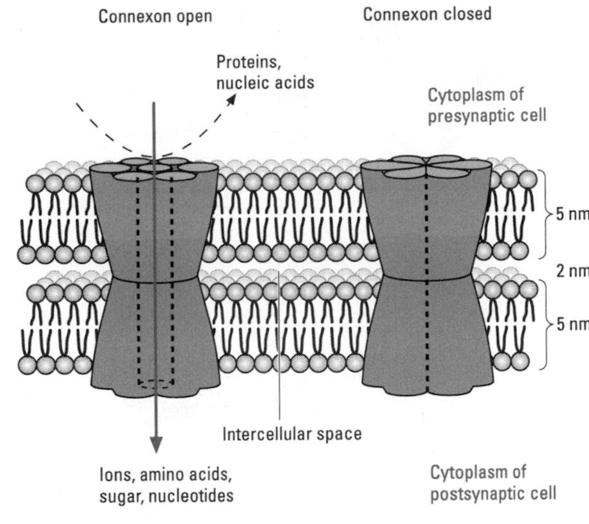

Fig. 9.25 The structure of electrical synapses. Connexons provide intercellular bridges, closely packed together and spanning across the two adjacent membranes. The subunits of the connexon protein form an aggregate with a central pore allowing ions and small molecules to pass and thus giving electrical continuity, but excluding proteins and nucleic acids. The pore can also be closed by configurational changes in the protein. (Adapted in part from Alberts *et al.* 1994.)

Electrical synapses are very useful where very fast conduction is important, as in some escape-response pathways, and are also excellent for accurately transmitting small changes in membrane potential from one cell to another. However, they have a major disadvantage in only working effectively between two similarly sized neurons, in a 1 : 1 fashion; the current in a tiny neuron can hardly have any effect on a giant neuron. Furthermore, unlike chemical synapses, there is no possibility of signal amplification.

9.4 Nervous systems

9.4.1 Patterns and evolution

Nerve nets

As we mentioned earlier, nerve cells originated very early on in animal evolution, and in modern animals at their first manifestation they were already highly organized into nervous systems showing characteristic abilities to summate, integrate, and conduct at high speed. The simplest nervous systems seen today are in the Cnidaria (sea anemones, jellyfish, and their kin). These animals have essentially only two layers of cell, ectoderm and endoderm, but in the jelly-like mesoglia between these cell layers there is a diffuse network of neurons, derived from the ectoderm during development and ramifying as a two-dimensional "**nerve net**" (Fig. 9.26a). There may be more than one net, and some cnidarian jellyfishes have a fast specific nerve net of large bipolar neurons with a preferential directionality, including pacemaker cells that produce the movements of the swimming bell. In highly active cnidarians there may also be a concentration of neurons into a simple nerve ring or ganglion, although there is nothing that could be described as an integrative center or "central nervous system". A variety of ectodermal sensory receptors feed into this system, and when the neurons relay signals to the contractile myoepithelial cells various coordinated behaviors (such as patterned swimming, and tentacle movement to grasp and manipulate food items) can result.

In the simplest of the triploblastic animals the nervous system is still in essence arranged as a nerve net, as shown in the primitive flatworm in Fig. 9.26b. In echinoderms too, there is usually a net-like arrangement in much of the body, just below the epidermis, although there is a ring of nerves round the mouth and a very simple nerve tract along each of the five arms.

Bilateral nervous systems

Figure 9.27a shows the transition from nerve net to bilateral nervous system with a distinct "front end" and longitudinal nerve cords. In *Notoplana*, a free-living and relatively active flatworm, there are two main nerve cords but also several minor ones, all linked to a simple **brain** anteriorly. These cords have long axons running along a central core of "**neuropil**", with an outer layer of neuronal cell bodies. However, the peripheral nervous system is still essentially net-like.

In most other worms the anterior neurons are organized as a ring of nerves around the gut (the circumesophageal ring), the dorsal portion of which may be complex enough to merit the term "brain", so giving the first signs of **cephalization**. In roundworms (Nematoda) there are usually ventral and dorsal nerve cords running backwards from this simple brain (Fig. 9.27b), but in most other worm taxa the cords become entirely ventral and there is usually just one pair. Thus we arrive at the classic "invertebrate" nervous system with a dorsal brain and paired ventral connectives (Fig. 9.28), often containing a few giant axons amongst the thousands of tiny interneurons. A distinction between motor neurons (**efferent** fibers leading away from the CNS and linking to effector organs, notably muscles), and sensory neurons (**afferent** fibers, bringing

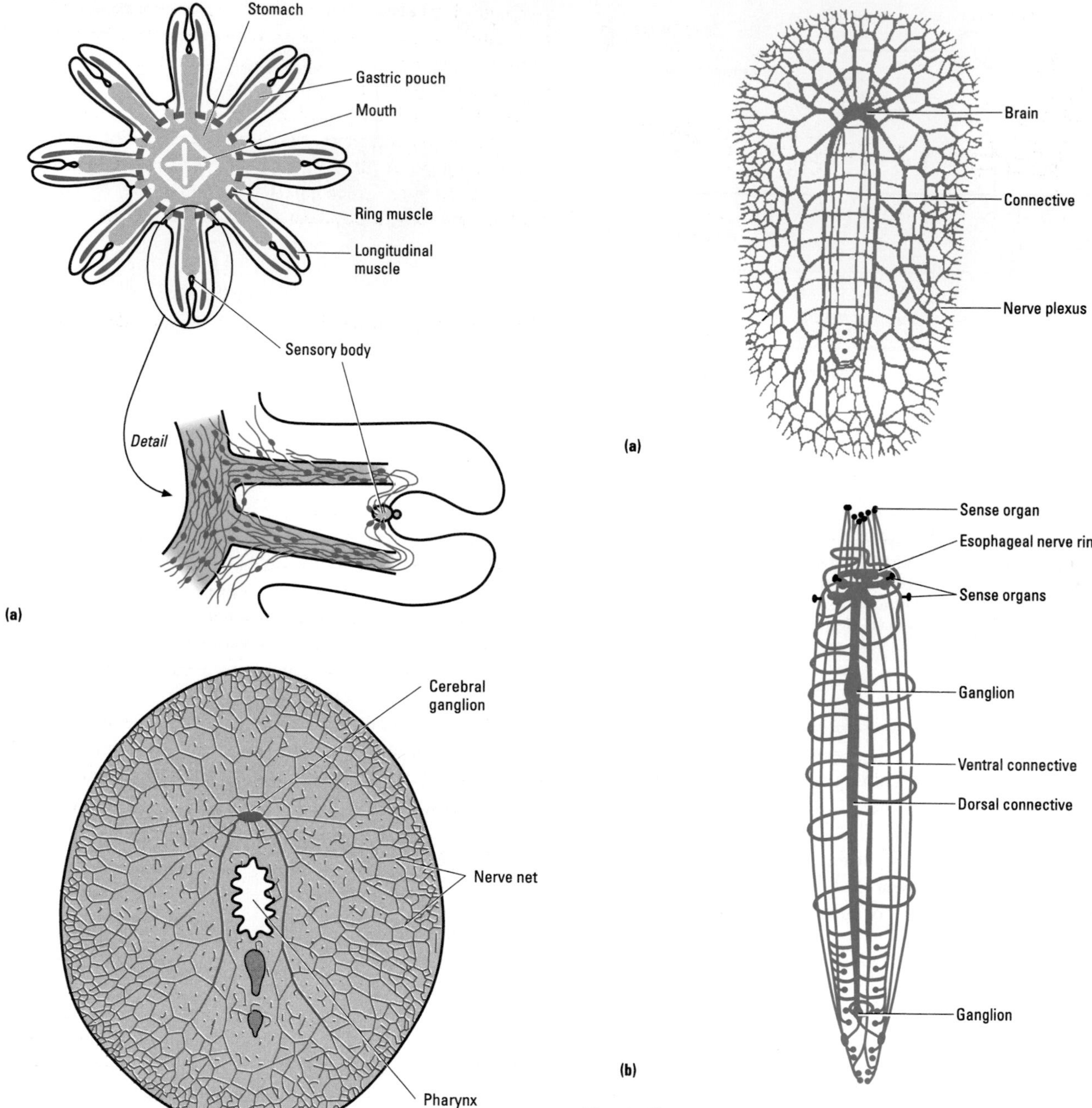

Fig. 9.26 (a) The anatomy and muscles of a cnidarian (a pre-adult jellyfish, *Aurelia*) with a close-up of the nerve net in one tentacle. (b) The nerve net of the simple flatworm *Planocera*, with a simple brain but otherwise an entirely net-like nervous system. (Adapted from Bayer & Owre 1968, and other sources.)

Fig. 9.27 The nervous system of (a) a higher flatworm (*Notoplana*), with several longitudinal connectives, and (b) a nematode roundworm (*Ascaris*) with a concentration of anterior sense organs, a circumesophageal nerve ring, and condensed ventral and dorsal connectives. (Adapted from Reichert 1992, after Hadenfledt and Goldschmidt & Voltenlogel.)

information from peripheral sense organs into the center) is set up for the first time.

In most animals the peripheral nervous system is organized into two functional parts: the **autonomic nervous system** whose nerves and ganglia regulate nonvoluntary functions such as heart contraction and gut motility (and split in higher tetrapods into the sympathetic and the parasympathetic system); and the **somatic nervous system**, processing sensory inputs and coordinating voluntary motor activity. In practice, many peripheral nerves carry neurons for both systems, in both invertebrates and vertebrates.

Segmentation

In several important invertebrate taxa the whole body plan is organized segmentally; this is especially noticeable in the annelids (earthworms and their kin) and in all the various forms of arthropods (insects, crustaceans, arachnids, etc.). Segmentation is evident

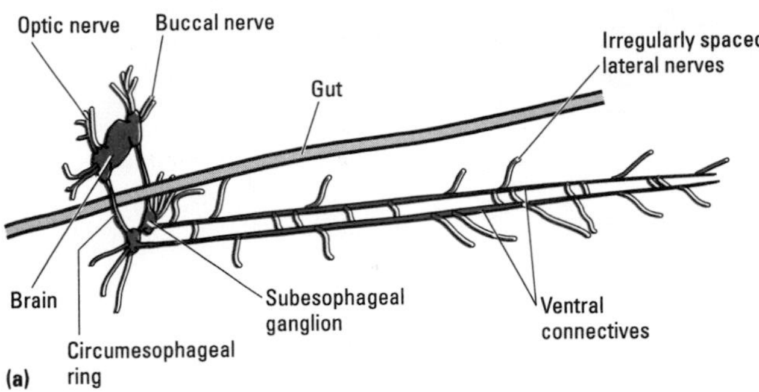

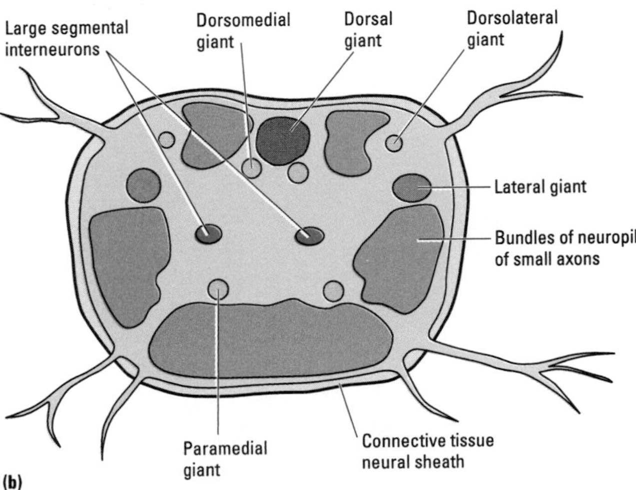

Fig. 9.28 (a) A typical invertebrate nervous system with brain and paired ventral connectives. (b) In transverse section there are commonly one or a few giant axons in the nerve tract.

both in the mesoderm (particularly the muscles) and in ectoderm (skin, cuticle, etc.). Not surprisingly, the nervous system also shows a segmental pattern, the ventral connectives becoming "ganglionized", with a swelling in each segment where there is a concentration of neurons (around 1000 per ganglion in annelids). If the two main ventral connectives are reasonably well separated, a ladder-like pattern results. A consistent arrangement of nerves emerges laterally from each segmental ganglion along the body (Fig. 9.29a). The consistency of organization is so great in annelids and many insects that specific neurons with specific functions can be mapped bilaterally in the ganglia of each segment of the body (Fig. 9.29b). Thus patterns of motor activity are controlled in identical fashion in each segment. Only in the first few segments are the innervation patterns different, though segmental ganglia do tend to become more complex and specialized in the thoracic regions of higher arthropods, as the locomotory appendages (legs and wings) become concentrated in the thorax. In the most advanced insects many of the abdominal ganglia are small and fused together and the "segmental" pattern of the CNS is heavily disguised (Fig. 9.29c).

Cephalization

In the segmental animals, and also in some nonsegmented invertebrates such as molluscs, we also see the appearance of real "cephalization". This often results from the fusion of several anterior ganglia into a brain. In other words the brain is merely a larger-than-usual ganglion, with the same basic structure of peripheral neuronal cell bodies and a central neuropil of axons and dendrites. Cephalization is crucially linked to a concentration of sensory inputs at the head end of a bilaterally symmetrical animal, so that the animal senses the portion of the world that it is moving into. Chemoreceptors, tactile sensors, and visual receptors ranging from simple light sensitive pits to complex eyes are arranged around the head end and close to the processing centers in the brain.

9.4.2 Invertebrate CNS

Most of the higher invertebrates have a fairly standard pattern of CNS organization, with a dorsal anterior brain, circumesophageal nerve tracts (commisures), and a paired ventral nerve cord that is usually ganglionated. The ventral nerve cord ganglia are themselves paired and bilaterally symmetrical, innervating the periphery via symmetrically arrayed nerves, and linked longitudinally by the paired ventral connectives. Cell bodies are peripheral throughout the CNS, with the central neuropil containing the axons and synapses.

In annelids and arthropods the increasing cephalization leads to the formation of a complex dorsal brain (Fig. 9.30) as a result of the fusion of initially separate cerebral ganglia, loosely corresponding to head segments (the exact correspondence of brain segments and head segments in these phyla is still controversial). Insects are probably the best studied animals within this assemblage; they normally have a brain made up of an anterior **protocerebrum**, receiving visual information from both compound eyes and ocelli (see section 9.7.2 below), a **deutocerebrum** mainly linked to the antennae, and a posterior **tritocerebrum** receiving inputs from the rest of the head surface. (A separate subesophageal ganglion below the gut, itself a fusion of several segmental ganglia, is the main neural center relating to the mouthparts.) In insects and crustaceans the three main brain areas may contain 0.5–1 million neurons, and make up around 90% of all the neurons in the body. Within the brain there are hierarchically organized regions of different complexity, some with cell bodies, others with axon bundles, and others again largely composed of neurosecretory cells.

In molluscs the CNS has a less obvious pattern, and shows no indication of segmental arrangement. The main processing center is a set of about five pairs of ganglia in a ring around the anterior gut, linked by connectives (Fig. 9.31). In simpler or less active molluscs, such as bivalves, these ganglia remain small and separate, and can be linked to particular anatomical regions (giving cerebral, buccal, pleural, pedal, and abdominal ganglia); but in advanced and highly active forms such as cephalopods they tend to fuse together into a brain (Fig. 9.31b). There may be additional small peripheral ganglia, and there is often also a peripheral nerve net highly reminiscent of that in flatworms (see Fig. 9.26b). The CNS of cephalopods such as squid, with highly active lifestyles, very rapid motor coordination, and excellent visual systems, may contain 100 million neurons and shows a significant ability to integrate and to learn.

9.4.3 Vertebrate CNS

The nearest invertebrate relatives of the vertebrates are not the arthropods or molluscs with their well-developed brains, but groups

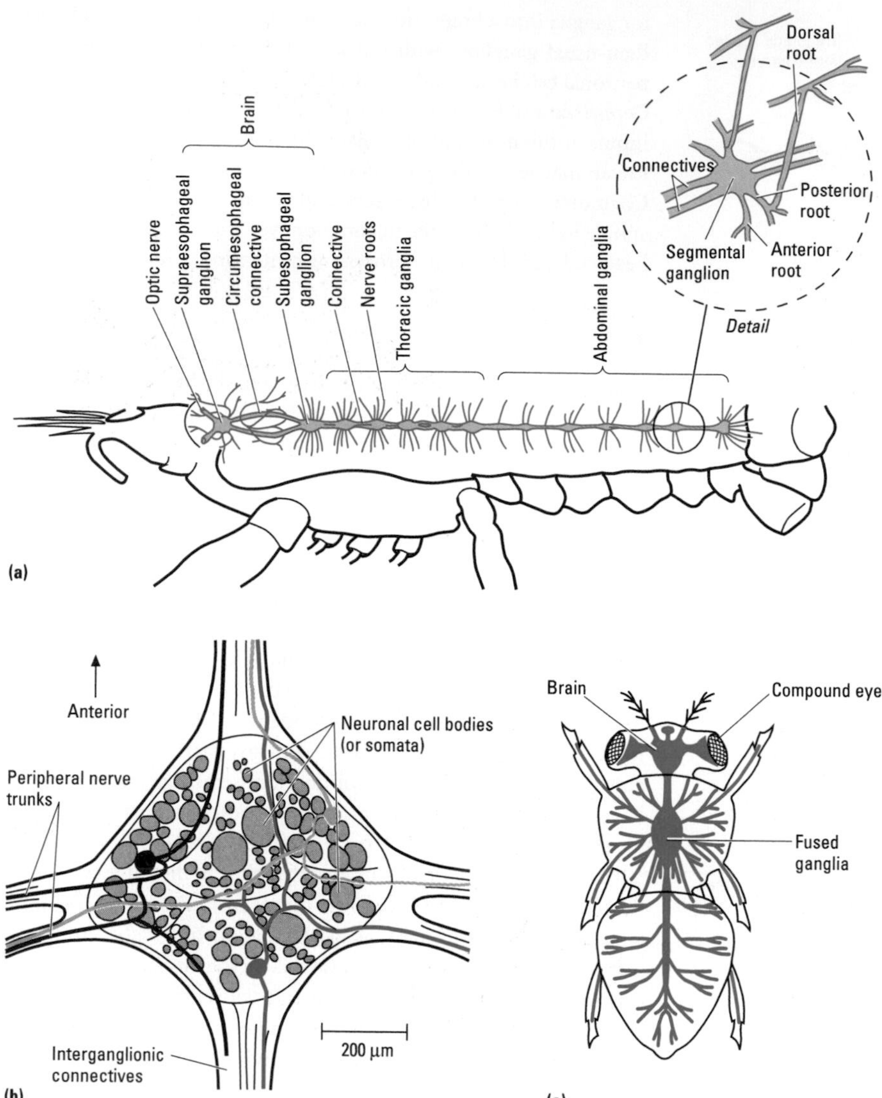

Fig. 9.29 (a) The segmental nervous system typical of annelids and arthropods. (b) The pattern of neurons in a ganglion is fixed, and in the main connective each ganglion would have this same repeated disposition of neurons. (c) The fusion and loss of obvious segmentation in an advanced insect such as a fly or bee.

of backboneless chordates such as the tunicates, which as larvae have a simple dorsal "neural tube"; this is their CNS, albeit very simple and with only a few hundred neurons, and it is largely lost in the adult forms which are sessile filter feeders. The related cephalochordates (best known as the lancelet or amphioxus, *Branchiostoma*) again have only a very simple neural tube and nothing that might really be called a brain. But the third class within the Chordata is the Vertebrata or Craniata, and the latter name draws attention to the vast evolutionary change that has taken place within the vertebrates—the evolution of a cranium and the brain that lies within it, which together with a prominent dorsal spinal cord makes up the CNS.

Although vertebrate brains appear rather diverse, there are clear underlying patterns of organization from the spinal cord up into the brain. The CNS has a segmental pattern, with a central dorsal (and primitively hollow) nerve cord arising by invagination of a simple epithelium derived from ectoderm (see section 9.6). The center of this cord contains the cell bodies, and the outer regions the neuropil. From this tubular structure, which in early development comes to be surrounded by the cartilaginous or bony vertebral column, peripheral nerves arise bilaterally and symmetrically, with one major pair of nerve tracts per segment. These lateral nerves arise from two "roots" (Fig. 9.32) before uniting as the spinal nerves. The **dorsal roots** (and the dorsal half of the spinal cord) largely process incoming sensory information, and the **ventral roots** largely contain motor neurons linking with the motor-processing centers of the ventral spinal cord. The cell bodies of the ventral root motor neurons are within the spinal cord, whereas the dorsal afferent sensory neurons have their cell bodies within small ganglia (the **dorsal root ganglia**) lying on the dorsal root just outside the spinal column. In the simplest vertebrates, the neurons of particular segments act largely as local reflex arcs (see section 9.14.2) with autonomous control over segmental activity; but within tetrapods, and particularly in birds and mammals, there is increasing dominance from ascending and descending connections to higher neural processing centers, with segmental ganglia becoming more subservient.

The anterior end of the tubular spinal cord is characterized by a series of vesicles expanded from the hollow core, and around these vesicles (known in the adult brain as ventricles) the cerebral ganglia are formed, which unite to form the brain within the protective

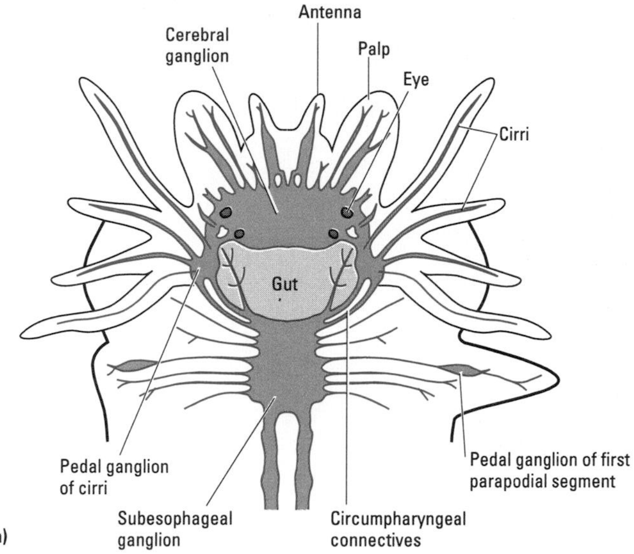

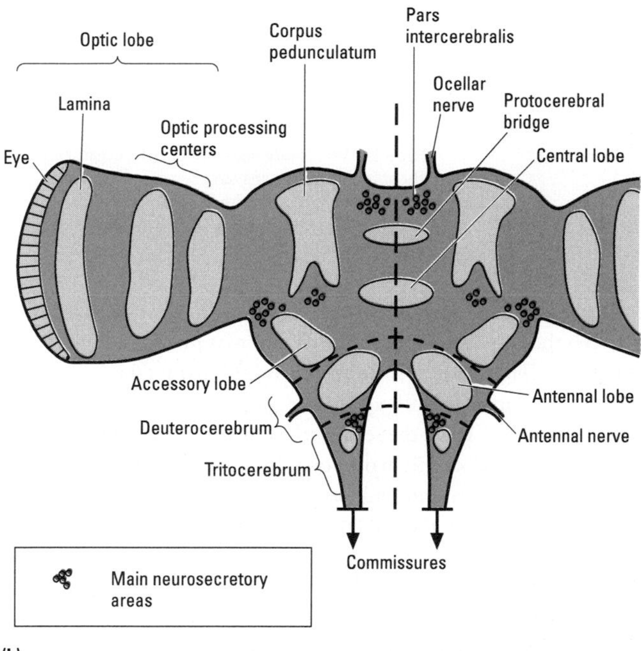

Fig. 9.30 Cephalization in the nervous systems of (a) an annelid worm, and (b) an insect, with the links from the brain to the sensory systems.

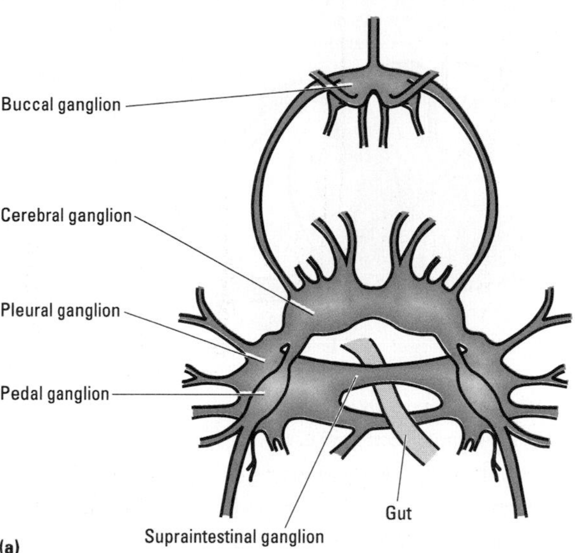

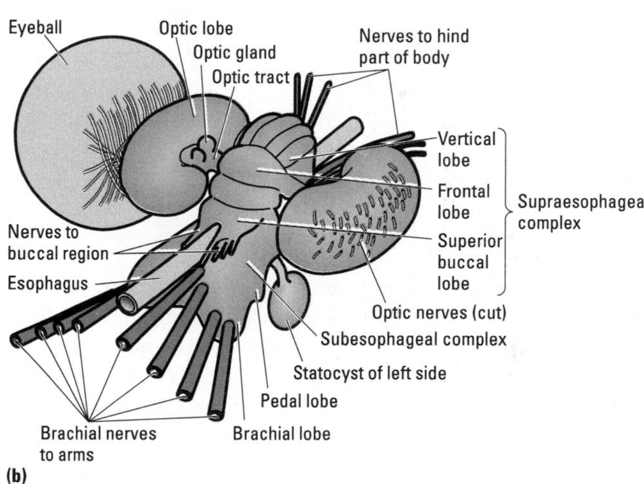

Fig. 9.31 Cephalization in the brains and associated ganglia of molluscs: (a) the simpler version seen in gastropods (such as sea slugs and snails), and (b) the very complicated brain of cephalopods (such as squid and octopus). (b, From Brusca & Brusca 1990.)

cranium (skull). The brain takes ever greater control of bodily processes through evolution, as it develops from a simple set of three small swellings on the spinal cord to a massively differentiated complex brain.

The basic structure of a vertebrate brain is shown in Fig. 9.33. In the most primitive fish, there is an anterior **prosencephalon**, an intermediate **mesencephalon**, and a posterior **rhombencephalon** barely distinguishable from the spinal cord that it runs into. Each of these three regions proliferates and becomes specialized (Table 9.6).

The most posterior (caudal) rhombencephalon is clearly segmental in origin, and is relatively constant in form and function in all vertebrates. It gives rise to the **medulla oblongata**, the **pons**, and (dorsally) the **cerebellum**. In fish the cerebellum is very large, and its primitive function was probably to integrate sensory and motor information relating to balance and position; but in higher vertebrates it becomes increasingly integrative and receives inputs from all the sensory modalities to give coordinated motor function.

The mesencephalon also retains some appearance of segmentation, with lateral spinal nerves arising from it (Fig. 9.33b). In fish its dorsal region is concerned with visual processing, whilst in mammals the same region splits into two nuclei, dealing with visual (superior colliculus) and with auditory (inferior colliculus) inputs. The ventral region of the mesencephalon has important motor centers.

The prosencephalon has two parts, an anterior unpaired region (telencephalon) and a more posterior paired region (diencephalon). The latter is primarily made up of the **thalamus**, which is the center for integrating all sensory inputs before transmission onwards to the most anterior part of the brain. Additionally, its ventral surface

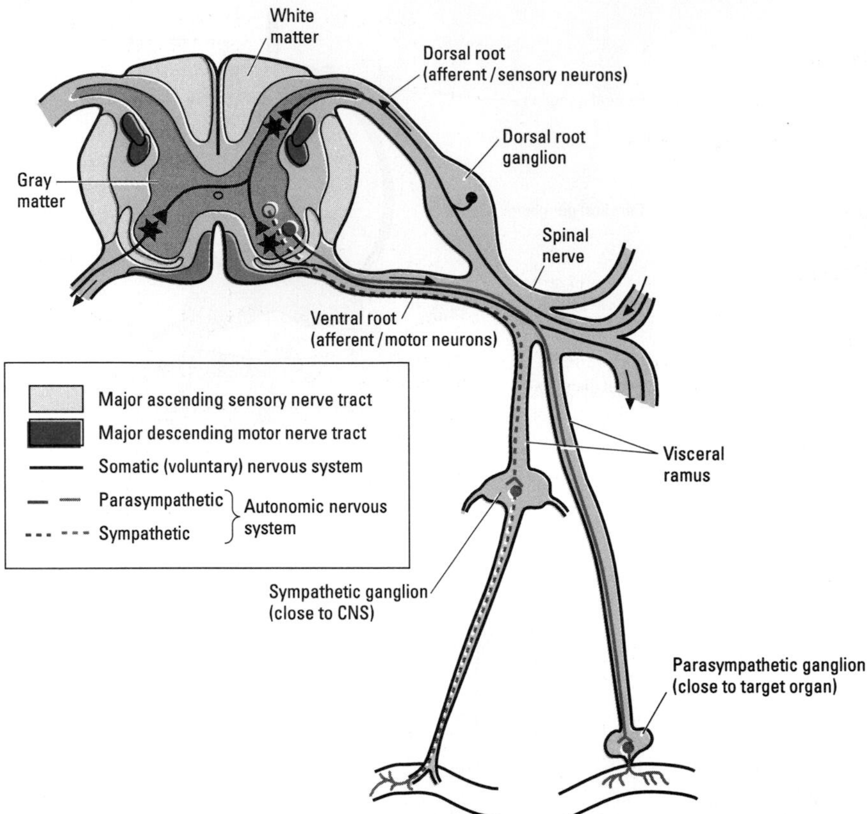

Fig. 9.32 Vertebrate segmental neural organization: the spinal cord in transverse section, with dorsal and ventral roots, and the normal paths of the sympathetic and parasympathetic nervous systems.

is elaborated into a separate region, the **hypothalamus**, through which the brain exerts control over the pituitary gland and thus influences most of the endocrine system and the body's visceral functioning (see Chapter 10). Finally, the telencephalon or **cerebrum** sits right at the front of the brain. In primitive vertebrates it was relatively small and only concerned with olfactory processing, but this is the region that in tetrapods has expanded vastly, growing as two "hemispheres", linked by the central corpus callosum. The much folded superficial hemisphere tissue (cortex) grows back over the surface of other brain regions, and comes to dominate the brain volume. One superficial region persists as an olfactory processing zone, and is often termed the paleocortex; but beyond and above this the **neocortex** develops.

The neocortex normally has six layers, five being neuronal cell bodies (gray matter) and the sixth most peripheral layer being the nerve fibers (white matter)—note that this is opposite to the disposition in the vertebrate spinal cord. There are several important neuronal aggregations within the cortex. On the outermost medial lip a single-celled layer rolls inwards and folds up on itself to form the **hippocampus**, important in learning processes. Deeper within the cerebral hemispheres near the thalamus are the basal ganglia (the **corpus striatum** plus some smaller nuclei), critical in motor control. Deeper still is the **amygdala** with links back to the hypothalamus. The amygdala and hippocampus are linked with other nuclei to form the **limbic system**, important in the control of emotions as well as in long-term memory.

The development of the neocortex is usually seen as the key to the success of the tetrapods. The size of the brain scales roughly with body size for a wide range of animals, but there is a clear discontinuity between the very large-brained birds and mammals, and the lower vertebrates (Fig. 9.34). This is almost entirely attributable to cortical growth, the cortex folding up to give greater surface and growing out as four major lobes that cover most of the rest of the brain. As the current "peak" of this evolutionary sequence, the human brain probably contains around 1 billion neurons, almost entirely made up of interneurons (only about 0.001% are sensory or motor neurons). Those interneurons are mostly neocortical and cerebellar. Collectively they may make about 10^{14}–10^{15} synapses with each other.

The neocortical neurons are arranged in columnar groups, perpendicular to the cortex surface, each group of neurons forming a functional unit, albeit with extensive lateral connectivity to other units. In some regions there is a clear morphological distinction between adjacent areas, resulting in **cortical fields**, and these fields (also termed "projections centers") may form a topographical representation of the areas they serve. The most familiar examples of this are the major sensory and motor fields that lie in a furrow across the lateral sides of the cortex in mammals (Fig. 9.35), and which are organized roughly as a "plan" of the body from the feet (in the cortical midline) to the head (at the sides of the cortex). As might be expected the areas most endowed with sensory inputs and most in need of fine motor control (hands, feet, and face) are over-represented topographically, while the trunk and viscera have little cortical surface devoted to them.

In most tetrapods a large part of the cortex is taken up with these sensory and motor fields. However, in higher mammals, and especially the primates, the majority of the cortex cannot be linked

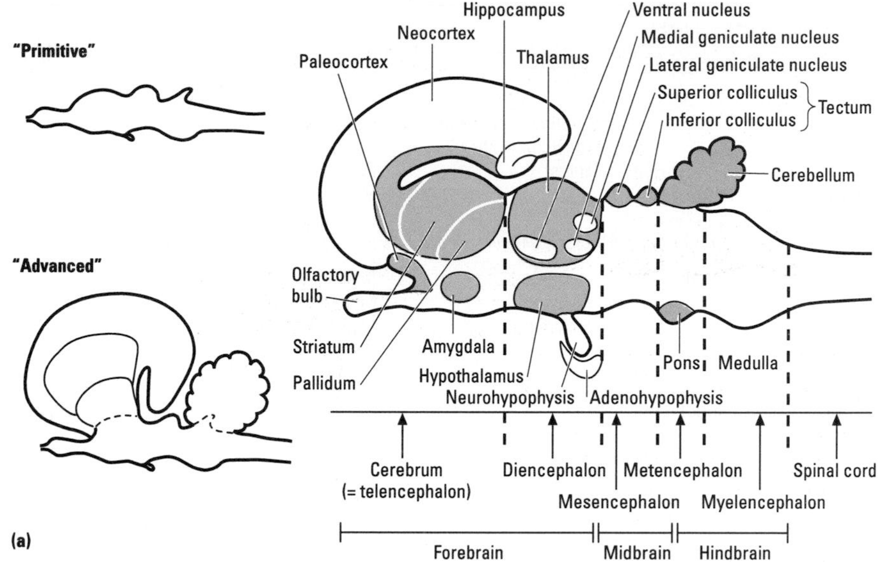

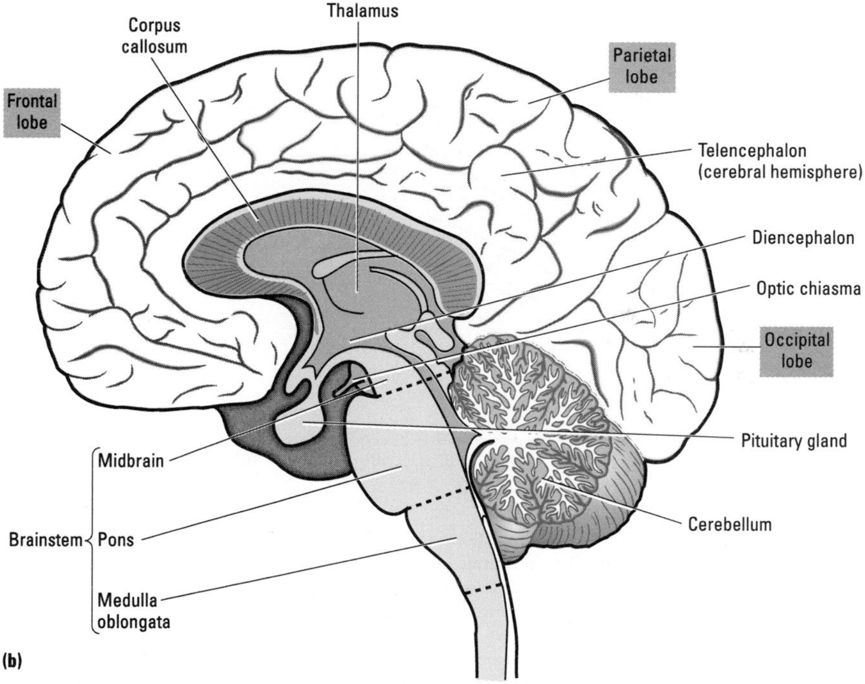

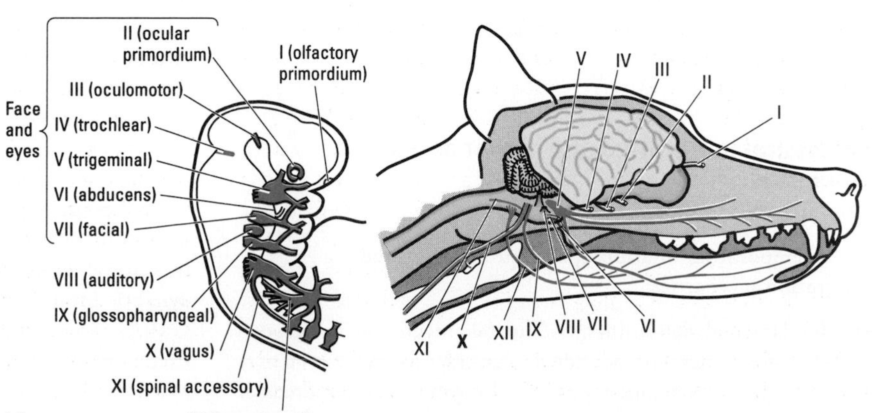

Fig. 9.33 (a) The basic anatomy of the vertebrate brain and its development from simple (e.g. frog) to more advanced (e.g. cat) forms. (b) A more detailed anatomy of the human brain. (c) Cranial nerves arising from the brain are clearly segmental in embryo mammals (left), but become spatially modified in the adult (right).

Table 9.6 The major parts of the mammalian brain.

Prosencephalon (forebrain)		
Telencephalon	→ Cerebral hemispheres	Higher mental functions, sensory processing, motor control
Diencephalon	→ Thalamus	Sensory processing
	→ Lateral geniculate nucleus	Visual relay
	→ Medial geniculate nucleus	Auditory relay
	→ Hypothalamus	Homeostatic regulation, link to pituitary gland
Basal ganglia	→	Motor coordination
Limbic system	→	Control of memory and emotion
(These last two systems exist as dispersed ganglia, but are predominantly in the forebrain)		
Mesencephalon (midbrain)		Sensory processing
	→ Superior colliculus	Oculomotor reflexes
	→ Inferior colliculus	Auditory relay
	→ Tegmentum	Orientation and auditory relay
	→ Red nucleus + substantia nigra	Posture, orientation, voluntary motor control
Rhombencephalon (hindbrain)		
Metencephalon	→ Cerebellum	Motor coordination, learning
	→ Pons	Respiratory control
Medulla oblongata	→	Control of respiration, heart rate, blood pressure, coughing, etc.

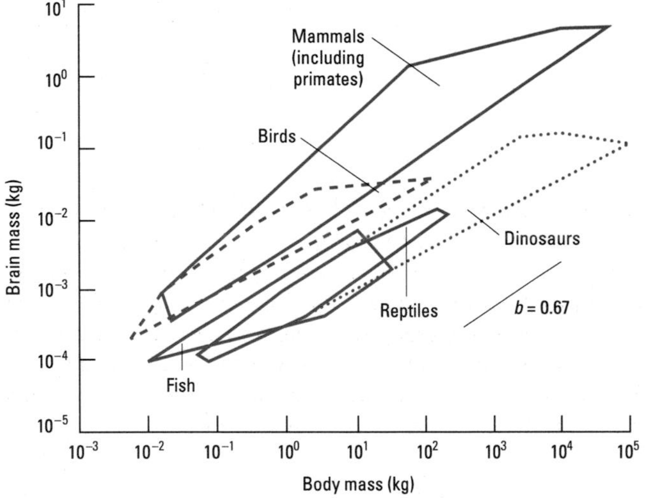

Fig. 9.34 Although the scaling between brain and body size in vertebrates shows a consistent exponent of about 0.67, there is a distinct discontinuity, with the endothermic birds and mammals having much larger brains predominantly due to the enlarged cortical areas. (Adapted from Jerison 1970.)

with specific anatomical areas, and instead is involved with higher cognitive processes—learning and memory, as discussed below. Cross-linking between fields becomes ever more important in these advanced cortical brains.

9.5 Neural integration and higher neural processes

9.5.1 Temporal summation, spatial summation, and integration at synapses

The fundamental determinant of neural integration in nervous systems lies at the level of individual synapses. Any one neuron may be receiving many synaptic inputs, resulting in tens or hundreds of separate PSPs at any one time. Some will be excitatory and some inhibitory, and most will be less than 1 mV in magnitude.

Spatial summation occurs when two or more inputs from different but more or less equidistant locations occur at the same time. Their EPSPs are individually below the threshold, but arriving simultaneously at the axon hillock they produce a depolarization that is well above the threshold and a spike results (Fig. 9.36a). However, if one is initiated a long way from the axon hillock they may not summate to produce an AP, because the more distant EPSP will have decayed more by the time it arrives. Only distant inputs from regions with larger length constants (thus suffering less attenuation) can effectively summate with inputs close to the hillock.

Temporal summation refers to two or more inputs that occur at different times, either multiple inputs at one synapse or single inputs at several synapses at different times (Fig. 9.36b). Here the crucial factor will be the time constant of the membranes involved; the larger the time constant, the longer it takes for an EPSP to decay, so the longer it is "available" for summation with other inputs. The squid giant axon has a time constant of only 1 ms and so undergoes very little temporal integration, whereas some axons have time constants of 20–30 ms and can respond strongly to repetitive weak inputs.

Synaptic integration is a more inclusive term embracing interactions of synaptic inputs of different types, both excitatory and inhibitory, but it involves the same principles as simple summation with both time and length constants being important. An early example of synaptic integration came from work on the crayfish claw-opener muscle and its neuromuscular junctions (NMJs), summarized in Fig. 9.37. This NMJ has an inhibitory neuron that releases GABA to the muscle, which responds via $GABA_A$ receptor chloride channels, producing an IPSP. However, the synaptic current in the normal excitatory (glutamate-transmitting) presynaptic cell is also seen to be reduced, often by 10-fold or more. This is because the inhibitory neuron also synapses with the excitatory input ("**presynaptic inhibition**"), where its effect is to open up presynaptic $GABA_A$ receptor chloride channels, tending to hyperpolarize the excitatory neuron and thereby depressing presynaptic excitability. We now know that presynaptic inhibition is very common, and is

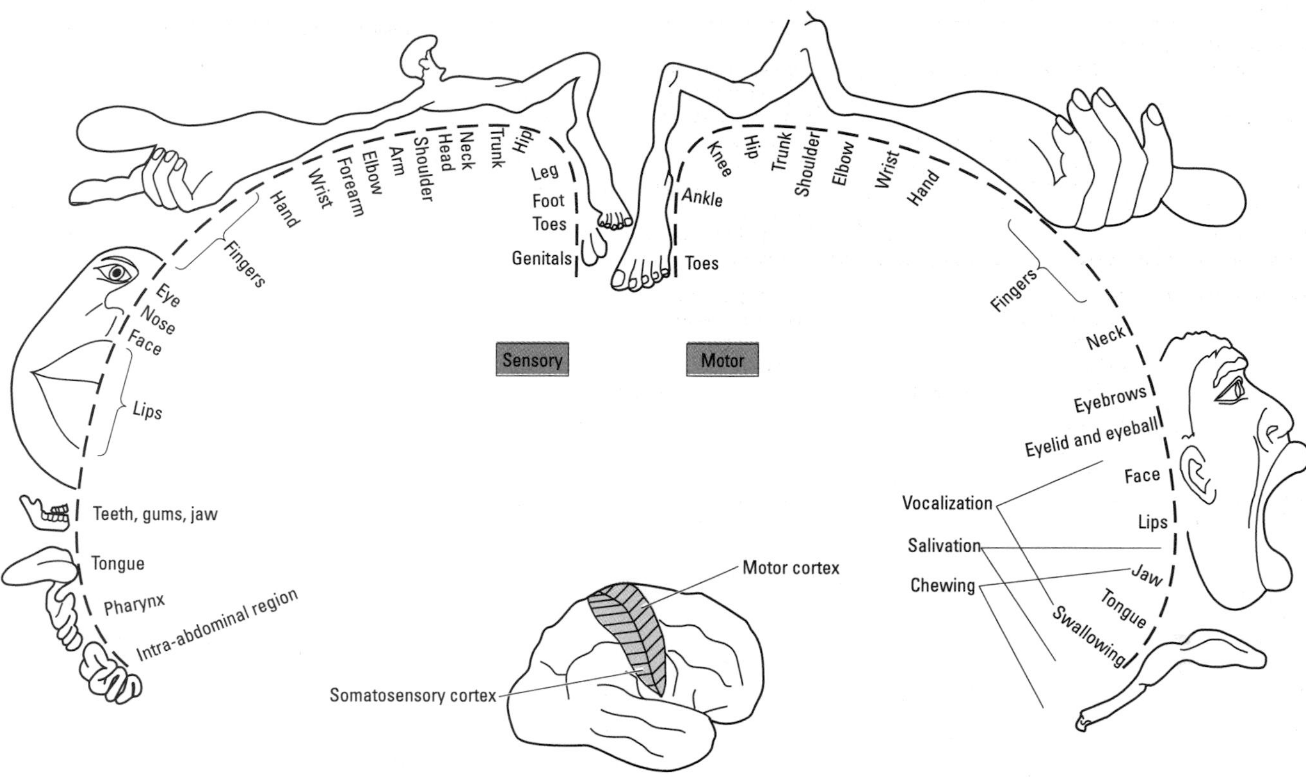

Fig. 9.35 Location of the main sensory and motor fields in the mammalian cortex (inset) and the spatial representation of the body within those fields.

often produced through the activation of metabotropic G-protein-coupled receptors. These can depress transmitter release either by "damping down" the rise in presynaptic calcium triggered by an incoming AP, or by direct modification of the vesicle-release mechanisms.

Similar examples relating to presynaptic summation are not hard to find. For example, presynaptic nicotinic receptors are known to enhance the release of both ACh and glutamate, whilst 5-HT enhances transmission at many invertebrate and vertebrate synapses. Again the mechanism may be either indirect effects on calcium levels or direct effects on the release process. Perhaps best known are the sensory motor neuron synapses of the mollusc *Aplysia*, where simple learning ("behavioral sensitization") occurs. Here a repeated noxious stimulation of the head or tail leads to sensitization by bringing about the release of 5-HT from many interneurons, increasing transmission at the sensory motor synapses; thus the animal withdraws from the stimulus more quickly. In this case the 5-HT both enhances the Ca^{2+} influx *and* modulates the vesicle-release processes. Its calcium-related effects (summarized in Fig. 9.38) operate via cAMP which then activates protein kinase A (PKA), which in turn closes a specific slow-acting K^+ channel (S channel). This makes the AP become broader and increases the sensory neuron excitability by decreasing potassium conductance. The AP also increases in duration due to reductions in yet another potassium current through the "delayed rectifier" K^+ channel, mediated by protein kinase C (PKC). We now know that this two-pathway system of facilitation is common in many animals, though not necessarily operating via the same targets. In most other known cases, effects via the vesicle-release system proteins are more important. Furthermore, the balance of effects on the two protein kinases PKA and PKC can vary at different times, according to whether a synapse is already depressed or not. At depressed synapses the PKC pathway dominates, and at nondepressed synapses the PKA pathway is more important. This indicates the complexity and subtlety of even the simpler types of synaptic interaction.

9.5.2 Neural plasticity

Nervous systems may alter their sensitivity and responses in relation to sensory experience, as we discussed earlier in relation to individual synapses. This is the phenomenon of plasticity, and it is particularly evident in the early development of an animal, when sensory inputs can act as organizers to the individual's developing neuronal network. Even in the adult animal most synapses are not completely stable, and may decrease or increase in importance in the light of experience and usage, giving the "long-term plasticity" that contributes to aspects of learning and memory.

Developmental plasticity is perhaps most evident in higher vertebrates, where we know that sensory experience is crucial to establishing specific structural connectivity in parts of the brain, leading to plasticity at a more behavioral level. Many species of mammals and birds require particular auditory experiences to develop acoustic recognition, and there may be quite brief "sensitive periods" in early development when these inputs are most important, with effects that last throughout adult life. So the neuronal network is initially plastic but gradually gets fixed by the acoustic inputs from the environment and by the properties of the receptors and

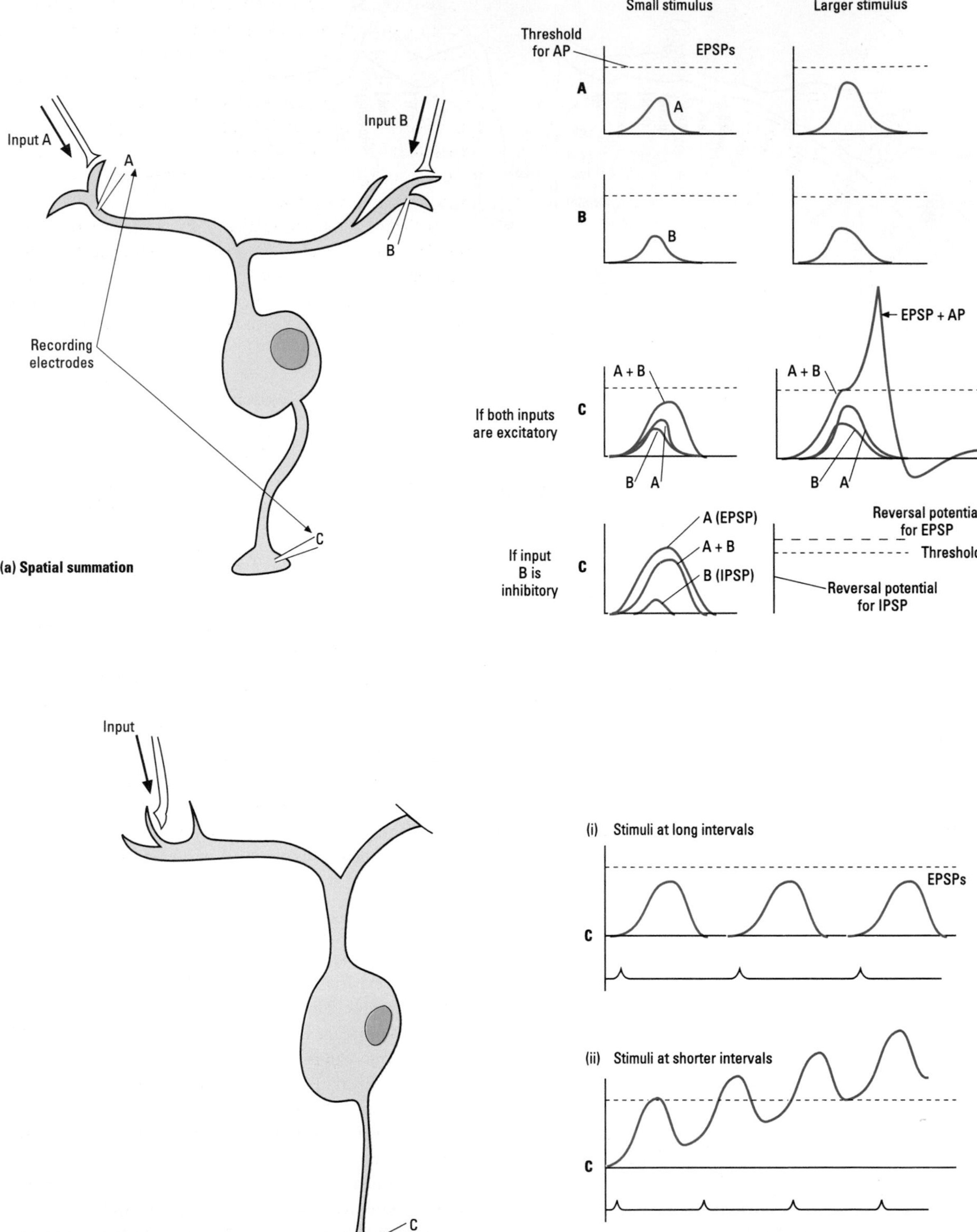

Fig. 9.36 (a) Spatial summation, where the incoming postsynaptic potentials (PSPs) from two (or more) neurons are summated at the onward synapses of the neuron, and can lead to a response exceeding the threshold even where each input neuron has subthreshold EPSPs on its own. (b) Temporal summation in response to repeated EPSPs in one input neuron. (Summation is reduced if the length constants in (a) are too small, or if the time constant in (b) is too short; see text.) AP, action potential; EPSP, excitatory PSP; IPSP, inhibitory PSP.

transmission characteristics of the ears. Plasticity is even more obvious in visual systems, where "what is seen" leads to major plastic changes in the visual cortex. Studies on cats and monkeys show that there are critical phases for establishing correct binocular vision, and if a monkey has one eye closed for the first 6 months of its life it never achieves vision in that (otherwise completely functional) eye because the visual cortex has not established the correct pathways. Hence sensory deprivation in a neonate can lead to major sensory disturbances in the CNS.

Plasticity of this kind clearly involves longer term changes than simple synaptic facilitation and depression, which normally only last a few seconds or minutes. Short-term effects (presynaptic facilitation and inhibition as described earlier, in section 9.5.1) usually use calcium as a second messenger, whereas in longer term plasticity the messenger is normally a G-protein-coupled receptor and/or a protein kinase. Where the synaptic transmission is permanently changed (as occurs with some learning processes) the mechanisms involve altered gene transcription and thus new protein synthesis. Additionally there are mechanisms for postsynaptic modification, affecting the magnitude of response to a quantal transmitter release.

One of the most thoroughly known examples concerns the nicotinic acetylcholine receptor (nAChR), which has a pentameric structure and is regulated by protein phosphorylation. The best study site for this is the electric fish *Torpedo*, whose electric organ provides a concentrated source of the receptor. Here PKA and PKC phosphorylate different identified serine residues on the receptor, increasing the rate at which the receptor desensitizes. The pentameric $GABA_A$ receptor undergoes similar phosphorylations. Even the glutamate receptors, from a different family of tetrameric molecules, undergo serine-residue phosphorylations by both PKA and PKC, though they are additionally regulated by phosphorylation with CaMKII. All these kinds of phosphorylation are thought to be very important in plasticity in the hippocampus of the vertebrate brain, where simple kinds of learning occur.

One further factor underlying plasticity during development is competition between neurons for synaptic sites. If two neurons both target a single postsynaptic site, the one that produces greater activation will gradually take over and become stabilized, while the weaker input neuron becomes inactive. Only if the two have inputs that are nearly or exactly matched in timing will both persist, as may happen in cases where bilateral inputs are crucial (binocularity or binaural hearing, for example). There may also be a role here for glial cells, which by covering and uncovering synaptic sites may influence the competition between synapses and thus the connections that become established.

In adult animals plasticity is reduced but is still present, perhaps

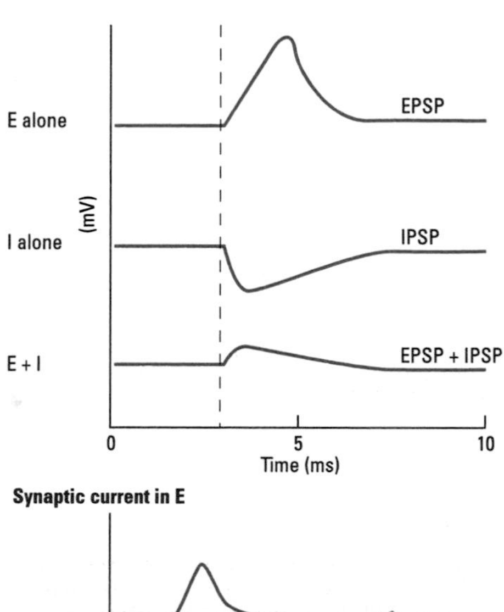

Fig. 9.37 (*left*) A simple synaptic integration at the crayfish claw-opener-muscle neuromuscular junction (NMJ) occurs by presynaptic inhibition, with a GABA-releasing inhibitory fiber not only synapsing on the muscle but also contacting the glutamate-releasing excitatory fiber (both at the NMJ and also within the central nervous system). The presynaptic inhibitory synapse, when triggered, increases the permeability of the excitatory terminal to chloride (and perhaps potassium) and thus hyperpolarizes it. This reduces the magnitude of excitatory action potentials and reduces the calcium influx per spike, so decreasing glutamate release. Claw opening can thus be graded by the balance of excitation in the two neurons. EPSP, excitatory postsynaptic potential; IPSP, inhibitory postsynaptic potential.

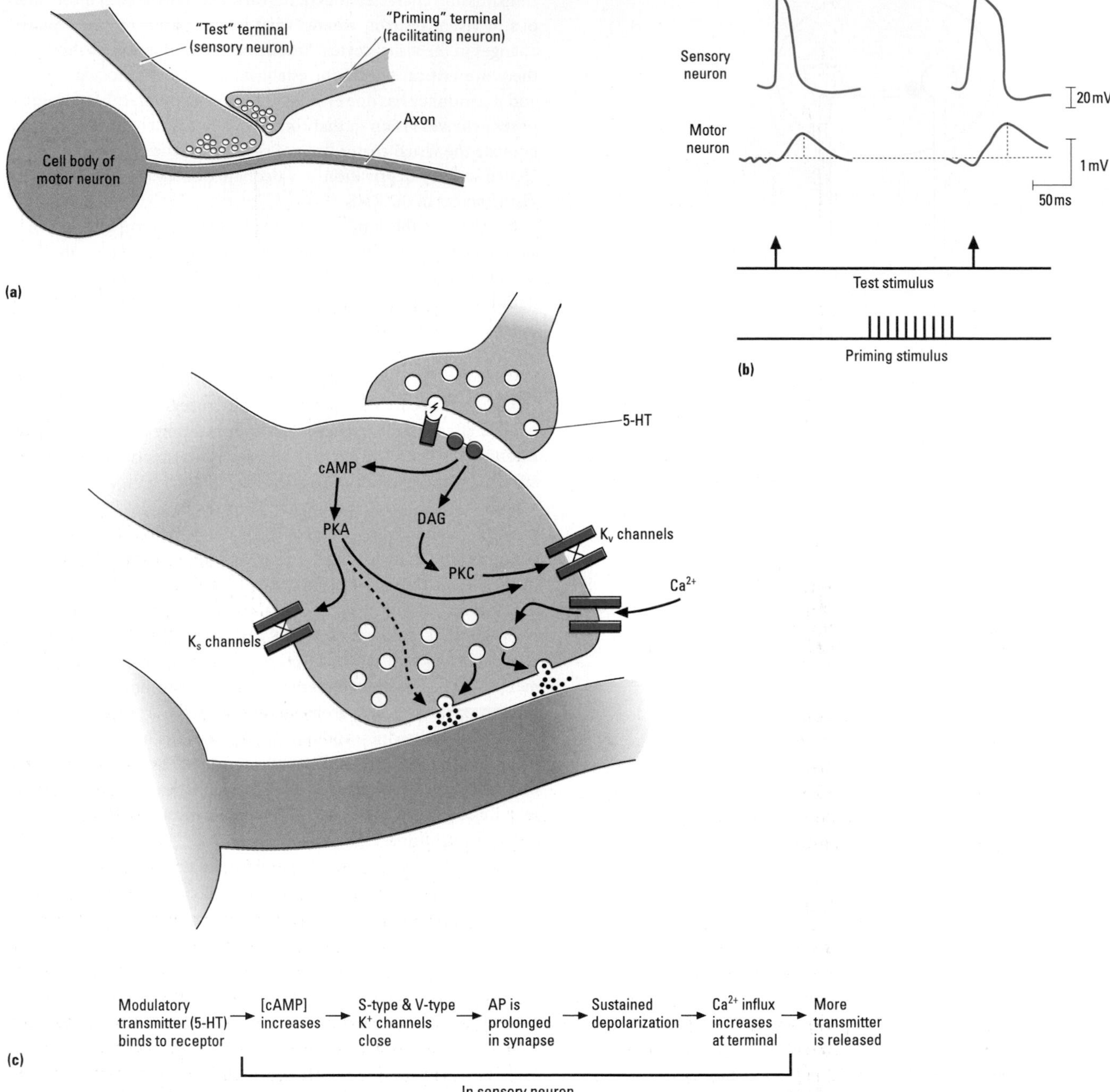

Fig. 9.38 Simple learning by heterosynaptic integration or facilitation in *Aplysia*. (a) A schematic for the required arrangement of neurons. If the priming neuron is activated, the amount of transmitter released by the sensory neuron is greater and the response is greater as shown in (b). (c) The underlying mechanism: 5-hydroxytryptamine (5-HT) is released from the facilitating neuron and binds to two types of G-protein-coupled receptors, one causing an increase in cAMP and hence of protein kinase A (PKA), the other via diacylglycerol (DAG) activating protein kinase C (PKC). These kinases phosphorylate and thereby close two types of K^+ channels (see text), prolonging the depolarization of the terminal and the calcium channel opening, so inducing the extra transmitter release. Direct effects on vesicle release (dashed line) are also possible; and there may also be longer term effects through the effect of PKA on protein synthesis.

particularly in the somatosensory networks. For example, a human may lose the nerve supply to one part of one hand, leading to changes in the motor cortex where that hand was represented; then sensory projections of nearby parts of the hand invade the denervated area of cortex over a few weeks or months. Conversely, if one or two fingers are repetitively stimulated their projection areas in the cortex increase. Effects of usage are clearly implicated, synapses altering according to both present and past usage. The degree or direction of plasticity is itself plastic, and this is sometimes termed "metaplasticity".

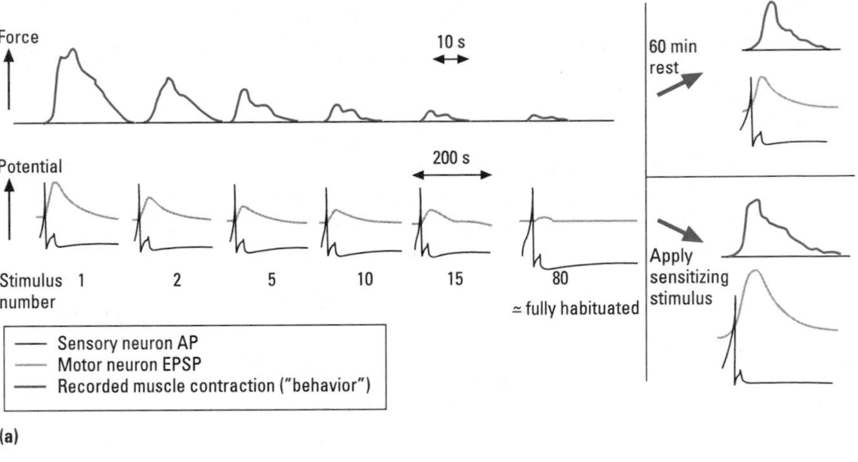

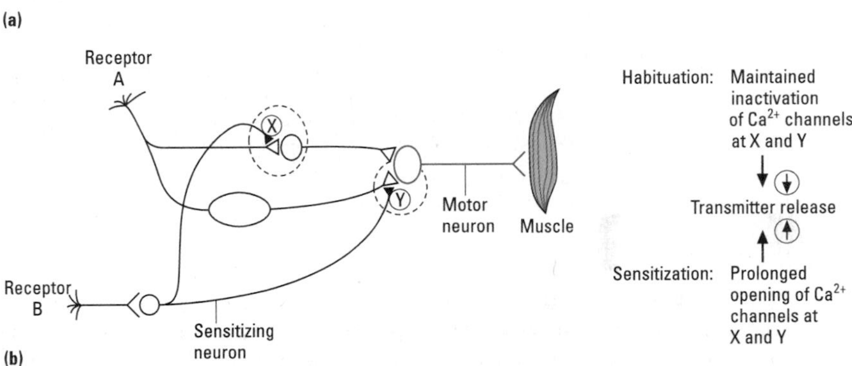

Fig. 9.39 Habituation and sensitization. (a) The motor neuron shows steadily declining responses to repeated stimulation, and the muscle contraction similarly reduces; but a period of rest can restore both, and a brief sensitizing stimulus can lead to restored or even enhanced responses. (b) A simple model where habituation can be overcome by a sensitizing neuron (operating in a similar presynaptic fashion to that shown in Fig. 9.38). In *Aplysia*, for example, this may involve strong stimulation in a different sensory pathway, such as a tap on the head end, which restores normal sensitivity to posterior tapping in the previously habituated gill withdrawal system. AP, action potential; EPSP, excitatory postsynaptic potential.

Adult plasticity becomes most critical when concerned with modifying the responses of circuitry at higher levels in the CNS; then it interacts with complex neuromodulatory inputs in multiple component neural networks to produce the richly varying communications that underlie learning and memory.

9.5.3 Learning

Learning, which can be reasonably defined as "an adaptive change in response or behavior, based on experience", also depends on changes in the effectiveness of synapses. Thus learning can be seen most simply as a form of neuronal plasticity and differentiation, happening later in life than developmental plasticity, but based on similar mechanisms. Largely based on work with vertebrates, learning is usually divided into **nonassociative learning** (where there is altered behavior in response to a single stimulus) and **associative learning** (where the animal learns and responds to a link between different stimuli). We now know that both types of learning can occur in a wide range of animals.

Nonassociative learning involves the relatively simple processes of habituation and sensitization, which can be readily understood at the level of single synapses or synaptic integration (Fig. 9.39). **Habituation** is a gradual reduction in response to (or "loss of interest" in) a signal that is repeatedly presented; the strength of response decreases without any real change in the qualities of that response. So an animal learns to ignore a stimulus that is no longer novel or particularly important to it, via an effect on a specific neuronal circuit. **Sensitization** is almost the opposite process of becoming more aware of a stimulus when it is long lasting or intense, though it is perhaps more complex than habituation and may involve several interacting neuronal circuits.

Neuronal networks underlying both of these systems have been well studied in invertebrates, especially in the marine mollusc *Aplysia* and in crayfish and insects. Here habituation involves the decrease in EPSP size (between sensory and motor neurons, and sensory neurons and interneurons) on repeated activation of sensory cells, giving a reduced frequency of APs. At the molecular level, a key factor is the maintained inactivation of calcium channels in the presynaptic sensory terminals, so that fewer quanta of transmitter are released in successive stimulations. Sensitization involves similar sets of neurons, but can also involve additional "facilitating interneurons" that "feedback" stimulation to the presynaptic terminals of the sensory neurons. Molecular cascades (via a G-protein and cAMP) are initiated there, increasing transmitter release and thus facilitating firing if there is repeat stimulation. Sensitization may also involve potentiation via the actions of glutamate, binding postsynaptically and leading to calcium influx and then phosphorylation of specific cAMP response element-binding proteins (CREBs) in the nucleus, eventually moderating gene expression patterns.

Associative learning includes the processes of **"classic" (Pavlovian) conditioning** where an animal learns to associate causes and effects, and **"operant" conditioning**, where it learns to solve particular problems (e.g. running through a maze) by trial and error, eventually gaining a reward. **"Aversion" conditioning** may be even more potent, with animals learning to avoid certain food items after single experiences of poisoning. This can be readily seen in birds and mammals, but is also quite common in much smaller-brained snails and insects. Associative learning also includes **imprinting** (recognition of parents or other key signals during a critical developmental phase), and **learning by observation**, where a more experienced animal shows the appropriate behavior and an inexperienced watcher learns it. Such processes in vertebrates are particularly linked with the hippocampus, cerebellum, and cortex.

Ultimately, associative learning must involve establishing a link between two pathways that were not previously connected. This may involve structural effects such as increases in numbers, or altered distribution, of presynaptic terminals in sensory neurons, or in interneurons. However, it may also be brought about by specific pathways such as the **long-term potentiation** (LTP) circuits found in the well-studied hippocampus of the vertebrate brain, a region dominated by glutamatergic synapses. Here there is a stereotyped microarchitecture of neurons, and a brief high-frequency input can lead to very long-term effects lasting a few hours in tissue slices, but days or weeks in intact animals, with PSPs becoming much larger than usual. Critically, LTP can show "associativity", such that a weak input to a postsynaptic cell only induces LTP when paired with a nearby strong input to the same cell (heterosynaptic LTP). LTP is currently the leading model for the synaptic changes that underlie learning and memory.

Molecular explanations for LTP are still emerging, but crucially involve the postsynaptic glutamate receptors that mediate most of the excitatory traffic in the brain. AMPA-type receptors open directly in response to glutamate, whilst NMDA receptors are functionally calcium channels triggered by glutamate released from the presynaptic side, but they only open when the postsynaptic membrane is also depolarized, resulting in a rise in Ca^{2+} (Fig. 9.40). The extent of calcium entry is crucial, and is linked to incoming stimulus strength and frequency because weak stimuli fail to dislodge Mg^{2+} ions that normally block the calcium channel entrance. A strong simultaneous input from another source dislodges the magnesium ions and facilitates calcium entry even to a weak local stimulus. The effects that follow the rise in calcium are complex, and multiple intermediates have been proposed, with CaMKII the most promising; synthesis of NO also occurs. Outcomes may include both increases in the release of glutamate presynaptically, perhaps triggered by the backward diffusion of the NO ("retrograde transmission"), and activation of messengers postsynaptically that leads to the unmasking of additional (probably AMPA) glutamate receptors. Deciding which of these pre- and postsynaptic mechanisms is operating is extremely difficult in the hippocampal system where

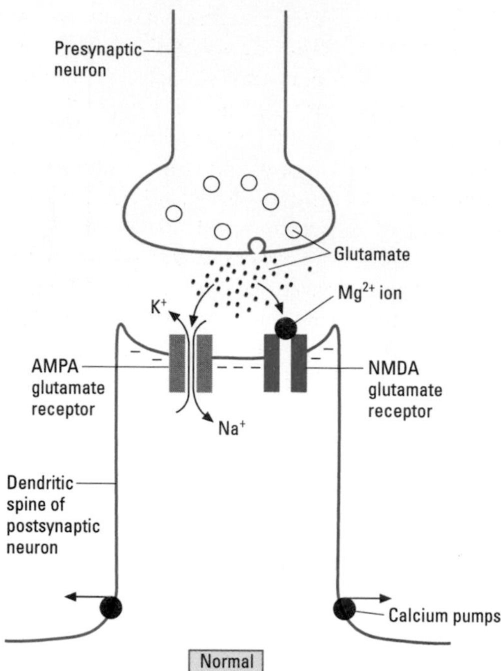

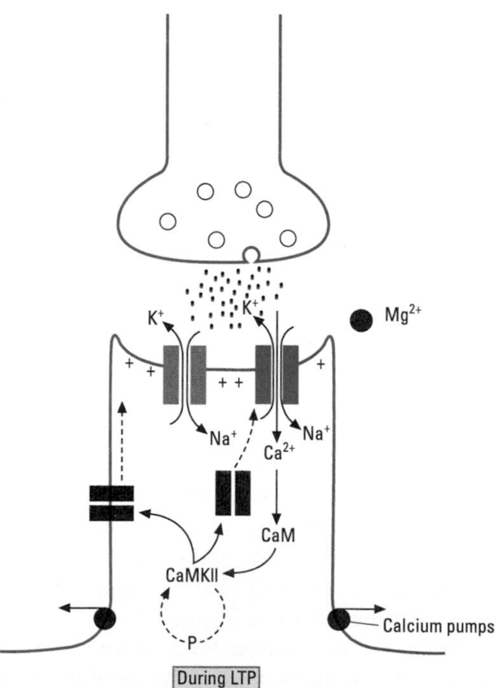

Fig. 9.40 (*right*) Long-term potentiation (LTP) as a mechanism for learning (seen at the glutamatergic synapses of the vertebrate hippocampus). With a normal stimulus, both AMPA and NMDA glutamate receptors are activated, but ions only flow through the AMPA receptors as the NMDA receptors are blocked by magnesium ions at normal membrane resting potential. However when the postsynaptic membrane is depolarized these Mg^{2+} ions are displaced and Ca^{2+} can flow through the NMDA receptor channel; this calcium binds to calmodulin (CaM) to activate calmodulin kinase II (CaMKII), which then maintains its own activity by autophosphorylation even when Ca^{2+} levels decline again. The CaMKII increases the levels of AMPA receptors in the postsynaptic membrane, either by phosphorylating and activating existing receptors there or by mobilizing extra receptors from cellular reserves. The net effect is an increased response, via these extra AMPA receptors, to any further incoming stimuli. Calcium pumps in the "neck" of dendritic spines may help to localize the change in calcium concentration.

each cell may have up to 30,000 synapses, but recent evidence seems to support a mainly postsynaptic modification, involving AMPA receptor function and localization. Overall, long-term sensitivity to glutamate is clearly enhanced, giving the desired LTP. Additional systems are certainly still to be found, and some cells (notably mossy fibers in the hippocampus, and perhaps parts of the cerebellum) are known to operate LTP independently of NMDA receptors.

In parallel with LTP, which is evoked by brief but strong stimulation, some cells in the hippocampus exhibit **long-term depression** or LTD, evoked by prolonged weak stimulation. LTD is probably produced by a small increment in postsynaptic calcium, in the same cells where larger increments produce LTP, and it may operate via the same receptor systems using phosphatases rather than kinases. LTD may in some sense represent a mechanism for "forgetting" information.

As yet we do not know how applicable such models may be to higher learning processes in general, though LTP has been observed at virtually every excitatory synapse in the mammalian brain that has been tested, and it clearly provides an excellent model system for memory storage processes.

9.5.4 Memory

Memory is the ability to both store and then recall information, and it is essential for learning. The central assumption in current neurobiology and behavior is that patterns of neural activity during a particular experience can induce changes in the strength of neural connections, which when reactivated provoke a memory of that experience. The great majority of animals have an ability to learn, and therefore even very simple invertebrates such as flatworms (phylum Platyhelminthes) can memorize simple tasks. For most animals that have been studied the process of memory has at least two stages. Firstly, in primary or "**short-term memory**", the information achieves some representation within the nervous system, presumably via the kinds of neuronal plasticity mechanisms already discussed, and at this stage the information is quite easily lost again if there is neuronal disturbance or sudden excessive sensory inputs. Indeed the link between synaptic plasticity and short-term memory (recently termed the "SPM hypothesis") is well established by a whole range of techniques. But then the stored information is transferred to a "**long-term memory**", assumed to involve long-lasting structural changes in the nervous system, giving more robust information storage, though by no means perfect storage or recall. In mammals, this transfer process seems to be centered on the hippocampus, where LTP may form a crucial part of the mechanism. Sometimes long-term memory in humans is split into secondary and tertiary components; the secondary memory involving things we learn by choice, which have a slower retrieval, and the tertiary memory giving very rapid retrieval of memories that are particularly critical, such as names, familiar locations, how to read, etc., and which are normally never forgotten except in certain pathologies.

Given that we do not understand learning in detail at a molecular level, we are clearly some way away from understanding memory. However, some features are becoming clear. Memory can sometimes be localized at single synaptic sites, so it is not necessarily dependent on complex and dispersed circuits. Nor does memory require a great deal of structural change, at least for short-term storage; existing synapses will do the job as we have seen. However, longer term storage does need new synapses and changes in dendrite patterning; hence it is inhibited by any drugs that reduce or block protein synthesis. Many models and experimental studies also suggest that second messengers such as calcium and cAMP are crucial to long-term memory storage, by providing the link from short-term membrane effects to these longer lasting effects at the gene and protein level.

9.5.5 Cognition

Cognition may be seen as a "higher" stage in nervous system functioning, beyond or above learning and memory. It can be viewed as an adaptive trait, incorporating the ability of animals to integrate perceptions of the environment, learn about their surroundings, and memorize where food, conspecifics (especially potential mates), and predators are located. Thus using an internal representation of the external world animals can make adaptive decisions according to the environmental circumstances. The neural and physiological bases of cognitive abilities are the evolutionary product of the interaction between the animal and the environment.

Cognition requires complex neurological interactions, particularly involving spatial memory and its use in predicting outcomes in varying environmental situations, so that the animal can solve a range of problems posed by its surroundings. Birds can store and later retrieve items of food such as seeds, and many insects can forage from a "central place" such as a nest, to which food is returned. Navigation over long distances also requires sophisticated cognitive abilities, as does the acquisition of "language", with bird song and primate calls having been intensively studied.

Studies of the spatial cognitive abilities of birds and mammals have established the central role of the hippocampus in these processes. In mice and in birds there are clear links between relative hippocampal volume and spatial learning ability, and indeed hippocampus volume can change seasonally within an individual, becoming enlarged during reproduction. The ability of some birds in the breeding season to learn new songs, or to store food, may parallel hippocampal cell birth (seasonal neurogenesis). However, our understanding of such complex processes is still very imperfect.

9.6 Neuronal development

9.6.1 Neurogenesis and axogenesis

The nervous system in animals is primarily an ectodermal tissue, and it develops gradually during embryogenesis. As with the development of many other tissues, it requires a highly organized developmental program involving cell proliferation, migration and navigation, recognition, synapse formation, and selective death. The proliferating neurons must follow some molecular signals that are genetic in origin and work at the level of gene regulation, and other molecular signals that are epigenetic, relating to current position within the tissue.

In invertebrates the differentiation into neurons occurs ventrally. In contrast, in vertebrates the first visible sign is a shallow groove on the dorsal surface of the embryo, which in a process termed

neurulation thickens into a long narrow neural plate whose edges arch upwards and meet and fuse, to form a **neural tube**. This is the precursor of the spinal cord, and it soon develops an anterior swelling that will become the brain. Along the lateral borders of the neural tube, another ectodermally derived tissue, the **neural crest**, gives rise to the vertebrate's peripheral nervous system.

Some of the key "segmentation genes" are involved in giving the positional information that triggers ectoderm cells into differentiating into neuroblasts, but once this **determination** has begun it is the incipient neuroblasts themselves that release chemical triggers to prevent other cells around them from becoming neural, to act as supporting cells and glial components. This has the advantage that if one neuroblast is damaged or destroyed during development the inhibition on its neighbors is relieved and another epidermal cell can differentiate into a neuron. Neuron generation begins anteriorly and proceeds backwards towards the tail, and also has a ventral–dorsal gradient so that the motor systems of the spinal cord differentiate (ventrally) before the more dorsally sited sensory systems. Differentiation of later generations of neurons depends on two major factors: firstly, the position of the cell in the dividing lineage (in effect where it lies in the family tree of that lineage), and secondly the position of the cell in space, relative to other cells and tissues around it.

Neurogenesis for peripheral nerves occurs in a fairly stereotyped sequence, with initial axon outgrowth followed by dendrite formation while still within the CNS, then axonal growth (axogenesis) from the CNS to the periphery to reach the target cell. At this point excitability appears, followed quickly by the formation of peripheral synapses and neurotransmitter vesicles. For those neurons that migrate long distances the determination of neurotransmitter type is not made until quite late on in their life, and is partly decided on the basis of the cellular environment in which they end up. Quite how neurons "know" where to go as they migrate and grow is still the subject of intensive research. In the cortex of the vertebrate brain they move along preordained pathways laid out for them by glial cells, particularly radially disposed glia that organize the cortex into its columnar units, acting as a cellular "scaffolding" to organize the neurons. However, this explanation just shifts the problem, and we have then to ask how the glial cells set up the initial structure. The problem is even more severe for neural crest neurons that have to migrate through long stretches of mesoderm to reach their targets. Here it seems that neurons are following signals from the various tissues they encounter, and are guided in part by extracellular matrix materials interacting with molecular triggers diffusing along the axon intracellularly. However, it should be noted that neuronal development is restricted, and beyond maturity it is strictly limited in many animals. The environment of the mammalian CNS is particularly hostile to axon growth, so that injured nerves do not readily regenerate; this is partly due to myelin, which incorporates growth-inhibiting proteins such as "Nogo" that effectively destroy axon growth cones.

The development of physiological neuronal function is perhaps the most pertinent issue here. It seems that neurons can form APs quite soon after they start differentiating, but initially the APs are carried by calcium ions; later they switch to a mixed calcium and sodium current, and only achieve "normal" Na^+-based spikes as they reach full differentiation.

9.6.2 Neuronal connectivity

Relatively long-range navigation by neurons involves diffusing chemical gradients of "growth factors", common throughout developmental biology. One important trigger is **nerve growth factor**, a polypeptide of 118 amino acids rather similar to insulin. It stimulates growth and differentiation of neurons in vertebrates, and has a clear trophic function, helping some neurons to survive. This polypeptide is released by target cells, so that neurons and glia grow towards the target along the concentration gradient. It is then taken up by the neurons and transported via the axon to affect gene expression back in the cell body. Many other neurotrophic factors are being identified in vertebrates, with varying degrees of specificity to motor neurons, sensory neurons, sympathetic neurons, and so on.

At relatively close range, the path taken and the connections made by an axon are specific **growth cones** (Fig. 9.41), located initially on the neuronal cell body and then later at the tips of the dendrites as the neuron advances. Growth cones undergo amoeboid movement, and push out extensions: fine tufted filopodia or flattened lamellipodia. These extensions serve as samplers of the local environment, contacting other cells (especially glial cells) and extracellular matrix, and then undergoing cell–cell adhesion when they receive the appropriate signals. This process of **"selective adhesion"** is important in establishing the morphology and connectivity of neurons. Two glycoproteins, **laminin** and **fibronectin**, are particularly involved in selective adhesion, and will bind to a class of receptors called **integrins** found on the neighboring cell surfaces. The final determination of connections may depend on specific recognition factors though; a class of glycoproteins termed "**fasciclins**" seems to determine ultimate axon pathways. Again, a large number of these are being identified (e.g. L1, F11, and TAG1), often with significant sequence homology, and surprisingly also with some similarity to immunoglobulins.

All these long- and short-range signals are particularly effective on so-called "**pioneer neurons**", which serve to lay out the primary structure of the nervous system. These then serve as guides for further axons to grow out in parallel bundles form the main nerve trunks. Again, selective adhesion is crucial, and neurons from various regions of the embryonic CNS show different patterns of preferential adhesion, relating to two key groups of cell adhesion molecules (CAMs), the cadherins (which are dependent on calcium and sometimes termed L-CAMs), and the calcium-independent N-CAMs. The patterns of CAMs vary in space and time as an axon grows, so determining which other cells it can bind to at any particular moment.

The final stage for a growing neuron is to achieve its own connections with its target cells—that is, **synaptogenesis**. Neurons acquire their identity partly from their position in the embryo, and become "imprinted" with a degree of specificity. However, the development of synapses involves only a limited degree of precision, and once in the right vicinity many neurons produce a number of synapses with adjacent cells. Selection of the appropriate synapses only comes rather later, through bidirectional communication by exchange of biochemical information; then only the "useful" synapses persist.

Synapses originate by conversion of the growth zones of neurons, with pre- and postsynaptic sites partially established in advance of a

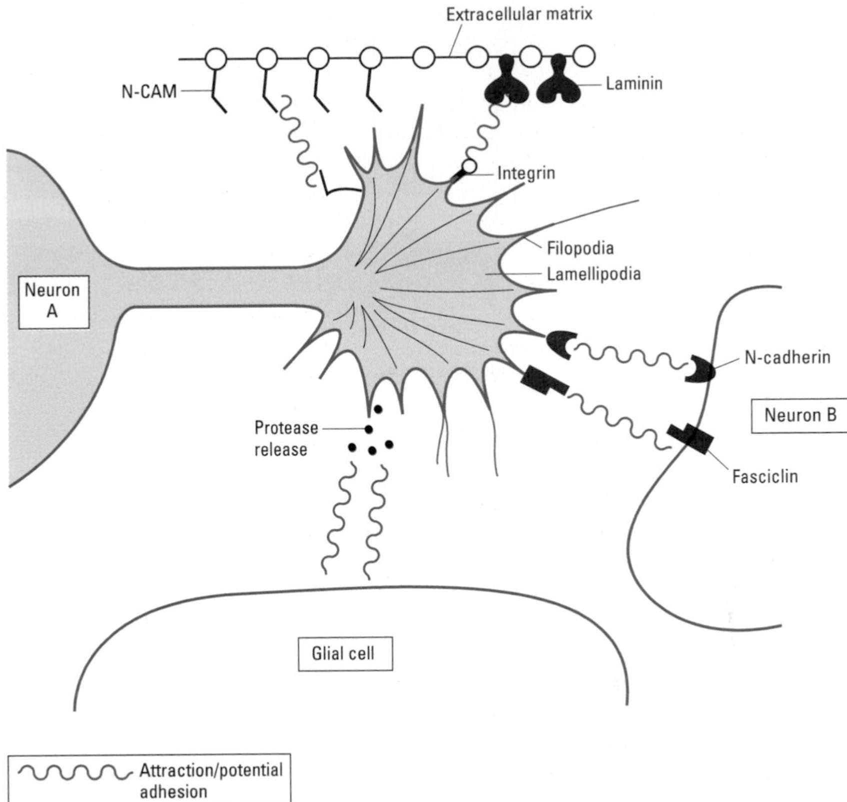

Fig. 9.41 Growth cones of axons and different possible systems for guidance to, and adhesion with, neighboring components of the nervous system. Note that all of these structures and systems are unlikely to occur in the same growth cone. CAM, cell adhesion molecule.

connection being made. It is unclear which side is the "initiator"; the presynaptic site can already achieve quantal release, and the postsynaptic site already has a degree of transmitter receptivity. Once physical contact is made the presynaptic membranes release a number of small polypeptide signaling molecules and this initiates reorganization of the postsynaptic membranes. Lateral migration of receptor molecules to the site occurs, so that the receptors instead of being present all over the target cell become concentrated in the region of synaptic membrane apposition. Synthesis of new types of receptor may then be invoked, and signals pass back to the presynaptic terminal that induce it to mature into a fully functional transmitter-release site. Thus there is "reciprocal induction"—the arrival of the nerve induces postsynaptic formation, which in its turn induces presynaptic differentiation. A key molecule in this, at least at vertebrate ACh neuromuscular synapses, may be the large glycoprotein **agrin**, which induces the clustering of dispersed transmitter receptor molecules. However, recent genetic evidence also indicates that these receptors are "preprogramed" to form patterned clusters even without innervation arriving.

The whole process of **synapse maturation** (summarized in Fig. 9.42) may take several days or even weeks. It involves only a minor change of shape and size, but a 4–10-fold increase in vesicle density. Receptors here may also gradually change; best known is the ACh receptor in mammals, which changes from a fetal to an adult type by replacing γ- with ε-subunits in the first postnatal week. Maturational changes of this type are often accompanied by changes in the reversal potential of the associated channel, which can even lead to switching between excitatory and inhibitory effects (for example GABA and glycine are excitatory in neonate synapses and only become inhibitory later as the chloride equilibrium potential becomes more negative).

Some synapses that form are later eliminated. This may include all the synapses that have been made with inappropriate target cells, and excess synapses made with the "right" target cells (for example many vertebrate NMJs start with 2–6 innervating neurons but end up with only one). Competitive interaction between the innervating neurons often results in just one neuron "winning" and linking to the target. Hence in some regions of the CNS whole axons, whole neurons, and even whole embryonic neural paths, may be laid down and then removed. Up to 75% of some groups of neurons die off completely, often according to a specific program of cell death (**apoptosis**). Conversely, after elimination of some synapses there may be a phase of plasticity when further "good" synapses with the correct cell–cell interactions are made. So the overall development of neuronal connectivity is established by a balance of cell death, synaptic elimination, and synaptic addition, and may persist well beyond the embryonic phase. It makes for a very plastic and adaptable nervous system.

9.6.3 Maintenance

Neurons, especially those with long axons and highly branched dendrites, are peculiarly susceptible to damage, yet serve a function that is peculiarly essential to normal functioning in animals. They should therefore be very demanding on maintenance and repair systems. Structural integrity of axons is largely assured by external protection from glial cells, but neurons also possess an interior "scaffold" of microtubules, microfilaments, and neurofilaments.

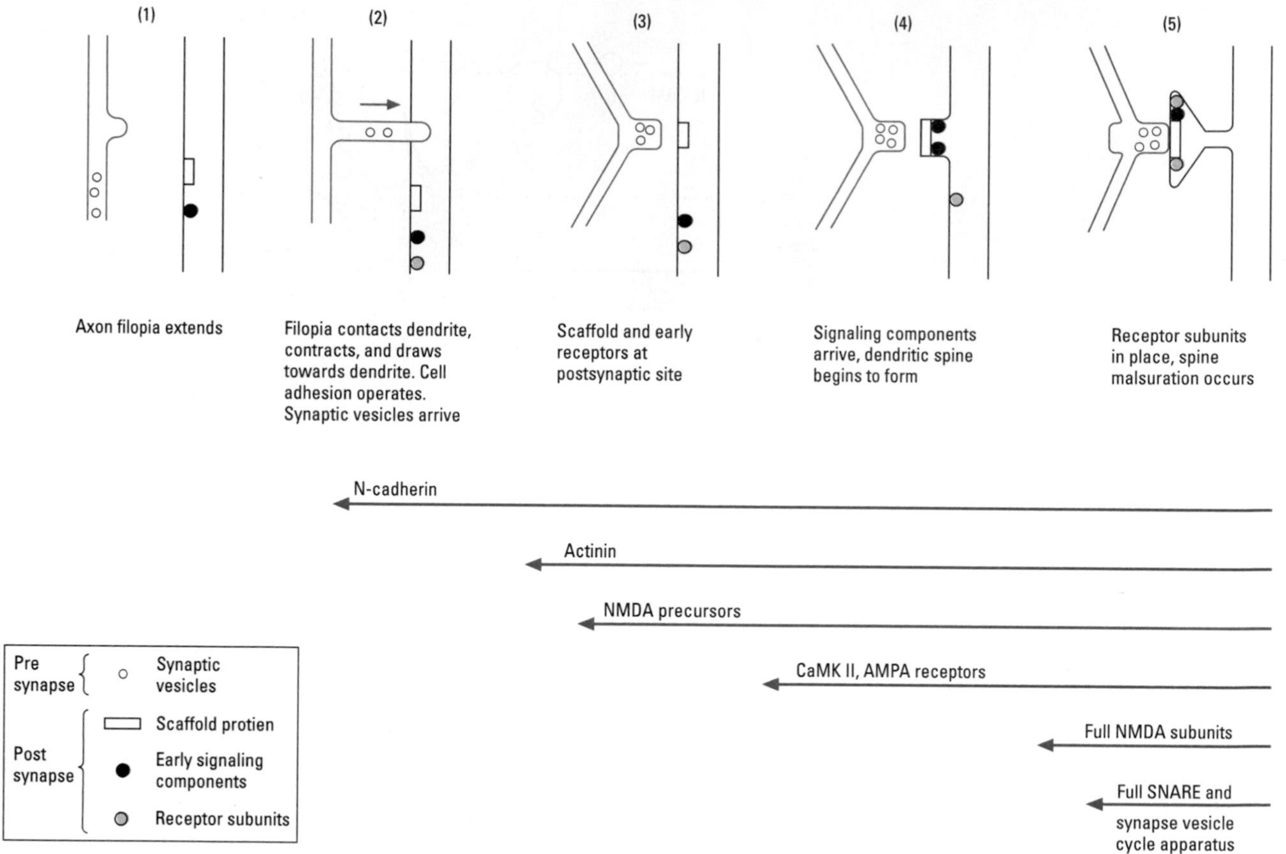

Fig. 9.42 Synapse maturation at a central axon–dendrite site with glutamate as the excitatory transmitter. At other sites, the dendrites may send out the filopodia, but the general sequence of events probably applies. (Adapted from Cowan 2001.)

Microtubules are made up of repeating tubulin elements giving hollow, tubular, 23–25 nm diameter "struts" which in a neuron run lengthwise along the axon. Microfilaments are much smaller (2–5 nm) and are made of actin monomers, and they link crosswise between the microtubules especially just below the cell membrane. Neurofilaments are intermediate in size (10 nm) and tend to occur mainly in the axons, where they again run parallel to the long axis. A major part of the maintenance system in neurons arises from the transport of key molecules along the axon, transport which must be fast and directed, and here again the cytoskeleton comes into play.

Intermediate and **slow axonal transport** both occur, moving enzymes and other proteins out to the periphery at 1–100 mm per day. The slowest movements may be driven by the growth of the microtubules themselves, since these add on monomers of tubulin at about the same rate. But **fast axonal transport** is particularly conspicuous for membrane-bound (e.g. transmitter-containing) vesicles, which move along an axon in a saltatory (discontinuous jumping) fashion. This transport can be in either direction, although it is usually faster outwards (around 400 mm per day in mammals) than inwards towards the cell body (200–300 mm per day). It relies on ATP-dependent binding of vesicles and their movement along the microtubules. A single microtubule can have vesicles moving along it in both directions, indicating that separate translocating motor proteins (see section 2.2) must be involved. Two key motor proteins here are **kinesin**, powering outward movements, and **dynein**, powering the inward movements; at least five others are also known, and are all part of the kinesin family of motor proteins. By these routes, membrane lipids, glycoproteins, and trophic factors can be sent to the periphery, while breakdown products from synapses can be returned to the cell body in lysosomes.

A third component of axon maintenance is supplied extrinsically, from the surrounding glial cells. As well as giving physical support to axons, these cells offer at least three other maintenance roles: a clean-up function, through phagocytosis of debris as neurons degenerate; a homeostatic function, regulating ion concentrations in the ECF and probably also transmitter concentrations; and a nutritional function, helping to organize the exchange of metabolites between neurons and blood vessels. This last function is particularly vital in the brains of vertebrates and some more complex invertebrates that have a "**blood–brain barrier**", whereby the nervous tissues are bathed in a special ECF that is highly controlled and inaccessible to many metabolites. Any challenges to the stability of the extracellular fluid can affect neuron functioning (e.g. from changes in environmental salinity, or from inputs in the diet), and these can often be offset by the presence of the blood–brain barrier. In vertebrates the blood–brain barrier of endothelial cells between the **cerebrospinal fluid** and the rest of the circulation plays a major role protecting the CNS from dietary or fluid-balance change. Blood capillaries supplying the CNS are lined by endothelial cells linked

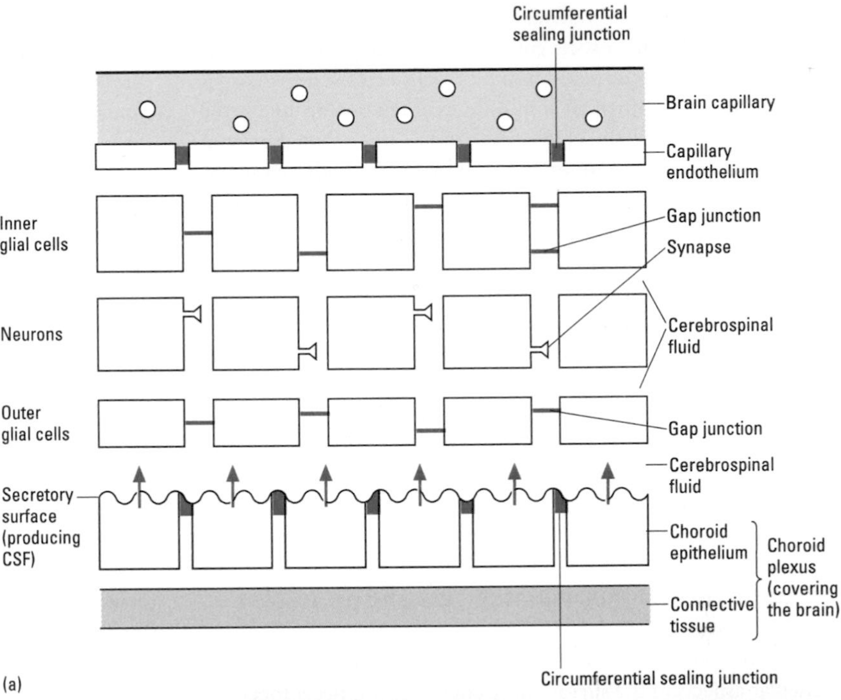

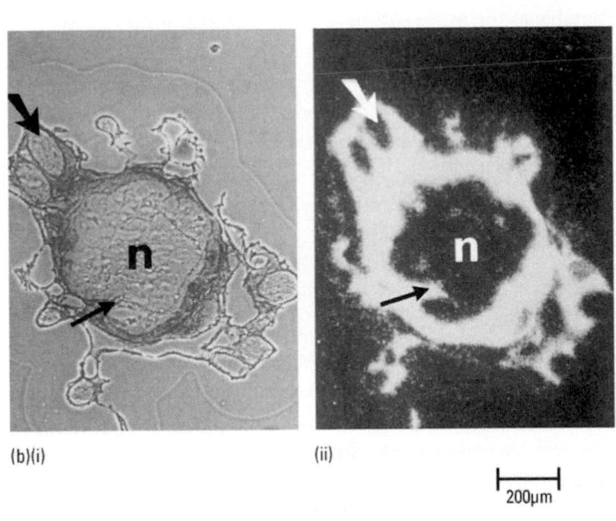

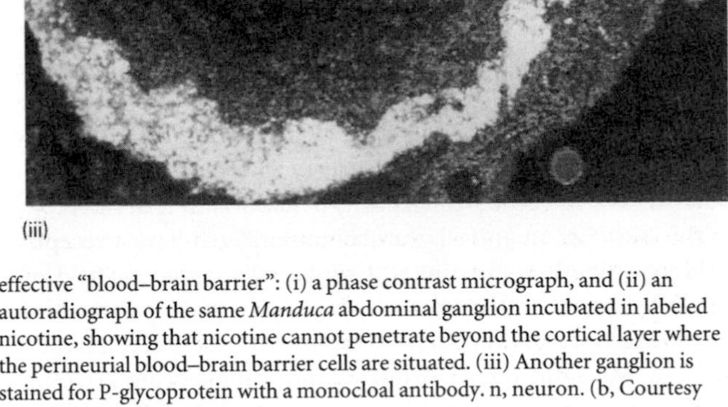

Fig. 9.43 (a) A schematic view of the vertebrate blood–brain barrier. The neurons are connected only by synapses, and glial cells by leaky gap junctions, but the endothelial cell walls of capillaries within the brain are tightly sealed together, as are the external epithelial cells making up the choroid layer. Hence the glia and neurons within the brain are bathed in a continuous cerebrospinal fluid (CSF) that is protected from free diffusion of molecules from the circulating blood. (b) The exclusion of nicotine from the nerve cord of *Manduca*, by an effective "blood–brain barrier": (i) a phase contrast micrograph, and (ii) an autoradiograph of the same *Manduca* abdominal ganglion incubated in labeled nicotine, showing that nicotine cannot penetrate beyond the cortical layer where the perineurial blood–brain barrier cells are situated. (iii) Another ganglion is stained for P-glycoprotein with a monocloal antibody. n, neuron. (b, Courtesy of Catherine Morris, Ottawa.)

together by continuous bands of tight junctions, preventing any uncontrolled exchange from blood to neurons (Fig. 9.43a). Thus hormones, drugs, antibodies, and many other crucial biochemical components of the blood have highly regulated access to the CNS, and the ionic concentrations of the ECF around the neurons is stabilized to insure constant signal-generating membrane properties. Only a few molecules, including the crucial fuel supply in the form of glucose (handled by the glucose transporter GLUT1), can pass across. Similar systems occur in some invertebrates; for example, herbivorous insects commonly have a high K^+ and low Na^+ diet, reflected in the blood, and in stick insects the nerve cord is surrounded by a sheath of fat body cells, maintaining the fluid in the extraneural space at slightly higher Na^+ and K^+ levels than the rest of the hemolymph, due to ionoregulation by the fat body cell membranes. In insects with unusual specialist food plants more specific measure may be needed. As an example, the neural sheath in the tobacco hornworm caterpillar (*Manduca*) is unusually impermeable to ingested nicotine, with tight junctions between the glial (cortical) cells, so that any nicotine entering the CNS must do so by a cellular route and thus runs the gauntlet of detoxifying enzymes (Fig. 9.43b). The nerves have a normal sensitivity to nicotine, and are therefore further protected by an alkaloid pump that removes nicotine rapidly from their vicinity; this pump is the same P-glycoprotein that operates to secrete nicotine in *Manduca* Malpighian tubules.

9.7 Sensory systems: mechanisms and principles

Sensory receptors are essentially filters, informing the body of particular aspects of the surrounding environment. Receptors each have their own "sensory modality", so that we can speak of mechanoreceptors (touch, vibration, and hearing), photoreceptors (vision), chemoreceptors (taste and smell), thermoreceptors, and so on. They respond to extremely weak stimuli within that modality, often ignoring all other information. But even within their own modality most sense organs are extremely selective, filtering out information of the "wrong" kind—too high a frequency, too strong a physical deformation, or the wrong wavelength of light. Sensory systems are not passive windows on the outside world; they actively shape each animal's perception of its environment.

9.7.1 Transduction processes and receptor potentials

Sensory systems are in essence mechanisms for transducing information from an animal's environment (both internal and external) into coded changes of electrical potential that can be transmitted into the nervous system and integrated with other inputs. Sensory cells, also termed **receptor cells**, have special sites on their membranes where this transduction can occur at a molecular level. Each type of receptor cell is specialized for extremely sensitive reception of a particular mode of energy (though it may also detect other kinds). Since the great majority of sensors are "exteroceptors", in the "skin" of an animal, they commonly each have a **receptive field**, covering some part of that surface. The receptive field of a single receptor is often extremely localized but may occasionally cover a large part of the body; and the size of the field depends on the stimulus strength, with a strong stimulus influencing a receptor much more distantly than a weak stimulus can do. Furthermore, receptive fields of separate neurons can overlap substantially, so that, for example, a point contact with the skin surface triggers many mechanoreceptors. This allows an increase in contrast and precise location of a stimulus by lateral inhibition (discussed in section 9.8.8).

Many of the sensory cell types operating in different "sensory modalities" have highly complex surfaces, elaborated from modified microvilli or cilia. Enlarging the surface area can help to lower the threshold for detecting a stimulus. Some receptor cells also have associated **accessory structures** derived from other cells (often made of cuticle in arthropods, or of connective tissue in vertebrates), used to conduct or amplify or increase the selectivity of the signal. By such means, a single type of receptor can be made sensitive to different "kinds" of trigger. For example a simple vertebrate hair cell (mechanoreceptor) can be made responsive to angular movement, to gravity, to water or air currents, or to sound waves, by modifying the accessory structures linking to it.

All the various kinds of receptor cells in animals work, fundamentally, by triggering a change in the properties of ion channels and thus inducing a current flow across the receptor membrane giving a **receptor potential** (sometimes called a generator potential if it is large enough to evoke APs, but many sensory receptors are nonspiking). However, as with synaptic membranes, the molecular pathways between the sensory trigger and the ion channel are variable and often highly complex. Due to the extremely small size, and often protected location, of many sensory endings, the details of these systems still remain relatively little known.

What is clear is that in many cases the transduction process also involves **amplification**, the potential change being far greater in energy equivalence (often by a factor of 10^3 or 10^4) than the initiating stimulus. This is partly due to the energy that is "stored" in the resting potential of all cells, and which is released by the permeability change to assist or boost the signal. But amplification is also often attributable to the intervention of biochemical cascades via second messengers, whereby a single activated receptor molecule can affect numerous channels. Wherever these additional biochemical stages intervene, though, the receptor inevitably loses reaction speed and the result is a decrease in temporal resolution for that sensory modality. There is thus a trade-off between sensitivity and speed that must be made in relation to the lifestyle of the animal and the characteristics of its environment.

9.7.2 Encoding the signal

All sensory receptors result in an opening (or sometimes a closing) of ion channels, and thus a flow of current. It is therefore easy to see how they can generate a receptor potential in the same manner that we have seen in synapses. Equally it is clear that there will normally be a quantitative relationship between the intensity of the stimulus, the number of channels responding, and hence the magnitude of the receptor potential. For the simplest kind of receptor the potential change will be related to the log of the stimulus intensity, because (from the Nernst and Goldman equations discussed in Chapter 4) the membrane potential varies with the log of the ion permeability. However, there are normally also a **subthreshold range** and a **saturation range** for any given receptor cell (Fig. 9.44a). The latter results when the receptor potential approaches the reversal potential for the particular ion that mediates it, and/or when all available ion channels are responding; in a spiking receptor it occurs when the spike frequency is 500–1000 s^{-1}, since each AP lasts 1–2 ms. Between threshold and saturation point the receptor is operating in its **dynamic range**, and here it can simply and directly encode the time and amplitude of a stimulus into its receptor potential.

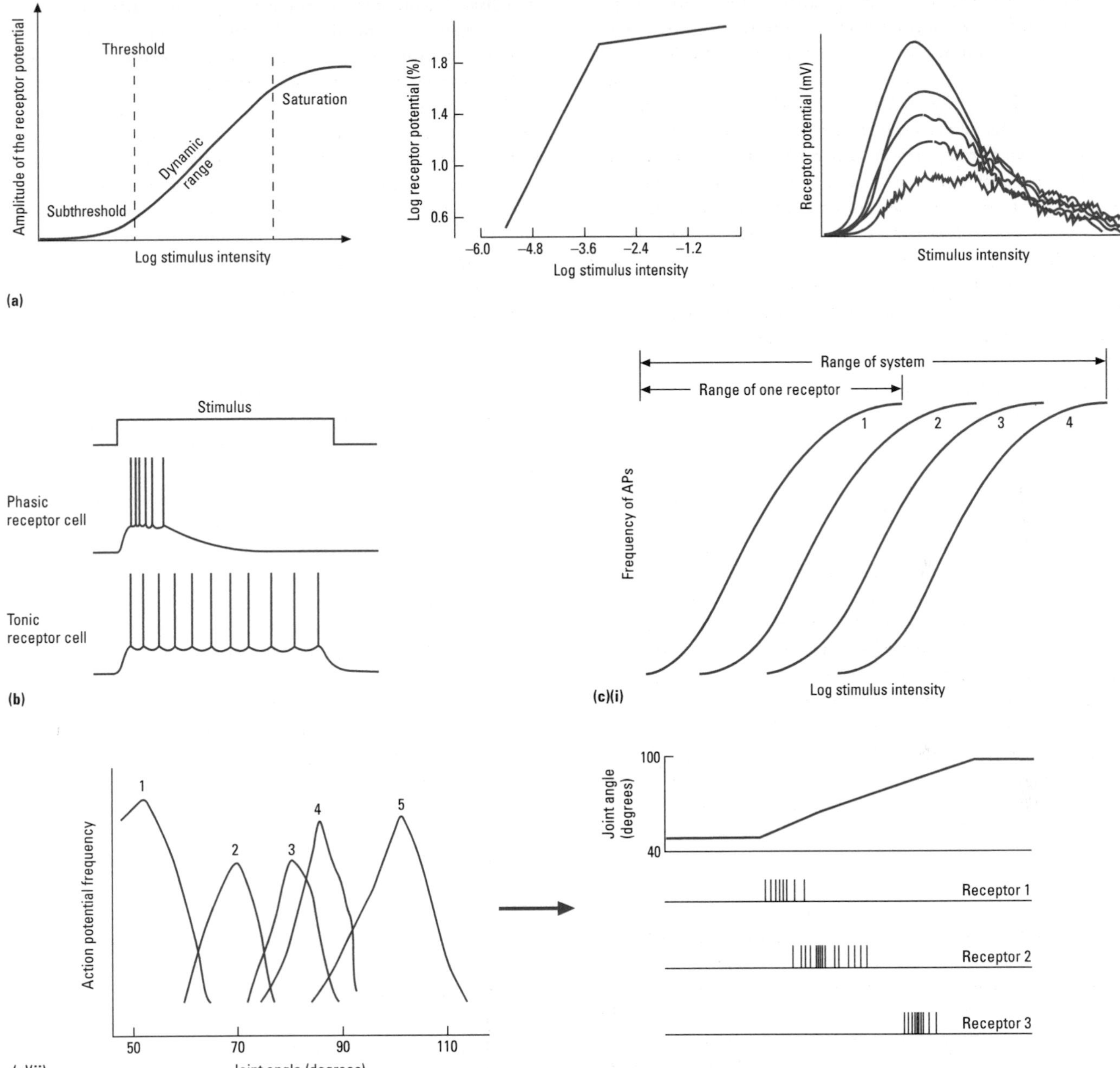

Fig. 9.44 (a) Subthreshold, above threshold (dynamic range), and saturation ranges in sensory receptors (left); in the dynamic range the amplitude of response is linearly related to stimulus on a log–log scale (middle), as shown in the recorded traces at different stimulus intensities on the right. (b) Tonic and phasic receptor responses to a similar stimulus. (c) The principle of "range fractionation" where different receptors respond maximally in different parts of the stimulus intensity range. (i) In a hypothetical example each of the receptors covers 3–4 orders of magnitude of stimulus intensity, but the dynamic range of the four receptors together is about seven orders of magnitude. (ii) Specific examples from five joint receptors in a vertebrate limb joint, where each receptor operates over a different angular range and therefore fires in a distinct phase of the full leg bending movement. (a, Adapted from Reichert 1992, after Uttal; cii, adapted from Reichert 1992, after Skoglund.)

As in other situations, the receptor potential spreads electrotonically from its site of origin. In some receptors, especially where there is no axon, the receptor potential generated at one end of the receptor cell may directly affect an area of transmitter release (a synapse) in another region and thus pass on its signal to neurons that then generate APs. For these kinds of receptor cells it is not uncommon to find that there is a low level of transmitter release all the time, giving the advantage that any small change in potential can move the level of transmitter release either up or down. This leads to a "no threshold" effect, directly encoding the magnitude of the receptor potential.

For receptors that have axons the electrotonic receptor potential must normally be recoded into APs for onward transmission. This process is similar to the summation of postsynaptic potentials into APs in other neurons, and again takes place in a special

AP-initiating zone. Usually, the graded nature of the receptor potential is reflected in the intervals between successive APs, which can vary almost continuously, limited only by the number of ion channels present (once they are all open the firing frequency clearly cannot continue to rise with stimulus strength), and by the duration of the absolute refractory period that must elapse between APs.

However, many receptors—probably the great majority—respond with decreasing strength to a stimulus of constant strength, reflecting the process of **sensory adaptation**. Thus constant pressure on a mechanoreceptor leads to a gradually decreasing perception of contact for the animal as a whole: constant input of bright light produces a reduction of visual acuity, and being in a room with a particular odor ceases to elicit any chemosensory response after a few minutes. This process of adaptation occurs at different rates in different kinds of receptor, and may have several different causes. An obvious possibility is simple fatigue in some part of the receptor system (as with bleaching of the visual pigment in a photoreceptor). More subtly, there may be processes built in at the membrane level, an example being a build up of intracellular calcium in some receptors when depolarized for long periods, which in turn activates calcium-dependent K^+ channels and partially repolarizes the membrane. A third mechanism may be due to changes in the accessory structures, either passively as when the lamellae of the Pacinian corpuscle (see Fig. 9.45b) deform under pressure, or in an active sense of adaptation as when the pupil of an eye closes in bright light. Finally, there may be influences from other parts of the nervous system ("efferent control"), discussed in section 9.8.8.

Some sensory cells adapt relatively little, and are described as **tonic** in that they alter their response minimally with maintained stimulation; others are **phasic**, and adapt very quickly (Fig. 9.44b), and of course there are intermediate phasi-tonic receptors. For phasic receptors, the encoded message reduces considerably in intensity, and the dynamic range of the sensory cell is effectively extended and improved, often giving a good sensitivity over many orders of magnitude. This is perhaps especially important in allowing animals to respond to changing conditions at a whole range of background levels of sensory input. Tonic receptors convey information about important sustained stimuli, such as the position of a limb, and are also crucial for internal homeostatic monitoring.

Transmission of information from sensory inputs, whether tonic or phasic, ultimately requires a train of frequency-modulated APs to be sent along axons to the CNS. However, the sequence of APs does not carry any information about the kind of stimulus that produced it. This can only come from the connectivity of the neuron, taking its message to the appropriate part of the brain and thus indicating its source.

A further important aspect of the encoding process occurs in situations where sensory cells are organized together into **sense organs**, giving a multireceptor that feeds into multiple neuronal channels. Information from each channel can be compared with that from its neighbors, to give a much more detailed information package for transmission onwards to the CNS. At its simplest a multireceptor system allows for "**signal averaging**", increasing the overall sensitivity by improving the signal to noise ratio (i.e. weak signals can be detected more clearly against a "noisy" background). But further improvements of information quality come where the receptors each have their own slightly different sensitivities, each cell being a specialist unit concentrating on its own relatively narrow range of stimuli ("**range fractionation**", illustrated in Fig. 9.44c). Better still, the receptors within a sense organ can be highly spatially organized so that they have their own **topography** relating to spatial effects in the world outside. An obvious example is the arrangement of photoreceptors in a single layer (the retina) across the inside of a vertebrate eye, where each part of that layer is "looking at" a separate piece of the outside world. Each photoreceptor therefore can signal news of what is happening in its own field, but in addition the temporal pattern of triggering of the receptors gives information on objects moving across the total visual field. The topographic information may be retained in the interneurons leading from such an organ, and even right through to the brain.

9.8 Specific senses and sense organs

9.8.1 Mechanical senses: touch, vibration, and balance

Mechanoreceptors work by detecting any kind of physical deformation of a rigid or semirigid structure, the deformation being linked to an ion channel in the receptor membrane. Transduction of a mechanical stimulus can be mediated by two different types of ion channels that are sensitive to stretch; one kind is activated (opened) by stretch, and the other kind is inhibited by it so that the channel closes and blocks an ion current. The response to a deformation is so fast that no biochemical processes can be present between the event and the potential change. The channels in a mechanoreceptor need to be very firmly anchored to the membrane and the underlying cytoskeleton, and sometimes also attached to a special extracellular skeleton of carbohydrate chains, to avoid spurious triggering. In at least some cases the external deformable structure is linked by very "springy" proteins to the channel protein, such that any movement of the cellular surface pulls on these springs and opens the channel gate.

Thus, at its simplest, mechanical sensitivity can reside in a simple unmodified nerve ending in the skin with the appropriate ion channels. This system persists from the simplest soft-bodied invertebrates (many of which use it to measure water currents past their body, so called "rheoreception") right through to the epidermis of mammals; though it is largely precluded in invertebrates with stiff exoskeletons such as the arthropods. Similar sensors occur internally, especially in the gut and in muscle, providing information on movement or position. A simple nerve ending linked to any larger scale deformable surface structure will also serve, and mammals achieve this with neurons linked to hair follicles, detecting any movement of the hair due to air movement or a light touch.

Mechanoreceptors can be equipped with a range of more complex accessory structures to give them sensitivity to varying kinds of mechanical disturbance—stretch, fluid movements, pressure, or vibration. In arthropods, mechanoreceptors are commonly enclosed in small sensory bristles (Fig. 9.45a) whose hinge arrangements and/or elliptical tip dimensions confer directional sensitivity. In this system epithelial cells adjacent to the receptor can modify its behavior; they secrete K^+ into the hemolymph, reducing the receptor's membrane potential and increasing its sensitivity. Arranged in groups, such bristles can become "hair beds", responding to wind

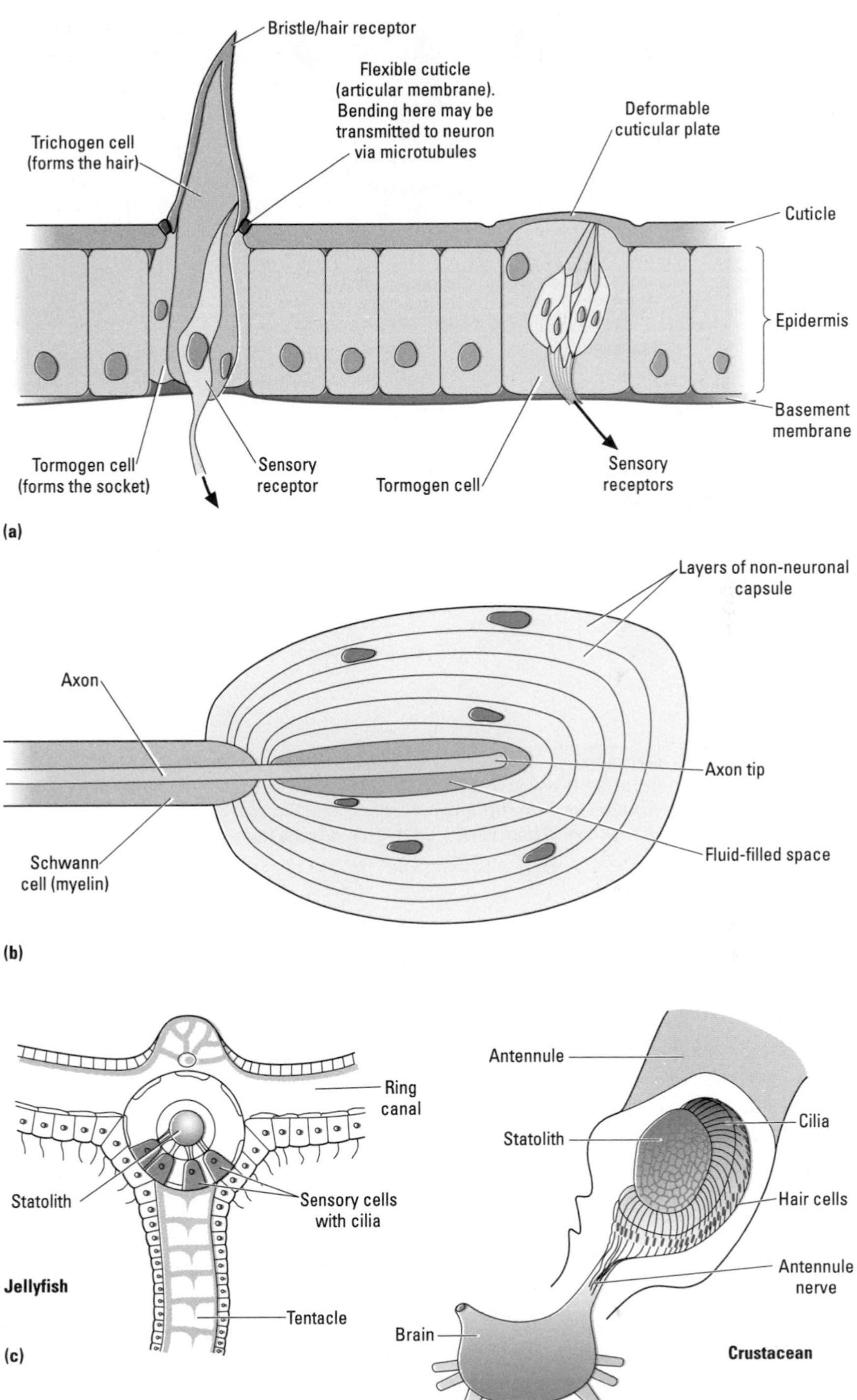

Fig. 9.45 A variety of mechanoreceptors: (a) two types of insect cuticular mechanoreceptor, (b) a Pacinian corpuscle from a vertebrate, and (c) statocysts from aquatic invertebrates (simple in jellyfish, more complex in crustaceans).

direction during flight or to gravity when arranged at a junction with another sheet of cuticle. An alternative system is to enclose sensory terminals in a multilayered mass of membranes derived from accessory cells, which reduces the sensitivity to mere touch but makes an effective pressure and vibration receptor, as seen in vertebrate Pacinian corpuscles (Fig. 9.45b). If the receptors protrude into a small fluid-filled sac that also contains a relatively dense sphere of fixed size, such as a ball of an insoluble inorganic material like calcium carbonate (known as a statolith), they become directional gravity receptors since this ball will rest upon them in different orientations according to the position of the body. This is the principle of the **statocyst** found in many aquatic animals (Fig. 9.45c), where each mechanoreceptor responds maximally at some particular static angle of the body.

Simple mechanoreceptors can also be used as pressure receptors, in water (where they necessarily become depth receptors) or in air.

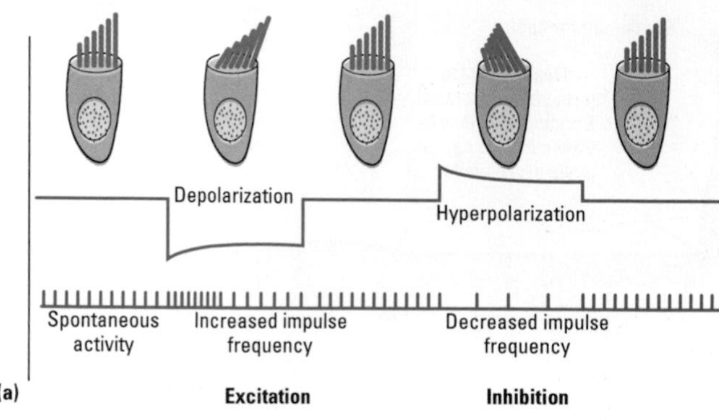

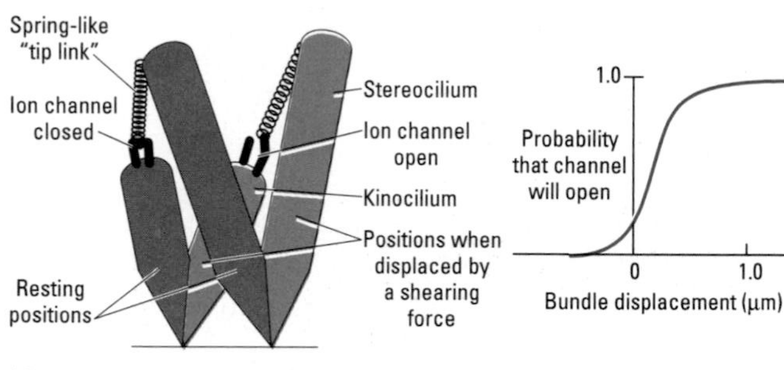

Fig. 9.46 Operation of a mechanoreceptor hair cell: (a) the basic structure, directionality, and responses to displacement of cilia in different planes, and (b) a model of the proposed mechanism involving tip links between adjacent cilia and the opening of ion channels. (a, Adapted from Flock 1965, 1967; b, adapted from Hudpseth 1989.)

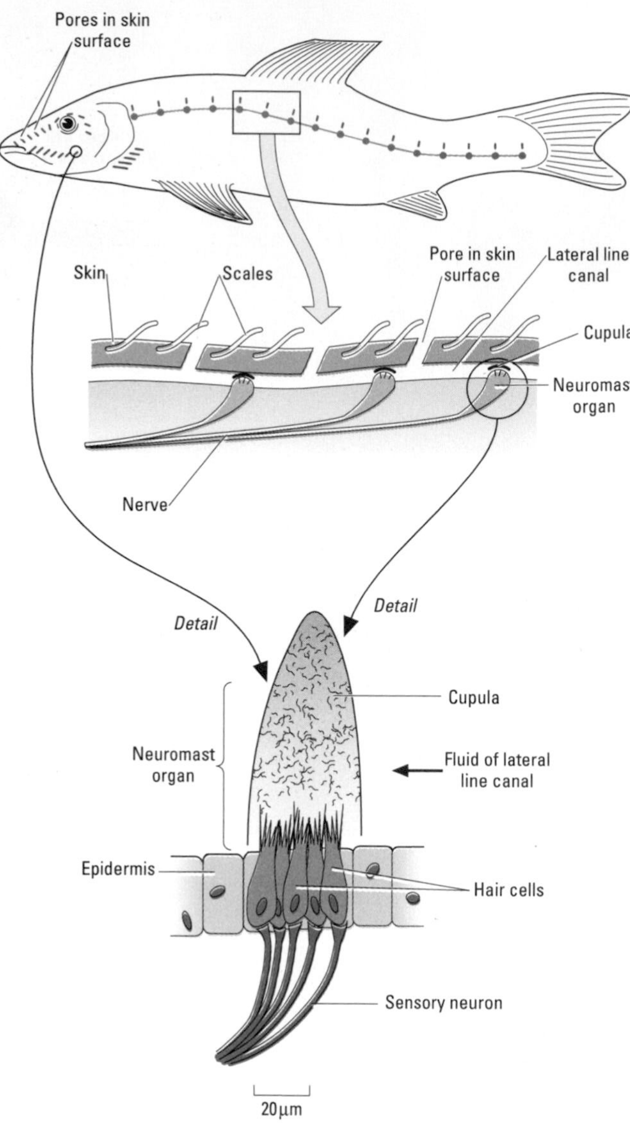

Fig. 9.47 The lateral line and neuromast sensory system in a schematic generalized fish. Most fish have a continuous lateral line with numerous separate neuromast organs lying within it, and many also have individual neuromasts located in pits, especially on the head. Note that the orientation of the neuromast and its hair cells (shown by the green dashes) changes in different parts of the body.

Aquatic insects often have water pressure detectors, for example using an air bubble trapped between sensory hairs, which bend as the bubble becomes compressed. Similar sensors attached to internal air sacs in different parts of the body can be used to sense orientation (whether the insect is head-up or head-down). Air pressure receptors are known to occur in some bats and birds; in pipistrelle bats the "Vitalli organ" in the middle ear can detect barometric changes of just 2–3 mbar. This aids both the avoidance of bad weather (especially for migrating birds), and the acquisition of prey; low air pressure tends to mean light rising air and warm cloudy weather, ideal for flying insects and so also ideal for catching them as food.

The best known example of a specific and specialized mechanoreceptor is the **hair cell** in vertebrates, of non-neural origin. The tip of a hair cell bears several "stereocilia", which are actually microvilli reinforced internally with some actin fibers, and stacked in a neat array of increasing length from one side of the cell to the other (Fig. 9.46). At one edge is a single true cilium (kinocilium). The tip of each stereocilium is linked to the side of the next (taller) one by a very fine elastic filament, termed a "tip link", and these links become stretched when the cilia are bent in the direction of the tallest ones. Each tip has a region with ion channels (probably only a few) that are directly sensitive to mechanical deformation and are pulled open by this stretching. When the stereocilia are bent the other way the tip links relax and the channels can close. Thus the whole receptor has directional selectivity. The ion channels themselves are relatively nonselective, allowing the passage of almost any small cations. When there is no deformation about 10–20% of the channels in the hair cell are open, giving a small depolarization, so they can give a quantitative (nonspiking) response in both directions by a change in the receptor potential; about 20 mV depolarization for movement towards the kinocilium and 5 mV hyperpolarization for movement the other way. The gating movement in a single transduction channel is estimated at about 4 nm, involving a force of just 2–3 pN (picoNewtons).

Hair cells can be used in various ways. Enclosing them in a channel of relatively viscous fluid so that they are influenced by the drag forces in that fluid will give information on fluid flow, and hence also on locomotory speed. This is the arrangement in the **lateral line** system of fish (Fig. 9.47), where a shallow canal runs along the side of the fish with openings to the surrounding water. Within the canal

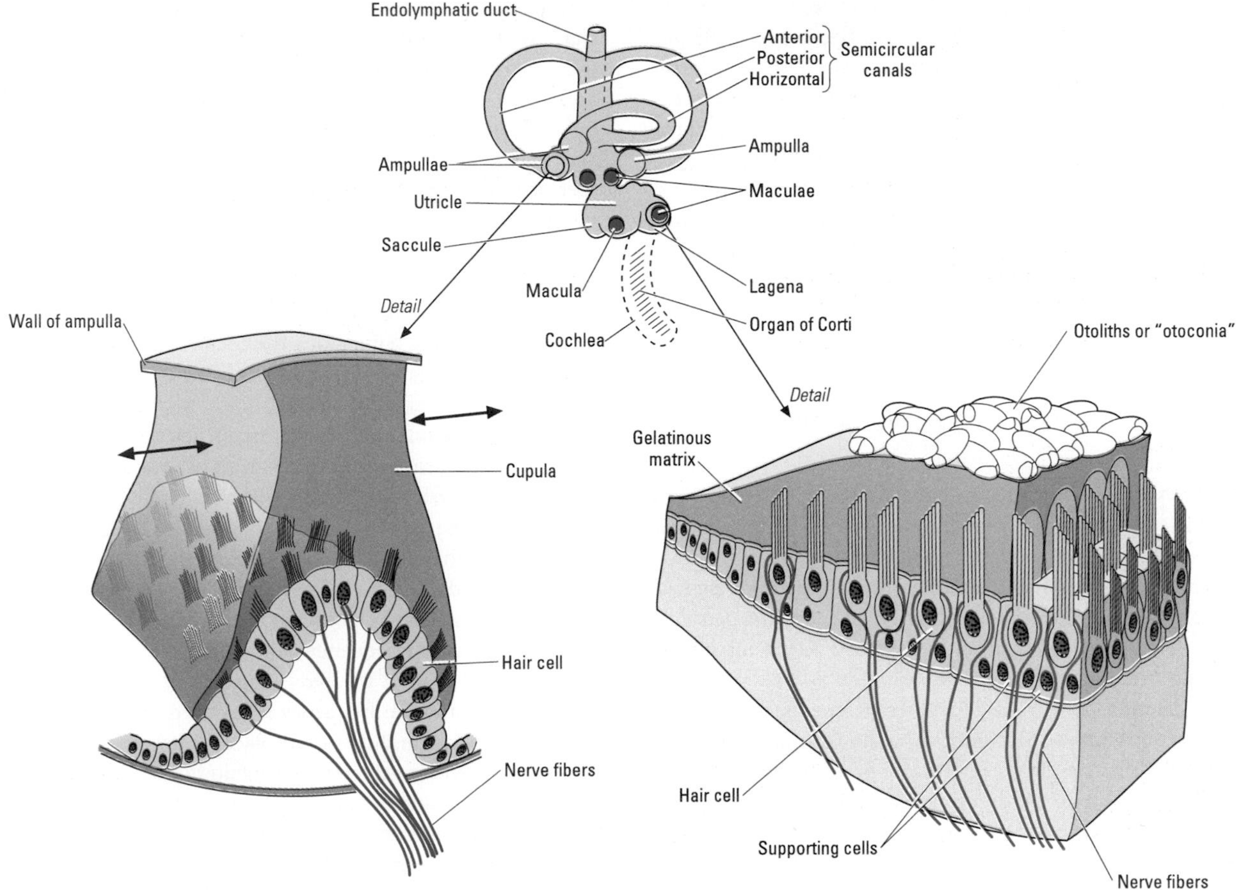

Fig. 9.48 The vertebrate vestibular system. A generalized view, with the three semicircular canals linked to neuromasts in their basal ampullae, and the otolith organs (utricle, saccule, and the smaller lagena) containing macula neuromasts. The hair cell stereocilia lie in varying directions in groups in the otolith ampullae to give sensitivity for all positions and movements.

(or sometimes free on the skin surface) are organs called **neuromasts**, each containing numerous (10–100) hair cells whose tips project into a small cap of a gelatinous material (the cupula). Any disturbance of the water around the fish creates a pressure wave through the lateral line canal and displaces the cupula, thus bending the hair cells. The axonless cells release transmitter at a low rate all the time; when the stereocilia are bent in one direction the cells depolarize and release more transmitter, thus increasing the firing of the interneuron, whereas bending the other way creates hyperpolarization and reduced transmitter release. The neuromast can thus signal the to-and-fro movements of water around the fish, allowing it to localize the source of a disturbance (e.g. a food item, or a predator).

The principles of the statocyst and the lateral line system are combined in the **vestibular system** of tetrapods (Fig. 9.48). The vestibular system lies adjacent to the cochlea in the inner ear, and is fluid-filled. One part consists of small chambers lined with a mass of hair cells forming the macula, each chamber also containing a gelatinous mass within which tiny mineralized particles are embedded, here termed **otoliths** rather than statoliths, and about the size of particles of dust. Movement of the otoliths relative to the macula causes deformation of the hair cells, either increasing or decreasing the rate of firing of their interneurons. One otolith organ (the utricle) lies horizontally and measures the direction of gravity; the other (the saccule) is vertical and mainly responds to linear acceleration of the head (and thus usually of the body as a whole); both have hair cells in organized arrays but with their stereocilia aligned in slightly different directions to give maximum sensitivity to any possible movement.

The second part of the vestibular system is the **semicircular canal** arrangement, essentially like three lateral line channels in different planes, so that a mammal or bird can achieve a fully directional sense of rotation (angular motion). This becomes the sense that we perceive as "balance". The three canals have a common opening leading to the otolith organs, with fluids shared throughout the vestibular system. One canal is horizontal to the axis of the head, the other two vertical but at 90° to each other. Each has at its base an enlarged ampulla, within which is a field of hair cells (a neuromast) projecting into a typically gelatinous cupula. Unlike the lateral lines of fish, though, there is usually no movement of the canal fluid due to external water currents; instead, the canals and ampullae work mainly by the effects of inertia. When the head is moved, the fluid tends to stay still so there is relative movement of the canal wall past the stationary fluid, causing relative deflection of the cupula and triggering the hair cells. The system is strongly phasic, reacting mostly to the initial acceleration as the head starts to move. This is because if a movement is maintained, the fluid itself is dragged into motion by the walls of the canal and the deflection of the cupula ceases, until the animal's movement stops again. With three canals, every possible angular movement in three dimensions produces a unique pattern of firing from the afferent neurons in the vestibular nerve. The most important function for this in most animals (apart from simply maintaining balance) is to integrate the movements of the head with the visual field. Head movements must be compensated for with appropriate eye movements if an animal is to be able to continue to look at a particular object while moving its own body, hence the vestibular system is strongly coordinated with eye movement control systems within the brain.

9.8.2 Mechanical systems: auditory systems and the sense of hearing

Sound waves can be generated in either water or air, whenever a solid vibrates and produces waves of compression and rarefaction of the adjacent fluid. The waves have a particular frequency (giving the pitch of the sound that is heard) and a particular amplitude depending on how vigorous the original vibration is (giving the perceived loudness of the sound). Hearing is therefore also a mechanical sense, being a response to the vibrations produced by sound waves (the term "phonon" is often used to represent the elementary energy of a single acoustic wave). Since vibrations, unlike other mechanical inputs, involve many cycles of a particular perturbation, the receptors that detect them are used to generate a mean (averaged across time) of the stimulus intensity. However, the fundamental transduction process in auditory sensors is the same as that described in the preceding section.

For vertebrates, the receptors are also the same, and are probably direct evolutionary descendants of the neuromasts described above. If hair cells are arranged in stacks on a relatively flexible membrane they can be made to vibrate in a resonant manner in response to mechanical vibrations of the membrane, thus becoming the basis of a hearing mechanism. Vertebrate ears each contain two sets of hair cells arranged in rows within the **cochlea**, each cell with 20–100 stereocilia (but lacking the kinocilium), mounted on the **basilar membrane** and projecting into (for the outer set), or very close to (for the inner set), a rather stiff gelatinous surface termed the **tectorial membrane** (Fig. 9.49). These two membranes and the hair cells between them form the **organ of Corti**, which is extended along the length of the spiraled cochlea. In a mammal, air vibrations entering the ear cause the **eardrum** (or **tympanic membrane**) to vibrate, and the vibration is transmitted via the three small bones (**ossicles**) of the middle ear to the oval window at the entrance to the cochlea (Fig. 9.50). The ossicles help to amplify the energy of the vibration about 15-fold, while decreasing its amplitude. Movements of the flexible oval window generate pressure waves in the fluids of the cochlea, eventually dissipated at the round window at its other end. But as they pass along the cochlea these pressure waves make the very flexible basilar membrane vibrate up and down, while the stiffer tectorial membrane, attached only at one side, moves less. The hair cells of the inner row, lying between the two membranes, are therefore subject to a shear force, deforming the tips of the stereocilia by only 0.2–0.4 nm, but enough to cause ion channel stretch effects as described in section 9.8.1, and thus leading to an oscillating receptor potential reliant on K^+ and Ca^{2+} currents.

Detection of frequency is critical, and depends on the varying character of the basilar membrane. At the beginning of the cochlea this structure is narrow and stiff, and it becomes steadily wider and more flexible along the length of the organ of Corti. Thus the wave traveling into the cochlea has different effects in different places, and any one frequency of vibration will have its greatest effect at a specific place (Fig. 9.51a). Low frequencies cause the broad flexible membrane at the far end of the cochlea to vibrate most, while high frequencies are maximally effective at the proximal narrower end; just a few hair cells are therefore stimulated for any given sound. The whole organ is effectively "**tonotopic**". In addition, the receptor hair cells at any one point along the cochlea are tuned, responding only to a narrow band of frequencies. This is partly due to differences in the stereocilia, which are shorter and stiffer at the base of the cochlea, so resonating at higher frequencies. But active processes also help to sharpen the tuning in at least two ways. Firstly the inner hair cells differ systematically at the level of their K^+ channel properties. Secondly the outer hair cells (which are effectors or mechanical motors rather than sensors) contract very slightly when stimulated, thus changing the tension in the tectorial membrane and helping to sharpen the frequency responses of the inner hair cells considerably. In effect the adjacent frequencies are suppressed, giving a result akin to lateral inhibition in the eye (see section 9.8.8). The end result is that information on frequency is encoded principally by which hair cells are firing, and at any one place the best frequency detected is approximately an exponential function of its distance from the eardrum. Amplitude is also coded, not in this case by firing frequency but by the number of cells firing, since a louder nose will trigger rather more hair cells with receptive fields adjacent to the peak frequency responders (Fig. 9.51b). Tetrapod ears have various ways of protecting themselves against excessive amplitude, and of increasing sensitivity to very quiet sounds. For example, small muscles adjust the coupling between the transmitting bones of the middle ear, and the epithelium lining the cochlea can secrete extra K^+ into the cochlear fluid to reduce the resting potassium current and increase the hair cells' sensitivity.

Hair cells, of which there are about 15,000 in the human ear, synapse onto neurons in the spiral ganglion, which also lies within the cochlea. From there the information passes via the cochlear nerve (part of cranial nerve VIII) to nuclei within the medulla, thence to the inferior colliculus of the thalamic midbrain and finally to the auditory cortex at the sides of the cerebral hemispheres. A tonotopic representation persists through much of this pathway, but with increasingly sharp frequency tuning in the interneurons (Fig. 9.51c). Lateral connectivity also exists, to focus the reception system on key auditory features of particular importance to the specific animal. Natural sounds are rarely "pure", containing multiple frequencies and harmonics, and are almost always heard against a noisy background. Thus each species may need mechanisms to pick out key sounds, a classic example being that species' own call or song. The cochlea itself can be adapted for particular frequency ranges (20 to 20,000 Hz in a young human, but above 100,000 Hz in some animals, and lower in many aquatic species), and it can be extraordinarily sensitive, detecting oscillations 10 times smaller than the diameter of a hydrogen atom. At a higher level some of the thalamic nuclei neurons are often specialized to pick up particular temporal patterns or "pulsed" sounds, while other neurons respond best to particular combinations of frequencies. By such means each

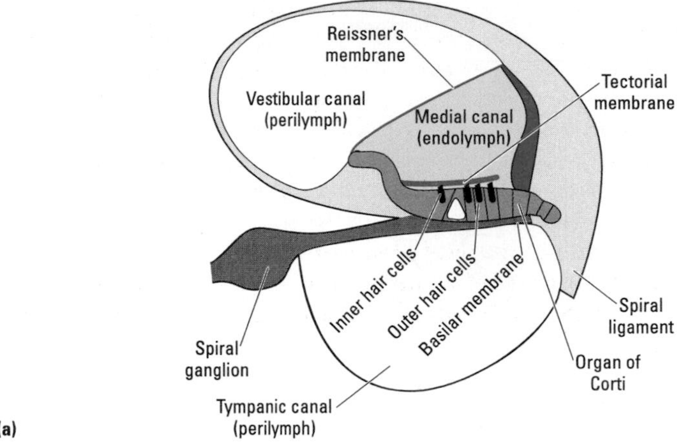

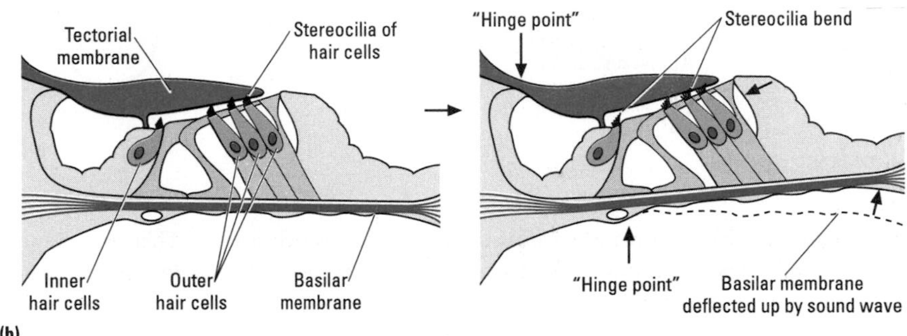

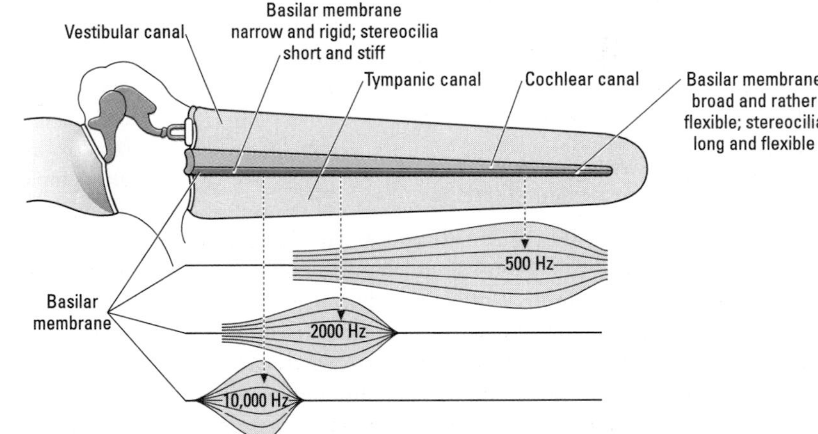

Fig. 9.49 The cochlea and organ of Corti. (a) A transverse section of the cochlea. (b) Close up of the organ of Corti, and the effect of a vibration on the position of the basilar membrane and hair cell cilia. (c) A schematic view of the localization of frequency in the basilar membrane, with high frequencies resonating near the oval window, and low frequencies traveling farther down the system before reaching maximum amplitude. Only the hair cells lying over the point of greatest membrane movement are stimulated.

animal can pick out the crucial sounds from its noisy environment. It can also localize them, since the same thalamic nuclei have a topographical map of the animal's surroundings, so the sounds picked up (by both ears) from a particular direction will fire particular neurons.

Outside the vertebrates, it is currently thought that only the insects could be described as having true "hearing". Many other invertebrates can detect low-frequency vibrations, in solids that they are touching or in the water or air they live in; but only some insects have proper "ears" that respond to high-frequency sounds. These ears are surprisingly diverse (Fig. 9.52), both in location and in mechanism. The simplest versions are **cerci**, which are hair-covered protrusions from the abdomen and which in most insects are ordinary mechanoreceptors, but which in some species have hairs so long and fine that they pick up air vibrations. Somewhat more sophisticated are the true vibration receptors, such as **Johnston's organs** at the base of the antennae of many insects (as wind detectors) but specialized in male mosquitoes to respond to sounds in air and particularly to the wingbeat frequency of the females. However, the best ears in insects are the **tympanic organs**, consisting of a membrane stretched over an air sac (part of the tracheal network), either in the sides of the thorax (many moths) or sometimes in the front legs (some crickets and grasshoppers). Simple sensory neurons are attached directly to the inside of the tympanic membrane as

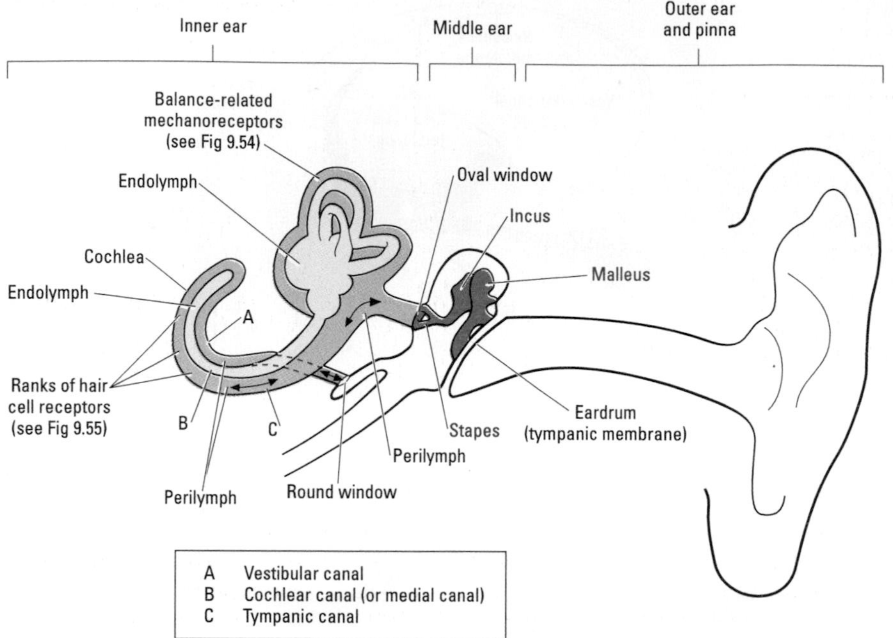

Fig. 9.50 A general view of the mammalian ear, including the transmission system of ossicles (in green lettering) in the middle ear. Note that the cochlea is shown as if "uncoiled" from the tight spiral it normally forms, and the relationship of the three parallel cochlear canals shown in Fig. 9.49 is clarified.

scolopidia, collectively forming "**chordotonal organs**", which fire when the membrane vibrates. Often there is little or no frequency discrimination, but in some species we find specific frequency tuning, with the larger more proximal scolopidia responding to lower frequencies (Fig. 9.52d). In these cases, as in vertebrates, the sensory projections are tonotopic and show increasing fine tuning at the interneuron level. This can allow a male insect to detect its own species' song, determine its direction, and roughly compute the distance to the source (important as there is likely to be another male with which it needs to compete).

9.8.3 Chemical senses: taste and smell

The ability to "sense" chemicals is widespread, and the skin surface of most animals can detect a variety of fluids (acids, etc.) and respond to them as noxious or "painful". Within the body, too, many chemicals can be sensed by "interoceptors", since it is crucial to detect the level of ions and pH, as well as gases such as oxygen and carbon dioxide, in the blood. For the detection of concentrations (osmolarity, or the balance of salt and water, whether in the environment or in encountered fluids) there is also a general phenomenon of "osmosensing" that is perhaps rather different in kind from most forms of chemical sensitivity. In principle osmosensing at the cell surface could involve specific solute sensors, or could be stretch activated, perhaps via phospholipase enzymes that respond to packing of the lipid bilayer. Genetic work with the nematode *Caenorhabditis* is also beginning to identify specific surface proteins involved in osmosensing, and there is recent evidence for an osmotically activated channel (osmoreceptor) in vertebrates. For example, some toads can "drink" through their skin at the ventral patch (see Chapter 5), and this skin is innervated by spinal neurons whose nerve endings have unusual Na^+ channels that respond to hyperosmotic solutions, allowing the toad to avoid such fluids. There may be additional intracellular osmosensors, possibly related to macromolecular packing or to cytoskeletal changes. Whatever the sensors (and there may be several), onward transduction of the osmotic signal may be via quite complex MAP (mitogen-activated protein) kinase pathways.

However, these kinds of chemical senses are not usually regarded as classic chemoreception, and they are processed in quite separate parts of the brain from the taste/smell modality. True chemoreceptors (olfactory receptor cells or ORCs) usually require a somewhat more complex transduction process than the basic mechanoreceptors described in the last section, although the ORCs are generally quite simple morphologically, involving small neurons with highly divided (ciliary or microvillar) tips. For a few other simple chemicals, such as salts and mild acids (and perhaps in the case of some biting flies, the CO_2 released from hosts), there is a direct effect on ion channels in the neuronal tip regions. But in most cases of chemoreception the chemical molecules that act as the signals must first bind with specific receptor molecules on the sensory cell membrane, and then the binding process must lead to a change in ion channel conductivity. Within the liquid that covers the tips of the chemoreceptor are specific **odorant-binding proteins**, which attach to the incoming scent molecules and move them to the receptor cell membrane, helping to concentrate the odorant at the sites of the odor receptors. Although only a few such receptor sites are known in detail, they commonly involve highly specific **receptor molecules** (usually proteins, though sometimes phosopholipids for salt detection) that are separate from the channel proteins, and the link between the two is usually a cAMP or cGMP second messenger operating via an adenylate (or guanylate) cyclase system (see Fig. 2.3). The system is thus not dissimilar to the chemical–electrical conversion occurring at some postsynaptic sites. In vertebrates another system has also been identified acting via phospholipase C and the IP_3/DAG route (see Fig. 2.19). For both routes, the second

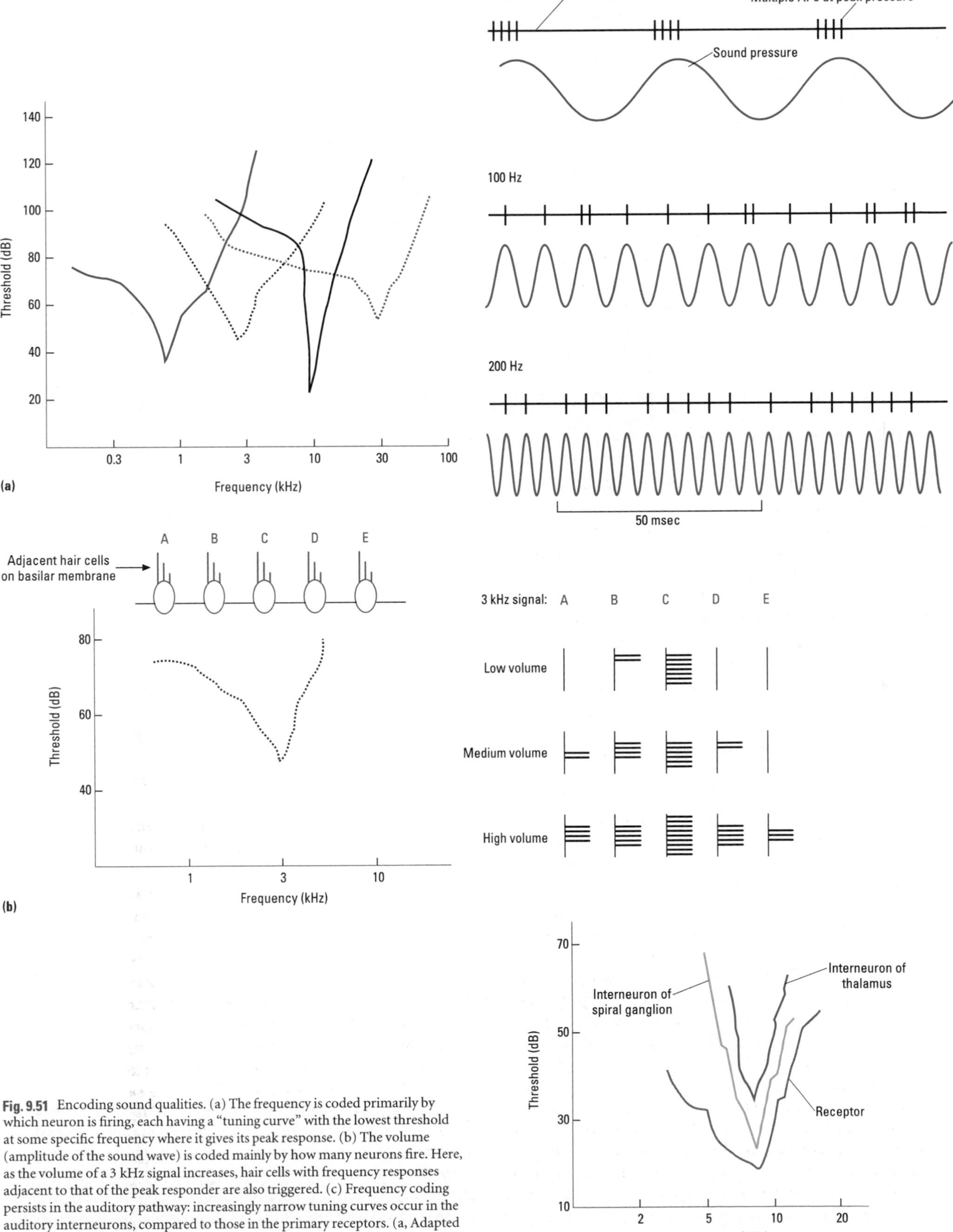

Fig. 9.51 Encoding sound qualities. (a) The frequency is coded primarily by which neuron is firing, each having a "tuning curve" with the lowest threshold at some specific frequency where it gives its peak response. (b) The volume (amplitude of the sound wave) is coded mainly by how many neurons fire. Here, as the volume of a 3 kHz signal increases, hair cells with frequency responses adjacent to that of the peak responder are also triggered. (c) Frequency coding persists in the auditory pathway: increasingly narrow tuning curves occur in the auditory interneurons, compared to those in the primary receptors. (a, Adapted in part from Delcomyn 1998.)

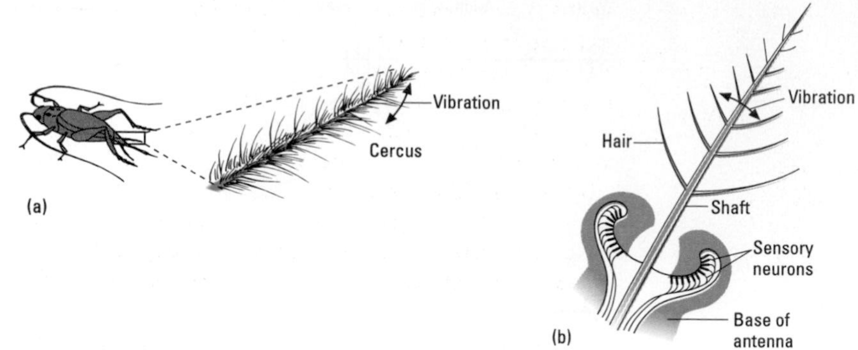

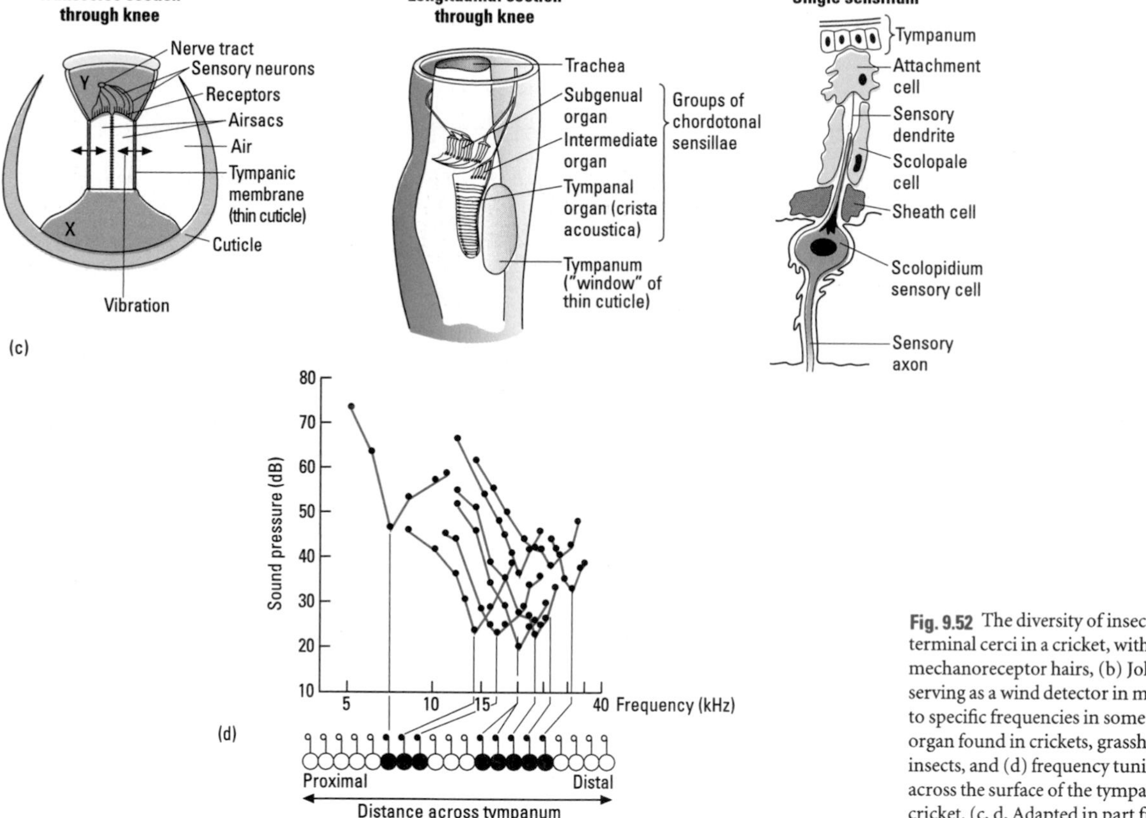

Fig. 9.52 The diversity of insect "ears": (a) the terminal cerci in a cricket, with many filiform mechanoreceptor hairs, (b) Johnston's organ, serving as a wind detector in many insects but tuned to specific frequencies in some flies, (c) the tympanic organ found in crickets, grasshoppers, and other insects, and (d) frequency tuning in scolopidia across the surface of the tympanic organ of a bush cricket. (c, d, Adapted in part from Withers 1922.)

messengers are the triggers for ion channel conformation changes. In at least some cases PKA and PKC are also produced by the internal biochemical cascade, and act to desensitize the receptor to the continued presence of an odor molecule by phosphorylating the receptor, thus providing a mechanism of sensory adaptation.

For aquatic animals there is really only one true chemical sense, and animals are often said to "taste" the water, having chemoreceptors of high sensitivity detecting chemicals from near and far (most spectacularly seen in the "homing" ability of fish such as salmon). But for land animals smell (olfaction) and taste (gustation) are clearly rather separate, sometimes termed contact and distance chemoreception, respectively (although the use of the term ORC is still common). Smell may be used to detect food, enemies, or mates, while taste is entirely related to checking food. Olfactory receptors respond to volatile chemicals, arriving on the wind from some distance away, while gustatory receptors require chemicals to be dissolved in fluid, usually in the mouthparts, and are generally orders of magnitude less sensitive. Vertebrates can generally taste a few molecules such as quinine at concentrations as low as 10^{-7} M, but (for members of the dog family and other taxa that rely heavily on scent) can smell distinctive odors that are 1 billion-fold more dilute than this (10^{-16}–10^{-17} M), equating to a response to a single molecule of the odor.

Chemoreceptors can occur in various regions of the body apart from the head area. Some fish have them all over the skin, especially on the gills and fins, and arthropods commonly have them on the

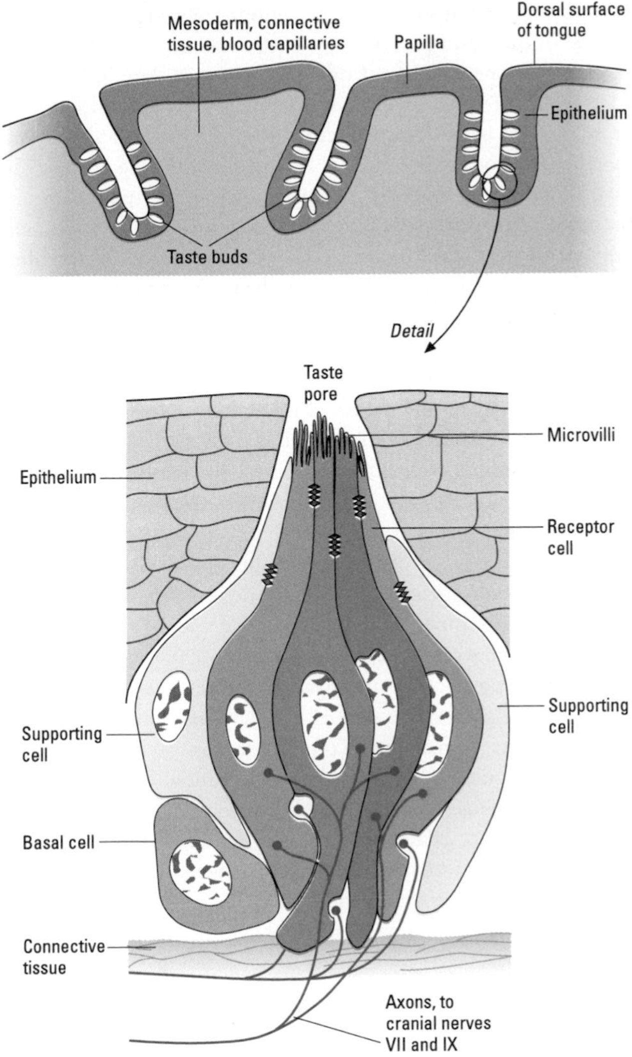

Fig. 9.53 The taste bud of a vertebrate, sited in crypts within the tongue.

feet and often also on the genitalia. But it is in terrestrial animals that the chemosensory system tends to be concentrated in a specific anterior organ (the nose and mouth in tetrapods, the antennae and mouthparts in insects), because on land olfaction can be a really important source of information.

Vertebrate taste is mediated mainly by receptors in the tongue, where each **taste bud** (Fig. 9.53) contains a few axonless receptor cells with an exposed microvillar tip, and a few supporting cells. Salt and sour receptors involve direct transduction via ion channels, whilst bitter and sweet stimuli provoke second messenger signaling cascades. Specific mechanisms also exist for the detection of glutamate and fatty acids. All the receptors synapse just below the tongue surface, and the afferent fibers then project via cranial nerves VII and IX to nuclei in the medulla and then the pons before reaching the cortex. Olfaction in these animals is via rather similar simple but ungrouped receptor cells protruding into the mucus of the **nasal epithelium** (and also in a special **vomeronasal organ** with a concentrated olfactory epithelium); these cells are associated with odorant-binding proteins in the lipocalin family. The scent receptors send their own axons to the **olfactory bulb**, just above the nose, which is a forward projection of the brain itself, and is massive in fish and some lower tetrapods. Here the interneurons are arranged in neatly organized packages termed olfactory **glomeruli**, each dealing with a particular set of inputs. The olfactory tract then projects straight to the cortex, without any mediation in the thalamus as with all other senses. It is possible that each olfactory ending contains just one specific receptor protein type, responding to one specific odor, since these particular receptor proteins belong to an especially diverse group with at least 1000 different versions expressed at the gene level.

In insects the chemoreceptors have to project through a cuticle-covered surface, and are therefore housed in cuticular bristles with tiny pores at the tip (Fig. 9.54). Contact (taste) receptors usually have just one terminal pore, whilst distance (olfactory) receptors have many pores. Within the liquid-filled shaft of the bristle are a few dendrites, often just four or five but occasionally up to 50, their tips protected by a plug of viscous mucopolysaccharide. Each neuron has a specific role; commonly there is one for salts, one for sugars, perhaps one for amino acids, even (perhaps) one for water in some flies. There may also be one or more with more specific functions, bearing receptors for chemicals associated with the diet; in herbivorous insects this normally means the host plant volatiles. Some neurons may respond to many such volatiles (i.e. they are generalists), but in insects that are monophagous feeders on just one or a few host plants there are often neurons that only fire in response to a specific chemical characteristic of that plant (for example there is a sinigrin or "mustard oil" receptor in cabbage-seeking *Pieris* butterflies; see Chapter 15), and in some dung and blood feeders there are specific responses to components of feces and to ammonia. Similarly, specialist neurons may be triggered only by the chemical(s) that act as the species' sex pheromone, so that in the silk-moth the male has receptors only sensitive to bombykol (see Chapter 10). Male moth antennae are often extremely large and feathery, and each may bear 70,000–100,000 receptors for the females' scent. Perhaps surprisingly, the olfactory receptor proteins in insects have very little sequence homology among themselves, or with the equivalent proteins in vertebrates or other phyla; there is also marked dissimilarity of odorant-binding proteins across taxa.

Moth antennae are convenient to record from with simple electrodes, and studies of the resultant "electroantennogram" outputs, together with mapping of the projections of the olfactory neurons into the relatively simple brain, have been crucial to understanding insect chemoreception. The axons from the receptors project to glomeruli in the brain's antennal lobes, which are remarkably similar in architecture to the glomeruli of the vertebrate olfactory bulbs, and where different types of axons segregate to specific glomeruli, so that individual glomeruli process information about particular scents. Information from the glomeruli is passed on to the forebrain and especially to olfactory neuropil in areas termed "mushroom bodies", implicated in olfactory learning and memory.

9.8.4 Vision: photoreceptors

Photoreception involves the transduction of light energy into electrical energy, and is a widespread phenomenon found in non-specialized cells (including ordinary neurons), very often giving a

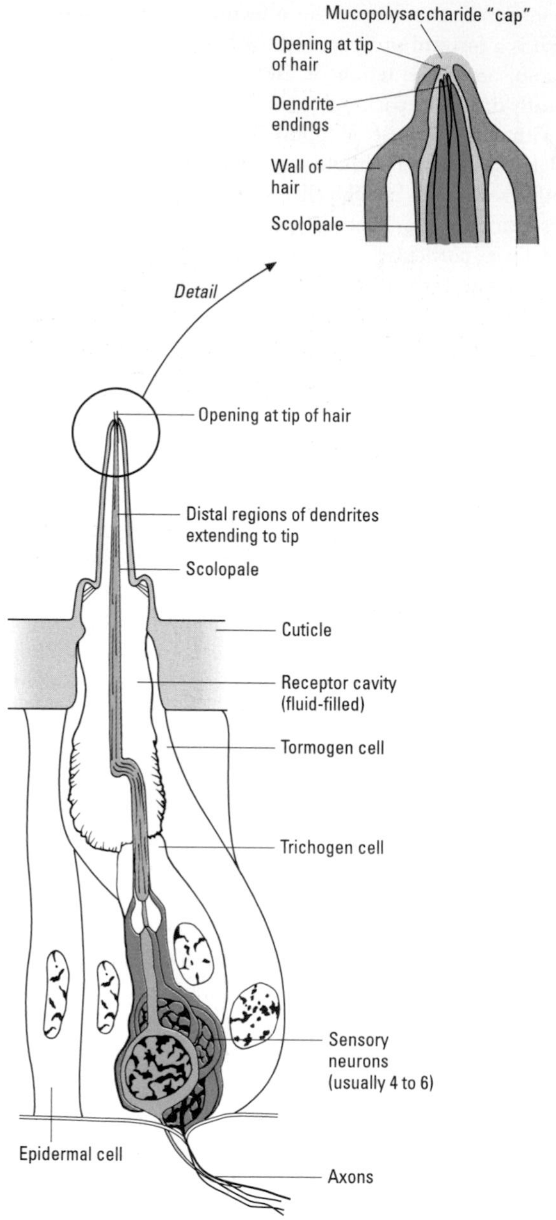

Fig. 9.54 A typical insect chemoreceptor, in this case a contact (taste) receptor with a single terminal pore (olfactory receptors normally have many such pores). The trichogen cell secretes the bristle and the thin protective scolopale (see inset), and the tormogen cell secretes most of the socket in which the receptor sits. (Adapted from Chapman 1998, and other sources.)

"dermal light sense" in animals sufficient to distinguish day and night and measure photoperiod. But in specific localized photoreceptors it is achieved via the absorption of photons of light by pigment molecules, which are membrane bound and composed of a **chromophore** group covalently bound to a lysine residue on the protein **opsin**, along with two polysaccharide chains. Opsins are ~40 kDa members of the same G-protein-coupled receptor family as the β-adrenergic and muscarinic membrane receptor proteins, with seven transmembrane α-helix segments (Fig. 9.55b; see also Fig. 2.17). When a photon is absorbed, the chromophore changes from one isomeric form to another, causing rapid changes in the opsin configuration so that the chromophore is expelled from the middle of the opsin protein, on which an enzyme-binding site is revealed. The production of this photo-excited state in the opsin takes less than 1 ms. The enzyme-binding site then links to a G-protein (here termed **transducin**), leading to a typical enzymic cascade via phosphodiesterase (PDE) and cGMP (Fig. 9.55c). The cascade is shut off again by a light-dependent phosphorylation at the opsin C-terminal, and the chromophore is converted back to the original isomer, which reassociates with the opsin. The eventual effect is a change in an ion channel in the receptor membrane, with cGMP being the ligand for a gated sodium channel in most eyes (but the effect differs in different animals). In invertebrates the cells usually depolarize, due to increased conductance as channels open in response to light (and mediated via a phospholipase rather than PDE). However, in vertebrates the effect is a hyperpolarization arising from a sodium conductance decrease as channels close. Most photoreceptors also have a particularly strong Na^+/K^+-ATPase activity, so that they can maintain the normal ion levels despite fairly continuous fluxes of ions in daylight.

A single quantum of light can be detected in many photoreceptors; for example, in the well-studied eyes of the horseshoe crab *Limulus*, one photon can open at least 1000 sodium channels. However, photoreceptors can also adapt to a very wide range of light inputs, allowing animal to see well both at dusk and at midday (at least 12 orders of magnitude of light intensity). **Photoreceptor adaptation** requires a change in the baseline number of open channels, so that changes in light are reflected in changes in membrane potential. Without adaptation, animals would become effectively blind temporarily when the light level changed. For example, in a relatively low light environment most of the sodium channels in a vertebrate are open, but a sudden bright light will close them all, and then further small changes in light are not detectable. Adaptation is best understood in vertebrate photoreceptors, where it seems to be mainly due to calcium effects (Fig. 9.56). A small amount of Ca^{2+} enters through the sodium channels, and within the receptor this calcium inhibits the synthesis of cGMP. In a dark-adapted eye with many channels open, calcium rises internally and keeps the cGMP level relatively low, so that even a moderate light will activate enough opsin to use up all the cGMP and close all the sodium channels, giving a maximum hyperpolarization. Further light input has no effect. However, over a few minutes the eye adapts; because the sodium channels are now closed, the free Ca^{2+} in the receptor declines, taking the brake off cGMP synthesis. As cGMP builds up some of the sodium channels reopen, restoring the membrane potential and allowing further responsivity to light inputs.

The overall **spectral sensitivities** of visual pigments are principally determined by the character of the chromophore, although specific and potentially adaptive sensitivities are also achieved by the expression of different opsin genes. The commonest chromophore is an aldehyde derivative of vitamin A, particularly 11-*cis*-retinal, which results in the classic visual pigment **rhodopsin**, with maximal absorption at about 500 nm. Rhodopsins vary (largely in their amino acid sequences) between and within species, many animals having more than one version. Where there is **color vision**, the different versions of the pigment conferring different spectral sensitivities are often termed **iodopsins**. Color vision is quite common, and may have originated from the ability to discriminate light with

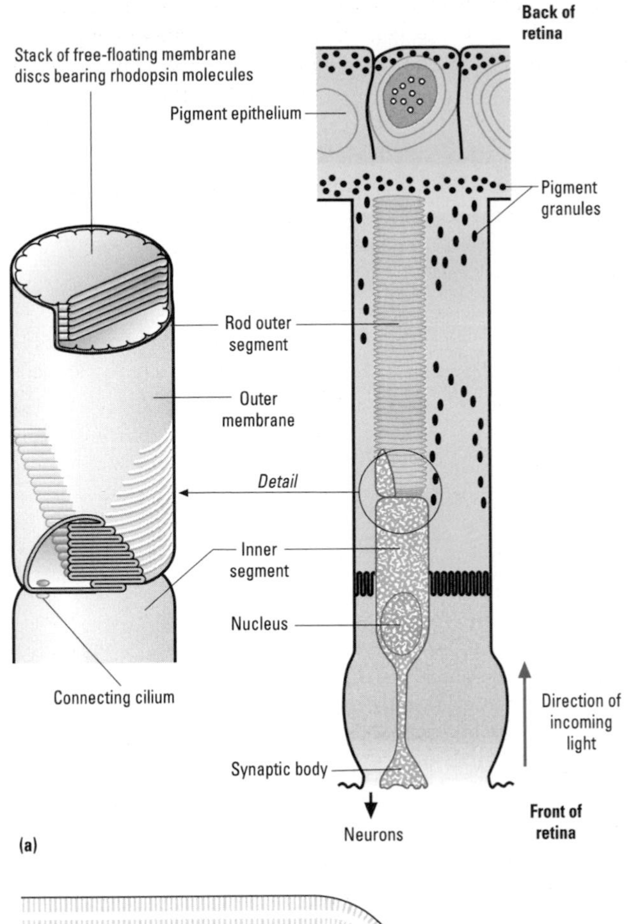

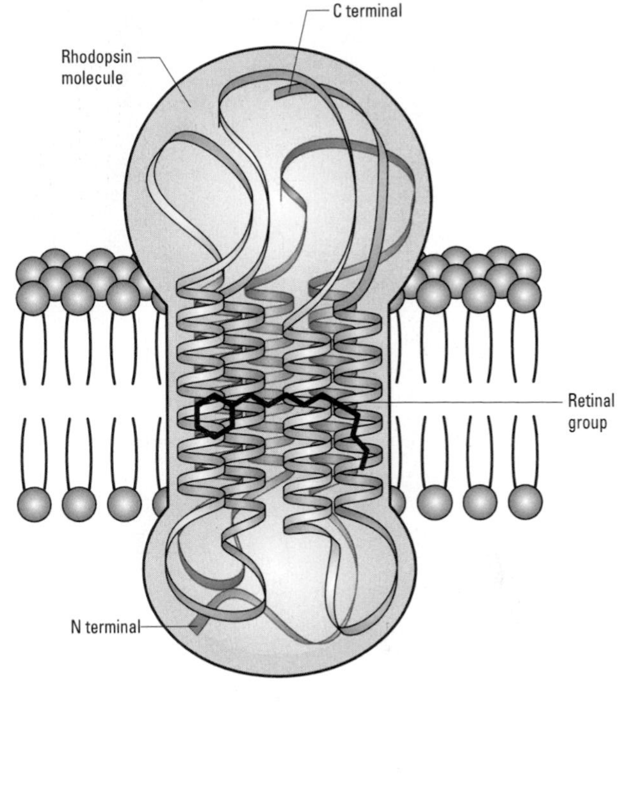

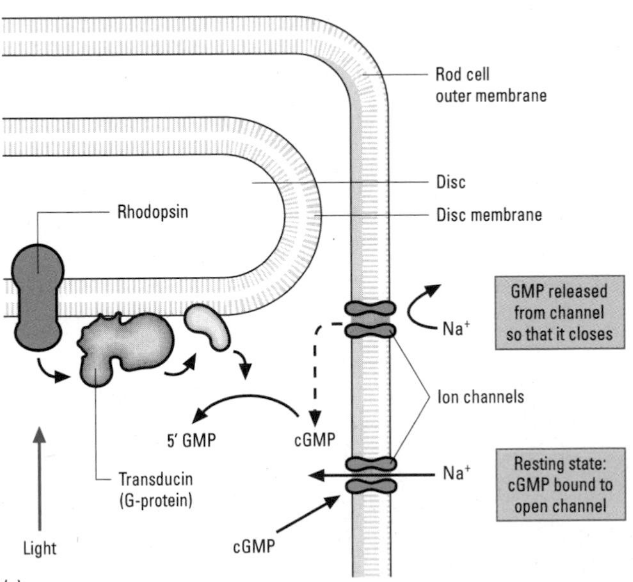

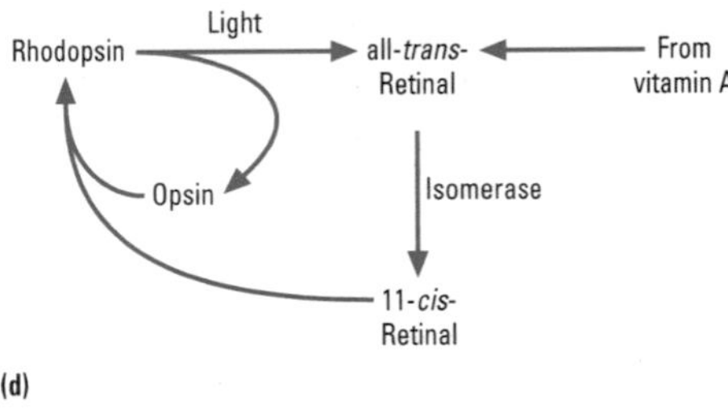

Fig. 9.55 (a) Structure of a vertebrate "rod" photoreceptor, embedded in the pigment epithelium of the retina. (b) Structure of rhodopsin. (c) The basic photoreception cascade in a rod, where rhodopsin is located in internal membrane discs. (d) The isomeric changes in the retinal chromophore when light is received. (b, Adapted from Reichert 1992, after Baehr & Applebury, and other sources.)

a high UV content (shorter wavelength, incoming from the sun, and in "open spaces") from low UV light (longer wavelength, as reflected from surfaces, when looking "down" at the substratum, or when enclosed in a complex habitat). In modern animals color vision is commonly trichromatic, involving blue, green, and red receptors in some fish reptiles, birds, and primates (though many mammals have only dichromatic and rather poor color vision). Each color

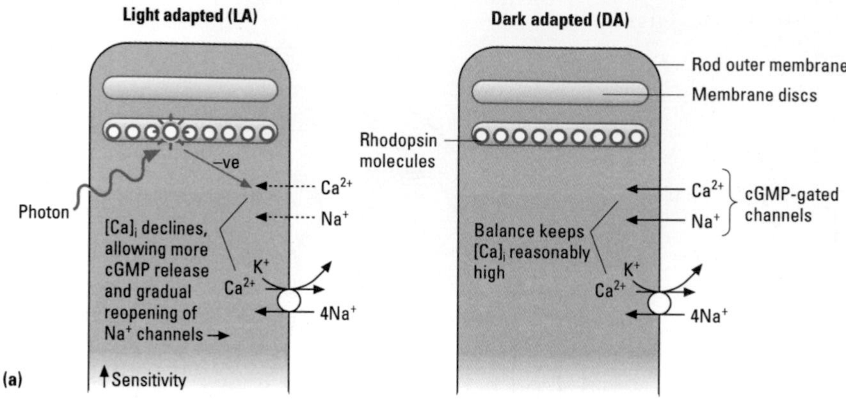

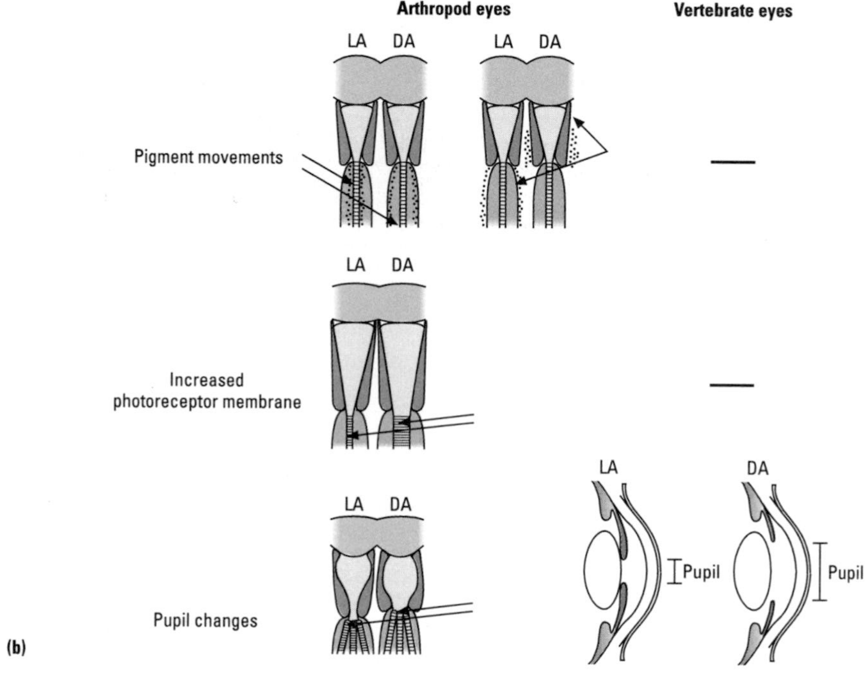

Fig. 9.56 Adaptation to maintain sensitivity at different light levels in eyes. (a) Adaptation at the molecular level in a rod photoreceptor: in the dark, the cGMP sodium channels are open and also allow some calcium entry, partly offset by a cation exchange system (4 Na^+ ions in exchange for 1 Ca^+ and 1 K^+ ion) keeping the internal calcium slightly raised overall. In the light the cation channels are initially mainly closed but the exchange is still operating, so intracellular calcium declines, cGMP levels rise, and some of the channels can reopen, restoring better sensitivity to additional light inputs. (b) The mechanisms of adaptation at the level of morphology, either by pigment movements or by altering the amount of photoreceptive membrane, or by altering the aperture (pupil) controlling light entry. (b, Adapted in part from Land & Nilsson 2002.)

is detected by the proportion of each of the types of cells that it stimulates; for example, orange light at 580 nm will stimulate red receptors 99% maximally, green ones by 42%, and blue ones not at all. Additional iodopsin types are occasionally found in vertebrates, and most diurnal birds have four or even five types. Arthropods can also detect color in a similar fashion, localized in individual cells, and honeybees (*Apis*) are probably the best known, with trichromatic peak responses in the green (540 nm), blue (435 nm), and ultraviolet (UV, 335 nm) ranges. Some butterflies are pentachromatic, and one mantis shrimp is known to have at least 10 different color photoreceptor types.

Photoreceptors are the elements that respond to light, and are nearly always specialized to have a large membrane surface with either a ciliated or a microvillar surface. If localized in groups they are termed **eyes**, giving some spatial and directional information about light. The most primitive animal eyes are merely small patches of photoreceptors, usually backed by a pigmented epithelium. Slightly more sophisticated organs occur when the sheet or receptor is invaginated to form a cup with a narrowed aperture, improving directionality. Adding a clear cornea, and better still a lens, with an orderly array of receptors internally, allows many animals to do better than merely receive and respond to light: they can form an image, in a structure that is then more recognizably a real eye. Image-forming eyes are confined to annelids, molluscs, and the various groups of arthropods, as well as vertebrates; the major optical types are summarized in Fig. 9.57. They can be divided into two main types: **simple eyes** (or vesicular eyes) with one lens directing all light to a common set of photoreceptors, and **compound eyes** (sometimes termed convex eyes) with multiple lenses each directing light to its own photoreceptors, found mostly in adult arthropods but also in some molluscs and annelids (Table 9.7). In either case the lenses used to direct the light by refraction can sometimes be replaced with mirrors to reflect it accurately onto the receptors. There is no doubt that eyes have evolved separately many times over (at least twice within each of these two "types"), even though the opsin photopigment seems to be homologous throughout all

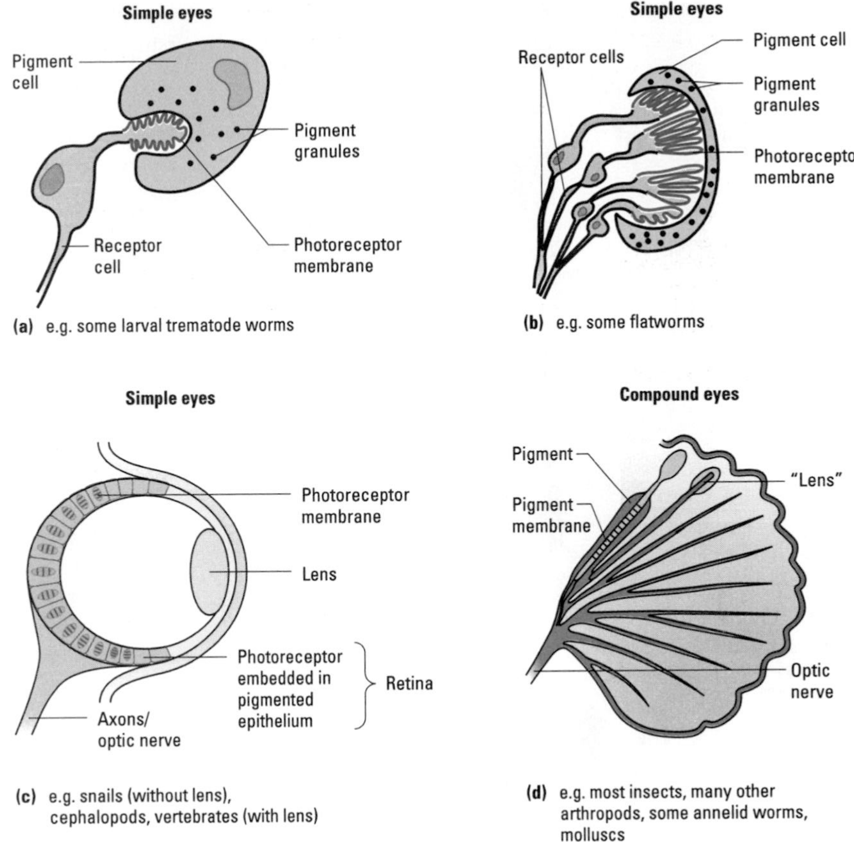

Fig. 9.57 The main eye types found in animals: (a) simple eye spots with a single receptor, (b) simple cup-shaped (pinhole) eyes with a few photoreceptors, (c) simple vesicular or chambered eyes, with an internal retina of photoreceptors and usually a lens, and (d) compound (convex) eyes, with the "retina" of photoreceptors external and in a convex array, often forming separate images.

Table 9.7 Major types of eyes in animals.

Photoreceptors with cilia	
Eye pits and pinhole eyes	Planarians
	Annelids and molluscs
	Nautilus
Simple chambered eyes	Vertebrates
	Some jellyfish
Compound eyes	Some clams
	Some tubeworms
Photoreceptors with microvilli (rhabdoms)	
Simple chambered eyes	
With lenses	Cephalopods
	Some snails
	Some insect larva
	Spiders
With mirrors	Scallops
Compound eyes	
With lenses	Trilobites
	Adult insects
	Crustaceans
With mirrors	Shrimps and lobsters

animals and the *Pax-6* gene that determines eye position and formation is also fairly widespread.

Simple eyes

Vertebrate eyes are the best known example of "simple" eyes, although unusual in their complexity. They have a protective cornea, a focusing lens, and an iris to control the amount of light entering. Many muscles act on the lens shape to alter its focus, on the surrounding iris to alter the pupil size (in humans the diameter can change four-fold, and thus the light entering the eye changes 16-fold), and on the whole eyeball to move the visual field. The visual pigments that transduce light energy into a nervous impulse are housed in a layer of photoreceptor cells forming a **retina** at the back of the roughly spherical eyeball, so that the image formed is inverted and left–right switched (see Fig. 9.59b). The photoreceptive cells in the retina are termed **rods** (by far the majority), or **cones** where they give color vision by having slightly different opsins with their differential wavelength sensitivities. A typical mammalian eye has about 100 million rods and 6 million cones, most densely packed at the central **fovea** where visual acuity is highest. Both rods and cones are elongate cells ($\sim 2 \times 300-600\ \mu m$) with inner and outer segments (Fig. 9.58). Transduction takes place in the outer segments, where the photopigment is bound to stacked discs of membranes (sometimes called a rhabdom). Oddly, though, these outer segments are not the first tissue encountered by incoming light; the processing neurons lie above the photoreceptors, closer to the light source.

Vertebrate **retinas** contain six main types of cell: the photoreceptors, some supporting pigmented cells that absorb stray light, plus four principal types of processing neuron interacting laterally and in layers (Fig. 9.59a). **Horizontal cells** only connect laterally between adjacent rods and cones, and have no onward connections. **Bipolar cells** link the photoreceptors to ganglion cells, but these links are

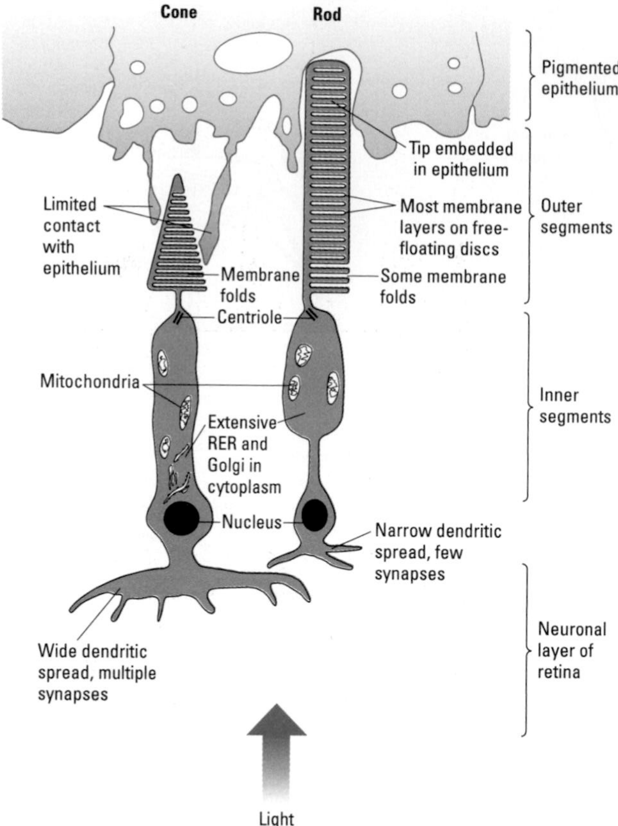

Fig. 9.58 Schematic typical structures of rods and cones from mammalian eyes. RER, rough endoplasmic reticulum.

strongly modulated by the **amacrine cells**. Only the final layer, the **ganglion cells**, have axons that project out of the eye via the optic nerve. Unlike any other sensory system, then, the vertebrate eye performs considerable peripheral processing before passing signals on to the CNS. The receptive fields of rods and cones are circular, but those of the crucial bipolar cells have two "rings", a central zone where light causes a response and a concentric surrounding ring where light has exactly the opposite effect. They may be either "off-center" or "on-center", respectively, according to whether the central region is inhibited (hyperpolarized) by light or stimulated by it. The difference in the central area response arises from variation in the effect of the neurotransmitter (glutamate) on receptor proteins in the bipolar cells; the difference in the surrounding area is largely due to the activity of the horizontal cells.

Ganglion cells are by necessity therefore also either off-center or on-center, according to their inputs from the bipolar cells, but their responses are shaped further by inputs from more than one bipolar cell and from the amacrine cells. They may have large or small receptive fields, altering visual acuity. **Small field cells** also respond to sustained stimulation (and in primates to color), while **large field cells** (Y cells) are less common but strongly phasic, so respond well to moving stimuli. Thus while the photoreceptors themselves are fantastically sensitive to light intensity but very poor at detecting contrast, pattern, or movement, by the time we reach the ganglion cells all these factors have been coded in to the signal that passes up the optic nerve. Much of this is achieved by convergence: in the cat retina, 1500 photoreceptors converge via about 100 bipolar cells on to a single ganglion cell projecting to the brain.

From the ganglion cells, visual information passes first to the lateral geniculate nuclei (LGN) of the thalamus, the main relay point before the visual cortex. In many mammals, some axons project to their own half of the brain (ipsilateral) and some cross to the other side (contralateral). For an animal with forward-looking eyes, the common arrangement is as in Fig. 9.59b, insuring that stimuli from the left visual field, seen by the right sides of each retina, project ultimately to the right side of the brain. In the LGN the ganglion cells' spatial patterning is maintained (a topographic representation) via a hierarchical series of six layers of neurons, and there seems to be little further processing; however a large part of the input to the LGN is "downwards" from the cortex, and there may be important modulation relating to "paying attention" to particular kinds of stimuli. Higher level processing is considered in section 9.8.8 below.

Simple eyes also occur in some invertebrate groups, from flatworms to cephalopods, and including most larval insects where they are termed **stemmata** (image forming) or **ocelli** (nonimage forming). In the squid and octopus the eyes are remarkably similar to those of vertebrates, though quite independently evolved. The retina contains only rod-like cells (with no color vision), but has a similar density of photoreceptors (20,000–50,000 mm^{-2}) and also a fovea of maximum sensitivity. Like vertebrates, the lens can move within the eye, and the eyes within the head. However, the receptors are pointed outwards, towards the lens and the outside world, with the relatively fewer interneuron cells and connections on the inside (an apparently more logical arrangement than in vertebrates).

Compound eyes

These are the main image-forming eyes in some immature insects and nearly all adult insects. They are formed from a large number of separate units called **ommatidia**, visible externally as distinct hexagonal facets on the surface of the eye. In fast predatory day-flying dragonflies there may be 10,000 ommatidia per eye, although in some ants there is only one! Each ommatidium consists of several parts (Fig. 9.60): a peripheral biconvex "**lens**" of cuticle, then a wedge of transparent material called the **crystalline cone** (which also helps to focus the incoming light), then a group of sensory neurons forming the **retinula**, typically made of eight sensory cells (though arrangements with 6–9 are possible), each contributing stacked membranes to a central area termed the **rhabdom**, where the pigment is housed (and which can be seen to be constituted from eight separate "rhabdomeres" of membrane, from the eight retinular cells; Fig. 9.60b,c). The rhabdoms of adjacent ommatidia are normally shielded from each other's incoming light by a layer of pigment, and an upright (i.e. noninverted) image is formed. The relative arrangement of rhabdom and pigment determines whether the eye works best at low or high light intensity. The pattern of microvilli stacking in the rhabdom is also important; many insect eyes can detect the plane of polarization of light (thus being able to navigate by the sun even when it is cloudy) because of the highly ordered directionality of the visual pigment on the membrane stacks in individual rhabdomeres, which can be precisely at right angles to each other.

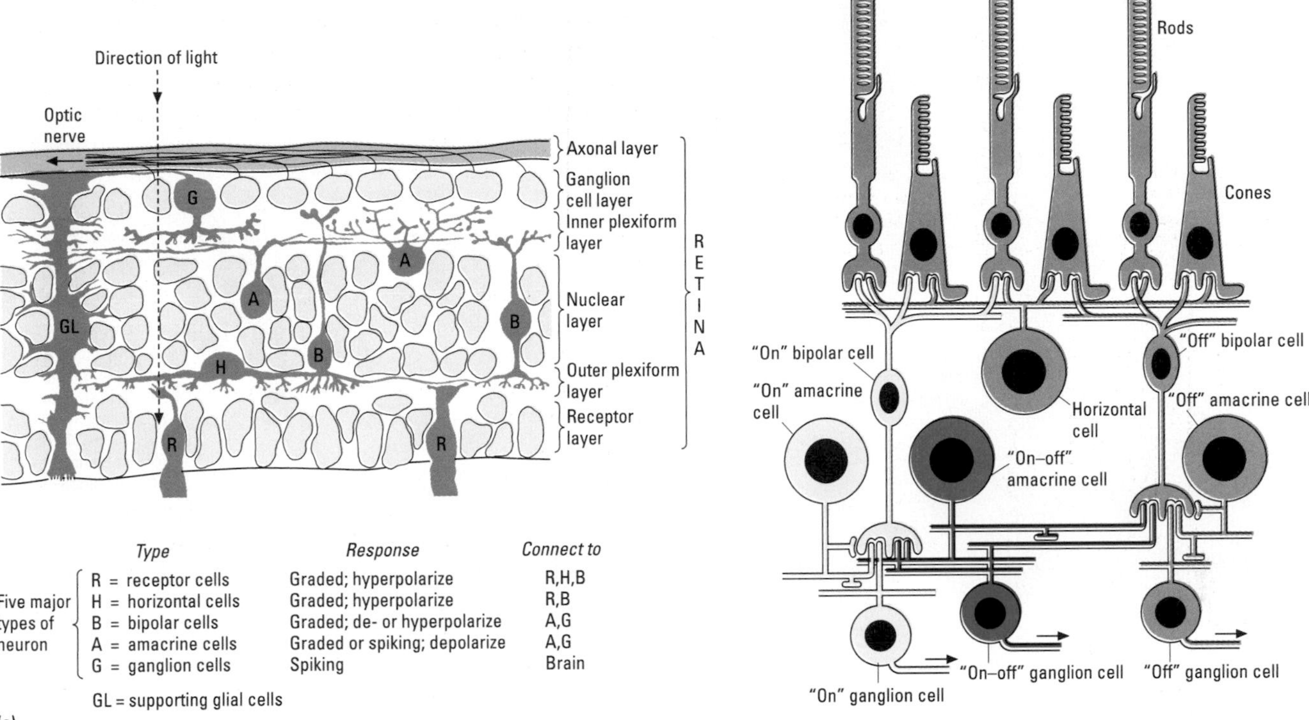

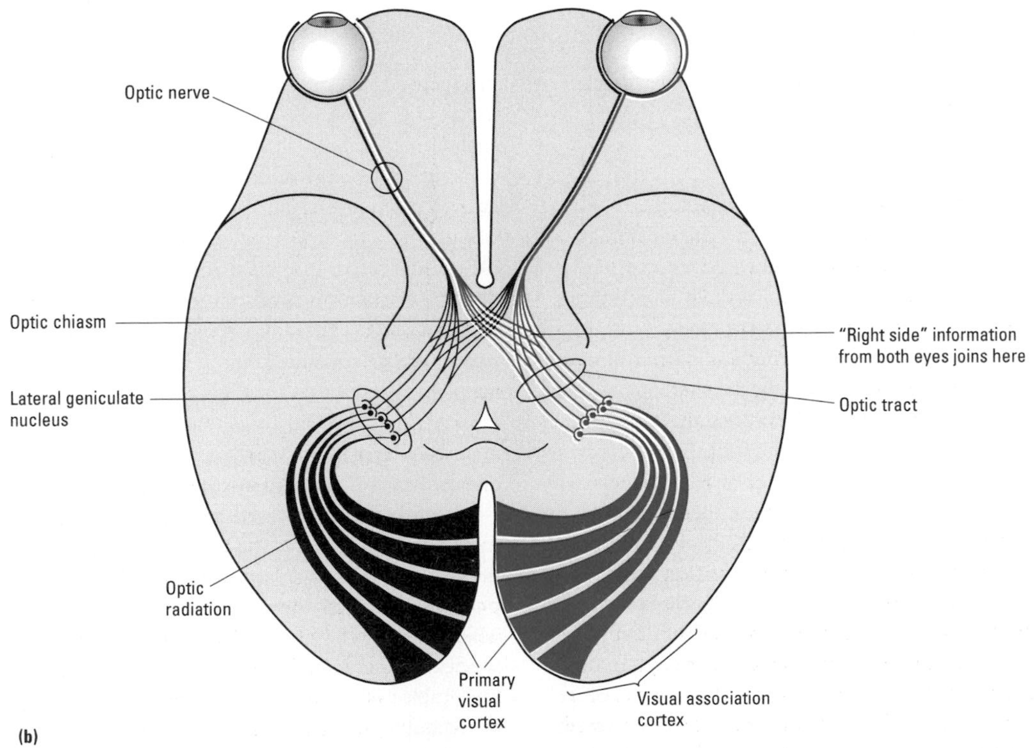

Fig. 9.59 The neural arrangements in vertebrate visual systems. (a) The five main neuronal cell types of the retina, and their typical connectivity, with a more diagrammatic representation to the right. (b) The path through the eyes to the brain in a mammal, allowing a topical representation and binocularity with left and right halves of the visual field combining.

Insect eyes contain rhodopsin or closely related pigments, and transduction operates very similarly to vertebrates, with cGMP as an important intermediate. However, IP_3 is also implicated, bringing about the release of Ca^{2+} from intracellular stores and thus providing an alternative way of regulating ion channels and the

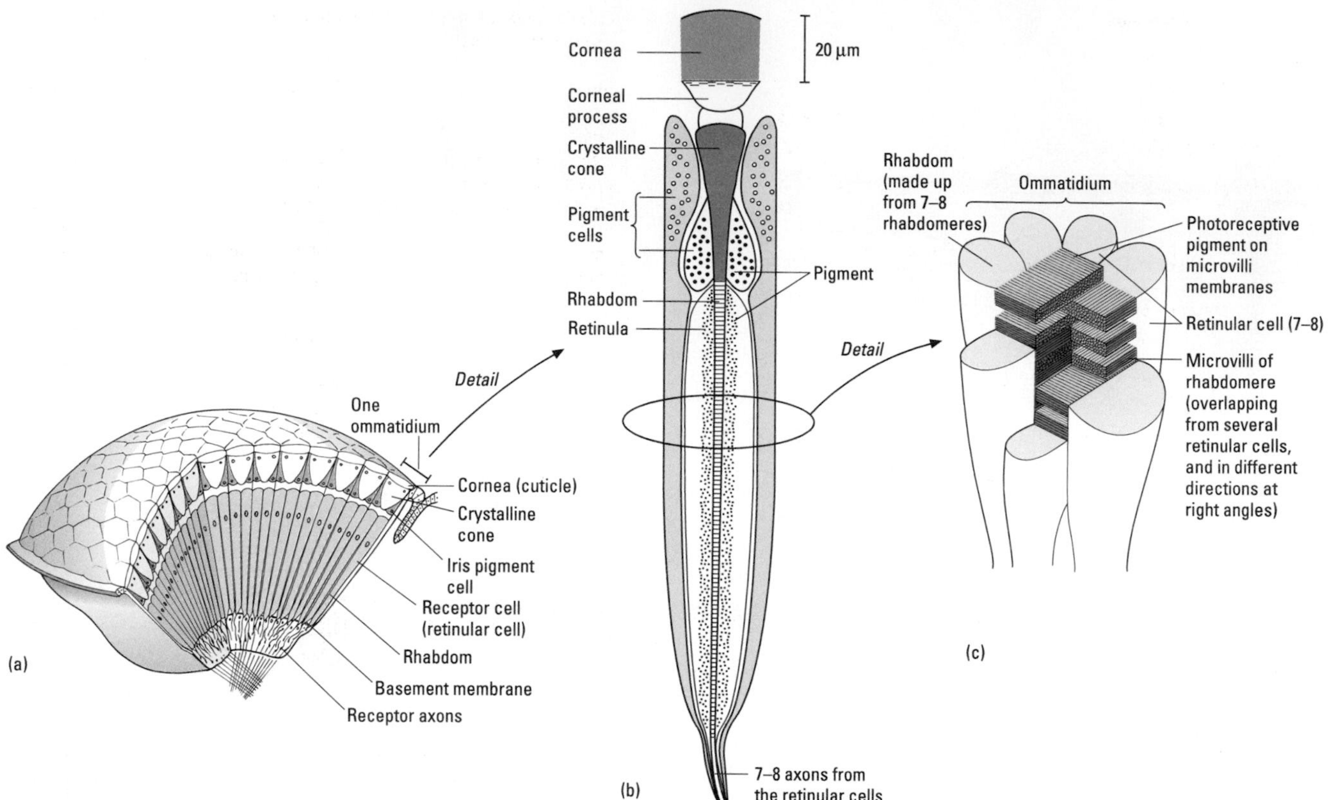

Fig. 9.60 Compound eye structure: (a) an overview, (b) a single ommatidium; (c) a transverse section of an ommatidium, showing the central rhabdom made up from overlapping microvilli from different retinular cells. (Adapted from Land & Nilsson 2002.)

polarization of the receptor membrane. And remember that, as in other invertebrate eyes, the end product of light reception is a depolarization, due to the opening of ion channels, in contrast to the hyperpolarizing response seen in vertebrate retinas.

The number, size, and spacing of the ommatidia determine the acuity of vision in any given insect eye. In the zones where vision is best (equivalent to a vertebrate fovea), the ommatidial surfaces are flat and large; elsewhere they are smaller and quite convex. A large flat surface collects more light from a particular part of the visual field and this improves resolution. In addition, there are different kinds of insect eye depending on how much the adjacent ommatidia interact with each other (Fig. 9.61). In **"apposition" eyes** each ommatidium acts as a single unit, responding to the light from the small arc at its own surface and forming its own tiny image. These occur in diurnal insect species, many of which also have an "iris" formed by pigment just below the cone to give adaptation to different light levels (see Fig. 9.56b). In **"superposition" eyes**, adjacent ommatidia are functionally overlapping, giving a single deeper lying image; this is common in crepuscular and nocturnal insects and deeper water crustaceans. The functional overlap ("superimposing") is achieved in two fashions. In some cases, the retinular cells in each rhabdom are slightly separated, and those in adjacent ommatidia that are in the same relative position (and thus "seeing" the same point) send convergent signals to the next layer of neurons, producing a "neurally superimposed" image. In other cases the retinular cells within one ommatidium remain closely linked but the rhabdom as a whole is rather long, having a clear zone between the cones and the receptive areas, each of which therefore receives light laterally from several cones and forms an "optically superimposed" image. Up and down movements of the pigment between the ommatidia determine how many cones send light to each rhabdom, giving a mechanism of adaptation in the nocturnal beetles, moths, and flies that use these superposition eyes (see Fig. 9.56). There are many further specializations possible, with mirrors replacing or supplementing lenses and with regional variations across the eyes. However, it should be noted that at best an insect with really good apposition eyes has a resolving power of only 1°, whereas a human eye resolves at 1′ and a bird of prey at perhaps 20″.

Whatever their type and origin, animal eye designs and sensitivities are strongly linked to the habitat in which they must perform. In deep-sea animals their sensitivity is high (i.e. they capture many photons per receptor when working at a standard level of radiance), whilst in midwater and coastal species and in terrestrial nocturnal species it is moderate, and for terrestrial diurnal species extremely low (Table 9.8), so that the overall differences cover five or six orders of magnitude. Deep-sea and nocturnal animals need even better sensitivity, which requires larger eyes (with large apertures and wide receptors) so it is not surprising that the largest recorded eye (40 cm diameter) comes from a deep-sea squid. This optical strategy to maximize photon capture can be assisted by spatial and temporal summation, but always at the cost of resolution. However, bigger eyes are not always associated with poor light and high sensitivity; they can also allow for greatly increased numbers of very narrow

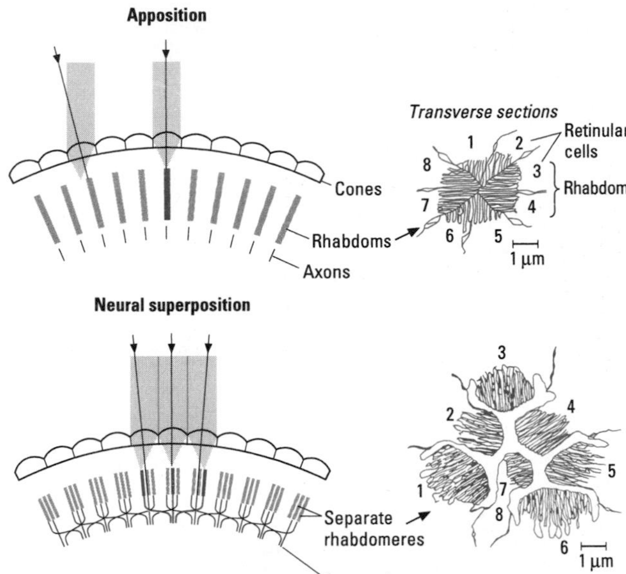

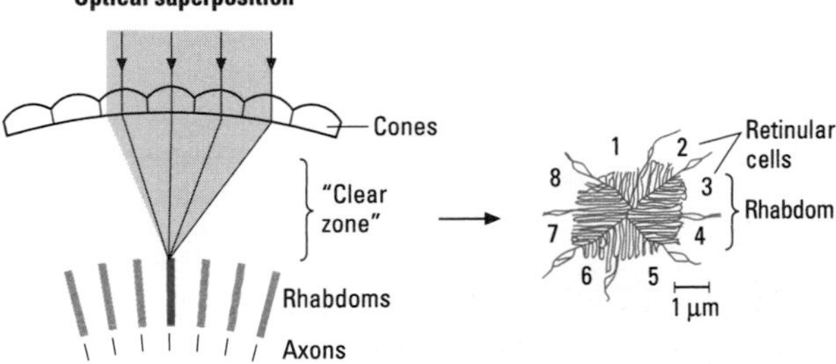

Fig. 9.61 A comparison of apposition and superposition compound eyes. In the apposition eye each rhabdom receives light from a different direction (shown by the arrows), and the light from any one direction (shaded) only hits one rhabdom. In transverse section, the rhabdom is made up from rhabdomeres from eight different receptors fused together and acting as a single light guide (although different rhabdomeres may be sensitive to different wavelengths). In the neural superposition eye, light from one direction hits several rhabdoms in neighboring ommatidia; here the rhabdomeres are spatially separate in transverse section, and the axons from each one go to different parts of the synaptic layer (the lamina) behind the eye (where axons from all the receptors receiving light from the *same* spot converge together). In the optical superposition eye the clear zone between the cones and rhabdoms allows light from several cones to converge on a single rhabdom. (Adapted from Land & Nilsson 2002, and other sources.)

receptors for high resolution and acuity, so terrestrial visually hunting predators (owls at dusk, eagles or dragonflies in daylight) also have relatively large eyes.

9.8.5 Thermoreceptors and hygroreceptors

Many sensory receptors respond to changes in temperature, with either a decrease or increase of firing frequency, and this in itself may convey some useful information to an animal. Indeed many neurons are inherently thermosensitive, and most hypothalamic neurons in mammals show this trait especially strongly. However, since temperature is so critical to proper cellular and physiological functioning, as we saw in Chapter 8, most animals also have specific

Table 9.8 Sensitivities (in relative units) of eyes in relation to habitat. (Data from Land & Nilsson 2002.)

Animal	Habitat	Sensitivity, *S*
Marine		
Isopod crustacean	Deep sea	4200
Decapod crustacean	Deep sea	3300
Amphipod crustacean	Midwater	38–120
Horseshoe crab	Coastal	83–317
Scallop	Sublittoral, benthic	4
Shore crab	Littoral, diurnal	0.5
Terrestrial		
Spider	Nocturnal	100
Moth	Crepuscular	38
Beetle	Crepuscular	31
Toad	Mostly diurnal	4
Beetle	Diurnal	0.35
Honey bee	Diurnal	0.32
Spider	Diurnal	0.04
Man (fovea)	Diurnal	0.01

sensory neurons responding to hot and cold, whether dispersed in the skin, concentrated in specific sense organs, or at internal sites. These are simple free nerve endings, with no special morphological features, and cannot easily be distinguished from the basic kinds of mechanoreceptors. However, their main responsiveness is to any *change* of temperature; that is, they always respond to relative, not to absolute, temperature. This may be mediated mainly by temperature-dependent changes in ion channel permeability, and perhaps modified by temperature-dependent changes in sodium pump activity. In vertebrates, and probably in all animals, there are separate detectors for rising ("hot") and for falling ("cold") temperatures, with response ranges that hardly overlap at all (Fig. 9.62a). Vertebrates have such detectors in the dorsal root ganglia and in the hypothalamus, and their sensitivity can be turned up by various factors, notably prostaglandins. For the cat, the cold receptor cells have axons 1–5 µm in diameter, myelinated, and with a conduction velocity of 5–30 m s^{-1}, and they show maximal firing at 25°C; when warmed, their firing frequency decreases. In contrast, the hot receptors are nonmyelinated, less than 2 µm across and conducting at less than 2 m s^{-1}, with maximum firing at 45°C; when warmed they show increased rates of depolarization in the receptor potential leading to faster firing rates. This is apparently mainly due to effects on a mechanically gated potassium channel, TREK-1, which is opened gradually and reversibly by heat. Since it is expressed in both central and peripheral mammalian sensory neurons this channel may be a true physiological thermoreceptor.

For both types of thermal neuron, the sensitivity can be exceptional, some neurons responding to a change of only 0.01°C. However both types also conduct much more slowly than touch receptor axons, so that humans and other animals are aware of contact with an object before they are aware of it being excessively hot or cold.

More specialized thermoreceptors are uncommon, but do occur in some invertebrates where nerve endings are sensitive to infrared radiation (acting much like a simple eye, but at much longer wavelength). Notable examples are the fire beetles such as *Melanophila*, where small pit organs containing 50–100 sensilla and sited on the legs give notice at up to 50 km of fires, on which the beetles depend to locate fresh burnt wood as mating and oviposition sites.

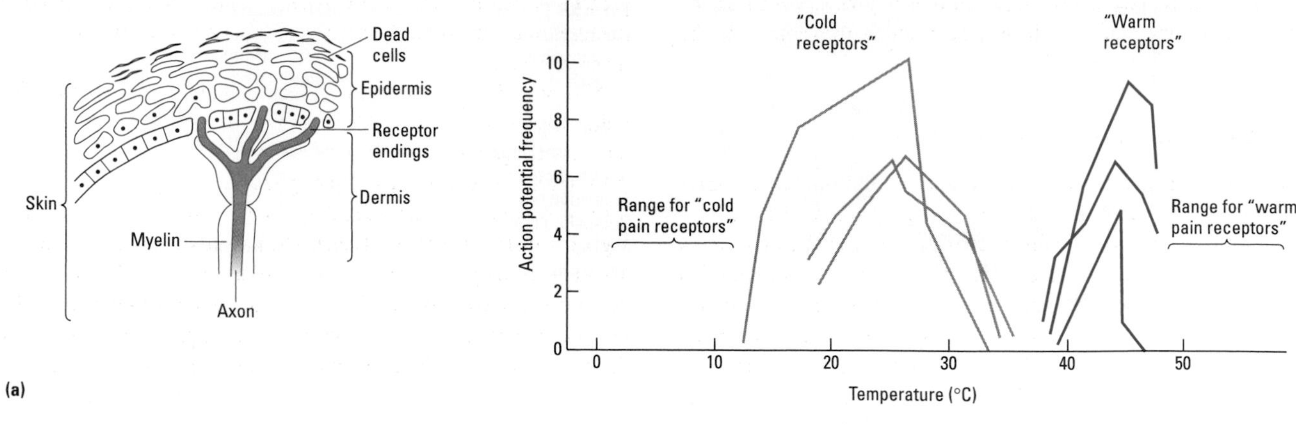

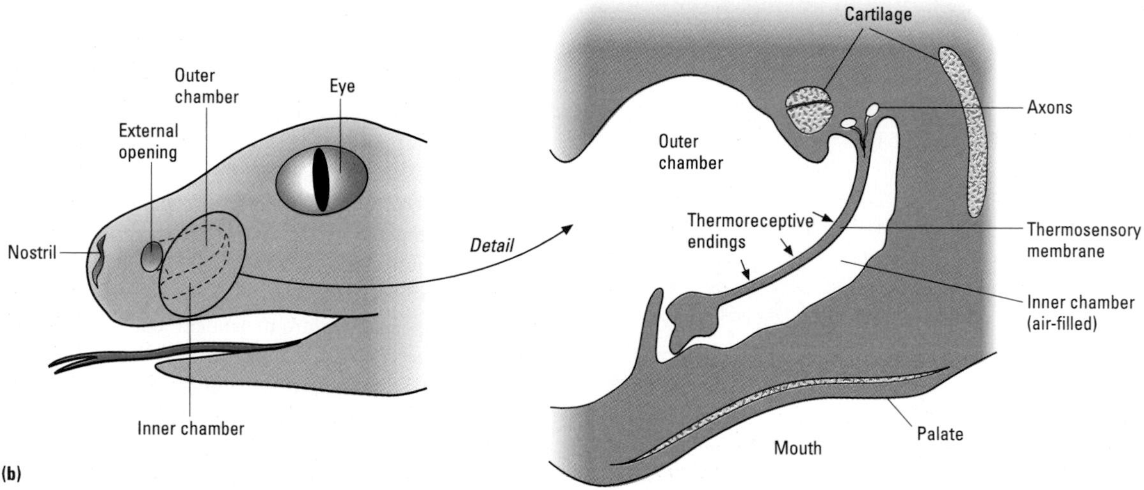

Fig. 9.62 (a) The typical structure of, and response curves from, single thermal receptors in a mammal. (b) The thermoreceptive "pit organ" in a snake.

Less spectacularly, some blood-sucking bugs can detect the presence and direction of potential endothermic hosts. Special thermoreceptors also occur in many snakes, where the "pit organs" on the face (Fig. 9.62b) can be used to detect the warm body or breath of nearby prey items. In pit vipers the organs are small air-filled depressions, covered with a thin membrane containing the thermoreceptors; infrared radiation warms the membrane, and the air below insulates it to prevent the heat being conducted away too quickly. The organs respond best to wavelengths of 700–1500 nm, and being paired they also give a "binocular" effect so that the snake can judge the direction and distance of the warm body.

Thermal receptivity can occasionally occur independent of ion channels. The electroreceptors of some sharks (see section 9.8.6) contain an extracellular glycoprotein gel that has properties similar to those of a semiconductor, developing a significant voltage in response to small temperature changes. Hence thermal information (for example slight warming of the water resulting from the presence of a nearby food source) can be converted directly into electrical information within the sensory endings, giving acute temperature sensitivity.

Hygroreceptors (humidity receptors, only relevant on land) remain somewhat controversial for most animals but are certainly present on the palps and antennae of many terrestrial arthropods, including millipedes, spiders, ticks, mosquitoes, and cockroaches. Ticks avoid any contact with liquid water, apparently using water sensors in the first pair of tarsi. In some beetles a single "sensilla styloconica" on the antenna may contain one hygroreceptor and one thermoreceptor neuron, together with 3–4 chemoreceptors. Within the larger hygroreceptive organs of other insects some receptors increase their firing in response to drier air, and others to moister air, and at best they can respond to a change of just 1–2% relative humidity in the ambient air. They permit the insect to seek moister microenvironments, and may also affect spiracle opening. However their transduction system remains unclear; it could involve a response to local evaporative effects on temperature, or to changed osmotic gradients, or to structural changes in hygroscopic materials swelling or shrinking within the receptor. The last of these is a currently favored model, but ultrastructural studies suggest that more than one mechanism of hygroreceptor may exist.

Some vertebrates also have hygroreceptors, for example toads

(*Bufo*) can detect changes in both barometric pressure and humidity in laboratory studies, which would allow movement in the direction of water in the field.

9.8.6 Electroreceptors and magnetoreceptors

Electroreceptors are perhaps the "simplest" kind of receptor since they involve no real transduction of energy from one form to another. They only work in aqueous media, and are best known in marine and freshwater fish (see Chapters 11 and 13), although also occurring in some amphibians and in one mammal, the platypus. They function because the receptor cell membranes are organized asymmetrically, with the surface contacting the water having a low resistance membrane, and the inner (basal) surface being of much greater resistance. Any electric current flowing through the cell therefore leads to a tiny but significant potential (usually a few microvolts) across the basal membrane, and this can serve as a receptor potential. The depolarization is usually mediated by calcium ions, and offset by a subsequent outward potassium current.

Most animals with an electric sense cannot generate any electric field of their own, but merely respond to existing fields (**passive electroreception**). However, some fish, such as the electric eel, can generate a strong electric field around themselves (they are **electrogenic**), and then detect changes in that field (**active electroreception**); these fish may generate either weak or strong fields (see Fig. 11.32). The weak fields are generated continuously and used for electrolocation and electrocommunication; strong fields are generated intermittently and used to stun prey or deter predators.

Electroreceptive sense organs are associated with the lateral line of fish, and derived from neuromast hair cells. The commonest type is the **ampullary organ**, a small invaginated cavity just below the skin surface, lined with receptors attuned to small low-frequency variations in electric fields, and with rather long entrance canals that help to maximize the voltage difference between different receptors and so improve discrimination. In elasmobranchs these structures are known as the **ampullae of Lorenzini**. In teleosts rather similar electroreceptors may be scattered over much of the body surface. In active-reception electric fish an additional type of receptor occurs, the **tuberous organ**, formed from sacs of epidermal cells but not connected to the skin surface, and responding to high-frequency voltage changes. Weak-field-generating fish possess an **electric organ**, usually near the tail, made from modified muscle or nerve, producing a high-frequency alternating current discharge (50–2000 Hz, species and sometimes sex specific), and their own electroreceptors respond maximally at this frequency so that they can readily detect changes in their electric field.

In all these electroreception organs, multiple receptors within the organ synapse with a single afferent neuron leaving it. Electrosensitivity is therefore high, and in sharks and rays it is extraordinary, sometimes to electrical gradients as low as 0.005 $\mu V\ cm^{-2}$. Thus the tiny muscle potentials in any other moving animal will be detectable; some sharks can detect a single muscle end-plate potential at 10–20 cm distance, and can discover a buried flatfish in sand from the electrical potential generated by its tiny gill movements. This level of sensitivity also allows an electric fish to navigate electrically, since by swimming through the Earth's magnetic field it will generate potential gradients of this magnitude (and these gradients will be greater when moving at right angles to, than when moving along, the lines of the magnetic field); in effect this is navigation by indirect magnetic reception.

A more direct form of **magnetic reception** is also possible. Behavioral work shows that bees, and some birds, have an ability to navigate magnetically by detecting where the north–south axis lies (though not which direction along this line is north), and some marine animals such as the sea slug *Tritonia* have also demonstrated a geomagnetic sensitivity. Usually this seems to be linked to the presence of magnetite (an oxide of iron) as tiny crystals somewhere in the body. However, detailed physiological understanding of this or other postulated mechanisms is still lacking. It seems likely that the magnetite crystals, by aligning themselves with the magnetic field, could provide a twisting force to trigger mechanoreceptors.

9.8.7 Overview of receptors

Sensory receptors in animals are about as sensitive as they theoretically can be, in some cases responding to a single chemical molecule, a vibration smaller than the size of a single atom, or a single light photon. The energy changes involved in these levels of sensitivity are of the order of just 10^{-18}–10^{-21} J per receptor cell. While sensory modalities do have much in common (particularly in their mechanisms of adaptation and the role of calcium in this) there are also important differences (Fig. 9.63) beyond those of purely cellular structure. In visual and olfactory systems the detection process is essentially quantal, with the binding of a single molecule or the absorption of a single photon providing an energy input significantly greater than the thermal "noise" in the receptor. However, for acoustic systems the energy of a phonon is too low for this, and a nonquantal or "classic" detection system is needed. The first two modalities rely on transduction via a G-protein cascade, while all acoustic systems use direct coupling of stimulus to ion channel, giving a significantly more rapid transduction (though without the amplification allowed by a second messenger system).

Individual receptors are more or less specific to one modality, and they do not in themselves code very much information; it is a common finding that there is just one type of receptor molecule per receptor cell. But by integration of their activities, receptors can produce sophisticated outputs about the nature and changing properties of the animal's environment, directed to sensory ganglia and to the brain, where further processing and interpretation occurs.

9.8.8 Sensory information processing

Messages from the sensory receptors or sense organs are sent to the brain, and their destination determines their effect. Thus they have to be routed through the correct transmission channels to make sense. Signals from mechanoreceptors in joints pass to the correct parts of the sensory motor areas and so are interpreted as giving information on the position or movement of that joint. Signals from visual receptors pass to the visual centers in the brain, and are thus interpreted as giving information about light intensity. (Hence when humans suffer brain damage and develop aberrant sensory pathways, they may perceive entirely inappropriate inputs—signals received by the fingers may be interpreted as coming from the

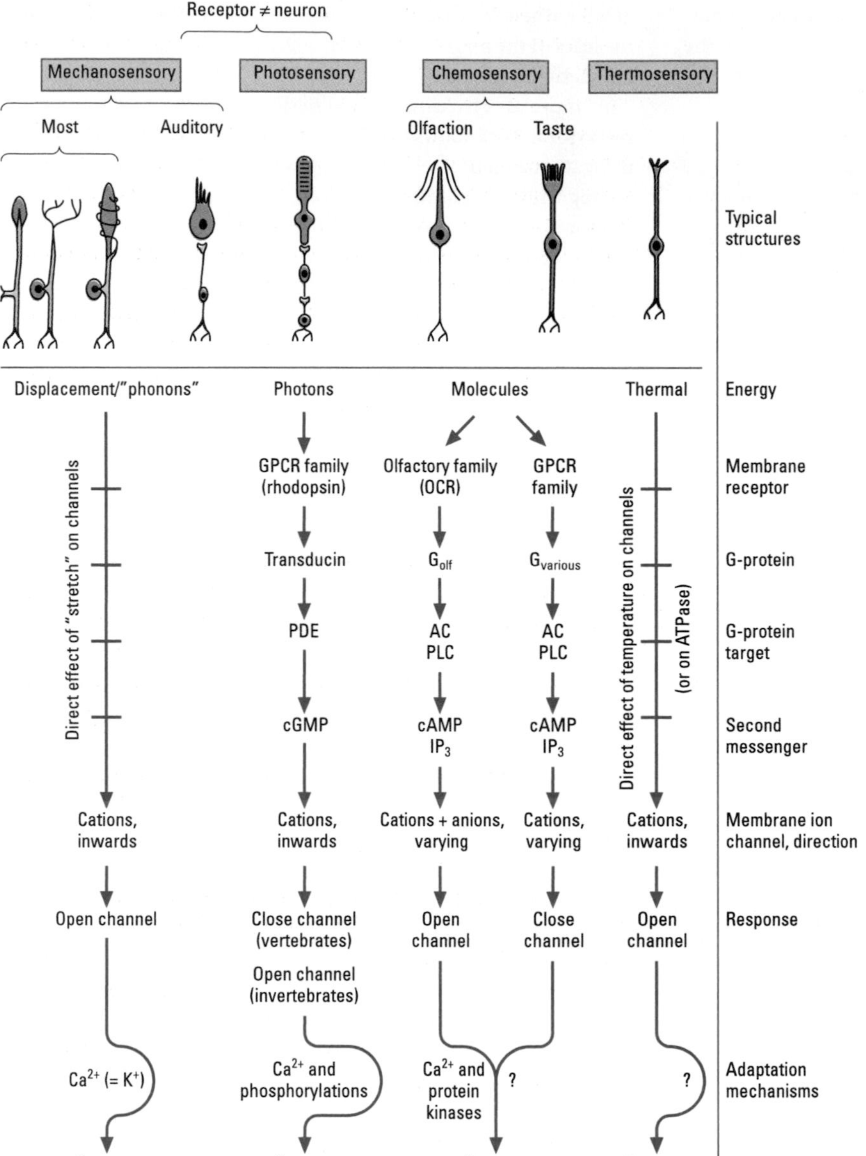

Fig. 9.63 A comparative overview of sensory receptors.

shoulder or chest, or even as being acoustic or visual in origin.) When signals are received centrally they therefore carry several kinds of information:

1 That a stimulus has been received.
2 That it has a particular strength and time course.
3 That it comes from a particular part of the body in a particular sensory modality.

In practice, a multitude of such stimuli are being received all the time, and central processing of this complexity has to be achieved. In many of the more complex animals some part of this processing is actually achieved more peripherally, by organizing sensory receptors into sense organs and having neural circuitry within the organs or in ganglia closely associated with those sense organs, as we have seen.

Sense organs that have built in spatial organization generally form connections with postsynaptic interneurons that maintain the spatial topography, as was described for the vertebrate eye. This spatial patterning may be modified in various ways. Firstly, the receptors may each link laterally to several interneurons (**divergence**), or several receptors may link to a single interneuron (**convergence**). The latter is particularly common (remember the example of the cat retina described above), and allows integration of several receptor inputs. Both convergence and divergence may occur in the same sense organ; again the cat retina provides an example since some photoreceptors synapse with two or more bipolar cells linking to at least five amacrine cells (Fig. 9.64a).

A second type of processing occurring peripherally is **lateral inhibition**. This arises where stimulation of a particular receptor results in reduced sensitivity in the neighboring receptors, due to the presence of reciprocal inhibitory synapses (Fig. 9.64b). This may help to increase the contrast between receptors, in space or in time ("contrast enhancement"), giving a very clear boundary effect. In visual

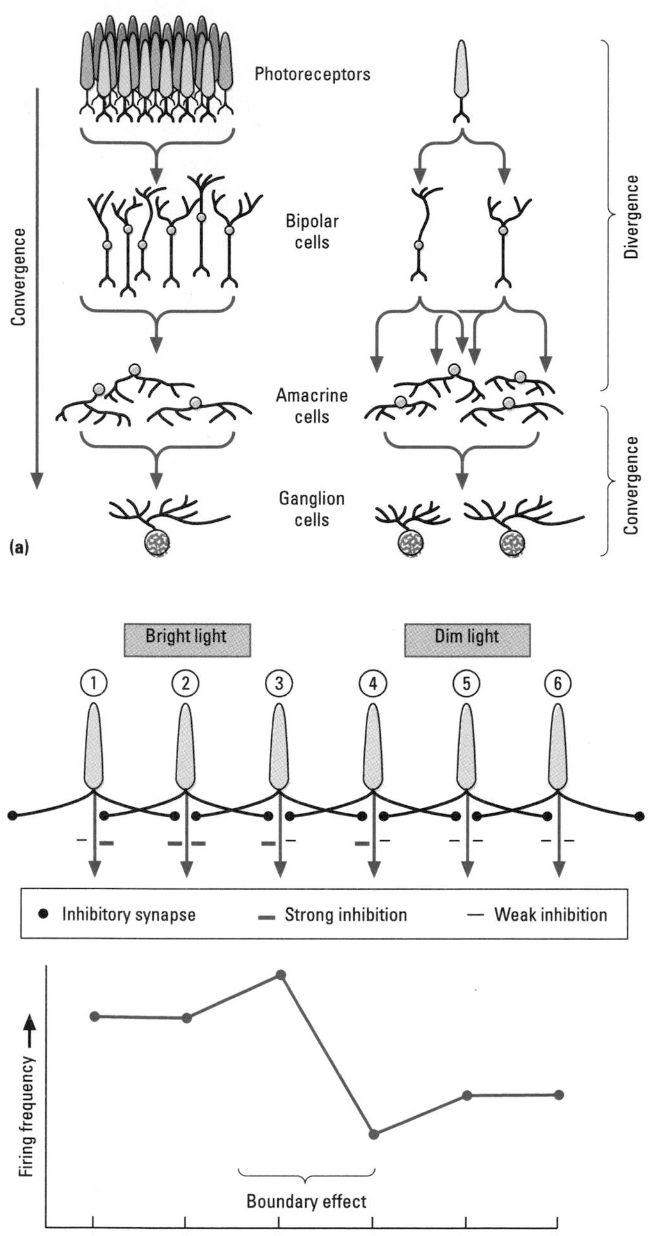

Fig. 9.64 Aspects of sensory processing. (a) Convergence and divergence of sensory neurons both occur in a typical vertebrate retina (and in many other sensory systems with spatial "tonotopic" organization) and allow integration of receptor inputs. (b) Lateral inhibition between adjacent receptors allows for contrast enhancement, here shown for photoreceptors. Each cell inhibits its neighbors in proportion to the light it is receiving, so receptor 3, although receiving the same light as 1 and 2, is less inhibited and produces more action potentials, while receptor 4 is more inhibited than the other dimly lit receptors 5 and 6 and has fewer action potentials, thus giving a sharpened "boundary effect".

systems lateral inhibition is particularly useful in detecting movements, because if it is organized asymmetrically then it can enhance the detection of directional movement for an object passing across the visual field: receptors ahead of the moving object are subject to less lateral inhibition from shaded neighbors while those just inside the shaded area are more inhibited than those that are fully shaded.

A third possibility is the use of **peripheral feedbacks**, so that signals flow back to the receptors from interneurons deeper within the sense organ or the associated ganglion. In some animals this feedback may come from the brain itself (**efferent control**), raising or lowering the sensitivity of a particular set of receptors. For example, there are feedback outputs going to the pigment cells of the insect compound eye, to the iris muscles in the vertebrate eye, and to the tympanal muscles in the mammalian ear. Efferent control can have several uses: most obviously it can suppress unimportant inputs, but it can also protect a sensory system against damage from excessive stimulation (for example our own hearing is dampened when we shout), and crucially it can help to compensate for the phenomenon of "reafference", sensory signals that are caused by an animal's own activities. Thus a fish can "blank out" the signals in its lateral line system that result from its own swimming activity, while still detecting signals due to another fish (potentially a predator or a food item) swimming nearby.

As we move higher up the sensory pathways though, we begin to find that processing systems get more sophisticated. Much of the initial filtering of information is done at the level of the peripheral ganglia in vertebrates. Here there may be neurons with relatively specialized and highly specific functions, and these are most clearly seen in visual systems. A classic example is the presence of "flying insect detecting neurons" in the frog retinal ganglia, responding only to small dark moving objects and ignoring all other visual inputs. In this manner, receptors can be finely attuned to the needs of particular animals and their specialist lifestyles and habitats.

Ultimately though it is the brain that has to sort out all the sensory information coming into it. As we saw in section 9.4.3 and Fig. 9.35, most sensory neurons connect with fairly discrete parts of the CNS that are termed **projection centers**, where they link with groups of interneurons. These interneurons are usually the primary processing systems, integrating the sensory inputs from different receptors or organs and transmitting signals onwards to other centers. The projection centers in many animal brains have a particular kind of fine structural organization, with localized concentrations of neurons (often visibly distinct in histological preparations) devoted to a particular kind of input and often termed "**glomeruli**" or "**sensory compartments**". For example, in the insect olfactory system we can locate a glomerulus of around 100 neurons that responds only to that species' sex pheromone, and in mammalian somatosensory cortex there are often identifiable sets of neurons, each group similarly disposed and each responding only to the inputs from one particular whisker on the animal's snout.

The spatial organization of such groups of compartmentalized neurons, in many kinds of animals, is often essentially columnar, with a series of layers of interacting neurons within each column of neurons. Each column of CNS neurons can be linked to a particular receptive field on the periphery of the body, and the receptive fields can vary in size from a few microns to several centimeters in diameter. Any information coming inwards that is "position coded" (which may be mechanical inputs from the body surface, or inputs from the highly spatially organized visual or olfactory systems) can be transmitted in a topographical manner to sequential neuronal layers within these sensory compartments of the brain. Where information relates to the whole body surface this is termed **"somatotopic" projection**—a spatially organized representation of the body is preserved at all levels and we can ultimately trace the outline of the body, albeit distorted by differential sensitivities, on the brain (as with the vertebrate somatosensory cortex; see Fig. 9.35). In the vertebrate auditory system there is a comparable "tonotopic" projection, where the processing centers are organized such that adjacent areas concentrate on neighboring frequencies (again there may be distortions where particular frequencies are important to a particular species, relating to mating or individual recognition calls).

Another important aspect of central sensory projections is that several representations of the sensory input can be built up at one time, existing "side by side". This means that the central processors are working both in series (as described above) and simultaneously in parallel, so that within one columnar compartment there may be subsystems of neurons processing the same information but in different ways. Some subsystems may be looking at the start and end of signals, others at their duration, others at the delay between similar kinds of receptors, and others again at spatial positioning in particular. As we move to higher order centers within the CNS the type of specialized features that neurons are responding to become increasingly complex—that is, the system is hierarchical. An early interneuron may respond to any type of touch on the skin, while at the next level we may find neurons only responding to a moving touch, and higher up again only to directionally moving touch in a fairly narrow band of possible directions. The same is true in visual systems, where the higher order neurons may be tuned to quite specific characters such as those that make up a hand or a face; and in auditory systems, where some human neurons in the median and temporal lobes only respond to particular spoken words. These higher order processors may link more than one sensory modality, as with specific movement detectors that tell an animal how it is moving with respect to its surroundings (linking mechanosensory and visual triggers); or how the visual information it is receiving is linked to the movements of its own eyes and head. These higher order detectors, whatever their modality, are sometimes termed "**feature detectors**", and they occur in insect visual systems as well as in vertebrates.

At a higher level still, the traditional distinctions between "sensory" and "motor" neurons begin to break down, and we find neurons that are both a part of a multimodal sensory unit and also have a direct influence on behavior. These are commonly termed "**sensorimotor integration neurons**" (SINs), and they work together in networks. Thus a flying insect such as a blowfly makes "automatic" small corrections to its flight path as a result of integrated visual inputs (eyes), aerodynamic information (halteres), and proprioceptive inputs (wing bases and muscle stretch receptors). Similarly a bat may act almost instantly in a fashion that indicates integration of inputs on the position, velocity, and size of its fast-moving prey. Any "target-oriented" movement must rely on continuing sensory inputs with modifying interactive feedbacks, detecting and correcting any deviations. In higher vertebrates a good deal of the SIN functioning of the body is concentrated in the cerebellum, where neurons are arranged in extraordinarily regular patterns (Fig. 9.65) with input being from "climbing fibers" and "mossy fibers", and output solely through the Purkinje fibers. The cerebellum contains more than half of all the brain's interneurons. Only from such systems can we get suitable adaptive behavior for a given animal facing changing environmental stimuli.

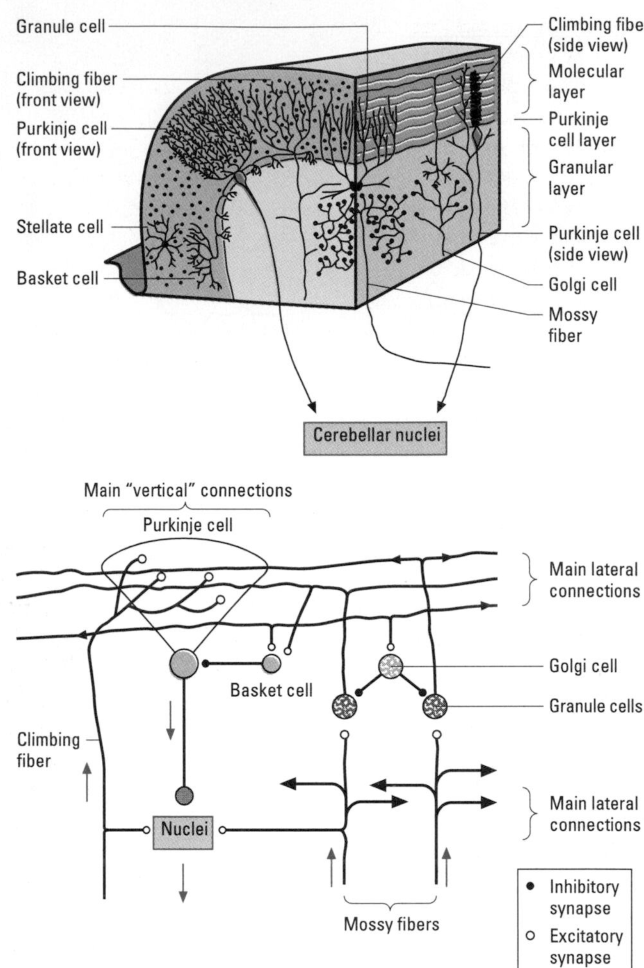

Fig. 9.65 The neuronal regularity shown in the vertebrate cerebellum. The cell types occur in organized columnar arrays perpendicular to the cerebellar surface. Note that the incoming climbing fibers and the outgoing Purkinje cells have two-dimensional "fan-like" dendritic trees, and that lateral connectivity between them is provided by parallel fibers and by colaterals of mossy fibers, providing a complex integrating network for sensorimotor functions. (Adapted from Reichert 1992.)

SECTION 2: MUSCLES

9.9 Muscles and movement: introduction

All movement by animals is generated by very small-scale cellular processes, in one of two ways: by the bending of cilia and flagella, or by the contraction of muscle. Muscular contraction is by far the more important of these mechanisms for most animals, producing both whole-body movements and many internal (visceral) movements. Muscle is the second major "excitable" tissue in animals, its function initiated (like that of nerves) by electrical activity in membranes. Muscles then act as biological engines, converting chemical energy into mechanical work. However, they can only act as pulling engines, and can never push, since they perform work only when actively shortening, and must relax again passively, either by contraction of an opposing muscle or by the effect of an elastic structure.

Muscle tissues are best known in vertebrates, where they can readily be subdivided (morphologically, and to a lesser extent functionally) into three or more types: **striated muscles** (further subdivided into skeletal and cardiac muscle) and **smooth muscle** (or nonstriated muscle). But their mode of action at the cellular level is essentially similar, and this similarity of action extends throughout the animal kingdom. It is found in the contractile tissues even of cnidarians, where muscle could be said to make its first evolutionary appearance as nonstriated contractile "tails" at the base of normal epithelial cells. However, it should be noted at the outset that the distinctions of muscle types are not as clear-cut as they may seem from a study of vertebrates. It is also increasingly apparent that muscle shows a remarkable range of specializations for different functions even within a single species. Thus individual muscles can vary considerably with respect to their shortening speed, strength, economy of action, and endurance. But since the basic cellular and molecular design of muscles is highly conserved in widely different animals, we can deal with the essentials first.

Muscles occur either as distinct elongate (roughly cylindrical) structures anchored at either end to the skeletal elements that they work against, or as sheets that often run around the walls of tubes. They are made up of cells that are generally referred to as **muscle fibers**; in fact the fibers are usually multicellular, and once mature are therefore multinucleate in many animals, because they arise from the fusion of many separate (embryonic) myoblast cells. The fibers normally run parallel to each other within the muscle, so that they all pull together in one direction during a contraction.

The key component of the muscle cell is a molecular motor protein called **myosin**, which occurs as long filaments running in parallel along the axis of the fiber. These interact with other axial filaments that are primarily composed of the protein **actin**. Contraction involves cross-bridges from the myosin molecules undergoing transient binding to sites on the adjacent actin molecules, which in turn makes the myosin molecule change its conformation and drag the attached actin past itself, the cross-bridge, and then detaching again. The interaction is in essence a process of the two filament types sliding over and past each other, using energy from ATP hydrolysis. The overall result is a shortening of the muscle, which produces a force that can move a load (usually part or all of the animal's own body).

The arrangement of the myosin and actin filaments determines the broad morphological category of the muscle, and the character of the various isoforms of the proteins that a muscle contains contributes to its particular specializations. The trigger for muscle activation is always electric and is always related to movements of Ca^{2+} ions, but there is diversity in the proteins that interact with the Ca^{2+} (the regulatory proteins), and in the means of achieving the changes in cytoplasmic Ca^{2+} concentration (using internal and external membrane systems) that are required to trigger bouts of activation and relaxation. Muscle fibers can also differ considerably in their metabolism, in the nature of their innervation, and in the way they respond to nervous inputs and chemical effectors. A summary of the nature and time courses of the component processes involved in typical muscle activity is shown in Fig. 9.66.

The primary function of a muscle is to contract or shorten, and so produce work against other muscles or other anatomical structures or the external environment. Thus muscles work in conjunction with each other and with structural components of an animal's skeleton, these components having tensile, elastic, and hydrostatic properties depending on the nature of the skeleton. Movement requires the coordinated function of all these elements, orchestrated by the nervous system.

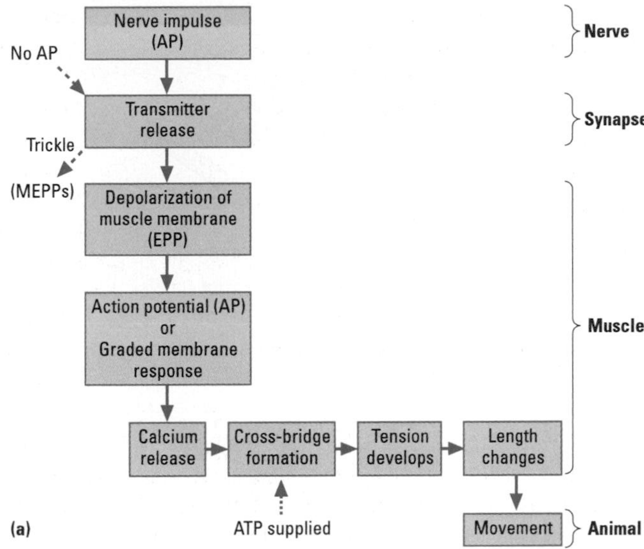

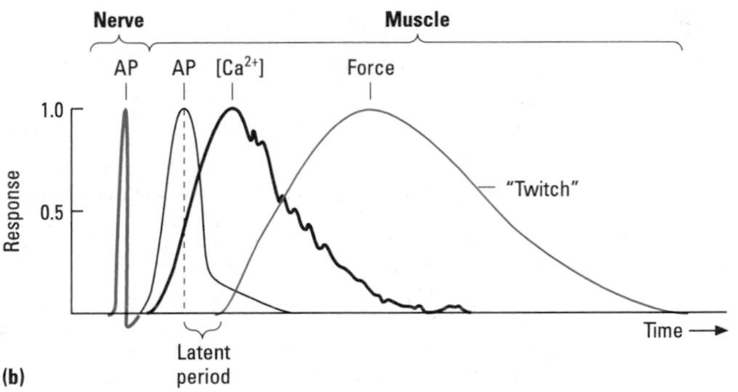

Fig. 9.66 (a) A summary of the basic processes leading to muscle activation, and (b) their relative time course. The action potential (AP) in the nerve is of shorter duration than that in the muscle; and the rise in calcium concentration in the muscle that is induced by the muscle AP is transient and has almost finished before any change in force or length is detectable in the muscle. EPP, end-plate potential; MEPP, miniature end-plate potential.

9.10 Muscle structure

The most widely studied types of muscle are the so-called "twitch fibers" found in the skeletal muscles of vertebrates, so we will use these to illustrate most of the basic principles of muscle design and contraction processes.

9.10.1 Gross anatomy

Each muscle in a vertebrate is a distinct organ (Fig. 9.67) bounded by connective tissue, and attached at either end to the vertebrate endoskeleton, usually through elastic tendons. The muscle is

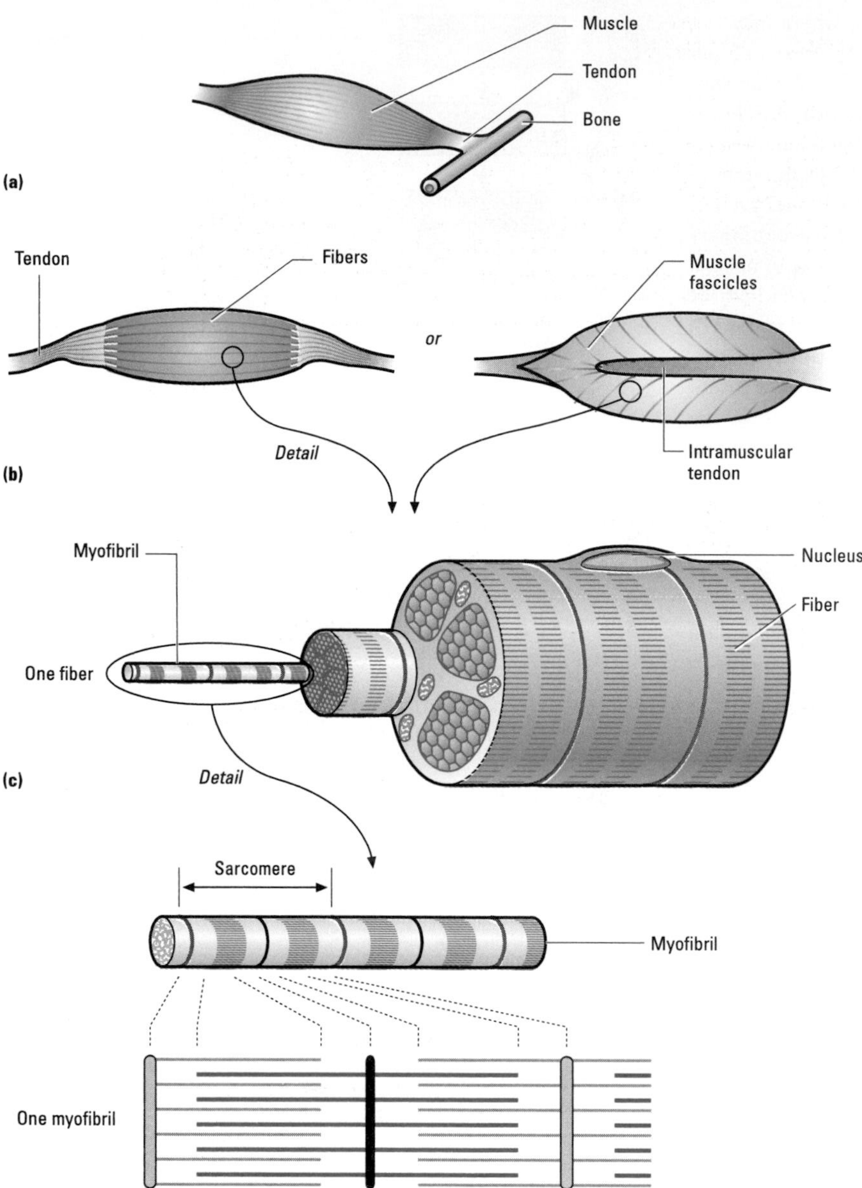

Fig. 9.67 Typical vertebrate skeletal (striated) muscle anatomy. (a) The anatomical arrangement of muscle, tendon, and bone. (b) The arrangement of muscle fibers within a muscle, either in parallel array or in the common "pennate" form where each fiber is shorter than the whole muscle and pulls at an angle to the main tendons. (c) A single muscle fiber and its constituent myofibrils. (d) A single myofibril and its sarcomeres, and the myofilaments within in highly schematic form.

innervated by motor neurons, and is always well vascularized, having a high demand for metabolites and oxygen. The more aerobic the muscle is, the greater the size of its capillary beds. Every muscle is made up from many fibers, and one fiber is the smallest unit of a muscle that is capable of a complete physiological response. Each muscle fiber receives one or more inputs from a motor neuron, at specialized synaptic structures called **neuromuscular junctions** (NMJs), also termed "end-plates"; and each fiber is also supplied by one or more blood capillaries. Muscle fibers are multicellular units contained within a common basal lamina, and are highly differentiated structures containing many nuclei. For example, the mouse extensor digitorum longus muscle has around 450 nuclei per fiber. The fibers may be several centimeters long but only fractions of a millimeter in diameter; in mammals, the range is from around 10 to 200 μm.

Vertebrate muscle fibers are incapable of division and growth. Repair from injury therefore requires inputs from a nearby population of highly proliferative muscle stem cells. The muscle stem cells and their progeny are located beneath the basal lamina of muscle fibers, and are collectively referred to as muscle satellite cells. Muscles also contain a variety of other cell types, including adipocytes, fibroblasts, and macrophages.

Each multicellular muscle fiber contains three main types of organelle:

1 The **myofibrils** (containing the actin and myosin filaments), organized in parallel axial arrays within the fibers, each about 1–2 μm in diameter.

2 The complex array of **sarcoplasmic reticulum** (SR) plus **T-tubules**, these being respectively the internal and external membrane systems that are involved in excitation–contraction coupling.

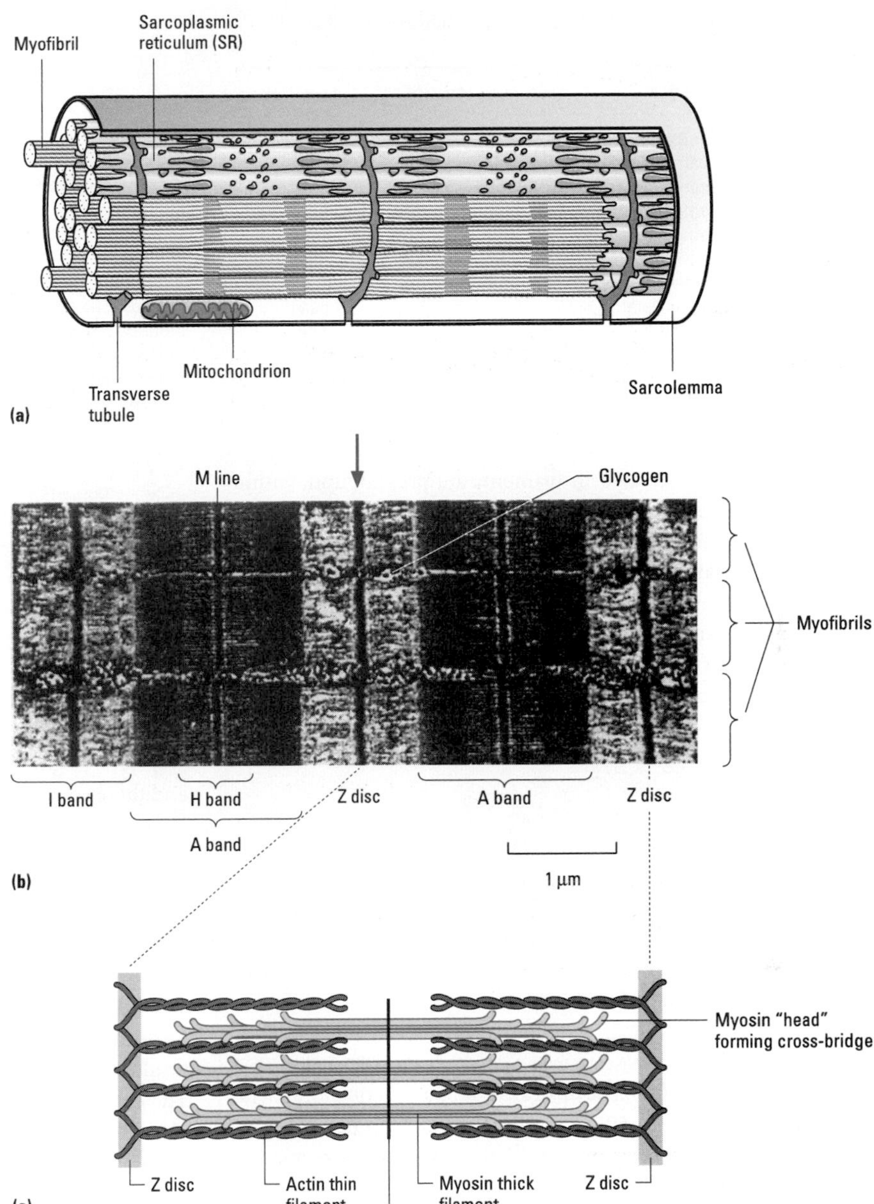

Fig. 9.68 (a) Schematic view of a single sarcomere from vertebrate striated muscle, with its characteristic banding pattern. (b) An electron micrograph of typical vertebrate striated muscle, showing two complete sarcomeres in three adjacent myofibrils, all aligned with each other. (c) The filament patterns that underlie the sarcomere banding seen in micrographs.

In frog sartorius muscle the T-system occupies 0.3% of the fiber volume, and the SR about 13%.

3 The **mitochondria**, which supply ATP using aerobic metabolic pathways.

9.10.2 Myofibrils and sarcomere structure

The myofibrils are the key effectors in muscle. Each myofibril appears to be cross-striated, with a periodicity of about 2 μm at rest, and this reflects the presence of numerous serially arranged contractile units called **sarcomeres** (Fig. 9.68). The sarcomere is the functional unit of striated muscle, and is bounded at each end by the so-called Z line, formed from the two closely apposed Z discs that terminate adjacent sarcomeres. Because the sarcomeres of adjacent myofibrils are in register, the whole muscle fiber has a striated appearance. When a muscle shortens, it is apparent with a light microscope view that each of the sarcomeres shortens to the same degree.

A further level of striation arises within the myofibrils because of the regular almost crystal-like arrangement of filaments inside each sarcomere (Fig. 9.68b). There are two types, the **thick or A filaments** (containing myosin), which lie relatively freely in the cytoplasm roughly in the middle of the sarcomere lengthwise, and the **thin or I filaments** (containing actin), which originate at either end of the sarcomere where they insert into the Z line. The two types of protein filament interdigitate in parts of the fibril as shown in Fig. 9.68c, giving rise to various different regions: the **I bands** at each end of the sarcomere where only actin filaments are present; the **A band**

Table 9.9 Composition of the protein content of myofibrils in vertebrate skeletal muscle.

Protein	Function	% wet mass of myofibril	Localization
Myosin	Contractile	43	A band
Actin	Contractile	22	I band
Tropomyosin	Regulatory	5	I band
Troponins	Regulatory	5	I band
M-protein	Regulatory	2	M line
C-protein	Regulatory	2	A band
Titin	Cytoskeletal	10	A and I bands
Nebulin	Cytoskeletal	5	I band
α-Actinin	Regulatory	2	Z line
β-Actinin	Regulatory	0.5	Z line

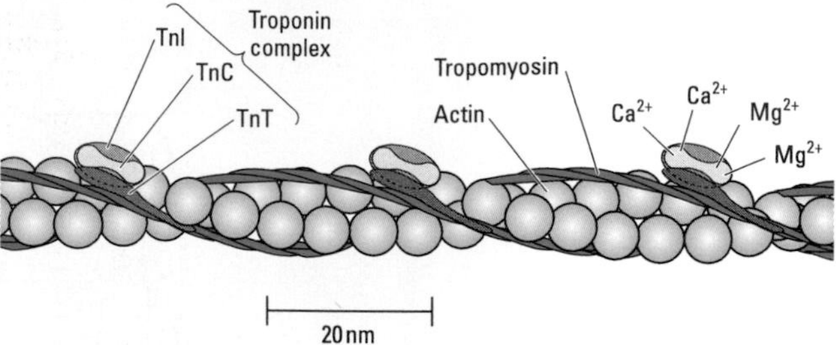

Fig. 9.69 The actin filament structure, showing schematically the actin monomers forming a double strand with a loose double helical coiling, the associated tropomyosin strands lying in the groove between the two actin strands, and the globular troponin (Tn) complexes at intervals along the filament.

centrally where denser myosin filaments are present; and, within the A band, are the two most dense zones where actin and myosin overlap and a central **H zone** where only myosin is present. The composition, relative abundance, and localization of the main proteins within the myofibril are shown in Table 9.9.

In electron micrographs another structural regularity can be seen, with the thick myosin filaments having small lateral projections at intervals along their length; these are termed **cross-bridges**, and are a crucial part of the muscular contraction mechanism.

Actin filaments

These filaments are typically 1 μm in length and about 8 nm in diameter, and they resemble a double string of beads twisted together in a loose helix (Fig. 9.69). Each bead in the chain is a **G-actin** monomer, 5.46 nm in diameter, and the whole filament is termed an **F-actin** polymer. When the actin monomers polymerize to form **F-actin**, they self-assemble in the loose two-stranded helix configuration, the two strands winding around each other with a "pitch" of 73 nm (thus crossing over each other every 36.5 nm).

In fact each filament is made up of around 350 G-actin monomers, but also contains about 50 molecules each of **tropomyosin** and three different **troponins** (troponin T, troponin I, and troponin C). The actin interacts with myosin to produce force, whilst the tropomyosin–troponin complex is involved in the Ca^{2+}-mediated regulation of contraction. The rod-shaped tropomyosin molecules also polymerize, end to end with a slight overlap, and they sit in the grooves between the actin strands (Fig. 9.69). There are 12–13 actin monomers per tropomyosin molecule. The troponins form loose globular complexes, attached to tropomyosin at intervals along the filament.

Myosin filaments and cross-bridges

Thick filaments in vertebrate muscle are about 1.6 μm long and around 12 nm thick. The main constituent of the thick filaments is myosin, a large protein with a molecular weight of around 480 kDa. It is large enough to be visualized in high-powered electron microscopes using shadow-casting techniques; each myosin molecule is then revealed as elongate, with two identical heads. Each thick filament contains around 300 of these myosin molecules, their tails overlapping along the axis of the filament (Fig. 9.70a).

Each myosin molecule is in fact made up of six polypeptide chains, in three pairs. There are two **heavy chains** of typical protein α-helix, which then supercoil together over most of their length to comprise the tail and neck of the molecule, but with the N-terminal end of each heavy chain forming a more globular "head" (so that we end up with the two-headed molecule). Each globular myosin head contains an ATPase site and an actin-binding site. Then there are two pairs of **light chains**, termed the "essential light chain" and the "regulatory light chain"; these both act as calcium-binding proteins and there is one of each kind per head. Both of them wrap around the neck of the myosin molecule (Fig. 9.70b). A minimum of three genes must be transcribed, twice each, to make a single myosin molecule: one for the heavy chain, one the regulatory light chain, and one the essential light chain. Molecular variants of all of the myosin chains are known, and these confer slightly different properties.

Myosin molecular structure is now known in some detail. The subunit composition has been determined by dissection using proteolytic enzymes, and the structures and periodicities of the resultant thick filament have been determined by X-ray diffraction. The myosin heads are about 17 nm in diameter, attached to a long (156 nm) tail, and with a flexible "hinge" region in the tail 43 nm from the heads (Fig. 9.71a). The molecule can be readily split into two types of fragment (Fig. 9.71b): the tail ("light" meromyosin) and the neck + heads ("heavy" meromyosin). Heavy meromyosin can then be split again, into fragments S1 and S2; S2 is the remaining α-helical coiled tail of the heavy chain, while two pieces of S1 result, each being a globular head of the heavy chain and its two attached light chains. The two binding sites (for ATP and actin) on the S1 heads are about 4 nm apart.

Each "neck" of the myosin molecule effectively becomes a "lever arm", formed by 73 amino acid residues, comprising a long single α-helix (Fig. 9.72) leading towards the globular head. The "wrapping" effect of the two light chains gives this neck region sufficient rigidity to act as a lever. The regulatory chain is at the base of the neck, and can be phosphorylated (with a specific kinase) and dephosphorylated (with light chain phosphatase), conferring its regulatory functions, while the essential chain is closer to the heavy myosin head. The whole lever arm is about 10 nm long. This lever arm, linking the head and tail regions of the molecule (Fig. 9.70), becomes the moveable "cross-bridge" that acts as the molecular motor during contraction.

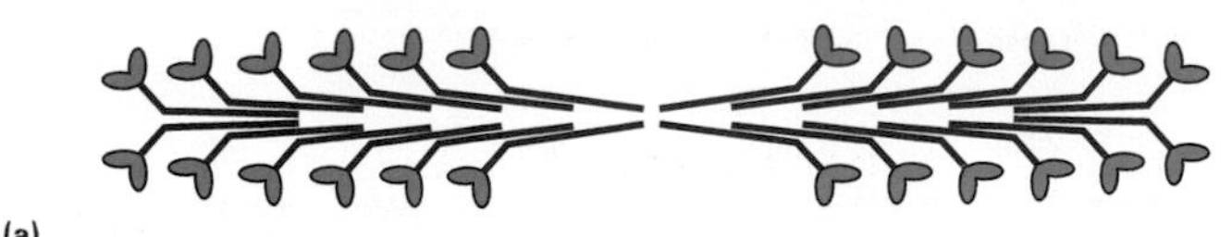

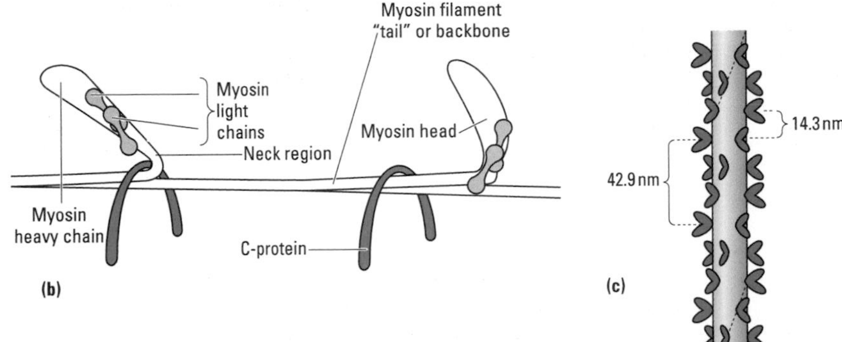

Fig. 9.70 Structure of the myosin filaments: (a) the whole filament formed from aligned myosin molecules (reversing their polarity about the midpoint) and with the double myosin "heads" protruding, (b) schematic close-up of a short length of filament with protruding heads and C-proteins, and (c) the helical periodicity of the myosin heads along the myosin filament.

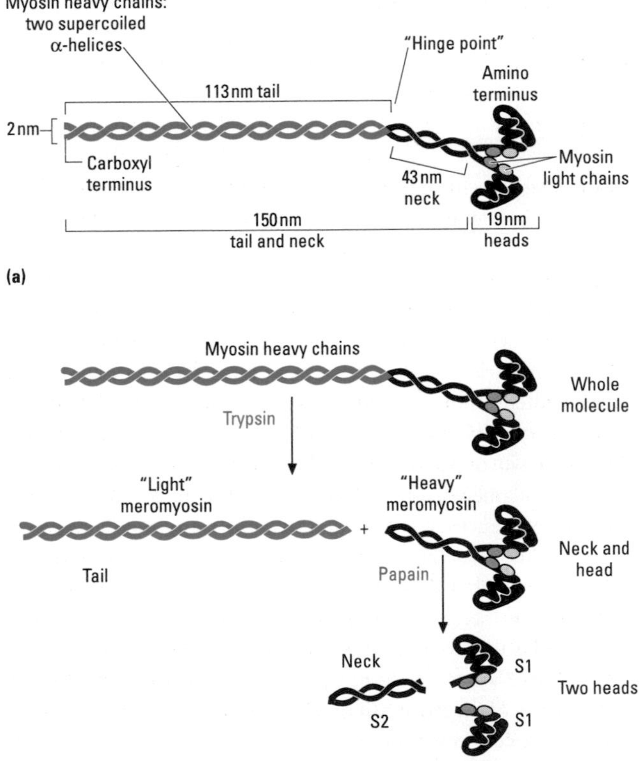

Fig. 9.71 (a) Structure and dimensions of a single myosin molecule, showing head, neck, and tail (or backbone) regions. (b) Effects of proteolytic dissection of myosin molecules, producing "light" meromyosin and "heavy" meromyosin fragments, corresponding to the backbone region and the head + neck region, respectively; then splitting the heavy meromyosin into three further fragments, two known as S1 (myosin heads) and one S2 (the coiled neck region).

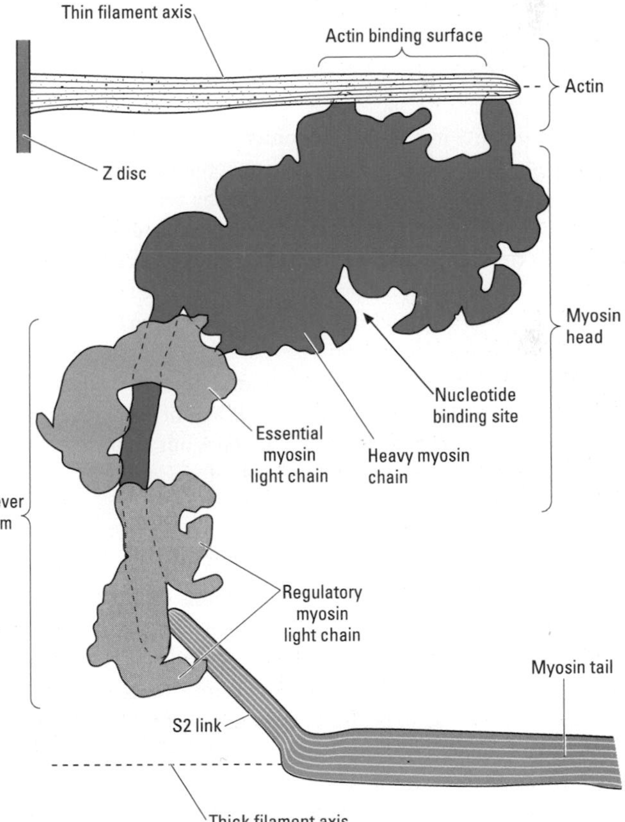

Fig. 9.72 Close-up view of the S1 fragment of the myosin molecule, representing a single "myosin head" and its adjacent neck region. Each head is made up of the amino-terminus of the main (heavy) myosin chain and the two myosin light or regulatory chains that wrap around the neck.

A whole thick filament can be assembled *in vitro* from its constituent myosin molecules, and the assembly is always initiated by a few molecules lining up axially but in opposite directions, with their head pointing outwards and their tail regions overlapping. More molecules then add on symmetrically at the ends of the line, always with the heads outwards. Hence a small region in the middle of the filament has only tails, and this will become the M line region of the sarcomere. All the heads point away from this area, and so have

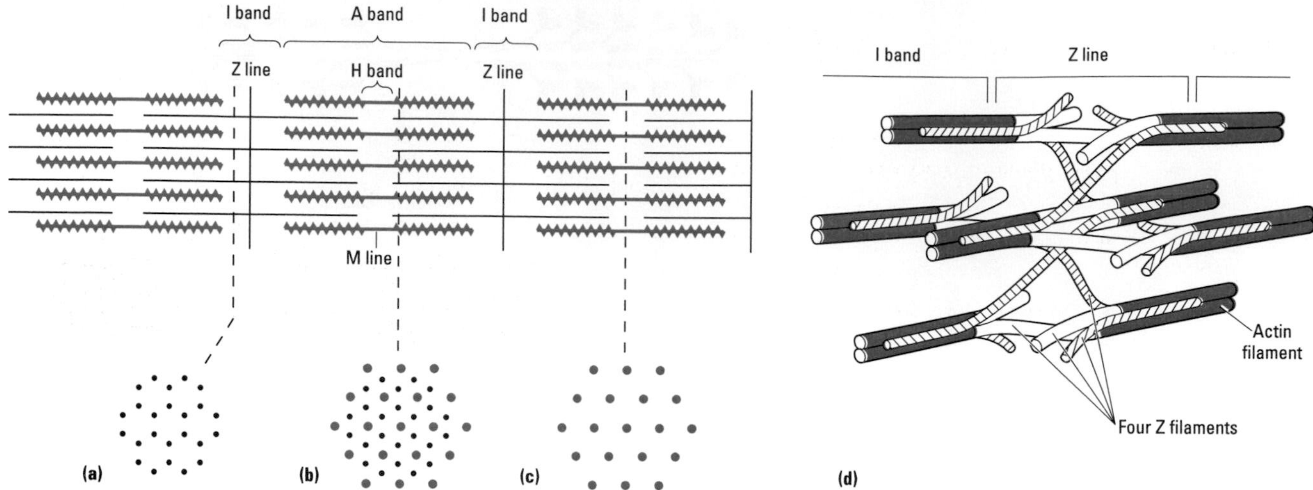

Fig. 9.73 Transverse section views of a myofibril showing the filament spacing: (a) in the I band, close to the Z line, where only actin is present, (b) in the A band where actin and myosin filaments overlap, and (c) midsarcomere in the H zone, with only myosin present. (d) The interactions of the Z filaments and actin filaments at the Z line.

opposite polarity in each half sarcomere (see Fig. 9.70a); hence the cross-bridges at either end of the filament oppose each other, and both ends can pull actin molecules in towards the center of the sarcomere. As the thick filament continues to assemble to its final length, the heads come to project sideways from its axis, at regular intervals and in different rotational positions, to form a helical array with a periodicity of 42.9 nm and an axial repeat of 14.3 nm (see Fig. 9.70c).

Thick filaments also contain a proportion of the less well-known **C-protein**, and in mammalian muscle antibodies to C-protein have revealed discrete binding regions in each half of the thick filament. The role of this component is still uncertain, but it may function to slow the contraction velocity at low levels of activation.

Patterning of filaments within the sarcomere

Electron micrographs of transverse sections at different points along the sarcomere can show the complex geometry of the filament proteins. Sections in midsarcomere, where thick and thin filaments overlap, show that myosin filaments here are arranged in a hexagonal lattice with the actin filaments at the trigonal points (Fig. 9.73a). Each actin filament is therefore surrounded by three nearest neighbor myosin filaments, whereas each myosin is surrounded by six nearest neighbor actin filaments. At the M line adjacent myosin filaments again show a hexagonal network structure, and the filaments here are cross-linked by M bridges (Fig. 9.73b), which are thought to contain creatine phosphokinase and some other protein(s), and which may serve to maintain the thick filaments in their hexagonal array.

The region next to the Z line only contains the thin actin filaments, and electron micrographs here reveal a characteristic "basketweave" pattern (Fig. 9.73c), where the actin I filaments form a square lattice such that each is connected by four Z filaments to four other I filaments. The protein **α-actinin** is an important component of these Z filaments and of the Z line; it is a rod-shaped molecule (194 kDa) with a diameter around 3–4 nm and a length of 30–40 nm. It is probable that other proteins are present here, since there is insufficient α-actinin to account for the mass of the Z line, but their nature and function are unclear.

Cytoskeletal proteins

Cytoskeletal proteins are present in striated muscle fibers and have a major role in stabilizing the sarcomere, thus allowing the correct interaction of the contractile filaments. At each end of a myofibril (i.e. at the terminal sarcomeres) the actin filaments are anchored to the basal lamina of the muscle fiber, and to the tendon. This occurs via a network of different cytoskeletal proteins including **dystrophin** and various **sarcoglycans**. The distribution of other cytoskeletal proteins thought to be involved in maintaining the stability and integrity of filaments in a myofibril is illustrated in Fig. 9.74. **Connectin** is an important cytoskeletal protein, linking the Z line to the myosin filaments, whereas **nebulin** (about 500 kDa) runs along the length of the actin filaments and gives them additional linkage to the Z line. Additionally, **desmin** is a filamentous protein forming a collar around each myofibril at the Z line and connecting two adjacent Z discs together, so producing a framework where myofibrils are aligned with their striations in register across the width of the muscle fiber. **Titin** is a very large molecule (about 1000 kDa) occupying about 9–10% of the myofibril mass (see Table 9.9), and apparently running as a continuous strand from the M line to the Z line. Its molecular structure may be a random coil, making it inherently elastic, but there are also local areas that fold and unfold particularly easily. The molecule may therefore act as a parallel elastic component to promote sarcomere stability.

9.10.3 Sarcoplasmic reticulum and the T-tubule system

Invaginations of the surface membrane of each muscle fiber in classic study tissue (frog gastrocnemius muscle) form a system of branched T-tubules, about 0.04 μm in diameter, running transversely into the fibers. A "ring" of these invaginated T-tubules occurs around each Z disc along the fibril; the T-tubule membranes

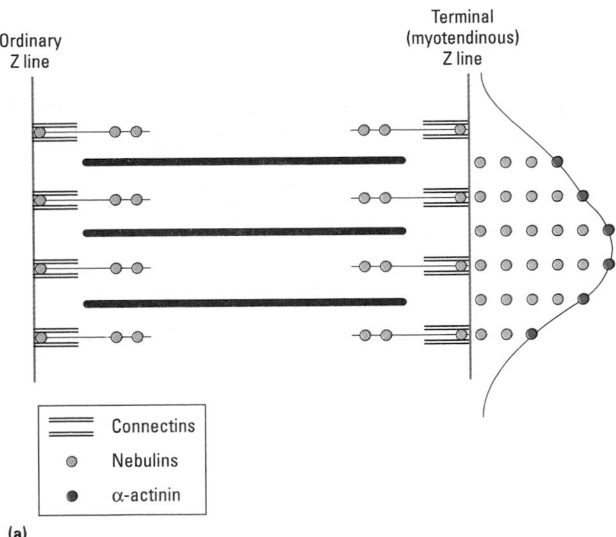

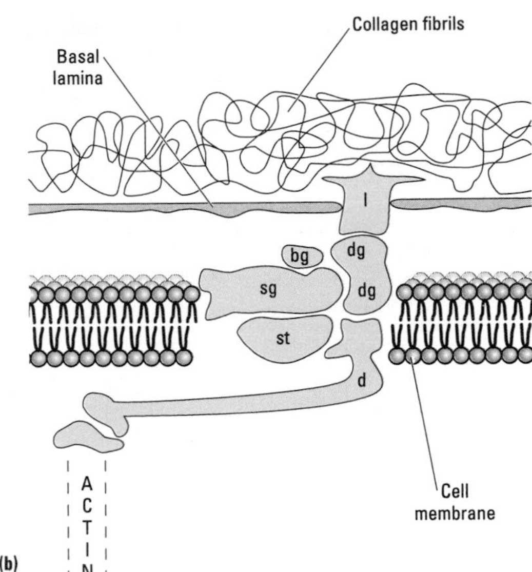

Fig. 9.74 A highly schematic view of cytoskeletal proteins linking actin filaments to the Z line, both at normal Z lines and at the terminal Z line where muscles join the tendon. (b) The possible method of anchorage of a single actin filament via dystrophin to a cell membrane; note that the anchorage is strengthened by going through the membrane and into the associated connective tissue collagen external to the cell. bg, biglycan; d, dystrophin; dg, dystroglycan; l, laminin; sg, sarcoglycan; st, syntrophin.

are continuous with the plasma membrane, and their lumen is continuous with the extracellular fluid outside the fiber. The T-tubules therefore provide the structural link between the surface membrane and the myofibrils in the fiber interior.

The SR is a further network of internal membrane vesicles, forming a collar of "terminal cisternae" around the myofibril on either side of the Z disc, so that the T-tubules are sandwiched between the cisternae (Fig. 9.75). Thinner longitudinal elements of SR also link between the cisternae of adjacent sarcomeres.

The SR is connected to the T-tubules at structures traditionally called "triads" because they appear in micrographs as three sacs: the two larger lateral SR cisternae and the central T-tubule (Fig. 9.76). The triad between the T-system and the SR may be regarded as an "intracellular synapse". In frog muscle there is a 15 nm gap between the T-tubule membrane and the membrane of the terminal cisternae, and there are electron-dense particles protruding from the SR membrane into this gap that have been termed "feet". These feet are seen in electron micrographs as having four subunits, and they form a regular tetragonal lattice, with about 800 "feet" per μm^2 of junctional membrane, so that the feet amount to 27% of the surface area of the terminal cisternae. The feet are now identified as membrane-bound protein complexes, called **ryanodine receptors**. Half of them are lined up with proteins on the T-tubule membrane, the **dihydropyridine receptors**. As we will see, together these two receptor proteins play a key role as calcium channels in linking excitation of the muscle to the contraction processes.

9.10.4 Neuromuscular junctions

The region of the muscle fiber where the neuron terminals make contact is called the end-plate, or neuromuscular junction, commonly referred to as the NMJ. Twitch muscle fibers in mammals each have one or a few discrete NMJs. An AP in the motor neuron produces an AP in the muscle plasma membrane via this NMJ, in an "all-or-nothing" fashion. A single stimulus produces a single "twitch" (a relatively rapid development of tension followed by a longer period of relaxation). The structure of a vertebrate NMJ can be seen in Fig. 9.77.

The gap between the opposing cell membranes at the NMJ is a synaptic cleft, which is continuous with other extracellular spaces in the body. The basic synaptic structure (between two nerves) was covered in section 9.3.1, and the NMJ synapse is essentially similar. However, the synaptic gap is wider, at about 50 nm, and there is an intervening basal lamina, which may house some of the signaling components. The presynaptic (acetylcholine-containing) vesicles are docked in two rows each side of the presynaptic density, and within a few nanometers of rows of calcium channels in the neuron membrane. There are also junctional folds postsynaptically (covered with a network of collagen fibrils), and in the frog NMJ these folds give a postsynaptic membrane area of around 1500 μm^2 area per end-plate. Voltage-gated sodium channels lie at the bottom of these postsynaptic folds, and acetylcholine receptors can be particularly dense there (up to 10,000 μm^{-2}), directly opposite the presynaptic active zones. Rapsyn is an important anchoring protein at the NMJ, occurring in a 1 : 1 ratio with AChR, and probably responsible for maintaining the ACh receptor clustering.

Initial neuromuscular synapse formation involves a trigger from the neuron, probably agrin (referred to earlier in section 9.3.1), and at least two muscle proteins, rapsyn and a transmembrane tyrosine kinase (MuSK), which together organize the initial patterned clustering of the acetylcholine receptors. There may also be a nerve-derived "dispersal factor" that eliminates any receptor clusters that have not been properly stabilized by the presence of agrin. In fact numerous signal molecules that are synaptogenic or antisynaptogenic are now known from genetic work on *Drosophila* and *Caenorhabditis*, and there may be significant functional overlap and redundancy here.

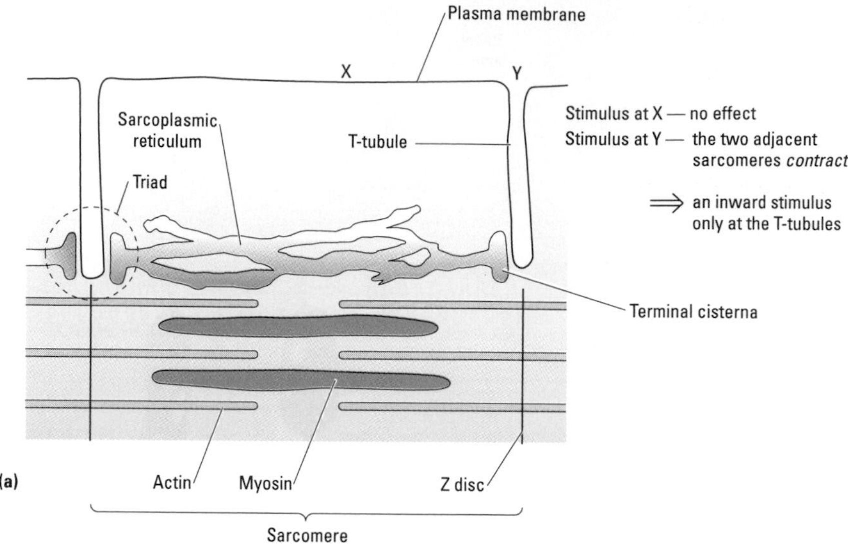

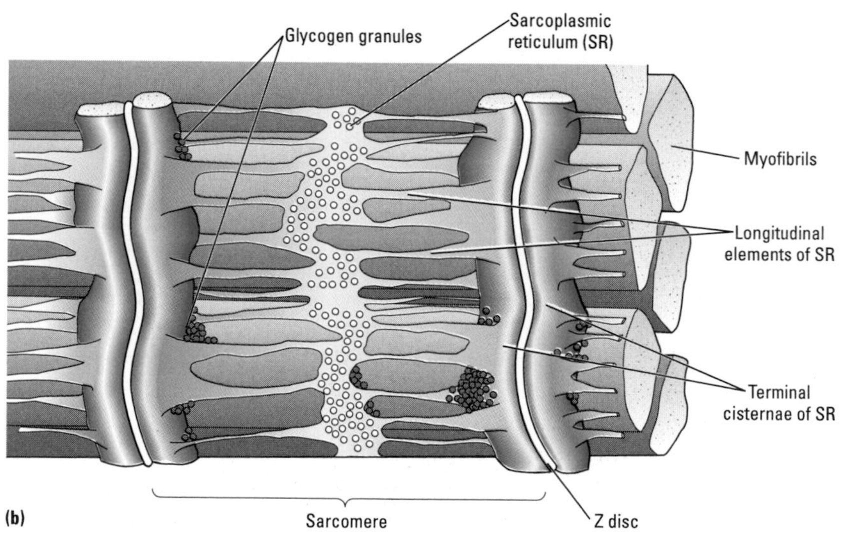

Fig. 9.75 The relationship of sarcoplasmic reticulum and T-tubules to individual sarcomeres, (a) in longitudinal section, and (b) in surface view.

The structure of an acetylcholine receptor in the postsynaptic membrane is shown in Fig. 9.78a. The receptors are pentameric proteins of 290 kDa, comprising two α and one each of the β-, γ-, and δ-chains. The subunits exhibit considerable sequence identity with each other, and they all have four hydrophobic segments, which are thought to form membrane-crossing helices (Fig. 9.78b).

Glial cells are present around the vertebrate NMJ, and neurotransmitter release also causes a G-protein cascade there, indicating some potential neuromodulatory role for the glia at a NMJ, as has been demonstrated in synapses between two nerve cells.

9.11 Muscle contraction

Muscle fibers can shorten by up to 25% of their resting length because each of the thousands of sarcomeres arranged in series in the myofibrils can itself shorten by about 25% (though in practice contractions are usually much less than this). When this happens it is apparent with light microscopy that the A bands (corresponding to the myosin filaments) do not change in length, whereas the I band and the H zone both become shorter (Fig. 9.79). From this it can be inferred that the thick filaments (A filaments) and thin filaments (I filaments) slide past one another, the thin filaments being pulled towards the center of the sarcomere. Neither type of filament changes in length at all, but the degree of overlap between them increases. This is the **sliding filament theory** of muscle contraction.

9.11.1 Sliding filaments and cross-bridge formation

Basic mechanism

The sliding filament mechanism is effected by the cross-bridges formed by the myosin heads attaching to and detaching from the

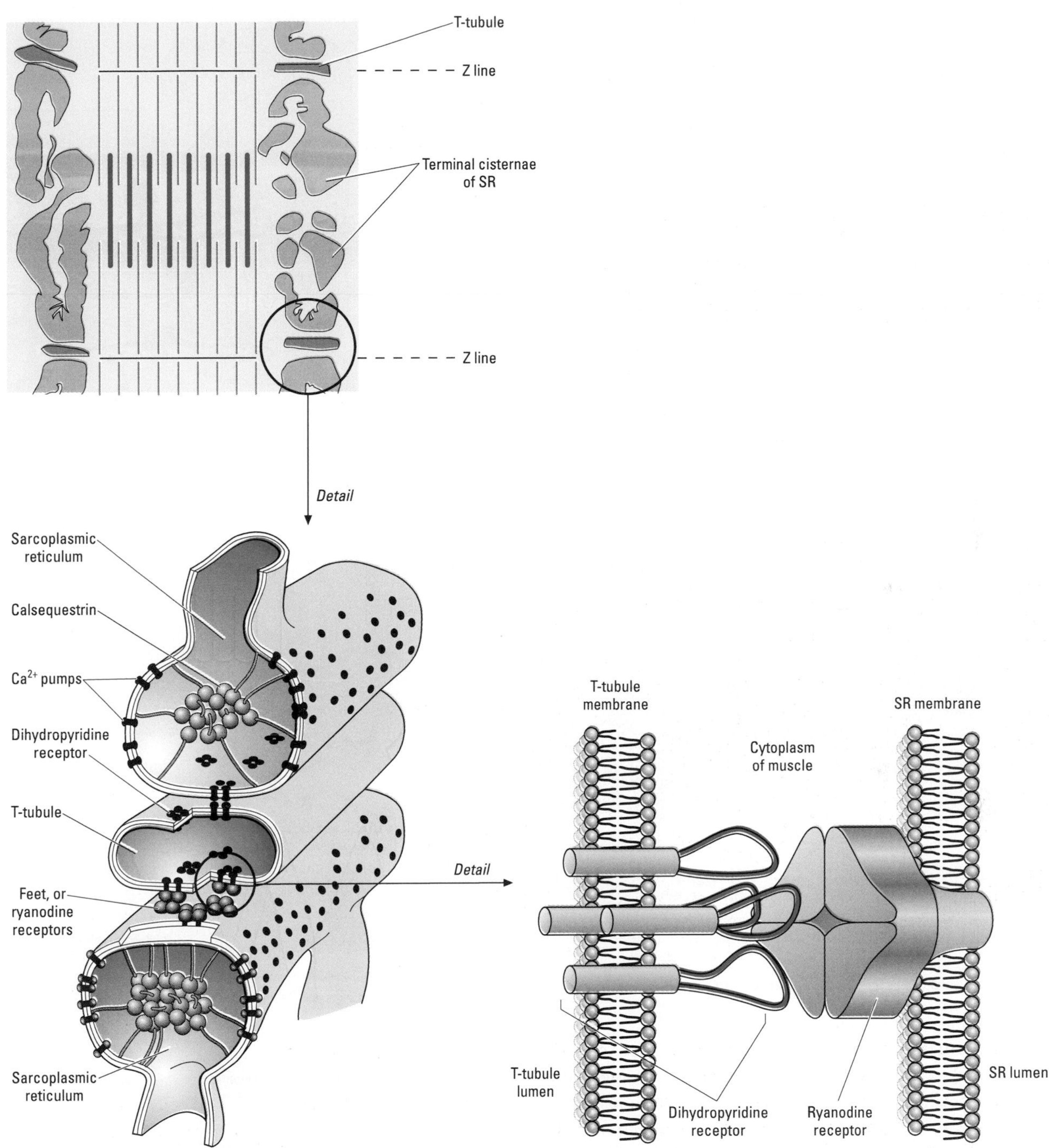

Fig. 9.76 The structure of a triad, where two terminal cisternae of the sarcoplasmic reticulum (SR) are in close contact with an invaginated T-tubule. The gap between the two systems is bridged by the "feet" on the SR membranes, which are actually the ryanodine receptors, many of these lining up exactly with dihydropyridine receptors on the T-tubule surfaces to form functional calcium channels.

adjacent actin filaments in a regular sequence. Each cross-bridge generates a tiny force that drags the actin along, and the total force generated is proportional to the number of cross-bridges that are forming at any time, which is in turn related to the degree of overlap of the two kinds of filament. The process can be usefully visualized as the myosin cross-bridges being like a series of oars on the sides of a boat, "rowing" themselves independently along the actin.

The essential elements of the cross-bridge mechanism are that the myosin heads bind to sites on the actin filaments forming cross-bridges; that the myosin heads generate a force that tends to move the actin filaments towards the center of the sarcomere, thus shortening the whole sarcomere; that the cross-bridges can attach and detach many times, so that the sarcomere can shorten more than the length of an individual cross-bridge; and that the energy to drive the cycle comes from ATP split during each cycle.

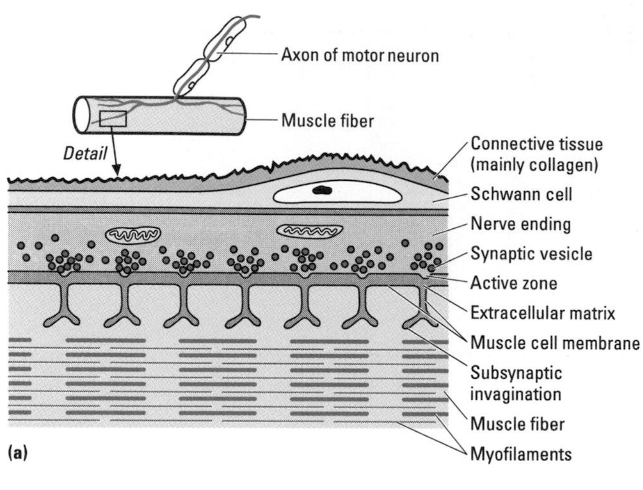

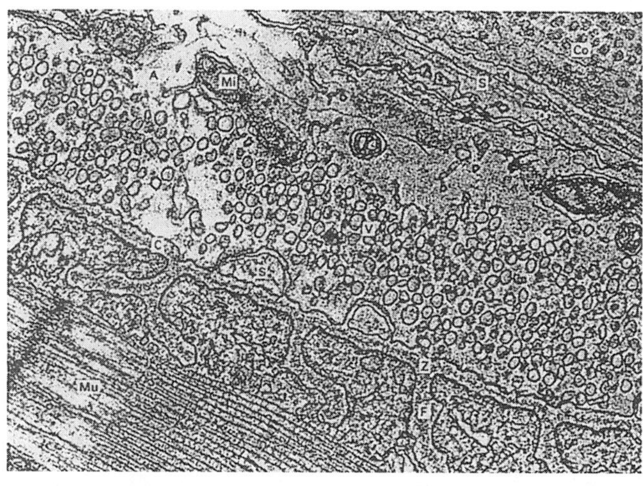

Fig. 9.77 The structure of a vertebrate neuromuscular synapse (or neuromuscular junction), (a) in diagrammatic section, and (b) as seen in an electron micrograph. C, synaptic cleft. (a, Adapted from Reichert 1992.)

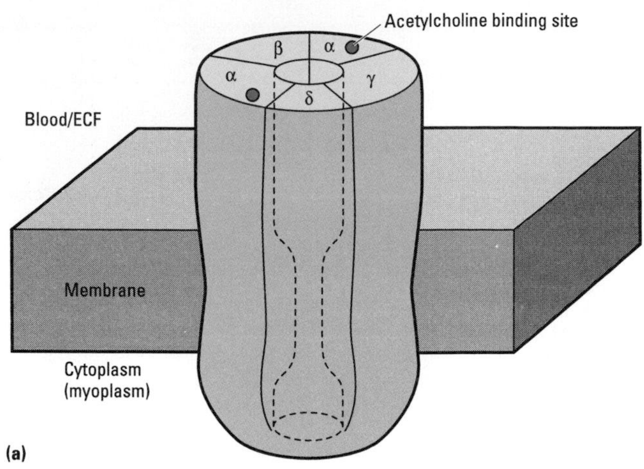

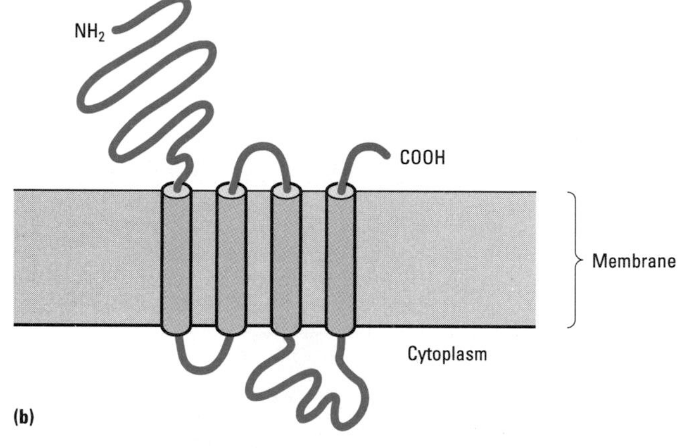

Fig. 9.78 Structure of the acetylcholine (ACh) receptor: (a) the five constituent subunits sited in the membrane, with two ACh-binding sites, and (b) the molecular structure of one subunit polypeptide, with four membrane-spanning segments. ECF, extracellular fluid.

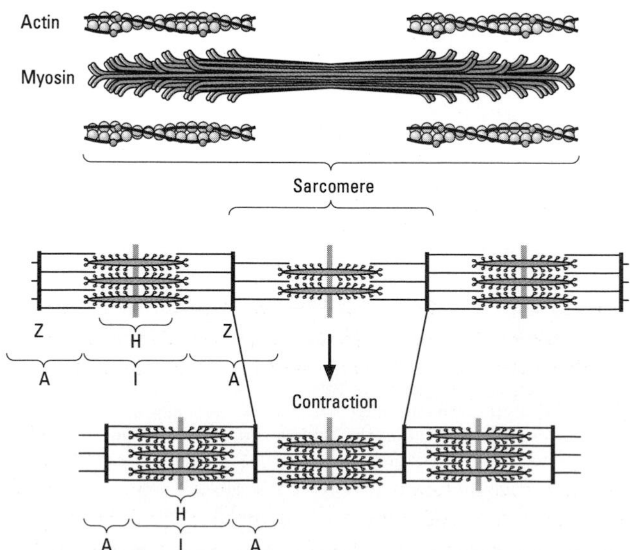

Fig. 9.79 Contraction effects on the fibril band pattern and its relation to myofilament overlap. Note that the A band and the H zone both shorten, but the I band stays at a constant length. (See also Box 9.2.)

Molecular sequence in cross-bridge formation

A highly simplified version of the **cross-bridge cycle** is shown in Fig. 9.80. One crucial point is that actin and myosin bind together readily, even *in vitro*; the main energy-requiring step is not to get them to bind, but to get them to dissociate. Another key observation is that when isolated myosin binds to ATP the myosin acts as an ATPase, but the resultant ADP and inorganic phosphate (P_i) dissociate from the myosin only very slowly (and this step is rate limiting). However, if actin is also present and fully bound to the myosin then the myosin is activated and becomes a "better" (faster) ATPase, and the breakdown products ADP and P_i dissociate much more quickly. Hence the binding of actin to a myosin–ADP–P_i complex is energetically favored. This in turn releases the P_i and the ADP, making room for another ATP; and this finally breaks the actin–myosin bonding.

We can therefore see the process as a series of steps:

- Step 1: hydrolysis of ATP to ADP and P_i, by the myosin-ATPase activity; the ADP and P_i remain tightly bound to the myosin. This [myosin–P_i–ADP] complex is in a state of low-affinity equilibrium with an actin, giving a loose [actin–myosin–P_i–ADP] complex.

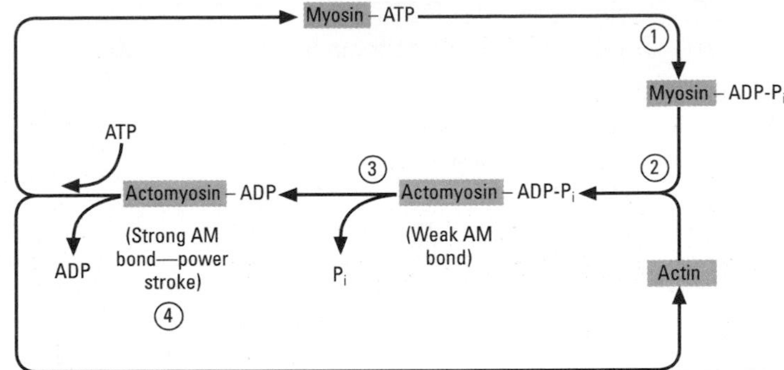

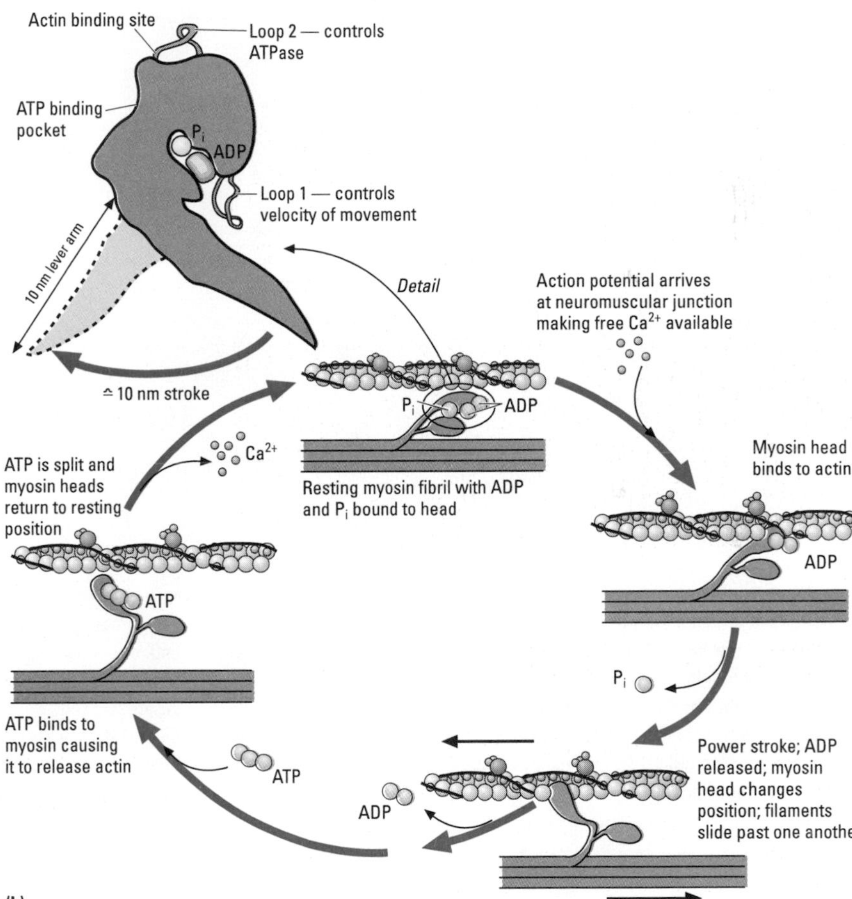

Fig. 9.80 The cross-bridge cycle, shown as (a) a highly schematic equation, and (b) as a structural cartoon of the actin filament and myosin head interactions. In the presence of ATP, a myosin head acts as an enzyme and splits the ATP to ADP and inorganic phosphate (P_i), but these products are released only very slowly from the nucleotide-binding site *unless* the myosin also binds to actin. The bond with actin is initially weak but as it forms it causes the release of P_i and this in turn makes the bond become stronger. Release of the P_i liberates energy, used to generate force as the myosin head rotates relative to the actin filament so that the filaments slide past each other. This rotation may involve the sequential formation of at least four bonds. ADP is then released and replaced by ATP, which breaks the strong actin–myosin complex again. Hydrolysis of the ATP then returns the myosin head to its "cocked" position. (Note that in (b) the second myosin head would normally be going through the same cycle of reactions with actin but for clarity it is shown as if disengaged.)

- Step 2: a slow transition to an activated, high-affinity or "tight" form of the same complex [*actin–myosin–P_i–ADP], this step probably determining the overall rate constant for the cycle. The step 2 conversion triggers step 3.
- Step 3: release of P_i, and the release of energy, resulting in a large conformational change in the myosin head, giving rise to the next step.
- Step 4: relative motion between the myosin head and the rest of the myosin molecule of about 10 nm, with the molecule bending at the hinge region; the conformational change allows release of ADP.
- Step 5: ATP binding to the globular myosin head in place of the ADP, which results in the rapid dissociation of myosin from actin and the myosin returning to its original "cocked" position.

The myosin head is then able to bind to a new site a little further along the actin filament, and the net result is a displacement of the actin filament relative to the myosin. This occurs repeatedly in these small incremental sequences of cross-bridge attachment, rotation, and release.

Although each myosin molecule is a dimer, each of its heads seem to operate quite independently, and only one head attaches to a myosin at any given moment. Since individual cross-bridges form and move independently (but all in the same direction), the outcome is a smooth sliding movement between the two types of filament, so that they increasingly overlap and there is a lengthwise contraction of the sarcomere.

Box 9.2 The length–tension relationship in striated muscle

Given the dimensions of the myofilaments in frog muscle (a), the characteristic shape of the length–tension relationship (b), as measured in a classic experiment on a single fiber, is neatly explained by the sliding filament mechanism. This is clear from (c), which shows the equivalent states of overlap of the actin and myosin filaments at different sarcomere lengths.

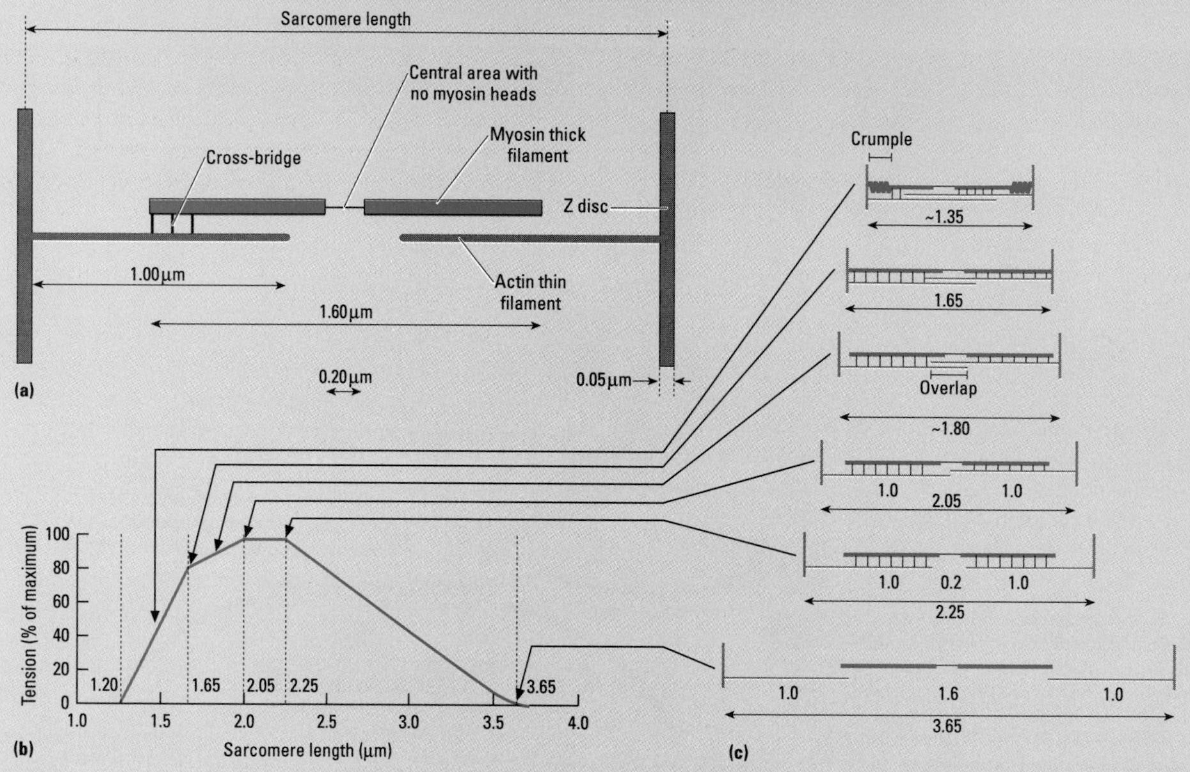

- The maximum force is produced when there is maximum overlap between the thick and the thin filaments, because the number of attached cross-bridges is then greatest; this corresponds to the "plateau" at sarcomere lengths between 2.05 and 2.25 μm.
- On the right-hand side of the curve (>2.25 μm) force declines because a part of each actin fibril does not overlap with a myosin fibril, so that the number of attached cross-bridges decreases, reaching zero at the point of no overlap (3.65 μm).
- On the left-hand side of the curve the force also declines because the attachment of cross-bridges is disrupted. Initially this is because the two sets of actin fibrils start to overlap with each other, and at even shorter sarcomere lengths (<1.65 μm) it is made worse because the myosin fibril length exceeds the sarcomere length and the myosin starts to "crumple". The force normally reaches zero at about 1.27 μm.

Thus *measured* length–tension curve exactly fits the *predictions* from the sliding filament theory based on the known filament dimensions for this muscle.

Evidence for the cross-bridge cycle

Our view of how muscle contraction works at the structural and molecular level is strongly supported by considering the relationship between length and tension in a striated muscle (Box 9.2), where the recorded tensions exactly follow the predictions of an actin–myosin sliding overlap system.

It is also possible to conduct *in vitro* assays, involving a myosin-coated surface and an actin filament attached to a glass needle, and these methods have allowed the force developed by the myosin head to be measured using "laser trap" technology. "Step sizes" of around 10 nm and forces of around 5 pN have been found. (Note that these values are similar to those for kinesin moving along a microtubule, and for dynein, the molecular motor that underlies the bending action of cilia and flagella, suggesting a common movement mechanism in all three of the molecular motors.)

The properties of the myosin motor clearly depend on the S1 head structure; this can be confirmed by the fact that this head contains two nonconserved surface loops, and any variability in these can produce molecular motors with different properties. Loop 1 (residues 204–216 in the structure) is near the ATP-binding pocket, and loop 2 (residues 627–646) is at the actin-binding site and is thought to interact with the negatively charged amino-terminal part of actin.

The functioning of the "lever arm" (Fig. 9.80) is also amenable to investigation. The S2 myosin portion that forms the lever arm of the myosin molecule is measured at about 10 nm, and the distance between the ATPase and actin-binding sites at 4 nm, which would together give a potential "swing" close to the step size of 10 nm measured in laser-trap experiments. The correspondence between the calculated movement of the myosin head lever arm and the actual movement of the actin fibril (10 nm per ATP split) has been confirmed by studies of a mutant myosin from the slime mold *Dictyostelium*, with a lever arm about 50% shorter, which moves actin at half the velocity of "normal" myosin, without affecting the ATPase activity.

The total cycle time of the actin-activated ATPase cycle is about 50 ms (for rabbit myosin, at 25°C), and for a 10 nm displacement the velocity along a single head would be about 0.2 $\mu m\ s^{-1}$. However, as the many cross-bridges on one filament go through their cycles asynchronously, more heads can be brought into play until the limit set by the duration of the "strongly bound state" (steps 3 and 4 of the cycle, together) is reached, giving a maximum velocity for filament sliding of about 0.5 $\mu m\ s^{-1}$.

Until recently there was some uncertainty as to whether the force-generating action was the "working stroke" of the actin-bound myosin head and lever, or whether force mainly came from fast-acting components of the attachment and detachment systems. However, evidence using new X-ray interference technology that can resolve changes well below 1 nm has indicated that there is indeed a force-producing working stroke. The first structural event is the movement of myosin heads from their resting helix on the myosin filaments and the shift of the myosin head towards actin; but substantial movement towards actin appears to occur well before any significant force is produced. This suggests that the myosin heads initially attach in the weak (nonforce-producing) state as described in the schema above. Only after a delay (10–15 ms) is force produced (strong binding state), this being accompanied by substantial changes in the X-ray diffraction pattern.

9.11.2 Initiation and control of contraction

End-plate potential

At the point at which a motor nerve makes contact with the muscle it loses its myelin sheath and splits up into a number of fine terminals, each contacting the muscle at a typical NMJ (end-plate). An incoming AP in the motor neuron opens voltage-gated calcium channels in the axon terminals at the end-plate, leading to the release of a large number of vesicles into the synaptic cleft. The postsynaptic potential change occurring at an excited neuromuscular junction is termed an **end-plate potential** (EPP). There is a short delay of <1 ms between the arrival of the AP and the resulting depolarizing EPP in the muscle membrane, largely due to events at the presynaptic terminal and to a lesser extent to the diffusion of transmitter across the synaptic cleft.

Acetylcholine is the transmitter for vertebrate end-plates, and binds to the α-chain of the acetylcholine receptor, causing an increase in membrane permeability to sodium and potassium ions by opening ion channels; two molecules of acetylcholine are needed to open one channel. When an NMJ is excited by an incoming AP, hundreds of quanta of transmitter are released. Each quantum (i.e. each presynaptic vesicle) normally contains nearly 10,000 molecules of acetylcholine, but the result is about 1000 ACh receptor channels opening in response to one vesicle being released, so that a full response involves a few hundred thousand channels being opened. However, in a normal response there is activation of less than 10% of the total receptive area of the NMJ, and the transmitter receptors are actually in considerable excess, and so are likely saturated at the site of a vesicle fusion.

Only the simultaneous release of many vesicles and the concomitant opening of a large number of channels results in an EPP. But even in resting muscle there are small fluctuations in membrane potential in the region of the end-plate, called **miniature end-plate potentials** (MEPPs). These are caused by the release of single vesicles; they can be seen as random "leakage", and are unlikely to be important in neuromuscular communication, but may have a role in guiding NMJ synapse formation.

In vertebrate twitch muscles the EPP crosses the threshold for excitation of the muscle membrane and so gives rise to an AP, which is propagated along the length of the muscle fiber in exactly the same manner as in neurons. Following a latent period of a few milliseconds, the muscle fiber then starts to generate tension. However, in many muscles with multiple innervation points from a single neuron (see section 9.13.1 below), nervous stimulation results in a graded depolarization of the muscle membrane, giving a slow contraction in the absence of an AP. These are referred to as tonic muscles. On repeated stimulation of a tonic muscle, the EPPs summate, resulting in a graded depolarization that depends on the frequency of stimulation.

Excitation–contraction coupling

One fundamental issue has still to be addressed: how is cross-bridge formation (and hence contraction) elicited by an electrical potential at the NMJ on the surface of the muscle fiber?

The initiation of the cross-bridge cycle inside the myofibrils can readily be shown to be due to calcium ions; isolated muscles will not contract in the absence of Ca^{2+} in the bathing medium, or if their internal calcium stores have been depleted. As with all cells, calcium levels in the cytoplasm of muscles are very low (usually much less than 10^{-6} M, and maybe as low as 10^{-9} M), and the relation between calcium concentration and contraction can be tested with "skinned" muscles in artificial ATP-containing solutions. The tension produced rises (nonlinearly) from zero (at 10^{-8} M) to be maximal as expected at about 10^{-6} M (Fig. 9.81).

The effect of calcium in striated muscle is primarily mediated by the two key "extra" proteins in the thin filaments: **troponin** and **tropomyosin** (see Figs 9.69 and 9.82). Tropomyosin is a long molecule lined up along the actin filaments, and when a muscle fiber is relaxed, the tropomyosin molecules block the myosin-binding sites on the actin filament with which they are associated. Troponin, in contrast, is a globular protein complex, and occurs at intervals of about 40 nm along the actin filaments; it has three subunits (Fig. 9.82) each with different functions. Troponin T is a relatively long chain and provides the connection to the tropomyosin molecules; troponin C binds Ca^{2+} with high affinity; and troponin I binds both actin and troponin C and is a specific inhibitor of the myosin-ATPase activity. When calcium binds to troponin C, the whole troponin complex attaches to tropomyosin, and the tropomyosin in

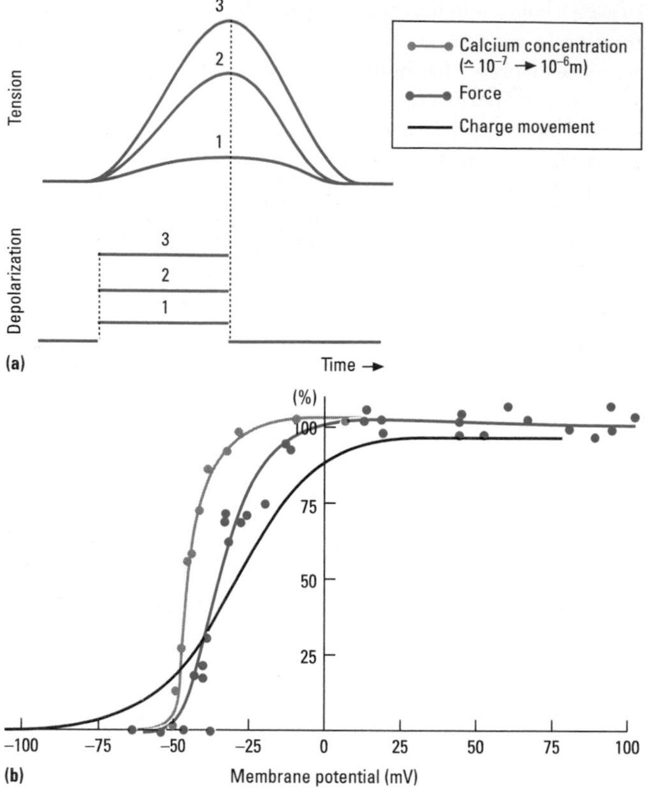

Fig. 9.81 (a) The tension developed in a muscle is directly related to the membrane potential, so that step depolarizations produce graded tension recordings. (b) The effect of membrane potential on the force, the charge movement, and the relative calcium concentration (measured as the arsenazo signal; see text). The threshold potential for a contraction is about −60 mV here.

turn changes its configuration and hence its links with the actin filament. The interactions of troponin C and troponin I thus seem to be central to the calcium regulation of muscle contraction, and different isoforms of the troponin complex can modify the concentration–tension relationship, thereby modifying the rate of relaxation. There is some evidence that the thick filament "C-protein" exerts its effect on contractility by altering the calcium affinity of troponin C, perhaps also having an effect on the mechanical properties of the head of the myosin molecule and tethering it more firmly to the thick filament backbone.

In effect, the presence and binding of calcium start a chain of events that releases the inhibition on actin being able to bind to the myosin cross-bridges, and hence calcium permits a contraction. Since there are two actin-binding sites on a single myosin head, a likely hypothesis that fits well with our earlier model is that when a myosin head interacts with an actin filament it binds initially through a weak interaction at one site. In the absence of Ca^{2+} the other and stronger binding site may be blocked or modified by tropomyosin, so that a full two-site attachment cannot occur. But in the presence of high Ca^{2+}, tropomyosin moves in towards the groove on the actin filament and this allows a second myosin–actin interaction (the step 2 transition, to [*actin–myosin–P_i–ADP]), removing the inhibition on phosphate release so that the step 3 transition can occur. This so-called "steric blocking mechanism" of thin filament regulation is the primary means of controlling contraction at the level of the filaments in vertebrate striated muscle (Fig. 9.83), though as we will see other mechanisms do operate in other types of muscle.

How then does an EPP or AP in the muscle fiber membrane affect calcium levels internally to start the contraction process? This is the essence of excitation–contraction coupling, and it requires a Na^+-dependent membrane potential change to be converted to substantial and distant movements of Ca^{2+} ions. The potential change at any excitable membrane is scarcely detectable in the cytoplasm, penetrating less than 1 μm into the interior, and will have no direct effect on internal myofibrils that may be up to 100 μm distant. Electrotonic spread via local currents cannot be an adequate explanation either, given the speed and magnitude with which excitation and contraction are coupled; nor is it possible for calcium ions to be simply diffusing into the muscle from the plasma membrane, as this would be many times too slow. Yet contraction can be induced rapidly and throughout a muscle by simply depolarizing the membrane; even a small "artificial" depolarization to about −25 mV (e.g. using extracellular potassium to change the cell's resting potential) gives a full tension in the muscle.

The key to this tight coupling of membrane potential and muscle contraction lies in the structure of the muscle fiber itself. To explain the speed of the coupling, the potential change must be getting "inside" the fiber, and must be triggering the release of calcium from some internal store close to the myofibrils. The structures that permit both of these events are, evidently, the T-tubule–SR complexes at the Z discs. The anatomical link between the fiber surface and the myofibrils provided by this deeply penetrating membrane system can be demonstrated in the frog by the fact that stimulating microelectrodes only work when applied over the Z discs. It is further confirmed because in some unusual crustacean striated muscle, and in some lizards, the T-system lies over the A bands rather than at the Z discs, and here microelectrodes only work when their tip lies over the A bands. Physically uncoupling the T-tubules from the plasma membrane is also possible (for example with an osmotic shock), and this too removes the ability of a depolarization to elicit contraction, although the muscle is still quite able to contract.

The AP is conducted deep into the fiber interior via the membranes of the T-system, and is thereby linked to the ramifying network of the SR. The SR is in essence a calcium-sequestering system; when an AP is conducted along the membrane of a T-tubule the neighboring SR releases calcium into the cytoplasm. The SR accumulates its calcium store via calcium pumps (Ca^{2+}/Mg^{2+}-ATPases) in the membranes, especially along the longitudinal elements. Within the SR the calcium is bound to the protein **calsequestrin**, so that the concentration of free calcium ions remains low even within the system and the gradient against pumping activity is relatively low. Various markers and dyes have been used to show the movements of calcium associated with this system. Originally the bioluminescent protein aequorin was used, which emits light in the presence of calcium; now more sensitive dyes such as furaptra or arsenazo III can be injected, which will fluoresce maximally only when free calcium is absent and so provide a concentration-dependent monitor of calcium levels. Such techniques show a brief increase of cytoplasmic calcium (the "calcium transient") each time a muscle is electrically stimulated (Fig. 9.84). The duration of

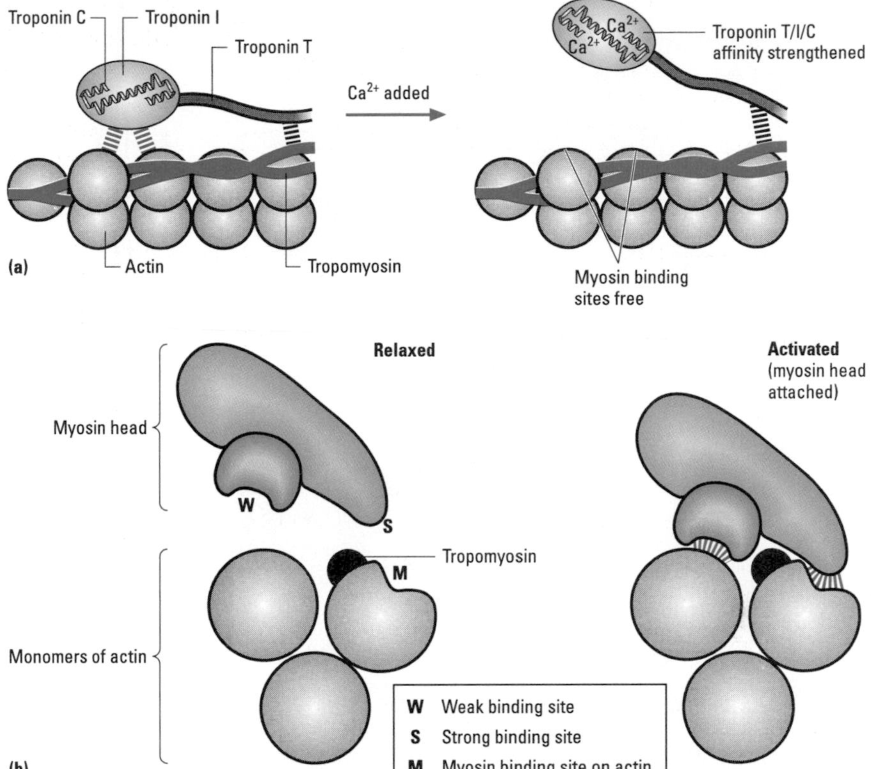

Fig. 9.82 (a) Details of the three troponin subunits and their binding to the actin filament, with and without calcium. When Ca^{2+} is present, it binds to troponin C and this strengthens both the troponin C–troponin I bond and the troponin C–troponin T bond, but breaks the troponin I–actin bond, which in turn exposes myosin-binding sites on the actin filament. The tropomyosin also shifts slightly in the presence of calcium; this is shown in more detail in (b). (b) An "end-on" view of the actin filament with a myosin head above it, where in the activated state the tropomyosin is deflected sideways by about 20° to expose the myosin-binding site.

the calcium transient sets the duration of the twitch response to a single stimulus (Fig. 9.84b); this duration is commonly several microseconds, but in the high-frequency, sound-producing muscles of the toadfish (*Opsanus tau*) it has a half width of only 1.5 ms at 25°C.

The link between the T-system and the SR at the "triads" is provided by the **ryanodine receptors** in the SR membrane, which function as intracellular calcium channels. These receptors are named after the plant alkaloid ryanodine, which locks open the calcium-release channels of the SR. The ryanodine receptors are large proteins (400 kDa), and these proteins aggregate automatically in artificial membranes into groups of four, to form particles with a characteristic clover-leaf appearance (Fig. 9.85), very like that of the "feet" seen at the triad (see section 9.10.3). These aggregates then function spontaneously as channels with high conductances.

The **dihydropyridine receptors** in the opposing T-system surface membrane are sensitive to voltage changes (i.e. they are voltage-gated L-type calcium channels), and in response to an AP (or any other imposed depolarization) they undergo a conformational change. This change of shape is transmitted, apparently purely mechanically, through the bulbous head of the ryanodine receptor, which spans the triad gap between the T-tubule foot and the SR, thus opening the calcium channel of the neighboring ryanodine receptor into the SR (Fig. 9.85). Hence an AP in the membranes of T-tubules triggers the receptor–protein complexes that lead through to the neighboring SR, and the SR releases a flood of calcium into the cytoplasm.

However, in frog muscle only about half the T-tubule ryanodine receptors are aligned with SR dihydropyridine receptors. The other half are activated differently, in direct response to the rising cytoplasmic calcium levels; thus there is a second rise in calcium as more Ca^{2+} channels open in the SR membrane ("calcium-induced calcium release"). Then, as the AP passes and the membrane potential returns to the resting level, all the calcium channels close again.

During activation, the calcium concentration inside the muscle fiber rises sharply, within a few tens of microseconds, from about 1 μM to 10–100 μM (the extent of this change is partially modulated by the calcium-binding protein **calmodulin**, which interacts with both of the key triad receptor proteins). This provides the trigger needed for calcium binding to troponin and the initiation of the contraction process. Troponin concentrations in vertebrate muscles are around 240 μm, so there is a large buffer or sink for the released calcium ions and they are bound very quickly. In some fast-contracting muscles an additional calcium-binding system is provided by cytoplasmic **parvalbumins**, insuring rapid relaxation. Any unbound calcium is pumped back into the SR lumen, reducing the cytoplasmic calcium levels and so favoring release of calcium from the troponin again.

A summary of the events mediating excitation and contraction is given in Fig. 9.86.

9.12 Muscle mechanics

9.12.1 Work, power output, and energy

Muscle performance is determined by three key variables, each determined by intrinsic mechanical features of the individual

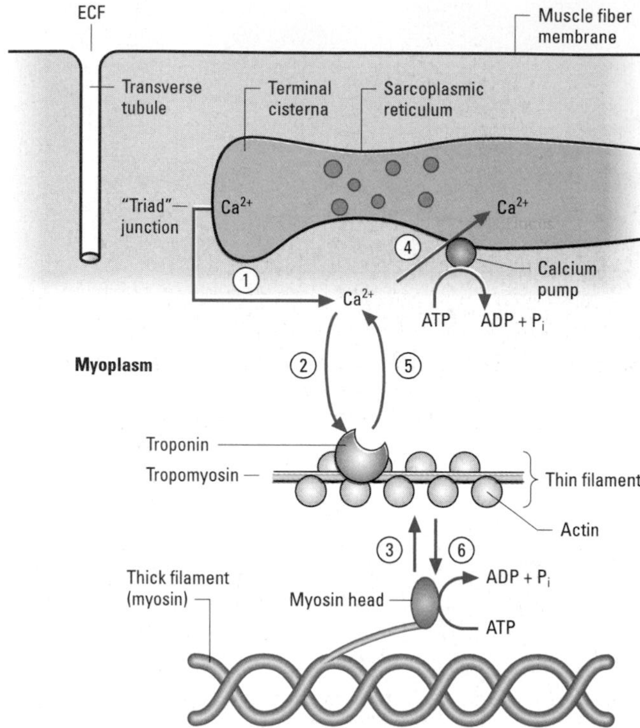

Fig. 9.83 Major steps in the troponin-based actin–filament control of striated muscle activation and relaxation. 1, Calcium is released from the sarcoplasmic reticulum (SR) into myoplasm via the triads, in response to membrane depolarization. 2, Calcium binds to troponin, so releasing inhibition of the actin filament. 3, Cross-bridges can then attach from the myosin heads to the actin. 4, Calcium is resequestered by calcium pumps in the SR. 5, As calcium concentration declines, calcium ions come off the troponin, preventing further cross-bridge formation. 6, Existing cross-bridges detach as ATP is split. ECF, extracellular fluid.

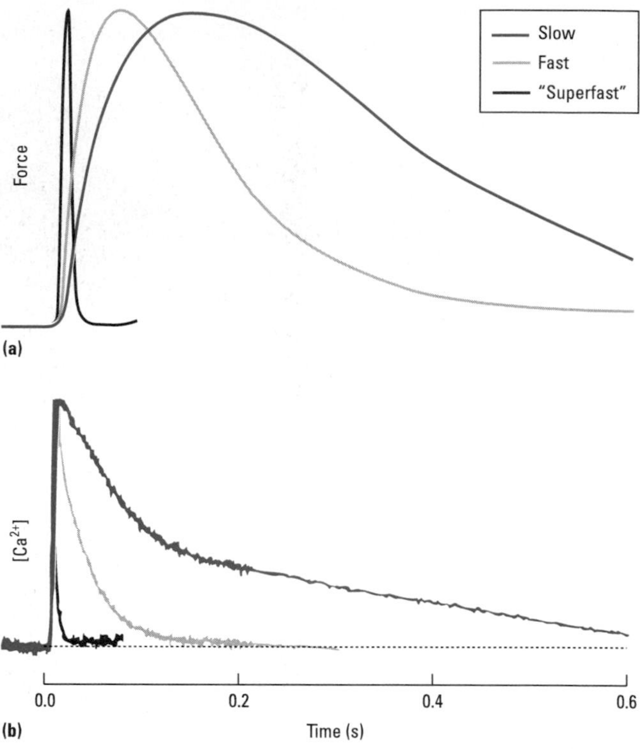

Fig. 9.84 The time course of the "calcium transient" (b) in three different types of muscle, in relation to the generation of tension shown in (a). All traces are normalized to their maximum value. The twitch time course and the calcium transients become shorter going from slow twitch to fast twitch fibers, and are exceptionally short in the "superfast" twitch fibers of the toadfish swim-bladder muscles used for high-frequency sound production.

muscle's design. The three factors are the **stress** (force per unit area, called tension) that the muscle can exert, the **strain** (the ratio of shortening distance to extended length, effectively the length through which the muscle can shorten), and the **frequency** with which it can contract.

To set the scene, typically a vertebrate skeletal muscle may have about 100 cross-bridges operating on one end of a myosin fibril at any one time, giving a maximum **tension** of about 530 pN at the middle of the filament, balanced by an equal tension from cross-bridges pulling on the other end of the fibril. More tension per filament could theoretically be supplied if more cross-bridges formed on a longer myosin filament, but that would put too much force on the filament and risk pulling it apart. There are about 5.7×10^{14} myosin fibrils per square meter in cross-section, giving a maximum stress for typical muscle tissue of about 300 kN m^{-2} (or kPa, kiloPascals). That same tissue also has a maximum **strain** of about 25% or 0.25, though this maximum is only reached in extreme exertion. The **rate of movement** of the filaments depends on the rate of attachment and detachment of the cross-bridges; though remember that, because of the sliding overlap system, the two ends of the sarcomere approach each other at twice the rate of filament movement.

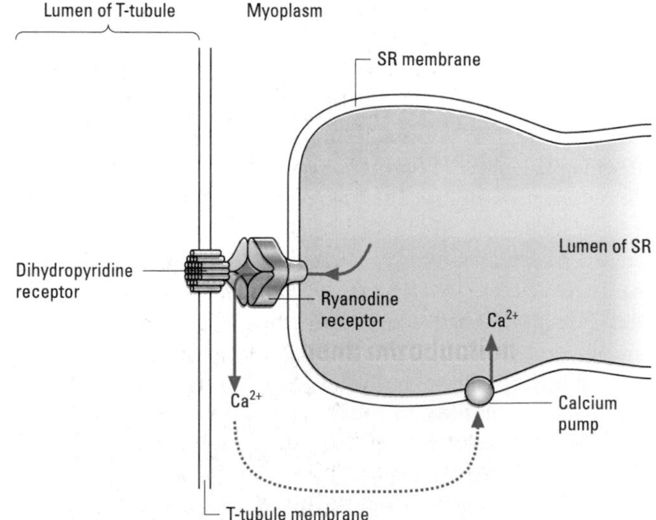

Fig. 9.85 Functioning of the ryanodine receptor + dihydropyridine receptor at the triad interface between the sarcoplasmic reticulum (SR) and T-tubules. When the T-tubule membrane depolarizes after receiving an action potential, the dihydropyridine receptor conveys this signal to the neighboring ryanodine receptor (with which it is aligned), and the calcium channels within the ryanodine receptor open. This allows calcium ions to flow out of the SR into the muscle cell cytoplasm (myoplasm), where they bind to troponin and allow cross-bridge formation (see Figs 9.80 and 9.82). When the T-tubule membrane potential is restored to normal the calcium channels close and excess calcium in the myoplasm is sequestered back into the SR by calcium pumps.

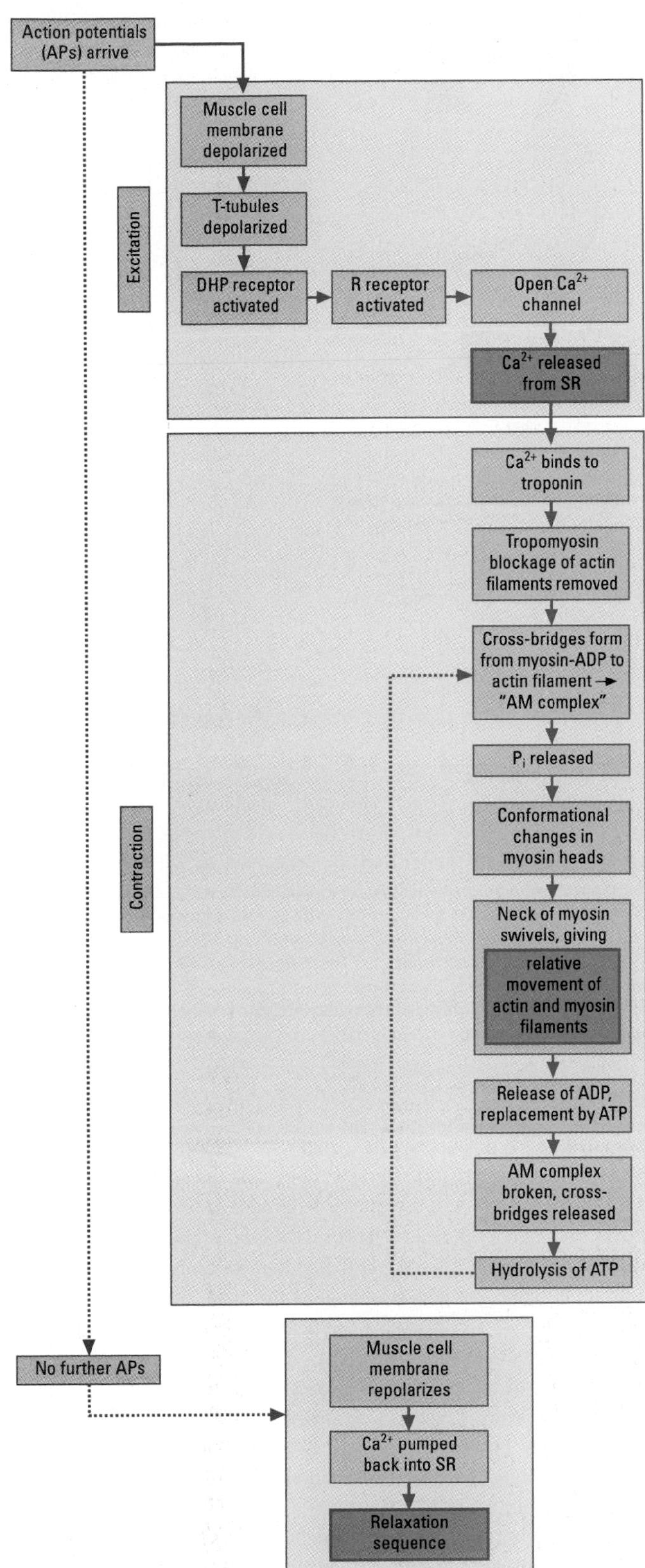

Fig. 9.86 A flow chart summary of the events involved in excitation/contraction coupling in striated muscle. ADP, adenosine diphosphate; ATP, adenosine triphosphate; DHP, dihydropyridine receptor; SR, sarcoplasmic reticulum.

Force and velocity of muscle twitch contractions

In early studies, muscle mechanics were studied largely by simple determinations of two fundamental mechanical properties of muscle: the **length–tension relationship** (see Box 9.2), and the **force–velocity curve**, which involves stimulating an isolated muscle to contract against a fixed load (or better, a servomotor system that regulates the apparent weight against which it is contracting) and then measuring its velocity of shortening.

Muscle contraction can be either **isometric** (meaning "same length"), when there is no shortening but work is being done (for example when you try and lift a load that is too heavy, the arm muscles are contracting very hard but not actually shortening at all); or they can be **isotonic** (meaning "same tension"), when a clear shortening is seen. During measurement of the force–velocity relation the muscle will initially contract isometrically, but once it is generating enough force (P) to move the attached mass it starts to contract isotonically at a constant velocity (Fig. 9.87), and a measurement of V is then taken (effectively at the plateau of the length–tension relationship shown in Box 9.2, with a maximum number of cross-bridges operating). With no load attached the muscle shortens at the maximum possible velocity (V_{max}); and at some high loading it will not contract at all. Between these two extremes the shortening velocity will decrease as the load increases. In fact, the force–velocity relation is parabolic (Fig. 9.88ci). Since the product of force and distance (in this case the contraction length ΔL) is work, it follows that:

$$\text{Power} = \frac{\text{work}}{\text{time}} = \frac{\text{force} \times \Delta L}{\text{time}} = \text{force} \times \text{velocity of contraction}$$

Thus we can work out the power output of the muscle at any one time, and plot this against velocity as in Fig. 9.88cii. The relation is bell-shaped, with maximum power at some intermediate velocity, indicating that muscles work optimally at "moderate" speeds. In fact this optimum power occurs where the speed is about 30–40% of V_{max}, and this ratio holds true for most muscles tested. Note also that overstretched muscles exert a negative power output, and instead absorb energy, thereby acting like a brake.

The metabolic rate of a muscle can be measured (as oxygen consumption or ATP consumption), and the relationship between metabolic rate and relative shortening speed for a mammal is shown in Fig. 9.88c; the increase in energy expenditure with shortening is known as the Fenn effect. The ratio of the mechanical power output to the metabolic rate provides one measure of the **efficiency** of a muscle. The maximum value is usually about 0.45, though it may be less in some muscles. (However, note that the overall efficiency is only about half this value, since the efficiency with which energy from combustion of foodstuffs is converted to ATP is itself only about 0.5.)

More recently the so-called "**work loop technique**" has become the main approach used to investigate the biomechanics of muscle under conditions more nearly simulating real life muscle action. In locomotion, muscles do not normally contract at a constant length or shorten under a constant load; rather, they undergo a series of cyclical contractions where activation state, length, and force are continuously changing, as wings or fins beat repetitively or legs swing backwards or forwards. To analyze a work loop, one end of

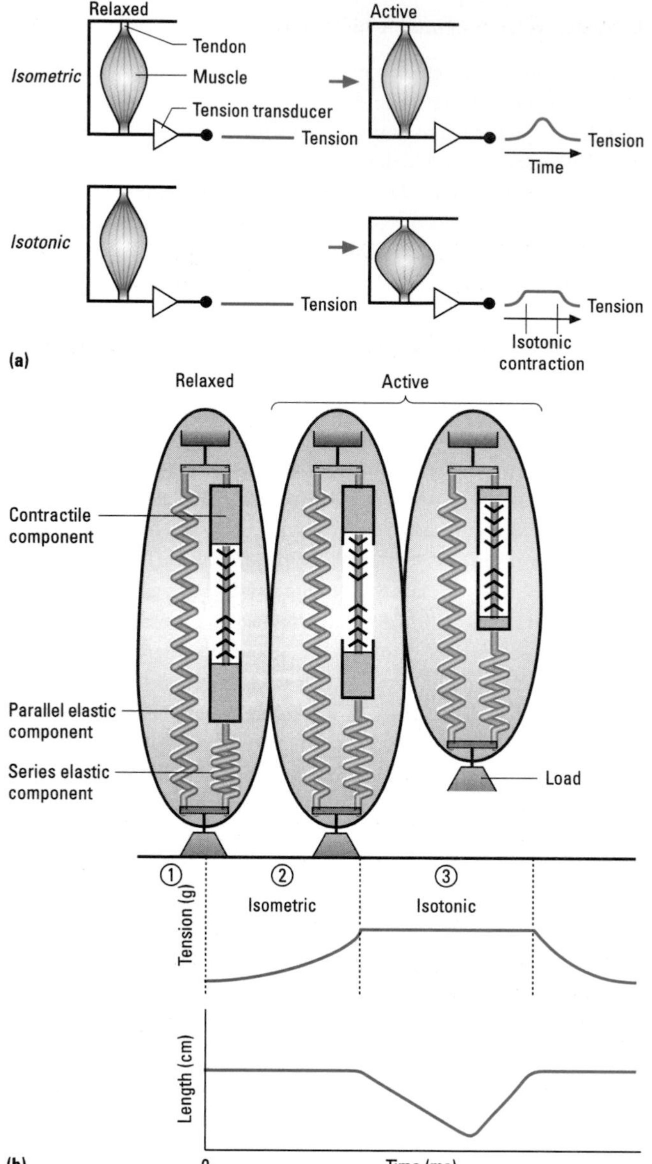

Fig. 9.87 A comparison of isotonic and isometric contraction in a muscle. (a) An idealized view—muscles can either contract with no change of length (top, an isometric contraction) or with change of length but constant tension (bottom, an isotonic contraction). (b) In practice, even an isotonic contraction begins and ends with an isometric phase. This is because muscles contain elastic components, both in series (e.g. tendons) and in parallel (e.g. connective tissues within and around the muscle) with the contractile system. At the start of a contraction as the filaments slide past each other the only effect is to stretch the series elastic components and there is no visible length change, hence giving an isometric phase; later the muscle has generated tension greater than the weight of the load and begins to lift it, changing its own length and working isotonically.

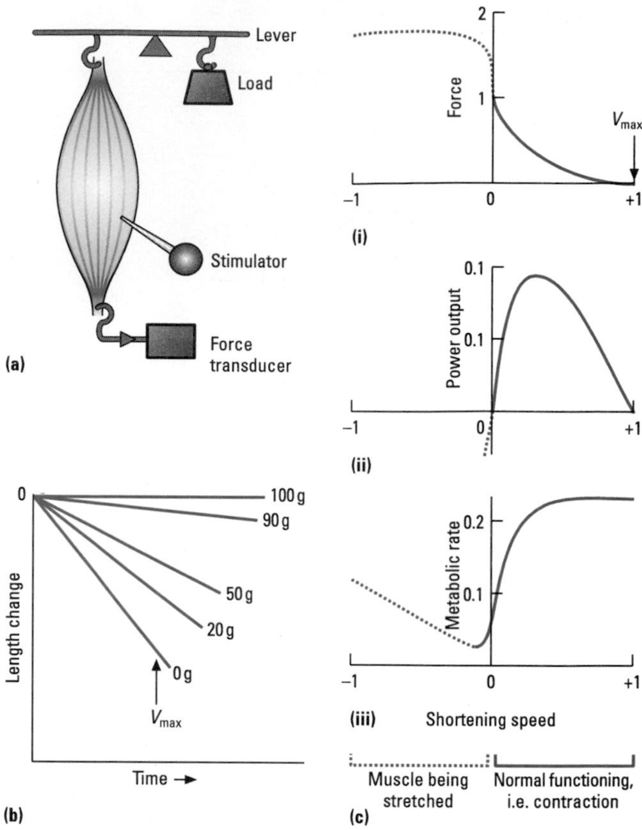

Fig. 9.88 (a) Measuring the force–velocity relation for a muscle. (b) The muscle works against a varying weight hung via a lever, and its rate of contracting is faster for small loads than for large ones. With no load it contracts at a maximum rate, V_{max}, and at some upper load it cannot contract at all (i.e. beyond 100 g in this case it works isometrically). (c) The resultant relationships for a typical vertebrate skeletal muscle: (i) the force–velocity relation, (ii) the equivalent power–velocity relation, obtained by multiplying force and velocity (see text), and (iii) the relationship between metabolic rate and shortening velocity for the same muscle.

the muscle is attached to a force transducer and the other end is attached to a servomotor. The servomotor applies sinusoidal length changes to the muscle at an appropriate amplitude and frequency and the muscle is phasically stimulated during each cycle. The resulting force generated is recorded and plotted against the muscle length to produce a "work loop" (Fig. 9.89c). At point A the muscle is extended and is developing near-maximal tension; it then shortens while still at high tension, doing work against the apparatus, until at B the tension is low and the muscle lengthens again as the apparatus works against it. As it returns around the loop to point A, the apparatus does work that is less in total than the work initially done by the muscle, because the length change is the same but the tension is lower. Hence we can calculate the net work done (per cycle) from the area contained within the loop. The muscle power output is the product of net work per cycle and cycle frequency (usually expressed as W kg^{-1} wet muscle mass). It is possible to determine the maximum power output of the muscle by optimizing the number and timing of stimuli, the strain applied, and the frequency of the movements.

However, muscles do not necessarily produce their optimum power output when acting inside a living animal. Hence another approach is to measure the duration and timing of muscle stimulation *in vivo* using electromyographic recordings, and at the same time to determine the muscle strain. This can be done either by calculation from an analysis of high-speed films of the behavior, or directly using sonomicrometry (where two piezoelectric crystals

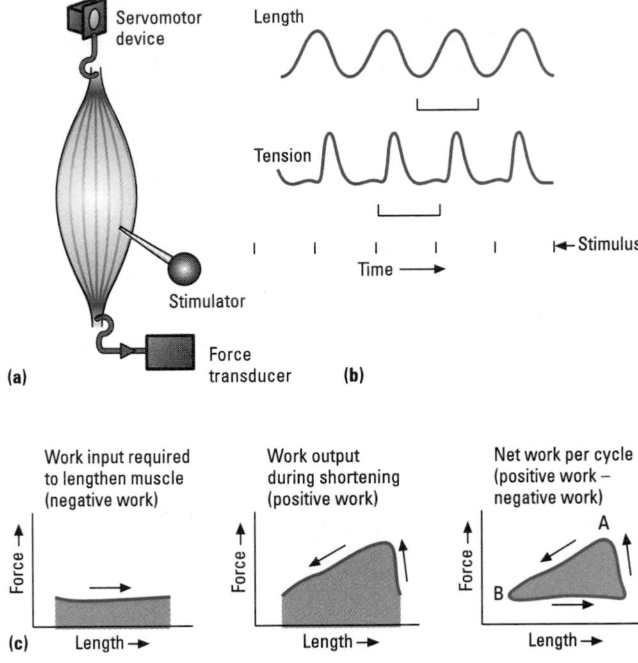

Fig. 9.89 (a) The measurement protocol of a "work loop" gives a more realistic view of the force–velocity relation for a muscle working cyclically, where the muscle is held at one length by a servomotor device. (b) The length and tension recorded during such an experiment, through several cycles. (c) The work loop that results: at first the muscle is being stretched by an outside force (usually an antagonistic muscle) and it performs "negative work"; then it shortens and does positive work. The net work is the difference between these two, represented by the area enclosed within the work loop. (Adapted from Josephson 1985, and other sources.)

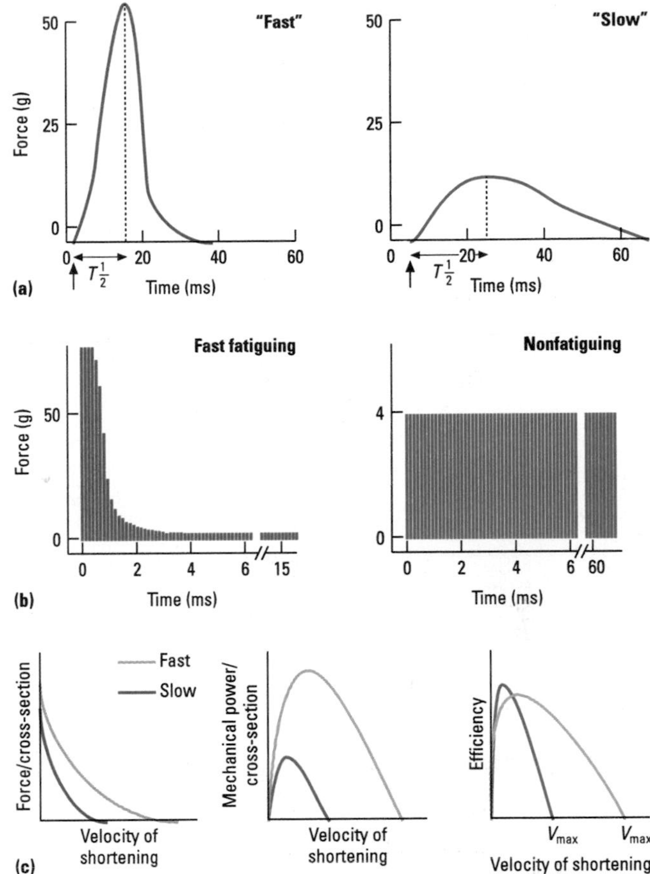

Fig. 9.90 A comparison of the properties of fast twitch and slow twitch muscle fibers in vertebrates; fast fibers have short contraction times and high tetanic force, but are also very prone to fatigue. (a) The basic twitch shape and duration. (b) Successive twitches in response to stimulation at 40 Hz. (c) The relationships between fiber speed and force, power and efficiency, all per unit cross-sectional area, plotted for fast and slow fibers operating at a range of speeds. Note that slow fibers are most efficient at their slowest speeds, while fast fibers have slightly lower peak efficiencies, and at only slightly faster speeds, with very low efficiency as they approach V_{max}.

are implanted into the muscle; sound is emitted from one crystal and the time required to reach the other crystal is a function of their distance apart). These parameters of "real" contractions can then be used to measure the power output of isolated muscles under the conditions observed *in vivo*.

On the basis of studies of this kind, it is estimated that at maximum contraction velocity only about 20% of the possible cross-bridges in a muscle are attached, rising to perhaps 30% during maximal isometric effort.

Contractile states: twitch and tetanus

The considerations in the last section mainly related to a single AP and a single contraction (twitch) in the muscle. The resultant "active state" is rapidly terminated because the SR quickly sequesters the released calcium, and the tension that the contractile system could achieve is not fully realized, although muscles can vary considerably in the speed of the twitch and are often referred to as "fast" or "slow" on this basis (Fig. 9.90). In fact in a single twitch much of the energy is wasted in stretching the elastic components of the muscle—the plasma membrane and the connective tissue that straps the muscle together.

In practice, of course, most muscles normally receive a train of APs, and each successive AP arrives before the SR has fully mopped up the calcium released by the previous AP. Hence the active state is prolonged, the contractile effects start to summate, and the overall tension builds up (Fig. 9.91). A maximum tension is reached when the force produced by the internal shortening of the sarcomeres and the stretching of the elastic components of the muscle is just enough to cause the cross-bridges to start slipping; at this point the muscle reaches the state called tetanus. The summation that leads to tetanus depends on both the inability of the calcium pumps to keep up with calcium release, and the inability of the elastic components of the muscle to relax between stimuli.

Energy consumption in muscle

Two major processes require energy when a muscle contracts. One is the hydrolysis of ATP by the ATPase of the myosin cross-bridges, and the other is the pumping of calcium ions back in to the SR, each Ca^{2+} ion requiring two ATP molecules. It has been calculated that the calcium pump activity accounts for about 25–30% of the total

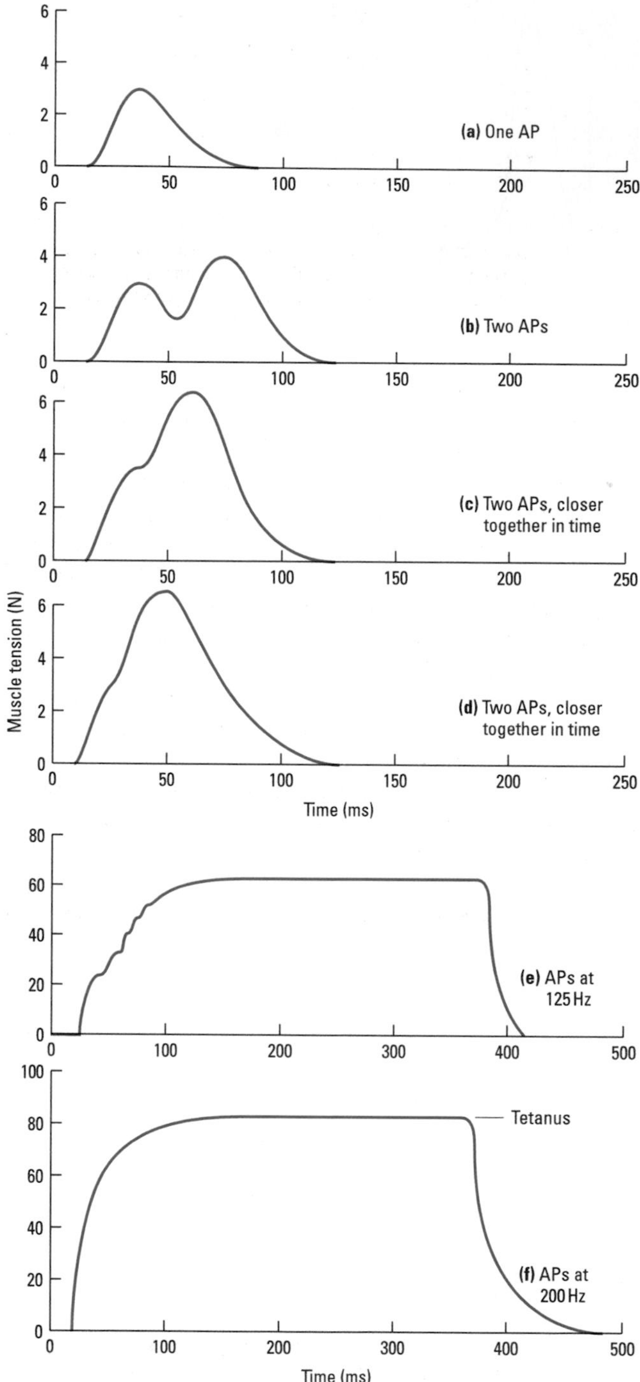

Fig. 9.91 The summation of twitches in a fast vertebrate skeletal muscle leading to a build-up of muscle tension and (if the action potential (AP) frequency is high enough) to a state of constant maintained high tension or "tetanus". At low frequencies the successive twitches begin before the fiber has had time to fully relax, so tension builds up; at high frequencies the twitches all fuse with one another. Note the different scales between (a–d) and (e, f).

ATP usage when a muscle contracts, and this value seems to be similar for all muscles regardless of their contraction velocity.

However, the ATP concentration in a muscle remains rather constant regardless of how hard it is working. This is largely because most muscles use one of the alternative phosphagens (see section 6.2.2) as a source of high-energy phosphate bonds. In most vertebrate muscle, **phosphocreatine** (PCr) is used, and this transfers a high-energy phosphate to ADP very rapidly, so that ATP is regenerated. PCr levels therefore fall substantially (from a high resting level of 20–40 mM) during activity, while the ATP levels stay constant at around 5 mM. Creatine phosphokinase (CPK) is the enzyme mediating this energy transfer, so it is generally found at high concentrations in muscles. This enzyme buffers the ATP supply and can supply crucial power to an animal in an emergency such as rapid escape movements.

Hydrolysis of PCr gives the fastest possible provision of ATP, occurring close to the sites of ATP utilization (though in some invertebrates, such as insects and molluscs, phosphoarginine (PAr) replaces PCr, and the enzyme responsible for ATP generation in the muscles is arginine phosphokinase). Nevertheless, the amounts of ATP and PCr available at any one time could only sustain very brief peak activity in the muscle of most vertebrates; for a human athlete, peak running with aerobic supplies is only possible for up to about 20 s, so that in any race longer than 200 m the athlete's muscles in part must operate anaerobically, allowing a 10-fold greater fatigue resistance. Likewise, in muscle fibers that have a high rate of contraction, or that generate a high power output for long periods, the requirement for ATP can exceed what can be generated using aerobic metabolic pathways. The next fastest method of providing energy is via anaerobic glycolysis, which involves 11 enzymes and has a molar yield of 2 ATP per glucose molecule, and which leads to the accumulation of lactate (see Chapter 6). The type IIB muscle fibers of mammals therefore contain a higher concentration of CPK and especially of glycolytic enzymes than either type IIA or IIAB (intermediate) fibers (see below, and Table 9.10). Just as with other characters, there is in practice a continuum of fuel supply and metabolic properties within a population of muscle fibers. When a muscle needs to contract most efficiently there is always an orderly and progressive recruitment of fiber types, starting with the slowest contracting and most fatigue resistant and finishing with the fastest and least fatigue resistant.

9.12.2 Variations in muscle performance

The mechanism underlying the conversion of chemical energy to mechanical energy is the same for all muscles in all animals. However, numerous factors can be varied to produce muscles with a range of very different mechanical properties, from very high-frequency operation (as often found in sound-producing systems) to slow sustained tensions. Factors that can vary include: the electrical properties of the plasma membrane (whether or not it produces an AP giving a twitch or phasic fiber, or just a graded potential change giving a tonic fiber); the density of ryanodine channels and calcium pumps in the SR (affecting how long free calcium levels stay high after a membrane potential change); the number of mitochondria in the muscle, the supplies of phosphagens and glycolytic enzymes, and the level of vascularization (all of which alter how quickly the muscle runs out of ATP and becomes fatigued). These factors are perhaps obvious, and some apply to other tissues as well, but in muscles two additional variables are particularly important.

Table 9.10 Types of twitch muscle fiber in vertebrate skeletal muscles.

Properties	Type I	Type IIA	Type IIB
Fiber diameter	Small	Medium	High
V_{max}	Low	High	High
Force per unit area	Low	Medium	High
Mitochondrial volume	High	High	Low
Myosin ATPase activity	Low	High	High
Fatigue resistance	High	Medium	Low
Anaerobic path enzymes	Low	Medium	High
Oxidative phosphorylation enzymes	High	High	Low
Occurrence	Mammalian postural muscles	Main leg/wing muscles, all vertebrate taxa	Wing muscle of flightless birds, leg muscle of amphibians and reptiles

Protein isoforms

Protein isoforms arise either through the expression of multigene families, or through different splicing of mRNA, or through some other post-translational modification of the protein (see Chapter 2). In muscles the types of myosin subunit present are particularly important. Different muscle fiber types can be recognized on the basis of histochemical staining reactions, especially using stains for myosin-ATPase activity, reflecting differences in isoform composition. Skeletal muscle fibers used to be subdivided with respect to their speed of contraction as "fast" or "slow", although this was always a relative classification, and slow muscles in small animals and in juveniles can actually have a comparable speed to many fast muscles in larger species and/or adult stages because of scaling effects. Use of myosin isoform analysis has led to an alternative distinction of three main muscle fiber types in mammalian skeletal muscle: slow type I, fast type IIA, and fast type IIB (see Table 9.10). Type I fibers contract slowly but also fatigue slowly as they are well endowed with mitochondria and can maintain oxidative metabolism for long periods, so they are used in mammals in postural muscles, and appear as "red muscle", which is relatively cheap to run. Type IIA fibers are larger, activate quickly and contract quickly, and fatigue moderately slowly; they predominate in muscles used for rapid repetitive locomotory movements. Type IIb fibers are larger again, and contract and fatigue quickly, with fast calcium kinetics and a dependence on anaerobic metabolism; they produce "white muscles" that cannot achieve sustained activity but are useful for occasional rapid movements. There are clear adaptive "trade-offs" between speed and cost operating here, though in reality there is a continuum of myosin isoforms, reflecting the different combinations of heavy and light chains.

Differences in isoform composition of thin filament Ca^{2+}-regulatory proteins and of the Ca^{2+} pump protein also contribute to muscle performance differences, particularly in terms of the relaxation rate.

Sarcomere and myofibril composition and organization

As we have seen, muscle power output is the product of force and velocity. The force produced per myosin head (5 pN) is very similar among different muscle fiber types, as is the packing of myosin molecules into the thick filament lattice. However, the fraction of the muscle fiber cross-section occupied by myofibrils (as opposed to mitochondria and SR) is much more variable. Thus, an important way to increase tension and hence power output is to have a higher volume density of myofibrils. The maximum packing of myofibrils (at about 90%) results in a maximum isometric tension of about 300 kN m^{-2} in a wide range of species.

However, filling the muscle with myofibrils leaves less space available for the mitochondria and SR, which are also, of course, important for aerobic capacity and rapid relaxation, respectively. Thus a high muscle power output is associated with a trade-off in other aspects of performance. Muscles that contract at high frequencies have a more abundant SR, specifically to shorten the average diffusion distance between the SR (the Ca^{2+} source) and the thin filaments (the Ca^{2+} sink). Muscles for which a high aerobic ATP synthetic rate is important have more mitochondria, at the expense of myofibrils and/or SR.

Typically, mitochondria comprise 6–8% of the volume of aerobic fiber types (e.g. in leg extensor muscles in mammals) but can reach 35% in the flight muscles of hummingbirds. Usually only a small volume of the muscle, under 3%, is devoted to fuel storage. This trade-off for space in the muscle in relation to different aspects of performance is illustrated in Fig. 9.92.

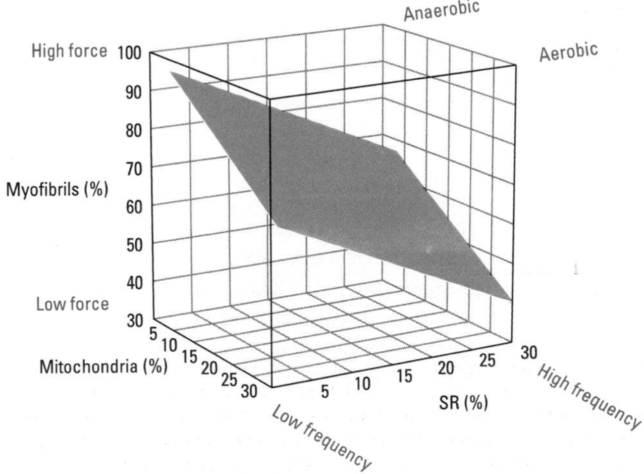

Fig. 9.92 The volume packing trade-offs that are possible in skeletal muscle. All muscles require a compromise between the proportions of myofibrils (determining the maximum force), of sarcoplasmic reticulum (determining the maximum frequency), and of mitochondria (determining the aerobic ATP production). Any increases in one component necessitate reductions in others, and all muscles may have to exist at a point somewhere on the depicted plane of what is feasible to maintain overall function. (Adapted from Lindstedt *et al.* 1998.)

Table 9.11 Comparison of major muscle types found in animals.

	Striated muscles				**Smooth muscles**		
	Vertebrate skeletal	**Cardiac**	**Invertebrate mollusc adductor**	**Invertebrate insect wings**	**Vertebrate multiunit**	**Vertebrate single unit**	**Invertebrate**
Myosin and actin filaments	+	+	+	+	+	+	+
Sliding filament system	+	+	+	+	+	+	+
Visible banding	+	+	+	+	–	–	–
SR abundance	High	High	Low	Medium	Low	Low	Low
T-tubule presence	+	+	–	–	–	–	–
Tropomyosin + troponin	+	+	–	+	–	–	–
Source of calcium	SR	ECF + SR	ECF + SR	SR	ECF + SR	ECF + SR	
Speed of contraction	I fast II slow	Slow	Variable	Fast	Very slow	Very slow	Slow?
Activation	Neurogenic	Myogenic	Neurogenic	Neurogenic*	Neurogenic	Myogenic	Neurogenic (± mechanical)
Gap junctions between fibers	–	+	+	–	+		

* Direct flight muscles are neurogenic, indirect flight muscles are neurogenic but asynchronous; see text for more details.
ECF, extracellular fluid; SR, sarcoplasmic reticulum.

9.13 Muscle types and diversity

A complete survey of the extensive literature on the comparative physiology of muscle in vertebrate and invertebrate animals is beyond the scope of this textbook. In this section, some of the diversity of muscle anatomy, fine structure, and physiology will be explored in relation to movement in water and in the air; here we look at the broad "types" of muscle found in animals. Note that all share a basic similarity of mechanism in that myosin and actin filaments slide over each other to produce the tension and movement. Their differences lie primarily in the microanatomy of the interrelationships of the myosin and actin filaments, and in the coupling and control systems. Many, perhaps most, muscles also differ form vertebrate skeletal muscle in their electrical conduction systems, relying on calcium APs rather than sodium APs. The voltage-gated sodium channel that we have discussed repeatedly in nerves and in striated muscle is not normally expressed in invertebrate muscles or in chordate smooth muscles. A review of all these kinds of muscle, in comparison with the vertebrate striated muscle that we have concentrated on until now, is shown in Table 9.11.

9.13.1 Smooth muscles

Muscles categorized as "smooth" include all those that lack obvious striation under the light microscope, although they represent an extremely heterogeneous grouping. The lack of striation arises because they are not organized into distinct sarcomeres. Instead there are small bundles of interdigitating actin and myosin filaments, with the thin actin filaments being anchored (again via α-actinin) to "dense bodies", either within the cell or just beneath the plasma membrane (in which case they are often called "attachment plaques"). They nearly always show a lower thick filament to thin filament ratio than is found in striated muscles. Smooth muscle fibers are present as individual cells, elongate and tapered at the ends, unlike the multicellular multinucleate fibers of vertebrate striated muscle.

Smooth muscles differ in other important fashions: they generally lack T-tubules and have very little SR, and they tend to lack some of the regulatory proteins, notably troponin. Contraction in these muscles is not mediated via troponin and tropomyosin, but instead the actin filaments often contain the protein **caldesmon,** which prevents actin binding to myosin cross-bridges. Caldesmon is removed from the actin (allowing cross-bridge formation) in one of two ways: either by being phosphorylated using protein kinase C, or by interacting with the calcium-binding protein **calmodulin.** Further regulation of contraction in some smooth muscles is provided by mechanisms that operate on the myosin filaments instead, via the regulatory light chains (LCs) in the neck region of the molecule. Direct binding of calcium to this regulatory chain is the simplest control system, so that when calcium levels are raised the neck changes its configuration and the head can bind to actin (often called "short LC" regulation). Alternatively the light regulatory chain can be phosphorylated by myosin light chain kinase which is itself activated by the calcium–calmodulin complex (this is the system common in vertebrate smooth muscle, "long LC" regulation). Finally, protein kinase C can affect the light chain too, phosphorylating it at a different site that prevents actin–myosin binding and so induces relaxation. There are thus a variety of different ways of regulating the actin–myosin interaction in smooth muscle (summarized diagrammatically in Fig. 9.93), via either the thick or the thin filament or both. In many taxa, including the arthropods, two or three of the calcium-regulatory mechanisms are present in the same animal; the known regulatory patterns among some major groups are shown in Table 9.12. Since each of these regulatory systems is itself part of an intracellular signaling cascade system, there is great scope for the modulation of smooth muscle tone. Note that smooth muscle relaxation is also very highly regulated by a considerable range of extrinsic modulators (including cyclic nucleotides, various neuropeptides, and nitric oxide as mentioned in the section on neural transmitters in section 9.3.1). Thus overall muscle tone depends on a balance of the signals that produce contraction and the signals that promote relaxation.

The calcium pulse in the cytoplasm of smooth muscles is significantly longer than in striated muscles. This presumably reflects the much reduced complement of SR (reduced to a scattering of rather flattened sacs just below the plasma membrane), most of the

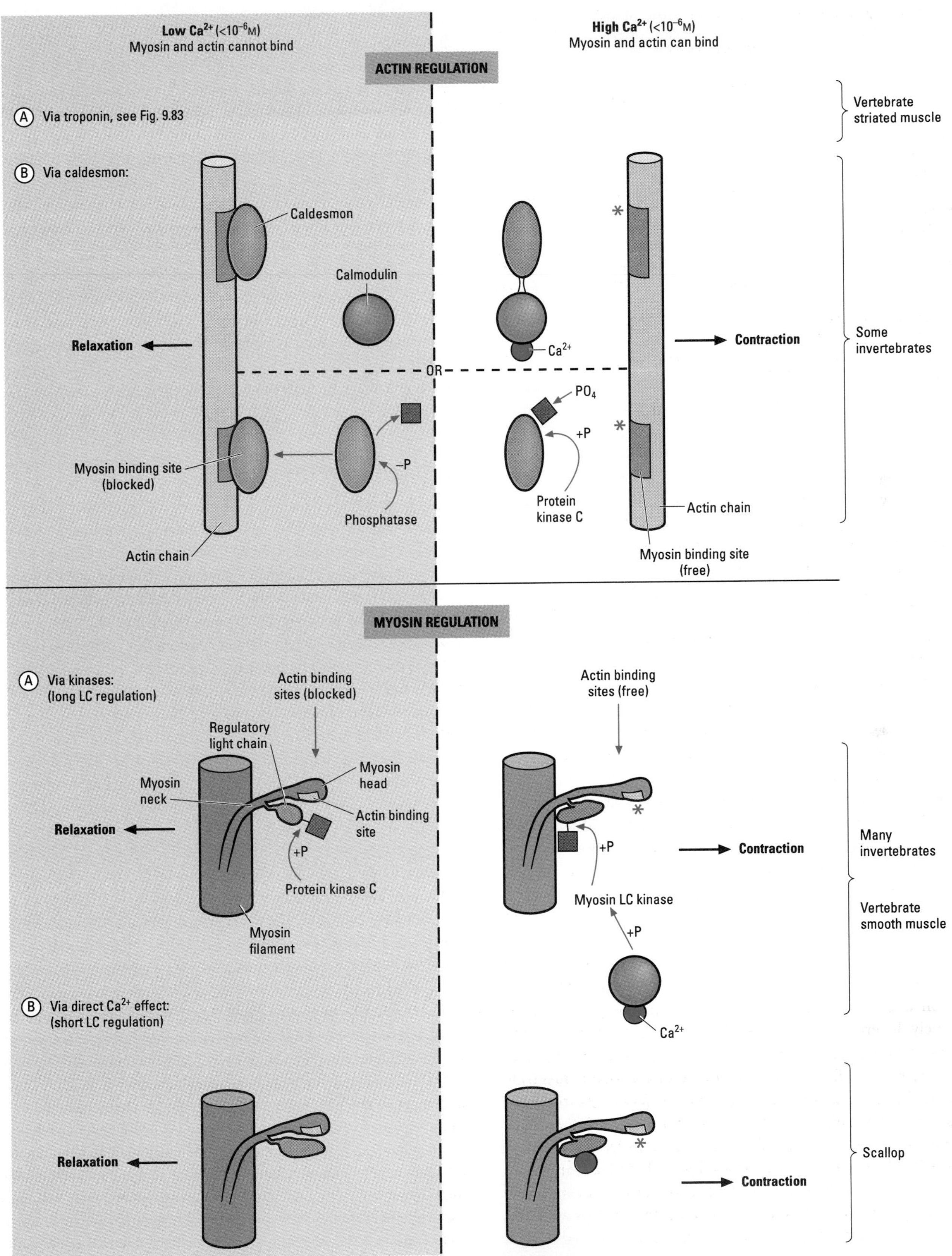

Fig. 9.93 A diagrammatic summary of different kinds of contraction control, via either the actin or the myosin filaments. On the left, the situations in low calcium conditions, preventing actin–myosin binding and leading to relaxation; on the right, high calcium conditions such that the actin- or myosin-binding sites are free (asterisks), cross-bridges can form, and contraction can occur. LC, light chain.

Table 9.12 Distribution of alternative mechanisms of regulation of calcium effects in muscles, via the actin (A) or myosin (M) filaments. See text for more details.

Phylum/class	Genus	Regulation
Nematoda	*Ascaris*	A+M
Mollusca	*Pecten*	M
Annelida	*Lumbricus, Glycera*	A+M
Sipuncula	*Golfingia*	A+M
Insecta	*Gryllus, Schistocerca*	A+M
Crustacea	*Balanus, Homarus*	A+M
Chelicerata	*Limulus*	A+M
Echinodermata	*Thyone*	M
Vertebrate	Various	
	striated	A
	smooth	M

calcium regulation and reuptake being done by the plasma membrane (never more than a few micrometers distant from the myofibrils in these cells). Because of this, and also because of the relatively slow actions of the protein kinases, contraction is generally rather slow, and relaxation even slower. However, the muscles are capable of very sustained contractions.

Commonly these kinds of muscle occur in the walls of tubular and very extensible structures. Thus many worm-like invertebrates with hydrostatic skeletons have nonstriated muscles in their body walls, which are their main locomotory effectors, as do echinoderms where hydrostatic principles operate the tube feet. In vertebrates the muscles around the tubular visceral organs such as the gut, the bladder, and most of the blood vessels are nonstriated.

Vertebrate smooth muscle

Vertebrate smooth muscles are rarely involved in voluntary locomotion, and instead are generally under either autonomic nervous control or hormonal control. Neural inputs arrive not at classic neuromuscular "end-plate" junctions but via multiple "varicosities", small swellings that are the terminals of autonomic nerves lying within and along the fibers. Released neurotransmitter from any one swelling may diffuse some distance and trigger a number of fibers, which have receptors scattered diffusely over their surfaces. The APs in the plasma membrane can be of the striated muscle type where the inward depolarizing current is due to sodium ions, but in most vertebrate smooth muscles the inward current is due to calcium ions.

The vertebrate smooth muscles can be subdivided into two main types, usually termed single unit and multiunit. The difference lies mainly in the control of contraction. **Single-unit smooth muscles** behave as if all the fibers were a single unit—they have their individual fibers linked by strong mechanical junctions and also linked electrically to each other through gap junctions. Hence depolarization of one cell is rapidly conducted to all others, so that the fibers are **myogenic** (as opposed to **neurogenic**, where a neural input is needed to trigger contraction). Thus a wave of contraction moves "automatically" through the muscle (particularly useful when it is being used for a wave-like rhythmic activity such as peristalsis in the gut). These muscles are found in the walls of the gut, the uterus, and the bladder, in most vertebrate taxa. **Multiunit smooth muscles** have fibers that act independently and need independent stimulation from neurons (i.e. they are neurogenic). There are no electrically coupling gap junctions. Such muscles mainly occur in the walls of blood vessels, and in addition to being controlled by their innervation they are often directly sensitive to mechanical stimulation, such that stretching them can induce depolarization and contraction. Hence many arterioles can autoregulate; when blood pressure rises the muscle gets stretched and automatically starts to contract, helping to maintain a fairly steady peripheral blood flow.

Some vertebrate smooth muscles can enter a state called "latch" (Fig. 9.94), where a contraction is maintained for long periods without much energy expenditure because the rate of cross-bridge cycling is greatly reduced. This is probably due to a balancing effect between the myosin kinase and the myosin phosphatase enzymes, both of which are reduced in activity during a latch so that many cross-bridges are formed and then persist; the calcium-binding proteins may also play a role. The outcome is sustained tension at less than 0.5% of the cost that would be incurred for the same job by a striated muscle.

Invertebrate smooth muscle

Many invertebrate muscles are nonstriated, especially in the soft-bodied taxa lacking exoskeletons, where locomotion is mainly achieved by hydrostatic effects (see section 9.13.2). Because these nonstriated muscles therefore appear in most of the supposedly "lower" invertebrates it is often assumed that nonstriated muscle is somehow more primitive. However, this is not the case, as some areas of striation occur even in very "simple" or "primitive" animals such as cnidarians. The pattern of occurrence seems instead to be related to the "action length" of a muscle. In an animal with a hard jointed skeleton the muscle is always pulling over a particular (usually short) length and tends to be striated, whereas in soft-bodied animals the body shape may change very substantially (a worm may be very short and fat or very long and thin at different times), and the length of a particular muscle therefore has to be much more variable. In that situation organized striation with sarcomeres does not work well, and a more extensible muscle arrangement is needed.

Most of the muscles in this category are tonic, having no propagated APs, but as might be expected they are in other respects extremely variable. Even within one animal there may be several different types of smooth muscle fibers, ranging from rapid all-or-nothing twitch fibers to slow fibers with fully graded depolarizations that do not propagate at all (Fig. 9.94). All these invertebrate muscles contain tropomyosin, which interestingly is a key cause of food allergy reactions in humans (especially to crustacean seafoods). The slower fibers have little or no T-tubule system and limited SR, and most of the fibers use caldesmon rather than troponin as the main regulator of contraction (see Fig. 9.93 and Table 9.12). It is also becoming clear that the tonic state of many invertebrate smooth muscles is carefully regulated by a vast range of neuropeptides, including many of the neurohormones covered in Chapter 10, showing their dual role as neuromodulators. Dopamine, octopamine, and allatostatin are classic examples of myoinhibitory and myostimulatory peptides in crustaceans, molluscs, and other invertebrates.

Some invertebrate smooth muscles can also enter states of sustained tension, in this case termed "catch"; the classic examples are

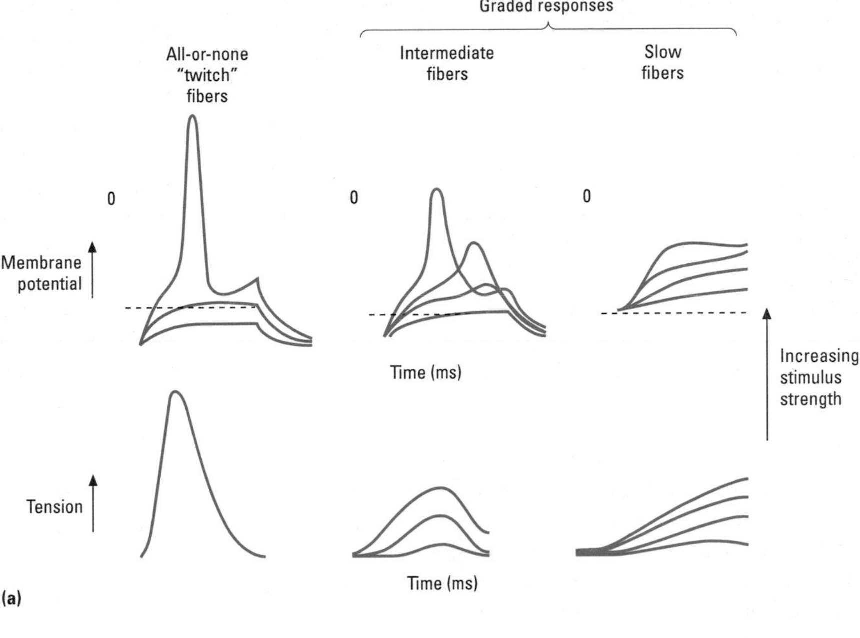

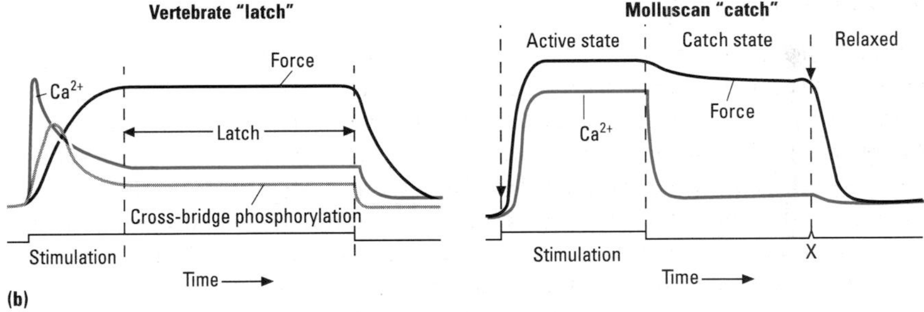

Fig. 9.94 (a) Properties of different kinds of smooth muscles: some have phasic twitch contractions, but many have the ability to produce graded tonic contractions varying with the strength of the stimulus. (b) These can summate to give a state of tetanus, which in a few rare cases can be maintained for long periods with little energy expenditure despite decreasing calcium levels. This state is known as "latch" in vertebrates, or "catch" in molluscs, where it persists without stimulation until a brief input (X) from another motor neuron releases it (see text).

the large adductor muscles of many bivalve molluscs, which hold the two shell valves together. The mechanism seems to be somewhat different from the latch state seen in vertebrates, and is not fully understood. The adductor muscle thick filaments contain **catchin** and **paramyosin** as well as myosin, and these extra proteins may somehow produce the very slow rate of cross-bridge cycling typical of catch. However, recent work shows that catch can be produced *in vitro* without paramyosin present—only actin, myosin, and catchin being crucial. The catch state induction depends on a calcium-dependent phosphatase, probably acting on the catchin, and catch termination depends on a protein kinase.

9.13.2 Invertebrate striated muscles

Many invertebrates have some striated muscle at appropriate points in their anatomy, following the pattern mentioned above of striation occurring where muscles have a fairly fixed action length. In arthropods, where the exoskeleton and jointed arrangements of limbs allow short contractions over fixed lengths, locomotory muscles in the swimming or walking legs are often fully or partly striated. At least in crustacean leg muscles the striation pattern and fibril packing are rather different from the vertebrate version, with longer and thicker myosin filaments having a different cross-sectional patterning. In some claw muscles there are giant sarcomeres (up to 8.3 μm long) which are also very extensible (up to 13 μm) due to a particularly elastic invertebrate version of connectin protein. In some of the larger earthworms the body wall musculature is also striated, but in an unusual pattern often termed "oblique striation"; a similar tissue occurs in the locomotory foot of snails. Many invertebrates also have striated muscle in their hearts.

Two further specific examples of invertebrate striated muscles are worthy of note.

Bivalve mollusc adductors

The adductors of most bivalves include a small portion of striated muscle that provides for an escape response of rapid shell closure (in addition to the large smooth muscle capable of sustained "catch"; see above). The striated muscle, seen most clearly in scallops and other large surface-dwelling bivalves, differs markedly from that of vertebrates. The muscle fibers are single cells, the resting sarcomere length is somewhat longer, and there is electrical coupling between the cells via gap junctions. The transverse tubular (T) system is absent and there is only a limited network of SR. The ratio of actin to myosin filaments is around 5.5 : 1, with an average of 11 actin filaments around each myosin filament, the latter forming a hexagonal lattice. Often a single myofibril occupies the center of the muscle cell and much of the remaining space is filled with glycogen stores.

The myofibrils contain little or no troponin, and the regulation of contraction is via the myosin filaments, where calcium binds to the light chains of the cross-bridges. Removing the regulatory light chain from molluscan myosin results in the loss of calcium sensitivity; however, the calcium-binding sites are not on the regulatory light chain itself, but rather the regulatory light chain is necessary for the proper folding of the cross-bridge to give a functional site. In the relaxed state (without calcium) the ATPase activity of scallop myofibrils is 600 times lower than in the activated state. A possible model whereby the binding of calcium to the myosin light chain in the neck region communicates via the regulatory light chain to control actin activation events at the ATPase site is included in Fig. 9.93.

Insect flight muscles

There are two different kinds of flight muscles in insects. Some taxa have **direct muscles** that pull against the flexible and elastic wing base to move the wing up and down. These direct muscles are **non-fibrillar**, a term that was coined because the myofibrils, although present, are thin and not readily recognized in the light microscope. They are also neurogenic muscles (sometimes called **synchronous**) where each beat of the wing is initiated by a nerve impulse leading to an AP and a twitch. However, this type of muscle activation clearly places a limit on the frequency at which the wings can flap, because to avoid tetany the time between nerve impulses needs to be longer than the time required for the SR to "mop up" Ca^{2+} in the cytoplasm back to a low level. To achieve the reasonably high flapping frequency required for flight, the SR in these muscles is very elaborate, allowing the rapid release and uptake of calcium; but even so, at frequencies above some critical value the calcium remains high enough between contractions to saturate the regulatory troponin, leading to a tetanic contraction. Hence insects with these synchronous direct muscles cannot fly at wingbeat frequencies above 100 Hz.

In more advanced insect orders we find **indirect flight muscles** (indirect because they pull on the walls of the thorax instead of on the wings; see section 9.15.2). This muscle is technically neurogenic in that APs are needed for activation, but it is highly **asynchronous**, with far less than one AP per contraction. Elevated calcium levels are needed for activation, but an isolated piece of muscle does not start to contract unless it is also given a quick stretch; and once it does contract, it will go on oscillating between contraction and relaxation if it is coupled to an oscillating mechanical system. So in some senses it behaves like a myogenic muscle, with APs merely acting as an on–off switch, and each contraction resulting automatically from the stretching of a muscle as its antagonist contracts ("shortening deactivation" and "stretch activation"). Each species has flight muscles that oscillate best at a particular frequency, generally close to the normal wingbeat pattern and inversely related to the insect's size: small flies can have natural wingbeat frequencies as high as 1000 Hz. Some flight muscles contract by only 1–2% of their own length, calculated as the equivalent of just one actin–myosin cross-bridge forming, and thus right at the limits of what can be achieved with normal biological materials. These muscles may also use fuel exceptionally quickly, usually being supplied with hemolymph containing 150 mM trehalose (equivalent to 300 mM glucose) to produce the massive inward fuel gradient required to keep the system going.

Asynchronous flight muscle is also structurally highly specialized (Fig. 9.95), and is sometimes called **fibrillar** flight muscle because regular fibril striation patterns are easily observed with light microscopy. The basic fibrillar structure is very similar to that of vertebrate striated muscle, and the myosin head structure is also conserved. However, some unusual proteins occur in association with the individual actin fibrils (for example **flightin**, and a novel **troponin H**), possibly helping to stiffen the myofibril matrix and/or improve the synchrony of events though the fiber.

Because of the odd activation pattern, in combination with the mechanical operation of the thorax, insect flight muscles can produce very high-frequency contractions even though calcium levels change only slowly, and they do not have a particularly large complement of SR. This in turn allows a very high percentage of the muscle volume to be devoted to myofibrils. Measurements of the power output per cross-bridge and the power generated in flight indicate that in this kind of muscle about 2 trillion myosin molecules must be working together. Some insects also have an unusual ability to regulate the cross-bridge power, via phosphorylations of the regulatory light chain on the myosin head, perhaps to allow the wings to be used both for flight (at maximum power) and also for wing-waving during courtship displays. Modulations of power on this scale seem to be a rarity amongst muscles.

9.13.3 Vertebrate cardiac muscle

The vertebrate heart is essentially a very large muscle, and is constituted from a special kind of tissue termed cardiac muscle. Cardiac muscle is present in all vertebrates, but in fish and amphibians it is less specialized, with smaller fibers and a reduced SR. Only in birds and mammals does it appear in really sophisticated forms. Most of the heart consists of **contractile fibers** with clear sarcomeres and an elaborate SR and T-tubule system that is sited over the Z discs as in normal striated skeletal muscle. The muscle is conspicuously striated, and the thick filaments show a strong axial regularity indicating ordered cross-bridge patterns. However, the isolated myosin filaments show a strong tendency to remain attached to actin under normally "relaxing" conditions, suggestive of rather different control systems. Cardiac muscle is also unlike skeletal muscle in being made up of small elongate and tapering cells with single nuclei. The fibers are connected by electrically conducting gap junctions, which

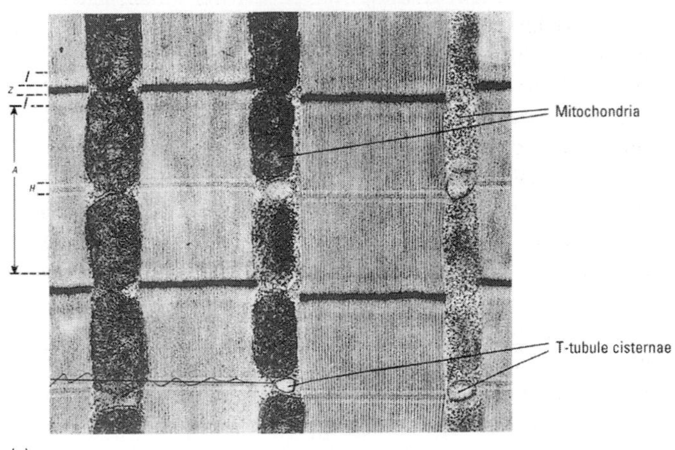

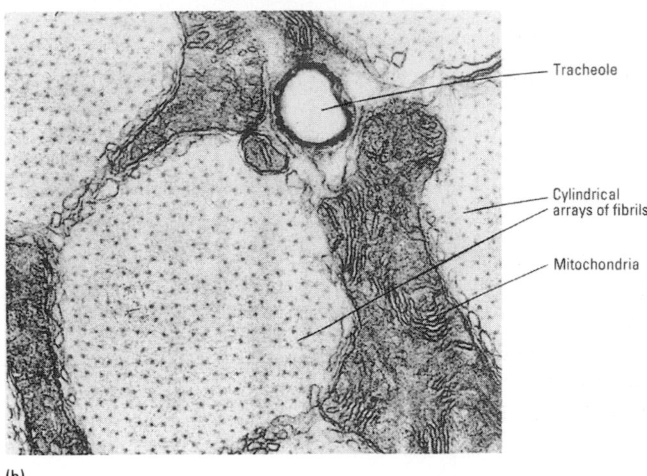

Fig. 9.95 The structure of fibrillar (asynchronous) insect flight muscle, in (a) longitudinal section (× 16,000), and (b) transverse section (× 70,000). Note the very prominent mitochondria, wrapping around the fibrils, and the very small proportion of sarcoplasmic reticulum although the T-system is still present. (From Smith 1975.)

are especially abundant where the ends of each cell abut on each other at "intercalated discs" (Fig. 9.96). The cells are also tightly anchored to each other by strong mechanical links termed desmosomes. Thus the muscle can act as an integrated mechanical and electrical unit.

In addition to contractile fibers, vertebrate hearts also contain a proportion of highly modified muscle fibers termed pacemaker fibers (or conducting fibers), which do not contract and lack the normal fibril proteins. These form the electrical signal-conducting system of the heart. They are crucial to heart functioning because cardiac muscle is a myogenic system, typified by the tight electrical coupling between the cells. An electrical signal is generated by these pacemaker fibers (without any neural input), and the resultant AP spreads all over the heart via gap junctions. This AP produces contractions in a fashion almost identical to that of vertebrate skeletal muscle. However, the mechanisms involved differ in two features: firstly, the calcium influx comes across the plasma membrane as well as from the SR, and secondly the SR influx fraction is dominated by calcium-dependent calcium efflux with a reduced role for the ryanodine receptor route (see section 9.10).

The AP in the plasma membrane of cardiac muscle is unusual, being prolonged into a plateau phase (Fig. 9.97a) due to the involvement of slow voltage-gated calcium channels; in other words, this is a calcium AP rather than a sodium AP. Each AP is followed by an even longer refractory phase. The plateau and the refractory phase together prevent tetanic contraction of the heart, and insure that the muscle relaxes between the APs (which are temporally widely separated). The contraction induced in the muscle has a rather similar time course to the AP and substantially overlaps it in time, very unlike the system in skeletal muscle (compare Fig. 9.66). In the four-chambered hearts of birds and mammals, the APs of the atrial muscle are shorter than those of the ventricles, and slightly precede them in time (Fig. 9.97b) because the atria are closer to the initiating point of the pacemaker system.

Vertebrate hearts require no neural input to initiate and sustain their rhythmic contractions. However, they do receive innervations, and these give important neuromodulatory inputs; the vagus nerve from the sympathetic nervous system is one of the most crucial influences. In addition, circulating catecholamine hormones (epinephrine and norepinephrine) have a regulatory effect by activating

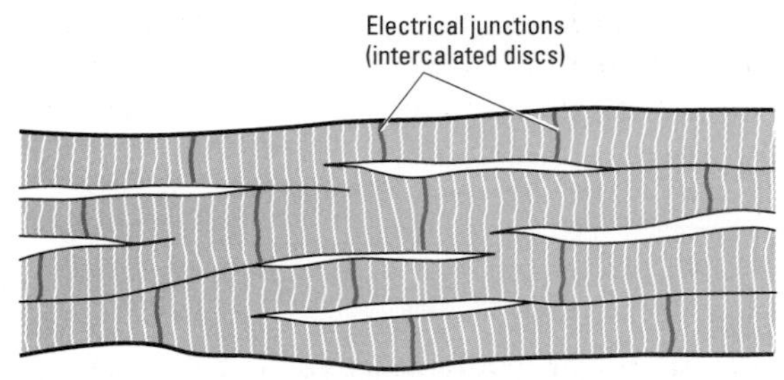

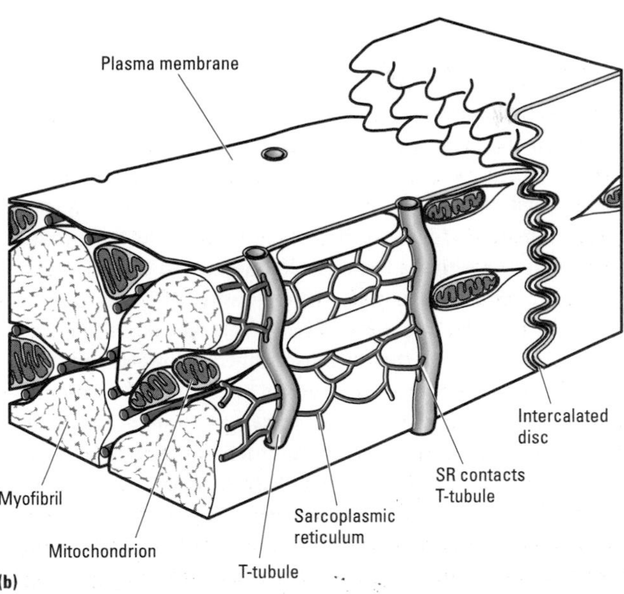

Fig. 9.96 Vertebrate cardiac muscle structure, (a) in diagrammatic view, and (b) as a close-up of two adjacent cells linked by intercalated discs where there are mechanical tight junctions as well as gap junctions that give electrical continuity.

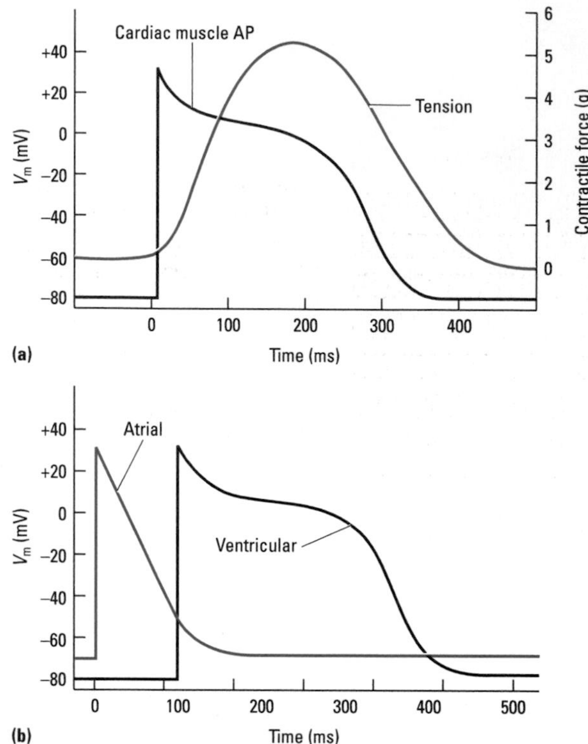

Fig. 9.97 (a) The action potential (AP) recorded from vertebrate cardiac muscle, and the associated tension. (b) Relative time course and shape of the AP in adjacent atrial and ventricular chambers, showing the much greater plateau effect in the ventricle muscle.

α- and β-adrenergic receptors (see Table 10.6); the α type produce increased calcium release from the SR, while the β type work via cAMP to increase calcium flux across the plasma membranes. Both produce increased contractile force and thus accelerate the blood flow round the body.

A combination of different gross morphologies and fibril arrangements, together with the key variables mentioned earlier—the proportions of myofibrils and SR, the expression of different protein isoforms, and the use of different fuels—produces enormous flexibility in the properties of muscle fibers. However, additional variability in performance can be added by changing the anatomical arrangement of the muscle within the framework of the whole body: changing the lengths and properties of tendons or apodemes, and incorporating lever systems formed with the skeleton or cytoskeleton. This leads us naturally to considering how entire muscles vary between animals, and how they work in whole animals to bring about locomotion.

SECTION 3: NERVES AND MUSCLES WORKING TOGETHER

9.14 Motor activity patterns

To be effective, for movement of internal organs and more particularly for locomotion, muscles must operate correctly in relation to each other, so that control of their contraction patterns is crucial. The nervous system generates the patterns of motor activity, in terms of both the timing of contractions and their force (since this so often depends on which muscle fiber, and how many of them, are recruited). Exactly how this is achieved varies quite substantially between animal taxa, though at present only the vertebrates and the arthropods have been investigated in very much depth.

In this section, after examining these different neuromotor systems, we will consider how they can be used to generate simple motor activities.

9.14.1 Motor innervation and control patterns

Vertebrates

The motor neurons of vertebrates leave the spinal cord via the ventral roots as part of large nerve bundles, and each muscle in the target area receives inputs from a proportion of the neurons (sometimes termed its "motor pool" of neurons). Each neuron branches repeatedly as it approaches its own target skeletal muscle, and a single motor neuron will innervate many muscle fibers. However, each muscle fiber only receives a neural input from one neuron (and usually only at one end-plate; see Fig. 9.98a). The neuron and the specific muscle fibers that it innervates are termed a **motor unit**.

Most motor units contain around 100–500 muscle fibers, though units smaller than 20 can occur in mammalian digits and tongues, while the units can be larger than 2000 in large trunk muscles where precise motions are not needed. A muscle with a large number of small motor units (that is, with many neural inputs each innervating relatively few muscle fibers) allows fine motor control but each motor unit delivers relatively little force and so force is developed relatively slowly (e.g. finger muscles). In contrast a muscle containing relatively few large motor units is better suited for the rapid development of force, but results in a relatively coarse degree of control (e.g. calf muscles).

All vertebrate motor innervations are excitatory, and all use acetylcholine as the transmitter at the end-plate. APs and twitch contractions have a 1 : 1 relationship, so there is an all-or-nothing effect of a single AP—the muscle fibers either give a full twitch or they do nothing. A somewhat more "graded" response appears, of course, because a series of APs can increase the contraction by inducing partial or full tetanus. But any degree of finer control can only be achieved by recruiting increasing numbers of motor units within a given muscle, and thereby varying the types of fibers (I, IIA, and IIB) that are recruited (since most motor units contain only one type of fiber). At the level of the whole animal yet more control is added by balancing contractions in more than one muscle; for example, producing tension in the two main muscles around any one joint will effectively hold the joint in a fixed position to maintain posture.

All of the above applies to the phasic motor fibers that predominate in most vertebrates. But tonic fibers also occur, especially in amphibians and reptiles where they have a postural role. Tonic fibers are unusual (as far as vertebrates are concerned) in having multiterminal innervations, i.e. many end-plates from a single neuron are distributed along the fiber, although the terminals are still always from just one neuron. These fibers have graded membrane potentials rather than APs, and so can generate graded contractions.

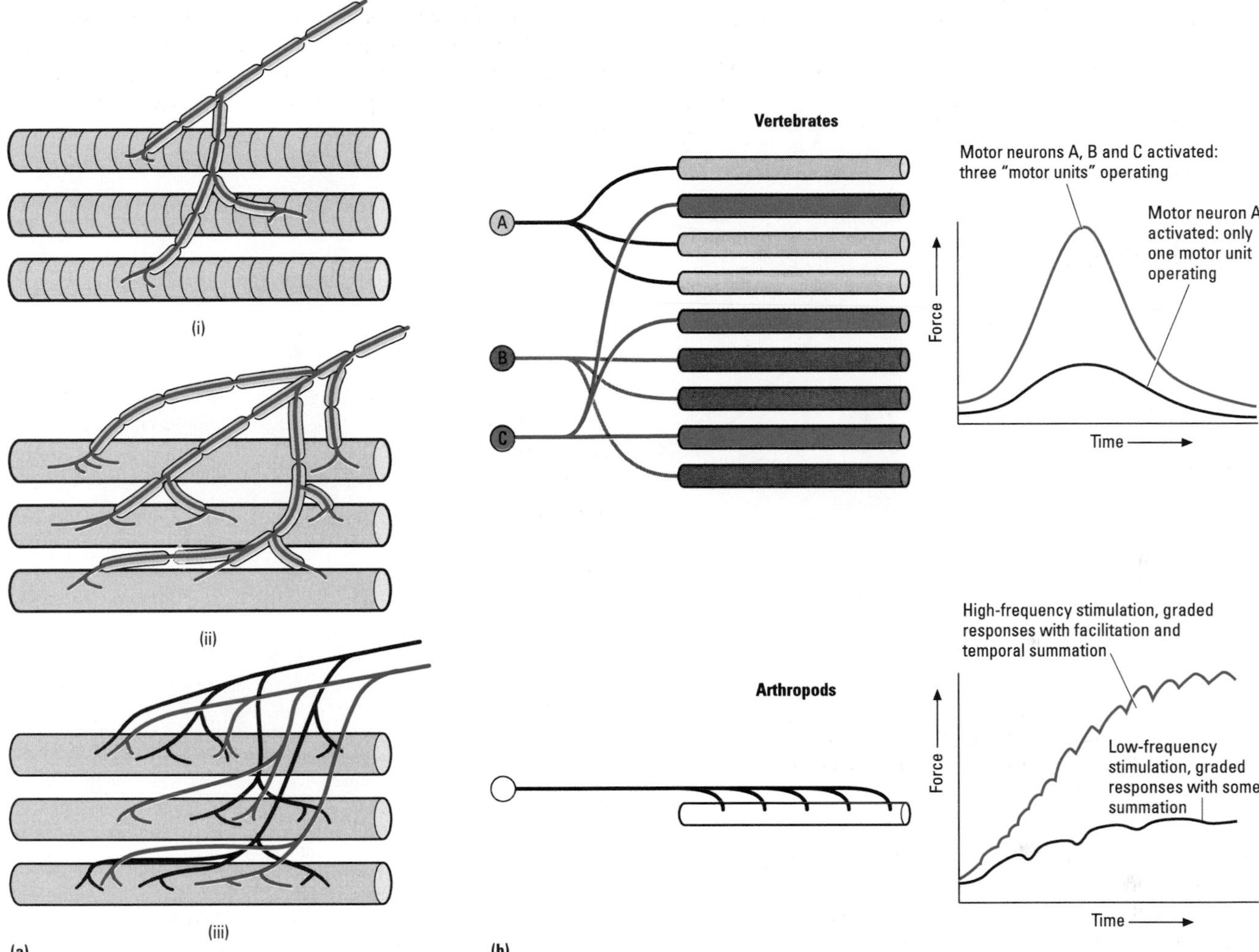

Fig. 9.98 (a) Multiterminal and multineuronal patterns of innervation for muscle: (i) uniterminal, with a single excitatory input, as in vertebrate skeletal muscle, (ii) multiterminal but unineuronal, found in vertebrate tonic fibers and in many invertebrate muscles, and (iii) multiterminal and multineuronal, the common pattern in arthropod muscles, where one or more of the neurons is normally inhibitory. (b) The resultant different patterns of motor control, by recruiting more phasic fibers grouped as motor units, or by varying the tonic response in relation to stimulus frequency (often in more than one exciter neuron).

Arthropods

Insects and other arthropods, and indeed most invertebrates, have muscles more like the vertebrate tonic fibers, which rarely or never produce APs. The muscles use graded potentials to produce graded contractions, rather than a system where APs summate. As in vertebrates, the fibers come in various types with differing combinations of structural, electrical, and contractile properties, and a few of them can "twitch" if stimulated very strongly. However invertebrates have relatively few neurons, and so need a system of motor control that allows varied muscle contraction with the minimum number of neural inputs. Muscle fibers in crustaceans and many insects are therefore **multiterminal**, having numerous synapses with the neuron occurring along their length (Fig. 9.98a), so that they contract synchronously without an AP having to be propagated along the fiber. (Note that here the neuromuscular synapses are not generally termed end-plates as their structures are less specialized.) Spatial summation occurs in the muscle fiber, so that the closer together two synapses and hence two EPPs are, the greater the depolarization of the muscle membrane. Hence the muscles can produce many different levels of tension, not just a twitch or a tetanus; and they do this by varied tension in single fibers, rather than by varied recruitment of multiple fibers.

The other crucial difference in many invertebrates is the presence of **multineuronal** innervations (Fig. 9.98a), which adds another level of flexibility. Each muscle fiber can have inputs from several motor neurons. Furthermore, these inputs can be inhibitory as well as excitatory; and these different inputs again summate at the plasma membrane. Many muscle fibers in crustaceans and insects are linked to two or three excitatory neurons (one of which gives particularly large EPSPs and is termed a "fast exciter") and also to one or two inhibitory neurons.

The essential differences between basic motor control in vertebrates and in arthropods are summarized in Fig. 9.98b.

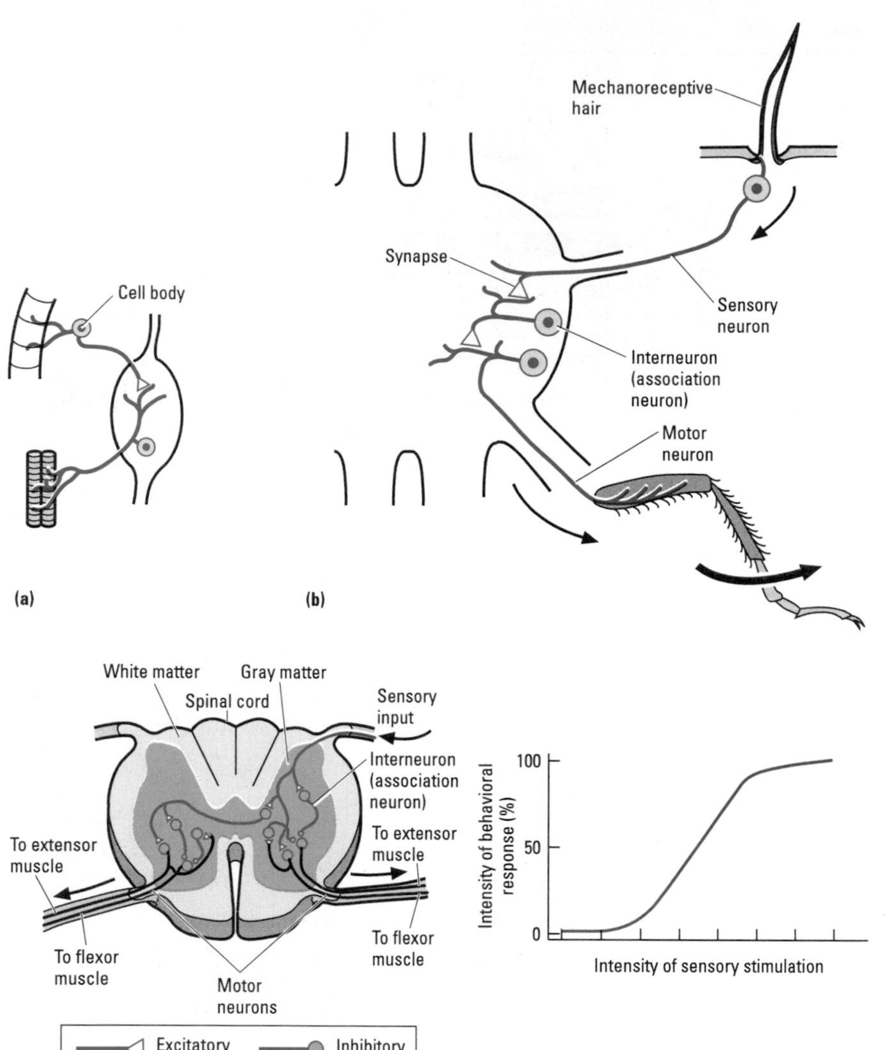

Fig. 9.99 Examples of reflex arcs. (a) A monosynaptic reflex arc. (b) A simple reflex with one interneuron in an insect, where stimulation of the bristle mechanoreceptor elicits a kicking response. (c) A polysynaptic reflex in the vertebrate spinal cord, incorporating reciprocal inhibition between an antagonistic muscle pair on one side as well as contralateral effects (the "crossed extension reflex"). (d) The effect of sensory stimulation intensity on the intensity of the behavioral response for a typical reflex; note that the response is *not* all-or-none.

9.14.2 Reflexes

All movement based on muscle systems depends on motor control systems, usually produced in neural networks formed from interacting neurons that are informed by inputs from sensors, and are linked by synapses that exhibit all the characteristics of plasticity described earlier in this chapter. However at their very simplest these controls may be extremely fixed and invariant, producing involuntary and stereotypical movements. These tend to be seen as reflex behaviors, and may be mediated by as few as two neurons, one a sensory receptor and the other a motor neuron, synapsing directly with each other (though one or a few interneurons may also be interposed without much affecting the stereotyped output). The whole system is commonly termed a **reflex arc** (Fig. 9.99), whether it is monosynaptic or multisynaptic. The reflex action may extend to larger and larger parts of the motor apparatus as the strength of the stimulus increases, in an almost linear fashion (Fig. 9.99b).

The most familiar example involves the knee-jerk reaction in humans. Here a sharp tap applied just below the knee stretches the main tendon of the knee-extending muscle (the quadriceps) and thus also stretches the muscle. This triggers sensory stretch receptors within the muscle that synapse onto spinal motor neurons, initiating a counteracting forward jerk of the knee. This reflex helps maintain upright posture, the muscle contraction countering any sudden stretching influence that might disrupt the balanced tonic states of the various knee muscles in a standing human. This is an example of a **myotactic** reflex, giving a reflex contraction in response to the stretching of a muscle, as with many other vertebrate "postural" reflexes that maintain muscle length and/or muscle tone. It relies on particular stretch receptors within the muscle that are known as **muscle spindles.** These sensory organs each contain a small bundle of specialized muscle fibers ("intrafusal fibers") with the terminals of sensory neurons known as Ia afferents wrapped around them centrally (Fig. 9.100). The intrafusal fibers are themselves innervated by small χ motor neurons, which cause them to shorten roughly in concert with the surrounding muscle tissue, so maintaining a constant tension in the spindle while the muscle is working, and insuring that the spindle Ia afferent nerves remain sensitive over a wide range of muscle lengths. Each of the Ia afferents normally synapse with all the motor neurons that innervate their

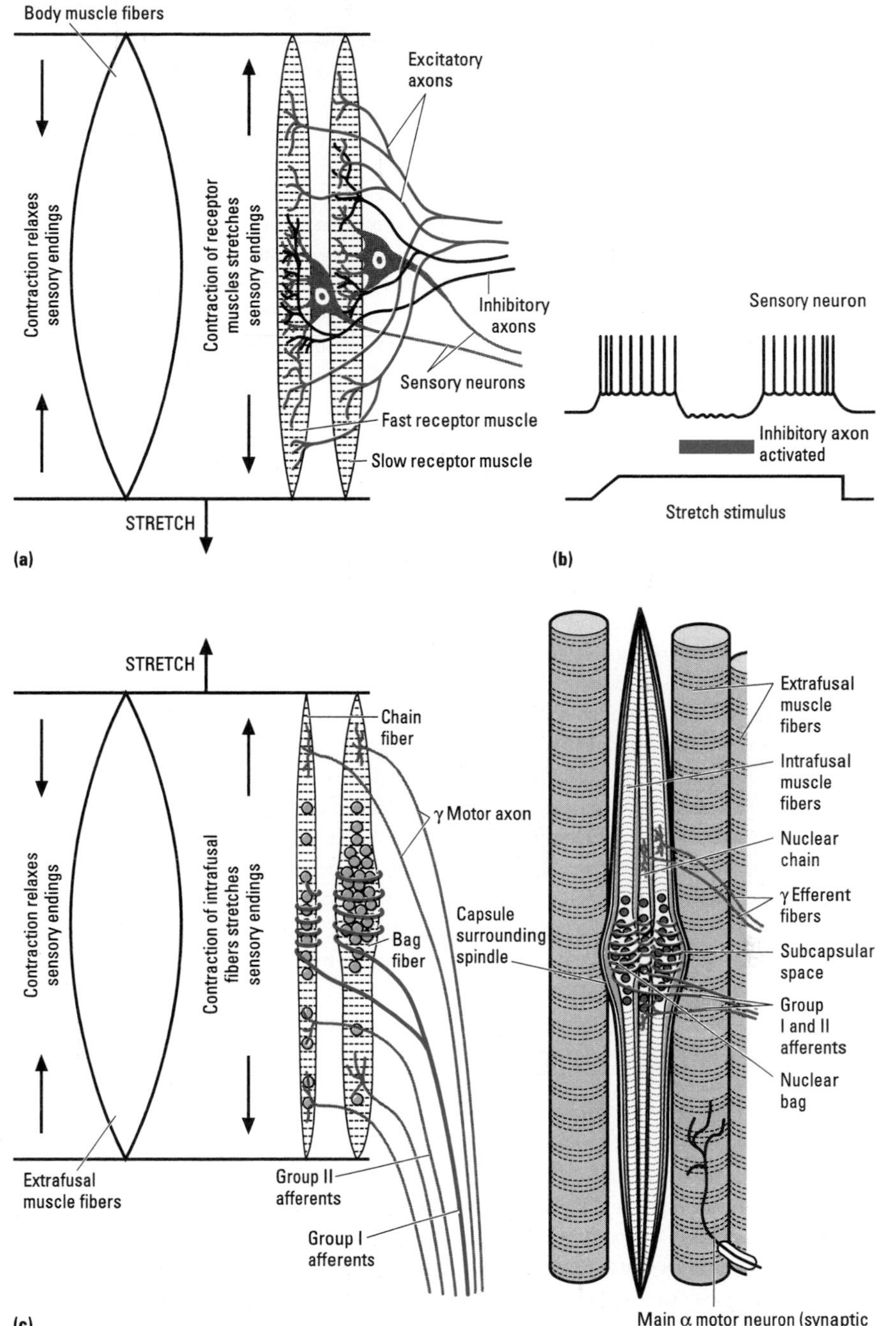

Fig. 9.100 Stretch receptors and reflexes. (a) A crustacean stretch receptor, where fast and slow stretch receptors operate, their contraction causing a relaxation of the sensory endings, but with additional influence from excitatory and inhibitory neurons to the receptor muscle fiber to reset the response range. (b) The recorded response of this receptor to stretch, and the effect of simultaneously triggering the inhibitory neuron. (c) A vertebrate muscle spindle, similar in principle (left) though with no inhibitory input, and as it appears in reality (right). In both cases, note that the contraction of the main muscle *relaxes* the sensory endings, whereas intrinsic contraction of the modified receptor muscle fibers *stretches* the sensory endings.

particular muscle, and may also synapse with inhibitory interneurons serving antagonistic muscles. Hence these Ia afferents can maintain the tone in their own muscle, and activate synergistic muscle, and inhibit antagonistic muscles, all these effects occurring simultaneously.

Arthropods also have well-developed muscle stretch receptors that produce myotactic or "resistance reflexes", helping to hold joints in a fixed position and so maintain posture. Indeed many other animals use simple reflexes based on stretch receptors, particularly for escape responses in lower animals, and for postural control in terrestrial animals.

Reflexes can operate on more than one set of muscles at once, for example in the "crossed extension reflex" in mammals, where pain in one foot leads to it being withdrawn and the contralateral foot being extended. They can also be linked (with slight temporal delays) between segments in vertebrate and arthropod nervous systems, to give phased responses along the axis; and they can be modulated in intensity by descending higher level neuronal signals, especially via spinal interneurons. Hence they can form the basis of quite complicated and graded movements and responses.

However, most of the behaviors we see in animals are rather more sophisticated, ranging from fixed multicomponent responses to particular stimuli, through repetitive patterns of movement giving locomotory or rhythmic movements, to highly complex voluntary or "conscious" movements. During such locomotory patterns the simple reflexes discussed here may be overridden or even reversed.

9.14.3 Fixed action patterns

Fixed action patterns (FAPs) are often seen (with reflexes) as the second most basic units of behavior, or as "elemental motor patterns". They are relatively complex motor sequences, always showing the same sequence of components and they differ from reflexes in that they are produced in full in response to very specific stimuli. Even if the stimulus stops, the motor pattern, having been "switched on", goes on through its complete sequence, and is therefore all-or-nothing. FAPs can be elicited even from an animal that has never seen the key stimulus (or "releaser") before, and has never watched the FAP being performed by another animal; thus it seems they are genetically inherited, not learnt. Indeed they are often so standardized that all members of a species, and often of a genus or even higher taxonomic category, show the same FAP, allowing their phylogeny to be traced. Furthermore, natural variants and hybrids can be used to investigate genetic control and heritability of different components of the behavior. These characteristics also make FAPs especially useful for study of the control of movement.

In invertebrates fixed behaviors of this type commonly occur in courtship, in mating, in aggressive displays or fighting, and in escape responses. The same is often true for vertebrates; the display behaviors of many fish show a very fixed pattern, as do various orientation and navigation mechanisms seen in the behavioral repertories of fish and tetrapods. More mundanely, higher vertebrate FAPs include vomiting, sneezing, and orgasm. However, the patterns become increasingly flexible and subject to modification by experience in these "higher" vertebrates.

9.14.4 Patterned activity and oscillators

Most repetitive behaviors, particularly those that underlie locomotion (swimming, walking, flying, etc., and also respiratory movements), cannot be explained in terms of simple reflex loops or entirely stereotyped FAPs, but instead rely on a network of neurons that together generate the basic pattern. These neural networks are termed **central pattern generators** (**CPGs**), also known as neuronal oscillators. The existence of such systems can often be demonstrated if the pattern of motor activity continues even after cutting sensory information from the muscle or joints. Sensory feedbacks can, of course, feed in to such a network to produce the continual small modifications needed in response to environmental changes such as currents, winds, and uneven terrain, but they do not in themselves cause the pattern.

In some cases the CPG relies in part on a single cellular oscillator, or "**pacemaker neuron**", where firing results endogenously from the interaction of ion currents within the cell. However, the pacemaker is always part of a neuronal circuit. Box 9.3 shows some ways of generating rhythmic firing in one or more cells.

CPGs have been shown to underlie the rhythmic oscillations involved in locomotion in a huge range of animals, from worms through slugs and insects to fish, birds and even advanced mammals such as cats. Classic examples can be seen in the flight systems of many insects with direct flight muscles, where the rhythmic activation of depressor and extensor wing muscles continues without any sensory inputs, but its frequency is altered (usually increased) when the sensors at the wing bases are allowed to feed into the neural circuit. Turning on the pattern generator is achieved by the insect jumping off the ground, often mediated via sensory hairs on the head that detect the relative air movement; and turning it off is usually related to some input from landing, through mechanical receptors in the feet.

Many invertebrate CPGs can be isolated to a single ganglion within the nerve cord. In both insects and crustaceans, and in segmentally organized worms, the same circuitry may occur in several ganglia but operate slightly out of phase in each, so giving a motor pattern where each successive segment has a slight incremental delay. This gives the familiar **metachronal rhythm** to many forms of locomotion, seen very clearly in walking centipedes, swimming shrimps, or peristaltic crawling earthworms. In many such systems the pattern generators in each segment are coordinated by **command interneurons**, located in ganglia within or close to the brain. Electrical stimulation of single identified command interneurons may elicit entire and complex behavioral sequences.

A well-known example of a CPG is the escape swimming mechanism in the nudibranch mollusc *Tritonia* (Fig. 9.101). Faced with an unpleasant stimulus this sea slug swims away by alternately flexing its body dorsally and ventrally, rising rapidly into the water column. The whole swimming sequence is controlled by just three kinds of neuron: a cerebral neuron, groups of dorsal and ventral swimming interneurons (which interconnect reciprocally with each other), and groups of dorsal and ventral swimming motor neurons. Most of the interconnections incorporate both excitatory and inhibitory synapses, in the normal invertebrate pattern. Once the network is activated it operates neurogenically, the dorsal and ventral interneurons generating alternate bursts of neuronal activity, so that the motor neurons produce alternate dorsal and ventral flexing.

Many aspects of vertebrate motor activity also rely on multiple coupled oscillators acting as spinal pattern generators, and the patterns persist without sensory inputs. Many fish can still swim, and frogs can still walk, when the sensory roots into their spinal cord have been cut; and respiratory rhythms centered on the brainstem persist under these conditions in all vertebrates, including the most sophisticated mammals. Flexibility in the performance of the responsible CPGs is conferred by modulatory neurons that control the cycle period and the duration and intensity of motorneuron spike bursts; these are commonly mediated by GABA and 5-HT acting both pre- and postsynaptically.

In many animals the pattern generators are subject to **command systems** formed by small networks of neurons, rather than single command neurons. For example in the decapod crustaceans there are commonly escape responses (involving tail flips) mediated by giant neurons, and there may be either forward or backward escape movements depending on which end of the animal is under threat. Backward escape is controlled by medial giant interneuron networks and forward escape by lateral giant interneuron networks (Fig. 9.102), both involving only three or four interacting command neurons.

Pattern generating systems are an economical and reliable way of

Box 9.3 Rhythmic neuronal patterns

Rhythmic motor activities can be produced either by the activities of single "pacemaker" (oscillator) neurons, or by the integrated actions of several neurons with specific connectivity.

Cellular pacemakers

Rhythmicity within one cell can be generated most easily by interacting ion channels and ion currents. Commonly this involves a **calcium** channel which is depolarizing, and a calcium-dependent **potassium** channel which is hyperpolarizing:

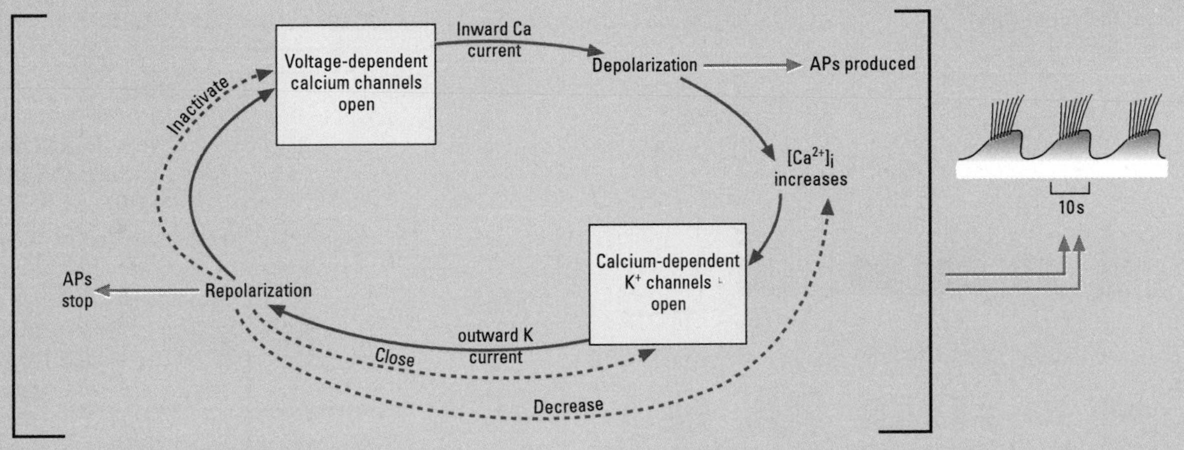

The output of such an intrinsic oscillator can be modified in several ways:

- Accommodation of the action potential (AP)-producing response.
- Postinhibitory rebound, where firing is triggered when the inhibition is removed.
- Stable membrane potentials other than the normal resting potential (RP)—some cells have "plateau potentials" above the threshold for APs, and switch rapidly between the RP and the plateau potential.
- Regulation by standard synaptic inputs.

Rhythmic neuronal networks

Three common types of rhythmically active circuits, or "network oscillators", occur.

1 Self-excitation. Here two neurons excite each other, giving positive feedbacks; a counterinfluence must be present to reset the system, and this can be achieved by an inhibitory neuron triggered at some critical discharge level.

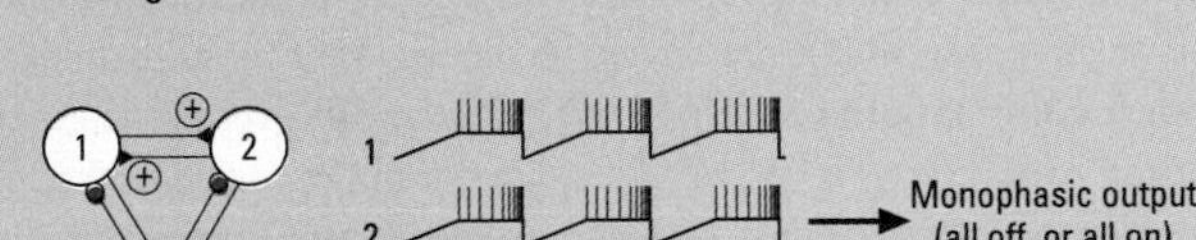

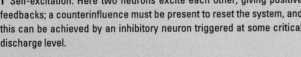

2 Reciprocal inhibition. Here two neurons inhibit each other and oscillate when both also show adaptation or synaptic depression; the counterinfluence is provided by a tonic input from an excitor neuron.

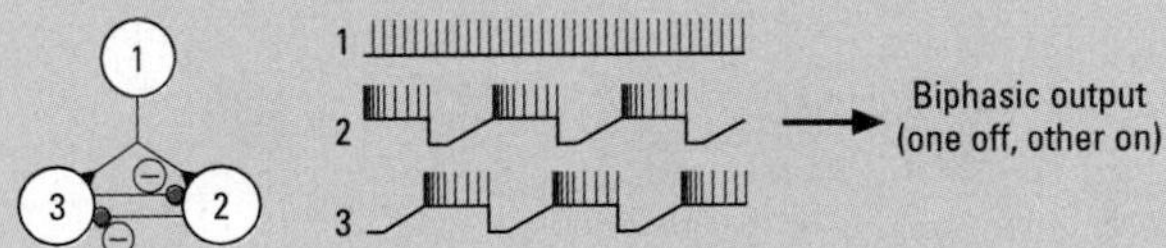

3 Cyclical recurrent inhibition. Here there is a complete circuit of inhibitory influence, and no need for a counterinfluence.

Polyphasic output

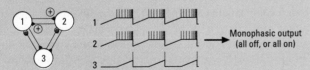

controlling locomotory rhythms and fast escapes. However, they are always subject to sensory inputs in living animals, to alter their expression in relation to circumstances, and generally they are also controlled by neuromodulatory inputs to reset both the properties and the sensitivity of the network. Instances where neuromodulators such as dopamine or 5-HT can change the effective connectivity within a command network by making some pathways more active and others more inhibited are well documented; in effect, the circuitry gets "rewired" by the modulating chemical.

9.14.5 Voluntary movements

In the absence of a forebrain many vertebrates can locomote fairly well, showing swimming, walking, or galloping patterns of limb activity. However the motions are rather robot-like, having no directionality or orientation, and no particular relation to the environment. These "extra" elements, together with initiation of the process, are the aspects of moving that are added by higher neural structures, which are only really known in vertebrates. The cortex

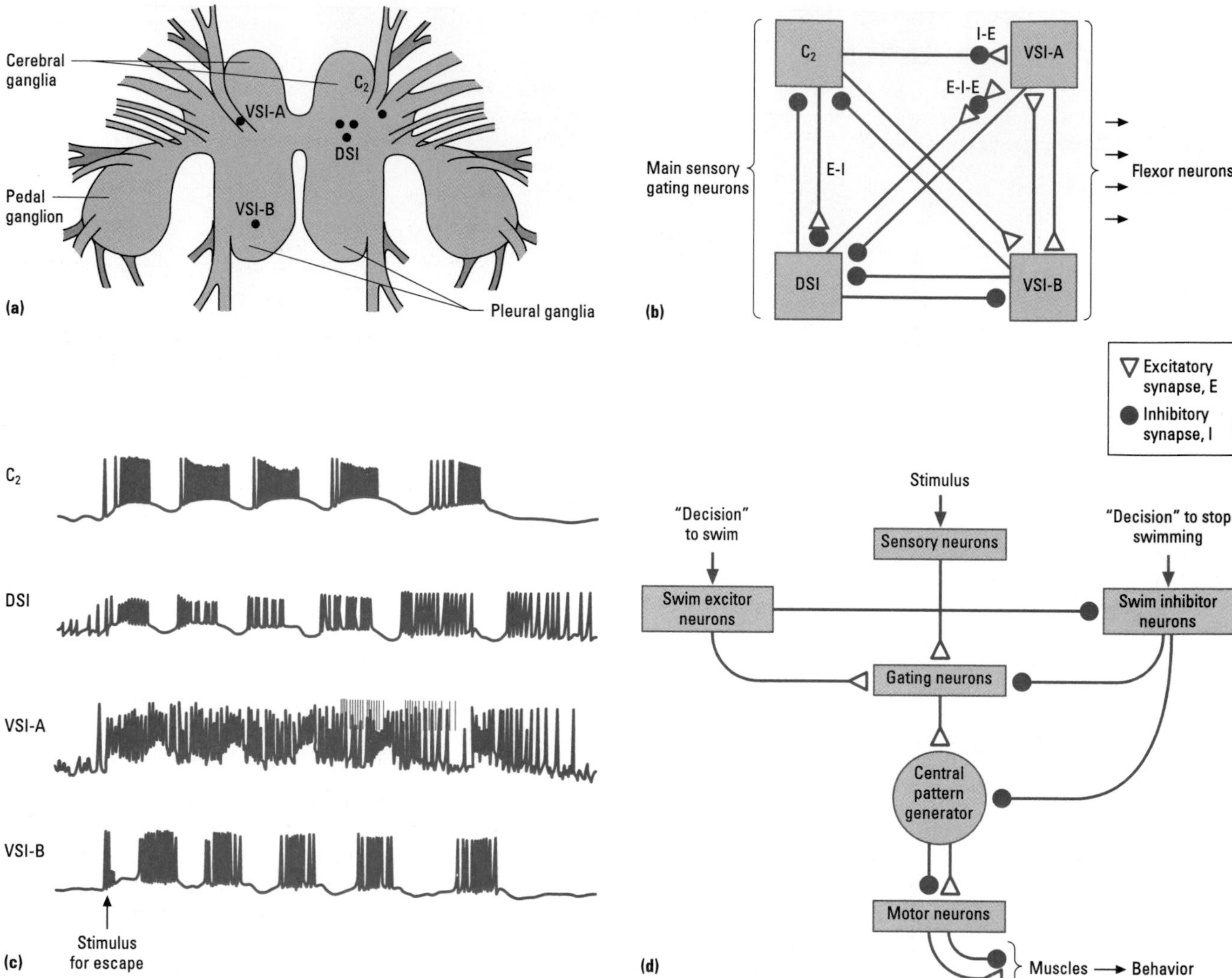

Fig. 9.101 The central pattern generator (CPG) system that mediates escape behavior in *Tritonia*: (a) the brain and location of the key neural components, which are four types of interneurons (VSI-A, VSI-B, C_2, and DSI), (b) the neuronal circuitry of these four interneurons, (c) the rhythmical patterns in the same neurons during an escape swimming response, and (d) how the CPG fits into the general control system. (Adapted from Reichert 1992, and other sources.)

is not particularly crucial, though; it is the brainstem that must be functional for coordinated and directed locomotion, and defects here lead to motor deficits such as tremor and poor balance. The basal ganglia and thalamic regions nearby (Fig. 9.103) receive inputs from the hippocampus and amygdala, and they project to a particularly important region, the **mesencephalic locomotor region** (MLR), which when stimulated can give rise to directed locomotion at speeds proportional to the intensity of the stimulation. The MLR projects directly to the lower brainstem, where the major spinal locomotory pathways begin and make descending connections to various CPG circuits.

MLR activity appears in its turn to be under tonic inhibitory control from levels higher still in the brain, perhaps with two separate inhibitory circuits, so that voluntary motion at least partly involves turning off this inhibition. How this is achieved is the subject of extensive and complex brain research; it involves interactions between many areas (Fig. 9.104), including the sensorimotor integration neurons (SINs) mentioned in section 9.8.8, and visuomotor coordination circuits. At the final hierarchical level are the main **motor cortex** (see Fig. 9.35) and, adjacent but anterior to that, the **premotor cortex** and the **supplementary motor cortex**, all these being organized topographically and with each vertical "column" of neural tissue devoted to the control of one or a few joints. The motor cortex areas are accessed by "conscious thought" from the frontal and parietal lobes. The motor cortex neurons begin to fire before the movement actually begins. Four main types of neurons occur here: dynamic neurons which code for the rate of development of force in the muscles; static neurons that code for the steady state force achieved; an intermediate set of neurons with elements of both these force parameters being encoded; and finally a set of direction-coding neurons where each neuron has a preferred direction. The premotor and supplementary motor areas begin to fire even earlier before the movement and seem to be crucial to the "decision to act". In essence they plan the movement in advance, and in humans these

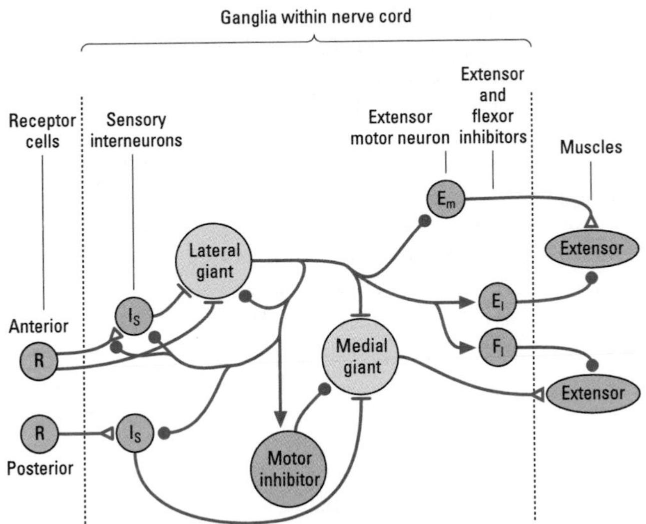

Fig. 9.102 The giant interneuron system that mediates escape in crustaceans such as the crayfish. The medial giants have an anterior receptive field, and so mediate escape backwards, while the lateral giants have a receptive field in the tail end and mediate escape forwards; they have strong inhibitory effects on other components of the motor system. The interconnections shown are repeated in each of the six main abdominal segments, with additional inputs from the muscle stretch receptors (not shown).

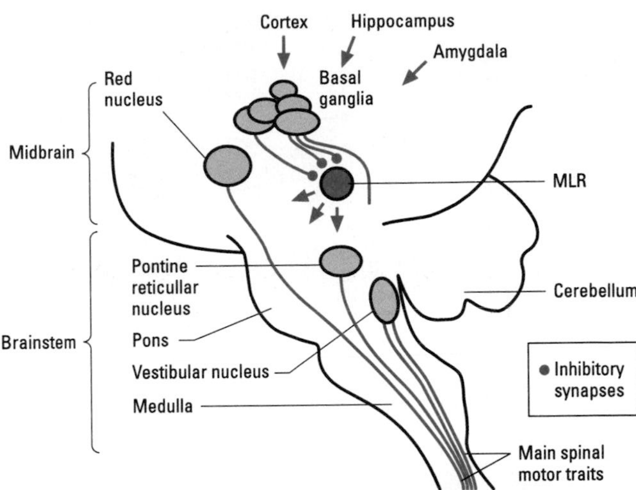

Fig. 9.103 The basal ganglia linking to the mesencephalic locomotor region (MLR) in vertebrates, involved in initiation of locomotion, and other nuclei forming part of the motor control system in the midbrain and brainstem. Note this does not show strict spatial relationships.

areas become active when someone is asked just to imagine making a particular movement of their hand, for example.

A further complication to the whole control system resides in the **cerebellum**, which is not directly part of the hierarchy of information transfer described so far (since it has no direct connections with the spinal cord) but which acts as a coordination center, comparing the ascending and descending signals and spotting discrepancies between the "intended" action and the "actual" motion performed. In this way it adds a "parallel" control component to the serial hierarchy of the vertebrate sensorimotor system. Parts of the cerebellum are also critical for the learning of complex motor tasks.

At this point we are getting closer to neuropsychology, and the details are somewhat beyond our present scope; they are also still a very long way from being understood.

9.15 Locomotion using muscles

Turning the intracellular processes of muscle contraction into locomotion of a whole animal sounds like a rather demanding task, not least because of the huge range of movement patterns that animals seem to exhibit. But the principles turn out to be rather simple, and common to nearly all the forms of animal locomotion: an animal contracts a muscle which moves a part of its body which exerts a force on the environment which makes the animal accelerate away in the opposite direction. Sustained locomotion results when a body undulates or undergoes peristaltic wave-like motions, or when an appendage moves in an oscillatory fashion. Locomotion thus relies

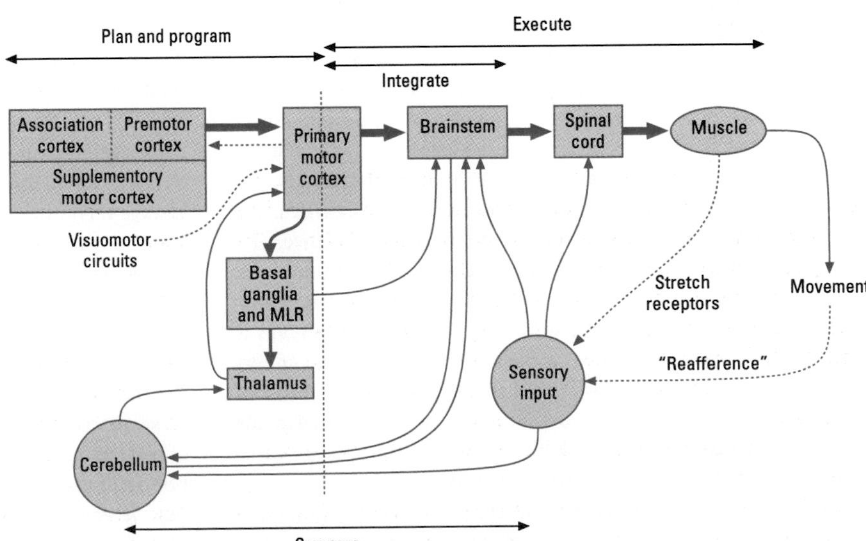

Fig. 9.104 A schematic view of overall motor control in vertebrates, with a hierarchical system "downwards" from the key cortical areas, plus a serial "comparator" component via the cerebellum (see text). MLR, mesencephalic locomotor region.

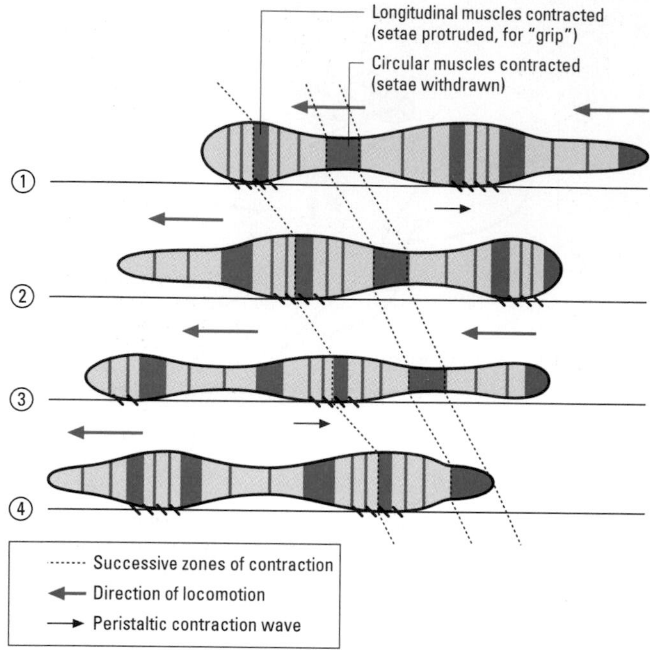

Fig. 9.105 Metachronal rhythm of muscle contractions and basic peristalsis, shown here for a segmented earthworm but similar in principle for all worms.

on the motor provided by the muscles, rapid mechanical reflexes with opposing (and usually rhythmical) stretches and contractions, and multimodal sensory feedbacks and feedforwards. But within this superficial simplicity, muscles are revealed as having a surprising range of functions, serving not only as motors but also as springs, as brakes, and as stiffening struts; and the force that they generate is used not only for propulsion, but also for achieving stability and maneuvrability in a complex and changing three-dimensional environment.

9.15.1 Hydrostatic movement

All animals have "soft" or hydrostatic skeletons within their bodies, because fluid pressure in internal cavities helps keep the animal in shape, and gives some degree of potential shape change when muscles act. To be effective as a skeleton for moving animals, a hydrostatic system needs three components: compressive, contractile, and tensile elements. The contained fluid, by definition incompressible, gives the first element; muscle of course provides the contractile element; and a strapping of fibers provides the tensile element to keep the whole system in shape. For many animals the body cavity is the hydrostatic skeleton, and the animal is roughly tubular (a "worm") with muscle-driven contractions acting on the fluid to produce movement of the whole body. To achieve this, the muscles must surround the incompressible fluid, and a two-layered arrangement is much the commonest, with one layer of circular muscle (fibers running around the circumference of the worm), and one layer of longitudinal muscle (fibers parallel to the body axis) arranged inside it. In these hydrostatic (soft-bodied) animals the muscles are commonly not striated and lack a clear fibrillar pattern.

Contraction of the muscles increases the pressure of the contained fluid, and the force involved is the product of pressure and area. The pressure is resisted by the body wall and its contained tensile lattice, so that at any point where the resistance is lower the animal will tend to bulge outwards and change shape. A long cylindrical animal can therefore perform a range of shape changes according to the state of tension in different parts of its surface. The change in shape performs mechanical work, which as always will be the product of the force applied and the distance moved.

Closed hydrostatic systems are found in all worms, many insect larvae, and many sessile tube- or box-dwelling aquatic animals. They operate at constant volume, with a restricted shape change. Most worms have a single undivided body cavity, or have a two- or three-segment system (e.g. phoronid and hemichordate worms) but where one segment is largest to form the trunk, largely responsible for locomotion (with the other more anterior segments specialized for feeding). Only one phylum, the Annelida, has a fully segmented body cavity, with the muscles also in segmental blocks, each segment operating as a separate hydraulic unit. Even there, rather a lot of annelids lose the partitions between segments internally.

In nearly all these hydrostatic animals movement depends on controlled **metachronal waves** of contraction in the two sets of antagonistic muscle. When the longitudinal muscles contract locally, that part of the body becomes short and fat; when the circular muscles contract, that part of the body becomes long and thin (Fig. 9.105). Successive waves of contraction give rise to **peristalsis**—the most familiar mode of locomotion in soft animals, and seen also in the wave-like contractions within tubular internal organs such as the gut. The peristaltic waves of successive contractions are easiest to see in a terrestrial earthworm moving over the surface of the ground. Here a contraction wave passes backwards along the body, with the edges of segments and the small bristle-like chaetae protruding to give purchase on the soil and prevent slippage backwards.

Based on this simple system, different patterns of waves of muscle contraction and antagonism permit a whole range of different types of locomotory patterns.

Surface gliding

This is used by many small aquatic and interstitial invertebrates, and in the smallest it is primarily achieved with cilia (e.g. small flatworms, only 1–2 mm thick). At such small sizes, viscous forces are the most important resistors to movement, and the animal merely glides along on the cilia of the lower epidermis and a film of mucus. Muscles only come into play in the larger flatworms, where inertial forces begin to dominate. Small amplitude waves of contraction then pass along the body. However, the overall body shape cannot change very much, and the flatworms cannot move fast because they have no real body cavity, instead having a relatively nondeformable parenchyma internally.

Creeping by pedal waves

Pedal waves can be seen readily in terrestrial slugs and snails as waves of contraction moving (usually forward) along the foot. In effect the snail lifts a part of the sole of its foot, using special obliquely orientated dorsoventral muscles, moves it forward, and puts it down again, and a wave of this activity moves forward. Hemocoelic spaces in the muscular foot help to transmit the forces

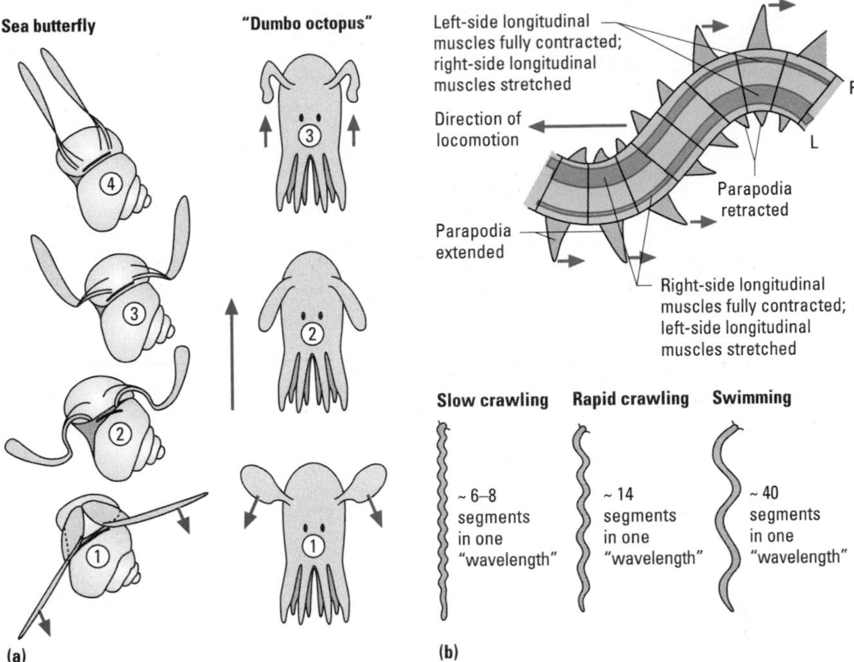

Fig. 9.106 (a) Flapping swimming in sea butterflies and in the "Dumbo octopus". (b) Sinusoidal swimming in an invertebrate, for example a ragworm, showing muscle contraction patterns and the increased wavelength and amplitude associated with swimming rather than crawling. (a, Adapted from Morton 1954.)

involved, and mucus secretion lubricates the movement. Recent evidence shows that this mucus has a key role, changing its properties at different parts of the contraction cycle to give alternating adhesion and slip, with the foot surface never actually lifting clear of the mucus layer.

In some species of mollusc there can be a variety of **gaits** with the pedal waves. A snail with a slow crawl may have many small almost undetectable waves traveling along its foot at once, whereas in a faster gait there are only one or two waves of larger amplitude. Some gastropod molluscs such as whelks have a so-called "gallop", where they pick up the entire right half of the foot to move forward and then the entire left half, requiring an additional left–right temporal asymmetry in the controlling neural circuitry.

Swimming and crawling

For a hydrostatic animal to achieve swimming, some part of the body is usually modified as a muscular undulating flap, and this alone undergoes the metachronal waves of contractions. This can be seen in the lateral mantle edge of cuttlefish (in their "slow" gait), in the two flapping "ears" of the deep-sea "Dumbo octopus", or in the edges of the body in large flatworms (Fig. 9.106a). Other animals also use a more extreme version of this kind of "**flapping swimming**"; for example the sea butterflies (pteropod molluscs) have extended mantle flaps that operate in a similar but more spectacular fashion, looking rather like a pair of wings. Waves of muscle contraction occur, again usually alternating in longitudinal and circular muscles; the flap has an aerofoil profile, and its gentle vertical flapping gives a combination of lateral and backward forces. The lateral forces cancel out on either side, and the backward force against the water propels the body forward.

In some small invertebrates a pattern of **helical swimming** occurs, similar to that seen in many unicellular organisms, where the body rotates on its axis as it moves along. In the "tadpole larvae" of some tunicates this has been analyzed kinematically and seems to result from an asymmetrical beat in the tail end of the animal, generating a "yawing" effect.

In larger and more elongate swimming animals, locomotion tends to involve an undulatory or **sinusoidal swimming** pattern (Fig. 9.106b). Occasionally (e.g. large leeches) this is achieved by a dorsoventral wave. More commonly the sinus wave is lateral, like the lateral wave-like bending seen in many fish (see below); forces are both forward and sideways, but again the lateral forces in either direction cancel each other out. This kind of swimming is often aided by lateral appendages, giving a greater degree of leverage against the fluid medium; the paddle-like parapodia in polychaete annelids are a classic example. Again the locomotion involves metachronal contraction waves in the body wall, with both front-to-back and left-to-right patterns of temporal offset. Waves of contraction pass from tail to head, and the left and right parapodia are exactly half a wavelength out of phase so that each paddle operates without interference from the one just behind it.

This system can also be used for crawling along the substratum in benthic animals, aided by some contact with the sand or rock. For polychaete annelids, in a slow crawl the body is kept fairly straight with little lateral undulation and the parapodia do most of the levering, controlled by their own intrinsic muscles. At faster speeds the body wall itself becomes more important, and patterned activity in the lateral longitudinal muscles means that the whole body is thrown into the typical sinusoidal wave, pushing against the water and substratum to give a net forward thrust.

Burrowing

Burrowing animals need both to penetrate into the substrate and then to retain an ability to move about within it. The principle for

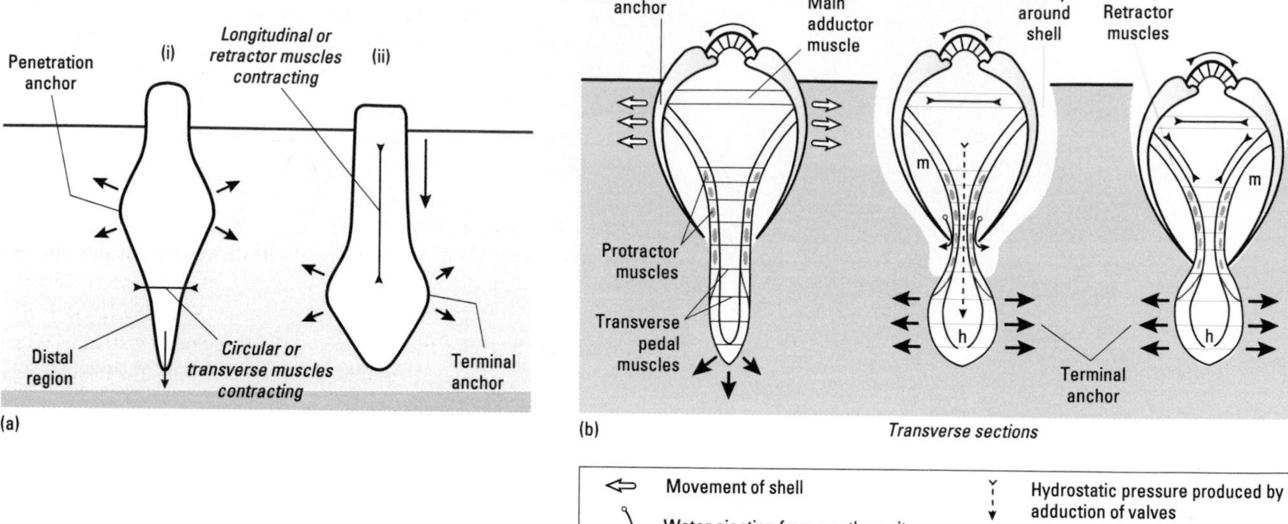

Fig. 9.107 (a) The principle of alternating anchors involved in basic burrowing: the penetration anchor holds the animal while the distal region is extended into the substratum, then the terminal anchor is expanded and holds the animal while the rest of the body is pulled downwards. (b) Successive stages in bivalve burrowing, where the shell forms the penetration anchor while the foot probes downwards, then the foot expands to form the terminal anchor as the shell is drawn down. h, hemocoel; m, mantle cavity. (From Trueman 1975.)

the main burrowing action is always the same for a whole range of animal designs, and is shown in Fig. 9.107. The idea is to have an alternation of two different kinds of anchorage. First the animal makes one part of the body expand widthways, to serve as the **penetration anchor**, from which the front end can push away into the substrate. Then it expands the advancing tip into a **terminal anchor**, which the rest of the body can then be drawn up to. This in effect creates a single (monophasic) wave along the body. In unsegmented worms, the unstriated longitudinal trunk muscles contract to form a fat penetration anchor, while the proboscis extends; then the proboscis is dilated with fluid by contracting the trunk circular muscles, to form the terminal anchor.

Figure 9.107b shows the same principle in a slightly more complicated form in a bivalve mollusc, many of which burrow in sand and mud. The shell itself is allowed to gape open to give the penetration anchor, and the foot pushes out from within this. Then the end of the foot is expanded by fluid pressure in the hemocoel as the retractors and other body muscles contract, to give a terminal anchor, and the shell is closed as far as possible and drawn down after the foot. Water expelled from the mantle cavity as the shell closes also helps to soften the substratum and ease penetration.

Looping and somersaulting

These systems represent further variants in which there is essentially only one wave along the body. However, now there are really only two points of attachment with the substratum. In effect one end is lifted, and placed down again some distance away; then the other end is detached and moved. In looping, normally the back is brought up to the front, then the front moves on again. This is familiar in leeches, and also in some terrestrial caterpillars, notably one group that are actually called "loopers". In somersaulting, the rear end is flipped over the head and placed down beyond it; this happens in the fast locomotion gait of *Hydra*, and in a few worms.

Jet propulsion

This represents the ultimate modification of a wave-based locomotory system; in effect it involves one wave at a time but made synchronous over the whole body. It requires a soft body with a high potential change of shape; an inflexible animal with hard skeletal parts would pull itself apart if all the muscles contracted at once. Crucially, jet propulsion also requires exceptional nervous coordination, with giant nerve fibers to transmit the instructions almost simultaneously to a large area of muscle.

The most obvious examples are the cephalopods, including squid, cuttlefish, and octopus, where the large open mantle cavity is the key. The mantle has thick layers of circular muscles that contract to increase the force at right angles to the mantle wall, resulting in an accelerated movement of water outwards via the siphon; then radial muscles contract to reinflate the mantle. The mass of water that is expelled is considerably smaller than the body mass, so it is expelled at a higher velocity than that of the animal's movement in the opposite direction. The circular muscles have a nerve supply from one massive giant neuron on either side of the body, insuring that the whole mantle contracts at once and water is ejected from the mantle cavity as a jet. The jet itself can be directed by a funnel, so the animal can dart off in an appropriate direction, usually backwards from danger.

Other examples do occur in invertebrates. Many jellyfish keep their station in the water column by alternately gently jetting upwards (by expelling water from the bell downwards), and then

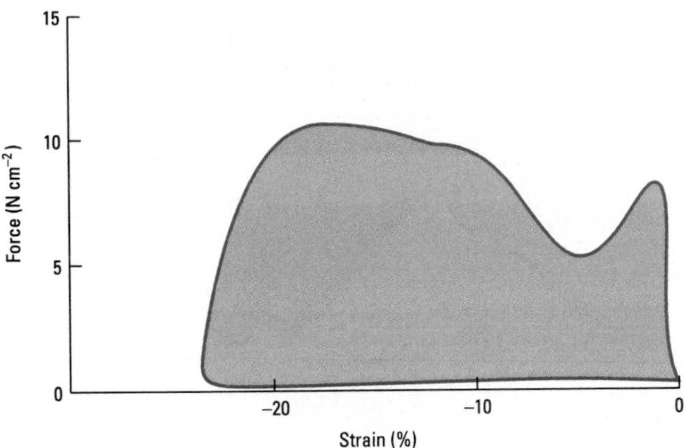

Fig. 9.108 The work loop recorded for the adductor muscle of a swimming scallop. At the lower right the shell has its maximum gape; as the adductor is activated the force rises and then the muscle shortens rapidly, providing the water jet effect. At the top left the muscle activation ceases, so force declines but shortening continues until at the lower left force is minimal and there is no further shortening. Along the base of the loop the shell opens by passive recoil of the elastic ligament, with no force involved. (From Marsh & Olson 1994.)

parachuting slowly down again. This jetting is most efficient in the bullet-shaped species, but it is traded off against better prey capture in the tentacles of flatter, more dish-shaped jellyfish.

Scallops and a few other bivalves can jet propel by clapping the two valves together to expel water, which causes them to jump upwards, mainly as an escape from the starfish that eat them. In scallops a small and striated part of the large adductor muscle (described in section 9.13.2) works in opposition to an elastic ligament. The rapid shortening of the adductor muscle snaps the two valves of the shell together, and water escaping through jets near the hinge propels the animal, gape side first, through the water. This is followed by a recovery phase as the muscle contraction stops and the shells open automatically under the action of the elastic ligament, drawing water into the animal. The jetting phase only occupies a relatively small part of the overall cycle, and relatively few claps are normally used to "jump" away from danger. The highest performance occurs during the first clap, and records using the work loop technique with *Argopecten irradians* (Fig. 9.108) show that the average muscle power output is 30 W kg^{-1} wet muscle mass, with a peak of 120 W kg^{-1} wet muscle mass just as the propulsive jet is formed.

Hydrostatic skeletons as "legs"

There are a few animal taxa in which hydrostatics are used to move leg-like structures and produce stepping movements. One obvious example is the echinoderm "tube feet"; these have various different functions, but they are the main locomotory effectors in starfish. They are linked to the circulating pipes of the water vascular system, but each foot has its own valve and its own water-storing ampulla. Water is forced into the foot by contraction of the ampulla muscles, and back out of the foot by longitudinal muscles in the foot itself, which executes a small step on each cycle by also bending slightly. Each foot operates independently from all others and with little coordination in terms of the timing of the steps, but all the feet in each of the five arms step in one direction, determined by the radial nerve. The extreme simplicity of the echinoderm nervous system has precluded any detailed analysis of the control systems here.

The "walking worm" *Peripatus* is a soft-bodied arthropod-like animal (Class Onychophora), endowed with paired stumpy soft legs that are operated hydrostatically. They give a fairly efficient "lever" action, extended mainly by muscular activity of the body wall musculature and retracted when they shorten under the influence of their own intrinsic muscles.

"Hydrostatic legs" can also be found in some primitive spiders, which of course are not really soft-bodied and have a true arthropodan exoskeleton, but where extension of the legs is hydrostatic and only retraction is by the more normal arthropodan system of direct muscle–joint action. Some spiders can even jump effectively using a rapid hydrostatic extension of the fourth pair of legs.

9.15.2 Locomotion with stiff skeletal systems and joints

Exo- and endoskeletons, muscles, and joints

Stiff skeletons occur in many animals, often with a primarily protective role (shells, boxes, tubes), but versions that are highly involved in locomotion and endowed with joints only occur in the arthropod taxa (insects, crustaceans, arachnids, and some minor groups) and in vertebrates. The arthropods have an exoskeleton, the vertebrates an endoskeleton; the merits of each system are strongly size dependent as summarized in Box 9.4.

In all these cases locomotion involves the use of limbs (and also of separately evolved wings in the case of most insects). Limbs are made up of a series of jointed elements, allowing extension and retraction (flexion) of the limb, often also with some basal rotation relative to the body axis. As the base of the limb is rotated the tip can trace a more or less linear path parallel to the motion of the body, acting as a mechanical lever, and the body itself need undergo very little lateral flexing. In vertebrates the limbs normally have two main longitudinal elements, but in arthropods there may be five or six. This difference arises from the greater flexibility of joints in an endoskeletal system, which can move in several planes by utilizing ball and socket principles, whereas in arthropods the movement between any two adjacent tubular segments is only in one plane so that more sections are needed to give a full range of position and motion at the tip.

Most forms of locomotion with legs (or, in water, with fins) involve energy storage and exchange systems such that energy in one half of each cycle is stored and then released, usually with very high efficiency, in the second half of the cycle. This leads to the possibilities of muscles acting as a **brake** (where in a running animal a contracting muscle is resisting the tendency of a limb to swing back to its starting point) or as a simple **spring** directing the force of a power muscle (as in insect flight), or as a **strut** transmitting forces through the body (as in some swimming fish). These principles are illustrated in the following sections.

Swimming

Swimming involves pushing against a deformable fluid, which distorts away from the animal and usually swirls into a complex wake

Box 9.4 Exoskeletons, endoskeletons, and locomotion

Both systems require **joints**, both within the body trunk and in the form of jointed legs, to move.

- **Exoskeleton**—muscles insert on the inside of the skeletal plates, and are connected to them via cuticular apodemes; rotation is only in one plane.
- **Endoskeleton**—muscles pull on the outside of the skeletal struts, via tendons, which are mechanically more efficient and more multifunctional, with rotation in different planes.

Advantages of joints

- They can function as pulley systems and lever systems, giving a ratio of distance moved to muscle distance shortened that is much greater than 1:1.
- They permit the local application of force, all exerted in one direction with no wastage as lateral components and with the rest of the body unaffected.
- They are useful on land because they give decreased friction against the substratum.
- Less energy is wasted in accelerating and decelerating the main body mass (only the limbs stop and start repeatedly, while the trunk moves relatively smoothly).
- Individual muscles can be smaller than in a hydrostatic system, and thus potentially faster with less impedance from connective tissues.
- There is positional sensing (proprioception) using a minimal number of receptors at each joint to specify where the limb is in three dimensions.

Effects of size

Endoskeleton

- Gives a better weight to volume ratio for larger animals. Beyond a certain size, a solid central strut gives greater support than the same amount of material arranged externally as a tube.

But for a small land animal:

- Thin solid struts are too prone to cracking and buckling (small tubes are safer weight for weight, and harder to damage).
- Relatively large soft surfaces exposed would give serious physiological problems.

Exoskeleton

- Gives better physiological protection, and so predominates in small animals. It reduces exchanges in an animal with a high surface area to volume ratio.

But a large land animal:

- Would need a disproportionate bulk of material, becoming cumbersome and eventually immovable.
- Would collapse under its own weight when unhardened following a molt.
- Would be hopelessly weak in collision, resulting in lethal surface damage.

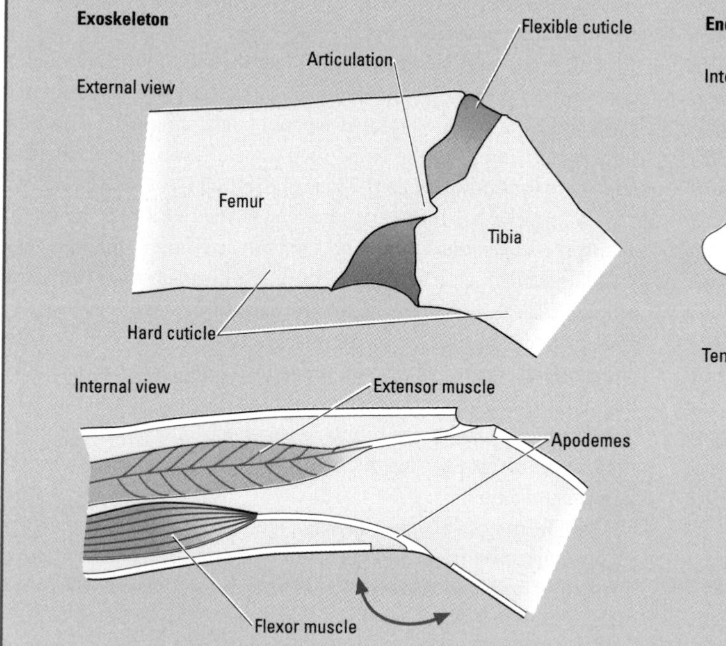

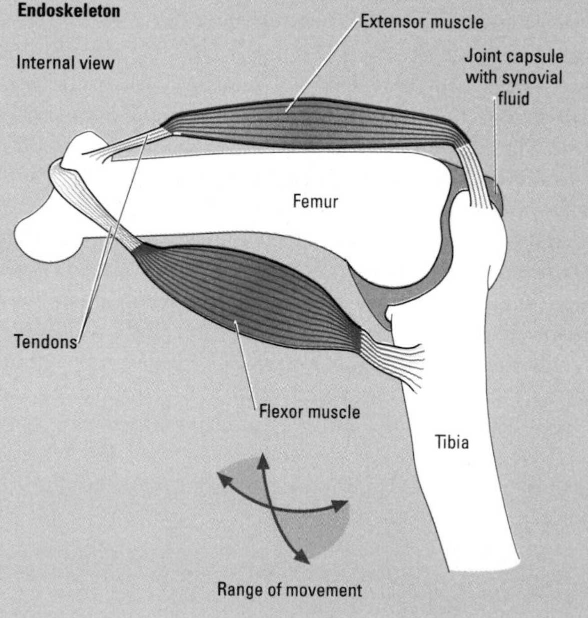

with spiraling vortices of water either side of the animal's path. In both swimming and flying, the vorticity tends to form a series of discrete "vortex rings", and techniques to reveal these have helped us to calculate the forces created by the animal as it moves.

AQUATIC ARTHROPODS

Swimming is a reasonably common form of locomotion in marine, brackish water, and freshwater crustaceans, as well as in many freshwater insects both as adults and larvae. Here the legs are used as paddles, sometimes many legs in metachronal patterns of beating (seen most clearly in the fairy shrimps which swim on their backs) and sometimes with most of the thrust provided by just one or two pairs of legs. In swimming crabs, and in many aquatic insects, the legs are flattened to provide a better paddle action, and often also have dense fringes of cuticular hairs (setae) to increase the effective surface (the legs being folded and the setae flattened during the

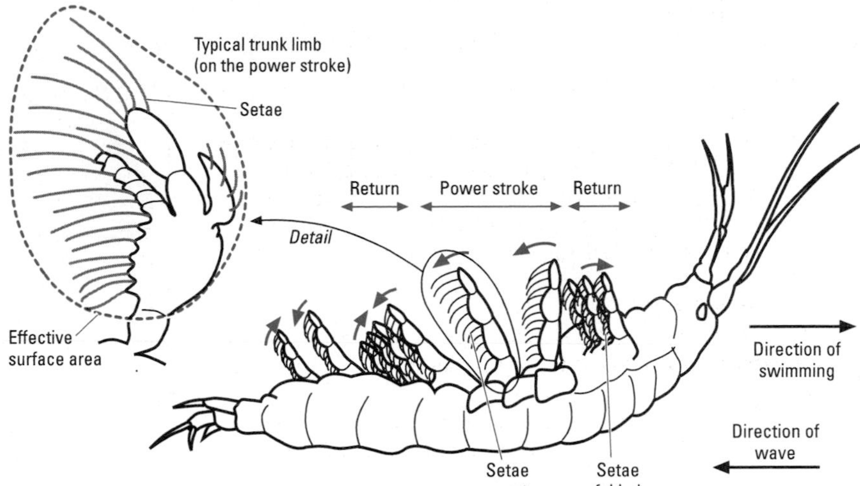

Fig. 9.109 Metachronal swimming pattern in a typical crustacean (here an anostracan that swims on its back). The long setae on each limb (see inset) help to extend the surface in the power stroke but the setae and the whole limb are partially folded down in the return stroke.

recovery stroke, Fig. 9.109). There is often a contribution from a "jetting" effect as well, with water drawn into the interlimb spaces as two successive limbs move apart and then forced rapidly out again laterally as the two limbs converge. Most of the muscle is concentrated in the upper part of the limb, with cuticular apodemes—the arthropod equivalent of tendons—making connections to the much thinner lower joints, reducing the mass of the most rapidly moving parts.

Some crustaceans and insects have an additional fast escape mode of swimming where most of the abdomen is rapidly flexed under the body (a "tail flip"), to propel the animal backwards away from sudden danger. In many shrimps up to 40% of the body mass may be devoted to the abdominal muscle required for this rarely used behavior, since it is so crucial for escaping predators. This is also a major reason why many crustacean tails make such excellent "seafood"!

FISH AND OTHER VERTEBRATES

Aquatic vertebrates have ways of achieving neutral buoyancy (see section 11.6.2) and do not have to swim to support their body mass against gravity. Most fish swim with a sinusoidal wave system (commonly termed **anguilliform motion** since it is most clearly seen in eels): waves of contraction pass down the body, out of phase along each side of the animal, and at a faster speed than the forward motion produced. This swimming uses a combination of paired and unpaired fins and, even more importantly, contractions of the trunk muscles, which are segmentally arranged as **myotomes**. Here the individual muscle fibers insert via short tendons into connective tissue sheets called myosepta. The myotomes have a complex shape, arranged helically (Fig. 9.110a) to form a series of overlapping cones, so that a transverse section through the trunk passes through several myotomes at different levels.

The anatomical separation of different fiber types in the myotomal muscles of elasmobranchs and teleosts is striking; the majority of the muscle is white, but red muscle fibers form a thin strip at the level of the major horizontal septum (Fig. 9.110b). The amount of red muscle is only 2–3% in sedentary benthic species but may reach 23% in active pelagic fish; some species also have an intermediate layer of pink muscle. Red muscle fibers have a well-developed capillary supply, a high concentration of myoglobin, and abundant mitochondria (15–50% fiber volume), while white muscle has only a sparse capillary supply, a low concentration of myoglobin, and a low volume density of mitochondria. White muscle is primarily recruited for movements associated with prey capture or escape behavior and is therefore used in short bursts. The energy supply to the white muscle is mainly provided by creatine phosphate, although anaerobic glycolysis is also important. Glycogen is usually at a lower concentration in white muscle than in red muscle, except in very active species such as tuna. The activation and relaxation speeds of the three muscle fiber types increase in the order red, then pink, and then white. This also reflects the order in which they become fatigued, and the order in which they are recruited as swimming speed increases.

Because of the density of water and the effects of drag, the power required to swim increases with approximately the cubed root of velocity. Thus for low speeds, which can be fully supported using aerobic respiration, only red muscle is required, and a small volume of it is adequate, operating close to its optimum velocity and sarcomere length (Fig. 9.111a). Such speeds can be maintained almost indefinitely (Fig. 9.111b) for normal foraging and during migrations. In contrast, a very high power output is required to swim at high speed, as in an escape response, and this requires a large recruitment of white muscle; but this starts to fatigue after a few seconds and the fish may become completely exhausted within a few minutes of high-speed swimming. However, carrying around a large amount of white muscle that is used relatively infrequently is not too much of a problem for a neutrally buoyant animal.

The red muscle fibers are arranged parallel to the longitudinal axis of the body, but the white muscle fibers always form complex helical patterns (see Fig. 9.110), giving significant "gearing", i.e. the body bends proportionally more for a given amount of shortening in the white than the red muscle. During escape behavior the white muscle length changes can reach ±20%, providing for considerable flexing of the trunk.

Figure 9.112 shows that the maximum power output of the white muscle is 4–5 times that of the red muscle, and occurs at a much higher frequency, though declining sharply at very high frequency.

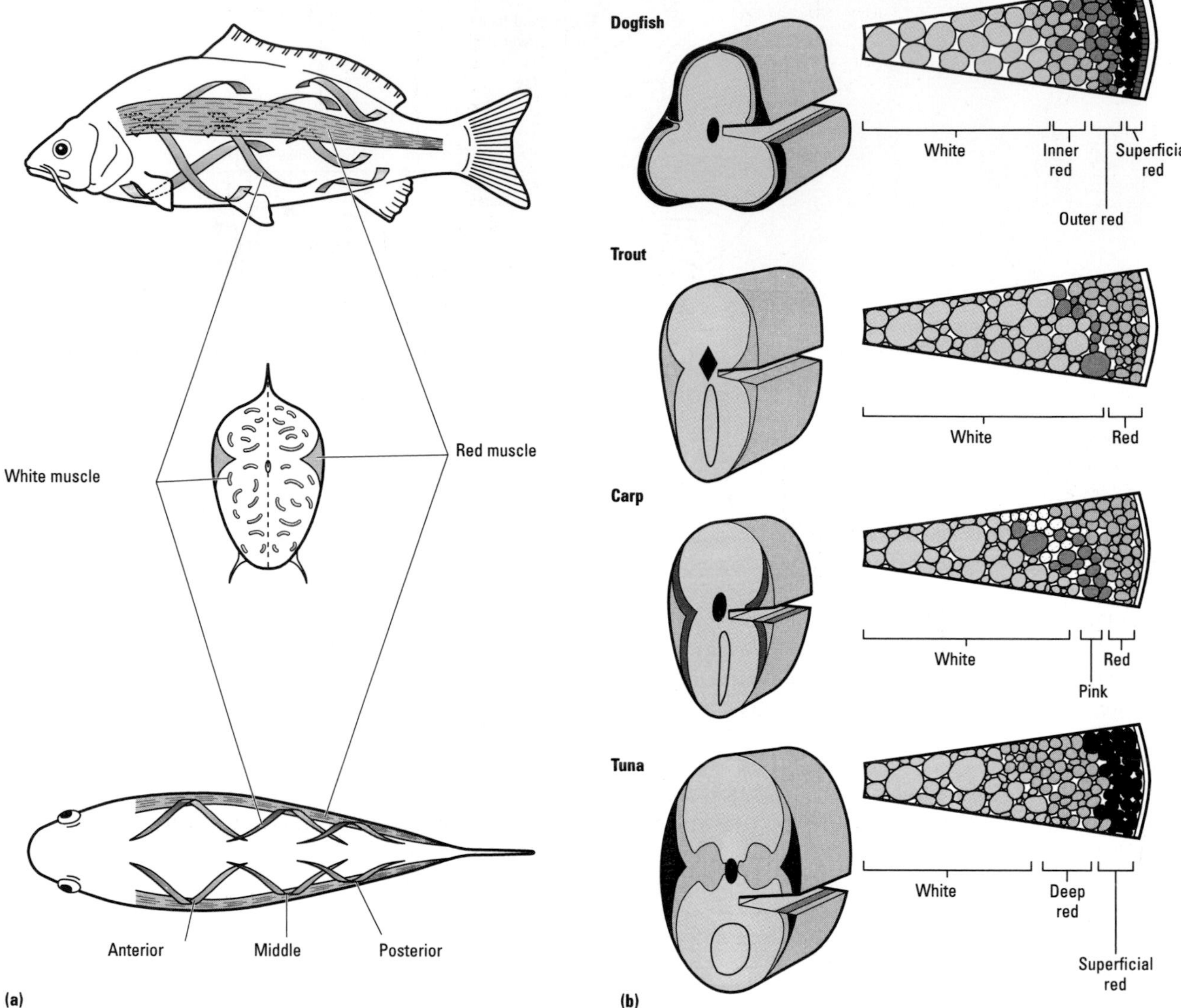

Fig. 9.110 (a) The orientation of fish muscles: the slow (red) muscle fibers lie in a thin layer beneath the skin, running parallel to the body axis, while the fast (white) fibers lie deeper and form helically arranged myotomes, which need to shorten only about 25–30% as much as the red fibers to produce a given curvature of the body. (b) The pattern of these two fiber types in various types of fish. (From Johnston 1981, with permission.)

In fact the red muscle contraction kinetics are too slow to generate any positive work above 8 Hz, and the highest power output recorded is only about 6 μJ compared with more than 30 μJ in white muscle. This is why it is necessary to have muscle fiber types with a spectrum of contraction kinetics for locomotion over a wide range of swim speeds.

In many fish the main propulsive force comes from the tail region and the tail fin; the anterior muscles still generate power, but this is transmitted to the tail fin by the stiffening "strut" action of more posterior trunk muscles. In other fish the fins mainly function to provide stability and some degree of buoyancy, and to permit change of direction. In a relatively few species the pectoral and/or pelvic fins provide significant thrust and are used to push the body along the sea bottom or over rocks, approximating a walk or crawl based on limbs. (However, we may also note the converse to this; eels use their fins very little, yet can use the basic anguilliform system for true terrestrial locomotion, gliding through grass for long distances when moving between streams.)

Other vertebrates that swim more or less continuously (marine mammals for example) use similar patterns though often with much more input from the flippers and less from the trunk muscles. Vertebrates that combine swimming with other locomotory modes tend to swim less efficiently; amphibious mammals never achieve the speed or maneuvrability of fish. For birds such as ducks that combine swimming, walking, and flying the trade-off effects on the muscle systems are even more noticeable, and swimming is relatively poor; the main gastrocnemius muscle in a duck leg generates only 37% of the work during swimming that it can achieve in walking mode, and there is little of the elastic storage in tendons seen in more cursorial birds, as explained in the next section.

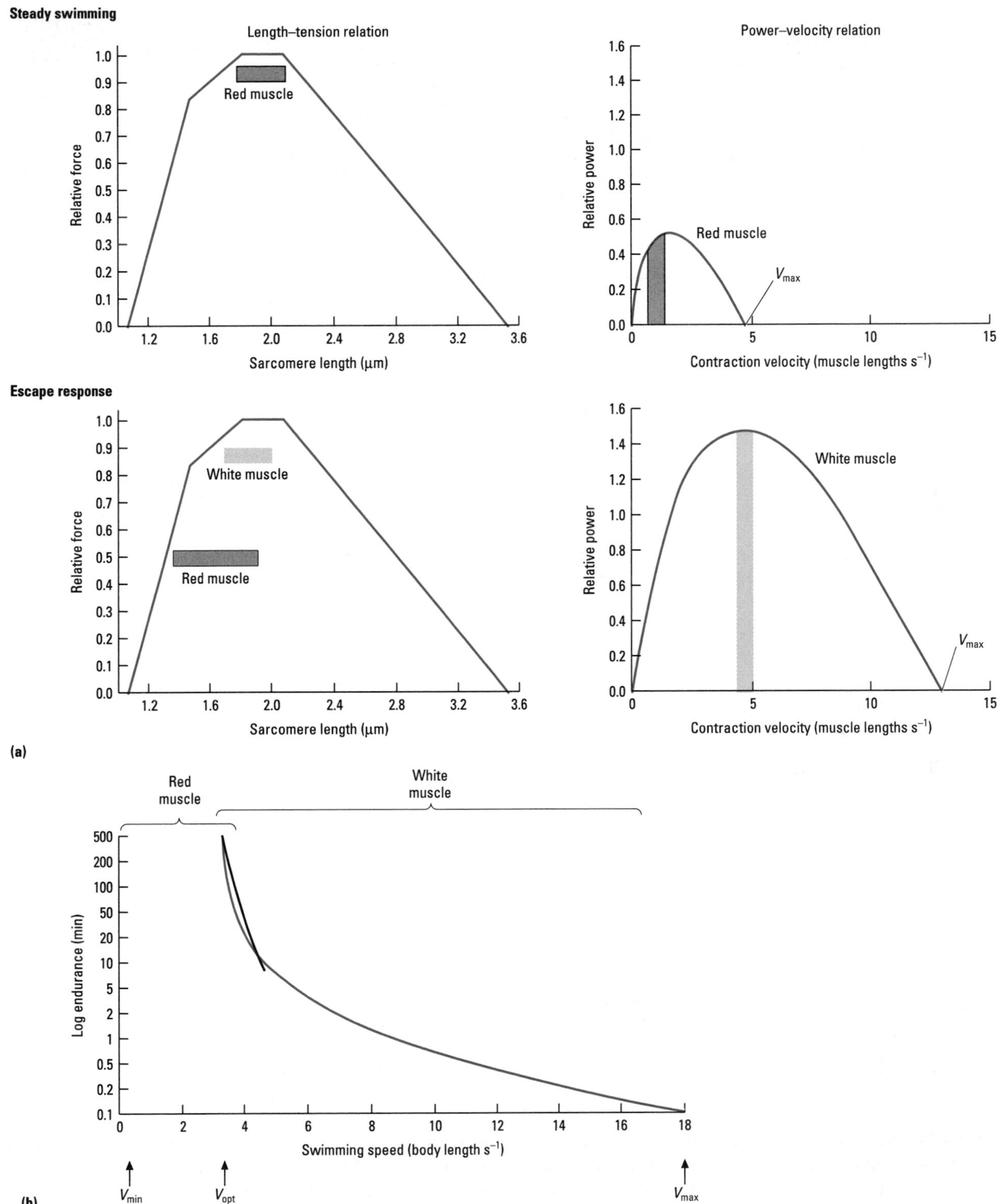

Fig. 9.111 (a) The relationships between force and length, and between power and velocity, for the two types of fish muscle. For the red muscle, the lengths of the sarcomeres are in the optimum range during slow swimming, and the contraction velocities are around 20–40% of V_{max}, close to the range of maximal power. The white muscle fibers are only operative in fast escape response, when their anatomy allows them to operate close to the optimal sarcomere length and at 38% of V_{max} at the peak of their power curve. In (b) the relation between swimming speed and endurance is shown, and the division of labor between red and white fibers. (a, Adapted in part from Rome & Sosnicki 1991.)

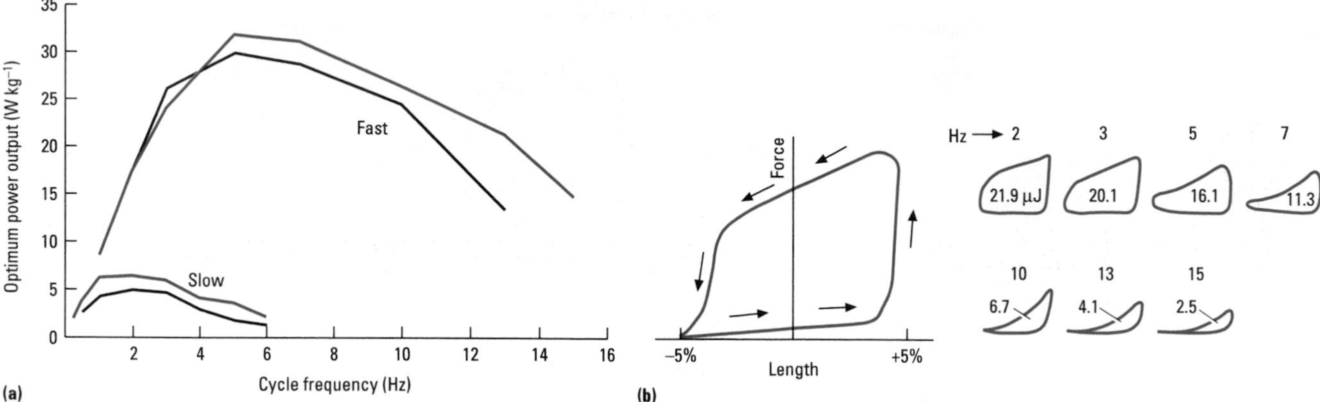

Fig. 9.112 The work loops in swimming fish. (a) Optimum power output for fast and slow fibers. (b) The relative shape of the work loop at different frequencies, for white muscle, showing the much lower work output for very fast escapes. (Adapted from Altringham & Johnston 1990.)

Crawling, walking, and running

On land, most animals with stiff skeletons use limbs, with a variety of walking and running gaits. A few crawl using only the trunk muscles (eels, snakes, some lizards), with movements akin to those of a swimming animal—indeed most of these animals can also swim, using a lower amplitude undulation.

When animals walk, the body effectively "vaults" over the stiff strut provided by the leg, so that kinetic energy in the first (rising) phase of the vault is transformed into potential energy, stored as elastic strain energy in muscle, tendons, or cuticle, which is then partly recovered as the body falls forwards and down again in the second phase.

ARTHROPODS

Swimming arthropods could fairly be said to be "preadapted" for terrestrial locomotion, with limbs that can walk as well as swim. In fact many amphibious decapods have thoracic legs suited for walking and abdominal limbs flattened for swimming. Many littoral crustaceans when submerged have a "walking" gait where just one or two legs give an occasional push against the substratum to keep them skimming along horizontally, but when the tide is out they walk on land using a six- or eight-legged gait similar to that of insects. For most animals, though, continuous terrestrial locomotion requires much more structural support for the body, which is no longer buoyant, and the legs need to become stouter and more cylindrical (less flattened) for maximum resistance to bending and twisting forces that might otherwise lead to breakage. They also tend to show the muscle mass concentrated more at the top of the limb (Fig. 9.113a).

Most walking arthropods use a metachronal rhythm of limb movements similar to that seen in swimming, and with two or more **gaits** at different speeds (Fig. 9.113b). At low speed only the limbs move to produce the power stroke, whereas at higher speeds the body may start to show lateral flexions, which increase the step length. For faster gaits a longer limb is useful with fewer limbs being involved; the decapod crustaceans have five pairs but often only use three or four of them, the arachnids have four pairs but may only use three, and the insects always only have three pairs (with at least three of the six limbs always kept in contact with the ground as a tripod for stability). In the faster gaits of insects such as cockroaches, some of the leg muscles have "negative" work loops and are actually serving as brakes, absorbing rather than producing net energy.

Most arthropods walk and run with a splayed-leg posture (Fig. 9.113c), and generate substantial lateral forces as they push against the ground. This sounds wasteful, but in practice the lateral forces too can be partly stored and recovered in the horizontal plane, and the posture also helps give robust and stable gaits at various speeds and over rough terrain. Many of these small land animals have to trade-off stability (whereby they resist winds or currents imposing external perturbations) against inherent maneuvrability (whereby they respond to voluntary steering commands from their own CNS).

VERTEBRATES

Tetrapods commonly walk quadripedally, and maintain each foot on the ground for more than half of each cycle, so that there is always at least two-footed contact with the ground. In contrast faster **cursorial gaits** allow successively less prolonged contact, until in the gallop all four feet may be off the ground at once and speeds of up to 25–29 m s^{-1} are possible for some antelope and large predatory cats (though only up to about 17–18 m s^{-1} for the more familiar horse, and 10 m s^{-1} for the fastest bipedal humans). The most important point about changing gaits is the energy saving incurred; Fig. 9.114a shows that each gait has its own optimal speed for economy in terms of energy per distance moved, and vertebrates in practice use each gait at the speed where it offers greatest savings. Faster locomotion is aided by longer legs and hence a long stride length, with the leg operating at close to the natural oscillation frequency of a similar-length pendulum. As in arthropods, the legs are thin with most of the muscle mass concentrated near the trunk and with tendons and elastic ligaments running down from each muscle (Fig. 9.114b). Speed is also improved by progressively reducing the area of foot contact with the ground so that the posture becomes unguligrade (weight borne on a hoof covering just one or two toes) instead of the primitive plantigrade state (weight borne on the whole foot flat on the ground).

Elastic energy-storage mechanisms play an important role in reducing the metabolic cost of fast-gait tetrapod locomotion. Running at constant speeds over level ground causes relatively little net work to be done against the environment; rather, the muscular work mainly lifts and accelerates the body and limbs. Before the limb extensor muscles shorten, the momentum of the animal

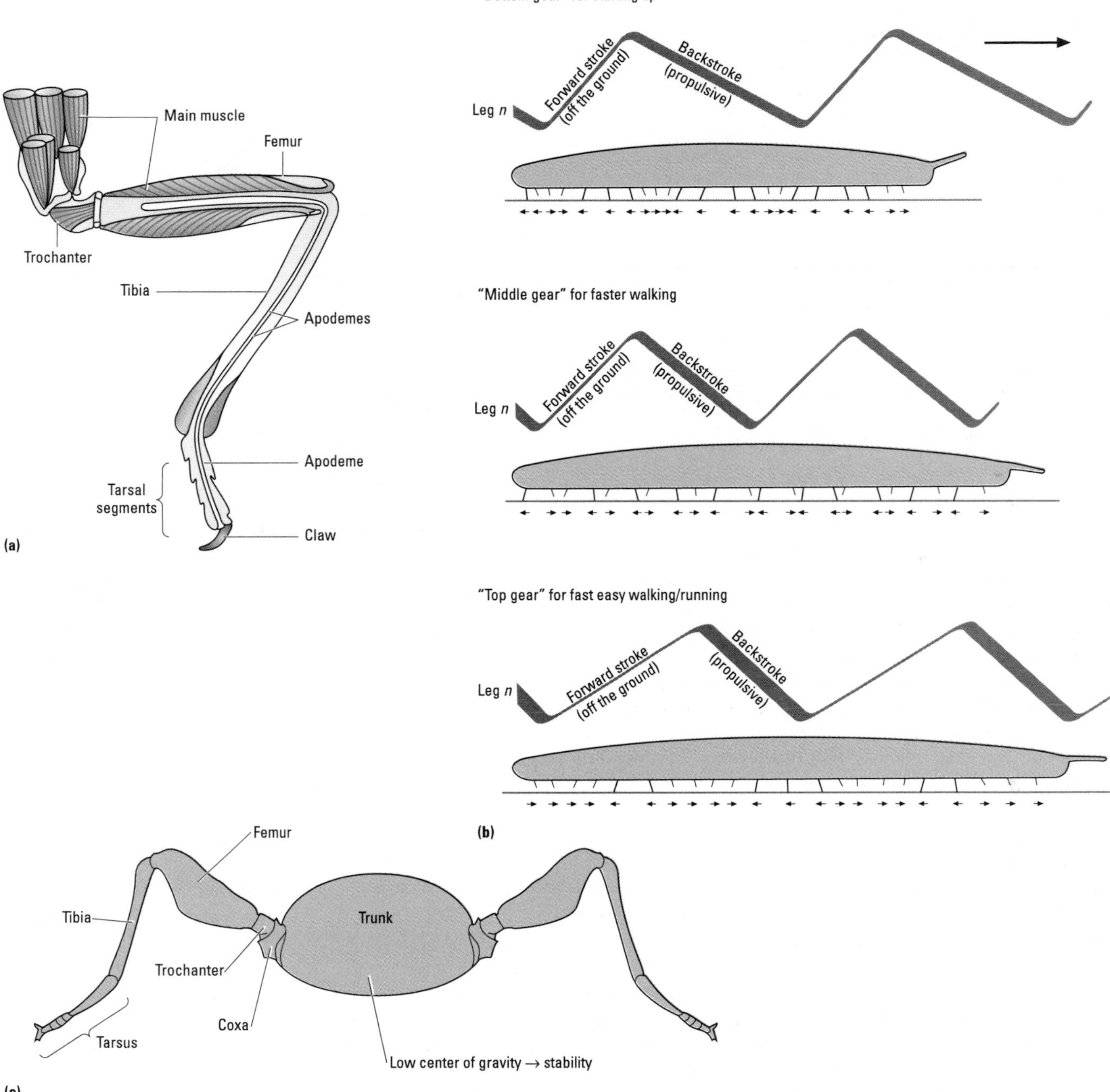

Fig. 9.113 (a) A typical arthropod limb structure with muscles mainly at the top of the limb in the coxa and femur, and linked to the apodemes extending distally. (b) Different gaits or "gears" in multilegged arthropods. (c) The splayed posture of an arthropod.

stretches them, so that some of the work done is stored (as elastic strain energy) and is then recovered when the muscle fibers shorten. The muscle–tendon system therefore functions like a spring that alternatively stores and releases this elastic strain energy as the animal moves along, leading to a considerable saving of metabolic energy. In running humans, more than half the energy supplied during each stride is derived from elastic energy storage. However the same muscle systems that are "merely" struts and springs in normal running mode for many animals can become a positive power generator when needed, for example when running up hills, and in such cases the work loop broadens significantly (Fig. 9.115).

The effectiveness of these systems can be increased by having short muscles and long slender tendons, readily seen in specialized runners such as many ungulates. In the plantaris muscle of the camel (a surprisingly effective and speedy runner) muscle fibers are only 1–3 mm long, while the tendon runs almost the whole length between the femur and the toes. Loading and unloading this tendon shows a very large recovery of energy, with the elastic resilience of the tendon measured as 93%.

Tendons in running animals are not always designed for maximum elastic efficiency, because there is a trade-off with strength and

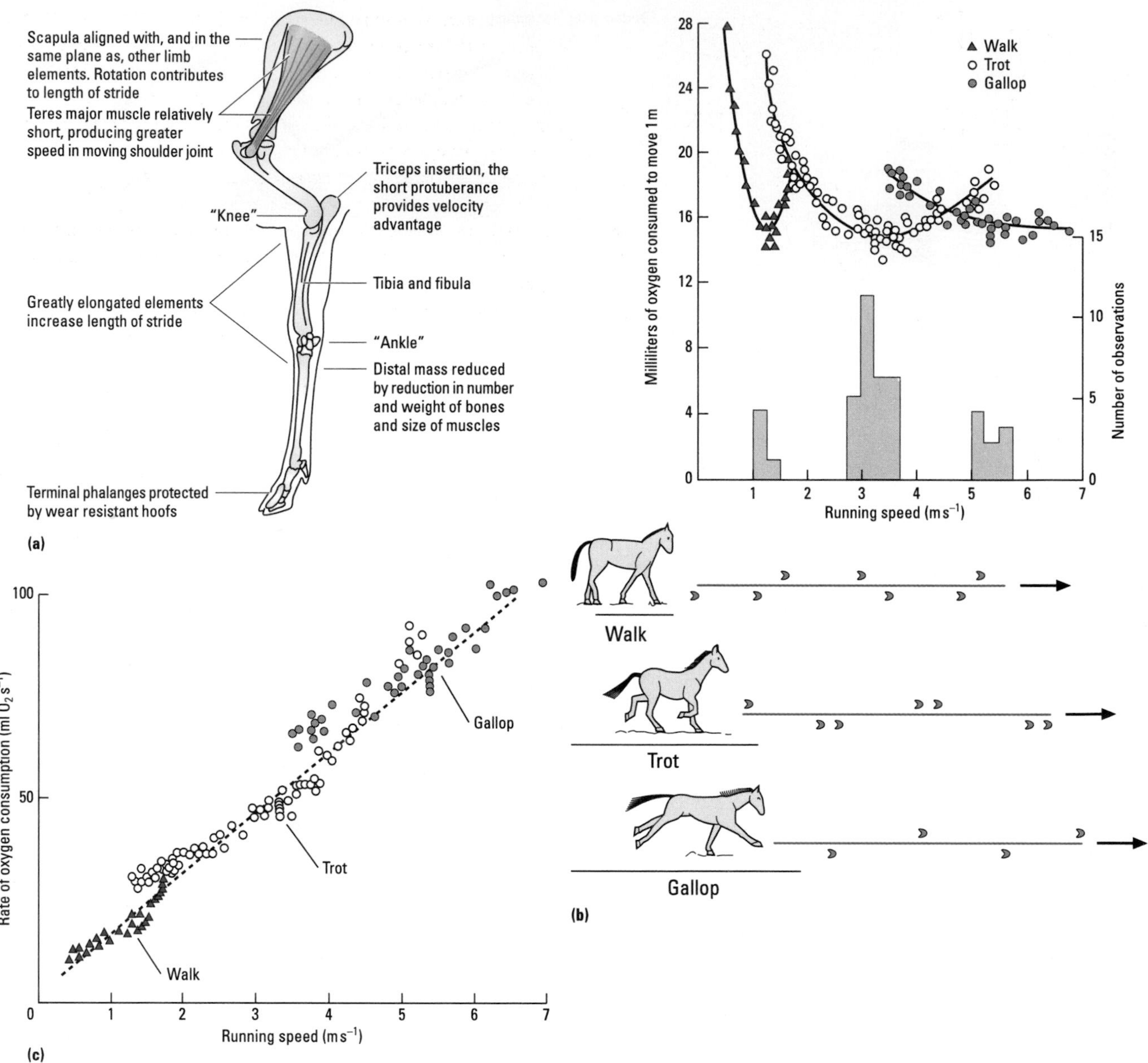

Fig. 9.114 (a) A typical vertebrate limb structure, with muscle concentrated at the top of the limb; note the specific adaptations for a cursorial (running) gait. (b) Different vertebrate gaits, the speeds and footfalls they produce, and (c) the associated energy costs.

acceleration. Halving the tendon thickness without changing its length results in four times more energy storage and 16 times more energy recovery per unit mass; but this extra efficiency comes at a price, since thinner tendons have a lower safety factor, and are very easily damaged.

Jumping, bouncing, and hopping

Jumping is a rather more radical solution to reducing friction with the substratum for land animals, and it has evolved repeatedly. Jumping (**saltatory**) systems are unlike any locomotory pattern seen in aquatic species; they are favored because they generally use highly efficient energy-storing systems, and these can work even at cold temperatures, allowing escape from fast-moving (possibly endothermic) predators.

ARTHROPODS

Jumping is reasonably common in insects, where the potential for inclusion of elastic proteins within the cuticle can be exploited. Collembolans (springtails) are apterygote insects that jump using a forked catapult-like springing organ (furca) on the ventral surface of their abdomen. Fleas and many orthopterans (Fig. 9.116) can jump to relatively enormous heights, using energy-storage systems in the "knee" joints of the metathoracic legs, where a rubber-like cuticle protein (**resilin**) stores energy from muscle contraction relatively slowly and then releases it very suddenly (by removing a restraining peg, effectively) to get remarkable acceleration. The

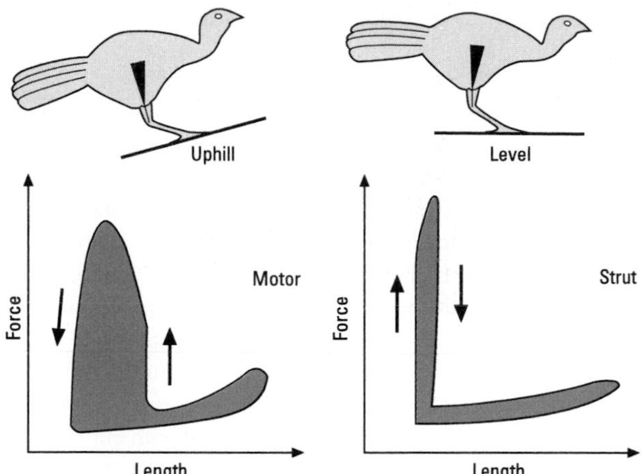

Fig. 9.115 The work loops for a bird running uphill (where the gastrocnemius muscle acts as a motor) and running normally on the flat (where its work output is much lower and it functions mainly as a strut, allowing the springy tendons to store and then release energy). (Adapted from Dickinson *et al.* 2000.)

elastic stores of a locust leg weigh only about 4 mg but can store and release the energy of a 70 mg muscle. Click beetles also use a jumping mechanism based on storing energy slowly and releasing it rapidly. A flea with a body mass of about 0.5 mg can jump 20 cm, a click beetle (40 mg) perhaps 30 cm, a locust (3 g) and a human (70 kg) both about 60 cm; these jump heights are all rather similar, yet for body sizes varying by eight orders of magnitude.

VERTEBRATES

Jumping locomotion has evolved several times in terrestrial vertebrates. It is common in anuran amphibians (see Chapter 15), and in some marsupial and primate mammals. Above all, jumping to any significant height again requires speed at take-off, which just as in most arthropods involves a very rapid straightening of the "knee" joints (usually with a preparatory bending of the legs), and employs built-in elasticity.

Jumping involves tendon and muscle changes, to allow for the elastic storage. The muscles can be operating isometrically, or can even be increasing in length, while the jump is in progress, so that the muscles are really acting as struts and springs to control the natural behavior of the skeleton's viscoelastic system. Scaling has crucial effects here. The red kangaroo can grow to 400 times the mass of the kangaroo rat although they both use a hopping bipedal gait and have a superficially similar hindlimb morphology. However, in the kangaroo rat relatively little strain energy is stored in the ankle extensor tendons and muscles during hopping, and only 14% of the work done on the muscles is recovered. This animal has disproportionately large muscles and tendons for its size, resulting in a high safety factor (at least 8–10) between the peak stress experienced in normal hopping and the stress that causes failure of the tendon. However, kangaroo rats are capable of remarkably high vertical jumps of 50 cm (10 times their own "hip height")—the muscle develops a peak stress of 350 kN m^{-2} and the safety factor drops to 2–3. The red kangaroo, on the other hand, has proportionately much thinner tendons and muscles of smaller cross-sectional area, and is able to store around three times more elastic strain energy during hopping than the kangaroo rat, which has a very narrow work loop in steady hopping (Fig. 9.117). These examples show different selective advantages accruing from bipedal hopping. The kangaroo rat lives in burrows in open terrain, evading predators by jumping and placing a premium on muscle designs with high strength and motor control. Having strong enough tendons for escape jumping has limited the efficient elastic energy storage for normal activity. The red kangaroo forages and migrates over long distances with impressive economy due to elastic energy savings, but with rather limited accelerative capacity, appropriate to a large animal with few natural predators.

Flying

In some animals, the ability to jump contributes to an additional locomotory possibility, providing the take-off for flight, whether by a jump-and-glide strategy (not uncommon in many vertebrates, from flying fish through to arboreal frogs, lizards, snakes, and primates) or by powered flight. As with swimming, animals generate vortex rings as they push against the air, and analysis of these vortices has been helpful in understanding forward flight dynamics. However, unlike swimming, flight also requires the generation of lift, and this is a rather substantial extra cost that restricts flight to specialists. Only two groups have really developed the capacity for powered flight using rapid muscular contractions, and this has contributed greatly to their terrestrial success.

Fig. 9.116 (a) Jumping in a grasshopper and a flea; in both cases the jump is preceded by contraction of the muscles that store energy in an elastic cuticular hinge (containing resilin) at the "knee", which is then explosively released. (b) The associated work loop. (b, From Dickinson *et al.* 2000.)

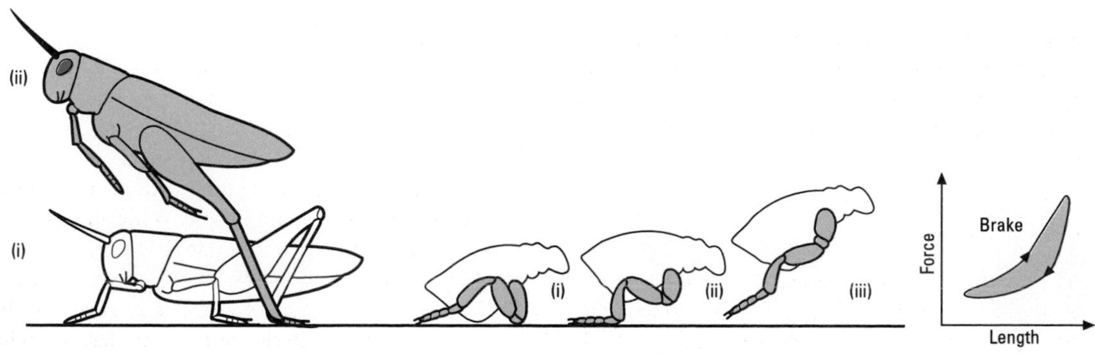

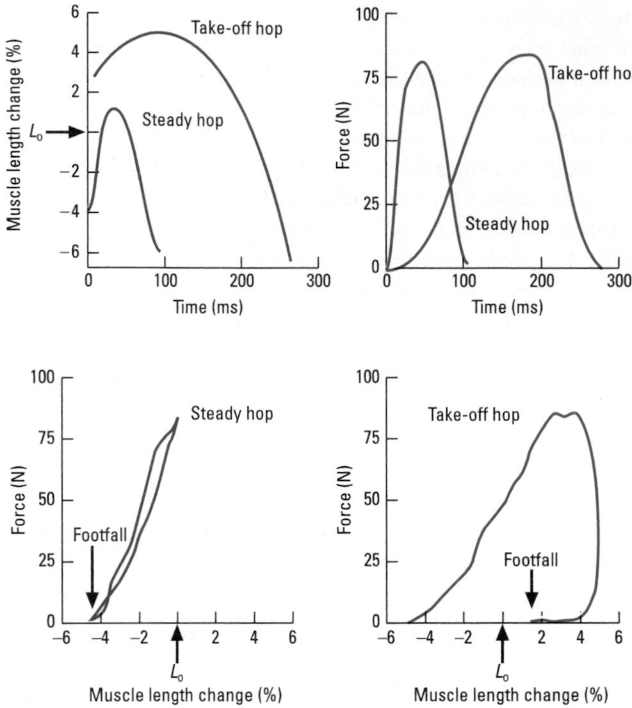

Fig. 9.117 The work loops for the leg muscles of a jumping kangaroo rat, comparing the steady hop and the take-off hop. (Adapted from Biewener & Blickhan 1988.)

INSECTS

Insects are unusual amongst animals in having wings that are not modified limbs, but instead evolved from lateral cuticular outgrowths that may have functioned originally as stabilizers or sails. Most of them launch their flights with a jump using the hind legs, but thereafter two different systems of flying and wing control are found in modern insects (Fig. 9.118), and we will deal with these in turn.

In section 9.13.2 the two kinds of insect flight muscle were described, and these are associated with two different kinds of muscle arrangement to move the wings. In insects such as dragonflies, cockroaches, locusts, butterflies, and moths, the downstroke is produced by **direct muscles** pulling against the flexible and elastic wing base. These direct muscles are nonfibrillar, and synchronous, with a one to one relationship between electrical and mechanical activity. However, insects with these synchronous direct muscles cannot fly at wingbeat frequencies above 100 Hz. The alternative is **indirect muscles,** not attached to the wings but pulling on the walls of the thorax instead, and these are usually of the asynchronous type. The side walls of the thorax are rigid, strengthened with cuticle ridges, while the dorsal "roof" (tergum) is quite flexible. Downward movement of the wings occurs when the longitudinal flight muscles in the thorax contract, reducing the thoracic volume and causing the thoracic roof to distort upwards; this automatically forces the wings down. The upstroke occurs when the vertical (dorsoventral) flight muscles contract and pull the tergum downwards again, flicking the wings back up against the fulcrum of the pleural process. Since the wing hinge area is again very elastic, with a high proportion of resilin in the cuticle, only the up or down positions of the wings tend to be stable, and the wing clicks very rapidly between these stable states. In the familiar fruitfly *Drosophila*, with a wingbeat frequency around 240 Hz, the wings move through 170° on each cycle. Since mechanical power is proportional to the cube of both wingbeat amplitude and wingbeat frequency, both of which are high, this fly (in common with many other insects) generates power at around 80 W kg^{-1} muscle mass, close to the maximum possible from an aerobic fast muscle. In these and other small insects the flight muscles make up at least 15% of the total body mass.

Wings must do more than move up and down. They must also be moved at an angle (pitch) to give forward thrust, so that a complete wing cycle involves the wing tip performing an elliptical or

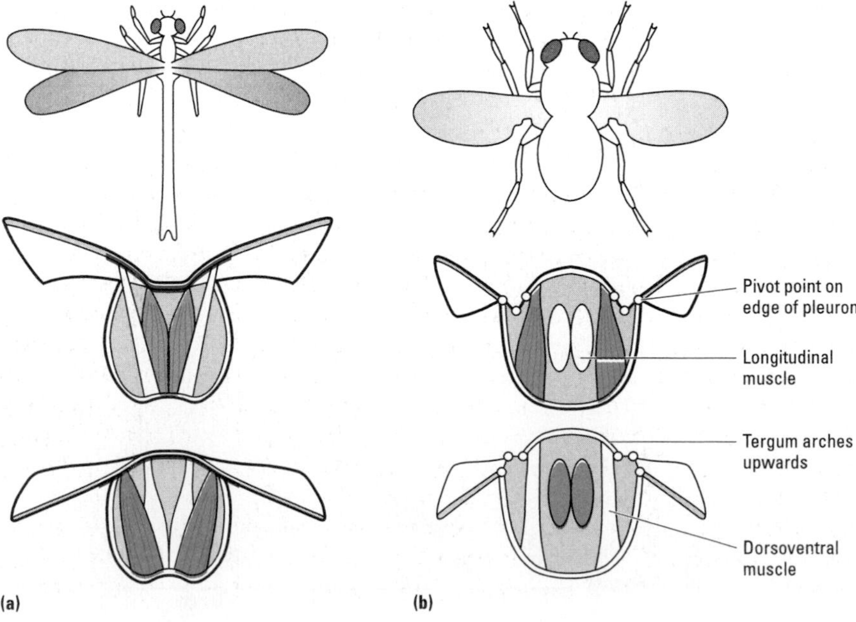

Fig. 9.118 The two alternative wing muscle systems in insects: (a) the direct system, as in dragonflies, and (b) the indirect system, as in bees and flies. Shading shows contracting muscles.

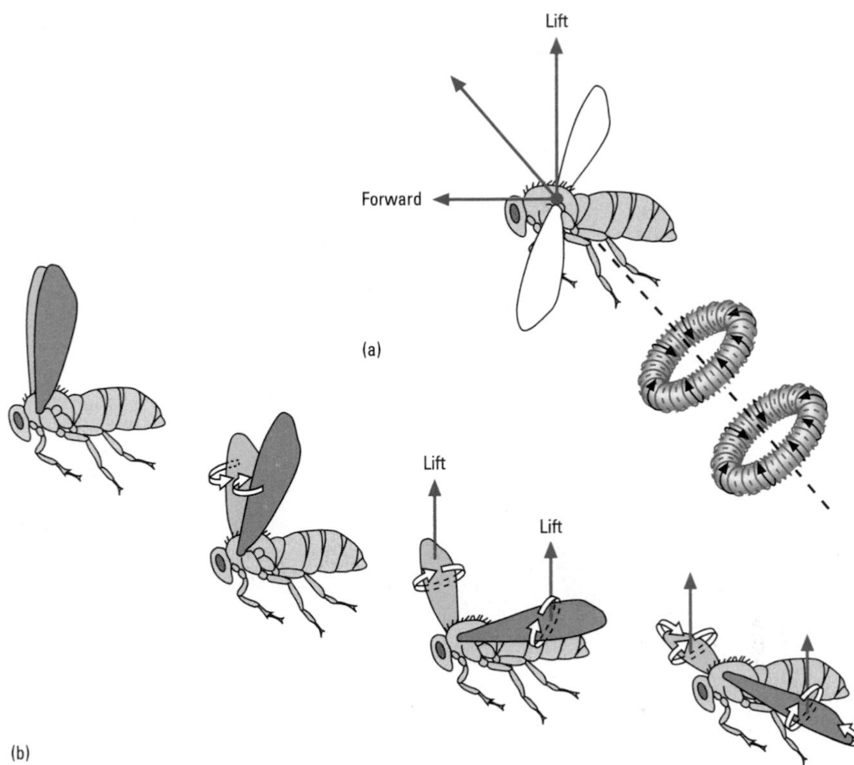

Fig. 9.119 (a) The spinning vortex of air that is shed from an insect in flight, which helps to give lift (see text). (b) In small insects the wings clap together on the upstroke and when peeled apart create an extra vortex effect.

figure-of-eight motion. The wing angles and contours must be varied to achieve directional control. These processes are controlled largely automatically in most flying insects by the mechanics of the wing hinge itself, with minor adjustments by smaller muscles acting directly on the complex sclerites of the wing base.

Despite all this, using conventional aerodynamic calculations, many insects should not be able to generate enough lift to fly. They fly with a high angle of attack, where a conventional aerofoil would stall. They must have ways of generating further downforce. Part of the answer is that they can use the vortices of moving air around the wing; the most recent evidence invokes the idea of "dynamic stall", showing that insects "deliberately" use their wings with a large angle of attack so that the leading edge vortex (Fig. 9.119) is prevented from spinning away from the wing and instead helps to draw air faster over the upper side of the wing, giving massive lift. By timing the rotation of the wing as it passes through the wake of a previous wingstroke, an insect can recapture some of the energy that was lost in that wake. Some very small insects also use a "clap, fling, and peel" system, whereby the two wings clap together in the midline at the top of the upstroke and then fling and peel away from each other so that air rushes between them and sets up a new vortex of air circulating around and under each wing (Fig. 9.119b). Even so, the "flight efficiency" of most insects (the ratio of mechanical power generated to metabolic power used) is only around 10%.

Insect flight involves wingbeat frequencies varying from around 4 Hz in large-winged butterflies through 100–200 Hz in bees and medium-sized flies, to above 1000 Hz in tiny midges. Remember that, as we discussed earlier, for insects with indirect flight muscles and wingbeat frequencies above about 25–30 Hz there is less than one AP to one contraction and the muscle in effect becomes myogenic. The faster wingbeats thus achieved allow rapid and forceful flight, with a honeybee able to cruise at 7 m s^{-1} (around 15 mph). Maneuvrability can also be exceptional, with flies able to hover, fly upside down and backwards, and turn within one body length.

BIRDS AND BATS

In birds two particular muscles become hugely dominant in relation to flight (Fig. 9.120a), both attaching to the keel of the pectoral girdle. The greatly enlarged pectoralis major muscle provides the downstroke and exhibits a particularly broad work loop (Fig. 9.120b), while the smaller supracoracoideus produces the less energy demanding upstroke. The supracoracoideus provides some lift in larger birds with fast steep take-offs, and in those that can hover, and in such species the ratio of the two muscle masses may be as low as 3 : 1. In most other species the supracoracoideus is relatively tiny, and the ratio of pectoralis to supracoracoideus may reach 20 : 1. In strong fliers both these muscles operate aerobically, with a high capillarity and dense mitochondrial packing, giving red muscle. In smaller and weaker birds that fly only as a transient escape response, where stamina is unimportant, the flight muscles (and the supracoracoideus in particular) are predominantly anaerobic with little myoglobin, and appear as white muscle. All birds also have small control muscles that control the stiffness of the wing joints at the elbow and wrist, and the shape of the wing surface. The same principles apply in flying bats, although here it is the elongated bones of the hand that provide the wing support and the wing shape is therefore more variable.

Birds exhibit a range of flying modes, ranging from **gliding** and **bounding** flight through normal **flapping** flight to the rare but impressive **hovering** flight shown especially in hummingbirds. The

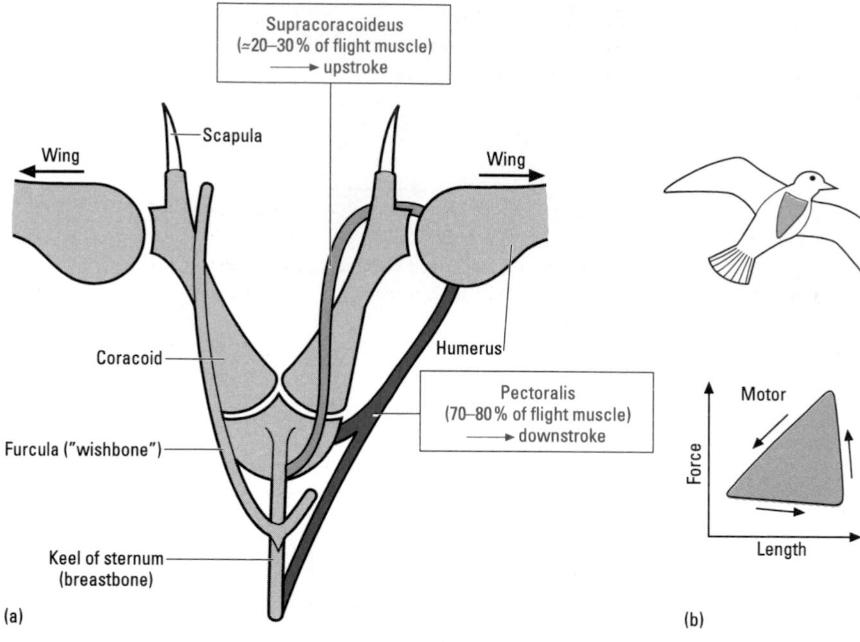

Fig. 9.120 (a) The major flight muscles of a bird. The pectoralis gives the downstroke, pulling on the sternum and upper forelimb, while the supracoracoideus provides the upstroke, with similar insertions but acting over a pulley formed by the coracoid and scapula bones. (b) A typical broad work loop from the pectoralis muscle of a flying bird. (b, From Dickinson *et al.* 2000.)

various types of flight can be seen as equivalent to gaits, and involve different patterns of neuromuscular recruitment, wingbeat frequency, and amplitude—though amplitude varies much more substantially than frequency, which is kept nearly constant at all speeds in level flight. Many species once aloft will show intermittent flight patterns, alternating bouts of flapping with gliding or bounding. Medium-sized birds tend to use flap/glide at lower speeds and switch to flap/bound at faster speeds in habitats where the wind allows this; both modes involve lower mechanical power output than continuous flapping. Smaller birds, usually having lower aspect ratio wings, may flap/bound at all speeds, while larger ones (above about 300 g) generally do not use the flap/bound mode at all, though the size limit on bounding can be somewhat altered by the scaling and composition of the pectoralis muscle. Birds can commonly fly at 50 km h^{-1}, with ducks and geese reaching 80–90 km h^{-1} and diving falcons touching 200 km h^{-1}.

9.16 Conclusions

A repeated theme in this chapter has been the presence of moderate to high phenotypic plasticity in excitable tissues. This allows such tissues to alter their properties substantially, in relation both to functional demand (whether to detect and respond, to learn, or to move) and to environmental variables. The plasticity operates at many levels, and may involve ion channels, pumps, cytostructural proteins, receptor proteins, or contractile protein isoforms; it may also on a longer timescale involve altered connectivity between excitable elements. There are also some intriguing signs of homeostatic controls operating between nerve and muscle to maintain function. For example, experimentally induced hyperpolarization of *Drosophila* muscles produces impaired muscle excitability, but increased presynaptic transmitter release then occurs that almost exactly compensates for this, suggesting some intrinsic monitoring of muscle membrane potential and intrinsic correction systems.

It is likely that such plastic and homeostatic changes underlie the ability of nerve and muscle to operate with reasonable efficiency in changing environments, but it is nevertheless evident that in many excitable tissues this ability is far from complete, and environmental variation can seriously disrupt neural, sensory, and contractile functioning. It may be that environmental plasticity is higher for muscles than for nerves, and that for neural tissue the use of physical and chemical barriers to protect the nerves and their immediate extracellular environment may be the better option.

A final important factor related to nerve and muscle functioning in relation to the environment may be the presence of fairly generous safety factors. Perhaps the most important safety factor occurs at synapses, including the NMJ, in that the amount of transmitter released per nerve impulse is usually considerably greater than that required to trigger an AP in the postsynaptic nerve (or muscle). For the mammalian NMJ the safety factor has been estimated at 3–5.

Given a degree of plasticity and a degree of protection, together with built-in safety factors, the excitable tissues in animals can then perform their key functions of detecting and responding to other environmental variations across a whole spectrum of habitats where conditions change on a moment-to-moment, daily, or seasonal basis.

FURTHER READING

Books

Aidley, D.J. (1998) *The Physiology of Excitable Cells*, 4th edn. Cambridge University Press, Cambridge, UK.

Alexander, R.McN. (1996) *Optima for Animals*. Princeton University Press, Princeton, NJ.

Alexander, R.McN. (1999) *Energy for Animal Life*. Oxford University Press, Oxford.

Ali, M.A. (ed.) (1984) *Photoreception and Vision in Invertebrates*. Plenum Press, New York.

Balda, R.P., Pepperburg, I.M. & Kamil, A.C. (eds) (1998) *Animal Cognition in Nature*. Academic Press, San Diego, CA.

Buser, P. & Imbert, M. (1992) *Vision*. MIT Press, Cambridge, MA.

Cowan, W.M., Sudhof, T.C. & Stevens, C.F. (2001) *Synapses*. Johns Hopkins University Press, Baltimore.

Hille, B. (2001) *Ionic Channels of Excitable Membranes*, 3rd edn. Sinauer, Sunderland, MA.

Junge, D. (1992) *Nerve and Muscle Excitation*, 3rd edn. Sinauer, Sunderland, MA.

Katz, P.S. (ed.) (1999) *Beyond Neurotransmission—Neuromodulation and its Importance in Information Processing*. Oxford University Press, Oxford.

Land, M.F. & Nilsson, D.-E. (2002) *Animal Eyes*. Oxford University Press, Oxford.

Matthews, G.G. (2003) *Cellular Physiology of Nerve and Muscle*, 4th edn. Blackwell Publishing, Oxford.

Reichert, H. (1992) *Introduction to Neurobiology*. Georg Thieme Verlag, Stuttgart.

Zigmond, M.J., Bloom, F.E., Landis, S.C., Roberts, J.L. & Squire, L.R. (eds) (1999) *Fundamental Neuroscience*. Academic Press, New York.

Reviews and papers

Altringham, J.D. & Johnston, I.J. (1990) Modelling muscle power output in a swimming fish. *Journal of Experimental Biology* **148**, 395–402.

Amara, S.G. & Kuhar, M.J. (1993) Neurotransmitter transporters: recent progress. *Annual Review of Neuroscience* **16**, 73–93.

Armstrong, C.M. (1992) Voltage-dependent ion channels and their gating. *Physiological Reviews* **72**, S5–S13.

Bajjalieh, S.M. (1999) Synaptic vesicle docking and fusion. *Current Opinion in Neurobiology* **9**, 321–328.

Breer, H., Raming, K. & Krieger, J. (1994) Signal recognition and transduction in olfactory neurons. *Biochemica et Biophysica Acta* **1224**, 277–287.

Brown, B.R. (2003) Sensing temperature without ion channels. *Nature* **421**, 495.

Cohen-Cory, S. (2002) The developing synapse: construction and modulation of synaptic structures and circuits. *Science* **298**, 770–776.

Davis, W.W., Eaton, B. & Paradis, S. (2001) Synapse formation revisited. *Nature Neuroscience* **4**, 558–560.

Dickinson, M.H., Farley, C.T., Full, R.H.J., Koehl, M.A.R., Kram, R. & Lehman, S. (2000) How animals move: an integrative view. *Science* **288**, 100–106.

Edwards, D.H., Heitler, W.J. & Krasne, F.B. (1999) Fifty years of a command neuron: the neurobiology of escape behaviour in the crayfish. *Trends in Neuroscience* **22**, 153–161.

Gillis, G.B. & Blob, R.W. (2001) How muscles accommodate movement in different physical environments: aquatic versus terrestrial locomotion in vertebrates. *Comparative Biochemistry & Physiology A* **131**, 61–75.

Herness, M.S. & Gilbertson, T.A. (1999) Cellular mechanisms of taste transduction. *Annual Review of Physiology* **61**, 873–900.

Hokfeld, T. (1992) Neuropeptides in perspective; the last ten years. *Neuron* **7**, 867–879.

Hudspeth, A.J. (1989) How the ear's works work. *Nature* **341**, 397–404.

Jin, Y. (2002) Synaptogenesis: insights from worm and fly. *Current Opinions in Neurobiology* **12**, 71–79.

Johnston, I.J. (1981) Structure and function of fish muscles. *Symposia of the Zoological Society of London* **48**, 71–113.

Josephson, R.K. (1993) Contraction dynamics and power output of skeletal muscle. *Annual Review of Physiology* **55**, 527–546.

Josephson, R.K., Malamud, J.G. & Stokes, D.R. (2000) Asynchronous muscle: a primer. *Journal of Experimental Biology* **203**, 2713–2722.

Lindemann, B. (1996) Taste reception. *Physiological Reviews* **76**, 719–726.

Malenka, R.C. & Nicoll, R.A. (1999) Long-term potentiation—a decade of progress? *Science* **285**, 1870–1874.

Marban, E., Yamagishi, T. & Tomaselli, G.F. (1998) Structure and function of voltage-gated sodium channels. *Journal of Physiology* **508**, 647–657.

Marden, J.H. (2000) Variability in the size, composition and function of insect flight muscles. *Annual Review of Physiology* **62**, 157–178.

Masson, C. & Mustaparta, A. (1990) Chemical information processing in the olfactory systems of insects. *Physiological Reviews* **70**, 199–245.

Matthews, G. (1996) Neurotransmitter release. *Annual Review of Neuroscience* **19**, 219–233.

Maughan, D.W. & Vigoreaux, J.O. (1999) An integrated view of insect flight muscles: genes, motor molecules and motion. *News in Physiological Sciences* **14**, 87–92.

Murad, F. (1998) Nitric oxide signaling: would you believe that a simple free radical could be a second messenger, autacoid, paracrind substance, neurotransmitter and hormone? *Recent Progress in Hormone Research* **53**, 43–60.

Nobili, R., Mammano, F. & Ashmore, J. (1998) How well do we understand the cochlea? *Trends in Neuroscience* **21**, 159–167.

Paradis, S., Sweeney, S.T. & Davis, G.W. (2001) Homeostatic control of presynaptic release is triggered by postsynaptic membrane depolarization. *Neuron* **30**, 737–749.

Piazzessi, G., Reconditi, M., Linari, M. *et al.* (2002) Mechanism of force generation by myosin heads in skeletal muscle. *Nature* **415**, 659–662.

Rutkove, S.B. (2001) Effects of temperature on neuromuscular electrophysiology. *Muscle & Nerve* **24**, 867–882.

Sakmann, B. (1992) Elementary steps in synaptic transmission revealed by currents through single ion channels. *Science* **256**, 503–512.

Scales, S.J., Bock, J.B. & Scheller, R.H. (2000) The specifics of membrane fusion. *Nature* **407**, 144–146.

Sheng, M. & Kim, M.Y. (2002) Postsynaptic signaling and plasticity mechanisms. *Science* **298**, 776–780.

Sillar, K.T., Kiehn, O. & Kudo, N. (1997) Chemical modulation of vertebrate motor circuits. In: *Neurons, Networks and Motor Behaviour* (ed. G.S. Stein, S. Grillner, A.I. Selverston & D.G. Stuart), pp. 183–194. MIT Press, Cambridge, MA.

Strausfel, N.J. & Hildebrand, J.G. (1999) Olfactory systems: common design, uncommon origin? *Current Opinion in Neurobiology* **9**, 634–639.

Tobalske, B.W. (2001) Morphology, velocity and intermittent flight in birds. *American Zoologist* **41**, 177–187.

Torre, V., Ashmore, J.F., Lamb, T.D. & Menini, A. (1995) Transduction and adaptation in sensory receptor cells. *Journal of Neuroscience* **15**, 7757–7768.

Vale, R.D. & Milligan, R.A. (2000) The way things move; looking under the hood of molecular motor proteins. *Science* **288**, 88–95.

Valtorta, F., Meldolesi, J. & Fesce, R. (2001) Synaptic vesicles: is kissing a matter of competence? *Trends in Cell Biology* **11**, 324–328.

Wood, S.J. & Slater, C.R. (2001) Safety factor at the neuromuscular junction. *Progress in Neurobiology* **64**, 393–429.

Yates, G.K., Johnstone, X.M., Patuzzi, R.B. & Robertson, D. (1992) Mechanical processing in the mammalian cochlea. *Trends in Neuroscience* **15**, 57–61.

Zucker, R.S. & Regehr, W.G. (2002) Short-term synaptic plasticity. *Annual Review of Physiology* **64**, 355–405.

10 Hormones and Chemical Control Systems

10.1 Introduction

All multicellular organisms require coordinating systems to link the activities of different kinds of cells. While the nervous system uses primarily electrical signals to transmit coordinating messages very quickly, the hormonal or endocrine system uses chemical messengers. A hormone is essentially a blood-borne messenger, released from one kind of cell and acting in a regulatory fashion on one or more other types. Whereas nerves achieve target specificity by the intimate physical proximity of cells, and regulate very fast coordination, hormones achieve their specificity only by the location of appropriate receptors on the relatively distant target tissues, and they play essential roles in "slower" phenomena such as growth, metabolism, and homeostasis, influencing behavior and annual or seasonal cycles, and mediating organism–environment interactions through the life cycle. In fact it is useful to consider three kinds of effects for hormones, with increasingly long timescales:

1 Trigger effects, where an irreversible event happens; a good example would be the triggering of an event such as spawning in many aquatic animals.

2 Homeostatic effects, controlling the levels of other chemicals; this may be achieved either by one hormone with a built-in negative feedback, or by pairs of hormones acting antagonistically. Examples here include the control of salt and water balance in many animals (often involving several hormones, though only one or two act at any one site), and the control of blood sugar in vertebrates by the antagonistic actions of insulin and glucagon.

3 Control of synthetic processes, especially development, growth, and reproduction, thus linking physiology to the life cycle.

However, as we will see, the differences between the neural and hormonal signaling systems are not as great as this simple overview suggests.

In this chapter we examine the properties and roles of hormones, especially in relation to the bigger issues of coping with environmental challenges, dealing first with the endocrine systems and component glands in different kinds of animals, then with the various functions (such as salt and water balance, metabolism growth and development, and reproduction) that are regulated by specific hormones. We end with a look at pheromones, functioning as important chemical mediators between animals in the control of sexual behaviors, group organization, and reactions to threatening environmental signals.

10.1.1 Types of chemical messengers

Table 10.1 summarizes the various different levels of chemical message that may operate within and between animals. The secreted chemical messages from a cell may also be characterized as in

Table 10.1 Different levels of chemical communication systems in animals.

Level	Types	Examples
Intracellular	Small molecules, within all living cells	Ca^{2+}, cAMP (see Chapter 2)
Intercellular, intraindividual	Neurotransmitters—very short range, one cell to another	ACh, GABA, 5-HT, dopamine, octopamine (see Chapter 9)
	Paracrine chemicals—very short range, one cell to many local cells	Histamine, wounding-related chemicals
	Hormones—endocrine chemicals, short to medium range, to more distant cells	
	(a) Neurohormones	Releasing hormones (CRH, TRH, LHRH, etc.)
	(b) Endocrine hormones, from specific glands	Insulin, prolactin, estrogen, ecdysone
	(c) Tissue-derived hormones (from heart, liver, gut, fat tissues, etc.)	Atrial natriuretic peptide (ANP), gastrin, leptin
Interindividual, intraspecies	Pheromones—medium to long range, between individuals, e.g. male to female	Bombykol
Interspecies	Allomones—medium to long range, between individual animals of different species	

See Abbreviations, p. xi, for definition of terms.

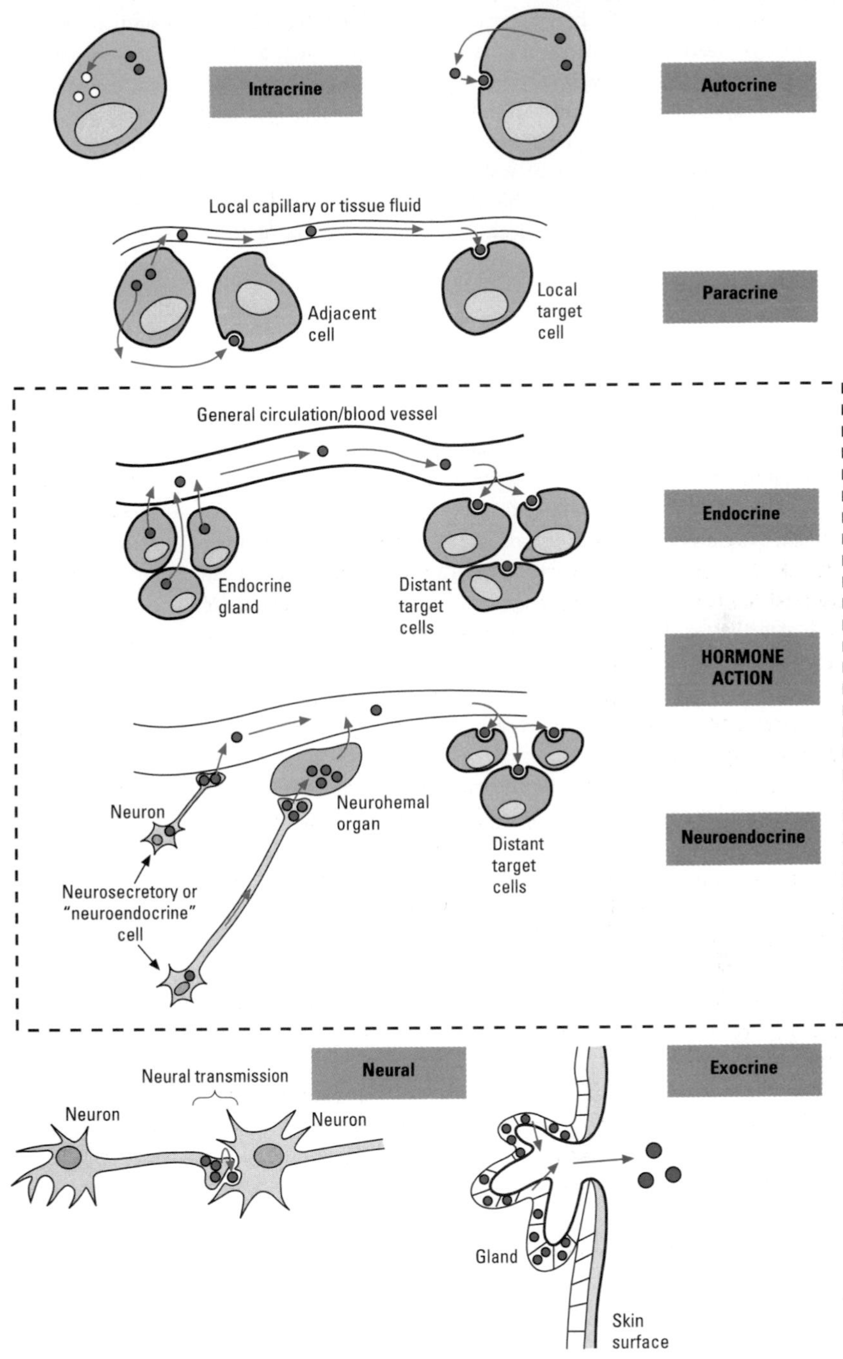

Fig. 10.1 Different types of communication between cells; only endocrine and neuroendocrine mechanisms are usually regarded as "hormonal".

Fig. 10.1, as **autocrine** (affecting that same cell), **paracrine** (affecting nearby cells), **endocrine** (affecting distant cells via the blood system), and **exocrine** (released onto the surface of the body, and potentially affecting other animals). However, this categorization can be confusing; the last group, for example, would include sweat and urine, mucus and silk, and chemicals like saliva released into the gut, as well as pheromones, allomones, and various defensive secretions or chemical "weapons".

At the "lower" intracellular end of the spectrum the chemical signals shown in Table 10.1 overlap, and interact, with the fast communication via neurotransmitters at chemical synapses described in Chapter 9. The distinction between types of target specificity also breaks down, as many neurohormones (hormones synthesized in and released from neurons) are delivered in very close proximity to their site of action. But at "higher" levels many of the chemical signals may be acting at long range and on a timescale of hours or days (or even months and years for some hormones).

We introduced the various types of internal chemical messenger in Chapter 2 (see sections 2.5.2 and 2.5.3) in relation to the control of gene expression; indeed intracellular messenger systems (via calcium ions, cAMP, G-proteins, and protein kinases) pervade many chapters of this book, reflecting increased understanding of physiologic processes at a molecular level. In this chapter we concentrate on hormonal systems on a larger scale: how they control processes in tissues, organs, and whole organisms, and influence the morphology, physiology, and behavior of animals. The first two types of message shown in Table 10.1 as "hormones" are particularly important in terms of large-scale physiologic regulation, and their activities form the core of this chapter (though we also deal briefly with interanimal communication systems in section 10.9).

10.1.2 Types of hormones

Patterns of release

Hormones are defined as chemical messengers, elaborated by source cells and influencing other target cells, and carried between source and target via the blood or some equivalent circulating fluid. As we saw above, hormones can be either neurosecretory products, i.e. the neurohormones that are released from nerve cell endings into the blood (especially in parts of the brain), or "true" hormones, released from special endocrine glands into the bloodstream. In each case, they are unlike neurotransmitters in that they travel in the blood and act at a *distant* target.

Neurohormones are released in exactly the same manner as neurotransmitters (see section 9.3) when the neuron is electrically stimulated. The product, however, passes into a blood vessel rather than into a synaptic cleft. The site of release may be one or a few neural endings in the brain, or may be a group of such endings forming a small gland (sometimes called a "neurohemal organ") closer to the effector (see Fig. 10.1). The secretory vesicles that contain neurohormones are typically larger (100–400 nm in diameter) than the presynaptic vesicles of ordinary neuronal synapses which may be only 30–60 nm. The terminals may also be unusually large, and often lie outside the central nervous system (CNS), beyond the restrictions of a "blood–brain barrier" (see section 9.6.3).

In contrast, true endocrine hormones are released by exocytosis from non-neural tissues, usually identifiable as distinctive small ductless glands that empty directly into blood capillaries.

Chemical type and mode of action

Hormones are of two main types, peptides and steroids, illustrated together with their typical mode of action in Fig. 10.2.

Peptide hormones (which can be as small as four amino acids, though most are much larger and usually termed polypeptides) are water soluble and can travel easily in the blood. But this means they are immiscible with lipid, so they do not actually cross cell membranes and enter their target cells. Instead they act rapidly at the cell membrane surface, binding to and triggering external receptors there and so inducing second messenger effects internally. Usually these reception systems involve adenyl cyclase and cAMP, but routes via inositol triphosphate (IP_3), diacylglycerol (DAG), or tyrosine kinase receptors are also found. A common outcome is phosphorylation of protein kinases, sometimes with calcium as a "third messenger", and with rapidly cascading amplification effects

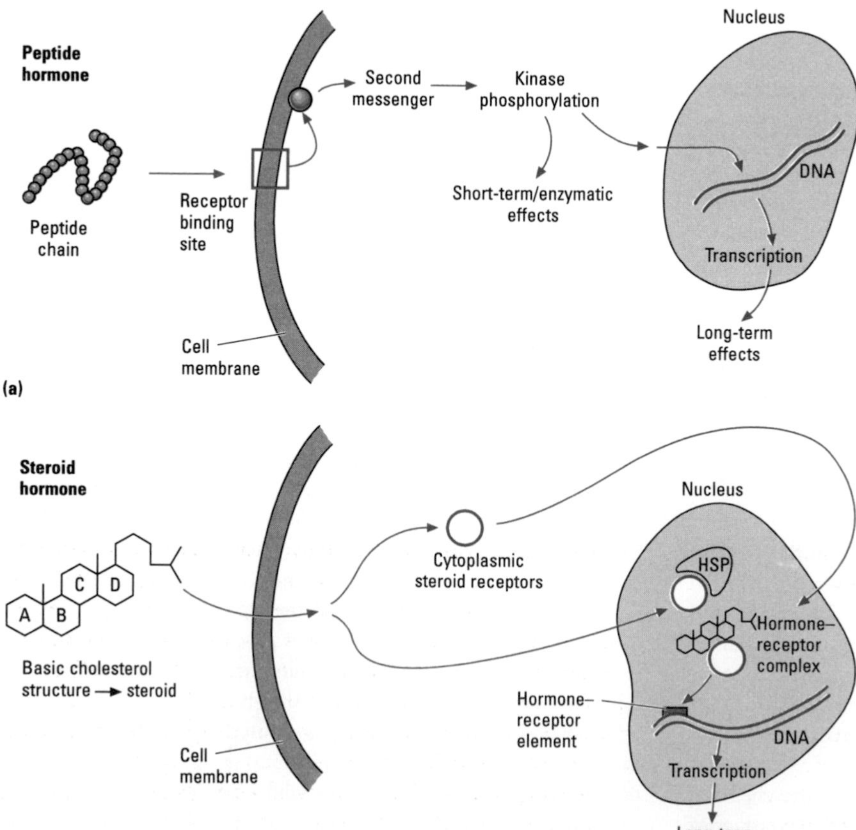

Fig. 10.2 The two main types of hormone and their differing modes of action: (a) peptide hormone, and (b) steroid hormone. HSP, heat shock protein.

(discussed more fully in Chapter 2). Thus these hormones usually mediate short-term effects.

Peptide hormones are synthesized on the ribosomes as "**prohormones**", which are rather longer polypeptide chains including the functional hormone (sometimes more than one of these) plus a terminal "signal" peptide group. This signal group is cleaved off when the peptide is in a secretory granule within the cell, to give the active hormone, ready for exocytosis. Sometimes the peptide incorporates carbohydrate groups at this stage, and is then termed a glycopeptide.

Nearly all the invertebrate "brain hormones", and all the neurally derived "releasing hormones" of vertebrates, are in the peptide hormone group; they are assembled in neurosecretory cell bodies (hence called **neuropeptides**). At least 40 peptide neurohormones have been identified in mammals. Many peptide hormones show significant sequence similarities across the whole animal kingdom, and even beyond it to chemical messengers in yeast and other microorganisms. Insulin-like molecules, for example, occur not only in vertebrates but also in many invertebrate groups, in protozoans and in slime molds. However, any one peptide hormone may not be constant in sequence wherever it occurs: as an example, the variations in a particular group of nonapeptides within the vertebrates are shown in Fig. 10.3. Homologous peptides may or may not work in other taxa—for example, human growth hormone (GH) has some effect in most vertebrates, including fish, but fish GH has almost no effect in humans.

Steroid hormones are lipophilic (i.e. they dissolve in lipid), so they normally act by crossing the lipid-rich cell membrane and binding to intracellular receptors. This binding usually occurs in the nucleus (but can be cytoplasmic, in which case a hormone–receptor complex then migrates into the nucleus). The nuclear receptors are linked to heat shock proteins (HSPs) when no hormone is present, thus being "chaperoned" and stabilized (see Chapter 8). When the hormone–receptor complex forms, the HSP vacates the receptor site so increasing its affinity for "hormone receptor elements"; these are specific DNA sequences in the promoter region of the genes whose expression is to be regulated (see Chapter 2). The hormone–receptor complex, once bound to the receptor elements, acts in effect as a transcription factor.

The main types of steroid are **estrogens** (C_{18}), **androgens** (C_{19}), **progestins** (C_{21}), and **corticosteroids** (C_{21} with extra hydroxylation). They are not always easy to distinguish in their effects because they interconvert readily. For example, the enzyme **aromatase** that converts testosterone directly to estrogen is widely distributed in vertebrate brains.

Being lipophilic, the steroid hormones cannot readily be stored in glands, as they so easily diffuse out of cells—they must be made as and when needed. Similarly they do not readily travel in free aqueous solution, and therefore must be carried in the blood in a bound form with plasma proteins, often albumins or globulins. This carriage system greatly extends their residence time ("half-life") in the blood, so they usually act rather slowly and over a long period, being particularly concerned with metabolism, growth, and reproduction. However, a few cases of quick-acting steroids are known; for example some steroids bind to the γ-amino butyric acid (GABA) receptor protein in the vertebrate brain, and these can affect mood in humans.

Other types of hormones also exist. Some are derivatives of modified amino acids, including the catecholamines such as epinephrine and the thyroid hormones (both of these types being derived from tyrosine), and melatonin which is synthesized from 5-hydroxytryptamine (5-HT) originating from tryptophan. Other hormones are derivatives of modified nucleotides (such as 1-methyl adenine). These are all lipid insoluble (excepting thyroid hormones), and so act by the second messenger system described for peptides. Invertebrates also have a range of **eicosanoid** hormones, especially those derived from arachidonic acid. These are lipid soluble and work intracellularly. In vertebrates the prostaglandins are of this type, synthesized from a C_{20} fatty acid, though prostaglandins tend to have paracrine rather than endocrine functions. Finally, **terpenes** (again derived from fatty acids) may also act as hormones, the classic example being the insect juvenile hormone (JH); these again are lipid soluble and act intracellularly.

Basic peptides
1 2 3 4 5 6 7 8 9
Arginine vasotocin (AVT)
Cys-Tyr-Ileu-Gln-Asn-Cys-Pro-Arg-Gly-NH2
Arginine vasopressin (AVP)
Cys-Tyr-Phe-Gln-Asn-Cys-Pro-Arg-Gly-NH2
Lysine vasopressin (LVP)
Cys-Tyr-Phe-Gln-Asn-Cys-Pro-Lys-Gly-NH2
Phenypressin (PP)
Cys-Phe-Phe-Gln-Asn-Cys-Pro-Arg-Gly-NH2
Neutral peptides
Oxytocin
Cys-Tyr-Ileu-Gln-Asn-Cys-Pro-Leu-Gly-NH2
Mesotocin
Cys-Tyr-Ileu-Gln-Asn-Cys-Pro-Ileu-Gly-NH2
Valitocin
Cys-Tyr-Ileu-Gln-Asn-Cys-Pro-Val-Gly-NH2
Isotocin
Cys-Tyr-Ileu-Ser-Asn-Cys-Pro-Ileu-Gly-NH2
Glumitocin
Cys-Tyr-Ileu-Ser-Asn-Cys-Pro-Gln-Gly-NH2
Aspartocin
Cys-Tyr-Ileu-Asn-Asn-Cys-Pro-Val-Gly-NH2
(a)

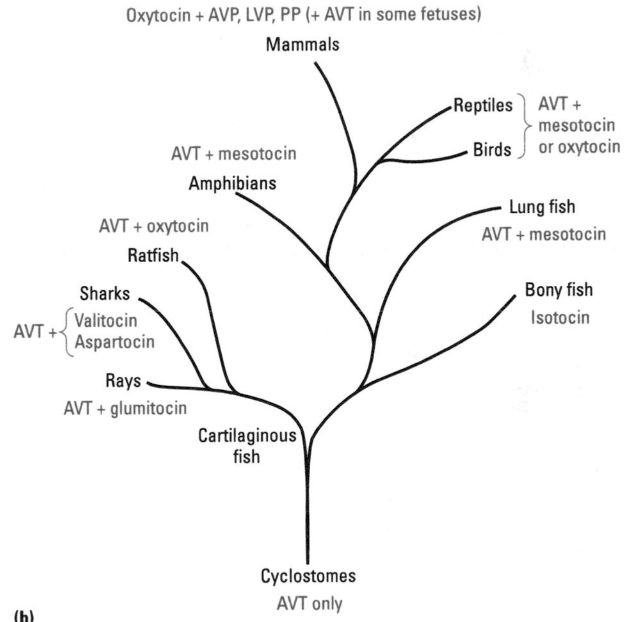

Fig. 10.3 (a) A comparison of closely related vertebrate nonapeptides, and (b) their occurrence in different vertebrate taxa.

Table 10.2 Mode of action of hormones.

Actions via an intermediate at the cell surface					
cAMP levels	**cAMP levels**	**Altered DAG levels**	**Altered tyrosine kinase levels**	**Altered IP_3 levels**	**Intracellular binding to intracellular receptors**
Catecholamines (β-adrenergic)	Catecholamines (α_1-adrenergic)	Catecholamines (α_2-adrenergic)	Insulin	Oxytocin	All steroid hormones
Glycopeptides	ACh (adrenergic)	ACh (muscarinic)	Vertebrate growth factors	ADH*	T_3 and T_4 thyroid hormones
LH					
FSH	Somatostatin	Angiotensin	GH		
TSH		GnRH	PRL		
Glucagon		ADH*	Erythropoietin		
PTH					
Calcitonin					
ADH*					
ACTH					
Most hypothalamic-releasing hormones					

See Abbreviations, p. xi, for definition of terms.
* Note that ADH can operate in at least three different ways (see p. 363).

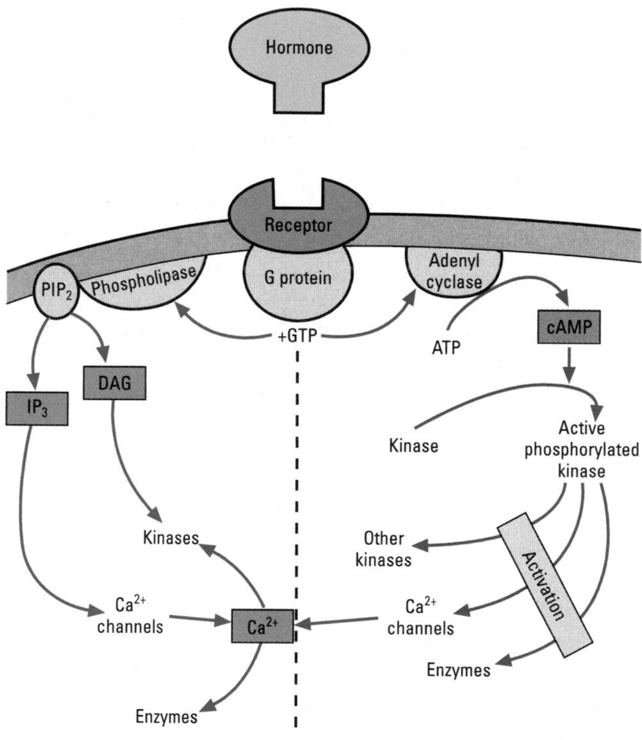

Fig. 10.4 A summary of the control pathways underlying typical hormonal actions via key second messengers (more detail is shown in Figs 2.18–2.20). (See Abbreviations, p. xi, for definition of terms.)

Table 10.2 summarizes the known modes of actions of some common hormones; note that some examples appear under more than one heading, as they have more than one kind of receptor. A reminder of the basic pathways involved is given in Fig. 10.4, though more details were included in Chapter 2.

Effects of hormones

At a cellular level, the outcomes of hormone actions in animals can usually be ascribed to one (or more) of five possible effects:

1 Synthesis of new proteins.
2 Activation or deactivation of enzymes.
3 Changes in cell membrane permeability.
4 Induction of mitotic cell division.
5 Induction of secretory activity.

This last effect is particularly common where the target of a hormone is another endocrine cell, in which case the initial hormone is referred to as a tropin, or **tropic hormone**. Obvious examples are the gonadotropins (usually from the brain) that trigger secretory activity in the gonads, from which the true and usually steroidal "sex hormones" are produced. Hence many of the hormones have common alternative names; for example, thyroid-stimulating hormone (TSH) from the anterior pituitary is also frequently termed thyrotropin.

Specificity of hormones

One main problem remains: if hormones are released into the general circulation and travel all round the body, how can they have specific effects on specific target tissues? Specificity clearly resides not with the hormone itself, but with the target cells, which are *only* those that carry (either on the cell membrane or internally) the protein receptors for a given hormone.

The degree of specificity may vary. For example, receptors for vertebrate adrenocorticotropic hormone (ACTH) occur only in a small proportion of the cells of the adrenal cortex, whereas receptors for vertebrate thyroid hormones that affect cell metabolism occur in nearly all the body's cells. Likewise the effect of the hormone on a target cell will vary not only with the hormone concentration, but also with the number of receptors present and with the affinity of those receptors. All three of these factors can vary, and indeed receptor number and affinity often vary in relation to the cells' recent exposure to hormone concentrations; receptor number may increase (upregulation) when the hormone level is low and decrease (downregulation) during chronically high exposure to hormone. Hence hormones are best seen as generalized triggers for preprogramed events in specific target cells; they are not in themselves information-carrying molecules.

10.1.3 Problems in studying animal hormones

Studying a hormone's activities used to be very difficult, partly because most hormones are present in the blood at very low levels (often only nanomolar or picomolar levels), and the normal or functional concentration was hard to assess. Early work relied on **bioassay** techniques, assessing the response of an isolated tissue, or the response of a whole animal, to artificially high or low hormone levels. This could be done by **ablation** (removing or destroying the tissue that produces the hormone), or by **implantation** of additional gland tissue (or injection of its products). But there were always uncertainties: whether all of the natural "source" had been removed, whether an implant was functioning normally, or whether neuronal connections were needed for full effect. Injection of "extra" hormone also made difficult assumptions: a hormone may have positive effects at one dose and negative at another, and injected hormone may be rapidly metabolized (usually by the liver, or the somewhat equivalent fat body in arthropods), or bound into an ineffective state, giving unrealistic results. Such techniques rarely proved that a hormone was *causing* a particular effect, and some of the early work on animal hormones is open to question.

However, endocrinology has been hugely aided by new technologies, especially **radioimmunoassays** (RIAs, using monoclonal antibodies against the hormone) and competitive binding assays. Analyses are now rapid, automated, and highly sensitive, using chemiluminescent immunometric systems developed from RIA technology. They are also very specific for the steroid or peptide of interest, accurately detecting the concentration of a hormone in the blood and thus its rate of secretion, its rate of clearance (by conversion or excretion), and hence its half-life. Hence a much more accurate analysis of specific hormonal effects is achievable. **Transgenic techniques** have added further insights, especially using "foreign" structural genes for particular hormones introduced into laboratory animals. Finally, we can now investigate endocrinology in a "reverse" direction, using molecular techniques to clone receptors, produce antisera, and then test for the presence of whole families of hormones and of receptors in different animals. The field of **reverse molecular endocrinology** is expanding rapidly, and we often now know what chemical messenger systems are present across a wide range of taxa without having fully disentangled their functions.

10.2 Endocrine systems

10.2.1 Invertebrates

Neurosecretory cells have been identified in virtually all animal phyla, and true endocrine cells and glands are probably also common in multicellular animals. In general, invertebrates tend to have far more neurohormones than endocrine hormones, implying that neurohormones are an evolutionarily earlier invention. Endocrine glands also tend to be much simpler in structure in invertebrates, with amorphous tissue releasing hormones straight into the open circulation, in contrast to the vertebrate pattern of complex follicular glands surrounding specific blood vessels. Invertebrates may also have simpler control systems, where neurosecretory axons release neuropeptides directly to their target tissues. Vertebrates commonly have two- or even three-stage links, with neuropeptides acting as releasing factors controlling other neurohormones and/or classic endocrine hormones. Finally, invertebrate hormones are more concerned with regeneration and growth (either continuous or discontinuous by moting), and reproduction (especially sex determination and gonad activity); they have rather limited roles in homeostatic systems.

Lower invertebrates and worms

Sponges do not really have nerves, and therefore cannot have neurosecretory hormones; neither do they have real "tissues", and so must lack endocrine glands. Such chemical regulation as does exist is probably by paracrine control within small groups of cells.

In cnidarians (e.g. *Hydra*) neurons within the diffuse nerve net produce a growth-promoting hormone influencing asexual budding and regeneration. Isolated neurosecretory granules can induce full regeneration of a head in a divided hydra, and the peptide they contain is often termed "Hydra head activator". Intriguingly, the same molecule (with 11 amino acids) turns up in mammals, with effects on the pancreas and uterus.

In most other invertebrate taxa the nervous system is more condensed. Nematodes (roundworms) have neurosecretory cells in the anterior ganglion and some axons, but lack specific neurohemal organs; the main function known for their neurohormones is control of molting. In platyhelminths (flatworms) specific neurohemal sites can be identified, and cerebral neurosecretory cells affect regeneration, reproduction, and (for euryhaline species) responses to lowered salinity.

The most "primitive" group so far shown to have true endocrine glands is the nemertines, with paired **cerebral organs** just behind (and linked with) the brain, which also has neurohemal areas. Products from the cerebral organs can inhibit gonad maturation, and regulate water balance.

The annelids have a fairly sophisticated circulatory system, and allied with this are both neurosecretory and endocrine hormones. Polychaetes have neurosecretory cells in the cerebral ganglia, the supraesophageal ganglion, and various ganglia in the nerve cord; and there is a neurohemal structure at the base of the brain (Fig. 10.5), receiving axons from the cerebral ganglion. Neurohormones from here (including annetocin, closely related to the vasopressin hormone of vertebrates) play key roles in growth, regeneration, and reproduction in annelids. They are also implicated in osmoregulation and glucose balance. Adjacent to the neurohemal organ is the **infracerebral gland**, probably a true endocrine structure. Polychaetes also have a true endocrine hormone derived from immature **oocytes**, and termed "feedback substance" because it prevents excessive egg production.

Echinoderms could also be included here; they have a very weakly developed central nervous system (CNS), and this releases some neurohormones from the **radial nerve** around the mouth. The **gonads** also produce at least two types of hormone, including a steroid in some species.

Molluscs

Molluscs lack an obvious neurohemal organ, but instead release neurosecretory peptides from many sites within the **brain** and the

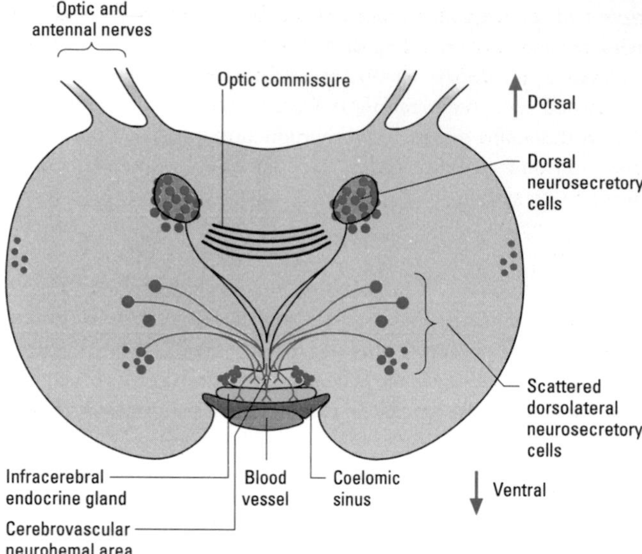

Fig. 10.5 The location of the main neurohormonal centers, and the infracerebral gland (at least partly endocrine), in the brain of a typical polychaete annelid worm. (Neurosecretory cells also occur in most of the ganglia of the nerve cord.)

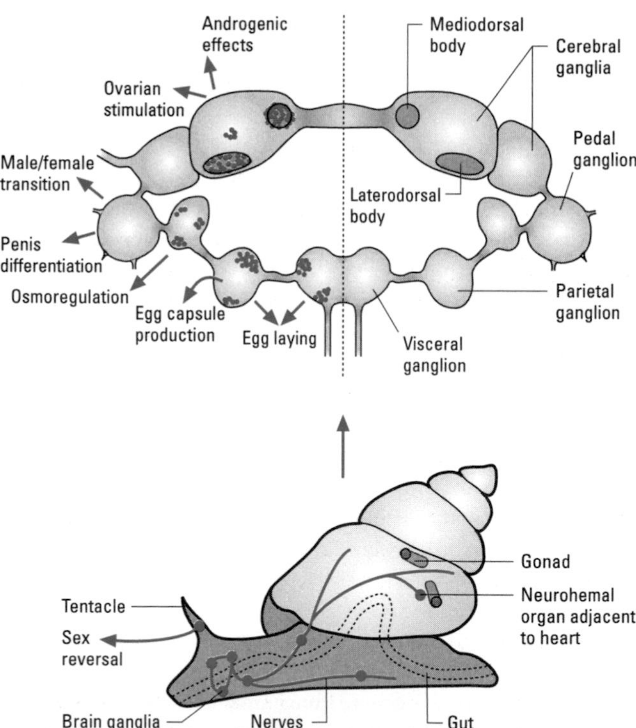

Fig. 10.6 The location of the main hormonal structures in a gastropod mollusc, with the "brain" enlarged to show the main neurosecretory centers (green dots) on the left, and the known neuroendocrine effects elicited from different ganglia (green arrows). (Amalgamated from work on various species.)

peripheral nervous system (Fig. 10.6). This may relate to their lack of a blood–brain barrier, with neural tissue in direct contact with the circulating hemocoel fluid. Some peripheral axons also penetrate directly to target organs and release neurohormones there, for example in the kidneys of gastropods. In gastropods most of the neurohormones are gonadotropic, or have other effects on reproduction. Cephalopods also have a gonadotropic neurohormone, derived from the **optic glands** adjacent to the brain. In some molluscs peptides controlling osmoregulation have been identified, with at least three variants (cephalotocin, conopressins) all having similarities with vertebrate vasopressin.

The **gonads** of some molluscs are also known to produce hormones, thus being true endocrine glands, although the hormones released are not necessarily related to the control of reproduction. However, in at least some cases they have been identified as steroids rather similar to those from vertebrate gonads.

Insects

Insect endocrine systems are easily the best known amongst the invertebrates, and again are dominated by neurohormones, with relatively few true endocrine glands. For a typical insect the known sources of hormones are summarized in Fig. 10.7. Neuropeptides are secreted by various groups of cells in the **brain,** including the median and lateral protocerebrum, and in some species also the tritocerebrum (see Fig. 9.30b). Axons from all these cells project to paired neurohemal glands behind the brain, known as the **corpora cardiaca**, where the brain products are stored. Additional neurosecretory cells occur in the **subesophageal ganglion**, and sometimes in ganglia further along the ventral nerve cord. These have axons passing along the segmental connectives to neurohemal organs on peripheral nerves, but their functions and products are largely unknown.

One of the brain's most important neurohormones, sent to the corpora cardiaca from the median neurosecretory cells, is a small protein originally just called "brain hormone" but more properly termed prothoracicotropic hormone (PTTH) or sometimes ecdysiotropin; in fact we now know there are several PTTH peptides. These are secreted from the corpora cardiaca under the integrated influence of various inputs from the brain, and initiate the process of molting. Another brain neuropeptide has been identified, eclosion hormone (EH), which acts via cGMP and specifically triggers the behavioral emergence of the new instar, rather than being involved in the whole molt cycle. EH targets specific neurons in the ventral nerve cord, and also cells in **epitracheal glands** near the abdominal spiracles which release the neuropeptide "ecdysis-triggering hormone". A further product from the brain and nerve cord is the relatively large protein bursicon, involved in later stages of metamorphosis and cuticle development.

The corpora cardiaca in some insects also have some secretory (endocrine) cells of their own, the probable source of other known products of the corpora cardiaca, including a cardioactive hormone that controls the heart beat, a hormone affecting ion transport, and hormones that influence carbohydrate and lipid metabolism via the fat body.

The target organ for the PTTH released from the corpora cardiaca is usually a paired endocrine gland in the first segment of the thorax, hence called the **prothoracic gland**. Once activated this is the principal source of the molting hormone steroid called ecdysteroid or more commonly ecdysone. Ecdysone production is upregulated by PTTH, and downregulated by a direct negative feedback of hemolymph ecdysone levels on the prothoracic gland cells.

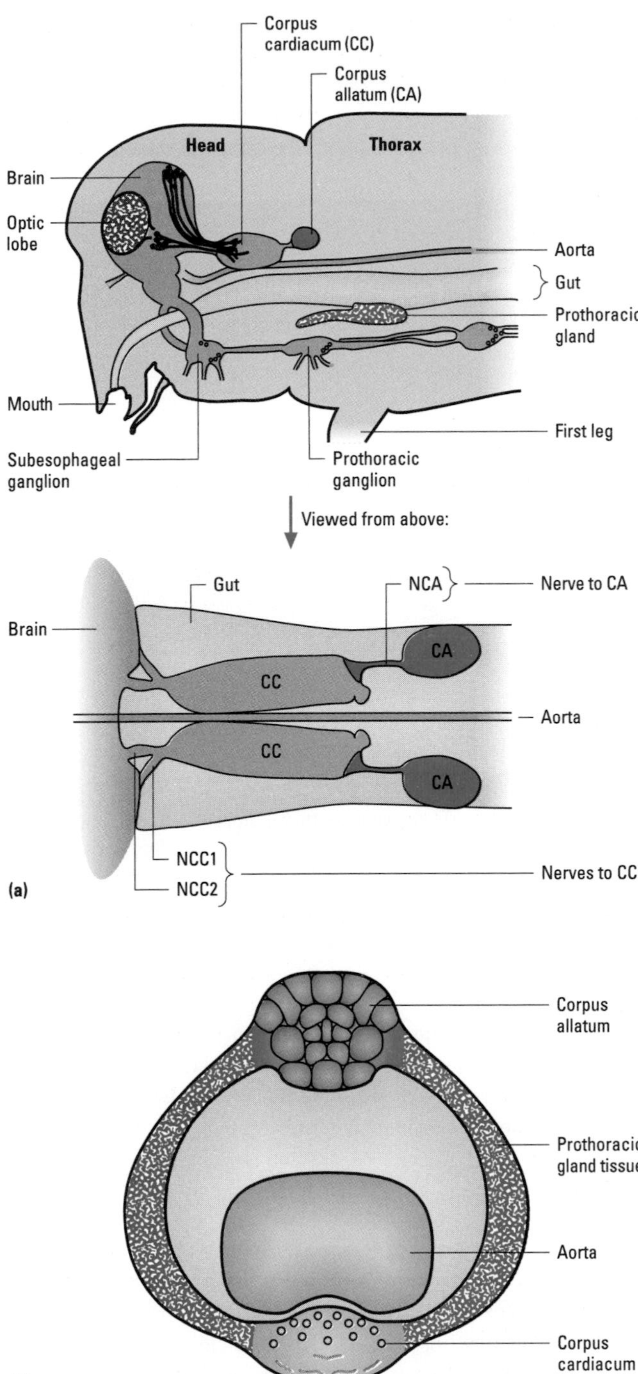

Fig. 10.7 (a) The location of the main hormonal structures in an insect, with the main neurosecretory cells in black. (b) The variant of the hormonal system in flies, with the main hormonal glands consolidated into a single "ring gland".

Adjacent to the corpora cardiaca lies another pair of glands, the **corpora allata**, which are true endocrine sources. (In many insects the corpora cardiaca and corpora allata are not really distinct, and in some of the most advanced dipteran flies the whole system of corpora cardiaca, corpora allata, and prothoracic glands are joined into a "ring gland" as shown in Fig. 10.7b). The corpora allata are again influenced by neuropeptides from the brain; in many insects there is both an allatotropin and an allatostatin (stimulating and inhibiting, respectively). The main endocrine product of the corpora allata is the terpene JH, derived from fatty acids, and occurring with various slightly different forms in different taxa of insects. It has a primary role in the control of metamorphosis in larvae. However, it also takes on an entirely different role as a sex hormone in adults, regulating ovarian activity (thus being a classic gonadotropin) and also stimulating male accessory organs.

Insects may also produce hormones from their **gonads**, like most other animals, but surprisingly they do not normally have a classic gonadal steroid as their main sex hormone. Instead the JH exerts most of the control of reproduction. However, the gonads often produce some ecdysone, and the ovaries may produce an oostatic factor feeding back on the brain to prevent excess egg production.

Crustaceans

Crustaceans have a relatively complex endocrine system, some parts common to all taxa but others present only in more advanced families (Fig. 10.8). The eyestalk contains the optic lobe of the brain, and this usually houses two organs, termed the **X-organ** and the **sinus gland** (Fig. 10.8b), that are similar in function to the median neurosecretory cells and corpora cardiaca of insects, respectively. The X-organ is a complex neurosecretory structure, and its products are released into the sinus gland for storage and release. The principal product is a molt-inhibiting hormone (MIH; a 78 amino acid peptide); contrast this with the molt promoter as secreted by the insect brain. Vitellogenesis-inhibiting hormones and hyperglycemic hormones are also produced in the X-organ, and its peptide products all share some sequence similarity.

There are two other neurohemal organs in more advanced crustaceans, a **postcommisure organ** near the commisures connecting the brain to the subesophageal ganglion, and a **pericardial organ** where the gill veins drain back into the pericardium around the heart.

Another structure, the small and paired **Y-organ**, is sited in the antennary or maxillary segments of the head; it is classically endocrine, with cells similar to steroid-secreting cells in vertebrates. It releases a molting hormone that is again ecdysone, involved in metamorphosis and reproduction. In many crustaceans a **mandibular gland** is present nearby, and this secretes methyl farnesoate (MF), which is a precursor of JH. It is now known that "JH-type activity" is present in many crustaceans, and either MF or JH is the main controller of metamorphosis, as in insects. A peptide that inhibits mandibular gland activity is also known, released from the X-organ.

Some crustaceans have additional endocrine glands. Males in higher taxa have an **androgenic gland** near the vas deferens, while the **gonads** in female amphipods and isopods secrete hormones that control secondary sexual characters.

Apart from the specific sources of invertebrate hormone identified above, there is a growing literature on the presence of hormones and hormone-like molecules identified from "reverse endocrinology" (see section 10.1.3). Many of these hormones are not yet associated with a precise endogenous location or function. For example tachykinin-related peptides have been identified in echiurans, molluscs, insects, and crustaceans, having varied effects on muscle contraction and on central neurons; natriuretic peptides are

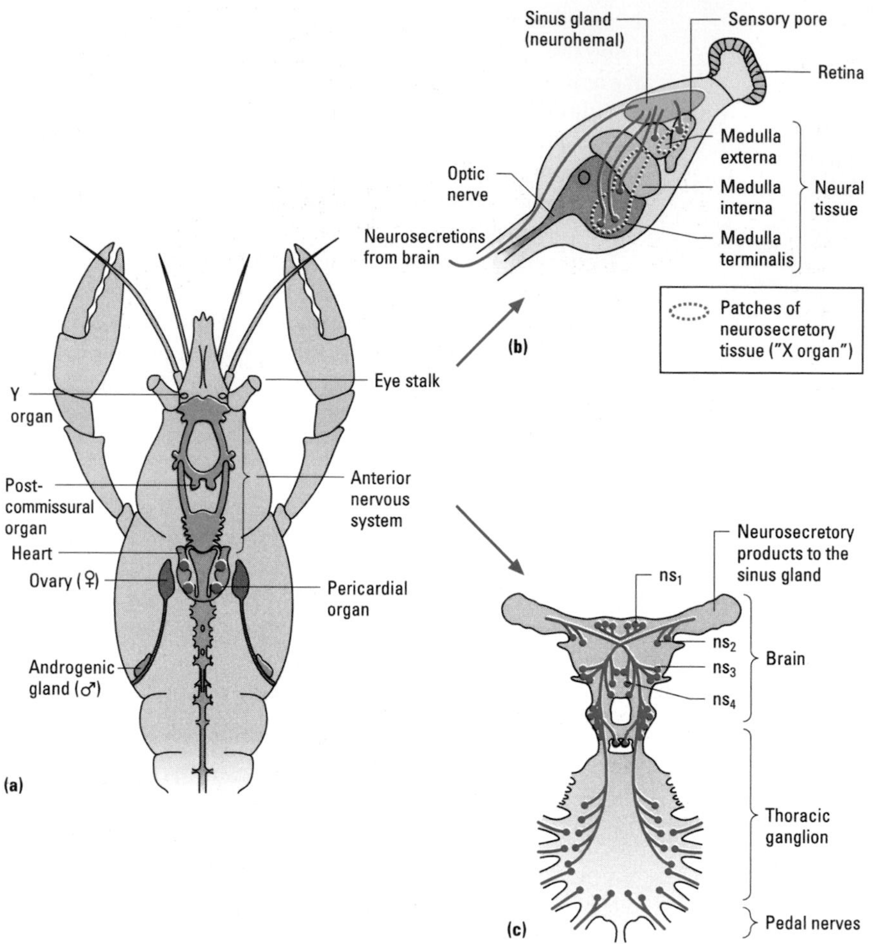

Fig. 10.8 (a) The location of the main hormonal structures in crustaceans. (b) A close up view of the hormonal sources and storage site (sinus gland) in the eyestalk. (c) Neurosecretory cells in the central nervous system, mainly feeding into the sinus glands.

increasingly being located in lower invertebrates; and insulin-like peptides are turning out to be almost ubiquitous. There is no doubt that a huge range of invertebrate hormones remains to be discovered and explained, and their hormonal complexity may come to rival that of the vertebrates.

10.2.2 Vertebrates

Compared with invertebrates, the vertebrate endocrine system has a more substantial component of true endocrine glands (often complex, and consolidated from several different types of cell) and less reliance on neurosecretory products. Indeed there is a trend away from mainly neurosecretory control in the more advanced vertebrates compared with those that are evolutionarily earlier, with primitive fish having many neurohormones and few true endocrine glands compared with tetrapods. There is also an increase in the role of hormones in control of physiological (homeostatic) processes in vertebrates, with a smaller proportion of endocrine control devoted to long-term growth and reproduction systems.

The major endocrine glands of a hypothetical ancestral vertebrate and of a typical modern mammal are shown in Fig. 10.9. Most of the glands are present throughout the vertebrates, although some of the hormones that they produce changed their function as fish evolved into tetrapods and then onto land. Similarly, some glands present in early vertebrates are no longer present in mammals. Table 10.3 summarizes the main hormones from each gland, which are then dealt with in sections 10.3–10.7 below. Here we briefly review the glands, their products, and their control systems.

Pituitary gland

The key endocrine gland in vertebrates is the pituitary (Fig. 10.10), also known as the hypophysis. It lies at the base of the skull, just below the part of the brain termed the hypothalamus. The pituitary and hypothalamus together form the principle central regulator of the endocrine system in all vertebrates, and link the neural and hormonal systems together.

The pituitary is formed from two quite distinct embryonic tissues, the upper part or posterior lobe being a stalked downgrowth from the brain, and the lower part or anterior lobe being an upward growth from the pharynx. Most vertebrates also have an intermediate lobe, formed similarly to the anterior lobe, but this is almost absent in higher mammals including humans. The origins of the pituitary can be traced back to protochordates, where the posterior lobe is part of the neural tube and the anterior lobe is a subneural gland (Hatschek's pit). Both tissues are directly exposed to water currents in the pharynx and so can respond to changes in temperature or water chemistry of the environment.

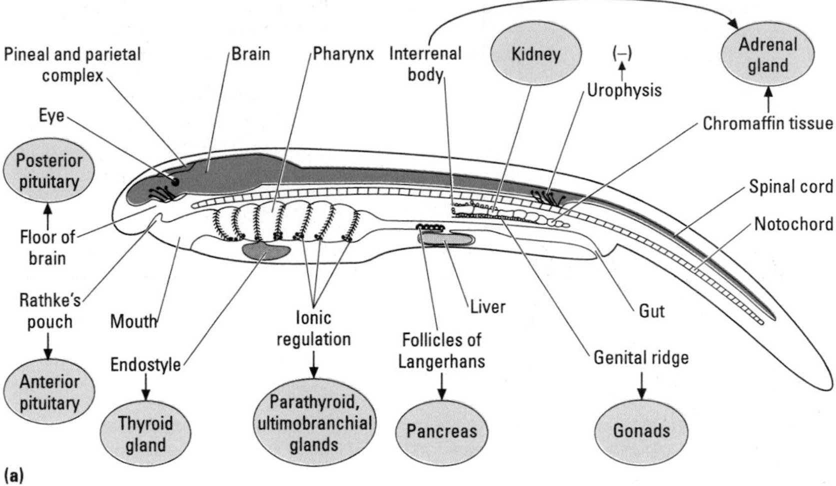

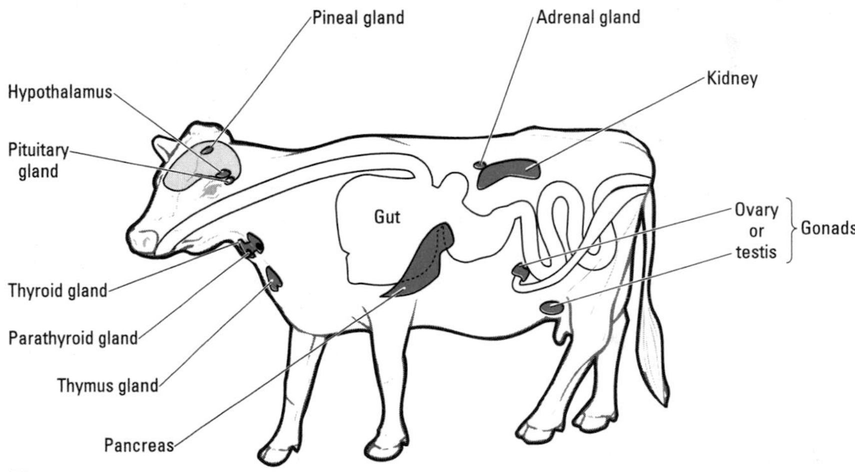

Fig. 10.9 (a) The endocrine system of a hypothetical ancestral vertebrate, with the modern status of each system shown in green. (b) Schematic location of the main endocrine glands in a modern mammal. (a, Adapted from Hoar 1965.)

POSTERIOR PITUITARY

The posterior lobe of the vertebrate pituitary is neural, and together with the adjacent "median eminence" tissue of the hypothalamus (with which it is intimately linked by neural tracts) it forms the **neurohypophysis**. This structure is a typical neurohemal organ, its neural terminals releasing two key hormones, **vasopressin** and **oxytocin** (Table 10.4). Both are nine amino acids in length and very similar, differing in just two of these amino acids (see Fig. 10.3a). Both operate via IP_3 as a second messenger (see Fig. 10.4). They are made up predominantly by neurons of the **supraoptic** and **paraventricular nuclei** of the hypothalamus, respectively (Table 10.5), and are derived from prohormones termed neurophysins.

"Vasopressin" is in fact a useful generic term for a small group of hormones with similar functions, varying in different vertebrate taxa (see Fig. 10.3b). In mammals the normal form is arginine vasopressin (AVP), whereas variants such as arginine vasotocin (AVT) and lysine vasopressin (LVP) occur in varying proportions in other vertebrates, with even more variety in the elasmobranchs. AVT is the dominant and probably the ancestral form. Vasopressin in all its versions plays a major role in regulating fluid balance, hence its alternative name of **antidiuretic hormone** (**ADH**). Its release is regulated mainly by changes in blood concentration and volume.

"Oxytocin" is also a generic term for some very similar hormones present in different vertebrates, the nonmammalian versions including mesotocin and valitocin. Oxytocin is released in response to afferent neural inputs, arising from various sources depending on the precise roles of this hormone in different vertebrate groups. For mammals, the main effects are on uterine contractions and milk ejection, and the presence of a fetus or of a suckling infant are the most potent triggers. Because the vasopressins and the oxytocins are so similar in structure, they also have some overlap in function; thus even in mammals vasopressin also has some effects on uterine contraction and on milk ejection.

ANTERIOR PITUITARY

The anterior and intermediate lobes of the pituitary are non-neural, and together with the pituitary stalk they form the **adenohypophysis**, a true endocrine gland, the main vascular supply to this being a portal blood vessel descending from the hypothalamus. The adenohypophysis is the core of the vertebrate endocrine system and

Table 10.3 Vertebrate endocrine glands and their products.

Gland	Protochordates	Elasmobranchs	Teleosts	Amphibians	Reptiles	Birds	Mammals
Pituitary							
Anterior	—	GnH? Prolactin GH ACTH TSH MSH	GnH? Prolactin GH ACTH TSH MSH	GnH? Prolactin GH ACTH TSH MSH	GnH? Prolactin GH ACTH TSH MSH	LH, FSH Prolactin GH ACTH TSH MSH	LH, FSH Prolactin GH ACTH TSH MSH
Posterior	AVT only —	AVT* Valitocin*	AVT* Isotocin*	AVT* Mesotocin*	AVT* Mesotocin*	AVT* Mesotocin*	AVP* (±LVP) Oxytocin*
Adrenal							
Cortex	—	Cortisol* *Interrenal glands*	Cortisol* *Interrenal glands*	Aldosterone* *"Head kidney"*	Corticosterone*	Corticosterone*	Aldosterone, cortisol*
Medulla	Catecholamines *Chromaffin cells*	Catecholamines *Chromaffin cells*	Catecholamines *Chromaffin cells*	Catecholamines *Chromaffin cells*	Catecholamines	Catecholamines	Catecholamines
Thyroid	Iodinated proteins *Endostyle*	—	—	Thyroxines	Thyroxines	Thyroxines	Thyroxines
		Calcitonin *C-cells*	Calcitonin *C-cells*	Calcitonin *C-cells*	Calcitonin *C-cells*	Calcitonin *C-cells*	Calcitonin
Parathyroid	—	—	—	—	—	—	Parathormone
Gut	—	CCK Secretin	CCK Secretin	CCK Secretin	CCK Secretin	CCK, gastrin Secretin	CCK, gastrin Secretin
Pancreas	*Follicles of Langerhans* (insulin, glucagons)	Insulin Glucagon	Insulin Glucagon	Insulin Glucagon	Insulin Glucagon	Insulin Glucagon	Insulin Glucagon
Kidney	—	—	—	Renin	Renin	Renin	Renin
Testis	?	Androgens	Androgens	Androgens	Androgens	Androgens	Androgens
Ovary	?	Estrogens Progesterone	Estrogens Progesterone	Estrogens Progesterone	Estrogens Progesterone	Estrogens Progesterone	Estrogens Progesterone
Placenta	—	—	—	—	—	—	Estrogens Progesterone Chorionic gonadotropin
Pineal	Melatonin	Melatonin	Melatonin	Melatonin	Melatonin	Melatonin	Melatonin
Urophysis	?	*Dahlgren cells* (urotensin)	Urotensin	—	—	—	—

See Abbreviations, p. xi, for definition of terms.
Equivalent/homologous tissues are shown in italics. Where several different versions of hormones exist the commonest version (*) is shown for each taxon.

Table 10.4 The hormones of the vertebrate pituitary gland.

	Families of hormone	Specific hormones	Amino acid chain length (in humans)
Posterior pituitary (neurohypophysis)	Nonapeptides	Oxytocin (OT)	9
		Vasopressin (arginine vasopressin (AVP) or antidiuretic hormone)	9
Anterior pituitary (adenohypophysis)	Somatotropic (single peptide chains)	Growth hormone (GH) or somatotropin	191
		Prolactin (PRL)	198–199
	Corticotropin related (excised from a single common peptide chain)	Corticotropin (ACTH)	39
		β-lipotropin (β-LPH)	91
		β-endorphin (β-LPH 61–91)	31
		α-melanocyte-stimulating hormone (α-MSH)	13
	Glycoprotein hormones (common α-peptide plus variable β-peptide)	Thyroid stimulating hormone (TSH) or thyrotropin	α 89, β 112
		Follicle-stimulating hormone (FSH)	α 89, β 115
		Luteinizing hormone (LH)	α 89, β 115

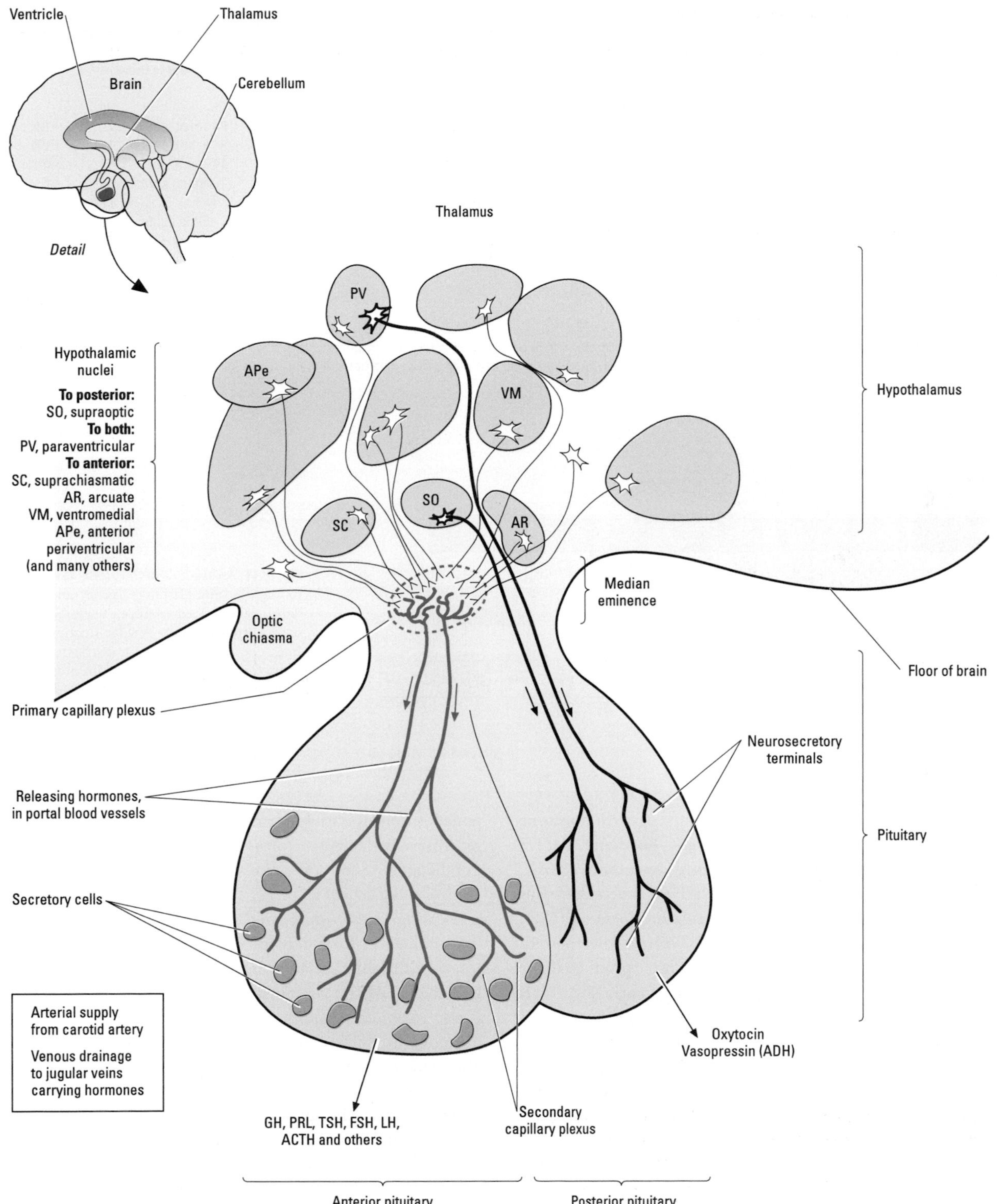

Fig. 10.10 The structure and products of the vertebrate pituitary gland, and the associated hypothalamic nuclei. The blood vessels are shown in green; the axons have a stellate cell body, and those to the posterior pituitary (from the supraoptic and paraventricular nuclei) are in bold. ACTH, adrenocorticotropic hormone; ADH, antidiuretic hormone; FSH, follicle-stimulating hormone; GH, growth hormone; LH, luteinizing hormone; PRL, prolactin; TSH, thyroid-stimulating hormone.

Table 10.5 Hormones from the vertebrate hypothalamus.

Name	Abbreviation	Action	Structure in humans	Source in hypothalamus
Releasing hormones (→ anterior pituitary)				
Growth hormone-releasing hormone	GHRH	Stimulates growth hormone (GH)	44 amino acids	ARN
Corticotropin-releasing hormone	CRH	Stimulates ACTH	41 amino acids	PVN
Growth hormone-inhibiting hormone (somatostatin)	GHIH	Inhibits GH and thyroid-stimulating hormone	14 amino acids	APeN
Gonadotropin-releasing hormone (formerly LHRH and FSHRH)	GnRH	Stimulates luteinizing hormone (LH) and follicle-stimulating hormone (FSH)	10 amino acids	ARN
Thyrotropin-releasing hormone	TRH	Stimulates TSH and prolactin (PRL)	3 amino acids	PVN
Prolactin-releasing factor	PRF	Stimulates PRL	Unknown (?TRH)	Unknown
Prolactin-inhibiting hormone	PIH	Inhibits PRL	Dopamine	ARN
Direct-acting hormones (→ posterior pituitary)				
Vasopressin, antidiuretic hormone	AVP, ADH		9 amino acids	SON (+ PVN)
Oxytocin	OT		9 amino acids	PVN (+ SON)

ApeN, anterior periventricular nucleus; ARN, arcuate nucleus; PVN, paraventricular nucleus; SON, supraoptic nucleus. (See Fig. 10.10 for locations.)

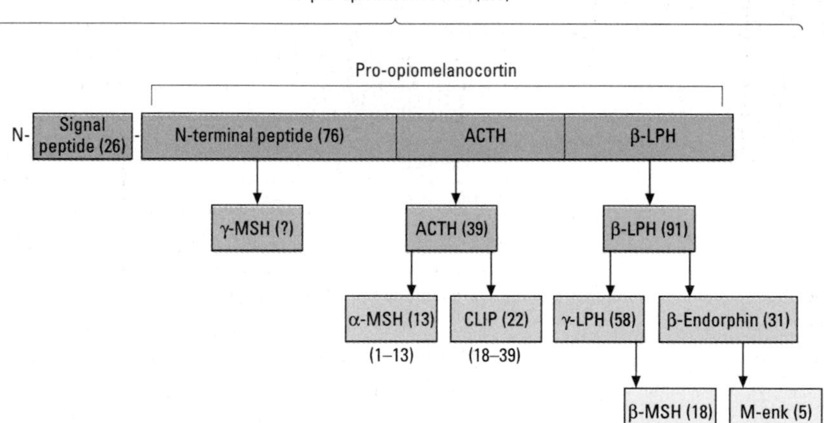

Fig. 10.11 An example of the production of multiple hormones from one gene transcript, here the transcribed vertebrate peptide pre-pro-opiomelanocortin. The smaller functional fragments result from post-transcriptional controlled enzymatic cleavage. The numbers in brackets show peptide lengths. ACTH, adrenocorticotropic hormone; CLIP, corticotropin-like intermediate peptide; LPH, lipotropic hormone: M-enk, metenkephalin; MSH, melanocyte-stimulating hormone.

secretes at least seven key "trophic" hormones: **growth hormone** (GH), **prolactin** (PRL), **adrenocorticotropic hormone** (ACTH or corticotropin), **melanocyte-stimulating hormone** (MSH), **thyroid-stimulating hormone** (TSH), and the two key gonadotropins **luteinizing hormone** (LH) and **follicle-stimulating hormone** (FSH) (perhaps occurring as a single gonadotropin (GnH) in lower vertebrates; see Table 10.4). Some of these are the multiple products of the splitting of a single prohormone into at least four parts (Fig. 10.11; note that in many nonhuman vertebrates the major product ACTH is itself split into two, giving an extra MSH). All the adenohypophysis hormones are peptides (sometimes glycopeptides) and act via cAMP on their target cells. Some are structurally variable between taxa, while others are remarkably conserved (e.g. all vertebrates have a common MSH with the same seven amino acids).

The adenohypophysis is itself under the direct control of **releasing hormones**, all of them neurohormone peptides, secreted from hypothalamic axons terminating on the hypophyseal portal blood vessels in the median eminence. They are thus carried down the pituitary gland's stalk to the anterior lobe and taken up there without ever entering the general blood circulation. These releasing hormones (Table 10.5) usually work via cAMP, and exert a simple and direct control over the rate of secretion of TSH, MSH, and ACTH (plus the other hormones related to ACTH). For FSH and LH there is normally a single releasing hormone, known as gonadotropin-releasing hormone or GnRH (though most vertebrates have at least two forms of this from slightly different cell populations). Some exceptions to the simple releasing hormone control path occur (Fig. 10.12). For example prolactin seems to have a releasing factor (unidentified) but is also controlled by inhibitory influences from dopamine, while GH is stimulated by a typical hypothalamic releasing hormone (growth hormone-releasing hormone (GHRH) or somatocrinin) and also by dopamine, but is also inhibited by another hypothalamic peptide termed growth hormone-inhibiting hormone (GHIH or somatostatin). There is also evidence that ACTH has a degree of dual control, since in addition to being stimulated by corticotrophin-releasing hormone (CRH) it can also be upregulated by vasopressin.

Most of the hypothalamic releasing hormone actions are subject to negative feedback control by the final hormone in the chain; for example, thyroid hormones in the blood can both reduce the output of thyroid-releasing hormone (TRH) and reduce the TRH sensitivity of the TSH-producing pituitary cells. All the releasing hormones, and hence the pituitary hormones that they control, show **pulsatile release** patterns, with short and regular bursts of release of hormones, probably controlled by intrinsic neural oscillators in the brain.

Although it was formerly called the "master endocrine gland", it should be clear from the above that the pituitary is in reality under strong control from the hypothalamus, either by direct neural con-

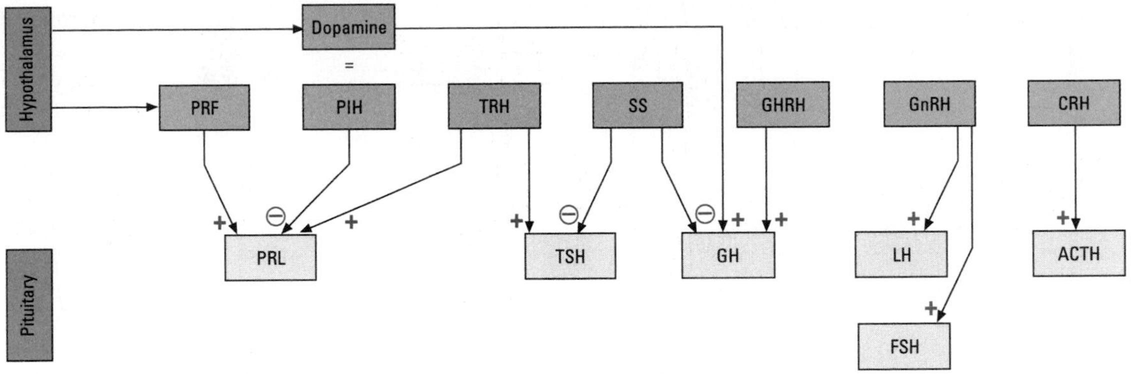

Fig. 10.12 Hormonal regulators (releasing hormones) provide control pathways between the hypothalamus and anterior pituitary. Most pituitary hormones are regulated by a single (positive) releasing hormone, but there are some unusual variants, including inhibitory controls. Standard abbreviations as elsewhere (see Tables 10.4 and 10.5); SS, somatostatin, also known as GHIH.

trol or via the hypothalamic releasing hormones, and it is therefore the hypothalamus that should be seen as the control center of the vertebrate endocrine system.

Adrenal glands

These are paired glands lying just above the kidneys, highly vascularized and each made up of an outer **adrenal cortex** and a central **adrenal medulla**. These two parts are essentially two different endocrine glands, and in lower vertebrates they may remain distinct as **the interrenal bodies** and the **chromaffin tissue**, respectively (Fig. 10.13a; see also Fig. 10.9). The cortex/interrenal tissue is mesodermal in origin and secretes steroid hormones. However, the medulla/chromaffin tissue is derived from the ectodermal neural crest (see section 9.6), and secretes catecholamines when stimulated via its sympathetic nervous supply; the medulla can be seen as part of the sympathetic nervous system.

The cortex is morphologically divisible into three zones (Fig. 10.13b), each making a different type of **steroid** hormone:

1 The outer zone makes mineralocorticoid hormones (**aldosterone** and some deoxycorticosterone).

2 The middle zone makes **cortisol** plus smaller amounts of corticosterone and cortisone.

3 The inner zone makes sex steroids, especially **androgens** (weak precursors of testosterone).

All these hormones are derived from cholesterol, most of which is taken up by endocytosis from lipoproteins in the blood. Conversion into active hormones particularly involves cytochrome P-450 enzymes in the mitochondria. After secretion the adrenal steroids are carried in the blood bound to albumins and globulins, especially transcortin. In lower vertebrates the cortical hormones mainly regulate sodium transport and hence salt balance. In mammals they have a more important role in metabolism and fuel supply.

Control of the cortical hormones is largely exerted by ACTH from the pituitary gland, in turn controlled by CRH from the hypothalamus. Cortisol probably feeds back onto both these sites. ACTH shows a pronounced daily rhythm of secretion related to the sleep–wake cycle, peaking around dawn in humans and other diurnal animals.

In the adrenal medulla the **catecholamine** hormones are synthesized in chromaffin cells, which are in effect specialized sympathetic ganglionic neurons. The cells contain granules of either **epinephrine** or **norepinephrine**, in the ratio of 4 : 1 in most mammals. The catecholamines act on different adrenergic receptors, epinephrine mainly on β-type receptors and norepinephrine mainly on α-type receptors; thus they can have slightly different effects. Both hormones have a very short half-life in the blood, usually less than 5 min, being taken up into other sympathetic nerve terminals or broken down in the liver, kidney, and brain.

Release of the catecholamines occurs as part of a general sympathetic stimulation, preparing the body to cope with an emergency ("fight or flight"). They promote a huge range of effects (Table 10.6), including glycogen breakdown, increased metabolic rate, and lipolysis. They can also produce vascular alterations (raised blood pressure and increased heart rate), and (in mammals and birds) bronchodilation and hair or feather erection. Unlike most other hormones they are not subject to complex feedbacks, but are directly controlled by nerves.

Thyroid and parathyroid glands

These two glands lie in the upper thorax or neck region, the much smaller parathyroid glands (usually 2–6 in number) being pressed against the surface of the large thyroid. The **thyroid** is an endodermal gland lying near the larynx and is derived from tissues that were originally part of the gill arches, whereas the **parathyroid** glands are derived from pharyngeal pouches in primitive chordate ancestors. All modern-day vertebrates possess thyroid glands, and their products always have similar functions in growth, maturation, and metamorphosis. However, only tetrapods have parathyroid glands; in teleost fish their role is performed by more diffuse groups of cells termed the **corpuscles of Stannius**, lying near the kidneys.

The thyroid is made up of many separate follicles around a central lumen, which contains the stored hormones. The follicular cells secrete two iodine-containing protein hormones, **thyroxine** (T_4) and the more active **tri-iodothyronine** (T_3), made from a precursor glycoprotein (thyroglobulin) by the incorporation of iodine. Unlike most endocrine glands, the thyroid can store enough hormones

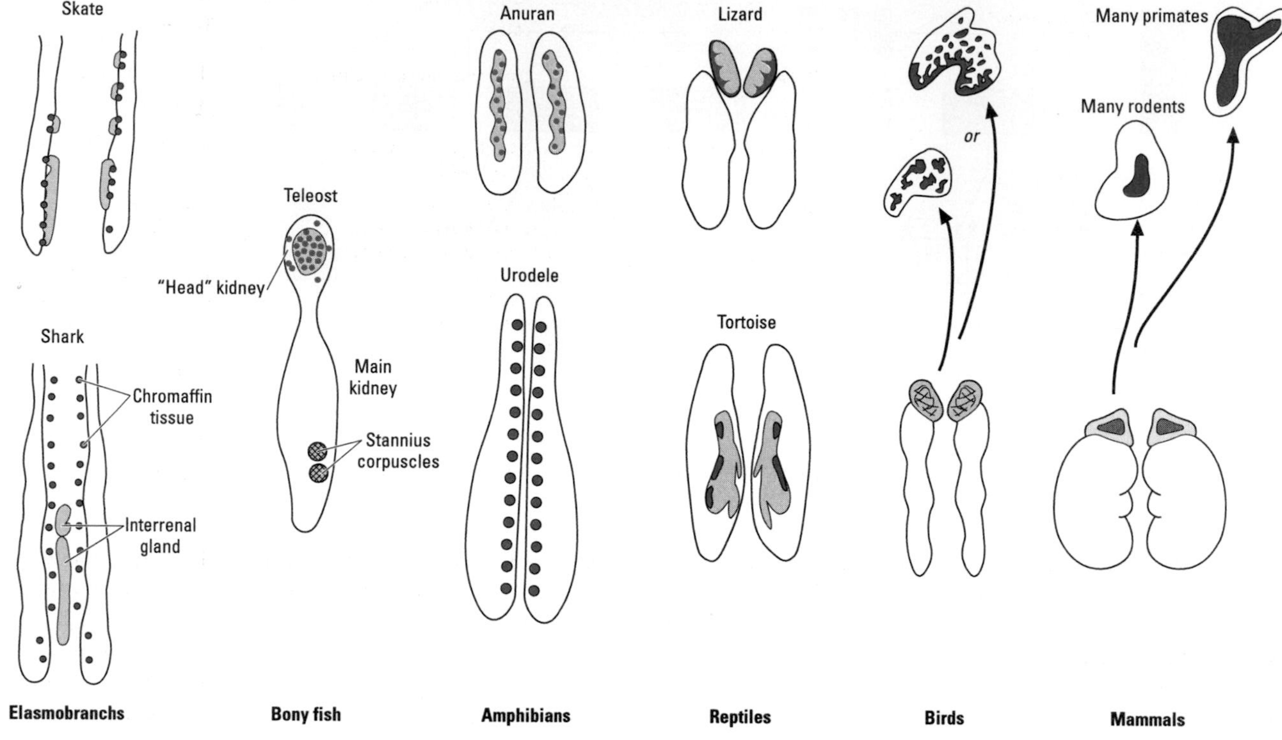

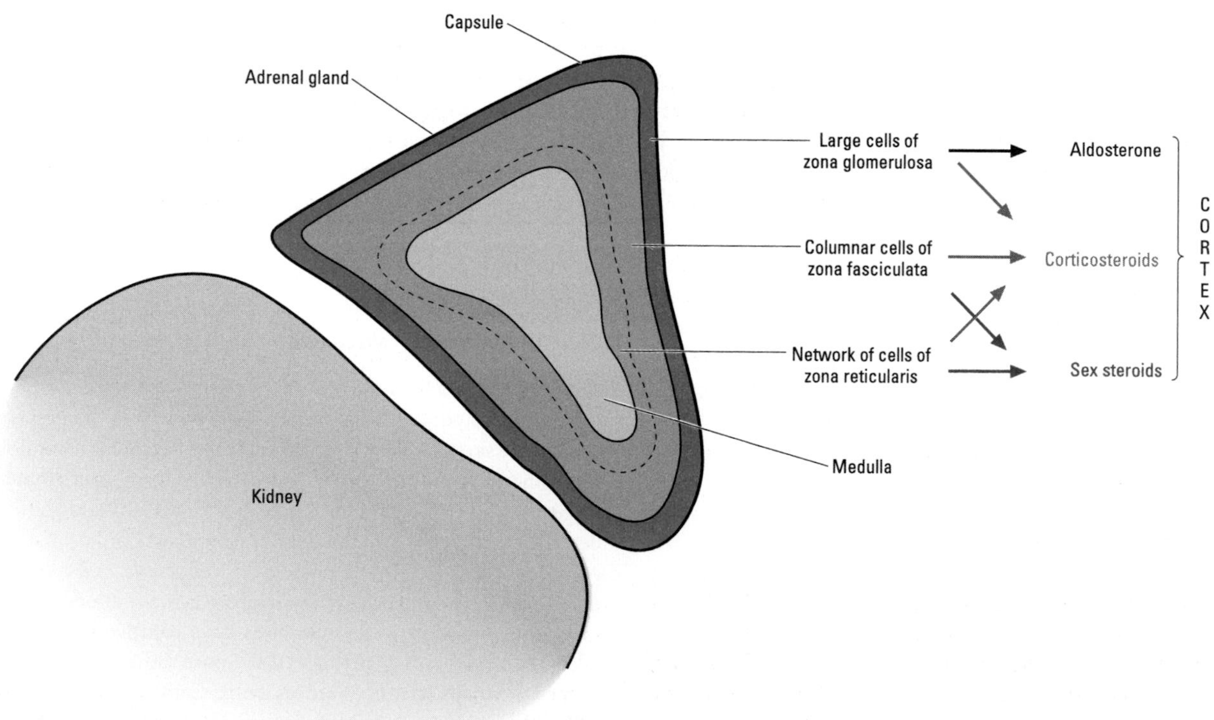

Fig. 10.13 (a) Different dispositions of the "adrenal tissues" in vertebrates, the interrenal glands becoming the adrenal cortex and the chromaffin tissue becoming the adrenal medulla. (b) Adrenal gland structure in a typical mammal, showing the three principal zones.

Table 10.6 Multiple effects of catecholamines.

	α effects	β effects
Effectors	Norepinephrine and epinephrine	Mainly epinephrine
Receptors	α1 and α2	β1 and β2
Second messengers	Increased IP_3 and DAG (α1) Decreased cAMP (α2)	Increased cAMP (β1 and β2)
Effects	Gluconeogenesis Decreased insulin secretion Vasoconstriction throughout Increased arterial pressure Tachycardia No effect on lungs Sweating (mammals) Piloerection (mammals) Ptiloerection (birds) Pupil dilation	Glycogenolysis Lipolysis Thermogenesis Increased insulin secretion Vasodilation in muscle and liver No change in blood pressure No change in heart rate Bronchodilation

cAMP, cyclic adenosine monophosphate; DAG, diacylglycerol; IP_3, inositol triphosphate.

to last several weeks, but the release of thyroid hormones depends on regulation by TSH from the anterior pituitary, in turn controlled by the hypothalamic releasing hormone TRH. Once released, the thyroid hormones are bound to globulins and albumins in the blood (including a specific thyroxine-binding globulin from the liver) for transport to the tissues. T_3 is usually about 10 times more active than T_4, and can have different effects at different sites: it may enter the cell and bind to intracellular receptors (like a steroid) when acting on some cells, but may have extracellular effects via cAMP on others.

The thyroid gland in mammals also has a small population of **parafollicular cells** (C-cells) that make a different hormone, **calcitonin**. However, the ancestral source of this hormone (as seen in all nonmammal vertebrates) is a separate gland, the **ultimobranchial body**, neurally derived and also located in the neck region. Calcitonin is a polypeptide that has a constant 32 amino acids in most species. In mammals it affects calcium levels, but in other tetrapods is more important in general salt balance.

The parathyroid glands are made up of "chief cells", secreting **parathyroid hormone** (PTH, also known as parathormone) a protein with 84 amino acids in humans, but rather variable between species. It bears the main responsibility for calcium regulation in all tetrapods (opposing calcitonin in mammals). Its release is triggered by falling blood calcium, via a specific calcium receptor on the chief cell membranes. PTH secretion is also increased by dopamine and by catecholamines.

Gut

The vertebrate gut is regulated by a number of chemicals, most of them polypeptides from the gut wall exerting a local effect (paracrine; see Table 10.1). These local effectors include histamine, somatostatin, and cholecystokinin (CCK). Other gut-regulating chemicals are truly endocrine, acting more distantly from their site of secretion. **Gastrin** and **secretin** are best known, but **cholecystokinin**, **gastric-inhibitory peptide** (GIP), **pancreatic polypeptide**, **motilin**, **neurotensin** and **enteroglucagon** have additional effects. The roles of these peptides vary in different parts of the gastrointestinal tract, and also vary with diet. The main targets in the gut wall are smooth muscle (changing gut motility and peristalsis), and secretory cells (altering rates of secretion of digestive enzymes and of pH-altering secretions).

Gastrin (a 17 amino acid peptide) is secreted by "G-cells" in the stomach wall, when stimulated by the vagus nerve as food is being eaten. It promotes secretion of acid from the stomach's parietal cells, and the release of enzymes and mucus from the gastric glands. Once the stomach pH falls below a fixed point (pH 2–3 in humans) further gastrin secretion is inhibited. Secretin (with 27 amino acids) is produced from the duodenal mucosa in response to acid, and reaches the stomach via the bloodstream where it helps to inhibit gastrin production. The duodenum also produces CCK and GIP in response to the presence of fat metabolites, and these factors again feedback on the stomach. Gastrin, secretin, and CCK all have additional influences in the duodenum, promoting secretion of an alkaline and mucus-rich fluid. Motilin is also produced there and interacts with neural inputs to help regulate peristalsis.

Pancreas

The pancreas, lying in the abdomen below the stomach, is both an exocrine gland serving as a source of digestive fluid (adding an enzyme-rich product to the gut) and a major endocrine organ, producing two hormones crucial to regulation of blood sugar levels. In lower vertebrates the exocrine and endocrine pancreas glands are separate, but in most tetrapods they are combined, the endocrine tissues lying within the main pancreas as "clumps" of hormone-secreting cells termed the **Islets of Langerhans**, comprising only about 2–4% of the pancreatic mass. The islets secrete the two main pancreatic hormones, **glucagon** from α-cells and **insulin** from β-cells. A third cell type (δ) produces small amounts of **somatostatin**, which has a limited inhibitory effect on the other hormones. These various hormones probably interact very directly with each other, since blood flow through each islet is from the central β-cells outwards to peripheral α- and δ-cells (Fig. 10.14). A pancreatic polypeptide is also produced, by a separate population of "F cells" (or "PP cells") mainly located in islets in the posterior part of the pancreas.

Glucagon is a small peptide, normally with 29 amino acids, and has a hyperglycemic effect, raising blood sugar levels. Its release

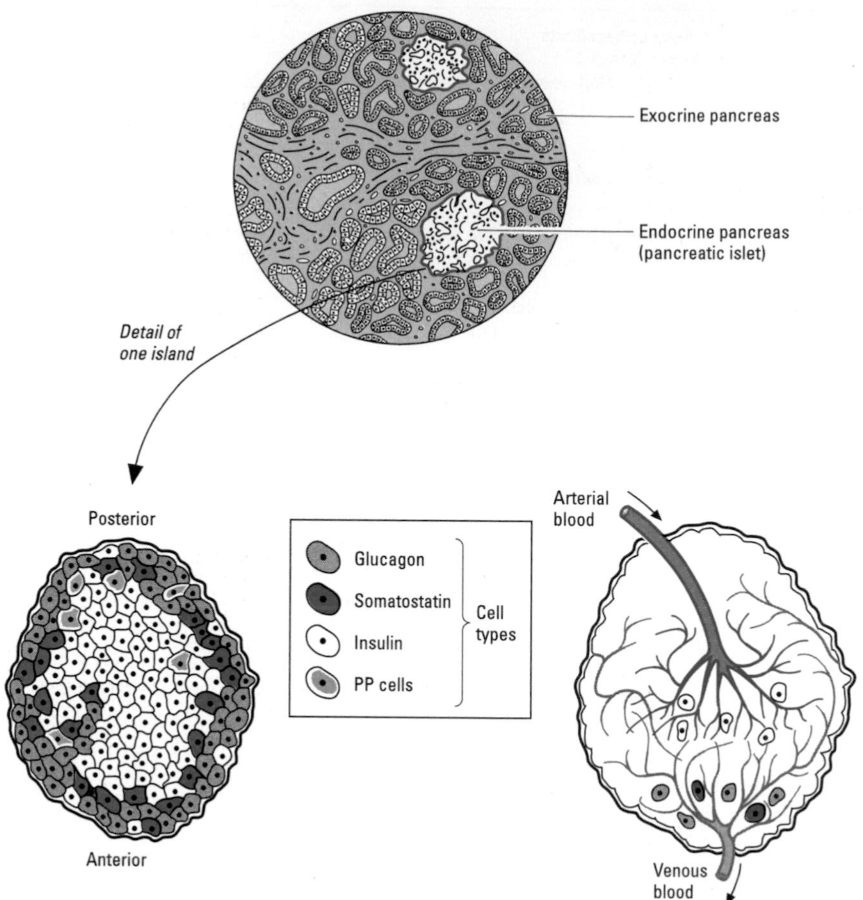

Fig. 10.14 A single pancreatic islet of Langerhans, with insulin-secreting β-cells concentrated centrally, and with the blood flow entering the core of the islet and draining peripherally.

from the pancreatic α-cells is primarily a direct response to circulating plasma glucose, but it is also stimulated by CCK from the gut, by certain amino acids, and by neural inputs (Fig. 10.15a). Insulin levels exert a negative feedback on glucagon release, perhaps very locally within each islet.

Insulin is again a small peptide, invariant at 51 amino acids in most mammals, made up from two peptide chains (21 in the α-chain and 30 in the β-chain). It is derived from a longer precursor (proinsulin), and carried in the blood bound to a β-globulin, where it has a half-life of only about 5 min because it is rapidly taken up into the liver, kidneys, fat cells, and muscle. It binds to glycoprotein receptors on cell surfaces that activate tyrosine kinase, inducing phosphorylation reactions that lead to insertion of glucose carriers into the membrane from intracellular reserves. Thus it can quickly lower blood sugar levels (a hypoglycemic action). Its release is promoted by increasing blood glucose, which causes a progressive depolarization of the β-cell membranes (where GLUT2 transporters are concentrated). Unusually, this leads to action potentials and thus the opening of voltage-gated calcium channels. Calcium therefore enters the cells, and elicits insulin release by exocytosis. The autonomic nervous system also has a regulatory effect, with catecholamines acting to reduce insulin release. A number of other hormones also have minor effects (Fig. 10.15b). Insulin secretion may increase 2–10-fold after a meal in a vertebrate, peaking anything up to an hour after feeding.

Kidney

The kidneys of vertebrates are important endocrine glands in terms of water balance regulation. The **renin–angiotensin system** is most important in tetrapod kidneys, with the hormone **renin** being released from the juxtaglomerular cells in the kidney cortex (Fig. 10.16) in response to various stimuli relating to blood concentration, composition, and volume. One key trigger is reduced stretch of these cells (a reflection of lowered blood pressure); another is direct stimulation by the renal sympathetic nerves. Renin is unusual in that it acts as an enzyme, breaking down the globulin angiotensinogen (made in the liver) into angiotensin I; this then enters the circulation and is converted to **angiotensin II** by an endothelial enzyme (especially in the lungs for mammals). Angiotensin II is a potent vasoconstrictor, activating smooth muscles and so causing blood vessels to narrow and blood pressure to rise. Hence enzymes that control angiotensin levels, including the "angiotensin-converting enzyme", ACE, are crucial in regulating the circulatory system (and ACE inhibitors are widely used therapeutically to control blood pressure). Angiotensin II also stimulates production of aldosterone in the adrenal cortex, causing renal tubules to begin resorbing more sodium from the urinary filtrate. The mechanism that starts with the kidney hormone renin thus primarily regulates blood pressure, but also stabilizes extracellular fluid volume and composition.

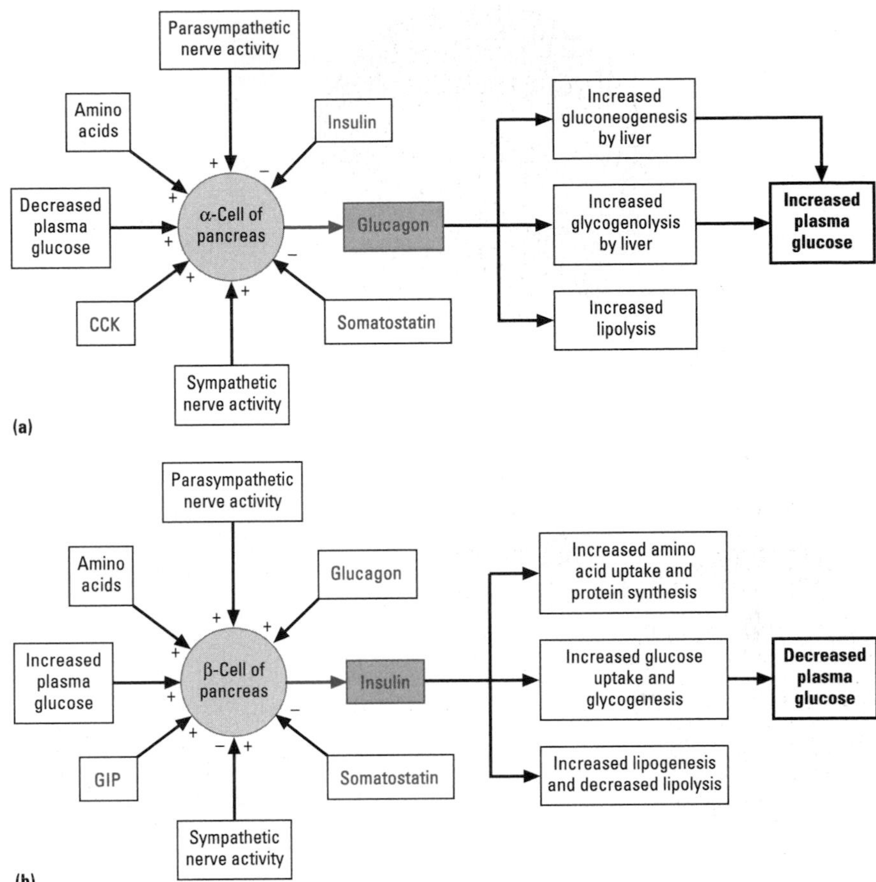

Fig. 10.15 Principal mechanisms for control of (a) glucagon, and (b) insulin, in mammals. CCK, cholecystokinin; GIP, gastric-inhibiting peptide.

Certain cells within the kidney also produce **erythropoietin** (Epo), a protein hormone that stimulates red blood cell production in bone marrow. Other kidney cells produce additional "hormones" with a very local (paracrine) influence, including **prostaglandins** that cause vasoconstriction or vasodilation, **nitric oxide** (a vasodilator), and **endothelin** which constricts renal blood vessels and inhibits renin release.

Gonads

Vertebrate gonads are invariably a source of hormones involved in the control of reproduction, and are themselves controlled by gonadotropic hormones from the anterior pituitary. In most vertebrates the ovaries and testes are the key sources of steroid sex hormones. All show the typical steroid mode of action described in section 10.1.2. Vertebrate gonads also produce various nonsteroidal factors, often glycoproteins, that help to regulate the pituitary–gonadal axis, and these include inhibins (as well as "activins" and various growth factors) although their functions are still not well known.

The ovaries in the female produce both estrogens and progestins. Estrogens are of several closely related kinds, and their relative frequency may vary in different parts of the reproductive cycle. They include estradiol 17β, estrone, and estriol, but we will refer to them collectively as **estrogens** for simplicity. By far the commonest progestin in most vertebrates is progesterone, and together with 17-OH progesterone this will simply be referred to as **progesterone**. Collectively all these hormones may control maturation of the ovaries, appearance of secondary sexual characters, egg development, and specific processes in higher tetrapods such as egg incubating in reptiles or birds or development of the mammary glands in mammals.

The testes produce androgens, primarily **testosterone**, which initiates male characters and sexual drive, maintains sperm production through adult life, and affects behavioral patterns such as aggression.

Another important source of hormones linked to reproductive function is the **placenta**, present only in gravid female mammals. During pregnancy this secretes both peptide and steroid hormones that maintain the fetus and influence pregnancy. Key products are again estrogens and progesterone, but also **chorionic gonadotropin** in primates.

Other endocrine glands

In elasmobranch fish there are giant neurosecretory cells in the spinal cord (termed Dahlgren cells) that are important in regulating water balance; in most teleost fish, cells in a similar region send out terminals to a nearby neurohemal organ called the **urophysis**. Neither of these structures is present in lungfish or in tetrapods. The urophysis secretes a number of peptides termed **urotensins** (UTs), having different effects: UTI regulates blood pressure, UTII affects smooth muscle contractility, UTIII is critical for sodium uptake at

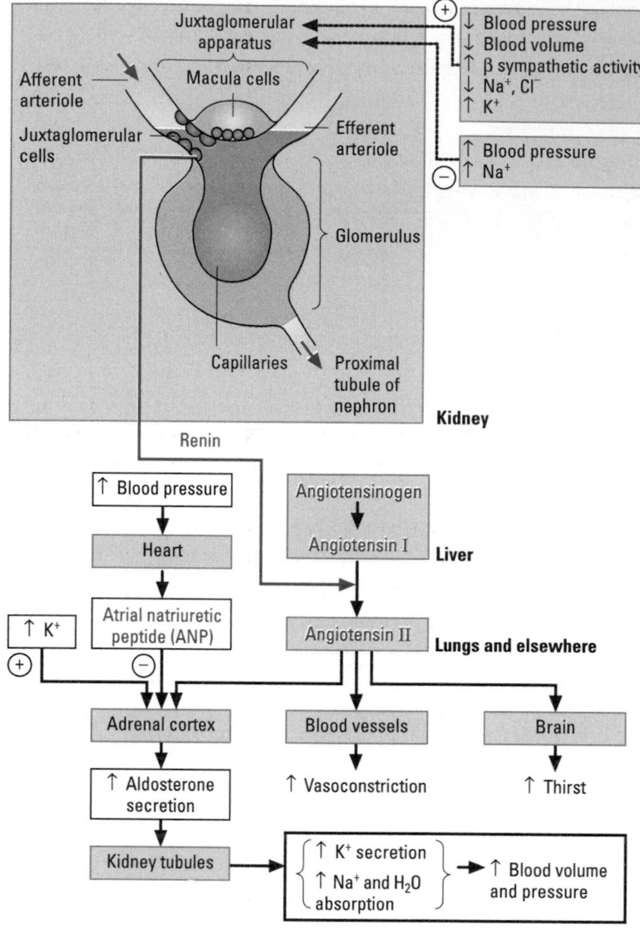

Fig. 10.16 The renin–angiotensin system based on the tetrapod kidney: from the site of production of renin in juxtaglomerular kidney cells, to the control of blood pressure and volume, and some components of salt balance.

the gills as part of the osmoregulatory response in nonmarine species, and UTIV has some antidiuretic effects.

The **pineal gland** (or epiphysis) lies in the brain, and in higher tetrapods is in the roof of the third ventricle. It produces the hormone **melatonin**, synthesized from tryptophan. In most vertebrates (excluding only mammals and snakes) the pineal gland includes photoreceptor units with neural connections to the brain, and is directly sensitive to light; indeed in some fossil vertebrates and a few living species it protrudes from the upper brain surface and is covered only by a thin layer of epidermis. However, in mammals it only receives information on the light cycle from the eyes, via neurons from the suprachiasmatic nucleus of the hypothalamus (an area often termed the "biological clock"); it does not have any outward neural connections. In response to the photoperiod, melatonin appears in the blood with circadian cyclicity, peak levels occurring in the inactive periods (night-time for diurnal animals). The secreted melatonin can mediate seasonal changes in gonad size and mating behavior, and probably also influences many other rhythmic physiological processes such as sleep, appetite, or body temperature. Its many roles are only just being discovered; it may have an inhibitory effect on gonadotropins in young animals (preventing precocious sexual maturation), it may alter ACTH and vasopressin secretion, and it has been implicated in stimulating the immune system.

The **thymus gland** lies in the neck or upper thorax, and in juvenile mammals it produces various peptide hormones (**thymopoietins, thymosins**). These are essential for the normal development and activity of T-cells, blood lymphocytes that mediate the immune response (see Chapter 17). The immature lymphocytes migrate from bone marrow to the thymus, where they mature under the influence of the thymic hormones and so become "immunocompetent". Then they migrate out to lymph nodes, spleen, and other lymphatic tissues. The thymus regresses as animals mature and is largely nonfunctional in older vertebrates.

The **heart** has one key product that acts as a hormone, a peptide termed **atrial natriuretic peptide** (ANP). This is an important downregulator of blood pressure and blood volume, released in response to stretch of certain atrial cells, and it promotes both water and salt secretion (i.e. diuretic and natriuretic effects) by opposing the release of all other kidney-affecting hormones. Other natriuretic peptides (notably VNP and CNP) may also be produced, and have important roles in fish and mammals.

Finally, several more diffuse tissues produce hormones. The **skin** (in terrestrial vertebrates, when receiving sunlight) produces **vitamin D**, an important prohormone for calcium regulation. The **vascular endothelium** produces several factors similar to those from the kidney that are important in blood pressure regulation, including **endothelins**, which are vasoconstrictors, **nitric oxide**, and various **prostaglandins**, which also influence smooth muscle in nonvascular sites. The **adipose tissue** also secretes the hormone **leptin**, mainly acting on receptors in the hypothalamus to regulate eating behavior and energy balance.

10.3 Control of water and osmotic balance

Many epithelia involved in osmoregulation are under hormonal control, with various possible effects: regulating permeability to particular ions or water, controlling the rate of transport of one or more ions, or changing the direction of this transport. Most vertebrates require particularly constant body fluid composition, not least because their hemoglobin is very sensitive and can easily suffer decrements in its oxygen-loading capacity if the blood alters in osmotic, ionic, or pH status. Hence they regulate both fluid volume and fluid concentration very closely and often independently.

Mechanisms of hormonal control of salt and water balance in most taxa involve induced changes in sodium pumps and sodium channels. Two distinct protein kinases are known that inhibit the activity of the crucial sodium pump (Na^+/K^+-ATPase) by phosphorylating its α-subunit, and these kinases are affected by such hormones as norepinephrine, angiotensin, dopamine, and insulin. A wide range of other molecules are known to affect the P-ATPase family (see Chapter 4), some directly and others (such as fatty acids and prostaglandins) via indirect effects on lipid–protein interactions in the membrane. Hence there are multiple possibilities for control pathways.

Most animals have a large part of their water balance control centered on their skin (for many marine and parasitic species) or in particular parts of it such as the gills (see Chapter 5). In contrast, some freshwater invertebrates and most terrestrial animals have rather impermeable skins and instead concentrate their water balance control on the excretory organs (kidneys), again discussed in Chapter 5. Gills primarily regulate salt uptake and secretion, while the kidneys regulate blood osmolality by varying the amount of water resorbed.

We therefore need to look at different taxonomic and environmental categories separately, even though similar hormones may be acting on the different sites.

10.3.1 Aquatic animals: water and salt exchanges at the skin and gill

Invertebrates

Many invertebrates have hormones that are called either diuretic hormones or antidiuretic hormones; these are respectively usually known as DHs and ADHs, but there is little evidence that they are related in different taxa. The names are often inappropriate anyway, as they may not affect diuresis at the kidney but instead alter skin uptake or loss of water.

Euryhaline flatworm species have some osmoregulatory neuropeptides, which are activated during exposure to lowered salinity; and a neurohormone regulates nemertine water balance, both in aquatic worms subject to low salinity and in semiterrestrial species. Currently we know of several osmoregulatory peptides in annelids: a neurohormone from the cerebral ganglion influences osmoregulation in earthworms, while in leeches several peptide hormones from other animals (including vertebrates) can alter skin sodium uptake, though endogenous hormones are not yet identified.

In molluscs, an ADH has been described for the marine sea slug *Aplysia*, allowing rapid water uptake through the skin in hyperosmotic media. In contrast, freshwater snails such as *Lymnaea* have a diuretic factor, secreted by "dark green cells" in the ganglia, which become inactive when the snails are in more concentrated media. The peptide hormone released from these neurosecretory cells has some sequence similarity with TRH in vertebrates, and the axons project mainly to the skin, where they presumably enhance water loss (though they may also act at the kidney to increase urine flow). Another neurohormone controls ion resorption in the renal tissue of these freshwater snails. Amongst terrestrial slugs, which still have rather permeable skins, a factor released from the pedal ganglion has a role in enhancing water permeability of the skin, allowing rapid water loss in excessively damp soils or after temporary submergence by heavy rains.

Crustaceans live in various different environments and thus require a range of salt-handling hormonal systems. Species that hyperregulate in brackish water have pericardial organs that secrete amines (including dopamine and 5-HT), working via cAMP to increase Na^+/K^+-ATPase and salt uptake at the gills. Freshwater species may have neurohormones in the eyestalks that control extracellular amino acid levels as part of the osmoregulatory response (see Chapter 5). Species exposed to very hyperosmotic media release a diuretic factor from the eyestalks that leads to elimination of excess water; the extreme case is the brine shrimp *Artemia* (see Chapter 14), where the brain also secretes a neurohormone promoting salt retention at the gills. Some semiterrestrial crabs, unable to use gills for breathing, nevertheless redirect their urine flow over the reduced gills (see Chapter 15). When fresh water is available the gills can resorb salt and the urine concentration is lowered (to only 5% of that of the hemolymph), whereas when the crab is drinking sea water, amines may trigger reduced inward sodium fluxes at the gills.

Vertebrates

In "lower" aquatic vertebrates (primarily fish and larval amphibians), control of water balance at the skin and gills remains important, in addition to thirst and drinking mechanisms when the environment makes this possible. Here an important role in water balance is played by prolactin, which occurs as a peptide of 185–188 amino acids in teleosts (up to 204 in nonteleost fish). This hormone reduces the losses of sodium and water at the gills and in the gut, with a minor effect on the kidneys. Adrenal cortex hormones also affect water balance, by altering sodium uptake systems, but their effects are differently located according to taxonomic status and environmental need. In elasmobranchs they mainly work on rectal glands, in teleosts on the gills and gut, in amphibians on the skin and urinary bladder. The posterior pituitary neurohormone AVT and its variants (equivalent to ADH) have relatively little influence in teleost fish, being only mildly antidiuretic by reducing the glomerular filtration rates. In elasmobranchs the AVT variants (of which there are many within the taxon) have almost no effect, perhaps related to the peculiar urea-retaining and rectal gland-based form of osmoregulation found in these animals (see Chapter 11). Here the rectal salt gland is stimulated by vasoactive intestinal peptide (VIP), which stimulates chloride secretion by activating Na^+/K^+-ATPase, enhancing the symport pathway handling Na^+ and Cl^-, and increasing the chloride permeability of the apical cell membrane. Additional VIP is released as a neurosecretory product of nerves directly innervating this salt gland, giving even faster control.

Many fish have a considerable capacity for osmoregulation and a high tolerance of brackish conditions; prolactin is often particularly important in these euryhaline species, and is sometimes described as the "freshwater-adapting hormone" (though not all freshwater and brackish fish respond to it). A few teleosts (e.g. salmon, eels, catfish) are even more euryhaline, some of them alternating between sea water and fresh water through their lives; here, additional hormonal control may be needed. In the catfish, analysis of mRNA levels during different salinity exposures indicates that GH is most important during adaptation to salty media, with a lesser role for cortisol, both mainly acting by stimulating the extrusion of salt at the gills by increasing Na^+/K^+-ATPase levels. However prolactin is still the main control during adaptation to fresh water, in this case increasing salt uptake at the gills. In eels and perhaps other migratory fish the natriuretic peptides (ANP, VNP, and CNP), mainly derived from the heart, are additional regulators of sodium extrusion. ANP is maximally expressed during adaptation to sea water, and CNP is more important when eels are exposed to fresh water. The natriuretic system may thus be a key factor in achieving euryhalinity.

10.3.2 Terrestrial animals: water balance via the excretory system

Invertebrates

Regulation via the excretory system is relatively unusual here, largely confined to the truly terrestrial arthropod invertebrates with urine production in various structures such as the Malpighian tubules, coxal glands, or antennary glands. Only the insect regulatory system is well studied.

Insects control their water balance mainly through the Malpighian tubule/rectal apparatus (described in detail in Figs 15.13–15.15). Isosmotic urine is produced by an active pump (V-ATPase) secreting cations into the tubule lumen, with water and other solutes following passively via cellular or paracellular routes, and the only real control here is on the *rate* of secretion and hence the rate of urine flow. Secretory activity is controlled by many hormones—in some insects at least seven classes of active regulators have been found. There are two main families of insect diuretic peptides: one group (related to vertebrate CRH) acts mainly in the lower Malpighian tubules via cAMP to cause changes in water permeability (P_w) and the V-ATPases, while another group contains kinins acting on the upper Malpighian tubule via calcium, raising chloride influx via cotransporters and/or increased apical P_{Cl}. In many insect species these peptides are released from the mesothoracic ganglion; in locusts two peptides, one of each family, act synergistically, while in other species one peptide acts synergistically with amines such as 5-HT to give a rapid-onset diuresis. Only two tubule-controlling peptides outside of these two families have so far been identified: a cardioactive and diuretic peptide in *Manduca*, and a calcitonin-like peptide in cockroaches. For all these hormones, release is triggered when a large watery meal is ingested, and the key effect is to increase tubule secretion. ADHs are also known in some species; for example the flour beetle *Tenebrio* has both DH and ADH peptides, acting via cAMP and cGMP, respectively, allowing for fine control of urine production. For at least a few species prostaglandins may also have some influence.

However, hormonal control is also needed at the resorption site, which is the rectum. Here an electrogenic chloride pump usually provides most of the gradient for the passive uptake of cations and for water resorption. Since primary urine is isosmotic, regulation of water balance must occur rectally. A peptide from the corpora cardiaca is known to control ion uptake in the rectal epithelium of locusts, so stimulating ion resorption and associated water retention. Similar effects are reported in insects with cryptonephridial rectal complexes that permit maximal water resorption. It is unclear how far rectal sites are regulated by the same DH and ADH factors that operate on the tubules or by separate hormones. However, in *Manduca* there are distinct tubule DH peptides and rectal ADH peptides to produce a drier excretory product.

In at least some insects (e.g. cockroaches and orthopterans), hormones from the brain that can reduce cuticle permeability have also been identified, and these will also have an influence on water balance.

Overall mechanisms to control water balance via these various sites remain largely speculative. Feeding can be a stimulant for DH release, and possibly also for ADH, so that the net effect is a faster cycling of fluid between tubules, hindgut, and hemolymph and a quicker clearance of potentially noxious compounds (hence the idea that DH is really acting as a "clearance hormone"). Possibly hygroreceptors also feed into the control circuitry, especially in relation to cuticular permeability.

Vertebrates

In the "higher", and mainly terrestrial, vertebrates prolactin, which was crucial for fish osmoregulation, takes on different roles mainly related to reproduction (see section 10.7), and to a lesser extent the same is true of the adrenal cortex steroids, although these do act at the salt glands in reptiles and some birds. The most important antidiuretic for terrestrial taxa, including the more terrestrial amphibians, is AVT (remember that the vertebrate ADH is usually arginine vasotensin, replaced by AVP in mammals, but collectively often termed vasopressin or ADH). This mainly acts by promoting nephron resorption of water (whereas in fish its main effect was decreasing the glomerular filtration rate). ADH prevents wide swings in water balance, avoiding excessive dehydration or water overload; low ADH levels result in copious watery urine, and high ADH levels lead to a small amount of concentrated urine. The release of ADH is triggered (Fig. 10.17) when osmoreceptors in the hypothalamus are depolarized by increasing blood osmolality, directly activating the neurons of the supraoptic nuclei that release the ADH. Additional triggers are reduced blood volume without change of blood concentration (e.g. due to hemorrhage after injury), and certain kinds of pain and stress.

The main physiological action of ADH, to facilitate water resorption from the collecting ducts (and the lower distal tubules) of the kidney, is effected in mammals by increasing the P_w of the cell walls. This is achieved by binding to receptors in the basal walls of the principal cells (P-cells), which are coupled to adenylyl cyclase, so increasing cAMP levels and activating a protein kinase. This phosphokinase then initiates fusion of "leaky" intracellular membrane vesicles containing extra water channels (porins) with the cell's apical membrane. Water then moves down its osmotic gradient from the lumen to tubule cell and thence to the interstitial fluid and ultimately to the blood, this water movement being independent of any active solute movements. The outcome is an increased urine concentration, and a decrease in flow and volume.

ADH also stimulates sodium resorption (by activating some quiescent Na^+ channels), and transport of urea (via urea channels) from the lumen of the collecting ducts to the interstitial fluid of the kidney, helping to maintain the crucial osmotic gradient from the cortex to medulla that permits urinary concentration by countercurrent multiplication (see Fig. 5.17 and Box 5.2). In addition, ADH has vasoconstricting effects on blood vessels (hence the alternative name vasopressin) and causes a reduction in heart rate with a reduced cardiac output. These effects can be very important during severe dehydration. In these roles, ADH binds to V_1-type receptors and acts via cAMP, whereas in its renal actions it interacts with V_2 receptors and via IP_3 and DAG pathways (see Table 10.2).

Blood volume is an important aspect of water balance, and is detected mainly by sensing alterations in blood pressure (mainly via receptors in the carotid sinus and aortic arch, and via arterioles in the kidney and pulmonary circulation). All these receptors can

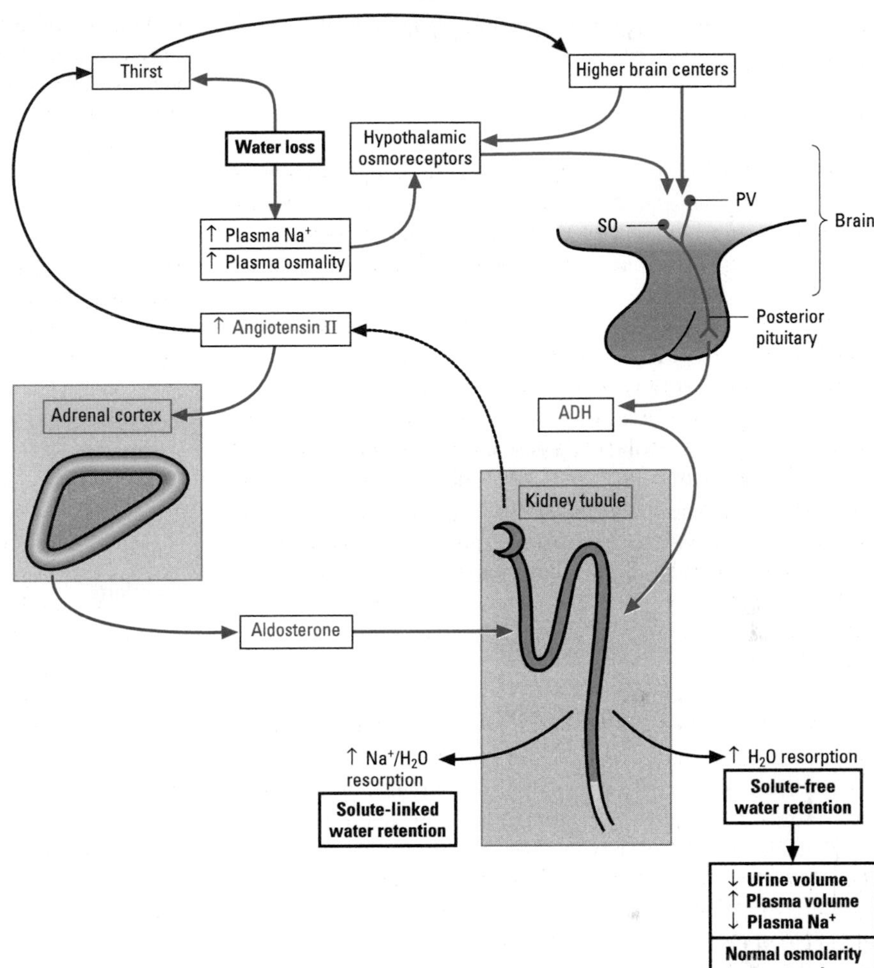

Fig. 10.17 Primary mechanisms for the triggering of vasopressin (antidiuretic hormone, ADH) release, and its role in the control of tetrapod water balance. Note the integration with other salt and water controlling hormones as in Fig. 10.16. PV, paraventricular nucleus; SO, supraoptic nucleus (both in the hypothalamus).

trigger increased activity in the sympathetic nervous system and especially in the renal nerve, leading to increased secretion of the hormone **renin** leading (see section 10.2.2) to increased production of **angiotensin II** and extra secretion of **aldosterone.** This latter hormone is also crucial to terrestrial vertebrate water and salt balance. It conserves salt in the body by increasing the P_{Na} of the apical membrane in the distal convoluted tubule, thus stimulating the resorption of sodium ions (in exchange for potassium ions) in the distal nephron. In the short term this involves activation of preexisting quiescent sodium channels from vesicles within the cell (see Chapter 4), but during prolonged exposure there is synthesis of new channel proteins and then of additional sodium pumps in the P-cells of the distal convoluted tubules and loops of Henle. Without aldosterone, sodium and chloride levels fall rather rapidly, leading to a reduction of extracellular fluid volume and thus of blood volume.

In addition, muscle cells in the heart atria secrete **atrial natriuretic peptide** as a response to increased blood pressure. ANP then acts on the kidney: it dilates renal arterioles to increase blood flow, it antagonizes the renin-angiotensin system to inhibit aldosterone, it directly inhibits sodium uptake, and it inhibits ADH secretion, so promoting water loss. Thus its actions all tend to reduce blood sodium and hence reduce blood volume.

Finally, some degree of regulation of water balance in terrestrial vertebrates is focussed on water *intake*, usually via thirst and drinking behavior. The hypothalamic osmoreceptors produce a desire for drinking, because they trigger vasopressin release and this regulates water-seeking behavior by its effects on the brain. In mammals, angiotensin II also stimulates thirst and drinking. In frogs and toads water uptake is achieved by "cutaneous drinking" (especially at the pelvic patch, see Chapter 5), also stimulated by angiotensin II and by AVT, which also alters the water permeability of the storage bladder. A further role is taken by a unique group of hormones, the **hydrins**, which are metabolites from pituitary provasotocin; these affect P_w *only* at the skin and bladder (probably again by incorporation of porins into the epithelial membranes) but have no antidiuretic effect on the kidney.

10.4 Control of ion balance and pH

Water balance is almost invariably intimately associated with salt balance, and much of the control discussed in section 10.3 is therefore relevant to control of sodium, potassium, and chloride. In primarily aquatic species, gill regulation is most important for both cations, with corticosteroids and natriuretic peptides being major control agents for sodium in fish.

For terrestrial animals, sodium is mainly regulated by the degree of its renal *resorption*, precisely tailored to need and controlled largely via aldosterone. Potassium is usually more invariant (90% is always reabsorbed early in kidney tubules), with regulation achieved in the collecting ducts by changing *secretion* into the final urine. For species with high potassium diets (omnivores and herbivores) there is some independent regulation of K^+ in the distal tubules, at the expense of sodium levels. Thus if blood potassium rises the collecting duct cells gradually accumulate the ion and begin to secrete more K^+ into the duct lumen. Additionally, aldosterone stimulation of Na^+ resorption inevitably enhances K^+ secretion, via Na^+/K^+-ATPase.

Other ions may require separate and more specific control systems. Little is known in invertebrates, and perhaps less control is required there for the "minor" cations. But vertebrates are unusual in having an internal calcitic skeleton, thus requiring special hormonal mechanisms for the control of calcium and phosphate levels. They also require close control of blood and tissue pH, again partly because of the susceptibility of hemoglobin to ionic disruption.

Calcium

About 99% of a vertebrate's body calcium is in the bones as stable calcium hydroxyapatites, with a tiny fraction near the bone surface as much more exchangeable calcium phosphate, and it is from this fraction that calcium can be mobilized if needed. Whilst intracellular calcium levels are extremely low (as in all animals), levels in the blood are perhaps 10^4 times higher and must be kept very constant, because calcium plays important roles in stimulus–secretion coupling, in excitability and muscle contraction, in cell adhesion, and in blood clotting. Low blood calcium leads to tetany or uncontrolled spasms in skeletal muscle. Yet calcium levels may have to vary with life cycle, for example to allow eggshell production, or antler and horn growth, so that a responsive regulatory system is essential.

In fish calcium regulation is achieved mainly at the gills, the corpuscles of Stannius (see Fig. 10.13a) secreting a hypocalcemic hormone (**hypocalcin**) that enhances calcium uptake at the gills. In most other vertebrates, Ca^{2+} is regulated at the gut, kidney, and bone mainly by **parathyroid hormone** (PTH) from the parathyroid gland. In mammals **calcitonin** (from the C-cells of the thyroid) acts antagonistically to this (though in other tetrapods calcitonin comes from the ultimobranchial glands, and has little effect on plasma calcium). In addition, metabolites of **vitamin D** (cholecalciferol) act as calcium-regulating hormones in many vertebrates. The control systems in a mammal are summarized in Fig. 10.18.

PTH works throughout an animal's life to increase calcium in the blood while decreasing phosphate levels, and its production is directly stimulated by low blood calcium. Normally it enhances the production of osteoblasts (bone-forming cells) so that more bone matrix is formed. But if calcium levels become unusually low three further effects occur: mobilization of the exchangeable surface calcium from bones, then increased bone resorption by osteoclasts, and finally increased calcium resorption in the kidneys.

In contrast, mammalian calcitonin is secreted in direct response to high blood calcium, and lowers this by targetting the skeleton, inhibiting osteoclast activity, and stimulating calcium uptake into bone matrix. It works very rapidly, but is often only effective in juvenile animals, having little effect beyond puberty.

Vitamin D is a prohormone, and is hydroxylated to form steroid hormones that alter calcium levels. A first hydroxylation occurs in the liver, and a second in the kidney to give "**calcitriol**". This reacts with intracellular receptors, attaching to a specific region of DNA to alter gene expression (as with other steroids; see section 10.1). The main effect is increased active uptake of ingested calcium from the gut (Fig. 10.18) at the intestinal brush border, and some stimulation of calcification in bone. The production of calcitriol in the kidney is stimulated by PTH, so it could be said that by this route PTH also acts indirectly on the gut to raise blood calcium, as well as on the bones and kidneys.

A number of other hormones affect calcium balance, but the most important extra factor appears to be estrogen level. In adult females, estrogen inhibits the PTH-mediated bone resorption process, and stimulates osteoblasts instead. Thus in postmenopausal mammals, including humans, the loss of the main estrogen supply can lead to excess bone resorption and bone fragility (osteoporosis).

Phosphate

Here close regulation is required for many animals, since phosphate is related to the concentrations of vital high-energy compounds and involved in pivotal phosphorylation reactions. As with calcium, vertebrate skeletons contain a large proportion of their total phosphate, and therefore act as a reservoir. However, regulation of blood phosphate is primarily via effects on gut absorption, in response to parathormone and calcitonin. Minor contributions come from the actions of these and other hormones on bone in concert with the calcium regulation systems already described, and from PTH effects on the kidney where phosphate excretion increases.

Hydrogen ions and pH

Another important aspect of ionic balance is the regulation of acid–base balance (**pH**) in the body. Chemical buffering in the blood of vertebrates (by bicarbonate, phosphate, and protein buffers) is the major rapid control system. The respiratory center in the brain is also crucial, regulating the depth and frequency of ventilatory movements at the gills or lungs and so the rate of removal of carbon dioxide (see Chapter 7). However the kidney also plays a role in regulating pH, on a scale of hours or days, and can remove phosphoric, uric, and lactic acids, and some ketones. The main strategy used by the kidneys is to change the level of excretion, resorption, or generation of bicarbonate ions (see Fig. 5.18), depending on the rate of secretion of protons (H^+ ions) into the kidney filtrate, mainly in the proximal convoluted tubule; this is therefore partly controlled by aldosterone effects on sodium transport.

10.5 Control of development and growth

10.5.1 Development and metamorphosis

All animals go through a process of development, largely controlled by growth factors and not strictly relevant here (although many growth factors share chemical similarity with hormones mentioned in this chapter). The extent to which animals also show metamorphosis during growth is highly variable, largely depending on

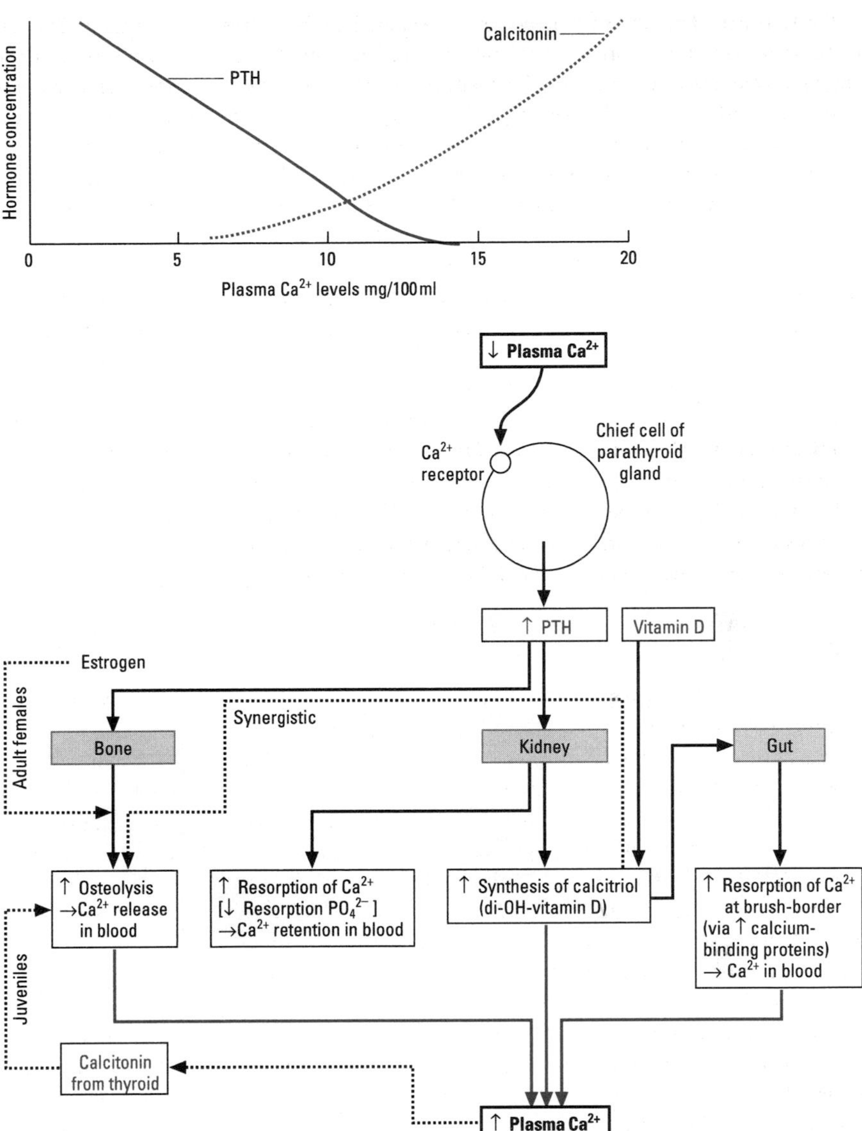

Fig. 10.18 The control and actions of calcium-regulating hormones in tetrapods, including the effects of vitamin D derivatives. The graph above shows the relative levels of the two main hormones in the blood of a mammal as calcium levels change.

changes in either habit or trophic relationships. Many aquatic invertebrates have pronounced metamorphosis, allowing an alternation of pelagic larval phases (feeding on plankton or on stored yolk) with sessile benthic filter-feeding adults. Many parasites also have morphologically very different larval phases to cope with finding and entering intermediate hosts. Terrestrial invertebrates tend to have a less drastic growth process, except in the flying insects where the mature animal is largely a reproductive and dispersive phase and less mobile larvae are the feeding phase. In vertebrates pronounced metamorphosis is rare, although an extreme version occurs in amphibious anurans to cope with their transition from aquatic herbivorous tadpole to terrestrial carnivorous frog or toad.

Insects

Arthropods are particularly amenable for study of metamorphosis because their morphology is almost entirely a function of their cuticle. Hormones act mainly on the epidermal cells to produce the relevant cuticle morphology at any particular stage in the life cycle. The insects are best studied because of their strongly metamorphic life cycle (Fig. 10.19), with a particularly drastic reorganization of the body where there is a pupal phase. Here metamorphosis can interact with reproductive processes, because mating often has to happen as soon as possible after the molt into the adult form; understanding it is therefore an area of huge applied importance in the development of control measures for pest species. The same applies to crustaceans because of their growing importance in aquaculture.

The insect molt is initiated from the brain, which produces the neuropeptide **PTTH** (prothoracicotropic hormone) from neurosecretory cells (see Fig. 10.7). These cells release their product to the corpora cardiaca. The stored hormone is then released in relation to various environmental stimuli (light intensity, photoperiod, temperature, food supply, and possibly population size), and in turn induces release of the classic molting hormone **ecdysone** from the prothoracic gland. In fact the main product released is α-ecdysone, an inactive prohormone, which is only transformed into the active β-ecdysone form when it reaches its targets at the epidermal cells.

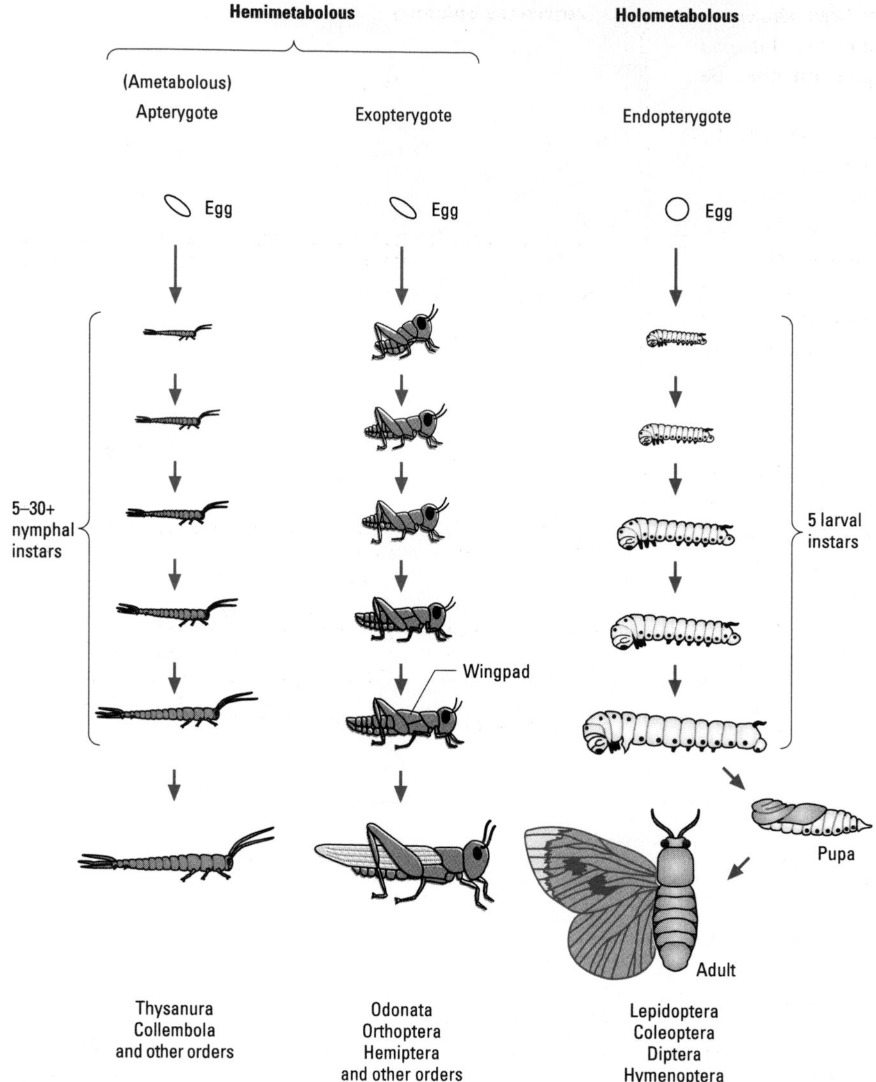

Fig. 10.19 The two main types of insect metamorphosis: hemimetabolous in exopterygotes (and also in apterygotes), with little change of form at each molt except for the growth of the exopterygote wing pads; and holometabolous in endopterygote groups, where larval instars are similar but there is a drastic change of form within the pupal stage to give a winged adult.

Molting itself is a complex process, involving separation of the old cuticle from the epidermis (apolysis), partial breakdown and resorption of the old cuticle (ecdysis) by hydrolase and protease enzymes, then enlargement of epidermal cells and secretion of the new cuticular layers. Final molting occurs as the next instar emerges, using stereotyped behaviors leading to splitting and shedding of the old covering. These behaviors must be fast, and are therefore brought about by a positive feedback loop between **eclosion hormone** (EH, a 62 amino acid peptide) released from the brain neurosecretory cells via the corpora cardiaca, and **ecdysis-triggering hormone** from epitracheal glands. EH itself is apparently triggered by falling levels of ecdysone. Following emergence the new skeleton is pale, crinkled, and flexible; it still needs to be stretched (by the uptake of air or water into the body), given a superficial protective layer, and finally hardened into a functioning exoskeleton. EH may be responsible for initiating secretion of the cement layer onto the surface of the cuticle. At about the same time, cardioactive peptides from the corpora cardiaca may increase in the blood, acting via cGMP to give rapid heart rate and hence faster circulation of blood into the expanding wings, along with a general expansion of the abdomen as air is taken in. Then the processes of hardening the cuticle by cross-linking the protein components (tanning) can occur. Hence, shortly after EH has acted, another hormone, **bursicon**, is released from neurohemal organs in the abdomen (at least in some species) to initiate tanning of the new cuticle. In flies (Diptera) at least three other neurohormones have been identified that control aspects of puparium formation and emergence from the pupal stage.

The developmental change through successive molts, in both hemimetabolous and holometabolous insects, is produced semi-automatically under the influence of ecdysone, which brings about programed cell death (apoptosis) in larval tissues, while enhancing the development of adult tissues. Ecdysone binding to receptors leads to the activation of specific sets of "**early response genes**" at each molt; pulses of ecdysone occur at intervals in the pupal stage and trigger prepupal to pupal and then pupal to adult transitions. Ecdysone-induced apoptosis has recently been shown to operate largely through a single gene (*E93*) in *Drosophila*, this gene product binding to other sites where gene expression must be regulated.

However, the *rate* of insect metamorphosis is controlled by the terpene **juvenile hormone**, JH, from the corpora allata. This is effectively a hormone that keeps the animal's cuticle (and hence its overall morphology) in the juvenile form; that is, it maintains the expression of juvenile genes and represses adult genes. Its level usually falls off progressively (Fig. 10.20) to allow increasing expression of adult characters. Removing the corpora allata in fifth instar larvae gives rise to a miniature adult, and adding extra JH gives an extra (sixth) and abnormally large larval stage and then a giant adult. However, JH cannot bring about "reverse metamorphosis"; its effect is only to inhibit further morphogenesis. There are complex feedbacks in the control of metamorphosis (Fig. 10.21) with release of JH regulated both by neuropeptides (allatostatin and allatotropin) and by direct neural inputs (possibly glutamatergic) to the corpora allata. JH exerts additional direct negative feedback, perhaps both onto the brain and onto the release site at the corpora allata; and ecdysone also has a reciprocal feedback with the brain on JH production. This level of complexity is crucial to achieving fast molts and orderly metamorphic progression.

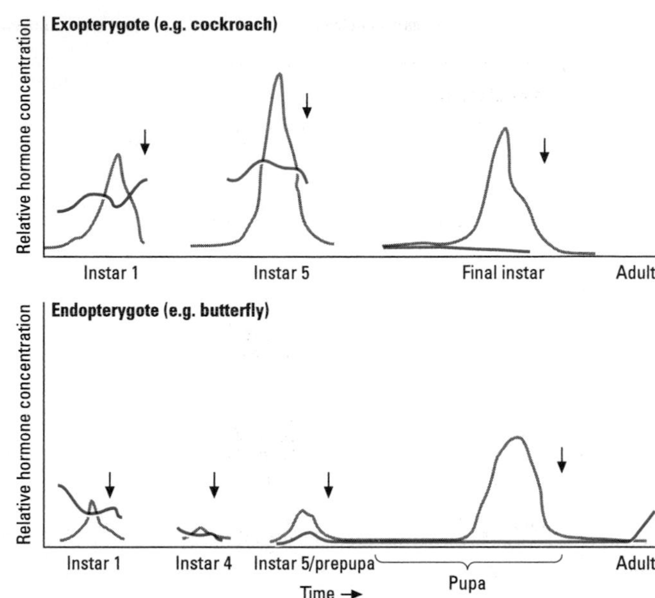

Fig. 10.20 A schematic view of the levels of hormones through typical insect molt cycles. The time of ecdysis (arrows) is always preceded by a surge of ecdysone (black), whereas levels of juvenile hormone (green) progressively decline, especially just before the ecdysis. In an exopterygote this juvenile hormone effect is often less clear-cut until the adult stage is reached.

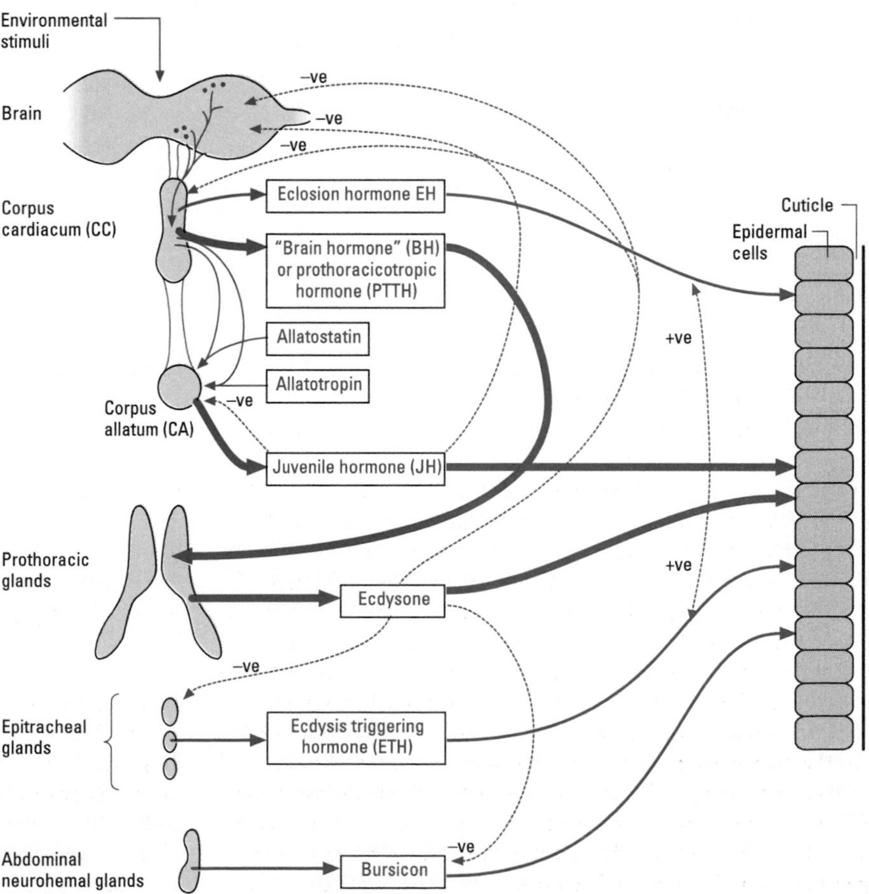

Fig. 10.21 The complex controls involved in the insect molting cycle, with multiple feedback loops between different hormones. The main axis involving prothoracicoctropic hormone, ecdysone, and juvenile hormone is shown with bold arrows, and feedbacks are shown with dashed lines. The other hormones shown probably do not occur in all insects.

Crustaceans

The crustaceans usually differ from insects in having many more larval stages, commonly of very different morphology, and with a much greater change of size. However, they too must molt their exoskeleton periodically, and molting goes on even through adult life. Furthermore the exoskeleton is often heavily calcified, so growth can *only* occur at the molt.

The crustacean eye stalk X-organ (see Fig. 10.8) releases molt-inhibiting hormone (MIH) into the sinus gland; MIH is stored and released gradually as needed, triggered by the neurotransmitter 5-HT. Normally, the X-organ is fully activated by 5-HT, so MIH is produced continually. Its main action is to inhibit the activity of the Y-organ in the thorax, so that there is no production of the molting hormone (which is again ecdysone, sometimes called crustic ecdysone or just **crustecdysone**). However, in the presence of appropriate stimuli (food, photoperiod indicating seasonal effects, presence of conspecifics), integrated via the CNS, the X-organ itself becomes inhibited and stops making MIH. Then the Y-organ can be active and produce α-ecdysone (activated in the epidermis to the β form) and the animal molts. Hence despite some real similarities with insects, crustaceans have as their "key" controlling hormone a neuropeptide that *inhibits* ecdysone action, whereas insects have a neuropeptide that *promotes* it.

Molting occurs by a similar process to that described for insects, though details of other neuropeptide controls are less well known. Recent evidence indicates that there is a separate **molt-accelerating hormone** in some crustaceans, acting on the Y-organ, and possibly an "**exuviation hormone**" that stimulates splitting of the old cuticle and emergence from it. The crustacean **hyperglycemic hormone** originating from parts of the gut wall also shows a surge at ecdysis in some species, likely linked to allowing the subsequent swelling needed to expand the new cuticle.

Control of metamorphosis is largely exerted by **methyl farnesoate** (MF), a precursor of JH and having very similar effects to those of JH in insects. In other words, MF maintains juvenile morphology, promoting early postembryonic stages but inhibiting later ones. As in insects, the later stages of ecdysis and emergence are in part dependent on a factor affecting heart activity, the **crustacean cardioactive peptide** from the pericardial organ, which stimulates the behaviors needed to break free from the old cuticle once the new one has been enlarged by massive water intake.

Other aquatic invertebrates

Very little is known as yet about the control of larval metamorphosis in aquatic worms and other aquatic invertebrates. However, interesting clues are emerging. Some indications of a role of JH in rotifer metamorphosis have been reported, where injected JH enhances production of mictic (sexual) females; however, it is unclear whether the animals have endogenous JH or whether the injected terpene is mimicking some other natural hormone. The trochophore larva of some polychaetes also responds to so-called "juvenoids", whose effect can be mimicked by MF derived from crustaceans or JH from insects, and also by eicosanoids. Eicosanoids such as arachidonic acid have been isolated from adult annelids (e.g. *Arenicola*), where they are described as sperm maturation factors. Hence it may be that annelids have eicosanoid hormones that have the same two functions in metamorphosis and reproduction as JH has in insects and MF has in crustaceans. It remains to be seen how far this pattern of control extends to other phyla.

A brain peptide hormone in some polychaetes may also control the proliferation of body segments, and seems to control "epitoky" (a peculiar form of adult metamorphosis; see Chapter 11) in marine species such as the palolo worm. Some worms have recently been shown to have ecdysones as well, active in the control of molting; nematodes are one such example, requiring regular ecdysis because they have particularly thick cuticles.

Vertebrates

The principal factors controlling development and metamorphosis in vertebrates are the iodinated protein hormones derived from the thyroid gland, **thyroxine** (T_4) and the more active **tri-iodothyronine** (T_3). In most vertebrates these hormones control developmental processes and act as tissue growth factors; they interact with intracellular receptors, then migrate to the nucleus and activate or repress gene transcription. Their effects include increased protein synthesis, and increased cellular respiration. Intestinal glucose absorption rises, muscle cells increase their glucose uptake, and fat cells undergo lipolysis so that fatty acids increase in the blood. The end-product is increased growth rate in the fetus for proper development of the skeleton and the nervous system. Other hormones regulate particular effects, for example cortisol, which stimulates production of the surfactants that allow lung expansion at birth.

Postnatal growth involves molting in many vertebrates, sometimes contributing to metamorphosis: mammals shed skin layers continuously, many birds and mammals have pronounced seasonal molts with change of form or color of the skin and its insulating layers, while reptiles and some amphibians and fish slough the whole skin at longer intervals. Such molting patterns are again mainly under the control of TSH and thyroid hormones, so that T_4 deficiency leads to poor hair or feather growth. There are also stimuli to epidermal growth from ACTH and vasotocin in some frogs and toads.

In these anuran amphibians there is also a much more substantial metamorphosis (Fig. 10.22) required for a tadpole to turn into an adult. This change is environmentally determined and can be accelerated by desiccation (e.g. a pond drying up), triggering a stress response via CRH. Then TSH from the pituitary promotes bursts of T_3 and T_4 release, which cause all the major effects, having very different effects on different tissues: initially relatively slow growth of the hindlimbs, then a sudden "metamorphic climax" involving resorption of the tail, reshaping of the head and mouth, changes in the gut (the diet switching from herbivory to carnivory), calcification of the skeleton, neuron growth and branching in the CNS, loss of gill arches and growth of lungs, and changes in the kidney, liver, and pancreas. Bursts of thyroid hormone secretion are often progressively larger, this hormone stimulating its own release by promoting a more extensive hypothalamus–pituitary portal blood supply. Prolactin also has a role in the metamorphic cycle, stimulating growth and inhibiting the metamorphic climax; but its inhibitory effects are overridden by the rapidly rising levels of thyroxine. The metamorphic changes must all be brought about very rapidly,

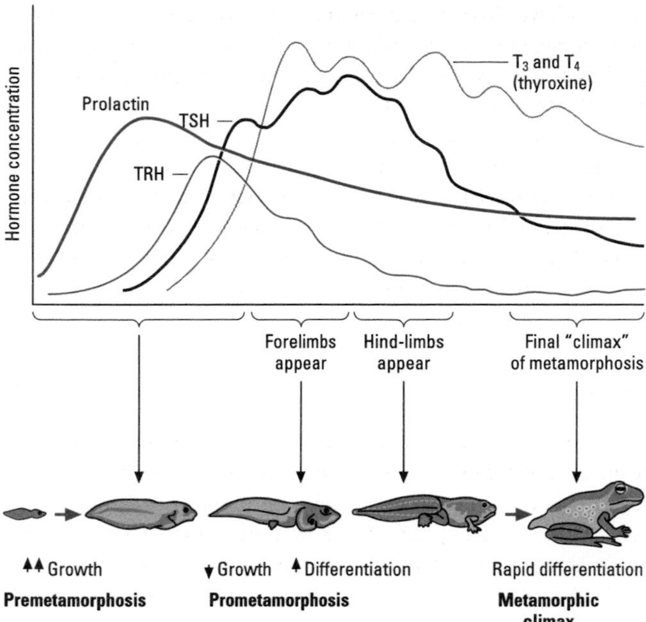

Fig. 10.22 The control of metamorphosis by hormones in anuran amphibians, shown schematically. Initially high prolactin levels stimulate early rapid growth, but as the hypothalamus and pituitary develop, thyroid-releasing hormone (TRH) production begins and thyroid-stimulating hormone (TSH) levels rise, causing thyroxine (T_3 and T_4) production from the thyroid, with levels peaking as the forelimbs appear. Additional vascularization of the median eminence then allows increased TSH production and further thyroxine surges, triggering final metamorphosis.

because the "half-way stages" are very vulnerable, suffering high predation rates. Hence this anuran metamorphosis is in many ways much more drastic even than insect metamorphosis, involving internal mesoderm and endoderm as well as surface ectoderm.

10.5.2 Growth hormones

Invertebrates

Invertebrates have relatively few known control systems for growth, and for many invertebrates that are roughly "worm like", growth is largely a process of elongation, sometimes achieved by adding extra segments to the body. The cerebral ganglia of annelids produce a general GH that maintains this elongation and addition of new segments. Molluscs also have a GH affecting both the soft tissues and also the shell via promotion of calcium transport into the mantle epithelium. For arthropods, having a rigid and restrictive exoskeleton, growth can only occur in coordination with regular molts, and the systems that control molting in insects and crustaceans, dealt with in the last section, are therefore the key to normal growth.

However, insects often also show a suspended growth (lowered metabolic rate) stage called **diapause**, which may be used for survival through adverse conditions (e.g. winter or drought). Diapause is the main "seasonal adaptive response" in many insects, and is usually an option rather then being obligate, often triggered by photoperiod and/or temperature. Normally it occurs in either the egg or pupal stages, but occasionally in the larva or adult. Again it is mainly hormonally controlled for both initiation and termination. In the moth *Hyalophora cecropia* pupal diapause occurs, and emergence is due to **ecdysone** triggering a molt, but a period of cooling is essential before diapause can be terminated. The effect is on the brain itself: implanting a brain kept at 20°C gives no emergence, but implanting one kept at 3°C for 6 weeks (and thereby "activated" to produce PTTH) switches on ecdysone production and thus adult emergence. However, in some other insects diapause in the egg or larva is largely controlled by **juvenile hormone**, which both induces and maintains the state. Unusually, in silk moths egg diapause is controlled in the previous generation, by the mother, who can lay either D eggs or non-D eggs (depending on temperature and food supply); temperature alters the levels of maternal dopamine, which controls a **diapause hormone** from the subesophageal ganglion acting on eggs in the ovarioles. In at least a few insects, including the common "greenbottle" *Calliphora*, larval diapause also seems to be controlled by the mother, though here the mechanisms are not clear.

Unlike other aspects of insect endocrinology, diapause control is highly variable, perhaps because it has to be so contingent on rather unpredictable environmental cues.

Vertebrates

The classic vertebrate **growth hormone** is a peptide, secreted from the anterior pituitary (see section 10.2.2). GH circulates in the blood bound to carrier proteins and its concentration varies with a circadian pattern, usually peaking during the early sleep hours. It has a wide range of actions (Fig. 10.23) affecting almost every kind of cell in the body, but especially the bones and skeletal muscles.

In essence, GH is an anabolic hormone—under its influence, rates of DNA and RNA synthesis, protein synthesis, and cell division all increase. These effects are especially evident as mitotic activity in the cartilage of juveniles, causing increasing bone length (aided by GH-stimulated uptake of the sulfur needed for cartilage deposition) and an associated increase in soft tissue mass. There is also increased hepatic glycogen breakdown and stored fats release free fatty acids to the blood. These actions provide more circulating glucose, but more importantly they promote the use of fats for ATP generation in muscles.

Various polypeptides from the liver and muscles (called **insulin-like growth factors**, IGFs, or "somatomedins") also affect growth, encouraging the division of cartilage cells and extra secretion of cartilage matrix. IGF-1 stimulates DNA production and cell division in fibroblasts, liver cells, and muscle, while IGF-2 mainly acts like insulin to increase glucose levels in the blood, especially in fetal and neonatal animals. GH and IGF-1 both appear as pulses during adolescence and puberty.

The thyroid hormones that act to control development (see section 10.5.1) also inevitably affect vertebrate growth, mainly by insuring regular and steady size increments from birth to puberty in birds and mammals. In fish the thyroid gland becomes particularly active during growth spurts (e.g. when migratory salmon transform into smolts *en route* to the sea). However in amphibians thyroxine has the unusual effect of arresting growth, while promoting metamorphosis as we saw earlier.

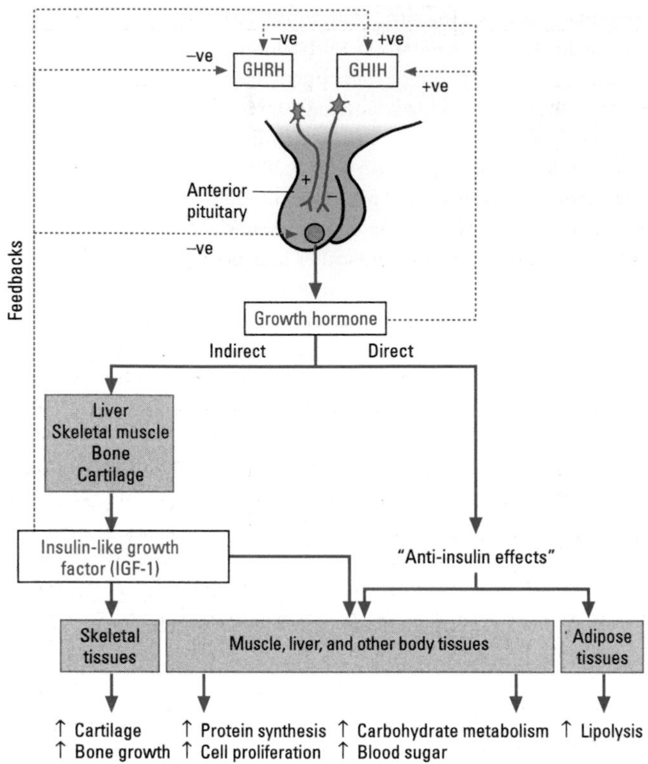

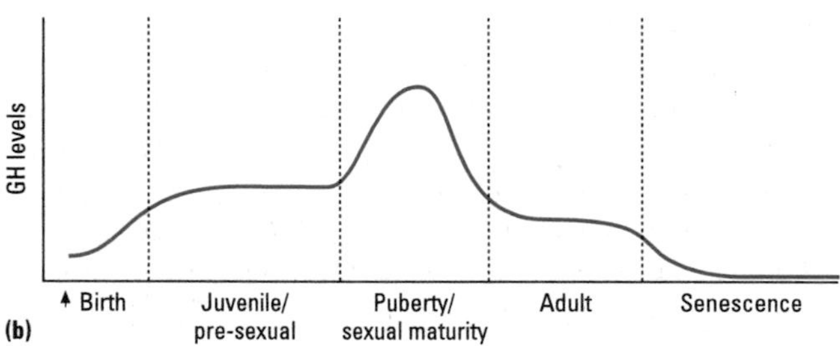

Fig. 10.23 (a) The effects of growth hormone (GH) in mammals, and its control. (b) The patterning of levels of GH through the lifetime of a mammal. GHIH, GH-inhibiting hormone; GHRH, GH-releasing hormone.

10.6 Control of metabolism, temperature, and color

10.6.1 Metabolism

Invertebrates

Very little is known about metabolic hormones in the simplest invertebrates. For marine species with a very stable habitat and the possibility of direct uptake of nutrients through the skin (see Chapter 11) it may be that little control is needed. However, bivalve molluscs such as *Mytilus* have a metabolic neurohormone from the visceral ganglia that increases storage of lipids, protein, and glycogen, and an insulin-like hormone from the cerebral ganglia also promotes glycogen storage. Earthworms also have a cerebral neurohormone that raises blood glucose. In crustaceans, neuropeptides from the eyestalk X-organ control some metabolic pathways, and crustacean hyperglycemic hormone from the sinus gland increases hemolymph glucose (some species also having a hypoglycemic factor). Perhaps such controls are more necessary for intertidal and terrestrial animals, where feeding is periodically precluded.

In insects there is a family of **adipokinetic hormones** (AKHs), released from the corpora cardiaca by neural triggers and acting on fat body cells. Insect hemolymph mainly transports lipid as diglycerides (especially DAG), bound to a lipophorin carrier, and high levels of hemolymph DAG inhibit the release of AKH. Additionally, low levels of sugar (especially trehalose) stimulate AKH release, and hence DAG is released from the fat body to give an alternative fuel supply. An antagonistic hormone promoting lipid uptake to fat body cells may also operate. Hypertrehalosemic hormone (HTH-II) from the corpora cardiaca affects sugar levels directly, breaking down glycogen into trehalose; while an octapeptide that increases levels of proline and some carbohydrates (so preparing the animal for flight) has been isolated in some beetles. The hormone bursicon, involved in the tanning processes, also helps to regulate amino acid levels through its effects on tyrosine (a precursor for the quinone-like tanning agents).

Vertebrates

Several of the hormones already discussed in relation to growth have important metabolic effects in vertebrates. **Catecholamines** elicit glycogen breakdown, gluconeogenesis (the synthesis of glucose from amino acid sources), and lipolysis. **Leptins** from adipose tissue regulate energy expenditure and metabolic rate in endotherms. **Thyroid hormones** and **growth hormone** cause an increase in protein synthesis, as well as raising blood glucose and lipid levels. Thyroid hormones are particularly important in many endothermic vertebrates as long-term regulators of basal metabolic rate. Under their influence, mitochondria increase in number and cristae density, the enzymes involved in oxidative phosphorylation increase in concentration, and ATPase activity rises.

The pancreatic sugar-regulating hormones, glucagon and insulin, are also crucial for short-term metabolic regulation (see Fig. 10.15). Both act on the liver to insure steady blood glucose, particularly crucial for brain functioning. **Glucagon**'s major targets are the liver cell surface receptors, via which it increases cAMP levels to trigger glycogen breakdown, inhibit glycogen synthesis, and increase the uptake of amino acids leading to gluconeogenesis. At adipose tissues glucagon produces lipolysis, thus giving alternative metabolic substrates and sparing glucose for use by the brain. Since an ability to increase blood glucose when it is falling is particularly vital, hyperglycemic effects also come from catecholamines, cortisol, thyroid hormones, and GH on a longer timescale. In contrast, **insulin** is the only hormone with the opposite effect of being hypoglycemic (reducing blood glucose); by stimulating the addition of extra glucose carriers into membranes it promotes the uptake of glucose, especially into muscle cells.

Some of the adrenal cortex hormones are also key players for longer term metabolic regulation. **Cortisol** and other glucocorticoids, triggered by pituitary ACTH (Fig. 10.24), bring about increased blood glucose by mobilizing protein from muscle to

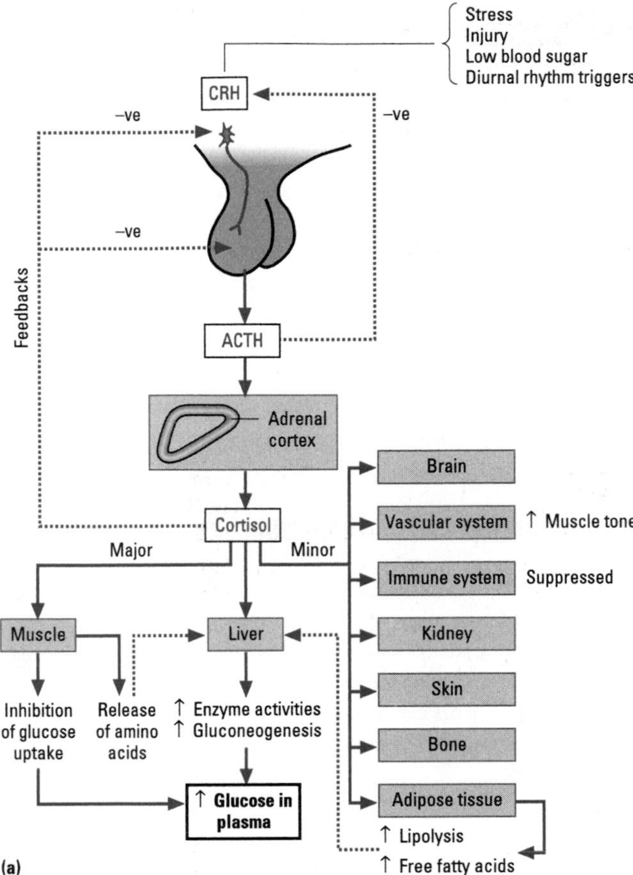

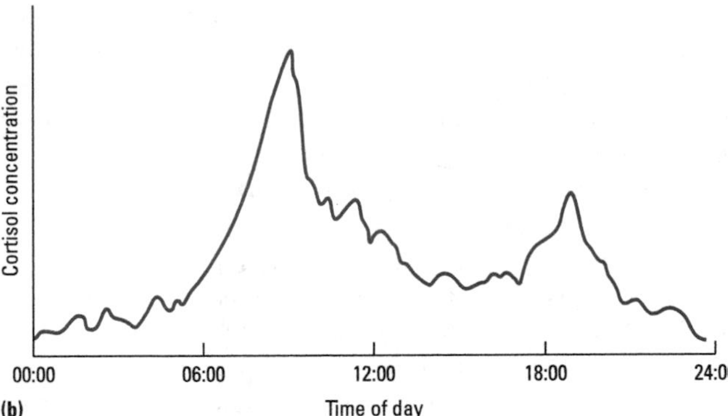

Fig. 10.24 (a) The effects of cortisol (the principal glucocorticoid hormone) in a mammal, and the control of its release via the hypothalamic–pituitary axis. (b) A typical cortisol secretion pattern in a diurnal mammal. ACTH, adrenocorticotropic hormone; CRH, corticotrophin-releasing hormone.

release amino acids, allowing gluconeogenesis in the liver. They also promote lipolysis, and stimulate erythropoietin to increase production of red blood cells. Finally, they trigger "appetite", allowing the replenishment of food reserves. These effects can all be seen as reactions to **stress** (e.g. injury, infection, extreme temperature changes, or emotional trauma); the glucocorticoids also tend to counter stress by having anti-inflammatory, antiallergic, and immunosuppressive effects, decreasing the levels of T-lymphocytes in the blood. In humans at least, an effect on the CNS also occurs, producing a feeling of euphoria that may be useful during periods of stress.

10.6.2 Temperature control

Hormones seem only rarely to have direct controlling effects on animal temperature balance. However, many of them affect metabolic processes in ways that influence body temperature, and these effects inevitably differ between ectothermic and endothermic animals.

Ectothermic animals: varying metabolic rates and effects of color

Metabolic rate in ectotherms is partly influenced by heart rate (altering the rate of circulation and hence supply of nutrients), and many invertebrates have **cardioactive peptides**. For example, cephalopods have one excitatory heart peptide, and the common snail *Helix* has two different ones, released from a neurohemal organ adjacent to the atria. Crustaceans and insects also have cardiac excitatory peptides, released from the pericardial organ and corpora cardiaca, respectively. Some of these factors have sequence similarities with vertebrate peptides such as vasopressin and MSH. In addition, calorigenic effects similar to those of thyroxine operate at the fat body in insects and crustaceans; prostaglandins may be the main controlling factors, as they are known to trigger behavioral hyperthermia in some arthropods.

Hormonal thermal effects may also occur in ectothermic vertebrates. **Leptins** can elevate T_b slightly in lizards, with increasing metabolic rates. In many diurnal reptiles **melatonin** decreases the preferred T_b so that behavioral thermoregulation patterns alter; this effect is lost in nocturnal species. Thyroid hormones may also act thermogenically in some lizards, and if injected there are increases in oxygen consumption, activity levels, and daily average T_b. Interestingly, the tissues are unresponsive at low T_b but respond close to the T_{pref} of around 30°C. In general, however, thyroid hormone concentrations are an order of magnitude lower in ectothermic reptiles than in endothermic mammals.

In many invertebrates, and in some reptiles and amphibians, hormones affecting body color also contribute to thermoregulation. Animals generally change color via **chromatophores**, epidermal cells in which the pigment can be either highly dispersed or highly concentrated; these are usually monochromatic but may be polychromatic with up to four different pigments. Chromatophores are stellate or dendritic in form; dispersion sends pigment to the extremities, imparting color to the epidermis, while concentration leaves a small central dot of color and the epidermis looks almost transparent. Dispersion is normally brought about by neurohormones. Many animals use color change to match their background and/or for display. But the chromatophore response may be modified by light and temperature, so giving the possibility of thermal regulation. Dispersing dark chromatophores when cold gives a dark color with high thermal absorbance, whereas at high temperatures the concentration of dark chromatophores (or dispersion of pale ones) can give a light color, increased reflectance, and a chance to cool. Thus a pale sea slater (*Ligia*) can be 3°C cooler than a dark one in the same microhabitat, and thermal needs can sometimes override camouflage constraints for this littoral animal. In crustaceans the control of epidermal color is exercised by at least two antagonistic hormones, acting on cell surface receptors that trigger Ca^{2+} and cAMP. For the shrimp *Crangon*, which can vary through white, gray, black, yellow, and red, the two main hormones are termed "Crangon darkening hormone" and "Crangon body lightening hormone"; but there are additional pairs of hormones regulating specific colors, two for white pigment, two for black, and two for red.

In insects color is altered by moving pigment granules within the epidermal cells, rather than within chromatophores. Here background-matching color change is under the control of a neurohormone from the median neurosecretory cells of the brain. But physiological color change can also be elicited by a tritocerebral neurohormone in many grasshoppers, which may be nearly black when cold and needing to absorb radiation, but turn brilliant reflective greens and blues in the middle of the day (see Chapter 15).

Interestingly, the most spectacular and rapid invertebrate color changes occur in cephalopods, for signaling purposes rather than thermoregulation, and here the changes are triggered directly by nerves with no hormonal involvement.

Some ectothermic vertebrates can change color using both epidermal and dermal melanophores. In the longer term they can alter the number of such cells, but short-term physiological color change is mediated by hormones. In frog skin there are three layers of pigment cells: an upper layer of yellow/green xanthophores, a middle layer of reflecting iridophores, and a lower layer of black melanophores. The absence of MSH causes melanin aggregation and expansion of the iridophores, giving a pale appearance, whereas MSH presence causes melanin dispersion and iridophore concentration giving a dark brown skin color. Teleosts have an additional posterior pituitary hormone (melanin-concentrating hormone, with 17 amino acids) that aids the speed and variation of their color changes.

Endotherms and the maintenance of body temperature

Body temperature is primarily regulated by the hypothalamus in vertebrates, especially by neurons in the preoptic area that receive inputs from both skin and body core receptors, including some receptors in the hypothalamus itself (see section 8.10). Responses to altered body temperature mainly involve behavior and circulatory adjustments, also dealt with in Chapter 8. However, hormones are often important because of calorigenic effects, raising the metabolic rate and thus the body temperature.

The **thyroid hormones** increase protein synthesis as we have seen, but also increase rates of cellular respiration via an increase in the size and number of mitochondria. There are also increments in the concentrations of key enzymes in the respiratory chain, and of the Na^+/K^+-ATPase, so increasing membrane ion transport. There is recent evidence of a direct effect of T_3 on the gene that transcribes the thermogenic uncoupling protein UCP-1 in brown fat (see section 8.7.2). The net effect is rising basal metabolic rate, leading to thermogenesis. This is important in thermal adaptation to cold environments (see Chapter 16); cold stress acts within 24 h to raise the circulating levels of T_4.

In addition, **epinephrine** from the adrenal medulla is thermogenic, giving a rapid transient warm-up in response to stress. **Insulin, glucagon** and **glucocorticoids** also have some thermogenic effect, the latter triggered by stress-related ACTH release. ACTH shows an interesting parallel with invertebrates, since it produces a skin-darkening effect in many birds and mammals. Ovarian **progesterone** has a slight T_b elevating effect in mammals, helpful in determining when ovulation occurs (the "fertile period").

Some of the vertebrate calorigenic hormones specifically promote brown fat metabolism (see Chapter 6), important in juveniles and also following periods of torpor, dealt with in Chapter 16. Futile cycling increases, with concomitant extra heat production, under the influence of thyroid hormones, because in some pathways T_3 and T_4 can stimulate both anabolic and catabolic enzymes.

One further thermal effect in birds and mammals comes in the form of antipyretic influences. Fever is used as a host defense mechanism, but an episode of fever must not cause T_b to get too high; hence endogenous antipyretics (MSH, AVT, AVP, and glucocorticoids) may work via the cytochrome P-450 enzymes in the mitochondria. Prostaglandins also have some regulatory effects.

10.7 Control of sex and reproduction

In most animals reproductive control involves two main types of hormone: **gonadotropins**, often from the brain, which are peptides and control the gonads, and **gonadal hormones**, usually steroids, from the gametes or gonads. The latter are what we generally call "sex hormones". This classic control pattern varies relatively little between taxa, but is modified crucially according to environment and life history.

10.7.1 Invertebrates

Worms

Very few types of worm reproduction are well studied, although medical concerns have led to an intensive analysis of a few parasitic taxa. Neurohormones play a role in platyhelminths, with active neurosecretory cells increasing in number in sexually mature flatworms, and being involved in the maturation and shedding of egg-laden segments (proglottids) in parasitic tapeworms. The cerebral ganglion of polychaete annelids produces a neurohormone that retards gamete maturation, and this kind of "**juvenile hormone**" (inhibiting reproductive development) seems to be common. In *Arenicola* brains there is a sperm maturation factor, identified as arachidonic acid, which also acts during metamorphosis as a JH; other species may also have eicosanoid hormones acting in this role. Another polychaete hormone released by the gametes inhibits the brain hormone at source, giving a classic negative feedback interaction. In a few species of polychaetes and in the related echiuran worms there is a sex reversal process, individuals transforming from female to male (see Box 11.1), again controlled by a cerebral neurosecretion.

Of all terrestrial worms *Lumbricus* is best studied. Here, removing the brain gives precocious maturation, indicating a classic gonadotropin "brain hormone" that sustains gamete maturation but inhibits full reproductive development. In another earthworm, *Eisenia*, a peptide termed "annetocin" is expressed only in neurons related to reproduction (mucus secretion and egg-laying behavior); this peptide is of the oxytocin/vasopressin family, suggesting interesting parallels with vertebrates. Since in earthworms reproduction involves direct development of protected eggs, hormones also affect vitellogenesis (yolk formation) and production of the clitellum in which fertilization occurs and where the protective cocoon is secreted.

Echinoderms

Starfish have a gonadotropin "spawning hormone" (or gonad-stimulating substance), which is a peptide released from the radial nerves. It induces gamete formation in both the testis and ovary, and subsequently also makes the ovarian follicle cells release a true gonadal hormone. This is a modified amino acid, 1-methyl-adenine, inducing meiosis in pre-oocytes and muscle contractions to expel the maturing gametes into the sea. It also causes the mature oocytes to produce a third hormone ("maturation-promoting factor"), which insures breakdown of the oocyte vesicle. These two oocyte-derived hormones possibly induce brooding in those few echinoderms that show it. There is also some evidence for steroid production in echinoderm gonads; estrogens occur in some species and can promote protein uptake into oocytes.

Molluscs

Most molluscs have two separate sexes, and in many marine species control is straightforward. Cephalopods generally have the simplest reproductive strategies, where males and females mate once and then die. Perhaps because of this (but also reflecting the complex neurological control of behavior in this taxon) there is only one known reproductive hormone, a neurosecretory gonadotropin from the optic glands. This controls spermatophore production and male accessory structures, as well as yolk formation and female accessory structures. The gonads do make steroids, but these appear to have metabolic rather than reproductive functions.

In contrast, gastropods are often hermaphrodite, either sequentially (with protandry and sex reversal from male to female; see Box 11.1) or simultaneously in some of the terrestrial gastropods. Eggs are often fertilized internally, and may be packaged in elaborate egg cases. Reproductive control is therefore complex and highly variable, and many neurohormones have been reported with specific effects on reproductive physiology and anatomy.

In the aquatic protandrous molluscs the juvenile gonad is bisexual and is "turned male" by an androgenic factor from the cerebral ganglion. However, the optic tentacles produce a hormone that cyclically suppresses maleness, resulting in seasonal bouts of sex reversal when sperm production is inhibited. At these times, the cerebral ganglion produces a factor that promotes oogenesis. Neurohormones also control development of female accessory structures (oviduct, vagina) and the male sperm duct, seminal vesicles, and penis. In aquatic molluscs such as sea slugs there are additional egg-laying hormones controlling the full sequence of ovulation, fertilization, egg packaging, and egg laying.

Terrestrial gastropod slugs and snails are hermaphrodites, and here the gonads become female unless influenced by an androgenic hormone from the cerebral ganglia and tentacles. In the female phase, vitellogenesis is promoted by a hormone from the dorsal body, in turn controlled by a brain hormone. A third influence comes from the gonads themselves, which release male and female hormones controlling secondary sexual characteristics. Finally, a hormone from the caudal ganglia controls ovulation and egg laying.

Insects

Insects seem to be rather unlike most other higher taxa, having gonads that do *not* necessarily produce crucial reproductive hormones. Most aspects of reproductive physiology and behavior are instead controlled more centrally by gonadotropins. Somewhat surprisingly the main gonadotropin is not a brain hormone, but instead it is again **juvenile hormone** from the corpora allata. There are indications that the control of reproduction was the "primitive" role of terpene hormones, with control of metamorphosis being a secondarily acquired function. Thus, having completed its role as metamorphic regulator in the juvenile insect, when its levels declined progressively to permit maturation, JH becomes the main sex hormone in adults. The source glands (having shrunk through successive larval molts) grow rapidly after the pupal–adult transition, so that hormone levels rise sharply (and, at least in some flies, additional JH may be produced from accessory sexual glands).

JH has multiple effects in adult females (Fig. 10.25), though these differ between species. Firstly it is gonadotropic, acting directly on ovarioles to promote oocyte maturation, and it has direct effects on protein and lipid synthesis and uptake here. It can also stimulate accessory glands that produce the constituents of the eggshell. It may act on the fat body, to promote production of key metabolites including the vitellogenins needed by the eggs. In species with particularly long extensible ovipositors, such as locusts, it affects the longitudinal muscles there, so that they tolerate remarkable lengthening. Furthermore, it affects both responsiveness and attractiveness in females, and causes them to secrete one or more pheromones for long-range attraction of males (see section 10.9). JH also acts directly on the brain in some insects to increase female receptivity; for example in the mosquito *Aedes atropalpus* there are two reproductive "morphs", one receptive at emergence and mating within 2 days, the other unreceptive at emergence and not mating for 5–6 days, linked to its JH secretion being much delayed (a difference which is directly heritable via a single gene). In yet another role, JH can also be involved in triggering and controlling oviposition behavior, making it certainly the most multifunctional hormone yet discovered in invertebrates.

However, in many insects JH does not on its own promote successful reproduction. Ecdysteroids may themselves be gonadotropic, and other neurosecretory peptides stored in the corpora cardiaca also have reproductive effects. One brain factor ("egg development neurosecretory hormone", EDNH, belonging to the F-peptide family) may work with JH to increase vitellogenin production and egg maturation. In locusts a functionally equivalent but chemically different peptide occurs, termed ovary maturating parsin, or OMP. In some flies the ovary produces its own hormone ("oostatic hormone") that prevents excess egg production. Thus the whole cycle can become rather complex, perhaps on a par only with terrestrial vertebrates. Mosquitoes can serve as an example. Here JH firstly promotes blood-feeding behavior in newly emerged females (to give the nutrients needed to make egg proteins), and then promotes the early development of eggs, but it also primes the gonad to become responsive to EDNH. The blood meal then triggers release

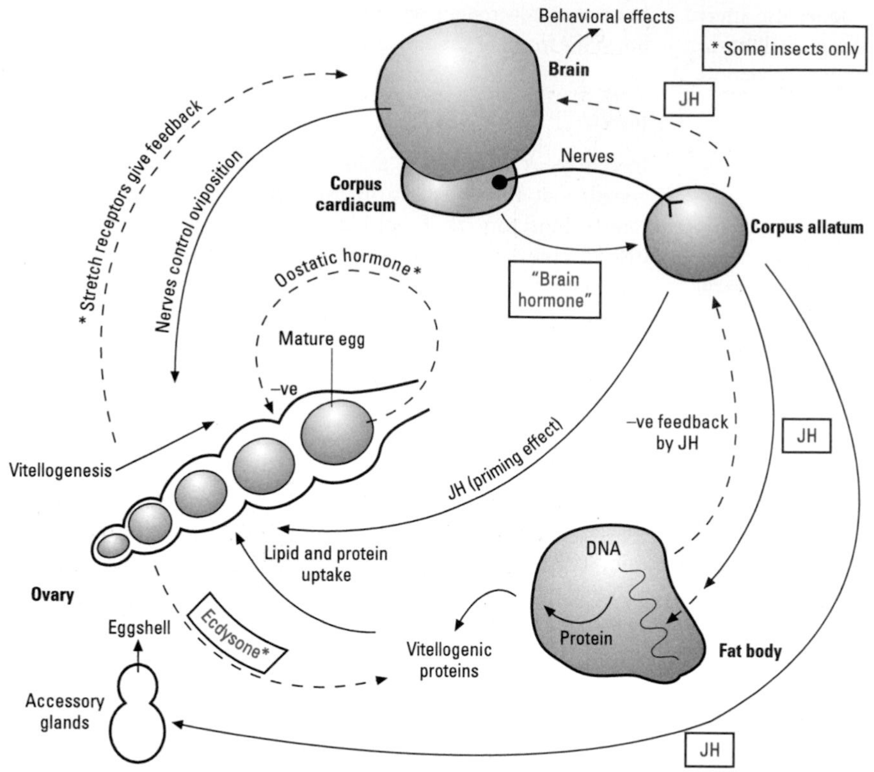

Fig. 10.25 Juvenile hormone (JH) effects in the control of female insect reproduction, with feedback loops and interactions with other hormones.

of EDNH, which controls the final stages of egg maturation and also stimulates the ovary to produce some ecdysone, which in turn promotes vitellogenin production in the fat body. Only then can fully mature eggs be laid; and if they are not laid, oostatic hormone intervenes to curtail further egg production. In a few insect species that are viviparous, such as the tsetse fly *Glossina*, another brain peptide affects regulation of the oviduct "milk glands" and changes the female behavior (to relative placidity and selection of high-protein diets) while she is carrying the young larvae in her enlarged oviducts.

Male insect reproductive systems are usually much simpler, but are again influenced mainly by JH. This can sometimes induce sperm production, but more often has its main effects on accessory organs producing the sperm packaging materials. Male sexual characters can be under the control of an androgenic hormone from the testis itself, and the testis may also produce some ecdysone that seems to influence sperm maturation (again having a very different role from that seen in juveniles).

Crustaceans

Reproduction in crustaceans has become better known in the last decade, largely due to the perceived need in aquaculture to promote early maturation and high reproductive output. It is clear that most crustaceans also use JH (or its near relative, MF) to control sexual maturation and gonad function. This control can be quite complex; for example, in the water fleas (*Daphnia*), which show a life cycle alternating between sexual reproduction and parthenogenesis, JH stimulates the production of male offspring later in the cycle while also reducing production of resting eggs and increasing the output of parthenogenetic offspring.

But crustaceans often also have a specific and independent sex hormone, unlike most of the insects. In the males of crabs, amphipods, and isopods, a glycoprotein hormone is secreted by the androgenic glands and promotes gametogenesis, secondary sexual characteristics, and male behavior patterns; it also inhibits female characteristics. In at least one species of mysid prawn, endogenous testosterone has been detected, although other androgens may also be active.

10.7.2 Vertebrates

It is difficult to generalize about reproductive endocrinology in vertebrates, since patterns vary so much from aquatic mass-spawning fish, to egg-laying reptiles or birds, to viviparous mammals. Mammals are particularly well studied, but fish, amphibians, reptiles, and birds are much less cyclical than mammals, and also have far more elaborate and stylized courtship patterns, so in many ways are more interesting.

Many vertebrates have a specific breeding season, timed to occur so that the vulnerable young appear in periods with plenty of food and reasonable environmental conditions. The endocrine system is often the mediator between changes in the external environment and triggering of internal changes, first in reproductive anatomy and physiology, and then in reproductive behavior (receptivity, attractiveness, nest building, maternal actions, etc.), whether by altering sensorimotor processes or higher motivational processes. The key hormones are summarized in Fig. 10.26. The initiator is GnRH, forming the key link between the brain and reproduction and delivered via the hypothalamic portal vessels to the pituitary gland; this then controls the two gonadotropins LH and FSH from

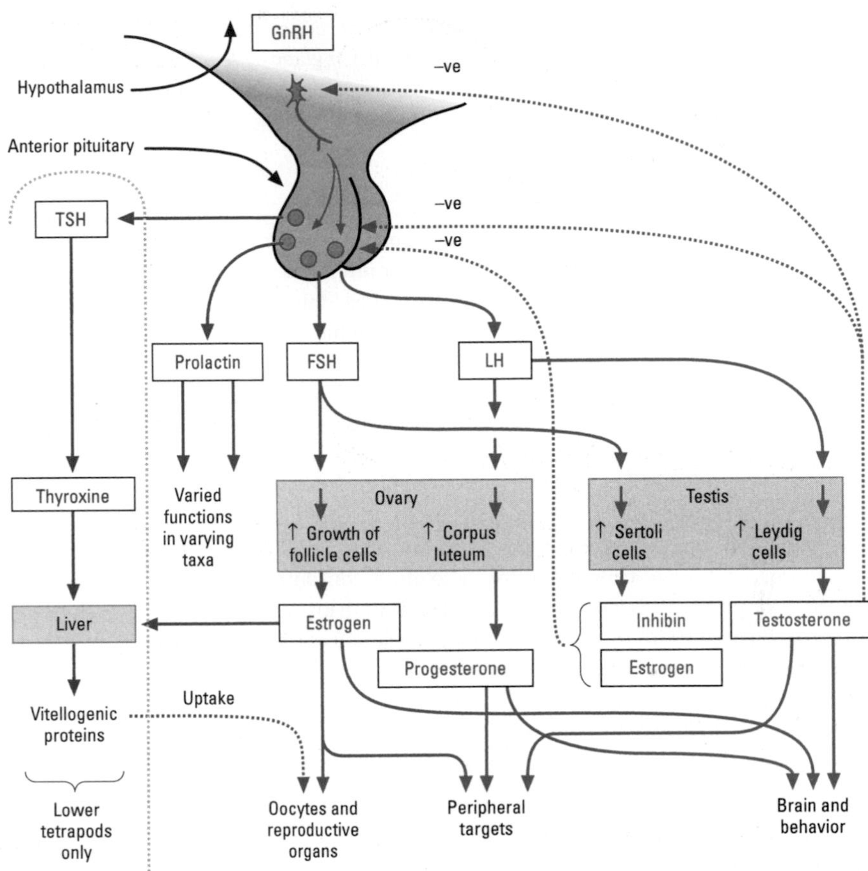

Fig. 10.26 An overview of the hormones involved in the control of sexual physiology and behavior in vertebrates. Standard abbreviations as elsewhere (see Tables 10.4 and 10.5).

the anterior pituitary. Other players are the anterior pituitary hormone prolactin, and the gonadal sex hormones, primarily testosterone in males (from the interstitial cells of the testes) and estrogen plus progesterone in females. However, the exact role played by each of these hormones, and the balance between them, is very different in the various vertebrate taxa.

Sex determination and maturation

In most vertebrates sex is determined genetically, usually such that two X chromosomes give a female, while one X and one Y give a male (but birds have homogametic WW males and WZ females). In some reptiles the temperature of the early embryo within the egg has a vital sex-determining effect (see Chapter 15). In most other vertebrates the development of sexual *characters* requires one or more hormones. In most taxa, testosterone is the main determinant of primary sexual characters; its absence (from about 6 weeks after fertilization in humans) leads to female genitalia. The female state, physically speaking, is therefore the basic or neutral one, which can be "turned male" by testosterone (though birds are again different in that steroids have an opposite effect and the male is the neutral sex requiring no hormone-dependent process). The sex-related hormones are thus active from the early embryonic stages, controlling sexual development. The early presence of testosterone is also an important key to the roles and effects of the releaser and gonadotropic hormones (mainly GnRH, LH, and FSH but also to a lesser extent prolactin) in later life. In other words, the gonadal hormones control the subsequent effects of the pituitary gonadotropins.

In young female birds and mammals, the balance of FSH from the pituitary and a constant dribble of estrogen from the ovary through early life keep each other in check, with a negative feedback of estrogen on the brain, so that the ovaries do not develop further. But at puberty the quantities of both hormones rise, and the sensitivity of the hypothalamus to negative feedback from estrogen declines, so that FSH and LH together cause ovarian growth, FSH stimulating gamete production and LH (in some taxa also prolactin) promoting secretion of gonadal hormones. There is therefore a further increase of estrogen output, which brings about female secondary characters, also starting the cyclicity of GnRH (and thus LH release that triggers each ovulation in humans and other periodic ovulators), and stimulating the cervix to produce copious alkaline mucus creating an ideal environment for sperm survival and motility. The adrenal sex steroids also seem to control some secondary characters at this stage.

Young males are rather similar initially, though in some mammals testosterone crucially triggers descent of the testes around the time of birth to a site outside the abdominal wall, where they can remain cooler and so mature the sperm better. But the increases of FSH and LH at puberty lead to a further rapid growth of the testes and hence more testosterone output, which largely determines the male secondary characters. These include a spurt of growth, changes in the vocal cords in many species, changes in hair growth, and special features such as antler growth or bright male plumage, plus systemic effects in increasing hematopoiesis and enhancing metabolic rate. Testosterone also inhibits female characters at this stage, notably development of the mammalian milk-producing glands. At puberty the hypothalamus becomes much less sensitive to negative inhibition by testosterone, thus upregulating the control system.

Controlling male physiology and behavior

In male vertebrates, sexual physiology is centered on the production and ejection of mature sperm (normally close to or into a receptive female). This is initiated by LH (sometimes also called "interstitial cell-stimulating hormone" in male mammals), which binds to the interstitial (Leydig) cells of the testes and stimulates them to make testosterone. However, FSH also has a key role in stimulating another group of testicular cells (Sertoli cells) to secrete an androgen-binding protein, which prompts the spermatogenic cells to bind testosterone and begin full-scale spermatogenesis. In effect, FSH acts by making the cells responsive to testosterone that was already present due to LH. To complete the control loop, testosterone inhibits hypothalamic release of GnRH, and may also affect gonadotropin release in the pituitary. Extra control is added (at least in mammals) by a protein termed **inhibin**, which is produced by Sertoli cells when the sperm count is high and has an extra negative feedback on both the hypothalamus and the pituitary.

In most vertebrates, if the testes are removed, sexual physiology is compromised and sexual behavior stops almost immediately, indicating a strong and perhaps primary effect of gonadal steroids. The brain-derived neurohormones (FSH, LH) alter this effect to varying degrees in different taxa.

The stickleback is the most studied of all fish, and in the spring breeding season (as day length increases) males develop nuptial colors with a conspicuous red belly, then build a nest, court the females and fight off other males, subsequently guarding the nest and fanning the eggs to maintain oxygenation. All these changes are regulated by testosterone, but with different thresholds: sand digging needs less hormone than does fanning, for example, suggesting steadily rising hormone levels and increasing thresholds for each successive element of the behavior.

Newts have a distinctive courtship by males of females, involving tail vibrations and pheromone production, and here the male behavior is controlled by prolactin and androgens, which also stimulate pheromone production. However, discharge of the pheromone is triggered unusually and directly by AVT.

In birds, male behavior is largely testosterone regulated via the preoptic lobe (a source of hypothalamic releasing hormones). But the hormone is also taken up in other areas of the brain, including areas involved in song. Castration effects vary according to sexual experience, and to the availability of social and vocal interactions. In budgerigars, hearing another bird singing causes the listening bird to sing and this in turn sets off gonadotropin secretion, so that here behavior may trigger hormone activity, rather than vice versa. The hormonal effects on male behavior relating to courtship, egg laying, and nesting are also reasonably well known in birds, especially from work on the ring dove. Male courtship is under testosterone control, and in the dove there are two male "types" (each with a different song type), one rather aggressive and a second that is nest-oriented, the latter requiring lower levels of testosterone. Male incubation behavior has some dependence on progesterone, but prolonged male incubation and nest building activity seem to be largely due to stimuli from the female rather than to any further male hormonal changes.

Male mammals show some hormonal effects on behavior, but the effects even of castration are reduced in the "higher" mammals, or in any mammal that is already sexually experienced, indicating a lesser role of gonadal steroids on sexual drive. However, with no testosterone being produced the accessory organs atrophy and most male mammals become both sterile and impotent. Almost invariably testosterone also has neural effects, promoting aggressiveness in sexually active male mammals. Effects of other hormones do exist though—recent evidence indicates that oxytocin is sometimes involved in arousal and sexual satiety.

Controlling female physiology and behavior

Female vertebrate sexual physiology is highly variable, and is much better known in the very complex and often cyclic mammals than in other taxa. In some lower vertebrates there may be only one pituitary gonadotropin hormone, rather than separate FSH and LH production, and then prolactin has some of the gonadotropic functions instead. In all cases, though, the cyclicity or seasonality of breeding may arise from direct neuronal inputs (via temperature or photoperiod) to the GnRH-releasing terminals in the hypothalamus, giving the key environmental control.

In the pattern of early egg development for birds and mammals (Fig. 10.27), GnRH stimulates increased production of FSH and LH, which act on different aspects of ovarian oogenetic function. FSH stimulates the follicle granulosa cells (a layer of flattened cells surrounding each oocyte) to produce estrogens. As the follicles mature, they develop an outer layer of thecal cells, and these respond mainly to LH instead, by producing androgens that diffuse inward and are converted into more estrogens by the granulosa cells. In mammals, the resultant rising levels of estrogens in the blood exert a negative feedback on the pituitary, inhibiting the release of gonadotropins. Rising estrogen levels also cause the ovary to produce yet more estrogens itself, by intensifying its response to FSH. The resulting "estrogen surge" passes some critical concentration and begins to have a *positive* feedback on the hypothalamus and pituitary, setting a cascade of events in motion. A burst of stored LH (plus some FSH) is released, and this "LH surge" stimulates the maturing follicles to complete meiosis and rupture out of the ovarian wall, the process termed **ovulation**. In most vertebrates AVT is important at this point in producing oviduct or uterus contractions to expel the eggs (paralleling the function of the closely related oxytocin in mammals). In fish and amphibians the eggs are usually released *en masse* from the oviduct into the environment where fertilization occurs. Whereas in reptiles and birds fertilization is internal, with the ovum meeting sperm from the male ejaculate. One or more eggs are then generally carefully laid in a nest; hormonal changes are implicated in stimulating accessory glands that make yolk proteins, albumen, and the egg cases or eggshells.

In mammals, the ovulated eggs remain tiny and are released in an unprotected state into the fallopian tubes, where estrogens induce ciliary activity in the tube walls to move the egg along. Eggs meet sperm in the uterus, and prostaglandin hormones in the male ejaculate facilitate this by causing contractions in the uterine muscle. The

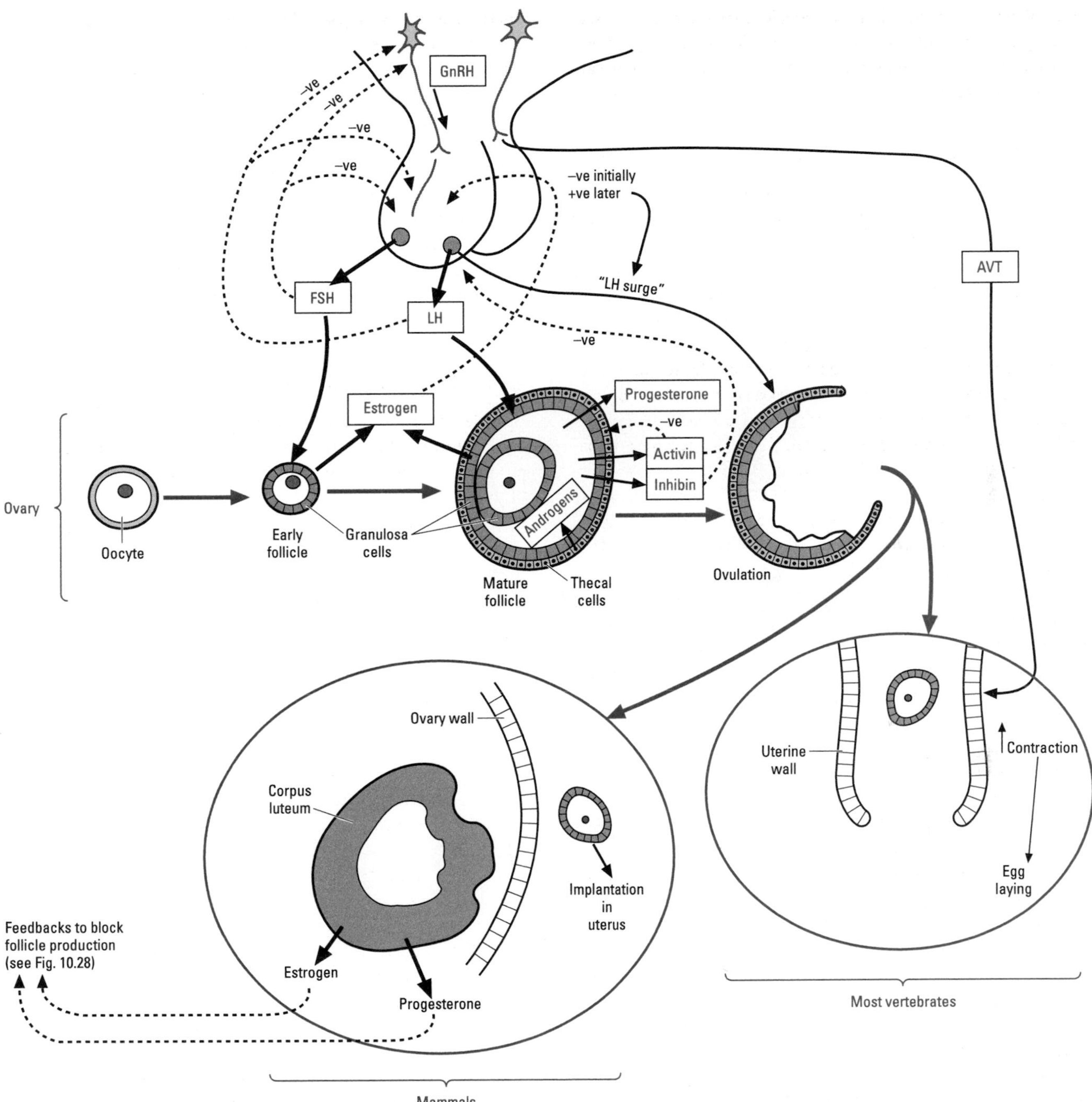

Fig. 10.27 The main reproductive control systems underlying egg development and release in a female vertebrate. Standard abbreviations as elsewhere (see Tables 10.4 and 10.5); AVT, arginine vastocin.

LH surge also transforms the ruptured follicle into a **corpus luteum** (Fig. 10.28), which effectively becomes an endocrine gland in its own right and begins to produce both estrogens and progesterone. These hormones have a strong negative feedback on the pituitary to lower the FSH and LH levels, so that the development of new follicles is inhibited. Estrogens also act on the uterine lining (endometrium) in preparation for embryo implantation; there is proliferation of epithelial cells, extra vascularization, increased secretory cell number, and the induction of progesterone receptors. Prolactin levels rise in concert with estrogen, because high estrogen can both promote prolactin-releasing hormone and block prolactin-inhibiting hormone (dopamine). If there is no fertilization or implantation of the released eggs, LH levels decline and the corpus luteum begins to degenerate, ceasing its output of steroids. Thus estrogenic inhibition of the pituitary stops, and gonadotropins are released again, to restart the whole cycle (of about 28 days in humans). However, if implantation of one or more fertilized embryos has occurred, the activity of the corpus luteum persists,

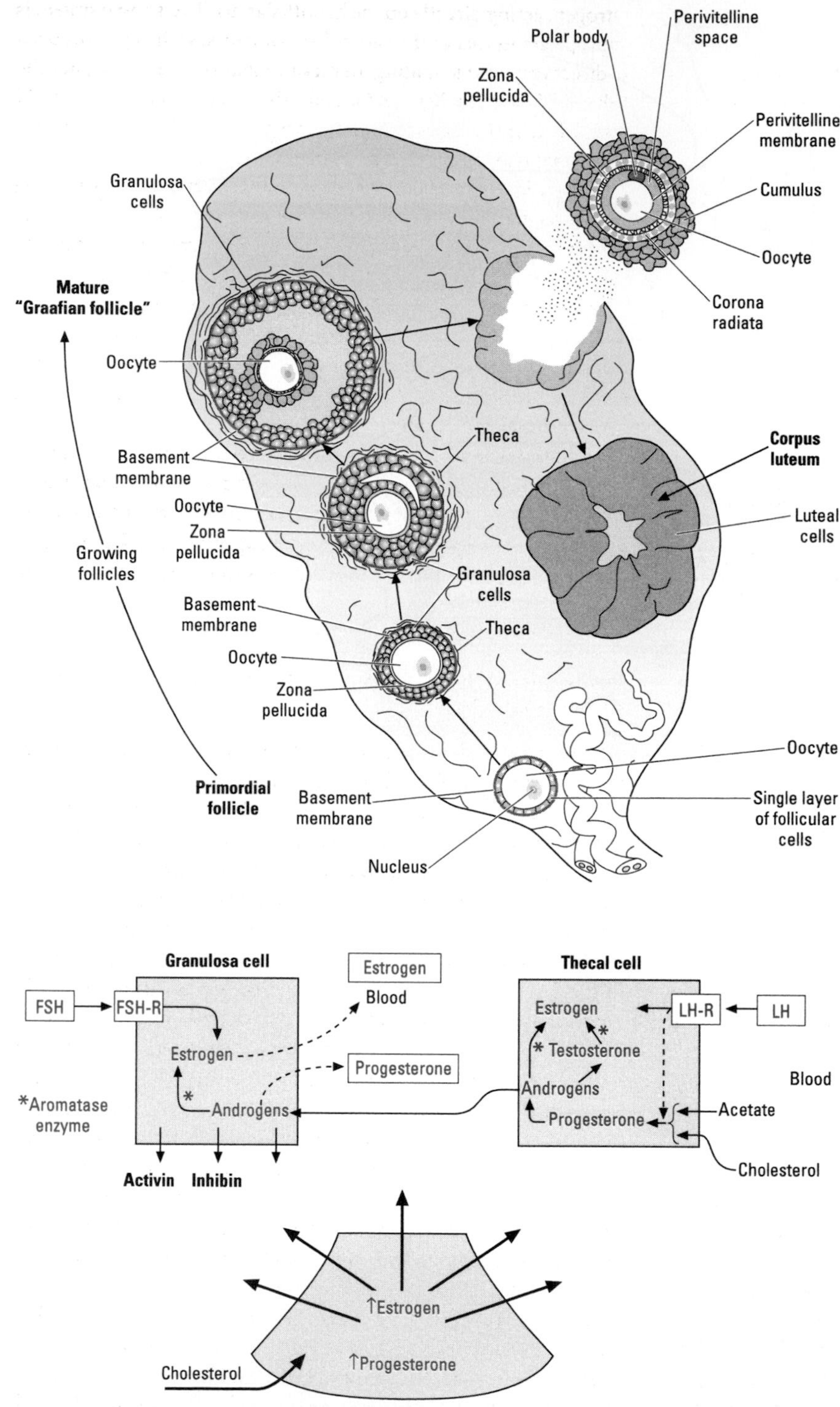

Fig. 10.28 Follicle maturation and corpus luteum formation in a mammal. FSH, follicle-stimulating hormone; LH, luteinizing hormone.

due to a hormone released from the embryo itself and pregnancy begins (dealt with below). An overview of hormone levels through an ovulatory cycle is shown in Fig. 10.29.

Hormonal control of female behavior is also complex in vertebrates, but removal of the ovaries usually causes sexual behavior to stop almost immediately, indicating the key role of the sex steroids. However, the effects get less in the "higher" mammals, so that ovariectomy after puberty in primates and some carnivores does not greatly alter behavior. The effects also depend on timing, and are much reduced in most adult mammals if they are already sexually experienced.

Available data from amphibians and reptiles indicate a primary role for estrogens, and some later effects from corticosterone. For example, in turtles, mating is linked to estrogen but nesting behavior is associated with corticosterone, which also helps to transfer

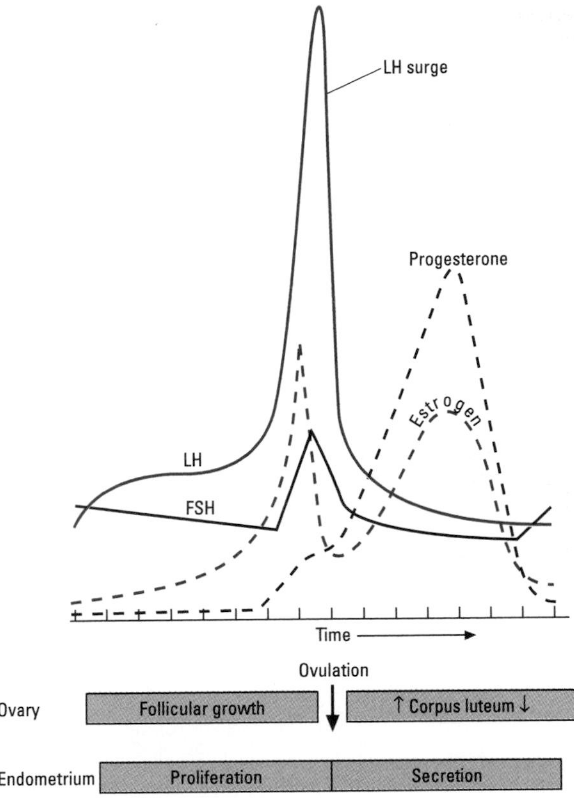

Fig. 10.29 Diagrammatic summary of the levels of key hormones through a mammalian ovulatory or estrous cycle. FSH, follicle-stimulating hormone; LH, luteinizing hormone.

lipid to the ovarian follicles prior to egg laying. Female birds are generally induced to ovulate by courtship, and estrogen is the main controller of their behavior, but with some effects of GnRH. The presence of eggs in the oviduct and then in the nest itself induces further changes, while behavioral and physical interactions with the male are also important in maintaining nesting behavior (see Fig. 15.47). Courtship with a male affects gonadotropins in the female, giving initial increases in estrogens that induce nest building, and then increases of progesterone that mainly affect receptivity. Incubation of the eggs brings about release of prolactin, which maintains broodiness and may (in doves, for example) initiate crop milk production ready for the hatchlings. Hence in birds similar behaviors in the two sexes (nest building, incubation, and care of the young) can be brought about by quite different hormonal paths.

A strange twist to the endocrinology of reproduction in birds comes with those rare species showing sexual role reversal, for example the phalaropes and dotterel, where females are brightly colored and males are dull, and the female takes the active role in courting, then abandons the nest for the father to incubate and raise the young. Here the hormones retain their "normal" functions, but the female secretes unusually high levels of testosterone which override other signals in determining her behavior.

In mammals, behavioral effects of hormones vary with species. Cats are best known, where a female "on heat" shows a characteristic "lordosis" posture (back bowed, rump raised, and tail to one side). This posture is lost in the absence of ovaries; the crucial controller is estrogen, acting directly on the hypothalamus. This simple system is appropriate in cats as they are reflex ovulators, with eggs produced in direct response to mating. In rats a similar system occurs, but the behavior is less marked, and an additive effect comes from GnRH (as in birds), released as a surge to give an LH peak. This allows copulatory behavior to be precisely timed to ovulation, appropriate for rodents, which are spontaneous or automatic ovulators producing eggs at regular intervals. In rhesus monkeys matters are more complex: estrogen makes the female behave only slightly more sexually, but males become more interested in her, because in the presence of extra estrogens bacteria present in her vagina secrete a scent that attracts males. The female monkey's receptive behavior is largely induced, unusually, by androgen hormones from the adrenal cortex. Oxytocin also plays some role in the primates, affecting sexual arousal and orgasm.

From this it is clear that even for the three well-studied mammals the control of female sexual behavior differs, with different hormones having different effects and acting on different sites. We can add to this complexity by noting that testosterone also has an effect on female behavior in many species, promoting sexual drive in the same fashion as in males.

Viviparous animals: the control of pregnancy, fetal physiology, and parental behavior

When a mammalian ova has been fertilized and an embryo arrives in the uterus, it floats freely for a few hours or days but then begins to implant in the prepared uterine wall. The egg is now a blastocyst, like a blastula but surrounded by an outer trophoblast layer. The trophoblast cells adhere to the uterine endometrial lining, and start to secrete digestive enzymes, cytokines, and growth factors. Effectively this provokes a localized inflammatory response, so the endometrium becomes swollen and permeable, with lymphocytes and macrophages invading the area. The trophoblast cells continue to invade, digesting more and more uterine cells, until the embryo is covered over by endometrium. The trophoblast cells may secrete a hormone similar to LH, termed **chorionic gonadotropin** (Fig. 10.30), though this seems only to occur in primates; it targets the corpus luteum in the ovary and causes it to persist rather than degenerate, so that it goes on making progesterone and estrogen and thus the uterine lining is not shed. After implantation the trophoblast becomes the **chorion** and continues with this function, although its activity declines during later gestation. In effect, the embryo has now taken over hormonal control of the ovary and uterus.

As the embryo grows, the **placenta** develops from both trophoblastic and endometrial tissues, the chorion being thrown into multiple finger-like folds (villi) that penetrate the uterine tissues and make intimate contact with lacunae filled with maternal blood. The placenta becomes the source of nutrients and oxygen for the fetus, and also serves as its excretory outlet. In addition it becomes the key hormonal tissue, responsible for secreting progesterone and estrogens through the remainder of gestation; typical patterns of these hormones through gestation are shown in Fig. 10.30. **Progesterone** levels rise steadily, mainly functioning to keep the uterine muscles quiescent and to prevent shedding of the uterine lining. It also promotes the secretion of viscous, acidic cervical mucus that is hostile to sperm, to prevent any further fertilizations.

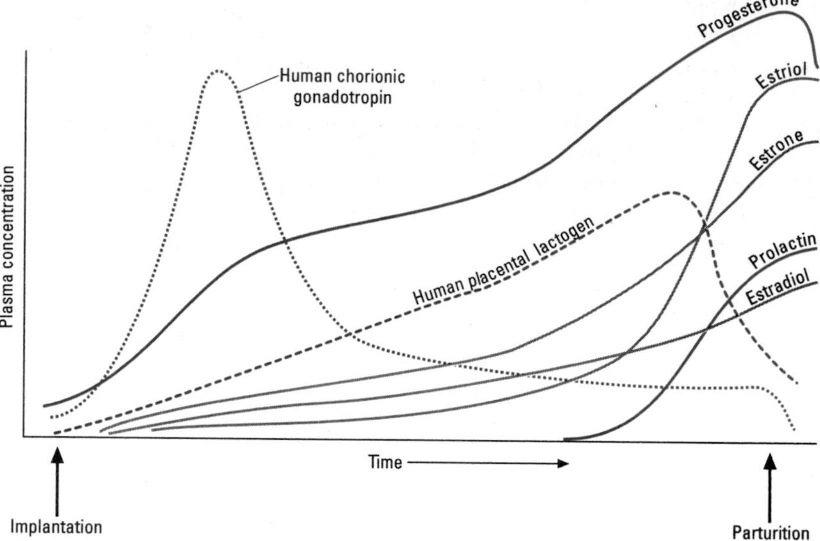

Fig. 10.30 Diagrammatic summary of the levels of key steroid hormones—including three different estrogens—through a mammalian pregnancy (additional hormones known mainly from humans are shown by dotted lines).

Estrogen levels (especially estriol) also rise, inducing maternal changes including the development of oxytocin receptors in the uterus, an increased irritability of the uterine muscles, and mammary gland differentiation. (The fetus may also promote maternal estrogen levels via its own rising levels of adrenal steroids, indicating that a fetus has some measure of control over its own time of birth.) **Prolactin** levels increase markedly near the end of gestation, triggering further swelling of the mammary glands and the beginnings of milk production. The placenta also produces other hormones needed for birth to occur and postnatal events to be successful: these include **placental lactogen** (mainly secreted into the maternal blood, stimulating breast maturation and faster fetal growth exploiting the maternal glucose supply), **chorionic thyrotropin** (increasing the maternal metabolic rate), and **relaxin** which causes the pelvic ligaments to relax and the pelvis to widen.

Quite what triggers **parturition** and birth is still unclear, although many factors are known to have some effect (Fig. 10.31). The fetus probably contributes via a spurt of CRH from its hypothalamus, triggering adrenocorticotropin release from its pituitary and thus cortisone release from its adrenal cortex. This affects enzyme activity in the placenta where progesterone converts to estrogens (helping to release the muscle-relaxing "progesterone block" on the uterus), and also affects the production of prostaglandin, causing uterine smooth muscle contraction and cervical dilation. Some fetal cells produce oxytocin, which causes the uterus or placenta to release prostaglandins, both of these being uterine muscle stimulants, leading to reflex contractions. Direct stretching of the uterus by the fetus may also be crucial, making the muscles more excitable and likely to go into contractions.The insufficiency of the placenta (becoming unable to supply the needs of the fetus) is also implicated, perhaps starting the cortisol stress response of the fetus. As the fetus pushes downwards in the uterus, stretch receptors in the cervix send afferent impulses to the brain, so that further maternal oxytocin is released giving a positive feedback and causing the very strong uterine contractions needed for labor and parturition. Oxytocin then acts as the trigger for milk ejection in breast tissues that are already under the milk-producing influence of prolactin.

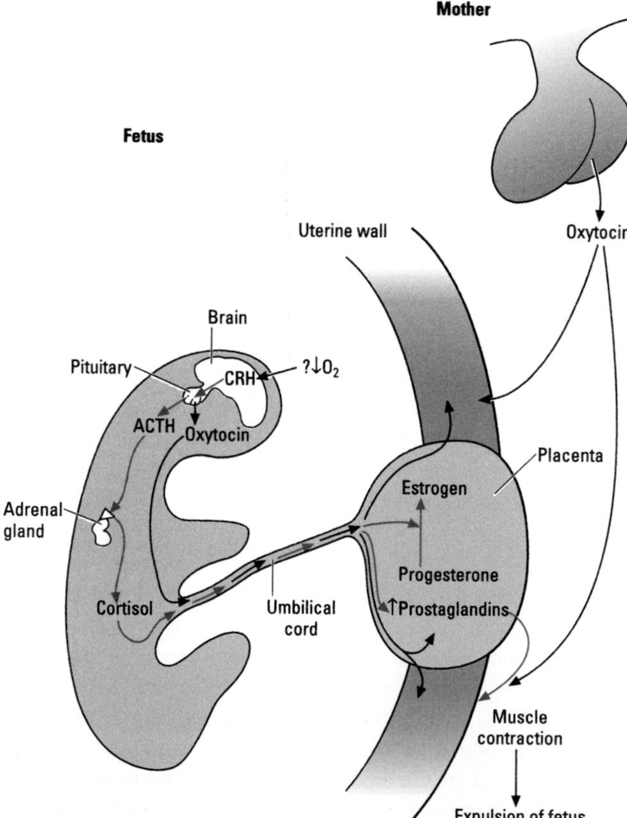

Fig. 10.31 The effects of hormones from the mother and fetus during a mammalian birth. ACTH, adrenocorticotropic hormone; CRH, corticotropin-releasing hormone.

After mammals have given birth, a whole new suite of maternal behaviors comes in; in a typical small mammal with a large litter, such as a mouse, this includes eating the placenta, cleaning the young, retrieving them, and adopting a spread-eagled posture so they can find the nipples. By then the main source of hormones during pregnancy (the placenta) has been lost, and prolactin becomes predominant in controlling maternal physiology, especially milk production. Oxytocin also often promotes nurturing behaviors. However, hormones surprisingly do not have much effect on higher mammal maternal behavior, as the components of such behavior are also shown by virgin females, or even by males when pups are presented.

Thus while reproductive *physiology* remains critically under the control of hormones, there may be a trend in "higher" taxa to an increasing freeing of all sexual and reproductive *behaviors* from direct gonadal control—the sex steroids have a lesser role and the brain hormones become more important in higher tetrapods. Indeed, within the mammals reproductive behavior is often almost completely released from hormonal control and is more under cortical influence, so that gonad removal has few behavioral effects.

It is also noteworthy that the relatively few available hormones in vertebrates can be used for a range of quite different triggering roles in reproduction, according to the environmental and life-history needs of the particular species. This is one of the clearest examples of evolution "making do" with what is available rather than starting from scratch with an "ideal" design for a control system.

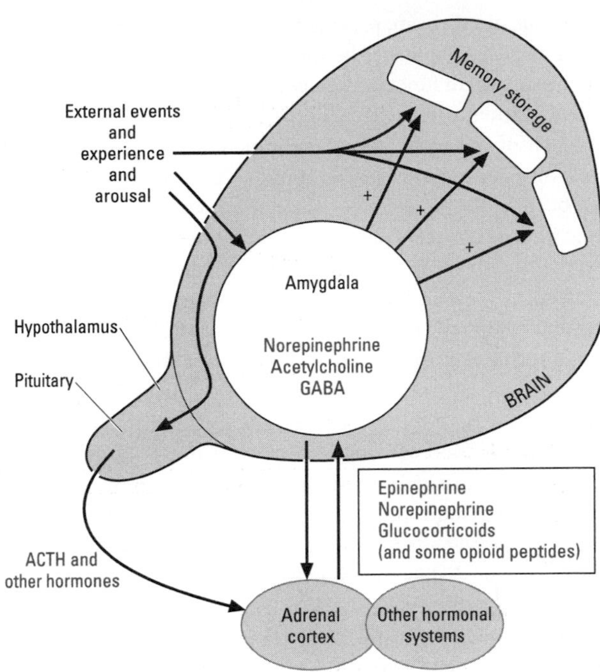

Fig. 10.32 The postulated hormonal effects on memory storage in the brain of mammals. ACTH, adrenocorticotropic hormone; GABA, γ-amino butyric acid.

10.8 Hormones and other behaviors: aggression, territoriality, and migration

Aggression between animals is not particularly widespread, sometimes involving contests for access to food or homes, but commonest in relation to the breeding season, when males in particular are competing for access to mates, or more indirectly are competing for territories to which mates may be attracted. Therefore perhaps a priori the hormones involved in **aggressive behaviors** are likely to be, or to be linked to, the sex hormones that we have already discussed. Indeed there are strong indications that sex hormones affect alternative "life-cycle states" in birds or mammals, so that (for example) birds may molt in certain hormonal states, may hold territories, fight other males, and reproduce in others, and have a third option of an "emergency" physiological state with low risk taking, reduced aggression and sex drive, and a high propensity to escape and to forage. Having the same hormones controlling all these three states insures that any one individual responds flexibly to its environment.

Undoubtedly in mammals and birds, **testosterone** is the main controller of male aggression (although in some species, and in some seasons, there are indications that testosterone is converted to estrogen by aromatase in the brain, with estrogen as the final trigger). In red grouse, territory size and vigor of fighting are increased after testosterone implants, while domestic chickens crow more and fight more if treated with testosterone. But in chickens, and also in mice, the area of brain affected by testosterone is not the preoptic nucleus (as is the case with sex), so the hormone must be acting via a different effector mechanism. Within the vertebrate CNS, circuits that mediate aggression often use serotonin as the neurotransmitter; this amine affects aggression levels in many invertebrates and vertebrates, up to and including humans. Increased male aggression may therefore result in part from upregulation of these serotonin pathways. By contrast, female aggression in birds and mammals tends to be greatest when they are sexually nonreceptive, and increasing estrogen and progesterone levels make a female less aggressive and more receptive.

Aggression also has effects on the recipients, who may be losing contests. Here the main consequence is a rising level of corticosteroids, the classic "stress hormones", and in social animals subordinate individuals may have more or less permanently elevated glucocorticoids in their blood, in turn making them more susceptible to a range of diseases and stress-related conditions.

Many insects and vertebrates also have hormonal control of **migration behaviors**. In the monarch butterflies (*Danaus plexippus*) this is linked with the control of aging; the migrating adults of the fall generation persist over winter for up to 8 months, while the summer generations survive for only 2 months. This increased longevity is attributed to suppression of **juvenile hormone**, which in effect puts the migrants into a state of reproductive diapause. In migrating birds and mammals migratory behavior is commonly related to dwindling food supplies rather than climate; it is often triggered by photoperiod and linked in to the "clock" mechanisms mediated by the pineal–pituitary links, with **prolactin** as a key controller.

In mammals, and presumably in other vertebrates, hormones also have complex interactions with higher neural functions, including memory storage and especially the memory of particularly emotionally arousing events. Such effects are especially mediated via the amygdala, as summarized in Fig. 10.32.

10.9 Pheromones and the control of behavior

Pheromones are chemicals that influence the behavior of conspecific animals. In a sense they should be regarded as the *original* communication system, present even in premetazoans, influencing aggregation and avoidance between unicellular organisms. Perhaps their most ancient function was as a gamete attractant, released into the environment from female cells and inducing the release and approach of male gametes, a phenomenon that still occurs quite commonly in mass-spawning marine invertebrates. Hormones then become the descendants of pheromones, acting between cells in the newly aggregated organisms. Pheromones later were adopted into the slightly more complex roles of mediating interactions between groups of cells that became independent individual animals. Certainly chemical signals between conspecifics do occur in most animal phyla, and equally certainly there are strong links between hormonal and pheromonal systems. Occasionally the same chemicals can have both roles, and often the receptivity to pheromones is under endocrine control with many animals becoming more responsive to mating-related pheromones when they are most fertile. Indeed some pheromones that act rather slowly to alter behavior ("primer effects") work via the endocrine system, whereas others act immediately and directly ("releaser effects").

Pheromones have to be picked out from amongst a complex chemical environment, and they have to give unambiguous signals. This requires novel chemicals, or mixes of chemicals, that can (in many cases) identify a species precisely. In aquatic environments the chemicals used need to be soluble and reasonably persistent in moving water; in the terrestrial environment, where transmission is through air, the chemicals need to be volatile.

Pheromones are best known in insects, where the chemoreceptors on the antennae are very accessible, so that recording from individual neurons allows the detection of thresholds for reception and response, and analysis of the receptor specificity to individual compounds. Furthermore, the behaviors elicited in the insect may be very clear-cut and stereotypical. Hence this is an area where we can establish the linkage between molecular stimulus, receptor response, subsequent CNS processing, and resultant behavior.

The first pheromone to be identified was **bombykol** (hexadecadien-10-*trans*-12-*cis*-ol; Fig. 10.33a) from the silk moth *Bombyx mori*, isolated in 1959 from 500,000 silkworms. Male *B. mori* antennal sensillae can respond to a single bombykol molecule, and just 200 molecules will give a full behavioral response. In fact the natural silkworm pheromone is a 10 : 1 mix of bombykol and its aldehyde form bombykal. Most terrestrial pheromones show similar "design principles": they are small organic volatiles (commonly C_5–C_{20}); they are acids, aldehydes, or ketones (Fig. 10.33b) derived from fatty acids; and they are commonly multicomponent, involving a small range of similar chemicals with species specificity residing in the exact mix of volatile components. Hence these kinds of communication systems can evolve and "speciate" quite fast, and can allow species recognition on the basis of relatively few chemicals.

Here we examine briefly how pheromones affect key aspects of behavior. More details are given in Chapters 11–16, in relation to specific environments.

10.9.1 Mating behavior

Pheromones are most commonly analyzed, and are perhaps most potent, in terrestrial nocturnal species such as moths, where scent is the most useful form of communication. Most insects probably

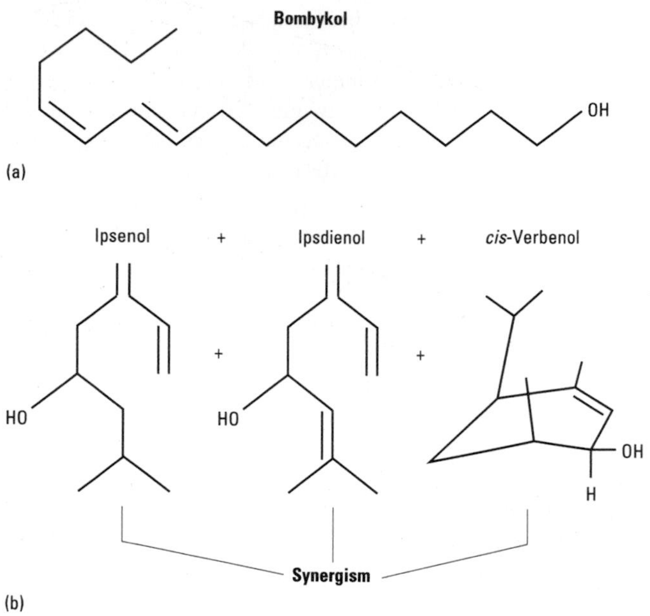

Bombykol

OH

(a)

Ipsenol + Ipsdienol + *cis*-Verbenol

HO HO OH H

Synergism

(b)

Fig. 10.33 Typical pheromone structures: (a) bombykol, the sex pheromone of the silk moth, and (b) three components used in varying mixtures by species of bark beetle from the genus *Ips*. (From Wilson 1970.)

use them, and they are prerequisites for successful mating in the majority of species studied.

Mate location

In most insects (and also terrestrial crustaceans and spiders) there are "sex attractant" pheromones emitted by the female. The pheromone here is a long-lasting volatile mixture, which disperses downwind and is picked up by males' chemoreceptors. The pheromone source produces a classic **odor plume** (Fig. 10.34) or "active space"; note that it is strongly directional, and the plume is smaller at faster wind speeds. Hence the system works optimally in gentle breezes, when the active space within which a male responds can be 3–5 km in length. Some insects improve targetting by emitting pheromone only at a species-specific time; for example the oakleaf moth *Antheraea polyphemus* always "calls" with its pheromone 3–4 h after dusk.

When the scent is picked up by the male it elicits a variety of behaviors: wing fluttering, abdominal movements, then upwind flight, usually with a side-to-side or "zig-zag" casting component. This is because the flying male simply increases his turning rate whenever he loses scent at the plume edge, so that he tends to get back into the plume. His behavior is simple upwind tracking (**anemotaxis**), and the pheromone concentration is not very important. However, the exact nature of the chemical mix released may convey quite subtle shades of meaning to the male, perhaps including cues on the age and health of the emitting female. In some species that can only mate once each night because they make only one spermatophore per day, it has recently been shown that after mating the pheromonal receptivity of the male's neurons is turned down, so he does not follow plumes again that night; his responsivity increases again after 16–24 h.

In aquatic worms and crustaceans the same kind of plume system

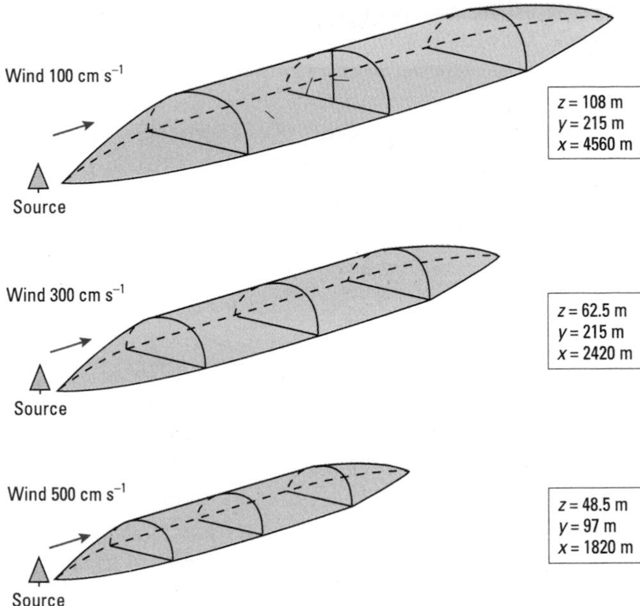

Fig. 10.34 The structure of a pheromonal odor plume in air (shown at three different wind speeds. The plume remains coherent in fairly still air, and can then travel up to 5 km. (From Wilson 1970.)

operates with mating pheromones in the water currents; however, the chemicals are of higher molecular weight and less volatile, often small polypeptides. Furthermore, in aquatic environments animals use **chemotaxis** as well as a simple upstream response. In response to a pheromone they can track across the plume to assess the chemical distribution, because turbulence is at least 100 times less disruptive in water, allowing better use of reliable spatial information in the plume.

Pheromones for mate attraction are also widespread in all land tetrapods, up to and including the primates. Here they quite often have primer effects, acting in the longer term and contingent on other stimuli, rather than having immediate releaser effects. They tend to act through the vomeronasal organ (VNO), close to the nasal cavity; drawing air over this to sample the smells in the air produces the raised upper lip commonly seen in cats and dogs. The VNO connects via the olfactory nerve to the hypothalamus, thus feeding into the neuroendocrine system.

Courtship

Once a pair of animals has located each other, closer range pheromones help to organize courtship. Some may be potent "arrestants" for both male and female; others specifically induce male copulatory actions. Some male beetles will repeatedly attempt to court and copulate with a piece of paper impregnated with female pheromone, while assiduously ignoring a real female nearby covered with a glass lid. Male butterflies and moths have scent patches ("hair pencils") on the wing that are held up to waft scent to the female. This is often part of a complex courtship sequence (a fixed action pattern of male behavior) that each male proceeds through once given an encouraging signal by a female. Newts also have a courtship pheromone, termed sodefrin, unusually released by the male to attract the female; the female responsiveness to this is affected by her hormonal state.

Copulation

Copulation often requires very close-range pheromonal interactions, with chemicals secreted from glands in the genitalia and operative over only a few millimeters; sometimes these are cuticular hydrocarbons, requiring contact, and are not volatile at all. Without these cues, which are genuinely chemotactic effects, either partner may disengage, or may position themselves wrongly so that insemination fails.

In some animals copulation results in the transfer of chemicals from one partner to the other, whether via the genitalia, the mouth (for some animals that bite during copulation), or in the curious case of terrestrial snails via the calcareous "love dart" that each hermaphrodite partner shoots into the other. These chemicals may induce changes in the recipient, such as increased egg production, improved sperm handling or storage, or the inhibition of further mating. Such chemicals are not strictly pheromones since they have a direct internal action, nor are they hormones since they are affecting another individual; instead the term "**allohormones**" has been proposed. They can be seen as covert chemical "gifts", usually from the male, and are likely to be subject to particularly strong sexual selection. In some insects (e.g. *Drosophila*), the male ejaculate contains "sex peptides" that diffuse into the female's hemolymph and can suppress her future mating behaviour, so insuring that the inseminating male's sperm fertilizes all her eggs—a somewhat dubious "gift" from the female's perspective!

Postnatal care

Pheromones influencing brooding after egg laying, or maternal behavior after giving birth, are relatively uncommon largely because such behavior is rather rare. In insects the eggs of some species release chemicals that elicit brooding by the mother; and true postnatal care occurs in some beetles and bugs as well as in the obvious cases of social insects, although chemical cues between egg or larva and mother are relatively unexplored.

In vertebrates pheromones are clearly operating in the period of maternal care, and there is some evidence that human mothers can detect the smell of their own infant on clothing whereas fathers cannot. Similarly infants can identify their own mother's clothing after it has had breast contact, and will preferentially turn towards the odor of their own mother rather than that of other lactating mothers.

Reproductive synchrony

Synchrony may be particularly useful in social animals, and is commonly mediated by pheromones operating between females. These are well known in rodents, where estrous synchrony between females is the outcome. The same pheromones influence mating preferences, and synchrony of weaning time. Similar pheromones may occur in social primates, including humans (though in humans we tend to define pheromones rather differently, as signals that regulate behavior without the recipient being consciously aware of having detected an odor). Compounds from female human armpits in the early stage of the menstrual cycle accelerate the preovulation LH surge in other women, whereas compounds from the underarm when close to ovulation have the reverse effect of lengthening other women's cycles. These two compounds have been termed "ovulatory pheromone" and "follicular pheromone".

10.9.2 Aggregation and group behavior

In many herbivorous insects, feeding individuals can sequester host-plant compounds and later use these as pheromones to attract groups of conspecifics. "Aggregation pheromones" also occur in some butterflies that aggregate to hibernate, and in tent caterpillars that aggregate to create a more equable microenvironment. Many marine invertebrate larvae also use pheromones to organize group settling behavior, insuring a reasonable aggregation of individuals to mate with in future. Aggregation into clusters in bees and some other social insects is elicited by a pheromone from the queen's Nasonov gland.

10.9.3 Alarm, caste, and trail pheromones

This range of pheromones is most conspicuous in social animals, as major organizers of the behavioral system of the whole colony as well as of individuals.

Alarm pheromones

These are not particularly common except in social insects, where they are produced by workers but never by drones. The same compounds recur in many species, but they are mainly used to warn other members of the colony (indeed in a few ant species they can be colony specific). They operate only over a few centimeters, and tend to be very volatile molecules (less than C_{12}). In ants they derive from mandibular glands or the abdominal Dufour's gland, but in bees and wasps they come from the sting shaft. At high concentrations these same compounds may become a defense, being highly irritable to other insects; the obvious example here is formic acid from many ants.

The effects are orientation towards the source at low concentrations, and furious activity (either attacking or fleeing) at higher levels; i.e. behavior is concentration dependent, as with many pheromones. In weaver ants, which make communally defended leaf nests, complex alarm behavior follows the secretion of pheromone by a single major worker; each of four components works over a different range, so that furthest from the source highly volatile hexanal is detected and causes alerting, next 1-hexanol acts as an attractant and brings in more workers, then undecanone causes orientation and biting, until finally (almost at the source) the ants meet the low volatility 2-butyl-2-octenal which elicits furious biting and fighting behaviors.

Outside the social insects a well-known example comes from the alarm pheromone of the sea anemone *Anthopleura* (see section 11.7.3); when a predatory sea slug eats these anemones it acquires the pheromone and releases it slowly from its own tissues, thus being forced to warn of its own approach! Earthworms also have an alarm pheromone, released in their mucus. Perhaps the most famous vertebrate example is the "schreckstoff" of fish such as minnows; if one fish is wounded, schreckstoff is released into the water and all others in the vicinity panic and flee (though whether this is really a "signal" effect has been questioned). Also in this category we might include "warning chemicals"; skunk musk is a classic example, which both warns conspecifics and warns off potential predators, and is so potent that humans can detect it 30 km away.

Territoriality, trail, and marker pheromones

Here the simplest pheromones are the chemical marks accompanying egg laying (for example by parasitoids), which effectively say "site occupied". Actual "trail markers" are rare, and best known in ants where locomotion is mainly overland and trails can be left from abdominal glands or feet. These are used as navigational aids over hundreds of meters, leading other workers to good food sources, to a site of alarm, or to new nests. Workers are poor at detecting directionality, and may follow the trail either way; though in some ants the trail is laid as "pear-shaped" smears, so another ant can assess the direction. Sometimes alarm pheromone is added to the trail, apparently to keep recruited workers highly excited.

Some bees use aerial odor trails for recruitment too; for example some tropical stingless bees (*Trigona*) put a drop of scent on leaves and twigs at strategic intervals along the way to a good food source. Some bumblebees may also do this, and thus follow consistent trails among their foraging plants.

Pheromones are also used in some vertebrates not as trails but certainly for territory and home-range marking; mice and rat marking pheromones are well known, and many species in the cat and dog families can produce complex and individually distinct and identifiable chemical marks.

Caste pheromones

The whole caste structure of a social insect colony may be held together by pheromonal secretions in bees and ants. Normally queen honeybees (*Apis*) secrete "queen substance", a mix of 9-oxydecanoic and 9-hydroxydecanoic acids. This inhibits gonad development of workers, and causes them to show brood-cell constructing behaviors (i.e. this is a pheromone affecting morphological and physiological change, as well as eliciting behavior). The same substance is the queen's sex attractant in swarming later in the season. Another queen pheromone inhibits building of new queen cells, an example of a pheromone directly *preventing* expression of a normal behavior pattern. When one queen dies the workers are released from this inhibition and start making new queens almost immediately.

Other social insects have complex variations in pheromone emission within the species or within one nest, allowing for caste recognition, giving cues on physiological state, or even on genetic strain and kin relatedness. Such pheromones may help in maintaining outbreeding.

Mammals may also use pheromones to organize "castes" in their groups, particularly well known in rodents such as hamsters and mice. In many pack-dwelling mammals pheromones in the urine or from special glands near the tail act as dominance signals to structure the group and its "pecking order".

10.9.4 Allomones and kairomones

These terms are used for behavior-altering signals between species. **Allomones** are chemical signals that are adaptively favorable to the emitter. They can be "one way", where the receiver suffers (e.g. defensive secretions of all kinds, often termed "chemical weaponry"), or can be "two way" where both emitter and receiver benefit (often the basis of symbiotic relationships, e.g. where ants tend and "milk" lycaenid butterflies, or cleaner fish feed on scraps around the mouths of large predatory fish).

Kairomones are chemical signals that are adaptively favorable only to the receiver (i.e. have negative effects on the emitting species). Kairomones are common if interpreted in a loose sense, since many animals pick up a chemical signal from another species as a way of finding prey or food generally, or of finding a settlement site in aquatic organisms. More precisely, though, kairomones should be chemical signals "deliberately" emitted by one organism for its own purposes, but picked up by another; these can usually be seen as pheromones that have "boomeranged" on the emitter and been hijacked by another species. For example sex or alarm pheromones from one species can become attractants for its parasites and predators. The spider *Habronestes* homes in on the alarm pheromone from injured ants, and both pike and predatory swimming beetles respond to the schreckstoff of minnows. In some parasites, the host *hormones* can be used as kairomonal cues for parasite breeding; in rabbit fleas, hormone levels are detected by fleas biting the mother rabbit's ears, and a rabbit pregnancy makes the fleas breed so that the next generation of fleas can disperse onto the baby rabbits. Thus more than behavior can be affected, with morphological and physiological change also brought into play. This is seen very clearly in some nudibranchs, whose drastic metamorphosis to adult form from veliger larva is only induced when they detect small organic compounds released by their coral prey.

Whether a chemical is an allomone or a kairomone may be time or context dependent. For example, lobsters produce chemicals in their urine that reveal their sex, aggression level, and perhaps unique identity, potentially very useful information to an antagonist, but they can time the release of the urine during a fight for

Table 10.7 Differing roles of prolactin in vertebrate taxa.

	Function	Action	Sites
Teleosts	Water and salt balance	Controls Na^+ flux	Gill, kidney
		Reduces water flux	Gill
		Increases urine flow	Kidney
		Increases mucus secretion	Skin, gut
	Reproduction	Increases growth, maturation	Gonad
		Increases secretion	Seminal vesicles
		Induces parental behavior	Brain
	Immunoregulation		Thymus, spleen
	Growth		Most tissues
	Pigmentation	Increases secretion	Melanocytes
Amphibians	Water and salt balance	Reduces Na^+ flux	Skin, gut, kidney, bladder
		Increases water uptake	Skin, bladder, kidney
	Reproduction	Reduces maturation	Sperm
		Stimulates secretion	Cloacal glands
		Induces parental behavior	Brain
	Metamorphosis	Cell proliferation	Most tissues
		Cell resorption	Gill, tail
		Limb regeneration	Muscle, bone
	Immunoregulation		Thymus, spleen
Reptiles	Water and salt balance	Regulates blood Na^+	Kidney
	Reproduction	Reduces maturation	Gonad
	Metamorphosis	Regeneration	Tail
		Induces molting	Skin, scales
	Immunoregulation		Thymus, spleen
Birds	Water and salt balance	Induces secretion	Salt glands
	Reproduction	Increases secretion of "milk"	Crop (some species)
		Induces maturation	Brood patch
		Alters behavior, e.g. singing	Brain
		Induces parental behavior	Brain
	Metamorphosis	Induces molting	Feathers
	Immunoregulation		Thymus, spleen
Mammals	Water and salt balance	Reduces Na^+ flux	Kidney
	Reproduction	Induces growth, maturation	Gonads
		?	Uterus
		Increases secretion of milk	Mammary gland
		Enhances secondary sex characters	Hair, glands, etc.
		Induces maternal behavior	Brain
	Immunoregulation (not primates?)		Thymus, spleen

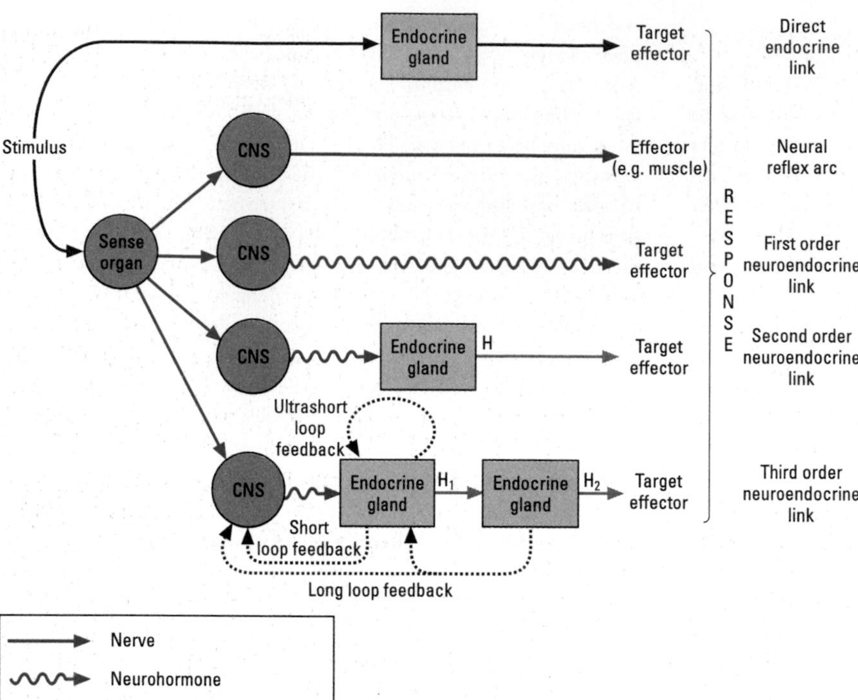

Fig. 10.35 An overview of the regulation of hormones with first, second, and third order links and with first and second order feedback loops, including both direct endocrine and neuroendocrine controls. A simple neural reflex arc is also included for comparative purposes. CNS, central nervous system.

maximum benefit to themselves, both within a fight and in terms of improving the losing opponent's future recognition of a dominant fight-winning opposing lobster.

Note also that chemical mimicry of pheromones is a possible way of generating effective kairomones, as a way of cheating the system. For example, the bolas spider impregnates its small web with pseudomoth pheromones and thus lures in male moths.

Thus the whole boundary between hormones, pheromones, kairomones, and allomones can get very blurred. The situation is made even more complex by all the possibilities of deceptive signaling, of jamming signals, and by the ability of many animals to "learn" about the reliability or otherwise of chemical signals over time.

10.10 Conclusions

It is evident that chemical control systems within animals have a long evolutionary history, originating in prokaryotic organisms. The very earliest cells probably used intracellular peptide-signaling systems mediated through cAMP, with families of peptides gradually diversifying. It is likely that steroidal signals must have been in place very early on too, since bacteria, algae, and plants all use them. Extracellular secretion of any of these signaling molecules would give scope for communication between individual unicells (equivalent to pheromonal signaling, as we noted above) and this could be readily modified to serve as a signaling system between the differing cell types of a multicellular organism, with particular peptides becoming expressed, secreted, and received only in particular kinds of cells. Peptide-secreting cells may have been commonplace and had common origins, but where the secretions came from cells that were also excitable a neuroendocrine capability was achieved, giving faster and more direct control. As we have seen, in many lower invertebrates it is the neurohormones that are most in evidence. In more complex animals the controlling neurosecretory peptides became concentrated in the brain, culminating in the vertebrate anterior pituitary, controlled in turn by releasing hormones from the hypothalamus. However, in the most sophisticated birds and mammals we see a further step taken, with higher brain centers often developing the ability to override hormonal influences, at least in the control of behavior. In these higher tetrapods there is also a trend to concentrating hormone sources into just a few glands (sometimes incorporating two or three separate tissues) rather than using diffuse tissues or scattered cells.

A general message that has often been drawn from study of hormones, and which relates to their likely evolutionary origins, is that "it is not the hormones that evolve, but the uses to which they are put". Hence the same molecules occur repeatedly; the insulin family is very ancient, with similar molecules occurring in vertebrates, insects, molluscs, and even sponges, and the oxytocin/vasopressin family also crops up in various invertebrates. Hence "old" hormones frequently get coopted and put to new uses in widely differing systems. Hormones that originally controlled slow processes of growth, maturation, and reproduction often become key players in shorter term homeostatic regulation, and the control of very short-term behavioral patterns. A classic example here is the multiple and changing uses of prolactin (Table 10.7); it has been estimated that prolactin has at least 300 different functions just in the vertebrates.

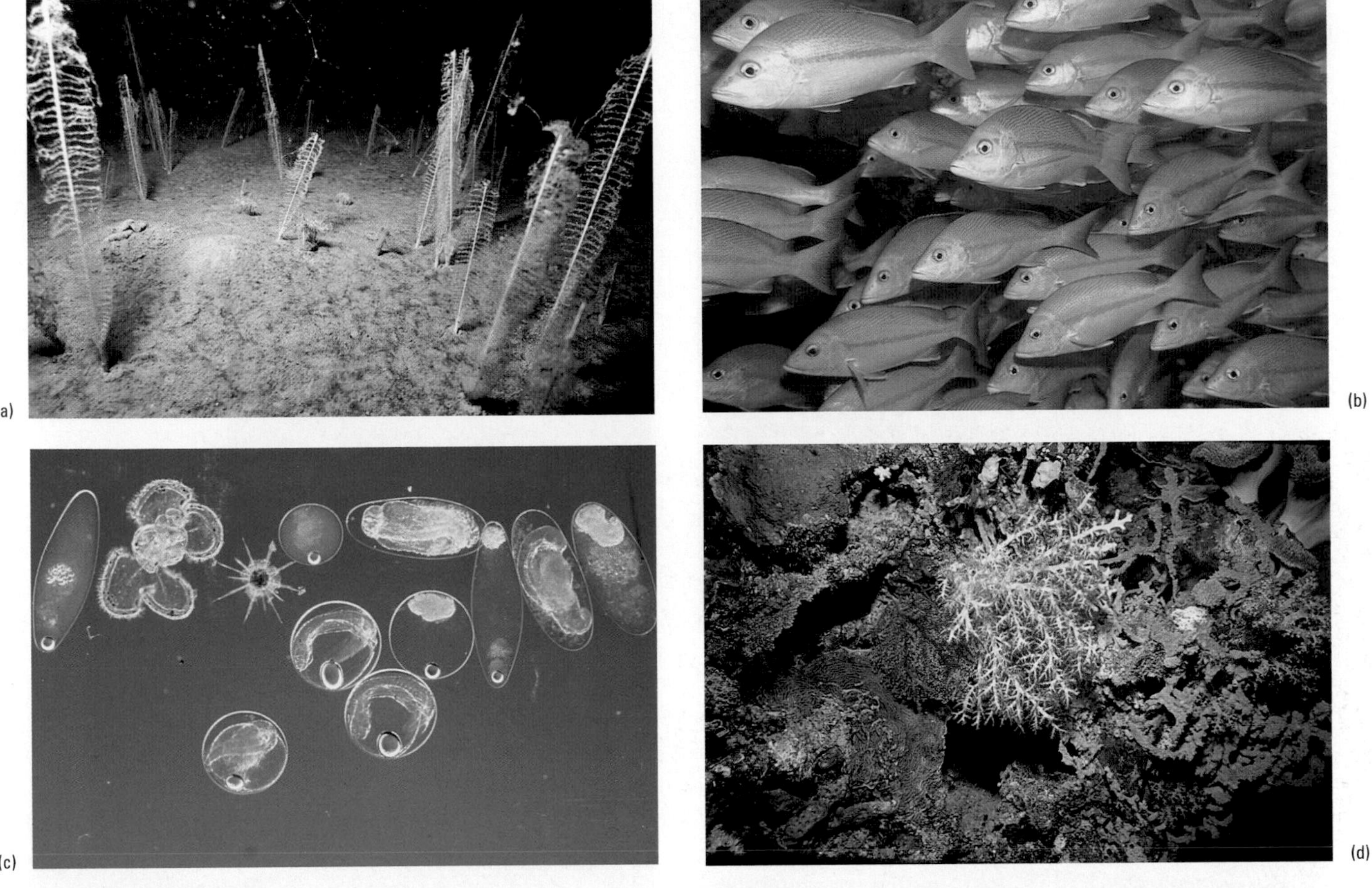

Plate 1 Marine environments
(a) Marine benthos. Shallow soft mud with sea pen *Virgularia mirabilis* and the anemone *Sagartiogeton laceratus*, Loch Laxford, Scotland, UK. (Courtesy of R. Holt/Joint Nature Conservation Council). (b) Marine nekton. A shoal of fish (hussars *Lutjanus amabilis*), Heron Island, southern Great Barrier Reef. (Courtesy of L. Newman & A. Flowers.) (c) Marine plankton, including invertebrate larvae, fish eggs, and phytoplankton. (Courtesy of P. Parks, Imagequest.) (d) Coral reef: underwater scene with soft corals, off Madang, Papua New Guinea. (Courtesy of L. Newman & A. Flowers.)

[*Facing p. 386*]

Plate 2 Littoral environments
(a) Sea cliffs, Lundy Island, UK. (Courtesy of A.S. Edwards.) (b) Vertical zonation on a moderately exposed temperate shore, Devon, UK. (Courtesy of P.G. Willmer.) (c) Mixed sandy/rocky shore with sea lions, Galápagos islands. (Courtesy of Lucy Crooks.) (d) Rock pools and rocky shore, Jervis Bay, southern Australia. (Courtesy of P.J. Sunnucks.) (e) Sand-dunes with marram grass, South Wales, UK. (Courtesy of P.G. Willmer.) (f) Volcanic lava shore, with basking iguanas, Fernadina, Galápagos. (Courtesy of Lucy Crooks.)

(a)

(b)

(c)

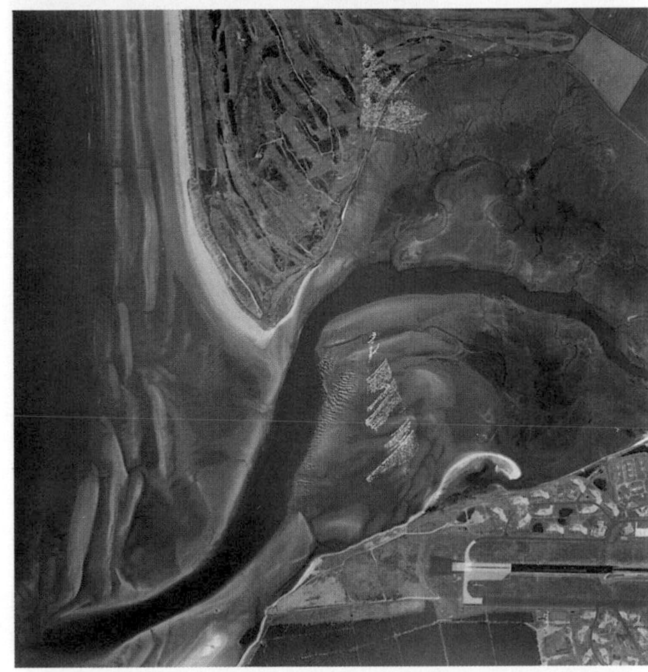
(d)

(e)

(f)

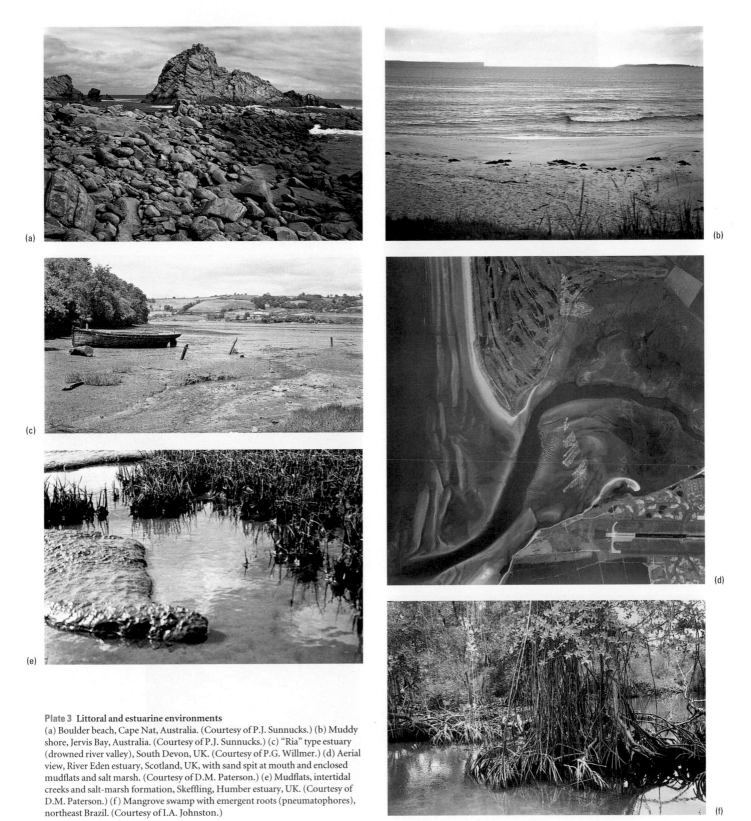

Plate 3 Littoral and estuarine environments
(a) Boulder beach, Cape Nat, Australia. (Courtesy of P.J. Sunnucks.) (b) Muddy shore, Jervis Bay, Australia. (Courtesy of P.J. Sunnucks.) (c) "Ria" type estuary (drowned river valley), South Devon, UK. (Courtesy of P.G. Willmer.) (d) Aerial view, River Eden estuary, Scotland, UK, with sand spit at mouth and enclosed mudflats and salt marsh. (Courtesy of D.M. Paterson.) (e) Mudflats, intertidal creeks and salt-marsh formation, Skeffling, Humber estuary, UK. (Courtesy of D.M. Paterson.) (f) Mangrove swamp with emergent roots (pneumatophores), northeast Brazil. (Courtesy of I.A. Johnston.)

Plate 4 Freshwater lotic environments
(a) Lowland tropical river, Trauna River, Mt Hagen, Papua New Guinea. (Courtesy of G.N. Stone.) (b) Dry river bed in summer in "Mediterranean"-type climate, Mootwingee, Australia. (Courtesy of P.J. Sunnucks.) (c) Upland stream in arid zone, southern Chile. (Courtesy of J.M. Lambert.) (d) Upland stream, cool temperate Kyrgyzstan. (Courtesy of J.M. Lambert.) (e) Tropical eutrophic stream, Java, Indonesia. (Courtesy of G.N. Stone.) (f) Humic river in rain forest, northern Queensland, Australia. (Courtesy of I.A. Johnston.)

(a) (b) (c) (d)

Plate 5 **Freshwater lentic environments**
(a) Upland oligotrophic lake, Mt Rugby, New Zealand. (Courtesy of P.J. Sunnucks.) (b) Tropical lowland seasonal eutrophic lake, Dindira, Tanzania. (Courtesy of G.N. Stone.) (c) Small mesic lake, Bosherston Ponds, Pembroke, UK. (Courtesy of D.M. Paterson.) (d) Crater lake, Rinjani, Lombok, Indonesia. (Courtesy of G.N. Stone.)

Plate 6 Unusual aquatic environments
(a) Pitcher plant, Mt Kinabalu, Borneo. (Courtesy of J.M. Lambert.) (b) Soda lake, hypersaline and sulfurous, northern Chile. (Courtesy of J.M. Lambert.) (c) Salt flat (resulting from completely dried out salt lake), northern Argentina. (Courtesy of J.M. Lambert.) (d) Rift Valley soda lake, Lake Magadi, Kenya, dominated by flamingos. (Courtesy of I.A. Johnston.) (e) Hot-water springs and small geysers feeding soda lake, with thermophilic algae, El Tatio, Chile. (Courtesy of J.M. Lambert.) (f) Deep-sea vent fauna: giant, tubicolous, pogonophoran worms, *Riftia pachyptila*, found abundantly around deep-sea fumaroles. (Courtesy of R.R. Hessler.)

Plate 7 **Semiterrestrial and terrestrial environments**
(a) Tropical flooded forest, Amazonas, Brazil. (Courtesy of I.A. Johnston.) (b) *Melaleuca* swamp (swamp cypress), southern Australia. (Courtesy of P.J. Sunnucks.) (c) Temperate southern beech (*Nothofagus*) forest, Tierra del Fuego. (Courtesy of J.M. Lambert.) (d) Tropical grassland, Mt Hagen, Papua New Guinea. (Courtesy of G.N. Stone.) (e) Temperate moorland, Isle of Hoy, Orkney, UK. (Courtesy of A.S. Edwards.) (f) Mediterranean "garigue", Mt Carmel, Israel. (Courtesy of B. Vulliamy.)

Plate 8 **Terrestrial environments**
(a) Tropical rain forest, New World, Ecuador. (Courtesy of J.M. Lambert.) (b) Monsoon forest, Kakadu, northern Australia. (Courtesy of P.J. Sunnucks.) (c) Temperate upland grassland ("paramo"), with llama, Ecuador. (Courtesy of J.M. Lambert.) (d) Subtropical savanna thorn scrub dominated by Acacia, Kenya. (Courtesy of G.N. Stone.) (e) Sand desert, windblown dunes, Namibia. (Courtesy of N.P. Ashmole.) (f) Semidesert scrub with goats, Turkana, northern Kenya. (Courtesy of G.N. Stone.)

(a)

(b)

(c)

(d)

(e)

(f)

Plate 9 Extreme terrestrial environments
(a) Cold mountain desert, Andes, northern Chile. (Courtesy of J.M. Lambert.) (b) Rock desert, with dry river bed (wadi), Sinai, Egypt. (Courtesy of S.A. Whiten.) (c) Arctic ice floes with seals, Spitsbergen. (d) Tundra with reindeer, Siberia. (e) Northern coniferous woodland (taiga), Canada. (Courtesy of A.S. Edwards.) (f) High-altitude mountain peaks, Mt Kinabalu, Borneo. (Courtesy of J.M. Lambert.)

(a)

(b)

(c)

(d)

(e)

Plate 10 **Manmade environments**
(a) Rice paddy, eastern Java, Indonesia. (Courtesy of G.N. Stone.) (b) Industrialized river with transport and heavy chemical influences, River Severn, UK. (Courtesy of D.M. Paterson.) (c) Intensive agriculture, ploughed field, southern Australia. (Courtesy of P.J. Sunnucks.) (d) Extreme urbanization, Athens. (Courtesy of B. Vulliamy.) (e) Arpley landfill site, Warrington, Cheshire, UK. (Courtesy of David Drain/Still Pictures.)

On an even broader taxonomic scale, the family of prostaglandins (all closely related) control uterine contractions in mammals, wound responses in other vertebrates, and hatching processes in crustacean barnacles, while the richest known source for them is gorgonian coral where their functions are largely undiscovered!

Linked to this conservatism of hormone types, it is also becoming apparent that the evolution of hormonal systems has often involved faster radiation in the receptors than in the hormones themselves. Thus sequence homology within taxa is often rather low for the receptors and their subtypes, and they have clearly increased in diversity rather readily through evolutionary time, often by gene duplications.

The regulation of hormonal systems (Fig. 10.35) also becomes increasingly sophisticated, both with an evolutionary pattern and with an environmental influence. Simple first and second order control loops are common and seemingly adequate in marine invertebrates, with second order control very common in terrestrial insects and lower vertebrates. However, there is an increasing tendency to find third order control systems in terrestrial vertebrates where homeostasis becomes both more difficult and more crucial. In this critical sense hormonal control is strongly influenced by environmental physiology, as well as being a major influence upon it. It is pertinent here to mention the very widespread occurrence of "endocrine disruptors" (or "endocrine-active compounds") in the modern world. Some of these cases are clearly the outcome of strong natural selection, as when plants produce estrogen mimics or JH analogs that affect their herbivores. But others are anthropogenic phenomena, with manufactured chemicals released into the environment that seriously disrupt normal hormonal function and can have major ecological effects on animal populations, especially in aquatic environments. Major concerns currently focus on xenoestrogens and antiandrogens, which can effectively castrate animals or even induce sex change.

Hormones are generally seen as regulators of internal physiological processes, but it should be evident that the endocrine system is hugely responsive to environmental stimuli and frequently serves to synchronize animals with their surroundings. Many animal processes and behaviors are strongly seasonal in both aquatic and terrestrial environments; the most obvious examples are reproduction, migration, and dispersal, and states of lowered metabolism such as hibernation, estivation, or diapause. The key environmental signals for such seasonalities are often temperature and photoperiod, and sometimes also the presence of conspecifics. The endocrine system is then a crucial intermediary between the brain and sense organs on the one hand, and physiological effector mechanisms on the other.

FURTHER READING

Books

Bolander, F.F. (1994) *Molecular Endocrinology*. Academic Press, San Diego.

Nijhout, F. (ed.) (1994) *Insect Hormones.* Princeton University Press, Princeton, NJ.

Norman, A.W. & Litwack, G. (1997) *Hormones*, 2nd edn. Academic Press, San Diego.

Reviews and papers

Coast, G.M. (1998) Insect diuretic peptides: structures, evolution and actions. *American Zoologist* **38**, 442–449.

DeLoof, A., Baggerman, G., Breuer, M. *et al.* (2001) Gonadotropins in insects; an overview. *Archives of Insect Biochemistry and Physiology* **47**, 129–138.

Gade, G., Hoffmann, K.H. & Spring, J.H. (1997) Hormonal regulation in insects: facts, gaps and future directions. *Physiological Reviews* **77**, 963–1032.

Hillyard, S.D. (1999) Behavioral, molecular and integrative mechanisms of amphibian osmoregulation. *Journal of Experimental Zoology* **283**, 662–674.

Hoyle, C.H.V. (1999) Neuropeptide families and their receptors: evolutionary perspectives. *Brain Research* **848**, 1–25.

Koene, J.M & ter Maat, A. (2001) "Allohormones": a class of bioactive substances favoured by sexual selection. *Journal of Comparative Physiology A* **187**, 323–326.

LaFont, R. (2000) The endocrinology of invertebrates. *Ecotoxicology* **9**, 41–57.

Laufer, H. & Biggers, W.J. (2001) Unifying concepts learned from methyl farnesoate for invertebrate reproduction and post-embryonic development. *American Zoologist* **41**, 442–457.

Lee, C.-Y., Wendel, D.P., Reid, P., Lam, G., Thummel, C.S. & Baehrecke, E.H. (2000) E93 direct steroid-triggered cell death in *Drosophila. Molecular and Cells* **6**, 433–443.

Manzon, L.A. (2002) The role of prolactin in fish osmoregulation; a review. *General & Comparative Endocrinology* **125**, 291–310.

Nassel, D.R. (1996) Peptidergic neurohormonal control systems in invertebrates. *Current Opinions in Neurobiology* **6**, 842–850.

Ricklefs, R.E. & Wikelski, M. (2002) The physiology/life history nexus. *Trends in Research in Ecology & Evolution* **17**, 462–468.

Stoka, A.M. (1999) Phylogeny and evolution of chemical communication; an endocrine approach. *Journal of Molecular Endocrinology* **22**, 207–225.

Truman, J.W. & Riddiford, L.M. (2002) Endocrine insights into the evolution of insects. *Annual Review of Entomology* **47**, 467–500.

PART 3

Coping with the Environment

Introduction

Our planet is very unusual. It is so, not just in the trivial sense that all planets are unique, but in several distinct and important ways. It is at just the right distance from its star (the Sun), and of just the right size, to experience a crucially significant range of temperatures. These are the temperatures that permit a molten core on which a thin crust of solid rock can float, giving us the "plates" of land making up the continents. These temperatures are also in the range that allow the existence of both liquid water and water vapor, giving oceans between the land masses and water vapor above them. Furthermore the planet is just large enough to have a sufficient gravitational pull to allow it to retain any atmosphere that forms. Our planet also has the only oxygen-rich atmosphere in the solar system (though this is primarily a consequence of the presence of life rather than a prerequisite for life to have evolved). Thus we have a lithosphere, a hydrosphere, and an atmosphere, as a result of which an evolving biosphere becomes possible.

All these components interact and change relative to each other over time; in particular, the continents move around continuously over their molten foundations, creating changing climates and biogeographic patterns, and precipitating major geological events such as earthquakes and volcanoes. All these effects have major impacts on the rates and patterns of evolutionary change and adaptation in the biota.

Within this biosphere, a multitude of habitats are thus made available, determined by the range of interactions of the solid, watery, and gaseous phases. Much of the living space on the planet is aquatic, and most of the aquatic phase is rather salty; this marine habitat is certainly where life first evolved, and where it is "easiest". Where evaporated water vapor precipitates back onto the land as rain it may collect to make freshwater habitats (always geologically short-lived, but giving rise to their own specialist fauna), and a range of "brackish" zones where sea water and fresh water meet at the seashores and estuaries. Finally there are the more strictly terrestrial animals, always exposed to the atmosphere, and only transiently in contact with external liquid water via their food and drink. Most of these live on the surface of the land, while a few are closer to being genuinely "atmospheric" inhabitants.

Cyclical effects on the planet's environments

Many cycles and rhythms are imposed upon planet Earth and its habitats by the physical behavior of the solar system, organizing the dimension of time as perceived by the biota, and affecting patterns of energy flow on Earth. Most of these rhythms originate with the orbital motions of the planets and planetary satellites, particularly of the Sun, Earth, and Moon.

The most important and immediate cycle is the circadian rhythm of sunlight and dark, resulting from the rotation of the Earth around the Sun once every 23.93 h. Secondary to this is the lunar cycle, with the cycle of moonlight being 29.5 days, or 1 lunar month. The Earth is unusual among small planets in having its own companion satellite, and the gravitational effects of the Moon are a major influence on the movements of the Earth's seas, giving rise to regular lunar tides with a period of 24.8 h (imposed on smaller solar tides due to the gravitational pull of the Sun). Thirdly, there is a seasonal cycle, mainly resulting from the fact that the Earth is tilted on its axis, with the northern hemisphere tilted towards the Sun in June (northern summer) and away from it in December (northern winter), the southern hemisphere having the opposite experience. These daily, monthly, and annual cycles have huge effects on virtually all the animals that inhabit the planet, principally through their effect on the pattern of solar energy received at any particular place. In addition, the life cycles of many coastal marine organisms must be organized to be in harmony with the complex tidal cycles.

There are also several longer term elements of periodicity in the Earth's heat budget that result from astronomical factors. Firstly the Earth has a slight wobble or "precession" in its spinning, so that we are not always in the same position relative to the Sun at a particular time of year. This has a periodicity of about 21,000–23,000 years. Secondly, the angle of tilt of the Earth's axis varies over time, between 22.1° and 24.5°, and with a periodicity of 41,000 years; at present we are in a period of decreasing tilt, which should give a slight cooling. Thirdly, we suffer from "orbit eccentricity", having an elliptical path around the Sun but slightly off center so that the ellipse varies, with a periodicity of 93,000 years. The elliptical variation makes a difference of up to 4.8 million km (3 million miles) in our distance from the Sun, and hence of up to 7% in the solar radiation we receive. All these long-term cycles certainly have effects on evolutionary patterns on Earth. We can use carbon isotopes in ancient deposits that reflect the temperature of the seas in which they were laid down, and so track how much of the world's water was locked up as ice in different periods. It is clear that large cyclical natural fluctuations in the Earth's temperature have been happening throughout the planet's history, and have probably been one of the major influences on glaciations, sea levels, plate tectonic movements, and hence the patterns of extinctions, radiations, and physiological biodiversity on Earth.

In this section, a series of habitats offering different challenges to their occupants are considered in turn. We look at the nature of each habitat and the particular problems it poses, and the types of animals that live there; then we deal with the major themes of adaptation in a standard format to allow easier cross-reference:

- ionic and osmotic adaptation and water balance;
- thermal adaptation;
- respiratory adaptation;
- reproductive and life-cycle adaptations;
- mechanical and sensory adaptations;
- feeding and being fed on;
- anthropogenic problems.

11 Marine Life

11.1 Introduction: marine habitats and biota

The seas occupy about 70% of the surface of the Earth, and because of their great depth (a mean of about 3800 m, with trenches to 11,000 m) they form at least 99.9% of the available living space on the planet. Marine animals are found throughout the sea, from benthic organisms in the soft oozes of the cold and profoundly dark ocean floors to drifting or swimming pelagic organisms in the photic zone near the surface.

The marine world is the obvious place to start a habitat-based review of environmental adaptation, because existence in the sea should be the "easiest" of all modes of life for animals on this planet. This simple truth arises because life originated in the sea, probably in shallow ocean fringes and pools, so that cells were originally adapted to function optimally in an ionic and osmotic environment probably not very different from that of modern oceans. The vast majority of contemporary inhabitants of the seas are therefore isotonic and isosmotic with their surroundings, and in most respects have rather limited needs for complex physiological strategies. Only at the fringes of the seas—shorelines and estuaries—do problems become more severe, and these habitats are dealt with separately in the next chapter.

However, as vertebrates we tend to lose sight of this situation, given our own difficulties in coping with the salty, undrinkable, and tissue-shrinking seas around us. The reasons for this, and the special adaptations needed by marine vertebrates and other "invaders" of the marine environment, will be dealt with as a special issue at the end of this chapter, although some of the basic biology of marine vertebrates is included in earlier sections where it corresponds closely with that of invertebrates.

11.1.1 Chemistry of the marine environment

The oceans of this planet are all continuous with each other, and are relatively consistent in chemical composition throughout, being well mixed by currents and tides. This constancy applies on daily and seasonal and even geological timescales, and on coastal or global spatial scales. Why should the seas have been stable through such long periods, and why is the stable point the apparently strange mixture we find today, dominated by sodium and chloride ions? After all, the oceans are continuously fed by rivers, which consist of fresh water plus dissolved crustal rocks and some nutrients. Therefore we might expect to see gradually increasing levels of several of the typical terrestrial mineral elements in the seas, but this does not happen.

Ocean chemistry is actually created and maintained by complex geochemical cycling processes, which have apparently always gone on and hence have kept the seas in a fairly steady state. Several different components are involved in maintaining this stability. The rocks that are eroded into fresh waters from the land are mainly feldspars: these are the potassium, sodium, and calcium salts of aluminum silicates. Thus rivers should gradually become rich in K^+, Na^+, Ca^{2+}, and silicates, and be relatively poor in Cl^-. However, much of the K^+ gets removed; in particular, particles of kaolin (china clay silicates) in the sediments of rivers take it up, and the result is the mineral "illite", which is deposited and held in muds. As the rivers disgorge into the seas, Ca^{2+} is progressively removed by the biota, and deposited after death as skeletal elements made largely of calcium carbonate and calcium phosphate. At the same time chloride ions are added to sea water, both via the land and rivers from terrestrial volcanoes and directly to the sea from mid-ocean ridge volcanic activity. Sodium is regulated as part of a complex cycle, being added to the sea via rivers, but replaced onto land via wind-borne sprays in the short term and via erosion of emergent sedimentary rocks on a geological timescale. Thus the sea does become, roughly speaking, dissolved NaCl, and it is maintained in that state indefinitely. Other minor elements get "swept up" within the seas such that their levels remain constant: some accumulate into clays, others into algae (this is especially true for phosphates and for copper ions) and thus are ultimately deposited as sediment when the algae die.

The longer term stability of the composition of sea waters is still a very contentious subject, though. Certainly on a geological timescale volcanic action, rock erosion, sedimentation rates, and the balance of precipitation and evaporation would all affect the marine environment, whilst biological processes themselves would change the oxygen and carbon dioxide levels and alter the distribution of ions such as calcium and carbonate. However, evidence as to what chemical features actually varied in the past, and to what extent, is incomplete and controversial. A gradual rise of dissolved oxygen through the Precambrian and early Palaeozoic is almost certain, and a gradual loss of potassium into clays and other deposits seems very likely, but all else remains contentious. A significantly lower overall concentration—perhaps about 60–70% of present levels—during early geological eras (Precambrian) has been inferred from data on

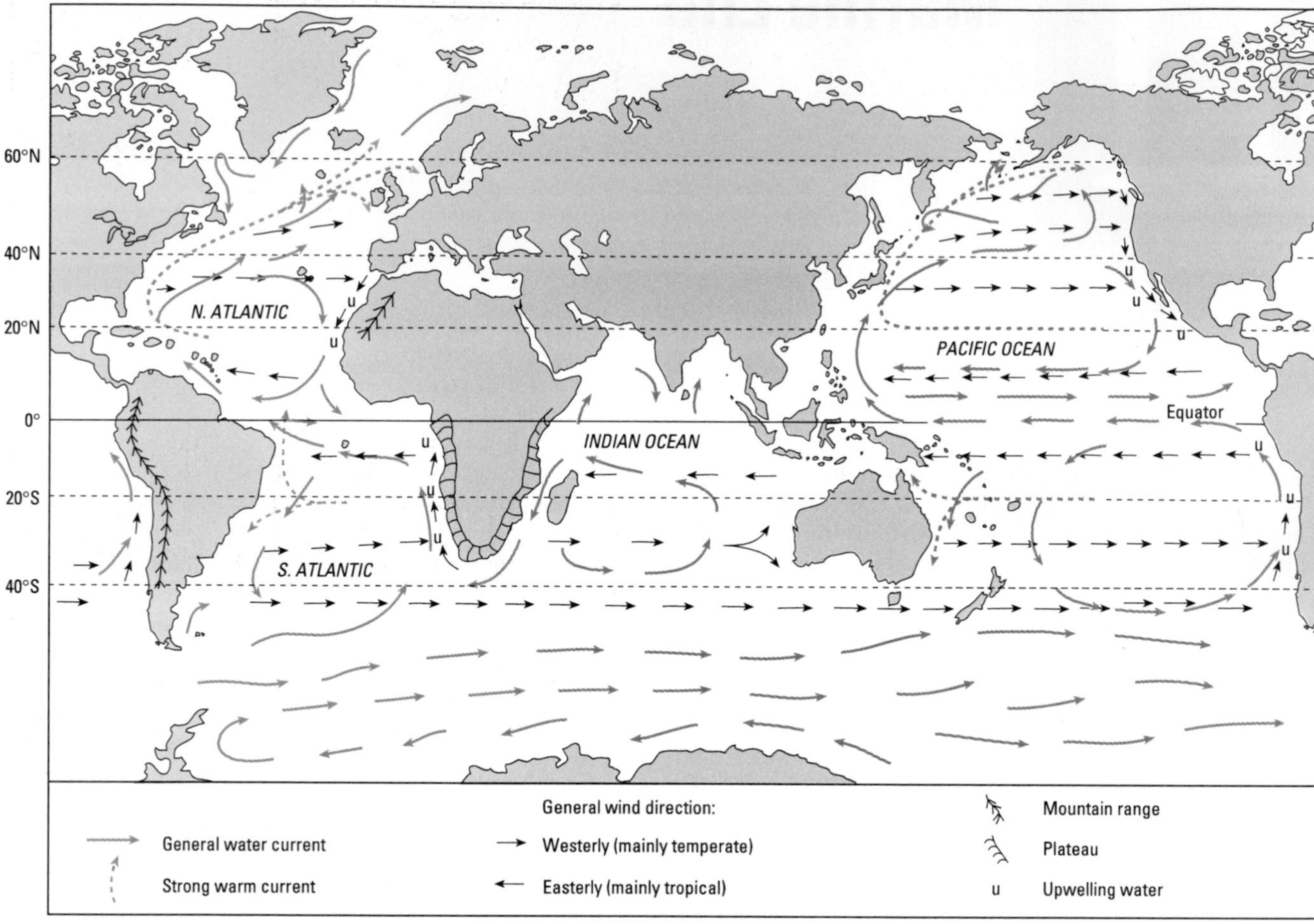

Fig. 11.1 Patterns of ocean currents and winds on the Earth. The winds tend to produce strong warm currents off eastern coasts and cold upwellings off western coasts. (Adapted from Crowley & North 1991.)

the regulatory ranges, and points of greatest stability, of modern animals. Most modern animals are better able to cope with moderate dilution to around 60–70% than with even slight concentration above 110–120%. It is also likely that an increase in salinity and in carbonate levels may have been accelerated from the Silurian era onwards, once life on land induced greater terrestrial rock weathering. However, the most severe changes were probably completed quite early in the story of the evolution of life, and modern sea water may therefore be rather a good model of what animal groups have had to contend with through most of their evolutionary history.

11.1.2 Modern-day oceans

The distributions and physical characteristics of the seas on Earth have varied continuously through geological time as the continental plates coalesced, drifted apart, and reformed in new configurations, giving rise to changing patterns of a few or many separate circulating oceanic systems. Today we have two vast ocean basins, the Atlantic and the Pacific, with their own circulating currents, together with a fairly small relatively enclosed Arctic Ocean and a number of smaller seas linked to the two main oceans. But this pattern is of rather recent origin, dating from the time about 4 mya when the two American continents joined at the Isthmus of Panama and for the first time in hundreds of millions of years separated the Atlantic waters from the larger Pacific area, so that there was no single equatorial ocean with a single circulation.

Each basin therefore now has its own circulation of major currents, shown in Fig. 11.1. These currents perturb the waters, promoting mixing and constant salinity but also bringing together cold and warmer waters. In some areas, large loops of water break off from the currents and form massive eddies or "rings", moving across the open ocean, with a core of either cold or warm water. In the North Atlantic there are commonly about 10 cold-core rings and three warm-core rings at any one time, broken off from the Gulf Stream. Marine waters of different temperature and salinity also mix and mingle to create "salt fingers" of interdigitating rising warm and falling cold waters. Thus animals adapted to different thermal regimes may be translocated *en masse* across apparently inhospitable marine zones.

11.1.3 Marine flora and productivity

Marine phytomass is dominated by thallophyte plants, especially the microalgae (phytoplankton) and macroalgae (seaweeds). There are only about 30 species of marine seed plants, and these are restricted to estuaries and coasts. Thus primary productivity derives

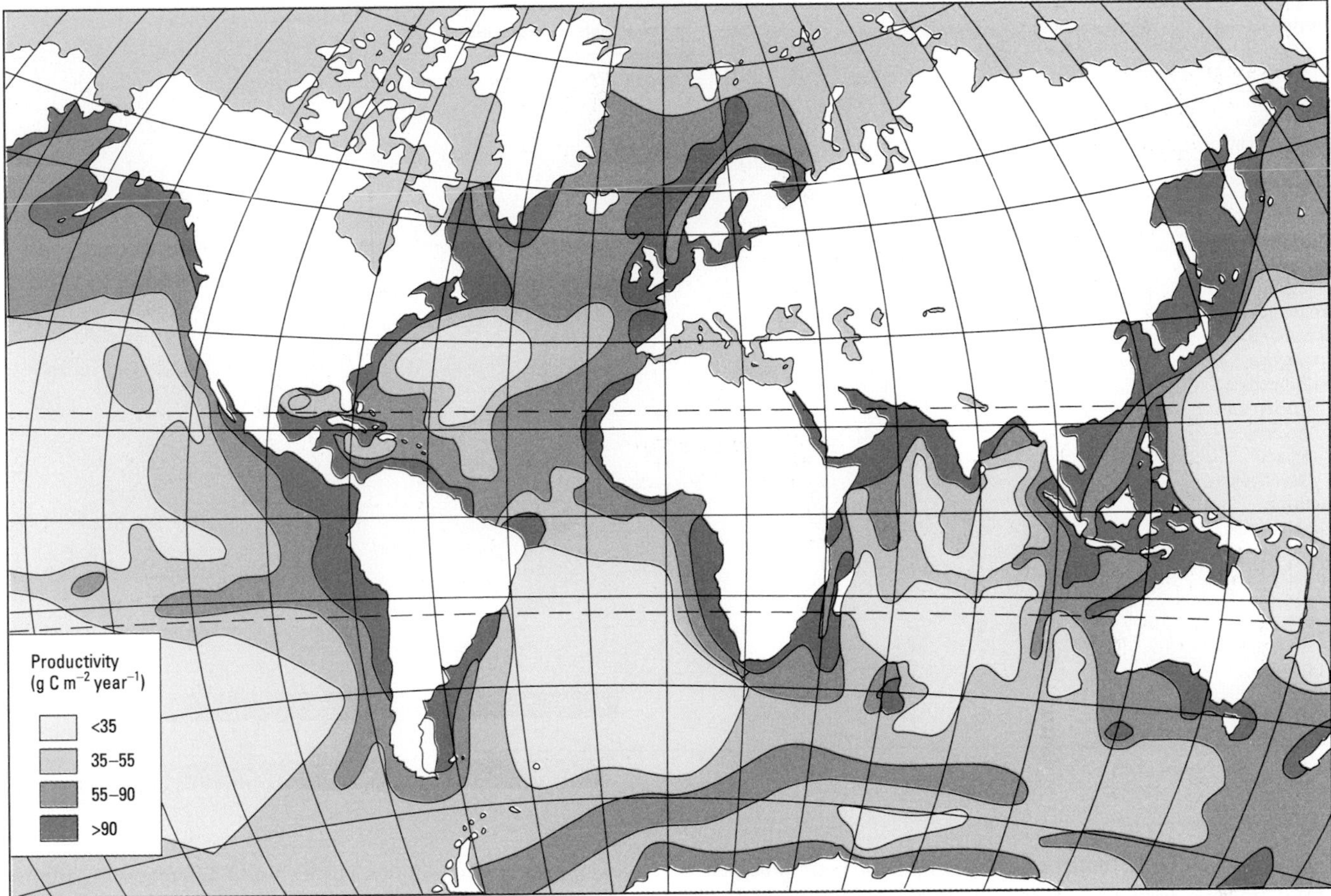

Fig. 11.2 Distribution of primary productivity in the world's oceans.

mainly from algal activities, and for most of the oceans this means floating phytoplanktonic algae (in addition to bacteria and other prokaryotic organisms).

Currents and eddies have major effects on marine productivity in every ocean. Productivity is rather low in vast areas, especially of the Pacific, with low iron levels often implicated as the limiting factor. An illustration of this causal link is provided by the leaching of iron from sub-Antarctic islands, resulting in locally enhanced productivity. By contrast there are areas of very high primary productivity (Fig. 11.2) usually also associated with abundant animal life. Such areas traditionally occur in the turbulent regions where warm and cold waters meet, a classic example being the (now almost depleted) fishing grounds off Newfoundland known as the Grand Banks, where the Gulf Stream and the Labrador Cold Current collide. Further areas of high production occur where cold nutrient-rich waters are forced into upwellings (see Fig. 11.1) as they meet the west coasts of continents, best exemplified by the coasts of Ecuador and Peru, where a large proportion of the world sardine catch originates. The polar seas represent another area of good productivity, at least in the polar spring and summer when glacial meltwater introduces nitrates and silicates, and promotes rapid growth in the algal communities formerly held in the ice pockets. The final major areas of production are the coastal and estuarine waters, especially in temperate latitudes where seasonal floodings bring in vast amounts of nutrients; these habitats come into the category of brackish waters, covered in Chapter 12.

11.1.4 Extant marine animals

Most animal phyla are represented in the marine habitat (Table 11.1); among the few absentees are the taxa that have evolved as terrestrial specialists (insects, myriapods, and arachnids) or as parasites in or on terrestrial organisms. Some of the deepest sea areas are only just being explored, and are proving to have much greater biodiversity than previously thought. However, large parts of the pelagic marine environment are relatively low in species numbers, perhaps largely because the habitat has no real structural complexity to allow niche subdivision and speciation. Only in relatively shallow seas where there are areas of reef or kelp forest (Fig. 11.3) do we find any real architectural complexity to promote very high diversity. Coral reefs (see Plate 1d, opposite p. 386) are largely restricted to areas where the surface temperatures are on average above 20°C even in the coldest month, whereas kelp forests thrive in much cooler waters. In both areas, and in some deep-water benthic communities, representatives of the majority of the phyla listed in Table 11.1 can be found.

Marine animals are probably best subdivided according to where they live in relation to depth. Fundamentally they may be either bottom dwelling (benthic) or living within the water column (pelagic). The **benthos** may be further subdivided into: **epifauna**

Table 11.1 Distribution of major phyla in marine habitats.

Phylum/group	Marine benthic	Marine pelagic	Fresh water	Terrestrial	Parasitic
Porifera	++	l			
Cnidaria	++	++	+		
Ctenophora		++			
Platyhelminthes	+	+	++	+	++
Nemertea	++	l	+	+	
Nematoda	++		++	++	++
Acanthocephala					++
Rotifera	+	+	++		r
Mollusca	++	+	++	++	r
Annelida	++	+	++	+	+
Pogonophora	+	l			
Sipuncula	+	l			
Priapulida	+	l			
Arthropods					
Crustacea	++	++	++	++	r
Arachnida	r		+	++	++
Insecta			++	++	++
Onychophora				+	
Bryozoa	++	l	+		
Brachiopoda	+	l			
Echinodermata	++	lr			
Urochordata	++	+			
Chordates					
"Pisces"	++	++	++		
Amphibia	r		+	+	
Reptilia		+	+	++	
Aves		(+)	(+)	++	
Mammalia		+	+	++	

++, common; +, present; r, rare; l, larvae only.

(or epibenthos), living (either attached, sessile, and nonmotile, or as errant crawling animals) on top of the substratum (see Plate 1a); and **infauna** living within the substratum, the latter including both sedentary macroinvertebrates that burrow and a host of microinvertebrates living between the substrate particles and often referred to as the **interstitial fauna** or **meiofauna**.

Within the pelagic zone animals are again divided largely by size linked to locomotory abilities. Pelagic animals that are relatively active swimmers (and usually fairly large) are referred to as **nekton**, a category currently dominated by fish and cephalopods (see Plate 1b), but with spectacular representations from the marine mammals (cetaceans, sirenians, and members of the Carnivora, the last of these being amphibiously marine and including pinnipeds, sea otters, and polar bears). In the Mesozoic these marine mammals were absent and various radiations of marine reptiles occupied the large carnivore niches.

The smaller animals that essentially float and drift with the currents (though they may also be good swimmers) are termed **plankton**, or more strictly zooplankton (see Plate 1c). The plankton of most oceanic areas are often extremely diverse, with many jellyfish (cnidarians), comb jellies (ctenophores), and arrow worms (chaetognaths), although usually numerically dominated by a variety of classes of Crustacea. It also includes a vast range of larval forms (including juvenile fish), varying in composition with the season, since many marine animals have a pelagobenthic life history with a relatively sessile adult form and a more or less prolonged planktonic larval stage.

Although the marine habitat is continuous, there are, of course, regional and latitudinal trends in faunal composition, in both benthic and pelagic communities. There are also differences in patterns of size, one striking effect being the occurrence of "gigantism" in polar marine invertebrates, with spectacularly large (and perhaps rather long lived) examples of isopods, jellyfish, and nemertine worms found in the seas fringing ice-bound polar islands and land masses.

11.2 Ionic and osmotic adaptation

11.2.1 Sea water

The chemical composition of "normal" open sea water is shown in Table 11.2; this gives the major components, but virtually every other element is also present in trace amounts. This mix of elements results in an osmotic concentration of 1000–1150 mOsm, a salinity of around 35–36 parts per thousand (ppt or ‰), and a freezing point depression of around 1.86°C. Below a depth of about 1000 m,

Table 11.2 Composition of normal sea water.

Component	mmol l^{-1} sea water	g l^{-1} sea water
Na	470	10.8
K	10	0.39
Mg	54	1.30
Ca	10	0.41
Cl	548	19.4
SO_4	28	2.7
HCO_3	2	0.14

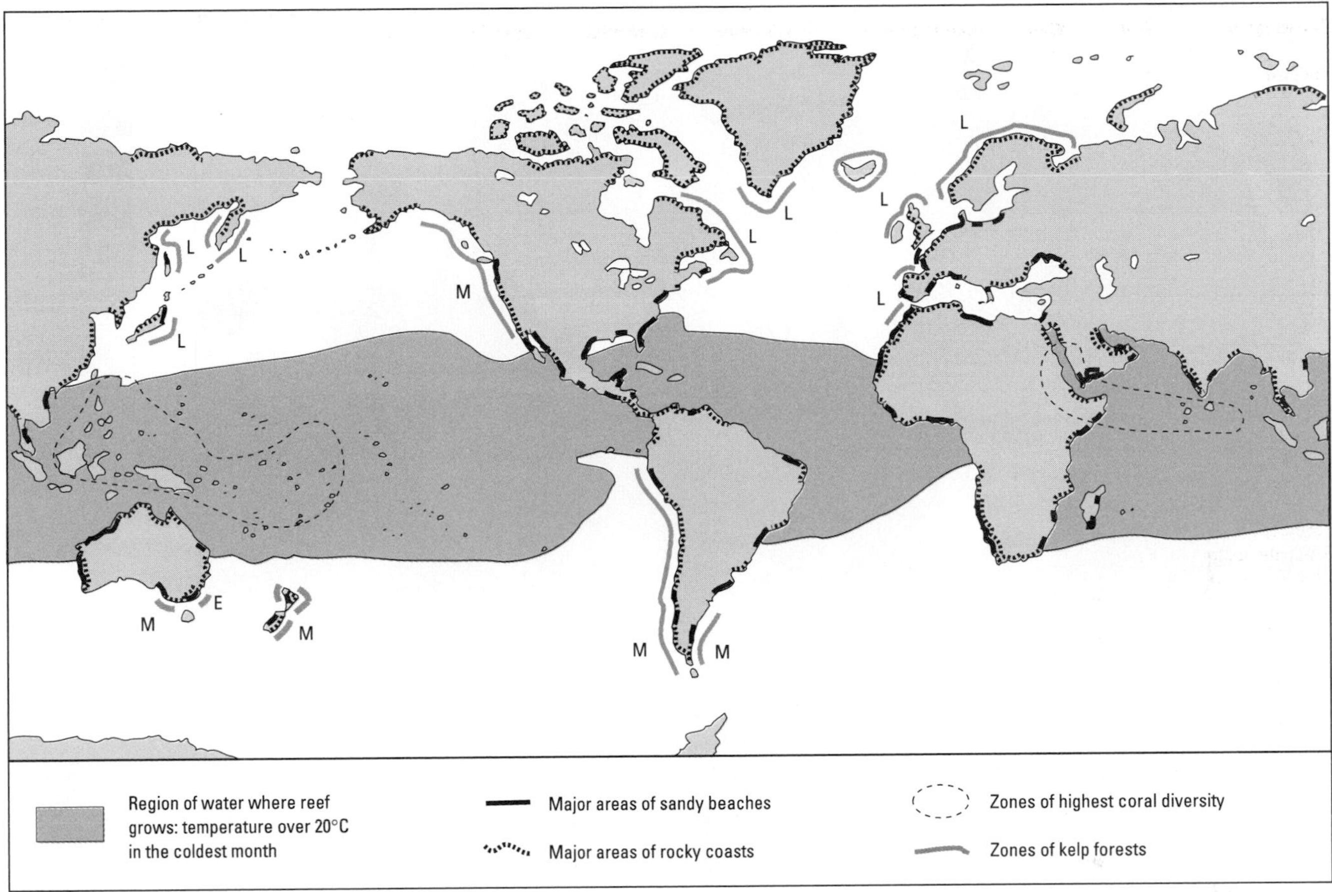

Fig. 11.3 Major regions of reef and of kelp forest in the world's oceans. For the kelps: E, *Ecklonia*; L, *Laminaria*; M, *Macrocystis*.

sea water is practically invariant, and always has a salinity of 34.5–35.0 ppt. Near the surface the two opposing effects of higher rates of loss of water by evaporation, and gain by precipitation, produce greater local variation in salinity. This property can therefore change somewhat with latitude (Fig. 11.4a), reaching 37.5 ppt in parts of the tropical seas. Only in a few unusual situations do the composition parameters change more substantially. Some very large enclosed "bays" such as the Hudson Bay and the Baltic Sea, with substantial freshwater inputs from surrounding mountains, may be moderately diluted, and polar seas may also be somewhat dilute for parts of the year due to melting ice. By contrast, enclosed areas such as the tideless Mediterranean Sea and the Red Sea, in high temperature zones and hence with particularly high evaporation, may become more concentrated. Variation of salinity with depth also occurs, with a cool and salty current running at depth southwards through the Atlantic and north again through the Pacific, as shown in Fig. 11.4b.

11.2.2 Composition of fluids in marine animals

Virtually all truly marine invertebrates are stenohaline, and their body fluids are nearly isosmotic with their surroundings; a range of examples is given in Table 11.3. However, it is evident that there is

Table 11.3 Osmotic concentration of marine animals' body fluids.

Phylum/genus	Osmotic concentration (mOsm)
Cnidaria	
Aurelia	1050
Mollusca	
Sepia	1160
Eledone	1061
Mytilus	1148
Annelida	
Arenicola	1120
Neanthes	1108
Crustacea	
Carcinus	1100
Ligia	1220
Nephrops	1108
Chelicerata	
Limulus	1042
Echinodermata	
Echinus	1065
Chordata	
Elasmobranch fish	
Scyllium	1075
Squalus	1096
Teleost fish	
Paralichthys	337
Gadus	308
Sea water	1000–1050

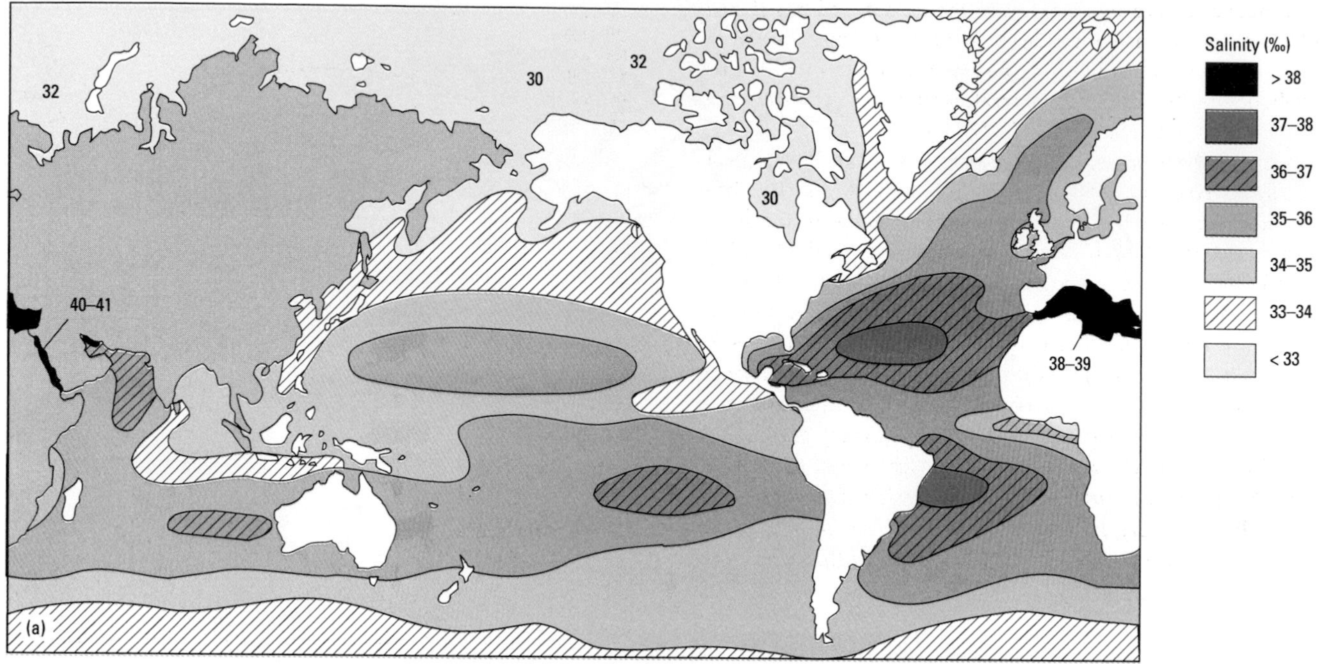

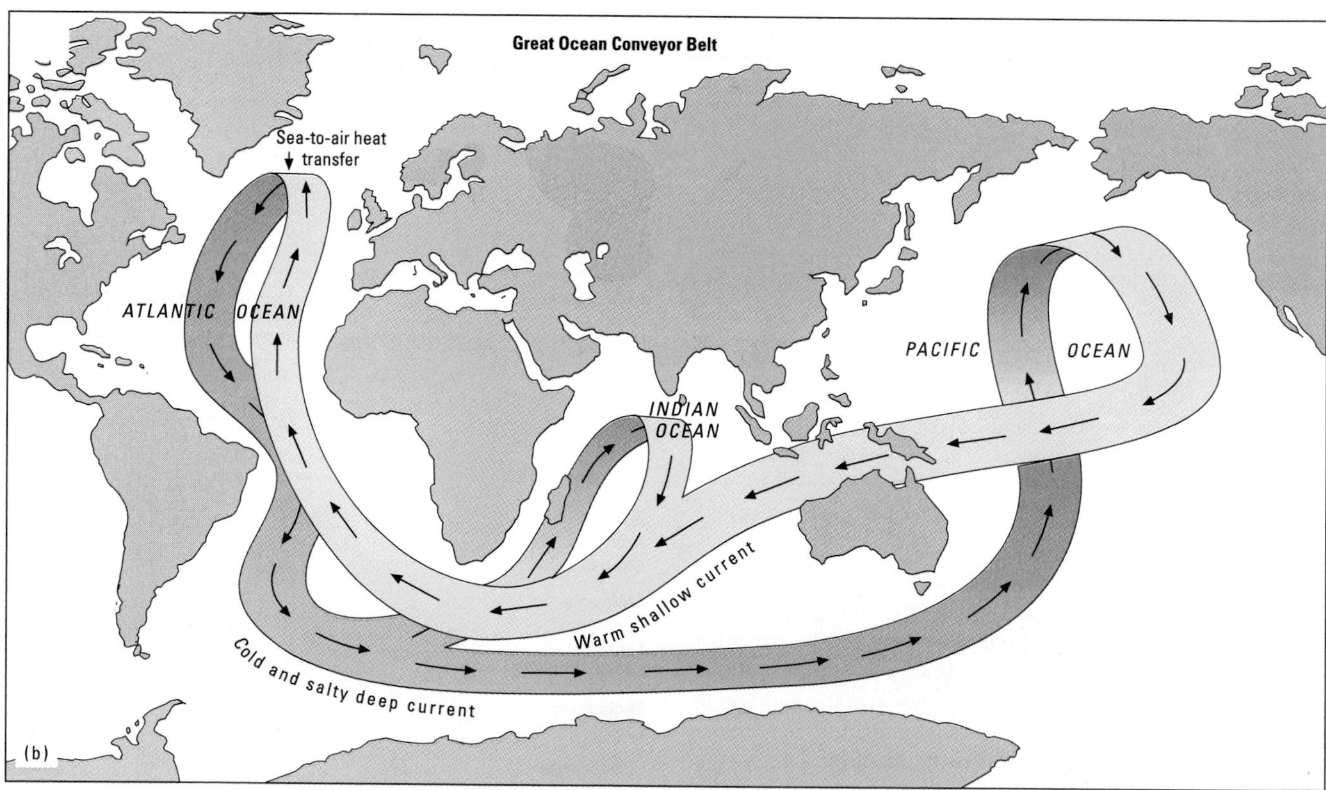

Fig. 11.4 Variation in salinity (a) in surface waters through the major oceans, and (b) in deep-water circulations, showing the deep salty water flowing out of the north Atlantic, around Africa and past Antarctica, then northwards into the Pacific. Here some of the water wells up and warms and there is then a shallower warm current of less salty water back through the Indian Ocean. This salt conveyor system compensates for water vapor transport through the atmosphere. (a, Adapted from Ingmanson & Wallace 1989; b, adapted from Kerr 1988.)

commonly a slight osmotic excess in the tissues, and this is attributable to the Donnan effect due to the proteins and other metabolites contained in all living organisms (see section 5.3.3). Since metabolic activity within the cell creates more of the relatively nondiffusible small molecules, any cell would tend to increase slowly in osmotic concentration, thus drawing in water and so gradually swelling up. Hence all cells, even in nearly isosmotic sea water, have a constant need for a minimum degree of volume regulation. A membrane-bound sodium-activated ATPase (the Na^+/K^+-ATPase or "sodium pump"; see section 4.4.2) actively contributes to this by extruding cations. At the same time, amino acid regulation systems also help to control the internal osmotic concentration and thus the cell volume (see section 4.6). In most stenohaline marine invertebrates volume control is limited and cellular swelling does occur in moderately dilute media; but some compensatory reduction in intracellular osmotic concentration is brought about by losses of intracellular constituents. The principal osmotic effectors that move out of the cell during hyposmotic stress are the common neutral amino acids (especially alanine and glycine, with proline also important in many crustaceans). However, a group of amino acid derivatives are also expelled from the cells according to need, and in molluscs and crustaceans, which have been best studied, these include betaine, taurine, and trimethylamine oxide (TMAO). Long-term isosmotic balance is thus maintained.

With respect to ionic balance and particular ionic concentrations in cells, the situation is rather different (see Table 4.4). The osmotic concentration in marine animals' cells may be very like that of sea water, but ions in fact make up only about half of this total intracellular concentration in most animals, the remainder being due to the cell's amino acids and other small organic molecules already mentioned. The ion fraction in the cell is dominated by potassium (150 mM, or rarely up to 400 mM) rather than sodium (25–80 mM), with chloride (25–100 mM) making up only a part of the anionic total as with any intracellular situation (see Chapter 4). A single cell in a marine organism therefore corresponds to the simple model of osmotic and ionic regulation shown in Fig. 11.5. Marine cells are indeed the archetype for all others, and they differ little in proportional composition from the cells of the freshwater or terrestrial animals that have evolved from them.

Moving outside the cell, examples of the ionic compositions of the extracellular body fluids in marine organisms were given in Table 5.1, and are shown relative to normal sea water in Table 11.4. The values there make it clear that many organisms, despite being nearly isosmotic, are specifically regulating at least some of their ionic levels at the "skin" interface between the sea and the body fluids. Sodium and chloride rarely deviate much from being isionic in marine bloods, but potassium is commonly hyperregulated, and calcium may also be high at times during the molt cycle of many arthropods. Sulfate ions may be roughly as in sea water or very low (in crustaceans and in pelagic animals), and magnesium may also be kept low in crustaceans. Some groups, notably echinoderms, show virtually no ionic regulation.

It should be noted that there are just a few marine invertebrates that are not isosmotic with their surroundings. Some grapsid crabs and some palaeomonid shrimps are always hyposmotic to sea water (often around 80–90% concentration), and these same animals are commonly fairly euryhaline, able to survive in both brackish and hypersaline habitats when necessary. Another oddity is the common octopus, which is always somewhat hyperosmotic to sea water at about 1300 mOsm, with a continuous need to produce a slightly dilute urine and to absorb salts from imbibed sea water.

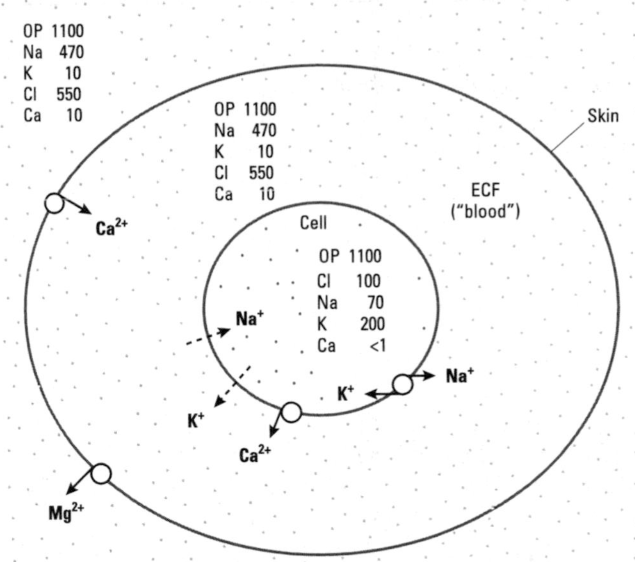

Fig. 11.5 A model of salt movements between compartments (cells and extracellular fluid (ECF)) in a typical marine animal. Circles with solid arrows indicate active transport; OP, osmotic pressure.

Table 11.4 Ionic composition of marine animals' body fluids given as percentages relative to sea water.

	Ion					
Phylum/species	Na^+	K^+	Ca^{2+}	Mg^{2+}	Cl^-	SO_4^{2-}
Cnidaria						
*Aurelia**	99	106	96	97	104	47
Mollusca						
Pecten	100	130	103	97	100	97
*Sepia**	93	205	91	98	105	22
*Loligo**	95	219	101	102	103	36
Annelida						
Arenicola	100	104	100	100	100	92
Aphrodite	99	103	100	101	99	92
Sipuncula						
Phascolosoma	104	110	104	69	99	91
Crustacea						
Maia	100	125	122	81	102	66
Carcinus	110	118	108	34	104	61
Nephrops	113	77	124	17	99	69
Echinodermata						
Marthasterias	100	111	101	98	101	100
Echinus	99	100	101	99	99	99
Urochordata						
*Salpa**	100	113	96	95	102	65

* Indicates a pelagic species.

11.2.3 Adaptive mechanisms

Marine invertebrates clearly require some ion-pumping capacity, and must retain some degree of control over the ion fluxes across their surfaces. Hence, although relatively permeable compared with close relatives living in other habitats (see Chapter 5), few seawater animals act as perfect osmometers, but instead have a somewhat lower water permeability at their outer surfaces than elsewhere in their bodies. Some examples of marine skin permeabilities were given in Table 5.3; the general range for water flux rate in invertebrates is around 5–20 mg cm^{-1} h^{-1} (1–10 mg cm^{-1} h^{-1} for vertebrates) corresponding to a skin resistance (see Chapter 5) of 6–15 s cm^{-1} (up to 35 s cm^{-1} for vertebrates). Marine animals also require impermeability to their own organic constituents, and restriction on the movements of divalent cations and anions in particular. In fact calcium permeability must be especially low in all the animals' membranes, to keep internal cellular levels correct, as this ion in turn affects all other membrane permeabilities.

But how and why are the blood ions regulated in marine animals? A basic mechanism for ion regulation was discussed in Chapter 4, but that largely accounted for the regulation of Na^+ and Cl^-, the two ions that are least varied in the standard marine animal's blood. The standard Na^+/K^+-ATPase enzyme that effects such regulation has a relatively high K_m value in marine animals (see Table 5.4), and achieves quite rapid sodium transport rates (cf. Table 12.5) out of the cells to the already high levels of sodium concentration in their surroundings. Beyond this, though, special mechanisms are clearly needed to raise calcium or potassium levels, or to extrude magnesium and sulfate, and though little information is available on these topics, ionic "pumps" requiring the splitting of ATP are inevitably involved, though probably indirectly via symport and antiport mechanisms (see Chapter 4). In some special cases, such as the floating organisms that reduce their buoyancy by eliminating heavy sulfate ions (see below), we know that exchange mechanisms (extruding sulfate in exchange for lighter chloride ions) are involved. In most marine mammals, ion-regulating mechanisms operate at the general body surface rather than in the excretory system; in a range of stenohaline invertebrates, the body surface is responsible for 75–95% of salt loss and uptake.

Thus the marine animal's body is not just a passive equilibrium system, but is involved continuously in minor regulatory adjustments to keep its cell volumes (and hence overall volume) constant, and to maintain the ionic gradients between blood and medium, and between cells and blood. This degree of regulation must have been instituted very early in evolution, probably as soon as the cell membrane was "invented" by the first marine cells in the Precambrian era.

Since most of the work of osmoregulation is performed at the level of individual cells, marine invertebrates require only very simple osmoregulatory/excretory organs, generating a gentle flow of fluid from which any nitrogenous wastes (those that have not diffused away from the gills or general body surface) can be lost. The marine mussel *Mytilus* and a marine ostracod crustacean provide simple examples in Fig. 11.6, with collecting areas leading via short unspecialized ducts to the discharge point. The octopus is somewhat unusual in having nitrogenous wastes added to the urine from a separate area distinct from the filtration site. Marine animals commonly produce a daily urine volume equivalent to 3–15% of their body volume, the volume being particularly low in stenohaline crustaceans (3–5%); this is in marked contrast to daily production of 15–50% in many euryhaline intertidal animals (see Chapter 12).

11.3 Thermal adaptation

11.3.1 Thermal environment of the seas

The seas are remarkably stable in thermal terms, due to their enormous volume, the high heat capacity of water, and the currents that tend to disperse any local cool or hot spots. Therefore, they again provide the easiest of all environments for their inhabitants. Heat can only be gained geothermally (important only in very localized areas near hot "vents") or by surface absorption from the sun, particularly at longer wavelengths. Sea temperatures therefore have a much narrower range than air temperatures. Salt water would freeze at −1.86°C, but formation of a surface ice layer in very cold habitats buffers the deeper waters and prevents substantial deep freezing. The deep waters where the sun's radiation cannot penetrate, below the levels of continental shelves and making up 90% or more of the total marine living space, are at a constant 1–4°C all year round. They are kept very cold because of inputs from oceanic bottom water. This cold water (always just above −1.86°C) is formed near the polar ice sheets, where it sinks due to its high density and flows towards the lower latitudes beneath the surface currents of warmer, less dense water that flow towards the poles.

Surface waters, especially over the shelves, may therefore be both warmer and thermally more variable. Since warmer water is less dense than cold water and tends to float on top of it, mixing between different thermal masses of water is reduced, which may give rise to stratification of the water and the existence of a **thermocline** (Fig. 11.7a). However, cycles of mixing and stratification can occur on a daily or seasonal basis to modify the thermocline. These cycles result from the interactions between heat input, night-time convection, and day-time wind-driven turbulence (Fig. 11.7b), the last of these being augmented in shelf areas by the extra turbulence generated by tides interacting with the seabed.

Thus in regions where radiation is strong all year round the thermocline is permanent and lies quite deep down, giving a large living space of fairly warm water in tropical seas; here the surface water is usually 24–28°C throughout the year. With increasing latitude the magnitude and depth of the thermocline decrease, although the density gradients associated with a stable thermocline do help to resist penetration by turbulent eddies of water. Nevertheless in temperate zones the seasonal permanence of a thermocline is usually partly lost, and there may be a "fall overturn" with fairly sudden mixing as the thermal and density gradients break down. In such zones we see a temporary "summer" thermocline superimposed on the permanent one (Fig. 11.7a). In polar seas there is usually no thermocline at all.

However, even with a thermocline the temperature range for many sea areas may be only about 5–8°C through a year, and it is normally only 2–5°C in the tropics and 4–6°C in Arctic waters or less than 1°C in the Antarctic. Only in some temperate zones with

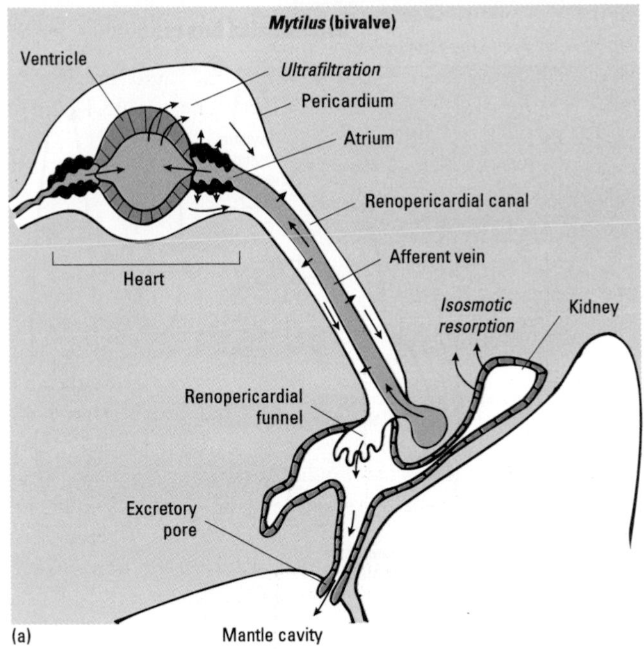

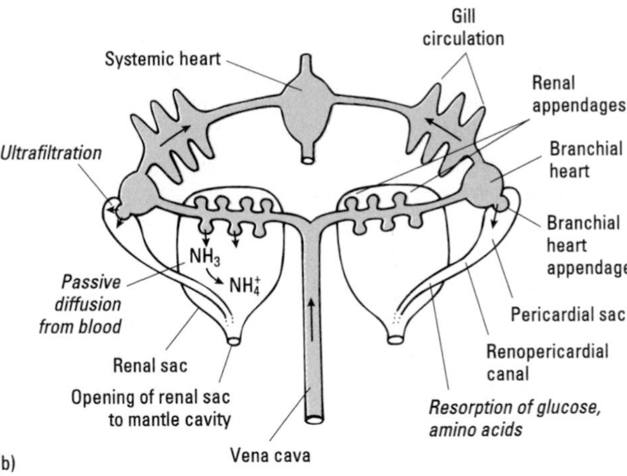

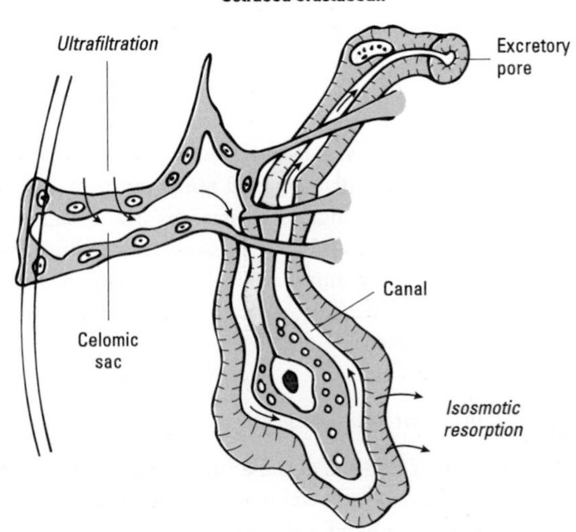

temporary thermoclines over continental shelves is there a really marked seasonality; the western shores of the UK are a good example of this, with sea temperatures of 2°C in January rising to as high as 18°C at the end of a good summer. In such zones the mixing in the warmed surface layers may be very limited.

A global survey of the mean annual sea temperatures for surface waters (conveniently measured using cameras from satellites these days) is shown in Fig. 11.8; values can vary from about 30°C in the tropics to −2°C around Antarctica. Note also that in enclosed areas of sea the recorded means and extremes can be exceptional: temperatures of up to 40°C are found in shallow seas such as the Red Sea and Arabian Gulf, whilst enclosed high-latitude seas such as the Baltic and Hudson Bay may freeze over substantially in winter.

Superimposed on fundamental latitudinal effects, the world's oceans also show pronounced local variations in temperature in both surface and deep waters due to particular currents determined by geomorphology and crustal plate positions. The North Atlantic Drift/Gulf Stream is a classic case, where the warm surface current from the southwest keeps the western shores of Europe milder than expected. In terms of marine animals' adaptations it is therefore important to know the local thermal situation and water current patterns through the year rather precisely.

11.3.2 Temperature in marine animals

The vast majority of marine animals live in water that is below 5°C all the time. Equally the great majority of them are ectotherms and have body temperatures that are within a degree or so of their surroundings at all times and thus essentially invariant. Most of these animals are also stenothermic, and cannot tolerate artificially imposed changes in temperature.

Indeed, virtually *all* the primary inhabitants of sea water are essentially ectothermic, partly because water is a sufficiently good conductor of heat that marine endothermy is particularly hard to maintain, with any internally generated heat being rapidly dissipated at the body surface, especially at the large surface area gills that are associated with active marine life. It is normally only some secondary invaders, with their previous history of life on land, that show true endothermy. Even then only relatively large endotherms such as the cetaceans (whales and their kin), with highly efficient insulation and no gills, can achieve permanent residency in the sea with a raised body temperature (T_b; see section 11.10); although it seems likely that large marine dinosaurs and other Mesozoic marine reptiles may also have achieved this.

Thus, for one reason or another, the great majority of marine animals have a marked constancy of T_b and little need for acclimatory or regulatory strategies. However, some pelagic ectotherms may experience variation in T_b in association with changing environmental temperature, especially in migrating species. Diurnal migrations from depths of 400–800 m by day to the top 100 m by night are

Fig. 11.6 (*left*) A variety of kidneys from marine invertebrates. In molluscs (a,b) and crustaceans (c) there is a filtration system (from the heart to pericardium in molluscs) followed by a short tubule where volume may be reduced by isosmotic resorption, but little or no change of concentration occurs. A terminal bladder or sac may be present. In the cephalopod (b) the renal appendages are a site for extra diffusion of NH_3 into the sac, where it is held by conversion to NH_4^+ by the pH conditions.

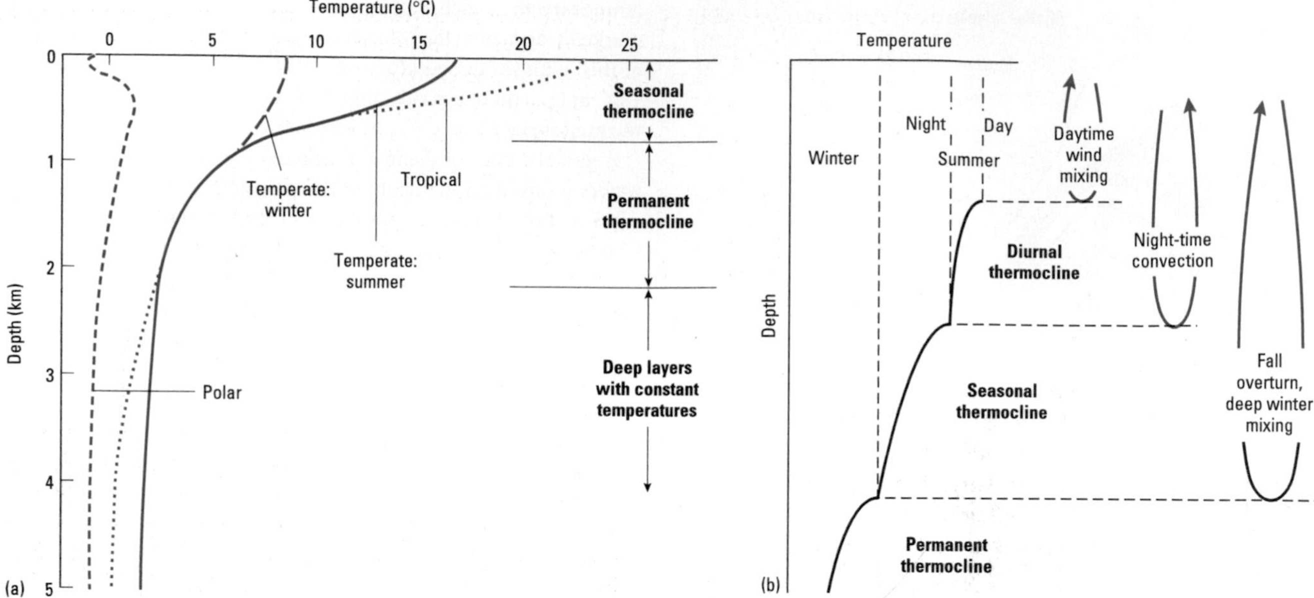

Fig. 11.7 (a) Patterns of temperature gradient in the upper ocean showing seasonal and permanent thermoclines. (b) The associated patterns of mixing and stratification for the open ocean. The strongly mixed layer extends only to the top of the diurnal thermocline, though there may be some deeper mixing where night-time temperatures are very low and at the onset of winter (the "fall overturn").

quite common for planktonic species, and in the tropics this may mean a temperature change of 15°C or more in less than 2 h. For larger nektonic crustaceans and fish there may be latitudinal migration, or movement from mid-ocean to offshore regions following food supplies, or seasonal migration through a thermocline. For example, some sculpin species and other fish migrate to deep water in winter to avoid freezing surface waters.

11.3.3 Adaptive mechanisms

For all those animals dwelling at a particular latitude and below the surface waters, thermal acclimation is not a problem, and cell machinery and enzyme biochemistry can be adapted to work optimally over a very narrow and rather cold temperature range, with little need for isozymic variation or for allosteric modulators (see section 8.2.1). Many cases of evolutionary adaptation of the cellular machinery to particular environmental temperatures are well documented. The myosins or "molecular motors" from a range of fish are perhaps the best known. The myosins from typical fast muscles can all generate similar stresses when operating at the normal habitat temperature for each species, although function is lost at unphysiological high or low temperatures. Some interspecific adaptations in the structure of the molecule are evident in relation to the ease at which enzymes can be thermally denatured. Figure 8.3 showed that the myosins from Antarctic fish have denaturation half-times about 100 times less than those recorded in fish from the Indian Ocean. This reflects the more "open" character of cold-adapted marine enzymes compared with the rigid, highly bonded structures of enzymes from animals adapted to other aquatic (and terrestrial) biomes. However, the maximal contraction speed and power output of the intact fast muscle fibers from these species is approximately a function of the water temperature at which they are tested, providing little evidence for interspecific adjustment in performance at different latitudes.

One clear example of adjustment of enzyme function across a latitudinal range is afforded by the lactate dehydrogenases in the eastern Pacific barracudas (genus *Sphyraena*). There are three species having similar morphologies and ecologies but different and overlapping temperature ranges. The turnover numbers (K_{cat}) of their muscle M_4 lactate dehydrogenase (LDH) enzymes at 25°C are inversely related to average habitat temperature (T_m), such that K_{cat} values are similar at the physiological temperature of each species (Fig. 11.9a). The figure shows that under physiological assay conditions, K_m values for pyruvate (the substrate for LDH) in these muscles are also highly conserved. DNA sequencing studies have shown that interspecific differences in the properties of the barracuda LDHs involve only a few base substitutions, all coding for amino acids remote from the substrate- and ligand-binding sequences, but nevertheless affecting the conformation of the ligand-binding pockets.

Membrane adjustments are also necessary to achieve efficient functioning at particular environmental temperatures, as we saw in Chapter 8. Studies of a range of crustaceans from polar waters indicate that both branchial and neural membranes are more fluid (less "ordered", with more unsaturated lipids) than those from temperate relatives, with the neuronal membranes being even more fluid than the gill membranes.

Acclimating to different temperatures

Relatively few cases of thermal acclimation in marine animals are described as yet. Most aquatic ectotherms prefer a thermal niche close to their normal ambient water temperatures. Large inshore

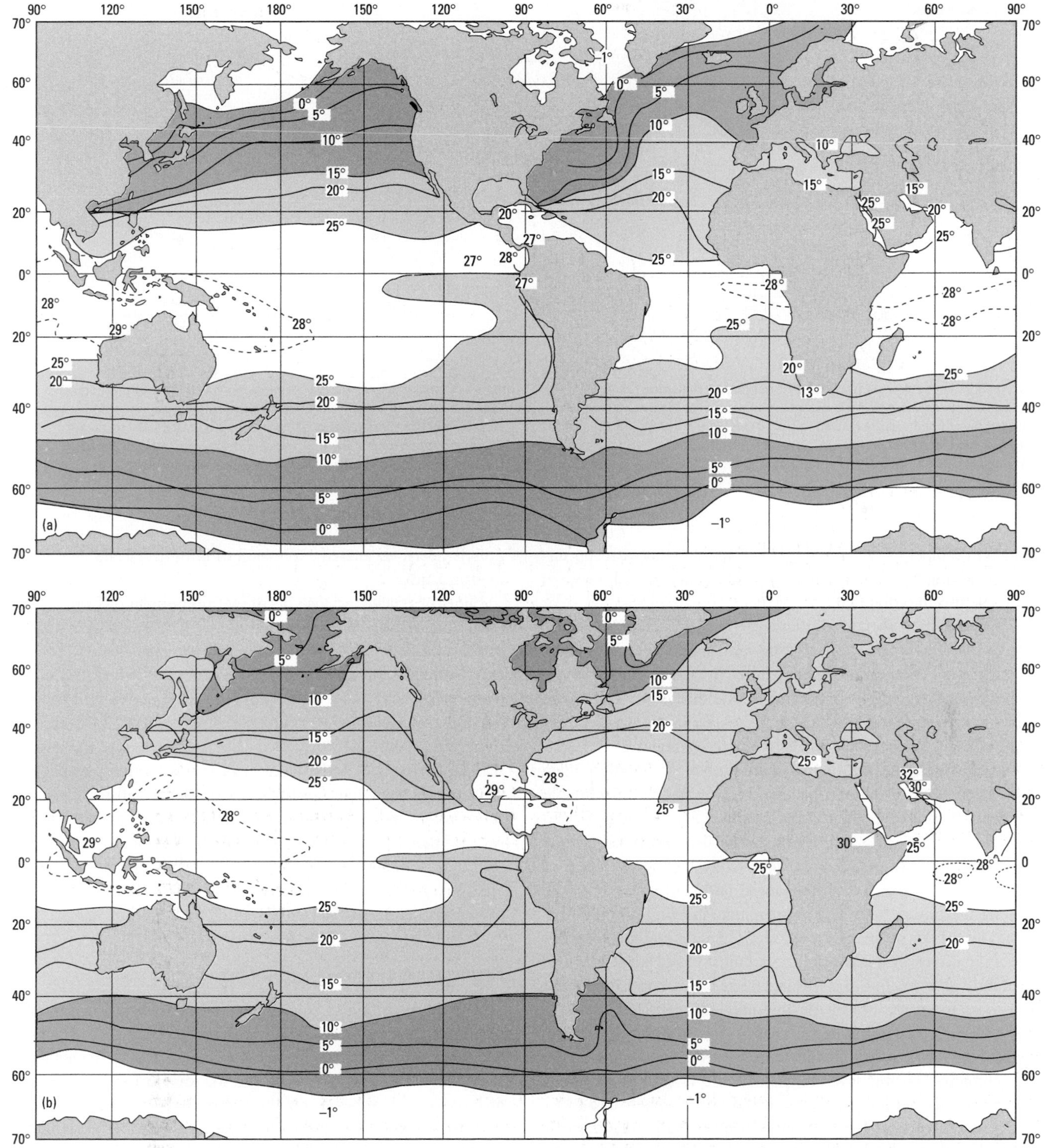

Fig. 11.8 Mean sea temperatures (°C) worldwide, for surface waters in (a) February and (b) August. (Adapted from Ingmanson & Wallace 1989.)

crustaceans, such as lobsters, show a degree of behavioral thermoregulation, selecting water temperatures of 13–19°C that are a little above their mean ambient temperature, with the exact preferred temperature varying by 1–2°C according to acclimation history. They may move into warmer waters in spring to enhance their growth and reproduction rates. Some fish are also known to exhibit seasonal acclimation of their general metabolic machinery, by increasing the concentration of many enzymes and of metabolites in cold periods. In at least some species the mitochondrial volume in cells also rises to insure an adequate supply of ATP. Such fish also show specific patterns of heat shock protein (HSP) synthesis, with

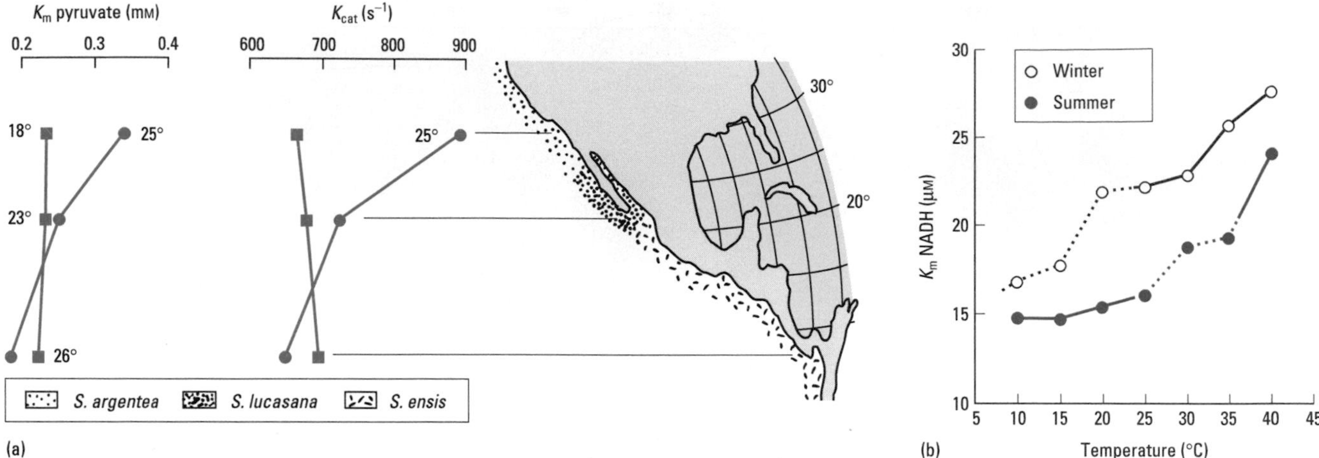

Fig. 11.9 Enzyme property variation with temperature in marine fish. (a) Latitude effects, shown here for three species of barracuda (*Sphyraena*) with different distributions in the eastern Pacific. Values of K_{cat} and K_m for M_4 lactate dehydrogenase are given, both at 25°C for all species (circles) showing latitudinal variation, and also at the normal environmental thermal midpoint for each (squares) showing much more consistent properties. (b) Seasonal effects. K_m values for malate dehydrogenase from the goby, *Gillichthys*, in winter and in summer. Dotted lines show the normal body temperatures of the fish in that season, indicating conservation of K_m at different seasons. In fact the fish contains two isozymes of this enzyme in both seasons, but the ratio of the thermolabile form to the thermostable form varies to produce the differences shown. (b, From Somero *et al.* 1996.)

induction temperature varying between species, and with concentration varying between species, individuals and, indeed, between tissues within an individual.

Ciliary function should also be under strong selective pressure since it underlies respiration, feeding, and larval dispersal in many invertebrates. Thermal latitudinal clines in ciliary function have been reported in some bivalves. In the common oyster, strains from colder temperature regimes have more active cilia at lower temperatures than do strains from warmer temperature regimes.

The clearest examples of the physiology of acclimation again come from work on muscle enzyme function in fish, operating at different acclimated T_b values. For example, Fig. 11.9b shows seasonal acclimation of K_m values in malate dehydrogenase from a goby. With any fish, initial exposure to reduced temperatures gives deleterious effects on swimming speed and endurance, but with prolonged exposure there is an improvement in swimming performance. This reflects numerous adaptations in nerve and muscle properties, including an increase in the amount of red muscle and in the density of mitochondria within muscle fibers. For example, the maximum speed attained by the long spine sea scorpion, a bottom-living ambush predator from the coastal areas of northern Europe, increases at both low and high temperatures following a period of acclimation lasting several weeks. At 15°C (this fish species' maximum summer temperature), the fast muscle fibers of a 15°C-acclimated fish are capable of generating more than three times the power of muscles from a 5°C-acclimated fish. This is sufficient to increase the rate of successful predation strikes on prey (shrimps) from 23 to 73%. The mechanism involves altered expression of myosin light chain subunits, which modulate the maximum contraction speed of muscle fibers. The composition of myosin heavy chains, and the ATPase activity, are unchanged by temperature acclimation in this species. This contrasts with the situation in freshwater cyprinid fish, where thermal acclimation modulates both the maximum tension generated and the contraction speed of muscle fibers, via altered expression of both heavy and light chain myosin genes.

Clines of genetic loci also occur, within populations where little individual acclimation can be observed, so that the proportions of different polymorphic forms at a given enzyme locus appear to vary with temperature and, consequently, often with latitude. An oft-quoted example (see Chapter 8) occurs in the common killifish, *Fundulus heteroclitus*; the occurrence of different M_4-LDH enzyme morphs is apparently correlated with latitude along the eastern seaboard of the USA, with the allozyme that is commonest at northern latitudes being the best catalyst at low temperatures. However, more recent work on mitochondrial DNA in this species has revealed that there are in fact two rather distinct races, northern and southern, with a boundary between the two lying off the coast of New Jersey. It seems likely that here the earlier allozyme work has been overinterpreted in terms of a neatly selected cline.

Regional endothermy in large ectothermic vertebrates

The major exception to ectothermy in primarily marine animals occurs in a few very fast active fish that come into the category of "regional endotherms" as described in section 8.4.1. The ability to maintain elevated body temperatures is known in just 27 species of large oceanic fishes, from the Scombroidei (mackerels, tunas, bonitos, and billfishes) among the teleosts and from the Lamnidae (mako, porbeagle, and white sharks) among the elasmobranchs. Conditions for regional endothermy are a large body size, a heat source, and heat exchangers to conserve heat. In all cases muscles or modified muscles provide the source of heat.

The tuna and some of its relatives can retain endogenous heat within the core of the swimming musculature, where the temperature may be more than 10°C above the surrounding water temperature (Fig. 11.10a). In contrast with most terrestrial endotherms, they do this by regulation of heat retention and not of heat production,

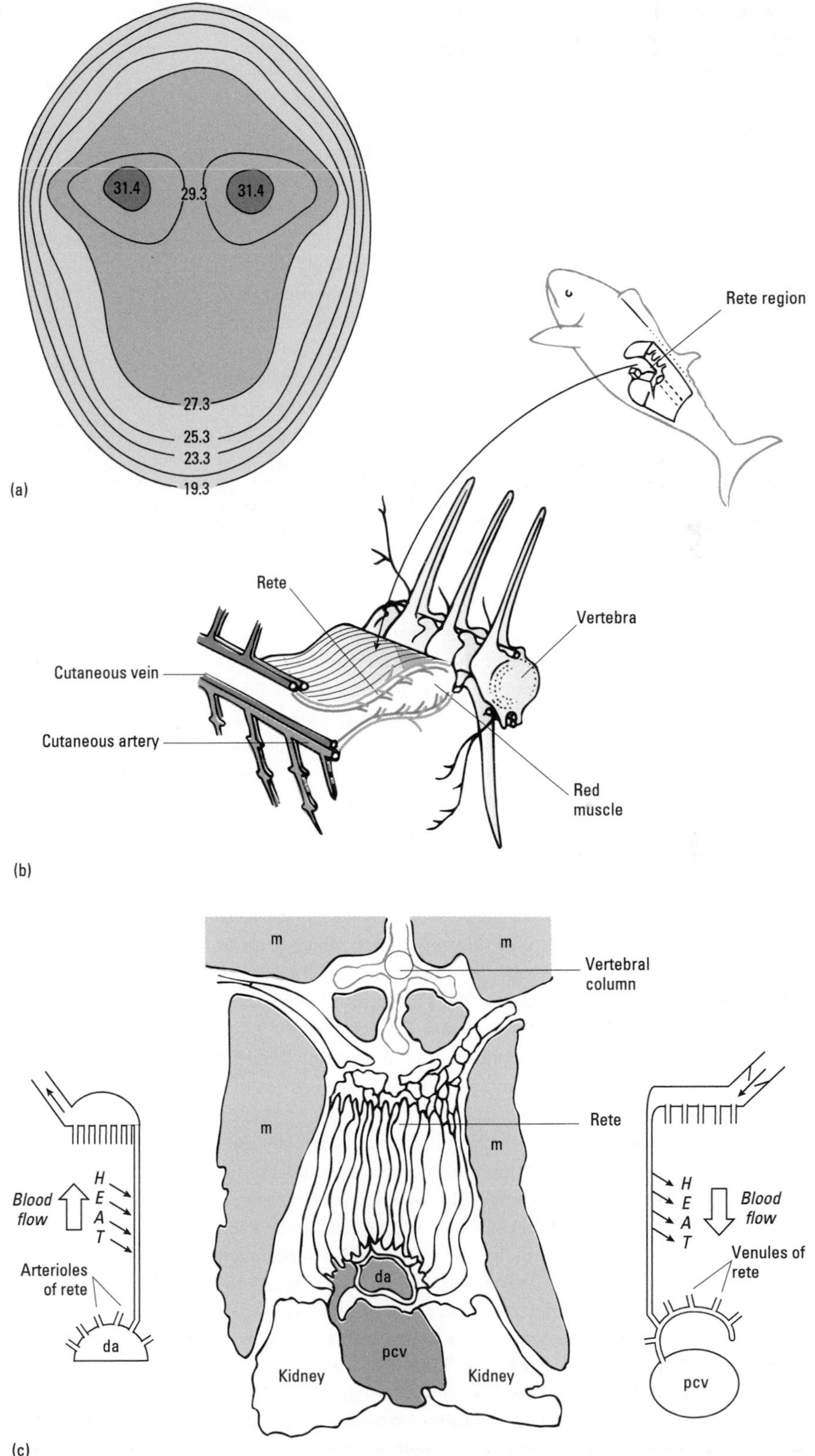

Fig. 11.10 Regional endothermy in tuna. (a) The temperature profiles across the trunk of a bluefin tuna, showing the "hot" swimming muscle at the core. (b) The paired lateral "cutaneous retes" of the big-eye tuna, where the cutaneous artery and vein run in parallel so that heat generated centrally is exchanged between the two vessels and retained at the core. (c) The central rete of the skipjack tuna, where a capillary bed beneath the vertebral column contains arterioles and venules running in opposite directions (up from the dorsal aorta (da), and down from the postcardinal vein (pcv)–see insets) between the main muscle blocks (m). (a, From Bone *et al.* 1982.)

the latter being independent of water temperature. The red muscle used for cruising in tunas is abundant and is internalized close to the body core. In most teleosts the blood supply to the red muscle comes from a central dorsal aorta and the venous blood is returned to a central postcardinal vein. In contrast, the blood supply to red muscle in most tuna is delivered via subcutaneous arteries, and the venous blood is also returned to subcutaneous vessels. Heat (resulting from the inefficiency of chemical energy conversion to muscle contractions) is therefore retained by a classic countercurrent heat exchanger system (Fig. 11.10b); warm blood leaving the muscle is used to heat the incoming cold blood. There is evidence that the internalization of red muscle preceded the evolution of vascular countercurrent heat exchangers and predisposed this taxon for conserving metabolic heat from muscle. The bonitos are the ectothermic sister group to the endothermic tunas, and some species have a degree of red muscle internalization intermediate between tuna and other scombroids.

The countercurrent rete design varies somewhat in different species. Larger forms, like the bluefin tuna, have a cutaneous rete as described above, where the peripheral artery and vein enter and leave the muscle mass as dense interdigitating sheets of arterioles and venules (Fig. 11.10b), whilst smaller species, like the yellowfin tuna, have multiple central retia where cool arterial blood from the gills meets warm venous blood draining inwards from the surrounding muscle (Fig. 11.10c). The exchanger is always arranged to keep heat in the core of the trunk muscles, insuring that up to 90% of the heat from the venous blood flowing out from the muscle is returned to the arterial blood (note that since heat exchange is markedly faster than oxygen exchange the arterial system can still maintain its normal oxygen delivery function). Thus, although this regional endothermy keeps these core locomotory muscles above ambient, the surface temperature of the tuna is close to that of its surroundings, with a sharp temperature gradient across the body. The raised temperature of the swimming muscles gives the tuna a higher sustainable speed than a nonendothermic fish of equivalent size, although maximum speeds remain similar and neither is there much improvement in time to exhaustion during a burst of activity.

There is little evidence that most tuna species can regulate their body temperature by physiological means; the temperature difference between the tuna and its surroundings does not change much as ambient temperature changes (Fig. 11.11). However, the large bluefin tunas do have considerable thermal inertia, which helps to stabilize T_b so that rapid changes in water temperature are not reflected in changes in body temperature. They may therefore thermoregulate behaviorally by operating in sea temperatures (T_a) that result in an acceptable excess temperature at routine cruising speeds. Tunas caught by anglers and played on fishing lines so that they cannot select a suitable T_a may actually overheat, producing what is known as "burnt tuna", of much reduced commercial value.

A variety of other marine fish from the mackerel and shark groups also use regional endothermy. Certain sharks have elevated visceral temperatures, routing the arterial blood via a rete close to the liver. Indeed, recent evidence indicates that the white shark, *Carcharodon carcharias*, is a remarkable regional endotherm with a stomach temperature up to 14°C above ambient and strongly regulated, allowing this and related species to continue as active predators in very cold waters. Brain heaters occur in the "billfishes", comprising the families Xiphiidae (swordfish) and Istiophoridae (marlins, sailfish, spearfish), and independently in a primitive member of the tuna family, the butterfly mackerel, *Gasterochisma melampus*. Endothermy in the billfish is restricted to the warming of the brain and eyes. Swordfish spend most of the night at the surface, but during the day undergo vertical excursions as deep as 600 m, thus encountering water temperature changes of up to 19°C for short periods (Fig. 11.12). Telemetry of free-swimming fish reveals that cranial

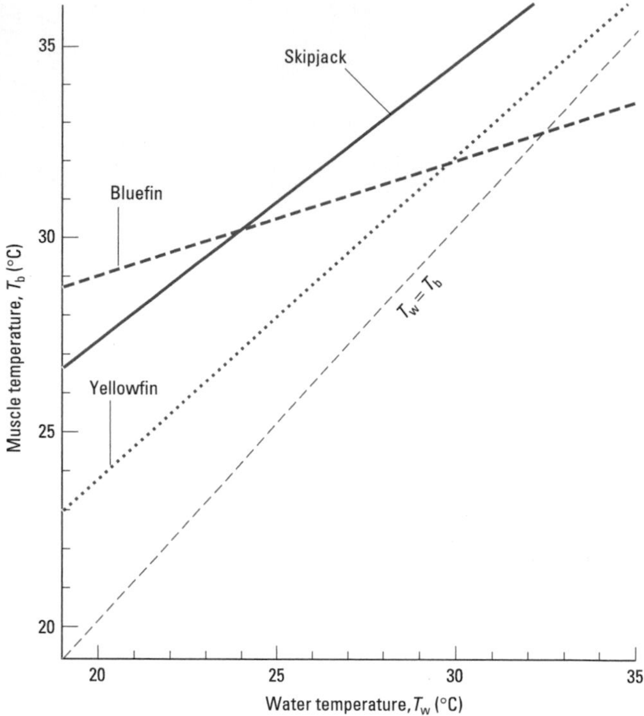

Fig. 11.11 The relation between water temperature and muscle temperature for some tuna species, showing their ability to maintain a high core T_b. Note that there is often little ability to thermoregulate since the lines are parallel to $T_w = T_b$. However, the bluefin tuna (large and with an efficient cutaneous rete arrangement) does regulate efficiently (see text). (From Carey *et al.* 1971.)

temperatures remain above, and buffered from, short-term decreases in water temperature. In the swordfish the brain temperature elevation is partly brought about by a carotid rete arrangement, but it also involves a so-called "heater organ". This is derived from a group of small muscles used to control eye movement in other fish (the superior rectus muscles). These heater muscles from each side of the head converge at the base of the braincase. The basisphenoid bone, which forms the lower braincase, is reduced to a membranous sheath, so that the brain is embedded on to the dorsal surface of the heater organ. Large amounts of adipose tissue surround the brain and insulate the brain, eyes, and heater organ. The amount of heater tissue varies between species and correlates with the ability to maintain elevated brain temperatures: a 40 kg swordfish has around 60 g of heater tissue. At the base of the heater a countercurrent heat exchanger, formed from the carotid artery and the venous return from the heater, prevents dissipation of the heat from the head region.

Heater cells in these fish are packed with mitochondria (63% by volume), contain high concentrations of myoglobin, and have an extensive internal smooth membrane system (tubules, cisternae, and membranous stacks, equivalent to the sarcoplasmic reticulum (SR) of normal muscle) making up 30% of the cell volume. Thermogenesis appears to involve the uptake and release of calcium from these SR membrane systems, providing a version of futile cycling and thus an energy "short circuit". Release of calcium through the SR Ca^{2+} channel promotes uptake back into the SR via the Ca^{2+}-ATPase, thus using ATP generated by the mitochondria

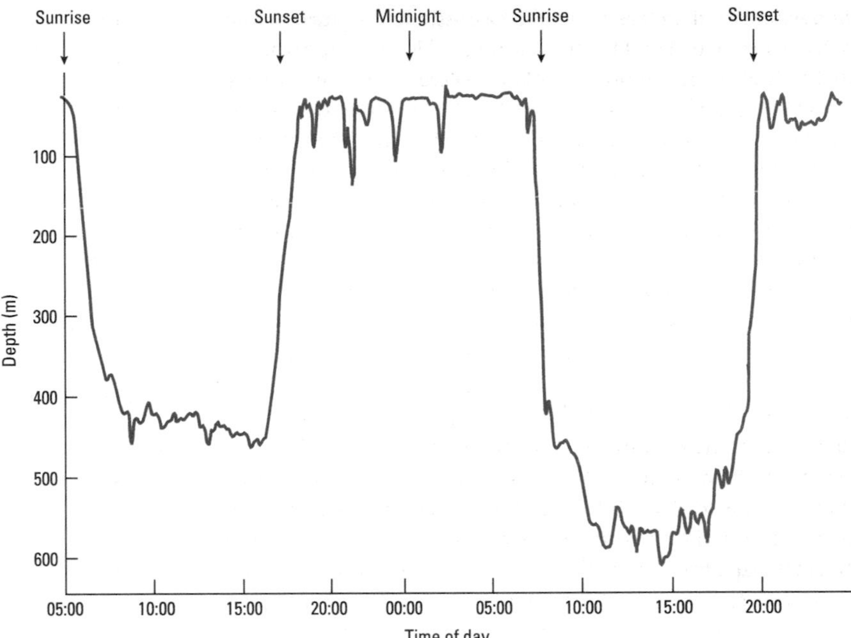

Fig.11.12 Telemetric recordings of depth in the swordfish *Xiphias gladius*, showing its diurnal movements, upward at sunset and downward at sunrise, exposing it to substantial water temperature change.

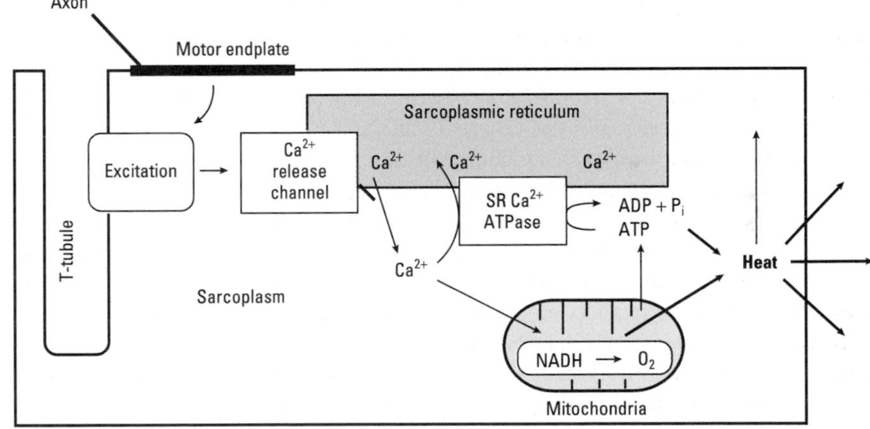

Fig. 11.13 A model for calcium-related thermogenesis in the heater cells of certain fish. Calcium is released from the sarcoplasmic reticulum (SR) (possibly in response to excitation from the axon, as in normal muscle) then recycled back into the SR by the calcium ATPase; thus ATP is split and heat released. Some calcium may also be taken into mitochondria, where it stimulates the calcium–proton transporter and generates heat via the electron transfer system. (Adapted from O'Brien & Block 1996.)

and producing heat as a by-product. Calcium may also stimulate the calcium/proton transporter of the mitochondria (Fig. 11.13). The calcium coupling ratio of the SR (i.e. the calcium ions pumped per ATP) is low compared with normal muscle, so that more heat is generated per calcium ion transported. Thus heater organ thermogenesis is rather different from other systems. In particular, the electron transport chain in the billfish heater organ mitochondria is tightly coupled to oxidative phosphorylation, contrasting with the uncoupled mitochondria of mammalian brown adipose tissue (see section 8.7.2).

Endothermy must have evolved at least three times within the Scombroidei, and cranial endothermy involving a thermogenic organ must have evolved twice (once in the billfish and once in the butterfly mackerel). The repeated evolution of endothermy suggests that there is a strong selection for this energetically expensive metabolic strategy. However, the advantages of regional endothermy in all these fish are not entirely obvious. It is not clear that muscle power or sustained swimming speeds are any greater in fish such as tuna than in any other tropical fish of the same size and habit. However, they can penetrate below the thermocline into cold water without suffering the same drop in temperature as other fish. Thus pursuit speed and predator–prey interactions may be fairly independent of rapid changes in water temperature, allowing these fish to be versatile predators at the top of the marine food chain. Recovery from exercise may also be quicker than in other fish, which may be an advantage for foraging in an environment where prey are patchily distributed. Tunas have relatively small guts, and a high visceral temperature may be advantageous in speeding the rate of food digestion and assimilation. The bluefin tuna and salmon sharks, the warmest of the endothermic fish, have ranges that extend far to the north, enabling them to exploit the food-rich seas of temperate and subpolar latitudes and to achieve rapid somatic and gonadal growth. Buffering of brain temperature in the billfish may also allow them to forage in relatively cold water, permitting extended periods of foraging beneath the thermocline, opening up the resources of the mesopelagic zone to exploitation. Thus in most cases the selective pressure driving the evolution of endothermy may well have been niche expansion.

Finally, some large aquatic turtles also have marked regional heterothermy. Metabolic heat production from the muscles is again conserved by countercurrent heat exchangers. Adult turtles have significant thermal inertia due to their large size, and the insulative properties of the enclosing carapace also aid heat retention. Leatherback turtles can generate local excess temperatures of up to 18°C. Again the advantages are not entirely clear, but the locally raised T_b may increase swimming speed or endurance, helping them to complete their long migrations.

11.4 Respiratory adaptation

Oxygen occurs at relatively constant concentrations in sea water; levels of 4–6 parts per million (equivalent to < 6 ml l^{-1}) are usually recorded, although the deep northern Pacific Ocean yields only 2–3 ml l^{-1}. Only in a few relatively enclosed seas does substantial deoxygenation of the bulk of the water ever occur. However, because oxygen has a somewhat lower solubility coefficient in warm than in cold waters, there may be rather less oxygen available for pelagic animals at the surface of the sea than there is at depth, especially at low latitudes. Furthermore, an **oxygen minimum layer** (OML) often occurs, somewhere between 400 and 1000 m, where biological oxygen usage is greatest in relation to the replenishment of oxygen by convection. Such layers are normal in areas of highly productive tropical oceans (especially the eastern Pacific and northern Indian oceans, where the oxygen concentration may fall as low as 0.1–0.5 ml l^{-1}), and in some temperate seas. At points where the OML abuts on continental margins, stable hypoxic conditions also exist for the benthic communities.

By contrast, carbon dioxide is exceedingly abundant in sea water, and forms about three-quarters of all the gas dissolved in the sea (compared to only 0.04% of the gases in air). Most of the carbon dioxide is dissolved as bicarbonate ions (HCO_3^-), which has an important role as a major buffer of sea water. Expired carbon dioxide can readily be added to this pool without creating any difficulties for the respiring organism.

Respiration is therefore not too much of a problem for most marine animals, in terms of acclimation to changing availability of gas. The exceptions are:

1 Pelagic (and some benthic) animals exposed to the oxygen minimum layers.

2 Benthic animals where decomposing sludge and unstirred waters can produce an almost anaerobic environment.

3 Animals that migrate, whether vertically or latitudinally, diurnally, or seasonally, between these problematic environmental zones.

4 Some fish and most endothermic vertebrates, where metabolic rate tends to be high, and for whom bursts of swimming can lead to temporarily anaerobic conditions at the muscle level.

The major physiological problem in all these cases is to avoid hypoxia, by establishing a respiratory system with enough exposed and/or ventilated surface area to extract the requisite oxygen by diffusion. The relation between this necessary surface area and body size was considered in general terms in Chapter 7 (see section 7.3.2). The respiratory surface area needs to be quite high for marine animals because water is viscous and thus quite difficult to move compared to air, tending to form a thick boundary layer next to the respiring surfaces. Large surfaces are also needed because sea water has a rather low oxygen content. These problems are partly compensated for by the possibility of having a unidirectional flow of water over the exchange surface, rather than the bidirectional tidal system used in lungs. Nevertheless a marine animal must move about 1 l of water to get 1 ml of oxygen, compared to only 25 ml air per 1 ml oxygen for an air breather.

The general body surface is certainly the primitive site of gas exchange, and may still be the only surface used in smaller animals. In benthic and moderately active marine invertebrates the larger requisite diffusive surface can be achieved by a highly perforated body structure (e.g. sponges, tunicates), or by frilled tentacles or gills (many worms and molluscs, but also many "simpler" animals such as sea anemones), or by dermal papillae (some anemones, many starfish). Within just one group, the echinoderms, different species can be found using papillae, tube-feet, invaginated respiratory trees, and even the gonad cavities as sites of oxygen uptake.

Examples of some of the more elaborate respiratory structures used in a range of marine invertebrates are shown in Fig. 11.14. In many molluscs and crustaceans very elaborate filamentous or lamellar gills occur. In crustaceans these are borne on the limbs, where the covering of the chitinous cuticle somewhat increases the diffusion path length (to 10–15 μm) and thus reduces rates of diffusion. However, the more active malacostracan crustaceans, such as prawns and crabs, have these gills tucked under a protective carapace, and ventilate the resulting branchial chamber either by gill movements or by using the scaphognathite baler on an anterior limb to give a one-way flow (see Fig. 7.9). The frequency of beating by this structure is the main determinant of water flow. In relatively active invertebrates, such as swimming crabs and cephalopods, the gills and their vascular supply are arranged to give countercurrent flow (e.g. Fig. 11.14b). Gill areas in most of these reasonably active animals generally scale with body mass to the power 0.7–0.8 (Fig. 11.15), giving a figure of about 5–10 cm^2 g^{-1}.

Marine fish may achieve some gas exchange through a cutaneous route (usually only about 5% of the total uptake), but invariably breathe principally with gills, sometimes external in embryos but always more or less internal in adults. Marine fish gills have short diffusion distances and therefore need a relatively lower surface area than many similarly sized crustaceans (Fig. 11.15), with a range of about 1–15 cm^2 g^{-1}, the higher values occurring in active forms such as tuna. These gills are ventilated with an organized one-way flow of sea water in most species, over a complex arrangement of filaments and lamellae (Fig. 11.16). Blood circulation is normally such as to give countercurrent flow, with water flow rates being substantially faster than blood flow rates to achieve maximum differences in oxygen partial pressure (Po_2) across the diffusing surface. Path lengths from water to blood may be only 2–5 μm in such gills. Ventilation may be through a spiracular opening, by active pumping from buccal and opercular muscles (see Fig. 7.10), or by "ram ventilation" achieved by fast forward swimming with the mouth open. Some fish use opercular pumping at slow speeds and ram ventilate when swimming faster and consuming more oxygen.

But achieving adequate oxygen uptake is still going to be particularly problematic for those species that are active swimmers with high demand, or those that inhabit the OML or experience marked temperature changes. Animals from the OML may show three key

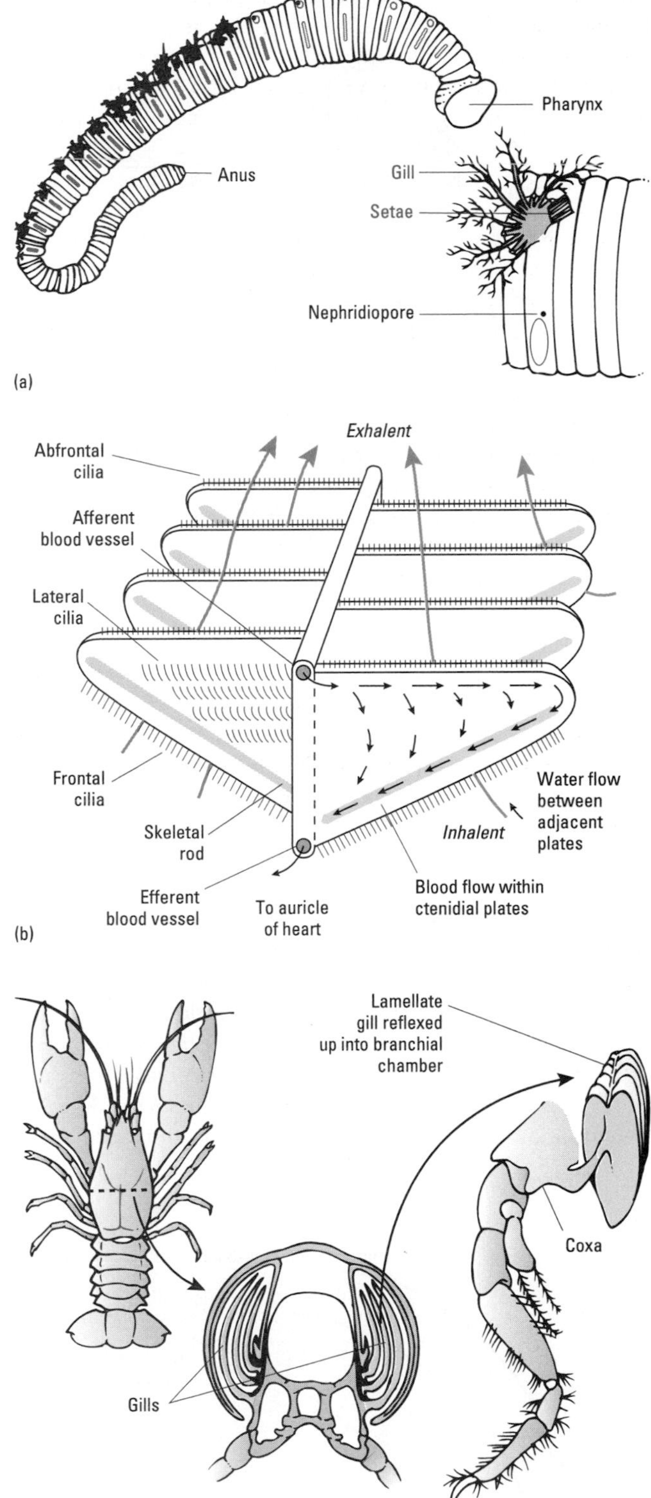

Fig. 11.14 A variety of respiratory structures in marine invertebrates: (a) polychaete, (b) gastropod mollusc, and (c) crustacean.

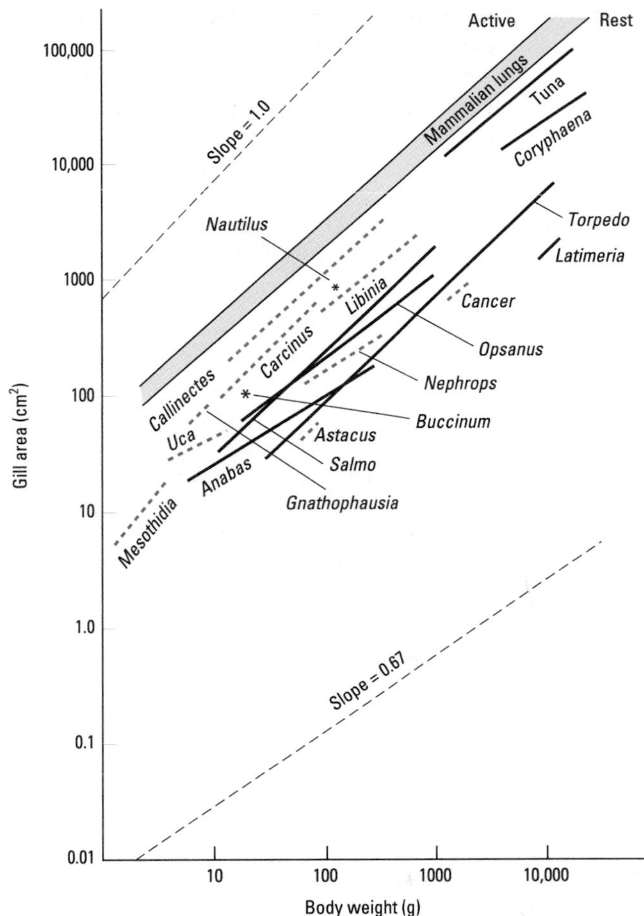

Fig. 11.15 The relationship between gill area and body mass for a range of crustaceans (dashed green lines), molluscs (*), and fish (solid lines). (From Withers 1992.)

approaches to cope with the reduced oxygen gradient into their tissues: increased oxygen extraction efficiency, reduced metabolic rate, or the use of anaerobic pathways. These may be used singly or in concert. For example, ctenophores in the OML probably cope because most of their oxygen demand is at the body surface in the comb-rows of cilia, and inward diffusion to the mitochondria is fast enough to supply them. Pelagic crustaceans from the OML off California show a more substantial suite of features that improve oxygen removal from the water: enhanced ventilation, larger gill surface areas, reduced diffusion distances across gill surfaces, and higher affinity hemocyanins. These crustaceans therefore rarely have to use sustained anaerobiosis. Pelagic crustaceans, such as euphausid shrimps ("krill"), that may experience moderate seasonal temperature change may have hemolymph with oxygen-binding properties suited to take up oxygen at lower Po_2 in summer than in winter (Fig. 11.17). For many marine fish modification occurs at a structural level, so that there are clear correlations between normal activity levels and various aspects of gill surface areas (Table 11.5). In relation to the fast-swimming fish referred to in the preceding section though, it is worth noting that periods of anaerobic activity are rather common. The marlin, for example, which often pursues and preys on tuna, has a 9 : 1 ratio of white to red muscle and can tolerate very high tissue acidity after a long chase and prolonged anaerobic work. Lactate is well buffered in the blood, and there is a remarkably pronounced Bohr effect (see section 7.5.2) to overcome this buffering and permit oxygen delivery when aerobic respiration is again possible. The tuna itself uses anaerobic activity effectively,

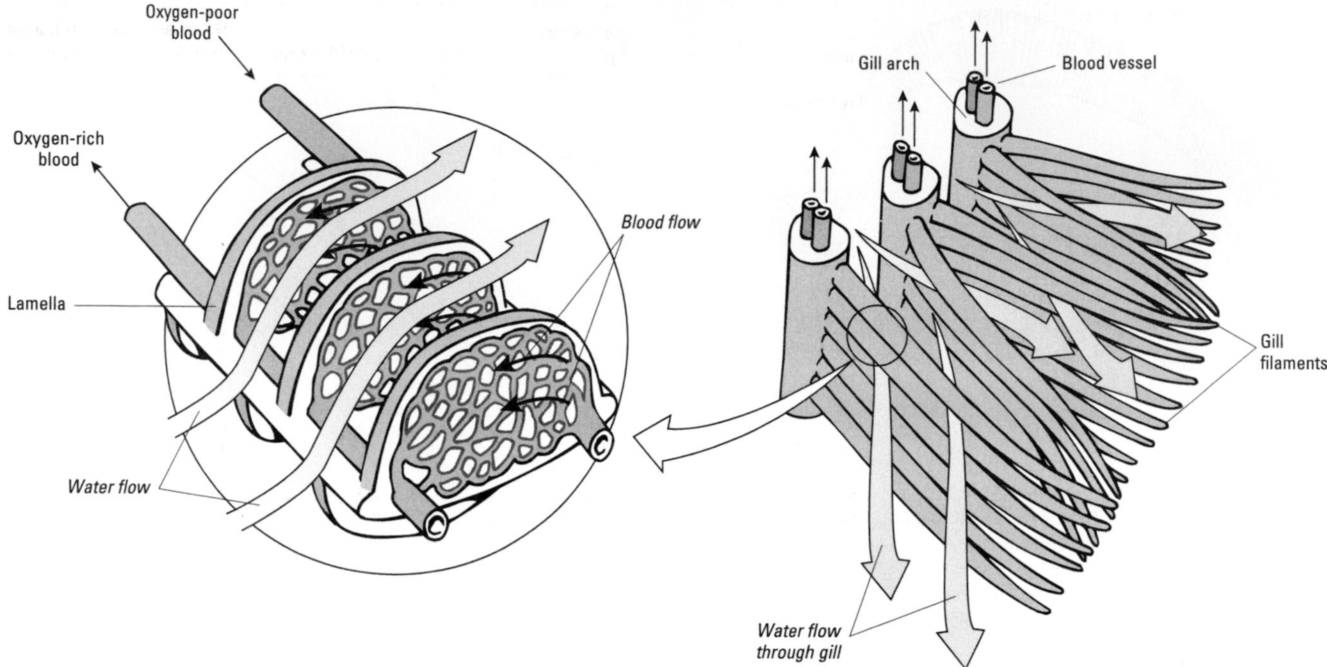

Fig. 11.16 The structure of fish gill lamellae showing the countercurrent flow of water and of blood.

and can recover from the associated acidosis in only 1–2 h, where other fish lacking warm musculature may take 8–24 h to clear lactate build-ups.

When structural sophistication is not enough and marine animals are faced with a degree of hypoxia, most (whether invertebrate or fish) will show two physiological responses: an increase in the ventilation rate, and/or a slowing of gill perfusion, usually achieved by slowing the heartbeat (bradycardia). In the molluscan sea slug *Aplysia*, for example, mild hypoxia produces an increase in respiratory pumping as the animal attempts to maintain its oxygen uptake (behaving as an "oxygen regulator", see Chapter 7). But if hypoxia persists or worsens its pumping rate declines and it becomes an oxygen conformer, entering a hypometabolic state. The capacity to control ventilatory or circulatory changes is greatest in species that routinely migrate through the regions of lower oxygen tension or that use bursts of sustained activity for predation or escape. One specialist at dealing with relative hypoxia is a continuous inhabitant of the OML, the crustacean *Gnathophausia ingens*, which has unusually large gill surfaces with very thin walls (hence low diffusion distances), an unusually fast circulation for a crustacean, and a

Fig. 11.17 Oxygen binding curves for the hemolymph of (a) euphausid shrimps (krill) at 5 and 10°C, and (b) various fish in relation to depth and activity. (a, Adapted from Brix *et al.* 1989; b, from Stewart 1991.)

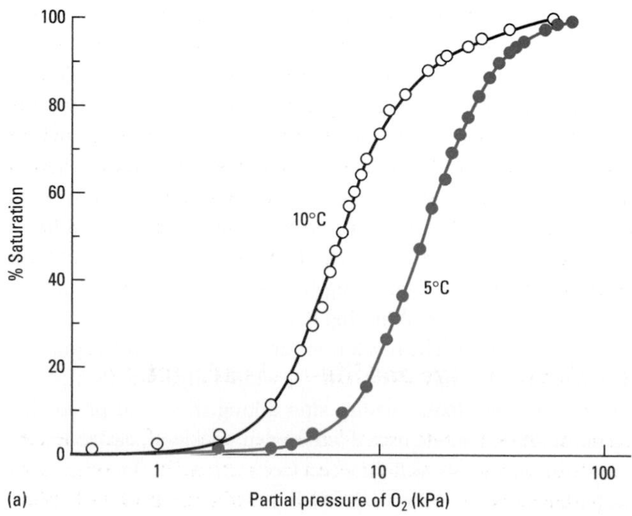

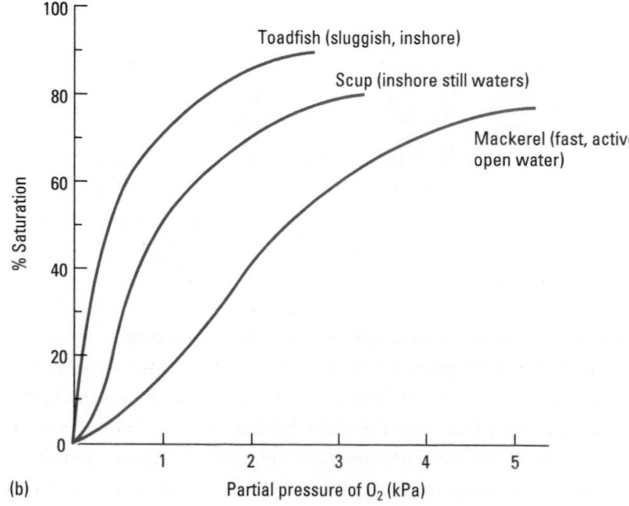

Table 11.5 Gill structure and activity levels in marine animals.

Genus	Body mass (g)	Activity	No. of filaments	No. of lamellae (mm^{-1})	Gill area ($cm^2\ g^{-1}$)
Crustaceans					
Callinectes		High			14
*Uca**		High			6
*Ocypode**		High			3
Panopeus		Moderate			9
Libinia		Low			6
Fish					
Lucioperca	70	High	1811	15	18
Salmo (salmon)	390	High	1606	19	2
Katsuwonus (tuna)	3258	High	6066	32	14
Thunnus (tuna)	26,660	High	6480	24	9
Pleuronectes (plaice)		Moderate			4
Callionymus (dragonet)	3	Low	478	16	2
*Saccobranchus**	42	Low	658	23	1
*Anabas** (climbing perch)	54	Low	567	21	1
Ictalurus (catfish)	239	Low	–	10	1
Opsanus (toadfish)	251	Low	660	11	2
Tinca (tench)	268	Low	1764	22	2

*Indicates fully or partially air-breathing species.

Table 11.6 A comparison of respiratory performance in marine animals.

Genus	Mass (g)	O_2 consumption rate (ml O_2 min^{-1})	Gill water flow (ml min^{-1})	Po_2 excurrent (kPa)	O_2 extraction efficiency (%)
Molluscs					
Octopus	10,000	2.8	2050	12.5	27
Crustaceans					
Callinectes (crab)	200	0.206	111	9.7	53
Homarus (lobster)	460	0.225	224	15.4	23
Cancer (crab)	900	0.504	288	13.6	34
Fish					
Callionymus (dragonet)	100	0.108	30	10.6	55
Scyliorhinus (dogfish)	200	0.134	24		40–70
Salmo (trout)	210	0.193	31	11.1	46
Acipenser (sturgeon)	900	0.86	380	14.0	30
Katsuwonas (tuna)	2250	13.3	4000	14.5	56

rather low routine metabolic rate. It also possesses a hemocyanin with high affinity and a large Bohr shift. This and other crustaceans from the same habitat can take up oxygen from water with a Po_2 of only 0.4–0.8 kPa. Other inhabitants of this layer, including some fish, are probably largely anaerobic for much of the day, and migrate to the surface at night. In common with other fish they may have a store of "extra" red blood cells in the spleen that can be mobilized if hypoxia is a problem.

While respiration in the marine environment may be fairly straightforward, and relatively unaffected by depth in a direct sense, it should be noted that there may be specific physiological problems associated with anaerobic zones and with the general sensitivity of respiratory pigments to high pressure and to low temperature. Animals living in low-oxygen benthic zones usually have oxygen dissociation curves well to the left on a standard plot (see section 7.5.1). Respiratory pigments in such animals can also show highly specific (adaptive) responses to temperature, pressure, and CO_2 and pH levels, according to the precise nature of their habitat. Deep-sea animals therefore often have hemoglobins or hemocyanins that are unusually temperature independent (e.g. tuna and a deep-sea vent crab, *Bythograea*), while others use oxygen-affinity modifiers such as organic phosphates or lactate ions to bind to the pigments and modify their properties as needed. The deep-sea shrimp *Glyphocrangon* has a hemocyanin that is virtually insensitive to temperature but is markedly influenced by pH, probably to allow a substantial Bohr shift for unloading oxygen to transiently active tissues.

In summary, the respiratory performance of a range of marine animals is compared in Table 11.6. The most telling information perhaps comes from the values of efficiency of oxygen extraction shown there, this factor varying from around 20% in some relatively inactive invertebrates and benthic fish to well over 50% in some crabs and in fast-swimming fish with good countercurrent organization.

11.5 Reproductive and life-cycle adaptation

Marine animals use both asexual and sexual modes of reproduction, with asexual processes being more abundant than in any other habitat, perhaps favored by the stability and predictability of the habitat, where the genetic variation arising from recombination is less vital. A vast range of "lower" invertebrates use systems such as budding to reproduce, either budding off free-living asexual larvae (e.g. in many anthozoans and scyphozoans) or budding incompletely to produce attached clonal organisms, as in so many colonial animals (e.g. hydroid corals, bryozoan sea mats, and tunicate sea squirts).

Sexual reproduction is present at some point in the life cycle of virtually all animals, however, and the process must include gametogenesis, gamete release and/or sperm transfer, and fertilization with a gamete from another individual. The majority of marine animals have simple uniflagellate sperm (presumed to be the "primitive" form) and relatively small buoyant eggs (oocytes). Rather large numbers of marine invertebrates (and a few fish, such as wrasse) are hermaphrodite. Some have an "ovotestis" that produces both male and female gametes at the same time or sequentially, although they tend to exchange gametes with another individual rather than self-fertilize. Other animals have a progression within their own lifetime (see Box 11.1) from young male to old female (protandrous hermaphroditism) or less commonly from young female to old male (protogynous hermaphroditism). The switch between sexual phases may be environmentally (seasonally) determined, or may depend on received signals (pheromones) from nearby potential mates.

11.5.1 Pelago-benthic alternation

There are a number of different reproductive options available to marine animals, ranging from broadcast spawning in massed populations to careful one-to-one copulations and viviparity, each of these having its own advantages and problems. Most marine animals have a larva (Fig. 11.18) as the pelagic stage of the life cycle, with relatively few hatching directly as juveniles (small adults) without metamorphosis. An alternating pelago-benthic lifestyle therefore predominates. This may be because it allows small larvae to feed on seasonal surface plankton blooms and grow rapidly, before becoming large and benthic. It also reduces competition between the generations, as well as allowing larval or juvenile dispersal. In addition, there are no particular "extra" physiological burdens on small immature stages as compared with larger adults, for this is the one habitat whose equable ionic, osmotic, and thermal characteristics make the surface area : volume ratios of most animals relatively unimportant. Osmotic and ionic control, respiration, and temperature relations are fairly straightforward whatever the animal's size, so that small larvae are perfectly feasible and may even have advantages in relation to buoyancy and dispersal.

Evolutionary choices as to the exact reproductive mode are therefore likely to be determined more by adult habit and habitat (benthic/interstitial vs. pelagic, sessile vs. active, and so on), and by phylogenetic constraints, than by physiological considerations.

Fig. 11.18 Marine larval forms.

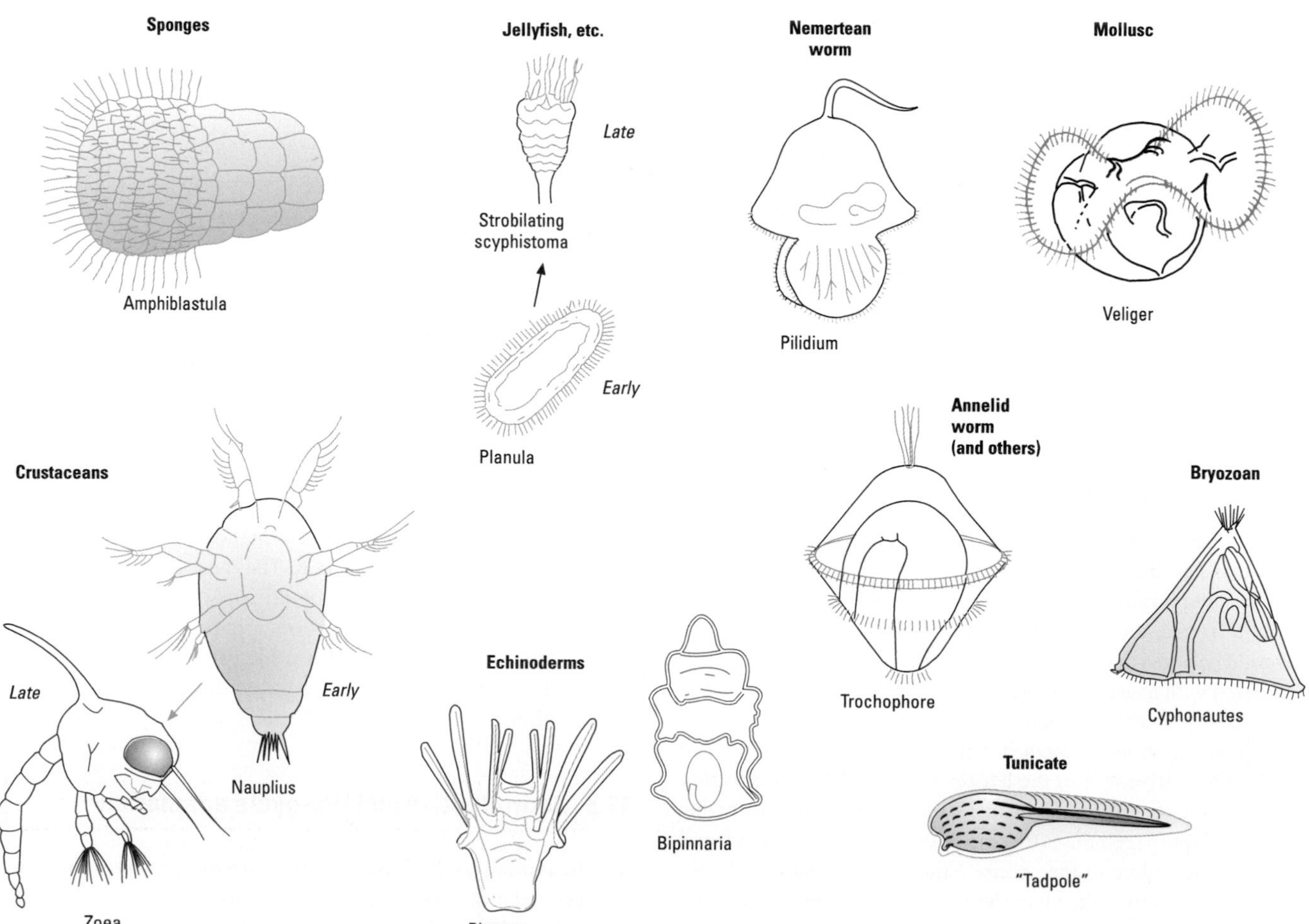

Box 11.1 Hermaphrodites and sex change in marine animals

Explanations of sex change in animals are based on the fact that the reproductive success achievable as a male or a female changes with body size.

Male success via sperm can begin in very small individuals, because tiny sperm are cheap to produce and easy to store in large numbers. Reproductive success for males increases steeply with reproductive investment, as sperm released into the sea can potentially reach many females. Beyond a certain level, however, further investment fails to produce more success, perhaps because the limited longevity of sperm puts an upper limit on the distance over which they can disperse and still retain fertility.

Females produce eggs which are much more expensive than sperm, so that a larger minimum body size is required for female reproductive function. Once this minimum size is reached, female function increases more or less linearly with body mass, as more eggs can be made and/or stored, and there is no clear upper limit to this.

The relationships of reproductive success to size for males and for females are therefore as shown below:

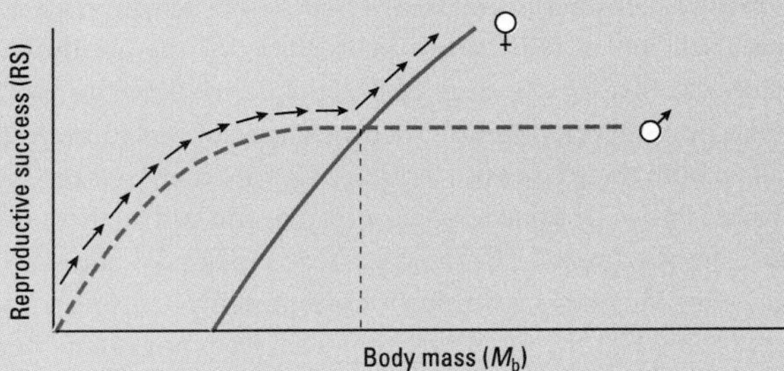

The lines for males and females intersect at some particular body size/investment value. For maximum reproductive success through a single lifetime it therefore pays off for an individual to follow the arrowed track in the diagram, changing from male to female at the body size where the intersection occurs. This is the most common direction of sex change, and is seen in animals such as pandalid shrimps, and clown or anemone fish on tropical reefs.

Cases of female-to-male reversal tend to occur when the relation between size and reproductive success is distorted. In certain wrasse, males are territorial and big males achieve many matings, while small males cannot maintain a big territory and get very little reproductive success. Hence the preferred track for sex change is modified, and individuals become males only when they are big enough to do well:

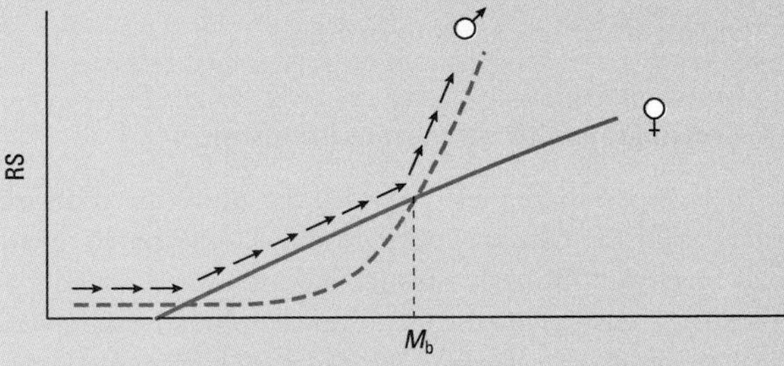

The size at which sex reversal occurs is known to depend on the make up of the population. For example, pandalid shrimps usually show male-to-female switches, and the size at which the switch occurs changes in at least two situations.

1 Invasion of lots of small shrimps during a good year for larvae, producing lots of small competing males. The value of being male at a given size falls, so it pays to become female at a smaller body size:

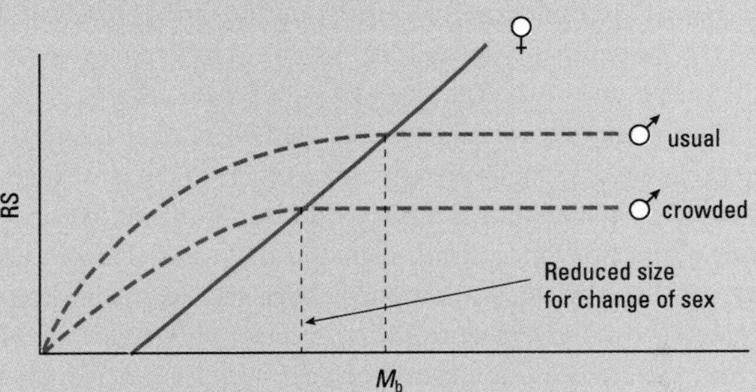

2 Enhanced mortality of large females (produced by selective fishing by man), where the success of smaller females is enhanced, again favoring the sexual switch at a smaller size:

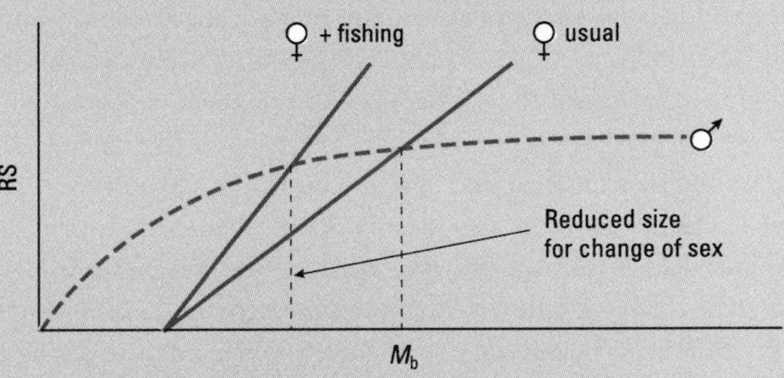

Given the advantages of hermaphroditism, why is it not more common? It may be that the competing requirements of male and female functioning in a single body are too expensive, especially where there is marked sexual dimorphism requiring substantial morphological change at the time of the sexual switch.

11.5.2 Planktotrophic and lecithotrophic larvae

A major distinction is usually made between these two modes of larval development. A planktotrophic larva feeds for itself on the abundant plankton, whereas lecithotrophic larvae feed off yolk reserves supplied by the mother.

This distinction has a limited range of physiological correlates. Most marine invertebrates adopt the pelagic planktotrophic option, producing very large numbers of simple gametes that are fertilized externally to give small larvae that then fend for themselves. This is the "easiest" reproductive/life-history option, and is not open to animals in most other habitats. The major control system needed, whether a species is hermaphrodite or with separate sexes (dioecious), is an ability to accumulate and then to release gametes *en masse*, preferably in synchrony with many others of the same species ("epidemic spawning") to insure good cross-fertilization, and if possible also in synchrony with planktonic blooms. This is commonly achieved by an integration of hormonal control systems (ultimately the major gonadotropic hormones) with environmental cues. Such cues may be limited in deep marine environments, where there is little temperature variation to give periodicity. However, recent evidence shows that deep-sea clams (*Calyptogena*) can spawn in response to very small thermal changes of just 0.1–0.2°C that occur with hydrographic water movements. For shallower waters more consistent cues are likely to be available relating to tide and to season, and it is no surprise that many marine animals use photoperiod and/or lunar cycles to trigger spawning. The best known example here is the South Seas palolo worm (*Palola viridis*), a polychaete where gametes accumulate and mature in the coelom and are then shed by rupture of the parental body wall. At a precise time of year (a night of full moon, usually September or October), the gravid individuals migrate upwards through the water column and form vast surface swarms where the gametes are shed and fertilized as milky clouds in the sea water. Mortality may be huge, as is usually the case with planktotrophic life cycles; survival rates of well under 1% are commonplace.

Lecithotrophic larvae have some inbuilt food supply, and tend to spend a shorter time in the planktonic phase, often settling out again within a few hours or days of being shed. They serve primarily to disperse the young away from the parents, without too much risk of long-range dispersal to entirely unsuitable habitats. The strategy is quite common in deep-water animals from cold and rather unproductive seas, where planktonic food organisms are relatively scarce. Fewer gametes are released, and each fertilized egg has a better chance of survival, rarely reaching the hazardous surface waters but settling again among the deep benthos. Species with lecithotrophic larvae require a period of vitellogenesis in the mother, to add yolk (lipid and protein) to the developing eggs; the female reproductive tract needs to be more complex, with accessory yolk glands, and more sophisticated physiological control systems involving hormonal signals.

Larvae that have matured enough to undergo metamorphosis, whatever the food supply used, are termed "competent", and may then respond to environmental cues that trigger settling and metamorphosis responses. Negative phototaxy and/or positive geotaxy bring the larva to a substratum, which is then assessed for suitability on the basis of physical, chemical, and biological information.

11.5.3 Direct development

Relatively few truly marine invertebrate animals use a degree of direct development. Those that do produce a fairly small number of large and yolky oocytes that are usually fertilized by direct insemination, the female often storing the male sperm or sperm packages (spermatophores). Vitellogenesis (yolk production) is essential, with follicle cells or nurse cells in the ovary providing nutrient transfer to the oocytes. Embryos are invested with some degree of parental care, either by encapsulation or by brooding in or on the parent's body. Most marine invertebrates that have direct development are either oviparous (laying the fertilized eggs externally, albeit in some sort of capsule) or occasionally ovoviviparous (retaining the eggs internally, but relying on the yolk to feed the young, with no additional feeding from the mother's body).

Direct development occurs in species where dispersal is not a problem, for example where the adult is pelagic, as in the squids and the arrow worms (chaetognaths), or where the environment is relatively harsh and/or unpredictable, which in the context of the seas often means in high-latitude zones. Examples include *Octopus*, where spermatophores are transferred to the female cloaca by a specialized arm of the mature male. Some of the opisthobranch molluscs known as sea slugs also copulate in elaborate fashions, though being hermaphrodite sperm transfer occurs in both directions. In both these cases the fertilized eggs are then packaged internally into fairly elaborate jelly-like capsules and laid on the substrate as discrete egg masses. The female octopus then takes the extra step of guarding the eggs and turning them repeatedly to insure adequate oxygenation. Her maternal care, however, does not extend beyond hatching, and in some species she normally dies soon after the eggs mature. Rather less elaborate direct development occurs in some echinoderms, where the fertilized eggs are brooded on the upper surface of the body, in the gonads or gonadal bursae, or even in the stomach.

In some cases direct development is followed by the release of a larval form that must still metamorphose to reach its adult morphology. This kind of "mixed life history" is particularly common in marine arthropodan groups, where both yolk production and a protective cuticular eggshell are characteristic features of the taxa, with elaborate mating behaviors and specially modified limbs (gonopods) for sperm transfer, but where larval dispersal and metamorphosis are also desirable for the predominantly benthic species. Mixed life histories also occur in many of the colonial tunicates, with eggs brooded in a common cloacal chamber until the relatively large "tadpole larvae" are ready to emerge.

11.5.4 Semelparous and iteroparous life histories

Semelparous species are those in which the adults breed once and then die, as in the *Octopus* example above, whereas iteroparous animals breed continuously through a season, or repeatedly over more than one season. This distinction affects physiological sophistication in the adult quite profoundly. Iteroparous animals must control rates of germ-cell maturation more carefully, and thus need to organize the partitioning of their own trophic resources in time and space; their endocrine systems, and the receptors that feed into them, have to be both more complex and more plastic. Semelparity is, perhaps not surprisingly, quite common among marine invertebrates of all taxa and size ranges. Iteroparity occurs more commonly in species that have direct development of a very few eggs in each batch.

11.5.5 Vertebrate reproduction

Fish can adopt life-history strategies that are similar to those of invertebrates. They often produce many thousands or even millions of tiny floating eggs for external fertilization, with a semelparous life cycle. They tend to endow their eggs with a small amount of food reserve, to allow the young to hatch slightly later and thus slightly larger, limiting the problems of hyposmotic regulation at extremely small sizes. The egg and larva may also be endowed with a small oil droplet to confer buoyancy (see below). At certain times of the year, fish larvae make up a high proportion of the zooplankton in temperate seas and provide a particularly rich food resource for the nekton.

The marine tetrapods are faced with very different problems, and cannot discharge large numbers of eggs and sperm into the sea as most invertebrates and fish do. Reptiles and birds are derived from essentially terrestrial egg-laying ancestors, and most of the species associated with the sea have no alternative but to return to land to breed, as the amniotic egg is designed for aerial rather than aquatic breathing. Many reptiles then run into the problems of finding thermally suitable nesting sites, given their temperature-sensitive sex-determination (TSD) systems (see Chapter 15). Today only sea snakes are viviparous and can bear live young while still submerged (though there is evidence that Mesozoic ichthyosaurs also achieved it). In this they parallel the mammals, some of which (cetaceans, including whales and dolphins) can achieve live birth at sea. The young of all the permanently marine tetrapods are inevitably relatively precocious, already able to swim, breathe air, and achieve homeostasis. The only essential difference from the adults' biology arises from the provision of an unusually concentrated and lipid-rich milk (see Chapter 15) from the parent mammals (matched by a milk-like uterine secretion in sea snakes).

Other than cetaceans, most marine mammals routinely come ashore and raise the pups terrestrially at first. This can put substantial strain on the maternal physiology, particularly with respect to temperature balance, as the adult is adapted for the lower temperatures associated with submersion and, encased in insulating blubber, can overheat rather quickly when hauled out on a sunny beach. Seals and walruses seek (and often compete fiercely for) shady patches and cool substrata, and those that achieve the most favorable microhabitats may rear young most successfully. These animals may also take rapid cooling dips in the sea whenever possible, but thereby often risk predation from sharks and other large carnivores waiting just offshore from the breeding sites.

11.6 Depth problems, buoyancy, and locomotion

The sea has an average density (in g cm^{-3}) of 1.025 (tropical surface waters) to 1.028 (colder and deeper polar seas). Cytoplasm has a density of 1.02–1.10 g cm^{-3}, and most organisms also have some heavy skeletal materials in their make-up giving an overall density markedly greater than their surroundings. All organisms therefore tend to sink, so must be able to cope with the adverse effects of depth and/or must have adaptations to maintain themselves at a particular depth, leading us on to the issue of buoyancy.

11.6.1 Effects of depth and pressure

The major problems of depth lie not in salinity, temperature, or gas availability, as we saw in preceding sections, but in the effects of pressure. For every 10 m below the surface the pressure increases by 1 atm (101 kPa). Since life is found at all depths of the sea, right down into the deepest trenches where depths can reach 11,000 m, some animals are clearly adapted to coping with pressures in excess of 1000 atm (101 MPa), and may be termed **barophilic**. Many more animals migrate vertically and must be able to cope with a wide range of pressures; for example, the pygmy shark moves up and down daily by about 1500 m to feed on bony fish and squid, while diving mammals and birds can reach depths of 500–1500 m (see section 11.10.3).

A pictorial summary of the kinds of life found living permanently at depth is shown in Fig. 11.19. Some rather "simple" animals, such as sponges, cnidarians (anemones and sea pens) and echinoderms (sea stars, sea cucumbers, and the like), survive very well at 5000–6000 m, together with more complex forms such as crustaceans and squids. Highly specialized teleost fish are particularly prominent, reaching depths of at least 7000 m (4.5 miles down), although those at greater depths lack the gas-filled swim-bladders described later.

Adaptations to high pressure occur at all the levels discussed in the introductory chapters of this book, but are most evident in the realms of physiology and biochemistry. All effects of pressure are in effect a consequence of volume change (ΔV). Pressure primarily affects molecular structure, altering the volumetric packing of enzymes and other proteins, and of lipid-based membranes. At a higher level of organization structures such as cilia and other microtubular structures may be disrupted, leading to problems with spindle formation in mitotic (and meiotic) cell division. Pressure can also be particularly disruptive for solute/solvent interactions. The hydration shells around individual ions change their effective size when pressurized, so that diffusion rates are affected and events dependent on permeation of cell surfaces—such as excitability and impulse transmission—can be grossly modified. Locomotor activity and behavior are therefore particularly sensitive to pressure in all kinds of animals, and the most critical adaptations often occur in nervous systems. When shallow-water organisms are exposed to gradually increasing pressures their motor activity tends to increase and they exhibit enhanced excitability, culminating in convulsions and spasms at about 100 atm. Humans show "high-pressure nervous syndrome" at depths greater than 200 m, with tremor, dizziness, and inattention. Thereafter immobilization occurs, which soon becomes irreversible. In contrast, animals collected from moderate depth in isobaric traps show no such responses to increasing pressures, but some are reversibly immobilized by normal shallow-water pressures ("decompression paralysis"); they have become obligate barophiles.

Only recently have the adaptations at the level of reaction rates, enzymes, and membranes attracted much attention. Hydrostatic pressure may perturb protein and membrane function wherever

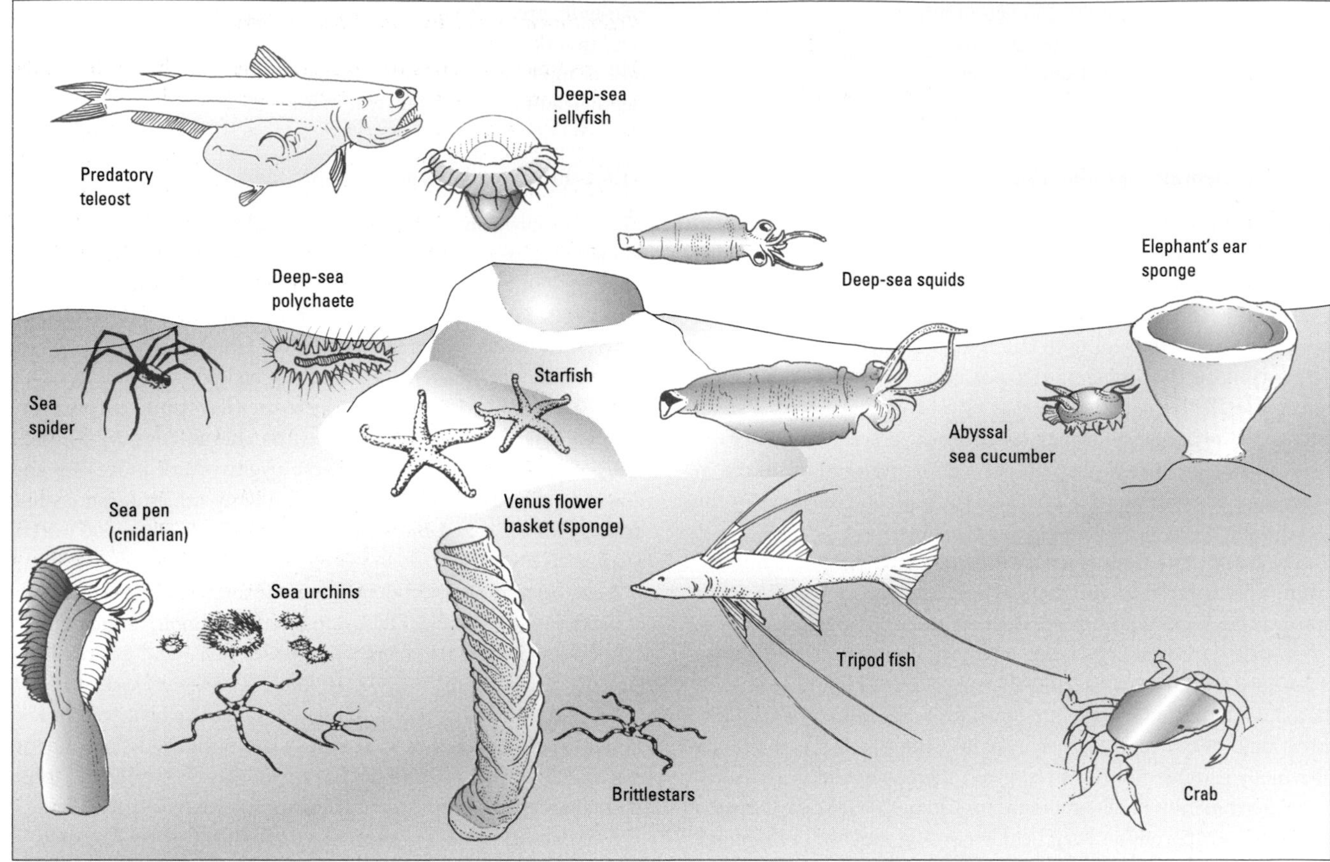

Fig. 11.19 A pictorial view of deep-sea fauna.

volume change occurs at specific molecular sites, and hence phenomena such as structural stability, ligand-binding affinity, and bilayer fluidity are all affected. Metabolic reactions can be accelerated, retarded, or remain unaffected by increasing pressure, according to the molecular geometry of enzymic reaction sites. Consequently, complex multistep enzyme pathways can be completely disrupted at depth as constituent reactions become unbalanced. Ion transport systems, and hormone or neurotransmitter binding, can also be disrupted. For example, membrane channel kinetics may change, probably by altered movement of the charged sensor regions that activate and inactivate the pore opening; though channel conductances are commonly unaffected.

Some level of protection can be afforded by the osmolytes such as methylamines that also protect against osmotic change (see Chapter 4). Deep-sea teleosts may have 250–300 mmol kg^{-1} of TMAO (about four times higher than in similar shallow-water species), helping to stabilize proteins. However, enzymes must also be modified to operate at high pressure by having reduced pressure sensitivity, probably achieved by increasing the hydrophobic (or in some cases electrostatic) bonds within the molecular structure, and by increasing bond strength between protein subunits. Where necessary, enzymes should ideally retain a fairly constant K_m value irrespective of pressure, and this has been shown to be true for LDH binding to its NADH cofactor in several deep-sea fish, whereas the same enzyme in shallow fish has its K_m value roughly doubled for a pressure equivalent to 700 m depth. In diving mammals and birds, however, LDH is apparently similar to that in nondiving species and is inherently pressure insensitive; this is perhaps because the LDH of endotherms is much less fluid in structure anyway, to allow it to function at a T_b of 37°C. Retention of sensitivity to modulators is also crucial, and has been demonstrated in G-protein-coupled receptors in many deep-sea fish (where receptor/protein interactions, agonist interactions, and function of the effector molecule adenylyl cyclase all vary with pressure).

Pressure-insensitive enzymes are particularly useful for nonendotherms that migrate vertically. For example, the fish *Sebastolobus* has larval and young stages that are pelagic but the adults are deep-sea specialists; its muscle enzymes can function without modification over a range of 130 atm, with no switching to alternative suites of isozymes during development.

For membranes, a classic strategy is the modification of the lipid composition, replacing standard phospholipids with less saturated forms to counteract the effects of pressure (and of the routinely low temperatures also experienced at depth) in making normal membranes more viscous. This is sometimes called "homeoviscous adaptation" (see Chapter 8). In fish, we know that monounsaturated lipids, such as phosphatidylcholine and phosphatidylinositol, become much more important in the cell membranes of deep-sea forms, in order to retain membrane fluidity. In the sunfish and other well-studied cases these effects are largely due to the desaturase enzyme systems discussed in Chapter 8, rather than to any change in dietary lipids or transfer enzyme activities.

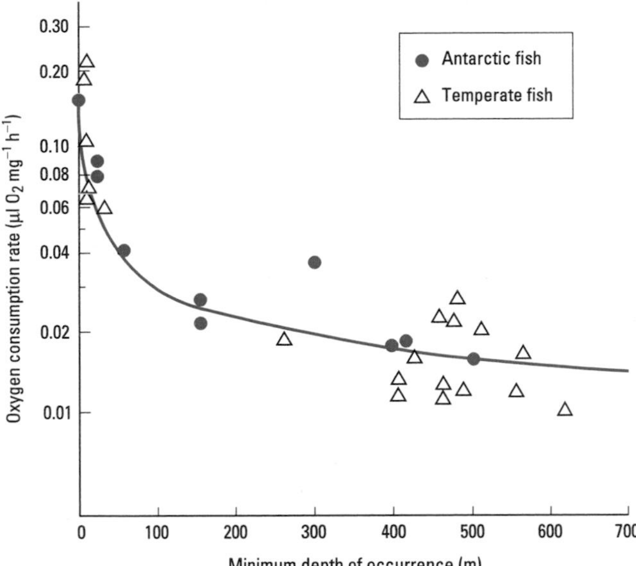

Fig. 11.20 Effects of depth (pressure) on oxygen consumption rates for a range of Antarctic and temperate fish. (Reprinted from *Comparative Biochemistry and Physiology B* **90**, Torres, J.J. & Somero, G.N., Vertical distribution and metabolism in Antarctic mesopelagic fish, pp. 521–528, copyright 1988, with permission from Elsevier Science.)

For whole animals, physiological performance can perhaps best be compared between species in relation to their **minimum depth of occurrence** (MDOC), so that permanent deep residents and vertical migrators can be considered together. For pelagic fish, cephalopods, and crustaceans, rates of oxygen consumption (metabolic rates) decrease rapidly with increasing MDOC (Fig. 11.20), and to an extent much more than can be explained by the decreasing temperature. Most of the decrease occurs in the first 200–400 m, so that at even moderate depth most animals have very low metabolic rates for their size. But the decrease with depth may be more to do with ecological factors favoring limited locomotion (low prey density, absence of light and therefore of plant-feeding opportunities, etc.), rather than with physiological constraint. This would accord with the particularly low activity of white muscle enzymes at depth.

11.6.2 Buoyancy mechanisms

Buoyancy can be seen as a means of avoiding the effects of variable pressure, by a combination of behavior and physiology. There may be three very general ways of achieving neutral buoyancy (Fig. 11.21): removing heavy material, adding very light material, or improving flotation by greatly increasing the surface area : volume ratio. For very small animals this last strategy is commonly found, as organisms of only a few millimeters in size are dominated by viscous drag forces, and having an elaborately frilled or spiky outline can in itself increase the drag sufficiently to prevent sinking. A great many marine larvae and small planktonic creatures exploit this property, and are strikingly different in form from the relatively streamlined larger animals in the same habitat. For these larger animals, there are various different mechanisms for achieving neutral buoyancy.

Reduction and substitution of heavier ions

This technique involves replacement of heavier (higher molecular weight) ions in the body fluids and tissues (such as magnesium and sulfate) with lighter ions, using some kind of exchange pump system, leading to some of the oddities of body fluid composition discussed above. The jellyfish *Aurelia* and the cuttlefish *Sepia* both show unusually low sulfate levels, with chloride making up virtually the entire anionic component of their body fluids. A range of comb jellies (ctenophores), jellyfish (cnidarians), salps (tunicates), and sea slugs (molluscs) are known to extrude sulfate in a similar fashion, and their measured lift correlates with the degree of this extrusion (Fig. 11.22).

Even more extreme is the deep-sea squid *Heliocranchia*, which has a large fluid-filled pericardial cavity (up to 65% of the body volume). This contains very little magnesium or even sodium, the cationic component being dominated instead by light ammonium ions, produced from protein metabolism and trapped in the pericardium by the low pH of that cavity, probably in turn resulting from a Na^+/H^+ exchange pumping system. Again, the anionic component is almost entirely chloride, with virtually no sulfate; the resultant fluid has a specific gravity of only 1.01, and is isosmotic with sea water. Other squid in the same family have similarly buoyant fluid vesicles containing NH_4Cl in their arms or mantle.

Many elasmobranchs use urea and TMAO as important balancing osmolytes in relation to maintaining their body fluid composition (see section 11.10.1), and recent calculations indicate that in species such as the Port Jackson shark these solutes also have a significant positive buoyancy effect. TMAO is also useful in this sense in some decapod crustaceans, and the effect may be widespread.

Losing ions without substitution

If the heavier ions are simply extruded actively from the body a degree of buoyancy can be achieved, although there will be inevitable osmotic costs as the body fluids will be relatively dilute, the animal becoming hyposmotic to the sea. This strategy is uncommon in invertebrates, but serves as an important buoyancy device in teleost eggs. It may in fact be a major advantage to the vertebrates of having retained hypotonic body fluids, giving a degree of buoyancy to all marine teleosts. It is taken to a greater extreme in some deep-sea teleosts, where skeletons and muscle mass are also reduced (see below).

Reduction of heavy minerals

In particular, buoyancy can be increased by the loss of calcium carbonate or phosphate in the skeleton. A number of pelagic gastropod molluscs (especially the heteropods and pteropods) that live at depth have substantially reduced their shells, to the extent that some have become the delicately buoyant animals known as "sea butterflies". Vertebrates also commonly exploit this strategy, especially in the elasmobranch fish. The same effect is found in marine mammals, which include the largest animals on Earth (the blue whales, at up to 4×10^5 kg); many of these cetaceans have limb bones and ribs with depressed skeletal maturation and no marrow cavity, the bone being very porous and containing significant quantities of

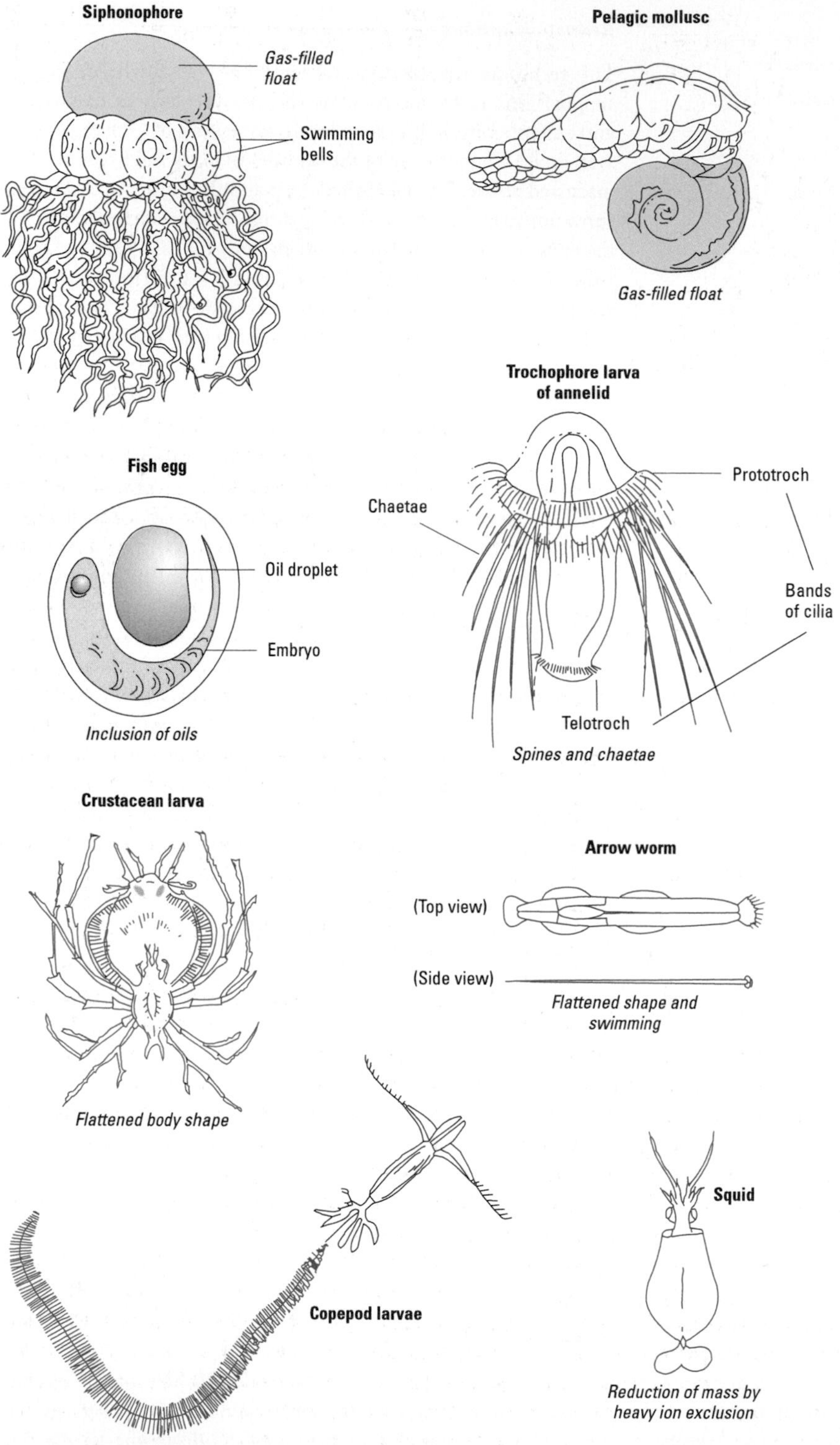

Fig. 11.21 Examples of methods to increase buoyancy in marine animals.

lipid (see below). Similar phenomena are found in marine reptiles and in penguins. Alternatively, skeletal materials in some parts of the body may be much reduced whilst others that are crucial to feeding are strengthened. Here, some deep-sea pelagic teleosts are the best examples, often having shrunken trunk musculature and reduced backbones with cartilage (density 1.1 g cm^{-3}) replacing bone (density 2.0 g cm^{-3}), the vertebral processes being reduced leaving just a narrow bony collar around a persistent gelatinous notochord. A vivid example is provided by a few Antarctic notothenioid fish that (following glaciations that reduced the pre-existing pelagic fauna) have become secondarily adapted to a pelagic mode of life from a benthic ancestor. They lack the normal teleost swim-

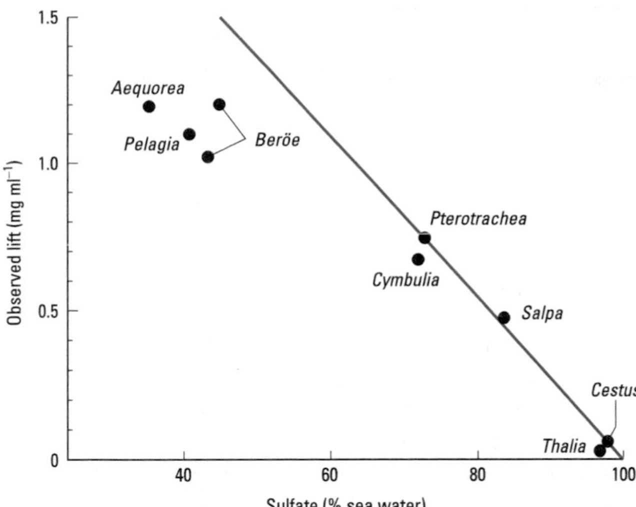

Fig. 11.22 The effects of extrusion of sulfate on buoyancy in a range of pelagic cnidarians, ctenophores, and tunicates. (From Denton & Shaw 1961.)

Table 11.7 Lipids providing upthrust in marine animals.

Lipid	Specific gravity	Upthrust (g upthrust per g lipid)
Triacylglycerol	0.93	0.103
Alkyldiacylglycerol	0.89	0.127
Wax ester	0.86	0.193
Squalene	0.986	0.193

bladder, and achieve buoyancy in the water column by reducing their mineralization and accumulating lipids.

Increasing the amounts of light fats and oils

Fats and oils commonly float on water, having densities less than 1 g cm^{-3}, so that if present in an animal's body they can produce an overall density less than that of sea water. This strategy is very common in planktonic crustaceans, where droplets of oily material may be visible within the body. It is also evident in many fish eggs, where oils may account for 50% of the organic matter present. More spectacular examples occur in some sharks in the family Squalidae (including the massive basking shark), which never have swim-bladders but may have greatly enlarged livers (up to 20% of the body weight) in which up to 75% by weight is oily material, particularly the unsaturated hydrocarbon squalene ($C_{30}H_{70}$: specific gravity 0.86). In other fish, and in marine mammals such as whales, a similar function is performed by a range of other lipid materials (Table 11.7). These lipid compounds are all rather inert, with very low metabolic turnovers.

Increasing the amount of gas in the body

Using gas "floats" by enclosing gas within the body is the "best" solution to the buoyancy issue, since compressed air has a density of only 0.0012 g cm^{-3} at sea level and even a small float will provide substantial lift (the swim-bladder of a marine fish is normally only 5% of its body volume). Many organisms exploit gases, including marine plants such as the kelps. In marine animals there are two major kinds of gas float:

1 The rigid incompressible float characteristic of various cephalopod molluscs such as *Nautilus* and the cuttlefish (and even more common in various extinct cephalopod groups).

2 The soft-walled chamber found in *Physalia* (the Portuguese man o' war) and other cnidarians, and also in a more complex form as the swim-bladder of fish. The disadvantage here is that the float is compressible so that its volume changes with depth, and lift reduces during a descent. Species that migrate vertically may therefore have problems with this system.

In cephalopods gas is retained within a shell—either in a complex internal shell, which may be coiled (*Spirula*) or straight (*Sepia*), or in an external coiled shell where the most distal chambers are given over to gas storage (*Nautilus*). A solid float like this has the merit of having a constant volume and is thus relatively independent of depth. These animals do not have specific gas-generating tissues, but rather the gas (principally nitrogen) just accumulates as a consequence of active salt transport. Initially the chambers of *Nautilus* shells are filled with a NaCl-rich isosmotic fluid, from which Na^+ ions are actively pumped by the mitochondria-rich cells of the central strand of living siphuncle tissue that extends up the middle of the shell. Chloride follows passively, and water then follows down the osmotic gradient, leaving a gas space. This equilibrates at about 0.8 atm N_2 (as in air), though the chambers maintain a rather lower than ambient oxygen pressure. The system in *Sepia* (cuttlefish) is rather similar (Fig. 11.23), except that the shell is tightly laminated with no clear chambers, and is internalized, making up about 10% of the body volume. The gas composition within the laminae is as in *Nautilus*, but the relative proportions of gas and of NaCl-rich fluid are varied in different parts of the cuttlebone and at different times. When the animal descends to greater depth, salts are extracted from the shell fluids by surrounding epithelial cells, so that the fluids become hyposmotic and the tendency of the increased hydrostatic pressure to force more water into the cuttlebone is counteracted. A freshly extracted cuttlebone therefore has a specific gravity varying between 0.55 and 0.65. In this way the cuttlefish varies its own density, and can live with neutral buoyancy at various depths from the surface to about 200 m (whereas complete withdrawal of fluid from the cuttlebone would theoretically permit penetration to about 240 m).

Other cephalopods using this kind of buoyancy system can descend to about 450 m (even down to 800 m in *Spirula*), suggesting that additional mechanisms operate, possibly to enhance blood concentration in the siphuncle with a countercurrent multiplier. But as the rigid flotation device has to be relatively rather massive, and strongly calcified to withstand the imposed pressures, it is also heavy, and at these depths around 80% of the gas is needed just to support the float itself.

Clearly a nonrigid float, although compressible as depth increases, would have the advantage of being much lighter to carry around. It would be particularly suitable over a range of depths if the quantity of enclosed gas could be regulated to counter the hydrostatic effects of pressure (otherwise the float volume would be halved for every 10 m of descent). Thus in floating cnidarians and in fish there are distinct gas-secreting tissues to give rapid control.

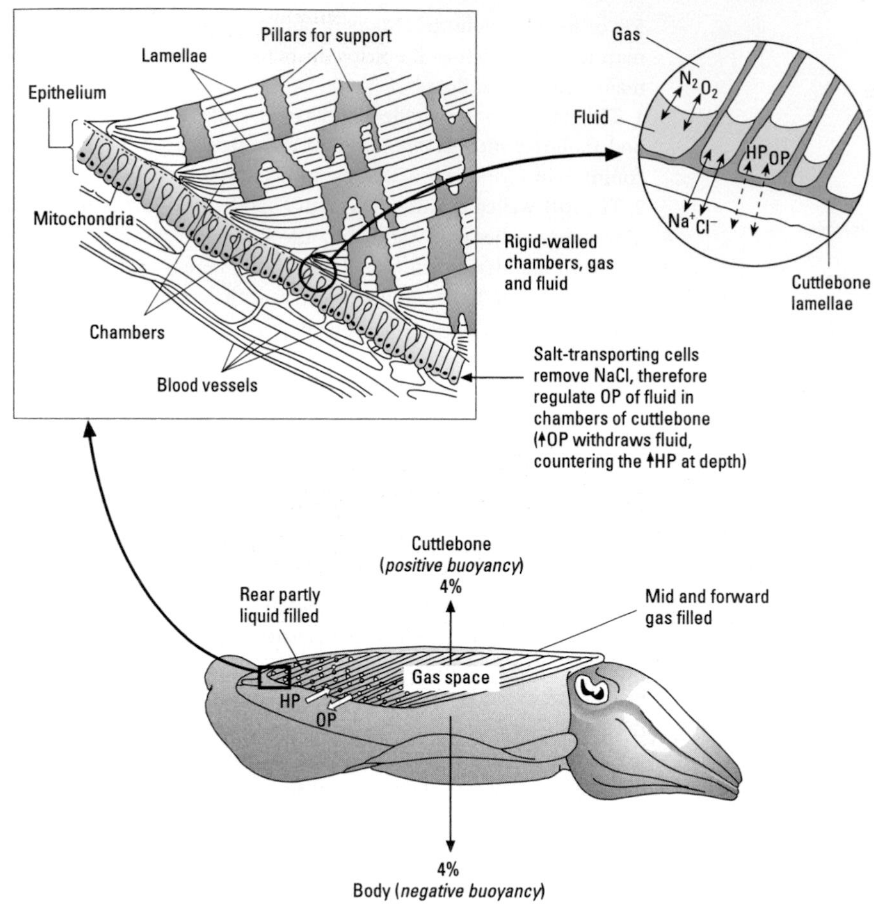

Fig. 11.23 Form and functioning of the *Sepia* cuttlebone; salt transport and gas accumulation. HP, hydrostatic pressure; OP, osmotic pressure. (Adapted from Denton & Gilpin-Brown 1961.)

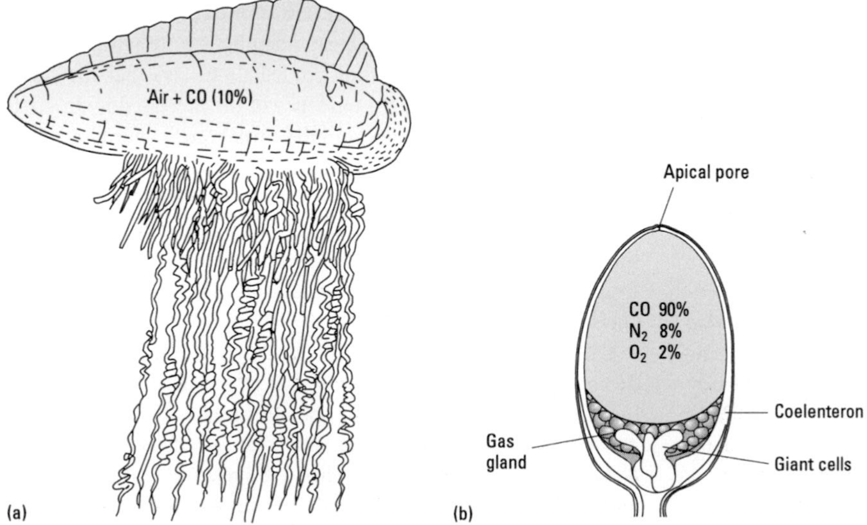

Fig. 11.24 (a) *Physalia* (the Portuguese man o' war) and its gas float. (b) Cross-section of the small gas float of *Nanomia*, with the gas gland and apical pore for gas release.

In the surface-dwelling colonial siphonophore *Physalia*, the chamber is up to 1 l in volume (Fig. 11.24a), filled with the normal components of air but also with 12–15% carbon monoxide, CO, derived from the amino acid serine. Other siphonophores live at variable depth and produce rather more CO, which is present at up to 90% in the tiny floats of *Nanomia*. In most of these animals there are specific gas-generating cells (the "gas gland") at one side of the float (with cells rich in folic acid, which appears to be the storage precursor for the gas), and an exit pore lying at the other end (Fig. 11.24b). The exit pore is important in any soft-walled float operating below the surface because a rising animal would have an expanding float and positive buoyancy, causing increased upward acceleration; gas must be released quickly to avoid unduly fast ascent and possible rupture.

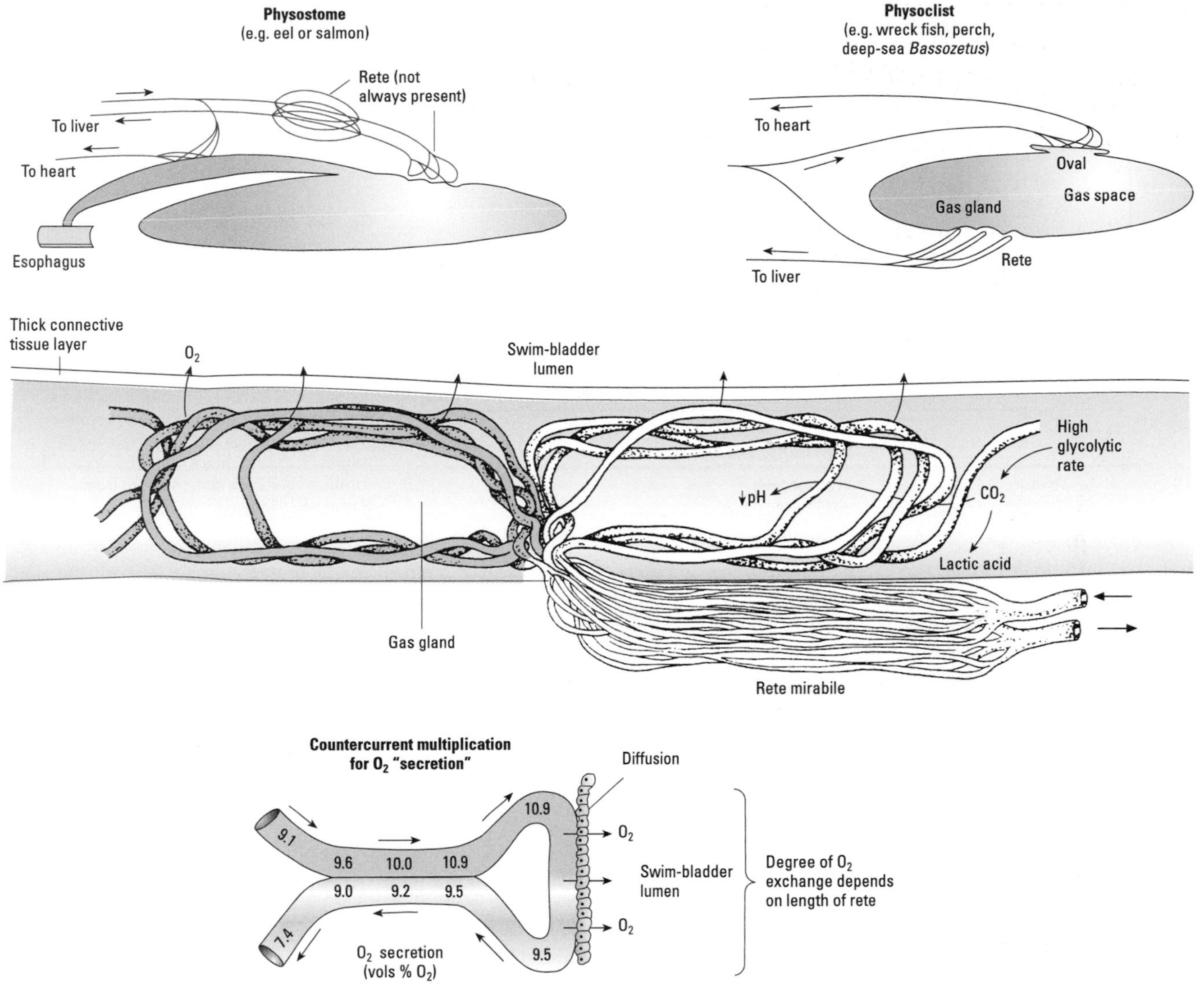

Fig. 11.25 The gas gland rete and its function in the swim-bladder of fish. (Adapted from Wittenberg 1958; Denton 1961.)

In teleost fish swim-bladders are the common, though by no means ubiquitous, buoyancy structures. These soft-walled sacs lie above the gut and just below the spinal column, taking up about 4–5% of the body volume, and there are two basic arrangements. In the **physostomes**, such as the salmonid fish, the bladder retains a connection with the gut, so that gas can pass in either direction, and this may be the sole means of buoyancy control. The system can be filled by gulping at the surface, and the intervening duct is controlled by the autonomic nervous system. However, some physostome fish also have a simple version of the rete and gas gland system described below.

The more "advanced" fish (**physoclists**) have a separated swim-bladder, which is filled with gas (oxygen and nitrogen principally) from the blood supply, via a **gas gland**. No gas is in itself actively secreted of course; gaseous movements are purely diffusive, even at great depth and apparently against enormous gradients. The gas gland has an associated rete of exceptionally long and unusually highly ordered blood capillaries (Fig. 11.25), the length of these capillaries being correlated with the depth to which the species can descend (they are up to 25 mm long in the deep-sea genus *Bassozetus*). In effect the gas gland is the distal end of a **countercurrent multiplier** (see Box 5.2) formed by this rete, so that gas from the blood accumulates there at high concentration and diffuses into the swim-bladder. The cells of the gas gland mainly function by undergoing intense glycolysis (an unusual occurrence in the presence of high O_2 levels), with lactic acid and CO_2 as the principal end-products. This creates a very low pH so that in the efferent venous blood the Root effect operates very effectively and there is also an enhanced Bohr effect (see Chapter 7). Hence oxygen is very rapidly unloaded from the venous hemoglobin and can diffuse across the rather permeable rete into the afferent arterial blood. In the gland this enriched arterial blood gets even more acidified and gives up its oxygen to the chamber. In the eel (*Anguilla*) the swim-bladder receives 83% of the oxygen removed from the blood, with just 17% being metabolized in the gas gland tissues. This countercurrent multiplier may be aided by a "salting out" effect, the gland adding ions (lactate and bicarbonate, and possibly NaCl) to the blood to

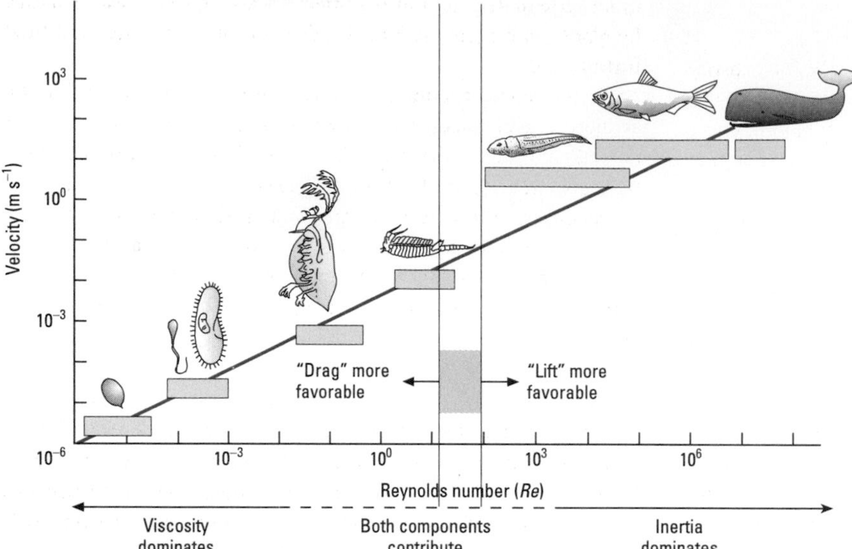

Fig. 11.26 Effects of size and Reynolds number (*Re*) on swimming patterns and the relative importance of viscous forces and frictional forces. *Re* is given by $\rho VL/\mu$, where V is velocity, L is dimension of moving surface, ρ is the density of the medium, and μ its viscosity, so giving a dimensionless ratio related to size. A high Reynolds number applies to a large or a very fast animal, where flow patterns are dominated by inertia and streamlining is useful. At low Reynolds numbers (small or very slow swimmers, with high surface areas relative to volume) viscosity is much more important and streamlining is irrelevant. (Adapted from Nachtigall 1983.)

free extra oxygen; this would also explain how nitrogen and other inert gases are accumulated in the swim-bladder.

The chamber of the swim-bladder retains this gas very efficiently, having highly impermeable walls with a thick layer of connective tissue, in some species impregnated with flat plate-like structures of crystalline guanine. The permeability of such a layer to oxygen is only about 10% that of normal connective tissues. However, in one particular area of the swim-bladder there is a window of vascularized tissue capable of resorbing gases from the chamber, often with a muscular valve so that it can be exposed to the gas or shut off from it. By balancing the activity of the gas gland and the resorptive window the volume of gas in the chamber can be regulated.

Pelagic fish with gas-filled swim-bladders normally only function at depths less than 1000 m, and very deep-sea fish in mid-ocean trenches have oil-filled bladders instead. However, a few species of near-bottom teleosts have been caught at depths of up to 4500 m with normal gas-filled bladders, and it may be that these fish have extra physiological specializations to maintain gas secretion beyond those already described.

11.6.3 Locomotion

It is clear from our consideration of buoyancy that some part of many marine animals' locomotory need is linked to station-keeping in relation to depth, and many achieve this mainly or partly by swimming mechanisms. Size plays an important role here (Fig. 11.26), since for very small animals (including most larvae) locomotion in water is dominated by viscous forces and a highly elaborated non-streamlined profile is possible (and, in terms of reducing sinking rates, appropriate). For larger animals frictional drag becomes increasingly important instead, and swimmers tend to converge on a relatively streamlined shape, culminating in the essentially torpedo-like shape of fast-swimming cephalopods, elasmobranchs, teleosts, and marine mammals, as well as the extinct reptilian ichthyosaurs. Here the power requirement to overcome drag increases very dramatically with speed, so that fast swimming is extremely expensive and streamlining is crucial.

Swimming mechanisms to propel animals through the water come in many varieties. Many of the smallest animals rarely swim actively but instead drift passively with the currents, relying on viscous forces to maintain stability and height in the water column. Small invertebrates, and most larvae, tend to rely on ciliary locomotory systems, though in ctenophores (sea gooseberries) there are complex cilia arranged as comb-plates and individuals can be up to 30 cm long. Normally, for animals above a few millimeters in diameter, cilia are inadequate and muscles are universally employed. The muscles may be used to operate a water jet system, giving jet propulsion, or may operate appendages or true limbs functioning as "paddles", seen in crustaceans and many tetrapods, such as turtles and some birds swimming underwater. A third option involves some part of the body modified as a muscular undulating flap—as in the fins or tails of many fish and marine mammals—acting as oscillating hydrofoils and giving lift-based swimming that is more energetically efficient than paddling systems. The waves of muscle contraction that underlie all such swimming motions are covered in sections 9.15.1 and 9.15.2. Most swimming aquatic animals have a considerable capacity to vary the speed and energy of their swimming, with a factorial aerobic scope of 5–7 (though varying with temperature); species with a capacity for long-distance endurance swimming (notably some crabs, lobsters, and fish) may have somewhat lower values.

Crawling and burrowing mechanisms also occur in benthic marine animals, and usually are variations on the undulatory swimming systems described above, aided by some contact with the substratum. The mechanics of these systems are dealt with in Chapter 10. Marine burrowers exploit the physical security and often also the food reserves of the benthic sludge. However, active burrowing is rather uncommon in fully marine animals; they may form a shallow burrow and live in it with one end intermittently protruding into the water, but they do not often burrow continuously for a living. By contrast, for animals in the littoral and estuarine zones dealt with in Chapter 12, this is the main strategy of a high proportion of the fauna, affording protection, physiological stability, and major food-gathering opportunities.

11.7 Sensory issues: marine signaling

Life in the deeper waters of the seas creates a number of sensory problems. Animals have no light to provide for normal vision, and high pressures create problems for sound and vibrational signaling systems. Modifications in the senses and in effectors are therefore major features of physiological adaptation in these habitats.

11.7.1 Light and vision

Water represents a far richer and more varied photic environment than that encountered by terrestrial animals, because of complex interactions between light and suspended or dissolved materials, varying with depth. Light in the sea varies in intensity, frequency, and polarization as it progressively penetrates to greater depths. The water molecules themselves scatter and attenuate light, particularly at the blue (short wavelength) end of the visible spectrum; and polarization of the light is also affected by scattering from the water. Dissolved salts have little effect on intensity, but suspended particulate matter attenuates all components of the transmitted light. Thus for typical sea water 50% of the light is absorbed within 2 m of the surface, and 90% within 8 m. At the same time, the frequency of transmitted light varies with depth in a predictable manner, with red light absorbed the most readily to give mostly blue wavelengths below 50 m (Fig. 11.27), though in coastal waters the greater amounts of dissolved matter shift this relation somewhat. The net result is a relatively predictable vertical gradient of light intensity, dividing the open ocean into the **euphotic zone** (epipelagic, where down to 200 m there may be enough light for photosynthesis), the **dysphotic zone** (mesopelagic, 200–1000 m, where blue light is still present but at insufficient intensities for photosynthesis to balance respiration), and the **aphotic zone** (bathypelagic, >1000 m, where all solar light has been absorbed; here most species in certain taxa are eyeless). Away from the open ocean the photic zones may be much attenuated, and also shifted towards greener wavelengths, by plankton blooms, sediment-laden run-off from land, and tidal turbulence.

The presence of so many distinct light regimes is reflected in the evolution of a great diversity of visual systems (see section 9.7). For animals that live in the euphotic and dysphotic zones, photoreception is always a major feature of the sensory repertoire, often residing in the epidermis, as with many corals where the tentacles have high sensitivity to blue light (around 480 nm) and can thus detect the timing of the full moon to which their spawning is synchronized. At a more sophisticated level, "eyes" of some sort occur in virtually every phylum of animals (with clearly convergent designs recurring across the animal kingdom). Most marine animals with eyes make use of simple and nearly spherical lenses, with the refractive index decreasing from the center to the periphery, and this type of eye must have arisen on at least six separate occasions. Multilayered lenses also occur (e.g. in some copepods), as do concave mirrors to channel incident light (e.g. some bivalves and ostracods). In a range of arthropodan animals compound eyes are also found, with at least three different designs, based on lens cylinders, mirrors, or a combination of lens and parabolic mirrors. For many of these animals, light (photoperiod) is a major influence on behavioral cycles, particularly for reproduction in polychaetes, starfish, and decapod crustaceans.

Visual pigments are the essential components of visual systems, as discussed in Chapter 9, and are the source of much habitat-related variation. In invertebrates and fish, absorption peaks are somewhat variable according to habitat (Fig. 11.27), many species showing two or more spectral peaks, with a trend to shorter wavelengths and broader sensitivity curves as one moves offshore. In general the peaks for invertebrate eyes occur at about 450–500 nm, and are never above 560 nm. This contrasts with the situation in some marine fish where 11-*cis*-3-dehydroretinal is used as the chromophore (instead of the normal 11-*cis*-retinal; see Chapter 9), and

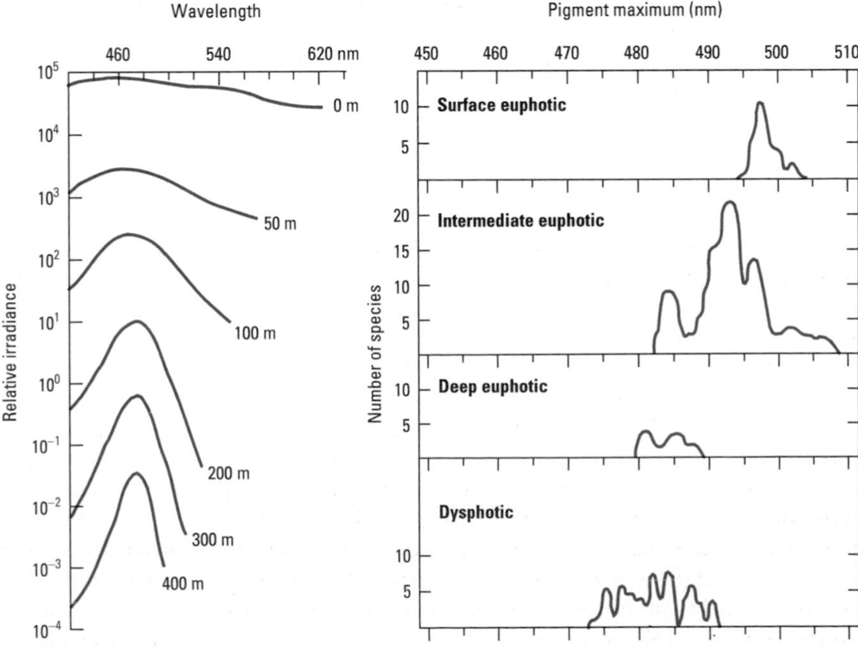

Fig. 11.27 Light attenuation and selective absorption in the ocean in relation to depth, and the associated trend towards blue-shifted rhodopsins in the eyes of fish from increasing depth.

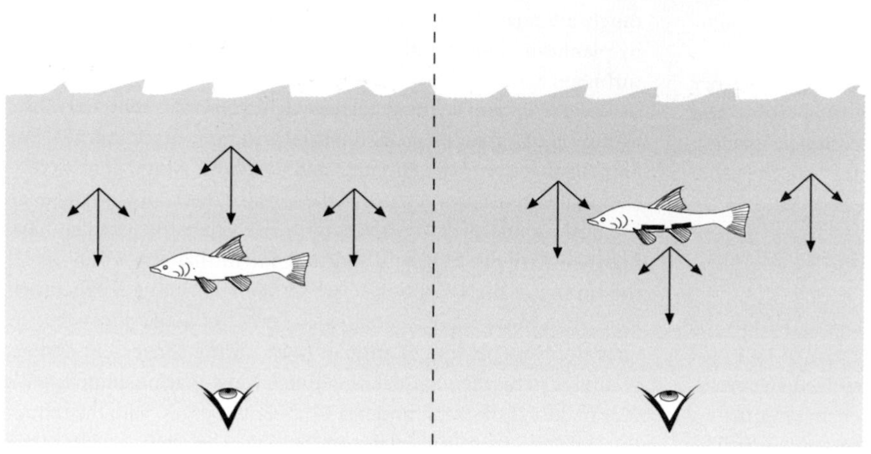

Fig. 11.28 The use of ventral camouflage light emission in marine animals. On the left, an observer from deeper water sees the animal because it blocks the limited daylight coming downwards. On the right, the animal itself emits light from its lower surface to replace the daylight and thus becomes camouflaged.

peak sensitivity may sometimes be towards 600 nm, perhaps giving better contrast vision. Fish in the euphotic zone have both rods and cones ("duplex retinas") with several different peak sensitivities. Sensitivity and resolution are adapted to different parts of the day–night cycle, with lower wavelength (green to blue) maxima as depth increases. It has been suggested that the fish's visual pigment absorbency peaks at the dominant wavelength occurring at the typical depth and the time of day most important to the active fish. For example, in reef fishes the rhodopsins match the spectrum of twilight underwater irradiance, since twilight is when visual sensitivity is most crucial for survival. Interestingly, most marine mammals have lost the typical bichromatic color vision of their near relatives (having "blue" and "green" cones) and are effectively color blind, with only the green cones present. Since this has occurred in both whales and in seals it is presumably a convergent pigment loss, though hard to explain as the animals live in an environment where any light present is strongly blue-shifted.

Some of the eyes found in cephalopods and crustaceans (and perhaps some fish) are also known to be sensitive to variations in polarization, though this has been little studied.

Marine animals may have particular problems with camouflage, given that threats may come either from above or below them. Many marine fish and squid exemplify the conflicting needs. To achieve camouflage from visual predators from above in daytime, the back of the animal needs to be very dark, so that the reflected light equals the light passing upwards on either side of its body. The sides of the body need to be near-perfect mirrors so that the light reflected back to a lateral observer is almost the same as if the animal was not there. The ventral surface can only be rendered invisible if the intensity and spectral distribution of light from it matches that of the surrounding sea. Below 250 m the spectral distribution of light is fairly constant, so that in practice this can be achieved by the animal sensing the intensity of light falling on its dorsal surface and matching it with emitted (bioluminescent) light from its ventral surface (Fig. 11.28).

11.7.2 Bioluminescence

The abyssal plains of the oceans average 3800 m in depth, well into the aphotic zone; so the majority of the Earth's surface receives no sunlight. Although vision would seem to be unimportant at aphotic depths where solar light cannot penetrate, in practice many animals there retain eyes and can see rather well. This is because **bioluminescence**—light (photons) produced by the biota itself—is a prominent feature of a wide range of deep-sea animals. At least 700 genera from 16 phyla are known to use bioluminescence.

Bioluminescence may occur as a continuous glow or in a great variety of flash patterns, and it is used for many different functions. Single cells (photocytes) or complex glands (photophores) can be used, and these may be sited deep within the animal tissues or superficially in the epidermis. Some bioluminescence involves intrinsic photocytes with intracellular reactions, while some originates from an extracellular process. Extracellular bioluminescence is particularly common in marine invertebrates, often involving two separate groups of cells within a gland. When both cellular products are released under the influence of muscular contraction, the reaction between them occurs in the gland lumen and light is emitted, with the other end-products excreted from the gland. A great deal of both intracellular and extracellular bioluminescence involves symbiotic bioluminescent bacteria. In any of these cases, accessory structures such as lenses, reflectors, light guides, and pigment "shutters" are commonly found (Fig. 11.29a). This diversity clearly indicates multiple convergent origins for the phenomenon, probably as a serendipitous acquisition of symbionts and/or as by-products of other metabolic pathways. Many different kinds of cell or tissue may be modified to emit bioluminscence, the most unlikely recent example being the suckers of some deep-sea octopuses.

There is considerable biochemical diversity in bioluminescent systems, though all are commonly referred to as using a **luciferin–luciferase** system, as in bioluminescent bacteria. The process involves the oxidative breakdown of a high-energy, complex organic molecule ("luciferin") by a specific enzyme ("luciferase") in a process that releases photons of light. The "true" luciferin of bacteria is a modified form of a reduced flavoprotein, complexed with an aldehyde. Under the influence of oxygen and the luciferase enzyme this complex breaks down to the flavoprotein and a carboxylic acid. Marine animals may also have a variety of more specific luciferins, of which the most famous are the calcium-activated molecule, aequorin, found in certain jellyfish and other cnidarians (but also in a squid, several crustaceans, and some fish), and the *Vargula*-type luciferin originally isolated from the ostracod *Vargula*, but also found in other crustaceans and some fish.

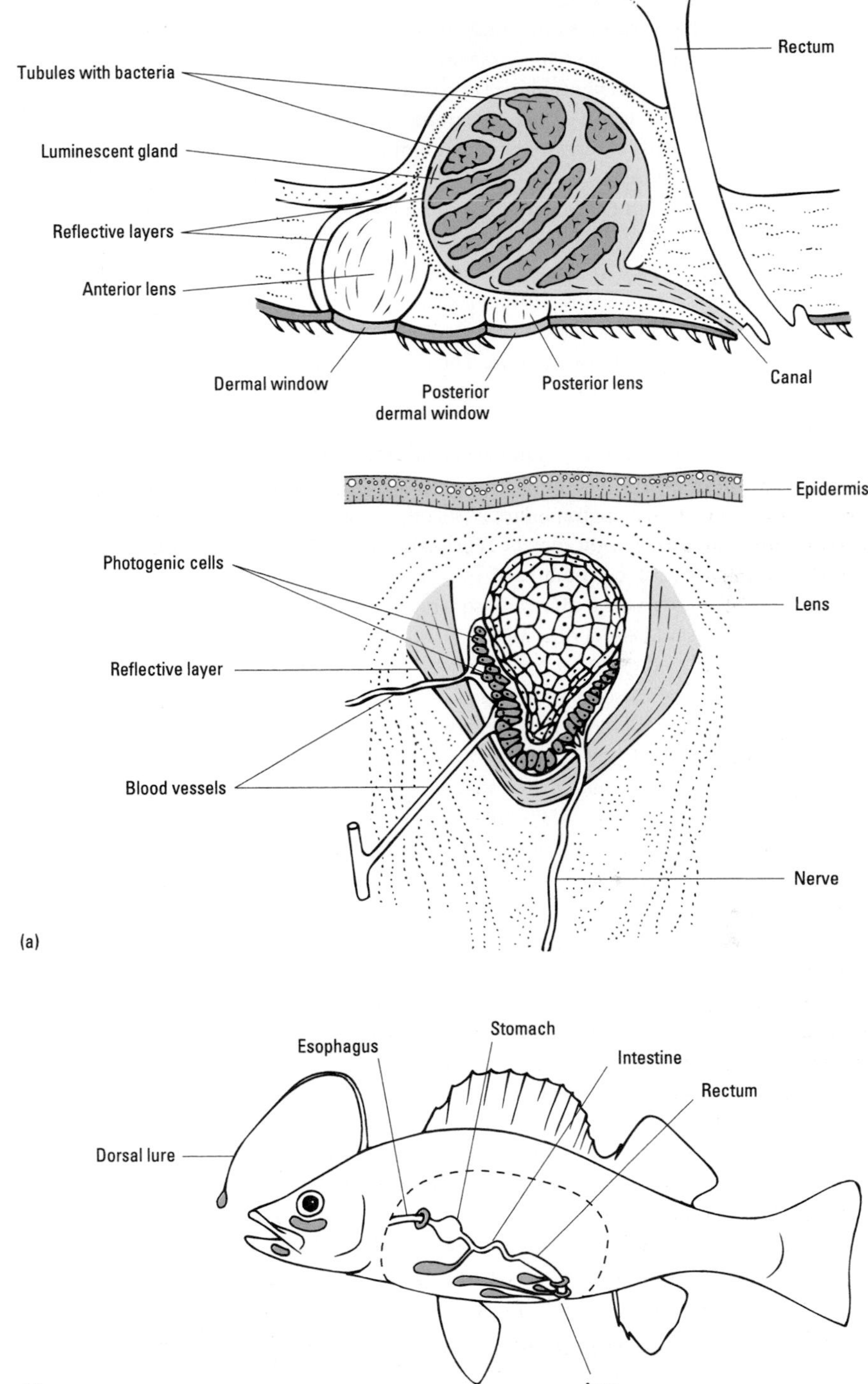

Fig. 11.29 Bioluminescence in fish. (a) Structures that emit and control light, including shutters, lenses, and light guides; and (b) the distribution of bacterial light organs in fish (shaded areas). (a, Adapted from Herring & Martin 1978.)

In the fish taxa, bioluminescence is a purely marine phenomenon, and can be taken to extravagant lengths. Again, it may arise from either the inherent tissue biochemistry or from symbiotic bacteria. Bacterial organs are commonest in fish from the aphotic zone, occurring in 20 families, nine of which are in the order of deep-sea anglerfishes (Ceratioidea). Each species has just one species of bacterium, although the same bacterium may occur in more than one fish species or even in more than one family. Inoculation of the light organs by the bacteria probably occurs from the surrounding sea water at the larval stage, since at least some species of the bacteria (e.g. *Photobacterium*, *Vibrio*) can be readily cultured outside the fish.

Fish from the dysphotic zone commonly use nonbacterial luminescence, where the luciferins are chemically variable but are never flavoproteins. Such fish may use either the cnidarian/aequorin-type luciferin or the *Vargula*-type. In some species, bioluminescence is supported by luciferin ingested in the diet, by swallowing either the cnidarian or the ostracod from which the luciferin originates. For example, the midshipman fish, *Porichthys notatus*, has several hundred superficial ventral photophores that contain *Vargula* luciferin only when the fish is in *Vargula*-containing habitats. A single dietary intake of *Vargula* confers a lifetime of luminescence due to an efficient recycling system.

Whatever the source of the luciferin, fish luminescent organs are widely distributed (Fig. 11.29b) and are often very similar to the eyes in construction, probably having coevolved with the fish eye. For example, in both the eyes and luminescent organs reflectors occur and are composed of purine platelets, while the pigmented layers contain high concentrations of melanin. The retinas of bioluminescent fish from aphotic depths have only rods, which achieve better photon capture than cones.

The light-emitting organs often contain bacteria that emit luminescence continuously, but there may be a covering mechanism or shutter (rich in melanin to give a dark absorptive cover), so that flashes can be produced. Light may be reflected from the internal swim-bladder, and may be modulated by superficial chromatophores or by filters of varying color. In fish with nonbacterial light organs the nervous and endocrine systems may also regulate light intensity.

Functions of bioluminescent organs in all taxa appear to be very varied. In pelagic marine animals they are most frequent on the ventral surfaces, and towards the periphery, thus presumably playing some role in ventral camouflage as described above. Emission is often in the blue–green range (450–490 nm), where transmission in sea water and visual sensitivity of most marine animals are both greatest. Benthic and deep-water animals may have slightly longer wavelength emissions (500–520 nm), suited to the more turbid waters they encounter (Fig. 11.30). For many animals the bioluminescence may serve to aid predation (e.g. as a "lure" for food gathering, as in the anglerfish), or as a deterrent to predation (e.g. as a "startle" effect in various small fish that are heavily preyed upon and may hope to escape while a predator is confused by their display, or as a directly off-putting signal especially to negatively phototactic predators). Intraspecific luminescent signaling is also occasionally found, in relation to recognition systems and prior to mating, and in such cases the signal may involve unusual and highly tuned wavelengths. For example, the rhodopsins of the deep-sea *Bathylagus* have peaks at 466 and 500 nm, detecting and discriminating between inter- and intraspecific bioluminescent red emissions.

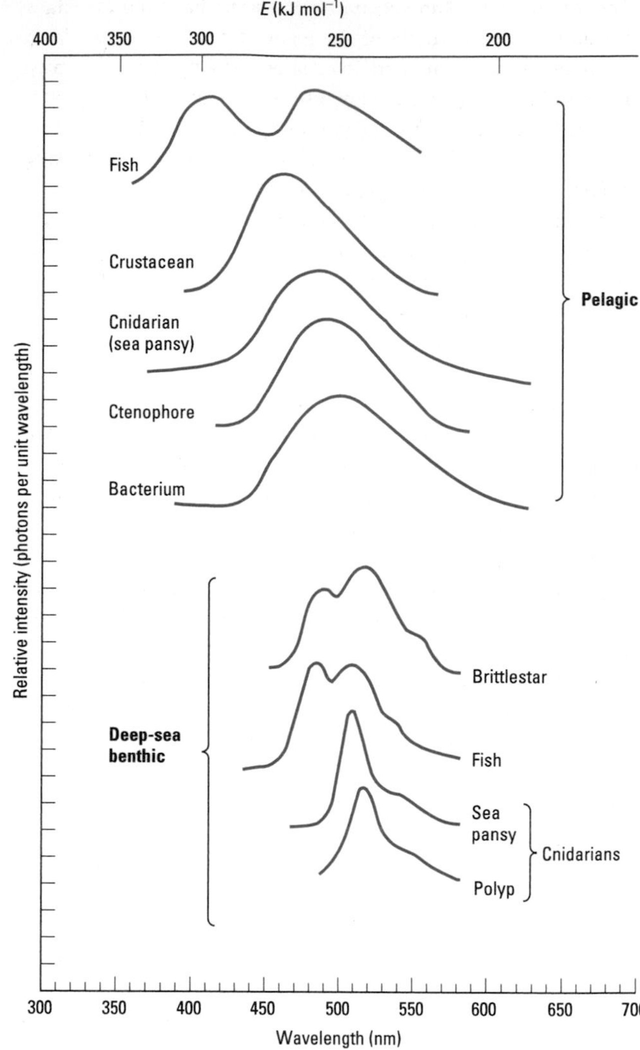

Fig. 11.30 Emission wavelengths of various bioluminescent marine animals. (Adapted from Wampler 1978.)

11.7.3 Chemical signaling and chemoreception

Chemoreception plays a major part in the lives of marine organisms. Chemicals can diffuse over relatively long distances in aquatic systems, retaining directional information where currents are of moderate and consistent velocity and with minimal turbulence. In inshore waters they may be more problematic, however, and are normally only useful over a short range or as chemical deposits on a substrate (recent work shows that variable shear velocities at turbulent boundary layers greatly modify chemical signal transfer over muddy substrata).

The detection of food by chemical cues is the most obvious case where chemical inputs are useful, and the signal is frequently an amino acid, or a mixture of several of these. This has been particularly well studied in fish, where "taste buds" occur on various parts of the body; in deep-sea fish such as rat-tails, these can be used to follow odor trails to the distant rotting remains of large fish and whales on which the fish feed. More localized chemical responses occur in marine cnidarians where chemical feeding signals (often in concert with a mechanical trigger) elicit the firing of specific nematocysts (stinging organelles). The thresholds for such chemicals can be extraordinarily low, perhaps only 10^{-9}–10^{-11} M.

More sophisticated cases of chemical signaling in the form of pheromones (see section 10.9) are also becoming increasingly well documented in the marine environment. Pheromones are used by animals in the sea for alarm purposes, as settling or metamorphosis cues, and perhaps most obviously as mating signals. Orientation to such chemicals is primarily by movement upstream ("up-current") when they are detected, but at least in crabs also involves sampling of attractant distribution across the stream to take account of turbulence.

The sea anemone *Anthopleura elegantissima* has been especially well studied as an example of the possession of a marine alarm pheromone. When wounded it releases "anthopleurine", which causes rapid contraction and convulsion in neighboring anemones within a few seconds. This pheromone is particularly intriguing because it can be carried by the anemone's chef predator, a sea slug (*Aeolidia*). When the slug eats some of the anemones it receives a dose of anthopleurine, which is stored in its tissues for several days and gradually diffuses out; thus when the slug approaches further anemones they are forewarned of its presence.

A few marine animals from the benthos of shallow waters also release trail pheromones, the most obvious examples here being molluscs that deposit a chemical signal with their mucus trail. The sea slug *Navanax inermis* uses intraspecific trails to locate mates and also to locate other slugs as (cannibalistic) food. However, when attacked by other predators the slug adds another component ("navenone") to its mucus, which serves as an alarm pheromone to conspecifics, and they then avoid rather than follow its trail.

Chemical regulatory systems are particularly important for reproduction in mass-spawning invertebrates such as sea urchins, ascidians, hydrozoans, and corals, since spawning needs to occur simultaneously and each sperm must find a conspecific egg or be wasted. The chemicals used are rather varied, including unsaturated fatty alcohols and modified terpenoids. In planktonic organisms the problems may be rather different, requiring two mature adults to come together and then release gametes. Some sea urchins, some crustaceans, and one marine rotifer have recently been shown to use surface glycoproteins as contact-mating pheromones. These glycoproteins are used in the crustacean *Tigriopus* as a cue for mate-guarding behavior; they arise mainly from the female's terminal appendages, the male having appropriate protein receptors on his antennules. However, there are good theoretical reasons why small pelagic animals should *not* use long-distance pheromones. Depending on the pheromone diffusibility and threshold concentrations, calculations show that efficiency of detection would be inadequate below a minimum size of between 0.2 and 5 mm. Microorganisms and small planktonic animals, such as rotifers and ostracods, are therefore unlikely to use sex pheromones. The more spectacular cases of reproductive pheromones can thus be found in larger and more complex marine invertebrates. For example, in certain gnathiid isopods males release an attractant from their burrow that spreads over a long range as a pheromone "plume" in the water and attracts a number of females to the burrow, where they are captured and kept as a harem. Another particularly well-studied marine pheromone working at close range occurs in barnacles (*Semibalanus balanoides*), which are inseminated by a highly extensible "pseudo-penis" from a neighbor and then brood the fertilized eggs over winter. In spring the onset of algal blooms permits filter feeding, which in turn promotes the production of a "hatching pheromone", a modified fatty acid derived from the algal/invertebrate diet. This is secreted into the egg cavity and causes the dormant eggs to hatch as free-swimming nauplius larvae. (It is noteworthy that this same barnacle species also has a well-known "settlement pheromone", in this case a protein that gradually leaches out from its cuticle and serves as an attractant to other barnacles to settle nearby, so allowing the close-range fertilization to operate as the barnacles mature.) The crab *Callinectes sapidus* uses a sex pheromone to elicit more specific and complex behaviors, with a compound from the almost mature female's urine exciting the male into his full courtship display, after which he grabs the female and carries her around until she molts to her final mature state, when copulation occurs. Sex pheromones in marine animals may also have even longer term effects; the sex and gross morphology of a marine echiuran worm such as *Bonellia* are determined by pheromones, the larvae developing into small and effectively parasitic males if they encounter a signal ("bonellinin") from a female, but growing into large females themselves if they settle on a substrate where no females and thus no pheromone are present.

Olfactory organs in marine fish are large with a high density of receptors, linking to further chemical sensors on gills, skin, and fins, and suggesting a more sophisticated chemical environment and signal-detection system than we are used to imagining in the seas. Fish certainly use pheromones as sexual attractants, and in some species (e.g. the lamprey, *Petromyzon*) the females can be attracted in their hundreds by fishermen who trap a single mature male as lure. Fish have also been shown to use pheromones in regulating pair interactions, in parent–young interactions, and in spawning. The fairly widespread phenomenon of *schreckstoff*, or alarm substance, in the skin of fish such as minnows that shoal (where a damaged fish becomes a source of alarm to its conspecifics), and the extraordinary "homing" ability of salmonids that detect their own particular spawning stream with extremely low "taste" thresholds, certainly also attest to the sophistication of aquatic chemoreception. Indeed, the limited evidence available as yet suggests that the timing, pulsing, and detection systems of pheromones in marine habitats are of similar complexity to the far better studied terrestrial systems of insects and of mammals.

11.7.4 Mechanical senses

In the sea mechanoreceptors can be used to detect gravity, pressure, water currents, and oscillatory particle movements (vibration), as well as the more obvious "sound" or far-field vibrations. All are essentially based on one of two fundamental designs, either statocysts or hair cells, both described in section 9.8. Statocysts are common in marine invertebrates, from radial arrangements in the umbrella margins of jellyfish to multiple arrays in the heads of crustaceans and cephalopods. Fish use ranks of hair cells in their lateral-line systems, either as distinct neuromasts or embedded within a complex subepidermal canal system. This system is sensitive enough to permit the monitoring of neighbors' movements in a shoal, and the precise coordination of maneuvers by all the constituent fish of that shoal. In some fish this lateral-line system is modified into the beginnings of an inner ear, and in some species its sensitivity and frequency range is enhanced by an association with the swim-bladder, whose walls magnify the oscillations via modified vertebrae known as Weberian ossicles. Curiously, coleoid cephalopods (squid, etc.) with a similar lifestyle have no real mechanosensory system and are effectively "deaf".

Acoustic energy is transmitted more efficiently through water than almost any other form of energy, and sound therefore persists without serious attenuation or interference to the signal over a long range, allowing submerged animals to use it as a primary method of communication. Sounds that seem rather diffuse and unfamiliar to

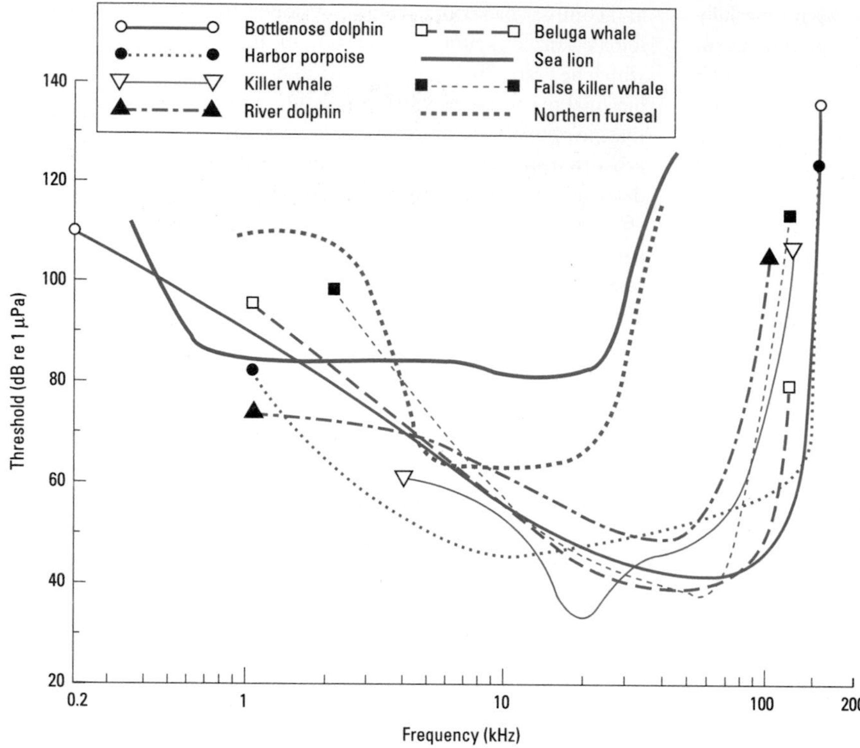

Fig. 11.31 Long-range sound and hearing in the sea: audiograms showing the frequency spectra of various cetaceans and pinnipeds using long-range sound communication under water.

us can give vital information to fish and mammal ears—the noise of surf, or of ice creaking, or of the faint disturbance created by the passage of a potential predator such as a shark. Many marine mammals go further than this and capitalize on the long-range potential of underwater sound by using it to communicate, with calls related to the presence of danger, of food, and of conspecifics, as well as giving information about their own identity, position, and territory, or their reproductive state. The toothed whales use calls of fairly high frequency (1–20 kHz), while baleen whales (when not echolocating) mainly use rather low-frequency calling noises (20–500 Hz, the wavelength being hundreds of meters at the lower frequencies). Behavioral audiograms for a range of species are shown in Fig. 11.31. Calls are variously described as grunts, moans, whistles, and barks, but there is no agreement on their classification or even on their function. For the toothed whales, which are often very gregarious, certain sounds appear to serve as individual recognition markers ("signature calls"); these are well documented in dolphins and their relatives. Some marine mammals can also recognize key characters of other species' calls. For example harbor seals can detect whether the call of a killer whale is that of a "safe" inshore individual (mainly salmonid feeding) or of one of the roaming individuals that prey on seals. Pinnipeds (seals, etc.), which are able to hear both in water and in air, often lack very long-range underwater communication systems, though they do call using barking noises within a group and between mother and pups.

Underwater calls are particularly effective at moderate depth, where sound can be channeled by the properties of water layering in the oceans. For example, humpback whale "singing" sounds are propagated halfway round the globe through the "deep sound channel", typically occurring at 600–1200 m depth in low and middle latitudes; this is the depth of minimum sound speed, where sounds are focused and propagated without loss from bottom and surface reflections. A similar phenomenon occurs at much shallower depths in polar waters. However, depth variation does present some problems for sound communication, since for example the whistle spectra of white whales alter markedly with depth and pressure (even though their hearing is unattenuated, suggesting that sound is mainly transmitted through the head rather than through the eardrum and ossicles as in land mammals).

11.7.5 Other senses

Marine organisms may also use more unusual senses to navigate and locate objects in the very dark or turbid parts of the oceans, particularly echolocation and electrical sensitivity. Marine mammals, such as dolphins, some whales, and a few seals, are well-studied exponents of aquatic **echolocation**, and demonstrate how useful a sense this can be in the seas where sound attenuation is much less than in air. Usually the system involves high-frequency sound (12–150 kHz), above the frequencies used for intraspecific calls, and commonly described as "clicks". Each click may be only 50–200 μs in duration, and can be useful over a range exceeding 100 m, possibly further for very large targets. Weddell seals can echolocate prey and also (crucially) find their breathing holes in the ice. Toothed whales echolocate at very high frequencies (20–150 kHz), and dolphins in particular have very complex clicking emissions at about 30–150 kHz (peaking in amplitude at around 50 kHz), using nasal and laryngeal structures to emit a highly directional "cone" of sound with a beam width of only 5–12°, according to species. Unusually, they appear to receive most of the returning sound through their bones, especially the lower jaw, which is filled with fatty material. From here it is transmitted to the middle ear and inner ear, which

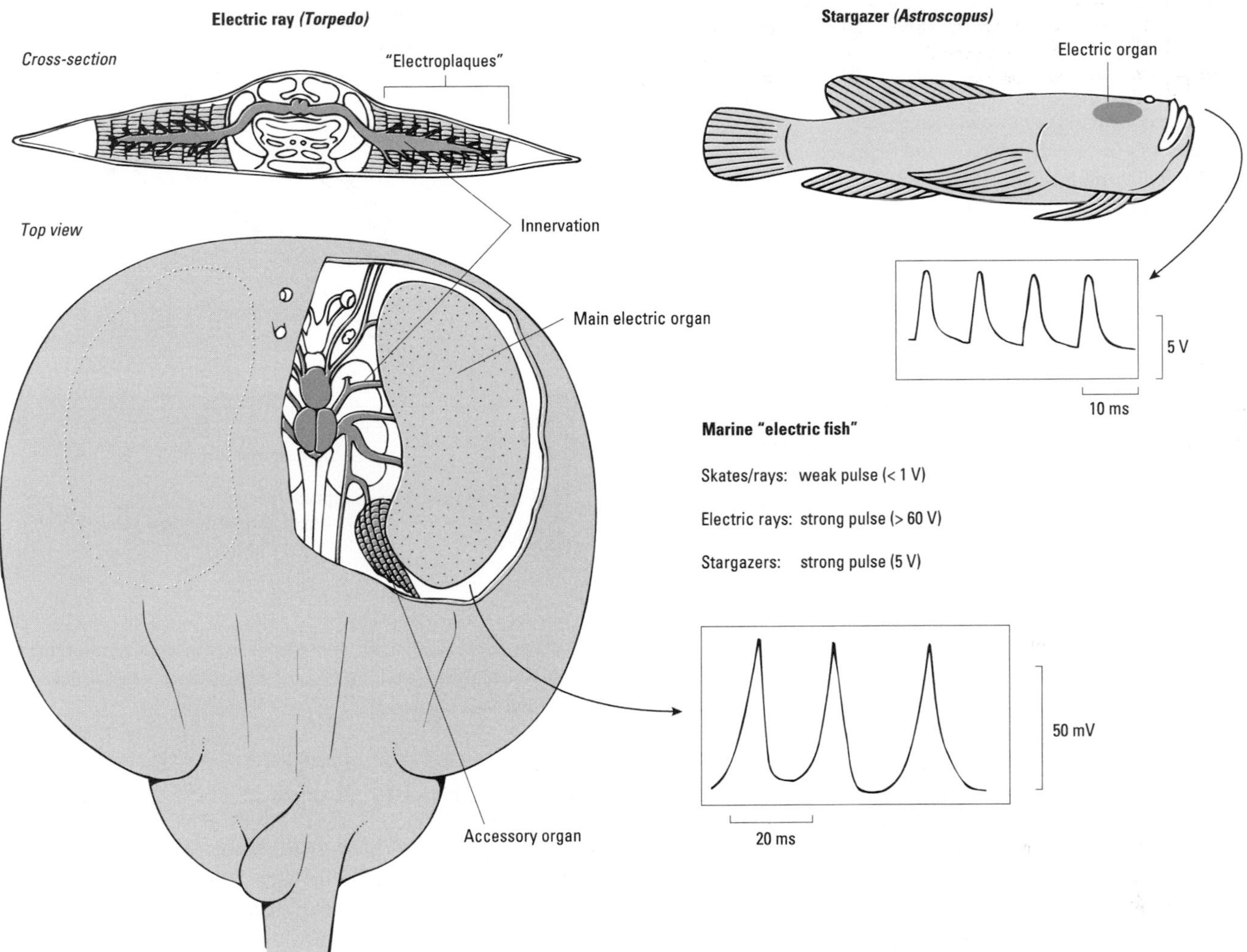

Fig. 11.32 The electric organs of marine fish: occurrence, structure, and discharge patterns.

are embedded in unusually dense bone. Dolphins can discriminate a difference of just 1 dB in the amplitude of echoes, allowing them to detect very subtle differences in targets.

Electroreception is one sensory modality that is confined to aquatic habitats and is reasonably common in marine fish. There are several alternative levels of sophistication (see section 9.8.6). Sharks and rays can detect very small electrical emissions (only a few microamps) from the hearts or gill muscles of other active animals that form potential prey, but they do not produce any external electrical discharge of their own. However, a number of fish take matters further and make their own electric field. The so-called "weakly electric" fish achieve an endogenous electric field of alternating current discharge at some species-specific (but often sexually dimorphic) frequency between 50 and 200 Hz, usually from an electric organ of modified muscle, sited near the tail (Fig. 11.32). They can electrolocate intruders by sensing the disturbance to this field, or can signal to conspecifics. In all these cases, the actual electroreceptors are small ampullary organs closely associated with the lateral-line systems. These electroreceptors may be sufficiently sensitive to detect the Earth's magnetic field (see Chapter 9), because current flow is induced in any conductor moving through a magnetic field, and sea water (unlike fresh water or air) provides a sufficiently good conducting medium to allow completion of an electric circuit. This opens up the possibility of **magnetoreception** for marine fish, and sharks and rays with their ampullae of Lorenzini have indeed been shown to use this detection mode.

However, representing the final level of development of exploitation of electricity, there are a few species of "strongly electric" marine fish, such as electric rays (*Torpedo*) and stargazer fish, which have very powerful discharges of up to 600 V, originating from "batteries" of modified muscle cells. Such discharges are used for actually stunning the prey, and in these animals there are usually no special electroreceptors.

11.8 Feeding and being fed on

In the pelagic marine environment, many feeding options are open to animals. They may filter feed on phytoplankton (though it is normally impractical to separate out phytoplankton and zooplankton, so there are no real "herbivores" at this level); or prey on other nektonic or planktonic animals, which are often present in huge numbers (but, as we saw earlier, not necessarily with great diversity of species); or in a few special areas, such as the Sargasso Sea, they may graze on floating seaweeds (in similar fashion to coastal algal grazers dealt with in the next chapter).

Filters may be provided by cilia and flagella, by mucus sheets (on their own or combined with cilia), or by cirri (cuticular bristles) in invertebrates. Vertebrates use various modifications of the jaw structures, such as gill-rakers, culminating in the impressive sieving baleen plates of various whales. Pelagic feeders vary from small passive floaters, doing little more than unselectively filtering out the surrounding life, to highly active swimmers, up to and including sharks and whales, which may be very selective and often fiercely predatory in their feeding. Taking the middle ground are many pelagic feeders, such as crustaceans, that select their food on size and palatability, and migrate to keep up with the best food supply. Planktonic species commonly show diurnal vertical migration, while many of the nekton have prominent seasonal latitudinal migrations (e.g. the whales that migrate from the poles to the equator and back, covered in section 8.9.3).

Nutrition in the benthic environment, at all depths, is rather different. Some animals in the shallower seas again operate as filter feeders, exploiting all the floating life forms. These benthic filterers are often colonial (tunicates, bryozoans, and many hydroid cnidarians), or in the case of molluscs may be individually massive (giant clams). Others feed on dead remains, as deposit feeders and scavengers; these more commonly occur at depth, and are usually solitary (many worms, crustaceans, and brittlestar echinoderms). There are also options for more active feeding, by preying on other benthic animals; but at depth there is only a very low-density biota, and food is scarce (e.g. one fish per 1000 m^3). Therefore deep-water marine predators have to be very specialized structurally, while being catholic in their tastes: many have very extensible mouths and guts, and a reduction of less vital skeletal components, with a very low metabolic rate. Fish that make their living in this way often also have reduced trunk muscles and skeletons as we saw earlier; but some of these species also have strengthened and highly ossified jaws, so that they become rather static feeding traps for smaller organisms. To escape moderately active predators, many of the smaller deep-sea inhabitants that might serve as prey tend to be highly transparent.

Many benthic marine animals have been found in recent years to rely at least partly on direct uptake of simple dissolved chemicals from the surrounding sea water. Some of the marine annelids are particularly good at this (Fig. 11.33). Better again are the pogonophorans, which have no gut at all and absorb some of their nutrients directly through their own skins; they often live in areas unusually rich in nutrients. Living in a tube probably helps them, by creating a local concentration of nutrient; furthermore, being immensely long and thin, they have a remarkably high surface area to volume ratio, which also helps.

We should not forget that some benthic marine animals live in and around deep-sea ridge crests and at subduction zones where there is geothermal activity, or around oil seeps, or in highly reducing sediments. These animals have very "odd" metabolisms where sulfur bacteria living as symbionts within their tissues or body cavities provide food from the unusual sulfurous cocktail of chemicals that occur. Here again the best examples are the pogonophorans; we will deal with the physiology of these hydrothermal vent organisms as a special case in Chapter 14.

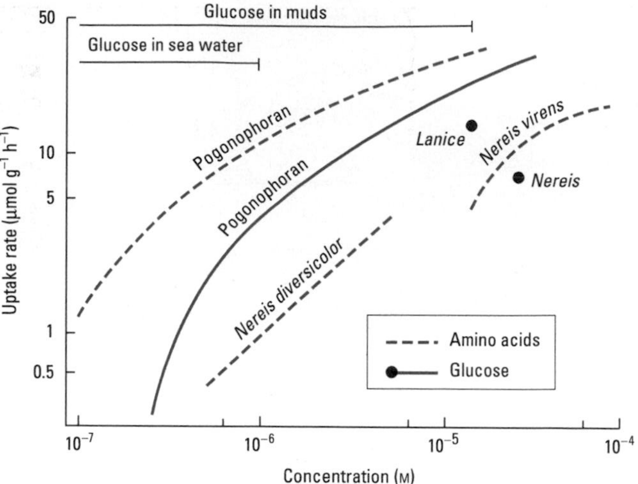

Fig. 11.33 Amino acid and glucose uptake rates through the cutaneous surface in various polychaete annelids and (at even lower concentrations) in the gutless pogonophoran worms.

11.9 Anthropogenic problems

There is no longer any part of the ocean system that does not contain pollutants, though in the deep oceans, beyond continental shelves, it remains true that most pollutants get diluted in such a huge volume of sea that they likely represent no real threat. Random dumping of "loose" chemicals at sea is undoubtedly unwise, although far too little is known of deep currents and sediment movements to predict how such things might get concentrated in food chains and where they might surface again. But waste disposal that is genuinely deep sea, and in properly sealed well-designed weighted containers, may indeed be a relatively cost-free solution to some of the disposal problems that can be so difficult on land.

However, there are other threats to the marine biosphere from man's activities. Some of these come from the various atmospheric changes that man precipitates: global warming caused by increased CO_2 and CH_4 levels, acidification of rainfall, and atmospheric heavy metals such as lead and copper. There is mounting evidence that warming sea temperatures are already having effects on coral reefs, which lose their photosynthetic symbiotic algae and become bleached as a result, with knock-on effects on all the associated biota. Even small changes in temperature can clearly be hazardous to marine animals that are almost inevitably stenothermic. Particular attention has been drawn to sea turtles, already under threat when they come ashore on tourist beaches to lay eggs, and for whom small temperature changes may have chaotic effects on sex ratios (due to their temperature-dependent sex determination; see Chapter 15), and hence on population dynamics.

Marine transport also causes problems, chiefly arising from oil lost from tankers, which can be especially dangerous in relatively

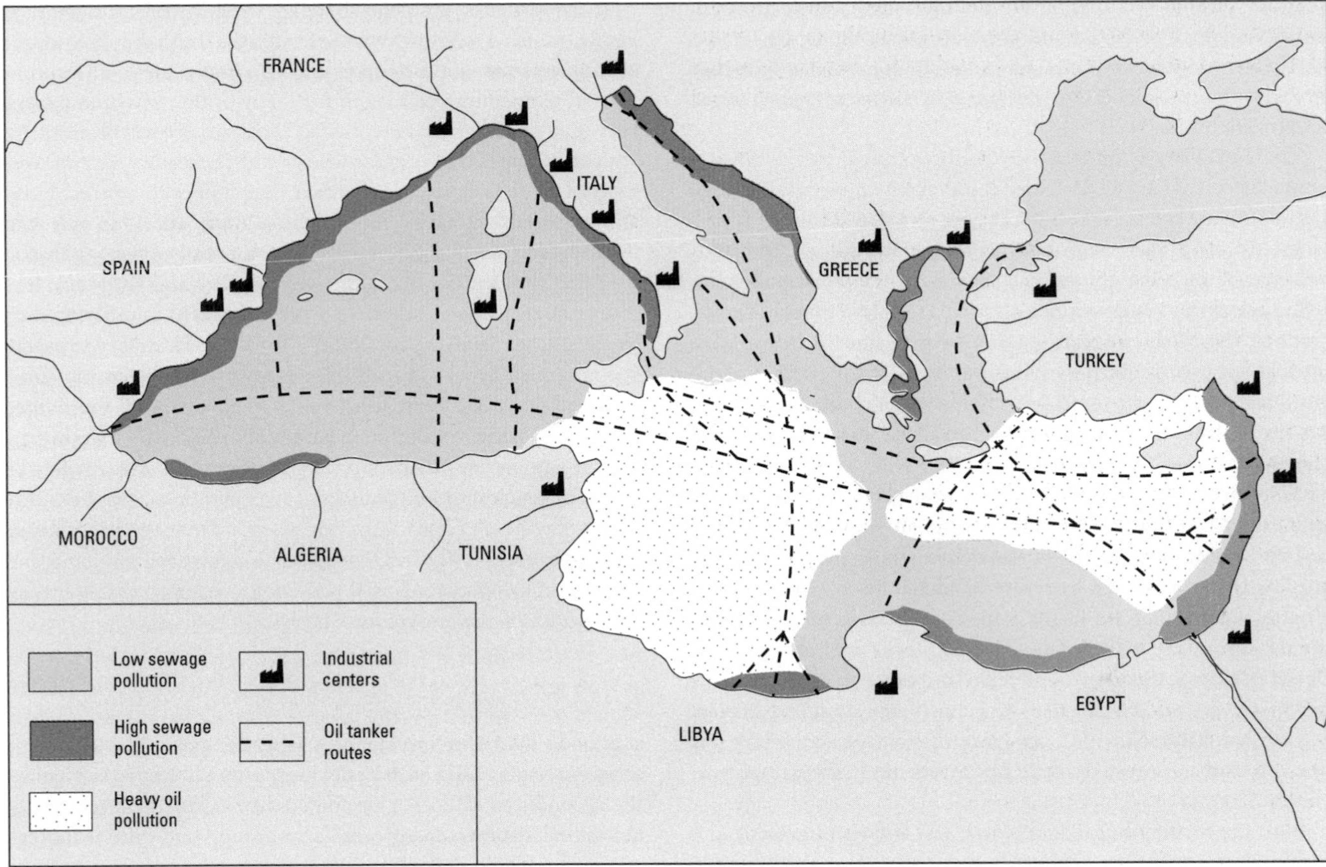

Fig. 11.34 Oil and sewage pollution zones in the Mediterranean.

enclosed areas such as the Red Sea, the Caribbean, areas around the Philippines, and the Mediterranean. The major causes are uncontrolled discharge, leakage, and tank cleaning; all this has declined somewhat since oil prices rose in the late 1970s, and with better policing. But still about 3 million tonnes of oil get into the seas each year. Figure 11.34 shows areas of the Mediterranean commonly found to be polluted with oil, corresponding closely to the oil-refining areas and shipping lanes. Then there are also periodic oil transport accidents, the most publicized recently being the grounding of the *Exxon Valdez* in Alaska in 1989 (this episode released around 36,000 tonnes, killing at least 35,000 birds and most of the sessile organisms in Prince William Sound), and of the *Sea Empress* in February 1996, when about 72,000 tonnes of light crude oil were released into the seas around the coast of southwest Wales, a region renowned for the diversity of its coastline, of which more than 100 km became seriously polluted.

The high mortalities accompanying such disasters result because oils are potentially directly lethal to all filter-feeding organisms and indirectly to the organisms higher in the food chains that feed on them. Oils are also commonly fatal to sea birds and some sea mammals that rely on their surface properties for buoyancy or for insulation, and that will therefore groom off and swallow surface deposits from fur or feathers. A summary of the effects of hydrocarbons on birds is shown in Table 11.8.

Nevertheless, natural degradation and wave action may often be the best option for cleaning up beaches, because many of the oil dispersants used are detergents with effects worse than the oil, disrupting the biological lipid layers on which so many animals depend for their intrinsic impermeability, and especially destroying the integrity of their respiratory surfaces. Bacterial remediation may be the long-term solution, whereby nitrogen and phosphorus are added to oil slicks to encourage bacteria to degrade the oil rapidly, although this has not really been shown to work well *in situ* on a large scale as yet.

Apart from oil, marine cargoes of toxic wastes and chemicals also get into the seas. Usually this does not matter too much because of

Table 11.8 Effects of hydrocarbons on marine birds.

Site of oil	Effects
Plumage	Reduced buoyancy Increased swimming activity and metabolic rate Blockage of nasal gland
Gut (ingestion)	Reduced gut absorption
Blood (after ingestion)	Hemolytic anaemia (damaged red blood cells); hypoxia Disruption of adrenal cortex function; reproduction disrupted Malfunction of nasal glands; salt accumulation, kidney failure
Eggs	Disrupted embryonic development; death or teratogenic effects

the vast volumes and diluting effect of the oceans, but very persistent toxins, such as dioxin and polychlorinated biphenyls (PCBs), can become locally concentrated, and PCBs in particular have been linked with the loss of disease resistance in marine mammals, birds, and invertebrates.

The North Sea is undoubtedly badly polluted; the volumes of waste deposited have all decreased (by about 40% overall) in the last decade but are still high. Fish stocks have seriously declined, though there are additional reasons for this (climate, silting, and overfishing). Sewage sludges and oils have been the worst pollutants; heavy metals and radionuclides are also a problem though. Conditions in the Mediterranean are even worse, since it is a shallow enclosed sea surrounded by industrial areas to the north and oil-producing areas to the south. Any foreign material dumped into this sea tends to stay in it; this also applies to all the waste coming down the rivers Rhône, Po, Danube, Don, and Nile, as well as dozens of lesser rivers. Each year the Mediterranean gets 1.5 million tonnes of industrial and agricultural effluent, and 30–50 million tonnes of untreated sewage. Figure 11.34 also demonstrates the huge areas of this sea that are affected by severe or moderate sewage pollution. Coastal resorts that are major tourist centers are affected, some almost terminally; marine life and fish stocks are reduced, with detectable contamination by mercury and other metals, and some key species are now endangered. An action plan was put together by the United Nations in 1982 involving 18 countries bordering this sea, and there have been resultant improvements in some respects in the last 15 years.

Both the Mediterranean and North Sea, and also the Baltic and localized sea areas around Japan and Indonesia, have recently become infamously prone to red algal blooms that can release toxins and deplete fish stocks. There have been incidents in Japan of multiple severe poisoning from eating affected fish, and the tourist industry along parts of the Italian Adriatic coast has been seriously affected in the last few years after great blankets of seaweeds grew offshore. This was probably a consequence of unbalancing the nutrients in the ecosystem due to pollutants combining with climatic changes and CO_2 levels; a salutary lesson in the way that differing environmental problems actually interact to make things even worse.

11.10 Secondary invasion of the seas: marine vertebrates

Much of the discussion so far in this chapter has dealt with invertebrate animals in the sea, for whom life is relatively straightforward in physiological terms. But marine vertebrates (and also the very rare marine insects) have a special set of problems when living in the sea. Both groups have arrived in the sea secondarily, after a long history of life in fresh water or on land. (Note that while there is an argument about the original habitat of the first vertebrates, it is clear that a large part of teleost history, and radiation from crossopterygian fish into the tetrapods, occurred in fresh water.) The modern animals in these taxa all have low blood concentrations, usually in the range 250–450 mOsm, and are effectively in a state of water shortage ("physiological drought") when they live in the seas, as their body water is continuously susceptible to leaking out down an osmotic gradient, causing tissue shrinkage.

Three particular problems therefore need special consideration. Firstly, we need to know how these animals deal with the osmotic and ionic stresses imposed upon them. Secondly, there will be problems of temperature regulation for many of the vertebrates, some having an endothermic inheritance. Thirdly, there will be problems with breathing in the sea in cases where the respiratory systems were evolved for use on land—a situation that applies to reptiles, birds, and mammals, all having air-breathing lungs, and to insects with their tracheal systems. But it is also worth noting in passing that all the other problems of life in the sea—pressure and buoyancy, life-history strategies, and sensitivity—will be affected in differing ways by the altered "starting conditions" for these secondary invaders. For example, the eyes of seals have to revert to a flat cornea and a spherical lens (like most fish), rather than retaining the domed cornea and flatter lens of most land mammals. Note that, in particular, any marine adaptation that involves ion pumping may be rendered much more difficult for a hyposmotic vertebrate in the sea. It may be no coincidence, for example, that vertebrates have adopted a buoyancy system that does not involve ion pumping, and that the cephalopods, endowed with such a system, have never been able to leave the sea and colonize freshwater habitats.

11.10.1 Ionic and osmotic regulation in marine vertebrates

The body fluid composition of a range of marine vertebrates is shown in Table 11.9. The most obvious feature is the marked hyposmotic condition of most of the animals listed, with these vertebrates almost invariably retaining osmolarity around one-third that of sea water (about 300–400 mOsm) that their ancestry gave them. They have blood dominated by Na^+ and Cl^-, with low potassium levels and relatively little organic content.

Ureo-osmoconforming

There are some clear "anomalies" in Table 11.9, notably the cyclostome hagfish and the various cartilaginous fish (Chondrichthyes). Figure 11.35 reveals that only the first of these isosmotic curiosities has body fluids that are also roughly isionic with its surroundings, at least in respect to the major sodium and chloride ions; in other words, the hagfish is unique amongst vertebrates in having blood very like sea water, being an osmoconformer and an ionic conformer. This may reflect a marine ancestry, although the levels of ions in the animal's cells are surprisingly like those of all other fish, and the hagfish also has fully glomerular kidneys, neither of these features being particularly helpful to a primitively marine animal.

In contrast to the hagfish, the cartilaginous elasmobranch fish although isosmotic actually have ionic levels more similar to those of other vertebrates (see Fig. 11.35, the dogfish and ray), and less than half of seawater levels. In these animals the osmotic deficit is restored in a very different fashion, by incorporating high levels of organic material into the blood, in particular the waste product urea, at about 300 mOsm, and its derivative TMAO, at about 100 mOsm (with small amounts of other osmolytes such as betaine). The methylamine derivative TMAO acts as a counterbalance to the perturbing effects of high urea concentrations on protein structure. TMAO levels are down to 40 mOsm in shallow-water species, and up to 400 mOsm in some very deep-water species, suggesting that it

Table 11.9 Body fluid composition in marine vertebrates.

Species	Osmotic concentration (mOsm)	mM Na	K	Ca	Mg	Cl	Urea	TMAO
Agnatha								
Myxine (hagfish)	1062	558	10	6	18	546	–	–
Petromyzon (lamprey)	317							
Chondrichthyes (elasmobranchs)								
Mustelus (dogfish)	1 011	288	8	5	3	270	342	~100
Raja (ray)	1050	289	4				——444——	
Osteichthyes								
Latimeria (coelacanth)	1181	181	51	4	14	200	355	~100
Teleosts								
Lophius (anglerfish)	452	180	5	3	3	196	<1	–
Opsanus (toadfish)	392	160	5				<1	–
Anguilla (eel, in sea water)	371	177	3				<1	–
Gadus (cod)	308	174	6	7	3	150	<1	–
Amphibia								
Rana cancrivora (crab-eating frog)	830	252	14				——350——	

TMAO, trimethylamine oxide.

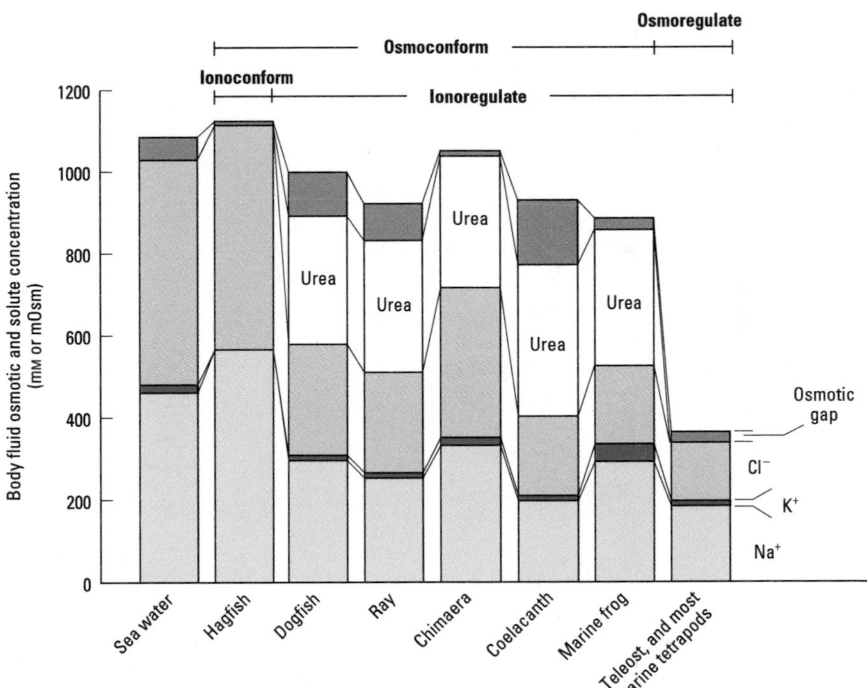

Fig. 11.35 Body fluid composition in marine fish and a marine frog, showing the ionic and osmotic conforming or regulating patterns. The "osmotic gap" is the blood solute concentration not accounted for by the listed components. (From *Comparative Animal Physiology*, 1st edn, by Withers © 1992. Reprinted with permission of Brooks/Cole, a division of Thomson Learning: www.thomsonrights.com. Fax 800 730-2215.)

may also help to counteract the destabilizing effects of pressure on proteins (and remember that it can also help with positive buoyancy; see section 11.6.2). These animals are therefore osmoconformers but have to be strong ionic regulators, and the inward ion diffusion is principally countered by secretion of NaCl by a rectal gland, which can secrete a fluid containing about 500 mM Na^+. The composition of different body fluids in the elasmobranch *Squalus*, including those of its rectal gland, are shown in Table 11.10.

Only three other marine animals are known to use this particular technique of "ureo-osmoconforming" (see Fig. 11.35 and Table 11.9). Two are fish—the chimaera (cartilaginous rat-fish) and the coelacanth *Latimeria* (a "living fossil", close to the root of the tetrapods and lungfish, and only distantly related to other "fish"). In addition there is the single species of modern amphibian that can live in sea water, the crab-eating frog, *Rana cancrivora*, of Southeast Asian mangrove swamps (oddly enough, the tadpoles of this frog can also live in sea water but do so without urea accretion, adopting instead the ionic and osmotic regulation patterns of teleost fish). In all these cases the level of urea plus TMAO in the blood may be around 350–400 mOsm, and the presence of these materials substantially limits the osmotic stress for the animal by preventing or greatly reducing the continuous outward leakage of water.

Component	Osmotic concentration (mOsm)	mM Na	K	Ca	Mg	Cl	Urea + TMAO
Blood/ECF	1096	296	7	3	4	276	504
Intracellular (muscle)	1189	18	130	2	14	13	905
Rectal gland fluid	1050	512	3	1	3	520	–

ECF, extracellular fluid; TMAO, trimethylamine oxide.

Table 11.10 Body fluids, intracellular concentrations, and rectal gland fluid concentrations in the elasmobranch fish, *Squalus*.

Coping with hyposmotic body fluids in teleosts

All the rest of the marine vertebrates, including the teleosts and the tetrapods, are substantially hyposmotic to their surroundings and will therefore tend to lose water and gain salts. They must regulate continuously, both osmotically and ionically. The water and ion leakages will occur especially through those parts of the body surface that are most permeable, such as the gill epithelia of teleosts; other parts of the outer surface generally have a rather reduced water permeability.

In teleosts the inward passage of salts and outward passage of water are countered by four main systems (Fig. 11.36).

1 Drinking the surrounding salty water to get water into the body.

2 Uptake (70–80%) of this water and of the univalent ions (Na^+, K^+, Cl^-) from the imbibed sea water through the gut wall, particularly the esophagus; but with minimal gut uptake of divalent ions (Ca^{2+}, Mg^{2+}, SO_4^{2-}), these being passed with the feces.

3 Elimination of most of the excess salt thus gained by active transport of sodium and of chloride across the gill surfaces, with varying levels of cutaneous transport also occurring. The gills contain

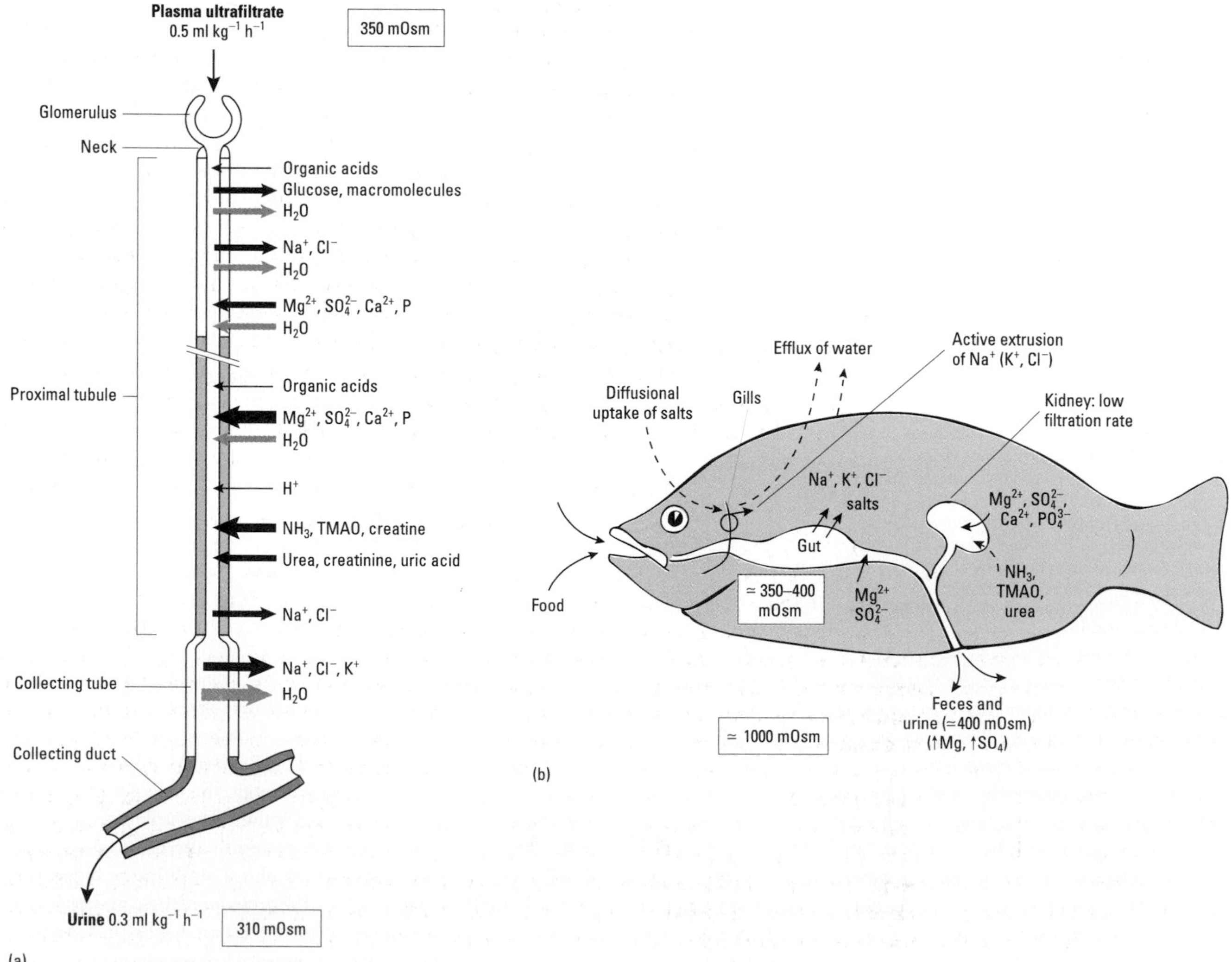

Fig. 11.36 Movements of salts and water: (a) in the nephron of a marine teleost, and (b) in the whole organism. TMAO, trimethylamine oxide. (a, From Hickman & Trump 1969.)

"chloride cells" (or "mitochondria-rich cells"; see Chapter 5), responsible for much of the Cl^- secretion, via sodium pumps, $Na^+/K^+/2Cl^-$ cotransporters, and chloride channels.

4 Excretion of excess divalent salts through the kidney, via an isotonic urine in which Mg^{2+} and SO_4^{2-} concentrations are nevertheless unusually high. Urine volume is generally rather low, and the glomeruli are reduced in number, sometimes even lacking (e.g. in the toadfish, *Opsanus*, and some goosefish and pipefish, where urine is produced by secretion rather than filtration). Figure 11.36a shows the major ion and water movements in the nephron of a marine teleost.

The net result is the retention of a fluid that (if it could be isolated) would be hypotonic both to the ingested sea water and to the urine, and which therefore restores the balance of the blood. In effect, these fish are distilling sea water in their bodies. The marine teleosts also use the gill epithelia as a major route for excretion of nitrogenous wastes, with NH_3 and NH_4^+ leaking passively outwards (perhaps with some Na^+/NH_4^+ exchange). This may be aided by NH_4^+ substituting for the similarly sized K^+ in basolateral sodium pumps.

The tissues are regulated relative to the blood in a rather similar fashion to the invertebrates considered earlier, i.e. by the manipulation of amino acid levels in the cells. Some teleosts use low levels of TMAO as an osmotic effector in addition, indicating that the strategy of urea/TMAO retention in elasmobranchs discussed above is perhaps just an extension of the general principle of using small organic osmotic effectors to maintain isosmotic conditions.

Using these kinds of mechanism, marine teleosts can maintain their blood and cells at remarkably constant concentrations. Indeed some can migrate into freshwater environments for part of their life cycle without substantial osmotic problems, by modifying the relative rates of these various processes through careful hormonal control. These anadromous fish (breeding stage in fresh water—lampreys and salmon) and catadromous fish (breeding stage in sea water—eels) have the regulatory patterns characteristic of the inhabitants of whichever environment they are in, thus being able to pump sodium ions outwards in sea water, and inwards when in fresh water, although their body fluids are fairly constant throughout (see Chapter 12).

Hyposmotic body fluids in marine tetrapods

Marine tetrapods have rather different strategies from those so far described, and may conveniently be considered in two groups: first, the reptiles and birds, and second, the mammals.

There are a number of reptiles and birds that live in or around sea water for most of their lives (Table 11.11). In the case of sea snakes this residency is permanent, whereas turtles must breed on land, while marine iguanas and all marine birds are essentially terrestrial animals that submerge themselves for varying periods. Like the teleosts they are unable to produce a significantly hyperosmotic urine to get rid of excess salts, yet they have no gill surfaces at which salts could be excreted. Instead they have a very low P_{Na} in their skin to slow inward salt leakage as far as possible, and they have specific **salt glands** (Fig. 11.37) where salt from drinking and from the slow leak into the body can be actively extruded. These glands can produce NaCl at very high concentrations (Table 11.11). For example,

Table 11.11 Salt gland performance in marine tetrapods.

Species	Na^+ concentration (mM)	Flow rate ($\mu l\ kg^{-1}\ h^{-1}$)
Sea snake	620	35
Terrapin	680	4
Turtle	690	19
Marine iguana	1000–1400	8
Cormorant	500–600	70
Pelican	600–750	100
Gull	600–900	30
Penguin	720–850	40–70
Petrel	900–1100	–

1170 mM Na^+ and 1330 mM Cl^- is found in the salty extrusions of the marine iguana, which can be sprayed out some distance from the head, whilst 1100 mM NaCl in Leach's petrel is the record for any fluid concentration achieved by a bird. They operate using standard sodium pumps across a rather impermeable epithelium so that there is little entrained osmotic water flow. In the avian salt gland this is an ouabain-sensitive Na^+ transporter, apparently sited on the basal membrane of these epithelial cells. The glands may be variously situated in reptiles: the nasal cavity (iguanas), eye sockets (turtles), mouth (sea snakes), or tongue (estuarine crocodiles). In birds they are usually situated on top of the skull, and invariably open to either side of the "nose", near the top of the bill. They seem to be modified mucoid or lacrimal glands, and probably have multiple convergent origins. These various salt glands are remarkably effective in removing salt from the body, and they allow their possessors to drink only salt water. They can be switched on and off, according to osmotic need, by cholinergic nerves, and their activity level is also affected by prolactin and steroid hormones. In birds they can become enlarged or shrunken within an individual's lifetime according to the extent of its encounters with a salty environment and diet. However, many partially marine birds and reptiles ameliorate their problems by taking as much water as possible from their food instead. The marine iguana gets about 58% of its water from food, only 38% from drinking sea water, and 4% from metabolism.

Secondly, there are several groups of marine mammals, in different families. Most obvious are the permanently resident cetaceans (whales, porpoises) and sirenians (dugongs), and the amphibious pinnipeds (seals and their kin). These have no salt-secreting glands, because they have inherited the benefits of the mammalian kidney (see Chapter 5), superior to that of other tetrapods in that by using the elongate loop of Henle it can produce a very markedly hyperosmotic urine, more concentrated than the animal's own blood and more concentrated even than the sea water in which the animal lives. Thus these animals get rid of excess salt directly from the blood. A whale drinking 1 l of sea water gets a net gain of about one-third of a liter of pure water once the kidneys have done their work. (A human can produce urine more concentrated than human blood, but less concentrated than sea water, so that drinking sea water is of no benefit to a thirsty human, producing a net water loss just to get rid of the salt load.) However, marine mammals do not normally drink much sea water and so do not get as much excess salt as the reptiles and birds anyway. This is at least partly because all marine mammals are predominantly carnivorous, and take in

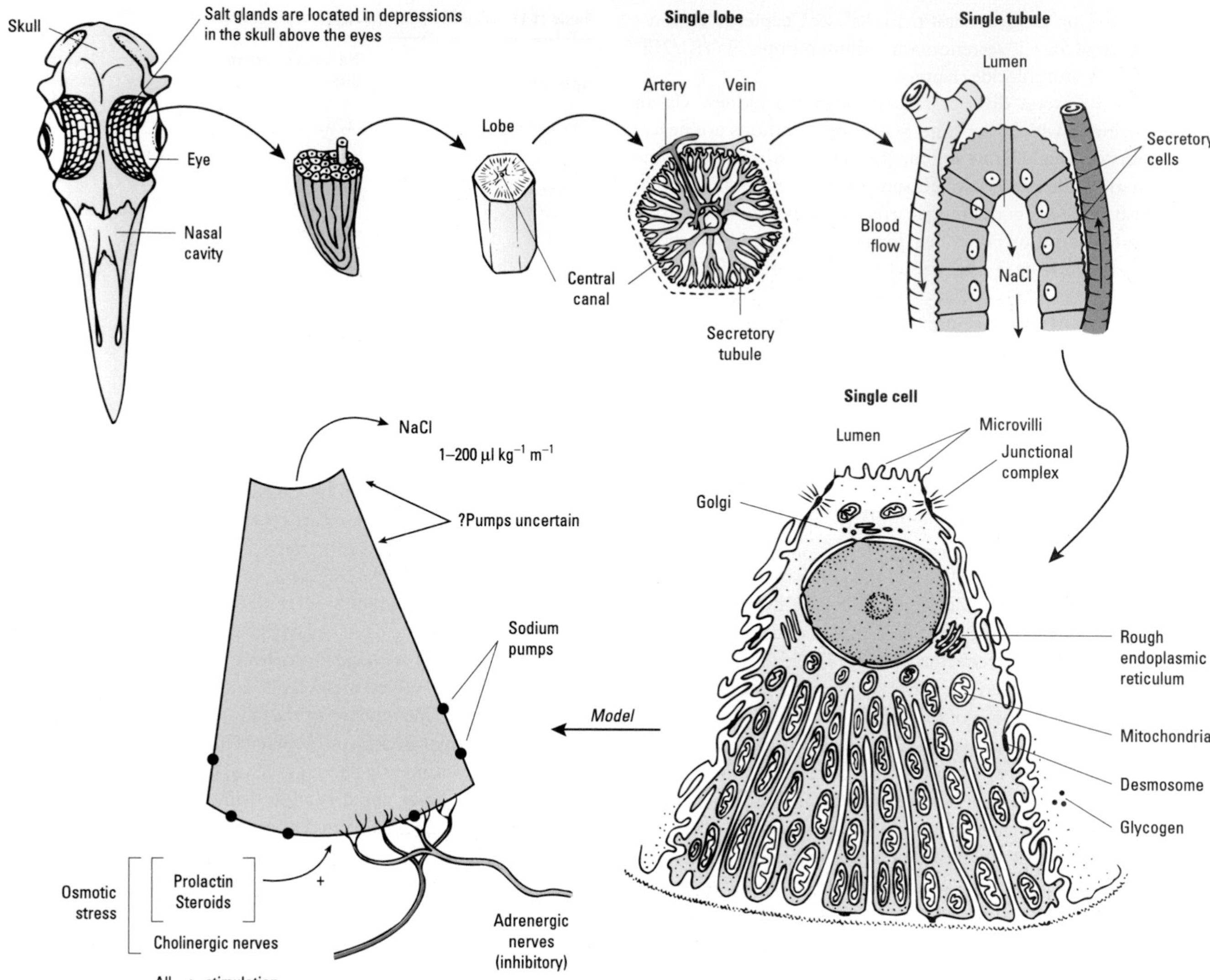

Fig. 11.37 Salt gland positions and functions in marine tetrapods. (Adapted from Peaker & Linzell 1975.)

dietary water only as a component of the flesh they get as food, having relatively dry stomach contents. In a sense then, the marine mammals are "parasitizing" the osmoregulatory abilities of their various prey, by taking in only these preregulated fluids.

A summary of the fluid compositions, and of the variety of osmoregulatory mechanisms operating, in a range of marine vertebrates is shown in Fig. 11.38.

The more complex ionic and osmotic regulatory strategies of marine vertebrates necessitate rather more complex control systems than are found in invertebrates. Prolactin from the anterior pituitary (in the brain) affects chloride uptake in general, and also more specifically controls mucus production by the skin and fish gills, and ion uptake in the salt glands. Vasopressin and other posterior pituitary peptides act on kidney glomeruli in most marine vertebrates (though on tubules in marine mammals), and the size of these peptides appears to correlate with the need for water conservation, being especially large in the marine mammals. In addition, a range of steroid hormones, including aldosterone, affect sodium secretion and resorption, in the kidney, gut, salt glands, and gills, thus regulating Na^+ and K^+ levels in plasma. These steroids are in turn affected by the renin–angiotensin hormone system (see Chapter 10), which is activated by changes in fluid volume and electrolyte levels.

11.10.2 Temperature in marine vertebrates

Surviving in cold seas with dilute body fluids

Marine fish, amphibians, and reptiles, like most of the invertebrates, are essentially ectotherms with T_b levels very close to the temperature of their surroundings. This can pose severe problems for those living in cold polar seas, as their body fluids (being more dilute) may freeze even while the surrounding sea water remains liquid. Furthermore, their surroundings are likely to contain small ice crystals that would precipitate rapid freezing in any unprotected animal. Hence polar fish, in particular, are very well endowed with the "standard" antifreeze mechanisms discussed in Chapter 8 (although as we pointed out there they have often lost the expression of heat

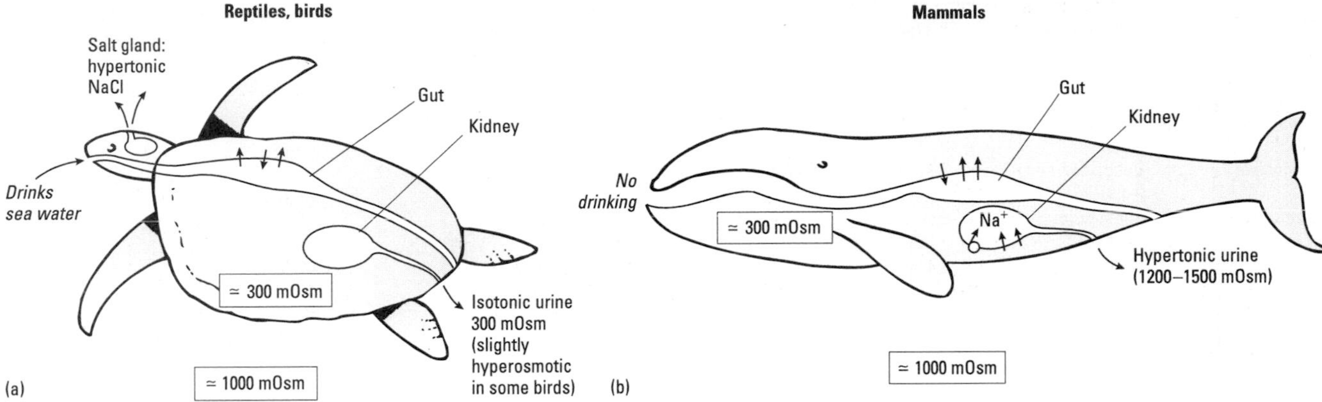

Fig. 11.38 Overview of salt and water balance in marine tetrapods.

shock proteins, since they never normally experience changes of temperature).

The body fluids of most marine fish are hyposmotic to sea water, freezing at −0.7°C. Since the temperature in shallow ice-laden water is close to −1.9°C, most Antarctic fish are supercooled by about 1°C. They may have at least three further adaptations that help them to avoid freezing. Firstly, the freezing point of sea water is depressed by 0.0075°C for each 10 m increase in depth, so that many polar fish avoid freezing by living in or migrating to sufficiently deep water. Deep-living Antarctic fish caught on hook and line through holes in the ice therefore have no real defenses against ice formation, and commonly freeze solid on being brought to the surface. Secondly, Antarctic fish have a higher plasma NaCl concentration than typical marine teleosts (Fig. 11.39), but this can only account for 40–50% of their recorded plasma freezing point depressions. The remainder is due to their third adaptation, the presence of particularly effective antifreeze peptides (AFPs) or glycopeptides (AFGPs), whose mode of action in preventing any existing ice crystals from growing was described in Chapter 8. These molecules are found in at least 11 different families of teleost fish (Table 11.12).

The structure of the antifreeze glycopeptide in the Antarctic notothenioid fish *Pagothenia borchgrevenki*, which lives in crevices in sea ice, is shown in Fig. 11.40a. This AFGP consists of many repeating units of a disaccharide molecule linked to the third amino acid of a tripeptide (alanine–alanine–threonine). In fact in this species a range of eight AFGPs have been identified, ranging in relative molecular mass from 2600 to 33,700. In a related species, *Notothenia coriiceps*, one of the AFGP structural genes has been shown to consist of 46 tandemly repeated segments encoding 44 copies of AFGP8 and two copies of AFGP7. This is one of the highest gene copy numbers presently known for any organism, and may reflect the need for high and seasonally constant levels of the AFGP.

Unrelated species of fish living at opposite extremes of the Earth (Antarctic notothenioids and Arctic cod) have almost identical AFGPs. However, while the original AFGP gene is thought to have

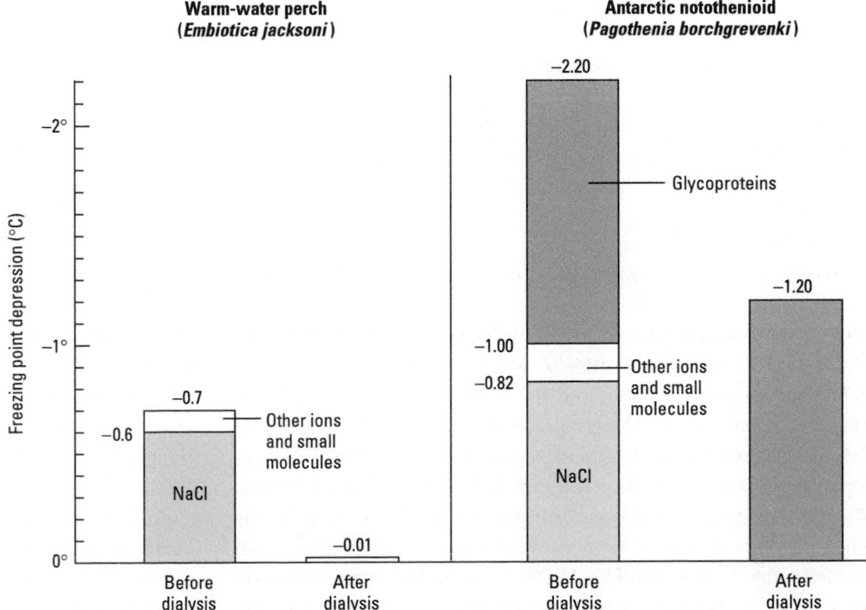

Fig. 11.39 Levels of NaCl in the blood of temperate and polar teleosts. In the warm-water perch, the freezing point of the blood before dialysis was −0.7°C, and after dialysis (removing small solutes) it rose to −0.01°C, indicating that NaCl and other small solutes are the main cause of freezing point depression (FPD). In the Antarctic notothenioid the NaCl levels are elevated, and their removal raises the freezing point by a full degree; however, the nondialyzable glycoproteins account for more than half the total FPD. (From Eastman 1993.)

Genus	Environment (temperature, °C)	FP, organism (°C)	FP, blood (°C)	MP, blood (°C)	Antifreeze molecules
Gadus	Deep, ice free	−1.0	−1.1	−0.7	AFGPs
Chaenocephalus	Shallow, ice free (−1.0)	−1.5	−1.5	−0.9	
Rhigophalia	Deep, ice free (−1.9)	−1.9	−2.0	−0.9	AFP 1–3
Notothenia	Shallow, icy (−1.9)	−2.0	−2.1	−1.1	AFGPs
Pagothenia	Shallow, icy (−1.9)	−2.4	−2.7	−1.1	AFGP 1–8

AFGP, antifreeze glycoprotein; AFP, antifreeze protein.

Table 11.12 Freezing points (FPs) and melting points (MPs) due to antifreeze glycoproteins in polar marine fish.

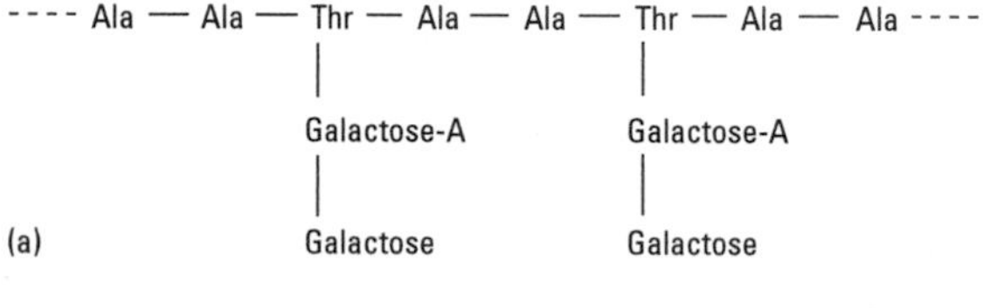

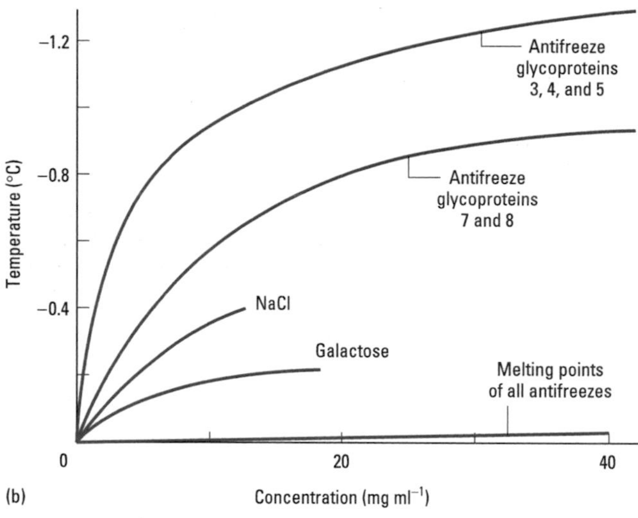

Fig. 11.40 The antifreezes of Antarctic fish. (a) The structure of the antifreeze glycoprotein (AFGP) from an Antarctic fish, with the typical Ala–Ala–Thr repeat and attached disaccharides. (b) The freezing point depressions produced by antifreeze proteins (AFPs) and AFGPs (NB 3, 4, and 5 are larger than 7 and 8), in comparison with the much lesser effects of NaCl and of galactose; note also that the AFPs of all kinds have little or no effect on melting points.

evolved from the trypsinogen gene, it seems that the Ala–Ala–Thr repeats in the cod antifreeze are derived from duplications in an unrelated ancestral coding element; thus the similarity of structure may be a rare example of protein sequence convergence.

When Antarctic notothenioid fish are placed in ice-free water, they do not freeze until their temperature drops to −6°C. Apparently they do not usually develop ice crystals unless ice first penetrates from the outside. The main roles of their AFGPs may be to keep ice from propagating across their integument, and to coat the edges of ice crystals that happen to get into their body along with the sea water they drink. The AFGPs have a very substantial effect on the freezing point but little or no effect on the melting point (Fig. 11.40b). The AFGPs are synthesized in the liver, and secreted into the blood. They appear in most body fluids except urine, ocular fluid, and the cytoplasm of cells (with the exception of the hepatocytes where they are made). They are excluded from the urine because most Antarctic fish have aglomerular kidneys: the urine is formed by secretion, so that the cells lining the kidney tubules leave the antifreeze in circulation. However, the tissues surrounding the urine and ocular fluids are fortified with antifreeze, constituting a barrier to ice propagation. It has recently been reported that AFGPs are found in the urine of the winter flounder, sea raven, ocean pout, and Atlantic cod, at a concentration that approaches that of plasma. Thus in some species AFGPs probably also increase the freezing resistance of urine. The other site at risk of ice formation is the gut, since its contents are often hyposmotic to plasma, and in most polar fish antifreezes (AFGPs 6, 7, and 8) are secreted into the gut lumen via the bile (translocated into the gall bladder from blood). Salt, water, and digested foodstuffs are absorbed during passage down the intestine, concentrating the AFGPs; they are eventually lost with the feces, representing an energetic cost of freezing avoidance. The loose anal sphincter is a potential danger spot for ice entry, but at this point the intestinal tract is quite concentrated and only freezes at −2.2°C.

At least five different classes of AFPs have also been identified in polar fishes, most being either alanine- or cysteine-rich. Since these types all have very different structures and amino acid sequences it seems likely that their mechanism of action is different in each case. In Arctic fishes, the production of AFP is seasonal (Fig. 11.41a). For example, the appearance of AFP in the plasma of the winter flounder is suppressed when day length exceeds 15 h, and it appears to be primarily under the control of photoperiod rather than low temperature (Fig. 11.41b). Exposure of the winter flounder to temperatures of 10–12°C and long day lengths results in a loss of blood antifreeze within 3 weeks. Synthesis is controlled at the level of DNA transcription and translation, as AFP mRNA appears in the blood about 4 weeks before the peptide is detectable. Control of transcription by photoperiod is hormonally mediated. Growth hormone represses transcription of the AFP gene in the summer, but as day length decreases this inhibition is reduced, resulting in higher levels of liver AFP mRNA. Control of translation then comes into operation, and is related to water temperature. Low temperatures reduce the rate of AFP mRNA degradation, which together with high levels of tRNA^{Ala} and of alanyl-tRNA synthetase favors the translation of the AFP mRNA and the appearance of AFP in the blood.

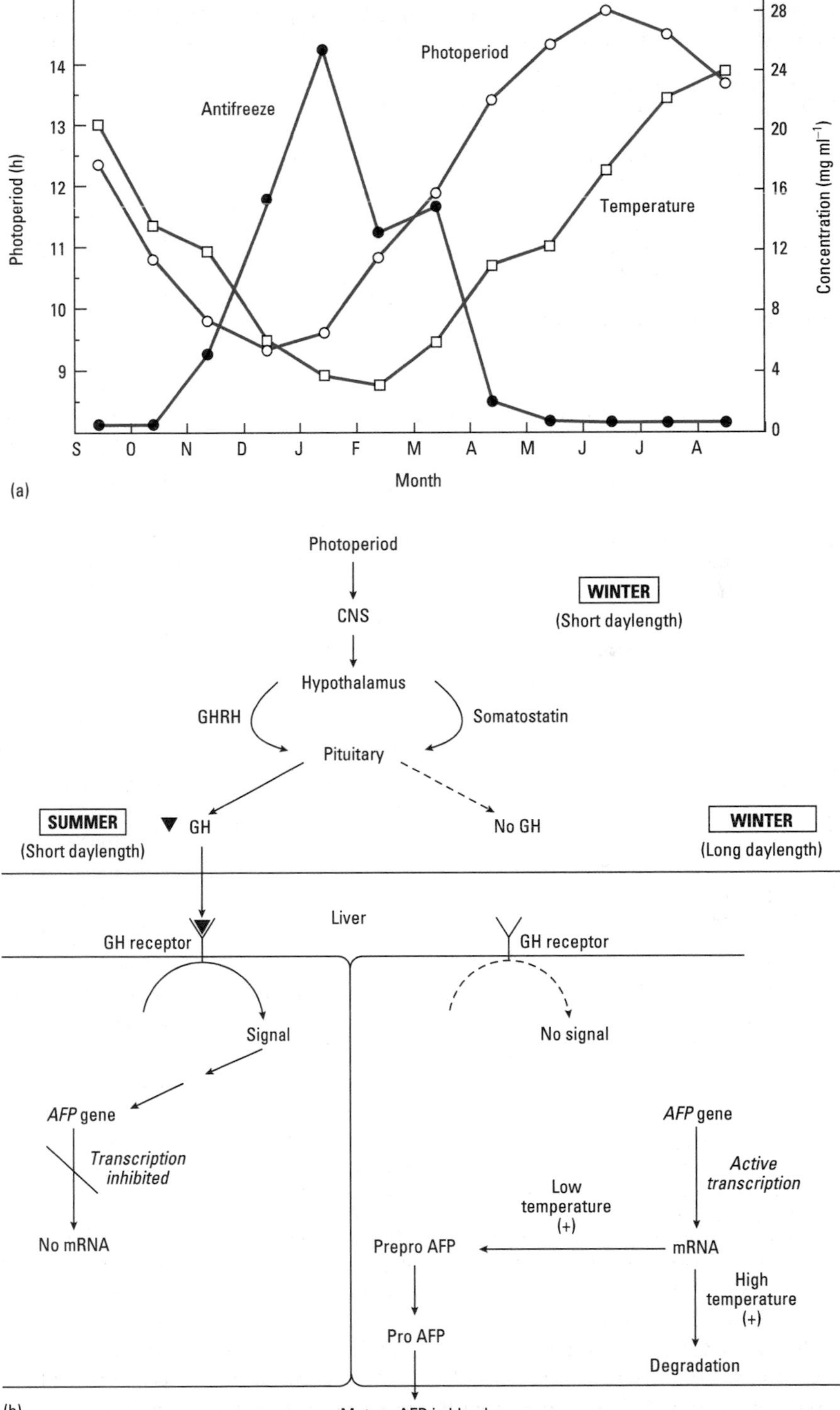

Fig. 11.41 (a) Seasonal production of antifreeze proteins (AFPs) in the winter flounder. (b) Regulation of this AFP production by photoperiod and the intermediary effects of growth hormone (GH) initiating transcription. Note that post-transcriptional control by temperature also operates, to insure that AFP is only present if temperatures are low. CNS, central nervous system; GHRH, growth hormone-releasing hormone. (a, Adapted from Cossins & Sheterline 1983; b, adapted from Chan *et al.* 1993.)

Marine endotherms

Birds and mammals that have adopted a marine habitat have very different thermal problems to those of marine ectothermic vertebrates. They are permanently endothermic and tachymetabolic, and must maintain constant T_b even when resting, despite the high thermal conductance of their surroundings. Permanent or semi-permanent residents of the sea therefore tend to be rather large animals (cetaceans, pinnipeds, penguins, etc.), with a reduced relative surface area. They have exceptionally effective insulating layers, whether as internal fat or external pelage (fur or feathers), or both. Where fatty "blubber" is used, the overlying skin is normally at roughly the same temperature as the sea, with a sharp temperature gradient over the outer 5–6 cm of fat and an almost constant

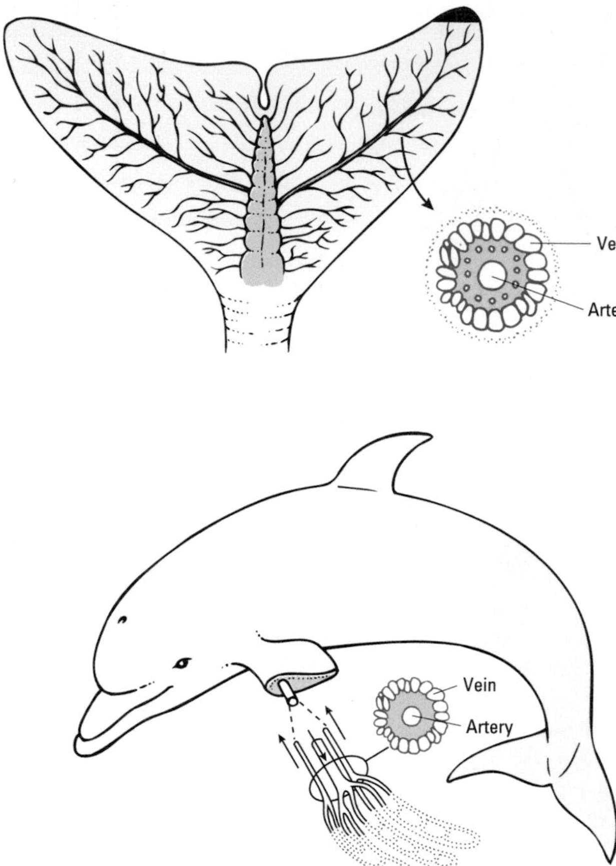

Fig. 11.42 Countercurrents in the tails and flippers of marine mammals.

temperature inwards of this. Most marine mammals and birds also use sophisticated countercurrent heat exchangers in their feet or flippers, with a central artery surrounded by 15–20 small veins (Fig. 11.42), to avoid heat loss at the extremities. The thermal consequences of these features when marine mammals are beached for pupping were considered earlier (see section 11.5). In the permanently aquatic whales, not only are the thermal losses through the fins and flukes controlled by countercurrent systems, but so too are the temperatures of the reproductive organs (which must not be allowed to overheat) and the mouth area, for it has been shown that the tongues of baleen whales, feeding fairly continuously with open mouths and in cold high-latitude waters, are well endowed with heat exchangers to reduce oral heat loss.

11.10.3 Respiration in marine vertebrates

In marine fish, breathing is a relatively straightforward issue comparable to that in other marine animals, and thus has already been dealt with in section 11.4. However, in the context of fish as secondary invaders it is worth mentioning that the very short diffusion distances and hence decreased areas and perfusion rates possible in many teleost gills have the additional advantage that the gains and losses of ions at the gill surfaces are reduced.

In tetrapods breathing is very different, as their terrestrial origins dictate a reliance on intermittent supplies drawn into the lungs from the air above, so that all marine tetrapods are intimately associated with the sea surface. Their "normal" respiratory modes and adaptations are therefore similar to those for lung-based respiration, discussed in Chapter 15.

Most of them do, however, submerge for long periods, deriving their food from below, and they show complete cessation of breathing (apnoea) during the dive, posing more interesting physiological problems. Even when using an artificial air supply, fully terrestrial tetrapods, such as man, experience blackouts when submerged for even short periods in shallow water; they exhibit severe nervous disruption in deeper waters, and suffer nitrogen narcosis. Yet emperor penguins can dive to at least 500 m, sperm whales have been recorded below 1000 m, and elephant seals penetrate down to 1500 m. Many sea mammals spend only about 10% of their time at the surface, and may be at 250–500 m the rest of the time. Weddell seals stay submerged for at least 60 min at a time, and sometimes up to 2 h. Green turtles actually seem to "hibernate" at 10–15 m depth for several months. During submersion, though, there must always be sufficient oxygen to maintain the animal, as the mammalian central nervous system in particular cannot survive even brief anoxia (more than 1 min can be fatal in humans). Dive durations in many animals appear to exceed the theoretical "aerobic dive limit" (the time they could stay submerged and still have enough oxygen to retain aerobic metabolism), so they were assumed to switch to anaerobic respiration. But in fact most tetrapod divers do not return to the surface for long enough to repay the kind of oxygen debt that this would imply. Instead of prolonged anaerobiosis, they use several linked adaptations to achieve maximum dive duration, collectively often termed the "**diving response**":

1 Apnoea, with suppression of the normal respiratory muscle action and its control systems.

2 Some system of oxygen storage, primarily the blood and its hemoglobin, and to a lesser extent muscle myoglobins (which occur at 10 times higher levels in sperm whales than in humans). Creatine phosphate is also unusually abundant in the muscles to provide a rapid energy supply (see Chapter 6). Blood volume may be enlarged in divers (in birds and mammals, though not apparently in marine reptiles). Hemoglobin also occurs at a raised concentration in diving mammals (30–40 ml per 100 ml blood, compared to 20 ml per 100 ml in most mammals), and the hemoglobin may have reduced temperature sensitivity. In mammals, red blood cells are also stored in the spleen, which undergoes regulated emptying in a dive.

3 Reduced blood flow (hypoperfusion) to all visceral organs other than the brain, with carefully controlled patterns of vasoconstriction (Fig. 11.43).

4 Reduced metabolic rate (hypometabolism), partly because organs are oxygen starved. This metabolic depression is the main factor reducing the need for anaerobic metabolism. Some seals show an associated reduction in body temperature, particularly in smaller species, and the abdominal temperatures of penguins have been recorded as low as 11°C, possibly due to extra heat loss from the exposed flippers and brood patch.

5 Anaerobic metabolic pathways in most visceral tissues and in muscles, with some lactic acid formation, but much less than expected, due to the reduced metabolic rate. In mammals lactate is not released to the blood until the dive ends, when there is a phase

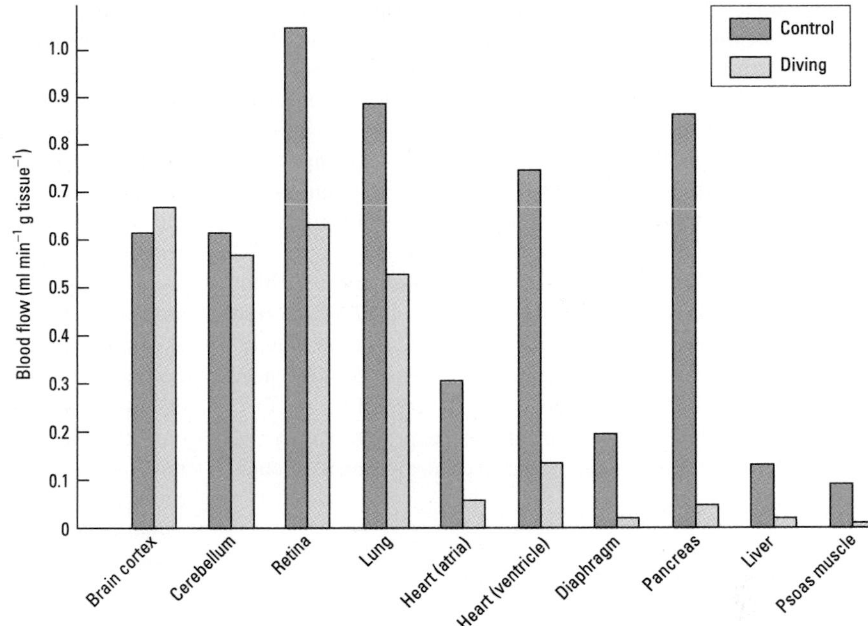

Fig. 11.43 Selective blood distribution in a diving seal. Blood flow to the brain cortex and cerebellum is maintained, and that to the lungs and eyes remains moderately high, but all other organs experience a very marked reduction in circulation. (Data from Zapol *et al.* 1979.)

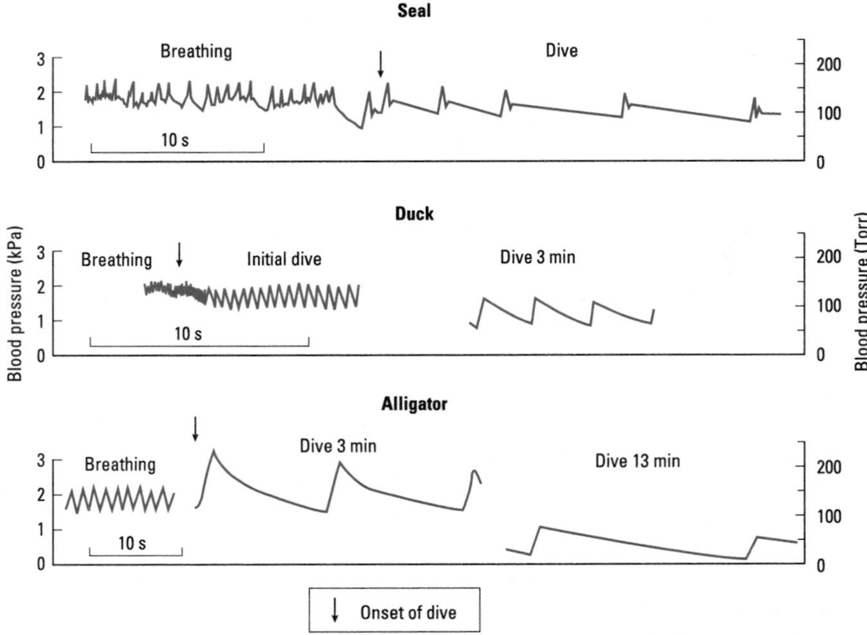

Fig. 11.44 Patterns of blood pressure recorded during diving in seals, ducks, and alligators, showing substantially reduced heart rate and in some cases also a decrease in overall blood pressure. (From Scholander 1964.)

of "lactate washout". By contrast, lactate concentrations can reach 200 mmol kg^{-1} in the muscles of diving turtles, five times normal levels in active vertebrates.

6 Reduced heart rate (bradycardia) and reduced cardiac output (Fig. 11.44). Bradycardia is triggered directly by chemosensitivity to oxygen levels.

7 Neuronal and hormonal control of cardiac and spleen action, mainly by catecholamines.

8 In marine reptiles, some cutaneous respiration—up to 30% of the oxygen requirement comes through the general skin in sea snakes, and a small proportion through the esophagus and rectum in some turtles. The incomplete separation of venous and arterial blood in reptiles also means that increasing proportions of the blood can bypass the lungs entirely during diving.

Of all these strategies, bradycardia was traditionally regarded as the centerpiece. But recent evidence from phylogenetically controlled comparisons in pinnipeds shows that bradycardia and hypoperfusion are ancestral characteristics, and it is the spleen and blood volumes, and the red blood cell changes, that are specific to divers.

Many diving animals also have surprisingly small lungs, especially in relation to the volume of the upper airways; and (counterintuitively) they exhale forcibly, rather than inhaling a good lungful of air, just before submerging. This is because a full lung would be a liability rather than a useful oxygen store. As the animal penetrates to greater depth, the increasing hydrostatic pressure causes lung compression, and air is forced back out of the alveoli into the (large) tracheoles and bronchi. This avoids diffusion of unwanted nitrogen through the alveolar surfaces into the blood supply, which would otherwise cause bubble formation and "the bends" (decompression sickness) when the animal resurfaced. Reptiles are again unusual in that they can avoid this problem by allowing some cutaneous escape of nitrogen.

Even oxygen diffusing from the lung to the blood under the high pressures during diving would be undesirable, as high Po_2 can cause convulsions in mammals. During the dive, the inspiratory muscles are inhibited via receptors that sense the presence of water near the mouth and nose; and the signals from the carotid O_2 and CO_2 receptors (see Chapter 7) are also overridden, to prevent the normal stimulus to ventilate as blood gas composition changes through the dive.

11.10.4 Vertebrate overview

Marine vertebrates, and air-breathing tetrapods in particular, clearly have a range of special features, relating to ion and water balance, temperature balance, and respiratory supply, that permit their survival in what is for them a rather hostile environment. Since the extinction of the large marine reptiles such as ichthyosaurs and plesiosaurs at the end of the Cretaceous (65 million years ago), the marine mammals in particular have proved adept at exploiting the marine world and are the only really large predators there apart from elasmobranch sharks. Most other tetrapods, those not primarily adapted for life in the sea, cannot survive the osmotic problems posed by marine conditions. This is clearly illustrated by the fact that humans are unable to live afloat without access to fresh drinking water; if we drink only sea water our bodies accumulate too much salt and lethal dehydration effects set in.

11.10.5 Marine insects

Insects have an entirely terrestrial evolutionary history, and though many are secondarily aquatic very few have ever invaded the sea. This seems very surprising given their overwhelming diversity in all other habitats. Those few species that do occur are essentially intertidal or surface dwellers, including some collembolans, heteropteran water striders (especially the genus *Halobates*), and beetles. They therefore avoid respiratory problems, and are unaffected by pressure or temperature regimes in the sea. Indeed, they can only really be considered "marine" in the sense that they face the osmotic and ionic problems of having access purely to salt waters; but the insect osmoregulatory and excretory system (based on Malpighian tubules and rectum working together) can readily cope with many worse problems than this, as we will see in Chapter 15. The major point of interest is perhaps *why* insects do not live in the sea. The likeliest answer is that the very features that make them particularly successful on land (a hydrophobic cuticle with very low permeability; reduced numbers of specialized legs; wings and flight; waterproof eggs; tracheal respiratory systems) are precisely those that will be of neutral value or even disadvantageous in the sea. Insect tracheal systems may be of much reduced efficiency at high hydrostatic pressures in the sea, yet insects would need to be able to dive deeper to escape fish predators. The insect cuticle may also make it very difficult to invade the sea, since penetrating the surface tension film is hard; amphibious freshwater insects largely make the transition by dragging themselves down stout vegetation stalks, but this is particularly difficult in disturbed waters such as the littoral and estuarine fringes of the sea. Indeed the prevalence of insect interactions with angiosperm plants may be important in a broader sense too: angiosperms are absent from open oceans, and insects therefore cannot benefit from their many coevolutionary links to such plants. It has also been argued that while insects are very good at osmoregulation they are in many ways poor competitors. The oceans already had a vast diversity of crustaceans, and these were endowed with most of the advantages of insects, with none of the potential disadvantages, and perhaps with a particular advantage of their own in the form of numerous complex cirri-bearing legs that could make excellent filter-feeding apparatus. These crustaceans would compete in the same size range for the same niches and resources, so it is perhaps not surprising that insects have not made the transition back to the seas.

11.11 Conclusions

Most phyla of animals live in the sea and have done so throughout their history. Indeed, most of the evolutionary history of the planet has been marine, and most of the really important evolutionary innovations in design occurred there. At the cellular level, physiological systems also evolved against a marine background, with fairly constant conditions of salinity, temperature, and gas availability. Because of this history marine environments do not impose a great physiological challenge to most components of an animal's living machinery. It is primarily those animals that have invaded the seas secondarily that have real difficulties with osmotic and ionic control, with respiration, and sometimes with their thermal biology, and that therefore require complex regulatory systems at the tissue and organ levels.

With an understanding of the essentially marine legacy of structure and functioning that all living cells have inherited, it should be easier to follow the progressively more difficult physiological challenges, and the appropriate adaptive changes, encountered in more dilute, more desiccating, and above all more variable environments, discussed in the following chapters.

FURTHER READING

Books

Atema, J., Fay, R.R., Popper, A.N. & Tavolga, W.N. (1988) *Sensory Biology of Aquatic Animals*. Springer-Verlag, New York.

Clark, R.B. (1992) *Marine Pollution*. Oxford University Press, Oxford.

Gage, J.D. & Tyler, P.A. (1991) *Deep-Sea Biology: a Natural History of Organisms at the Deep-Sea Floor*. Cambridge University Press, Cambridge, UK.

Herring, P.J., Campbell, A.K., Whitfield, M. & Maddock, L. (1990) *Light and Life in the Sea*. Cambridge University Press, Cambridge, UK.

Hochachka, P.W. & Somero, G.N. (2002) The diving response and its evolution. In: *Biochemical Adaptation: Mechanism and Process in Physiological Evolution*, pp. 158–185. Oxford University Press, Oxford.

Jobling, M. (1995) *Environmental Biology of Fishes*. Chapman & Hall, London.

Kooyman, G.L. (1989) *Diverse Divers: Physiology and Behaviour*. Springer-Verlag, Berlin.

Laverack, M.S. (1985) *Physiological Adaptations of Marine Animals*. Society for Experimental Biology, Cambridge, UK.

Rankin, J.C. & Jensen, F.B. (eds) (1993) *Fish Ecophysiology*. Chapman & Hall, London.

Vogel, S. (1994) *Life in Moving Fluids*. Princeton University Press, Princeton, NJ.

Reviews and scientific papers

Block, B.A. (1986) Structure of the brain and eye heater tissue in marlins, sailfish and spearfishes. *Journal of Morphology* **190**, 169–189.

Boyd, I.L. (1997) The behavioural and physiological ecology of diving. *Trends in Ecology & Evolution* **12,** 213–217.

Childress, J.J. (1995) Are there physiological and biochemical adaptations of metabolism in deep-sea animals? *Trends in Ecology & Evolution* **10**, 30–36.

Childress, J.J. & Seibel, B.A. (1998) Life at stable low oxygen levels: adaptations of animals to oceanic oxygen minimum layers. *Journal of Experimental Biology* **201**, 1223–1232.

Croll, D.A., Nishigushi, M.K. & Kaupp, S. (1993) Pressure and lactate dehydrogenase function in diving mammals and birds. *Physiological Zoology* **65**, 1022–1027.

Crossin, G.T., Al-Ayoub, S.A., Jury, S.H., Howell, W.H. & Watson, W.H. (1998) Behavioural thermoregulation in the American lobster, *Homarus americanus*. *Journal of Experimental Biology* **201**, 365–374.

Davenport, J. (1997) Temperature and the life history strategies of sea turtles. *Journal of Thermal Biology* **22**, 479–488.

DeVries, A.L. (1982) Biological antifreeze agents in coldwater fishes. *Comparative Biochemistry & Physiology A* **73**, 627–640.

Dietz, T.J. & Somero, G.N. (1993) Species- and tissue-specific synthesis patterns for heat shock proteins HSP70 and HSP90 in several marine teleost fish. *Physiological Zoology* **66**, 863–880.

Dusenbery, D.B. & Snell, T.W. (1995) A critical body size for use of pheromones in mate location. *Journal of Chemical Ecology* **21**, 427–438.

Fedak, M.A. & Thompson, D. (1993) Behavioural and physiological options in diving seals. *Symposium of the Zoological Society of London* **66**, 333–348.

Fish, F.E. (1996) Transitions from drag-based to lift-based propulsion in mammalian swimming. *American Zoologist* **36**, 628–641.

Fujiwara, T., Tsukahara, J., Hashimoto, J. & Fujikura, K. (1998) In situ spawning of a deep sea vesicomyid clam: evidence for an environmental cue. *Deep Sea Research* **45**, 1881–1889.

Gillett, M.B., Suko, J.R., Santoso, F.O. & Yancey, P.H. (1997) Elevated levels of trimethylamine oxide in deep sea gadiform teleosts: a high pressure adaptation? *Journal of Experimental Zoology* **279**, 386–391.

Goldman, K.J. (1997) Regulation of body temperature in the white shark, *Carcharodon carcharias*. *Journal of Comparative Physiology* **167**, 423–429.

Handrich, Y., Bevan, R.M., Charassin, J.-B. *et al.* (1997) Hypothermia in foraging king penguins. *Nature* **388**, 64–67.

Heyning, J.E. & Mead, J.G. (1997) Thermoregulation in the mouths of feeding gray whales. *Science* **278**, 1138–1139.

Johnston, I.A. (1990) Cold adaptation in marine organisms. *Philosophical Transactions of the Royal Society B* **326**, 655–667.

Kirschner, L.B. (1993) The energetics of osmotic regulation in ureotelic and hyposmotic fishes. *Journal of Experimental Zoology* **267**, 19–26.

LaBarbera, M. (1984) Feeding currents and particle capture mechanisms in suspension feeding animals. *American Zoologist* **24**, 71–84.

MacDonald, A.G. (1997) Hydrostatic pressure as an environmental factor in life processes. *Comparative Biochemistry & Physiology A* **116**, 291–297.

Maddrell, S.H.P. (1998) Why are there no insects in the open sea? *Journal of Experimental Biology* **201**, 2461–2464.

Newton, C. & Potts, W.T.W. (1993) Ionic regulation and buoyancy mechanism in a pelagic deep sea crustacean. *Journal of the Marine Biological Association of the UK* **73**, 15–23.

Pelster, B. & Scheid, P. (1992) Countercurrent concentration and gas secretion in the fish swim bladder. *Physiological Zoology* **65**, 1–16.

Rasmussen, A.D. & Anderson, O. (1996) Apparent water permeability as a physiological parameter in crustaceans. *Journal of Experimental Biology* **199**, 2555–2564.

Sanders, N.K. & Childress, J.J. (1988) Ion replacement as a buoyancy mechanism in a pelagic deep sea crustacean. *Journal of Experimental Biology* **138**, 333–343.

Sebert, P., Simon, B. & Pequeux, A. (1997) Effects of hydrostatic pressure on energy metabolism and osmoregulation in crab and fish. *Comparative Biochemistry & Physiology A* **116**, 281–290.

Siebenaller, J.F. & Garrett, D.J. (2002) The effects of the deep sea environment on transmembrane signalling. *Comparative Biochemistry & Physiology B* **131**, 675–694.

Siebenaller, J.F. & Somero, G.N. (1989) Biochemical adaptation to the deep sea. *Aquatic Science* **1**, 1–25.

Snyder, G.K. (1983) Respiratory adaptations in diving mammals. *Respiratory Physiology* **54**, 269–294.

Somero, G.N. (1990) Life at low volume change: hydrostatic pressure as a selective factor in the aquatic environment. *American Zoologist* **30**, 123–135.

Somero, G.N. (1992) Adaptation to high hydrostatic pressure. *Annual Review of Physiology* **54**, 557–577.

Suchanek, T.H. (1993) Oil impacts on marine invertebrate populations and communities. *American Zoologist* **33**, 510–523.

Toulmond, A. (1985) Circulating respiratory pigments in marine animals. *Society of Experimental Biology Symposia* **39**, 163–206.

Van der Hage, J.C.H. (1996) Why are there no insects and so few higher plants in the sea? New thoughts on an old problem. *Functional Ecology* **10**, 546–547.

Wilkie, M.P. (1997) Mechanisms of ammonia excretion across fish gills. *Comparative Biochemistry & Physiology A* **118**, 39–50.

Withers, P.C., Morrison, G. & Guppy, M. (1994) Buoyancy role of urea and TMAO in an elasmobranch fish, the Port Jackson shark. *Physiological Zoology* **67**, 693–705.

Wright, S.H. & Manahan, D.T. (1989) Integumental nutrient uptake by aquatic organisms. *Annual Review of Physiology* **51**, 585–600.

Zimmerfaust, R.K., Finelli, C.M., Pentcheff, N.D. & Wethey, D.S. (1995) Odor plumes and animal navigation in turbulent water flow: a field study. *Biological Bulletin* **188**, 111–116.

12 Shorelines and Estuaries

12.1 Introduction: brackish habitats and biota

Life in the sea is, as we have seen, a relatively easy affair for most animals. But problems start increasing as organisms move towards the edges of the sea, for both benthic and pelagic fauna. There are particular difficulties (and advantages) for life at the shifting interfaces of water, land, and air; animals here are in a cyclically changing habitat, with new challenges but also with new kinds of strategies opening up to them.

Seashores and estuaries therefore constitute the subject matter for this chapter. Shores occur where the sea meets up with the land–air interface, and estuaries are where the sea meets a freshwater interface as rivers flow from the land masses. These habitats collectively involve what are termed **brackish waters**. They share certain key characters:

1 They are not very common, in terms of surface or volume.
2 They are not at all easy to live in because cyclical variation is a continuous problem.
3 They are variable over quite short periods of evolutionary time, as the interface between media advances and recedes due to erosion, deposition, or changes in sea level.

These habitats are nevertheless zones of quite high diversity and of careful patterning among the biota. This arises from the multiplicity of problems of life in these habitats (Fig. 12.1). Most obviously, a shore is not just at the interface of air, land, and water, but has complex spatial and temporal variations in the balance of these three media, with environmental gradients much in evidence. The rise and fall of water level due to tides brings problems particularly affecting breathing patterns, osmotic balance, temperature balance, and predation pressure. Wave action may lead to crushing or dislodgment, and may increase water turbidity with sand and mud becoming suspended in the water column. Freshwater inputs lead to large variations in water and ion balance. Coastal boundaries also exert strong effects on the air and water movements of the planet as a whole, and oceanic air masses often move laterally along shores rather than inland over rough or steep terrain. Heat exchange between air and land is also complex and may affect the movements of maritime air masses. Thus shores have their own "weather", and

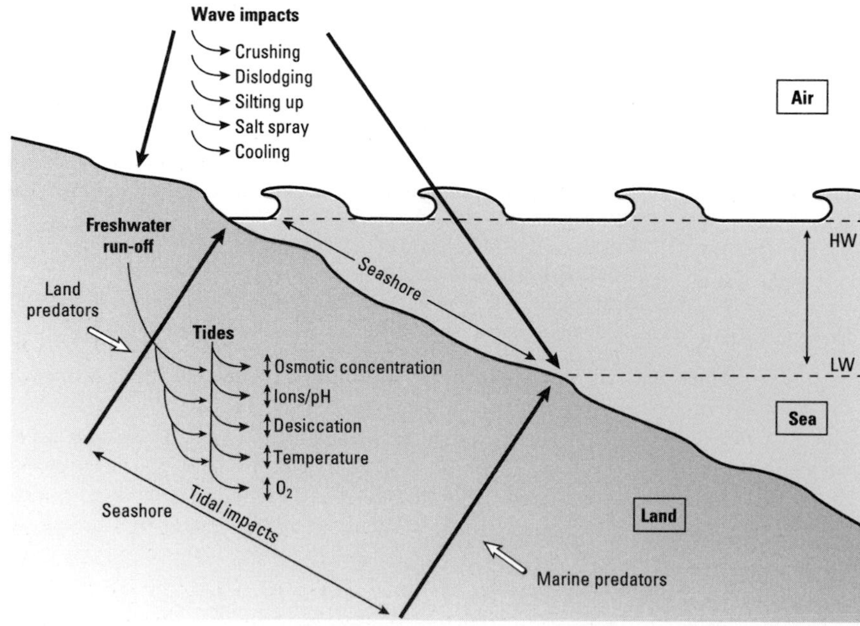

Fig. 12.1 A summary of the problems experienced by animals living on seashores. HW, high water; LW, low water.

Table 12.1 Productivity of brackish ecosystems (in comparison with marine and freshwater systems).

Ecosystem	Productivity (g C m^{-2} $year^{-1}$)
Open ocean water	5–50 (cf. Fig. 11.2)
Coastal waters (temperate)	
Shallow continental shelf	30–150
Zones of upwelling	50–220
Coastal bays	50–120
Subtidal	
Seaweeds	800–1500
Coral reefs	1700–2500
Sea grass	120–350
Intertidal	
Sandy beaches	10–30
Rocky beach seaweeds	100–250
Estuarine mudflats	500–750
Supratidal	
Salt marsh	700–1300
Mangrove swamp	350–1200
Sand dunes	150–400
Eutrophic freshwater lake	400–600

storm fronts often track along them (e.g. in the English Channel or the Gulf of St Lawrence) rather than over land. Oceanic water masses impinging on land also have effects on water mixing, often with downwelling close in and upwelling further offshore, which may help to promote and localize inshore productivity.

But there are also some key advantages of life on the seashore. Above all these relate to food supply. The sea water is penetrated by light and there is a substratum for marine plants to attach to, so that a wide range of multicellular algae (seaweeds) can thrive and produce a strong base to the food web. These seaweeds also stabilize the substratum in shallow waters, allowing invasion by some specialist angiosperms, notably the sea grass *Zostera*. Coastal benthic habitats are therefore zones of high primary productivity (Table 12.1). With abundant algae to eat, most herbivorous marine animal taxa do extremely well either just offshore or on the shore itself. There will also be plenty of suspended matter, primarily marine plankton but also larger organic flotsam, which can be filtered or gathered by sessile animals and which is continually renewed by tide and wave action. The herbivores and filterers in turn provide opportunities for predators, both marine animals (especially crustaceans and fish) coming in to feed, and terrestrial animals (especially birds) combing the shore at low tide. Finally, with damage and death resulting from waves, dislodgment, and other hazards, there is a ready supply of food for a host of scavengers, either on the surface or amongst the sandy or muddy substrata. The abundance of resources leads to periods of rapid depletion and strong interspecific competition for food, so that population densities of many animals are cyclical.

All the organisms on a seashore are faced with the same problems and are exploiting the same resources within a narrow strand. Inevitably, therefore, there is a great deal of competition for space as well as for food. This is a major reason why there is also very precise patterning on most seashores. They provide the best possible example of where severe physiological constraints and severe ecological competition come together to give observable and definable structure to a biological community.

Box 12.1 Classification of coastlines

Geological tectonic categories

1 Collision coasts (leading-edge coasts, active coasts): abrupt and steep coasts, eroding cliffs, narrow continental shelves.
- (a) Continental collision coasts, e.g. Pacific coasts of North and South America.
- (b) Island arc coasts, e.g. Philippine islands and Caribbean islands.

2 Trailing-edge coasts (passive coasts): flat shores, with lagoons and wide continental shelves.
- (a) Recent rifting coasts, e.g. Red Sea.
- (b) Spreading basin coasts, e.g. southeast Africa, eastern USA.
- (c) Basins with one trailing and one leading edge, e.g. North Pacific.

3 Marginal coasts.
- (a) Back-arc basins, e.g. China Sea.

Geological/biological categories

1 Mountain coast.

2 Narrow shelf coast.
- (a) Headlands and bays.
- (b) Coastal plains.

3 Wide shelf coast.
- (a) Headlands and bays.
- (b) Coastal plains.

4 Deltaic coast.

5 Reef coast.

6 Glaciated coast.

12.1.1 Littoral habitats

Seashores are in many ways extremely diverse in character. Coastlines may be classified very broadly according to their geological nature (Box 12.1: geological tectonic categories), imparting a fundamental character to long stretches of shore. Within this, matters of scale impose a more varied outcome and a classification into six main categories (Box 12.1: geological/biological categories) is more useful. Worldwide distributions of these major categories are shown in Fig. 12.2. At a local level coasts may include not only the traditional sandy, muddy, and rocky shores that surround all continents and islands, but also sea cliffs and sand dunes where wave splash may impinge on more terrestrial zones (see Plate 2a, between pp. 386 and 387). In some respects even the coral reefs (see Plate 1d) are "shores", where an essentially marine zone providing dwelling space on living coral is periodically exposed around the littoral margins of tropical land masses (see Fig. 11.3 for reef zones).

However, all these littoral fringes also have some key similarities, making the littoral zone as a whole a particularly fascinating place to study environmental adaptation. First of all, and most obviously, all three of the major habitat zones—air, land, and water—come together, each imposing its own stresses. Secondly, the balance between the three habitats changes cyclically and with several interacting rhythms relating to the sun and moon, giving daily tides and

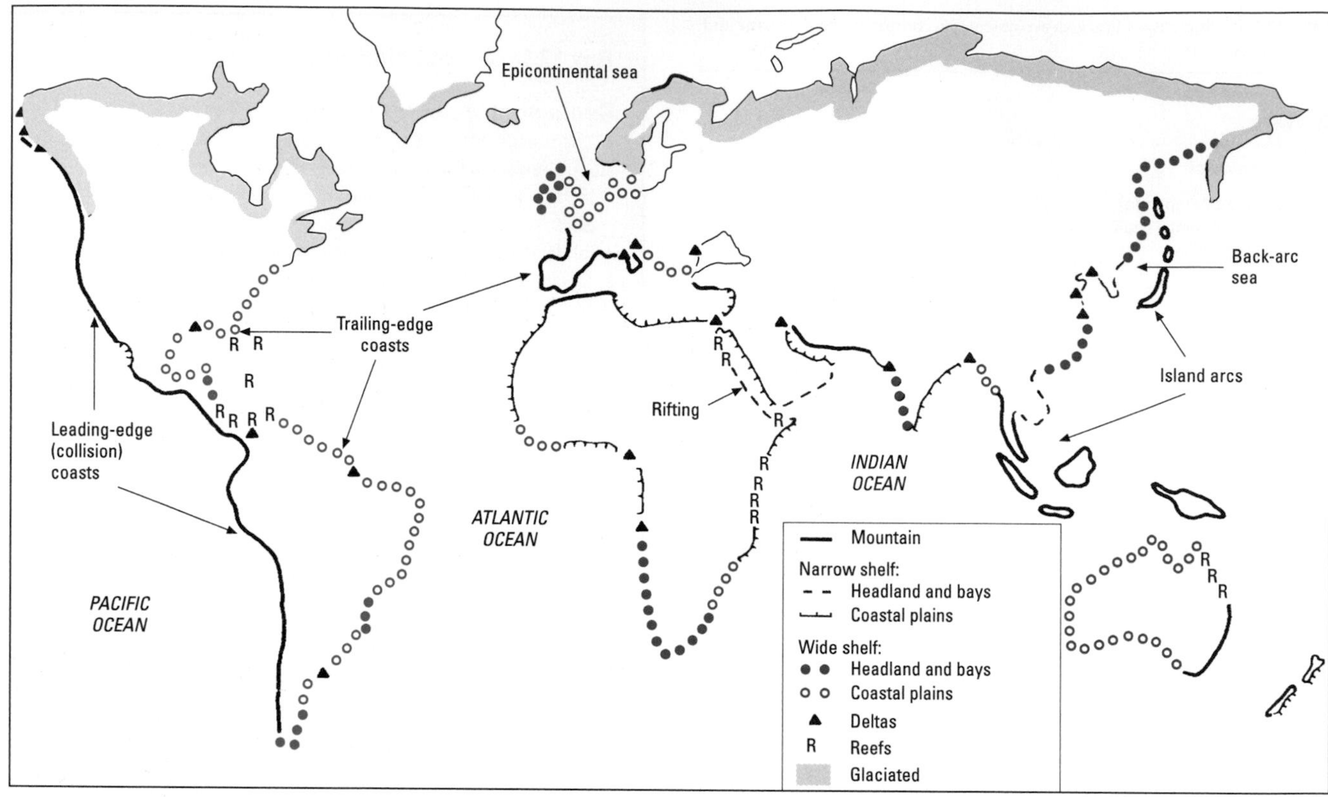

Fig. 12.2 Major types of coastlines around the world's continents. (From Carter 1988.)

predictable variations in emersion (exposure to the air) and immersion (return of a water covering) for the biota. Thirdly, and as a consequence of this, the organisms on a seashore are all laid out in a neat two-dimensional array, with distinct "layers" of animals and plants, and with the precise nature of the layering varying according to the characteristics of the particular shore.

Tides

Tides are primarily the results of astronomical phenomena, with the gravitational forces of the moon and sun pulling on the Earth's surface water. Since the moon is much closer to us, it has a much greater effect, and in effect it pulls an ellipse of water around the Earth as it moves on its own rotational path, creating a "bulge" of water pointing towards wherever the moon is, and a reciprocal bulge on the other side of the Earth as the planet's rotation adds a centrifugal effect. This means that each site on the planetary surface gets a surge of water—a high tide—every half lunar day, where the lunar period is 24 h and 50 min, thus cycling through low–high–low every 12 h and 25 min. Hence shores normally experience two high tides a day, each delayed by a fixed period of around 25 min on each successive day.

The effect of the sun modifies this basic pattern by superimposing a lesser gravitational pull with a different periodicity. At new and full moons, the sun and moon are in line and the overall pull is therefore a bit greater, so that we get particularly extreme high and low tides known as **spring tides**. At half moons the sun and moon are working against each other so the tidal ranges are reduced, producing **neap tides**. Thus springs and neaps occur twice in each lunar cycle, or roughly twice a month (Fig. 12.3). Spring-tide low waters are particularly good times for ecological work on beaches, as a long stretch of the shore will be exposed and low-tide animals will be accessible.

A third periodicity affects tides, since they also vary with the declination of the sun, which runs on a yearly cycle. Declination is greatest at the equinoxes, so the most extreme spring tides always come in March and September.

Thus far we have only been generalizing on a global scale about tidal patterns. The actual degree to which tides rise and fall on any shore is highly variable and must be computed for each location. Likewise the precise time and daily progression of high and low tide varies with the constraints of geography. Tides are greatly modified by the shapes of the land masses around which the seas are pulled by gravitational forces, by the shape of local coastlines, and by the effects of prevailing winds. For example, average tide ranges for Great Britain are about 3.5 m (neaps) and 4.7 m (springs), but the east coast has smaller tides as tidal motion is constrained by the relatively small basin of the North Sea, while in enclosed seas, such as the Mediterranean, there is effectively no tide at all. On a smaller scale, long, thin inlets pointing upstream from prevailing winds can have massive tidal ranges, often experiencing a "tidal bore" when the fast incoming tide is increasingly channeled into a narrowing profile; the Bristol Channel in England is famous for a massive tidal range (12–13 m) and periodic large bores. Tides are also modified from day to day by weather, particularly by strong winds and the after-effects of storms out at sea.

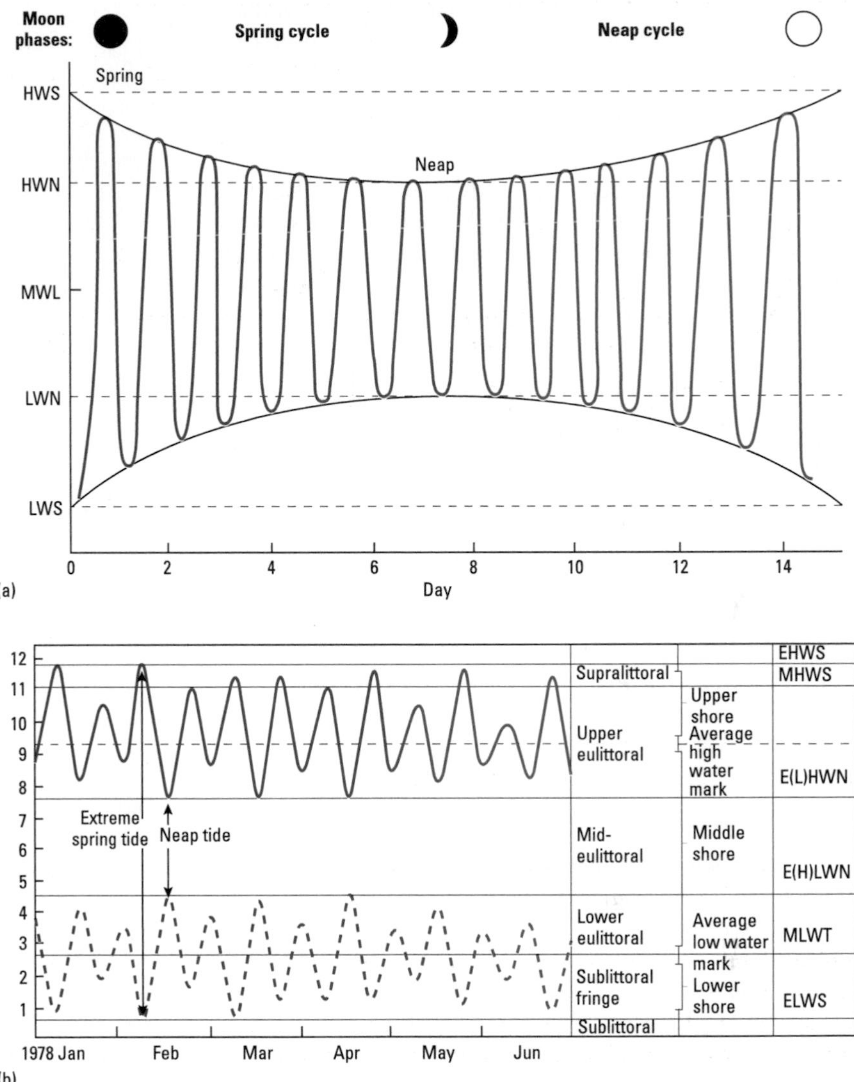

Fig. 12.3 Basic patterns of tidal cycles (a) with twice-daily and twice-monthly cycling related to the phase of the moon, and (b) records over 6 months and the relation of tide to terminology used on beaches. (See Abbreviations, p. xi, for definition of terms.)

Waves

Waves form whenever the water surface is disturbed, whether by winds, earth tremors, or underlying planetary rotation. Most wave action is due to the effects of wind, and it is therefore restricted to surface water, having little effect on most marine organisms. It is also variable seasonally in mid-latitudes, with wave effects being much greater in winter when storms occur. In the littoral zone, and especially on beaches, waves are often the most important component affecting the interplay of marine and terrestrial conditions and impacting upon shoreline organisms.

Waves appear to move smoothly for long distances across the ocean, often for hundreds of miles (but remember that the water mass itself is *not* moving laterally, it is only the wave form that moves). In light winds the waves are seen only as slight ripples, but in very strong winds waves may reach over 50 m in height and 1000 m in length. When they reach shallow water, the waves become even higher and shorter and eventually unstable, "breaking" onto the shore itself either by a gradual "spilling" or by more dramatic "plunging" especially on steeper beaches. Wave height is often critical to shore biology, affecting the energy that has to be absorbed by the shore. Where wave action is strong the effects of saltwater splash are also considerable and may occur for quite some distance inland. Wave height depends on three main factors: the wind velocity, the wind duration, and the "fetch" of open water seawards of a particular coastline (effectively the distance over which the waves have traveled uninterrupted before reaching the land). On west coasts of Britain (see Plate 2a, between pp. 386 and 387) the fetch can be the whole of the Atlantic, and for the surf beaches of California it is most of the Pacific. But in the English Channel (see Plate 2b) it may be only a few miles.

Waves are largely responsible for the shape and the character of beaches. As they hit a coastline they undergo complex processes of reflection and refraction, so that their direction is reorientated. Although resultant water movements are extremely complex, and interact with local currents, a net longshore drift is commonly created and material is normally moved laterally. The waves therefore cause continuous erosion and deposition, leading to all the characteristic geological and geographic formations of coastlines (cliffs and caves, blowholes, stacks and coves, wave-cut platforms, dunes, spits, etc.).

Sediments

Coastal sediments, whether rocky, sandy or mixed (see Plate 2c), represent a relatively short-term "staging post", where material accumulates briefly in transit from the mountain tops to ocean benthos. Some sediments may persist for thousands or even millions of years as part of the coastline, while others last for only a few seconds or minutes before being moved on by currents, tides, storms, or waves.

The coastal sediments may be divided into primary and secondary sources. Primary sediment comes from cliffs, platforms, or offshore submerged rock outcrops. Where gravel and sand predominate, much of the material comes from the erosion of relatively loose glacial deposits. Additional primary material is from biogenic deposits, i.e. from the shelly remains of marine organisms. Secondary sediment arrives from rivers or glaciers and as wind-blown matter. Here the sediment is usually "sorted" in some fashion (commonly by particle size or density) before arriving at the coast.

Sediment type has an enormous influence on the biota of shorelines, so we will often deal with the two major types (rocky and sandy/muddy) separately in this chapter.

Patterns of biota on rocky shores

Vertical **zonation**, where dominant species occur in distinct horizontal bands, can be readily observed on shores, often from a long way away (see Plate 2b), and it is surprisingly similar in its basic elements for all shores worldwide. It is often described mainly in terms of upper and lower limits for particular organisms (Fig. 12.4), with an upper "black lichen zone", then a "periwinkle zone", then a "barnacle zone", and lowest of all a fourth band where the dominant species are more variable. The width and number of distinct zones, and the diversity within these zones, then depends on local effects: geography, climate, magnitude of the tides, the timing of low and high tides, the substratum, and above all **exposure**. This is a useful

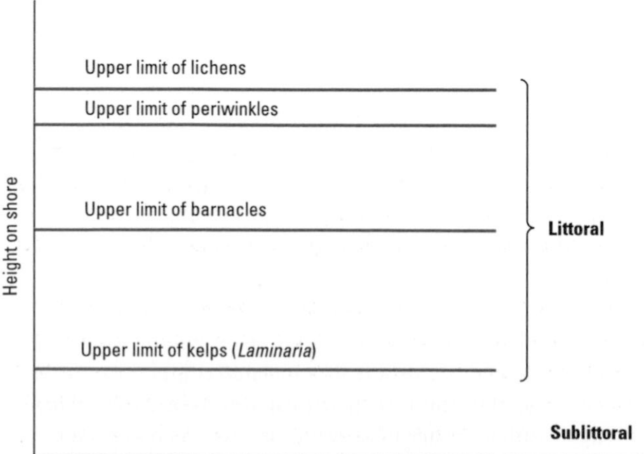

Fig. 12.4 Fundamental patterns of zonation described by the limits of key types of biota.

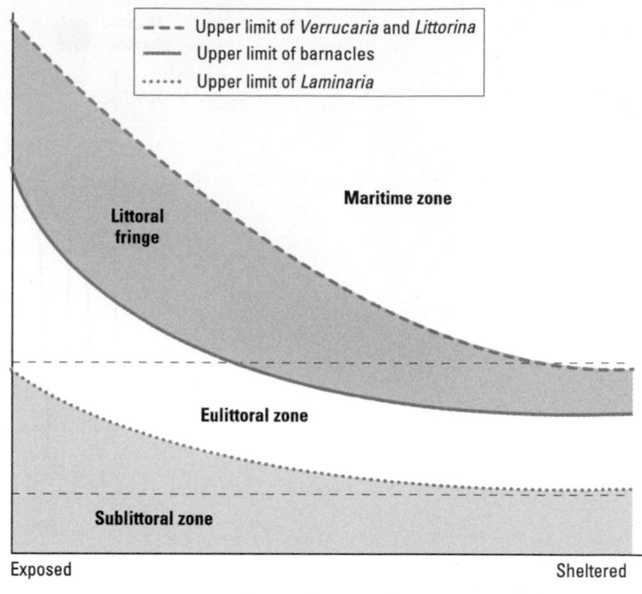

Fig. 12.5 A more realistic description of beach zonation where the degree of wave action modifies the height of the beach over which the patterns in the biota are spread: highly compressed on a sheltered beach, and greatly spread (often well up the cliff face) on an exposed shore.

composite term involving beach aspect, fetch, wave force, and winds as influenced by all the other astronomical and geographic factors; it is the most widely used term when comparing shores and their fauna and flora.

An exposed shore has a long fetch and heavy wave action, with wave splash occurring to a considerable height and with all the zones fairly broad and spread out. By contrast a sheltered shore, such as occurs in an enclosed bay with reduced wind action, gentle waves, and a short fetch, may have only limited wave splash upwards, and the zones are compressed together in terms of height. So a simple representation has to include the element of exposure, related to degree of wave action, as in Fig. 12.5; high-exposure shores are those on the left, with the indicator species' zones all wide, while low-exposure sheltered shores are on the right, their zones compressed together. On a very exposed shore the lichens and the small winkles (gastropod snails in the genus *Littorina*) may extend a long way up cliffs behind the seashore. This figure also shows some of the terminology used in studying seashores: the maritime zone, the littoral fringe, the eulittoral "shore", and the sublittoral zone.

In terms of actual species composition, we can identify particular zones quite clearly wherever shores of relatively rocky substratum occur. There is an upper zone of gray-green lichens and maritime dicotyledonous plants and grasses, seen at the tops of cliffs worldwide. Below this is the black lichen zone, where the animals include isopods such as *Ligia*, collembolan springtails such as *Anurida*, and a range of small to medium-sized periwinkles (gastropod molluscs) that can fit into cracks and crevices. Further down the winkle species become somewhat larger, more diverse and more dominant, and merge into a zone dominated by barnacles (sessile crustaceans), usually with a high percentage covering of one or a few species of barnacle (*Balanus*, *Semibalanus*, *Chthamalus*). Seaweeds include

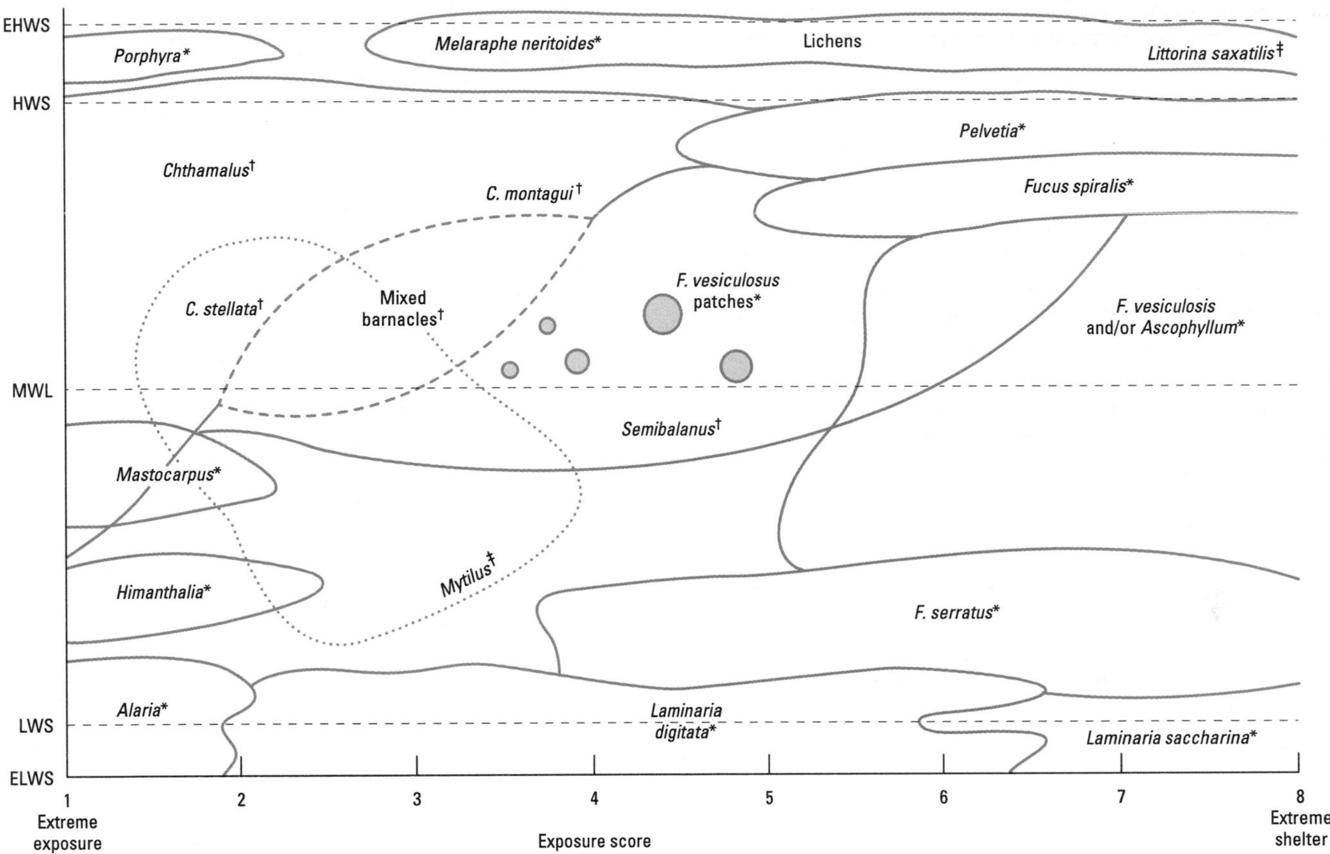

Fig. 12.6 A biotic analysis of zonation in terms of key genera and species encountered at different heights in relation to exposure (using a scale of 1–8 from exposed to sheltered). *, Seaweeds; †, barnacles; ‡, molluscs. (See Abbreviations, p. xi, for definition of terms.) (Adapted from Ballantine 1961.)

the tougher fucoid species. Amongst the weeds and barnacles are gastropod limpets (*Patella*) and whelks (*Nucella*), some sea anemones (*Actinia*) and often very dense colonies of mussels (*Mytilus* or *Modiolus*). The common brown seaweeds, such as bladderwrack, abound. Finally we meet the red algal and kelp zone, with dense beds of long strappy seaweeds such as *Laminaria*. Here animals are often very diverse, including crabs, top-shells and cone-shells, anemones, tunicates, sponges, echinoderms, bryozoans, and polychaete annelids—note that the list includes representatives of many of the soft-bodied phyla of animals. These will be particularly abundant and diverse on more sheltered shores with a long shallow slope and plenty of rocky crevices (see Plate 2d).

It is thus possible to build up a more complex view of a zonation scheme for a typical rocky shore (Fig. 12.6). A scheme such as this can be used to assess exposure on shores and to compare them with all other shores worldwide. The patterns of occurrence therefore give us a way of *quantifying* exposure, usually given on a scale of 1–8. In other words, if we get information on species' abundances and their heights on the shore, we can easily describe that shore and can make directly useful comparisons between shores. The exact species may differ on shores in Europe, Australia, or in North America, but the principles will be the same. The major exceptions are rocky shores in Antarctica where ice-scouring almost entirely eliminates intertidal macrofauna.

Patterning of biota on sandy shores

Sandy shores (here taken broadly to include muddy beaches and shingle shores) have very different characteristics from rocky shores. Sand is predominantly silica (SiO_2), together with $CaCO_3$, both deriving from rock and from shelled animals, and in varying proportions. Beaches in areas where deposition is predominantly from coralline material are dominated by calcium carbonate and tend to be very white (see Plate 2e, between pp. 386 and 387), while those on volcanic islands (see Plate 2f) may be of very dark sand with rock erosion products dominating. There is also usually some complex chemical clay material present, and a small component of organic matter. Sands, of course, also contain water, because the "packing factor" between roughly spherical grains means that there is normally a minimum volume of fluid (calculated as 26% for perfect spheres) that is hard to remove.

The main variable for sands is the **grain size**, which in turn affects a number of other features. Drainage is the most important of these, being the outcome of gravity working against capillarity, so that fine grains are more water-retentive when wetted. Oxygenation is also affected, with well-drained large-grain sands being more highly oxygenated. Small-grain fine sand may retain considerable volumes of water for long periods and so readily become anoxic. This condition merges into even finer grained muds, which can get

Table 12.2 Wentworth classification of surface deposits.

Grade	Description	Particle size (mm)
1	Boulder	>256
2	Cobble	64–256
3	Pebble	4–64
4	Granule	2–4
5	Very coarse sand	1–2
6	Coarse sand	0.5–1
7	Medium sand	0.25–0.5
8	Fine sand	0.125–0.25
9	Very fine sand	0.062–0.125
10	Silt	0.031–0.062
11	Clay/mud	<0.031

unpleasantly smelly as anaerobic decomposition products build up. Grain size is a parameter commonly measured and frequently quoted for beaches. One of the schemes most widely used is Wentworth's classification, shown in Table 12.2, listing types from boulders (see Plate 3a, between pp. 386 and 387) to fine clay and mud (see Plate 3b). Grain size correlates rather well with the exposure factor discussed above. In exposed sites the fine-grained material gets removed, leaving rather coarse or gritty sands or shingles ("clastic" beaches). The smaller grains are deposited again in areas of gentle wave action, producing shallow sheltered beaches of soft sands, often with some intrinsic patterns of low ridges, cusps, and bars resulting from waves interacting with local topography. Grain size therefore directly affects animal distribution in sandy shores, as shown for crab species in Fig. 12.7a.

Another factor differing between beaches is that of **sorting**. Some beaches have a very uniform particle size and are said to be well sorted, while others are highly variable in different areas, with a poor sorting factor. Thus two beaches can clearly have the same mean grain size but very different sorting factors, in turn affecting their drainage and oxygenation. Note also that the presence of animals in itself can affect sorting; the processes involved are **bioturbation**, **biodeposition**, and **biosecretion**. These are brought about by burrowing and feeding animals that pass sediments through their guts, such as polychaete worms, thalassid shrimps, sea cucumbers, and many bivalves. These animals may move finer egested particles up to the surface, giving a layer of fine material on top of rather coarser sands; but where the sand particles are bound together with fecal pellets the effective grain size may increase. Such processing of the sediments often shows distinct seasonal patterns (Fig. 12.7b). For the lugworm *Arenicola*, high summer rates of bioturbation are associated with the lowest sediment shear strength, the surface mud having a higher water content and low bulk density. The presence of tube structures made by animals can also have a marked effect on

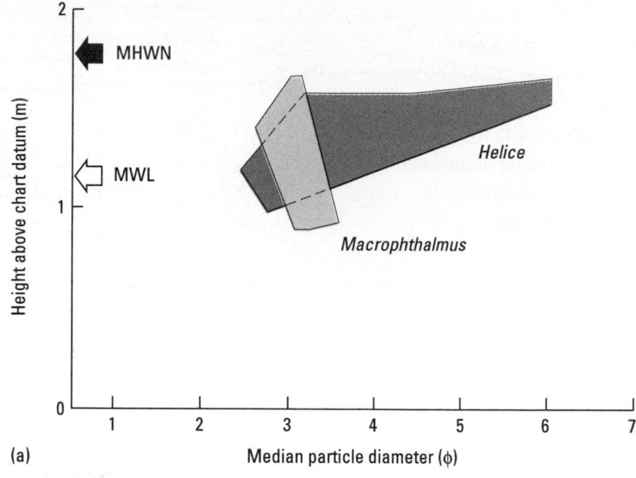

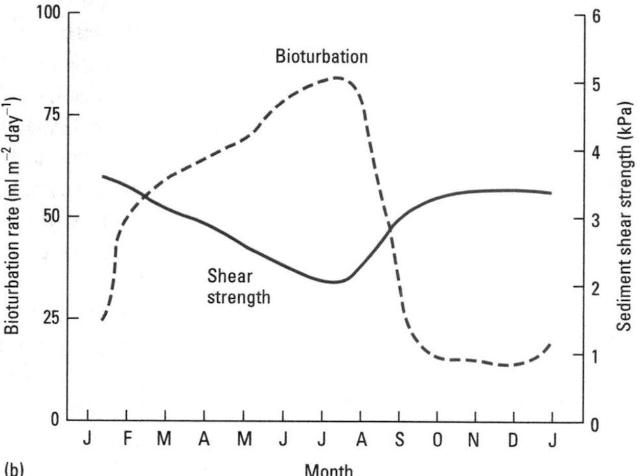

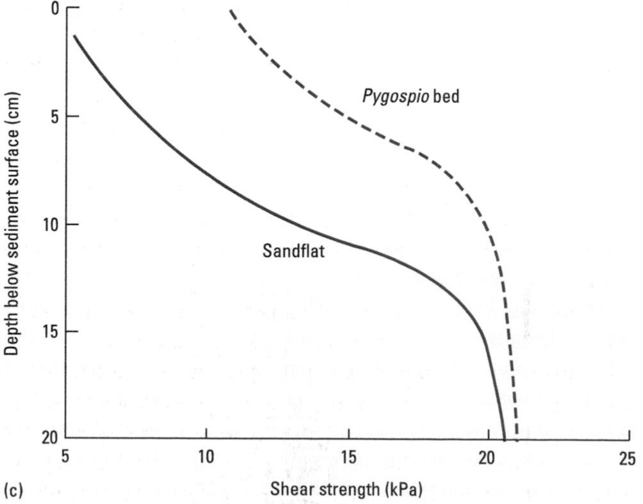

Fig. 12.7 (*right*) Interactions of animals and particle size on beaches. (a) Effects on the distribution of crabs; *Helice* is found higher on the beach and in finer sediments, while *Macrophthalmus* occurs lower down and in a more restricted habitat of rather coarse sand grains (particle diameter is given as $\log_2$ of the particles in millimeters). (b) Seasonal bioturbation changes brought about by burrowing lugworms (*Arenicola marina*), mirrored by changes in the sediment shear strength. (c) Increased shear strength due to sediment accumulation between tubes of *Pygospio* worms in a sandflat. MHWN, median high water (neap); MWL, median low water. (a, From Little 1990, courtesy of Cambridge University Press; b and c, from Lockwood *et al.* 1996.)

the stability of sediments. High tube densities can result in a reduced flow regime between the tubes, often leading to sediment accumulation. Figure 12.7c shows that high densities of the worm *Pygospio* (tens of thousands of tubes per square meter) on an estuarine sand flat can greatly increase the shear strength of the sediment, and this may over time raise the worm bed well above the surrounding sand.

The small-grain beaches provide the kind of material that animals can dig in, where they will be able to keep wet fairly reliably, so they can make a living without too much physiological stress. Foods will be available, because organic material and bacteria are usually associated with the surfaces of sand grains, and fine particulate sediments have relatively much greater surfaces available. Inevitably it is sheltered beaches that we are mainly concerned with here; the exposed sandy beaches do not have a great deal of organic material or metazoan life associated with them. But note that in sheltered beaches only the upper sandy part is oxygenated, and there is usually a sharp transition vertically to deoxygenated mud. This transition layer is termed the **redox potential discontinuity** (RPD), where the sand changes from an oxygenating to a reducing environment. This is often visible as the interface between sand and black mud and it may be at various depths. In fact its depth affects how much mixing of oxygen occurs across it, and therefore how sharp it is. The lower layer below the RPD is a highly deoxygenating (reducing) environment; ferric ions are converted to ferrous, nitrates turn to nitrites and ultimately ammonia, and sulfates end up as hydrogen sulfide, producing the characteristic "bad egg" smell. Some anaerobic fermentation by microbes also occurs, yielding alcohols and fatty acids. Animals and aerobic bacteria therefore occur only in the upper part of sheltered beaches and are always at some risk of anoxia. However, the activities of animals themselves affect the mixing a little, and in particular the tube-dwelling animals may penetrate into the anoxic mud by keeping a circulation of reasonably oxygenated fluid through their burrows or tubes; when excavated, the edges of such tubes often appear as a zone of altered coloration because the substratum is locally oxygenated.

In effect there are three main habitat options for animals in sandy shores: living amongst the sand grains, living on top of the sandy layer, or projecting downwards within tubes or burrows. For practical purposes the biota of sandy beaches is divided into a number of categories largely related to size and how the organisms can be collected, rather than to taxonomic categories:

1 **Microfauna** are organisms with dimensions less than about 0.1 mm, living amongst sand grains. The category includes protists and some very tiny invertebrates; these can only be collected by fine sieving of the sand.

2 **Meiofauna** are around 0.1–0.5 mm, and are also called the **interstitial fauna**, again living within the sand. Included here are some hydroids, small platyhelminths, a range of pseudocoelomates (especially nematodes), some of the curious rather simplified groups of annelid worms termed "archiannelids", and a few small crustaceans. These animals tend to be rather convergent in body form, all being vermiform (worm-like) with bristles or stubby "legs" to help push through the sand.

3 **Macrofauna**, which is divided into two types:

(a) **Infauna**, which are the visible large creatures living in burrows and tubes within the sediment, often leaving characteristic traces on the surface. Annelids and molluscs are most dominant, especially the filtering bivalve molluscs like cockles and clams (*Cardium*, *Mya*, *Mercenaria*, *Donax*), and polychaetes such as the lugworm *Arenicola* and the sand mason *Lanice*. The carnivore niche is filled by a few specialist polychaete worms, such as *Glycera*, a few nemertean worms, and boring gastropods such as *Natica*. There are also normally some echinoid and asteroid echinoderms (urchins and starfish), a number of other worm groups, and a profusion of small burrowing crabs, amphipods, and isopod crustaceans scuttling or hopping about amongst and over the sand grains. Towards the top of the beach more terrestrial amphipod sand-hoppers such as *Talitrus* may be dominant. At the lowest levels, especially where the beach merges into mud, more fragile filterers such as the peacock worms (*Sabella*) and burrowing shrimps (e.g. *Callianassa*), and rarer deposit feeders, such as peanut worms (sipunculans) and sea cucumbers (holothurian echinoderms), occur patchily.

(b) **Epifauna**, which are the visible surface dwellers, including everything from seagulls to bivalves to crabs, with some essentially marine animals in the shallow water (including stingrays and small sharks, various flatfish, and gobies) and some essentially terrestrial animals on the upper shore such as insects (beetles and flies in particular). There is actually rather little epifauna (or epiflora) on most sandy beaches.

12.1.2 Estuarine habitats

Estuaries are semienclosed water bodies linked to the sea within which the water is measurably diluted by freshwater inputs. The category of estuaries could be said to include all zones where we get sea water and fresh water coming together cyclically, such that as water depths go up and down (varying the interaction with the aerial environment) so too does the salinity vary. This would therefore include salt marshes, mudflats, coastal swamps, and mangrove forests, and the upper reaches of sea lochs and fjords. The three-way interaction of sea water, fresh water, and terrestrial habitats (dictated by tides and waves as above) means that life in estuarine conditions is particularly difficult. Plant diversity (in terms of number of species) is usually low. There are also far fewer animal species in such habitats than in either sea water or fresh water (Fig. 12.8); the critical salinity range is around 10–25% sea water, where few essentially marine animals survive and equally few freshwater species can remain active, in both cases mainly due to a breakdown of the normal cell volume regulatory systems (see below). Whether this is the primary reason for the low species number of brackish-water animals remains unclear. It may also partly be brought about by the transience of estuaries through evolutionary time, so that there is never a long enough period for a large number of species to establish themselves (except perhaps in the very largest estuaries, such as the Amazon). A combination of physiological stress and evolutionary ephemerality is likely to explain the paucity of estuarine species; it is certainly reflected in a rather similar fashion in every taxon, as shown in Table 12.3.

Habitat characteristics and biota

Estuaries are inevitably very variable in type and in terms of key characteristics such as salinity regime, depth and width, water flow

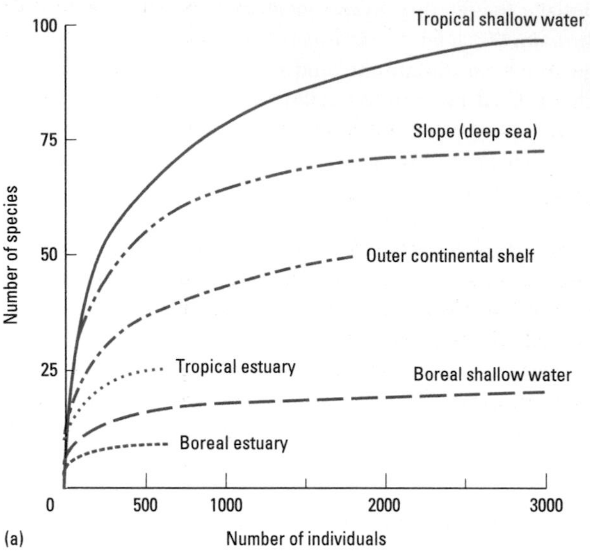

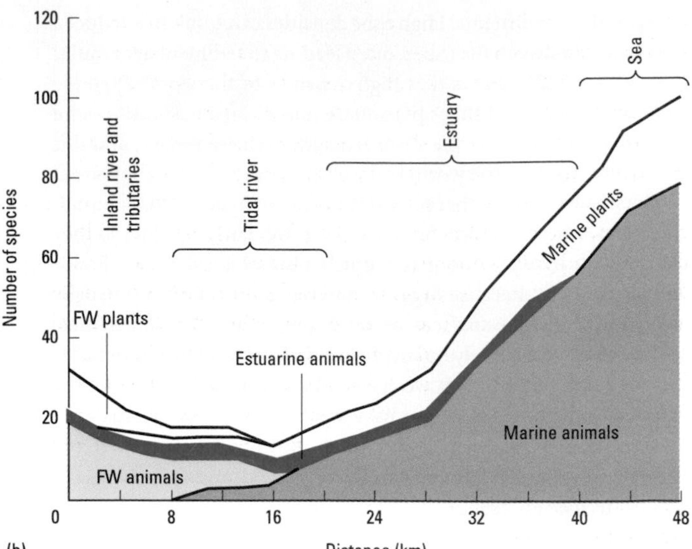

Fig. 12.8 The species diversity of estuaries. (a) Species number plotted arithmetically against number of individuals for different marine and estuarine habitats. (b) Number of species plotted against distance along a typical estuary showing the seaward decline in freshwater (FW) species and increase in marine species, with truly estuarine species always in a small minority. (a, Adapted from Sanders 1969; b, adapted from Alexander *et al.* 1935.)

patterns, and substratum. Most of the "true" estuaries result from drowned river valleys (rias; see Plate 3c, between pp. 386 and 387), and they are therefore also geologically impermanent and greatly altered in conformation by small changes in sea level. Their average lifespan may be less than 10,000 years, and many of those observed today resulted from marine flooding immediately after the last Ice Age. Others arise from the gradual occlusion of a river mouth by spits and bars (see Plate 3d).

But despite their ephemeral nature estuaries are biologically very rich, with a high biomass. They can act as substantial nutrient traps, with nutrient run-off from the land and nutrient cycling occurring just within the bay, being retained by flow characteristics from both ends. Thus they can be areas of very high primary productivity (cf. Table 12.1), especially in tropical latitudes, limited usually by nitrogen availability. This productivity is reflected in high faunal density, despite the low faunal diversity; some estuaries have become major sites of fisheries (notably the large Chesapeake Bay area and the huge Hudson Bay), supporting large populations of molluscan and crustacean "shellfish" as well as fish. The resident animals present tend to be specialists, particularly drawn from these three groups (molluscs, crustaceans, and fish such as mullet, flounder, and bass). There is also a great range of "amphibious" animals, dominated by wading birds, for whom coastal mudflats and other brackish wetlands are crucial stopovers during migrations, for winter feeding, or for summer breeding. Also in this category would be tetrapod vertebrates such as otters, living on the rocky shores of inlets and sea lochs. In the more freshwater parts of estuaries many of the crustaceans and annelids are replaced by insect larvae and adults, especially midges, caddis flies, and beetles, often able to survive a few hours of submersion in brackish water. Few animals are able to tolerate the full range of estuarine conditions; only those relatively rare fish and crustaceans that can regulate across the entire salinity range are able to migrate right through estuaries.

Mudflats and salt marshes (see Plate 3e) commonly form on the sheltered sides of estuaries in temperate zones, where wave energy is low and there is a surplus of fine sediment, and these areas may retain water through almost all the monthly and annual tide cycles. Mudflats may be bare of vegetation or show a patchy covering of sea grass. In the tidal saline creeks some green algae survive. But the tops of mudflats may get desiccated and very warm, thus there are often dense populations of filmy green algal species through the winter, which get burnt off in summer and replaced by tougher brown algae with denser growth forms. In these creeks, certain mussels, crabs, and gastropods thrive, and the habitat may be an important nursery ground for fish. Waters in the creeks may become unusually rich in sulfides and nitrites leaching out of the adjacent anaerobic mud, so that resident animals require tolerance or regulation of these compounds.

Mudflats usually occur in various successional stages that lead eventually to marsh and then solid ground. The original bare mud gradually forms into a more solid marsh as plants colonize and

Table 12.3 The numbers of species in various taxa of molluscs as salinity decreases from the North Sea into the Baltic Sea.

Region	Approximate salinity (ppt)	Scaphopods	Bivalves	Gastropods	Cephalopods
North Sea	35	11	189	351	32
Kattegat	26	1	92	162	14
Danish Belts	20	1	42	68	5
Outer Baltic	8	0	11	19	0
Gulf of Bothnia	4	0	4	4	0

begin to reduce the flow rates, their roots trapping and holding the finer sediments. Salt marshes are initially dominated by relatively few specialist salt-tolerant plants, notably *Spartina* and other cord-like grasses with extensive rhizomes, gripping into the mud and spreading asexually over vast areas. Succulent species, such as glass-worts, also invade and act as pioneer stabilizers. When the sediment surface rises above the normal neap high tide level, more terrestrial species such as reeds, rushes, sea blites, and sea lavenders may colonize. Mud and peat continues to accrete amongst the marsh plants, giving a steady rise in height, until eventually the marsh is normally above the high tides and it becomes an essentially terrestrial habitat, colonized by insects and birds. This whole process may take perhaps 150–350 years if the system is left undisturbed, again stressing the ephemeral character of estuarine habitats.

In tropical areas, salt marshes are usually replaced by mangrove swamps (see Plate 3f)—intertidal zones where emergent mangrove plants are rooted in an essentially marine anoxic sediment. There are just a few mangrove genera (e.g. red mangrove, *Rhizophoa*, with surface prop roots, and black mangrove, *Avicennia*, with protruding pneumatophore roots for aeration), with a succession of species from seaward to landward. Small offshore islands may be formed from the root mats of mangroves. Each mangrove swamp has its own associated submerged fauna, usually again dominated by small crustaceans, annelids, and gastropods that are essentially detritivores. Many mangroves are important fish nurseries and may provide day-time resting places for oceanic species such as sharks. But unlike the temperate salt marshes, these mangrove swamps also have a strong above-ground fully terrestrial community, which may include birds, bats, and monkeys as well as the inevitable insects. Tropical mangroves are exposed to normal salinity sea water at around 28°C, but with variation due to intense evaporation, torrential tropical storms, and run-off from shallow freshwater pools inland.

Patterning of biota in estuaries

Estuaries are most obviously characterized by a gradient of decreasing salinity as we move upstream from the mouth, giving a zonation of less tolerant to more tolerant species. But the pattern is complicated by two other factors.

Firstly, many estuaries also have a vertical gradient (stratification), resulting from the different densities of fresh and salt waters. Salinity will be greater at the bottom than in the surface water, with the fresh water moving over the top of a "wedge" of saline water, while the underlying sediment may contain water of relatively stable intermediate salinity giving a **halocline** (Fig. 12.9). Thus burrowing and benthic animals can often penetrate further into an estuary than planktonic animals can (and adults may survive better than their planktonic larval forms). Flow is usually seaward in the surface water, except briefly at the peak of an incoming tide; animals can exploit this to achieve station-keeping in an estuary, adjusting their depth within the water column to come inwards near the surface on the landward flood tide and then becoming benthic and drifting only a little way out again on the seaward ebb.

Secondly, the flow patterns of the river may override the expected mixing pattern within the apparent estuarine zone. For large tropical rivers, like the Amazon and the Congo, the flow coming out from the river is usually so forceful that the "zone of mixing" is actually well out to sea. For moderately large rivers, with limited tides, flow progressively forces silts and mud out into a fanning "birdfoot-type"

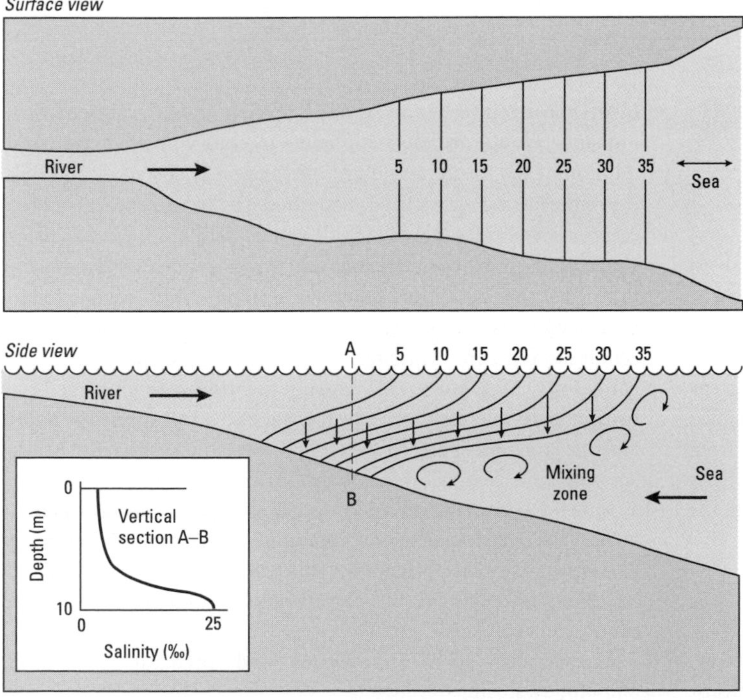

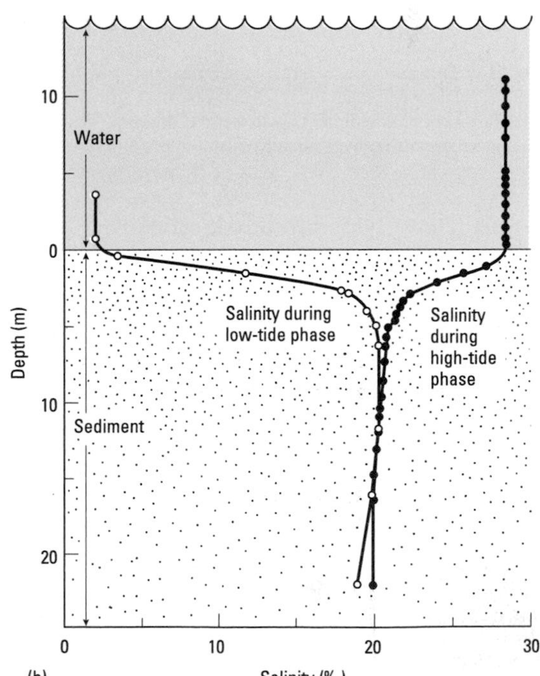

Fig. 12.9 Zonation in estuaries. (a) The distribution of salty and fresh water in top and side view, with a deep "salt wedge". (b) The variation in salinity in the water column and in the sediment at low tide and high tide, with the deep sediment salinity constant at all phases of the tide cycle.

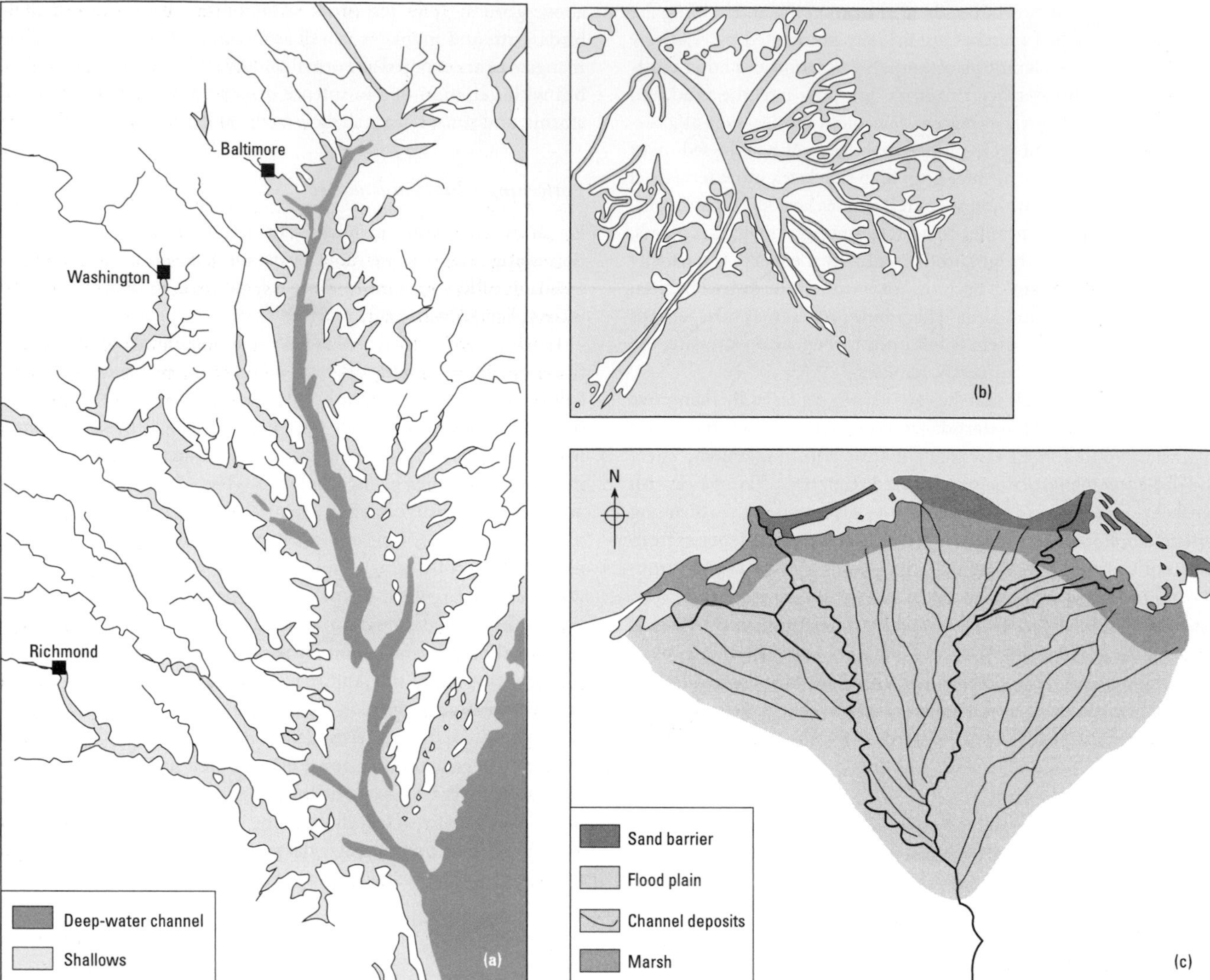

Fig. 12.10 Different types of estuary and mixing patterns. (a) An example of a traditional drowned valley or ria (Chesapeake Bay). (b) A birdfoot estuary where high flow rates push sediment out to sea (the Mississippi delta). (c) A classic triangular river delta, again mixing well out to sea (Nile).

estuary (Fig. 12.10b) where mixing occurs. Only with smaller rivers does the mixing happen strictly within the river mouth. Thus true estuarine conditions do not occur exclusively within the apparent geographic confines of an estuary.

There may also be strong seasonal patterning in brackish waters. Temperate estuaries in particular commonly show seasonal changes in biota, with a spring bloom of phytoplankton when inputs from the land and river run-off peak. This bloom of primary productivity is followed by successive peaks of animal species that feed on the phytoplankton and on each other (Fig. 12.11).

12.1.3 Fauna of brackish waters

Brackish-water faunas, as we have seen, are not normally very diverse or speciose, but can be of very high biomass and density. Most marine invertebrate groups achieve some representation, but some groups are too stenohaline to cope with even moderate dilution of the medium and are conspicuously absent, found only below the littoral zone and so encountered by beachcombers only on extremely low tides—these include cephalopod molluscs, tunicates, sipunculans, most echinoderms, and scyphozoan cnidarians (jellyfish). There are occasional examples of sponges, anthozoan cnidarians (anemones), bryozoans ("sea mats"), and minor groups such as echiurans and hemichordates. A few amphibians tolerate brackish conditions (*Rana cancrivora*, already mentioned in Chapter 11 as being able to tolerate full sea water, but also a few species of *Bufo*), and one crocodile species lives in estuaries in Australasia. But brackish waters are clearly dominated by four major groups:

1 Annelid worms, including deposit-feeding sand masons and terebellids, filter-feeding sabellids, and omnivorous mobile ragworms.

2 Gastropod molluscs (including the grazing winkles, limpets, etc.) and filter- or deposit-feeding bivalve molluscs (including mussels, clams, and razor shells).

3 Crustaceans, both omnivorous and predatory, including amphipods, shrimps, crabs, etc.

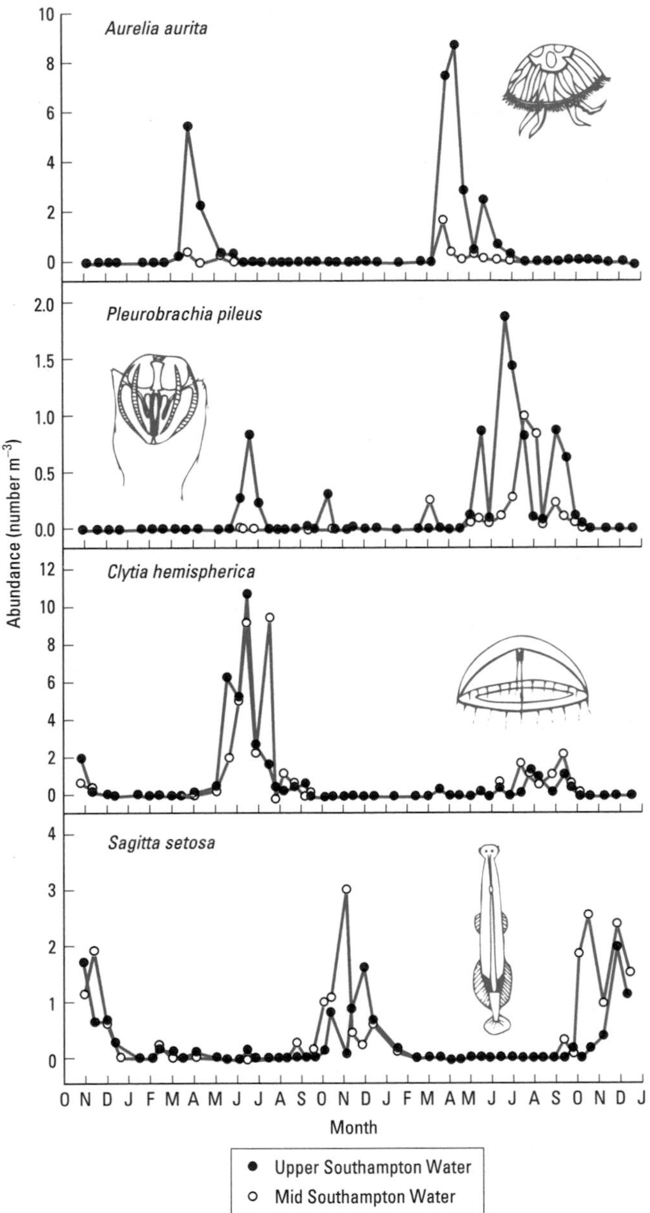

Fig. 12.11 Seasonal patterns of abundance in a temperate estuary; spring peaks of phytoplankton occur and are followed by successive peaks of animals that eat the plankton and are themselves eaten by other animals such as the predatory planktonic arrow worm, *Sagitta*. (From Lockwood *et al.* 1996.)

4 Fish, especially the gobies, blennies, killifish, etc. on shores and in pools, and larger fish migrating in and out of estuaries (including sawfish, stingrays, bull sharks, etc.).

12.1.4 Central problems and strategies

The problems of living on the fringes of seas largely arise due to ongoing **temporal change**, mainly due to tidal patterns and wave actions. Most obviously there are the effects of being at the habitat interface, affecting osmotic and thermal balances, respiratory exchange, and feeding strategies for both predators and herbivores. Tidal variation adds another layer of hazards, affecting all those same parameters on a complex rhythmic time-base, and also altering ions, pH, and light levels. Waves produce the further problem of crushing (with a pressure of up to 22 t m^{-2}), and a major risk of dislodging and displacement. They also bring in suspended material and cause turbidity, which may silt over and bury animals, clogging up gills and filtering surfaces, and reduce the rate of photosynthesis for algae. Freshwater run-off down a beach may add further variability in osmotic concentrations, in ionic levels, and in pH; this can be critical in cold regions where there may be sudden and prolonged seasonal exposure to meltwaters. In estuaries there is the extra variation of cyclical water concentration changes as the sea retreats and the river adds fresh water.

In the face of all these problems, the simple strategy of avoidance is extremely important, and we will meet many examples of this. Where change is cyclical, and conditions can be guaranteed to reverse and ameliorate within an established timeframe, it makes very good adaptive sense to avoid the predictably short period of adverse conditions by simply shutting down activity, hiding, closing a shell, or burrowing deeper. This requires a detection system, and many littoral animals have chemoreceptors (osmoreceptors) in peripheral sites suited to sampling the changing medium. Crabs use their antennules, other crustaceans their feet, molluscs the osphradium (near the gills) or the mantle edge generally, and polychaetes the cirri or tentacles on their heads. Avoidance can then take many forms. For example, some lower shore organisms may only get exposed to air briefly twice a day at low tide, and closing up valves or lids, with some sea water within to provide an osmotic buffer and an oxygen store, may be the simplest and almost cost-free strategy. However, this may not be enough at times of the year where avoidance would need to work on quite a long timescale. Some animals at the extremes of a beach or estuary may experience cycles not just involving the 12 h or so between tides, but of monthly recurrence, where tides only cover them (or uncover them) for a few hours on a couple of days each month. For these animals, as well as for many unprotected softer bodied organisms that cannot readily avoid conditions, some degree of tolerance and regulation is essential, involving physiology and biochemistry at the cell or whole organism levels. These organisms need to be **euryhaline**, **eurythermal**, and **euryoxic** to have any real chance of long-term survival in an estuary or seashore.

12.2 Ionic and osmotic adaptation and water balance

We know that all animals regulate their ionic balance to some extent, and that the necessary pumps and channels are a basic property of cells. However, on shores and in estuaries all animals need to cope with more external variation. Shores experience seasonal and diurnal/tidal variations; there will be periods of hyperosmotic conditions when the tide is out, due to evaporation of water from rock pools and water in crevices, etc., and periods of hyposmotic stress due to rain and freshwater run-off. Seasonally the interstitial water in a beach may vary markedly in salinity. In the example shown in Fig. 12.12a, this water is hyposmotic in most winter months but roughly as concentrated as the sea water in the summer, when it also becomes much more anoxic. Superimposed on this, of course, is a

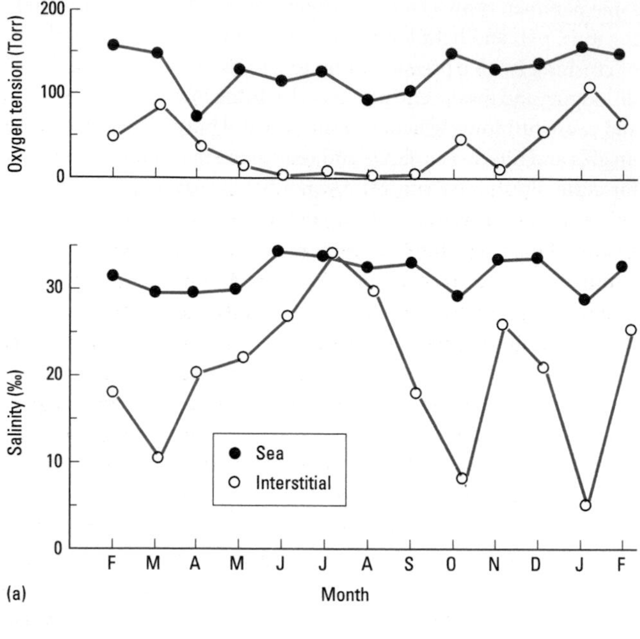

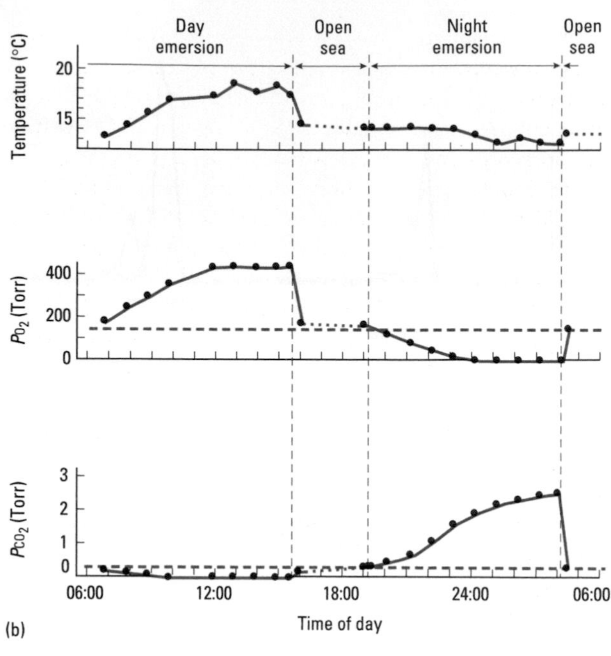

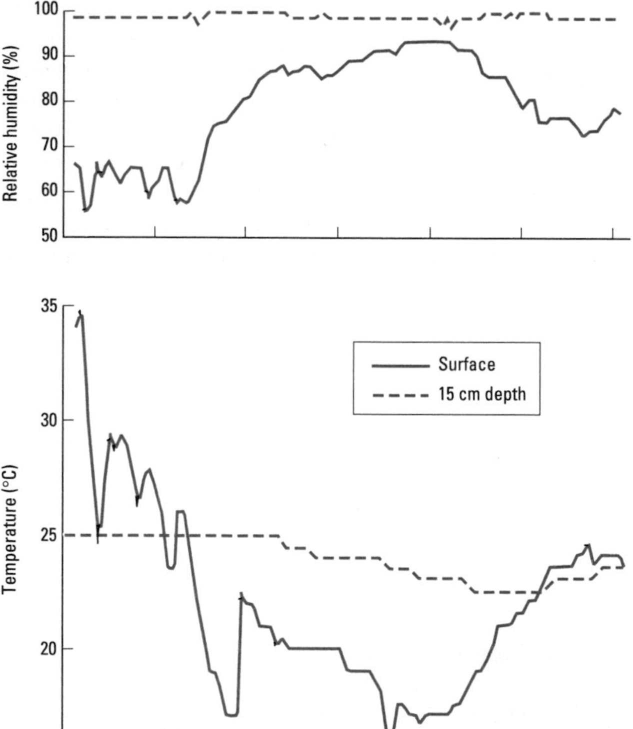

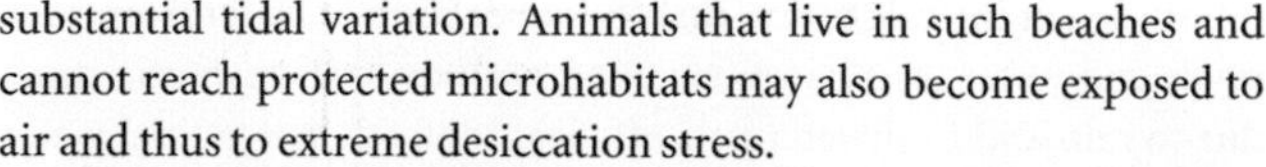

Fig. 12.12 Microenvironmental variation in shore habitats. (a) Annual patterns of salinity and oxygenation in interstitial water at the high-water mark (MHWN) on a boulder beach. (b) Daily temperature, oxygen and carbon dioxide in tidepools in summer (dashed line shows air). (c) Daily variation in temperature and humidity in a pile of seaweed at the high-water mark on a sand beach, with constancy at about 15 cm depth. (a, Adapted from Agnew & Taylor 1986; b, adapted from Dejours 1981; c, adapted from Moore & Francis 1985.)

substantial tidal variation. Animals that live in such beaches and cannot reach protected microhabitats may also become exposed to air and thus to extreme desiccation stress.

Estuarine waters range from almost fully marine to substantially hyposmotic, but alter to a varying extent through time and with distance upstream, to be nearly fresh water at low tide close to the river outlet or after prolonged rainfall.

12.2.1 Strategies

Avoiders

It is possible to avoid osmotic stress and/or actual drying out of tissues to a large extent, and a great many animals in brackish waters do this. They can easily stay covered in water by retreating into crevices, remaining in rock pools, or burrowing deeper in sand or mud. However, these options may give extra problems of oxygen depletion or of exposure to unusual pH levels or (in pools) to raised water temperatures, as shown in Fig. 12.12b. For shoreline animals that are more terrestrial, excessively dry air may be avoided by burrowing into the piles of seaweed and debris that accumulate at the strandline, where temperature and humidity stay relatively constant (Fig. 12.12c); this niche always has a thriving community of small invertebrates associated with it.

Another solution is to carry a water store ("exosomatic water"), perhaps within a gill chamber as in many crustaceans, or in the mantle cavity for molluscs. This provides exposure of the soft tissues to water whose salinity differs from that in the external environment, and may allow the animal to "smooth out" the variations it actually experiences. Many isopods and some amphipods also carry an external water source in their ventral groove, a fluid-conducting system using capillarity to move water over the lower surfaces and allowing "aquatic" methods of gas, water, and ion exchange in otherwise semiterrestrial beach dwellers.

A third possibility is to have an intrinsic resistant covering. This may be mucus, of limited value but present in certain animals such as the common beadlet anemone, *Actinia*; or more specifically it may involve tubes or boxes, manufactured by the animal, and usually capable of being sealed off from the outside world temporarily with a lid or sphincter. Obvious examples include calcareous shells or cuticles in molluscs, brachiopods, and crustaceans, chitinous boxes in bryozoans, and tubes of sand, chitin, consolidated mucus, or calcium carbonate in a range of annelid worms. It is no coincidence that so many shores are dominated by animals with intrinsic covers—most spectacularly by barnacles, limpets, and mussels.

Conformers and regulators

The basic pattern of conformers and regulators within some common euryhaline organisms is shown in Fig. 12.13 (cf. Fig. 5.2). In essence, following the concept of four different "spatial levels" of control that we considered in Chapter 1, animals that are conforming have no control at the skin–blood interface, but are to some extent regulating (in terms of maintaining reasonable cell volume) at the blood–cell interface. In contrast, regulating animals concentrate their controls at the skin–blood interface and are thus maintaining a relatively constant body fluid concentration.

Animals like *Carcinus maenas* (the common shore crab) are very good ("strong") hyperosmotic regulators, keeping their blood osmotic concentrations quite high and stable over a wide range of external concentrations. But there is intraspecific variation here: shore crabs from the North Sea are less tolerant of dilution than those from the more brackish waters of the Baltic Sea, and also regulate rather less strongly at any given dilution. *Cancer* (the edible crabs) and *Nereis* (the polychaete ragworms) are moderate ("weak") regulators, allowing their body fluids to fall in conformity with external dilution (while using moderate volume regulatory measures) down to some critical point, and then instituting a degree of skin- or gill-based regulation to cope with any further dilution. Animals such as *Actinia* and *Anemonia* (the commonest shoreline anemones) and *Mytilus edulis* (the common mussel) are essentially conformers, with only a little regulation when in moderately or (for *Mytilus*) very dilute habitats, to give an internal osmotic concentration slightly above that of the medium. Over most of the concentration range they do not attempt to control their water balance but act as simple osmometers, incurring substantial tissue volume changes. (However, remember that *Mytilus* is a bivalve mollusc, so does have the

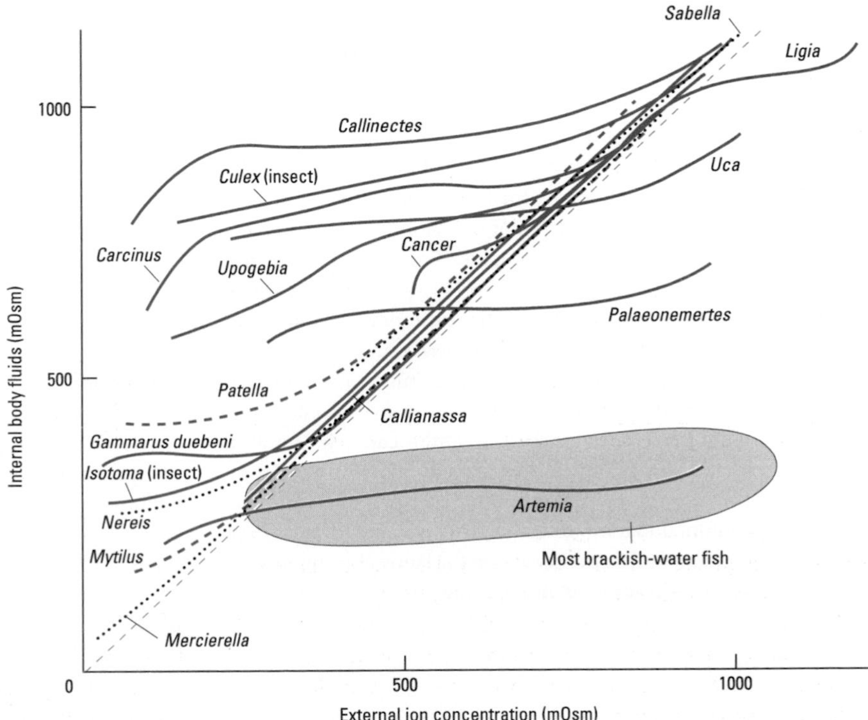

Fig. 12.13 Variation of internal body fluid concentrations with external salinity for a range of brackish-water species. Arthropods are shown by solid lines, molluscs by dashed lines, and worms by black dotted lines, with teleost fish represented by the tinted area.

option of temporary avoidance by shutting the shell, and will probably only need to conform on higher shore levels during periods of spring and neap tides.) At the extreme, there are a few invertebrates that come close to being perfect conformers, with body fluids that even in tap water are scarcely elevated: the tiny reef-forming polychaete worm *Mercierella* (also known as *Ficopomatus*) is a well-known example, surviving both in tap water and in 200% sea water.

The kind of plot shown in Fig. 12.13 clearly reveals that there is a continuum of strategies, not a simple dichotomy. Several points emerge when we look at specific examples. Firstly, there are no perfect regulators. Organisms exhibit varying degrees of maintaining their blood fluid concentration, but most brackish-water animals that survive dilution seem to settle at around 30–70% of original seawater concentration when in dilute media. Secondly, note that crustaceans usually regulate more than worms and molluscs, which is probably due to their exoskeleton. Since they have a very hard covering they are unable to swell significantly and therefore cannot help but regulate somewhat; their cuticle gives an inbuilt physical aid to being a regulator, which also makes the process cheaper. Thirdly, conforming *can* be almost perfect, and is a very cheap solution as it avoids the need for using up substantial energy in active transport, so reducing the metabolic rate and food requirement. Fourthly, not many animals hyporegulate (i.e. keep their blood concentration lowered when the medium gets more concentrated); amongst the invertebrates there are just a few crustaceans able to do this (some crabs and prawns, branchiopods, and isopods). Most upper shore animals instead just "conform upwards" when their pools or crevices get hyperosmotic on warm days. Finally, as we should by now expect, vertebrates do not fit the general pattern, as they have dilute blood anyway and they regulate it fairly constantly at a moderately dilute level, whether the medium is above or below normal sea water.

It is amongst the crustaceans (such as the blue crab, *Callinectes*, and the mitten crab, *Eriocheir*) and the vertebrates (such as salmon, *Salmo* spp., and eels, *Anguilla*) that we find the animals with the greatest osmoregulatory powers, capable of migrating right across the estuarine zones as they move from fresh water to sea water and back cyclically (the anadromous and catadromous species, respectively, spawning in fresh water and in sea water but moving regularly between the two media). However, their osmotic ability is matched or exceeded by some of the "amphibious" shore animals that are essentially terrestrial rather than aquatic. For example, the sea slaters in the genus *Ligia* are good hypo- and hyperregulators when immersed but achieve even better osmotic constancy when in air, as do supralittoral beach-hoppers, such as *Orchestia*.

It is also worth noting that osmoregulatory ability varies through ontogeny, many crustacean larvae being almost osmoconformers with poor salinity tolerance where the adults can regulate and survive at much lower salinities. The changes arise in part from maturation of the hormonal system, with hyperglycemic hormone only becoming functional in postlarvae, JA: style and in part from changes in the osmoregulatory site (gills in the larvae, but epipodites and branchiostegites in adults as gills become purely respiratory).

12.2.2 Mechanisms

For all brackish-water animals, there are three main mechanisms that may be involved in the adaptation to osmotic changes: alterations of external permeability, varying salt uptake, and cellular osmoregulation.

Table 12.4 Values of water permeability (P_w) and sodium permeability (P_{Na}) in brackish-water animals in comparison with marine and freshwater species and races.

Genus and habitat	P_w (10^{-4} cm s^{-1})	P_{Na} (10^{-6} cm s^{-1})
Bivalves		
Brackish water		
Mya	3.8	5.7
Geukensia	1.3	4.0
Crustaceans		
Sea water		
Libinia	12.8	13.2
Porcellana	4.7	1.0
Eupagurus	12.4	1.6
Maia	3.4	12.9
Carcinus	9.0	5.0
50% sea water		
Carcinus	2.8	5.0
Cancer	1.3	2.7
Fresh water		
Eriocheir	0.22	0.87
Astacus	0.88	0.10
Vertebrates		
Fresh water		
Carassius	3.9	0.4
Salmo	1.6	0.3
Anguilla	1.2	0.6

Permeability

For any animal living in an uncertain environment, it makes obvious sense to reduce surface permeability as far as possible, since this will reduce the rate of change of concentration, and the rate of change in blood volume, to levels where they can be manageable. This is potentially even more useful in brackish forms than in true freshwater organisms. Notice how many of the common shore and estuary animals are preadapted in this regard, especially the molluscs and crustaceans, by having "built-in", low-permeability external surfaces as part of their body plan.

As we saw in Chapter 4, there are various separate components when considering permeabilities, and values for both water permeability (P_w) and salt permeability (P_{Na}) for some crabs and fish are summarized in Table 12.4. P_w is clearly reduced in most brackish-water groups. For example, nonmarine crabs show lower fluxes for water than marine crabs, and they may also have measurably thicker cuticles. The value of P_w can sometimes be at its lowest in the brackish species, below that of freshwater congeners, underlining the point that this zone is most difficult and it is really worthwhile to invest in low permeability, thus "smoothing out" the environmental changes. However, the trend is not universal; in some brackish-water bivalves that burrow deeply (e.g. the clam *Mya*) P_w is quite high, whilst in others that are fairly exposed (e.g. the mussel *Geukensia*) it is lowered. For P_{Na} the brackish species are not normally the most impermeable, having values that lie between sea water and fresh water; examples of sodium loss for the polychaete worm genus

Table 12.5 Rates of sodium loss in annelids and crustaceans from aquatic habitats.

Taxon	Habitat	ECF* [Na+] (mM)	Sodium loss rate†
Annelida—Polychaeta			
Nereis succinea	SW/BW	200–440	54
N. diversicolor	BW	150–350	13
N. limnicola	FW		8
Crustacea—Amphipoda			
Marinogammarus finmarchicus	SW	300	80
M. obtusus	SW	–	57
Gammarus tigrinus	BW	220	20
G. duebeni	BW	273	11
G. zaddachi	BW	252	14
G. lacustris	FW	135	5
G. pulex	FW	135	4

BW, brackish water; FW, fresh water; SW, sea water.
* Sodium concentration in the extracellular fluid (ECF).
† Sodium losses are expressed as μmol g^{-1} h^{-1} for annelids, and as μmol ml^{-1} blood h^{-1} for crustaceans.

Nereis and the amphipod crustaceans allied to *Gammarus* are shown in Table 12.5.

Animals in these zones may also be able to modify P_w within their own lifetime by acclimatory change, as shown in Fig. 12.14 for *Gammarus duebeni*; here the rate of exchange of labeled water depends on the salinity, changing at best from about 8.3 h^{-1} (at 35 ppt) to 1.7 h^{-1} (at 1 ppt). It should be noted that some animals that also undergo periods of aerial exposure, such as the bivalves *Mya* and *Geukensia*, do *not* change P_w with decreasing salinity, having, respectively, a deep burrow or a routinely low P_w value to reduce water loss during such exposure. Where it does occur, permeability acclimation has a very rapid time course, probably due to hormonal action. Certainly in vertebrates there are a series of hormones, such as antidiuretic hormone (ADH), that control P_w variations in the skin and/or the kidney tubule cells (see Chapter 10).

Salt loss through the skin is also very important for most invertebrates; it usually makes up 80–95% of the total salt loss, with the "kidneys" not being very important. A few shoreline animals, such as sea anemones, are surprisingly permeable to ions at their epithelial surfaces, which allow rapid diffusion in either direction; but in general the P_{Na} of skin is as low in brackish-water animals as in freshwater species (Fig. 12.15a). There is little evidence that this can be acclimated in invertebrates, except in a few hyposmotic regulators such as the prawn *Palaeomonetes serratus*, which has decreased P_{Na} in 120% sea water. However, in vertebrates aldosterone and prolactin hormones control P_{Na}; this occurs in fish gills and frog skins, for example, as well as in the kidney (see Chapter 10). P_{Na} also varies genotypically, where populations are exposed to different media concentrations, as with the isopod *Mesidotea* shown in Fig. 12.15b.

Changes in permeability may be accompanied by changes in cellular morphology in some brackish-water animals. In the hydroid *Cordylophora*, and in some crab gills, cells in the exposed epithelia become more rounded and have a reduced total surface area in estuarine conditions, presumably helping to reduce net fluxes.

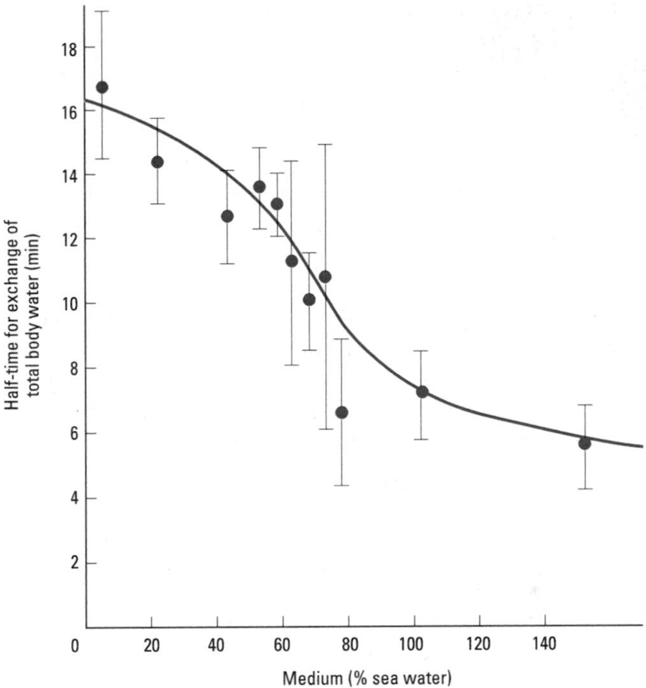

Fig. 12.14 Changes of water permeability (P_w) in *Gammarus duebeni* acclimated to different salinities. Vertical axis shows half-time for exchange of tritiated water, so that high values indicate reduced permeability. (From Lockwood 1961.)

Reduced surface permeability is the first line of defense, and it can be varied and controlled. But it is only a delaying tactic. Living organisms can never be totally impermeable and completely isolated from the variations in the world outside. Thus the next part of the total strategy must be a mechanism to keep the internal fluids at acceptable levels.

Ion transport mechanisms

For an animal with reduced ion and water permeability, but in water more dilute than sea water, ions and water will still move passively down their gradients—water moving in, and ions leaking out. This will happen especially at parts of the surface that have to be permeable for gas exchange, which for aquatic animals normally means the gills. Thus it is frequently the gills that are also used as ion-exchange sites, allowing control to be localized. In crustaceans the gills commonly have two cell types: "thin" cells important in gas exchange, and "thick" cells assumed to be the main site of ion exchange. In many developing embryos the ability to osmoregulate independently coincides with gill development, although in some crustaceans (e.g. *Gammarus*) regulation can be detected before the adult gills grow (on the coxal leg-bases) and seems to be a function of the vitelline membrane of the developing egg and a special "dorsal organ" in the larva, which later regresses. Salt uptake in the midgut area is also likely in juvenile and adult crustaceans. In some adult aquatic animals the kidneys may also be used for some salt uptake, tubule cells resorbing ions from the filtrate into the body, producing slightly hyposmotic urine (5–15% body volume

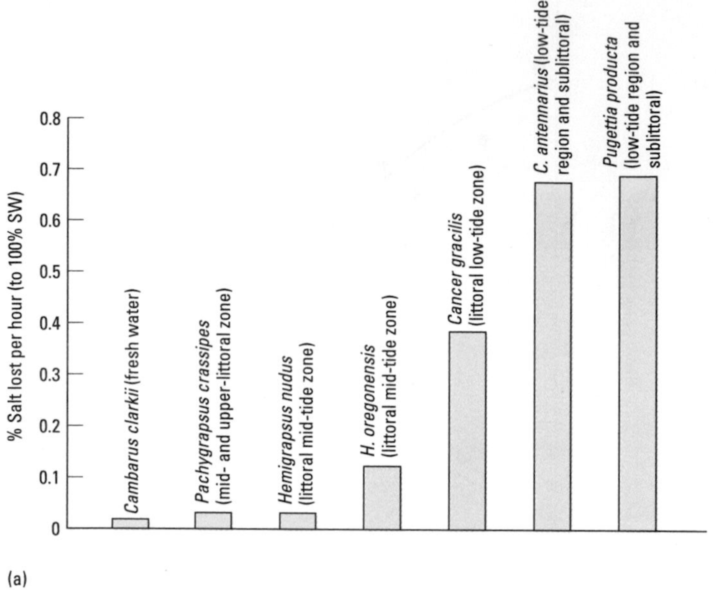

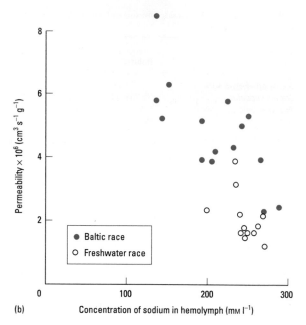

Fig. 12.15 (a) Ionic permeability (P_{Na}, expressed as salt loss per hour into 100% sea water (SW), measured with sections of isolated cuticle) for a range of crustaceans from different habitats, showing the reduced values in freshwater and brackish-water species. (b) Adaptive changes in sodium permeability; a comparison of P_{Na} in the Baltic race and freshwater race of the isopod *Mesidotea entomon*. (Adapted from Croghan & Lockwood 1968.)

per day). This is fairly uncommon in brackish-water invertebrates, reported only for a few annelids (e.g. *Nereis diversicolor*) and some gammarid crustaceans. Salt uptake at the kidneys does occur in most brackish vertebrates, and is well studied in the estuarine crocodile, where urine is routinely about 80% of the concentration of the plasma with around 95% of all the salts being resorbed in the tubules after glomerular filtration. Some further processing may occur in the cloaca, particularly of potassium levels.

The major contributor to blood osmolarity in all animals is sodium chloride, so the regulation of the fluxes for Na^+ and Cl^- ions, whether at the gills or kidneys, is bound to be at the center of any osmoregulatory response. The basic principle for salt uptake by sodium transporters was described in Chapter 4 (see Fig. 4.10), and the details of the system found in most aquatic animals are shown in Fig. 12.16. It is based on a classic sodium ion pump, located on the inner (basal) epithelial membrane. This involves an exchange system, which though based on the transporter (adenosine triphosphatase, ATPase) that normally exchanges Na^+ for K^+, in practice often appears to utilize a Na^+ for NH_4^+ exchange. (Direct evidence for such active exchange is controversial, and it may be that NH_3 diffusion accounts for most of the ammonia excretion; a dependence of NH_4^+ loss on Na^+ uptake is perhaps caused secondarily, by boundary-layer effects on acidification.) There is commonly also a separate basal Cl^-/HCO_3^- exchange in salt-uptaking epithelia. This gives the animals the two ions they most want in exchange for two waste products (ammonia and dissolved carbon dioxide). Details of what happens at the outer or apical membrane to move sodium into the cell are more controversial, as Fig. 12.16 shows: Na^+/H^+ antiporters, $Na^+/K^+/2Cl^-$ cotransporters, and simple Na^+ channels have all been proposed. The first of these is strongly supported by recent molecular evidence showing the presence of an appropriate mRNA sequence in the gill tissues.

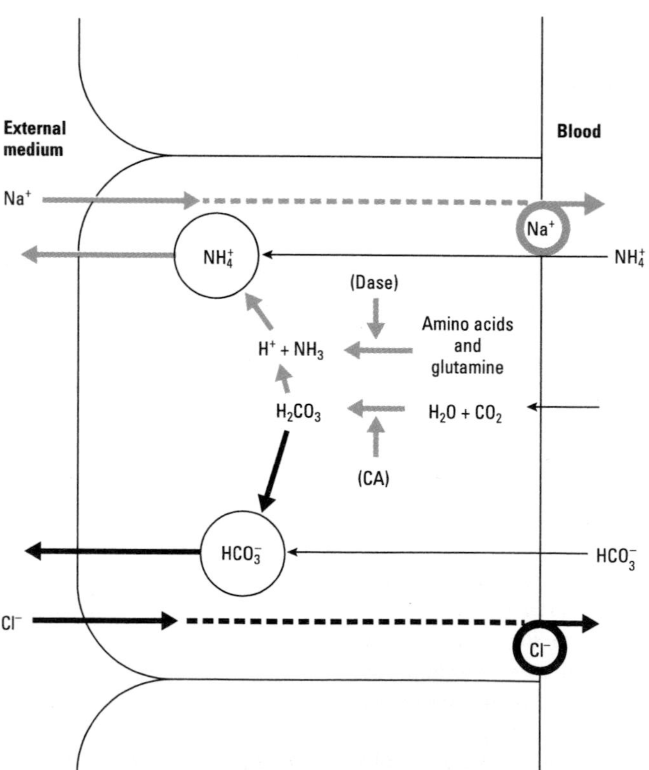

Fig. 12.16 Ion uptake in brackish-water species; the basic mechanism for Na^+/NH_4^+ exchange and Cl^-/HCO_3^- exchange in uptake epithelia. CA, carbonic anhydrase; Dase, deamination enzymes. (Adapted from Maetz 1973.)

Table 12.6 The K_m and J_{max} values for sodium influx in brackish-water osmoregulating animals.

Species	Habitat	K_m (mM)	J_{max} (mM kg^{-1})
Carcinus maenas (crab)	Brackish	20	
Mesidotea entomon (isopod)	Brackish	9	8.5
Gammarus duebeni (amphipod)	Brackish	1.5	15
Rana cancrivora (frog)	Brackish	0.4	5
Freshwater species (general range: see Table 13.9)		0.1–2 (lower)	0.1–20 (similar range)

Cells involved in these transport processes always have a characteristic structure (see Fig. 4.12) with highly infolded apical membranes to give a large diffusive surface, a somewhat indented basal membrane on which the main cell/blood Na^+ transporters are found, and very large mitochondrial volumes. This architecture allows the pumps to work fast, with a high rate of fluid flow.

Ion transport normally increases with environmental concentration up to a limiting value (J_{max}), determined at least in part by the number of pump proteins, and it shows saturation kinetics, with the K_m value (the "affinity") being the external concentration for half-maximal transport (see Box 4.5). Thus the lower the value of K_m, the lower the range of concentration over which the transport system can achieve its maximum rate. It turns out that K_m (ultimately a function of pump protein structure) is well correlated with habitat, as Table 12.6 shows. Rates of salt uptake can thus be varied in the short term by acclimation, and also by evolutionary changes at the genotypic level, although the ability to survive low salinity is in fact a complex function of changes in J_{max}, K_m, and permeability parameters. Many species are known to increase the number of ATPase pump proteins in their membranes during hyposmotic acclimation (e.g. in *Mytilus* nerve membranes, and in crab gills), and these changes can often be seen as an increase in membrane-bound particles in freeze–fracture studies. Adaptation may also be exhibited at a higher morphological level by adaptive growth of the regulatory organs or surfaces, giving a greater membrane area for active pumping of ions. For example, in the brackish mosquito larva *Culex*, the anal papillae where salt-uptake processes are concentrated grow very markedly in more dilute media (Fig. 12.17).

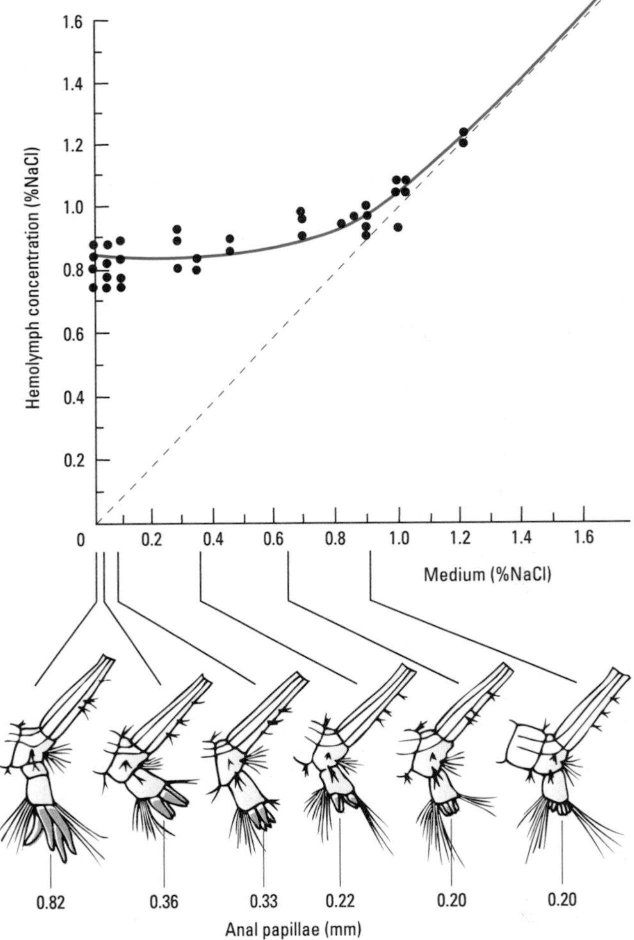

Fig. 12.17 The relationship between hemolymph and medium NaCl concentrations for the brackish-water mosquito *Culex*, also showing the associated growth of the anal papillae (where uptake occurs) in dilute media. (Adapted from Wigglesworth 1938.)

These changes must be triggered by sensors that detect changing external salinity. This aspect of adaptation is relatively little studied (see Chapter 9) but in some crabs it is known that the main sensors are specific "hair-peg" organs on the legs, which may set in motion a hormonal cascade that involves a small peptide hormone and/or dopamine.

Ion uptake processes and their regulation have been particularly studied in anadromous and catadromous fish, such as salmon and trout, which migrate through estuaries, and a range of other fish, such as tilapia and guppies, that tolerate a wide range of conditions. When adapting to fresh water, these fish show increases in gill Na^+/K^+-ATPase, and also increases in a bicarbonate-dependent ATPase and a Ca^{2+}-ATPase. An additional and unusual mechanism operates in the gills of these fish (and many other freshwater fish, considered in Chapter 11), whereby the chloride cells (see Chapter 5) can be physically covered or uncovered by volume changes in the nearby pavement cells, effectively turning the chloride cells on or off. In fish that can live in fresh or salt water, the gill chloride and pavement cell systems can take up NaCl in fresh water and secrete it in salt water; they also vary in function according to the pH of the medium in freshwater systems. But the change from one functional mode to another takes some time, with morphological change in the chloride cells and a change in the molecular components of the pumps, primarily under the control of the hormones prolactin and cortisol. Indeed many of the adaptive changes in moderately euryhaline fish are under tight hormonal control, with calcitonin and calcitonin gene-related peptide also implicated in gill membrane changes, and with arginine vasotocin and isotocin levels, as well as growth hormone, operating in longer term adjustments (see Chapter 10). In salmonids growth hormone and cortisol also act to promote long-term seawater adaptation via actions on the differentiation of the chloride cells towards their marine condition. One signal for this change of function appears to be the shrinkage of the chloride cell itself during high salinity exposure. Recent evidence indicates that there may be two kinds of chloride cells in these fish, termed α and β: the α-cells play a key role in chloride-dependent sodium absorption in fresh water and sodium excretion in sea water, while the β-cells have some other unknown function in fresh water and are transformed into additional α-cells on exposure to higher salinities. In conditions where ion regulation is particularly difficult, chloride cell populations proliferate, but this also tends to thicken the gill tissue and reduce the respiratory gas transfer. A further trade-off may arise from the effects of temperature on salinity tolerance; this has been particularly neatly shown in chum salmon migrating off Japanese coasts, where the salinity tolerance is reduced (and the expression of the vasotocin gene proportionately affected) in years where cold currents rather than a warm surface current dominate the migration route.

A brackish-water animal thus has a means of taking up the extra salts it needs to counter passive outflow, and the blood concentration is taken care of by a regulatory system. Osmoregulation is not perfect as we have seen; the extracellular fluid (ECF; usually blood) is regulated but in a state that is still hyperosmotic to the medium, reducing the gradients for further exchange to a level where they can be balanced without excess expenditure. In crustaceans that are hyperregulating, the ECF concentration may be reduced to roughly half that of sea water and the P_w is perhaps three-fold reduced, the net effect being a reduction in osmotic water loading of about 1 order of magnitude.

This provides the first level of regulation as we discussed in Chapter 1 (between the outside world and circulating fluid). But there remains the problem of keeping the cell concentrations under control—how do these animals maintain the primary level of regulation, between relatively diluted circulating fluids and more concentrated individual cells?

Cellular osmotic adaptation

Cells need to hold on to ions, even when the world outside them, and their own immediate surroundings (i.e. the blood), are becoming dilute. This is because ions affect so many internal cellular processes dramatically, especially via allosteric effects on enzyme activities and DNA–histone interactions (see Chapter 2). Thus when faced with dilute surroundings, cells tend to keep their ions but lose instead some other small osmotically active solutes, known as **osmotic effectors** or **compensatory osmolytes** (see Chapters 4 and 5). These are a fundamental characteristic of cells, and are predominantly amino acids, which make up a substantial proportion of the normal cellular osmotic concentration (see Table 4.4).

Amino acid reduction during acclimation to dilute media could theoretically be achieved equally effectively in two ways: by increasing protein synthesis to turn the small osmotically effective amino acids into a few large, osmotically less effective, proteins; or by having a net export of the amino acids from the cell into the blood. Both systems do occur, but the latter is more important, i.e. when blood ion concentrations fall, many cell types in brackish-water animals respond by exporting amino acids. The control of this process is complex, involving transaminations and deaminations, *de novo* synthesis controlled by MAP kinases, and variations of efflux and influx rates. Some of the enzymes that control the synthesis of key amino acids (see Fig. 4.20) turn out to be very sensitive to Na^+ concentration, and cease to function at low Na^+ levels. Transmembrane movement of amino acids is also Na^+ dependent, so that at low external Na^+ concentration the net outward movement of amino acids is facilitated. Therefore, as Na^+ in the blood varies, the amino acid levels are automatically adjusted. Cell osmotic concentration gradually gets balanced downwards, and blood osmotic concentration is also corrected upwards slightly by receiving the amino acids. The system also works in reverse, whereby in hyperosmotic stress amino acids are synthesized and retained in the cells. This system has been particularly well studied in mammalian cells, where it appears that genes coding for enzymes and transporters that produce accumulation of organic osmolytes contain "osmotic response elements" that are directly implicated in the activation of the transcription process. Overall, the system of osmolyte regulation clearly forms a complex and neatly controlled automatic regulatory feedback mechanism.

The key osmotic effectors differ in different animals. In crustaceans, the amino acid loss is particularly of alanine, with some proline and glycine; in the lugworm it involves alanine and glycine; in molluscs the modified amino acids taurine and betaine are particularly important. Inevitably the key enzyme in the mobilization of amino acids as osmotic effectors therefore also varies from the scheme shown in Fig. 4.20. In the copepod *Tigriopus*, inhabiting splash pools on the Pacific coast of North America, it is glutamate-pyruvate transaminase, which leads to alanine synthesis. This enzyme shows a two-allele polymorphism, where fast (F/F) homozygotes have around 50% higher specific activity for alanine synthesis than slow (S/S) homozygotes. In laboratory tests, hyperosmotic stress therefore induces faster alanine accumulation in F/F and F/S individuals, and the S/S individuals die more quickly and are less successful in more exposed sites. In the common mussel, *Mytilus edulis*, the L-aminopeptidase-I enzyme (LAP) is crucial and is polymorphic, with the LAP^{94} allele, having higher kinetic capacity and showing a cline between open ocean and estuarine conditions. However, each year some carriers of this allele invade very low-salinity habitats as larvae and suffer weight loss and eventually high mortality; their disadvantage seems to be an excessively rapid use of their nitrogenous reserves, leaking away as excess amino acids, in these low salinities.

The regulatory system is even neater than we have so far described. Remember that the salt-uptake systems that are working to restore blood concentration are using a Na^+/NH_4^+ exchange system. Therefore the very amino acids that have been shed by the cells into the blood can now be used (usually at the gills) to make the NH_4^+ by deamination. Commonly some alanine is converted to proline, and proline oxidase in the gills then completes the deamination and so can "fuel" the regulation of Na^+ uptake. Low Na^+ levels have thus stimulated high amino acid levels in blood, and this in turn stimulates the uptake of Na^+ from the medium into the blood, giving a classic feedback regulation circuit (Fig. 12.18).

Therefore, all shore and estuarine animals have built-in biochemical and physiological circuitry allowing them to maintain blood Na^+ and keep their cells at a reasonable osmotic concentration and volume, despite regular and drastic changes in their environment.

Volume regulation

The net effect of exposure to lowered salinity for all these animals, and for tissues within animals (muscles having been particularly well studied), is the process of volume regulation, characterized

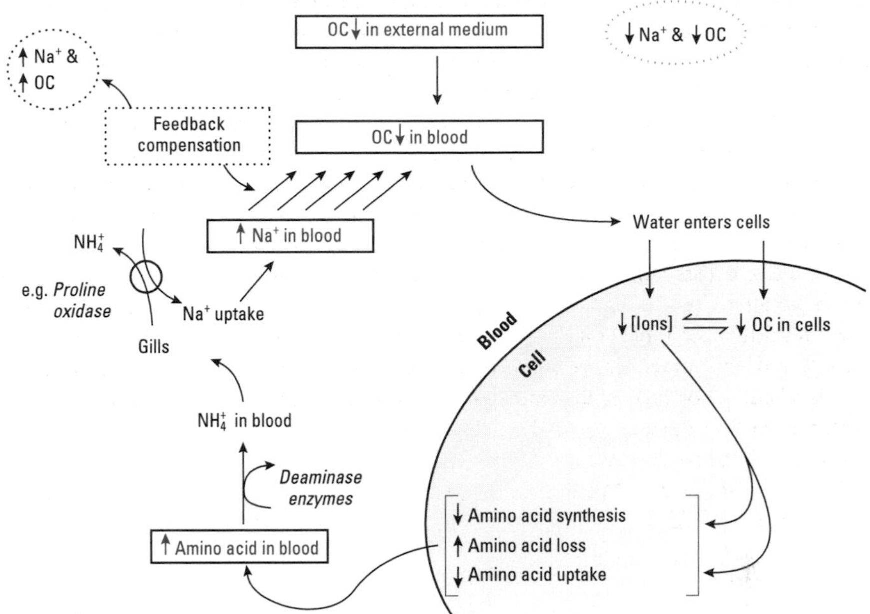

Fig. 12.18 A summary of the regulatory cycle whereby amino acid levels are adjusted in response to changing ionic (Na^+) levels in a brackish-water animal. Reduced external Na^+ levels automatically trigger events that produce an efflux of amino acids from cells to blood and thus a reduction in cellular osmotic concentration (OC) without much deleterious ion loss (see text for details; cf. Fig. 4.20).

by an initial swelling, and a subsequent gradual restoration of a less swollen (though rarely completely unswollen) state, known as the **regulatory volume decrease** (RVD), mentioned in Chapter 4. Figure 12.19 shows examples of the magnitudes and timescales of this response; it varies with taxon, but also with the pattern of exposure to osmotic stress (being largely eliminated if the change in concentration is introduced gradually enough).

We can now understand what underlies the phenomenon of RVD. During initial exposure the animal is gaining water into (and losing salt from) its blood through its finitely permeable surfaces, and the speed of volume increase will largely reflect the effectiveness of permeability control (it may be exceedingly slow in bivalve molluscs where the closed valves reduce the rates of exchange even further, as seen in Fig. 12.19). Volume control of the whole animal may be aided by reduced drinking, especially in fish, and by increased urine production, but these options cannot entirely eliminate tissue swelling. The resulting osmotic gradient between the blood and cells causes the cells in turn to take up water, a process which is often very fast but which is also corrected very quickly. This correction can only be due to solute loss, since we know that there can be no active water extrusion. The main effect is activation of K^+ and Cl^- channels, via calcium-dependent routes. The amino acid content of the cells also drops very sharply in this RVD phase, until the blood (or other ECFs) and cells are approaching equilibrium again at some new level. While this is going on the active uptake of extra solute is also instituted (from the medium or sometimes from the urine) and the blood levels are corrected on a somewhat slower timescale by this uptake from outside the animal, until a final equilibrium volume level is attained, which may be little different from the starting condition. Fluid-filled body cavities such as the coelom may be used to buffer volume change, so that cell volume returns to normal even when total body volume remains elevated.

In some mid-shore bivalves, such as mussels, where the external concentration rises and falls predictably with a relatively short time course, there is little or no RVD in many of the cells, which are allowed to swell and shrink cyclically, presumably thereby saving energy that might be used in trying to regulate their volume. However, heart cells do show an RVD response, as do nerves.

Volume regulation may be aided by circulatory adjustments in crustaceans and other animals with arterial blood systems. In the crab *Cancer magister*, which suffers relatively large volume changes in dilute media, exposure to low salinity produces an increased heart rate, with a reduced stroke volume. Blood flow is redistributed, with anterior flow (mainly to the gills) reducing and with an increase through the posterior aorta. This may help to reduce the average exchange gradient for the inward movement of water at the gill surfaces.

Maintaining ionic levels for nerve and muscle functioning

The mechanisms so far discussed allow cells to continue to function in dilute media, with relatively little loss of internal ionic content. However, certain kinds of cell are especially reliant on ionic gradients across membranes for their functioning; excitable cells depend upon axonal, synaptic, and neuromuscular transmission systems that require inwardly directed concentration gradients of sodium, calcium, or chloride ions and outwardly directed potassium ion gradients (see Chapter 9). Thus it might be expected that muscles and nerves in particular would need special adaptations in euryhaline brackish animals, above all in osmoconformers where concentrations of these ions in the body fluids vary widely. In muscles the primary adaptations relate to maintaining cellular volume by regulation of osmotic effectors, thus preventing damage to membranes caused by swelling. In both muscles and nerves the main osmotic effectors are often aspartate and glutamate, since their net negative charge helps to maintain the normal cell resting potential (see Chapter 4). Maintaining cell volume using amino acids also insures adequate ionic gradients for the conduction of sufficient

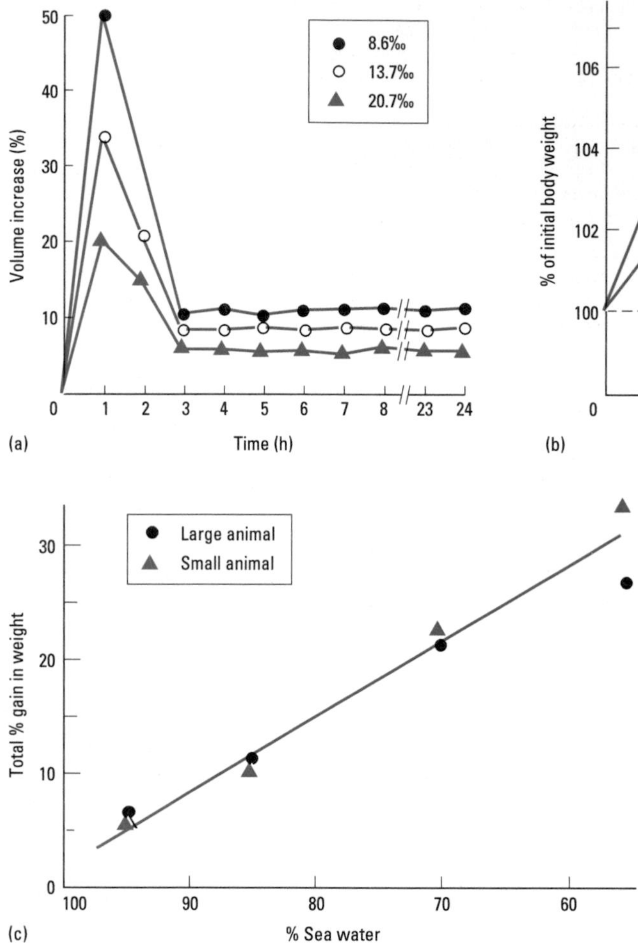

Fig. 12.19 Volume regulation in dilute media in brackish-water species. The regulatory response is fast in a polychaete (a), moderate in a mussel (b), where the gain of volume is also much less, and almost nonexistent in a starfish (c), where water gain and swelling are very marked and the animal hardly deviates from the behavior of a perfect osmometer. (Adapted from Binyon 1961; Pierce 1971.)

electrical potential changes to trigger the contractile or conductive systems.

The problems of nerve functioning in estuarine animals are even more complex, since the size and frequency of the action potentials, which depend upon the magnitude of ion gradients, also carry information. Regular changes of the ion gradient due to changing body fluid concentrations risk subverting the supply of information. The strategies adopted have been studied in relatively few cases. In the more stenohaline animals nerves show a rapid hyperpolarization and reduced action potential amplitude when in hyposmotic media, apparently due to increased potassium leakage as P_K increases to assist volume regulation. The hyperpolarization is reversible on return to normal sea water if exposure has not been for too long or too severe. However, conduction block (no further propagation of action potentials) often sets in irreversibly at 60–70% sea water.

In the moderately euryhaline polychaete worm *Sabella*, which copes with a 50% dilution of its medium, changes in neuronal responses occur much more slowly, primarily because hyposmotic conditions produce a swelling of the protective glial cells that surround the nerve cords, thus reducing the access of ions to the extracellular spaces around the axons and helping to maintain their function. This purely physical response may be enough to sustain the nerves through many tidal cycles, insuring that in the natural habitat hyposmotic conditions are experienced only gradually at the axon membrane surface. Meanwhile, a slow decline in internal K^+ concentration is matched to the declining external K^+ levels, so that the resting potential is relatively unaltered and conduction can continue. Axons also show an increased sodium selectivity in hyposmotic media, compensating for the reduced sodium gradient; hence the overshoot of the action potential can be reasonably well maintained.

The common mussel, *Mytilus edulis*, can osmoconform down to about 25% sea water, and shows another suite of responses that allow it to maintain nearly full-sized action potentials. During acute exposure to low salinity the axons, all of which are of small diameter, swell up and lose both sodium and potassium, together with nonionic solutes such as amino acids, and they therefore depolarize and lose their ability to conduct. However, this effect is reversible and the axons rapidly recover in normal sea water. In longer term and more gradual exposures (more typical of these animals' normal experience) axonal swelling is resisted by an onset of thickening (with increased collagen production) in the acellular neural sheath that surrounds the bundles of axons, allowing the nerve cells to maintain a relatively normal volume. Over a similar time course the axon membranes also synthesize extra sodium pump proteins

(Na^+/K^+-ATPases), with an increment of about 70% for a dilution to 25% sea water, which will help to maintain an adequate Na^+ gradient and also sustain internal potassium levels.

Nerve function has been examined in the even more spectacular conformer *Mercierella enigmatica* (*Ficopomatus*), an annelid worm that can function even in tap water. Here there is no obvious neural sheath around the axons, yet they again seem to have a mechanism to counter swelling in the form of extensive hemidesmosome systems in the axon membranes, which are the anchor points for an elaborate internal network of neurofilaments, strapping the axons together from within and preventing much change of volume. The giant axons in this species also show reduced intracellular electrolyte concentrations in hyposmotic media, but there is a disproportionate retention of potassium so that the nerve hyperpolarizes markedly, maintaining a large overall action potential amplitude despite the reduced overshoot resulting from sodium depletion.

Overall, the nerves of osmoconformers thus seem to have various options to cope with osmotic dilution, and by some combination of all these techniques, operating over appropriate timescales, they can avoid losing electrical conduction and can adapt their nerves to varying osmotic regimes.

1 They have systems for limiting damage due to axonal swelling: a volume increase of the surrounding glial tissues, or intracellular strapping within particularly large diameter cells, or synthesis of extra neural sheath as a physical protection.

2 The kinetics of responses to hyposmotic stress are approximately matched to ecological requirements. The polychaetes cope with very rapid change, whereas *Mytilus* nerves can only maintain excitability if dilution is gradual, normally being protected from more rapid change by the simple trick of closing their shell valves.

3 All the invertebrate osmoconformers tested show a reduction in intracellular potassium, but the effect of this on the cell resting potential varies. In *Mercierella* intracellular loss is not proportionate to extracellular dilution, so there is a hyperpolarization, whilst in *Sabella* it is in proportion to reducing external K^+ so that the resting potential is maintained, and in *Mytilus* the accompanying losses of other solutes lead to a net depolarization.

4 The resultant effects on the crucial action potential spike amplitude are equally varied: in *Sabella* an increased sodium selectivity allows the action potential size to be maintained, whilst in other species the internal sodium level is reduced to help maintain a Na^+ gradient, and hyperpolarization may insure that the total spike amplitude is largely unaltered. Over a longer timescale extra synthesis of sodium pumps may be a common strategy to assist in sodium gradient maintenance.

By some combination of all these techniques, operating over appropriate timescales, osmoconforming invertebrate animals avoid losing electrical conduction and can adapt their nerves to varying osmotic regimes. Preliminary studies on euryhaline fish show similar effects, with particularly rapid and marked increases in brain Na^+/K^+-ATPase and creatine/creatine kinase levels.

Table 12.7 Lower salinity tolerance of brackish-water species.

Species	Salinity tolerance (% sea water)	Strategy
Annelids		
Sabella	~50	Conformer
Nereis diversicolor	15–20	Moderate regulator
N. limnicola	~5	Moderate regulator
Mercierella enigmatica	<1	Conformer
Molluscs		
Nassarius	25–30	Weak regulator
Mytilus edulis	15–20	Conformer
Patella vulgata	10–15	Moderate regulator
Crustaceans		
Cancer	~50	Weak regulator
Gammarus obtusatus	~30	Moderate regulator
Callianassa	25–30	Conformer
Ligia oceanica	40–50	Hypo-hyper-regulator
Palaeomonetes varians	20–30	Hypo-hyper-regulator
Upogebia	15–20	Moderate regulator
Callinectes	10–20	Strong regulator
Carcinus maenas	10–15	Strong regulator
Artemia salina	~10	Hypo-hyper-regulator
Gammarus duebeni	<1	Strong regulator

Salinity tolerance and water loss tolerance

The ability to tolerate a range of salinities, resulting from some mix of all the above strategies, is often seen as the key adaptation for brackish-water animals. Their tolerances may be listed by species, as in Table 12.7, reflecting the lower point of all the lines in plots such as Fig. 12.13. It is worth remembering three key points here, however. Firstly, this lower limit is not absolute but is exposure time dependent—most species will survive considerable periods below this "threshold", although they may be sluggish and unable to maintain a positive energy balance. Secondly, there will be considerable individual variation in this tolerance limit within many species, affected in part by size but also by the variability in all the other factors that interact with salinity tolerance (oxygen uptake and varied acid–base balance, nitrogenous balance, thermal tolerance, continued ability to feed or to locomote, etc.). For example, the whelk *Nassarius*, listed as having a species lower limit of 11.5 ppt, may have individual limits varying from 7 ppt to as high as 25 ppt, and the mean tolerance will also vary somewhat between populations. Thirdly, and perhaps most crucially, tolerance may be very different in juvenile stages. For example, in supratidal grapsid crabs the larval stages have some euryhaline tolerance from the moment of hatching (15–35 ppt), but tolerance of low salinities gradually increases through development to 5–45 ppt in the third (zoea) larva and 5–55 ppt in the final (megalopa) stage and the adult. This may be mainly a reflection of the decreasing surface area to volume ratio of the growing individual.

Tolerance of water loss is also highly developed in many estuarine and littoral animals (see Table 5.2): the loss of body water without fatality can be 35–60% in limpets, and up to 75% in chitons. As desiccation sets in, many of these intertidal animals can use metabolic depression (see Chapters 6 and 7) to reduce their oxygen need and thus limit the need for any shell-gape for respiratory exchange, thereby reducing their water loss rates further. In a few species, such as the copepod *Tigriopus*, from supralittoral splash zones where pools of water may be very ephemeral, there is a capacity to be almost

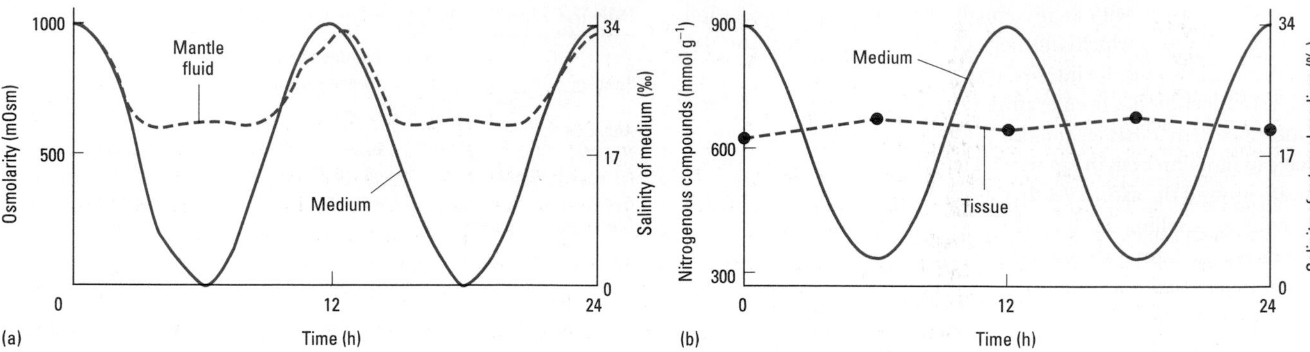

Fig. 12.20 The effects of "smoothing out" environmental change in a bivalve mollusc (here the mussel, *Mytilus edulis*), where (a) mantle fluid osmolarity changes much less than the external salinity, and (b) tissue amino acid concentrations can remain almost unchanged through a double tidal cycle.

completely dried out and yet to recover when wetted, an example of the phenomenon of "cryptobiosis", covered in Chapter 14.

Finally we should return to the key point, which is that many littoral and estuarine animals eliminate the potential problems of osmotic change and dehydration not by complex physiological or biochemical means but by simple evasion. Thus even animals that have the machinery to adapt may rarely need to use it. The prawn *Palaeomonetes*, though a good regulator, moves around an estuary to minimize its need to regulate, following the salinity displacement pattern and being commoner higher up river in the summer and fall when inwards salinity encroachment is greater. Figure 12.20 shows that for *Mytilus*, which as we have seen also has a suite of possible mechanisms to protect its tissues, the strategy of shell closure at low tide is common enough to make these other mechanisms unnecessary. The mantle fluid remains relatively concentrated even in dilute media, and the tissues show hardly any change in amino acid level through a normal 24 h tide cycle. Similarly for many crabs with good osmoregulatory physiology, such as *Uca*, much of their normal "regulation" is in practice achieved by the use of the burrow during periods of environmental stress, conditions inside the burrow being thermally and osmotically much more stable. Physiological change is a last line of defense here (perhaps rarely called upon except when very low tides coincide with warm dry weather); it is not the mainstay of the adaptive response. An interesting demonstration of the interactions and trade-offs of physiology and behavior comes from work on semiterrestrial crabs (*Sesarmops*) in Taiwan. Here the large and small individuals live in burrows near the waterline, whilst only medium-sized ones live further up the shore. The small crabs lose body water very rapidly when exposed but can regain it very quickly from the substratum or when immersed, while the large crabs have a low water loss rate but cannot readily access interstitial water and take a long time to rehydrate and so cannot afford to be too far from water. Only the medium-sized crabs benefit from both moderate water loss rates and an ability to rehydrate quickly from the interstitial supplies, and can dig their burrows further from the water creeks.

12.3 Thermal adaptation

Essentially, life is thermally very much more variable away from the buffering of constant cover by the seas, so that every animal living in the seashore zone and in estuaries needs to be eurythermal to cope with increased thermal variation. This variation can be manifested in several ways and over differing timescales. Tidal effects mean that on shores there are marked differences over the 12 h cycle as the water advances and retreats (e.g. Fig. 12.12b). Seasonally the temperature of both water and air will vary, especially in temperate and high-latitude estuaries. Figure 11.7 showed the thermocline effects in surface sea waters, so that the waters lapping temperate shores and estuaries may be 7–10°C higher than most of the sea. Some kinds of shoreline animals are therefore required to be more tolerant of thermal extremes than almost any other animals. For example, the rock pool goby, *Gillichthys seta*, is one of the most eurythermal of animals, tolerating body temperatures (T_b) from 8 to 40°C. But in addition to the tidal and seasonal variations there are also globally varying interactions of warm surface currents and cold deeper currents, causing substantial intermittent thermal overturns on some shores. Of these, the El Niño events in South America are best known (Box 12.2). Finally there are long-term global changes in climates that may need to be accommodated by genotypic change within a species; and on a similar magnitude and timescale there may be effects arising from animals invading along shores and colonizing new thermal habitats.

Thermal changes on the scale of tidal and seasonal effects require some level of phenotypic adaptation. Usually we find behavioral change first, and then an intervention of physiology or biochemistry; remember the mechanisms for thermal acclimation and adaptation of enzymes and membranes in animals discussed in Chapter 8. Hence, amongst more sedentary littoral invertebrates there are clear zonational patterns in thermal traits, including tolerance limits, membrane fluidity, mitochondrial respiration, heat shock protein (HSP) expression, and protein stability, many of which impose adaptive maintenance costs that may contribute to limiting a species' upward penetration of the shore environment. But what further choices does a shoreline organism have when it comes to temperature regulation? There are various sources of heat loss and gain, and ways of using and controlling these; for essentially aquatic organisms the main problem is often to avoid getting excessively hot when the tide is out.

Box 12.2 The El Niño Southern Oscillation (ENSO) and its effects on coasts

The phenomenon of El Niño is a periodic reversal of the main currents across the Pacific Ocean, occurring every few years and having substantial consequences for world weather patterns.

It arises in an oscillatory (periodic) manner when tropical Pacific water warms up (by up to 2°C in places) and instead of remaining local (and causing intense tropical rains) the surface waters from the western Pacific, around the Indonesian area, migrate eastwards, accompanied by storm systems. Recent ENSO events have occurred in 1982 and 1987 (for 1 year), 1991 (with effects lasting for up to 3 years), and 1997 (even more prolonged). There is evidence that the effects are becoming more frequent, more intensive, and more prolonged, probably because of an interaction with global atmospheric warming.

On land, the meteorological consequences commonly include risk of drought and wildfire in Australia, Brazil, East Africa, and parts of India, and risk of torrential rain and flooding in normally desert regions of North America, the Peruvian coast of South America, and the western side of the Indian subcontinent.

Such effects clearly also impact upon coastal ecosystems, and in the eastern Pacific these changes have been well documented:

1 There may be a substantial deepening of the thermocline, with the phytoplankton redistributing downwards.

2 Californian fish have occurred well north of their normal latitudes, as have crabs and squid; commercial catches in southern California are often greatly reduced.

3 More northerly species have been displaced, with salmon fisheries in Oregon much depleted and knock-on effects on sea birds and on sea lions and fur seals.

4 Coastal planktonic communities have changed, with correlated decreases in filterers, predators, and top predators; Peruvian anchovy stocks have therefore crashed, again affecting sea lions and pelicans, with many sea birds abandoning attempts to breed.

5 Relatively stenothermal low-shore animals have declined on sensitive beaches.

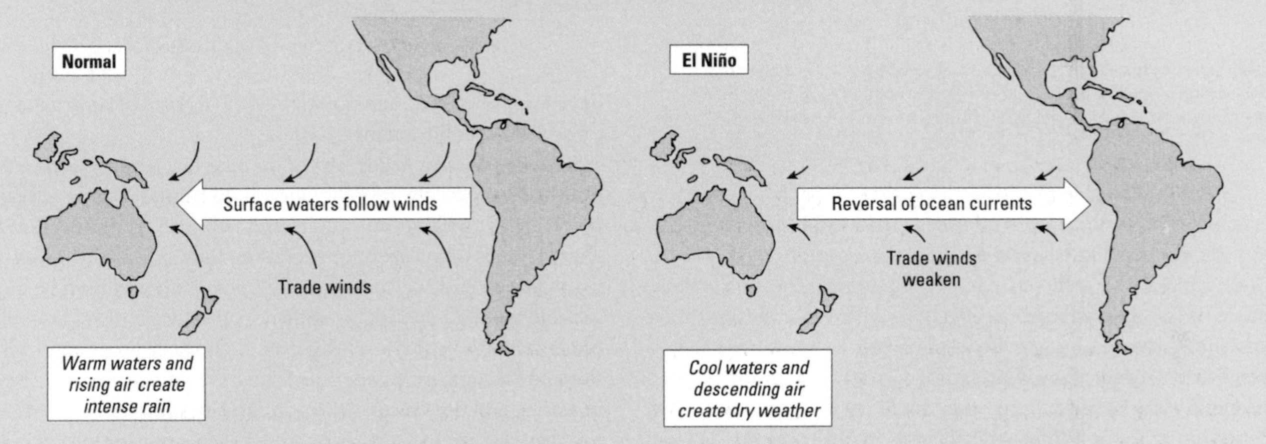

12.3.1 Problems of overheating

Heat stress is likely to be a cyclical problem, overlaid with a seasonal component on all but tropical shores. Summertime low tides will bring the greatest risk of overheating for most littoral and estuarine animals; large proportions of intertidal mussels can be killed or suffer severe tissue damage in unusually hot dry summers.

Being the right size, shape, and color is part of the solution to thermal stress, so that we often find intraspecific clines of size or shell height or shell color on beaches, particularly within the molluscs. Small size gives a rapid heat gain or loss, but also a rapid desiccation rate; larger size helps to reduce heat exchange rates in very variable conditions such as the middle shore, but may lead to overheating through prolonged exposure to the sun on the upper shore. Thicker shells in dog whelks (*Nucella*) appear to help reduce thermal stress (though they were probably selected principally to reduce predation). Shells with elaborate ridges appear to help increase convective heat loss in some beach gastropods, and such shells are much commoner in warmer latitudes. Paler colors help to reflect some of the radiation so that heating rates are reduced. For example, the dark shells of winkles have a reflectivity of 2–10%, whereas bright yellow or white shells may give a value of 30–50%. Even in mussels the difference between black and brown morphs is enough to produce a significant thermal difference, so that populations are nearly 100% black in eastern Canada and only about 40% black in the southeastern USA (Fig. 12.21). An ability to change color may also be particularly useful to seashore animals; it has been demonstrated in a wide range of isopod and amphipod crustaceans, as well as in certain fish. The isopod *Ligia* changes color using cells (melanophores) in which dark pigments can be either widely dispersed or tightly condensed, producing a reflectance varying from 2% (almost black) to 10% (very pale beige). Here the cuticle color is determined both by body temperature needs and by the requirements of background matching for camouflage. In addition there are underlying circadian and circannual rhythms of color change. As these animals are dawn and dusk scavengers the main priority is to insure crypsis at such times, but this can be adjusted according to hygrothermal need (paler to stay cool or darker to warm up).

In addition to morphological adaptation, increased tolerance in the animal's cells and tissues is also required, exploiting the

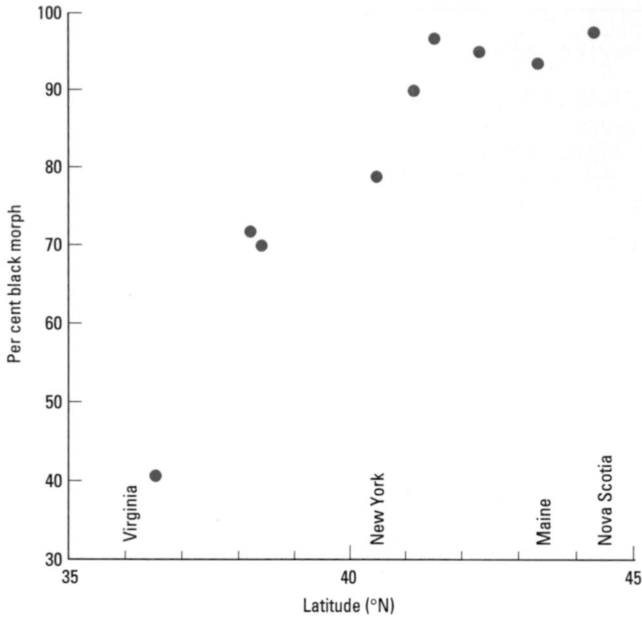

Fig. 12.21 A cline in coloration in the mussel *Mytilus edulis* along the eastern shores of North America, with progressively more of the black morph (better at absorbing solar radiation) at higher latitudes. (From Innes & Haley 1977; Mitton 1977.)

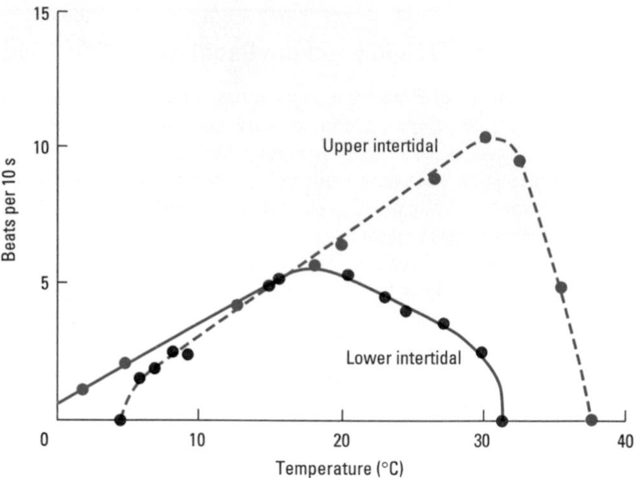

Fig. 12.22 Effects of height on the beach on thermal responses. The high-tide barnacle species show increasing cirri beat frequency with rising temperature up to and beyond 30°C, whereas low-tide species progressively lose cirri activity above just 18°C. (From Southward 1964.)

mechanisms for protecting membrane function and enzyme activity that were discussed in Chapter 8. The mussel *Mytilus californianus* experiences a 20°C cyclic change in body temperature on a daily basis, as well as marked seasonal effects. Its gill membrane phospholipids show seasonal homeoviscous adaptation, optimizing membrane fluidity (order) for a winter mean T_b and for a summer upper extreme T_b despite the fact that they are likely to experience water submersion at about 10°C on a daily basis in both seasons. At least in high-shore populations they also show an ability to alter their membrane order within hours, to counter the thermal effects of the tidal cycle. Mussels also show local and seasonal variation in their levels of HSPs (see Chapter 8), with ubiquitin and HSP70 increasing in gill tissue in summer months, but more so in intertidal than subtidal populations.

These phenomena of increased membrane and tissue tolerance are manifested in maintained or increased activity levels in many animals. For example, high intertidal barnacles are able to keep their filter-feeding cirri beating at markedly higher temperatures than individuals from further down the beach (Fig. 12.22). Mean upper critical temperatures (UCTs) and thermal niche width—the difference between UCT and lower critical temperature (LCT)—are generally positively correlated with the maximum height of distribution for a range of species on any particular shore. However, some common shore animals such as the winkles (*Littorina*) show a lack of the normal temperature-related changes in oxygen consumption expected in most ectotherms. This may reflect their experience of marked diurnal temperature change, where any benefits from longer term temperature compensation would be negated by extreme daily fluctuations. Instead, the littorinids have an unusual ability for nearly instantaneous suppression of metabolic rate and entry into short-term metabolic "diapause" at high temperature (20–35°C, depending on species and population). This kind of strategy may be present in other littoral invertebrates.

Behavior is also a major part of the adaptive strategy to cope with heat stress, especially for invertebrates. Examples include retreating to a burrow, seeking out cooler shaded crevices, hiding beneath dense clumps of seaweed, or excluding the warming air by clamping tightly to a rock. The pattern of burrowing may itself be highly variable within a species, as shown in the supralittoral amphipod beach-hoppers, which may dig deep high-shore burrows in winter to avoid freezing, shallower downshore burrows nearer the tide line in spring, and deeper burrows again in late summer to avoid overheating. Aggregation may also be used to increase the effective thermal mass (whether to keep warmer in winter or cooler in summer), and/or produce a more humid microclimate. Mussels in aggregations have T_b values 4–5°C lower than similarly sized loners in short hot spells of weather, whilst winkles often aggregate more when on dry and cool substrata and then achieve significantly higher body temperatures over a longer term. Behavioral thermal regulation is also in evidence in seashore vertebrates. Blennies, for example, seek out deeper colder water in rock pools in the summer months. But probably the best known example is on a larger scale, with the Galápagos iguana, *Amblyrhynchus cristatus*, living intertidally and subtidally on cool rocky shores. The Galápagos islands are equatorial, therefore receiving strong solar inputs, but they are bathed by cold currents, so that diving for their seaweed food cools the iguanas substantially. After a dive, when T_b falls despite peripheral vasoconstriction (see Chapter 8), they use a complex basking regime: they initially orientate sideways on to the sun on fully exposed rocks to regain their preferred T_b of 36°C, or they may warm even faster by crouching down on to the black lava rocks in the middle of day, when these surfaces may reach 50°C. Once T_b reaches about 40°C, the iguanas need to avoid overheating, and so turn head on to the sun, with the head and neck raised so that the rest of the body is in shade, and with the thoracic region raised off the substratum, so that breezes coming off the sea will cool the chest region.

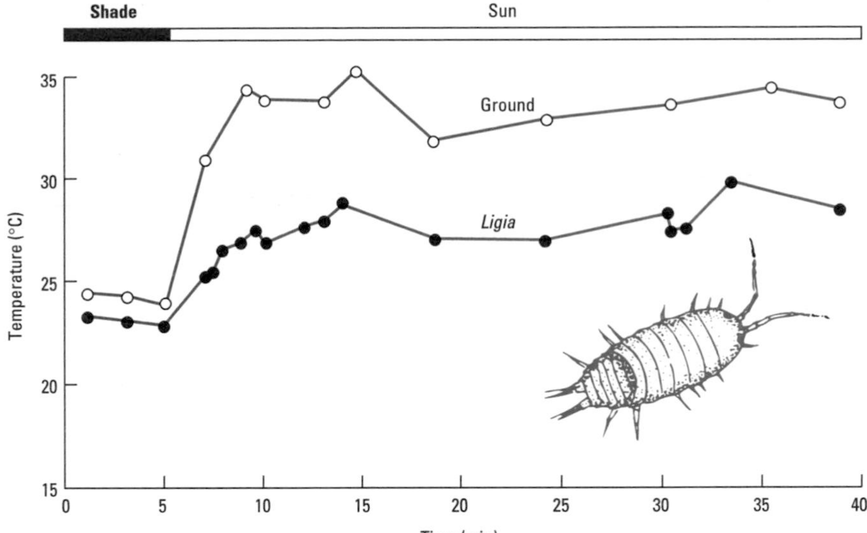

Fig. 12.23 The body temperatures (T_b) of the isopod *Ligia oceanica* (the sea slater), measured before and during exposure to the sun, in comparison with nearby ground, at a humidity of 45–52%. The insolated animals are cooled by as much as 7°C relative to their surroundings by the effects of evaporative water loss, keeping their T_b below 30°C. (Adapted from Edney 1953.)

For seashores, though, there is a whole new area open for possible thermal regulation, and that is **evaporative cooling**. This is not available to fully aquatic animals (except in very rare cases of floaters that emerge above the surface, such as the Portuguese man o' war jellyfish), and it is usually too risky to be used except in emergencies for fully terrestrial animals without reliable water supplies. But it is the perfect solution on shores; animals can use up some of their water to keep cool, with the security that the water will come back again to replenish the stores with the next tide. Remember that evaporative cooling is also very efficient, using up 2500 J g^{-1} of water lost. For example, a littoral isopod scavenging in the sun at low tide may be able to keep its body temperature 5–8°C below the air temperature by use of evaporative cooling (Fig. 12.23), and a crab such as *Uca* achieves a similar reduction, with its burrow providing a source of renewable standing water when the tide is out.

Seashore and estuarine animals show a new range of strategies to take advantage of this opportunity. They may take up an extracorporeal water store (in the mantle cavity of molluscs, or the gill chambers of crabs, etc.), and they may control access of this water to air, for example by a judicious gaping of the shell or dribbling of water from the gill space, to get their temperature balance right. Gaping and the use of evaporative water loss (EWL) are clearly adaptively exploited according to the environment experienced, and are not just a reaction to stress. For example, tropical oysters growing high up on the shore have a UCT of 40–42°C, roughly corresponding to the T_a during midday exposure, and they remain tightly closed during such exposure. In contrast, mangrove oysters in similar latitudes, which are exposed to less extreme ambient temperatures and remain in more humid air when open, have a lower UCT and use gaping to increase their EWL and stay cool. Similarly, tropical limpets are more prone to use "mushrooming" behavior, lifting the shell off the substrate, than their temperate relatives, and this would appear to be a strategy for increasing EWL.

Shoreline animals able to gape and regulate EWL may also withstand temporary severe desiccation if necessary, and desiccation resistance tends to increase up a beach. The values of 60–75% water loss quoted in section 12.5 are far greater than would normally be needed for the 12 h tidal cycle, i.e. these animals are built to withstand the worst possibility of being exposed for a full 13–14 days between extreme high tides once or twice a year. They are also greater than the desiccation tolerances of most terrestrial animals (see Table 5.2), and are really only bettered by truly cryptobiotic creatures, dealt with in Chapter 14. Hence it is fair to say that water loss is a fundamental strategy for regulating T_b in shoreline animals, unlike almost any other habitat.

Because of this role for evaporative cooling on seashores, clearly we cannot isolate temperature studies from water balance studies for littoral species. Surviving high temperature depends on an ability to be euryhaline and to cope with varying blood concentrations. For example, in various shore isopods the ability to withstand lower salinity decreases as the temperature rises, and the two systems are intimately linked. But again the settings for temperature tolerance and desiccation tolerance vary with species, and with the position on shore that each species occupies, as well as varying intraspecifically.

Thus far we have dealt mainly with animals on rocky shores and in estuaries, but of course temperature is also a problem for animals living in and on sandy shores. Once it is exposed and dried, sand can get very hot (often too hot for humans to walk on). It is usually not protected by weeds or other surface cover as on rocky shores. Problems of overheating are particularly bad on dark-substrate sands such as on volcanic islands, where the absorptivity is high, so that these kinds of beaches tend to have a different and rather specialist fauna, deeply buried during low tide. Deep burrowing can help solve most of the ecophysiological problems on shores for those animals that can actively move downwards reasonably fast. Intertidal animals tend to burrow deeper than subtidal ones (Fig. 12.24); they need to get down to about 250 mm in temperate climates to avoid heat and drought stress effectively. Some (e.g. razor shells) can migrate vertically through the sand with the tide, which helps to maintain a fairly constant T_b. Animals with shells are also in a sense preadapted to reduce heating stress in sandy beaches, though they may lose out on the option of keeping cool by evaporation through the gape if they also become too water-stressed.

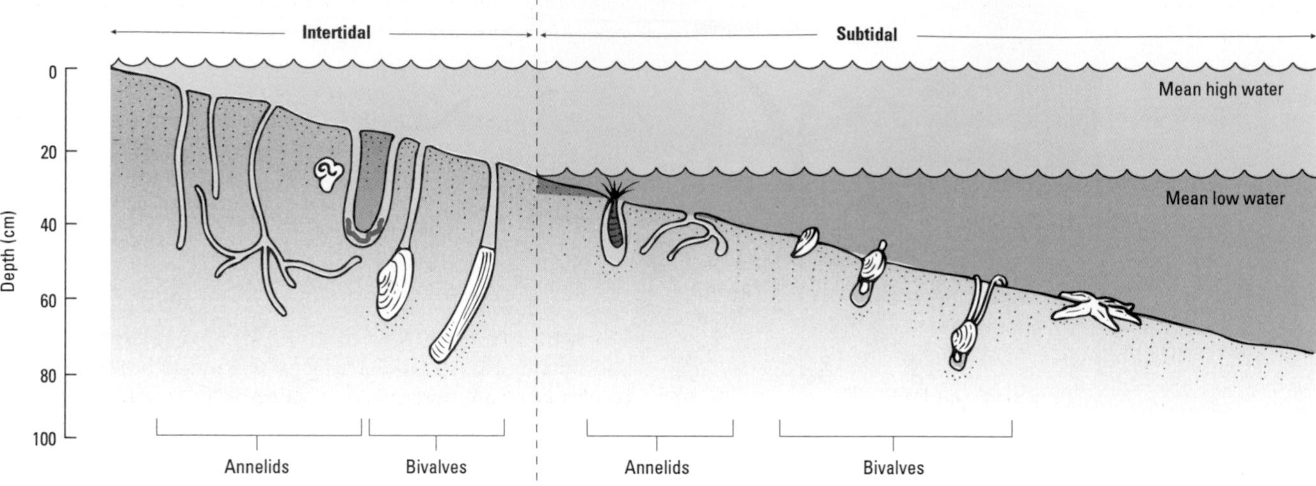

Fig. 12.24 Burrow depth vs. shore height in a soft sediment beach. Depth is greater in the intertidal zone, where the burrow provides microclimatic amelioration at low tide. (From Rhoades 1966.)

We should also note here the "special" shoreline cases of coral reefs and mangrove swamps. Coral reefs constitute a "manufactured" sublittoral and littoral zone, and are at their maximum development in up to 10 m of water at unusually high sea temperatures of 25–29°C. In some areas the reef may be almost exposed at low tide, giving even higher temperatures due to direct solar heating. Hence coral reef organisms (especially the coral cnidarians themselves) are all adapted to high thermal regimes, and have very high growth rates and hence enormous productivity. The same is true of many mangrove specialists, including the snapping shrimp, *Alpheus viridari*, which can survive 25–45 ppt salinity (regulating its body fluids fairly constant) and up to 35°C water temperature. However, animals within the mangrove forest will be more protected from the sun because of shading by foliage. Nevertheless, all these warm-water animals may be at risk when temperatures rise beyond their limit. Many are at special risk since this may also be linked with increased surface evaporation, causing local sea-level lowering and an increased exposure of the coral or mangrove root habitats to direct unfiltered radiation from the sun and thus to UV damage.

12.3.2 Coping with cold

Rock pools and interstitial waters in temperate and cool biomes may freeze for some part of the tidal cycle during winter months. Shore and estuarine animals in higher latitudes are therefore mostly freeze tolerators, with their body fluids able to tolerate freezing and thawing on a regular basis; this is true of most of the barnacles, bivalves, and gastropods tested. The minimum temperatures survived by most intertidal invertebrates with freeze tolerance are not particularly low, often around −10°C; with reimmersion in sea water above −2°C always imminent with the returning tide, body tissues will rarely cool below this level. Indeed, in deeper rock pools animals may not freeze at all, since the surface ice both protects the water beneath and withdraws fresh water from it, leaving the pool water strongly hypersaline and therefore unfrozen to as low as −8°C (Fig. 12.25). Hence there is often no clear relationship between freezing resistance and position on the shore.

Cold acclimation for shore animals involves the classic strategies of membrane and enzyme adaptation discussed in Chapter 8. For example, crabs undergo membrane lipid changes allowing increased membrane fluidity, specifically decreasing the cholesterol to phospholipid ratio but leaving the saturated to unsaturated fatty acid ratio largely unchanged. The stenothermal crab, *Cancer*, has slightly greater membrane fluidity and a lower cholesterol: phospholipid ratio than the eurythermal *Carcinus*. Enzyme adaptation has been particularly studied in the mussel *Geukensia*, in which freezing causes coordinated suppression of multiple cytosolic and mitochondrial enzymes involved in carbohydrate metabolism, primarily by reversible protein phosphorylations, so that biochemical pathways remain in balance.

For more extreme high-latitude seashores, air temperatures of −30°C are not uncommon, and resident animals on the beach may have 90% of their body water frozen. The cells must then be

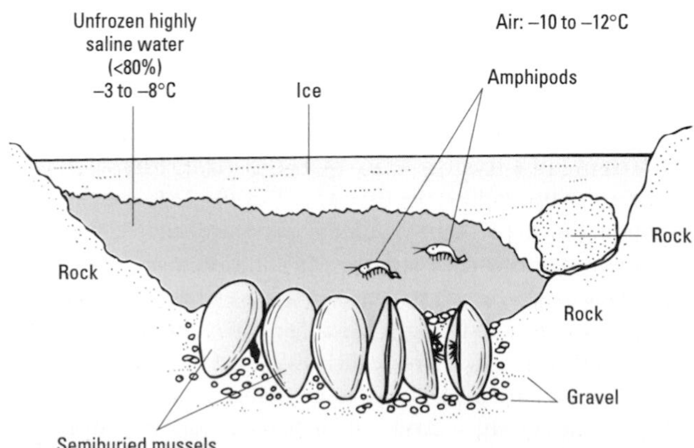

Fig. 12.25 The microenvironment of a high-shore rock pool in winter on a cold Norwegian shore; the upper pool is frozen, withdrawing fresh water from the water below, which becomes substantially hypersaline and therefore does not freeze even at −3 to −8°C, allowing amphipods also to remain unfrozen. (From Davenport 1992.)

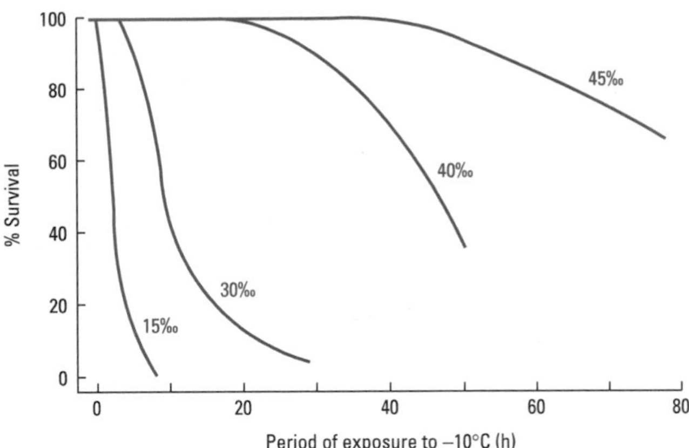

Fig. 12.26 Freezing resistance in the common cockle (*Cerastoderma edule*), showing acclimatory effects such that pre-exposure to high salinity substantially improves survival of freezing, presumably due to intracellular accumulation of amino acids. (From Davenport 1992; adapted from Theede 1972.)

tolerating remarkably high osmotic concentrations (incidentally rendering them even less likely to freeze intracellularly). The tissues and cells become very distorted by the surrounding ice formation, cells becoming shrunken and morphologically altered. Nevertheless, many of these animals, including the common upper- and mid-shore winkles and limpets, can survive repeated freeze–thaw cycles and have fully restored tissue function shortly after each thaw. Note too that salinity and freezing resistance are interactive; animals that are already acclimated to hypersaline conditions are substantially more freeze resistant (Fig. 12.26), since both phenomena involve increased concentrations intracellularly.

Polar inshore and shoreline invertebrates that are essentially marine but may experience transient aerial exposure are exceptions to the general rule of low UCTs in animals from cold seas. For example, Antarctic copepods may have UCT values of 14–18°C compared with only 4–6°C in fish that come from similar habitats but are mobile enough to guarantee staying submerged.

12.4 Respiratory adaptation

Oxygen is usually fairly freely available in littoral waters, but estuaries can become both significantly hypoxic (especially so where man's activities supervene) and significantly hyperoxic on sunny days when primary productivity is at its peak. In these habitats hypoxia tends to be accompanied by hypercapnia (elevated CO_2), and vice versa. Additionally, many of the animals on shores have respiratory problems because they are endowed with a system that only works well either during submersion or during exposure. The gills of molluscs, crabs, annelids, and fish will commonly collapse and adhere together when unsupported by water, reducing the exchange surface area and producing hypoxia even though oxygen is freely available in the air. At the other extreme, lungs and tracheae are unable to function if filled with water, and their possessors rapidly drown. Relatively few animals are truly amphibious in this respect, and therefore many shore dwellers are facultative anaerobes. In particular, primarily aquatic forms may get stranded in pools or in mud as the tide goes out: in either site they can get short of oxygen quite quickly, because of rising temperatures and evaporation, leaving little or no O_2 in solution. Marine muds are particularly impermeable to gases, so rapidly go black and anoxic below the surface, with the resident bacteria converting nitrates and sulfates to nitrites and sulfides. We therefore need to consider both aerobic and anaerobic respiration for most shore and estuarine animals.

12.4.1 Aerobic metabolism

Whenever oxygen is fairly freely available we find both regulators (i.e. animals that can maintain their consumption rate regardless of conditions), and conformers (i.e. those whose consumption varies in line with oxygen availability). Animals from more consistently highly oxygenated environments are more likely to be conformers (e.g. the fully marine swimming crabs, and the coastal polychaete worm *Marezelleria viridis*), and those from variable sites not surprisingly tend to be regulators (e.g. spider crabs from tidepools, shore crabs from estuaries, and the littoral polychaete worm *Nereis diversicolor*). The latter can genuinely be described as "euryoxic" animals, although their respiratory responses may be modified in some conditions by other interacting stresses such as lowered salinity or raised temperature.

Conformers

Conforming strategies generally involve a reduction of metabolic rate (adaptive hypometabolism) and/or a gradual inclusion of the anaerobic pathways dealt with in the next section. The high-shore limpets, *Siphonaria*, can reduce their metabolic rate to just 18% and incur little or no oxygen debt, while in the lower shore limpet, *Patella*, anaerobic respiration is commoner and tolerance much lower. Even within a genus, both regulators and conformers in this respect may occur according to height on the shore. The porcelain crab genus (*Petrolisthes*) has high-shore species that retain a reduced level of aerobic metabolism at low tide, and low-shore species that shift to anaerobic metabolism whenever they are in air. Oxygen usage limitation by slowing metabolism has also been demonstrated in shore crabs and in oyster larvae, amongst others, and in many intertidal animals it is greatly aided by reducing activity when exposed to air. Many species lower their heart rates and ventilatory rates, and bivalves reduce their heart rate and stop all filter-feeding activity when the tide goes out.

Regulators

By contrast, oxygen regulators can achieve a steady oxygen uptake by a greater variety of means, different species adopting a different combination of these.

CHANGES TO THE VENTILATION RATE OR DEPTH

This may be the simplest response, with increased rate or depth in low-oxygen conditions. For example, many crabs can alter their ventilation rate by varying both the frequency and the "stroke volume" of the scaphognathite (the flattened plate whose beating forces water through the gill chamber; see Fig. 7.9).

CHANGES IN HEART FUNCTION

In animals with closed or partly closed circulations, including the littoral crabs, there is the option of quite subtle changes in heart rate, or in the stroke volume of the heart. The crab *Callinectes* elevates its heart rate by up to 200% during aerial exposure.

USE OF PIGMENTS

A further adaptation that assists the maintenance of aerobic respiration is the predominance of respiratory pigments in littoral and estuarine animals. This is particularly common for annelids in sandy and muddy shores, where the lower layers are substantially deoxygenated. For example, some worms, such as *Arenicola* and *Sipunculus*, have enough hemoglobin to keep them oxygenated through the whole of the normal low-tide cycle, and estuarine fish can mobilize extra red blood cells from the spleen during periods of hypoxia. In the longer term, regulation may require changes to the oxygen carriers, switching gradually to higher affinity pigments; for example, the annelid tube-dwelling worm *Pista pacifica*, from anaerobic mudflats, has three different hemoglobins with different oxygen affinities. Acclimation may occur within an individual's lifetime; for example, in the blue crab, *Callinectes*, a higher affinity hemocyanin is produced during prolonged hypoxia. Evolutionary adaptation tends to produce higher affinity pigments (ones having a lower P_{50}, the concentration for half-maximal loading; see Chapter 7) in the fully water-breathing species from shores and salt marshes. For example, aquatic littoral crabs with gills have much lower P_{50} values for their hemocyanin pigment than upper littoral and fully terrestrial crabs, where P_{50} is higher and arterial blood readily becomes fully saturated in air (Fig. 12.27).

Pigments in some animals from muddy shores and salt marshes, such as the salt marsh killifish (*Fundulus parvipinnis*), may also need to be unusually tolerant of sulfides, which can interact with hemoglobin by binding with the central porphyrin groups, thus making the pigment nonfunctional. The sulfide may be detoxified by reactions that also yield useful sources of energy, the S^{2-} entering the electron transport chain to be oxidized. The unusual echiuran worm (*Urechis*), found on sulfide-rich mudflats, appears to oxidize sulfides in its coelomic fluid, and can maintain aerobic respiration even when exposed between tides in mud with up to 30 μM sulfide. We will meet more elaborate versions of this strategy in deep-sea thermal vent animals (see Chapter 14).

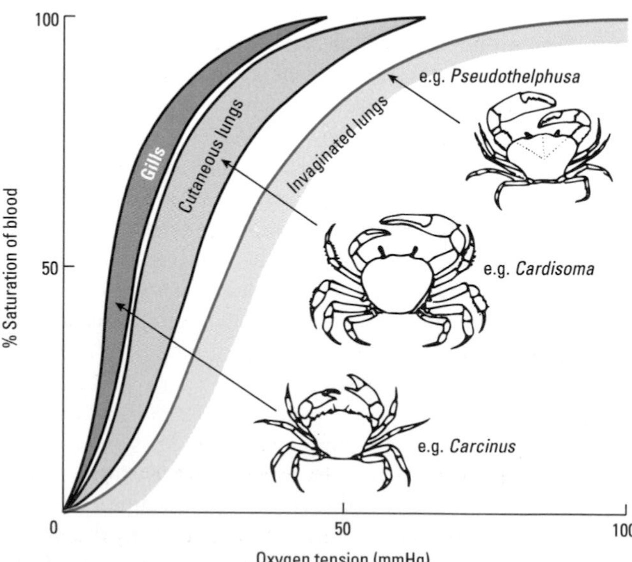

Fig. 12.27 Differences in oxygen-affinity curves and types of respiratory surface for air-breathing littoral crabs. *Carcinus* is fully aquatic with gills and has high-affinity hemocyanin (low P_{50}) whereas *Cardisoma* is upper-shore dwelling with vascularized gill chambers, while *Pseudothelphusa* is a tropical crab living inland with well-ventilated lungs, both these latter species having higher P_{50}s. (From Little 1990; adapted from Innes & Taylor 1986, courtesy of Cambridge University Press.)

USE OF A TEMPORARY OXYGEN STORE

This is an alternative (or supplementary) strategy for many regulators. Water stores are used by primarily aquatic animals (which as we have seen may use the stores as a source for maintaining water balance as well) and air stores are used by primarily terrestrial air-breathers. Many intertidal molluscs retain a store of water in the mantle cavity when the tide is out, and crustaceans may use the branchial chamber in a similar fashion. Many burrowing animals on sandy shores can draw in fresh oxygenated water when the tide is in, and store it in the burrow (perhaps aided by the mantle cavity or gill chambers). Most spectacularly, the mudskipper fish of tidal mudflats can accumulate air in the horizontal side-shafts of their burrows, apparently by air gulping at the surface followed by bubble-release underground, providing safer conditions for their eggs in the roofs of the air-filled spawning chambers during potentially hypoxic periods. Littoral mites and some of the rare littoral insects use air films or trapped air bubbles (discussed in more detail in Chapter 13).

Switching between different sites of oxygen uptake

Maintaining aerobic respiration in regulating littoral animals may involve a more elaborate strategy that entails genuinely "bimodal" respiratory systems that can function, at least to some degree, in both air and water, usually by altering the site of uptake.

Water-breathers in air

On shores, animals that normally breathe from water have to cope with periods of air-breathing when the tide is out, despite their collapsed gills. Thus many prosobranch molluscs have parts of the mantle edge and mantle floor vascularized to take up oxygen once the gills become useless. These molluscs, and many amphibious crustaceans, have reduced gill size and/or number, in parallel with the development of the accessory respiratory surfaces. One of the best adapted examples is the high-tide and supralittoral crab *Leptograpsus*, which shows fully bimodal breathing, maintaining oxygen delivery in both air and water equally, and regularly remaining emersed for 8 h. Switching respiratory mode raises the associated problem of varying acid–base balance. In *Leptograpsus* Po_2 varies markedly during submersion–emersion cycling, producing alkalosis, which is apparently controlled largely by the behavioral trick of air–water shuttling, with a transient period of rapid ventilation and tachycardia on return to water.

Some quite unexpected animals can switch to taking oxygen directly from air to allow continued aerobiosis. For example, some

North American mussels will gape to achieve gaseous exchange across the mantle tissue (though they avoid doing this in very dry air), while grass shrimps (*Palaeomonetes pugio*) can be observed jumping up into the air when the sea water becomes hypoxic. Some littoral crabs have invaginated "lungs" within the branchial chamber, while others have vascularized membranous areas on the outer surfaces of their upper limb segments where they take up oxygen from air. Even within a genus the respiratory surface may vary; for example, the high-shore porcelain crab, *Petrolisthes cinctipes*, has leg membranes, and suffers high lactate accumulation when these are experimentally obscured, whilst the lower shore *P. eriomerus* has no leg membranes and breathes solely with gills.

One estuarine dweller that appears to "prefer" air-breathing despite its aquatic ancestry is the salt marsh mudskipper fish, *Periophthalmus*. This feeds just above the tideline at low tide, and then clings to the creek walls just above the rising high tide, breathing in air for much of each tidal cycle, especially in periods of warmer weather.

Air-breathers in water

Animals that have air-breathing apparatus (lung books, lungs, or tracheae) may be forced to shut these down during periods of submersion, most of them then getting their supplies through the skin instead. Small mites, and strandline insect larvae, probably achieve this readily for short periods, but it is uncommon in other arthropods. Frogs exposed to brackish waters in the supralittoral zone certainly use cutaneous respiration.

12.4.2 Anaerobic metabolism

Intertidal habitats, including mudflats and salt marshes, can be extremely reducing in character, and have some of the highest known rates of sulfide production. Although we have seen that a few animals can remain aerobic, anaerobic metabolism is much the commoner response. In a typical salt marsh up to 12 times as much of the primary productivity is degraded via anaerobic sulfate reducers as by aerobic processes. The hydrogen sulfide concentrations vary with water flow, organic content, and season but, in general, sulfide is near zero at the sediment surface and increases with depth to the low micromolar range. The most outstanding anaerobic abilities are found in certain burrowing annelids living in reducing sediments where there is no possibility of the pigment storage of oxygen being adequate to sustain aerobicity, and in some intertidal bivalves during aerial exposure. These abilities are accentuated in temperate and boreal habitats, where extreme temperatures also force animals into anaerobiosis.

When an organism becomes cut off from its normal source of O_2, three critical needs must be satisfied (see Chapter 6):

1 Provision of suitable substrate(s) for fermentation.

2 Provision to store, recycle, excrete, or minimize the production of noxious waste products.

3 Provision to re-establish metabolic homeostasis in the recovery phase when oxygen is again available, without undergoing damage from rapid increases in oxygen radicals.

Glycolysis is a highly conserved pathway, as we saw in Chapter 6. Reducing equivalents (NADH) formed by glyceraldehyde phosphate dehydrogenase are oxidized in the terminal step of the pathway, catalyzed by another dehydrogenase. This enzyme is most commonly lactate dehydrogenase (LDH), because for most animals pyruvate is the terminal electron acceptor, producing lactate as the end-product. Arthropods, echinoderms, and vertebrates all rely on classic glycolysis for anaerobic energy production, resulting in the accumulation of large amounts of lactate. However, the anaerobic pathways incorporating amino acids (see Chapter 6) are particularly common in littoral and estuarine species. Such pathways result in the accumulation of a different "imino" end-product: arginine yields octopine, glycine yields strombine, and alanine yields alanopine. LDH is therefore replaced by an analogous "imino dehydrogenase" enzyme. All of these imino dehydrogenases are functionally equivalent to LDH, and thus the ATP output per glycosyl unit is the same. The ATP yields of the different substrates of fermentation were shown in Table 6.2.

There is a distinct preference for octopine formation in many of the brackish-water species studied, and this appears to be linked with high resting levels of phosphoarginine (PAr) (see Chapter 6); the pathway is especially common in swimming bivalves and cephalopods, where PAr levels range from 20 to 30 μmol g^{-1} wet weight. The advantage of octopine production over lactate production may lie with the fact that arginine has serious effects on the catalytic properties of enzymes, whereas octopine is fairly benign. Hence, it is preferable to maintain arginine at a low level and convert it to octopine. However, this advantage does not apply to alanopine or strombine, as the amino acid precursors alanine and glycine seem to have no deleterious effects on enzymes. In annelids, octopine dehydrogenase is usually absent, and alanopine dehydrogenase is more common. Thus prolonged digging activity in the lugworm *Arenicola* results in the formation of relatively large amounts of alanopine. Aspartate and certain branched-chain amino acids can also be fermented in some invertebrates; in intertidal molluscs, aspartate fermentation is stoichiometrically coupled to glucose fermentation, resulting in the production of succinate and/or propionate (see Fig. 6.6).

The metabolic rate is greatly depressed during anoxia. This can be calculated from the rate of ATP utilization and rate of end-product formation. The ATP turnover rates for long-term environmental anaerobiosis vary. In molluscs the rates are rather low, at 5–20 mmol g wet wt^{-1} min^{-1}; in annelids and crustaceans they are higher, around 20–60 mmol g wet wt^{-1} min^{-1}. When compared with the corresponding aerobic resting rates it appears that molluscs can reduce their metabolic rate by as much as 75-fold, and both annelids and crustaceans by about 10-fold. However, since crustaceans seem to use only the lactate pathway they are probably less well adapted than annelids to long-term anaerobic survival. A case study of anaerobic patterns in the common lugworm is shown in Box 12.3.

Bivalves and other molluscs turn out to be peculiarly resistant to anoxia, and in part this is because they have good antioxidant defenses for use in the recovery phase; enzymes such as superoxide dismutase (see Chapter 7) and high levels of the antioxidant glutathione have been recorded in mussels and periwinkles. Their survival of anoxia is also due to their inactive lifestyles, linked to their capacity for much reduced metabolic rates, and is probably further enhanced because the carbonates in their shells act as buffers for the

Box 12.3 Anaerobic lifestyles: a case study

Arenicola marina (the lugworm) lives in burrows up to 300 mm deep in the sand and mud sediments of tidal flats. At low tide it may be exposed to several hours of anoxia. The switch from aerobic to fully anaerobic metabolism occurs in two phases.

During the first phase, or transition period, various metabolic pathways that are immediately available are used; the energy produced comes from utilization of the unusual phosphagen phosphotaurocyamine, and from glycogen. The glycogen is metabolized to pyruvate and then further converted to strombine or alanine. At the same time aspartate is catabolized, yielding oxaloacetate and then malate. This reaction reoxidizes the NADH originating from the earlier stage of glycolysis. Malate is further metabolized to succinate, acetate, and propionate in the mitochondria. However, through the transition phase the energy content of the body wall musculature falls, indicating that energy production is failing to keep pace with requirements.

After about 3 h a second phase of prolonged anaerobiosis is initiated. The affinity of pyruvate kinase for the substrate phosphoenolpyruvate (PEP) is reduced, involving a phosphorylation of the enzyme. PEP is therefore converted to oxaloacetate instead of pyruvate, and feeds into the mitochondria via malate. The energy yield of glycogen catabolism by this direct route to propionate production is more than double that of conventional anaerobic glycolysis, though only about 15% of that gained by aerobic glycogen catabolism. At this stage phosphagen reserves are depleted but the ATP supply and demand pathways have come into closer balance, with the energy content of the cells falling only very slowly.

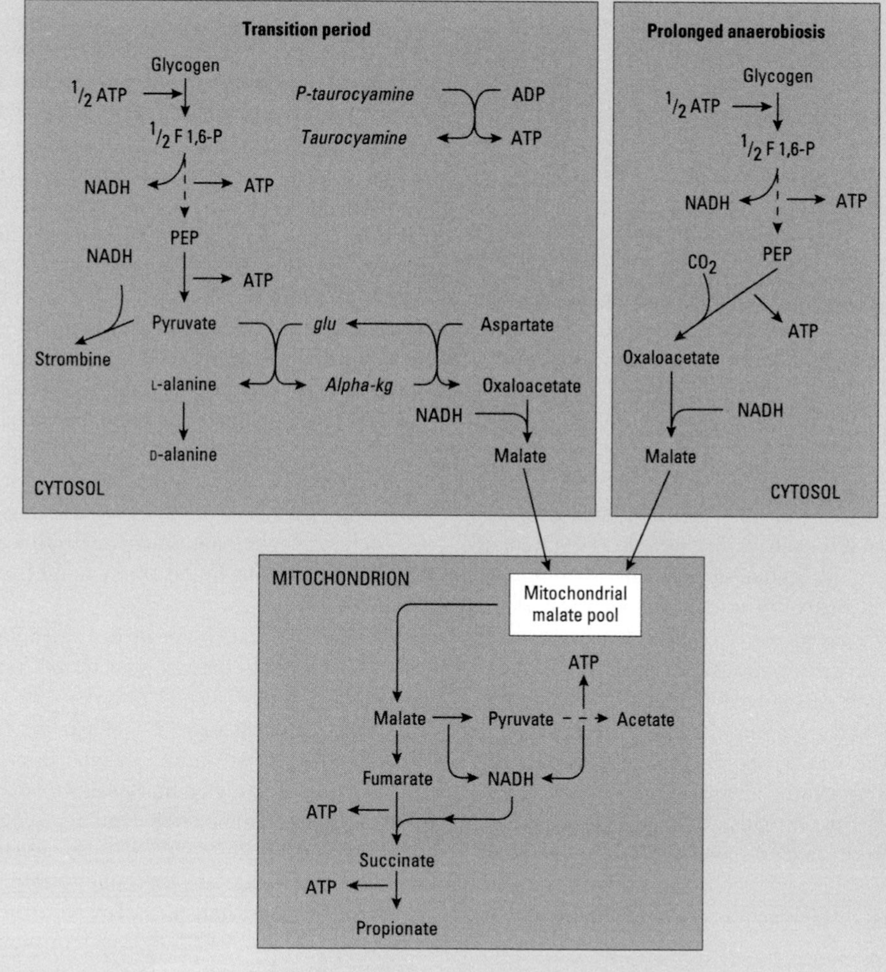

acidic end-products of anaerobiosis. This anoxia tolerance may be one more reason for the preponderance of molluscs in sand and mud beaches.

Restoring metabolic homeostasis after a period of anoxia involves the clearance of anaerobic end-products and the recharging of glycogen stores. Intertidal species that regularly have to sustain low oxygen partial pressures need a quick re-establishment of normoxic cellular conditions, so that ATP and phosphagen levels are rapidly recharged. This recharging, as well as the resynthesis of aspartate and glycogen, the clearance of acidic end-products (by excretion or metabolism), and the resumption of normal activities such as feeding, all increase the energy demand. This is met by a temporarily increased O_2 consumption rate, but also by a small ongoing anaerobic component in the first 2–4 h of recovery, resulting in the accumulation of various pyruvate derivatives.

12.5 Reproductive and life-cycle adaptation

All the problems that are faced by adults in brackish waters, already discussed in the preceding sections, are much worse for the young (and thus almost invariably smaller) stages of animals. Physiological problems are accentuated by the higher surface area to volume ratios, and some studies indicate that desiccation is the greatest physiological threat to juveniles, particularly juvenile molluscs, on shores. Mechanical problems of holding on or station-keeping are also problematic for larval morphs lacking heavy protective shells or fully developed locomotory systems. Nevertheless it is worth noting that some otherwise oceanic vertebrates use inshore waters as breeding grounds, while many semiterrestrial crabs come back to the shore to breed, so that shorelines can be seen as reproductive refugia for various species that have perhaps not fully adapted to a secondarily adopted adult habitat that is in some sense even more "difficult".

Many littoral and estuarine animals avoid some of the problems for their young by adopting more complex reproductive behaviors (Fig. 12.28) compared with marine species. Broadcast spawning is either carefully synchronized with tide patterns to achieve high fertilization rates, or is replaced with direct copulation. A startling example of the latter is found in the sessile barnacles, where a greatly elongated penis is used to transfer sperm to neighbors. Most of the invertebrates lose or greatly attenuate their pelagic larval stages; any larvae that are produced tend to be already rather large and advanced, with a reasonable ability to swim, and they will settle quickly and relatively locally. They are timed to coincide with periods of reasonably benign weather conditions. Thus in high-shore grapsid crabs the eggs are yolky and substantially enlarged, and the larvae are relatively large and highly salinity tolerant, with their stages reduced from four to three; they are able to survive for at least two stages just by lecithotrophy. Many invertebrates similarly switch to laying much yolkier eggs (Fig. 12.29), often laid in protected sites on seaweeds or in rock crevices and overhangs. Sometimes these eggs are protected in toughened or gelatinous cases as in whelks and other gastropods; these egg cases give significant though by no means complete protection against desiccation, osmotic stress, and UV radiation. In the gravel beach-dwelling gammarids, where no such protection is offered, egg size varies seasonally, being greater in winter when disruptive storms are likely. Other invertebrates use a brood pouch to protect the eggs, retaining the embryos until they are relatively large and potentially independent, as in many littoral crustaceans. An alternative strategy is to lay and then

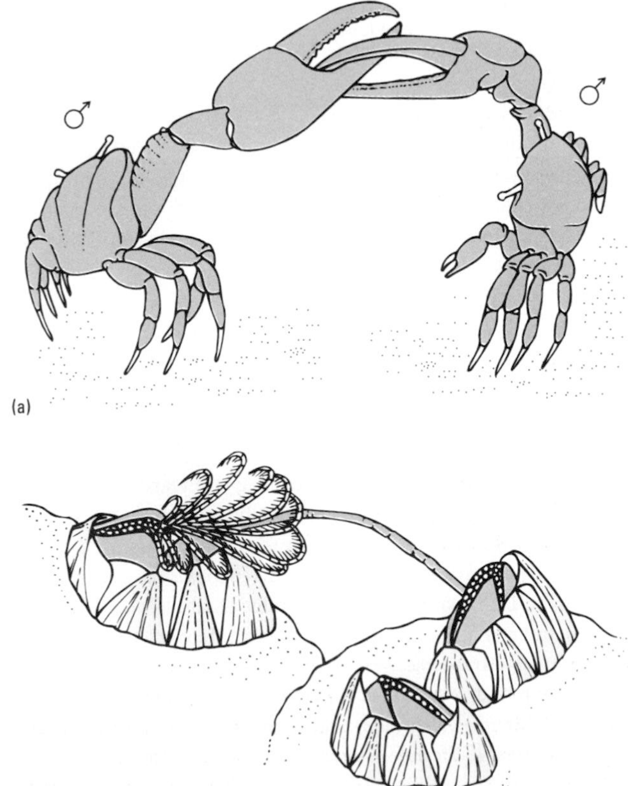

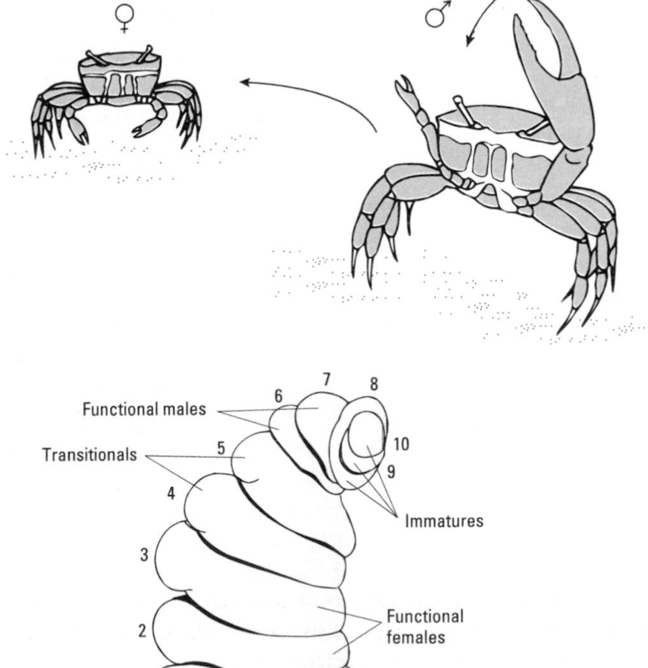

Fig. 12.28 Reproductive strategies in seashore and estuarine animals. (a) Courtship competitions and dancing in fiddler crabs. (b) Extended penis for mating in barnacles. (c) Piled-up slipper limpets (*Crepidula fornicata*) where the sex of individuals changes as the pile builds up.

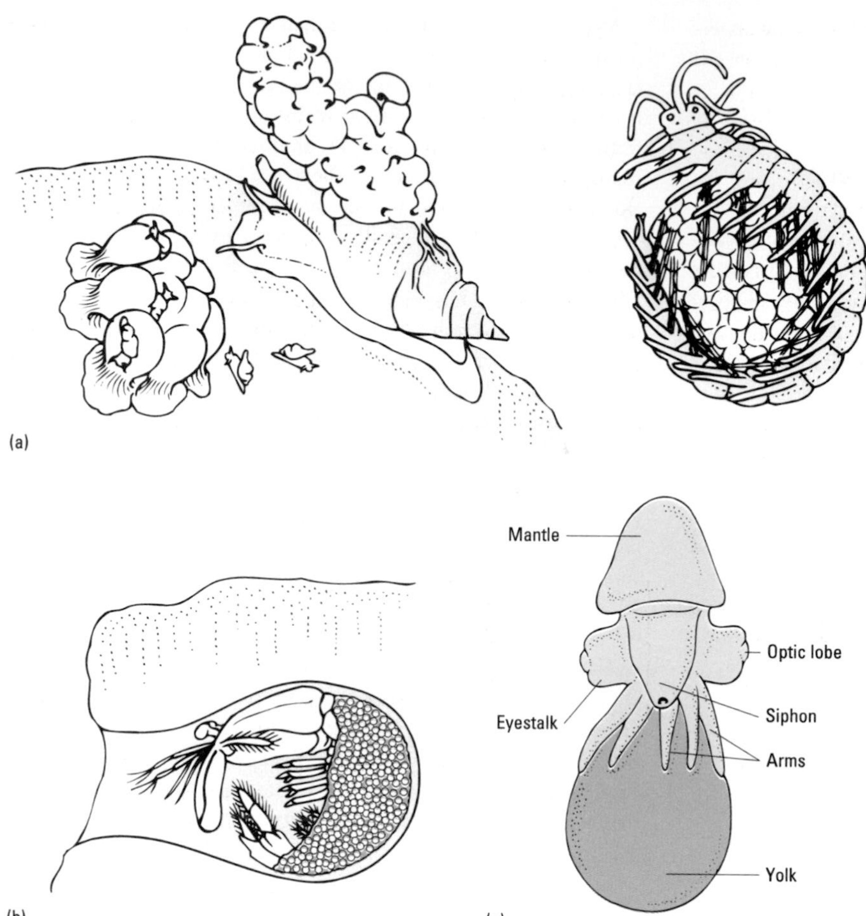

Fig. 12.29 Provision for the care of eggs and juveniles in littoral animals. (a) Protected egg masses in whelks and polychaetes. (b) Mantis shrimp with egg masses in the burrow. (c) Embryo with massive yolk sac in a sublittoral octopus.

guard eggs carefully, a phenomenon found in many shoreline fish such as gobies.

Some species of littoral molluscs and some fishes can change sex within the lifetime of an individual (see Box 11.1). This may help in relatively dispersed mollusc populations by allowing any encounter between conspecifics to become a mating opportunity. In the slipper limpets (*Crepidula*) early arrivals in a location are young males, and they in turn become a settlement site for later arrivals. Those at the bottom of the heap then gradually switch from being male to being intersexual and then female as numbers build up (Fig. 12.28c). The sex of each individual is controlled by the sex ratio of the whole aggregation, probably by the action of pheromones.

Life-cycle patterns vary with the exposure and extremity of shore conditions. On very "difficult" shores such as some mangrove systems, where salinity change and heat stress are unusually marked and currents may be fierce, some specialist invertebrates adopt a life-history pattern dominated by extremely slow growth. For example, the mangrove gastropod *Thais kiasquiformis* grows at only 1 mm year^{-1}. Such species also show very high plasticity for shell shape and thickness, depending on their exact position on the shore, and during their lifetime may migrate according to the levels of desiccation and predation they experience.

The age of a particular shore also affects life cycles. Shores that are of relatively recent origin (where there have been marked coastal level changes, or on shores at high latitude covered by ice until the end of the last Ice Age around 10,000 years ago) tend to be dominated by brooders or direct developers that have arrived by rafting. This is presumed to be due to isolation from feasible sources of larval colonization. The south Atlantic island of South Georgia, for example, lacks both mussels and barnacles, though the environment could certainly support them.

12.6 Mechanical, locomotory, and sensory systems

12.6.1 Physical damage

Waves produce very considerable crushing and tearing forces. Not surprisingly, littoral zones require substantial investment in systems for protecting soft tissues against abrasion and pressure damage, and against drag forces as waves recede. Behavioral options may suffice (deep burrowing, sheltering under stable large rocks or in crevices); but a very high proportion of littoral invertebrates use the option of protective shells.

The design and mechanical properties of molluscan shells have been fairly extensively studied, and there is a relationship between microstructure and lifestyle. The mollusc shell is formed from small crystalline blocks of calcium carbonate each wrapped in a protein matrix ("conchiolin") and assembled together in different geometric arrays. Where the crystals are in parallel rows like flat bricks in a wall, in the plane of the shell, the material is termed nacre or mother-of-pearl; this gives maximum mechanical strength, at 35–115 N m^{-2}. The crossed lamellar structure is a much lower strength material (9–60 N m^{-2}) where the layers of crystals alternate in direction rather like a plywood; however, this structure appears to be the most resistant to chemical attack from boring animals. A third alternative is the prismatic structure, with stacks of prisms perpendicular to the shell surface; this gives reasonable mechanical strength against crushing (around 60 N m^{-2}). Rarely, molluscs use a fourth option of a disorganized chalky structure. Many mollusc shells have a multilayered composition with more than one of these microstructures incorporated, presumably achieving the preferred balance of properties. Thus on very exposed sites where strong resistance to waves is crucial, this may mean an inner nacreous layer and a thick band of prismatic or crossed-lamellar shell distal to this, as in many mussel species. Whilst in more sheltered habitats where predation is strong, the crossed-lamellar format dominates, and in sheltered sublittoral zones a thin inner layer of nacre may be covered only with a thick chalky layer, as seen in oysters.

The macrostructures of shells are also potentially adaptive. Animals from outer coastal regions commonly have thicker coverings than conspecifics from inner, more sheltered waters; this applies both to shelled animals, such as mussels, and to endoskeletal animals, such as starfish, where the dorsal surface can be greatly strengthened by stronger calcareous ossicles. Ornamentation of shells is also much reduced in individuals from very exposed sites, though it may be useful in areas of moderate wave action where spines and ridges can help to anchor an animal on the beach.

Wave shock is also a major problem on sandy shores, where the impact of the water itself, and the debris the waves may carry, can crush animals. Because wave action is somewhat seasonal, usually fiercer in winter, the shape and slope of beaches change through the year and they normally become steeper in winter. The constant seasonal and daily shifting of the substratum is the main reason for the lack of macroalgae and of surface invertebrates on sand. The eastern surf clam, *Spisula solidissima*, is a rare example of a bivalve strong enough, and able to burrow fast enough, to live on exposed western US shores, while the rapidly burrowing streamlined crustacean *Haustorius* survives on many exposed beaches of western Britain. It should be noted in passing that tidepools on sandy beaches provide something of a refuge from damage for macroalgae and for many animals. They are often more common on "groyned" beaches (where manmade breaks have been installed in an attempt to retain sand and stabilize a beach for human purposes). Some kinds of sheltered mudflats also provide refugia, where the substratum is not battered by strong waves and rarely dries out completely.

12.6.2 Dislodgment

Currents, tides, and waves all combine to produce powerful forces tending to dislodge shoreline animals and so displace individuals from their adaptive niche. It should be evident by now that small displacements up or down or along a beach may produce a substantial change in the conditions experienced, especially the regime of emersion and immersion, circumventing all the carefully balanced adaptations we have already discussed relating to osmotic and thermal survival and tolerance of anoxia.

Thus virtually all animals in these brackish-water ecosystems need adaptations for holding on and avoiding displacement from the shore. Algae have elaborate and very tough holdfasts, which help to create local zones of stability that a range of animals can live on or in. Anemones have adhesive pedal discs, and some worms, such as the sand mason *Lanice*, with filter-feeding crowns emerging above the surface, have radiating anchors of hardened mucus and sand grains. Gastropod molluscs use strong suction of the foot to the substratum, and in the case of limpets have the additional safeguard of a homebase where the rock is abraded to fit exactly the contours of an individual shell and improve the efficiency of the suction-pad effect. An absent limpet leaves a clear abrasion scar, and a limpet caught by waves when away from "home" is much more likely to be dislodged. Mussels and some other bivalves have strong byssus threads cementing them to a rock or to other mussels in the bed. Many intertidal and rock pool fish have suckers anteriorly or ventrally. Burrowing again provides a solution for worms.

In the absence of systems for attachment, the alternative is a system of careful station-keeping to insure that an individual stays at about the right level on a shore. The well-known limpet "homing" responses are a way of achieving this; the animal leaves its circular rock homebase to graze over the rocks but returns to its own site reliably, mainly using chemical cues. Fixed behavioral responses may also help to keep an animal in the right site. The winkle *Littorina neritoides* uses negative geotaxis when submerged but switches to negative phototaxis when exposed and drying out, lacking any clear directional preferences when exposed but still moist. This suite of responses automatically maintains it in its high intertidal zone, after local feeding bouts or even after dislodgment by waves. Some sandy beach dwellers may stay in the "right" zone in a rather different sense, leaving the sediment as they detect an incoming wave and thus riding the waves up the beach on an incoming tide then burrowing rapidly at the higher tide mark. This keeps them in the ideal balance of a moist but not extensively waveswept habitat for a greater proportion of each tide cycle. This "swash-riding" technique is found in certain crabs and in some of the faster burrowing clams.

12.6.3 Locomotion

With problems of damage and dislodgment, littoral animals have both to move around rather carefully, usually just above or within the substratum, and to time their locomotory activities appropriately.

Rhythmic locomotory activities

Rhythmic locomotory activities are extremely common on seashores, regulated by **circatidal** and **circadian rhythms**. Those animals that are essentially aquatic and that swim and crawl in search of food or mates tend to do so as close as possible to the substratum, in the hour or two either side of high tide. Grazing molluscs and a host of creeping detritivores move slowly over and around rocks or across the sand surface throughout the period of emersion, but in many cases time their return to a homebase to occur just before the retreat of the tide. Some of the more motile animals, particularly predators such as crabs, have regular patterns

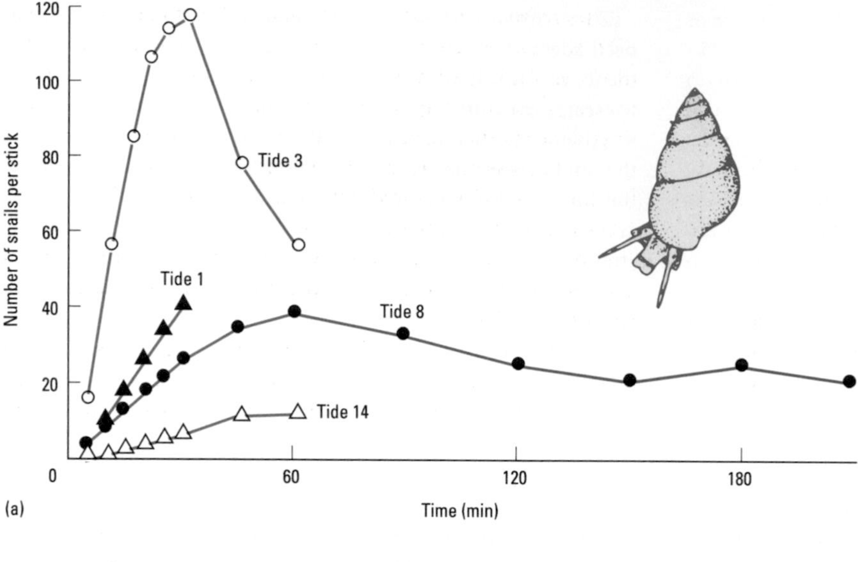

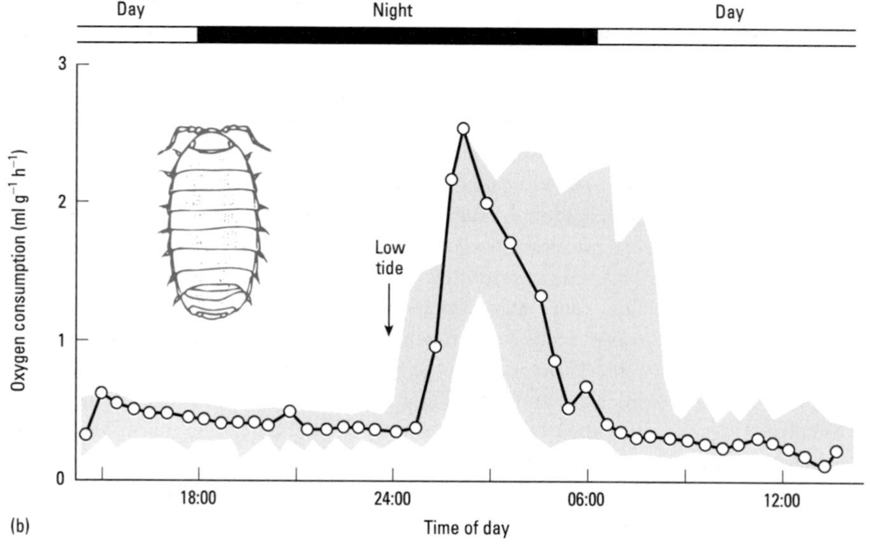

Fig. 12.30 Rhythmic behavioral strategies in brackish-water animals. (a) Patterns of climbing in the estuarine snail *Hydrobia*. Given sticks above mud in a laboratory simulation, snails climbed at each tide, especially in early tides (1 and 3) when they were starved, but less so at later tides once they had restored their nutritional balance. (b) Respiratory rhythms in the isopod *Tylos*, showing an oxygen consumption peak just after low tide at night, with inactivity from dawn onwards (the line shows one individual, and the tint shows the range for six animals). (a, From Barnes 1981; b, from Little 1990; adapted from Marsh & Branch 1979, courtesy of Cambridge University Press.)

of moving up and down the beach, and the crab *Carcinus maenas* appears to have at least three circatidal neural oscillators setting its behavioral rhythms, cued by salinity, temperature, and pressure. Regular movements through beach levels also occur in some of the essentially terrestrial animals that inhabit sandy beaches and mudflats (Fig. 12.30). Predatory birds invade shores and estuaries and move down with the retreat of the water. Insects and mites move up and down the vegetation on salt marshes, or up and down the strandline zone on beaches, to stay in proximity to the water surface (and thus potential food sources) without drowning. Certain shore insects are known to use upwind orientation, invoked by humidity changes, to stay in the "safe" zone as the tide level changes.

Burrowing

Burrowing is probably the single most important locomotory mode for animals in the littoral and estuarine zones. As we have seen, it helps to solve nearly all their problems: osmotic, thermal, and respiratory hazards may be avoided when the tide goes out, and predation is also reduced. It also opens up new food sources, either by poking one end out above the substratum to filter feed or deposit feed, or by eating the substratum itself to extract organic debris. The basic mechanism was covered in section 9.15.1. It is worth noting here that animals with unsegmented bodies can burrow more effectively in softer littoral sediments such as sand and mud than can segmented worms, because there has to be substantial lateral expansion of the body to allow any kind of purchase on the deformable substratum.

Boring animals do something rather similar to burrowing, but in very hard substrates such as rock and wood on beaches. *Teredo*, a bivalve that bores into wood (and often known as the "shipworm"), uses its foot and shell as anchors as before but with the extra trick of rasping with the serrated edge of the shell to soften the substrate. The bivalve piddock *Pholas* works its way into rocks, adding chemical secretions to soften and dissolve the substratum.

Crawling and swimming

Most worms and many crustaceans use a crawling mode for foraging and dispersing amongst the rocks and weeds of the littoral zone, modified in some of the more amphibious crustaceans into a scuttling or hopping mode when moving over dry sand or rock between tides. The sideways walk of crabs is a special case, where the leading legs pull and the trailing legs push; the adaptive value of this is not entirely clear, though it may reduce tripping and allow a longer stride length.

Swimming is only appropriate in rock pools, or when the tide is well in, whether in littoral or estuarine zones. Animals swimming in the shallow rapidly moving surf zones have little control of their own direction, and in estuaries they may be swept too far inland to areas of low salinity if they enter the main water column on an incoming tide. It is essentially the marine animals of the sublittoral zone that retain a free-swimming mode, together with the specialist rock pool dwellers (including fish) that can enter a crevice or hold on to a rock with suckers at times when swimming is too dangerous. However, some of the relatively specialized invertebrate larval forms do become free swimming in the main water body, allowing currents to disperse them more widely.

12.6.4 Sensory systems

The littoral zone and estuarine habitats are both rather disturbed habitats with many physical and chemical factors undergoing cyclical change, so that sensory detection systems face particular problems. Perhaps the most fundamental of these is that all sensory neurons operate on the basis of ionic gradients, so that as we saw earlier (with nerves in general) there will be problems in maintaining a constant level of sensitivity in environments of changing salinity. This will apply to all internal sensors for all but the most precise of osmoregulators; and it will also apply to external chemoreceptors for all animals. Some key issues here were covered in section 9.16.

Visual senses

Visual sensitivity and responses to light will be more important for littoral animals than for most species dwelling in the ocean, where 70% of the living space receives no light at all. Most of the living space in brackish water is shallow enough to receive light, though this may be reduced by vegetation growth in summer. Animals that remain water-covered in the sublittoral zone and around coral reefs are perhaps particularly able to benefit from good vision and in particular from color (polychromatic) vision. These zones are home to fish with up to five types of photoreceptors giving pentachromatic color vision, and to some mantis shrimps with 10 or more types of photoreceptor. In contrast, animals from muddy shores and estuaries rarely have good visual acuity or color discrimination.

Animals in the littoral zone are effectively amphibious and thus have the particular problem of focusing in two different media. Any lens system that focuses light accurately on a retina under water (refractive index (RI) = 1.33) will be severely myopic in air (RI = 1.00), focusing light in front of the retina. Similarly, eyes that work well on land lack sufficient focusing power to form an image on the retina when under water. Thus many littoral animals merely have eyes that give information on light and dark, with little acuity. The problem can be overcome though by combining the properties of the lens and the cornea; a flat cornea and a very spherical lens (having high focusing power) is a common solution. The mudskipper represents a good example of this. Another alternative is to have two pupils with different properties, and a few of the shoreline blennies do have paired side-by-side pupils. Amphibious shore birds that hunt under water tend to vary their focus using very strong accommodation instead, with powerful ciliary muscles that squeeze the lens so that it is more convex and bulges forwards almost through the iris.

Bioluminescence is relatively uncommon in littoral animals relative to the marine fauna, perhaps because it risks attracting attention from two lots of predators (from above and from below).

Chemical senses

Species that can survive and feed over a wide range of salinities must have olfactory receptors capable of working at highly variable Na^+ levels. There is little information on how this is achieved, although in salmonid fish the olfactory systems are similar in both pond water and sea water, the receptors themselves somehow providing inbuilt tolerance. Permitted reductions in internal K^+ levels at low salinity are likely to be part of the response, as with nerves, so that a reasonable resting potential is maintained. In crabs there is evidence that the exposed chemosensory dendrites show changes in length positively correlated with salinity, with shorter dendrites and reduced sensitivity in freshwater-acclimated individuals, also accompanied by sodium efflux, the dendrite length perhaps being constrained by the distance over which a suitable ionic microenvironment can be sustained. However, in many brackish animals proper functioning is lost in artificial salinity media lacking adequate calcium, so that Ca^{2+} ions also appear to be essential for maintained sensor responsivity.

Chemoreception in littoral and estuarine animals serves similar functions and faces similar problems to those described for marine animals in Chapter 11. Many species detect their food chemically, or detect and respond to incoming predators, sometimes with defensive chemicals. Commensal and parasitic species locate their hosts chemically, and many species locate their mates with sex pheromones. However, the disturbed pattern of flow inevitably disrupts any pheromone plume and thus reduces the range over which chemical cues can act, and on exposed beaches it is unlikely that any chemical signaling system can be of use in the surf zone between tides. Nevertheless, the fact that salmon can home in on their spawning grounds through estuaries should caution us that very precise chemoreception remains a potent source of information.

Mechanical senses

We have already stressed that brackish habits pose a number of mechanical problems for animals, so it is no surprise that mechanoreception is important to many of them. Detection of flow in the external medium may be a crucial component of the response to tide and to waves, and can be readily signaled by stretch receptors in the skin, by deformable projections such as microvilli on hair cells (see Fig. 9.46), or by sensory bristles in arthropods and many fish.

Most phyla of soft-bodied invertebrates have ciliated neuroepithelial cells that act as direct "rheoreceptors" for flow and pressure. Any animal dislodged by wave action is also greatly aided by statocysts (see Fig. 9.45) that act as gravity receptors, allowing it to right itself and find the substratum again. In fish the lateral line system containing neuromast cells (modified hair cells) provides the main sensor for flow and pressure, with otoliths giving orientation and gravity information.

Other senses

Electroreception in littoral fish is rare, since stable electric fields are unlikely and varying salinity also alters the field parameters. However, electric senses are particularly useful in turbid but otherwise more stable waters in the river mouth, and are important in some teleosts and sharks that penetrate large tropical rivers. Echolocation also occurs, with both dolphins and seals regularly fishing inshore and moving up estuaries. The fish and mammals using these sensory modes often occur very far up river in the case of turbid large estuaries such as the Ganges, Amazon, and Yellow River; details were considered in Chapter 11.

12.7 Feeding and being fed on

All animals in the littoral and estuarine habitats have the potential benefit of two sources of food (from the sea and from the land) and the potential hazard of two parallel sets of predators. Predation may be the single most important population determinant on many shores, especially in taking out large proportions of the juvenile prey items.

Rocky shores contain an abundance of food: plankton for filterers and detritus for scavengers, both renewed on each tide; macroalgae (seaweeds) providing a dense growth for herbivores at various depths; and a wide range of invertebrates for predators to pick off. Estuaries too contain an abundance of food: all the sources of the seashores, but also the local growths of phytoplankton supported by nutrients from the river. By contrast, at first glance there seems to be little food around on a sandy beach. However, animals can feed on the sand and mud itself, thereby extracting mostly bacteria, as with lugworms, fiddler crabs, and mole crabs, and many flatworms and nematodes. They can also feed at the mud surface on the detritus deposited there; examples include *Hydrobia* snails (especially on salt marshes), the sand mason *Lanice* (an annelid), and many bivalves, like the clam *Scrobicularia*, whose feeding tube (the inhalant siphon) comes up to lie on the surface. Other animals may live in the sand but feed from the water: some achieve this using structures projecting upwards, such as the tentacles of epifaunal hydroids or of tube-dwelling annelids like *Sabella* and other fanworms; others draw water though a burrow and filter from it with a mucous net (the annelid worm *Chaetopterus* being the classic example here; see Fig. 7.9a). Obviously these options depend on the duration of water immersion, which is often the limiting factor in feeding and growth rates; high-tide individuals feed less and grow more slowly than low-tide individuals.

It is worth dealing specifically with the problems of herbivory in these tidally influenced habitats. Marine plants are almost exclusively seaweeds, and varying proportions of their productivity are diverted into grazing animals—rather more in sheltered and low-latitude communities, and from the mid-shore downwards, but less towards the high-water mark and on exposed or high-latitude beaches, where decomposition is more important. Where herbivory is dominant, the key grazers are arthropods, molluscs, and some polychaetes and echinoderms (sea urchins), together with specialist fish. In general, small urchins and gastropod molluscs are more important in temperate systems, with larger urchins and fish dominating on tropical shores, and with tropical reefs showing the highest herbivory rates of any known communities. Microalgal grazers, such as limpets and winkles, scrape or sweep over the substratum, eating diatoms and recently settled juvenile algae, whereas macroalgal grazers, such as sea urchins, scrape or bite off pieces of seaweed. All face substantial problems: how to find appropriate food (which may be widely dispersed or extremely patchy), when and how often to feed on it, and crucially when to stop feeding to avoid the onset of unfavorable or dangerous conditions as the tides or currents alter—inherent rhythmicity of behavior is again a useful common feature. In addition they may face problems of predation while feeding, forcing them to select more protected or sheltered sites of poorer food quality. But above all they may have to overcome problems posed by the food itself, since many littoral weeds are well defended against herbivory. Like terrestrial plants (see Chapter 15), seaweeds possess a range of chemicals that may serve to deter grazers. Polyphenolics and terpenes are commonplace, and certain aromatic or amino acid-derived compounds are also used, often with chloride, bromide, or iodide incorporated. Only the alkaloids are seemingly absent from the defensive repertoire compared with land plants. Most of the chemicals are broad-spectrum defenses active against a wide range of herbivores; but herbivore size and mobility are often correlated with resistance to seaweed chemicals, and small sedentary grazers may eat weeds that are extremely deterrent to more active fish, for example. Just as with insect–plant interactions in the terrestrial world, some littoral grazers may use the compounds from their seaweed foodstuffs to their own advantage: as cues to locate and feed on these weeds, or as chemicals in their own bodies to act as defensive/deterrent antipredator agents. Coevolution between specific weeds and specific herbivores is relatively unlikely compared with terrestrial systems, since most marine and littoral herbivores are fairly unspecialized in their choice of diet. However, there may be morphological and/or chemical "mimicry rings" among seaweeds, some harmless mimic species resembling toxic model species (Batesian mimicry) or various toxic species converging on a common phenotype (Müllerian mimicry). Seaweeds may also improve their chances of surviving herbivory by having the ability to pass through guts without being completely destroyed. Pieces of vegetative tissue may regenerate after passing out with fecal pellets, and in some species algal propagules may also germinate after passage through a fish gut (a direct parallel to some kinds of seed dispersal on land).

One other form of nutrition on shores should be mentioned, as quite a range of animals in shallow waters have microalgae within their bodies, in a symbiotic relationship. Corals are the most obvious example (Box 12.4); these get the benefits of some carbon fixed by the algal cells' photosynthesis, as well as indirect assistance with deposition of their calcium carbonate skeletons. Certain clams

Box 12.4 Corals and their microalgal symbionts

Many corals (including all the reef-building types), and some forms of jellyfish, harbor small algal cells (from two groups of protistans, the cryptomonads and the dinoflagellates) which are collectively termed zooxanthellae. In corals these occur in dense populations that may exceed 30,000 algal cells mm^{-3} of host tissue. Some freshwater cnidarians, including the familiar *Hydra*, also house green algal cells, termed zoochlorellae. All animals that house such symbionts must live in the photic zone where the algae can receive some light; thus corals are only present to a maximum depth of about 90 m.

The benefits to the coral are probably variable in different habitats. Some may derive relatively limited nutrition, or help in eliminating metabolic wastes (nitrogen and some phosphorus, which get used and recycled by the algal cells). Others appear to get very substantial nutritional gain, with a large part of the algal photosynthetic production being passed on to the host as glucose, glycerol, or amino acids.

An additional advantage to the host comes in the process of depositing the calcareous coralline exoskeleton. By using up CO_2 in photosynthesis the zooxanthellae help to drive equilibria in favor of $CaCO_3$ formation as shown below.

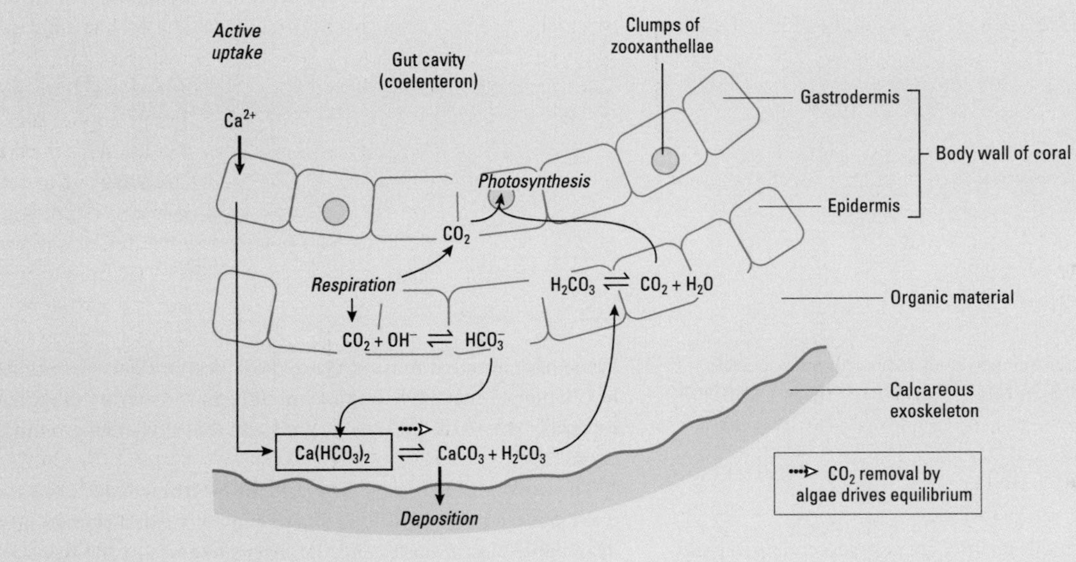

and anemones in littoral waters also house microalgae; and perhaps the most spectacular example is the green flatworm, *Convoluta*, found on beaches around the English Channel, where it migrates in large numbers to the surface at low tide on sunny days to give the appearance of green sheets across the sand.

A large proportion of littoral animals run the risk of becoming food in their turn, and must possess strategies to avoid potential predators. Some invertebrates are assisted in this by chemical cues. The littorinids (winkles) appear to respond to "alarm substances" in the presence of nearby crabs, migrating up shore or hiding in crevices—chemicals leaching from the crushed remains of conspecifics are the probable cue. Not surprisingly, many other littoral animals avoid predation primarily by hiding or by burrowing. This is not an absolute guarantee of safety, however, as some common beach predators have adaptations to reach down into muds and sands, or into crevices, or amongst small stones, to extract prey items. Figure 12.31 shows the adaptations of a variety of wading birds' beaks. Curlews and godwits have long bills able to reach deep-burrowing and softer shelled clams; oystercatchers delve for burrowing worms and can open cockles; redshank, sandpipers, and similar birds consume shallower burrowers, with plovers and turnstones taking the superficial amphipods, etc. At high tide when the terrestrial predators have departed, and all these potential prey animals are active, they face the additional hazard of predators such as crabs and fish, especially flatfish and gobies, nipping off their extended siphons and tentacles. A study of one estuarine fish, called the spot, showed that the two sets of predators have effects that are not simply additive; spot avoid the aquatic predators (mainly other fish) by moving into shallower water, where they might be expected to be more vulnerable to birds, but in fact since they also then tend to aggregate more they reduce their susceptibility to bird predation, their behavior insuring that they suffer lower mortality in the presence of both predators than would be predicted by adding the mortalities due to each predator acting alone.

12.8 Anthropogenic problems

Man has always tended to settle along coastlines and thus to have an unusually forceful impact on shore environments, through food-related, industrial, and leisure activities. Most of the current problems of coastline management bring up conflicts of interest, between developers and ecologists, landowners and economists; in many of these conflicts, the outcome for the biota is particularly hard to predict due to the very complexity of the brackish environment in time and space.

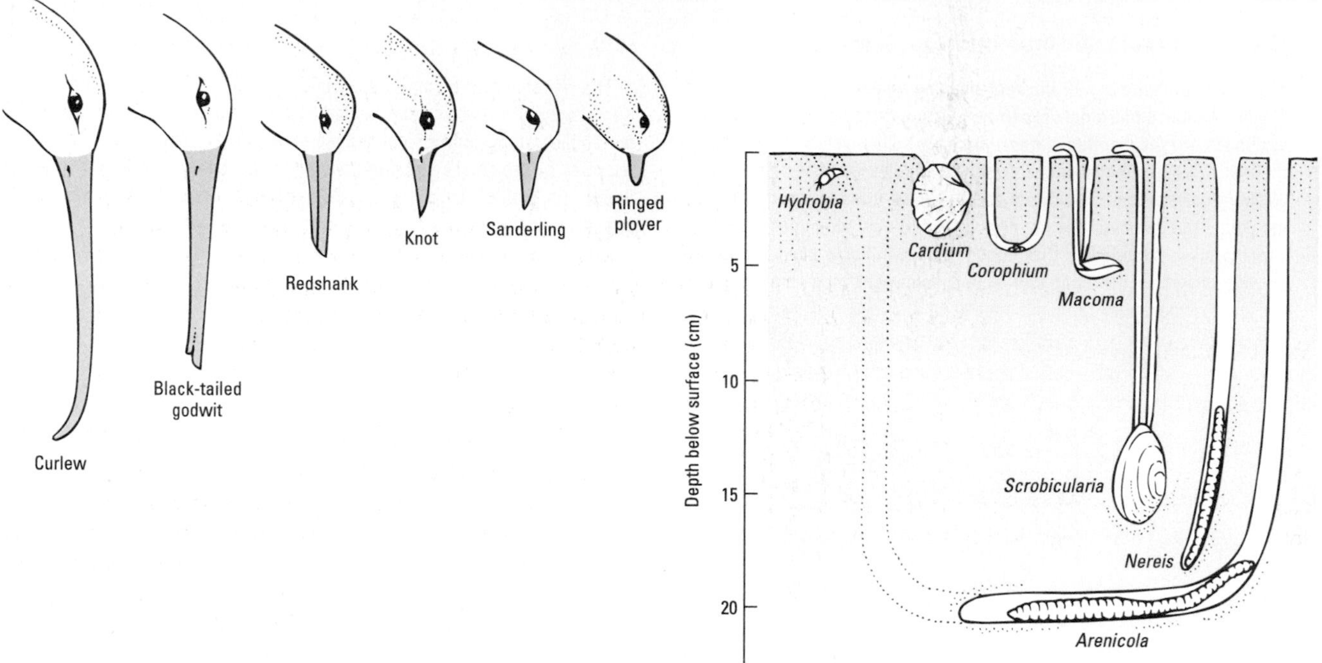

Fig. 12.31 Depths of burrows for common sandy and muddy beach animals, and the corresponding lengths of beaks of common waders that feed upon them. (Adapted from Green 1968.)

12.8.1 Pollution

All the habitats covered in this chapter are very susceptible to pollution. Most of the oil and chemical spills from the seas accumulate on shores. Because humans are apt to live on rivers and estuaries their sewage is often also deliberately fed out there, at or below low-tide level, leading to increased coliform bacteria counts and zones of oxygen depletion. Deposition of solid wastes from pipes or from barges may also lead to choking of kelp forests and their holdfast communities; this occurred in Santa Monica Bay off Los Angeles, where the outcome was decreased diversity and a population explosion in sea urchin communities. The accumulating deposits of agrochemicals (such as nitrates, phosphates, and DDT) and industrial wastes (including hydrocarbons and heavy metals) coming down from rivers to the estuaries and nearby shores also add to the general problems of eutrophication and possible toxicity. Tributyl tin is used in antifouling paints and is a specific problem in some estuaries used for marinas and boat moorings; it causes sexual dysfunctions in dog whelks and perhaps in fish. Some coastal species detoxify heavy metals by incorporating them into nonreactive complexes such as metallothioneins, which can reach quite high levels in some of their tissues. Robust species such as the shore crab, *Carcinus*, can acclimate to some extent, gradually reducing their levels of copper uptake and more efficiently transferring it from their hemolymph to their tissues during prolonged exposures.

Apart from visible filth, the evidence of plankton blooms ("red tides") reveals an increasing problem, and excess macroalgal growth is also now regularly appearing on the shores of the Mediterranean, the Black Sea, and other areas of relatively closed-off inshore sea. Remember also that most of the oil pollution problems mentioned in Chapter 11 inevitably impinge mainly on the marine shorelines, especially where oil spillage results from the accidental grounding of tankers.

The worst problems at present and for the immediate future arise from polluted river outflows, inadequately treated sewage, non-degradable plastic waste, and the insensitive use of the nearshore seabeds for disposal. Inland deforestation is also becoming a problem as it often allows erosion and increases river sediment loads, leading to increased offshore deposition. Important areas of shore are being lost completely, especially coral reefs, wetlands, and mangrove forests. In some parts of the world long sections of beaches are so polluted they become devoid of life and no longer usable for human leisure interests. Prime indicators of coastline pollution are the shellfish, particularly filter-feeding bivalves, and these often become too contaminated to be safely consumed by humans (or by the birds, crabs, and fish that normally prey upon them).

Many coastal communities also rely on pumped groundwater for their fresh drinking-water supplies, and shallow coastal freshwater aquifers are easily damaged by abstraction for domestic and industrial use (often with induced seepage of salt water inwards), as well as by polluting influences.

12.8.2 Coastline alterations and tourism

Another problem impinging on many littoral zones and their fauna is man's tendency to alter and "manage" coastlines, to reduce erosion of cliffs, stabilize dunes, drain wetlands, or alter the movement of deposits along the coastline. These management practices can affect the balance of recycling processes whereby organisms normally recycle nutrients back to the land—especially from mudflats and wetlands. Most of the traditional methods of "holding" the shoreline with groynes, seawalls, etc. are now being questioned due to their down-coast repercussions. Protecting shorelines against erosion or flooding is commonly more environmentally sensitive now, and may involve manipulation of the natural agents that give stability, with lower cost and better long-term outcomes.

Drainage schemes to produce tourist resources such as marinas may also destroy some rather important chemical cycling, and rebound on the land ecosystems. They may also destroy the breeding areas of fish and the important relatively sheltered zones where young fry mature, so damaging exploitable fish stocks. (Similar effects occur in certain areas from the use of shark nets, however desirable these may otherwise be!) The loss of estuarine land may also lead to unforeseen modification of tidal and river channels and thus have effects far inland.

At the extreme, use of coastlines for tourism in areas such as the Mediterranean, with managed "cleansed" beaches and offshore sporting activities, can produce complete destruction of the fragile ecosystem and also effectively destroy local human communities formerly dependent on fishing and local agriculture.

12.8.3 Invasions

There have been some spectacular cases of coastlines being affected by invasions of **nonindigenous species** (NIS) from other biogeographic areas, brought deliberately or accidentally (often as larval or adult settlers on ship hulls, or perhaps most commonly in ballast water) by man, or given access to new areas by the building of canals between seas (notably at Panama and Suez).

In Britain new barnacle species have been introduced, and the invader *Elminius modestus* is still spreading around the coastline from its first foothold in Southampton Water, outcompeting some native barnacles. The Chinese mitten crab, *Eriocheir sinensis*, is also invading up many European rivers. In North America at least 400 NIS have been recorded along the Pacific and Atlantic coasts, of which perhaps 15% are classed as nuisance species. The non-native periwinkle, *Littorina littorea*, is now the single commonest winkle in most of New England, after an introduction to Nova Scotia about 100 years ago. The European shore crab, *Carcinus maenas*, is also now a common predator on these shores. Some such introductions are relatively benign, but others can seriously damage local communities. In San Francisco Bay the introduced snail *Ilyanassa obsoleta* has almost completely outcompeted the local mud-snails, and the clam *Potamocorbula amurensis* has displaced a wide range of other endemic invertebrates. In reciprocal fashion, the introduction of the North American ctenophore *Mnemiopsis ledyi* into the Black Sea and Azov Sea has been implicated in a massive loss of fisheries revenue.

12.8.4 Coastal energy

Not only are relatively remote coasts commonly regarded as suitable sites for oil refineries and nuclear power plants, but there are significant plans for the use of wave power and tidal power as renewable energy sources for the future. Such ideas are superficially attractive but coastal energy is so far proving very difficult to harvest, not least because of the extraordinary variability of waves. Wave-energy devices are mainly at the testing stage, with limited success as yet. Tidal power is perhaps easier to harness, and small-scale tide mills have been in operation for centuries. Some small tidal power stations do now exist, and the impact of large tidal barrages in suitable rocky channels has been intensively studied in France, Canada, China, and the UK, but with no very large projects in place as yet. However, these developments are bound to impact on coastal environments and biota in the future, perhaps especially on sessile species close to their environmental limits, and possibly on those more conspicuous fish and birds feeding (and breeding) at higher trophic levels.

12.9 Conclusions

Given all the problems covered here, it is clear that littoral communities are amongst the most finely patterned and perturbation-sensitive to be found anywhere. Every organism has its own very tightly defined niche, often within a vertically zoned habitat. Several of the figures in this chapter have shown the *interactions* of different environmental variables—notably salinity, desiccation, temperature, and oxygen—stressing the point that these factors are strongly interdependent and that simplistic studies looking at just one aspect of the littoral environment rarely work. Figure 12.32 provides a particularly clear demonstration of this, with the mortality due to salinity and temperature change being interrelated and in turn modified by changes to oxygen availability. Reduced salinity exposure in any one species may lead to changes in oxygen consumption, altered nitrogenous excretion, and altered acid–base balance, all by relatively simple mechanisms. But it may also produce reduced feeding, reduced activity levels, and reduced reproductive output. Thus individuals of one species in different parts of a highly zoned shore may be operating very different physiological and life-history strategies; and apparently straightforward interindividual differences in salinity tolerance produce very subtle effects on the ability to penetrate estuarine conditions.

Zonation between species is in part *caused* by the highly variable interacting physiological tolerances of desiccation, heat stress, freezing, oxygen deprivation, and reduced feeding time. It is also inevitably influenced by larval preferences and tolerances and larval gregariousness. But above all it is also induced by competition, whether direct (mussels may overgrow other species, while limpets may undercut others) or indirect by more effective resource use; and it may be supplemented by differential predation, especially on the lower shore where fish and crustacean predators are most common.

The limits to distribution and success in brackish-water systems are therefore set by a mixture of intrinsic factors and biotic interactions. A few examples should underline this point. Firstly, Fig. 12.33 shows a general picture of the interacting factors in estuaries. Each species is limited at the upper shore by temperature and desiccation, and especially at the upstream end by hyposmotic tolerance; while its lower limit is set by competition with other species. Secondly, a specific case from a shore is illustrated in Fig. 12.34 with an analysis of the factors controlling mussel distribution. The upper shore limits are again determined primarily by physiology, and the lower shore limits by competition and predation, but with the exact limits and constraints depending on individual size. Small/young individuals are limited at the top by water loss and at the bottom by

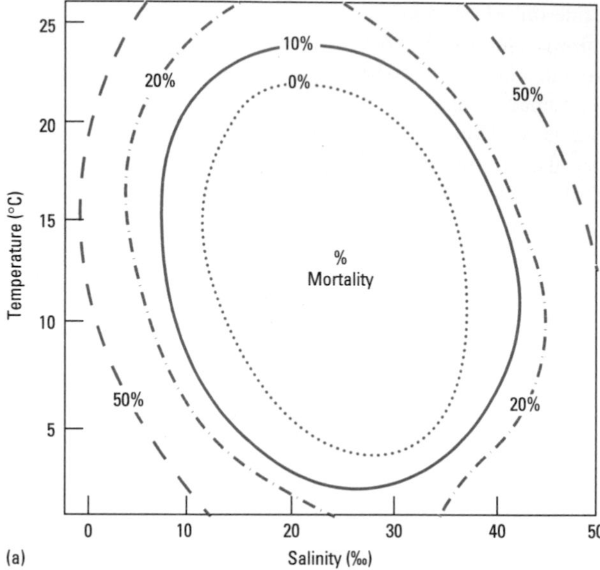

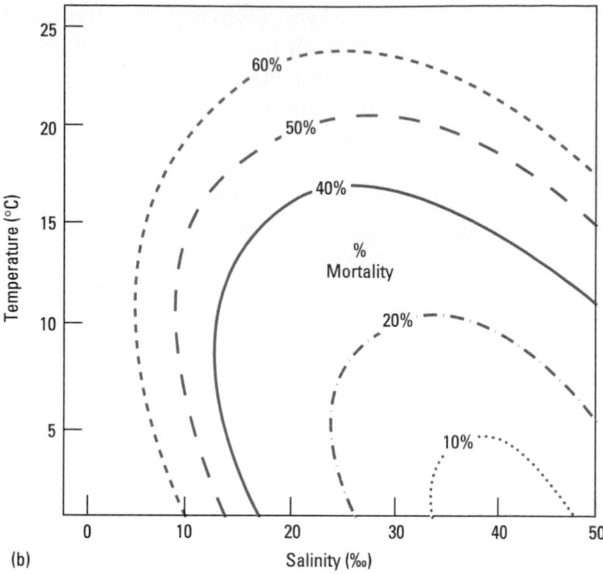

Fig. 12.32 Interactions between temperature, salinity, and oxygenation in determining mortality of the common shrimp *Crangon*. In (a) where the water is fully oxygenated mortality is minimal at moderate temperatures and with salinities close to and somewhat below full sea water. In (b), with poorly oxygenated water, mortality is much higher but animals survive somewhat better with colder water and at a higher salinity range. (Modified from Haefner 1969, 1970.)

predation (especially by starfish on many beaches); larger/older ones are limited at the top by their risk of overheating out of water, and at the bottom by competition with other mussels. The situation on some shores may also change seasonally with a strong impact from transitory populations of migrating predatory shore birds. A third example is the well-known interaction in Europe of two barnacle genera, *Chthamalus* and *Semibalanus*, shown in Box 12.5.

As a final comment it is worth noting in particular the extraordinary number of cases where quite different littoral and estuarine animals arrive at the same designs and solutions to their problems, a point amply made by a simple comparison of barnacles, mussels, winkles, and limpets, the most dominant taxa on so many shorelines. Convergent evolution is spectacularly apparent in these fringe

Box 12.5 Biotic and abiotic interactions: patterns of occurrence of barnacles

Two barnacle genera (each with several species) occur in the littoral zone around much of Europe and exemplify the interactions of physiological tolerances and biotic factors in determining shoreline zonation (Connell 1961).

Chthamalus is more heat tolerant with a higher upper critical temperature (UCT), and occurs alone on southern European shores, whereas *Balanus* or *Semibalanus* are relatively intolerant of heat and do better in higher latitudes, occurring on their own on the beaches of northern European countries such as Sweden.

At intermediate latitudes the two species co-occur, but with *Chthamalus* on the upper beach and *Balanus* lower down. The tidal height where one takes over from the other shows a latitudinal cline; even within Great Britain we find southern and western beaches that are dominated by *Chthamalus* with a narrow belt of *Semibalanus* low down, while colder beaches in the north and east have a broad zone of *Semibalanus* and a narrow high-tide strip of *Chthamalus*.

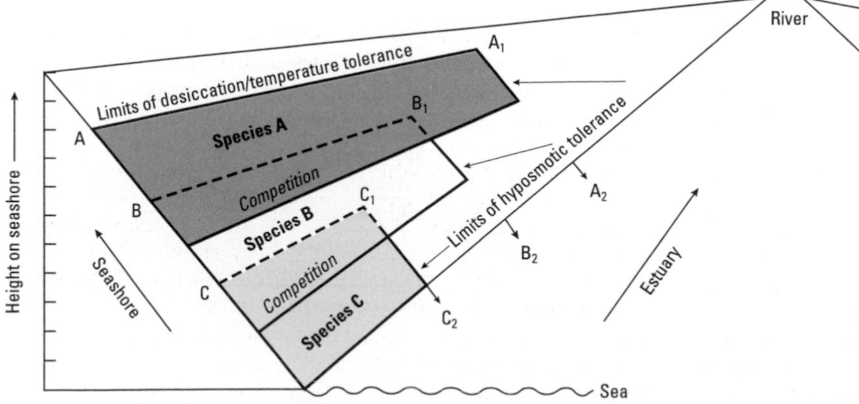

Fig. 12.33 Limiting factors on animal distribution in shores and estuaries, for three representative species. Species A has high osmotic and thermal tolerance and could survive physiologically in the whole rectangle specified by A–A_1–A_2, but is in practice limited on the lower shore by competition with species B. B has a potential range B–B_1–B_2 but is only found well within its physiological limits, constrained upwards by this competitive interaction with A and downwards with C. C is able to live downshore to the limits of its physiological tolerance, but cannot extend up the shore where B outcompetes it. All species have ranges into estuarine conditions limited by their osmotic tolerance in line with their potential upshore ranges. (From Newell 1979.)

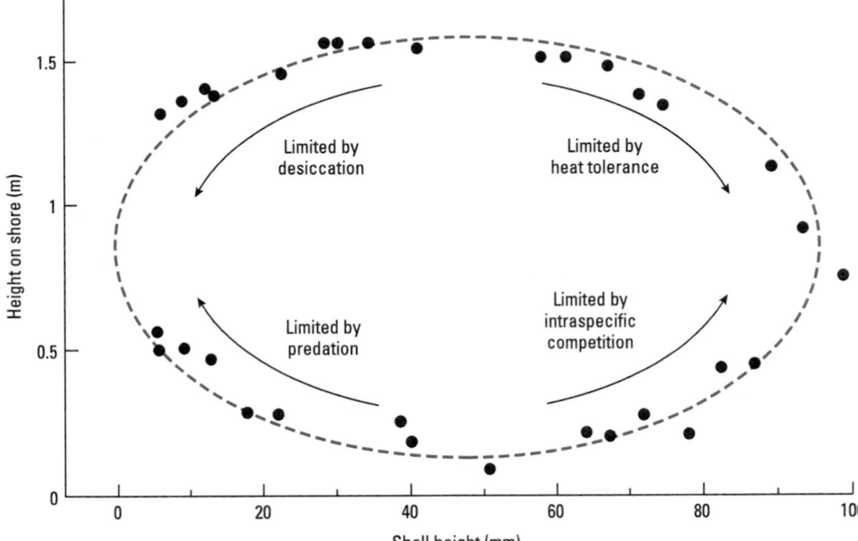

Fig. 12.34 Limiting factors for *Mytilus edulis* on seashores in relation to height of shore and size of shell (approximating age of individual). The dashed line represents the overall distribution limits, and the main limiting factor in each part of the range is shown; thus small mussels are limited on the lower shore mainly by predation, and so on. See text for more details. (From Newell 1979.)

environments; this is an issue that should become even more apparent as we go on to look at more terrestrial environments.

FURTHER READING

Books

Baker, J.M. & Wolff, W.J. (eds) (1987) *Biological Surveys of Estuaries and Coasts*. Cambridge University Press, Cambridge, UK.

Boaden, P.J.S. & Seed, R. (1985) *An Introduction to Coastal Ecology*. Blackie, Glasgow.

Carter, R.W.G. (1988) *Coastal Environments*. Academic Press, London.

Denny, M.W. (1988) *Biology and the Mechanics of the Wave-swept Environment*. Princeton University Press, Princeton, NJ.

Hawkins, S.J. & Jones, H.D. (1992) *Rocky Shores*. Immel Publishing, London.

John, D.M., Hawkins, S.J. & Price, J.H. (eds) (1992) *Plant–Animal Interactions in the Marine Benthos*. Clarendon Press, Oxford.

Little, C. & Kitching, J.A. (1996) *The Biology of Rocky Shores*. Oxford University Press, Oxford.

Moore, P.G. & Seed, R. (eds) (1985) *The Ecology of Rocky Shores*. Hodder & Stoughton, London.

Newell, R.C. (1979) *Biology of Intertidal Animals*. Marine Ecological Surveys Ltd, Faversham, UK.

Pequeux, A., Gilles, R. & Bolis, L. (eds) (1984) *Osmoregulation in Estuarine and Marine Animals*. Springer-Verlag, Berlin.

Reviews and scientific papers

Borgatti, A.R., Pagliarini, A. & Ventrella, V. (1992) Gill (Na^{+}+K^{+})-ATPase involvement and regulation during salmonid adaptation to salt water. *Comparative Biochemistry & Physiology A* **102**, 637–643.

Bridges, C.R. (1993) Ecophysiology of intertidal fish. In: *Fish Ecophysiology* (ed. J.C. Rankin & F.B. Jensen), pp. 375–386. Chapman & Hall, London.

Brinkhoff, W., Stockmann, K. & Grieshaber, M. (1983) Natural occurrence of anaerobiosis in molluscs from intertidal habitats. *Oecologia* **57**, 151–155.

Burnett, L.E. (1997) The challenges of living in hypoxic and hypercapnic aquatic environments. *American Zoologist* **37**, 633–640.

Charmantier, G., Haond, C., Lignot, J.H. & Charmantier-Daures, M. (2001) Ecophysiological adaptation to salinity throughout a life cycle: a review in homarid lobsters. *Journal of Experimental Biology* **204**, 967–977.

Connell, J.H. (1961) The influence of interspecific competition and other factors on the distribution of the barnacle *Chthamalus stellatus*. *Ecology* **42**, 710–723.

Cronkite, D.L. & Pierce, S.J. (1989) Free amino acids and cell volume regulation. *Journal of Experimental Zoology* **251**, 275–284

Crowder, L.B., Squires, D.D. & Rice, J.A. (1997) Non-additive effects of terrestrial and aquatic predators on juvenile estuarine fish. *Ecology* **78**, 1796–1804.

Deaton, L.E. (1992) Osmoregulation and epithelial permeability in two euryhaline bivalve molluscs: *Mya arenaria* and *Geukensia demissa*. *Journal of Experimental Marine Biology & Ecology* **158**, 167–177.

Deaton, L.E. & Greenberg, M.J. (1991) The adaptation of bivalve molluscs to oligohaline and fresh waters: phylogenetic and physiological aspects. *Malacological Reviews* **24**, 1–18.

Denny, M.W. (1995) Predicting physical disturbance: mechanistic approaches to the study of wave-swept shores. *Ecological Monographs* **65**, 371–418.

Gilles, R. (1987) Volume regulation in cells of euryhaline invertebrates. *Current Topics in Membrane Transport* **30**, 205–247.

Gleeson, R.A., Wheatly, M.G. & Reiber, C.L. (1997) Perireceptor mechanisms sustaining olfaction at low salinities; insights from the blue crab *Callinectes sapidus*. *Journal of Experimental Biology* **200**, 445–456.

Gosselin, L.A. & Chia, F.S. (1995) Characterising temperate rocky shores from the perspective of an early juvenile snail—the main threats to survival of newly hatched *Nucella emarginata*. *Journal of Marine Biology* **122**, 625–635.

Graham, J.B. (1990) Ecological, evolutionary and physical factors influencing aquatic animal respiration. *American Zoologist* **30**, 137–146.

Heble, D.K., Jones, M.B. & Depledge, M.H. (1997) Responses of crustaceans to contaminant exposure: a holistic approach. *Journal of Estuarine & Coastal Shelf Science* **44**, 177–184.

Helmuth, B.S. (1998) Intertidal mussel microclimates: predicting the body temperature of a sessil invertebrate. *Ecological Monographs* **68**, 51–74.

Hofmann, G.E. & Somero, G.N. (1995) Evidence for protein damage at environmental temperatures: seasonal changes in levels of ubiquitin conjugates and hsp70 in the intertidal mussel *Mytilus trossulus*. *Journal of Experimental Biology* **198**, 1509–1518.

Ishimatsu, A., Hishida, Y., Takita, T. *et al.* (1998) Mudskippers store air in their burrows. *Nature* **391**, 237–238.

Levinton, J.S. (1995) *Marine Biology*. Oxford University Press, New York.

Lockwood, A.P.M., Sheader, M. & Williams, J.A. (1996) Life in estuaries, salt marshes, lagoons and coastal waters. In: *Oceanography* (ed. C.P. Summerhayes & S.A. Thorpe). Manson, London.

McCormick, S.D. (2001) Endocrine control of osmoregulation in teleost fish. *American Zoologist* **41**, 781–794.

McGaw, I.J. & McMahon, B.R. (1996) Cardiovascular responses resulting from variation in external salinity in the Dungeness crab *Cancer magister*. *Physiological Zoology* **69**, 1384–1401.

McMahon, R.F. (1988) Respiratory response to periodic emergence in intertidal molluscs. *American Zoologist* **28**, 97–114.

McMahon, R.F. (1990) Thermal tolerance, evaporative water loss, air–water oxygen consumption and zonation of intertidal prosobranchs: a new synthesis. *Hydrobiologia* **193**, 241–260.

Morritt, D. & Spicer, J.L. (1995) Changes in the pattern of osmoregulation in the brackish water amphipod *Gammarus duebeni* during embryonic development. *Journal of Experimental Zoology* **273**, 271–281.

Murphy, D.J. (1983) Freezing resistance in intertidal invertebrates. *Annual Review of Physiology* **45**, 289–299.

Paine, R.T. & Levin, S.A. (1981) Intertidal landscapes: disturbance and the dynamics of pattern. *Ecological Monographs* **51**, 145–178.

Pannunzio, T.M. & Storey, K.B. (1998) Antioxidant defenses and lipid peroxidation during anoxia stress and aerobic recovery in the marine gastropod *Littorina littorea*. *Journal of Experimental Marine Biology & Ecology* **221**, 277–292.

Ruiz, G.M., Carlton, J.T., Grosholz, E.D. & Hines, A.H. (1997) Global invasions of marine and estuarine habitats by non-indigenous species: mechanisms, extent and consequences. *American Zoologist* **37**, 621–632.

Somero, G.N. (1986) From dogfish to dogs: trimethylamines protect proteins from urea. *News in Physiological Sciences* **1**, 9–12.

Somero, G.N. (2002) Thermal physiology and vertical zonation of intertidal animals: optima, limits and costs of living. *Integrative & Comparative Biology* **42**, 780–789.

Towle, D.W. (1997) Molecular approaches to understanding salinity adaptation of estuarine animals. *American Zoologist* **37**, 575–584.

Treherne, J.E. (1980) Neuronal adaptations to osmotic and ionic stress. *Comparative Biochemistry & Physiology B* **67**, 455–463.

Tsai, M.L., Li, J.J. & Dai, C.F. (2000) Is large body size advantageous for terrestrial adaptation? A study of water balance in a semi-terrestrial crab, *Sesarmops intermedium*. *Evolutionary Ecology* **14**, 61–78.

Warman, C.G. & Naylor, E. (1995) Evidence for multiple, cue-specific circatidal clocks in the shore crab *Carcinus maenas*. *Journal of Experimental Marine Biology & Ecology* **189**, 93–101.

Willmer, P.G. (1978) Electrophysiological correlates of ionic and osmotic stress in an osmoconforming bivalve (*Mytilus edulis*). *Journal of Experimental Biology* **77**, 181–205.

Willmer, P.G., Bayliss, M. & Simpson, C.L. (1989) The roles of colour change and behaviour in the hygrothermal balance of a littoral isopod *Ligia oceanica*. *Oecologia* **78**, 349–356.

Zanders, I.P. & Rojas, W.E. (1996) Osmotic and ionic regulation in the fiddler crab *Uca rapax* acclimated to dilute and hypersaline seawater. *Marine Biology* **125**, 315–320.

13 Fresh Water

13.1 Introduction: freshwater habitats and biota

13.1.1 Nature and occurrence of fresh water

"Fresh water" is not really a strictly defined concept, but is usually taken to be any water body of very low salt content, such that it is not detectably brackish. Commonly this corresponds to salt concentrations of between 0.01 and 0.5 ppt, i.e. usually less than 1% of sea water. It makes up only a tiny proportion of the water on Earth (Table 13.1); about 3% of all water is fresh, and the habitable volume of liquid fresh water is less than 1%, because two-thirds of the total 3% is permanently frozen in polar ice caps and glaciers. Of that 1% free fresh water, much is bound up in underground aquifers or within soils, so that only about 0.1% of all the Earth's water is "visible" liquid fresh water, as lakes, ponds, and rivers. An even smaller percentage of the planetary water is within the biosphere at any one time.

Yet as a habitat fresh water is of great biological interest, for a number of reasons:

1 The high overall water availability, and the continual input of geochemically freshwater run-off from the surrounding land, makes floodplains and river deltas exceptionally productive areas. Although they make up only 3% of the terrestrial surface of the planet, they may account for as much as 12% of the "land-based" (nonoceanic) productivity.

2 The habitats concerned are highly variable, encompassing a great range of types and of tremendous chemical variability, more so than any other type of habitat, so that no two freshwater bodies are ever quite the same.

3 Fresh water is an important driving force, cycling minerals and nutrients around the terrestrial environment, via the familiar hydrological cycle (evaporation and precipitation).

4 The habitats impinge on human activity very considerably, since settlements have always been concentrated alongside rivers and lakes.

5 Humans also impinge on the freshwater zones very considerably, so that they are a central cause for concern in environmental study and conservation.

Natural waters come in many forms with very varied physical and chemical characteristics. Figuer 13.1 shows patterns of temperature and pressure and of ionic strength and pH, for atmospheric, surface, and subterranean (groundwater and hydrothermal) sources. In essence the surface fresh waters can be divided into moving, or **lotic** waters (rivers, streams, and temporary trickles), and still, or **lentic** waters, such as lakes, pools, puddles, and rain drops, but also including bogs, fens, marshes, swamps ("wetlands"), and even damp soils, moss cushions, etc. Note that these habitats also differ greatly in permanence (Table 13.2), and that they form a continuum in time and in space through to what we normally think of as "terrestrial" habitats. Indeed, all surface freshwater habitats have a littoral zone where terrestrial influences increase, and in many cases this undergoes a periodicity of emersion and immersion as water levels change.

Table 13.1 Distribution of water sources on Earth.

Source	% of total water	Volume ($km^3 \times 10^6$)	Renewal time (vol/vol added time^{-1})
Oceans	97.2	1350	300–11,000 years
Saline lakes	0.008	0.1	1–4 years
Fresh water (total)	2.8	37	
Glaciers and ice caps	2.15	29	12,000 years
Aquifers/ground water	0.62	8.2	60–300 years
Soil water	0.005	0.07	–
Lakes	0.009	0.12	1–100 years
Rivers	0.0001	0.001	2–10 days
Atmospheric water	0.001	0.013	~7 days
Water in biota	0.0001	0.001	–

13.1.2 Flowing (lotic) waters

Rivers and streams vary considerably along their length, with the discharge (volume per unit time) and current (distance per unit

Table 13.2 Types and permanence of freshwater habitats.

Moving (lotic) waters	
Rivers Streams Trickles	→ *Decreasing permanence*
Still (lentic) waters	
Lakes Ponds Puddles Rain drops	→ *Decreasing permanence*
Wetlands: bogs, fens, swamps	↓ *Increasing terrestriality*
Interstitial zones: soils, moss cushions, etc.	

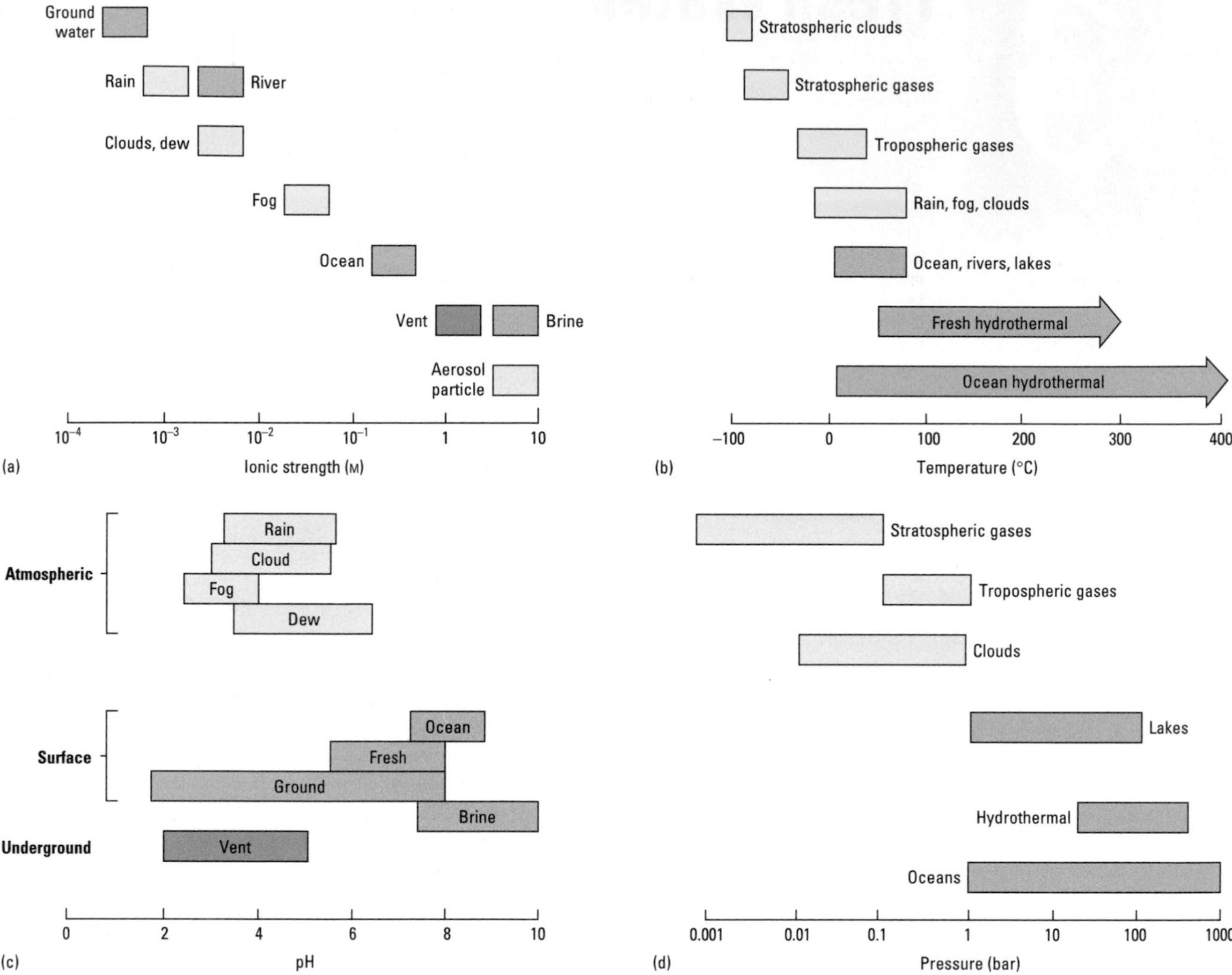

Fig. 13.1 A survey of the chemical characteristics of natural waters (atmospheric, surface, and underground), showing (a) ionic strength, (b) temperature, (c) pH, and (d) pressure ranges. (From Graedel, T.E. & Crutzen, P.J. *Atmospheric Change: an Earth System Perspective*, copyright 1993 by AT&T, used with permission by W.H. Freeman and Company.)

time) interacting with the slope and local geology to determine the water and bed characteristics. It is helpful to divide lotic waters into two categories:

1 Permanent (or "reservoir") rivers, normally with permanent channels and without large flooding events outside the main channel or large seasonal reductions in water level (except in some very large tropical rivers).

2 More temporary "streams".

Reservoir rivers show a fairly clear pattern of alternating deep pools and shallow riffles and bars (see Plate 4a, between pp. 386 and 387), giving a range of benthic habitats that are consistently present even though their distribution may be altered by storms and varying erosion patterns. Streams lack this predictability of habitat and may lack any deep pools for most of their lifetime, even drying up completely (see Plate 4b).

These flowing waters also vary greatly according to their climatic zone and topography, from torrential cold upland streams (see Plate 4c,d) (high current, low discharge) through to warm and sluggish lowland tropical rivers (low current, high discharge). They also differ enormously in the area that they drain. Some of the largest rivers are compared in Table 13.3, where it is evident that large rivers in the tropics with massive basins achieve vastly greater discharge volumes than any others. Table 13.4 shows the variation in nutrient status that is possible, from rivers rich in nitrate and phosphate and often choked with weeds to others that have almost undetectable nutrient levels or are dominated by humic acids, in either case being largely lifeless. Indeed, the same river basin may have areas of entirely different nutrient status, as with the whitewaters, clearwaters, and blackwaters of the Amazon.

This underlines the fact that in nearly all rivers we tend to find considerable variation along the length of the river bed, a **zonation** phenomenon (Fig. 13.2). This is traditionally classified by the commonest conspicuous animals, inevitably types of fish. In Europe the classic four zones are usually listed as follows:

1 The "trout zone" in the upper reaches, with steep gradients, a rapid current, rocky bed, and well-aerated cool water, where the

Table 13.3 Major river basins of the world and river flow rates. Note that rivers in tropical and subtropical forest areas have much higher discharges than rivers draining similar areas of temperate or dry habitat.

River	Drainage area ($km^2 \times 10^3$)	Discharge ($km^3\ year^{-1}$)
Amazon (Brazil)	7000	5500
Mississippi (USA)	4800	560
Congo (Africa)	4000	1800
Parana (Argentina)	3200	730
Nile (Egypt)	3000	90
Mackenzie (Canada)	1800	333
Volga (Russia)	1300	238
Niger (W. Africa)	1100	220
Murray–Darling (Australia)	1100	22
Mekong (SE Asia)	780	4800
Orange–Vaal (S. Africa)	650	12
Colorado (USA)	600	18
Rhine (Europe)	220	70

Table 13.4 Nutrient levels of important rivers. Values are high relative to most lakes, and European and North American rivers are generally higher in nitrate levels than African and South American rivers.

River	Nitrate ($\mu g\ l^{-1}$)	Phosphate ($\mu g\ l^{-1}$)
Amazon		
Whitewater	4–15	10–15
Clearwater	<1	<1
Blackwater	35	6
Mississippi	700–3000	40–440
Parana	500+	50–100
Nile	10–1000	1–40
Mackenzie	600	16
Volga	50–4000	1–250
Niger	1100–6300	500–3100
Orange–Vaal	300–1400	30–100

main inputs of organic matter are external (allochthonous), from leaves falling in from terrestrial vegetation.

2 The "grayling and minnow zone", with moderate gradients, some pools, and gravelly surfaces.

3 The "barbel and chub zone", with shallow gradients and plenty of quiet water, and with increasing plant growth giving internal (autochthonous) organic sources.

4 The "bream zone" in the lower reaches, with slow currents, high summer temperatures when oxygen levels may be low, turbid flow, and deeper water with substantial plant growth.

Other fish associated with each zone are shown in Fig. 13.2. Along with this there are associated zonations of other common animals, notably of flatworms, of caddis flies, and of mayflies.

However, this simplistic view has been largely superseded by the concept of a **river continuum** (Fig. 13.3), where zones are classified by the relative abundance of different types of benthic invertebrate, from shredders through grazers to collectors, linked with the coarseness of particulate organic matter present in the water. Associated with this is a 12-point scheme of river "order". Upper reaches (low order) are dominated by inputs of entire leaves (plus twigs, fruits, etc.); further down these are gradually shredded into fragments and recycled into smaller particles (fine particulate organic matter in the figure), to be collected by filter and deposit feeders in the lower reaches, where they are often sheltered in the gravels of riffles (Fig. 13.4) away from the predators that are important in all parts of the river system. Plant zonation runs from predominantly bryophytes at the upper end, through the first angiosperm pondweeds to appear downstream, then abundant water lilies, rushes, and green algae. In broad, slow rivers, vegetation may decrease again at high orders as light is blocked out by dense growths of riverside trees.

The primary cause of this vertical zonation is of course that lotic waters are fundamentally running downhill at all times, through a succession of altitudinal, climatic, and vegetational zones, usually with a decreasing gradient, decreasing flow rate, and increasing

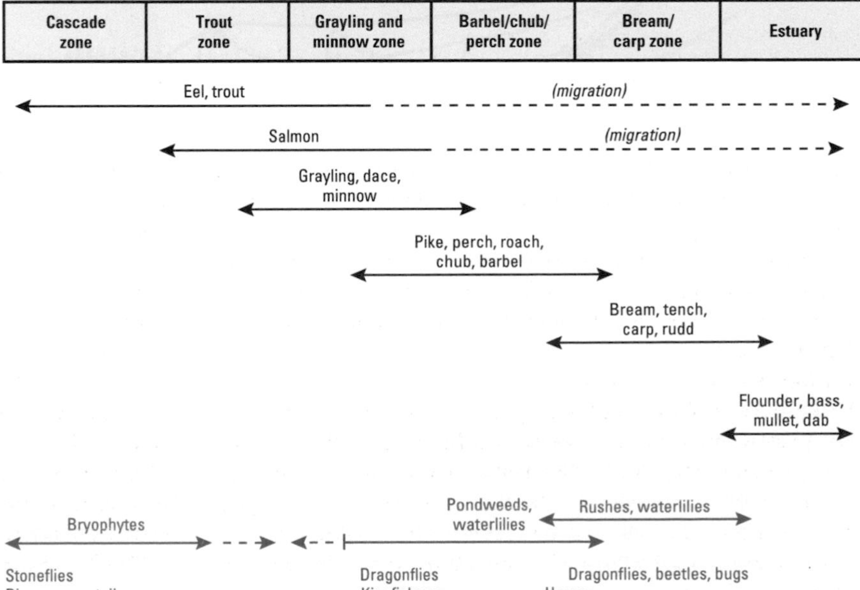

Fig. 13.2 A traditional view of river zonation by major fish zones. Species apply for palaearctic rivers.

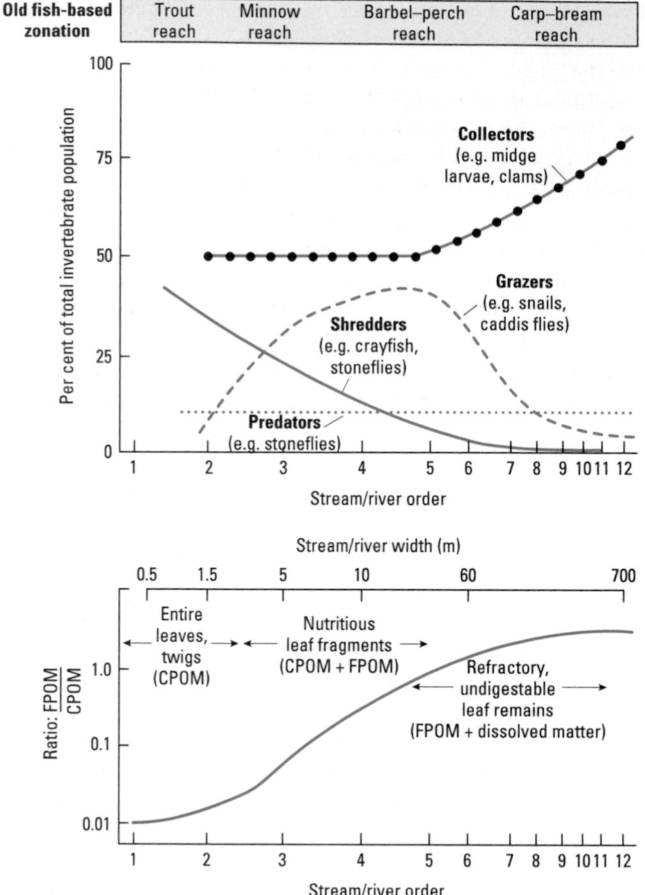

Fig. 13.3 The river continuum concept, where zones are seen in terms of the relative abundance of various types of benthic invertebrates that collect, shred, or graze on vegetation or act as predators. In (a) the patterns for each category are plotted against river order from origin to mouth (scale of 1–12); in (b) the resulting pattern of particulate matter (from coarse particulate organic matter (CPOM) to fine (FPOM)) is shown, reflecting twigs and leaves being shredded and recycled by invertebrates and microorganisms. (From Horne & Goldman 1994.)

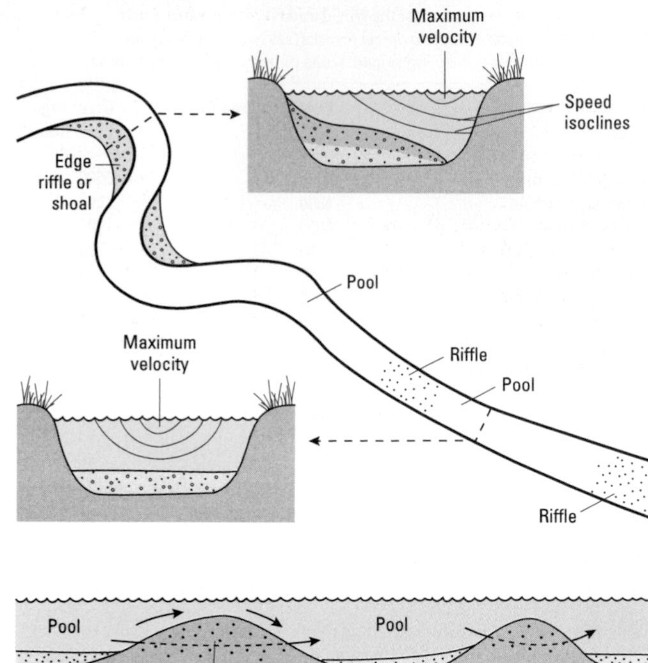

Fig. 13.4 Water flow and the structural patterns of small rivers and streams, with alternating pools and riffles where water passes through and over gravel bars. (From Horne & Goldman 1994.)

volume from source to sea, as hills flatten out into plains and small stream tributaries coalesce into larger rivers. Linked to this, the primary problem for life in all lotic waters is that they ultimately lead down to the sea; any animal living there has a continuous need to work against the current and maintain its position to achieve a degree of stability of conditions.

Nevertheless, animals in lotic fresh water are going to experience variation in many characters of their environment. Temperature will change, both seasonally (especially at higher latitudes) and daily if the water volume is small or the flow rate slow. Ion and pH levels will change according to the rate of input from rainfall and patterns of run-off over surrounding soils and rocks, with pH values of 9–10 in chalk uplands and 4–5 in lowland humic forests. Oxygen levels will vary with temperature, with mixing patterns and flow rate, and with the amount of submerged plant growth and of decay. Food availability is likely to be strongly seasonal and floating or swimming food may be difficult to trap. The physical structure of rivers and streams in itself insures a wide range of differing niches; there are areas of fast current contrasting with slow eddies behind sheltering rocks or trunks, well-lit sunny pools and cool, deep gravel beds, warm transparent upper surface waters and dark protected muddy corners, even completely underground (hypogean) streams. The animals that exploit different parts of lotic water systems therefore tend to be rather specialist in character.

13.1.3 Still (lentic) waters

In lentic waters there may no longer be the problems resulting from continual flow, ultimately to the sea, but there are new sets of problems posed by being in a closed volume of water, each pond or lake effectively being an ecological "island" surrounded by inhospitable land. Problems also arise from the habitat being relatively ephemeral. Table 13.5 shows the area and volume of the world's biggest lakes, but even these are relatively transient on a geological timescale. Again there is also great chemical variability; most lakes have a pH between 6 and 9, but there are also small calcareous upland ponds and highly acid peat bogs well outside these limits, and some naturally acid and alkaline salt lakes (dealt with in Chapter 14).

No particular "zonation" occurs lengthwise in lakes, but instead there may be other kinds of patterning known as **stratification** in the communities, largely depending on the size of the water body. It is therefore easier to deal with these habitats in an orderly size-related sequence.

Table 13.5 The world's largest lakes. Note that Tanganyika and Baikal have very large volumes due to great depth rather than area.

Lake (location)	Area (km^2 × 10^3)	Volume (km^3 × 10^3)	Salinity
Caspian Sea (Russia/Iran)	374	78.2	Marine/brackish
Lake Superior (USA/Canada)	82.1	12.2	Fresh
Lake Victoria (E. Africa)	68.5	2.7	Fresh
Aral Sea (Russia/Kazakhstan)	64.1*	1.0	Brackish
Lake Huron (USA/Canada)	59.5	3.5	Fresh
Lake Michigan (USA)	57.7	4.9	Fresh
Lake Tanganyika (E. Africa)	32.9	18.9	Fresh
Lake Baikal (Russia)	31.5	23.0	Fresh
Great Bear Lake (Canada)	31.3	3.4	Fresh
Great Slave Lake (Canada)	28.6	2.1	Fresh

* Aral Sea now much reduced from this area (see text).

Lakes

Lakes are tremendously variable in character, and three main factors determine this, though they interact heavily: depth, climate, and nutrient status.

Firstly, the depth (or shape, or morphometry) of the lake basin most obviously affects how far from the shore rooted angiosperms can develop. Shallow lakes may have plant growth throughout, with floating and emergent leaves giving total cover; deep lakes have only a tiny marginal fringe of visible plants (macrophytes). Lake depth also affects water mixing patterns and water retention times (the time for all the water in a lake to be replaced); this is commonly 5–10 years for a moderately sized lake, but up to 100 years for Lake Victoria in the East African Rift Valley and much higher still for very deep lakes, such as Tanganyika and Baikal.

The second important factor is climate, especially temperature. In some areas lakes may be well mixed vertically by winds, but particularly in seasonal (mid-latitude) climates the lake separates into two zones demarcated by a sharp thermocline (thermal stratification), with the **epilimnion** above and **hypolimnion** below (Fig. 13.5). The **thermocline** normally forms in spring and disappears again in fall (the "fall overturn"); the level at which it forms is sometimes termed the **metalimnion.** There may also be an associated **oxycline,** with the deep hypolimnion being unmixed and gradually losing all its oxygen. This is also linked to a distinction between the photic zone (where light penetrates and photosynthesis is possible) and the deeper aphotic zone (where only respiration can occur). The compensation depth, at which the boundary between these two zones occurs, varies with the strength of the sun (hence with both latitude and season) and with the water transparency, determined largely by nutrient inputs. Because of all these climatic factors, many lakes can be clearly separated into a pelagic zone where variation is substantial and a profundal zone where conditions are physically and chemically uniform, each zone having its own characteristic biota.

Figure 13.6 shows the resultant different patterns of mixing. In seasonal climates where stratification patterns change predictably once a year, lakes are described as **monomictic** (mixing once). In seasonal continental biomes, there may be inversions of stratification in winter, with the warmest and densest water at 4°C at the bottom, covered by colder water grading up to the surface ice layer. Such lakes have both spring and fall mixing periods, in opposite directions, and are termed **dimictic**. Very high-latitude lakes may be permanently ice-covered and therefore **amictic,** whilst tropical forest lakes mix rarely but rather unpredictably and are termed **oligomictic** or **polymictic.**

Climatic region, and hence the prevalence of winds, is also the main factor producing variation between lakes by the effects of wave action. The extent of wave scouring affects the horizontal patterning of the littoral communities in particular (Fig. 13.7), with the variation from muds and silts on sheltered shores to gravel and bare rock on windswept shores paralleling that found on marine shores (see Chapter 12).

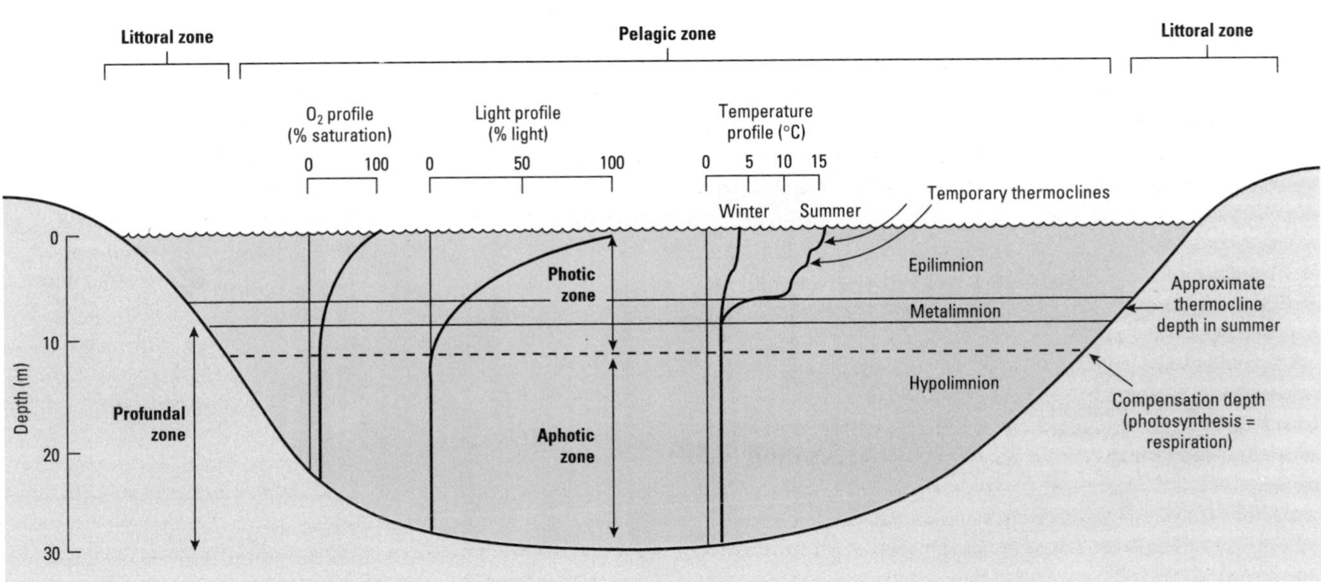

Fig. 13.5 Patterns of stratification in lakes, showing the deep hypolimnion and superficial epilimnion in relation to temperature profiles and oxygenation.

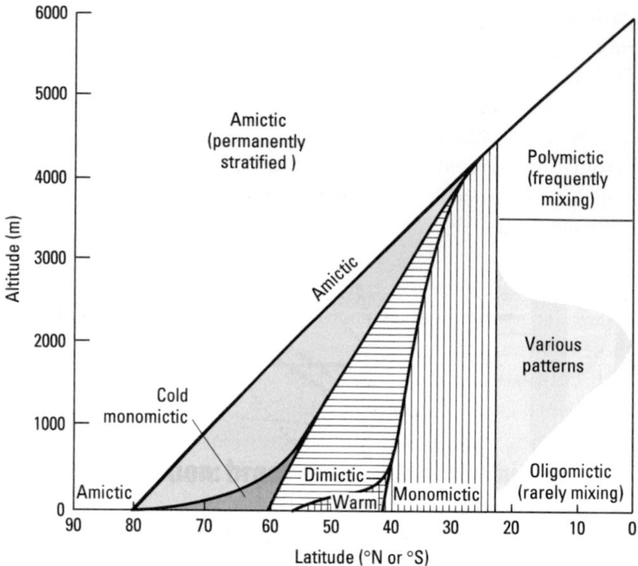

Fig. 13.6 Types of stratification and mixing in lakes in relation to latitude. High-latitude lakes are usually amictic, with progressively more mixing events occurring annually at lower latitudes. Note that increasing altitude usually also reduces mixing except in the tropics where montane lakes are often polymictic. (From Moss 1980.)

The third key factor determining lake biology is that of nutrient level. Two main categories are recognized, termed oligotrophic (very low in nutrients) and eutrophic (very high nutrient levels), with mesotrophic as an intermediate category.

Oligotrophic lakes are usually deep and clear, have low nutrient levels, low ionic conductivity, and few algae. Characteristic examples include most of the upland waters in northern Europe and North America: Wast Water and Buttermere in the English Lake District are well-studied cases, as are Lake Tahoe and Lake Superior in North America, and Lake Baikal (the oldest and deepest freshwater lake) in Siberia. Such lakes often arise due to glacial action, forming a long deep bed between hard rocks such as granites (see Plate 5a, between pp. 386 and 387). Their plant life is primarily floating plankton; but where even this is very sparse there may be some rooted vegetation at surprising depth, because of good light penetration through the clear water, giving an extended "photic zone". Biological production in an oligotrophic lake is usually mainly in the top 5–10 m. The hypolimnion usually stays oxygenated, so benthic mud-dwellers are characterized by *Orthocladius* chironomid midges (these are green rather than red, lacking the hemoglobin pigment found in midges from hypoxic habitats), and the characteristic fish are salmonids. In such lakes, as in no other freshwater systems, the physiological problems of depth (see Chapter 11) may become relevant; for example, in the extreme case of Lake Baikal some animals such as gammarid crustaceans occur at a depth of 1300 m. There may be a high diversity of shallow-water animals, but these are restricted to the lake fringes. The lake water is cold in winter, but then as the day length and warmth increase there may be a spring burst of phytoplankton, before the lakes stratify in late spring. These changes will occur later if the lake surface has frozen (which is rare in Britain, but common in Scandinavia, Russia, and Canada). In even more Arctic regions, there is a sufficiently long freeze to reduce the winter oxygen level substantially, so that such lakes cannot support any fish stocks at all.

Eutrophic lakes are often shallow, have high nutrient levels (especially high phosphate and nitrate concentrations), high ionic

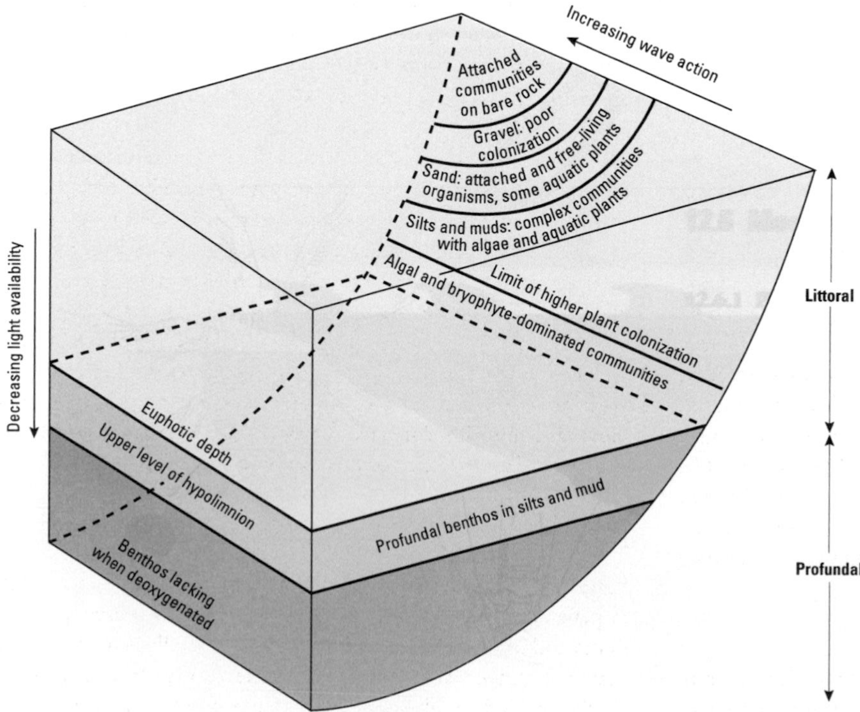

Fig. 13.7 Littoral and profundal habitats in lakes, and the effect of increasing wave action on the littoral characteristics. (From Moss 1980.)

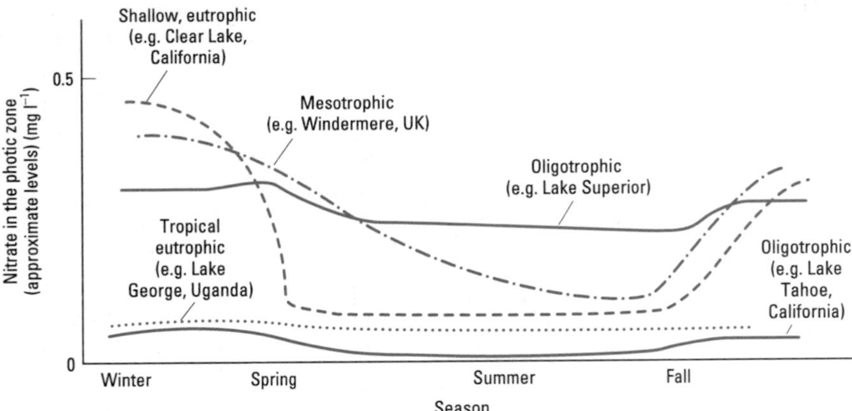

Fig. 13.8 Typical patterns of nutrient status (as nitrate availability) in different kinds of lakes through the annual cycle. (From Horne & Goldman 1994.)

conductivity, and abundant phytoplankton and/or macrophytes (see Plate 5b). This condition is often associated with a large drainage area. In Britain many of the Norfolk Broads are classic examples, as is Derwentwater in the Lake District; in North America, California's Clear Lake and Cayuga Lake in New York State are well known. These lakes are often not deep enough to show much stratification, but deeper examples do get a deoxygenated hypolimnion in the summer; the classic indicator organism is therefore *Chironomus*, a red midge larva containing hemoglobin and thus surviving hypoxia. Often there is a high biomass of animals, but with low diversity. Such lakes may tend to be rather transient unless they are managed, gradually giving way either to more terrestrial wetland habitats or (especially when they have become eutrophic due to artificial addition of nutrients over a relatively short timescale) to unpleasantly smelly anoxic and effectively dead wastelands.

Obviously there are many intermediate types of lakes (see Plate 5c), and an overview of the seasonal changes in nutrient status for classic lakes in each major category is shown in Fig. 13.8. There are also a few special types of lakes that are worth considering.

LARGE LOWLAND TROPICAL LAKES

Many of these occur in the Rift Valley of Africa, and were formerly known as "Victoria", "Albert", etc. but they now have more appropriate African names. These show permanent stratification into upper warm and lower cool regions, because of high temperatures and low wind so that mixing is minimal. Anything that dies falls through the thermocline and never gets recycled, so there is a huge nutrient-rich sediment. Thus wherever the water is moderately shallow the lakes may be dominated by flamingos, filtering the rich mud. There is some recent and current debate about forcibly mixing these lakes to release this nutrient and improve fish stocks. But the lakes already have highly speciose fish communities; they are high-stability environments, so that groups such as cichlid fish (*Tilapia*, etc.) have had time to partition their niches very finely. Lake Malawi has 200+ species, most of them endemic. Mixing the lakes up could give short-term benefits, but would risk destroying all this diversity in the longer term.

TROPICAL MOUNTAIN LAKES

These are especially found in the Andes (such as Lake Titicaca), and in the high plateau of Ethiopia. They show a continuous cycle of stratification, layers forming by day when the sun heats the surface water and mixing at night when it is cold and windy; they are truly polymictic (cf. Fig. 13.6). Often therefore these lakes are very productive indeed, and the fish are an important food source for many surrounding communities.

CRATER LAKES

These occur throughout the volcano belts of the world, when water fills the often doughnut-shaped cavities in the top of extinct eruption sites (see Plate 5d). The volcano walls insure that the water is very protected, therefore the waters do not mix much. The water itself is nominally "fresh" but often with high levels of sulfur and other unusual chemicals, commonly within a layer of denser, quite salty water at the bottom, so that there may be intriguing and unique communities of animals present. Crater lakes have varying patterns of stratification and may exhibit seasonal or sporadic overturns. These can occasionally be disastrous, releasing sulfurous fumes to the surroundings, as in the case of Lake Nyos (in Cameroon, West Africa) in 1986, when fumes from the overturning lake killed hundreds of people and thousands of cattle locally, as well as most life in the lake itself, largely due to CO_2 and SO_2 release.

UNDERGROUND LAKES AND AQUIFERS

Hypogean lakes occur in many parts of the world where caves are common, especially in limestone regions, and aquifers accumulate above the relatively impervious rock strata. The waters here may be strongly mineralized, and of course are usually lightless with no real plant growth. They may undergo marked seasonal drops of level and become hypoxic, stranding some animals on underground shores and in channels that become aerial. Inhabitants of these waters are invariably rather specialist, often eyeless and completely unpigmented, often with elongated sensory appendages, and sometimes with a trend to persistent larval morphology (pedogenesis).

Ponds

Ponds are defined loosely, only by their smaller size than lakes; in general, they are subject primarily to convective mixing rather than wind stirring. They are invariably rather shallow, but may have highly stratified waters. Biological patterns therefore tend to be similar in kind to lakes, but much more rapidly changing—small ponds are notoriously prone to become eutrophic and anoxic very quickly. In temperate zones such small water bodies usually need management (especially periodic weed removal) to stay in a constant and productive state.

Ponds tend to have a much more obvious surface fauna than lakes, because they are usually more sheltered from wind. Animals such as pond-skaters, water bugs, and beetles occur, as do floating duckweeds and larger macrophytes.

There is a much greater likelihood of a pond totally drying up in summer droughts, so the life cycles of resident species tend to be shorter, for both plants and animals, with each individual only experiencing a part of the total climatic cycle. There is nearly always a resistant stage in the life cycle, encysted and able to withstand extreme conditions, and commonly also suited for dispersal to another pond.

Wetlands

Wetlands lack the structured character of most lentic waters, being characterized by shallow fresh water with little thermal or density stratification and little current. The water and underlying soils are often anoxic, permitting the growth of only specialist plants. The wetlands therefore tend to be classified by the main vegetation present, though this may be spatially heterogeneous. Temperate examples include habitats such as: **fens**, with reed and sedge and willow floras; **swamps**, with emergent trees; and **bogs**, with *Sphagnum* moss and cottongrass often dominant. Strictly, fens are dependent on underground water and tend to be alkaline, whilst mires or bogs depend only on rainfall and are acidic, though the terminology is often confused in the vernacular. Productivity (Table 13.6) is usually in the order marshes → swamps → fens → bogs; in all cases most of the primary production is not directly eaten by herbivores but instead cycled into the detritus pathway.

These temperate wetlands are perhaps best known and valued as vital areas for migrant birds and native coastal birds, and as spawning grounds for fish. Recently they have tended to be well studied, both for ecological and hunting interests. There are also tropical wetlands, including papyrus swamps in Africa and Asia, and tree swamps such as the Florida Everglades and the wetlands of northern Australia, with areas of open water interspersed with dense stands of swamp-cypress, freshwater mangroves, and sedges, plus a range of insectivorous plants, vast numbers of insects, and turtles and alligators, together with important populations of grazers such as manatee, dolphins, tapirs, and capybara. However, the ecology of many such areas is almost unknown; the hotter swamps are impenetrable and difficult to monitor and are also major centers for diseases such as bilharzia, malaria, and yellow fever. In some parts they have been exploited for turtle, caiman, and crocodile hunting (for the carapace and skin trades).

The term "wetlands" also encompasses a range of more specialist conditions, notably:

• Riverine forests of the upper Amazon, particularly the "blackwaters" colored by high humus contents, with low pH and low nutrient levels, and a highly specialist fauna and flora.

• Floodplains of very large rivers, producing a highly movable littoral zone; the floodplain of the lower Amazon may be up to 100 km across, and is dotted with small lakes.

Table 13.6 Productivity of freshwater habitats and freshwater plant communities. (Compare Table 12.1 for brackish and marine examples.)

Plant or habitat type	Primary production ($g\ C\ m^{-2}\ year^{-1}$)
Plant types	
Open-water phytoplankton	10–3000
Submerged macrophytes	
Tropical	500–1700
Temperate	300–1300
Floating macrophytes	
Tropical	4000–6000
Temperate	100–1500
Lakes	
Oligotrophic	1–50
Eutrophic	400–600
Wetlands	
Marsh	
Oligotrophic	1000
Eutrophic	6000
Polluted eutrophicated	22,000
Swamp	
Alder/ash	570–650
Reed	1500–3700
Papyrus	6000–9000
Bog	
Oligotrophic	100
Eutrophic	1800
Fen	
Eutrophic	350

Table 13.7 Relative success in freshwater habitats for different taxa (in terms of number of species).

	Very abundant	Moderate	Absent
Algae	Chlorophytes	Other algae	
Plants	Angiosperms	Bryophytes	Conifers
Animals	Crustaceans (ostracods, cladocerans) Rotifers Nematodes Oligochaetes Gastropods Insects Teleosts	Planarians Bryozoans Bivalves Tardigrades Other vertebrates	Echinoderms Cephalopods

• Rice paddies in Asia (see Plate 10a, between pp. 386 and 387), totally managed communities with monoculture vegetation, but vast enough in many parts of the world to have a strong influence on the fauna, and particularly the birdlife, of surrounding areas.

Freshwater biota

Table 13.7 summarizes the taxa that are abundant in fresh water, and those that are largely or entirely absent (cf. Table 11.1, which also includes a few rare taxa). In all fresh waters, invertebrates tend to dominate the benthos and fish dominate the open water. Rotifers (in huge numbers and with tens or even hundreds of species) and small crustaceans (especially copepods and cladocerans or "water fleas") are usually numerically dominant in the filtering zooplankton, though in streams blackfly larvae are important specialist filterers. Other insects are common, the majority having only the immature phases in the aquatic habitat, with relatively short-lived aerial adults (mayflies, dragonflies, caddis flies, alder flies, mosquitoes, blackflies, etc.), but amongst the bugs and beetles there are many permanent freshwater species with raptorial diving adults. Also common are planarian, nematode, and oligochaete worms (the latter especially in colder regions), amphipod crustaceans, chironomid fly larvae, and bivalve and gastropod molluscs; these groups tend to dominate the deep benthic ("profundal") communities, and tubificid annelid worms and chironomid larvae may reach densities of the order of 10^4 per m^2. Fly "midge" larvae (chironomids and chaoborids) dominate in muddy and sandy/gravelly sediments, with particular species occurring in specific sediment types and at specific depths (varying predictably even between successive riffles and bars in riverine habitats). Some specialist groups also live in the substrata of transient water bodies that may dry up completely; these animals (especially rotifers, tardigrades, nematodes) can undergo "cryptobiosis" during drought or freezing, and are dealt with as a special case in the next chapter.

The higher trophic levels in fresh water are mainly occupied by fish, especially teleosts; the salmon, pike, and perch taxa provide some of the larger species worldwide, with minnows and their kin in smaller size ranges, while cichlids are particularly important in more tropical biomes. Some sharks and rays also penetrate far up rivers, as we saw in Chapter 12. Amphibians routinely live in and around fresh water, as do some reptiles (the alligators and their kin, some water snakes and turtles) and many waterfowl, from flamingos to dippers and kingfishers. Beavers, muskrat, mink, coypu, otters, rats, shrews, and voles are the most familiar temperate amphibious freshwater mammals, with less familiar specialists such as desmans and platypus occurring more locally. There are also some riverine specialists, such as hippopotamus, in warmer biomes. All of these are strongly associated with rivers yet they inhabit a macroenvironment effectively remote from the typical river variables such as dissolved oxygen, thermoclines, and water turbulence. However, a few mammals are more strictly fresh water in their lifestyles; manatees and river dolphins are perhaps the most truly aquatic freshwater mammals, and Lake Baikal (Siberia) and Lake Iliamna (Alaska) both have species of freshwater seals.

Surprisingly, there is no very clear pattern here. Certainly the reasonably large animals endowed with hard coverings (arthropod groups, molluscs, vertebrates) do very well, but then so do some conspicuously soft creatures, such as flatworms, and some very small ones, such as rotifers. There are some obvious freshwater specialists, notably amongst crustaceans, rotifers, and teleosts. Only about 3–5% of all insect species are aquatic, but that still represents a considerable abundance and diversity. The main "failures" are echinoderms (starfish are marginal survivors in estuaries, but no echinoderm seems ever to have lived in fresh waters), tunicates, and cephalopods (again never found in fresh or even brackish water throughout their known evolutionary history). Most sponges and cnidarians are excluded, but there are a few successful species such as *Spongilla* and the hydroids *Craspedocusta* and *Hydra*. It is quite hard to find good general reasons for these exclusions (see section 13.9).

13.1.5 Overall strategies for life in fresh water

In principle, fresh water poses the same kinds of problems as brackish water but with the hazards more constant through time (lacking a tidal influence) and somewhat accentuated in degree. The medium is always very dilute and puts a great strain on osmoregulatory mechanisms; animals can no longer be true osmoconformers, since cell contents can *never* be as dilute as fresh water or the cells will die. Nor will behavioral and avoidance techniques be of much use, since there is nowhere to go where the osmotic stress is reduced, and no possibility of short-term shut down until "better" osmotic conditions return. However, on the positive side the limited need to adapt to osmotic *variation* means that enzyme polymorphisms can be reduced, and a single suite of adaptive responses to give adequate efficiency at ion uptake, volume regulation, and cellular osmoregulation may suffice.

In large lakes, and especially for animals living within the hypolimnion, other factors are likely to be relatively constant within the lifetime of an individual. But in smaller bodies of lentic water, and in flowing waters, variation in other parameters does become a real problem. Obviously the temperature can be highly variable with the season, and there may be strong thermoclines. Transience of the liquid water as a habitat therefore also becomes a problem; temperatures may vary to the point of evaporating away the water body completely in summer and freezing it solid in winter. Oxygen levels can similarly be very variable, and the water may go completely stagnant and anoxic in parts. Ion levels and pH will vary drastically with run-off patterns and with freak inputs. Light levels and hence photosynthetic rate and food availability can vary drastically with the season (especially due to shading from plants on the margins). For all these reasons, reproduction also becomes very tricky; the gametes and young stages, with higher surface area to volume ratios, will find life even harder than adults. It is perhaps not very surprising that freshwater life is dominated by relatively few and rather specialist taxa.

13.2 Ionic and osmotic adaptation and water balance

Freshwater animals face the central problem of a permanently dilute external medium, with sodium, potassium, and calcium levels often below 1 mM, with a permanent gradient for ion loss out of their bodies and a net inward osmotic flux of water, so that they must continuously counteract a tendency to become diluted and to swell up. They may also face "extra" osmotic problems if they live in the littoral lentic or shallow lotic freshwater zones, where periodic emersion is likely on unpredictable timescales (unlike the regular emersion–immersion cycles in the marine littoral zone, dealt with in Chapter 12).

Figure 13.9 shows the classic internal medium versus external medium plot, this time for a range of fully freshwater invertebrates. Freshwater lake and river dwellers are all capable of osmotic and ionic regulation, as it is impossible to maintain functioning tissues at these continuously low concentrations. Two factors do vary, and tend to be interrelated: the level of body fluid concentration maintained in fresh water (itself commonly 0.1–5 mM), and the tolerance

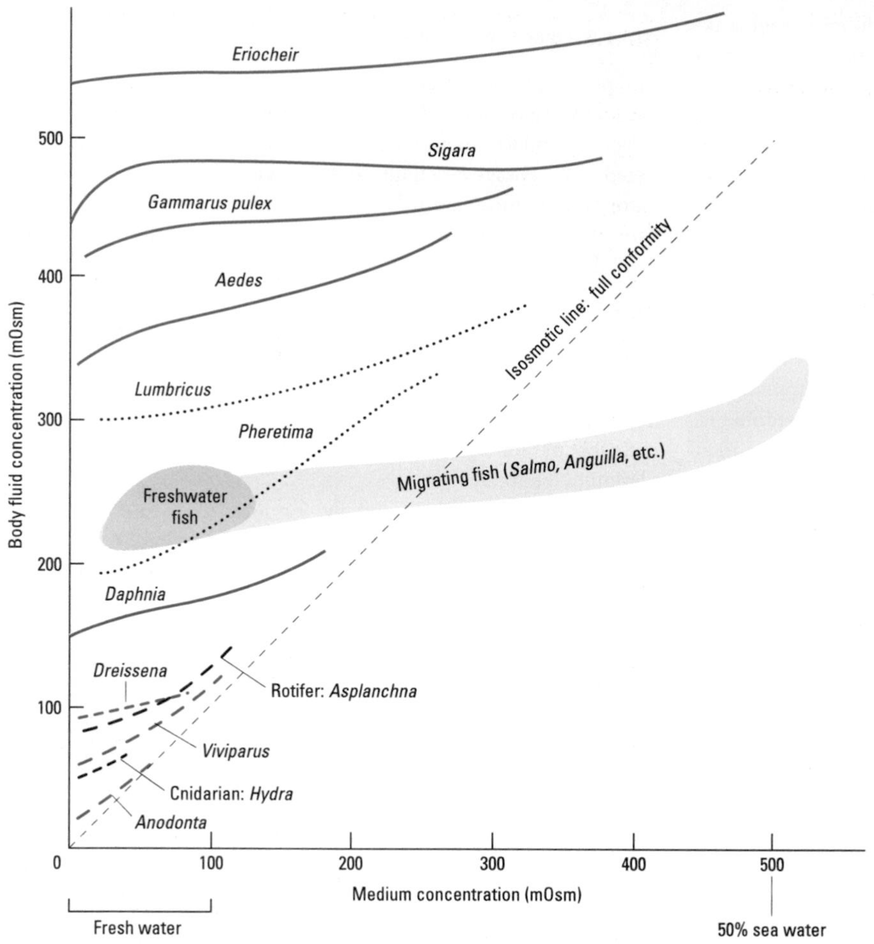

Fig. 13.9 The relationship between internal body fluid concentration and external medium, for various freshwater animals. Arthropods are shown by solid lines, molluscs by green dashed lines and worms by black dotted lines, with teleost fish represented by the tinted area. Compare Fig. 12.13.

range. Thus the Chinese mitten crab, *Eriocheir*, is very euryhaline, as it can penetrate deeply up rivers but can also live in the sea, to which it must return to breed; even in its freshwater phase it maintains a blood concentration at least two-thirds that of sea water. In contrast, *Anodonta*, a river mussel, is a strictly stenohaline freshwater inhabitant and will not survive above about one-tenth of seawater concentration, having its blood concentrations a little less than one-tenth sea water. The bivalve *Dreissena* is similarly restricted, dying at around 50 mM external NaCl concentrations. Many freshwater bivalve molluscs, rotifers, sponges, and cnidarians show this pattern, while gastropods, annelids, and insects tend to be intermediate between this and the crustacean mode. Body fluid compositions of typical examples are shown in Table 13.8.

Despite the variation in degrees of hyperregulation, all freshwater animals (and all terrestrial animals making a secondary return to fresh water) are clearly regulators to some degree, and all have continuous water balance problems because water will always tend to flow into their bodies from the very hyposmotic surrounding fluid.

For all these animals, all the mechanisms described in relation to shores and estuaries in Chapter 12 still apply; i.e. they exhibit:

- Reduced permeability.
- Ion uptake mechanisms.
- Cellular osmoregulation with small osmotic effectors.

But there is also a new mechanism in most freshwater animals:

- Regulated hyposmotic urine.

13.2.1 Permeability

The permeabilities to water and to sodium (P_w and P_{Na}) for some freshwater animals were included in Table 12.4. The values for both invertebrates and vertebrates tend to be much lower than for marine animals, but there will still be substantial osmotic gain of water and diffusional loss of ions, due to the relatively large gradients involved. The larva of a mosquito (*Aedes*) may gain 3% of its total body water volume per day osmotically, and for the crayfish *Astacus*, with more concentrated blood, the figure may be nearer 5%.

Amphibians are well known for their relatively high permeability, but also show an unusual ability to control P_w, particularly in the region used for water uptake known as the "pelvic patch", a section of ventral abdominal skin. When dehydrated, or when the bladder is empty, toads in the genus *Bufo* produce the hormone arginine vasotocin (see Chapter 10), increasing P_w in this area. A second hormonal axis, the renin–angiotensin system, controls the water uptake rate in the same patch.

13.2.2 Ion uptake

Ion uptake is commonly centered on the skin, or on the gills in fish and in invertebrates such as crustaceans and insects (these often having secondary abdominal or anal gills). Rates (as J_{max}) can be quite high (Table 13.9) compared with marine and brackish-water

Table 13.8 Composition of extracellular fluids in freshwater animals.

Animal group/genus	Concentration (mM)			Osmotic concentration (mOsm)
	Na^+	K^+	Cl^-	
Sponges				
Spongilla				55
Cnidarians				
Chlorohydra				45
Rotifers				
Asplanchna	21	7		81
Molluscs				
Anodonta	16	0.5	12	66
Margaritifera	14	0.3	9	40
Pomacea	56	3	52	139
Viviparus	34	1	31	80
Annelids				
Lumbricus	76	4	43	300
Pheretima	43	7	54	152
Hirudo	136	6	36	200
Crustaceans				
Astacus	208	5	250	477
Eriocheir	309	6	280	636
Potamon	259	8	242	522
Asellus	137	7	125	–
Insects				
Aedes				266
Sialis	109	4	199	436
Ephemera				237
Vertebrates				
Carassius	142	2	107	280
Salmo	161	5	120	320
Rana	92	3	70	210
Alligator	140	4	111	278
Anas	138	3	103	294

Table 13.9 Maximum sodium uptake affinities (K_m) and rates (J_{max}) in freshwater animals.

Animal group/genus	K_m (mM)	J_{max} (mM kg^{-1})
Crustaceans		
Mesidotea entomon	2–3	21
Gammarus duebeni	0.40	20
G. lacustris	0.15	
G. pulex	0.15	
Astacus	0.07	0.05
Annelids		
Lumbricus	1.3	0.1
Hirudo	0.14	
Molluscs		
Lymnaea	0.25	
Margaritana	0.04	
Fish		
Carassius	0.26	0.2
Salmo	0.50	0.4
Amphibians		
Ascaphus	0.07	0.05
Hyla	0.14	0.09
Rana	0.20	0.27
Bufo	0.25	0.48
Xenopus	0.05	
Brackish species general range (see Table 12.6)	0.4–20 (higher)	5–15 (similar range)

fauna, with affinities (K_m values) substantially lower (cf. Table 12.5). The rate of salt uptake tends to be variable according to the acclimation history; thus the freshwater crustacean *Gammarus zaddachi* is capable of faster uptake after exposure to 0.3 mM Na^+ than to 10 mM Na^+ solutions (Fig. 13.10a). The freshwater crayfishes have been studied in some details, their gill epithelia showing what is now regarded as the typical pattern for freshwater uptake epithelia (Fig. 13.10b). An apical V-ATPase pumps H^+ ions out of the cell, providing a steeper electrochemical gradient for Na^+ entry (so that Na^+/H^+ exchanges are indirectly coupled), and a more direct Cl^-/HCO_3^- exchange also occurs; overall the effect is an apical electroneutral (1 : 1) ion exchange. This is followed by active basal ion transport via the sodium pump, the counter-ions (H^+ and HCO_3^-) being provided by CO_2 from the body fluids. There may be some interrelation with ammonia levels, but this appears to work via acidification effects on gill permeability to NH_3 rather than via direct involvement of NH_4^+ ions.

The essential processes of uptake and regulation in fish gills were considered in Chapter 12, for brackish/freshwater species. The linkage between CO_2 and ammonia excretion by fish gills is also complex (Fig. 13.11). Ammonia probably leaks passively from the gill down favorable blood-to-water gradients, but this is augmented by the carbonic anhydrase-catalyzed hydration of CO_2, generating H^+ ions that again help to trap the ammonia as NH_4^+. There are thought to be H^+-ATPases in this epithelium rather than Na^+/H^+ exchangers, and these would assist ammonia excretion by actively extruding more H^+ ions. Note that this is somewhat different from the situation in marine fish (see section 11.10.1), where direct diffusion of NH_4^+ (as Na^+/NH_4^+ exchange) is more likely given the much higher cationic permeability of marine gill surfaces.

Adaptation may involve not only changes in the abundance or rate properties of uptake sites in the membranes, but also, on a grosser level, morphological change in the uptake surfaces, as we saw for the anal gills of brackish-water mosquitoes in Fig. 12.17. Morphological modification to the ion uptake system may also occur "in reverse", where freshwater invertebrate taxa have some capacity to invade more saline habitats. For example, in *Aedes* mosquito larvae the species that are obligate freshwater dwellers (about 95% of the genus) have a single rectal segment, whereas euryhaline species have a more complex two-segment pattern. In fact a degree of secondary salt tolerance has evolved at least twice in the mosquitoes: some species osmoregulate using this two-part rectum with both resorptive and secretory cells, while other lineages tolerate salt as osmoconformers.

13.2.3 Osmolytes and cellular regulation

Strictly freshwater invertebrate animals cannot afford to accumulate large amounts of intracellular organic osmolytes, in the manner of brackish/estuarine animals (see section 12.2.2), so that cellular osmotic regulation by variation of levels of free amino acids is relatively unimportant to them. Instead they tend to regulate their volume using movements of potassium ions from cytoplasm to extracellular fluid, thus reducing osmotic intake and resultant

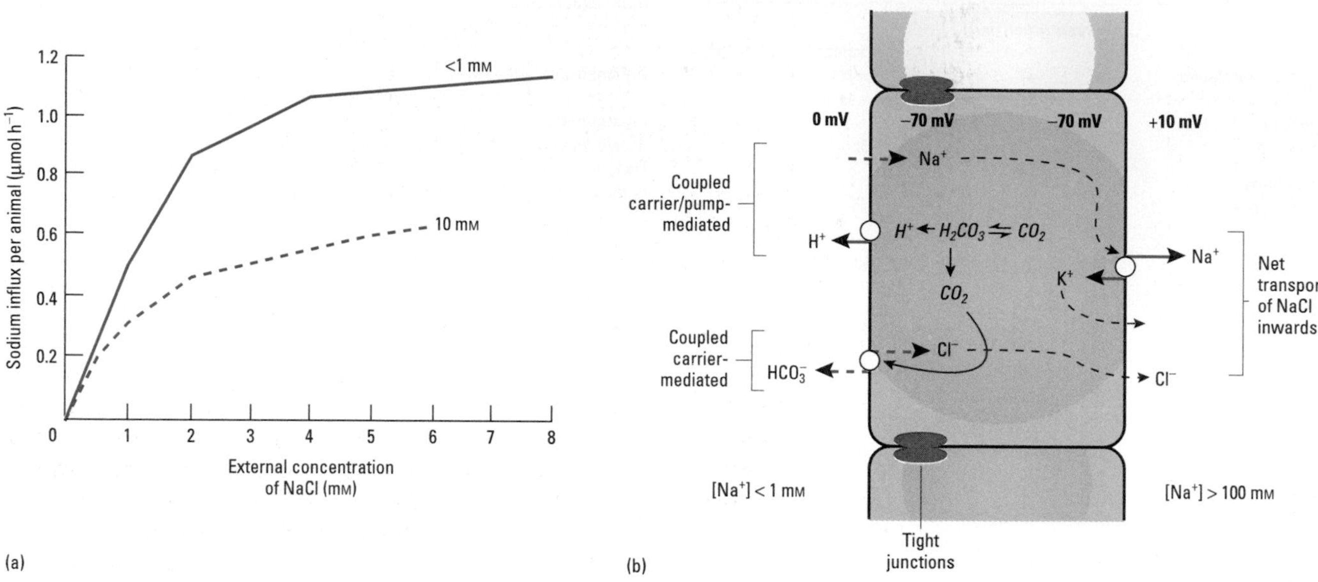

Fig. 13.10 Freshwater ion uptake. (a) Acclimation effects: the rate of uptake in the freshwater shrimp, *Gammarus zaddachi*, previously acclimated to very low salinity and to slightly higher salinity. (b) The basic pattern of ion uptake in freshwater gills or skins. ECF, extracellular fluid.

swelling, allowing a suitable electrochemical balance to be maintained. For example, volume regulation in the bivalve *Dreissena* fails if potassium is unavailable, and the ideal ratio of K^+ to Na^+ in the surrounding medium is around 0.01.

However, freshwater animals that can tolerate a degree of increased salinity tend to synthesize amino acids as intracellular effectors in brackish media; in the freshwater shrimp *Macrobrachium*, amino acids appear in the hemolymph at 21 or 28 ppt, while gill Na^+/K^+-ATPase levels decrease as the rates of ionic uptake are downregulated. The levels of amino acids are regulated by endocrine factors from the eyestalks and from neurohormonal cells in the thoracic ganglion. The osmoconforming mosquitoes mentioned earlier also use the trick of accumulating compensatory

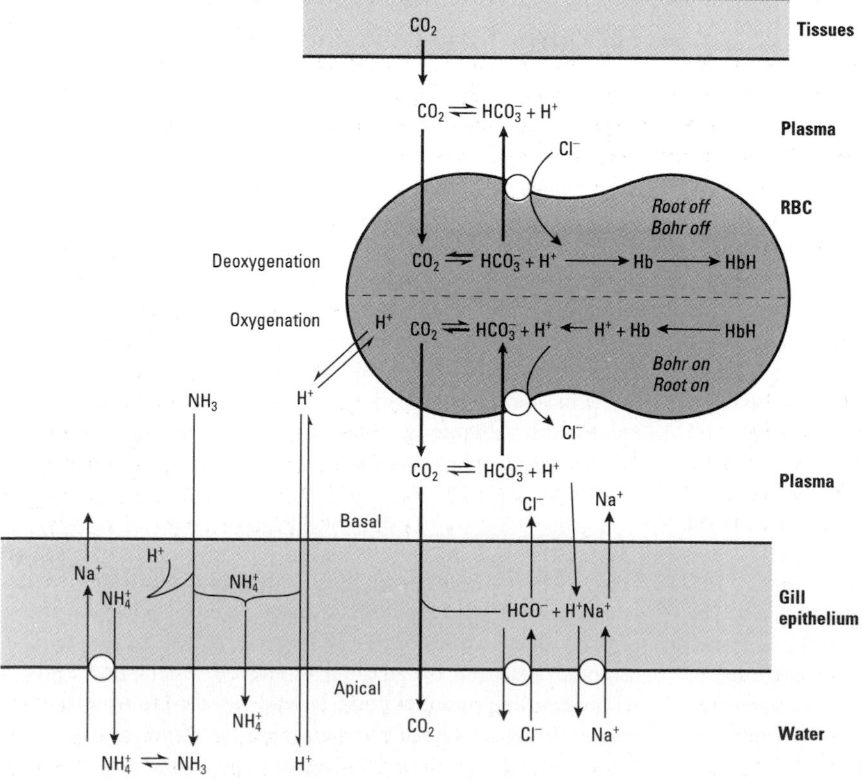

Fig. 13.11 The interaction of ion transport, CO_2 movement, and NH_3 excretion across a freshwater fish gill. RBC, red blood cell. (Adapted from Randall & Daxboeck 1984.)

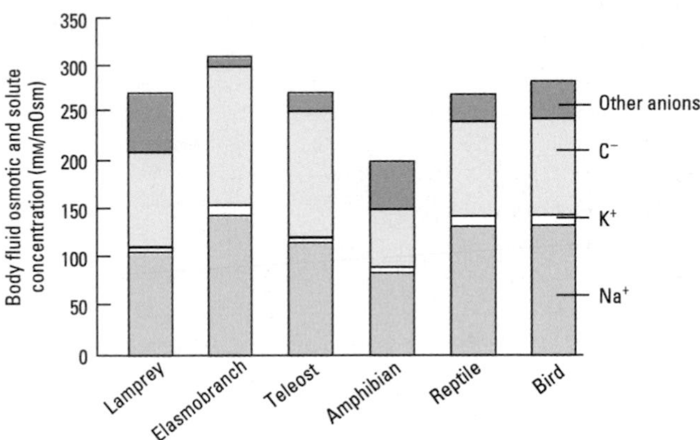

Fig. 13.12 Extracellular fluid composition in a range of freshwater vertebrates.

osmolytes in high-salinity media, in this case incorporating trehalose and proline especially (both these osmolytes also being used in insects for other purposes).

Vertebrates in fresh water routinely have a total osmotic concentration of around 250–400 mOsm, and Fig. 13.12 shows how the extracellular fluid is made up in various taxa. In teleosts and elasmobranchs Na^+ and Cl^- are dominant, while in lungfish and in the "primitive" fish, such as sturgeon and lampreys, there is a moderate organic component. Intracellularly the patterns are similar but with K^+ inevitably replacing sodium as the main cation. Freshwater reptiles, birds, and mammals operate in similar fashion.

13.2.4 Hyposmotic urine

Most marine and brackish-water invertebrates have unmodified, isosmotic urine. But freshwater forms, with a continuous influx of water through their external surfaces and also usually an unavoidable input of fresh water as they feed, often show the additional strategy of recovering ions from their urine to leave it distinctly **hyposmotic** to themselves (urine : blood ratio < 1), as shown in Table 13.10.

Table 13.10 Urine concentrations and flow rates in freshwater animals.

Species	Flow rate (ml kg^{-1} h^{-1})	Concentration (mOsm)	Urine : blood ratio
Annelids			
Pheretima		45	0.19
Crustaceans			
Orconnectes	1.3		
Pseudotelphusa	0.13	120	
Eriocheir		800	1.04
Gammarus	21		
Molluscs			
Viviparus	6	30	
Anodonta	19	25	
Fish			
Carassius	14		
Salmo	5		
Lampetra	7		
Esox		30	0.09

The mechanism is essentially the same as ion uptake in gill or skin epithelia, but involves ion resorption from the urine filtrate back into the body. Classic examples of the structures involved are the **flame cells** (or, when collected together in groups, **protonephridia**) shown in Fig. 13.13. These are found in various forms in most "lower" freshwater invertebrates, including flatworms, rotifers, and nemerteans. A single flame cell has an intracellular invagination into which 30–90 flagella project. The surrounding wall has slit-like perforations, where the flame cell meets the duct cell, and these slits act as the filter. Flagella draw the isosmotic fluid through these slits and then drive it on down the duct, a narrow fast-flow zone where further cells have typical fluid transport structure (see Fig. 4.12), and effect the resorption of ions sufficiently quickly that the water cannot follow proportionately and a progressively hyposmotic fluid is left in the duct lumen.

Other examples of **freshwater kidneys** are shown in Fig. 13.14. The exact classification of these in anatomical or evolutionary terms is still uncertain, but gastropods have a nephridium whose functioning is similar to the basic protonephridial system, though with initial fluid formation by ultrafiltration from the heart cavity (pericardium) or coelom. The crustacean antennal gland (Fig. 13.14b) involves similar filtration and resorption systems, with up to 95% of the initial filtrate being resorbed in the distal tubule in a crayfish (though in amphibious crabs further urine reprocessing occurs in the gill chamber when they are living in air; see section 15.2.4). Leeches are an unusual case, living predominantly in fresh water but also occasionally moving around on land and having the additional problem of intermittent blood meals. Their primary urine is formed in multiple paired nephridia tubules by a combination of ultrafiltration from the blood system and a secretory process based on chloride transport from special canalicular cells in the upper tubules (analogous to the secretory urine production system of insects), driving transcellular K^+ flow and paracellular Na^+ flow. The lower tubule then normally resorbs 85% of Na^+ and 97% of K^+, to leave a very hyposmotic urine. However, after a blood meal the flow rate increases due to an upregulation of paracellular flow, switching the fluid from a K^+-rich to an Na^+-rich balance, and ion resorption is reduced, to give an 80-fold increase in NaCl output and clear excess volume as quickly as possible. Control of this switching is exerted via nephridial nerve cells, releasing a small peptide; these are directly inhibited after blood feeding by the presence of extra Cl^-, so that the nerves stop firing and peptide release is blocked.

Freshwater insects retain their terrestrially adapted Malpighian tubule system, but often switch to excretion of high percentages of ammonia rather than uric acid. In mosquito larvae (*Aedes*) the secretion rates of fluid, and of sodium and potassium, are controlled by 5-hydroxytryptamine (5-HT), and elevation of hemolymph K^+ also causes increased fluid secretion rates, as do increased salinity and increased feeding rates; but in all these cases urine composition is relatively unaffected.

In vertebrates the main uptake sites are the kidney tubules, and in some cases (e.g. turtles) also the urinary bladder. The urine is invariably hyposmotic to the blood (Table 13.10). In fish and amphibians, the kidney usually contains some nephrons with a glomerulus and some that are aglomerular, gathering fluid from the coelomic cavity (though the proportion of aglomerular nephrons is usually less than in marine fish). In most cases about half the primary filtrate is

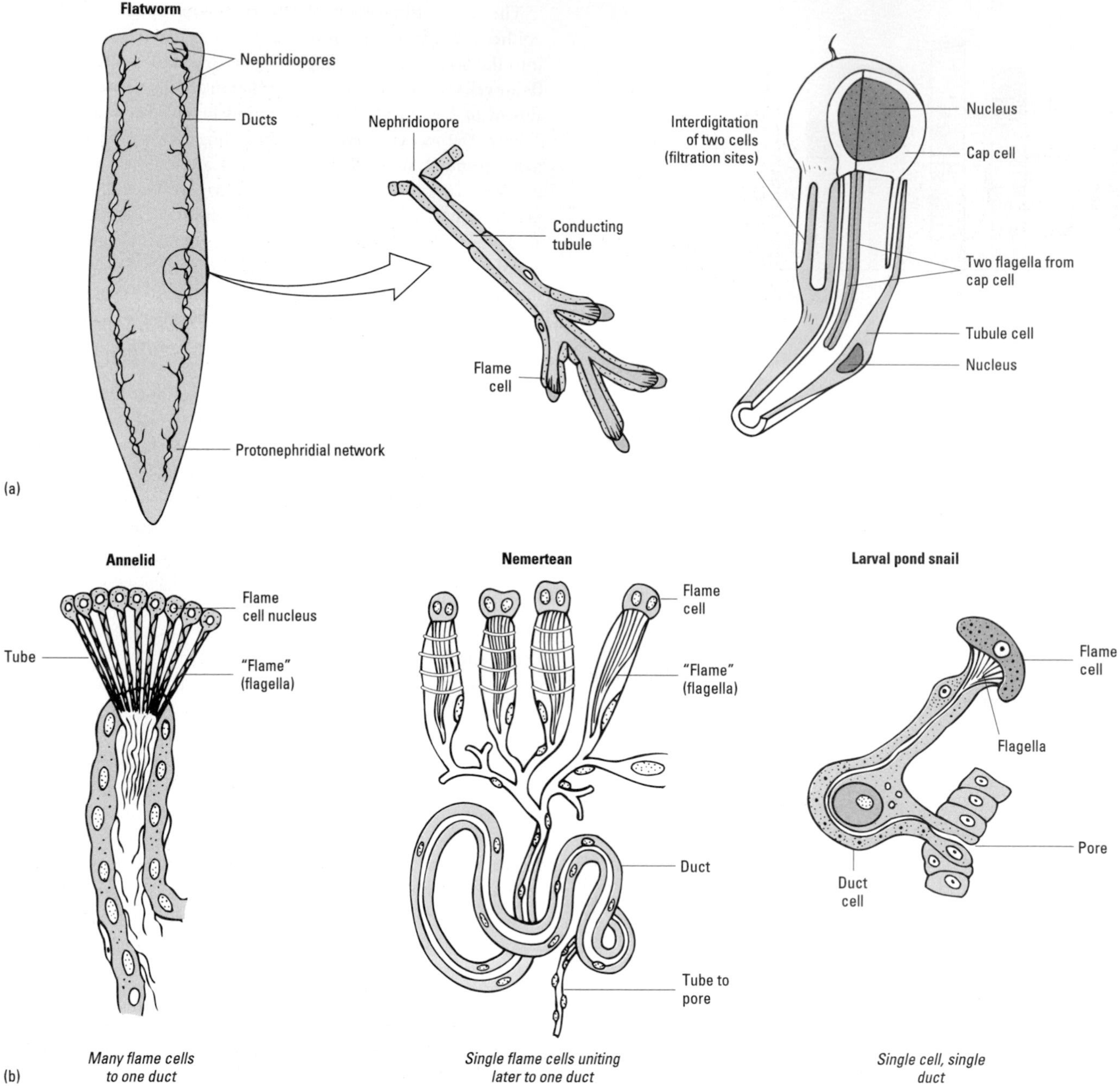

Fig. 13.13 Flame cells, the excretory organs of many smaller freshwater invertebrates: (a) shows the general structure, and (b) shows examples from major freshwater taxa.

resorbed and only half therefore reaches the bladder; but as much as 99% of the filtered ions may be resorbed. Figure 13.15 shows the functioning of a freshwater fish nephron, and gives an overview of the salt and water balance for the whole animal.

For the relatively few fish that can survive in both sea water and fresh water, such as eels, salmon, and blennies, urinary water loss is much increased in fresh water. Glomerular nephrons are more abundant and the glomerular filtration rate rises sharply to permit this flow control. For amphibians, the filtration rate is highly controlled in relation to the hydration state of the animal, largely by the hormone arginine vasotocin. In all animals, **urine volume** (flow rate, also shown in Table 13.10) must be equivalent to the daily osmotic uptake of water, and is therefore high in freshwater fauna. Marine invertebrates may have urine production as low as 0.05 ml kg^{-1} min^{-1}, but freshwater species may achieve 2–10 ml kg^{-1} min^{-1} (see above in relation to permeability), although in the relatively few animals that continue to produce isosmotic urine, such as the mitten crab, *Eriocheir*, this volume can be much lower. The same pattern applies in vertebrates, though here the freshwater range is 2–20 ml kg^{-1} min^{-1}.

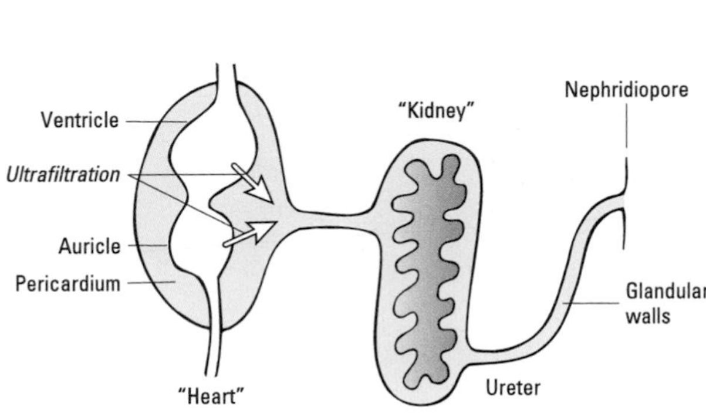

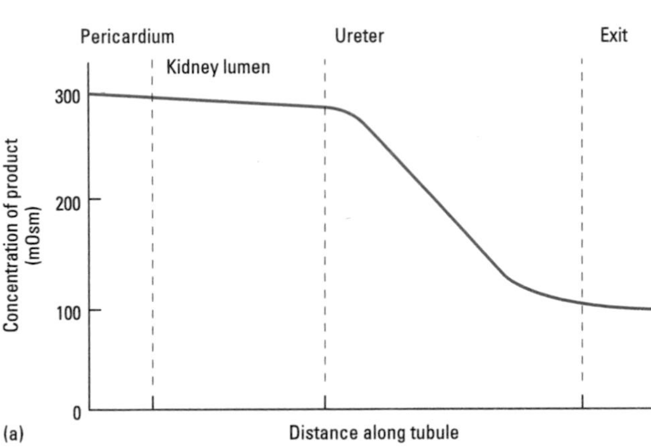

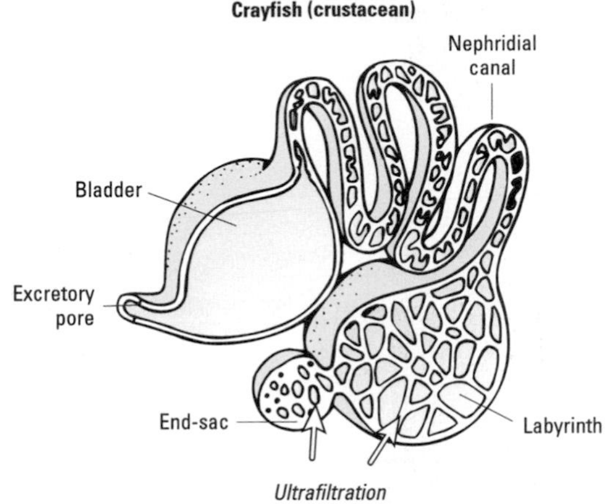

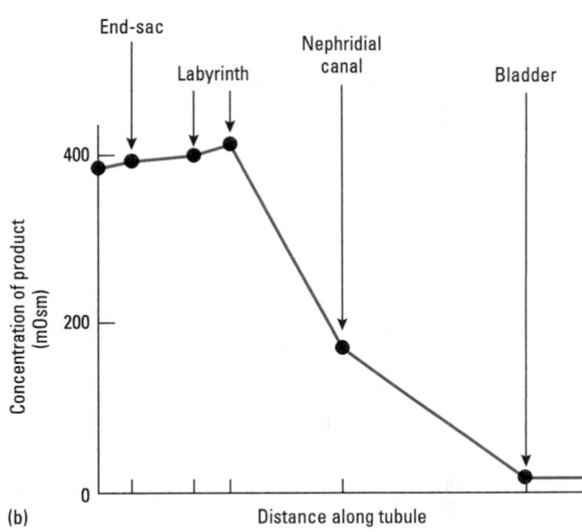

Fig. 13.14 Excretory organs in (a) molluscs and (b) crustaceans from fresh water.

13.2.5 Other ionic problems

Regulation of ions other than sodium and chloride can be crucial in freshwater bodies, which may be individually peculiar in composition. In particular, exposure to high or low pH regimes causes new problems for water and ion regulatory systems, and here analysis has been advanced in view of the problems posed by anthropogenic acidification.

In general, acidic habitats cause problems by disrupting electrolyte balance in both invertebrates and vertebrates, primarily by inhibiting active sodium uptake and increasing diffusional sodium loss. In freshwater clams (*Anodonta*) acute exposure to acidic low pH water produces both low pH and much reduced Po_2 in the blood, but this is redressed over a few days, apparently by mobilizing bicarbonate from the shell. Control of acid–base balance in more active animals can be achieved by varying ventilation patterns to change the availability of CO_2, which provides the counter-ions for NaCl uptake, giving a degree of automatic feedback. The Australian crayfish, or "yabby" (*Cherax*), can survive in both acid and alkaline water, but becomes hypometabolic in both, with O_2 uptake reduced by at least 40–55%.

In Chapter 12 we noted that in fish, varying salinities lead to changes in the degree of exposure of the branchial "chloride cells". Similarly in freshwater species, alkalosis causes the surface area of these cells to increase by withdrawal of the surrounding pavement cells, enhancing the rate of Cl^-/HCO_3^- exchange and so removing basic products from the blood. In acid waters, the chloride cells are covered over by the pavement cells and HCO_3^- is retained.

Magnesium levels may also need particular regulation in freshwater animals, and Mg^{2+} is mainly obtained from food, with gill uptake as a secondary route. Fish can reabsorb Mg^{2+} in the kidney; but in magnesium-deficient fresh waters they can also minimize their losses, increase their intestinal uptake, and mobilize some Mg^{2+} from hard tissues (a reservoir containing around 50% of the body's Mg^{2+} pool).

It should be noted that crustaceans and insects face the additional problem of having to molt and yet maintain osmotic and ionic homeostasis. In the crayfish, freshwater uptake generates the physical force needed to shed the old cuticle and expand the new one, leaving the animal with abnormally dilute hemolymph and calcium deficiency. There is therefore a period of intensive post-molt branchial (gill) NaCl uptake, with additional branchial uptake of calcium and bicarbonate using a Ca^{2+}-ATPase and Ca^{2+}/Na^+ exchange. In insects, the gut helps as a reserve for ions and water during the molt.

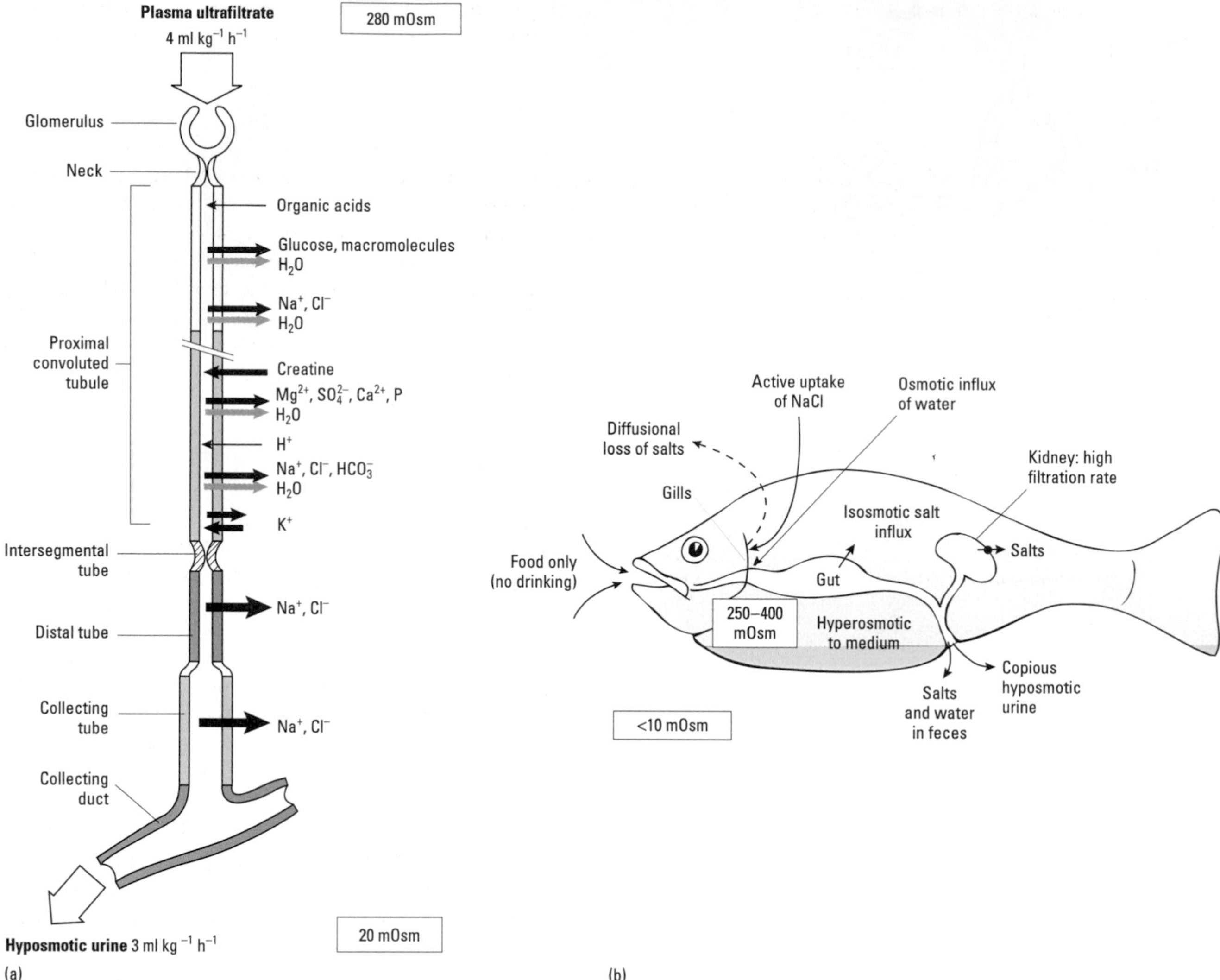

Fig. 13.15 (a) Functioning of a single nephron in a freshwater fish; and (b) an overview of salt and water balance in the whole animal. (a, Adapted from Hickman & Trump 1969.)

13.2.6 Problems of aerial exposure

The remaining water balance problem for some freshwater animals is that of coping with periodic loss of water cover (emersion), which in seasonal droughts may last for weeks or even months. Where water levels merely become lowered, all but the most sessile of animals can retreat down the shore; but where ponds dry up completely, more drastic moves (migration to other ponds, deep burrowing and a period of anaerobic dormancy, or use of a resistant life-cycle stage) may be needed. Only rarely do freshwater animals have to face real drought. Freshwater bivalves are one example, where motility is not an option and dispersal is difficult. Many in this taxon have extensive capacities to withstand long emersions, controlling the rate of water loss using valve movements; during emersion Na^+ and Cl^- levels in the blood are tightly regulated. All ammonia production ceases, suggesting they rely on nonprotein metabolism, and respiratory acidosis is again compensated using carbonate stores from the shell.

Aerial exposure may also compromise the normal aquatic pattern of nitrogenous excretion using ammonia. Freshwater insects, with a terrestrial ancestry, tend to use allantoin and allantoic acid, apparently unable to switch "back" to ammonia excretion but at least finding a moderately soluble alternative to uric acid. In lungfish, the African genus *Protopterus* excretes ammonia when in water, and switches to urea when in its terrestrial cocoon phase, while the much more aquatic Australian genus *Neoceratodus* never achieves a high rate of urea production and though it gulps air when necessary it cannot survive in air for any substantial period.

13.3 Thermal adaptation

The patterns of temperature profiles and of heat cycles in lakes are summarized in Fig. 13.16, indicating that problems are very different for animals in the stable hypolimnion compared to those in the seasonally variable epilimnion. In flowing waters thermal problems

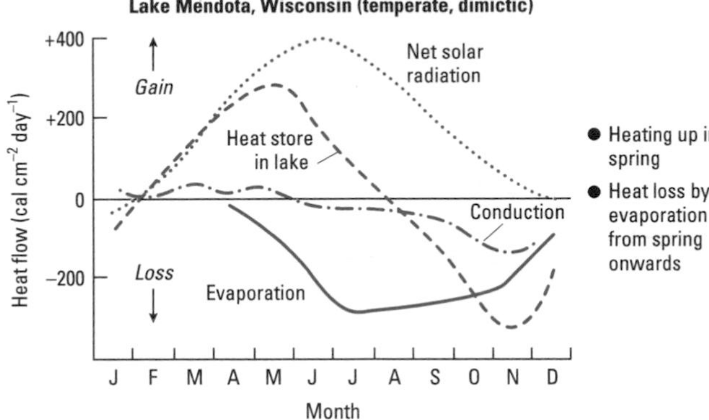

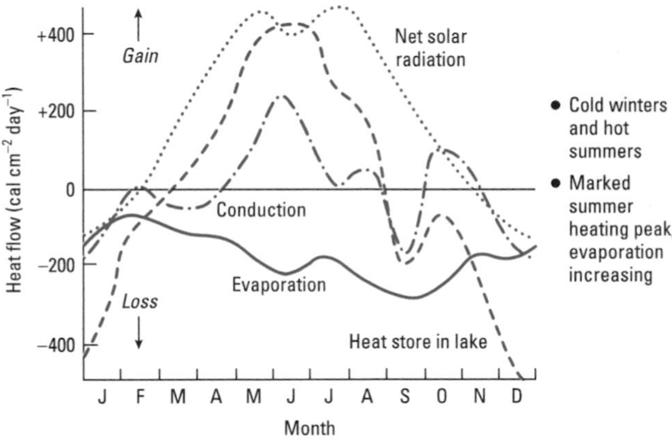

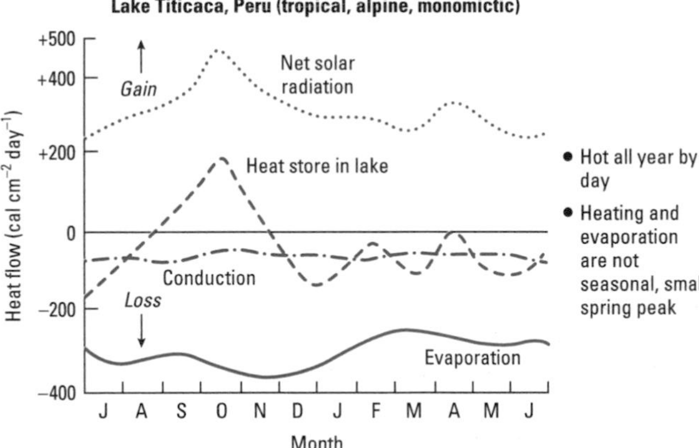

Fig. 13.16 Heat cycles in three contrasting lakes. The temperate lake has a single annual cycle of temperature, while tropical lakes have more irregular heat storage patterns, and higher and more continuous evaporation. (From Horne & Goldman 1994.)

may be compounded on a short timescale by variation in discharge and volume, hence thermal capacity; and on a longer scale by seasonality and (potentially) latitudinal/altitudinal movement of organisms with the water flow. Thermal strategies therefore depend critically on the character of a particular freshwater habitat. An additional indirect problem related to thermal variation for all freshwater animals, whether ectotherm or endotherm, is that higher water temperatures promote bacterial action, so there can be drastic and rapid chemical changes in the water, compounding the problems for the rest of the biota. This is considered further in section 13.8.1.

13.3.1 Freshwater ectotherms

The thermal relationships of freshwater ectotherms, as with all aquatic ectotherms, are greatly constrained by the high thermal conductivity and high specific heat capacity of water, the lack of radiative heat gain through much of the habitat, and the impossibility of evaporative heat loss. Heat dissipation at any large surface area such as the gills is so rapid that the animal inevitably has a T_b virtually identical to water temperature. Moreover, there can be much more variety of water temperature within the lifetime of a freshwater organism than for organisms in the marine and estuarine habitats considered in earlier chapters.

Biochemical adaptations

Given this variation, the central biochemical adaptations for a freshwater ectotherm, as discussed in Chapter 8, are likely to be: (i) broad-spectrum enzymes whose properties are little affected by thermal change, and/or (ii) several morphs of each enzyme, with a capacity to switch between them seasonally or as appropriate.

In many freshwater animals, different enzyme morphs are expressed in different seasons, and for key enzymes such as the sodium pump Na^+/K^+-ATPase there may be a substantial change in protein density in the uptake epithelia and in the nerves and muscles. For example, in a teleost fish, the roach, ouabain-binding sites increase nearly two-fold in winter. Swimming performance in goldfish and carp is modified following acclimation to both low and high temperatures. Myosin-ATPase activity increases after cold acclimation, and the ratio of different isoforms of the myosin light and heavy chains is also altered, influencing the muscle-shortening speed. Thus muscles from cold-acclimated fish can activate, shorten, and relax more quickly at low temperatures. In other animals, particularly the salmonid fish, enzyme changes are more limited and control seems to be exercised more by altering membrane fluidity, with an exaggerated homeoviscous response (see Chapter 8).

Behavioral adaptations

The only exceptions to the general rule that T_b matches the water temperature occur for some motile animals that can exploit local thermal niches in shallow waters, using a limited degree of behavioral thermoregulation. Tadpoles may select warmer water at the edges of ponds, and may shuttle in and out of the warmer waters to maintain a fairly constant T_b. Certain small freshwater fish, and *Hyla* tadpoles, clearly show "basking" behaviors in these shallow water edges, orientated to the sun; and the latter may also aggregate to get warmer. Many semiaquatic animals also use basking as a sophisticated control of T_b; alligators will generally move into shallow water to warm up, but may also expose part of their backs to the air to allow drying, and if T_a is high enough they will leave the water and bask on the river bank. Water snakes also bask out of the water, especially in the early morning, returning to water around midday; when basking, the species *Nerodia sipedon* has a T_b of 26.3 ± 0.7°C,

Table 13.11 Preferred body temperatures (T_{pref}) for freshwater ectotherms.

	T_{pref} (°C)
Fish	
New Zealand species	16–27
Bullhead	26
Carp	22–28
East African tilapia	20–32+
Amphibians	
Salamander (juvenile)	25
Bullfrog (juvenile)	24–30
Bullfrog (adult)	22–28
Reptiles	
Box turtle	21–25
Painted turtle	29–35
Alligator	32–35

much lower than expected for this body mass by passive warming in the sun, indicating some thermoregulation to avoid overheating.

Although most aquatic ectotherms cannot have a T_b deviating from the surrounding water temperature, most do still show thermal preferences (T_{pref} or "eccritic temperatures") wherever there is a gradient of available temperature, for example from deep water to shallow or from midstream to unstirred edges. Preferred body temperatures for a range of freshwater ectotherms appear in Table 13.11, showing clear relations with habitat temperatures. Very subtle changes in preferred temperature occur in some fish, the tropical catfish showing daily patterning (but rather surprisingly to a higher tolerance nocturnally), apparently arising from control via the light-sensitive pineal gland.

Given the availability of choice of temperatures, and an ability to exploit this, it is not surprising that overall thermal tolerance polygons are generally larger for freshwater animals than for marine ones (cf. Table 8.10). Variations in the size of the polygon for different activities for a freshwater trout are shown in Fig. 13.17. The comparison of upper lethal temperatures against time of exposure in Fig. 13.18 also shows a clear correlation with latitude and habitat for seven species of freshwater fish.

Amphibious animals face varied rates of cooling according to whether they are in air or water. Small invertebrates that become exposed in summer droughts commonly enter estivating states, and the temperature balance of a snail in this condition is substantially alleviated by partial burial in pond mud (Fig. 13.19). For larger amphibious animals, there may be a variation between the rate of heating and cooling within the aerial environment, for example in the freshwater turtle *Mauremys*, suggesting cardiovascular changes similar to those in the iguana (see Fig. 15.25). The turtle raises its T_b by basking in water or in air, and lowers it by immersion.

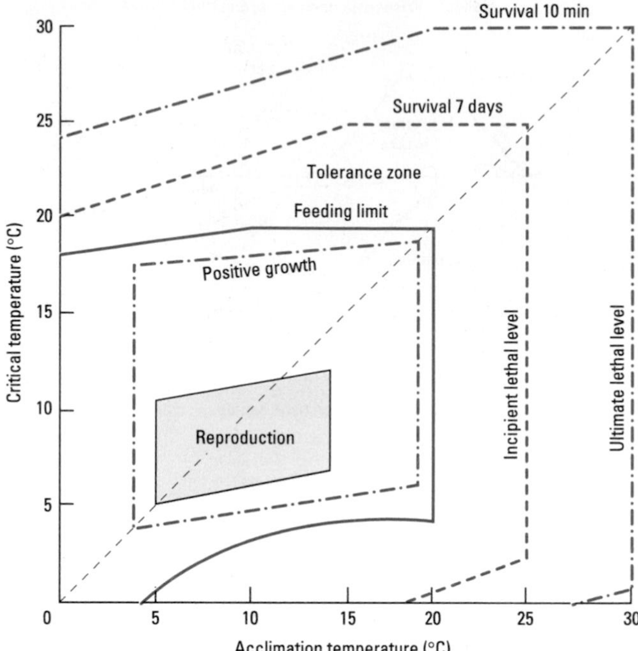

Fig. 13.17 Thermal tolerance polygons for the brown trout (*Salmo trutta*) for various activities. (Adapted from Elliott 1994, and other sources.)

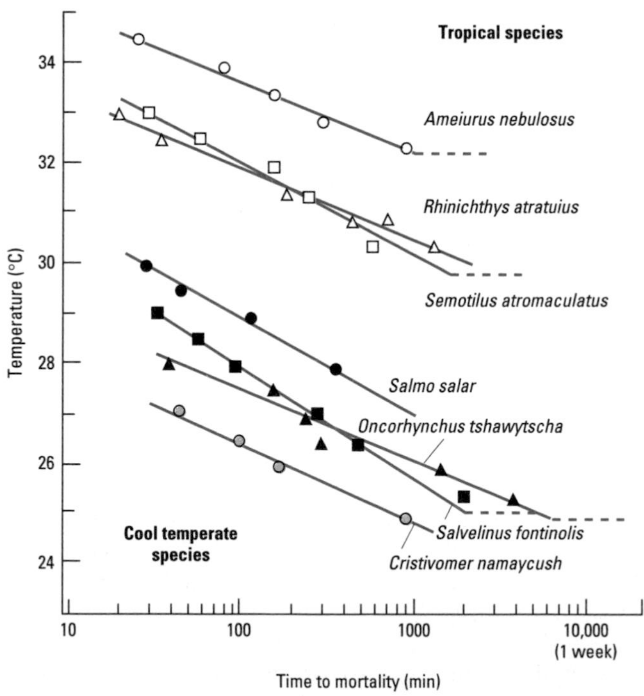

Fig. 13.18 Upper lethal temperature for varying times of exposure in freshwater fish species. Fish were acclimated at 20°C. (Adapted from Brett 1956.)

Coping with freezing

Freezing is a significant problem for temperate freshwater ectotherms, especially in small lotic bodies, though it is worth remembering that even in moderate- to high-latitude lakes the bottom water probably never freezes and may stay at 4°C where water has its greatest density, protected by insulating snow and ice above. Freshwater animals tend to show little accumulation of colligatively active solutes, and supercool only moderately (−5 to −7°C compared with −10 or −20°C in many terrestrial invertebrates from similar latitudes). They are susceptible to inoculative freezing and must therefore rely on not encountering any ice. This is why some species in temperate latitudes seek out larger ponds as winter approaches. Ponds may

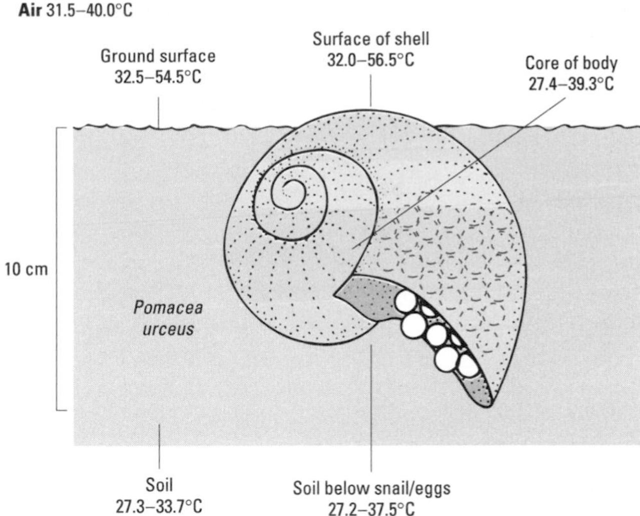

Fig. 13.19 Temperature balance in a freshwater swamp snail, during estivation at the mud surface. The snail's body and its eggs remain below 40°C though the shell surface may reach 56°C. (From Little 1990; adapted from Burky *et al.* 1972, courtesy of Cambridge University Press.)

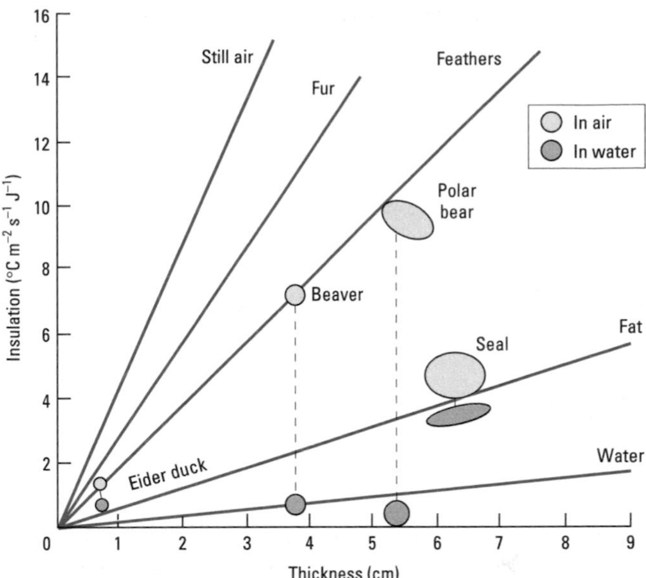

Fig. 13.20 Insulation and the effects of water immersion on fur properties in some aquatic endotherms. (Adapted from Scholander *et al.* 1950.)

also be safer than soils in temperate zones; hatchling turtles leave their soil burrows where they nested after birth and seek a pond, since in the soil they are more at risk of inoculative freezing. Certain species of turtle can survive a degree of freezing, however, at least for a few hours.

Freshwater invertebrates in high latitudes may suffer seasonal freezing, and whilst many may migrate away from a freezing front in a stream, or burrow deep enough to avoid actual freezing, whole communities of insect larvae (especially midges) do become frozen within the gravel in Canadian and Alaskan streams, using freeze-tolerance mechanisms to survive (see Chapter 8).

13.3.2 Freshwater endotherms

Freshwater endotherms are limited to secondarily invading birds and mammals. Most of these are amphibious, swimming on the water surface (ducks, etc.) or diving beneath the water in search of food, though the manatees (sea cows) and river dolphins are permanently aquatic. There are no partially endothermic freshwater fish as we met in the marine environment (see section 11.3.3), nor are there any known cases of freshwater insect endotherms (heterotherms) as with the terrestrial bees, beetles, and moths covered in section 15.3.2.

Endothermic mechanisms for freshwater tetrapods are essentially unmodified from the classic patterns used on land, dealt with in principle in Chapter 8. The main problem for an aquatic endotherm is not with heat generation but with heat retention, the animal being surrounded by a much more conductive medium that is below the body temperature, into which heat is rapidly dissipated. Insulation layers, in the form of fur and feathers, therefore become critical. This raises the problems of wetting, which normally reduces the insulating value of such layers substantially by replacing the trapped air with water as well as compressing the fur or feathers (Fig. 13.20). The common solution in birds is to possess oil glands whose product is spread over the feathers to reduce their wettability, so that an air layer remains trapped against the skin; this has a substantial buoyant effect, useful during surface swimming but increasing the energetic cost of diving somewhat. Mammals may achieve a similar insulating effect with an underfur that retains an air layer, or may use the normal dense fur to trap a layer of stagnant water that at least reduces convective heat loss (the "wet suit" effect). However, insulation does almost invariably decrease during a dive: in the eider duck, insulation due to feathers in air is about 1.6°C m^{-2} s J^{-1}, but only 0.83°C m^{-2} s J^{-1} in water.

Most freshwater endotherms use countercurrent exchangers in their extremities (feet or flippers) to reduce heat loss, and in cold climates the membranes of tissues in these extremities in animals such as ducks and waders incorporate lipids of short chain length to maintain fluidity—these have melting points up to 30°C lower than the lipids in the body core. Because of the cooling effect of a dive, some semiaquatic mammals, such as muskrats, also use brief periods of nonshivering thermogenesis (NST) in the body core (see section 8.7.2) to warm up after diving or prolonged swimming.

13.4 Respiratory adaptation

Although water contains relatively little oxygen compared with air, in principle fresh waters hold more oxygen than salty ones (see Table 7.2). Often, therefore, small freshwater animals have little problem in gaining oxygen. Equally, they can lose carbon dioxide readily because of its high solubility. It is common to find O_2 and CO_2 levels as mirror images of each other in lakes (Fig. 13.21), the pattern of each depending strongly on temperature.

But the difficulties for respiration arise from the variability of freshwater habitats. Whilst large oligotrophic lakes may stay nearly saturated, several factors can reduce the available O_2 in the water column of rivers and eutrophic lakes, including:

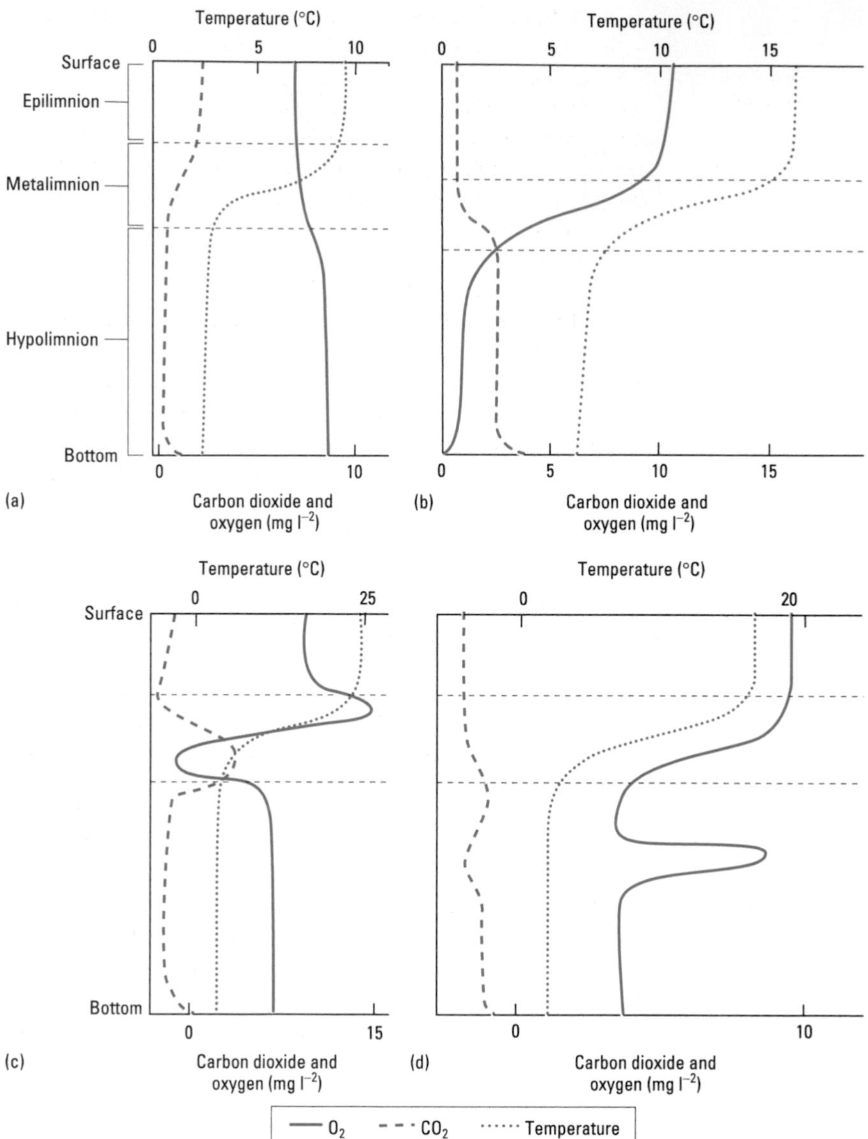

Fig. 13.21 Oxygen and carbon dioxide profiles in different kinds of lake. (a) An oligotrophic lake. (b) A temperate eutrophic lake. (c) A heterograde curve in a relatively unstirred lake where algae just above the thermocline raise oxygen levels and bacteria just below it reduce oxygen. (d) The anomalous pattern that results locally from the inflow of dense, cool oxygen-rich stream water that forms a discrete layer. Note that carbon dioxide levels are often inversely related to oxygen levels. (From Horne & Goldman 1994.)

1 Seasonal cycles of productivity and vegetation decay, producing anoxia in the hypolimnion and potentially supersaturation in the epilimnion. For streams, ponds, and swamps there may even be a diel pattern of productivity-related oxygen levels.
2 Raised temperature (although somewhat compensated for by increased speed of diffusion at high temperature): in spring and summer lakes may lose up to 50% of their oxygen simply due to temperature change.
3 Prolonged freezing: below the ice all the oxygen may get used up and it is not renewed until the spring melt.
4 Build up of water weeds in high-nutrient zones.
5 Flow patterns where rainfall is highly seasonal.
6 Human interference.

In addition, conditions in the benthic muds of still waters (where many animals burrow), and in most wetlands such as marshes and swamps, tend to be rather anaerobic (Table 13.12), so that there are whole communities of specially adapted animals there.

Table 13.12 Composition of gases in freshwater bodies.

	O_2 (ppm)	CO_2 (ppm)
Normal fresh water (surface)	8–10	0.02–0.04
Normal fresh water (100 m depth)	8–10	0.02–0.04
Reed swamp	1–6	3–18
Guinea-grass swamp	0.2–1.2	8–9
Water-hyacinth swamp	2–7	0–9

Just as with estuarine animals, coping with the variation in oxygen levels can be done at several levels. A simple response shown by many invertebrates and by amphibians is to move to cooler water when hypoxia threatens (the "behavioral hypothermia" response). Thus frogs preacclimated to 4°C shift their preferred temperature from 6.8 to 1.9°C when also exposed to hypoxia. If hypoxia cannot

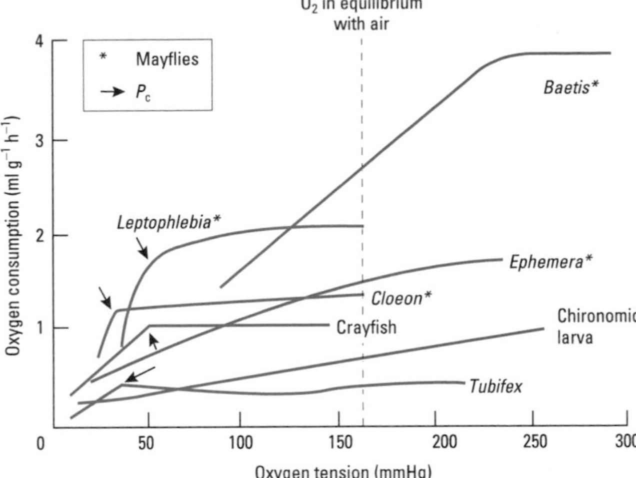

Fig. 13.22 Oxygen consumption in relation to environmental Po_2: conformers and regulators amongst freshwater annelids, insects, and crustaceans.

be avoided by behavior, animals may respond by either **conforming** or **regulating**. Figure 13.22 includes examples from the mayflies: *Baetis* is a classic conformer, and *Cloeon* a near-perfect regulator over most of the Po_2 range, largely achieving this by variations in the frequency of gill movement. Both it and the crayfish represent the common condition of regulation above some fixed point (P_c; see Chapter 7) and conformity below it, and the values of P_c are lower for most freshwater animals than for either open-water marine or terrestrial species (see Table 7.1; cf. Fig. 7.3).

The most obvious and widespread solutions to variable oxygen supply are three-fold:

1 To expand or elaborate the respiratory surface.

2 To use higher affinity oxygen-storing pigments.

3 To modulate the ventilatory and/or circulatory rates.

Some of these features may be varied on an ontogenetic, seasonal, or daily basis. Alternatively, some freshwater animals opt out of the problem and instead (either intermittently, or permanently if they have terrestrial origins) breathe from the more reliable aerial oxygen supply. A relatively small number of freshwater animals are able to withstand quite prolonged periods of anoxia, as might be experienced when ponds dry out or freeze over.

13.4.1 Respiratory surfaces

In freshwater invertebrates gills are the most common adaptation to insure oxygen uptake, and these can be elaborated from almost any part of the body. Most freshwater soft-bodied invertebrates are derived from marine ancestors already endowed with aquatic respiratory systems, and most annelids and molluscs retain the cutaneous exchange surfaces, or serially repeated filamentous gills or tentacular crown gills, or enclosed lamellate gills, of their relatives. In specialist benthic animals, such as oligochaetes (e.g. *Tubifex* worms), the head is buried in mud and the skin of the posterior half of the body is particularly well vascularized. The wetland oligochaete *Alma* has a particularly deeply grooved tail with dense vascularization. Use of different exchange sites in different conditions also occurs. For example, the common pond snail, *Lymnaea stagnalis*, has 25% cutaneous uptake at low Po_2, but this can rise to 50% as Po_2 rises. The snail *Biomphalaria* has both gills and lungs.

Freshwater crustaceans have equally inherited an efficient gill system from marine ancestors, and the branchial surfaces are generally borne on the legs, either free in more primitive forms, where they may be ventilated by leg movements (as in cladocerans and amphipods), or as complex feathered structures enclosed within a branchial chamber and ventilated by the scaphognathite ("baler") in the decapods (crabs and crayfish).

For freshwater insects that are fully aquatic (as larvae or throughout the life cycle), some modification to the aerially adapted tracheal system is needed, and this usually means the addition of cuticular gills. Many larval forms have tracheal gills (Fig. 13.23a) where extensions of the body surface contain a dense network of tracheae (seen on the abdomens of stonefly, mayfly, and caddis fly larvae, and as long tails on many damselfly larvae). Other aquatic insects, particularly fly larvae, have spiracular gills (Fig. 13.23b), where a tube-like structure extends from a spiracle, or rectal gills (Fig. 13.23c), as in dragonfly larvae, with tracheoles investing folds of the rectal surface. Other insects, especially those that are amphibious, have achieved yet other respiratory devices to breathe from air stores when under water (see section 13.4.4).

Freshwater vertebrates may use cutaneous or gill-based respiration, or be reliant on aerial supplies via lungs. Skin-breathing is quite common in freshwater fish larvae, and many eels and catfish rely on it extensively to sustain their resting metabolic rate. Swamp-dwelling fish often develop a form of air-breathing; the bichir (*Polypterus*) and the true lungfish in the genera *Protopterus*, *Lepidosiren*, and *Neoceratodus* all show the development of lungs from the swim-bladder, while less specialized developments of vascularized swim-bladders or gut regions occur in some catfish and bowfins, and in the giant Amazonian fish *Arapaima*. Some salamanders also show a high dependence on cutaneous oxygen uptake, and certain adult frogs have highly vascularized and folded areas of skin, or even areas of "hairy" skin, that contribute a large part of their O_2 requirement, only using their lungs when active or if exposed to hypoxic surroundings. But most fish, and juvenile or neotenous amphibians, use gills, and nearly all adult amphibians and reptiles, birds, and mammals (being secondarily aquatic) rely on air-breathing through lungs. Figure 13.24 shows that while alveolar surface area for mammals is strongly related to oxygen uptake rate (scaling with a mass exponent of 1.0), the surface area in freshwater mammals (porpoise, manatee, dugong) produces a rather higher uptake rate than predicted.

13.4.2 Respiratory pigments

Increased exploitation of pigments as oxygen carriers and stores is very apparent in common freshwater invertebrate communities, where many more of the species are red or green in general color than is evident in marine communities. Obvious examples are the tubificid annelids and the chironomid midge larvae ("bloodworms") living in the muddy benthos. Pigment characteristics, as always, vary with habitat and lifestyle. For example, in the surface-active amphipod *Gammarus*, oxygen consumption rises linearly with oxygen availability, whilst for benthic *Chironomus* worms the shift to full O_2 consumption (full activity) occurs at less than 20% O_2

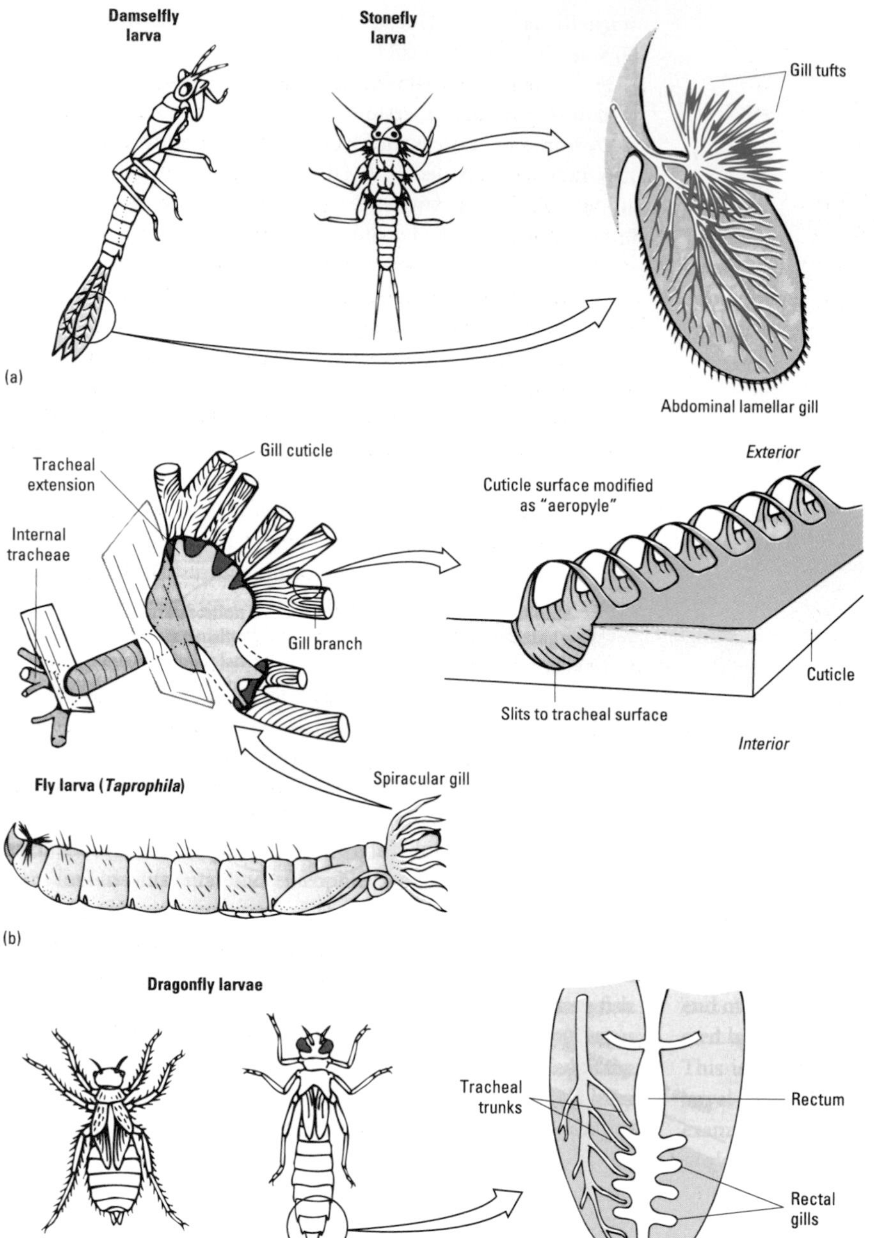

Fig. 13.23 Respiration in freshwater insects. (a) Tracheal gills in the damselfly and stonefly. (b) Spiracular gills in an aquatic fly larva. (c) Rectal gills in the dragonfly. (Adapted from Hinton 1957, and other sources.)

saturation, its hemoglobin having a characteristically left-shifted saturation curve (Fig. 13.25). In the water flea (*Daphnia*) hypoxia leads to an increase in hemoglobin concentration, which is directly correlated with its swimming ability in oxygen-depleted waters.

Pigment adaptation is particularly evident in the hemoglobins of larger crustaceans and fish with active lifestyles, as might be predicted. In crayfish regulation is achieved by increased hemocyanin affinity. In the blue crab, *Callinectes*, hypoxia is accompanied by an increase in the amount of circulating hemocyanin, but also a shift in the hemocyanins in the blood, with more of the high-affinity 1×6-meric oligomer and less of the normal 2×6-meric oligomer. The properties of the hemocyanin also vary in relation to both temperature and ionic strength of the blood (see Chapter 7). Fish species from rivers where seasonal flow changes may result in very different oxygenation regimes often show rather similar patterns, with seasonal adjustments in the expression of different hemoglobin fractions, and also in hematocrit or red blood cell (RBC) abundance. Goldfish acclimated to different temperature regimes show complex responses involving erythropoiesis (formation of new RBCs), loss of existing RBCs, and division of circulating juvenile RBCs, thus adjusting the abundance of hemoglobin isomorphs without greatly affecting overall hematocrit and blood viscosity. Mobilization of stored RBCs from the spleen (or in the case of some cave-dwelling fish from the liver as well) provides another safety valve.

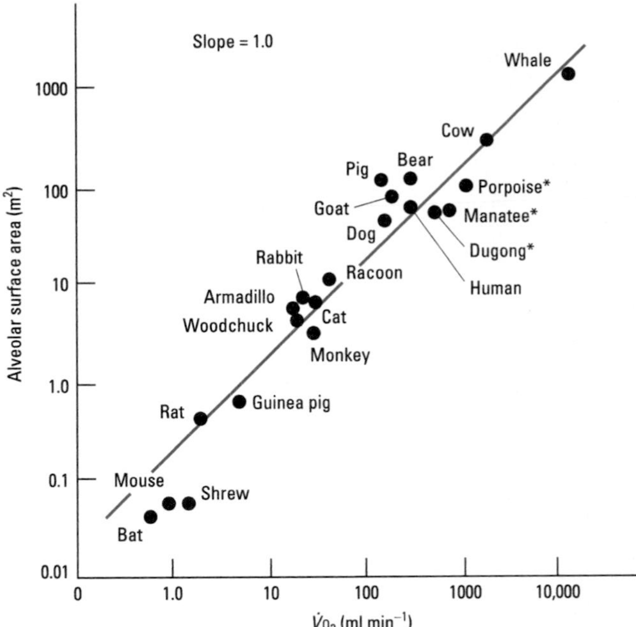

Fig. 13.24 Relationship between alveolar surface area and oxygen uptake in mammals; note that the three freshwater species (*) have somewhat higher uptake rates than predicted from their lung structure.

13.4.3 Ventilation and circulation

Increased ventilation rate is usually the first and quickest response to changing oxygen demand in freshwater animals, whether this arises from environmental hypoxia or from metabolic activity. In small animals, such as sponges and rotifers, ventilation is mainly by cilia or flagella, whose activity increases as needed. In invetebrate bottom-dwellers it may involve wiggling of the whole body, as in tubificid worms, or localized bursts of gill activity in many insect larvae. Mayflies can keep their burrows above 75% saturation at all times by continuous gill-pumping, whereas alder fly larvae ventilate only intermittently, dependent on temperature and water quality. Molluscs and crustaceans tend to have baling systems and use either an increase in rate or in stroke volume, or both, as Po_2 declines. Crayfish show clear hyperventilation initially in low Po_2, but as hypoxia sets in the animals show bradycardia, and the circulatory flow patterns alter to give increased flow anteriorly over the brain.

Freshwater vertebrates show relatively sophisticated ventilatory responses to hypoxia. Fish employ modulation of both stroke volume and frequency (by varying the buccal and opercular pumping patterns) to match oxygen uptake at gills to demand, as detected by oxygen receptors that are usually sited in the brain and aorta. In resting snakes and turtles ventilation may be intermittent (with short bursts of breaths between periods of complete rest), but becomes continuous during steady swimming; the green turtle shows a seven-fold increase in mean ventilation frequency and increases in both pulmonary and aortic blood flow.

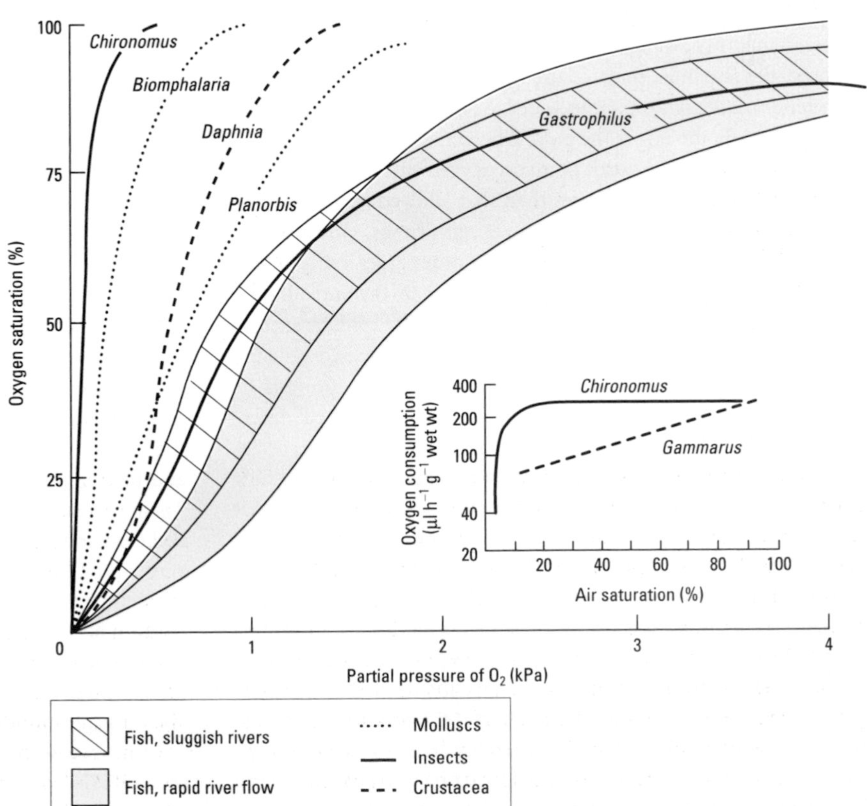

Fig. 13.25 Hemoglobin oxygen-binding curves in freshwater invertebrates and riverine fish. The inset shows the effect on oxygen consumption, the chironomid from the stagnant benthos achieving maximum oxygen uptake at very low Po_2. (Data from Weber 1980, and other sources.)

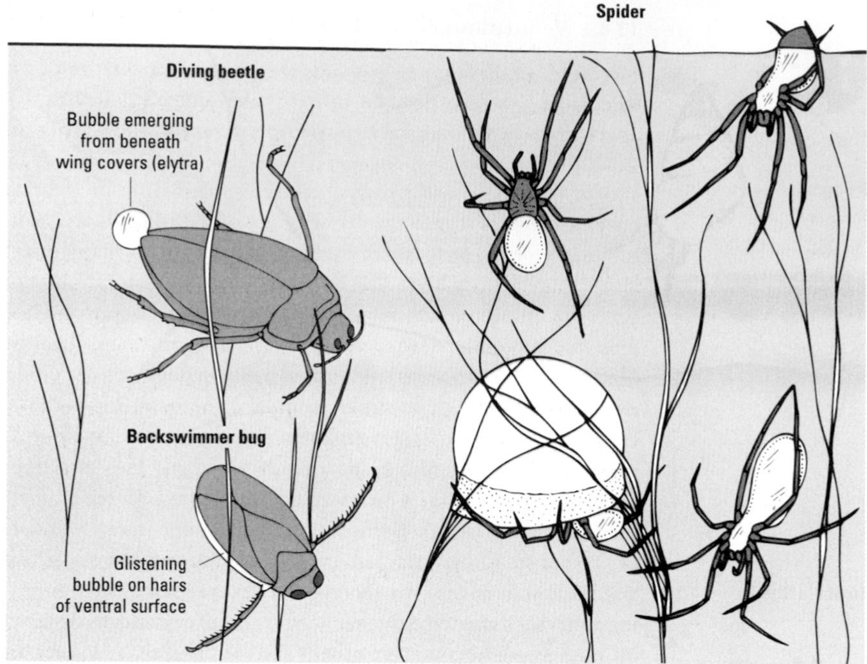

Fig. 13.26 Air bubbles used in freshwater air-breathing insects and spiders (as a diving bell in the spider *Argyroneuta*).

13.4.4 Respiration in air-breathers and bimodal breathers

Many aquatic animals are secondary invaders from land, using strictly *aerial* supplies. In arthropods two main air supply systems are used underwater: air bubbles (defined as where the gas phase is compressible) and plastrons (where the gaseous phase is somehow rendered incompressible). With air bubbles (Fig. 13.26), as used in many beetles and bugs, the pressure rises as the animal dives and the gases tend to diffuse from air to water. As the animal consumes the oxygen, though, the Po_2 in the bubble becomes less than in the surrounding water, and from this point on oxygen enters the bubble, so that the total dive time is greater than that allowed by the original oxygen content of the bubble, by about eight times. However, the system cannot continue indefinitely, as the other gases in the bubble increase in partial pressure and diffuse faster to the surrounding water, reducing the bubble's lifespan. In the water spiders, such as *Argyroneuta*, an air bubble is held under a silken framework strung across submerged vegetation, and this may help to prevent its shrinking somewhat. The spider returns to its "diving bell" intermittently between foraging expeditions. There are also a few species of insect that live in fast-flowing streams where the water flow around the bubble decreases the pressure in the bubble and forces it into an elliptical shape with increased surface area to volume, so that the bubble survives indefinitely as a good permanent "gill".

Plastrons, found in water bugs and water beetles, are essentially a specialist adaptation on the air-bubble theme, where an air supply is maintained at constant volume and reduced pressure by being contained in a dense mat of fine water-repellent cuticular hairs (Fig. 13.27) that prevent bubble collapse and counteract the tendency for nitrogen to leave. The cuticular hairs that trap the film of air are about 5–10 μm long, usually with a pronounced kink near the tip. In the bug *Aphelocheirus* they are about 0.2 μm in diameter and 0.6 μm apart, giving an amazing 2–3 million hairs per mm^2. As the volume of air beneath the hairs tends to contract (as oxygen is used up and nitrogen begins to leave), it is resisted by a high surface tension effect from the increasingly curved meniscus at the air–water junction between each pair of hairs. Further nitrogen loss is prevented and the system reaches a constant volume/reduced pressure equilibrium at a particular depth. The hairs are dense enough in most diving insects to resist a pressure of 3 atm (a depth of 40–50 m) without collapsing. This mat of hair may cover just the ventral surface, or most of the body, but it always covers the areas where the spiracular openings occur.

Special diving adaptations are also found in freshwater reptiles, birds, and mammals, and in lentic fresh water the problem of diving may be exacerbated by the presence of a substantial unstirred hypoxic boundary layer. Hence, for example, diving times for frogs may be 2–3 times shorter in still waters compared to running waters. For the endotherms, specific diving adaptations differ little from those described in marine tetrapods (see section 11.10.3), including blood-shunting tricks during submersion, enhanced oxygen-carrying capacity, and the use of bradycardia, etc. But for ectothermic freshwater tetrapods, such as amphibians, snakes, and turtles, there is an additional complication in that the water temperatures and hence their own body temperatures are much more variable. To some extent they can solve this problem by their pronounced ability to undergo long periods of apnoea interspersed with bursts of ventilation, but the ventilation rate during such bursts varies with temperature. For example, in the turtle *Mauremys*, lung ventilation per unit of O_2 uptake declines linearly with increasing temperature, and alveolar Pco_2 therefore increases with temperature. However, at very high temperatures (40°C) there is an increased breathing response, possibly related to evaporative heat loss.

Many essentially freshwater animals are able to use both aerial and aquatic oxygen, and may switch to air-breathing when their aquatic habitat begins to dry up or overheat and become hypoxic, or if it becomes too rich in H_2S due to decomposition. This bimodal

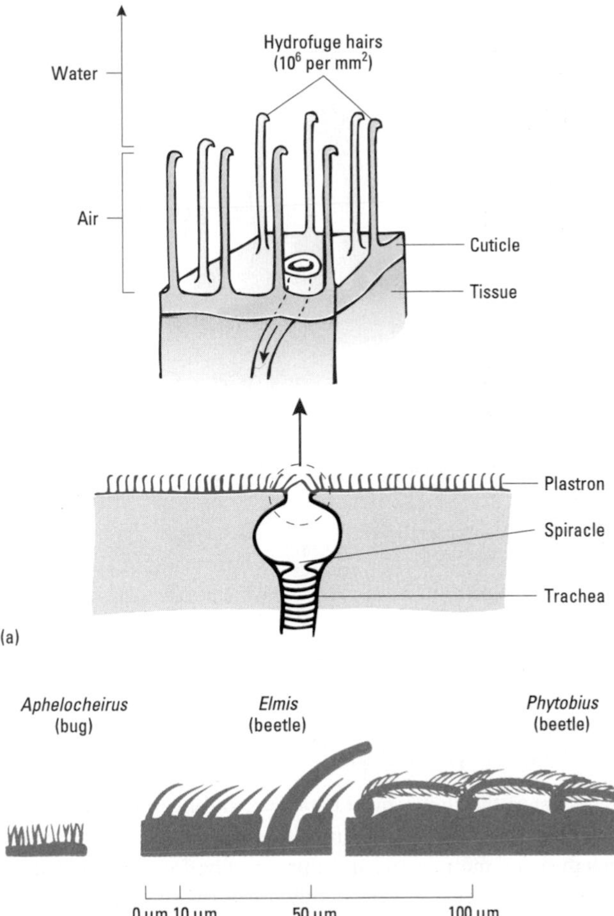

Fig. 13.27 Cuticular plastrons in freshwater insects. (a) The principle of plastron design, and (b) the structure and dimensions of cuticular plastron hairs in three species. (a, From Randall, D. *et al. Animal Physiology: Mechanisms and Adaptions*, 1997, 1988, 1983, 1978, copyright by permission of W.H. Freeman & Co.)

breathing represents a complex physiological problem, since air has nearly 30 times as much oxygen as water, whereas water can contain about 28 times as much CO_2 as air, giving two very different respiratory environments. Crustaceans, fish, and amphibians are most notable as bimodal breathers, although some soft-shell turtles also achieve substantial (cutaneous) aquatic oxygen uptake. All experience the problem that while oxygen becomes easier to acquire on land, carbon dioxide excretion is problematic.

In crustaceans the presence of a high-affinity hemocyanin is common, and switching between sites of oxygen uptake may occur. Many bimodal crabs show the reduced gill area typical of air-breathers (see Chapter 15), as well as an elaboration of the branchial chamber epithelia where gas exchange can be effected. The chamber normally contains a limited store of water such that gas exchange can take place from both air and water simultaneously, and in some species gas exchange seems to be partitioned so that O_2 uptake occurs from air into the chamber lining while CO_2 excretion occurs mainly across the gills into the stored water. (Note that fully terrestrial crustaceans lose this mixed pattern and can excrete CO_2 directly into air from the gills, perhaps due to the incorporation of extra carbonic anhydrase enzyme into their gill epithelia.)

In fish and tetrapods, bimodal breathing obviously requires a marked degree of remodeling of the vascular plumbing to allow the perfusion of gills, lungs, or cutaneous sites in various parallel or series arrangements. But at the physiological end it is mainly related to strategic pigment use. Two adaptations are common during the transition to air: a decrease in oxygen affinity and an increase in pigment concentration. The change in pigment affinity may arise intrinsically in some amphibians, or it may be achieved by increases in the organic phosphate : hemoglobin ratio. Behavioral effects also occur, for example in the papyrus swamp fish *Barbus neumayeri* where surface respiration increases in frequency as hypoxia increases and dispersal within the swamp only occurs in the wet season. In the amphibian *Xenopus*, surfacing behavior depends on temperature, the frequency increasing sharply between 5 and 20°C, but with surfacing duration kept very short (10–40 s) in daylight and only increasing with temperature at night, when the animal may stay at the surface for up to 60 min (probably due to lower predation pressure rather than physiological need). Amphibians suffer from a substantial hypoxic boundary layer in still waters, and can dive for significantly longer periods where the water is well stirred; when diving in still waters they are observed to undergo more voluntary movements, which may serve mainly to stir up the water.

Certain water snakes, such as the garter snake *Thamnophis*, have an acute seasonal problem with respiration in that they overwinter for several months in submerged water-filled hibernacula at around 5°C. In this period they switch to nonpulmonary ventilation, and their oxygen uptake is only around 50% of that in snakes induced to hibernate in air. At such low temperatures they can survive without anaerobicity using just cutaneous uptake. Turtles that overwinter for 3 months or more under water with persistent anoxia do become anaerobic but cope largely by accumulating some of the resultant lactate in the shell; in the genus *Chrysemys* about 44% of the lactate is buffered in this way. These turtles produce very little of the normal oxidative damage products such as dienes, having high levels of key enzymes such as superoxide dismutase and catalase (see Chapter 7), which decline rapidly during anoxia and rise again during reoxygenation. Thus they are minimizing the potential damage by reactive oxygen species (free radicals) during the reoxygenation of organs after anoxic bouts. They also reduce the Na^+/K^+-ATPase activity in their brain during anoxia, with an associated reduction of electrical activity, concomitant with a substantial drop in metabolism and a complete lack of movement.

13.5 Reproductive and life-cycle adaptation

In freshwater habitats there are several key problems that are bound to influence reproductive strategies:

1 The transience and changeability of the habitat, which may require opportunistic breeding carefully tied to seasonal change, and a protected stage in the life cycle.

2 The physiological difficulty of maintaining ion and water balance, made much more difficult in small animals such as larvae and juveniles with a high surface area to volume ratio.

3 In streams and rivers there is the additional problem, especially for small or immature individuals, of countering continuous downstream flow.

There is therefore a general tendency in freshwater invertebrates to have very short life cycles with a rapid turnover of generations, and to reduce larval forms, with more direct development involving larger and yolkier eggs (many molluscs) or brood pouches (e.g. the crustacean water flea *Daphnia*). Where larvae persist they tend to be either unusually large, or to adopt a crawling rather than a pelagic habit (e.g. the planula larva of some cnidarians), or to have protected surfaces (e.g. a chitinous covering in that same planula), or in a few special cases to become parasitic on larger animals such as fish (examples include the cnidarian *Polypodium* living on sturgeon, and the bivalve *Unio* whose hooked veliger larvae grow in the gills of various fish). Also rather more species use direct insemination, or protected spermatophores, to protect their gametes prior to fertilization. Protective "dormant" stages other than shelled eggs include sponge gemmules, bryozoan statoblasts, the durable encysted eggs or larvae of flatworms and nematodes, and the cocoons of certain oligochaete worms, all of which are resistant to drought or freezing or any other environmental stress. Dormancy in all its manifestations is much commoner in freshwater than in marine species, though less frequent than on land. These resting stages get widely distributed by winds or water spray or by animal means (such as on the feet of birds), ready to colonize further ponds when conditions are favorable again. In one unusual case a nonresting ostracod actively insures its own dispersal to better ponds by hitching a lift on migrating toads.

Commonly there is very high annual reproductive output in freshwater invertebrates. One well-studied case is the zebra mussel, *Dreissena*, in which one female may release over 1 million oocytes per year. The mussel is dioecious and fertilizes its eggs externally, therefore using coordinated maturation and spawning behaviors. Spawning occurs in late spring, with the hormone serotonin acting as a trigger, causing oocytes to undergo accelerated meiosis. Spawned oocytes then contain species-specific sperm attractants, increasing the chances of rapid fertilization.

13.5.1 Short life-cycle strategies

Many of the zooplankton exemplify the short life-cycle strategy, allowing them to take best advantage of seasonal algal blooms. Rotifers and cladocerans may complete their lives in just a few days, producing many generations per year (multivoltine); they mature quickly and put most of their assimilated energy into gamete production. In these taxa the time-wasting process of finding a mate is commonly avoided by the use of parthenogenesis, where the ova develop without fertilization and only females are produced (often brooded in sacs within the carapace; Fig. 13.28). Generations of the asexual female morphs can thus be produced in vast numbers by one colonizing female finding a suitable pond away from predation pressure, and laying many thin-shelled rapidly developing eggs, with a generation time of only 1–4 weeks. A brief phase of production of sexual morphs usually occurs in the fall, or as drought sets in; these mate with each other and produce a further generation, often with heavily shelled eggs that are resistant to desiccation or to low temperatures.

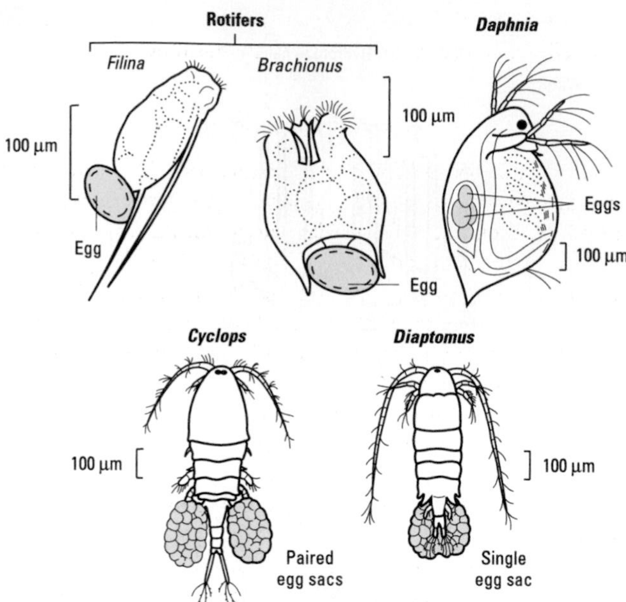

Fig. 13.28 Brooding in freshwater invertebrates.

13.5.2 Long life-cycle strategies

In contrast to the above, most copepods and insects, and most of the benthic invertebrates, are univoltine and grow relatively slowly through many molt cycles, and they usually do not show parthenogenesis and rather less frequently use dormant stages. Indeed a long adult lifespan is usually a direct alternative to the long dormant or diapausing phases described earlier; out of 167 crustacean species surveyed, not one had an adult life of greater than 1 year *and* a diapausing stage exceeding 1 year. Some copepods use a period of summer diapause, either avoiding fish predation or avoiding food bottlenecks in competition with other grazers; this usually occurs at the egg stage, and the diapausing egg may be capable of extended cryptobiosis (see Chapter 14).

Annual reproductive output in these longer lived animals is lower (though it may still be high compared to other habitats). It often involves iteroparity, and each egg or embryo is more carefully protected. Freshwater crabs and crayfish have larger eggs than marine/brackish relatives, and higher C : N ratios in the eggs and young larvae, indicative of high lipid content. Copepods carry their egg sacs around on the abdomen, while insects tend to lay them in protected sites stuck to vegetation or buried in mud. In tropical and temperate habitats most of the insect species have an annual life cycle, emerging as adults in summer after the spring burst of food availability, mating and laying eggs within a few days. In higher latitude ponds and lakes the life cycle may be greatly prolonged though, with 5–7 years between egg production and adult emergence. In all cases there may be mass emergence, with clouds of midges or mayflies issuing from the lake or river surfaces on just one or a few days of the year.

Lotic waters are particularly hazardous for free gametes and for pelagic larval stages, which would move downstream freely, so these are *very* rarely found. Instead lotic invertebrates use internal fertilization, and either the young are then retained in the body, or large yolky eggs are firmly cemented to the substrate.

13.5.3 Phenotypic plasticity and polymorphism

Either of these broad life-history strategies may be accompanied by a pronounced degree of phenotypic plasticity, allowing the reproductive output to vary with environmental conditions. In *Daphnia*, water temperature has a clear effect, with larger size at maturity (measured as the age when eggs are first laid into the brood pouch) when water temperatures are lower. This operates via an effect on maturation threshold body length two instars before the eggs are actually produced, since ovarian development is not initiated until a threshold body size is reached. This may be because higher temperatures and rapid metabolism produce an energy deficit for growth and molting, or because there is selection against larger size in warmer weather via an external agency such as predation.

The well-known phenomenon of **cyclomorphosis** occurs in many of the parthenogenetic freshwater animals, where generations are polymorphic in form, physiology, or behavior, or all of these. A classic case is again the water flea *Daphnia* (Fig. 13.29a), and something similar occurs in some rotifers. Explanations for this are controversial (see section 13.7).

There are also many examples of resource-based or trophic polymorphisms involving differences in life-history strategy (as well as behavior and morphology) in freshwater animals. The ecological conditions that promote resource polymorphisms probably include the relaxation of interspecific competition providing unfilled niches that can be exploited. A striking example occurs with Arctic charr (*Salvelinus alpinus*), which have invaded "new" freshwater lakes within the last 15,000 years since the retreat of northern ice sheets. In a single volcanic lake in Iceland, four morphs of these charr are found (Fig. 13.29b), differing in size, fin shape, body and head depth, and jaw structure. Two are benthic, the smaller morph exploiting laval tubes, a third is planktivorous, and a fourth piscivorous. The phenotypic differences between the morphs include age at maturity, egg size, and reproductive investment; they are partly genetic, but also have a strong environmental component, with triggers acting during early embryonic development.

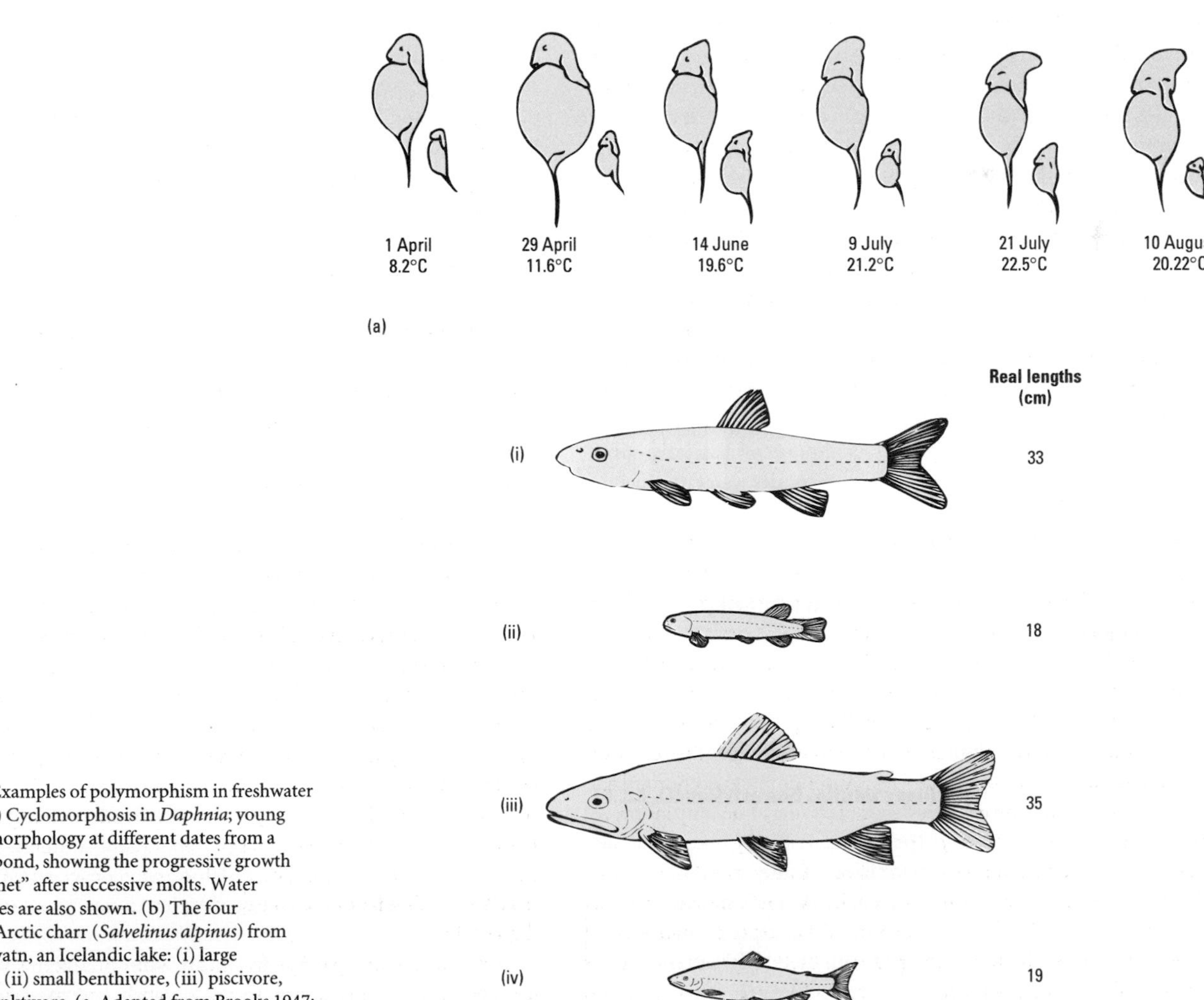

Fig. 13.29 Examples of polymorphism in freshwater animals. (a) Cyclomorphosis in *Daphnia*; young and adult morphology at different dates from a temperate pond, showing the progressive growth of the "helmet" after successive molts. Water temperatures are also shown. (b) The four morphs of Arctic charr (*Salvelinus alpinus*) from Thingavallvatn, an Icelandic lake: (i) large benthivore, (ii) small benthivore, (iii) piscivore, and (iv) planktivore. (a, Adapted from Brooks 1947; b, adapted from Skulason & Smith 1995.)

13.5.4 Life cycles of freshwater vertebrates

Reproduction in freshwater fish shows similar patterns to those found in invertebrates, often with the eggs serving as a protected stage for seasonal endurance. Freshwater fish eggs are generally larger (1–30 mm diameter) than marine fish eggs (0.8–2 mm), and may be protected in mucus froth "nests", or in vegetation; where a pond or stream dries up the eggs may survive in the bottom mud. Some of the lungfish from swampy habitats can survive as adults encased in a cocoon of hardened mud reinforced with secreted slime, with a narrow air-tube to the surface. Lake species of fish, especially the many endemic cichlid species in the Rift Valley lakes of eastern Africa, have often evolved particularly strange reproductive habits that provide isolating mechanisms between species, including elaborate nest constructions and even "mouth brooding" of the juveniles. Some lotic fish have also acquired careful nesting patterns, as in sticklebacks and guppies, to avoid the young being swept downstream.

Secondary freshwater vertebrates that are essentially land animals usually resort to a terrestrial site for reproduction. For freshwater reptiles, such as turtles, this is usually the upper shore. Some turtles manage to lay eggs in mud under shallow water, but these stay in developmental arrest until the water recedes and oxygen is able to diffuse into the embryo. This phase may last up to 12 weeks, and the egg albumen loses up to 90% of its Na^+, while its water diffuses slowly into the yolk but at a very slow rate (the vitelline membrane having a very low P_w) compared with other turtles. Freshwater amphibious mammals also choose a birth site very close to the water, which may be an air-filled burrow in the river bank with an underwater access.

However, some extra problems arise, particularly with the rare ectothermic species that are viviparous and are at extra risk while carrying the embryos. For example, tropical water snakes when pregnant have a higher eccritic T_b (27–28°C rather than the normal 22–24°C), and thus choose to spend longer at higher temperatures, losing their normal diel pattern of T_b and so both increasing and stabilizing the embryonic developmental rate. Endothermic viviparous species (manatees, lake seals, river dolphins) also have some problems in giving birth whether onshore or under water, paralleling those of the marine cetaceans and pinnipeds.

13.6 Mechanical, locomotory, and sensory adaptations

13.6.1 Depth, buoyancy, and locomotion

Fresh water is rarely very deep, the largest lakes (Baikal, Tanganyika) being only 1000–1500 m, so that freshwater animals do not normally have to cope with great pressures. However, buoyancy is more difficult in fresh water due to the reduced specific gravity of the medium giving very little lift. Hence invertebrates need more exaggerated versions of the buoyancy mechanisms discussed in Chapter 11; either the body must be even more spiky or frilly to increase the drag forces, or more oil or gas must be included within the tissues. Cephalopods do not occur in fresh water, so the problem of achieving enough salt transport from the fluids of the rigid float system does not arise. In fish the swim-bladder is present in a much higher percentage of species and is larger, about 6–9% of body volume in freshwater teleosts compared with only 4–5% in marine species.

Fresh water provides few problems or opportunities for locomotion not found in the marine world (see Chapter 11), except in lotic systems where animals may need to locomote continuously to counter the tendency to flow down river and ultimately out to sea. Station-keeping is a continuous problem for any species that cannot cope with slightly salty conditions. Suckers and hooks for adhesion may be appropriate for bottom-dwellers; certain fish have ventral suckers, and blackfly (*Simulium*) have hooks to hold on to rocks while filter feeding with elaborate fans of bristles. Adhesive systems, such as the byssus threads of freshwater bivalves, may also be used. In moderately fast flow, a streamlined shape flattened against the substratum may be enough to insure stability, as the fast water flow over the upper surface effectively presses the animal down and keeps it in place. Flatworms often rely on this, and many freshwater bivalve shells are very flattened for similar reasons.

13.6.2 Senses

Vision

In general lentic environments are well lit through most of their depth, and only deep lakes extend beyond the photic zone. However, in both rivers and eutrophic lakes light is depleted by vegetation in summer, so that the light environment can be very variable. Eyes may therefore need to work in very dim conditions, especially for predators in weed-filled ponds or in stretches of river overhung by dense trees. Freshwater beetles and bugs, and fish such as the pike (*Esox*) have eyes with properties similar to those of terrestrial nocturnal species, with large facets or large pupils. Rhodopsins are sensitive to green wavelengths in particular, filtering through the floating or shading foliage.

Amphibious animals share the same problem as some littoral species (see Chapter 12), in that they need to be able to focus in both air and water. An unusual solution is found in certain surface swimmers—that of having two sets of eyes. The four-eyed fish *Anableps* can focus simultaneously on a terrestrial and an aquatic image, having in fact only two eyes but with a pear-shaped lens and two sets of pupils above and below the water meniscus. Any animal that hunts across the air–water interface also has to be able to make allowance for the refraction at the surface. This applies to archer fish shooting water out at insect prey in the air, as much as to herons or kingfishers or bears dipping or diving in to spear fish or invertebrates spotted from above. Diving mammals and birds that hunt by swimming beneath the water, such as otters or dippers, must be able to accommodate their eyes by altering the lens–cornea relationship using the ciliary muscles.

Some teleosts, and the crayfish *Procambarus*, can detect polarized light, an ability that seems to be rare in aquatic animals. Polarized light detection has also been demonstrated in some surface-dwelling beetles and bugs that need to find a new pond when their habitat dries out. The "glare" reflected from a water surface is highly polarized and provides a potent cue for water boatmen (*Notonecta*) and other surface predators. In the crayfish and these insects, as in fully terrestrial arthropods, detection is due to the precisely orientated microvilli in adjacent retinular cells in a single ommatidium being interdigitated at 90° to each other (see Chapter 9).

Chemoreception

Chemoreception is important to many freshwater animals for prey location, predator avoidance, and the location of hosts or mates. In rivers it is particularly favored because the currents will help to keep chemical plumes directional and reasonably intact. Thus both general and specific sensitivity are appropriate. For example, teleosts have good acid, alkali, and salt receptivity, and the minnow can detect 10^{-5}M sucrose or 4×10^{-5}M NaCl. But in these and many other shoaling fish there are also specific sensors for intraspecific cues, such as the alarm substance ("*schreckstoff*"; see Chapter 10) released from the skin of damaged individuals; these chemicals are commonly polypeptides or proteins.

Mechanoreception

Many aquatic insects have water pressure (= depth) receptors. For example, the bug *Aphelocheirus* has a group of long cuticular hairs that trap an air bubble, the hairs then compressing around the bubble and detecting its shape change with depth. Other bugs with relatively long bodies use pressure sensors adjacent to their abdominal air spaces to detect differences in tracheal pressure between each segment, so getting information on whether they are swimming head-up or head-down.

Flow receptors are also critical for lotic species, and are designed from deformable neuroepithelial cells or from hair cells (see Fig. 9.46) as in other aquatic animals. In larger animals, such as crustaceans and fish, the receptors may be associated with cephalic appendages such as antennae or vibrissae.

Other senses

Electroreception occurs in a range of freshwater fish (Fig. 13.30), with both weak and strong field producers (see section 11.7.5). Weak electric fish sometimes use the system for mate recognition, the male having a discharge about an octave lower than the female. The electric eel is one of the most forceful field generators, and can both detect and stun its prey in very murky waters. Some electric elasmobranch fish, including rays and sawfish, penetrate up large rivers. Freshwater amphibians retain the electroreceptors of their ancestors, and the system even persists into a few diving mammals, notably the platypus. Echolocation by contrast is largely lost in freshwater species, although it persists in freshwater dolphins.

13.7 Feeding and being fed on

An overview of the trophic cascade in a typical temperate lake is shown in Fig. 13.31, emphasizing the importance of planktonic and microphagous feeding systems in most freshwater systems.

13.7.1 Microphagous feeding

Most of the freshwater zooplankton are filter feeders, exploiting the algae, bacteria, and detritus in the water column. Some are generalists, but many are highly selective, taking particular kinds of green algae, diatoms, or flagellates. Cyanobacteria are generally avoided. Rotifers feed mainly on particles of 1–20 μm, using their ciliated corona. Cladoceran "water fleas" (including the extremely common herbivorous genera *Daphnia* and *Bosmina*) take slightly larger algae up to 50 μm, and the larger crustacean copepods feed on a larger range again, taking phytoplankton and zooplankton of 5–100 μm. Within each group, some prefer live food, others detritus, while yet others take both types of food material. Surprisingly large numbers of freshwater animals, including cnidarians and some flatworms, have found an additional way of "exploiting" algae for feeding purposes, adopting a symbiotic association with photosynthetic algae termed zoochlorellae. In *Hydra*, for example, green algae within the animal's cells provide a proportion of the animal's carbon needs (and probably receive some nitrogen in return).

Other benthic invertebrates feed on the sinking remains of algal blooms, particularly the diatoms, which remain fairly intact as they sink and are a rich source of fatty acids. True detritus feeding (swallowing the mud, in effect) is practiced by oligochaete worms (*Tubifex*, etc.). Thus all the vegetation that dies within a freshwater body eventually becomes a food source, although this occurs relatively slowly compared with land habitats due to the exclusion of fungi and other key decomposers. In the benthic communities grazers are also abundant, scraping microbial communities (the epiphyton) off rocks and plants fairly nonselectively. Mayfly and caddis larvae, gammarid shrimps, and most snails are classic microbial grazers.

13.7.2 Herbivory

Herbivory on a scale larger than algal grazers is relatively rare in temperate systems, since there are few macrophytes in many lakes, and they are relatively nondiverse in lotic waters. Herbivory is commoner in parts of the tropics, where papyrus and water hyacinth can provide high rewards, and where large grazers such as hippopotamus, water deer, capybara, and dugong are important.

Where temperate macrophytes do occur, they may be very heavily exploited as living spaces; a single plant of the pondweed *Potamogeton* has been recorded as a habitat for 500–600 individuals, including chironomid midge larvae, oligochaete worms, water mites, ostracods, *Hydra*, and small snails. Animals that actually eat the weed are much rarer; pondweeds and water lilies do get attacked by some insects (especially small leaf- and stem-mining larvae) and many molluscs, as well as amphipods, crayfish, and some fish. There are also a few herbivorous turtles, and some of the freshwater mammals take aquatic plants as part of their diet—the dugongs and manatees are entirely herbivorous. Defenses against herbivory in freshwater plants are relatively poorly studied.

13.7.3 Carnivory

Carnivores operating in fresh water may be usefully subdivided into lurkers and hunters. Stationary lurking is exemplified by *Hydra*, taking any small prey that brush against and trigger the stinging cells (cnidoblasts) in its tentacles. Leeches also tend to operate by lurking in the concealing weeds, as do stonefly and alder fly larvae. Predatory

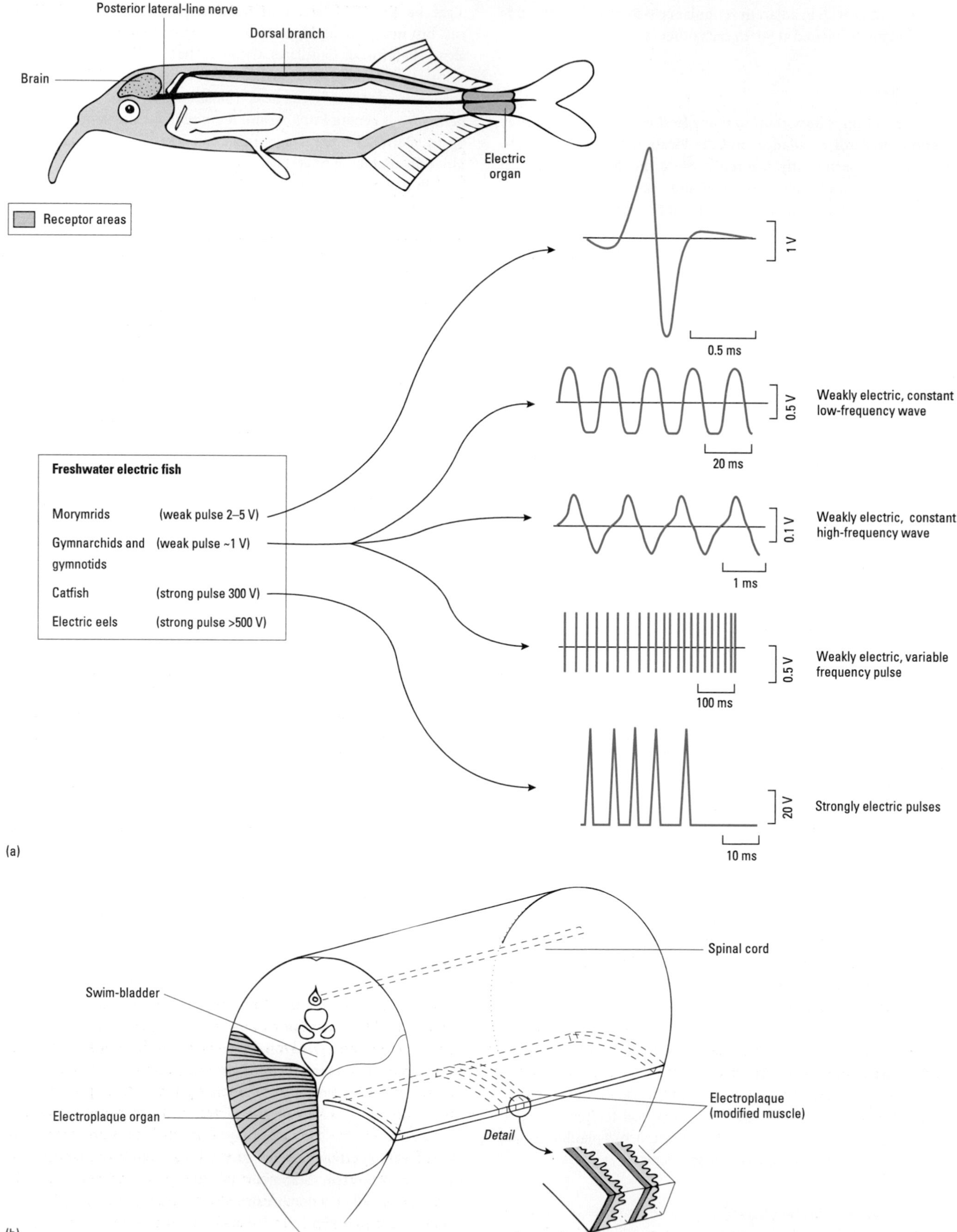

Posterior lateral-line nerve
Dorsal branch
Brain
Electric organ
Receptor areas
1 V
0.5 ms
0.5 V
Weakly electric, constant low-frequency wave
20 ms
Freshwater electric fish
Morymrids (weak pulse 2–5 V)
Gymnarchids and gymnotids (weak pulse ~1 V)
Catfish (strong pulse 300 V)
Electric eels (strong pulse >500 V)
0.1 V
Weakly electric, constant high-frequency wave
1 ms
0.5 V
Weakly electric, variable frequency pulse
100 ms
20 V
Strongly electric pulses
10 ms
(a)
Spinal cord
Swim-bladder
Electroplaque (modified muscle)
Electroplaque organ
Detail
(b)

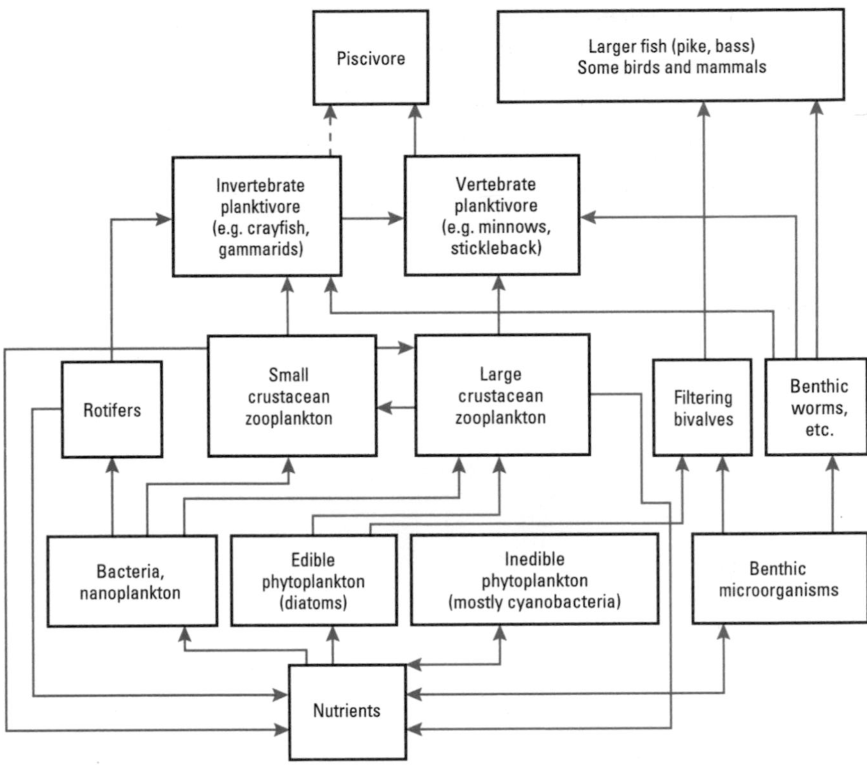

Fig. 13.31 The trophic web in a typical temperate lake.

hunting is also common amongst insects (bugs, beetles, dragonfly larvae with their extensible "mask", and some midge larvae) and almost ubiquitous at some point in the life cycle of fish, since larger zooplankton form an important part of most fish diets. The larger insects, crustaceans, worms, and snails, and also many fish, in turn are eaten by hunting reptiles (e.g. crocodiles, anaconda), mammals (e.g. otters, river dolphins) and birds (e.g. herons, dippers). The larger crocodilians perhaps represent the most ferocious freshwater carnivores, and some species of alligator have the most acidic stomach contents yet discovered, and are able to break down and digest even the bones of their prey.

There is an additional possibility for carnivores in and around fresh water, taking advantage of the surface film. Many insects have a specialist lifestyle relying on the surface tension of the water film for support, preying upon other small insects that become entrapped by that same surface tension when they fall into the water, or catching small crustaceans just below the surface. The water striders (gerrid bugs) and the pirate spiders live above the surface, standing on tiptoe on the water film, while the water boatmen or backswimmers (notonectid bugs) live in the water suspended upside down from the surface film by the tip of their abdomens. Exploiting the surface film fauna is rarely an option in marine or estuarine habitats where the film is constantly disrupted by wave actions.

Fig. 13.30 (*opposite*) Electric fish in fresh water. (a) The types of electric fish and their pulse patterns. (b) The large electroplaque organ in the trunk of the electric eel, formed from modified myotomes (muscles).

13.7.4 Feeding and growth patterns

Diel and seasonal patterns of food abundance in fresh water have a major effect on growth patterns, which in some parts of a season may be strongly negative. The freshwater flatworms can undergo extraordinary degrees of starvation, reducing their body mass to as little of 1/300 of its maximum and resorbing most of their gut and parenchyma. Seasonality is also very strongly influential on patterns of feeding activities. Many components of the zooplankton show vertical diurnal migrations (Fig. 13.32), and planktivorous carnivores must move with them. In temperate lakes there are commonly "spring blooms" (Fig. 13.33), sometimes in two phases, where a build up of diatoms and flagellates allows bursts of herbivorous cladocerans and copepods, accompanied by carnivorous copepods. Fish populations inevitably are trophically linked to these blooms.

13.7.5 Avoiding predation

Diurnal movements may be partly related to limiting losses to predators. Other predator avoidance strategies are much in evidence in freshwater plankton. Here cyclomorphosis may be relevant. The small and very transparent summer generations of many cladocerans may be selected for by intense pressure from predatory fish populations. This view is supported by their greater abundance inshore where the fish take refuge from their own predators, compared with greater numbers of the larger and more conspicuous forms in the open water. Similarly the spines, helmets, and other protuberances produced seasonally by cyclomorphic rotifers and cladocerans (see Fig. 13.29a) may be a defense against invertebrate predation, allowing extra food reserves to be stored in a transparent

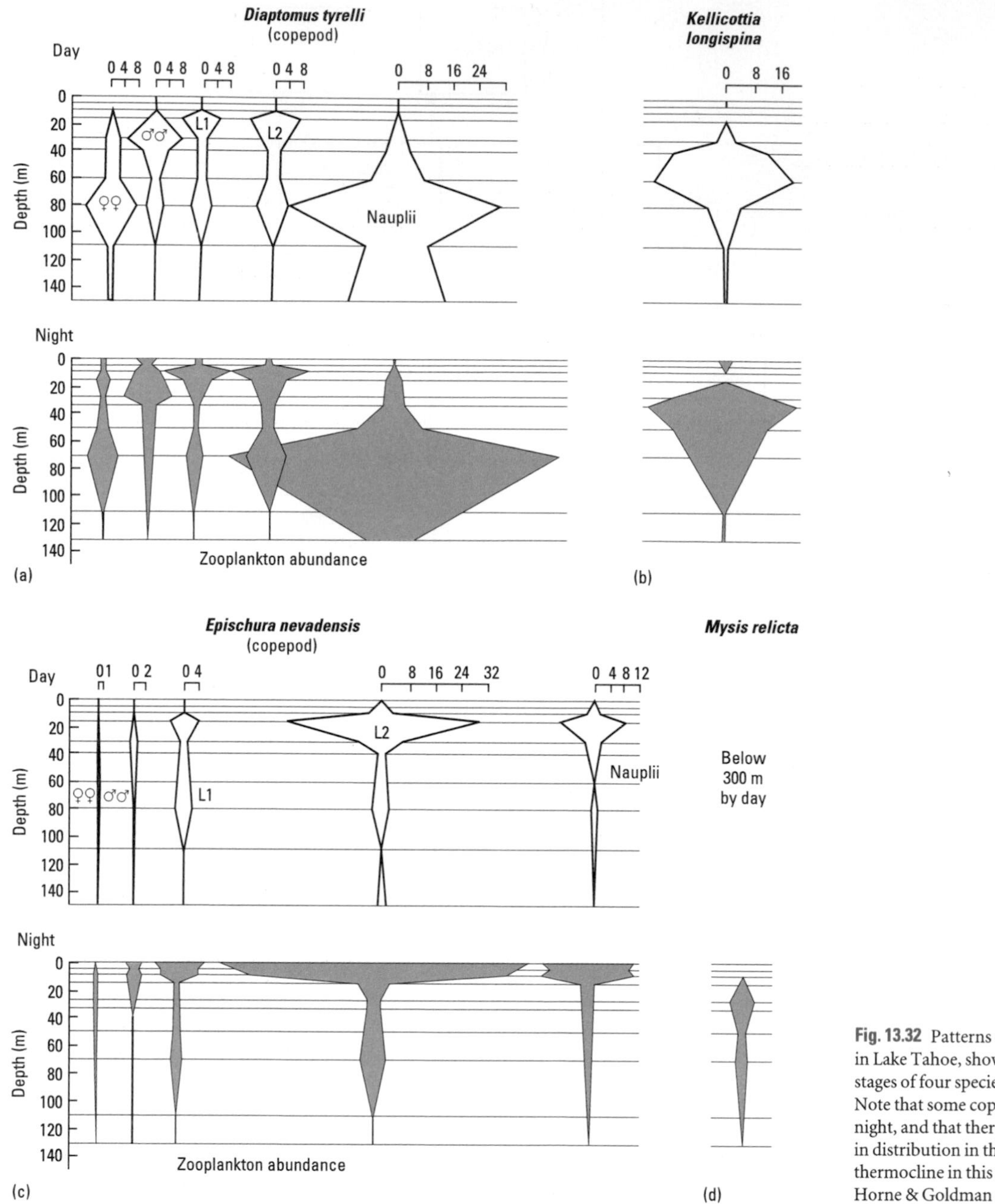

Fig. 13.32 Patterns of diurnal plankton migration in Lake Tahoe, showing the abundance of various stages of four species by day and by night (in green). Note that some copepods migrate to the surface at night, and that there are clear sexual differences in distribution in the copepod *Diaptomus*. The thermocline in this lake is at about 20 m. (From Horne & Goldman 1994.)

structure not easily seen by visual hunters and making the animal less easy to swallow. In larger animals, good attachment systems, strong molluscan shells with appropriate resistance to boring mouthparts, and surface prickliness, as seen in many smaller insects, may also serve as antipredator devices.

Predation may also be linked to physiological constraints, perhaps especially for arthropods that must undergo periodic molting. In the genus *Gammarus*, the species *G. pulex* is strictly freshwater, while *G. tigrinus* prefers water of somewhat higher ionic concentration; where the two species co-occur they prey upon each other's molting stages. In fresh water the predation is differentially in favor of *G. pulex*, whereas in slightly more saline conditions it is balanced. Thus the better physiological adaptation of *G. pulex* to very dilute media is probably the reason for its exclusion of its congener. Snails also have a problem with predators, since in very fresh waters they cannot accumulate enough minerals to make dense calcareous shells and must rely on rather less resistant proteinaceous coverings.

13.8 Anthropogenic problems

13.8.1 Wastes and pollution

Human settlements have always tended to develop on river banks, as rivers provide natural defenses, natural drinking water, potential power for mills, and above all channels of communication. Most

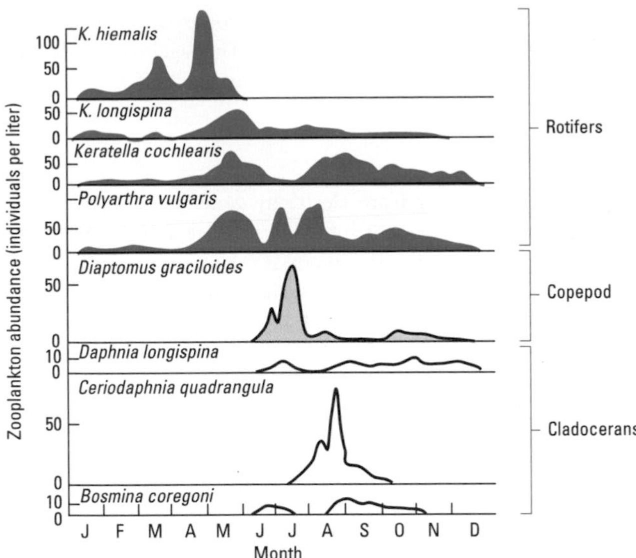

Fig. 13.33 Seasonal bloom patterns in Lake Erken (Sweden); in spring for various rotifers, followed by herbivorous cladocerans and carnivorous copepods as algae build up. (From Horne & Goldman 1994.)

Table 13.13 Categories of wastes discharged into fresh waters.

Type	Source
Acids and alkalis	Industrial
Anions (S^{2-}, SO_3^{2-}, CN^-, etc.)	Industrial, mining
Detergents	Industrial, domestic
Sewage	Domestic
Silage and farm manures	Agricultural
Food wastes	Domestic, agricultural
Gases (Cl_2, NH_3, etc.)	Industrial
Heat	Industrial, power generation
Metals (Cd, Zn, Pb, Hg)	Mining, industrial
Nutrients (nitrates, phosphates)	Agricultural
Oil and oil dispersants	Industrial
Organic toxins (C_6 residues)	Industrial
Pathogens	Various
Pesticides, herbicides	Agricultural
Polychlorinated biphenyls	Shipping, tourism
Radionuclides	Industrial, power generation

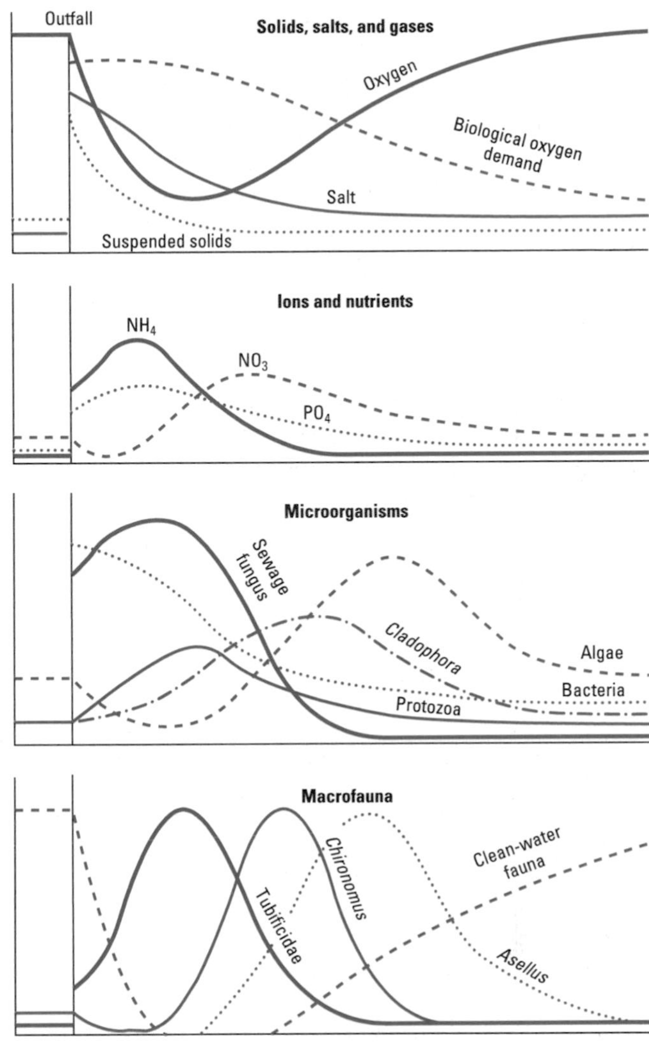

Fig. 13.34 Effects of effluent discharge in a river: downstream patterns of physical and chemical change, and the associated changes in microorganisms and macroinvertebrates. (Adapted from Hynes 1960.)

of the big cities of today still straddle the world's major rivers. Unfortunately fresh waters are among the most fragile of all the Earth's habitats and this juxtaposition does not enhance freshwater communities (see Plate 10b, between pp. 386 and 387). These riverside cities have always tended to discharge their sewage into the river; more recently they have discharged all their other wastes as well (Table 13.13). Five main categories of anthropogenic waste can be recognized.

Sewage and silage

Whether from human settlements, intensive farming units, or fish farms, untreated or poorly treated effluents produce a burst of biological oxygen demand in fresh waters, creating an "oxygen sag" downstream in lotic systems. The **biological oxygen demand** (BOD) of raw sewage may be as high as 120 mg l^{-1} day^{-1}, whilst that of well-treated sewage may be reduced to 6 mg l^{-1} day^{-1}, whereas clean fresh water is only around 1 mg l^{-1} day^{-1}. This oxygen usage encourages fungal bacterial communities, and animals such as tubificid worms that tolerate low oxygen; it may take a long stretch of river (and/or time) for recovery (Fig. 13.34). Indeed, accelerated (or cultural) **eutrophication** is a well-recognized process in both rivers and lakes (Fig. 13.35). The high algal growth may produce "blooms", so that the water becomes unpleasant, potentially toxic, and often smelly. Reduced oxygen levels may alter the solubility of nutrients and some metals. Submerged plants gradually disappear as light is cut out, and animal casualties inevitably follow. First the submerged animals change (*Tubifex* worms replace crustaceans, salmonids lose out to coarse fish like pike and perch), and then surface dwellers decline. Table 13.14 summarizes the major categories of pollution susceptibility in freshwater animals, showing the key role of highly susceptible (anoxia-intolerant) groups, such as mayflies, stoneflies, and caddis flies, as indicator species for environmental damage.

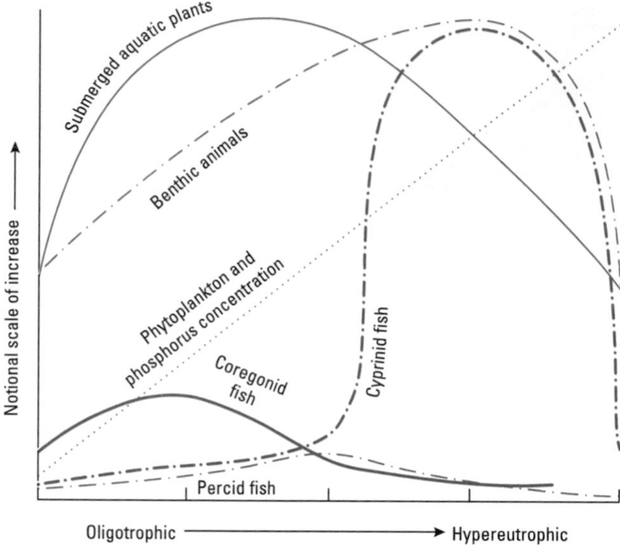

Fig. 13.35 Changes in a temperate lake as it becomes eutrophicated (as indicated by the steady build-up of phosphorus and phytoplankton). Yields of fish may increase overall until the lake is highly eutrophicated, but the cyprinid fish are much less desirable as a commercial resource than are coregonids and perch. (From Moss 1980.)

Table 13.14 Pollution susceptibility in freshwater animals (1 = resistant, 10 = highly susceptible).

Category	Animal taxa
1	Oligochaeta
2	Chironomid larvae
3	Most pond snails; most leeches; water louse *Asellus*
4	*Baetis* mayflies; alder flies; fish leeches
5	Most bugs and beetles; crane flies; blackflies; flatworms
6	*Viviparus* pond snails; mussels; gammarids; some dragonflies
7	Some mayflies; some stoneflies; most caddis
8	Crayfish; most dragonflies
9	Some dragonflies
10	Most mayflies and stoneflies; most caddis; river bug *Aphelocheirus*

Damage can be reversed by managing the effluent, but this has normally only been done where the lake is a valuable drinking-water source, or valuable as a leisure resource. It can be reduced by putting animal wastes back on the land rather than into rivers; and improved treatment and recycling of human waste has also got to be part of the long-term solution. However, sewage treatment commonly involves surfactants and detergents, and the breakdown products of these may include estrogenic substances that are known to affect fertility (especially of fish) and which are environmentally persistent.

Mining wastes

Usually there are acid drainage waters from mining activities, resulting from sulfides in the ores being converted to sulfuric acid, partly by bacterial action. The acid water then dissolves out metals, like copper, iron, and zinc, adding to the pollution problem—copper is especially toxic to fish. A little way downstream, minerals such as iron may subsequently redeposit as red-brown sludge, killing stream biota completely. Mineral deposits may become particularly serious in subtropical rivers in the dry season. This again needs on-site remedial technology, which is expensive but not "difficult".

Thermal waste

This arises particularly from electricity generation, where power stations discharge hot water into rivers. Some animal species die directly due to their own low thermal tolerance, others succumb because bacteria proliferate rapidly and reduce the oxygen levels. Heat is a valuable resource, and it is certainly wasteful to offload it in rivers. Some cities are therefore developing recycling systems whereby the heat is piped to buildings. In the shorter term, it is better to use the heat for some purpose near to the generator, perhaps for farming some freshwater species that tolerates or prefers warmer water, such as eels or carp.

Agricultural chemicals

The most important discharges reaching rivers and lakes are nitrates and phosphates. Levels of these ions may be five-fold greater in streams flowing through agricultural land compared with forest. When these ions enter a watercourse, the water tends to get overgrown with algae, whose decay may kill fish by reducing oxygen and/or by increasing alkalinity. This is principally a problem in causing long-term eutrophication, of slow-flowing rivers and especially of lakes. In the UK, East Anglia and parts of the West Country fruit-growing areas are badly affected; many lakes of Europe and in the crop belts of North America are also threatened; and in China one-fourth of all lakes are thus contaminated. On a large scale Lake Erie and the Aral Sea are the most notable, near terminal, casualties, though the latter has been exacerbated by diverting inflows for irrigation leading to a massive loss of water volume and salinification (Fig. 13.36).

A second agriculture-related problem is the input of herbicides and pesticides, such as organochlorines, all directly toxic to freshwater animals, even ones in the least susceptible categories; levels of all these are now controlled by international directives. DDT, which was once widely used and is still applied in some countries, is directly toxic and also has an estrogenic effect, causing sexual dysfunction in many organisms but especially noticeable as feminization and reduced fertility in male fish.

Industrial outflows

Industrial discharges commonly involve organic and inorganic chemicals, including oils, bleaches, dyes, detergents, formaldehydes, and phenols, and a variety of heavy metals, in varying degrees of pretreatment (see Plate 10b, between pp. 386 and 387). These factory effluents may be almost untreated from older style industrial plants in Eastern Europe and parts of the Third World, and again tend to be directly toxic. For heavy metals this toxicity can occur at very low doses, often because metals such as cadmium, mercury, and zinc are inhibitors of key enzyme activities, while copper is an inhibitor of hemoglobin function.

A particular hazard to freshwater areas used for boating and recreation are the tributyl tins (TBT) and polychlorinated biphenyls

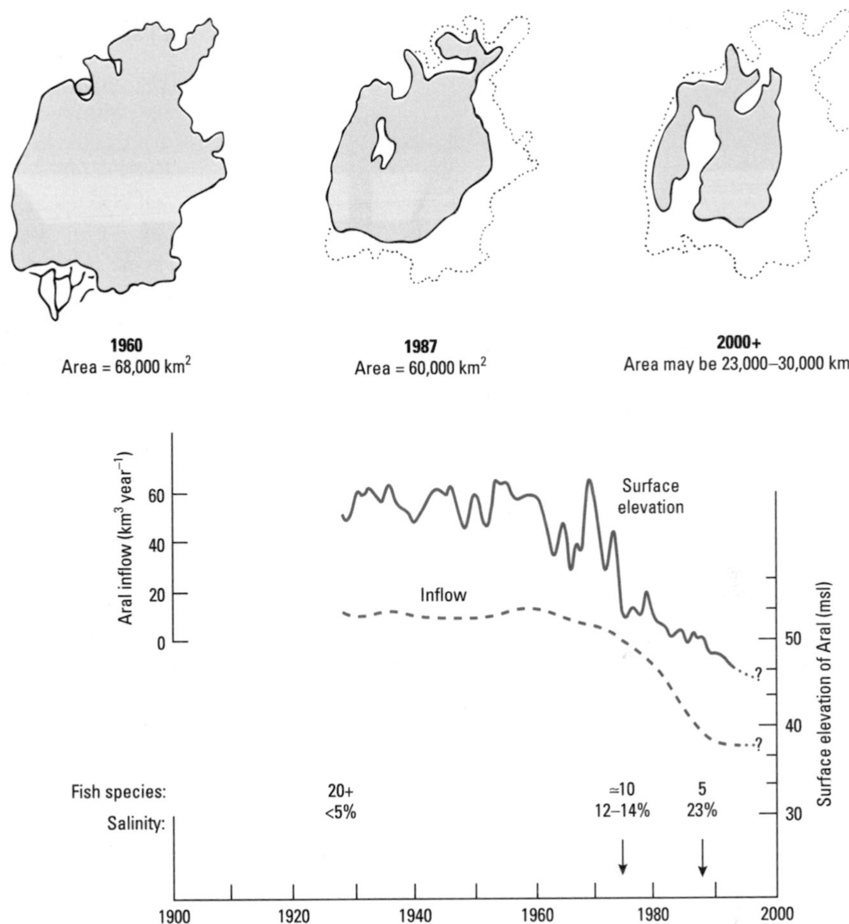

Fig. 13.36 Progressive loss of surface area in the Aral Sea, largely resulting from diversion of inflows for agricultural irrigation in adjacent cotton-growing areas. An elevation drop of just 10 m reduced the area by 60%, and what was formerly the fourth largest freshwater lake in the world is now rising in salinity and much reduced in diversity (with brackish-water species outcompeting freshwater species). (Adapted from Micklin 1988, and other sources.)

(PCBs) used to clean boat hulls and other underwater structures in order to render them unsuited to larval settlement, but they are also lethal to other animal life where they leak into the lake or river. PCBs also get into the atmosphere, and have recently been recorded precipitating out in the cold air of the polar regions, where once ingested they may seriously affect the growth and fertility of animals as large as polar bears.

These last two categories of contaminant are interactive, with eutrophication impacting on concentrations of industrial pollutants. PCB and DDT concentrations in pike, in lakes of southern Sweden, have been shown to be inversely correlated with the total phosphorus, chlorophyll, and humic content of the water. Otter populations are also reported to persist in eutrophic PCB-contaminated lakes but not in oligotrophic ones. One suggestion is that increased productivity in the lakes reduces the pollution burden in fish and mammals due to higher growth rates, quicker turnover times, and more rapid sedimentation of the pollutants into an inactive form adsorbed to particles (Fig. 13.37).

Agricultural and industrial chemicals are also longer lasting than other pollutants, and many affect the groundwater system more than the rivers. Once there they are nearly impossible to purify, as all the cleansing microbial agents need some oxygen source not available underground. However, the rock through which the ground water filters does actually slow the cycling sufficiently so that most organic pollution degrades into a relatively harmless state. Eventually though, these groundwater aquifers recycle into the drinking water, and long-lived organic pollutants can sometimes get through the system.

Polluted waters and disease

Contaminated natural fresh waters are the single biggest source of human disease in the world today. Some of the diseases result from parasites and pathogens naturally present in fresh water, such as river blindness (carried by blackflies and caused by the nematode *Onchocerca*) and malaria (due to the protozoan *Plasmodium*, carried by mosquitoes). But a huge additional hazard is added by pathogens deriving from sewage and animal wastes, made worse by erosion/silting after deforestation. Infantile diarrhea is the single largest killer in areas where these anthropogenic effects are greatest. In the Western world pathogenic contents of water are usually controlled by chlorination, which is not ideal but is certainly a relatively acceptable risk compared to the pathogens themselves. The cleaning up of rivers is also well under way in the West, with the Thames and the Rhine now both in reasonable condition compared to the 1960s and early 1970s. However, the Elbe, running through the Czech Republic and former East Germany, is still very heavily polluted and lacking in fauna and flora, as little effort was formerly put into ecological planning in major industrial centers such as Leipzig and Dresden.

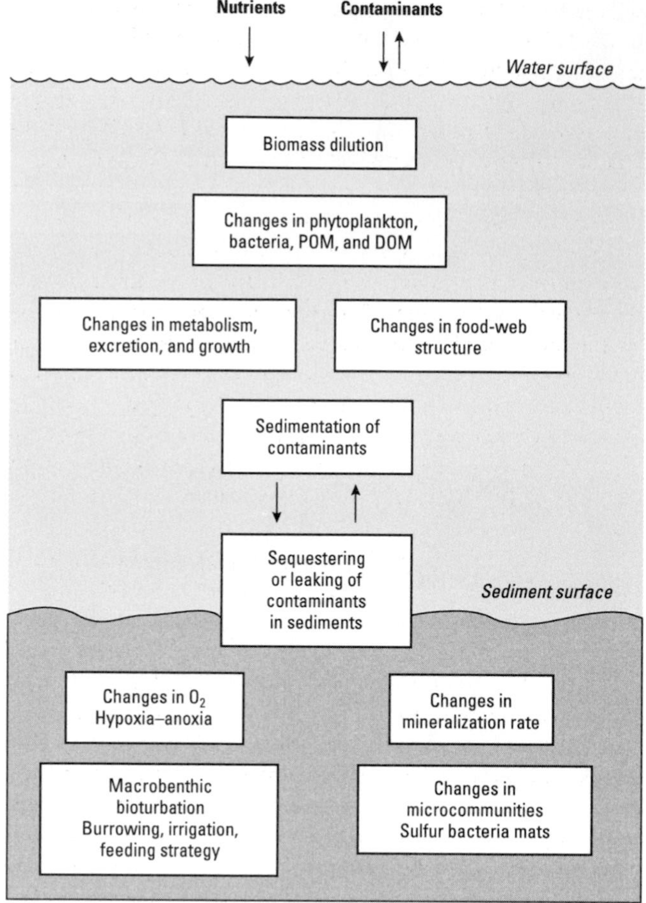

Fig. 13.37 Interactions between contaminants and eutrophication in freshwater environments. Nutrients may increase the primary productivity and thus growth rates of larger animals, increasing turnover rates, and sedimentation of contaminants. DOM, dissolved organic matter; POM, particulate organic matter. (Adapted from Park 1991.)

13.8.2 Acidification

The problem of acid deposition was first noted at least 100 years ago, with the observation that rainfall was sometimes acidic and that this did harm to vegetation, stonework, and metals. But in the 1960s Scandinavian lakes had become seriously acidified, with fish and invertebrates dying, and the damage was linked with pollution blown from Britain and north Europe; similar problems began to appear in North America. By the early 1980s, 8–10% of lakes in Sweden, and a full 20% in the USA, were too acid to support any fish life. In 1997, 16,000 out of a total of 85,000 Swedish lakes were recorded as anthropogenically acidified.

Causes

Rain is naturally acidic, because the CO_2 in air dissolves to give weak carbonic acid, having a pH of usually about 5.5. But rain in Europe and the USA can now be regularly recorded at 4.0–5.0, with even lower figures in Central Europe. Some rainstorms have scored 3.0, and some smogs lower still, with a pH equivalent to vinegar. The obvious clue to the source of this problem is that acid rainfall is a regional rather than a global problem (Fig. 13.38); northern Europe, northeastern USA, and southeastern Canada are much the worst areas, all lying to windward of heavy industry. Sulfur dioxide emissions, either unconverted as a gas or dissolving as sulfuric acid, are therefore targeted as the prime cause of the acidity; 60% of this gas comes from power stations (especially coal-fired ones), and 20% from other industrial plants.

In fact, recent evidence from the analysis of foraminiferan shells in lake sediments dates the onset of acidification to around 1800, when coal burning became more common; particles of soot appear in the sediments at the same time. This was really the conclusive evidence as to the cause of acid rain. However, emissions from coal burning have been going down in the UK, USA, and most of Western Europe since the early 1970s, and SO_2 has therefore been declining, yet damage has still been worsening. This is suggestive of long-term

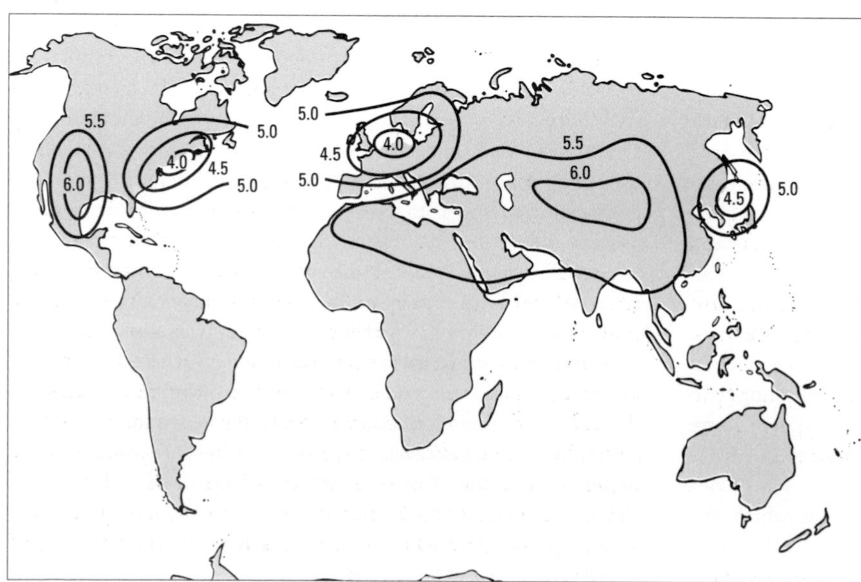

Fig. 13.38 Regional patterns of acidification of rainfall and of lakes. The black lines define pH isoclines.

cumulative effects and/or interactive effects with other pollutants or stress factors. Several gases are therefore now implicated, but the evidence is especially damning against interactive effects of sulfur dioxide and nitrogen oxides (both from power plants, and the latter also from car exhausts) and ozone (mainly from car exhausts). There are some recent suggestions that climate modification is making matters worse; for example, lakes in the US state of Michigan have become rapidly and seriously acidified only during recent drought summers. Thus if global warming really sets in it may hasten changes in lake chemistry.

Consequences for animals

There are a number of consequences of acidification for freshwater animals:

1 Reduced fecundity and increased mortality occur, operating through a variety of different physiological effects on ion and acid–base balance with knock-on behavioral implications. Animals show reduced body weight in acid waters; this has particularly been recorded for fish and for amphibians, and in the latter its proximate cause is mainly reduced rates of prey capture.

2 Acid rain solubilizes minerals from the river or lake sediment. Of these, aluminum is the most important, solubilizing below pH 4.5. In many acidified lakes, the death of fish is now clearly attributable to aluminum poisoning, affecting mucus production and clogging the fish gills (mucus has a similar oxygen diffusion coefficient to water, but its presence creates a nonconvective and thus effectively stagnant boundary layer around the gill surfaces). It can be reasonably assumed that mucus production in other (noneconomic and therefore less studied) freshwater animals, such as flatworms, snails, and annelids, is also seriously affected, causing mortality. Mobilized aluminum also leads to the precipitation of phosphate and thus an oligotrophication effect, resulting in reduced algal growth and very clear water.

3 There are direct effects on physiological processes such as cuticle formation in insects and crustaceans (where calcium uptake is especially affected), so reducing these animals' subsequent osmoregulatory abilities. Their resultant mortality upsets the food chain for fish, and these food-chain effects also impact on birds (e.g. dippers and other stream dwellers)—one particular result is thinning of their eggshells and thus increased egg mortality in some species.

4 Particularly acid conditions often occur in spring at snowmelt, coinciding with the hatching time of many salmonid fish. Many fail to hatch at low pH, apparently because the enzyme chorionase, which allows the larva to escape from the chorionic membrane, can only function fully at a pH of 6.5–8.5. Thus fish stocks suffer very considerably.

The abundance of many species therefore declines, particularly snails, amphibians, small crustaceans contributing to the zooplankton, and salmonid fish. Other fish such as eels, and insects and rotifers, are relatively unaffected at first, but in the long term pH levels below 5.0 will kill off most of the fauna, with only some green algae and sphagnum moss flourishing in the littoral fringes.

Dealing with acidification

Some of the "curative measures" proposed to deal with acidification are also treated with some scepticism by environmental biologists. In the 1970s it was thought that the situation could be helped by dumping ground-up lime into lakes. Norway and Sweden have been attempting to promote lake recovery in this way, until their neighbors reduce their emissions sufficiently. About half of all Sweden's acidified lakes are now treated with lime powder, and some significant increase of species richness has been recorded. The amount of lime needed is very large and very hard to calculate, as it depends on flow patterns and the turnover time of each lake, and liming must be repeated regularly; it also does not affect the inflow streams, which are where fish often breed. "Target liming", going for streams and bogs in the catchment area, may be the best compromise, and this has been done recently with reasonable success in some lochs in southwest Scotland. But liming tends to kill off sphagnum and other mosses in the catchment zone, which may have other long-term effects on bog formation and stability, and which certainly badly affects birds and invertebrates associated with upland moors and bogs.

There are some recent and encouraging indications that lake and water acidifications can be naturally reversed reasonably quickly when acid inputs stop. Many Scottish lochs have been regaining a higher pH in the 1990s, though no fish are returning yet. A particularly detailed study has been made at Whitepine Lake in Canada. In 1980 the lake was showing severe acid stress, with many acid-resistant perch but a loss of numbers and/or reproductive ability in trout and burbot, plus a decline in diversity of most other taxa. Over the next 10 years, SO_2 emissions were halved and the pH rose from 5.4 to 5.9, with aluminum levels dropping. The trout recovered very effectively, as did the diversity of benthic invertebrates. At least two species of formerly common fish did not recover though, and must be regarded as extinct in that lake. It is also apparent that as yet there is not much improvement of lake quality in Norway or Sweden.

13.8.3 Dams and irrigation systems

The construction of dams and irrigation systems drastically alters flow regimes in river systems, and while this must not be seen as necessarily "bad", since there may be proliferation of some new and valuable species downstream of management areas, nevertheless it is very difficult to predict the outcome of damming a watercourse, and it is therefore biologically speaking a "chancy business". Damming directly affects all migratory species, especially anadromous fish, and both damming and other management schemes tend to remove rapids, waterfalls, and shallows, which can greatly reduce a river's ability to re-aerate following any unusual oxygen demand. On the other hand, water in the spills below high dams can become supersaturated with oxygen at up to 140% saturation, and this too can be damaging, causing lethal "gas bubble disease" in migrating fish.

Damming a river can also have drastic knock-on effects on downstream communities, especially by causing silt to be deposited in the lake above the dam, so being lost to the downstream systems. Both fisheries and natural silt fertilizations of the Nile lowlands were seriously disrupted after construction of the Aswan High Dam. Downstream effects on human communities are also potentially serious, particularly where rivers cross national boundaries and countries compete for the water supply.

13.8.4 Invasions

Human activities have often brought new species (nonindigenous species) into freshwater ecosystems, which as with all biological invasions can have complicated and far-reaching effects. A number of spectacular invasions of the Great Lakes system in North America have been recorded. In the 1830s the sea lamprey invaded the Lakes via the Hudson river, and seriously disrupted fish stocks. More recently (1986) the European freshwater zebra mussel, *Dreissena polymorpha*, arrived, probably in a ship's ballast water, and has now spread to every one of the Lakes, occurring at densities of at least 0.5 million per square meter in some areas, displacing local clam beds and seriously affecting fish-spawning areas, as well as fouling many manmade structures. These *Dreissena* apparently came from populations in the Black Sea with a rather high upper thermal limit, so that in America the species is spreading into much warmer lakes and rivers than it normally tolerates in northern Europe.

13.8.5 Effects of global warming on freshwater systems

Direct effects of carbon dioxide build up are expected to be limited, in that aquatic animals usually respond more to $P\text{O}_2$ than $P\text{CO}_2$ (see Chapter 7), and changes in external $P\text{CO}_2$ in the water will in any case be very small due to the high solubility of this gas. Temperature effects are of more concern, but will perhaps be less drastic than for the less buffered terrestrial environments (see Chapter 15).

However, the timing and distribution of rainfall and run-off will certainly alter, according to most of the global circulation models attempting to predict future weather changes. There may be 3–15% more rain overall on IPCC (Intergovernmental Panel on Climate Change) models. This is likely to result in increased river currents and discharges, with potential bouts of flooding in vulnerable river delta areas such as Bangladesh. It will also be likely to alter animal zonation patterns in many key rivers; and it could again be especially critical for humans where resources are shared between nations.

13.8.6 Deforestation and afforestation

The presence of forests or other dense vegetation around both lakes and rivers affects the light received and hence flora and fauna patterns, and resultant water chemistry, as well as affecting the allochthonous inputs. Thus whole river systems can be greatly affected by deforestation, with effects even out to sea on vulnerable communities such as coral reefs. There may be a complete loss of the humus layer in the tropics, for example, which runs into the rivers and creates transient "blackwaters". On the other hand, replanting naturally or anthropogenically deforested areas with dense stands of conifers, as is occurring in many regions of European and North American uplands, can greatly alter the acidity of river systems running through the area.

13.8.7 Uses of fresh water

The worldwide use of freshwater resources by humans is such that an amazing 73% goes on crop irrigation, usually being used extremely inefficiently; 21% of the usage is industrial; and just 6% involves domestic use. However, these figures are somewhat misleading. Most of the irrigation water gets recycled through the agricultural system many times, so it is not actually "lost"; and much of the industrial water is purified within the plant now, in Western countries at least, and so is returned to the cycle.

13.9 Conclusions

It should be very clear that freshwater life is so varied that it is virtually impossible to generalize. Given the problems and the many axes of environmental variations, it is perhaps quite surprising that so many animals live so successfully in these habitats. In fact, most animal taxa achieve some freshwater representation, with many very abundant, and it is not particularly easy to account for this.

Can we end instead by deciding why some groups do not succeed? We can recall that the chief exclusions are echinoderms, tunicates, most sponges and cnidarians, and the cephalopods (see Table 13.7). For all but the last of these, the answer may be reasonably simple:

- All these animals are built with large, soft exposed surfaces that could not be made impermeable, either because they are used as food filters (most) or as uptake surfaces (echinoderms and cnidarians);
- All these animals are from groups that are not already endowed with osmoregulatory organs.

Where an occasional sponge (*Spongilla*) or cnidarian (*Hydra*) does live in fresh water, we know very little about how it is achieved, but it is evident that energy-consuming pumping across the exposed epithelial surfaces must be going on.

For the cephalopods, however, these arguments do not really work. This taxon is entirely marine now, and furthermore always has been. There is no immediately obvious reason why cephalopod skin could not be more impermeable, and the animals have perfectly good kidneys; after all, other molluscs do well in fresh water with the same basic apparatus. Several suggestions have been advanced to explain the cephalopod's restriction to saline waters:

1 They cannot toughen the skin to give greater impermeability because they use this surface so much for delicate, rapid, and complex chromatophore signaling, and thicker impermeable layers would mask or slow down this system.

2 Being fast active predators, with quite high metabolic rates, they cannot cope with a variation of oxygen because they have a relatively poor pigment (hemocyanin, like nearly all molluscs) with little ability for potentiated loading.

3 Their buoyancy mechanisms depend on active pumping of ions, and this is harder in fresh water against increased gradients.

4 In combination with reasons 2 and 3, the extra cost of osmoregulation would be just too much for them.

5 Their central nervous system complexity and neuromuscular coordination need constant conditions to work properly.

It may well be that a mixture of all these—again stressing the interaction of environmental and physiological factors—makes freshwater life just too difficult for cephalopods; perhaps they could cope with any one problem, but not with all of them at once.

However, there is an additional possibility. Cephalopod history has been a continuing saga of coevolution with fish; the cephalopods were dominant in the early Palaeozoic, but when teleosts returned to the sea, the cephalopods slumped, and recovery has only really occurred in those groups (squid) that are most like fish and could compete with them. So perhaps with teleosts already established in fresh water with low blood concentrations, the cephalopods just could not compete in potentially similar predatory niches, so leaving the fish (and subsequently a few other vertebrates) as the supreme actively swimming big predators of fresh waters.

This point neatly underlines some of the messages that should have emerged from this chapter, as a summary of freshwater biology. Firstly, it is crucially important to understand the interplay of *all* factors in environmental adaptation, including their effects at physiological and behavioral levels and on morphology, biochemistry, and life history. Studying the effects of single environmental variables in isolation is particularly inappropriate. Secondly, freshwater organisms give us yet more insight into the critical role of coevolution and convergence between groups of organisms in shaping present environments.

FURTHER READING

Books

Adams, S.M. (1990) *Biological Indicators of Stress in Fish.* American Fisheries Society, Bethesda, MD.

Horne, A.J. & Goldman, C.R. (1994) *Limnology.* McGraw-Hill, New York.

Mason, B.J. (1992) *Acid Rain: its Causes and its Effects on Inland Waters.* Oxford University Press, Oxford.

Mason, C.F. (1991) *Biology of Freshwater Pollution.* Longman, London.

Morris, R., Taylor, E.W., Brown, D.J.A. & Brown, J.A. (1988) *Acid Toxicity and Aquatic Animals.* Cambridge University Press, Cambridge, UK.

Moss, B. (1988) *Ecology of Freshwaters: Man and Medium*, 3rd edn. Blackwell Science, Oxford.

Rankin, J.C. & Jensen, F.B. (eds) (1993) *Fish Ecophysiology.* Chapman & Hall, London.

Reviews and scientific papers

Byrne, R.A. & Dietz, T.H. (1997) Ion transport and acid–base balance in freshwater bivalves. *Journal of Experimental Biology* **200**, 457–465.

Byrne, R.A. & McMahon, R.F. (1994) Behavioural and physiological responses to emersion in freshwater bivalves. *American Zoologist* **34**, 194–204.

Chaui-Berlilnck, J.G. & Bicudo, J.E.P.W. (1994) Factors affecting oxygen gain in diving insects. *Journal of Insect Physiology* **40**, 617–622.

Deaton, L.E. & Greenberg, M.J. (1991) The adaptation of bivalve molluscs to oligohaline and fresh waters: phylogenetic and physiological aspects. *Malacological Review* **24**, 1–19.

Dietz, T.H., Neufeld, D.H., Silverman, H. & Wright, S.H. (1998) Cellular volume regulation in freshwater bivalves. *Journal of Comparative Physiology B* **168**, 87–95.

Feder, M.E. & Booth, D.T. (1992) Hypoxic boundary layers surrounding skin-breathing aquatic amphibians: occurrence, consequences and organismal responses. *Journal of Experimental Biology* **166**, 237–251.

Fryer, G. (1996) Diapause: a potent force in the evolution of freshwater crustaceans. *Hydrobiologia* **320**, 1–14.

Goss, G.G., Perry, S.F., Wood, C.M. & Laurent, P. (1992) Relationships between ion and acid–base regulation in freshwater fish. *Journal of Experimental Zoology* **263**, 143–159.

Henry, R.P. (1994) Morphological, behavioural and physiological characterisation of bimodal breathing crustaceans. *American Zoologist* **34**, 205–215.

Kirschner L.B. (1995) Energetics of osmoregulation in fresh water vertebrates. *Journal of Experimental Zoology* **271**, 243–252.

McKee, D. & Ebert, D. (1996) The effect of temperature on maturation threshold body length in *Daphnia magna. Oecologia* **108**, 627–630.

McMahon, R.F. (1996) The physiological ecology of the zebra mussel *Dreissena polymorpha* in North America and Europe. *American Zoologist* **36**, 339–363.

Morris, S. & Bridges, C.R. (1994) Properties of respiratory pigments in bimodal breathing animals: air and water breathing by fish and crustaceans. *American Zoologist* **34**, 216–228.

Okland, J. & Okland, K.A. (1986) The effects of acid deposition on benthic animals in lakes and streams. *Experientia* **42**, 471–486.

Olson, K. (1994) Circulatory anatomy in bimodally breathing fish. *American Zoologist* **34**, 280–288.

Perry, S.F. (1997) The chloride cell: structure and function in the gills of freshwater fish. *Annual Review of Physiology* **59**, 325–347.

Pynnönen, K. (1994) Hemolymph gases, acid–base status and electrolyte concentration in the freshwater clams *Anodonta anatina* and *Unio tumidus* during exposure to and recovery from acidic conditions. *Physiological Zoology* **67**, 1544–1559.

Ram, J.L., Fong, P.P. & Garton, D.W. (1996) Physiological aspects of zebra mussel reproduction: maturation, spawning and fertilization. *American Zoologist* **36**, 326–338.

Reiber, C.J. (1995) Physiological adaptations of crayfish to the hypoxic environment. *American Zoologist* **35**, 1–11.

Schindler, D.W. (1988) Effects of acid rain on freshwater ecosystems. *Science* **239**, 149–156.

Schwarzbaum, P.J., Wieser, W. & Cossins, A.R. (1992) Species-specific responses of membranes and the Na^{+}+K^{+} pump to temperature change in the kidney of two species of freshwater fish, roach (*Rutilus rutilus*) and Arctic char (*Salvellinus alpinus*). *Physiological Zoology* **65**, 17–34.

Wenning, A. (1996) Managing high salt loads—from neuron to urine in the leech. *Physiological Zoology* **69**, 719–745.

Wheatley, M.G. & Gannon, A.T. (1995) Ion regulation in crayfish: freshwater adaptations and the problem of molting. *American Zoologist* **35**, 49–59.

Wilkie, M.P. (1997) Mechanisms of ammonia excretion across fish gills. *Comparative Biochemistry & Physiology A* **118**, 39–50.

Zerbst-Boroffka, I., Bazin, B. & Wenning, A. (1997) Chloride secretion drives urine formation in leech nephridia. *Journal of Experimental Biology* **200**, 2217–2227.

14 Special Aquatic Habitats

14.1 Introduction

In this chapter we cover a range of "aquatic" habitats that are in various ways not strictly within the definitions of marine, littoral, estuarine, or freshwater habitats. Most of these are "extreme" on one or more of the normal scales: so transient as to share more characteristics with the terrestrial world; or so osmotically or thermally or barically extreme, or so lacking in aqueous content, that they demand very special adaptations. In each case, the animals to be found there are real specialists, and are part of very strange communities often excluded from consideration in classic physiological textbooks.

14.2 Transient water bodies

A range of very specialist habitats occur where fresh water is only intermittently present, whether as puddles or small pools, or in interstitial habitats such as moss cushions and crevices in rocks, or as temporary pools of water trapped within vegetation or animal remains.

14.2.1 Puddles

Puddles occur in shallow depressions on slow-draining soils and are rapidly colonized by inocula of phytoplankton such as the spores of *Euglena* and *Chlamydomonas* (flagellate protistans), and by adult insects with aquatic larvae, most notoriously the mosquitoes and other flies. Larval stages of chironomids and ceratopogonids (nonbiting and biting midges, respectively) are often hugely abundant, usually achieving success by having very short larval life cycles, often lasting only a few days. Amongst the chironomids, a famous puddle specialist is *Polypedilum vanderplanki*, whose larvae are longer lived and can survive virtually complete drying out and baking by the sun by using anhydrobiotic mechanisms as described below. In the biting midge *Dasyhelea* an alternative strategy occurs with the larvae constructing protective cocoons as their puddle dries out. Some water fleas also occur in puddles, from eggs dispersed on birds' feet, etc., and are known to absorb organic nutrients from the water directly through their cuticle. The same may be true of other small invertebrates.

Ground puddles in some parts of Africa and tropical America are also invaded by certain frogs as a tadpole habitat. With luck the tadpoles metamorphose before the puddle dries out. In East Africa the savanna ridge frog (*Ptychadena*) lays its eggs in vast numbers in small pools after rain, and tadpoles occur at densities of up to 1000 m^{-2} of water surface. Where tadpoles occur, their excretions enrich the puddles and help to speed up development of the insect larvae they share the habitat with.

Puddle specialists must be relatively eurythermal and also tolerant of anoxia, but we have little detailed knowledge of their physiological adaptations.

14.2.2 Water pools in plants and animals

Water may be trapped within the leaf bracts of many kinds of plants, but the most conspicuous examples occur in the strong spiky bromeliad plants of the tropics, where the base of each leaf may trap a small tank of water for relatively long periods (Fig. 14.1a). Water is also retained in tree holes produced by natural damage or decay or by certain kinds of burrowing or nesting animals. Water pools also occur in various carnivorous pitcher plants, in the capsules of various nuts and seeds, and even in shelly animal remains. Where such habitats occur in living plants they may be well oxygenated by day but have periods with high levels of carbon dioxide and low pH; in animals or plant remains the water is commonly strongly eutrophicated and may be anoxic.

In these kinds of habitat a range of insects and other invertebrates can be found, together with larval tree frogs in the tropics. Frogs may deposit eggs and hence tadpoles individually in bromeliad leaf bracts. In some species the tadpoles are nonfeeding, living off their yolk reserves; others are predatory, consuming larval insects and even other tadpoles, and in a few species the adults revisit their offspring and provide food for them. Some tree frogs that routinely use bromeliad pools or tree-hole pools have strangely ossified cranial structures, which they use to seal up the entrance to their shelters, and these ossified patches may have unusually low evaporative water loss (EWL) rates so that once sealed into its hole the frog is resistant to drying out. Jamaican bromeliads also constitute a home for a specialist genus of crabs (*Metopaulias*), which undergo abbreviated larval development within the tanks provided between the leaf bracts. The larval stages are completed in the low pH water of the tanks (usually pH $< 5-6$), but not in normal river water from the same habitat (pH 8). Recent evidence indicates that the crabs can achieve some degree of pH and Ca^{2+} control by importing fragments of calcareous snail shells into the tanks.

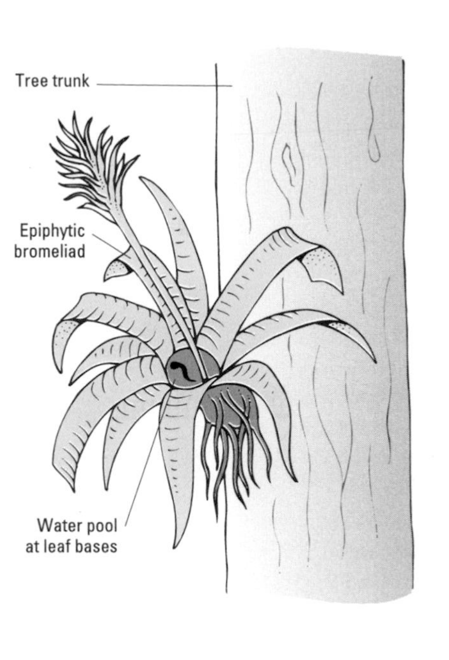

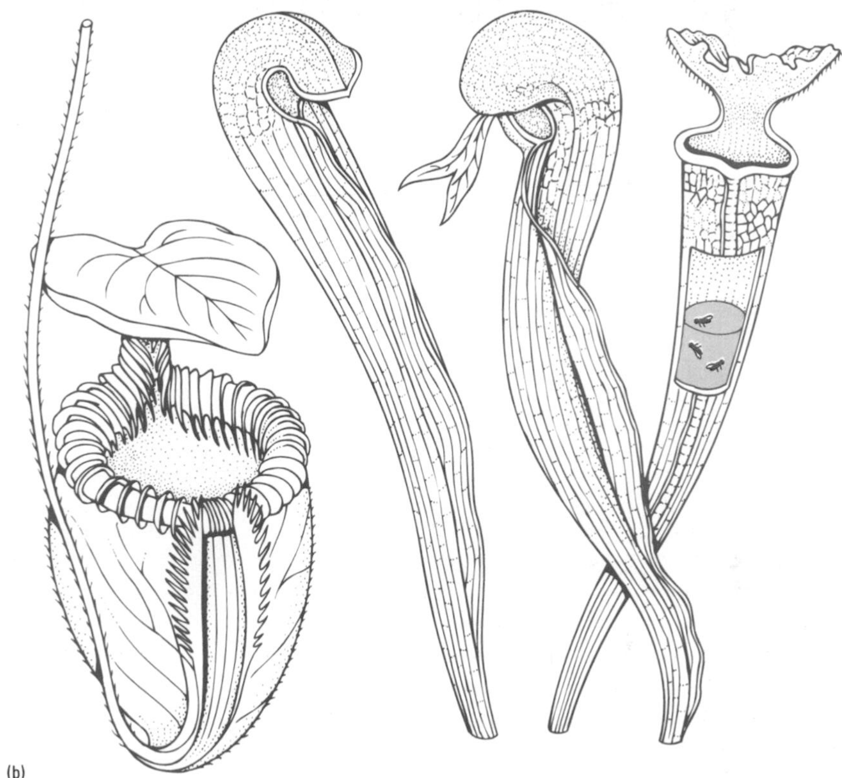

Fig. 14.1 Transient water pools used by various invertebrates and frogs in (a) leaf bracts of bromeliads and (b) pitcher plants.

Pitcher plants provide another kind of puddle (Fig. 14.1b; see also Plate 6a, between pp. 386 and 387), although here the organisms living in the water may be even more specialist, often being carnivores exploiting the trapped animals that form the plant's "prey", and thus perhaps best seen as inquilines. Pitcher plant mosquitoes are a well-studied group, able to live and breed in the fluid.

The vacated shells of land snails provide yet another small enclosure that can trap water and provide a temporary aquatic home. In Jamaica the crab *Sesarma* uses these: females carrying eggs find a snail shell of suitable size and fill it with up to 5 ml of water using a hairy patch of cuticle to carry the liquid by capillarity. Juveniles hatch and develop within the snail shell, guarded by the female for up to 3 months.

In these small aquatic habitats within plant tissues or shells, drying out is unlikely but temperatures may still fluctuate markedly. This leads to substantial differences in larval mortality and developmental rate according to the size and exposure of the initial water pool, so that the degree of physiological stress on the larval animal is largely determined by pre-emptive maternal choices.

14.2.3 Extreme transience: moss cushions, cracks, and crevices

Some of the most transient of all aquatic habitats are sites such as moss cushions, cracks, and crevices. Such habitats attract a unique fauna dominated by very small and mostly unfamiliar invertebrates (Fig. 14.2), such as nematodes, rotifers, collembolans, tiny crustaceans, and also the tardigrades (commonly known as "water bears"). When active these moss dwellers are covered by fresh well-oxygenated water (either bulk water or at least a covering film of water) and operate much like other freshwater life. But they must also be able to survive when the habitat dries up; this may occur in summer on an hourly or daily basis and with no very predictable pattern, relating more to sporadic rainfall than to clear-cut seasonality, and it may also occur in winter if the microhabitat suddenly freezes.

For these animals, many of which also survive in other extreme habitats such as high-latitude gravels, small crevices on glaciers, and sands of desert wadis, the only possible survival strategy is to be able to cope with almost total desiccation, and all therefore exhibit the phenomenon of **cryptobiosis**, or "hidden life", also known as **anhydrobiosis** (life without water), appearing to be dead for long periods and then apparently miraculously coming back to life days, weeks, or even years later when liquid water is supplied.

A number of groups can achieve this in their juvenile stages as eggs or cysts, but others (the rotifers, nematodes, and protists) do it as adult organisms. Perhaps the most famous and peculiar examples are the tardigrades, amongst the most endearing of all animals. Figure 14.3 shows an active tardigrade in its hydrated state, and then two further stages showing what happens as its habitat dries out until eventually it is in the form called a "tun" (meaning barrel), with all the limbs withdrawn, and the intersegmental cuticle tucked in, leaving just a smooth surface of the thicker cuticular plates exposed. The animal can be reduced to just a small percentage of its initial water content and yet it stays alive.

Three main mechanisms are known to be involved in the process of entering cryptobiosis, whether it is initiated by high temperature, by drought, or by freezing, and the mechanisms involved seem to be the same in all the groups that can achieve this "playing dead" trick:

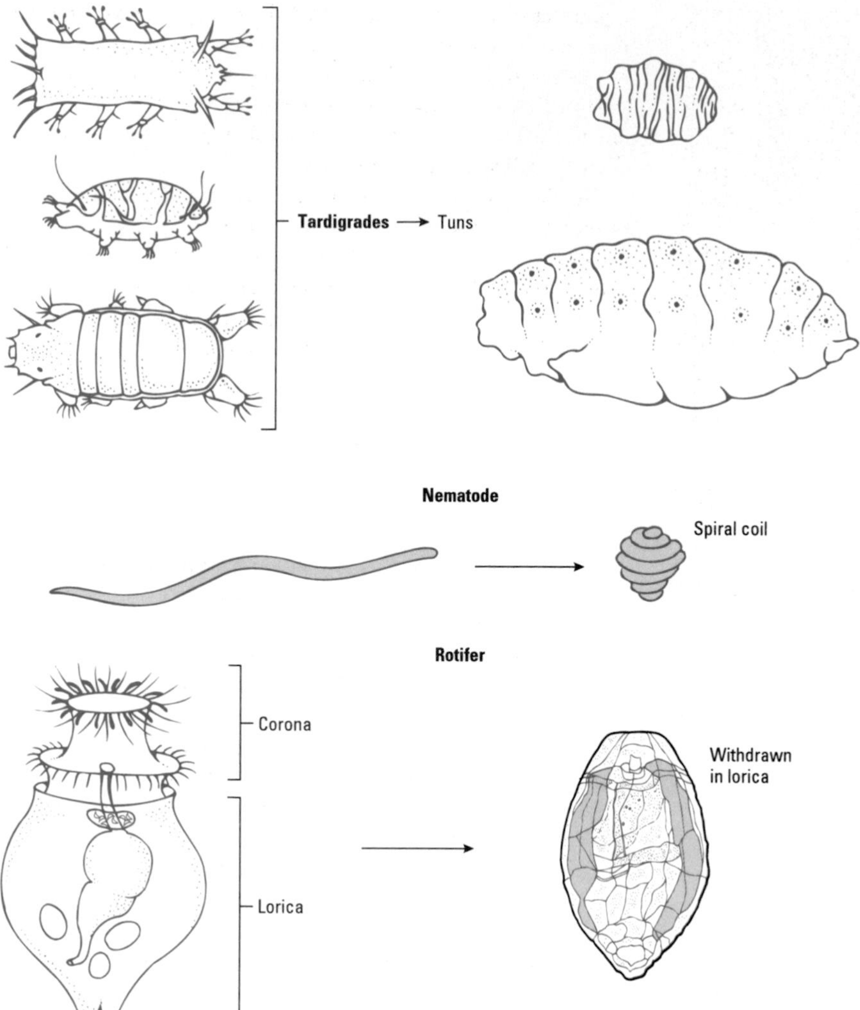

Fig. 14.2 Fauna that use cryptobiosis as a survival strategy in transient aquatic habitats.

1 Changing the morphology of the surface, by tucking in more permeable areas of the cuticle and changing the relative thickness or chemistry of cuticle layers. Nematodes cannot contract into a tun, so they either spiral themselves up into a neat ball instead, sometimes in aggregations, or they enter cryptobiosis within the shed but not discarded skin from the previous molt. Rotifers withdraw into their lorica (the casing that protects them), and tuck in their high-surface-area feeding structure, the ciliated corona. In nematodes the hyaline layer of the cuticle is visibly altered, and in tardigrades wax extrusion may occur.

2 Reducing the permeability of the cuticle, by lipid phase changes and/or wax production, so insuring that a minimum amount of water is retained. The "permeability slump" has a time course of tens of minutes in the relatively complex tardigrades such as *Echiniscus*, and of only about 2 min in the anhydrobiotic nematode *Ditylenchus*. The final water content of the animal may be only 5–15% of the initial value.

3 Metabolic production of a large quantity of particular sugars, such as trehalose and glycerol, that protect cell membrane phospholipids and proteins from damage during drought by binding into the membranes in place of the water and stabilizing their structures. This process is accompanied by "cytoplasmic vitrification", turning the intracellular fluid into a glass-like material. Trehalose is an α_1-linked nonreducing disaccharide of glucose, formed in many microorganisms as well as in cryptobiotic animals and in insects. It is probably a stress protectant for two main reasons: firstly because it interacts with and directly protects both lipid membranes and proteins, probably by hydrogen bonding to phosphates and other polar residues within the dried out macromolecular assemblages; and secondly because it naturally forms a "glass" material, resulting in vitrification of the whole cytoplasm. In a cryptobiotic nematode subjected to either freezing or desiccation it occurs in all tissues, but especially in muscle and in reproductive organs. Its production is catalyzed by two key enzymes, trehalose 6-phosphate (T6P) synthase and T6P phosphatase, and its degradation to glucose is mainly controlled by trehalase. However, the regulation of these enzymes is largely unknown. Likewise the reasons for the particular success of trehalose as a stabilizer relative to other sugars are not fully elucidated. In part, it may be that trehalose has a high glass-transition temperature that is relatively unaffected by small amounts of water. Cytoplasmic vitrification due to trehalose has to be maintained throughout the cryptobiotic phase, and if it is lost then phase

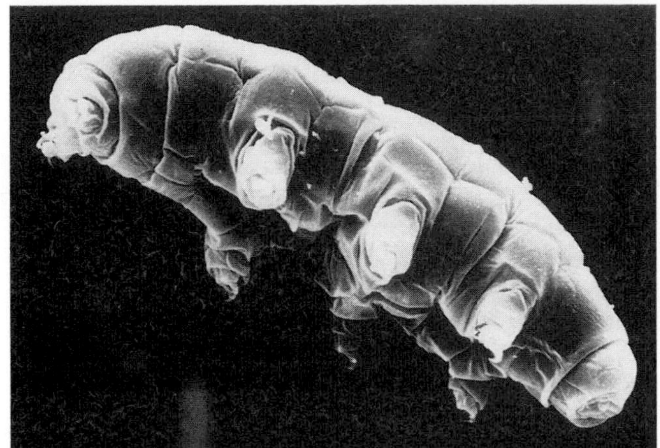

(a)

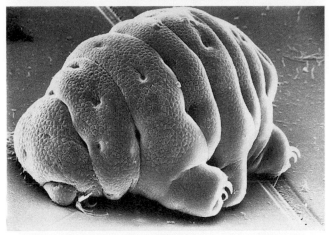

(b)

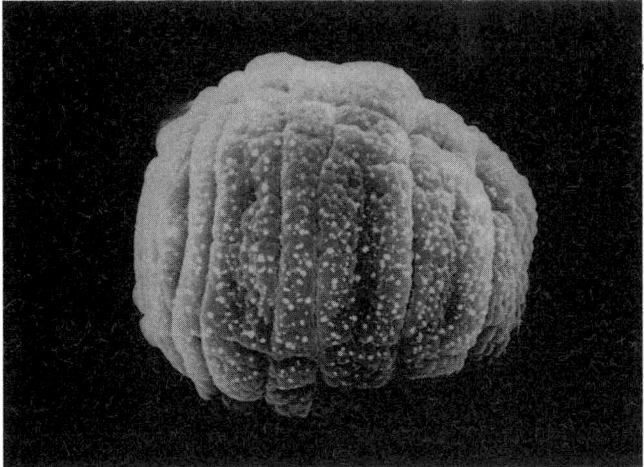

(c)

Fig. 14.3 SEMs of a tardigrade (a) in its normal hydrated state, (b) shortly after beginning entry to the cryptobiotic state, and (c) when fully into this state (the "tun"). (Courtesy of J.C. Wright.)

separations occur in membranes, free radical oxidation begins, and the cytoplasm may instead undergo a process akin to crystallization, which irreversibly damages the cells.

A summary of the control of anhydrobiosis is shown in Fig. 14.4; note that there is a need for relatively slow drying in moist air to give a good chance of survival, so that these three mechanisms have time to be set in place, and that if resuscitation begins in dry air it can be reversed. The moisture-retaining properties of mosses tend to insure humid air in the phases before and after rainfall, so that the animals in this habitat have a reasonable guarantee of successful entry to and exit from cryptobiosis.

The ability to enter cryptobiosis seems to be age dependent, and it cannot be achieved in very young eggs or embryos in most species where it occurs. However, when it occurs in more mature animals it modifies the timing of subsequent life-cycle events but not the reproductive output (the pattern of age-specific fecundity), indicating merely a resetting of an internal clock system.

14.3 Osmotically peculiar habitats

14.3.1 High-salinity lakes, brine ponds, and brine seeps

There is almost as much water contained in salt lakes (including inland "seas") in the present world as there is in freshwater lakes (see Table 13.1). Wherever freshwater lakes occur without an outflow they tend over time to turn into salt lakes (see Plate 6b, between pp. 386 and 387), or **endorheic lakes**, and often eventually dry up completely into salt flats (see Plate 6c). The resulting inland brackish waters will normally have a salt composition similar to that of the sea, but where they are produced mainly by evaporation from springs they can have a very different ionic composition, and are described as **athalassic** (meaning not sea-like). The compositions of some salt-lake water are shown in Table 14.1; note that in extreme cases the total osmotic concentration may be 6–8 times that of sea water.

Such lakes mainly form in three conditions (leading to the world distribution shown in Fig. 14.5):

1 In the rain-shadow areas of high mountains—to the north of the Himalaya, in Nevada and Utah, and central Canada beyond the Rockies, and in parts of eastern Australia. Lake Mono in California and the Great Salt Lake of Utah are particularly well studied.

2 As the relatively undrained and long-lived lakes formed by geological rifting in hot climates where evaporation is intense, notably in East Africa (Table 14.2) and the eastern Mediterranean. The Dead Sea has formed primarily by this process of longstanding evaporative concentration, but receives a small surface input from spring waters and river water, so that periodically it undergoes a substantial mixing event (most recently in 1979).

3 At higher latitudes where lakes that formed at the end of the last Ice Age have become isolated from the sea by isostatic rebound (rising land levels as the heavy ice retreats). These lakes are quite common in Iceland, Greenland, and parts of Canada, and have very variable degrees of hypersalinity. Often they have become freshened from the top down, and have saline and hypersaline water only at depth.

High-salinity conditions also occur as smaller "brine ponds" in many areas, often where water accumulates temporarily on top of dried salt flats (Plate 6c) or where pools and coastal lagoons become isolated just above the littoral zone, especially on low-latitude coasts. Here concentrations of up to four times sea water are not uncommon.

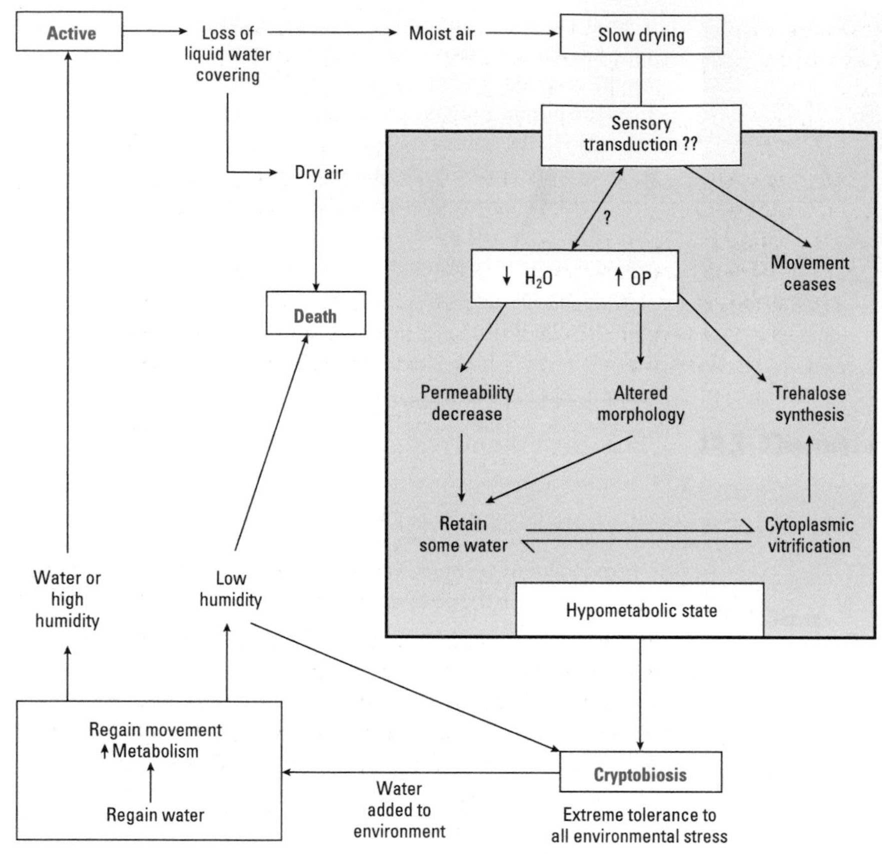

Fig. 14.4 The control of cryptobiosis: the effects of water presence, and humid or dry air externally, and of endogenous controls within the animal. OP, osmotic pressure.

Table 14.1 Compositions of some salt lakes.

Lake (location)	Osmotic concentration (mOsm)	Ionic concentrations (mM) Na	K	Mg	Ca	Cl	SO_4	Salinity (ppt)
Great Salt Lake (USA)	~6000	3000	90	230	9	3100	150	210
Mono Lake (USA)	~2000							70
Dead Sea (Middle East)	6200–6500							226
Lake Koombekine (Australia)	~7500	4070	17	110	32	1670	72	263

Salt lakes and ponds obviously pose a major physiological problem to animals in terms of osmotic and water balance. However, they also tend to be very strongly stratified, due to high water viscosity, needing much stronger wind action to bring about an overturn (i.e. they are usually amictic or monomictic). Thus marked oxygen differences and thermal gradients must also be dealt with for mobile animals to be successful colonizers.

Biota in the lake water

The biota of endorheic lakes and ponds is usually very low in diversity. Extreme cases are dominated by just one or a few species of blue–green algae, flagellates, or halobacteria; but where water is reasonably persistent above the salt pans, the brine shrimps *Artemia* and *Parartemia* are almost ubiquitous colonizers. The species *Artemia salina* may achieve several generations per year in salt lakes as these commonly occur at lower latitudes. In Lake Mono there is a spring generation and a much faster developing summer generation, when warmer temperature outweighs reduced food availability (but mortality is rather higher with many failures to develop). Other crustaceans also do well in high-salinity pools above tropical shores: the fiddler crab, *Uca rapax*, is common in South American lagoons. In the most extreme salinities all fish are absent, but in salty (up to 140 ppt) high-temperature lakes the desert pupfish (*Cyprinodon*) can occur. The larvae of several species of brine fly (*Ephydra*, *Ephydrella*, and *Hydropyrus*) also occur in vast numbers in brine pools in many parts of the world, together with some chironomid midges, such as *Cricotopus*, and mosquitoes, such as *Aedes*. All the insects and crustaceans are extremely good osmoregulators (Fig. 14.6), maintaining rather constant blood concentrations in both dilute and very salty media; they are sometimes termed "hypo-hyper-regulators".

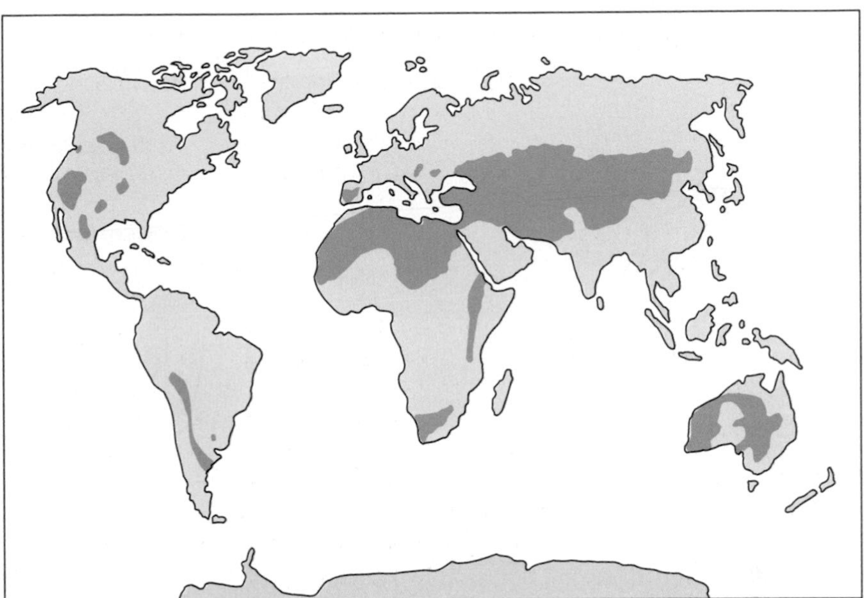

Fig. 14.5 Dark shading shows areas in which endorheic salt lakes are commonly found. (From Moss 1980.)

Table 14.2 East Africa soda lake characteristics.

Lake	Area (km^2)	Maximum depth (m)	Mean pH	Alkalinity (mEq l^{-1})
Turkana	8000	120	9.6	19–24
Manyara	400	1.5	9.5	78–800
Baringo	150	8	9.0	4–10
Magadi	95	0.6	10.1	400–3600
Nakuru	43	1.3	10.4	120–1500

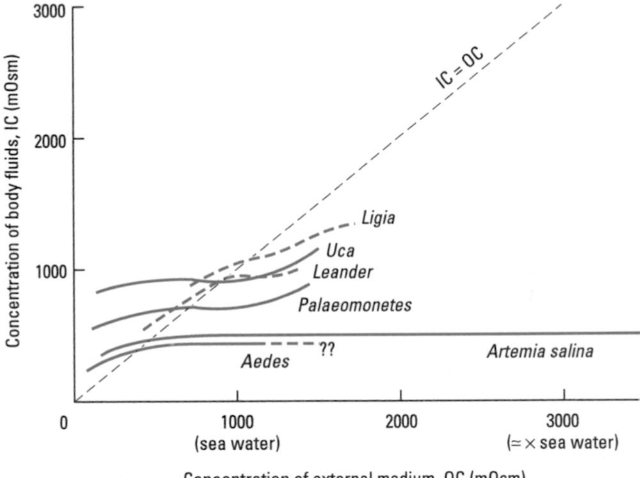

Fig. 14.6 The patterns of hypo-hyper-regulating in various brine-dwelling animals, comparing internal and external media.

For the various invertebrates faced with these problems, the main response seems to be to increase the drinking rate several fold, this being the only way to acquire the essential water. Adult *Artemia* can drink 5–8% of their own body mass per day, with water and ions then gained across the gut epithelium. Surface permeability is relatively low compared to most crustaceans (P_{osm} only 0.1 μm s^{-1}, cf. Table 4.1), so that the outward leak of the water thus gained is reduced, but the inevitable consequence of drinking is a salt load, which must be excreted. In adult *Artemia* (Fig. 14.7b) the salt glands on the gills perform this function, though in the larvae it is achieved by a specialized area of tissue at the back of the head, known as the "neck organ" (Fig. 14.7a). At both sites the epithelium concerned has very high sodium pump (Na^+/K^+-ATPase) activity, and chloride pumps basally or apically (or both) are also proposed. In the very small larvae of mosquitoes in the genus *Aedes*, hyporegulation is achieved by extraordinary levels of drinking (up to one-third of the body volume per day in *A. campestris* and even double the body volume per day in *A. taeniorhynchus*), with rapid and strongly hyperosmotic salt secretion from the anal papillae projecting from the rectum (the normal site for ion and water regulation in insects).

Many brine-dwelling animals (especially the insects) are able when necessary to excrete urine strongly hyperosmotic to their own blood, with concentrations of up to 8700 mOsm in *Ephydrella*. The crustaceans achieve this via their antennal glands, which can resorb or excrete ions into the urine as necessary, though to only a limited extent. Similarly in the insect's rectum there are large numbers of specialized secretory cells, as well as the more normal resorptive cells that produce dilute urine in dilute media. The Malpighian tubules of insects from hypersaline lakes rich in Mg^{2+} and SO_4^{2-} are also unusual in being able to pump these ions very rapidly to the lumen. It is worth noting that hyposmotic regulation in the brine shrimp and mosquito larva is relatively much more costly (22–33% of the metabolic budget) than hyperosmotic regulation in any animal yet investigated (cf. Table 5.10), which may help explain why it is a strategy adopted by so few animals.

In very salty water the ion levels may be enough to reduce dissolved oxygen content markedly. It is therefore not surprising that in *Artemia* the concentration of hemoglobin (Hb) and the proportions of different Hb morphs vary with salinity; hypoxia and high salinity both cause an increase in levels of Hb_3, a β-homodimeric

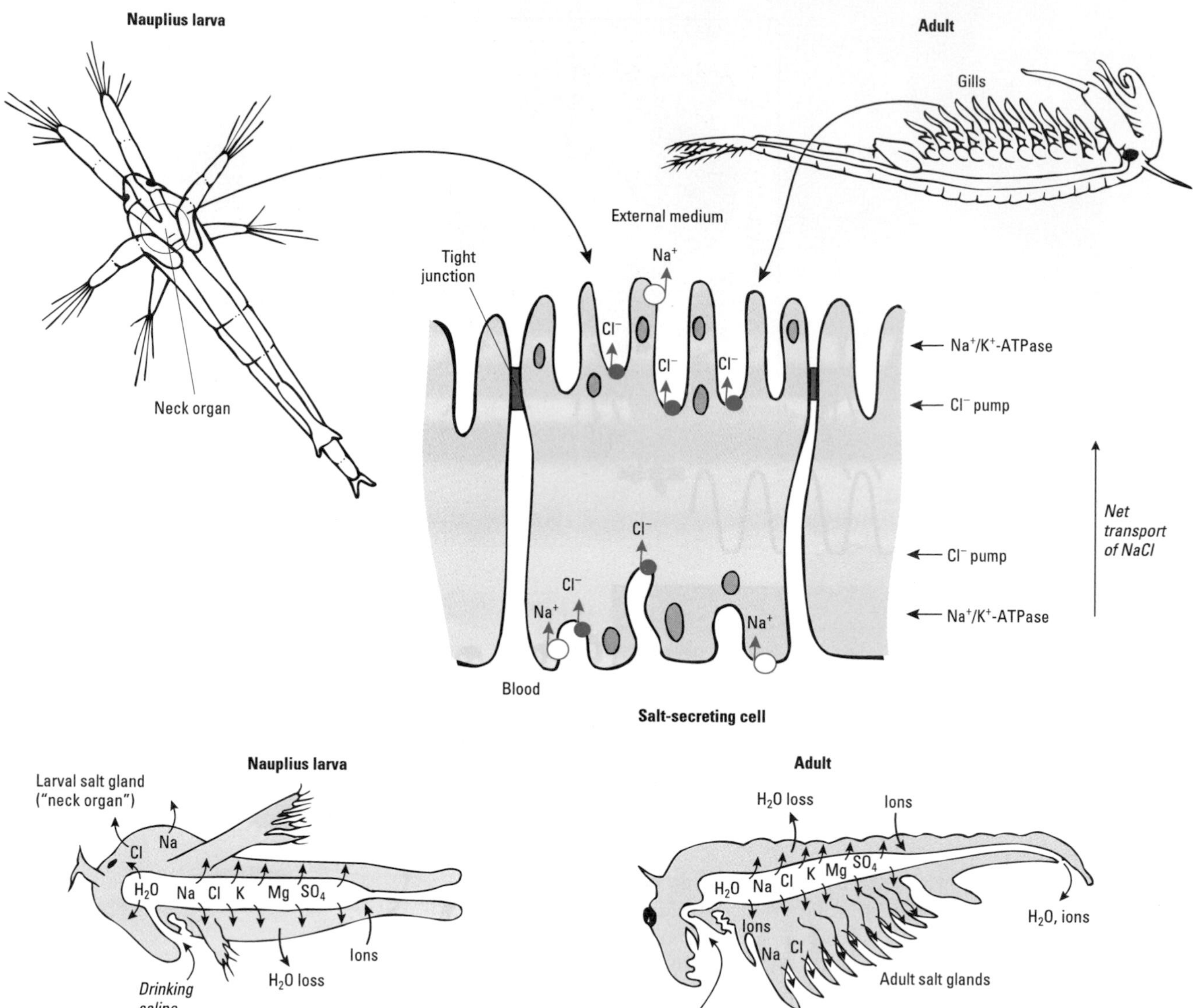

Fig. 14.7 The structures that regulate salt elimination during hyporegulation in the brine shrimp, *Artemia salina*.

form of the pigment that is structurally relatively unaffected by high ionic strength.

Artemia also links us back to the issue of cryptobiosis, since the brine shrimp uses this strategy to withstand drying out in the shallow fringes of salt lakes. The resistant stages are usually called "eggs" but are in fact encysted gastrula embryos; they may have a water content of less than 0.15 ml g^{-1} dry matter, and have no measurable metabolism.

Fauna associated with salt lakes

There are a number of larger animals found in association with salt lakes, although not living entirely within the salty water. Most obvious of these are the flamingos, which are spectacularly abundant on the salt lakes in central Africa such as Lake Nakuru, in some Andean lakes, and in some salt lagoons of southern Europe. They survive by filtering out the unicellular algae. It might be supposed that they would thus be bound to accumulate a high salt load and should therefore require special kidney adaptations, but in fact they solve the problem behaviorally rather than physiologically, by flying off to freshwater lakes regularly for a drink.

14.3.2 Deep-sea brine seeps

A variety of "brine seeps" occur in deep cold marine sites, often accumulating above undersea oil deposits, though these are scarcely explored. In some cases they are surrounded by vast beds of mussels, probably relying on the methane released from the hydrocarbons, and being fixed by bacteria within the mussel tissues. Shrimps, squat lobsters, and polychaetes also occur in these habitats; although their physiology has not been explored, they presumably share some features of the brine-lake inhabitants discussed above.

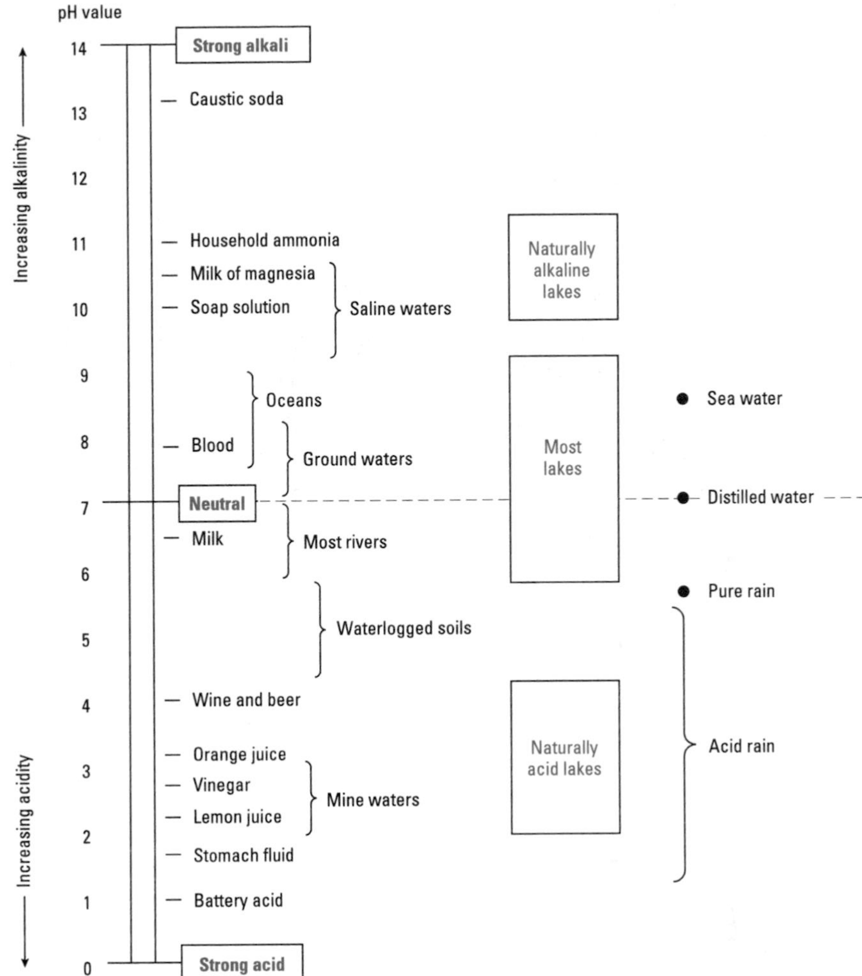

Fig. 14.8 The range of pH found naturally in lakes, in comparison with other natural fluids. Remember that the pH scale is logarithmic.

14.3.3 Acid and alkaline waters

Figure 14.8 summarizes the range of pH found in naturally occurring lakes, in comparison with normal lakes and various other natural substances. Both high and low pH pose particular problems for animal life.

Alkalinity: soda lakes and crater lakes

Moderately alkaline waters occur in chalk and limestone areas worldwide, and lakes with inputs of high pH and few outlets may become increasingly alkaline through their life. This poses real problems for animals' acid–base balance, given the linkage between H^+ ions and key transport processes in many tissues.

Problems arise especially in soft-water areas (with run-off from noncalcareous rocks), and mortality in nonspecialists can be very high. Where the water is "hard", Ca^{2+} from the environment (chalk and limestone) appears to protect animals against excessive disruption of their ionic and acid–base balance. The main effect operating may be a calcium stimulation of ammonia excretion, via a Na^+/NH_4^+ exchange mechanism, helping to keep the blood from experiencing raised pH. Certainly fish die if there is no calcium present, and do better in hard waters, which are perhaps fortunately the commonest kind of alkaline aquatic systems.

Moderately alkaline lakes have similar fauna to normal freshwater lakes at similar latitudes, but usually with greatly reduced species diversity. For example, there are just 16 species of chironomid ("bloodworm") flies in the Neusiedler See in Austria, one of Europe's largest shallow alkaline lakes, compared with 40–50 in many other central European lakes. The chironomid biomass is also markedly reduced.

More extreme alkaline conditions occur in "soda lakes" in various parts of the world, especially in areas of rifting and volcanic activity. One of the best studied examples is Lake Magadi in Kenya (see Plate 6d, between pp. 386 and 387), routinely having a pH of 10 and a temperature of 30–42°C with a "salinity" equivalent to about 50–65% sea water. Pyramid Lake in Nevada, with a pH of 9.4, and Lake Van in Turkey (pH 9.8), are also homes to well-known alkaline-adapted fish. Certain alkaline crater lakes in Central America lack fish species but are dominated by chironomid larvae and oligochaetes; they are often warmed by peripheral "geysers" (see Plate 6e).

One of the more successful species from African alkaline lakes (and the only fish to survive in Lake Magadi) is *Oreochromis*

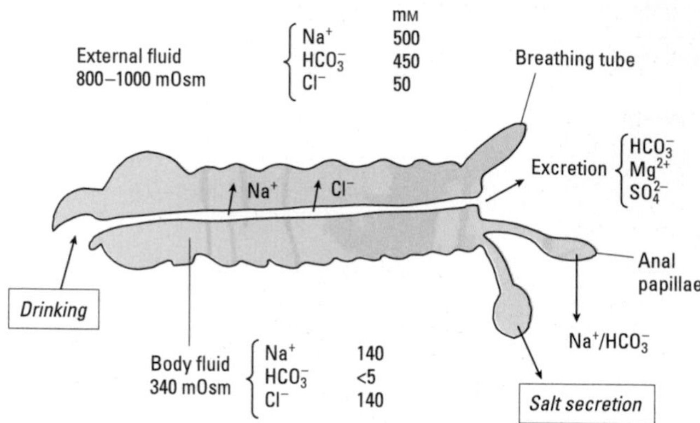

Fig. 14.9 Regulation of body fluids by *Aedes campestris* in strongly alkaline water. (Data from Phillips *et al.* 1971.)

grahami, a cichlid that excretes at least 90% of its nitrogen as urea so avoiding toxic ammonia accumulation. Trout from Pyramid Lake and the unusual tarek fish from Lake Van achieve a lesser degree of switching to ureotely, and tolerate relatively high levels of ammonia in their tissues.

Oreochromis suffers acidosis and ion balance disruption if transferred to water of normal pH. This species has apparently standard seawater-type chloride cells in its gills that can be activated or covered over by pavement cells as in their estuarine and freshwater relatives (see section 12.2.2), and which presumably carry out "salt" excretion as in seawater teleosts, with carbonate/bicarbonate replacing chloride in the main outward transport system. Exposure to nonalkaline waters tends to deactivate these cells in the short term (2–3 h), although this is reversed within about 1 day of acclimation. However, in the other alkaline lake fish, exposed to relatively dilute media, the gill chloride cells are probably more like the freshwater teleost type (see Chapter 13).

Various species of mosquito larvae are also found in soda lakes, with *Aedes campestris* (Fig. 14.9) being typical. Here again there is excretion of a hyperosmotic urine, with HCO_3^- accumulated from the environment replacing the normal chloride.

As with other osmotically peculiar habitats, the raised salt concentration in these lakes tends to produce respiratory problems as well as osmotic stress. Typically there are low oxygen concentrations, and this may explain the relatively thin diffusion barrier in the gills of alkaline lake fish, rather than the thickened protective gill surface that might have been expected. In fact Lake Magadi is highly oxygenated during the day due to cyanobacterial photosynthetic activity, but becomes nearly anoxic at night as bacterial respiration dominates, so *Oreochromis* fish have to cope with extraordinary diurnal variation in oxygen availability. Hence at night they commonly show surface skimming behavior, with mouths agape, to give passive ventilation of the gills and air bladder with relatively aerated water.

Lake Natron in East Africa is an extreme example of a large soda lake, fed by hot springs from nearby volcanoes and having a solid soda crust. This particular lake has almost no signs of life, except that it is used by lesser and greater flamingos for breeding, and the young are hatched on raised nesting platforms above the almost crystalline soda. The fledglings rapidly die of heat exhaustion if they fall off the nest; but so long as they stay on the nest platform they survive, fed on regurgitated crop "milk" from the parents, which has been desalted from their own food. It is presumed that the flamingos migrate in from other lakes such as Nakuru to breed at Lake Natron because there is no predation pressure on the fledglings.

Acidic waters

Acid waters occur as lakes, but also in semiterrestrial conditions such as peat bogs, and even in some manmade situations such as vinegars. All of these may support animal life, with varying degrees of specialization to regulate the organisms' acid–base balance in extreme H^+ surroundings. We looked at some of the consequences of anthropogenic acidification in Chapter 13, and the deleterious effects on species not adapted to cope with low pH. It follows that the same problems must have been solved by acid-water residents. The adaptations needed can therefore be summarized as follows:

1 Control of the physiological effects of low pH on ion uptake and internal acid–base balance.

2 Tolerance or detoxification of solubilized minerals, of which aluminum is most important, aluminum poisoning causing mucus production and clogging of the gills, etc. (with iron and copper also potentially important depending on the local geology).

3 Limitation of the effects on physiological processes, such as cuticle formation in insects and crustaceans, which would otherwise reduce osmoregulatory abilities.

4 Changes to the control of hatching in many invertebrates, and especially in vertebrate eggs, because the enzyme chorionase, which allows the larva to escape from the chorionic membrane, can normally only function fully at a pH of 6.5–8.5.

Adaptation of key membrane and enzyme parameters must underlie most of these issues, with proteins functioning under unusual ionic conditions. Many of the acid-water dwellers appear to have slightly to moderately acidic blood, representing some degree of hydrogen ion regulation at the gills, kidney, or gut that will limit the stress on the cellular constituents. The crustacean *Allanaspides*, from Tasmanian peat bogs with a pH as low as 3–4, can keep its blood at neutrality (pH 7), which represents very considerable hydrogen ion regulatory powers. There is probably net H^+ secretion at the gills, and also at the "neck organ", by modified Na^+/H^+ exchange processes.

14.3.4 Living in oils

In certain parts of the world natural waters become readily contaminated by seepage from underground oil deposits, and the resultant "hydrocarbon seeps" still support an array of metazoan life. The fluid in these seeps may contain dissolved sulfides and volatile potentially toxic hydrocarbon compounds, but also provide an enriched source of carbon and nitrogen. Nematodes are particularly well represented, and their numbers often increase with oil content, whereas other taxa such as polychaetes, oligochaetes, gammarids, and copepods tend to reduce in number in the more oil-rich seeps. A peculiarly resistant cuticle with a dense inert hyaline layer, and a pronounced tolerance of anoxia, are assumed to contribute to the success of nematodes in such strange habitats.

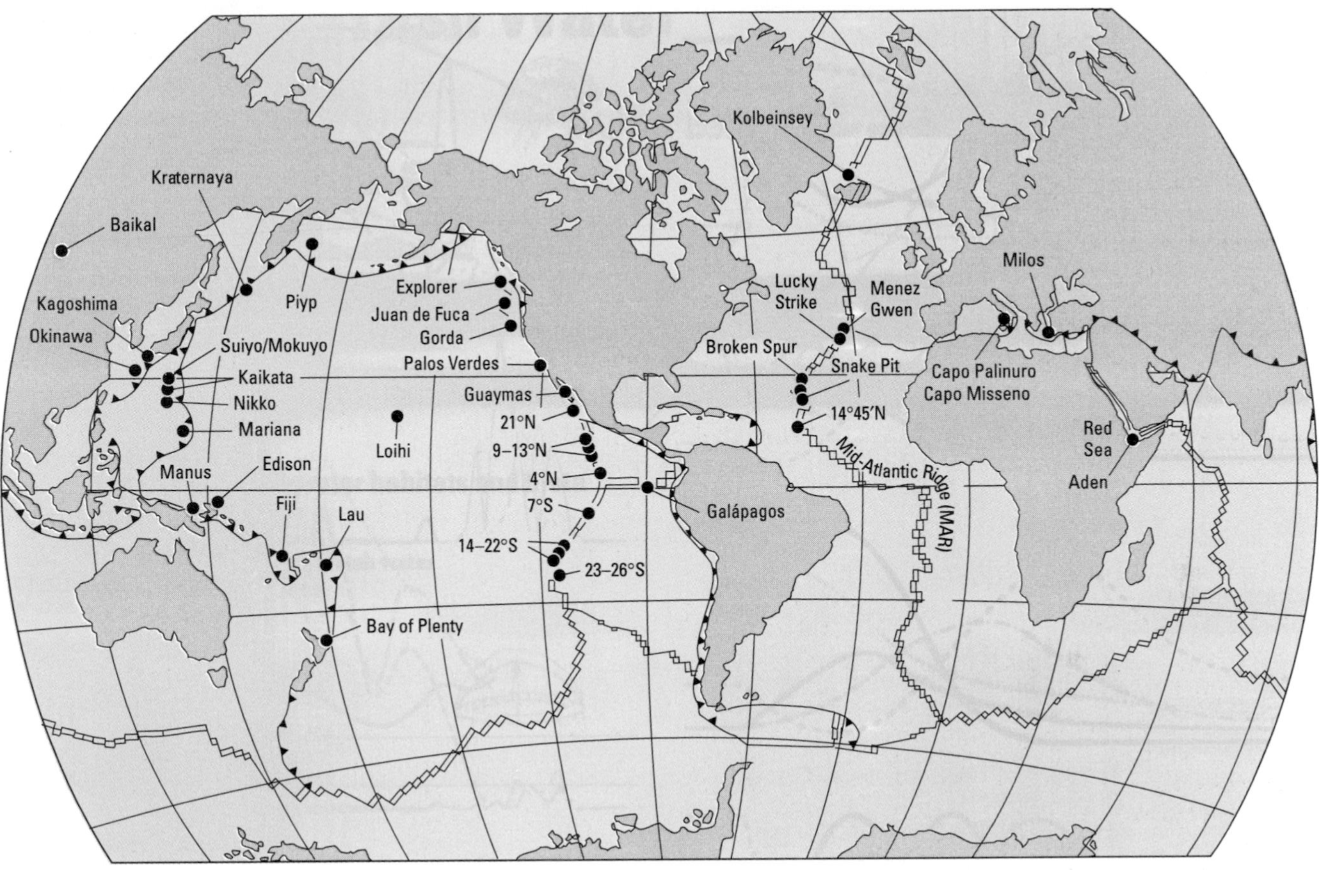

Fig. 14.10 Distribution of hydrothermal vents, primarily around mid-ocean ridges. (From Desbruyeres & Segonzac 1997, copyright Pierre Chevaldonné, IFREMER.)

14.4 Thermally extreme waters

14.4.1 Deep-sea thermal vents

Throughout the world's ocean depths, where tectonic plates meet, a mid-ocean ridge is formed from material rising up from the mantle as the continental plates move apart. Sea water seeps downwards through numerous crevices and fractures and becomes heated by the hot mantle magma. The superheated pressurized water (up to 350°C) then forces up through the crust to emerge as "hydrothermal vents" (Fig. 14.10). The water is rich in silicates, hydrogen sulphide, and various sulfide minerals incorporating iron and magnesium, which crystallize as the hot water meets the cold oceanic water, forming "black smokers", "chimneys", and other mineral deposits. At the typical depth of 2000–3000 m the ambient water is at about 2°C, with an ammonium concentration of less than 0.5 μM, a nitrate concentration of 40 μM, silicate at 150 μM, and no hydrogen sulfide. In contrast the rising vent water contains zero oxygen due to reaction with high sulfide levels, ammonium concentrations can be high, and there may be natural radioactivity. In the zone of mixing, conditions become such that life is possible. Prokaryotes can occur in water almost up to boiling point (which at depth may mean 120°C); the enzymes in such organisms ("extremophiles") are highly specialized, and stabilized with many S–S bonds. However, eukaryotes do not appear until the water is cooled to 55°C and below, while metazoan animals generally need temperatures below 50°C. In practice, most animal life is associated with areas where the vents interact with the surrounding water to produce temperatures below 30°C and hydrogen sulfide concentrations of less than 400 μM. Vent animals are particularly abundant in and on the "lava" fields that surround the vents, immediately above which the water temperatures may still be 12°C or more; their very presence alters the mineralogy of the chimney structures.

Discovered in 1977, this fauna (Fig. 14.11) is dominated by annelid and pogonophoran tubeworms and bivalve molluscs, together with some crustaceans. Some 250–300 species of vent macrofauna have been identified, of which 200 or more are new to science. The fauna is endemic and taxonomically diverse. Some vent faunas are also very ancient relative to the surrounding deep sea.

Some vent structures pose an additional problem since they are very unstable, showing major changes in structure and water flow over weeks or months. *In situ* recordings have shown that the temperature and water chemistry at sites where tubeworms and other fauna live are also highly variable as the waters are not well mixed. The vents at 13°N in the east Pacific typically last for only a few years. Furthermore, vent fields are often separated by tens to hundreds of kilometers of "normal" ocean. Thus the composition of the fauna at particular vents is not fixed, but may be predictably dictated by their age and site in relation to recruitment and larval dispersal.

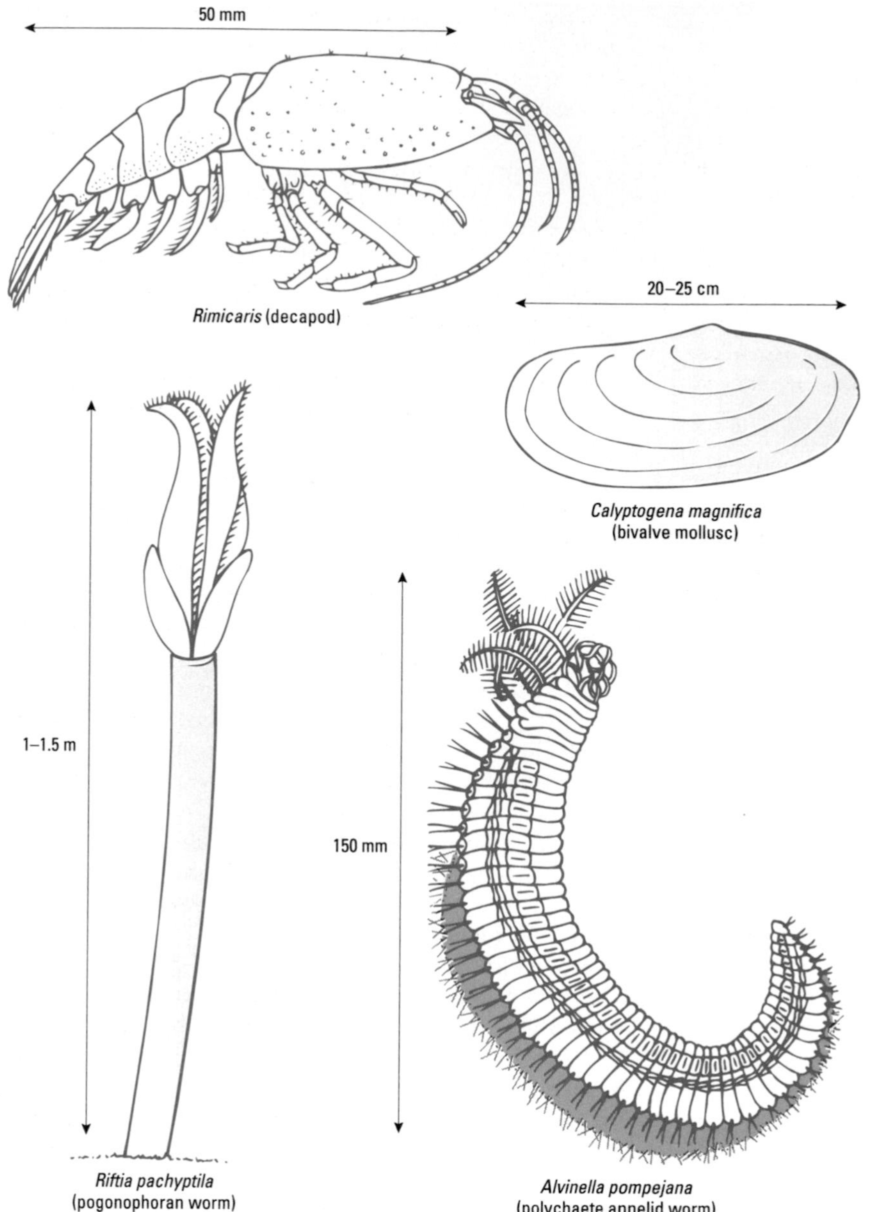

Fig. 14.11 Fauna found at deep hydrothermal vents in the oceans.

Coping with high sea temperatures and pressures

Vent animals live in temperatures that may range from nearly ambient (2°C) to 50°C and above. One of the most spectacular cases is the aptly named "Pompeii worm" (*Alvinella pompejana*), a tube-dwelling polychaete that lives on vent chimneys. The water in its tubes has been recorded at 68°C, with occasional peaks up to 81°C. At the tube mouth the water is only around 22°C, giving a possible gradient of 60°C along the length of the worm's body. This makes *Alvinella* potentially the most eurythermal animal on record. As yet we have no real information on the temperature of its tissues, or whether the proteins and membranes of these worms can function under such conditions, although there are clear differences in the thermostabilities of some of the key enzymes in different species.

Detecting high temperatures may also be an issue for more mobile animals in these habitats. The deep-sea shrimp *Rimicaris exoculata*, which occurs around vents, lacks normal eyes on its head but has a thoracic "eye" adapted for vision in very low light that is probably mainly used to detect the dim black-body radiation emitted by the very hot vent water emissions.

High pressures also create problems, with pressures of around 30 MPa being commonplace at the vents of the Pacific Ridge. Studies of pressure effects on vent crabs reveal, as expected from physical principles, a reduced level of damage when high pressures are accompanied by high temperatures (the former causing reduced membrane fluidity, the latter increasing this fluidity). Thus effects on key enzymes and on membrane bilayers are relatively mild when the conditions at the vent are reproduced in laboratory studies.

Nutrition and respiration: coping with hydrogen sulfide

At the Galápagos hydrothermal vents, hydrogen sulfide concentrations vary from 0 to 300 μM. Hydrogen sulfide can only accumulate in those environments that are low in oxygen, as it is readily oxidized. Sulfide is potentially toxic, binding to the heme site of cytochrome *c* enzymes to inhibit cellular respiration. Sulfide can also potentially reduce the disulfide bridges in proteins, inhibiting normal protein function. Passive resistance to sulfide (having insensitive enzymes) appears to be rare, and the tubes or mucous layers of the fauna provide no real protection, so most animals living in high sulfide environments have one or more detoxification mechanisms. In addition, nutritional specializations are required to cope with the environment, which is dominated by sulfide rather than by carbon and oxygen.

One possibility is to use sulfide-binding proteins, as found in the pogonophoran tubeworm, *Riftia pachyptila* (see Plate 6f, between pp. 386 and 387), and the hydrothermal vent clam, *Calyptogena magnifica*. These proteins draw free sulfide into the general circulation where it is strongly bound, and thus achieve three important functions:

1 The preservation of aerobic respiration.
2 The prevention of sulfide precipitation in the blood (sulfide granules might impede circulation).
3 The transport of sulfide to internal centers where symbionts' enzymes can exploit it for energy production.

These blood-borne binding proteins have probably evolved separately in each group. The sulfide-binding protein in *R. pachyptila* is the hemoglobin itself, present in high concentration in both vascular and coelomic blood; it can bind oxygen and sulfide at different sites. The same is true for the polychaete *Paralvinella*. In contrast, in the clam *C. magnifica* the hemoglobin occurs in erythrocytes and the sulfide-binding protein occurs quite separately in the serum.

Many of these animals are also symbiotic with bacterial chemoautotrophs that use a sulfur-based metabolism and release carbon compounds to their hosts. The animals provide oxygen and carbon dioxide from the sea water, as well as the necessary dissolved sulfides. Representatives from many phyla incorporate specialized structures containing sulfide-oxidizing bacteria. The nematode, *Eubostrichys dianeie*, houses sulfur-oxidizing bacteria in a mucus web over the exterior of its body. The worm, *A. pompejana*, maintains an epithelial "fur" of fine projections densely packed with sulfur-containing filamentous bacteria. Bivalves have specific intracellular (endosymbiotic) strains of sulfur bacteria housed in their gills in modified cells called bacteriocytes; gastropods also have bacteriocytes in their gills. In pogonophoran worms, sulfide oxidation occurs in the "trophosome", an internal organ that fills much of the coelomic cavity and contains extremely high concentrations of sulfide-oxidizing bacteria. It is worth looking at the strategies of some of these animals in more detail.

BIVALVES

Vesicomyid bivalves live in many deep-sea reducing environments, including hydrothermal vents, hydrocarbon seeps, and anoxic sediments. In the vent clam, *Calyptogena magnifica*, a mouth and gut are present although the feeding and digestive apparatus are reduced. The enlarged gills contain sulfur-oxidizing chemolithoautotrophic bacteria that provide the major energy source of the clams and may average 17% of the animals' wet weight. *C. magnifica* lives on rocks and extends its highly vascularized foot down into cracks, where the hydrothermal fluids rich in hydrogen sulfide (up to 40 μM) are slowly venting. Sulfide appears to be taken up by the foot and is transported bound to the serum sulfide-binding factor via the bloodstream to the gills. The clam's serum can reversibly bind up to 8 mM sulfide, resulting in a significant concentration above ambient seawater levels. The serum-binding component in *Calyptogena elongata* has a large molecular mass and contains Zn^{2+} at the active site.

Any unbound sulfide is oxidized to the nontoxic compound thiosulfate in the foot and transported to the gills. The animal therefore has moderate concentrations of circulating thiosulfate but virtually no free sulfide. This suggests that in these high-sulfide environments, the clams have evolved the ability to oxidize sulfide to thiosulfate for themselves, and that the incorporation in the gills of chemoautotrophic bacteria that utilize thiosulfate is a secondary adaptation to exploit what would otherwise have been a waste product. The gill bacteriocytes are exposed to the sea water on one side and the blood of the clam on the other, so that nutrients can be obtained from both sources. The siphon of the clam extends into the ambient water such that the animal effectively forms a bridge between the reducing environment of the substrate and the oxygenated sea water. Some of the energy released during sulfur metabolism in the bacteria is trapped as ATP and NADPH, which are used in part to drive net CO_2 fixation. The clams then digest a fraction of the bacteria to satisfy their nutritional needs. As a consequence, the gut can be fairly degenerate (an adaptation found in many of these vent animals).

Other vesicomyid clams, such as *C. elongata*, which lives on muddy substrates in other parts of the deep sea, appear to have a broadly similar type of symbiosis, with the foot penetrating into the anoxic sulfide-rich sediments and the siphon extending into the water above. Mussels with chemoautotrophic bacteria are also common on hydrothermal vents and in the sulfide-rich oozes. They are much less specialized for symbiosis than the vesicomyid clams, in that they have retained the ability to ingest particulate matter for themselves, and have no significant concentrations of sulfide binding in their gills and hemolymph. The mussel symbionts utilize thiosulfate rather than sulfide. Some species do have methanotrophic (instead of sulfur-oxidizing) symbionts, but must still deal with the potentially toxic sulfide.

POGONOPHORAN WORMS

These have a very different strategy for sulfide metabolism. Most of their sulfide oxidation occurs in the "trophosome". The worms lack a mouth and gut, and the plume of tentacles serves as the primary site of exchange of gases and solutes with the ambient sea water. The abundant hemoglobin (up to 26% wet mass of the worm) in the vascular and coelomic spaces carries both oxygen and sulfide, as we saw earlier, and the sulfide is delivered to the trophosome from the hemoglobin. The bacteria are then able to fix CO_2 by involving the enzyme ribulose 1,5-biphosphate carboxylase, providing reduced carbon compounds, which in *Riftia* sp. supply almost 100% of the worm's requirements. The bacterial symbionts in the pogonophoran trophosome have several adaptations to protect the animal from sulfide toxicity. They have a V_{max} for sulfide that is at least an order of magnitude greater than in free-living sulfur-oxidizing bacteria, and they can shunt excess sulfide into nontoxic elemental sulfur deposits. Elemental sulfur (which can constitute up to 20% dry weight of the trophosome) is further oxidized when blood sulfide levels decrease. The symbionts utilize only sulfide (*not* thiosulfate, as in many other vent animals), and the worm is highly specialized to minimize the interaction of its own metabolism with sulfide. The use of hemoglobin as the sulfide-binding protein enables the worm to concentrate sulfide from the medium by almost two orders of magnitude. However, because the hemoglobin binds sulfide, and the bacteria use it up, the free concentration of sulfide in the hemolymph is an order of magnitude below external H_2S concentrations. The Hb has a very high affinity for oxygen (P_{50} = 0.1–0.3 Torr at pH 7, equivalent to about 2 mM).

Recent evidence suggests that the bacteria in *Riftia pachyptila* can use nitrate in addition to oxygen, producing ammonia and nitrite as end-products. The presence of sulfide stimulates nitrate respiration up to 500 μM sulfide, whereas thiosulfate has no effect, again supporting the idea that the symbionts are sulfide specialists. Nitrate respiration increases with decreasing oxygen concentrations. The use of nitrate respiration presumably enables the symbionts to function in a very low-oxygen environment whilst still gaining energy through respiratory pathways.

CRUSTACEANS

The crab *Bythograea thermydron* occurs at hydrothermal vents, and shows high tolerance for hydrogen sulfide and for raised temperatures. Like the molluscs described above, it has been shown to oxidize H_2S to thiosulfate, this time in the hepatopancreas. In this species the thiosulfate has an additional adaptive function in increasing the affinity of the respiratory pigment (hemocyanin) for oxygen. In contrast, the vent copepod *Benthoxynus* contains significant levels of hemoglobin with a very high oxygen affinity, suited to the low-oxygen environment where some O_2 acquisition is useful to support a minimal level of aerobic respiration.

Coping with unusual minerals

Several of the common vent animals exhibit bioaccumulation of unusual elements in their tissues (notably cadmium, copper, and in some cases zinc), with levels fluctuating as the vent emissions change. These elements are stored as mineral compounds or bound with insoluble ligands of high chemical stability. Some are also detoxified as more soluble metallothioneins (low molecular weight proteins), whose half-life varies according to the associated cations.

Coping with reproduction and dispersal

Hydrothermal vents are widely separated and often rather short-lived, so that vent organisms face regular bouts of extinction unless they can produce effective dispersive propagules. These may be either larvae or motile adults, which can be carried by seafloor currents supplemented by their own swimming activities. There is little doubt that dispersion is effective and rapid—new vents have been shown to be colonized within a year of their formation. It remains unclear how the colonizers find the vents though. Recent evidence indicates that sulfide itself may serve as an attractant, at least to highly motile vent shrimps. Late larval stages of these shrimps have unusual deposits of lipid in their thorax and abdomen, comprising 75–82% wax esters, these same compounds being absent in the adults. These reserves presumably provide the food reserve for a relatively prolonged bathypelagic existence in which opportunities for planktotrophy are rare. A similar explanation may underlie an observation of "giant" gametes in vent bivalves, the enlarged oocytes having extra yolk to allow an extended lecithotrophic stage.

14.4.2 Hot springs and thermal ponds

Hot springs occur due to geothermal action in terrestrial habitats in areas of volcanic activity. Ground water becomes superheated and forced to the surface through cracks, emerging as steam or as nearly boiling water. In its passage through the laval rocks the water becomes very modified in composition, often heavily charged with hydrogen sulfide, carbonates, or silicates. Springs may produce their own outflow streams, or may feed into other natural streams, part of whose flow system is then heavily affected.

Close to the geothermal source only bacteria survive, but downstream of the discharge point algal mats appear and are colonized by specialist protozoans (at about 60°C). As the waters cool below 50°C, some ostracods and larval flies may colonize, and inevitably some nematodes and rotifers appear, albeit with low species diversity. In many hot-spring effluents midge larvae are also early colonizers, with chironomids and ceratopogonids becoming abundant, the latter being especially thermophilic at some Colorado hot springs.

Where geothermal water enters otherwise "normal" streams, some freshwater animals may drift into the heated and potentially sulfurous regions. Some caddis flies in Californian streams show raised upper critical temperatures (UCT values) in such zones, though unable to penetrate the most strongly affected waters.

Anthropogenic activities also create a new kind of hot thermal environment for animals to exploit in the form of thermal ponds near industrial and nuclear plants. Inevitably, a range of similar highly dispersive opportunist species colonize these. In some areas existing ponds have become heated, and here some fish and invertebrates have proved able to adapt to near-lethal temperatures. The mosquito fish, *Gambusia*, occurs in some thermal ponds, and within about 60 generations has acquired a much higher thermal tolerance (raised UCT), with increased heterozygosity and an apparent heritability as high as 32% for this raised lethal temperature.

14.4.3 Sea-ice dwellers

Polar sea ice

Although polar ice may look completely abiotic, it has fairly recently been appreciated that within the seasonal ice fields there is a maze of fine channels that house a surprisingly complex ecosystem. This is dominated by single-celled algae, but there are also some protozoans and bacteria, and an assortment of animals: amphipods, copepods, nematodes, and flatworms in particular.

The ice is colonized by its specialist biota as it forms in the polar fall, gradually spreading from its summer minimum of 4 million km^2 to its peak of 20 million km^2 around Antarctica. Small ice crystals mix with plankton, which become caught up in the "frazil" ice as it forms and rises above the sea surface by up to perhaps 2 m. The plankton-containing crystals form "ice pancakes", stacked on top of each other. As ice is pure water, the sea salt is concentrated in the intervening water, which becomes very hypersaline. Most of the trapped plankton die, and larger animals are crushed by the ice, but the small-bodied, very cold hardy, and euryhaline species can survive, effectively having a habitat that switches from nearly solid to liquid as the annual temperature cycle progresses.

In fact there are separate distinct habitats within the ice. At the upper surface all the water is frozen and tightly packed together at a temperature as low as −15°C; the channels have brine at a concentration at least seven times that of sea water (250 ppt) and little can survive. But lower down the ice sheets contain more abundant pockets and channels, filled with concentrated very cold brine (perhaps −1.5°C) at about 2–4 times the concentration of sea water; these channels range from a few micrometers to several centimeters across. Here algae are reasonably abundant, accumulating K^+ as an osmolyte and producing proline and dimethylsulfoniopropionate (DMSP) as antifreezes. Lower still the sea permeates the ice and it is even more pitted and channeled, though the population of algae is lower as light penetration is much reduced.

In the middle and lower reaches of the ice floes, zooplankton can make a living. Harpacticoid copepods may occur at densities of up to 30 per liter in the lower ice channels, with nematodes also abundant. The occurrence of flatworms, with no protective cuticle, is rather more surprising and their physiological strategies need investigating. These sea ice communities also provide food to the animals in the water beneath the ice, with krill (euphausid shrimps) surviving the winter almost entirely by grazing on the biota at the under surface of sea ice.

In spring as the ice melts, the sea ice communities are released and seed a huge bloom of algal production in the seas around the ice fringe, again affecting the larger marine organisms. The spring surge of diatoms and algae is also a mixed blessing on a larger scale; the organisms release bromine gas, which causes ozone depletion, and dimethyl sulfide (from DMSP), which promotes cloud formation and may help reduce global warming.

Seafloor methane ice

In some parts of the world lumps of "ice" form under the conditions of low temperature and high pressure that occur in the marine benthos. In fact the ice is a mixture of frozen water, methane (CH_4), and other hydrocarbons, often termed "methane hydrate". Lumps of this rock-hard methane ice material are not uncommon in the Gulf of Mexico at depths of 600–800 m and temperatures around 6°C. They form in the relatively anoxic sediment, and can then emerge to lie just above the sea floor. They are extremely unstable, and can evaporate into bubbles almost explosively if the temperature rises even slightly—some believe that these upwelling methane bubbles underlie the "Bermuda Triangle" phenomenon!

Recent undersea explorations have shown that even these inhospitable habitats are teeming with life—most commonly a species of polychaete now named *Hesiocaeca methanicola*. The worms appear to "burrow" into the ice surface by gently wafting water currents at it with their parapodia. They then survive with the assistance of bacteria that colonize their burrows (perhaps finding conditions there particularly amenable) and that can metabolize the methane hydrate into usable organic molecules. As yet it is uncertain whether the worms colonize the "ice" before or after it emerges from the deoxygenated benthic sludge, but it seems likely that coping with hypoxia is also part of their physiological repertoire.

FURTHER READING

Books

Ashcroft, F.M. (2000) *Life at the Extremes. The Science of Survival.* Harper Collins, London.

Desbruyeres, D. & Segonzac, M. (eds) (1997) *Handbook of Deep-Sea Hydrothermal Vent Fauna.* IFREMER, Plouzane, France.

Van Dover, C.L. (2000) *The Ecology of Deep Sea Thermal Vents.* Princeton University Press, Princeton, MA.

Wharton, D. (2002) *Life at the Limits.* Cambridge University Press, Cambridge, UK.

Reviews and scientific papers

Airriess, C.N. & Childress, J.J. (1994) Homeoviscous properties implicated by the interactive effects of pressure and temperature on the hydrothermal vent crab *Bythograea thermodron. Biological Bulletin* **187**, 208–214.

Brauner, C.J., Ballantyne, C.L., Randall, D.J. & Val, A.L. (1995) Air breathing in the armoured catfish as an adaptation to hypoxic, acidic and hydrogen sulphide rich waters. *Canadian Journal of Zoology* **73**, 739–744.

Cary, S.C., Shank, T. & Stein, J. (1998) Worms bask in extreme temperatures. *Nature* **391**, 545–546.

Cavanaugh, C.M. (1983) Symbiotic autotrophic bacteria in marine environments from sulfide-rich habitats. *Science* **302**, 332–333.

Cosson, R.P. (1996) Bioaccumulation of mineral elements within the vestimentiferan tube worm *Riftia pachyptila*—a review. *Oceanologica Acta* **19**, 163–176.

Cosson, R.P. (1997) Adaptations developed by hydrothermal vent organisms to face the stress of heavy metals. *Bulletin du Societe de Zoologie de la France* **122**, 109–126.

Crowe, J.H., Carpenter, J.F. & Crowe, L.M. (1998) The role of vitrification in anhydrobiosis. *Annual Review of Physiology* **60**, 73–103.

Crowe, J.H., Hoekstra, F.A. & Crowe, L.M. (1992) Anhydrobiosis. *Annual Review of Physiology* **54**, 579–599.

DeWachter, N., Blust, R. & Decleir, W. (1992) Oxygen bioavailability and haemoglobins in the brine shrimp *Artemia franciscana. Marine Biology* **113**, 193–200.

Ellis, B.A. & Morris, S. (1995) Effects of extreme pH on the physiology of the Australian yabby (*Cherax destructor*). *Journal of Experimental Biology* **198**, 409–418.

Grieshaber, M.K. & Volkel, S. (1998) Animal adaptations for tolerance and exploitation of poisonous sulfide. *Annual Review of Physiology* **60**, 33–53.

Grueber, W.B. & Bradley, T.J. (1994) The evolution of increased salinity tolerance in larvae of *Aedes* mosquitoes: a phylogenetic analysis. *Physiological Zoology* **67**, 566–579.

Juniper, S.K. & Martineau, P. (1995) Alvinellids and sulfides at hydrothermal vents of the eastern Pacific—a review. *American Zoologist* **35**, 174–185.

Laurent, P., Maine, J.N., Bergman, H.L., Narahara, A., Walsh, P.J. & Wood, C.M. (1995) Gill structure of a fish from an alkaline lake: effects of a short-term exposure to neutral conditions. *Canadian Journal of Zoology* **73**, 1170–1181.

Marshall, A.T., Kyriakou, P., Cooper, P.D., Coy, R. & Wright, A. (1995) Osmolality of rectal fluid from two species of osmoregulating brine fly larvae (Diptera: Ephydridae). *Journal of Insect Physiology* **41**, 413–418.

McLachlan, A. & Ladle, R. (2001) Life in the puddle: behavioural and life cycle adaptations in the Diptera of tropical rain pools. *Biological Reviews* **76**, 377–388.

Narahara, A., Bergman, H.L., Laurent, P., Maia, J.N., Walsh, P.J. & Wood, C.M. (1996) Respiratory physiology of the Lake Magadi tilapia (*Oreochromis grahami*), a fish adapted to a hot, alkaline and frequently hypoxic environment. *Physiological Zoology* **69**, 1114–1136.

Nelson, J.A. & Mitchell, G.S. (1992) Blood chemistry response to acid exposure in yellow perch (*Perca flavescens*): comparison of populations from naturally acidic and neutral environments. *Physiological Zoology* **65**, 493–514.

Pond, D., Dixon, D. & Sargent, J. (1997) Wax-ester reserves facilitate dispersal of hydrothermal vent shrimps. *Marine Ecology Progress Series* **146**, 289–290.

Randall, D.J., Wood, C.M., Perry, S.F. *et al.* (1989) Urea excretion as a strategy for survival in a fish living in a very alkaline environment. *Nature* **337**, 165–166.

Scott, K.M. & Fisher, C.R. (1995) Physiological ecology of sulfide metabolism in hydrothermal vent and cold seep vesicomyid clams and vestimentiferan tube worms. *American Zoologist* **35**, 102–111.

Somme, L. (1996) Anhydrobiosis and cold tolerance in tardigrades. *European Journal of Entomology* **93**, 349–357.

Steichen, D.J., Holbrook, S.J. & Osenberg, C.W. (1996) Distribution and abundance of benthic and demersal macrofauna within a natural hydrocarbon seep. *Marine Ecology Progress Series* **138**, 71–82.

Sun, W.Q. & Leopold, A.C. (1997) Cytoplasmic vitrification and survival of anhydrobiotic organisms. *Comparative Biochemistry & Physiology A* **117**, 327–333.

Tunnicliffe, V. (1992) Hydrothermal vent communities of the deep sea. *American Scientist* **80**, 336–349.

Tunnicliffe, V. & Fowler, C.M.R. (1996) Influence of sea-floor spreading on the global hydrothermal vent fauna. *Nature* **379**, 531–533.

Wharton, D.A. (1996) Water loss and morphological changes during desiccation of the anhydrobiotic nematode *Ditylenchus dipsaci*. *Journal of Experimental Biology* **199**, 1085–1093.

Wilke, M.P. & Wood, C.M. (1996) The adaptations of fish to extremely alkaline environments. *Comparative Biochemistry & Physiology B* **113**, 665–673.

Yesaki, T.Y. & Iwama, G.K. (1992) Survival, acid–base regulation, ion regulation and ammonia excretion in rainbow trout in highly alkaline hard water. *Physiological Zoology* **65**, 763–787.

Zierenberg, R.A., Adams, M.W.W. & Arp, A.J. (2001) Life in extreme environments: hydrothermal vents. *Proceedings of the National Academy of Science USA* **97**, 12961–12962.

15 Terrestrial Life

15.1 Introduction

The earliest organisms to colonize land were probably prokaryotic cyanobacteria, living in fine silt deposits and screes even back into the Precambrian; certainly more than 550 million years ago (mya), and perhaps in excess of 800 mya. These "primitive" microorganisms still survive today, as cyanobacterial mats, often in association with terrestrial algae and higher plants. Plants were probably present on land by the Ordovician period, but only begin to appear as fossil deposits from the Silurian (about 430 mya) when they had become fairly widespread. We can be reasonably sure that animal (metazoan) life on land did not predate the appearance of at least some of these multicellular land plants. This is because the plants must first have produced decayed matter to serve as food and as shelter, and once they had a moderately upright growth form they would also give a certain amelioration of conditions at the ground surface, with shade, reduced wind speeds, and a higher, more stable humidity. The presence of burrow traces in Ordovician soils suggests that invertebrates were established by that time, but again fossils are lacking. A range of small myriapod fossils (forerunners of modern millipedes) do appear in the late Silurian, and wingless insects (Apterygota) and some arachnids are present soon after this.

It is nearly impossible to determine what were the first "real" land animals though. This is because of the complex intergrading of terrestrial habitats with neighboring semiaquatic systems. Certainly there would have been ample scope for gradually acquiring characters more appropriate to terrestrial life in a range of small worms, molluscs, and arthropodan groups that inhabited the littoral zones, the freshwater marshes, and the interstitial spaces of moist fringing soils through the early Palaeozoic. As we will see, the true land environment, where animals essentially live in air rather than in water, presents so many challenges to physiology and lifestyle that successful invasion must have depended on gradually increasing adaptation through a series of intermediate environments. No animal group could have achieved all the necessary modifications of form and function to make the transition to land quickly. However, many animal groups *have* achieved some or all of the necessary adaptations to become partially or fully terrestrial. Some groups did it during the early spurt of terrestrialization that accompanied movements of plant groups onto land in the Silurian, Devonian, and Carboniferous. A further range of animals invaded the land later on, when the flowering (angiosperm) plants radiated (perhaps originating in the Cretaceous, around 120 mya), supplanting many types of conifer and seed fern, and opening up new niches for seed eaters, pollinators, and specialist coevolved herbivores. It is probably no coincidence that these two key phases of land invasion occurred in periods of equable climate across most of the world.

In this chapter we cover the essential strategies of the broad range of animals that live in the majority of terrestrial habitats, particularly in temperate zones and the humid tropics where thermal extremes are rarely encountered and where water balance, though difficult to achieve, is not pushed to the limits for survival. In such habitats, large endothermic animals are more or less always in their thermoneutral zone, and humidities are rarely so low, or food and drinking water so scarce, as to create grave water loss problems. Small animals have a huge variety of microclimates to choose from to extend their active periods.

However, there may still be some severe short-term problems. Most of these areas do have significant daily temperature fluctuations, so that small vertebrate endotherms may get chilling problems at night, and tend to exhibit "heterothermy", as also do large insects (see Chapter 8). There may also be transient droughts or transient waterlogging, so that good avoidance tactics, especially short-range evasive movements and migrations, are a good idea. For similar reasons resistant stages in the life cycle will often be required, especially by smaller animals. Only rarely is excess heat a severe problem in the wet tropics, although a neat example of problematic "wet heat" does occur in the hot caves of the Neotropics, favored as roosts by several species of bats, where relative humidity (RH) exceeds 90% and ambient temperature (T_a) levels of up to 40°C year round give rise to basal metabolic rates (BMRs) in the bats that are substantially lower than expected (only 48–66% of the predicted levels). On the whole, though, the inhabitants of tropical areas can be regarded as showing "normal" terrestrial adaptations.

Chapter 16 will deal with some special cases of terrestrial life: hot arid deserts, where the hygrothermal endurance limits of animal residents may be severely tested; polar regions, tundra, and northern coniferous forests, where extreme cold is superimposed on the generality of terrestrial problems; and montane habitats, where altitude effects may parallel the latitudinal effects at the poles.

Table 15.1 Properties of water and air, and of aquatic and terrestrial environments.

Property	Water/aquatic	Air/terrestrial	Approximate ratio of water : air
Density (g ml^{-1})	1.00	0.0012	850
Viscosity (kg m^{-1} s^{-1})	1.00	0.02	50
Thermal capacity (J ml^{-1} $°C^{-1}$)	1.00	0.0003	3300
Velocity of sound (m s^{-1})	1485	343	4.33
Refractive index	1.33	1.00	1.33
Oxygen content (ml l^{-1})	4–7	210	1/30
Oxygen diffusion ratio			1/300,000
Carbon dioxide content (ml l^{-1})	0.4	46	1/115
Salts	Freely available	Not directly available	
Water	Abundant, but may be osmotically unavailable	Rare, always hard to find and keep	
Food for plants	Only at surface	At all levels	
Food for animals	Throughout, all mechanisms possible	Throughout, but often hard to catch or eat	

15.1.1 Problems and advantages of life on land

For any animal, there are a series of distinct problems and equally distinct advantages of life on land and thus in air. Many of these arise directly from the different physical and chemical properties of air and of water (see Table 15.1).

Firstly, water has approximately 1000 times greater **density** than air, and provides a supportive fluid for all the pelagic organisms that float in it. Support of the body (itself essentially watery) is much harder in all terrestrial habitats, and the problems of self-weight and coping with gravity really arise for the first time. The composition and design of skeletons and structures concerned with motility therefore become more critical, and more energy must be expended in maintaining the extra skeletal weight. However, water is also much more **viscous** than air (approximately 50-fold), so that on land the medium actually impedes movement to a much smaller degree and makes acceleration considerably easier. Consequently, on land a specifically adapted animal can achieve faster speeds and an altogether quicker approach to life. The fastest speeds achieved by any animal in water are about 10–20 m s^{-1}, maintained for only very short bursts, whereas animals running on land can achieve around 25 m s^{-1} in short bursts (66 mph, recorded in a cheetah, is 29 m s^{-1}) and steady speeds of 15 m s^{-1}. Birds may fly in bursts at up to 40 m s^{-1} and steadily at 20 m s^{-1}. Flying is clearly the fastest option, and in fact, given the enormous numerical dominance of insects, it would be fair to say that the majority of land animals have adopted flying as their main way of getting around.

This effect of a speeded-up lifestyle is accentuated by the differing **thermal** properties of air. Its thermal capacity is 3000 times less than that of water, so that temperatures may change very much more rapidly and on a much finer spatial scale. Terrestrial environmental temperatures also reach substantially higher levels, as the air heats up from direct and reflected sunlight during the daylight hours, so that terrestrial ambient temperatures can exceed 50°C. Since temperature affects so many other aspects of metabolism and physiological activity (see Chapters 7–8), we have an added reason why life often proceeds much faster on land. Equally though, polar land habitats can be substantially colder than the neighboring seas, with no mass of slowly circulating ocean to act as a heat sink; the air over the South Pole can reach temperatures below −70°C towards the end of the long Antarctic winter when the sun has not been above the horizon for nearly 6 months. This variability of temperature links to a generally much greater instability and unpredictability of terrestrial environments. A great many factors vary much more widely and more quickly for a land animal, so that a faster response and broader sensitivity are likely to be needed.

Two other changing physical properties then become relevant. Watery media have a 4–5-fold higher **velocity of sound** conduction, and most aquatic animals can perceive only low-frequency stimuli, with only the marine mammals using extensive sound communications systems. By contrast, land animals perceive higher frequency noises, and have a much more widespread and sophisticated capacity for sound production, tying in with their rapid lifestyle. Water also has a higher **refractive index** (RI) than air, with terrestrial eyes needing more refraction from the cornea to supplement the lens. Whilst this difference in RI affects visual perception across air–water interfaces markedly, it has only limited effects on the visual systems of animals living entirely within one medium.

Then there are differences in the **chemical properties** of air and water, as solvents and as carriers of other molecules. Most obviously, the oxygen that makes up a substantial part (21%) of the atmosphere has only a limited solubility in water, with seas and fresh waters containing only around 4–7 ml l^{-1} compared to about 210 ml l^{-1} of oxygen in air, at least a 30-fold difference. The reduced density of air compared to water also means that the diffusion of gases is very much faster in air, so that in practice oxygen can be supplied to static (nonventilated) respiring surfaces about 300,000 times faster in air than in water. When ventilation is also taken into account, being much more difficult in viscous water than in air, the business of supplying oxygen to an animal's body is clearly much easier in terrestrial habitats. This again makes a difference to the rate at which aerobic metabolism (and hence many other linked processes, including locomotion) can proceed in the two media, even in the absence of sophisticated respiratory and circulatory systems.

Carbon dioxide is often rather abundant in sea water, though largely in the form of carbonic acid (HCO_3^- ions), the total concentration being about 46 ml l^{-1}; whereas in clean fresh waters and in air it is at quite low concentrations (about 0.3 ml l^{-1}). Simple diffusive loss is nearly always adequate for living organisms in any of these media. However, on land most animals regulate the CO_2 levels in the intermediary circulatory fluid, as this is important in controlling ventilation (see Chapter 7). The availabilities of oxygen and carbon dioxide, together with the much more limited support provided by air for complex respiratory structures, have an enormous impact on the design of gas-exchange surfaces in terrestrial animals.

However, the most profound differences in chemical properties between aquatic and terrestrial habitats clearly reside in the **water and salt contents** of the media. Salts are at least theoretically readily available in solution in all watery environments (unless they are exceptionally fresh). On land there are normally no salts in solution, and they must be obtained from food, often requiring complex dietary choices or specific behaviors such as salt-licking or lapping at excreta. But of course those same salts may easily diffuse out of the body of any animal living in an aquatic habit, whilst they are fairly readily retained by terrestrial animals. Similarly, water may seem abundant in all aquatic habitats, but as has repeatedly been seen in earlier chapters it may be osmotically unavailable where the animal's own fluids are markedly hyposmotic (e.g. vertebrates in sea water). Or it may cause problems by a tendency to enter too readily if the animal's tissues are hyperosmotic. On land, for all animals, water loss is always inevitable, but rates may be variable as the ambient water vapor pressure (RH) varies; only when the RH is over 99% does a terrestrial animal come close to achieving water equilibrium.

As a general rule then, for land animals both salts and water have a continuous tendency to leak from the body, and so must normally be regulated by careful control of skin, respiratory, and excretory surfaces. The smaller the animal, the more critical the control of these regulatory surfaces becomes, making land life for young stages particularly hazardous (and thus entraining a whole complex series of reproductive adaptations). This salt and water regulation could in theory be paid for easily by the greater availability of oxygen, but in practice gaining oxygen always means losing water, so the benefit is at best minimal. Ultimately salt and water balance must be maintained by eating and drinking.

This raises a final category of differences between aquatic and terrestrial habitats relating to feeding habits and community **trophic structures**, linked to the different character of photosynthetic life in the two habitats. In the oceans, where the medium is in constant motion, plants are small and usually short-lived (phytoplankton and benthic macroalgae), whereas on land, with a relatively stable soil to provide anchorage and a degree of longer term stability, plants may be both larger and more persistent, with specialized multicellular structures both below and above ground level. This leads directly to a difference in the type of organic compound that predominates in plant bodies: in the seas proteins are all important, allowing rapid growth but little energy storage, whereas on land carbohydrates predominate, allowing a slower growing organism but one with the ability to store energy over several seasons, decades, or even centuries, and with large deposits of purely structural materials. This dichotomy is to a large degree also reflected in the chemistry of the animals of the two environments, and overall there are reckoned to be about 14,000 times greater reserves of carbohydrates on land than in the seas. There may also be a less efficient conversion of carbohydrate to protein by land animals than by their marine counterparts, partly because of the energy requirements of land animals (especially the high proportion of terrestrial birds and mammals); but this is partly offset by the very rapid conversions achieved in terrestrial soils by bacteria and fungi.

There are also inevitably differences in the variety of feeding systems that are practicable in the different media. In the seas nutrients are kept on the move by currents but show a gradual tendency to sink, so that the surface waters where most animals live may become depleted of food, while the benthic zones benefit from a rain of food that can be garnered by a sessile animal. On land the same inevitable effects of gravity mean that foods (of both plant and animal origin) tend to accumulate at the soil surface, which is just where most animals make their living. However, land animals cannot usually adopt a sessile strategy, waiting for their food to come along and be filtered out; the fluid low-density air cannot support a suspended biomass, with only bacteria, very small spores, and the small resistant egg stages of a few animals occurring. Thus the air does not really provide a supply of food to "filterers" (the obvious exception to this being the extraordinary sophistication shown by orb-web spiders as aerial filter feeders on flying insects).

Instead, food must generally be actively sought. The layer of dead and decaying detritus is the one easy food source, and probably the starting point for many animal groups in their transition to land. True herbivory (macroherbivory) on land becomes a much more difficult way of life than in water, as land plants are very much tougher (reflecting their own need for support), rather high in carbohydrate, and nutritionally not very well balanced as a diet for animals. However, while herbivorous grazing is rare in aquatic communities, terrestrial grazing is a major way of life for many groups, despite the poor quality of food obtained and the presence of strong antiherbivore defenses (see section 15.8.1). At the same time carnivory may be harder on land because passive sit-and-wait strategies (e.g. suspension-feeding carnivory) are rare, and specific targets must always be involved; most of the prey items are also likely to be moving and responding faster.

Overall, then, the properties of air mean that the successful evolution of animal life on land needs a more specific adaptation of the skeletal framework and motile systems; more physiological control of hygrothermal balance, with an overriding need for integration of water balance, temperature balance, and respiration at all stages of the inevitably more specialized life cycle; and sensory, neurological, and behavioral adaptations to permit a more active mode of life for animals generally.

15.1.2 Terrestrial climates and microclimates

The terrestrial environment provides a balance of solids, water, and air that is highly amenable to plant growth, with an almost unimpeded supply of light for photosynthesis. The resultant pattern of **primary productivity** over the planet's surface (Fig. 15.1) therefore coincides closely with the supply of light, heat, and water; in other words, with meteorological patterns. Ecosystems are generally classified largely in terms of their floral components (Fig. 15.2), giving categories such as tundra, grassland, tropical rain forest, or desert. These patterns of climatically determined plant communities inevitably affect animal distributions, resulting in biomes with

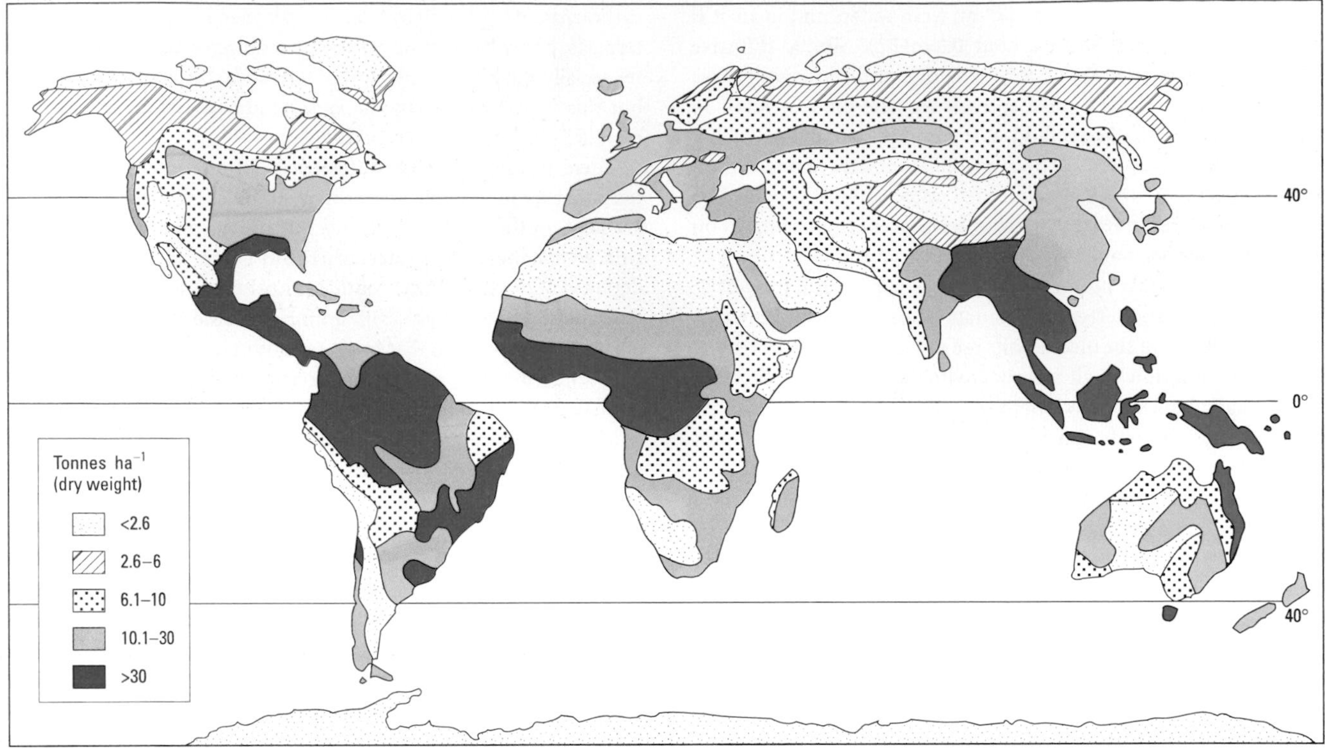

Fig. 15.1 Patterns of primary productivity across the land masses.

distinctive kinds of fauna repeated in different geographic locations around the globe.

In addition, terrestrial **weather** has obvious direct effects on animals' physiology and hence on their behavior, ecology, and distribution. Terrestrial temperatures and rainfall regimes are dramatically variable. Maximum temperatures of 58°C have been recorded, with a record minimum of −89°C. At the same time rainfall may vary from effectively zero in deserts that are in rain-shadow areas, to above 3000 mm year^{-1} in mountain ranges within the tropical rain forests.

The average global surface temperature is only about 9°C, taking all habitats, all seasons, and all latitudes into account. However, a more apt figure for terrestrial habitats is a mean ground level temperature of 15°C. There are somewhat higher averages around coasts at all latitudes than in the center of continents, due to the thermally stable seas giving a "maritime" rather than a "continental" weather pattern. However, terrestrial temperatures and climates obviously vary greatly at any one place, largely in relation to all the cyclical patterns outlined in the Introduction to Part 3 (p. 389).

For most animals with short life cycles it is diurnal temperature change that has the greatest impact on their existence. The most dramatic diurnal changes in terrestrial temperatures, irrespective of habitat, are found close to the ground. This is because during the day solar energy warms this surface faster than the energy can be transmitted downwards or reradiated away; while during the night the surface loss of infrared radiation is faster than the replacement of heat up through the soil, causing rapid cooling. Since air has a low thermal conductance, heat from the ground will not easily be conducted away from the irradiated surface in the absence of convection, giving rise to steep thermal gradients. Such considerations, plus the presence of vegetation, make terrestrial habitats very varied and complex, particularly for very small animals. Thus it is always more appropriate to talk in terms of **microclimate**, defining the conditions actually experienced by the animal of interest on an appropriate spatial and temporal scale, rather than the macroclimate that would be experienced by a human (with the head some 2 m above the ground—the height at which our routine meteorological instruments tend to be sited). Most animals live on or in the ground, or closely associated with vegetation emerging from (and intimately connected with) the ground. Microclimatic conditions can be very different 50 mm above a ground surface, where a rodent's body might be, than just 5 mm above that same surface, where an insect or worm might rest. Similarly, conditions 5 mm above the surface may vary from second to second with small-scale air movements and the shifting of shadows, whereas larger animals, with their main body mass and their receptors much further above the ground, will experience a much more stable environment through time.

Effects of soil

The soil provides an important refuge from temperature extremes in terrestrial animals; at depths of 0.2 m all diurnal temperature change is smoothed out in most soil types. Soils are a mixture of minerals derived from the underlying bedrock (and/or by erosion and deposition from other rocks), together with dead organic material (humus). The temperature regime of a particular soil is largely controlled by the thermal conductivity, thermal capacity, and

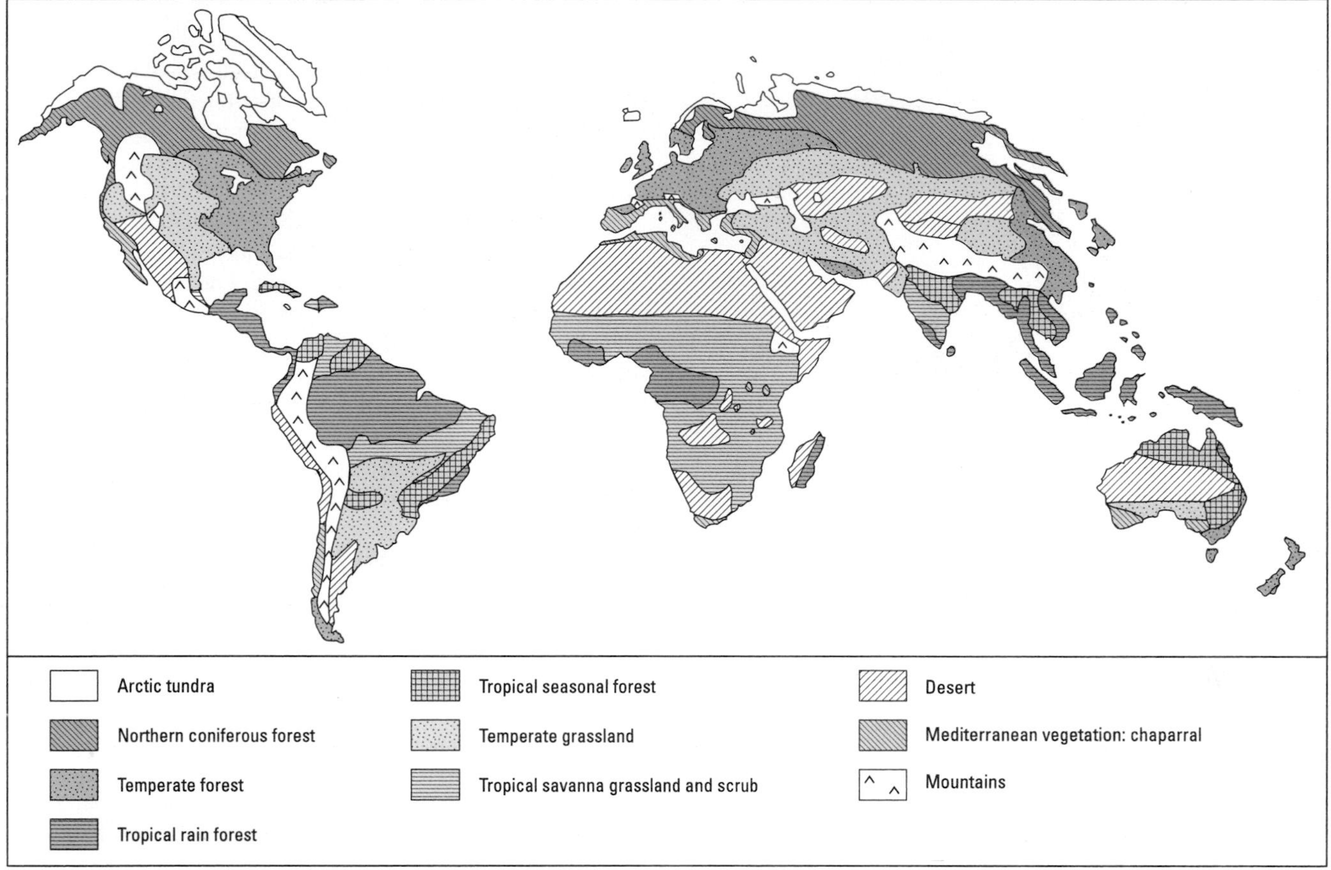

Fig. 15.2 Occurrence of the major terrestrial biomes.

diffusivity of its particular mixture of components (Table 15.2). For dry sandy soils **conductivity** may be five times greater than for a dry peat, so the peaty soils heat up much more quickly at their surfaces. However, conductivity alone is misleading, since the presence of water (rather than air) in the soil interstices affects the rate at which soil temperature changes. The most useful measure is therefore the soil **diffusivity**, which indicates more exactly how a particular soil responds to a thermal input, by measuring the time taken for heat to travel within that soil. The lowest values of diffusivity occur for peat soils, whether wet or dry, and the highest values are for saturated sands. Sands and clays permit heat to penetrate quite rapidly, but peats allow only shallow heat penetration and their surfaces are consequently subject to extremes of temperature.

Table 15.2 Properties of terrestrial soils.

	Conductivity (W m^{-1} °C^{-1})	Diffusivity (m s^{-1} × 10^{-6})
Soil type		
Sandy		
Dry	0.3	0.24
Wet		0.74
Clay		
Dry		0.18
Wet		0.51
Peat		
Dry	0.06	0.10
Wet		0.12
Water at 10°C		0.14
Air at 10°C		20.50

However, soils also vary in their ability to hold on to water, which in turn affects gaseous diffusion rates. Wet sands have an open texture and can be dried out rather easily; wet clays have a high soil moisture tension and dry out only slowly; wet peats dry fairly easily but can be difficult to re-wet. Thus the distribution of animals in soils is not solely governed by thermal considerations, and for those animals that rely on soil moisture—either to give a high or stable burrow humidity, or as a source of drinking water—assessment of soil moisture retention may override thermal diffusivity.

Examples of temperature and humidity profiles in soils, and in burrows within soils, are shown in Fig. 15.3. Burrowing into the relatively stable soil layer is an exceptionally useful way to escape, at least temporarily, from many of the problems of life on land. Many land animals spend a considerable part of the day in burrows: this time may coincide with, for example, the hottest and driest hours in summer, with the animal only coming out onto the surface at dawn, dusk, or at night; or alternatively in winter the coldest hours may be spent in the burrow, with the animal emerging only in the few warm hours around midday. Whilst in the burrow, the metabolic rate may be reduced or at least may be more constant, with the animal often resting in conditions that are particularly favorable for digestive functions.

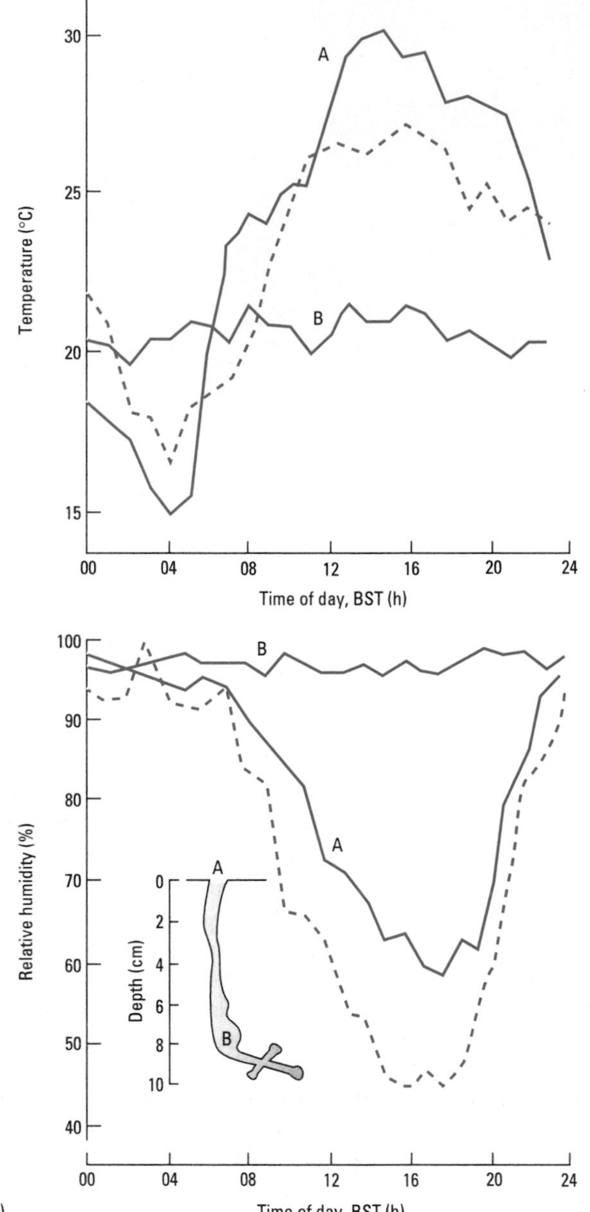

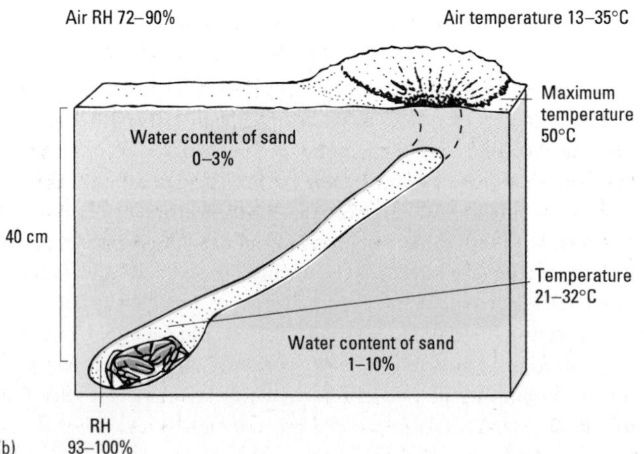

Fig. 15.3 (*left*) Profiles of temperature and relative humidity (RH) or water content in terrestrial soils and in burrows. (a) The burrow of a solitary wasp (*Cerceris*) in sandy soils. Dashed lines show ambient conditions. BST, British Summer Time. (b) The burrow of the land crab *Gecarcinus* in the upper beach zone. (a, From Willmer 1982; b, from Bliss 1979.)

Effects of vegetation

Wherever life occurs, the environment changes, and often in ways that make local conditions more suitable for further life. This is particularly true on land, where plants create ecospaces for animals, and both then become habitats for a huge range of symbionts and parasites.

Plant life becomes the focus of **energy exchanges**, raising the main exchange site above the ground surface. Wherever there is vegetation, the plants reduce the amount of radiation reaching the ground below, and temperature changes at the soil surface and within the plant layer tend to be much reduced. The main temperature gradient occurs instead at the top of the plant "canopy", whether this is a few millimeters or many tens of meters above the ground (Fig. 15.4a). For all animals living in burrows or in the litter layer, then, plant cover gives a great amelioration of conditions, and a new range of animals can live on and amongst the plants themselves, in a reasonably equable microclimate.

Plants also increase the **humidity** above the ground, however. This effect is quite complex, arising partly from the retention of rain water on the plant surfaces (plants actually reduce the amount of precipitation reaching the ground by anything up to 50%), and partly from the slower evaporative and diffusive dispersion of this precipitation brought about by reduced wind speeds within the vegetation. But most importantly, this moisture is continually added to by evaporation and transpiration from the plants themselves, via their stomata, giving a regulation of the water content within the vegetation. In effect, the vegetated area becomes a thick boundary layer of relatively still, stable, moist air.

More drastic effects of plants on the microclimate surrounding them and in the soil below can also be seen. In dry climates where water is scarce plants may deplete the soil moisture around their roots. In wetter climates the opposite effect may occur: plants delay the arrival of rain onto the ground, causing it to drip steadily through rather than arriving in bursts, so that the run-off from the ground is much reduced and water is more likely to be absorbed into the soil. This effect is especially striking in tropical forests, where run-off is very low and almost no soil erosion occurs. When such an area is deforested, sudden storms can erode the soil very quickly and surrounding areas may experience flash flooding from the run-off.

Plants also produce very local stable microclimates within their own structures due to their own transpiration, with distinct regimes created either side of a leaf (Fig. 15.4b) or within buds and open flowers (Fig. 15.4c); all these zones are exploited by small terrestrial animals.

Effects of animals

Life modifies and very often improves its own environment. Not only do plants modify ecospaces for animals, but the animal itself has effects on its own environment. In fact the presence of an animal in a small enclosed space cannot fail to modify the conditions, as the

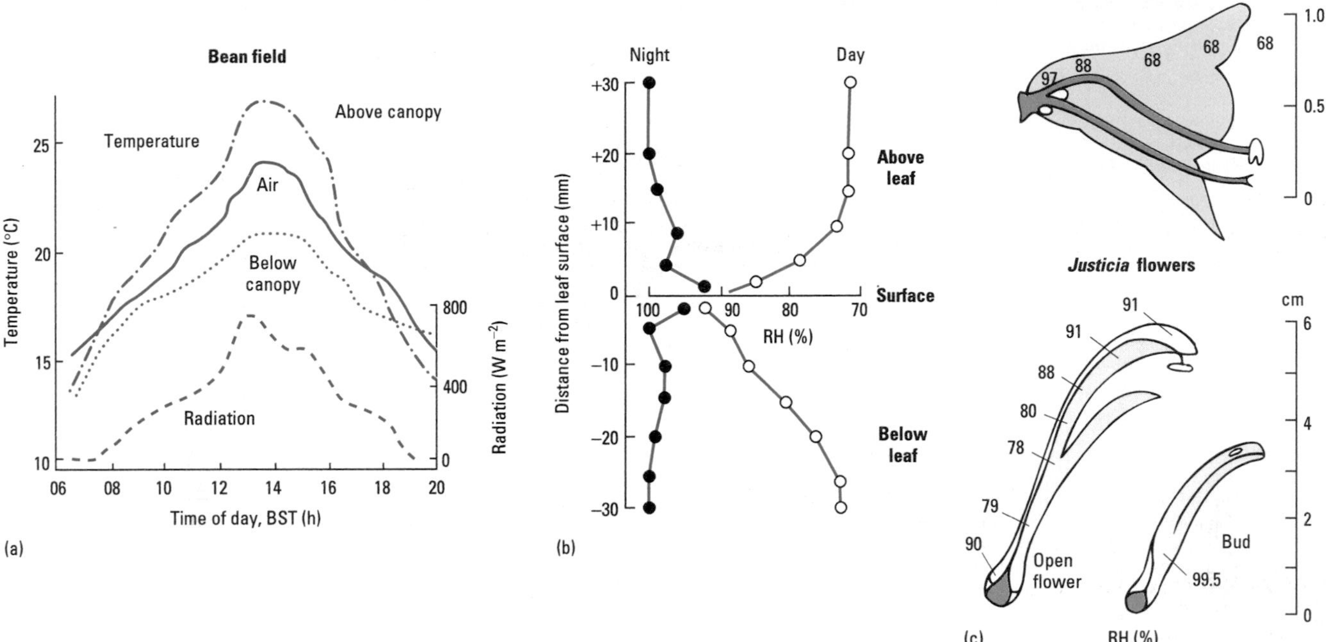

Fig. 15.4 Microclimates associated with plants: (a) the stabilized conditions below the plant canopy; (b) the microclimate around a single leaf in sunshine; and (c) the microclimate within flowers. BST, British Summer Time; RH, relative humidity. (b & c, From Willmer 1982.)

animal emits both heat and water vapor, and may also produce excreta that will equilibrate with the surrounding air. The nest tunnel of a solitary bee is both warmer and slightly more humid when the female bee is present and digging a new cell than when she is absent foraging. On a larger scale, occupied badger setts have higher levels of CO_2 (up to 0.6%) and lower O_2 levels (around 19.5%) than unoccupied setts. Enclosed bird nests also have their own microclimate, with humidity rising when the owners are present.

In social animals the effects can be more dramatic, with "social thermoregulation" occurring. Tropical stingless bees will use endothermic incubation to warm up the brood areas of their nest, and cover cooler areas with insulating secretions; overheating is also controlled, using accelerated and directed wing fanning. The nests of ants and termites provide more extreme examples of controlled architecture for stable microclimate.

15.1.3 Categories of terrestrial animals

"Terrestrial animals" is a rather unhelpful term to a physiologist. Animals that can be broadly considered to live on land cover a very wide range of lifestyles, and it is useful to distinguish between these at the outset. A taxonomic summary of the animals in these various groups is given in Table 15.3 as a reference point for this chapter.

Interstitial fauna

Small animals that are encountered in soils often experience a habitat that is really no different from that of a freshwater animal: they are normally permanently surrounded by a film of fresh water, with which they exchange ions and gases by osmosis and diffusion in a typically aquatic fashion. If the soil dries out too much they become inactive and may enter a cryptobiotic state (see Chapter 14) or may desiccate and die. In this category are all small soil organisms whose dimensions are such that the soil moisture will usually be adequate to give a film of moisture over their bodies; in particular it includes soil nematodes, oligochaetes, mites, and rotifers.

Cryptozoic fauna

Soil-dwelling organisms of a larger size will not experience a continuous covering of moisture, and though they may live in exactly the same places as the interstitial fauna described above, these larger animals are more properly terrestrial (e.g. earthworms that burrow in soil are terrestrial in many aspects of their physiology, and may risk drowning in very waterlogged soils). These soil animals, and all others that live in zones of continuously high humidity, are termed **cryptozoic** ("hidden animals"), rarely being seen "out in the air". The grouping includes animals from leaf litter and detritus, from burrows and boreholes inside trees and other plants, and from the debris in other animals' nests. Some worms, woodlice, centipedes, and many larval insects are familiar animals that belong in this category; so too (stretching a point) do various lizards and snakes, and mammals such as gophers and moles. Many of these "soil dwellers" have cylindrical or rather flattened bodies, and are often extremely flexible in three dimensions.

Hygrophilic fauna

There is a further group of animals that still need a water supply and high humidity for activity, but that can tolerate some degree of desiccation. They can best be described as **hygrophilic** ("wet-loving").

Table 15.3 Taxa of terrestrial animals.

Worms
"Planarians", or flatworms
- Cryptozoic, interstitial, in soils and damp litter
- Most in humid ecosystems such as tropical rain forest
- Some triclad species in temperate and Mediterranean biomes, under stones and logs
- Nocturnal, carnivorous on other worms and small molluscs
- Genera: *Bipalium, Rhynchodemus,* and *Microplana*

Nemertean worms, or ribbon worms
- Probably direct from the marine habitat
- Cryptozoic, humid places, supralittoral zone or in deep litter and among rotting wood
- Nocturnal predators
- Genera: *Geonemertes, Argonemertes*

Nematodes, or roundworms
- Strictly interstitial animals, living an essentially freshwater life
- Extraordinarily abundant in soils, with up to 20 million m^{-2} in humic woodland soils

Annelids
- Cryptozoic oligochaetes, commonly known as earthworms
  - Deposit-feeding burrowers
  - Genera: *Lumbricus.* Tropical and subtropical genera: *Lampito, Allolobophora,* also "giant earthworms", more xeric, family Megascolecidae: *Megascolides* and *Pheretima*
- Leeches in tropical forests, damp vegetation
  - Predatory or parasitic, usually on vertebrate hosts

Molluscs
Only from class Gastropoda
Order Prosobranchia: some Littorinacea (littoral route) and Architaenioglossa (freshwater route); persist only in tropical forests
- Order Pulmonata: variety of shell forms, closure by flaps and ridges; or shell loss (slugs)
- Hygrophiles, active at high humidity, rapid onset of estivation in drought
- Herbivores, with chitinous radula; or sometimes carnivorous
- Air-breathers with lungs
- Hermaphrodite with complex reproductive systems
- Genera: supratidal: *Ovatella* and *Pythia*; slugs and snails: *Helix, Pomatia, Arion, Limax* and in deserts *Sphincterochila*

Arthropodan groups
Crustacea
- Copepoda: few, interstitial, and cryptozoic in soils
- Amphipoda, family Talitridae: beach-living sandhoppers (*Orchestia, Talitrus*), plus true land dwellers only in the tropics and Japan (e.g. *Arcitalitrus*)
- Isopoda, suborder Oniscoidea (woodlice) from damp litter (*Oniscus*), grasslands, and soils (*Porcellio, Armadillidium*), and in desert burrows (*Venezillo, Hemilepistus*)
- Decapoda (crabs, lobsters, and prawns):
  - Anomurans (hermit crabs and their kin): e.g. coconut crab *Birgus* and *Coenobita*
  - Brachyuran ("true" crabs): Grapsidae, shingle shores, beaches, mangrove swamps; Ocypodidae, beach burrowers, *Ocypode* (fiddler crabs), *Uca* (ghost crabs); Mictyridae, soldier crabs; Gecarcinidae, land crabs, but return to water to breed

Chelicerata
- Predominantly terrestrial and xerophilic
- Mainly predatory on insects, using trapping and ambush techniques and venoms
- Araneae (true spiders) and Acari (mites and ticks). Both widespread
- Scorpiones (true scorpions), Uropygi, and Amblypygi (whip scorpions and whip spiders), and Solpugida (sun spiders)—all in warm and often dry habitats

Myriapoda
- Entirely terrestrial
- Chilopoda (centipedes): predators in litter and soils, sometimes in caves
- Diplopoda (millipedes): browsers or detritivores
- Also Pauropoda and Symphyla, tiny and specialist litter dwellers, cryptozoic

Onychophorans
- All terrestrial, genera *Peripatus* and *Peripatopsis*, "velvet worms" or "walking worms"
- Cryptozoic, moist forest habitats, often in rotting wood. Predatory

Insecta
- Wingless insects, Apterygota: litter and cryptozoic habitats
- Winged insects, Pterygota: all possible terrestrial niches (+ fresh water)
- 27–29 orders, nine include herbivores, five predominantly carnivores, parasites, parasitoids; also omnivores and detritivores
- Many orders into arid zones and true deserts, notably beetles (Coleoptera), ants and bees (Hymenoptera), termites (Isoptera), and grasshoppers and locusts (Orthoptera)

Vertebrates
Four classes of tetrapods:
- Amphibia, hygrophilic
- Reptilia (perhaps three separate classes of tortoises, crocodilians, and lizards + snakes)
- Aves
- Mammalia
- (These last three groups all mesophilic and xerophilic)

This would include snails and slugs, which are active and conspicuous in damp habitats and at night (**nocturnal**) or around dawn and dusk (**crepuscular** for dusk activity, **eocrepuscular** if both dawn and dusk are used). Larger animals with permeable skins, such as amphibians, also belong here. Most of these hygrophilic animals have strategies for survival through water shortages. In temperate and arid zone species this may involve diapause or estivation (summer hibernation); in cold habitats it may involve torpor when all water is frozen. Again some of the smaller species use the more extreme form of suspended animation known as cryptobiosis (see Chapter 14).

Xerophilic fauna

The final category of terrestrial animals is the most familiar one, and includes all those animals that can be active in dry conditions. These are usually described as **xerophilic** ("dry-loving"), although many do not survive in real drought conditions and might be better termed **mesophilic**, implying a preference for moderate conditions. These are the animals we see above ground and active in all the sorts of terrestrial places where humans live, tolerating the macroclimate of the terrestrial world. The most notable examples are the insects, arachnids, reptiles, birds, and mammals.

15.1.4 Littoral and freshwater routes onto land

There are two major alternative routes for the evolution of life on land, and both have certainly occurred several times in different groups. Terrestrial life could arise from animals from the seashore and rock pools moving via the splash zone and onto land from an essentially **marine** source; these animals would be endowed with the kinds of adaptation discussed in Chapter 12. Furthermore, terrestrial life could equally easily arise from animals living in **fresh water**, via the marginal swamps and bogs (see Plate 7a,b, between pp. 386 and 387) that inevitably appear as freshwater bodies shrink and dry up, and with the specifically freshwater adaptations outlined in Chapter 13.

In general, invertebrate animals with a marine littoral evolutionary background could be said to do better as "true" land animals (i.e. as the xerophilic types discussed above), with a higher ability to tolerate desiccation and osmotic variation. Freshwater groups usually have an ability to produce dilute urine (not helpful to a real land animal) and so they do well as partially terrestrial hygrophiles. Vertebrates could be seen as an exception—the one group that has done well terrestrially from an essentially freshwater background, and which has evolved new ways of coping with the physiological stresses on land.

There is probably a third route onto land, taken by some small soft-bodied worms and perhaps by the insects. This involves an **interstitial** habitat, living entirely within a sediment, where there is buffering against changing salinity and temperature. Sediments can readily serve as a transition zone between aquatic and land life, whether the water within the sediment is salty or fresh, and such habitats probably require less tolerance of osmotic change in their occupants. Any animals adapted to life in sediments are likely to acquire mechanisms for regulating their salt and water balance, and it may be impossible to decide whether this fauna—including the interstitial and cryptozoic animals—moved via intertidal sands or via freshwater muds.

15.1.5 Terrestrial biomes

It is fairly conventional to divide up the terrestrial habitat into the nine "biomes" shown in Fig. 15.2, roughly determined by their positions along intersecting temperature and rainfall axes (Fig. 15.5). Many other categorizations exist, but this system serves reasonably well in a review of the physiological consequences for animals. There are three major categories of biome that might be considered representative of "normal" or nonextreme terrestrial life, supporting animals with "generalist" terrestrial adaptations. These are:

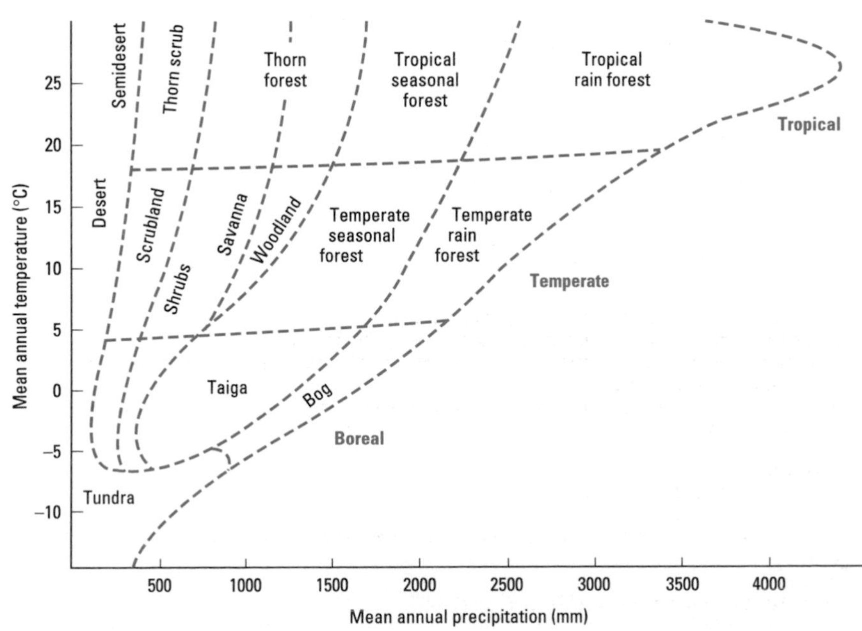

Fig. 15.5 Temperature and rainfall as determinants of the major terrestrial biomes. (Adapted from Holdridge 1947.)

(i) temperate (usually deciduous) forests (see Plate 7c); (ii) temperate and semitropical grasslands (steppe, pampas, savanna, and prairie) (see Plate 7d,e) and the not dissimilar scrub forests and "Mediterranean" zones (see Plate 7f); and (iii) tropical seasonal and monsoon tropical rain forests (see Plate 8a,b, between pp. 386 and 387). Each has certain repeatable features in terms of climate, flora, and fauna, regardless of which continent is considered.

Temperate forests

Forests once covered much of the temperate land mass of the planet, in a wide belt at latitudes between the boreal coniferous forests and the temperate and subtropical grasslands. However, they have become increasingly restricted over the last 10,000 years largely through human activities, and in many countries less than 1% of their original area remains. Temperate deciduous forests are rather varied in appearance throughout the globe, with trees from many taxa, shrubs, and bulbs, and a dense litter layer. Within such forests, small deer, bears, wild boar, squirrels, chipmunks, rabbits, mice, and shrews may thrive, with a great abundance of passerine birds, woodpeckers, and owls. Ectothermic vertebrates and many invertebrate groups are also strongly represented in association with the continually renewed supply of dead and dying wood and with the litter; there is also an enormous pterygote insect and molluscan fauna.

Temperate grasslands and scrublands

The grasslands include the steppes of Eurasia, the prairies in North America, the pampas and paramo (see Plate 8c) in Argentina, some parts of Australia, and the savanna and veldt of East and South Africa (see Plate 8d). These kinds of habitats may have been partly brought about by human activities (felling and burning forests) over the last few thousand years, and thus may not be really "natural" at all (though elephants and mammoths may have felled woodland naturally in prehuman times). Today the grasslands tend to exist as flat or rolling landscapes, and most experience periodic drought when the dry soil is eroded by wind. Since they are often sited towards the middle of large land masses, the grasslands can be fairly extreme in terms of climate, with hot summers and cold winters, but the reasonably continuous perennial vegetation provides amelioration of conditions for a wide range of animals. Soils are usually quite shallow and moderately acidic in character. The vegetation is of grasses with many bulbous flowering plants, often with scattered stands of relatively small spiny trees. Grazers are inevitably abundant, with larger animals dominated by equids (horses and zebra), antelopes, and huge numbers of colonial rodents such as marmots (prairie dogs) and gophers; elephants, rhinoceros, buffalo, or bison may be less numerous but important in biomass terms. There is frequently a large flightless bird, also evolved to fill the grazing niche: ostriches, rheas, and emus occur on different continents. Birds of prey are usually the commonest large carnivores, together with various felids and canids (the cat and dog families). Reptiles and amphibians are fairly sparse. Insects are not hugely diverse, although grasshoppers are common and ants and termites can be incredibly abundant, the ants using grass seeds as food. There is often an ant-eating mammal to exploit these: edentates, pangolins, aardvarks, and echidnas have all evolved convergently in the grasslands of different continents. A very large proportion of the fauna is actually based underground, thereby limiting the seasonal temperature stress.

Towards coastlines, and in areas around 40° latitude, especially where human influence has been longstanding, the grasslands are commonly replaced by what may be termed "Mediterranean" habitats, as on the European and North African coasts of the Mediterranean Sea (see Plate 7f), in California and the western USA ("chaparral" communities), and in parts of South America and Australia. These are characterized by hot summers and mild wet winters, and often have thin and rocky soils. Many of the original grassland animals remain, but the fauna may also become dominated by rodents, small lizards, and snakes, and a vast range of insects of which orthopterans (grasshoppers, crickets, and cicadas) are often especially conspicuous.

Tropical rain forest

Where rainfall exceeds 200 cm year^{-1} and is evenly distributed, evergreen forest results (see Plate 8a,b); in areas where there is some variation in rainfall patterns the forest is more seasonal. Mean monthly temperatures may be as high as 24–28°C. Such tropical forests are often said to be more speciose and of higher productivity than any other communities on Earth, although the reality may be more complicated than this. Tropical forest soils are diverse in character on different continents and also within large land masses such as Amazonia. Sometimes they are deep but well-weathered soils, forming clays; but where rainfall is particularly high, and where forests lie over sandstone, the soils may get severe weathering and only a thin layer of organic soil persists over meters of whitened quartz.

Tropical forests are characterized by a pronounced vertical stratification of vegetation from the tallest trees (>50 m) down to woody shrubs and large herbs and creepers (up to 1000 species of plant per hectare in total). This automatically creates a huge range of habitats; yet the animal biomass in rain forests, at least on the visible scale, is surprisingly low. Many herbivores and frugivores are arboreal and consequently small; primates are especially effective in this role. The main large forest floor dwellers are pigs, bovids, and insectivores. There are few large predators, the main ones being felids (tigers, jaguars, and leopards). However, birds and insects are incredibly abundant throughout the three-dimensional architecture of the forest, as also are reptiles and amphibians. All the marginally terrestrial hygrophilic and cryptozoic litter invertebrates are well represented, though the lack of a deep litter layer in many forests keeps their biomass in check. The preponderance of invertebrates and amphibians underlines the lack of physiological problems; the "soft-skinned" animals do well.

15.1.6 The success of animals in land habitats

Different groups of animals have very different patterns of success in the various available terrestrial habitats. Quite how these patterns appear depends on whether "success" is viewed in terms of species diversity, numerical abundance, or total biomass. With species diversity it is difficult to disentangle ecosystem effect from latitudinal

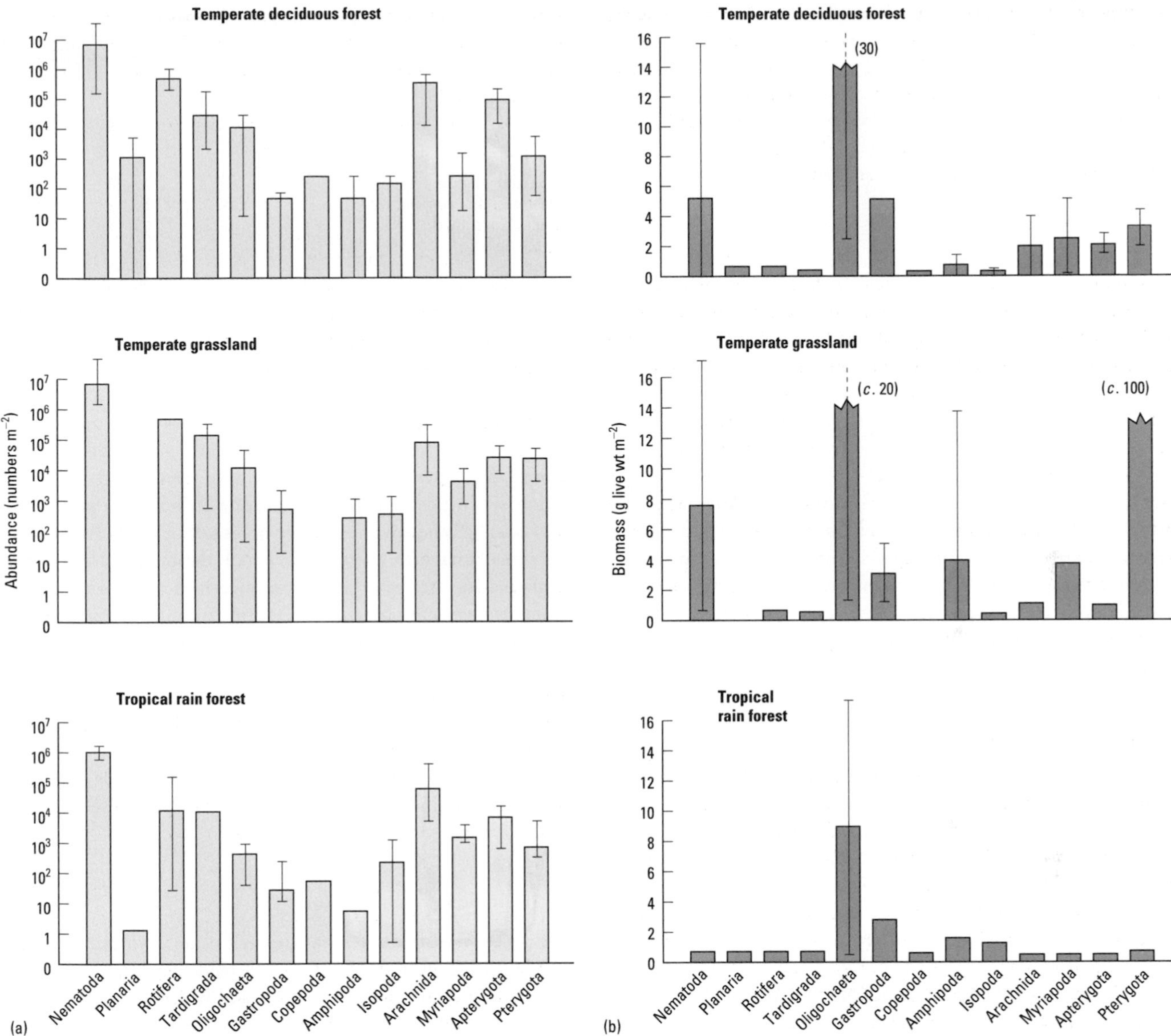

Fig. 15.6 (a) Abundance and (b) biomass of soil animals in temperate and tropical biomes; note that earthworms and winged insects may go off the scale in terms of biomass. Error bars are ±1 standard error; numbers above bars show off-scale values. (Adapted from Little 1990, courtesy of Cambridge University Press.)

effect, with species number in virtually all taxa decreasing towards the poles; clearer habitat effects can be seen with abundance and biomass.

On the basis of the relative abundances of the major terrestrial animal groups in the soils of temperate habitats, shown in Fig. 15.6a, it is clear that the small interstitial and cryptic animals dominate the scene, with nematodes occurring by the millions in virtually all sediments and outnumbering other groups in all ecosystems. Tardigrades, rotifers, and mites also do well. On a larger scale, oligochaetes and collembolan springtails are also abundant in soils. Planarian flatworms are surprisingly common in cooler ecosystems, while various groups of crustaceans and myriapods generally increase in abundance in warmer soils, as do salamanders. The same is true of gastropod molluscs, with the added complication of more shelled forms in tropical and desert zones and a predominance of slugs in cooler deciduous and coniferous forests. Pterygote insects are common in all soils (particularly as dipteran and beetle larvae in temperate soils, and as ants and termites in more tropical zones) but they do not dominate in the spectacular way that they generally do above ground. In general, invertebrate soil communities strongly correlate with vegetation type, with the largest overall abundances in the temperate forests and grassland zones where there is good vegetation cover and a deep protective litter layer. Note that tropical rain forest soils are *not* especially rich in fauna; in these areas the soil is rapidly leached, often waterlogged, and sometimes anoxic, and most of the diversity and abundance of life is above ground here.

Figure 15.6b shows the biomass of different taxa in soils, and a rather different picture emerges. The larger body mass of

Table 15.4 Estimates of number of terrestrial species in major groups of animals.

Animal group	Number of terrestrial species
Platyhelminthes (triclad flatworms)	500
Nemertines	20
Gastropod prosobranchs	4000
Gastropod pulmonates	20,500
Oligochaetes	2000
Onychophorans	70
Crustacean isopods	1000
Crustacean amphipods	50
Crustacean decapods	50+
Arachnids	65,000+
Myriapods	11,000
Insects	5–30 million
Amphibians	2000
Reptiles	5000
Birds	9000
Mammals	4000

oligochaetes immediately elevates them to dominance. In temperate soils, they are almost always from the family Lumbricidae, the true earthworms, consistent with their well-known role in soil formation and turnover; while in the northern coniferous forests, where the soil is impoverished below the litter layer, the oligochaetes are represented instead by the smaller enchytraeids. Other taxa play only minor roles in most soil habitats; the exceptions are mites in coniferous forests and pterygote insects in temperate grasslands.

If the same plots were to be made including above-ground fauna (the truly terrestrial xerophiles) several major differences would be apparent. Pterygote insects would dominate all the habitats in terms of abundance, while vertebrates, though always low in abundance, would make a significant appearance in the biomass stakes. Looking at species diversity gives us yet another slant on "success". Table 15.4 summarizes the species diversity of major land groups and underlines the insects' dominance.

15.2 Ionic and osmotic adaptation and water balance

Maintaining water balance can be seen as the key problem for land animals, interacting with all other aspects of their life, and above all requiring extensive adaptations of the skin and/or cuticle, the excretory systems, and the respiratory systems. The problems are always worse for smaller animals (see Chapter 5), but almost invariably for any truly terrestrial animal water loss must be minimized. In the aquatic habitats described in previous chapters, animals had very variable problems: to take up water and extrude ions (vertebrates in sea water), to try to limit all water exchanges and hold on to ions (in brackish zones), or to take up and conserve ions and try to get rid of water (in freshwater systems). But on land the problem is always clear and constant: to take in and hold on to ions and to take in and hold on to water.

Analysis of how this is achieved is complicated by the existence of the two quite different evolutionary routes onto land. Animal groups that have invaded the land via a littoral route have usually inherited a very high tolerance of changing osmotic conditions and of desiccation, relatively little osmoregulatory ability via skin or kidney, and a respiratory system able to cope with alternating supplies of water or air as the respiratory medium. Animals that arrived from a freshwater ancestry inevitably have pronounced osmoregulation with hyposmotic urine and strong capacities for salt uptake, and essentially aquatic respiratory organs. It could be argued that, because of their high tolerance of change, the animals derived from a marine ancestry have been more successful on land than those from a freshwater background, inheriting an emphasis on expensive osmotic regulation. All of this points to a fundamental and quite surprising point: most land animals are indeed **osmotic tolerators** rather than strong osmotic regulators. It is only some vertebrates (birds and mammals in particular) and perhaps some insects that are particularly good at **osmotic regulation**. Strategies for life on land based on physiological regulation have only been conspicuously successful when the animals concerned have also evolved a relatively impermeable outer covering.

Table 15.5 Tolerance of water loss in temperate land animals.

Animal group	% water loss tolerated
Flatworm (*Bipalium*)	>50
Earthworm (*Lumbricus*)	70–75
Snail (*Helix*)	50
Slug (*Limax*)	80
Crab (*Gecarcinus*)	15–18
Caterpillar (*Manduca*)	50
Beetle (*Coccinella*)	35
Fly (*Eristalis*)	40
Frogs	28–45
Birds	5–10
Mammals	10–12
Camel	25–30

15.2.1 Ionic and osmotic balance

A simple guide to a terrestrial animal's osmotic tolerance can be gained by looking at its ability to withstand changes in **water content** (Table 15.5). Many cryptozoic and hydrophilic animals are fairly intolerant in this respect, especially when they are also small. However, hygrophiles such as earthworms, slugs, and snails can often cope with 40–80% water loss, the latter figure being enough to cause very substantial shrinkage of the body. Many terrestrial insects from temperate habitats will survive 30–50% loss of body water, while anurans (frogs and toads) from the same environments can survive around 30% water loss. But most terrestrial birds and mammals, although xerophilic, cannot survive more than 5–10% loss of water; the camel is really exceptional in tolerating up to 30%.

Examples of the osmotic and ionic concentrations of the blood of land animals are shown in Table 15.6. A number of features stand out. Firstly, most land animals have osmotic concentrations in the range 200–500 mOsm, roughly one-fourth to one-half that of a marine animal, and more concentrated than most freshwater animals. Secondly, the differences between species are striking, with some invertebrates (mostly those that are thought to have a freshwater background) having osmotic concentrations below 100 mOsm, and some with a clearly marine littoral background yielding values of 500–900 mOsm, approaching the typical values for a fully marine animal.

Table 15.6 Blood composition in land animals.

Animal group/species	Osmolarity (mOsm)	Ionic concentrations (mM)					Ancestry
		Na	K	Ca	Mg	Cl	
Nemertea							
Argonemertes dendyi	145						Interstitial
Annelida							
Lumbricus terrestris	150–370	76	4	3		43	Freshwater/ interstitial
Mollusca							
Poteria lineata	74	31	2	5	2	25	Freshwater
Pomatias elegans	254	110	6	16	3	106	Littoral
Helix pomatia	183						Littoral
Agriolimax reticulatus	345						Littoral
Arion ater	97–231						Littoral
Limax maximus	140–200						
Crustacea							
Amphipoda							
Arcitalitrus dorrieni	400						Littoral
Isopoda							
Porcellio scaber	700	227	8	15	11	279	Littoral
Oniscus asellus	577	230	8	17	9	236	Littoral
Decapoda							
Holthuisana transversa	517	270	7	16	5	266	Freshwater
Sudanonautes africanus	~400	207	6	12	11		?
Cardiosoma armatum	744						Littoral
Gecarcinus lateralis	~900	468	12	17	8		Littoral
Insecta	200–510						? Interstitial
Locusta	220–350	60	12	17	25		
Periplaneta	320–380	161	8	4	6	144	
Chelicerata	360–480						? Littoral
Onychophora	180–200						
Diplopoda	150–240						
Chilopoda	320–500						
Vertebrata							Freshwater
Amphibia							
Usual range	180–270						
Bufo bufo	205						
Bufo americanus	310						
Scaphiopus couchii	294–606						
Mammalia							
Homo	295	142	4	5	2	104	
Rattus	320	145	6	3	2	116	

Thirdly, Table 15.6 reveals the variability within species, particularly for relatively permeable land animals, reinforcing the point that land animals are good osmotic tolerators. Where a single figure for osmotic concentration is given in the table, it usually means that only a few individuals have been tested and only under one set of conditions. In practice, when exposed to a range of environmental conditions, or when either deprived of food and water or given a large meal, most terrestrial animals show a wide range of tolerated osmotic concentrations in the blood, reflected in the table for some groups. In the insect *Trichostetha*, hemolymph concentration may increase from 437 to 742 mOsm during desiccation. Figure 15.7 shows the possibilities for variation in blood concentration in various caterpillars in relation to ambient humidity and desiccation. Other animals' blood may vary drastically in relation to feeding: the housefly, *Musca domestica*, shows a blood concentration increasing from about 160 to over 760 mOsm after a sugary feed.

15.2.2 Behavioral regulation of water balance

A great deal can be done to alleviate water problems on land by using appropriate behavior: in other words, as we stressed in the introductory chapters of this book, animals can use behavior as a first line of defense against most of their environmental problems, and may only need to use complex and expensive physiological adaptations when the behavioral strategy fails.

This is already evident from our categorization of land animals described above. Cryptozoic and hygrophilic fauna survive entirely successfully on land because of their ability to find and remain within habitats of high humidity—most of them have inbuilt behavioral patterns that will take them to areas of darkness and high RH, and often also into intimate contact with wet substrates. Once within such zones they may show reduced activity, or activity with increased rates of turning, both behaviors that will tend to keep

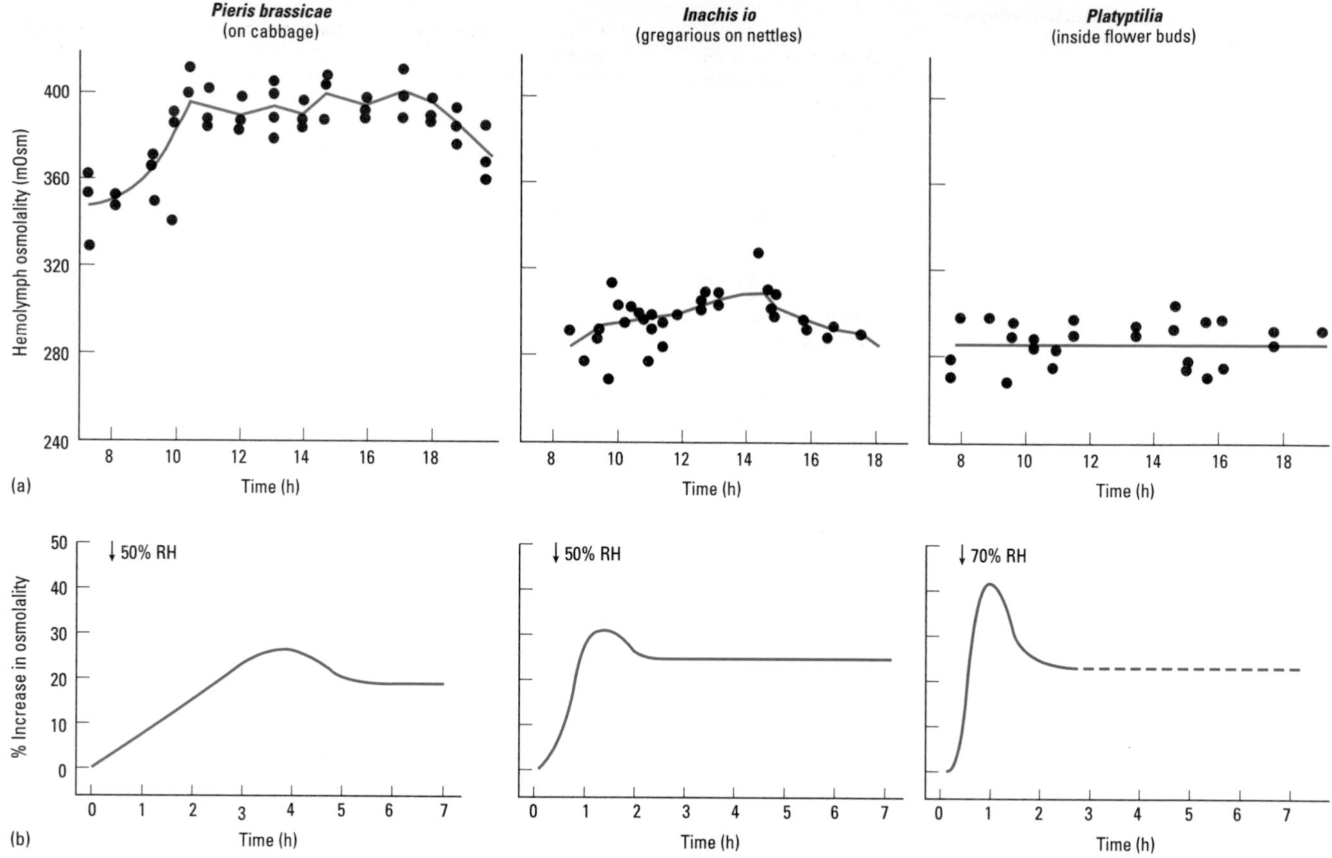

Fig. 15.7 Changes of hemolymph concentration in caterpillars measured within (a) and measured when removed from (b) their normal microclimates. Species that live in very enclosed spaces (such as *Platyptilia* inside flower buds) suffer excessive water loss and lethal hemolymph concentration increases away from their normal microclimate, whereas species that live in the open (*Pieris* on cabbage leaves) are better able to osmoregulate and survive lower humidities. (Adapted from Willmer 1980.)

them in the favored conditions. Very often they have circadian activity rhythms that peak in the night hours, again to insure that they encounter only dark (= cooler) and humid conditions. Hygrophiles, such as slugs, may only emerge when they detect a wet substrate and a falling temperature.

All these animals, but also the more xerophilic land animals, also have important behavioral mechanisms that assist their water balance. The choosing of moister rather than drier foods, or of drinking fluids of appropriate osmolarity, are obvious possibilities. Most animals also tend to reduce their activity with prolonged exposure to unfavorable conditions. On a short timescale this may mean retreating to a burrow or other favorable microclimate during the hottest and driest parts of the day. Taken to extremes on a seasonal scale this leads to winter hibernation or summer estivation. These responses may be coupled with aggregation, which helps to reduce water loss both in actively feeding larval insects (such as caterpillars) and in diapausing adults, such as fungus beetles and ladybirds; similar effects are to be expected in vertebrates too.

15.2.3 Skins and cuticles

Techniques for estimating the permeability of skins were considered in Chapter 5, and the values of permeability to water (P_w) and of cutaneous evaporative water loss (CEWL, given by various kinds of measurements) were compared across the whole spectrum of animals in Table 5.3. A more detailed terrestrial data set is shown in Table 15.7.

Rates of water loss comparable to those for aquatic animals are found in the terrestrial cryptozoans and hygrophiles—all small and thus with a high surface area to volume ratio—including flatworms, nemerteans, and oligochaetes, and also in larger land animals such as the molluscan slugs and snails and the anuran and urodele amphibians. These are all essentially "soft-skinned" animals, where the skin often retains a respiratory function and most of them have surfaces continuously lubricated with mucus from epidermal glands. In many of these animals, as the skin dries out on exposure to low humidities the CEWL decreases quite markedly, so that the figures given in Table 15.7 represent the maximum rate of water loss. But in practice exposure to drought will normally produce a rapid behavioral response, insuring the return to a humid or thoroughly aquatic microhabitat where water reserves can be replenished. If a watery habitat is not located, for example during long, hot summer months, some of these animals go into a period of estivation and the outer skin dries up almost completely. Here the snails have a particular advantage as they can retreat fully into their shells and may seal off the opening with a calcareous or mucoid lid

Table 15.7 Measures of permeability in temperate terrestrial animals.

Animal group/species	Resistance (s cm^{-1})	Water flux (mg cm^{-2} h^{-1})	Transpiration rate (μg cm^{-2} h^{-1} $mmHg^{-1}$)	Water turnover (ml g^{-1} day^{-1})
Molluscs				
Helix				
active	1.6		2500	
inactive	46			
Limax (slug)				0.96
Otala (estivating)			16	
Annelids				
Lumbricus (earthworm)	8.8			0.53
Onychophorans				
Peripatopsis				0.91
Crustaceans				
Porcellio (woodlouse)	3.1		110	
Oniscus (woodlouse)			165	
Armadillidium (woodlouse)			20–80	
Ocypode (crab)	32	2.6		
Gecarcinus (crab)	74	1.2		
Arachnids				
Pandinus (scorpion)			80	
Euscorpius (scorpion)			31	
Uroctonus (scorpion)			7.7	
Pinata (spider)	83			
Lycosa (spider)			33	
Ixodes (mite)			60	
Myriapods				
Lithobius (centipede)		3.8	270	
Scolopendra (centipede)			34	
Mastigona (millipede)			234	
Paradesmus (millipede)		1.9		
Glomeris (millipede)			200	
Insects				
Tomocerus (collembolan)			700	
Coptotermes (termite)			37	
Reticulotermes (termite)			28	
Hepialis (caterpillar)			190	
Pringleophaga (caterpillar)				0.24
Bibio (fly)			76	
Eristalis (dronefly)				0.48
Calliphora (blowfly)			51	
Acheta (cricket)			59	0.36
Locusta (locust)		0.8	22	0.13
Periplaneta (cockroach)	199		55	
Diploptera (cockroach)			21	
Blatta (cockroach)			48	
Cicindela (beetle)			24–49	
Rhynchophorus (beetle)			39	
Solenopsis (ant)			25	
Vertebrates				
Caiman (crocodilian)			70	
Gehyra (lizard)	198	0.22		
Amphispiza (sparrow)		1.48		
Columba (dove)	56			
Zenaida (dove)		0.73		
Poephila (finch)		2.69		
Honeyeater				3
Rattus (rat)		0.75		
Cow		9.96		
Merino sheep				0.07

("epiphragm"); the small area of exposed mantle collar in a snail such as *Otala* or *Helix* becomes fairly impermeable (see Table 15.7; the transpiration rate in estivating *Otala* is 16 μg cm^{-2} h^{-1} $mmHg^{-1}$). The mechanism involved is not clear, but live estivating snails lose water less quickly than dead snails, suggesting some active regulation. Certain arboreal frogs also have additional tricks to conserve water: the so-called "waterproof frogs", in the genera *Phyllomedusa*, *Chiromantis*, *Hyperolius*, and *Litoria*, can make themselves much

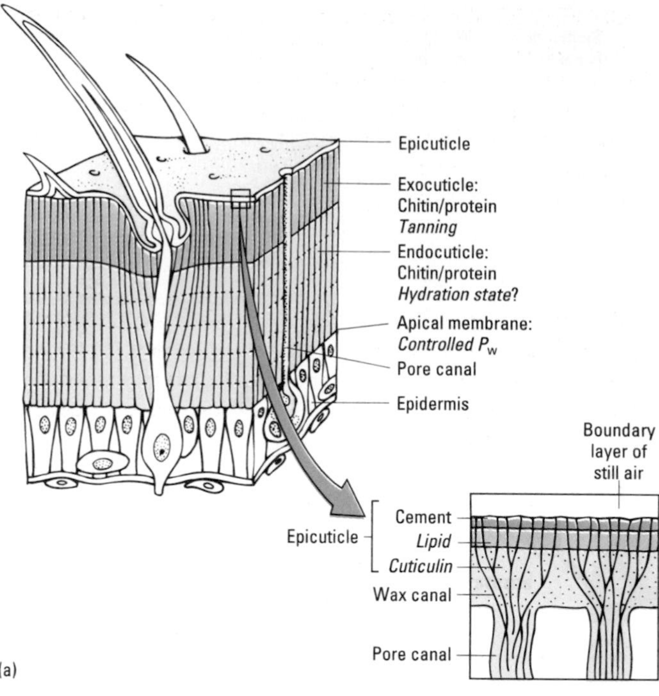

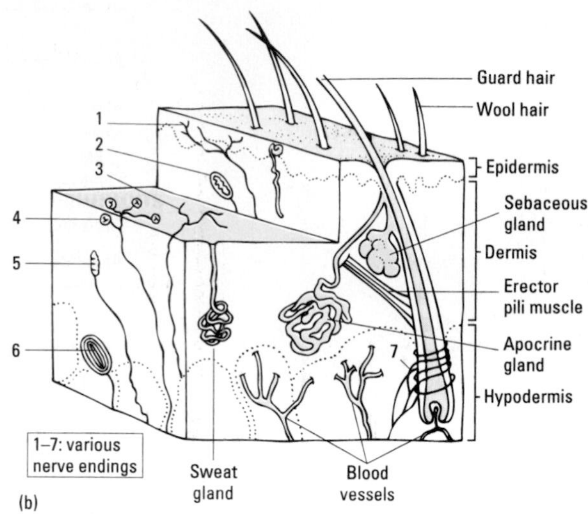

Fig. 15.8 (a) The structure of a typical insect cuticle; note that at the joints and in growing larval stages it may be much thinner with little sclerotized exocuticle. The main control points of resistance to water loss are shown in *italic*. (b) Structure of mammalian skin (not to scale).

more impermeable in dry spells when they become relatively inactive, due to the addition of a cutaneous lipid layer, derived from a specific posterior lipid gland and "wiped" over the body using the limbs. Some other treefrogs have a lesser degree of waterproofing related to lipoid secretions from their normal mucous glands.

For the various groups that have supplemented the soft skin with some form of cuticle, rates of water loss are clearly reduced by one or two orders of magnitude. Within each taxon there is still considerable variability, but even the most permeable are always more waterproofed than the soft-bodied animals. However, the terrestrial representatives of some taxa are not necessarily much more impermeable than their aquatic relatives; land crabs and most land isopods, for example, are still subject to rather high rates of water loss.

The waterproofing effect of a cuticle is invariably due to the structure and chemistry of the **epidermis** and/or the **cuticle** itself. The common thread in all the more xerophilic animals is the acquisition of **lipid barriers** at some site (see Fig. 5.4). This is most clearly seen in arthropod cuticles (Fig. 15.8a), where several features contribute to a low P_w. Firstly, the epidermis itself exerts some control over P_w; the cuticle of hydrated cockroaches is substantially more permeable than that of desiccated individuals, and the difference is attributed partly to the epidermal layer. Secondly, the bulk exocuticle and endocuticle material of chitin and protein is inherently relatively impermeable, and its water resistance may alter according to the density of its molecular packing as it becomes more or less hydrated, so that an insect under water stress is automatically more impermeable. (These two effects may be related, since the epidermis, influenced by hormones, could be responsible for withdrawing water via dermal gland canals, producing this reduction in P_w.) Thirdly, and usually much the most important of all, the outer layer, or epicuticle, confers very low permeability due to its high lipid content: C_{10}–C_{37} fatty acid derivatives (especially the even-number chain length C_{12}–C_{18} series), together with some branched and unsaturated hydrocarbons. Lipids within the cuticle may be augmented by a thin layer of surface lipid, often predominantly wax esters.

In arachnids the cuticle is traversed by fine ducts from the epidermal glands, and the nature of the epicuticular lipids is also often rather different, with a greater sterol component. In fact in these animals the epicuticular lipids may change in nature through the life of an individual, which is probably the reason for glands leading directly to the surface.

Following from this, there is an increasing appreciation of the possibilities of adaptive variation in the properties of the arthropod cuticle. Summer-active species may contain more long-chain hydrocarbons than closely related winter-active species, an effect demonstrated both in beetles and in scorpions. The difference occurs even between field-caught winter- and summer-adapted individuals within one species, most clearly demonstrated for the beetle *Eleodes armata*. Moreover, pupae of *Drosophila* laboratory reared from eggs at 17°C have more permeable cuticles than those reared at 24°C, with concomitant changes in constituent lipid chain lengths. The cast-off cuticles (exuviae) of the grasshopper *Melanoplus sanguinipes* contain lipids of significantly higher melting point when the individual is acclimated to higher temperatures, largely due to the incorporation of more straight-chain hydrocarbons relative to branched forms; and analysis of siblings shows that the natural genetic variation of their cuticle permeability largely also resides in the straight-chain lipid class. Changes in the cuticle clearly may occur during the lifetime of an individual land arthropod, above and beyond the differences between hydrated and dehydrated

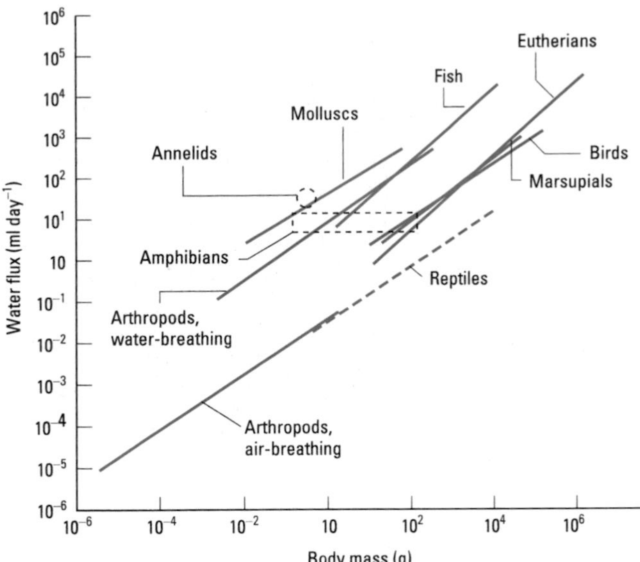

Fig. 15.9 Water flux rates in relation to body size for land animals; all show similar gradients, though arthropods have routinely lower fluxes than vertebrates, and soft-bodied animals much higher fluxes. (Adapted from Nagy & Peterson 1987.)

individuals referred to earlier. The changes may be related to hormonal control; for example, in cockroaches a hormone from the brain operates at the epidermis to slow desiccation, and water loss rates in the lubber grasshopper (*Romalea*) have been shown to increase with rising humidity, indicating a relaxation of some controlled barrier mechanism. More impressively, in the cattle tick, *Boophilus*, the amount of wax increases when a tick drops off its host, and the thermal properties of the lipids also seem to alter, indicating fine control over the waterproofing properties of the surface layers. Some insects undergo a progressive increase in cuticle permeability through their adult life, perhaps due to increasing abrasion and loss of surface waxes. More spectacularly, some homopteran bugs and coleopteran beetles have superficial wax "blooms" on their cuticles, which may be replenished seasonally or even daily; and a few have cuticular pores through which they can achieve controlled "sweating" (see Chapter 16).

In xerophilic tetrapod vertebrates the basic epidermal type, with high concentrations of the protein keratin, is modified by the inclusion of some extra phospholipid in the outer layers of the epidermis (e.g. in the stratum corneum of reptiles). This arrangement works most effectively as a waterproofing layer in the lizards and snakes, where the skin is not interrupted by glandular pores, such as those of sweat glands. Nevertheless, CEWL in reptiles is normally 60–85% of the total evaporative water loss (EWL; higher range in water snakes, and lower in desert tortoises). In the skin of endothermic mammals (Fig. 15.8b), where such glands often have to be present to provide a cooling mechanism, CEWL may be much more variable according to levels of activity and environmental temperatures, though even its basal level is set higher than in a similarly sized reptile.

Reptiles, insects, and arachnids, with thick keratinized or sclerotized cuticles having a high lipid content, therefore hold the records for animal impermeability. This is reflected in Fig. 15.9, showing daily water flux always increasing with animal body mass but with the setting lower for arthropods and reptiles, higher for birds and mammals, and of course higher still for annelids and molluscs.

Given reasonably good defenses at the external barriers (as in these groups), and/or a moderately large body size to slow the rate of exchanges (as in the vertebrate endotherms), regulation of the internal fluids becomes a viable option.

15.2.4 Regulatory organs

All animals living on land need a site for the excretion of waste materials, and most excretory materials, being potentially toxic, have to be lost quickly and diluted in a moderate volume of fluid. Therefore most animals also use this **excretory system** ("kidney") as a site where they can conveniently regulate their water content, and consequently also as a site for regulating salt balance. In fact most of the so-called excretory systems of animals probably had osmoregulation as their primary function.

Soil animals tend to have osmoregulatory/excretory systems similar to their aquatic relatives, as they will often have an excess of water available and need to excrete it quickly. Animals such as earthworms may need systems that can expel excess water but that can also act as water conservers during drought. Xerophilic animals almost always need to conserve as much water as possible, and (with the exception of mammals) most have combined their "excretory" opening with their gut opening, so that all the excreta are regulated in one site (the rectum) with minimal water loss.

Terrestrial animals therefore show a great variety of mechanisms of excretory water loss. Chapter 5 dealt with the principles of urine formation and regulatory reabsorption; here we look at the form and functioning of the different organs found in land animals.

Cryptozoic and hygrophilic land animals

Most of these retain an essentially aquatic approach to osmoregulation and excretion. Flatworms and nemertines retain the **flame cell** protonephridial system (Fig. 15.10a) discussed in Chapter 13. The lower parts of the flame-cell tubules likely have some resorptive function, but normally allow the primary urine to pass with minimal salt resorption, flushing excess water out of the body as quickly as possible. The product is therefore usually **hyposmotic**, with ammonia as the main nitrogenous waste, both features reflecting the more or less freshwater nature of the habitat. However, some terrestrial flatworms are unusual (for invertebrates) in producing a large percentage of their nitrogenous output as urea.

Larger soil dwellers, such as earthworms, also have a nephridial system, organized segmentally and exiting via nephridiopores, although the organs (Fig. 15.10b) have a more substantial coiled resorptive tubular portion than in the flatworms and nemertines. The product is clearly hyposmotic in most oligochaete species. However, in some of the larger earthworms, such as *Pheretima* from India and the Australasian megascolecids (Fig. 15.10c), the nephridia are "enteronephric", opening into the gut rather than to the outside world, and here some drying of the feces + urine is achieved within the rectum. In most terrestrial annelid species the main excretory product is ammonia, although urea levels may increase in worms that are not feeding. In addition a substantial amount of

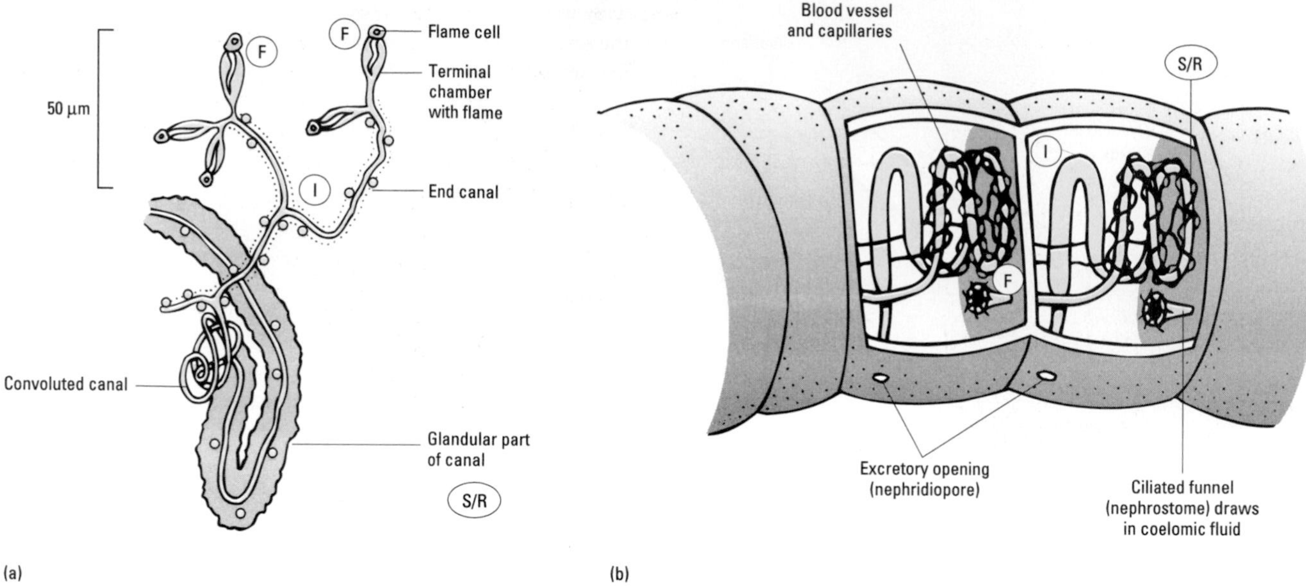

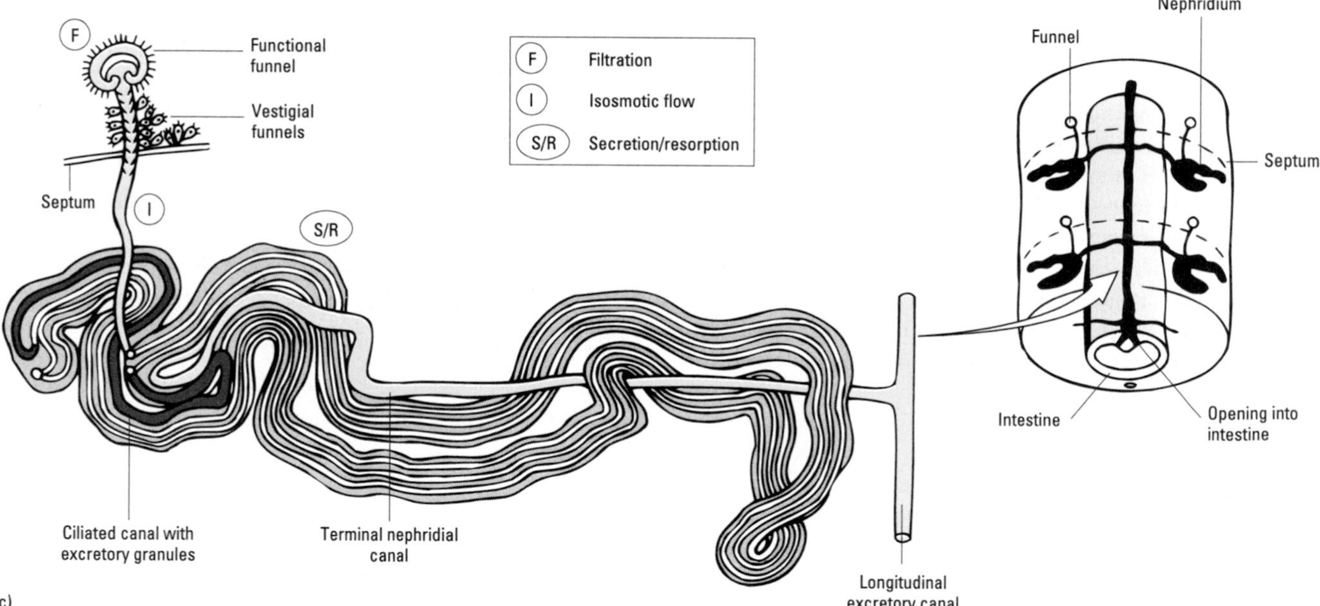

Fig. 15.10 (a) Flame cells from terrestrial nemertines; (b) the more elaborate nephridia of earthworms; and (c) the nephridia of xeric megascolecid earthworms, which empty into the gut allowing greater resorption of water.

nitrogen is lost as mucoprotein in the copious mucus secretions from the skin.

Land snails and slugs

Terrestrial prosobranch snails have a kidney essentially similar to their littoral and marine kin. Fluid flows to the kidney from the pericardium, and drains directly from there to the outside, with no special absorptive area. Excretion always involves the loss of considerable quantities of water, because there is no spatial separation of the sites of nitrogenous excretion and of water balance regulation.

Terrestrial pulmonates (the more familiar slugs and snails) do rather better, and show a wide range of adaptations to the kidney. The generalized structure from a temperate snail is shown in Fig. 15.11. The kidney has a classic **ultrafiltration** system to form primary urine, with the fluid supply coming either from the pericardium surrounding the heart or from the blood surrounding the kidney epithelium. The kidney then opens into an elongate ureter, which turns back on itself to give a primary and secondary arm, then opening to the outside at a renal pore in the mantle cavity. Uric acid is secreted into the urine together with some guanine and xanthine, usually within the kidney but sometimes in the upper part of the

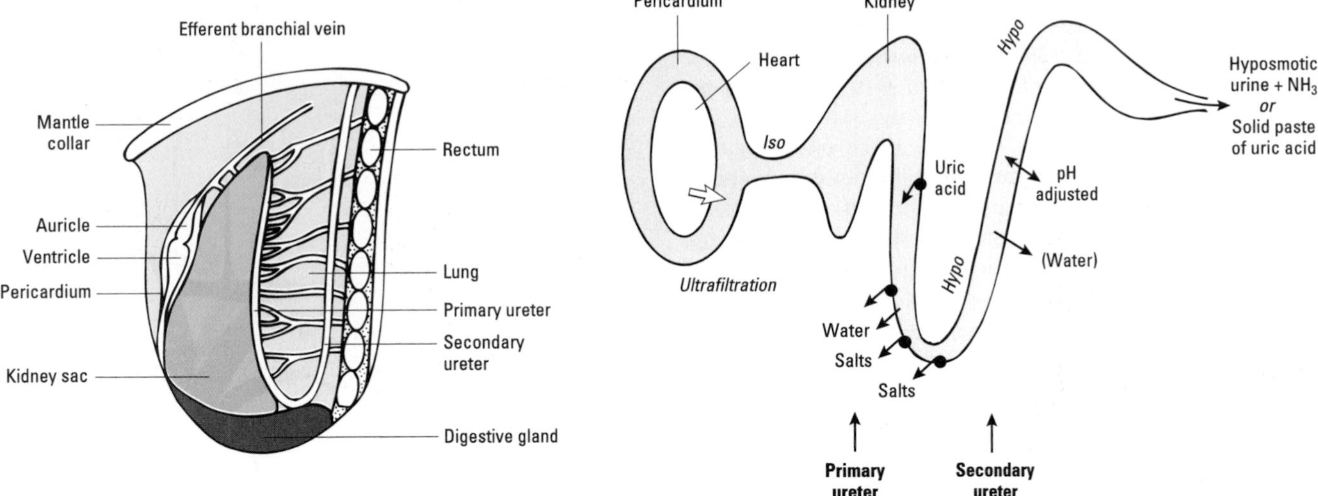

Fig. 15.11 The kidney of pulmonate snails (*Helix*), receiving a filtrate from the pericardium and normally producing a hyposmotic secondary urine in the primary and secondary ureter. (Adapted from Little 1983.)

ureter. The lower ureter is absorptive and regulatory, taking some salts and water back into the body and regulating the urine pH. In damp surroundings, there is a copious flow of hyposmotic urine, but in drought conditions a dry paste emerges from the renal pore. Curiously, there is no intermediate of a hyperosmotic liquid urine. In other words, either the filtration system is turned on and a dilute watery product emerges, or the filtration is shut down and dry waste is produced independently, from the (spatially separated) uric acid secretion site. In this way pulmonates can safely excrete their nitrogenous wastes in a wide range of climatic conditions. Indeed, when undergoing estivation they can survive without any excretion by storing nitrogenous products, especially urea, within the body (though some also lose gaseous ammonia by diffusion through their shell in this phase of their life). Estivating snails show elevated blood osmolarity, sufficient to depress neuronal activity and reduce the heart rate to around 60% of normal.

Hygrophilic arthropods

Semiterrestrial crabs, living at the top of beaches and in damp mangrove forest zones of the tropics, show a relatively strong osmoregulatory ability (see Table 15.6). However, this is not achieved by osmoregulation at the **antennal glands** (Fig. 15.12a), the crustacean version of the kidneys—these can only produce isosmotic urine.

Fig. 15.12 (a) Structure and function of the antennal glands in land crabs (note that the branchial chamber often does some "urine reprocessing"). (b) Coxal gland of a typical arachnid.

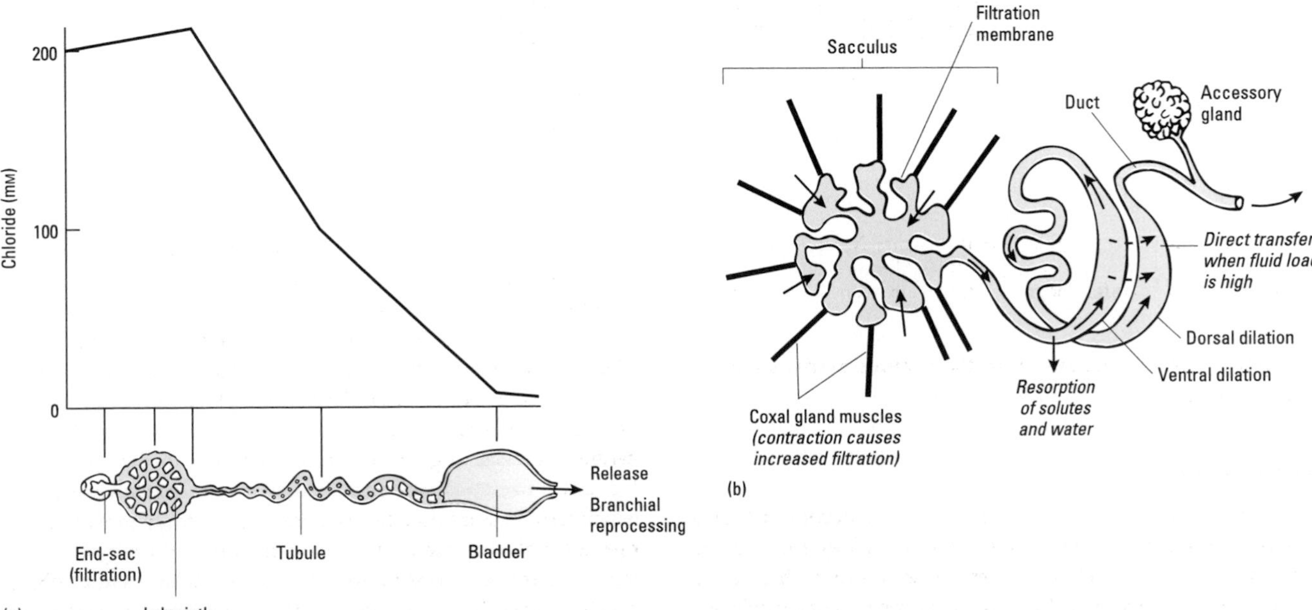

These glands do modify the ionic composition of the filtrate they receive, to regulate blood ion levels, but they cannot produce hyperosmotic urine to aid water conservation. Instead, salt is secreted at other sites, especially at the gill epithelium as in so many aquatic invertebrates (see Chapter 5), but perhaps also in the rectum. The only water balance "regulation" achieved at the antennal glands concerns the overall rate of **urine flow**, so that in dry conditions flow stops entirely, with no primary ultrafiltration occurring. Land crabs are thus physiologically similar to their aquatic relatives, although the balance between physiological prowess and behavioral strategies to avoid water loss and achieve adequate water input is very variable between species. The only possible extra "tricks" that some may have are to reingest some of their urine or, quite commonly, to divert some of it into the branchial chamber, from where extrarenal resorption of ions may occur through the gills, resulting in a branchially modified urine known as the "P-product". This **gill reprocessing** system is under the control of dopamine, which can increase branchial Na^+/K^+-ATPase by around 67%. In the blue crab, *Cardisoma*, around 90% of the urinary ions are reclaimed in the gills, and the two tricks of urine reingestion and reprocessing together allow the robber crab, *Birgus*, to reclaim 70–99% of the ions from its initial urine, especially important when living in particularly dry inland sites where only rain water is available and ions are in short supply. This phase in urine production also allows the urine to be augmented with extra ammonia.

Systems that are structurally and functionally similar to crustacean antennal glands occur in other hygrophilic arthropods, including flightless insects and myriapods (where the organs involved are maxillary glands, opening near the mouthparts), and in onychophorans (where there are segmentally repeated excretory organs). Most of these glands have two regions (Fig. 15.12b). Firstly there is a sacculus, where ultrafiltration occurs across thin-walled cells termed podocytes (somewhat similar to cells in the vertebrate kidney glomerulus). The sacculus then leads on to a tubular section in which salts and water are resorbed.

Woodlice (pillbugs) have an extra trick. Urine from their maxillary glands is released via capillary channels to flow over the flat ventral respiratory flaps called pleopods, keeping them moist for gas exchange (and allowing ammonia to diffuse away, disposing of the nitrogenous waste from the woodlouse's body). At least some of this excreted water is then returned to the body via the resorptive rectum, giving an unusual involvement for the gut in that fluid arrives in it via an external route. Species from arid habitats, such as *Hemilepistus*, can resorb significant quantities of water via this rectal route.

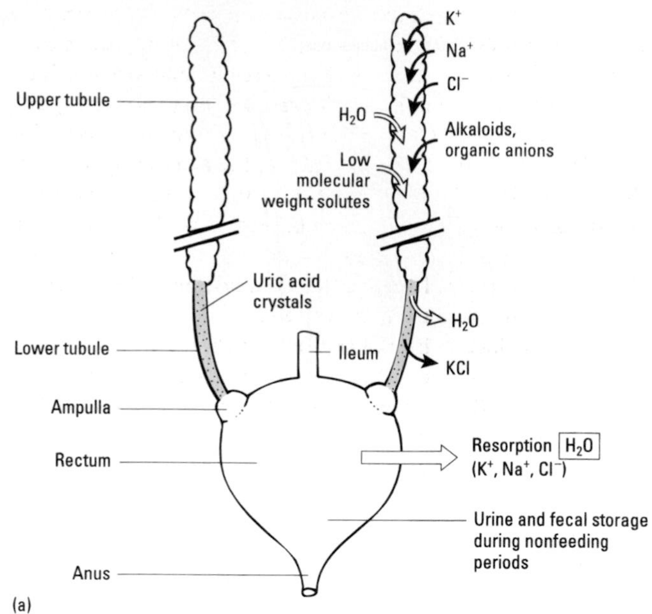

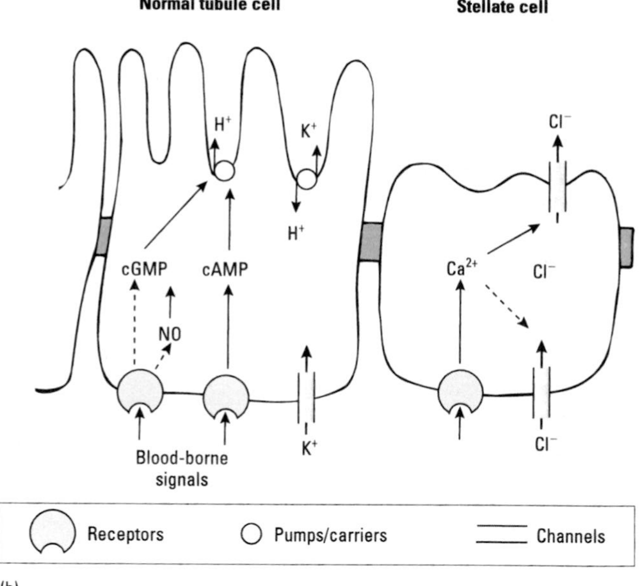

Fig. 15.13 The functioning of Malpighian tubules in insects: (a) an overview of the system; and (b) details of the ion movements in normal tubule cells and stellate cells. (Adapted from Phillips 1981; Maddrell & O'Donnell 1992; O'Donnell *et al.* 1998.)

Xerophilic arthropods

Two rather different systems occur in truly terrestrial arthropods. Most of the arachnids have a system analogous to the maxillary gland system described above; the structures are termed coxal glands, and although they have a different origin they function in a similar manner, though in some species ultrafiltration is speeded up by muscular stretching of the walls of the filtration chamber. Unusually, the excretory coxal fluid is often delivered into the prey rather than to the outside world. All other xerophilic arthropods (insects and many myriapods) have an excretory–osmoregulatory system based on the "Malpighian tubules plus rectum" composite, where the stages of urine formation and resorption are in widely separated structures. Urine is produced by secretion in the **Malpighian tubule** cells (see Chapter 5); details of this tubule secretion system, which has been very thoroughly studied, are shown in Fig. 15.13. The tubule wall is composed mainly of standard cation-handling cells, with pumps (V-ATPases) effectively moving K^+ in most insects with herbivorous or omnivorous diets, but able to transport Na^+ very rapidly in blood feeders with high-sodium diets. The entrained flow brings most other solutes passively through into the lumen, alhough specific urate carriers to insure excretory clearance are present in many species. There are usually also smaller numbers of "stellate cells" in the Malpighian tubule wall, which are now known to act mainly as chloride-transporting cells. Malpighian tubules are controlled by hormonal triggers in many insects, discussed in section 10.3.2. The net result is a fast and usually isosmotic flow across from the hemolymph into the tubule lumen. Some resorption of salts and water may take place in the lower reaches of the tubules themselves, where P_w is probably reduced; but usually the primary urine is discharged, unmodified, into the gut at the point where the midgut joins the rectum.

The insect hindgut (ileum + rectum) then has a variety of specializations to resorb water effectively and to regulate the combined composition of the feces and urine. The first part of the hindgut mainly executes a substantial resorption of fluid so that the ileum contents reduce in volume with little change in concentration (except that some molecules specifically secreted into the system earlier on are unable to diffuse out, and so become more concentrated). The **rectum** is the more crucial area, where there is a substantial change in the concentration of all the urine's components. The first step in this is to draw water from the rectal lumen into the gut cytoplasm; it is assumed that the cells are endowed with osmolytes that draw this water in. Then the rectal cells effectively secrete (to the hemolymph) a hyposmotic fluid. The simplest system that can achieve this is the rectal pad, an intracellular resorption system essentially as described for the model system outlined in Figs 5.14 and 5.15. Slightly larger scale systems operate, using the same principle, in the **rectal papilla** (Fig. 15.14), found in orthopterans and in some flies, such as the common "greenbottle" or blowfly (*Calliphora*). Spaces between the rectal cells create a compartment separated from both the blood and the gut contents, and ions are transported into these spaces from the cells, with water flowing from the gut across water-permeable membranes, down the osmotic gradient thus created. The fluid that accumulates is then forced out into the hemolymph under the ever-increasing hydrostatic pressure created by the secretory processes, along channels where most of the ions in the fluid can be resorbed and recycled into the cells, but where water cannot readily follow due to a high flow rate and relatively low P_w. This system is a further example of the "concentration by compartmentalization" mechanism described in Chapter 5; it can also recover useful compounds, such as amino acids and sugars, from the gut to blood.

In some beetles and larval lepidopterans a yet more complex system is found, with the blind-endings of the Malpighian tubules reflexed back against the wall of the rectum, to create a **cryptonephridial** system (Fig. 15.15). The spaces around the tubules now form an additional compartment (the perinephric space), which can be used in a similar fashion to the intercellular spaces described above. In addition, the flow forward in the Malpighian tubules is now running in the opposite direction to the flow backward along the rectum, so we have a countercurrent created. The whole complex is enclosed in a relatively impermeable membrane, punctuated with "leptophragma" cells where intense ion movement from the hemolymph into the tubules occurs (Fig. 15.15d). At the anterior end of the system concentrations are quite low, though always with a water gradient from gut to hemolymph. Near the anal end of the rectum, there is still a gut-to-blood gradient but by now concentrations in both are much higher and water is still withdrawn from the gut contents, which may reach 4000–5000 mOsm (up to five times seawater concentration). Hence a very concentrated dry paste of uric acid and some fecal waste results.

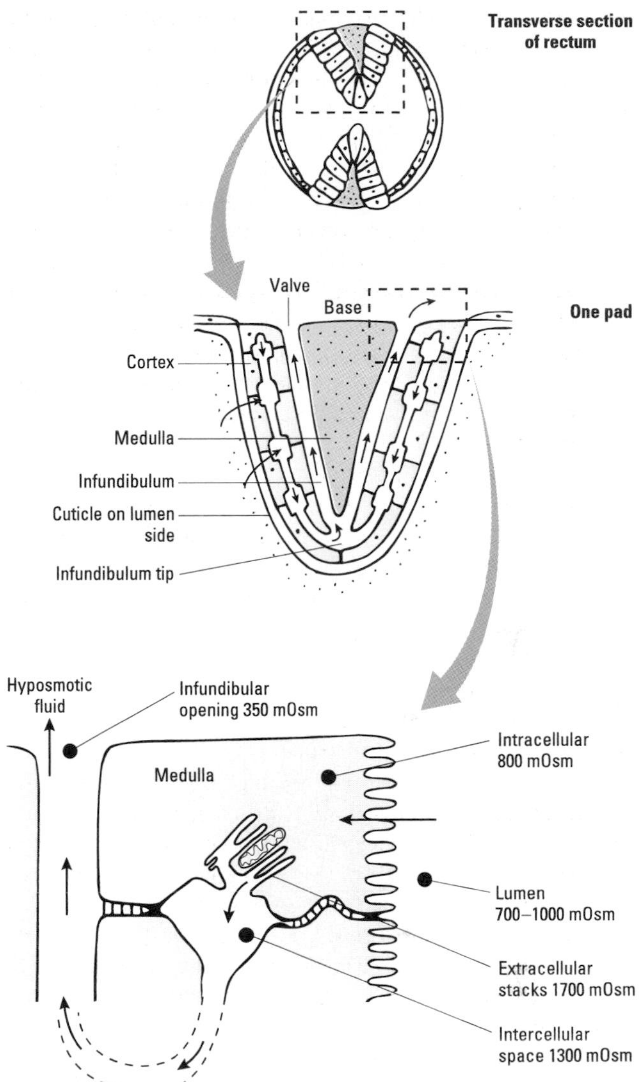

Fig. 15.14 The rectal papillae of the blowfly *Calliphora*, showing water uptake from the gut by solute recycling (cf. Fig. 5.14 for the principle involved). (Adapted from Maddrell 1971; Gupta *et al.* 1980.)

Tetrapods

The key features of the vertebrate kidney and excretory products were dealt with in Chapter 5; here we need only look at the variation in different land tetrapods. Amphibians certainly inherited an essentially freshwater-adapted kidney from their ancestors, and can only make **hyposmotic** urine. They can decrease the rate of urine formation and increase urine resorption from the bladder by the action of the amphibian antidiuretic hormone (arginine vasotocin), and can also correct their osmotic balance with active salt uptake through the skin.

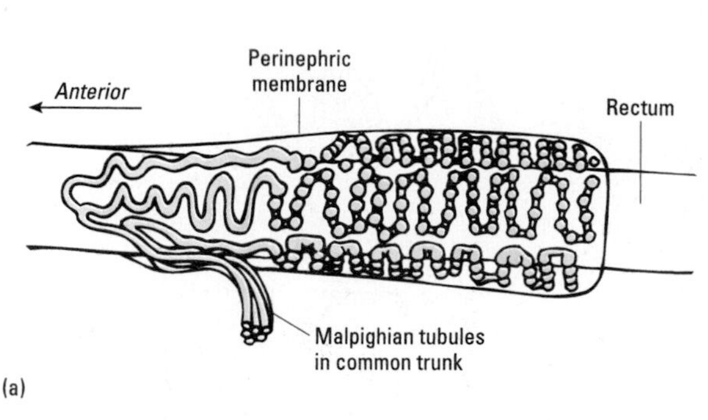

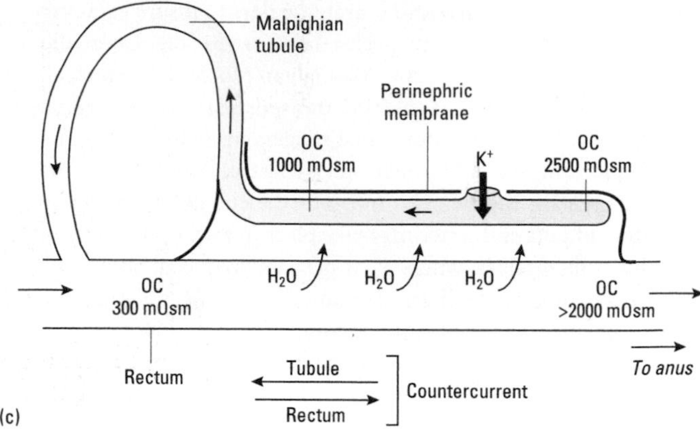

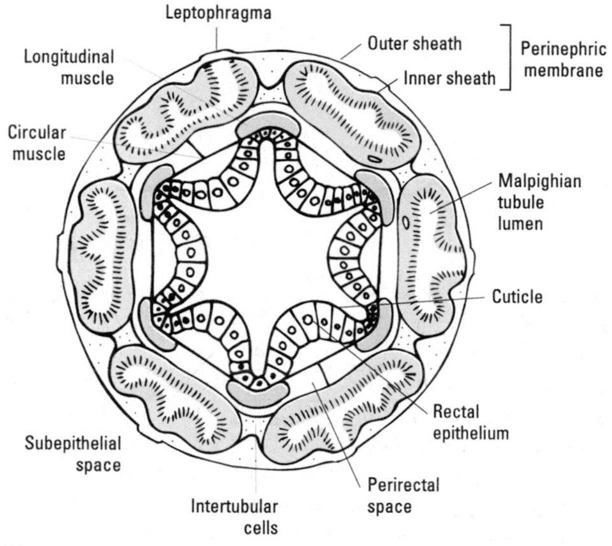

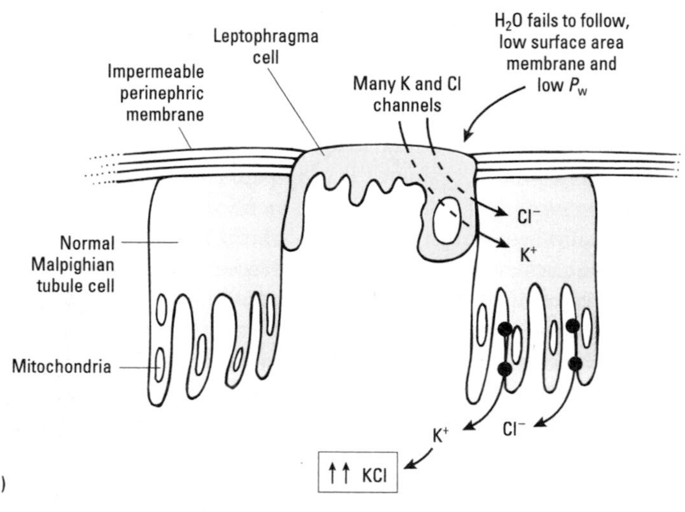

Fig. 15.15 The cryptonephridial system found in a variety of xeric insects, here shown in the flour beetle *Tenebrio*. (a) The whole system with Malpighian tubules folded back against the gut. (b) In transverse section, showing six tubules around the rectum. (c) Ion and water movements and concentration gradients along the system. OC, osmotic concentration. (d) Ion transport at leptophragma cells. (Adapted from Ramsay 1964; Maddrell 1971.)

All other tetrapods have rather impermeable skin and cannot use these systems. Reptiles nevertheless remain unable to produce hyperosmotic urine, even when they live in extremely xeric habitats. Their urine is **isosmotic** and excess salts are excreted through their nasal salt glands (see Chapter 5). Most birds and mammals also do not normally produce hyperosmotic urine. Instead, when drinking water is freely available many of them release a hyposmotic fluid; this also occurs where the diet is very watery (as in some fruit bats, which may excrete around 15% of their body mass each day, and in some nectar-feeding hummingbirds). Alternatively, when water stressed they commonly produce a roughly isosmotic fluid. Some birds (and also some reptiles) discharge this urine into the cloaca but it then flows back up into the rectum where additional (approximately isosmotic) resorption of fluid occurs (Fig. 15.16).

However, some birds and nearly all mammals have a **loop of Henle** interposed between the descending and ascending portions of their nephrons, to give a site of countercurrent multiplication (see Chapter 5). They can then produce a significantly **hyperosmotic** urine (Table 15.8; U : P (urine : plasma) ratio > 1) when needed. Here the cells in the hairpin tip of the countercurrent loop are exposed to very high levels of urea and are almost anoxic; they cope by incorporating high levels of betaine and sorbitol into their own cytoplasm as counteracting osmotic effectors (see Chapter 4). Remember that the absolute length of the loop is not necessarily the key factor in relation to urine concentration. For example, the desert mouse *Notomys* has loops up to 5.2 mm long and can produce urine of 9370 mOsm, while the horse has loops up to 36 mm long and can achieve urine of only (at best) 1900 mOsm. Indeed, in birds urine-concentrating ability commonly decreases with increased loop length, perhaps partly because many of their loops lie crosscurrent rather than countercurrent to blood flow. Urine concentration

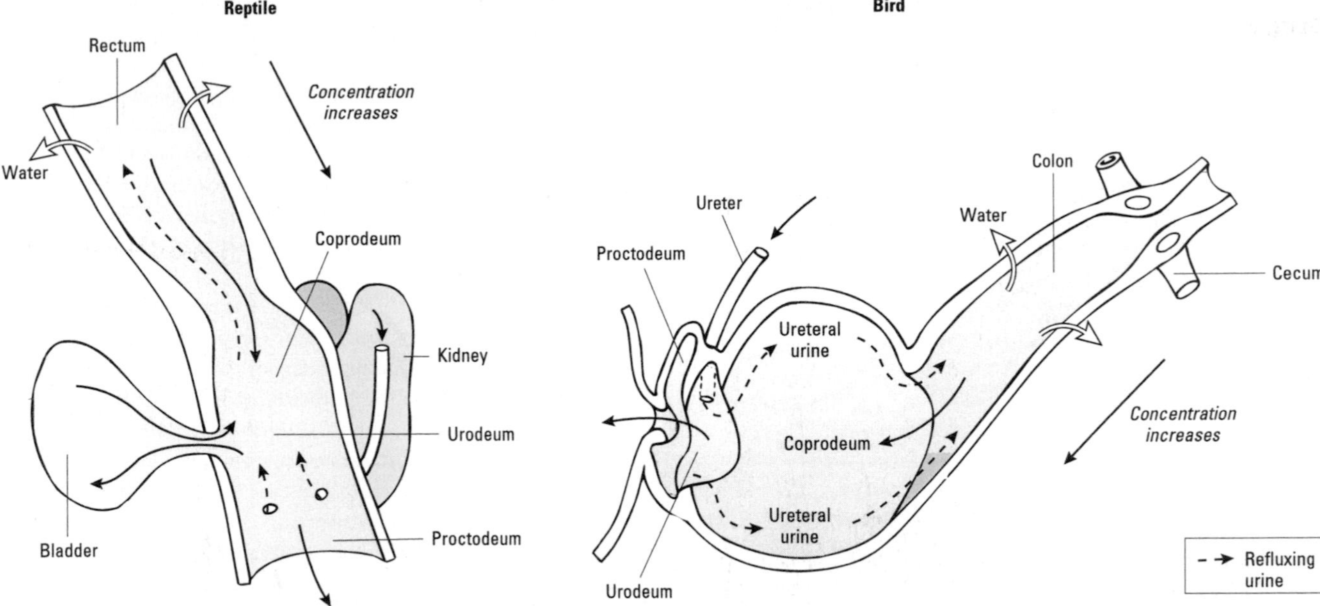

Fig. 15.16 Distal portions of the gut and excretory systems of a reptile and a bird, showing the reflux of urine up into the rectum via the common "coprodeum". The hindgut contents are substantially more concentrated than the blood. (Adapted from Skadhauge 1981; Minnich 1982.)

is determined more by the specific metabolic rate of the tubule tissues (higher in smaller animals, with a much higher density of actively transporting membrane surfaces) than by length of tubule (see Chapter 5).

Most tetrapods excrete urea or (in most birds and reptiles) uric acid, but diet also introduces some variation here; hummingbirds, for example, are able to switch to ammonia secretion when feeding on dilute nectars at low temperatures.

Table 15.8 Urine concentrations and urine : plasma ratios in temperate land animals (values are maxima from dehydrate animals).

Group/species	Urine concentration (mOsm)	Urine : plasma ratio
Insects	~1000–5000	~3–15
Reptiles		
Tuatara	270	1.0
Gecko	325	0.7
Iguana	362	1.0
Birds		
Chicken	538	1.5
Pigeon	655	1.7
Pelican	700	2.0
Mammals		
Rat	2900	9
Domestic cat	3100	10
Vampire bat	4650	14
Beaver	520	1.7
Pig	1100	3
Domestic cow	1160	3.7
Human	1400	4–5

15.2.5 Acquiring water

Liquid uptake

Most land animals gain their water by drinking, from streams, rainwater puddles or rain drops on vegetation, or from water vapor condensed as dew. The other important source is the water content of the diet, and a surprising range of terrestrial animals rely on this alone, requiring no free water. Many species will select wetter food sources over drier ones when water stressed. Some may even take more food than they need just to get enough water, although the evidence for this is controversial. Animals that live in burrows and other enclosed microhabitats may store dry foods, such as seeds, for some time before eating them, and the humid microclimate then allows the foodstuff to hydrate substantially at no cost to the animal. This phenomenon is an important contributor to many desert rodent's water balance (see Chapter 16), eliminating the need for free water in the diet.

However, a number of animals have specific adaptations to acquire water from the environment independently of drinking water or eating food. The possibilities were listed in Chapter 5, and are particularly obvious amongst the various arthropodan groups on land.

Some arthropods have evolved behavioral mechanisms to make use of dawn **dews and fogs**. Simple versions of this involve drinking dew as it forms on plants or on the sand, but some desert species again have more sophisticated ways of using heavy dawn fogs in coastal deserts, which we will deal with in Chapter 16.

Standing water can also be acquired by **osmosis**, in ways that do not involve traditional "drinking". This is usually only an option for animals that retain a strong link with water sources anyway, and often retain fairly permeable skin surfaces. Crabs therefore exploit a number of these possibilities. Species with gills that live around high-tide level on beaches can immerse the gill chambers in pools of

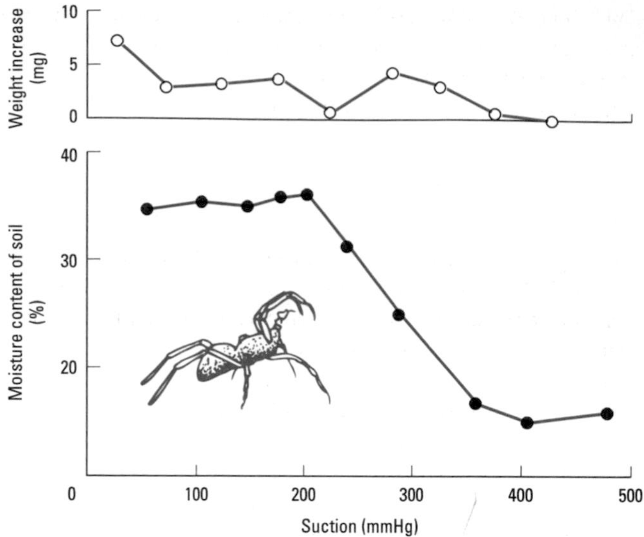

Fig. 15.17 The ability of spiders to suck water from soil of different hydration states; this ability is still present at only 20% soil water content, though weight gain is then very small. (From Little 1990; adapted from Parry 1954, courtesy of Cambridge University Press.)

fresh or slightly brackish water so that water enters the body. The hermit crab, *Coenobita*, can select the osmolarity of the pool it uses according to its own osmotic needs. One species of ghost crab (*Ocypode*) can actually suck water from the sand into its gill chamber, creating a pressure of about 40 mmHg by muscular action on the chamber walls. Many other semiterrestrial crabs can take up water using a patch of densely hairy cuticle on the ventral body surface. When the crab is squatting down on damp sand, the hairs draw up water by capillarity, and conduct the water to the leg bases, where it is drawn into the gill cavity.

Uptake of water from wet soils and surfaces can also be achieved osmotically in the more terrestrial arthropodan groups. Quite a range of wingless insects and myriapods can absorb water using structures at the top of the legs called **coxal sacs**. These can be everted and placed onto damp substrates, and an active salt-pumping process is used to create a large osmotic gradient that will rapidly take in water. A xeric millipede, *Orthoporus*, can extrude a pad of rectal tissue from its anus that works in a similar fashion. Isopods, such as the littoral *Ligia*, and several of the more terrestrial woodlice, can also absorb water osmotically from wet soils using permeable surfaces at the mouth or anus.

Getting interstitial liquid water from relatively unsaturated soils generally requires both a very good muscular pump, and modifications of the skin surface to produce a strong **capillarity** that will overcome the capillary attraction of water to soil. Some spiders can extract water from soils (Fig. 15.17) against a suction pressure of 450 mmHg, where the soil has only 20% water content. They are presumably aided by their strong pharynx muscles (normally used for sucking fluid from prey), and so could be said to be preadapted for extracting moisture from soil.

Thus most of the specialist mechanisms for acquiring water are found amongst the arthropodan groups of land animals. Nevertheless, a few vertebrates also achieve similar tricks. Some frogs and toads have patches of abdominal skin ("pelvic patches") specialized to absorb water osmotically; they can sit in puddles or on very damp sand and take water into their bodies, storing a surplus in the bladder. Local sensors within the skin deter them from taking up the water-absorption posture on salty soils or in brackish puddles. This cutaneous drinking response also appears to be under the control of angiotensin II (see Chapter 10). It can be modified by barometric pressure, and so is attuned to the perceived future availability of water. Toads therefore show "anticipatory cutaneous drinking", before they are dehydrated and with the water-storing bladder still moderately full.

However, most other vertebrates have to rely on drinking liquid water, and may have very much more elaborate physiological controls to regulate their drinking behavior. Control of **drink rate** is mainly via osmolarity of the extracellular fluid (ECF), with osmoreceptors feeding information to a "thirst center" in the hypothalamus. Decreased ECF volume resulting from water loss also acts as a trigger, causing renin secretion and thus the release of angiotensin II, which acts on the "subfornical organ" close to the brain ventricles, again feeding information to the hypothalamic control center.

A final possibility sometimes mentioned as a source of water for animals is the **metabolic water** derived from oxidative processes within all the cells of the body, when carbohydrates and other large molecules are broken down. Certainly these processes do produce water as a by-product (see Chapter 6) and animals retain and use this. It is usually only a small component of overall water intake. But in animals from dry habitats, or with very high metabolic rates (e.g. in flight), it may be a significant proportion of the water budget. However, it cannot normally be regarded as a mechanism for gaining "extra" water. Extra oxidation requires extra ventilation, and in most animals the extra respiratory water loss exceeds the water gained internally. Only in special cases could there be a net gain of water; these may include certain vertebrates with an ability to reduce the water content of their expired air using a nasal countercurrent exchanger, or large flying insects that may also be using countercurrent water saving and that do not use much evaporative cooling in flight (unlike bats and birds).

Vapor uptake

Some representatives of the insects, crustaceans, and chelicerates on land can use the technique of **water vapor uptake**, briefly discussed in Chapter 5. This is an unusual and specialist adaptation found only rarely (see Table 5.5): in a few isopods, some mites and ticks, some apterygote insects, and some flightless insects in the orders Anoplura (lice), Siphonaptera (fleas), Dictyoptera (cockroaches), and Coleoptera (beetles, though they show uptake only in the larval stages). An example of uptake rates at different humidities in a flea was shown in Fig. 5.7. Uptake seems to be virtually always associated either with the mouthparts or with the rectum.

In mites and ticks the ability to take up water may vary in different life stages. For example, in deer ticks it is most evident in the larvae, which emerge on the ground and may have to survive for months before encountering a host. In these animals, and also in lice, blocking the mouthparts prevents vapor uptake. A clear **hygroscopic fluid** accumulates near the palps, derived from the salivary glands, which in *Dermatophagoides* (the infamous house-dust mite) produce a secretion that is rich in potassium and chloride, and probably also contains some organic material. At low RH this dries to a crystalline deposit, but at higher humidities water vapor condenses into it and it is then withdrawn by a muscular suction into the pharynx. As yet we do not know how the very high concentrations of solute needed for uptake at low RH can be achieved; for example, uptake at 45% RH (see Table 5.5) would require fluid with a concentration of at least 50 Osm. In the tick *Amblyomma* it is thought that a **hydrophilic cuticle** plays a role in water condensation, with the hyperosmotic secreted fluid then altering the water affinity of this cuticle to release the adsorbed water into the sucking pharynx. In the particularly well-studied desert cockroach, *Arenivaga*, the concentration of the salivary fluid alone is certainly inadequate to explain vapor uptake, and other cuticle-based mechanisms are proposed; these are discussed in relation to desert adaptations in Chapter 16.

Other vapor-absorbing arthropods use a rectal rather than a salivary site. The rectum is the normal site for uptake of liquid water from the gut anyway, and its tissues may have very high concentrations of salt to allow osmotic uptake, particularly where the gut has a cryptonephridial arrangement. Calculations show that this is adequate to account for the extraction of water from air above the 88% CEH (critical equilibrium humidity; see Chapter 5) of *Tenebrio* larvae. In effect, during vapor absorption the gut contents have a reversed fluid flow, and ongoing K^+ transport in the Malpighian tubules allows the build up of very high osmotic concentrations there, so extracting water from the rectal tissues and ultimately from the air within the rectum. However, the wingless insect *Thermobia* (commonly known as the firebrat) can take up water anally from air at 45% RH, while the rat flea, *Xenopsylla*, can operate at 65% RH, again via the rectal surfaces—yet neither has a cryptonephridial system. In the firebrat there are three anal sacs, with a much-folded and mitochondria-rich (presumably ion pump-rich) epithelium, and these may be filled with air and some hygroscopic material. The sacs open to the outside via anal valves, which open and close rhythmically, suggesting a cyclic process perhaps involving **pressure increases**, which would raise the humidity, making uptake easier. This observation links neatly to recent work on water vapor uptake in certain woodlice, where hyperosmotic fluid in the cavity between the pleopods and the abdomen is implicated, with uptake again augmented by pressurizing this cavity cyclically.

15.3 Thermal adaptation

For any animal on land, all the avenues of heat exchange discussed in Chapter 8 are available; a summary is shown in Table 15.9. Heat can readily be gained by conduction and convection from heated surfaces, or by direct radiation from the sun. Heat can be lost by these same three processes for an animal that moves into cooler air, away from heated surfaces and out of direct sunlight; crucially, it may also be lost by evaporation.

Remember the issues of terminology (see Chapter 8). Both **ectothermic** and **endothermic** strategies are favored in different circumstances. Ectotherms may be thermal **conformers**, but more commonly both strategies on land involve a fairly high degree of more or less expensive **thermoregulation**. Land animals may also be either **stenotherms**, operating over only a narrow range of body temperature (as in most endothermic tetrapods and at the opposite extreme in animals from very protected stable microhabitats) or **eurytherms**, coping with very variable body temperatures as necessary (as in many ectothermic insects and vertebrates). Different strategies and solutions tend to operate in different size ranges. Small animals (mostly eurythermic ectotherms) may use a considerable suite of behavioral adaptations to avoid thermal stress, while large animals (mostly stenothermic endotherms) may have elaborate behavioral and physiological techniques to keep their bodies at the preferred constant temperature. Life on land, as far as temperature is concerned, is not just a matter of increasing physiological or biochemical sophistication, as sometimes portrayed. It is more often a case of finding new balances of physiological and behavioral regulation, with the common endpoint being a raised and stable body temperature whenever an animal is active.

Table 15.9 Mechanisms of temperature regulation on land.

	Source	Route	Influences
Heat gain	The sun	Radiation Conduction Convection	Color, surface properties Size, and surface area exposed Behavior (posture, orientation, movement)
	Metabolism		Size and surface area Enzyme concentrations, mitochondrial density Special thermogenic tissues Superficial blood flow, vascular shunts Hormones, nerves Activity
Heat loss		Radiation Conduction Convection	Size/surface area, color, special exposed surfaces, behavior (posture, orientation, movement) Superficial blood flow, vascular shunts Hormones, nerves Activity
		Evaporation	Size and surface area Panting, sweating, licking, urinating Superficial blood flow, vascular shunts Hormones, nerves Activity

15.3.1 Terrestrial ectotherms

As we stressed in Chapter 8, ectotherms are not necessarily "cold blooded", or "at the mercy of their environments" with respect to body temperature, and these traditional terms are particularly inappropriate for many terrestrial ectotherms. Air has a lower thermal conductivity and specific heat than water, therefore it is relatively easy for a land animal to maintain a gradient between body temperature (T_b) and ambient temperature (T_a). In the cryptozoic and hygrophilic fauna a gradient is rarely observed, and the habitats have fairly equable conditions anyway, with an ambient temperature buffered from the more extreme changes in the neighboring freely moving air; animals living here are moderately eurythermic. However, for xerophilic animals, T_b may be regulated in a variety of ways, usually to keep it above T_a. This difference is related not only to habitat, but also to **size** (itself linked to habitat, since xerophiles can be, and usually are, much bigger than the cryptozoic and hygrophilic fauna). Small animals inevitably change temperature more rapidly than larger ones, losing and gaining heat on a per gram basis much more quickly, so they cannot maintain a gradient between body and air easily. But a reasonably large xerophilic ectotherm has every chance of regulating its T_b at high and fairly steady levels, principally using behavioral techniques to be active only at appropriate ambient temperatures, to sunbathe or seek shelter by day, and (perhaps less obviously) to seek warmer microniches at night; it can be regarded as stenothermic, at least during all periods of actual or imminent activity. Note also that preferred body temperatures, achieved T_b ranges, and normal metabolic rate values vary with habitat. For example, tropical species within the lizard genus *Sceloporus* have higher T_b values, metabolic rates about 1.6-fold higher, and higher water flux rates, than do closely related temperate relatives.

Where ectotherms are concerned, avoidance, tolerance, and regulation tend to merge into each other as strategic approaches to varying metabolic rate and achieving temperature balance. Thus animals may use behavioral techniques to avoid very cold microhabitats and avoid the risk of freezing, but could also thus be said to be regulating their T_b above ambient. Nevertheless, it might be useful to deal with avoidance and tolerance first, and then concentrate separately on adaptations that specifically tend to regulate T_b at particular levels.

Thermal avoidance and thermal tolerance

AVOIDING THE COLD

Behavioral choices constitute the most important, and the cheapest, ways of maintaining thermal balance, especially by timing an activity appropriately. Here we return to the crucial issue of microclimates. Staying in a warm burrow at night or in winter is the simplest of all strategies. But even when active, relatively small movements can take an animal away from cold spots and into a strikingly different thermal environment, hence bringing about a change in T_b.

Short-term movement effects can be especially striking for small terrestrial ectotherms such as insects, for whom shelter is readily available. Figure 15.18 gives examples of the benefits of climbing up a grass blade, rotating around the stem of a plant as the sun passes over, or moving around, under, or within a shrubby plant. Slightly more substantial movements may be needed for something the size of a lizard to avoid the cold, although choices are still available (Fig. 15.18d). However, garter snakes can achieve a better stability of T_b by picking a rock of just the right size and thickness to shelter under than by sun–shade shuttling through a day. This also makes the point that burrows are not the only form of shelter; crevices and gaps within rocks will serve, and in some cases a shelter may even be manufactured, in which case we usually call it a nest. Nests that are not also burrows are relatively uncommon amongst invertebrates or ectotherms generally, but examples may be seen in froghopper insects, which produce a frothy ball of foam ("cuckoo spit") around themselves, in "tent" caterpillars, in the use of rolled up leaves by insects and spiders, and of course in the social insects.

Figure 15.19 shows a few examples of insect behavior patterns in relation to ambient conditions (temperature, radiation, RH, and wind). Not only overall active periods and time above ground, but also orientation, basking postures, and even the whole hierarchy of feeding and reproductive behaviors shown are crucially dependent on the hygrothermal environment.

Movement in and out of shelter is usually a very short-range strategy. Longer range and longer term movements to avoid cold are generally termed migration, although the immediate stimulus to move may be food supply or photoperiod (see Chapter 8). The alternative way of effectively avoiding cold weather is "avoidance in time", insuring that during the coldest periods the body is "shut down" or dormant. Many insects achieve this process of dormancy by diapause (see Chapter 8), while other animals may use torpor (or "hibernation"), dealt with in Chapters 8 and 16. Dormancy in all its forms is conspicuously more common in terrestrial species than in any of the aquatic habitats covered in earlier chapters.

All that said, remember that many terrestrial ectotherms do nevertheless achieve activity and complete their lives with very low T_b values. This can even be true for some flying animals; classic examples are the winter moths (e.g. *Operophtera*), which can fly in freezing air temperatures with their T_b within 1°C of T_a, using a very gentle flapping locomotion.

COPING WITH FREEZING

Many terrestrial animals can survive freezing temperatures by freeze-tolerance or freeze-avoidance strategies. However, this is largely a phenomenon of colder climates, so we leave consideration of details to Chapter 16. Note, though, that there may be cold winter nights in many temperate habitats, and that invertebrates, amphibians, and reptiles in these biomes may have the same kinds of adaptations to cope with freezing as those discussed in section 16.3.2.

AVOIDING OVERHEATING

Ground-dwelling or plant surface-dwelling animals are particularly vulnerable to potential overheating since they live in the boundary layer of rapidly heated air. Caterpillars may lift their head and thorax up off the substrate (or allow them to hang down from a leaf undersurface) when overheated. But longer legs are a particular advantage in avoiding boundary layers, and many adult insects and arachnids will raise their trunk well above the surface by "stilting", effectively standing on tiptoe. Lizards may show similar behavior, with successively raised legs seen in animals moving over hot sand. Such behaviors not only lift the trunk out of the heated boundary

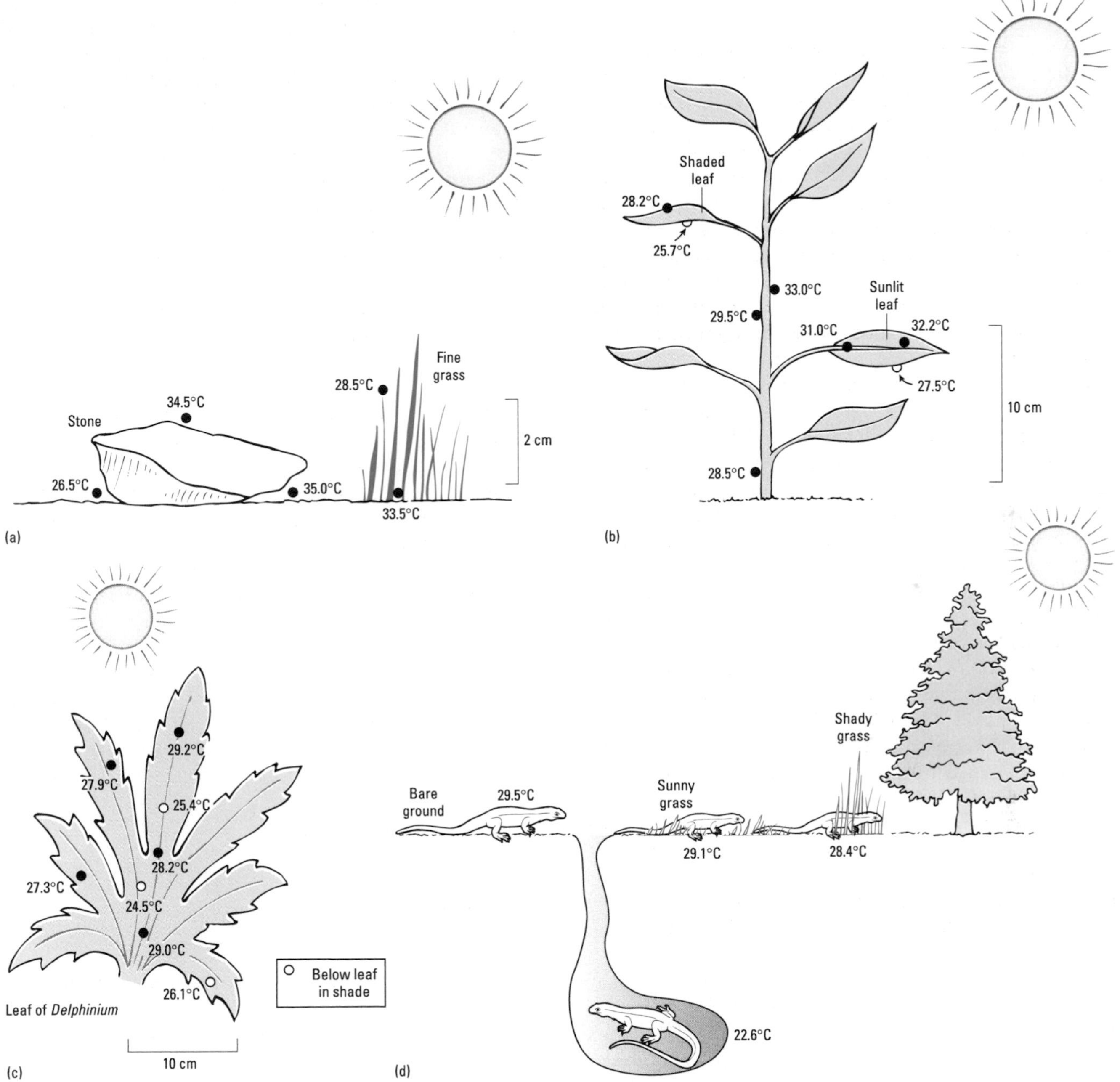

Fig. 15.18 (a–c) Microclimates and body temperatures of small insects (assessed from the temperature of a dead 50 mg insect with an inserted thermocouple): effects of climbing grass blades, moving around small rocks, around plant stems, and around leaves. (d) The effects are still present, though less extreme, for a much larger lizard.

layer but also increase convective cooling by placing most of the body mass in the air moving at a higher velocity above the still surface layer. This can be enhanced by orientating the body side-on to the direction of the wind. On a larger scale, moving to the top of a large rock or up a hillock will increase convective cooling.

Posture may also be used to avoid overheating, usually by orientating the body for minimum exposure to the sun (commonly "head-on"). Any extremities may also be withdrawn or tucked in, so that heat is not absorbed through their surfaces; in particular, the wings of insects are folded in.

Thermal regulation

The extent to which any animal can thermoregulate is roughly indicated by the slope of the relation between T_b and T_a. With a slope of 1, the dependence of T_b on T_a is total, indicating thermoconformity; something approaching this value may be observed in nocturnal ectotherms, and in many cryptozoic animals. However, values of 0.5–0.7 can still indicate a lack of thermoregulation, as such figures can be obtained even in passive systems such as water-filled metal

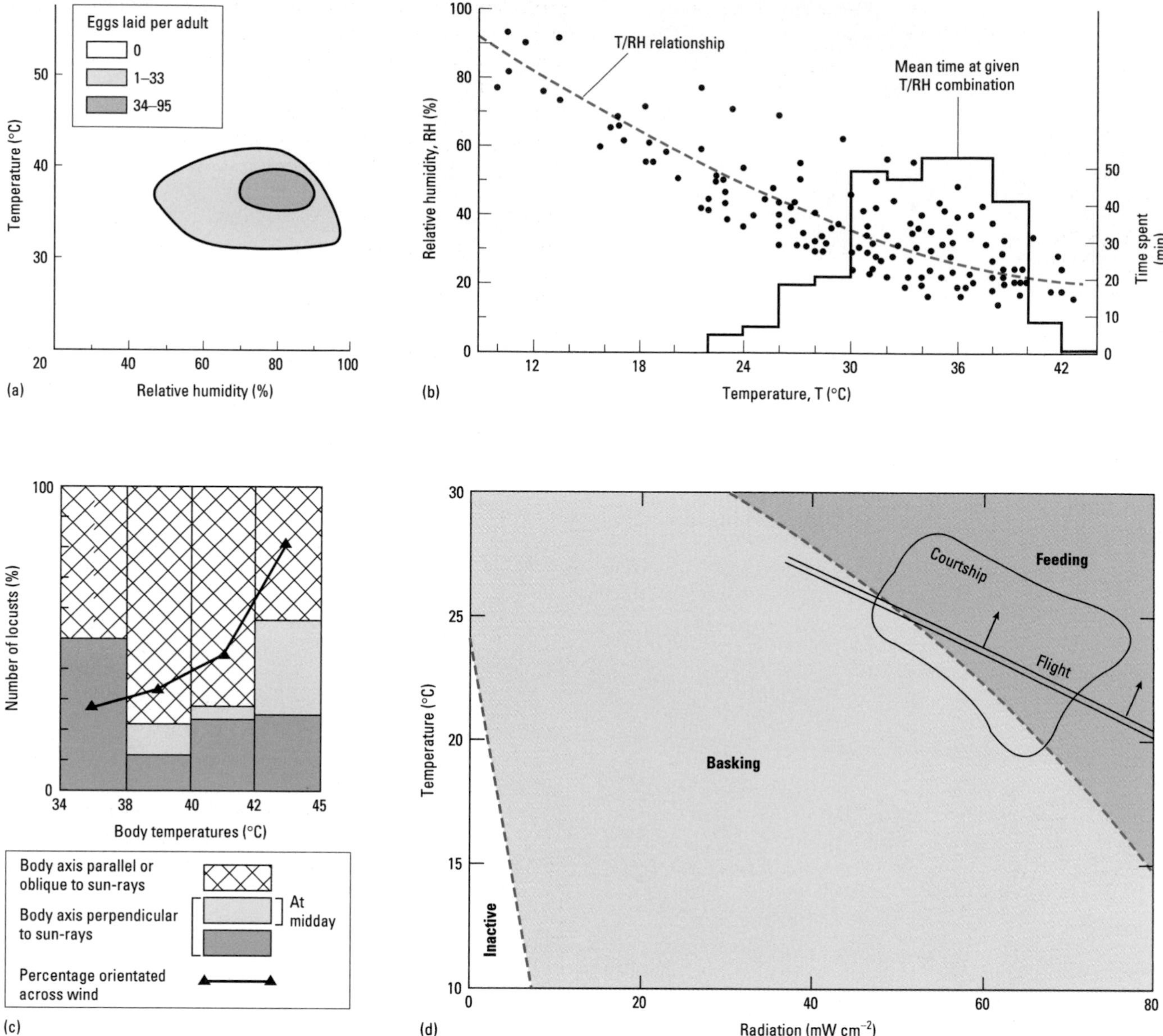

Fig. 15.19 Activities of insects in relation to microclimatic constraints. (a) Oviposition by the apterygote *Thermobia* in relation to temperature and humidity. (b) Active periods of the tiger beetle, *Cicindela*, with the advantages of high temperature overriding the problems of low humidity. (c) Orientation of locusts with respect to sun and wind at different body temperature (T_b) values. (d) Hierarchy of activities in the butterfly *Heodes* as temperature and radiation increase (note the wide range of conditions for basking). (Data from Sweetman 1938; Waloff 1963; Douwes 1976; Dreisig 1980.)

cans lying in the sun. With a slope of 0, though, an animal is clearly showing total independence of environmental temperature, indicating full thermoregulation. Examples of the values for this coefficient are shown in Table 15.10.

However, such data can be misleading for ectotherms. If the body temperatures are collected carefully in the field and only during active periods (usually in daylight) some species will properly appear to be good regulators with a steady high T_b, but where data are gathered over full 24 h cycles the picture will usually be rather different, with very low T_b values at night when the animal is quiescent. Furthermore, very misleading data may be obtained from laboratory studies where animals lack a normal range of choices of microhabitat or of behavioral repertoires. Hypothetical illustrations of these three kinds of assessment are shown in Fig. 15.20. Actual examples of plots of T_b versus T_a are shown in Fig. 15.21, for a range of reptiles and insects in reasonably natural conditions. Here we consider the ways in which terrestrial ectotherms can gain, lose, and conserve heat, to achieve this moderate thermoregulation.

BEHAVIORAL REGULATION TO GAIN HEAT

Moving to a warmer microclimate is the simplest and most obvious way to warm up, although on its own it only allows T_b to be raised to the new higher T_a. All the movements described above as part of the strategy of "avoiding the cold" could also be viewed as ways of

Table 15.10 Gradients of regression of body temperature (T_b) on ambient temperature (T_a) in terrestrial animals.

Animal	Gradient
Ectotherms	
Anolis lizard (forest)	1.17
Tenebrionid beetle	1.00
Salamander	1.00
Python (non-brooding)	0.48
Cicada (non-singing)	0.44
Anolis lizard (grassland)	0.43
Sphinx moth caterpillar	0.40
Heterotherms	
Honey-bee	0.81
Solitary bee (foraging)	0.65
Honey-bee (foraging)	0.55
Dragonfly	0.43
Naked mole rat	0.41
Bumble-bee (foraging)	0.23
Python (brooding)	0.12
Bumble-bee (brooding)	0.07
Endotherms	
Pocket mouse	0.08
Possum	0.07
Finch	0.05
Parrot	0.05
Weasel	0
Human	0

gaining heat. In addition, crouching against a heated soil or rock surface, within the still boundary layer, will aid warm-up, and this strategy (sometimes termed **thigmothermy**) is used by reptiles and by surface-dwelling insect types such as ground beetles and tiger beetles.

Basking (heliothermy) is one of the commonest additional mechanisms for gaining radiative heat, to raise T_b well above T_a. It is particularly effective in small ectotherms able to use limited patches of incident sunlight within vegetation, and in animals with wings that can be spread out to gather radiation; in other words, in pterygote insects. Temperatures in moderately sized insects can be raised at least 15°C above ambient by basking alone, and basking behavior may be carefully controlled by sensors on the wing itself, with a slow wing closure when the local temperature gets too high. The effect of wings as thermal absorbers (and reflectors) may in part explain why adult winged insects are often much better thermoregulators than larval wingless forms, even though the latter may have a higher body mass (see Fig. 15.21).

Basking postures may also be crucial. Insects exhibit three different basking modes (Fig. 15.22), usually termed dorsal basking, lateral basking, and reflectance basking. These are particularly evident in large-winged butterflies, which have a rather narrow thermal window for flight and to really exploit basking opportunities. The merits of the three possible postures are somewhat controversial; dorsal basking only works well when the insect is flattened against a warm surface, and the benefit may come principally from the warm trapped air. Body basking, or "reflectance basking", is supposed to involve the wings reflecting heat down onto the body, and occurs in mainly white-winged insects, but its effectiveness has been seriously doubted, and it may merely involve wings being partially raised to reduce convective cooling. Lateral basking may work partly by trapping a "greenhouse" of warmed air above the thorax and around the abdomen.

Basking also occurs in vertebrate ectotherms. It is not particularly common in amphibians, since if they achieve a raised T_b they also increase water loss through their permeable skins and cool down again evaporatively—basking achieves no thermal gains. But some species of *Bufo* and *Rana* have preferred T_b values in the range 25–30°C, and as long as water is freely available they will bask to achieve a T_b at least 15°C higher than in the shade. Reptiles are generally much more regular and predictable in their basking behaviors, and are not so restricted by water availability. The common European lizard, *Lacerta vivipara*, lives in habitats with yearly mean T_a values of around 9°C yet has a preferred T_b of 30°C and an active temperature range of 28–32°C; it can warm from 15 to 25°C within about 5 min in good sunshine. Many reptile species from fairly open habitats such as grasslands and heaths, or from deserts, are

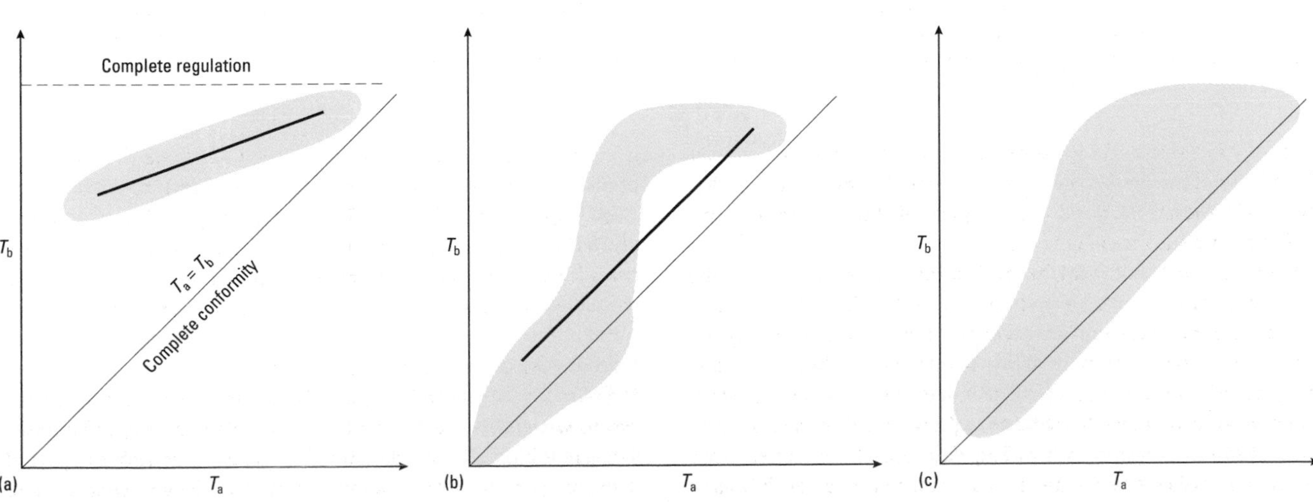

Fig. 15.20 Hypothetical relationships between body temperature (T_b) and ambient temperature (T_a) for a small ectotherm on land. (a) Field measurements, daytime only, showing good regulation (behavioral). (b) Field measurements for 24 h cycle, with regression line concealing the daytime regulation and night-time conformity. (c) Laboratory studies may give various combinations of these results depending on the choices available to the animals.

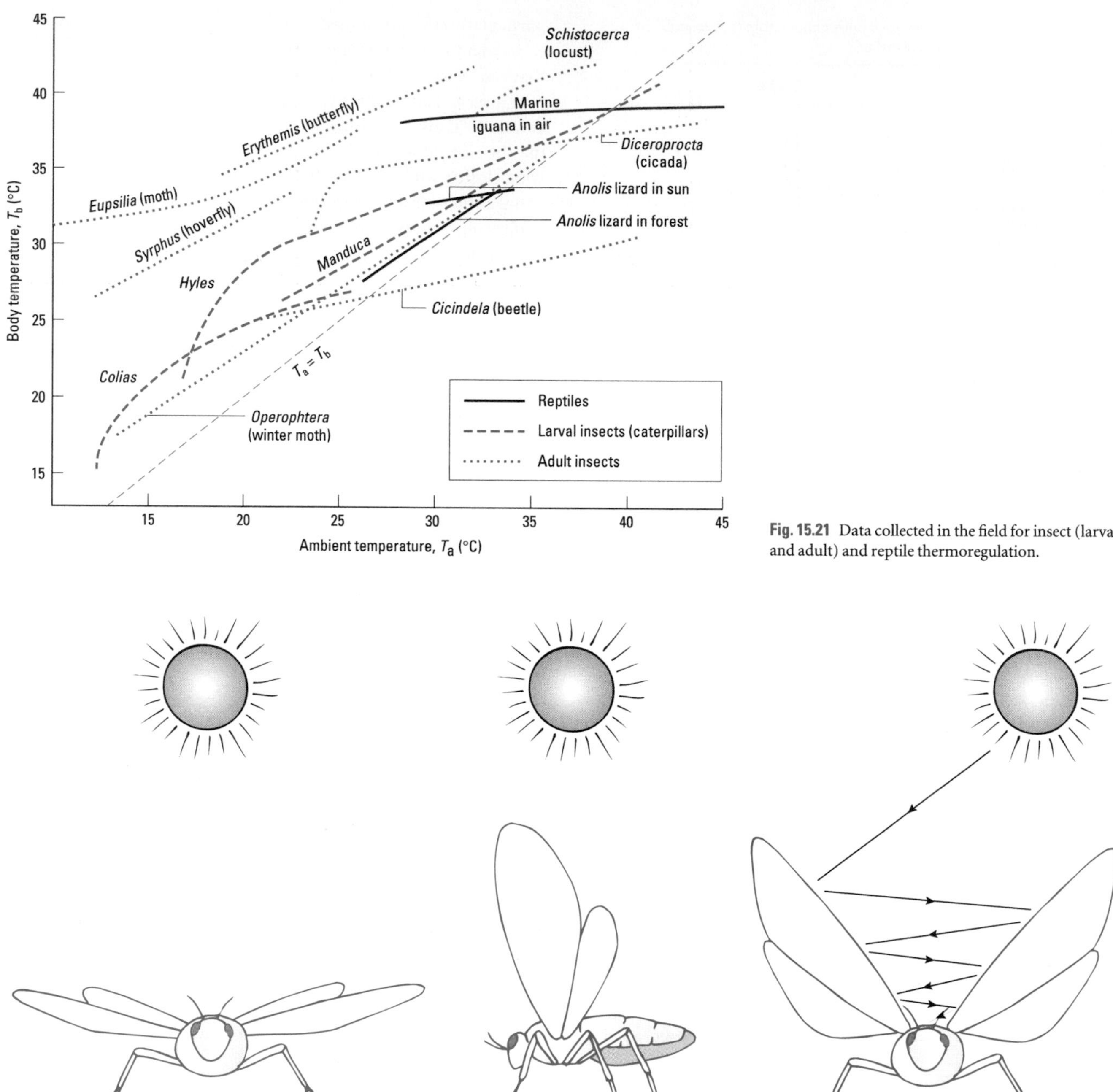

Fig. 15.21 Data collected in the field for insect (larva and adult) and reptile thermoregulation.

Fig. 15.22 Three postures that may be used for basking in winged insects (see text for more details).

sophisticated basking heliotherms, but forest species rarely find sun-spots large enough to accommodate their lengths. The precision of thermoregulation at a particular preferred temperature can be exceptional in a medium-sized reptile (Fig. 15.23), alternating between side-on basking (often with crests erected) for maximal heat gain and head-on postures to reduce heat uptake. But the preferred temperature is often variable within a species, for example being reduced in poor-quality habitats where food is sparse or predation risks high. The iguanid *Dipsosaurus* maintains 39.1 ± 2.0°C in favorable habitats, but only 32.9 ± 4.0°C (lower and with greater variance) in poor environments where thermoregulation bears a high cost. Within each habitat, reptile T_b values are strongly correlated with success, measured in the short term as locomotory speed or rate of prey capture, or in the longer term as growth rate or reproductive output.

Color is particularly important for basking species from many taxonomic groups. Since about 50% of the radiant energy from the sun is in the visible wavelengths of the spectrum, visible reflectance (the color seen) affects radiative heat gain significantly (see

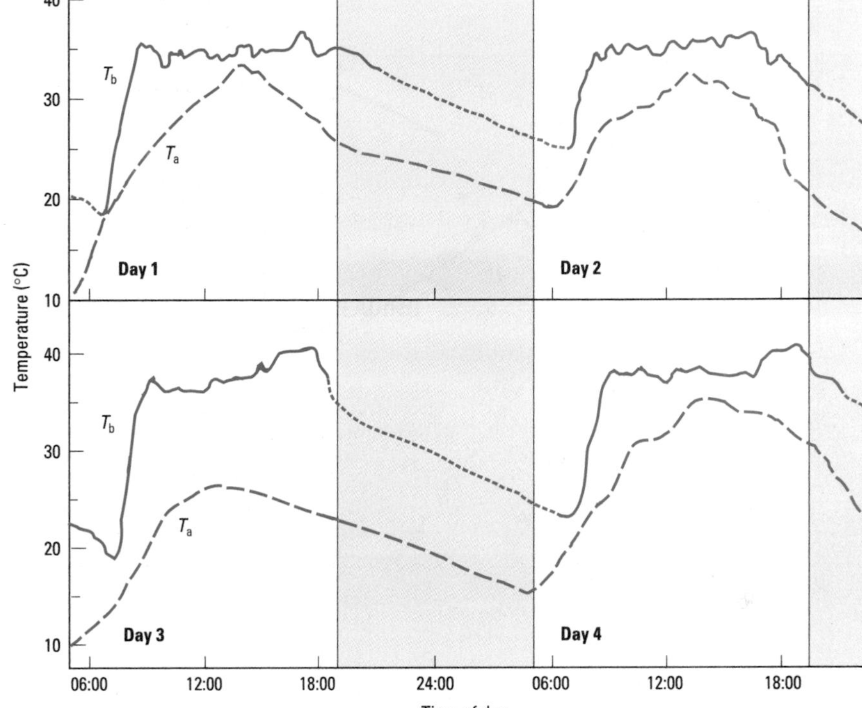

Fig. 15.23 Patterns of temperature through a day for a *Varanus* monitor lizard, measured by telemetry, showing the rapid rise in body temperature (T_b) (to well above ambient temperatures, T_a) due to early-morning basking and subsequent effective thermoregulation through the daylight hours by sun–shade shuttling. Note that the lizard has high thermal inertia curled in its treehole den at night so that T_b falls quite slowly. (From Cossins & Bowler 1987, with kind permission from Chapman & Hall.)

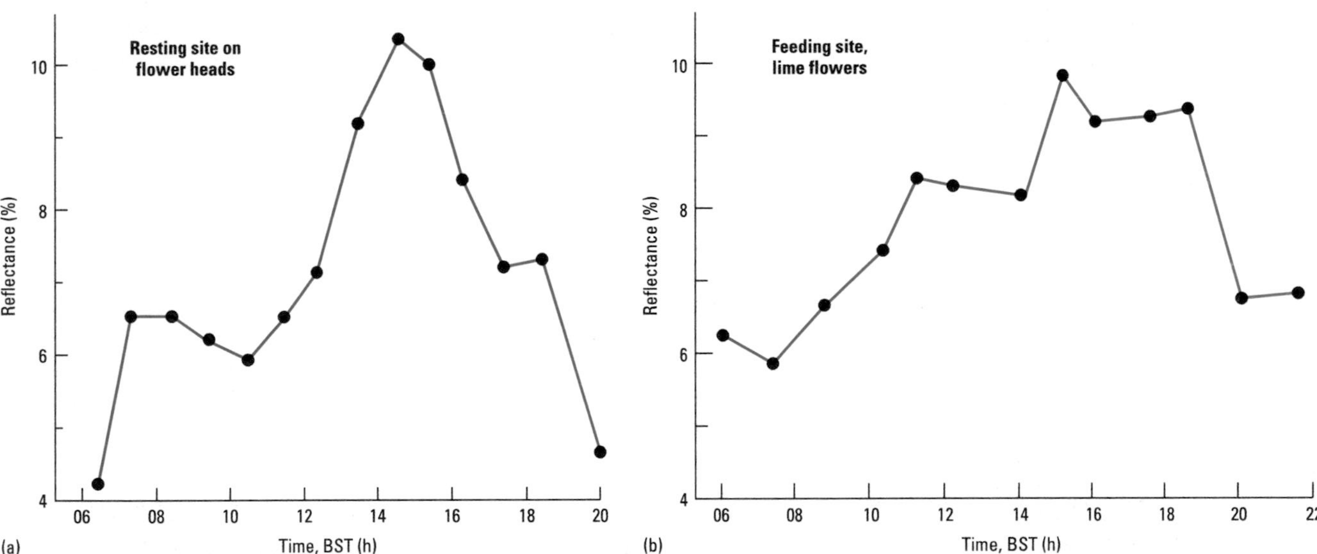

Fig. 15.24 Patterns of insect color (reflectance) through a day at (a) a basking site, and (b) a feeding site on a flower. In both cases dark species are commoner towards dawn and dusk (having higher thermal absorption rates) while only light-colored highly reflective insects are active around midday in full sun. BST, British Summer Time. (From Willmer 1983.)

Chapter 8). Insects with white and pale bodies warm more slowly than dark ones; small insects active at dawn and dusk are more likely to be dark, while the middle of the day may be dominated by insects with bright or pale colors (Fig. 15.24). However, insects that live close to the ground and gain heat largely through reradiated longwave radiation will be less affected by their color; they are commonly brown or black irrespective of their diurnal activity pattern. Some insects have both white and dark surfaces and may change posture according to thermal need. Vertebrates may use similar strategies; many savanna antelope have white faces, rumps, and ventral surfaces, minimizing the absorption of shortwave reflected radiation from the pale sandy soils and when orientated head-on to the sun.

Given the effects of color on radiative gain it is no surprise to find many ectotherms able to change color as a thermoregulatory strategy. Some species of frog can blanch when heat stressed, changing their reflectance from 35 to 60%. Chameleons are particularly famous for their ability to alter their coloration and indeed their

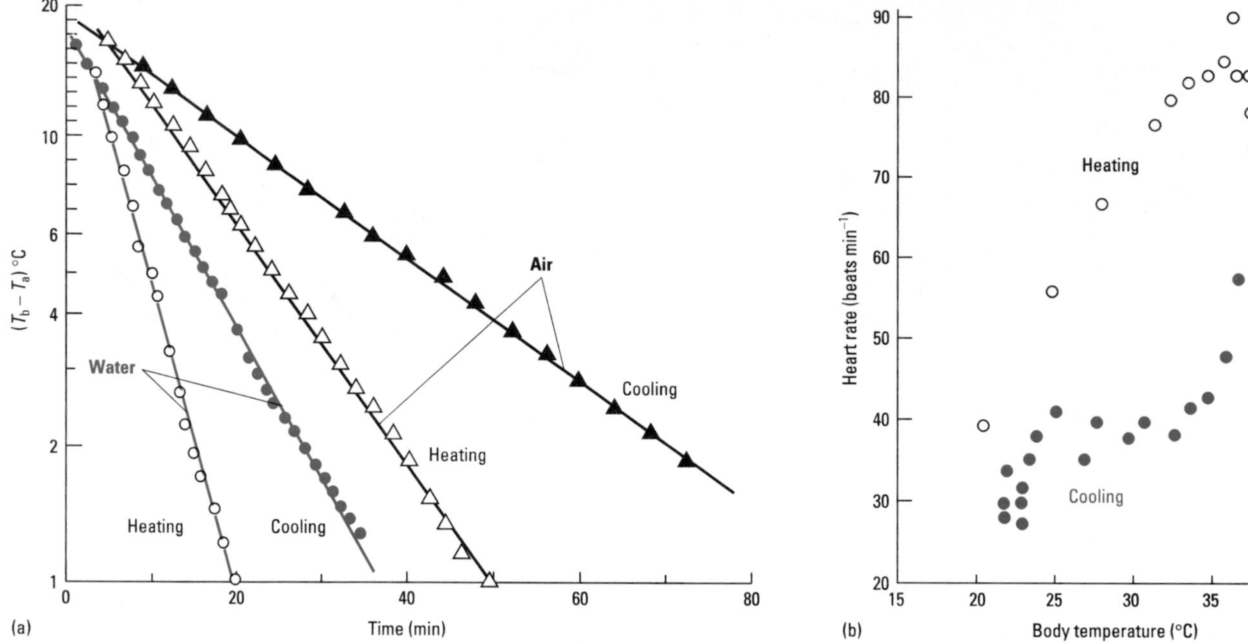

Fig. 15.25 (a) Heating and cooling rates of the marine iguana *Amblyrhynchus cristatus* in water and in air, showing that heating is always faster than cooling. T_a, ambient temperature; T_b, body temperature. (b) This is due to blood shunting and altered heart rate: when cooling there is a marked bradycardia, so blood moves slowly and stays away from the periphery, allowing heat to be retained in the core and reducing the overall cooling rate. (Adapted from Bartholomew & Lasiewski 1965.)

shape, balancing camouflage and thermal need. A cold chameleon darkens and basks, irrespective of other requirements for concealment or aggressive coloration. A warming *Chamaeleo dilepis* can increase its reflectance from 31% at 20°C to 46% at 35°C, with the major changes incurred in the visible and near infrared range (600–1000 nm). This ability is calculated to give a mean change of 0.7°C in its equilibrium temperature averaged over a whole year. Other lizards can also alter between a sandy color and dark brown, or through green and blue shades. The lizard *Urosaurus ornatus* can change parts of its belly and throat from light green to intense blue using the thermally sensitive iridophore system described in Chapter 8. Some insects too can darken or lighten according to need; for example, certain grasshoppers and dragonflies alter from bright metallic blue or green when warm to almost black when cooled.

BEHAVIORAL REGULATION TO CONSERVE HEAT

Postural effects to conserve heat in terrestrial ectotherms can be used to retain the heat gained by basking and heliothermy. General "balling" postures, with extremities tucked in, are very common, and perhaps best exemplified by the coiling response of snakes, which reduces the effective surface area by 50–70% in the early part of the night, retains heat gained by day, and probably speeds up digestion significantly.

Huddling and aggregation also serve as excellent methods for conserving body heat in ectotherms. Several larger snakes (pythons, anacondas, and boas) can be found clustered together at night in the cooler parts of their ranges, slowing down the loss of heat gained by basking. Insects aggregated on their food plants may gain a similar advantage, for example in the tent caterpillars (where the silken tent serves as a reasonably windproof greenhouse) or in various butterflies that mass together in the evening, thereby reducing convection within the group.

PHYSIOLOGICAL REGULATION TO CONSERVE HEAT OR LOSE HEAT

Regulatory conservation or loss of heat in ectotherms by nonbehavioral means is relatively rare, but not impossible. This underlines again the important distinction between the ability to generate heat by endothermy, and thermoregulatory ability to maintain a stable T_b (see Chapter 8). Ectotherms can achieve regulation by vascular adjustments (blood shunting), or by using evaporation. **Blood shunting** to control rates of heat loss occurs quite widely in reptiles, and an example is shown in Fig. 15.25. Rates of cooling are substantially lower than rates of heating, allowing the iguana to retain its acquired heat for longer, because heart rate is slowed and peripheral circulation decreased, reducing heat loss through the skin. A few ectothermic insects appear to achieve a similar conservation of heat by preventing hot blood in the thorax from losing its heat to the cooler abdomen, using a countercurrent exchanger at the "waist".

Evaporative cooling is an excellent way of dissipating heat if, and only if, the consequent loss of water can be tolerated. So it is only common in larger terrestrial ectotherms as a specific thermal strategy. In many small or hygrophilic ectotherms, such as worms, snails, and amphibians, it occurs unavoidably whenever the animal gets warmed by the sun, and in some woodlice the T_b can be depressed by as much as 4–8°C as a result. This could be termed passive evaporation, and the T_b reduction and water loss are often a problem for the animal rather than an advantage. However, some ectotherms, especially reptiles and insects, are able to enhance their evaporative cooling when at risk of overheating, by increasing either cutaneous or respiratory water loss (CEWL or REWL). Insects that can "sweat"

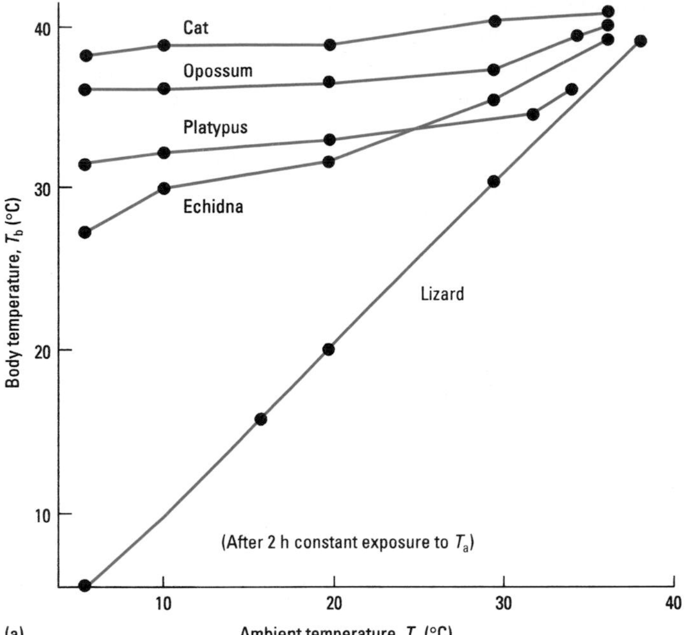

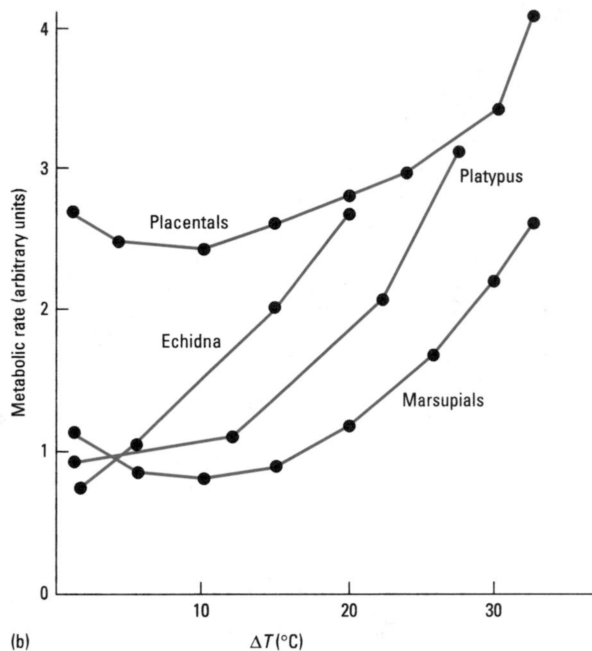

Fig. 15.26 (a) Plots of body temperature (T_b) against ambient temperature (T_a) for endothermic mammals and a lizard in laboratory studies, showing the better regulation in the eutherian cat compared with monotremes and marsupials. (b) Metabolic rate (MR) against ($T_b - T_a$), showing the thermoneutral zone where MR is roughly constant (for marsupials and placentals), rising at either extreme. (Data from Martin 1980.)

through pores in their cuticles are now reasonably well known, and include desert cicadas such as *Diceroprocta* (see Chapter 16); these have access to ready water supplies from their sap-feeding habit, and can maintain a temperature of 37–38°C at a T_a of 42°C. A few insects also use excretory fluid to achieve cooling—the larvae of the sawfly *Perga* can raise their rear segments vertically and allow excreta to trickle down their exterior surfaces. Two genera of the so-called "waterproof frogs", *Phyllomedusa* and *Chiromantis*, are also able to increase their normally low CEWL when their body temperatures rise above 37–38°C. In both genera this is due to a rapid onset of watery secretion from epidermal glands, but in *Phyllomedusa* it additionally involves an apparent melting of an epidermal waxy layer. The iguanid lizard *Dipsosaurus* pants whenever T_b exceeds 40°C, and the agamid lizard *Amphibolurus* can use panting to reduce its T_b from about 42 to 39°C. Other lizards use liquids (saliva) regurgitated from the mouth over the throat region to achieve cooling.

15.3.2 Terrestrial heterotherms and endotherms

Endotherms derive much of their body heat from metabolic processes (endogenous heat) rather than from the sun or from geological sources. They can therefore maintain a fairly constant T_b by moderate thermoregulation, though they do not necessarily regulate very precisely. Some animals can be endothermic for just part of their lives; they are more properly termed heterothermic. These include both traditional ectotherms, such as some insects, which can turn on endothermic processes (often in just part of their bodies) when they need it to initiate activity; and traditional endotherms, such as hummingbirds and small rodents, which can turn down or turn off endothermic processes when they become too costly to maintain. Other animals, notably the great majority of birds and mammals, are endothermic throughout their lives.

Types and abilities

The basic physiology of a typical endotherm can best be appreciated with a display of the relations between T_a and T_b, and between temperature excess ($T_b - T_a$) and metabolic rate (MR), shown in Fig. 15.26 for various mammals. T_b is more or less constant for all ambient temperatures, with a slight upturn occurring as T_a rises beyond about 35°C. At the same time, for placental and marsupial mammals MR is reasonably constant for moderate ambient temperatures but rises at either extreme; at low T_a it rises as the animal compensates for high rates of heat loss, and at high T_a it rises as the ability to thermoregulate begins to break down. Between these two phases of rising MR the animal is in its **thermoneutral zone** (TNZ), and is maintaining its "normal" T_b. More explanatory details of the relevant terminology were given in Chapter 8.

The values of normally maintained T_b for bird and mammal families were shown in Table 8.11, revealing the generally higher fixed values in birds (about 40°C) compared to mammals (37–38°C), and the rather low values for ratite birds and for monotreme, marsupial, and insectivore mammals. Certain taxa also have unusual levels of energy expenditure that affect their thermoregulatory responses; again the Insectivora are notable, where the smallest members (shrews) possess very high mass-specific MRs and the largest members (e.g. hedgehogs) have lower than predicted values, giving an unusually low slope to the classic BMR/M_b (basal metabolic rate/body mass) plot (see Chapter 6). There is no obvious overall trend of T_b with size in the mammals as a whole, however. Nor do mammals and birds from cold climates maintain body temperatures any different from their temperate and tropical relatives, although they have a broader TNZ and may briefly tolerate greater extremes of T_b. The differences between birds and mammals, and between temperate and cold-adapted species, seem to be largely explained by variations in thermal conductance (insulation). Indications of thermoregulation are again best given by the coefficient relating T_b to T_a, and values are included in Table 15.10, showing examples of the very tight regulation in birds and mammals ($k < 0.1$). However, Table 15.10 also shows some rather loose thermoregulation in a few birds and in the marsupials, and a lack of thermoregulation in the naked mole rats (the only mammals that hardly use endothermy, living more or less permanently in subterranean social nests where T_a may be constant, for example at 26–28°C in *Cryptomys* nests from East Africa). Table 15.10 also shows that some insect heterotherms are nearly as good at thermoregulating as the more strictly endothermic vertebrates.

Avoidance and tolerance

Behavioral avoidance techniques for endotherms are similar to those in ectotherms, for avoiding cold areas, avoiding freezing, and avoiding overheating. However, the use of behavior as a way of achieving equable microclimates and T_b values is limited by the inherently large size of endotherms. With body masses usually in excess of 1 g, and linear dimensions nearly always at least tens of millimeters, it becomes much harder to use the fine-grained variation in climatic conditions. Most of the body is inevitably above the still boundary layer of air above a soil or rock surface, so that stilting and other postures make little difference to convective exchanges. Endotherms may have a choice between a patch of incident sunlight and the shade under a bush, but they cannot normally choose between the two sides of a leaf.

Nevertheless, some aspects of behavioral avoidance are still very effective. Behaviors that minimize thermal conductance can help animals to limit cooling; assuming a roughly spherical shape, with the body curled up and paws and nose tucked in, is the ideal for a mammal, whilst a fluffed-up squatting posture with the legs covered by feathers and the head drawn down is characteristic of a chilled bird. Many endotherms enhance these insulation/conductance effects either with burrows or by gathering extrinsic materials; nests built by their own efforts or borrowed from other animals may reduce effective conductance and metabolic rate. For example, a lemming surrounded by cotton wool or pieces of collected fur and feathers decreases its own thermal conductance by 40–50%. Similarly a vole maintained in the laboratory at 4°C without nesting materials may have a daily energy expenditure of 3.5 kJ g^{-1}, compared with a value of 2.3 kJ g^{-1} when nest materials are freely available. Huddling behaviors give further benefits; the same vole given no nest but the company of three other voles can lower its daily expenditure to 3.0 kJ g^{-1}. All these effects can be demonstrated under natural conditions as well. For example, shrews in winter have an overall oxygen consumption 20% lower when they have access to a nest than when their nest is sealed off.

Behavioral mechanisms in general tend to keep an animal within its thermoneutral zone and reduce or eliminate the need for raised metabolic rates or more complex physiological regulation.

Thermal regulation

ENDOTHERMIC HEAT GAIN

As we saw in Chapter 8, the principle mechanisms of increasing heat production involve increased activity of skeletal muscles. This may be achieved by exercise, but the drawback is increasing blood flow to the extremities and dishevelling the pelage of birds and mammals, so that heat-loss routes are also increased. The better option is to uncouple the muscles from locomotory effects and simply shiver. Endothermic vertebrates thus show a clear inverse relation between rate of **shivering** (as recorded by an electromyograph) and metabolic rate (Fig. 15.27). These animals are regulating the extent of their shivering to compensate for cold and to generate internal heat, though the slope of the relation is much shallower for larger animals than for small ones. Shivering tends to have a specific onset temperature (shivering threshold temperature or STT), varying between species and also variable according to acclimation state. In king penguins the STT is about −18°C when cold-acclimated but as high as −9°C after prolonged exposure to 25°C.

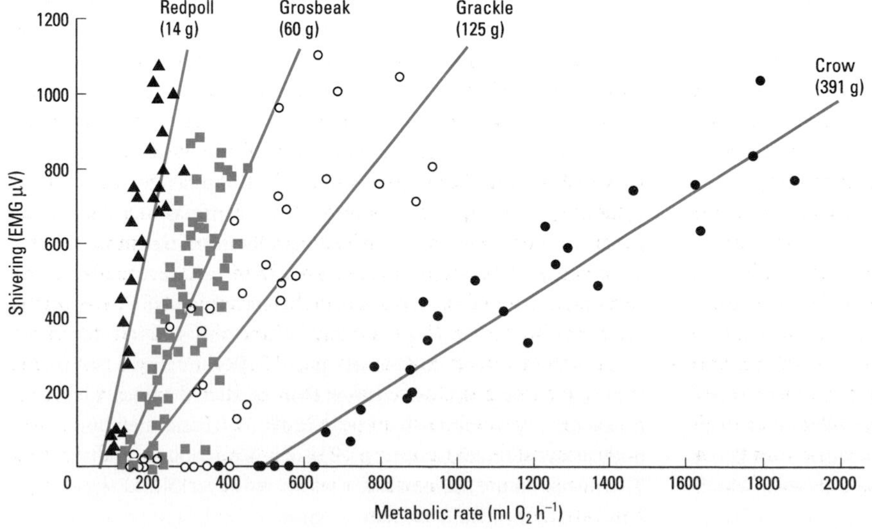

Fig. 15.27 The linear relation between the degree of shivering in various birds and their metabolic rate. The relation is flatter for larger birds, but in all cases shivering substantially elevates metabolic rate and allows warming. (Shivering was quantified from electromyograms, EMGs.) (From West 1965, courtesy of University of Chicago.)

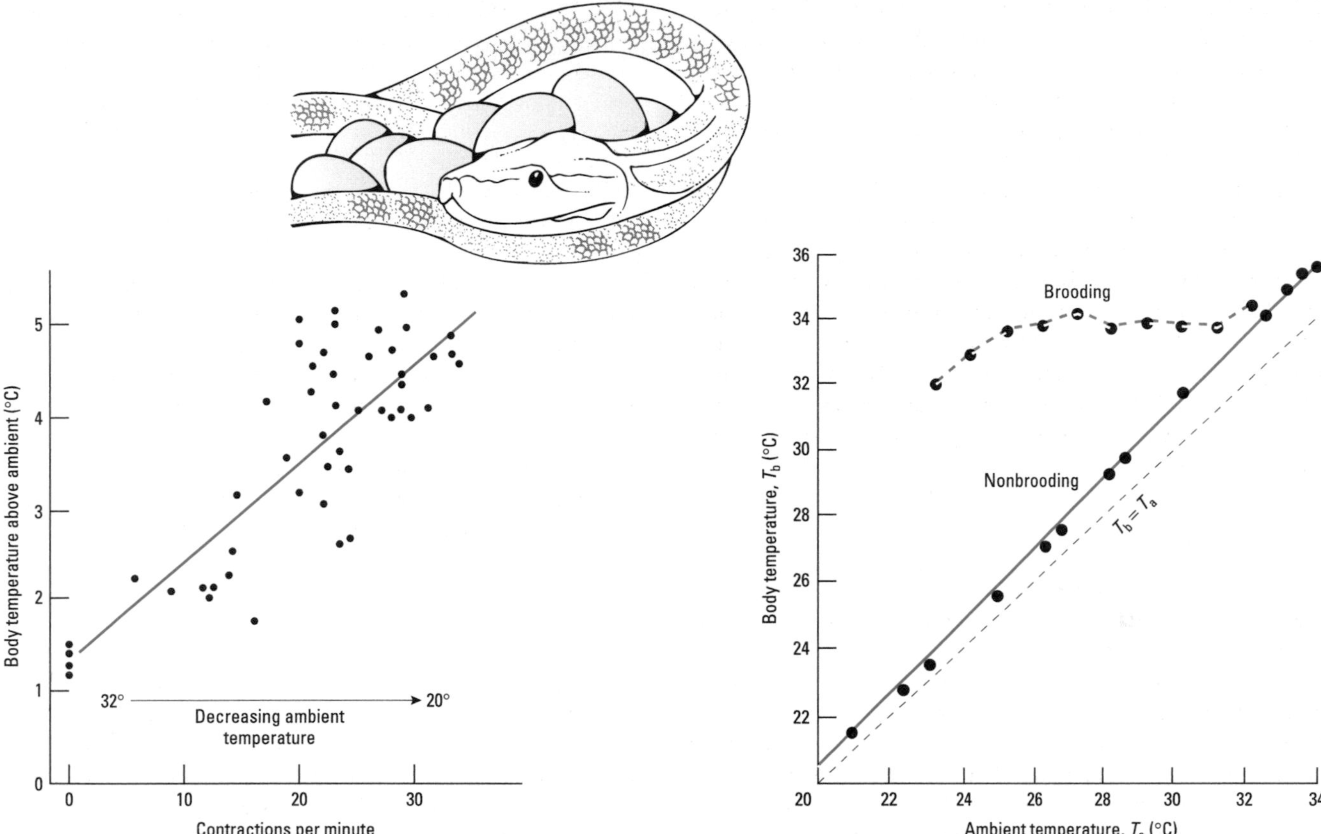

Fig. 15.28 Thermogenic incubation in a python; curled around her eggs, the female contracts her trunk muscles rhythmically to raise metabolic rate and increase T_b. (Adapted from Hutchinson *et al.* 1966; van Mierop & Barnard 1978.)

A few snakes, such as the Indian pythons (*Python* spp.), use similar shivering mechanisms to achieve endothermic warm-up, during the periods when they are incubating eggs (Fig. 15.28). The female wraps her body tightly round the clutch and produces low-frequency but powerful spasmodic shivering in the trunk muscles for long periods, raising her T_b and thus the egg temperatures to around 30–33°C, perhaps 7–8°C above ambient, allowing faster development.

Certain insects can also shiver. Some moths, beetles, dragonflies, and many bees need to warm their muscles before flying, to achieve an adequate power output (about 100 W) for take-off, i.e. they are obligate heterotherms (see Chapter 8). For example, the saturniid moth *Hyalopohora* requires a muscle temperature of at least 35°C to reach the required power output for flight, but once airborne it is able to maintain its thorax at 35°C and fly and feed at air temperatures as low as 10°C. Some winter-flying moths, such as *Eupsilia*, can fly with a thoracic temperature of 30°C at subzero ambient temperatures. Bumble-bees (*Bombus*) operate best at a muscle temperature around 40°C with a wingbeat frequency of about 50–100 Hz, and must commonly use endothermy to achieve this temperature before take-off. Retention of heat during the warm-up period is aided in all these insects by long scales or dense fur covering the thorax, or internally by elongated air sacs. More details of insect endothermy and other uses for it are given in Table 15.11.

The ability to warm up in insects has been most extensively studied in the bees, where activity patterns must coincide with times of nectar and pollen availability. Since these resources often peak in the early morning there may be a strong selection for flight at low T_a. There is a significant positive correlation between warm-up rate and body size (Fig. 15.29), probably resulting from the greater surface area to volume (*SA*/*V*) ratio and higher heat-loss rates of the smaller species. Tropical species may have lower warm-up rates than cool-temperate species of similar size; however, this is complicated by a taxonomic difference between the bees present in temperate and tropical/desert fauna. Species active at low temperatures have higher metabolic rates, and even within a species this trend can be detected between sea-level and high-altitude (cooler) populations. The ability of individual flying bees to vary thermogenic output is unclear, although honey-bees (*Apis*) certainly seem able to vary metabolic heat production to maintain thermal stability while in flight.

Nonshivering thermogenesis (NST) is another possible way of achieving warm-up (see Chapter 8). It occurs in many placental mammals and some marsupials, especially in young animals, but it is rare in birds. It can produce a 2–4-fold increase in metabolic rate in a small mammal such as a rodent, bat, or rabbit. NST involves the release of heat from "futile cycling" of metabolic substrates in the cytoplasm or mitochondria (see Fig. 8.35). NST may occur in the liver and some muscles, but in mammals it is commonly concentrated into special brown adipose tissue (BAT). This tissue is laid down around the shoulder region of many juvenile and hibernating mammals, where it provides a warm-up site close to the crucial heart and respiratory muscles.

Taxon	Mass range (mg)	Preactivity thermogenesis?	Biochemical thermogenesis?	Controlled blood shunting?
Odonata				
Libellulids	70–600	Yes		No?
Aeshnids	120–1200	Yes		Yes
Orthoptera				
Tettigonids (katydids)		Yes (sing)		?
Gryllotalpids (mole crickets)		Yes (sing)		?
Neuroptera				
Ascalaphids	60–250	Yes		Yes
Lepidoptera				
Sphingids	200–3000	Yes		Yes
Saturnids	200–2500	Yes		Yes?
Lasiocampids	90–120	Yes		No
Lymantrids		Yes		?
Hesperiids		Yes		?
Coleoptera				
Scarabeids				
Scarabeus	1300–5400	Yes (fight, roll dung)		No
Cotinus	600–2000	Yes		No
Megasoma	10,000–35,000	Yes	No	?
Trichostetha		Yes		?
Diptera				
Syrphus	18–40	Yes		No?
Eristalis	70–150	Yes		Yes?
Sarcophaga	30–130	No		Yes
Gasterophilus	30–240	Yes		No
Calliphora	20–100	Yes		?
Glossina	20–80	Yes		?
Hymenoptera				
Apids (bees), most, e.g.				
Apis	80–120	Yes (+ brood)	No?	Yes
Bombus	240–600	Yes (+ brood)	Yes?	Yes
Xylocopa	400–2000	Yes		No?
Euglossines	60–700	Yes		?
Centris	100–200	Yes		No?
Andrena	30–50	Yes		?
Anthophora	90–600	Yes		Yes
Vespids (wasps), some, e.g. *Vespa*, *Vespula*	50–500	Yes (+ brood)	Yes	?

Table 15.11 Insect heterothermy and thermoregulation. Activity after warm-up is flying unless otherwise stated. Blanks indicate that information is unavailable. Fighting and dung-rolling need a high/controlled body temperature to give a competitive edge; singing needs it to maintain a good output at the correct (species-specific) frequency.

NST may also occur in bumble-bees, although not in other endothermic insects (and some regard it as unproved even in the bumble-bee). It involves futile cycling between fructose 6-phosphate and fructose 1,6-diphosphate, the thoracic muscles having unusually high concentrations of the two necessary enzymes phosphofructokinase (PFK) and fructose biphosphatase.

BEHAVIORAL HEAT GAIN AND LOSS

Endotherms can of course also use behavioral techniques to gain heat as their "cheapest" option, particularly at times of day when ambient temperatures are low but sunlight is available. They may crouch against east-facing sun-warmed rocks to gain heat by conduction from the ground in the hour after dawn. They may also employ more traditional basking postures, just as in ectotherms, to gain heat. Mammals as large as camels will bask in the early morning sun, while small rodents and insectivores may get a significant heat input in this way, reducing their need for metabolic thermogenesis. Birds will also bask, and a herring gull basking in the sun may have a measurably lower metabolic rate than a nearby gull in the shade. Perhaps most spectacularly, the North American roadrunner, *Geococcyx californianus*, allows its T_b to drop by as much as 4°C overnight and then exposes as much of its body as possible to the sunrise to regain its "normal" T_b.

Large animals may also use behavior to dissipate heat. Erection of large and well-vascularized surfaces such as ears (in elephants or desert foxes) will give a significant heat loss, especially with postures and orientation adjusted to the height and direction of the sun. The spreading tails of some desert rodents may be used as radiators in some circumstances, but may also have a parasol effect; and birds may spread their wings open in shade and into a breeze to radiate excess heat.

Heat loss also comes from a quick dip in water or wallow in mud, as seen in many African savanna mammals such as hippos and elephants. Bouts of temporary submersion followed by moderate

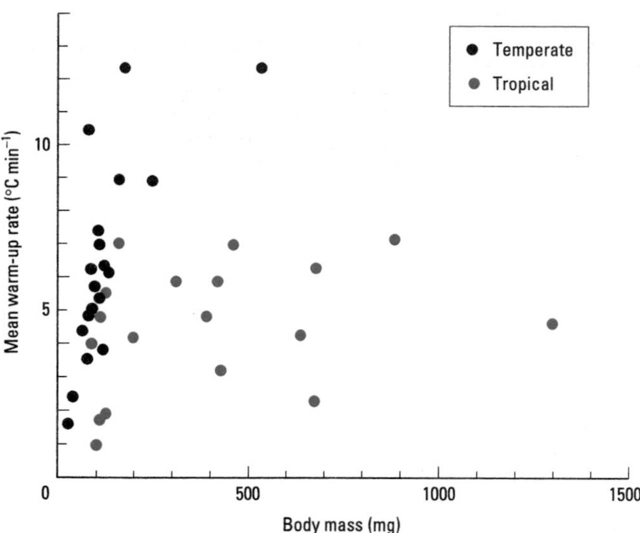

Fig. 15.29 Mean warm-up rates in a range of bee species from temperate and tropical habitats, showing an apparently faster warm-up at lower body mass in the temperate species. (In fact this is partly due to a phylogenetic effect, certain families of bee with large bodies being more common in tropical zones.) (Adapted from Stone & Willmer 1989.)

sunbathing give evaporative cooling that may be particularly useful for a very large endotherm where internal heat generation may overload the ability of dry surfaces to dissipate heat.

Huddling may be used by endotherms to reduce heat loss, particularly in small mammals and birds, such as bats and wrens, that overwinter in cooler climates where food is scarce. It is particularly useful in flying animals unable to accumulate large lipid reserves to see them through the winter at "normal" endothermic metabolic rates. Bats may aggregate in groups of several hundred, not to maintain normal body temperatures but to insure that the T_b values of individuals, all of which are torpid, do not fall below 5–10°C, so avoiding any risk of freezing. Wrens also huddle in winter, with up to 60 entering a single manmade nest box.

Heterotherms also use huddling, conspicuously in the overwintering social honey-bee, *Apis mellifera*. Whereas in most temperate bees only reproductive females overwinter, in *Apis* the whole colony persists, as an almost spherical dense cluster on the central nest combs, with a temperature gradient from core to periphery (Fig. 15.30). The cluster forms when the T_a falls to around 15°C, and inserted thermocouples show a steady 20°C within the mass of bees throughout the ensuing winter, rising to over 30°C as brooding begins in spring. This colony homeostasis is probably largely due to the much reduced *SA*/*V* ratio of the cluster, rather than to extensive thermogenesis by individual bees, at least in the larger clusters. Indeed, in very large groups of perhaps 5000 bees there is little evidence of sustained endothermic muscular activity. Individuals do move around within the cluster, with a positional cycling such that cooler bees migrate in from the periphery; and the temperature of the cluster is controlled by the degree of packing, becoming denser as the ambient temperature falls.

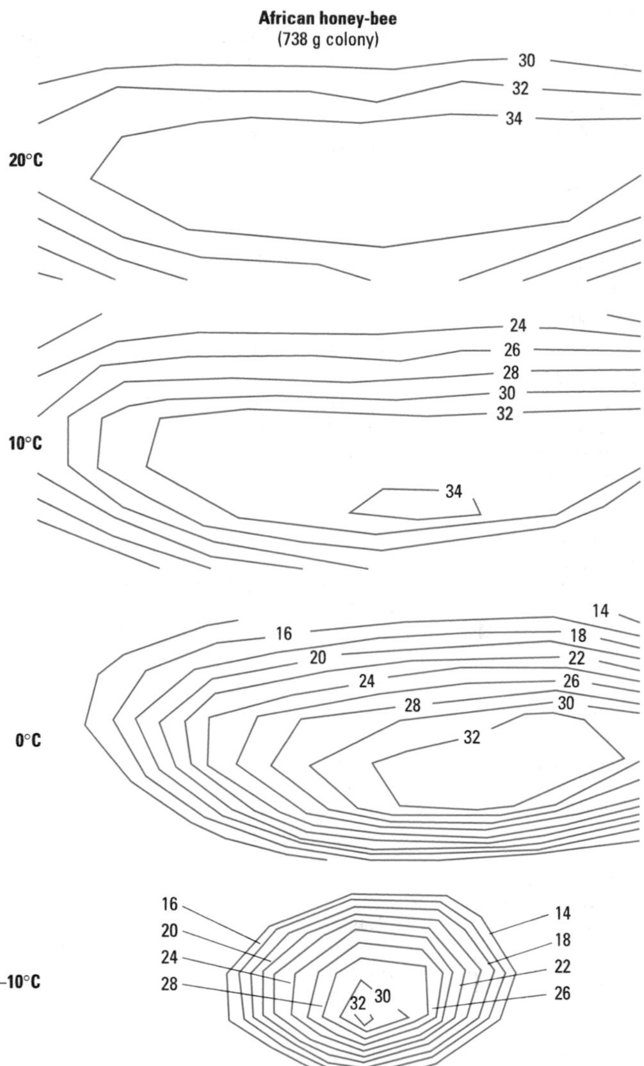

Fig. 15.30 Thermoregulation in honey-bee swarms. Profiles are shown at four different ambient temperatures, with the bees much more tightly aggregated as external temperatures fall, giving a much greater thermal gradient across the cluster.

HEAT CONSERVATION BY VASCULAR CONTROL

Most endotherms have a capacity to operate vascular shunts to regulate their heat-loss rate. In heterothermic insects, endothermic warm-up allows flight and other kinds of activity at low T_a, but once activity begins most of the regulation of T_b is due to blood shunting—in particular the control of heat loss from the thorax—rather than to changes in metabolic rate. The best studied examples are large sphingid moths and large bees, shown in Fig. 15.31. Bumble-bees have a countercurrent shunt at the narrow petiole between the thorax and abdomen, which keeps heat largely within the thorax (where it is generated by the flight muscles). However, this shunt can also be "turned off", by pulsing the blood flow at the petiole. In an overheated bee, the heated hemolymph from the thorax passes backwards through the petiole in a pulse that is out of phase with cold hemolymph coming forwards from the abdomen. Thus heat is discharged into the abdomen and can be lost from the relatively uninsulated "thermal window" of the ventral abdominal surface.

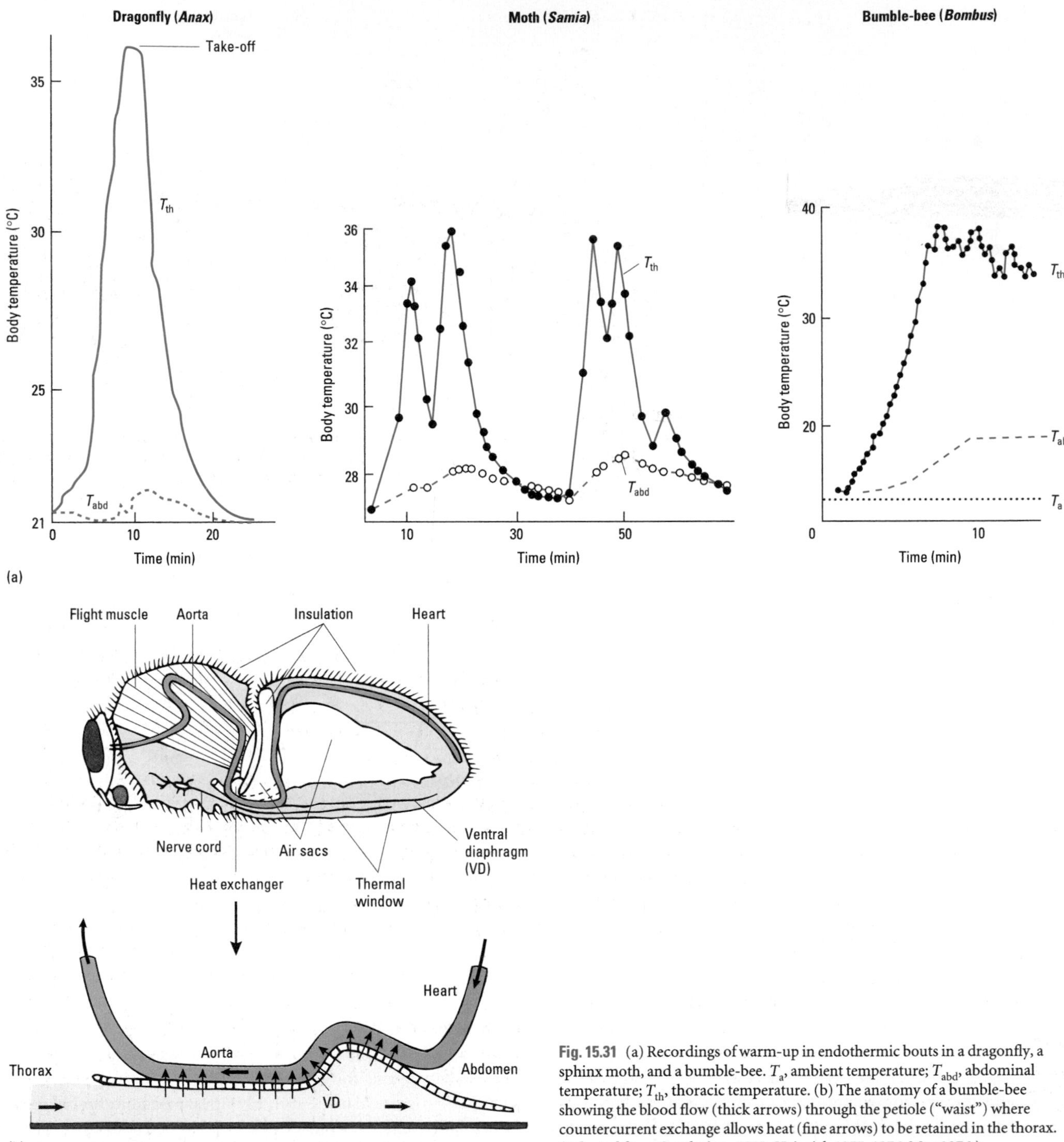

Fig. 15.31 (a) Recordings of warm-up in endothermic bouts in a dragonfly, a sphinx moth, and a bumble-bee. T_a, ambient temperature; T_{abd}, abdominal temperature; T_{th}, thoracic temperature. (b) The anatomy of a bumble-bee showing the blood flow (thick arrows) through the petiole ("waist") where countercurrent exchange allows heat (fine arrows) to be retained in the thorax. (Adapted from Oosthuizen 1939; Heinrich 1975, 1976; May 1976.)

In birds and mammals, short-term regulation of heat loss may be achieved by adjustments to the blood flow in peripheral capillary beds; we saw in Chapter 8 that these may be bypassed by arteriovenous shunts. In a normal mammal the blood flows via the skin capillaries, but in a chilled mammal where the maintenance of T_b is under threat, flow to these capillaries is shut off, with all the blood passing through the deeper shunt vessels, so reducing heat loss at the skin surface.

Many endothermic vertebrates from cool climates operate longer term vascular shunting so that parts of their body are regularly at a lower temperature than the core T_b. Extremities such as feet, flippers, tails, noses, or ears, and even keratinized areas such as horns, are "at risk" sites for an endotherm: they have a high *SA*/*V* ratio, are usually poorly insulated, and would allow severe heat loss in cool conditions if their temperatures were maintained. These areas may have an internal vascular countercurrent (rete) associated

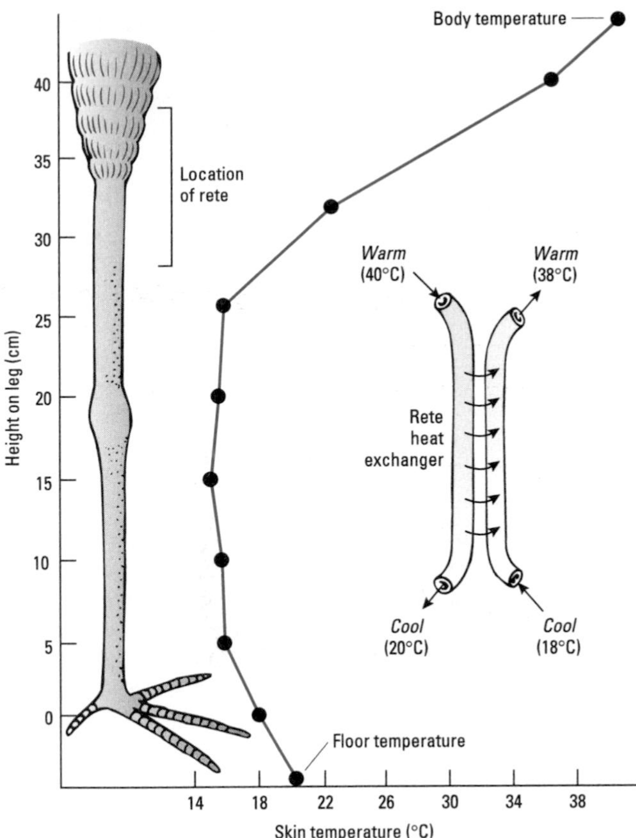

Fig. 15.32 Temperatures recorded from skin surface along the leg of a wood stork, showing the heat-conserving effect of an upper limb countercurrent rete. (From Kahl 1963, courtesy of University of Chicago.)

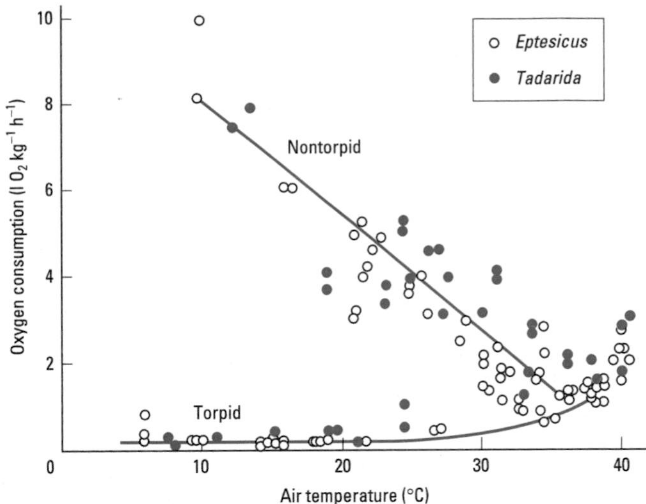

Fig. 15.33 Daily torpor in two species of small bat, with greatly reduced metabolic rate as the temperature drops at night (a very well-fed bat avoids torpor and increases its metabolic rate at night to maintain body temperature). (Adapted from Herreid & Schmidt-Nielsen 1966.)

with them so that warm blood in the afferent arterioles passes close to the efferent venules and heat is transferred across, so that much of the heat never reaches the tip of the extremity. The associated vasculature comes in three different forms (shown in Fig. 8.37).

An example from a temperate bird is shown in Fig. 15.32; a thermal gradient is set up along the bird's limb, reducing the gradient for heat loss to the environment from the foot. Intermittently, pulses of blood are sent to the foot by bringing into play nonrete vessels (e.g. superficial veins or venules) to insure that just enough warmth is supplied to prevent freezing damage to the tissues. These bypass routes also mean that the extremities can be used for heat dissipation if the animal ever gets overheated. This is an example of **regional heterothermy** (see Chapter 8), and it is common in endotherms from all habitats.

The ability to keep the brain relatively cool using a countercurrent heat exchanger is also important in many endothermic vertebrates, but as it is particularly well developed in animals from hotter climates we cover it in detail in Chapter 16.

HEAT CONSERVATION BY CHANGING CONDUCTANCE

A decrease in thermal conductance (C) can substantially reduce the lowest ambient temperature that an endotherm can survive, and can also reduce the oxygen consumption (metabolic rate). Increased insulation is thus a "cheap" way to improve survival in cool climates. Larger animals have lower mass-specific C-values than smaller species, because of their lower SA/V ratio, but they usually have a higher absolute thermal conductance. Larger animals therefore have more scope for decreasing their thermal conductance, usually by increasing the density or length of their fur or feathers (pelage). Smaller endotherms tend to be very well insulated anyway, and cannot increase the length of their pelage greatly without hampering the operation of their limbs; a small mouse cannot have fur more than a few millimeters in length or its legs would not emerge to meet the ground. Thus in small endotherms, increasing conductance by storing some subcutaneous fat is a better option. This not only insulates the surfaces, but also gives a small increase in body mass with only a limited increase in metabolic rate, and simultaneously provides a store of energy for the colder periods when food supplies may be reduced.

Many medium- to large-sized birds and mammals show acclimatory seasonal adjustments in pelage quality (with a spring molt to lose the thicker winter coat) and sometimes also in pelage color. This is particularly common in polar and tundra animals and is dealt with in Chapter 16.

The other possible conservation strategy is hypothermia and torpor, also dealt with in detail for animals from cold climates in Chapter 16. But note here that some temperate animals do undergo seasonal torpor, and that many very small endotherms, even in the tropics, undergo transient nocturnal torpidity. Figure 15.33 shows an example for a Neotropical bat. In bats and in small marsupials this kind of torpor is used to adjust energy expenditure to match food availability, and only occurs when they cannot feed enough to maintain their T_b throughout the 24 h cycle.

HEAT STORAGE AND DISSIPATION

When faced with an excessive heat load, endothermic animals have a number of possible physiological methods for dealing with the heat. If T_b is raised due to extra endogenous heat production (e.g. from bouts of exercise), they may use postural changes, pilo or ptilo

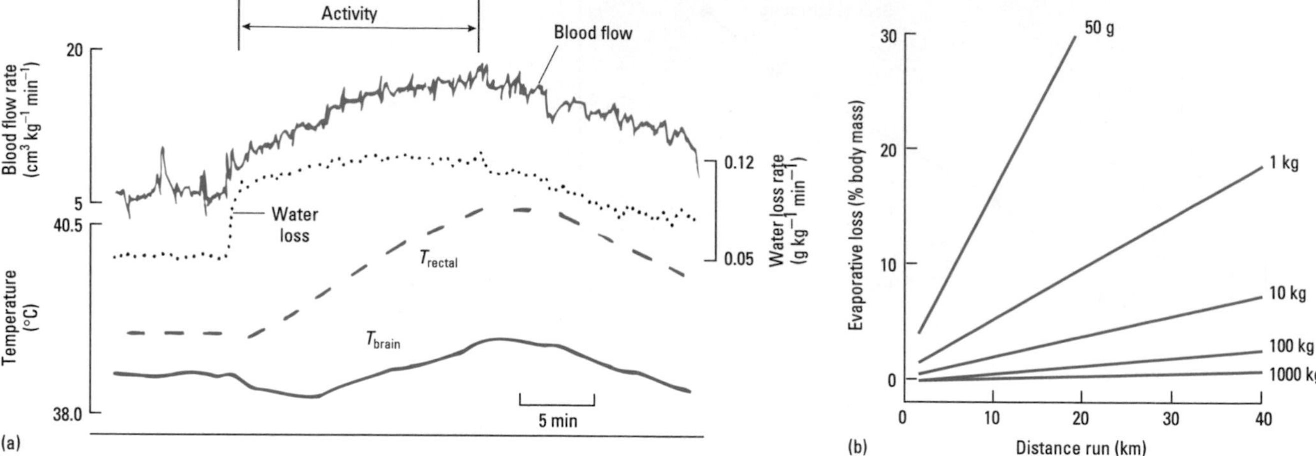

Fig. 15.34 (a) Blood flow rate and water loss rate in an exercising panting dog, also showing increments of rectal and brain temperature (T_{rectal} and T_{brain}) but with T_{brain} much lower due to the cooling effect of panting. (b) Rates of water loss as percentage body mass in different sized panting mammals; in very small species rates are too high to survive for long, and exercise with panting must be avoided as far as possible in daylight hours. (a, From Baker 1982; b, adapted from Mitchell *et al.* 1987.)

depression (flattening fur or feathers) and vascular adjustments (extra blood flow to the skin, producing flushing) to lose excess heat. Even if T_b is rising in concert with warm ambient conditions they will still normally be able to conduct, convect, or radiate away the heat, this being one of the main advantages of maintaining a high level of body temperature. But if the T_a approaches or exceeds T_b the options for regulation become limited. Temporary heat storage (adaptive hyperthermia, with raised T_b) may be worthwhile, to insure a $T_b - T_a$ gradient, but this can normally only be allowed for short periods. In a small rodent it may be tolerable for just long enough to give time to get back to a cool burrow, for example. In larger animals, heat storage has more substantial and longer term uses, as raising the T_b by 1°C takes much longer and represents a much larger quantity of heat stored, so that some large desert animals use this option on a daily cycle (see Chapter 16). However, for temperate animals it is rare; ultimately the animals must cool down again, and they normally resort to evaporation to achieve this.

Evaporative cooling is, as with ectotherms, an excellent way of dissipating heat if the consequent loss of water can be tolerated. This is more likely to be feasible in endotherms because of their necessarily large size and low *SA*/*V* ratio. So it is a common mechanism for regulating T_b in terrestrial endotherms, evolving as a specific thermal strategy.

Evaporative water (and therefore heat) loss may be achieved by either cutaneous (CEWL) or respiratory (REWL) routes. The relative merits of sweating and panting were considered in Chapter 8. In general, panting is a better option, and it is extensively used by birds and by some groups of mammals when exercising (Fig. 15.34). Note that in small mammals the EWL can be excessive, and exercise is only possible at night. Birds also have a problem, and for them CEWL may be more important (about 50% of the total water loss) at moderate air temperatures; at higher temperatures REWL usually dominates, and in pigeons the switch between the two modes is largely controlled by adrenergic signals that can increase cutaneous blood flow. In many canids the effectiveness of panting is increased by bypassing the nasal countercurrent system, taking air in at the nose but exhaling entirely through the mouth. Birds and mammals both tend to use a very high frequency of panting, perhaps 10 times the normal respiratory frequency (e.g. from 32 to 320 min^{-1} in a medium-sized dog). In birds the panting frequency is often close to that of their wingbeat, at around 600 min^{-1} in pigeons. In each case this also gives a panting frequency that is close to the resonant frequency of the lungs. In many birds REWL is further enhanced by a high-frequency "gular flutter" moving air in and out of the throat region (see Fig. 8.44b).

As with ectotherms, other sources of water may be used when necessary to achieve emergency cooling. Some mammals, particularly marsupials and rodents, salivate and lick their fur to spread the fluid; a few birds, such as vultures, will urinate over their own legs.

Size limits for terrestrial endotherms

Since the *SA*/*V* ratio of animals decreases as they become larger, small endotherms have a relatively large surface area and a very real problem: the metabolic heat generated in the smallest species is lost as fast as it is produced even under the most favorable circumstances. Thus the T_b of small dipterans (weighing, say, 1 mg) is close to the ambient environmental temperature even during long flights with very high rates of heat generation. Insects such as bees require a minimum size of around 20–30 mg before significant heat gains are observed during muscular activity, and in less highly insulated insects the size limit may be much higher, perhaps 100 mg in many taxa. However, once above this size limit, periods of endothermy and temperature regulation with T_b above ambient become a real possibility, as we have described.

The smallest vertebrate endotherms, the hummingbirds and shrews, have a body mass which is at least 10 times greater than that of the smallest endothermic insect. Why are there not even smaller birds and mammals? Are the conditions for heat loss different in insects? The answer to this question is apparently no, since the conductance of large moths and bees falls on a direct extension of the regression line for birds and mammals (Fig. 15.35).

The reason why insects can be successful endotherms at a smaller

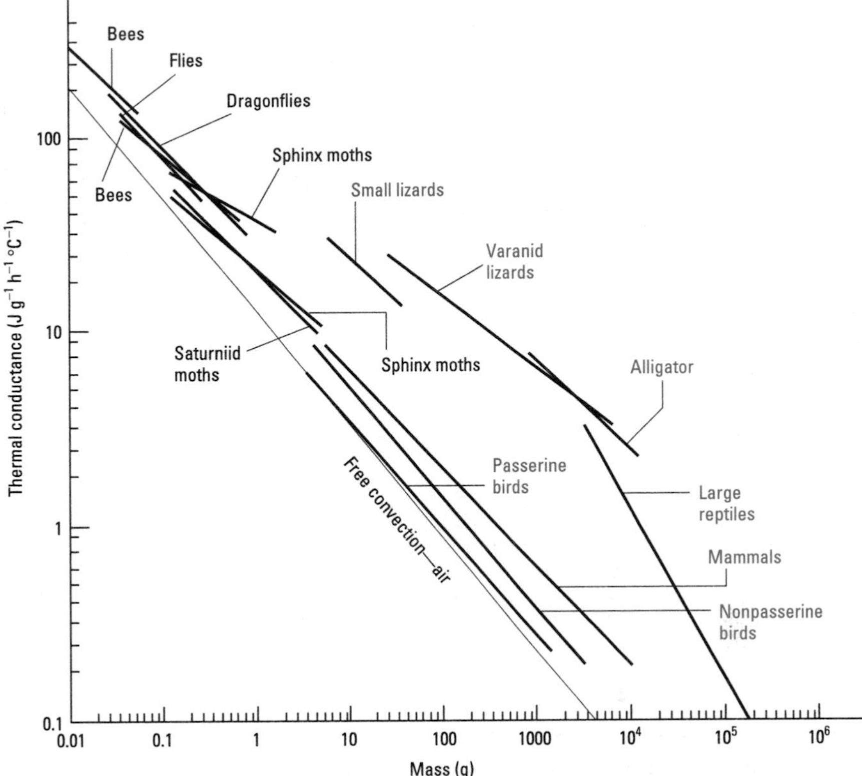

Fig. 15.35 Allometric relations for thermal conductance against mass in insects and vertebrates. Overall the slopes for insects (0.47) and for birds and mammals (0.52) are very similar. (From *Comparative Animal Physiology*, 1st edn, by Withers © 1992. Reprinted with permission of Brooks/Cole, a division of Thomson Learning: www.thomsonrights.com. Fax 800 730-2215.)

body size than birds and mammals is therefore probably related to differences in energy output and consequently heat production. In particular, differences in oxygen supply may well account for the constraints on the minimum size of birds and mammals. The cardiovascular system of shrews and hummingbirds is tuned to the limit. The hearts in these animals are two to three times larger than expected from the general scaling of heart size in birds and mammals. The hemoglobin content of the blood (hematocrit) in hummingbirds is also as high as that found in any other animal. It is probably not possible to increase hematocrit further without inordinately raising the viscosity of the blood and therefore the energy needed to circulate it around the body. Even with their large hearts and high hematocrits, the heart rate needed to sustain the maximum activity in shrews and hummingbirds is around 1200–1400 beats min^{-1}. Thus, each heartbeat lasts only 40–50 ms, which probably represents an irreducible minimum for the design of a muscular pump. By contrast, insects are able to sustain higher levels of energy output using their novel and radically different oxygen supply system —the tracheae, described in Chapter 7—allowing oxygen to diffuse directly in the vapor phase to the tissues where it is consumed.

15.4 Respiratory adaptation

Air contains about 21% oxygen, so for the great majority of land animals there is essentially no problem in obtaining the oxygen that they need. Only when constrained within a poorly ventilated space, such as a burrow or cave, is an animal likely to experience a degree of hypoxia.

However, all terrestrial animals must have a respiratory exchange site permeable to oxygen, so they inevitably also have a site permeable to water, from which EWL is bound to occur. Striking the right balance between the conflicting demands of a fast, active lifestyle demanding high aerobic metabolism and the need to conserve water by reducing REWL is a dominant aspect of terrestrial respiratory design. Inevitably the nature of this compromise is affected by the lifestyle and the size of each land animal, as well as by its phylogenetic legacy.

The smallest and soft-bodied land animals are able to breathe cutaneously. A few larger and moderately xerophilic animals retain a system based on the gill, for example, isopod pleopods, though these develop internal "pseudotracheae" in the more fully terrestrial species. All other land animals have fully internalized gas-exchange sites, either **lung books** or **lungs** or **tracheae**.

In Chapter 5 we considered the general balance of oxygen uptake and water loss from different respiratory surfaces operating in air (see Fig. 5.26), and concluded that for a simple cutaneous surface the ratio of water loss to oxygen uptake might be 2.1 (mg H_2O lost per ml O_2 taken up), whereas for lungs it could theoretically be only 0.2 and for tracheae as low as 0.1. Any terrestrial system of respiration that improves on the poor ratio for skin alone will be strongly selected for, so it is not surprising that both the systems of invaginated surfaces appear to have evolved many times over.

15.4.1 Lung breathing and water loss

Lungs can either be simply **diffusional** or can be **ventilated**, but always involve a substantial **vascularization** of the exchange surface

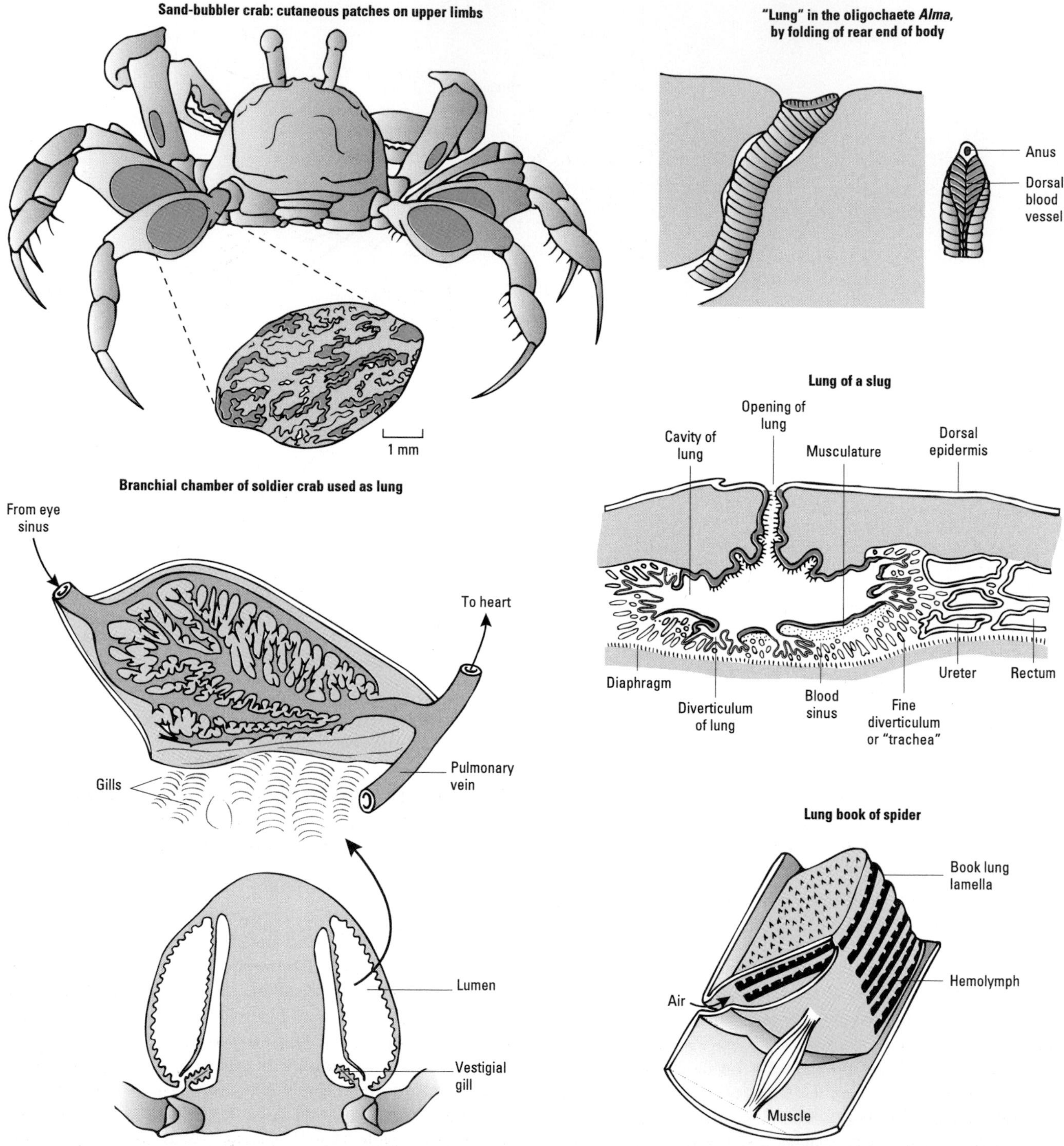

Fig. 15.36 Respiratory surfaces in a range of land invertebrates: cutaneous patches, lung books, and lungs.

(evident even in the simple terminal "lung" of some giant earthworms and in the modified branchial chamber of land crabs, where a rather complex double portal system of blood vessels has been described). Diffusional and ventilated lungs represent identical situations in terms of controlling water loss. The lungs of crabs and pulmonate snails, and the lung books of many arachnids (Fig. 15.36), all constitute water-saving devices because by invaginating the exchange surface they help to reduce Po_2 at the point of uptake. This effect is particularly striking in vertebrate alveolar lungs, where the Po_2 at the internal alveolar surfaces is reduced to around half of its atmospheric level, giving the favorable ratio of water loss to oxygen uptake mentioned above. EWL is therefore inherently rather low, around 15–40% of total EWL in a medium-sized reptile.

However, extra savings of water can be made in the nasal exchange systems of some birds and mammals, where, by reducing

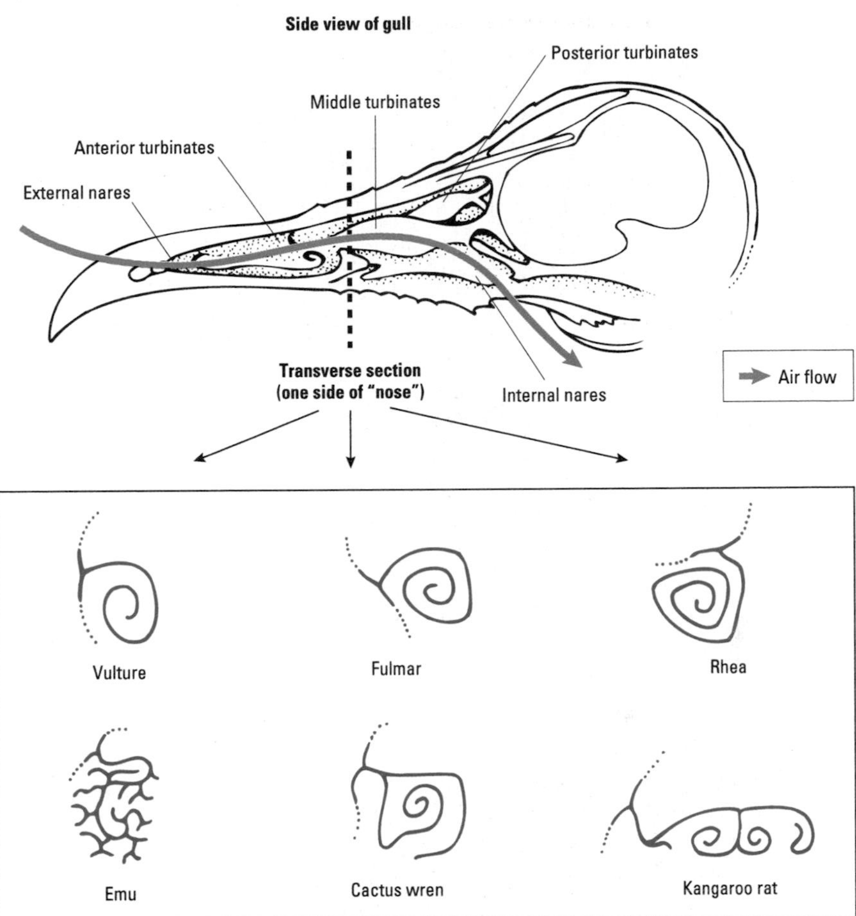

Fig. 15.37 Cross-sections of the nasal passages (turbinates) in various birds and mammals, showing the greater development of the turbinates in more xeric species. (Adapted from Schmidt-Nielsen *et al.* 1970b; Hillenius 1994.)

the temperature of expired air in the turbinals of the "cold nose", some water recondenses into the nasal mucosa (see Fig. 5.27 for the general principle). Examples of the nasal turbinal systems in vertebrates are shown in Fig. 15.37. The extent of cooling achieved depends on the precise anatomy and length of the system. For example, in humans the expired air has a temperature range of 28–34°C, well below the core temperature of 38°C, and representing a moderate water saving, while in birds the expired air may be as cool as 20°C. However, long noses should not be assumed to be adaptations for water saving during evaporative cooling. A theory that this was the "reason" for long noses in baboons was largely discarded when it was pointed out that baboons lack the characteristic Steno's gland (permeable tissue producing water for the cooling effect) and lack a valve on the epiglottis to allow the nose to be bypassed during panting. Here the long nose probably has more to do with sexual display and flaunting the impressive canines during threat behavior than it does with physiology!

15.4.2 Tracheal systems and water loss

Tracheal systems by their very nature provide a site where Po_2 can be substantially lowered at the exchange site, giving a favorable oxygen uptake to water loss ratio. In the more hygrophilic land arthropods this is the limit of their water-conserving design features; the spiracles cannot be closed, and REWL must be considerable, but this matters less in the litter habitat of these animals. By contrast, insects have fully closable spiracles, and their respiratory system can have additional specializations to improve ventilation as well as to reduce water loss.

Ventilation and condensation

In at least some insects ventilation of the tracheal system occurs directionally, with most of the intake at the front end of the animal (thoracic spiracles) and most of the exit of spent gases at the rear (Fig. 15.38). In locusts, for example, air enters at spiracles 2 and 3, and leaves mainly through spiracles 5–10; in some beetles only the thoracic spiracles are shaped so as to scoop in air during forward movement. This raises the possibility of operating condenser systems with the expired air (as in vertebrate noses), and some active insects probably exploit this. A locust may have a thoracic temperature of 30°C with the abdomen at only 20°C; and air that is 10°C cooler, while still saturated, holds substantially less water. Certain desert beetles have exploited this option rather more explicitly, as we will see in Chapter 16.

Spiracular control

The onychophorans have open access to their tracheal system, as do most centipedes and millipedes (though a few centipedes do achieve

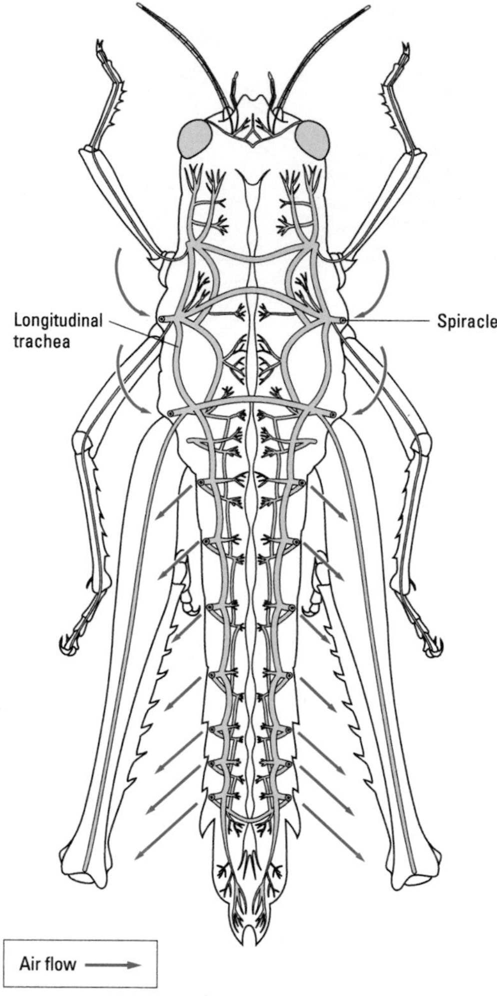

Fig. 15.38 Directional flow of air from thorax to abdomen through the body of a locust, potentially allowing some water conservation by condensation from exhaled air in the cooler abdomen.

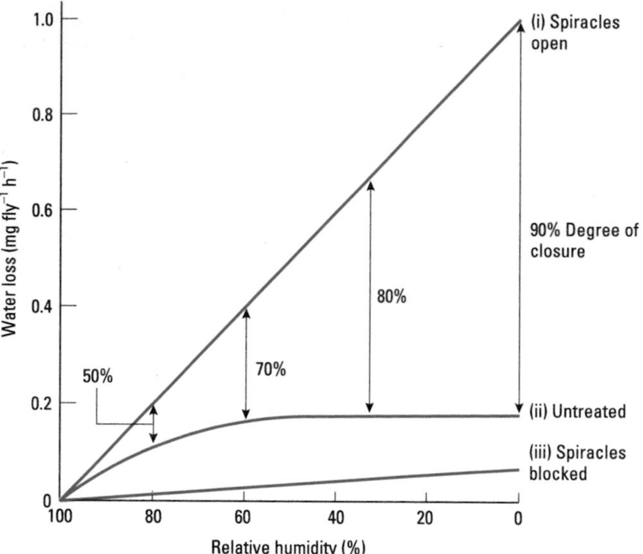

Fig. 15.39 Water loss in a normal tsetse fly, and in one where the spiracles are kept open by raised (15%) CO_2. Respiratory water loss is normally well controlled and little greater than cutaneous evaporative water loss (CEWL; cuticular loss, with all spiracles blocked). (Adapted from Bursell 1957.)

some muscular contraction around the spiracle) and most apterygote insects. In some mites the spiracles have a perforated cuticular covering, and the scale of the perforations is such that they aid water conservation by creating a very humid environment in the enclosed atrium and reducing vapor loss above. But most of the pterygote insects and some arachnids have control systems to open and close cuticular valves over the spiracles and thus manage their water loss. Many have closely interlocking bristles to prevent dust and external water entering the spiracles, and in addition they have two sets of muscles (openers and closers) to control the size of the aperture (see Fig. 7.31). This control may be exercised directly via CO_2 effects on the spiracular closure muscle (Chapter 7), but temperature and humidity, and probably wind speed, also have effects on spiracular closure, and the state of hydration of the insect modulates these effects. The best known example is the African tsetse fly, *Glossina*, where a direct effect of humidity on spiracular closing has been demonstrated (Fig. 15.39). Thus the respiratory demand and the water balance of the animal can be closely integrated. Fully closed spiracles in a resting insect can reduce REWL by 70–90% in many insects; but in flight the spiracles have to be fully open to oxygenate the thoracic flight muscles, and water balance needs are secondary.

Discontinuous respiration

At the extreme, in some insects, including a range of pupae and some adults, the spiracles remain closed for all but occasional brief bursts of opening. This is termed the **discontinuous ventilation cycle** (DVC, or discontinuous gas-exchange cycle), and each cycle is made up of a closed spiracle phase, a flutter-spiracle phase, and an open phase. In fact the insect continues to use oxygen at a constant level throughout (Fig. 15.40), but it is able to store carbon dioxide and release it only intermittently, thus also avoiding continuous water loss. When the spiracles are closed, accumulating CO_2 is retained in solution in the hemolymph as bicarbonate, and this means that the air initially taken in and depleted of oxygen has not been replenished with an equivalent gaseous volume. Thus inside the tracheal tubes there is a small but measurable negative pressure as the O_2 is used up. This negative pressure is countered by the spiral thickening of the tracheae, and is enough to insure that air continues to leak in at the spiracles without significant outward gaseous or water vapor loss, so oxygen continues to be supplied to the animal's tissues ("passive suction ventilation"). After long periods of closed spiracles and gentle inward leak of air, the bicarbonate levels in the hemolymph become too highly elevated and the spiracles open up. Over a period of a few minutes there may be a series of brief spiracular openings (the F phase) where inward oxygenated flow greatly exceeds outward leaks, and then a more substantial open period of real "flush out", until the hemolymph pH and $[HCO_3^-]$ are back to normal levels (though it is unclear how such massive outward movements of CO_2, from solution, are achieved so quickly). Then the spiracles close again and the cycle repeats, with no apparent "breathing" occurring for another few minutes or (in pupae) several hours.

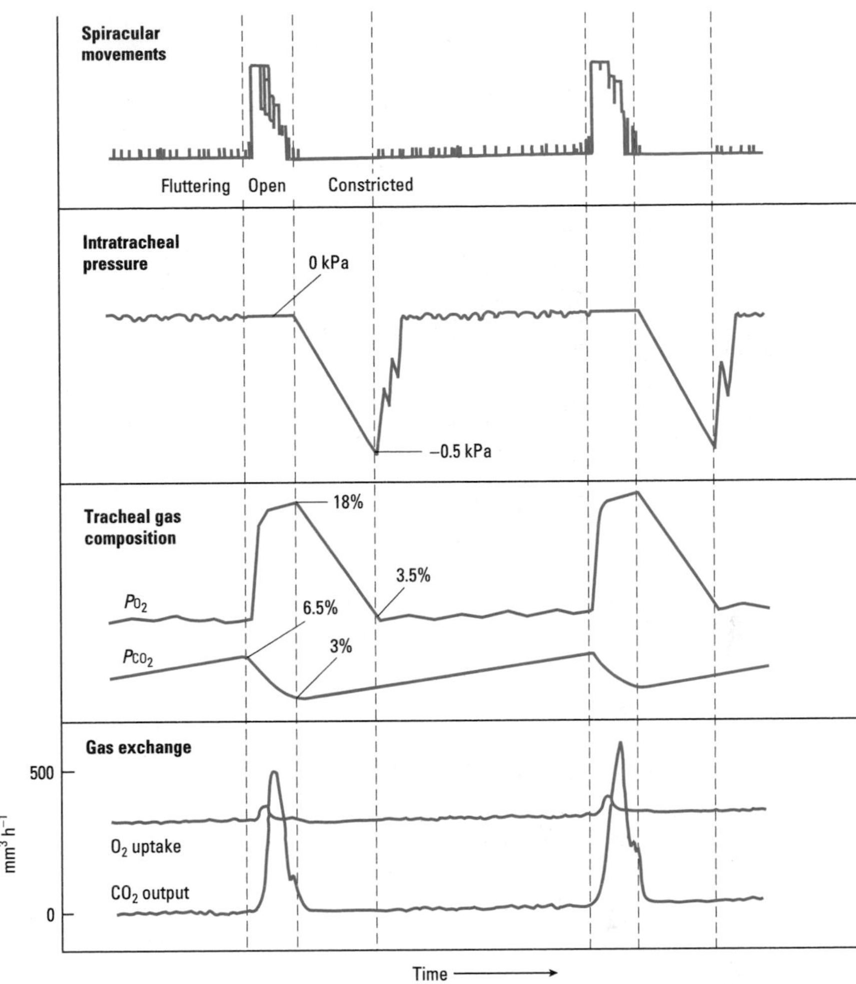

Fig. 15.40 Discontinuous respiration in a silkmoth pupa, showing spiracle movements and the patterns of pressure and gas exchange in the tracheae. (Reprinted from *Journal of Insect Physiology* **12**, Levy, R.I. & Schneiderman, H.A., Discontinuous respiration in insects. IV. Changes in intratracheal pressure during the respiratory cycle of silkworm pupae, pp. 465–492, copyright 1966, with permission from Elsevier Science.)

This pattern of respiration has now been shown in a range of quiescent adult insects as well as in fairly active ants and various insect pupae, and something very similar occurs in ticks and other chelicerates, in some centipedes (all of these having tracheal systems), but also in land snails and even in torpid bats. The water-conserving benefits of discontinuous breathing have been disputed on both practical and theoretical grounds, however. Water loss is certainly very low while the spiracles are closed, but it rises greatly at the start of the ventilatory burst. Furthermore, dehydrated grasshoppers tend to reduce rather than increase their use of discontinuous breathing; and some other species exhibiting this pattern have REWL as only a small component of water balance anyway. However, elaborate tests comparing EWL with and without spiracular control in the more xeric animals, such as ants and moth pupae, do indicate a several-fold saving in water loss during DVC, and xeric ants also show lower frequency DVC and higher flutter phase CO_2 emission volumes than mesic relatives. *Drosophila* reared under desiccating stress also show more pronounced cyclic respiration. All these findings suggest that discontinuous breathing is a genuine adaptive response to reduce REWL; its presence in many burrowing mesic species suggests it may have evolved ancestrally partly in relation to anoxia in underground habitats, but it has been markedly enhanced in xeric conditions.

Despite all these adaptations in both lung and tracheal systems, there is ultimately no way of avoiding substantial water loss from respiratory surfaces in a fully active terrestrial animal. The behavior of the animal in choosing microhabitats and appropriate food and drink, together with its cutaneous and renal adaptations, must compensate for its inevitable respiratory losses.

15.4.3 Oxygen carriage, circulation, and respiration

In insects the tracheal system delivers oxygen in the gaseous phase almost to its point of use, and no circulatory fluid is needed, the hemolymph containing no respiratory pigment. In contrast, most other land animals use the blood system to carry the gas to the tissues. Most also use a blood pigment to carry oxygen: molluscs and crustaceans use hemocyanins, while earthworms and tetrapods use hemoglobin. In many vertebrates there is a "store" of extra pigment available as unused erythrocytes in the spleen, which can be mobilized if any hypoxia is experienced (or if there is blood loss following wounding). However, the pigments of terrestrial animals are unexceptional—they do not appear to be significantly different in chemistry or in properties from those of closely related aquatic species (see Chapter 7), although they load to a higher percentage oxygen content (see Fig. 7.23).

In contrast, the circulatory patterns of all terrestrial animals except insects undergo substantial modification relative to aquatic animals. Blood flow has to be organized so that oxygenated blood from the new respiratory sites (in effect, the lungs) is quickly distributed to the most demanding tissues, commonly the locomotory muscles and the brain. In invertebrates the blood always flows from the heart to the systemic organs and then to either the gills or lungs (see Fig. 7.17), so that there is a delay in supplying freshly oxygenated blood where it is needed. However, the rate and pattern of circulation can be modified considerably in relation to demand; even in earthworms, the degree of dilation of cutaneous vessels is under hormonal control and the O_2 level in the dorsal vessel can rapidly increase on contact with moist oxygenated soils.

The system in vertebrates such as fish is perhaps better organized, with blood flowing from the heart to the gills to the systemic organs (see Figs 7.17 and 7.18). However, in air-breathing vertebrates the situation is complicated by the progressive division of the circulation into two parallel systems, with blood returning to the heart twice in each full circuit of the body. The disadvantage of the "delay" this introduces in getting freshly oxygenated blood to the respiring tissues is presumably offset by the extra pressure boost given by the second passage through the heart.

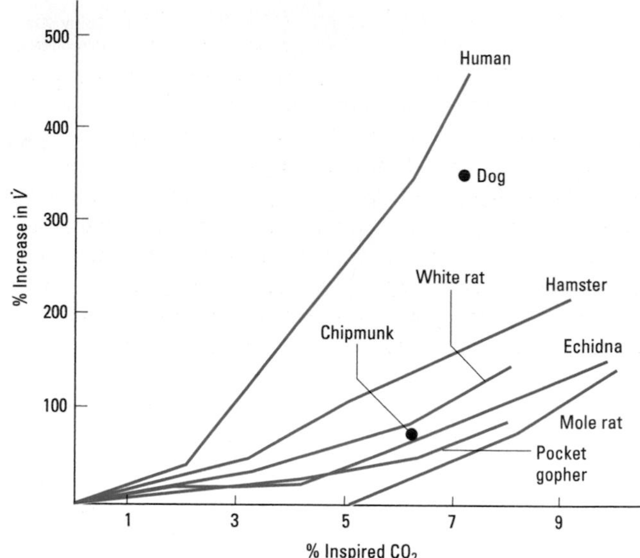

Fig. 15.41 Reduced CO_2 sensitivity in various burrowing animals compared with nonburrowing species.

15.4.4 Carbon dioxide loss

Carbon dioxide loss in terrestrial animals is much like that in aquatic animals—loss is essentially diffusive, from dissolution in the blood across the permeable surfaces direct into the medium. However, whereas aquatic animals lose most of their CO_2 at the gills, many soft-bodied terrestrial animals use their general body surface. Most worms use this route, as do amphibians. In fact a moderately active amphibian with typical bimodal (cutaneous + lungs) breathing may gain 50–78% of its oxygen from its lungs, but lose 57–84% of its carbon dioxide from its skin.

In land animals with relatively impermeable skins and cuticles (the arthropods and tetrapods) there is little alternative but to lose CO_2 from the respiratory system. Since raised CO_2 levels (hypercapnia) can represent a significant hazard, especially in terms of reduced pH levels and thus acidosis, the tracheae and lungs have to be better ventilated to flush out the carbon dioxide. This is especially important in birds and mammals again, where increases in blood $P\text{CO}_2$ compromise oxygen carriage and delivery, and upset the acid–base balance, with knock-on effects on homeostasis throughout the body. CO_2 becomes an important blood buffer, and there is therefore also a potential problem with losing too much CO_2 too quickly (hence the dangers of hyperventilation) as acid–base balance is again compromised.

Adaptation to land also involves a change in the site of CO_2 sensors, from an external site in fish and aquatic invertebrates to an internal one (usually in the arterial blood system) in the land animals (see Chapter 7). This allows a direct response of increased ventilation when blood $P\text{CO}_2$ builds up. However, this direct response varies somewhat with habitat; for example, CO_2 sensitivity is substantially reduced in most burrowing land animals compared to nonburrowers (Fig. 15.41), reflecting the raised CO_2 levels that must occur within the confines of most burrows and nests. There are a few exceptions to this rule, notably in the burrowing owls (*Athene*) where cardiopulmonary responses are normal but there is a substantially increased blood buffering capacity so the owl can tolerate acidosis.

15.5 Reproductive and life-cycle adaptation

The essential requirement for terrestrial reproduction is to keep the eggs and embryos from desiccating (Table 15.12). In terms of types and degrees of reproductive adaptation, we again need to

Table 15.12 Problems and solutions for reproduction on land.

Problem	Solution	Examples
Protect the gametes	Return to water	Most crabs, amphibians
	Use sperm droplets	Myriapods
	Use protective spermatophores	Most arachnids, some insects
	Internal insemination	Most insects, spiders, amniote vertebrates
Protect the young	Shelled yolky eggs	Worms, insects, arachnids, snails, amphibians, reptiles, birds, monotremes
	Viviparity	Few insects, all eutherian mammals
	Postnatal parental care	Few insects, some arachnids, many vertebrates

distinguish between animals that achieved terrestriality by a freshwater route and those that invaded across the littoral zone. Littoral animals show a considerable range of reproductive strategies (see Chapter 12) reflecting their varied experiences of immersion and transient desiccation; many produce small pelagic eggs and larvae, but some have internal fertilization and curtailed larval development, often with parental brooding. Groups such as amphipods and isopods retain this latter habit on land, while other arthropodan groups (insects and myriapods) commonly use sperm droplet or spermatophore transfer systems that may have originated in marine interstitial habitats. Freshwater animals often have a rather different approach, again often fertilizing internally but commonly producing fewer and larger eggs, as dense egg masses attached to the substrate or vegetation, with substantial yolk as a nutritional endowment; they only infrequently adopt brooding habits. Thus the young are released quite early from any association with the mother but are already relatively large to avoid the osmotic problems associated with a very large surface area to volume ratio in a dilute medium; traditional aquatic larval forms such as trochophores, veligers, and nauplius are abandoned. Either route may allow the development of an ovoviviparous or fully viviparous habit, although the marine route with its relatively greater environmental constancy may less often exert selective pressures to abandon oviparity.

15.5.1 Reproductive strategies

General life-history patterns

We dealt with the general theories of life-history strategy briefly in Chapter 1. Although the terminology has been criticized and has sometimes been overdone, animals and plants alike can usefully be seen as occupying particular positions on the *r*–*K* continuum (see Table 1.1). From the brief discussion above it will be evident that on land there might inevitably be a general tendency to the *K*-selected end of these possible strategies, with fewer, larger, better resourced, and better protected gametes and young. But the counterargument is that *r*-selection usually occurs with less predictable habitats and less stable resources, as expected on land. Within the terrestrial animals there is therefore a wide range of possibilities, from the rather profligate egg-laying habits of some amphibians and insects to the extreme of investment in a single offspring once a year or even once every few years in some birds and mammals. It is commonplace to speak of some of the insects as being extremely *r*-selected, but important to bear in mind that within the animal kingdom as a whole their reproductive strategies are actually rather modified towards fewer and higher investment offspring, albeit produced with great (and sometimes alarming!) abundance and regularity.

Remember that the energy that females invest in reproduction usually scales with a mass exponent (b) between 0.5 and 0.9 (see Chapter 6), which means that larger species tend to invest relatively less in their offspring per unit time. But larger animals also tend to live longer, and lifespan scales with a mass exponent of 0.15–0.29. Thus over a female's lifetime, these two factors together mean that the energy invested in reproduction varies more or less isometrically with respect to body mass ($b = 1.0$). Larger animals on land tend to invest proportionally less of their resources in each breeding cycle, but will have a longer reproductive life and produce more broods. However, this effect is much more evident within a taxon (where reproductive constraints and techniques are comparable) than when comparing across widely differing taxa such as insects and vertebrates.

A further generalization about terrestrial animals is their greater tendency towards seasonal breeding. In most invertebrates breeding occurs annually (more rarely there are multiple times at roughly fixed intervals, in a particular season of the year) but its precise timing is very dependent on climatic conditions. For example, a spring breeding norm may be delayed after a cold winter. However, in many vertebrates there are one or a few breeding seasons each year that are fairly rigidly fixed by hormonal triggers, dependent either on the invariant signal of the photoperiod or perhaps on endogenous timing systems.

Terrestrial invertebrates

Soft-bodied animals living on land are often highly dependent on mucus at many stages of their reproduction. Both flatworms and nemertines may use mucoid sheaths to enclose a copulating pair, and in flatworms there are often multiple eversible cirri each acting as a penis to inject sperm into the general body tissues of any other worm encountered (each being hermaphrodite). In nemertines the gametes may be shed into the mucoid sheath, where fertilization then occurs and the eggs remain (though at least one species of nemertine is ovoviviparous). However, most flatworms and a few nemertines achieve internal fertilization. Flatworms then lay rather well-protected shelled eggs, which hatch directly as small worms with no larval stage.

Earthworms too secrete a mucoid sheath during copulation. Two worms come close together ventrally, lying in opposite directions; each (hermaphrodite) worm makes its own sheath and deposits into it a package of its spermatozoa. The sperm then pass backwards along a seminal groove and enter the sheath of the other worm, migrating into that worm's spermathecae (sperm storage organs). Each worm sloughs off its sheath and they separate. Some time later the worm secretes a further sheath of protein-based material and some albumen proteins, and the eggs are deposited into it, together with some of the stored sperm, at which point fertilization occurs. The sheath slips off the worm, and later hardens and darkens to form the cocoon protecting the fertilized eggs. The young earthworms hatch from the eggs within the cocoon and feed on the stored albumens.

In terrestrial molluscs, many of which are simultaneous hermaphrodites (see Box 11.1), more complex reproductive behaviors and anatomies come into play. Courtship is well developed, with two snails or slugs coming together in an entwining "dance" of tentacular and oral stroking. In some genera, after a short period of dancing each partner everts a structure called the dart sac and projects a calcareous dart into the other's body, where it stimulates the recipient into the next phase of courtship, culminating in penis protrusion and cross-copulation. Slugs are particularly adept at conducting their copulatory exchanges in mid-air, suspended from vegetation on the end of a mucus thread. Many of the slugs and snails then lay small groups of rather large eggs in a sheltered damp place; even the very large and strongly calcified eggs of the genus *Achatina* (the giant African land snail) are still relatively permeable and desiccate easily. There are even a few pulmonate species that are ovoviviparous and retain the eggs in the oviduct prior to hatching. Note that in both annelids and molluscs there is a trend from jelly-covered eggs in aquatic species to shelled protected eggs in terrestrial species, paralleling the amphibian–reptile transition in the vertebrates.

Crustaceans on land generally show brooding and maternal care. In isopods and amphipods brood care is present in marine species, and is simply somewhat extended in terrestrial species so that the brooded eggs hatch as miniature adults. In the amphipods another aspect of reproduction is carried over from the marine to the land species, that of precopulatory mate carriage. An aquatic amphipod male carries the female around for some time, then copulates with her and releases her very quickly. In the terrestrial forms (e.g. sandhoppers) the male carries the female around only briefly, and copulation is often a lengthier process, but again the male then departs. The female retains the fertilized eggs in her brood pouch for a variable time, depending on her stage in the molt cycle.

Land crabs show rather more sophisticated behaviors associated with mating, but with their essentially marine ancestry they are actually more reliant on water than the two groups described above. In many aquatic crabs mating has to occur at the point when the female has just molted and her carapace is soft, since at other times the genital operculum is inconveniently hardened in the closed position. The newly molted female is therefore transiently very vulnerable and may have little choice in who she mates with. In more terrestrial species, though, the operculum may be hinged, so that both copulation and oviposition can occur at any stage of adult life. So in land crabs mating behaviors become much more complex, and the male may have to perform a prolonged courtship ritual to secure a female. The fiddler crabs (*Uca*) and ghost crabs (*Ocypode*) are well known for elaborate mating dances, with visual and acoustic signals, although the complexity and duration of these varies widely. Copulation often occurs inside a burrow (sometimes built specially for the occasion by the male), and the female crab may remain in that burrow until the eggs are mature. Eggs are large, as are any young larval stages that persist, and their carbon : nitrogen ratio is particularly high due to a high content of lipid in the yolk. In some species the hatching of young juveniles occurs within the burrow and the young are held against the maternal abdomen, below the pleopods, prior to release. However, most of the grapsid and ocypodid females return to the sea to release their young for a brief marine larval phase, and even the highly terrestrial gecarcinids have to return briefly to sea, often from sites many miles inland.

In the insects and myriapods a different approach to terrestrial reproduction appears, probably derived from an interstitial lifestyle. Male chilopods (centipedes) deposit sperm droplets; at its simplest the drop is laid directly onto the soil, but in many species a pad or web of silk, or a small pillar of gelatinous material, is laid down first. The female centipede then takes up the sperm straight into her genital opening. In another group of myriapods, the symphylans, a sperm droplet is laid by the male and then eaten by the female; as with the flatworms mentioned earlier, the sperm must then migrate within the female body to her sperm storage organ. The more primitive apterygote insects again use sperm droplets, without copulation, but in many species there is a behavioral ritual whereby the male leads the female into position over his sperm. In all these cases the sperm droplet is an appropriate means of transfer in the humid and protected habitat of cryptozoic fauna.

In most of the arachnids and winged (pterygote) insects, which have emerged fully from the interstitial and litter zones into the drier, fully terrestrial world, droplets are replaced with more elaborate spermatophores (Fig. 15.42), transferred to the external female genitalia during normal mating behavior (most arachnids and many insects), or directly inseminated through an intromittent organ into the female (most pterygote insects). Spermatophores may become extremely elaborate, and in the orthopteran groups, such as katydids, the basic sperm packaging is augmented with a gelatinous mass, the spermatophylax (Fig. 15.42c), which the female eats after the male has left. This is thought to have a nutritive function and thus forms a kind of "courtship feeding".

Other kinds of nuptial gift come into play with insects during precopulatory behavior in the higher orders, where direct insemination is the norm; for example, many male flies bring small packages of food to a female. Other complex behaviors are also common: males may wait for (or even dig for) females at emergence sites, or show territorial defense at known feeding sites such as flowers or carrion, or they may form "leks", male swarms to which females are attracted visually and by scent (pheromones) and where they may choose the most competitive male as a partner. In many insects the males show alternative mating tactics, with some males being territorial and others adopting a "sneaky" tactic, for example. In many bees and wasps this behavioral variation has been shown to be due to size differences and to the directly related differences in thermal tolerance of larger and smaller (or paler and darker) males. Other female and male behaviors may also affect the reproductive process of course—female insects commonly attract males with pheromones, while either sex may attract the other acoustically with song, or visually with bioluminescent systems, as in fireflies.

Some insects have adopted unusual reproductive modes incorporating asexual phases, and these can usually be related to the particular characters of their terrestrial habitat. Many aphids (Homoptera) reproduce parthenogenetically for much of the year, and some of them produce nymphs directly rather than eggs, with the adults being wingless. Only in the late summer are sexual (usually winged) males and females produced, and these may be nonfeeding stages without mouthparts; they mate in protected crevices on the host plant, and the female then disperses to a new habitat and lays just one or a few giant fertilized eggs that overwinter to produce the foundress of the spring (parthenogenetic) generation. Even more remarkably, some midges and some parasitoids regularly omit an adult phase entirely, with larvae becoming sexually mature and laying eggs, or retaining cells that develop internally into new larvae and eventually kill the mother larva. These kinds of life cycles are probably largely determined by food (host plant) availability rather than by temperature or humidity *per se*.

Insects and most other arthropods produce eggs with a stiff chitinous cuticular covering, often with a waxy layer incorporated to give low permeability (see Chapter 5). This "eggshell" always bears pores to allow gas exchange, though the total area of such pores may be only 1% of the total egg surface. In many insects the outer surface becomes very complex, with an aeropyle layer of reticulate sponge-like tissue (see Fig. 5.28), so that the eggshell acts as an air-filled plastron in the manner already described for some freshwater insects

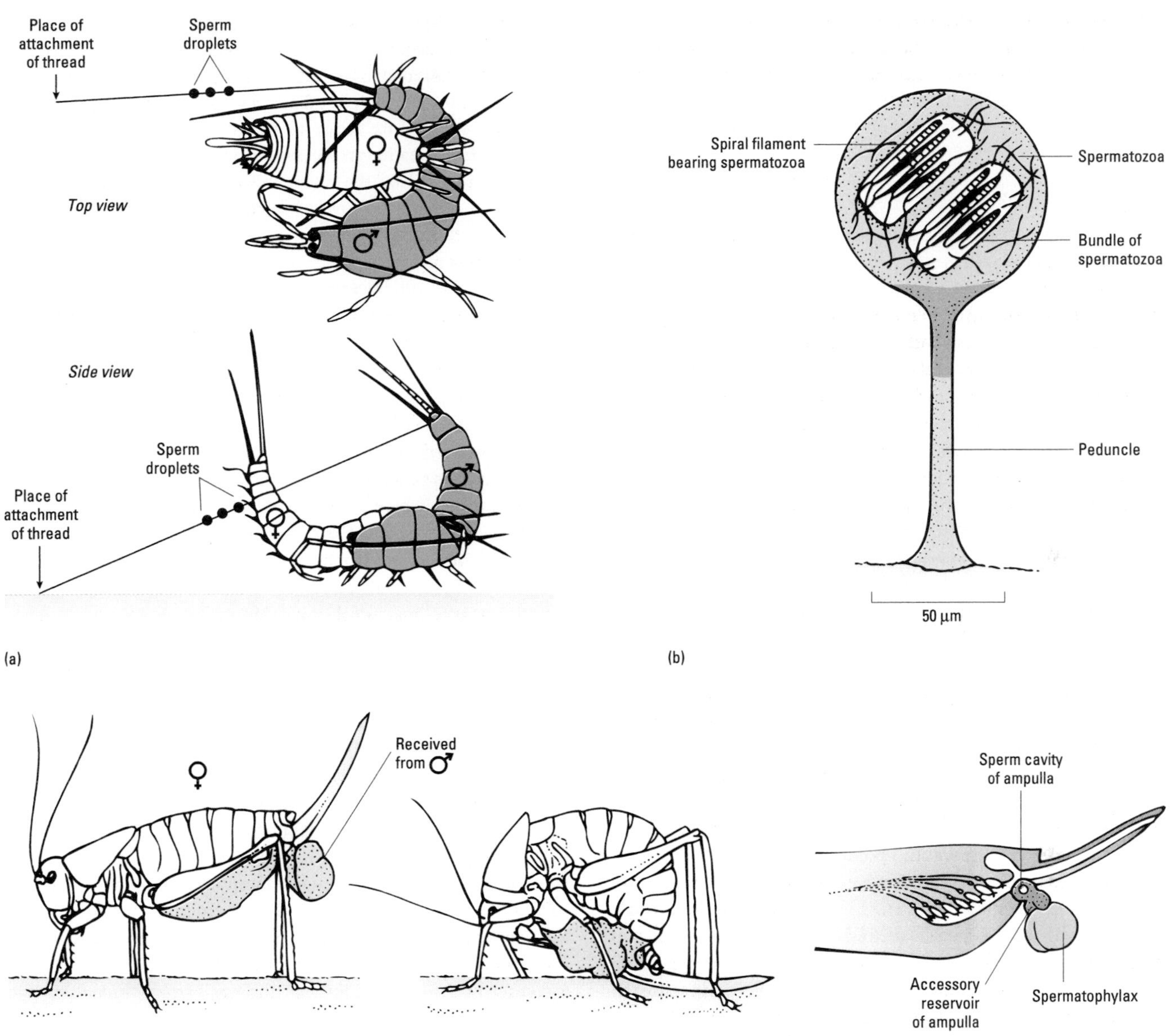

Fig. 15.42 (a) Sperm droplets in an apterygote insect, (b) spermatophores in a myriapod, and (c) the elaborate nutritive spermatophylax of an orthopteran, all serving as ways of transferring male gametes on land. (a, From Little 1990, courtesy of Cambridge University Press.)

(see Chapter 13). There is also evidence that air spaces in the eggs of some insect species are structured so that they act as reflectors, keeping the eggs cool in sunlight.

Most insects lay their eggs in specific protected microclimates where desiccation will be minimized. This often involves a site associated with the food source used by the developing larvae or nymphs; eggs may be laid on the leaves or flowers or bark of the host plant, or in the flesh of another animal that will be a host for the parasitic or parasitoid juvenile stages. Many terrestrial insects lay their eggs underwater and have an aquatic larval stage. Adult dragonflies and mayflies, and some beetles and true flies, must struggle to pierce the surface meniscus of ponds and streams to permit oviposition in a suitable place. Relatively few insects provide any after-care to their young; these include a few bugs ("parent bugs"), together with the social insects. However, there are also some species that retain the eggs within the body, usually in the lower ovarioles, and provide nutrition to the developing offspring; for example, tsetse flies (*Glossina*) "lay" very advanced larvae.

In contrast to insects, many arachnids lay their eggs in prepared nests or carry them around with them. Scorpions are typically ovoviviparous or fully viviparous, and these and other arachnids commonly provide brood pouches for their developing embryos or young. Certain spiders also effectively brood their eggs to speed up the development rate; in the genus *Pirata*, the preferred T_b of the female rises when she is carrying eggs, from around 21°C to around 27°C, and she will bask on the upper surface of her moss habitat to achieve this T_b. Other spider genera lay the eggs in a silken nest but will move them around the nest to keep their temperature constant, or even out into the cooler moving air if they get too hot. Arachnids also quite commonly exhibit maternal care after hatching, with juvenile scorpions riding on the parental back and young spiderlings living on within the parental nest for some time after birth.

Vertebrates

Vertebrates, with their freshwater ancestry, inherited typically large eggs and larger offspring size, with more advanced precocious hatchlings that are at such an advantage in osmotically stressful and variably provisioned ponds and streams. Hence freshwater amphibians have distinctly larger eggs than estuarine species, and terrestrial amphibians lay eggs that are even larger—in some species each egg approaches 10% of the mother's linear dimensions. But most amphibians still have unprotected, essentially aquatic eggs, and lay a great many of them (commonly hundreds at a time, and up to 20,000 throughout the lifetime of the average female in some species). The jelly that surrounds each egg expands enormously by taking up water, and the resultant capsule may help to trap heat and keep the eggs warm; however, the maximum diameter of modern amphibian eggs is restricted to about 9 mm, as beyond this diffusion alone would not permit adequate oxygenation. The main problem is that a large egg mass of soft jelly-like material inevitably suffers a high mortality to fish, insect larvae, ducks, etc.

In urodele amphibians (salamanders and their kin) the female generally picks up a package of sperm into her cloaca and fertilization occurs internally just before laying. But fertilization is virtually always external in anurans, even though each male may have courted the female carefully and may grasp her tightly during egg laying. These common amphibians need to insure hydration of the sperm, and many therefore mate within a pond.

An additional obvious disadvantage in amphibian reproduction is that since the eggs are not waterproof, and the male and female both normally have to return to fresh water to lay them and to insure sperm safety, the amphibians have to be metamorphic, with aquatic young preceding the more terrestrial adults. In practice the young of apodans and urodeles are relatively like the adults except for having gills, and the juveniles are only very different in modern anurans, for whose young the term "tadpole" is usually reserved. These young stages have to cope with a range of freshwater predators throughout their development.

Faced with all these problems, some frogs have adopted different strategies; they lay far fewer eggs and protect them better, keeping them away from open water for as long as possible. In some cases this involves forms of brooding behavior, with egg mortality cut down or even eliminated. For example, the Surinam toads (*Pipa*) have extraordinary broad, flat bodies, and the male spreads newly fertilized eggs over the female's back, where the skin swells to embed the eggs, which disappear entirely within 30 h. The mother remains in the water, and about 3 weeks later the young tadpoles break out and swim off. Alternatively, gravid female frogs may find their own predator-free swimming pool, and in tropical forests this commonly involves pools of water that collect within the leaf bases of bromeliad plants (see Chapter 14). At least one Brazilian species makes its own "birthing pools" with raised mud walls. However, some anurans come closer to a truly terrestrial mode of reproduction. The midwife toad, *Alytes*, is a well-known example occurring in Europe, Majorca, and Morocco, where the male wraps up the eggs in stringy mucus around his own back legs for several weeks while he pursues a terrestrial lifestyle; then he stands with his back end in the water when the tadpoles are ready to emerge. In some of the South American dendrobatid (arrow poison) frogs, eggs are laid in moist earth and guarded by the male, and after hatching the tadpoles climb onto his mucus-covered back. Some African and Asian treefrogs mate in tree canopies, and then choose a branch overhanging water on which they make a ball of mucus; this is lathered up into an airy froth by both the male and female, after which the female lays her eggs in it. Either it then hardens to form a protective desiccation-resistant nest, or in a few species it is kept moistened by the female's urine, until the tadpoles drop out when mature into the pond below. Anuran foam nests reduce desiccation, but also provide insulation, the eggs being up to 8°C warmer (and therefore developing faster) than if unprotected.

A few species of anuran have gone a little further and produce large yolky eggs where the whole development can occur. In the Caribbean whistling frog there may be only 20–30 eggs, each with 50 mm^3 of yolk, and the froglets emerge in about 3 weeks. Occasionally the whole process can be retained inside the adult body. In *Gastrotheca* (the "marsupial frog") the females have brood pouches on the back, from which emerge either tadpoles or froglets, depending on the species. In *Rhinoderma* ("Darwin's frog" from Chile and Argentina) the male waits for the eggs to hatch and then "eats" the young, which develop in his enlarged vocal sacs and are subsequently spat out as froglets. The extreme case of terrestrialization is *Nectophrynoides*, a West African frog, where males introduce sperm into the female vent using their cloacal "tail", so that fertilization occurs inside, and the eggs are retained in the oviduct. The tadpoles hatch there and feed off flaky material from the walls of this duct, which are also very well oxygenated by an arterial supply.

All other vertebrates—collectively the Amniota—have overcome the limitation of egg size and egg water balance in a quite different way, with the evolution of the **cleidoic egg** (see Fig. 5.28). This has accessory breathing structures in the egg surface, linked via the chorioallantoic membrane to the embryonic blood system inside the egg, so that reptile and bird eggs can be very large indeed and the young can be highly advanced before hatching. Reptiles are commonly oviparous, with a few excursions into viviparity, while birds are exclusively oviparous. The eggs produced may be leathery and quite flexible, as in many turtles and lizards, or calcified, as in crocodiles and birds (which often have an outer coating over the eggshell to reduce bacterial invasion). Both types are still relatively permeable to water, but the calcified eggs are sufficiently rigid to resist volume change, and water lost during development, which is around 14–16% of egg mass in birds, is compensated by a growing air space within the shell. This air space is important in facilitating gas exchange in the developing embryos. However, eggs increase their metabolic rate as they mature, and run into oxygen shortage; they compensate by using inositol pentaphosphate to shift the hemoglobin dissociation curve to the right (decreased P_{50}), with a rapid return shift to normal after hatching (Fig. 15.43). The egg air space also allows for some accumulation of wastes, as the aging embryo switches progressively from producing ammonia to producing urea and then uric acid. The eggshell itself is perforated by

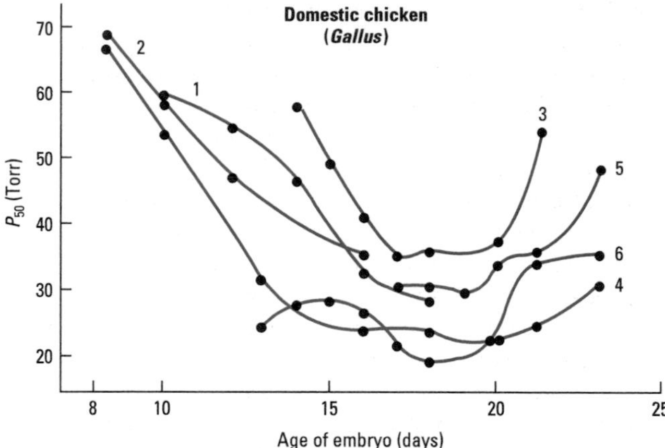

Fig. 15.43 The value of P_{50} for hemoglobin in maturing bird embryos. For six separate individuals the greatest oxygen affinity occurs at about 15–20 days (and is due to changing levels of inositol pentaphosphate, IPP—see text). (From Lutz 1980.)

pores (see Fig. 5.28) through which both gases and water vapor can diffuse, these pores having a complex anatomy usually constricted towards the inner surface. Though numerous, the pores are tiny and commonly give perhaps only 1 mm^2 of "open" exchange surface, though this is adequate to supply the egg's oxygen demand by simple diffusion. These pores inevitably allow rather substantial water loss during incubation, particularly in the latter stages when the shell becomes increasingly eroded.

The cleidoic egg frees its parents from any dependence on liquid water supplies at the site of mating or oviposition, and the amniotes can therefore become truly emancipated terrestrial animals. However, their eggs still require careful siting, either in a nest in a protected environment or retained within the parental body. These techniques serve several functions: for example, preservation against predators (some monitor lizards visit their nest periodically after laying, primarily for defensive reasons) or maintenance of equable humidities (to limit the egg water loss). But their commonest function is control of egg temperature, which must be high enough to achieve a reasonably rapid development. Tuataras in New Zealand provide a nice example, leaving their residential burrows in woodland and producing special nest burrows in open pasture where it is warmer.

There may also be other reasons why temperature control of the egg is important. In many semiaquatic and terrestrial reptiles a new phenomenon appears, that of **temperature-dependent sex determination**, or TSD. In three of the five major reptilian lineages sex is determined by environmental temperature, and many of these reptiles lack heteromorphic sex chromosomes (i.e. X and Y versions, or some equivalent). TSD is found in all crocodiles and many turtles, but is rare in lizards, and is absent in the tuatara, the amphisbaenians, and all snakes that have been examined so far. In general, TSD is more prevalent in long-lived species.

Three general patterns of TSD have been found. In type A (most crocodilians and lizards), females are produced at low temperatures and males at high temperatures ("FM"). Type B (many turtles) gives males at low temperatures and females at high temperatures ("MF"). Finally, some crocodiles, one lizard, and three turtles have been reported to produce females at high and low temperature extremes and males at intermediate temperatures (type C or "FMF"). More recent evidence, though, suggests that some cases of FM are actually FMF when sufficiently high temperatures are tested. Patterns tend to be associated with the direction of sexual dimorphism, i.e. which sex is larger.

Figure 15.44 shows the effects of constant temperature on the sex ratio of hatchling turtles. The gonadal sex of the embryo is determined not by the temperature at any particular moment but by the cumulative effects of temperature through a critical phase in development, which in turtles commonly begins shortly after laying and extends through the first half of development. In crocodiles the thermosensitive period is somewhat later, roughly in the third quarter of development (days 30–45), and it coincides with the timing of gonadal differentiation. In the field, the sex ratio of hatchlings is

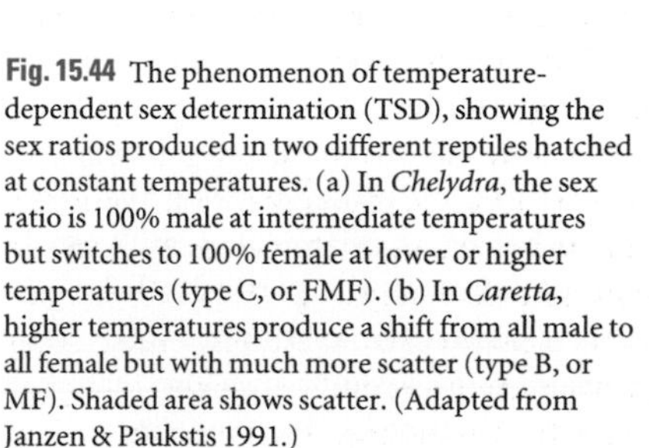

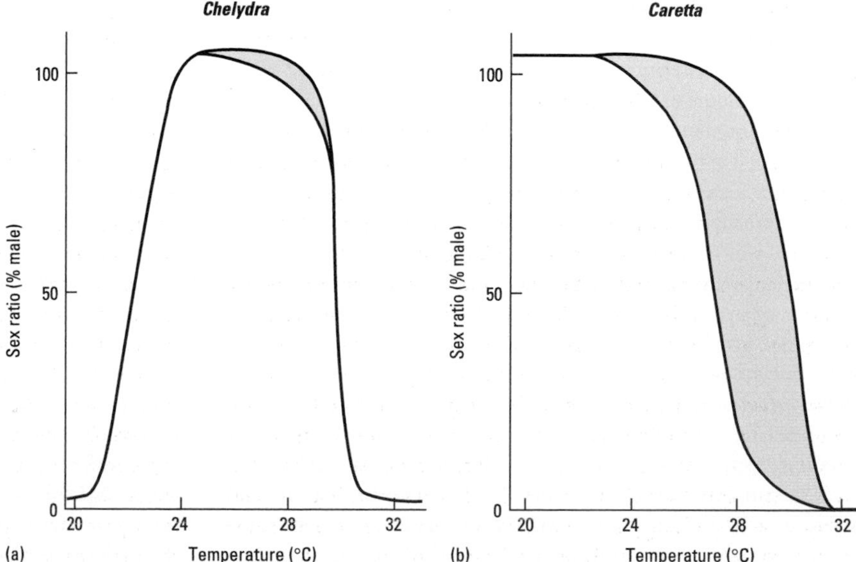

Fig. 15.44 The phenomenon of temperature-dependent sex determination (TSD), showing the sex ratios produced in two different reptiles hatched at constant temperatures. (a) In *Chelydra*, the sex ratio is 100% male at intermediate temperatures but switches to 100% female at lower or higher temperatures (type C, or FMF). (b) In *Caretta*, higher temperatures produce a shift from all male to all female but with much more scatter (type B, or MF). Shaded area shows scatter. (Adapted from Janzen & Paukstis 1991.)

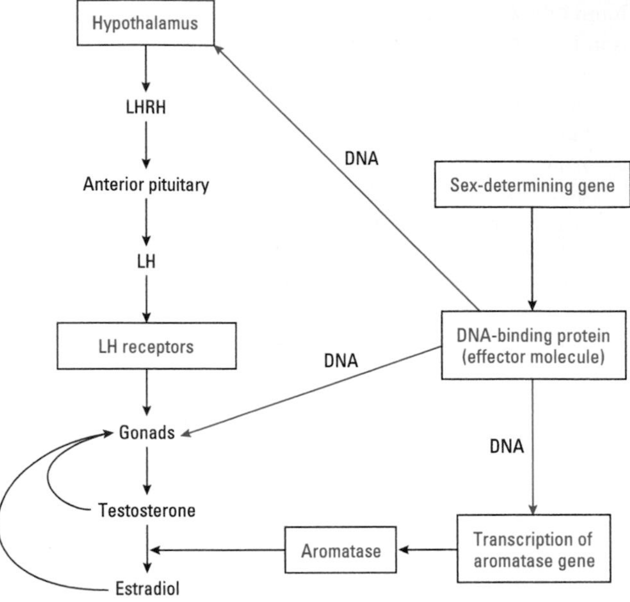

Fig. 15.45 Possible mechanisms controlling temperature-dependent sex determination (TSD). LH, luteinizing hormone; LHRH, LH-releasing hormone. (From Janzen & Paukstis 1991.)

therefore affected by the location of the nests, and the resultant thermal microclimate of the nest interior. Some nests produce both sexes whilst the majority produce only one sex.

Species with TSD are found in thermally patchy environments that allow the production of both sexes. Various hypotheses have been advanced to explain the mechanisms of TSD. The gonadal ratio of androgenic to estrogenic steroids is known to be important; females are produced by a low ratio and males by a high one. This ratio is controlled by the enzyme P-450 aromatase, which converts testosterone to estrogen. One idea is that aromatase itself exhibits temperature-dependent activity, or is produced by temperature-dependent gene transcription. Another hypothesis is that incubation temperature affects the secretion of luteinizing hormone (LH; see Chapter 10) or the density of LH receptors on the gonads. This could involve the regulation of gene expression by a temperature-sensitive effector molecule (a DNA binding protein) which is the product of a sex-linked gene. Some of these ideas are illustrated in Fig. 15.45.

In birds and mammals, sex determination is controlled independently of the environment, with heteromorphic sex chromosomes. In mammals two similar chromosomes (XX) makes a female and the heteromorphic condition (XY) makes a male, due to the effects of a gene (*SRY*) on the Y chromosome that initiates testis formation. In birds the situation is reversed and the heteromorphs are females. In both groups temperature has no effect on genetic sex, although both share some later operating sexual character-determining genes with the reptiles.

Nevertheless, in birds there is still a need to control the egg temperature, to achieve a smooth and rapid development so that nestlings appear at times when the adult can both feed them adequately and insure that their body temperatures are maintained. Hence nearly all birds use nests, and prolonged brooding behavior by one or both parents occurs. All birds also lay eggs whose shape is tailored for particular nesting sites and whose size reflects different degrees of development (precocious or altricial) at hatching time. Birds have never switched to viviparity, and this is presumed to be mainly because the female could not carry the weight of developing young around in her body during flight.

About 90% of all birds are monogamous (at least in theory, though DNA fingerprinting is increasingly revealing a surprising degree of infidelity in supposedly lifelong partners!). This is unusual in any animal group, but links to the needs of the young birds before and after hatching. The eggs need to be kept warm by incubation, as the embryo is also "warm blooded" but incapable of endothermy and thermoregulation. Then there is a long period of feeding and protecting the young, one or preferably both parents foraging intensively, giving time for behavioral patterns to be learnt, endothermic regulation to be established and stabilized, and flight systems to mature and become manageable.

Mammals have adopted a different solution to reproduction on land from their reptilian and avian relatives. They almost exclusively use viviparity, with very much smaller nonshelled eggs (the only exceptions being the egg-laying monotremes of Australasia, which are reproductively very similar to reptiles until after birth, when mammalian-style lactation occurs). The strategies of retaining the embryo internally and feeding it first from the maternal bloodstream and then from special mammary glands could be seen as the ultimate reproductive adaptations to life on land. Lactation may be one of the most important components of this strategy in terms of selective value. It allows the provision of a highly nutritious and digestible fluid even in mammals with poor-quality diets (e.g. coarse fibrous plants), where the mother may store reserves as fat in times of plenty and then breed in times of dearth; it allows the delivery of large meals in a short time, freeing the mother to leave the nest and forage; it permits the feeding of a litter of many altricial young with underdeveloped jaws and no teeth; and it promotes a strong bond between parent and offspring, allowing the development of complex learned behaviors. Mammalian milk can be delivered at a concentration suited to the particular habit and lifestyle (Table 15.13): very concentrated in marine mammals, as we saw, but very dilute in many large savanna-dwelling mammals where water balance in the young may be hard to maintain.

Table 15.13 Milk composition in mammals.

	Percentage composition				
Species	**Water**	**Carbohydrate**	**Protein**	**Fat**	**Energy content ($kJ\ g^{-1}$)**
Human	88	6.5–8	1–2	3–5	
Cow			3.4	4	
Seal				50–60	
Rhino	95		1.2	0.3	
Arctic fox	72	3	8	16–18	8
Red fox	82	3	6	9	4–5
Polar bear	50–70			20–35	12–20
Sun bear	73				7
Canids	80–95			2–6	4–9
Felids					5–10
Marsupials					
early	9	3–8	~1		
late	~1	11–15	17–20		

Birds and mammals share the problem of newborn young that may not yet be fully capable of endothermy and thermoregulation. In both groups a spectrum of abilities occurs, related to whether the young are altricial or precocial. Many altricial young are virtually naked (lacking fur or feathers) and are effectively ectothermic at first, their T_b being dependent on brooding by the mother. Chicks of boobies are good examples, having no inherent thermoregulation until they grow to about 200 g, but with a constant T_b of 38°C due to parental behavior. The same is true of most marsupial young, and of some eutherian offspring such as rabbits. It may take days or weeks for normal endothermy and regulation to be achieved. In essence these young animals are ectothermic behavioral regulators, but it is someone else's behavior that does the work!

However, the altricial strategy has its advantages; it allows shorter gestations and smaller birth sizes, putting less physiological strain on the mother, and it also normally allows larger clutch or litter sizes. In marsupials, and perhaps in other groups living in relatively variable and risky habitats, it allows the mother some choice in whether to continue investing in a youngster. In very harsh conditions, it may pay her to abandon a weak 1 cm long juvenile and invest in a new pregnancy instead, rather than spending up to 7 months (in kangaroos) with an ill-fated offspring in her pouch. Many marsupials can even "choose" to abort a fetus within the womb if conditions and food availability deteriorate. Average investment in reproduction for a marsupial is therefore less, lactation being a little cheaper and safer to the adult than placentation. In hard times, a eutherian fetus feeds through the placenta even at the expense of the mother's health. Within the eutherians, altricial young usually occur in taxa that have very short seasonal "windows" for successful breeding, so that more than one litter can be squeezed into the time available (e.g. desert rodents) or in taxa that have dens or protected homebases to return to (e.g. canids, primates, felids); precocial young are the rule in antelopes and many other large mammals that breed in the open.

15.5.2 Reproductive control systems

With land animals in general requiring more careful production and packaging of gametes, more complex mating and maternal behaviors, and much longer maturation times, there are inevitable consequences for the complexity of control of all stages of reproduction. The simple strategies of a gonadotropic hormone and a spawning hormone or pheromone in aquatic animals become entirely inadequate. Each stage of the reproductive process must be regulated both intrinsically and in relation to the changing environment to insure that young are produced at a time when their survival chances are optimal. However, many groups have been very inadequately studied and most of our knowledge comes from the insects and the vertebrates (especially birds and mammals). The control systems in these two groups were extensively described in section 10.7, detailing the roles of insect **juvenile hormone** (JH), and of the main tetrapod gonadotropins (**follicle-stimulating hormone**, FSH, and **luteinizing hormone**, LH) and gonadal hormones (**estrogens**, **progesterones**, and in males **testosterone**). A reminder of the essential control systems in mammals is shown in Fig. 15.46. Here we deal mainly with the cues for reproduction, and differences from aquatic systems.

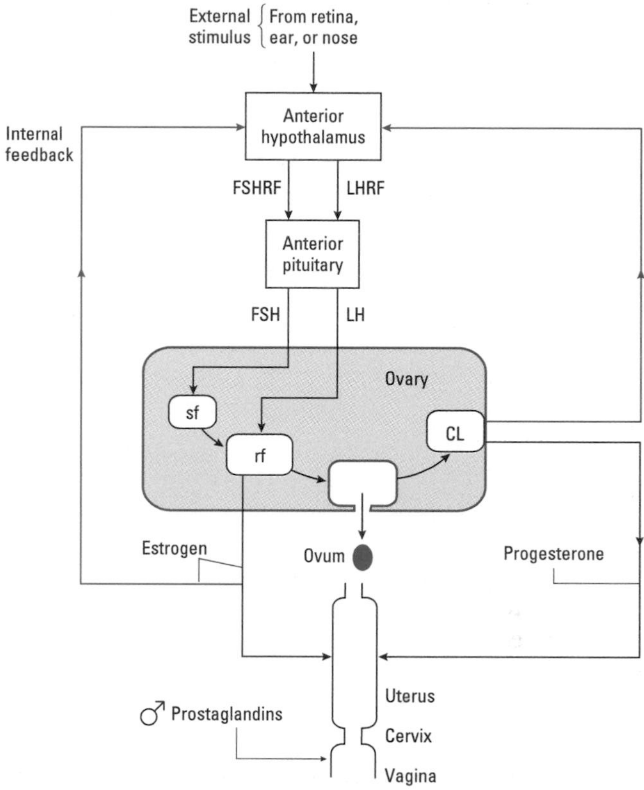

Fig. 15.46 Hormonal control of reproduction in female mammals: ovarian function. CL, corpus luteum; FSH, follicle-stimulating hormone; FSHRF, FSH-releasing factor; LH, luteinizing hormone; LHRF, LH-releasing factor; sf, small follicle; rf, ripe follicle.

Control of gamete production

The controlling influence of gonadotropic hormones is one of relatively few cases where parallels between terrestrial and aquatic animals can be drawn. Most land animals have a female **gonadotropin**, and many also have a male one (though a few species are born in a relatively advanced state, with eggs and sperm already maturing). Land animals may use different cues to initiate gamete production, however. Lunar triggers are rare, while a photoperiodic response is very common, often interacting with temperature or nutritional state, and sometimes with specific auditory, visual, or even tactile cues from potential mates.

In insects, JH initiates yolk production in the ovarioles, and hence egg development. A brain hormone stimulates release of ecdysone ("molting hormone") from the prothoracic gland, and the ecdysone triggers egg maturation. This in turn sets in train an automatic sequence of events through eggshell production, mating, and oviposition, likely involving secondary secretions from the ovariole tissues.

In terrestrial vertebrates gametes need to be produced at precise times, either in a regular cycle or at specific points in the seasonal cycle. Gonadotropic hormones from the pituitary are released under the influence of releasing hormones from the hypothalamus; in annual breeders photoperiod is the most commonly used hypothalamic cue. However, in many social and herd mammals seasonal breeding is abandoned to avoid the production of many vulnerable

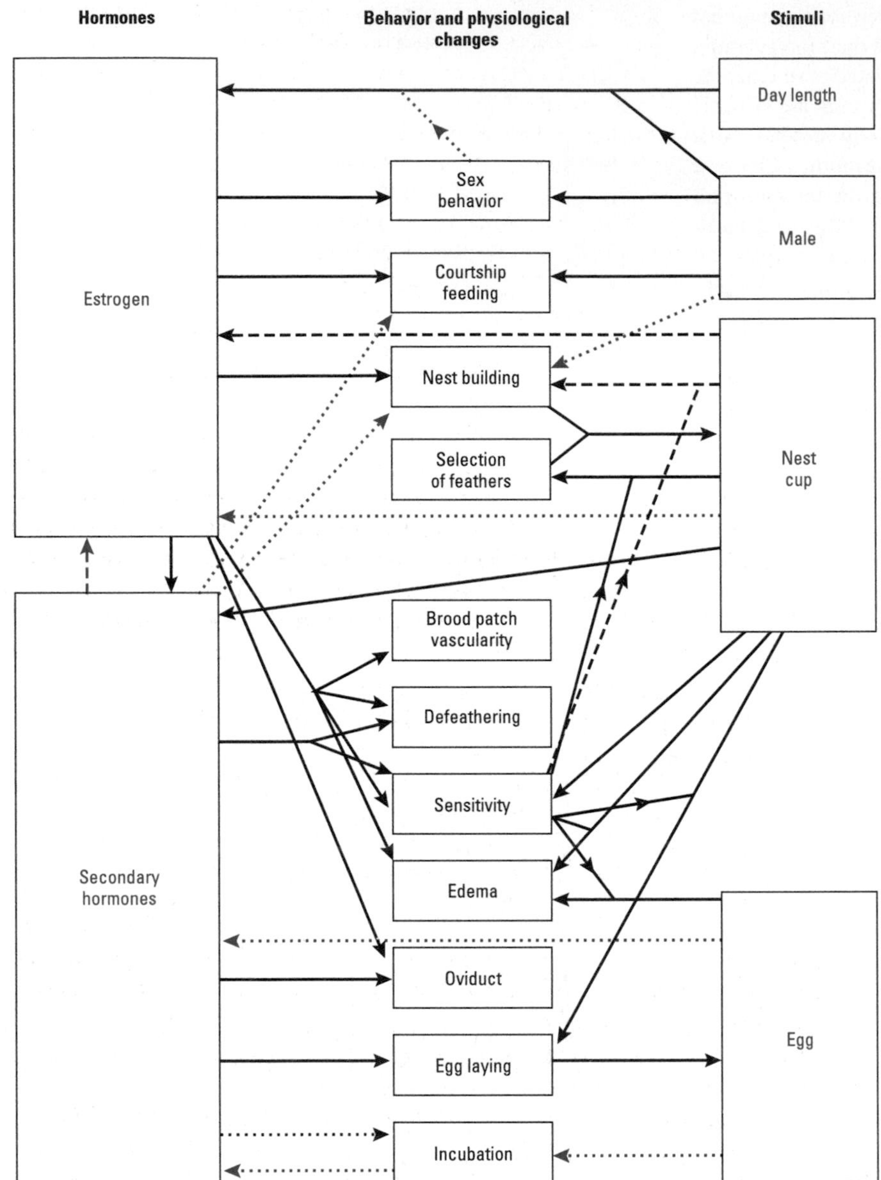

Fig. 15.47 Effects of hormones in controlling bird breeding behaviors and accompanying physiological changes. (From Hinde 1982.)

young at the same time, so females produce gametes and become receptive at random.

Control of mating behavior

In soft-bodied terrestrial animals mating behavior tends to be limited, although we have seen that snails and slugs have complex courtships, probably initiated at a distance by the scent of a potential partner and maintained by sense organs primarily in the tentacles. In insects mating behavior is often triggered seasonally by hormones, but is very much controlled by pheromones (see section 10.9) in most species that have been studied.

In birds and mammals, female mating behavior (termed **estrus** in mammals) is initiated by the effects of estrogen on the central nervous system; it is normally a short period of intense sexual receptivity. In males circulating testosterone brings about changes in external genitalia and development of secondary characters such as breeding coloration or breeding plumage, and it may also stimulate specific territorial behaviors and mate-attracting behaviors, with both visual components and acoustic song components. As a reminder of the potential complexity, some of the behaviors and their controls in breeding birds are shown in Fig. 15.47.

Control of gamete release and fertilization

The stimulus for gamete release in insects is often indirect, because of the ability of many insects to store sperm and to release them as required when an egg is mature and an oviposition site has been located. Many insects, notably dragonflies and fleas, have highly complex sperm storage organs, where the sperm donations of several different males may be kept. In some cases the last sperm to enter are the first to be used in fertilizations, and in these species males may try to scoop existing sperm out of the female before inseminating her themselves. In other cases the first sperm to enter take precedence. But recent evidence from DNA fingerprinting suggests that certain female dragonflies may exercise some control over which sperm fertilize which eggs, and they will behave rather differently according to which sperm they have "chosen". Most of the control here is probably nervous rather than hormonal; stretch receptors may signal when a mature egg is present in the lower ovariole, and nerves will control the release of sperm from the spermatheca.

There is also some evidence of the control of sperm release in the hermaphroditic land snails, and in genera such as *Helix* this may be related to the stimulus from the calcareous "love darts" that each partner fires into the other.

In land vertebrates, ovulation is triggered when the level of circulating estrogens interacts with the pituitary to affect the response to the releasing hormones from the hypothalamus. This may be achieved cyclically by complex internal feedbacks as in humans, or may occur in response to an external stimulus such as scent or the sound of a male calling; in some cases the act of copulation itself triggers ovulation. Details are given in section 10.7. Briefly, instead of continuing to produce FSH, the pituitary produces a surge of LH. This acts on the ovary to release the ovum from its follicle (the remnants of which become the corpus luteum and take over production of progesterone). One or more eggs are released into the fallopian tubes, and normally encounter spermatozoa from the male ejaculate in the upper portion of the uterus.

Fertilization is controlled in a similar manner in most animals, but details at the molecular level are still somewhat unclear. Mammals are inevitably best known; here the spermatozoa, as many as 250 million per milliliter of seminal fluid, begin to approach the egg by dissolving the sticky cumulus cells that surround it, using hyaluronic acid derived from the sperm tip (acrosome). They then penetrate the zona pellucens and the vitellogenic membrane of the ovum itself. Normally only one sperm achieves complete penetration and fertilizes the ovum nucleus, all other sperm then being prevented from entry. However, internal fertilization for land animals poses some extra problems. Firstly, infective agents such as bacteria and fungi may gain access to the body through this route, encountering relatively unprotected surfaces. In higher vertebrates this problem is partly solved by the presence of specific commensal bacteria (*Lactobacillus vaginalis*) within the vagina that react with glycogen (common in cells in the walls of the reproductive tract, especially at the high levels of estrogen preceding ovulation) to produce a highly acidic environment (pH 5 in humans), in which most foreign bacteria and fungi cannot grow. Seminal fluid therefore has to be well-buffered and alkaline (pH 7.2–7.8). A further problem is that immunologically foreign material is introduced deep within the body, where it may contact highly vascularized surfaces and will provoke an immune response. In mammals there is an exudation of neutrophil white cells from the uterine wall, which quickly agglutinate and begin to destroy the sperm. Fertilization must therefore occur quickly and efficiently, and the male gamete of land animals has only a very short life outside its owner's body.

Control of egg maturation and development

In insects and arachnids, hatching of the eggs (whether as nymphs, larvae, or tiny adults) is usually highly attuned to environmental conditions. Where the eggs have overwintered or diapaused, temperature is the commonest cue for hatching, although cumulative "day degrees" above a particular temperature are often needed rather than a simple threshold effect. If the eggs have been laid in water, for example in dragonflies, mosquitoes, and some hoverflies, rising temperature and decreasing oxygen levels together may serve as a cue.

After hatching, time to maturity (the final molt to adult form and/or size) is determined in insects by the interplay of two main hormones, JH and ecdysone. As the level of JH in the hemolymph falls progressively, each molting episode triggered by ecdysone results in a less juvenile morphology until finally the adult genes are fully expressed and the adult body architecture appears (see section 10.5 for more details). However, there may also be maternal effects on maturation determined across generations via the physiology of the egg-laying adult, so that eggs laid as winter approaches have an inherently greater tendency to enter diapause (see Chapter 10).

In mammals the initial control of fetal development is largely brought about by the gonadal steroids, estrogen, and progesterone, and full details were given in section 10.7. Estrogen causes the uterine wall to prepare for implantation by the fertilized embryo, and once fertilization has occurred the presence of the embryo brings about the persistence of the corpus luteum. From this point, the fetus has therefore begun to exert some control over the maternal physiology, a process that will continue for the weeks or months of pregnancy. Progesterone is secreted throughout this period, either from corpora lutea in the ovary or from the uterine placental tissues themselves. Estrogen is also present at low levels, and begins to function in preparing the mammary glands.

When the fetus reaches maturity, in many species it partly triggers its own birth as described in section 10.7.4. Thus pregnancy and birth involve a complex array of interactions between fetal and maternal hormones, with little intervention from environmental stimuli.

Post-hatching care and maternal behavior

Little is known about the control of parental care in insects, except for the social insects, particularly the honey-bees. Here pheromones from the queen control the behavior of all other females (the sterile workers), who are full- or half-sisters of the new eggs. Workers construct and provision the egg cells, tend and clean the larvae, and gather provisions from outside the nest to maintain the young and the queen, all under the influence of "queen substance".

In mammals, birth is followed by a rapid decline of both estrogen and progesterone in the mother once the placenta is shed, and when the young then begin to suckle the physical stimuli promote a surge of prolactin from the pituitary, which together with oxytocin initiates and maintains milk production and ejection.

Overall, we can see that reproductive problems on land are solved by an increased investment of time or energy or both, by one or both parents; whether in finding a mate and achieving efficient sperm transfer, in protecting and provisioning the eggs, or in either preparing for or directly overseeing the postnatal needs of the juveniles. All of these changes necessitate a far more elaborate behavioral repertoire, and substantially greater sophistication of the neural and hormonal control systems.

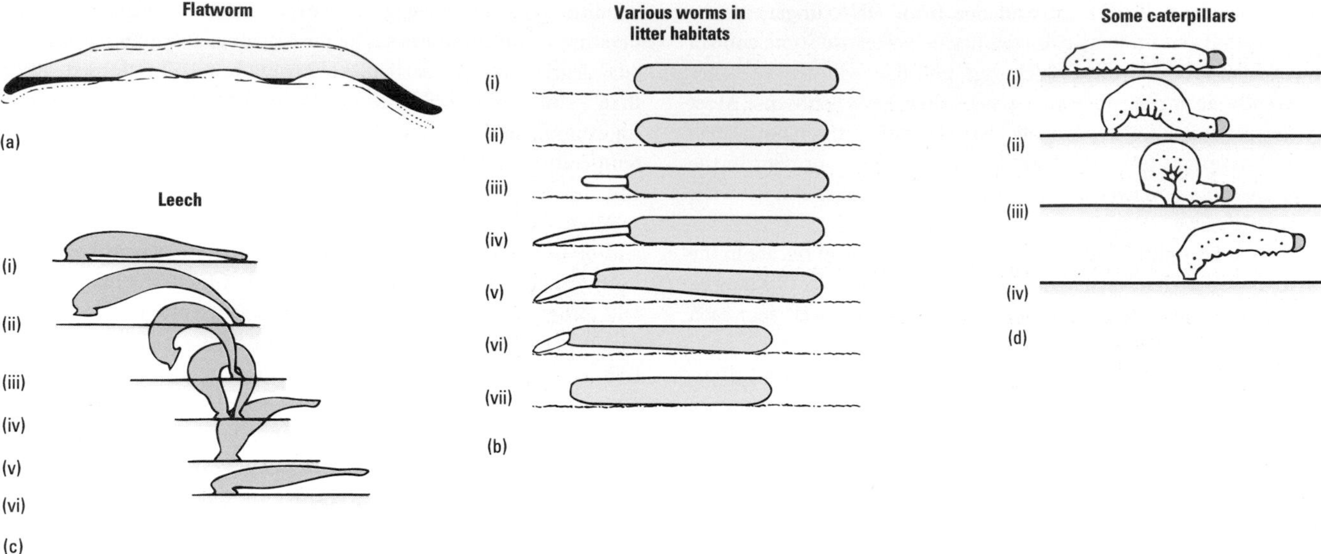

Fig. 15.48 (a) Ventral "pedal waves", (b–d) looping.

15.6 Locomotion and mechanical adaptations

15.6.1 Soft-bodied animals

The basic principles of hydrostatic locomotion in soft-bodied animals were dealt with in section 9.14.1. Such animals are at a serious disadvantage on land. They will have less support from the medium and a tendency to flatten under their own weight; thus they need relatively thicker muscle layers to maintain tone and shape compared to related aquatic forms. Their entire anatomy and functional integrity relies on maintained hydration, and any degree of desiccation will compromise their normal activity—resisting water loss is crucial. They will also suffer severely from the effects of friction in crawling over or burrowing through the relatively hard substrata prevalent on land, and may be damaged by such contacts. They therefore require more copious production of mucus as a lubricant and as a protectant, which entails extra water loss. Clearly this design will only work adequately for small animals in the humid microhabitats of soil or litter layers.

Most nematodes are so small (20–50 μm diameter) that they remain within the water films around soil particles; indeed, their design is such that they locomote most efficiently when within such films. Thus they cannot be considered terrestrial in terms of locomotion, any more than in terms of their physiological relations with their environment. However, land flatworms and nemertines are somewhat larger and do experience a less than saturated world, and have certain locomotory adaptations to ease their passage through it. Land flatworms move with contractile muscular waves, which are of longer wavelength than their aquatic kin, so that the body touches the ground only intermittently (Fig. 15.48a), leaving broad mucoid "footprints" rather than a continuous trail. Terrestrial gastropod molluscs (snails and slugs) can also modify their "gait" to limit ground contact, and often leave a discontinuous "footprint" trail indicating the reduced points of contact. Some nemertines probably use the same trick of limited ventral contact points, but many of them have an additional "looping" strategy to achieve fast escape movements: the proboscis is shot out ahead of the worm and its tip adheres to the ground, so that the rest of the body can be rapidly drawn up to it using longitudinal muscle contractions (Fig. 15.48b). Many leeches (Hirudinea) are found moving about well away from water in tropical forests, using suckers to achieve a similar looping motion (Fig. 15.48c) that reduces contact with the substrate, thus reducing frictional costs and potential damage; "looper" caterpillars also use the same trick (Fig. 15.48d). Earthworms, though, moving mainly underground, retain more obvious repetitive longitudinal and circular waves of contractions, accentuated by their pronounced segmentation (see Fig. 9.105). Their movements occur almost entirely within their soil burrows, the expanded segments (plus chaetae) acting as anchors; fundamentally their locomotion differs little from that of marine annelids (see Chapter 11).

In evolutionary terms the real innovators in soft-bodied land locomotion have been the tropical onychophorans (velvet worms), with the body wall protruded into many pairs of stumpy legs. The body operates essentially hydrostatically, undergoing tremendous shape changes (up to 10-fold variation in cross-sectional area), and dehydration causes loss of tone and a disrupted gait. But the legs act as "real" limbs: protrusions and retractions due to extrinsic leg muscles give a genuine lever action and greatly reduce frictional contact with the ground.

15.6.2 Arthropods

First of all, remember that a great many land arthropods spend most of their life cycle as soft larval stages (the maggots and caterpillars of many endopterygote insects), often completely lacking legs, and moving very similarly to the hydrostatic animals already discussed. Peristaltic muscular waves drive the hemolymph forward to dilate the anterior end and penetrate the substratum, be it soil or the tissues of another animal or of a plant. The cuticle is usually com-

Table 15.14 Advantages of legs (contrasted with soft-bodied locomotion).

Lever action—small movement of muscle gives large movement at limb tip
Limited ground contact—reduced friction
Legs stop and start, body moves forward smoothly—reduced acceleration and deceleration of large masses, energy costs reduced. Aided by having main mass of leg (muscle, etc.) at the top, close to body
Muscles can be small instead of sheet-like, so with less connective tissue strapping and increasing speed of contraction
No lateral sinusoidal components that waste energy
Largely independent of hydration state
Effects of muscle contraction are localized, do not affect other body-wall muscles
Increases number of gaits and gait/speed/energy trade-offs
Permit good proprioception
Legs can be diversified for other uses

pletely unsclerotized, excepting only the head capsule in forms such as soil beetles. Only in the adults of such insects do we find innovative kinds of land locomotion.

Walking and running

Fully functioning legs are present in myriapods, arachnids, juvenile and adult exopterygote insects, and adult endopterygotes. In all of these, the same principles are used for crawling and walking as their aquatic ancestors used for crawling and swimming. They exploit to the full the merits of a cuticle that can be secreted in a rather fluid state and then "set" (with very variable degrees of hardness) into any form that is needed. The evolution of a proper leg, with flexors and extensors at each joint, brings immediate advantages on land (Table 15.14). For arthropods, the particular benefits include a lever action, reduced frictional contact with the substrate, and a smooth forward action for the main body mass with only the legs themselves alternately accelerating and decelerating. Efficient locomotion is therefore achieved with a reduced number of legs, each containing minimal muscle mass (see section 9.15.2). The major changes in locomotion in walking land arthropods as compared with aquatic arthropods are the reduced numbers of legs engaged in locomotion, and altered proportions of parts of the legs; there are few fundamental changes in actual mechanisms.

Speeds in land arthropods can be substantially higher than in marine species, largely due to the reduced viscosity of the medium. Land crabs can achieve 1–2 m s^{-1}, with only eight of the 10 legs in use and with only three in contact with the ground at any time to form a tripod support, where two legs are thrusting and the third is merely providing balance. Centipedes such as *Scutigera* can also run fast, up to about 0.5 m s^{-1}, each leg having a quick backstroke and a slow recovery phase (see Fig. 9.113). These faster speeds tend to be enhanced by longer limbs, but to avoid these tripping over each other their lengths are staggered (Fig. 15.49).

Equally, the forward force of a land arthropod can exceed that of a marine species of the same size; millipedes in particular are renowned for their "pushing power", moving slowly but very forcefully when burrowing, and often described as "bulldozing" in their action, with the blunt head and shield-like collar region thrusting into the soil. Burrowing beetles, with heavy limbs, a narrow anterior, and a very heavily sclerotized cuticle, achieve similar forces.

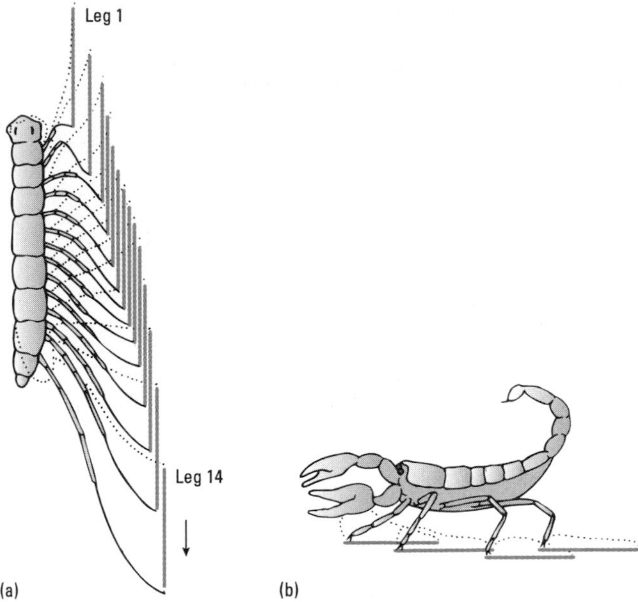

Fig. 15.49 (a) Top view of the gait and stance of a running centipede and (b) side view of a scorpion, both organized such that multiple pairs of legs are different lengths and do not step on each other. Solid bars show stride length; dotted lines show leg positions at end of stride. (Adapted from Manton 1952; Herreid & Fourtner 1981.)

Jumping and flying

These types of locomotion offer more radical solutions to reducing friction with the substratum. Various jumping (saltatory) arthropods use energy-storing systems that can allow rapid escape, and sometimes a continuous hopping gait. Collembolans, fleas, and many orthopterans can jump to relatively enormous heights, using the energy-storage systems as discussed in Chapter 9. Some spiders use a different system, jumping by using a rapid hydrostatic extension of the fourth pair of legs.

Flight is restricted to the pterygote insects among the arthropods (though aerial locomotion is achieved by parachuting on silken threads in spiderlings). The wings appear to have evolved initially for other purposes; perhaps as stabilizers, or as areas of increased surface area to aid in thermoregulation, or possibly as "sails" to assist propulsion over water surface films. The initially stubby wing-buds may have expanded enough to become useful in gliding, and only then became jointed to the thorax and capable of being flapped. Once evolved, and making full use of the advantages of the arthropod chitinous cuticle as a light but immensely strong material, wings became a huge part of the insect success story from the Carboniferous onwards, giving access to new habitats within the canopy of the radiating angiosperm forests and allowing insects to disperse more rapidly, and potentially perhaps to speciate ever more quickly.

Insect flight primitively involved two pairs of large wings with a complex net-like venation, veins being formed where the two apposed cuticular surfaces (with their epidermal cells jettisoned at the molt) are separated by thickened channels carrying hemolymph, nerves, and tracheae. These wings have tended to become smaller and/or more easily folded away, allowing winged insects to climb

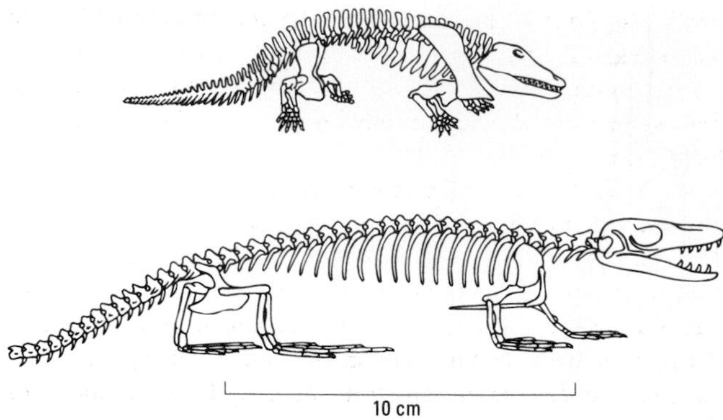

Fig. 15.50 The sprawled gait of early land vertebrates (here a labyrinthodont amphibian from the Permian) contrasted with the upright gait of most modern forms, the legs rotating to lie under the body, holding the belly aloft with much reduced friction.

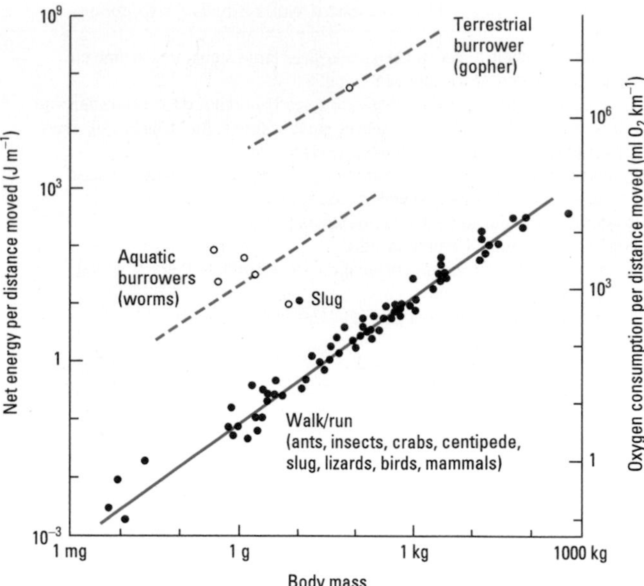

Fig. 15.51 Cost of terrestrial burrowing is substantially higher than either aquatic burrowing in softer substrates, or terrestrial walking and running. (Adapted from Alexander 1982.)

through vegetation, and in some cases to burrow. The wings have also tended to become coupled so that the two wings on each side of the body operate together, efficiently supported by just a few strut-like veins, especially along the leading edge. Strong terrestrial fliers therefore have rather rigid short wings, with mainly longitudinal veins, levering against a dorsal process on the insect thorax like an off-center see-saw. Two different systems of wing control are found in modern insects, and the principles and mechanisms involved were discussed in section 9.16. In the most advanced fliers, including bees, flies, and many beetles, the two sets of indirect flight muscle take up most of the volume of the thorax; flight muscle may be 60% of total mass in males of the dragonfly *Plathemis lydia*. These highly developed wing muscles are the main source of internal heat generation in the active insect (see section 15.3), so that locomotory mode is strongly interlinked with other aspects of insect physiology.

15.6.3 Vertebrates

Walking and running

Proper "walking" appeared rather gradually amongst the first land vertebrates. The early labyrinthodont amphibians used a fish-like belly crawl based on the trunk muscles, only slightly augmented by a pulling action from the front limbs and hardly affected by hindlimb action. As skeletal elements fused to form the pectoral and pelvic girdles (each forming strong links to the limbs and spine and becoming freed anteriorly from the back of the skull), so the force provided by the short amphibian limbs could increase, and they became rotated somewhat inwards to lie below the animal rather than sprawled out to the side (Fig. 15.50). Modern salamanders still use sinusoidal movements of the body to complement their limb action, but in most other tetrapods the limbs have become greatly modified to support the body well off the ground, giving all the advantages of limbs discussed above. Inevitably there have been concomitant increases in speed, agility, and maneuvrability in three dimensions.

Walking and running have therefore become the primary locomotory modes for land vertebrates. On four legs the common **gaits** are usually walking (slow) and running (fast), but for animals larger than cats there may be one or more additional gaits, usually termed the trot and (for horses and similar animals) the canter, with the fastest gait then termed galloping. For bipedal animals there is a slow walk and either a hop or a run to go faster. The principles and mechan-isms of these gaits were discussed and illustrated in section 9.15.2.

Burrowing and digging in soil or litter remain important locomotory modes even for quite large vertebrates, with some toads, lizards, birds, and many mammals dwelling underground in burrows they have excavated. Figure 15.51 shows that terrestrial burrowing is an expensive occupation, but of course it has huge benefits in terms of hygrothermal physiology and concealment. In amphibians and reptiles a subterranean or litter-centered life may be accompanied by the loss of limbs, as in salamanders, caecilians, skinks, and amphisbaenians. However, birds and mammals always retain girdles and limbs and use the limbs as the digging tools. Life may be permanently centered on a burrow system, with marmots, mole rats, moles, and badgers exemplifying the required adaptations. Forelimbs are generally short and very heavy with stout claws, heads are broad and rounded with the nostrils well protected, and prominent incisor teeth may play an important role in burrow excavation.

Jumping and flying

Jumping has evolved several times in vertebrates. It is used as a continuous gait in anuran amphibians, in many birds when on the ground, and in some marsupial mammals. Moreover, leaping between trees (and sometimes across the ground too) is common in many primates, and jumping is employed as a transient escape or attack response in many other taxa. Above all, jumping to any significant height again requires speed at take-off, which in vertebrates involves a very rapid straightening of the "knee" joints, usually with

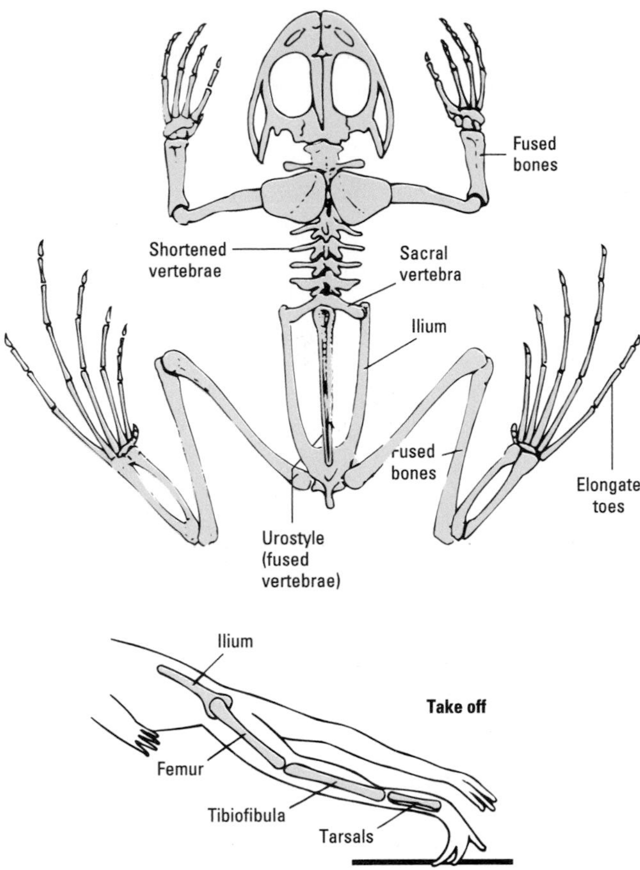

Fig. 15.52 The highly modified skeleton of a frog, showing specializations for jumping.

a preparatory bending of the legs, and built-in elasticity. The principles were discussed in section 9.15.

Anurans are spectacularly specialized for jumping (Fig. 15.52), with several key features:

1 Long back legs, with an elongated ilium bone in the pelvis plus elongate tarsals in the foot effectively adding two joints to the hind leg for the actual jump.

2 The second elements in both limbs fused, as the radio-ulna and tibio-fibula, giving greater strength and reduced twisting.

3 The loss of the tail, since a symmetrical hind kick leaves the tail with no function as a stabilizer, as in walkers and runners.

4 A shortened backbone, comprising only about eight vertebrae, with the last few fused as the rod-like urostyle.

5 A shoulder girdle modified as a partial shock-absorber.

Jumping in anurans may have evolved primarily to assist escape back to the water, since nearly all anurans tend to jump towards higher humidity whenever startled. The Goliath frog of West Africa can jump 3 m, although smaller frogs do proportionately better, possibly using a catapult mechanism to amplify muscle power production. For example the Cuban treefrog, *Osteopilus septentrionalis*, weighing around 14 g, is able to jump 1.4 m, and calculations show that the power required for take-off is more than 800 W kg^{-1} muscle, with peak power production perhaps twice this, exceeding the maximum output possible from the hindlimb muscles by about seven-fold. Thus a substantial part of the work must be performed before take-off and stored in the elastic mechanisms. Some tree frogs can jump and then glide 15–18 m, having flaps between the fingers and toes and along the body, and sucker-like tips to their toes for safe landing.

However, as far as vertebrates are concerned, flying in the strict sense occurs only in birds and in bats (and in extinct pterodactyls). The principles and mechanisms of different kinds of vertebrate flight were covered in section 9.15. Wing surfaces have been produced in a variety of ways: using membranes between the fingers (bats), or between the forelimbs and hindlimbs ("flying" squirrels, etc.), or feathered extensions from the long bones of the arms (birds). Flight muscles are modifications of the thoracic/forelimb muscles, and in birds two particular muscles have become hugely dominant (see Fig. 9.120), with the larger pectoralis muscle powering the downstroke and the smaller supracoracoideus muscle producing the recovery upstroke, the two together being up to 35% of the body mass in a strong flier. Flapping flight is extremely expensive, especially for small birds and bats (though recent evidence indicates that it may be somewhat more economical in bats), but of course this is offset by the advantage of an extremely economical means of transportation per unit of mass for a given distance (see Chapter 3). In relatively few very small birds an even more expensive hovering flight is used; this is especially well known in hummingbirds, whose unusual flight muscles were considered in Chapter 9, where the fast wingbeats occur at the resonant frequencies of the wing/thorax system. Gliding flight is much more economical, using a slow muscle to hold the wing in position; it is widely used for soaring in rising thermals or along the air currents over the slopes of land or waves, and can permit cheap travel over great distances, as in petrels and albatrosses.

15.7 Sensory adaptations

Changes in motility and agility, and in the mode and speed of locomotion, have inevitably entailed marked adaptation in the senses and coordinating systems in all the land animals, especially the arthropods and vertebrates.

15.7.1 Mechanoreception and hearing

A sense of touch via mechanoreceptors works well in both water and air, and mechanoreceptors are not much modified on land as a result. However, they may become more numerous, or concentrated on protrusions around vulnerable areas, to guard against the more forceful impacts that can occur on land. Mammals improve their sense of touch with specific vibrissae (whiskers), particularly around the muzzle and lower legs; arthropod feet and antennae have become crowded with mechanoreceptors.

Vibration signals at low frequency work well in aquatic media, and can be used for very long-range signaling (as in the whales); but true aquatic "hearing" is rare, and most marine animals use only a localized sense of vibration. In contrast, land animals continue to use a vibrational sense, but also routinely use the sense of hearing as part of their main detection and signaling repertoire, with high-frequency airborne sounds both produced and perceived. Vibrations through the ground are important to many snakes, and various arthropods (e.g. scorpions) also have ventral receptors that pick up vibration through the sand. Land crabs, such as *Uca*, signal by tapping the sand with their claws. However, the related crab *Ocypode* can stridulate using ridges and tubercles on the limbs, producing sounds through the air that attract females to the mating burrows (see above). Similar systems occur in many insect groups, with crickets and cicadas particularly noted for their species-specific "calling" songs, produced by serrated areas of the legs, abdomen, or wings and often amplified by air-filled chambers or even by carefully constructed resonant burrows (e.g. in the mole crickets). Detection in both crabs and insects usually involves chordotonal organs (see Fig. 9.52), sometimes incorporated in ear-like structures on the legs or abdomen.

Sound production amongst the vertebrates usually involves internalized organs. The calling notes of frogs originate from the lungs, and are often amplified by throat sacs; the larynx at the top of the windpipe has a pair of vocal cords and accessory cartilages, to produce the actual noise. From this kind of arrangement also come the intricate songs of passerine birds, and the squeaks, honks, and trumpetings (and ultimately speech) of mammals. This is accompanied by increasing sophistication of the ear apparatus in the land tetrapods, with progressive recruitment of "spare" gill-arch bones and jaw bones into the ear as sound-transmitting ossicles, and the extension of the cochlear system of frequency-tuned hair-cell receptors (see Fig. 9.50). Sensitivity is improved by variation in the size ratio of the eardrum and the internal oval window; this ratio is commonly 10–25 but reaches 40 in owls, where nocturnal hunting requires acute hearing. Some birds can also hear extremely low-frequency noises (infrasound), which may be used in navigation in pigeons and other species; and elephants are now known to communicate over long distances using infrasound.

15.7.2 Vision

Air has a very different refractive index from water, so the physical properties of visual systems need to alter. Eyes are usually fitted with relatively flattened lenses compared with the spherical lenses found in aquatic animals, although the cornea may be more domed to give it some refracting power. The lens also tends to be deformable, so that its focal length can be changed by the surrounding ciliary muscles ("visual accommodation").

Visual communication is more effective over long distances in air than it is in water, light traveling much further without attenuation by absorption. So again it is no surprise that visual signals become more complex and the eyes of land animals more sophisticated. Many land crustaceans respond to visual stimuli over a range of several meters, where their aquatic relatives rarely see more than half a meter; land crabs hunt largely by visual cues whereas aquatic crabs use chemical and tactile cues. Land animals also tend to respond to much higher frequency visual changes, aided by sharper and faster lateral inhibition between adjacent receptors (see Chapter 9). Smaller land animals also tend to live in a rather two-dimensional surface world, where most events take place roughly "on the horizon", and many insects therefore have a horizontal band of particular sensitivity across the middle of their eyes (seen as a band of larger flatter ommatidia).

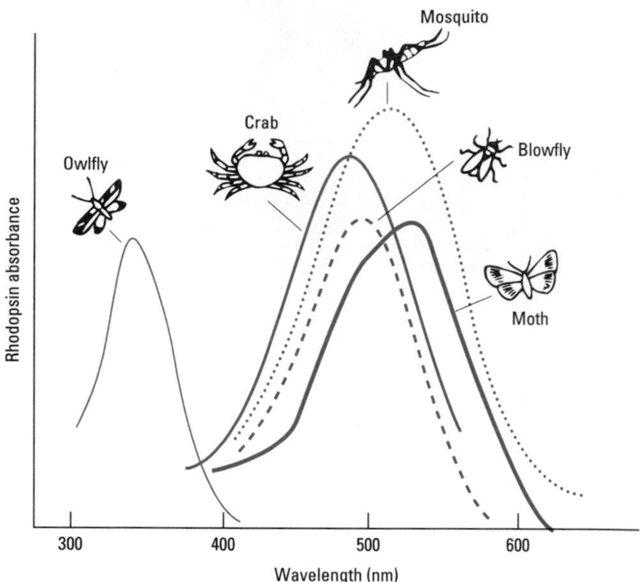

Fig. 15.53 Rhodopsin spectral sensitivities in a range of terrestrial insects and crabs. In man the peak sensitivity is at 498 nm, with cones tuned to 440, 535, and 575 nm. (Data from Hoglund *et al.* 1973.)

Most obviously, visual receptors can only respond to those wavelengths that reach an animal's surface unattenuated, so that in the atmosphere there is a wider range of spectral sensitivity than in the hydrosphere and color vision becomes a key component of many animals' sensory repertoire, with varying pigments in the photoreceptors giving a wide range of possible spectral sensitivities (Fig. 15.53). Truly diurnal species have particularly high visual acuity and excellent color vision. However, crepuscular and nocturnal species have "scotopic" eyes, with high sensitivity to low light levels but low acuity and little or no color vision, some vertebrates and insects also having a reflecting layer (tapetum) behind the eye to reflect back any photons that are not initially absorbed. Land animals (some insects and some birds, at least) are also able to use the plane of polarization of light as a cue to direction, which relatively few aquatic animals can do. A few land animals produce and respond to bioluminescence, notably fireflies and glow-worms, and also such unlikely creatures as earthworms; but on the whole it is less common than in the seas.

15.7.3 Chemical senses

Chemical communication also becomes more dominant in land animals, with pheromones playing an important part in the lives of most insects and many vertebrates. Chemical signals have relatively poor resolution and their transmission may be rather slow (depending on wind speed), but they operate over considerable range, by day or night, and are not impeded by intervening physical obstacles in the way that vision is. They can also be deposited on soils or vegetation or on other organisms (e.g. a mated female, an already occupied host, or a recently visited food source) where they will persist as markers. They can potentially form a "trail", whereas in water a mark or trail would be quickly dissolved or disrupted.

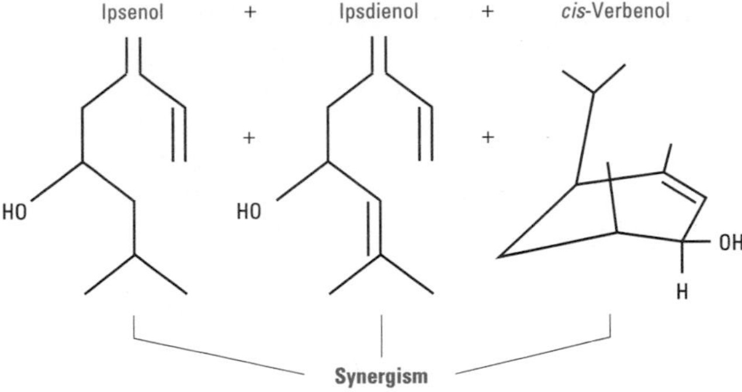

Fig. 15.54 The three components used in pheromone mixes in bark beetles in the genus *Ips*, each species using a different mixture.

Pheromones in terrestrial environments have to be detected against a complex background of competing chemicals in the environment, and must give unambiguous signals to the receiver, requiring a novel volatile chemical or more probably a unique mix of small volatiles (see section 10.9). Further specificity may be imparted by the timing of release of the chemical, with many species only emitting their mating pheromones at particular times of day (and/or of season) in still weather where the odor plume will remain reasonably intact and directional (see Fig. 10.34). For example the oak-roller moth, *Archips*, has an attractant pheromone consisting of 21 components, 17 of which are effective individually, although less so than the complete mixture. In the bark beetle genus *Ips* there are three main components of the mating pheromone (Fig. 15.54), with the proportions varying in different species. Hence some male moths and beetles can respond specifically to their females' signals from many hundreds of meters away.

Pheromones in mammals may also achieve specificity from mixing components, perhaps even to the level of individual recognition in the scent marks left by territorial species (in these highly olfactory vertebrates the surface of the much-folded nasal olfactory epithelium may be greater than the entire body surface!). Table 15.15 summarizes the functions which pheromones can serve in terrestrial animals, stressing the prevalence of marking and trail systems; more details appear in section 10.9.

Table 15.15 Functions of pheromones in land animals.

Activity	Specific function
Mating and reproduction	Long-range attraction Short-range arrestment Courtship behavior maintenance Copulation induction/maintenance Postnatal care elicitation Fertility suppression (S) Nesting behavior induction (S) Larval settling Cycle synchronization
Feeding and foraging	Food-source marking (S) Trail marking (S) Aggregation
Alarm and defence	Conspecific alarm (S) Fight induction (S)
Social organization	Hierarchy maintenance (S) Caste recognition (S) Colony recognition (S) Territorial marking Dominance marking

S, mainly or solely in social insects, i.e. ants, wasps, bees.

Table 15.16 Frequencies of echolocation sounds in land animals.

Group/species	Frequency (kHz)
Cave-dwelling birds	
Swiftlet	4–8
Oilbird	2–7
Nocturnal insectivores	
Tenrec (*Centetes*)	5–17
Shrew (*Sorex*)	18–60
Shrew (*Crocidura*)	70–110
Nocturnal bats	
Fruit bat (*Rousettus*)	13–40
Greater horseshoe (*Rhinolophus*)	35–90
Mouse-eared (*Myotis*)	25–100
False vampire (*Megaderma*)	30–125
Painted (*Kerivoula*)	230–243

15.7.4 Other senses

As might be expected, land animals generally lack electric senses as the conductivity of the medium is too low to allow the passage of signals, whether for detecting disturbances of field due to nearby objects (electroreception) or for delivering defensive discharges (electrogenesis). The electrocyte system of fish apparently disappeared very quickly in the first land crossopterygians, although use of electrical sensitivity does remain in aquatic vertebrate lineages right up to mammals such as the duck-billed platypus.

Echolocation can have a place on land though, occurring in bats and some shrews active at night and in a few birds that live in caves (Table 15.16), either for prey detection or to avoid obstacles. Bats emit clicks either from their larynx or with their tongues, through the mouth or via an elaborately folded nose structure. They receive the echo with greatly enlarged ear pinnae, and their ears are highly frequency-tuned. Sounds must be high energy when emitted (up to 120 dB) but on receipt are attenuated down to just a few decibels; how then can a bat avoid deafening itself and still hear the echo? The answer lies with contractions of muscles in the middle ear that dampen the tympanic membrane just as a click is emitted, relaxation occurring exactly at the end of the click so that full sensitivity is restored in time to hear the echo.

One other sense that occurs in a few land animals is the detection of heat. In a sense this is a special case of vision, as it involves picking up wavelengths (infrared radiation) that are outside the normally visible range. It is impossible in water, but useful on land where marked temperature gradients can occur on a small scale. The best known cases are the heat sensors on the heads of pit vipers (see Fig 9.62b), although of course many vertebrates have thermal sensors in their skin that will detect and give warning of dangerous heat sources before contact is made.

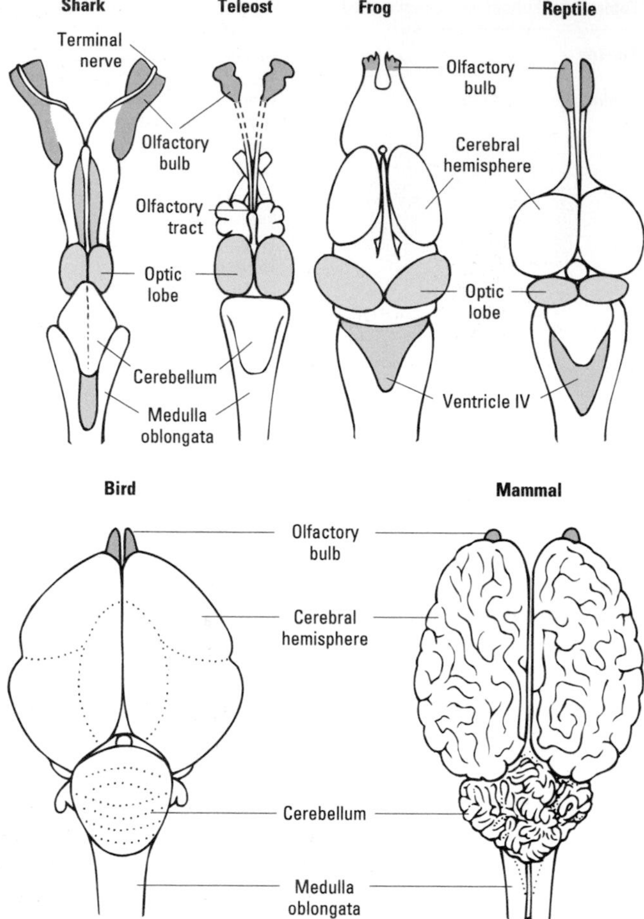

Fig. 15.55 The increasing size of sensory areas in terrestrial groups compared with fish (with the cerebrum and cerebellum also becoming dominant in the endothermic groups). (From Pough *et al.* 1999.)

15.7.5 Sensory coordination

Given all these sensory modalities, and their very considerable complexity in many land animals where elaborate intraspecific signaling systems have evolved, it is inevitable that the sensory coordination systems have also tended to become more intricate than those of aquatic animals. This is clearly expressed in increasing overall brain size (see Fig. 9.34), but with particular growth of the areas that integrate and pass on the information from the dominant senses (Fig. 15.55). The antennary lobes of arthropods and the olfactory bulbs of vertebrates are enlarged in terrestrial taxa, as are the optic lobes (and in higher vertebrates the visual cortex). The telencephalon at the front of the brain is the major sensory integrating center in lower vertebrates, and enlarges markedly to reach its zenith in birds, where it dominates the brain (along with the cerebellum, controlling motor performance and stability and therefore linked with the problems of flight and balance). Only in mammals is this trend overridden, with the massive growth of the neocortex dominating the other integrating centers. In birds and mammals, operating at the faster sensory and motor pace demanded by an endothermic lifestyle, the brain is roughly 15 times larger than in an equivalent ectothermic fish, amphibian, or reptile.

15.8 Feeding and being fed upon

In the introduction to this chapter we pointed out that, in comparison with watery habitats, on land the plants are tougher and better defended, while animals are faster. Also the supply of aerial "plankton" is limited and cannot support many filter-feeding animals (really only the web-building spiders). Land animals can therefore be detritivores, herbivores, or active predators, or some omnivorous combination of these; or they can be parasites, although they perhaps then cease to be really terrestrial (see Chapter 17). In fact, many of the major terrestrial groups are fundamentally predatory, including the flatworms, nemertines, leeches, crabs, centipedes, most arachnid and many insect orders, onychophorans, amphibians, and reptiles. Set against this, earthworms, gastropods, amphipod and isopod crustaceans, millipedes, many acarines (mites), and many insect orders are essentially herbivorous. Birds and mammals probably had carnivorous (insectivorous) origins but have diversified into other modes. Indeed amongst the insects, birds, and mammals in particular, a huge diversity of trophic roles have evolved, often in concert with radiations in other groups such as the angiosperms. Referring back to Table 15.4, it is noteworthy (though not very surprising) that the most speciose groups are precisely those that have been able to diversify into the whole range of trophic niches rather than sticking to their ancestral feeding mode.

15.8.1 Herbivores

Terrestrial plants are not very easy things to eat. Their support systems make them inherently tough, hard to penetrate, and hard to digest. These support systems are formed from the carbohydrate cellulose, and the complex and variable chemical lignin, based on a phenyl-propane polymer. Table 15.17 shows the content of these relatively indigestible components in common terrestrial plants. Vegetation is also hard to hold on to for a small land animal—it is faced with vertical moving windblown surfaces, often with a waxy and shiny cuticle. Land plants are also inherently poor nutritionally for animals, having a low nitrogen content (Fig. 15.56), the wrong balance of amino acids, insufficient sodium content, and an absence of necessary steroids. There may also be excess water and/or excess sugar for some sap feeders and fruit feeders, or excess salt for some desert herbivores. Finally, and again very obviously, land plants are well defended against herbivores, either physically or chemically or both.

Despite being poor food for animals, plants are the primary producers of all trophic systems and are therefore bound to be eaten. Some of the groups mentioned above only eat them once they are dead or decaying, and thus beginning to soften. Woodlice, for

Table 15.17 Fiber contents of different foodstuffs (as % dry matter).

Plant	Cellulose	Lignin
Potato	1.8	0.2
Apple	4.4	0.4
Carrot	8.7	0.3
Wheat bran	9	4
Lettuce	14	2
Alfalfa	27	8

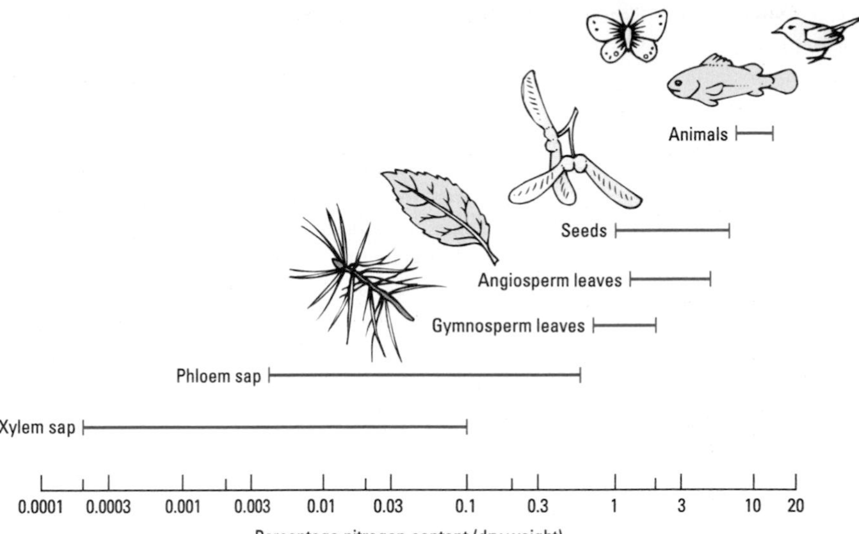

Fig. 15.56 Nitrogen contents of different types of plant tissue; note that almost all are very low compared with animal tissues. (From Strong *et al.* 1984; adapted from Mattson 1980.)

example, perceive litter as food only when it begins to produce odors of the metabolites that arise from decomposing microorganisms. Others feed only at a cellular level, with stylets used for piercing individually the relatively soft cells of roots and young foliage (many nematodes and some mites). But all groups that eat the tissues of living land plants have to be equipped to get over the first hazard of toughness, so that there is an absolute prerequisite for **hard mouthparts**. The land herbivores are therefore dominated by just three groups:

1 Some terrestrial slugs and snails using the chitinous molluscan radula, feeding on leaves.

2 A vast array of insects, using their cuticular mouthparts as chewers for feeding on leaves, roots, seeds, flowers, or fruits.

3 A range of vertebrates, using teeth to chew all plant solids.

In addition to hardened mouthparts, the terrestrial herbivores that ingest large quantities of plant material tend to show five other key characteristics. Firstly, they must have **elongated guts**, where the bulky food can be processed adequately. All land herbivores show markedly greater overall gut length than closely related carnivores. Some birds and mammals can show acclimation by increasing the overall gut length by as much as 50% when fed on particularly poor diets; some birds change gut length seasonally as the diet alters.

Secondly, they must also have guts of unusual structure, as measured by the **coefficient of gut differentiation** (the ratio of stomach and large intestine to small intestine, or of the equivalent areas). In carnivores this is around 0.1–0.4, but in herbivores it may be 2–6, with a relatively much reduced absorptive small intestine and much enlarged "processing" areas.

Thirdly, they nearly all require some form of anaerobic **cellulase-providing symbionts** in their gut to provide the initial fermentation process whereby cellulose is broken down to usable constituents. A fully effective cellulase in fact requires at least three separate kinds of carbohydrase enzyme, and very few animals possess these, the exceptions being a few crustaceans, molluscs, and wood-eating beetles.

Fourthly, they need mechanisms to achieve **pH homeostasis** in the gut in the face of vegetation with very different pH characteristics (especially if they take in some decomposing leaves as part of their diet). Many invertebrate herbivores achieve reasonable pH stability in the midgut with foods ranging between pH 4.0 and 7.5, and in vertebrates homeostasis is even more marked.

Finally, all land herbivores need to supplement their **salt intake**, with sodium being much less available in the diet than for a carnivore or for any aquatic animal. Herbivorous vertebrates usually meet their sodium needs by resorting to salt licks, or selecting some plants of unusually salty composition. Some insects use the alternative of "puddling" behavior; moths and butterflies visit small puddles and drink for many minutes or even hours on end, voiding excess water as repetitive anal jets in order to get enough salt. The males of the butterfly *Gluphisia* are especially adapted for puddling, with wide oral slits; much of the acquired sodium is transferred to the female at mating, for incorporation into eggs—a novel form of nuptial gift. Where puddles are contaminated by fecal droppings from other animals they may be particularly favored as their salt content will be higher. Even rain drops that have been in contact with bird droppings on foliage are used in this fashion.

As we will see, most terrestrial herbivores also have special problems with particular plant chemicals that have a defensive function, and therefore need good **detoxification systems**.

Molluscs and other invertebrates

Earthworms eat terrestrial vegetation mainly as already decaying material. They possess a gizzard where food is partially crushed, and a typhlosole (a dorsal longitudinal fold in the gut wall) to increase the absorptive surface area. There are also calciferous glands in the gut epithelium, probably helping to regulate internal pH levels, although it remains unclear whether this is related to a build-up of CO_2 in the burrows or an excess of calcium in the food.

Snails and slugs have inherited the chitinous radula from their aquatic ancestors, and this serves as an effective rasping tool for penetrating the cuticle of leaves and scraping plant tissue into the mouth. Land pulmonates generally have a radula that is wider and flatter than normal, with multiple rows of very similar simple small teeth, functioning rather like sandpaper. They will eat a range of land plants fairly indiscriminately, although they may reject some genera that are heavily chemically defended (see below). They have no problems with calcium, using any excess in shell construction; noncalcareous soils where Ca^{2+} is in short supply may be a major selective force promoting shell loss in slugs.

Some land crabs have become herbivorous, mainly selecting fallen leaves that are beginning to decay but also grazing on living foliage that is within their reach. On some isolated sites, such as Christmas Island, crabs have become the dominant litter recyclers.

Insect herbivory

Within the Insecta, nine out of the 29 orders are entirely or mostly herbivorous, notably the lepidopterans, orthopterans, and true bugs (heteropterans and homopterans). Insects are the major terrestrial defoliators in most habitats, with caterpillars in particular eating relatively huge amounts, and in some species growing by a factor of 5000 in just 20 days. Many of the defenses found in the plant world are related to deterring these insect herbivores.

Table 15.18 Chemicals used as plant defenses against herbivory.

Chemical type	Examples	Sources
Alkaloids	Nicotine Atropine Solanine	Tobacco Deadly nightshade Potato
Thiols, glucosinolates, and mustard oil glycosides	Sinigrin	Crucifers
Cardenolides (cardiac glycosides)	Ouabain	Asclepiadaceae, Apocynaceae
Cyanogenic glycosides (releasing cyanide)	Cassavin Prunacin	Cassava Seeds of Rosaceae
Phenolics, terpenes, steroids, coumarins, tannins	Cucurbitacin Juvabione Ecdysones Tannin	Cucumber Firs Bracken Oaks, other trees
Other	Silicates Odd amino acids Polypeptides	Grasses

SELECTION AND INGESTION OF FOOD PLANTS

The term "coevolution" was first coined when it was noted that plants possessing "odd" chemicals and unusual tastes were only ever fed on by a few genera of relatively specialist insects. Many of the chemicals found in plants have metabolic effects in animals; many, perhaps most of them, have been selected for because they interfere with insect and/or mammalian metabolisms or behaviors, so preventing herbivores from utilizing the plant. Familiar examples include: alkaloids, such as atropine in deadly nightshade, solanine in parts of potatoes, and above all nicotine in tobacco; the mustard oils (glucosinolates) that give a "hot" taste to various crucifers, mild in cabbage but very concentrated in mustard and horseradish; terpenes from many shrubs and trees; and pyrethrum, widely used as a broad-spectrum insecticide. These curious plant chemicals are now generally termed secondary plant compounds or **allelochemicals**. Presumably these plant by-products had serendipitous deterrent effects on insects, and were retained by natural selection. But some insects have subsequently specialized to eat those particular plants, becoming monophagous or oligophagous on them, and leaving the less well-defended plants to be competed for by the mass of generalist polyphagous feeders. Over evolutionary time, the defended plants might increase their investment in chemical defenses, so the "coevolutionary arms race" between herbivorous insects and plants sets in.

It is difficult to classify the chemicals used as defenses against insects, a huge range being involved. Table 15.18 gives an overview of chemical types. They can be classified according to effects on the insect herbivore as follows:

1 Development inhibitors—often found in woody plants, and including the tannins (which can be up to 5% dry weight in many trees, and which work by complexing with dietary proteins so the herbivore cannot digest them) and mimics of insect hormones (JH or ecdysones). These are broad-spectrum dosage-dependent deterrents, present in high quantities—"quantitative defenses"—causing decreased growth in a range of herbivores. They are expensive to produce, but hard to counter for the herbivore.

2 Lethal chemicals—alkaloids and glucosinolates are frequently lethal to nonspecialist insects. They are produced in only very small doses—"qualitative defenses"—and in relatively short-lived weeds. They are cheap to make, and relatively easy to detoxify if eaten by specialists, but most insects never encounter them and are not under selective pressure to develop detoxification systems.

3 Chemicals with physical effects—resins, latexes, and gums, oozing from wounded plants and gumming up the feet or mouthparts, or silicates in grasses, which are sharp and painful to eat for mammals (there may be more silicon than nitrogen in a mature maize plant).

4 Feeding or oviposition deterrents—chemicals that modify insect behavior, e.g. wild potatoes release an aphid alarm pheromone mimic from their trichomes when bitten.

5 Attractants—bringing in predatory and parasitic insects that feed on the plant's herbivores.

Many toxins are concentrated in the most valuable and susceptible parts of plants, like seeds and young leaves. They can sometimes be induced by damage, and they often occur in higher concentrations in populations more exposed to attack. For example, the wild ginger plant has two morphs, one inhabiting sites where there are few slugs and beetles and another (with more chemical defenses) where there are many.

In addition to chemical defenses against herbivores, many land plants invest in **physical defenses**. Spines, hairs, and trichomes all make it hard to hold on: sometimes hairs actually puncture and kill invading caterpillars. Alternatively, slippery surfaces, using exudates from glands, can be used to form barriers against attack by insects. As another alternative, some plants can "cheat" and use **biotic defenses**, employing other animals as guards. *Acacia* is the classic case, where the plant provides both nesting space (in expanded thorns) and food (via extrafloral nectaries) to resident ants, who aggressively deter potential insect herbivores and even some large grazing mammals.

Many plants have defenses; how then do herbivorous insects eat them at all? **Behavioral strategies** are often the first possibility, allowing the animals to avoid or to limit the effects of the defense, in space or time. They may specifically feed from nondefended parts, such as the phloem, or from inside a leaf-mine (avoiding the most toxic palisade layer of the leaf), or even (for very small larvae) from the leaf surface but avoiding the glandular trichomes in which the toxins are stored. For example, the first instar of *Heliothis* (the boll-worm) feeding on cotton can avoid the toxin gossypol by eating around the base of individual hairs. Another way to avoid toxins is shown by grasshoppers that feed on cassava: the animals take a bite and then jump off the leaf, avoiding the sudden release of cyanide from the damaged surface. These grasshoppers are gregarious, and many together seem to use the "bite and retreat" technique to produce wilting of the cassava leaf, so that all can then move in and eat from a no longer defended, limp and floppy leaf. Yet another strategy is to cut off the inducible chemical defense supply route: some caterpillars cut a shallow trench in the host-plant leaf, through all the veins, before feeding, so preventing any mobilization of feeding deterrents. A recent further example of behavioral manipulation of the host plant involves leaf-rolling; many insects roll up leaves to make a shelter within which they eat in relative safety and in an equable microclimate, but the rolling also reduces the light falling on the leaf surfaces by 95% in some neotropical shrubs, which in turn means a 31% decrease in physical toughness and a 15% decrease in leaf tannins.

If behavioral tricks fail, many insects resort to physiological and biochemical techniques against ingested toxins. The most obvious strategy is excretion. For example, the principal defense against nicotine for most tobacco-feeding insects is to excrete this compound very rapidly. *Manduca* (the tobacco hornworm) has a specific active nicotine pump in the walls of its Malpighian tubules. Some such pumps are inducible, for example in milkweed bugs where an ouabain pump appears after the insects have been exposed to this natural cardiac glycoside in the diet. However, if a compound cannot be excreted, it must be dealt with internally by a detoxification or storage system. Most herbivores have **polysubstrate monooxygenase** (PSMO) enzyme systems designed to detoxify a range of unfamiliar chemicals. These are found in the liver of vertebrates and in the midgut and fat body of insects. The amount of PSMO varies with the diet in a predictable manner, with more present in polyphagous species. (This poses a massive problem for insect control, since insecticide resistance often involves acquiring higher levels of PSMO, so that the "pest" insect tends to become resistant to all insecticides, not just the one used!) Specific detoxification mechanisms also exist, more commonly in the monophagous herbivores that are dealing with small amounts of highly toxic compounds. **Sequestration** is a further option, best understood for the tannins. These tend to be countered in habitual tree feeders by changing the pH of the foregut or midgut to strong alkalinity, so inhibiting the tannin's complex-forming abilities. The tannin is then tolerated and becomes the equivalent of roughage, accumulated in a sequestered inactive form and then either defecated or simply retained in the gut and lost at the next molt. But this in turn could make the normal H^+-related uptake systems (see Chapter 4) less rapid, and the processes of normal sugar and amino acid uptake would be slowed down. In some species the symport proteins that mediate uptake in the gut wall are therefore unusually insensitive to pH, suggesting a long association between the tannin-producing trees and the insect's alkaline guts.

Overcoming physical defenses principally involves making the mouthparts, whether piercing, sucking, chewing, or cutting, sufficiently hardened by strong tanning and cross-linking of the protein/chitinous structures to overcome the basic problem of cellulose-toughened materials. Avoiding penetrating hairs and trichomes is also crucial for softer bodied arthropods such as caterpillars and aphids. Some caterpillars spin a raft of silk over the leaf and feed from it to get round this defense. Feet adaptations also occur: some aphids have long tarsal tips that allow them to hold on between hairs without damage, while suckers and circlets of fine hairs allow caterpillars to hold on to very slippery surfaces.

As well as circumventing the plant's defenses, there are various ways in which the insect can turn the tables back on the plant. One is to use the plant chemical as a feeding attractant, and this has turned out to be extraordinarily common, the supposedly deterrent chemical becoming the insect's **phagostimulant**, initiating and maintaining feeding behavior. For example, mustard oils are used by *Pieris* butterflies, and gossypol is used by the cotton boll-worm. Even the "quantitative" defenses, such as tannins, are phagostimulants to gypsy moths on oak trees. Alternatively (or often additionally), insects may use the plant chemical as an oviposition cue to identify their preferred host plant, or may use it as a pheromone to attract further insects to feed on the plant and mate with the founder.

There is one further possibility in relation to plant chemicals that is widely exploited by insect herbivores: that of using the plant chemical as the herbivore's own defense. Most caterpillars feed on mildly or moderately poisonous plants, and most "get by" using either detoxification or excretion. However, their butterflies are largely unprotected, which is a problem as butterflies are conspicuous daytime fliers. Some lepidopterans therefore accumulate significant amounts of plant toxin in larval tissues outside the gut, and pass these chemicals to the adult. They also usually show **aposematism** (warning coloration), so that predators rapidly learn to associate unpleasant taste or illness with the bright (red/yellow/black) colors. The herbivore often adopts an "aposematic lifestyle", warding off attack before the predator even gets hostile, reinforcing a strong negative search image due to the colors with other obvious visual cues, or conspicuous behaviors such as slow flight or undulating walking. There may be a tendency to aggregate, and perhaps to have tougher body structures to allow handling and learning by predators. The phenomenon of plant-derived chemical defense is particularly common on milkweeds (*Asclepias*), which produce irritant and toxic milky sap but nevertheless attract a range of specialist feeders. Most of these use the plants' cardiac glycosides to achieve unpalatability, and they are yellow/black or red/black. The most familiar examples are the monarch butterflies (*Danaus plexippus*), where larvae store the poison and pass some of it on to the adults, which are very toxic to birds, perhaps explaining why the butterflies can overwinter in such conspicuous aggregations. At least one grasshopper and a few moths actually spray the milkweed poisons out as a defensive secretion when attacked by birds.

A final possibility for insect herbivores is to overcome the plant's defenses by releasing chemicals that make the plant form galls. Here plant morphology is altered to the insect's advantage. Galls are pathologically developed cells or tissues, arising by hypertrophy (increased cell size) or hyperplasia (increased cell number). About 2% of all described insects are gall makers or "cecidozoa", mainly from the orders Hemiptera, Diptera ("gall-midges"), and Hymenoptera ("gall-wasps"). Gall formation involves two phases: initiation and then growth and maintenance. Initiation is usually in actively growing tissues, young leaves or buds, and involves an exaggerated wounding response sometimes interacting with plant hormones, especially cytokinins. Oral and anal secretions from the insect are also sometimes implicated in gall production. Hemiptera produce saliva containing amino acids and salts, but also low levels of auxins and phenolic compounds probably derived from the plant. There are some recent suggestions that gall formation is also related to a variety of semiautonomous genetic entities (viruses, plasmids, etc.) transferred from insect to plant during feeding, with the insect thus acting as an RNA or DNA donor. The gall insects get all of their food from the galled tissues, away from potential predators and pathogens, and protected from the weather. But even more important, the gall tissues are sinks for plant assimilates; they tend to be relatively rich in sugars, proteins, and lipids. In at least some cases the gall insect also manipulates the plant chemical defenses; outer layers of the gall are very heavily endowed with the defensive secondary plant compounds, while the inner nutritive tissues are dedifferentiated cells and almost free of toxins.

DIGESTION OF FOOD PLANTS

Once insects have taken plant material into their guts and overcome any toxic effects, they still have the problem of digesting the cellulose and releasing the more nutritious cellular contents. The "best" insect herbivores achieve 60–65% fiber digestion, as good as that in many mammals. Many do this using a flourishing gut fauna of symbionts, almost always in the hindgut. Some have a relatively small hindgut bacterial fermentation chamber, but primitive termites (having an almost pure cellulose diet) have a very large rectal sac ("paunch") housing flagellate protistans and bacteria. They repeatedly regurgitate and re-swallow the wood they eat to achieve maximum breakdown, which can be up to 89% efficient. The symbiotic gut fauna of arthropods are relatively unstudied, but are known to provide certain key elements in the diet (the B vitamins, and some sterols) in addition to breaking down the cellulose. In termites and some plant-feeding bugs, the gut fauna aids in nitrogen balance, metabolizing a proportion of the uric acid waste.

The apparent exceptions to the necessity of symbiotic fauna are certain anobiid and cerambycid wood-eating beetles, and probably also the locust *Schistocerca*, which have low numbers of intestinal bacteria and appear to make their own (endogenous) cellulases, although there are contradictory indications that these derive from intracellular yeasts. Certain woodlice may also have endogenous hindgut cellulases. Another clear exception comes with the more advanced termites, which use a rather special kind of "extracorporeal symbiont"; these are fungi especially cultivated in their underground nest "gardens", to provide the initial breakdown of woody material. The termites then consume this partially digested material, and may also produce some foregut cellulase of their own.

EXCRETORY PROBLEMS FOR INSECT HERBIVORES

Some insects feed on plant material that gives them too much water or too much sugar in the diet, and they often have systems to bypass the midgut epithelium (see Chapter 5 and Fig. 5.25). This applies particularly to sap feeders: xylem (fed on by cicadas, for example) is very dilute with an enormous water content relative to its energetic value, whilst phloem (the food of aphids and many leafhoppers) is excessively rich in sugar and sometimes also nitrogenous foods, but very deficient in inorganic salts. The "filter chamber" in the gut brings the anterior midgut, Malpighian tubules, and anterior hindgut close together, and much of the water taken into the gut thus bypasses the main midgut epithelium and passes directly from the anterior midgut to the rectum. Phloem feeders adopt the strategy of rapid excretion instead, with a sugary fluid ("honeydew") emerging from the anus or from dorsal abdominal glands called cornicles, and a certain amount of excess water being lost as salivary secretions. They may also be able to combine simple sugars into oligosaccharides, to reduce the osmotic effect of the latter in drawing water out of the aphids' tissues. Certain fruit-feeding flies, such as the tephritid *Rhagoletis*, exhibit "bubbling" behavior after a large dilute meal, repeatedly regurgitating and re-swallowing some of the fluid, presumably to concentrate it by evaporation. Many bees do the same thing with floral nectar, although this may also serve as a mode of evaporative cooling.

Vertebrate herbivores

Vertebrates that are predominantly herbivorous include anuran tadpoles (mainly eating relatively soft aquatic vegetation), some reptiles, relatively few birds (except in the sense of specialist feeders on seeds and fruits, which are very common), some marsupials, and many eutherian mammals, especially rodents (specializing on seeds and/or woody tissues), ruminants (leaf eaters), and primates (leaf and fruit eaters). The practice of bulk feeding on green leafy material is rather rare, restricted to the ruminant mammals and a few large species in the other groups (e.g. geese, tortoises), where the necessarily large guts can be accommodated.

SELECTION AND INGESTION OF FOOD PLANTS

Plant defenses operate against vertebrates as well as against insects, although the scale of the defense may be different. Smaller scale physical features such as trichomes may become irrelevant and are replaced by sharp and solid thorns, deterring all but the most thick-lipped of browsers. Chemicals are more likely to be an irritant or unpalatable rather than toxic, since the larger quantities needed to kill a vertebrate can rarely be achieved. However, there may be enough toxin to have significant physiological effects: for example, ringtail possums feeding on different species of *Eucalyptus*, all with toxic oils, produce urines of different pH, and where the eucalypt is particularly high in terpenes they have to take in more food and then excrete up to 40% of the metabolizable energy, compared with only 10–15% on a low-terpene food plant. Koalas feeding on eucalypts face similar problems and are only just able to detoxify enough leaf to survive, rarely accumulating any fat stores. In droughts, when the gum trees stop producing new leaves, koalas may die of nitrogen deficiency even though they have full stomachs.

Inaccessibility of the most important plant parts is another important aspect of defense against grazing mammals. This is exemplified by the low growth of grasses and of many pasture weeds, where the growing meristems and often also the important reproductive organs may be at ground level; and by the raised canopy of many savanna trees, where the lowest flowering branches are just out of reach of many of the antelope and other herbivorous mammals.

Again plant defenses, both physical and chemical, may be "used" by the herbivore to find and choose its food, producing relatively narrow dietary niches. Primates are particularly fussy about the nature of the plants they ingest, having different and quite specific techniques of manual dexterity to deal with the variety of prickly and stinging vegetation that they encounter. Fruits may be particularly valued, partly because they often rely on being eaten and so are usually sweet and lacking in secondary defenses. Monkeys and apes often seek out favorites over large distances, revisiting known trees at particular times of year. However, the larger size and necessarily large food volume of most vertebrate herbivores make them fairly unselective in what they eat, and monophagy and oligophagy are rather rare, occurring only in a few (often rather threatened!) cases, such as giant pandas and koalas. Diet selection may be more influenced by what can be found easily and gathered in relatively benign thermal and hygric conditions; small rodents, for example, have been shown to balance nutrition with thermoregulation.

DIGESTION OF FOOD PLANTS

Vertebrates are incapable of digesting cellulose and related plant polymers, but once again they make use of the many species of microorganisms that have the ability to ferment cellulose. The amount of fermentation in vertebrate guts is hindered by a high degree of molecular cross-linking and lignification in ingested plant material. The space available for fermentation in the gastrointestinal tract (GIT) increases in direct proportion to body mass, whereas we know that metabolic rate scales with $M_b^{0.75}$. Thus in small animals the rate of food passage through the GIT is too rapid for extensive microbial growth, and substantial fermentation is impossible. To achieve enough fermentation, several techniques occur. The Australian wood duck, a rare example of obligate foliar herbivory in a bird, has no obvious adaptation to retain or recycle fiber, but selects soft grasses and herbs rich in hemicellulose, with little assimilation of lignin or cellulose. Rabbits and other grass eaters may reingest a proportion of their own feces (coprophagy) so that the digestion products pass through the GIT several times; the rabbit feeds by night, and while resting by day it produces two kinds of fecal pellets, only reingesting the softer kind (25–65% of total fecal production) for recycling. This gives more time for fermentation and leads to a highly efficient utilization of nitrogen. An alternative approach is to have voluminous intestines, and this adaptation is found in the herbivorous reptiles such as iguanas, with valve-like folds that increase the time food takes to pass through the gut. The iguanas also bask in the sun, which stimulates fermentation but at the expense of higher maintenance costs. In the green iguana, transit time in the gut can be decreased from 10 days to 3 days by raising the T_b from 30 to 36°C.

In larger animals there are two additional strategies to achieve good fermentation:

1 **Hindgut fermentation**: nonruminant herbivores (many mammals and some birds) have an enlarged area of hindgut, usually a large cecum, where fermentation occurs long after initial gastric digestion. Hindgut fermentation provides the host animal with volatile fatty acids as a source of energy, but the lack of subsequent digestive processes and the limited opportunity for the absorption of other products (especially vitamins) puts the nonruminants at a relative disadvantage. Efficiency of fiber digestion is variable, between 20 and 65%.

2 **Foregut fermentation**: usually termed **rumination**, and seen as characteristic of ruminant mammals, although also found in a few birds, such as the hoatzin, which eat leaves. Ruminant herbivores provide accommodation for their microbial partners in a modified stomach, where fermentation precedes gastric digestion. Here efficiency is greater, at 52–80%.

Vertebrate ruminants The Ruminantia includes bovids (cattle, sheep, goats), cervids (deer, antelope, giraffe, pronghorn), and camelids. Sheep, cattle, and goats are the best known ruminants physiologically, and they share three characteristics. Firstly, they lack upper incisor teeth, having instead a dental pad against which plant material is crushed. Secondly, they have a complex four-chambered stomach (Fig. 15.57). Thirdly, they regurgitate the food from the first stomach chamber and subject it to a lengthy mastication (i.e. they ruminate, or "chew the cud"). The newly ingested fodder is attacked by a mixed population of anaerobic bacteria, protozoa, and fungi. Carbohydrates, proteins, and some fats are metabolized (only saturated fatty acids are not subject to fermentation), to produce a foregut mixture of volatile fatty acids, ammonia, and microbial cells, containing 50% protein. As the ingesta is reduced to small particles it passes with ruminal fluid to the reticulum (second chamber) and omasum (third chamber), where further physical maceration occurs, together with the absorption of much fluid. True gastric digestion occurs in the abomasum (fourth chamber), which is the equivalent of the stomach of other animals. Since large numbers of microorganisms accompany the ingesta and are killed and digested in the abomasum they represent an appreciable proportion of the solid part of the animal's real food ration.

The particular advantages of pregastric or ruminal digestion are three-fold:

1 Cellulose and other structural plant polymers are solubilized. Of the total ingested energy, about 46% is lost as waste (feces and methane) and 18% is utilized by the microorganisms for their metabolism and population growth. This extra microbial growth is passed on to the host's gut, however, so that 54% of the total energy is made available to the ruminant. The 3% that is converted into heat of fermentation is also potentially useful for temperature regulation by the ruminant if it is cold stressed.

2 The microorganisms can use nonprotein nitrogen for growth, so that inorganic nitrogen or urea from the host's protein metabolism may be converted to microbial protein, which eventually becomes available to the host animal.

3 The vitamin content of food is increased by microbial synthesis. Ruminants are independent of all dietary vitamins, other than A and D.

Metabolism of the food produces acetic, propionic, and butyric acids, roughly in the proportions 70 : 20 : 10, together with some lactate, formate, valerate, isovalerate, ammonia, methane, carbon dioxide, and hydrogen. The rumen has very particular and unusual conditions: its temperature is about 39°C, maintained by heat produced during fermentation; its pH varies between 7.4 and 5.5, with a buffer system provided by the bicarbonate from saliva and by the volatile fatty acids; and it has a gas phase consisting of about 27% CH_4, 65% CO_2, 7% N_2, and a trace of H_2, and the redox potential can reach as low as −350 to −400 mV at pH 7, reflecting profoundly anaerobic conditions. Methane present in the rumen represents a

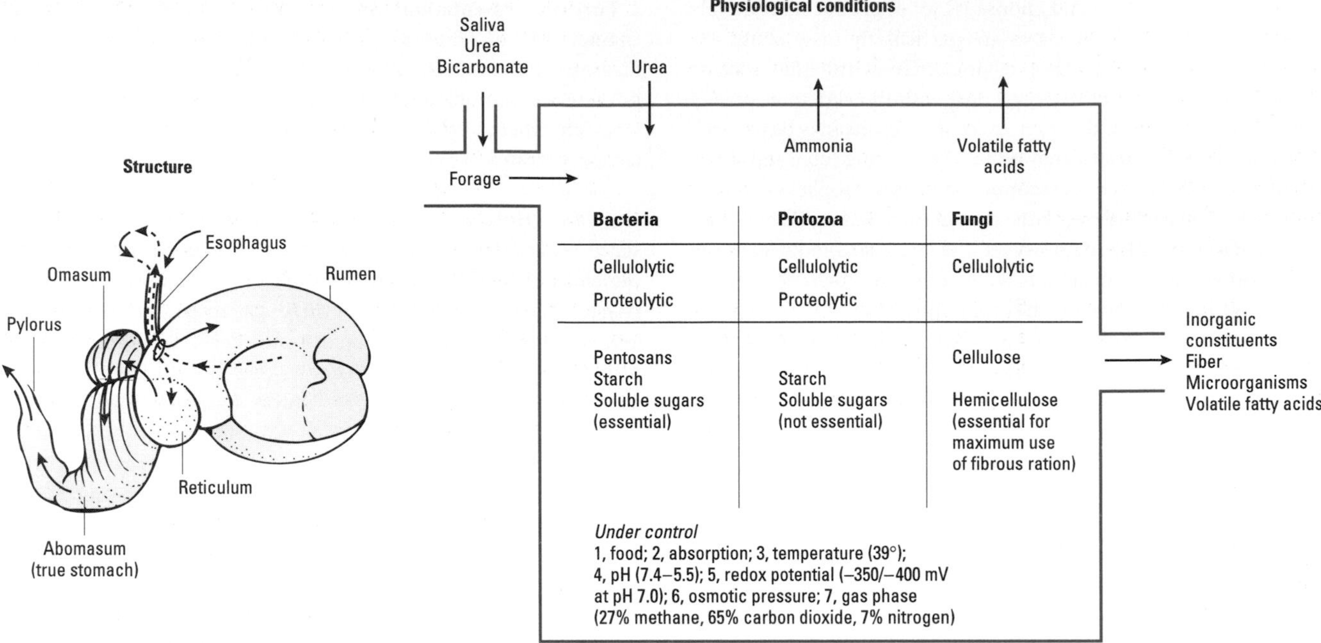

Fig. 15.57 The complex stomachs of a ruminant and their physiological conditions. (From Strong *et al.* 1984; adapted from Mattson 1980.)

sink for the considerable amount of hydrogen produced in fermentation reactions.

Rumen microorganisms consist of bacteria, protists, and fungi. At least 12 genera of **bacteria** can be regarded as true rumen organisms, and they act symbiotically such that the end-products of one class are the substrates of another. Two groups of ciliated **protists** are also found in the rumen, and these can ferment carbohydrates with the production of volatile fatty acids, CO_2, and H_2. In sheep fed a grain diet there are about 2.5 million protists and 10 billion bacteria per milliliter of rumen fluid, reflecting the ready availability of carbohydrate; whereas on a hay and grass diet with less available carbohydrate, populations of protists are about an order of magnitude lower. Rumen **fungi** are also higher in animals on fibrous diets and are much reduced in animals on a high grain diet. The fungi actively ferment cellulose to acetate, formate, lactate, ethanol, CO_2, and H_2. The latter is converted to methane by bacteria.

Volatile fatty acids (acetate, propionate, and butyrate) are absorbed directly from the stomach and utilized immediately. Acetate is especially important: it passes through the rumen epithelium and liver unchanged but is transported in the blood and then oxidized (via acetyl-CoA) in the tissues to produce energy, supplying up to 50% of the carbon expired as CO_2 in a ruminant. The importance of acetate as a source of energy is emphasized by the low blood sugar levels of ruminants (only about half that of adult humans). Ruminants are also unusual in using acetate to fuel milk production and to produce body fat stores.

Other species with ruminant-like digestion of cellulose A number of nonruminant placental mammals use ruminant-like digestive processes with a stomach modified for microbial fermentation. These include leaf-eating tree sloths, colobid and langur monkeys, and some rodents. Sloths are noteworthy for their low metabolic rate and very large gut volume, which may be up to 37% of the body mass; fermentation occurs within the foregut, but at an extremely low rate compared with normal ruminants. The macropod marsupial mammals (kangaroos and wallabies) also have a ruminant-like digestive system. The stomach of the grey kangaroo, *Macropus*, is highly specialized as a fermentation chamber but is more tubular than the four-chambered ruminant stomach, and there is a more continuous flow of material through it. Relatively little material is regurgitated, rechewed, and reingested. The forestomach again contains bacteria, ciliate protists, and fungi, and the overall digestive efficiencies of ruminants and macropods are strikingly similar.

Postgastric fermenters Postgastric fermentation occurs in many placental mammals (perissodactyls, such as horses; the elephants and hyraxes; some rodents and lagomorphs) and in many marsupials (koalas, and many herbivorous possums). The site of fermentation in postgastric fermenters may vary. In koalas the **cecum** predominates ("cecant digestion"), whereas in wombats the colon is crucial and the cecum is vestigial. In the horse, bacterial fermentation occurs in the cecum and large intestine, and the volatile fatty acids are absorbed across the wall of the large intestine. In pigs fermentation takes place solely in the large intestine, while starch, sugars, proteins, and fat are digested in the small intestine; but fermentation accounts for no more than 10–20% of the feed for pigs. The capybara is particularly interesting, being the largest living rodent at around 50 kg, and possessing both a cecum and a habit of coprophagy, eating its own feces especially in the early morning.

However, the microorganismal production of the cecant mammals is not normally digested and assimilated, as occurs in ruminants, because the microbial production always occurs after the stomach and small intestine. Consequently, cellulose digestion is less effective in the horse than in a ruminant such as a cow, and horses compensate for this by having a higher rate of food intake. Postgastric fermentation also does not confer the advantage of nitrogen recycling. Nor does it allow the microorganisms to detoxify plant allelochemicals, so that in hindgut fermenters most of these get into the bloodstream and must be dealt with by the liver. On the other hand it does work well with diets high in fiber or tannins or silica; these pass on through the gut without entering the cecum, whereas in ruminants they cannot leave the rumen until ground to tiny particles, impeding the movement of other materials to the absorptive surfaces.

During the Eocene epoch (57–36 mya), when forests covered most of the available land, the dominant herbivores were nonruminant perissodactyls, now represented by the horses, tapirs, and rhinoceroses (a mere six genera of the 158 known to have existed). By the end of the Eocene, however, the artiodactyls (the group that includes the ruminants) had increased, allied to the appearance of grasslands during the Miocene. The ruminants have now come to dominate the artiodactyls, with over 70 genera. Even today, ruminants and horses or zebras often compete in terms of dietary intake and digestive strategies, and horses do better on low-quality forage such as straw. Thus they still do well on grasslands when there is drought, and might perhaps do better in the future in areas of anthropogenic desertification.

Overall effects of terrestrial herbivory

Plants do get eaten, but rarely to excess. They have undergone a prolonged period of coevolution with their herbivores, especially the insects, evolving ever more sophisticated ways of avoiding being eaten, which has resulted in complex chemical niche separation with specialists and generalists on both sides of the competitive arena. Animal–plant trophic relations may have generated much of the present species richness on land, especially in the tropical forest systems where so many of the coevolutionary stories have been worked out. Herbivores can therefore be seen as largely responsible for maintaining genetic diversity and polymorphisms in the land plants, and plant–insect relations have made enormous contributions to shaping present-day environments.

15.8.2 Carnivores

Carnivory is generally no more difficult on land than elsewhere, since animal tissues almost by definition provide the perfect balanced diet for other animals. In terms of nutrition and access to the digestible food there may nevertheless be a few difficulties. As with plant feeders, a major problem lies with skeletal material, although in animals the relatively indigestible parts are discreet, with internal rods or external coverings, rather than forming a protective shield around each individual cell. Land fauna tend to be dominated by groups with stiff skeletons, of course, so most carnivores either have to crush the chitinous exoskeletons of small arthropods or pull the soft flesh off the bones of vertebrates. Chitin is indigestible to many animals, so that the exoskeleton is commonly left behind as a husk. Specialist feeders on small arthropods, such as the ant- and termite-eating edentate mammals and the insectivorous bats, tend to have rather low metabolic rates, reflecting their relatively poor-quality diet where a substantial proportion of the captured biomass is discarded. However, some earthworms and crustaceans, and a few vertebrates such as bats and lizards have a chitinase, and others use bacterial chitinase in the gut. Bone is also fairly indigestible to most carnivores (most merely crushing it to extract the marrow); but there are always a few specialists around to eat even this material, the most famous examples being hyenas.

There are two main strategies for land carnivores: active searchers, and sit-and-wait predators. Either may impose physiological demands on the predator due to the speed of land prey, making any kind of capture difficult and very often intermittent. Many land carnivores therefore have a very high resistance to starvation (notable in some spiders and worms), or if they are active hunters they have unusual stamina, as in various dogs, so that they can pursue prey for long periods. Soft-bodied fauna attack prey with simple jaws and often some extruded enzymes to produce extracellular predigestion, sucking up the resultant soup. Similar tactics are used by most of the arachnids, using powerful sucking pharynx muscles to deal with their liquefied diet. Many of these arachnids use venoms, allowing them to attack prey larger and often more powerful than themselves. Leeches use a modification of this approach in feeding on the blood of much larger animals, using the proboscis to make an incision, then injecting anticoagulants to insure a steady blood flow. In these animals, and in the abdomens of other blood feeders such as mosquitoes and certain bugs, the body wall has to be very extensible to permit the intake of large meals at very long intervals. Many blood feeders have a problem simply of getting too much at once, perhaps feeding only at intervals of weeks or even months. Many leeches, and insects such as the bug *Rhodnius* and the tsetse fly *Glossina*, have extremely rapid excretory outputs controlled by hormones, to eliminate most of the volume taken in as quickly as possible as dilute urine (up to 50% of the volume being passed within 3 h).

Overall, carnivores on land have populations of lower density and there are fewer of them at any particular body mass compared with herbivores (Fig. 15.58), reflecting the inefficiency of transfer of energy through successive trophic levels and the traditional "ecological pyramid". Remember, though, that land animals are not only prey for the predators dealt with in this section, but in turn become hosts for a huge range of other animals that feed in and on their bodies: parasitic groups that as endoparasites lead essentially aquatic lives insulated from the vagaries of the terrestrial environment. Many of these animals are from the same groups of terrestrial animals already considered here, notably the insects, and are dealt with in Chapter 17.

15.9 Anthropogenic problems

The effects of human populations on terrestrial ecosystems represent one of the greatest hazards for other animals, a relatively "new" selective agent determining the survival or extinction of other species. The varied threats that we are imposing on land habitats set complex physiological problems that a few animals survive and even thrive on, where others are stressed beyond their acclimatory or adaptive limits. Here we review some of the major changes being brought about by human activities, and the effects on major groups of land animals.

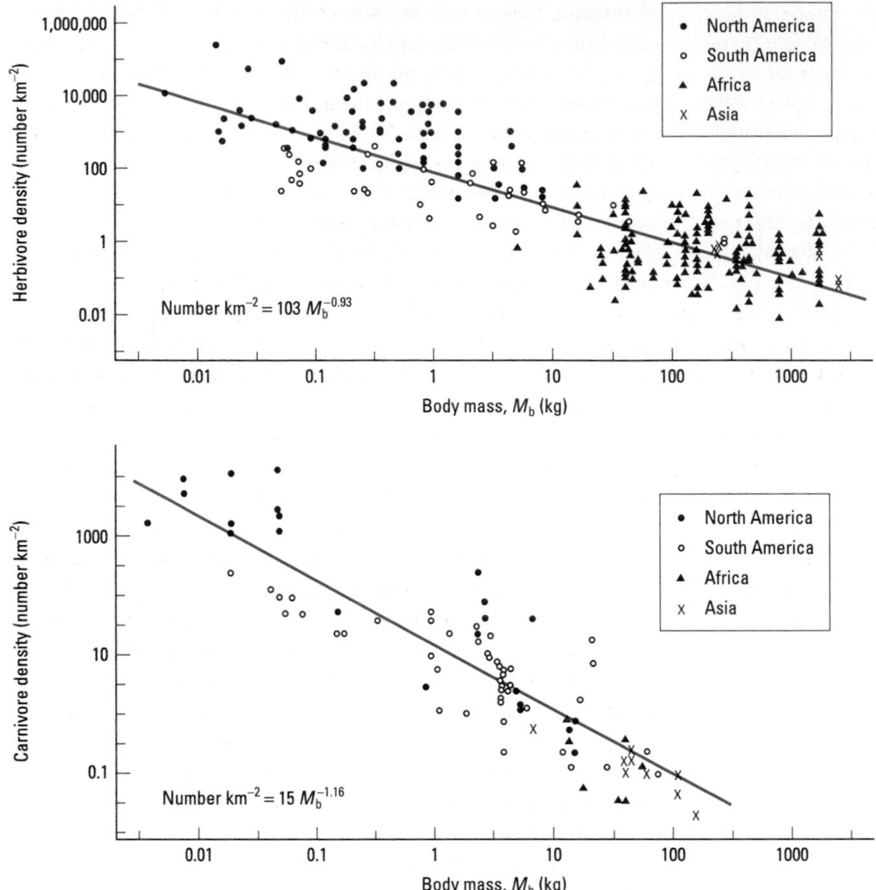

Fig. 15.58 The relation between body size and abundance for mammalian carnivores compared with herbivores. Note that both absolute numbers and slope are different. (From Peters 1983, courtesy of Cambridge University Press.)

The most obvious effect of humanity is wholesale and direct destruction of habitats, either to make way for human constructions or to provide raw materials. Even where this is on a small scale—for example pollution leaking from a waste site, removal of a few trees to construct a new road, or building a few houses—the effect on the biota is obvious and usually fatal. Large animals may migrate out, but thousands of microhabitats for the small and unglamorous fauna are destroyed. On the much greater scale of mass rain forest destruction, any potential for moving out may be abolished, and the whole community is extinguished. No adaptation of physiology or life history makes very much difference here. But humans also have more subtle and yet potentially more serious effects on the global ecosystem, and these anthropogenic effects may be more selective between species.

15.9.1 Greenhouse effect and temperature rise

Carbon dioxide is in many ways at the root of the atmospheric and climatic regulation problems that have occurred since the beginning of the Holocene period (i.e. since the last Ice Age, 10,000 years ago). If there were no CO_2 the Earth's surface temperature would be only −18°C, instead of the present 15°C average; we would have a frozen planet, ice-bound and with high reflectivity (albedo) giving no chance of warming and no possibility for the evolution of life. But CO_2 is continually produced by volcanoes erupting, and if unregulated by life it would therefore soon smother and poison all living organisms. In fact, because of life the atmospheric levels of CO_2 have been declining since the Precambrian. Living plants continuously absorb CO_2 from the air, using it as raw material for photosynthesis, releasing it again as carbon compounds underground when they die, and so keeping the atmosphere suitable for other life. The geosphere and the biosphere have interacted, as in so many other ways, maintaining conditions suitable for life.

Some of the CO_2 still persists in the air and acts as a blanket. This kept the Earth warm in the past when the sun was cooler, but as the sun grew gradually hotter the CO_2 also became thinner and acted less effectively as a thermal blanket. On balance this has led to rather stable planetary temperatures overall. In a sense we should therefore see the long-term "problem" for the planet as that of progressive lowering of CO_2 and potential cooling. Looking at the temperature levels through time confirms this: the planet has been getting on average colder for many thousands of years.

The problem is that humanity has now caused the CO_2 level to rise by 7–10% in three decades (Fig. 15.59), so increasing its thermal blanket effect. Carbon dioxide levels are naturally cyclic (higher in spring, lower in fall and winter) but the net levels are undoubtedly rising. The preindustrial average level was 260 parts per million (ppm) and in the 1990s we reached levels of up to 360 ppm, rising by at least 1.5 ppm year^{-1}. At the same time the amplitude of the annual cyclical changes are now somewhat greater.

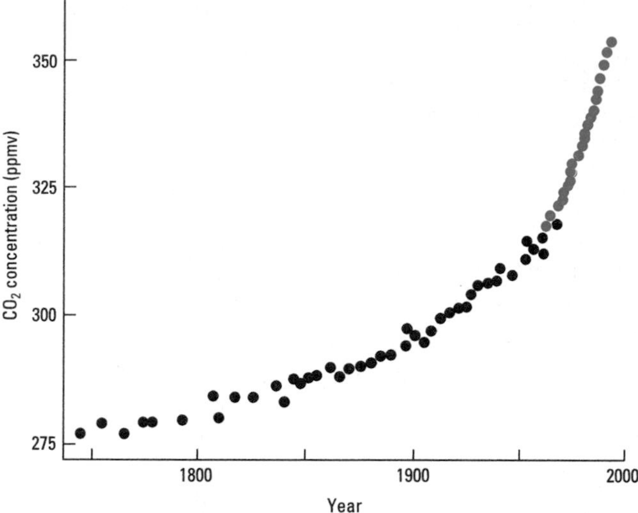

Fig. 15.59 Patterns of atmospheric carbon dioxide recorded since 1750; green symbols are the data from the recording center at Mauna Loa (Hawaii). (Reprinted from *Nature* **324**, Friedli, H. *et al.* 1986, copyright 1986, with permission from Macmillan Magazines Limited.)

In terms of temperature rise, a doubling of CO_2 levels is calculated to cause a 3°C rise in temperature. So far we have had a clear 0.5–0.7°C temperature rise globally since 1900, with the 1980s and 1990s being exceptionally warm in Europe (Fig. 15.60). The eight hottest years on record (as a global average) have all been in the last two decades. The projections, from the Intergovernmental Panel on Climate Change (IPCC), are for progressively higher temperatures that go beyond anything seen in the time since man evolved as a species (Fig. 15.61). Thus the biota is having to cope not only with increased CO_2 levels but also with steadily rising average temperatures.

The causes of CO_2 increase are not hard to find. The industrial and domestic burning of fossil fuels currently releases about 5.6 thousand million tonnes (gigatonnes, or Gt) of carbon per year, forming a large part of the problem (Fig. 15.62). Deforestation also has major effects; it leads in the long term to less fixing of CO_2 in vegetation, and the trees themselves are often destroyed by burning, which also releases CO_2 directly (up to 2 Gt carbon year^{-1}).

The effect of CO_2 as a pollutant greenhouse gas is probably responsible for about 50% of the total warming effect. However, there are other gases that are increasing due to human activities and

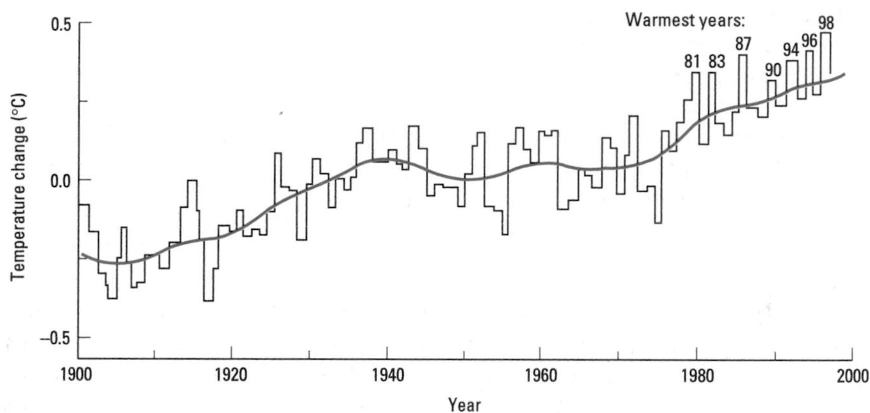

Fig. 15.60 Globally averaged surface (land and ocean) temperatures for the last century, expressed as differences from the mean in 1950–80. (Data from *Climate Change IPCC Report 1995.*)

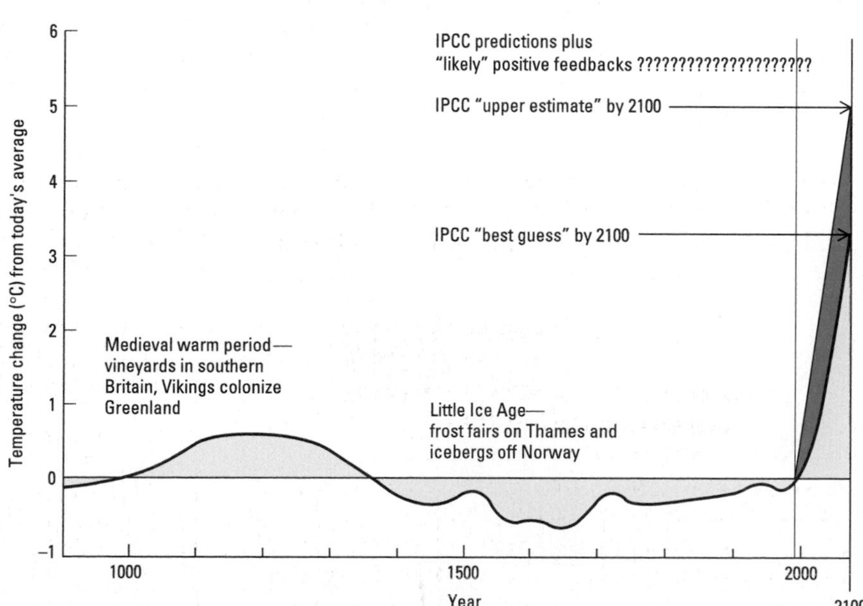

Fig. 15.61 Scenarios for temperature change based on different models of climate sensitivity, from the Intergovernmental Panel on Climate Change (IPCC). (Adapted from *Climate Change IPCC Report 1995.*)

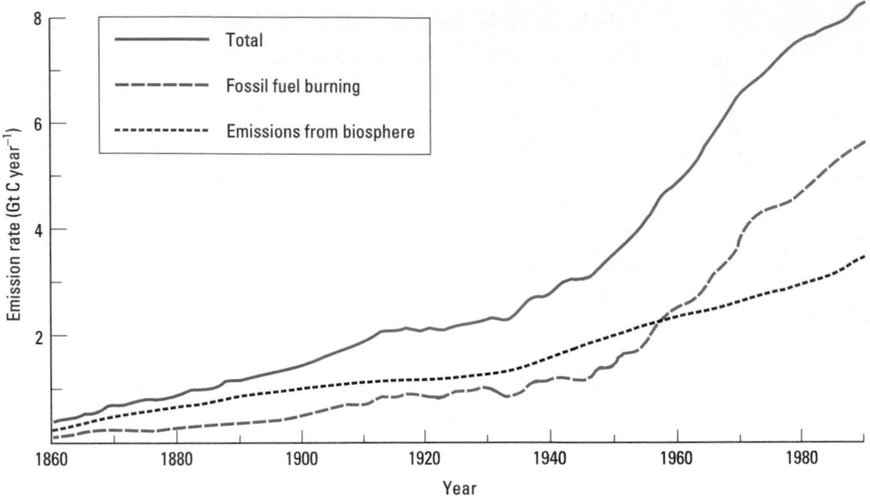

Fig. 15.62 Carbon emission rates increasing through the twentieth century, largely paralleling fossil fuel burning but still increasing beyond the 1970s (when fuel use leveled off somewhat), probably reflecting increasing deforestation rates.

that also affect atmospheric temperatures (Fig. 15.63). One major culprit is methane (CH_4), and in many cases deforestation also affects methane levels because forest is replaced by ranches where ruminants are grazed, and ruminants are very good at producing methane from their guts as a by-product of fermentation. Methane is also released from rice paddies in tropical zones and from bogs and other natural wetlands in temperate zones; perhaps also by dredging up oil, coal, and gas deposits. Methane fluxes into the atmosphere are made worse by extensive nitrogenous fertilization as part of human agricultural practice.

The levels of all of the other gases shown in Fig. 15.63 are also rising. The worst culprits for such emissions tend to be the more industrial countries, with power stations, smelting plants, and heavy industrial centers particularly implicated. Figure 15.64 summarizes the main ways in which human activities have contributed to global warming.

An additional problem in terms of potential temperature rise is that all these kinds of emission to the atmosphere may interact. Interactions may be positive or negative (Table 15.19), but on balance the IPCC group have estimated that positive feedbacks are more likely than negative ones, so that estimates of the magnitude of temperature change and of its timing may actually be on the low and slow side. There is no doubt that CO_2 is rising very fast and abnormally; and while there is some doubt about temperature rise (how much, how fast) there is every likelihood that there will be further changes in the upward direction. Best predictions are for a rise to 600 ppm CO_2 by a date between 2030 and 2080, which might mean a 3–5°C global rise, with worse effects at higher latitudes and in the northern hemisphere, perhaps up to 12°C warmer at the North Pole. This does not sound like a spectacular change, but any rise above the average has enormous implications, especially when it occurs very quickly.

Fig. 15.63 Proportional contribution of atmospheric gases (excluding natural water vapor) to greenhouse effect warming: (a) up to 1980 and (b) since 1980. CFC, chloroflurocarbon. (Adapted from Mintzer 1992.)

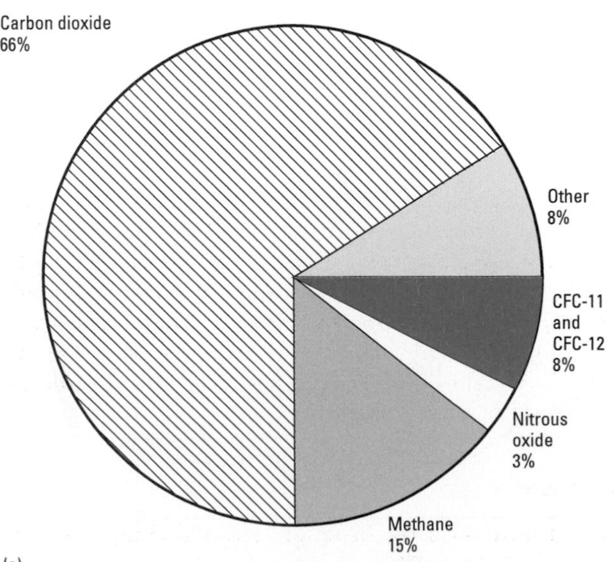

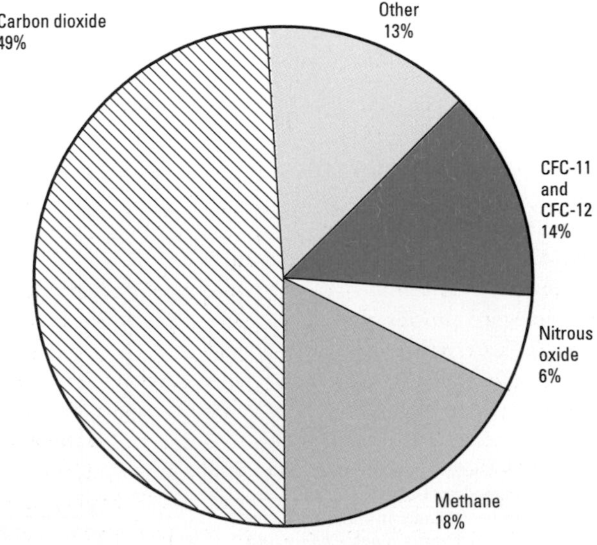

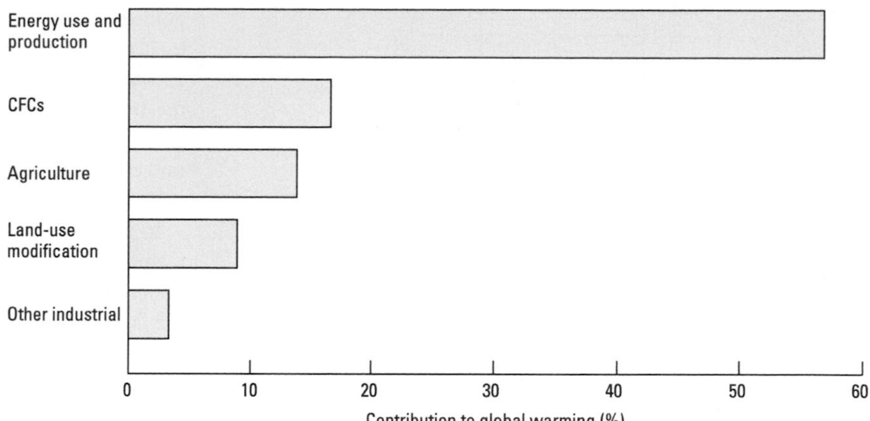

Fig. 15.64 Major influences of human activities on greenhouse emissions and global warming. CFC, chloroflurocarbon.

Table 15.19 Feedbacks that may alter the greenhouse effect.

Negative feedbacks
Rising temperatures lead to more cloud formation; this increases the reflectivity of the atmosphere and reduces solar inputs
Rising productivity of plants (using high CO_2 levels to fix more carbon) may use up the excess CO_2 and bring things back into balance (but plant responses are highly variable: weedy plants do well, others are often limited by nutrient resources, perhaps because the enhanced soil microflora are doing better and sequestering the nutrients)
The seas will serve as a sink and mop up excess CO_2 (but present evidence suggests this can only occur where CO_2 levels rise much more slowly; the oceans could perhaps mop up an "extra" 1–3 Gt per year, whereas 7–8 Gt per year are being released)
Positive feedbacks
Increasing temperature may release more of the carbon held in soils, the dead organic matter being released as CO_2 and CH_4, especially from temperate wetlands and from reserves currently trapped below northern permafrosts
Increasing temperature may enhance sea stratification, with a warm unmixed upper layer having decreased primary production so that it gets depleted of nutrients, giving decreased CO_2 uptake by the seas
Increasing temperatures could cause some polar ice melt, thus exposing new areas of land, giving areas of dark color with lower albedo and higher heat absorptivity than the ice, and so accelerating the process of warm-up

Direct effects

Land animals are unlikely to be directly affected at a physiological level by the projected changes in CO_2 levels, since they excrete the gas readily as a waste product of respiration, with very substantial gradients from their tissues to the outside world. Terrestrial animals are usually responsive to $P\text{CO}_2$ (more so than aquatic species), but they have their sensors internally to detect blood concentration, so that direct effects on respiration are unlikely.

However, the effects of rising environmental temperatures on physiological processes are potentially more serious. We have explored the various biochemical and physiological mechanisms that underlie thermal adaptation and phenotypic plasticity in animals (see Chapter 8), and it is clear that those species with a suite of enzyme and membrane adaptations that render them eurythermal should be able to cope with moderate changes in ambient temperatures. Some eurythermal species may be able to exhibit phenotype plasticity, i.e. switching to slightly different phenotypes as a direct response to different thermal conditions. Others may be able to effect genetic changes in traits with a fast enough time course; this would be especially true for animals with high reproductive outputs and thus rapid evolutionary rates (insects, crustaceans, some worms, but probably only a few small vertebrates), and for traits that show high heritability. The relatively low heat tolerance of metazoans compared with unicells and bacteria may indicate limits due to complex systemic factors rather than simple molecular processes, and for "higher" animals the whole-animal aerobic scope seems to be the first process to become stressed in a warming environment, as circulation and ventilation suffer with oxygen becoming limiting. But interactions of physiology and behavior inevitably complicate matters; for example, recent studies indicate that for the common mouse, *Mus domesticus*, body weight and "nesting score" show high heritability but body temperature and weight of brown fat (BAT) do not, so that more northern mice with larger bodies and an inherent tendency to build good nests might fare better when challenged with climate change. It is also possible to use behavioral alterations to cope with warming. There is accumulating evidence for "laying dates" in temperate birds becoming earlier in the spring, probably because food is available earlier, and similar effects are documented for some insects where the earlier start may allow an extra generation to be produced in each annual cycle (potentially serious with "pest" species such as aphids). Indeed some animals may be able to migrate quickly enough to keep pace with the changing climate (after all some species certainly did live further north in previous interglacial periods and may be able to recolonize lost ground northwards). There are already indications of poleward movements in the ranges of some insects, including butterflies and bees.

But stenotherms may be at real risk of experiencing ambient temperatures outside their tolerance ranges. This may be especially problematic towards the poles, where temperature changes may be greatest. Remember also that cold-adapted proteins are generally inherently more stenothermal (and likely to denature rapidly) than warm-adapted ones. Thus larger, slowly reproducing, and/or less mobile, polar and high-altitude stenotherms may be at particular risk in a warming world of the future.

Indirect effects

First and foremost a rise in temperature will change the patterns of global climate. There are many alternative computer global

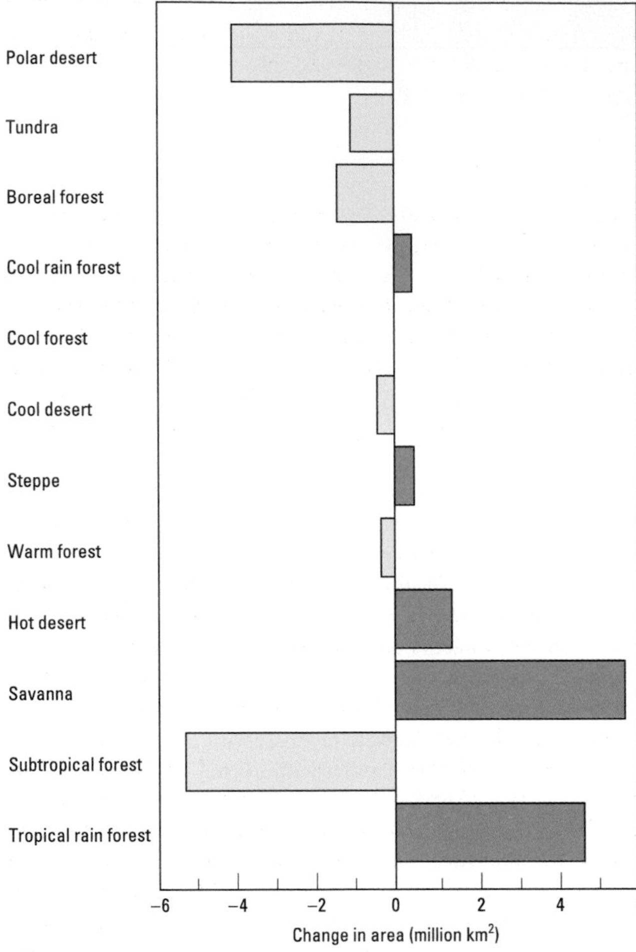

Fig. 15.65 Predicted patterns of change in the land biomes due to global warming.

circulation models, though most are reasonably in agreement and most predict average temperature rises of 1°C already, and up to 5°C in 100 years (varying with latitude). The best guesses for climatic changes accompanying global mean surface warming are large-scale stratospheric cooling, global mean precipitation increases, reduced sea ice with polar winter warming, and continental summer warming and drying with increased cyclones and storms.

Climatic changes of this magnitude must have major effects on vegetation distribution, pushing successive biome types towards the poles. Globally, the current projections of overall changes in biomes (Fig. 15.65) suggest a loss of polar and tundra systems especially, with some increase in savanna, desert, and (if it is not cut down) tropical rain forest. The forest belts have been moving northwards at 1 km year^{-1} since the last Ice Age anyway; they probably cannot sustain much faster migration, although they may still have hundreds of kilometers of potentially habitable land still to recolonize. More seriously, different maximal migration rates would surely apply to different species of both plants and animals, so that the ecosystems would change and potentially become very unbalanced. Coevolved species could become mismatched in time or in space: pollinators and seed dispersers with the flowering and fruiting of their relevant plants, monophagic predators with their prey, parasites with their normal hosts, and so on. This is especially likely in the tropics, where such relations are often finely tuned. There is growing concern that vector-borne parasites might expand their ranges in space and through the season, exposing immunologically naïve human populations to new threats.

In general, fast growing and fast reproducing (*r*-selected) "weed" plants and "pest" animals should almost inevitably do better due to their high dispersal powers, higher reproductive rates, and potential for faster adaptive change. Animal species with eurythermic and euryhaline tolerances should out-compete all other species; polyphagous species should thrive at the expense of monophages; and those with good locomotory/dispersive powers in at least one stage of the life cycle should do well. Again, the potential future problems created by parasites and pathogens are rather obvious.

15.9.2 Ozone depletion

As a gas, ozone (O_3) is harmful low down in the troposphere. It is an oxidant and a pollutant, and where it collects in cities as part of the "photochemical smogs" it causes respiratory problems and skin rashes. But ozone is absolutely necessary higher up in the atmosphere, because it helps protect living organisms from potentially harmful UV-B radiation (280–320 nm) in the stratosphere. It is the only atmospheric gas that absorbs strongly around 300 nm. And in the upper layers of the planet's atmosphere, ozone levels have been changing quite drastically in some places and at some times of year.

The first suspicions of a manmade threat to the ozone layer came in the mid-1970s, and a "hole" was detected between 1977 and 1984 by Antarctic survey teams. The hole now develops each spring in the Antarctic (through September and October) and at its worst grows to be about the size of the USA, spreading over the tip of South America and a moderate area of southern Australia. The chlorofluorocarbon gases (CFCs, especially $CFCl_3$ and CF_2Cl_2) that cause this were "wonder compounds" when invented in the 1920s, due to their extreme chemostability and low toxicity, and so were widely used in refrigeration, bubble-foamed plastics, aerosols, and air conditioners. But photodissociation of these CFCs can occur under the influence of short-wavelength UV light (200–350 nm); they then break down and release Cl^- ions, which react with ozone to produce oxygen. Chlorine is particularly effective because it precipitates a chain reaction with ozone in which the active principle, ClO (chlorine monoxide), is continuously regenerated; a single ClO molecule, once formed, can break down 20,000–100,000 ozone molecules.

The CFCs are very long-lived molecules (15–100 years) and are therefore bound to reach the stratosphere when released at the Earth's surface. The link between the resultant rising levels of manmade chlorine and the seasonal loss of ozone is now quite clear (Fig. 15.66). In the mid-1980s the annual production of CFCs was nearing 1.2 million tons and the total stratospheric content of halogens had risen from 0.6 ppb (parts per billion) to around 3 ppb chlorine and 0.02 ppb bromine. The UN Montreal Protocol of 1987 agreed to halve production of CFCs, and many governments now plan their complete elimination. But chlorine levels are still rising by about 5% per year, the "hole" is still enlarging each year to cover

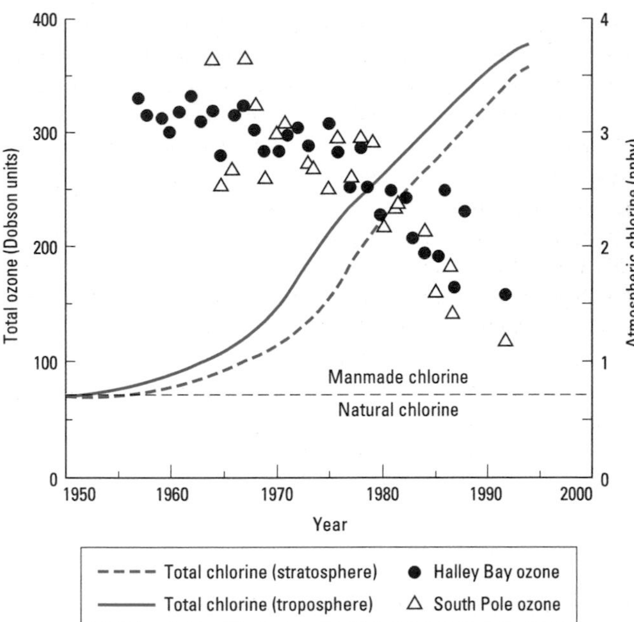

Fig. 15.66 Rising anthropogenic chlorine levels, and associated declines in stratospheric ozone at two sites in Antarctica as the seasonal ozone hole develops.

large parts of the southern continents, and some perturbations are also occurring over the northern polar region. In any event, with the longevity of the molecules involved, recovery of the ozone layer is unlikely until atmospheric chlorine levels fall back below 2 ppb, currently projected for the end of the twenty-first century if strict controls are adhered to.

Thinning of the ozone layer is potentially serious to life below because organisms become exposed to an excess of UV-B. This has several adverse effects. Firstly, UV-B slows photosynthesis in many plants (not all, but at least 50% of all species tested). It might also decrease productivity substantially in the phytoplankton; marine plants are not well defended against UV-B, and through evolutionary history their main defense has been to live at a depth where it does not penetrate, so they can be harmed by small increases. These effects on land and sea plant productivity would also of course enhance global warming by reducing CO_2 sinks. UV-B may also directly affect some secondary plant chemistry pathways, and there are indications that it reacts directly with DNA in plants. This may relate to its effects in animals, which have mostly been studied in man: a 1% decrease in ozone, by allowing 2% more UV-B through, may cause a 5% increase in benign human skin cancers, of which some unknown percentage will turn malignant. But a 10% decrease in ozone would mean a 50–90% increase in mild skin cancers, and probably a small increase in the much more serious melanomas too. There is also an increasing prevalence of eye cataracts and potential blindness, and some suggestions of effects on the human immune system. Projecting these kinds of effects into the rest of the animal kingdom suggests a serious indirect effect for certain kinds of mono- and oligophagous herbivores, whose food plants could be seriously reduced or out-competed, and a direct risk for those animals that lack a good protective skin or that spend most of their time out in the open. Animals with thicker cuticles or shells, and those that spend much of their life in protected burrows or within the soil or litter layers, may be at little risk.

15.9.3 Acidification of the atmosphere and biosphere

As we noted in Chapter 13 in relation to lake acidification, the problem of acid deposition was first noted at least 100 years ago, with the observation that rainfall was sometimes acidic and that this did harm to vegetation, stonework, and metals. By the 1960s the phenomenon of "Waldsterben" (tree sickness) was noted, with a general forest decline in southern Germany and Czechoslovakia. Trees, especially the silver fir (*Abies alba*), showed a premature increase in old-age disorders, particularly showing fungal infections and thickening branches. By the late 1970s other trees, including the Scots pine, spruce, and even beech, were noticeably affected, with bleached leaves, and by the mid-1980s around one-fourth of all trees were affected in Western and Central Europe. Figure 15.67 summarizes the position throughout Europe from the 1989 tree damage survey: both deciduous and coniferous forests were damaged, the former slightly more so, and matters were especially bad in Central Europe (Bulgaria, Czechoslovakia, Poland, Greece, and eastern Germany). In 1995–96, 50% of all trees in Germany were assessed as badly damaged. Acid rainfall can be detected very frequently in all these areas, with the rain in Europe and the USA now commonly recorded at pH 4.0–5.0 in areas lying to windward of heavy industry. Sulfur dioxide and nitrogen oxide emissions, either unconverted as gases or dissolving as sulfuric and nitric acids, are the prime causes of the acidity (see Chapter 13). Soils are directly affected, even in isolated areas; the average pH of soils in Great Britain over the last 100 years showing a decrease from about 7 to 4.5–6 according to depth (Fig. 15.68).

Acidified rainfall has several effects for land animals:

1 It removes nutrients from the soils. This is relevant to trees, because the acid rain washes out minerals, like magnesium, that are needed for growth, leading to yellowing of conifer needles. Thus tree feeders suffer. However, acid conditions also solubilize aluminum in soil, and aluminum-poisoning affects mucus production in invertebrates such as earthworms and flatworms.

2 The gaseous components of acid rain affect plant growth directly. The gases have interactive negative effects: SO_2 and ozone cause additive damage to leaf growth and synergistic damage to root growth and leaf appearance, while they are antagonistic in their effects on fungal infections. With spruce, SO_2 opens the stomata and lets ozone in, also allowing more drying out and frost susceptibility. Thus the very dry summer of 1976 was one of the worst ever recorded for tree damage in Germany, as the acid gases made drought unusually hard to bear. Folivorous animals certainly experienced associated stresses.

3 The dissolved acid destroys useful fungal (mycorrhizal) associations in soils and so indirectly reduces plant growth and affects the associated animals.

4 Reduced diversity in streams and soils (see Chapter 13) affects dietary breadth for associated land fauna such as birds and insects.

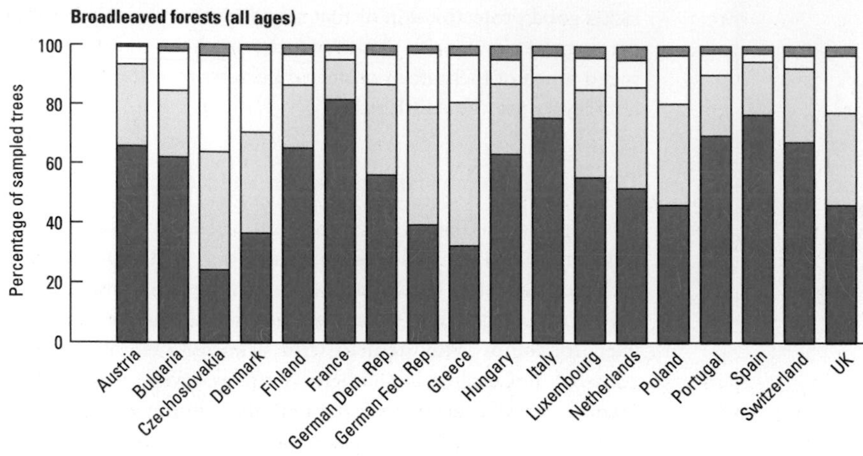

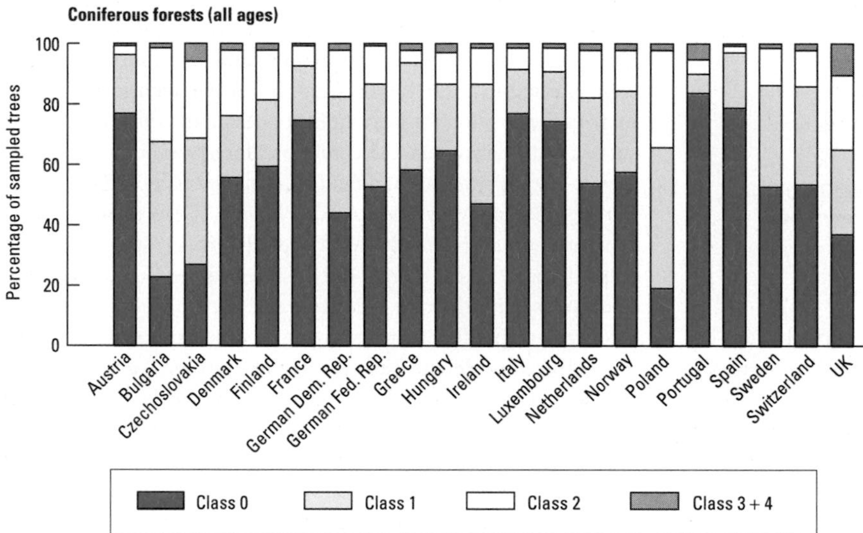

Fig. 15.67 Results of the comprehensive 1989 forest survey in Europe, showing the percentage of trees in different damage (defoliation) classes by country; 1–4 = increasing defoliation. Many show less than half the trees with no damage (class 0). (Adapted from *Climate Change IPCC Report 1995.*)

5 Soil pH directly affects many soft-bodied animals, including amphibians; salamanders, for example, suffer disrupted sodium balance and weight loss in acid soils.

6 Very low pH rainfall almost certainly kills small soft-bodied invertebrates in leaf litter directly.

Note that any or all of these effects can leave trees, tree herbivores, and litter animals alive but under stress, and thus more susceptible to other pollutants, to pathogens, and to parasites.

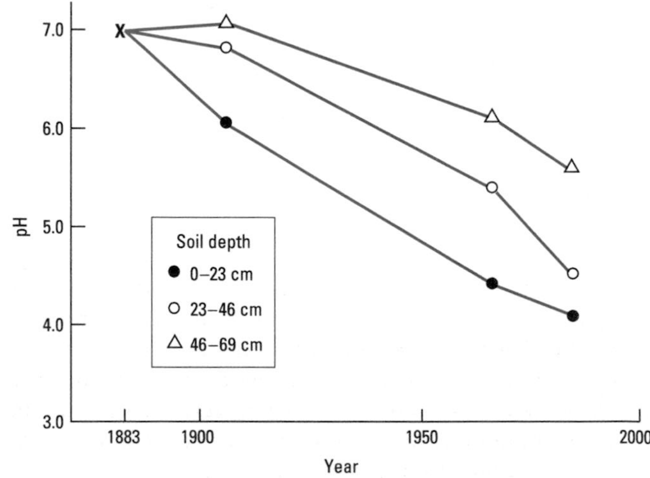

Fig. 15.68 Changes in pH at a site in Britain over the last century, linked to acid precipitation.

15.9.4 Urbanization

Wastes and pollutants

Wastes can be divided into four major categories: municipal, industrial, agricultural (including mining and dredging substrates), and nuclear. The amount of each varies for different human communities: agricultural wastes dominate in Europe and most of the USA, whereas in Japan industrial waste is hugely dominant. More than 75% of the waste is recyclable in ways that would pose little threat to the rest of the biota, but *at best* only about 30% does get recycled. Instead, extensive and ever-increasing use of land-fill sites destroys habitats for flora and fauna, while incineration contributes further CO_2 to enhance global warming.

Chemical waste is particularly problematic for the biota. About 80,000 chemicals are now in common use by humans, by far the majority being organic carbon-based compounds, mostly derived ultimately from natural gas and oil. We generate about 250 million tonnes of organic chemical wastes annually. The two main groups are aromatic hydrocarbons used in numerous industrial processes, and organochlorines and organophosphates, especially used as pesticides and herbicides and thus directly threatening to the biota. In addition there are inorganic wastes, with various heavy metals being particularly difficult to dispose of and potentially hazardous.

About 75% of this chemical waste is disposed of in land-fill tips (see Plate 10e, between pp. 386 and 387), mostly around urban areas, but older and/or badly managed tips may threaten the ground water or leak noxious gases, and toxins, gases, and microorganisms can all become hazards. The alternative of incineration, which decreased immediately after clean air legislation was put in place in many industrialized countries, is increasing again now. Wastes can generate useful energy, but also greenhouse gases; and the potentially toxic ash still has to go to land-fill. Bacterial remediation has the most promising outlook, in association with modern biotechnology.

Nuclear waste disposal represents a very specific extra problem, because the major effects of radiation are at the level of the genome itself. Reactors, however well managed, contaminate everything nearby with radiation, including clothes, dust, and the cooling water, producing low- and intermediate-level wastes (LLW and ILW). They also generate spent fuel, which is still very highly radioactive and constitutes the high-level waste (HLW). This is primarily still uranium, with 1–2% plutonium and 5% other fission products. Plutonium is often described as the most toxic substance on Earth, partly because it has a half-life for decay of 25,000 years. It causes excess mutations in all organisms, potentially leading to teratogenic deformities at birth and carcinogenic effects later in life. A typical reactor may produce 200 kg of it per year, and this is enough in theory to cause cancer in the entire vertebrate population of the world. If it were uncontrolled (or if large quantities of equivalent radiation were released through nuclear warfare or terrorism) it could destroy much of the Earth's animal and plant life; with the somewhat depressing proviso that animals with tough exoskeletons, and particularly a few species such as cockroaches that have resistant highly dispersed genetic material, might be most able to survive.

Nuclear material cannot be disposed of by controlled burning, which would just generate radioactive smoke instead. Most of the LLW was formerly dumped at sea, but now international conventions insure it is put into shallow pits on land, together with MLW. The problem of HLW is still quite unresolved, and it is all currently in store at reactor sites. Geologists are not yet agreed on *any* sites, on land or under the sea, as being stable enough for deep burying, even if accompanied by initial stabilizing vitrification.

Manmade habitats

Man has not only destroyed habitats, but has also, on various scales, created new ones (see Plate 10, between pp. 00 and 00), and certain kinds of animals have exploited these. We mentioned earlier that savanna grasslands are probably a very long-term "creation" of human activities. An even more obvious recent example is intensively farmed agricultural land (see Plate 10c), especially where crops are grown as virtual monocultures. Here the diversity of the endemic fauna and flora is very severely reduced, though there can be "weed" proliferation and severe plagues of "pest" insects, largely due to the removal of natural enemies and parasites, but also in part due to the physiological toughness and rapid powers of recovery of such organisms. The native vegetation and all the associated animals get squeezed into tiny islands of hedgerows and small woods, and they are often at risk of local extinction, having to deal with gross chemical disturbances to the soils and surroundings. This leads to intense ecological stress, with disruption of the normal biotic balance between predators, prey, and parasites, and between herbivores, pollinators, and plants. It underlines once again the point that for all these relatively equable climatic zones the most important shaper of the community is often the range of competitive and cooperative biotic interactions, rather than any particular physiological pressure.

A second kind of manmade habitat is associated with agriculture, and involves all the resultant stored products, in grain silos, in processing centers and at points of sale. All provide high concentrations of certain kinds of foods that are by definition attractive to animals, and most also provide very stable (but often very dry) conditions. Grains and seeds are particularly susceptible, and for "pests" of these the main physiological problem is drought, as these products are stored at low humidity to prevent fungal invasions. The adaptations shown therefore tend to be rather like those of desert organisms, and such habitats are attractive to many species of mites (often those species capable of water-vapor uptake), to insects such as beetles, and to rodents.

The third and most obvious kind of manmade habitat is, of course, houses (see Plate 10d). In most parts of the world these are deliberately built to have stable, mild, and reasonably humid interiors—ideal living conditions for a great many insects. Thus our dwellings attract carpet and clothes eaters, wood eaters, and even dry skin eaters (the house-dust mites, especially *Dermatophagoides*); and of course generalist detritus eaters such as cockroaches and rodents.

Notice that all these manmade habitats attract species with very high reproductive rates and high evolutionary adaptation rates, and therefore are stretching our control measures to the limit. Pesticides increasingly do not work, and this situation is certainly not going to get any easier.

15.10 Conclusions

A great many groups of animals have achieved some degree of terrestriality, and it is a mistake to underestimate the success and importance of the small and less "exciting" soil dwellers, which may form a huge but invisible part of the land fauna. However, the real success stories are restricted to just a few taxa whose impact on land has been more striking. Molluscs do well in all biomes, but it is nearly always the arachnids and insects and the vertebrates that are dominant, the arthropods taking the small-size niches and vertebrates the larger ones. Above all, this emphasizes the crucial factor of controlling surface properties: the arthropod cuticle and vertebrate dermis that have underlain so much of their owners' success. It is worth remembering that there is a real paradox in relation to "skin". It has to serve many purposes at once: it is a barrier keeping some things inside the body and keeping others out, but it must also let some things in (notably oxygen and information) and let some out (waste products). It is just about impossible to do all these things equally well, and compromises must therefore be made. In the light of this it is perhaps no surprise that the outstanding success story has been that of the arthropod exoskeleton. This represents the best possible compromise for land life: inbuilt defense and protection against unwanted physiological exchanges, coupled with versatile properties residing in the different layers of the structure, which can be varied between and within species and stadia, and a relative ease of making gated entry and exit points. The highly interactive problems of water balance, temperature balance, and respiration can be solved in diverse ways using the cuticle; potential modes of locomotion are greatly increased by it; every conceivable trophic possibility can be realized with cuticular mouthparts; and reproduction on land is made easy by protective egg cases and by the elaboration of the genitalia and of signaling devices to provide species-recognition cues. Arachnids and insects, endowed with this material, dominate the terrestrial abundance, biomass, and species diversity in almost all habitats.

FURTHER READING

Books

Burggren, W.W. & McMahon, B.R. (eds) (1988) *Biology of Land Crabs.* Cambridge University Press, Cambridge, UK.

Cossins, A.R. & Bowler, K. (1987) *Temperature Biology of Animals.* Chapman & Hall, London.

Dejours, P., Bolis, L., Taylor, C.R. & Weibel, E.R. (eds) (1987) *Comparative Physiology: Life on Water and on Land.* Liviana Press, Padova, Italy.

Hadley, N.F. (1994) *Water Relations of Terrestrial Arthropods.* Academic Press, San Diego, CA.

Heinrich, B. (1993) *The Hot-Blooded Insects.* Harvard University Press, Cambridge, MA.

Herreid, C.F. & Fourtner, C.R. (1981) *Locomotion and Energetics in Arthropods.* Plenum Press, New York.

Little, C. (1983) *The Colonisation of Land.* Cambridge University Press, Cambridge, UK.

Little, C. (1990) *The Terrestrial Invasion; an Ecophysiological Approach to the Origin of Land Animals.* Cambridge University Press, Cambridge, UK.

Louw, G.N. (1993) *Physiological Animal Ecology.* Longman, Harlow, UK.

Rosenthal, G.A. & Berenbaum, M.R. (1991) *Herbivores, their Interactions with Secondary Plant Metabolites.* Academic Press, San Diego, CA.

Skadhauge, E. (1981) *Osmoregulation in Birds.* Springer-Verlag, Berlin.

Reviews and scientific papers

Alexander, R.McN. (1989) Energy-saving mechanisms in terrestrial locomotion. In: *Energy Transformations in Cells and Organisms* (ed. W. Wieser & E. Gnaiger), pp. 170–174. Georg Thieme Verlag, Berlin.

Beuchat, C.A., Calder, W.A. & Braun, A.J. (1990) The integration of osmoregulation and energy balance in hummingbirds. *Physiological Zoology* **63**, 1059–1081.

Blickhan, R. (1989) Running and hopping. In: *Energy Transformations in Cells and Organisms* (ed. W. Wieser & E. Gnaiger), pp. 183–190. Georg Thieme Verlag, Berlin.

Cazemier, A.E., Op den Camp, H.J.M., Hackstein, J.H.P. & Vogels, G.D. (1997) Fibre digestion in arthropods. *Comparative Biochemistry & Physiology A* **118**, 101–109.

De Vries, M.C. & Wolcott, D.L. (1993) Gaseous ammonia evolution is coupled to reprocessing of urine at the gills of ghost crabs. *Journal of Experimental Zoology* **267**, 97–103.

Full, R.J. (1989) Mechanics and energetics of terrestrial locomotion: bipeds to polypeds. In: *Energy Transformations in Cells and Organisms* (ed. W. Wieser & E. Gnaiger), pp. 175–182. Georg Thieme Verlag, Berlin.

Full, R.J., Zuccarello, D.A & Tullis, A. (1990) Effect of variation in form on the cost of terrestrial locomotion. *Journal of Experimental Biology* **150**, 233–246.

Geiser, F. (1988) Reduction of metabolism during hibernation and daily torpor in mammals and birds: temperature effect or physiological inhibition? *Journal of Comparative Physiology B* **158**, 25–37.

Hayes, J.P., Speakman, J.R. & Racey, P.A. (1992) The contribution of local heating and reducing exposed surface area to the energetic benefits of huddling by short-tailed field voles. *Physiological Zoology* **65**, 742–762.

Hillard, S.D., von Seckendorff Hoff, K. & Propper, C. (1998). The water absorption response: a behavioural assay for physiological processes in terrestrial amphibians. *Physiological Zoology* **71**, 127–138.

Hoffmann, A.A. & Blows, M.W. (1993) Evolutionary genetics and climate change; will animals adapt to climate change? In: *Biotic Interactions and Global Change* (ed. P.M. Kareiva, J.G. Kingsolver & R.B. Huey), pp. 165–178. Sinauer Associates, Sunderland, MA.

Hofman, R.R. (1989) Evolutionary steps of ecophysiological adaptation and diversification of ruminants: a comparative view of their digestive system. *Oecologia* **78**, 443–457.

Lance, V.A. (1997) Sex determination in reptiles: an update. *American Zoologist* **37**, 504–513.

Lovegrove, B.G., Heldmaier, G. & Ruf, T. (1991) Perspectives of endothermy revisited: the endothermic temperature range. *Journal of Thermal Biology* **16**, 185–197.

Minnich, J.E. (1982) The use of water. In: *Biology of the Reptilia*, Vol. 12 (ed. C. Gans & F.H. Pough), pp. 325–396. Academic Press, London.

Oliviera, M.F., Silva, J.R., Dansa-Petretski, M. *et al.* (1999) Haem detoxification by an insect. *Nature* **400**, 517–518.

Parmesan, C., Ryrholm, N., Stefanescu, C. *et al.* (1999) Poleward shifts in geographical ranges of butterfly species associated with regional warming. *Nature* **399**, 579–583.

Parsons, P.A. (1990) The metabolic cost of multiple environmental stresses: implications for climatic change and conservation. *Trends in Ecology & Evolution* **5**, 315–317.

Patz, J.A & Reisen, W.K. (2001) Immunology, climate change and vector-borne diseases. *Trends in Immunology* **22**, 171–172.

Phillips, P.K. & Heath, J.E. (1995) Dependency of surface temperature regulation on body size in terrestrial mammals. *Journal of Thermal Biology* **20**, 281–289.

Portner, H.O. (2001) Climate change and temperature-dependent biogeography: oxygen limitation of thermal tolerance in animals. *Naturwissenschaften* **88**, 137–146.

Quinlan, M.C. & Hadley, N.F. (1993) Gas exchange, ventilatory patterns and water loss in two lubber grasshoppers: quantifying cuticular and respiratory transpiration. *Physiological Zoology* **66**, 628–642.

Schmitz, H. (1994) Thermal characterization of butterfly wings. 1. Absorption in relation to different color, surface structure and basking type. *Journal of Thermal Biology* **19**, 403–412.

Smedley, S.R. & Eisner, T.A. (1995) Sodium uptake by puddling in a moth. *Science* **270**, 1816–1818.

Walton, B.M. & Bennett, A.F. (1993) Temperature-dependent color change in Kenyan chameleons. *Physiological Zoology* **66**, 270–287.

Warburg, M.R. (1994) Review of recent studies on reproduction in terrestrial isopods. *Invertebrate Reproduction & Development* **26**, 45–62.

Willmer, P.G. (1980) The effects of a fluctuating environment on the water relations of larval Lepidoptera. *Ecological Entomology* **5**, 271–292.

Willmer, P.G. (1982) Microclimate and the environmental physiology of insects. *Advances in Insect Physiology* **16**, 1–57.

Wright, J.C. & O'Donnell, M.J. (1992) Osmolality and electrolyte composition of pleon fluid in *Porcellio scaber* (Crustacea, Isopoda): implications for water absorption. *Journal of Experimental Biology* **164**, 189–203.

Wright, J.C. & Machin, J. (1993) Energy-dependent water vapor absorption (WVA) in the pleoventral cavity of terrestrial isopods: evidence for pressure cycling as a supplement to the colligative uptake mechanism. *Physiological Zoology* **66**, 193–215.

Zimmerman, L.C. & Tracy, C.R. (1989) Interactions between the environment and ectothermy and herbivory in reptiles. *Physiological Zoology* **62**, 374–409.

16 Extreme Terrestrial Habitats

16.1 Introduction

Extreme terrestrial habitats classically include those that are unusually hot in low latitudes, or those that are unusually cold towards the poles and at high altitude. Hot habitats are normally only a real problem to animals when compounded by aridity, with the high ambient temperatures and low humidity together putting excessive strain on thermal and water (hygric) balance. Where heat is accompanied by high rainfall and freely available water it is less of a problem, as both ectotherms and endotherms can regulate their temperature without compromising their water balance, and many use evaporative cooling to this end. In this chapter we therefore concentrate on the really biotically "difficult" areas of dry heat, exemplified by deserts. For cold regions, the distinction between arid and humid zones is less important to the fauna, since water loss is much less of a problem at low temperature and evaporative cooling is most unlikely to be an important strategy. Cold zones can therefore be treated together in terms of the physiological adaptations needed. All of these habitats tend to be of low productivity as well as climatically extreme, so that animals require specialist adaptations on several fronts, over and above the range of "normal" terrestrial strategies covered in Chapter 15. For these reasons the extremes of the land masses have always been especially attractive to environmental physiologists, and many of the classic studies come from deserts and ice-fields. This chapter deals with these specialist adaptations, where animals are pushed to their limits. We particularly look at the realms of temperature, water balance, food and energetics, and related effects on life histories. Adaptations to respiratory, reproductive, sensory, and locomotory systems are not covered in detail; essentially these systems are similar in equable and extreme terrestrial habitats.

16.1.1 Common characteristics of life in extreme conditions

Size

It is often said that animals in general are larger towards the poles, and for endothermic animals this idea is formalized as Bergmann's rule. Certainly there are examples that fit well with this rule: penguins, for example, where tropical species are quite small while the much larger king and emperor penguins are found deep in Antarctica;

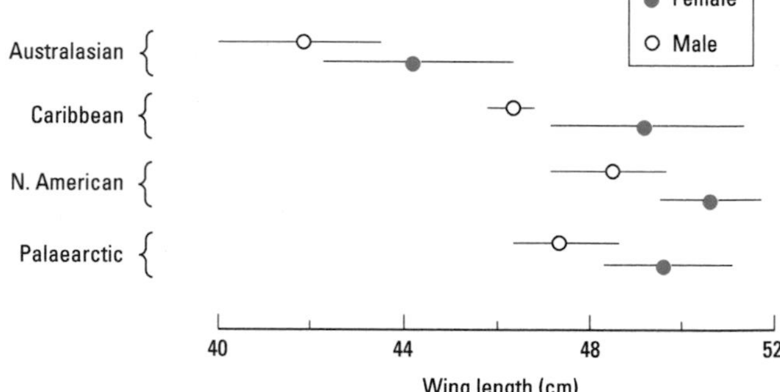

Fig. 16.1 Body size vs. latitude for ospreys (*Pandion haliaetus*) from Australasia to northern Europe, showing mean and standard error for wing length. The data are suggestive of a trend to smaller size at low latitude (though only the Australasian birds are statistically distinct). (Adapted from Prevost 1983.)

bears, where the large kodiak and polar bears occur further north than other species; and ospreys, as illustrated in Fig. 16.1. But at the same time many small birds and rodents occur towards the poles, escaping the rigors of their environment by burrowing. In fact if we look at extreme environments of all kinds, whether polar, desert, or at high altitude, it might be fairer to say that there is a general rule that being *either* unusually large *or* unusually small is a good strategy. (Inevitably, proving this kind of generalization is difficult as it is confounded by taxonomic artefacts, and a strict phylogenetic comparison has not been attempted; but it is noteworthy that even within a taxon such as the trogid beetles a tendency towards larger and hence more desiccation-resistant species occurs in the Kalahari desert relative to neighboring more temperate zones.) The logic of being either large or small is obvious: large species get all the advantages of a lower water loss rate and a high thermal inertia so that they can stay warm longer (at the poles) or stay cool longer (in a desert, as with the camel). The small species get the option, unavailable to larger relatives, of being able to exploit fine-grained microhabitats and escape from the harshest conditions. In deserts in particular, there is a striking paucity of medium-sized animals, which are denied either of these advantages; mammals of the size range that includes the dog, cat, and goat only really do well around the fringes of deserts or at oases.

In considering the strategies for effective life in extreme conditions, it is therefore often helpful to split the animals into a small number of (nontaxonomic) groups. It could be said that there are often only two possible broad strategies, which are strongly related to the size of the animals; these are the small "evaders" and the large "endurers". In the particular case of deserts, there is perhaps a third group comprising the rarer middle-sized animals, which might be termed the "evaporators".

Lifestyle

In any extreme environment, certain kinds of lifestyle seem to be favored. Many animals cope using a strategy of **storing energy** during periods of relative abundance to maintain themselves through periods of scarcity, whether in a daily, seasonal, or temporally random fashion. Others use a period of reduced metabolic need (dormancy, torpor, or estivation) through periods of scarcity. Others again have a nomadic lifestyle, abandoning each area as it becomes depleted of resources or enters a period of unendurable climatic extreme.

There may also be particular behavioral needs, so that in some extreme habitats there appears to be a trend towards more **gregariousness** or **eusociality**. This may have several advantages: it allows a colony to store food in periods of plenty as a buffer against intermittent and unpredictable scarcity; it allows a relatively small predatory animal, such as an ant, to attack, subdue, and store a much larger range of prey sizes; and it may involve shared underground nests that are strongly buffered against climatic extremes. Overall, the population size of a colonial or social species may be much less variable than that of a solitary epigean species.

Range and biogeography

For extreme habitats there appears to be a tendency for animals to have a **greater range** than occurs in more equable biomes. For high-latitude species this has sometimes been formalized as "Rapoport's rule", indicating that such species range over a broader latitudinal belt than occurs in the tropics. The same seems to be true for elevational range at altitude. These observations are not particularly surprising given that tropical lowland species are likely to be closely adapted to stable and predictable climatic regimes whereas polar or highland species must cope with much greater climatic variation at any one site. The issue is, of course, compounded by the much greater species richness at low latitude and altitude.

Selective regime

Harsh environments produce populations living on the edge of their abilities to cope, and may involve a particular kind of **adversity selection** (see Chapter 1), species being selected for their tolerance of the physical conditions and for an ability to reproduce successfully even with the small numbers remaining after mass mortalities. Adversity selection allows species to survive through periodic population bottlenecks. At the same time the periodic extinction events keep diversity so low that there are few interactions between the species and little coevolutionary selection.

16.2 Hot and dry habitats: deserts

16.2.1 Types and characteristics of deserts

Occurrence

Arid zones occur between 15 and 40° latitude, either side of the warm and wet equatorial zone. The boundaries of deserts were set quite recently in geological terms; the massive Pleistocene glaciers tied up free water, leading to aridity at low latitudes so that desertification occurred, and in the 10,000 years since the last glaciations these areas have never recovered. We generally think of deserts as exceptionally hot and dry areas with sweeping sand dunes (see Plate 8e, between pp. 386 and 387), but in fact a variety of biotopes are included within this general category and they are surprisingly varied. Their most important shared characteristic is aridity; the three subdivisions of semiarid (see Plate 8f), arid, and hyperarid deserts, as defined by international agreements, cover one-third of the Earth's land surface, about half of this being true desert as traditionally perceived by the layman (Fig. 16.2).

Natural aridity arises for three main reasons, varying in importance for different desert areas. Deserts in parts of North and South America arise from a rain-shadow effect, lying on the leeward side of high mountain ranges (sierras), and where no rain clouds occur as the rising air has precipitated all its water on the mountains (see Plate 9a, between pp. 386 and 387). Deserts in the center of large continents are arid principally due to their sheer distance from the sea; the Gobi and Turkestan deserts are examples. However, the largest deserts on Earth arise from the third factor, occurring in latitudes (25–35°N or S) where there are dry, stable air masses of high pressure, resistant to invasion by storm systems to the north or south. The Saharan and Arabian deserts, and the Australian desert, are of this type.

Hyperarid deserts, where high-pressure weather dominates, normally receive less than 25 mm of rainfall per annum; within this there are areas that effectively receive no rain at all. Even when rain does fall it may be as violent convective showers, causing flash-flooding in dry river beds (wadis; see Plate 9b) and very fierce run-off, with little of the rain becoming available to plants or animals in the area; that which does remain local is subject to very rapid evaporation due to the high temperatures. Thus water availability is not only low but highly unpredictable, and all organisms must be opportunistic in responding to its presence. These deserts usually also have high wind speeds, leading to even more fierce evaporation, and they have cloudless skies, so that it can be very cold at night as heat radiates away from the soil. Their soils are hardly leached at all by any passage of water through them, so they are often sandy or stony (leading to high heat penetration and very poor water retention; see section 15.1.2) and also have considerable deposits of salts very near the surface. As this is eroded and stirred by the high winds life becomes even harder for any colonizing plants. Drought alone is difficult, but when combined with hot and cold extremes, and saltiness, and the instability of the terrain, living here is impossible for most kinds of organism. These desert communities are therefore dominated by specialists, usually occurring at low biomass.

Less extreme areas, the arid and semiarid deserts, may have up to 600 mm of rain per year, more evenly and predictably distributed,

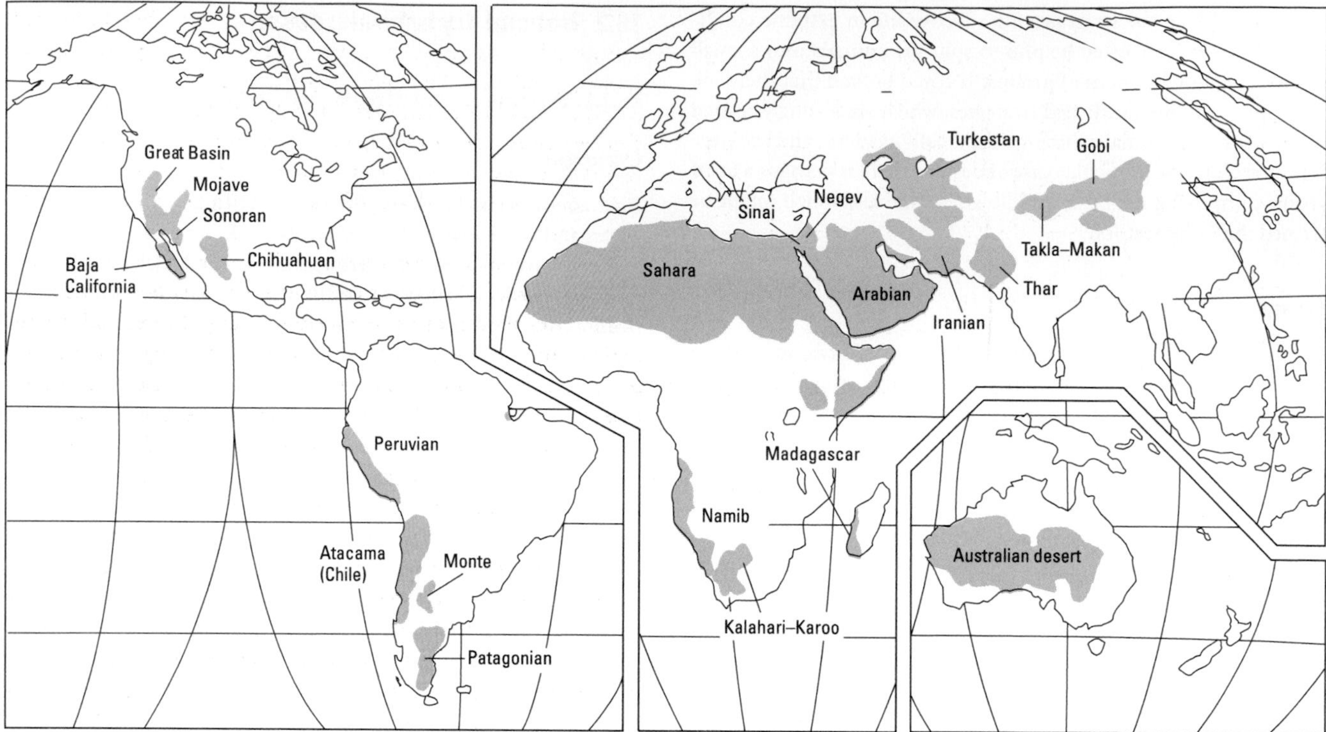

Fig. 16.2 The occurrence of deserts on a worldwide basis, with two broad latitudinal belts north and south of the equator.

and they are often much cooler than the hyperarid areas. Coastal deserts, such as the Namib (southwest Africa), the Atacama (Chile), parts of southern Israel, and the Baja California coast of USA/Mexico, are characterized by cooling fogs, especially after dawn. Cooler inland deserts, such as the Gobi and large areas of Patagonia, also occur, with limited fog but prolonged periods of winter cold.

Here we are concerned particularly with the problems faced by the biota in hot arid areas, where terrestrial adaptations are pushed to extremes. We will therefore concentrate on animals from the large sand and rock deserts of Saharan north Africa, Arabia, and the Middle East, southwestern states of the USA, and Australia. By far the most detailed work to date has derived from fieldwork in the deserts of Arizona and Colorado, the Negev, and the Australian outback, all areas where high temperatures and aridity are compounded by high winds and continuously disturbed wind-blown sandy and/or salty soil.

Desert microclimate and vegetation

Microclimatic niches make survival possible for many terrestrial animals, as we saw in Chapter 15. In the Libyan desert a daytime air temperature of 58°C has been recorded, and other deserts can be below freezing at night; but within desert sands conditions just 100 mm deep approach constancy, with diurnal variation virtually eliminated at 200 mm (Fig. 16.3). This degree of "fine-grained" microclimatic variation becomes paramount in determining animal distributions and lifestyles. Any abiotic irregularity can cause additional microclimatic variation; pebbles and rocks on sand create local shadows and potential shelters, while any vegetation creates not only shade but a local increase in humidity and a reduction of wind speed. Figure 16.4 shows microclimates recorded around and below a desert shrub in terms of the body temperatures of beetles

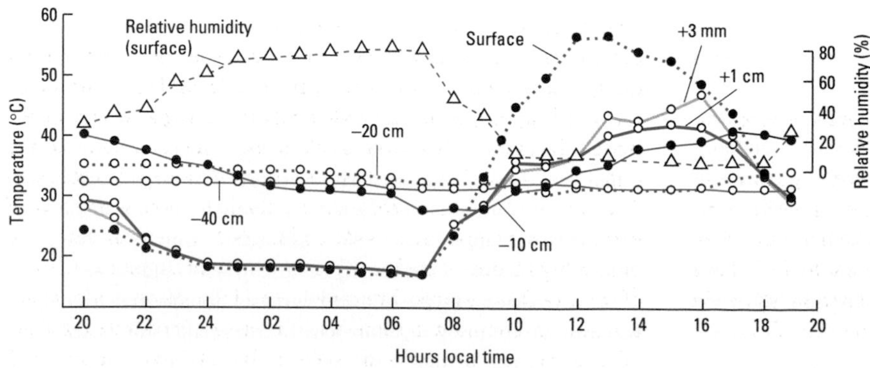

Fig. 16.3 Patterns of temperature and humidity against depth for dune slopes in a desert, through the 24 h cycle. Notice that there is reasonable constancy at –10 cm and complete stability at –20 and –40 cm depth, while the surface shows a temperature range of nearly 40°C. (From Holm & Edney 1973.)

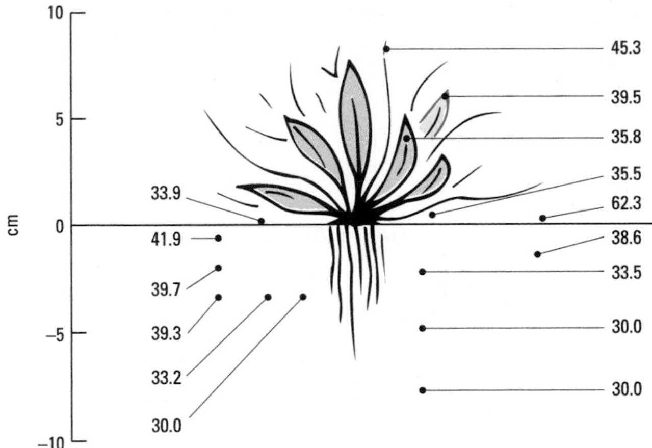

Fig. 16.4 The microclimate (shown by temperatures, °C) as experienced by small beetles in, under, and around a desert shrub. (Adapted from Hamilton 1975.)

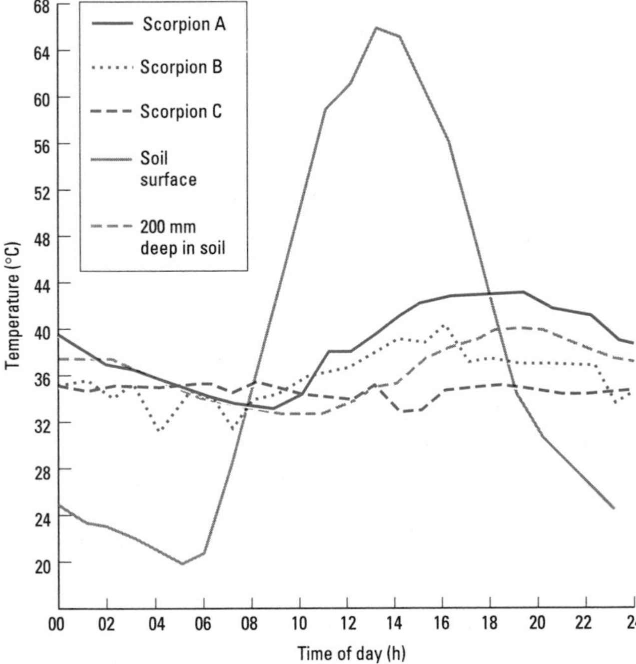

Fig. 16.5 Temperatures in the burrow of the scorpion *Hadrurus*, compared with temperatures at the soil surface and at a depth of 200 mm. (Adapted from Hadley 1970.)

found in those sites. Desert arthropods commonly occur clustered around vegetation in this fashion.

A great many species also live cryptically, hidden away in microclimatic conditions that allow survival. Burrows 200–300 mm below the surface may be at an almost constant temperature of 35–40°C, and at humidities always above 80–85%, allowing comfortable living for arthropods and small mammals alike even when the surface temperature varies from below freezing to well above 45°C. A specific example for a 200 mm scorpion burrow is shown in Fig. 16.5. Only at dawn and dusk, or after brief periods of rain, are these animals to be seen active on the desert surface.

The particular microclimatic problems of immobile stages such as eggs and pupae have to be solved in other ways. Eggs are laid deep underground or within other organisms by many desert invertebrates. Pupal microclimates can only be regulated indirectly by choices made in the larval stage, and this again may involve locating a sheltered spot on a host plant, or for parasites a site within a much larger host animal, which will itself be maintaining reasonable homeostasis.

Desert vegetation inevitably has characteristic specializations. There is usually a "crust" of modified sand densely occupied by a community of microorganisms including fungi, lichen, and mosses, perhaps only 1 mm in depth and held together by mucilaginous secretions. Hardly any visible perennials exist, except at watercourses (see Plate 9b). Small plants that do survive are mostly evaders in time, almost invisible for most of their life as seeds or dried-out prickly husks, undergoing sudden synchronous bursts of growth, flowering, and seed production after occasional rains. Such plants are often termed ephemerals; they are technically "arido-passive", being inactive when dried. Some can complete their life cycle in only 14 days. There are also some arido-passive desert plants that endure by virtual cryptobiosis; these are the macroscopic lichens, often called cryptogams. Plants of the genus *Ramalina* from the Negev can survive near-total desiccation for at least a year, with photosynthesis shut right down, and ambient temperatures up to 65°C. They derive their water from dew in the hour or two after dawn, and can even take water from unsaturated air above about 80% relative humidity (RH).

On a larger and more obvious scale, "normal" plants occur only at oases. These include a variety of palms, a small range of annual flowers, and the few specialist perennial "arido-active" forms that keep photosynthesizing even in drought. Cacti are the classic examples, with 2000+ species in the Americas. They show loss of leaves, succulent stems, waxy surfaces, deep roots, sunken stomata, C_4 metabolism with reduced temperature sensitivity, and fearsome herbivore deterrents so that all the physiological and morphological specialization is not wasted by surfaces getting damaged. There are also small prickly trees such as the creosote bush, mesquite, and many eucalypts and wattles; most of these have enormous roots, penetrating to depths of 50 m or more and vastly in excess of the aerial biomass, and some of them will lose all their leaves and even drop whole branches "deliberately" when they are particularly severely desiccated.

Desert fauna

Desert faunas have very different taxonomic compositions from other terrestrial habitats, as Fig. 16.6 shows. The microfauna within the soil is, as always, dominated by nematodes and some groups of microarthropods, especially collembolans (springtails) and mites. But in the fauna living some or all of the time above ground the differences are striking. Most of the taxa with more permeable surfaces are underrepresented, with a very low species richness of earthworms, millipedes, isopods, and snails. However, there may be quite a high biomass made up from a few specialist representatives of some of these groups; for example, specialist desert woodlice are surprisingly common in some areas. There are moderate numbers of cockroaches, orthopterans such as locusts and crickets, and

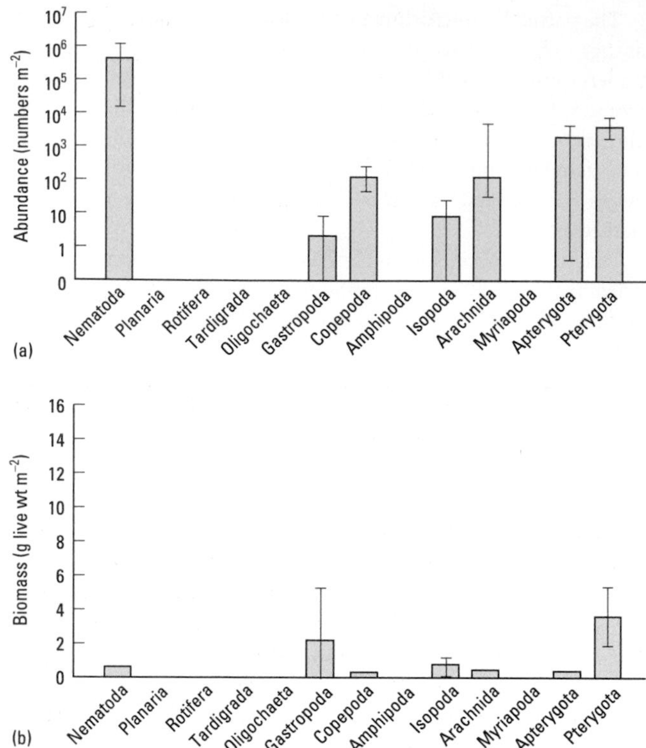

Fig. 16.6 Desert faunal composition (for invertebrates only) in terms of (a) abundance and (b) biomass. (Adapted from Little 1990; courtesy of Cambridge University Press.)

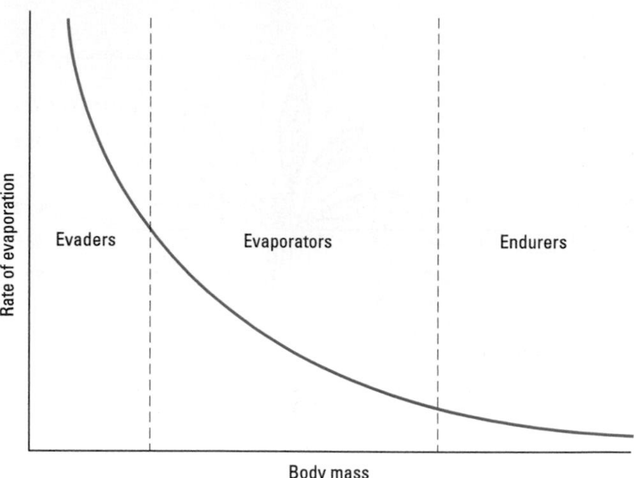

Fig. 16.7 Three main strategies for desert animals, in relation to size and rate of evaporation.

surprisingly high numbers of hemipteran bugs such as aphids, hoppers, and cicadas, small enough to live in the microclimates around desert plants. Relatively few lepidopterans occur, because the caterpillar stages are rather large and permeable plant feeders. However, taxa with grub-like juvenile stages do survive where the larvae are underground root feeders (e.g. beetle larvae) or are parasitic within other animals (dipterans and some wasps) or are laid in underground nests (most notably bees, ants, and termites). There may also be relatively high diversity and biomass for a few particularly adapted groups, notably the beetles, spiders, and scorpions from the invertebrate spectrum, and the reptiles and rodents from the vertebrate classes. In some deserts a single group may be particularly dominant. For example, one family of beetles, the Tenebrionidae, or "darkling beetles", may account for up to 15% of the species diversity in some deserts, and may be more than 50% of the total insect species diversity in the Namib desert, while in the western Sahara ants have been calculated to represent 75% of the entire faunal biomass, but with the sand cockroach *Heterogamia* being the single most abundant species.

Larger vertebrate animals also of course occur in deserts, and seem most obvious to casual visitors, although they are of low species diversity and low overall biomass. This category would include desert anurans (frogs and toads especially, but also tiger salamanders in American deserts), plus many lizards, snakes, and tortoises; and a range of endotherms, dominated by relatively huge numbers of rodents, moderate numbers of antelope, and some specialist birds.

Following the theme outlined in the first section of this chapter, Fig. 16.7 shows how these components of the fauna may be broadly split up into three strategic categories on the basis of their size. We will look at the strategies of each group in turn, although it should be borne in mind that many of the techniques used by the small evaders are also relevant in modified form to many of the larger desert animals.

16.2.2 Evaders and their strategies

All the small desert animals, of sheer necessity, come in this category. Four different "grades" can perhaps be identified:

1 Strict evaders active only at night, otherwise resting deep underground in humid, reasonably equable microclimates: the "softer" invertebrates like woodlice, apterygote insects, etc., and also nematodes, which may occur at 1 million m^{-2} even in shallow desert soils. Many of these groups use cryptobiosis (see Chapter 14) when conditions are particularly harsh.

2 Evaders, but rather less restricted: insects, spiders, and scorpions, commonly of small size and using daytime evasion with nocturnal/crepuscular behavior.

3 Small vertebrate ectothermic evaders: the desert amphibians and reptiles, having somewhat different problems.

4 Small vertebrate endotherm evaders, dominated by rodents.

Collectively, all these animals exhibit a wide range of adaptations, most of these being extensions of characteristics found in other terrestrial animals in less severe climates.

Burrowing

A fossorial (digging) lifestyle is characteristic of all members of the first group of animals described above, with woodlice, collembolans, mites, and nematodes living within the soil rather than in specific burrows. Many of these migrate through the sandy deposits on a regular cycle, moving up and down each day or through the seasons. The cockroach *Arenivaga* moves up to within 20–50 mm of the soil surface at night, but down to 200–600 mm depth in the day.

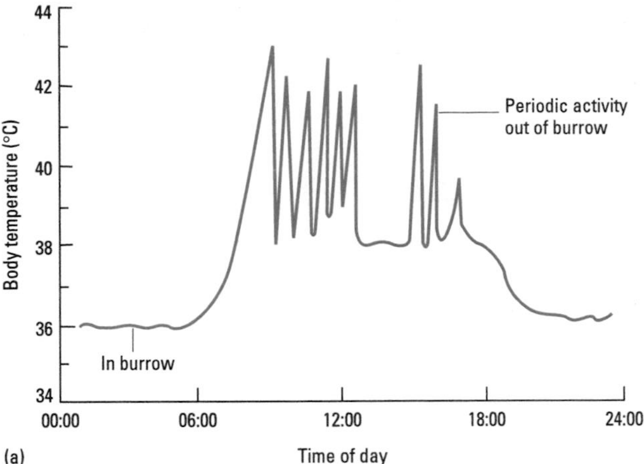

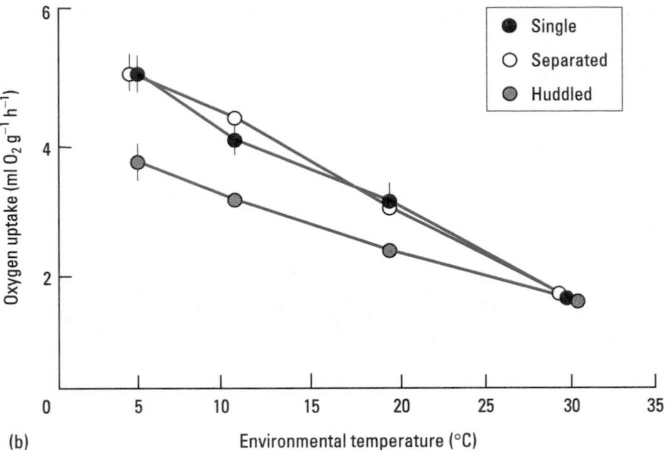

Fig. 16.8 Temperature and activity in small desert mammals. (a) In relation to the use of a burrow, for the antelope ground squirrel, where periods of absence from the burrow are brief and heat gained is then unloaded in the cool burrow depths. (b) In relation to the presence of conspecifics, in the Mongolian gerbil from relatively cool deserts, where huddling with two other gerbils reduces the need for increased metabolism as temperature falls. (a, Adapted from Bartholomew 1964; b, adapted from Contreras 1984.)

On a smaller scale many mites move from the shallow litter layer into the soil during the hotter parts of the day.

The use of properly constructed **burrows** combined with fossorial habits is exceptionally common in deserts, being exhibited by virtually all the larger invertebrate desert dwellers and by the ubiquitous rodents. Quite short burrows, even in extreme desert conditions, give reasonably **equable microclimates** (see Fig. 16.5), and burrows may be used by species as large as birds (burrowing seasonally) and even tortoises. Most of these animals show a degree of modification of the limbs to allow burrowing, from the modified forelimbs of scorpions to the stout digging front legs of desert tortoises. However, given the ubiquity of burrows, and the fact that constructing one has costs in itself, there are also those who "cheat" by using a hole created by another animal and are ill-equipped to burrow for themselves.

Burrows become an important component in the regulatory strategies of their owners. For example, desert spiders (which are normally representatives of the burrowing families or are nomadic hunters, not the web-spinners of more temperate habitats) use burrows to shuttle in and out of to keep their body temperature (T_b) reasonably constant. The Australian burrowing species *Geolycosa* emerges soon after sunrise and basks in the sun to raise its T_b to 40°C, then basks and burrows in turn to maintain it at this level through much of the day. Desert tortoises in North America use both shallow summer burrows to avoid overheating, and deeper winter dens (often enormous and reused by succeeding generations over decades or centuries) to avoid winter freezing, spending the majority of their life underground. To give a mammalian example, the air within deer mice burrows in Nevada is usually close to 26°C at about 1 m depth throughout the 24 h cycle, while the external temperatures vary between 16 and 44°C. Small rodents such as the antelope ground squirrel (Fig. 16.8a) can thus be briefly active even in the hottest periods of the day, using their cool burrow to unload the heat they pick up on quick forays, so that their T_b never goes above 42–44°C. Most of the nocturnal desert rodents have high and narrow thermoneutral zones, with burrow temperatures always in the range 25–35°C. Indeed, some small desert mammals and birds are gregarious in burrows at night, the clustering habit helping to reduce each individual's metabolic rate (e.g. by 11–22% in mousebirds) while keeping the T_b elevated to normal levels; Fig. 16.8b shows the effect in gerbils.

Burrows may also give high **humidity** that aids osmotic regulatory strategies. For naked mole rats the burrow 700 mm below ground provides not only a constant 26–28°C temperature but also a constant high humidity that helps minimize evaporative water loss (EWL). The spadefoot toads (*Scaphiopus*) only survive in deserts because of an interaction of their permeable skins and their burrowing habit. They live in deep burrows, constructed at the end of the rainy season in moist soil, and initially further water enters the body osmotically from the soil. As the soil dries the gradient is reversed, and the toad then stops secreting any urine and switches to urea retention (see Control and tolerance of water loss, below), raising the concentration of its body fluids to reduce this gradient. The toads can remain buried for 9–10 months without access to further liquid water, and may emerge at the end of the dry season with the blood still only at 600 mOsm.

Burrows of course also have advantages beyond the creation of a favorable microclimate. They allow **food storage** (especially for social and gregarious animals), and provide protection from predators and parasites. Use of a burrow also allows a sit-and-wait type of foraging strategy, for example in the mygalomorph spiders, and in insects such as larval ant-lions and tiger beetles, where passing prey fall into or are grabbed from the opening of a small pit in the sandy soil. On a larger scale, sidewinder adders and golden moles also conceal themselves in sand burrows and ambush passing prey.

However, burrows also have their problems. There may be an accumulation of ammonia, less from adult excretion (normally restricted to outside the burrow) than from urination and defecation from juveniles. Furthermore, the restricted entrances of many burrows tend to limit aeration, and when the owner is resident they become zones of reduced Po_2. This is especially important for small mammals and birds with high endothermic metabolic rates; the Po_2

values in the nests of small desert rodents are often only 10–15 kPa, instead of the normal 21 kPa, and this may link to cases of rather low basal metabolic rates (BMRs) in desert rodents discussed below. (Note, however, where animals burrow continuously through sand, rather than constructing a fixed burrow, there is little respiratory problem; the Po_2 near the nose of a sand-swimming desert mole is very similar to that in the free atmosphere, the sand being dry and porous.) We also noted in Chapter 7 that mammals and birds with burrowing habits tend to have reduced sensitivity to CO_2, allowing them to overcome the related problem of the burrow air having elevated Pco_2 (up to 6 kPa). Some burrowing rodents even excrete much of their CO_2 as bicarbonate, reducing the build-up of the gas in the burrow.

High mobility and navigational ability

For a surface-active animal in a hot desert, it is particularly useful to be able to locomote at high speeds in short and very effectively directed bursts, allowing darting movements between patches of shade or shelter, and rapid retreat when physiological tolerances are pushed close to their limit. Species of *Onymacris* and *Stenocara* (both tenebrionid beetles) utilize rapid sun–shade shuttling as the primary means of maintaining T_b. *Onymacris plana* runs very rapidly between shady spots with the body raised about 15 mm above the substratum, "stilting" on its long legs. A comparative phylogenetic analysis has confirmed that desert species in this genus do indeed have longer legs than coastal species.

This point emphasizes that there may also be a range of special **locomotory tricks** to move through or over hot sand while minimizing contact of the substrate with the bulk of the body, to reduce conductive heat gain (Fig. 16.9). The sand-swimming moles of the Namib desert hardly achieve the latter, but do move more rapidly through the hottest sand. Lizards run on tiptoe, and when briefly stationary will raise alternate feet high off the ground in turn to avoid superficial burning. Sidewinder snakes slide rapidly sideways across dune slopes, leaving only an intermittent track as parts of the trunk are always off the substratum. Even more curious is the cartwheeling strategy of a dune spider, adopted during rapid descent of dunes (Fig. 16.9c).

Rapid and darting modes of locomotion are also associated with active foraging, often from a central place such as a nest or burrow. In rodents this commonly involves short, darting forays, using the characteristic hopping gait of the small gerbils and kangaroo rats. In social insects such as ants and bees it may involve predetermined trails, again helping to reduce time on the surface seeking food. The preferred trails lead quickly to known sources of seeds, pollen, and nectar, or to prey such as the nests of other ants or termites. These animals often become highly **territorial** and defend their trails and food sources. Desert ants are also noted for particularly well-developed detection of the plane of polarization light, so they can **navigate** in an almost featureless environment.

Rhythmic activity patterns

Foraging is rarely continuous in desert animals, but commonly shows a **diel pattern** with the number of active individuals peaking either at night or at dawn and/or dusk. Scorpions, centipedes, and spiders from deserts are far more likely to be nocturnal than are their temperate counterparts, foraging between particular maximum and minimum temperatures. Figure 16.10 shows an example of foraging ants, with the amount of time spent foraging on the surface linked to surface temperature, and the respite length between foraging bouts concomitantly increasing as the temperature rises. Like many desert species, this ant has an upper tolerable limit of sand surface temperature in the range 47–50°C, and the tight relation between activity and sand surface temperature results in a sharply

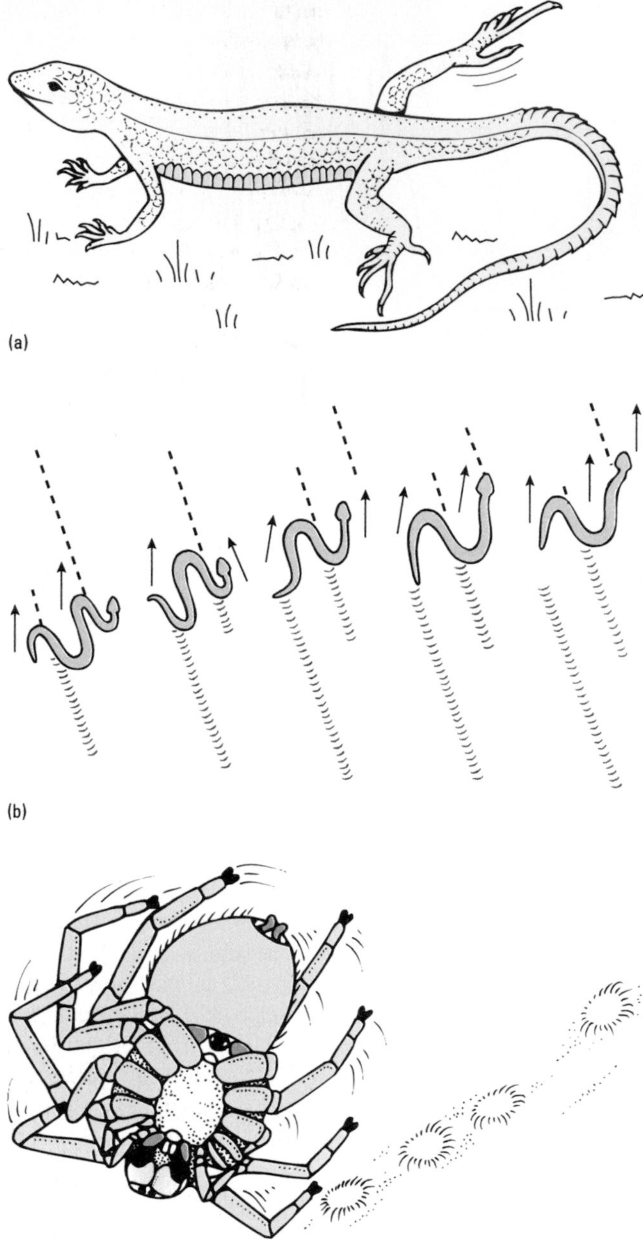

Fig. 16.9 Methods of locomotion on hot sand. (a) Tiptoe stance in lizards (also shown in long-legged insects). (b) Sidewinding snakes, allowing reduced contact with sand. (c) Cartwheeling down dunes in the spider *Carpachne*. (b, Adapted from Pough *et al.* 1999; c, adapted from Henschel 1990.)

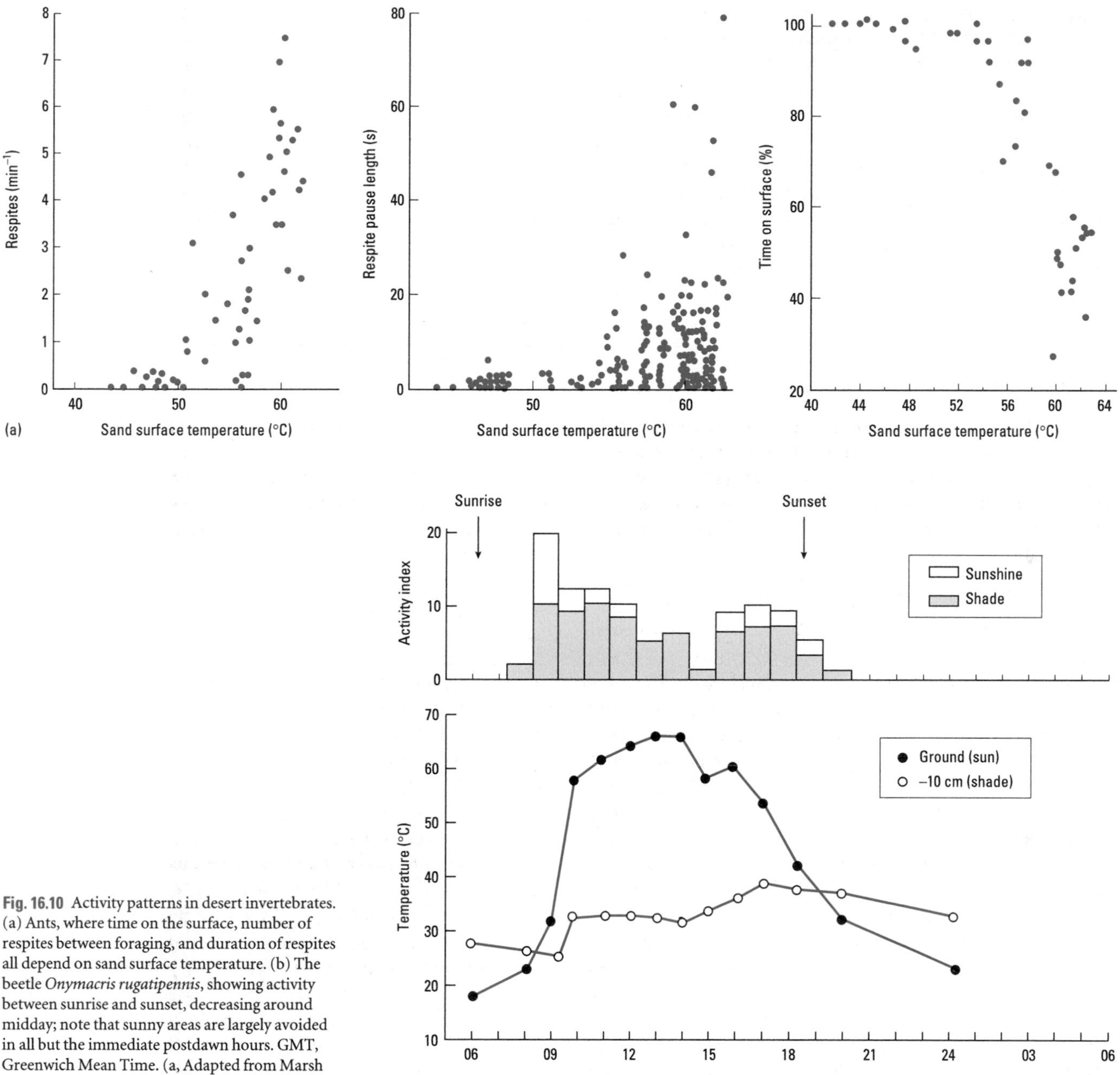

Fig. 16.10 Activity patterns in desert invertebrates. (a) Ants, where time on the surface, number of respites between foraging, and duration of respites all depend on sand surface temperature. (b) The beetle *Onymacris rugatipennis*, showing activity between sunrise and sunset, decreasing around midday; note that sunny areas are largely avoided in all but the immediate postdawn hours. GMT, Greenwich Mean Time. (a, Adapted from Marsh 1985; b, from Holm & Edney 1973.)

defined diurnal activity pattern. Other insects may have much lower limits; desert bees may avoid ambient temperatures above about 36°C since their body temperature in flight cannot be greater than around 45°C, while desert caterpillars cease activity above about 34°C and retreat to the shady undersides of leaves or stems.

Many of these desert invertebrates have **endogenous rhythms** that persist under constant laboratory conditions and serve to get the animals away from the harshest surface conditions automatically. Some evidence suggests that the endogenous rhythms of desert animals are particularly likely to be entrained by direct temperature or humidity cues, rather than by photoperiod as is commoner in most aquatic and many terrestrial animals. Perhaps this is because in deserts the link between seasonality and productivity is often broken, with spurts of productivity linked instead to sporadic rainfall events. Certainly, endogenous rhythms can often be overridden when sudden rainfall produces a period of cool damp surfaces and a subsequent flush of green vegetation, resulting in an abundance of flowers, seeds, and ephemeral invertebrate prey items. In some tenebrionid beetles, though, the circadian rhythm that promotes narrow windows of activity at dawn and dusk seems to have no reliance on external triggers, as the inactive beetle is underground in the dark at constant temperature and humidity. At another extreme, daily rhythms may be almost entirely suppressed, as in the desert snail *Sphincterochila*, which is active on only a few days each year when the rains have wetted the ground. It may be that endogenous biological clocks have been lost in some desert species,

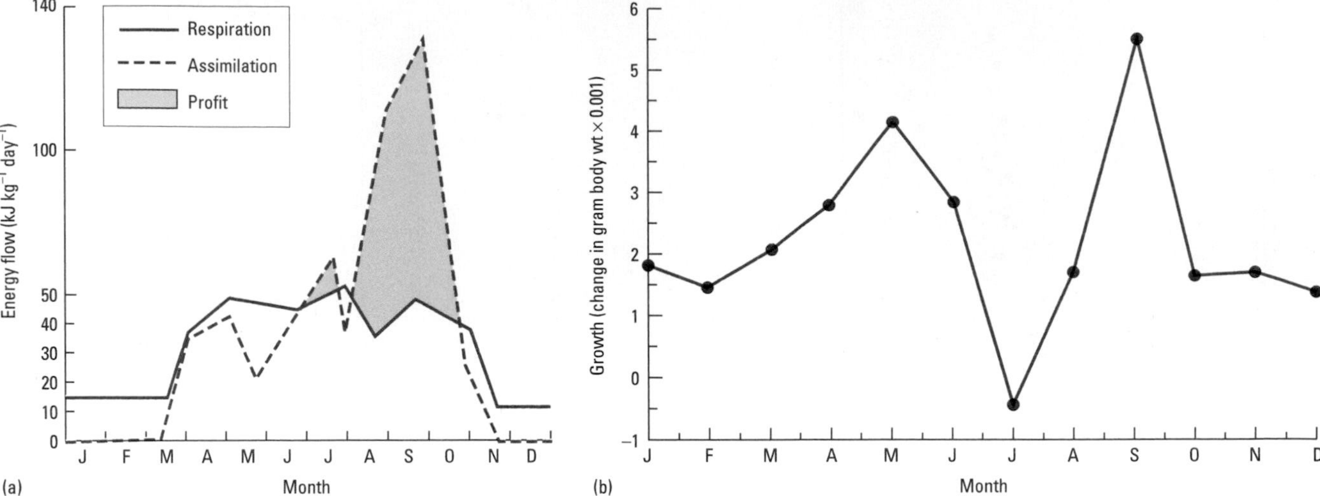

Fig. 16.11 Seasonal energy budgets: (a) for a desert tortoise, with uneven and relatively limited periods of profit when moist vegetation is available, and (b) for a desert scorpion. (a, Adapted from Nagy & Medica 1986; b, adapted from Polis 1988.)

to allow them this more opportunistic approach to organizing their activity.

Where there is a mutualistic relation between food source and feeder, as with pollinating insects such as bees, the plants with which they associate usually also show appropriate diel patterns. Many desert shrubs, such as the caper (*Capparis*), open their fragile flowers at dusk and they last only a few hours, during which small nocturnal carpenter bees visit and transfer pollen between individuals.

In some deserts there are also marked seasonal patterns of activity; this depends on the geographic location of each desert, since some experience marked summers and winters and others are nearly aseasonal. **Seasonality** may promote endogenous rhythms entrained by photoperiod. The deserts of Nevada are seasonal with plant growth concentrated in spring and early summer; here desert tortoises show marked seasonal behavior patterns (Fig. 16.11a). They hibernate in winter, and in spring are active for only a few hours on some days. At this time the vegetation has too high a water content to give the tortoise a net energy gain; body weight does increase but this is due to water stored in the bladder, and body solids actually decline. In the summer, when the vegetation has dried out somewhat, the tortoises achieve positive energy balance, but incur water loss; there is a depletion of the stores in the bladder coupled with an increase in plasma and urine concentration. In high summer the animals estivate, emerging briefly only after rainstorms in July; they stay mainly underground until the August/September rains rehydrate the vegetation and give a period of water storage and energy gain prior to hibernation in November. Thus these animals have a seasonal cycle where there is rarely genuine homeostasis; energy and water budgets are balanced over a whole year, but are highly varied on a daily or weekly basis.

Seasonal patterns of behavior relating to vegetation quality may also result in complex seasonal patterns of growth rate in invertebrate predators, as shown for a desert scorpion in Fig. 16.11b. Clearly, it is always important to consider the long-term productivity of the whole ecosystem in relation to climate; short-term studies will not do.

Raised thermal tolerance, lowered metabolic rate, and patterns of Q_{10}

Some desert animals are diurnal, and these must inevitably have high thermal tolerances, i.e. **raised upper critical temperature** (UCT). Desert arthropods commonly have upper limits of 45–47°C. Desert ants, as we have seen, can often tolerate surface temperatures in excess of 50°C for short periods. One species, *Ocymyrmex barbiger*, has a UCT measured at 51.5°C. Possibly even more extreme is an Australian species of *Melophorus*, which can survive for an hour with a T_b of 54°C; it does not become active until surface temperatures rise to 56°C, and has been recorded on sand in excess of 70°C for very brief periods, with no cessation of colony activity even at midday, albeit with long respite periods for individuals amongst cooler vegetation. Some of these animals are clearly on a "thermal tightrope" in pursuing their activities in a desert, and some do die during foraging treks. Probably such risky tactics are only possible in eusocial animals, where the nest will persist despite some individual losses. There is a trade-off between heat tolerance and "risk-prone" foraging in Mediterranean ant species, with the more tolerant species foraging "riskily" closer to their UCT and achieving high foraging efficiency, while less tolerant species forage in the evening and night for lower rates of return.

Such species may exhibit thickened cuticles and contain lipids with relatively high melting points, but many of them must also have unusually tolerant enzymes and membranes, as yet little studied but assumed to incorporate the kinds of biochemical changes discussed in Chapter 8. There is good comparative evidence that desert species of beetle show a particularly good match between their preferred T_b and their achieved field temperatures, which should allow the fine-tuning of membrane and enzyme properties.

It has often been claimed that a **reduced metabolic rate** is a possible mode of energy conservation for an ectothermic desert animal, either allowing permanently lowered energy expenditure or a tem-

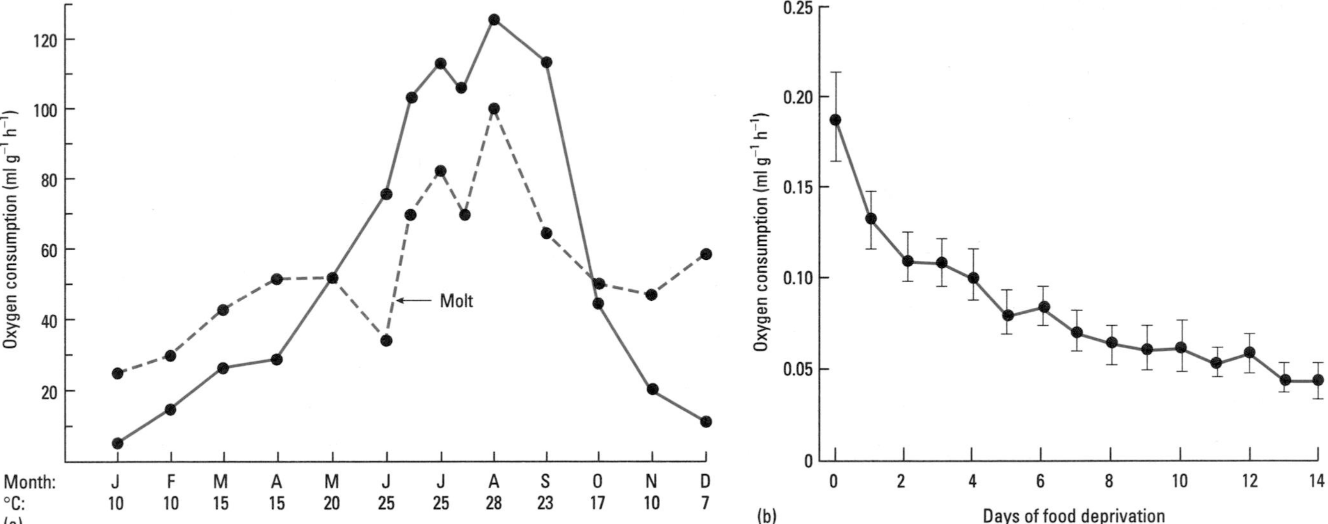

Fig. 16.12 (a) Seasonal differences in metabolic rate for two individuals in the desert millipede *Orthoporus*, measured at burrow temperature; decreases occur during winter food shortage and at the molt. (b) The direct effect of food shortage on metabolic rate in the ant *Camponotus*. (a, From Wooten & Crawford 1975; b, from Lighton 1989.)

porarily lowered rate during food scarcity, and thus also reducing water losses. The same could be true for small desert endotherms. Some taxa that are pre-eminently desert-adapted, such as scorpions and sun-spiders, have notoriously low metabolic rates. Desert orthopterans, ants, and tenebrionid beetles also seem to show a general lowering of metabolic rate compared with mesic or hygric relatives; among the tenebrionids, desert species of *Onymacris* show particularly low metabolic rates. Desert lizards (*Uromastyx*) have a lower metabolic rate than predicted, as do desert golden moles (only about one-fifth that predicted for their size), probably in both cases linked to low (and poor-quality) food availability. Desert rodents generally have lower resting metabolic rates than mesic species, with indications of somewhat higher rates of nonshivering thermogenesis to compensate for this when necessary. Some small ungulates (e.g. steenbok) and some birds (e.g. pin-tailed sandgrouse) are also reported to have unusually low resting metabolic rates. Perhaps most convincingly, intraspecific effects are well documented. For example, in the brown hare, populations from the Israeli Negev desert have resting metabolic rate values just 61% of those recorded in populations from France. Thus while the data overall are somewhat limited and have not been subject to careful phylogenetic analysis, they do strongly suggest that a general lowering of metabolic rate is an adaptive feature. (Nevertheless in some species an opposite effect has been documented; some desert grasshoppers, such as *Taeniopoda*, appear to have unusually high metabolic rates, perhaps compensating for the exceptionally short growing season they experience.)

The ability of desert animals to withstand starvation for months (or even years in a few cases) is also very marked and is accompanied by a further reduction of metabolic rate to an extent greater than is possible in mesic congeneric species. Many desert invertebrates that are strongly reliant on plant productivity show a very marked seasonal reduction of metabolic rate when food supplies are low (Fig. 16.12a), with a rapid onset of lowered metabolic rate when starved of food artificially (Fig. 16.12b). Of course the same effect is often manifest in desert endotherms, where in some cases exceptional reduction in metabolic rate is achieved during summer dormancy or estivation (see below).

Many desert invertebrates also show a particular **pattern of temperature quotients** (Q_{10}) as indicated in Fig. 16.13 for oxygen consumption (metabolic rate) in an ant. The Q_{10} is relatively high at low temperatures, but depressed (i.e. below the expectation of a straight-line fit) in the normal environmental temperatures, rising again at very high temperatures. Remember that a higher Q_{10} means a more substantial response to temperature, so that when ambient temperatures are low and the animals are inactive their metabolic rate drops quite substantially with even a small further decrease in T_a. This allows reduced energy expenditure in inactive periods, such as in the desert night. During normal foraging, in contrast, the Q_{10} is relatively low and there is a more limited response to changing ambient temperatures so that the animals avoid undue "thermal acceleration". This kind of pattern has been shown in various ants, orthopterans, and beetles, as well as in desert spiders and scorpions; in the active range the values of Q_{10} are commonly only 1.5–2.2.

Color, shape, and posture

Invertebrates encountered in deserts are commonly almost black, almost white, or of a nondescript sandy brown color. Reasons for all three color schemes are not hard to find. Black is characteristic of some eocrepuscular species, since dark surfaces allow the body to warm up more quickly in the limited daylight at these otherwise rather cold times of day. White surfaces of very high reflectivity are found in some species active or exposed in the heat of the day, and can be shown to be effective. For example, the beetle *Onymacris brincki*, with white elytra having a reflectance of 35%, has an abdominal temperature of only 37°C in full sunlight when the ground temperature is 47°C, and when the abdomen of a similar beetle with dark elytra is at 43–45°C (Fig. 16.14a). The estivating snail *Sphincterochila*, which has one of the most highly reflective of

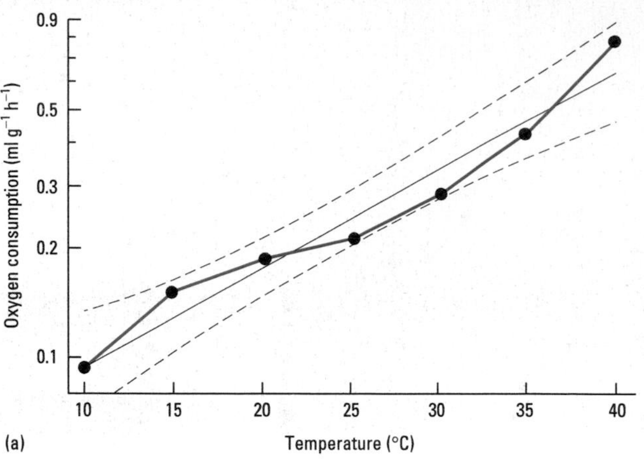

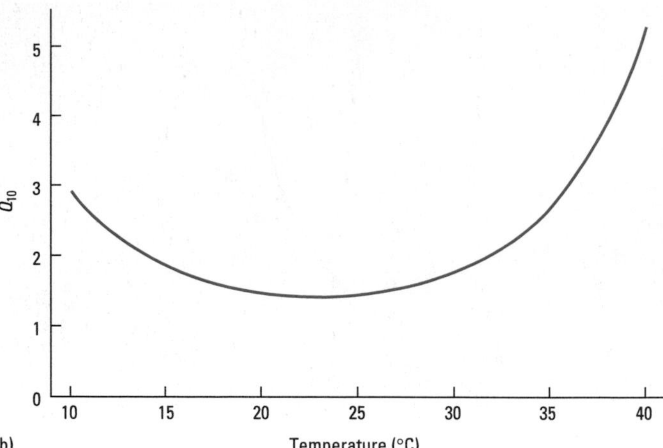

Fig. 16.13 Q_{10} adaptation for a desert ant (*Camponotus fulvopilosus*). (a) Declining temperature produces declining oxygen consumption, but the regression line obscures a consistent pattern of relative increases at low temperature and relative reductions close to the normal environmental temperatures (30–35°C). The Q_{10} vs. temperature plot is therefore shown as in (b). (From Lighton 1989.)

all biological surfaces (95%), also demonstrates the efficacy of white surfaces (Fig. 16.14b). However, it is equally true that many insect species active in daytime in deserts are black, which appears to rather spoil the argument. In fact for species with short legs that live very close to the ground color may be relatively unimportant in thermal terms, since most of their heat gain will come from long-wave radiation reflected off the soil, and absorption of this is unaffected by color. It may be no accident that the more spectacularly long-legged insect species, gaining heat more from direct shortwave radiation, are more likely to be "adaptively" white or black, whereas

Fig. 16.14 (a) Color effects in desert beetles, where a species with black elytra has substantially higher temperatures than a similar species with white elytra. (b) The effectiveness of the highly reflective white shell of the snail *Sphincterochila* in reducing heat stress for the soft tissues within. (a, Adapted from Henwood 1975; b, from Schmidt-Nielsen *et al.* 1971, courtesy of Company of Biologists Ltd.)

creatures such as scorpions that are flattened closer to the sand are usually drab-brown or dull-black (even those species active by day).

Sandy brown surfaces may also be the best compromise where the animal is essentially burrowing, emerging only briefly and perhaps subject to intense predation pressure, so that the need for crypsis dominates over thermal needs; this is particularly true in the smaller invertebrates. Similar pale, sandy colors are found in most small diurnal mammals in hotter climates (though often with counter-shading for additional camouflage purposes). Many desert rodents have sandy colored fur, although in some this is unusually sparse (e.g. in the desert squirrel *Spermophilus*, where fur is sparse everywhere but the tail, which may be used as a shading parasol).

Color change is also reasonably common in desert animals. Quite a number of desert arthropods show a special cuticular feature in having substantial "wax blooms" on their surface, especially at low humidity. These may be formed from a meshwork of extruded wax filaments (Fig. 16.15), which can be rapidly regenerated if damaged. In some tenebrionid and buprestid beetles they affect the color of the animal, with *Onymacris rugatipennis* being whitish-blue when the bloom is present but black when it is removed. The amount of wax may vary with the season, being greatest in the hottest months, and it probably helps to protect the animals from solar radiation,

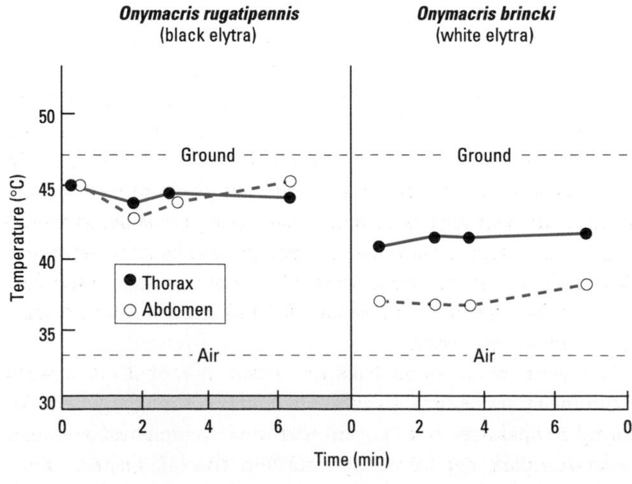

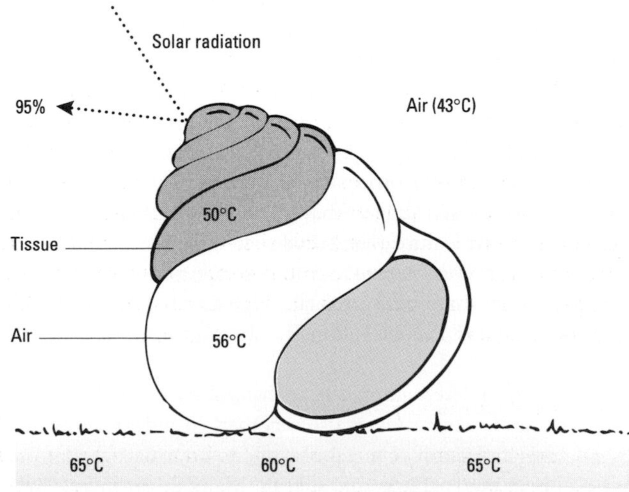

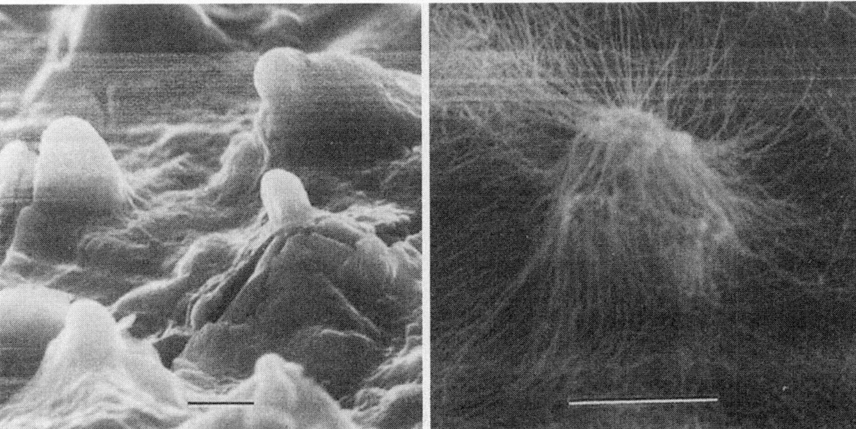

Fig. 16.15 Wax blooms extruded on the surface of desert beetles, increasing reflectance and reducing water loss. (From Louw 1993; photographs courtesy of N.F. Hadley.)

having a high reflectivity. Certain desert grasshoppers also change color, from bright blue-green when warm to almost black when chilled; some desert agamid lizards may use the same trick, being dull dark brown at dawn and dusk and a bright turquoise around midday. A reed-frog, *Hyperolius viridiflavus*, from very dry African savanna regions, has been shown to change color dramatically in the dry season as it begins estivation. Initially the skin turns from yellow-brown to a brilliant white; but as dehydration sets in it turns from white through coppery iridescence and finally to green iridescence when more than 25% of the body mass has been lost. This change is brought about by a layer of iridiophores in the skin, with crystals causing multilayer interference reflection and hence iridescence; purine deposition in the skin accompanies this process. In dry-adapted skin there are 4–6 layers of iridiophores in a layer up to 60 μm thick and with a reflectance of around 65%.

Changes to the fur are also recorded in endotherm evaders, with summer coats usually paler than winter ones. In the Sonoran desert rock squirrels the fur is renewed seasonally and the animals absorb 33–71% less solar heat load in summer as a result (though there is no obvious visible change, suggesting rather subtle changes in structural and optical properties of the hairs).

Color is highly interactive with shape and with postural adjustment, to get maximum benefit from particular pigmented surfaces and from areas of large surface area. The reed-frog *Hyperolius*, mentioned above, shows "body shape optimization", with a half-cylinder shape rather than the more common half-sphere, allowing a "head-on" posture where a large area for conductive and convective heat loss can be shaded by a small area exposed to the direct solar radiation. This frog can survive 3 months of estivation without water at temperatures of up to 45°C. Heat gain in many desert animals is avoided in similar fashion, usually involving orientation into the sun's rays; the desert beetle shown in Fig. 16.16 illustrates this point. Ideally, and whatever the taxon, the shape should be elongate rather than squat, with the front or rear, and the midline dorsum, as pale as possible.

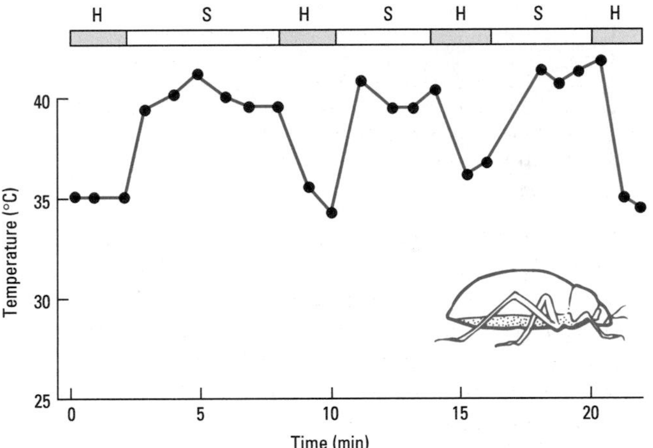

Fig. 16.16 Orientation effects for a desert beetle; thoracic temperatures are reduced when head-on to the sun (H) and rise in the side-on (S) position. (Adapted from Edney 1971.)

Control and tolerance of water loss

Many desert animals are exceptionally **impermeable**, and an overview is given in Table 16.1 (see also Table 5.3 and Table 15.7). The lowest values of water loss rate have been recorded for certain scorpions and ants, at 0.02 mg cm^{-1} h^{-1} (around 0.03–0.04% fresh weight per hour), with certain tenebrionid beetles (e.g. *Onymacris plana*) only slightly higher at around 0.1 mg cm^{-1} h^{-1}. This impermeability is largely due to the normal cuticular layers described in Chapter 15, which may be enhanced with more epicuticular lipids. However, there are also special cases of "extra" impermeability; the wax blooms in tenebrionids, described above, certainly help to reduce water loss as well as limiting thermal gain, so that in field-caught *Cauricara* beetles the water loss rate is about 40% higher when the wax bloom is at its minimal thickness in August. In *Onymacris* the occurrence of wax blooms is significantly commoner in desert interior species than in coastal species. For desert insects in general, cuticular water loss (cutaneous evaporative water loss, CEWL) has been so much reduced that respiratory water loss (respiratory evaporative water loss, REWL) is the dominant component of water balance.

However, while undoubtedly important, relationships between cuticular properties and water loss rates can be oversimplified. For example, the desert fruit fly, *Drosophila mojavensis*, does have more cuticular lipid, and longer hydrocarbon chain lengths in these lipids, than more mesic fruit flies; but since it is also smaller when it

Table 16.1 Water loss rates and permeabilities in desert animals.

Animal group/genus	Resistance (s cm^{-2})	Water flux (mg cm^{-2} h^{-1})	Transpiration rate (μg cm^{-2} h^{-1} mmHg^{-1})	Water turnover (ml g^{-1} day^{-1})
Arthropods				
Isopods				
Hemilepistus (woodlouse)			14–23	
Venezillo (woodlouse)			15–32	
Arachnids				
Scorpio (scorpion)	1317			
Androctonus (scorpion)	4167		0.6	
Hadrurus (scorpion)		0.02	1.2	
Buthus (scorpion)			1	
Leiurus (scorpion)			0.8	
Latrodectus (spider)			1.2	
Ornithodorus (mite)			2–4	
Myriapods				
Orthoporus (millipede)	433		8	
Alloporus (millipede)		0.02		
Insects				
Thermobia (apterygote)			15	
Ctenolepisma (apterygote)			0.7	
Rhodnius (bug)			12	
Geocoris (bug)			14	
Arenivaga (cockroach)			12–80	
Diceroprocta (cicada)			100	
Locusta (locust)		0.70		
Onymacris (beetle)	5030	0.10	3.1	
Centrioptera (beetle)			6.3	
Cryptoglossa (beetle)			8.4	
Lepidochora (beetle)			3.1	
Tenebrio (beetle larva)			5	
Tenebrio (beetle pupa)			1	
Eleodes (beetle)		0.20	17	
Glossina (tsetse fly)			8	
Glossina (pupa)			0.3	
Manduca (caterpillar)			40	
Anaphes (wasp)			4–8	
Pogonomyrmex (ant)			26	
Messor (ant)			18	
Vertebrates				
Pternohyla (frog, cocooned)	457			
Gopherus (tortoise)	120	1.56		0.003
Sauromalus (lizard)	1360			
Dipsosaurus (lizard)		0.13		0.03
Gehydra (lizard)		0.22		
Pitnophis (snake)		0.23		
Struthio (ostrich)	158	1.73		0.09
Dipodomys (kangaroo rat)		0.53		0.03
Gerbillus (gerbil)				0.13
Peromyscus (cactus mouse)		0.66		
Capra (desert goat)		2.10		0.09
Oryx		3.2–5.9		0.03
Megaleia (red kangaroo)				0.09
Camelus (dromedary)				0.06

develops at high ambient temperatures, its water loss rate per unit body mass is in fact higher than predicted, and it has to rely on avoidance of really desiccating conditions. Routinely higher values of cuticular water loss occur in many species from all taxa that are more inclined to seek protected microclimates or moister foods, giving a good inverse correlation between moisture content of the microhabitat and/or diet and the desiccation resistance of a species. For example, the nocturnal honeypot ant, *Myrmecocystus mexicanus*, is more vulnerable to water stress than its diurnal relative *M. mendax*. There is also the complication of evaporative water loss ("sweating") in certain insects as a strategy to avoid lethal overheating. This occurs in some desert cicadas and grasshoppers through cuticular pores, since they have ready access to water in their food (see below). As always, then, it is important not to overgeneralize, since there are a range of possible adaptive strategies within and between species, with different balances between physiological and behavioral solutions to problems.

Some desert animals may also be very tolerant of **water loss** (see

Table 5.2). Insects from arid habitats may tolerate over 50% water loss, with 75% recorded for some tenebrionids. In such cases dehydration is largely at the expense of the hemolymph, which may be reduced to less than 5% of its normal volume in a very dehydrated beetle. Amino acids contribute to the regulation of osmolarity during dehydration (as in estuarine animals; see Chapter 12), with glycerol also playing a role as an osmotic effector. The snail *Sphincterochila* survives 50% loss, and values up to 48% have been recorded in desert toads such as *Scaphiopus*, where again loss of blood volume occurs. Dehydration inevitably accompanies estivation in many animals, and with no urination occurring there may be a substantial accumulation of the excretory product. For some estivating Australian frogs, urea levels rise to 100–300 mmol l^{-1}. However, this does not appear to be accompanied by substantial co-accumulation of balancing osmolytes such as methylamines (e.g. trimethylamine oxide, TMAO) or polyols (e.g. sorbitol) as in other urea accumulators, like elasmobranch fish (see Chapter 11). It may be that the amphibian enzymes are insensitive to perturbation by urea, or even that the perturbing effects are used as an aid to metabolic depression while estivating. Urea retention linked to having a bladder as a water/urea reservoir was perhaps a "preadaptation" to becoming terrestrial in amphibians.

Tolerance of water loss in endothermic birds and mammals is a rather different phenomenon, in that the blood is necessarily a much more highly controlled fluid. Many of the small desert mammals and birds are characterized by a particular ability to maintain an almost constant plasma volume even when losing substantial body mass and total body water, with the water being lost from other compartments, including the cells, where concomitant adjustments of amino acid and other osmotic effector levels must be involved (see Chapter 5).

Water loss may also produce **salt balance** problems, and many desert-dwelling reptiles use nasal salt glands to offload excess salts, producing a potassium-rich fluid if they are herbivorous but a sodium-rich fluid if the diet is animal flesh. However, concentration of the excretory product into very **hyperosmotic urine**, with minimal fluid excretion, is the more obvious desert adaptation to lose excess salt, and may be allied to water-storage systems. Many desert tenebrionid beetles exhibit the cryptonephridial system for concentrating their urine and feces (see Fig. 15.15) and therefore also potentially gain a method of **taking up water vapor**. Desert woodlice (*Hemilepistus*) can resorb extra water from fluid passing up into the rectum after flowing over the pleopods where ammonia is lost. Desert toads use the bladder as a massive water reserve during drought. In contrast, some desert tortoises lack a water-storage bladder (always found in their temperate relatives) and instead have powerful nasal salt glands and cloacal resorptive tissues to regulate their salt and water balance. Desert rodents are famous for having long loops of Henle, in proportion to their size; as we saw in Chapter 5, this is linked to concentrating their urine such that the cells lining the loop are particularly well endowed with mitochondria to support high salt transport rates. The same is true of desert birds, where the resorptive rectal epithelium is also strongly developed. Desert populations of the brown hare form urine up to 4470 mOsm, compared to 2500 mOsm in European populations.

Changing the excretory product also helps; desert toads have switched to excreting urea rather than ammonia, while many other small desert vertebrates (reptiles and birds) excrete most of their nitrogen as **uric acid**. This latter product is ubiquitous among the most successful desert invertebrates, and is commonly excreted as an almost solid paste or crystals of uric acid and urates, with virtually no accompanying water.

The net effect of all these water-retention adaptations can be quite remarkable, and has been best studied in the small desert rodents such as the kangaroo rat (*Dipodomys*) and the hopping mouse (*Notomys*). These animals feed on very dry seeds, and in 1 month may gain less than 1 ml of water from that contained directly in their food; by far the greater gain comes indirectly from the food as metabolic water, perhaps 50 ml. In the same period they lose the same total water, in the ratio of roughly 75% by evaporation, 5% in the feces, and 20% in the urine (at 5000–9000 mOsm). Plasma volume is maintained almost constant under all conditions, and no drinking water is needed. However, the strategies involved are labile even within a species, and kangaroo rats from more xeric areas have significantly lower water loss rates than their mesic conspecifics, with their REWL contributing only 40–50% of total water lost.

Note that water conservation systems cannot be divorced from other problems arising in deserts, notably thermal stress, with similar behavioral strategies alleviating both problems. For example, relatively permeable evaders, such as desert toads, will choose cooler habitats when they are particularly water stressed, and may show a greater propensity to burrow and to huddle together.

Water uptake and use of condensation

In deserts, the condensation at dawn as the overnight temperature profiles invert can be very substantial, but the dew that forms would normally dissipate into the sand and evaporate very quickly. However, in certain deserts, such as the Namib, dense morning fogs may occur where the water vapor from adjacent cool oceans rolls in over the sands, and condensation may then persist long enough to be really important to the fauna. Some desert tenebrionid beetles have ways to exploit this. *Onymacris unguicularis* ascends to the tops of dunes during dense dawn fogs, and stands with its hind legs fully stretched up and its head down ("fog basking", perhaps enhanced by complex alternating hydrophobic and hydrophilic architecture on the cuticle surface). Water condenses on the body and runs down towards the head where it is drunk, giving up to 34% increase in body weight. *O. plana* selects dunes facing in different directions at different times of day on which to bask (east-facing at dawn, west-facing at dusk, retreating to its burrow at midday). Yet other beetles (genus *Lepidochora*) dig short trenches in the sand to gather the condensing dew. It is not only beetles that benefit from desert fogs; water will condense on any surface in these conditions, and many animals lick it off the vegetation. Some geckos and snakes also lick it from their own bodies, and spiders gather it from the hygroscopic silk of their webs.

Water acquisition from sources other than free liquid is the only other option. We have seen that some desert amphibians extract water from soil through their permeable skin when there is a brief rainy period. Some desert arachnids may achieve the same trick at such times, using their suctorial apparatus (see Fig. 15.17). But this is a rare adaptation in deserts, as soils are normally far too dry for any osmotic or capillary uptake to be possible.

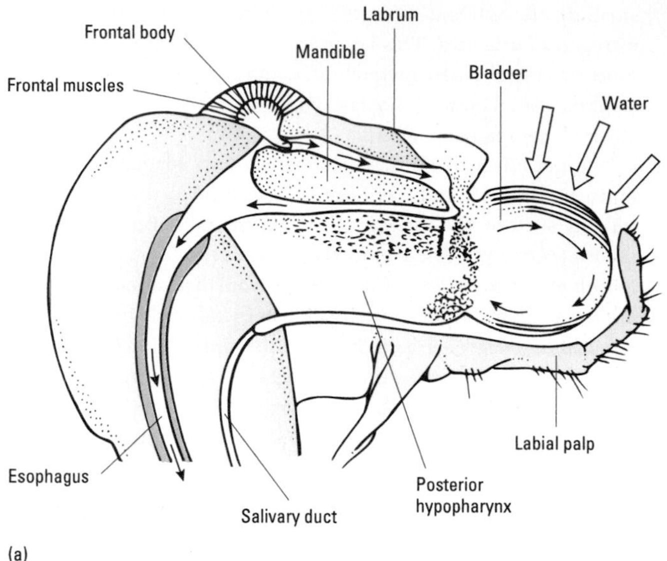

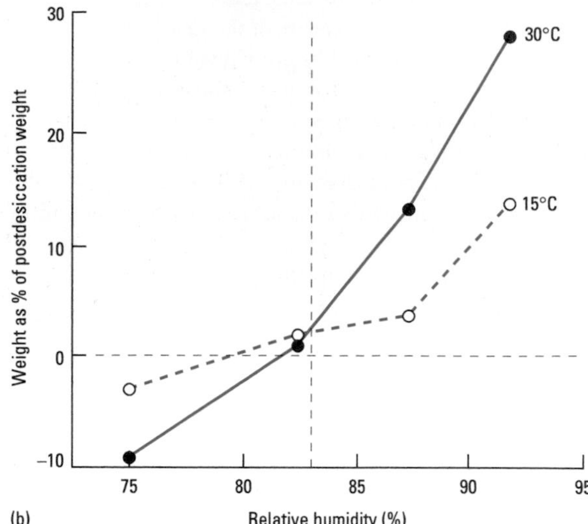

Fig. 16.17 The water uptake mechanism in the desert cockroach *Arenivaga*. (a) Fluid is produced in the frontal bodies and flows onto the extruded bladder surface. (b) Water vapor is absorbed onto this surface at humidities above 83%, and is then drawn back into the esophagus. (a, Adapted from O'Donnell 1987; b, adapted from Edney 1966.)

The alternative of using active **water vapor uptake** mechanisms (see Chapters 5 and 15) is of obvious utility in deserts. Certainly some species do it via the mouth and salivary glands, including the millipede *Orthoporus*, some desert mites, and the North American desert cockroach, *Arenivaga*. In this well-studied animal, uptake is observed above 83% RH (Fig. 16.17b) in association with the protrusion of two salivary sacs (Fig. 16.17a) from below the mouthparts. The concentration of the salivary fluid alone is inadequate to explain vapor uptake, and the secreted fluid is not hygroscopic. Instead a partially physical explanation involving the mat of cuticular hairs on the exposed sacs has been invoked. These hairs appear to undergo cyclical swelling; they take up condensed water when they are exposed and shrunken, releasing this water osmotically to a new flush of salivary fluid that is then resorbed into the mouth along fine capillary channels, allowing the hairs to shrink again. Other desert animals acquire vapor at the alternative site of the rectum, and many tenebrionid beetles achieve vapor uptake using the cryptonephridial chamber. At least one case of vapor uptake by eggs has also been demonstrated, in the Australian desert stick insect *Exatostoma*, which can take up vapor at 40% RH. The "sneaky" tactic of indirect uptake of water vapor, by storing seeds and consuming them after they have taken in vapor from the air of a humid burrow, could also be mentioned here (it is dealt with fully below).

Respiratory control

By a combination of spiracular control and reduced overall respiratory activity, desert insects are commonly able to maintain a low REWL. Early gravimetric studies were criticized for showing apparently very tight control; for example, at 25°C resting beetles of the genus *Eleodes* appeared to lose only 3% of their total EWL through the spiracles. With more sophisticated techniques using radioactively labeled water most studies have found much variation but in general somewhat higher values. For example, REWL is about 4–5% of total water loss in the grasshopper *Taeniopoda*, and 8% in the ant *Cataglyphis*, but these values are still notably low (see Chapter 15). In larger insects with very impermeable cuticles the value may be higher, so that in the tenebrionid genus *Onymacris* up to 45% of total water is lost through its spiracles. As temperatures rise the control of respiratory water loss also generally becomes less effective, and at 40°C (not uncommon in the Namib desert) REWL in beetles may be 40–70% of total EWL. Large-scale comparative analysis has recently confirmed that respiratory transpiration usually constitutes a greater proportion of EWL in xeric than in mesic insects.

Beetle elytra are thick, and as we noted above in some tenebrionid species they are white and highly reflective to reduce radiative heat gain. The air within the **subelytral cavity** (below the wing cases, and sheltering the spiracles) is therefore somewhat cooler than ambient air, so that some of the water vapor in the expired air will condense onto the beetle's dorsal abdominal surface, and is recovered via the spiracles or anus (Fig. 16.18). (The system is also found in species with black elytra, however, where the subelytral air is actually hotter than ambient; therefore it may be that the major function of the cavity is more mundanely to reduce air flow over the spiracular openings and so limit water loss.)

Many desert insects use **discontinuous respiration** (discontinuous ventilation cycle, DVC, or discontinutous gas-exchange cycle; see Fig. 15.40), with periods of apnoea, as discussed in Chapter 15. This is particularly common in burrowing species, including many tenebrionids and most ants. Compared with similar mesic species, these ants and beetles show a longer cycle of spiracle closure, and a relatively long flutter phase with a shorter fully open ventilatory phase (Fig. 16.19), presumably giving a saving on water loss. The water saving in cycling animals may be rather limited; it only occurs when the animals are motionless, and the REWL of small desert insects is rather low anyway. Thus savings might be quite limited,

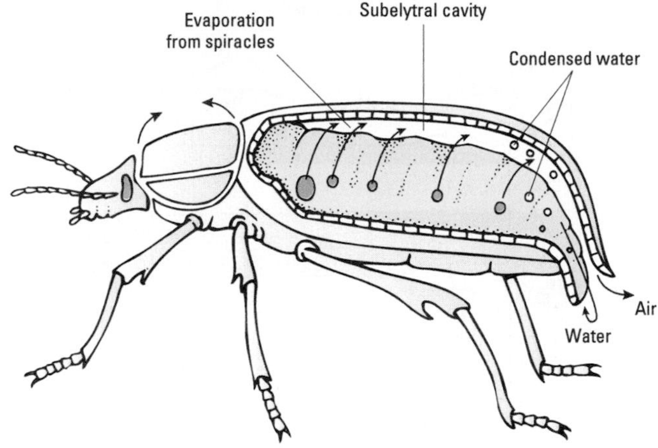

Fig. 16.18 The subelytral cavity of a tenebrionid beetle, into which the spiracles open, allowing the cavity to act as a condensation chamber, water being returned to the body via the anus. (Adapted from Ahearn 1970.)

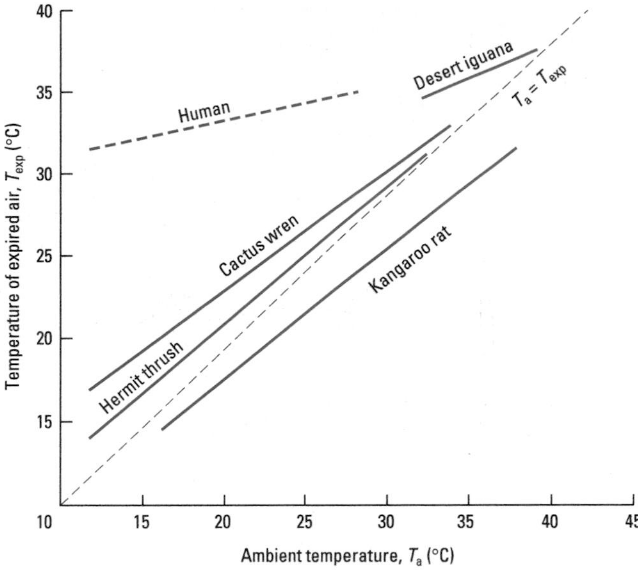

Fig. 16.20 Temperature of expired air (T_{exp}) in desert vertebrates; it is substantially below air temperature (T_a), allowing significant water saving (water condensing in the nose). (Data from Murrish & Schmidt-Nielsen 1970; Schmidt-Nielsen *et al.* 1970b; and other sources.)

but in a desert they may be large enough to make the difference between a positive and a negative net water balance.

Respiratory control of water loss is very difficult in the small endothermic desert vertebrate evaders, with their very much higher metabolic rates. However, **nasal heat exchange** is improved (the temperature of expired air is reduced, closer to the temperature of inspired air, or T_a) in many species of small desert mammals and birds that have elongated and elaborately coiled nasal passages (turbinals; see Fig. 15.37). In the kangaroo rat, *Dipodomys*, the expired air is actually cooler than the inspired air (Fig. 16.20). The important point here is that any cooling of the expired air relative to its temperature at the respiratory surface of the lung will cause it to give up moisture as it leaves the body, since saturated cold air holds less water vapor than saturated warm air. Hence water condenses in the elongated nose, which is a site of countercurrent recovery both for heat and water.

Fig. 16.19 Discontinuous respiration in desert species. (a) Oxygen uptake and highly discontinuous carbon dioxide loss in the beetle *Psammodes*. (b) Carbon dioxide loss ($\dot{V}CO_2$) and water loss rate (WLR) in the grasshopper *Romalea*. (Cf. Fig. 15.40.) F, flutter phase; O, open phase; C, closed phase. (Adapted from Lighton 1988; Hadley & Quinlan 1993.)

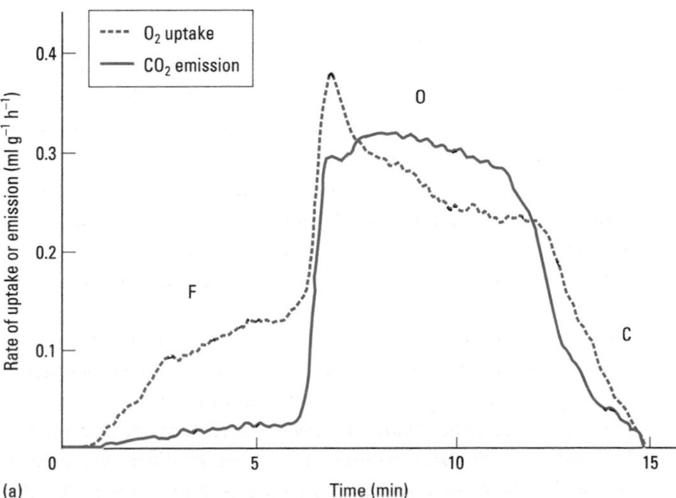

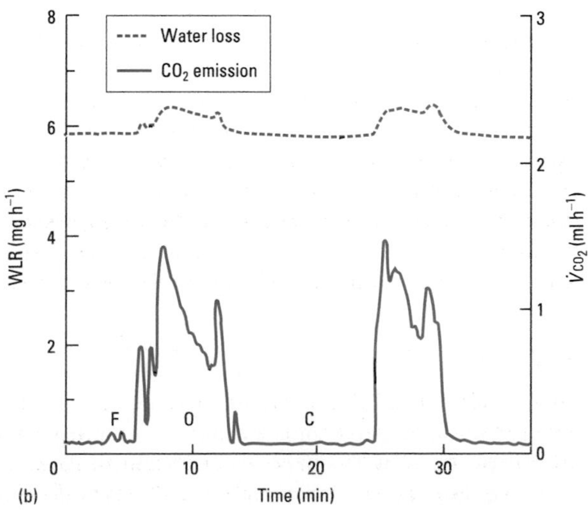

Manipulating the microclimate

Many arthropods make their own microclimates, by building burrows (dealt with above) or more specifically by making nests, usually designed to keep the occupant cool. Spiders' nests in the Negev are an interesting exception, being sited on the hottest side of shrubs apparently to increase prey capture while minimizing disturbance from possible browsers. The nest can become very hot indeed, and the adult spiders may move out to the nest entrance during the midday hours; their nests are clearly not a thermal refuge. In fact these spiders are exemplary "maxitherms", living close to their thermal

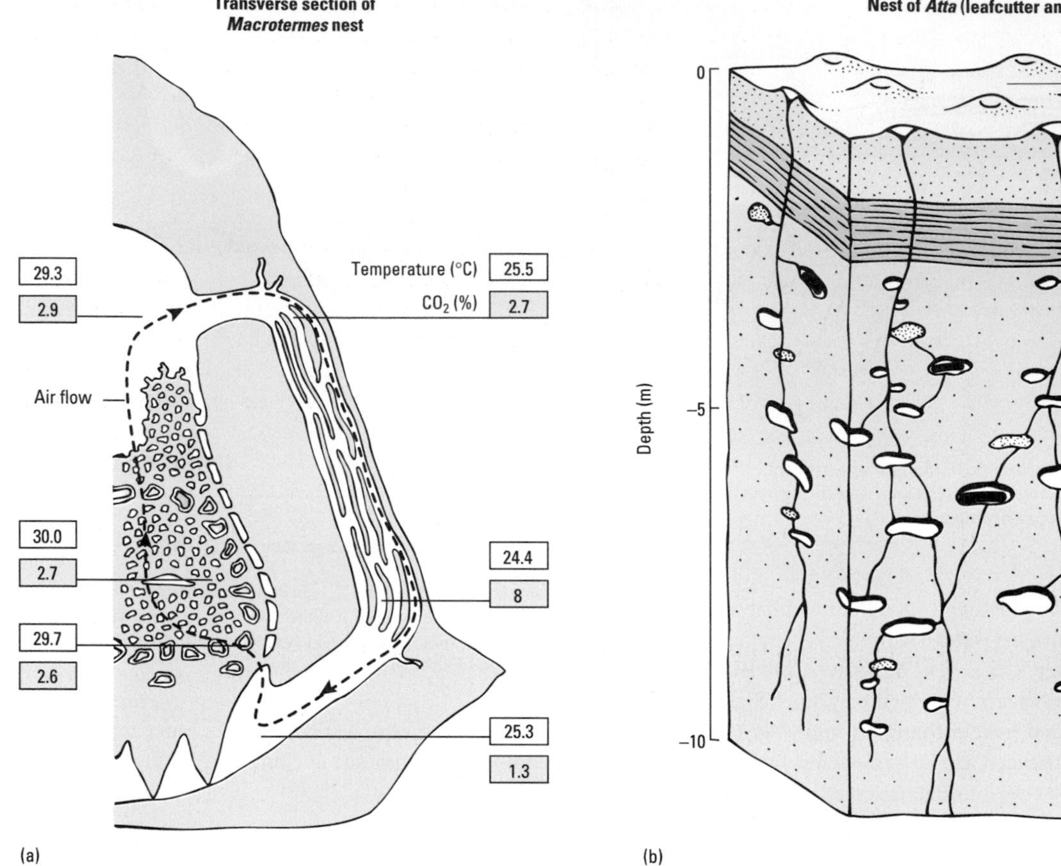

Fig. 16.21 (a) Termite and (b) ant nest architecture and temperatures.

limits, and not surprisingly the females are a very pale cream color, cream models being shown to stay cooler than black models.

However, the nests of ants and termites are the most spectacular examples of managed microclimate, with the insulation, ventilation, and orientation of the nest all attended to. Termites, for example, are mound builders in most of Africa and in Australia; in forest areas their nests are compact domes, whereas in semiarid areas the nests may appear as complex turreted and ridged mounds. But in the hot and/or dry deserts of central Asia termites are mostly subterranean, while in the Sahara there are just a few species and they live almost entirely underground, feeding on remnant semifossilized tree trunks from more humid Pleistocene periods. The best known examples of sophisticated termite nest architecture are the "compass termites", their nests always being built with the long axis running north/south, so that the structure warms up quickly after dawn but keeps relatively cool in the middle of the day with minimum surface exposed to the sun. Elaborate patterns of vertical shafts provide ventilation systems (Fig. 16.21a).

Ants' nests are usually underground, and may extend down for several meters (Fig. 16.21b), providing different microclimates at different levels for different functions. For example, *Pogonomyrmex rugosus*, a seed-harvesting species, constructs its storage chambers in the upper 0.7 m of the soil profile, but has its workers (mainly) and larvae (almost entirely) living in galleries 1.5–2 m down.

Getting the right food

Feeding in deserts is necessarily opportunistic, and many species live close to starvation for large parts of their lives. Invertebrates are particularly subject to long periods when there is no food available, between intermittent rains. Desert spiders, for example, may make a capture on fewer than 5% of all nights, and on only 1.5% of all nights during drought periods.

Most of the desert insects are herbivores, many feeding on seeds, which survive for very long periods between rains because the rate of decomposition in a desert is very low. Rodents and some reptiles use the same food resource. Relatively few insects (aphids and hoppers, orthopterans, and a few caterpillars) and a few small mammals use the other more succulent parts of plants (often below ground). A small guild of nectarivores usually occurs, often dominated by solitary or semisocial bees from the family Anthophoridae in particular ("flower bees" and "carpenter bees", particularly diverse in deserts) with some bee-flies (Bombylidae) and small sphinx moths (Sphingidae).

The selection of plant food in deserts may perhaps be more relaxed than in other terrestrial habitats. Granivores certainly collect seeds that are higher in protein content and lower in secondary metabolites by preference, but may be forced to take a wider range in xeric climates than in mesic zones. Foliar feeders may face particularly well-defended plants, since no desert plant can easily afford the desiccating effects of being damaged; physical defenses such as spines are ubiquitous and require careful manipulation by any intending herbivore. Chemical defenses are also very common, and it is perhaps no coincidence that so many desert plants are used by local human populations as medicines. It might be a fair generalization to say that desert herbivores are more likely to be *either* very generalist, taking whatever is available at any one time, *or* very specialist, having a coevolved dependence on a very few species of relatively abundant well-defended plants. It is the oligophagous species that are quite rare.

Generalist adaptations for desert herbivory probably involve large-capacity but relatively simple guts, able to provide flexibility in dealing with whatever may be available. This is exemplified by the desert tortoise, *Xerobates agassizii*, which eats very fibrous grasses or much softer herbaceous plants according to the season, utilizing colonic fermentation (see Chapter 15). Its gut constitutes up to 21% of the body mass, and is heavily endowed with mucous glands to provide some protection from abrasion, but is otherwise relatively unmodified.

Certain desert plants are halophytic (growing on very salty soils) and attract only specialist feeders because of their high electrolyte content. A classic case is the saltbush, *Atriplex*, common in deserts of the Middle East. Fat sand rats (*Psammomys obesus*) are diurnal rodents that feed specifically on saltbush tissues, but they always scrape off the outermost saltiest layers of the plant leaves before ingesting the underlying tissues.

Conspicuous guilds of specialist feeders may also occur on highly toxic plants such as asclepiads (milkweeds) and solanaceous herbs (the deadly nightshade family). These specialist herbivores can be picked out as they are often aposematically colored, warning of their own acquisition of some of the plant's chemicals. In some cases desert herbivores can get around this problem of secondary plant chemicals by selectively feeding on plants at times when such defenses are reduced. For example, in North American deserts the creosote bush (*Larrea*) has reduced levels of translocated defensive chemicals in the evenings and is then much more strongly grazed. It has also been suggested that after rains many desert plants are bound to invest more in growth than in defense and so may be particularly favored by herbivores at such times, many animals taking in plants that they would have avoided during droughts when chemical defenses were stronger.

Given a range of generalist and specialist herbivores, there are clearly niches for a limited number of predators, and in most deserts the arachnids and the snakes are pre-eminent. They survive largely by eating the plant-feeding insects, but with the use of poisons they can also take on much bigger creatures such as lizards and rodents. By eating animal flesh, they automatically acquire a well-balanced and relatively moist diet (about 70% water content), so may need little or no extra drinking water. Spiders have very low metabolic rates and can recycle their silk and recoup its energy, giving very high starvation resistance. Sand-vipers locate their prey in an unusual manner partly by detecting ground-borne vibrations, and can thus feed even if their visual and olfactory systems are blocked. Likewise scorpions are extremely sensitive to vibrational information from prey, transmitted through the sand.

For all desert animals it is advantageous to take in as much food as possible during the transient periods of plenty. An active desert snail (*Sphincterochila*) may take in nearly 50 times as much energy as it uses in 1 day of foraging activity. Many social insects and also most of the desert rodents will store food at times of plenty to use later. Sometimes this involves storing within the body itself: laying down fat reserves prior to estivation is one example, but a more extraordinary version of the same principle is found in the honeypot ants (*Myrmecocystus* and other genera), which use a few members of the colony as living underground storage jars, their abdomens hugely distended with gathered nectar.

Extracorporeal stores are more normal though. Here, storing seeds is a particularly sensible strategy given the low rate of attack by fungi and other decomposing agents, and is also favored because many seeds will take in water hygroscopically when stored in a burrow at raised humidity, so giving the prudent gatherer an extra source of water in the future. For example, dry seeds collected by gerbils above ground at 10% RH contained 3.7 g water per 100 g of seed; stored at 75% RH in a relatively short burrow, their water content increases to 18.0 g per 100 g. Rodents with cheek pouches for gathering seeds should be best equipped to exploit this effect. However, it has recently been realized that a "normal" rodent with internal pouches actually loses a significant amount of salivary water to dry seeds when these are held in the pouch for just a few minutes, and cannot recover this water as it mainly stays in the indigestible seed coat. Only some specialist desert rodents with external fur-lined cheek pouches can avoid this. This phenomenon allows many desert heteromyid rodents, such as kangaroo rats, pocket mice, and kangaroo mice, to survive on little else but seeds. The kangaroo rats (*Dipodomys*) of the Mojave desert may eat nothing but dry *Larrea* (creosote bush) seeds for 9 months of the year, without needing any drinking water.

The sheer volume of stored food in a desert can be astounding: a single termite colony can accumulate 0.5 kg of dried grass, and a nest of harvester ants may contain at least 100 g of seeds, more than four times the total biomass of the ants. One nest of *Messor* ants was calculated to contain around 170,000 seeds!

Estivation

Many semiarid and desert animals that are essentially evaders use a period of physiological inactivity as part of their survival repertoire, and this is conventionally termed **estivation** (summer sleep) as it is usually timed to avoid the hottest and driest periods. Metabolic rates are reduced, and usually the thermal tolerance limits are expanded. Growth and reproduction cease, and the animal becomes relatively unresponsive to external stimuli.

As with winter dormancy and torpor, terminology is confusing and different terms are used in different fashions. A period of behavioral inactivity without much physiological adjustment is often termed **quiescence**, whereas deeper states of withdrawal including modified physiological states are termed estivation, and the deepest forms of metabolic suppression (with no visible sign of life) are called **cryptobiosis** (see Chapter 14). Where the dormant state is triggered internally and occurs as a regular and defined part of the life cycle, particularly in arthropods, it is usually termed **diapause**. These states also differ in the method of arousal, since quiescence normally ends automatically when conditions improve whereas diapause requires specific signals and internal control mechanisms.

Many burrowing desert invertebrates show periods of quiescence, usually plugging up their burrows, both at the individual level as in spiders and sun-spiders, and at the colony level as in ants. More prolonged dormancy (estivation) is very characteristic of desert snails, which may spend a large part of their life in this state (up to 98% of their life for *Sphincterochila* in the Negev). In snails such as *Helix*, metabolic rate is suppressed by 84% after a month in estivation, and in the more drought-resistant snail *Rhagada* this rises to over 90%, with a 97% reduction in evaporative water loss (and thus relatively little lowering of body water content). In the snail *Otala*, estivation-specific proteins are synthesized in the hepatopancreas during the process of reducing the metabolic rate. This state, with the shell sealed off by a calcareous lid (epiphragm), can prolong survival time to several years. Different species of snail arrange themselves with the body suspended in air on the coolest side of a desert shrub, or on the least insolated side of a rock; the air space within the shell serves as extra insulation for the tissues (see Fig. 16.14b). A few species estivate in large aggregations, with those at the base of the clump having the best survival chances as they are protected from the sun.

Diapause, as we saw in Chapters 8 and 10, is a complex phenomenon of arrested growth and development occurring in arthropods in response to environmental adversity, and it may affect physiology, behaviour, and even morphology. It is normally triggered well in advance of the onset of really harsh conditions (see Fig. 8.45). Desert species often lack the photoperiodic cues commonly used by mesic species, relying instead on triggering by food scarcity or climatic factors, always acting cumulatively some time in advance of the onset of the diapause state and so allowing extra control of its occurrence. Thus diapause may persist over many months or even several years in the absence of desert rains.

Estivation in ectothermic vertebrate evaders occurs in many amphibians and reptiles, which burrow into the soil and in the case of amphibians may form a cocoon. Estivation is particularly well studied in the genus *Scaphiopus* and in several frog genera from Australia, with the metabolic rate declining by 50–67%. *Scaphiopus* can spend 10 months of every year in estivation utilizing stored lipid reserves, with reduced hemolymph volume and increased hemolymph concentration accompanied by urea accumulation, as we saw above. In this species the antioxidant enzymes (needed to protect the tissues during arousal and reoxygenation; see Chapter 6) are both abundant and unusually insensitive to urea.

Estivation also occurs in desert endotherms in drought periods, with behavioral lethargy and a reduction of T_b to a level close to ambient. It is well known in desert ground squirrels, gerbils, and some possums, where T_b values during the "summer rest" are commonly around 25–30°C. Gerbils and other burrowers may also show circadian patterns of torpor during daylight, allowing T_b to drop to the burrow temperature of about 30°C (occasionally as low as 19–20°C), thereby saving considerably on both food and water.

Reproduction: protecting the eggs and young

Life-history strategies consisting of short lifespan, high reproductive output, and frequent breeding are usually debarred by the scarcity of resources in deserts; infrequent but regular cycles are also difficult given the unpredictability of the habitat. Many desert animals are therefore long-lived and breed opportunistically and very quickly whenever there is rain, being effectively sterile in some years. Certainly the more traditional terrestrial patterns of seasonal breeding are rare. Different species may be more to the *r*-selected or *K*-selected end of the spectrum within these broad generalizations, and in some ways desert animals therefore represent a mixture of the classic *r*- and *K*-characteristics, conflicting with the normal demarcations of life-history strategies.

Cryptobiotic nematodes living in the sands can become active and lay eggs within minutes of their first wetting, and certain collembolans, mites, and isopods can complete a generation within a week or two. At least some invertebrates appear to be able to oviposit in anticipation of rain, and the juveniles therefore appear just as the flush of green plants and associated insects occurs. Many of these examples are underground cryptic species, and they produce very large numbers of offspring with every onset of rain, most of which do not survive.

Producing relatively few young in each breeding episode, and investing heavily in those few offspring, is also favored, particularly among surface dwellers, where maternal care is therefore rather common. In some special cases the mother may control the microclimate of the juveniles herself: ants move the brood around the nest to the most favorable microhabitats, and many spiders carry their eggs around in an egg sac. Spiders also take care of their spiderlings, and the Namib dune spider (*Leucorchestris*) is known to feed its young within the burrow. Scorpions are particularly renowned for prolonged care of juveniles, these being carried around on the mother's back.

Most of the arthropod evaders burrow and therefore also lay eggs underground. Even the large migratory locust, unable to burrow, achieves underground eggs with its very long ovipositor at the end of a telescopic abdomen. Alternatively eggs may be laid on or in the host food plant, taking advantage of microclimates on the underside of leaves or within stems or flowers.

The spadefoot toads (*Scaphiopus*) are well known for their specialized tricks for reproducing in deserts, although they retain the aquatic egg-laying habit of most mesic amphibians. They depend on careful timing and very rapid development for their success. During the seasons when rain might be "expected" their gonads mature and they move each night close to their burrow surface, retreating down again by day if there is no rain. But when rain does come (even a light shower, <0.5 mm of rain) individuals emerge almost instantly and seek out a temporary surface pool of water, arriving at this "breeding pond" on the same day as the first rain and laying eggs that night. By the next morning the embryos are well developed and have acquired moderate thermal tolerance (unusual in amphibian eggs); they complete their metamorphosis in 2–3 weeks instead of the 10–12 weeks of "normal" frogs and toads. The pools they live in may start to dry up and become very crowded, and this triggers a further response, as some of the tadpoles then turn cannibalistic and develop hypertrophied jaws, speeding up their own development even further with a high-quality tadpole diet!

The desert tortoises, whose seasonal cycle was referred to earlier, manage to reproduce every year despite the variation in water and food availability, reducing their field metabolic rate in drought years and forfeiting body condition to produce a few eggs. This "bet-hedging" life-history strategy is probably typical for a relatively large and long-lived desert ectotherm.

Desert rodents use a variety of different reproductive patterns. Some are "pulse averagers", with slow prolonged reproductive effort matching the historical probability of rainfall timing; some are "pulse matchers", responding directly to rain and food availability; some, which store foods or use completely reliable foods, are "pulse ignorers"; and yet others are "pulse gamblers", producing large litters at the end of their hibernation period irrespective of conditions, some of which fail completely.

The most striking factor in many desert-breeding species is their ability to detect and respond to the very smallest quantities of rain almost instantly. Spadefoot toads appear to use the low-frequency sound of rain drops on the ground as their cue. Some desert birds use even the distant sight of rain, or the accompanying slight rise of humidity, as their cue for initiating nesting and egg laying.

Sociality

We have already noted that social insects are particularly common in deserts, exploiting the advantages of stable nest microclimates in providing thermal refugia, bulk storage of ephemeral resources, and defense against predators and parasites. The utility of sociality has been realized in other taxonomic groups as well. Very arid habitats house a few species of social isopods, of social spiders, and of at least one social tenebrionid beetle. Southern African deserts are also the home of a now famous rodent, the truly social mammal known as the naked mole rat, *Heterocephalus glaber*. This has a queen and nonreproductive workers living together in a burrow network, and its lifestyle is ideal as a strategy for exploiting limited and patchy food resources and sharing the costs of tunneling in hard, baked soils. It is also noteworthy for being the only mammal that is approximately ectothermic, having noninsulated surfaces with very high heat-loss rates and thus a potentially variable T_b (though it can produce heat endothermically in the laboratory). Its preferred T_b is 33°C, which is also the temperature at which its gut flora, performing the cecal fermentation of the very high fiber diet, work fastest.

16.2.3 Evaporators and their strategies

The evaporators are the "middle-sized" animals of the deserts, including some birds, plus dogs, cats, smaller antelope, foxes, etc. The group could also be said to include domesticated goats and sheep, and man himself where he lives on desert fringes.

These animals are all dependent on a reasonable water supply to allow them to cool down by the use of evaporation, at ambient temperatures where other cooling modes are impossible. Evaporators are therefore inevitably few and far between, mostly living only at the desert fringes and where there is access to water at oases. They do rather better in the margins of rocky deserts, where there is more microclimatic variation to exploit on a scale that is useful to them (e.g. sheltering behind boulders). Moderately sized creatures, such as the rock wren and the elf owl, can survive by sheltering in crevices in such habitats.

Evaporative cooling

As with all other animals, either **panting** or **sweating** may be used as the means of evaporation; as we saw in Chapter 8, the latter is less controllable and also loses salts, but it uses up less energy. Panting is normally preferable for desert animals, and it can be combined with countercurrent systems in the nose to save water as described for the smaller evader animals. Desert sandgrouse show a very pronounced gular flutter (see Fig. 8.44b) to supplement normal panting, and have markedly enhanced evaporative thermoregulation compared with less arid-adapted relatives. The pin-tailed sandgrouse, *Pterocles alchata*, dissipates 89% of its metabolic heat production by evaporation at 40°C, which is 152% of the allometric prediction. Some xeric birds are particularly well suited to be evaporators by virtue of feeding on nectar. The Australian honeyeater *Manorina flavigula* has a very high daily water turnover and a high and variable urine concentration, reflecting the water inputs it gets from plants.

One group of desert invertebrates deserve honorable mention as being part of the "evaporator" cohort; these are the insects that can use evaporative water loss for cooling, again because of a reliable watery food supply. The desert cicadas are best known, their transpiration rate increasing at least five-fold when exposed to temperatures above a specific setpoint. Species such as *Diceroprocta apache* release water from pores on their dorsal cuticular plates, allowing them to maintain a body temperature of only 39–45°C when (in the case of males) singing to attract a mate in ambient temperatures of up to 48°C. This particular cicada feeds on the abundant mesquite plants of the North America deserts. The cicada *Okanagodes gracilis* is even more impressive, maintaining activity even through the hottest days ($T_a > 50$°C) while feeding and evaporatively cooling. Control of sweating in the cicada *Diceroprocta* is turning out to be rather complex, involving internal and surface sensors and at least two regulatory pathways involving prostaglandin-like chemical signals.

A less familiar example of the evaporative strategy in desert insects is the grasshopper *Poekilocerus bufonius*, which gets enough water from its asclepiad food plant to be able to afford some evaporative cooling. It seems probable that other desert xylem feeders may use a similar technique. We should also mention some social bees, particularly in the genus *Apis*, that must cool their bodies and also their whole colony using gathered water when temperatures rise too high, inevitably causing them to be restricted to oases and desert fringes.

Other thermal and water balance strategies

Inevitably the desert evaporators use most of the behavioral and postural strategies already described above for their smaller counterparts; behavior continues to be a major part of the overall adaptive pattern. **Huddling** is perhaps more common than burrowing, which can be difficult for these larger animals. The wood hoopoes of the sub-Saharan savannas huddle in response to low ambient temperatures, with metabolic rates 30–60% lower at night than for solitary birds, saving up to 29% of the daily energy expenditure.

Desert birds such as babblers, finch larks, and sandgrouse show some particular tricks to survive quite deep into the deserts. The mobility endowed by flight enables them to gather seeds over a very wide area, and more importantly to get water from oases. In the sandgrouse genus (*Pterocles*) the males are adapted to **carry water** back to the nest and to the young; they wade chest high into the water and return with it soaked into and below their specialized breast feathers, carrying up to 40 ml per trip over astonishing distances of up to 80 km. Other desert birds use a special kind of "shading behavior" at the nest, squatting just above the eggs rather than nestling down on them. This may be more important for cooling the parent's lower body by raising it above the boundary layer, than for allowing convective cooling from the eggs themselves.

However, there are some additional physiological components in the repertoire of these animals. The carotid rete system for **selective brain cooling** (see Chapter 8) appears to be relatively common in these medium to large mammals. It is best developed in those that pant, especially artiodactyls, canids, and felids, where the necessary cool blood is that returning from the nasal area, cooled by evaporation. The carotid rete linked to the cold nose in these panting animals can cool the blood to the brain by 2°C or more, which in a heat-stressed desert species is very important in that it avoids triggering the brain-based receptors that normally turn on extra panting mechanisms to cool the body down, and which if activated could lead to disastrous evaporative water loss.

Similar brain cooling occurs in other mammals (even in man), where there is no panting mechanism, and then seems to involve cool blood from the facial veins entering the cavernous sinus at the base of the skull where the carotid artery, though not subdivided into a rete, follows a fairly tortuous course so that its blood can be slightly cooled. In desert birds brain cooling has also been demonstrated, with a countercurrent operating in the ophthalmic rete and involving cooled venous blood from the nose, mouth, and eye areas. Brain cooling by heat exchange occurs in at least some reptiles, too, where the brain tissues can be 2–3°C lower than the core body temperature or rectal temperature. This is achieved in desert lizards by a rapid and shallow breathing through the open mouth, allowing the pharynx area to cool evaporatively. While this does not actually achieve much body cooling in the sense that true panting can do for a mammal or bird, it does provide a brain temperature-regulating system, for the carotid artery runs very superficially through this pharyngeal area and thus the arterial supply into the brain is again cooled.

Some of the desert birds may also demonstrate a **lowered metabolic rate** (e.g. 26% below the allometric prediction in Houbara bustards, and 46% lower in the double-banded sandgrouse *Pterocles bicinctus*). There may also be a tolerance of some **hyperthermia**, and a high upper critical temperature permitting survival at somewhat elevated T_b. In the Negev the great gray shrike (*Lanius excubitor*) shows an increase of T_b with rising T_a even within its normal thermoneutral zone (30–36°C), and this controlled hyperthermia gives it a saving of water by allowing extra dry heat loss rather than evaporative water loss. However, this and other desert birds also exhibit a rather high rate of increase of EWL with temperature, a puzzling observation not yet adequately explained.

Many of these medium-sized animals from hot climates seem anomalous in having very dark surfaces (e.g. desert goats, sheep, and ravens). The phenomenon of black desert mammals and birds may be explained by several factors. The individual black feathers and hairs can become exceptionally hot at the tip (80°C is not uncommon), enough to actually encourage heat loss to cooler air by convection and radiation, while at the same time the trapped insulating air layer prevents the heat of the pelage being relayed to the skin surface. Black color may also be appropriate in the winter months for animals such as desert goats, when nights are very cold and food is rare. The metabolic rate of a black goat may be 25% lower than that of a white one, because they can absorb more heat from the sun and have less need to shiver. Furthermore, remember that in all these cases we are considering environments where reradiated longwave radiation from the bare ground, the absorption of which is largely independent of color, may be a particularly large component of the total heat spectrum.

Table 16.2 Urine concentrations and urine : plasma ratios in desert vertebrates.

Taxon	Urine concentration (mOsm)	Urine : plasma ratio
Reptiles		
Desert tortoise	337	1.0
Birds		
Struthio (ostrich)	900	2.7
Kookaburra	944	2.7
Zebra finch	1005	2.8
Savanna sparrow	2020	5.8
Mammals		
Eland	1880	6
Bedouin goat	2200	7
Megaleia (kangaroo)	2700	8
Camelus (camel)	3200	8
Dipodomys (kangaroo rat)	5500	16
Notomys (hopping mouse)	9370	25

Desert goats also show an unusual ability to reduce metabolic rates when food is scarce and maintain their body weights on less than half their normal food intake. Unusually, this involves reduced metabolism mainly in the muscles, whereas in most mammals it is the gut that is "turned down" most conspicuously.

Medium-sized desert animals also show a pronounced ability to produce **hyperosmotic urine** (Table 16.2), well beyond the levels seen in temperate species of similar body mass (cf. Table 15.8). Mammals with more numerous long loops of Henle inevitably succeed best, with high urine : plasma solute ratios.

Migration

Many evaporators are essentially **nomadic**, using migration as part of their strategy to track the patchy resources, moving between water sources as they fluctuate and become exhausted. This may involve regular movements outward from a fairly permanent central refugium, or a more random wandering between chance-found resource foci. However, migration does impose substantial metabolic costs, and is a risky strategy since location of future supplies can rarely be guaranteed.

Following on from this point, it could also be said that the famous examples of migratory desert insects come into the "evaporator" category rather than being evaders. Many species of locusts and grasshoppers are neither particularly impermeable nor particularly cryptic in lifestyle, by desert standards. Instead they move in vast numbers between patches of water or areas of recent rainfall that are sustaining food resources, ranging over grassland and into deserts, their ability to fly allowing them access to the occasional areas of plenty within profoundly arid areas. Indeed, the true desert locust, *Schistocerca gregaria*, is almost entirely restricted to the hot Palaearctic deserts, moving continuously between wadis and oases.

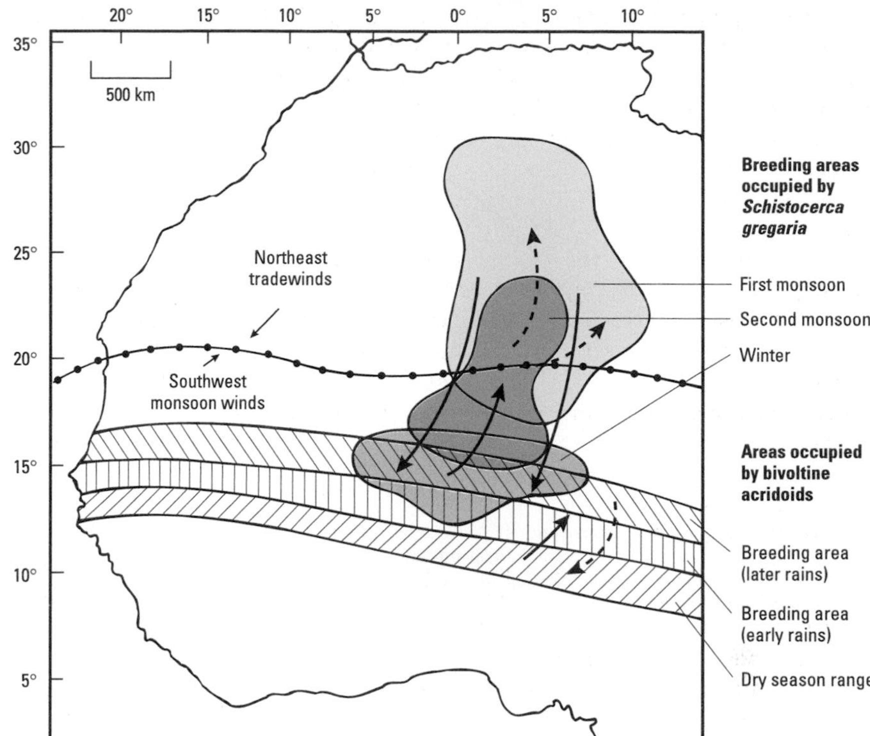

Fig. 16.22 Locust migrations around the Sahara desert. (Adapted from Farrow 1990.)

Its movements within the Sahara are shown in Fig. 16.22, and are largely determined by the seasonal movements of air masses.

16.2.4 Endurers and their strategies

Camel, oryx, and other large mammals

The camel and oryx are the famous examples of enduring desert animals, being large endothermic mammals. The camel of the Arabian and Sahara deserts is *Camelus dromedarius*, which in some parts of its range routinely endures air temperatures up to 55°C in the daytime and below freezing at night. The oryx (*Oryx beisa*) is almost equally famous as an Arabian desert gazelle with an exceptional ability to withstand aridity. In the same functional category, although less highly specialized, come some of the large gazelles of dry grasslands and deserts, including the eland and the springbok, and perhaps also the two species of African rhinoceros and the African elephant.

Because they are large, such mammals have real difficulties in losing excess heat through their surfaces. Much of their body is conspicuously lacking in thick fur, as insulation would compromise heat loss even further, but many do have large appendages (ears and tails, or long necks and legs) where heat can be dissipated. Elephant ears viewed with infrared thermography show a remarkable rate of heat dissipation of up to 70 W at a T_a of 32°C, allowing them to shed almost 100% of the animal's total heat-loss needs when maximally vasodilated and flapped gently.

These large mammals also tend to be very inactive in the heat of the day, avoiding any metabolic overload; and they will use behavioral strategies to exploit any source of cooling, such as spread-eagling against cooler rocks, seeking the shade under any desert scrub vegetation, orientating head-on to the sun, locating the slightest breeze on a hillock, or wallowing in patches of mud.

However, the great compensation offered by large size is that of high thermal inertia, opening up the possibility of substantial heat storage and the strategy of **adaptive hyperthermia**. Here the body temperature is allowed to fluctuate rather widely on a daily basis, with heat stored during the day giving a barely tolerable peak T_b just before sunset, and then dissipated during the night so that there is a just tolerable minimum T_b around dawn. Using this technique the camel is able to survive even when dehydrated and in the hottest of deserts over a 24 h cycle (T_b range 34–41°C, exceptionally up to 45°C briefly; Fig. 16.23). Effectively it is acting as a storage radiator, and this is particularly obvious when the camel T_b is compared with the much more rapidly fluctuating T_b of a small rodent in the same environment (see Fig. 16.8a). Storing heat in the daylight hours can reduce the daily water loss of a camel by up to 5 l, since it need not use EWL to regulate T_b tightly.

This type of thermal behavior is aided by a light color and thin curly coat, especially dorsally. In stark contrast to the situation in cold climates, the fur of mammals from hot arid habitats reduces in thickness as body size increases (Fig. 16.24; cf. Fig. 8.17). Most desert gazelles are pale brown, but commonly have a darker flank or an almost black strip along the flank (e.g. Thomson's gazelle, Grant's gazelle, and springboks; Fig. 16.25), which is known to facilitate heat gain when the animals stand side-on to the sun in the early morning. For exactly opposite reasons many also have a white rump and will orientate parallel to the sun's rays at midday (Fig. 16.26), paralleling the behaviors of ectotherm evaders. Having long ears, long necks, and long legs may also be adaptive in providing a range of postural options for controlling heat gain and loss. Long horns also serve as heat dissipators, and the horns of desert wild sheep are

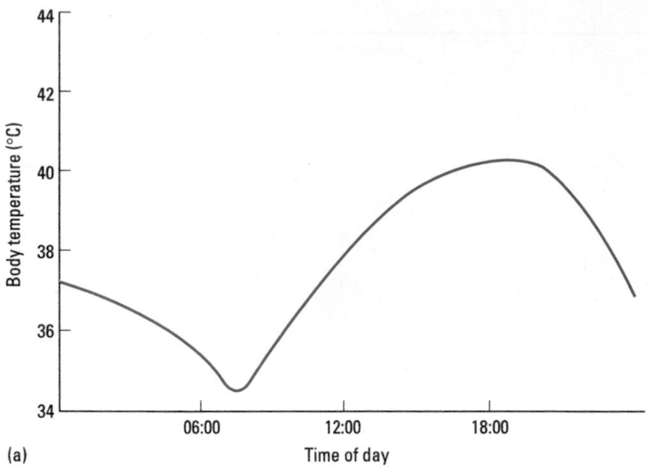

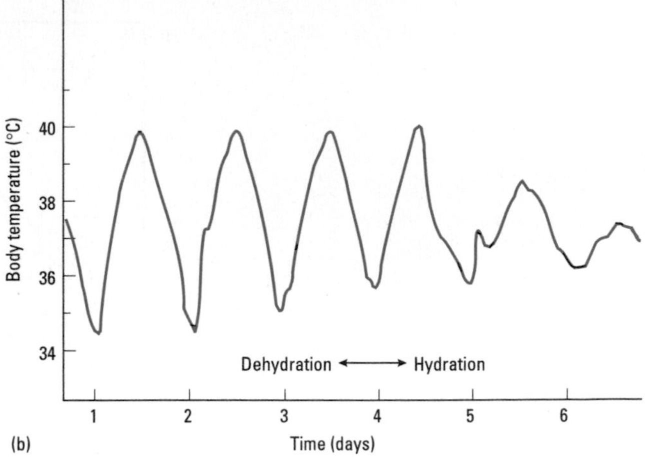

Fig. 16.23 Daily patterns of body temperature (T_b) in a camel: (a) a single cycle with daytime heat storage and nocturnal heat dissipation, and (b) the effects of hydration, allowing cooling by evaporation and so reducing the need for the heat-storage effect. (Adapted from Schmidt-Nielsen 1964.)

notable for large internal vascularized cores useful for shedding heat from the blood.

A camel is also substantially aided by an ability (unusual amongst large mammals) to endure up to 30% water loss, without losing its ability to behave normally and to feed. For comparison, humans and other mesic mammals cannot tolerate more than 10–12% losses (cf. Table 15.5). A camel's special ability is that of **drinking** very rapidly to replenish this deficit, up to 200 l or around one-third of its own body weight in one drinking bout lasting just 3 min. It can dehydrate without compromising its blood viscosity; the fluid lost comes from the tissues rather than blood, so that the blood composition and volume remain reasonably constant, the hemoglobin functioning remains normal, and explosive heat death due to reduced blood circulation is avoided. During rehydration, large inputs of drinking water are stored temporarily (up to 24 h) in the gut to avoid an equally dangerous rapid dilution of the blood. Camels and other arid-dwelling vertebrates, including kangaroos, also have unusually robust red blood cells that are resistant to potential osmotic shock accompanying changes in body water content.

There are many special physiological tricks in the camel and the antelopes to help all this endurance. Some assistance comes from good water conservation mechanisms. They have relatively low water loss rates for their size (around 3–4 mg cm^{-2} h^{-1} in the dehydrated oryx and Grant's gazelle), and can produce very dry fecal pellets by

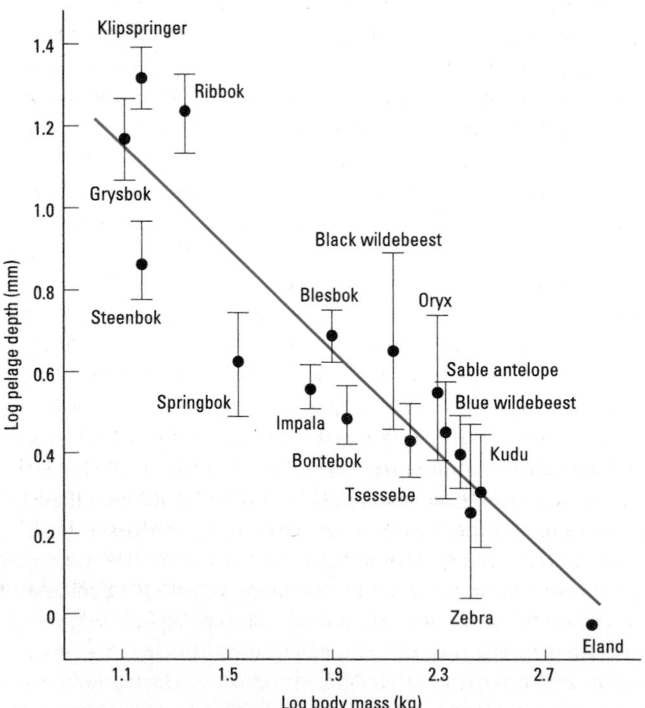

Fig. 16.24 The negative relation between fur depth and body size for various African desert and savanna ungulates (cf. Fig. 8.17). (From Louw 1993.)

Fig. 16.25 Gazelle color pattern, typical of desert antelopes. (From Louw 1993.)

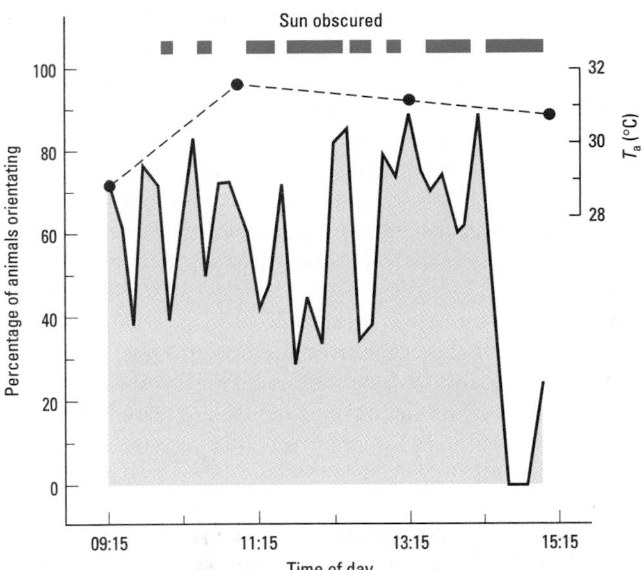

Fig. 16.26 The percentage of gazelles showing an orientation head into or away from the sun (white face or rump exposed); in cool periods when the sun is more obscured they are more likely to turn sideways, when the dark flanks absorb heat better. T_a, ambient temperature. (From Louw 1993.)

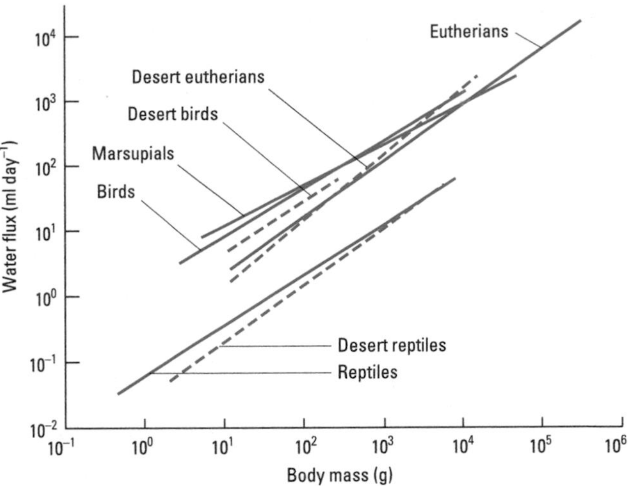

Fig. 16.27 Water flux in desert animals compared with nondesert species. (From Nagy & Petersen 1987.)

mammalian standards (e.g. about 40–50% water in the camel and the dik-dik, compared with 70–80% in temperate ungulates). As might be expected many of them also have very concentrated urine: about 4000 mOsm in the dik-dik, and over 3000 mOsm in the springbok and camel. Of all these large enduring mammals, the dik-dik and the oryx are perhaps the most spectacular in their ability to survive for very long periods with no drinking water, but of the two only the much larger oryx does so in real desert conditions. The camel, in similar conditions, can survive without drinking for 14 days. These mammals have a series of highly regulated hormonal responses to dehydration that maintain sodium and other ionic balances and water compartmentalization.

In desert antelopes and other ungulates the occasional need to run in the daylight hours may involve a 30–40-fold increase in metabolic rate, and thus an enormous heat load. While hyperthermia may be possible for most of the tissues, there is a vital need to keep the brain functioning at a normal temperature. The carotid rete system was first described in desert gazelles, and soon afterwards in the camel, where it is very well developed. This serves the "brain cooler" function, where cool blood from the nasal heat exchanger is used to reduce the temperature of the carotid artery blood passing to the brain. Blood cooled in the nasal cavity is diverted to the brain sinuses via the nasal and angular veins. The muscular tone in these veins is particularly temperature sensitive, with relaxation occurring during small thermal increments in the range 33–45°C; under the same conditions the facial veins vasoconstrict, so that as much cool blood as possible is directed to the brain sinuses and carotid rete. A few animals can use the nasal heat-exchange system in a particularly sophisticated manner by expiring nonsaturated air, thus achieving even greater water economy in the "long nose" respiratory system (see Fig. 16.20). The camels are the best known examples, as they can exhale air at only about 75% RH, thereby saving about 60% of the water unloaded to the air in the lungs.

The net effect of all these features that reduce water loss is a steeper regression between water flux and body mass for larger desert animals as compared with the regressions for all species taken together (Fig. 16.27). This steepened regression is observed especially in eutherian mammals, but also in birds and in reptiles.

Reproduction also produces some constraints in these large desert endurers. Most of them produce rather **dilute milk** (e.g. rhinoceros milk has only 0.3% fat and 1.2% protein, compared with 4.0 and 3.4%, respectively, in a cow; see Table 15.13), presumably helping to offset evaporative water loss in the juvenile. Also, antelopes and camels need very precocial young as the habitat gives no chance to hide and the young must escape predation and be able to seek shade. Allied with precocity they are often also rather large proportional to the adult at birth, so that they can experience the advantages of thermal inertia as soon as possible.

In addition to physiological features, remember some of the detailed morphological factors that assist the largest mammals in deserts. They have large flattened hooves to avoid sinking in the sand; dense protective eyelashes and strong eyelids to avoid sand damage during sandstorms; pelage that is dense on the back but sparse elsewhere; and a trunk shaped to give a minimal surface area to volume ratio when resting head-on to the sun in the heat of the day.

Finally, of course, we come to the famous camel's hump. Hopefully, the old wives' tales scarcely need to be refuted: it is not full of water, nor is it full of fat as a source of water (remember that fats give more energy per gram, and more metabolic water per gram, but produce *less* water per kilojoule of energy, and with extra evaporative water loss also incurred when providing the respiratory oxygen needed for their full metabolism). The camel's hump is simply fat as a store of food as in most other mammals, but for a large desert animal it is sensible to concentrate the fat only on the dorsal surface since there it also has a localized insulating effect, helping to reduce the uptake of solar radiation from above in the day and allowing the flanks to be used for thermal emission at night.

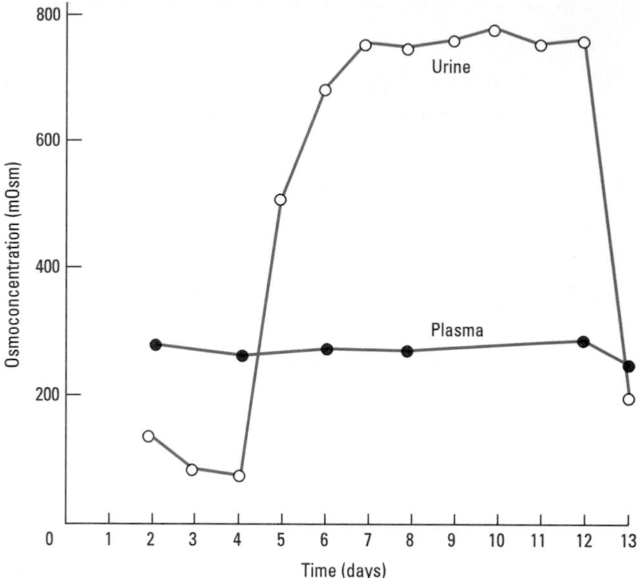

Fig. 16.28 Ostrich urine variability; the bird was dehydrated from day 4 to day 12. Note the minimal variation in blood concentration. (From Louw 1993.)

Ostrich and rhea

The ostrich, *Struthio camelus*, does particularly well in deserts and dry grasslands, as do some other large flightless birds (ratites, including the rhea, emu, and cassowary). Their general build—long legs, long necks, and often long beaks—in some ways parallels that of the large desert mammals.

Their strategies are rather similar to those outlined above, relying particularly on adaptive hyperthermia and countercurrent cooling. For example, when *Rhea americana* runs for 20 min at 10 km h^{-1} about 75% of the total heat generated is stored, being lost again by radiation and convection after the bird stops running. In these ratites the surface temperatures of all the extremities appear to be well regulated. Most obviously, these birds again have characteristic long noses and elaborate nasal turbinals (see Fig. 15.37). Ostriches are one of the very few animals, apart from camels, that can exhale unsaturated air (Fig. 16.20); for example, in ambient air at 36°C the exhalant air from an ostrich is at 85% RH, so recovering about 35% of the water initially evaporated into the air during inspiration and saving up to 500 g water per day. And again the cooling effect at the nose can be used to achieve brain cooling; as in other birds this is largely via an elaborate ophthalmic rete just below the brain.

An ostrich at risk of severe overheating also uses a combination of panting (at about 40 breaths min^{-1}) and alterations of the plumage, erecting the rather sparse dorsal feathers to give a thicker insulating layer but one in which air moves freely to give some convective cooling. The birds also orientate face-on to the sun and droop their wings laterally away from the thorax, giving the rather bare sides of the thorax some shade and allowing them to act as thermal windows.

Ostriches can reduce their urine volume dramatically when dehydrated (Fig. 16.28), with the concentration rising (though only to around 800–1000 mOsm). In such conditions they also become more selective in food choice, eating plants with a higher water content.

Large desert birds face a particular problem in keeping their eggs cool; they cannot build large protective insulated nests in such an environment. The ostrich lays its enormous eggs in a shallow scrape, and relies on its own high thermal inertia and thermoregulatory mechanisms of panting and ptiloerection to keep the clutch cool beneath it.

The Australian emu (*Dromaius*) is also worth mentioning as its habitat ranges from high snowy mountains to the central continental desert, and it can maintain a constant T_b at ambient temperatures ranging from −5 to +45°C despite being too big to exploit any microclimatic variation. Its large thermal inertia certainly helps it; in the deserts it uses techniques similar to those of the ostrich and rhea, while on snow slopes it maintains T_b largely by reduced conductance and increased heat production, spending much of its time sitting as a huddled ball with all appendages tucked in.

16.2.5 Desertification and anthropogenic effects

In a sense a desert is the one type of habitat that is not really fragile and vulnerable to human interference in the way that the rest of the globe is: for deserts are the "endpoint" of our interference in all low-latitude zones (Table 16.3). Existing deserts are spreading because of human influence, agricultural practices, and grazing livestock herds. In some areas entirely new deserts are being created, for example where forests are felled on hill slopes and the land quickly erodes into a dustbowl (Fig. 16.29). These processes are unlikely to be halted; in fact desertification is almost certainly getting worse now, and will continue to get worse into the future, because of pressure from growing human populations, especially in sub-Saharan Africa, and because of projected anthropogenic climatic changes (cf. Fig. 15.65).

Once deserts have formed they are of no use to man, and so may suffer relatively little interference, persisting as zones of low but specialized biodiversity. There have been attempts to resurrect certain desert areas by grand irrigation schemes, ongoing particularly in Israel and projected in other parts of the Middle East (e.g. diverting water from the Nile to irrigate the Sinai Peninsula). However, such

Table 16.3 Estimated causes of desertification (as percentage of desertified land area).

Regions	Overcultivation	Overstocking	Fuel/wood gathering	Salinization	Urbanization	Other
Northern China	45	16	18	2	3	16
North Africa/Near East	50	26	21	2	1	–
Sahel/East Africa	25	65	10	–	–	–
Middle Asia	10	62	–	9	10	9
Australia	20	75	–	2	1	2
USA	22	73	–	5	–	–

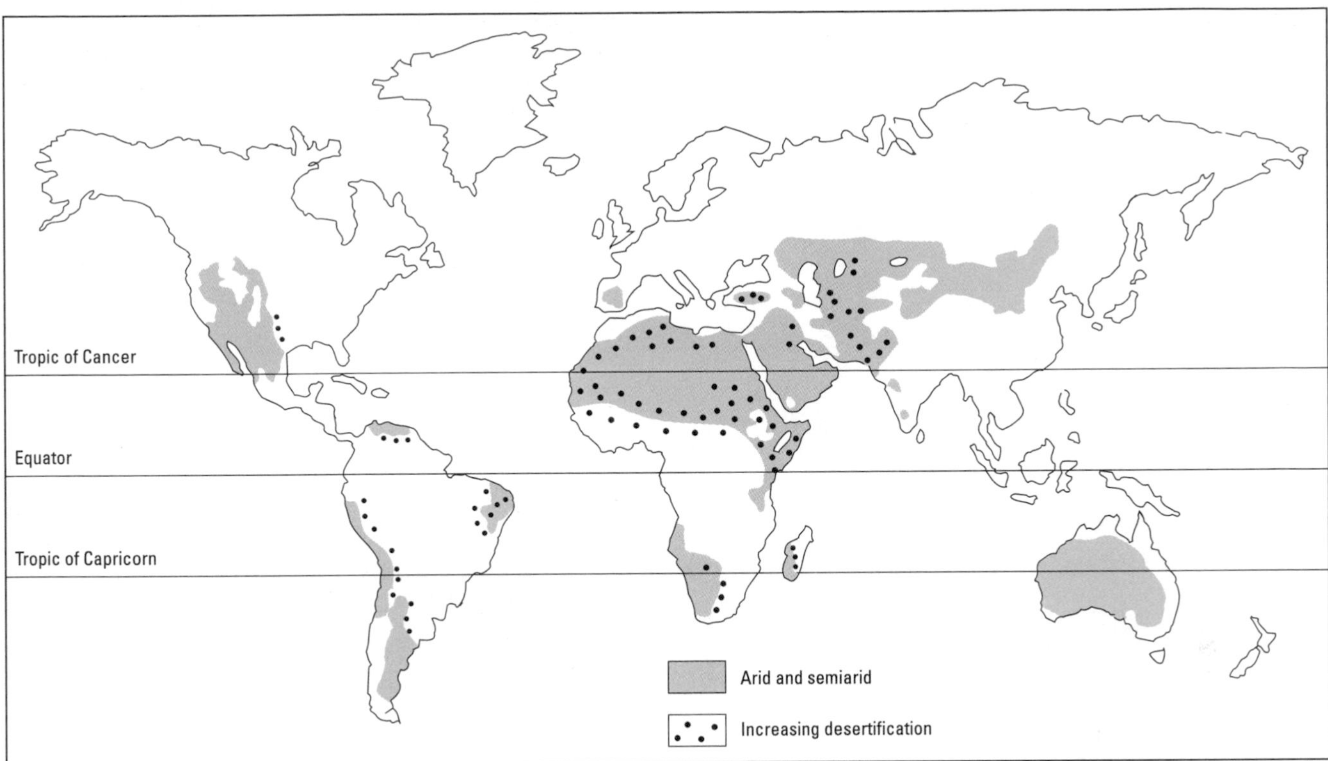

Fig. 16.29 Regions where desertification has increased in the last three decades. (Adapted from Williams 1986.)

schemes tend to produce localized areas of highly specialized and artificial agricultural productivity rather than returning the land to a natural savanna or subtropical status.

16.2.6 Overview of hot arid environments

Before we move on, it is worth reiterating that while we have presented a series of adaptations that desert animals may show, these are very strongly interactive and can never be treated in isolation. To take a simple example, desert insects may have very low values of cuticular water loss (CEWL) such that REWL is a larger proportion of their water balance. They have a number of possible tricks to reduce REWL, but in addition to these they may also show a lowering of metabolic rate, which lowers their need for oxygen and hence also lowers REWL. But lowering metabolic rate means a reduction in cellular ATP production, which leads to a reduced cellular sodium pumping and a reduced sodium gradient across membranes, in turn leading to a reduced sodium-coupled amino acid transport into cells. The hemolymph of these insects may therefore be low in sodium and high in amino acids. In most insects this would lead to substantial urinary loss of amino acids, but since desert insects also have a very effective rectal resorptive system they are able to sustain their internal conditions. In this fashion many of the key physiological adaptations are linked together. Similar arguments clearly apply in relation to integrated temperature and water balance in camels and other large endurers.

The biology of deserts is strongly dominated by the presence or absence of water, setting the dynamics of resting, foraging, and reproducing for all animals and imposing an unusually unpredictable regime requiring opportunistic adaptive strategies. Life appears to proceed as a series of pulses of activity with long periods of relative quiescence in both flora and fauna. For animals that depend on the ephemeral parts of plants—leaves, flowers, sap—these pulses have a direct and immediate impact, and the animals' population densities may thus be directly related to rainfall. For animals that rely on more persistent plant parts such as seeds, or which store foods in nests, and for those that are omnivorous or carnivorous, the pulses can be smoothed out somewhat, and more stable (albeit small) population densities can be maintained.

16.3 Very cold habitats

16.3.1 Types and characteristics of the cold biomes

Most of the living space on this planet is rather cold, since the majority of it is deep cold seas; the animals that live there were dealt with in Chapter 11. But even restricting discussion to terrestrial habitats, the world is predominantly a cold place, especially by human standards as we are essentially a tropical species by origin.

However, land-based temperatures that could reasonably be termed "very cold" are mainly encountered with higher latitude, where the radiant energy from the sun has to travel through a greater thickness of atmosphere, so that it is partially absorbed, and what does get through must spread out over a larger area of ground (Fig. 16.30a). The circumpolar ecosystems, sometimes collectively

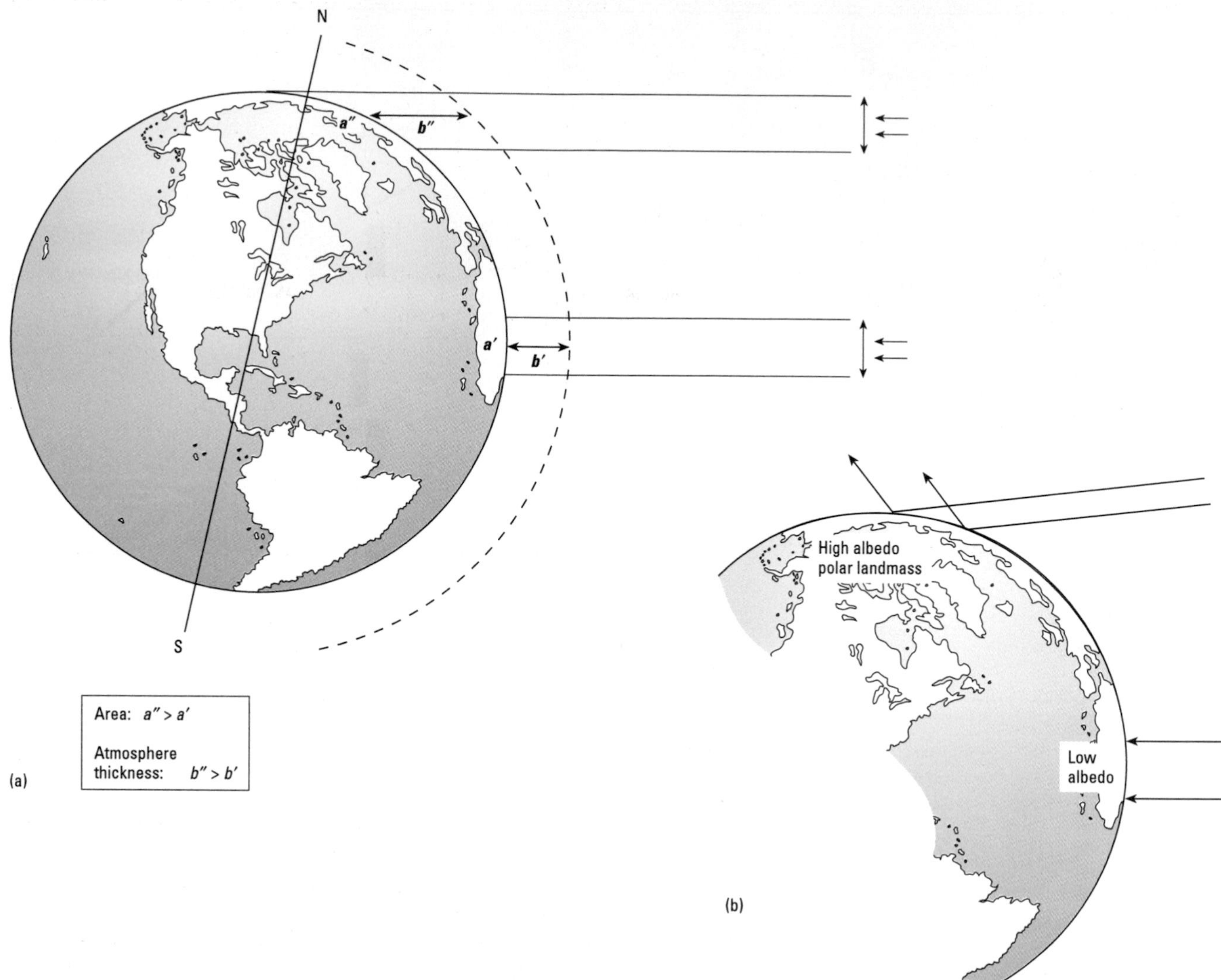

Fig. 16.30 Causes of reduced radiation load at high latitude: (a) input spread over a larger surface area, and (b) reflected by ice (high albedo) over the seas or land masses.

termed the boreal systems, are therefore marked by extreme cold. The records for the lowest temperatures in overland climates are held by Siberia in the northern hemisphere (–68°C) and by central Antarctica in the southern hemisphere (–89°C). These zones also show a negative winter energy balance (more heat is lost than is gained from the sun), and a very short growing season.

In most of these cold habitats there is also ice or lying snow for large parts of the year, creating additional and unusual problems. Intense direct and reflected radiation can cause sunburn to unprotected animal surfaces, and "snowblindness" by dazzling and damaging the eyes. There is also a more general problem with visual sensations, as land, sea, and sky merge to give a "white-out" with no shadows to assist detection of contour or of cracking unstable surfaces.

The cold biomes comprise three main biotic types: (i) the polar zones themselves, with continuous ice or snow cover; (ii) the tundra zones, with snow and minimal low vegetation in the short summer; and (iii) the taiga, where coniferous forests survive. Cold biomes could also be said to include some areas of "cold desert", often found on relatively high-altitude plateaus in the interior of large continents (e.g. the Tibetan plateau).

Polar zones

Towards the poles, low temperatures lead to snow and ice formation, creating a surface of **high albedo** (high reflectivity), so that some of the sparse incoming solar energy is reflected at a tangent and lost to the system (Fig. 16.30b). Each of the polar regions becomes a giant low-pressure center, with unmixed cold air at its center hardly able to mix with the peripheral circulation of milder westerly winds (the "circumpolar vortices").

Temperatures of course do vary seasonally, with a "warmer" summer, although the growing season may last only 1 month. The poles experience periods when the planetary tilt means a complete absence of direct radiation, and by definition beyond the polar circles (roughly 66°N and S) there are days when the sun never rises.

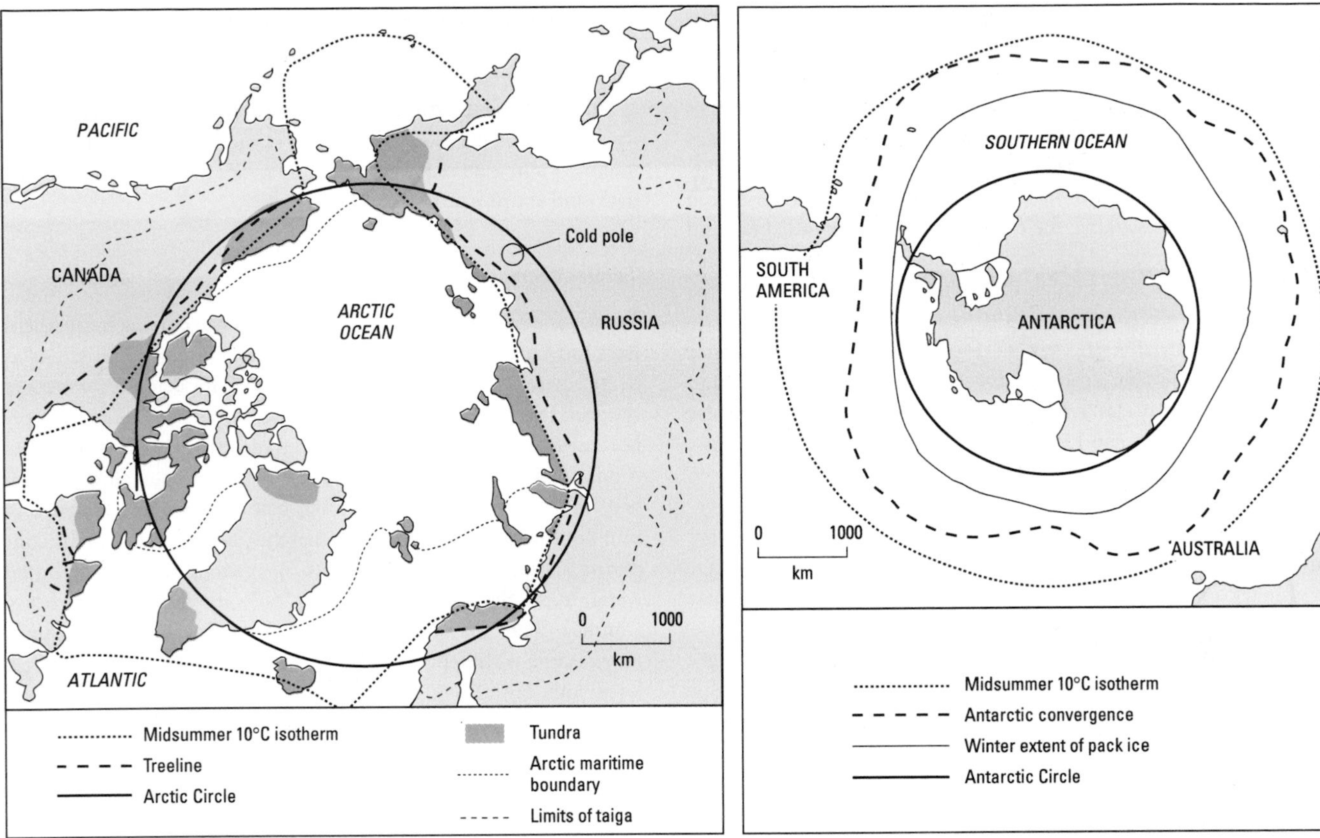

Fig. 16.31 Defining the Arctic and Antarctic, and regions of tundra and of taiga. (Adapted from Davenport 1992.)

But there is a marked difference between the two hemispheres. The Arctic is merely a frozen sea (see Plate 9c, between pp. 386 and 387), but is fringed by land masses, whereas the Antarctica is a real continent, 3000–4000 m above sea level in parts and with another few hundred or thousand meters of ice on top. The south polar region is therefore always colder for a given latitude, because of the presence of a large icy polar continent instead of a thermally buffering polar sea. The coldest northern hemisphere temperatures are recorded around the so-called "cold pole" of eastern Siberia, and the northern fringes of the land masses of Russia, Canada, and Greenland are routinely colder than the true North Pole, again due to the ameliorating effects of the thermally stable Arctic oceanic mass, which is never less than −1.86°C. However, the coldest temperatures in the south occur deep into the continent of Antarctica itself, and may be 20°C lower than in the north.

The polar zones are conventionally defined as the areas within the Arctic and Antarctic circles (and thus confusingly excluding a small part of the land mass of Antarctica). However, it is more useful to a biologist to define these zones in terms of actual temperatures or in terms of vegetational zones. Figure 16.31 shows the possibilities. The Arctic may be defined by the treeline on the three surrounding land masses, although this has been substantially modified by human activities. The Antarctic may be defined by the extent of pack ice in winter. Alternatively each polar zone could be delimited by the midsummer 10°C isotherm (the line joining all places having a mean air temperature of 10°C in the warmest summer month), which has the advantage of applying to both land and sea.

Polar areas are in many senses best seen as "cold deserts", suffering in effect from the same kind of aridity that we met earlier in the hot deserts. This is because precipitation rate is low (zero in some parts of Antarctica), and any snow falls largely onto ice, so that any water present is unavailable to either plants or animals, which exist in a "physiological drought". The Antarctic probably contains 90% of the planet's fresh water, but in a form that is almost entirely useless for supporting life. In many valleys within Antarctica, at temperatures below −50°C where there is no snowfall, the air becomes extraordinarily dry, giving the most arid places on Earth with fiercely desiccating winds. On the other hand, when snow does fall it creates a covering for overwintering animals where the enclosed air is saturated, in equilibrium with the vapor pressure of the ice and snow; thus animals burrowing within the snow are protected against desiccation.

POLAR MICROCLIMATES AND VEGETATION

Ground temperatures towards the poles are greatly affected by seasonal and daily changes in solar altitude, and small variations of slope can have marked effects on soil temperatures. In the Antarctic the generally colder air and extensive ice cover produce more extreme conditions, where microclimatic amelioration may be irrelevant over large areas. However, the Antarctic fringes do provide areas where soil is protected by a thermal blanket of snow, with relatively stable conditions 50–100 mm below the surface. In the gravelly substrates at the melt line in summer, many small invertebrates do survive, with summer temperatures rising just above freezing for a few hours a day even though T_a may remain below zero. The Antarctic has just two species of flowering plant, together with an abundant fringe of lichens and algae. It might be thought that there is little scope for microclimatic variation in such a shallow layer of vegetation, but in fact thermal gradients on a very small scale can be most striking: for example, the surface of lichens and mosses in the maritime Antarctic may reach 50°C at noon on cloudless summer days even though the air temperature 1 m above the ground is only 5°C.

In the terrestrial Arctic fringes, topography and the influence of the sea have a substantial influence on local climate. Hills and mountains lead to cloud formation and may interrupt the solar insolation since the sun may be close to the horizon. Hill slopes also give rise to "katabatic winds", with warmer air flowing down the slopes and giving favorable conditions; by contrast, low temperatures occur where the air is stagnant in valleys. Temperatures are always highest in summer within the reasonably substantial plant boundary layer, where there may be a temperature excess of at least 20°C above ambient amongst small plants. Temperatures may also be raised by several degrees in the soils up to 10 cm below such vegetation and also below lichens and moss turf. Wind speeds may limit this temperature augmentation, however, especially in areas of very patchy discontinuous vegetation.

POLAR FAUNA

In the true Antarctic there are no entirely terrestrial animals except a few mites and two species of midge; the mite *Nanorchestes antarcticus* is the most southerly invertebrate known, occurring within 5° latitude of the South Pole. Even on the fringing maritime Antarctic islands the species diversity is remarkably low, with a limited soil fauna of nematodes, tardigrades, and rotifers, and further mites and insects, including some lice and fleas associated with semiaquatic birds and mammals. This probably reflects the geographic isolation from sources of invading species and the unfavorable direction of winds for dispersal of small species (most of which are wingless). The paucity of animal species is also partly because the Antarctic has nothing like a continuous cover of plants, so that there can be no real herbivores. All the larger animals present are semiaquatic and instead depend on the sea for food, with seals, whales, and penguins dominant, subsisting on the fish and the often very large (see section 11.1.4) marine invertebrates. Of all animals, penguins are the most profoundly adapted for polar life, the emperor penguins (*Aptenodytes forsteri*) breeding in midwinter in continuous dark at temperatures down to −70°C. Antarctic birds in general, and penguins in particular, are the most important members of their ecosystem, in terms of both biomass and of interactions with other components of the environment; they also import organic material from the sea onto the land.

The Arctic, by contrast, having a fringe of plant life on the enclosing land masses, possesses a surprisingly abundant invertebrate fauna, notably spiders, mites, springtails, biting flies, lice, and a reasonable range of butterflies. There is also a substantial burrowing fauna, including enchytraeid oligochaete worms but surprisingly lacking in beetles (at least when compared with their abundance in other habitats). Above ground there is again a wide range of semiaquatic seals, but only one truly terrestrial carnivore, the polar bear (*Thalarctos maritimus*), although in the Arctic fringes both wolves and Arctic foxes are to be found.

Since in all these habitats the species number is low, the community structure is inevitably very simple, the individual community members above all requiring great capacities for endurance of cold, dark, and at times near-starvation.

Tundra zones

Tundra is a biome of low treeless vegetation, occurring in northern Asia and Canada, mostly within the Arctic Circle, often defined as being where temperatures remain below 0°C for at least 7 months of the year. It does not really occur in the southern hemisphere, simply because there are no large land masses at appropriate latitudes. The tundra climate is an extreme version of continental climate, with very short summers between long, cold, dry winters. Tundra in mid-continents is so cold all year that the deep ground is in the condition known as **permafrost**, with no penetration of free liquid water. In the shallow layer of "active" soil, above the permafrost in summer, there may be a negative thermocline, i.e. temperatures may decline with depth down to the permanently cold permafrost layer; this layer may also be almost permanently waterlogged. Alternating freezing and thawing also make this layer physically unstable, and it may become highly fractured.

Nearer the coasts the soil and climate can be much more variable, depending on the extent of snow cover. Here an abundance of snow keeps conditions *less* cold, but also delays the onset of spring growth.

TUNDRA MICROCLIMATES AND VEGETATION

The characteristic vegetation is abundant moss and lichen; at best there may also be very sparse and very low-growing trees (especially dwarf willow), and some coarse grasses (see Plate 9d, between pp. 386 and 387). No large trees or shrubs can survive, because the permafrost precludes taproot penetration, and tall plants with shallow roots would inevitably be blown down. The majority of the living plant tissue exists underground as roots and extensive rhizomes, and in fact most of the organic carbon is in the form of undecomposed dead remains, with only 2% of it "alive" (compared with perhaps 50% in most of the forests of the world). Low cushion-like growth helps to trap this dead material and allow local, slow nutrient recycling.

At these high latitudes thick snow may cover the ground to a depth of several meters during most of the winter months, and this provides a whole range of possible habitats. Thick snow has poor thermal conductivity and acts as an excellent insulating layer between cold air and the soil surface, which in many areas is warmed to around 0°C by geothermal energy so that it remains unfrozen. In fact the lowest soil temperatures through most of the tundra zones are typically recorded in the fall rather than in winter. Even in extreme inland regions the minimum winter temperatures at the soil surface under leaf litter and snow are only −4 to −7°C.

Tundra lichens are so well protected biochemically (usually by glycerol) that they never freeze, providing year-round food to anything that can find them. Many of the larger evergreen plants have small leaves with thickened and hairy cuticles, and most have vegetative budding from runners with protected ground-level buds to allow rapid growth in spring.

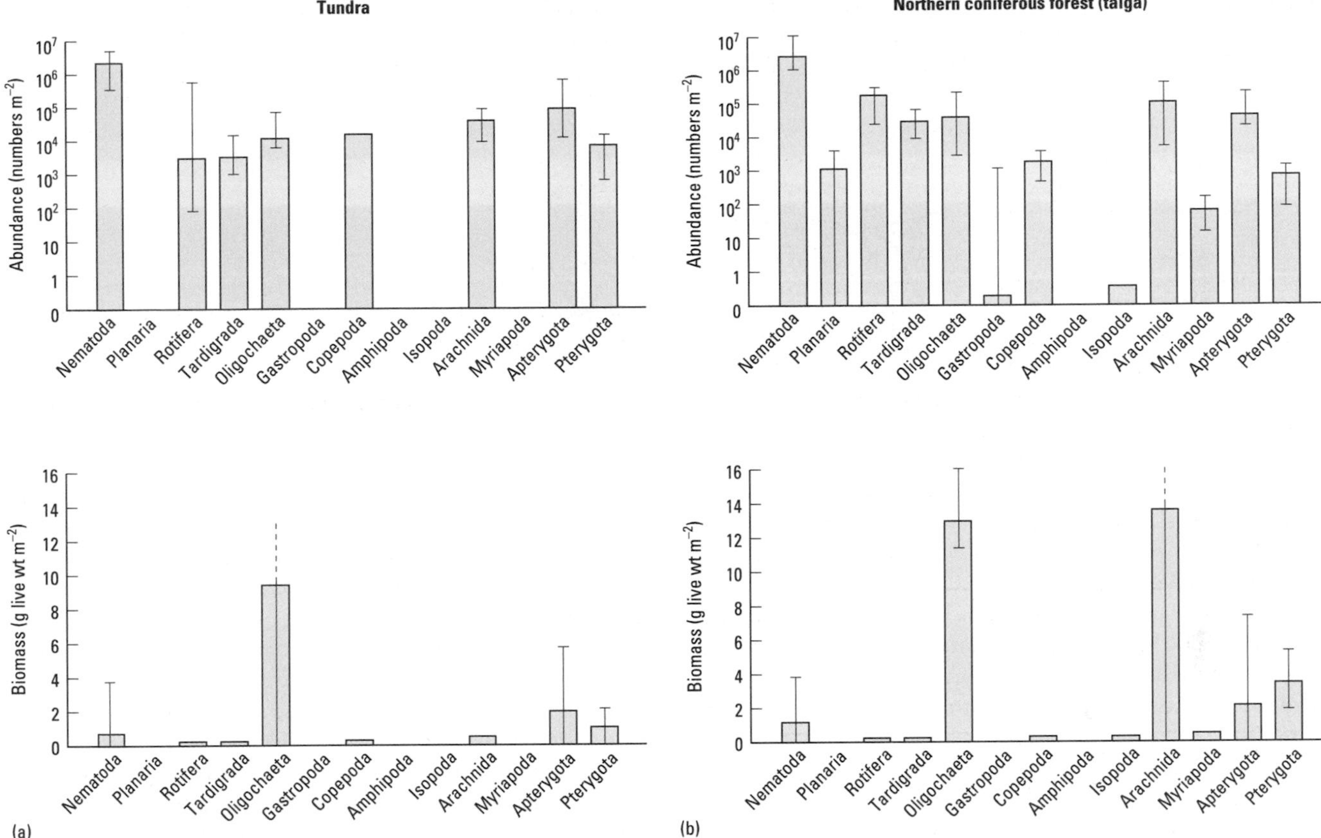

Fig. 16.32 (a) Abundance and (b) biomass of tundra and taiga fauna. (Adapted from Little 1990, courtesy of Cambridge University Press.)

TUNDRA FAUNA

The fauna is inevitably sparse, but there is scope for a reasonable range of rather specialist herbivores. The large herbivores, such as reindeer or caribou (*Rangifera*), migrate south to the timberline in winter and move back to the tundra to breed. Medium-sized herbivores such as musk ox (*Ovibos*), dall sheep, and snow sheep (*Ovis*) stay put and survive the winter feeding on lichens. In addition there are small vertebrate herbivores such as Arctic hares and above all lemmings (microtine rodents). In summer, the tundra is invaded by waterfowl, especially geese, using the long daylight hours to feed. There are relatively few large predators, the commonest being various mustelids and wolves; avian predators include the snowy owl and the gyr falcon. Many of these mammals and birds turn white in winter, giving a high degree of crypsis.

The conspicuous fauna is therefore almost entirely composed of endotherms. Towards the southern edge of the Arctic tundra the species diversity increases though, with a few reptiles surviving; the common adder (*Vipera berus*) occurs throughout northern Norway well into the Arctic Circle, and lizards of various kinds are also found in Norway, as well as in the southern tip of South America within the Antarctic Circle.

There is, of course, also a cryptic invertebrate fauna living in and on the vegetation and thin decomposing layer (Fig. 16.32a), with mites, springtails, and nematodes present in particular abundance, and with a surprisingly high biomass of enchytraeid earthworms. Some gall-forming insects occur on dwarf willow at most latitudes.

The tendency described in the introduction to this chapter to be large or small, but relatively rarely in the middle size ranges, therefore clearly applies (except perhaps for the special case of migrating bird species). There are also many examples in the tundra illustrating "Allen's rule", where animals have relatively small extremities compared with their temperate relatives, to avoid undue heat loss. The tundra rodents and mustelids generally have small ears and toes and short tails, as do many polar birds and mammals, particularly in their juvenile stages. The rule is particularly well illustrated by foxes, where the relative sizes of ears, noses, and paws decrease progressively from desert species (kit fox and fennec fox) through the ubiquitous temperate red fox to the Arctic fox. The same rule may apply to ectotherms: legs, antennae, and mouthparts may be shortened in northern insects, and the incidence of brachyptery (reduced or absent wings) is also higher in Alaskan dipterans than in other faunas.

Boreal forests (taiga)

The more northerly forests of the world are often labeled "taiga", from a Russian word for "swamp forest". They stretch in a wide belt across Canada (see Plate 9e), Scandinavia and northern Europe, and most of Russia, from the northernmost treeline down to a gradual merger with more deciduous temperate forests (see Fig. 16.31). These northern boreal coniferous forests are not really matched in

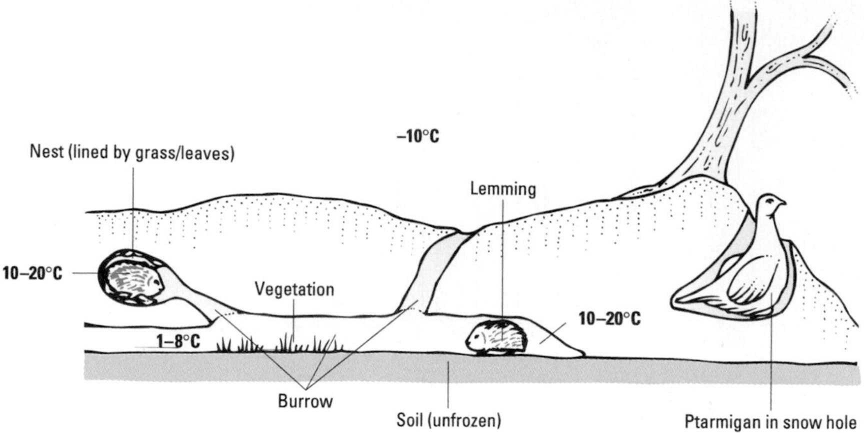

Fig. 16.33 Microclimate within snow—the subnivean habitat. (Adapted from Davenport 1992.)

the southern hemisphere, again because land masses are much smaller at the relevant latitudes, although some of the southern beech forests of southern Chile and Argentina do present some substantial parallels.

The climate allows only a short growing period (2–3.5 months), since it is too cold for much of the year and the fairly warm summers have limited rainfall. Soils are nutrient poor and strongly acidic, commonly called "podzols" and turning with time into peats; remember that peats warm up only slowly but dry out rather easily (see section 15.1.2).

TAIGA MICROCLIMATES AND VEGETATION

Conifers always dominate, usually in low diversity, so that these regions have a dense layer of decomposing conifer needles up to 100 mm deep, producing a very acid effect but providing a thermally buffered zone. There may be larger areas of peaty swampland, dominated by mosses. These swamps are in part created by intermittent bouts of tree loss in natural wildfires that sweep through the forests, feeding on the resinous undergrowth and litter and sometimes clearing thousands of hectares at a time. Within snow-covered areas a **subnivean** habitat results where temperatures in burrows may be quite mild (Fig. 16.33) especially when gregarious endotherms are present.

TAIGA FAUNA

Apart from the moose, herbivores tend to be of only moderate size (usually smaller than those of the tundra): marmots, squirrels, and voles are common, many relying on seeds and berries and also taking invertebrates as food. Small carnivores are drawn especially from the mustelid group (martens, sables, ermine—these zones are the center of the fur trade). Larger carnivores include bears, wolves, snow leopard, Siberian tiger, and lynx. There are many owls and raptors, and specialist conifer-feeding birds such as grouse and crossbills. Few ectotherm vertebrates live in these forests, although there are rare amphibians such as the wood-frog (*Rana sylvatica*) and a few cold-tolerant reptiles (again including the adder) right up to the Arctic Circle.

Invertebrates in the soils are dominated again by nematodes, springtails, and a particularly high abundance of mites (Fig. 16.32b). Sawflies, gall-midges and weevils (insect groups often specialized in feeding on conifers), and a plethora of biting flies, are evident above ground.

The fauna of cold biomes is clearly dominated by small invertebrate ectotherms and by relatively large bird and mammal endotherms, each of which have their special adaptations. Although we will deal with these separately, it is important to remember that some of the adaptations discussed apply to both groups. In particular, all the animals that survive at high latitude use behavioral strategies as a first line of defense against the cold, and this is true for both ectotherms and endotherms. The major behavioral strategies to be found are exploitation of favorable microhabitats, migration, and the development of gregarious habits. Similarly all of them are likely to share biochemical strategies of enzymes with lower temperature optima and membranes with homeoviscous adaptation. For each category of animal we cover behavioral strategies first and then move on to physiological adaptations.

16.3.2 Small ectotherm evaders and their strategies

Use of microhabitats

As always the utility of behavioral tricks is much enhanced for smaller animals. In the polar zones their choice, and exploitation, of somewhat less severe and more sheltered microhabitats is particularly evident. The winter temperature range beneath the snow cover (subnivean zone) is markedly less than the range of air temperature above snow, and this microhabitat is exploited by a whole range of animals (see Fig. 16.33). The advantages of this way of life are greatly enhanced by the penetration of light through snow, allowing plants to stay green and edible. Many invertebrate herbivores exploit this resource, and some are specialized as gall-formers, getting added protection from the modified plant tissue they induce (see Chapter 15).

To create and manipulate microclimates, burrows can be dug into the snow itself, or below it into the litter or soil layers. Some toads in the taiga zones can burrow down more than 1 m to get below the frost line; large numbers of beetles and other ground-dwelling insects also occur at this depth in winter.

Huddling and aggregation

Huddling behavior is the commonest form of gregariousness used as a thermal strategy, and is very obvious in a whole range of animals from cold habitats. Its effectiveness increases as the difference between air temperature and preferred T_b increases, and with the

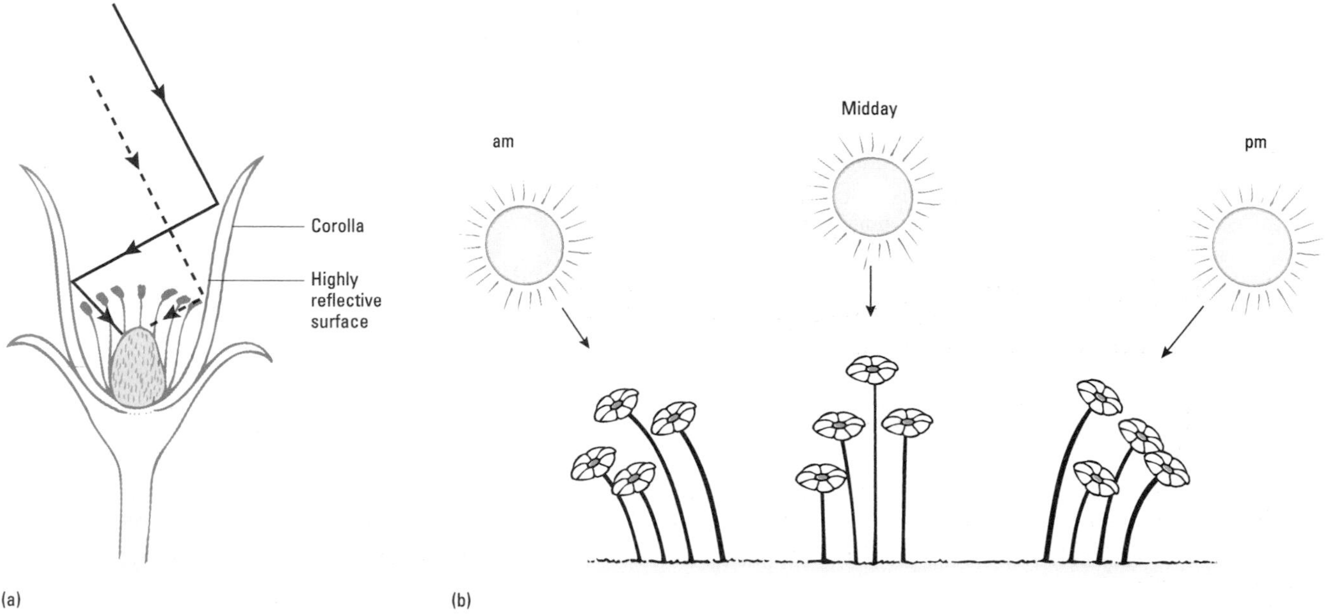

Fig. 16.34 Arctic flowers acting as (a) parabolic reflectors and (b) sun trackers to give a warm microclimate around their ovules, exploited by many small insects. (Adapted from Kevan 1975.)

number of participating individuals. Examples among the smaller invertebrate ectotherms are not particularly well studied. The mite *Alaskozetes* aggregates in clusters under rocks, and other mites and some collembolans also aggregate, although it is debatable whether this is real thermally advantageous huddling or just a shared preference for the same microhabitat. Amongst the reptiles several cases are well known, with winter aggregations occurring in many snakes and lizard. The adder is a familiar example, with dens of up to 80 individuals having been found in Finland. Similarly the red-sided garter snake, which occurs in western Canada (where the winter temperature is often –40°C), shelters in huge groups of thousands at a time in sink-holes beneath the snow.

Migration

Long-range migration is not a very common strategy for ectotherms from cold climates. There are, however, just a few spectacular examples: the monarch butterflies (*Danaus plexippus*) that move *en masse* from sites in Canada and the northern USA each year to overwinter in Mexico are the best studied case, though only some of them come from really cold habitats. These butterflies then spend the winter in huge aggregations, subsisting largely on their stored lipid reserves (triglycerides, derived from summer feeding on nectar) as there are inadequate nectar supplies to maintain them in the Mexican forests. *Colias* butterflies also migrate seasonally from genuinely Arctic habitats in northern Scandinavia to the southern Baltic regions.

Thermoregulation

Small terrestrial animals from cold climates must have highly cold-adapted enzymes and membranes. For example, the mite *Alaskozetes* shows low enzyme activation energies and elevated metabolic rates at low temperatures, relative to temperate species. But they must also be good thermoregulators of necessity, and must make any possible use of the minimal heat inputs available. Most boreal animals are highly adept at **behavioral thermoregulation** by microhabitat choice and aggregation, as described above, but can also utilize sophisticated basking techniques, as with the examples we looked at in temperate habitats (see Chapter 15).

A small ectotherm may raise its T_b at least 25°C above ambient by **basking** in the polar summer. Basking butterflies from Greenland and Canada have been studied in some detail, and they can warm themselves very substantially above ambient using their wings as heat gatherers. Many are melanic, the darker color improving absorption. Arctic fritillaries are dorsal baskers (see Fig. 15.22), the dorsum of the wings being dark, while *Colias* butterflies are lateral baskers—the Arctic species (e.g. *C. hecla*) are unusually dark in color especially on the ventral sides of the hind wings; their extra melanism allows them to achieve an excess temperature up to 80% greater than lighter forms. Note that, in addition to basking, contact with the substrate is also used to warm the body in many of these Arctic insects. Butterflies use bare ground for contact warming, seeking warm and sheltered spots.

Basking also occurs in wingless insects, the best example being the "woolly bear" caterpillars, *Gynaephora groenlandica*, which may spend up to 60% of their active time in basking and only 20% in feeding. This species may spend as long as 14 years in the caterpillar stage, reflecting its very slow growth rate.

There are also some specialist cases where insects use particular species of flowers as basking warm-up sites (e.g. *Dryas integrifolia* and the Arctic poppy *Papaver radicatum*). These flowers are shaped like bowls, and they rotate by phototropism so that the corolla always points towards the sun as it passes across the sky each day (solar tracking). Their shape acts as a parabolic reflectoı to concentrate radiation into the center of the bowl of the flower, raising it by 5–8°C above ambient (Fig. 16.34). Mosquitoes, hoverflies, blowflies, and danceflies all exhibit a raised T_b in association with sitting in these flowers.

Larger polar ectotherms may also use basking as a key component of their thermal strategy. The adders and lizards of northern Norway may bask for large parts of the 20 or more hours of daylight in the summer, feeding between basking bouts to permit lipid stores to be built up for the winter.

In the case of a few large insects, although nominally ectothermic, remember that a component of **endothermy** also comes into play in achieving thermoregulation. The Arctic bumble-bee, *Bombus polaris*, found in northern Canada, is able to fly at low temperatures due to its inherent endothermic warm-up (see Chapter 15), and the same almost certainly applies to several *Bombus* species found in northern Scandinavia. These species also use endothermic warm-up at times of egg production, the queen pumping warm blood to her abdomen to speed up egg maturation, and then staying at an elevated T_b to brood the larvae within the nest.

Coping with freezing

Diverse taxa have adopted the "freeze tolerance" strategy discussed in Chapter 8, involving some tolerance of extracellular ice formation, with cryoprotectant polyols or sugars to prevent damage. Examples include many terrestrial insects, a few other arthropods such as centipedes, a range of gastropods, annelids, and nematodes, and several species of frogs, lizards, and turtles that overwinter on land. Recent evidence shows that in polar nematodes freezing may also occur intracellularly, an otherwise unheard of phenomenon—the worms freeze in all body compartments, and can subsequently melt, grow, and reproduce apparently normally.

Among the insects, **freeze tolerance** is the exception rather than the rule, but is widespread in some families from the Coleoptera, Diptera, Hymenoptera, and Lepidoptera, especially in larvae and pupae but rarely in adults and apparently never in eggs. In many insects low temperature is the immediate trigger for **polyol synthesis**, in conjunction with other environmental cues, which might include photoperiod, humidity, food availability, and to a lesser extent endogenous factors such as hormones and the physical changes of diapause. Among the best studied insects are a parasitic wasp, *Bracon cephi*, the high Arctic caterpillar *Gynaephora*, and the larvae of the goldenrod gall-fly, *Eurosta solidaginis*. In *Bracon* larvae, which overwinter in frozen Canadian prairies, glycerol is present at up to 25% body weight in winter, giving a supercooling point as low as −47°C. *Gynaephora* caterpillars can spend as long as 10 months every year in a frozen state, at −50°C or lower. In field populations of *Eurosta*, which survive a mere 12 weeks in the frozen state, long-term chilling at the right time of year is required to initiate glycerol production: at least six consecutive days with daily average temperatures of less than 5°C in mid-November. Laboratory experiments show that glycerol is synthesized first, followed by sorbitol as the temperature falls (Fig. 16.35). Glycogen is clearly the source of these polyols, and the glycogen phosphorylase enzyme in the fat body is cold-activated, with low temperature causing a rapid conversion from the inactive to the active form. From September to November the larvae also accumulate monounsaturated fatty acids at the expense of saturated forms, suggesting activation of a Δ^9-desaturase enzyme (see Chapter 8). However, traditional ice-nucleating agents (INAs) have never been identified in *Eurosta*, and recent evidence indicates that calcium phosphate performs this role, existing as

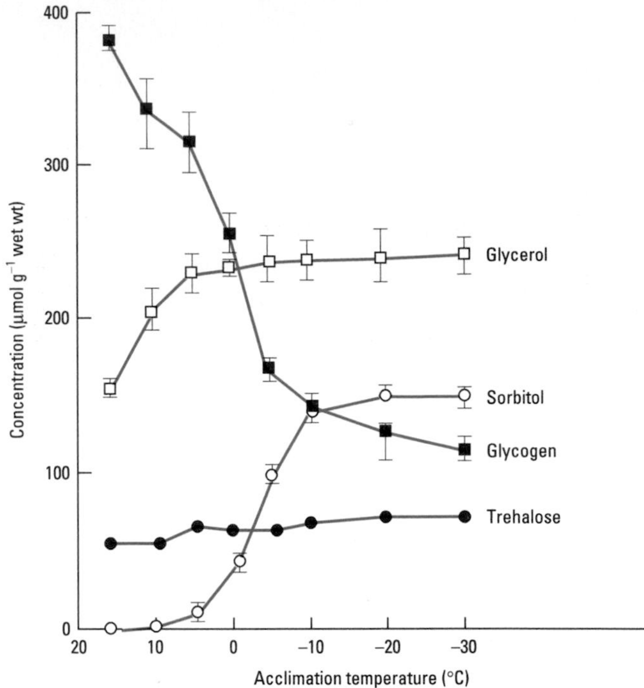

Fig. 16.35 Sorbitol and glycerol synthesis in the gall-fly *Eurosta* at decreasing acclimation temperatures, using up glycogen stores to make the cryoprotectants. (Adapted from Storey *et al.* 1981.)

small crystalline particles within key tissues such as the Malpighian tubules.

Many animals show **seasonal patterns** of glycerol content, of **supercooling point**, and of melting and freezing points (Fig. 16.36). Examples of the supercooling points (SCPs), and the solutes used to produce them, are shown in Table 16.4 for a range of terrestrial ectotherms that are freeze tolerant; the values are not exceptionally low, with a slow controlled freezing at moderately high subzero temperatures being preferable. Values of −5 to −25°C are clearly quite common, and there is a general correspondence between SCP and the normal site of overwintering: species that survive in vegetation have lower values than those that spend the winter in caves or other more protected locations. Remember that polyols also give protection against desiccation, which may be an important factor in insects that overwinter in very cold dry air. For example Fig. 16.37 shows the variation in glycerol content in relation to humidity in the polar mite *Alaskozetes*. Cold and drought are both potential problems for high-latitude animals, but the biochemical and physiological strategies to deal with both problems are often complementary.

Many of the polar arthropods can tolerate substantial ice formation in their extracellular fluids. In some species up to 90% of this water may freeze, particularly when the animal is in contact with external ice and undergoes "inoculative freezing" from the outside inwards. Precautions against contact with ice are often present to limit the risk; examples include the use of silken cocoons in many lepidopterans, or galls as in *Eurosta*, or specially impregnated winter burrow walls in some beetles. Where ice contact is unavoidable, slow, safe freezing is promoted by INAs in the hemolymph, and protein nucleators are fairly common in insects as part of the protective

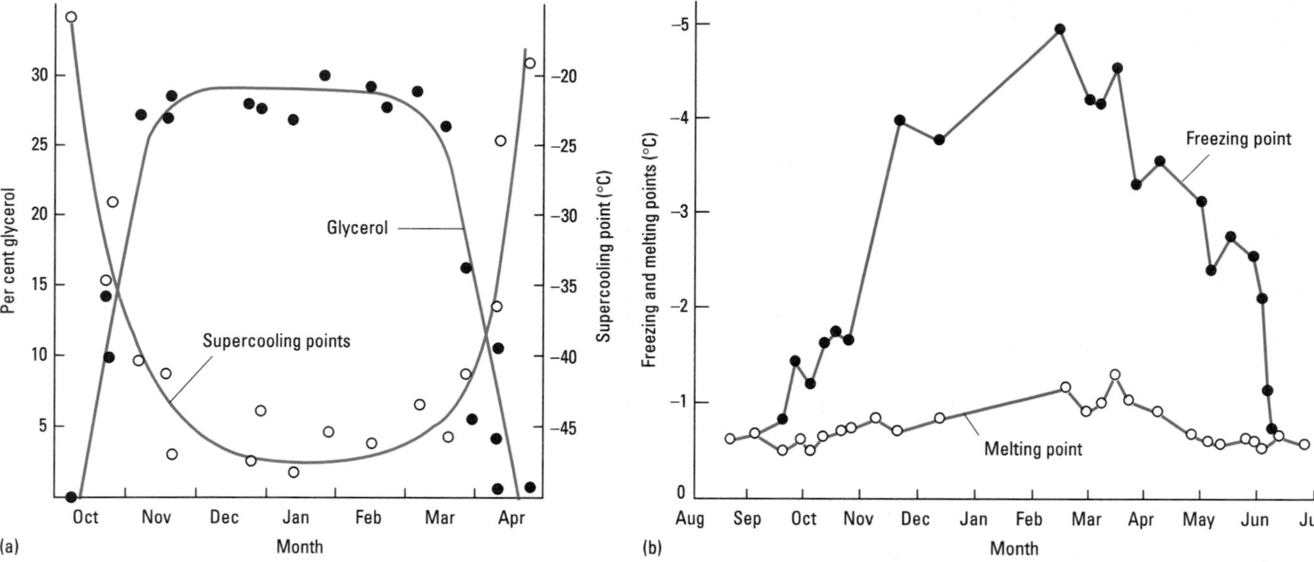

Fig. 16.36 (a) Seasonal changes in glycerol contents and supercooling points in the larvae of a caterpillar, and (b) in the melting point and freezing point of a larval beetle. (From Hansen 1973.)

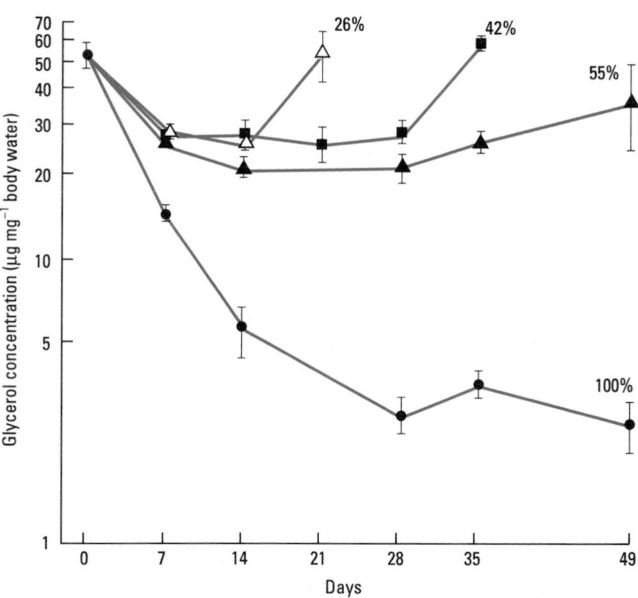

Fig. 16.37 Glycerol contents (means ± standard errors) of the mite *Alaskozetes* at 4°C are moderated in relation to humidity, with glycerol gradually disappearing at high humidities. (From Cannon & Block 1988.)

mechanism. For larvae of the crane fly *Tipula trivittata*, both protein and lipoprotein nucleators have been identified in the hemolymph. The ice-nucleating protein from the wasp *Vespula maculata* has been purified and is a protein of relative molecular mass 74,000, with an amino acid content that includes 20% glutamate/glutamine, 12% serine, and 11% threonine.

Amongst vertebrates, at least four species of North American frog, the Siberian salamander (*Salamadrella keyserlingii*), and even some reptile species can also tolerate freezing. Either glycerol (in the

Table 16.4 Supercooling points (SCPs) in species living at subzero temperatures.

Species	SCP (°C)	Known solutes
Spiders		
Clubiona	−15.4	Glycerol, AFP
Philodromus	−26.2	Glycerol, AFP
Mites		
Alaskozetes	−30	Glycerol, polyols
Lepidoptera		
Pringleophora	−5.0	Glycerol
Isia	−18.2	Glycerol, sorbitol
Nemapogon	−26.1	Trehalose, amino acids
Pieris	−26.2	Trehalose, amino acids
Laspeyresia	−31.5	Glycerol, trehalose
Alsophila	−44.6	Glycerol
Beetles		
Pterostichus	−10.0	Glycerol
Meracantha	−10.3	AFP
Dendroides	−12	Glycerol, sorbitol, AFP
Ips	−32.4	Ethylene glycol
Dendroctonus	−34	Glycerol
Hymenoptera		
Trichiocampus	−8.6	Trehalose
Megachile	−27.7	Glycerol
Camponotus	−28.7	Glycerol
Bracon	−41.2	Glycerol
Eurytoma	−49.2	Glycerol
Flies		
Xylophagus	−6.0	Sugars, amino acids
Tipula	−7	Sorbitol, INP
Eurosta	−10.3	Glycerol, sorbitol, trehalose
Diplolepis	−32.7	Glycerol
Rhabdophaga	−49.1	Glycerol
Vertebrates		
Hyla	−2.0	Glycerol, glucose
Pseudacris	−2.0	Glucose
Rana sylvatica	−3.0	Glucose
Turtle	−3.3	Glucose, amino acids
Northern ground squirrel	−2.9	?None

AFP, antifreeze protein; INP, ice-nucleating protein.

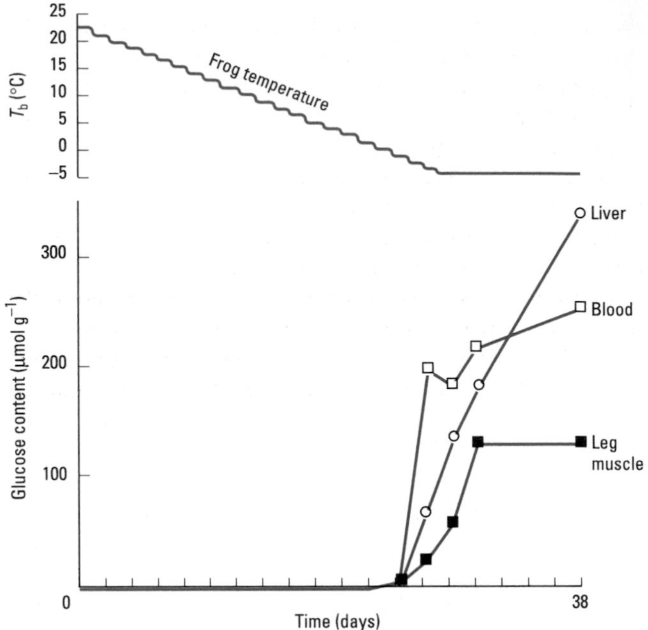

Fig. 16.38 Levels of glucose in the tissues of the freeze-tolerant frog *Rana sylvatica* as its body temperature (T_b) declines. (From Storey & Storey 1984.)

tree-frog *Hyla versicolor*) or glucose (in some wood-frogs and in the lizard *Lacerta vivipara*) is used as a cryoprotectant in the few cases that have been studied. The wood-frog *Rana sylvatica*, which is found north of the Arctic Circle, is by far the best studied vertebrate freeze tolerator. Here the levels of glucose in the muscle, liver, and blood begin to rise about 14 days after T_b reductions are induced in laboratory studies (Fig. 16.38). But at least some of the frogs do not back up their cryoprotectants with nucleating agents, and instead appear to rely on their relatively large size to keep the cooling rate slow enough to insure that ice only forms extracellularly. Frog skin is highly water permeable, and an important part of the strategy for winter survival is the choice of a well-protected and humid hibernation site where temperatures will fall only slowly.

R. sylvatica is famous for its ability to survive repeated freezing of its whole body, with all breathing and circulation stopped and the nerves unresponsive to stimulation. It will survive for 3–14 days in any one freezing cycle, and when frozen a single frog may contain 7–8 g of ice in its coelom and beneath the skin, indicating that 65% of its body water is tied up in the frozen state and all of its organs must be substantially dehydrated. Nevertheless, on thawing it recovers all its functions, with excitability of the peripheral nerves and simple reflexes being reinstated in 5–14 h. There is an in-built seasonality in the frog's cryobiology, with fall frogs better able to survive longer and harsher freezing than summer frogs; the fall frogs have higher osmotic concentrations and glucose concentrations in their plasma. Glucose is produced from massive glycogen stores (up to 180 mg g^{-1} are built up in the liver before hibernation), and synthesis of the cryoprotectant glucose is triggered by the beginnings of ice nucleation in the body (often when ice first begins to form on the skin of the frog), via an adrenergic and cAMP-mediated signal transduction system. As Fig. 16.38 shows, glucose concentrations may reach 150–300 μmol g wet wt^{-1} in the core organs of frozen wood-frogs (liver, heart, and brain) compared with 1–5 μmol g^{-1} in unfrozen frogs. The rapid distribution of glucose from the liver is critically aided by the ability to move this sugar across cell membranes, out of the liver and into other organs via glucose transporters (see Chapter 4); cAMP and protein kinase A are also mobilized during this process. For wood-frog liver the $\dot{V}_{max}$ value for glucose transport by membrane vesicles is 70 nmol mg^{-1} s^{-1} at 10°C, compared with only 8 nmol mg^{-1} s^{-1} for the common temperate frog *Rana pipiens*. The total number of transporter sites is 4.7-fold higher in the liver of the freeze-tolerant species, and varies seasonally in that species by about eight-fold. There is also a gradient of glucose concentration from the core to the peripheral tissues, which enables the frog to thaw uniformly, thereby avoiding tissue damage in the spring. The thawing of frogs can now be studied by proton magnetic resonance imaging, since the signal undergoes a large change during the transition from ice to water; protons in ice are invisible to this technique and frozen areas therefore show up as black. From such studies it is clear that the liver and heart thaw first during recovery from freezing, so that their vital functions can begin while the rest of the frog is still thawing out. We are also beginning to understand the processes that occur at a gene transcription level during freezing in this species.

Freeze-tolerant animals must survive and maintain some degree of homeostasis despite a lack of circulating fluids. Their enzymes must be able to tolerate very low temperatures and very high salt concentrations. For many of them, in the frozen state energy production switches to the hydrolysis of phosphagens, phosphoarginine (PAr), or phosphocreatine (PCr) (see Chapter 6), followed by anaerobic glycolysis with lactate and alanine as end-products. The metabolic rate is usually greatly depressed (to 5–10% of normal resting values), giving a typical hypometabolic state.

Avoiding freezing

Many animals that encounter subzero temperatures are freeze intolerant but are able to survive by virtue of the phenomenon of supercooling alone. Such animals must eliminate or mask all INAs as winter approaches. The small terrestrial mites and insects that cannot tolerate ice formation use additional strategies to assist their cold-hardiness: they allow the body to become very dehydrated, to reduce the availability of freezable water, and they may void their guts as winter sets in, to clear any potential ice nucleators. As a result of all these mechanisms, a number of Arctic species have lower critical temperatures of −55 or even −70°C.

Springtails and mites, among the most important components of polar soil fauna, are almost invariably freeze intolerant and rely on their supercooling properties for survival. The supercooling points of circulating fluids in these cold-hardy but freeze-intolerant animals are usually (though not invariably) very low compared with those in freeze-tolerant species. Antarctic mites and Arctic insects can generally supercool down to −30 or even −40°C, with eggs and pupae showing even lower values. Some animals can remain almost fully active at subzero temperatures. For example, *Alaskozetes antarcticus* (body mass 150–200 μg) shows peak locomotory activity at 16–24°C but does not enter cold stupor until about −5°C. Another Antarctic mite, *Nanorchestes antarcticus*, has even been observed moving at −11°C, but it has a relatively high temperature requirement for optimum activity, reflecting the high temperatures found at the radiative surface on the mosses, algae, and lichen on which this species lives.

In some of the polar insects and mites glycerol acting as an antifreeze can reach concentrations of 25% of fresh body mass during the winter. In addition, due to their small size these animals undergo substantial dehydration, which reduces their content of freezable water and elevates the effective antifreeze concentration. Antifreeze proteins (AFPs) are also now known to occur in some mites, spiders, and insects (as well as in the polar fish where they were first identified, described in Chapter 11). The associated antifreeze mRNA has been studied in some northern temperate insect species, and its activity is inducible at around 2–4°C in the laboratory, but induction only occurs in winter, and cannot be achieved after about February in these tissues. This suggests that AFP synthesis normally involves an interaction of low temperature priming with an endogenous clock system. AFPs tend to be recycled in spring as amino acids for growth and egg production.

Although freeze tolerance may be a "cheaper" strategy (see Chapter 8), recent evidence suggests that freezing intolerance may be the better strategy in low temperature but thermally quite variable environments, such as the maritime Antarctic margins. Here small terrestrial springtails and mites are very common, and switching rapidly from inactive to active states may be the ideal way to exploit winter thaws and avoid summer frosts.

Life cycles

Almost all the small "evader" animals from cold climates have specialized life cycles, in which multivoltinism (having many generations per year) is very rare indeed. Usually the invertebrate fauna show very brief active periods during summer, otherwise being in a resting state as eggs, pupae, etc. They may achieve fairly "normal" growth rates during the brief summer, but inevitably have lower annual growth increments. This results in two main patterns of life history, each requiring substantial flexibility and opportunism.

The first strategy involves **univoltinism** and adult lifespans that are greatly reduced, so that in some insect species the eggs mature within the pupal phase, with emergence, mating, and oviposition squeezed into just a few days of suitable weather. On the other hand, while retaining univoltinism and an opportunistic burst of reproductive activity, there may be a **life-cycle extension** to 2 or more years in other taxa that normally complete the cycle annually, so that there is much overlap of generations. Mites and springtails have particularly long lifespans (5 years in *Alaskozetes*), as does the caterpillar *Gynaephora* referred to above. *Pardosa* wolf-spiders may have a 6-year life cycle, whereas temperate relatives take only 1 year; the high-latitude populations mature at a much later age. Even within a species the effect can be striking; for example, the springtail *Hypogastrura tullbergi* lives 3–5 years in polar regions and just 2–12 months in temperate regions. Water-striders (bugs living on the surfaces of fresh water) have univoltine cycles with an obligate winter diapause in Quebec but show a southwards cline to a bivoltine cycle; interestingly, total reproductive output is similar at all latitudes.

In many Arctic insects "hibernation" can occur at any stage except the egg, to facilitate this lengthened life. In some cases this may be a true diapause, but more commonly it may be a less specific form of torpor/dormancy, probably because full diapause could be inappropriately triggered by a period of poor summer weather. Torpor may also occur in larger animals, including the relatively rare vertebrate ectotherms in cold climates; in reptiles such as the cobra, living on cold plateaus, it leads to a drop in blood sugar and pyruvate and an increase in blood lactate, but also to a rise in nonprotein nitrogen and particularly uric acid levels, indicating protein catabolism. This is in contrast to endothermic vertebrates, where torpor is accompanied mainly by lipid catabolism.

Reproduction itself may also be highly specialized. In insects the females are often wingless and almost immobile (e.g. in crane flies and midges) and males seek them out, presumably using pheromonal cues. The red-sided garter snake provides another example of specialized reproductive behavior; the males emerge synchronously from the winter dens in spring, then the females come out in small groups and are immediately set upon by males, forming massive "mating balls". Each female then bears up to 30 live young. The triggers for emergence are unknown, and androgens in both sexes seem to be at very low levels.

Viviparity in the vertebrate ectotherms is rather common (e.g. *Vipera* and *Lacerta* amongst the reptiles). This may be because retention in the mother's body, which is undergoing typical shuttling heliothermy to maintain its T_b, is the best way of keeping development of the young at a fast and steady rate.

Feeding and predation

As in the extreme desert environments, foodstuffs may pose a problem. Many of the small invertebrates are omnivorous with little diet specialization, taking only detritus for much of the year. Macrovegetation may be unavailable or deeply frozen, and is heavily physically defended, so that animals with food storage habits are favored. However, lichens remain unfrozen so that caribou, using their hooves to dig and tough lips to forage, can survive year-round. Green plants may also remain in a minimally active state within the snow layer, with just a little light penetrating, providing enough food for the burrowing rodent populations. Small populations of carnivores are found in tundra and taiga, and large populations of biting ectoparasites are also to be found, benefitting both from blood meals and from the microclimates created by their hosts.

16.3.3 Large endotherm endurers and their strategies

Use of microclimates and burrows

The subnivean vegetation is a food resource for some small specialist herbivorous mammals such as lemmings and shrews. They in turn become a food source for small carnivores like foxes and raptors. Thus a whole submerged community exists through the winter, and some of the mammals even choose to reproduce in winter because of the benefits of snow cover.

These animals all exhibit a form of microhabitat selection, but for the large ones it is more specifically expressed as a **burrowing** behavior. Perhaps the most famous mammalian examples are the lemmings, found in most of the northern tundra areas throughout the year. They dig winter burrows within the snow, lined with grass, where they live gregariously and do not enter torpor; in fact these

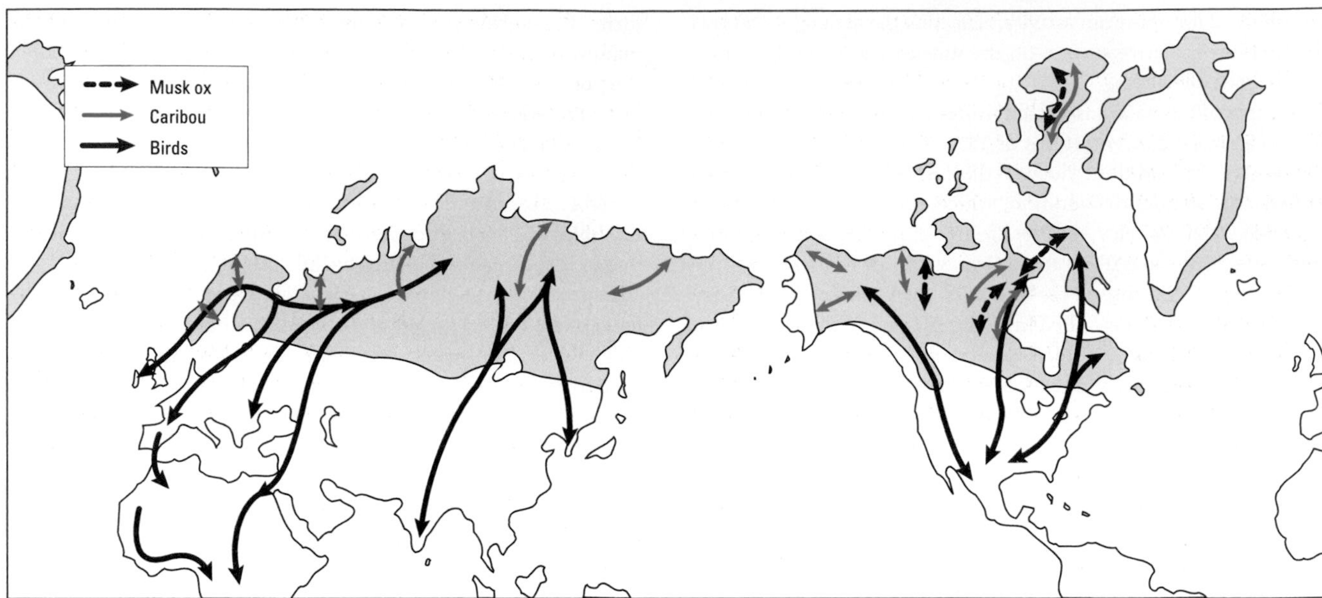

Fig. 16.39 Migration patterns in tundra animals.

winter burrows may be warmer places than the lemmings' summer holes dug in the earth. Rodents in the tundra and taiga biomes have a particular problem in being the major prey for raptorial birds, so their burrows provide an important escape from predation as well as a climatic refuge (though owls can hear lemmings even under a dense snow layer). Some raptors can also detect rodent abundance by perceiving the ultraviolet emission from deposits of urine and feces. Therefore lemmings in Scandinavia and Canada have elaborated the architecture of their summer burrow, unusually for a rodent, to include a latrine area so that no clues are left above ground. In winter, when predation risk is much lower, their snow burrows have no latrine and they defecate freely away from the nest. The burrows of ground squirrels, used during winter torpor, are also decidedly elaborate, and are thought to act as biological condensing towers, water evaporating from the warm lower levels and recondensing in the upper cold levels, where it may be available as an important water source for the squirrels when they arouse.

Permanent Arctic residents among the birds also use snow burrows, although rarely gregariously; the classic example here is the burrowing rock ptarmigan, rather larger than a lemming. Burrowing is also employed quite extensively by seabirds at high latitudes. Petrels and shearwaters dig into rock crevices, while puffins excavate burrows in the turf above sea cliffs in many cold maritime areas of the northern hemisphere. Even the largest terrestrial animal found at the poles, the polar bear, may dig temporary snow dens in the harshest periods of the winter, with suckling females spending many months in deep snow dens.

Polar and tundra endotherms also use **basking** in association with microclimatic variation, which may have little effect on their core temperatures but will certainly help to maintain peripheral temperatures and thus reduce the overall cost of thermoregulation. Semiaquatic animals such as penguins and seals bask extensively, to assist in the replacement of heat lost during dives; they may rest facing into the sun with the wings (or in the case of seals the tail flippers) raised and splayed out.

Migration

Migration is the most striking behavioral strategy shown by the bigger polar animals, and could be seen as a larger scale version of the movements between microhabitats discussed above. It is also perhaps the equivalent of a life cycle involving resistant stages in terrestrial polar invertebrates, providing an escape in space rather than in time. It is infrequently encountered in ectothermic tetrapods, the reptiles and amphibians, since their locomotory speeds and endurance are limited. But there are many endothermic examples, classic cases being the caribou and reindeer, migratory mammals that trek overland to the forested taiga zones as winter approaches, and the lemmings with their fall migrations southward. Polar bears also migrate south across the frozen Hudson Bay in winter. Among the birds migration is especially common since movement by flight is energetically favorable. Penguins of course miss out on this benefit; many of the smaller penguin species merely move away from the Antarctic mainland and spend the winter at sea or on small islands such as the Falklands and South Georgia. But high latitudes in summer provide areas of high productivity and thus serve as good feeding grounds for flying birds, with ducks, terns, and many waders moving into littoral zones, and martins and swallows moving onto exposed land areas to exploit the copious insect life. Most of the summer tundra and taiga fauna retreats to more temperate areas for the winter months (Fig. 16.39). Some then show quite profound changes of lifestyle; for example, the knot (*Calidris canutus*) switches from being a rufous-feathered, arthropod-eating land bird in summer in the high Arctic tundra where it breeds, to being a gray-feathered, mollusc-eating temperate shorebird in the nonbreeding winter season. But some birds may avoid this need for a changed lifestyle by moving to the opposite hemisphere to enjoy another "summer". At the extreme, the Arctic tern (*Sterna paradisaea*) flies from pole to pole every year, spending the northern summer in

Greenland or Norway and the southern summer around the shores of Antarctica.

Migration may have multiple functions and advantages, and may be triggered by multiple cues, but it nearly always results in taking the animals to more equable thermal environments. However, migrations in the endothermic birds are probably less to do with avoiding very low temperatures and rather more to do with leaving areas where foraging will become unprofitable due to the imminent snow cover. The same can be said of the large mammals, whose lichen food would be inaccessible under a thick snow cover. In semiterrestrial mammals such as elephant seals (*Mirounga leonina*), the situation may be rather different in that migration by swimming may be unrelated to food but necessitated by the requirements of breathing through ice holes, so that the seals move north from Antarctica to areas where the ice is thin enough to maintain such a hole; but again temperature itself is not the main problem. It is normally only in small birds and mammals that cold alone would be likely to be lethal.

Migration may be set in train by temperature cues in a few cases, but is probably more commonly brought about by changing **photoperiod** or by food supply. In most cases, the start of migration is inevitably preceded by a period of laying down energy stores as fat, to fuel the journey.

Reproduction and life-cycle adjustment

Endotherms that migrate into high latitudes for the summer commonly breed there, exploiting the transiently highly productive ecosystem. This exposure to prolonged summer day lengths necessitates a "resetting" of the response to photoperiod, to insure that gonadal development is initiated at the correct time, avoiding an inappropriately early stimulation of the reproductive system and associated behavioral repertoires. Primarily, the relation between day length and gonadotropin secretion must be readjusted. The effect of decreased sensitivity to photoperiod and hence a shorter reproductive cycle can be seen in birds that have been introduced to new latitudes, and in species such as the house sparrow, with a wide latitudinal distribution, where the period of active spermatogenesis is shortened by about 2 days per degree of latitude the further north the birds breed.

High-latitude reproduction may also require a higher degree of **parental investment** from one or both parents. In the hamster *Podopus*, from cold deserts, the reproductive cycle is extremely compressed, requiring rapid milk production, which puts a real strain on the mother's water balance, and also maternal hyperthermia. Species from the more arid and cold parts of the desert show biparental care, the male's presence increasing humidity in the burrow and so alleviating thermoregulatory and water balance stresses for the female. Whereas species from areas with more predictable rainfall rear the young without male assistance.

Huddling and aggregation

Huddling in endotherms has been a favorite topic for polar researchers, as it is so spectacular in some seals and penguins. Smaller penguins probably huddle largely to insure group cohesion and aid mating access in the spring. The thermal advantage for these penguins, in the face of horrendous wind-chill factors in the Antarctic winter, may be very limited. But penguins such as the emperor penguin, *Aptenodytes forsteri*, and the king penguin, *A. patagonica*, have the additional advantage of large size and compact shape, lacking protruding "eartufts" and tails. Emperor males may spend the winter in the Antarctic interior incubating eggs, and fasting for up to 115 days, when they certainly do gain thermal benefit from huddling. At moderately cold temperatures the emperor penguins form fairly loose semicircular huddles, all with their backs to the wind. When temperatures drop to the typical –50°C of an Antarctic winter, with wind speeds over 160 km h^{-1} (100 mph), the birds form much denser tight huddles of several thousand individuals, giving a biomass of perhaps 100 tonnes. Most of each bird's surface is then in contact with other penguins rather than with the cold air, reducing weight loss per individual (from metabolism of stored fat) by 50% relative to a nonhuddling penguin, and saving as much as 80% of the heat loss that would occur from one bird alone. There is a 25% reduction in metabolic rate, perhaps in part because the birds sleep for long periods.

Polar mammals such as seals and walruses exhibit a less dramatic but also thermally effective huddling when hauled out on beaches in cold weather. In many species, individuals that get too cold on land will return to the sea at –1.86°C to "warm up". However, there are a few species of pinniped, such as elephant seals, that have molt periods when they have to remain out of water and fairly immobile for several weeks. In colder periods, these seals will huddle together as pods of up to 30 individuals, staying warm enough to melt surrounding snow.

Burrowing combined with huddling may give energetic benefits in cold climates, but it may be effective at least as much due to the local heating of the burrow that a huddle produces, as to the reduced surface area of each participating individual. For example, in short-tailed voles (*Microtus agrestis*) the local heating effect is estimated as 55% of the total energy saving.

Insulation

When active (i.e. in less cold periods) the regulation of T_b in cold-climate endotherms is largely at the heat *loss* end. The larger polar mammals show an obvious tendency to have long, fine, and densely packed fur, commonly 30–70 mm thick in the larger species (see Fig. 8.17) and without any particular correlation with body size. However, in small species such as lemmings or weasels the fur is much shorter, and usually not significantly longer than in an equivalent tropical species, although it is denser, and is spread more completely over the body including the head and extremities.

Fur in boreal mammals also varies with season (Fig. 16.40). Arctic foxes and the stoat (ermine) are also classic examples of sub-Arctic tundra species that grow a denser and highly camouflaged **winter coat**. The advantages of increased pelt thickness are strictly size related though: mammals of less than 1 kg show negligible improvements in winter pelt depth, whereas larger mammals such as the red fox and the timber wolf show depth and insulation increments almost in proportion to body size. Phenotypic acclimatory responses also occur; domestic dogs, cats, and rabbits may all increase their coat length when kept in colder conditions. The fur of large boreal mammals is often so effective that the complete thermal gradient

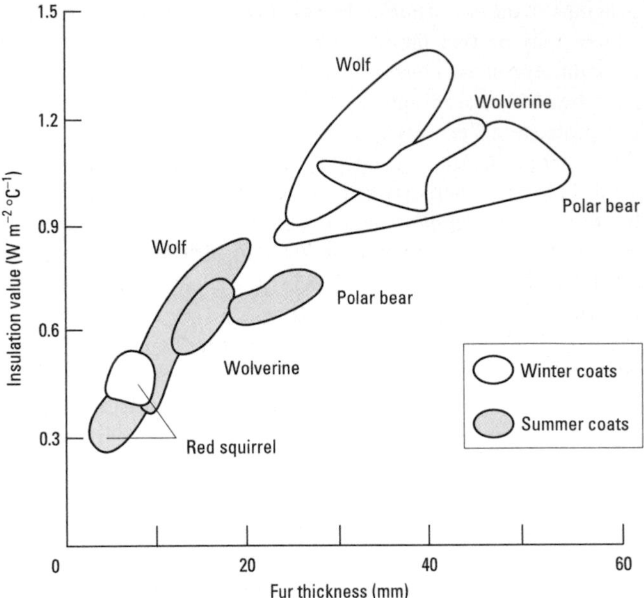

Fig. 16.40 Winter coats of four species of mammals that are thicker and better insulating than summer coats; note that changes are greater for larger species. (Adapted from Hart 1956.)

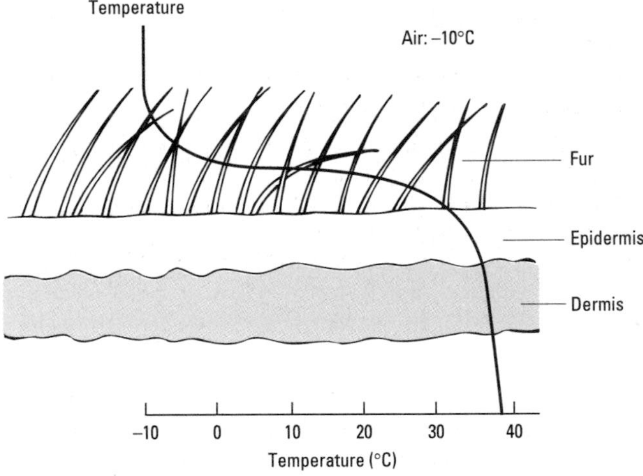

Fig. 16.41 The temperature gradient across the surface of a high-latitude mammal; the gradient is largely across the highly effective fur, and skin surface temperatures are close to core temperatures.

from air temperature to normal T_b is contained across the fur and skin (Fig. 16.41). The polar bear is something of an exception in having coarse fur, probably because it frequently dives in the water between ice floes and must shake water quickly out of the fur again when it surfaces. There is an apparent paradox in animal coloration here, in that the polar bear and many other polar animals are white, when black fur would appear to be beneficial in absorbing maximum solar radiation. For example, the light coat of the Siberian hamster in winter has a thermal reflectance of 43% compared with the summer coat of 18%, and for the snowshoe hare the relevant figures are 61% (white in winter) and 23% (brown in summer). An obvious explanation is that white pelage makes the animals less conspicuous against a bright ice or snow background, whether they are predators stalking their food or prey trying to avoid being eaten. But this explanation has recently been supplemented for polar bears by a demonstration that the hairs are very unusual in having hollow shafts, which has two effects: firstly the air space reflects visible light so that the hair appears white (while it is in fact translucent), and secondly the hairs act like optical fibers, with total internal reflection of shortwave UV radiation so that it passes in from the hair tips to the bear's skin, allowing radiative heating through an apparently "reflective" pelt.

High-latitude birds such as penguins have dense, smooth layers of feathers, extending over all their extremities, and particularly effective in resisting ruffling and loss of insulation at high wind speeds. Beneath the plumage is a substantial adipose layer in winter, providing insulation during diving and a massive food store for the incubating and often nonfeeding adults. King penguin males fast for a month or more at the start of their breeding cycle, and 93% of their energy is then provided by the subcutaneous fat stores. The insulation in penguins is so effective that they can survive in icy water for days on end, with their feet continuously vasodilated to maintain the thermal window effect for heat loss to prevent overheating; and can breed on land in air well below freezing point while maintaining high levels of REWL even when motionless, again to avoid overheating.

The terrestrial high-latitude mammals do not have particularly thick subcutaneous **adipose layers**, as these are too heavy and cumbersome. However, aquatic and amphibious mammals such as whales, seals, and polar bears match the penguins in investing in a massive "blubber" layer, since for them fur would lose most of its value through being wetted (wet fur is up to 50-fold more prone to heat loss than dry fur). The relative utility of fur to terrestrial and aquatic animals is neatly demonstrated by many of the seals, where the essentially terrestrial pups have much denser fur than the primarily aquatic blubber-bearing adults. Baby seals have similar problems to penguins and can die of overheating rather readily; indeed sea lions in the Galápagos have to breed in cool caves or behind boulders away from the sun. Some amphibious diving mammals, such as beavers and mink, have a pelt of dense air-trapping underfur beneath a coarser wettable outer fur, but these animals normally only immerse themselves for short periods. Species that dive in sea water rather than fresh water, such as the sea otter (*Enhydra lutra*), have an extra problem of salt crystals forming in the pelt, which compromises its insulating role both on land and on return to the water. These species spend unusually large proportions of their day in grooming the fur to remove the salt.

Control of heat loss at extremities

Many polar animals suffer from potentially extreme rates of heat loss at their extremities, which usually have less dense fur to aid maneuvrability and/or sensitivity. In seals and whales, for example, it is calculated that 10–30% of heat production in a resting animal is lost through the flippers, fins, and flukes, rising to 70–80% during moderate exercise. Most of these animals use **countercurrents** at their extremities to prevent these losses being even higher. Seal and dolphin flippers are among the best studied cases of this regional heterothermy, with elaborate rete systems (see Fig. 11.42). Each artery is surrounded by a ring of small veins, normally allowing a

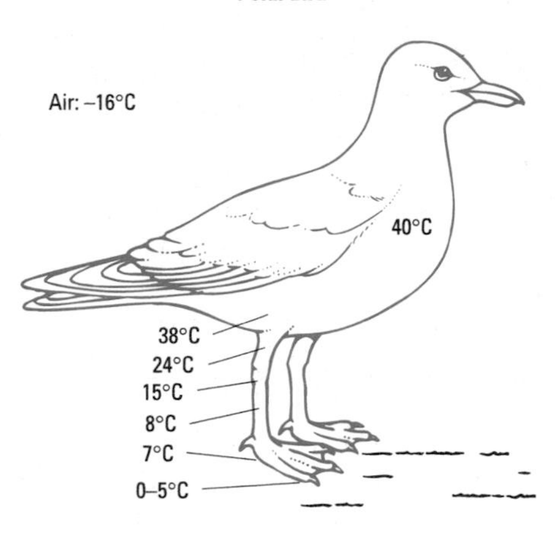

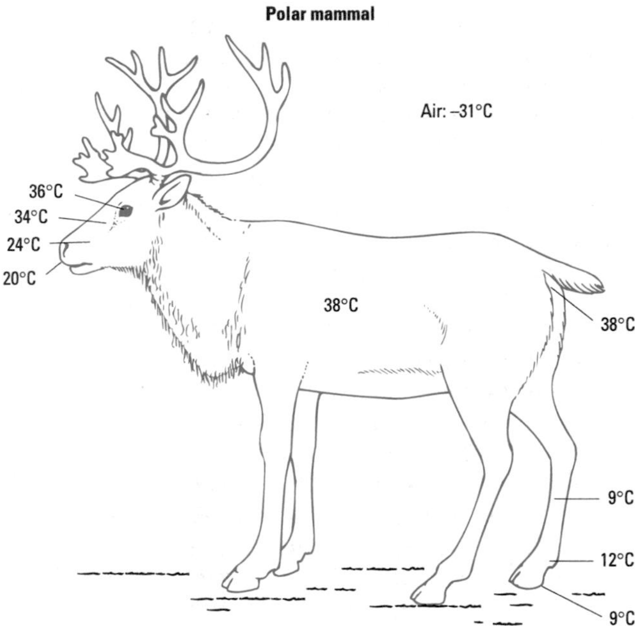

Fig. 16.42 Peripheral countercurrents producing cold extremities in polar mammals and birds. (Adapted from Irving & Krogh 1955.)

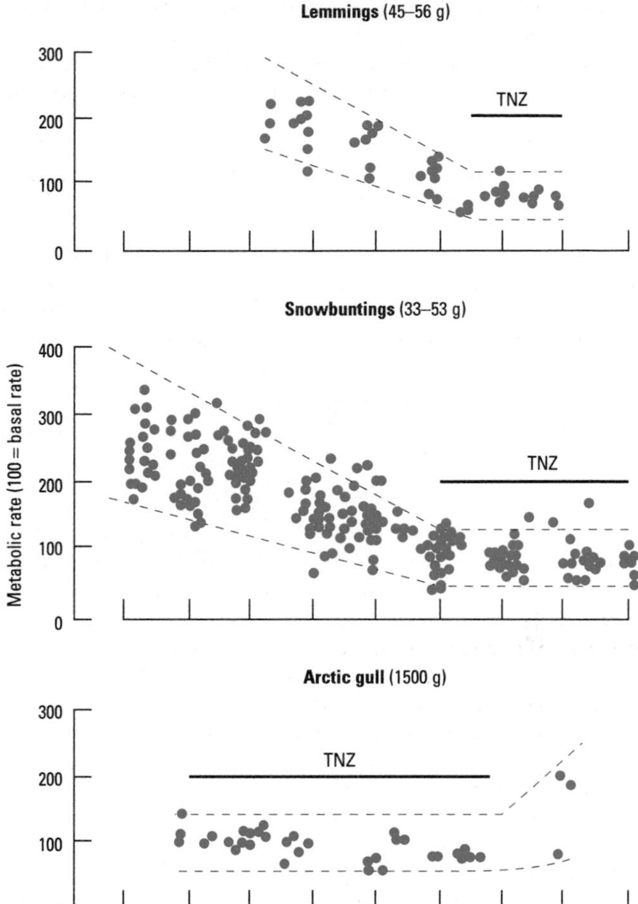

Fig. 16.43 Metabolic rates (MR) vs. declining temperature in three polar endotherms. The thermoneutral zone (TNZ) is relatively broad, especially for the Arctic gull, in which MR does not increase even at −30°C. (From Davenport 1992.)

countercurrent exchange to reduce heat loss through the uninsulated flippers. But when the animal is overheating due to prolonged exercise, the anatomy of the system is such that the raised blood flow and higher blood pressure cause the central artery to dilate and the surrounding veins to collapse, bypassing the heat exchanger and making blood return from the flipper by alternative more peripheral veins, losing heat to the surrounding water in the process.

Similar heat exchangers normally keep hot blood away from gull and penguin feet, from caribou noses and hooves, from beaver tails, and probably even from the small ears of many polar mammals (Fig. 16.42). Since so many of these animals are standing on or swimming in ice-cold environments, they particularly need to insure that "frostbite" (cold-induced tissue damage) does not set in, so they must also supply the peripheral tissues with occasional pulses of warmer blood to give some protection. At very low temperatures the heat transfer to the tip becomes small but continuous (see Fig. 8.40).

Control of heat production

Some high-latitude animals have **higher basal metabolic rates** (BMRs) than predicted from the simple allometric equation (see Chapter 3). In rodents such as lemmings it is 200–240% greater than predicted. Metabolic rate can also be modified in many cold-adapted animals if food supplies permit. But note that, on the whole, metabolism is less sensitive to temperature than in tropical animals (Fig. 16.43). Polar animals have a much broader zone of environmental temperature where they keep "ticking over" at much the same rate regardless of ambient temperature (i.e. a **broader thermoneutral zone**).

Hypothermia and torpor

Hypothermia involves an adaptive decline in core T_b, to a new value that may be just a few degrees below normal (perhaps 30–35°C) or may be very low indeed, often below 10°C and within 1 or 2°C of T_a. Slight reductions in T_b are best described just as mild hypothermia; many birds and mammals use this in cold spells, but these animals can normally become active again more or less instantaneously. The term **torpor** is generally used for the phenomenon of a relatively deep hypothermia involving a pronounced naturally occurring decline in T_b where metabolic rate, respiration, and circulation are strongly depressed. Its occurrence in birds and mammals was outlined in Chapter 8. Torpor requires the specific and energetically expensive phenomenon of **arousal** to permit subsequent activity.

Remember that "**hibernation**" is a much looser term, sometimes thought of as synonymous with prolonged and profound torpor but widely misused in relation to such animals as badgers and bears, both of which sleep for long periods in the winter but readily can and do become active when hungry; the descriptive terms "winter sleep" or "carnivore lethargy" are often preferred in modern literature. Brown bears (*Ursus arctos arctos*), for example, show "hibernation" at only 4–5°C below normal T_b, with very minor modifications to circulatory pattern and a slightly increased red blood cell count, probably triggered by a fall decrease in production of thyroid hormones.

Torpor is not especially common in polar or tundra mammals, because rates of heat loss are too high for it to be sustainable at all, the habitats being too unproductive to allow adequate accumulations of fat reserves. Polar species that do exhibit torpor, such as the Arctic ground squirrel (*Spermophilus parryi*), use it for only brief periods. It is the small eutherian mammals of the boreal forests that yield the best examples of prolonged deep torpor.

Torpor normally has three phases: that is to say, the rapid entry phase and the rather slower arousal phase can be physiologically distinguished from the middle period of relatively steady, low metabolic rate, when the metabolic rate is much reduced, representing a very large energy saving. In these boreal mammals there is no single trigger for going into a torpid hypothermic state; its onset may vary with food supply and with the imposed ambient temperature regime, although where there is a yearly cycle the onset of torpor is generally regulated in part through photoperiodic changes acting via the endocrine system. The onset of torpor may be quite gradual, and it largely results from failure to increase metabolic rate as T_a declines, with ventilation slowing rather than increasing. This leads to an inevitable decrease in T_b and a further decrease in metabolism, so that the animal slides rather gently, often over 8–24 h, into a torpid state without initiating any special effects such as increased conductance. Entry into this state sometimes involves a short series of transient drops in T_b (Fig. 16.44), with "test runs", followed by reinstatement of regulation, then a single smooth rapid drop to the new low level during which thermoregulation is temporarily abandoned and the body cools largely passively. For a bird or mammal of about 2 g, the torpid body temperature of 15°C would be reached in only about 40 min, whereas for a 100 kg bear to become fully torpid would take a week (Table 16.5). Simple calculations nevertheless show that for any endotherm, despite the rapid rate of warm-up required for arousal and its cost in energy and food, even short periods of torpor are energetically worthwhile.

In the fully torpid state, the metabolic rate may be only 2–20% of its normal resting value, and it is primarily body fat that is catabolized. Metabolism is suppressed mainly by a reversible phosphorylation

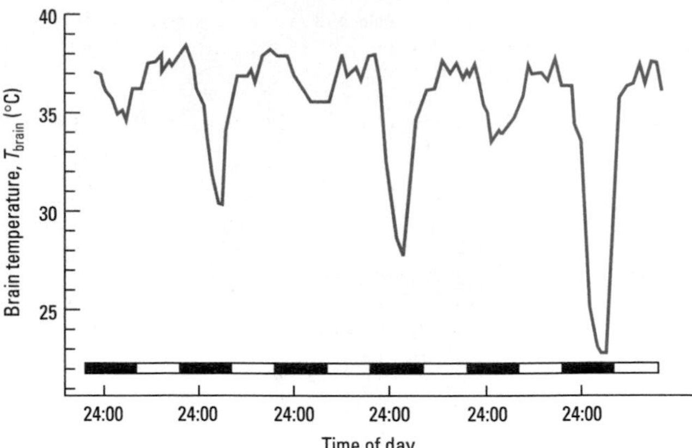

Fig. 16.44 The onset of torpor in a ground squirrel, with progressive lowering of brain temperature (T_{brain}) on successive nights. (Adapted from Strumwasser 1960.)

of key glycolytic enzymes, especially pyruvate dehydrogenase (although some of the mitochondrial enzymes are surprisingly upregulated in torpor, perhaps to minimize damage to the electron transport chain during cold exposure). Many enzymes also become relatively temperature insensitive, and levels of phosphagens are decreased. As a result, for example, the oxygen consumption of a torpid bat is only 2.5% that of the active rate (Fig. 16.45), so that its fat reserves will last 40 times longer than "normal" under cold conditions. In at least some torpid mammals the degree of saturation of the dietary fats affects torpor, with an unsaturated diet giving lower minimum T_b and longer torpor bouts, possibly via interactions with the energy-regulating protein leptin. But this enhanced torpor is at the expense of greater lipid peroxidation during the torpid period. In ground squirrels the intake of polyunsaturated fatty acids such as linoleic and linolenic acids is therefore reduced during the fall months.

In full torpor, ventilation may be as low as 1–2 breaths min^{-1} in a small mammal, with periods of total apnoea of up to 5 min. Torpid animals appear unresponsive and uncoordinated, and indeed their sensory apparatus may deteriorate. In the Siberian ground squirrel (*Citellus undulatus*) the taste bud cells have much reduced ribosome, endoplasmic reticulum, and Golgi contents, indicating reduced protein synthesis and reduced sensory function during torpor. However, torpid animals still have some functioning neural control of their own condition. If the T_b is in danger of dropping too low, with the animal approaching freezing, there may be spontaneous arousal or a gradual increase in heat production to keep the body at 4–5°C without actually initiating expensive full warm-up. This is reflected in a rise in oxygen consumption as the air temperature falls too low, and again indicates that the torpid animals are not entirely going over to ectotherm-like thermal biology.

Arousal from torpor may occur quite rapidly (1–6 h) relative to the process of entering torpor. Arousal is also very size dependent (see Table 16.5). However, the arousal rate also inevitably depends on the starting point of T_b, since metabolic rate and temperature are so closely related. In fact at a very low body temperature it is impossible for chilled tissues to use oxygen rapidly and metabolic rate may not be able to get high enough to raise T_b at all. In other

Table 16.5 Time to enter and arouse from torpor for animals of different body mass. Calculated from allometric equations, for cooling and arousing at 15°C, body temperature (T_b) changing from 17 to 37°C.

Species	Body mass	Entry time	Arousal time
Shrew	2 g	35 min	13 min
Hummingbird	4 g	59 min	17 min
Honey possum	10 g	80 min	24 min
Poorwill	40 g	224 min	41 min
Nightjar	86 g	350 min	55 min
Vulture	230 g	39 h	3.2 h
Echidna	3.5 kg	27 h	3.8 h
Marmot	4.0 kg	29 h	4.0 h
Badger	9.0 kg	45 h	5.4 h
Bear	80 kg	138 h	12.3 h

words, for each size of animal there is a theoretical body temperature threshold below which it cannot fall or arousal becomes biochemically impossible. Each species also in practice has a minimum temperature from which it will initiate arousal, known as the **critical arousal temperature**, and this parameter seems relatively unrelated to body size or to phylogeny, perhaps reflecting natural habitat temperatures (e.g. it is 2°C for the boreal pocket mouse *Perognathus longimembris*, but as high as 20°C for the pygmy mouse *Baiomys taylori*). It may also be linked to habitat humidities, since progressive desiccation may be a real problem for small hibernating mammals such as the little brown bat, where EWL during torpor would exceed metabolic water production at any humidity less than 99.3% RH.

Arousal in most boreal mammals is a two-stage process, triggered by day length and/or rising temperature. At first it is not accompanied by obvious shivering or muscular activity, because a large part of the initial thermogenesis in mammals is achieved not by the muscles but by **nonshivering thermogenesis** (NST; see Chapter 8) in the specific tissue known as brown fat or brown adipose tissue (BAT). The BAT where NST occurs is found particularly around the neck and shoulder areas of torpid mammals (and of many newborn mammals too, where it may be up to 5% of the birthweight), although it is absent in birds. Its activation results in a very rapid warming of the thoracic and head area, including the heart, spine, and brain. Rectal temperature lags many minutes behind, indicating active control of blood distribution during arousal. Tracer studies show a high blood flow to the heart, to the brown fat itself, and then to skeletal muscles, so that the key organs can be supplied with oxygenated blood. Only then are corticosteroids and insulin released into the blood so that the muscles can initiate shivering thermogenesis, to give the second phase of speeded up arousal. Thus in arousing hedgehogs blood insulin secretion does not begin until T_b has risen to about 25°C, showing the importance of NST in the early warming stage. Indeed, NST accounts for about 40% of the heat needed for full arousal in hamsters, and somewhat less in bats (Fig. 16.45).

Figure 16.46 summarizes the role of reduced winter metabolism and torpor in mammals of different sizes. In large species seasonal acclimation (with the metabolic rate reduced to about 50% of summer levels) is largely achieved by extra fur insulation and torpidity is avoided. In smaller mammals, there may be prolonged deep torpor in some cases, but in other species winter activity is maintained by a combination of weight reduction, improved thermal conductance,

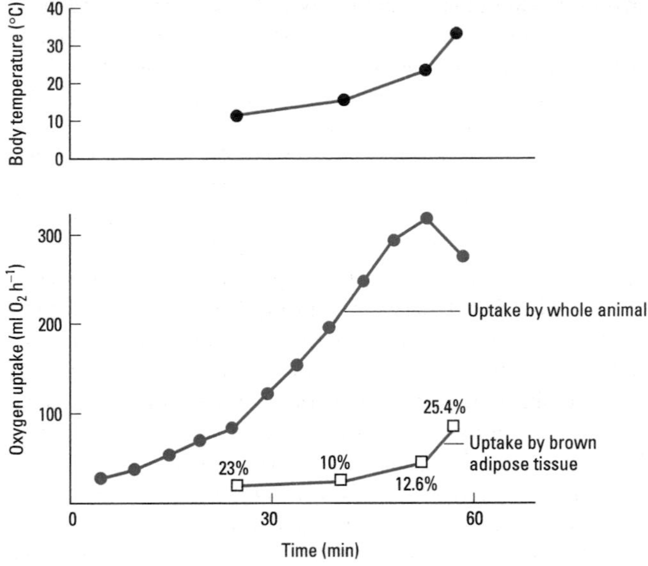

Fig. 16.45 Temperature and oxygen consumption in a bat during arousal from torpor, showing that brown adipose tissue (BAT) metabolism is important but forms only part of the whole thermogenic response. (From Davenport 1992; adapted from Jansky 1973.)

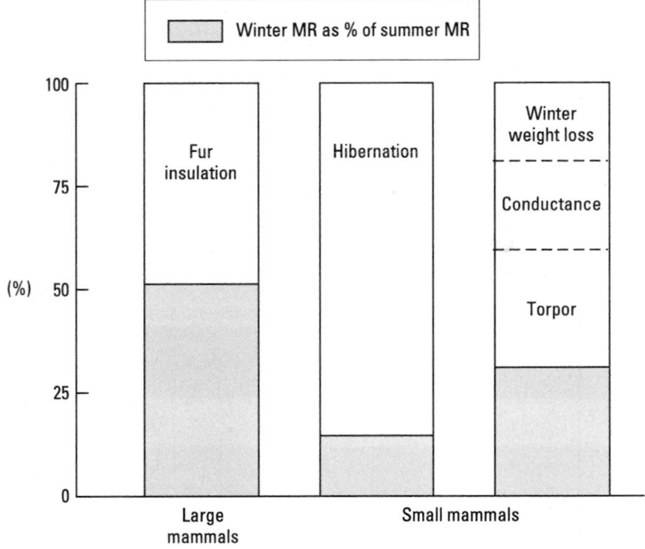

Fig. 16.46 A comparison of the seasonal thermal strategies of large and small mammals; large species mainly respond to winter with improved insulation, while small mammals use either prolonged torpor (hibernation) or a mixture of weight loss, improved insulation, and short periods of torpor. MR, metabolic rate. (Adapted from Heldmaier *et al.* 1989.)

and a much reduced degree (length and/or depth) of torpor. The winter metabolic rate of these species is about 30–40% of that in summer, compared with a drop to 15–20% of summer levels in small mammals with deep prolonged torpor.

Coping with freezing

Because of their endothermic and thermoregulatory abilities, most of the "endurer" animals are not required to cope with freezing and cannot do so if very low temperatures are forced upon them in laboratory studies. However, there is at least one endotherm, the Arctic ground squirrel, mentioned above as a true polar user of torpor, that has the ability to supercool to −2.9°C to avoid freezing.

Food supplies

As we saw for ectotherms, foodstuffs may pose a problem. Many tundra and taiga species that do not migrate are food hoarders, hiding caches of seeds or nuts for future use, although they normally combine this with periods of torpor. Herbivores tend to face rather nutrient-poor and high-fiber diets, and gut mass may be larger than normal and may be subject to acclimation, since it has been shown to increase markedly as a proportion of body mass in lemmings fed on low-quality high-fiber diets. Polar mammals and birds that are herbivorous show more distinct plant preferences than related temperate species (e.g. willow ptarmigans select willow, rock ptarmigans select birch).

Permanent active residents such as polar bears, penguins, and pinnipeds tend to be carnivores, or more especially piscivores; fish and other large marine fauna are the only rich dietary items available year-round that are not themselves frozen. Polar mammals have the additional problem of providing adequate nutrition to their offspring, often in the depths of winter. It is no coincidence that cetacean and pinniped milks are unusually rich in fats and proteins, and polar bears too produce a milk richer than that of other Carnivora (see Table 15.13). Hence Weddell seal pups, receiving milk with 60% fat content, can double their weight in the first 10 days after birth.

16.3.4 Anthropogenic effects in the cold biomes

The low overall productivity in boreal zones makes the plants, and hence the communities as a whole, particularly susceptible to human interference. Humans have occupied the Arctic tundra since at least the Neolithic, but had little impact before the last century except in reducing some populations of vertebrates through whaling, hunting, and fur-trapping activities.

In the last 40 or so years, however, exploration and the accompanying **exploitation** have boomed, with oil, gas, and minerals much sought after and with pipelines, roads, and railways all invading the landscape. The problem is that localized damage to these environments is only very slowly repaired by regenerative growth, and soil turnover is even slower, so that we can still see extensive surface damage from the early gas and oil pipelines installed across Alaska and Siberia up to 40 years ago, with damage accentuated by activities causing localized scree flows and the creation of deep gullies. Such lines may seriously interrupt animal migrations.

Global warming also poses particular problems for high-latitude systems. Most of the climate models predict substantially faster and greater changes in temperature at the poles, commonly as much as 6–12°C within the next century, leading to warmer polar winters. An early proposal that this might lead to more snowfall and hence a thicker, more extensive area of snow and ice field, with greater albedo and greater reflection of radiation, has not been upheld by the evidence. There is now extensive documentation of the thinning and **break-up of ice sheets**. In Antarctica several ice shelves have been monitored carefully and show marked disintegration and retreat of the ice front (Fig. 16.47), while in Greenland ice thinning has also been reported. Since many polar invertebrates appear to

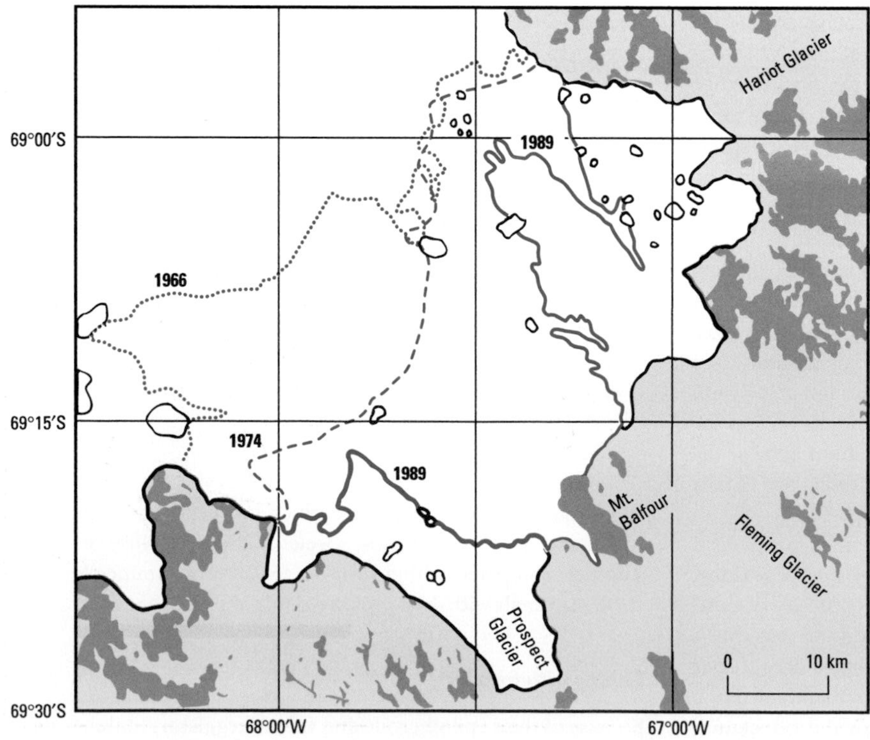

Fig. 16.47 Documented ice-sheet thinning in part of Antarctica since 1966. (Reprinted from *Nature* **350**, Doake, C.S.M. & Vaughan, D.G., copyright 1991, with permission from Macmillan Magazines Limited.)

have upper lethal temperatures rather close to their current maximum microclimate temperatures, there may be casualties from overheating.

Ozone depletion is currently a phenomenon very much centered on the polar zones, with polar stratospheric clouds implicated in the reactions with atmospheric halogens (chlorine and bromine) that lead to ozone destruction. The seasonal Antarctic ozone "hole" is still expanding every year, and some annual thinning in the Arctic is also documented. The ozone-depleted air has on occasions moved out to lie over the southern tips of Australia and South America, and routinely affects the Antarctic continent and fringes. The expected effects of increased frequencies of skin cancers, and of cataracts, will certainly be most likely in the polar mammals and birds, with penguins perhaps most profoundly at risk.

All of these problems can be accentuated by increased predation risks for weakened animals faced by introduced predators; feral mice in the sub-Antarctic are already a problem for insects. Hence in some ways these cold terrestrial environments are every bit as "fragile" as the tropical forests and temperate wetlands that are more common areas of concern, but as yet they have received much less research effort.

16.4 High-altitude habitats

Life at altitude in many ways presents a special case of low-temperature habitat, where height above sea level parallels the effects of latitudinal distance from the equator. However, there are also additional problems relating to temperature, water availability, respiration, and pressure, so these habitats are given a section of their own here.

16.4.1 Montane habitats

Occurrence

Mountains occur in all continents (Fig. 16.48), and in a wide range of forms, both as long mountain chains and as relatively isolated peaks. The process of **orogenesis** (mountain building) is primarily related to plate tectonic movements, with the land mass rising up where two plates collide and forming a mountain chain along the perimeter of one plate. This may be either at the edge of a continent where a submerged plate undergoes subduction below a land mass, as with the Andes in South America, or it may be in the center of a continent formed from two land masses abutting on each other, as with the Ural mountains between Europe and Asia, or the Himalayas between Asia and India. However, every mountain range is unique, since the processes underlying orogenesis are very complex.

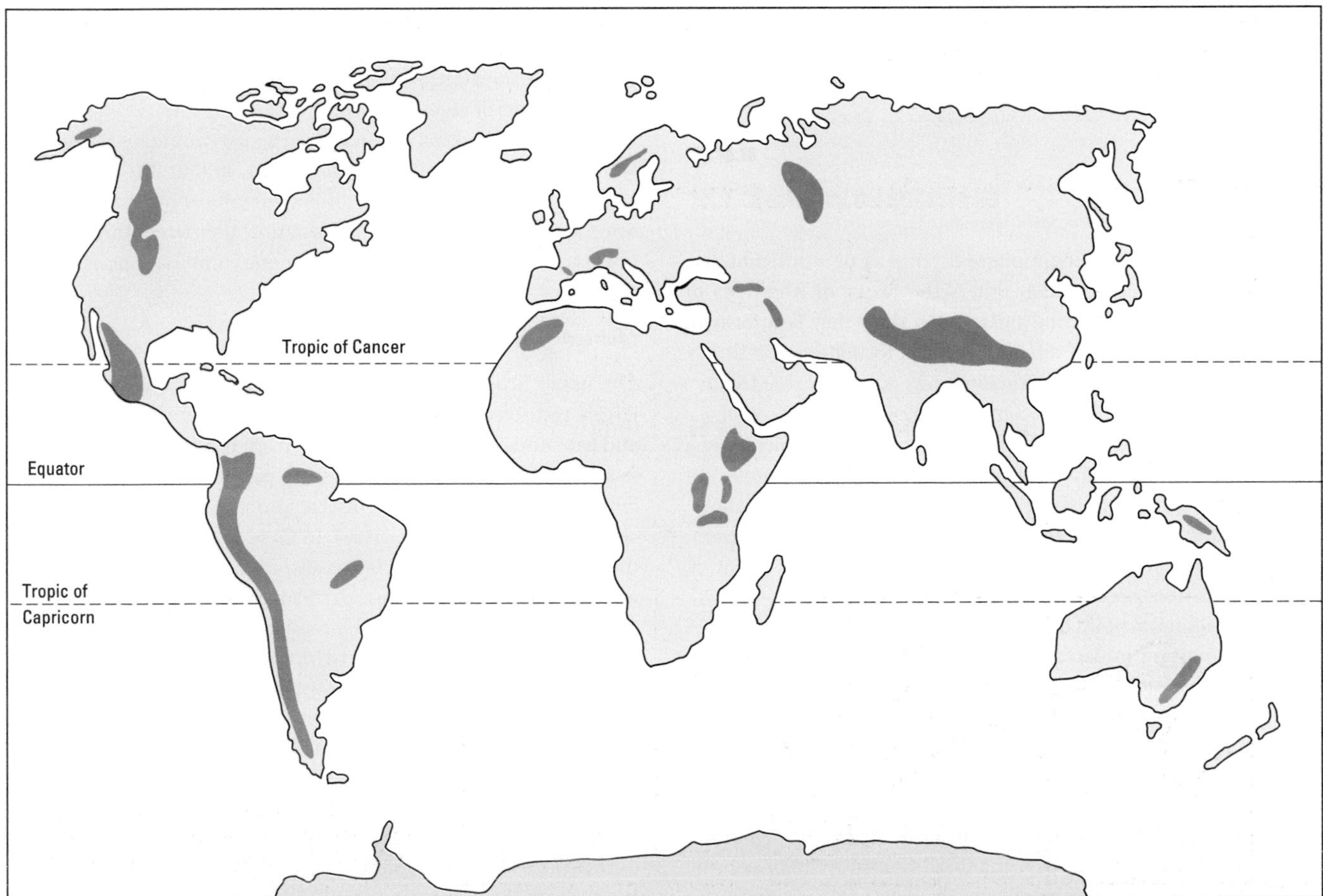

Fig. 16.48 The world's mountain ranges.

Table 16.6 Characteristics of mountain ranges.

Mountain range	Highest peak (m (ft))	Latitude
Alps	Blanc 4807 (15,771)	45–47°N
Andes	Ojos del Salado 7034 (23,241)	35°S–7°N
Antarctica	Vinson Massif 5140 (16,863)	
Appalachians	Mitchell 2037 (6684)	34–42°N
Atlas	Toubkal 4165 (13,664)	30–33°N
Australian Alps/ Great Dividing Range	Kosciusko 2230 (7316)	28–38°S
Caucasians	Elbrus 5642 (18,510)	42–43°N
Ethiopian plateau	Ras Dashen 4620 (15,157)	5–15°N
Himalayas	Everest 8848 (29,028)	28–35°N
Iran/Turkish plateau	Ararat (Büyük Agri) 5165 (16,945)	30–42°N
Kenyan plateau	Kilimanjaro 5895 (19,340)	5°S–3°N
Pyrenees	Pico de Aneto 3404 (11,168)	42°N
Rockies	Elbert 4399 (14,431)	35–62°N
Sierra Madre	Citlaltepetl 5699 (18,697)	35–40°N
Sierra Nevada	Whitney 4418 (14,495)	36–39°N
Southern Alps (NZ)	Cook 3764 (12,346)	43–45°S
Urals	Narodnaya 1894 (6213)	53–68°N

Furthermore, every existing mountain range is of a different age, some still rising and forming, others the result of hundreds of millions of years of fracturing and erosion since they first formed, thus giving a vast range of different types of formations. Some key characteristics of major mountain ranges are given in Table 16.6.

Being areas of marked recent geological upheaval, younger montane areas also tend to suffer from other geological phenomena such as earthquakes and vulcanism. The Andes, for example, are part of the Pacific "ring of fire", with numerous active volcanoes. Mountainous areas may also be sufficiently elevated to be frozen, thus giving rise to glaciers, and the scouring effects of recent or current glaciations are often very obvious.

From a biological point of view, several of these geological factors have an important impact. Elevation in itself is perhaps the most obvious, with effects on temperature, pressure, and oxygen availability. In mountain ranges daily average temperatures are reduced by about 1°C for every 150 m of altitude. This effect is roughly similar at all latitudes, so that even tropical mountain ranges sitting astride the equator can be snow-covered (see Plate 9f, between pp. 386 and 387). They are therefore profoundly cold at night even though highly irradiated by day, and these diurnal cycles of extreme heat and cold are accentuated during the dry season. A classic example is the isolated peak of Mt Kilimanjaro in Tanzania, almost on the equator yet with snow and glaciers at the summit all year round even in blazing midday sunshine. Other tropical mountain ranges may be extremely arid, with the northern Andes having exceptionally low-humidity air since the precipitation has been shed from air rising up the west-facing slopes, giving a rain-shadow effect. These dry mountain peaks are again always profoundly cold by night; many animals living there must show expanded thermal tolerances. In more temperate latitudes, mountain ranges may show a lesser diurnal temperature range but a very drastic seasonal variation, so that temperate mountain ranges such as the Alps and Rockies have an extremely variable snowline.

Aside from elevation, mountains also may pose challenges in terms of geological instability, with periodic fracturing, quaking, or volcanic emissions. While potentially lethal for local populations in the short term, such effects may in the longer run produce increased diversity (by separating subpopulations on isolated peaks or plateaus, leading to potential allopatric speciation), or in the case of vulcanism to increased local soil fertility and biotic productivity.

Montane vegetation

Vegetational patterns at altitude parallel those with latitude, but may show the same sequences compressed into a much smaller linear dimension (Fig. 16.49a). In ascending from the lower slopes one may pass from deciduous forest and chaparral or grassland into coniferous forest, above which is a distinct treeline; above that again there may be regions of shrubby vegetation or ericaceous plants, sages, and junipers, with some thinning grass, then bare rock with lichens. The snowline in winter descends downwards through these successive tiers of vegetation.

But the actual vegetation on any particular mountain range varies with continent and latitude (Fig. 16.49b), so that it is very hard to generalize beyond this. In addition, mountain ranges have often acted as refugia and as centers of isolation at different times in their history, so that many of them have unique combinations of flora.

Montane fauna

The upper lichen zones of mountains have an invertebrate fauna largely restricted to the soil layers and including springtails, mites, and nematodes. Spiders, particularly from the genus *Pardosa*, occur in most mountain ranges, subsisting partly on windblown insect fauna such as aphids; some insects also occur as residents (up to about 6000 m in the Himalayas). In these upper rocky areas, where there is patchy vegetation there is also a range of relatively large and quite specialized vertebrates, particularly camelids, with the bactrian camel occurring in the Himalayas, the dromedary in central Asia, and the llama, alpaca, guanaco, and vicuña in South American mountain ranges. There are also various sheep and goats, including the alpine ibex, mouflon, and chamoix, the Himalayan goral, tahr, and markhor, and the Rocky Mountain bighorn. Amongst the crags there may be resident raptorial birds (eagles, etc.) and a range of passing migrant birds. Just a little lower down, where the soil is deeper and vegetation more prolific, there is a wide diversity of burrowing rodents, providing prey for the raptors. Where there are lakes within the mountain ranges, particularly in the long chain of

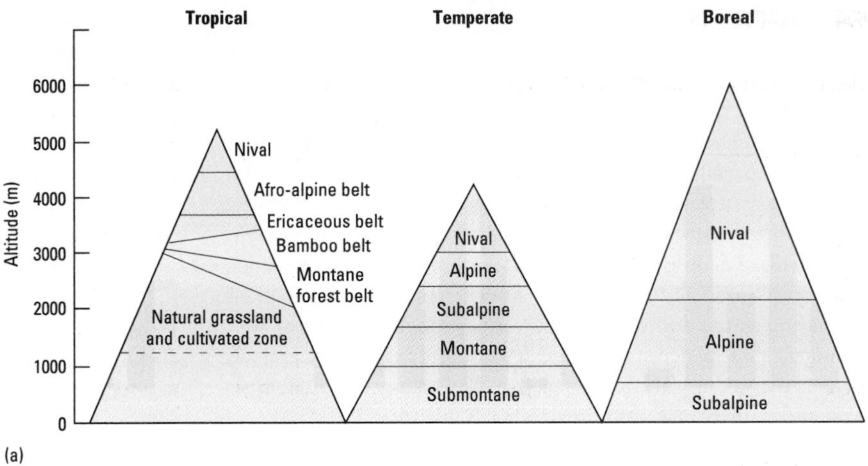

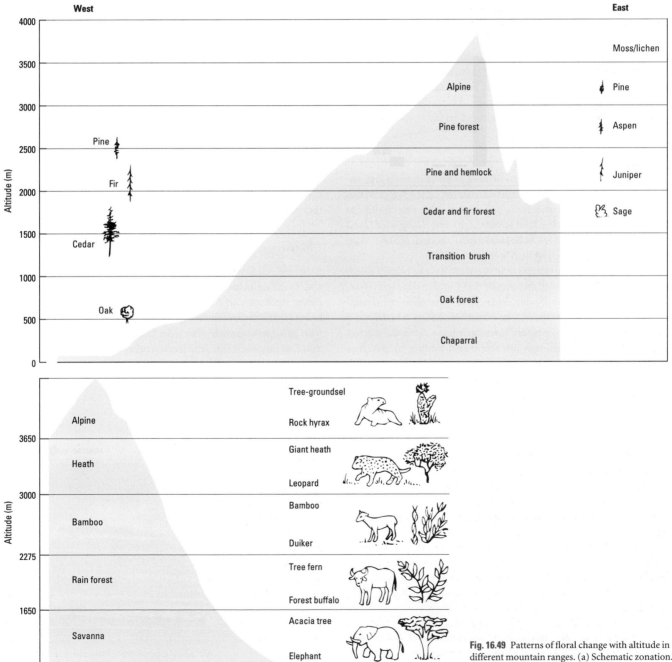

Fig. 16.49 Patterns of floral change with altitude in different mountain ranges. (a) Schematic zonation. (b) Patterns in the Sierra Nevada (above) and in East Africa (below).

the Andes, quite a range of waterfowl also occurs, especially geese and coots but also including flamingos.

Further down the mountains faunal diversity inevitably increases and largely parallels the fauna of the equivalent latitudinal belt, with a particularly strong representation of insects, such as grasshoppers, and some pollinators, such as bees, butterflies, and flies. Amphibians and reptiles become rather common, each species often having a quite substantial altitudinal range. Rodents are again common, as are various large deer and antelope (elk, pronghorn, taruga, klipspringer, etc.) in dense bush and forests, and very bulky animals such as the yak in plateau areas of Asia. Thus many taxa of animals have surprisingly high altitudinal limits (Table 16.7).

To make the point that every mountain range is different, an example of the faunal distribution up a tropical mountain is shown in Fig. 16.49b. Here hyrax are a speciality in the upper slopes, with leopards making homes in isolated crags and only descending at night to feed; lower down the typical African bush fauna appears.

It is generally found that species diversity decreases with altitude, and that species ranges increase with altitude. The habitats thus become dominated by relatively few, relatively specialist, species extending over broad elevational bands. An example of elevation ranges for bats in the Andes is shown in Fig. 16.50.

Table 16.7 Altitudinal limits for various vertebrates.

Taxon	Altitude (m)	Location
Fish		
Trout	2800	Alps
Trout	3800	Andes
Nemachilus	4700	Asia
Amphibians		
Salamander	3000	Alps
Eleutherodactylus	4500	Andes
Toad	5000	Himalayas
Reptiles		
Lizard	3400	Rockies
Skink	4000	Kilimanjaro
Iguanid	4900	Andes
Lizard	5500	Himalayas
Birds		
Several	4000–6500	Rockies
Lammergeier	5000–6000	Himalayas
Barheaded goose	8800+	Himalayas
Mammals		
Deer mouse	4000	Rockies
Human (permanent)	4500+	Andes, Himalayas
Llama, alpaca	4800–5400	Andes
Chinchilla	5000	Andes
Yak	5800	Asia
Taruca deer	6000	Andes

16.4.2 Thermal problems

The main environmental effect of altitude, as far as animals are concerned, is the **declining temperature.** However, on most mountains there is in addition a particularly stark contrast between conditions in sun (since the radiative load is very high) and in the shade, and often a stark contrast between day and night. Thermal problems can also be greatly magnified by high winds.

Animals in mountain ranges may therefore have to exhibit a high thermal tolerance, to avoid freezing by night, and to cope with overheating and desiccation by day. In tropical mountains the change between the two conditions may be extremely sudden at sunrise and sunset, with no time to "prepare" on a daily basis. To make matters worse, since smaller animals are active and feeding by day they cannot empty their guts and are bound to be full of INAs, so that they are particularly vulnerable to bodily freezing.

Heliothermy: basking, sun–shade shuttling, and color

Amongst ectotherms, the insects, lizards, and snakes use basking and shuttling **heliothermy** to a marked degree. There have been classic studies on an Andean lizard species, *Liolaemus multiformis*, living at 4000 m, which emerges from its burrow at dawn (0700 h) with an air temperature of only −5°C and basks on heaps of vegetation that insulate it from the ice, warming up to a T_b of 35°C within

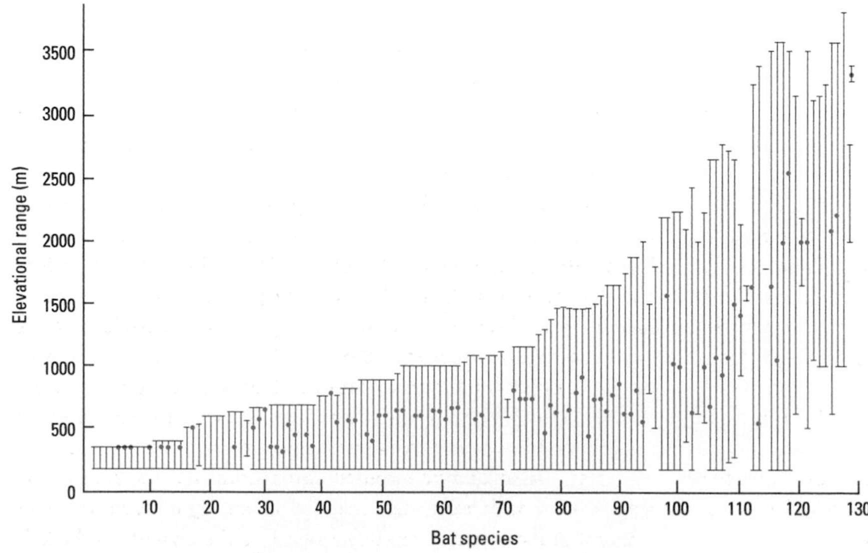

Fig. 16.50 Elevational ranges of 129 bat species in the Andes; species that occur at a higher mean altitude (solid circle) also have substantially greater altitudinal ranges. (From Patterson *et al.* 1996.)

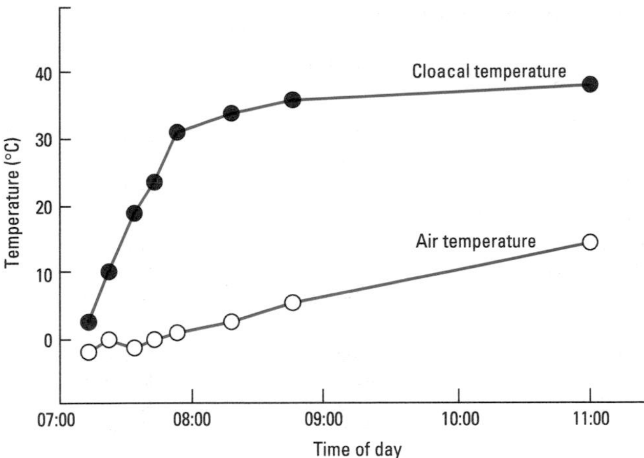

Fig. 16.51 Basking in a Peruvian mountain lizard allows cloacal temperature to rise sharply after dawn and then be maintained at a steady 35–38°C, well above air temperature. (Adapted from Pearson 1954.)

2 h (Fig. 16.51). However, life at altitude complicates the issue of thermal balance and patterns are not straightforward. Within the genus *Liolaemus* there is no clear association between thermal tolerance and elevation, because varying body masses affect thermal inertia; with mass factored out, high-elevation species heat more slowly. High-altitude Andean frogs have a broader thermal tolerance than congeners from neighboring lowlands, and greater aerobic metabolic scopes at low temperatures. The lizard *Sceloporus variabilis* achieves a mean T_b of only 28.9°C at 1000 m in Mexico, whereas the same species at 45 m can achieve 32.4°C, despite the two populations having similar metabolic rates. Another closely related lizard, *S. jarrovii*, is winter-active in mountains, and it can darken rapidly when cold, with the melanophores directly responding to melanophore-stimulating hormone. Skin samples show a differing dose–response curve to this hormone at different temperatures, and at 35°C the darkening response is dramatically reduced compared with lower temperatures. Pigment cell sensitivity thus appears to be acclimated at the cellular level, to allow appropriate thermoregulatory responses.

Postural thermoregulation is also very obvious in montane fauna. Basking and ground-hugging is common in ectothermic montane butterflies such as *Colias* species, and crouching postures side-on to the sun are conspicuous in montane grasshoppers such as *Melanoplus sanguinipes*, which can raise T_b up to 20°C above ambient. These high-altitude grasshoppers often exhibit higher resting metabolic rates than their low-altitude congeners, which may compensate for lower temperatures and shorter seasons. Amongst endotherms, the Andean guanacos bask with their flanks and axillary regions exposed to the sun, but adopt curled up postures where these "thermal windows" are shut off towards sunset, achieving a maximum 22% reduction in thermal window heat loss. They then bed down in groups, with their hindquarters into the wind, these postural changes together giving a 67% energy saving.

For many montane animals there are indications that the metabolic rate rises and the critical thermal maximum decreases, inter- and intraspecifically, with increasing altitude.

Use of vegetation and microclimate

Because of the combination of freezing and desiccation dangers referred to above, many montane animals from the tropics cannot be either particularly cold-hardy or particularly desiccation resistant, and there is a strong tendency to use behavioral rather than physiological techniques to survive. Hiding in crevices and under rocks, and seeking shelter within vegetation, are almost ubiquitous strategies amongst invertebrates. Thus, for example, insects sampled from above 4000 m on Mt Kenya showed almost no special adaptations to low temperature, but instead sheltered in the rosettes of large *Senecio* plants and in other dry flower heads. The predominance of large rosette plants in these habitats may be particularly important in providing stable microclimates amongst the bulk of decaying older leaves.

Certain insects are particularly adept at exploiting the sharp temperature gradients around mountain vegetation. One example is the checkerspot butterfly, *Euphydryas gillettii*, in the mountains of Wyoming, USA, which preferentially lays its eggs on the southeast-facing leaves of certain honeysuckles, to gain maximum warming from the early morning sun. Eggs laid in such sites mature more quickly than those from other zones of the plant, and so are more likely to produce caterpillars that will have fed enough to survive when the early montane winter sets in.

Microhabitat selection certainly also matters for vertebrates, but the physical nature of the terrain at high elevations may limit the options for these larger animals. For example, the microclimate choices are relatively similar for the Andean and lowland frogs mentioned above, although the times of day (and air temperatures) at which each kind of habitat is sought out do vary markedly.

Burrowing and huddling

Burrowing in general serves as a way of avoiding very cold temperatures for montane animals. The Andean lizard *Liolaemus* spends on average 82% of its time in deep burrows, with 15% basking and crouching on warm surfaces and just 2–3% actively foraging or interacting with other animals. Deer mice in Wyoming mountains similarly use burrows to avoid freezing; with night temperatures of −15°C, the burrow never falls below 0°C. Note that this deer mouse is the same species that can be found in the Nevada desert further south in the USA, where, as we saw, the burrows have the opposite role of providing a buffer against overheating.

Montane animals also frequently combine the benefits of **burrows** and **aggregations**. The alpine marmot (*Marmota marmota*) overwinters in large underground nests, containing 5–8 animals of the same family (whereas its close relative the woodchuck does not live at altitude, and hibernates singly). The alpine marmot undergoes torpor with a T_b of only 7–8°C for periods of about 2 weeks, interspersed with short arousals when T_b rises to normal levels for a day or so. Within the nest all the individuals are synchronized, and huddling occurs both during torpor and during the high T_b phases. When one individual begins to arouse, some of its body heat is transferred to the other marmots, helping them to warm passively. The first one to be fully aroused also grooms its nest mates and covers them with extra nest material to aid their warming. The marmot burrows are commonly around 12°C internally in the fall, but this has dropped close to 0°C by early spring. When burrow temperatures fall below 5°C, the marmots raise their metabolic rate somewhat, but because of the burrow this is only required for rather less than half the total period of torpor. The more animals there are in the burrow, the longer it takes for the burrow temperature to fall to 5°C, giving a real benefit to "social hibernation".

Even in the larger mountain mammals a gregarious habit is common, with sheep and goats living in small herds and huddling together at night and in stormy weather.

Insulation, heat distribution, and metabolic rate

Montane mammals show some interesting adaptations of furs. The duck-billed platypus, *Ornithorhynchus anatinus*, lives in mountain streams in eastern Australia and may dive for prolonged periods in water close to freezing point. It has one of the most effective fur insulations known, with a dense underfur and flattened guard hairs to trap air, combined with countercurrent heat exchangers at the base of its tail and all four limbs. *Oreamnos americanus*, a wild goat from the Rocky Mountains, has evolved the same strategy as we met earlier in the polar bear, with a rather coarse pelt of hollow hairs that appears white (and presumably it gets the same advantages). The large camelids, deer, and antelope from mountain areas are also famous for the depth and quality of their pelage. Many are used as providers of high-quality wools (e.g. chinchillas, alpaca, and cashmere goats). The individual hairs may be very long, most spectacularly in the wild yak where the rather coarse fur almost reaches the ground. During the short summer periods, when these animals might risk overheating, blood can be shunted to the less insulated extremities to dissipate excess heat.

In several invertebrates, notably insects, mass-specific **metabolic rates** tend to increase with elevation, and it has been argued that this is an adaptation to accelerate development in species that remain annual (see below). However, in the grasshopper genus *Xanthippes*, population differences in mass explained most of the effects (high-altitude grasshoppers were smaller), and when mass was factored out little effect of elevation remained. In heterothermic bees T_b and warm-up rate increase with altitude, although it is unclear whether this results from decreased thermal conductance or higher mass-specific heat-generation rate.

Coping with cold and freezing

Small ectotherms in mountains may have real problems with achieving activity in continuously cold conditions. In insects this is sometimes circumvented by a degree of endothermy, as for example in the rainbeetles (*Plecoma*) of the Sierra Nevada of the USA, which achieve a T_b of 35°C even in air temperatures around freezing. Bumble-bees in the Himalayas and Andes achieve similar feats.

Many small animals from seasonal temperate mountains have classic physiological adaptations to achieve cold-hardiness in winter, with both freeze tolerance and freeze intolerance exhibited. Smaller species (mites, springtails, etc.) tend to be **freeze intolerant**, voiding their guts to eliminate nucleating agents and accumulating glycerol or other polyols or sugars. Larger invertebrates tend to be **freeze tolerant**, accumulating both nucleating agents and cryoprotectants, and surviving repeated freezing of their extracellular fluids. As in other habitats, levels of protection against freezing tend to show seasonality; Fig. 16.52 shows the strongly seasonal variation in supercooling points for alpine beetles.

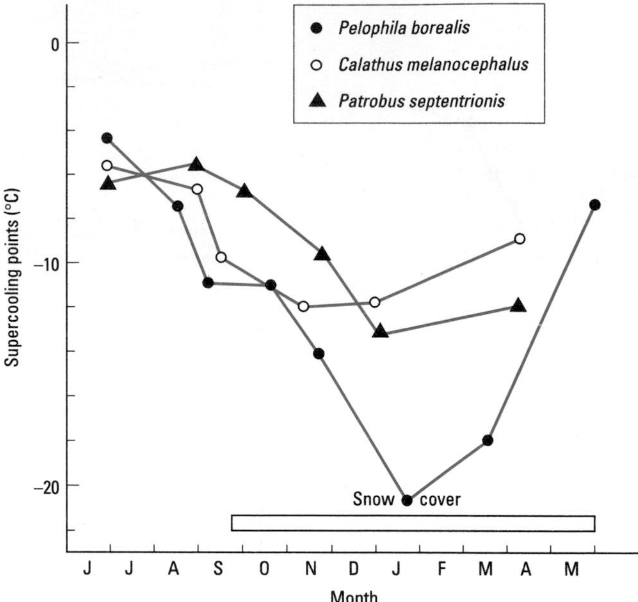

Fig. 16.52 Seasonal variation in supercooling point in three species of alpine ground beetles. (From Sømme 1995.)

One of the more spectacular examples of freezing tolerance is the New Zealand weta (*Hemideina maori*), a very large montane grasshopper that survives in the frozen state for much of the winter, and which can be seen freezing and thawing on a daily basis in spring and fall. The weta may have more than 80% of its body water converted into ice, promoted by an ice-nucleating protein, though the intracellular contents remain unfrozen through osmotic dehydration effects and a build-up of proline. A few cases of freeze tolerance in species from tropical mountains have also been recorded, including beetles from Mt Kenya, *Meridacris* grasshoppers from the northern Andes, and *Agrotis* caterpillars from Mauna Loa on Hawaii.

16.4.3 Dehydration and water balance

Animals suffer an increased tendency to dehydrate at altitude. At low ambient temperatures the water vapor content of the ambient air is reduced, and any animal with a T_b above T_a will lose water very quickly in its warm exhaled saturated breath. Furthermore the diffusion constant for all gases, including water vapor, increases with altitude. Increased wind speeds may add to the dehydrating effect. Where there is standing water or snow, larger animals, including most of the vertebrates, may have little problem, but small invertebrates are more at risk.

Invertebrates from mountains may therefore have relatively high **dehydration tolerance**. For example, tenebrionid beetles from exposed habitats on the slopes of Mt Teide (Tenerife, 2000–2500 m) can survive at least 40 days at 5% RH, with only 15–30% weight loss, matching or outdoing some more xeric lowland species for rates

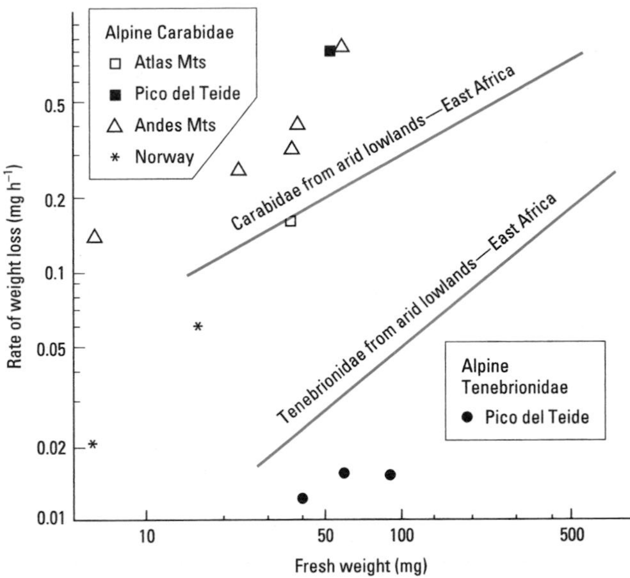

Fig. 16.53 Rates of water loss (cutaneous evaporative water loss, CEWL) in montane beetles compared with those from arid lowlands (measured at similar temperature and humidity). There is a lowered evaporative water loss (EWL) in tenebrionids from Pico del Teide (Tenerife), but for carabids (living in the litter layer) there is little effect except in the relatively exposed Norwegian beetles. (From Sømme 1995.)

of cuticular evaporative water loss (CEWL). Likewise grasshoppers (*Melanoplus*) from montane populations also show lower water loss rates, and the amount and melting point of cuticular lipids are correspondingly higher. However, in groups such as carabid beetles that normally live within the litter layer the montane species are no more desiccation resistant than similar lowland species; only a few species that survive in rather exposed zones above the treeline in Norway have lowered CEWL values (Fig. 16.53). In addition to cuticular adaptations, there may be **excretory adaptations** to minimize water loss in invertebrates. Montane tenebrionids normally have a cryptonephridial system, and the gut lumen (and therefore the urine/feces) may reach 2500–2750 mOsm. However respiratory adaptations to reduce water loss are not clearly documented, so that although *Melanoplus* grasshoppers use discontinuous ventilation at times, this behavior shows no good relationship with altitude or water loss rates.

For endotherms the issues of water balance are somewhat more complicated. Many mammals are reported to have **lower total body water** values at altitude, and in humans this effect can be marked (29 kg in Andean populations compared with 39 kg in weight- and age-matched lowland humans). High-altitude populations can also clear a drinking-water load much more quickly, mainly because the collecting ducts of the kidney become relatively insensitive to the hormone arginine vasopressin. It is still unclear whether other aspects of water loss rates are altered in mammals. Careful analysis of deer mice water loss rates with altitude show a clear increase, but this was an indirect effect, with mass and metabolic rate as confounding variables. Differences in the thermal environment are probably the main cause of higher overall water-flux rates in small mammals at higher altitude.

Table 16.8 Altitude and pressure effects in humans.

Altitude (m)	Atmospheric pressure (kPa)	Ambient Po_2 (kPa)	Alveolar Po_2 (kPa)	Alveolar Pco_2 (kPa)
0	101	21.1	13.8	5.3
3100	71	14.6	8.9	4.8
4340	62	12.8	6.0	–
5300		Ceiling for unacclimated humans		
5500		Highest human habitation		
6200	46	9.7	5.3	3.2
6000–7000		Ceiling for acclimated humans		
8500		Human breathing possible for a few hours		
8848	33	6.9	4.0	1.5
9200	30	6.3	2.8	–
12,300	19	3.9	1.1	–
14,460		Ceiling for humans with pure oxygen supply		
15,400	12	2.4	0.1	~0
20,000	6	1.3	0	0

16.4.4 Pressure and respiration

Another important effect of altitude is the change in barometric pressure, which dramatically alters the **partial pressures of gases** in the atmosphere. The relationship between altitude and pressure is exponential, pressure decreasing by 50% for every 5500 m of elevation. Thus at the top of Mt Everest (8848 m above sea level) the barometric pressure is only 250 Torr or 33 kPa, about one-third of the sea-level value, with Po_2 at 6.9 kPa.

The additional physiological problems of altitude are therefore mainly respiratory, due to low pressure and low oxygen. This is unlikely to be a significant problem for animals in high-altitude ponds and streams, and indeed the amphibious African crab *Potamonautes warreni* shows little specialization in ponds at 1400 m elevation. Nor is it a substantial problem for terrestrial arthropods using a tracheal respiratory system, which is rarely at the limits of its capacity to supply oxygen. But for vertebrates with lungs it represents an additional constraint on performance, and especially for the maintenance of a high BMR in endotherms. Table 16.8 shows the effects of altitude on partial pressures of gases for a human.

For nonadapted birds and mammals, **hyperventilation** is the normal response, in an attempt to maintain alveolar Po_2, but arterial oxygen levels inevitably decline, and for humans loss of consciousness results at about 50% arterial oxygen concentrations, normally occurring at around 7000 m above sea level. Hyperventilation also tends to lead to reduced alveolar Po_2, which in turn inhibits respiration for terrestrial vertebrates (see Chapter 7), giving an automatic brake on the acclimatory response. This can be partly overcome by excreting excess carbonate at the kidneys leading to metabolic acidosis and thus an improved central chemoreceptor response to increase the respiratory rate. The need to oxygenate the tissues conflicts sharply with the need for acid–base homeostasis. Inorganic phosphate and 2, 3-diphosphoglycerate in the red blood cells play some role in changing the hemoglobin affinity to assist in this balance, overriding the Bohr shift (Fig. 16.54). Nevertheless both arterial blood concentrations and blood acidity are reduced, which may lead firstly to mild cerebral edema and secondly to ever-increasing pulmonary edoma as the pulmonary capillaries constrict in some parts of the lung leading to increased pulmonary pressure

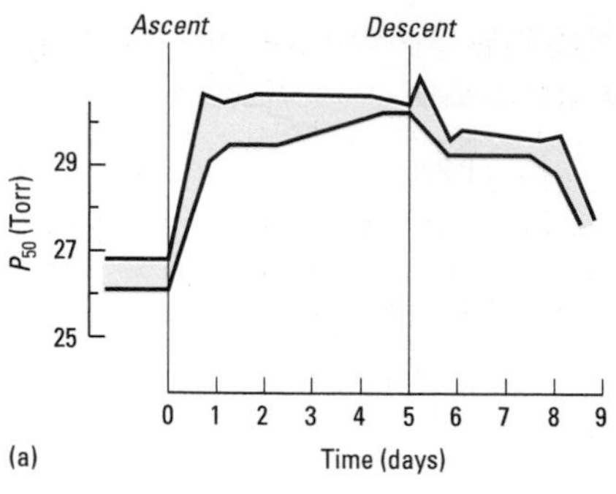

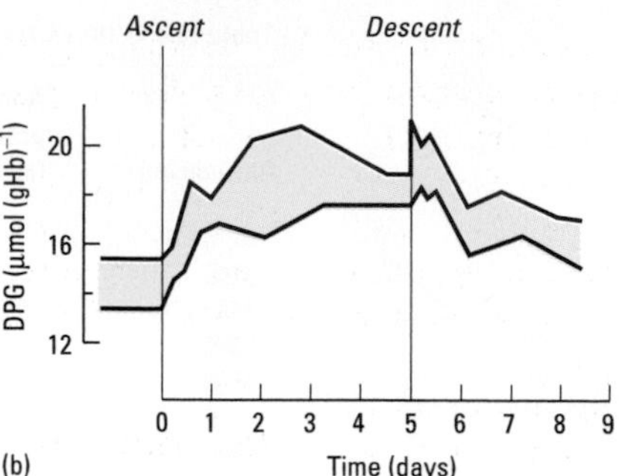

Fig. 16.54 (a) Acclimatory changes in hemoglobin affinity in humans at altitude (4510 m) for 5 days, and (b) the associated change in erythrocyte diphosphoglycerate (DPG) concentrations. (Adapted from Bouverot 1985.)

and patchy leakage of fluid into the alveoli, leading to crackly breathing as a symptom of the "altitude sickness" experienced by humans.

For resident adapted animals, the respiratory and circulatory systems become modified. In vertebrates this means that enhanced lung volume, and extra pigment in the blood (increased hematocrit) with high-affinity properties, are the commonest responses. **Enhanced lung volume** is hard to prove in endemic species, but is readily seen in comparisons within species, even in humans, where high altitude natives have larger barrel-shaped chest with larger lungs, and may also have larger hearts, allowing exertion levels that can never be matched by incoming humans however long they stay and attempt to acclimate. An **increased hematocrit** is useful during acclimation, but is rare in permanent residents such as camelids, since it also increases blood viscosity. **High-affinity hemoglobin** (at normal concentration) is a better response in permanently resident mammals and birds, with the oxygen dissociation curve well to the left. This is particularly apparent in camelids (Fig. 16.55); but do the camelids have this characteristic due to residence at high altitude over hundreds of generations, or have they come to inhabit high altitudes because they were already endowed with suitable hemoglobin? The fact that the two camel species (essentially lowland animals) have high-affinity hemoglobin, albeit less spectacularly than in the four montane species (llama, guanaco, alpaca, and vicuña) suggests that the latter explanation is more likely. However a comparison of carnivores from high altitude in Peru shows P_{50} values of 31.1 Torr (puma) and 18.5 Torr (fox) compared with 36.3 and 26.2 Torr, respectively, in the same genus or species from sea level, suggesting that in this group the change in respiratory pigment affinity and structure occurred after invasion of the montane habitat. The same is true for geese, where the high-altitude migrators (the barheaded goose in Asia and the Andean goose) have specially modified pigment relative to close lowland relatives. In every case, the changed affinity can be traced to just one or a few amino acid substitutions within the protein, and for the two geese species the point mutations are quite distinct, confirming their separate evolutionary origins.

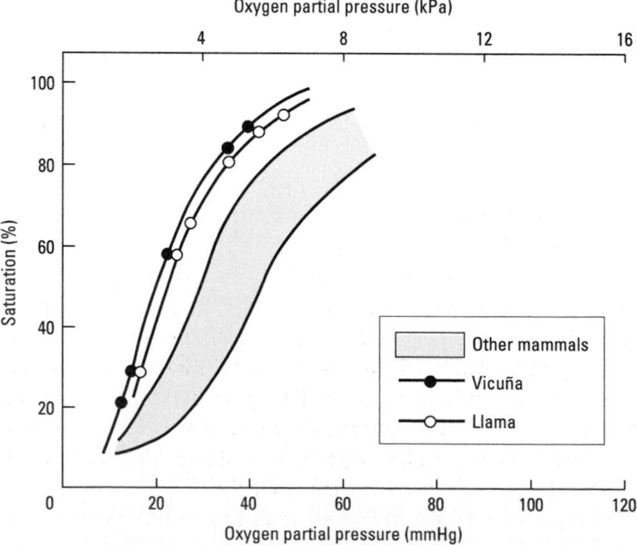

Fig. 16.55 Oxygen dissociation curves and lowered P_{50} values in high-altitude llama and vicuña compared with sea-level mammals. (Data from Hall *et al.* 1936.)

A high-altitude rodent from China, the pika, reveals further circulatory adaptations, having reduced pulmonary vasoconstriction responses with a larger right ventricle giving high pulmonary perfusion; in this species again there is no increase in hematocrit. Comparisons of the circulatory systems of high-altitude birds (Andean coots) show further circulatory changes. **Capillarity** is increased in muscles, and mean muscle fiber diameter is decreased. These changes appear to be sufficient to allow regulated energy production in muscles without increases in enzymatic activity, since oxidative enzyme function shows no significant differences from sea-level coots.

Birds' eggs also have a particular respiratory problem at altitude, the shell needing larger than usual pores to compensate for reduced rates of gas exchange and therefore incurring rapid EWL. Most birds that nest at altitude must compensate by providing more protected nest holes with a humid microclimate.

16.4.5 Life cycles

As with polar species, montane invertebrates often have **extended life cycles**, needing two or more seasons to complete a cycle that in related lowland congeners would take only 1 year. Carabid beetles from the Austrian Alps routinely take 2–3 years; even at the relatively limited heights of Snowdonia in Wales, carabids seem to be biennial, whereas their lowland conspecifics are annual. Wolf spiders of the genus *Pardosa* have a 2-year cycle above the timberline in the Rockies, whereas species below the timberline always complete the cycle in one season; the same applies for high-altitude *Pardosa* from Norway (3 years) and from Austria (2–3 years). Similar effects have been reported in mites, springtails, moths, and crane flies. **Diapausing** stages may contribute to the lengthened cycle, and as in other extreme habitats this phase is often triggered by temperature rather than by photoperiod, allowing greater phenotypic plasticity. However, there is also evidence that on reaching adulthood high-altitude populations undergo more rapid senescence. In *Melanoplus* grasshoppers montane populations aged and died more quickly than lowland populations when both were kept in a common equable environment, perhaps due to selection on their reproductive schedules.

An alternative strategy at somewhat lower elevations, or where microclimatic variation can be exploited to speed up development, is to **curtail the life cycle** somewhat so that it can be fitted into one season. A Rocky Mountain grasshopper, *Aeropedellus*, has one fewer nymphal stage than normal, and many insects at similar elevations are rather smaller as adults than in lowland populations. But patterns of change with altitude are not simple. In lizards, growth rates and reproductive potentials can either increase or decrease with elevation, depending on rather subtle balances of behavioral adjustments and local climate patterns.

In mammals, the most conspicuous adaptation is that of extremely **precocial young**, which on high mountain slopes can move around and look after themselves almost from the first day after birth. Birds appear to show an alternative pattern of increased biparental care in montane species (linked with less sexual dimorphism and relatively drabber males). This may be associated with another problem for many high-altitude species, that of attracting a mate for reproductive purposes. Data on this issue are limited, but at least in Andean frogs there is evidence that the more montane species show less energetically demanding vocal behaviors.

16.4.6 Other problems and adaptations in mountains

A major problem in high mountain crags is that of locomotion, and of keeping a grip and maintaining balance on unstable difficult terrain that is often very windswept. For the large mammals that make a living in these habitats, some show careful slow gaits, picking their way amongst rocks, while others have a bouncing jumping locomotion achieved at amazing speeds. The latter group have major adaptations, particularly of the feet, to give this surefootedness. The bighorn sheep of the Rockies have soft elastic pads in their feet that absorb the shock of impacts and provide an effective grip on slippery rock.

Communication across craggy mountain expanses may be difficult, with many obstructions to long-range visual signals, so that auditory communication has become particularly common in amphibians, rodents, and the larger mammals. Honking and barking sounds that carry well in thinner cool air are used to communicate between and among the herds in larger mammals, and to signal danger among marmots and other rodents.

Some higher-altitude populations may be among the most susceptible to potential **global climatic change**. Most obviously, they almost by definition occur in isolated small populations, relatively close to the edge in terms of local extinction. More specifically, analyses of locomotor performance and stamina in a widespread North American lizard *Urosaurus ornatus*, across a range of habitats, indicate that the high-elevation populations in ponderosa pine forests may be the most at risk, lacking an ability to modify their geographic distribution on an appropriate timescale.

The overall pattern of life on mountain slopes is likely to change markedly in response to continued global warming (Fig. 16.56). An analysis of likely changes in vegetation distribution on the east and west slopes of the Cascade Range in USA (Fig. 16.56b) shows the eastern faces lose much of their interesting alpine upper zones and become dominated by sagebrush scrub, while the west loses alpine, hemlock, and fir habitats and gains further oak savanna.

16.5 Aerial habitats

Aerial life is in some senses a special case of life at altitude, where the animal has usually generated the altitude by its own activity. This "altitude" may be very limited in small insects, but can exceed that of the highest mountains for some birds, who are able to fly at heights greater than those achieved by human mountaineers using accessory oxygen (see Table 16.7). All of these animals may therefore have altitude-related problems as described above, relating to temperature, water balance, pressure, and respiration; but in addition they have to solve the more mechanical problems of keeping a low body weight and good buoyancy, of having appropriately shaped aerodynamic surfaces and, of course, of fueling. The mechanical and physiological problems often become coupled, since flying entails substantial power output, potentially producing unusually high heat loads and compounding the problems of water balance by increasing convective and evaporative cooling and/or increasing metabolic water production. In flying animals, temperature, water balance, and respiratory problems are particularly closely tied together.

Many organisms live continually in the air for short periods, often in dispersive phases, and often involuntarily. A great many more use flight and other kinds of airborne locomotion in short bursts, for prey capture, escape from predators, or competition for and acquisition of mates. Just a few insects and some birds (e.g. swifts) stay aloft for extended periods, and could be said to lead truly aerial lives. It is worth remembering that *most* animals fly; the majority of insects of course, but also most birds (of which there are twice as many species as there are of mammals) and even about one-fourth of all mammals (the bats). Remember that though flight is energetically expensive, it is also very efficient in terms of cost per unit distance traveled.

Flight, whether of short or long duration, needs particularly **high-energy foods** such as nectar, fruit, or very good quality flesh, especially in the form of fish for the birds that spend long periods aloft gliding and flying over the sea. Very few leaf-eating herbivores are good fliers, since the diet is too poor and the gut needs to be too large. Fueling within the body has to be critically adjusted to keep the wing musculature adequately supplied. In insects, proline and lipids are used, while in vertebrate fliers glycogen stores remain the primary source of blood glucose. However, small passerine migratory species have particularly high levels of plasma triglycerides, and this may be an adaptive system of fuel supply.

Wing muscle design also varies according to lifestyle and time spent aloft. Birds that fly only in short bursts have predominantly **white anaerobic muscle** fibers in their flight muscles, while prolonged fliers have predominantly red aerobic fibers in the main pectoralis (downstroke) muscles. Flying birds have the highest

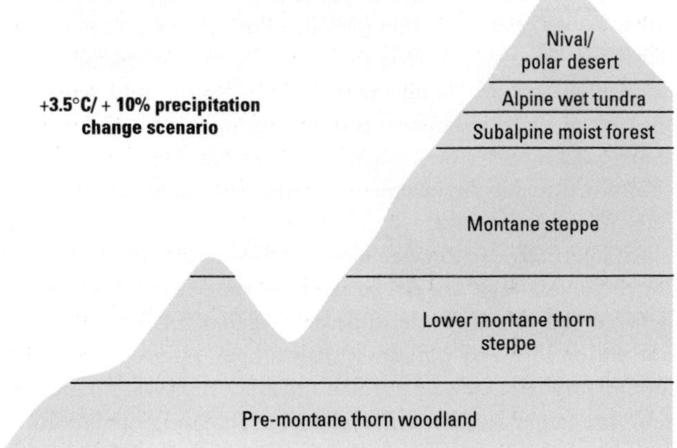

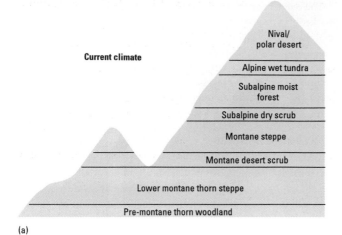

(a)

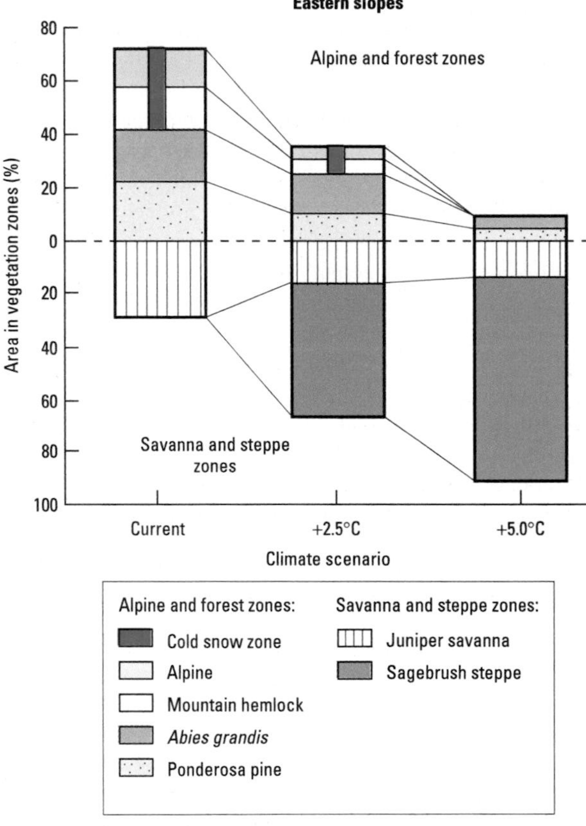

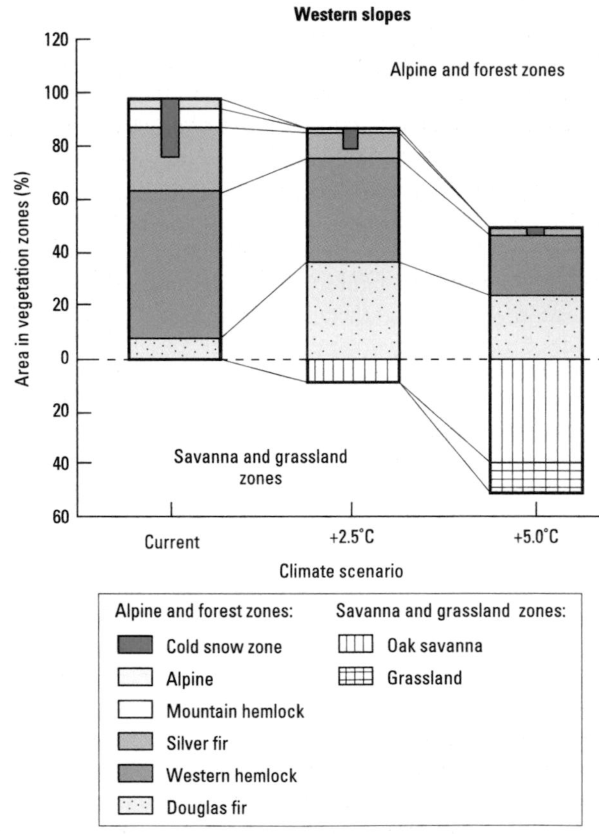

(b)

Fig. 16.56 Effects of global warming on ecosystems at altitude. (a) A warming of 3.5°C is predicted to move floral zones upwards and severely reduce the high alpine communities in general. (b) Predictions for the east and west faces of a northern hemisphere mountain range, for two different levels of warming. (a, Adapted from Beniston 1994.)

mass-specific energy turnover, and the highest core body temperatures, of any animals. The design features of the smallest hummingbirds, which allow this power output, were considered in Chapter 6.

Because they are generally small, birds cannot store large amounts of energy and may have fewer options for patterns of **energy allocation** than any other vertebrates. Their power output in sustained flight is 8–12 times that of the resting metabolic rate, reflecting a much higher factorial scope than is found in exercising mammals of similar size. In bats especially, there is considerably respiratory specialization to give fast uptake and transfer of oxygen: a thin blood–gas barrier, a vast alveolar surface, and a large lung capillary volume. The first two features are also present in birds, with the addition of a small and rigid lung aided by cross-current in the parabronchi (see Fig 7.11).

Respiratory cycles are often coupled to wingbeat frequency in flying animals, with thoracic musculature (in both insects and birds) generating both effects and so giving a saving on muscular effort. The insect tracheal system is probably always adequate to cope with supplying oxygen fast enough, but for vertebrates it is a different story, particularly when flying at any height. At very high altitudes, where the partial pressure of oxygen is low, the driving force for diffusion of oxygen across the pulmonary surfaces is very low, and here only the cross-current blood flow system of the parabronchial bird lung will suffice, by insuring that partially depleted air continually encounters new deoxygenated blood. Hence high flying is only possible for birds, and is never seen in bats.

The water balance problems of aerial animals are rather poorly understood. Flapping flight in birds certainly increases **respiratory water loss** and may also increase cutaneous losses leading to rather rapid dehydration. However, all flying birds appear to be excellent plasma-volume regulators, even when undergoing substantial dehydration, an ability that in mammals tends only to occur in desert species. It appears that this goes with an ability to **tolerate high heat loads** during flight. In insects, small species appear to suffer net water loss in flight, which may constrain their flight duration and require modified dietary choices (e.g. more dilute nectar for bees). But large insect species, for example bees of more than 200 mg, appear to generate more metabolic water than they can readily cope with due to their exceptionally high metabolic rates, and they have been observed jettisoning excess water while in flight, both by loss of saliva or crop fluid from the mouth and by copious dilute urination.

Finally, it should be noted that flying is an expensive business, with costs ranging from 50 to 120 W kg^{-1} in birds and bats, and up to 300 W kg^{-1} in insects. There may therefore be design trade-offs between flight apparatus and other organ systems. For example, in territorial insect species where flight determines mating success, such as in the dragonfly *Plathemis*, extra investment in flight muscle may mean reduced gut tissues and fat reserves in the males, and perhaps reduced longevity.

16.6 Conclusions

Extreme environments pose a series of linked problems for their inhabitants, and some of these problems are rather similar whether we are looking at deserts, polar regions, or mountains. Not only are the problems similar, but the convergence of adaptation between quite different groups is striking. Animals living at these extremes have repeatedly acquired the same kinds of solutions: from behavior (use of burrows and other refugia, and gregariousness) through to features at the gross morphological scale (overall size and shape, and the reduction or elaboration of appendages) down to physiological systems (peripheral countercurrent exchangers, altered insulation, changes in metabolic rate, and periods of hypometabolism) and the fine details of biochemical endowments (altered membrane and enzyme properties, and the use of osmolytes that protect against freezing and desiccation damage). Convergent evolution is extremely commonplace where there are only limited possible solutions to common physical problems.

FURTHER READING

Books

Bouverot, P. (1985) *Adaptations to Altitude-Hypoxia in Vertebrates.* Springer-Verlag, Berlin.

Bradshaw, D.S. (1986) *Ecophysiology of Desert Reptiles.* Academic Press, Sydney.

Carey, C., Florant, G.L., Wunder, B.A. & Horowitz, B. (1993) *Life in the Cold: Ecological, Physiological and Molecular Mechanisms.* Westview Press, Boulder, CO.

Hadley, N.F. (1994) *Water Relations of Terrestrial Arthropods.* Academic Press, San Diego, CA.

Heller, H.C., Musacchia, X.J. & Wang, L.C.H. (eds) (1986) *Living in the Cold: Physiological and Biochemical Adaptations.* Elsevier, New York.

Hochachka, P.W. & Guppy, M. (1987) *Metabolic Arrest and the Control of Biological Time.* Harvard University Press, Cambridge, MA.

Lee, R.E. & Denlinger, D.L. (1991) *Insects at Low Temperature.* Chapman & Hall, New York.

Louw, G.N. (1993) *Physiological Animal Ecology.* Longman Scientific & Technical, Harlow.

Louw, G.N. & Seely, M.K. (1982) *Ecology of Desert Organisms.* Longman, London.

Reviews and scientific papers

Addo-Bediako, A., Chown, S.L. & Gaston, K.J. (2001) Revisiting water loss in insects: a large scale view. *Journal of Insect Physiology* **47**, 1377–1388.

Ancel, A., Visser, H., Handrich, Y., Masman, D. & LeMaho, Y. (1997) Energy saving in huddling penguins. *Nature* **385**, 304–305.

Badyaev, A.V. (1997) Altitudinal variation in sexual dimorphism: a new pattern and alternative hypotheses. *Behavioural Ecology* **8**, 675–690.

Baker, M.A. (1982) Brain cooling in endotherms in heat and exercise. *Annual Review of Physiology* **44**, 85–96.

Bale, J.S. (1993) Classes of insect cold hardiness. *Functional Ecology* **7**, 751–753.

Block, W. (1995) Insects and freezing. *Science Progress* **78**, 349–372.

Butterfield, J. (1996) Carabid life cycle strategies and climate change; a study on an altitude transect. *Ecological Entomology* **21**, 9–16.

Caceres, C.E. (1997) Dormancy in invertebrates. *Invertebrate Biology* **116**, 371–383.

Cerda, X., Retana, J. & Cros, S. (1998) Critical thermal limits in Mediterranean ant species: trade off between mortality risk and foraging performance. *Functional Ecology* **12**, 45–55.

Christian, K.A. & Morton, S.R. (1992) Extreme thermophilia in a central Australian ant *Melophorus bagoti. Physiological Zoology* **65**, 885–905.

Convey, P. (1996). Overwintering strategies of terrestrial invertebrates in Antarctica—the significance of flexibility in extreme seasonal environments. *European Journal of Entomology* **93**, 489–505.

Danks, H.V. (1992) Long life cycles in insects. *Canadian Entomology* **124**, 167–187.

DeLamo, D.A, Sanborn, A.F., Carrsco, C.D. & Scott, D.J. (1998) Daily activity and behavioural thermoregulation of the guanaco (*Lama guanicoe*) in winter. *Canadian Journal of Zoology* **76**, 1388–1393.

Duman, J.G. (2001) Antifreeze and ice nucleator proteins in terrestrial arthropods. *Annual Review of Physiology* **63**, 327–357.

Elkhawad, A.O. (1992) Selective brain cooling in desert animals: the camel (*Camelus dromedarius*). *Comparative Biochemistry & Physiology A* **101**, 195–201.

Faraci, F.M. (1991) Adaptations to hypoxia in birds—how to fly high. *Annual Review of Physiology* **53**, 59–70.

Fletcher, G.L., Hew, C.L. & Davies, P.L. (2001) Antifreeze proteins of teleost fish. *Annual Review of Physiology* **63**, 359–390.

Florant, G.L. (1998) Lipid metabolism in hibernators: the importance of essential fatty acids. *American Zoologist* **38**, 331–340.

Gaede, K. & Knülle, W. (1997) On the mechanism of water vapour sorption from unsaturated atmospheres by ticks. *Journal of Experimental Biology* **200**, 1491–1498.

Gibbs, A.G., Louie, A.K. & Ayala, J.A. (1998) Effects of temperature on cuticular lipids and water balance in a desert *Drosophila*: is thermal acclimation beneficial? *Journal of Experimental Biology* **201**, 71–80.

Haim, A. & Izhaki, I. (1995) Comparative physiology of thermoregulation in rodents: adaptations to arid and mesic environments. *Journal of Arid Environments* **31**, 431–440.

Hainsworth, F.R. (1995) Optimal body temperatures with shuttling—desert antelope ground-squirrels. *Animal Behaviour* **49**, 107–116.

Henen, B.T. (1997) Seasonal and annual energy budgets of female desert tortoises (*Gopherus agassizii*). *Ecology* **78**, 283–296.

Jenni-Eiermann, S. & Lukas, J. (1992) High plasma triglyceride levels in small birds during migratory flight: a new pathway for fuel supply during endurance locomotion at very high mass-specific metabolic rates? *Physiological Zoology* **65**, 112–123.

Klok, C.J. & Chown, S.L. (1997) Critical thermal limits, temperature tolerance and water balance of a sub-Antarctic caterpillar *Pringleophora marioni*. *Journal of Insect Physiology* **43**, 685–694.

Kuhnen, G. (1997) Selective brain cooling reduces respiratory water loss during heat stress. *Comparative Biochemistry & Physiology A* **118**, 891–895.

Machin, J. & O'Donnell, M.J. (1991) Rectal complex ion activities and electrochemical gradients in larvae of the desert beetle *Onymacris*: comparisons with tenebrionids. *Journal of Insect Physiology* **37**, 829–837.

Maina, J.N. (2000) What it takes to fly: the structural and functional respiratory refinements in birds and bats. *Journal of Experimental Biology* **203**, 3045–3064.

Nagy, K.A. & Gruchacz, M.J. (1994) Seasonal water and energy metabolism of the desert-dwelling kangaroo rat (*Dipodomys merriami*). *Physiological Zoology* **67**, 1461–1478.

Navas, C.A. (1996) Metabolic physiology, locomotor performance and thermal niche breadth in high-elevation anurans. *Physiological Zoology* **69**, 1481–1501.

O'Donnell, M.J. (1982) Hydrophilic cuticle; the basis for water vapour absorption by the desert burrowing cockroach *Arenivaga investigata*. *Journal of Experimental Biology* **99**, 43–60.

O'Donnell, M.J. (1987) Water vapour absorption by arthropods: different sites, different mechanisms. In: *Comparative Physiology: Life in Water and on Land* (ed. P. Dejours, L. Bolis, C.R. Taylor & E.R. Weibel), pp. 155–179. Liviana Press, Padova, Italy.

Patterson, B.D., Pacheco, V. & Solari, S. (1996) Distribution of bats along an elevational gradient in the Andes of south-eastern Peru. *Journal of Zoology* **240**, 637–658.

Reinertsen, R.F.E. & Haftorn, S. (1986) Different metabolic strategies of northern birds for nocturnal survival. *Journal of Comparative Physiology* **156**, 655–663.

Rourke, B.C. (2000) Geographic and altitudinal variation in water balance and metablic rate in a California grasshpopper, *Melanoplus sanguinipes*. *Journal of Experimental Biology* **203**, 2699–2712.

Samaja, M. (1997) Blood gas transport at altitude. *Respiration* **64**, 422–428.

Schmidt-Nielsen, K. (1981) Counter-current systems in animals. *Scientific American* **244**, 118–128.

Schmidt-Nielsen, K., Kanwisher, J., Lasiewski, R.C., Cohn, J.E. & Bretz, W.L. (1969) Temperature regulation and respiration in the ostrich. *Condor* **71**, 341–352.

Schmidt-Nielsen, K., Schroter, R.C. & Shkolnik, A. (1981) Desaturation of exhaled air in camels. *Proceedings of the Royal Society of London B* **211**, 305–319.

Sherbrooke, W.C., Castrucci, A.M. de L. & Hadley, M.E. (1994) Temperature effects on *in vitro* skin darkening in the mountain spiny lizard *Sceloporus jarrovii*: a thermoregulatory adaptation? *Physiological Zoology* **67**, 659–672.

Sinclair, B.J., Vernon, P., Klok, C.J. & Chown, S.L. (2003) Insects at low temperatures: an ecological perspective. *Trends in Research in Ecology & Evolution* **18**, 257–262.

Sinclair, B.J. & Wharton, D.A. (1997) Avoidance of intracellular freezing by the New Zealand alpine weta *Hemideina maori* (Orthoptera; Stenopelmatidae). *Journal of Insect Physiology* **43**, 621–625.

Stone, G.N. & Purvis, A. (1992) Warm-up rates during arousal from torpor in heterothermic mammals—physiological correlates and a comparison with heterothermic insects. *Journal of Comparative Physiology B* **162**, 284–295.

Storey, K.B. (1997) Metabolic regulation in mammalian hibernation: enzyme and protein adaptations. *Comparative Biochemistry & Physiology A* **118**, 1115–1124.

Storey, K.B. & Storey, J.M. (1990) Metabolic rate depression and biochemical adaptation in anaerobiosis, hibernation and estivation. *Quarterly Review of Biology* **65**, 145–174.

Storey, K.B. & Storey, J.M. (1996) Natural freezing survival in animals. *Annual Review of Ecology & Systematics* **27**, 365–386.

Strathdee, A.T. & Bale, J.S. (1998) Life on the edge: insect ecology in arctic environments. *Annual Review of Entomology* **43**, 85–106.

Tatar, M., Gray, D.W. & Carey, J.R. (1997) Altitudinal variation for senescence in *Melanoplus* grasshoppers. *Oecologia* **111**, 357–364.

Tracy, R.L. & Walsberg, G.E. (2000) Prevalence of cutaneous evaporation in Merriam's kangaroo rat and its adaptive variation at the subspecific level. *Journal of Experimental Biology* **203**, 773–781.

Wagner, P.D. (2000) Reduced cardiac output at altitude—mechanisms and significance. *Respiratory Physiology* **120**, 1–11.

Walsberg, G.E., Weaver, T. & Wolf, B.O. (1997) Seasonal adjustment of solar heat gain independent of coat coloration in a desert mammal. *Physiological Zoology* **70**, 150–157.

Ward, D. & Seely, M.K. (1996) Adaptation and constraint in the evolution of the physiology and behaviour of Namib Desert tenebrionid beetles. *Evolution* **50**, 1231–1237.

Wharton, D.A. & Ferns, D.J. (1995) Survival of intracellular freezing by the Antarctic nematode *Panagrolaimus davidi*. *Journal of Experimental Biology* **198**, 1381–1387.

Willmer, P.G. & Stone, G.N. (1997) Temperature and water relations in desert bees. *Journal of Thermal Biology* **22**, 453–465.

Withers, P.C. (1983) Energy, water and solute balance of the ostrich *Struthio camelus*. *Physiological Zoology* **56**, 568–579.

Zachariassen, K.E. (1996) The water conserving physiological compromises of desert insects. *European Journal of Entomology* **93**, 359–367.

17 Parasitic Habitats

17.1 Introduction

A great many animals live in an intimate association with another organism, either on its outer surface or within its body; these "commensals" range in their effect on the host from harmful parasites to beneficial symbionts (see Box 17.1), although here we are concerned primarily with parasites having an intimate and negative effect on their host. Living in or on other organisms is phylogenetically extremely widespread in the Metazoa, and it is clear that invasions of the body surfaces and the interiors of other species have happened many times. Here we deal mainly with parasites living on other animals, though of course many land animals are notionally "parasitic" on plants.

Parasites are known in most of the common phyla of animals (Table 17.1), but certain groups are extremely rich in parasitic species, particularly the tapeworms, flukes, acanthocephalans, and nematodes (collectively often termed **helminths**), and the insects. Nematodes are extraordinarily abundant and highly speciose, while the insect group contains huge numbers of species (especially parasitic wasps and flies) that develop as larvae inside other insects or

Box 17.1 Interactions between species

It is traditional to view interactions between species in terms of positive and negative impacts for each partner, resulting in an "interaction grid" as follows:

		Species 1		
		+ve	0	−ve
Species 2	+ve	Mutualism	Commensalism	Contramensalism
	0		Neutralism	Amensalism
	−ve			Competition

However, the terminology is often confused in existing literature, particularly in relation to the terms "commensalism" and "symbiosis". **Commensalism** is often used in a wider sense than is shown in the matrix above, to include all kinds of situations where two species live together (usually one on or within the surfaces of another). **Symbiosis** is sometimes used as a catch-all term for species living and interacting together, but is perhaps better reserved for a more restricted usage where two species are both benefitting from an interaction, making symbiosis synonymous with mutualism.

Thus zooxanthellae within corals are an example of mutualism or symbiosis, where both plant and animal benefit; bromeliad plants that live on a tree trunk but take nothing from the tree are true commensals; and herbivory and carnivory are both "contramensalisms". But the boundaries are not always clear-cut, and in almost every case both partners may over evolutionary time gradually change the relationship in ways that benefit themselves.

On such a matrix, **parasitism** is usually classified as a **contramensalism**; the parasite benefits and the host is harmed.

Note, however, that the boundaries between parasitism and predation become rather difficult, and in effect there is a continuum of kinds of relationship. Predators kill and eat many items of prey that are usually smaller than they are; parasites do not normally kill their single much larger host but eat it (or its resources) while it lives. A biting fly, or a vampire bat, that alights on a larger animal and sucks its blood is normally seen as predatory (but sometimes as parasitic, especially if it transmits disease); a fly that punctures the skin to feed and then lays eggs in the wound is more clearly a parasitic species. The term parasitism therefore needs to be reserved for associations that not only have a negative impact on the host but that also involve a relatively lengthy association between the two species (so excluding the "hit-and-run" biting flies).

There still remain problems with many arthropod "parasites" (commonly wasps and flies) where one or more eggs are laid on a host (usually a larval insect, not dissimilar in size from its attacker) by the adult female. The resulting offspring hatch and feed on the host, which is usually immobilized by their presence but still alive (hence the relation is parasitic) but which dies when they mature into adults and leave the carcass (hence the outcome is effectively predation). The term "**parasitoid**" is commonly used for these intermediate kinds of relationship.

For our purposes, the parasitoids are living for most of their life in an internal enclosed environment created by the host and with physiological problems like those of a parasite rather than those of a predator; therefore we will include them in this chapter without further distinguishing them from "true" parasites.

Table 17.1 Occurrence of parasite members in animal groups.

Animal group	Importance as parasites
Poriferans (sponges)	A few "ectoparasitic", boring into shells, etc.
Cnidarians (sea anemones, hydroids, etc.)	A few ectoparasites on fish, etc.
Ctenophores (comb-jellies)	Very few
Platyhelminths (tapeworms and flukes)	All groups (except most turbellarians) are ecto- and endoparasites on invertebrate and vertebrate hosts
Nemertines	A few parasitic/symbiotic? in other invertebrates
Rotifers	A few ectoparasites
Nematodes (roundworms, filarial worms)	Huge numbers of endoparasites in invertebrate and vertebrate hosts
Nematomorphs (horsehair worms)	All endoparasitic in arthropods as larvae
Acanthocephalans (spiny-headed worms)	All endoparasitic in vertebrate guts
Molluscs	Some larvae are parasitic on fish; a few adult ectoparasites and endoparasites
Annelids	Leeches are ectoparasitic/predators; a few polychaetes are parasitic
Insects	Very numerous ectoparasites (e.g. lice, fleas, tsetse flies, mosquitoes); also parasitoids on other insects (e.g. ichneumon wasps, flies); some larvae endoparasitic in vertebrates (e.g. bot flies)
Crustaceans	Uncommon; some barnacles on crabs, some copepods and branchiurans mostly on fish or on other crustaceans, pentastomids in vertebrate lungs
Arachnids (spiders, mites, etc.)	Mites and ticks are ectoparasites
Chordates	Some jawless fish are ectoparasites/predators

inside plants. From consideration of these two groups alone, which dominate the animal kingdom numerically, it is probably fair to conclude that *most* animal species are parasites.

This chapter surveys the departures from a free-living physiology that are associated with a range of types of parasitism. There are two main aspects to cover. Firstly, being a parasite involves coping with the *abiotic* conditions to be found when living in or on a particular part of the host. For many parasites, homeostatic mechanisms within the host provide the parasite with energy and nutrients, regulate temperature and water balance, and supply the dissolved gases for aerobic respiratory exchange while removing metabolic wastes. In this sense, the parasitic life is physiologically "easy" most of the time. However, some parts of the host environment (e.g. the acidic or anoxic parts of the gut) are less suitable for normal life; thus parasites may show adaptations to unusual pH, redox potential, dissolved gas levels, and so on. Some of these adaptations parallel features found in free-living organisms that occupy similarly extreme environments—for example, some of the adaptations found in gut-inhabiting nematodes and platyhelminths are similar to those found in metazoans inhabiting anoxic muds.

However, being a parasite also involves interactions with a host, which constitutes a *biotic* environment, and which creates a unique set of problems. Most obviously, the host commonly seeks to evict the parasite, and is therefore an aggressive environment that "fights back"; and just as hosts are part of the parasite environment, so parasites are an ancient part of the host environment. Many hosts have developed a range of defenses against parasite attack, and these in turn have led to an equally diverse array of parasite countermeasures. Whereas physiological solutions to a given abiotic environmental challenge should have long-term selective value in that environment, the same cannot be said of the host–parasite arms race, which is an ongoing evolutionary struggle with no predefined optimum or endpoint. Defense and counterdefense are therefore a dominant part of the biology of many parasites, and a large part of this chapter deals with the physiology of host defenses and the biochemical, physiological, and behavioral tricks used by parasites to cope with these.

There are other ways in which a biotic environment is unusual. Animals normally control their own metabolic resource allocation using hormones as control agents, making "decisions" on allocations to growth and to reproduction, eventually undergoing programed senescence followed inevitably by death. But many parasites are not passive bystanders in the control of host resource allocation, but instead actively reroute resources to favor their own development, by releasing hormones and neuromodulators that operate directly on host receptors. Effects include prevention of host reproduction, alteration of host size, induction of entirely new tissues for the parasite's benefit, and considerable changes to host longevity.

To be successful a parasite must also reproduce or be transmitted within its host's lifetime, posing a further set of physiological hurdles to be overcome, and mortality of propagules is often high. Parasites show many adaptations that improve transmission between hosts, including methods to synchronize parasite and host reproduction, and manipulation of host behaviors. These mechanisms operate through intimate chemical interactions between the parasite and host, giving an added dimension to parasite environmental physiology.

Clearly, many of the more dramatic metabolic adaptations in the environmental physiology of endoparasites are responses to a host environment that is both hostile and capable of coevolutionary responses. Indeed host and parasite genomes may become intimately linked, and recent evidence indicates convergent changes in a range of parasite genomes, including reduced genome size and lowered G + C content, while genes for surface proteins or cell wall constituents divergently multiply. It may be that the parasites should not really be seen as having "invaded" hostile environments in an evolutionary sense, but as having "created" them by driving the evolution of the hosts' protective and immune systems. Certainly these evolutionary interactions are a major part of what makes parasite environmental physiology unique.

17.2 Parasite environments

The environment of parasitic organisms is unusual in consisting of two components:

1 The macrohabitat in which the host organism lives.

2 The microhabitat, resulting from modification to the free-living environment brought about by living in or on a particular region of the host.

The importance of these two components in determining the environment experienced by a parasite varies enormously across the spectrum of host–parasite interactions, differing between parasite species and between stages in the same parasite life cycle. At one extreme, some parasites living on the external surface of their host (termed **ectoparasites**) may experience much the same environment as their host, and so require similar physiological tolerances and adaptations to free-living organisms in the same habitat. At the other extreme, parasites living within host tissues (termed **endoparasites**) commonly experience a relatively stable but often highly unusual internal microhabitat, and they may have their thermal, osmotic, respiratory, and nutritional needs met entirely by the host. The term "endoparasite" covers a wide range of different niches within a host's body (especially if the host is a vertebrate), and differences between these niches have major consequences for the environmental physiology of parasites. In addition to the ectoparasitic environment on the skin, we therefore need to consider three major categories of endoparasitic niche—the respiratory passages, gut, and blood—in what follows.

17.2.1 Living on the skin

For parasites in aquatic environments there is little "microclimatic" amelioration around the surfaces of the host animal, whereas for terrestrial ectoparasites, especially on birds and mammals, the environment close to the host skin may be much less thermally variable and much less desiccating than the macrohabitat. Skin ectoparasites on a land vertebrate may need only fairly narrow thermal tolerances and many will have relatively little problem with water balance. This will be especially true if the parasite is "plugged in" permanently or intermittently to the capillaries or tissues of its host, as with many ticks and fleas, and thus endowed with a freely available water supply.

The physiological tolerance of particular groups of skin parasites may mean that they are capable of living in some environments and not others. This is again seen most clearly when comparing terrestrial and aquatic habitats. For example, while platyhelminth flukes are major ectoparasites of vertebrates in marine habitats, they are far less abundant and speciose in terrestrial hosts, presumably being unable to cope with the desiccation associated with a terrestrial environment given their permeable outer surfaces (see Chapter 5). Some monogeneans survive in terrestrial habitats by generally avoiding the most exposed outer surface of the host, and limiting any free-living phase in their life cycle to periods when the host is in water. A good example of this is the common fluke *Polystoma*, which as an adult inhabits the bladder of frogs.

17.2.2 Living in respiratory passages

Many parasites inhabit the buccal cavities or respiratory chambers of larger animals; most notable are the great range of flukes and leeches found in the gills and buccal cavities of marine and freshwater crustaceans and fish. These locations are still essentially ectoparasitic, though less exposed to some aspects of the external conditions. Many parasites have taken this route further, and inhabit deeper respiratory passages as endoparasites. These environments are moist enough to allow essentially aquatic taxa to survive; for example, the lungs of terrestrial vertebrates support crustaceans (e.g. pentastomids in snakes and birds) and platyhelminths (e.g. the human lung fluke). Parasites derived from terrestrial ancestors have also invaded respiratory passageways, for example the mites whose infestation of the tracheae of bees causes the disease varroa, and the maggots of the nostril fly, *Oestrus ovis*, which live within the nasal passages of sheep and goats. Here most aspects of the parasite's physical environment (such as temperature, levels of respiratory gases, and water content) are dominated by the host, and physiological problems are relatively limited. The entry and exit of eggs or infective larval stages is also uncomplicated, via the respiratory openings. A dramatic example of this is provided by the larvae of the nostril fly, which when mature migrate towards the nostril openings and are expelled into the environment in an explosive sneeze! However, this also points to a major drawback of life in respiratory passages; many hosts, and especially vertebrates, have defenses such as coughing and sneezing that serve to eject parasites.

Many of the parasites in respiratory passages feed by penetrating the wall of the host's body to feed on blood. They then become vulnerable to attack by the various blood-mediated host defenses discussed below, and their environment is best regarded as somewhat intermediate between respiratory and blood-based.

17.2.3 Living in the gut

Parasites inhabiting the gut are surrounded by host food material, which provides nutrition for many of them (particularly cestode tapeworms). The gut lining is also richly vascularized, providing a further food source for those gut parasites that penetrate the gut wall and feed on blood (nematodes and some trematode flukes) or tissue fluids (acanthocephalan worms). Entry via the mouth is easily achieved, and the exit of eggs or infective larval stages is also uncomplicated, via the anus or possibly via the mouth if the parasite can move itself or its propagules upwards against the gut movements.

However, there are severe problems as well. The gut of most potential hosts is a relatively anoxic environment, especially in its lower reaches, and many gut parasites are therefore at least facultatively capable of anaerobic respiration. Also, gut parasites are exposed to the digestive processes of the host, including extremes of pH, high levels of activity of a range of enzymes, and the mechanical action of peristalsis. Conditions are most extreme in the stomach, with a very low pH, abundant proteolytic enzymes, and a high turnover of the stomach lining making attachment difficult; relatively few metazoan parasites survive here. At a given location in the gut, conditions may also change dramatically and cyclically during the processing of a single host meal. Species feeding on blood may be attacked by blood-borne host defenses, and some host defenses are also known to reach nonblood feeders in the gut lumen. Furthermore, the gut again has ways of physically removing parasites at either end, by vomiting or by the copious fecal emissions (diarrhoea) induced by some parasitic infections.

17.2.4 Living in tissues and blood

Host tissues include not only the muscles, body cavity, and organs of the host, but also host circulatory systems containing cells (e.g. blood and lymph). In some ways, these constitute ideal parasite habitats: as long as the parasite tissues have similar general requirements to the host tissues, they are provided with nutrients and oxygen, and waste products and carbon dioxide are removed for them. In other ways, inhabiting host tissues is the most demanding of all parasitic habitats. Firstly, although tissue parasites are protected from physical removal, they are exposed to the full range of host defenses based on host recognition of nonself tissue (described below). Such exposure has given rise to a spectacular diversity of parasite countermeasures, including the exploitation of a number of specific parts of the host's body—**immunologically privileged sites**—where host defenses are reduced. Second, there is often no direct aperture in the host's tissues through which the parasite can infect the host or release eggs or larvae. Tissue-inhabiting parasites thus tend to have complex life cycles and infective strategies. Eggs or larvae may escape by release into the gut or lungs (thus encountering, albeit briefly, the problems already outlined for such areas), or may be transmitted by a blood-feeding vector such as a mosquito. Entry into a second host may involve boring into host tissues by larvae, boring out of the host's gut after being swallowed, or being injected by blood-sucking insect vectors.

17.2.5 Microenvironments in space and time

A feature of parasites inhabiting each of these habitat types is that the region of the host body they exploit is extremely specific (e.g. a particular part of the respiratory tract, or of the gut, or a particular set of blood vessels). Characteristics of internal host environments that confer this specificity not only include general aspects such as pH, salt concentrations, and levels of oxygen and carbon dioxide, but many more intimate fluctuations in host physiology. Recognition of these is important in many aspects of parasite biology, ranging from parasite migration within the host to the timing of parasite reproduction.

Host physiology may alter on a variety of temporal scales. Changes may be unidirectional and long term (such as those associated with sexual maturation), or medium term and cyclic (such as the changes in levels of mammalian reproductive hormones in the blood, or changes in levels of insect molting hormones), or operate on even shorter timescales (such as changes in the chemical composition of the gut associated with the processing of a meal, or changes in the levels of metabolites associated with host activity–sleep cycles), and these short-term changes can be particularly important in synchronizing the release of infective stages in some parasites (e.g. the release of nematode filariae into the blood). In addition, many hosts are able to deploy mechanisms that have evolved specifically to impede parasites, including a wide range of relatively specific blood-borne (humoral) and cell-mediated responses, as well as some more widespread defensive systems.

The tapeworm *Hymenolepis diminuta* provides the best studied example of adult migration between microenvironments within the host gut. The adults show a daily (circadian) migration in response to the position of food along the gut. When the host's stomach is full after feeding, the scolex ("head" region) of the worm is found within the intestine close to its exit from the stomach (the pyloric sphincter). With the passage of food into the intestine, the worm shifts back down the intestine, and finally moves forward again to await the next host meal. An adaptive explanation for this change in position is that the cestode is using the balance between the surrounding host gut contents and the physiological environment to achieve adequate absorption of nutrients. Indeed there is evidence that some parasites adjust their position in the host's gut when the host *expects* to be fed, even if the host does not actually feed. This occurs in the digenean fluke *Bunoderina encoliae*, which inhabits the rectum of sticklebacks, and it suggests that parasites can detect stimuli more subtle than the mere presence/absence of gut contents.

It is also important to realize that parasites (particularly endoparasites) may move between very different environments during their life cycle. This is well illustrated by the life stages of the digenean fluke *Schistosoma mansoni* (Fig. 17.1), which causes the disease bilharzia in humans. Schistosomes generally live as adults in the blood system of vertebrates, and adults of *S. mansoni* prefer the fine mesenteric blood vessels surrounding the human gut. Schistosome eggs are spiny, and with the help of the host's immune system (see section 17.7) make their way out of the host's body with the feces or urine. The eggs hatch into free-swimming ciliated miracidium larvae, which locate and penetrate a freshwater snail as the second host organism. The miracidium larvae give rise to a different type of larva, the sporocyst. These reproduce asexually in the snail, producing first a second generation of sporocysts and then a third larval type, the cercaria. The cercariae leave the snail and swim actively until they encounter human skin; they then bore through the skin and shed their ciliated larval epidermis, metamorphosing very rapidly into a smaller version of the adult worm, called a schistosomula. These immature adults migrate to the blood supply, and travel in it until they reach the hepatic portal vein. This life cycle thus involves two distinct dispersive free-living freshwater stages, which must cope with the vagaries and variability of freshwater bodies (see Chapter 13), and also must cope with abrupt changes in their external environment as they penetrate the tissues of their next host. There are also two stages that develop inside host tissues, but the two hosts are extremely disjunct taxonomically, differing in such basic variables as body temperature and levels of dissolved gases and metabolites, and also in the types and magnitudes of defenses they are able to mount against invading organisms. The different environments encountered by the life stages of *S. mansoni* have been inevitably selected for stage-specific physiological adaptations; successful completion of the life cycle requires appropriate sense organs at each stage to detect which environment the current life-cycle stage is in. These sensory inputs must be linked to switches able to trigger the expression of genes relevant to each particular environment. This example shows that the modifications to the environmental physiology of a free-living organism—even a relatively "simple" platyhelminth—may be extremely complex.

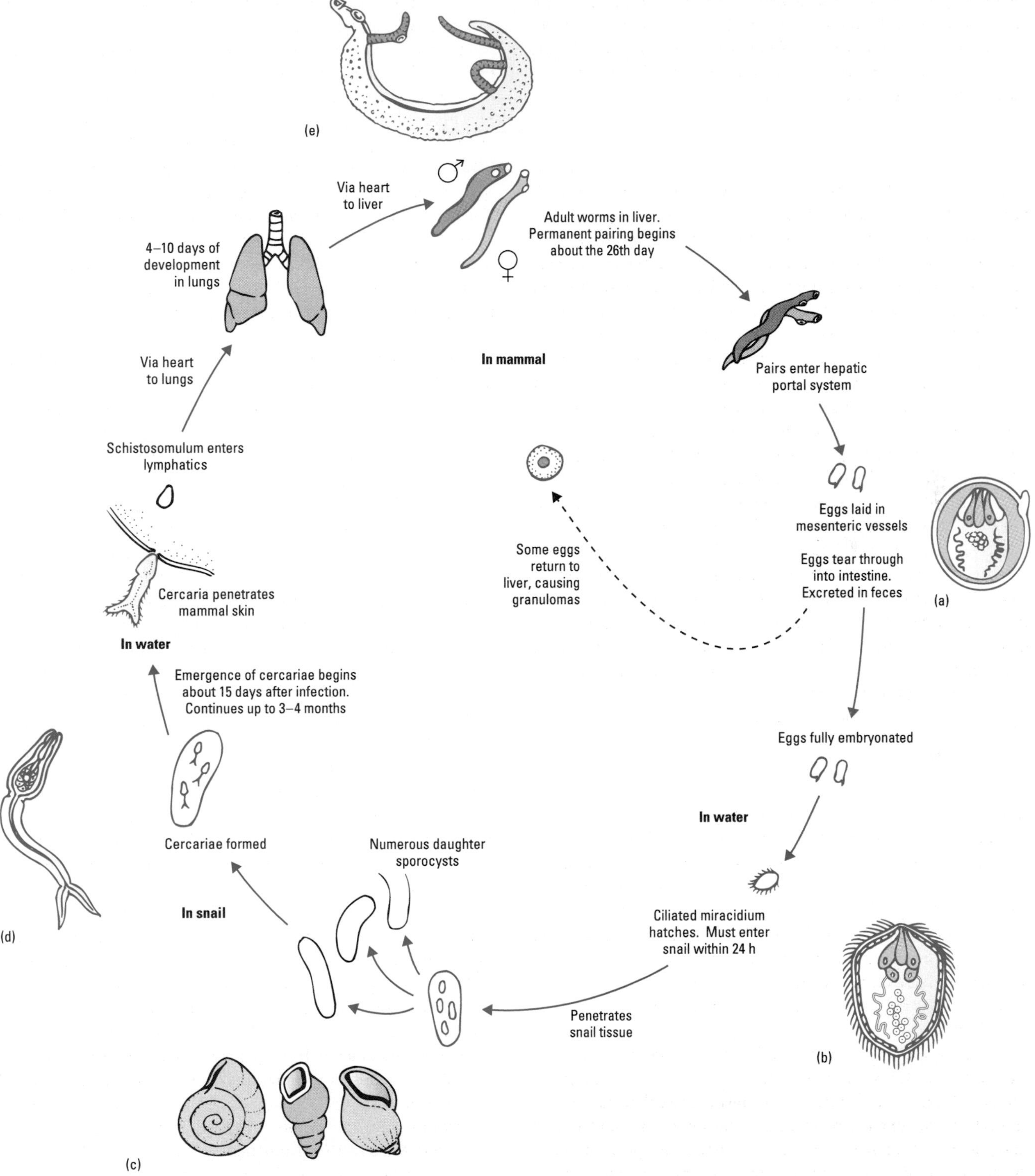

(e)
Via heart to liver
Adult worms in liver. Permanent pairing begins about the 26th day
4–10 days of development in lungs
In mammal
Pairs enter hepatic portal system
Via heart to lungs
Schistosomulum enters lymphatics
Eggs laid in mesenteric vessels
Some eggs return to liver, causing granulomas
Eggs tear through into intestine. Excreted in feces
(a)
Cercaria penetrates mammal skin
In water
Emergence of cercariae begins about 15 days after infection. Continues up to 3–4 months
Eggs fully embryonated
In water
Cercariae formed
Numerous daughter sporocysts
In snail
Ciliated miracidium hatches. Must enter snail within 24 h
(d)
Penetrates snail tissue
(b)
(c)

Fig. 17.1 The life cycle of the human blood fluke, *Schistosoma mansoni*. Enlarged life-cycle stages are: (a) a fertilized egg, (b) a miracidium larva, (c) alternative snail hosts, (d) a cercaria larva, and (e) a pair of adult flukes, with the female lying in a groove in the male. (Adapted from Trager 1986.)

A feature of any complex life cycle is that the behavior and environmental physiology of any one life cycle stage may limit the distribution of the parasite. Thus while adult *Schistosoma* can probably survive in human beings anywhere, they also require freshwater habitats containing the snail intermediate host. Furthermore, snail and human hosts must be abundant enough for the larvae involved

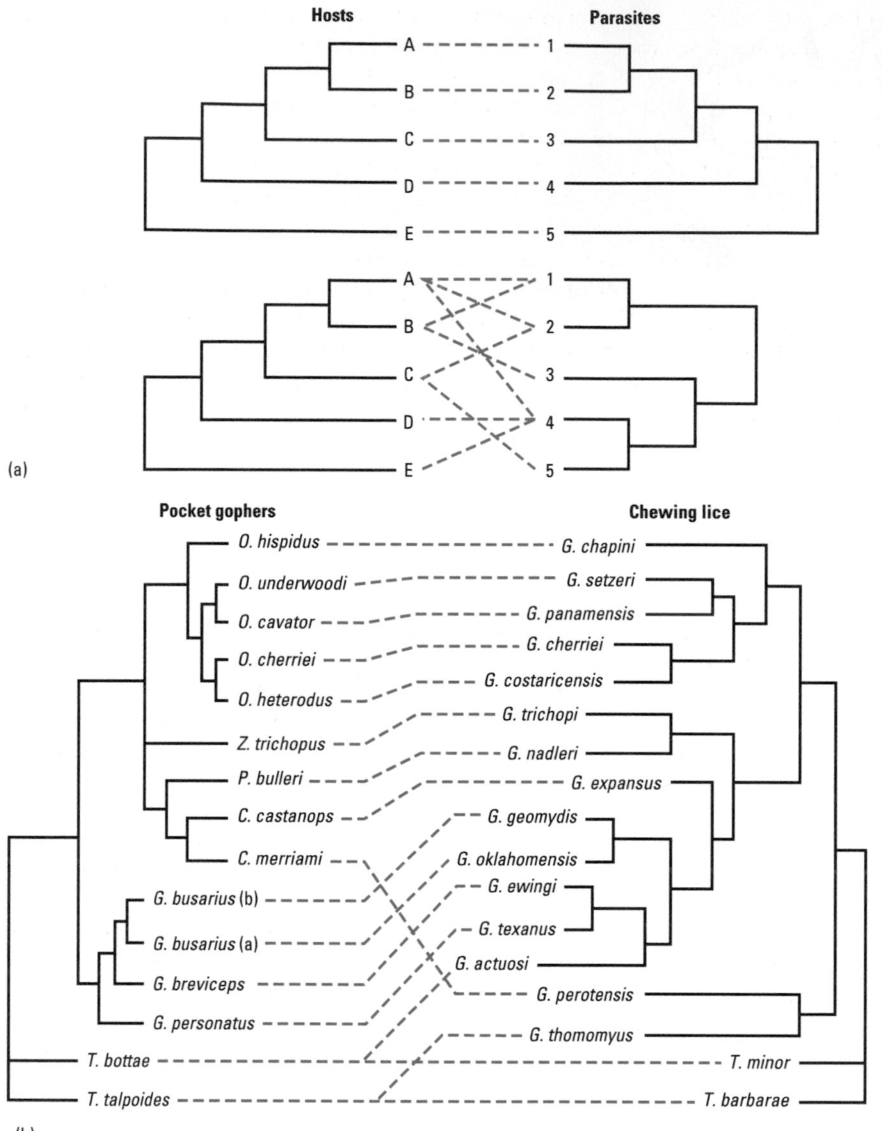

Fig. 17.2 (a) Alternative patterns of speciation in parasites and their hosts, with host–parasite relationships indicated by dashed lines. In the upper example, speciation of host-specific parasites exactly tracks speciation of hosts. Such parallel evolution results in Fahrenholz's rule: i.e. taxonomically related hosts have taxonomically related parasites. The lower example shows a case in which host and nonspecific parasite phylogenies do not match. (b) A clear example of parallel evolution, demonstrated between pocket gophers (rodents of the family Geomyidae) and their chewing lice (order Mallophaga, family Trichodectidae). Gophers: O, *Orthogeomys*; Z, *Zygogeomys*; P, *Pappogeomys*; C, *Cratogeomys*; G, *Geomys*; T, *Thomomys*. Lice: G, *Geomydoecus*; T, *Thomomydoecus*. (a, Reprinted from *International Journal for Parasitology* **23**, Page, R.D.M., Parasites, phylogeny and cospeciation, pp. 499–506, copyright 1993, with permission from Elsevier Science; b, from Hafner & Page 1995.)

to find them in their very limited lifetimes—a matter of hours in each case. Considerations of this kind mean that many parasites have a limited distribution even in regions where they are generally endemic.

17.2.6 Parasite environment and parasite speciation

The importance of host microenvironments in the evolutionary radiation of parasite groups is encapsulated in what has become known as Fahrenholz's rule, which states that the evolutionary radiations of parasites tend to parallel those of their hosts (Fig. 17.2). Parasite groups that show this pattern tend to be highly adapted to their host environment and to infect only a small diversity of host species; for example, 78% of monogenean parasites of fish and 70% of parasitic copepod crustaceans are restricted to a single host species. However, not all parasite groups follow Fahrenholz's rule, and some parasitic taxa are characteristic of a given habitat rather than a group of closely related hosts; a good example is the fluke *Benedenia hawaiiensis*, which is an exception to the monogenean generalization made above, being able to infect 24 fish species distributed in a range of families but all living in the same coral reef habitat. The study of the relative importance of parasite–host relationships and external macroenvironments in determining speciation patterns in parasites is very much in its early stages, though, because the sampling of host species for their parasite faunas is still very incomplete.

17.3 Basic parasite physiology

17.3.1 Ionic and osmotic adaptation, and water balance

In general, parasites are either bathed in, or are taking in fairly continuously, a fluid that is already being regulated by their host. In this sense they have fewer problems with osmotic physiology than almost any other kind of animal. Once inside the host many of them are highly permeable, are fairly intolerant of water loss (because they never experience it), and have weak or absent kidney function. However, it must always be remembered that the larval or transmission phases may have very different problems, surviving outside the host in what may be very different osmotic conditions. Endoparasitic nematodes that infect insect larvae are known to remain within the host cadaver for variable times dependent on the external humidity; at low relative humidity (RH) they may stay within the corpse for up to 50 days rather than emerge and risk desiccation. Other nematode parasites adopt the cryptobiotic response (see Chapter 14) when exposed to desiccation during transmission.

Many aquatic ectoparasites are surprisingly sensitive to variation in the salinity of the environment inhabited by their host, and monogenean and trematode platyhelminths are generally rare in low-salinity environments. Where the host is better able to tolerate variation in salinity than its ectoparasite, the ectoparasite may be absent from part of the host's range; this applies for the monogenean fluke *Benedenia melleni*, which has a lower tolerance of high salinity than its fish hosts.

Terrestrial ectoparasites may have very little amelioration of water balance or ionic problems across their external surfaces relative to a free-living animal in the same environment, but they do have the advantage of being plugged in to a reliable regulated fluid source. Some species with terrestrial hosts may have the option of excreting into the host (as in the tick shown in Fig. 5.23) rather than to the outside world. Some endoparasites of desert ectotherms may face significant problems, however; the monogenean *Pseudodiplorchis* inhabits the bladder of the spadefoot toad *Scaphiopus* (see section 16.2.2) and must therefore survive greatly increased osmotic concentrations for many months at a time.

Endoparasites are probably usually in osmotic balance with their animal host's tissues, though this is rarely studied. For any host of reasonable size, whether in sea water, fresh water, or on land, this will give the parasite a stable habitat. In land and freshwater animals, and in marine vertebrates, a stable osmotic concentration of 100–400 mOsm is likely, while in marine invertebrates the concentration is around 1000 mOsm. The interesting experiences will come for parasites on migratory aquatic hosts, for example eels, salmon, and crabs such as *Eriocheir* that move between seas and rivers. Little is known of such parasites, but it is very likely that the host does most of the regulatory work and the parasite conforms with the resulting osmotic concentration of the host tissues.

One interesting example of host habitat effects on parasite physiology is provided by some endoparasitic wasps that develop inside larvae of species of fruit fly (*Drosophila*). These fruit flies live in fermenting fruit, and particular species differ in their tolerance of the ethanol generated by fermentation in this habitat. Parallel adaptations are found in the parasitic wasps: strains and species that attack *Drosophila* in alcohol-rich habitats show greater tolerance of ethanol.

17.3.2 Thermal adaptation

Small aquatic ectoparasites effectively experience the thermal vagaries of the macrohabitat, since water is usually well stirred and of quite high thermal conductivity. Thus temperature may be important in constraining the environment occupied by these kinds of parasite. For example, the trematode fluke *Derogenes varicus* is absent from hosts in warm marine areas, preferring cool temperate seas; and it is found near the surface at cooler high latitudes but in deeper waters at lower latitudes.

For terrestrial ectoparasites, temperature balance is also usually fairly easy, as they exist mainly within the stable boundary layer created around a reasonably sized host (only during transmission between hosts are cooler temperatures potentially encountered, so that egg or larval stage may need greater cold tolerance or even freeze tolerance). Ectoparasites are therefore common on land vertebrates, and especially on endothermic hosts where the trapped, still air within fur or feathers may be at a rather constant temperature (close to the host's body temperature, T_b), and at a high humidity. Thus an ectoparasite here may need a rather high but narrow-band thermal tolerance, and can use enzymes and membranes with rather limited acclimatory ranges. In cold climates, ectoparasites tend to congregate in the warmest places: the "armpit" or "groin" regions of mammals, or beneath the wings on birds. On marine mammals they may be fairly evenly dispersed in the aquatic phase, but tend to accumulate on the tail flipper when seals and walruses haul out to give birth, this being the warmest spot and used for basking; for example, the lice on elephant-seal flippers stay at 27–34°C throughout a diurnal cycle.

For endoparasites, problems are again effectively handed over to the host's regulatory systems for large parts of the life cycle. But many endoparasite life cycles involve transitions between different thermal environments. This is particularly true for those life cycles (such as the infection of human hosts by schistosome cercariae, or the injection of nematode filaria larvae into mammalian blood by an insect vector) where a free-swimming larva enters an endothermic host. Such abrupt changes in thermal environment can be very damaging (as discussed in Chapter 8), and parasites make extensive use of heat shock proteins (HSPs) to protect their metabolism. HSP manufacture can occupy a major proportion of a parasite's metabolism during such transitions; in adult schistosomes, production of HSP-70 represents more than 1% of the total adult protein production, and levels are higher again in the postinvasion schistosomula larva. The transmission stages of vertebrate endoparasites may also need much broader thermal tolerance; nematode parasites have a marked degree of cold tolerance that is also susceptible to rapid selection. They have also been shown to change their membrane lipid saturation in similar fashion to other ectotherms exposed to colder conditions. Most nematodes that exploit insect hosts also have a suite of inducible HSPs that can protect them from excess heat (e.g. in a heated decomposing corpse until conditions are favorable for leaving). There are indications that nematode parasites of vertebrate hosts living in extremely arid and hot conditions, such as those of the dromedary of the Sahel, have rather specialist suites of enzymes with reduced isozymic variation compared with nematodes from more mesic hosts.

While intricate temperature regulation may not be a notable feature of parasitic adaptation, sensitivity to temperature certainly is. For many parasites that develop in or on endothermic hosts, temperature is a major cue for larval penetration of the host, or for egg hatching within the host. For example, schistosome cercariae move up a temperature gradient when locating human skin, and fleas locate new hosts by moving towards warmth. Hatching of a range of eggs, including taeniid tapeworms, is triggered by temperature.

Nematodes that have snails as intermediate hosts have an unusually high developmental temperature threshold, and this probably helps to insure that larvae remain in their first instar while the snail undergoes winter hibernation, the first instar being more cold resistant than later larvae. In general, a very wide range of parasites associated with endothermic hosts show high sensitivity to temperature gradients as infective stages.

While parasites may show adaptation to high body temperatures in endotherms (37–41°C), a range of host organisms react to infection by producing short bursts of exceptionally high body temperature—generally termed fever (see Chapter 8). Fever in vertebrate endotherms is induced in response to chemicals termed **pyrogens**, which can be produced either by parasitic organisms (exogenous pyrogens) or by host blood cells (endogenous pyrogens). In mammals both operate, as endogenous pyrogens are produced by leucocytes in response to exogenous pyrogens, and they act directly on the hypothalamic temperature control center to elevate the setpoint about which temperature is regulated. Some ectothermic animals (including some caterpillars and some lizards) also react to infection by elevating their body temperature. In these cases elevation is usually achieved by a change in behavior—increased basking duration, and altered posture to increase the surface over which heat is absorbed. In these ectothermic cases, fever has a demonstrated advantage in conferring protection from bacterial infections, but in endotherms the role of fever is less clear. High temperatures may slow infections down, but can also cause harm or even death to the host.

Temperature also has marked effects on parasite–host interactions, altering relative levels of susceptibility, latency, and virulence; the net effects are rarely linear and there can be complex outcomes at the population level.

17.3.3 Respiratory adaptation

Ectoparasites

In general, ectoparasites exchange gas with the macroenvironment, whether water or air, and show little specialization. However, some aquatic ectoparasites show a range of behavioral responses when exposed to hypoxia by the behavior of their host. For example, *Entobdella soleae*, a monogenean ectoparasite of the sole, shows increased body undulations to maintain movement of water over the body surface through which gas exchange takes place. Similarly fish lice increase the ventilatory movements of their pleopods (modified abdominal limbs). Parasites that occupy the outer respiratory passages of their hosts are supplied with oxygen by the host's ventilatory currents (e.g. flies living in the nasal passages of ruminants, or marine leeches in the branchial cavities of fish).

Some tissue endoparasites are able to be "ectoparasitic" in terms of respiration, reaching the exterior for gas exchange by causing lesions in the host body wall. For example, the larvae of many tachinid flies have posterior spiracles armed with hooks that they use to penetrate either the skin of their host, or an air sac or major tracheal airway. The host responds by forming a funnel of wound tissue around the parasitoid larva's spiracles, providing permanent access to gas exchange with the exterior. The jigger flea has a similar trick, the adult living beneath the skin as little more than an egg-producing bladder, but with the posterior segments of the flea containing a narrow airway that penetrates the host's skin through a small pore.

Endoparasites

All endoparasites need the physiological scope to cope with the particular levels of oxygen and CO_2 (and the associated pH) of their niche within the host. However, in any one particular spot these conditions should be relatively constant, so in some sense the adaptation is not too difficult. Some parasitic sites, such as the bloodstream and lungs, are unambiguously aerobic, while the lumen of the small intestine and the bile duct may have consistent but extremely low oxygen tensions. The hemoglobin of the fly *Gasterophilus* can even, apparently, take up O_2 from the stomach contents of the horse it lives in, having a P_{50} of only 0.003 kPa. Parasitic worms such as the nematode *Ascaris lumbricoides* or the fluke *Fasciola hepatica* are oxygen conformers, i.e. the rate of oxygen uptake varies with the partial pressure.

But the greatest physiological problem by far is usually that respiration must necessarily be anaerobic, particularly in the large intestine.

RELATIVE USE OF ANAEROBIC RESPIRATION

In a number of ways, respiration in parasites inhabiting relatively anaerobic habitats (such as the lower gut of vertebrates) parallels pathways seen in free-living metazoans inhabiting environments such as anoxic surface muds in aquatic habitats, and the pathways must often have been inherited from free-living ancestors. Indeed, parasites sometimes show metabolic pathways that seem unsuited to their habitat in the host. For example, anaerobic respiration is common in adult nematodes and platyhelminths, and is found even in species inhabiting relatively aerobic habitats, such as the blood of vertebrates. Use of relatively inefficient anaerobic metabolism even where oxygen is available suggests that the efficient exploitation of high-energy metabolites has not been important in these species. Perhaps this is because these materials are so abundant in the host environment that parasites can "afford" to be wasteful. In explaining the metabolic pathways used for energy generation by parasites, we therefore need to separate genuine adaptations to selective pressures from ancestral mechanisms characteristic of particular taxa that may have little or nothing to do with a parasitic way of life. Remember also that the use of entirely anaerobic respiration does not necessarily mean that a particular species can thrive in the total absence of oxygen. Schistosomes, for example, largely respire anaerobically as adults, but require a certain minimal level of oxygen to carry out the synthetic reactions associated with the manufacture of egg shells. In the total absence of oxygen, adult schistosomes survive well but cannot reproduce.

Aerobic respiration is now known to occur at a low level in many adult parasites formerly thought to rely entirely on anaerobic respiration. While the activity of the aerobic respiration pathways is limited, the far greater yield of ATP per glucose molecule generated aerobically means that even low activity can make a significant contribution to the parasite's energy budget. In the filarial nematode *Litomosoides carinii*, adults pass only 2% of their carbon through aerobic respiration, and 98% is fermented; yet starting with 100 mol of glucose, the energy yield of the 98 mol processed via fermentation is 196 mol of ATP, and the 2 mol of glucose respired aerobically generate 72 mol of ATP, or 27% of total energy storage.

The large intestinal nematode *Ascaris lumbricoides* exhibits transitions in energy metabolism during its life cycle that are characteristic of a number of parasites. The eggs leave the host with feces, and survive in the soil. Larval development within the egg is aerobic, with a high activity of enzymes involved in the Krebs cycle. After ingestion of the egg, a second-instar larva emerges within the host's gut. It burrows through the gut mucosa to enter the bloodstream, reaching the liver 2–3 days after infection, where it molts into a third-instar larva. This then switches to anaerobic metabolism as it migrates via the heart to the lungs, reaching them after 7–9 days. The larva then moves up the trachea and is swallowed, reaching the intestine again after 8–10 days where it molts into a fourth instar before finally reaching adulthood. Both the fourth-instar larva and the adult can continue their development in the complete absence of oxygen.

In *Schistosoma*, the transition from aerobic respiration in the cercaria to anaerobic metabolism in the schistosomula (see Figs 17.1 and 17.8) is very abrupt. When a cercaria penetrates human skin there are two rapid morphological changes: the propulsive tail is lost, and the outer layer of cuticle (the glycocalyx) surrounding the remaining anterior part of the body is shed. The loss of the glycocalyx results in a massive increase of the permeability of the parasite epidermis to host metabolites, some of which are thought to be the cause of the metabolic shift from aerobiosis to anaerobiosis. The key host metabolite may be serotonin (5-hydroxytryptamine, 5-HT), which among other effects it has in vertebrates, activates the enzyme adenyl cyclase and promotes the formation of cAMP. This has a direct effect on respiration through its ability to activate a number of enzymes involved in glycolysis.

In other parasites, shifts between mechanisms may be gradual rather than abrupt. In large parasitic worms, mitochondria with cristae on the inner membrane are concentrated in outer tissues, while mitochondria in deeper tissues lack cristae. Cristate mitochondria generally indicate aerobic respiration, and the absence of cristae usually indicates anaerobiosis, leading to the conclusion that deeper tissues predominantly use anaerobic respiration. This has been explained as the result of limited diffusion of oxygen into the tissues of larger parasites. It may also explain in part the general shift from aerobic to anaerobic respiration through the life cycle of parasitic platyhelminths, being an effect of increasing size as well as of changes in habitat.

ANAEROBIC PATHWAYS

Respiration involves the oxidative breakdown of a small organic molecule, and this is most typically glucose, but raw materials can include amino acids, glycerol, or fatty acids (see Chapter 6). It seems that parasites have an almost complete dependency on carbohydrate (either as glycogen or exogenous glucose) as their sole energy source. In adult helminths there is no active β-oxidation sequence, so there is no catabolism of fatty acids, and amino acid catabolism is very limited. Similarly, there is no evidence of the cofermentation of amino acids and carbohydrate or of fatty acids and carbohydrate; this is found in the free-living larvae of some endoparasites, but is absent from the adults.

There are two other important differences relating to parasite anaerobiosis compared with free-living versions. Firstly, for parasites it is a semipermanent state, not an emergency measure until oxygen returns. Thus helminths do not accumulate anaerobic end-products for later resynthesis into glycogen. Secondly, in parasitic helminths anaerobic metabolism persists even if some oxygen is available, and the anaerobic pathways are not inherently inhibited by the presence of oxygen.

In general, anaerobic parasitic helminths can be divided, on the basis of their end-products, into two groups:

1 Those that rely essentially on glycolysis alone and produce lactate or some other reduced derivative of pyruvate as end-products of carbohydrate breakdown ("homolactic" fermentation).

2 Those that go beyond glycolysis to fix carbon dioxide and have what is often referred to as an "*Ascaris*-type metabolism".

In parasitic helminths where carbohydrate is broken down glycolytically, the amount of lactate varies from 30% under aerobic conditions to 80% under anaerobic conditions. Acetate and acetoin, the other end-products, are produced by the decarboxylation of pyruvate via the pyruvate dehydrogenase complex. In the acanthocephalan *Moniliformis*, the main end-product is ethanol, with small amounts of lactate, succinate, and volatile fatty acids. These parasitic helminths do not produce alanine, octopine, alanopine, etc. as anaerobic end-products, unlike free-living anaerobic invertebrates.

Ascaris adults illustrate the alternative anaerobic energy-generation mechanism that is also found in other (unrelated) gut parasites, such as the tapeworm *Hymenolepis*, as well as in some free-living invertebrates. Glycolysis terminates with phosphoenolpyruvate (PEP), rather than pyruvate, and the enzyme PEP carboxykinase then converts the 3-carbon PEP to a 4-carbon product, oxaloacetate, a step that generates 1 ATP. The enzyme malate dehydrogenase is then used to reduce oxaloacetate to malate. In the process, the NADH formed in glycolysis is oxidized to NAD^+, maintaining the $NADH/NAD^+$ balance.

The malate produced enters the mitochondrion, and rather than entering the normal Krebs cycle it acts as the substrate for two other linked pathways. These pathways together are termed **malate dismutation**; malate is reduced in one series of reactions, and oxidized in the other. In the oxidation route it is converted to pyruvate and CO_2, while in the reduction route it is converted to fumarate and then succinate (Fig. 17.3). These two reactions are coupled in that the oxidation route generates NADH from NAD, and this NADH is then responsible for the reduction of malate in the second route. As in normal aerobic respiration, the NADH generated is the raw material for an electron transport system. The cytochrome involved in the anaerobic electron transfer system (cytochrome b_{558}) is different to those used in the usual aerobic chain (cytochromes a, a_3, b, c, and c_1), and the terminal electron acceptor is fumarate rather than oxygen. The cytochrome and fumarate reductase enzyme make up a complex that is the major component of adult *Ascaris* mitochondria, and rhodoquinone is used as an electron carrier, rather than the ubiquinone characteristic of the aerobic electron transfer chain (see Chapter 6). The redox potential of ubiquinone is more positive (+113 mV) than that of the fumarate/succinate couple (+33 mV), such that electron transfer to fumarate would not be favored. In contrast, the redox potential of rhodoquinone is strongly negative (−63 mV), favoring electron transfer to fumarate. In *Ascaris* and

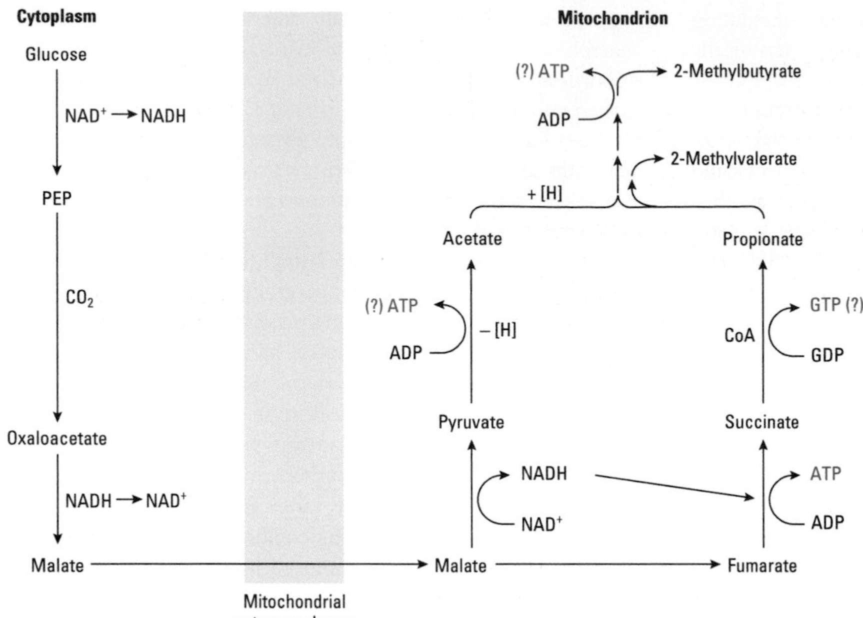

Fig. 17.3 The malate dismutation reaction in the mitochondria of the nematode *Ascaris lumbricoides*. Note the principal stage resulting in ATP generation (between fumarate and succinate), together with three other possible ATP- or GTP-generating stages (marked with ?). PEP, phosphoenolpyruvate. (Adapted from Trager 1986.)

many other parasites the fumarate, succinate, and pyruvate are usually further metabolized to propionate, acetate, and volatile fatty acids such as 2-methylvalerate and 2-methylbutyrate.

Energy is thought to be stored in ATP at several stages in the dismutation reaction. The best known ATP generation point is during the electron transfer chain in the reduction route. The enzyme fumarate reductase, which converts fumarate to succinate, is coupled to an electron transfer system in the mitochondrial inner membrane, phosphorylating ADP to ATP. This is the major known energy storage reaction in *Ascaris*. Three other possible stages at which ATP is generated are also indicated in Fig. 17.3. The endpoint of the metabolism of malate is the fatty acid 2-methylbutyrate. In contrast to the 36–38 ATP molecules generated from a single glucose molecule in traditional aerobic metabolism, the *Ascaris* mechanism generates between three (two in glycolysis, one additionally in the mitochondrion) and five (if the additional ATP-generation steps in the mitochondrion actually occur).

This pathway is unusual in the role played by PEP carboxykinase, which differs radically from its usual role in animals. In mammals this enzyme is present in the liver, and is involved in the manufacture of glucose from lactic acid (gluconeogenesis; see Chapter 6). In contrast to the carbon fixation seen in *Ascaris*, in mammals the enzyme catalyzes the same reaction but in the opposite direction! In effect the animal is using the second span of the tricarboxylic acid cycle, from succinate to oxaloacetate, but operating in the reverse direction. It has what has been described as a partial, reversed tricarboxylic acid cycle. How is this reversal achieved? Three aspects of the situation in *Ascaris* favor this kind of CO_2 fixation:

1 Levels of the substrate PEP are far higher in *Ascaris* tissues than they are in mammalian tissues; pyruvate kinase, which normally breaks PEP down, is present only in very low concentrations in *Ascaris*.

2 *Ascaris* PEP carboxykinase has a seven-fold higher K_m (i.e. lower affinity) for oxaloacetate than for PEP.

3 Oxaloacetate is rapidly metabolized by the enzyme malate dehydrogenase, favoring further CO_2 fixation by the PEP carboxykinase.

Similar transitions in energy metabolism between life-cycle stages are seen in other parasites. *Schistosoma mansoni* has two free-living larval stages, the miracidium that locates the snail intermediate host, and the cercaria that penetrates the human host. Cercariae are definitely aerobic, and miracidia are probably so, but the adults rely to a large extent on anaerobic respiration, even in the presence of oxygen. In contrast to the metabolic pathways used by adult *Ascaris*, schistosome adults generate lactic acid in much the same way as anaerobically functioning invertebrates described in Chapter 6, yielding only two molecules of ATP per molecule of glucose. The same energy-storage system is also important in filarial nematodes. Thus across parasites as a whole, a range of end-products of anaerobic glycolysis are found, most yielding only 2 ATP per molecule of glucose (Table 17.2).

17.3.4 Feeding

Parasite feeding strategies can be broadly divided into two very different types:

1 Parasites feeding on host tissues and fluids via the mouth. This category includes those feeding on blood or hemolymph, and on host exudates such as mucus on the skin.

2 Parasites absorbing nutrients from host body fluids or gut contents, across their body wall.

Feeding via the mouth

Where the parasite takes in food via the mouth, it is possible for the parasite's body to be enclosed in a cuticle or sheath that is metabolically relatively inactive, conferring protection against attack by host immune responses or digestive enzymes. Most nematodes, for example, feed via the mouth and have metabolically inactive cuticles

Table 17.2 Pathways used in parasite anaerobic respiration.

Respiration type	End-product	mol ATP generated per mol glucose	Organism
Aerobic	CO_2, H_2O	36	Almost all aerobes
Anaerobic	Lactic acid	2	Very common; e.g. *Schistosoma mansoni*, filarial nematodes
Anaerobic	Ethanol	2	Some helminths
Anaerobic	Alanine	2	Many invertebrates
Anaerobic	Acetate + succinate	3.7	Many helminths
Anaerobic	Acetate + propionate	5.4	Many helminths
Anaerobic	Acetate + propionate, methylbutyrate, methylvalerate	2–5	Some nematodes, e.g. *Ascaris lumbricoides, Hymenolepis diminuta*

that are thick and multilayered. The same is true of many blood-feeding ectoparasites, whose outer body layers must protect them from conditions in the external environment. The mouthparts of all blood feeders need to be modified for attachment and/or suction (Fig. 17.4).

To generalize, the main problems for this group, and hence the principal feeding adaptations found in members of this group, are three-fold:

1 coping with large volumes of fluid.

2 Coping with the metabolic breakdown products of blood (particularly iron).

3 Coping with possible blood-borne infective agents such as bacteria.

Many parasites from a diversity of animal groups feed on blood, including platyhelminths (monogeneans and trematodes), nematodes, leeches, crustacean fish-lice, arachnids such as ticks, and insects such as lice, fleas, mosquitoes, and tsetse flies. Problems may result from very intermittent meals of large volume; adaptations are best known in leeches and in biting insects, where in both cases diuretic elimination of water is initiated by hormones (often triggered by body-wall stretch receptors) to promote very rapid fluid loss after a meal. Initial blood digestion occurs in response to enzymes released by pharyngeal or "salivary" glands. Homogeneous acellular products of this digestion, including hemoglobin, are usually digested intracellularly. Iron is discarded back into the gut lumen by exocytosis and shed with the normal feces, or in the case of schistosomes iron-rich waste products are voided by regurgitation (flukes usually have a blind-ending gut, with no anus). To cope with the third problem, it is known that antibacterial peptides are upregulated in response to ingestion of a blood meal in the blood-feeding fly *Stomoxys calcitrans*, and such a system may be widespread.

Some parasites with normal guts and mouths that could feed on blood or other tissues nevertheless use the skin as an uptake surface for key nutrients. For example, the monogenean *Diclidophora merlangi* feeds mainly on blood, but supplements its diet by the absorption of amino acids through its outer tegument.

Most parasites can use their metabolic pathways to synthesize the majority of the amino acids they require, often by the described routes relating to carbon dioxide fixation, and most can synthesize pyrimidines; but few if any can manufacture the purine ring, which must be obtained from the host. Salvage pathways therefore exist in many parasites to allow recycling of certain host molecules from the blood before they are broken down in normal catabolic paths.

Direct absorption across cuticles

Where direct absorption occurs, the outer tissues of the parasite are typically highly metabolically active, and experience rapid fluxes of nutrients, ions, and water; it is not possible for the parasite to isolate itself from the host by a barrier layer. Direct absorption of host nutrients across the body wall is found in a wide diversity of parasites, particularly among the platyhelminths. It is also found in acanthocephalan worms, in the first larval stage of some insect endoparasitoids, and in a small number of highly specialized nematodes.

Parasitic platyhelminth teguments consist of several layers (Fig. 17.5a,b). The outer layer is composed of syncytial living tissue, the nuclei lying more centrally within the animal below layers of circular and longitudinal muscle. The outer surface of the tegument is covered in a thin glycocalyx of mucopolysaccharides and glycoproteins. This layer has a net negative charge, and is able to bind both inorganic and organic charged molecules, including host enzymes. These may then be used to digest surrounding food material, which is absorbed by the parasite. The surface area of the tegument is enlarged (by up to 10 times) by microvilli (called microtriches in the case of cestodes), which are particularly well developed in cestode tapeworms (Fig. 17.5b) and are presumably adaptations for large-scale secretion and absorption across the tegument. The outer surface of acanthocephalan worms also has an enlarged surface area (Fig. 17.5c), in this case through high densities of surface pores, which lead to blind-ending spaces ("crypts"), enlarging the surface area by about 60-fold. The nematodes that absorb nutrients across their body wall lack the usual thick and nonabsorptive cuticle, and develop microvilli all over their surface, through which it is assumed nutrient absorption takes place.

The major physiological adaptations in nutrition are associated with digestion and absorption of host material, both inside and often also immediately outside the parasite's body, while protecting the same tissues from attack by host defenses. Cestodes and acanthocephalans especially are in competition with their hosts for nutrients in the gut lumen. If they are to compete successfully for these, their ability to bind and absorb nutrients must be at least as good as or better than the host's. Adult cestodes, for example, are able to absorb all detectable glucose from their surrounding medium, indicating a very high-affinity (low K_m) active uptake for this compound.

Living in the gut clearly has real hazards, and gut parasites show a

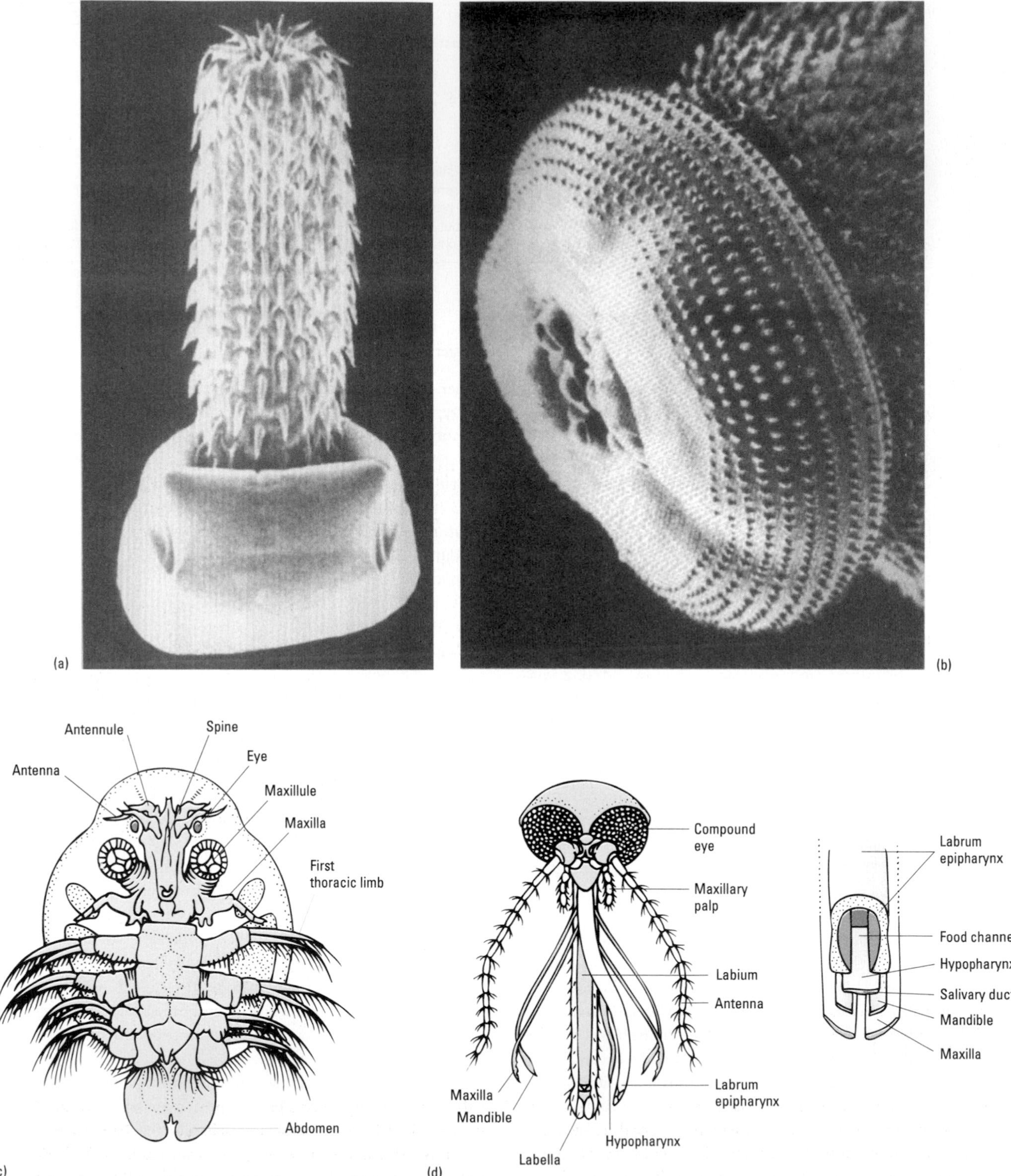

Fig. 17.4 Mouthparts of blood-feeding parasites. (a) The proboscis of an acanthocephalan worm. (b) The spiny head collar of the nematode *Gnathostoma*, with the mouth at the center. (c) Ventral view of a crustacean fish louse, where the enlarged maxillules (limbs modified for feeding) hold the parasite to the fish's skin by suction. (d) Piercing mouthparts of a mosquito, with the components spread out of their natural alignment; to the right are the components in their functioning arrangement. (a, From Matthews 1998, photograph courtesy of H. Melhorn; b, from Matthews 1998, photograph courtesy of L. Gibbons, CAB International; c, adapted from Kaestner 1980; d, from Brusca & Brusca 1990.)

Digenea

Outer body surface

Pinocytotic vesicle

Mitochondria

Spine

Basement membrane

Circular muscle

Longitudinal muscle

Mitochondria

Nucleus

(a)

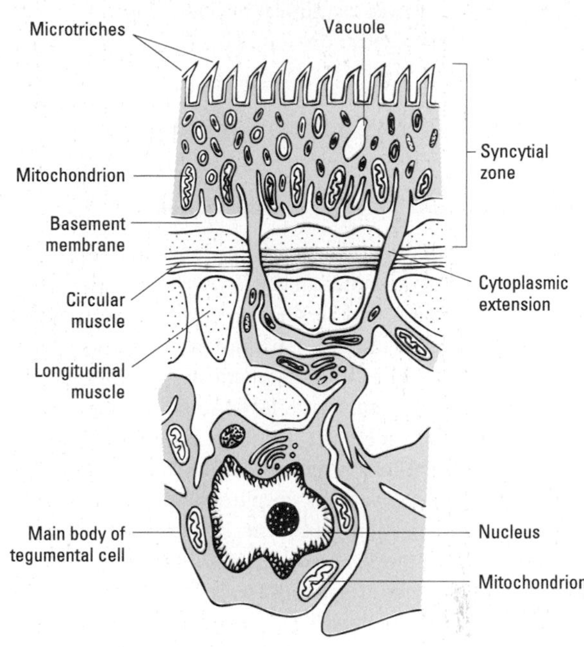

(b)

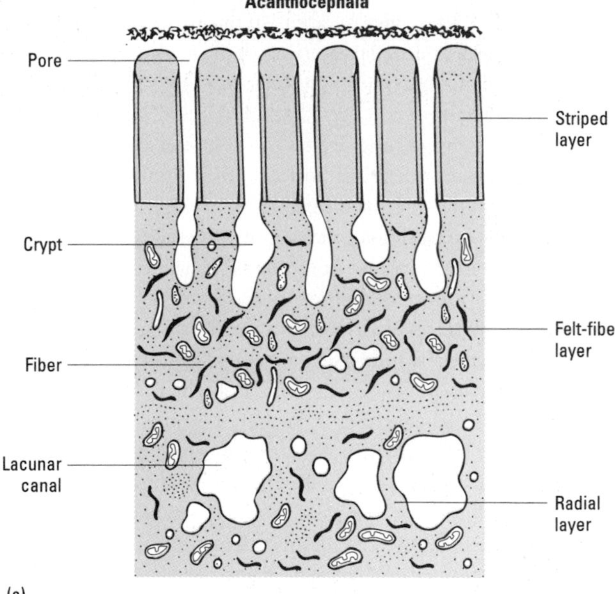

(c)

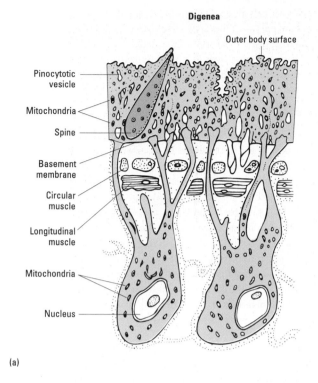

Fig. 17.5 Highly modified cuticles in gut parasites: (a) tegument of a digenean fluke, (b) tegument of a cestode tapeworm, and (c) cuticle of an acanthocephalan worm, showing internal crypts. (Adapted from Matthews 1998.)

variety of tricks to enhance their own competitive success. Many gut parasites coat themselves in mucus, which is a good ion-binding matrix, to limit the effects of extreme pH in the host gut. Some intestinal parasites inhibit digestive activity of the hosts and indirectly inhibit vitamin and blood-sugar metabolism. For example, the tapeworm *Diphyllobothrium latum* has very high affinity for vitamin B_{12}, and can cause deficiency disease (pernicious anaemia) in humans. In the nematode *Nippostrongylus*, adults secrete large quantities of acetylcholinesterase from their excretory pore, perhaps to produce local inhibition of peristaltic activity in the host gut (which will also help prevent them being dislodged). Insect parasitoids may disrupt host absorption of materials from the hemolymph, which increases resource availability for themselves.

Many parasites secrete enzymes associated with feeding. A number of gut-inhabiting parasites are able to reduce the risk of digestion by their host by releasing "antienzymes"—compounds that neutralize the host enzymes. Survival may also be augmented by the tolerance of high levels of toxic compounds, as illustrated by tapeworms in the guts of rabbits. Rabbit bile contains very high levels of deoxycholic acid: 36 mg ml^{-1}, in comparison with 1.77 mg ml^{-1} in dog bile. Rabbit gut parasites that live below the point where the bile duct enters the intestine are able to cope with this level of deoxycholic acid, while larvae of the dog tapeworm, *Echinococcus granulosus* (which lives in a similar position in dogs), are destroyed by the bile acid concentrations found in rabbits.

17.4 Reproduction and transmission

17.4.1 Goals and challenges of transmission

All hosts have a finite lifetime, and the long-term survival of parasites requires the transmission of infective stages from one host to another. In marine and freshwater environments, transmission between hosts can take place, as we have seen, via ciliated free-living larval stages or (in the case of ectoparasites in particular) by direct adult migration. In many cases larval dispersal is passive, without direct attraction to the hosts, although larvae may concentrate in areas more often visited by the host. An example is larval concentration in surface waters at times when intermediate hosts feed.

The situation is quite different in terrestrial habitats. Adult tapeworms, nematodes, and other parasitic worms, though able to locomote and survive in very moist terrestrial habitats, are unable to survive outside the host in drier environments. Movement from one host to another thus cannot be direct. However, direct transmission is often possible for ectoparasites with a more recent terrestrial free-living ancestry, such as biting flies, fleas, or ticks. Endoparasites are generally unable to survive as adults in terrestrial habitats and rely on resistant eggs and/or larvae released from the host to achieve infection of further hosts. One parasitic strategy is to produce long-lived, desiccation-resistant eggs that are released into the environment in large numbers and are taken up by the same host species (usually via the mouth). The host occupied by the adult parasite (usually termed the **definitive host**) may be extremely patchily distributed in the environment, making the possibility of direct infection of another host very low. Some parasites have evolved life cycles that exploit an **intermediate host** species, usually one that is more abundant in the larval environment than the definitive host. Many parasitic life cycles include **asexual reproduction** by larvae in the intermediate host, increasing the probability that at least one larva will successfully reach either a further intermediate host or the definitive host. In many life cycles, the intermediate host is a prey item for the next intermediate host or the definitive host, so that the limited ability of the parasite to move between hosts is vastly enhanced by the ability of a predator to detect and locomote towards its prey. The same advantage applies for parasites transmitted by vectors. For endoparasites such as tapeworms that infect terrestrial hosts, complex life cycles of this type can be seen as a necessary consequence of the limited ability of the parasite to survive and locomote on land. What a blood-feeding tsetse fly (*Glossina morsitans*) can achieve by flying between two cattle, a tapeworm must achieve by apparently tortuous and indirect means.

Parasite environmental physiology shows adaptations associated with several steps in the transmission process. Firstly, there are parasite traits associated with different methods of movement between hosts. These include adaptations to particular environments, and the dramatic changes that often take place when moving from one to the other. Secondly, a number of parasites have evolved mechanisms allowing reproductive synchronization with their host, and some show daily migrations in the host blood system that enable them to make use of blood-feeding insect vectors. Third, and most elegantly, parasites that reach their definitive host by moving up a food chain are sometimes able to manipulate the behavior of their intermediate host such that they are more likely to be eaten.

17.4.2 Movement between hosts

Reinfection requires that a suitable host is located, and usually a specific part of the host's body. In cases where the parasite habitat is essentially external, the movement of individual adults or larvae through the medium may be possible. In marine ectoparasitic platyhelminths such as flukes, the detection and settling of larvae on host external surfaces is analogous to detection of settling sites by the planktonic larvae of many other marine organisms. In many endoparasites, the situation is rather more complicated. Reinfection involves not only the release of eggs or larvae from adults that may be located deep within host tissues, but also the subsequent re-entering of hosts. This is achieved via the mouth and digestive tract, or via the host's skin. In the latter case, location of the adult site in the host usually also involves highly complex larval migration through the host's body.

The physiological tolerances of particular parasite groups have a major bearing on the infection routes possible for free-living adults or larvae. Adults and larvae of many parasites can survive in aquatic environments but are less tolerant of terrestrial ones. In platyhelminth parasites in particular, infection by free-living life stages tends to take place only when the host is aquatic or is a terrestrial animal that intermittently comes to water (as in the life cycle of *Schistosoma*). Some platyhelminth adults can survive for short periods outside the host, even in terrestrial environments. For example, adults of cestode tapeworms release their eggs packed into reproductive segments called proglottids. These leave the host with the feces and can survive and disperse eggs for a while outside the host. Even large proglottids, however, probably do not disperse far. Nematodes are better able to survive in free-living terrestrial environments, and several serious parasites of humans have larvae that spend time in the soil. In the case of the nematode hookworms *Necator americanus* and *Ancyclostoma duodenale*, the parasite life cycles include free-living larval stages that actually feed on bacteria in the soil (Fig. 17.6). Dispersal by larvae in terrestrial environments can also be achieved if they are collected from one host and delivered to another without exposure to the external environment. A number of parasites (particularly nematodes) survive by being spread via blood-feeding vectors, notably insects and ticks. Parasite larvae enter the vector as it feeds from one host, and (sometimes after intermediate larval stages) enter the next host as the vector feeds again.

The alternative to free-swimming larvae or adults is the use of eggs or larvae that are protected from environmental conditions by egg shells or other structures. Parasite eggs leave the host either through a natural aperture (most often with host feces), or through a hole in the host's body wall induced by the parasite, as in the nematode commonly known as the guinea worm, *Dracunculus medinensis*, or the jigger flea, *Tunga penetrans*. Once released, these resistant stages are unable to locomote themselves, but there is a probability that if they survive long enough they may encounter

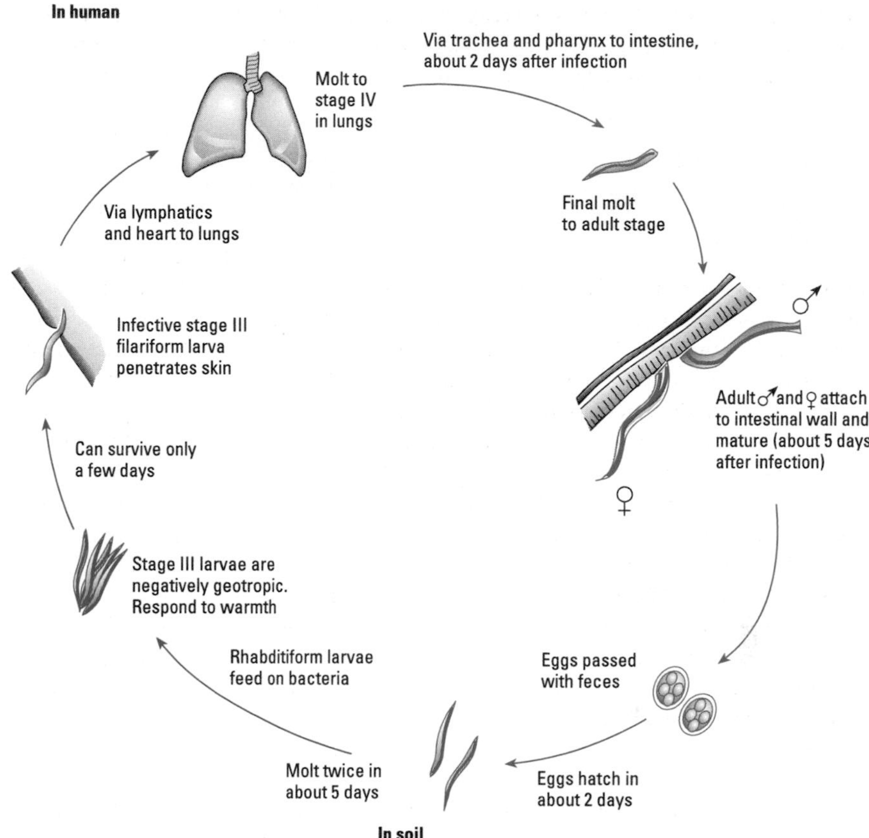
In human
Molt to stage IV in lungs
Via trachea and pharynx to intestine, about 2 days after infection
Via lymphatics and heart to lungs
Final molt to adult stage
Infective stage III filariform larva penetrates skin
Adult ♂ and ♀ attach to intestinal wall and mature (about 5 days after infection)
Can survive only a few days
Stage III larvae are negatively geotropic. Respond to warmth
Rhabditiform larvae feed on bacteria
Eggs passed with feces
Molt twice in about 5 days
Eggs hatch in about 2 days
In soil

Fig. 17.6 Life cycle of the hookworm nematode *Necator americanus*. (From Trager 1986.)

another host. Sometimes the larva is dispersive for a while, but forms a resistant cyst before reaching the limits of its environmental tolerance. Here the liver fluke *Fasciola hepatica* is a good example; its life cycle is similar to that of *Schistosoma*, except that the cercaria larvae do not actively bore through the skin of their definitive host (a ruminant mammal). Instead, they form resistant cysts on grass leaves, and rely on the ruminants to eat them. Dispersive stages of this type typically maintain a very low metabolic rate, are very desiccation resistant, and able to survive unfavorable osmotic conditions for long periods. Eggs of the gut nematode *Ascaris* possess a lipid layer inside the egg shell that is extremely impermeable to both water loss from the egg and entry of other chemicals from the external environment. In some nematodes, the dispersive stage is a larva that resists desiccation rather differently, by retaining the cuticle of a previous molt as an extra layer around the worm. These adaptations clearly parallel those seen in other essentially aquatic Metazoa—such as rotifers, tardigrades, and many free-living nematodes—that disperse through terrestrial environments using a cryptobiotic stage (see Chapter 14) between phases in aquatic habitats.

Infection by free-living adults or larvae

In aquatic ectoparasites, dispersal by free-swimming adults is possible. The adult must already be able to tolerate conditions experienced in the external environment, and needs only to be able to release its attachment to one host and then locate another. This is certainly possible for some monogenean flukes, but is probably impossible for many species that, though ectoparasitic, form complex and permanent attachments to the host. This is the case with many monogeneans that grow into the tissues of fish gills.

Free-swimming larvae are found in many parasitic groups, particularly the platyhelminths. The longevity, and sensory and locomotory abilities, of platyhelminth larvae are relatively limited, and without the direct inoculation of another host by a vector, only a tiny proportion reach the next host in the life cycle. Host location by free-swimming larvae involves two general steps. First, responses to general stimuli such as light or gravity allow the larvae to concentrate in regions of the environment where contact with the next host has a better than random probability. For example, larvae of flukes that attack bottom-living fish tend to stay near the bottom, while species attacking surface-swimming fish are positively phototropic and maintain themselves in upper water layers. Second, responses to characteristic host cues stimulate egg hatching, larval orientation, and attachment behaviors. Thus in the monogenean *Entobdella soleae*, the sole fluke, which parasitizes a bottom-living fish, eggs hatch in response to urea in the mucus of the host's skin, and detection of this cue stimulates the release of a proteolytic hatching fluid and emergence of the larva. The larvae show a diurnal rhythm in their hatching activity, which peaks during the middle of the day when their hosts are lying inactive and accessible. Having located a host, the larvae use adhesive glandular secretions to maintain contact until the main attachment organ (the hook-bearing opisthaptor) can be applied to the fish's skin.

The situation becomes more complex in endoparasites, many of which have complex life cycles. The miracidium larvae of schistosomes (see Fig. 17.1) locate the intermediate snail host, and the

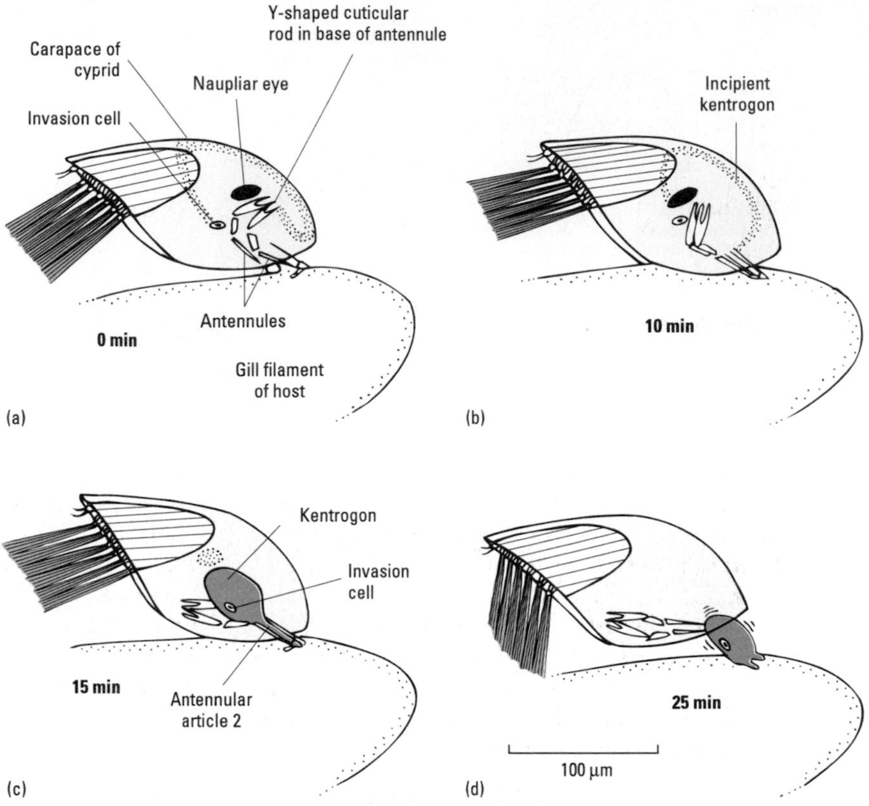

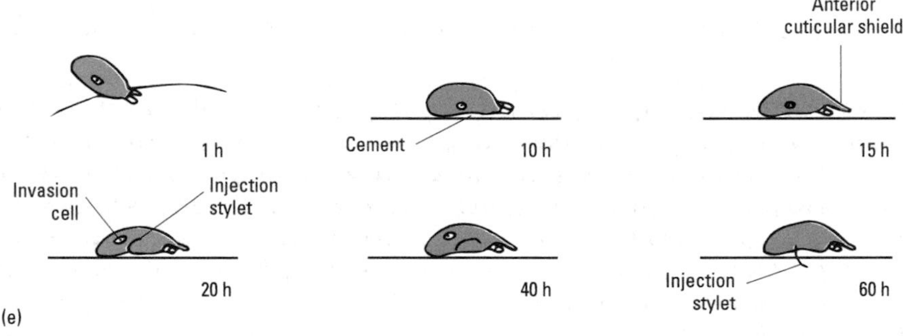

Fig. 17.7 Rapid body reorganization during host infection in the rhizocephalan barnacle *Lernaeodiscus porcellanae.* (a–d) The free-swimming cyprid larva locates and settles on the gills of its host, a crab, attaching by secretions from a large cement gland. Within 25 min the cyprid larval body is extensively reorganized, forming a sac-like structure termed the kentrogon. Then the carapace of the cyprid is shed. (e) The kentrogon develops a stylet, and injects a specialized "invasion cell" into the crab between 40 and 60 h after attachment; this divides to provide a network of parasite tissues, more similar to a fungus in structure than to a typical crustacean. (Adapted from Brusca & Brusca 1990.)

cercaria larvae re-enter the definitive host. There is a dramatic and rapid reorganization of the larval body on entering a host from the free-living aquatic environment. In the case of the schistosome transitions a motile outer layer (ciliated in the miracidium, and smooth with a muscular swimming tail attached in the cercaria) is rapidly shed to leave a new outer layer. An even more extreme example is shown in Fig. 17.7 for *Lernaeodiscus porcellanae*, a rhizocephalan barnacle that is parasitic on crabs.

In life cycles involving more than one host, stage-specific host location behaviors and recognition cues must be used. Larval recognition of the host is normally only effective over short distances, and for schistosomes it involves movement up a temperature gradient, and response to lipids characteristic of the host skin. Once contact is made, the cercaria uses secretions from two specialized acetabular glands whose ducts open into an anterior sucker (Fig. 17.8). Secretions from these glands are sticky, and while holding the larva in place help to soften the outer layers of the host's skin. Once the larva has gained entry into a small break in the host's skin, proteolytic secretions break down the host cells and allow the larva to move inwards. Then the larva undergoes a dramatic transformation into the next larval stage, the schistosomula; the free-swimming cuticle and tail are shed (Fig. 17.8c), leaving the highly specialized absorptive outer covering (tegument) characteristic of the adult. While the body wall of the cercaria was relatively impermeable, though probably with salt uptake regions to maintain larval metabolism in the hyposmotic freshwater environment, the new tegument is a permeable surface through which all metabolic exchange with the host takes place for the rest of the parasite's life. An important early function of the tegument is the rapid acquisition of host surface antigens, rendering the parasite invisible to the host immune system (see below).

Infection by free-living larvae in nonaquatic terrestrial environments is less common. In the nematode parasites of mammals that

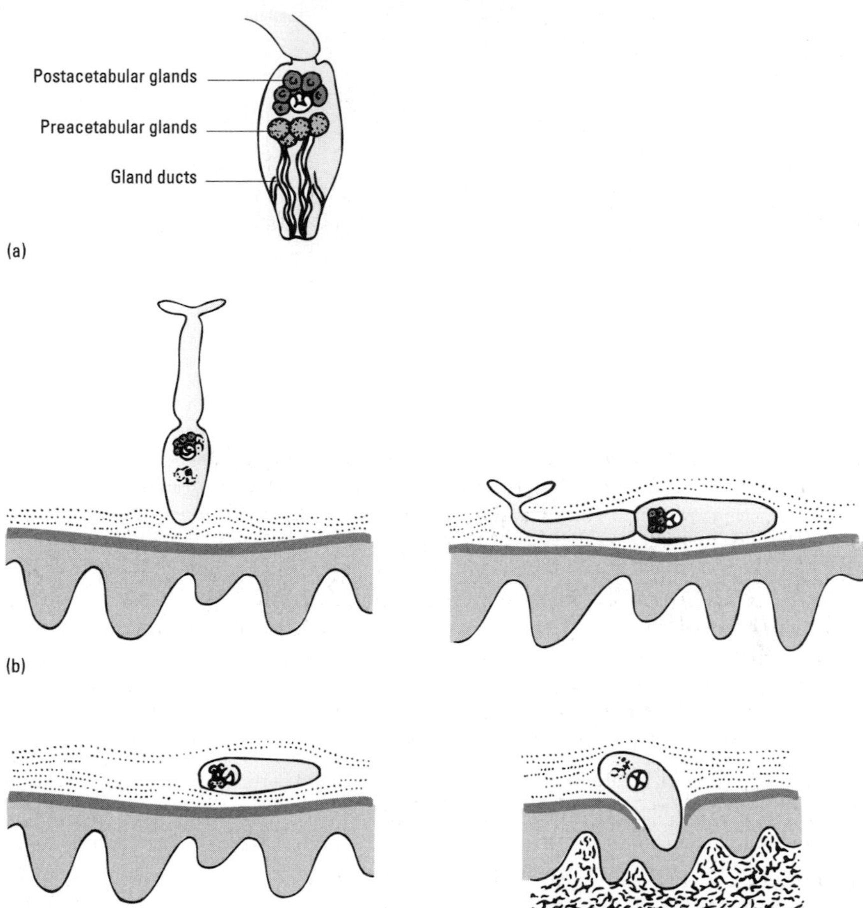

Fig. 17.8 Penetration of human skin by a cercaria larva of *Schistosoma*: (a) enlargement of anterior portion of a cercaria, showing glands and ducts; (b) adhesion to, and penetration of, the skin surface; (c) shedding of the larval tail; (d) migration of the metamorphosed schistosomula larva into deeper tissues and then to the human's circulation. (Adapted from Matthews 1998.)

have free-living larvae surviving and feeding in the soil, infection of a host is preceded by a molt to a nonfeeding stage, which must locate a host within a few days. Host location is brought about by negative geotropism (keeping the larvae at the soil surface) and by rapid movement towards a source of warmth. Once contact is made, the filariae of the nematode hookworm *Necator americanus* also bore through the skin of their vertebrate host, using proteolytic secretions from their esophageal glands to break down host cells in their path.

Vector-mediated parasites

In a sense, parasites utilizing vectors avoid some of the problems associated with entering the external environment. They do not need sense organs to find the next host, since the vector does that for them, and they are not exposed to the risks of desiccation. However, they encounter other challenges. First, the larvae must reach a part of their host in which they become accessible to their vector. For parasites occupying internal tissues, this requires migration towards the host body surface in the peripheral circulation. Although this brings the parasites to within reach of the mouthparts of blood-sucking vectors, traveling in the bloodstream may expose the parasite larvae to intensified host defense. A partial solution to this problem is for the parasite to show a very brief and precisely timed diurnal migration between the deeper tissues and the blood vessels just beneath the skin, as observed for the microfilariae of the nematode *Wuchereria bancrofti*. The vector exploited by the parasite varies geographically, and parasites emerge in the peripheral circulation at the right time for their local vector. Thus numbers of *Wuchereria* microfilariae just beneath the skin peak during the night in regions where the main vectors are night-biting *Culex* and *Anopheles* species, and during the day where the main vector is the day-biting *Aedes*. The parasites must detect some stimulus in the host metabolic cycle that has a daily rhythm, and experimental confirmation of this comes from studies of *Wuchereria* filarial migrations in people working night shifts, with biorhythms 12 h out of phase in comparison with the normal human cycle, who show filarial migrations that are 12 h out of phase with filariae in people working day shifts (unfortunately for the parasites, this results in larval presence in the peripheral host blood system at a time when their mosquito vectors are not active).

In addition to getting the timing right, vector-borne parasites must also be present in the blood in an area of the body favored by their vector. *Onchocerca* is another nematode parasite of humans, being the agent of the disease river blindness. It is transmitted by several species of biting blackflies that show species-specific preferences for biting particular parts of human hosts. Thus the nematode larvae aggregate in the legs in regions where their vector preferentially bites legs, and in the upper body when this region of the host is preferred. This clearly requires very specific larval detection of—and response to—host stimuli.

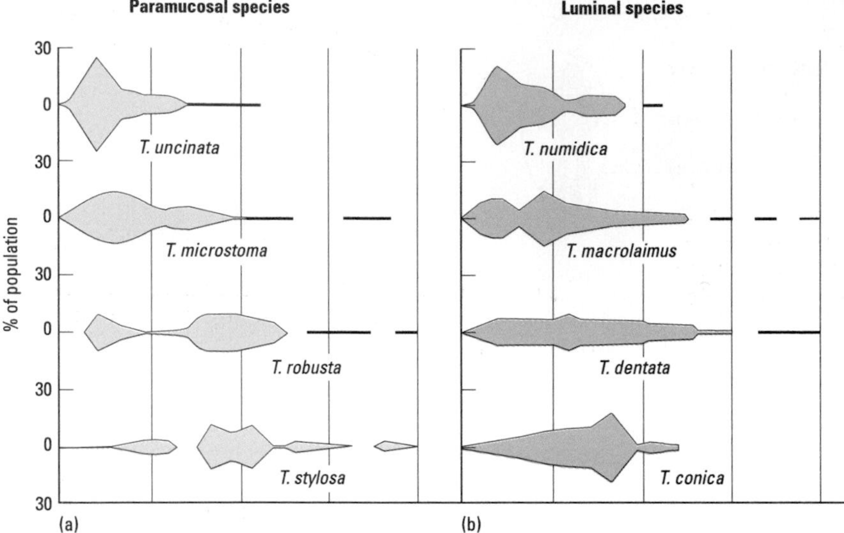

Fig. 17.9 Distribution of nematode species of the genus *Tachygonetria* in the intestine of the Greek tortoise. The eight species are divided into two groups: (a) those living close to the gut wall (paramucosal) and (b) those inhabiting the lumen of the gut. For each group, the distribution of the parasites along the intestine is shown from left to right. (Adapted from Schad 1963.)

A second set of challenges involves parasite responses to the environments present in the vector. Vector-transmitted parasites are initially swallowed into the gut of the vector, and thence move to the blood and on into the tissues. Survival requires parasite resistance to host digestive processes, and defenses in host tissues, which are extensive even in invertebrates (see below). Eventually (sometimes after intermediate larval stages) the larvae reach the vector's salivary glands or proboscis, and are reinjected into another host. This requires the parasite to cope with three very different vector environments, often within a very short timeframe.

Reinfection via the gut

Infection via the gut typically involves eggs or encysted larvae that emerge in response to the conditions in a certain part of the host gut. Protective coats around the egg or larva, coupled with relatively specific hatching or excysting stimuli, allow the parasite to reach the appropriate part of the host gut without being digested on the way. Protective coats include egg shells, cysts formed by larvae, and larval sheaths in some nematodes. They also include cases where the larva is surrounded by tissue laid down by an intermediate host. In the tapeworm *Taenia solium*, the cysticercus larvae are surrounded by cysts of pig tissue in the muscles, and emergence is triggered when the flesh of the pig is consumed and then digested by the human definitive host.

A wide range of stimuli are known to trigger emergence from eggs or cysts in the gut. Generally, stimuli in invertebrate hosts are less specific than those in vertebrates, even where these are both present in the same parasite life cycle. For reasons that are not always obvious, gut parasites usually occupy a relatively specific position in the host's alimentary canal (Fig. 17.9). This may represent the region of the gut in which resistance to host defenses and levels of nutrient acquisition are optimum for a given parasite. Host guts show characteristic gradients in chemical and physical conditions along their lengths (Fig. 17.10), and there may be triggers corresponding to particular combinations of conditions that allow parasites to emerge at the correct position. Conditions required for the emergence of larvae can be divided into two types. First, some stimuli indicate that the infective stage has passed unsuitable regions of the gut; thus the emergence of several intestinal parasites is initiated by low pH, characteristic of the host stomach, although hatching does not progress further until more favorable conditions are reached. Second, some stimuli indicate more directly that the egg or larva has reached the right part of the gut for hatching or metamorphosis. For example, the increase in temperature associated with entry into an endothermic host provides a very simple trigger for a few parasites. Alternatively, conditions in the host gut may stimulate secretion of enzymes by the egg or larva, leading to rupturing of the egg/cyst wall. *Ascaris* eggs hatch in response to a critical concentration of CO_2, an appropriate pH (slightly alkaline; see Fig. 17.11) and a high temperature (close to 37°C). Within the eggs, larvae release lipase and chitinase, and the lipid-rich inner layer of the egg shell increases in permeability. The enzymes diffuse through and dissolve a small area, through which the larva emerges. As another option, digestive enzymes present in the host gut may break down the egg/cyst wall; here examples are provided by hatching of the eggs of tapeworms, and excystment of digeneans and cestodes. Host enzymes remove the outer layers of the egg or cyst, and then bile causes changes in the permeability of the remaining membranes.

17.5 Parasite sensory abilities

It is usually suggested that in comparison with their free-living relatives, many parasites have reduced sense organs—particularly those associated with stimuli absent from the interior of a host's body, such as light. This is true of some parasites, such as the rhizocephalan barnacle *Sacculina*. This parasite, which in the adult stage looks nothing like a barnacle and quite like a mushroom, forms a network of tissue within the body of its crab host, extending externally as a formless mass of tissue beneath the crab's abdomen. Unlike the free-living barnacles from which it is derived, the adult *Sacculina* lacks obvious sense organs.

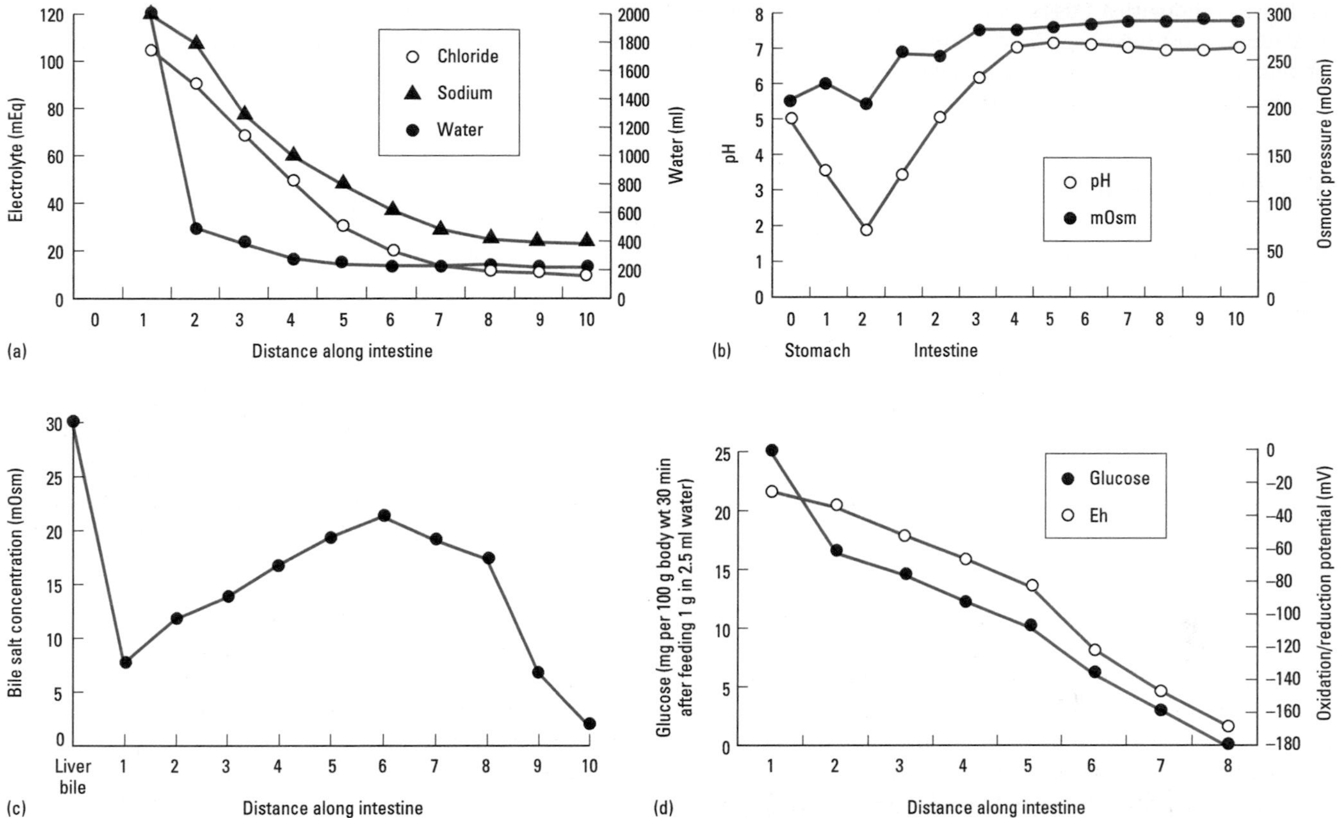

Fig. 17.10 Gradients in conditions along the mammalian gut, for: (a) water and salts; (b) pH and osmotic pressure; (c) bile salts; and (d) glucose and redox potential (Eh). (From Matthews 1998.)

However, to extend the example of *Sacculina* over all parasites may be unwise. Many parasites are able to successfully locate hosts as larvae, and furthermore to locate amazingly specific microhabitats within their hosts. Within the human blood system, young larvae of *Schistosoma* are able to recognize and congregate in the hepatic portal vein, while other flukes are able to migrate to a wide diversity of highly specific sites in other parts of the body, including the brain, heart, and bladder.

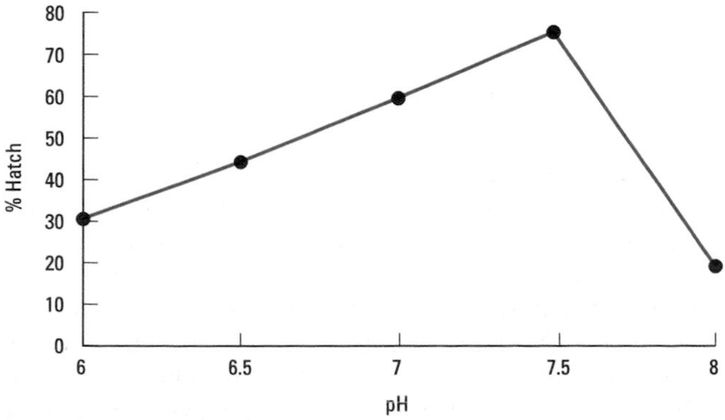

Fig. 17.11 The pH sensitivity of hatching by eggs of the gut nematode *Ascaris lumbricoides*. (From Matthews 1998.)

It could be that parasites infect hosts at random, but die in the "wrong" hosts, or in the wrong part of the right host—generating specificity without sensory selection of hosts by the parasite. Alternatively, parasites could possess sense organs allowing specific selection and rejection of alternative hosts and sites. Site recognition could in part be attributed to a small number of highly specific recognition cues, perhaps mediated by recognition of specific host cell types at a molecular level. Detailed studies are now revealing that parasite sense organs and nervous systems are far more complex than previously appreciated. Superficially simple endoparasites such as trematodes and nematodes have highly complex sensory systems. As an example, adults of the marine aspidogastrid fluke *Lobatosoma manteri* have a probable total of 20,000–40,000 surface sensory receptors of at least 14 different types. Such a complex sensory system may be required to monitor the physiology of the host, and to prevent overexploitation leading to host death (and so parasite death).

The ability of parasites to find such specific locations in their hosts suggests an ability (across a range of parasite species) to detect extremely precise combinations of pH, osmotic concentration, oxygen tension, patterns of fluid pressure in water currents, and temperature. In addition, the exploitation of different habitats through the life cycle also implies that different subsets of receptors or activation thresholds for given behaviors are present in different life-cycle stages. Remember also that parasite behaviors are often very flexible, and able to take into account variations in the parasite's environment. Thus the adults of a number of intestinal worms are able to adjust their location in the gut in response to the digestive cycle of their host, while others adjust their spatial patterning in the host in response to whether it is well fed or fasting. These alterations of position again suggest a sophisticated ability to detect and respond to changes in environmental stimuli.

17.6 Parasite regulation of host physiology

Parasite manipulation of the way hosts invest their resources is widespread, and is thought to result in greater resource availability for the parasite. The scale of the manipulation can range from a small area of host tissue immediately around the parasite to reorganization of the host's entire body. The mechanism of parasite modification is rarely known, but in at least some cases it is through active secretion of chemical signals.

17.6.1 Modification of host tissue growth

On the most local scale, parasites may modify the physiology and development of individual cells. The nematode *Trichinella spiralis* is able to transform individual cells in the muscles of its pig host: secretory/excretory products of the worm transform muscle cells into nurse cells, probably through direct modulation of host genomic expression.

Some of the clearest examples of parasite control of host tissue differentiation come from insects that form galls in plants (Fig. 17.12). Galls are plant tissues induced by another organism, providing that

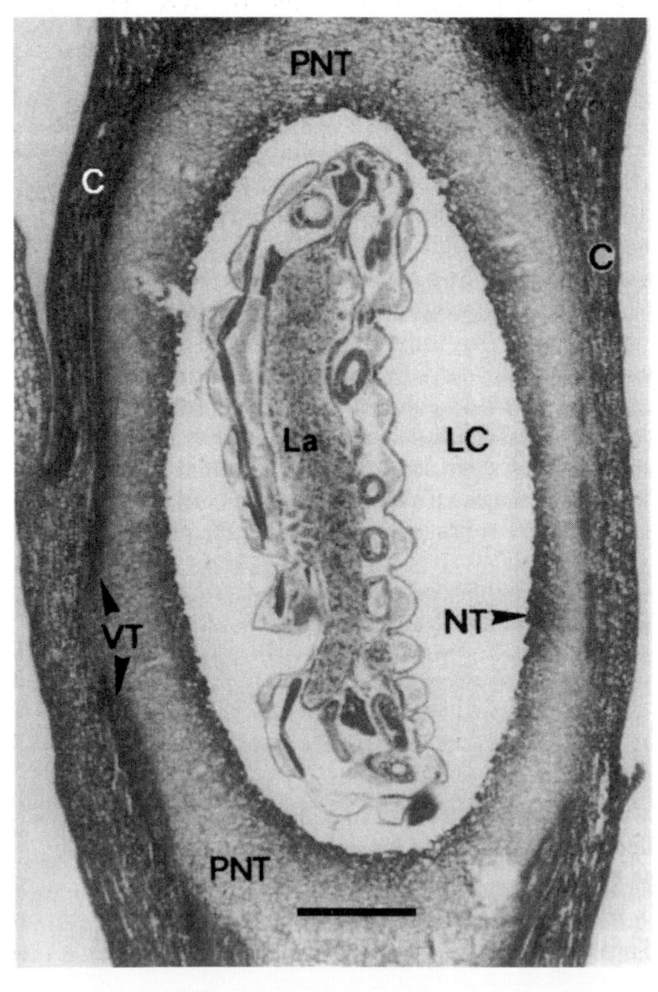

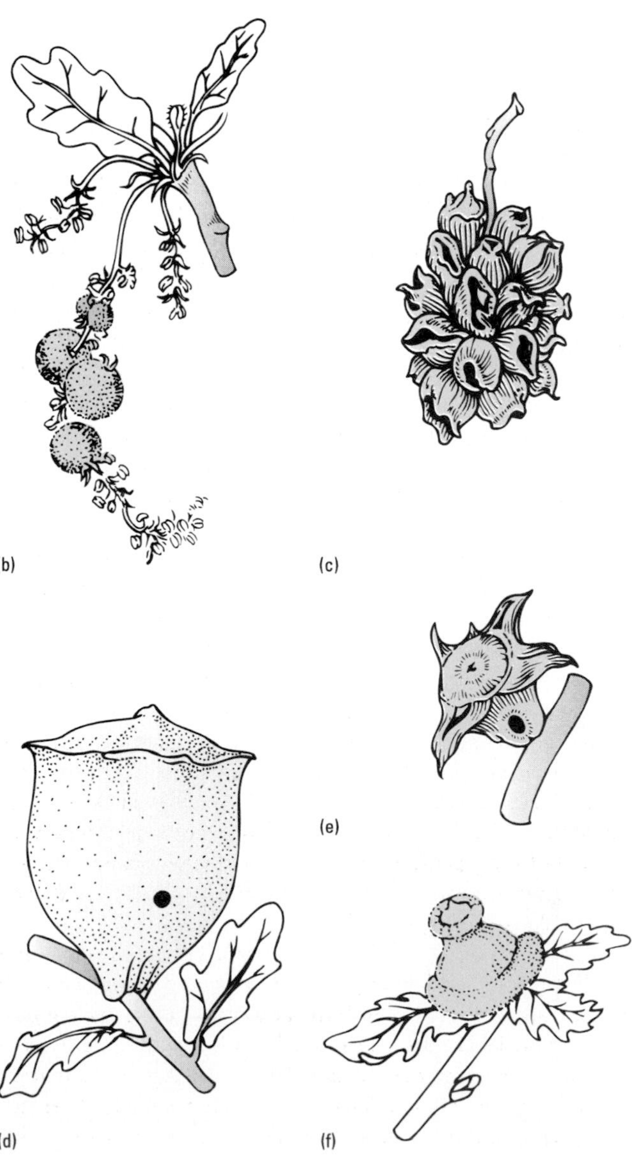

Fig. 17.12 Galls formed by cynipid gall-wasps. (a) Longitudinal section through a gall induced on a rose, showing the larva (La) in its hollow cell (LC), and the surrounding plant tissue layers on which the larva feeds (C, outer plant tissues; NT, nutritive tissue; PNT, parenchymatous nutritive tissue; VT, vascular tissue supplying the gall). (b–f) Examples of galls induced on oak trees by gall-wasps: (b) *Neuroterus quercus baccarum*, (c) *Andricus aestivalis*, (d) *A. quercus tozae*, (e) *A. coronatus*, (f) *A. tomentosa*. (a, From Brooks & Shorthouse 1998.)

organism with food and a measure of physical protection (see Chapter 15). Typically, galls consist of nutritive inner tissues, on which the galling organisms feed, and outer tissues that are often hardened. The gall-forming organism is often able to induce the plant to produce abnormal cell types and structures, and to direct resources such as nitrogen preferentially to the gall tissues. A group of small wasps (commonly termed gall-wasps, in the hymenopteran family Cynipidae) induce the most structurally complex and diverse galls known. At the center of the gall is a layer of endosperm-like tissue on which the gall-wasp larva browses. A series of outer tissue layers give the gall of each species a characteristic form, and show a wide diversity across gall-wasp species, including layers of woody or spongy tissue, complex air spaces within the gall, and surface coats of sticky resins, hairs, or spines. It is thought that these outer tissues protect the gall-former from attack by insect parasites and vertebrate predators. In addition to physical protection, gall tissues are often rich in toxic secondary plant compounds (see Chapter 15). Some oak-feeding gall-wasps are even able to induce oak tissues to secrete a sucrose-rich nectar, attracting ants to the surface of the gall, which further protects the gall-wasp from insect enemies.

17.6.2 Modification of host reproduction and life cycles

Rather than initiating the development of novel tissues, parasites may prolong existing phases of the host's life cycle to favor their own development. One way of diverting resources into parasite growth is to suppress sexual maturation of the host, such that host resources are never directed into eggs or sperm. This occurs in chaetognaths and snails infected with trematode larvae, and in crabs infected by *Sacculina*. In the snail case, trematode larvae interfere with the endocrine control of host development, by causing the host snail's central nervous system to manufacture a protein ("schistosomin") that interferes with several of the snail's usual hormones involved in the control of egg development. A vertebrate example is given by tapeworm larvae that suppress sexual behavior and egg maturation in many fish, probably by preventing the production of the host's gonadotropic hormones.

Sometimes the parasite does not completely inhibit reproduction, but inhibits or delays it. For example, male hamsters infected by schistosomes show decreased testosterone levels and so reduced investment in male behavior and secondary sexual characters. The change is thought to be under parasite control, through release of chemicals termed opioid peptides. These have many effects on host behavior and immune responses, and in the hamster suppress immune defenses as well as altering host investment in reproductive effort. Some parasites can directly inhibit mating ability via morphological or behavioral effects; one example occurs in the schistosome bilharzia parasites infecting freshwater snails, where a parasite-derived neuropeptide directly inhibits development of the host snail's copulatory organ.

Another parasite-induced effect is host gigantism, sometimes associated with host castration. Again, the suggestion is that such a manipulation of resource investment by the host results in greater resource availability for the parasite. Host gigantism in response to infection is best known for cestode larvae of the genus *Spirometra*, whose larvae live in mice. The larvae secrete a hormone ("plerocercoid growth factor") that mimics the effects of pituitary growth hormone (see Chapter 10), causing infected mice to reach a spectacular size.

Some parasites are also able to prevent molting in their host, for example in crabs infected by *Sacculina*. Suppression of molting is required because the parasite has external parts that disrupt the usual form of its host's exoskeleton, and an attempt to molt would probably kill both host and parasite. In other cases, the suppression of host molting maintains feeding and growth by a particular host life-cycle stage attacked by the parasite, so enhancing parasite growth.

The clearest examples of this type of host manipulation are to be found among insect **parasitoids**, representing a special case of parasitism in which the parasite always kills the host. **Idiobiont parasitoids** kill or paralyze their host at the time of egg laying, and no further host growth takes place. **Koinobiont parasitoids** represent a more subtle strategy in which the parasite allows the host to continue to develop after egg laying; parasitoid larvae avoid damaging organs essential to host survival and growth, and may also manipulate the host's endocrine physiology, nutritional biochemistry, and/or behavior. The eggs or larvae of koinobiont parasitoids typically delay their own development until the host reaches a large size, and may also prolong the life cycle of the host, both strategies resulting in greater resource availability for the growing parasite. For example, the parasitoid *Cotesia congregatus* is able to prevent the larva of its caterpillar host *Manduca sexta* (the tobacco hornworm) from metamorphosing into a pupa, and can induce up to six extra host larval instars. Metamorphosis of the host is usually brought about by a drop in the levels of juvenile hormone and an increase in the level of ecdysone in the host's hemolymph. Parasitoid larvae are able to prevent the drop in levels of juvenile hormone, and some may also block the conversion of inactive ecdysone into the active 20-hydroxyecdysone. However, it is worth noting that delay in parasitoid development as the host grows may come with a cost; most koinobiont insect parasitoids are endoparasites, and they are thus exposed to host defenses for rather longer.

Synchronization between host and parasite life cycles is also an option that many parasites exploit, insuring that parasite young are produced at the same time as host young. Molting in insect parasitoids appears to occur in response to host titers of juvenile hormone and ecdysone. More spectacularly, rabbit fleas time their own reproduction by sensing the changes in circulating hormones in their pregnant female host's blood.

17.6.3 Parasite manipulation of host behavior

Many host behaviors have direct impacts on parasites, ranging from obvious examples such as grooming behavior, coughing, sneezing, or vomiting, to less obvious cases such as predator avoidance. It is in the interests of parasites to increase the probability of host behaviors that benefit the parasite, and to suppress those which do not. Host behaviors result from the interactions of stimulus detection and assimilation systems (sense organs and central nervous system), communication mechanisms (nervous and endocrine systems), and effector mechanisms (muscles). A number of cases are known in which parasites interfere with one or more of these components (though it has recently been argued that the associated evidence is not always particularly strong). In the best known cases, much of the

Polymorphus paradoxus

Infested amphipods

Vegetation

Mallard

Uninfested amphipods

(a)

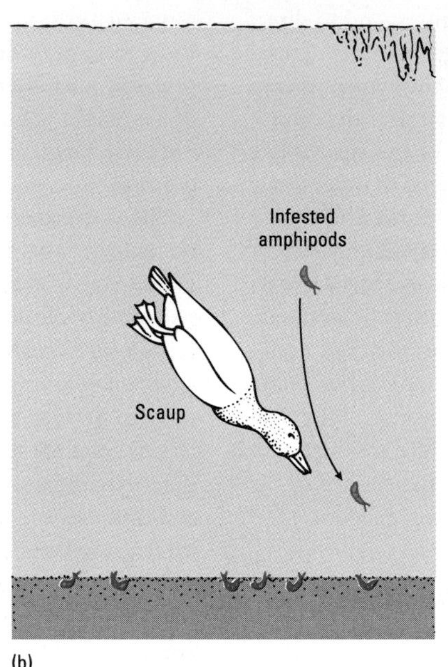

(b)

Fig. 17.13 Behavioral manipulations by parasites: the interactions of duck predators and gammarid amphipod crustaceans infested by different species of acanthocephalan worms (*Polymorphus*). One species causes gammarids to stay at the surface where they are eaten by dabblers such as mallard (a), while another causes them to swim at depths where divers such as scaup eat them (b).

manipulation is achieved by chemical interventions, and thus represents components of the environmental physiology of the parasite.

Parasite-induced host behaviors can be divided into three types:

1 Behaviors that help maintain the parasite in the host. An example is provided by the behavior of portunid sand crabs infected with *Sacculina* (similar to the parasite in Fig. 17.7), which show increased cleaning behavior, tending the parasite as if it were a crab egg mass stored in the same location. This behavior is shown even by male crabs, which do not usually carry egg masses.

2 Behaviors that increase the probability of parasite transmission. For example, healthy individuals of the estuarine snail *Ilyanassa obsoleta* avoid being exposed by tidal flow. In contrast, individuals infected with the trematode *Gynaecotyla adunca* follow high tides and become stranded on beaches and sandbars, increasing the probability that larvae will reach the semiterrestrial crustaceans that are the next host. In the acanthocephalan worm *Polymorphus paradoxus*, for which surface-feeding ducks are the definitive host and the freshwater amphipod *Gammarus* is an intermediate host, the parasite alters the amphipod behavior from benthic scavenging well away from ducks to a suicidal pattern of skimming along the surface and clinging to surface weed (Fig. 17.13a), where ducks readily eat them. The related parasite *P. marilis* causes the gammarids to swim in the lower waters of the pond, insuring that they are picked up by a different required host, in this case diving ducks such as scaup (Fig. 17.13b). *Polymorphus* is thought to alter *Gammarus* behavior by somehow interfering with 5-HT metabolism.

3 Behaviors that favor parasite survival after host death. Insect parasitoids may pupate within or on their dead host, and a number of species manipulate the host so that before it dies it reaches a location suitable for pupal survival. For example, conopid flies that parasitize bumble-bees insure that just before death the bee burrows into soil (a highly unusual behavior for a worker bumble-bee), providing the fly larva with a protected location in which to pupate.

17.7 Biotic interactions: host–parasite conflicts

17.7.1 Host defenses

Host defense mechanisms and parasite responses have presumably been evolving since one species first invaded another, hundreds of millions of years ago. Thus parasites are part of the host environment, requiring the evolution of defense mechanisms. Similarly, as soon as hosts begin to evolve defenses, there is strong selection on parasites to evolve countermeasures. Defenses and countermeasures thus both fall into the category of environmental physiology. This is particularly true for endoparasites, although several host defense mechanisms are able to reach ectoparasites on the outer body surface.

It is a common misconception that host defenses are essentially limited to vertebrates, but this is certainly not the case. It is now known that a number of host defense systems are widespread phylogenetically, and probably of very ancient origin (Fig. 17.14 and Table 17.3), while others appear to be limited to vertebrates. As well as withstanding host defenses, some parasites have evolved mechanisms that directly suppress host defenses, or utilize them in their own development. The control of host immune responses involves the complex interplay of a variety of chemical messages, including hormones that influence host behavior; the secretions of some parasites, perhaps primarily aimed at suppression of host immune responses, thus also modify host behaviors to the benefit of the parasite.

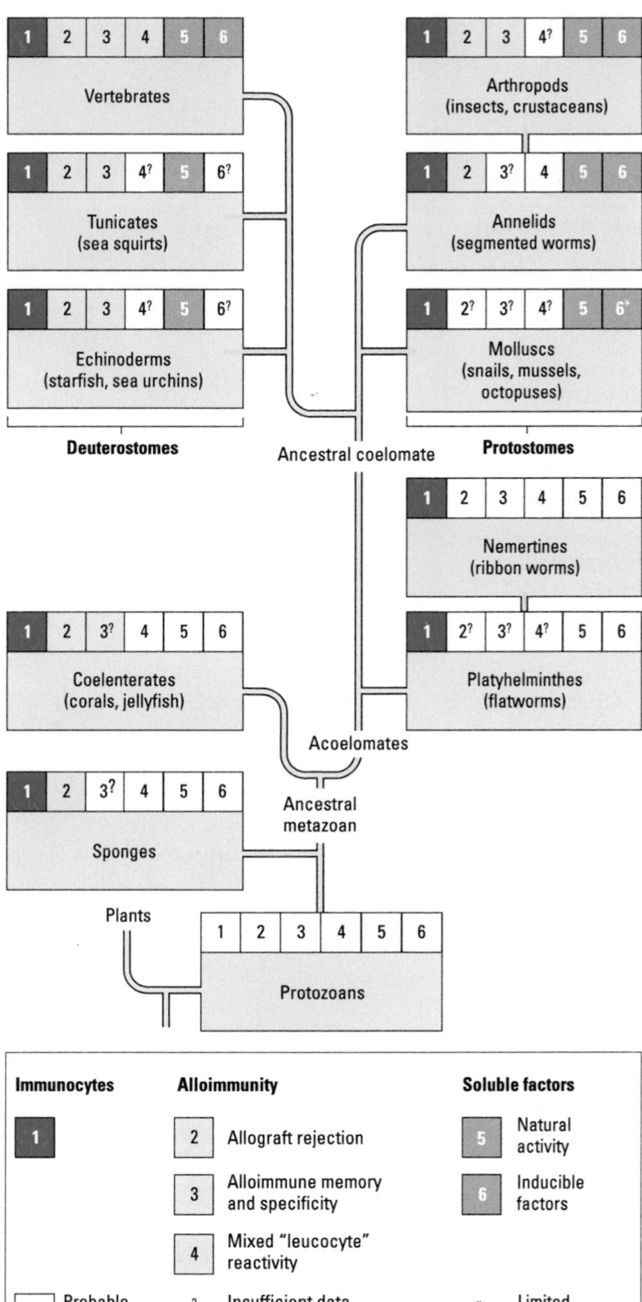

Fig. 17.14 Phylogenetic patterns in immune-type responses and defense systems in the Metazoa. Cells attacking nonself tissues (immunocytes) are ubiquitous, while more complex cellular immune responses and humoral responses (natural and inducible soluble factors) have a more limited distribution. See also Table 17.3. (From Roitt *et al.* 1998.)

There are a number of defense systems that parasites have to withstand, and what follows is a very brief review of the environmental physiology of host defense.

Distinguishing "self" from "nonself", and the initiation of defense

Defense in an organism relies on the ability to distinguish self cells from nonself cells. Recognition normally takes place in response to complex glycoproteins, termed **antigens**, present on the surface of cell membranes. A primitive recognition system in all animal cells is mediated by a number of proteins, including agglutinins, cecropins, and lectins, in tissue fluids such as the blood. A second recognition system results from detection of nonself antigens by a range of receptors on the surface of host white blood cells (found in invertebrates as well as vertebrates; see Table 7.8). Vertebrates possess an additional highly complex self-recognition system, involving cell surface markers of the **major histocompatibility complex** (MHC). This involves incorporation into host cell membranes of gene products coded for by two groups of polymorphic loci—MHC class 1 (on all adult cells) and MHC class 2 (on a subset of adult cells). The presence of these surface markers gives cells a highly characteristic signature recognized by host defenses. Either the presence of other surface antigens or the absence of the correct MHC signature can lead to attack by host defense systems.

Once tissues have been recognized as foreign, hosts may mount an **immune response** against them. Such a response can take several forms, broadly divided into those involving blood and other cells (cell-mediated responses), and those relying on the presence of compounds existing freely in host blood and tissue fluids (humoral responses). The initial reaction to foreign tissue is often a local swelling of host tissue, and local aggregation of host defensive cells. This defensive effort is achieved through signaling between host cells; avoidance of host defenses by parasites can therefore be brought about by manipulating these signals.

Signaling between host cells during the immune response

Both vertebrates and invertebrates possess tissues, particularly white blood cells, whose primary function is the control of the behavior and production of particular types of host defensive cell. In vertebrates this control includes instructions for the manufacture of highly specific defensive **antibodies** by host cells that never actually confront parasites directly. On a more local scale, signals between cells dictate the type of response mounted in the immediate vicinity of a particular parasite. Communication is achieved via cell surface receptors to specific messenger molecules, and differential stimulation of sets of receptors can produce very different responses by the same host cell. The most important chemical signals can broadly be divided into two types: cytokines and hormones. The known diversity of these signals is far higher in vertebrates than in invertebrates.

CYTOKINES

These are polypeptides released by a variety of activated immune and nonimmune cells (including endocrine glands), regulating host defenses in both invertebrates and vertebrates. There are four basic sets of cytokines (Table 17.4), with a range of effects summarized in Fig. 17.15. They are best studied in vertebrates, but at least three types are already known in invertebrates, and the invertebrate interleukin IL-1 shows a marked diversity of functions: (i) stimulation of phagocytosis and proliferation in white blood cells; (ii) promotion of leucocyte aggregation leading to encapsulation of nonself tissue; (iii) cytotoxicity to some cell types; and (iv) induction of increased vascular permeability allowing greater accessibility of nonself tissue to host defenses.

Table 17.3 Evolution of the immune system. (Adapted from Rowley & Ratcliffe 1988.)

Evolutionary step or selection pressure	Immunological implications
Single-celled animals	Recognition and discrimination
Multicellularity (including colonial forms)	Histocompatibility system, recognition, and short-term memory
Mesoderm and circulatory system, nutrition and defense as separate functions	Freely circulating and more diverse blood cell types, cellular immunity, and erythrocytes
Cancer and viral infections associated with increasing complexity and longevity	Immunosurveillance of own cells for those that are infected or cancerous
Ancestral protovertebrates	Increased recognition and discriminatory powers?
Lower vertebrates: increased size, longer lifespan, and reduced reproductive potential compared with invertebrates	True lymphocytes, lymphoid tissue, and antibody production (IgM), longer term memory
Emergence onto land, exposure to irradiation, and development of high-pressure blood vascular systems	Bone marrow, additional antibody classes, T- and B-lymphocytes, lymphoid organs with increased complexity
Amniotes (reptiles, birds, mammals) with loss of free-living larval form	Advanced differentiation of immunocompetent cells allowing increased diversity and efficiency of immune system
Endothermy provides a more favorable environment for pathogens	Increased efficiency of immune system, integrated cellular and humoral responses, germinal centers in secondary lymphoid organs, lymph nodes
Viviparity with maternal–fetal interactions	Additional fine-tuning of immune system to avoid rejection by mother

Table 17.4 Types of cytokines in animals.

Type	Functions
Interferons	Defense of the body against viruses
Interleukins (17 different types)	Directing other cells to divide or differentiate Up- or downregulation of cellular processes
Colony-stimulating factors	Division and differentiation of bone marrow, hence levels of specific white blood cell types
Tumor necrosis factors α and β, and transforming growth factor	Mediating inflammation and cytotoxic responses

OTHER CHEMICAL MEDIATORS OF IMMUNE RESPONSES

Vertebrate white blood cells possess surface receptors for many hormones, neuropeptides, and neurotransmitters, including steroids, epinephrine and norepinephrine, enkephalins, endorphins, and others. Corticosteroids, endorphins, and enkephalins are of particular importance for parasite environmental physiology because they suppress host immune responses. Cytokine- and hormone-mediated systems can interact in complex ways; for example, interleukins from blood cells can elevate or depress the production of corticosteroids, while interleukins from pituitary and adrenal glands affect the numbers and activities of white blood cells.

Defense systems

The defenses mobilized following the detection of nonself tissue can be broadly divided into two types. Firstly, there are nonspecific, innate immune defenses (also termed "natural", "nonadaptive", and "nonanticipatory"). These tend to occur in response to a broad spectrum of parasites, with relatively general defensive properties. The defenses have little or no "memory", so prior exposure to a parasite does not result in a more effective defense next time the host is attacked.

Secondly, there are induced, adaptive defenses (also termed "specific" or "anticipatory"). These defenses are more specific to particular parasites, and include those with a "memory" that confer enhanced protection of the host after an initial exposure to the parasite.

The division between these types is in part taxonomic. A range of nonspecific systems are common to all Metazoa, while long-term memory in the immune system and the manufacture of highly antigen-specific antibodies (immunoglobulins) are known only in vertebrates.

NONSPECIFIC IMMUNITY

Both cell-mediated and humoral nonspecific immunity systems are widespread in both invertebrates and vertebrates. Cell-mediated defenses include three reactions mediated by host cells—phagocytosis, natural killer (NK) cells, and encapsulation—which may be ubiquitous in the Metazoa. All are mediated by white blood cells (leucocytes; see Fig. 17.16), which may have their evolutionary roots in the phagocytic cells responsible for intracellular digestion and transport of food in basal metazoans. Phagocytosis occurs in invertebrates and vertebrates alike, when phagocytes move towards and engulf invading material, which is digested intracellularly. NK cells are large granular white blood cells that recognize and destroy nonself cells by releasing cytotoxins, possibly pore-forming proteins that become incorporated in the membrane of the nonself cell so that it either swells and explodes through osmotic inflow of water, or ceases to function as a result of metabolite leakage. The third process of encapsulation involves surrounding nonself material with tightly adhering layers of host cells. Parasites may be killed when walled in by encapsulation; or this may be merely the first stage in a sequence of cell-mediated host responses resulting in the destruction of the parasite, including the action of lysosomal enzymes such as peroxidases and lysozymes. The cell-mediated defenses of invertebrates are now known to involve a wide diversity of cell types (Table 17.5).

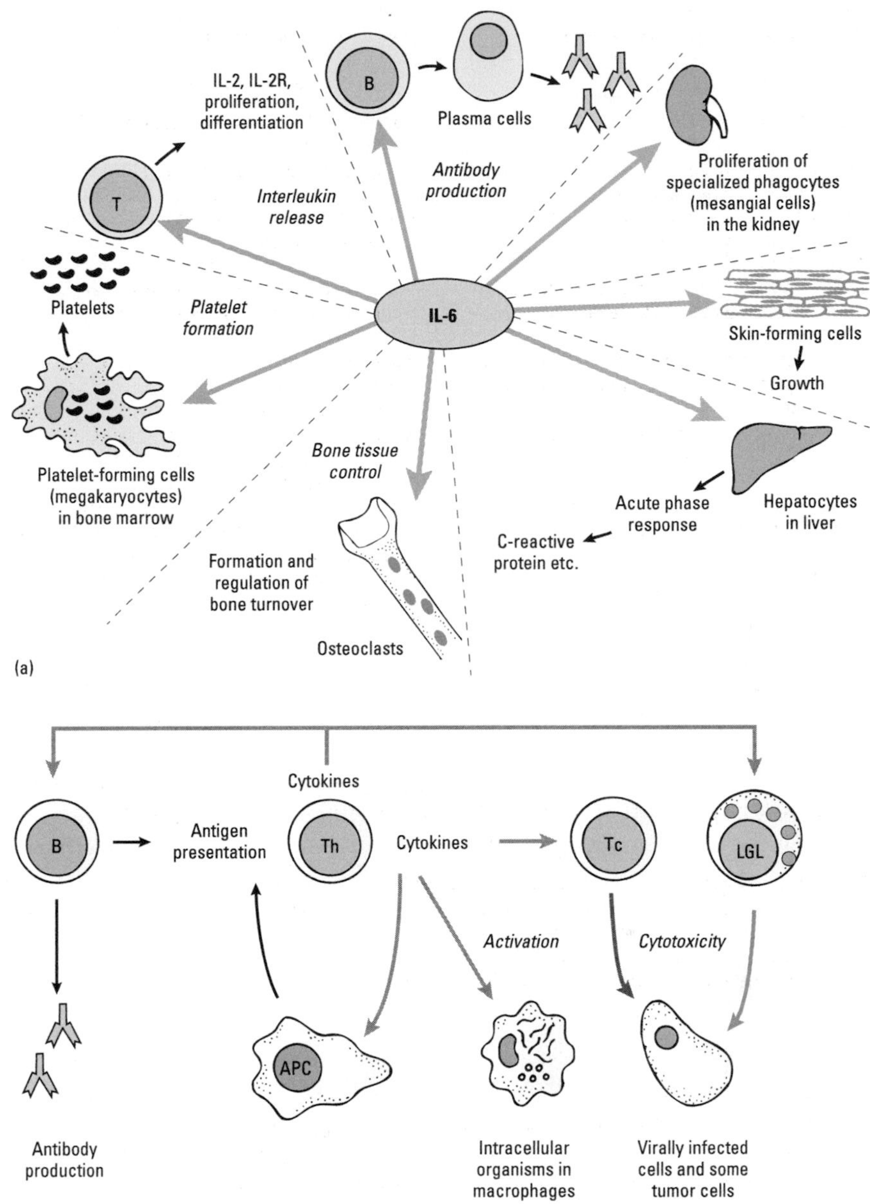

Fig. 17.15 (a) The range of effects of a cytokine, interleukin-6 (IL-6). (b) An example of the role of cytokines in mediating interactions between populations of white blood cells: here, between T-helper cells (Th), B-cells (B), cytotoxic T-cells (Tc), and large granular lymphocytes (LGL). Th-cells are stimulated by antigen-presenting cells (APC) and B-cells to produce cytokines; macrophages are activated to kill intracellular microorganisms; and Tc-cells and LGLs recognize and kill target cells. See text for more details. (From Roitt *et al.* 1998.)

Humoral defenses include toxins that are permanently present in the tissues of some metazoans as defenses against predators and parasites alike. These occur in many sponges and tunicates, and in those herbivorous insects that store toxic secondary plant compounds in their tissues (see Chapter 15). More generally, humoral defenses common to vertebrates and invertebrates include proteins that adhere to and aggregate foreign material (e.g. agglutinins and lectins), antibacterial peptides (e.g. tachyplesins and defensins), and relatively broad-spectrum digestive enzymes (lysozymes). Although many of these broad-spectrum humoral defenses are described as principally "antibacterial", some confer effective protection against metazoan parasites. Humoral defenses can be rapidly elevated in response to invasion (termed an "induced response"), but the duration of response is generally only a few days. Agglutinins may show some specificity to markers on particular types of invading cell, and short-term memory in humoral immunity (up to several weeks) has been demonstrated in molluscs and insects.

ADAPTIVE IMMUNITY

Adaptive immunity involves the development of highly specific host responses to particular parasite antigens. This is coupled with an ability of the host immune system to maintain a small population of cells specific to a given parasite antigen even when the infection is cleared. This immunological "memory" allows later reinfections to be met with a rapid and highly specific defensive response.

As far as is known, such adaptive immunity is only present in vertebrates, and it represents a major breakthrough in controlling parasites. If rats are infected with the nematode *Nippostrongylus brasiliensis*, nematode larvae reach the gut and mature, beginning to release eggs about 5 days after infection. Egg production reaches a

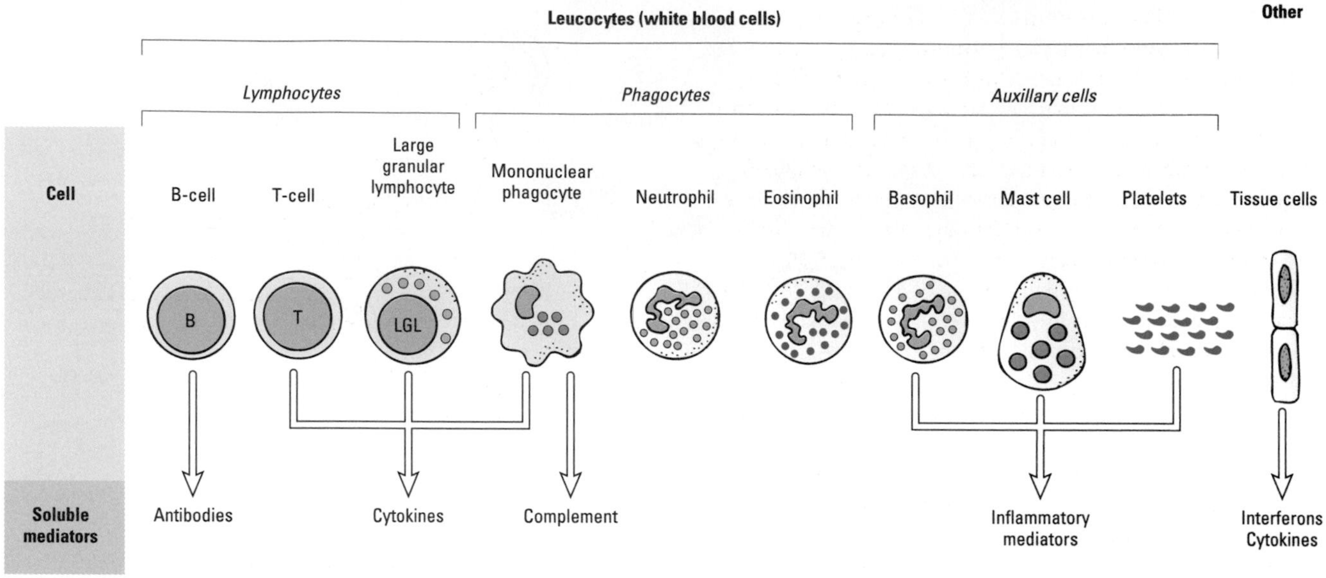

Fig. 17.16 Summary of the main cell types involved in the immune system, the majority belonging to one of the classes of white blood cells (leucocytes). (From Roitt *et al.* 1998.)

peak about 9 days after infection, and then decreases rapidly. After 21 days, all the adult worms have been expelled from the intestine. But if these exposed rats are then reinoculated with infective nematode larvae, very few adult worms reach maturity, and infections are cleared in 10 days rather than 21.

Table 17.5 Cells and tissues of invertebrate immune systems. (From Roitt *et al.* 1998.)

Cells/tissues	Role(s) in immunity/physiology
Mucus, cuticle, shells, tests, and/or gut barrier	Physicochemical barriers to invasion
Five groups of free and sessile white blood cells	Mediate cellular and many of the humoral defense reactions
Progenitor cells	May act as stem cells for other cell types
Phagocytic cells	Phagocytosis, encapsulation, clotting, wound healing, and killing
Hemostatic cells	Plasma gelation and clotting by cell aggregation; nonself recognition, lysozyme, and agglutinin production
Nutritive cells	Encapsulation reactions and wound healing? Nutritive role?
Pigmented cells	Role in defense (if any) unknown; respiratory function
Fixed cells such as pericardial cells, nephrocytes, or pore cells, etc.	Pinocytose colloids and small particulates; synthesize lysozyme (pericardial cells) and other antimicrobial factors?
Hemopoietic organs—well organized in some invertebrates	Hemopoiesis and phagocytosis; synthesize antimicrobial factors in a few animals
Fat body (insects), midgut, and sinus lining cells (molluscs, crustaceans)	Synthesize immune proteins and agglutinins (fat body), phagocytosis (midgut cells), clearance of foreign particles

The specific leucocyte membrane receptors and antibodies involved contain a common structural motif, the **immunoglobulin domain**. In vertebrates there is a wide diversity of similar genes with this motif, almost certainly derived by multiple gene duplication events and sequence divergence from a single ancestral gene. Several cell types are involved:

1 Mononuclear phagocytes (**"monocytes"**). A range of different forms of these cells exists in different vertebrate tissues (Fig. 17.17),

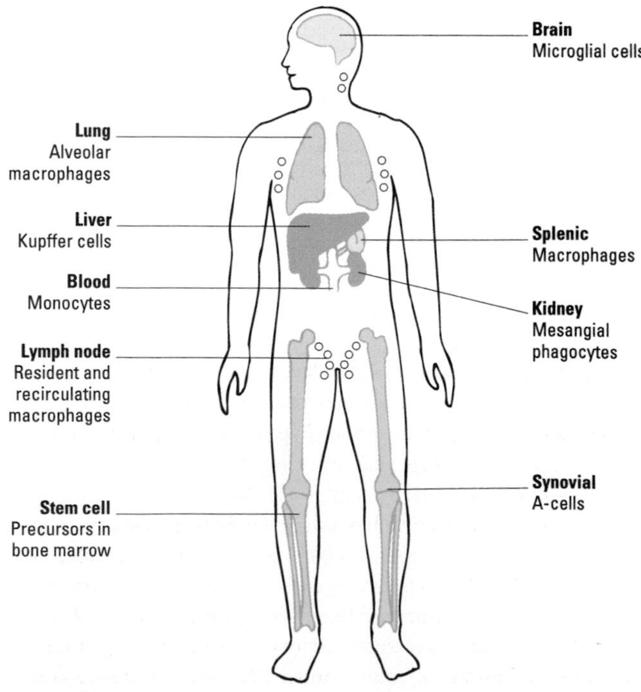

Fig. 17.17 Distribution of different types of phagocytic cells (collectively termed mononuclear phagocytes, or monocytes) in vertebrates. These are manufactured in the bone marrow, and pass out of the blood to form a range of tissue-specific morphs. (From Roitt *et al.* 1998.)

including mammalian monocytes, which become tissue macrophages. They recognize particular portions of antigen molecules, and if these are not compatible with the MHC signature of host cells, then the structure bearing them is attacked. In addition to their phagocytic ability, macrophages can destroy other cells by secreting a range of toxic compounds. Cytotoxicity can be through the release of proteolytic and other cytolytic enzymes, or through the release of toxic compounds termed "reactive oxygen intermediates" and "reactive nitrogen intermediates". Oxygen intermediates are formed by enzymes that reduce oxygen to the superoxide anion, OO^-. This in turn can give rise to toxic hydrogen peroxide and hydroxyl radicals. Monocytes in the blood use peroxidase to create more toxic bleach-like oxidants. Nitrogen intermediates are formed by the enzyme nitric acid synthase, which combines oxygen with nitrogen from the amino acid arginine to produce toxic nitric oxide. Under certain conditions, nitrogen intermediates and oxygen intermediates may interact to produce even more toxic peroxynitrites. These compounds are all highly oxidizing, and their application in defense is sometimes termed the "oxidative burst"; it can destroy multicellular parasites such as nematode microfilaria larvae and the cercariae of schistosomes.

2 **Neutrophils.** These are the most common type of leucocyte in the blood, and are shorter lived as after ingesting the target material they die. Neutrophils are able to mount cytotoxic responses using toxic proteins and reactive O_2/N_2 intermediates, against targets labeled with antibodies. This defense is also effective against various metazoan parasites.

3 **Eosinophils.** These are a specialized group of leucocytes that can recognize and damage large extracellular parasites. Although capable of phagocytosis, they function principally by cytotoxic responses.

4 **Lymphocytes.** These are wholly responsible for the specific recognition of parasite antigens, central to adaptive immunity. There are two types, derived from bone marrow stem cells: T-cells develop in the thymus gland, while B-cells develop in the bone marrow in mammals. Both B- and T-cells accumulate in the lymph nodes and spleen, from where they respond to parasite antigens in the blood. T-cells exist in two types with different surface markers; T-helper (Th) cells and cytotoxic T (Tc) cells. Tc-cells are involved mainly in recognizing antigens resulting from intracellular parasites, and killing infected cells, while Th-cells are more important in responses to multicellular parasites. Th-cells divide to produce a clonal population of antigen-specific cells, which then release cytokines to activate other cells in the immune system (see Fig. 17.15b). B-cells also possess highly specific antigen-recognition abilities, and once activated they manufacture and release large quantities of antibody.

ANTIBODIES AND COMPLEMENT

Immunoglobulin (Ig) antibodies are proteins composed of four peptides—two heavy (≈ 400 amino acids) and two light (≈ 200 amino acids) chains (Fig. 17.18) bound together to give a Y-shaped structure. The arms of the Y contain highly variable regions on both the heavy and light chains, and this is the part of the immunoglobulin involved in binding to a specific antigen. The base of the Y has a more constant amino acid sequence characteristic of particular immunoglobulin types, and is the part of the molecule recognized by host cells. In mammals, antibody molecules can be divided into four types, with characteristically different roles in host defense.

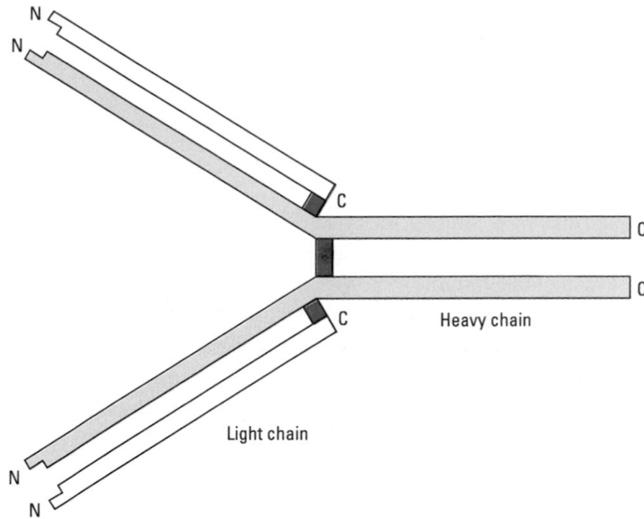

Fig. 17.18 The basic structure of immunoglobulins. The Y-shaped molecule consists of two identical light polypeptide chains and two identical heavy polypeptide chains linked together by disulfide bonds (dark green). N is the amino terminal and C the carboxy terminal of the chains. The variable regions are at the N-terminal ends of the heavy and light chains. (From Roitt *et al.* 1998.)

1 **IgM.** This is the first immunoglobulin secreted in response to infection. It is a five-molecule polymer that stimulates phagocytosis and also binds cells bearing the correct antigen together. It is secreted across the wall of the intestine, and is thus active against gut parasites.

2 **IgG.** After 4–5 days, IgM is replaced by IgG, which plays a major role in activating cytotoxicity by macrophages, neutrophils, and eosinophils. It also activates a complex cascade of reactions termed the **complement system**. This consists of at least eight factors that interact to result in a membrane attack complex, forming pores in bacterial cell walls. Complement components also attract macrophages, enhance phagocytosis, and increase local vascular permeability.

3 **IgA.** This is secreted in body fluids such as tears, saliva, and milk, and boosts the immunity of young infants. It is also secreted into the gut, and is resistant to enzymatic attack.

4 **IgE.** This is bound to cells termed mast cells. In response to antigens and cytokine messages from activated T-cells, it causes the mast cells to release chemicals including histamines into the surrounding tissues and stimulates cytotoxicity by macrophages and eosinophils. The histamines cause direct damage to parasites, or disrupt the usual parasite microhabitat by tissue inflammation, or increase exposure of the parasite to other attacking agents as a result of increased vascular permeability. Mast cells and IgE are important in defense against helminth infections, particularly for worms that live in contact with the gut lining. Activated mast cells cause shedding of the intestinal epithelium, aiding removal of the worm.

Antibodies are able to reach parasites in a wide diversity of locations in or on the host. While tissue and gut parasites are most obviously exposed to attack, parasites on the body surface can also be reached if they feed on blood. Fish produce antibodies in response to attack by ectoparasitic monogenean flukes, targeting them either via the blood they ingest or via the mucus on the fish skin. Mammals produce antibodies to anticoagulants in the saliva of arthropod blood feeders such as ticks, leading to infiltration of the bite site by mast cells. Ticks attempting to feed from immune hosts may fail to engorge, fail to molt normally, or die of desiccation.

Coordination of host defenses

In parasites with complex life cycles, each host provides a different defensive environment, and a single parasite species may be attacked by several of the immune responses described above. Often particular components of the immune system are of differing importance in different tissues, and where parasite larvae migrate through the host they may be exposed to a sequence of defenses. This can be illustrated by the diversity of defenses brought to bear by the human immune system on the larvae of *Schistosoma mansoni*. Attacks are directed against the larval tegument, whose antigens induce Th-lymphocytes to initiate production of IgG and IgE antibodies by B-cells. The antibodies bind to the larval tegument. Direct humoral attack on the worm is carried out by the complement system, both alone and in concert with worm-bound antibodies. The cytokine, interferon, secreted by Th-lymphocytes, activates macrophages, which in turn secrete another cytokine, tumor necrosis factor. This activates neutrophils, eosinophils, and platelets. All four blood cell types respond to the antibody-labeled worm tegument with cytotoxins. Macrophages and neutrophils attack the host tegument using toxic oxygen and nitrogen metabolites, while eosinophils attack using toxic proteins. Worm antigens also bind to specific IgE receptors on mast cells, and these in turn are involved in activating the eosinophils.

17.7.2 Parasite countermeasures

There are two kinds of protection available to parasites in or on their hosts. One type is associated with resisting chemical attack in extreme environments, particularly the gut. The second is associated with avoiding specific antiparasite aspects of host physiology. Some aspects of parasite protection are common to both types of environmental challenge.

Physical protection

Isolation of the parasite from external conditions confers protection from all unfavorable aspects of the surrounding environment. Some parasite larvae resist host defenses in their intermediate hosts by constructing thick-walled protective cysts, as in the nematode *Trichinella spiralis*. These cysts are broken down when the parasite enters the next host in its life cycle (normally by predation of its intermediate host), either by host- or parasite-generated enzymes. Adult parasites in their definitive host may avoid immune responses by inducing host tissues to form a cyst around them. This can perhaps be regarded as parasite adaptation to survive within a cyst following detection of a host encapsulation response.

Immunologically privileged sites

These are regions of the host body in which immune responses are reduced or lacking, usually due to the absence of host white blood cells. In vertebrates such sites include the eye lens (occupied by larvae of some digenean fish flukes) and parts of the central nervous system. In insects certain sites (including nerve ganglia, salivary glands, gonads, muscle fibers, and the fat body) show far lower risks of encapsulation than others, and these too are exploited by parasites. Life-cycle stages that migrate through host tissues are more at risk from host immune defenses than those inhabiting the gut lumen, and it is possible that exploitation of the gut as an adult site may represent avoidance of host immune responses.

After parasites enter the host, the speed with which they reach immunologically privileged sites may be crucial to their survival. The larvae of some endoparasitic tachinid flies hatch in the gut of their insect host where they are relatively safe from host defenses; they then migrate rapidly from the gut wall towards the anterior nervous system of their host, a protected site. On the way they initiate a vigorous encapsulation response from their host, and only the most rapidly moving are able to reach safe sites in the nervous system. Tachinids also illustrate the fact that safe sites are sometimes exploited only for a while immediately after the parasite enters the host. After a period, generation of more effective defenses by the tachinid larvae allows them to re-emerge into the better defended host hemocoel without being encapsulated.

Detoxification of host humoral defenses

Parasites are able to neutralize a very wide range of toxic host compounds, including toxins generally present in the body fluids, the complement system, toxic reactive oxygen and nitrogen intermediates, and antibodies. Where hosts sequester compounds to deter attackers, specialist parasites are usually able to detoxify these compounds, although generalist parasites may be susceptible. For example, larvae of moths of the genus *Zygaena* sequester cyanogenic compounds from the food plants and are avoided by most parasitoids, but a few specialist parasitoid wasps possess enzymes that detoxify these compounds.

Parasites show a range of responses to humoral products of the immune system. Metazoan parasites are often able to resist attack by the host complement system through the possession of surface glycoproteins that mimic host molecules that inhibit the complement cascade. A number of parasites show enzyme-based defenses against the reactive oxygen and nitrogen intermediates from cytotoxic phagocytes. The enzymes are the typical metazoan antioxidants (see Chapter 7) and may either be secreted from the parasite's body or bound to its surface. Thus the nematode *Onchocerca* secretes superoxide dismutase, filarial nematodes inhabiting the lymph system use a surface-bound glutathione peroxidase, and schistosomes use a surface-bound glutathione *S*-transferase.

Detoxification of antibodies is achieved in some nematodes and trematode flukes by cleaving the antibody molecules with proteases. The antibody molecules are split in such a way that the part of the heavy chains essential for detection of the antibody by host cells is removed.

Hiding and molecular mimicry

Intracellular parasites such as protistan plasmodia or coccidia are able to "hide" inside host cells, such that their antigens are not exposed to host immune systems. Relatively few metazoan parasites are small enough to hide inside cells, but the filarial nematode *Trichinella* is an example, hiding inside the large cells of pig skeletal muscle.

Parasites are sometimes able to avoid detection by the host immune system in other ways, either coating their surface with host tissues, or disguising their tissues with host antigens (molecular disguise). *Schistosoma mansoni* absorbs host antigens soon after invading the human body, and so effectively disappears from host detectors of "nonself". The larva is able to absorb surface antigens corresponding to blood groups and the MHC. An immune response is mounted to the antigens presented immediately after invasion, but the larvae have "disappeared" by the time the host response becomes effective. However, later invasions by cercariae are often destroyed by the activated defenses during penetration of human skin. Immunity to reinfection established by the first wave of parasites is termed "concomitant immunity", and may help reduce overexploitation of hosts by a given parasite species. Schistosomes are able to incorporate host antigens in both their snail and human hosts; their ability to take up host antigens is highly host specific, and cercaria larvae invading the incorrect host are rapidly destroyed. For example, the cercariae of marine schistosomes, normally living in sea birds, occasionally try to penetrate the skin of human swimmers or fishermen. They are halted by host defenses and die in the skin, leading to a dermatitis condition known variously as "swimmer's itch", "clam-digger's itch", "seabather's eruption", and "weed itch".

There is also some evidence that parasites can synthesize their own host-like molecules (molecular mimicry). For example, a few insect parasitoids appear able to coat their eggs with virus-like particles that mimic host proteins. Likewise, pentastomids living in vertebrate lungs evade the hosts' immune defenses and reduce inflammation by coating their cuticles in their own stage-specific lipid surfactants, very similar to the host surfactants that line the alveoli.

Changing the antigens presented to the host

If a particular set of surface antigens on a parasite have been changed by the time the host can produce antibodies to them, then the parasite can stay one step ahead of the host's immune system. Changing parasite antigens is sometimes associated with molts or metamorphoses in the parasite life cycle. This strategy is illustrated by the nematode *Nippostrongylus* in rats. Infective larvae penetrate the skin, and migrate to the lungs in less than 24 h. In the lungs they molt, presenting a different set of stage-specific antigens. The next larval instar moves from the lungs to the intestine, where it molts again to present yet another set of surface antigens. In each case the change is quick enough to prevent effective host immune responses. However, the effect of memory in the Th- and B-cells in vertebrate immune systems is that antibodies against each larval instar can be produced far more rapidly in subsequent infections, leading to levels of acquired host immunity.

A number of parasites are able to maintain a rapid turnover of surface antigens within a single life stage. The best known example of this is provided by protozoan trypanosomes, the agents of sleeping sickness. By the time the host has mounted an immune response against one set of antigens, the parasite population consists of individuals with a new type of glycoprotein forming their surface coat. This type of strategy has been suggested to explain the low immune responses mounted against some nematodes, such as *Wuchereria bancrofti* (the agent of elephantiasis), which can survive in the lymph system even though this is the heart of the human immune system.

Active suppression of host defenses

Some parasites reduce host defenses by interfering with the structural and/or functional integrity of host leucocytes. For example, bilharzias schistosomes produce neuropeptides that directly suppress their snail host's hemocytes. Other parasites are able to divert the recruitment of attacking host blood cells. For example, the complement system activates neutrophils and macrophages through a protease enzyme, elastase, and tapeworms are able to secrete an elastase inhibitor, preventing the generation of characteristic complement fragments following elastase digestion of targets and so hiding the parasite from host defensive cells.

Suppression of detection may also be involved in the defense of parasitoid eggs inside insect hosts. Within its egg, the parasitoid embryo is surrounded by a layer of cells called teratocytes, and in many species these continue to surround the young larva when the egg shell is lost. The surface of the teratocytes is covered in microvilli, and the cells are also supplied with extensive endoplasmic reticulum, suggesting a secretory function. Some property of teratocytes (as yet unidentified) suppresses host abilities to encapsulate the parasitoid; embryos from which these cells have been removed are quickly encapsulated.

A fascinating mechanism of host defense suppression in insect parasitoids involves the release into the host of polyDNA viruses by the egg-laying female. These viruses have a characteristic structure, either rod-shaped or spindle-shaped, and always contain double-stranded circular DNA molecules coiled into superhelices. The origin of polyDNA viruses remains a subject of debate. One attractive theory is that viruses are in fact parts of the parasitoid genome that are injected into the host to provide local delivery of venom. Evidence for this is the similarity between the polyDNA viral sequence and other parasitoid genes expressed in the venom glands. After injection by the female parasitoid, the viruses rapidly infect a whole range of host cells (transcription products of the viral genome are detectable within 2 h of infection), and suppress the host encapsulation of parasitoid eggs.

Parasites in vertebrates show a further set of immunosuppressant adaptations. Some are able to cause direct damage to leucocytes through the release of toxic secretions. In others, the ability of the immune system to produce suitable antibodies is disrupted by the release of large quantities of materials that interfere with antigen processing by macrophages. Alternatively the activity of defensive cells in the immune system may be reduced by parasite-induced release of immunosuppressants, such as prostaglandins and other hormones. This may be achieved either through parasite manipulation of host secretions or (in the case of filarial nematodes and tapeworms) by secretion by the parasites themselves. *Schistosoma mansoni* secretes endorphins, which probably have immunosuppressant effects in both invertebrate and vertebrate hosts. Similar disruption of host defenses may also be achieved by altering the balance of cytokines; filarial nematodes induce human hosts to release high concentrations of an IgG which inhibits protective IgE-mediated responses. Similarly, patients infected with schistosomiasis produce IgM and IgG antibodies that inhibit normal cytotoxic responses by neutrophils and eosinophils.

17.7.3 Parasite exploitation of host defenses

Parasites are adept at turning the tables on their hosts' attempts at defense. For example, they may feed on the tissues produced as a host defense in response to them, or induce such complex and costly defenses that the host is weakened and is more likely to be preyed upon by a predator that is itself a host in the parasite life cycle. Some parasites actually need part of the host immune response to survive. Here an example is seen in remoras (marine fish) infected with the fluke *Dionchus remorae*, since the egg bundles of this ectoparasite induce the gill epithelium to which they attach to grow around the egg bundles, thus usefully anchoring them in place. Similar immune response-based attachment of parasites is seen in the reaction around the heads of acanthocephalan worms in vertebrate guts, which holds the worms firmly in place. The nematode *Strongyloides*, parasitic on rats, appears to use the host's immune system as a trigger for its own sexuality, only producing sexual phases from larvae that encounter hosts with acquired immune protection; here sex is perhaps being used as an adaptation to counter the rapid evolution of vertebrate immunity.

17.7.4 Chemical warfare between parasites

Parasites not only have to contend with host defenses, but also with competition and attack from other parasites. Sometimes this occurs within species (even between siblings) and sometimes between species. Suppression of host defenses by one parasite may lead to increased susceptibility to attack by others, an outcome which may well be detrimental to the first parasite.

Perhaps in response to this, parasites may deploy secondary host defenses (termed "parasite-mediated internal defenses") to prevent damage to the host or to the first parasites by later opportunists. Such defenses may be very general; the bactericidal anal secretions from some insect parasitoids are thought to protect their hosts (which are in effect living carcasses) against bacterial decay, prolonging the "shelf-life" of the larval food supply. Other examples are more specific; some insect parasitoids lay eggs in previously parasitized hosts that inhibit the development of the resident parasitoids. Competition among insect parasitoid larvae is often determined by direct combat, but differing physiological tolerances for limiting resources, particularly oxygen, may also influence the outcome of these fights. For example, in ichneumonid wasps the eggs and young larvae have a much lower tolerance of oxygen depletion than older larvae, and older larvae are able to deplete oxygen levels in the host such that younger competitors die. Competition is also recorded for nematode and platyhelminth parasites in the gut. Two parasites that normally occupy the same region of intestine when alone may be spatially separated if both invade one host, the competitively inferior species being displaced to a new region. The acanthocephalan *Moniliformis* causes the cestode *Hymenolepis* to move further back in the rat intestine, but if the acanthocephalans are killed the cestodes migrate anteriorly. Physiological interference by the acanthocephalan, perhaps through depletion of essential nutrients or the release of alcohol as a toxic excretory product, may be involved. However, just as so many parasites are able to overcome host defenses, some parasites (regarded as specialized opportunists) are able to cope with other parasite species in the same host, and consistently parasitize previously infected hosts.

17.8 Conclusions

Parasites may be the single most important cause of reduced fecundity and/or early mortality in natural populations of animals, thus regulating populations and ecological balances at the most fundamental level. There has been an increasing realization that parasites may be one of the major factors shaping all host populations, both in terms of ecological distribution (ranges occupied) and cyclical behavior of population numbers. This is linked to models of how host–parasite interactions may evolve. Are they in a sense mutualistic, the parasite and the host both seeking to minimize the damage done, or entirely aggressive, both host and parasite always seeking to outdo the other, and selection always operating to maximize individual success?

Early work on parasitism tended to stress mutualisms, giving examples like trypanosomes in Africa that are relatively benign to native cattle (with whom they are coevolved) but lethal to introduced cows and to humans. "Older" parasites (in the evolutionary sense) do seem to be generally less virulent and have slower population increases. However, more recent studies have concentrated on aggressive models, seeing host–parasite relations as a coevolutionary arms race, in which each organism always tries to keep one step ahead of the opposition and have maximal growth rate and maximal virulence. Perhaps a balanced view now is that either model (or an intermediate model of a "prudent parasite", doing only as much damage as the host can take) may apply in different circumstances. Whether a parasite evolves towards greater or lesser virulence will depend on its genetics and its epidemiology, including factors such as its lifespan relative to its host, its reproductive rate, its transmission mode, the host's population size and dispersion, and so on. In very simple terms, microparasites that have already infected a few hundred other hosts will not "care" at all about killing off an original host, and nor will a parasite that achieves transmission by one host being eaten by the next. However, many macroparasites—helminths and insects, generally—may have relatively low reproductive rates, longer lifespans relative to their host, and more secure transmission systems, so that they need and benefit from much lesser virulence.

Given these various possibilities, different types of parasites can clearly have wildly different ecological consequences for their hosts. For highly virulent, strongly "*r*-selected" microparasites, epidemics and mass mortalities are always possible, and such organisms invading new habitats or new hosts are often lethal. For more benign parasites, effects may be more subtle. For example, the ranges of hosts may be controlled by their susceptibilities to endemic parasites. One example here is the grazers of North America, where the bighorn sheep range is largely determined by the prevalence of lungworms (*Protostrongylus*). Similarly, rinderpest virus in Africa seems to control distribution of native ungulates: it is usually fatal to kudu and eland, has a moderate effect on giraffe and wildebeest, and is fairly harmless to gazelle and hippopotamus.

Parasitism has almost certainly had widespread ecological and evolutionary effects in terms of speciation. There are many examples now known where host and parasite phylogenies can be fairly precisely matched (see Fig. 17.2), a strong indication that speciation in one has led to speciation in the other. There may thus be positive feedbacks whereby overall speciation rates are greatly increased; and the effect is so clear that specific parasite distributions are now sometimes used to trace the phylogenetic relationships of their hosts.

On a different kind of evolutionary scale, parasitism may also be the underlying reason for the evolution of sex. If this is true then parasitism has had one of the most vital of all effects on evolutionary rates and processes. Put simply, it is suggested that it was (and is) to escape the pressures of rapidly evolving parasites that other organisms had to use sexual reproduction, despite its rather heavy costs relative to asexual reproduction, just to maintain enough genetic diversity to insure some resistant strains could keep up with or ahead of the coevolving parasites. The nematode example *Strongyloides*, referred to in the last section, gives some support to this idea.

Sexual selection and the strong divergence between morphologies and colors in the sexes may be a consequence of parasitism too. At a very basic level, evolution of very small sperm in one sex, where the genetic material is stripped of all cytoplasm, may have been selected for partly as a way of reducing possible parasite transmission. Furthermore, at the organism level it is suggested that males (usually) are bright and often gaudy not just to inform females of their quality, but perhaps specifically to indicate their resistance to parasitic loads and give females a way of assessing this resistance. So female choice is related to selecting good genetic resistance to parasites as a desirable feature in a partner.

Some of these ideas are still controversial, but there is no doubt that parasites are indeed coming to be seen as perhaps the major reason for much of the genetic polymorphism seen in host populations. It is hard to imagine what modern environments would be like *without* parasitism, since their effects are immense. Every organism in existence probably has a suite of various different kinds of parasites acting upon it, and to which it is reacting in contemporary time and over evolutionary time. Thus each organism's current status—its chemistry, behavior, physiology, and ecology—are all consequent upon its parasitological history, and its very genome is a product of coevolution with other infecting organisms. Host–parasite relations could well be the single most important factor affecting modern environments and their occupants.

FURTHER READING

Books

Brooks, D.R. & McClennan, D.A. (1991) *Phylogeny, Ecology and Behaviour.* Chicago University Press, Chicago.

Bryant, C. & Behm, C. (1989) *Biochemical Adaptation in Parasites.* Chapman & Hall, London.

Cox, F.E.G. (1994) *Modern Parasitology: a Textbook of Parasitology.* Blackwell Scientific Publications, Oxford.

Douglas, A.E. (1994) *Symbiotic Interactions.* Oxford University Press, Oxford.

Godfray, H.C. (1994) *Parasitoids: Behavioural and Evolutionary Ecology.* Princeton University Press, Princeton, NJ.

Marr, J. & Müller, M. (1995) *Biochemistry and Molecular Biology of Parasites.* Academic Press, London.

Matthews, B.E. (1998) *An Introduction to Parasitology.* Cambridge University Press, Cambridge, UK.

Poulin, R. (1998) *Evolutionary Ecology of Parasites; from Individuals to Communities.* Chapman & Hall, London.

Rohde, K. (1993) *Ecology of Marine Parasites.* CAB International, Wallingford, UK.

Roitt, I., Brostoff, J. & Male, D. (1998) *Immunology.* Mosby, London.

Rollinson, D. & Simpson, A.J.G. (1987) *The Biology of Schistosomes.* Academic Press, London.

Shorthouse, J. & Rohfritsch, O. (1992) *Biology of Insect-Induced Galls.* Oxford University Press, Oxford.

Tinsley, R.C. (ed.) (1999) *Parasite Adaptation to Environmental Constraints.* Cambridge University Press, Cambridge, UK.

Williams, M.A.J. (1994) *Plant Galls: Organisms, Interactions, Populations.* Clarendon Press, Oxford.

Reviews and scientific papers

Baudouin, M. (1975) Host castration as a parasitic strategy. *Evolution* **29**, 335–352.

Bayne, C.J. (1990) Phagocytosis and non-self recognition in invertebrates. *Bioscience* **40**, 723–731.

Beckage, N.E. (1985) Endocrine interactions between endoparasitic insects and their hosts. *Annual Review of Entomology* **30**, 371–413.

Brown, I.M. & Gaugler, R. (1997) Temperature and humidity influence emergence and survival of entomopathogenic nematodes. *Nematologica* **43**, 363–375.

Combes, C. & Morand, S. (1999) Do parasites live in extreme environments? Constructing hostile niches and living in them. *Parasitology* **119**, S107–S110.

De Jong-Brink, M., Bergamin-Sassen, M. & Soto, M.S. (2001) Multiple strategies of schistosomes to meet their requirements in the intermediate snail host. *Parasitology* **123**, S129–S141.

Dunn, P.E. (1990) Humoral immunity in insects. *Bioscience* **40**, 738–744.

Gemmill, A.W., Viney, M.E. & Read, A.F. (1997) Host immune status determines sexuality in a parasitic nematode. *Evolution* **51**, 393–401.

Lawrence, P.O. (1986) Host–parasite hormonal interactions: an overview. *Journal of Insect Biochemistry &Physiology* **32**, 295–298.

Loker, E.S. (1994) On being a parasite in an invertebrate host—a short survival course. *Journal of Parasitology* **80**, 728–747.

Moore, J. (1984) Parasites that change the behaviour of their host. *Scientific American* **250**, 82–89.

Page, R.D.M. (1993) Parasites, phylogeny and cospeciation. *International Journal of Parasitology* **23**, 499–506.

Patz, J.A. & Reisen, W.K. (2001) Immunology, climate change and vector-borne diseases. *Trends in Immunology* **22**, 171–172.

Poulin, R. (2000) Manipulation of host behaviour by parasites: a weakening paradigm? *Proceedings of the Royal Society of London B* **267**, 787–792.

Slansky, F. (1986) Nutritional ecology of endoparasitic insects and their hosts: an overview. *Journal of Insect Physiology* **32**, 255–261.

Tamas, I., Klasson, L.M., Sandstrom, J.P. & Andersson, S.G.E. (2001) Mutualists and parasites; how to paint yourself into a (metabolic) corner. *FEBS Letters* **498**, 135–139.

Thomas, M.B. & Blanford, S. (2003) Thermal biology in insect–parasite interactions. *Trends in Research in Ecology & Evolution* **18**, 344–350.

References

Agnew, D.J. & Taylor, A.C. (1986) Seasonal and diel variations of some physico-chemical parameters of boulder-shore habitats. *Ophelia* **25**, 83–95.

Ahearn, G.A. (1970) The control of water loss in desert tenebrionid beetles. *Journal of Experimental Biology* **53**, 573–595.

Aidley, D.J. (1998) *The Physiology of Excitable Cells*, 4th edn. Cambridge University Press, Cambridge, UK.

Alberts, B., Bray, D., Lewis, J., Raff, M., Roberts, K. & Watson, J.D. (1994) *Molecular Biology of the Cell*. Garland Publishing, New York.

Aleksiuk, M. (1971) Temperature-dependent shifts in the metabolism of a cool temperate reptile, *Thamnophis sirtalis parietalis*. *Comparative Biochemistry & Physiology A* **39**, 459–503.

Alexander, R.McN. (1982) *Locomotion of Animals*. Blackie, Glasgow.

Alexander, R.McN. (1999) *Energy for Animal Life*. Oxford University Press, Oxford.

Alexander, W.R., Southgate, B.A. & Bassindale, R. (1935) Survey of the River Tees—the estuary, chemical and biological. *DSIR Water Pollution Research Technical Paper 5*.

Altringham, J.D. & Johnston, I.J. (1990) Modelling muscle power output in a swimming fish. *Journal of Experimental Biology* **148**, 395–402.

Baker, M.A. (1982) Brain cooling in endotherms in heat and exercise. *Annual Review of Physiology* **44**, 85–96.

Ballantine, W.J. (1961) A biologically-defined exposure scale for the comparative description of rocky shores. *Field Studies* **1**, 1–19.

Barnes, R.S.K. (1981) An experimental study of the pattern and significance of the climbing behaviour of *Hydrobia ulvae* (Pennant). *Journal of the Marine Biological Association of the UK* **61**, 285–299.

Bartholomew, G.A. (1964) The roles of physiology and behaviour in the maintenance of homeostasis in the desert environment. *Symposia of the Society for Experimental Biology* **18**, 7–29.

Bartholomew, G.A. & Lasiewski, R.C. (1965) Heating and cooling rates, heart rate and simulated diving in the Galapagos marine iguana. *Comparative Biochemistry & Physiology* **16**, 575–582.

Bayer, F.M. & Owre, H.B. (1968) *The Free-Living Lower Invertebrates*. Macmillan, New York.

Bayne, B.L. & Hawkins, A.J.S. (1997) Protein metabolism, the costs of growth, and genomic heterozygosity: experiments with the mussel *Mytilus galloprovincialis* Lmk. *Physiological Zoology* **70**, 391–402.

Behan-Martin, M., Bowler, K., Jones, G. & Cossins, A.R. (1993) A near-perfect temperature adaptation of bilayer order in vertebrate brain membranes. *Biochimica et Biophysica Acta* **1151**, 216–222.

Beniston, M. (ed.) (1994) *Mountain Environments in Changing Climates*. Routledge, London.

Benzinger, T.H. (1961) The diminution of thermoregulatory sweating during cold reception at the skin. *Proceedings of the National Academy of Sciences of the USA* **47**, 1683–1688.

Biewener, A.A. & Blickhan, R. (1988) Kangaroo rat locomotion—design for elastic energy storage or acceleration. *Journal of Experimental Biology* **140**, 243–255.

Binyon, J. (1961) Salinity tolerance and permeability for water of the starfish *Asteria rubens* L. *Journal of the Marine Biological Association of the UK* **41**, 161–174.

Bliss, D.E. (1979) From sea to tree: saga of a land crab. *American Zoologist* **19**, 385–410.

Bond, T.F., Kelly, C.F., Morrison, S.R. & Pereira, N. (1967) Solar atmospheric and terrestrial radiation received by shaded and unshaded animals. *Transactions of the American Society of Agricultural Engineers* **10**, 622–625.

Bone, Q., Marshall, N.B. & Blaxter, J.H.S. (1982) *Biology of Fishes*. Blackie, Glasgow.

Bosch, I., Beauchamp, K.A., Steele, M.E. & Pearse, J.S. (1987) Development, metamorphosis and seasonal abundance of embryos and larvae of the Antarctic sea-urchin *Sterechinus neumayeri*. *Biological Bulletin* **173**, 126–135.

Bouverot, P. (1985) *Adaptation to Altitude-Hypoxia in Vertebrates*. Springer-Verlag, Berlin.

Boyd, I.A. & Martin, A.R. (1956) The end-plate potential in mammalian muscle. *Journal of Physiology* **132**, 74–91.

Brett, J. (1956) Some principles of the thermal requirements of fish. *Quarterly Review of Biology* **31**, 75–87.

Brix, O., Borgund, S., Barnung, T., Colosimo, A. & Giardino, B. (1989) Endothermic oxygenation of haenocyanin in the krill *Neganyctiphanes norbegica*. *FEBS Letters* **247**, 177–180.

Brooks, J.L. (1947) Turbulence as an environmental determinant of relative growth in *Daphnia*. *Proccedings of the National Academy of Sciences of the USA* **33**, 141–148.

Brooks, S.E. & Shorthouse, J.D. (1998) Developmental morphology of stem galls of *Diplolepis nodulosa* (Hymenoptera: Cynipidae) and those modified by the inquiline *Periclistus pirata* (Hymenoptera: Cynipidae) on *Rosa blanda* (Rosaceae). *Canadian Journal of Botany* **76**, 365–381.

Brusca, R.C. & Brusca, G.J. (1990) *Invertebrates*. Sinauer Associates, Sunderland, MA.

Burky, A.J., Pacheco, J. & Pereyra, E. (1972) Temperature, water and respiratory regimes of an amphibious snail *Pomacea urceus* from the Venezuelan savannah. *Biological Bulletin* **143**, 304–316.

Bursell, E. (1957) Spiracular control of water loss in the tsetse fly. *Proceedings of the Royal Entomological Society (London)* **32**, 21–29.

Cannon, R.J.C. & Block, W. (1988) Cold tolerance of microarthropods. *Biological Review* **63**, 23–77.

Carey, F.G., Teal, J.M., Kanwisher, J.W. & Lawson, K.D. (1971) Warm-blooded fish. *American Zoologist* **11**, 137–145.

Carter, R.W.G. (1988) *Coastal Environments*. Academic Press, London.

Chan, S.L., Fletcher, G.L. & Hew, C.L. (1993) Control of antifreeze protein gene expression in winter flounder. In: *Biochemistry & Molecular Biology of Fishes*, Vol. 2 (ed. P.W. Hochachka & T.P. Mommsen). Elsevier, Amsterdam.

Chapman, R.F. (1998) *The Insects, Structure and Function*, 4th edn. Cambridge University Press, Cambridge, UK.

Cheung, W.W.K. & Marshall, A.T. (1973) Water and ion regulation in cicadas in relation to xylem feeding. *Journal of Insect Physiology* **19**, 1801–1816.

Connell, J.H. (1961) The influence of interspecific competition and other factors on the distribution of the barnacle *Chthamalus stellatus*. *Ecology* **42**, 710–723.

Contreras, L.C. (1984) Bioenergetics of huddling: test of a psychophysiological hypothesis. *Journal of Mammalogy* **65**, 256–262.

Cossins, A.R. & Bowler, K. (1987) *Temperature Biology of Animals*. Chapman & Hall, London.

Cossins, A.R. & Sheterline, P.S. (1983) *Cellular Acclimatisation to Environmental Change*. Cambridge University Press, Cambridge, UK.

Cowan, W.M., Südhof, T.C. & Stevens, C.F. (eds) (2001) *Synapses*. Johns Hopkins University Press, Baltimore, MD.

Croghan, P.C. & Lockwood, A.P.M. (1968) Ionic regulation of the Baltic race and freshwater races of the isopod *Mesidotea (Saduria) entomon* (L). *Journal of Experimental Biology* **48**, 141–158.

Crowley, T.J. & North, J.R. (1991) *Palaeoclimatology*. Oxford University Press, New York.

Damuth, J. (1991) Ecology—of size and abundance. *Nature* **351**, 268–269.

Daniel, P.M., Dawes, J.D.K. & Pritchard, M.M.L. (1953) Studies on the carotid rate and its associated arteries. *Physiological Transactions of the Royal Society of London B* **237**, 173–208.

Davenport, J.A. (1992) *Animal Life at Low Temperature*. Chapman & Hall, London.

Davenport, J., Blackstock, N., Davies, D.A. & Yarrington, M. (1987) Observations on the physiology and integumentary structures of the Antarctic pycnogonid *Decalopoda australis*. *Journal of Zoology* **211**, 451–465.

de Fur, P.L. & Mangum, C.P. (1979) The effects of environmental variables on the heart rates of invertebrates. *Comparative Biochemistry & Physiology A* **62**, 288–300.

de Vries, A.L. (1980) Biological antifreeze and survival in freezing environments. In: *Animals and Environmental Fitness* (ed. R. Gilles), pp. 583–607. Pergamon Press, Oxford.

de Vries, A.L. (1988) The role of antifreeze glycopeptides and peptides in the freezing avoidance of Antarctic fishes. *Comparative Biochemistry & Physiology B* **90**, 611–621.

Dejours, P. (1981) *Principles of Comparative Respiratory Physiology*. Elsevier/Holland Biomedical, Amsterdam.

Delcomyn, F. (1988) *Foundations of Neurobiology*. W.H. Freeman, New York.

Denton, E.J. (1961) The buoyancy of fish and cephalopods. *Progress in Biophysics & Biophysical Chemistry* **11**, 178–233.

Denton, E.J. & Gilpin-Brown, J.B. (1961) The distribution of gas and liquid within the cuttlebone. *Journal of the Marine Biological Association of the UK* **41**, 365–381.

Denton, E.J. & Shaw, T.I. (1961) The buoyancy of gelatinous marine animals. *Journal of Physiology* **162**, 14P–15P.

Desbruyeres, D. & Segonzac, M. (eds) (1997) *Handbook of Deep-Sea Hydrothermal Vent Fauna*. IFREMER, Plouzane, France.

Dickinson, M.H., Farley, C.T., Full, R.J., Koehl, M.A.R., Kram, R. & Lehman, S. (2000) How animals move: an integrative view. *Science* **288**, 100–106.

Doake, C.S.M. & Vaughan, D.G. (1991) Rapid disintegration of the Wordie Ice Shelf in response to atmospheric warming. *Nature* **350**, 328–330.

Douwes, P. (1976) Activity in *Heodes virgaureae* in relation to air temperature, solar radiation and time of day. *Oecologia* **22**, 287–298.

Dreisig, H. (1980) Daily activity, thermoregulation and water loss in the tiger beetle *Cicindela hybrida*. *Oecologia* **44**, 376–389.

Eastman, J.T. (ed.) (1993) *Antarctic Fish Biology, Evolution in a Unique Environment*. Academic Press, London.

Edney, E.B. (1947) Laboratory studies on the bionomics of the rat fleas *Xenopsylla brasiliensis* Baker and *X. cheopsis* Roths. II Water relations during the cocoon period. *Bulletin of Entomological Research* **38**, 263–280.

Edney, E.B. (1953) The temperature of woodlice in the sun. *Journal of Experimental Biology* **30**, 331–349.

Edney, E.B. (1966) Absorption of water vapour from unsaturated air by *Arenivaga* sp. (Polphagidae, Dictyoptera). *Comparative Biochemistry & Physiology* **19**, 387–408.

Edney, E.B. (1971) The body temperature of tenebrionid beetles in the Namib desert of southern Africa. *Journal of Experimental Biology* **55**, 253–272.

Edney, E.B. (1980) The components of water balance. In: *Insect Biology in the Future* (ed. M. Locke & D.A. Smith). Academic Press, New York.

Elliott, J.M. (1994) *Quantitative Ecology and the Brown Trout*. Oxford University Press, Oxford.

Else, P.L. & Hulbert, A.J. (1985) An allometric comparison of the mitochondria of mammalian and reptilian tissues: implications for the evolution of endothermy. *Journal of Comparative Physiology* **156**, 3–11.

Farlow, J.O. (1976) A consideration of the trophic dynamics of a late Cretaceous large dinosaur community. *Ecology* **57**, 841–857.

Farrow, R.A. (1990) Flight and migration in acridoids. In: *Biology of Grasshoppers* (ed. R.F. Chapman & A. Joern). Wiley, New York.

Feder, M.E. (1987) The analysis of physiological diversity: the prospects for pattern documentation and general questions in ecological physiology. In: *New Directions in Ecological Physiology* (ed. M.E. Feder, A.F. Bennett, W.W. Burggren & R.B. Huey), pp. 38–75. Cambridge University Press, Cambridge, UK.

Flock, A. (1965) Transducing mechanisms in the lateral line canal organ receptors. *Cold Spring Harbor Laboratory Symposia on Quantitative Biology* **30**, 133–144.

Flock, A. (1967) *Lateral Line Detectors* (ed. P. Cahn). Indiana University Press, Bloomington, IN.

Friedli, H., Lotscher, H., Oeschger, H., Siegenthaler, U. & Stauffer, B. (1986) Ice core record of the $^{13}C/^{12}C$ ratio of atmospheric CO_2 in the past two centuries. *Nature* **324**, 237–238.

Fry, F.E.J. & Hochachka, P.W. (1970) Fish. In: *Comparative Physiology of Thermoregulation* (ed. G.C. Whittow). Academic Press, New York.

Geiser, F. (1988) Reduction of metabolism during hibernation and daily torpor in mammals and birds: temperature effect or physiological inhibition? *Journal of Comparative Physiology B* **158**, 25–37.

Gilles, R. (1979) Intracellular organic osmotic effectors. In: *Mechanisms of Osmoregulation in Animals* (ed. R. Gilles). John Wiley & Sons, New York.

Goldsworthy, G.J., Robinson, J. & Mordue, W. (1981) *Endocrinology*. Blackie, Glasgow.

Gracey, A.Y., Logue, J., Tiku, P.E. & Cossins, A.R. (1996) Adaptation of biological membranes to temperature: biophysical perspectives and molecular mechanisms. In: *Animals and Temperature* (ed. I.A. Johnston & A.F. Bennett), pp. 1–21. Cambridge University Press, Cambridge, UK.

Graedel, T.E. & Crutzen, P.J. (1993) *Atmospheric Change: an Earth System Perspective*. W.H. Freeman, New York.

Green, J. (1968) *The Biology of Estuarine Environments*. University of Washington Press, Seattle, WA.

Gupta, B.L., Wall, B.J., Oschmann, J.L. & Hall, T.A. (1980) Direct microprobe evidence for local concentration gradients and recycling of electrolytes during fluid absorption in the rectal papillae of *Calliphora*. *Journal of Experimental Biology* **88**, 21–48.

Hadley, N.F. (1970) Micrometeorology and energy exchange in two desert arthropods. *Ecology* **51**, 434–444.

Hadley, N.F. & Quinlan, M. (1993) Discontinuous carbon dioxide release in the Eastern lubber grasshopper *Romalea guttata* and its effect on respiratory transpiration. *Journal of Experimental Biology* **177**, 169–180.

Haefner, P.A. Jr (1969) Temperature and salinity tolerance of the sand shrimp, *Crangon septemspinosa* Say. *Physiological Zoology* **42**, 388–397.

Haefner, P.A. Jr (1970) The effect of low dissolved oxygen concentration on temperature–salinity tolerance of the sand shrimp *Crangon septemspinosa*. *Physiological Zoology* **43**, 30–37.

Hafner, M.S. & Page, R.D.M. (1995) Molecular phylogenies and host–parasite co-speciation: gophers and lice as a model system. *Philosophical Transactions of the Royal Society of London B* **349**, 77–83.

Hall, F.G., Dill, D.B. & Barron, E.S.G. (1936) Comparative physiology in high altitudes. *Journal of Cellular & Comparative Physiology* **8**, 301–313.

Hamilton, W.J. (1975) Colouration and its thermal consequences for diurnal desert insects. In: *Environmental Physiology of Desert Organisms* (ed. N.F. Hadley). Dowden, Hutchinson & Ross, New York.

Hansen, T. (1973) Variation in glycerol content in relation to cold-hardiness in the larvae of *Petrova resinella* L. (Lepidoptera, Torticidae). *Eesti Teaduste Akadeemia. Toimetised. Bioloogia* **22**, 105–112.

Hardy, J.D. (1949) Heat transfer. In: *Physiology of Heat Regulation and the Science of Clothing* (ed. L.H. Newburgh). Saunders, Philadelphia, PA.

Hart, J.S. (1956) Seasonal changes in insulation of the fur. *Canadian Journal of Zoology* **34**, 53–57.

Haugaard, N. & Irving, L. (1943) The influence of temperature upon the oxygen consumption of the cunner (*Tautogolabrus adspersus*) in summer and in winter. *Journal of Cellular & Comparative Physiology* **21**, 19–26.

Hawkins, A.J.S., Bayne, B.L., Day, A.J., Rusin, J. & Worrall, C.M. (1989) Genotype-dependent interrelations between energy metabolism, protein metabolism and fitness. In: *Reproduction, Genetics and Distributions of Marine Organisms* (ed. J.S. Ryland & P.A. Tyler). Olsen & Olsen, Fredensborg.

Heinrich, B. (1975) Thermoregulation in bumblebees II. Energetics of warm-up and free flight. *Journal of Comparative Physiology* **96**, 155–166.

Heinrich, B. (1976) Heat exchange in relation to blood flow between thorax and abdomen in bumblebees. *Journal of Experimental Biology* **64**, 561–585.

Heldmaier, G., Steinlechner, S., Ruf, T., Wiesinger, H. & Klingenspor, M. (1989) Photoperiod and thermoregulation in vertebrates—body-temperature rhythms and thermogenic acclimation. *Journal of Biological Rhythms* **4**, 251–265.

Henschel, J.R. (1990) Spiders wheel to escape. *South African Journal of Science* **86**, 151–152.

Henwood, K. (1975) A field-tested thermoregulation model for two diurnal Namib Desert tenebrionid beetles. *Ecology* **56**, 1329–1342.

Herreid, C.F. & Fourtner, C.R. (1981) *Locomotion and Energetics in Arthropods*. Plenum Press, New York.

Herreid, C.F. & Schmidt-Nielsen, K. (1966) Oxygen consumption, temperature and water loss in bats from different environments. *American Journal of Physiology* **211**, 1108–1112.

Herring, P.J. & Martin, J.G. (1978) Bioluminescence in fishes. In: *Bioluminescence in Action* (ed. P.J. Herring). Academic Press, London.

Heusner, A.A. (1982) Energy metabolism and body size: is the 0.75 mass exponent of Kleiber's equation a statistical artefact? *Respiration Physiology* **48**, 1–12.

Hickman, C.P. & Trump, B.F. (1969) The kidney. In: *Fish Physiology*, Vol. 1 (ed. W.S. Hoar & B.J. Randall). Academic Press, London.

Hillenius, W.J. (1994) Turbinates in therapsids—evidence for late Permian origins of mammalian endothermy. *Evolution* **48**, 207–229.

Hinde, R.A. (1982) *Ethology*. Fontana Press, London.

Hinton, H.E. (1957) The structure and function of the spiracular gills of the fly *Taphrophila vitripennis*. *Proceedings of the Royal Society of London B* **147**, 90–120.

Hinton, H.E. (1981) *Biology of Insect Eggs*. Pergamon Press, Oxford.

Hoar, W.S. (1965) The endocrine system as a chemical link between the organisn and its environment. *Transactions of the Royal Society of Canada* **3**, 175–200.

Hochachka, P.W., Emmett, B. & Suarez, R.K. (1988) Limits and constraints on the scaling of oxidative and glycolytic enzymes. *Canadian Journal of Zoology* **66**, 1128–1138.

Hochachka, P.W. & Somero, G.N. (1973) *Strategies of Biochemical Adaptation*. Saunders, Philadelphia, PA.

Hodgkin, A.L. & Huxley, A.F. (1952) A quantitative description of membrane current and its application to conduction and excitation in nerve. *Journal of Physiology* **117**, 500–544.

Hofmeyr, M.D. & Louw, G.N. (1987) Thermoregulation, pelage conductance and renal function in the desert-adapted springbock *Antidorcas marsupialis*. *Journal of Arid Environments* **13**, 137–151.

Hoglund, G., Hamdorf, K. & Rosner, G. (1973) Trichromatic visual systems in an insect and its sensitivity control by blue light. *Journal of Comparative Physiology* **86**, 265–279.

Holdridge, L.R. (1947) Determination of the world's plant formations from sample climate data. *Science* **105**, 367–368.

Holm, E. & Edney, E.B. (1973) Daily activity of Namib Desert beetles in relation to climate. *Ecology* **54**, 45–56.

Hoppeler, H. & Weibel, E.R. (1998) Limits for oxygen and substrate transport in mammals. *Journal of Experimental Biology* **201**, 1051–1064.

Horne, J. & Goldman, C.R. (1994) *Limnology*. McGraw Hill, London.

Hoyt, D.F. & Taylor, C.R. (1981) Gait and the energetics of locomotion in horses. *Nature* **292**, 239–240.

Hudspeth, A.J. (1989) How the ear's works work. *Nature* **341**, 397–404.

Huey, R.B. (1987) Phylogeny, history and the comparative method. In: *New Directions in Ecological Physiology* (ed. M.E. Feder, A.F. Bennett, W.W. Burggren & R.B. Huey), pp. 76–101. Cambridge University Press, Cambridge, UK.

Humphreys, W.F. (1979) Production and respiration in animal populations. *Journal of Animal Ecology* **48**, 427–453.

Hutchinson, V.H., Dowling, H.G. & Vinegar, A. (1966) Thermoregulation in a brooding female Indian python *Python molurus bivittatus*. *Science* **151**, 694–695.

Hynes, H.B.N. (1960) *The Biology of Polluted Waters*. Liverpool University Press, Liverpool, UK.

Ingmanson, D.E. & Wallace, E.J. (1989) *Oceanography*. Wadsworth Publications Co., Belmont, CA.

Innes, D.J. & Haley, L.E. (1977) Inheritance of a shell-colour polymorphism in the mussel. *Journal of Heredity* **68**, 203–204.

Innes, A.J. & Taylor, E.W. (1986) The evolution of air-breathing in crustaceans: a functional analysis of branchial, cutaneous and pulmonary gas exchange. *Comparative Biochemistry & Physiology A* **85**, 621–637.

Irving, L. & Krogh, J. (1955) Temperature of skin in the Arctic as a regulator of heat. *Journal of Applied Physiology* **7**, 355–364.

Janzen, F.J. & Paukstis, G.L. (1991) Environmental sex determination in reptiles: ecology, evolution, and experimental design. *Quarterly Review of Biology* **66**, 149–179.

Jerison, H.J. (1970) Brain evolution: new light on old principles. *Science* **170**, 1224–1225.

Jobling, M. (1993) Bioenergetics: feed intake and energy partitioning. In: *Fish Ecophysiology* (ed. J.C. Rankin & F.B. Jensen). Chapman & Hall, London.

Jobling, M. & Davis, P.S. (1980) Effects of feeding on metabolic rate and the specific dynamic action in plaice, *Pleuronectes platessa*. *Journal of Fish Biology* **16**, 629–638.

Johnston, I.A. (1981) Structure and function of fish muscles. *Symposia of the Zoological Society of London* **48**, 71–113.

Johnston, I.A. (1990) Cold adaptation in marine organisms. *Philosophical Transactions of the Royal Society of London B* **326**, 655–667.

Johnston, I.A. & Battram, J. (1993) Feeding energetics and metabolism in demersal fish species from Antarctic, temperate and tropical environments. *Marine Biology* **115**, 7–14.

Johnston, I.A. & Walesby, N.J. (1977) Molecular mechanisms of temperature adaptation in fish myofibrillar adenosine triphosphatases. *Journal of Comparative Physiology* **119**, 195–206.

Josephson, R.K. (1985) The mechanical power output of a tettigoniid wing muscle during singing and flight. *Journal of Experimental Biology* **117**, 357–368.

Jusiak, R. & Poczopko, P. (1972) The effect of ambient temperature and season on total and tissue metabolism of the frog *Rana esculenta* L. *Bulletin of the Academy of Science. Polar Science Series B* **20**, 523–529.

Kaestner, A. (1980) *Invertebrate Zoology*, Vol. 3, *Crustacea*. Wiley, New York.

Kahl, M.P. (1963) Thermoregulation in the wood stork with special reference to the role of the legs. *Physiological Zoology* **36**, 141–151.

Kaufmann, W.R. & Phillips, J.R. (1973) Ion and water balance in the ixodid tick *Dermacentor andersonii*. *Journal of Experimental Biology* **58**, 523–547.

Kerkut, G.A. & Gilbert, L.I. (1985) *Comprehensive Insect Physiology, Biochemistry and Pharmacology*, Vol. 8. *Endocrinology II*. Elsevier Science, Amsterdam.

Kerr, R.A. (1988) Linking earth, ocean and air. *Science* **239**, 259–260.

Kevan, P.G. (1975) Sun-tracking solar furnaces in high arctic flowers: significance for pollination and insects. *Science* **189**, 723–726.

Kilgore, D.L., Bernstein, M.H. & Hudson, D.M. (1976) Brain temperatures in birds. *Journal of Comparative Physiology* **110**, 209–215.

Kilgore, D.K. & Schmidt-Nielsen, K. (1975) Heat loss from ducks' feet emersed in cold water. *Condor* **77**, 475–478.

Kleiber, M. (1932) Body size and animal metabolism. *Hilgardia* **6**, 315–333.

Kleiber, M. (1961) *The Fire of Life: an Introduction to Animal Energetics*. Wiley, New York.

Kram, R. & Taylor, C.R. (1990) Energetics of running: a new perspective. *Nature* **346**, 265–267.

Land, M.F. & Nilsson, D.-E. (2002) *Animal Eyes*. Oxford University Press, Oxford.

Lasiewski, R.C. & Snyder, G.K. (1969) Responses to high temperature in nestling double-crested and pelagic cormorants. *Auk* **86**, 529–540.

Lees, A.D. (1947) Transpiration and the structure of the epicuticle in ticks. *Journal of Experimental Biology* **23**, 397–410.

Levinton, J.S. (1995) *Marine Biology*. Oxford University Press, New York.

Levy, R.I. & Schneiderman, H.A. (1966) Discontinuous respiration in insects. IV. Changes in intratracheal pressure during the respiratory cycle of silkworm pupae. *Journal of Insect Physiology* **12**, 465–492.

Lighton, J.R.B. (1988) Discontinuous CO_2 emission in a small insect, the formicine ant *Camponotus vicinus*. *Journal of Experimental Biology* **134**, 363–376.

Lighton, J.R.B. (1989) Individual and whole-colony respiration in an African formicine ant. *Functional Ecology* **3**, 523–530.

Lindstedt, S.L., McGlothlin, T., Percy, E. & Pifer, J. (1998) Task-specific design of skeletal muscle: balancing structural composition. *Comparative Biochemistry & Physiology B* **120**, 35–40.

Little, C. (1983) *The Colonisation of Land*. Cambridge University Press, Cambridge, UK.

Little, C. (1990) *The Terrestrial Invasion; an Ecophysiological Approach to the Origin of Land Animals*. Cambridge University Press, Cambridge, UK.

Lockwood, A.P.M. (1961) The urine of *G. duebeni* and *G. pulex*. *Journal of Experimental Biology* **38**, 647–658.

Lockwood, A.P.M., Sheader, M. & Williams, J.A. (1996) Life in estuaries, salt marshes and coastal waters. In: *Oceanography* (ed. C.P. Summerhayes & S.A. Thorpe). Manson, London.

Louw, G. (1993) *Physiological Animal Ecology*. Longman, Harlow, UK.

Lutz, P.L. (1980) On the oxygen affinity of bird blood. *American Zoologist* **20**, 187–198.

Maddrell, S.H.P. (1971) The mechanisms of insect excretory systems. *Advances in Insect Physiology* **8**, 199–331.

Maddrell, S.H.P. & O'Donnell, M.J. (1992) Insect Malphigian tubules—V-ATPase action in ion and fluid transport. *Journal of Experimental Biology* **172**, 4117–4429.

Maetz, J. (1973) Na^+/NH_4^+, Na^+/H^+ exchanges and NH_3 movements across the gill of *Carassius auratus*. *Journal of Experimental Biology* **58**, 255–275.

Malenka, R.C. & Siegelbaum, S.A. (2001) Synaptic plasticity. In: *Synapses* (ed. W.M. Cowan, T.C. Südhof & C.F. Stevens), pp. 395–453. Johns Hopkins University Press, Baltimore, MD.

Manton, S.M. (1952) The evolution of arthropodan locomotory mechanisms. II. General introduction. *Journal of the Linnean Society of London* **42**, 93–117.

Marsh, A.C. (1985) Thermal responses and temperature balance in a desert ant, *Ocymyrmex barbiger*. *Physiological Zoology* **58**, 629–636.

Marsh, B.A. & Branch, G.M. (1979) Circadian and circatidal rhythms of oxygen consumption in the sandy-beach isopod *Tylos granulatus* Krauss. *Journal of Experimental Marine Biology & Ecology* **37**, 77–89.

Marsh, P.L. & Olson, J.M. (1994) Power output of a scallop adductor muscle. *Journal of Experimental Biology* **193**, 139–156.

Martin, R.D. (1980) Body temperature, activity and energy costs. *Nature* **283**, 335–336.

Matthews, B.E. (1998) *An Introduction to Parasitology*. Cambridge University Press, Cambridge, UK.

Mattson, W.J. Jr (1980) Herbivory in relation to plant nitrogen content. *Annual Review of Ecology & Systematics* **11**, 119–161.

May, M.L. (1976) Warming rates as a function of body size in periodic ectotherms. *Journal of Comparative Physiology* **111**, 55–70.

Micklin, P.P. (1988) Desiccation of the Aral Sea—a water management disaster in the Soviet Union. *Science* **241**, 1170–1175.

Minnich, J.E. (1982) The use of water. In: *Biology of the Reptilia*, Vol. 12 (ed. C. Gans & F.H. Hough). Academic Press, London.

Mintzer, C. (1992) *Confronting Climate Change*. Cambridge University Press, Cambridge, UK.

Mitchell, D., Laburn, H.P., Nijland, M.J.M., Zurovsky, Y. & Mitchell, G. (1987) Selective brain cooling and survival. *South African Journal of Sciences* **83**, 598–604.

Mitton, J.B. (1977) Shell color and pattern variation in *Mytilus edulis* and its adaptive significance. *Chesapeake Science* **18**, 387–389.

Moore, P.G. & Francis, C.H. (1985) On the water relations and osmoregulation of the beach-hopper *Orchestia gamarellus*. *Journal of Experimental Marine Biology & Ecology* **94**, 131–150.

Morton, J.E. (1954) *Journal of the Marine Biological Association of the UK* **33**, 297–312.

Moss, B. (1980) *Ecology of Fresh Waters: Man and Medium*. Blackwell Scientific Publications, Oxford.

Murrish, D.E. & Schmidt-Nielsen, K. (1970) Water transport in the cloaca of lizards: active or passive? *Science* **170**, 324–326.

Nachtigall, W. (1983) The biophysics of locomotion in water. In: *Biophysics*. (ed. W. Hoppe, W. Lohmann, H. Markl & H. Ziegler). Springer-Verlag, Berlin.

Nagy, K.A. & Medica, P.A. (1986) Physiological ecology of desert tortoises in southern Nevada. *Herpetologica* **42**, 73–92.

Nagy, K.A. & Petersen, C.C. (1987) Water flux scaling. In: *Comparative Physiology: Life in Water and on Land* (ed. P. Dejours, L. Bolis, C.R. Taylor & E.R. Weibel). Liviana Press, Padova, Italy.

Newell, R.C. (1979) *Biology of Intertidal Animals*. Marine Ecological Surveys Ltd, Faversham, UK.

O'Brien, J. & Block, B.A. (1996) Effects of Ca^2 on oxidative phosphorylation in mitochondria from the thermogenic organ of marlin. *Journal of Experimental Biology* **199**, 2679–2687.

O'Donnell, M.J. (1987) Water vapour absorption by arthropods: different sites, different mechanisms. In: *Comparative Physiology: Life in Water and on Land* (ed. P. Dejours, L. Bolis, C.R. Taylor & E.R. Weibel). Liviana Press, Padova, Italy.

O'Donnell, M.J., Rheault, M.R., Davies, S.A. *et al.* (1998) Hormonally controlled chloride movement across *Drosophila* tubules is via ion channels in stellate cells. *American Journal of Physiology* **42**, R1039–R1049.

Oglesby (1968) Some osmotic responses of the sipunculid worm *Thermiste dyscritum*. *Comparative Biochemistry & Physiology* **26**, 155–177.

Oldfield, B.P. (1985) The role of the tympanal membranes and the receptor array in the tuning of auditory receptors in bushcrickets. In: *Acoustic and Vibrational Communication in Insects* (ed. K. Kalmring & N. Elsner). Paul Parey Verlag.

Oosthuizen, M.J. (1939) The body temperature of *Samia cecrophia* Linn. (Lepidoptera, Saturnidae) as influenced by muscular activity. *Journal of the Entomological Society of South Africa* **2**, 63–73.

Page, R.D.M. (1993) Parasites, phylogeny and cospeciation. *International Journal for Parasitology* **23**, 499–506.

Palmiter, R.D., Gagnon, J., Ericsson, L.H. & Walsh, K.A. (1977) Precursor of egg-white lysozyme: amino acid sequence of NH_2 terminal extension. *Journal of Biological Chemistry* **252**, 6383–6393.

Park, C.C. (1991) Trans-frontier air pollution—some geographical issues. *Geography* **76**, 21–35.

Parry, D.A. (1954) On the drinking of soil capillary water by spiders. *Journal of Experimental Biology* **31**, 218–227.

Patterson, B.D., Pacheco, V. & Solari, S. (1996) Distribution of bats along an elevational gradient in the Andes of south-eastern Peru. *Journal of Zoology* **240**, 637–658.

Peaker, M.J. & Linzell, J.L. (1975) *Salt Glands in Birds and Reptiles*. Cambridge University Press, Cambridge, UK.

Pearse, B. (1980) Coated vesicles. *Trends in Biochemical Sciences* **5**, 131–134.

Pearson, O.P. (1954) Habits of the lizard *Liolaemus multiformis* at high altitudes in southern Peru. *Copeia* **1954**, 111–116.

Peters, R.H. (1983) *The Ecological Implications of Body Size*. Cambridge University Press, Cambridge, UK.

Phillips, J.E. (1981) Comparative physiology of insect renal function. *American Journal of Physiology* **241**, R241–257.

Phillips, J.E., Bradley, T.J. & Maddrell, S.H.P. (1971) Mechanisms of ionic and osmotic regulation in saline-water mosquito larvae. In: *Comparative Physiology: Water, Ions and Fluid Mechanics* (ed. K. Schmidt-Nielsen, L. Bolis & S.H.P. Maddell). Cambridge University Press, Cambridge, UK.

Phillipson, J. (1981) Bioenergetic options and phylogeny. In: *Physiological Ecology: an Evolutionary Approach to Resource Use* (ed. C.R. Townsend & P. Calow), pp. 20–45. Blackwell Scientific Publications, Oxford.

Pierce, S.K. (1971) Volume regulation and valve movements by marine mussels. *Comparative Biochemistry & Physiology A* **39**, 103–117.

Pierce, S.K., Warren, M.K. & West, H.H. (1983) Non amino acid mediated volume regulation in an extreme osmoconformer. *Physiological Zoology* **56**, 445–454.

Piiper, J. & Scheid, P. (1982) Models for a functional analysis of gas exchange organs in vertebrates. *Journal of Applied Physiology* **53**, 1321–1329.

Polis, G.A. (1988) Foraging and evolutionary responses of desert scorpions to harsh environmental periods of food stress. *Journal of Arid Environments* **14**, 123–134.

Pough, F.H., Janis, C.M. & Heiser, J.B. (1999) *Vertebrate Life*, 5th edn. Prentice Hall, Englewood Cliffs, NJ.

Prevost, Y.A. (1983) Osprey distribution and subspecies taxonomy. In: *Biology and Management of Bald Eagles and Ospreys* (ed. D.M. Bird). Macdonald Raptor Research Centre of McGill University, Quebec.

Purvis, A. & Harvey, P.H. (1997) The right size for a mammal. *Nature* **386**, 332–333.

Ramsay, J.A. (1964) The rectal complex of the mealworm *Tenebrio molitor* L. *Philosophical Transactions of the Royal Society of London B* **248**, 279–314.

Randall, D., Burggen, W. & French, K. (2002) *Eckert Animal Physiology*, 5th edn. W.H. Freeman, New York.

Randall, D.J. & Daxboeck, C. (1984) Oxygen and carbon dioxide transfer across fish gills. In: *Fish Physiology* (ed. W. Hoar & D. Randall). Academic Press, Orlando, FL.

Redmond, J.R. (1955) The respiratory function of haemocyanin in Crustacea. *Journal of Cellular & Comparative Physiology* **46**, 209–247.

Reichert, H. (1992) *Introduction to Neurobiology*. Georg Thieme Verlag, Stuttgart.

Reinertsen, R.F.E. & Haftorn, S. (1986) Different metabolic strategies of northern birds for nocturnal survival. *Journal of Comparative Physiology* **156**, 655–663.

Rhoads, D.C. (1966) Missing fossils and palaeoecology. *Discovery (Yale Peabody Mus. Nat. Hist.)* **2**, 19–22.

Richards, O.W. & Davies, R.G. (1977) *Imm's General Textbook of Entomology*. Chapman & Hall, London.

Roitt, I., Brostoff, J. & Male, D. (1998) *Immunology*, 5th edn. Mosby Publishers, London.

Rome, L.C. & Sosnicki, A.A. (1991) Myofilament overlap in swimming carp. II. Sarcomere length changes during swimming. *American Journal of Physiology* **260**, C289–C296.

Ross, M.H. & Romrell, L.J. (1989) *Histology*. Williams & Wilkins, Baltimore, MD.

Roughton, F.J.W. (1964) Transport of oxygen and carbon dioxide. In: *Handbook of Physiology*, Vol. 3.1. *Respiration* (ed. W.O. Fenn & H. Rahn). American Physiological Society, Bethesda, MD.

Rowley, A.F. & Ratcliffe, N.A. (eds) (1988) *Vertebrate Blood Cells*. Cambridge University Press, Cambridge, UK.

Ruppert, R.F. & Smith, P.R. (1988) The functional organisation of filtration nephridia. *Biological Reviews* **63**, 231–358.

Sanders, H.L. (1969) Benthic marine diversity and the stability-time hypothesis. *Brookhaven Symposium in Biology* **22**, 71–81.

Santos, C.A.Z., Penteado, C.H.S. & Mendes, E.G. (1989) The respiratory responses of an amphibious snail *Pomacea lineata* to temperature and oxygen tension variations. *Comparative Biochemistry & Physiology A* **86**, 409–415.

Schad, G.A. (1963) Niche diversification in a parasitic species flock. *Nature* **198**, 404–406.

Schmidt-Nielsen, K. (1964) *Desert Animals—Physiological Problems of Heat and Water*. Oxford University Press, Oxford.

Schmidt-Nielsen, K. (1972) *How Animals Work*. Cambridge University Press, Cambridge, UK.

Schmidt-Nielsen, K. (1984) *Scaling: Why is Animal Size so Important?* Cambridge University Press, Cambridge, UK.

Schmidt-Nielsen, K. (1997) *Animal Physiology*, 5th edn. Cambridge University Press, Cambridge, UK.

Schmidt-Nielsen, K., Bretz, W.L. & Taylor, C.R. (1970a) Panting in dogs: unidirectional air-flow over evaporative surfaces. *Science* **169**, 1102–1104.

Schmidt-Nielsen, K., Hainsworth, F.R. & Murris, D.E. (1970b) Countercurrent heat exchange in the respiratory passages; effects on water and heat balance. *Respiration Physiology* **9**, 263–276.

Schmidt-Nielsen, K., Taylor, C.R. & Shkolnik, A. (1971) Desert snails: problems of heat, water and food. *Journal of Experimental Biology* **55**, 385–398.

Scholander, P.F. (1964) Animals in aquatic environments: diving mammals and birds. In: *Handbook of Physiology*, Section 4 (ed. D.B. Dill, E.F.V. Adolph & C.G. Wilbur). American Physiological Society, Bethesda, MD.

Scholander, P.F., Walters, V., Hock, R. & Irving, L. (1950) Body insulation of some arctic and tropical mammals and birds. *Biological Bulletin* **99**, 225–236.

Skadhauge, E. (1981) *Osmoregulation in Birds*. Springer-Verlag, Berlin.

Skulason, S. & Smith, T.B. (1995) Resource polymorphisms in vertebrates. *Trends in Ecology & Evolution* **10**, 366–370.

Smith, D.S. (1975) The flight muscles of insects. In: *The Insects* (ed. T. Eisner & E.O. Wilson), pp. 41–49. W.H. Freeman, New York.

Smith-Gill, S.J. (1983) Developmental plasticity: developmental conversion versus phenotypic modulation. *American Zoology* **23**, 47–55.

Somero, G.N., Dahlhoff, E. & Lin, J.J. (1996) Stenotherms and eurytherms: mechanisms establishing thermal optima and tolerance ranges. In: *Animals and Temperature* (ed. I.A. Johnston & A.F. Bennett), pp. 53–78. Cambridge University Press, Cambridge, UK.

Sømme, L. (1995) *Invertebrates in Hot and Cold Arid Environments.* Springer-Verlag, Berlin.

Southward, A.J. (1964) The relationship of temperature and rhythmic cirral activity in some Cirrepedia considered in connection with their geographic distribution. *Helgolander Wissenschaftliche Meeresuntersuchungen* **10**, 391–403.

Staehelin, L.A. (1974) Structure and function of intercellular junctions. *International Review of Cytology* **39**, 191–283.

Stewart, M. (1991) *Animal Physiology.* Hodder & Stoughton Educational, Sevenoaks, UK.

Stone, G.N. & Willmer, P.G. (1989) Warm-up rates and body temperature in bees: the importance of body size, thermal regime and phylogeny. *Journal of Experimental Biology* **147**, 303–328.

Storey, K.B., Baust, J.G. & Storey, J.M. (1981) Intermediary metabolism during low temperature acclimation in the overwintering gallfly *Eurosta solidaginis. Journal of Comparative Physiology B* **144**, 183–190.

Storey, K.B. & Storey, J.M. (1984) Biochemical adaptations for freezing tolerance in the wood frog *Rana sylvatica. Journal of Comparative Physiology B* **155**, 29–36.

Strong, D.R., Lawton, J.H. & Southwood, T.R.E. (1984) *Insects on Plants.* Blackwell Scientific Publications, Oxford.

Strumwasser, F. (1960) Mammalian hibernation. XV. Some physiological principles governing hibernation in *Citellus beecheyi. Bulletin of the Museum of Comparative Zoology* **124**, 285–320.

Südhof, T.C. (1995) The synaptic vesicle cycle—a cascade of protein–protein interactions. *Nature* **375**, 645–653.

Sweetman, H.L. (1938) Physical ecology of the firebrat, *Thermobia domestica* (Packard). *Ecological Monographs* **8**, 285–311.

Taylor, C.R. & Lyman, C.P. (1972) Heat storage in running antelope: independence of brain and body temperatures. *American Journal of Physiology* **222**, 114–117.

Taylor, C.R., Maloiy, G.M.O., Weibel, E.R., Langman, V., Kamau, J.M.Z., Seeherman, J.H. & Heglund, N.C. (1981) Design of the mammalian respiratory system. III. Scaling of maximal aerobic capacity to body mass: wild and domestic animals. *Respiration Physiology* **44**, 25–37.

Taylor, C.R., Schmidt-Nielsen, K. & Raab, J.L. (1970) Scaling of the energetic cost of running to body size in mammals. *American Journal of Physiology* **219**, 1104–1107.

Tenney, S.M. & Temmers, J.E. (1963) Comparative quantitative morphology of the mammalian lung: diffusing area. *Nature* **197**, 54–57.

Theede, H. (1972) Vergleichende ökologisch-physiologische Untersuchungen zur zellulären Kälteresistenz mariner Invertebraten. [Some ecophysiological studies on the cellular cold resistance of marine invertebrates.] *Marine Biology* **15**, 160–191.

Torres, J.J. & Somero, G.N. (1988) Vertical distribution and metabolism in Antarctic mesopelagic fish. *Comparative Biochemistry & Physiology B* **90**, 521–528.

Trager, W. (1986) *Living Together: the Biology of Animal Parasitism.* Plenum Press, New York.

Trueman, E.R. (1975) *The Locomotion of Soft-Bodied Animals.* Edward Arnold, London.

van Mierop, L.H.S. & Barnard, S.M. (1978) Further observation on thermoregulation in the brooding female *Python molurus bivattatus* (Serpentes; Boidae). *Copeia* **1978**, 615–621.

Villee, C.A. (1977) *Biology*, 7th edn. Saunders, Philadelphia, PA.

Virkar, R.A. & Webb, K.L. (1970) Free amino acid composition of the softshell clam *Mya arenaria* in relation to salinity of the medium. *Comparative Biochemistry & Physiology* **32**, 775–783.

Wall, B.J. & Oschman, J.L. (1970) Water and solute uptake by rectal pads of *Periplaneta americana. American Journal of Physiology* **218**, 1208–1215.

Walliman, T., Wyss, M., Brdiczka, D., Nicolay, K. & Eppenberger, H.M. (1992) Intracellular compartmentation, structure and function of creatine-kinase isoenzymes in tissues with high and fluctuating energy demands—the phosphocreatine circuit for cellular energy homeostasis. *Biochemical Journal* **281**, 21–40.

Waloff, Z. (1963) Field studies on solitary and transient desert locusts in the Red Sea area. *Anti-Locust Bulletin* **40**, 1–93.

Walsberg, G.E. & King, J.R. (1978) The energetic consequences of incubation for two passerine species. *Auk* **95**, 644–655.

Wampler, J.E. (1978) Measurements and physical characteristics of luminescence. In: *Bioluminescence in Action* (ed. P.J. Herring). Academic Press, London.

Wayne, R.P. (1991) *Chemistry of Atmospheres.* Clarendon Press, Oxford.

Weber, R.E. (1980) Functions of invertebrate haemoglobins with special reference to adaptations to environmental hypoxia. *American Zoologist* **20**, 79–101.

Weigert, R.G. (1964) Population energetics of meadow spittlebug (*Philaenus spumarius* L.) as affected by migration and habitat. *Ecological Monographs* **34**, 217–241.

West, G.C. (1965) Shivering and heat production in wild birds. *Physiological Zoology* **38**, 111–120.

Wigglesworth, V.B. (1938) The regulation of osmotic pressure and chloride concentration in the haemolymph of the mosquito larvae. *Journal of Experimental Biology* **15**, 235–247.

Wigglesworth, V.B. (1945) Transpiration through the cuticle of insects. *Journal of Experimental Biology* **21**, 97–114.

Wilkie, D.R. (1977) Metabolism and body size. In: *Scale Effects in Animal Locomotion* (ed. T.J. Pedley). Academic Press, London.

Williams, M.A.J. (1986) The creeping desert: what can be done? *Current Affairs Bulletin* **63**, 24–31.

Willmer, P.G. (1980) The effects of a fluctuating environment on the water relations of larval Lepidoptera. *Ecological Entomology* **5**, 271–292.

Willmer, P.G. (1982) Microclimate and the environmental physiology of insects. *Advances in Insect Physiology* **16**, 1–57.

Willmer, P.G. (1983) Thermal constraints on activity patterns in nectar-feeding insects. *Ecological Entomology* **8**, 455–469.

Wilson, E.O. (1970) Chemical communication within animal species. In: *Chemical Ecology* (ed. E. Sondheimer & J.B. Simeone). Academic Press, New York.

Withers, P.C. (1992) *Comparative Animal Physiology.* Saunders College Publishing, Fort Worth, TX.

Wittenberg, J.B. (1958) The secretion of inert gas into the swim bladder of fish. *Journal of General Physiology* **41**, 783–804.

Woledge, R.C., Curtin, N.A. & Homster, E. (1985) Energetic aspects of muscle contraction. *Monographs of the Physiological Society 4.*

Wooten, R.C. & Crawford, C.S. (1975) Food, ingestion rates and assimilation in the desert millipede *Orthoporus ornatus* (Girard). *Oecologia* **20**, 231–236.

Zapol, W.M., Liggins, G.C., Schneider, R.C. *et al.* (1979) Regional blood flow during simulated diving in the conscious Weddel seal. *Journal of Applied Physiology* **47**, 968–973.

Index

Page numbers in **bold** indicate pages with tables or boxes; page numbers in *italic* indicate pages with figures